S0-BBU-881

ENCYCLOPEDIA OF

FOOD SCIENCE AND TECHNOLOGY

VOLUME 2

WILEY ENCYCLOPEDIA OF FOOD SCIENCE AND TECHNOLOGY, Second Edition

Editor-in-Chief
Frederick J. Francis
University of Massachusetts, Amherst

Associate Editors
Christine M. Bruhn
Center for Consumer Research
Pavinee Chinachoti
University of Massachusetts, Amherst
Fergus M. Clydesdale
University of Massachusetts, Amherst
Michael P. Doyle
University of Georgia
Kristen McNutt
Consumer Choices, Inc.
Carl K. Winter
University of California, Davis

Editorial Staff
Publisher: **Jacqueline I. Kroschwitz**
Associate Editor: **Glenn Collins**
Managing Editor: **John Sollami**
Editorial Assistants: **Susan O'Driscoll, Hugh Kelly**

ENCYCLOPEDIA OF

FOOD SCIENCE AND TECHNOLOGY
Second Edition

VOLUME 2

Frederick J. Francis
University of Massachusetts
Amherst, Massachusetts

A Wiley-Interscience Publication
John Wiley & Sons, Inc.
New York / Chichester / Weinheim / Brisbane / Singapore / Toronto

This book is printed on acid-free paper. ♾

Published simultaneously in Canada.

For ordering and customer service, call 1-800-CALL-WILEY.

Library of Congress Cataloging-in-Publication Data:

Wiley encyclopedia of food science and technology.—2nd ed. / [edited by] Frederick J. Francis.
p. cm.
Rev. ed. of: Encyclopedia of food science and technology / Y.H. Hui, editor-in-chief. c1992.
Includes bibliographical references.
ISBN 0-471-19285-6 (set : cloth : alk. paper).—ISBN 0-471-19255-4 (v. 1 : cloth : alk. paper).—ISBN 0-471-19256-2 (v. 2 : cloth : alk. paper).—ISBN 0-471-19257-0 (v. 3 : cloth : alk. paper).—ISBN 0-471-19258-9 (v. 4 : cloth : alk. paper)
1. Food industry and trade Encyclopedias. I. Francis, F. J. (Frederick John), 1921- . II. Encyclopedia of food science and technology.
TP368.2.E62 2000
664′.003—dc21 99-29003
CIP

Printed in the United States of America.

10 9 8 7 6 5 4 3 2 1

EVAPORATION

The primary objectives of evaporation, as a unit operation in food processing, are to reduce the volume of the product by some significant amount with minimum loss of nutrient components, and to preconcentrate liquid foods such as fruit juice, milk, and coffee before the product enters a dehydration process, thus saving energy in subsequent operations and reducing handling (transport, storage, and distribution) costs. Evaporation is accomplished by boiling liquid foods until the desired concentration is obtained, as result of the difference in volatility between the solvent and solutes. In the food industry, the solvent to withdraw is almost always water (1). Evaporation increases the solids content of a food and hence preserves it by a reduction in water activity; however, the flavor and color of a food may be changed during the process. The technical simplicity of evaporation gives it an obvious advantage compared to other methods such as reverse osmosis and freeze concentration. The removal of unpleasant volatile substances, for instance, during evaporation of milk, may also be an advantage. Evaporation of heat-sensitive materials is often accomplished under relatively high vacuum (low pressures), lower temperatures, and short residence time.

The first commercial use of evaporating equipment was made by Borden in 1856 (2). Since then, equipment for evaporation has evolved from open kettles through rotating steam coils to the modern natural or forced circulation evaporators. A wide range of different types of evaporator are available and evaporation can take place under different processing conditions. Commercial evaporation capacities range from 100 kg/h for pilot plants to over 200,000 kg/h for industrial installations (for instance, FMC, San Jose, Calif.; APV, Crawley, UK; Alfa-Laval, Lund, Sweden, and Dedert, Olympia Fields, Ill.).

Evaporation involves a sequence of processes. Computers can be used in the design of individual components and the overall system; however, to compromise between vastly opposed factors, the choice of the optimum design solution must be based on acquired experience and sound judgment. The intensive energy use of the evaporation process poses a continuing challenge in minimizing energy costs while maintaining product quality, especially at times of escalating energy costs. Multiple-effect evaporators and vapor recompression are measures that can be employed to improve the energy efficiency of the evaporation plant.

THEORY

The complete design of evaporator systems requires engineering calculations that involve combined heat and mass balances, and is essentially an application of the first law of thermodynamics. In an evaporator, the steam condenses at constant pressure, and the condensate leaves at this pressure and corresponding equilibrium temperature. The heat from the condensing steam boils water from a liquid that is being concentrated. The water vapor leaving the evaporator is at a lower temperature and pressure than the original steam supply, and can give up its latent heat in a double-effect evaporator to evaporate a lower boiling temperature liquid. This necessary lower boiling temperature can be maintained by operating a second effect of evaporation at a lower pressure than the first, usually under a vacuum. The same principles can be extended to more than two effects in industrial applications.

Steady-state mathematical models with equations that represent heat and mass transfer may be used for design purposes. Steam requirement, heat transfer surface area, and system capacity (production rate) can be determined for a given degree of concentration and processing time. A double-effect evaporator, with feed and steam or vapor streams moving in counterflow arrangement (Fig. 1) will be used here to illustrate the theory.

For the first effect, mass balance that relates solids entering and leaving the evaporator gives

$$\dot{m}_2 X_2 = \dot{m}_1 X_1 \tag{1}$$

and the rate of heat transfer to the evaporator from the steam is

$$\dot{m}_s \lambda_s = U_1 A_1 (T_s - T_1) \tag{2}$$

(The terms in equations are defined in the notation list at the end of the article). Assuming that there are negligible heat losses from the evaporator, the heat balance states that the amount of heat given up by the condensing steam equals the amount of heat used to raise the feed temperature to boiling point and then to boil off the vapor. In other words, the input energy of condensing vapor and feed equals the output energy of vapor and product, thus

$$\dot{m}_2 h_2 + \dot{m}_s \lambda_s = (\dot{m}_2 - \dot{m}_1) H_1 + \dot{m}_1 h_1 \tag{3}$$

Similarly, for the second effect, mass and energy balances are represented by the equations

$$\dot{m}_F X_F = \dot{m}_2 X_2 \tag{4}$$

$$(\dot{m}_2 - \dot{m}_1)\lambda_1 = U_2 A_2 (T_1 - T_2) \tag{5}$$

$$\dot{m}_F h_F + (\dot{m}_2 - \dot{m}_1)\lambda_1 = (\dot{m}_F - \dot{m}_2) H_2 + \dot{m}_2 h_2 \tag{6}$$

The overall heat-transfer coefficient U in equation 2 is defined by the sum of three thermal resistance terms, as represented by

$$\frac{1}{UA_m} = \frac{1}{h_s A_s} + \frac{1}{kA_m} + \frac{1}{h_p A_p} \tag{7}$$

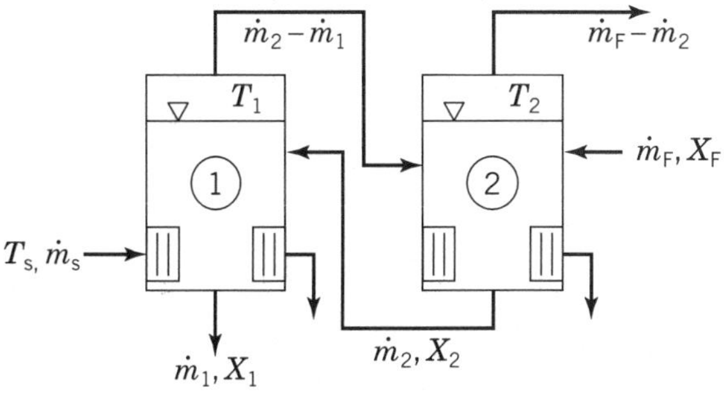

Figure 1. Schematic of double-effect evaporator, backward feed.

These resistances include resistance in the boundary layer (heat-transfer film) on the heating medium side, resistance to heat conduction in the material making up the heat-transfer surface, and resistance in the heat-transfer film on the product side.

In most evaporators, the heating medium will be steam or some other condensing vapor. The resistance to heat transfer on the heating medium side of the heat-transfer surface is normally created by a condensation film, and an expression for condensation heat-transfer coefficient as presented in one study (3) was

$$h_c = 0.94\left(\frac{k_f^3 \rho_f^2 g \lambda}{L\mu_f(T_s - T_w)}\right)^{1/4} \tag{8}$$

for vertical tubes and

$$h_c = 0.725\left(\frac{k_f^3 \rho_f^2 g \lambda}{D\mu_f(T_s - T_w)}\right)^{1/4} \tag{9}$$

for horizontal tubes. Both expressions have accounted for heat transfer involved in phase change. The heating-surface material is usually stainless steel, and it is relatively easy to compute the corresponding resistance knowing the thickness and the thermal conductivity of the material. The most complex resistance to heat transfer is due to the resistance film on the product side.

The heat transfer from a heated surface to a boiling liquid is by convection, but the heat flux varies with the temperature difference between the surface and the bulk liquid, as illustrated in Figure 2. When the temperature difference is very small (less than 5°C), the heat transfer is by natural convection in a regime called pool boiling. By increasing the temperature difference, vapor bubbles tend to form at selected locations on the surface, and the regime is called nucleate boiling, which continues until the temperature difference is near 55°C. Heat-transfer coefficient from surface to boiling liquid is increased. Beyond 55°C, a film of vapor bubbles forms and collects near the heat-transfer surface, resulting in a reduced heat flux. At temperature differences greater than 55°C, the film becomes stable; however, radiation heat-transfer mode begins to contribute to the heat flux and leads to greater heat-transfer coefficients. Either pool or nucleate boiling will probably describe the majority of physical situations that may exist in liquid food evaporation (4).

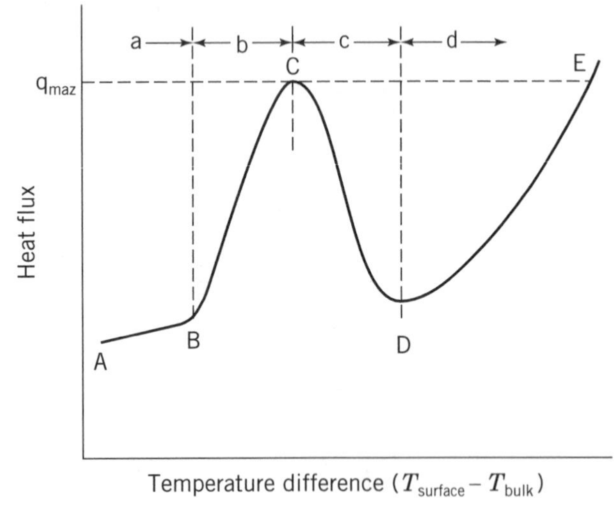

Figure 2. Typical boiling curve illustrating various regimes of boiling near heated surfaces. a, natural convection; b, nucleate boiling; c, transition boiling; d, film boiling. *Source:* Reprinted with permission from Ref. 4.

Heat transfer to boiling liquids is normally expressed in terms of dimensionless groups in the same manner as other types of heat transfer (5). Experimental studies with evaporation systems have resulted in a number of empirical expressions for convective heat-transfer coefficients to the product. For the case of natural circulation evaporators, the following correlation was developed (6):

$$h = 0.0086\,\frac{k_f}{D}\left(\frac{u_m D \rho_f}{\mu_f}\right)^{0.8}\left(\frac{c_f \mu_f}{k_f}\right)^{0.6}\left(\frac{\sigma f}{\sigma}\right)^{0.33} \tag{10}$$

For the case of forced convection systems, the following expression was suggested (7):

$$h_b = 3.5h\left(\frac{1}{Z}\right)^{0.5} \quad \text{for } 0.25 < \frac{1}{Z} < 70$$

$$\frac{1}{Z} = \left(\frac{y}{1-y}\right)^{0.9}\left(\frac{\rho_f}{\rho_v}\right)^{0.5}\left(\frac{\mu_v}{\mu_f}\right)^{0.1} \tag{11}$$

where h_b and h are for convective heat transfer with and without boiling, respectively.

In falling-film evaporation three different flow regimes normally exist, depending on the Reynolds number (Re) of the film: laminar, wavy laminar, and turbulent. The truly laminar regime takes place for Re < 20–30, the wavy laminar regime starts at Re = 30–50, and at 1,000–3,000 a fully turbulent flow is established. The Reynolds number for a film is calculated according to the following equation (8):

$$\mathrm{Re} = \frac{4G}{\mu} \tag{12}$$

where G is the mass flow rate per unit width of the wall.

For the laminar region, the local heat-transfer coefficient may be written as (8):

$$h = a\phi\left(\frac{G}{\mu}\right)^{-0.33}$$

$$\phi = \left(\frac{k_f^3 \rho_f^3 g}{\mu^2}\right)^{0.33} \tag{13}$$

Table 1 gives formulas for local heat-transfer coefficients according to different investigators.

Experimental product-side heat-transfer coefficients h_p were determined (9) for thin-film wiped-surface evaporators, which are extensively used for concentrating high-viscosity or heat-sensitive products. Product degradation is minimized by vigorous agitation and continuous removal of liquid films on the heat-transfer wall. The correlation obtained for the transition-flow regime (Re = 100–1,000) was

$$\mathrm{Nu} = 0.0483\mathrm{Re}^{0.586}\mathrm{Pr}^{1.05}\mathrm{Fr}^{0.118}\left(\frac{\mu}{\mu_w}\right)^{-2.93} \tag{14}$$

Table 1. Local Heat-Transfer Coefficients

Flow regime	A	ϕ_1	Exponent for Re	Notes
Laminar (Re < 20–30)	0.69	ϕ	−0.33	Nusselt theory
Wavy laminar (Re < 1,000–3,000)	0.61	ϕ	−0.223	
Turbulent (Re > 1,000–3,000)	0.067	$\phi \cdot$ Pr	0.2	Colburn theory based on pipe friction
	0.030	$\phi \cdot$ Pr	0.33	
	0.19	$\phi \cdot$ Pr	0.067	
	0.0078	$\phi \cdot$ Pr	0.4	Pr ≈ 5
	0.0122	$\phi \cdot$ Pr	0.33	Pr ≈ 5

Source: Ref. 8.

which is valid for liquid feed rate of 58–87 kg/h and agitation speed of 150–500 rpm. The correlation obtained for the turbulent-flow regime (Re > 1,000) was

$$\mathrm{Nu} = 4.137\mathrm{Re}^{0.283}\mathrm{Pr}^{3.325}\mathrm{Fr}^{0.032}\left(\frac{\mu}{\mu_w}\right)^{-0.753} \tag{15}$$

which is valid for liquid feed rate of 87–133 kg/h and agitation speed of 500–1,300 rpm. The authors found that unstable operation of the evaporator occurred when the speed of the wiper blades was below 150 rpm, and rapid increase in h_p was observed for rpm range 250–350. Under similar conditions of evaporation, heat-transfer coefficients were lower for high-viscosity liquids.

A detailed expression for the overall heat-transfer coefficient in multiple-effect falling-film evaporators was worked out (10). This coefficient U was given as a function of the enthalpy and temperature of the liquid and vapor phases, local heat-transfer coefficients, and mass fractions of solute in the inlet and outlet liquid streams. The fundamental mathematical model of multiple-effect evaporators was further applied (11) along with an accurate estimate of U and fouling factor R_d to design a five-effect evaporation unit capable of treating the input citrus juice flow rates indicated in (Figure 3). Results of the calculations are shown in Table 2 for orange and lemon juices that were concentrated from 11.7 to 65°Brix, and from 9.3 to 40°Brix, respectively.

According to one study (12), factors influencing the rate of heat transfer can be summarized as

1. Temperature difference between the steam and the boiling liquid.
2. Fouling of evaporator surfaces.
3. Boundary film thickness.

Factors influencing the economics of evaporation include

1. Loss of concentrate due to foaming and entrainment.
2. Energy expenditure.

The total evaporation depends on the steam consumption and the specific surface area of the evaporator. Capital costs increase with the size of the evaporator whereas energy cost decrease with it. If the fouling of the heating surface is expressed as specific product losses and is reduced from 1.0 to 0.2 kg thin milk/m^2, for modern evaporators with a specific surface area of 0.12–0.14 m^2(kg/h), the production costs will decrease by about 40% for whole milk evaporation and 30% for skim milk evaporation (13). Increased fouling is caused by a nonuniform distribution of liquid to individual heating tubes in falling-film evaporators such that some groups of tubes have more deposit than the others.

Steam economy, defined as the mass of water evaporated per unit mass of steam utilized, is often used as a measure of evaporator performance. A similar definition could be used if another condensing vapor (ammonia, Freon, or diphenyl) is employed as the heating medium.

Steam requirement for double-effect is computed directly from solving the heat and mass balances presented earlier. Equations 1–6 can be used to compute the heat-transfer surface area after the overall heat-transfer coefficients U_1 and U_2 have been evaluated for both effects. The usual procedures for computing steam requirement involve assumptions, such as equal heat flux, equal heat-transfer area, or equal temperature gradient in each effect of the system.

For plants with increasing number of effects resulting in decreased potential temperature difference in each effect, the boiling point rise is of importance when calculating the temperature program of an evaporation plant. For well-defined solutions such as water, the elevation of boiling points is proportional to the molar concentration of the solute, thus

$$\Delta = \frac{RT_{A0}^2 X^1}{\lambda} \tag{16}$$

Food liquids are normally more complicated and the boiling point rise has to be determined experimentally. For milk, the following values have been given (8):

Concentration	% DM	16	27.5	39	49	62	69	73
Temperature	°C	0.5	1	1.5	2	3	4	5

The Duhring plot for sucrose for various conditions are presented in Figure 4.

EQUIPMENT

Evaporators are basically heat exchangers, that transfer heat from steam or other heating medium to the food. Vapors produced must be separated. The selection of an evaporator should include the following factors (12):

1. Production rate (kilograms of water removed per hour).

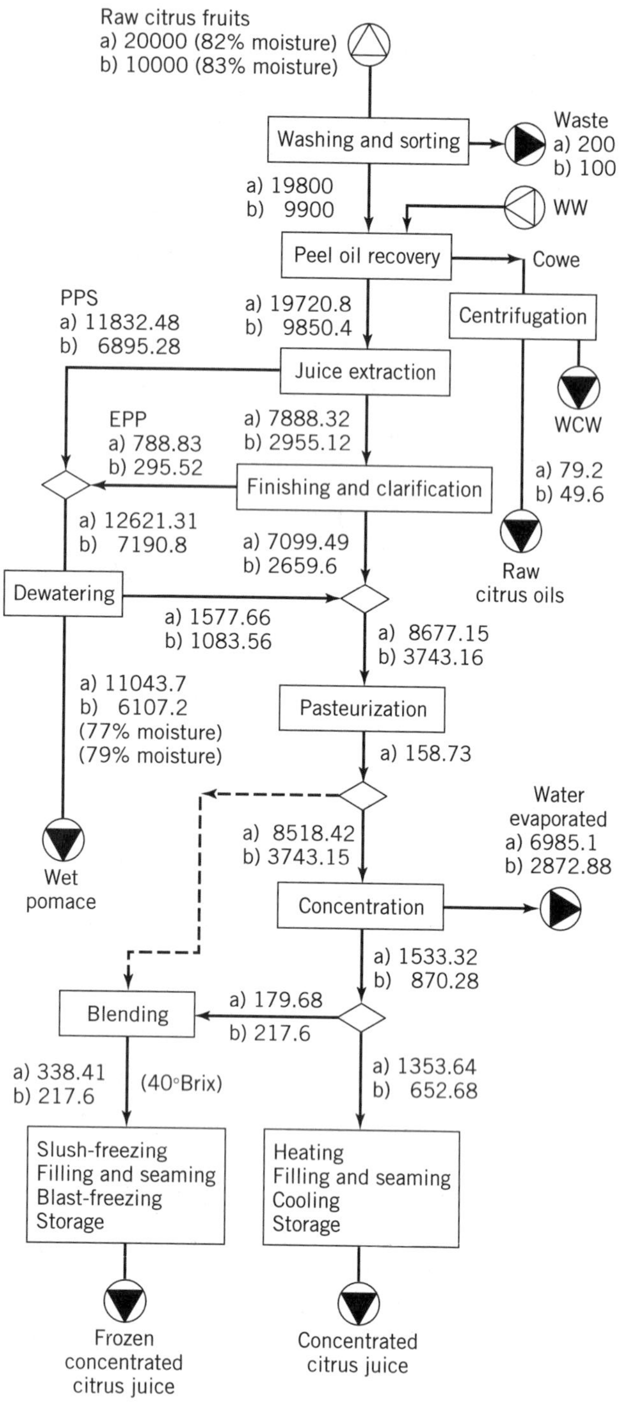

Figure 3. Flow diagram and material balance flow sheet for a concentrated citrus juice production plant with working capacity of 20,000 kg of raw oranges per hour and 10,000 kg of raw lemons per hour. The data reported here are expressed as kgh^{-1}; symbol (a) refers to orange, while symbol (b) to lemons. Stream identification symbols: COWE, citrus oils and water emulsion; EPP excess pulp and peel; PPS, peel, pulp, and seeds; WCW, water and citrus waste; WW, washing water. *Source:* Reprinted with permission from Ref. 11.

2. Degree of concentration required (percentage dry solids in the product).
3. Heat sensitivity of product in relation to the residence time and temperature of evaporation.
4. The requirement for volatile recovery facilities.
5. Ease of cleaning.
6. Reliability and simplicity of operation.
7. Size of evaporator.
8. Capital and operating costs.
9. Product quality.

Evaporators are classified as natural circulation evaporators and forced circulation evaporators.

Natural Circulation Evaporators

These may be open or closed pan evaporators, short-tube evaporators (also known as calandria vacuum pans in the food industry) (Fig. 5), or long-tube evaporators. Natural circulation evaporators operate by the thermo-syphon principle. The density difference between the boiling liquor and the circulation leg produces the driving force for liquid circulation. Typical applications include beet sugar, low to moderately viscous liquors, and nonsalting materials.

Pan evaporators are heated directly by gas or electrical resistance wires or indirectly by steam. They have relatively low rates of heat transfer, low energy efficiencies, and cause damage to heat-sensitive foods. However, they have low capital costs and are easy to construct and maintain. They have found wide applications in the preparation of sauces, gravies, and jam or other preserves (12).

Short-tube evaporators consist of a vessel or shell that contains a bundle of tubes. The feed solution is heated by steam condensing on the outside of the tubes. Liquor rises through the tubes, boils, and recirculates by flowing back down through a wide central bore. The vertical arrangement of tubes promotes natural convection currents and, therefore, higher heat transfer rates. The flow velocity is typically 0.3–1 m/s (15). These evaporators have low capital costs and are flexible, although generally unsuited to high-viscosity solutions. They are widely used for concentrating syrups, salt, and fruit juices.

The long-tube evaporators consist of a vertical bundle of tubes, each up to 50 mm in diameter and 5–15 m high. The length-to-diameter ratio is 70:130 (15). Liquid is preheated almost to boiling point (evaporation temperature) before entering the evaporator. For low-viscosity (less than $0.1\ N \cdot s/m^2$) foods such as milk, the thin film of liquor is forced up the evaporator tubes and this arrangement is known as the climbing-film evaporator (Fig. 6). A great deal of foaming does not matter, but they are susceptible to caking (15). As steam is admitted to the chest of the evaporator, the liquid reaches the boiling point and evaporation starts; a lot of small vapor bubbles appears in the liquid and starts the two-phase flow. The bubbles expand and push the mixture of liquid and vapor upward, further heat is applied resulting in more vapor formed (plug flow). Farther upward the remaining liquid is maintained as a film on the tube wall with vapor flowing as a core (annular flow). In the final part of the tube, the film is bursting and the liquid appears as droplets in a flow of vapor (mist flow).

Table 2. Material and Energy Balance for the Evaporation Unit of the Concentrated Citrus Process

Parameter		1st effect	2nd effect	3rd effect	4th effect	5th effect	Unit
Internal pressure	a[a]	73.5	63.8	54.2	44.0	6.0	kPa
	b	57.0	39.2	25.0	14.2	6.0	
Juice temperature	a	91.7	88.1	84.2	79.6	41.2	°C
	b	85.9	76.9	66.9	55.5	40.8	
Boiling point rise	a	0.44	0.55	0.77	1.3	4.9	°C
	b	1.21	1.45	1.84	2.57	4.5	
Clean heat-transfer coefficient	a	3,764	3,416	2,971	2,259	130	$Wm^{-1}\,°C^{-1}$
	b	4,061	3,470	2,902	2,188	799	
Input liquid flow rate	a	8,518	7,140	5,765	4,392	3,014	kgh^{-1}
	b	3,743	3,233	2,694	2,123	1,517	
Input liquid concentration	a	11.7	14.0	17.3	22.7	33.1	°Brix
	b	9.3	10.8	12.9	16.4	23.0	
Output liquid flow rate	a	7,140	5,765	4,392	3,014	1,533	kgh^{-1}
	b	3,233	2,694	2,123	1,517	870	
Output liquid concentration	a	14.0	17.3	22.7	33.1	65.0	°Brix
	b	10.8	12.9	16.4	23.0	40.0	
Recirculation ratio	a	0	0	0	0	6	
	b						
Thermal loss	a	0.03	0.03	0.03	0.03	0.03	
	b						
Heat-transfer surface	a	193.3	193.3	193.3	193.3	193.3	m^2
Fouling factor ($\times 10^3$)	a	0.43	0.43	0.43	0.43	0.43	$m^2\,°CW^{-1}$
	b	4.5	4.5	4.5	4.5	4.5	
Live steam consumption (95°C)	a			1,484			kgh^{-1}
	b			625			
Cooling water consumption (21–33°C)	a			72.3			m^2h^{-1}
	b			29.7			

Source: Adapted with permission from Ref. 11.
[a]Orange juice = a; lemon juice = b.

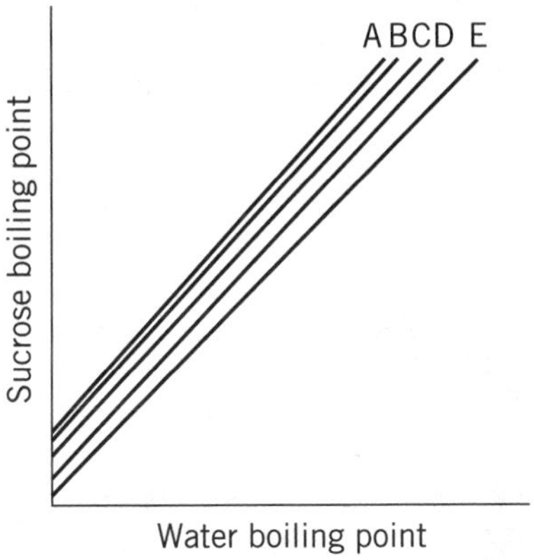

Figure 4. Duhring plot for sucrose for various sucrose concentrations per 100 g of water. A, 1,000 g; B, 800 g; C, 600 g; D, 400 g; E, 200 g. *Source:* Reprinted with permission from Ref. 12.

Pressure drop increases as more vapor is formed, and hence the evaporation temperature drops. This temperature drop must be as small as possible because it leads to higher energy consumption. The flow regime and temperature profile are depicted in Figure 7.

For more viscous foods, or those that are very heat sensitive, the feed is introduced at the top of the tube bundle, and the concentrate and the vapor are leaving the tubes at their bottom. Normally the liquid is superheated when entering the tube, thus the liquid is already at its boiling point and no part of the heating surface is used for preheating. This type of equipment is called the falling-film evaporator (Fig. 8). Only a small amount of product is in the tube, compared with the column of liquid in the climbing-film evaporator. Hydrostatic head loss found in climbing-film evaporators is absent in falling-film evaporators. Thus little temperature drop is seen during the passage through the tube (14). The gravitational force enhances the forces due to expansion of the steam, to produce high flow velocities of up to 200 m/s at the end of the tubes (12). Falling-film evaporators were developed to increase the rate of removal of water and decrease the danger of localized overheating.

A falling-film evaporator consists of the following major components: vertical heating tube bundle with surrounding heating jacket, calandria upper section with product distribution device, and calandria base section with separator. The liquid to be concentrated is fed into the upper part of the calandria where it is distributed evenly on the tube sheet, from which it flows down the interior walls of the vertical tubes as a thin boiling film. The most important factor in the trouble-free operation of this evaporator is the liquid coverage of the heating tubes, that is, the liquid flow rate required to cover each heating tube. Without enough liquid, the liquid film will break near the bottom of the tube. Dry spots develop on the tube, the heating surface becomes coated with a deposit, and eventually the tube may get clogged with burned-on product. The minimum liquid coverage depends on the type of product, required concentration, and operating time. If the coverage

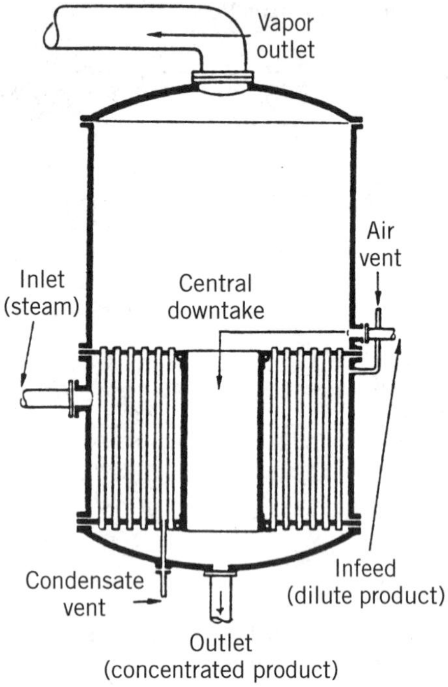

Figure 5. Calandria evaporator. *Source:* Reprinted with permission from Ref. 14.

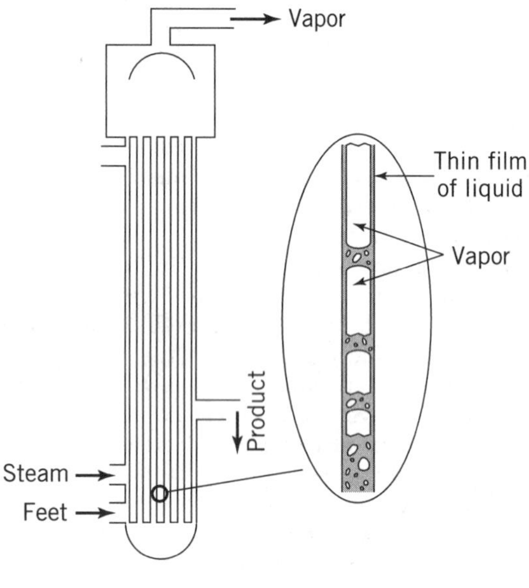

Figure 6. Climbing-film evaporator. *Source:* Reprinted with permission from Ref. 12.

falls short of a minimum value, fewer but longer heating tubes may be used, thus giving more liquid per tube while maintaining the same total heating surface area. The next choice is to subdivide the calandria to create more passes on the product side; recirculation of product within a calandria also improves coverage, however, it increases the residence time. Tube length design shall be a compromise between adequate liquid coverage and pressure drop. Short tubes with larger diameters have smaller pressure loss, whereas small diameter tubes lead to high velocity of

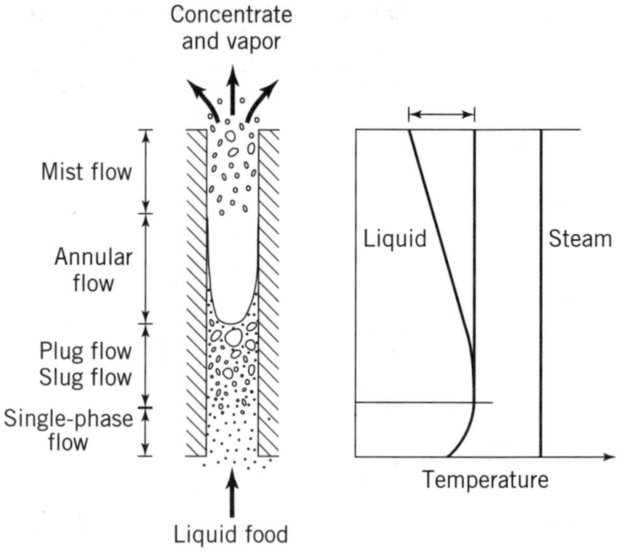

Figure 7. Flow regimes and temperature distribution in a climbing-film evaporator. *Source:* Reprinted with permission from Ref. 16.

the two-phase flow, causing high pressure loss and boiling point elevation. To this effect, it has been recommended that tubes between 8 and 12 m in length and with a diameter between 30 and 50 mm be used (16).

Product distribution in the upper part of the calandria can be done in two ways; dynamic distribution or static distribution. Dynamic distribution may be done by a full-cone nozzle in which the kinetic energy of the liquid is used to atomize the fluid into droplets and distribute them on the upper tube sheet. This distribution device, although simple and economic, does not guarantee even distribution. Static distribution devices (Fig. 9) are generally preferred. Even distribution is achieved by having uniform liquid static pressure over all openings in the base of a predistribution bowl, giving uniform wetting of all tubes. The system adapts readily to varying feed conditions by varying the liquid level. The falling-film evaporator is widely used for sugars and syrups, yeast extracts, fruit juices, dairy products and starch processing.

It has been pointed out that both types of long-tube evaporator are characterized by short residence times, high-heat transfer coefficients, and efficient energy use (steam economy = 2.5 to 3.3) in multiple-effect systems (12).

Forced Circulation Evaporators

Forced circulation evaporators have a pump or scraper assembly to move the liquor, usually in thin layers, to enhance heat-transfer rates and shorten residence times. The forced circulation design is generally used in applications where a solid is precipitated during evaporation or a scaling constituent is present.

Recirculated falling-film configurations are used when insufficient feed liquor is available to use the heat-transfer surface with single-pass operation (natural circulation method). A portion of the product liquor is combined with the feed stream and is pumped to the upper liquor cham-

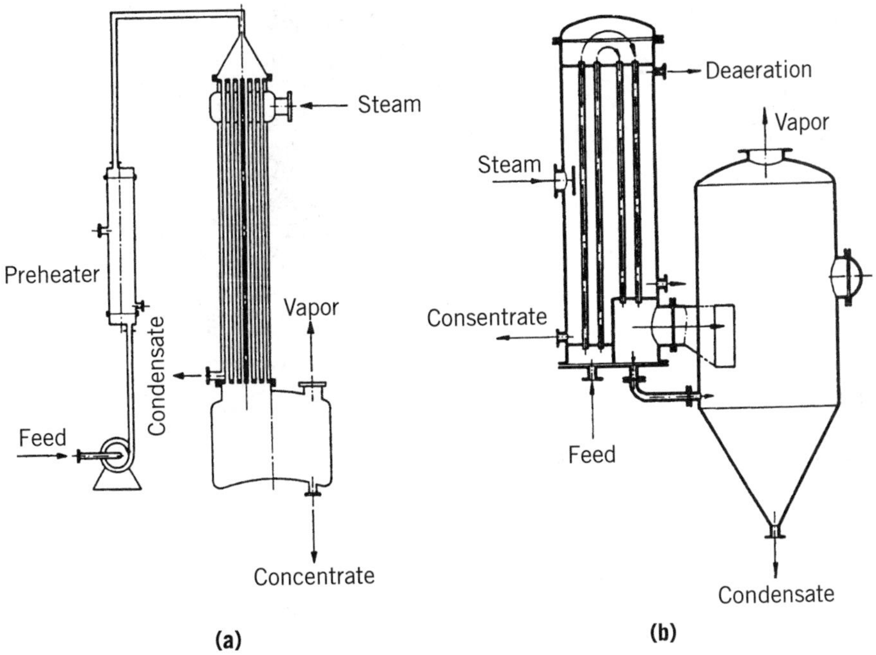

Figure 8. (**a**) Falling-film evaporator. (**b**) Falling-film evaporator, with falling and climbing sections. *Source:* Reprinted with permission from Ref. 15.

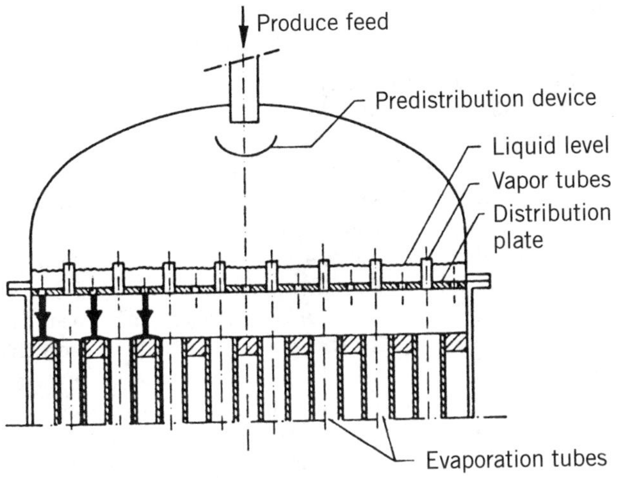

Figure 9. Liquid distribution in a falling-film evaporator. *Source:* Reprinted with permission from Ref. 8.

ber. Product retention time is greater than for the single-pass evaporator but is still relatively short as the operating volume is small. These evaporators find applications in moderately heat-sensitive foods.

The plate type or AVP evaporator, developed in the UK in the 1950s, is suitable for viscous liquids (0.3–0.4 N · s/m^2), because the food is pumped through the plate stack, consisting of a series of climbing-film and falling-film sections. The heating medium is steam. They are compact and are advantageous in having a small amount of product in the system at any one time. They have the additional advantage of ease of maintenance and inspection.

An example of a high-temperature, short-time (HTST or flash) evaporator is one with rotating heating surface and is shown in Figure 10. It is a mechanical thin-film evaporator. Feed flows to the underside of a stack of rotating hollow cones from a central shaft. The feed liquor is kept in rapid motion, and the centrifugal force (750–3,000 N) causes the formation of a very thin film (about 0.1 mm thick) in which the liquid remains for only a very short time of 0.6 to 1.6 s (17). Steam condenses on each cone, and heat is conducted rapidly through the thin metal to evaporate the liquor. High rates of heat transfer are obtained as steam condensate is flung from the rotating cones as soon as they are formed. A high degree of flexibility can be achieved by changing the number of cones. This evaporator is used for coffee and tea extracts, meat extract, fruit juices and enzymes for use in food processing.

A schematic cross section of an agitated-film (also called scraped-surface or wiped-surface) evaporator is shown in Figure 11. The product film near the heat-transfer surface is continuously agitated by a rotor to increase the heat-transfer rate on the product side. The evaporator may also consist of a steam jacket surrounding a highspeed rotor, fitted with short blades along its length (Fig. 12). The liquid enters the wide end and is forced through the precise space between the rotor and the wall (heated surface) toward the outlet for the concentrate. The blades agitate the thin film of liquid vigorously, thus promoting high heat-transfer rates and preventing the product from burning onto the hot surface. If the viscosity increases greatly during evaporation, it is better to feed the liquid at the narrow end. This type of equipment is highly suited to viscous (up to 20 N · s/m^2), heat-sensitive foods, or to those that are liable to foam or foul evaporator surfaces (fruit pulps and juices, tomato paste, meat extracts, honey, cocoa mass, coffee, and dairy products). Only single effects are possible, thus giving low steam economy. It is more intended for finishing highly viscous products after preconcentration in other equipment (12).

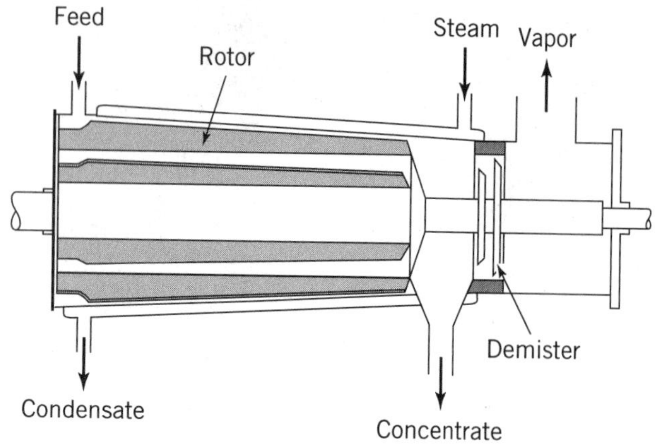

Figure 10. Mechanical thin-film evaporator. *Source:* Reprinted with permission from Ref. 15.

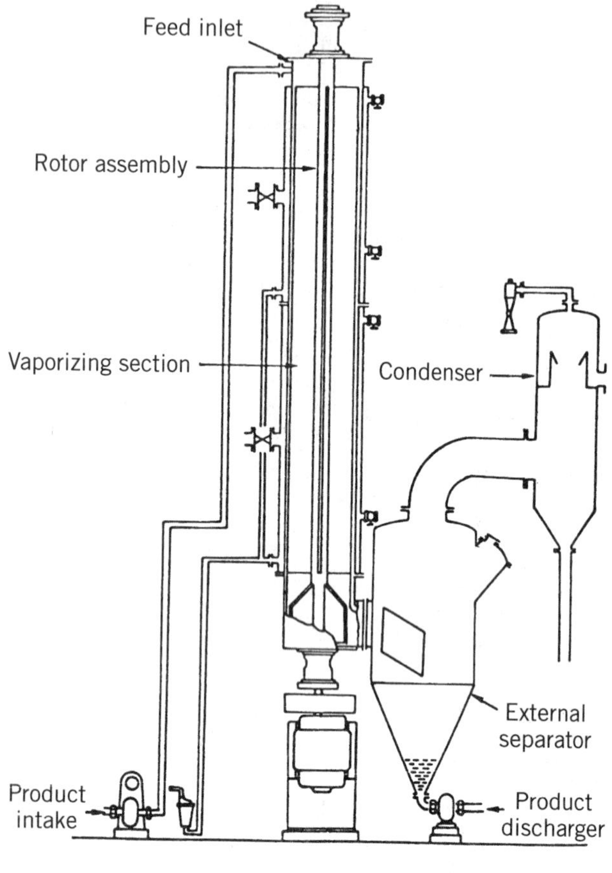

Figure 11. Schematic cross-section of agitated film evaporator. *Source:* Reprinted with permission from Ref. 4.

Energy Efficiency

The energy efficiency of evaporation systems may be improved by the following approaches (19): multiple-effect, vapor recompression, or refrigerant cycle evaporators.

Multiple-Effect Evaporating System. For the evaporation of temperature-sensitive materials it is important to keep the temperature as low as possible. This is achieved by

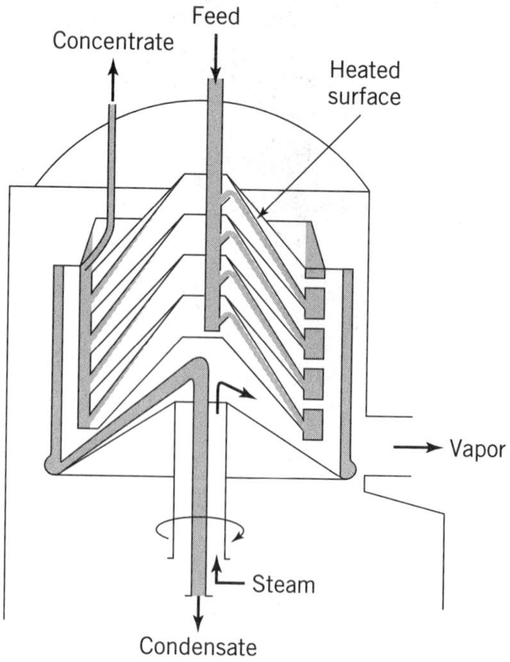

Figure 12. Centri-term evaporator. *Source:* Reprinted with permission from Ref. 18.

working at a lower pressure. Practically all the evaporators mentioned above can be used at pressures below atmospheric.

To make use of the large amount of heat in the vapor driven off the liquid food, the water vapor can be used as the heating medium in a following evaporator. This second evaporator must then operate at a lower pressure than the first. This is called double-effect evaporation. Repetition of this step produces multiple-effect evaporation. The effects then have progressively lower pressures to maintain the temperature difference between the feed and the heating medium. A multiple-effect evaporating system can be designed in several ways in regard to the direction of flow of the solution. Figure 13 illustrates, with a triple-effect evaporator, the arrangements of forward feed, backward feed, parallel feed, and mixed-feed system.

The liquid in the forward feed system enters the highest temperature evaporator first and exits as a concentrated liquid from the third evaporator that operates at the lowest temperature and pressure. The liquid is usually preheated by steam or vapor from one of the effects before it enters the first evaporator. The liquid leaving effect 1 has a higher temperature than the liquid in effect 2, and in crossing the throttle valve to effect 2, a small amount of the liquid water will flash to water vapor and the temperature of the liquid will drop to the operating temperature of effect 2.

In the backward feed system, the liquid flows in opposite direction of the forward feed evaporator. This system allows the concentrated liquid to exit at the highest temperature, which is advantageous for products having a high viscosity in the concentrated form. Because the liquid leaving effect 3 has a lower temperature than the liquid in effect 2, a pump is needed to provide additional energy to increase its temperature to the saturation temperature

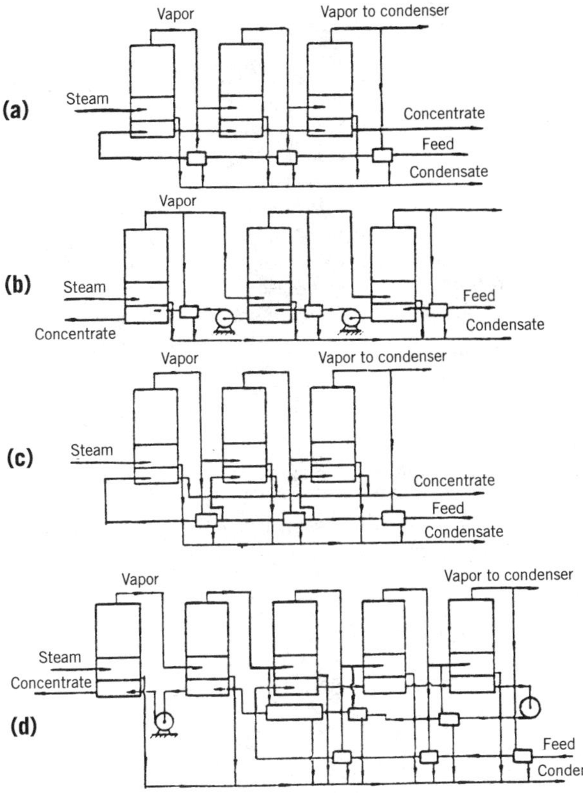

Figure 13. Multiple-effect evaporation. (**a**) Forward feed; (**b**) backward feed; (**c**) parallel feed; (**d**) mixed feed. *Source:* Reprinted with permission from Ref. 15.

corresponding to effect 2. A similar situation exists between effect 2 and effect 1.

The parallel feed evaporator essentially acts like three separate evaporators, each operating at a different pressure. It has the advantage of simple operation. However, it is the most expensive arrangement, because extraction pumps are required for each effect. Mixed-feed evaporators are useful for viscous food, such as distillery by-products (20). They have the simplicity of forward feed and economy of backward feed. The merits and shortcomings of each arrangement are described in Table 3.

In each evaporator of a multiple-effect installation it is necessary for the temperature of the heating medium to be higher than the boiling temperature; the number of stages is, therefore, limited. Savings in energy consumption decreases with every effect added. Ideally, 1 kg of steam vaporizes 1 kg of water in all effects, so that the steam economy would be 1, 2, 3, and 4, respectively, for one to four effects. In practice, the steam consumption is somewhat higher because of heat losses. Actual steam economy for evaporating milk is 0.8, 1.7, 2.5, and 3.3 (2). The steam economy for evaporating tomato paste was calculated (22); the values were 0.95, 1.91, and 2.60 compared to measured values of 0.84, 1.45, and 2.20 (on average), respectively, for one, two, and three effects.

Many HTST evaporators, under the trade name TASTE (thermally accelerated short time evaporator) have been installed in Florida and elsewhere (23). The initial feed is preheated to about 90°C in a conventional heat exchanger and flashed through a series of evaporators with progressively higher vacuums (15). A typical unit has seven stages plus a vacuum flash cooler arranged so there are four effects. The word *effect* indicates vapor flow in the evaporator, while the word *stage* indicates the flow of product (24). Temperature in the various stages range from 41 to 96°C. These units are small for their capacity and lack the vapor–liquid separation chambers of the low-temperature evaporators. There is more tendency for hesperidin to build up on the heat exchange surface of the TASTE evaporator in comparison with low temperature units, but because of the low retention time, it can be emptied rapidly. It can be cleaned in place, without dismantling, in minutes.

The steam economy of two tubular TASTE evaporators with four effects and seven stages (Fig. 14) and with nominal evaporation capacity of 27,211 kg (60,000 lb) and 45,352 kg (100,000 lb) of water per hour is 0.85 N; N being the number of effects of the evaporator (24). More stages are incorporated in the last effect because of reduction in product volume as it is concentrated. Similarly, it was found that the steam economy for two plate evaporators (four effects, five stages) (Fig. 15) with a capacity of 13,605 kg (30,000 lb) per hour was 0.82 N.

Vapor Recompression. Thermal recompression with a steam-jet compressor (25) is illustrated in Figure 16. A fraction of the vapor produced is combined with steam entering the heat exchanger in the low-pressure section of the Venturi. The remaining vapor is used in subsequent effects. This system is sometimes advantageous when the temperature of steam supplied by the boiler is too high for the foods. By regulating the size of the Venturi the pressure can be controlled and hence condensing temperature of the steam. Part of the kinetic energy associated with the high-velocity flow in the Venturi is used to compress the recirculated vapor. Vapor recompression will decrease steam consumption and increase the steam economy. When steam costs are moderate and electricity costs comparatively high, thermal compression is recommended. This simple method is relatively inexpensive, has a long operating life, and requires little maintenance. However, its characteristic operating curve has only one optimum operating point, thus restricting flexibility between design and actual operating conditions (16).

In a mechanical vapor recompression system (Fig. 17) all of the vapor produced is compressed to a slightly higher pressure by a mechanically driven compressor. The compressed vapor then enters the heat exchanger and on condensation provides heat to drive the evaporation process. This method can virtually eliminate steam requirement, as only a small amount of make-up steam is needed. Yet both the high capital and energy costs to operate the compressor need to be considered; it is about 10 times more expensive than the thermal type. Also, when evaporation pressures become too low, the volume of vapor to be handled increases dramatically, and mechanical compression is not economical. An evaporator operating at 5.0 kPa (33°C) would produce almost four times as much vapor volume as an evaporator operating at 20 kPa (60°C) on the same production rate basis (19). Nevertheless, mechanical

Table 3. Advantages and Limitations of Various Methods of Multiple-Effect Evaporation

Arrangement of effects	Advantages	Limitations
Forward feed	Least expensive, simple to operate, no feed pumps required between effects, lower temperatures with subsequent effects and, therefore, less risk of heat damage to more viscous product.	Reduced heat-transfer rate as the feed becomes more viscous, rate of evaporation falls with each effect, best quality steam used on initial feed, which is easiest to evaporate; feed must be introduced at boiling point to prevent less of economy (if steam supplies sensible heat, less vapor is available for subsequent effects)
Reverse feed	No feed pump initially, best-quality steam used on the most difficult material to concentrate, better economy and heat-transfer rate as effects are not subject to variation in feed temperature and feed meets hotter surfaces as it becomes more concentrated thus partly offsetting increase in viscosity	Interstage pumps necessary, higher risk of heat damage to viscous products as liquor moves more slowly over hotter surfaces, risk of fouling
Mixed-feed	Simplicity of forward feed and economy of backward feed, useful for very viscous foods	More complex and expensive
Parallel	For crystal production, allows greater control over crystallization and prevents the need to pump crystal Murries	More complex and expensive of the arrangements, extraction pumps required for each effect.

Source: Adapted with permission from Ref. 21.

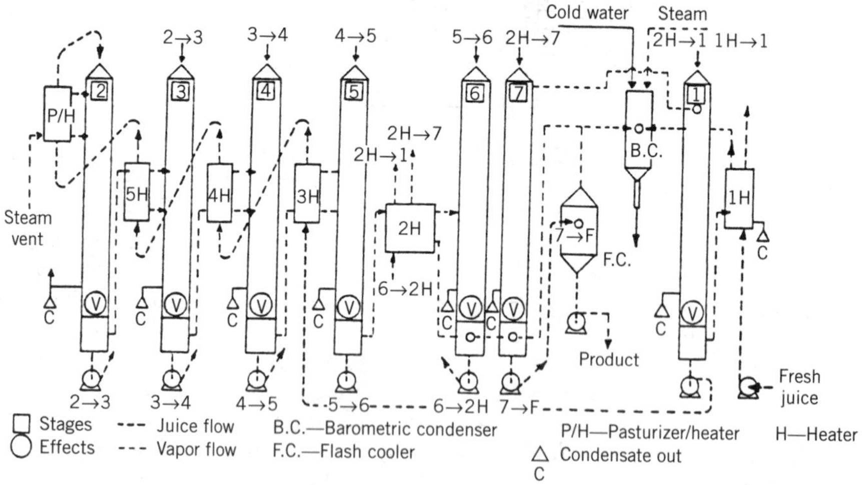

Figure 14. Schematic of a 5-effect, 7-stage TASTE Evaporator. *Source:* Reprinted with permission from Ref. 24.

vapor recompressors have more or less the same limited flexibility as thermal recompressors. This applies particularly to the high-speed radial turbo compressor, which reaches its surge limit if the actual vapor flow is too small, thus rendering the system unstable (16). Mechanical recompressors are complicated in design, and need careful maintenance and supervision of mechanical vibration.

Refrigerant Cycle Evaporators. Figure 18 shows a schematic system of juice concentration using a technique that combines a vapor compression refrigeration cycle with the evaporation system (25). Condensing water vapor boils liquid refrigerant to form a refrigerant vapor, which is then compressed to a higher temperature level and acts as a heat source to boil the liquid being concentrated (19). The evaporator, for simplicity, is shown as a single-effect design with means for recirculating the juice. Variations may employ multiple effects. The pressure in the system can be maintained by a relatively small steam ejector or a mechanical vacuum pump. Its advantage over a mechanical vapor recompression system is that the volume of vapor being compressed is much reduced at a given evaporator temperature. However, the refrigerant must be compressed through a wider temperature range because an additional heat exchanger is required.

Auxilliary Equipment

Preheaters. After feeding the product into the evaporation plant via a float balance tank and feed pump, the first process step is to preheat the product. In general, liquid food is stored at between 5 and 20°C for quality preservation; therefore, it must be preheated to at least the evaporation temperature of the first effect, most commonly between 40 and 110°C. Three types of preheater are used: plate, spiral tube, and straight tube heat exchangers (16). If there are solid particles in the product, a plate heat exchanger should not be used, because the particles may block the narrow product passages. The same problem may

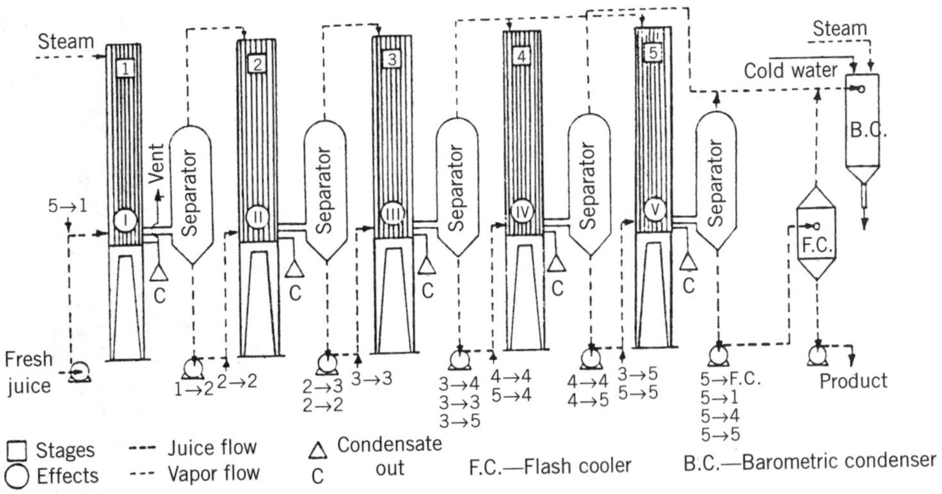

Figure 15. Schematic of a 4-effect, 5-stage plate evaporator. *Source:* Reprinted with permission from Ref. 24.

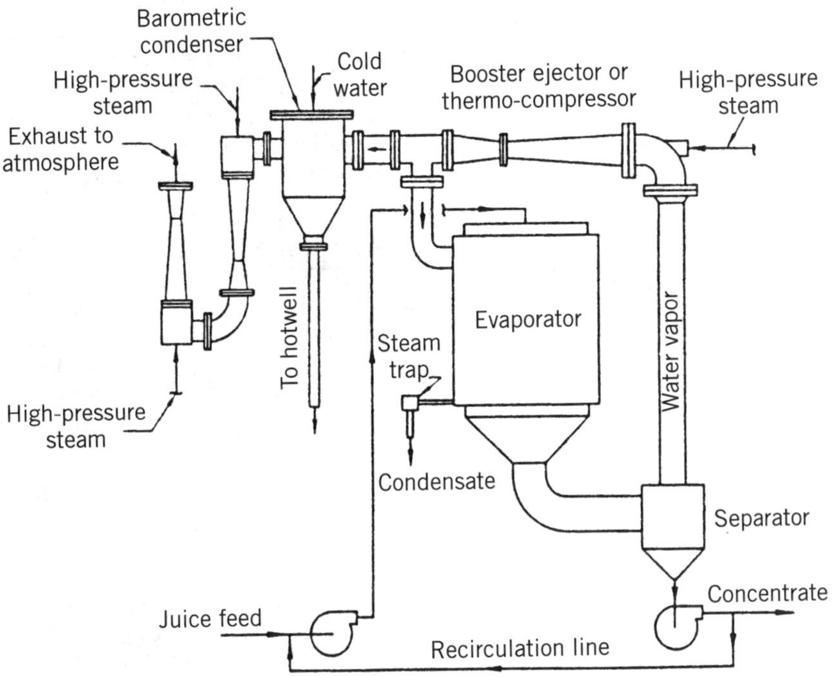

Figure 16. System of concentration using thermal recompression with a steam-jet compressor. *Source:* Reprinted with permission from Ref. 25.

also occur with protein-containing products, due to the formation of deposits if the preheating temperature exceeds the denaturalization temperature. Secondary flows developed in the spiral tube leads to higher heat-transfer coefficients than straight tube, and a reduced tendency to foul. The disadvantages are that inspection and replacement of the tube is very difficult. Straight tube preheaters are especially useful as intermediate preheaters following first step preheating with plate heat exchanger, where product of high concentration is required, and where equipment inspection is essential.

Separators. Separators are used to separate the mixture of vapor and concentrated product produced in the calandria. A separator should have high separation efficiency, small pressure loss, and short residence time. Because a fine mist of concentrate is produced during vigorous boiling and carried over in the vapor, separators should be designed to minimize entrainment. Large pressure loss is to be avoided lest extra heating surface must be installed in the next effect. The many different separators in use can be classified according to three operating principles: gravity separator, baffle separator, and centrifugal separator.

Condensers. Surplus vapor produced in evaporation can be condensed and reused where possible. Condenser cooling water is normally provided from a cooling tower or the city main; in certain cases, it may be extracted from rivers, lakes, or the sea. The tasks of a condenser are to maintain the heat balance of an evaporation plant, to stabilize the

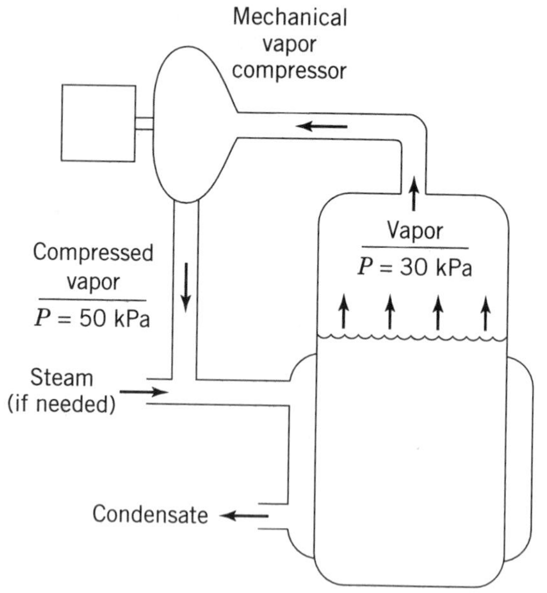

Figure 17. A mechanical vapor recompression system. *Source:* Reprinted with permission from Ref. 19.

temperature in each effect, and to produce and maintain vacuum.

EFFECT ON FOODS

Nutritional changes take place during processing, the extent varying with the type of food, the process, the plant in use, and the degree of control exercised. Many losses are inevitable, particularly if the process involves heating. Part of the water-soluble nutrients such as the B vitamins, together with lesser and less-important amounts of mineral salts, protein, and even carbohydrate will be precipitated out. Vitamin C is oxidized in air and accelerated by heat, whereas vitamins A and D and niacin are unaffected (20). A comparison of nutrient losses in milk preserved by evaporation and UHT sterilization is shown in Table 4.

Aroma compounds that are more volatile than water will be lost during evaporation. With some products such as fruit juices, the retention of taste and aroma is important, yet in other foods such as cocoa and milk, the loss of unpleasant volatiles improves the product quality. The color of foods darkens, partly due to an increase in the solids content and partly because the reduction in water activity promotes chemical changes (for example, Maillard browning) (12). As these changes are time and temperature dependent, short residence times and low boiling points produce concentrates with higher qualities. For in-

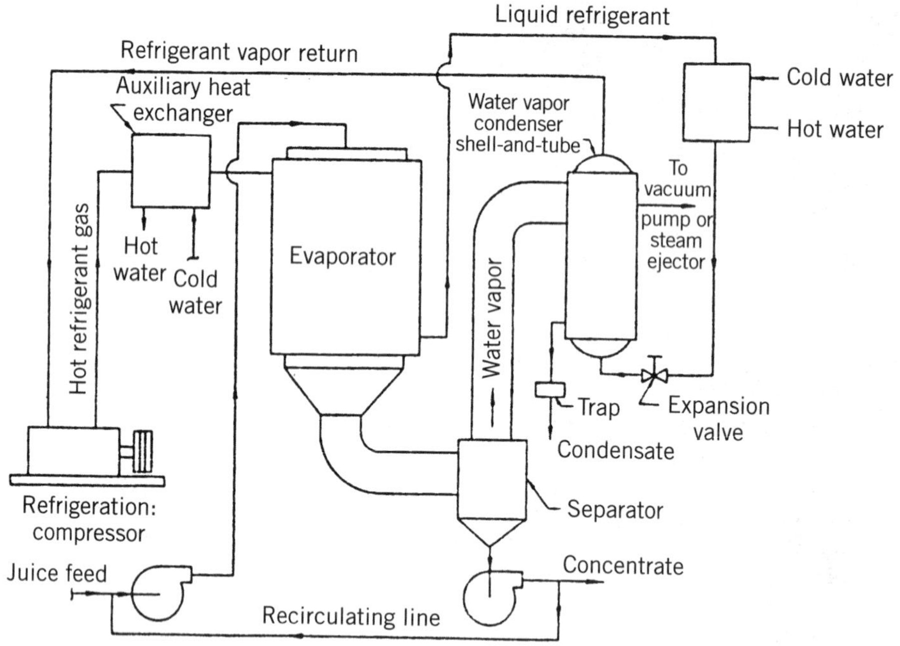

Figure 18. System of concentration using refrigerant cycle evaporators. *Source:* Reprinted with permission from Ref. 25.

Table 4. Vitamin Losses in Concentrated and UHT Sterilized Milk

Product	Loss (%)				
	Thiamin	Vitamin B_6	Vitamin B_{12}	Folic acid	Ascorbic acid
Evaporated milk	20	40	80	25	60
Sweetened condensed milk	10	<10	30	25	25
UHT sterilized milk	<10	<10	<10	<10	<25

Source: Adapted with permission from Ref. 26.

stance, the Centri-therm mechanical thin-film evaporator produces a concentrate that, when diluted, has sensory and nutritional qualities that are virtually unchanged from those of the feed material (12).

NOMENCLATURE

A	Area, m^2
D	Diameter, mm
DM	Dry matter content, %
H	Enthalpy, J/kg
G	Mass flow rate per unit width of wall, kg/m · s
R	Gas constant = 8,314 N · m/g · mol °K
T	Temperature, °C
U	Overall heat transfer coefficient, W/m^2 °C
X	Mass fraction of solute in the liquid
X^1	Molar concentration
Fr	Froude number
Nu	Nusselt number
Pr	Prandtl number
Re	Reynolds number
a	Coefficient used in eq. 13
c	Specific heat, J/kg °C
g	Gravitational constant, m/s^2
h	Enthalpy of liquid, J/kg
h_c	Convective heat-transfer coefficient, W/m^2 °C
k	Thermal conductivity, W/m °C
$\dot{m}$	Mass flow rate for evaporator, kg/s
u	Liquid–vapor velocity, m/s
x	Thickness, mm
y	Vapor quality
Δ	Boiling point elevation, °C
λ	Latent heat of vaporization, J/kg
μ	Viscosity, kg/m · s
ρ	Density, kg/m^3
σ	Surface tension, N/m

Subscripts

1	First effect in a double-effect evaporator
2	Second effect in a double-effect evaporator
A	Solvent component of product (water)
F	Feed
b	Boiling
c	Concentration
f	Fluid, or liquid state
m	Log-mean value
p	Product
s	Steam (or other heating medium)
v	Vapor
w	Wall condition
0	Pure state

BIBLIOGRAPHY

1. H. A. Leniger, "Concentration by Evaporation," in T. Hoyem and O. Kvale, eds., *Physical, Chemical and Biological Changes in Food Caused by Thermal Processing*, Elsevier Applied Science Publishers, Ltd., Barking, UK, 1980, pp. 54–76.
2. A. W. Farrall, *Food Engineering Systems. Vol. 1, Operations*, AVI Publishing Co., Inc., Westport, Conn., 1976.
3. F. Kreith and W. Z. Black, *Basic Heat Transfer*, Harper & Row, 1980.
4. D. R. Heldman and R. P. Singh, *Food Process Engineering*, 2nd ed., AVI Publishing Co., Inc., Westport, Conn., 1980.
5. *Handbook of Fundamentals*, American Society of Heating, Refrigeration and Air Conditioning Engineers, 1989.
6. E. L. Piret and H. S. Isbin, "Natural Circulation Evaporators," *Chemical Engineering Progress* **50**, 305–310 (1954).
7. J. M. Coulson and J. F. Richardson, *Chemical Engineering*, Vol. 2, 3rd ed., Pergamon Press, Elmsford, N.Y., 1976.
8. B. Hallstroem, "Heat Exchange," in *Evaporation, Membrane Filtration and Spray Drying in Milk Powder and Cheese Production, North European Dairy Journal*, 1985, pp. 37–41.
9. K. Stankiewicz and M. A. Rao, "Heat Transfer in Thin-Film Wiped-Surface Evaporation of Model Liquid Foods," *J. Food Process Eng.* 10, 113–131 (1988).
10. S. Angeletti and M. Moresi, "Modeling of Multiple-Effects Falling-Film Evaporators, " *Journal of Food Technology* **18**, 539–563 (1983).
11. M. Moresi, "Economic Study of Concentrated Citrus Juice Production," in B. M. McKenna, ed., *Engineering and Food*, Vol. 2, *Processing Applications*, 1984, pp. 975–991.
12. P. Fellows, *Food Processing Technology—Principles and Practice*, VCH Publishers, New York, 1988.
13. S. Bouman, D. W. Brinkman, P. de Jong, and R. Waalewijn, "Multistage Evaporation in the Dairy Industry: Energy Savings, Product Losses and Cleaning," in S. Brun, ed., *Preconcentration and Drying of Food Materials*, 1988, pp. 51–60.
14. M. A. Joslyn and J. L. Heid, *Food Processing Operations—Their Management, Machines, Materials, and Methods*, Vol. 3, AVI Publishing Co., Inc., Westport, Conn., 1964.
15. H. A. Leniger and W. A. Beverloo, *Food Process Engineering*, Reidel Publishing, Boston, 1975.
16. G. Hahn, "Evaporator Design," in D. MacCarthy, ed., *Concentration and Drying of Foods*, 1987, pp. 113–132.
17. P. P. Lewicki and R. Kowalczyk, in P. Linko, Y. Malkki, J. Olkku and J. Larinkari, eds., *Food Processing Systems*, Elsevier Applied Science Publishers, Barking, UK, 1980, pp. 501–505.
18. C. H. Mannheim and N. Passy, "Aroma Retention and Recovery During Concentration of Liquid Foods," Proceedings of the Third Nordic Aroma Symposium, Hemeelinna, 1972.
19. J. C. Batty and S. L. Folkman, *Food Engineering Fundamentals*, John Wiley & Sons, Inc., New York, 1983.
20. A. E. Bender, "Food Processing and Nutritional Values," in S. M. Herschdoerfer, ed., *Quality Control in the Food Industry*, Academic Press, Inc., Orlando, Fla., 1984.
21. J. G. Brennan, J. R. Butters, N. D. Cowell and A. E. V. Lilly, *Food Engineering Operations*, Applied Science, London, 1976.
22. T. R. Rumsey, T. T. Conant, T. Fortis, E. P. Scott, L. D. Pedersen and W. W. Rose, "Energy Use in Tomato Paste Evaporation," *J. Food Process Eng.* **7**, 111–121 (1984).
23. B. S. Luh and J. G. Woodroof, *Commercial Vegetable Processing*, 2nd. ed. van Nostrand Reinhold, N.Y, 1988.

24. J. G. Filho, A. A. Vitali and F. C. P. Viegas, Energy Consumption in a Concentrated Orange Juice Plant. *J. Food Process Eng.* **7**, 77–89, 1984.
25. *HVAC Systems and Applications*, Amer. Soc. Heating, Refrigeration and Air Conditioning Engineers, 1987.
26. J. W. G. Porter and S. Y. Thompson. "Effects of Processing on the Nutritive Value of Milk," Vol. 1 *Proc. Fourth International Conference on Food Science and Technology*, Madrid, 1976.

A. K. Lau
University of British Columbia
Vancouver, British Columbia
Canada

EVAPORATORS: TECHNOLOGY AND ENGINEERING

TYPES OF EVAPORATORS

In the evaporation process, concentration of a product is accomplished by boiling out a solvent, generally water, so that the end product may be recovered at the optimum solids content consistent with desired product quality and operating economics. It is a unit operation that is used extensively in processing foods, chemicals, pharmaceuticals, fruit juices, dairy products, paper and pulp, and both malt and grain beverages. It also is a unit operation, which, with the possible exception of distillation, is the most energy-intensive.

Although the design criteria for evaporators are the same regardless of the industry involved, the question always arises as to whether evaporation is being carried out in the equipment best suited to the duty and whether the equipment is arranged for the most efficient and economical use. As a result, many types of evaporators and many variations of processing techniques have been developed to take into account different product characteristics and operating parameters.

The more common types of evaporators include the following:

1. Batch pan
2. Natural circulation
3. Rising film tubular
4. Falling film tubular
5. Rising-falling film tubular
6. Forced circulation
7. Wiped film
8. Plate equivalents of tubular evaporators

Batch Pan

Next to natural solar evaporation, the batch pan as shown in Figure 1 is one of the oldest methods of concentration. It is somewhat outdated in today's technology but still is used in a few limited applications such as the concentration of jams and jellies where whole fruit is present and in processing some pharmaceutical products. Up until the early 1960s, it also enjoyed wide use in the concentration of corn syrups.

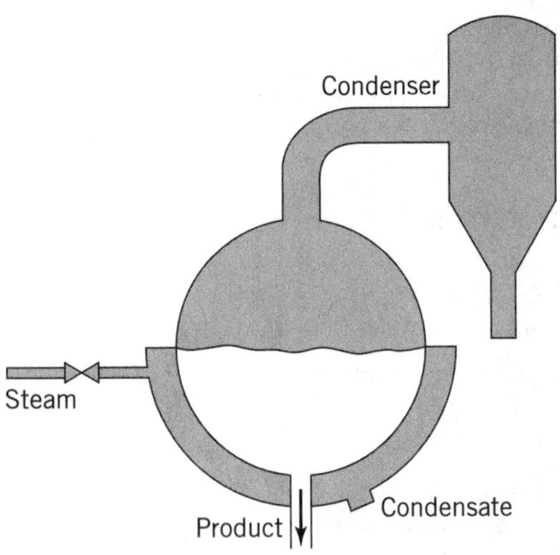

Figure 1. Jacketed batch pan.

With a batch pan evaporator, product residence time normally is many hours. Therefore, it is essential to boil at low temperatures and high vacuum when a heat-sensitive or thermodegradable product is involved. The batch pan is either jacketed or has internal coils or heaters. Heat-transfer areas normally are quite small as a result of vessel shapes, and heat-transfer coefficients tend to be low under natural convection conditions. Heat transfer is improved by agitation within the vessel. Low surface areas together with low HTCs (heat-transfer coefficients) generally limit the evaporation capacity of such a system. In many cases, large temperature differences cannot be used for fear of rapid fouling of the heat-transfer surface. Relatively low evaporation capacities, therefore, limit its use.

Tubular Evaporators

Natural Circulation. Evaporation by natural circulation is achieved through the use of a short tube bundle within the batch pan or by having an external shell and tube heater outside the main vessel as illustrated by Figure 2.

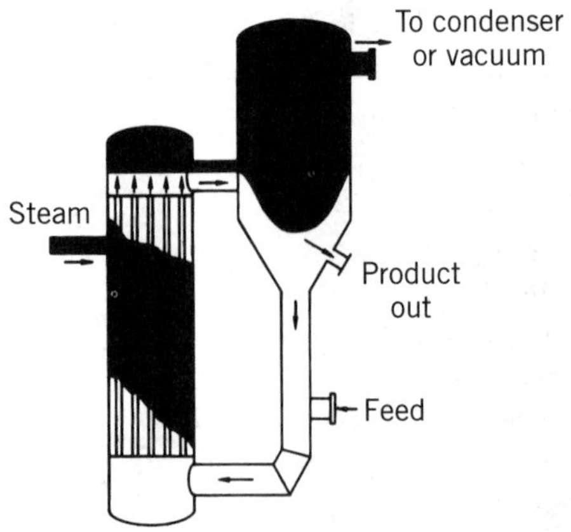

Figure 2. Natural circulation.

The external heater has the advantage that its size is not dependent on the size or shape of the vessel itself. As a result, larger evaporation capacities may be obtained. The most common application for this type of unit is as a reboiler at the base of a distillation column.

Rising Film Tubular. The first of the "modern" types of evaporators is the rising film unit, although its use by industry goes back to the early 1900s. The rising film principle was developed commercially by using a vertical tube with steam condensing on its outside surface (Fig. 3). Liquid inside the tube is brought to a boil, the vapor generated forming a core in the center of the tube. As the fluid moves up the tube, more vapor is formed, resulting in a higher central core velocity that forces the remaining liquid to the tube wall. Higher vapor velocities, in turn, result in thinner and more rapidly moving liquid film. This provides higher HTCs and shorter product residence time.

The development of this principle was a giant step forward in the evaporation field, particularly from the viewpoint of product quality. Its further advantage of high HTCs resulted in reduced heat-transfer area requirements and consequently, in a lower initial capital investment.

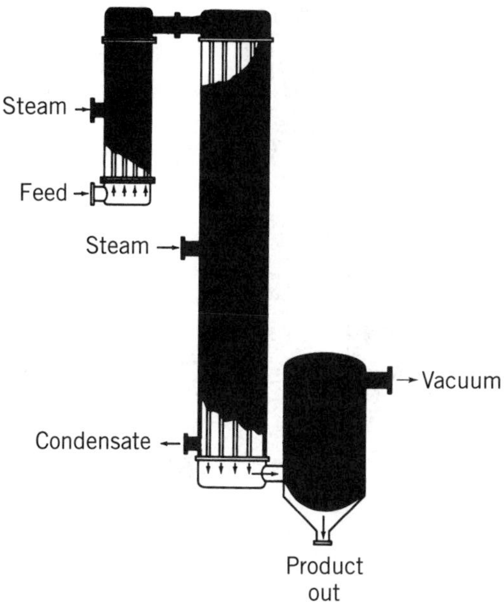

Figure 4. Falling film tubular.

Falling Film Tubular. Following development of the rising film principle, it took almost a further half century for a falling film evaporation technique to be perfected (Fig. 4). The main problem was to design an adequate system for the even distribution of liquid to each of the tubes. Distribution in its forerunner, the rising film evaporator, was easy as the bottom bonnet of the calandria always was pumped full of liquid, thus allowing equal flow to each tube.

While all manufacturers have their own techniques, falling film distribution generally is based on the use of a perforated plate positioned above the top tube plate of the calandria. Spreading of liquid to each tube sometimes is further enhanced by generating flash vapor at this point. The falling film evaporator does have the advantage that the film is "going with gravity" instead of against it. This results in a thinner, faster moving film and gives rise to even shorter product contact time and a further improvement in the value of HTC.

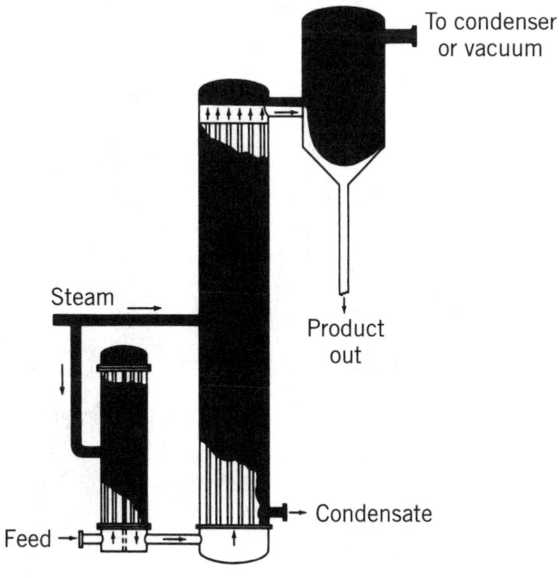

Figure 3. Rising film tubular.

The rising film unit normally needs a driving force or temperature difference across the heating surface of at least 25°F to establish a well-developed film, whereas the falling film evaporator does not have a driving force limitation. This permits a greater number of evaporator effects to be used within the same overall operating limits, in other words, if steam is available at 2.5 psig corresponding to 220°F and the last effect boiling temperature is 120°F, the total available ΔT is equal to 100°F. This would limit a rising film evaporator to four effects, each with a ΔT of 25°F. It would be feasible, meanwhile, to have as many as ten or more effects using the falling film technique.

Rising-Falling Film Tubular. As illustrated by Figure 5, the rising-falling film evaporator has the advantages of the ease of liquid distribution of the rising film unit coupled with lower head room requirements. The tube bundle is approximately half the height of either a rising or falling film evaporator and the vapor-liquid separator is positioned at the bottom of the calandria.

Forced Circulation. The forced circulation evaporator (Fig. 6) was developed for processing liquors that are susceptible to scaling or crystallizing. Liquid is circulated at a high rate through the heat exchanger, boiling is prevented within the unit by virtue of a hydrostatic head maintained above the top tube plate. As the liquid enters the separator, where the absolute pressure is slightly less than in the tube bundle, the liquid flashes to form a vapor.

The main applications for a forced circulation evaporator are in the concentration of inversely soluble materials,

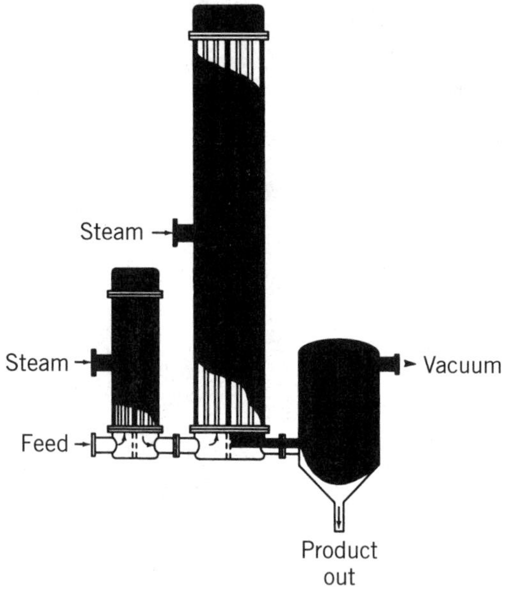

Figure 5. Rising-falling film tubular.

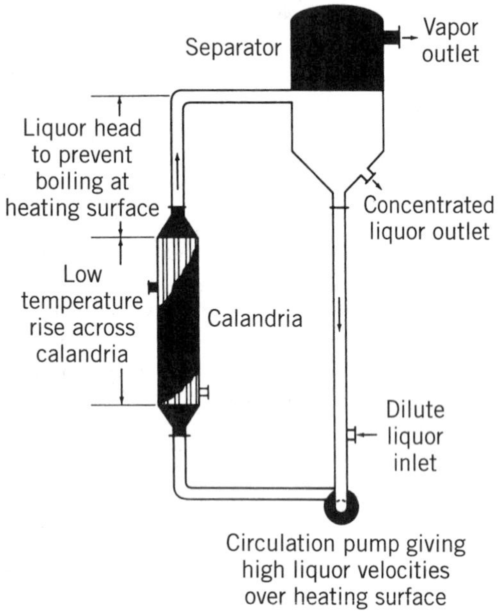

Figure 6. Forced circulation.

which results in the deposition of solids. In all cases, the temperature rise across the tube bundle is kept as low as possible, generally in the region of 3–5°F. This results in a recirculation ratio as high as 200–330 lb of liquor per pound of water evaporated. These high recirculation rates result in high liquor velocities through the tubes, which helps to minimize the buildup of deposits or crystals along the heating surface. Forced circulation evaporators normally are more expensive than film evaporators because of the need for large-bore circulating pipework and large recirculating pumps generally of the axial flow type. Operating costs of such as unit also are considerably higher.

Wiped Film. The wiped or agitated thin film evaporator depicted in Figure 7 has limited applications and is confined mainly to the concentration of very viscous materials and the stripping of solvents down to very low levels. Feed is introduced at the top of the evaporator and is spread by wiper blades on to the vertical cylindrical surface inside the unit. Evaporation of the solvent takes place as the thin film moves down the evaporator wall. The heating medium normally is high-pressure steam or oil. A high-temperature heating medium generally is necessary in order to obtain a reasonable evaporation rate since the heat-transfer surface available is relatively small as a direct result of its cylindrical configuration.

The wiped film evaporator is very satisfactory for its limited applications. However, in addition to its small surface area, it also has the disadvantage of requiring moving parts such as the wiper blades, which, together with the bearings of the rotating shaft, need periodic maintenance. Capital costs in terms of dollars per pound of solvent evaporated also are very high.

Plate Evaporators

The plate equivalent of the tubular evaporator is available in four configurations: rising-falling film, falling film, Parvap, and Parflash plate evaporators. All have been developed by APV to handle the concentration of products that have varying characteristics.

Plate-type evaporators now have been sold commercially for nearly 30 years, and during that time over 1600 units manufactured by APV have been installed for concentrating hundreds of different products.

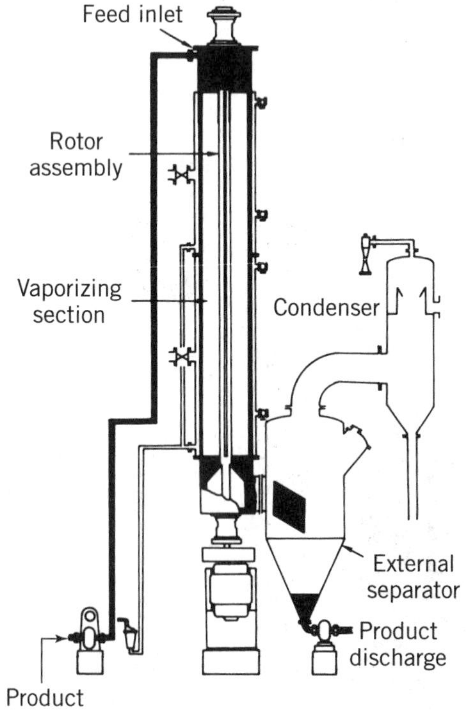

Figure 7. Wiped film.

Rising/Falling Film Plate. The original plate type evaporator, the rising-falling plate system basically is a flexible, multiduty unit engineered for medium production runs of heat-sensitive products under sanitary conditions and at the lowest possible capital investment.

Operating on a single-pass rising and falling film principle, as shown in the exploded view of Figure 8, the evaporator consists of a series of gasketed plate processing units within a compact frame. As a thin layer of feed liquor passes over the rising and falling film section in each evaporation unit, it is vaporized on contact with adjacent steam heated plates and is discharged with its vapor to a vapor-liquid separator. All evaporation is accomplished in a matter of seconds within the plate pack. The product then is extracted and the vapor passed to a condenser, the next evaporator effect or a mechanical compressor. Product quality is given maximum protection against thermal degradation by means of high heat-transfer rates, low liquid holding volume, and minimum exposure to high temperatures.

The plate evaporator has a number of advantages over its tubular counterpart. Since it is designed to be erected on a single level with minimum headroom requirements, in many cases it will fit within an existing building with overhead restrictions as low as 12 or 13 ft. It also can be arranged as single or multiple effects without extensive building modifications or structural steel supports and can handle expanded duties by the addition of more plate units. Systems are available for evaporation rates to 35,000 lb/h with efficient in-place cleaning of all stainless, steel product contact surfaces.

Falling Film Plate. The success of the rising-falling film plate evaporator prompted the development of the falling film plate system. This design has several advantages over its predecessor, including even lower residence times and higher evaporation capabilities. While the rising-falling film plate unit is restricted to a maximum of 30,000–35,000 lb/h of water removal, the falling film plate evaporator with its larger vapor ports can accommodate 59,000–60,000 lb/h of evaporation. One of this unique design features as illustrated in Figure 9 is that each side of the plate may be used independently of the other, thus permitting two pass operation within the same frame for even shorter residence time and improved product quality.

Paravap. Especially designed to concentrate foaming liquids or those with high solids content or non-Newtonian viscosity characteristics, the APV Paravap replaces the wiped film evaporator in many cases. It successfully concentrates corn syrups and soap to better than 97–98% total solids or strips hexane and other solvents from vegetable oils and similar products. Figure 10 shows this type of unit in a simple schematic.

Note that the key element in the system is a plate heat exchanger (Fig. 11). Although specifically designed for liquid-liquid applications, it has been found that if a fluid is allowed to vaporize within the plate pack, the small plate gap and corrugated plate pattern create high vapor velocities (Fig. 12). This causes atomization of the liquid within the high-velocity vapor stream, resulting in a greater liquid surface area for mass transfer and enabling low residual solvent concentrations to be realized. Although the final product may be extremely viscous after separation from the vapor, the apparent viscosity within the plate pack is very low since only droplets are being transported in the vapor.

Paraflash. Similar to its tubular counterpart, the APV Paraflash (Fig. 13) is a suppressed boiling forced-circulation evaporator used mainly for concentrating products subject to fouling that is too excessive for a film evaporator. While this system uses a plate heat exchanger, vaporization is not allowed to take place within the exchanger itself. Boiling is suppressed either by a liquid static head above the heat exchanger or by the use of an

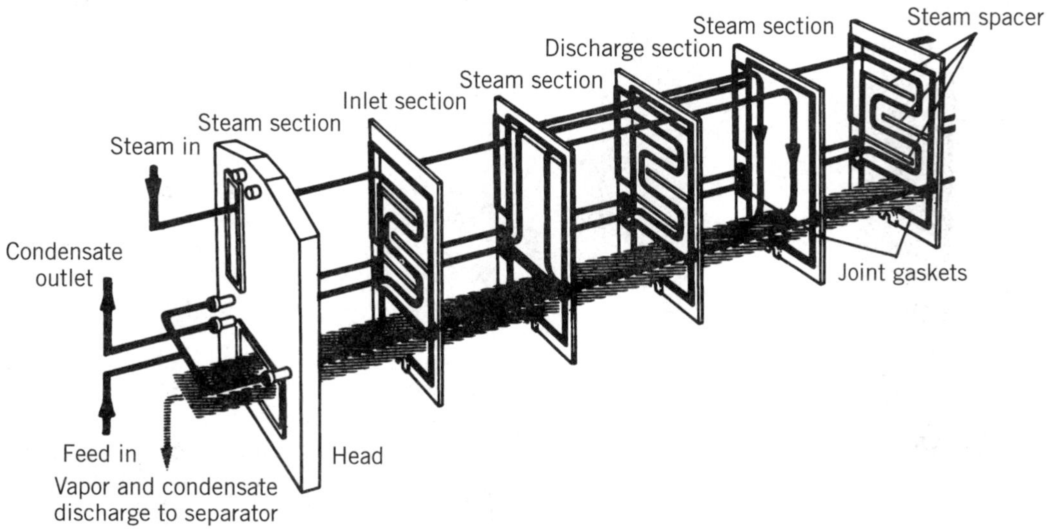

Figure 8. Rising-falling film plate evaporator.

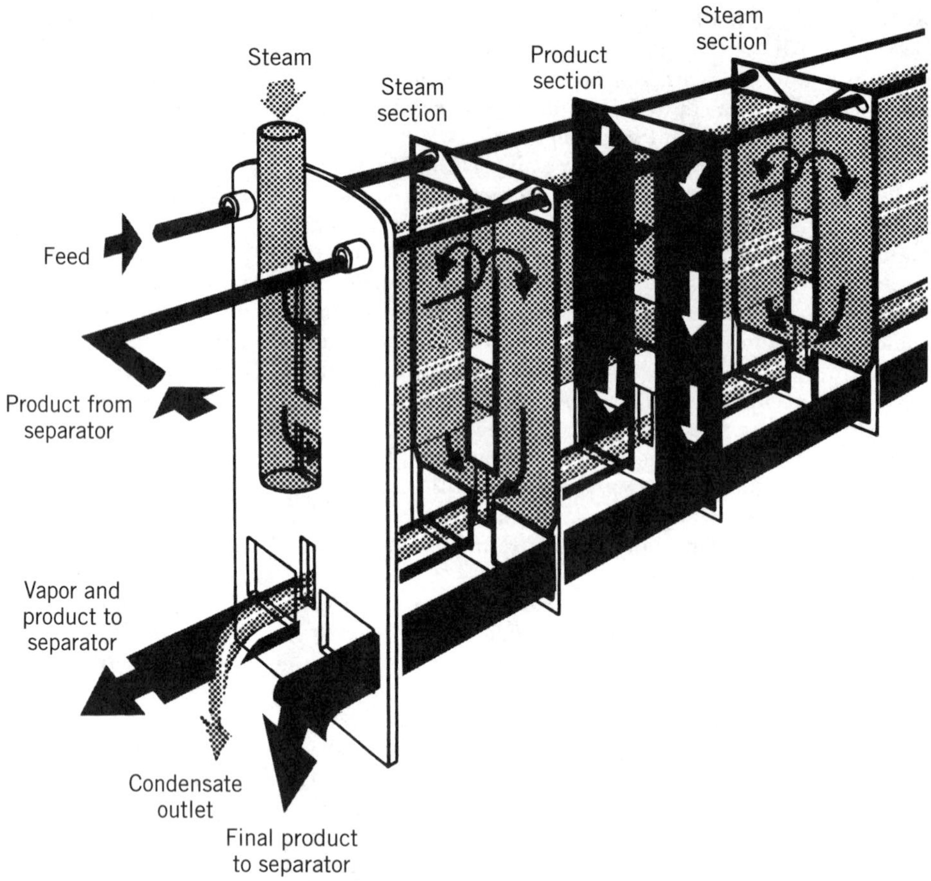

Figure 9. Falling film plate evaporator.

orifice piece in the discharge line. From a heat-transfer standpoint, the forced-circulation Paraflash is far more efficient than a tubular unit but suffers from a smaller equivalent diameter when large crystal sizes are present.

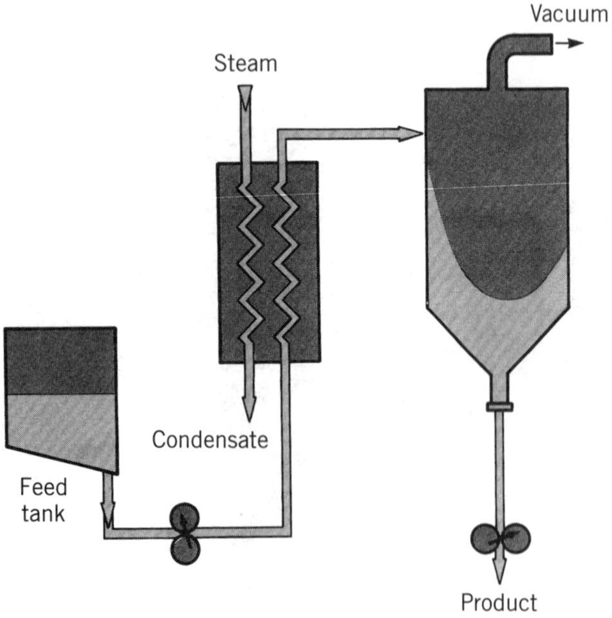

Figure 10. Paravap evaporator.

SELECTION OF EVAPORATORS

In choosing the evaporator best suited to the duty on hand, the following factors should be weighted.

Heat-Transfer Coefficients

The evaporator with the highest heat-transfer coefficients (HTCs) generally is the falling film type. Higher HTCs lead to smaller surface areas, resulting in minimum capital cost. Wherever possible, therefore, a film evaporator should be considered first.

While the biggest single disadvantage of the film evaporator is its susceptibility to scale buildup on the heat-transfer surface, its lower cost offsets this inconvenience on mildly scaling duties, particularly when high evaporative capacities are involved. A typical example of this is spin bath liquors. This solution, which results from the production of rayon and cellulose, consists mainly of sulfuric acid, sodium sulfate and water plus various impurities such as zinc sulfate. Scale builds up over a period of about 1 week and the evaporator then is boiled out.

In food related industries where the nature of most products is heat sensitive, it generally is desirable or nec-

Figure 11. Plate heat exchanger.

Figure 12. Corrugated plate pattern.

essary to clean out an evaporator at least once per day for sanitary reasons. In the chemical field, however, a minimum run of a week normally is required, and, in many cases, an evaporator will run for months.

Capacities

For evaporation capacities of up to about 60,000 lb/h of water removal, the possibility of using either a plate or tubular system should be considered. The capital investment is relatively close, but the lower cost of installing a plate evaporator together with its compactness, expendability, and reduced building requirements generally makes it extremely attractive. Above 60,000 lb/h evaporation, the

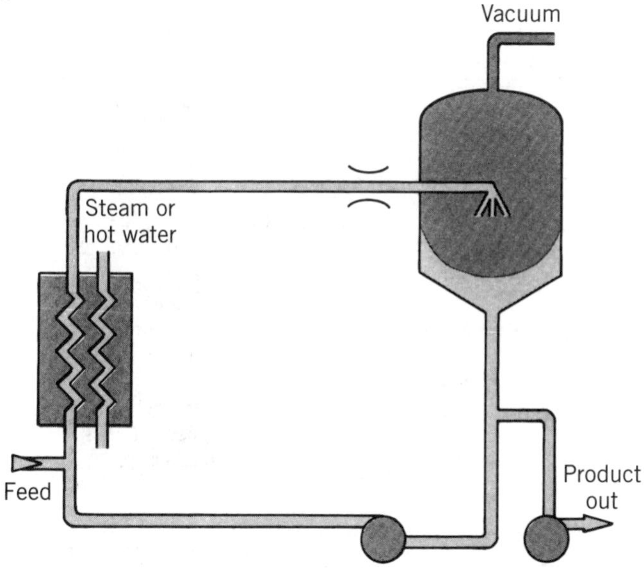

Figure 13. Suppressed boiling Paraflash evaporator.

Figure 14. Single-effect R56 Paraflash concentrates brewers yeast from 13.

choice lies among various types of tubular evaporators. Available configurations are multieffect or mechanical vapor recompression.

Materials of Construction

In many cases, the choice of evaporator is determined by the necessary materials of construction (Fig. 15). An example would be a sulfuric acid evaporator, where, for product concentrations of up to 50%, the heating surface normally would be graphite. This immediately indicates use of a tubular evaporator as opposed to plate unit. With satisfactory temperature and concentration conditions, Incoloy and Hastelloy could be employed using either a tube or plate but would be much more expensive than graphite. Separators and piping would be either a fiber-filled phenolic resin or rubber-lined carbon steel. Sulfuric acid between 50 and 72% concentration again would normally use graphite as the heating surface providing temperatures are kept below 230–240°F, but piping and vessels would be lead-lined steel, tantiron, or cast lead. For construction reasons, a graphite tube bundle has to arranged as a rising film or forced-circulation evaporator with the vapor-liquid separator positioned at the top.

Multiple Evaporators

More than one evaporator may be used to concentrate a particular product. For example, there may be a main or bulk evaporator followed by a finishing unit. The main evaporator would be used to remove the majority of the water or solvent to a point where the liquid concentration gives rise to restrictions. Among these are boiling-point elevation, which reduces the overall temperature difference of a multieffect or MVR evaporator, product viscosity resulting in low HTCs, or saturated liquor resulting in crystallization. A forced-circulation crystallizing evaporator very often is the last product stage of a multieffect evaporator.

Consider the example of concentrating a solution from 4 to 60% total solids, a 15:1 concentration ratio. At the 40% total solids level, the viscosity is too high for a film evaporator and the product has to be processed in a Paravap or wiped film unit. However, in concentrating from 4 to 40% or a 10:1 ratio, 90% of the feed has been evaporated in the initial evaporator. The finishing unit now has to concentrate only from 40 to 60% or 1.5:1 ratio. This means that 96.43% of the total evaporation has been achieved in a multieffect or mechanical vapor recompression bulk evaporator at high thermal economy and probably while obtaining good HTCs. Only 3.57% of the duty is confined to a small single-effect evaporator, which handles the most difficult part of the concentration.

Crystallization

When the requirement is for a crystallizing evaporator, there never is any doubt that the stage where crystalliza-

Product		Materials of Construction
Most dairy products		304/316 stainless steel
Most food products		304/316
Most fruit juices		304/316
Most pharmaceuticals		304/316
if NaCl in any of above		Monel or titanium
Sulfuric acid	<50%	FFPR/graphite
		RLCS/graphite
	<72%	Lead lined/tantiron or cast lead
Hydrochloric acid		FFPR/graphite/RLCS
Phosphoric acid	(dilute/pure)	316
	(with impurites)	FFPR/graphite/RLCS
Caustic soda	<40%	Stress relieved C.S.
Caustic soda	(with NaCl)	Monel/nickel/NiLCS
Ammonium nitrate		304/304L
Ammonium sulfate		316

FFPR—fiber-filled phenolic resin; RLCS—rubber-lined carbon steel; NiLCS—nickel-lined carbon steel.

Figure 15. Guide to materials of construction.

tion takes place has to be a forced-circulation evaporator and, in most cases, a tubular unit. However, when the problem is not one of crystallization but a general scaling of the heat-transfer surface instead, the Paraflash is much more efficient. As shown in Fig. 16, the circulation rate through the plate is considerably less than through a tube and results in much lower horsepower requirements.

If crystals are formed and are too large for a plate evaporator to handle, the designer has to determine whether the tubular forced circulation unit should be a single- or multiple-pass configuration. Normally, forced-circulation tubulars are designed with tube velocities of 6–8 ft/s or higher to minimize fouling. In the Figure 17 comparison between single- and three-pass configurations, three-pass units are shown to have lower circulation rates directly resulting in higher temperature rises, lower log mean temperature differences, and, therefore, greater surface area. Even though the circulation rate is less, the pressure drop on the three-pass versus one-pass is more because of greater overall tube length. This results in larger absorbed horsepower. It generally is concluded, therefore, that a single-pass arrangement is less expensive from both the capital and operating cost standpoints.

HTCs, Forced Circulation, and Film

Estimation of HTCs from physical properties is easier and more reliable for forced-circulation units than for film evaporators. As a general rule, HTCs used for the latter are determined by experience or test work.

EVAPORATION CONFIGURATION AND ENERGY CONSERVATION

Engineers can help cut energy consumption even when evaporation requirements seem to become greater every year. This can be done by

Solution properties:

Specific gravity	1.35
Specific heat	0.62
Thermal conductivity	0.26 Btu units
Viscosity	100 cps

	Tubular Unit	Plate Unit
Heat duty, Btu/hr	5×10^6	5×10^6
Element	1¼ in. OD, 16 SWG, 25-ft tube	APV R10 plate
Total surface, ft^2	2620	2622
Fluid temperature in, °F	160	160
Fluid temperature out, °F	161.08	170
Service temperature, °F	190	190
Recirculation rate, gpm	10,975	1610
Total ΔP, psi	16.9	10.8
BHP absorbed	107.7	10.1

SWG = standard wire gauge; OD = outer diameter; gpm = gallons per minute.

Figure 16. Comparison between forced circulation tubular and plate evaporators.

Duty: Triple effect forced circulation evaporator
Feed: 102,000 lb/h, Cornsteep 6.14% → 51%

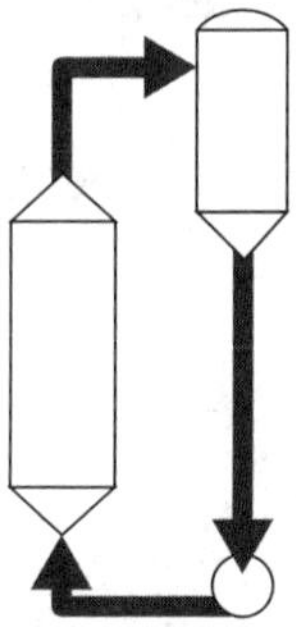

	1st Effect		2nd Effect		3rd Effect	
	1 Pass	3 Pass	1 Pass	3 Pass	1 Pass	3 Pass
Circulation rate (GPM)	12,275	6,190	12,275	6,190	14,500	4,800
Heat transfer area (ft)	4,434	5,541	4,434	5,541	4,434	4,656
Absorbed BHP	50	91	66	113	156	192

	One Pass	Three Pass
Equipment	$491,280	$540,960
Absorbed BHP	272	396

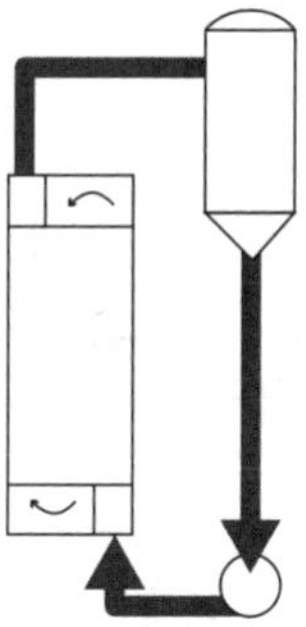

Figure 17. Comparison between single- and triple-pass calandria.

1. Installing a greater number of steam effects to an evaporator.
2. Making use whenever possible of steam thermocompression or mechanical vapor recompression.
3. Making sure that the feed to an evaporator is preheated to the boiling point in the most efficient manner.
4. Using low-grade heat from the evaporator whenever possible.
5. Insulating equipment to keep losses to a minimum.

Some of these points are illustrated by the Figure 18 schematic of a three-effect evaporator with steam thermocompression that uses evaporator condensate and intereffect vapor for feed preheating. Note that even this vapor has been used efficiently in prior effects and that condenser water also may be used if the feed temperature is low. Figure 19, meanwhile, outlines a four-effect evaporator with thermocompression and the use of a slightly different technique with a spray heating loop.

While a greater number of effects increases initial capital investment, savings in operating costs will justify the expenditure if evaporation capacities are large enough and the operating period long enough. A thermocompressor generally adds the equivalent of one or more effects at relatively little capital cost, although the surface area in the effects between thermocompressor suction and discharge must be increased. The HTC of these effects generally is very good so the increase in surface area is minimized. Care should be taken, however, to design with a low ΔT across the thermocompressor to ensure high entrainment ratios of vapor to steam.

It is also important when using a steam thermocompressor not to have a high-boiling-point elevation in those effects. This will cut down the (ΔT) available for heat transfer. Thermocompressors are somewhat inflexible and do not operate well outside their design conditions. They should not be used when the product may foul the evaporator surface and cause the design pressures and pressure boost across the thermocompressor to rise. Under these conditions, the amount of vapor entrained is reduced and a fall-off in evaporator capacity results.

Effective preheating of the feed also can reduce energy consumption considerably. The capital cost of preheaters is fairly small with the plate heat exchangers being particularly effective owing to their capability for realizing a very close temperature approach to the heating medium, whether it is condenser water, waste evaporator condensate, or intereffect vapor.

Although all of these features reduce energy consumption, still better methods are available. In less than a decade and probably within the next 5 yr, the majority of new large-capacity evaporators installed, particularly for bulk duties, will be of the mechanical vapor recompression (MVR) type. The mechanical vapor compressor can be driven by electricity from coal, hydropower, or nuclear

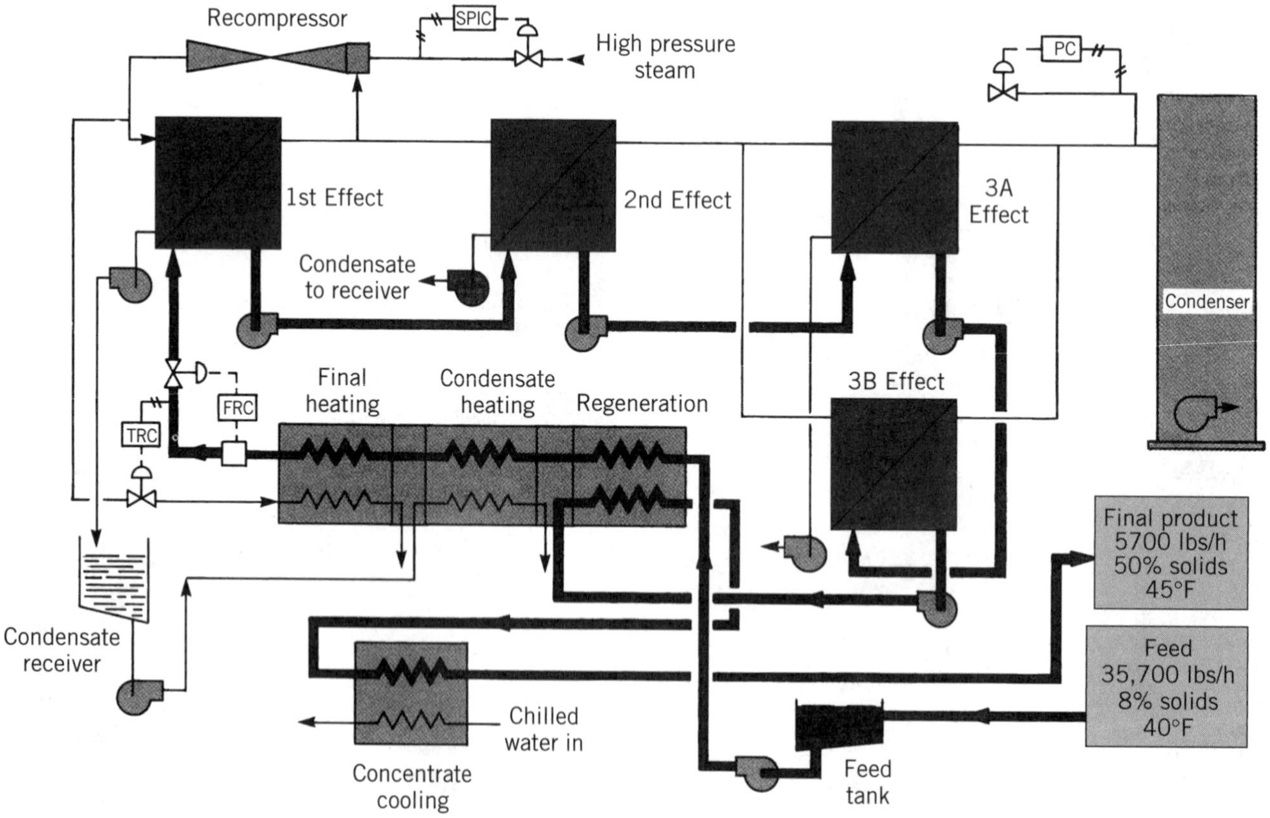

Figure 18. Three-effect steam jet recompression.

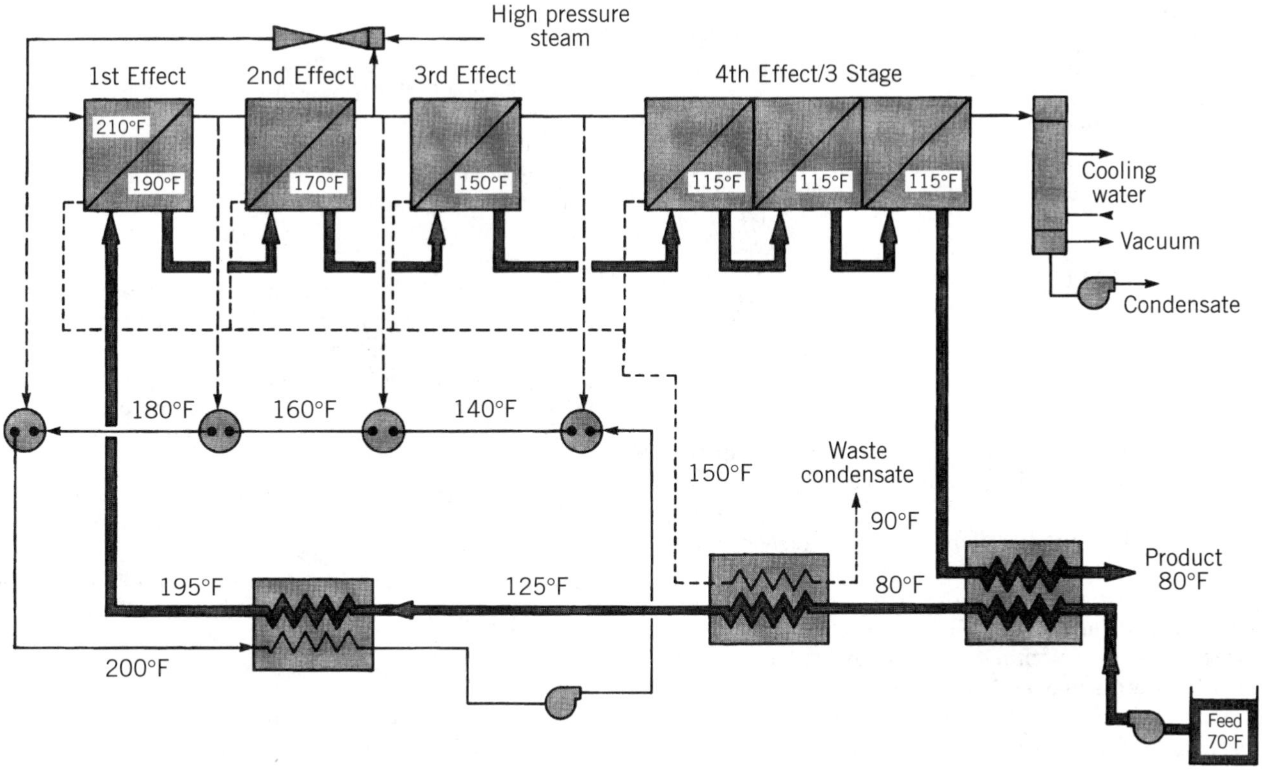

Figure 19. Four-effect thermocompression.

power even when all the gas and oil has been exhausted or is too expensive to use.

Briefly examining the thermodynamics involved with MVR, the Figure 20 schematic shows an evaporator with a liquid boiling point of 212°F (atmospheric pressure). All of the water vapor that is boiled off passes to a compressor. In order to keep the energy input to the system as low as possible, the pressure boost across the compressor is limited. In the majority of cases, this pressure boost will correspond to a saturated temperature rise in the region of 15°F or less.

In this example, there is a pressure boost of 4.5 psi across the compressor. Assuming that there is a pressure low of 0.5 psi in the system, the effective pressure on the

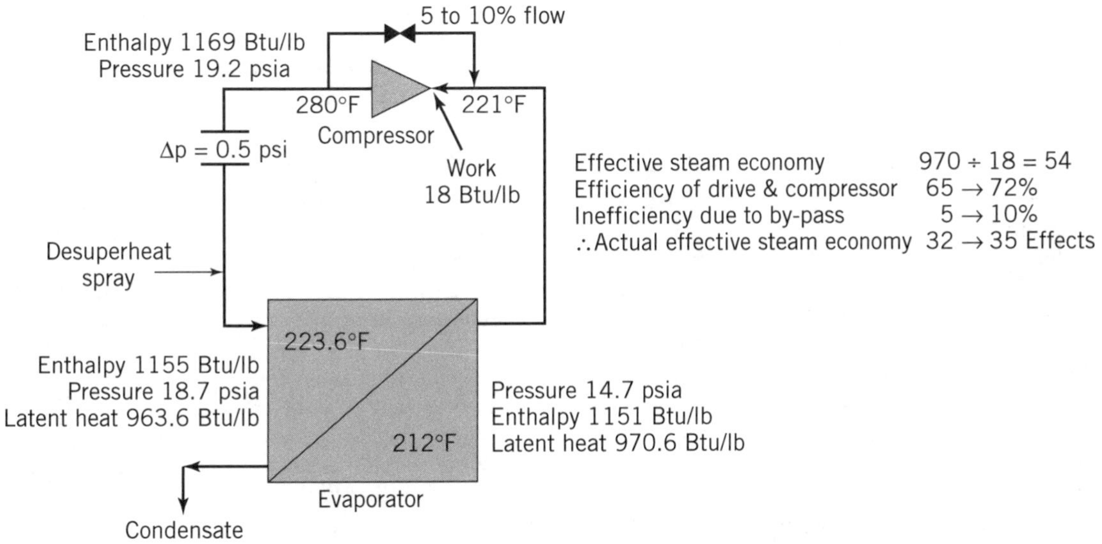

Figure 20. Mechanical recompression evaporator.

steam side of the evaporator is 18.7 psia. This compressed water vapor condenses and gives up its latent heat, thus vaporizing more water from the liquid that is being concentrated. The latent heat, thus vaporizing more water from the liquid that is being concentrated. The latent heat of vaporization of water at atmospheric pressure is 970 Btu/h. Note that it requires a theoretical energy input of only 18 Btu/lb to raise the water vapor from 14.7 to 19.2 psia. The theoretical steam economy, therefore, is 970/18 = 54. When compressor efficiency is taken into account, this figure is brought down to between 32 and 35, which is another way of saying that the MVR system is equivalent to a 32–35-effect steam evaporator. Related to energy costs of 4 cents per kilowatt-hour for electricity for the compressor drive and $6.00 per 1000 lb of steam for a conventional steam evaporator, the MVR system then becomes the economic equivalent of just under a 19-effect evaporator.

The MVR has another definite advantage over steam. The condensate is available at high temperature and is ideal for evaporator feed preheating, particularly if the condensate rate is as high as 90% of the feed rate, ie, a 10:1 concentration ratio within the evaporator. There are many such evaporators in operation where the sole energy input to the system is through the compressor with steam requirements limited to approximately 15 min during startup.

Although Figure 20 illustrates an evaporator system that uses a compressor, it will be shown that the use of a turbo fan in place of the centrifugal compressor not only yields lower boost but also provides more surface area and better operating economics.

RESIDENCE TIME IN FILM EVAPORATION

Since many pharmaceutical, food and dairy products are extremely heat-sensitive, optimum quality is obtained when processing times and temperatures are kept as low as possible during concentration of the products. The most critical portion of the process occurs during the brief time that the product is in contact with a heat-transfer surface that is hotter than the product itself. To protect against possible thermal degradation, the time—temperature relationship therefore must be considered in selecting the type and operating principle of the evaporator to be used.

For this heat-sensitive type of application, film evaporators have been found to be ideal for two reasons: (1) the product forms a thin film only on the heat-transfer surface rather than occupying the entire volume, and thus residence time within the heat exchanger is greatly reduced, and (2) a film evaporator can operate with as low as a 6°F steam-to-product temperature difference. With both the product and heating surfaces close to the same temperature, localized hot spots are minimized.

As described previously, there are rising film and falling film evaporators as well as combination rising—falling film designs. Both tubular and plate configurations are available.

Rising Film Evaporators

In a rising film design, liquid feed enters the bottom of the heat exchanger, and when evaporation begins, vapor bubbles are formed. As the product continues up either the tubular or plate channels and the evaporation process continues, vapor occupies an increasing amount of the channel. Eventually, the entire center of the channel is filled with vapor while the liquid forms a film on the heat-transfer surface.

The effect of gravity on rising film evaporator is twofold. It acts to keep the liquid from rising in the channel. Further, the weight of the liquid and vapor in the channel pressurizes the fluid at the bottom and with the increased pressure comes an increase in the boiling point. A rising film evaporator therefore requires a larger minimum ΔT than does a falling film unit.

The majority of the liquid residence time occurs in the lower portion of the channel before there is sufficient vapor to form a film. If the liquid is not preheated above the boiling point, there will be no vapor and since a liquid pool will fill the entire area, the residence time will increase.

Falling Film Evaporators

As liquid enters the top of a falling film evaporator, a liquid film formed by gravity flows down the heat-transfer surface. During evaporation, vapor fills the center of the channel and as the momentum of the vapor accelerates the downward movement, the film becomes thinner. Since the vapor is working with gravity, a falling film evaporator produces thinner films and shorter residence times than does a rising film evaporator for any given set of conditions.

Tubular and Plate Film Evaporators

When compared to tubular designs, plate evaporators offer improved residence time since they carry less volume within the heat exchanger. In addition, the height of a plate evaporator is less than that of a tubular system.

ENGINEERING CONVERSION

Table 1 provides data for engineering conversion applicable to the use of an evaporation in food processing.

Estimating Residence Time

It is difficult to estimate the residence time in film evaporators, especially rising film units. Correlations, however, are available to estimate the volume of the channel occupied by liquid. Table 2 describes a number of applicable equations. Equation 1 is recommended for vacuum systems.

For falling film evaporators, the film thickness without vapor shearing can be calculated by equation 2.

Since the film is thin, this can be converted to liquid volume fraction in a tubular evaporator by equation 3.

For a falling film plate evaporator, equation 4 is used. As liquid travels down the plate and evaporation starts, vapors will accelerate the liquid. To account for this action, the rising film correlation is used when the film thickness

Table 1. Engineering Conversions

To convert from	to	Multiply by
	Heat capacity	
Calories/(gram)(°C)	Calories/(Gram)(mole)(°C)	Molecular weight
	Btu/(pound)(°F)	1.0
	Density	
Gram/milliliter	Pounds/gallon	8.33
	Pounds/ft^3	62.42
	Thermal conductivity	
Kilocalorie/(hr)(m)(°C)	Btu(h)(ft)(°F)	0.6719
W/(m)(°C)		0.5778
	Viscosity	
Centistokes	Centipoise	Specific gravity
Dynamic viscosity pound-mass (ft)(s)	Centipoise	1488.2
Kinematic viscosity cm^2/s	Centistokes	100
	Pressure	
Kilopascal	psi	0.14504
Bar		14.504
Inches Hg absolute	psia	0.4912
Atmosphere		14.696
Torr		0.01908
mmHg		0.01908
	Enthalpy	
Calorie/gram	Btu/pound-mass	1.8
	Work / Energy	
(Kilowatt)(hr)	Btu	3412.1
(Horsepower)(hr)		2544.4
Calorie		0.003968
	Heat-transfer coefficient	
Kilocalorie/(h)(m^2)(°C)	Btu(h)(ft^3)(°F)	0.2048
W/(cm^2)(°C)		1761.1

Table 2. Estimating Residence Time

1.[a]	$R_L = 1 - \dfrac{1}{1 + \left(\dfrac{(1-y)}{y}\right)\left(\dfrac{2\rho_v}{\rho_L}\right)^{1/2}}$
2.	$m = \left[\dfrac{3\Gamma\mu}{g\rho_L(\rho_L - \rho_v)}\right]^{1/2}$
3.	$R_L = 4m/d$
4.	$R_L = 2m/z$
5.	$t = AL/R_L q_L$

[a]Symbols: A = cross-sectional area, ft^2; d = tube diameter, ft; g = 32.17 ft/s^2, L = tube length, ft; m = film thickness, ft; R_L = liquid volume fraction (dimensionless); q_L = liquid rate, ft^3/s; t = time, s; z = plate gap, ft; Γ = liquid wetting rate, lb/s · ft; ρ_L = liquid density, lb/ft^2, ρ_v = vapor density; lb/ft^3; μ = liquid viscosity, lb/ft s; y = local weight fraction of vapor (dimensionless).

Source: Refs. 1 and 2.

falls below that of a falling film evaporator. In practice, the film thickness may be less than estimated by either method because gravity and vapor momentum will act on the fluid at the same time.

Once the volume fraction is known, the liquid residence time is calculated by equation 5. In order to account for changing liquid and vapor rates, the volume fraction is calculated at several intervals along the channel length. Evaporation is assumed to be constant along with channel length except for flash due to high feed temperature.

Table 3 shows a comparison of contact times for typical four-effect evaporators handling 80 GPM of feed. The tubular designs are based on 2-in.-diameter by 30 ft. Designs using different tube lengths, incidentally, do not change the values for a rising film tubular system.

The given values represent total contact time on the evaporator surface, which is the most crucial part of the processing time. Total residence time would include contact in the preheater and separator as well as additional residence within interconnecting piping.

Table 3. Residence Time Comparison

Contact time	Rising film tubular	Rising film plate[a]	Falling film tubular	Falling film plate
1st effect	88	47[b]	23	16[b]
2nd effect	62	20	22	13
3rd effect	118	30[b]	15[b]	9
4th effect	236[b]	78[b]	123[b]	62[c]
Total contact time	*504*	*175*	*183*	*100*

[a]Plate gap of 0.3 in assumed.
[b]Two stages.
[c]Three stages.

While there are no experimental data available to verify these numbers, experience with falling film plate and tubular evaporators shows that the values are reasonable. It has been noted that equation 2 predicts film thicknesses that are too high as the product viscosity rises so, in actuality, four-effect falling film residence times probably are somewhat shorter than charted.

Summary

Film evaporators offer the dual advantages of low residence time and low temperature difference that help assure a high product quality when concentrating heat-sensitive products. In comparing the different types of film evaporators that are available, falling film designs provide the lowest possible ΔT and the falling film plate evaporator provides the shortest residence time.

Figure 21. Rising-falling film plate evaporator in final stages of fabrication.

PLATE-TYPE EVAPORATORS

To effectively concentrate an increasing variety of products that differ by industry in such characteristics as physical properties, stability, or precipitation of solid matter, equipment manufacturers have engineered a full range of evaporation systems. Included among these are a number of plate-type evaporators.

Plate evaporators initially were developed and introduced by APV in 1957 to provide an alternative to the tubular systems that had been in use for half a century. The differences and advantages were many. The plate evaporator, for example, offers full accessibility to the heat-transfer surfaces. It also provides flexible capacity merely by adding more plate units, shorter product residence time (resulting in a superior quality concentrate), a more compact design with low headroom requirements and low installation cost.

These APV plate evaporation systems are available in four arrangements—rising-falling film, falling film, Paravap, and Paraflash—and may be sized for use in new produce development or for production at pilot plant or full-scale operating levels.

Rising-Falling Film Plate

The principle of operation for the rising-falling film plate evaporator (RFFPE) involves the use of a number of plate packs or units, each consisting of two steam plates and two product plates. As shown in Figure 21, these are hung in a frame resembling that of a plate heat exchanger. The first product passage is a rising pass and the second, a falling pass. The steam plates, meanwhile, are arranged alternately between each product passage.

The product to be evaporated is fed through two parallel feed ports and is equally distributed to each of the rising film annuli. Normally, the feed liquor is introduced at a temperature slightly higher than the evaporation temperature in the plate annuli, and the ensuing flash distributes the feed liquor across the width of the plate. Rising film boiling occurs as heat is transferred from the adjacent steam passage with the vapors that are produced helping to generate a thin, rapidly moving turbulent liquid film.

During operation, the vapor and partially concentrated liquid mixture rises to the top of the first product pass and transfers through a "slot" above one of the adjacent steam passages. The mixture enters the falling film annulus where gravity further assists the film movement and completes the evaporation process. The rapid movement of the thin film is the key to producing low residence time within the evaporator as well as superior heat-transfer coefficients. At the base of the falling film annulus, a rectangular duct connects all of the plate units and transfers the evaporated liquor and generated vapor into a separating device. Steam and condensate ports connect to all the steam annuli (Fig. 22).

The plate evaporator is designed to operate at pressures extending from 10 psig to full vacuum with the use of any

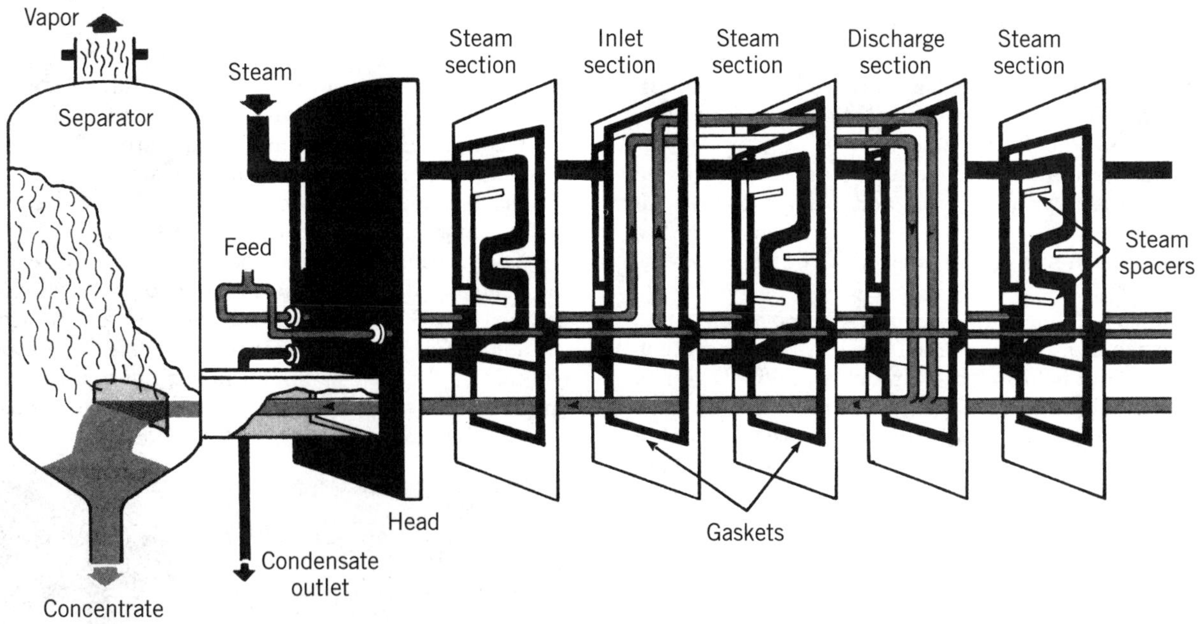

Figure 22. Rising-falling film plate evaporator arranged for one complete pass.

number of effects. However, the maximum pressure differential normally experienced between adjacent annuli during single effect operation is 15 psig. This, and the fact that the pressure differential always is from the steam side to the product side, considerably reduces design requirements for supporting the plates. The operating pressures are equivalent to a water vapor saturation temperature range of 245°F downward and thus are compatible with the use of nitrile or butyl rubber gaskets for sealing the plate pack.

Most rising-falling film plate evaporators are used for duties in the food, juice, and dairy industries where the low residence time and 100–200°F operating range temperatures are essential for the production of quality concentrate. An increasing number of plate evaporators, however, are being operated successfully in both pharmaceutical and chemical plants on such products as antibiotics and inorganic acids. These evaporators are available as multi-effect and/or multi-stage systems to allow relatively high concentration ratios to be carried out in a single pass, nonrecirculatory flow.

The rising-falling film plate evaporator should be given consideration for various applications.

- That require operating temperatures between 80 and 210°F
- That have a capacity range of 1000–35,000 lb/h water removal
- That have a need for future capacity increase since evaporator capabilities can be extended by adding plate units or by the addition of extra effects
- That require the evaporator to be installed in an area that has limited headroom
- Where product quality demands a low time-temperature relationship.
- Where suspended solid level is low and feed can be passed through 50 mesh screen

A "junior" version of the evaporator is available for pilot plant and test work and for low-capacity production. If necessary, this can be in multieffect-multistage arrangements such as the system illustrated in Figure 23.

Falling Film Plate

Incorporating all the advantages of the original rising—falling film plate evaporator system with the added bene-

Figure 23. Three-effect, four-stage, "junior."

fits of shorter residence time and larger evaporation capabilities, the falling film plate evaporator has gained wide acceptance for the concentration of heat sensitive products. With its larger vapor ports, evaporation capacities typically are up to 50,000–60,000 lb/h.

The falling film plate evaporator consists of gasketed plate units (each with a product and a steam plate) compressed within a frame that is ducted to a separator. The number of plate units used is determined by the duty to be handled.

As shown in Figure 24, one important innovation in this type of evaporator is the patented feed distribution system. Feed liquor first is introduced through an orifice (1) into a chamber (2) above the product plate where mild flashing occurs. This vapor/liquid mixture then passes through a single product transfer hole (3) into a flash chamber (4), which extends across the top of the adjacent steam plate. More flash vapor results as pressure is further reduced and the mixture passes in both directions into the falling film plate annulus through a row of small distribution holes (5). These ensure an even film flow down the product plate surface where evaporation occurs. A unique feature is the ability to operate the system either in parallel or in series, giving a two-stage capability to each frame. This is particularly advantageous if product recirculation is not desirable.

In the two-stage method of operation, feed enters the left side of the evaporator and passes down the left half of the product plate, where it is heated by steam coming from the steam sections. After the partially concentrated product is discharged in the separator, it is pumped to the right side of the product plate where concentration is completed. The final concentrate is extracted while vapor is discharged to a subsequent evaporator effect or to a condenser.

Paravap

The operating principle of the Paravap represents an application of the thin film, turbulent path evaporation process and in many cases, replaces the wiped or agitated film evaporator. Since it has no mechanical moving parts within the heat transfer area, costly maintenance repairs common to wiped film systems are eliminated. It is especially designed to concentrate liquids with high solids contents or non-Newtonian viscosity characteristics (Fig. 25).

With feed liquor and the heating medium of steam, hot water or hot oil directed into alternate passages within a plate heat exchanger, boiling begins as the liquid contacts the heated plates. The small plate gap and high-velocity flow pattern (Fig. 26) atomizes the feed and provides a greater liquid surface area for mass transfer. Since the vapor carries the feed in the form of minute particles, the apparent viscosity within the heat exchanger plate pack is

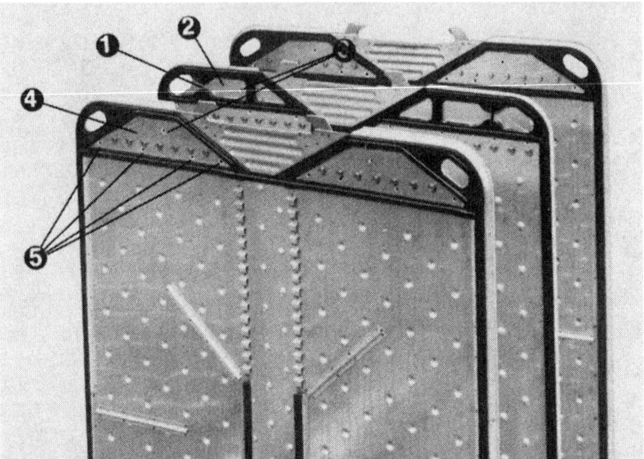

Figure 24. Typical product and steam plate unit for falling film plate evaporator.

Figure 25. Single-effect R86 Paravap.

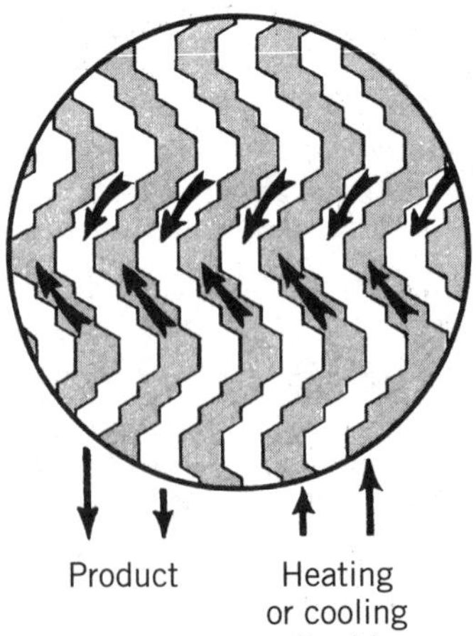

Figure 26. Small plate gap and high-velocity flow pattern.

very low. The final product, however, may be extremely viscous after separation from the vapor.

Advantages include low residence time to minimize thermal degradation, low rates of fouling, and economical operation.

Typical applications: concentrating soap to final product consistency; concentrating apple puree from 25 to 40° brix, grape puree to 50° brix, cherry puree to 60° brix, fruit juices to 95% solids, and sugar or corn syrup solutions to over 97%; solvent stripping duties such as hexane from oil (See also Figs. 25 and 27.)

Paraflash

Operating under a suppressed boiling, forced circulation principle, the APV Paraflash is used to concentrate products subject to fouling too excessive for film evaporators. Unlike the Paravap, vaporization does not occur within the heat-exchanger plate pack. Instead, liquor flashes as it enters a separator, crystallization takes place, and a suspended slurry results. Suppressed boiling combined with high liquid velocities deters scaling on the heat-transfer surface, minimizes cleaning downtime, and promotes longer production runs.

The Paraflash can be used in single or multiple effects for such products as grape juice (tartrate crystals), coffee, wheat starch, distillery effluent, and brewer's yeast (suspended solids). (See Fig. 27).

Figure 27. Single-effect, forced circulation Paraflash.

MECHANICAL VAPOR RECOMPRESSION EVAPORATORS

In recent years, MVR technology has been introduced and widely accepted as an effective approach to powering medium to large-capacity evaporators for both chemical and sanitary applications. While experience has shown that the higher capital cost of this equipment relative to that of steam-driven evaporation systems has been offset by significant energy savings, advances in MVR technology now have reduced energy requirements even further.

Definition

Simply stated, with mechanical recompression the water vapor boiled off in the evaporator is passed to an electrically powered compressor and is compressed through 1–3 psi. This raises the temperature of the vapor, which then is used as the heating medium. The difference in enthalpy between the vapors on the heating and process sides is comparatively small, with a resultant reduction in energy consumption and, depending on regional steam and power costs, an operating cost equivalent to at least an 8–30-effect evaporator. Theoretical thermal efficiency may exceed that of a 60-effect steam evaporator.

Typically, Figure 28a shows a single-effect steam evaporator operating at atmospheric pressure (14.7 psia) with vapor being produced at 212°F and sent to a condenser. The heat source for this system is steam at 17.2 psia, condensing at 220°F.

In Figure 28b, the same evaporator is shown operating with mechanical vapor compression. In this case, the vapor at 212°F, 14.7 psia, is sent to a compressor, where its pressure is raised to 17.2 psia. It then is charged to the steam side of the evaporator as the heating medium, condensing at 220°F. Savings are realized in two areas: no steam is required for evaporation, although nominal amounts are required for startup and compressor seals, and since the vapor from the evaporator is not sent to a condenser, cooling water requirements are dramatically reduced.

It will be shown later that excess vapor is produced gas a result of the inefficiencies in the compression cycle. This excess vapor is used to compensate for vent and radiation losses and at times, as an assist to preheating.

Development

To a large extent, the design of MVR evaporators has been dictated by the capabilities of the compressors available at the time. The primary limitation has been the compression ratio (absolute discharge pressure divided by absolute suction pressure) that can be achieved since this determines the temperature difference available for the evaporator. For example, if an evaporator is run at 212°F and atmospheric pressure and a compression ratio of 1.4 is used, the steam pressure at the discharge of the compressor is 1.4 × 14.7 psia or 20.58 psia. Since water vapor at 20.58 psia condenses at 229.5°F, the allowable temperature difference ΔT for the evapo-

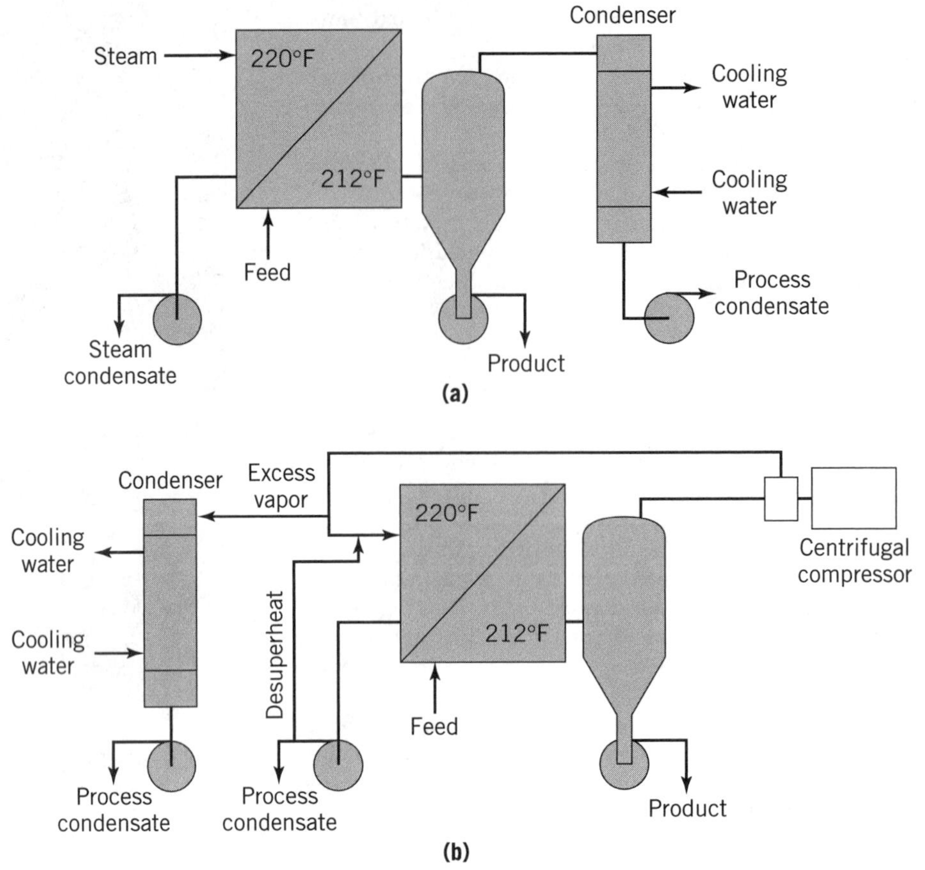

Figure 28. (**a**) Single-effect evaporator. (**b**) Single-effect MVR evaporator.

rator is 17.5°F before losses are considered. Table 4 shows the total temperature differences for a number of conditions.

Centrifugal Compressors

The relatively high volume of vapor encountered in evaporators requires the use of centrifugal compressors in most cases. Initially, centrifugal compressors on steam were limited to a compression ratio of 1.4, which, in turn, limited the available temperature difference to between 13 and 17.5°F depending on boiling temperature. Only film evaporators could use MVR since there was not sufficient temperature difference to consider forced-circulation designs. Furthermore, products having significant boiling point elevation were excluded for this technique.

Two trends developed in recent years, however, have increased MVR capabilities and applications.

First improved design allows centrifugal compressors to run at higher speeds, increasing compression ratios approximately to 2.0 to 1. Consequently, a (ΔT of 27–36°F now can be obtained and thus, MVR can be used with forced-circulation systems and for the evaporation of products with boiling-point elevations.

Secondly, advances in evaporator design allow the operation of film evaporators with a lower (ΔT), thereby minimizing energy requirements. For some products, film evaporators can be calculated with a (ΔT (excluding losses) in the range of 6°F.

By increasing the ΔT available and decreasing the ΔT required, it also became possible to design multiple—effect MVR evaporators. Figure 29 illustrates a double-effect MVR system with a forced-circulation finisher operating across all effects. With this arrangement, the flow to the compressor is reduced and the finisher, operating at the higher concentration, takes advantage of the full ΔT available.

Fans

The next development in MVR design was an apparent reversal of prior advances.

Table 4. Saturated ΔT(°F) at Various Compression Ratios

	Boiling temperatures		
Compression ratio	130°F	170°F	212°F
1.2	6.9	8.0	9.3
1.4	12.9	15.0	17.5
1.6	18.2	21.2	24.7
1.8	23.0	26.8	31.2
2.0	27.3	31.9	37.2

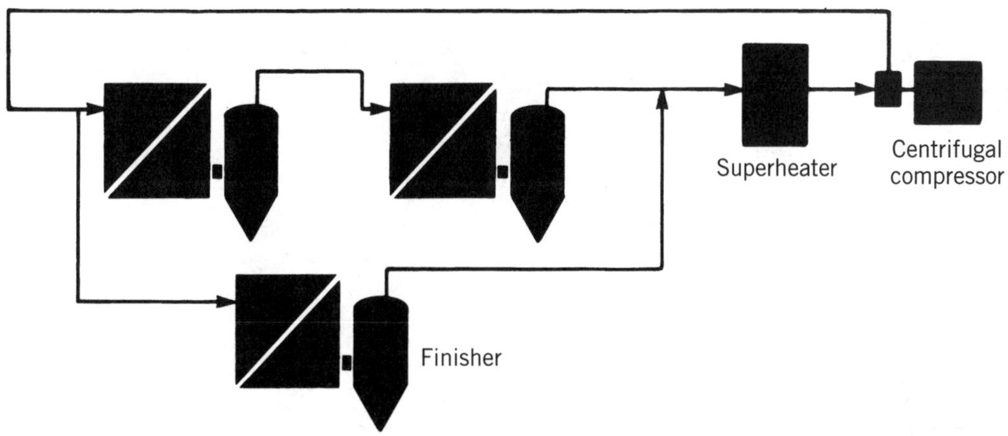

Figure 29. Multiple-effect MVR, centrifugal compressor with finisher.

As the required ΔT across a film evaporator decreased, it became possible to provide that ΔT by using a fan. This produces a compression ratio on the order of 1.2 to 1, providing approximately 7 or 8°F ΔT (before losses) for evaporation. The MVR system previously discussed and shown in Figure 30 is typical of fan use. It should be noted that the fan MVR technique can be used only on single-effect film evaporators handling product without significant boiling point elevation. Where the heat-transfer coefficient of the product is low, it sometimes is better to make use of the higher ΔT available from a centrifugal compressor.

Where fans can be used, however, the horsepower requirement usually is less than in centrifugal compressor designs. Furthermore, fans do not require that the inlet vapor be superheated. Instead, this is done either by a separate heat exchanger using steam or by recycling some of the compressor discharge vapor to the compressor suction.

It is, incidentally, a common practice to use a fan evaporator to concentrate a product up to a point where a large ΔT is required and then to switch to a small steam or MVR finishing evaporator for final concentration.

Other Designs

Rotary blowers are low-capacity positive compressors that occasionally are used with small MVR evaporators of up to approximately 15,000-lb/h capacity. This generally is a finishing-type evaporator.

In some cases, two compressors will be arranged in series as is shown in Figure 30. Here, the bulk of the evaporation is done in a single effect MVR system having a centrifugal compressor. Some of the compressor discharge vapor then is further compressed in a blower in order to operate a finishing evaporator.

Estimating Compressor Power Requirements

To calculate estimated power requirements for an MVR compressor power and ΔT values for both centrifugal compressors and fans have been plotted in Figures 31 and 32, respectively. These values compare reasonably well with installed MVR systems.

Table 5, meanwhile, compares the power requirements for different MVR designs. Note that it is possible to calculate an equivalent steam economy by converting horsepower to Btu/h and multiplying by 2545. This value then is divided by the latent heat to arrive at the equivalent steam used.

Thermodynamics

To comprehend the potential of MVR evaporation fully, it is important that the thermodynamics of this technique be

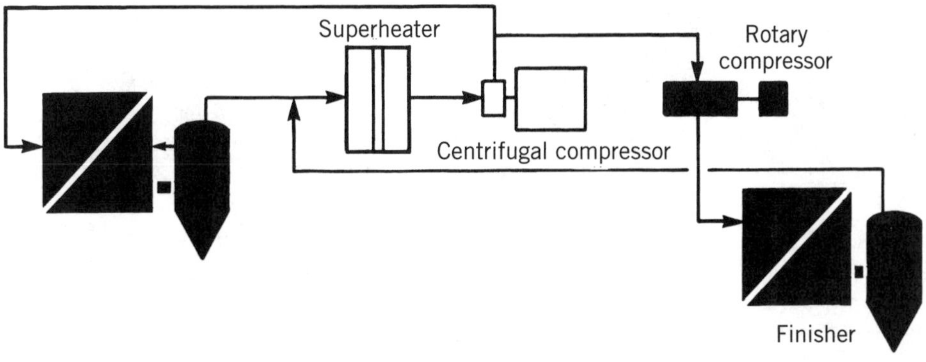

Figure 30. Single-effect MVR evaporator with centrifugal compressor and rotary-blower-driven finisher.

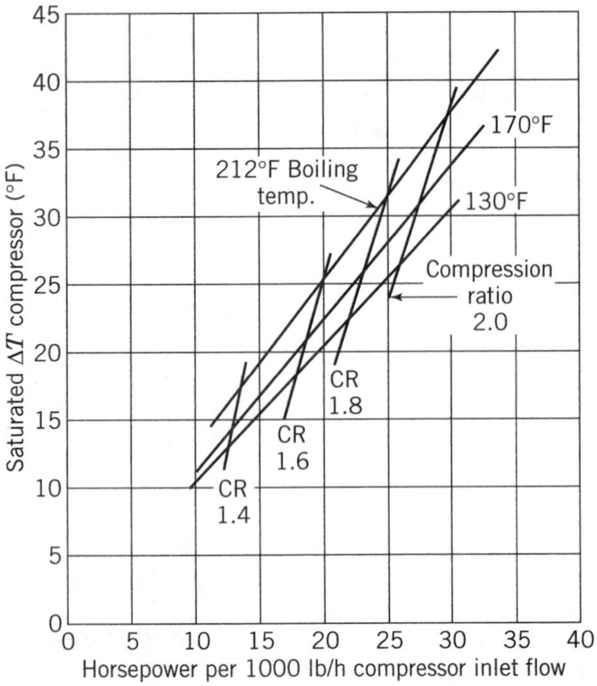

Compressors								
Boiling temperature 130°F			Boiling temperature 170°F			Boiling temperature 212°F		
CR	ΔT	HP	CR	ΔT	HP	CR	ΔT	HP
1.3	10.00	9.39	1.3	11.65	10.01	1.3	13.53	10.67
1.4	12.91	12.18	1.4	15.03	12.99	1.4	17.47	13.85
1.6	18.21	17.38	1.6	21.23	18.54	1.6	24.69	19.76
1.8	22.98	22.15	1.8	26.81	23.63	1.8	31.20	25.18
2.0	27.31	26.57	2.0	31.89	28.34	2.0	37.16	30.20
2.2	30.30	30.69	2.2	36.57	32.74	2.2	41.31	34.88

Figure 31. Horsepower versus ΔT for centrifugal compressors.

examined. This analysis involves Table 6 as well as the steam and MVR evaporator schematics shown in Figures 28a and 29b.

The first two columns of Table 6 give the properties of saturated water vapor 14.7 psia, 212°F and at 17.2 psia, 220°F. These values can be used to estimate the energy requirements for an evaporator operating at these conditions.

Looking first at the steam evaporator shown in Figure 29a, water at 17.2 psia, 220°F has an enthalpy (heat con-

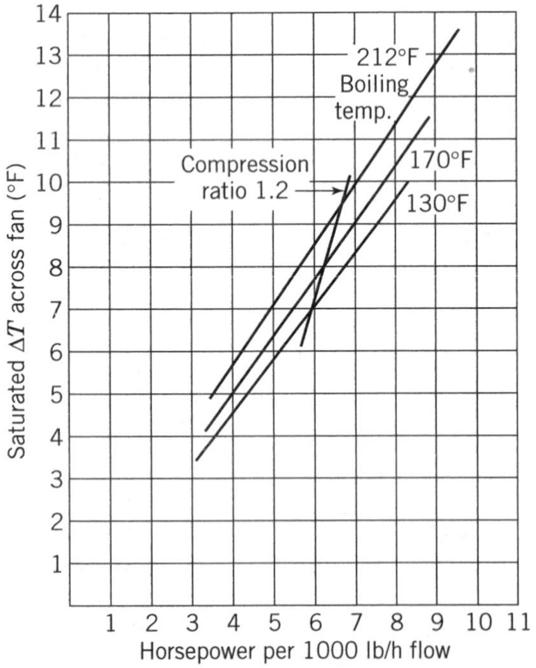

Fans								
Boiling temperature 130°F			Boiling temperature 170°F			Boiling temperature 212°F		
CR	ΔT	HP	CR	ΔT	HP	CR	ΔT	HP
1.1	3.59	2.99	1.1	4.17	3.19	1.1	4.85	3.41
1.2	6.91	5.79	1.2	8.04	6.18	1.2	9.34	6.60
1.3	10.00	8.43	1.3	11.65	9.00	1.3	13.53	9.60

Figure 32. Horsepower versus ΔT for fans. *Note:* Vapor assumed to be superheated 10°F by external steam.

Table 5. Comparison of Typical MVR Designs (Approximate)[a,b]

	Single effect fan	Double effect centrifugal	Triple effect centrifugal
Compression ratio	1.2	1.6	2.0
Total ΔT available, °F	6.9	18.2	27.3
Vapor to compressor, lb/h	60,000	30,000	20,000
Horsepower/1000 lb h vapor flow	5.79	17.38	26.57
Total horsepower	374.4	521.4	531.4
Equivalent Btu	884,133	1,326,963	1,352,413
Equivalent steam, lb/h	870	1315	1347
Equivalent steam economy	69	45.6	44.5
Average ΔT per effect before losses, °F	6.9	9.1	9.1

[a]Boiling temperature 130°F; evaporation rate 60,000 lb/h.
[b]*Note:* In this example, the fan horsepower is lower than either of the centrifugal designs, but the lower ΔT required the greater the surface area.

Table 6. Properties of Water Vapor

Pressure, psia	14.7	17.2	17.2
State	Saturated	Saturated	Saturated
Temperature, °F	212	220	243
H-enthalpy vapor, Btu/lb	1150.5	1153.4	1164.6
Latent heat, Btu/lb	970.3	965.2	—
H-enthalpy liquid, Btu lb	180.2	188.2	—
S-entropy Btu/lb	1.7568	1.7442	1.7596

tent) of 188.2 Btu/lb while steam at the same conditions has an enthalpy of 1153.4 Btu/lb. The difference between the vapor and liquid is the latent heat or 965.2 Btu/lb. In other words, in order to produce one pound of steam at 220°F, 17.2 psia, from water at 220°F, 17.2 psia, the addition of 965.2 Btu/lb of energy is required.

In the case of the MVR evaporator, however, the vapor at 17.2 psia, 220°F (enthalpy 1153.4 Btu/lb) is produced from vapor at 212°F, 14.7 psia (enthalpy 1150.5 Btu/lb). Theoretically, only 2.9 Btu/lb of energy must be added.

The compressor, a fan in this case, operates such that the entropy of the discharge vapor be at least as high as the entropy of the inlet vapor. Since inefficiencies in the compression cycle will result in an exit entropy above that of the inlet entropy, the temperature of the existing vapor and energy input to the compressor must be increased.

Figure 33 is an enthalpy—entropy diagram for water vapor. The vapor at 14.7 psia, 212°F, is shown at point *A*. During compressions the entropy remains constant (ideally) or increases (actually). With typical efficiencies for this duty, the discharge temperature may be expected to rise to 243°F where the enthalpy is 1164.6 Btu/lb. The energy input from the compressor is 14.1 Btu/lb.

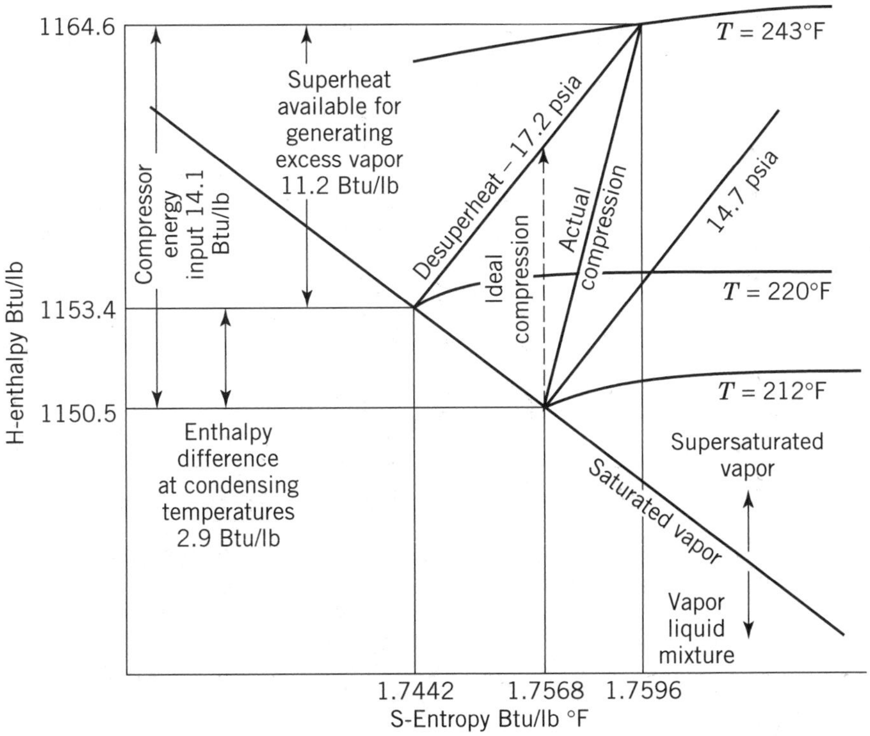

Figure 33. Enthalpy/entropy diagram.

In order to cool the vapor to its condensing temperature of 220°F, 11.2 Btu/lb of heat is removed. This can be done by introducing condensate at 220°F from the steam side of the evaporator. The 11.2 Btu/lb of heat removed from the vapor is absorbed into the condensate, some of which will vaporize and give off what is known as "excess vapor." For every pound of vapor cooled, 11.2 Btu of heat is absorbed in the condensate, which requires 965.2 Btu/lb to boil. Therefore, for each point of vapor leaving the compressor, 11.2/965.2 or 0.0116 lb of excess vapor is available.

This excess vapor is used in several ways. Since there is a slight difference in latent heat between the steam and the vapor, slightly more steam is required than the vapor generated. Other excess vapor is used to cover losses due to radiation and venting, is made available in some instances for preheating, or is sent to a condenser. It is significant to note that the condenser on an MVR evaporator is responsible only for vent and excess vapors. This results in a much lower cooling requirements than is necessary for steam evaporators.

It is possible to calculate an equivalent steam economy for an MVR system. In this example, for every pound of water evaporated, 970.3 Btu is absorbed. The compressor supplies 14.1 Btu but with motor and gear losses, probably requires 14.5 Btu of energy. The equivalent economy (970.3/14.5) is 67 to 1. Since one horsepower is equivalent to 2545 Btu/h, the compressor in the example requires 14.5/2545 or 0.0057 hp/lb · h of evaporation.

It should be noted that pressure losses through the evaporator that must be absorbed by the compressor have not been considered in this example. These losses would be taken into account by either higher compressor horsepower or lower ΔT over the heat-transfer surface.

BIBLIOGRAPHY

Adapted from:

P. Worrall, "Selection and Design of Evaporators," *CPI Digest* **4**(4), 2–10 (Feb. 1977); **5**(4), 2–9 (March 1979).

P. Worrall, "Residence Time in Film Evaporation," *CPI Digest* **8**(3), 17–19 (Dec. 1986).

P. Worrall, "Plate Type Evaporators," *CPI Digest* **8**(4), 12–15 (July 1987).

P. Worrall, "Tubular Evaporators and 3A Sanitary Standards," *APV Evaporator Handbook*, Nov. 1977, pp. 24–26.

P. Worrall, "Advances in MVR Evaporator Technology," *APV Evaporator Handbook*, Nov. 1989, pp. 24–26.

1. HTR1 report, May 1978, p. 7.
2. *Perry's Chemical Engineer's Handbook*, 5th ed., pp. 5–57.

APV Crepaco, Inc.
Lake Mills, Wisconsin

EXTRACTION

All engineering studies of food processes are related to the basic unit operations of mass transfer and/or heat transfer. These principles are applied to the numerous procedures of transporting, preparing, processing, packaging, and distributing of basic and value-added foods. Secondary unit operations include techniques of separating components, reducing particle size, mixing ingredients, and concentrating desired components. The unit separation operations, distillation, evaporation, dehydration, and filtration all involve the extraction of a component from a liquid, gas, or solid by physical or chemical means. However, each category listed has such different applications of scientific and engineering principles that it is too unwieldy to place them all under a combined unit operation. Thus, due to these complexities, the procedures have been divided into numerous categories related to the mechanisms of separation. This greatly simplifies the scientific and engineering studies of various unit operations and the subsequent engineering design, manufacture, and integration of the process facilities into an overall food process. Various separations of components in a basic harvested food or a partially processed product are the most important sequences of processing a food to a finished product for the market. The unit operation of extraction is considered to be the removal or separation of a component by material that has greater affinity for the component being removed.

Extraction, the separating of a component from a liquid or solid by another liquid is handled as a separate unit operation since the process is based on diffusion of one component from the base material to an extracting liquid. Since the mechanics of these processes are based on diffusion, study of the processes and design of extracting equipment is based on the diffusivities of the solute being removed from the base material. This differs from other separation processes such as distillation that involve a change of phase in one or more components requiring different scientific principles. The original applications of extraction processes began many years ago with application of gas absorption and solvent extraction in the industrial chemical industry. It has been in relatively recent years that the technology has been extensively applied to separating components in food (1).

Extraction as practiced in the food industry is essentially the operation of removing or separating a component from the food to ensure food safety or to alter the properties. A rudimentary form of extraction begins with the basic separation or removal of components from an as-received food. With the growing problems of microbiological contamination in harvested fruits, vegetables, and animals, the extraction operation of washing as-received raw materials has received much emphasis over the past few years. The simple washing of a vegetable or fruit is necessary to ensure that contamination from soil, fertilizers, living organisms, pesticides, and so on is removed or at least reduced to an acceptable level, making the vegetable safe to eat. Proper washing is particularly important in slaughtered animals since extremely dangerous microorganisms such as *Escherichia coli* and *Salmonella* contaminate many of them. A much more complex procedure for removing or extracting a component is found in the process for producing vegetable oils. For example, many vegetable and seed oils are extracted from the base material by solvent extraction. A solvent in which the corn oil is soluble contacts a product, such as ground corn. This results in an oil-rich solvent with little of the solvent-insoluble products remaining in the oil fraction. The oil is then removed from the solvent by other separation operations such as distillation. Extraction processes can be carried out on a batch basis or as continuous steady-state operation. As is the case with most unit operations, the continuous operation is the easiest to analyze and design. Continuous operations are particularly important to maximize the operating costs and efficiency of extraction processes used in the food industry (1–3).

The basis for success of extraction processes is the difference in affinity for one component or material over another. For example, water has little affinity for vegetable oil, but the oil is completely soluble in an organic solvent. Hence, when an oil-containing food raw material, such as soybeans or corn, is placed in contact with water, the water will not absorb any oil. Oil will only be released if the food is in hot water, such as during cooking, when the structure of the food is changed so that oil is released. In this case the oil is not extracted by water but the water acts as a vehicle for washing out the oil and floating it to the surface.

Extraction during food processing is involved with the diffusion mass transfer of one component of the food being extracted, or leached, in the solvent phase. Although the extraction is enhanced by thorough contact or mixing, the controlling factor is the diffusion as measured by the diffusivity of the component being removed from the base material. There are two primary unit operations to consider under extraction of foods. The first is a liquid–solid extraction in which a soluble component is removed from the solid by a liquid. This process is known as leaching. The second, liquid–liquid extraction involves the removal of a component in a liquid by another liquid.

There are several special extraction processes whereby the extraction liquid changes phase during the process. However, the basic diffusivity of the condensing extracting material is still the basis for the component transfer rate. For example, steam is used for stripping volatile materials from both solids and liquids. Thus, the condensing steam is actually the solvent phase that either remains a vapor or changes from a vapor to a liquid during the extraction process. Some extractions, such as the use of hexane in extracting oil from soybeans, utilize a vapor solvent that condenses during the extraction and is recovered as an oil-rich liquid solvent.

PRINCIPLES OF EXTRACTION

The extraction rate at which a component (solute) is transferred from the material (solid or liquid) phase being treated to a solvent phase is the controlling factor involved in extraction. This is related to the mass-transfer coefficient (K_LA) as determined by the following conditions of the process and properties of the raw material, solute, and solvent:

- The properties of the solvent as related to its affinity for the solute
- The immiscibility of the raw material with the solvent
- The particle size of the solid and thus the exposed surface area and depth of solvent penetration required (for leaching only)
- Temperature of the liquids
- Degree of contact between liquids, which is a function of the agitation or equipment design

The driving force is the difference between the saturation concentration of the solute in the raw material and the solvent ($C_s - C$). Hence, the rate of extraction (amount per unit time, dN/dt) would be

$$dN/dt = K_LA(C_s - C) \quad (1)$$

Extraction involves three separate operations: (*1*) mixing the raw material and the solvent to bring them into intimate contact, (*2*) separating the two phases following contact, and (*3*) removing the solute from the solvent so that the solvent can be recycled. Often the economic viability of a process depends on the third step of removing the solute and reclaiming the solvent. In many cases the solvent is relatively expensive (eg, organic solvent) and good separation and recycling is imperative. When the solvent in leaching is water and because the solute is often a waste material, the solvent would not be recycled.

LIQUID–SOLID EXTRACTION (LEACHING)

As a solvent comes in contact with the solid material, the solvent becomes richer in the solute that is being removed from the solid. The rate at which the solvent becomes

Table 1. Examples of Commercial Leaching Operations

Raw material	Solvent	Solute	Product
Sugar beets	Water	Sugar	Refined sugar
Sugar beets	Water	Sugar	Refined sugar
Corn, rapeseed, cottonseed, soybeans	Organic solvent, nonpolar (hexane)	Oil	Edible oil (vegetable)
Vanilla beans	Supercritical carbon dioxide	Vanilla	Vanilla extract
Coffee beans	Supercritical Carbon Dioxide	Caffeine	Decaffeinated coffee
Roasted coffee beans	Water	Insoluble fraction	Soluble coffee constituents
Vegetables	Steam	Debris	Cleaned vegetables
Fish and meat flesh	Nonpolar solvent (dichloroethane)	Oil	Refined fish oil
Fish and meat flesh	Polar solvent (isopropyl alcohol)	Water-oil	Oil, protein concentrate
Fish and meat flesh, enzyme hydrolysis	Water plus enzymes	Proteins	Water-soluble proteins

Table 2. Examples of Commercial Liquid–Liquid Extraction Operations

Raw material	Solvent	Solute	Product
Oil seed for oil recovery (ie, cotton seed, soybeans)	Organic solvent (ie, propane, hexane)	Fatty acid (oleic)	Refined vegetable oil
Fish oil	Hot water	Low-molecular-weight odor and taste components	Semirefined fish oil
Fish oil	Supercritical carbon dioxide	Cholesterol	Low-cholesterol high omera-3 refined fish oil
Butterfat in milk or cream	Supercritical carbon dioxide	Cholesterol	Low-cholesterol milk or cream

richer in the solute depends on the factors in equation 1: contact time, mass-transfer coefficient for the specific system, area of the solid exposed to the solvent, and concentration driving force. Diffusion in the solid is most often the rate-controlling factor. Table 1 shows examples of typical commercial processes in which a food is leached by a solvent to remove a specific solute. Note that the product or products can be the removed solute (eg, leaching sugar beets, vegetable oil, etc), the solid material remaining after extraction (eg, decaffeinated coffee, cleaned vegetables, etc), or both the remaining solid and solute (eg, fish oil and protein concentrate).

A diagram of a semicontinuous solid–liquid extraction system is shown in Figure 1. The pure solvent enters the extractor and leaves rich in solute while the solid that has been placed in the extractor decreases in solute soluble components. The rate of reaction decreases during the process as the solid is depleted in solute. Because fresh solvent is passing over the solid, the extraction can continue until essentially all solute is removed from the food. If the process is made continuous by conveying the solid into and out of the reactor in one stage, the extraction normally will not be complete because the solid–solvent exposure time is reduced and equilibrium is not reached. Throughput can be increased and extraction can be more complete when two or more extractors are used in series. Usually, a series of batch-continuous reactors are connected and fresh solvent is used for each reactor while the solid is moved down the reactor series. Figure 2 shows this multiextraction process in which three reactors are in series. If the remaining solid is one of the end products, it is necessary to remove the residual solvent. Often this secondary process turns out to be another extraction process such as steam stripping. If the solvent is water, the normal process is to remove it from the solid by a drying operation.

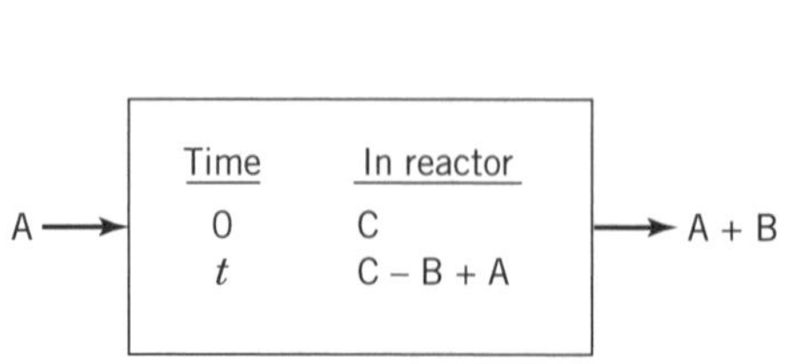

Figure 1. Batch solid–liquid extraction (leaching). A, Incoming pure solvent; A, outgoing solvent rich in solute (B); B, solute leached from solid (C); C, original solid rich in solute (B); C, extracted solid depleted of solute with retained solvent (A); t, extraction time.

A process whereby components of a food material are concurrently extracted while being made soluble by microbial and chemical action complicates leaching. A good example of this type of process is the batch hydrolysis of minced fish flesh processing waste by enzyme action. The water-insoluble components are changed into soluble (digested) materials. The resultant water-based protein solution can be dried to produce a high-quality protein product for both animal and human consumption. This technique is a good means for using the tremendous amounts of unused industrial fish, by-catch, underutilized edible species, and waste from processing whole fish (4).

LIQUID–LIQUID EXTRACTION

Liquid–liquid extraction operations are like solids leaching except the food, solvent, and solute are all in the liquid state. Most commercial processes are normally operated on a continuous basis. Table 2 shows the solvent, solute, and product produced from extracting some liquid raw materials. As shown in Figure 3, the ordinary extraction process involves countercurrent flow to maintain the maximum driving force. However, the relationship between the liquids sometimes makes a cocurrent operation more advisable.

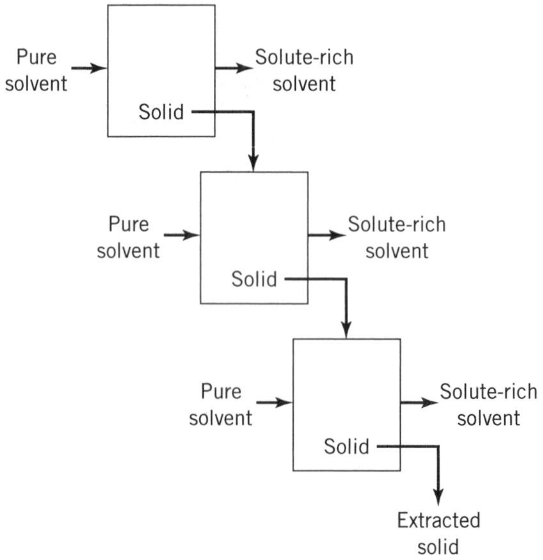

Figure 2. Batch-continuous three-phase solid–liquid extraction.

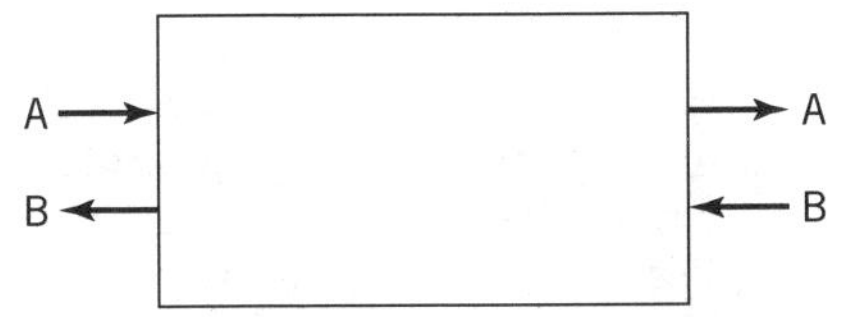

Figure 3. Continuous liquid–liquid extraction. A, Incoming food product rich in solute (S); A, outgoing food depleted of solute (S); B, incoming pure solvent; B, outgoing solvent rich in solute (S); S, solute originally in food.

The amount of solvent remaining in the food stream is dependent on the partial miscibility of the two liquids. Often the solute is recovered from the solvent as a viable product, or economics and environmental factors dictate that the solvent must be recycled. These goals are accomplished by steam stripping, evaporation, distillation, adsorption or chemical means, depending on the requirements and the materials.

SPECIAL EXTRACTION OPERATIONS

Over the past two decades, a considerable amount of scientific and applied research effort has been applied to the development of supercritical extraction. Many uses have been found for application of this technology (see the article SUPERCRITICAL FLUID TECHNOLOGY). Processes have been developed for extraction using supercritical fluids as the solvent. Carbon dioxide has been chosen as the supercritical fluid because it is nontoxic, less expensive than many organic solvents, leaves no residue of chemicals, and has a molecular weight and reaction behavior that is desirable. The extraction process using CO_2 is carried out at approximately 35 to 49°C and several thousand pounds per square inch (psi).

Decaffeination of coffee and tea were the first commercial processes carried out using CO_2 in a supercritical state, reducing initial caffeine content of 0.7 to 3% to as low as 0.02%. This was followed by processes for extracting hops and high-quality flavorings such as vanilla, ginger, paprika, rosemary, sage, and celery. Spices from supercritical extraction have been found to have superior antioxidant properties to some of the synthetic products in current use (5).

There is much interest in replacing hexane for extracting and refining oils from plant, animal, and seed sources. Interest ranges from eliminating the possible carcinogenicity of hexane residues to concentrating desired omega-3 fatty acids in fish oils. As development of critical extraction proceeds, new uses and improved economics of operation should see considerable increase in the commercial application of this technology.

BIBLIOGRAPHY

1. D. R. Heldman, *Food Process Engineering*, AVI Publishing, Westport, Conn., 1977, pp. 325–337.
2. G. M. Pigott, "Fish Processing," in *Encyclopaedia Britannica*, Vol. II, Encyclopaedia Britannica, Chicago, Ill., 1996.
3. G. M. Pigott, "Marine Oils," in Y. H. Hui, ed., *Baileys Industrial Oil and Fat Products*, Vol. III, 5th Ed., John Wiley & Sons, New York, 1996, pp. 225–254.
4. G. M. Pigott, "The Status and Future of Aquatic Food Research in the United States," *Proceedings of the International Seafood Research Meeting*, Dept. of Chemistry of Fishery Resources, Mie University, Tsu, Mie, Japan, September 30, 1994.
5. D. D. Duxbury, "High-Quality Flavor and Color Extracts Derived by Supercritical Extraction," *Food Processing* **50**, 50–54 (1989).

GEORGE PIGOTT
Sea Resources Engineering, Inc.
Kirkland, Washington

EXTRUSION

DEFINITION AND HISTORY OF USAGE

Extrusion is a process that combines several unit operations that include mixing, kneading, shearing, heating, cooling, shaping, and forming. This process also involves compressing and working a material to form a semisolid mass under a variety of controlled conditions, and then forcing it to pass through a restricted opening such as a shaped hole or slot at a predetermined rate (1). The first known record of extrusion was in 1797 when Joseph Bramah used a piston-driven device to manufacture seamless lead pipes (2). This innovation was adapted later by the food industry for the production of macaroni products. The first twin-screw extruder was developed in 1869 by Follows and Bates for sausage manufacture (2). A single-screw extruder, with shallow flights and rotating in a stationary cylindrical barrel, was developed for the Phoenix Gummiwerke A.G. in 1873 (3).

In the mid-1930s, forming extruders were used to mix semolina flour and water to form pasta products. A few years later, extrusion technology that combined transport, mixing, and shaping operations was used to produce the first ready-to-eat breakfast cereal from precooked oat-flour dough (4). In the late 1930s, Robert Colombo and Carlo Pasquetti developed a corotating and intermeshing twin-screw extruder for mixing cellulose without the use of a solvent (5). In the middle of the 1940s, the first extrusion-cooked, expanded food products, corn snacks, were commercially produced using single-screw extruders (6). Dry expanded pet food, dry expanded ready-to-eat breakfast cereals, and textured vegetable protein, produced by single-screw extruders, were introduced in the 1950s, 1960s, and 1970s, respectively (4,6). In the early 1970s, Creusot-Loire had developed the twin-screw plastics extruder for food applications. Finally, the 1980s have brought rapid commercialization of the production of feeds for aquatic species using the process of extrusion (6).

FOOD APPLICATIONS

Many different food products have been produced directly or indirectly by extrusion. Applications of extrusion have been classified into two categories: semifinished products and finished products (7). For semifinished products, extrusion cooking has been shown to offer economic advantages over the traditional process such as drum drying for producing pregelatinized cereal flours, potato starch, and cereal starches. By controlling the processing conditions to achieve the desired balance of gelatinization and molecular degradation, starches, and chemically modified starches with a wide range of cold water solubility values can be produced (8). Extensive patent literature has been covered on extruded breakfast cereals (9), snack foods (10) and textured foods (11–13). Other applications include dry and soft-moist pet foods, precooked and modified starches, flat bread, breadings, croutons, full-fat soy flour, precooked noodles, beverage bases, soup and gravy bases, and confections such as licorice, fruit gums, and chocolate (4,14).

GENERAL DESCRIPTION OF EQUIPMENT

Extrusion equipment can be classified thermodynamically or by pressure development in the extruder (15). According to the former classification, there are (*1*) autogenous extruders, which convert mechanical energy into heat energy during the flow process; (*2*) isothermal extruders where constant temperature is maintained throughout the extruder; and (*3*) polytropic extruders that operate between the extreme conditions of (*1*) and (*2*). If classified by the manner in which pressure is developed in the extruder, there are (*1*) direct or positive displacement types, include the ram or piston-type extruders and the intermeshing counterrotating twin-screw extruders; and (*2*) indirect or viscous drag types, which include roller extruders, single-screw extruders, intermeshing corotating twin-screw extruders, and nonintermeshing multiple-screw extruders.

The components of an extruder have been well described (4,16). They consist of drive, feed assembly, screw, barrel, die head assembly, cutters, and take-away system. Figure 1 shows a twin-screw extruder system manufactured by Werner & Pfleiderer with various components. Either AC or DC types of motors can be used to rotate the extruder screw in the barrel. Depending on the capacity, the size of the motors can be as small as a few horsepower for a laboratory-type extruder or several hundred horsepower for a commercial extruder used for full-scale production of extruded foods. Except for simple extruders such as the collet type, most extruder drive motors have a transmission and a speed variation device for controlling the screw rotation speed. Since the screw pushes food forward against the die, which builds pressure that could exceed hundreds of atmospheric pressure, a thrust bearing is required to support and center the screw and absorb the thrust exerted by the screw.

The feed assembly consists of dry ingredient and liquid ingredient feeders. Preblended dry ingredients are held in hoppers or bins above feeders, which can be vibratory, variable speed auger, or loss-in-weight type. Liquid ingredients are metered through positive displacement pumps, variable orifice, variable head, or water wheels. If a preconditioner is used, dry ingredients are mixed with water, steam, or other ingredients in a closed vessel, which can be operated under pressure if needed. The uniform feeding of dry and liquid ingredients is imperative for the consistent operation of an extruder.

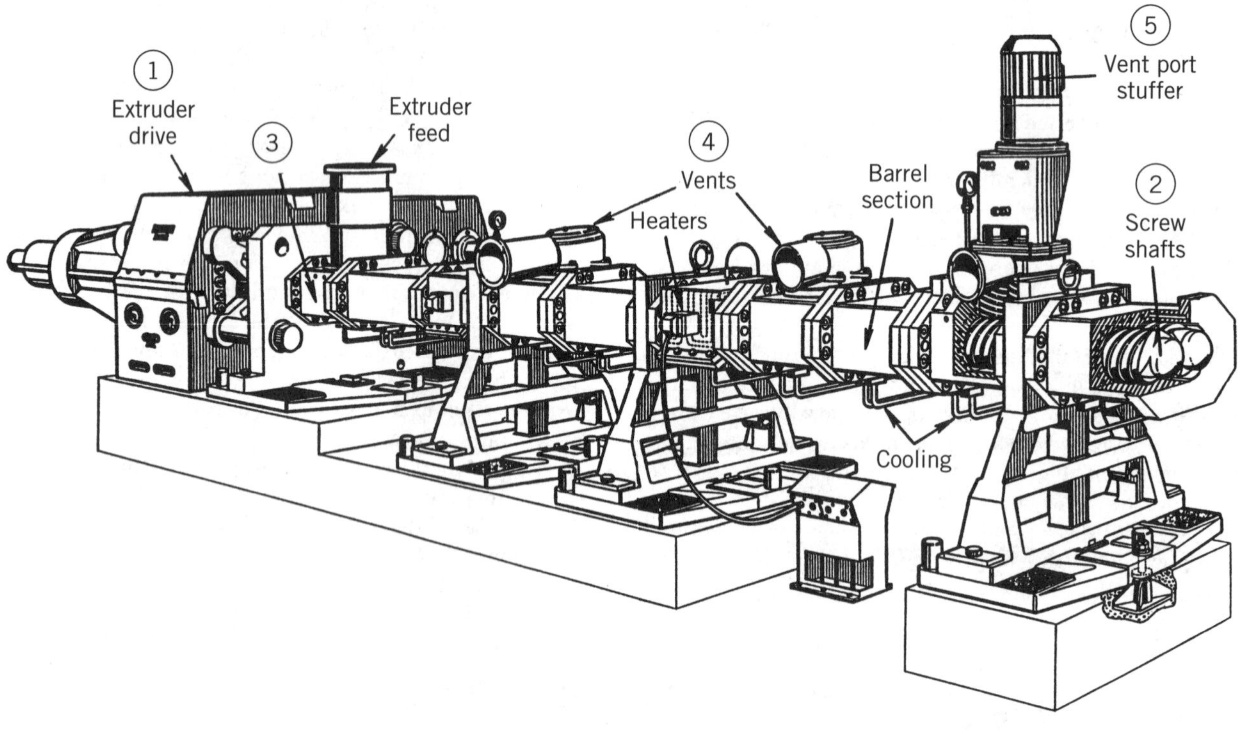

Figure 1. A Werner-Pfleiderer corotating, intermeshing twin-screw extruder. *Source:* Courtesy Krupp Werner & Pfleiderer, Ramsey, N.J.

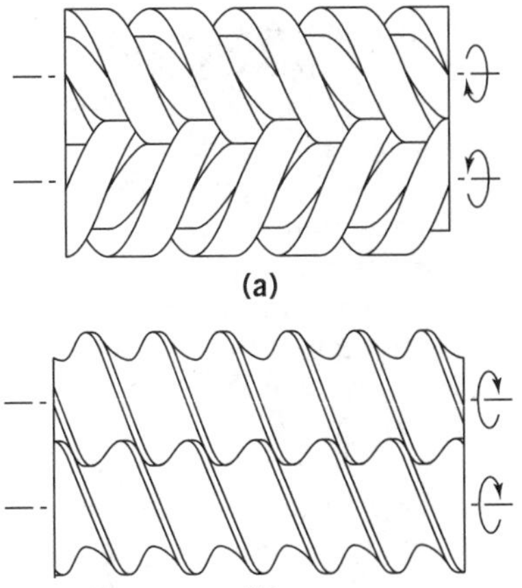

Figure 2. Schematic diagram of fully intermeshing counter- and corotating extruder screws. (**a**) Counterrotating screws and (**b**) corotating screws.

Figure 3. Four pairs of kneading disks are shown with corotating twin-lead screws.

The screw can be one piece or consist of many individual screw elements assembled in a screw shaft. The screw is usually divided into feed, transition, and metering sections for single-screw extruders. The feed section has deeper flights to accept the food ingredients from the feeder. In the transition section, the food ingredients are mixed, heated, and worked into a continuous mass. Thus, the transition section is also called the compression section because the materials are changed from a loose, powdery state into a plasticized dough. This is accomplished by (*1*) gradually decreasing flight depth or pitch in the direction of discharge, (*2*) heating from the barrel, and (*3*) working of the feed material to generate frictional heat. The metering section has shallow flights that increase the shear rate; therefore, the temperature of material increases rapidly in this section.

Twin-screw extruders are classified by the direction of screw rotation, corotating or counterrotating, and by the depth of screw engagement—fully intermeshing, partially intermeshing, and nonintermeshing (16,17). The schematic diagram of fully intermeshing counter- and corotating extruder screws is shown in Figure 2. For fully intermeshing twin-screw extruders, the flight depth is always constant and equal to the distance of screw overlap. The screw design consists of single-, twin-, and triple-lead types, which refers to the number of helical-shaped flights parallel along the length of the screw. To improve mixing and increase the conversion of mechanical energy into heat, another type of screw design, kneading disks (Fig. 3), is usually used.

The barrel fits tightly around the rotating screw and is assembled by bolting or clamping several segments together. An important parameter for the specification of the extruder is its L/D ratio, which is the extruder barrel length divided by the extruder barrel bore diameter. The barrels and their liners, in particular, are usually constructed of special hardened alloy to become wear resistant. The interior surface of barrel is either smooth or grooved. The presence of grooves increases the pumping capability of the extruder against high back pressure (4).

The die head assembly is located at the end of the extruder barrel and holds the extruder die plate and sometimes serves as support for the cutter. The die plate can hold many dies, which shape the food product before it emerges from the extruder. The dies are small openings that can be round, annular, slit, or of specially designed shape, such as alphabets or animals, and so on.

BIBLIOGRAPHY

1. J. D. Dziezak, "Single- and Twin-Screw Extruders in Food Processing," *Food Technol.* **43**, 164 (1989).
2. L. P. B. M. Janssen, *Twin Screw Extrusion*, Elsevier Science, New York, 1978.
3. P. Linko, P. Colonna, and C. Mercier, "High-Temperature, Short-Time Extrusion Cooking," in Y. Pomeranz, ed., *Advances in Cereal Science and Technology*, Vol. IV, AACC, St. Paul, Minn., 1981, pp. 143–235.
4. J. M. Harper, *Extrusion of Foods*, Vol. I & II, CRC Press, Boca Raton, Fla., 1981.
5. F. G. Martelli, *Twin-Screw Extruders*, Van Nostrand Reinhold, New York, 1983.
6. B. W. Hauck and G. R. Huber, "Single Screw vs Twin Screw Extrusion," *Cereal Foods World* **34**, 930 (1989).
7. W. Wiedmann, and E. Strobel, "Technical and Economic Advantages of Extrusion Cooking," *Technische Mitteilungen Krupp* **2**, 95 (1987).
8. M. G. Fitton, "Extruded Starches—Product Analysis, Structure and Properties," in G. O. Philips, D. J. Wedlock, and P. A. Williams, eds., *Gums and Stabilizers for the Food Industry 3*, Elsevier Applied Science, New York, 1985, pp. 213–220.
9. R. Daniels, *Modern Breakfast Cereal Processes*, Noyes Data Corporation, Park Ridge, N.J., 1970.
10. M. H. Gutcho, *Prepared Snack Foods*, Noyes Data Corporation, Park Ridge, N.J., 1973.
11. M. H. Gutcho, *Textured Foods and Allied Products*, Noyes Data Corporation, Park Ridge, N.J., 1973.

12. M. H. Gutcho, *Textured Food Products*, Noyes Data Corporation, Park Ridge, N.J., 1977.
13. M. H. Gutcho, *Textured Protein Foods*, Noyes Data Corporation, Park Ridge, N.J., 1977.
14. O. B. Smith, "Extrusion Cooking of Corn Flours and Starches as Snacks, Breadings, Croutons, Breakfast Cereals, Pastas, Food Thickeners, and Additives," in G. E. Inglett, ed., *Maize: Recent Progress in Chemistry and Technology*, Academic Press, New York, 1982, pp. 193–219.
15. J. L. Rossen, and R. C. Miller, "Food extrusion," *Food Technol.* **27**, 46 (1973).
16. J. M. Harper, "Food Extruders and Their Applications," in C. Mercier, P. Linko, and J. M. Harper, eds., *Extrusion Cooking*, AACC, St. Paul, Minn., 1989, pp. 1–15.
17. J. Fichtali and F. R. van de Voort, "Fundamental and Practical Aspects of Twin Screw Extrusion," *Cereal Foods World* **34**, 921 (1989).

FU-HUNG HSIEH
University of Missouri
Columbia, Missouri

EXTRUSION COOKING

BACKGROUND

Extrusion cooking generally refers to the combination of heating food products in an extruder to create a cooked and shaped product. Raw materials, such as flours, starches, proteins, salt, sugar, and other minor ingredients, are mixed, kneaded, cooked, and worked into a plasticlike dough mass. Heat can be applied directly by steam injection or indirectly through heated barrel or through the conversion of mechanical energy. The final process temperature in the cooking extruder can be as high as 200°C, but the residence time is relatively short, 10 to 60 s (1). Thus, extrusion cooking is also called a high-temperature, short-time (HTST) process.

The first cooking extruder was developed in the late 1940s and led to a great expansion of the application of extruders in the food field (2). Cooking extruders have many different sizes and shapes, which allow a wide range of moisture contents (10–40%), feed ingredients, cooking temperatures, and residence times. In addition, the ability for extruders to vary the screw, barrel, and die configurations for the cooking and shaping requirements of different food product makes extrusion cooking one of the most versatile food processes.

EFFECTS OF PROCESSING PARAMETERS

The process parameters for extrusion generally include screw speed, feed moisture, feed rate, and barrel temperature. Changes in these processing parameters will change the motor torque, die pressure, and product temperature. Visual product characteristics such as expansion ratio, specific length (length of extrudate per unit mass), and bulk density also will change.

Increases in screw speed at constant feed rate will cause a quick initial increase in die pressure and motor torque, followed by a sharp decrease to the new equilibrium value (3). The initial quick increase in die pressure and motor torque is due to the effort of the motor drive to overcome the inertia of the screw shaft and the momentary increase in the output. Because of a rapid decrease in the degree of fill due to screw speed increase, both torque and die pressure fall sharply after the initial increase (4,5). The specific energy input increases with increasing screw speed at constant feed rate; therefore, the product temperature increases. Extrudates are usually lower in expansion ratio, higher in specific length, and lower in bulk density with increasing screw speed (4,5).

Increasing feed moisture lowers the torque, specific mechanical energy input, and product temperature (3). Die pressure drops initially due to a lower dough viscosity caused by moisture increase. However, as the product temperature gradually becomes lower, the dough viscosity rises, causing die pressure to recover. Final equilibrium die pressure will depend on the relative effect of feed moisture and product temperature on the dough's viscosity. Extrudates are generally lower in expansion ratio and specific length, but higher in bulk density with a higher feed moisture (6).

An increase in feed rate causes rapid increases in motor torque and die pressure and a slower decrease in product temperature (3,7). Increases in feed rate while maintaining constant screw speed increase the degree of fill in the extruder barrel (8) and, thus, raise motor torque and die pressure (9). Since the specific energy input is decreased with increasing feed rate, the product temperature decreases. Increasing feed rate usually increases both expansion ratio and specific length, resulting in a lower bulk density (10).

Decreasing the barrel temperature causes the product temperature to decrease. A lower product temperature corresponds to a lower dough temperature or a higher dough viscosity and, therefore, a higher motor torque and die pressure (3). The physical characteristics of extrudates from lower barrel temperatures are usually similar to those from higher feed moistures.

CHANGES IN FOOD AFTER EXTRUSION

Since extrusion cooking is an HTST process, it changes the physical and chemical properties of food. Many methods have been developed and applied to characterize extruded products. Expansion of extrudates in both longitudinal and lateral directions can be evaluated by direct physical measurements. Apparent bulk density is determined by weighing the extrudates held in a container with a fixed volume. Bulk density or specific volume can be obtained using sand or rapeseed displacement methods. Scanning electron micrographs have been used to examine the extrudate's internal porous structure. Crispness, crunchiness, hardness, and other texture attributes can be determined objectively with a texture analyzer such as Instron or subjectively with human sensory evaluation methods. The color of extrudates can be assessed with a colorimeter. Flavor compounds formed can be determined with a GC-MS.

Many analytical tests have been developed to evaluate extruded products. For tests on cereal or starchy-type products, there are the water absorption index (WAI), water solubility index (WSI), amylograph, and gelatinization tests. The WAI is the weight of gel obtained per gram of dry product after dispersing the dry product into water, centrifuging, and decanting the supernatant. The WSI is the amount of dried solid recovered by evaporating the supernatant from the WAI test. The amylograph test determines the viscosities of a slurry from the finely ground, extruded product through a preprogrammed heating and cooling cycle.

Gelatinization of starches is one of the most important changes that occurs in cereal types of foods during extrusion. Starch, in the raw or native state, exists in the form of granules with many different shapes, ranging from round, oval to irregular, with sizes between 1 and 100 μm depending on the sources of starch. As water and heat are added during extrusion, the starch granules swell and the amylose in the granules begins to diffuse out. Eventually, the starch granules, consisting mostly of amylopectin, collapse and become a colloidal gel held in a matrix of amylose. For different types of starch, the temperature range at which gelatinization occurs differs according to the sources of starch. Many methods have been developed to determine the degree of starch gelatinization after extrusion. Some commonly used methods are (*1*) loss of birefringence (crystallinity) in polarized light, (*2*) X-ray diffraction patterns, (*3*) differential scanning calorimetry (DSC), and (*4*) enzyme susceptibility.

For proteinaceous-type products, two indices are of interest: the protein dispersibility index (PDI) and the nitrogen solubility index (NSI). While the latter measures the percentage of total nitrogen in extrudate that is soluble in water, the former determines the amount of protein in the extrudate that is dispersible in water; both are conducted under controlled conditions of extraction (2). As the amount of heat treatment during extrusion increases, more protein in the extrudate will be denatured, which will result in decreases in both the PDI and NSI. Similar to a cereal-type product, texture is also an important quality attribute for an extruded proteinaceous product. Both sensory and instrumental methods have been developed. The extrudate can also be examined by utilizing scanning electron microscopy to reveal the structural integrity of the protein matrix. Changes in proteins during extrusion can be briefly summarized as follows: (*1*) physically, extrusion converts protein bodies into a homogeneous matrix; and (*2*) chemically, the process recombines storage proteins in some way into structured fibers (11).

Besides starch gelatinization and protein denaturation, many other changes may occur during an extrusion process, such as nonenzymatic browning, enzyme inactivation, destruction of antinutritional factors, and microorganisms.

HTST EXTRUSION COOKERS

HTST extrusion cookers can be classified into two main categories: single-screw and twin-screw cooking extruders. The latter is further divided into corotating and counterrotating twin-screw extruders. Within the same type of cooking extruders, the mechanical features and operating characteristics may differ substantially from each other since there are many manufacturers supplying each type of cooking extrusion equipment. The following is a partial list of manufacturers that supply the HTST extrusion cookers:

Single-screw cooking extruders: Anderson, Bonnot, Buss, Egan, Johnson, Mapimpianti (Mapa), Sprout-Waldron, and Wenger

Corotating twin-screw cooking extruders: APV Baker, Buhler, Clextral, Wenger, and Werner-Pfleiderer

Counterrotating twin-screw cooking extruders: Brabender, Cincinnati Milacron, and Textruder

A more complete list can be found in a recent article (12).

METHODOLOGY FOR HTST EXTRUSION COOKING

Before beginning the HTST extrusion cooking process, the extruder and its related equipment must be assembled first. All fasteners should be tight and all parts aligned. If there is a torque specification for the retaining bolts, it must be closely followed using a torque wrench. After the barrel and screw(s) are assembled, the screw should be turned manually to ensure that it moves freely. The die plate with its die inserts is then bolted to the end of the extruder barrel. The rotating cutter assembly is installed and the cutting knives are adjusted to provide a cleanly cut product and an extended life for the blades. Water and steam lines are then connected to the extruder barrel as required. Condensate in the steam line must be removed. All feeders and regulators should be checked. If a volumetric feeder is used for dry ingredients, it should be calibrated first for each batch of feed materials. A typical Wenger extrusion cooking system is shown in Figure 1.

To minimize the amount of waste produced during the start-up, the cooking extruder is usually brought to operating condition and equilibrium as quickly as possible. This can be accomplished by preheating the barrel and die plate to the desirable operation temperature. The water pump is then started at a rate about 1.5 times the target operating rate. Once the water is in the extruder, start the extruder and the feed stream at a low rate. Increase the screw speed and feed rate, but reduce the water rate to operating levels gradually. When the barrel and die have reached stable operating temperatures, adjust the water rate, feed rate, and screw speed to obtain the desired product at a moderate torque level (90% or less motor load).

Similar to the start-up procedures, shutdown procedures for HTST cooking extruders must also be followed in sequence. Water is gradually increased first to allow uncooked material to purge the cooked material from the system. Barrel temperatures are reduced to stop cook reactions, which can prevent burning on after shutdown. When the percent load meter readings are below 25%, decrease the screw speed and shut off the extruder; in addition, shut off the feeder and water pump. The die plate is then removed by gradually and carefully undoing the bolts because some pressure may still exist behind the die plate.

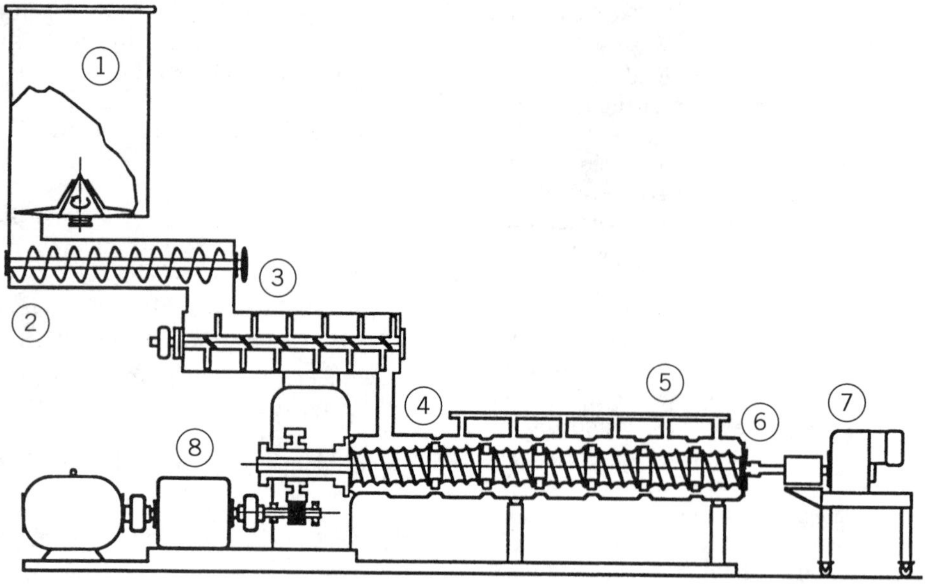

Figure 1. A Wenger extrusion cooking system. 1, Bin discharger; 2, feeder; 3, blender/preconditioner; 4, extruder, 5, steam injection ports; 6, die and variable speed cutter; 7, variable speed drive for cutter; 8, gear drive. *Source:* Courtesy of Wenger, Sabetha, Kansas.

APPLICATIONS AND DESCRIPTION FOR EXTRUSION COOKING OF DIFFERENT CATEGORIES OF PRODUCTS

Breakfast Cereals

There are four types of breakfast cereals that can be produced by HTST extrusion cooking process: (*1*) extrusion flaked cereals, (*2*) extrusion gun-puffed cereals, (*3*) extrusion shredded cereals, and (*4*) extrusion expanded cereals (13). Traditional flaked cereals are produced by flaking cooked grits. Extruded flaked cereals are different from traditional flaked cereals in that the grit for flaking is formed by extruding the mixed ingredients with a feed moisture of 25 to 35%. After exiting from the die holes, the product is cut off at the desired size and forms pellets. The pellets are then dried slightly and tempered before flaking. This process allows additional flexibility in the ingredients, such as multigrains or brans, used for the manufacture of flaked breakfast cereals.

Similar to extrusion flaked cereals, the first step of manufacturing extrusion gun-puffed and extrusion shredded cereals is to form pellets with the HTST cooking extruders. The pellets are dried and tempered before gun-puffing or tempered only before shredding. Thus, the pellet moisture content for flaked and gun-puffed cereals is usually lower than that for shredded cereals. After flaking, gun-puffing, or shredding, cereals may be sugarcoated if desired and then dried or toasted in an oven to impart cereals with a desirable toasted color and crisp or crunchy texture.

Extrusion expanded cereals differ from the other three types of extruded cereals in that the cereals are expanded into a porous matrix during the HTST extrusion cooking process. A typical moisture content in feed is about 15 to 25%, depending on the type of extruder and the extrusion conditions used. The extruded cereals are then sugarcoated, if desired, and then dried or toasted in an oven as with the other three types of extruded cereals.

Snacks

Many advances have been made over the years since corn collet was commercially developed in the 1940s. For example, extruded snacks now come in many different shapes other than the common collets or curls and in many different flavors other than basic cheese/oil such as products from potato, wheat, rice flour, or brans along with starches or modified starches (2).

Collet snacks can be divided into two types, baked or fried, depending on the moisture of the collets (14). Baked corn collets are extruded using a collet extruder at low moisture (<15%) to produce highly expanded products from degerminated corn grits. After extrusion, the collets are dried (baked) in an oven and coated with flavor and oil. Fried collets are extruded from corn meal that is moistened to a higher moisture (20–30%). Product expansion takes place during the deep fat frying step instead of during extrusion. The latter partially cooks and forms the product shape.

In addition to corn meal or corn grits, potato and other grain-based extruded snacks have been developed. Some examples are French fried potato stick, flatbread, rice cake, and multigrain snacks. Similar to collet snacks, they can be divided into either baked or fried types. Although the collet extruder can be used in some cases, other snacks, such as crispbread, may require more sophisticated extruders such as the HTST single-screw or twin-screw cooking extruders.

Extruded half-products are another type of snack products. Potato and other starches or modified starches are extruded to form very dense and thin pellets that are dried at a low temperature and controlled humidity environment for storage stability. The half-products need to be expanded by deep fat frying or baking and then enrobed with flavorings before consumption.

The latest development in the extruded snacks involves dual or coextrusion of two different components to form a single piece of snack using a specially designed die. Two extruders may be needed although, occasionally, one of the extruders may be substituted by a high-pressure gear pump if the second component is pumpable. This allows the development of many new snacks, for example, dual-textured snack products. The flow rates of the two extruders and the viscosity of the two components must be well regulated.

Texturized Protein Meatlike Products

According to Harper (2), two types of texturized meatlike products are derived from extrusion processes. The first is the meat extender, which can be produced by a single HTST extrusion cooking step using defatted soy flours or soy grits mixed with a variety of additives. The resulting product is highly expanded and shows distinct fiber formation. Once hydrated, the products can be used to extend ground meat or meat products. This product has found extensive applications in pizza toppings, meat sauces, and fabricated food formulations.

The second type of texturized meatlike product is a meat analog that can be used to replace meat. The finished product must be dense and have a layered fiber formation similar to that found in real meat. Additional requirements include maintaining a meatlike character after extensive cooking or retorting and giving the appearance and taste of real meat. The manufacture of meat analogs usually requires multiple extrusion processes or specially cooled dies (15).

Breadings and Croutons

The manufacture of breadings by either single-screw or corotating twin-screw extruders has been described (16,17). The starting materials are corn flour, wheat flour, or the second clears of wheat flour. Leavening agents, dry skim milk, salt, emulsifier, and other food additives may be added. Processing conditions during extrusion can be varied to obtain breadings that are functionally similar to traditional breadings. The extrudate is wet milled and sized prior to drying. Croutons are made in a similar manner. The small cubes are sized at the die of the extruder using a variable speed knife. Thus, the wet milling stage for the extrudate is not needed.

Pasta

The raw materials used in pasta products is semolina, the purified middlings of durum wheat, or durum flour, water, eggs and other ingredients (18). Commercially, pasta products are formed by the extruder or continuous press with a deep flighted screw. The screw not only forces the dough through the die, but it also kneads the dough into a homogeneous mass. A very accurate control of semolina flour and water are needed. Traditional pasta product uses low-temperature extrusion (<50°C). If the temperature of the dough exceeds 60°C, the cooking quality of the finished product will be damaged (2,18). Thus, excessive heat generated from mechanical energy input must be removed by circulating cooling water around the extruder barrel.

BIBLIOGRAPHY

1. E. W. Schuler, "Twin-Screw Extrusion Cooking Systems for Food Processing," *Cereal Foods World* **31**, 413 (1986).
2. J. M. Harper, *Extrusion of Foods*, Vol. I and II, CRC Press, Boca Raton, Fla., 1981.
3. Q. Lu et al., "Model and Strategies for Computer Control of a Twin-Screw Extruder," *Food Control* **4**, 25 (1993).
4. F. Hsieh, I. C. Peng, and H. E. Huff, "Effects of Salt, Sugar and Screw Speed on Processing and Product Variables of Corn Meal Extruded With a Twin-Screw Extruder," *J. Food Sci.* **55**, 224 (1990).
5. F. Hsieh et al., "Twin-Screw Extrusion of Rice Flour With Salt and Sugar," *Cereal Chem.* **70**, 493 (1993).
6. B. W. Garber, F. Hsieh, and H. E. Huff, "Influence of Particle Size on the Twin-Screw Extrusion of Corn Meal," *Cereal Chem.* **74**, 656 (1997).
7. M. K. Kulshreshtha and C. A. Zaror, "An Unsteady State Model for Twin Screw Extruders," *Trans. I. ChemE., Part C, Food and Bioproducts Proc.* **69**, 189 (1992).
8. G. Della Valle, J. Tayeb, and J. P. Melcion, "Relationship of Extrusion Variables With Pressure and Temperature During Twin-Screw Extrusion Cooking of Starch," *J. Food Eng.* **6**, 423 (1987).
9. F. Martelli, *Twin-Screw Extruders*, Van Nostrand Reinhold, New York, 1983.
10. M. Liang et al., "Barrel-Valve Assembly Affects Twin-Screw Extrusion Cooking of Corn Meal," *J. Food Sci.* 59, 890 (1994).
11. D. W. Stanley, "Protein Reactions During Extrusion Processing," in C. Mercier, P. Linko, and J. M. Harper, eds., *Extrusion Cooking*, American Association of Cereal Chemists, St. Paul, Minn., 1989, pp. 321–341.
12. "Extruder Directory," *Feed Tech* **2**, 32 (1998).
13. R. B. Fast, "Manufacturing Technology of Ready-to-Eat Cereals," in R. B. Fast and E. F. Caldwell, eds., *Breakfast Cereals and How They Are Made*, American Association of Cereal Chemists, St. Paul, Minn., 1990, pp. 15–42.
14. S. A. Matz, "Extruding Equipment," in S. A. Matz, ed., *Snack Food Technology*, AVI, Westport, Conn., 1984, pp. 203–230.
15. A. Noguchi, "Extrusion Cooking of High-Moisture Protein Foods," in C. Mercier, P. Linko, and J. M. Harper, eds., *Extrusion Cooking*, American Association of Cereal Chemists, St. Paul, Minn., 1989, pp. 343–370.
16. O. B. Smith, "Extrusion Cooking of Corn Flours and Starches as Snacks, Breadings, Croutons, Breakfast Cereals, Pastas, Food Thickeners, and Additives," in G. E. Inglett, ed., *Maize: Recent Progress in Chemistry and Technology*, Academic Press, New York, 1982, pp. 193–219.
17. P. L. Noakes and W. A. Yacu, "Extrusion Cooking of Wheat Flour to Process Breadings," *Cereal Foods World* **33**, 687 (1988).

18. O. J. Banasik, "Pasta Processing," *Cereal Foods World* **26**, 166 (1981).

FU-HUNG HSIEH
University of Missouri
Columbia, Missouri

See also EXTRUSION PROCESSING: TEXTURE AND RHEOLOGY.

EXTRUSION PROCESSING: TEXTURE AND RHEOLOGY

In recent years extruders have been widely used in the human food and animal feed industry to manufacture products such as textured proteins, snack foods breakfast cereals, pasta, confectioneries, and pet foods. A list of products presently produced using the extrusion process is given in Table 1. The history of food extrusion has been reviewed in Ref. 2.

The two major functions of a food extruder are cooking and forming. In the cooking process, the product is heated through the transfer of heat energy, which is applied by steam injection or heaters and/or by the dissipation of viscous energy through shearing action between the rotating screw and the material. As a result of the cooking process, starches are gelatinized, proteins are denatured, undesirable enzymes are inactivated, and antinutritional substances (eg, trypsin inhibitors in soybeans) are destroyed. Because the temperature reached during the process can be quite high (ca 200°C) and the material takes a relatively short time to travel through the extruder (5 s to a few min), the cooking process is often referred to as high-temperature, short-time (HTST) extrusion. The forming extruder is used to produce special shapes from precooked material. Typically, the operating temperature of forming extruders are lower than that of cooking extruders.

Two major types of extruders are used in the food industry: single-screw and twin-screw extruders. The twin-screw extruders can be further classified according to the type of screws, ie, nonintermeshing or intermeshing and the type of rotation, ie, corotating or counter-rotating. Use of single-screw extruders began in the early years of the industry and they are still used today, though the twin-screw machines are becoming increasingly popular. Twin-screw extruders are comparatively more expensive than single-screw machines of the same production capacity. The flow of material in single-screw extruders depends on the friction between the barrel wall and the material. Hence the type of material that can be used is limited by its properties; proper conditioning of the material is necessary to reduce or eliminate process instability. The corotating, intermeshing twin-screw extruder uses the positive pumping characteristics created by the two intermeshing screws to convey the material forward. This type of mechanism reduces the need to precondition material. Other potential advantages of the corotating twin-screw extruder over the single-screw extruder include (1) narrow residence time distribution, leading to uniform processing, (2) better mixing capabilities, and (3) wide formulations of product, particularly low bulk density and high fat material that cannot be processed using a single-screw machine. The twin-screw extruder provides better control of the process and more uniform product characteristics than does the single-screw extruder. These advantages of the twin-screw machine compensates to some extent for the

Table 1. Extrusion Application in the Food and Feed Industry

Human food		Extrusion processes	Animal feed	
Starch-based	Protein-rich	Agrochemicals	Cattle and pig feed	Fish feeds
Pasta (2 types)	Animal protein	Fertilizers	Partial precooked cereals/seeds	Precooked flours (starch and protein) with different bulk densities
Tortilla flours	Sausages	Vegetables/soup	Enzyme-engineered feed	
Bakery products	Scrap meat/fish	Bagasse and other cellulosic material	Complete gelatinized cereals	
Confectioneries	Blood protein	Cocoa waste material	Yeast wastes	
Candies	Milk protein	Sucrose-based chemicals	Breaking up of cellulose	
Licorices	Germ extrusion	Spices	Protein hydrolysates	
Caramels	Co-extruded meat/starch	Maltodextrins	Pet foods	
Chocolate	Soft cheese	Pulp for paper production		
Starches	Vegetable protein	Proteins chemically modified		
Partly cooked	Soya/cottonsee/rapeseed	Cross-linked starches		
Gelatinized	Oilseed meals	Lactose-based chemicals		
Malt	Health foods			
Cereal snacks				
Health foods				
Breakfast cereals				
Flat breads/cookies				
Bran extrusion				
Crumb				
Instant-cooked foods				
Potato products				
Baby foods				

Source: Ref. 1. Courtesy of Elsevier.

high initial capital investment required. The advantages and disadvantages of different types of extruders have been reviewed in Refs. 2–4. The engineering principles of extruders have also been reviewed (5).

During the HTST process, material typically enters the extruder in a powdered form. The heat, shear, and pressure generated during the process melts the material into a dough-like mass. The material flows from the barrel to a die of a smaller cross section. The pressure at the barrel exit, in the extruder-die setup, can be as high as 3000 psi, depending on the product and process conditions. In the extruder, no water vapor formation occurs at the elevated temperatures attained inside because of high pressures. The material expands as it exits from the die to the atmosphere. This expansion is caused partly by the stresses that are free to relax and partly by the moisture in the material that is flashed-off as vapor, owing to the pressure drop between the die exit and the atmosphere. The expanded product imparts textural and functional characteristics that determine the acceptability of the product. This article reviews the existing literature regarding the textural and rheological properties during extrusion cooking and addresses the relationship between the two.

TEXTURE

The term texture describes a wide range of physical properties of the food product. A product of acceptable texture is usually synonymous with the quality of a product. Table 2 is a glossary of terminology used to describe textural attributes. Textured has been defined as "the attribute of a substance resulting from a combination of physical properties and perceived by senses of touch (including kinaesthesis and mouthfeel), sight and hearing (6)." Texture as defined by the International Organization for Standardization is "all of the rheological and structural (geometric and surface) attributes of a food product perceptible by means of mechanical, tactile and where appropriate, visual and auditory receptors (7)." The following terms have been used to describe product characteristics of extrudates: tender, chewy, soft, tough, brittle, crunchy, crisp, smooth, fine, coarse, puffed, flaky, fibrous, and spongy. One or more of these terms may describe the same behavior.

A variety of properties have been used to quantify or measure the texture and functional properties of extrudates. Some of these are:

1. Bulk density, or density, defined as the mass per unit volume of the extrudate.
2. Breaking strength, defined as the stress required to shear a material. This stress is the ratio of the breaking force to the area of the material. In certain instances it is referred to as the force required to break or shear the sample. The breaking strength is an indirect measure of the stress–strain relationship of the extrudate. It is an important parameter that determines the final quality of the extrudate and can be related to such textural attributes as toughness, chewiness, crunchiness, and brittleness.

Table 2. List of Terms and Groups

Structure	Texture	Consistency
	Terms of relating to the behavior of the material under stress or strain	
Firm	Rubbery	Brittle
Hard	Elastic	Friable
Soft	Plastic	Crumbly
Tough	Sticky	Crunchy
Tender	Adhesive	Crisp
Chewy	Tacky	Thick
Short	Gooey	Thin
Springy	Glutinous	Viscous
	Glutenous	
	Terms relating to the structure of the material	
Smooth	Gritty	
Fine	Coarse	
Powdery	Lumpy	
Chalky	Mealy	
	Relating to shape and arrangement of structural elements	
Flaky	Cellular	Glassy
Fibrous	Aerated	Gelatinous
Stringy	Puffed (puffy)	Foamed (foamy)
Pulpy	Crystalline	Spongy
	Terms relating to mouthfeel characteristics	
Mouthfeel	Watery	Creamy
Getaway	Juicy	Mushy
Body	Oil	Astringent
Dry	Greasy	Hot
Moist	Waxy	Cold
Wet	Slimy	Cooling

Source: Ref. 6.

3. Viscosity of extrudate paste is measured using amylographs.
4. The extrudate dimension, particularly for snack foods, is another important characteristic. It is often referred to as the expansion ratio or the puff ratio and is defined as the ratio of the area of cross section of the extrudate to the area of cross section of the die. In this article, the swelling of extrudate due to elastic and moisture effects is referred to as extrudate expansion. Extrudate swell is referred to as swelling from all effects other than moisture. Puffing is referred to as the swelling believed to be predominantly from the moisture effect.
5. Water absorption index (WAI) is the weight of sediment formed per unit mass of sample, after the sample is centrifuged and the supernatant is removed. Water absorption capacity (WAC) is the ratio of the weight gained over the original weight after a known mass of powdered extrudate is soaked in water for a fixed period of time at a given temperature. Water solubility index (WSI) is the percentage of the initial sample present in the supernatant obtained from the WAI test.

The methods used to obtain these properties are not standardized. Because different methods have been used to obtain a property by the same name, procedures used to determine the properties should be reviewed before drawing conclusions.

The textural properties of the extrudate are affected by the process variables, including temperature, screw speed, screw and barrel dimensions, and product variables (eg. moisture content, composition, and particle size). The properties are evaluated as a function of different parameters, the most common of which includes product moisture, process temperature, and extruder screw speed. Other parameters, some of which may be interrelated, include residence time, die dimensions, screw dimensions, shear rate or shear stress, and product composition. When a number of process conditions are to be evaluated at different levels, the experimental size could be very large. To reduce the experiments to a manageable scale without increasing the experimental error, the response surface methodology (RSM) has been used to develop relationship between dependent and independent variables. According to Meyers (8), "response surface procedures are a collection involving experimental strategy, mathematical methods, and statistical inference which, when combined, enable the experimenter to make an efficient empirical exploration of the system in which he is interested." The RSM helps in finding an approximate function to enable one to predict future response and to determine conditions of optimum response.

In the RSM approach, the entire system is modeled as a black box. The mathematical functions that are developed to correlate the dependent and independent variables are empirical in nature. An advantage of this approach is that a detailed understanding of the changes that occur during the process is not necessary. The disadvantage is that a general relationship is not available, which requires studies for different machines and different formulations of raw material.

Proteins

The basic constituents of all proteins are amino acids. The linear protein chain consists of amino acids linked together by peptide bonds between amino and carboxylic sites. Proteins can have different levels of structure—primary, secondary, and tertiary—that result in a three-dimensional shape. Texturization of proteins can be used to produce two types of products (9). The first are meat analogues that are sold in place of meat. Such products have found acceptance in some countries, such as Japan (10). The second type of textured plant proteins (TPP) is as a meat extender. Several reviews on protein texturization using extruders can be found (9,11–15). During the extrusion process in the presence of water, globular proteins and aluerone granules unravel and align themselves in the presence of the shear field that exists in the extruder (14). Thermal denaturation (loss of native structure) is also expected to occur and is thought to be necessary for texturization (9), although some investigators contend that denaturation is not a necessary step in the texturization process (16).

The mechanism of texturization has been the subject of intense research. It is generally agreed that intermolecular disulfide bonding of proteins is responsible for the structure formation of spun soy fiber (17). It has been reported that the presence of heat and pressure cause the protein molecules in defatted soy meal to disassociate into subunits and become insoluble (18). The mechanism of bond formation has been investigated during extrusion (19). Results indicated that disulfide bonding had no role in extrusion texturization, suggesting that intermolecular amide bonds are instead responsible for texture formation during extrusion processing. The importance of amide bonds in protein texturization was also reported (20).

However, other researchers disagree with these findings. One study reported the formation of intermolecular disulfide bonding when defatted soy meal was extruded at temperatures between 110–150°C and found no evidence of intermolecular peptide bond formation (21). Another study used a single-screw extruder without a die and reported the formation of intermolecular disulfide bond for blends of defatted corn gluten and defatted soy flour extruded at 145°C (22). Others have also refuted the idea that intermolecular peptide bonds are responsible for texturization of extruded soy protein (23). The reason for this disagreement is not clear. The temperatures used by the first investigators (20) were higher than 175°C whereas the latter investigators (21,23) used lower temperatures. It was reported that thermal polymerization of peptide bond formation requires a minimum temperature of 180°C (24). This could be one possible reason for the conflicting results obtained by these researchers.

Most of the work evaluating product quality has used soy proteins. One study extruded defatted soy meal of moisture content 30% (dry basis) at a constant screw speed of 100 rpm with temperatures ranging from 107 to 200°C (25). The product density decreased with temperature (Fig. 1). Shear force and work increased with increasing temperature, but breaking strength increased as temperature increased to 160°C and then decreased with further increase in temperature. Another study used a three-

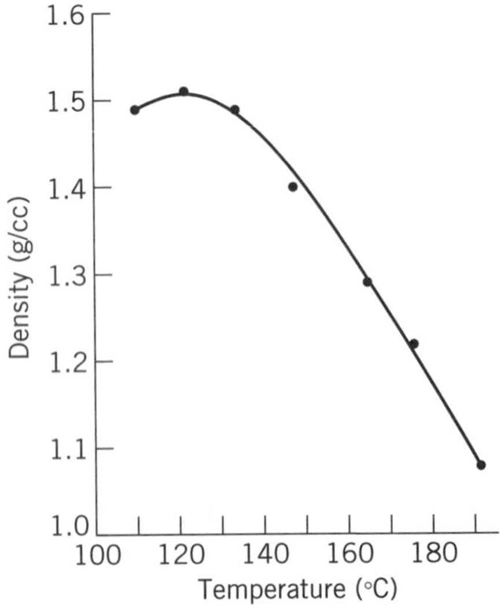

Figure 1. Effect of process temperature on product density. *Source:* Adapted with permission from Ref. 25.

variable, three-level fractional factorial design to measure the effect of extrusion of defatted soy flakes on residual trypsin inhibition activity (TIA), Warner Bratzler shear (WBS), and water absorption capacity (WAC) (26). Lower product moisture decreased residual TIA. Higher temperatures and lower moistures increased WAC, but increased feed moisture content decreased WBS.

A third study used both sensory and mechanical means to evaluate the texture of an extruded soybean meal product (27). A five step temperature range of 135–180°C was used in this experiment. A high correlation between the instrumental and sensory parameter was obtained. With increases in temperature, the product became less compact and spongy in appearance and had increased aligned fibers as reflected by cohesiveness. In a subsequent work, these investigators used a four-factor (temperature, screw speed, product moisture, and protein content) three-level RSM to evaluate the textural properties of extruded defatted soybean meal (28). Results from this study indicated that protein level and extrusion temperature were the most important factors affecting WBS values.

The effect of pH (5.5–10.5) of defatted soybean meal and soybean isolate on WBS values (20) indicated that at the extremes of the pH values, the extrudate appeared to have suffered a loss of structural integrity. The maximum shear force and sensory values were obtained at about a pH of 8, as indicated in Figure 2a and b. At pH < 5.5, the extrusion of the product becomes quite difficult, as pH is increased from 5.5 to 8.5 the product becomes tender, less chewy, and rehydrates rapidly (29). These effects have been attributed to the lower solubility of the proteins at acidic pH, which produced a denser product due to the formation of aggregates in the extruder (14).

Independent variables such as moisture content, temperature, and screw speed affect shear strain, stresses, and shear rate. The effect of shear environment on the textural properties of unroasted defatted soy flour was reported (30). Increasing strain enhanced cross-linking in the dough, as indicated by a higher work required to shear the sample. Higher product moisture reduced the density. A high shear rate (or shear stress) in the die achieved by increasing flow rate led to a denser and less absorbant product, caused by the disruption of cross-linking. When low levels of hydrocolloid (sodium alginate or methylcellulose) were added to defatted soy flour, sodium alginate increased chewiness, maximum force, WAC, and bulk density, whereas the addition of the same amount of methylcellulose decreased maximum force and chewiness (31).

A seven-factor, five-level experiment using RSM to optimize the process variables of temperature, screw speed, screw compression ratio, die diameter, and product moisture content during extrusion cooking of defatted soy grits studied the individual and interactive effects of the independent variables (32). Results indicated that high screw compression ratio and maximum temperature decrease between the barrel and the die favored good texture. In an investigation of the effects of product moisture, barrel temperature, and die temperature on the properties of extruded soy isolate and soy flour, it was found that soy isolate required higher pressure to extrude and had higher expansion and a narrower range of textures than did soy flour (33). The peak force measured using a Ottawa texture machine was found to correlate well with protein solubility, as illustrated in Figure 3. The higher barrel and low die temperature resulted in a better texture for both materials.

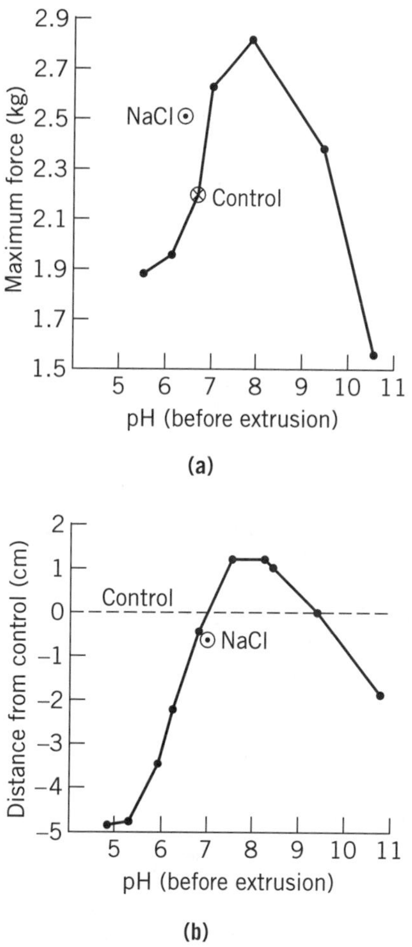

Figure 2. (**a**) Influence of pH adjustment on Warner-Bratzler shear values of extruded soy protein; (**b**) influence of pH adjustment on sensory response of extruded soy protein. *Source:* Adapted with permission from Ref. 20.

Although most research has been on soy proteins because of the potential to simulate meat products, some research has evaluated the properties of other high-protein materials after extrusion. These included studies of the effect of extrusion on glandless cottonseed flour and soy meal mixtures (34,35) and on cowpea meal (36) and the effect of process temperature, screw speed, moisture content, and pH on the WAC and nitrogen solubility index of cowpea, mung bean, defatted soybean, and air-classified mung bean (37). Protein type and amount were found to affect properties. Results from extrusion of blends of corn and soy proteins have been reported (22,38–40). Increasing the amount of soy protein in the corn–soy mixture led to products with better textural and functional properties.

Starch

Much literature is available on extrusion cooking of starch-based materials. Reference 41 is an excellent review of

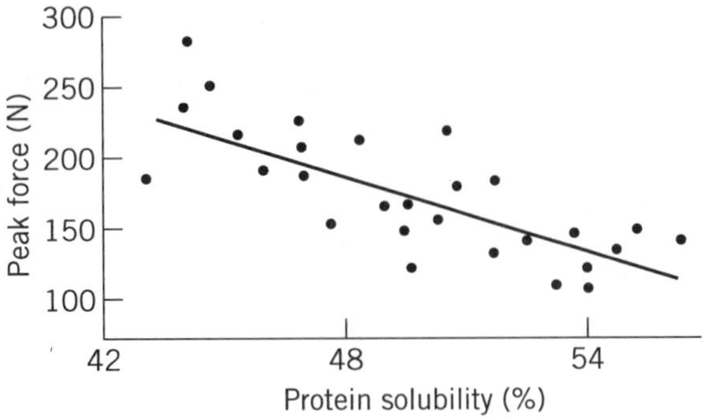

Figure 3. Correlation between OTMS texture and protein solubility for soya isolate. *Source:* Adapted with permission from Ref. 33.

starch extrusion. Starch consists of α-D-glucose units that are linked to form large macromolecules. Native starch is a mixture of two different glucose polymers—the linear amylose and the branched amylopectin. Under the severe shear environment inside the extruder, the starch molecule loses its granular and crystalline structure (42) and undergoes macromolecular degradation (43–48). The insoluble native starch is partially solubilized after extrusion at room temperature. The magnitude of the transformation is a function of multiple processing parameters and their interactions as well as the type of starches.

The extrudate characteristics of several varieties indicated that wheat starch had the highest expansion ratio at 135 and 225°C (Fig. 4) (49). Increasing product moisture content decreased the expansion ratio (49–54). Increasing temperature increased the expansion ratio up to a point after which it decreased with further increase in temperature (49,55). This decrease can be attributed to the degradation of starch molecules at higher temperatures. Expansion is affected by the dimensions of the die or nozzle (56,57) and product composition. Increasing the length-to-diameter ratio of the nozzle increased the expansion ratio. Increasing the amylose content decreased the expansion ratio (49,53,58). Presence of lipids, proteins, salts, and

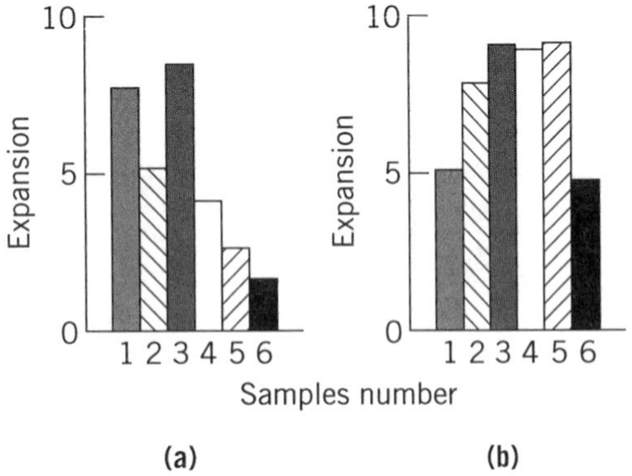

Figure 4. Expansion of products extruded at (**a**) 135°C and (**b**) 225°C. Starches are (1) waxy corn; (2) corn; (3) common wheat; (4) rice; (5) amylon 5 (6) amylon 7. Initial moisture content was 22% by weight. *Source:* Adapted with permission from Ref. 49.

sugar can also affect the puff ratio (50,54,59–61). Decreasing the lipid content increased the expansion ratio of the product whereas increasing the salt and sugar concentration increased the expansion ratio.

Increasing moisture content increased the breaking strength of extrusion-cooked cornmeal (53). Increasing temperature was found to decrease the breaking strength for nonwaxy corn, but the reverse was true for waxy corn. Increasing the concentration of salt and sugar in the extrudate reduced the breaking strength (54). Results of two studies indicated an inverse relationship between expansion ratio and shear strength (54,55).

Native starches are insoluble in water at room temperature. However, on extrusion these materials are soluble in water and could be added as ingredients in other food mixes to provide an acceptable consistency. The solubility of starches increase with increasing severity of treatment. At lower moisture content, the solubility index is higher (49,62–64). Higher temperature also increases solubility.

A small portion of a considerable body of literature that addresses the effect of process and product variables on properties of extrudates has been cited. Some researchers have used the RSM to locate optimum conditions for the best product texture. Empirical predictive models have been developed to relate product quality to independent parameters. These models or equations suffer from two major drawbacks: (1) they are machine-dependent in that the data obtained from one extruder cannot be extrapolated to other machines, and (2) they do not explain the physical phenomena. Understanding of how the basic physical properties control texture formation is limited.

RHEOLOGY

Although, rheology is defined as the science of deformation and flow of matter, it typically is used to describe the flow of non-Newtonian fluids (65). The flow properties of the fluid can affect important extrusion parameters such as velocity and pressure profiles in the extruder, heat transfer between the walls and the fluid, pressure drop in the die, and energy requirements. These parameters determine the product quality, extruder and die design, and production rate.

During the extrusion process the flow patterns are complex (Fig. 5) and involve six major flow regions (66). These regions are:

1. Metering section, where the material is fully plasticized.
2. The entrance region to the die, where a converging flow field is developed generating large stresses.
3. The region within the die, where the disturbance caused by the entry flow gradually disappears.
4. The steady flow region in the die, where a fully developed flow exists.
5. The exit region, where the vapor bubble nucleation and growth and velocity profile rearrangement occur within the die followed by the expansion of the extrudate outside the die.

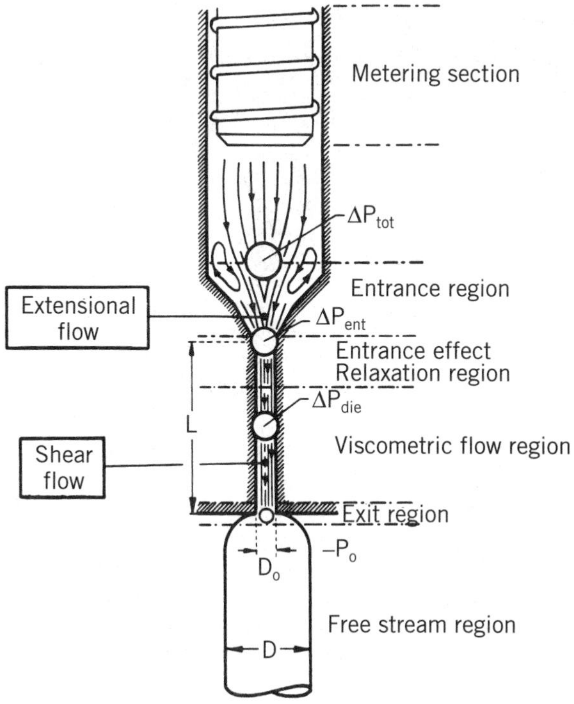

Figure 5. Flow pattern in an extruder barrel and die system. *Source:* Adapted with permission from Ref. 66.

6. The free stream region, where the extrudate expansion reaches an equilibrium value.

These flow situations of viscoelastic fluids necessitates the measurement of rheological parameters such as steady shear viscosity, primary and secondary normal stress differences, and elongational viscosity.

Two types of flows occur during extrusion cooking. The first is the shear flow that occurs owing to the presence of the walls, as in the screw channel and the die. The second is the extensional flow that is present farther away from the wall where the stream lines converge or diverge, such as at the entrance and exit to the die. The stress (σ) components in shear flow, are given by:

$$\sigma_{12} = \eta_s\dot{\gamma},\ \sigma_{23} = \sigma_{13} = 0,\ \sigma_{11} - \sigma_{22} = N_1,\quad \sigma_{22} - \sigma_{33} = N_2 \qquad (1)$$

where 1 is the direction in which the fluid flows, 2 is the direction of the velocity gradient, and 3 is the remaining neutral direction. The shear viscosity (η_s) and the normal stress differences (N_1, N_2) are functions of shear rate. For Newtonian fluids the shear viscosity is a constant and under simple shear flow N_1 and $N_2 = 0$.

For extensional flows, extensional viscosity is defined by:

$$\sigma_{11} - \sigma_{22} = \eta_e\epsilon \qquad (2)$$

Extensional viscosity is constant for a newtonian fluid, but it may be a function of extensional rate for non-Newtonian fluids.

The three types of extensional flows are (1) uniaxial, (2) biaxial, and (3) planar. A cylindrical sample when stretched along its axis results in uniaxial extension. Compression of a cylindrical sample along its axis results in biaxial extension. Stretching a rectangular sheet of a sample along one direction, while keeping one of the remaining dimensions constant, results in planar extension (there are two viscosities in planar extension; see Ref. 67 for details). The ratio of the elongational viscosity (η_e) to the shear viscosity (η_s) is defined as the Trouton ratio. For Newtonian fluid, this ratio is a constant and equal to 3 for uniaxial extension, 4 for planar extension, and 6 for biaxial extension. For non-Newtonian fluids, this ratio is generally much higher and may vary with extensional rate.

The measurement of these rheological parameters during extrusion processing poses special problems. The rotational instruments used in the measurements of rheological parameters of liquids foods cannot be used because the shear rates obtained in these rheometers are several orders of magnitude below those encountered during extrusion and it is difficult to reproduce extrusion-like conditions in the rheometer. Thus on-line measurements of rheological parameters are necessary to avoid these problems.

Shear Viscosity

The rheological property that has received the most attention by scientists is shear viscosity. Some of the published viscosity studies are summarized in Table 3. A die or a viscometer (capillary or slit cross section with multiple pressure transducers located along the wall) attached to the extruder is often used to measure the viscosity. Alternatively, a single pressure transducer mounted near the exit of the barrel, along with capillary dies of different lengths but having the same radius and entry geometry, can be used to obtain the shear viscosity (77).

The first investigators to evaluate the viscosity of doughs during extrusion found that doughs exhibited shear thinning behavior (68). The viscosity decreased exponentially with temperature and was found to follow Arrhenius kinetics. Moisture was found to act as a plasticizer. Increased moisture content (M) decreased the dough viscosity. An empirical model of the form below was proposed:

$$\eta_s(\dot{\gamma}, T, M) = \eta^*\dot{\gamma}^{n-1} \exp(\Delta E/RT) \exp(KM) \qquad (3)$$

This model is one of the simplest and the most popular. The following model was proposed for constant moisture dough (69):

$$\eta_S(T, \dot{\gamma}, t) = \eta^*\dot{\gamma}^{n-1} \exp(\Delta E/RT) \cdot \exp \int_0^t k \exp(\Delta E_k/RT(t))dt \qquad (4)$$

In equation 4 it is assumed that the temperature is a function of time is known. Reactions such as gelatinization and denaturation can lead to network formation and affect viscosity (parameters k and ΔE_k control the reactions). A log polynomial model was used to express viscosity as a function of moisture content, shear rate, and temperature (71).

Table 3. Summary of Viscosity Studies Conducted During Extrusionlike Processes

Product	Method	Moisture (% w.b.)	Temperature (°C)	Reference
Cooked cereal dough	CDV	25–30	67–100	68
Defatted soy flour	CR	32	35–60	69
Pregelatinized corn flour dough	CD	22–35	90–150	70
Soy fluff	CD	25.5–35.5	100–160	71
Soy flour	CR	25–60	25–120	72
Corn grits	SDV	20–32	100–140	73
Potato grits	SDV	18–42	80–140	73
Corn starch	CDV	20–40	100–170	74
Corn grits	CDV	15–30	130–180	75
Corn grits	SDV	25–45	150–180	76

Note: CD, capillary die attached to the extruder; CR, capillary rheometer; CDV, capillary die viscometer is capillary die with multiple transducers along its length; SDV, slit die viscometer is slit die with multiple transducers along its length.

To account for the shear history in the extruder, an equation was suggested of the form (73).

$$\eta_S(\dot{\gamma}, T, M, N) = \eta^* \dot{\gamma}^{n-1} \exp(\Delta E/RT) \exp(KM) N^{-\alpha} \quad (5)$$

The flow curves obtained using a capillary rheometer (9) and slit-die viscometer were significantly different for food doughs (corn grits and potato flour), whereas for low-density polyethylene the flow curves were the same whether obtained using a capillary rheometer or slit-die viscometer. The flow behavior index was found to be affected by screw speed and temperature. This would indicate that shear history is varying, due to different screw speeds. Hence, the Bagley procedure for obtaining the viscosity during extrusion cooking is invalid because the fluid entering the die is rheologically not the same for each condition. A model has been developed that accounts for the thermal and mechanical energy imparted to the product (shear history) before the viscosity is measured (78).

In addition to these viscosity models, two recent models are worth mentioning. They present almost identical expression for starch and protein doughs based on reaction kinetics (79,80). The models are cumbersome because of the large number of constants (10 or more). One notable feature of these models is the inclusion of yield stress. Another study indicated that soy doughs exhibited yield stress (72). The magnitude of yield stress was found to be a function of temperature.

There are several drawbacks to the existing viscosity models. The model of Harper and co-workers (68) assumes that the decrease in viscosity is due to shear thinning only. Most plastic polymers are modeled by a network of macromolecular entanglements. Shear thinning flow is associated with the decrease of the entanglement density under the influence of deformation of the polymer (81). For food doughs the decrease in viscosity is due to the sum of shear thinning process history (screw configuration, residence time), and molecular degradation. A complete model that quantifies the effect of parameters other than product moisture and temperature on shear viscosity is still lacking. These effects are discussed in Ref. 82, though no attempt has been made to quantify them. Another drawback of these models are that the fluid is assumed to be inelastic, ie, normal stress effects are neglected. Because doughs are viscoelastic fluids, inelastic models are inadequate. Evidence has been presented that the pressure drop experienced at the entrance of the die is much greater than that predicted by shear viscosity alone, indicating that elasticity is important (79). As will be discussed later, elasticity is important in the manufacture of products at temperatures below 100°C.

Extruded starches are used as thickening agents or as ingredients in instant foods. The viscosity of powdered extrudate in solvent (water) has also received attention. Typically, shear thinning behavior is observed in all cases. Increasing the concentration of wheat starch from 5 to 9% in solution resulted in the shifting of flow behavior from almost Newtonian to shear thinning (83). For plastic polymers it is known that when the molecular weight is greater than a certain critical molecular weight, non-Newtonian flow behavior is observed (81). Extruded cornstarch in solution exhibited a constant viscosity at high and low shear rates (Fig. 6) (84). Increasing barrel temperature decreased the values of the constant viscosities, whereas increasing moisture content was found to increase the values of constant viscosity, ie, depended on the severity of the extrusion environment.

The effect of emulsifiers on dough viscosity has also been studied (70). The presence of sodium stearoyl-2-lactylate (SSL) was found to affect the viscosity, whereas diacetyltartaric acid ester of monoglyceride did not affect the viscosity. Fat did not affect the viscosity when SSL was present, but it was found to increase the viscosity in the absence of additives.

Normal Stress Difference

It is seen from equation 1 that when N_1 and N_2 are not equal to zero the normal stresses are unequal. The presence of unequal normal stresses can create some interesting phenomena (85). The reason for nonzero values of N_1 and N_2 can be attributed to the anisotropy of fluid microstructure in the flow field and is observed in elastic fluids (86). The ratio of primary normal stress difference to the wall shear stress is an indirect measure of the elasticity of the fluid, ie, the higher the ratio, the more elastic is the fluid.

Measurement of normal stress differences on-line has been a subject of intense research among polymer engi-

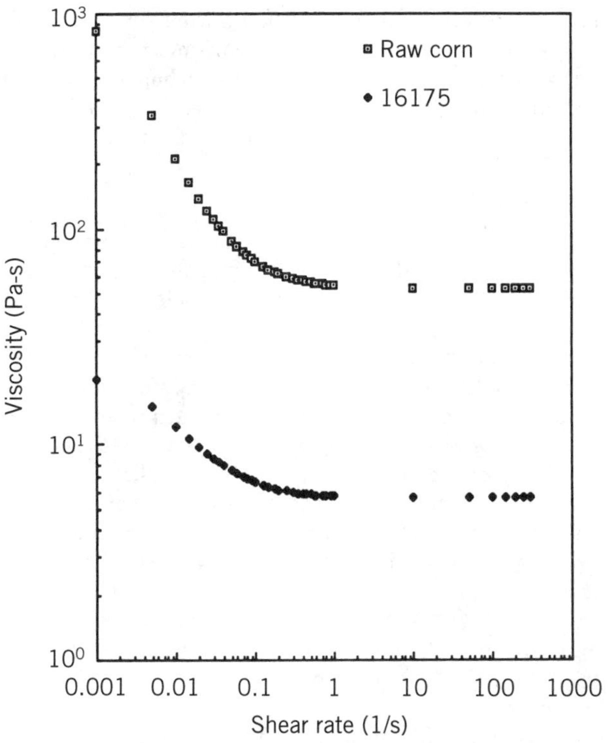

Figure 6. Flow behavior of extruded (16% moisture, 175°C and 150 rpm) and unextruded corn starch at 6% concentration and 60°C (84).

neers and rheologists. The two methods used to measure primary normal stress difference are the exit pressure method and the hole-pressure method. Exit pressure is the residual stress at the die exit. For inelastic fluid this pressure is zero, but positive for a viscoelastic fluid. Hole pressure is defined as the difference between the pressure measured by a flush-mounted transducer and a transducer located at the bottom of a slot directly opposite to it (Fig. 7). For a Newtonian fluid, under creeping flow conditions (Reynolds number ~ 0), the hole pressure is zero.

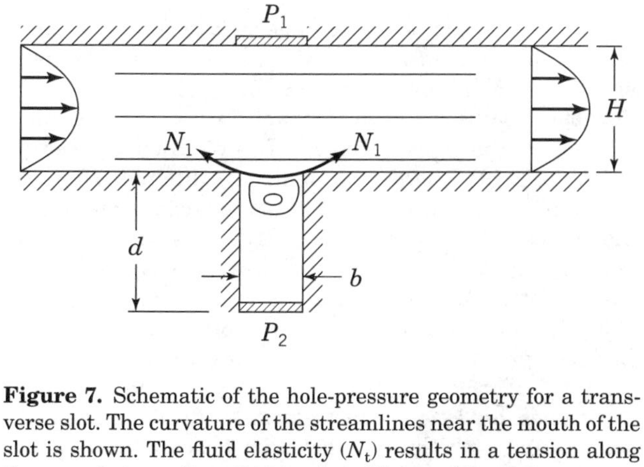

Figure 7. Schematic of the hole-pressure geometry for a transverse slot. The curvature of the streamlines near the mouth of the slot is shown. The fluid elasticity (N_t) results in a tension along the curved streamlines that tends to lift out of the slot.

The exit pressure technique has been used to calculate the normal stress differences (87,88). As it is difficult to measure the pressure at the die exit directly, exit pressure is usually obtained by linear extrapolation of the pressure profile from the region of fully developed flow in the die. The expressions for the normal stress differences for a slit die are (89):

$$N_1 = P_{ex}\left[1 + \frac{d \ln P_{ex}}{d \ln \tau_w}\right] \quad (6)$$

However, this method is controversial. Critical analyses of the method have been done (88–90). The origin of the controversy lies in the fact that for polymers, positive (87,88,91), zero (92), and negative (90,93–97) exit pressures have been obtained. There has been very few published studies using this method on foods. Both positive and negative exit pressures were reported for corn grits, but positive exit pressures only for potato grits (98). Another study examining this method for corn grits showed that products having lower moisture (25–35% db) content gave mostly negative exit pressures, whereas higher moisture values (40 and 45% db) gave positive exit pressures (99). Positive exit pressure was obtained using a capillary die (100). As it is impossible to flush-mount transducers on the walls of the capillary, the presence of hole pressure could lead to erroneous exit pressure values, so the use of a capillary die for this procedure is not recommended.

One assumption in obtaining the expression for the normal stress differences from the exit pressure theory (eq. 6) is that the flow remain fully developed till the die exit. There is evidence that for polymers, the flow does not remain fully developed till the exit (101,102), owing to the presence of exit disturbance. The exit disturbance will lead to a rearrangement of the velocity profile. The velocity rearrangement would lead to an overestimation of the exit pressure (89). Beyond a critical shear stress, the exit disturbance is negligible (88). The value of this stress would depend on the material. For low-density polyethylene this value is 25 kPa. However, there are data in the literature where negative exit pressures were obtained for experiments conducted above the recommended 25 kPa value for wall shear stress. A problem during the extrusion cooking of foods is that at temperatures greater than 100°C the presence of moisture flash at the die exit could cause an additional exit disturbance. Other problems with the exit-pressure method are viscous dissipation in the die and pressure dependence of viscosity of the material. Both these effects could lead to a concave pressure profile and would result in erroneous estimates of exit pressure.

The hole pressure was originally observed as an error in the measurement of normal stress difference using standard geometry rheometers (103). Studies subsequently established the presence of significant hole-pressure errors (104). The interest in the hole pressure arises from the claim that this property can be used for reliable prediction of the elastic properties of viscoelastic fluids. As the fluid passes over a hole or a slot, the stream line for a viscoelastic fluid tends to dip and this leads to the development of normal stresses that tend to lift the fluid (Fig. 8), thus creating a lower pressure at the bottom of the slot or hole

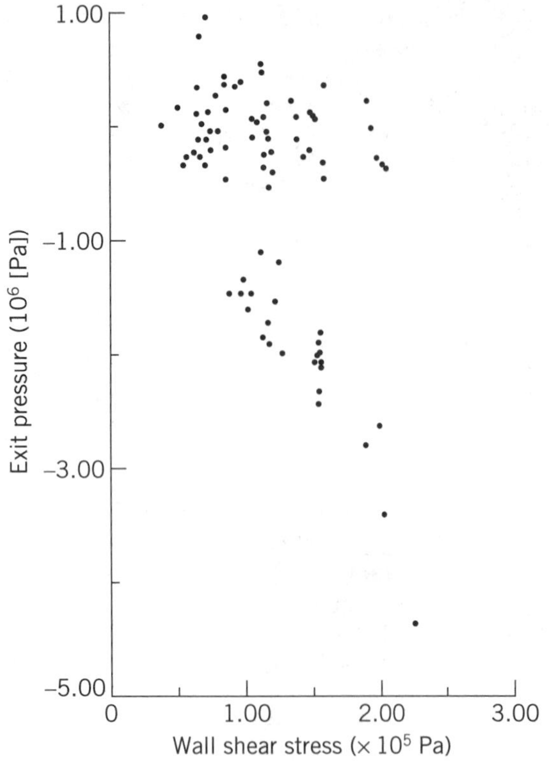

Figure 8. Exit pressure values for corn grits extruded under different conditions. *Source:* Ref. 76.

(105). This method has been used successfully to measure the normal stress difference for polymer solutions and melts (90,96,106–111). Typically, a slit die with holes or slots of different geometries is used to estimate the normal stress differences. The common hole geometry is a rectangular slot with its length transverse to the direction of the flow. This geometry is known as the transverse slot.

The original equations relating hole pressure and the normal stress difference (112) were later modified by Baird (109) to give the HPB equations. The HPB equations were reinterpreted to give the HPBL equations (90). Depending on the geometry of the hole, the HPBL equations are given by:

$$N_1 = 2P_{HE} \frac{d \ln (P_{HE})}{d \ln (\tau_w)} \quad \text{(Transverse slot)} \qquad (7a)$$

$$N_2 = -P_{HE} \frac{d \ln (P_{HE})}{d \ln (\tau_w)} \quad \text{(Parallel slot)} \qquad (7b)$$

$$N_1 - N_2 = 3P_{HE} \frac{d \ln (P_{HE})}{d \ln (\tau_w)} \quad \text{(Circular hole)} \qquad (7c)$$

where P_{HE} signifies the elastic contribution to the hole pressure. Despite violations of some assumptions, reasonably good comparisons of N_1 obtained from standard geometries and slit die were obtained (90,96,111). However, recent numerical studies have shown the HPBL equations to be valid (113,114).

This method is presently under evaluation in the authors' laboratory and has yielded promising results. A potential problem with this method is that for a fluid with yield stress, an error in pressure transmission could occur. For a fluid that has yield stress, higher hole pressure values will be obtained (95). Thus the effect of yield stress on hole-pressure values will have to be accounted for.

Extensional Viscosity

When a fluid flows from a larger to a smaller diameter tube (eg, from the barrel of the extruder to the die), a pressure drop is encountered. For a polymer melt or solution the magnitude of this pressure drop is significantly higher than that obtained for a Newtonian fluid of the same viscosity (87,115). Similar results were reported for corn grits during extrusion cooking (75). The excess pressure drop at the entrance was originally believed to be due to the elastic properties of the fluid. Recent studies for plastic polymers indicate that the flow in the die entry region cannot be explained by melt elasticity alone (116) and that the extensional viscosity is an important parameter that must be considered.

Cogswell (117) was the first to propose a method for obtaining extensional viscosity from the entrance pressure drop method. The entrance pressure drop is calculated by linear extrapolation of the pressure readings along the walls from the fully developed flow region of the capillary or slit to the die entrance. This reading is subtracted from that of the transducer located at the barrel exit or a reservoir. The basis for Cogswell's analysis is that the large entrance pressure drop in the converging flow region is due to the extensional nature of the flow. In addition, the presence of the wall introduces a shearing component. Thus, the entrance pressure drop is assumed to consist of two components: due to shear flow and due to extensional flow. The expression for the shear component of the pressure drop due is given by:

$$\Delta P_{\text{Shear}} = \frac{2B}{3n \tan \alpha} \left(\frac{1 + 3n}{4n}\right)^n \dot{\gamma}^n \left(1 - \left(\frac{r_1}{r_0}\right)^{3n}\right) \qquad (8)$$

and the extensional component for the pressure drop is:

$$\Delta P_{\text{Ext}} = \frac{2A}{3m} \left(\frac{\tan \alpha}{2}\right)^m \dot{\gamma}^m \left(1 - \left(\frac{r_1}{r_0}\right)^{3m}\right) \qquad (9)$$

These equations have been derived assuming a power law behavior in both shear and extensional flow. Expressions for sudden contraction in cylindrical and rectangular flow can also be found in Cogswell's paper. Several other expressions have been suggested using differing assumptions (118,119). There is no significant difference from the values of strain rate obtained from these expression for low die angles (<30°). A spherical coordinates system was assumed to arrive at a slightly modified expressions for equations 8 and 9 (120). An alternative analysis for determining the extensional viscosity in planar and uniaxial extension has been proposed from entrance pressure-drop data (121–123).

Studies on extensional viscosity of food doughs are scarce. A stretch-thinning behavior for corn grits using a capillary rheometer has been reported (124). Similar behavior for corn grits has been noted during extrusion cooking (76). Using Gibson's analysis, these authors found that the extensional viscosity decreased with increasing moisture content. The shear component of the entrance pressure was less than 5%. The Trouton ratio for the doughs ranged from 80–100. Not all materials exhibit shear thinning and stretch-thinning behavior. Several polymeric liquids exhibit stretch-thickening behavior while displaying shear thinning behavior in shear flows.

The entrance pressure-drop method for obtaining extensional viscosity has not been universally accepted. The shear viscosity parameters (B and n in eq. 8) are assumed to be known a priori. These parameters can be calculated from the pressure profile and flow-rate data in the die. At large die entrance angles, poor data fit is observed. The assumption used in the calculation of the shear component of the entrance pressure drop can lead to error. However, the magnitude of this error is expected to be small because extensional flow dominates. Also, only average quantities are computed, whereas in reality the stress and the rate of strain will change with position. It should be noted that this method gives only qualitative comparisons and not definitive data. Hence it is important that a comparison be made between the magnitude of the extensional viscosity obtained from entrance pressure calculations with those obtained by other methods. Such a comparison has been made for polymers, and the values obtained were reported to be of the same order of magnitude (125,126).

Dynamic Oscillatory Flow

The dynamic properties of rice dough have been studied using an on-line rheometer (127). Storage and loss moduli increased with increasing frequency and decreasing moisture content. However, these authors reported significant problems with the equipment and recommended several modifications.

TEXTURE AND RHEOLOGY

The relationship between product structure and its texture is well recognized (9). The structure achieved is dependent on the product composition, processing condition, and molecular configuration. Rheological properties are sensitive to changes in product composition, processing condition, and molecular configuration. Thus, rheological properties present themselves as a measure of product structure and hence its texture.

One of the most important characteristics of an extrudate is the expansion ratio. To some extent, extrudate expansion is indirectly related to other textural properties such as breaking strength, bulk density, and WAI. The expansion phenomena can be characterized in part from rheological parameters. Two recent articles discuss the expansion of extrudate from a rheological point of view (100,128).

During HTST (>100°C) extrusion, the moisture that is unable to escape due to the high pressures in the die is free to vaporize at or near the die exit and is assumed to be responsible for the expansion of the extrudate. Expansion ratios as high as 15 have been obtained. Experiments where the extrudates were cooled to below 100°C (as in the forming or half-product process) before exiting the die were reported (100). Extrudate swells (diameter of extrudate to that of the die) of 1.5, 2.27, and 3.00 were obtained for wheat flour, wheat starch, and manioc starch, respectively. When temperatures are below 100°C, no water vapor is released. In such instances the expansion is due to the elasticity of the material. A Newtonian fluid flowing out of a tube would expand at low Reynolds numbers (<16) but shrink at higher Reynolds numbers (85).

An approach commonly used to explain the extrudate swell is the elastic recoil mechanism. When an elastic fluid is flowing through the tube, it is subjected to extra tension along the stream lines. Once the fluid exits, the walls are no longer present to constrain the flow, which causes the fluid to contract along the stream lines, resulting in radial expansion. The magnitude of this tension is represented by the primary normal stress difference. Expressions for obtaining the extrudate swell have been derived (129). Other factors affecting extrudate swell include surface tension, gravity, inertia, wall slip, temperature effects, flow behavior index, and extensional viscosity. In general, the effect of inertia, surface tension, gravity, and wall slip reduce the extrudate swelling, whereas temperature effects, flow behavior index, elastic recovery, and extensional viscosity increase extrudate swell.

The rheological phenomena involved during extrusion cooking of foods have received some attention. The first attempt to incorporate the effect of elasticity on extrudate expansion qualitatively used pore volume as an indication of elasticity, ie, if the pore volume was small, elastic effects was responsible for the expansion (58). Product morphology was used to estimate dough elasticity (130). These studies, along with that of Alvarez-Martinez and coworkers (131), have speculated that dough elasticity is an important parameter that contributes to the magnitude of the radial expansion, although no quantification has been attempted.

At temperatures above 100°C the extrudate expansion contains contributions from both the elasticity and the moisture flash. A two-stage expansion occurs under these conditions. This phenomenon has been observed and photographed (132). The first stage happens immediately after the extrudate emerges from the die and is due to the elastic effect. A few seconds after the first effect and further downstream, bubble formation due to moisture vapor trying to escape can be observed. For high forces due to moisture effects, the moisture vapor quickly finds its way out. At these conditions the swelling occurs almost instantaneously. Depending on the process conditions, elastic or moisture effects may dominate.

Several investigators (58,131,133) have speculated that radial expansion alone is an insufficient parameter to quantify extrudate expansion. It was proposed that expansion occurs both radially and longitudinally and that a negative correlation exists between the two (58). Longitudinal expansion was defined as the mean length of unit mass of extrudate, or as the ratio of extrudate velocity to that of the velocity of fluid inside the die (58,131). Both these researchers report that longitudinal expansion is inversely proportional to the shear viscosity. However, these authors do not give a reason why shear viscosity should be important in longitudinal expansion.

Consider the flow inside the die, where the melt adheres to the die but flows at a maximum velocity at the center (parabolic profile for a Newtonian fluid). Downstream of the die exit the velocity profile approaches that of a plug flow, with all points having a uniform velocity. This means that the surface layer must be accelerated from rest, and it has been shown that conservation of mass and momentum demands that this acceleration cause an extensional flow in the surface layer while the center is decelerated and compressed (134). This could be a possible reason for the longitudinal expansion. Owing to the extensional nature of the flow, this effects to account for extrudate swelling are necessary (135). Another reason for longitudinal expansion may be the moisture vapor formed in the extrudate. As the vapor tries to escape from the extrudate, it stretches the dough. The magnitude of the expansion due to moisture loss would depend on the extensional viscosity (biaxial extension) of the fluid. This phenomenon is similar to the rise of bread during baking. The vapor pressure of the water will stretch the filament of dough that is cooling in the ambient temperature. If the vapor pressure is sufficient to rupture the extrudate, open structures on the extrudate surface are produced. On cooling, the material dries and solidifies to its final shape.

One must caution against the use of shear viscosity to explain the expansion phenomenon. It has been speculated that lower shear viscosity is favorable for greater expansion (41,136). Data contrary to the findings of these researchers have been presented (132). An increase in dough moisture causes a decrease in shear viscosity, but also causes a reduction in extrudate expansion. Shear viscosity is the least sensitive to molecular weight changes. From a rheological point of view, primary normal stress difference and elongational viscosity are probably far more important. These parameters may be able to better explain some of the conflicting results that exist in the literature and provide an insight as to why lipids, salts, and sugars behave the way they do. Moreover, both primary normal stress difference and extensional viscosity are significantly more sensitive to average molecular weight and molecular weight distribution (and hence to the structure of the material) than shear viscosity and could serve as an effective tool in process control.

CONCLUSION

Food texture as practiced in the industry has remained more of an art than a science. The ingredients of dough components, ie, proteins, starches, and lipids interact in a manner that is not fully understood. Evaluating the rheological parameters of the dough during extrusion may help in providing an understanding of the effect of product and process conditions on the texture of the extrudates. Techniques such as the hole pressure and entrance pressure drop for the measurement of elastic and extensional material functions could help provide a better understanding of the relationship between rheology and texture. These parameters could also be useful in developing a complete rheological equation of state (one that will predict both the viscous and elastic properties). Such equations of state would be useful in numerical modeling of the food extrusion process and help in developing a better understanding of the flow patterns in the extruder and aid in process control and product development.

NOMENCLATURE

A, B	Constants in equations 8 and 9
b	Slot width
d	Slot height
ΔE	Activation energy of flow
ΔE_k	Activation energy for cooking reaction
H	Slit height
k	Specific reaction velocity constant
K	Coefficient for moisture in equations 3 and 5
M	Moisture content
m	Power law index in extensional flow
N_1	Primary normal stress difference
N_2	Secondary normal stress difference
n	Power law index in shear flow
P_{ex}	Pressure drop in the exit
ΔP_{Shear}	Pressure drop due to viscous effects
ΔP_{Ext}	Pressure drop due to elongational flow
ΔP_{tot}	Total pressure drop
ΔP_{ent}	Entrance pressure drop
ΔP_{die}	Pressure drop across the die
P_{HE}	Elastic hole pressure
Q	Volume flow rate
r_0	Radius of the barrel
r_1	Radius at the die
T	Temperature
t	Time
α	Half-cone angle of the die
$\dot{\epsilon}$	Elongational strain rate
$\dot{\gamma}$	Shear rate
η_s	Shear viscosity
η_e	Elongational viscosity
η^*	Reference viscosity
σ_{ij}	Total stress
τ_{ij}	Extra stress components

BIBLIOGRAPHY

1. D. J. Van Zuilichem and W. Stolp in C. O'Connor, ed., *Extrusion Technology for the Food Industry*, Elsevier Science Publishing, New York.
2. J. D. Dziezak, "Single and Twin-Screw Extruders in Food Processing," *Food Technology* **43**, 164–174 (1989).
3. J. M. Harper, "Food Extruders and Their Applications," in C. Mercier, P. Linko, and J. M. Harper, eds., *Extrusion Cooking*, AACC, St. Paul, Minn., 1989, pp. 1–5.
4. B. W. Hauck and G. R. Huber, "Single Screw vs Twin Screw Extrusion," *Cereal Foods World* **34**, 934–939 (1989).

5. L. P. B. M. Janssen, "Engineering Aspects of Food Extrusion," in Ref. 3, pp. 17–38.
6. R. Jowitt, "The Terminology of Food Texture," *Journal of Texture Studies* **5**, 351–358 (1974).
7. International Organization for Standardization, "Sensory Analysis Vocabulary, Part IV," Geneva, Switzerland, 1981.
8. R. H. Meyers, *Response Surface Methodology*, Allyn & Bacon, Boston, 1976.
9. J. M. Harper, *Extrusion of Foods*, Vols. 1 and 2, CRC Press, Boca Raton, Fla., 1981.
10. N. Noguchi, "Extrusion Cooking of High-Moisture Protein Foods," in Ref. 3, pp. 343–370.
11. F. F. Huang and C. K. Rha, "Protein Structures and Protein Fibers—A Review," *Polymer Engineering Science* **14**, 81–89 (1974).
12. J. P. Clark, "Texturization by Extrusion," *Journal of Texture Studies* **9**, 109–123 (1978).
13. D. W. Stanley and J. M. deMan, "Structural and Mechanical Properties of Textured Proteins," *Journal of Texture Studies* **9**, 59–76, 1978.
14. J. E. Kinsella, "Texturized Proteins: Fabrication, Flavorings and Nutrition," *CRC Reviews in Food Science and Nutrition* **10**, 147–207 (1978).
15. D. A. Ledward and J. R. Mitchell, "Protein Extrusion—More Questions Than Answers," in J. M. V. Blanshard and J. R. Mitchell, eds., *Food Structure—Its Creation and Evaluation*, Butterworths, Boston, 1988, pp. 219–229.
16. P. R. Sheard et al., "Macromolecular Changes Associated with the Heat Treatment of Soy Isolate," *Journal of Food Technology* **21**, 55–60 (1986).
17. J. J. Kelley and R. Pressey, "Studies with Soybean Protein and Fiber Formation," *Cereal Chemistry* **43**, 195–205, 1966.
18. D. B. Cumming et al., "Fate of Water Soluble Soy Protein During Thermoplastic Extrusion," *Journal of Food Science* **38**, 320–323 (1973).
19. L. D. Burgess and D. W. Stanley. "A Possible Mechanism for Thermal Texturization of Soybean Protein," *Journal of the Institute of Canadian Science and Technology—Alimentation* **9**, 228–231 (1976).
20. R. W. Simonsky and D. W. Stanley, "Texture-Structure Relationships in Textured Soy Protein. V. Influence of pH and Protein Acylation on Extrusion Texturization," *Canadian Institute of Food Science Technology* **15**, 294–301 (1982).
21. D. F. Hager, "Effects of Extrusion upon Soy Concentrate Solubility," *Journal of Agricultural and Food Chemistry* **32**, 293–296 (1984).
22. P. E. Neumann et al., "Uniquely Textured Products Obtained by Co-Extrusion of Corn Gluten Meal and Soy Flour," *Cereal Chemistry* **61**, 439–445 (1984).
23. P. R. Sheard et al., "Role of Carbohydrates in Soya Extrusion," *Journal of Food Technology* **19**, 475–483 (1984).
24. P. Melius, "Thermal Polymerization of Amino Acids Mixtures," *Federation Proceedings, Federation of American Society of Experimental Biologist* **34**, 573, (1975).
25. D. B. Cummings et al., "Texture–Structure Relationships in Texturized Soy Protein. II. Textural Properties and Ultrastructure of an Extruded Soybean Product," *Journal of the Institute of Canadian Science and Technology—Alimentation* **5**, 124–128 (1972).
26. J. M. Aguilera and F. W. Kosikowoski, "Soybean Extruded Product: A Response Surface Analysis," *Journal of Food Science* **41**, 647–651 (1976).
27. T. J. Maurice et al., "Texture—Structure Relationship in Textured Soy Protein. III. Textural Evaluation of Extruded Products," *Canadian Institute of Food Science Technology* **9**, 173–176 (1976).
28. T. J. Maurice and D. W. Stanley, "Texture–Structure Relationships in Texturized Soy Protein, IV. Influence of Process Variables on Extrusion Texturization," *Canadian Institute of Food Science Technology* **11**, 1–6 (1978).
29. O. B. Smith, "Extrusion and Forming: Creating New Foods," *Food Engineering* **47**, 48–50 (August 1975).
30. S. H. Holay and J. M. Harper, "Influence of the Extrusion Shear Environment on Plant Protein Texturization," *Journal of Food Science* **47**, 1869–1874 (1982).
31. G. Boison et al., "Extrusion of Defatted Soy Flour-Hydrocolloid Mixtures," *Journal of Food Technology* **18**, 719–730 (1983).
32. P. J. Frazier et al., "Optimization of Process Variables in Extrusion Texturing of Soya," *Journal of Food Engineering* **2**, 79–103 (1983).
33. P. R. Sheard et al., "Comparison of the Extrusion Cooking of a Soya Isolate and a Soya Flour," *Journal of Food Technology* **20**, 763–771 (1985).
34. M. V. Taranto et al., "Textured Cottonseed and Soy Flours: A Microscopic Analysis," *Journal of Food Science* **43**, 767–771 (1978).
35. M. V. Taranto et al., "Morphological, Ultrastructural and Rheological Evaluation of Soy and Cottonseed Flours Texturized by Extrusion and Nonextrusion Processing," *Journal of Food Science* **43**, 973–979, 984 (1978).
36. M. B. Kennedy and co-workers, "Effects of Feed Moisture and Barrel Temperature on the Rheological Properties of Extruded Cowpea Meal," *Journal of Food Processing and Engineering* **8**, 193–212 (1986).
37. C. B. Pham and P. R. Del Rosario, "Studies on the Development of Texturized Vegetable Products by the Extrusion Process," *Journal of Food Technology* **19**, 535–547 (1984).
38. J. A. Maga and K. Lorenz, "Sensory and Functional Properties of Extruded Corn-Soy Blends," *Lebensmittel Wissonschalf Technologie* **11**, 185–187 (1978).
39. Y. C. Jao et al., "Evaluation of Corn Protein Concentrate: Extrusion Study," *Journal of Food Science* **50**, 1257–1259, 1288 (1985).
40. M. Bhattacharya et al., "Textural Properties of Plant Protein Blends," *Journal of Food Science* **51**, 988–993 (1986).
41. P. Colonna and co-workers, "Extrusion Cooking of Starch and Starchy Products," in Ref. 3, pp. 247–319.
42. P. Colonna et al., "Flow, Mixing and Residence Time Distribution of Maize Starch Within a Twin Screw Extruder with a Longitudinally-Split Barrel," *Journal of Cereal Science* **1**, 115–125 (1983).
43. C. Mercier, "Effect of Extrusion Cooking on Potato Starch Using a Twin Screw French Extruder," *Staerke* **29**, 48–52 (1977).
44. P. Colonna and C. Mercier, "Macromolecular Modifications of Manioc Starch Components by Extrusion Cooking With and Without Lipids," *Carbohydrate Polym* **3**, 87–108 (1983).
45. P. Colonna et al., "Extrusion Cooking and Drum Drying of Wheat Starch. I. Physical and Macromolecular Modifications," *Cereal Chemistry* **61**, 538–543 (1984).

46. V. J. Davidson et al., "Degradation of Wheat Starch in a Single Screw Extruder: Characteristics of Extruded Starch Polymers," *Journal of Food Science* **49**, 453–458 (1984).

47. V. J. Davidson et al., "A Model for Mechanical Degradation of Wheat Starch in a Single Screw Extruder," *Journal of Food Science* **49**, 1154–1157 (1984).

48. L. L. Diosady et al., "Degradation of Wheat Starch in a Single Screw Extruder: Mechanico-kinetic Break Down of Cooked Starch," *Journal of Food Science* **50**, 1697–1699, 1706 (1985).

49. C. Mercier and P. Feillet, "Modification of Carbohydrate Components by Extrusion-Cooking of Cereal Products," *Cereal Chemistry* **52**, 283–297 (1975).

50. J. M. Faubion and R. C. Hoseney, "High Temperature Short-Time Extrusion Cooking of Wheat Starch and Flour. I. Effect of Moisture and Flour Type on Extrudate Properties," *Cereal Chemistry* **59**, 529–533 (1982).

51. M. Gomez and J. M. Aguilera, "A Physicochemical Model for Extrusion of Corn Starch," *Journal of Food Science* **49**, 40–43, 63 (1984).

52. J. Owusu-Ansah et al., "Textural and Microstructural Changes in Corn Starch as a Function of Extrusion Variables," *Canadian Institute of Food Science Technology* **17**, 65–70 (1984).

53. M. Bhattacharya and M. A. Hanna, "Textural Properties of Extrusion Cooked Corn Starch," *Lebensmittel Wissenschaft und Technologie* **20**, 195–201 (1987).

54. F. Hsieh et al., "Effects of Salt, Sugar and Screw Speed on Processing and Product Variables of Corn Meal Extruded with a Twin-Screw Extruder," *Journal of Food Science* **55**, 224–227 (1990).

55. R. Chinnaswamy and M. A. Hanna, "Optimum Extrusion-Cooking Conditions for Maximum Expansion of Corn Starch," *Journal of Food Science* **53**, 834–836, 840 (1988).

56. R. Chinnaswamy and M. A. Hanna, "Die-Nozzle Dimension Effects on Expansion of Corn Starch," *Journal of Food Science* **52**, 1746–1747 (1987).

57. A. L. Hayter et al., "The Physical Properties of Extruded Food Foams," *Journal of Materials Science* **21**, 3729–3736 (1986).

58. B. Launay and J. M. Lisch, "Twin-Screw Extrusion Cooking of Starches: Flow Behavior of Starch Pastes, Expansion and Mechanical Properties of Extrudates," *Journal of Food Engineering* **2**, 259–280 (1983).

59. D. Paton and W. A. Spratt, "Component Interactions in the Extrusion Cooking Process. I. Processing of Chlorinated and Untreated Soft Wheat Flour," *Cereal Chemistry* **55**, 973–980 (1978).

60. J. M. Faubion and R. C. Hoseney, "High Temperature Short-Time Extrusion Cooking of Wheat Starch and Flour. II. Effect of Protein and Lipid on Extrudate Properties," *Cereal Chemistry* **59**, 533–537 (1982).

61. M. Bhattacharya and M. A. Hanna, "Effect of Lipids on the Properties of Extruded Products," *Journal of Food Science* **53**, 1230–1231 (1988).

62. R. A. Anderson et al., "Gelatinization of Corn Grits by Roll and Extrusion Cooking," *Cereal Science Today* **14**, 4–7, 11–12 (1969).

63. R. A. Anderson et al., "Roll and Extrusion Cooking of Grain Sorghum Grits," *Cereal Science Today* **14**, 372–376 (1969).

64. H. F. Conway, "Extrusion Cooking of Cereals and Soybeans," *Food Product Development* **5**, 14–17, 27–29 (1971).

65. S. Middleman, "Advances in Polymer Science and Engineering: Application to Food Rheology," in C. K. Rha, ed., *Theory, Determination and Control of Physical Properties of Food Materials*, Riedel Publishing, Dordrecht, Holland, 1975, pp. 39–53.

66. J. L. Leblanc, "Recent Progress in Understanding Rubber Processing Through a Rheological Approach," *Progress and Trends in Rheology* **2**, 32–43 (1988).

67. C. J. S. Petrie, "Some Asymptotic Results for Planar Extension," *Journal of Non-Newtonian Fluid Mechanics* **34**, 37–62 (1990).

68. J. M. Harper et al., "Viscosity Model for Cooked Cereal Doughs," *American Institute of Chemical Engineering Symposium Series*, No. 108, 67, 40–43 (1971).

69. C. H. Remsen and J. P. Clark, "A Viscosity Model for a Cooking Dough," *Journals of Food Process Engineering* **2**, 39–63 (1977).

70. N. W. Cervone and J. M. Harper, "Viscosity of an Intermediate Moisture Dough," *Journal of Food Processing Engineering* **2**, 83–95 (1978).

71. Y. C. Jao et al., "Engineering Analysis of Soy Dough Rheology in Extrusion," *Journal of Food Process Engineering* **2**, 97–112 (1978).

72. L. A. Luxenburg et al., "Background Studies in the Modeling of Extrusion Cooking Processes for Soy Doughs," *Biotechnology Progress* **1**, 33–38 (1985).

73. A. Senouci and A. C. Smith, "An Experimental Study of Food Melt Rheology. I. Shear Viscosity Using a Slit Die Viscometer and a Capillary Rheometer," *Rheologica Acta* **27**, 546–554 (1988).

74. L. S. Lai et al., "On Line Rheological Properties of Amylose and Amylopectin Based Starch. The Role of Viscosity on Extrudate Expansion," *Proceedings of the Xth International Congress on Rheology* **2**, 55–57, (1988).

75. M. Padmanabhan and M. Bhattacharya, "Analysis of Pressure Drop in Extruder Dies," *Journal of Food Science* **54**, 709–713 (1989).

76. M. Bhattacharya and M. Padmanabhan, "Elongational Viscosity During Extrusion Cooking Using a Converging Flow Analysis," Paper 243, Presented at the Institute of Food Technology Annual Meeting, Anaheim, Calif., 1990.

77. E. B. Bagley, "End Corrections in the Capillary Flow of Polyethylene," *Journal of Applied Physics* **28**, 624–627 (1957).

78. B. Vergnes and J. P. Villemaire, "Rheological Behavior of Low Moisture Molten Maize Starch," *Rheologica Acta* **26**, 570–576 (1987).

79. R. G. Morgan et al., "A Generalized Viscosity Model for Extrusion of Protein Doughs," *Journal of Food Process Engineering* **11**, 55–78 (1989).

80. K. L. Mackey et al., "Rheological Modeling of Potato Flours During Extrusion Cooking," *Journal of Food Process Engineering* **12**, 1–11 (1989).

81. G. V. Vinogradov and A. Y. Malkin, *Rheology of Polymers*, Mir Publishers, Moscow, 1980.

82. B. van Lengerich, "Influence of Extrusion Processing on In-Line Rheological Behavior, Structure, and Function of Wheat Starch," in H. Faridi and J. M. Faubion, eds., *Dough Rheology and Baked Product Texture*, Van Nostrand Reinhold, New York, 1989, pp. 421–471.

83. J. L. Doublier et al., "Extrusion Cooking and Drum-Drying of Wheat Starch. II. Rheological Characterization of Wheat Starch," *Cereal Chemistry* **63**, 240–246 (1986).

84. B. Launay and T. Kone, "Twin-Screw Extrusion Cooking of Corn Starch: Flow Behavior of Starch Pastes," in P. Zeuthen et al., eds., *Thermal Processing and Quality of Foods*, Elsevier Applied Science, London, 1984, pp. 54–61.

85. R. B. Bird and co-workers, *Dynamics of Polymeric Liquids*, Vol. 1, John Wiley & Sons, New York, 1987.

86. H. A. Barnes et al., *An Introduction to Rheology*, Elsevier Science Publishing, Amsterdam, The Netherlands, 1989.

87. C. D. Han, *Rheology in Polymer Processing*, Academic Press, New York, 1976.

88. C. D. Han, "Slit Rheometry," in A. A. Collyer and D. W. Clegg, eds., *Rheological Measurement*, Elsevier Applied Science, New York, 1988, pp. 25–48.

89. D. V. Boger and M. M. Denn, "Capillary and Slit Methods of Normal Stress Measurements," *Journal of Non-Newtonian Fluid Mechanics* **6**, 163–185 (1980).

90. A. S. Lodge and L. de Vargas, "Positive Hold Pressures and Negative Exit Pressures Generated by Molten Polyethylene Flowing Through a Slit Die," *Rheologica Acta* **22**, 151–170 (1983).

91. C. Rauwendaal and F. Fernandez, "Experimental Study and Analysis of a Slit Die Viscometer," *Polymer Engineering Science* **25**, 765–771 (1985).

92. R. Eswaran et al., "A Slit Viscometer for Polymer Melts," *Rheologica Acta* **3**, 83–91 (1963).

93. J. L. Leblanc, "New Slit Die Rheometer. Some Results with a Butadiene-Styrene Block Copolymer," *Polymer* **17**, 235–240, (1976).

94. L. Choplin and P. J. Carreau, "Excess Pressure Losses in Slit," *Journal of Non-Newtonian Fluid Mechanics* **9**, 119–146 (1981).

95. H. M. Laun, "Polymer Melt Rheology with a Slit Die," *Rheologica Acta* **22**, 171–185 (1983).

96. D. G. Baird et al., "Comparison of the Hole Pressure and Exit Pressure Methods for Measuring Polymer Melt Normal Stresses," *Polymer Engineering Science* **26**, 225–232 (1986).

97. N. Y. Tuna and B. A. Finlayson, "Exit Pressure Experiments for Low Density Polyethylene Melts," *Journal of Rheology* **32**, 285–308 (1988).

98. A. Senouci and A. C. Smith, "An Experimental Study of Food Melt Rheology. II. End Pressure Effects," *Rheologica Acta* **27**, 649–655 (1988).

99. M. Bhattacharya and M. Padmanabhan, "On-line Rheological Measurements of Food Dough During Extrusion Cooking," in *Proceedings of the Second International Conference in Extrusion and Rheology of Foods*. Rutgers University, New Brunswick, N.J., 1990.

100. R. C. E. Guy and A. W. Horne, "Extrusion and Co-Extrusion of Cereals," in Ref. 15, pp. 331–349.

101. B. A. Whipple and C. T. Hill, "Velocity Distributions in Die Swell," *American Institute of Chemical Engineering Journal* **24**, 664–678 (1978).

102. M. Gottlieb and R. B. Bird, "Exit Effects in Non-Newtonian Liquids. An Experimental Study," *Ind. Eng. Chem. Fund.* **18**, 357–368 (1979).

103. J. M. Broadbent et al., "Possible Systematic Errors in the Measurement of Normal Stress Differences in Polymer Solution in Steady Shear Flow," *Nature* **217**, 55–57 (1968).

104. A. Kaye et al., "Determination of Normal Stress Differences in Steady Shear Flow. II. Flow Birefringence, Viscosity and Normal Stress Data for Polyisobutene Liquid," *Rheologica Acta* **7**, 368–379 (1968).

105. R. I. Tanner and A. C. Pipkin, "Intrinsic Errors in Pressure-Hole Measurement," *Transactions of the Society of Rheology* **13**, 471–484 (1969).

106. A. S. Lodge, "Low-Shear Rate Rheometry and Polymer Quality Control," *Chemical Engineering Communications* **32**, 1–60 (1985).

107. A. S. Lodge, "A New Method of Measuring Multigrade Oil Shear Elasticity and Viscosity at High Shear Rates," *SAE Technical Paper*, Series 872043 (1987).

108. A. S. Lodge et al., "Measurement of the First Normal-Stress Difference at High Shear Rates for a Polyisobutylene/Decalin Solution D2," *Rheologica Acta* **26**, 516–521 (1987).

109. D. G. Baird, "A Possible Method for Determining Normal Stress Differences from Hole Pressure Error Data," *Transactions of the Society of Rheology* **19**, 147–151 (1975).

110. D. G. Baird, "Fluid Elasticity from Hole Pressure Error Data," *Journal of Applied Polymer Science* **20**, 3155–3173 (1976).

111. R. D. Pike and D. G. Baird, "Evaluation of the Highashitani and Pritchard Analysis of the Hole Pressure Using Flow Birefringence," *Journal of Non-Newtonian Fluid Mechanics* **16**, 211–231 (1984).

112. K. Higashitani and W. G. Pritchard, "A Kinematic Calculation of Intrinsic Errors in Pressure Measurements Made with Hole," *Transactions of the Society of Rheology* **16**, 687–696 (1972).

113. R. I. Tanner, "Pressure-Hole Errors—An Alternative Approach," *Journal of Non-Newtonian Fluid Mechanics* **28**, 309–318 (1988).

114. M. Yao and D. S. Malkus, "Error Cancellation in HPBL Derivation of Elastic Hole-Pressure Error," *Center of the Mathematical Science*, Technical Summary Report 90-18, University of Wisconsin, Madison (1989).

115. W. Philippoff and F. H. Gaskins, "The Capillary Experiment in Rheology," *Transactions of the Society of Rheology* **2**, 263–284 (1958).

116. S. A. White and D. G. Baird, "The Importance of Extensional Flow Properties on Planar Entry Flow Patterns of Polymer Melt," *Journal of Non-Newtonian Fluid Mechanics* **20**, 93–101 (1986).

117. F. N. Cogswell, "Converging Flow of Polymer Melts in Extrusion Dies," *Polymer Engineering and Science* **12**, 64–73 (1972).

118. D. R. Oliver, "The Prediction of Angle of Convergence for the Flow of Viscoelastic Liquids into Orifices," *The Chemical Engineering Journal* **6**, 265–271 (1973).

119. A. B. Metzner and A. P. Metzner, "Stress Levels in Rapid Extensional Flows of Polymeric Fluids," *Rheologica Acta* **9**, 174–181 (1970).

120. A. G. Gibson "Converging Flow Analysis," in Ref. 88, pp. 49–82.

121. D. M. Binding, "An Approximate Analysis for Contraction and Converging Flows," *Journal of Non-Newtonian Fluid Mechanics* **27**, 173–189 (1988).

122. D. M. Binding and K. Walters, "On the Use of Flow Through a Contraction in Estimating the Extensional Viscosity of Mobile Polymer Solution," *Journal of Non-Newtonian Fluid Mechanics* **30**, 233–250 (1988).

123. D. M. Binding and D. M. Jones, "On the Interpretation of Data from Converging Flow Rheometers," *Rheologica Acta* **28**, 215–222 (1989).

124. A. Senouci et al., "Extensional Rheology in Food Processing," *Progress and Trends in Rheology* **2**, 434–437 (1988).

125. R. N. Shroff et al., "Extensional Flow of Polymer Melts," *Transactions of the Society Rheology* **21**, 429–446 (1977).

126. H. M. Laun and H. Schuch, "Transient Elongational Viscosities and Drawability of Polymer Melts," *Journal of Rheology* **33**, 119–175 (1989).

127. J. F. Steffe and R. G. Morgan, "On-Line Measurements of Dynamic Rheological Properties During Food Extrusion," *Journal of Food Process Engineering* **10**, 21–26 (1987).

128. M. Padmanabhan and M. Bhattacharya, "Extrudate Expansion During Extrusion Cooking of Foods," *Cereal Foods World* **34**, 945–949 (1989).

129. R. I. Tanner, "A Theory of Die Swell," *Journal of Applied Polymer Science* A-2, **8**, 2067–2078 (1970).

130. R. C. Miller, "Low Moisture Extrusion: Effect of Cooking Moisture on Product Characteristics," *Journal of Food Science* **50**, 249–253 (1985).

131. L. Alvarez-Martinez et al., "A General Model for Expansion of Extruded Products," *Journal of Food Science* **53**, 609–615 (1988).

132. R. Chinnaswamy and M. A. Hanna, "Relationship Between Viscosity and Expansion Properties of Variously Extrusion-Cooked Corn Grain Components," *Food Hydrocolloids* **3**, 423–434 (1990).

133. K. B. Park, "Elucidation of the Extrusion Puffing Process," Ph.D. Thesis, Department of Food Science, University of Illinois, Urbana, Il., 1976.

134. S. Richardson, "The Die Swell Phenomenon," *Rheologica Acta* **9**, 193–199 (1970).

135. R. J. Tanner, "Recoverable Elastic Strain and Swelling Ratio," in Ref. 88, pp. 93–117.

136. B. Vergnes et al., "Interrelationships Between Thermomechanical Treatment and Macromolecular Degradation of Maize Starch in a Novel Rheometer with Preshearing," *Journal of Cereal Science* **5**, 189–202 (1987).

GENERAL REFERENCES

D. Q. Ada and S. S. Wang, "Modeling Shearing Resistance of Powdery Starch for Simulation Studies of Extrusion Cooking Processes," *Starch / Staerke* **50**, 147–153 (1998).

E. S. M. Aal Abdel et al., "Effect of Extrusion Cooking on the Physical and Functional Properties of Wheat, Rice and Fababean Blends," *Lebensm.-Wiss. Technol.* **25**, 21–25 (1992).

E. D. Beecher and M. S. Starer, "Direct-Expanded Cereals Produced on High-Speed, High-Torque Twin Screw Extruders," *Cereal Foods World* **43**, 753–757 (1998).

H. H. Ben et al., "Extrusion-Cooking of Pea Flour: Structural and Immunocytochemical Aspects," *Food Structure* **10**, 203–121 (1991).

S. Bhajmohan and S. J. Mulvaney, "Modeling and Process Control of Twin-Screw Cooking Food Extruders," *Journal of Food Engineering* **23**, 403–428 (1994).

S. Bhatnagar and M. A. Hanna, "Starch-Stearic Acid Complex Development Within Single and Twin Screw Extruders," *J. Food Sci.* **61**, 778–782 (1996).

M. Bhattacharya, M. Padmanabhan, and K. Seethamraju, "Uniaxial Extensional Viscosity During Extrusion Cooking From Entrance Pressure Drop Method," *J. Food Sci.* **59**, 221–226, 230 (1994).

J. C. Cheftel, M. Kitgawa, and C. Queguiner, "New Protein Texturization Processes by Extrusion Cooking at High Moisture Levels," *Food Rev. Int.* **8**, 235–275 (1992).

Q. Di and S. S. Wang, "Kinetics of the Formation of Gelatinized and Melted Starch at Extrusion Cooking Conditions," *Starch / Staerke* **46**, 225–229 (1994).

W. Fangzhi, "Study of Single-Screw Extruders for Continuous Feeding of Canola Paste for Supercritical Extraction," *Dissertation Abstracts International B* **55**, 512 (1994).

S. Ilo et al., "The Effect of Extrusion Operating Conditions on the Apparent Viscosity and the Properties of Extrudates in Twin-Screw Extrusion Cooking of Maize Grits," *Lebensm.-Wiss. Technol.* **29**, 593–598 (1996).

A. S. Jaber Ba et al., "Texturization of Hand- and Mechanically-Deboned Poultry Meat Combinations With a Soy Protein Isolate by Extrusion Cooking," *Lebensm.-Wiss. Technol.* **25**, 153–157 (1992).

J. L. Kokini, "The Effect of Processing History on Chemical Changes in Single- and Twin-Screw Extruders," *Trends Food Sci. Technol.* **4**, 324–329 (1993).

M. K. Kulshreshtha, C. A. Zaror, and D. J. Jukes, "Simulating the Performance of a Control System for Food Extruders Using Model-Based Set-Point Adjustment," *Food Control* **6**, 135–141 (1995).

L. Levine and E. Boehmer, "The Fluid Mechanics of Cookie Dough Extruders," *Journal of Food Process Engineering* **15**, 169–186 (1992).

S. Lue, F. Hsieh, and H. E. Huff, "Extrusion Cooking of Corn Meal and Sugar Beet Fibre: Effects on Expansion Properties, Starch Gelatinization, and Dietary Fiber Content," *Cereal Chem.* **68**, 227–234 (1991).

K. L. Mackey et al., "Rheological Modeling of Potato Flour During Extrusion Cooking," *Journal of Food Process Engineering* **12**, 1–11 (1990).

M. Padmanabhan and M. Bhattacharaya, "Rheological Measurement of Fluid Elasticity During Extrusion Cooking," *Trends Food Sci. Technol.* **2**, 149–151 (1991).

G. J. Rockey, "RTE Breakfast Cereal Flake Extrusion," *Cereal Foods World* **40**, 422–424, 426 (1995).

W. Seibel and H. Ruguo, "Gelatinization Characteristics of a Cassava/Corn Starch Based Blend During Extrusion Cooking Employing Response Surface Methodology," *Starch / Staerke* **46**, 217–224 (1994).

M. Thiebaud, E. Dumay, and J. C. Cheftel, "Influence of Process Variables on the Characteristics of a High Moisture Fish Soy Protein Mix Texturized by Extrusion Cooking," *Lebensm.-Wiss. Technol.* **29**, 526–535 (1996).

S. M. Wang, J. M. Bouvier, and M. Gelus, "Rheological Behaviour of Wheat Flour Dough in Twin-Screw Extrusion Cooking," *International Journal of Food Science and Technology* **25**, 129–139 (1990).

S. Wang, J. Casulli, and J. M. Bouvier, "Effect of Dough Ingredients on Apparent Viscosity and Properties of Extrudates in Twin-Screw Extrusion Cooking," *International Journal of Food Science and Technology* **28**, 465–479 (1993).

W. Yacu, "Process Instrumentation and Control in Food Extruders," *Cereal Foods World* **35**, 919–926 (1990).

MRINAL BHATTACHARVA
MAHESH PADMANABHAN
University of Minnesota
St. Paul, Minnesota

See also EXTRUSION COOKING.

F

FATS AND OILS: CHEMISTRY, PHYSICS, AND APPLICATIONS

Fats and oils are the commercially important group of substances classified as lipids. Lipids are compounds usually associated with solubility in nonpolar solvents. They are mostly esters of long-chain fatty acids and alcohols and closely related derivatives. The most important aspect of lipids is the central position of the fatty acids. The scheme of Figure 1 illustrates this (1). The basic components of the lipids (also known as derived lipids) are listed in the central column. In the left-hand column are the phospholipids, most of which contain phosphoric acid groups. The right-hand column includes the compounds most important from a quantitative standpoint in fats and oils. These are mostly esters of fatty acids and glycerol. Fats are solid at room temperature, oils are liquid.

Food fats can be divided into visible and invisible fats (2). More than half of all the fats consumed are in the latter category, that is, those contained in dairy products (excluding butter), eggs, meat, poultry, fish, fruits, vegetables, and grain products. The visible fats include lard, butter, shortening, frying fats and oils, margarines, and salad oils.

Fats and oils may differ considerably in composition, depending on their origin. Fatty acid as well as glyceride composition will influence many of the properties of fats and oils. These properties may be further modified by appropriate processing methods. The fats and oils can be classified broadly into the following groups: animal depot fats, ruminant milk fats, vegetable oils and fats, and marine oils. The processing of fats and oils serves several purposes: to separate the oils or fats from other parts of the raw materials, to remove impurities and undesirable components, to improve stability, to change physical properties, and to provide desirable functional properties.

COMPONENT FATTY ACIDS

Even-numbered straight-chain saturated and unsaturated fatty acids make up the largest proportion of fatty acids in fats and oils. Minor amounts of odd-carbon-number acids, branched-chain acids, and hydroxy acids may also be present. Processed fats, especially hydrogenated fats, may contain a variety of geometric and positional isomers. The division of fatty acids into saturated and unsaturated groups is important because it generally reflects on the melting properties of the fat of which they are a part. There are some exceptions to this rule. Short-chain saturated fatty acids such as those present in milk fats and lauric fats have

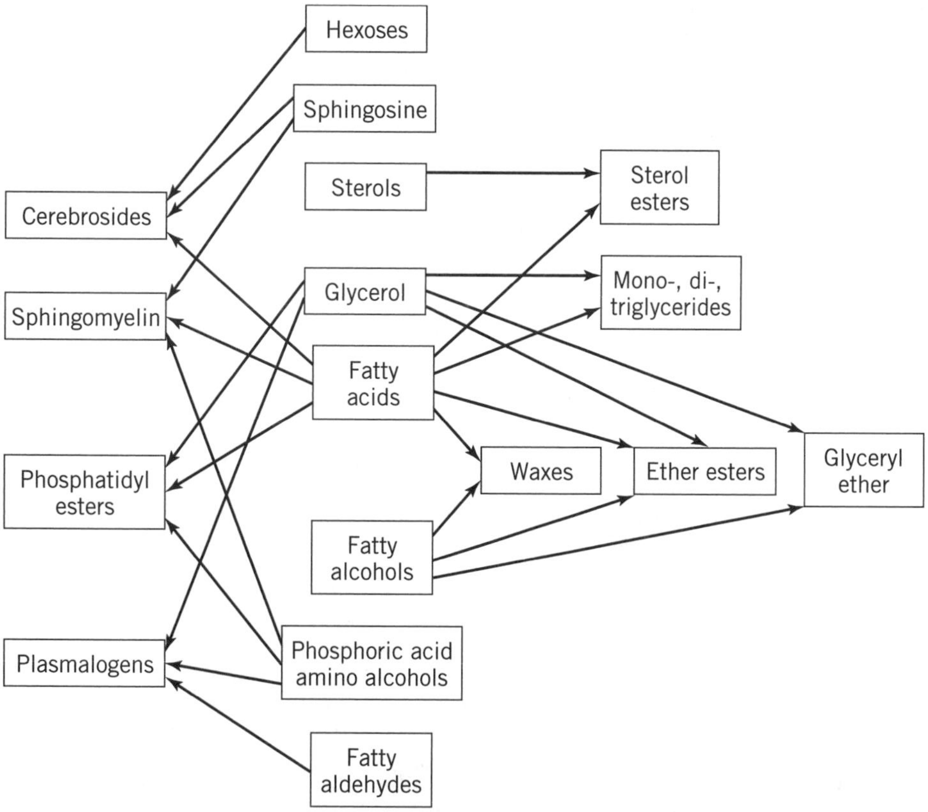

Figure 1. Interrelationship of the lipids.

very low melting points; unsaturated trans isomers have much higher melting points than the cis isomers and are therefore comparable to saturated fatty acids in their effect on melting characteristics.

Some of the important saturated fatty acids are listed in Table 1 with their systematic and common names. The unsaturated fatty acids are listed in Table 2. The naturally occurring unsaturated fatty acids are almost exclusively in the cis form. Trans acids are abundant in ruminant milk fats and in hydrogenated fats.

The depot fats of higher land animals consist mainly of palmitic, oleic, and stearic acids and are high in saturation. The fats of birds are somewhat more complex. The fatty acid compositions of the major food fats of this group are listed in Table 3. The kind of feed consumed by the animals greatly influences the fatty acid composition of the depot fats. For example, the high linolenic acid content of the horse fat in Table 3 is the result of pasture feeding. Animal depot fats are generally low in polyunsaturated fatty acids. The iodine value of beef fat is about 50 and of lard about 60.

Ruminant milk fat has an extremely complex fatty acid composition. The following fatty acids are present in cow's milk fat (3): even and odd saturated acids from 2:0 to 28:0; even and odd monoenoic acids from 10:1 to 26:1 with the exception of 11:1, and-including positional and geometric isomers; even unsaturated fatty acids from 14:2 to 26:2 with some conjugated trans isomers; monobranched fatty acids 9:0 and 11:0 to 25:0; some iso and some anteiso (iso acids have a methyl branch on the penultimate carbon, anteiso on the next to penultimate carbon; multibranched acids from 16:0 to 28:0, both odd and even with three to five methyl branches; and, finally, a number of keto, hydroxy, and cyclic acids.

Marine oils also contain a large number of different fatty acids. As many as 50 to 60 fatty acids have been reported (4). The 14 major ones consist of few saturated fatty acids (14:0, 16:0, and 18:0) and a larger number of unsaturated fatty acids with 16 to 22 carbon atoms and up to 6 double bonds. This provides the possibility for many positional isomers. It is customary to number carbon atoms of

Table 1. Saturated Even- and Odd-Carbon-Numbered Fatty Acids

Systematic name	Common name	Formula	Shorthand description
n-Butanoic	Butyric	$CH_3 \cdot (CH_2)_2 \cdot COOH$	4:0
n-Hexanoic	Caproic	$CH_3 \cdot (CH_2)_4 \cdot COOH$	6:0
n-Octanoic	Caprylic	$CH_3 \cdot (CH_2)_6 \cdot COOH$	8:0
n-Decanoic	Capric	$CH_3 \cdot (CH_2)_8 \cdot COOH$	10:0
n-Dodecanoic	Lauric	$CH_3 \cdot (CH_2)_{10} \cdot COOH$	12:0
n-Tetradecanoic	Myristic	$CH_3 \cdot (CH_2)_{12} \cdot COOH$	14:0
n-Hexadecanoic	Palmitic	$CH_3 \cdot (CH_2)_{14} \cdot COOH$	16:0
n-Octadecanoic	Stearic	$CH_3 \cdot (CH_2)_{16} \cdot COOH$	18:0
n-Eicosanoic	Arachidic	$CH_3 \cdot (CH_2)_{18} \cdot COOH$	20:0
n-Docosanoic	Behenic	$CH_3 \cdot (CH_2)_{20} \cdot COOH$	22:0
n-Pentanoic	Valeric	$CH_3 \cdot (CH_2)_3 \cdot COOH$	5:0
n-Heptanoic	Enanthic	$CH_3 \cdot (CH_2)_5 \cdot COOH$	7:0
n-Nonanoic	Pelargonic	$CH_3 \cdot (CH_2)_7 \cdot COOH$	9:0
n-Undecanoic		$CH_3 \cdot (CH_2)_9 \cdot COOH$	11:0
n-Tridecanoic		$CH_3 \cdot (CH_2)_{11} \cdot COOH$	13:0
n-Pentadecanoic		$CH_3 \cdot (CH_2)_{13} \cdot COOH$	15:0
n-Heptadecanoic	Margaric	$CH_3 \cdot (CH_2)_{15} \cdot COOH$	17:0

Table 2. Unsaturated Fatty Acids

Systematic name	Common name	Formula	Shorthand description
Dec-9-enoic		$CH_2{=}CH \cdot (CH_2)_7 \cdot COOH$	10:1
Dodec-9-enoic		$CH_3 \cdot CH_2 \cdot CH{=}CH \cdot (CH_2)_7 \cdot COOH$	12:1
Tetradec-9-enoic	Myristoleic	$CH_3 \cdot (CH_2)_3 \cdot CH \cdot CH \cdot (CH_2)_7 \cdot COOH$	14:1
Hexadec-9-enoic	Palmitoleic	$CH_3 \cdot (CH_2)_5 \cdot CH{=}CH \cdot (CH_2)_7 \cdot COOH$	16:1
Octadec-6-enoic	Petroselinic	$CH_3 \cdot (CH_2)_{10} \cdot CH{=}CH \cdot (CH_2)_4 \cdot COOH$	18:1
Octadec-9-enoic	Oleic	$CH_3 \cdot (CH_2)_7 \cdot CH{=}CH \cdot (CH_2)_7 \cdot COOH$	18:1
Octadec-11-enoic	Vaccenic	$CH_3 \cdot (CH_2)_5 \cdot CH{=}CH \cdot (CH_2)_9 \cdot COOH$	18:1
Octadeca-9:12-dienoic	Linoleic	$CH_3(CH_2)_4 \cdot (CH{:}CH \cdot CH_2)_2 \cdot (CH_2)_6 \cdot COOH$	18:2ω6
Octadeca-9:12:15-trienoic	Linolenic	$CH_3 \cdot CH_2 \cdot (CH{=}CH \cdot CH_2)_3 \cdot (CH_2)_6 \cdot COOH$	18:3ω3
Octadeca-6:9:12-trienoic	Gamma linolenic	$CH_3 \cdot (CH_2)_4 \cdot (CH{=}CH \cdot CH_2)_3 \cdot (CH_2)_3 \cdot COOH$	18:3ω6
Octadeca-9:11:13-trienoic	Elaeostearic	$CH_3 \cdot (CH_2)_3 \cdot (CH{=}CH)_3 \cdot (CH_2)_7 \cdot COOH$	20:3
Eicos-9-enoic	Gadoleic	$CH_3 \cdot (CH_2)_9 \cdot CH{=}CH \cdot (CH_2)_7 \cdot COOH$	20:1
Eicosa-5:8:11:14-tetraenoic	Arachidonic	$CH_3 \cdot (CH_2)_4 \cdot (CH{=}CH \cdot CH_2)_4 \cdot (CH_2)_2 \cdot COOH$	20:4ω6
Eicosa-5:8:11:14:17-pentaenoic acid	EPA	$CH_3 \cdot CH_2 \cdot (CH{=}CH \cdot CH_2)_5 \cdot (CH_2)_2 \cdot COOH$	20:5ω3
Docosa-13-enoic	Erucic	$CH_3 \cdot (CH_2)_7 \cdot CH{=}CH \cdot (CH_2)_{11} \cdot COOH$	22:1
Docosa-4:7:10:13:16:19-hexaenoic acid	DHA	$CH_3 \cdot CH_2(CH{=}CH \cdot CH_2)_6 \cdot (CH_2) \cdot COOH$	22:6ω3

Table 3. Component Fatty Acids of Animal Depot Fats

	Fatty acids, % of total									
Animal	12:0	14:0	16:0	18:0	20:0	16:1	18:1	18:2	18:3	20:1
Cow		6.3	27.4	14.1			49.6	2.5		
Pig		1.8	21.8	8.9	0.8	4.2	53.4	6.6	0.8	0.8
Sheep		4.6	24.6	30.5			36.0	4.3		
Goat	3.5	2.1	25.5	28.1	2.4		38.4			
Horse	0.4	4.5	25.9	4.7	0.2	6.8	33.7	5.2	16.3	2.3
Chicken	1.9	2.5	36.0	2.4		8.2	48.2	0.8		
Turkey	0.1	0.8	20.0	6.4	1.3	6.2	38.4	23.7	1.6	

fatty acids starting from the carboxyl end; however, for biological activity it is interesting to number from the methyl carbon, which is done by using the symbol ω. Three different types of fatty acids can be distinguished, the oleic type with one double bond removed nine carbons from the methyl end (18:1ω3). Three different types of unsaturated fatty acids can be distinguished: the oleic type with one double bond removed nine carbon atoms from their methyl end (18:1ω9 or oleic acid type); the linoleic type with two double bonds removed six carbon atoms from the methyl end (18:2ω6 or linoleic acid type); and the linoleic type with three double bonds removed three carbon atoms from the methyl end (18:3ω3 or linoleic acid type). The end structure is usually retained even if additional double bonds are introduced or if additional carbon atoms are added. Thus, linoleic acid (18:2ω6) may be changed into arachidonic acid (20:4ω6) while retaining the ω6 structure which confers essential fatty acid character to the molecule. The latter two types are now often referred to as n-6 and n-3 fatty acids. The high content of polyenoic fatty acids makes fish oils highly susceptible to autoxidation. The component fatty acids of some marine and freshwater fish oils are listed in Table 4 (5).

Considerable interest has developed recently in the health effect of certain n-3 fatty acids, especially eicosapentaenoic acid (EPA) (20:5ω3) and docosahexaenoic acid (DHA) (22:6ω3). These fatty acids can be produced slowly from linoleic acid by herbivore animals, but not by humans. EPA and DHA occur in major amounts in fish from cold deep waters, such as cod, mackerel, tuna, swordfish, sardines, and herring (6,7). Arachidonic acid is the precursor in the human system of prostanoids and leukotrienes.

The vegetable oils and fats can be divided into three groups on the basis of their fatty acid composition. The first group comprises oils containing mainly 16- and 18-carbon fatty acids and includes most of the seed oils—cottonseed oil, peanut oil, sunflower oil, corn oil, sesame oil—as well as palm oil. The second group comprises seed oils containing erucic acid (docos-13-enoic), and includes rapeseed and mustard oil. The third group is that of the vegetable fats, comprising coconut oil and palm kernel oil, which are highly saturated and also known as lauric fats, as well as cocoa butter. The component fatty acids of some of the common vegetable oils and fats are listed in Table 5. Palmitic is the most common saturated fatty acid. Oils containing high levels of linolenic acid are susceptible to rapid oxidative deterioration.

The Crucifera seed oils, including rapeseed and mustard oil, are characterized by high levels of erucic acid (docos-13-enoic) and smaller amounts of eicos-11-enoic acid. Plant breeders have succeeded in replacing virtually all of these fatty acids by oleic acid, resulting in what is now known as canola oil (8).

Cocoa butter is unusual in that it contains only three major fatty acids, palmitic, stearic, and oleic, in approximately equal proportions.

COMPONENT GLYCERIDES

When a fat or oil is characterized by the determination of its component fatty acids, there still remains the question of how these acids are distributed among and within the glycerides. The stereospecific numbering (sn) of glycerol for a triacid glyceride is as follows:

Table 4. Major Component Fatty Acids of Some Marine and Freshwater Fish Oils

	14:0	16:0	18:0	16:1	18:1	20:1	22:1	18:2ω6	18:3ω3	18:4ω3	20:4ω6	20:5ω3	22:5ω3	22:6ω3
Herring	6.4	12.7	0.9	8.8	12.7	14.1	20.8	1.1	0.6	1.7	0.3	8.4	0.8	4.9
Turbot	6.5	12.0	0.9	15.4	17.4	18.6	17.8	0.5	0.2	0.8	0.1	3.0	0.6	1.9
Sablefish	6.7	11.1	1.9	6.6	29.0	18.1	14.8	0.7	0.2	0.3	0.3	1.4	0.5	1.0
Cod	1.4	19.6	3.8	3.5	13.8	3.0	1.0	0.7	0.1	0.4	2.5	17.0	1.3	29.8
Sole	4.3	16.5	2.4	14.4	12.2	3.9	Tr	0.3	2.0	1.6	4.0	11.9	10.6	7.0
Halibut	0.8	9.6	9.0	2.5	12.3	4.0	5.0	Tr	Tr		1.4	13.0	2.5	37.6
Carp	3.1	16.8	4.3	17.1	28.3	3.9		13.2	2.3		2.5	3.2		
Trout	2.7	20.9	8.3	3.9	18.4			7.3	1.6	3.2	1.7	5.8	Tr	7.0
Catfish	1.0	15.2	3.9	2.9	29.7	0.9		10.0	0.5	0.4	0.8	0.2	0.2	0.6

Source: Ref. 5.
Note: Values are percentage of total: Tr = trace.

Table 5. Component Fatty Acids of Some Vegetable Oils

	Fatty acid									
Oil	14:0	16:0	18:0	20:0	22:0	16:1	18:1	18:2	18:3	Total C_{18}
Cottonseed	1	29	4	Tr		2	24	40		68
Peanut	Tr	6	5	2	3	Tr	61	22		88
Sunflower		4	3				34	59		96
Corn		13	4	Tr	Tr		29	54		87
Sesame		10	5				40	45		90
Olive	Tr	14	2	Tr		2	64	16		82
Palm	1	48	4				35	9		51
Soybean	Tr	11	4	Tr	Tr		25	51	9	89
Safflower	Tr	8	3	Tr			13	75	1	92

Source: Ref. 10
Note: Values are weight %. Tr = trace.

$$CH_2O_2CR^1 \quad (1)$$
$$|$$
$$R^2CO_2CH \quad (2)$$
$$|$$
$$CH_2O_2CR^3 \quad (3)$$

The molecule is shown in the Fisher projection with the secondary hydroxyl pointing to the left. The location of fatty acids in the various positions on the glycerol molecule can be determined by stereospecific analysis (9). Several theories of glyceride composition have been proposed, such as even distribution, random distribution, and restricted random distribution. The distribution of fatty acids in the glycerides is of utmost importance for the physical properties of a fat. This is illustrated by pig fat and cocoa butter, which have similar fatty acid composition. In pig fat most of the unsaturation is located in the 1- and 3-positions, in cocoa butter it is in the 2-position (Table 6).

PHOSPHOLIPIDS

All fats and oils and fat-containing foods contain a number of phospholipids. The lowest amounts are present in animal fats such as lard and beef tallow. In some crude vegetable oils, such as cottonseed, corn, and soybean oils, phospholipids may be present at levels of 2 to 3%. Phospholipids are surface active, because they contain a lipophilic and a hydrophilic portion. Since they can be easily hydrated, they can be removed from fats and oils during the refining process. The structure of the most important phospholipids is given in Figure 2. The phospholipids removed from soybean oil are used as emulsifiers in foods. Soybean phospholipids, also known as soy lecithin, contain about 35% lecithin and 65% cephalin. The acyl groups in phospholipids are usually more unsaturated than those of the triglycerides in which they are present. Saturated fatty acids are found mostly in position 1 and unsaturated fatty acids in position 2.

UNSAPONIFIABLE COMPONENTS

The unsaponifiable portion of fats consists of sterols, terpenic alcohols, squalene, and hydrocarbons. In most fats the major unsaponifiable component is sterols. Animal fats contain cholesterol, and in some cases, minor amounts of the other sterols, such as lanosterol. Plant fats and oils contain phytosterols, usually at least three and sometimes four (10). The predominant phytosterol is β-sitosterol; others are campesterol and stigmasterol. The sterols are solids with high melting points; part of the sterols in natural fats are present as esters of fatty acids, part in free form. Cholesterol makes up 99% of the sterols of fish. The sterol content of some fats and oils is given in Table 7.

PROCESSING

In the commercial production of fats and oils, processing is used to separate, purify, and modify the oils and fats to make them suitable for the various functions they fulfill in the food system. A large portion of the seed oils produced in the temperate regions of the world are used in the form of solid fats: margarine, shortening, and frying and baking

Table 6. Positional Distribution of Fatty Acids in Pig Fat and Cocoa Butter

		Fatty acid, mole %					
Fat	Position	14:0	16:0	16:1	18:0	18:1	18:2
Pig fat	1	0.9	9.5	2.4	29.5	51.3	6.4
	2	4.1	72.3	4.8	2.1	13.4	3.3
	3	0	0.4	1.5	7.4	72.7	18.2
Cocoa butter	1		34.0	0.6	50.4	12.3	1.3
	2		1.7	0.2	2.1	87.4	8.6
	3		36.5	0.3	52.8	8.6	0.4

$$CH_2OCOR$$
$$|$$
$$CHOCOR$$
$$|$$
$$CH_2O-PO_2{}^--OCH_2CH_2N^+(CH_3)_3$$

Phosphatidylcholine
(lecithin)

$$CH_2OCOR$$
$$|$$
$$CHOCOR$$
$$|$$
$$CH_2O-PO(OH)-OCH_2CH_2NH_2$$

Phosphatidylethanolamine
(cephalin)

$$CH_2OCOR$$
$$|$$
$$CHOCOR$$
$$|$$
$$CH_2-PO(OH)-OCH_2CH(COOH)NH_2$$

Phosphatidylserine

$$CH_2OCOR$$
$$|$$
$$CHOCOR$$
$$|$$
$$CH_2O-PO(OH)-O-$$

Phosphoinositides

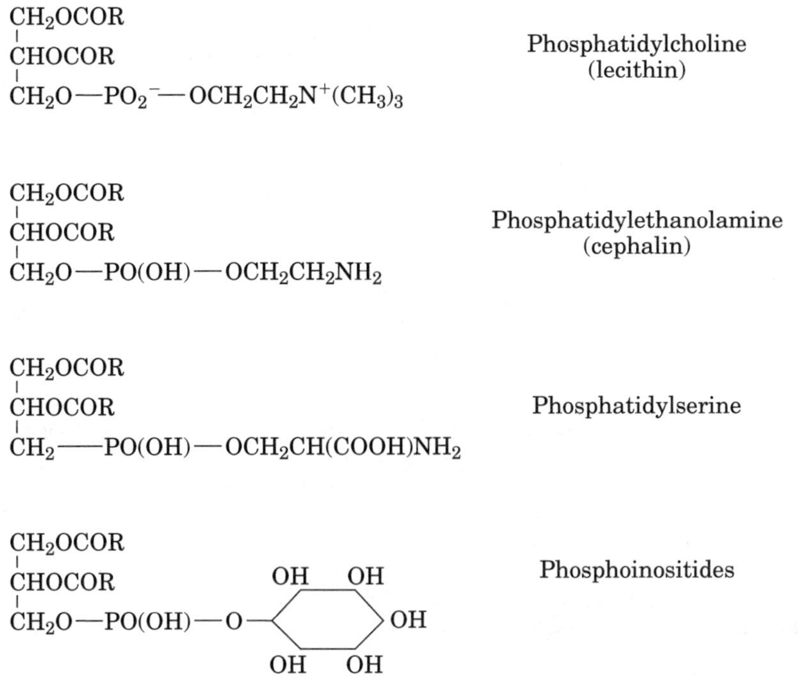

Figure 2. Structure of the major phospholipids.

fats. Hydrogenation is most often used to change oils into fats. Another use for hydrogenation is to improve the oxidative stability by partial hydrogenation to remove most of the linolenic acid.

The separation of oils and fats from animal tissues is done by rendering, either dry rendering or steam rendering. The separation of oils from oilseeds usually involves a pretreatment, crushing or flaking of the seeds, followed by pressing (11). This is usually followed by solvent extraction to remove the remainder of the oil and yield a residue with less than 1% residual oil (12).

The crude oils obtained by rendering, pressing, and/or extraction are purified by a series of operations designed to remove impurities that may detract from the quality of the oil. Removal of phospholipids is achieved by degumming (13). The crude oils are treated with steam, which hydrates the phospholipids and makes them settle out. Degumming can also be achieved by using solutions of phosphoric or organic acids. The soybean "gums" are purified and used as food emulsifiers, known as soy lecithin.

Free fatty acids in crude oils are removed by alkali refining (13). Solutions of caustic soda are used to reduce the level of free fatty acid to 0.01 to 0.03%. Care is required to prevent saponification of neutral oil. Removal of free fatty acids can also be achieved by physical refining. This involves treatment of the oils under vacuum with steam. The advantage of physical refining is that the process is similar to deodorization and these processes can be combined. A possible disadvantage of physical refining relates to the high temperatures (up to 270°C) employed in this process. This may cause randomization of the glyceride structure, formation of dimers and conjugated fatty acids (positional isomerization) and *cis–trans* isomerization (14).

Bleaching is used to remove colored impurities, such as carotenoids and chlorophyll. In the bleaching process the

oils are treated with bleaching earth or activated carbon. The yellow-red color of most vegetable oils, mostly carotenoids, is easily removed by bleaching earth. The green and brown pigments are more difficult to remove.

After refining and bleaching, vegetable oils are further processed into margarines, shortenings, and frying and baking fats. Two-thirds of all liquid oils produced in North America are used in the form of fats. Hydrogenation is used to change oils into fats, and involves the reaction of gaseous hydrogen, liquid oil, and solid catalyst under pressure and at high temperature (15). The catalyst used for edible oil hydrogenation is invariably of the activated nickel metal type (16). The hydrogenation reaction can be represented by the following scheme, in which the reacting species are the olefinic substrate (S), the metal catalyst (M), and hydrogen:

$$S + M \rightleftharpoons \left[S{-}M \right]$$
$$(1)$$

$$\left[S{-}M{-}H_2 \right] \longrightarrow SH_2 + M$$
$$(3)$$

$$M + H_2 \rightleftharpoons \left[M{-}H_2 \right]$$
$$(2)$$

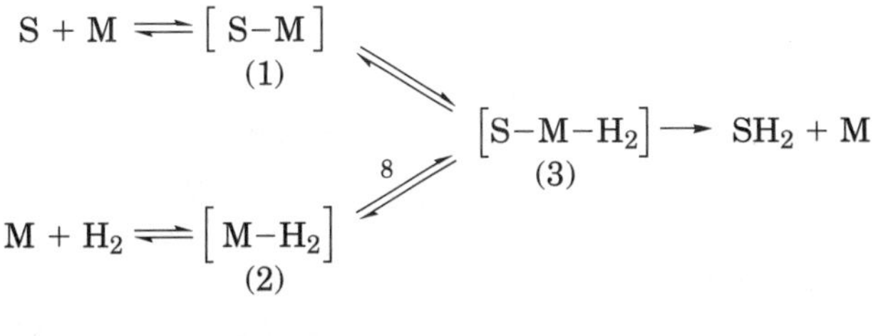

The intermediates **1**, **2**, and **3** are organometallic species and are labile and short-lived and cannot usually be isolated. In heterogeneous catalysis the metal surface performs the catalytic function. In theory, the finer the particle size, the more active the catalyst will be. In practice, however, particle size has to be balanced against filterability, since removal of the catalyst at the end of the process should not be too difficult.

When hydrogen is added to double bonds in natural fats and the reaction is not carried to completion, a complex

Table 7. Sterol Content of Fats and Oils

Fat	Sterol, %
Lard	0.12
Beef tallow	0.08
Milk fat	0.3
Herring	0.2–0.6
Cottonseed	1.4
Soybean	0.7
Corn	1.0
Rapeseed	0.4
Coconut	0.08
Cocoa butter	0.2

mixture of reaction products results. Hydrogenation may be selective or nonselective. Selectivity means that hydrogen is added first to the most unsaturated fatty acids. Selectivity is increased by increasing hydrogenation temperature and decreased by increasing pressure and agitation. Selectively hydrogenated oil is more resistant to oxidation because of the preferential hydrogenation of the linolenic acid.

Another important factor in hydrogenation is the formation of positional and geometric isomers. Formation of trans isomers is rapid and extensive. The isomerization can be understood by the reversible character of chemisorption. When the olefinic bond reacts, two carbon-metal bonds are formed as an intermediate stage. The intermediate may react with an atom of adsorbed hydrogen to yield the half-hydrogenated compound, which remains attached by only one bond. Additional reaction with hydrogen would result in formation of a saturated compound. There is also the possibility that the half-hydrogenated olefin may again attach itself to the catalyst surface at a carbon on either side of the existing bond, with simultaneous loss of hydrogen. Upon desorption of this species a positional or geometric isomer may result. The proportion of trans isomers is high because this is the more stable configuration. Double bond migration occurs in both directions, but probably more extensively in the direction of the terminal methyl group. The hydrogenation of oleate can be represented as follows:

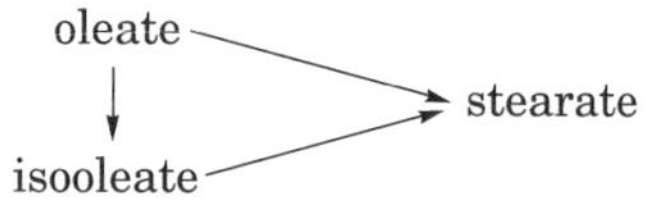

The change from oleate to isooleate involves no change in unsaturation but does result in a considerably higher melting point. This is why the hardening effect of hydrogenation is only partly the result of saturating double bonds; trans-isomer formation has a major effect on hardness.

For example, olive oil with an iodine value of 80 is liquid at room temperature. When soybean oil is hydrogenated to the same iodine value it is a fat with the consistency of lard.

Hydrogenation of linoleate first produces some conjugated dienes, followed by the formation of positional and geometric isomers of oleic acid, and finally stearate:

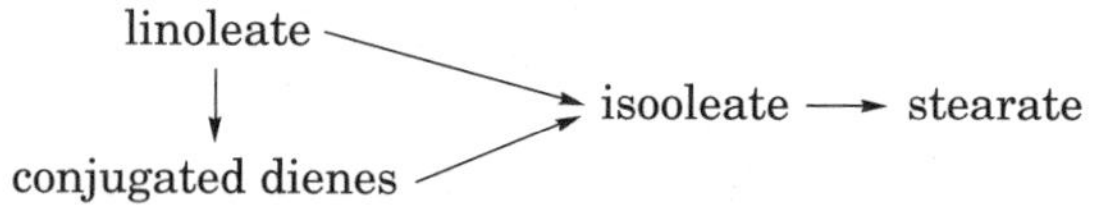

Hydrogenation of linolenate is more complex and is greatly dependent on reaction conditions. The possible reactions can be summarized as follows:

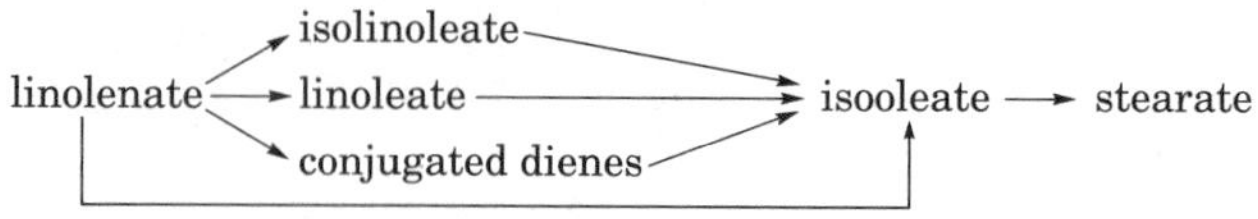

In the nonselective hydrogenation of seed oils, polyunsaturated fatty acids are rapidly isomerized or reduced and trans isomers increase to high levels (Fig. 3).

Interesterification is a process whereby fatty acid radicals can be made to move from one hydroxyl of a glycerol moiety to another one, either within the same glyceride or to another glyceride. The reaction pattern has been described (17):

$$RCOOR^2 + R^1COOR^3 \rightarrow RCOOR^3 + R^1COOR^2$$

The reaction is used in industry to modify the crystallization behavior and the physical properties of fats. The catalysts are usually alkaline and consist of sodium methoxide or alloys of sodium and potassium. At temperatures above the melting points of the reactants, several raw materials may be interesterified together so that new products are produced. If the reaction is carried out below the melting point, so that only the liquid fraction reacts, the process is called directed interesterification. Industrially, lard has been interesterified to improve its properties. Lard has a narrow plastic range, creams poorly, and gives poor cake volume. After interesterification these properties are greatly improved. Acetoglycerides can be prepared

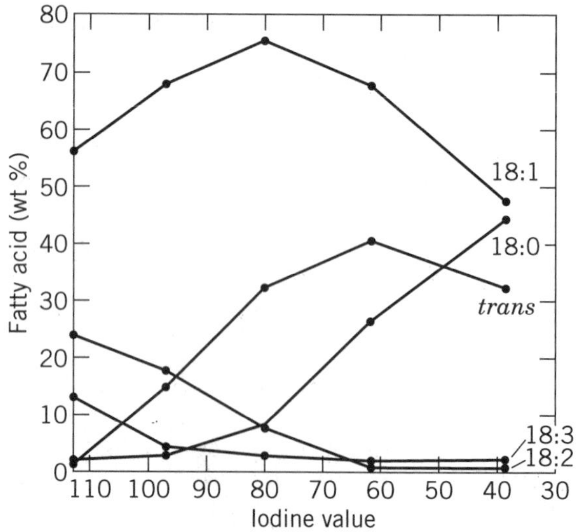

Figure 3. Change in fatty acid composition during hydrogenation of canola oil.

by interesterification of natural fats with glyceryl triacetate. The resultant products are waxy, translucent materials that can be used as edible coatings and plasticizers.

Interesterification may provide an alternative to hydrogenation for the production of margarine and shortening fats. Hydrogenation has the disadvantage of forming trans isomers and of losing essential fatty acids. Interesterification of liquid oils with highly saturated fats, obtained by complete hydrogenation or by fractionation, may result in nutritionally more desirable fats.

Ester interchange of fats with a large excess of glycerol, at high temperature, under vacuum, and in the presence of a catalyst, results in an equilibrium mixture of mono-, di-, and triglycerides. After removal of excess glycerol, the mixture is called technical monoglyceride. Technical monoglycerides are used as emulsifying agents in foods. Molecular distillation yields products with well over 90% 1-monoglycerides; these are also widely used in foods.

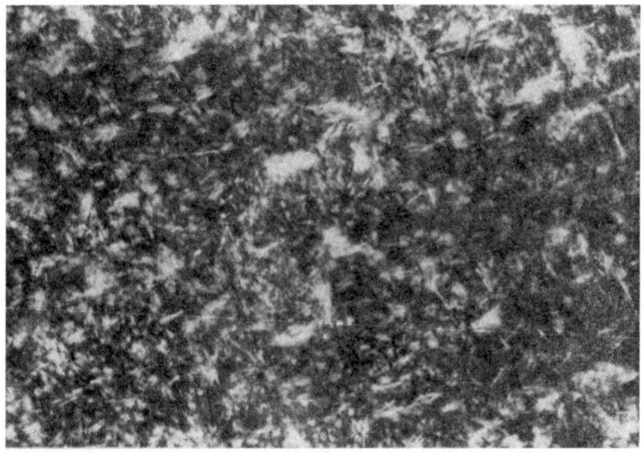

Figure 4. Fat crystals in a fat as seen in the polarizing microscope.

PHYSICAL PROPERTIES

Fats and oils are long-chain compounds that have particular physical properties of importance in processing and final use. Properties of surface activity, viscosity, solubility, and melting behavior are important in formulating emulsions and fat products such as shortenings and margarines.

The most important aspect of the physical properties of fats and oils is related to the solid-liquid and liquid-solid phase changes; in other words, melting and solidification. In fact, fats can be defined as partially solidified oils. When oils solidify they form crystals, usually of a size in the range of 1 to 10 μm. These fat crystals form a three-dimensional network that lends solid properties to the fat. The nature of this crystal network determines the rheological or textural properties of the product (18). The amount and size of the crystals in a fat determine its physical properties by influencing the density of junction points and, therefore, the strength of the crystal network. The crystals are held together by weak van der Waals forces and these can be disrupted by working or kneading. Upon resting, many of these bonds may be reestablished and the product is said to be thixotropic. Not all bonds are reformed after working, leading to what is known as work softening. The crystals in fats can be observed by polarized light microscopy as shown in Figure 4 (19).

The proportion of solids in a fat is of major importance in determining the rheological properties of a product. Fats may retain their solid character with solid fat content as low as 10%. Desirable spreadability occurs in a narrow range of solid fat content, roughly 15–35%. This is called the plastic range of fats.

Solidification of liquid oil results in a volume contraction and a positive (exothermic) heat effect. Melting of a fat results in volume expansion and a negative (exothermic) heat effect. Traditional methods of determination of solid fat depend on specific volume measurement at a given temperature (dilatometry). The heat effect is used in differential scanning calorimetry to determine melting and solidification properties. Modern methods of solid fat determination are based on nuclear magnetic resonance. Protons in solid fat behave differently in a magnetic field after being excited by radio-frequency energy than protons in liquid fat. This enables rapid and accurate estimation of solid fat content. The dilatometric method is empirical and the results are expressed as solid fat index (SFI).

In addition to crystal content and crystal size and shape, an important factor in the solidification of fats is polymorphism (the existence of different crystal modifications). Polymorphism results from different patterns of molecular packing in fat crystals. Triglycerides may occur in three main forms named α, β', and β, in order of increasing stability. After crystals of a lower melting form have been produced, change into a higher melting form may take place. The change is monotropic, that is, it always proceeds in the sequence from lower to higher stability. When a fat is crystallized in an unstable form and heated at a temperature slightly above its melting point, it may resolidify in a more stable form (18). The melting point of tristearin in the α form is 54.7°C; in the β' form, 64.0°C; and in the β form, 73.3°C. The cross-sectional structures of long-chain compounds in the different polymorphic forms is shown in Figure 5. In the α form the chain axes are randomly oriented and the crystals are of the hexagonal type. In the β' form alternate rows are oriented in opposite directions and the crystal type is orthorhombic. In the β form the rows are oriented in the same direction and the crystal type is triclinic. The polymorphic forms are also distinguished by differences in the angle of tilt of the unit

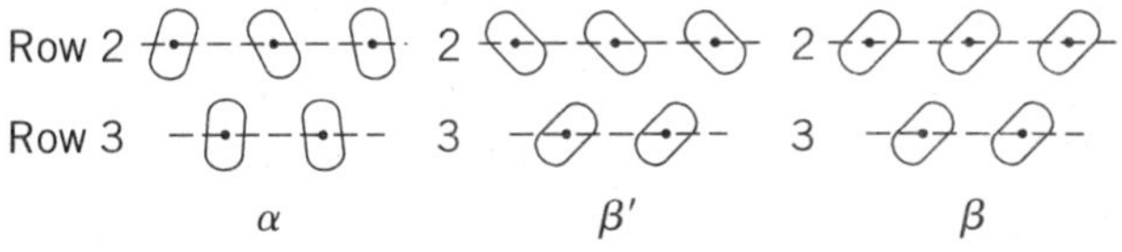

Figure 5. Cross-sectional structures of long-chain compounds. *Source:* Ref. 18, courtesy of *Journal of the American Oil Chemists' Society*.

cell in the crystal. X-ray diffraction analysis enables measurement of the short spacings and long spacings of the unit cell. Principal short spacings of the polymorphic forms are: α, 4.15 Å; β', 4.2 and 3.8 Å; β, 4.6, 3.85, and 3.7 Å (20). Fat crystals of different polymorphs have different crystal habit. That is, they appear to have different shapes when seen in the microscope: α crystals are fragile, transparent platelets of about 5-μm size; β' crystals are tiny needles about 1 μm long; β crystals are large and coarse, averaging 25 to 50 μm in size (21). These crystals can now be made visible by scanning electron microscopy (22).

For the production of margarines and most shortenings, the fat should be present in the β' form. Some fats have a tendency to convert slowly from the β' to the β form and this results in a coarsening of the crystal structure, making the product unacceptable. Fats which contain a very high level of fatty acids with identical chain lengths are especially prone to this effect. Canola and sunflower oils are notorious in this respect and have well over 90% of C_{18} fatty acids (Table 5).

The melting point of a fat is basically determined by the melting points of its constituent fatty acids. When we speak of the melting point of a fat we mean the end of the melting range, since only simple substances have a sharp melting point. The end of the melting range of a fat is dependent on the method used. Chain length and unsaturation of fatty acids determine their melting point. In addition, the configuration around the double bond is important, as is the arrangement of fatty acids at the different positions of the glycerides. Trans-unsaturated fatty acids have much higher melting points than their cis counterparts, making them more comparable with saturated fatty acids. For example, oleic acid (*cis*) has a melting point of 13°C and elaidic acid (*trans*) has a melting point of 44°C.

AUTOXIDATION AND ANTIOXIDANTS

The unsaturated bonds present in all oils and fats represent active centers, which may react with oxygen. This reaction leads to the formation of primary, secondary, and tertiary oxidation products which may result in the fat or fat-containing food becoming unfit for consumption. Oxidation of fats and oils may occur by a free-radical chain reaction known as autoxidation. Another process of oxidative deterioration involves the presence of a sensitizer and exposure to light. This is known as photosensitized oxidation (23,24).

The process of autoxidation and the resulting deterioration in flavor of fats and fatty foods are often described by the term rancidity. Lundberg (25) distinguishes several types of rancidity. When fats are exposed to oxygen, common oxidative rancidity will result in sweet but undesirable odors and flavors which will progressively become more intense and unpleasant. Flavor reversion is the term used for the objectionable flavors that develop in oils containing linolenic acid, especially soybean oil. This type of oxidation is produced with considerably less oxygen than common oxidation.

Among the many factors which affect the rate of oxidation are the following: amount of oxygen present, degree of unsaturation of the oil, presence of pro- and antioxidants, presence of heme-containing molecules and lipoxidase, light exposure, nature of packaging material, and storage temperature.

The autoxidation reaction can be divided into the following three parts: initiation, propagation, and termination. During the initiation, hydrogen is abstracted from an olefinic compound to yield a free radical:

$$RH \rightarrow R^{\cdot} + H^{\cdot}$$

The removal of hydrogen takes place at the carbon atom next to the double bond. The dissociation energy of hydrogen in various olefinic compounds has been listed by Ohloff (26). The value for an isolated double bond is 103 kcal/mol; for two double bonds separated by a methylene group it is only 65 kcal/mol. Once a free radical is formed, it will combine with oxygen to form a peroxy free radical, which can in turn abstract hydrogen from another unsaturated molecule to yield a peroxide and a new free radical, thus starting the propagation reaction. This may be repeated up to several thousand times and has the nature of a chain reaction:

$$R^{\cdot} + O_2 \rightarrow RO_2^{\cdot}$$

$$RO_2^{\cdot} + RH \rightarrow ROOH + R^{\cdot}$$

The propagation reaction can be followed by termination if the free radicals react with themselves to yield nonactive products:

$$R^{\cdot} + R^{\cdot} \rightarrow R\text{-}R$$

$$R^{\cdot} + RO_2^{\cdot} \rightarrow RO_2R$$

$$n RO_2^{\cdot} \rightarrow (RO_2)_n$$

The hydroperoxides formed in the propagation part of the reaction are the primary oxidation products. The hydroperoxide mechanism of autoxidation has been described (27) and reviewed (23). The primary oxidation products are unstable and decompose into secondary oxidation products, mostly carbonyls. The peroxides are not important in flavor deterioration, which is caused wholly by secondary oxidation products. In the initial stages of the reaction there is a slow increase in the amount of hydroperoxides formed; this is the induction period. At the end of the induction period there is a sudden and rapid increase in peroxide content. The induction period is measured in accelerated tests to determine the storage stability of a fat or oil.

The rate of oxidation depends greatly on the degree of unsaturation. In the series of 18-carbon-atom fatty acids 18:0, 18:1, 18:2, 18:3, the relative rates of oxidation have been reported to be in the ratio of 1:100:1200:2500. In addition to the degree of unsaturation, the position of the double bonds in a polyunsaturated fatty acid may affect its oxidation rate. Zhan and Chen (28) found that conjugated linolenic acid oxidized considerably faster than linolenic acid. The reaction of unsaturated compounds proceeds by the abstraction of hydrogen from a carbon α to the double bond, resulting in a free radical stabilized by resonance. These free radicals are then transformed into a number of isomeric hydroperoxides of the general structure

$$\begin{array}{l}\text{-CH}_2\text{-CH-CH=CH-}\\ \qquad\quad |\\ \qquad\text{OOH}\end{array}$$

In addition to the changes in double bond position, there are isomerizations from cis to trans and 90% of the peroxides formed may be in the trans configuration (25).

The decomposition of hydroperoxides has been described (29). It involves decomposition to alkoxy and hydroxy aldehydes, which are in great part responsible for the oxidized flavor of fats. They are powerful flavor compounds with very low flavor thresholds. For example, 2,4-decadienal has a flavor threshold of less than one part per billion.

Trace metals, especially copper, and to a lesser extent iron, will catalyze fat oxidation; metal deactivators, such as citric acid, can be used to counteract the effect of metals.

Antioxidants may be present naturally or added to fats and oils. Many foods contain natural antioxidants; the tocopherols are the most important of these. They are present in greater amounts in vegetable oils than in animal fats, which may explain the greater stability of the former. Antioxidants react with free radicals, thereby terminating the chain reaction. The synthetic antioxidants are phenolic compounds and limited in number. The four most widely used are BHA (butylated hydroxy anisole), BHT (butylated hydroxy toluene), PG (propyl gallate), and TBHQ (tertiary butyl hydroquinone). BHT is volatile with steam, which makes it less suitable for use in frying oils and baked products. Nonvolatile antioxidants are said to have carry-through properties for these applications.

EMULSIONS AND EMULSIFIERS

An emulsion is a heterogeneous system of two immiscible liquids, one of which is intimately dispersed in the other. The droplets of the dispersed phase generally have a diameter of over 0.1 μm. The emulsions are stabilized by the presence of a third component, the emulsifier, a surface-active agent which is partly soluble in both phases (30). Food emulsions usually contain water and oil as the immiscible phases, giving rise to emulsions of the oil-in-water (O/W) type or water-in-oil (W/O) type. The action of emulsifiers can be enhanced by the presence of stabilizers. Emulsifiers are surface-active compounds that have the ability to reduce the interfacial tension between liquid-liquid and air-liquid interfaces. This ability is the result of an emulsifier's molecular structure: the molecules contain two distinct sections, one having a polar or hydrophilic character, the other having nonpolar or hydrophobic properties. Most surface-active agents reduce the surface tension from about 50 dynes/cm to less than 10 dynes/cm when used in concentrations below 0.2%.

The relative sizes of the hydrophilic and hydrophobic sections of an emulsifier molecule mostly determine its behavior in emulsification. To select the proper emulsifier for a given application, the so-called HLB system was developed (HLB = hydrophile-lipophile balance). It is a numerical expression for the relative simultaneous attraction of an emulsifier for water and for oil. The HLB of an emulsifier is an indication of how it will behave but not how efficient it is. Emulsifiers with low HLB tend to form W/O emulsions, those with intermediate HLB form O/W emulsions, and those with high HLB are solubilizing agents. The HLB value can be either calculated or determined experimentally (31). The scale goes from 0 to 20, at least in theory, since at each end of the scale the compounds would have little emulsifying activity. The HLB value of some commercial nonionic emulsifiers is given in Table 8 (32).

Foods contain many natural emulsifiers, of which phospholipids are the most common.

Emulsions are stabilized by a variety of compounds, mostly macromolecules such as proteins, starches, and gums.

NOVEL FATS AND OILS

The modification of the composition of fats and oils to improve or change their functional or nutritional properties is a well-established practice. The process of hydrogenation was invented in 1897 and is still used widely in the fats and oils industry. Other means of modifying the properties of fats and oils include interesterification, fractionation, blending, and combinations of these techniques.

Plant breeding has been used to not only produce improved yields of oilseeds but also to change the composition of many oilseeds, the first example of which was low–erucic acid rapeseed oil (LEAR), or canola oil; it is now one of the

Table 8. HLB Values of Some Commercial Nonionic Emulsifiers

Trade name	Chemical designation	HLB
Span 85	Sorbitan trioleate	1.8
Span 85	Sorbitan trioleate	1.8
Span 65	Sorbitan tristearate	2.1
Atmos 150	Mono- and diglycerides from the glycerolysis of edible fats	3.2
Atmul 500	Mono- and diglycerides from the glycerolysis of edible fats	3.5
Atmul 84	Glycerol monostearate	3.8
Span 80	Sorbitan monooleate	4.3
Span 60	Sorbitan monostearate	4.7
Span 40	Sorbitan monopalmitate	6.7
Span 20	Sorbitan monolaurate	8.6
Tween 61	Polyoxyethylene sorbitan monostearate	9.6
Tween 81	Poly(oxyethylene sorbitan monooleate)	10.0
Tween 85	Polyoxyethylene sorbitan trioleate	11.0
Arlacel 165	Glycerol monostearate (acid stable, self-emulsifying)	11.0
Myrj 45	Polyoxyethylene monostearate	11.1
Atlas G-2127	Polyoxyethylene monolaurate	12.8
Myrj 49	Polyoxyethylene monostearate	15.0
Myrj 51	Polyoxyethylene monostearate	16.0

Source: Ref. 30.

world's most widely used vegetable oils. In LEAR oil, the level of erucic acid is less than 1%, and erucic acid has been replaced by oleic acid. This oil is still about 10% linolenic acid by weight, similar to the level in soybean oil. Plant breeders have developed low-linolenic-acid varieties of canola and soybean oil. These oils have improved oxidative stability and are better suited for the purpose of frying (33). Both sunflower and safflower oils have been modified by plant breeding to attain high levels of oleic acid. These are commercially available as Sunola (about 85% oleic acid) and Saffola (about 75% oleic acid) (34).

Genetic engineering has developed into a powerful technique to modify oilseed plants. In addition to changing agronomic properties such as yield and resistance to pests and herbicides, major changes in fatty acid and glyceride composition have been achieved (35). The rapeseed plant is easily amenable to genetic manipulation and has been the basis for the development of laurate canola, which can be used as a replacement of the tropical oils obtained from coconut and palm kernel. Because of its special glyceride structure, with no lauric acid in the 2-position, this is a truly novel oil. When partially hydrogenated, this oil may have many applications in the food industry (confectionery coatings, baked goods, coffee whitener, icings, toppings, etc.).

Fat replacers are substances that are meant to partially or completely replace the calories provided by fats (9 kcal/g). They may consist of fat-like substances that are not or only partly absorbed by humans. One example is Olestra, a sucrose polyester with six to eight acyl groups. The fatty acids are derived from soybean, corn, cottonseed, or sunflower oil. Olestra is not absorbed in the human digestive system and therefore yields no calories. Other fat replacers are based on the fact that the level of 9 kcal/g does not apply when short-chain fatty acids are present in the triglycerides. It is also known that long-chain saturated fatty acids, such as stearic acid, are incompletely metabolized and yield less than 9 kcal/g. By combining these two types of fatty acids into glycerides, fats are obtained that have energy values of 5 kcal/g. A product range of this type is Salatrim (an acronym for short and long acyl triglyceride molecules). It contains at least one stearic acid and one short-chain acid in each of the glycerides. It is produced by fully hydrogenating a liquid oil (canola or soybean) and then interesterifying with short-chain fatty acids (acetic, propionic, or butyric). By varying the composition, a family of products is obtained that range from solid to liquid at room temperature (36).

Medium-chain triglycerides (MCTs) are another type of structured fat made by transesterification of caprylic and capric fatty acids derived from coconut or palm kernel oils (37).

BIBLIOGRAPHY

1. J. M. de Man, *Principles of Food Chemistry*, 3rd ed., Aspen Publishers Inc., Gaithersburg, Md., 1999.
2. L. M. Smith, "Introduction to the Symposium on Milk Lipids," *Journal of the American Oil Chemists' Society* **50**, 175–177 (1973).
3. R. G. Jensen, "Composition of Bovine Milk Lipids," *Journal of the American Oil Chemists' Society* **50**, 186–192 (1973).
4. R. G. Ackman, "The Analysis of Fatty Acids and Related Materials by Gas-Liquid Chromatography," in R. T. Holman, ed., *Progress in the Chemistry of Fats and Other Lipids*, Vol. 12, Pergamon Press, Oxford, 1972.
5. R. G. Ackman, "Marine Lipids and Fatty Acids in Human Nutrition," *FAO Technical Conference on Fishery Products, Japan*, 1973.
6. R. G. Ackman, "The Year of the Fish Oils," *Chemical Industries* March, 139–145 (1988).
7. A. P. Simopoulos, "ω-3 Fatty Acids in Growth and Development and in Health and Disease," *Nutrition Today*, 10–19 (March/April 1988).
8. J. K. Daun. "Composition and Use of Canola Seed, Oil, and Meal," *Cereal Foods World* **29**, 291–293 (1984).
9. A. Kuksis. "Newer Developments in Determination of Structure of Glycerides and Phosphoglycerides," in R. T. Holman, ed., *Progress in the Chemistry of Fats and Other Lipids*, Vol. 12., Pergamon Press. Oxford, 1972.
10. E. Fedeli and G. Jacini, "Lipid Composition of Vegetable Oils," *Advances in Lipid Research* **9**, 335–382 (1971).
11. D. K. Bredeson. "Mechanical Oil Extraction," *Journal of the American Oil Chemists Society* **60**, 211–213 (1983).
12. P. L. Christensen, "Solvent Extraction: Recent Developments," *Journal of the American Oil Chemists' Society* **60**, 214–215 (1983).
13. G. Haraldsson, "Degumming, Dewaxing and Refining," *Journal of the American Oil Chemists' Society* **60**, 251–256 (1983).
14. G. Hoffmann, *The Chemistry and Technology of Edible Oils and Fats and their High Fat Products*, Academic Press, London, 1989.
15. R. Larsson, "Hydrogenation Theory: Some Aspects," *Journal of the American Oil Chemists' Society* **60**, 275–281 (1983).
16. M. J. Beckman, "Hydrogenation Practice," *Journal of the American Oil Chemists' Society* **60**, 282–290 (1983).
17. M. W. Formo. "Ester Reactions of Fatty Materials," *Journal of the American Oil Chemists Society* **31**, 548–549 (1954).
18. J. M. de Man and A. M. Beers, "Fat Crystal Networks: Structure and Rheological Properties," *Journal of Texture Studies* **18**, 303–318 (1987).
19. E. S. Lutton. "Technical Lipid Structures." *Journal of the American Oil Chemists' Society* **49**, 1–9 (1972).
20. J. M. de Man, "Microscopy in the Study of Fats and Emulsions," *Food Microstructure* **1**, 209–222 (1982).
21. C. W. Hoerr, "Morphology of Fats, Oils and Shortenings," *Journal of the American Oil Chemists' Society* **37**, 539–546 (1960).
22. A. N. Mostafa, A. K. Smith, and J. M. de Man, "Crystal Structure of Hydrogenated Canola Oil," *Journal of the American Oil Chemists' Society* **62**, 760–762 (1985).
23. E. N. Frankel. "Lipid Oxidation: Mechanisms. Products and Biological Significance," *Journal of the American Oil Chemists' Society* **61**, 1908–1918 (1984).
24. A. Sattar and J. M. de Man, "Photo-oxidation of Milk and Milk Products: A Review," *Critical Reviews in Food Science and Nutrition* **7**, 13–38 (1975).
25. W. O. Lundberg, *Autoxidation and Antioxidants*, John Wiley & Sons, Inc., New York, 1961.
26. G. Ohloff, "Fats are Precursors," in J. Solms, ed., *Functional Properties of Fats in Foods*, Forster Publishing Ltd., Zürich, 1973.

27. E. H. Farmer, "Peroxidation in Relation to Olefinic Structure," *Transactions of the Faraday Society* **42**, 228–236 (1946).
28. A. Zhang and Z. Y. Chen, "Oxidative Stability of Conjugated Linoleic Acids Relative to Other Polyunsaturated Fatty Acids," *Journal of the American Oil Chemists' Society* **74**, 1611–1613 (1997).
29. M. Keeney, "Secondary Degradation Products," in H. W. Schultz, E. A. Day and R. O. Sinnhuber, eds., *Lipids and Their Oxidation*, AVI Publishing Co., Inc., Westport, Conn., 1962.
30. P. Becher, "Emulsions—Theory and Practice," Van Nostrand Reinhold Co., New York, 1965.
31. S. Friberg, *Food Emulsions*, Marcel Dekker Inc., New York, 1976.
32. W. C. Griffin, "Emulsions," in *Kirk-Othmer Encyclopedia of Chemical Technology*, 2nd ed., Vol. 8, 1965, pp. 117–154.
33. M. D. Erickson and N. Freg, "Property Enhanced Oils in Food Applications," *Food Technology* **48**, 63–68 (1994).
34. F. D. Gunstone, "Movements toward Tailor-Made Fats," *Progress in Lipid Research* **37**, 277–305 (1998).
35. M. Lassner, "Transgenic Oil Crops: A Transition from Basic Research to Product Development," *Lipid Technology* **9**, 5–9 (1997).
36. R. Kosmark, "Salatrim: Properties and Applications," *Food Technology* **50**, 98–101 (1996).
37. C. C. Akoh, "Making New Structured Fats by Chemical Reaction and Enzymatic Modification," *Lipid Technology* **9**, 61–66 (1997).

GENERAL REFERENCES

T. P. Hilditch and P. N. Williams, *The Chemical Constitution of Natural Fats*, 4th ed., John Wiley & Sons Inc., New York, 1964.
A. Kuksis, ed., *Handbook of Lipid Research*, Vol. 1. *Fatty Acids and Glycerides*, Plenum Press, New York, 1988.
V. C. Mehlenbacher, *The Analysis of Fats and Oils*, The Garrard Press, Champaign, Ill., 1960.
H. B. W. Patterson, *Hydrogenation of Fats and Oils*, Applied Science Publishers, London, 1983.
C. Ratledge, P. Dawson, and J. Rattray, eds., *Biotechnology for the Oils and Fats Industry*, American Oil Chemists' Society, Champaign, Ill., 1984.
D. M. Small, *Handbook of Lipid Research, Vol. 4. The Physical Chemistry of Lipids*, Plenum Press, New York, 1986.
D. Swern, ed., *Bailey's Industrial Oil and Fat Products*, 4th ed., 3 Volumes, John Wiley & Sons Inc., New York, 1979.

J. M. de Man
University of Guelph
Guelph, Ontario
Canada

FATS AND OILS: FLAVORS

Fats and oils comprising the bulk of foods such as shortenings, mayonnaise, and cooking oils often contribute unique flavors to the product. They are chemically referred to as lipids and occur in food mainly in the form of triglycerides. Fats and oils can either increase or decrease the flavor quality of foods, depending on the chemical reactions taking place during processing and storage. Under the right conditions fats and oils can be susceptible to oxidation, which produces undesirable volatile compounds and causes detrimental flavor effects to foods. The rate of oxidation in fats and oils is influenced by a wide variety of factors including oxygen, light, trace metals, antioxidants, sensitizers, enzymes, temperature, and fatty acid composition. Although thermal oxidation in deep-fat frying may provide good flavors, the volatile flavor compounds produced through autoxidation are mainly responsible for the deterioration of the flavor quality of oils during storage. Some of the most common volatile flavor compounds produced by fats and oils include esters, aldehydes, alcohols, ketones, lactones, and hydrocarbons. Unsaturated aldehydes and ketones, which have the lowest sensory threshold, are principally responsible for the undesirable oxidized flavor of food. The separation and identification of these volatile flavor compounds by gas chromatography and mass spectroscopy have increased the knowledge base for fats and oils flavor chemistry.

The isolation, separation, and identification of volatile compounds from fats and oils is difficult due to their chemical complexity, variations in concentrations and volatilities, interactions between the volatile compounds themselves, and also reactions between the volatile compounds and food components such as proteins and carbohydrates (1). The mechanisms responsible for the production of volatile flavor compounds during oxidation must be thoroughly studied and understood to improve the flavor quality of fats and oils.

MECHANISMS OF AUTOXIDATION

Autoxidation is a free-radical chain reaction that includes initiation, propagation, and termination steps:

Initiation	$RH \rightarrow R\cdot + H\cdot$
Propagation	$R\cdot + O_2 \rightarrow ROO\cdot$
	$ROO\cdot + RH \rightarrow ROOH + R\cdot$
Termination	$ROO\cdot + R\cdot \rightarrow ROOR$
	$R\cdot + R\cdot \rightarrow RR$
	$ROO\cdot + ROO\cdot \rightarrow ROOR + O_2$

Unsaturated fatty acids undergo oxidation by the above free-radical chain reaction to form hydroperoxides. The hydroperoxides are decomposed to form off-flavor compounds, which are called secondary oxidative products (2,3). Though the reaction of singlet-state fatty acids with triplet-state oxygen is thermodynamically difficult, fatty acid free radicals can react with triplet-state oxygen molecules fairly easily. Heat, metal catalysts, and both ultraviolet and visible light can accelerate the formation of free radicals of unsaturated fats during the initiation step of autoxidation. Typically, a hydrogen atom is removed from the methylene group alpha to the double bond of the fatty acid to form a fatty acid free radical, which has a short lifetime due to its tendency to achieve a stable electron pair. The alkyl radical reacts with oxygen to form a peroxy radical. This peroxy radical then reacts with a hydrogen atom that was abstracted from another unsaturated fatty acid to form a hydroperoxide. The formation of hydroperoxides in this manner is usually accompanied by a shift in the position of the double bonds, due to resonance stabilization of the fatty acid free radical and resulting formation of isomeric hydroperoxides. A mixture of four isomeric hydroperoxides that have -OOH groups at the 8, 9, 10, and 11 carbon positions are produced in the autoxidation of oleic acid. Hydroperoxides themselves are generally tasteless, colorless, and odorless; however, they are relatively unstable and therefore readily decompose to form hydrocarbons, alcohols, esters, acids, aldehydes, and ketones, each having differing molecular weights and flavor thresholds. The principal aldehyde compounds formed from the decomposition of hydroperoxides at the 8, 9, 10, or 11 carbon atom positions are 2-undecenal, 2-decenal, nonanal, and octanal, respectively. Oxidized linoleic acid produces hydroperoxides at either the 9 or 13 position, and a new conjugated double bond system is formed at either the 10 or 11 position (4). The autoxidation of linolenic acid typically results in the formation of hydroperoxides at either the 9, 12, 13, or 16 position on the hydrocarbon.

MECHANISMS OF PHOTOSENSITIZED OXIDATION

Although free-radical autoxidation is the primary mechanism for the formation of volatile flavor compounds from fats and oils, photosensitized oxidation initiated by sensitizers such as chlorophyll in the presence of light should be included as oxidation mechanisms that contribute to the formation of off-flavor volatile compounds (2,5). The reaction of fatty acids with oxygen to produce hydroperoxides requires a change in their electron spin states because the unsaturated fatty acid and hydroperoxide are in singlet-states, whereas oxygen is in a triplet-state. The reaction between singlet-state fatty acids and triplet-state oxygen to form singlet-state hydroperoxides can only occur if enough energy is present to overcome the spin barrier. Electrophilic singlet-state oxygen can directly react with the electron-rich, singlet-state unsaturated fatty acids to form singlet-state hydroperoxides. Naturally occurring pigments such as chlorophyll in vegetable oils can absorb ultraviolet and visible light and act as photosensitizers (sens) (6). The sensitizer transfers energy to triplet-state oxygen to form singlet-state oxygen, which reacts with unsaturated fatty acids to form singlet-state hydroperoxides:

$$\text{sens} \rightarrow {}^1\text{sens}^* \rightarrow {}^3\text{sens}^*$$

$${}^3\text{sens}^* + {}^3\text{O}_2 \rightarrow {}^1\text{O}_2 + {}^1\text{sens}$$

$${}^1\text{O}_2 + \text{RH} \rightarrow \text{ROOH}$$

Singlet-state oxygen reacts directly with singlet-state double bonds of unsaturated fatty acids by an "ene" reaction to form hydroperoxides:

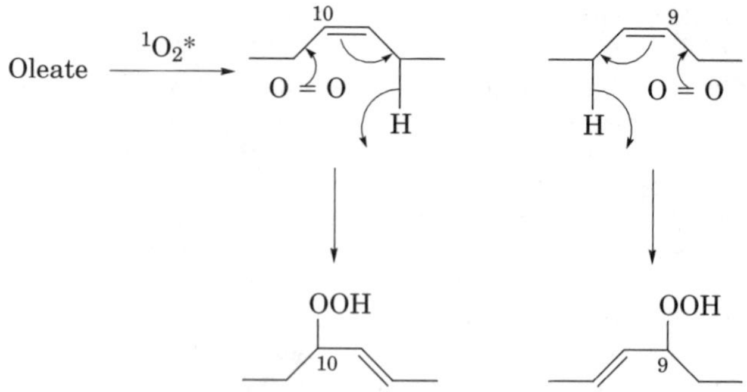

The singlet-state oxygen molecule is joined onto one end of a double bond simultaneously as it abstracts an allylic proton, and a new double bond is formed between the allylic position and the other end of the original C = C bond (7). A mixture of conjugated and nonconjugated hydroperoxides is formed by the reaction of singlet oxygen with linoleic acid. Because the flavor stability and quality of vegetable oils can be decreased by singlet oxygen oxidation, attempts have been made to inhibit singlet oxygen oxidation. Antioxidants such as β-carotene, lycopene, and tocopherols, which have been found to be able to deactivate singlet-state oxygen back to its low-energy triplet-state (6,8,9), are effective at minimization of singlet oxygen oxidation.

THERMAL OXIDATION OF FATS AND OILS

Although autoxidation is the principal mechanism for the development of off-flavors in fats and oils, foods can retain desirable flavors from fats and oils during deep-fat frying (10). Thermal oxidative decomposition of fats and oils occurs as foods are continuously heated in oil at about 185°C in the presence of air. The lipid decomposition mechanism involved is virtually the same as in autoxidative processes, where a hydrogen atom is abstracted from fatty acids to form a fatty acid free radical, after which molecular oxygen reacts with the fatty acid free radical to form first the peroxy radical, and then the hydroperoxide. The hydroperoxide eventually decomposes to form volatile flavor compounds. Flavor compounds formed at deep-fat frying temperatures are different from those produced at room temperature, where each particular pathway of flavor compound formation has its own activation energy. When the reaction rate ($\ln K$) versus temperature ($1/T$) is plotted for specific flavor compounds, one compound might be produced at a higher rate at deep-fat frying temperatures, but when the temperature is decreased to room temperature, another flavor compound might be produced at a higher rate. Therefore, the formation of flavor compounds from fats and oils is strongly dependent on the processing temperature (11). Different fats and oils produce different volatile compounds during deep-fat frying due to differences in fatty acid composition. Corn and soybean oils produce relatively large quantities of decadienals and unsaturated aldehydes, whereas coconut oil produces more saturated aldehydes, methyl ketones, and γ and δ lactones during deep-fat frying (12). Coconut oil, which contains 90 to 94% saturated fatty acids, has extremely high thermal oxidative stability. It has been reported that γ lactones with unsaturation at the 2 or 3 position are of particular significance in deep-fat-fried flavors (13).

FACTORS AFFECTING OXIDATION

The rate of autoxidation in vegetable oils is affected by many factors including degree of unsaturation of fatty acids, metals, and antioxidants. For example, linoleic acid, which contains two double bonds, is oxidized at a faster rate than oleic acid, which contains only one double bond. The relative rates of autoxidation of oleic to linoleic to linolenic (3 double bonds) acid has been reported as 1:40–50:100 on the basis of oxygen uptake (14). Reasons for differences in autoxidative susceptibility include the fact that the bond strength of linoleic acid (52 kcal/mole) is much lower than oleic acid (77 kcal/mole), and linoleic acid has a higher reactivity with oxygen than oleic acid. Metals such as copper and iron that exist in two different valence states play an important catalytic role in the oxidation of fats and oils. Metals differ in their ability to catalyze the autoxidation of fats and oils depending on the concentration, reaction temperature, and polarity of the reaction medium. The addition of chelating agents such as ethylenediaminetetraacetic acid (EDTA), phospholipids, and/or free fatty acids to oils can improve their oxidative flavor stability by sequestering the potentially catalytic metal atoms (15).

Natural and synthetic compounds that directly reduce the oxidation of fats and oils in one of three ways: (*1*) by becoming oxidized themselves, (*2*) by donating hydrogen to fatty acid free radicals to terminate free-radical chemical chain reactions, or (*3*) by forming a complex between the lipid radical and the antioxidant radical. Examples of each type of reaction can be seen here:

$$\text{Antioxidant (A)} + O_2 \rightarrow \text{Oxidized Antioxidant} \quad (1)$$

$$R\cdot + AH \rightarrow RH + A\cdot \quad (2)$$

$$R\cdot + A\cdot \rightarrow RA \quad (3)$$

Antioxidants such as butylated hydroxyanisole (BHA), butylated hydroxytoluene (BHT), tocopherols, and propyl gallate are commonly used in the food industry to minimize the oxidation of fats and oils. The presence of light, oxygen, and high storage temperatures should be controlled to help antioxidants most effectively minimize oxidative deterioration in fats and oils.

FLAVOR FORMATION BY ENZYMATIC REACTIONS

Flavors associated with fats and oils sometimes originate through the activity of enzymes that first create fatty acid hydroperoxides and then volatile secondary products (16). The release of free fatty acids through lipolytic processes usually occurs before lipoxygenase, a major lipolytic enzyme, can catalyze the oxidation of the polyunsaturated fatty acids linoleic and linolenic acid to hydroperoxides. Lipoxygenase is limited to reactions with polyunsaturated fatty acids because of its substrate specificity for a *cis,cis* penta-1,4-diene unit with a methylene group at the ω-8 position, a structure that is only present in fatty acids containing at least two nonconjugated double bonds. Therefore, the higher the polyunsaturated fatty acid content in a vegetable oil, the more susceptible the oil is to enzymatic oxidation and degradation.

Vegetable oils may also contain the enzyme hydroperoxide lyase, which is responsible for catalyzing the formation of aldehydes directly from the fatty acid hydroperoxides formed by lipoxygenase (17). The "grassy" or "beany" flavors often attributed to fatty acid autoxidation in vegetable oils may also originate from the formation of such volatile compounds as *cis*-3-hexenal and hexanal through the cleavage of hydroperoxides by hydroperoxide lyase.

FLAVOR PROPERTIES OF FATS AND OILS

The types of volatile compounds produced from the oxidation of fats and oils are strongly influenced by the composition of the hydroperoxides formed as well as the extent of oxidative cleavage of double bonds contained in the fatty acids (18–20). Due to an abundance of possible reaction pathways, a variety of volatile compounds such as hydrocarbons, alcohols, furans, aldehydes, ketones, and acids are formed during oxidation. Most of the compounds responsible for oxidized flavors, however, are the aliphatic carbonyl compounds (21). The flavor characteristics and corresponding threshold values of aldehydes vary as shown in Table 1.

CONTRIBUTION OF SPECIFIC OILS TO FLAVOR

Soybean oil, the most abundant vegetable oil in the United States, contains approximately 7% linolenic acid. The relatively unsaturated hydrocarbon of linolenic acid results in it being highly susceptible to oxidation and often

Table 1. Flavor Characteristics and Threshold Values of Aldehydes

		Threshold value in oil (ppm)	
Aldehyde	Description	Odor	Taste
3:0	Sharp-irritating	3.60	1.00
5:0	Sharp-bitter	0.24	0.15
6:0	Green	0.32	0.08
7:0	Oily-putty	3.20	0.05
8:0	Fatty-soapy	0.32	0.04
9:0	Tallowy-soapy	13.50	0.20
10:0	Orange peels	6.70	0.70
5:1	Sharp-paint-green	2.30	1.00
6:1	Green	10.00	0.60
7:1	Putty-fatty	14.00	0.20
8:1	Woodbugs-fatty	7.00	0.15
9:1	Tallowy-cucumber	3.50	0.04
10:1	Tallowy-orange	33.80	0.15
7:2	Frying odor	4.00	0.04
9:2	Fatty-oily	2.50	0.46
10:2	Deep-fried	2.15	0.10

responsible for the oxidation of oil to produce off-flavor compounds. Soybean oil also contains photosensitizers such as chlorophyll that produce singlet oxygen in the presence of light (22). Singlet oxygen can directly react with fatty acid double bonds and oxidize the fatty acids at a rate 1,500 times that of autoxidation (23,24). Analysis of headspace has shown that 2,4-decadienals, 2-heptenal, 2,4-heptadienal, pentane, and hexanal are the principal volatile compounds formed from oxidized soybean oil (25). The flavor characteristics of oxidized soybean oil have been described as grassy or fishy, but a single volatile flavor compound that gives unique grassy or fishy flavor in oxidized soybean oil has not been identified (26). However, *trans*-2-hexenal and 2,6-nonadienal have been reported to give general grassy flavors, whereas deca-*trans*-2,cis-4,*trans*-7-trienol and oct-1-en-3-one have been linked to fishy flavors (13).

Vegetable oils such as corn, cottonseed, peanut, coconut, safflower, and olive oil, as well as animal fats such as lard and beef tallow provide unique flavor characteristics to foods. Oil processing, which involves the three steps of refining, bleaching, and deodorization, removes impurities that contribute to the development of undesirable flavors. Olive oil, which contains 50 to 83% oleic acid, is popular in the United States as a result of the positive nutritional implications of monounsaturated fats. The flavor compounds of olive oil have been separated and identified by gas chromatography–mass spectrometry. Hexanal, *trans*-2-hexenal, 1-hexanol, and 3-methylbutan-1-ol, which are formed by autoxidation, are the principal volatile compounds of olive oil. Its flavor can be influenced by many factors such as climatic and soil conditions, the maturation process of the olive, and storage conditions (27). Although many volatile compounds have been identified, specific chemical compounds that are responsible for olive oil's flavor have not been determined. Safflower oil, which contains 78% linoleic acid, possesses the highest content of polyunsaturated fatty acid of all the commercial oils. The high content of linoleic acid increases the susceptibility of the oil to oxidation, resulting in the formation of off-flavor compounds (28). Cottonseed oil, which contains 18% oleic acid and 53% linoleic acid, is mainly used in restaurants for frying potato chips, seafood, and snacks. The bland flavor of cottonseed oil reportedly does not mask the flavor of the product nor does it revert during deep-fat frying (29).

DETERMINATION OF FLAVOR COMPOUNDS

The isolation, separation, and identification of volatile flavor compounds is important to determine the chemical mechanisms of flavor compound formation from fats and oils during storage, to identify chemical compounds that are responsible for specific desirable and undesirable flavors, and to develop optimum processing and storage methods that can minimize the undesirable oxidation of fats and oils. Analysis of the flavor compounds present in fats and oils is difficult because of their volatility, presence in low concentrations, and likelihood of undergoing chemical alteration during isolation. Food flavors are extremely complex mixtures of volatile compounds that act individually or in combination to produce unique tastes and aromas. Any modification of the chemical composition of these compounds could drastically change the flavor characteristics of food. Solid phase microextraction (SPME) is a relatively new technology that holds great promise for aiding in the evaluation of flavors by gas chromatography (30). SPME is rapidly gaining popularity as the technique of choice in the isolation and identification of volatile compounds because of its ability to integrate sampling, extraction, concentration, and sample introduction in a single step. Both the cost and time involved with sample analysis is reduced, while at the same time efficiency and selectivity are increased (31).

Threshold values are significantly different among flavor compounds, thus the presence of a small quantity of a compound with a low threshold value could have a more significant influence on the overall flavor than the presence of a large quantity of a particular compound with a high threshold value (Table 2). Hydrocarbons, for example, have a threshold value of 90 to 2,150 ppm; alkanals, 0.04 to 1.0 ppm; and vinyl ketones, 0.00002 to 0.007 ppm (2). This wide range of threshold values for flavor compounds illustrates how one specific flavor compound could have a significant influence on the overall flavor regardless of the concentration. Several volatile compounds have been proven to be responsible for specific described flavors in oxidized fats, as shown in Table 3 (13).

SENSORY EVALUATION

Sensory evaluation is considered the most important test to determine the flavor quality and stability of oils and is therefore used to evaluate the effects of processing conditions, storage time and periods, and packaging environments on the flavor of fats and oils. The most common sensory evaluation method involving fats and oils is a hedonic scale of 1 to 10; with 1 indicating a "repulsive" flavor and 10 a "completely bland" flavor, as shown in Table 4 (32).

Sensory evaluation can be expensive, subjective, difficult to reproduce, time-consuming, and not readily available. As a result, chemical and physical methods such as peroxide value, thiobarbituric acid (TBA), conjugated diene determination, fluorescence, Schaal oven test, and active oxygen methods have been developed to more efficiently evaluate the flavor quality and stability of fats and oils. The identification of volatile compounds that are responsible for specific flavors such as fishy or grassy has been made possible by correlating sensory evaluations with instrumental analysis such as gas chromatography.

FLAVOR EVALUATION BY GAS CHROMATOGRAPHY

With recent advancements in isolation techniques for flavor compounds, gas chromatography, and statistical analyses, several evaluations of oil flavor quality by gas chromatographic methods have been published (32–36). These papers have reported excellent correlation coefficients of better than 0.9 between sensory scores and predicted sensory scores by gas chromatography of corn oil, soybean oil, sunflower oil, and hydrogenated soybean oil. When com-

Table 2. Effect of Threshold Values on Flavor Significance of Soybean Oil Volatiles

Major volatiles	Relative %	Threshold value[a] (ppm)	Weighted % (1-octen-3-ol)	Relative order
t,t-2,4-decadienal	33.7	0.10	2.5	2
t,c-2,4-decadienal	17.9	0.02	6.7	1
t,c-2,4-heptadienal	11.1	0.04	2.1	3
2-heptenal	5.6	0.2	0.21	8
t,t-2,4-heptadienal	4.5	0.1	0.34	7
n-hexanal	4.5	0.08	0.42	6
n-pentane	3.1	340	6.8×10^{-5}	16
n-butanal	1.5	0.025	0.45	5
2-pentenal	1.2	1.0	0.009	13
1-octen-3-ol	0.9	0.0075	0.9	4
2-pentyl furan	0.8	2.0	0.9	14
n-pentanal	0.7	0.07	0.075	10
2-hexenal	0.7	0.6	0.009	13
n-nonanal	0.7	0.2	0.026	11
n-heptanal	0.6	0.055	0.082	9
1-penten-3-ol	0.5	4.2	8.9×10^{-4}	15
2-octenal	0.5	0.15	0.025	12

Source: Ref. 23.
[a]Ref. 19.

Table 3. Compounds Responsible for Specific Flavors in Oxidized Fats

Flavor	Source compound(s)
Cardboard	Non-*trans*-2,-*trans*-6-dienal
Oily	Aldehydes
Painty	Pent-2-enal, aldehydes
Fishy	Deca-*trans*-2,cis-4,*trans*-7,trienol, oct-1-en-3-one
Grassy	*Trans*-hex-2-enal, nona-2,6-dienal
Deep-fried	Deca-*trans*-2,*trans*-4-dienal

Table 4. Sensory Evaluation of Oil

Flavor score		Description of flavor
10		Completely bland
9	(good)	Trace of flavor, but not recognizable
8		Nutty, sweet, bacony, buttery
7	(fair)	Beany, hydrogenated
6		Raw, oxidized, musty, weedy, burnt, grassy
5	(poor)	Reverted, rubbery, butter
4		Rancid, painty
3	(very poor)	Fishy, buggy
2		Intensive flavor and objectionable
1	(repulsive)	

pared with sensory evaluation, gas chromatography has many advantages including rapid analysis time, convenience, low cost, and high reproducibility. A preliminary gas chromatographic analysis of flavor compounds of mayonnaise has been done to predict the sensory quality of mayonnaise during storage (37). Various flavor compounds have been identified in mayonnaise including allyl isothiocyanate, acetic acid, ethyl acetate, pentane, and isomers of 2,4-decadienal. As storage time increased, levels of pentane and 2,4-decadienal increased, whereas other volatile concentrations stayed the same. The flavor compound allyl isothiocyanate, the major flavor constituent of mustard, and acetic acid and ethyl acetate, the flavor compounds in vinegar, did not change during the six-month storage period. There was a correlation coefficient of greater than 0.9 between actual sensory scores of mayonnaise and predicted sensory scores by gas chromatography. Therefore, the shelf life of mayonnaise can be accurately predicted by analyzing specific compounds associated with off-flavors such as pentane and 2,4-decadienal during storage periods.

Gas chromatography can also be used to study the effects of storage conditions and processing methods on the oxidation of fats and oils. Soybean and corn oils were exposed to different periods of light to produce a wide range of flavor quality (38). As the period of light storage increased, the amount of 2,4-decadienal (as measured by gas chromatography) increased and the sensory scores decreased. The correlation of sensory evaluation and instrumental analysis using gas chromatography was greater than 0.95.

BIBLIOGRAPHY

1. M. Supran, *Lipids as a Source of Flavor*, American Chemical Society, Washington, D.C., 1978.
2. E. N. Frankel, "Chemistry of Autoxidation: Mechanism, Products and Flavor Significance," in D. B. Min and T. H. Smouse, eds., *Flavor Chemistry of Fats and Oils*, American Oil Chemists' Society, Champaign, Ill., 1985, pp. 1–37.
3. H. W.-S. Chan, "The Mechanism of Autoxidation," in H. W.-S. Chan, ed., *Autoxidation of Unsaturated Lipids*, Academic Press, Orlando, Fla., 1987, pp. 1–16.
4. H. W. Chan and G. Levett, "Autoxidation of Methyl Linoleate. Separation and Analysis of Isomeric Mixtures of Methyl Linoleate Hydroperoxides and Methyl Hydroxylinoleates," *Lipids* **12**, 99–104 (1977).
5. I. Lee et al., "Quantitation of Flavor Volatiles in Oxidized Soybean Oil by Dynamic Headspace Analysis," *J. Am. Oil Chem. Soc.* **72**, 539–546 (1995).

6. M. Y. Jung and D. B. Min, "Effects of Quenching Mechanisms of Carotenoids on the Photosensitized Oxidation of Soybean Oil," *J. Am. Oil Chem. Soc.* **68**, 653–658 (1991).

7. H. W.-S. Chan and D. T. Coxon, "Lipid Hydroperoxides," in H. W.-S. Chan, ed., *Autoxidation of Unsaturated Lipids*, Academic Press, Orlando, Fla., 1987, pp. 17–50.

8. S. Kaiser et al., "Physical and Chemical Scavenging of Singlet Molecular Oxygen by Tocopherols," *Arch. Biochem. Biophys.* **277**, 101–108 (1990).

9. H. Sies and W. Stahl, "Vitamins E and C, β-carotene, and other Carotenoids as Antioxidants," *Am. J. Clin. Nutr.* **62**, 1315S–1321S (1995).

10. W. W. Nawar, "Chemistry of Thermal Oxidation of Lipids," in D. B. Min and T. H. Smouse, eds., *Flavor Chemistry of Fats and Oils*, Champaign, Ill., 1985, pp. 39–60.

11. S. S. Chang, R. J. Peterson, and C.-T. Ho, "Chemistry of Deep Fried Flavor," in M. K. Supran, ed., *Lipids as a Source of Flavor*, ACS Symposium Series 75, American Chemical Society, Washington, D.C., 1978, p. 18–22.

12. M. W. Formo et al., "Composition and Characteristics of Individual Fats and Oils," in D. Swern, ed., *Bailey's Industrial Oil and Fat Products*, Vol. 1, John Wiley & Sons, New York, 1979, p. 289–478.

13. W. Grosch, "Lipid Degradation Products and Flavour," in I. D. Morton and A. J. Macleod, eds., *Food Flavours, Part A, Introduction*, Elsevier, Amsterdam, The Netherlands, 1982, pp. 325–398.

14. R. T. Holman and O. C. Elmer, "The Rates of Oxidation of Unsaturated Fatty Acids and Esters," *J. Am. Oil Chem. Soc.* **24**, 127–129 (1947).

15. J. Pokorny, "Major Factors Affecting the Autoxidation of Lipids," in H. W.-S. Chan, ed., *Autoxidation of Unsaturated Lipids*, Academic Press, Orlando, Fla., 1987, pp. 141–206.

16. H. Brockerhoff and R. G. Jensen, *Lipolytic Enzymes*, Academic Press, New York, 1974.

17. R. Tressl and F. Drawert, "Biogenesis of Banana Volatiles," *J. Agric. Food Chem.* **21**, 560–565 (1973).

18. E. Selke, W. K. Rohwedder, and H. J. Dutton, "Volatile Components from Triolein Heated in Air," *J. Am. Oil Chem. Soc.* **54**, 62–67 (1977).

19. E. N. Frankel, J. Nowakowska, and C. D. Evans, "Formation of Methyl Azelaaldehydrate on Autoxidation of Lipids," *J. Am. Oil Chem. Soc.* **38**, 161–162 (1961).

20. A. C. Noble and W. W. Nawar, "Identification of Decomposition Products from Autoxidation of Methyl 4,7,10,13,16,19-Docosahexaenoate," *J. Am. Oil Chem. Soc.* **52**, 92–95 (1975).

21. D. A. Forss, "Odor and Flavor Compounds from Lipids," *Prog. Chem. Fats and Other Lipids* **13**, 177–258 (1972).

22. Y. A. Tan, C. L. Chong, and K. S. Low, "Crude Palm Oil Characteristics and Chlorophyll Content," *J. Sci. Food Agric.* **75**, 281–288 (1997).

23. E. N. Frankel, "Soybean Oil Flavor Stability," in D. R. Erickson, et al., eds., *Handbook of Soy Oil Processing and Utilization*, American Soybean Association, St. Louis, Mo., and American Oil Chemists' Society, Champaign, Ill., 1980, pp. 229–244.

24. H. R. Rawls, and P. J. Van Santen, "A Possible Role for Singlet Oxygen in the Initiation of Fatty Acid Autoxidation," *J. Am. Oil Ch. Soc.* **47**, 121–125 (1970).

25. E. N. Frankel et al., "Comparison of Gas Chromatographic Methods for Volatile Lipid Oxidation Compounds in Soybean Oil," *J. Am. Oil Chem. Soc.* **62**, 620 (1985).

26. T. Mounts, "Processing of Soybean Oil for Food Uses," *Cereal Foods World* **34**, 268 (1989).

27. G. Montedoro, M. Bertuccioli, and G. Anichini, "Flavor of Foods and Beverages," Academic Press, Orlando, Fla., 1972.

28. T. J. Weiss, *Food Oils and Their Use*, AVI Publishing, Westport, Conn., 1970.

29. J. D. Dziezak, "Fats, Oils, and Fat Substitutes," *Food Technol.* **7**, 65–74 (1989).

30. C. L. Arthur and J. Pawliszyn, "Solid Phase Microextraction with Thermal Desorption using Fused Silica Optical Fibers," *Anal. Chem.* **62**, 2145–2148 (1990).

31. Z. Zhang, M. J. Yang, and J. Pawliszyn, "Solid Phase Microextraction," *Anal. Chem.* **66**, 844A–853A (1994).

32. D. B. Min, "Analyses of Flavor Qualities of Vegetable Oils by Gas Chromatography," *J. Am. Oil Chem. Soc.* **60**, 544–545 (1983).

33. H. P. Dupuy, S. P. Fore, and L. A. Goldblatt, "Direct Gas Chromatographic Examination of Volatiles in Salad Oils and Shortenings," *J. Am. Oil Chem. Soc.* **54**, 340–342 (1973).

34. J. L. Williams and T. H. Applewhite, "Correlation of the Flavor Scores of Vegetable Oils with Volatile Profile Data," *J. Am. Oil Chem. Soc.* **54**, 461–463 (1977).

35. K. Warner, E. N. Frankel, and K. J. Moulton, "Flavor Evaluation of Crude Oil to Predict the Quality of Soybean Oil," *J. Am. Oil Chem. Soc.* **65**, 386–391 (1988).

36. K. Robards et al., "Headspace Gas Analysis as a Measure of Rancidity in Corn Chips," *J. Am. Oil Chem. Soc.* **65**, 1621–1626 (1988).

37. D. B. Min and D. B. Tickner, "Preliminary Gas Chromatographic Analysis of Flavor Compounds on Mayonnaise," *J. Am. Oil Chem. Soc.* **59**, 226–228 (1982).

38. D. B. Min, "Correlation of Sensory Evaluation and Instrumental Gas Chromatographic Analysis of Edible Oils," *J. Food Sci.* **46**, 1453–1456 (1981).

DAVID B. MIN
DONALD F. STEENSON
The Ohio State University
Columbus, Ohio

FATS AND OILS: PROPERTIES, PROCESSING TECHNOLOGY, AND COMMERCIAL SHORTENINGS

Fats and oils are key functional ingredients in a large variety of food products. Common product applications are baked goods, snacks, icings, confections, fried food, and imitation dairy products. Fats and oils are also key parts of the human diet providing concentrated sources of energy and essential nutrients. This article briefly covers the chemistry, processing, physical characteristics, formulation, nutrition, and consumption of fats and oils in food applications.

CHEMISTRY AND STRUCTURE

Fats and oils are predominately mixtures of triglycerides. Triglycerides are the triesters of fatty acids and glycerol formed as a condensation product.

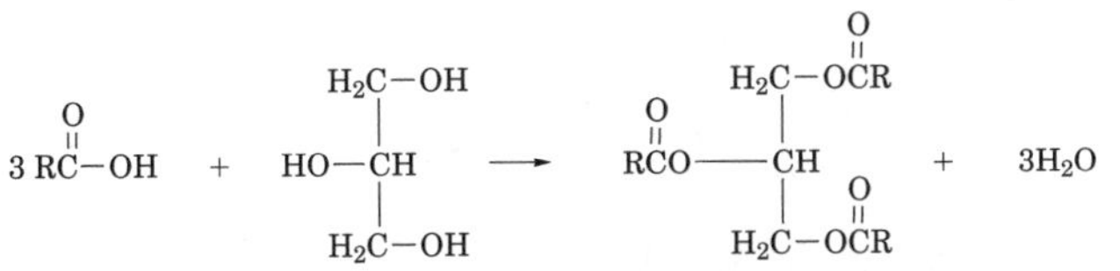

Fatty acid Glycerol Triglyceride

Triglycerides may range from solid to liquid consistency at room temperature depending on the fatty acid esters they contain. Solid triglycerides are commonly called fats, and liquid triglycerides, oils. If all three fatty acids that comprise the triglyceride are the same, it is called a single triglyceride; conversely, if the fatty acids differ, the triglyceride is mixed.

Triglycerides typically comprise about 95% of fat or oil. The other components include free fatty acids, monoglycerides, diglycerides, phosphatides, sterols, vitamins, natural antioxidants, color-causing pigments, minerals, flavor-odor compounds, and other substances. Free fatty acids are the uncombined or unesterified fatty acids present in a fat or oil. Crude oils may contain several percent fatty acids, which result from hydrolysis. The formation of free fatty acids via hydrolysis leaves the partial glycerides (mono- and diglycerides) behind. Monoglycerides and diglycerides are frequently used in foods as emulsifiers. They are often found in functional cake and icing shortenings. They are produced commercially by the alcoholysis reaction of triglycerides or fatty acids with glycerol or by esterification of glycerol with fatty acids (1).

Phosphatides are found in fat and oil from plant and animal sources. These materials consist of a polyhydric alcohol (usually glycerol) that is esterified with fatty acids, phosphoric acid, and a nitrogen-containing compound. The most common phosphatide is a family of compounds found in soybean oil referred to as lecithin, which includes chemical lecithin (phosphatidyl choline), cephalin (phosphatidyl ethanolamine), lipositols (phosphatidyl inositol) (2). Sphingomyelins are a group of phosphatides that contain no glycerol (3). Lecithin is typically used as a food additive for its emulsification properties. Phosphatides are essentially removed in the refining process.

Colored materials occur naturally in fats and oils and generally are a result of two classes of pigments. Carotenoids and chlorophylls are the yellow-red and green colors, respectively, found in fats. Level and type of pigment vary greatly with the source, from the orange color of crude palm oil due to its carotene content to the dark green color of canola (or rapeseed) oil due to its high chlorophyll content. Vegetable oils also contain tocopherols that serve as natural antioxidants, retarding oxidation, and as a source of the essential nutrient vitamin E. Other antioxidants may be added to processed fats to preserve freshness (4,5). In contrast, meat fats contain negligible levels of tocopherols.

Sterols are common minor components of all natural fats and oils. Cholesterol, the infamous sterol associated with heart disease, is the principal sterol found in animal fats. Vegetable oils only contain trace quantities of cholesterol. Vegetable oils contain other sterols that vary in type and quantity with the vegetable oil source (6–13).

Fats and oils may contain a number of other minor constituents such as waxes, gums, hydrocarbons, and fatty alcohols. These materials are generally of little importance and are typically handled during processing.

Fatty Acids

The key to understanding the chemical and physical characteristics of a fat is understanding the component fatty acids that comprise the triglyceride. These can vary in chain length (number of carbon atoms), type of carbon–carbon bonding (saturated or unsaturated with hydrogen), and stereochemistry both within the fatty acid and its position in the triglyceride. The most common edible fatty acids are aliphatic, straight chain, saturated and unsaturated, containing an even number of carbon atoms and ending in a carboxyl group.

$$\underline{H_3C-(CH_2)_n}-\underline{\overset{O}{\overset{\|}{C}}OH}$$

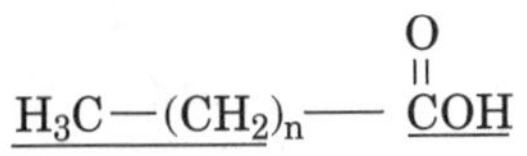

Saturated aliphatic carbon chain — Carboxyl group

Small amounts of odd-numbered, straight-chain acids, branched chains, and cyclical acids are also present in edible vegetable oils (9,14).

The length of the chain can vary in edible fatty acids from as few as 4 carbons to as many as 24. Most fats predominately contain fatty acids that are 16 or 18 carbons long, although there are some exceptions like coconut oil, palm kernel oil, and butterfat, which contain high percentages of shorter (8–14 carbons) chain fatty acids.

The fatty acids are classified by structure:

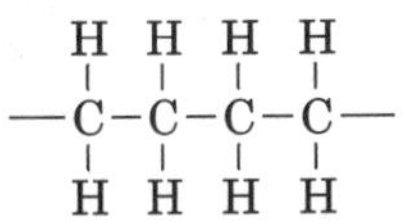

Saturated

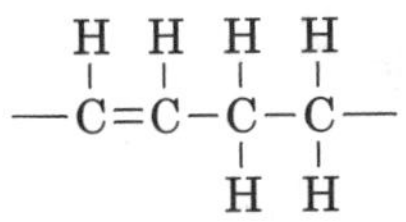

Monounsaturated

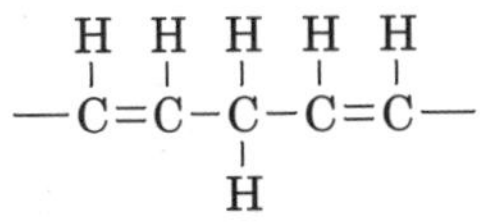

Polyunsaturated

The more unsaturated the fatty acid, the more chemically reactive a fatty acid will be. Saturated fatty acids have a higher melting point than the corresponding unsaturated fatty acids of the same chain length. The more unsaturated, the lower the melting point. Fatty acids with double bonds may differ in geometric form. Orientation of the hydrogen atoms around a double bond determines whether the fatty acid is *cis* or *trans*.

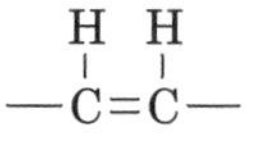

$$-\overset{\overset{\displaystyle H}{|}}{C}=\overset{\overset{\displaystyle H}{|}}{C}-$$
Cis

$$-\overset{\overset{\displaystyle H}{|}}{C}=\underset{\underset{\displaystyle H}{|}}{C}-$$
Trans

The *trans* form of a fatty acid is more stable and has a higher melting point. The double bond may also vary by position on the chain, although most natural fatty acids have double bonds that commonly occur in fixed positions.

There are two accepted systems of nomenclature to identify the position of double bonds in the chain. The Geneva system numbers the carbons in the chain consecutively starting from the carbon in the carboxyl group, which is numbered 1. The carbon with the lower number is used to identify the location of the double bond, that is, the carbon closest to the carboxyl group.

Biochemists typically identify the position of the double bond by the omega or n minus classification. This system refers to the position of the carbon in the double bond closest to the methyl end of the chain; the carbon of the end methyl group is number 1. Linoleic acid, a fatty acid 18 carbons long with two double bonds may be identified as 9,12-octadecenoic acid by the Geneva system. This same fatty acid would be an omega-6 or n-6 fatty acid by the alternate system. The second double bond would be an omega-9 or n-9. All of these structural variations can affect the chemical and physical properties of a fatty acid and the resulting triglyceride. The effects of chain length, unsaturation, and geometric isomerism on the melting point are shown in Table 1.

Polyunsaturates may also be described by the relative position of the double bonds.

Table 1. Fatty Acid Melting Points

Common name	Number of carbon atoms	Number of double bonds	Melting point, °F
Caproic	6	—	26
Caprylic	8	—	62
Capric	10	—	89
Lauric	12	—	112
Myristic	14	—	130
Myristoleic	14	1	65
Palmitic	16	—	145
Margaric	17	—	142
Stearic	18	—	157
Oleic	18	1	61
Elaidic	18	1 (*trans*)	111
Linoleic	18	2	20
Linolenic	18	3	9
Arachidic	20	—	168
Behenic	22	—	176
Erucic	22	1	92

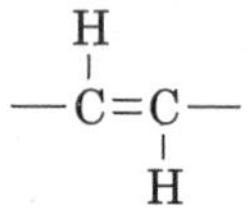

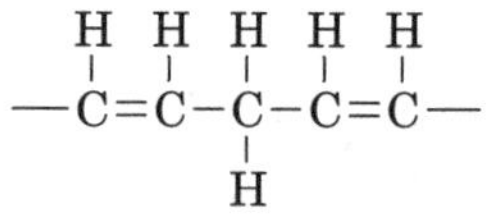

$$-\overset{\overset{\displaystyle H}{|}}{C}=\overset{\overset{\displaystyle H}{|}}{C}-\underset{\underset{\displaystyle H}{|}}{C}-\overset{\overset{\displaystyle H}{|}}{C}=\overset{\overset{\displaystyle H}{|}}{C}-$$
Nonconjugated
(*cis-cis*)

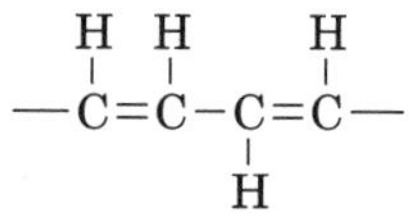

$$-\overset{\overset{\displaystyle H}{|}}{C}=\overset{\overset{\displaystyle H}{|}}{C}-\underset{\underset{\displaystyle H}{|}}{C}=\overset{\overset{\displaystyle H}{|}}{C}-$$
Conjugated
(*cis-trans*)

The double bond locations in unhydrogenated fats and oils usually occur in characteristic positions. Processing, particularly hydrogenation, may alter these locations forming conjugates and other positional isomers.

The physical properties of a triglyceride will be affected by the individual fatty acids it contains and the location of these fatty acids on the glycerine backbone. Some typical source oil fatty acid compositions are shown in Table 2. A triglyceride composed of a single fatty acid type will have uniform, distinct physical properties such as a sharp melting point. Fats and oils are mixtures of triglycerides. These mixtures result in a less distinct, broader range for a property, e.g., a melting point range. The molecular configuration gives rise to characteristic physical properties typical of a given source, which may be quite unique like cocoa butter. Monoglycerides and diglycerides have higher melting points than the corresponding triglycerides (Table 3).

Solid fats can also orient themselves in different crystal forms or polymorphs depending on how the fat crystals orient themselves in the solid state (15,16). Fats containing a diverse mixture of triglycerides are more likely to remain in lower melting crystal forms. The more singular triglycerides will readily transform successively from lower-melting to higher-melting forms. The extent and rate of transformation are a function of the triglyceride structure and mixture, temperature of crystallization, and the duration of crystallization at a given temperature. Mixtures of fats with different polymorphic tendencies can be directed to a particular polymorphic form by control of the composition and crystallization conditions. β prime crystal formation of partially hydrogenated soybean oil in bakery shortenings and margarines is promoted by inclusion of small amounts of β prime tending, partially hydrogenated palm oil (17–23).

Processing

Animals or oilseeds are the sources of edible fat in the United States. Animal fats are obtained by separating the fat from the tissues in a process called rendering. Heat is applied to separate the fat from the tissue protein. Vegetable fats are obtained by crushing the oil source and extracting the fats by direct compression and filtration. Modern processing utilizes solvent (typically hexane) extraction to improve efficiency and yield. The solvent is recycled after recovery and reused. Oils and fats obtained from this process are called crude because they contain small amounts of nontriglyceride components. Crude fats and oils are processed to remove the undesired material and modify the physical properties to obtain useful, functional products.

Two types of refining processes are commonly in use: chemical or physical (steam) refining. Both employ similar steps, but with different objectives. Most vegetable oils and

Table 2. Approximate Fatty Acid Composition, Percent Total Fatty Acids

Composition	Canola oil	Coconut oil	Corn oil	Cottonseed oil	Palm oil	Soybean oil	Tallow
C 6:0		<1					
C 8:0		8					
C10:0		6					
C12:0		47					
C14:0		19		<1	1		3
C14:1							1
C16:0	4	9	11	22	44	11	24
C16:1				<1			4
C17:0							2
C17:1							<1
C18:0	2	3	2	3	5	4	19
C18:1	61	6	27	19	40	24	43
C18:2	21	2	59	54	10	54	3
C18:3	9		1	1		7	<1
C20:0	1						
C20:1	1						
C22:0							
C22:1	<1						
C24:0							

Table 3. Fatty Acid Ester Melting Points, °F

Fatty acid	Monoglyceride	Diglyceride[a]	Triglyceride
C12:0 Lauric	145	136	116
C14:0 Myristic	159	152	135
C16:0 Palmitic	171	169	146
C18:0 Stearic	179	175	164
C18:1 Oleic	95	71	42
C18:1 Elaidic	137	131	108
C18:2 Linoleic	54	27	8
C18:3 Linolenic	60	10	−12

[a]1,3 diglyceride

meat fats are chemically refined, but for some vegetable oils, notably coconut and palm, physical refining is more economical. The choice is generally determined by the convenience and economics of the process.

Chemical, or caustic refining, is intended to remove most of the free fatty acids present in the crude oils. The process consists of treating the crude with an alkali solution, which converts the free fatty acids to soap. These soaps are removed by centrifugation and water washing. Phospholipid, mucilagenous, and proteinaceous materials are also removed in this step. Caustic strength, temperature, mixing, contact time, and removal are critical process parameters.

The residual moisture is removed by vacuum drying prior to bleaching. The natural pigments, carotenes and chlorophylls, are removed by physical absorption on acid-activated clay or bleaching earth. Other absorbent materials such as activated carbon may also be used in specific situations. The critical variables are absorbent level, contact time, temperature, and removal of the spent absorbent. Typically, hydrogenation follows bleaching. Hydrogen is added to unsaturated fatty acids in this heterogeneous reaction with gaseous hydrogen and catalyst, usually nickel although other precious metals may be used. Critical parameters are reaction time, temperature, pressure, catalyst type and amount, and gas purity. Vessel design to provide effective mass and heat transfer is also critical. Less commonly, the oil may be bleached again after hydrogenation.

Hydrogenation raises the melting point of the triglyceride mixture, thus altering physical properties. It also improves oxidative and thermal stability by reducing the number of labile, polyunsaturated reaction sites in the fatty acids of the triglyceride. Hydrogenation allows the processor to differentiate products for their various applications. A processor may make a full range of products, from salad oils to solid frying shortenings, margarines to bakery shortenings, by hydrogenating and blending various oils.

Deodorization typically follows hydrogenation or bleaching if natural liquid oil is desired. The process is a high-temperature, vacuum, steam distillation that removes residual fatty acids as well as undesirable flavor and odor compounds. This step will also remove a small portion of the natural antioxidants. The high-temperature exposure also reduces the color due to transformation of the red and yellow pigments into nonchromophores. Key elements are temperature, vacuum, steam rate, and exposure time. Again, vessel design is critical to provide effective and efficient heat and mass transfer.

Liquid oils may be packaged directly after deodorization, but fluid or plastic shortenings are usually votated, or crystallized, after deodorization. The blended, melted product is fed under pressure to a scraped-surface heat exchanger at slightly above its melting point. The product is chilled creating crystal nuclei, which are allowed to grow slowly for a controlled time in a mildly agitated vessel, commonly referred to as a B unit. Nitrogen may be added to gasify a solid shortening, providing a whiter, more plastic product. Proper crystal conversion may require tempering. Liquid products may be tempered in process, whereas solids are typically stored for 24 to 48 h at their tempering temperature. Key parameters are flow rate, temperature, and tempering conditions. Control of crystal structure is essential for functional bakery shortenings.

Other Processes

There are other specialized processing steps. The most notable are winterization, fractionation, transesterification, interesterification, and directed interesterification. Winterization and fractionation separate melting fractions by cooling the shortening and separating the solids that crystallize. Processors use winterization to provide stable salad oils. Fractionation consists of cooling the oil and separating the solids a number of times to obtain various melting fractions. These are used to create hard butters (cocoa butter substitutes) for coating applications and high stability oils. These processes may employ a solvent (hexane or acetone) to improve the separation, which is commonly referred to as solvent winterization or solvent fractionation (24).

Transesterification, a type of interesterification, is a process that rearranges or redistributes the fatty acids on the glycerine backbone of the triglyceride. Processors may select conditions to distribute the fatty acids more randomly, thus modifying the physical properties of the shortening. This process does not create a significant level of fatty acid isomers (*trans*) like hydrogenation. Hydrogenation may be combined before or after the process to allow further modification of the shortening.

Interesterification involves two or more oil sources, randomizing the fatty acids of the source triglycerides. By cooling and removing the higher-melting triglycerides the reaction can continue with the remaining components redirecting the interesterification. This process, used to provide relatively pure triglycerides, is called directed interesterification (25–27). If excess glycerol is present, monoglycerides and diglycerides are formed. The relative amounts of each at equilibrium are a function of the ratio of fatty acids available to glycerol (28). This is commonly called glycerolysis.

PROCESSING AIDS AND ADDITIVES

A number of food-approved materials may be added at low levels to improve processing or to protect the oil quality during processing (Table 4). These materials are removed or reduced to minimal levels in the final product. Other additives may also be used to provide a functional effect in the final product. These uses must comply with Food and Drug Administration (FDA) regulations covering addition levels, methods of addition, permissible residual levels, and labeling.

FATS AND OILS FUNCTIONALITY

Fats and oils provide several functional characteristics in a variety of food products. Shortening products can provide lubricity, structure, a heat-transfer medium, a moisture barrier, and aeration. Fat systems can also act as a carrier for flavor, color, and vitamins. Oils themselves provide vital nutrition.

Table 4. Additives Used in Fats and Oils

Additive	Effect
Antioxidants	
Tocopherols	Retard oxidation
Butylated hydroxyanisole (BHA)	
Butylated hydroxytoluene (BHT)	
tert-Butylhydroquinone (TBHQ)	
Propyl gallate	
Colors	
β-carotene	Provides yellow color to finished product.
Tumeric and annatto	
Flavors	
Various acids, aldehydes, and ketones	Provide flavor and odor to the finished product
Phosphoric acid	Metal chelators, process hydrating agents, neutralizing agents.
Citric acid	
Dimethylpolysiloxane (methyl silicone)	Antifoaming agent, high-temperature antioxidant
Polyglycerol esters	Crystal inhibitors, crystal modification
Oxystearin	
Emulsifiers	
Monoglycerides, diglycerides	Emulsification, aeration, dough conditioning, wetting, antispattering, antisticking, separation and viscosity control
Lecithin	
Polyglycerol esters	
Calcium/sodium steroyl lactylate	
Lactylated monoglycerides and diglycerides	
Sorbitan esters	
Propylene glycol esters	

Fat is multifunctional in most applications. Frying and cooking fats provide lubricity, a heat-transfer medium, and a flavor medium in some applications. Fats provide lubricity and aeration in cakes and additionally structure in icings, cream fillers, and the like.

SALAD AND COOKING OILS

Salad and cooking oils are usually prepared from refined, bleached, and deodorized vegetable oils, or they are lightly hydrogenated and winterized for stability. The oil may be used for flavor (eg, peanut and olive oils), nutrition (eg, corn and canola oils), and stability. Antioxidants and crystal inhibitors may be added to salad oil (29,30). Oils may be blended for flavor and stability, if they are intended for heating and reuse, and dimethylpolysiloxane may also be added. Oils used in salad dressings typically have light color and bland flavor and must resist clouding. Salad dressing must contain a minimum of 30% oil by federal regulation (31). Mayonnaise is an emulsified, semisolid dressing that must contain 65% vegetable oil and whole eggs or egg yolks (32). Reduced-calorie dressings must contain at least one-third fewer calories than the conventional product. This is usually accomplished by reducing the fat content and stabilizing the increased water content with gums and thickeners to provide emulsion stability.

FRYING SHORTENINGS

Frying fats must maintain stability under abusive conditions that include high temperature, extractable materials from the fried food, and oxygen from the air. Stability of fried products (snack foods) is also important. The shortening imparts textural and flavor effects to the fried food. The key is the type of oil used and its chemical and physical properties. There are now new varieties of common vegetable oils with modified fatty acid compositions through breeding and genetic modification to expand the variety of properties available.

Frying shortenings are available as meat fats, meat fat–vegetable blends, and solid or liquid vegetable shortenings. Meat fat and solid vegetable products provide excellent stability, particularly under abusive conditions. But these products have poor nutritional attributes (cholesterol in meat fat, saturated fat, and *trans* fatty acids). Pourable, vegetable, liquid shortenings offer convenient handling with good frying stability. There is some compromise on oxidative stability, but fry stability is acceptable for most applications. Liquid products have lower saturated and *trans* fatty acid levels than solid products.

Fry life is a function of the chemical and physical characteristics of the frying shortening. These depend on the source oil(s) and processing. Fry life is also a function of many operational factors such as the type and amount of food fried, the size of the fryers, the time and temperature of frying, frequency of filtration, and cleaning procedures. Analytical characteristics vary with the product (Table 5).

Table 5. Typical Fly Shortening Analysis

	Solid		Liquid
Analysis	Meat fat–vegetable blend	Partially hydrogenated soybean	Partially hydrogenated soybean
Iodine value	52	69	101
Mettler dropping point, °F	117	103	95
Solid fat Index			
at 50 F	40	50	5
at 70 F	27	36	4
at 80 F	23	30	3
at 92 F	17	14	2
at 104 F	12	2	2
Fatty acid composition, %			
C16	24	11	11
C18	18	11	7
C18:1	41	70	46
C18:2	8	2	28
C18:3	1	trace	3

Lovibond color (1.5 red maximum), free fatty acid (0.05 wt% maximum) and peroxide value (1.0 meq/kg maximum) are identical for all products.

GRIDDLE AND PAN FRYING SHORTENINGS

Specialty colored and flavored shortenings are designed to replace butter or margarine in grilling, pan frying, and other applications (soups, sauces, gravies, basting, and bun wash). These may be liquid or solid, resist darkening, reduce waste (no water), and eliminate scorching (no milk solids). Other ingredients may be added. These include salt (flavor enhancement), flavorings (garlic), and even small amounts of butter itself (33–36). Coconut oil may be added for abusive, high-temperature applications (37).

BAKERY SHORTENINGS

A variety of general and specialized shortenings are available for baking applications (24,38). The principal products are bread, rolls, biscuits, cakes, icings, cookies, pastry, doughnuts, and pies. All-purpose shortening is intended to function adequately in a broad variety of applications from baking to frying. It is formulated to have stability, a wide plastic range for proper mixing, and some structure for applications like icings.

Cake and icing shortening is designed to meet the dual needs of cake and icing applications. The shortening must entrap and disperse air as an emulsion that is a cake batter or icing. In a cake, the fineness and amount of the air bubbles determine the finished height and cell structure. The shortening also lubricates the batter components and tenderizes the cake. In icings, the shortening must provide the structural properties (body and firmness) as well as its creaming or aerating ability. Monoglyceride and diglyceride emulsifiers are typically added to cake and icing shortening to improve the emulsifying properties. Table 6 shows the typical analytical data for all-purpose and cake and icing shortenings.

The development of emulsifier technology has allowed the formulation of specialized cake or icing shortenings and the development of prepared mixes. Their formulation is a function of their application: continuous versus batch

Table 6. Typical Shortening Analyses

Analysis	All-purpose (soy base)	Cake and icing (soy base)
Iodine value	74.00	74.00
Mettler dropping point, °F	120.00	120.00
Solid fat index		
at 50 F	24.00	24.00
at 70 F	20.00	20.00
at 80 F	19.00	19.00
at 92 F	16.00	16.00
at 104 F	10.00	10.00
Lovibond color, red, maximum	1.00	2.00
Free fatty acid, wt%, maximum	0.05	0.15
Peroxide value, meq/kg, maximum	1.00	1.00
α-monoglyceride, wt%, maximum	—	3.50

mixed, rich or lean formulation, liquid content, and sugar content. The shortening base is formulated and processed to obtain optimum emulsifier performance. Emulsifiers generally permit a richer, moister cake due to their capability to incorporate high levels of moisture in the batter (Table 7). The specialized mix shortenings contain even higher levels of emulsification and are used for consumer cake mixes, leaner snack-type cake applications, or cake products where long shelf life is required (39–41).

Similarly, emulsifier technology has allowed incorporation of high levels of moisture and air into the icing. The shortening must provide structure to allow the icing to hold this moisture and air as well as retain its shape. Processing of the shortening, particularly formulation and votation, is critical to provide this structure. Generally, cooler votation conditions improve the icing's ability to aerate, as measured by specific gravity, and resistance to separation and slumping.

Different levels and types of emulsification are required for formulations with increased moisture and aeration requirements (Table 8). Flat or thick fudge icings require little, if any, aeration. However, butter cream and some decorator icings require some emulsification ability, usually accomplished with monoglycerides and diglycerides or low levels of polysorbates. Specialty icing products are formulated to produce fluffier and higher moisture-type icings. Polyglycerol esters or combinations of monoglyceride and diglyceride, poylsorbate emulsifier systems are used. As the level of emulsification increases, the structure provided by the shortening becomes more critical to icing performance.

Breads and Rolls

Shortening is added at relatively low levels to breads for lubricity, softening, and shelf life. Structure is primarily obtained from starch and protein complexes. Emulsifiers are employed as dough conditioners to resist staling (drying) (Table 9). These may be added either as part of the shortening or hydrated first by heating in water until a liquid crystalline phase (mesophase) is formed and maintained by cooling the mixture. Biscuit shortenings are formulated to blend as small, discrete chunks into the cold dough, yet still provide sufficient solids for a flaky, tender texture.

Pie Crust Shortening

Pie shortenings are also mixed loosely into the cold dough to maintain separate fat layers for a flaky crust. Overmixing, or mixing too warm, incorporates the shortening intimately, which reduces flakiness, increases toughness, and increases shrinkage. The shortening system contains a loosely bound oil fraction to provide lubricity and tenderness to the crust.

Lard is the classic choice for pie shortening due to its solids content, plasticity at cool temperatures, and crystal structure. Vegetable shortenings have been formulated to provide similar functional properties with an improved nutritional (cholesterol and saturates) profile.

Puff Pastry and Danish Shortening

Puff pastry and Danish shortenings are referred to as roll-in products because they are formed from a laminated dough sheet, consisting of alternating layers of dough and shortening, that is traditionally formed by rolling the coated dough sheet and folding it to form layers.

Danish has a higher fat content than puff pastry, and softer-consistency products are used to achieve lubricity. Long refrigerated retarding (dough-relaxing) steps are used between folds to minimize toughness and maximize lubricity while maintaining structural integrity of the fat film. Puff pastry shortening contains higher solids for flakiness due to the lower fat content used. This shortening must be carefully formulated and processed so it remains workable, not stiff, to properly apply to the dough sheet. Dough and shortening temperature are critical. Puff pastry has larger expansion because of the number of layers and the water vapor sealed by the shortening between the dough layers.

Butter was originally used for these applications. Puff pastry and Danish are formulated to have similar composition (fat, moisture, and salt) and functionality and may be butter flavored as well.

OTHER APPLICATIONS

There are a number of other specialized shortening applications. Confectionery fats are primarily intended to be substitutes or extensions for cocoa butter in confections and coatings. These consist of lauric-based products (coconut and palm kernel oils), which may be hydrogenated, fractionated, or interesterified to better match the solids profile and melting point of cocoa butter. Domestic substitutes may be hydrogenated or fractionated as well (18–42–45). These do not generally have the same crystal type as cocoa butter, and they do not require tempering. Eating quality, snap, mouth melt, and glossy appearance are key performance factors. Coatings provide a moisture barrier

Table 7. Typical Cake Shortenings

Shortening	Emulsifiers	Moisture incorporation	Fat content
Cake and icing	Monoglyceride and diglyceride	Low	Moderate
Specialty cake			
Plastic solid	Propylene glycol monoesters and monoglycerides	Moderate	Moderate
Liquid	Lactylated monoglycerides and monoglycerides	Moderate	Moderate
Specialized mix shortenings	Higher levels of propylene glycol monoesters and monoglycerides	High	High

Table 8. Typical Icing Shortenings

Shortening type	Emulsifier	Icing applications
All-purpose	None	Flat, fudge icings
Cake and icing	Monoglycerides and diglycerides	Butter cream and decorator icings, and cream fillers
Specialty icings	Monoglycerides, diglycerides, and polyglycerol esters	Butter cream, fluffy decorator icings, and cream fillers

Table 9. Bread Emulsifiers and Dough Conditioners

Monoglycerides and diglycerides	Sodium stearoyl lactylate
Distilled monoglycerides	Calcium stearoyl lactylate
Ethoxylated monoglycerides	Lactylated stearate
Succinylated monoglycerides	Diacetyl tartaric acid esters

and are typically high-solids, low-melting-point products. These are often referred to as hard butters.

Similar shortenings may also be used in imitation dairy products like sour cream, dips, nondairy creamers, and whipped toppings. Texture and stability are keys to successful shortening formulation for these products.

Peanut butter stabilizers are partially hydrogenated, high-melting fats, or emulsifiers, which are added to fluid peanut butter to prevent oil separation.

BUTTER AND MARGARINES

Butter contains at least 80 wt% butterfat (46), which serves as the matrix sponge for the aqueous phase. The aqueous phase consists of water, casein, milk solids, and minerals. The solids comprise about 1 to 2 wt% of the butter, and salt may typically be added at concentrations from 1.5 to 3 wt%. Margarines are butter replacements made from shortening or shortening and oil blends with other ingredients to match butter's aqueous phase. Coloring, flavoring, and vitamins A and D may also be added. Margarine must contain 80% fat by federal regulation (47). Margarines are available in solid form, whipped, and as pourable liquids. They are also available in reduced-calorie, diet, or imitation spreads, which have a lower fat content.

U.S. FAT CONSUMPTION

The U.S. Department of Agriculture provides data on the availability of food fat for consumption per capita in the United States. These data measure the amount of fat available and do not consider the quantities wasted or discarded. Fats are separated into visible and invisible categories. The visible portion consists of easily identified fat like butter, margarine, shortening, and salad oils. The invisible category is ingested as part of other foods and is not easily recognized as fat such as dairy products.

Visible fat consumption has increased 16 lb per capita from 1970 to 1994 (Table 10). Visible fat usage accounts for this increase. This is to some degree a reflection of the increase in fast-foods and convenience foods, meals not made from scratch at home. Fat consumption appears not to be affected by dietary and health concerns, except that saturated fat consumption is down (48). Meat consumption is up, but the fat content contributed by meat has declined due to the use of leaner meats.

The amount of vegetable fat consumption has increased from 1970 to 1996 while animal fat consumption has decreased (Table 11). The amount of fat consumed in the diet

Table 10. Edible Fat Availability per Capita

Source	1970	1980	1990	1994
	Visible fats			
Butter, fat content	4.3	3.6	3.5	3.9
Lard and tallow	4.6	3.7	2.4	4.7
Margarine and spreads	8.7	9.0	8.7	7.9
Shortening and other fats	17.3	18.2	22.2	24.1
Salad and cooking oils	15.4	21.2	24.8	26.3
Other	2.3	1.5	1.2	1.6
Animal source	14.1	12.3	9.7	11.4
Vegetable source	38.5	44.8	53.1	57.2
Total visible fats	*52.6*	*57.2*	*62.8*	*68.6*
	Invisible fats			
Dairy products, excluding butter	15.6	14.9	15.4	15.7
Eggs, meats, poultry, fish	46.3	42.1	36.1	34.0
Fruits, vegetables, legumes, nuts, soy, and grains	7.2	6.7	8.8	8.7
Other	2.1	2.0	3.0	2.9
Animal source	61.9	57.0	51.5	49.7
Vegetable source	9.3	8.7	11.8	11.6
Total invisible fats	*71.2*	*65.7*	*63.3*	*61.4*
Total fats and oils	*123.8*	*122.9*	*126.2*	*129.0*

Source: Ref. 48.

Table 11. Fats and Oils Used in Foods, mm Pounds

Source	Year			
	1970–7.1	1980–81	1990–91	1995–96
	Vegetable oils			
Soybean	5,780	8,610	10,772	11,877
Corn	403	624	1,143	595
Cottonseed	805	555	777	497
Palm	160	291	98[a]	Not avail.
Coconut	220	338	169	221
Peanut	181	119	129[a]	129[a]
Sunflower		79	166[b]	90[b]
	Meat fats			
Tallow	508	740	501	341
Lard	1,612	415	314	297
Total fats	*9,669*	*11,908*	*14,491*	*14,706*

[a]Incomplete data.
[b]Total edible and inedible.
Source: Ref. 49.

has decreased as a percentage of total calories, yet total fat consumption is higher. This is somewhat misleading since total caloric intake (carbohydrate and protein) has been increasing at a greater rate than fat consumption, thus reducing the relative amount of fat consumed when expressed as a percentage of total calories. The reality is that fat consumption remains high with the trend of a lower proportion of saturated fatty acid and a higher proportion of polyunsaturated fat intake.

NUTRITION

Fats are an important part of the human diet along with proteins and carbohydrates. Fats are a concentrated source of energy (9 kcal/g vs 4 kcal/g for proteins and carbohydrates), a source of essential fatty acids, and a source of fat-soluble vitamins (50).

Certain fatty acids cannot be synthesized by the body and are essential for our health. These acids, linoleic and linolenic, must be supplied by the diet. About 2% of linoleic acid is necessary in the human diet to prevent a deficiency, and 3% is considered to be a more adequate intake. The minimum requirement for linolenic acid has been estimated at a little over 0.5% of total calories (50,51).

Estimates of total fat in the U.S. diet have dropped to around 35% of calories and will vary depending on the source of the data. Reduction in fat intake is an important part of any weight loss-control diet because fat is such a dense source of calories. Most recent dietary health concerns center on cardiovascular disease, including heart attack and stroke. Diet is just one, and not necessarily the major, risk factor identified from epidemiological studies. Other factors include smoking, high blood pressure (hypertension), obesity, elevated serum cholesterol, male sex, sedentary lifestyle, diabetes, excessive stress, and a positive family history of cardiovascular disease. No direct cause-and-effect relationship has been established for these risk factors; only an increased statistical risk has been shown.

Most of the recent studies involving dietary fat focus on the specific type of fatty acid (saturated, monounsaturated, polyunsaturated, and *trans* isomers) and its relationship to serum cholesterol and lipoproteins (cholesterol carriers). There have been studies that have shown a relationship between carbohydrates, proteins, fiber, and trace minerals and cardiovascular disease (52).

Vegetable fats do not directly contain a significant amount of cholesterol. Animal and dairy fats do contain cholesterol and thus may directly affect serum cholesterol. Much recent work has been done relating specific fatty acid types with their effect on the lipoprotein carriers and, indirectly, serum cholesterol (53,61). Two general classes of lipoprotein carriers are largely involved. Low-density lipoproteins (LDL) contain the largest portion of total cholesterol in the blood. High levels of LDLs are associated with increased risk of cardiovascular disease, and they can be thought of as the bad cholesterol. High-density lipoproteins (HDL) transport cholesterol from the blood, and elevated levels of HDL are associated with decreased risk of coronary heart disease. Thus HDLs can be viewed as the good cholesterol (62).

Increased levels of saturated fat in the diet have been shown to increase total serum and LDL cholesterol, thus increasing the risk of heart disease (46). Much negative publicity has been associated with tropical fats (coconut, palm kernel, and palm oil) due to their high saturated fat content. Recent studies show palm oil may be an exception, despite its high saturated fat content (56,60,61,63–65). Similarly meat and dairy fats suffer because of their relatively high saturated fat content as well as their cholesterol content.

Diets high in monounsaturated fat and low in saturated fats, the so-called Mediterranean diet, have been shown to decrease total and LDL cholesterol while not affecting HDL cholesterol (53,55). Studies suggest this type of diet may be equal to, or better than, a low-fat diet for lowering blood cholesterol. Olive oil and canola oil are common sources of monounsaturates. Diets high in polyunsaturated fats beneficially decrease total and LDL cholesterol, but also decrease HDL cholesterol (53,55). Soybean and corn oil are common polyunsaturated fat sources.

Some animal studies have suggested a relationship between diets high in polyunsaturates and certain types of cancer, particularly breast cancer. Other animal studies indicate a high caloric diet due to increased fat may be related to increased cancer incidence, particularly colon and breast cancer (66,67). Any direct relationship between total caloric intake, fat unsaturation, and cancer incidence has not been proved.

Specific fatty acid isomers have also been studied. *Trans* isomers, which result from the hydrogenation process, are present at significant levels, 10 to 35%, in most processed shortenings. Most are present as monounsaturated, *trans* fatty acids. Early concerns about the safety of *trans* fatty acids in the U.S. diet were not supported by reliable epidemiological and animal data (68,69). Subsequent studies indicate that moderate *trans* fatty acid intake results in cholesterol response similar to saturated fat (70–73). Higher *trans* intake levels may also reduce HDL (70). *Trans* fatty acids do not appear to be associated with increased cancer risk (74).

Diets high in omega-3 polyunsaturates, like fish oils, have shown a reduced incidence of coronary heart disease. However, other data show an increased risk of stroke and bleeding tendency with this type of diet (54). More studies are needed to determine the long-term effects of diets supplemented with fish oil. Some vegetable oils (soybean and canola) are sources of omega-3 fatty acids without the cholesterol and saturated fatty acids associated with fish oil.

There is still controversy concerning nutritional and health effects of fat in our diets. Given the current state of knowledge, prudence would dictate moderation in the diet.

BIBLIOGRAPHY

1. M. Gupta, "Manufacturing Process for Emulsifiers," in Y. H. Hui, ed., *Bailey's Industrial Oil and Fat Products*, Vol. 4, 5th ed., John Wiley & Sons, New York, 1996, pp. 569–572.
2. O. L. Brekke, "Oil Degumming and Soybean Lecithin," in D. R. Erickson et al., eds., *Handbook of Soy Oil Processing and Utilization*, American Soybean Association, St. Louis, Mo., and

American Oil Chemists' Society, Champaign, Ill., 1980, pp. 77–79.

3. W. W. Nawar, "Chemistry," in Y. H. Hui, ed., *Bailey's Industrial Oil and Fat Products*, Vol. 1, 5th ed., John Wiley & Sons, New York, 1996, pp. 401–402.
4. *Oils & Fats International* **2**, 31 (1986).
5. D. F. Buck, "Antioxidant Applications," *Manufacturing Confectioner* **64**, 45–49 (1984).
6. N. A. M. Eskin et al., "Canola Oil," in Y. H. Hui, ed., *Bailey's Industrial Oil and Fat Products*, Vol. 2, 5th ed., John Wiley & Sons, New York, 1996, pp. 53–54.
7. L. R. Strecker et al., "Corn Oil," in Y. H. Hui, ed., *Bailey's Industrial Oil and Fat Products*, Vol. 2, 5th ed., John Wiley & Sons, New York, 1996, pp. 131–133.
8. L. A. Jones and C. C. King, "Cottonseed Oil," in Y. H. Hui, ed., *Bailey's Industrial Oil and Fat Products*, Vol. 2, 5th ed., John Wiley & Sons, New York, 1996, pp. 194–195.
9. D. Firestone, E. Fedeli, and E. W. Emmons, "Olive Oil," in Y. H. Hui, ed., *Bailey's Industrial Oil and Fat Products*, Vol. 2, 5th ed., John Wiley & Sons, New York, 1996, pp. 258–259.
10. Y. Basiron, "Palm Oil," in Y. H. Hui, ed., *Bailey's Industrial Oil and Fat Products*, Vol. 2, 5th ed., John Wiley & Sons, New York, 1996, pp. 280–282.
11. C. T. Young, "Peanut Oil," in Y. H. Hui, ed., *Bailey's Industrial Oil and Fat Products*, Vol. 2, 5th ed., John Wiley & Sons, New York, 1996, p. 386.
12. F. T. Orthofer, "Rice Bran Oil," in Y. H. Hui, ed., *Bailey's Industrial Oil and Fat Products*, Vol. 2, 5th ed., John Wiley & Sons, New York, 1996, pp. 404–406.
13. E. F. Sipos and B. F. Szuhaj, "Soybean Oil," in Y. H. Hui, ed., *Bailey's Industrial Oil and Fat Products*, Vol. 2, 5th ed., John Wiley & Sons, New York, 1996, pp. 499–502.
14. D. Kitts, "Toxicity and Safety of Fats and Oils," in Y. H. Hui, ed., *Bailey's Industrial Oil and Fat Products*, Vol. 1, 5th ed., John Wiley & Sons, New York, 1996, pp. 221–226.
15. C. W. Hoerr, *J. Am. Oil Chem. Soc.* **37**, 539–546 (1960).
16. E. S. Lutton, "Lipid Structures," *J. Am. Oil Chem. Soc.* **49**, 1–9 (1972).
17. U.S. Pat. 2,530,596 (November 21, 1950), N. W. Zills and W. H. Schmidt (to Lever Bros. Co.).
18. U.S. Pat. 2,625,478 (January 13, 1953), K. F. Mattil and F. A. Norris (to Swift & Co.).
19. U.S. Pat. 2,521,242 (September 5, 1950), P. J. Mitchell, Jr. (to Procter & Gamble Co.).
20. M. M. Chrysam, "Margarines and Spreads," in Y. H. Hui, ed., *Bailey's Industrial Oil and Fat Products*, Vol. 3, 5th ed., John Wiley & Sons, New York, 1996, pp. 79–82.
21. D. J. Metzroth, "Shortening: Science and Technology," in Y. H. Hui ed., *Bailey's Industrial Oil and Fat Products*, Vol. 3, 5th ed., John Wiley & Sons, New York, 1996, pp. 122–123.
22. R. D. O'Brien, "Shortenings: Types and Formulations," in Y. H. Hui, ed., *Bailey's Industrial Oil and Fat Products*, Vol. 3, 5th ed., John Wiley & Sons, New York, 1996, pp. 181–184.
23. U.S. Pat. 4,127,597 (November 28, 1978), J. C. Craig, M. F. Kozemple, and S. Elias (to U.S. Secretary of Agriculture).
24. R. Krishnamurthy and M. Kellens, "Fractionation and Winterization," in Y. H. Hui, ed., *Bailey's Industrial Oil and Fat Products*, Vol. 4, 5th ed., John Wiley & Sons, New York, 1996, pp. 314–315.
25. U.S. Pat. 3,232,961 (February 1, 1966), W. Stein, H. Rutzen, and E. Sussner (to Henkel & Cie).
26. U.S. Pat. 2,875,066 (February 24, 1959), G. W. Holman and L. H. Going (to Procter & Gamble Co.).
27. U.S. Pat. 2,875,067 (February 24, 1959), G. W. Holman and L. H. Going (to Procter & Gamble Co.).
28. M. Naudet, "Composition of Partial Esters of Glycerol and Fatty Acids," *Rev. Fr. Corps Gras* **17**, 97–104 (1970).
29. E. R. Sherwin, "Antioxidants for Vegetable Oils," *J. Am. Oil Chem. Soc.* **53**, 430–436 (1976).
30. *U.S. Code of Federal Regulations*, Title 21, Part 172, Sections 172.818 and 172.854, U.S. Government Printing Office, Washington, D.C., 1994.
31. *U.S. Code of Federal Regulations*, Title 21, Part 169, Section 169.150, U.S. Government Printing Office, Washington, D.C., 1994.
32. *U.S. Code of Federal Regulations*, Title 21, Part 169, Section 169.140, U.S. Government Printing Office, Washington, D.C., 1994.
33. U.S. Pat. 4,385,076 (May 24, 1983), T. G. Crosby (to Procter & Gamble Co.).
34. U.S. Pat. 4,384,008 (May 17, 1983), D. H. Millisor (to Procter & Gamble Co.).
35. U.S. Pat. 4,960,606 (October 2, 1990), T. G. Crosby (to Bunge Foods Corporation).
36. U.S. Pat. 4,962,951 (October 9, 1990), T. G. Crosby (to Bunge Foods Corporation).
37. U.S. Pat. 4,359,482 (November 16, 1982). T. G. Crosby (to Procter & Gamble Co.).
38. E. J. Pyler, *Baking Science and Technology*, Vols. 1 and 2, 3rd ed., Sosland Publishing Co., Kansas City, Mo., 1988.
39. U.S. Pat. 4,310,556 (January 12, 1982), J. L. Suggs, D. F. Buck, and H. K. Hobbs (to Eastman Kodak Co.).
40. U.S. Pat. 4,320,557 (January 12, 1982), J. L. Suggs, D. F. Buck, and H. K. Hobbs (to Eastman Kodak Co.).
41. U.S. Pat. 4,242,366 (December 30, 1980), J. E. Morgan, A. J. Del Vecchio, B. L. Brooking, and D. M. Laverty (to Pillsbury Co.).
42. U.S. Pat. 4,201,718 (May 6, 1980), J. T. Marsh.
43. U.S. Pat. 4,205,905 (May 27, 1980), M. Pike, I. G. Barr, and F. Tirtaux.
44. U.S. Pat. 4,268,534 (May 19, 1981), T. Kawada, and Y. Tanaka (to Kao Soap Co., Japan).
45. U.S. Pat. 4,108,879 (August 22, 1978), S. Minowa, Y. Tokoshima, N. Yasuda, and T. Tanaka (to Ashai Denko Kogyo KK, Japan).
46. *U.S. Code of Federal Regulations*, Title 21, Chapter 9, Section 321a (Butter Act), U.S. Government Printing Office, Washington, D.C., 1994.
47. *U.S. Code of Federal Regulations*, Title 21, Part 166, Section 166.110, U.S. Government Printing Office, Washington, D.C., 1994.
48. Economic Research Service, *Food Review*, U.S. Department of Agriculture, Washington, D.C., September–December, 1997.
49. Economic Research Service, *Oil Crops Situation and Outlook*, U.S. Department of Agriculture, Washington, D.C., October 1998.
50. J. E. Kinsella, "Food Lipids and Fatty Acids: Importance in Food Quality, Nutrition, and Health," *Food Technol.* **42**, 124–145 (1988).
51. Food and Nutrition Board, National Research Council, *Recommended Daily Allowances*, 10th ed., National Academy of Sciences, Washington, D.C., 1989.
52. E. H. Ahrens, Jr., and W. E. Connor, "Report of the Task Force on the Evidence Relating Six Dietary Factors to the Nation's Health," *Am. J. Clin. Nutr.* **32** (Suppl.), 2621–2748 (1982).
53. S. M. Grundy, "Cholesterol and Coronary Disease. A New Era," *J. Am. Med. Assoc.* **256**, 2849–2858 (1986).

54. A. Leaf and P. C. Weber, "Cardiovascular Effects of N-3 Fatty Acids," *N. Engl J. Med.* **318**, 549–557 (1988).
55. F. H. Mattson and S. M. Grundy, "Comparison of Effects of Dietary Saturated, Monounsaturated, and Polyunsaturated Fatty Acids on Plasma Lipids and Lipoproteins in Man," *J. Lipid Res.* **26**, 194–202 (1985).
56. A. Bonahome and S. M. Grundy, "Effect of Dietary Stearic Acid on Plasma Cholesterol and Lipoprotein," *N. Engl J. Med.* **318**, 1244–1248 (1988).
57. N. B. Carter, H. J. Heller, and M. A. Denkar, "Comparison of the Effects of Medium-Chain Triacylglycerols, Palm Oil, and High Oleic Sunflower on Plasma Triacylglycerol Fatty Acids and Lipid and Lipoprotein Concentrations in Humans," *J. Clin. Nutr.* **65**, 41–45 (1997).
58. K. Sundram, K. C. Hayes, and O. H. Sirn, "Both Dietary 18:2 and 16:0 May Be Required To Improve the Serum LDL/HDL Cholesterol Ratio in Normocholesterolemic Men," *Nutr. Biochem.* **6**, 179–187 (1995).
59. K. C. Hayes, "Saturated Fats and Blood Lipids: New Slant on an Old Story," *Can. J. Cardiol.* **11** (Suppl. G), 16–86 (1995).
60. K. Sundram, K. C. Hayes, O. H. Siru, "Dietary Palmitic Acid Results in Lower Seum Cholesterol Than Does Lauric-Myristic Acid Combination in Normolipemic Humans," *Am. J. Clin. Nutr.* **59**, 841–846 (1994).
61. P. Kholsa and K. C. Hayes, "Dietary Palmitic Acid Raises Plasma LDL Cholesterol Relative to Oleic Acid Only at High Intake of Cholesterol," *Biochim. Biophys. Acta* **1210**, 13–22 (1993).
62. B. Dolan and J. M. Nash, "Searching for Life's Elixir," *Time*, 62–66 (Dec. 12, 1988).
63. J. T. Anderson, F. Grande, and A. Keys, "Independence of the Effects of Cholesterol and Degree of Saturation of Fat in the Diet on Serum Cholesterol in Man," *Am. J. Clin. Nutr.* **29**, 1184–1189 (1976).
64. Anonymous, "New Findings on Palm Oil," *Nutr. Rev.* **45**, 205–207 (1987).
65. M. Sugano, *Lipids* **40**, 48–51 (1987).
66. C. Ip et al., eds., *Dietary Fat and Cancer*, Alan R. Liss, New York, 1986.
67. W. C. Willet and B. MacMahon, "Diet and Cancer—Overview," *New Engl. J. Med.* **310**, 697–703 (1985).
68. J. E. Hunter, C. Ip, and E. J. Hollenbach, "Isomeric Fatty Acids and Tumerigenesis: A Commentary on Recent Work," *Nutrition and Cancer* **7**, 199–209 (1985).
69. F. R. Senti, ed., *Health Aspects of Dietary Trans Fatty Acids*, Federation of the American Society for Experimental Biology, Bethesda, Md., 1985.
70. J. T. Judd et al., "Dietary Trans Fatty Acids: Effects of Plasma Lipids and Lipoproteins on Healthy Men and Women," *Am. J. Clin. Nutr.* **59**, 861–868 (1994).
71. P. H. Nestel et al., "Plasma Lipoprotein and Lp[a] Changes With Substitution of Elaidic Acid for Oleic Acid in the Diet," *J. Lipid Res.* **33**, 1029–1036 (1992).
72. D. Kromhout et al., "Dietary Saturated and Trans Fatty Acids and Cholesterol and 25-Year Mortality From Coronary Heart Disease: The Seven Countries Study," *Prev. Med.* **24**, 308–315 (1995).
73. ASCN/AIN Task Force on Trans Fatty Acids, "Position Paper on Trans Fatty Acids," *Am. J. Clin. Nutr.* **63**, 663–670 (1996).
74. C. Ip and J. R. Marshal, "Trans Fatty Acids and Cancer," *Nutr. Rev.* **54**, 138–145 (1996).

GENERAL REFERENCES

E. Bernadini, *Vegetable Oils and Fat Processing*, Vol. 2, B. E. Oil, Rome, Italy, 1983.

E. W. Eckey, *Vegetable Fats and Oils*, Reinhold Publishing, New York, 1964.

M. Gupta, "Manufacturing Process for Emulsifiers," in Y. H. Hui, ed., *Bailey's Industrial Oil and Fat Products*, Vol. 4, 5th ed., John Wiley & Sons, New York, 1996, pp. 569–601.

T. P. Hilden and P. N. Williams, *The Chemical Constitution of Natural Fats*, 4th ed., John Wiley & Sons, New York, 1964.

Institute of Shortening and Edible Oils, *Food Fats and Oils*, 7th ed., Institute of Shortening and Edible Oils, Washington, D.C., 1994.

H. W. Lawson, *Standards For Fats and Oils*, AVI Publishing, Westport, Conn., 1985.

K. S. Markley, *Fatty Acids*, 2nd ed., John Wiley-Interscience, New York, 1960.

T. J. Wiess, *Food Oils and Their Uses*, 2nd ed., AVI Publishing, Westport, Conn., 1983.

THOMAS CROSBY
Bunge Foods Corporation
Bradley, Illinois

FATS AND OILS: SUBSTITUTES

A growing public awareness about the relationship between food and wellness has kept dietary fat at the top of the list of consumer nutritional concerns for at least a decade (see Table 1 Ref. 1). According to the U.S. Department of Agriculture's (USDA's) Continuing Survey of Food Intakes by Individuals (CSFII) (1994–1996), almost half of Americans think their diets are too high in fat (2). Research by Susan M. Krebs-Smith and colleagues, reported in the *American Journal of Clinical Nutrition*, supports the CSFII findings. Of the six most prevalent nutrient intake patterns identified by these researchers, not one meets the 30% calories-from-fat guideline. In fact, only 1% of the population actually fulfills all of the U.S. government's Food Guide Pyramid recommendations (3).

At 9 kcal/g, fat is the most concentrated source of energy among the macronutrients. According to the *Surgeon General's Report on Diet and Health*, high intake of dietary fat is "associated with increased risk for obesity, some types of cancer, and possibly gallbladder disease. Epidemiologic, clinical and animal studies provide strong and consistent evidence for the relationship between saturated fat intake, high blood cholesterol, and increased risk for coronary heart disease. Excessive saturated fat consumption is the major dietary contributor to total blood cholesterol levels" (4). Reflecting national health policy, the Surgeon General, the National Academy of Sciences, the American Dietetic Association, the American Heart Association, the National Cholesterol Education Project, the American Cancer Society, the National Institutes of Health, the USDA, and the U.S. Department of Health and Human Services are among the many health and government authorities that recommend limiting dietary fat to 30% of calories.

ROLE OF FAT REPLACERS IN FOOD

In its April 1998 position paper on fat replacers, The American Dietetic Association stated:

Table 1. Nutritional Content of Food: Top Five Concerns 1994–1998

	Percentage of consumers concerned				
Concern	1994 (%)	1995 (%)	1996 (%)	1997 (%)	1998 (%)
Fat content, low fat	59	65	60	56	59
Salt/sodium content, less salt	18	20	28	23	24
Cholesterol levels	21	18	26	20	20
Sugar content, less sugar	14	15	12	11	12
Food/nutritional value	4	8	6	11	12

Source: Ref. 1.

> Fat replacers may be used for multiple purposes. One motivation is to facilitate a reduction in total and saturated fat consumption by maintaining the appeal of foods reduced in fat content. The goal of fat reduction is to decrease the incidence of obesity and chronic diseases. A second purpose may be to reduce total energy consumption to improve health and enhance physical appearance and self-concept. A third option entails using fat replacers to increase the volume of palatable foods that may be consumed without increasing fat or energy intake. (5)

According to Calorie Control Council research, the primary reason people use reduced-fat foods is to stay in better overall health (see Table 2). Not surprisingly, a food's fat content is the first item shoppers look for when reading a nutrition label (6). Yet even though consumers seem increasingly committed to lowering their dietary fat intake, they continue to rank taste as their top consideration when selecting food (see Table 3). And few nutrients affect the taste and overall sensory experience of food more than fat.

Fat not only adds aroma and flavor to food, but also influences palatability, flakiness, creaminess, and crispness. And to most people, the texture—or mouth-feel—of food ranks alongside taste as an important selection criterion. From the creaminess of ice cream, to the moist tenderness of cake, to the crispness of potato chips, fat is a critical component. Thus, removing fat from a food can affect far more than taste. Research has shown that most consumers, when given a food from which the fat has been removed and no replacement added, find the food unacceptable (7). The delicate balance between fat and a pleasant sensory experience can be seen in the home-cooking arena as well. Home bakers can replace *some* of the fat or oil in muffins and cakes with applesauce or pureed prunes. But removing all the fat will result in unacceptable appearance, texture, and taste.

Responding to a market that is ripe for low- and reduced-fat foods (see Table 4 and 5), manufacturers have made fat modification an increasingly important aspect of their research and development. Their objective is threefold: to create food products that qualify for nutrient content/health claim status (see Table 6); to capture the sensory attributes and functionality of fat; and to achieve all of this cost effectively. In 1990 Healthy People 2000 called on the food industry to create 5000 reduced-fat foods over the next decade. By 1992 manufacturers had exceeded that goal by 600 (8). In 1990 alone, the food industry rolled out more than 1400 products bearing a reduced- or low-fat label claim—up 70% from the number of such products the year before. By 1995 the number of fat-modified products had risen to 8500 (9,10).

Table 2. Why People Used Reduced-Fat Products

Reason	Percentage of adult Americans citing
To stay in better overall health	77
To eat or drink healthier food and beverages	71
To reduce fat	68
To reduce cholesterol	61
To maintain current weight	57
To reduce calories	56
To maintain an attractive physical appearance	52
To reduce weight	43
For refreshment or taste	39
To help with a medical condition	31

Source: Calorie Control Council 1998 National Survey.

COMPOSITION OF FAT REPLACERS

Focus groups conducted by the International Food Information Council suggest that there is a lack of knowledge concerning fat replacers among consumers as well as among health professionals. For example, very few consumers realized that the fat removed from low- and reduced-fat foods is replaced with another ingredient. And although health professionals understood that product reformulation replaced the taste and texture of fat, they typically had very little specific information (11).

Usually a fat replacer is not itself a fat. Some fat replacers, however, are lipids, but because of their structure, they may be undigestible or only partially digestible. Most fat replacers are limited in their use. Some are good thickening agents; others provide creaminess; still others make food moist. Having a variety of fat replacement options available allows manufacturers to use a "systems approach" to formulating low-fat and fat-free foods with the functional and sensory characteristics of their full-fat counterparts. The combination of fat replacers used in a product depends on which properties of that product are being duplicated. Combinations may include proteins, starches, dextrins, maltodextrin, fiber, emulsifiers, and flavoring agents. Some fat replacers are themselves blends of various ingredients (see Table 7).

Types

Fat replacers fall into three functional categories (5,7,12):

Table 3. Importance of Various Factors in Food Selection: 1994–1998

	Percentage of consumers saying "very important"				
Factor	1994 (%)	1995 (%)	1996 (%)	1997 (%)	1998 (%)
Taste	90	90	88	87	89
Nutrition	76	74	78	77	76
Product safety	69	69	75	73	75
Price	70	69	66	66	64

Source: Ref. 1.

Table 4. Percentage of Consumers Who Eat Various Reduced-Fat Foods

		Among shopper segment	
Food	% Total who eat	% Healthy eaters	% Nutritionally apathetic
Salad dressing	68	78	49
Mayonnaise	55	60	39
Crackers	53	61	35
Ice cream	52	60	35
Chips	51	57	40
Cookies	51	54	40

Source: Ref. 6.

Table 5. Most Popular Reduced-Fat Products

Product	Percentage of adult Americans who consume
Skim or low-fat milk	62
Salad dressings/sauces/mayonnaise	56
Margarine	44
Chips/snack foods	40
Meat products	39
Ice cream/frozen desserts	36
Cakes/baked goods	32
Dinner entrees	30
Candy	18

Source: Calorie Control Council 1998 National Survey.

Table 6. Food Label Nutrient Content Claims

Claim[a]	Criterion
Fat free	Less than 0.5 fat
Low fat	3 g or less of fat
Reduced or less fat	At least 25% less fat[b]
Light	One-third fewer calories or 50% less fat[b]

[a]The Food and Drug Administration and the U.S. Department of Agriculture have established specific regulation concerning allowable nutrient content claims. The following claims are defined for one serving.
[b]As compared with a standard serving size of the traditional food.

1. Fat mimetics replace the bulk, body, and mouth-feel of fat. They reduce calories both because they are less calorically dense than fat and because they contain water, which itself replaces some fat. Mimetics based on gums, cellulose, and seaweed are noncaloric; those based on starch and protein provide 4 kcal/g.
2. Low-calorie fats include products such as caprenin and salatrim, both of which provide 5 kcal/g. They are metabolized more like carbohydrate and are rapidly used as energy rather than stored as fat.
3. Fat replacers, substitutes, or analogs are similar to fats and oils in function and in sensory properties. Theoretically, they can replace fat on a one-to-one weight basis. Generally, they are heat stable and can be used for frying. Fat replacers provide 0 to 3 kcal/g. Olestra is a zero-calorie fat replacer.

Derivations

Fat replacers are also classified according to the substances from which they are derived (see Table 8). Carbohydrate- and protein-based fat replacers are fat mimetics; fat-based fat replacers are low-calorie fats or fat replacers (5,7,13):

- *Carbohydrate-based.* Carbohydrates such as starches, fibers, cellulose, and gums are commonly used in low-fat and fat-free foods as texturing agents, bulking agents, moisturizers, and stabilizers. Carbohydrate-based ingredients provide 0 to 4 calories per gram.
- *Protein-based.* Milk, egg, or whey protein is heated and blended at high speed to produce tiny particles that feel creamy to the tongue. This process is called microparticulation. Protein-based ingredients also help to stabilize emulsions in sauces, spreads, and salad dressings. These ingredients are fully absorbed by the body and range in calories from 1 to 4 kcal/g, depending on degree of hydration.
- *Fat-based.* Some fat-reduction ingredients are actual fats designed to contribute fewer calories and less fat to foods. Others, such as olestra, are structurally modified to be calorie-free.

NUTRITIONAL IMPACT OF FAT REPLACERS

Some observers have speculated that the availability of reduced-fat and fat-free foods is contributing to the increasing rate of obesity among adults and children. To characterize fat replacers as a magic formula for weight loss is to misrepresent their optimal dietary role. To prevent or slow weight gain, calories eaten cannot exceed cal-

Table 7. Selected Applications and Functions of Fat Replacers

Application	Fat replacer	General functions[a]
Baked goods	Lipid based	Emulsify, provide cohesiveness, tenderize, carry flavor, replace shortening, prevent staling, prevent starch retrogradation, condition dough
		Retain moisture, retard staling
	Carbohydrate based	Texturize
	Protein based	
Frying	Lipid based	Texturize, provide flavor and crispness, conduct heat
Salad dressing	Lipid based	Emulsify, provide mouth-feel, hold flavorants
	Carbohydrate based	Increase viscosity, provide mouth-feel, texturize
	Protein based	Texturize, provide mouth-feel
Frozen desserts	Lipid based	Emulsify, texturize
	Carbohydrate based	Increase viscosity, texturize, thicken
	Protein based	Texturize, stabilize
Margarine, shortening, spreads, butter	Lipid based	Provide spreadability, emulsify, provide flavor and plasticity
	Carbohydrate based	Provide mouth-feel
	Protein based	Texturize
Confectionery	Lipid based	Emulsify, texturize
	Carbohydrate based	Provide mouth-feel, texturize
	Protein based	Provide mouth-feel, texturize
Processed meats	Lipid based	Emulsify texturize, provide mouth-feel
	Carbohydrate based	Increase water-holding capacity, texturize, provide mouth-feel
	Protein based	Texturize, provide mouth-feel, water holding
Dairy products	Lipid based	Provide flavor, body, mouth-feel and texture; stabilize
	Carbohydrate based	Increase viscosity, thicken, aid gelling, stabilize
	Protein based	Stabilize, emulsify
Soups, sauces, gravies	Lipid based	Provide mouth-feel and lubricity
	Carbohydrate based	Thicken, provide mouth-feel, texturize
	Protein based	Texturize
Snack products	Lipid based	Emulsify, provide flavor
	Carbohydrate based	Texturize, aid formulation
	Protein based	Texturize

Source: Food Technology, March 1998.
[a]Functions are in addition to fat replacement.

ories expended. Fat-modified foods are but one tool that can be used to create this energy balance.

The theory is that people may be cutting down on fat consumption, but by overindulging in fat-modified foods, they are consuming extra calories and dramatically tipping the energy balance toward obesity. But the fact is that despite the proliferation of low-fat and fat-free foods, Americans are actually eating *more* fat than in the past. Although between the 1970s and the 1990s, Americans decreased their intake of fat from approximately 37% of calories to approximately 34% of calories, this drop in *percent of calories from fat* represents neither a decrease in *calories* nor a decrease in *fat grams*. Over the same 20-year time span, the average adult's daily calorie intake increased by 300, and total fat gram consumption increased from about 81 g/day to 83 g/day (13). At 9 calories per gram, this 2-g/day increase in consumption adds up to about 540 extra calories per month, or almost 6500 extra fat calories per year.

In addition, the 1996 National Institutes of Health Consensus Conference Statement on Physical Activity and Cardiovascular Health identified physical inactivity as a major U.S. public health crisis. Inactivity reduces muscle mass, which translates into fewer calories burned, even at rest. When physical activity is low, a reduction in dietary fat is needed simply to *prevent* weight gain, even if fat intake is moderate. But increased fat consumption coupled with stagnant-to-falling exercise rates is the magic formula for positive energy balance and weight gain.

Another often overlooked fallacy in the theory that fat-modified foods have led to obesity concerns timing. Recently released population figures on obesity were compiled as of 1990—when 33% of the American public was identified as overweight. The market influx of fat-modified foods, however, didn't peak until 1995 (14). In addition, first-generation fat-modified foods of the 1980s were far less palatable than today's low-fat and fat-free products. A proliferation of good-tasting fat-modified foods in the 1990s can't be blamed for obesity statistics based on 1980s data.

Satiety

Bell and colleagues have demonstrated that people eat to a certain *volume*, rather than to a calorie count (15). Diets high in fat are calorically dense and highly palatable, but they are generally *less* satiating than diets high in carbohydrates. Although once in the intestine, fat seems to create strong signals of satiety, the "I'm full" feeling is delayed because of the time it takes time for fat to reach the intes-

Table 8. Example Fat Replacers: Ingredients and Applications

Type	Brand name(s)	Ingredients	Applications
		Protein-based fat replacers	
Microparticulated protein	Simplesse K-Blazer	Milk and/or egg/white proteins, sugar, pectin, citric acid	Imparts fatlike creaminess and richness but no fat flavor; not heat stable; 1–2 kcal/g; digested as protein; used in ice cream, butter, sour cream, cheese, yogurt
Modified whey protein	Dairy-Lo	Whey or milk protein concentrate	Improves texture, flavor, stability of low-fat foods; 4 kcal/g; used in frozen dairy desserts, hard and processed cheese, sour cream, dips, yogurt, sauces, baked goods
Isolated soy protein	Supro ProPlus Supro Plus	Used in foods for 35 years; added to meat to reduce fat	
		Fat-based fat replacers	
Sucrose polyester (olestra)	Olean	Sucrose core with 6 to 8 fatty acids	Looks, tastes, and feels like fat; calorie-free; passes through body unabsorbed; may lower cholesterol; only approved fat replacer that can be used for frying; used in savory snacks
Salatrim	Benefat	Soybean and canola oil	Used in confections, baked goods, dairy; 5 kcal/g
Caprenin	Caprenin	caprylic, capric and behenic acids, and glycerine	Good substitute for cocoa butter; can be used in confections; 5 kcal/g
Mono- and diglycerides	Dur-Em, Dur-Lo		Used to replace all or part of the shortening in cake mixes, cookies, icings, and some dairy products; emulsifiers that disperse fat in watery mediums so less fat is used; 9 kcal/g
		Carbohydrate-based fat replacers	
Carrageenan	Carrageenan	Extract of red seaweed	First used as an emulsifier, stabilizer, thickener; provides gel-like mouth-feel; used to replace part of the fat in ground beef, hot dogs, processed cheese, low-fat desserts
Microcrystalline cellulose	Avicel (cellulose gel), Methocel		Used in salad dressings, mayonnaise, processed cheese, frozen desserts; forms gel in water; acts like fat and has the mouth-feel and appearance of fat; used as a stabilizer; calorie-free
Hydrolyzed oat flour	Oatrim, Beta-Trim, TrimChoice	Enzyme-treated oat flour containing β-glucan soluble flour	Used in baked goods, fillings and frostings, frozen desserts, dairy beverages, cheese, salad dressings, processed meat, confections; 1 to 4 kcal/g
Powdered cellulose	Solka-Floc		Often used in fried foods; causes less oil to be absorbed
Maltodextrins	CrystalLean; Lorelite; Lorelite; Lycadex; MALTRIN; Paselli D-LITE, EXCEL, SA2; STAR-DRI	Gel or powder extracted from tapioca, corn, potato, rice	Mimic fat's mouth coating, melting sensation, richness; used in salad dressings, pudding, spreads, dairy desserts, meat products; can be used in some cooked foods; 4 kcal/g
Polydextrose	Litesse, Sta-Lite	Water-soluble polymer of dextrose containing minor amounts of sorbitol and citric acid	Used as fat replacer and bulking agent; used in baked goods, chewing gum, confections, salad dressings, frozen dairy desserts, gelatins, puddings; 1 kcal/g
Gums	KELCOGEL, KELTROL, Slendid		Have creamy mouth-feel; provide thickening effect; pass through body virtually unmetabolized; common gums include guar gum, gum arabic, locust bean gum, modified carbohydrate gum/vegetable gum; used in a wide variety of foods

Table 8. Example Fat Replacers: Ingredients and Applications *(continued)*

Type	Brand name(s)	Ingredients	Applications
Inulin	Raftiline, Fruitafit, Fibruline	Chicory root extract	Used in yogurt, cheese, frozen desserts, baked goods, icings, fillings, dairy products, fiber supplements, processed meats, 1 to 1.2 kcal/g
Pectin	Splendid	Citrus peel and table sugar	Forms a gel that can replace up to 100% of fat in some foods; mouth-feel and melting sensation of fat; used in soups, sauces, gravies, cakes and cookies, dressings and spreads, frozen desserts, frosting
Z-Trim	Z-Trim	Processed hulls of oats, soybeans, peas, rice, or from bran of corn or wheat	Provides smooth mouth-feel of fat; can be used in a variety of foods; calorie-free

Sources: Refs. 5, 17, 13, and Fat Reduction in Foods, Calorie Control Council, 1997.

tinal tract. Large amounts of energy-dense fat can be ingested before satiety signals kick in. Thus, eating high-fat foods leads to what researchers call "passive consumption" or high-fat hyperphagia—that is, a tendency to consume *excess* energy (16).

Because of the weak effect it has on satiety, removing some fat or substituting for it does not substantially decrease the satiating effect of a food, but it can result in fewer calories being eaten. Consequently, lower-fat diets including use of fat-modified foods, rather than contributing to obesity, actually have the ability to *reduce* the passive consumption of excess calories. CSFII data from 1995 show that adults who used more fat-modified foods also consumed less energy overall (17). In addition, research by Arsenault and colleagues and by Sigman-Grant demonstrates that individuals who consume fat- and calorie-reduced foods have a more nutrient-dense diet than those who do not (18–20).

Compensating Behaviors

Probably the most common argument used to support the notion that overindulging in fat-modified foods is responsible for obesity is the compensation theory—that people equate "fat-free" with "low-calorie" and therefore eat more. Most of the research on dietary compensation and fat replacers has been done on olestra, available under the brand name Olean®.

Among individuals *who did not know* a fat replacer had been added to their diet, researchers have observed two behaviors, depending on participants' ability to regulate calories. The research has shown that children and young adult males, who generally regulate calories well, did compensate for the energy deficit created by fat replacement (21,22). In both groups, however, compensation was not 100% nor was it fat-specific. Consequently, although these children and lean young men did compensate for calories, they nevertheless consumed a diet that was lower in fat.

Researchers also studied a group of 105 individuals who do not regulate calories well. Among these participants, they observed both a decrease in the percentage of calories from fat and in total calories consumed. These preliminary findings indicate that for people who need to control body weight, fat replacers may be a useful tool for helping manage both fat and calorie intake (23–25).

When research participants *did know* that a fat replacer had been added to their food, compensation behavior remained negligible. In a 10-day study of 96 habitual potato-chip eaters, some individuals who knew they were eating fat-free potato chips ate a few more chips. Most participants however, ate the same amount of chips regardless of whether or not they knew they were eating full-fat chips or chips made with a fat replacer. It is important to note that all participants eating chips made with a fat replacer benefited from both a fat and a calorie reduction (26).

In another study examining the behavioral aspects of compliance with a low-fat diet, two groups totaling 44 women attempted to maintain a diet having less than 30% of calories from fat. Women who were allowed to use products with fat replacers reported feeling less deprived because they had a wider range of foods from which to choose (27). A six-month study conducted by Westerterp and colleagues tracked 217 free-living, nonobese men and women who received either reduced-fat or full-fat products. At the end of the intervention, study participants in the reduced-fat group had consumed fewer calories and had not gained weight. The control group, however, showed both an increase in calorie intake and weight (28). Similarly, in a recent three-month trial, conducted by Kelly and colleagues, study participants consuming a full-fat diet gained weight, whereas those consuming foods made with a fat replacer did not (29).

REGULATING FAT REPLACERS

The 1958 Food Additives Amendment to the Federal Food, Drug and Cosmetic Act requires that the Food and Drug Administration (FDA) approve all food additives prior to their use. Manufacturers must submit a food additive petition (FAP) to FDA, detailing the additive's production, its expected consumption level (average as well as 90th percentile); exposure by age group; and its safety profile, including for potentially sensitive subpopulations. When Congress enacted this legislation, it exempted (with some qualifications) two groups of substances:

- "Prior-sanctioned substances" are those which the FDA or the USDA had determined prior to 1958 as safe for use in food—for example, sodium nitrate and potassium nitrate used to preserve luncheon meats.
- GRAS substances are substances in use before 1958 that are "generally recognized as safe" by scientific experts based on the substances' history of use in food or on scientific procedures. If a substance was not in use in foods before 1958, its safety must be determined based on scientific procedures. Common GRAS substances—among the several hundred used in food—are sugar, baking soda, and vitamins.

Establishing GRAS status and filing a FAP require similar scientific support. One key difference between seeking GRAS status for a food component and filing a petition for approval of a food additive, however, is that the safety information about a GRAS substance is available in the scientific literature and readily accessible to the manufacturer, whereas safety information for the food additive may not yet be published. Acting independently from the FDA, manufacturers can convene a panel of experts to review the scientific information and determine whether or not a substance should be considered GRAS for a specific use.

The majority of fat replacers have garnered approval through the GRAS procedure (eg, various carbohydrate polymers, gums, gels and starches; microparticulated proteins; whey proteins; and fat emulsifiers). Procter & Gamble's caprenin offers a good example of an independent

Table 9. Dietary Impact of Fat Replacers

Regular diet				Diet with fat replacers			
Food	Portion	Kcal	Fat g	Food	Portion	Kcal	Fat g
Dietary impact of fat replacers: breakfast							
Coffee	10 oz	6	0	Coffee	10 oz	6	0
Creamer	1 t	16	1	Low-fat creamer	1 t	8	0
Bagel	1	195	1	Bagel	1	195	1
Cream cheese	1 oz	100	10	Light cream cheese	1 oz	60	5
Orange	1	65	0	Orange	1	65	0
Total		*382*	*12*	*Total*		*334*	*6*
Dietary impact of fat replacers: lunch							
Lean ground beef	4 oz	286	18	Lean ground beef	4 oz	286	18
Roll	1	114	2	Roll	1	114	2
Cheese	2 oz	120	9				
Low-fat cheese	2 oz	90	2				
Lettuce	1 oz	5	0	Lettuce	1 oz	5	0
Tomato	1 slice	4	0	Tomato	1 slice	4	0
Mayonnaise	1T	57	5	Fat-free mayonnaise	1T	10	0
Potato chips	1 oz	150	10	Fat-free potato chips	1 oz	0	0
Water	12 oz	0	0	Water	12 oz	0	0
Total		*736*	*44*	*Total*		*509*	*22*
Dietary impact of fat replacers: dinner							
Tossed salad	1-1/2 c	35	0	Tossed salad	1-1/2 c	35	0
Salad dressing	2 T	138	14	Low-fat salad dressing	2 T	70	6
Chicken breast	6 oz	284	6	Chicken breast	6 oz	284	6
Baked potato	1	220	0	Baked potato	1	220	0
Sour cream	1 oz	61	6	Fat-free sour cream	1 oz	17	0
Green beans	1/2 c	25	0	Green beans	1/2 c	25	0
Butter	2 pats	71	8	Low-fat margarine	2 pats	36	4
Ice cream	1/2 c	150	12	Low-fat ice cream	1/2 c	100	1.5
Iced tea	12 oz	99	0	Iced tea	12 oz	99	0
Total		*1,083*	*46*	*Total*		*886*	*19.5*
Dietary impact of fat replacers: snack							
Tortilla chips	1 oz	141	7	Fat-free tortilla chips	10 oz	98	0
Salsa	1 T	25	0	Salsa	1T	25	0
Soda	12 oz	148	0	Soda	12 oz	148	0
Total		*314*	*7*	*Total*		*271*	*0*
Grand Total		*2515*	*109*	*Grand Total*		*2000*	*45.5*

Source: Adapted from *The Weight Control Digest*, May/June 1996.

GRAS determination. Caprenin is a triglyceride formed of three fatty acids that are natural components of the diet. Procter & Gamble asked an independent expert panel of scientists to review caprenin for GRAS status. In 1991 the panel determined that caprenin should indeed be generally recognized as safe.

If the FDA chooses to disagree with a GRAS determination, it may take regulatory action. Companies can also petition FDA to affirm their independent GRAS determinations. In these cases, FDA publishes a notice in the *Federal Register* requesting public comment.

The FDA approval process for a *new* food ingredient begins with a food additive petition in which the manufacturer must supply extensive scientific data documenting intended use and safety. A food additive is defined as "an ingredient not previously found in food whose intended use results, or may reasonably be expected to result, directly or indirectly, in its becoming a component or otherwise affecting the characteristics of any food" (5). FDA may recommend intake limits and may require postmarket surveillance. Procter & Gamble's sucrose polyester fat replacer, olestra (Olean), went through the FAP process.

CONCLUSION

Currently, there is no single fat replacer that "does it all." The most successful approach to fat replacement may prove to be the "systems approach" that combines the features of various fat replacers to achieve the optimal taste and function in a particular food. In the future, the ideal fat replacer—whether a single entity or a blend of ingredients—will be one that captures the functionality and sensory qualities of fat; is calorie-free, inexpensive, and suitable for frying and cooking; and that does not interfere with drug utilization or with the absorption of other nutrients. Manufacturers may seek to promote other health benefits of certain fat replacers. For example, Oatrim and the soon-to-be-available Nu-trim, both developed by the USDA, contain β-glucan, the soluble fiber in oats and barley that lowers blood cholesterol.

Although an impressive number of fat-modified products are introduced every year, relatively few succeed. Until consumers are content with the taste and cost of fat-modified foods, their use will not be widespread, and their long-term impact on everyday eating habits and the health of the U.S. population will be difficult to measure. It is clear, however, that when used properly, fat-modified foods have the potential to play an important role in reducing both calories and fat (see Table 9). As the American Dietetic Association notes in its 1998 position paper on fat replacers:

> "Reduced-fat products can . . . be used to expand the diversity and volume of foods that can be consumed while maintaining a given level of energy consumption. . . . Safe and effective use of products containing fat replacers should be facilitated through efforts to improve consumer understanding of labeling information, the importance of food preparation practices, portion size and eating frequency control, and the relationship between energy expenditure and total energy balance." (5)

BIBLIOGRAPHY

1. *Trends in the United States: Consumer Attitudes and the Supermarket 1998*, Food Marketing Institute, Washington, D.C., 1998.
2. *What We Eat in America Survey 1994–1996. Continuing Survey of Food Intake in America (CSFII)*, U.S. Department of Agriculture, Washington, D.C., 1997.
3. S. M. Krebs-Smith et al., "Characterizing Food Intake Patterns of American Adults," *Am. J. Clin. Nutr.* **65**, 1264S–1268S (1997).
4. *U. S. Surgeon General's Report on Nutrition and Health*, U.S. Department of Health and Human Services, Washington, D.C., 1988.
5. "Position of the American Dietetic Association Position: Fat Replacers," *J. Am. Dietet. Assoc.* **98**, 463–468 (1998).
6. *Shopping for Health*, Food Marketing Institute, Washington, D.C., 1997.
7. K. Napler, *Fat Replacers: The Cutting Edge of Cutting Calories*, American Council on Science and Health, New York, 1997.
8. *Healthy People 2000 Review: 1995–1996*, U.S. Department of Health and Human Services, Hyattsville, Md., 1996.
9. Calorie Control Council home page, URL: *www.caloriecontrol.com* (last accessed November 1998).
10. L. Dornblaser, "New Product News," *Prepared Foods* **4**, 37 (1996).
11. Internal memo to International Food Information Council Task Force on Fat Replacers, April 24, 1995.
12. E. E. Ziegler and L. J. Filer, eds., *Present Knowledge in Nutrition*, 7th ed., ILSI Press, Washington, D.C., 1996.
13. *IFIC Review: Use and Nutritional Impact of Fat Reduction Ingredients*, International Food Information Council, Washington, D.C., 1997.
14. International Food Information Council, "Fat Facts and Fiction," *Food Insights* Sept./Oct., 2 (1997).
15. E. A. Bell et al., "Energy Density of Foods Affects Energy Intake in Normal-Weight Women," *Am. J. Clin. Nutr.* **67**, 412–420 (1998).
16. J. E. Blundell and J. I. Macdlarmid, "Fat as a Risk Factor for Overconsumption: Satiation, Satiety, and Patterns of Eating," *J. Am. Dietet. Assoc.* **97**, S63–S69 (1997).
17. A. H. Lichtenstein et al., "Dietary Fat Consumption and Health," *Nutr. Rev.* **56**, S3–S18 (1998).
18. M. Sigman-Grant, "Can You Have Your Low-fat Cake and Eat It Too? The Role of Fat Modified Products," *J. Am. Dietet. Assoc.* **97**, S76–S81 (1997).
19. J. E. Arsenault, "Nutritional Intake of Women Consuming Fat- and Calorie-Reduced Foods," poster session, American Dietetic Association Annual Meeting, Oct. 23, 1997.
20. M. Sigman-Grant, S. Poma, and K. Hsieh, "Update on the Impact of Specific Fat Reduction Strategies on Nutrient Intake of Americans," *FASEB J.* **12**, 3079A (1998).
21. L. L. Birch et al., "Effects of a Nonenergy Fat Substitute on Children's Energy and Macronutrient Intake," *Am. J. Clin. Nutr.* **58**, 326–333 (1993).
22. B. J. Rolls et al., "Effects of Olestra, a Noncaloric Fat Substitute, on Daily Energy and Fat Intakes in Lean Men," *Am. J. Clin. Nutr.* **56**, 84–92 (1992).
23. G. Bray et al., "Effects of Two Weeks of Fat Replacement by Olestra on Food Intake and Energy Metabolism," *FASEB J.* **9**, 2 (1995).

24. J. O. Hill et al., "Effects of 14 Days of Covert Substitution of Olestra on Spontaneous Food Intake," *Am. J. Clin. Nutr.* **67**, 1178–1185 (1998).
25. C. deGraff et al., "Nonabsorbable Fat (Sucrose Polyester) and the Regulation of Energy Intake and Body Weight," *Am. J. Physiol.* **270**, R1386–R1393 (1996).
26. D. L. Miller et al., "Effect of Fat-Free Potato Chips With and Without Nutrition Labels on Fat and Energy Intakes," *Am. J. of Clin. Nutr.* **68**, 282–290 (1998).
27. M. P. Bolton, P. W. Pace, and R. S. Reeves, "Use of Olestra Products in a Nutrition Education Program," *Journal of Nutrition in Recipes and Menu Development* **2**, 33–42 (1996).
28. K. R. Westerterp et al., "Dietary Fat and Body Fat: An Intervention Study," *International Journal of Obesity* **20**, 1022–1026 (1996).
29. S. M. Kelly et al., "A 3-Month, Double-Blind, Controlled Trial of Feeding With Sucrose Polyester in Human Volunteers," *Br. J. Med.* **30**, 41–49 (1998).

SUZETTE MIDDLETON
Procter & Gamble Co.
Cincinnati, Ohio

FIBER, DIETARY

Dietary fiber is a general term for plant polysacharides and lignin that are not broken down by the digestive enzymes of human beings. Dietary fiber is found in plant cell walls, mucilages, gums, and algal polysaccharides. Synthetic polysaccharides used in the food industry are also considered dietary fiber. Resistant starch, starch that is not digested in the small intestine, is chemically and physically similar to other nondigestable polysaccharides and also may be included in the definition of fiber. Many compounds with a wide range of physical and chemical properties are referred to as dietary fiber. In a single plant, the fiber composition of the stem or leaves is very different from the fiber in the fruit or seeds. Dietary fiber is not a uniform substance and every plant has a unique dietary fiber makeup. Because dietary fiber includes many compounds with very different physicochemical properties, it is difficult to characterize. Division of fibers by their solubility in water is the most common method of classification, and is convenient because many of the physiological properties of fiber seem to be based on this physical property. As a rule, water-soluble fibers form viscous solutions and are used in the food industry as emulsifiers, food thickeners, and gelling agents. In the body, soluble fibers are associated with lipid and carbohydrate metabolism. Insoluble dietary fibers, such as those found in wheat bran, tend to contribute to fecal bulk and decreased intestinal transit time.

COMPONENTS

Complex polysaccharides from plants are divided into several distinct groups based on chemical structure. The most common structural component of the cell wall is *cellulose*. Cellulose is composed of long linear chains of glucose molecules (approx. 10,000) with beta 1–4 links, which cannot be broken down by digestive enzymes. The *hemicelluloses* found in plants are a heterogeneous group of compounds that contain a variety of different sugars arranged in a primary chain or backbone, with numerous side chains. The specific names of the hemicelluloses are based on their sugar content, such as xyloglucans (xylose and glucose), glucomannans (glucose and mannose), and arabinoxylans (arabinose and xylose). Solubility in water is dependent on the sugar components and the physical structure of the individual hemicellulose, but most compounds in this group are soluble in dilute alkali. *Pectin* from plants refers to a complex mixture of substances that is composed of a galacturonic acid core esterfied with methyl groups on the uronic acid residues. Rhamnose is also found in the backbone of pectins. Pectin is water soluble and used to produce highly viscous gels, whereas protopectins are insoluble. The only nonpolysaccharide included in the definition of dietary fiber is *lignin*. The lignins consist of chains of phenyl propane residues that cement and anchor the cell wall matrix and stiffen the plant tissue. As plants grow and age they undergo a process of lignification, and the dietary fiber makeup changes over time.

ANALYSIS

The different chemical properties of the various components of dietary fiber make it quite challenging to determine the fiber content of various plants and foods. Early methods for isolating dietary fiber commonly destroyed many of the water-soluble components of the fiber matrix. The term crude fiber is used to describe fiber content determined by an acid and alkali extraction. A rapid, reliable method of determining total dietary fiber content (water soluble and insoluble) has been recommended by the Association of Official Analytical Chemists (AOAC). This methodology combines gravimetric and enzymatic processes and is appropriate for measuring dietary fiber content of foods for labeling purposes. Analysis of the individual components of dietary fiber is much more complex. The best known procedure, developed by Southgate (1), removes the individual fractions of dietary fiber through a series of extractions. The addition of gas liquid or liquid chromatography allows for the identification of individual sugars present in each fraction.

PHYSICOCHEMICAL PROPERTIES

Every fiber source has unique physicochemical properties that elicit variable physiological responses in the body and allows for a wide range of uses in the food industry. These properties are variable and are influenced by factors such as particle size (coarsely ground wheat bran versus finely ground bran), cooking or freezing, and breakdown of the matrix during digestion. Properties such as water-holding capacity or the hydration of dietary fibers to form a gel matrix influence the viscosity of the gastrointestinal contents and can increase fecal bulk and stool softness. Adsorption and binding of organic molecules is also an important trait of dietary fibers. Polysaccharides with an electrical charge, such as pectin and lignin, are known to sequester or bind bile acids and increase their excretion.

It is hypothesized that in the intestine dietary fiber is responsible for binding or adsorbing toxic and mutagenic substances, thus enhancing their removal from the body. Dietary fiber also has the capacity to act as a cation exchanger in the intestine. Minerals and electrolytes can bind to fiber, potentially decreasing their absorption in the digestive tract. An additional characteristic of fiber is the ability to undergo degradation or fermentation by microflora of the large intestine. Water-soluble fiber sources such as pectin or gums may be totally degraded in the colon, whereas cellulose, being a tightly packed linear insoluble polysaccharide, is almost totally resistant to fermentation by the human colonic flora. In plant tissues a mixture of polysaccharides and lignin is produced making it difficult to accurately predict their physicochemical properties and their potential physiological effects in the body. This explains why the effects of oat bran are so different than those of wheat bran and why the overall characteristics of a fiber source may not be related to the physicochemical properties of the individual polysaccharides it contains.

PHYSIOLOGIC AND METABOLIC EFFECTS

Burkitt and Trowell (2) were the first to report the physiological importance of dietary fiber consumption. Based on epidemiological studies, they showed associations between low-fiber diets and chronic disorders such as constipation, diverticulosis, colon cancer, diabetes, and cardiovascular disease. Since the 1970s research has been carried out that, for the most part, confirms the role of dietary fiber in disease prevention. Normal laxation is an important health benefit of dietary fiber consumption. Certain varieties of dietary fiber have been shown to increase stool weight and frequency, soften feces, increase fecal bulk, and reduce gastrointestinal transit times. This is particularly true of insoluble dietary fibers such as cellulose, found in large quantities in wheat bran, and of soluble but nonfermentable fibers such as psyllium gum. Constipation may be prevented, or successfully treated, by increasing dietary fiber intake. Various hypotheses have been suggested as to how dietary fiber affects the composition of fecal material and the frequency of defecation. A likely explanation is the sponge theory. Dietary fiber in the gastrointestinal tract acts as a "water-laden sponge" (3). The sponge entraps or adsorbs water, ions, and nutrients, altering intestinal contents and changing the rate of peristalsis. The presence of fiber also affects the microfloral activity of the colon, the end result being decreased intestinal transit time with increased fecal bulk and water. Stools tend to be softer and larger accompanied by increased frequency of defecation.

Consumption of soluble dietary fibers has been shown to lower blood cholesterol levels in both laboratory animals and humans. Fibers such as oat bran, guar gum, and pectin all have a hypocholesterolemic effect. Several mechanisms of action appear to be responsible for this physiological response. Soluble fibers increase the viscosity of intestinal contents, which alters the mixing and diffusion of nutrients in the intestine and changes rates of nutrient absorption. Impaired bile acid and/or cholesterol absorption from the intestine is thought to lead to lowered blood cholesterol levels. In the colon, most soluble fibers are degraded by the microflora and produce short chain fatty acids (SCFA). It has been hypothesized that SCFA absorbed from the colon may directly inhibit cholesterol synthesis in the liver. High-fiber diets also alter the structure of the intestine. In experimental animals, intestinal length is increased and there is a marked increase in mucosal and muscle mass following fiber consumption (4). It is assumed that morphological changes in the intestine alter absorption patterns thus influencing metabolic processes in the body.

Many varieties of soluble fibers influence the rate of carbohydrate digestion and absorption. It has been shown that when certain fibers are added to test meals, postprandial glucose and insulin levels are decreased (5). In addition, intake of high-fiber foods tends to modulate glucose excursions following meal consumption. This helps to maintain glucose homeostasis and is of particular importance to individuals with impaired glucose tolerance or diabetes.

Prevention of cancer is also considered a health benefit of high-fiber diets. Epidemiological studies indicate that dietary fiber, and in particular a diet rich in cereals and vegetables, helps to prevent colon cancer (6). It is thought that the insoluble fibers that promote laxation and decrease intestinal transit times are responsible for this effect. If toxins and mutagens in foods are moved rapidly through the intestine, exposure of the mucosa to these harmful materials is minimized. Although it is thought that dietary fiber may have a role in preventing numerous cancers including mammary and prostate, this hypothesis has yet to be proven.

RECOMMENDATIONS

The numerous types of dietary fiber and their many components have led to difficulties in establishing recommended daily allowances. However, several professional groups have issued guidelines for levels of dietary fiber intake that should promote good health and have little or no negative side effects. The United States Department of Agriculture (USDA) has suggested a guideline of 11.5 g dietary fiber per 1,000 calories consumed ranging from approximately 20 to 35 g/day for adults (7). Recent recommendations of the American Health Foundation suggest a goal of minimal intake of dietary fiber for children and young adults 3 to 20 years of age be the equal to age plus 5 g of dietary fiber per day (age plus 5). This recommendation suggests that 3-year-old children consume no less than 8 g/day of dietary fiber and a 20-year-old consume 25 g/day (8).

NUTRITIONAL VALUE

Dietary fiber has been called a nonnutritive bulking agent. This is not an accurate description because dietary fiber does indeed contribute energy to the body. In cases where the fiber undergoes fermentation by the intestinal microflora, the SCFAs that are formed and absorbed into the blood contribute kilocalories to the body. Not all fibers are degraded by bacteria, so an average caloric value of 2 kcal/g fiber has been suggested by the British Nutrition Foundation Task Force. Actual values for different fiber sources range from 0 to 4 kcal/g.

USES IN THE FOOD INDUSTRY

The physicochemical properties of individual fiber sources determine their usefulness in the food industry. Many synthetic or modified fibers are used as thickening agents, texturizers, or stabilizers. Gums, also called hydrocolloids, are considered food additives. To achieve desired properties in food systems, they are used at levels of less than 2%. Some examples of gums include seaweed extracts such as alginates, agar, and carrageenan. Gums are also isolated from seeds (locust bean, guar) or dried resins that are expelled from trees (gum arabic, gum ghatti, gum karaya, gum tragancanth).

Derivatives of cellulose are also used in the food industry. Chemical modification of cellulose produces water-soluble fibers such as carboxymethylcellulose, methylcellulose, and hydroxypropylmethylcellulose. These compounds have a wide variety of uses such as incorporation into baked goods to increase shelf life and to improve crumb texture. Pectin, which is commercially derived from citrus peels and apple pomace, is used as a thickener, gelling agent, and suspending agent. The degree of methylation of the main polygalacuronic acid backbone determines the gel strength and gelling time of the pectin. Numerous other fibers are used in industry to improve food quality and produce new products.

Table 1. Dietary Fiber Content of Commonly Eaten Foods

Product	Dietary fiber (g/100 g)
Fruits	
Apple (fresh)	1.9
Banana	2.4
Orange	2.4
Grapes	1.0
Peach	2.0
Watermelon	0.5
Vegetables	
Broccoli	3.0
Cucumber	0.7
Corn	2.7
Lettuce	1.4
Tomato	1.1
Potato	1.6
Peas	2.3
Grains and legumes	
Wheat flour (white)	2.7
Wheat bran	42.8
White bread	2.3
Whole wheat bread	6.9
All Bran (Kellogg's)	32.3
Lentils (cooked)	7.9

Source: USDA Nutrient Data Base.

DANGERS OF HIGH-FIBER DIETS

A major concern regarding the safety of high-fiber diets is the effect of dietary fiber on the bioavailability of vitamins and minerals. This remains one of the few possible adverse effects of increasing dietary fiber consumption. Dietary fiber can bind or adsorb ions and act as a physical barrier, potentially interfering with absorption of important micronutrients from the intestine into the bloodstream. In young children and the elderly, where only small quantities of foods are consumed and variety may be limited, dietary fiber consumption should conform to the recommended daily intakes. Despite studies that suggest that high-fiber diets may have deleterious effects, many researchers firmly believe that there is no convincing scientific evidence that dietary fiber, even when consumed in large amounts, has any adverse effects on nutrition in humans. Deficiencies have not developed in long-term vegetarian adults who consumed more than 50 g of dietary fiber per day (9).

BIBLIOGRAPHY

1. D. A. T. Southgate, "The Chemistry of Dietary Fibre," in G. A. Spiller and R. J. Amen, eds., *Fiber in Human Nutrition*, Plenum Press, New York, 1976; pp. 31–72.
2. D. P. Burkitt and H. C. Trowell, *Refined Carbohydrate Foods and Disease. Some Implications of Dietary Fibre*, Academic Press, London, 1975.
3. M. A. Eastwood and E. R. Morris, "Physical Properties of Dietary Fiber that Influence Physiological Function: A Model for Polymers Along the Gastrointestinal Tract," *Am. J. Clin. Nutr.* **55**, 436–442 (1992).
4. A. Stark, A. Nyska, and Z. Madar, "Metabolic and Morphometric Changes in Small and Large Intestine in Rats Fed a High Fiber Diet," *Toxicol. Path.* **24**, 166–171 (1996).
5. A. Stark and Z. Madar, "Dietary Fiber," in I. Goldberg, ed., *Functional Foods: Designer Foods, Pharmafoods, Nutraceuticals*, Chapman & Hall, New York, 1994, pp. 183–201.
6. M. J. Hill, "Cereals, Dietary Fibre and Cancer," *Nutr. Res. (N.Y.)* **18**, 653–659 (1998).
7. U.S. Department of Agriculture. *Food Guide Pyramid: A Guide to Daily Food Choices.* U.S. Department of Agriculture, Washington, D.C., 1992.
8. C. L. Williams, M. Bollella, and W. L. Wynder, "A New Recommendation for Dietary Fiber in Childhood," *Pediatrics* **96**, 985–988 (1995).
9. J. Rattan et al., "A High Fiber Diet Does Not Cause Mineral and Nutrient Deficiencies," *J. Clin. Gastroenterol.* **8**, 390–393 (1981).

ALIZA STARK
Hebrew University of Jerusalem
Rehovot, Israel

FILTH AND EXTRANEOUS MATTER IN FOOD

THE FEDERAL FOOD LAWS

Soon after the new century began in 1900, the efforts of that great champion of food purity, Dr. Harvey Wiley, began to pay off. The U.S. Congress passed the Pure Food Law in June of 1906 (1–3). A stronger and more detailed law emerged from Congress in 1938 (4). Since then the food law has been amended and expanded on several occasions. Today, the Federal Food, Drug, and Cosmetic Act (FD&C Act) stands as a major achievement in the protection of consumer health (5).

As great as that achievement was, however, it should be noted that protection of consumer health was not the only objective of the Pure Food Law. The courts have consistently ruled that it was the intention of Congress to keep filth, whether harmful or not, out of food. Congress wanted food held to the highest standards of hygiene and aesthetics (6,7). The hygienic standard has, however, received the greater amount of attention by the Food and Drug Administration (FDA) in its mission to protect the public's health (8).

Two Sections on Filth

The FD&C Act (9) contains two paragraphs that are especially pertinent to the topic of filth. The first, 402(a)(3), states that a food shall be deemed to be adulterated if it consists in whole or in part of any filthy, putrid, or decomposed substance or if it is otherwise unfit for food. The second, 402(a)(4), states that a food shall be deemed to be adulterated if it has been prepared, packed, or held under insanitary conditions whereby it may have been contaminated with filth, or whereby it may have been rendered injurious to health. There have been numerous challenges to FDA's interpretation of the "filth sections" of the Act.

ENFORCEMENT OF THE ACT

Enforcement actions fall roughly into two categories: actions against products and actions against persons (10).

The Butter Oil Case

In a precedent-setting case, the "butter oil case," decided in 1947, federal agents seized a large quantity of butter because the cream from which it had been made was contaminated with insect and rodent filth. The owner of the butter wished to recoup his losses by making strained butter oil out of the butter. FDA objected to this procedure on the grounds that the resulting butter oil would still contain soluble insect and rodent filth. The court upheld this objection, thus making soluble (invisible) filth just as much an adulterant as particulate (visible) filth (11).

Reconditioning

What the owner of the butter had proposed is a process called reconditioning; by this procedure a substandard product may be brought into compliance if the offending filth in the product is removed. Reconditioning processes are often approved by FDA (12–14).

"Too Small to See" Defense

Another landmark case involved some tomato paste that contained bits and pieces of larval corn earworms (*Helicoverpa zea*). The owner of the tomato paste claimed that the insect fragments could not be considered filth—and the consumer would not so perceive them—because the fragments were too small to see. The court rejected that argument (15).

Actions Against Persons: The *Park-Acme* Case

A series of inspections (1970–1972) of Acme Markets warehouses in Philadelphia and Baltimore resulted in charges against both the company and its president (Mr. Park) for violations of Section 402(a)(4) of the FD&C Act. The president claimed to be innocent on the grounds that the violative acts were committed by subordinates to whom he had delegated authority and that he had no knowledge of or responsibility for the violations. When the Supreme Court decided the case, it vitiated the excuse of "delegation" and determined that top executive officers do indeed bear complete responsibility for violations that occur in their organizations (7,13,16–18).

Filth: "Too Little to Matter"

The bold statements in the two principal filth sections of the FD&C Act would seem to totally preclude the presence of any filth in or around food. The courts, however, have never held to the letter of the law; in other words, there is such a thing as de minimis filth—too little to matter (12,13,15). Some judges have been remarkably lenient in their interpretations of what constitutes significant filth in or around foods, whereas other judges have been less forgiving. In any case, it must be admitted that, without the unconscionable waste of raw food materials, it is impossible to produce foods that are 100% free of filth. The Act, on the other hand, prohibits filth.

Food Defect Action Levels

To bridge the gap between this practical reality and the absolute ban on filthy food called for in the Act, the FDA has established a regulatory tool not mentioned in the FD&C Act, namely, the food defect action levels (DALs) (19,20). The DALs, which apply to about 90 categories of foods, mainly refer to field infestations and other unavoidable contamination. Infestations of stored foods and related insanitary conditions get no relief from the DALs.

The food DALs are administrative guidelines that are set on the basis of no hazard to human health (21). Any products that might be harmful to consumers are acted against on the basis of their hazard to health, whether or not they exceed the action levels. DALs are resorted to because it never has been possible to grow, harvest, and process foods that are totally free of natural defects. Insanitary manufacturing practices will prompt regulatory action whether or not the resulting product is above or below the pertinent DAL.

Recalls

To maintain their good reputations, many food firms are willing to act promptly to initiate a recall to protect both themselves and the public. Even though recalls are not mentioned in the FD&C Act and in spite of the fact that recalls pertaining to food are technically voluntary on the part of industry (10), the FDA can still be instrumental in prompting a recall when circumstances warrant it (7,22–24).

Extraneous Matter

The courts have always defined filth in its ordinary sense (ie, the dictionary definition) rather than giving the term a technical meaning (19). The same can be said for extraneous matter, that is, things that do not belong in food such as filth or any foreign matter that comes to be in food as a result of objectional conditions or practices in production, storage, or distribution of food. Included within the meaning of extraneous matter are filth (any objectionable matter contributed by animal contamination such as rodent, insect, or bird matter), decomposed material (decayed tissues due to parasitic or nonparasitic causes), and miscellaneous matter such as sand, soil, glass, rust, or other foreign substances (but not bacteria) (25).

The more common kinds of filth encountered during food inspections and analyses are whole and fragmented insects and mites, mammalian hairs and feces (mostly of rodent origin), feather barbules, urea (from mammalian urine), and uric acid (from avian excretions). Other kinds of filth and extraneous matter include molds and other fungi and weed seeds (26,27).

Light Filth

Since the 1930s the principal method for separating light filth such as rat hairs, feather barbules, and bits of insect exoskeleton from food has been by flotation. When mineral oil is added to a sample of food mixed in water, the oleophilic particles (ie, the light filth) float to the surface while the hydrophilic particles (ie, most foods) settle to the bottom. When the oily portion is passed through filter paper, the filth elements remain on the surface of the paper; the particles of filth are then identified and enumerated by a food analyst (28). Analysts often use reference manuals to assist them in identifying filth elements (26,29–35).

Although flotation methods are still the mainstay of food analysts, other kinds of analytical techniques have been and continue to be developed. These newer methods use biochemical analyses to search out contaminants or indicators of contamination that cannot be detected by direct visual inspection (36–42).

Heavy Filth

Most hard foreign objects (eg, pebbles, sand, glass, pieces of metal or plastic, ball bearings, pits), strictly speaking, are not filth; nevertheless they are often referred to as "heavy filth." These hard foreign objects may be considered adulterants if, when chewed, they cause a gritty mouthfeel (21), or if the particles are large enough or sharp enough to be hazardous to chew or swallow (43,44). The food DALs permit very low levels of natural hard objects (usually pit or shell fragments) in a few kinds of foods (21). When hard foreign objects occur in foods for which there is no applicable DAL, each alleged violation is treated on a case-by-case basis.

MODES OF CONTAMINATION

On the Farm

The circumstances by which foods may become adulterated by filth and extraneous matter are legion. The place where the food is grown may be the first source of filth. All field-grown crops are to some degree infested with or affected by insects, mites, rodents, birds, and molds (45). The process of harvesting or gathering the food from the field may inadvertently involve also bringing along live or dead animals, molds, weed seeds, pebbles, and particles of soil. The feces of animals or humans that live in or frequent the field may also be accidentally collected.

This kind of contamination is becoming increasingly important in crops such as lettuce, strawberries, and raspberries that are marketed with minimal after-harvest cleaning (46–48). Foods grown and harvested from aquatic habitats may also be subject to adulteration by a wide array of contaminants (49,50). Foodborne disease outbreaks associated with cross-contamination at retail outlets or in the home are at least as common, or perhaps more common, than outbreaks linked to contamination that occurs where the food is harvested.

In Storage

If the crop is temporarily stored on the farm or placed in longer-term storage off the farm, it may be attacked by insects, mites, molds, rodents, or birds (51,52). These adulterating entities produce mostly surface contamination, but some insects and molds are capable of penetrating food materials and living inside the seeds of beans, corn, wheat, and other crops.

Many kinds of manufactured foods that are stored in warehouses are very susceptible to molds and pests (insects, mites, rodents, birds). Food items stored in metal, glass, or plastic containers are generally safe from the penetrating sorts of infestations or contaminations (eg, urine), but these containers and all other kinds of packaging are susceptible to surface contamination by the urine and feces of rodents and birds. Rats and mice have little difficulty chewing their way into most kinds of plastic, paper, foil, or cloth packaging; some insects also have great penetrating capabilities (53).

In Processing

Standards of sanitation are supposed to be maintained at a high level in food manufacturing facilities, and this is often the case. However, it is difficult under even the best of circumstances to keep all pests out of food processing plants; any laxity of sanitary standards can lead to infestations by pests, and some of these pests may find their way into the foods being manufactured.

In Transportation

Opportunities for pests to infest foods during transport by railcars, trucks, and ships are numerous. The longer the journey in terms of time, the greater the likelihood that pest populations, especially of stored-grain pests (beetles, moths, mites), will dramatically increase.

In Retail

Every retail food outlet (eg, restaurants, grocery stores, sidewalk food stands) is to some extent the target of attack by flies, ants, and cockroaches and rats and mice, too, when the habitat is right for rodent pests. When these insects and rodents come into contact with food or food-contact surfaces, they may add to the food the bacteria and molds that they are carrying around with them. Occasionally a whole insect or rodent fecal pellet becomes entrapped in or falls into the food; the insect body or fecal pellet may remain largely intact or it may be broken up into small pieces as the food material is stirred during preparation. If the insect or fecal pellet had been carrying pathogenic bacteria, the food thus contaminated becomes hazardous to health if it is not subsequently heated to a high enough temperature to kill the pathogens. The Retail Food Code (54) provides guidance to managers on how to prevent foodborne diseases in retail food outlets.

At Home

Homes are no less a target for attack by pests than are retail food outlets. Unlike retail food managers, home occupants are under no legal obligation to keep pests under control. Many householders, if not for sanitary reasons, at least for aesthetic ones, attempt to keep pests under control. Both the obvious pests, such as mice, ants, cockroaches, and flies, and the more cryptic ones, such as stored-food mites, beetles, and moths, are capable of carrying pathogens and spoilage molds and leaving them behind on foods or food-contact surfaces. When homes and restaurants as sources of foodborne illness are compared, the home kitchen is the far more common source. Sanitation in the home kitchen is at least as important as sanitation in the commercial kitchen (55,56).

KEEPING FILTH OUT OF FOOD

Integrated Pest Management

Ever since DDT (a pesticide) came to be widely used after World War II, pesticides have been used as the major (sometimes, the only) answer to pest problems. In recent years, a somewhat different approach, integrated pest management (IPM), has come to be applied to pest problems (57–59). The IPM system first identifies the pest, then tries to suppress the pest population by capitalizing on weaknesses in the lifestyle of the pest. IPM may be applied to crops in the field; to stored, raw food grains; to transportation systems; to food facilities of all kinds (eg, warehouses, food factories, groceries, restaurants); and to the home kitchen.

Pesticides may play a greater or lesser role in IPM, depending on local circumstances. The goal of IPM is not to eliminate the use of pesticides (although the trend is in that direction), but to suppress or exclude pest populations (60). This goal requires a comprehensive approach that involves an array of techniques, often including application of pesticides and often emphasizing pest-exclusion practices. IPM merges with and becomes indistinguishable from those aspects of the Hazard Analysis and Critical Control Points (HACCP) system that deal with pest prevention and pest control (61).

For the food-facility manager, pest management is but one link in the chain of food sanitation that begins at the receiving door (62) and ends with the consumer. The important operational guidelines for the manager are to buy only from reliable suppliers, inspect all incoming materials, reject any substandard items, and maintain pest-free premises.

No matter how closely standard cleaning and sanitizing protocols are followed, if pests are tolerated in a food facility, they will inevitably deposit spoilage microorganisms and, occasionally, pathogens on food and food-contact surfaces. A major goal of the food-facility manager must therefore be: zero tolerance for pests. IPM makes it possible to reach this goal.

Hazard Analysis and Critical Control Points

Development of the HACCP system began in the early 1970s. The system has been widely adopted by the worldwide food industry; eventually, it will be mandatory for the entire food industry in the United States. The HACCP system emphasizes the identification, prevention, and elimination of microbiological hazards that could lead to foodborne illnesses (63–67). Extraneous materials such as filth and hard foreign objects do not figure prominently in HACCP, it is true. However, any HACCP program that proposes to be successful must take place in an environment where preventive maintenance protocols and sanitary standards are set at very high levels.

Food-manufacturing protocols often call for a biocidal step such as pasteurization or commercial sterilization that inactivates microorganisms and parasites. Cans or bottles that are aseptically sealed after the biocidal step may still contain hard foreign objects (eg, metal fragments, glass shards) that may be hazardous to health. Foods that go through additional handling after the biocidal step may be exposed to food-contact surfaces contaminated by pests, or the pests may actually come into direct contact with, and sometimes become incorporated into, the food. Such events would be of concern in a HACCP program because the pest may have introduced parasites, pathogens, or spoilage fungi that may proliferate and cause foodborne illness or spoilage of the food material.

Even if very high sanitary standards are maintained in a food-manufacturing facility, those standards will be compromised unless comparable standards extend backward along the food chain to the origin of the food on farms or in aquatic habitats. A system for the protection of the food from adulteration by filth must be in operation at all the steps from farm to table—harvesting, storing, transporting, manufacturing, warehousing, retailing. Defects in sanitation present challenges to the success of the HACCP system wherever it is applied along the food chain.

HOW DOES FILTH IN FOOD AFFECT HUMAN HEALTH?

No Effect

If the amount of filth in a food is so small that it falls within the limits set by the DALs, then that filth is assumed to have no adverse effect on human health (21). This conclusion is admittedly an assumption; the fact that this assumption is based on extensive practical experience rather than on scientific experimentation does not make it any less valid.

Indirect Effects

The presence of food-associated pests—insects, mites, rodents, birds—in a food facility is an indicator of unacceptable sanitation practices. The HACCP system works well only when high standards of environmental sanitation are maintained. The presence of food-associated pests compromises the gains in food safety that would otherwise be achieved by the HACCP program in a food facility.

There may be another indirect effect on human health that involves the production of an unhealthy or unstable emotional state when one discovers recognizable insects in or around one's food. This situation may occur anywhere, but it is especially significant if it occurs in a hospital or other patient-care facility when a person already debilitated from injury, illness, surgery, chemotherapy, or advanced age experiences revulsion and loss of appetite upon sighting obvious filth in his or her food.

Direct Effects

The accidental ingestion of insects and mites infesting food may result in the production of antibodies to the proteins of the ingested arthropods (68,69), but this rarely results in overt indications of disease. Matsumoto et al. (70) report on two patients who experienced anaphylaxis after eating foods made from ingredients contaminated with storage mites of a common species, *Tyrophagus putrescentiae* (the mold mite), and they comment on a third case involving anaphylaxis after ingestion of doughnuts made from ingredients infested by American house dust mites, *Dermatophagoides farinae* (71).

The potential for direct physical injury resulting from ingestion of insect-infested food was pointed out in the report of a case of colitis in an infant who had eaten baby cereal infested by dermestid beetles (72).

FILTH IN FOOD IN PERSPECTIVE

A Different World

Over the past 30 years or so, some very significant changes have transpired in the area of public health, especially as it relates to foodborne diseases (73). Widespread resistance to antibiotics by pathogens poses a major threat to human health. The population of surgical patients who undergo invasive procedures has burgeoned (such surgeries provide greater opportunities for colonization by pathogens, even some of those that may more usually be associated with food). The number of immunocompromised persons has increased enormously, not just because of infections by the human immunodeficiency virus, but also because more people are receiving chemotherapy and radiation treatments, and because that portion of the general population considered to be elderly (and therefore more susceptible to infections) is steadily increasing. All of these immunocompromised populations are regarded as being more susceptible to foodborne pathogens and toxins than are immunocompetent persons (74). Some kinds of opportunistic pathogens that never stood a chance of replicating in healthy hosts now find no constraints in immunocompromised hosts.

Emerging Pathogens

The classic foodborne pathogens that have long been known and well studied are quite capable of causing disease in immunocompromised persons and, to a lesser extent, in the immunocompetent. But now there are new pathogens and more virulent strains of classic pathogens to which virtually everyone is susceptible. Not only are the immunocompromised gravely challenged by the newer pathogens, but so also are the immunocompetent. No one is exempt from the threat of foodborne pathogens and toxins. Names and strains that were rarely or never heard of in the 1960s, are now in the news media on a daily basis. Enteropathogenic *Escherichia coli, E. coli* O157:H7 (and other highly virulent serotypes of this species), *Giardia lamblia, Cryptosporidium parvum, Salmonella typhimurium* Definitive Type 104, *Cyclospora cayetanensis*, and others are listed among dangerous emerging pathogens (75,76).

THE PARADOX OF FILTH IN FOOD

The Caricature

Some food safety experts and even at least one government have taken the position that looking for filth in food is simply not worth the required effort and expense. The small benefit to the public gained from occasionally finding a food item that is disgustingly filthy cannot, they contend, justify the expenditure of the time and money needed to make that discovery. To justify this position, these experts make a caricature of people's aversion to filth in food—according to them, everyone in a lifetime is going to "eat a peck of dirt" and, besides, it is harmless anyway. Apparently, even in this camp of experts there is still some interest in the maintenance of sanitation around food, if not in food, and that position continues to yield modest benefits to the consumer.

The Promise

The position just described reflects a misunderstanding of the significance of filth in food, especially if the filth was added to the food after the final biocidal step in food processing. The occasional rat hair that is extracted from a food sample probably does not represent the shedding of a single hair from a rat moving about above a food processing line. More likely that single rat hair represents two things—it is probably just one of a much larger number of hairs, most of which escaped detection; and it more probably represents a remnant of a fecal pellet, now disintegrated and dispersed throughout the food material.

When rats and mice groom themselves or each other, they remove and swallow hairs which then become incorporated into the feces of the rodent. Such pellets are never sterile, but only a small percentage of them carry pathogens that cause human disease. But some do, and here is where risk enters the picture, especially if the contamination occurs after the last biocidal step in processing and especially if the consumer of the food is immunocompromised.

What is said of the rat hair can be said in principle of all the filth elements that make their way into food—they all carry spoilage organisms, some of them carry pathogens, rarely do they emerge intact from the food production line, they are all more dangerous if they enter the food manufacturing process after the last biocidal step, and their danger to human health is related to the virulence of any pathogens they might carry and to the susceptibility status of the consuming host.

It is admirable for food safety monitors to say that action will be taken against a food manufacturer or warehouser or retailer if food pests or filthy conditions are found in a food facility, but it makes no sense to suggest that food emerging from that facility does not need to be examined for filth. The experience of countless food safety inspectors and food analysts testifies to the fact that food emerging from a filthy, pest-infested facility will bear filth far above the standard action levels and stands a good chance of becoming associated with a foodborne disease outbreak (77–79).

Early Warning

The analysis of food for adulteration by filth may serve as an early warning of lapses in sanitary protocols in the food facility and the nearly inevitable associated compromise of the HACCP program. Examination of food for light filth is a rapid, cost-effective, and efficient way to evaluate the status of sanitation in a food facility. This kind of ongoing quality assurance program gives the consumer confidence that the food as it leaves the production line is safe for consumption by virtually all persons, regardless of physical or emotional health, except for the small percentage of the population that suffers from specific food allergies. In view of the widespread occurrence of newly emerging, highly virulent pathogens, a laissez faire attitude toward filth in food is no longer tenable.

The reader may find a more detailed discussion of this topic in the first edition of this encyclopedia (80).

BIBLIOGRAPHY

1. R. C. Litman and D. S. Litman, "Protection of the American Consumer: The Congressional Battle for the Enactment of the First Federal Food and Drug Law in the United States," *Food Drug Cosmetic Law J.* **37**, 310–329 (1982).
2. R. A. Merrill and P. B. Hutt, *Statutory Supplement to Food and Drug Law—Cases and Materials*, The Foundation Press, Mineola, N.Y., 1980.
3. J. H. Young, *Pure Food—Securing the Federal Food and Drugs Act of 1906*, Princeton University Press, Princeton, N.J., 1989.
4. J. E. Hoffman, "FDA's Administrative Procedures," in *Seventy-fifth Anniversary Commemorative Volume of Food and Drug Law*, Food and Drug Law Institute, Washington, D.C., 1984.
5. W. F. Janssen, "Golden Anniversary of the FD&C Act: Consumers 'Never Had it so Good,'" *J. Assoc. Food Drug Off.* **52**, 59–60 (1988).
6. P. B. Hutt, "The Basis and Purpose of Government Regulation of Adulteration and Misbranding of Food," *Food Drug Cosmetic Law J.* **33**, 505–540 (1978).
7. J. T. O'Reilly, *Food and Drug Administration* (Regulatory Manual Series), Shephard's, Colorado Springs, Colo., 1979.
8. J. P. Hile, "New Theories of Enforcement," *Food Drug Cosmetic Law J.* **41**, 424–428 (1986).
9. U.S. Food and Drug Administration, *Federal Food, Drug, and Cosmetic Act, as Amended, and Related Laws*, HHS Publication No. (FDA) 89-1051, U.S. Department of Health and Human Services, Rockville, Md., 1989.
10. E. M. Pfeifer, "Enforcement," in *Seventy-fifth Anniversary Commemorative Volume of Food and Drug Law*, Food and Drug Law Institute, Washington, D.C., 1984.
11. V. A. Kleinfeld and C. W. Dunn, *Federal Food, Drug, and Cosmetic Act, Judicial and Administrative Record 1938–1949*, Commerce Clearing House, New York, 1953.
12. T. W. Christopher and W. W. Goodrich, *Cases and Materials on Food and Drug Law*, 2nd ed., Commerce Clearing House, New York, 1973.
13. R. A. Merrill and P. B. Hutt, *Food and Drug Law—Cases and Materials*, The Foundation Press, Mineola, N.Y., 1980.
14. Office of Enforcement, "Reconditioning of Foods Adulterated under 402(a)(4)," *Compliance Policy Guide 7153.04*, U.S. Food and Drug Administration, Washington, D.C., 1989.
15. V. A. Kleinfeld and C. W. Dunn, *Federal Food, Drug, and Cosmetic Act, Judicial and Administrative Record 1949–1950*, Commerce Clearing House, New York, 1951.
16. Y. H. Hui, *United States Food Laws, Regulations, and Standards*, John Wiley & Sons, New York, 1979.
17. J. S. Kahan, "Criminal Liability under the Federal Food, Drug, and Cosmetic Act—The Large Corporation Perspective," *Food Drug Cosmetic Law J.* **36**, 314–331 (1981).
18. J. A. Levitt, "FDA Inspections and Criminal Responsibility," *Food Drug Cosmetic Law J.* **36**, 469–477 (1981).
19. P. M. Brickey, Jr., "The Food and Drug Administration and the Regulation of Food Sanitation," in J. R. Gorham, ed., *Ecology and Management of Food-Industry Pests*, Association of Official Analytical Chemists, Arlington, Va., 1991, pp. 491–495.
20. J. M. Taylor, "Establishment and Use of Defect Action Levels," in *Association of Food Industries 82*, Association of Food Industries, Matawan, N.J., 1982.
21. Industry Activities Staff, *The Food Defect Action Levels*, U.S. Food and Drug Administration, Washington, D.C., 1998.
22. M. H. Bozeman, "Recalls—On Making the Best of a Bad Thing," *Food Drug Cosmetic Law J.* **33**, 342–359 (1978).
23. J. R. Phelps, "Actions in the Courts," *Food Drug Cosmetic Law J.* **35**, 502–510 (1980).
24. J. Bressler, "What FDA Expects During a Recall," *Food Product Development* **14**, 64–65 (Apr. 1980).
25. P. A. Cunniff, ed., *Official Methods of Analysis of AOAC International*, 16th ed., 3rd rev., 2 vol., AOAC International, Arlington, Va., 1997.

26. J. R. Gorham, ed., *Principles of Food Analysis for Filth, Decomposition, and Foreign Matter*, FDA Technical Bulletin 1, 2nd ed., Association of Official Analytical Chemists, Arlington, Va., 1981.
27. J. R. Gorham, ed., *Training Manual for Analytical Entomology in the Food Industry*, FDA Technical Bulletin 2, Association of Official Analytical Chemists, Arlington, Va., 1978.
28. E. C. Washbon and K. R. Halcrow, "Laboratory Procedures," in A. R. Olsen, T. H. Sidebottom, and S. A. Knight, eds., *Fundamentals of Microanalytical Entomology. A Practical Guide to Detecting and Identifying Filth in Foods*, CRC Press, Boca Raton, Fla., 1996, pp. 241–260.
29. O. L. Kurtz and K. L. Harris, *Micro-analytical Entomology for Food Sanitation Control*, Association of Official Analytical Chemists, Arlington, Va., 1962.
30. D. McClymont-Peace, *Key for Identification of Mandibles of Stored-food Insects*, Association of Official Analytical Chemists, Arlington, Va., 1985.
31. J. P. Sutherland, A. H. Varnam, and M. G. Evans, *A Colour Atlas of Food Quality Control*, Wolfe Publishing, London, 1986.
32. J. W. Gentry, K. L. Harris, and J. W. Gentry, Jr., *Microanalytical Entomology for Food Sanitation Control*, 2 vol., published by J. W. Gentry and K. L. Harris, Melbourne, Fla., 1991.
33. J. R. Gorham, ed., *Insect and Mite Pests in Food: An Illustrated Key*, Agriculture Handbook 655, U. S. Department of Agriculture, Washington, D. C., 1991.
34. D. H. Ludwig and J. R. Bryce, "Hairs and Feathers," in A. R. Olsen, T. H. Sidebottom, and S. A. Knight, eds., *Fundamentals of Microanalytical Entomology. A Practical Guide to Detecting and Identifying Filth in Foods*, CRC Press, Boca Raton, Fla., 1996, pp. 157–216.
35. G. Domenichini, ed., *Atlante delle Impurità Solide negli Alimenti—Manuale per il Riconoscimento dei Materiali Estranei*, Chiriotti Editori, Pinerolo, Italy, 1997.
36. J. J. Thrasher, "Detection of Metabolic Products, in J. R. Gorham, ed., *Principles of Food Analysis for Filth, Decomposition, and Foreign Matter*, FDA Technical Bulletin 1, 2nd ed., Association of Official Analytical Chemists, Arlington, Va., 1981, pp. 201–216.
37. H. R. Gerberg "Chemical Test for Mammalian Feces in Grain Products: Collaborative Study," *J. Assoc. Off. Anal. Chem.* **72**, 766–769 (1989).
38. P. E. Kauffman and D. B. Shah, "Enzyme Immunoassay for Detection of *Drosophila melanogaster* Antigens in the Juice of Various Foods," *J. Assoc. Off. Anal. Chem.* **71**, 636–642 (1988).
39. F. A. Quinn, W. Burkholder, and G. B. Kitto, "Immunological Technique for Measuring Insect Contamination of Grain," *Journal of Economic Entomology* **85**, 1463–1470 (1992).
40. B. Brader, "Are Food Sanitation Assays Meaningful?," *Cereal Foods World* **42**, 759–760 (1997).
41. P. A. Valdes, G. C. Ziobro, and R. S. Ferrera, "Use of Urease-Bromothymol Blue-Agar Method for Large-scale Testing of Urine on Grain and Seeds," *J. AOAC Int.* **79**, 866–873 (1996).
42. P. Valdes-Biles and G. C. Ziobro, "The Identification of the Source of Reagent Variability in the Xanthydrol/Urea Method," *J. AOAC Int.* **81**, 1155–1161 (1998).
43. F. N. Hyman, K. C. Klontz, and L. Tollefson, "Food and Drug Administration Surveillance of the Role of Foreign Objects in Foodborne Injuries," *Public Health Reports* **108**, 54–59 (1993).
44. J. R. Gorham, "Hard Foreign Objects in Food as a Cause of Injury and Disease: A Review," in Y. H. Hui et al., eds., *Food-borne Disease Handbook*, Vol. 3, Marcel Dekker, New York, 1994, pp. 615–626.
45. D. T. Wicklow, "Preharvest Origins of Toxigenic Fungi in Stored Grains," in E. Highley et al., eds., *Stored Product Protection, Proceedings of the 6th International Working Conference on Stored-product Protection*, Vol. 2, CAB International, Wallingford, United Kingdom, 1994.
46. Office of Enforcement, "Recall of Frozen Strawberries Linked to Outbreak of Hepatitis A," in *The Enforcement Story Fiscal Year 1997*, U.S. Food and Drug Administration, Rockville, Md., 1998, pp. 113–115.
47. Office of Enforcement, "Import Detention of Raspberries from Guatemala," in *The Enforcement Story Fiscal Year 1997*, U.S. Food and Drug Administration, Rockville, Md., 1998, pp. 116–117.
48. L. R. Beauchat and J-H. Ryu, "Produce Handling and Processing Practices," *Emerging Infectious Diseases* **3**, 459–465 (1997).
49. Office of Enforcement, "Import Detentions for Salmonella in Seafood," in *The Enforcement Story Fiscal Year 1997*, U.S. Food and Drug Administration, Rockville, Md., 1998, p. 127.
50. E. S. Garrett, C. Lima dos Santos, and M. L. Jahnacke, "Public, Animal, and Environmental Health Implications of Aquaculture," *Emerging Infectious Diseases* **3**, 453–457 (1997).
51. W. Stein, *Vorratsschädlinge und Hausungeziefer—Biologie, Ökologie, Gegenmaßnahmen*, Eugen Ulmer GmbH & Co., Stuttgart, 1986.
52. J. R. Gorham, ed., *Ecology and Management of Food-Industry Pests*, Association of Official Analytical Chemists, Arlington, Va., 1991.
53. H. A. Highland, "Protecting Packages Against Insects," in J. R. Gorham, ed., *Ecology and Management of Food-Industry Pests*, Association of Official Analytical Chemists, Arlington, Va., 1991, pp. 345–350.
54. *Food Code*, U.S. Food and Drug Administration, Washington, D.C., 1999.
55. J. E. Collins, "Impact of Changing Consumer Lifestyles on the Emergence/Reemergence of Foodborne Pathogens," *Emerging Infectious Diseases* **3**, 471–479 (1997).
56. R. L. Hall, "Foodborne Illness: Implications for the Future," *Emerging Infectious Diseases* **3**, 555–559 (1997).
57. E. G. Thompson, "The Integrated Pest Management Approach to Food Protection," *Cereal Foods World* **29**, 149–154 (1984).
58. J. V. Osmun, "Insect Pest Management and Control," in F. J. Baur, ed., *Insect Management for Food Storage and Processing*, American Association of Cereal Chemists, St. Paul, Minn., 1984, pp. 17–24.
59. D. F. Jones and E. G. Thompson, "Integrated Pest Management for the Food Industry," in J. R. Gorham, ed., *Ecology and Management of Food-Industry Pests*, Association of Official Analytical Chemists, Arlington, Va., 1991, pp. 551–556.
60. J. A. Gibson, "Review of Rodent and Insect Damage to Stored Products and Non-pesticidal Methods of Control," in D. R. Houghton, R. N. Smith, and H. O. W. Eggins, eds., *Biodeterioration 7*, Elsevier Applied Science Publishers, Barking, United Kingdom, 1988.
61. J. R. Gorham, "HACCP and Filth in Food," *J. Food Prot.* **52**, 674–677 (1989).
62. "The Right Way to Accept Deliveries," *Best Practices* **2**, 10 (1998).
63. O. P. Snyder, "HACCP in the Retail Food Industry," *Dairy, Food and Environmental Sanitation* **11**, 73–81 (1991).

64. E. J. Rhodehamel, "Overview of Biological, Chemical, and Physical Hazards," in M. D. Pierson and D. A. Corlett, Jr., eds., *HACCP—Principles and Applications*, AVI Book/Van Nostrand Reinhold, New York, 1992, pp. 8–28.
65. O. P. Snyder, Jr., "HACCP—An Industry Food Safety Self-Control Program, Part III," *Dairy, Food and Environmental Sanitation* **12**, 164–167 (1992).
66. F. K. Käferstein, Y. Motarhemi, and D. W. Bettcher, "Food-borne Disease Control: A Transnational Challenge," *Emerging Infectious Diseases* **3**, 503–510 (1997).
67. National Advisory Committee on Microbiological Criteria for Foods, "Hazard Analysis and Critical Control Point Principles and Application Guidelines," *J. Food Prot.* **61**, 762–775 (1998).
68. H. M. Johnson et al., "Antigenic Properties of Some Insects Involved in Food Contamination," *J. Assoc. Off. Agric. Chem.* **56**, 63–65 (1973).
69. R. A. Wirtz, "Food Pests as Disease Agents," in J. R. Gorham, ed., *Ecology and Management of Food-Industry Pests*, Association of Official Analytical Chemists, Arlington, Va., 1991, pp. 469–475.
70. T. Matsumoto et al., "Systemic Anaphylaxis after Eating Storage-mite-contaminated Food," *Int. Arch. Allergy Immunol.* **109**, 197–200 (1996).
71. A. M. Erben et al., "Anaphylaxis after Ingestion of Beignets Contaminated with *Dermatophagoides farinae*," *J. Allergy Clin. Immunol.* **92**, 846–849 (1993).
72. G. T. Okumura, "A Report of Canthariasis and Allergy Caused by *Trogoderma*," *Calif. Vector Views* **14**, 19–22 (1967).
73. E. M. Foster, "Historical Overview of Key Issues in Food Safety," *Emerging Infectious Diseases* **3**, 481–482 (1997).
74. J. G. Morris, Jr., and M. Potter, "Emergence of New Pathogens as a Function of Changes in Host Susceptibility," *Emerging Infectious Diseases* **3**, 435–441 (1997).
75. J. Lederberg, "Infectious Disease as an Evolutionary Paradigm," *Emerging Infectious Diseases* **3**, 417–423 (1997).
76. R. V. Tauxe, "Emerging Foodborne Diseases: An Evolving Public Health Challenge," *Emerging Infectious Diseases* **3**, 425–434 (1997).
77. A. R. Olsen, "Introduction," in A. R. Olsen, T. H. Sidebottom, and S. A. Knight, eds., *Fundamentals of Microanalytical Entomology. A Practical Guide to Detecting and Identifying Filth in Foods*, CRC Press, Boca Raton, Fla., 1996, pp. 1–10.
78. R. A. Baldwin, "Quality Assurance for Regulatory Science," in A. R. Olsen, T. H. Sidebottom, and S. A. Knight, eds., *Fundamentals of Microanalytical Entomology. A Practical Guide to Detecting and Identifying Filth in Foods*, CRC Press, Boca Raton, Fla., 1996, pp. 11–20.
79. J. R. Gorham, "Reflections on Food-borne Filth in Relation to Human Disease," in A. R. Olsen, T. H. Sidebottom, and S. A. Knight, eds., *Fundamentals of Microanalytical Entomology. A Practical Guide to Detecting and Identifying Filth in Foods*, CRC Press, Boca Raton, Fla., 1996, pp. 269–275.
80. J. R. Gorham, "Filth and Extraneous Matter in Food," in Y. H. Hui, ed., *Encyclopedia of Food Science and Technology*, 1st ed., Wiley-Interscience, New York, 1991, pp. 847–868.

J. RICHARD GORHAM
Uniformed Services
University of the Health Sciences
Bethesda, Maryland

FISH AND SHELLFISH MICROBIOLOGY

Humans consume over 1,000 species of fish and shellfish that grow in diverse habitats and geographic regions all over the world (1). These fish and shellfish carry a variety of microorganisms from both aquatic and terrestrial sources. The high levels of moisture, rich nutrients, including free amino acids, other extractable nitrogenous compounds, digestible proteins, and psychrophiles, render seafood easily perishable, often spoiling in a short period of time even under refrigeration. In addition to spoilage microorganisms, seafood may contain various potential pathogens that can threaten the public health.

It is often difficult to maintain the quality of seafood products because there is a considerable distance between consumers and the harvesting areas, which provides opportunities for microbial growth and recontamination. To process fish and shellfish into stable products, low temperature, heat, curing, fermentation, and irradiation can be applied. This article covers the quantitative and qualitative aspects of microorganisms found in fish and shellfish and the factors affecting seafood quality. Organisms involved during the seafood processing are also described and discussed. Emphases are placed on spoilage bacteria, which cause the degradation of products and organisms that present risks to the public health.

MICROORGANISMS IN FINFISH

In healthy fish, muscle tissue or flesh is generally considered sterile. However, the fish surface and certain organs contain various levels of microorganisms: skin, 10^2–10^7/cm^2; intestinal fluid, 10^3–10^8/mL; and gill tissue, 10^3–10^6/mL (2,3). Cold marine water fish mainly carry psychrophilic gram-negative bacteria including *Moraxella, Acinetobacter, Pseudomonas, Flavobacterium*, and *Vibrio* (4,5). Both *Moraxella* and *Acinetobacter* were designated as *Achromobacter* in the past (6–8). The levels of these bacteria vary somewhat depending on season and food ingested (2,5,9). Fish intestines normally contain *Vibrio, Moraxella, Acinetobacter, Pseudomonas*, and *Aeromonas*, in addition to a small number of anaerobic bacteria, including *Clostridium* and *Bacillus*. Warm-water fish carry large numbers of gram-positive, mesophilic bacteria such as *Corynebacterium, Bacillus, Micrococcus*, and sometimes *Enterobacteriaceae* or even *Salmonella* (4,5).

The flora of fish depends on intrinsic factors (season, fish ground, and species) and extrinsic ones (fishing method, fish handling on board, storage condition, sampling technique, medium, and incubation temperature). Trawled fish usually carry bacterial loads 10–100 times greater than those of lined fish, because fish are dragged for a long time along the sea bottom (4). The physiological condition of fish prior to death has an effect on postharvest quality. When tuna, the fastest swimming fish, are captured in a highly stressful state, the buildup of lactic acid combined with elevated muscle temperature degrade the muscle quality, although the tuna is still acceptable for canning. Salmon harvested by gill netting die after an exhausting struggle, resulting in a shorter period of rigor mortis and deterioration during icing (10). The fish should be handled as soon as possible after being landed on the vessel. Careful handling of fish with gaff hook or forks and avoiding severe physical damage are crucial. Any breaks in the skin and flesh quickly introduce spoilage bacteria, which deteriorate fish quality. Fish should be carefully cleaned, not exposed to sunlight or to the drying effects of wind, and cooled down to the temperature of melting ice (0°C) as quickly as possible (11,12).

Gutted and well-cleaned fish contain fewer bacteria than whole fish. Bleeding and gutting should be done as soon as fish arrive on deck (13). However, bleeding and gutting are not helpful in all fish harvesting operations; blue fish and dogfish are not benefited by this operation (13) if in-plant processing occurs shortly after harvest.

The method of stowing fish on the vessel can have different impacts on fish quality (14,15). Boxing, commonly used in Asia, Norway, and Iceland, provides good fish quality and quicker unloading at dockside, although it results in more labor needed in handling fish on the deck and less storage capacity. Bulking, commonly used in North America, allows for quick operation and maximum storage capacity; however, the pressure often causes poorer quality and decreases shelf life (11,16).

During the transport of fish from fish ground to fish pier, bacteria grow in the fish pen or hold board at various rates depending on fish handling and storage temperature. In the bulking stow in cold weather, a slime accumulates on the bottom of the fish hold. This fish hold slime is constituted mainly of various bacteria with a level ranging from 10^9 to 10^{10}/g (17). The bacterial flora includes *Moraxella, Acinetobacter, Flavobacter*, pseudomonads, and heavily mucoid corynebacteria, which are the major organisms responsible for the slimy deposits and are critical problems in boat sanitation (17,18). Thorough sanitation of the fish hold after unloading the catch is necessary to insure high quality of raw fish on future trips.

FISH SPOILAGE AND CHANGES IN BACTERIAL FLORA OF FISH DURING COLD STORAGE

The fish's regulatory mechanisms, which prevent invasion of the tissues by bacteria, cease to function after death. Bacteria then invade the fish body through the skin, enter the body cavity and belly walls via intestines, and penetrate the gill tissue and kidney by way of the vascular system. The low molecular substances and soluble proteins yielded from fish body during autolysis after rigor mortis provide rich nutrients for bacterial growth.

Various proteases and other hydrolytic enzymes secreted by psychrophilic and psychrotrophic organisms can act on the fish muscle even at low temperatures (17,19). The factors that influence microbial contamination and growth include fish species and size, method of catch, onboard handling, fishing vessel sanitation, processing, and storage condition (12,20,21). Fish are subject to rapid microbial spoilage if fish handling and storage are inadequate. It is estimated that about 10% of the total world catch is lost due to bacterial spoilage (22). Various microorganisms involved in spoilage are listed in Table 1 in descending order of spoilage activity. Some organism cause spoilage in different degrees depending on the total microbial flora, fish quality, handling and packaging methods, and storage temperature.

Table 1. Microorganisms Associated with Spoilage of Fresh Seafood

Spoilage activity	Microorganism
High	*Pseudomonas, (Alteromonas), putreijaciens, Pseudomonas, (Alteromonas), fluorescens,* other fluorescent pseudomonads, and other pseudomonads
Moderate	*Moraxella, Acinetobacter,* and *Alcaligenes.*
Low or active only in specific conditions	*Aerobacter, Lactobacillus, flavobacterium, Micrococcus, Bacillus,* and *Staphylococcus*

Refrigerated fresh haddock fillets contain about 10^5 g of initial bacteria, predominated by *Moraxella-Acinetobacter* and *Corynebacterium*. After storage at 1°C for 14 days the bacterial number reaches 2.1×10^8/g and seafood enters the spoilage stage. *Pseudomonas* (*Alteromonas*) *putrefaciens* and fluorescent pseudomonads are organisms responsible for the spoilage of haddock at refrigerated temperatures (23). These spoilers account for only about 1% of the total count at the beginning but increase to at least 30% at the stage of spoilage. In other words, whenever *Pseudomonas putrefaciens* and fluorescent pseudomonads reach 30% of the total bacterial count, fish spoilage will result regardless of total bacterial level. When cod is stored at 20°C *Alteromonas* and *Vibrionaceae* will cause spoilage in one day (24).

As *Pseudomonas putrefaciens*, fluorescent pseudomonads, and other potential spoilers increase rapidly in initial spoilage stage; they produce vast amounts of proteolytic and other hydrolytic enzymes (5,25). Various macromolecules of fish body are degraded. Proteins are decomposed by proteases to peptides and amino acids and then further broken down to indole, amines, acids, sulfide compounds, and ammonia (26). Lipases break down lipids to form fatty acids, glycerol, and other products. Nucleotides are decomposed into nitrogenous compounds. Many enzymatic tests can determine microbial spoilage activity in fish, including hydrogen sulfide, gelatin hydrolysis. DNase, RNase, amylase, lipase, and trimethylamine oxide reductase tests and inoculation test on fish juice or fillets (20).

MICROORGANISMS IN SHELLFISH

Shellfish is composed of crustaceans (shrimp, crabs, lobster, crawfish, etc) and mollusks (bivalves, squids, snails, etc). Shellfish normally contains more moisture, greater amount of free amino acids, and more extractable nitrogenous compounds than finfish. These biochemical characteristics facilitate bacterial growth and deteriorative reactions resulting in the rapid spoilage of shellfish (27). Many shellfish grow in estuarine, coastal waters, and aquacultural ponds near residential areas and are hence susceptible to contamination by potential pathogenic organisms. Deterioration of shellfish quality results from enzymatic action from both the tissue and the contaminating organisms. Microorganisms that spoil shellfish are similar to those responsible for finfish spoilage. However, such organisms as *Moraxella-Acinetobacter* and *Lactobacillus* are more active in shellfish than in finfish.

Crustaceans

Shrimp. Among shellfish, shrimp ranks first in value and second in quantity next to crabs (28). Immediately after shrimp death, the tissue enzymes phenolases become active oxidizing tyrosine to bluish black zones or spots at the edges of the shell segments. The dark color is produced by melanin pigments that form on the internal shell surfaces on the underlying shrimp meat. At the same time a variety of bacteria start to proliferate and the growth can be accelerated if the storage temperature is not kept low enough. Removing the heads can reduce 75% of the bacteria. Gulf of Mexico shrimp contain mainly. *Moraxella-Acinetobacter, Bacillus, Micrococcus*, and *Pseudomonas* (29). Most of these bacteria produce hydrolytic enzymes: 62% proteolytic, 35% lipolytic, 18% TMA-O reductive, and 12% indole positive (29–31). Shrimp unloaded from the trawlers have an average bacterial load of 6.0×10^5/g and market shrimp, 3.2×10^6/g. Bacterial counts used for indicating shrimp quality are 1.3×10^6/g, acceptable; 4.5×10^6/g, good; 1.1×10^7/g, fair; and 1.9×10^7/g, poor (32). During iced storage for 16 days, *Moraxella-Acinetobacter* increase from 27 to 82% of the total bacterial count while *Flavobacterium* decrease from 18 to 1.5%; *Micrococcus*, from 34 to 0%; and *Pseudomonas*, from 19 to 17% (33).

There are two putrefactive types of spoilage in shrimp. One is the production of indole, presumably from tryptophan by bacterial action before icing when exposed at a temperature favorable for bacterial growth. After commencement, the decomposition proceeds fairly rapidly even under ice. Indole is heat resistant and is a reliable spoilage indicator of raw material prior to processing. The other type of ammoniacal decomposition is slow and is characterized by an odor of free ammonia (34–36). The reaction is attributed to both microbial and tissue enzymatic activities depending on storage temperature and bacterial composition.

Crabs. The dominant crabs harvested in the United States are blue crab, king crab, Dungeness crab, and tanner-snow crabs. The bacterial flora of freshly caught crabs reflect that of the growing water, season, and geographic location. The hemolymph of healthy blue crabs from Chincoteague Bay, Va., is about one-fifth sterile according to 290 freshly caught crabs tested (37). The organisms found in the hemolymph of blue crabs are *Acinetobacter, Aeromonas, Pseudomonas, Flavobacterium, Vibrio, Bacillus*, coliforms, and *Clostridium* (38).

Greater numbers of bacterial species are found in Dungeness crabs from Kodiak Island and the Columbia River, waters close to human habitation, whereas the least number of species are found in the tanner crab from the Bering Sea, an area far from human habitation (39). The highest levels of bacteria occur in the gills, 10^3 to 10^7/g, as compared to 1×10^1 to 4×10^2/g in muscle tissue. Gills of Dungeness crabs from the Bering Sea contain *Moraxella, Acinetobacter, Alcaligenes, Micrococcus*, and *Staphylococcus*, whereas the muscle carries *Micrococcus* and *Staphylococcus* (39)

Crawfish. Spoilage of crawfish is caused by the potential spoilers similar to those detected in other crustaceans. In a total of 280 isolates found from spoiled crawfish, 22.1% were shown to be rapid spoilers; 16.4%, low spoilers; and 61.5%, nonspoilers (40). In the group of rapid spoilers over half were pseudomonads and less than half were *Moraxella-Acinetobacter*. Slow spoilers include *Pseudomonas, Moraxella-Acinetobacter, Alcaligenes, Flavobacterium, Aerobacter, Lactobacillus, Micrococcus*, and *Staphylococcus*. Organisms considered as nonrapid spoilers are *Aerobacter, Bacillus, Flavobacterium, Micrococcus, Sarcina*, and *Staphylococcus* (40,41). It is clear that organisms belonging to the same genus have different activities in spoilage.

Bivalves

Bivalves are mollusks, including oysters, clams, and mussels. They are soft-bodied animals that are enclosed by two rigid, bilaterally symmetrical shells. Bivalves are filter feeders and pass a large volume of water through their gills to obtain oxygen and food. Particulate matter, including microorganisms, from the water is trapped in mucus on the gills, then conveyed to the mouth, and finally to the digestive system. Bivalves, particularly oysters, ingest many microorganisms that can survive the digestive process and accumulate in the animals (42–44). The concentration of microorganisms in bivalves can be tens to hundreds of times as high as that in their growing water (42,45). Consumers and public health regulatory agencies are concerned about the pathogenic organisms found in bivalves that are affected by sewage pollution.

Fecal coliforms are generally used as indicators for bivalve quality and for domestic pollution in shellfish-growing waters (46,47). Following harvest, two microbiological guidelines are applied to determine the acceptability of shellfish meats. Bivalves at wholesale market level should have a 35°C standard plate count (SPC) of $<500{,}000$ g and a most probable numbers (MPN) fecal coliforms of $\leq 230/100$ g (46,47).

The microflora of bivalves at harvest is composed of both organisms that are symbiotic with the bivalves and organisms that are filtered from the water and ingested as food. These microorganisms vary qualitatively and quantitatively, depending on the nutrient level, salinity, temperature, and water quality. The commensal microflora include *Cristispira pectineus*, which colonizes the crystalline style of oysters, and spirochetes such as *Saprospira*, found in the crystalline style, stomach, and intestine of eastern oysters. These commensal organisms are difficult to culture and have no pathological significance to humans (48). Bivalves at harvest normally carry a total plate count of 10^3 to 10^5/g. Soft-shell clams harvested from different growing areas in the Chesapeake Bay contain a geometric mean of SPC, 2.0×10^4–7.2×10^4/g: total coliforms, 1.5×10^3–6.3×10^3 100 g: fecal coliforms, 29–62/100 g: and *E. coli*, 14–27 100 g (49).

The common microflora of bivalves at harvest consists primarily of gram-negative rods including *Pseudomonas* and *Vibrio* species (50,51). The *Flavobacterium–Cytophaga* group occasionally exists in a certain level in oysters. Other organisms in oysters include *Acinetobacter, Corynebacterium, Moraxella, Alcaligenes, Micrococcus*, and *Bacillus*. The microflora of Gulf of Mexico's oysters is dominated by *Vibrio, Aeromonas, Moraxella*, and *Pseudomonas* (52). Low levels of yeasts such as *Rhodotorula rubra, Trichosporon, Candida*, and *Torulopsis* are also frequently encountered in eastern oysters. Several potential pathogenic strains of *Vibrionaceae* are naturally occurring in nonpolluted estuarine waters and may be encountered in bivalves. Coliforms, fecal coliforms, and *E. coli* are the most common contaminating organisms in bivalves at harvest (45,46,49).

Reducing a high microbiological load in bivalves can be accomplished by placing the animals in clean water that is free of undesirable microorganisms and under conditions in which the bivalves will actively feed. This process is called relaying and usually requires about 15 days to reach the satisfactory microbiological quality (53,54). The relaying process is often applied by transferring bivalves harvested from moderately polluted water into approved shellfish-growing water until the animals clean themselves. A second approach is called depuration in which the shellfish are maintained in tanks of clean water with controlled salinity and temperature (55). The water is often recycled through a biofilter to control water quality. The bivalves in this condition can reach the satisfactory microbiological quality within two to three days (53,55). Nevertheless, removal of viruses often does not correlate with fecal coliform elimination even if fecal indicators have a similar reduction rate as do enteric bacterial pathogens (56,57). Ultraviolet irradiation can facilitate the reduction of contaminating bacterial flora in oysters (56). Because hepatitis A may not be eliminated as readily as other enteroviruses during dupuration, hepatitis may still occur by consuming raw depurated shellfish (58).

MICROBIOLOGY AND QUALITY

Fish Products

Fish carries a variety of organisms and should be handled adequately and processed as soon as possible. Refrigeration, freezing, canning, pasteurization, salting, drying, fermentation, curing, and a combination of these methods are commonly used to process seafood into relatively stable, marketable products. Other methods such as irradiation and modified atmospheric preservation (59) have been extensively studied and proved to have potential application.

Refrigerated and Frozen Fish. Refrigeration at 5°C ceases the growth of the mesophiles and as the temperature is further lowered, psychrophiles are eliminated. During refrigeration storage a gradual killing off of the microorganisms occurs. Gram-negative asporogenous pseudomonads are cold sensitive, whereas gram-positives such as micrococci, lactobacilli, and streptococci are more resistant (60,61).

Cooling rate affects the survival of microorganisms. The maximum survival of *E. coli* is at a cooling rate of about 6°C/min, and the minimum at about 100°C/min (62). A similar minimum survival rate has been found for *Streptococcus faecalis, Salmonella typhimurium, Klebsiella aerogenes, Pseudomonas aeruginosa*, and *Azotobacter chroococcum*, but these organisms have an optimum survival rate varying from 7°C/min for *A. chroococcum* to 11°C/min for *P. aeruginosa* (63).

Frozen fish should be stored at or below −20°C and preferably at −30°C. During freezing the number of cells that are inactive ranges from 50 to 90% of the initial bacterial population. Gram-positive bacteria are more resistant to freeze injury than gram-negative ones. Some pathogenic organisms can also have a certain survivability at freezing temperatures (65). Spores are the most resistant microbial entity to freeze damage. Poliovirus inoculated into oysters showed a gradual decline in plaque-forming units during frozen storage at −36°C (64).

Reasons for the cryoinjury of cells are thermal shock, concentration of extracellular solutes, toxic action of concentrated intracellular solutes, dehydration, internal ice formation, and attainment of a minimum cell volume (62). Although cryoinjury results in the cell death, survival of microorganisms occurs and is greater in a supercooled environment than in a frozen one. Some *V. parahaemolyticus* cells, inoculated into oysters, sole fillets, and crabmeat, can persist at −15 or −30°C with a greater survival at −30°C although there is a sharp reduction in viability during freezing (66). Some pathogens such as *Listeria* can survive in freezing temperatures (60).

Cryoprotective agents are substances that can protect bacterial cells during freezing and thawing. Glycerol, dimethyl sulfoxide, egg white, carbohydrates, peptides, serum albumin, meat extract, milk, glutamic acid, malic acid, diethylene glycol, dextran, Tween 80, glucose, polyethylene glycol, and erythritol have cryoprotective functions probably due to reduction of damage to the cell wall and membrane (60).

Canned Fish. Most canned fish are fully processed products such as canned tuna, salmon, sardines, mackerel, fish balls, and other fish. These canned fish are commercially sterile with 12D process to destroy all pathogenic and other organisms, allowing a satisfactory shelf life at room temperature. However, some problems may arise due to the presence of heat-resistant spore formers in the underprocessed products or can leakage from improper seam closure and cross-contamination through cooling water. In oil pack, the oil may protect bacterial spores against heat resulting in nonsterile canned products (67). Flat sour spoilage may occur due to thermophiles, such as *B. stearothermophilus*, which survive processing and multiply during slow cooling and storage at high temperatures. Swollen cans are occasionally encountered when clostria such as *C. sporogenes* survive inadequate processing.

Dried, Salted, and Smoked Fish. Dehydration is an old process that reduces the water activity (A_w) of the fish products below that required for the growth of microorganisms. The process involves drying with or without other preservatives to form dried, salted, or smoked fish.

Dried salted fish, fish bits (fried shredded fish), katsuobushi (dried and smoked skipjack stick), dried shark fins, and dried mullet roe are popular in Asia (68). Smoked salmon, herring, dogfish, and other fish are common fishery products in Europe. Growth of halophilic bacteria or molds may occur, resulting in the spoilage of salted fish. *Aspergillus* and *Penicillium* are major species associated with the color deterioration of salted round herring (68). Halophilic bacteria such as *Halobacterium* and *Halococcus* commonly present in solar salts are most troublesome during the salting and drying process. These bacteria cause red and pink discoloration and induce softness in salted fish.

Surimi-Based Products. Surimi is a mechanically deboned, water-washed frozen fish paste containing cryoprotectants. This high-protein, gel-forming material can be chopped and then mixed with salt, starch, and flavor compounds. The surimi mix is colored, textured, and cooked in two stages to set the gel to process into seafood analogues such as imitation crab, shrimp, lobster, or scallop. Freshly processed crab leg and flaked crab leg analogues contain only 10^2–10^3/g bacteria. During storage bacteria grow and SPC reaches 2×10^8/g at 10°C in 25 days and 10^4–10^6/g at 5°C in six weeks (69). SPC in flaked crab leg increases rapidly to 10^9/g after two and four weeks at 5 and 0°C, respectively. Spoilage and quality deterioration of crab analogues are indicated by number of bacteria (10^7/g), visible slime, odor, and appearance. Slime formation, softened texture, sour odor, and discoloration are consequences of spoilage. *Bacillus* is predominant initially but *Pseudomonas* gradually grows and finally outnumbers other genera at two weeks of storage at 0–5°C. *Bacillus*, which is possibly derived from the ingredient starch, is the major organism throughout the six-day storage at 15°C (69). The spoilage of other fish cake products such as kamaboko can be attributed to *Streptococcus, Leuconostoc*, and *Micrococcus* (70,71).

Shellfish Products

After being harvested, shellfish should be kept refrigerated or at low temperature and processed as soon as possible. Shrimp should be beheaded, peeled, or left unshelled and frozen.

Crabmeat. Blue crabs rank first in U.S. crab landings, and their major products are fresh and pasteurized meat. Other crabs caught are king, Dungeness, and tanner-snow crabs from which frozen section, claws and meat, and canned meat are commonly made (72,73).

To process blue crabmeat, live crabs are steam cooked at 121°C for 10 min. Cooked crabs are refrigerated overnight and the meat is removed by hand or machine. The meat is packed in plastic cups for fresh crabmeat or sealed in tin cans to process for pasteurized crabmeat (74). Three kinds of meat are available: lump meat taken from the back fins, claw meat extracted from claws, and regular meat collected from main body (72). In good plant sanitation, fresh crabmeat usually has a geometric mean SPC of 1.5×10^4–4.5×10^4/g, which increases to 1.4×10^5–3.2×10^5/g under poor plant sanitation (75). Cooked crabs should be stored in refrigeration (<2°C), separated from the live crabs. Refrigerated cooked crabs before picking usually contain bacteria of $<10^4$/g while cooked sponge crabs (gravid females carrying an egg mass) taken from the picking table contain bacterial levels as high as 10^6/g of whole crabs (76,77). Cooked sponge crabs have consistently been found to harbor greater numbers of bacteria than crabs without a sponge (77). Similarly, cooked green crabs, blue crabs that have recently molted and contain a higher level of water than fat crabs, carry higher levels of bacteria than normal crabs into the picking room. They contain higher moisture, which encourages bacterial growth during overnight refrigerated storage (76).

The commercial machine Quik Pik, which mechanically removes the body meat of blue crabs was started on a trial basis in 1978 and now operates successfully in Maryland under proper procedures (76,78). One quick Pik machine can pick 150 lb meat per hour, a rate equal to the work capacity of 30 hand pickers. The cooked crabs are placed in a round, rotating slotted cage to remove legs, fins, and claws, which are dropped through the slots. The crabs then pass through the debacking machine and cleaning device. The cores are loaded on racks for steam heating and then placed in the quik Pik shaker, which vigorously vibrates at 70 oscillations/s for 4 s. All the meat from cores will fall on the collecting belt for further inspection for broken shells and packaging (76).

Machine picking requires constant attention to cleanliness and sanitation to produce a meat product with a satisfactory bacterial quality. The machine should be disassembled and thoroughly cleaned and sanitized at the end of the day. Liquid on the machine during operation creates an aerosol that can greatly contaminate the meat (79). The meat conveyor belt needs continual washing with a tap water spray and sanitation in a chlorine (>200 ppm) bath. A comparison of the bacterial levels indicates that both the SPC and coagulase positive *S. aureus* for machine-picked meat are lower than for the hand product. However, a higher *E. coli* count is found in machine-picked meat than in hand-picked meat. This is not surprising because machine picking is processed under wet and warm environment whereas hand picking is operated in cool and dry conditions.

The normal shelf life of fresh crabmeat is 7–10 days and may last up to 14 days if meat with a low initial bacterial count (80) is stored under optimum refrigeration temperature. To extend the shelf life of crabmeat, the pasteurization process of holding crabmeat for 1 min at 171°–210°F, depending on the desired shelf life from 1 to 12 months, was patented (81). A process of 185°F for 1 min in the center of a 1-lb can (401 × 301) was found to sufficiently reduce an inoculated 10^8/100 g of *C. botulinum* type E spores to <6/100 g and to keep the meat nontoxic for 6 months at 40°F (82,83). A recommendation has been made to increase the time at 185°F to 3 min to provide for 12D cook based on the thermal death time studies of type E *C. botulinum* (84–86). Table 2 shows the thermal resistance characteristics of *C. botulinum* type E. For a complete process, an *F*-value of 31 based on an *F* 16/185 value and a cooling

Table 2. Decimal Reduction Time (D_{10}) for Heating *C. botulinum* Type E Spores

Strain	Heating medium	Temperature range (°C)	D_{10}-value minimum at 82.2°C	*Z*-value (°C)	Reference
Beluga	Blue crab meat	73.9–85.0	0.75	6.5	85
Beluga	Crabmeat	73.9–85.0	0.84	6.5	85
Sarotoga	Crabmeat	80.0–82.2	1.90	6.3	87
Alaska	Whitefish chubs	73.9–85.0	2.21	7.6	88
Sarotoga	Sardines in tomato sauce	76.7–82.2	2.9	6.3	87
Sarotoga	Tuna in oil	72.2–82.2	6.6	6.1	87
Strain 202	Crabmeat	76.7–85.0	1.16	6.38	85

meat temperature of 55°F within 180 min of heat process following storage at 35°F have been recommended (87).

If storage temperature of pasteurized crabmeat rises above 38°F, surviving bacteria may grow and bacterial spores may germinate leading to spoilage. Any leakage due to can defects can introduce psychrotrophic bacteria from cooling water and result in spoilage during refrigerated storage (89). Poor meat quality with a high level of initial bacteria will increase the chance of survival of potential spoilage bacteria. The spoiled pasteurized crabmeat usually contain both aerobic and anaerobic bacteria ranging from 4.0×10^2 to 5.0×10^8/g. Spoiled crabmeat prepared from machine picking has more anaerobic organisms than hand-picked crabmeat (79).

Bivalves. Heat shock has been attempted to facilitate oyster shucking by quickly passing the stocks through a steam tunnel to slightly open the shells. Although this process increases the shucking rate, disadvantages are yield reduction and high risk of bacterial contamination (90). Canning bivalves serves as a long-term preservation method. When recommended processing times and temperatures are followed, few microbiological problems are encountered (88). However, loss of texture and economical infeasibility are problems with this processing.

Pasteurization of oysters at 72–74°C for 8 min in a flexible pouch results in a relatively stable product with organoleptic properties similar to raw oysters. The products stored at 0.5°C for three months have a low level of both aerobic and anaerobic bacteria (91). Immediately after pasteurization, all the surviving bacteria in the oysters are *Bacillus* sp. At storage of five months *Bacillus* continues to dominate the aerobic bacteria while *Clostridium, Corynebacterium, Listeria, Peptostreptococcus*, and *Staphylococcus* constitute the facultatively anaerobic bacteria (92). Combining safe preservatives with the pasteurization can provide a safe and longer shelf life of oysters.

SEAFOOD IRRADIATION

Irradiation is a process extending seafood shelf life by exposing seafood to a certain level of ionizing radiation. Studies on the utilization of ionization radiation in food processing began in the early 1940s, and an extensive program implementing this process was launched in the early 1950s, in the United States (93,94). Ionization radiations are a group of corpuscular and electromagnetic radiations of extremely short wavelengths that can cause ejection of electrons from atoms or molecules. Only radiation with strong penetration, including gamma rays, x-rays, and electron beams, is useful in food preservation. Gamma rays from radioactive isotopes such as Co-60 have been the most acceptable radiation source due to its availability, price, and properties. One rad, a unit of irradiation dosage, is equal to 100 ergs of radiation energy absorbed per gram of substrate. A megarad (Mrad) is a million rads, and a kilorad (krad) is a thousand rads.

There are several advantages in using ionization radiation for seafood preservation. The extension of shelf life of seafood by using low-dosage irradiation allows the expansion of markets farther inland where fresh seafood is otherwise unavailable. The long shelf life of irradiated seafood can adjust the seasonal production providing a year-round supply and unfluctuating prices. Irradiation of high-quality catches produces top-quality seafood thus reducing the need to discard or process deteriorating fish to fish meal (94,95). Ionization radiation directly affects living cells by causing breaks in DNA and indirectly affects other cell components. The presence of water increases the degree of the DNA damage. Complex cells are more sensitive to irradiation than simple ones. The death or injury of microorganisms is due to the direct hit on DNA and the indirect effects from ionization and diffusion of free radicals and peroxides produced around the cells.

Food irradiation can be classified into three classes depending on the level of microorganisms to be destroyed (60). Radappertization or Radiation sterilization is the use of high dose with 12D usually more than 2 Mrad for complete destruction of all or practically all of the organisms. Radurization or radiation pasteurization is using low dose of radiation to destroy a sufficient number of organisms to enhance the shelf life of foods. This process usually uses 100–1,000 krad to destroy 90–99% of the organisms. Radicidation is a low-level irradiation treatment that kills nonspore-forming pathogens to reduce or eliminate the food-borne illness problem. This kind of irradiation treatment normally uses 400–600 krad dose for destruction of salmonellae from poultry and red meat or feed but not for spores of *C. botulisum* or *C. perfringens*.

In seafood irradiation, radiation resistance is expressed as the decimal reduction dose (D_{10}), the dose (krad) required for a 90% or 1-log reduction in bacterial count. The radiation resistance of microorganisms varies, depending on strains and species and the D_{10}-values of various organisms are shown in Table 3. The resistance in descending order is yeast, *Micrococcus, Moraxella-Acinetobacter, Flavobacterium*, and *Pseudomonas* (60). The gram-positive

Table 3. Decimal Reduction Doses (D_{10}-Values) for Irradiation of Some Microorganisms and Toxins

Organism or toxin	Suspending medium	D_{10} (krad)	Ref.
Vibrio parahaemolyticus	Crabmeat	5–12	96
Proteus vulgaris	Oysters	20	95
	Crabmeat	10	95
Aeromonas hydrophila	Blue fish	14–22	96
Escherichia coli	Soft-shells clams	39–42	96
	Mussel	41–48	96
	Oysters	35	95
	Crabmeat	14	95
Salmonella typhimurium	Soft-shell clams	60	96
	Mussel	58–63	96
S. paratyphi A	Oysters	75	95
	Crabmeat	50	95
	Shrimp	85	95
S. paratyphi B	Shrimp	61	97
Shigella flexneri	Mussel	24–34	96
S. sonnei	Oysters	27	95
Staphylococus aureus	Oysters	150	95
	Crabmeat	80	95
	Shrimp	190	95
Adenovirus, echovirus	Eagles medium	410–490	60
Toxins, botulinum type E			
Washed cells	Buffer	1,700–2,100	60
Purified	Buffer	40	60
Staphylococcal			
Enterotoxin B purified	Buffer	2,700	60

bacteria are more resistant than gram-negative ones. Coccoid forms are more radioresistant than rod-shaped cells. Bacterial spores exhibit the most resistance to radiation with the exception of some gram-positive cocci. Compared to bacterial spores, viruses are more or equally resistant to irradiation. Although the amount of radiation needed to inactivate microbial toxins is similar to that for bacterial spores, the radiation level for inactivation of botulinum toxin is much higher than that needed to kill the bacterial cells (98) (Table 3).

In general, D_{10}-values for bacterial vegetable cells (excluding cocci), bacterial spores, *M. radiodurans*, viruses, and bacterial toxins are 20–100, 150–450, 200–600, 400–800, and 200–2,000 krad, respectively (99–104).

Low-dose irradiation treatment of seafoods usually does not induce detectable changes in flavor, texture, and appearance with the exception of a slight loss of flavor in a few products (105,106). The level of irradiation is one factor in determining the type of microorganisms remaining that cause spoilage of irradiated seafood. The spoilage microflora of irradiated fish after a low dose of $\leq$100–150 krad treatment consists mainly of pseudomonads. At dose levels higher than 100 krad, *Maraxella-Acinetobacter* represented the dominant group of organisms in fish spoilage. Irradiation of packaged fish favors the growth of lactic acid bacteria that will be the major spoilers. In general, nonirradiated fish spoils at an aerobic plate count (APC) of 10^6/g whereas low-dose irradiated fish spoils at about 10^8/g (94).

Under the optimum dose of 100–300 krad, fish and shellfish can be maintained three to seven weeks at refrigeration without altering the fresh-product characteristics (Table 4). Compared with the regular 10- to 14-day shelf life of the unirradiated products, low-dose irradiated products have a shelf life two to three times longer (94,109). To insure efficiency and successful processing, the seafood should initially be of high quality and irradiated products should be maintained as near 0°C as possible without freezing.

The safety of irradiated seafood must be evaluated based on the absence of microorganisms and microbial toxins harmful to man, the nutritional contribution of the product, and the absence of any significant amount of toxic compounds formed in the irradiated products (101). Low-dose irradiated seafood, particularly lean fish species, have been found to pose no potential health hazards (94,110). Despite the approval of over 35 countries irradiation in foods and the USDA clearance of low-dose application for poultry irradiation, seafood irradiation has not been permitted. This process can become a successful technique for seafood preservation once it is approved in the United States.

SEAFOOD MICROORGANISMS OF PUBLIC HEALTH SIGNIFICANCE

Seafood, particularly shellfish, may contain a variety of pathogenic microorganisms that impose a threat to the consumers' health. These potential pathogens include both indigenous organisms and contaminating organisms. Pathogens may contaminate the seafood after harvest or during processing. Some indigenous pathogens found are *Vibrio, Clostridium botulinum* type E, *Aeromonas*, and poisonous phytoplankton such as dinoflagellates (111,112). Extraneous pathogens include *Salmonella, Shigella, Listeria, Campylobacter, Staphylococcus aureus, E. coli, Bacillus cereus*, Hepatitis A, and Norwalk virus (64,112).

The consumption of bivalves is of greatest concern to the public due to several factors. Bivalves, especially oysters, are frequently consumed raw. The whole animal is consumed rather than just the muscle tissue, as in the case

Table 4. Shelf Life at 0°C of Selected Seafood after Optimal Dose Irradiation Treatment

Seafood	Shelf life (weeks)	Optimal dose (Mrad)	Ref.
Haddock fillets	3–5	0.15–0.25	107
Cod fillets	4–7	0.10	108
Ocean perch fillets	4	0.15–0.25	107
Mackerel fillets	4–5	0.25	107
English sole fillets	4–5	0.20–0.30	107
Smoked chub	6	0.10	107
Petrale sole fillets	2–3	0.20	107
Shrimp	4	0.15	107
Lobster meat (cooked)	4	0.15	107
King crab meat (cooked)	4–6	0.20	107
Dungeness crab meat (cooked)	3–6	0.20	107
Oyster meat	3–4	0.20	107

Note: Fish or shellfish were packed aerobically in hermetically sealed cans or oxygen-impermeable plastic bags.

of other raw crustaceans and raw fish. Bivalves are filter feeders that can trap microorganisms from the growing water into their bodies. Because the animals grow in near-shore estuarine waters that may encounter sewage pollution, bivalves may become a vector for disease transmission. Based on past records of bivalve shellfish borne disease between 1900 and 1986, there were 12,376 cases of food-borne illnesses in the United States excluding cases of paralytic shellfish poisoning (112). This has been estimated to be only 5–10% of the actual number of cases. These documented cases include 43% gastroenteritis, 26% typhoid, 11% infectious hepatitis, 11% Norwalk virus, and 2% *Vibrio*. The specific etiologic agents have been identified as *Salmonella typhi, Vibrio cholera* 01, *V. cholera* non-01, *V. vulnificus, V. parahaemolyticus, V. fluvialis, V. mimicus, V. hollisae, V. furnissii, E. coli, Salmonella* sp., *Shigella* sp. *Bacillus cereus, Staphylococcus* sp., *Campylobacter* sp., *Aeromonas hydrophila. Pleisomonas shigelloides*, hepatitis A, Norwalk virus, and other viral agents (112–115).

Food poisoning outbreaks reportedly have often been due to the consumption of seafood. The most common agent involved is *V. parahaemolyticus* (114,116). This species was first isolated in 1950 in Japan from seafood (117,118). Since then it has been implicated in more than 1,000 outbreaks per year in Japan and accounts for over 50% of that country's bacterial food poisoning. *V. parahaemolyticus* strains require a low concentration of salt for growth, but some cultures isolated from fresh water have been reported (119–124). Raw seafood is the major vehicle for the organism in Japan; cooked seafoods that have been recontaminated are the source of implication in the United States (117). Kanagawa reaction is commonly used to differentiate between virulent and avirulent isolates by testing whether a strain can produce a heat-stable hemolysin to lysis a blood agar containing 7% NaCl and mannitol. Cultures isolated from stools of patients are always Kanagawa positive (99%). On the contrary only about 1% of isolates from waters or seafoods is Kanagawa positive (125,126). The mechanism for this discrepancy is still unknown. One of the hypotheses is that there may be a transformation of Kanagawa-positive strains on passage through the intestines. There is a possible competitive advantage for Kanagawa positive strains to proliferate more readily in the intestines. The current method of isolating and identifying Kanagawa-positive strains in seafoods and waters may also leave some undetected (116,126).

The infective dose of *V. parahaemolyticus* for humans ranges from 10^5 to 10^7 viable cells, and a decrease in stomach acidity may lower the infective dose. The incubation period for symptoms is 5–90 h and the duration of the illness is normally 2–10 days. The frequencies of symptoms are diarrhea, 98%; abdominal cramps, 82%; nausea, 71%; vomiting, 52%; headache and fever, 27%; and chills, 24% (116,126). The organism has three biologically active hemolysins, substances that can lyse the animal's blood cells: a heat-stable peptide with 45,000 mol. wt., a heat-labile hemolysin, and a phospholipase (127,128). *V. parahaemolyticus* are heat sensitive and most food-poisoning cases result from cross-contamination or poor sanitation (129–131).

V. cholerae causes a gastrointestinal illness in humans called cholera (132). This species is usually divided into two groups, serotype 01 and non-01: both are found in aquatic environment. The 01 serotype contains two biotypes: classical and El Tor. The classical biotype prevailed worldwide until the 1960s and the El Tor biotype presently predominates, including in the United States (111,133,134). Cholera in the United States is relatively rare but occasional outbreaks occur in southern states such as Texas, Louisiana, and Florida due to consumption of raw or partially cooked molluscs, cooked crabs, and other shellfish (133,134). The infective dose is estimated to be about 10^6 cells. Taking antiacids or medication to lower gastric acidity will lower the infective dose (115). The incubation time varies from 6 h to 5 days. Severe symptoms include profuse watery diarrhea, dehydration, and death in the absence of prompt treatment. In the beginning, the stool is brown with fecal material and quickly acquires the classic rice-water appearance. Enormous amounts of fluid leave the body resulting in dehydration and difficulties in circulation. The stool is high in potassium and bicarbonate (111,133). Besides severe diarrhea, victims suffer from thirst, leg cramps, weakness, hoarse speech, and rapid pulse (132,133). In emergency treatment, prompt replacement of fluid and electrolyte losses by intravenous injection is often used. After an initial recovery, oral intake of glucose and electrolytes can improve the condition. In general, *V. cholerae* is sensitive to heat and cold, but it can survive at low temperatures for a certain time (135).

Non-01 cholera is a group of nonagglutinable (NAG) cholerae, which exist naturally in estuarine and coastal waters and also in rivers and brackish waters (128). Only about 5% of this group from seafood and patients isolates in the United States produce cholera toxin. However, the nontoxigenic strains cause gastrointestinal illness with principal symptoms of abdominal cramps, fever, bloody stools, nausea, and vomiting (133). Some human isolates of this group in the United States are from extraintestinal sources including wound infection, ear infection, and primary and secondary septicemia (133). Eating raw oysters has been found to be the major cause of most non-01 *V. cholerae* infection cases in the United States. Other seafood, egg and asparagus salad, or potatoes can occasionally be a vehicle for these organisms (134).

V. vulnificus is widely distributed in the environment and has been found in most U.S. estuarine and coastal waters (136–138). This water and seafood organism is found most frequently at water temperatures >20°C and low salinity of 0.5–1.6%. Environmental isolates are phenotypically identical to clinical isolates. Some strains show bioluminescence and may also be pathogenic (138). This organism can cause illness and infection through the consumption of contaminated raw or undercooked seafood, particularly mollusks such as oysters and clams. The incubation period for this illness is 16–48 h after ingestion. Symptoms include weakness, chills, fever, hypotension, and fatigue with occasional vomiting and diarrhea. Infection occurs, progresses rapidly, and may cause death in 40–60% of patients (128). Patients may have greater risk of infection if they have skin cuts or suffer chronic liver disease, gastric disease, or hemochromatosis (111,128).

Other potential *Vibrio* pathogens such as *V. hollisae, V. mimicus* and *V. furnissi* have been implicated in seafood-borne illnesses (111). Vehicles for these organisms are shellfish including oysters, clams, shrimp, and crawfish. Common symptoms are diarrhea, nausea, vomiting, and abdominal cramps. Both toxigenic and nontoxigenic strains of *V. mimicus* have been isolated but food-poisoning cases have mostly occurred with nontoxogenic ones. Some strains of these organisms have been newly isolated in the last few years, and more information will be available after further studies (111).

Clostridium botulinum type E is another intrinsic pathogen in seafood causing botulism type E intoxication. Based on the serological classification of the neurotoxin, *C. botulinum* is composed of eight types: A, B, C_1, C_2, D, E, F, and G (139). These types are divided into four groups according to proteolytic activity (140,141). Groups I and II are of particular importance for causing botulism in humans. Group I includes type A and proteolytic strains of types B and F. Group II contains all type E and nonproteolytic strains of types B and F. Type E is distributed in sediments of marshes, lakes, and coastal ocean waters and is common in the intestines of fish and shellfish. The organism does not proliferate in living fish but it may multiply in bottom deposits and in aquatic vegetation when the growth of algae reduces the oxygen level of the water to conditions suitable for *C. botulinum*. Type E can grow and produce toxin at 4°C and is heat sensitive (142,143). It is inhibited by water activity of <0.975 (5% NaCl) and pH <5.3 (144). Most outbreaks of botulism associated with fishery products have been with semipreserved products. Smoked, salted, fermented, and canned products are eaten without further cooking and can involve the risk of botulism due to inadequate processing (89,143). Satisfactory heat processing is critical in assuming quality safety of pasteurized crabmeat described above.

Plesiomonas shigelloides is found in fresh surface water and possibly in seawater and is more often isolated in the summer months. This organism has been implicated in human gastroenteritis for 40 yr (145). Seafood that may carry this organism are cuttlefish, raw oysters, salt mackerel, and undercooked oysters. Incubation time for symptoms is one to two days after ingestion of the food. Symptoms include diarrhea, abdominal pain, nausea, chills, fever, vomiting, and headache (145).

Enteric viruses are those animal viruses that are excreted in feces and discharged into domestic sewage. These viruses are more resistant than enteric bacteria to sewage treatment process and various environmental stresses. Several human enteric viruses may be implicated in the viral disease by seafood consumption, such as hepatitis, fever, diarrhea, paralysis, meningitis, and myocarditis. Shellfish may play a significant role as vectors in the transmissions of viral diseases. A study of cockles-borne disease indicated that about one-quarter of the hepatitis A cases in the southeast UK could be caused by shellfish consumption (146,147). In Frankfurt, Germany, about one-fifth of infectious hepatitis was associated with the consumption of contaminated oysters (148). Molluskan shellfish is the major seafood involved in the outbreak of hepatitis A incidence (58) and also attributes to 12.5% of non-A, non-B hepatitis in the United States (114). Outbreaks of Norwalk virus gastroenteritis by shellfish ingestion must be documented in Australia and the United States (97).

According to the Centers of Disease Control annual documentation, nearly 90% of all reported outbreaks of seafood-borne illness are associated with consumption of molluskan shellfish and a very few species of fish (148,149). In geographic distribution, Hawaii has the most cases accounting for 35% of all these seafood-borne illnesses. About half of all these seafood-borne illness cases are found in four states and territories: Hawaii, Puerto Rico, Virgin Islands, and Guam. Another third of these seafood-related illnesses occur in the continental states of New York, California, Washington, Connecticut, and Florida (149).

Chemical intoxication resulting from the harmful metabolites of microorganisms are often implicated in seafood. An allergic food poisoning after eating deteriorated scombroid fish is caused by formation of histamine by microbial enzyme histidine decarboxylase when fish have been exposed to environments favorable to bacterial growth. The organisms involved in histamine poisoning include *Pseudomonas putrefaciens, Aeromonas hydrophila, Proteus vulgaris, Clostridium perfringens, Enterobacter aerogenes*, and *Vibrio alginolyticus* (150).

MICROBIOLOGICAL CRITERIA AND INSPECTION

Microbiological criteria are an important but controversial issue in establishing a regulatory food-control program, particularly for seafoods in the United States. The purposes of microbiological criteria are to insure food safety, adhere to good manufacturing practice, and provide established measure for food inspection and marketing. According to the degree of compliance, microbiological criteria include three categories: standard, guideline, and specification (151). A standard is part of a law or ordinance and is a mandatory criterion. A guideline is a criterion for assessing microbiological quality during processing, distribution, and marketing of foods. A specification is used in purchase agreements between buyers and vendors (152,153).

As more countries are recognizing the need to assess the safety and quality of foods due to food-poisoning outbreaks and international trade disputes, several agencies have formed organizations to formulate microbiological criteria. The International Committee on Microbiological Specifications for Foods (ICMSF) was established by the International Association of Microbiological Society (IAMS) in 1962 for this and related purposes (154). The committee assists in establishing microbiological standards implementing method of examination, promoting safe movement of foods in international trade, and mediating disputes caused by disparate criteria. The committee also publishes books of its reports and recommendations (153–155). Many of the ICMSF recommendations were incorporated into the work of Code Alimentarius Commission to implement the Joint FAO/WHO Food Standard Program of the United Nations World Health Organization. The goal of the Joint Food Standard Program is to protect the health of consumers, ensure fair practices in the food trade, guide the preparation of draft standards and codes of practice, and promote the incorporation of all food standards by different governments. The program has also published recommended international codes of practice for many common seafood products such as shrimps or prawns and frozen fish (156,157).

To ensure consumer safety, inspection systems are enforced by the regulatory agencies in most of the countries. In the United States, the inspection program is carried out by different levels: local, state, and federal government. The federal regulatory seafood control systems include the Food and Drug Administration (FDA), U.S. Department of Agriculture (USDA), U.S. Department of Commerce (USDC), U.S. Department of Defense (USDD), and the Environmental Protection Agency (EPA). The FDA is in charge of imported seafood inspection, a mandatory surveillance system. The National Marine Fisheries Service of USDC operates the voluntary Fishery Product Inspection Program mainly for domestic seafood industry on a fee-for-service basis. The USDA and other federal agencies are responsible for some inspection systems related to their domain of authority. Much of the domestic seafood products are inspected by the states and municipalities in their respective territories. In recent years, the public concern in seafood safety and confusion among various inspection systems have led to the proposal of a national seafood surveillance system. At the time of this writing the USDA will likely be chosen to head this mandatory seafood inspection program.

Microbiological criteria have been controversial and confusing because of the variation in agency objectives, legislative authorities, unsatisfactory definitions, questions in sampling procedures, methodology, quality requirements, and uninsured consumer protection. In recent years a hazard analysis critical control point (HACCP) approach has been regarded as an effective, practical, and economic program in microbiological monitoring and safety of the products (152,158). The industry should develop the actual details of a HACCP program and insure the implementation of the complete process. In conducting hazard analysis of the products each step of the processing operation should be defined to specify the hazards associated with each step. The preventative measures at each processing step will be identified for elimination of the hazards. The critical control points are thus defined, bringing the hazards under control. The theory behind the HACCP concept, if properly implemented, is to reduce the number of governmental inspections, and overcome many of the weaknesses inherent in traditional inspection schemes (158). When plant personnel have been well trained in HACCP concepts, the quality and safety of products can be enhanced with efficient cost inspection.

Seafood microbiology is a difficult subject because of the many obstacles in this area. A broad spectrum of microorganisms is involved, including indigenous microbial flora organisms that contaminate the catch after it is landed and those that multiply during the process from the water to the dining table. The ambiguous taxonomy of many seafood microorganisms is confusing and impedes the study of these organisms. The problems with growth conditions, such as inadequate growth temperature and media composition, often result in disputable results of seafood sample analysis. The large number of nonculturable viable cells in marine bacteria greatly limits the study of the role of these organisms in the natural aquatic ecosystem and the relation to fish and shellfish.

The fast growth of psychrophilic and psychrotrophic bacteria cause fish and shellfish to spoil rapidly even in refrigeration. In addition to spoilers the potential pathogenic microorganisms present in seafood can proliferate expeditiously and threaten consumer safety. Prompt seafood handling and adequate processing are critical in the control of both spoilage and pathogenic organisms to provide safe and stable products. Good quality control and inspection, particularly the HACCP program, can ensure a high quality of seafood products.

Application of modern basic microbiology, biochemistry, and biotechnology can advance fish and shellfish microbiology. DNA probe, immunoassay, monoclonal antibody, and other rapid-detection methods provide accurate and expeditious procedures for determination of seafood pathogens. The study of bacterial cell envelope, a new field that has arisen in the last two decades, can provide information about the functions and physiochemical properties of cell envelope components, which are important in understanding bacterial survival and interaction with other organisms in the aquatic ecosystem. This field of knowledge will also be helpful in establishing new technology to control bacterial growth during processing. Study of nonculturable viable or dormant cells may lead to a new field of study: bacterial survival in the natural environment and its impact on seafood. This can provide another avenue in exploring the key factors behind many nondetectable food pathogens. The progress in seafood microbiology will greatly benefit the seafood industry, providing safe, better quality seafoods.

BIBLIOGRAPHY

1. R. E. Martin, W. H. Doyle, and J. R. Brooker, "Toward an Improved Seafood Nomenclature System," *Mar. Fish. Rev.* **45**, 1–20 (1983).
2. D. C. Georgala, "The Bacterial Flora of the Skin of North Sea Cod," *Journal of General Microbiology* **18**, 84–91 (1958).
3. J. Liston, "Qualitative Variations in the Bacterial Flora of Flatfish," *Journal of General Microbiology* **15**, 305–314 (1956).
4. J. M. Shewan, "Some Bacteriological Aspects of Handling, Processing and Distribution of Fish," *J Roy. Sanit. Inst.* **59**, 394–421 (1949).
5. J. M. Shewan, "The Microbiology of Sea-Water Fish," in G. Borgstrom, ed., *Fish as Food*, Vol. 1. Academic Press, Inc. Orlando, Fla., 1961.
6. P. Baumann, M. Doudoroff, and R. Y. Stanier, "A Study of the *Moraxella* Group. I. Genus *Moraxella* and the *Neisseria catarrhalis* Group," *Journal of Bacteriology* **95**, 58–73 (1968).
7. P. Baumann, M. Doudoroff, and R. Y. Stanier, "A Study of the *Moraxella* Group. II. Oxidative-Negative Species (Genus *Acinetobacter*)," *Journal of Bacteriology* **95**, 1520–1541 (1968).

8. G. L. Gilardi, "Morphological and Biochemical Differentiation of *Achromobacter* and *Moraxella* (DeBord's Tribe Mineae)," *Applied Microbiology* **16**, 33–38 (1968).
9. J. Liston, "A Quantitative and Qualitative Study of the Bacterial Flora of Skate and Lemon Sole Trawled in the North Sea," Ph.D. dissertation. Aberdeen University, Aberdeen, U.K. 1955.
10. B. R. Botta, B. E. Squires, and J. Johnson, "Effect of Bleeding, Gutting Procedures on the Sensory Quality of Fresh Raw Atlantic Cod (*Gadus morhua*)," *Canadian Institute of Food Technology Journal* **19**, 186 (1986).
11. R. M. Brian and D. R. Ward, "Microbiology of Finfish and Finfish Processing," in D. R. Ward and C. R. Hackney, eds., *Microbiology of Marine Food Products*, Van Nostrand Reinhold, Co., Inc., New York.
12. D. R. Ward and N. J. Baj, "Factors Affecting Microbiological Quality of Seafoods," *Food Technology* **42**, 85–89 (1988).
13. W. K. Rodman, "On Board Fish Handling Systems for Offshore Wetfish Trawlers, Work Smarter Not Harder," in W. T. Otwell, compiler, *Proceedings of the First Joint Conference of the Atlantic Fisheries Technology Society*, Florida Sea Grant, University of Florida, 1987.
14. J. J. Connell, *Control of Fish Quality*, Fishing News Books Ltd., Surrey, UK, 1975.
15. R. Nichelson II, *Seafood Quality Control—Boats and Fish Houses: Processing Plant*, Marine Advisory Bulletins, Texas A & M University, Texas, 1972.
16. J. A. Dassow, "Handling Fresh Fish," in M. E. Stansoy, ed., *Industrial Fishery Technology*, Robert Krieger, New York, 1976.
17. T. Chai, "Studies on the Bacterial Flora of Fish Pen Slime," M. S. Thesis, University of Massachusetts, Amherst, 1970.
18. T. Chai, "Usefulness of Electrophoretic Pattern of Cell Envelope Protein as a Taxonomic Tool for Fish Hold Slime *Moraxella* Species." *Applied Environmental Microbiology* **42**, 351–356 (1981).
19. V. Venugopal, "Extracellular Proteases of Contaminant Bacteria in Fish Spoilage: A Review." *J. Food Prot.* **53**, 341–350 (1990).
20. T. Chai, *Fishery Bacteriology*, Joint Commission Rural Reconstruction, Taipei, Taiwan, 1979.
21. H. C. Chen and T. Chai, "Microflora of Drainage from Ice in Fishing Vessel Fish Holds," *Applied Environmental Microbiology* **43**, 1360–1365 (1982).
22. D. G. James. "The Prospects for Fish for the Undernourished Food and Nutrition," *FAO* **12**, 20–27 (1986).
23. T. Chai et al., "Detection and Incidence of Specific Species of Spoilage Bacteria in Fish. II. Relative Incidence of *Pseudomonas putrefaciens* and Fluorescent Pseudomonads on Haddock Fillets," *Applied Microbiology* **16**, 1738–1741 (1968).
24. E. M. Ravesi, J. J. Licciardello, and L. D. Racicot, "Ozone Treatments of Fresh Atlantic Cod, *Gadus morhua*," *Mar. Fish. Ref.* **49**, 37–42 (1987).
25. J. M. Shewan, G. Hobbs, and W. Hodgkiss, "A Determination Scheme for the Identification of Certain Genera of Gram-Negative Bacteria, with Special Reference to the *Pseudomonadeceae*," *Journal of Applied Bacteriology* **23**, 379–390 (1960).
26. J. Liston, "Microbiology in Fishery Science," in J. J. Connel, ed., *Advances in Fish Science and Technology*, Fishing News Books Ltd., Surrey, UK, 1980.
27. E. A. Fieger and A. F. Novak, "Microbiology of Shellfish Deterioration," in Ref. 5.
28. National Marine Fisheries Service, *Fisheries of the United States 1988*, NMFS, U.S. Department of Commerce, Washington, D.C., 1989.
29. O. B. Williams and H. B. Rees, Jr., "The Bacteriology of Gulf Coast Shrimp. III. The Intestinal Flora," *Texas J. Sci.* **4**, 55–58 (1952).
30. O. B. Williams. "Microbiological Examination of Shrimp," *J. Milk and Food Technol.* **12**, 109–110 (1949).
31. O. B. Williams et al., "The Bacteriology of Gulf Coast Shrimp. II. Qualitative Observations on the External Flora," *Texas J. Sci.* **4**, 53–54 (1952).
32. M. Green, "Bacteriology of Shrimp. II. Quantitative Studies of Freshly Caught and Iced Shrimp," *Food Res.* **14**, 372–383 (1949).
33. L. L. Campbell, Jr., and O. B. Williams, "The Bacteriology of Gulf Coast Shrimp. IV. Bacteriological, Chemical, and Organoleptic Changes with Ice Storage," *Food Technology* **6**, 125–126 (1952).
34. B. F. Cobb et al., "Effect of Ice Storage upon the Free Amino Acid Contents of Tails of White Shrimp (*Panaeus satiferus*)," *Journal of Agriculture and Food Chemistry* **22**, 1052–1056 (1974).
35. C. Vanderzant, E. Roz, and R. Nickelson, "Microbial Flora of Gulf of Mexico and Pond Shrimp," *J. Milk Food Technol.* **33**, 346–350 (Aug. 1970).
36. G. Finne, "Enzymatic Ammonia Production in Penaied Shrimp Held on Ice," in R. E. Martin and co-eds., *Chemistry & Biochemistry of Marine Food Products*, AVI Publishing Co., Inc., Westport, Conn., 1982.
37. H. S. Tubiash, R. K. Sizemore, and R. R. Colwell, "Bacterial Flora of the Hemolymph of the Blue Crab, *Callinectes sapidus*: Most Probable Numbers," *Applied Microbiology* **29**, 388–392 (1975).
38. R. K. Sizemore et al., "Bacterial Flora of the Hemolymph of the Blue Crab. *Callinectes sapidus*: Numerical taxonomy," *Applied Environmental Microbiology* **29**, 393–399 (1975).
39. M. A. Faghri et al., "Bacteria Associated with Crabs from Cold Waters with Emphasis on the Occurrence of Potential Human Pathogens," *Applied Environmental Microbiology* **47**, 1054–1061 (1984).
40. N. A. Cox and R. T. Lovell, "Identification and Characterization of the Microflora and Spoilage Bacteria in Freshwater Crayfish, *Procambarus clarkii*," *Journal of Food Science* **38**, 679–681 (1973).
41. R. J. Miget, "Microbiology of Crustacean Processing: Shrimp, Crawfish, Prawns," in Ref. 11.
42. D. W. Cook, "Microbiology of Bivalve Molluskan Shellfish," in Ref. 11.
43. D. W. Cook, "Fate of Euteric Bacteria in Estuarine Sediments and Oysters Feces," *J. Miss. Acad. Sci.* **29**, 71–76 (1984).
44. A. R. Murchelano and J. L. Bishop, "Bacteriological Study of Laboratory-Reared Juvenile American Oysters (*Crassostrea virginica*)," *J. Muertebr. Pathol.* **14**, 321–327 (1975).
45. T. Chai et al., "Comparison of Microbiological Quality between Soft-Shell Clams and Growing Waters," *Journal of Applied Environmental Microbiology*, submitted.
46. Food and Drug Administration, *National Shellfish Sanitation Program Manual of Operation. Part II. Sanitation of the Harvesting and Processing of Shellfish*, FDA U.S. Department of Health and Human Services, Washington, D.C., 1989.
47. J. D. Clem, *Status of Recommended National Shellfish Sanitation Program Bacteriological Criteria for Shucked Oysters*

at the Wholesale Market Level, FDA. Washington, D.C., (1983).

48. V. T. Dimitroff, "Spirochaetes in Baltimore Market Oysters," *Journal of Bacteriology* **12**, 135–177 (1926).
49. T. Chai et al., "Microbiological Studies of Chesapeake Bay Soft-Shell Clams *(Mya arenaria)*," *J. Food Prot.* **53**, 1052–1057 (1990).
50. R. R. Colwell and J. Liston, "Microbiology Shellfish: Bacteriological Study of the Natural Flora of Pacific Oysters (*Crassostrea gigas*)," *Applied Microbiology* **8**, 104–109 (Feb. 1960).
51. T. E. Lovelace, "Quantitative and Qualitative Commensal Bacterial Flora of *Crassostrea virginica* in Chesapeake Bay," *Proc. Nat. Shellfish Assoc.* **58**, 82–87 (1968).
52. C. Vanderzant et al., "Microbial Flora and Level of *Vibrio parahaemolyticus* of Oysters *(Crassostrea virginica)*, Water and Sediment from Galveston Bay," *J. Milk Food Tech.* **36**, 447–452 (Sept. 1973).
53. G. P. Richards, "Microbial Purification of Shellfish: A Review of Depuration and Relaying," *J. Food Prot.* **51**, 218–251 (Mar. 1988).
54. D. W. Cook and R. D. Ellender, "Relaying to Decrease the Concentration of Oyster-Associated Pathogens," *J. Food Prot.* **49**, 196–202 (Mar. 1986).
55. G. H. Fleet, "Oyster Depuration—A Review," *Food Tech. Aust.* **30**, 444–454 (1978).
56. G. J. Vasconcelos and J. S. Lee, "Microbial Flora of Pacific Oysters *(Crassostrea gigas)* Subjected to Ultraviolet-Irradiated Seawater," *Applied Microbiology* **23**, 11–16 (1972).
57. J. G. Metcalf et al., "Bioaccumulation and Depuration of Enteroviruses by the Soft-Shelled Clam, *Mya arenaria*," *Applied Environmental Microbiology* **38**, 275–282 (1979).
58. G. P. Richards, "Outbreaks of Shellfish-Associated Interic Virus Illness in the United States: Requisite for Development of Viral Guidelines," *J. Food Prot.* **48**, 815 (1985).
59. K. L. Parkin and W. D. Brown, "Preservation of Seafood with Modified Atmospheres," in Ref. 36.
60. G. J. Banwart, *Basic Food Microbiology*, AVI Publishing Co., Inc., Westport, Conn., 1979.
61. K. Schroder et al., "Psychrotrophic *Lactobactllus plantarum* from Fish and Its Ability to Produce Antibiotic Substance," in J. J. Connell, ed., *Advances in Fish Science and Technology*, Fishing News Books Ltd., Surrey, UK, 1980.
62. P. H. Calcott and R. A. Macleod, "Survival of *Escherichia coli* from Freeze-Thaw Damage: A Theoretical and Practical Study," *Canadian Journal of Microbiology* **20**, 671–681 (1974).
63. P. H. Calcott et al., "The Effect of Cooling and Warm Rates on the Survival of a Variety of Bacteria," *Canadian Journal of Microbiology* **22**, 106–109 (1976).
64. S. W. Weagant et al., "The Incidence of *Listeria* species in Frozen Seafood Products," *J. Food Prot.* **51**, 655–657 (1988).
65. R. Digirolamo et al., "Survival of Virus in Chilled, Frozen, and Processed Oysters," *Applied Microbiology* **20**, 58–63 (1975).
66. H. C. Johnson and J. Liston, "Sensitivity of *Vibrio parahaemolyticus* to Cold in Oysters, Fillets and Crabmeat," *Journal of Food Science* **38**, 437–441 (1973).
67. N. Neufeld, "Influence of Bacteriological Standards on the Quality of Inspected Fisheries Products," in R. Kreuzer, ed., *Fish Inspection and Quality Control*, Fishing News Ltd., London, 1971.
68. B. S. Pan, "Low Moisture Fishery Products," in J-L. Chuang, B. S. Pan, and G-C. Chen, eds., *Fishery Products of Taiwan*, JCRR Fisheries Series **25B**, Taipei, Taiwan, 1977.
69. I. H. Yoon, J. R. Matches, and B. Rasco, "Microbiological and Chemical Changes of Surimi-Based Imitation Crab during Storage," *Journal of Food Science* **53**, 1343–1346 (1988).
70. E. L. Elliot, "Microbiological Quality of Alaska Pollock Serimi." in D. E. Kramer and J. Liston, eds., *Developments in Food Sciences: Seafood Quality Determination*, Elsevier, Amsterdam, The Netherlands, 1987.
71. J. R. Matches et al., "Microbiology of Surimi-Based Products," in Ref. 70.
72. R. R. Cockey and T. Chai, "Microbiology of Crustacea Processing: Crabs," in Ref. 11.
73. J. S. Lee and D. K. Pfeifer, "Microbiological Characteristics of Dungeness Crabs *(Cancer magister)*," *Applied Microbiology* **30**, 72–78 (1975).
74. J. W. Duersch, M. W. Paparella, and R. R. Cockey, *Processing Recommendation for Pasteurization Meat from the Blue Crabs*, Marine Products Laboratory, University of Maryland, Crisfield, Md., 1981.
75. F. A. Phillips and J. T. Peeler, "Bacteriological Survey of the Blue Crab Industry," *Applied Microbiology* **24**, 958–966 (1972).
76. M. W. Paparella, *Information Tips 78-4*, Marine Products Laboratory, University of Maryland, Crisfield, Md., 1978.
77. J. Pace, R. R. Cockey, and T. Chai, "Sources of Spoilage of Pasteurized Crabmeat," *Proceedings of the Interstate Seafood Seminars*, 314–323 (Oct. 1989).
78. R. R. Cockey and T. Chai, "Quik-Pik Machine Processing of Crabmeat and Quality Control," *Proceedings of the Interstate Seafood Seminars 1990.*
79. R. R. Cockey, "Bacteriological Assessment of Machine Picked Meat of the Blue Crabs," *J. Food Prot.* **43**, 172–174 (1980).
80. M. A. Benarde and R. A. Littleford, "Antibiotic Treatment of Crab and Oyster Meats," *Applied Microbiology* **5**, 368–372 (1957).
81. U.S. Pat. 2,546,428 (1951), G. C. Byrd.
82. R. R. Cockey and M. C. Tatro, "Survival Studies with Spores of *Clostridium botulinum* Type E in Pasteurized Meat of the Blue Crabs *Callinectes sapidus*," *Applied Microbiology* **27**, 629–633 (1974).
83. Maryland State Department of Health, *Regulations Governing Crabmeat*, Maryland State Department of Health, Baltimore, Sept. 3, 1957.
84. D. A. Kautter et al., "Incidence of *Cl. botulinum* in Crabmeat from the Blue Crabs," *Applied Microbiology* **28**, 722 (Oct. 1974).
85. R. K. Lynt et al., "Thermal Death Time of *Clostridium botulinum* Type E in Meat of the Blue Crabs," *Journal of Food Science* **42**, 1022–1025 (1977).
86. R. K. Lynt, D. Kautter, and H. Solomon, "Differences and Similarities among Proteolytic and Nonproteolytic Strains of *Clostridium botulinum* A, B, E, and F: A Review," *J. Food Prot.* **45**, 466–474 (Apr. 1982).
87. D. R. Ward et al., *Thermal Processing Pasteurization Manual*, Department of Food Science and Technology. VPI State University, Blacksburg, Va., 1982.
88. E. Tanikawa and S. Poka, "Heat Processing of Shellfish," in G. Borgstrom, ed., *Fish as Food*, Vol. IV. Academic Press, Inc., Orlando, Fla., 1965.

89. M. W. Eklund, "Significance of *Clostridium botulinum* in Fishery Products Preserved Short of Sterilization," *Food Technology* **36**, 107–115 (Dec. 1982).

90. F. Huand and C. E. Hebard, *Proceedings of Engineering and Economics of the Oyster Steaming Shucking Process*, VPI & SU, Hampton, Va., 1980.

91. T. Chai, J. Pace, and T. Cossaboom, "Extension of Shelf-Life of Oysters by Pasteurization in Flexible Pouches," *Journal of Food Science* **49**, 331–333 (Feb. 1984).

92. J. Pace, C. Y. Wu, and T. Chai, "Bacterial Flora in Pasteurized Oysters after Refrigerated Storage," *Journal of Food Science* **53**, 325–327 (Feb. 1988).

93. G. G. Giddings, "Radiation Processing of Fishery Products," *Food Technology* **38**, 61–6, 94–97 (1984).

94. J. J. Licciardello and L. J. Ronsivalli, "Irradiation of Seafoods," in Ref. 36.

95. J. D. Quinn et al., "The Inactivation of Infection and Intoxication Microorganisms by Irradiation in Seafood," in FAO/IAEA, eds., *Microbiological Problems in Food Preservation by Irradiation*, IAEA, Vienna, 1967.

96. J. J. Licciardello, D. L. Dentremont, and R. C. Lundstrom, "Radio-Resistance of Some Bacterial Pathogens in Soft-Shell Clams *(Mya arenaria)* and mussels *(Mytilus edulis)*," *J. Food Prot.* **52**, 407–411 (1989).

97. C. W. Hung and B. S. Pan, "Effect of Gamma Irradiation Dosage on Frozen Shrimp," *Proceedings of the Atlantic Fisheries Technology Conference and International Symposium of Fisheries Technology*, Boston, Mass., 1985.

98. G. Hobbs, "*Clostridium botulinum* in Irradiated Fish," *Fd. Irrad. Inf.* **7**, 39–54 (1977).

99. A. Hobbs, "Toxin production by *Clostridium botulinum* Type E in Fish," in Ref. 97.

100. T. Miura et al., "Radiosensitivity of Type E Botulinum Toxin and Its Protection by Proteins, Nucleic Acids and Some Related Substances," in Ref. 97.

101. *Wholesomeness of Irradiated Food*, Joint FAO/IAEA/WHO Committee Report, WHO Technological Report Series **604-FAO**, Food and Nutrition Series 6, World Health Organization, Geneva, 1977.

102. R. B. Read, Jr., and J. G. Bradshaw, "γ-Irradiation of Staphylococcal Enterotoxin B," *Applied Microbiology* **15**, 603–605 (1967).

103. R. Sullivan et al., "Inactivation of Thirty Viruses by Gamma Radiation," *Applied Microbiology* **22**, 61–65 (1971).

104. T. Miura et al., "Radiosensitivity of Type E Botulinum Toxin and Its Protection by Proteins, Nucleic Acids, and Some Related Substances," in M. Herzberg, ed., *Proceedings of the First United States—Japan Conference on Toxic Microorganisms*, USDI, Washington, D.C., 1970.

105. E. F. Reben et al., "Biological Evaluation of Protein Quality of Radiation-Pasteurized Haddock, Flounder and Crab," *Journal of Food Science* **33**, 335–337 (1968).

106. R. O. Brook et al., "Preservation of Fresh Unfrozen Fishery Products by Low-Level Radiation. V. The Effects of Radiation Pasteurization on Amino Acids and Vitamins in Haddock Fillets," *Food Technology* **20**, 99–102 (1966).

107. J. W. Slavin et al., "The Quality and Wholesomeness of Radiation-Pasteurized Marine Products with Particular Reference to Fish Fillets," *Iso. Radiat. Technol.* **3**, 365–381 (1966).

108. L. J. Ronsivalli et al., *Study of Irradiated Pasteurized Fishery Products, Maximum Shelf Life Study. B. Radiation Chemistry*, U.S. Atomic Commission Contract No. AT(49-112)-1889, Bureau of Commercial Fisheries Technological Laboratories, Gloucester, Mass., 1970.

109. B. L. Middlebrooks et al., "Effects on Storage Time and Temperature on the Microflora and Amine Development in Spanish Mackerel *(Scomberomorus maculatus)*." *Journal of Food Science* **53**, 1024–1029 (1988).

110. J. H. Skala, E. L. McGown, and P. P. Waring, "Wholesomeness of Irradiated Foods," *J. Food Prot.* **50**, 150–160 (Feb. 1987).

111. C. R. Hackney and A. Dicharry, "Seafood-Borne Bacterial Pathogens of Marine Origin," *Food Technology* **42**, 104–109 (Mar. 1988).

112. S. R. Rippey and T. L. Verber, *Shellfish Borne Disease Outbreaks*, FDA. Shellfish Sanitation Branch. Northeast Technical Services Unit, Davisville, R.I., 1986.

113. J. M. Janda et al., "Current Perspectives on the Epidemiology and Pathogenesis of Clinically Significant *Vibrio* spp.," *Clinical Microbiology Reviews* **1**, 245–267 (1988).

114. M. J. Alter et al., "Sporatic Non-A, Non-B Hepatitis: Frequency and Epidemiology in an Urban U.S. Population," *Journal of Infectious Diseases* **145**, 886–893 (1982).

115. P. Blake, "Vibrio on the Half Shells; What the Walrus and the Carpenter Didn't Know," *Annals of Internal Medicine* **99**, 558–559 (1987).

116. L. R. Beuchat, "*Vibrio parahaemolyticus*: Public Health Significance." *Food Technology* **36**, 80–92 (1982).

117. T. Fujino. "International Symposium on *Vibrio parahaemolyticus*" Saikon Publishing Co., Tokyo, Japan, 1974.

118. J. Sakurai, A. Matsuzaki, and T. Miwatani, "Purification and Characterization of Thermostable Direct Hemolysin of *V. parahaemolyticus*," *Infection and Immunity* **8**, 775–780 (1973).

119. L. R. Beuchat, "Interacting Effects of pH. Temperature, and Salt Concentration on Growth and Survival of *Vibrio parahaemolyticus*," *Applied Microbiology* **25**, 844–846 (May 1973).

120. D. Covert and M. Woodburn. "Relationship of Temperature and Sodium Chloride Concentration to the Survival of *Vibrio parahaemolyticus* in Broth and Fish Homogenate," *Applied Microbiology* **23**, 321–325 (Feb. 1972).

121. D. Golmintz, R. C. Simpson, and D. L. Dubrow, "Effect of Temperature on *Vibrio parahaemolyticus* in Artificially Contaminated Seafood," *Developments in Industrial Microbiology* **15**, 288 (1974).

122. C. Vanderzant and R. Nickelson, "Survival of *Vibrio parahaemolyticus* in Shrimp Tissue under Various Environmental Conditions," *Applied Microbiology* **23**, 34–37 (Jan. 1972).

123. K. Venkateswaran et al., "Characterization of Toxigenic *Vibrio* Isolated from the Freshwater Environment of Hiroshima, Japan," *Applied Environmental Microbiology* **55**, 2613–2618 (Oct. 1989).

124. R. R. Colwell and J. Kaper, "*Vibrio cholerae, V. parahaemolyticus*, and Other Vibrios: Occurrence and Distribution in Chesapeake Bay," *Science* **198**, 394–396 (1978).

125. G. Spite, D. Brown, and R. Twedt, "Isolation of One Enteropathogenic, Kanagawa-Positive Strain of *V. parahaemolyticus* from Seafood Implicated in Acute Gastroenteritic," *Applied Environmental Microbiology* **35**, 1226–1227 (1978).

126. R. M. Twedt and D. F. Brown, "*V. parahaemolyticus*: Infection or Toxicosis?" *J. Milk Food Technol.* **36**, 129–134 (1973).

127. T. Honda et al., "Identification of Lethal Toxin with the Thermostable Direct Hemolysin Produced by *V. parahaemolyticus*, and Some Physiochemical Properties of the Purified Toxin," *Infection and Immunity* **13**, 133–139 (1976).

128. J. Oliver, "Vibrio: On Increasingly Troublesome Genus," *Diagn. Med.* **8**, 43–49 (1985).
129. R. Delmore and P. Crisley, "Thermal Resistance of *Vibrio parahaemolyticus* in Clam Homogenate," *Journal of Food Science* **41**, 899–902 (1979).
130. L. R. Beuchat and R. E. Worthington, "Relationships Between Heat Resistance and Phospholipid Fatty Acid Composition of *Vibrio parahaemolyticus*." *Applied Environmental Microbiology* **31**, 389–394 (Mar. 1976).
131. R. P. Delmore, Jr., and F. D. Crisley, "Thermal Resistance of *Vibrio parahaemolyticus* in Clam Homogenate," *J. Food Prot.* **42**, 131–134 (1979).
132. R. Sakazaki, "Vibrio Infections," in H. Rieman and F. Bryon, eds., *Foodborne Infection and Intoxication*, Academic Press, Inc., Orlando, Fla., 1979.
133. J. Morris and R. Black, "Cholera and Other Vibrioses in the United States," *New England Journal of Medicine* **312**, 343–350 (1985).
134. Centers for Disease Control, "Cholera in Louisiana—Update," *Morbidity Mortality Weekly Report* **35**, 687 (1986).
135. L. Shultz et al., "Determination of the Thermal Death Time of *Vibrio cholerae* in Blue Crabs *(Callinectes sapidus)*," *J. Food Prot.* **49**, 4–6 (Jan. 1984).
136. J. Oliver, R. Warner, and D. Cleland, "Distribution of *Vibrio vulnificus* and Other Lactose Fermenting Vibrios in the Marine Environment," *Applied Environmental Microbiology* **45**, 985–987 (1983).
137. R. Tilton and R. Ryan, "Clinical and Ecological Characteristics of *Vibrio vulnificus* in the Northeastern United States," *Diagnostic Microbiology Infectious Disease* **6**, 109–117 (1987).
138. J. Oliver et al., "Bioluminescence in a Strain of the Human Pathogenic Bacterium *Vibrio vulnificus*," *Applied Environmental Microbiology* **52**, 1209–1221 (1986).
139. G. Sakagrechi, "Botulism," in Ref. 132.
140. L. Smith, *Botulism: The Organism, Its Toxins, the Disease*. Charles C. Thomas, Springfield, Ill., 1977.
141. J. Simunovic, J. Oblinger, and J. Adams, "Potential for Growth of Nonproteolytic Types of *Clostridium botulinum* in Pasteurized Meat Products. A Review." *J. Food Prot.* **48**, 265–276 (Mar. 1985).
142. NCA. *Thermal Destruction of Type E Clostridium botulinum*, NCA Research Foundation. Washington, D.C., 1973.
143. G. Hobbs, "*Clostridium botulinum* and Its Importance in Fishery Products," *Adv. Food Res.* **22**, 135–156 (1976).
144. A. Emodi and R. Lechowich, "Low Temperature Growth of Type E. *Clostridium botulinum* Spores. I. Effect of Sodium Chloride and pH," *Journal of Food Science* **34**, 78–81 (1969).
145. M. Miller and S. Koburger, "*Plesiomonas shigelloides*: An Opportunistic Food and Waterborne Pathogen," *J. Food Prot.* **48**, 449–457 (May 1985).
146. C. P. Gerba, "Viral Disease Transmission by Seafoods," *Food Technology* **42**, 99–103 (1988).
147. M. C. O'Mahony et al., "Epidemic Hepatitis A from Cockles," *Lancet* **1**, 518–520 (1983).
148. W. Stille et al., "Oyster-Transmitted Hepatitis," *Deutsche Medizinische Wochenschrift* **97**, 145 (1972).
149. Centers for Disease Control, *Annual Summary of Foodborne Diseases*, CDC. U.S. Department of Health and Human Services, Atlanta, Ga., 1979–1986.
150. S. L. Taylor, "Marine Toxins of Microbial Origin," *Food Technology* **42**, 94–98 (1988).
151. MCCFP. National Research Council. *An Evaluation of the Role of Microbiological Criteria for Foods and Food Ingredients*. National Academic Press, Washington, D.C., 1985.
152. E. S. Garrett III. "Microbiological Standards, Guidelines, and Specifications and Inspection of Seafood Products," *Food Technology* **42**, 90–93 (1988).
153. International Committee on Microbiological Specifications for Foods. *Microorganisms in Foods. 2. Sampling for Microbiological Analysis: Principles and Specific Applications.* ICMFS, University of Toronto Press, Toronto, 1978.
154. International Committee on Microbiological Specifications for Foods, *Microorganisms in Foods. 1. Their Significance and Methods of Enumeration*, ICMFS. University of Toronto Press, Toronto, 1978.
155. International Committee on Microbiological Specifications for Foods, *Microorganisms in Foods. 2. Sampling for Microbiological Analysis: Principles and Specific Applications.* ICMFS, University of Toronto Press, Toronto, 1986.
156. Codex Alimentarius Commission, *Recommended International Code of Practice for Shrimps or Prawns*, Joint FAO/WHO, Rome, Italy, 1980.
157. Codex Alimentarius Commission, *Recommended International Code of Practice for Frozen Fish*, Joint FAO/WHO, Rome, Italy, 1980.
158. National Academy of Sciences, *An Evaluation of the Role of Microbiological Criteria for Foods and Food Ingredients*, National Academy Press, Washington, D.C., 1985.

TUU-JYI CHAI
University of Maryland
Cambridge, Maryland

See also MICROBIOLOGY OF FOODS.

FISH AND SHELLFISH PRODUCTS

THE SEAFOOD CHAIN

It is known that humans have used rather advanced techniques for obtaining and processing seafood throughout recorded history. A main source of protein for the ancient Egyptians was fish from the Nile, Mediterranean, and pond culture. Fish was consumed fresh and salted for preservation by the Greeks. Dried fish became a major source of animal protein in Europe when the Roman church banned meat consumption on Fridays and during Lent.

It is known that American Indians used fish as an important part of their diets 10,000 years ago. The early European settlers coming to the New World brought processing and preservation practices such as drying, salting, pickling, and cooling. During colonial times salted dried cod was exported back to England. A major part of the diet of American northwest Indians and Eskimos was salmon that was sun dried and smoked over a wood fire.

It was not until the 19th century in the Great Lakes region that fish were frozen by using the combination of salt and ice to lower temperatures below freezing. The growth of commercial freezing processes over the last 100 years has made freezing the leading means of preserving seafood for human consumption. As will be seen, the early fresh fish handling and freezing technology was not developed with good knowledge of the biological and physical factors that must be considered. Hence, the fishy odor of fish along with certain rancid and spoilage off-tastes were considered normal in fresh, frozen, and dried products.

Today it is realized that quality control of fish, as much or more than any other food, must be practiced from harvest to table. The seafood chain (Fig. 1) begins with the boat builder or hatchery designer and ends with the consumer. All must practice good techniques of sanitation, temperature control, and packaging protection. As the fish story unfolds, it will be seen that any break in this chain can be disastrous to the quality of seafood. If a processing plant, market, restaurant, or home smells fishy, then there has been a break in the chain. Good fish looks good, has a neutral fresh odor, and has firm flesh. Good fish is a healthful, highly nutritious protein food that has an unlimited market.

U.S. FISHERY

There are more than 2,000 species of finfish within the U.S. coastal marine waters, of which about ca 2% are consumed as food fish (1). In fact, the U.S. consumer is familiar with only a few of the some 200 species of finfish and 40 species of shellfish that are consumed throughout the world. This is not surprising when one compares the geographical annual per capita consumption of fish and shellfish as shown in Table 1 (2). Based on live weight of the landed animal, North America consumed 85.6 lb while consumption in other areas of the world ranged from 5 to 15 times that amount. There was little change in the amount of fish consumed by the average U.S. citizen from the early 1900s to 1970s (Table 2). Over the last 15 years, although still considerably below that of the rest of the world, the U.S. per capita consumption has risen about 30%. Most of this increase has been during the last 5 yr. With the increasing emphasis on eating lighter and leaner foods, this U.S. consumption is continuing to rise.

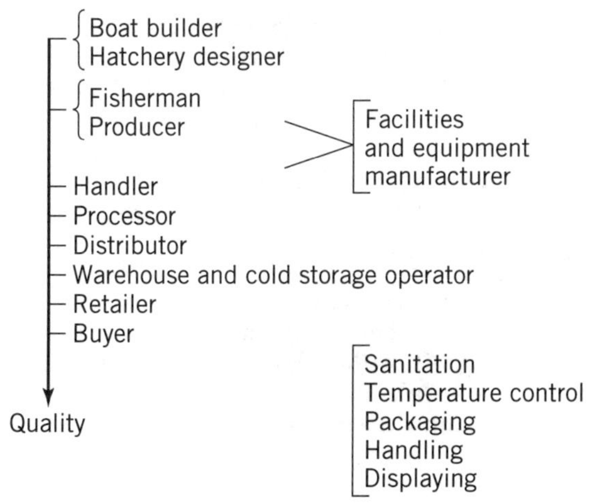

Figure 1. The seafood chain from the harvest to the table.

Table 1. Annual Per Capita Consumption of Live Weight Fish and Shellfish for Human Food, Selected Countries, 1982–1984

	Estimated live weight equivalent	
Country	Kilograms	Pounds
North America		
Canada	21.4	47.2
United States	17.4	36.4
Caribbean		
Barbados	33.6	74.1
Dominican Republic	6.8	15.0
Grenada	19.4	42.8
Haiti	3.7	8.2
Jamaica	17.0	37.5
Latin America		
Argentina	5.7	12.6
Bolivia	1.6	3.5
Brazil	6.2	13.7
Chile	18.7	41.2
Equador	14.2	13.1
Mexico	9.9	21.8
Panama	14.4	31.7
Europe		
Denmark	22.0	48.5
France	24.8	54.7
Iceland	88.4	194.9
Norway	46.0	101.4
USSR	27.3	60.2
Near East		
Egypt	5.5	12.1
Israel	14.2	31.3
Lebanon	0.5	1.1
Oman	35.5	78.3
Far East		
China	4.9	10.8
India	3.1	6.8
Japan	74.5	164.2
Philippines	35.7	78.7
Africa		
Congo (Brazaville)	33.4	73.6
Ethiopia	0.1	0.2
Kenya	4.7	10.4
Republic of South Africa	9.6	21.2
Oceania		
Australia	16.0	35.3
New Zealand	12.2	26.9

Source: Ref. 2.

Table 2. U.S. Annual Per Capita Consumption of Commercial Fish and Shellfish, 1953–1985

Year	Civilian resident population, July 1[a]	Fresh and frozen[b]	Per capita canned[c]	Consumption cured[d]	Total
1909[e]	90.5	4.3	2.7	4.0	*11.0*
1910	92.4	4.5	2.8	3.9	*11.2*
1915	100.5	5.8	2.4	3.0	*11.2*
1920	106.5	6.3	3.2	2.3	*11.8*
1925	115.8	6.3	3.2	1.6	*11.1*
1930	122.9	5.8	3.4	1.0	*10.2*
1935	127.1	5.1	4.7	0.7	*10.5*
1940	132.1	5.7	4.6	0.7	*11.0*
1945	128.1	6.6	2.6	0.7	*9.9*
1950	150.8	6.3	4.9	0.6	*11.8*
1955	163.0	5.9	3.9	0.7	*10.5*
1960	178.1	5.7	4.0	0.6	*10.3*
1965	191.6	6.0	4.3	0.5	*10.8*
1970	201.9	6.9	4.5	0.4	*11.8*
1975	213.8	7.5	4.3	0.4	*12.2*
1980[f]	225.6	8.0	4.5	0.3	*12.8*
1981[f]	227.7	7.8	4.8	0.3	*12.9*
1982[f]	229.9	7.7	4.3	0.3	*12.3*
1983[f]	232.0	8.0	4.8	0.3	*13.1*
1984[f]	234.4	8.3	5.0	0.3	*13.6*
1985[f]	237.0	9.0	5.2	0.3	*14.5*
1986[f]	239.4	9.0	5.4	0.3	*14.7*
1987[f]	241.5	10.0	5.1	0.3	*15.4*
1988[f]	243.9	9.6	5.1	0.3	*15.0*

Source: Ref. 2.
[a]Resident population for 1909–1929 and civilian resident population for 1930 to date.
[b]Fresh and frozen fish consumption from 1910 to 1928 is estimated. Beginning in 1973, data include consumption of artificially cultivated catfish.
[c]Canned fish consumption for 1910–1920 is estimated. Beginning in 1921, it is based on production reports, packer stocks, and foreign trade statistics for individual years.
[d]Cured fish consumption for 1910–1928 is estimated.
[e]Data for 1909 estimate based on the 1908 census and foreign trade.
[f]Domestic landing data used in calculating these data are preliminary.

Fishery Landings

The 1988 commercial domestic landings, by species, of marine fish and shellfish in the United States is shown in Table 3 (2). Near-shore landings (0–3 mi offshore) totaled 1,762,059 t (value $2,026,147,000), offshore landings (3–200 mi offshore) totaled 2,953,034 t (value $1,671,279,000), and international water catches totaled 259,735 t (value $313,642,000), for a grand total of 4,974,828 t (value $4,010,068,000). Almost 60% of the total fish and shellfish landings by the United States are caught 3–200 mi offshore. Much of the harvest in these waters is in the Fisheries Conservation Zone, which extends fishing jurisdiction beyond that normally considered international waters.

There is a large variation in the value of different species based on the consumer acceptance and the economics of harvest. Two relatively low-cost species, menhaden and pollock, accounted for 48.3% of the total fish and shellfish landed but had a value amounting to only 8.2% of the total. Menhaden is reduced to fish meal for animal consumption, whereas pollock is the basis for the rapidly growing surimi and pollock fillet industry. Conversely, high-value salmon accounted for only 5.5% of the catch but represented 22.7% of the total value, Shellfish also accounted for a relatively small volume of the total landings in the United States (less than 12%) but represented 37.5% of the total value.

It is important to include the production of fish and shellfish from the rapidly growing aquaculture industry when considering total fish and shellfish availability. Although long established in other parts of the world (particularly Asia) aquaculture is a relatively new and rapidly growing industry in the United States. As shown in Table 4, aquaculture-raised fish accounted for 281,800 t (620 million lb) in 1986 (3).

For many years trout was the leading aquaculture finfish grown in the United States. However, catfish became the dominant species in the 1970s and salmon production passed trout in 1985. Catfish farming in the United States has increased from 1,004 t (22.1 million lb) in 1977 to 134,090 t (295 million lb) in 1988 (4). This represents about 53% of the total aquaculture fish and shellfish and 5% of the total combined wild and aquaculture fish and shellfish.

Imports and Exports

Tables 5 and 6 show the imports and exports of fishery products to and from the United States (2). In 1988 1,350,000 t (2,971 million lb) of edible fishery products were imported while only 36% of that amount, 483,550 t (1,064 million lb), were imported.

Harvesting Gear

Static Gear. Although early fishing depended on spear fishing and later was the basis for sophisticated harpoon

Table 3. Commercial Landings of Fish and Shellfish by U.S. Fishing Craft, 1988

Species	Total	
	Metric tons	Thousand dollars
	Fish	
Alewives		
Atlantic and Gulf	2,561	626
Great Lakes	4,856	191
Anchovies	5,636	2,615
Bluefish	7,644	3,012
Bonito	4,178	1,827
Butterfish	2,468	3,407
Cod		
Atlantic	34,506	42,941
Pacific	232,727	68,858
Croaker	4,810	4,596
Cusk	1,072	1,021
Flounders		
Atlantic and Gulf:		
Blackback	8,211	21,533
Fluke	16,334	44,345
Yellowtail	5,041	13,187
Other	11,616	27,022
Pacific	392,612	87,818
Total flounders	*433,814*	*193,905*
Groupers	5,547	21,703
Haddock	2,916	7,030
Hake		
Pacific (whiting)	142,919	15,820
Red	1,739	618
White	5,446	3,639
Halibut	37,017	72,718
Herring, sea		
Atlantic	41,003	5,229
Pacific	59,617	57,431
Jack mackerel	10,227	1,685
Lingcod	2,965	2,385
Mackerel		
Atlantic	12,377	2,722
King	1,954	5,043
Pacific	45,514	7,498
Spanish	1,922	1,479
Menhaden		
Atlantic	360,462	35,162
Gulf	638,721	73,259
Total menhaden	999,183	108,421
Mullets	14,806	11,218
Ocean perch		
Atlantic	1,066	1,467
Pacific	5,406	2,546
Pollock		
Atlantic	14,992	11,071
Alaska	1,396,833	211,354
Rockfish	58,360	39,368
Sablefish	48,817	91,793
Salmon, Pacific		
Chinook or king	20,716	117,551
Chum or keta	66,436	134,689
Pink	80,053	127,297
Red or sockeye	86,199	437,630
Silver or coho	21,539	93,506
Total salmon	*274,943*	*910,673*
Scup or porgy	6,513	9,572
Sea bass		
Black	2,188	5,144
White	49	218
Sea trout		
Gray	9,314	7,948
Spotted	1,403	3,169
White	168	229
Sharks		
Dogfish	4,568	975
Other	6,622	8,454
Snapper		
Red	1,884	9,496
Other	3,083	11,319
Striped bass	185	517
Swordfish	5,814	42,703
Tilefish	2,066	7,222
Tuna		
Albacore	8,643	16,278
Bigeye	2,472	15,149
Bluefin	1,699	17,305
Skipjack	136,077	149,052
Yellowfin	127,140	182,202
Unclassified	354	546
Total tuna	*276,385*	*380,532*
Whiting	16,134	8,621
Other marine finfishes	119,697	74,953
Other freshwater finfishes	13,271	18,534
Total fish	*4,389,185*	*2,505,515*
	Shellfish	
Clams		
Hard	5,616	67,818
Ocean quahog	21,006	14,921
Soft	3,091	18,717
Surf	28,824	29,183
Other	1,225	4,142
Total clams	*58,757*	*134,781*
Crabs		
Blue, hard	99,184	84,357
Dungeness	21,518	54,771
King	9,513	84,153
Snow (tanner)	66,372	137,052
Other	10,082	23,227
Total crabs	*206,669*	*383,560*
Lobsters		
American	22,064	145,236
Spiny	3,250	23,030
Oysters	14,466	78,498
Scallops		
Bay	258	3,414
Calico	5,383	12,462
Sea	13,860	128,243
Shrimp		
New England	3,078	7,497
South Atlantic	10,997	51,667
Gulf	102,416	414,469
Pacific	33,590	32,401
Total shrimp	*150,081*	*506,034*
Squid		
Atlantic	21,200	16,220
Pacific	36,481	7,689

Table 3. Commercial Landings of Fish and Shellfish by U.S. Fishing Craft, 1988 *(continued)*

Species	Total	
	Metric tons	Thousand dollars
Other shellfish	52,174	65,386
Total shellfish	*585,643*	*1,504,553*
Grand total, 1988	*4,974,828*	*4,010,068*

Source: Ref. 2.
Note: Landings are in round (live) weight for all items, except clams, oysters, and scallops, which are reported in weight of meat.

systems, modern commercial fishing methods use either static gear or moving gear when a vessel is used for towing or dragging. There are three basic types of static gear used in the commercial fisheries. The hook-and-line technique is familiar to all sports fishermen. The fish is attracted to the hook by an edible bait or by an attractive device such as a feather. As the fish tries to take the food from the bait or attacks the lure, it becomes attached to the barbed hook and cannot shake the hook from its mouth. A technique called long-lining is the major commercial use of hook-and-line fishing. A series of baited hooks are suspended from a buoyed horizontal line. This use of multiple hooks greatly increases the chances of catching fish and makes the fishing a viable commercial venture.

There are two basic types of trap used to catch fish and shellfish. One is an enclosure placed in the pathway of a moving or migrating animal while the other is a container or pot in which bait is placed. An animal crawls or swims into the tunnel entrance and then cannot escape because it is unable to find its way out of the narrow opening. Traps usually have a one-way wire door that can be easily pushed aside to enter but prevents regression.

Static nets, called gill nets, are a form of static gear closely related to traps. Nets are suspended from a few fathoms up to 50 fathoms into the water from a buoyed line. They are placed in the water where fish are known to be passing. The moving fish hits and penetrates the net. As it realizes that it has hit an obstruction, it tries to pull back. This causes its gills to become entangled in the webbing where the fish is securely held until the fishermen pull the net to the surface. Figures 2–7 show the types of static gear described above (5).

Towed or Dragged Gear. It is often necessary to sweep large areas of the ocean to catch sufficient volumes of fish or shellfish. This is done by trawling, in which large nets are towed through the water, or dredging, in which gear is towed along the bottom.

A trawl net is constructed like a large net bag that has a restricted end called the cod end. As a fishing vessel pulls the funnellike nets, fish are swept into the nets and collected in the bottom or cod end. The vessel must travel at sufficient speed to insure that the fish being collected into the net cannot swim out of the opening. The goal of a trawl operation, lasting from less than an hour to several hours is to fill the cod end. A typical trawl net and a dredge are shown in Figures 8–10 (5).

Encircling Gear. Encircling gear is used for catching large amounts of fish that are schooled or densely concentrated. One end of the net, called a seine net is pulled from the fishing vessel and completes the operation by encir-

Table 4. U.S. Private Aquaculture Production and Value

Species	1980	1983	1984	1985	1986
Baitfish	22,046	22,046	23,598	24,807	25,247
	(44,000)	(44,000)	(47,045)	(51,280)	(51,522)
Catfish	76,849	220,000	239,800	271,357	326,979
	(53,572)	(132,000)	(191,840)	(189,194)	(228,886)
Clams	561	1,689	1,689	1,588	2,506
	(2,295)	(9,500)	(4,178)	(4,717)	(8,307)
Crayfish	23,917	60,000	59,400	64,999	97,500
	(12,951)	(30,000)	(27,700)	(32,500)	(48,750)
Freshwater prawns	300	275	317	267	178
	(1,200)	(1,500)	(1,698)	(1,540)	(893)
Mussels	—	775	917	928	1,206
	—	(1,500)	(1,584)	(1,248)	(1,725)
Oysters	23,755	23,300	24,549	22,473	24,090
	(37,085)	(31,500)	(38,970)	(39,997)	(42,797)
Pacific salmon	7,616	20,600	45,086	84,305	74,398
	(3,400)	(6,800)	(17,252)	(25,439)	(32,751)
Shrimp	—	255	528	440	1,354
	—	(874)	(1,566)	(1,687)	(3,408)
Trout	48,141	48,400	49,940	50,600	51,000
	(37,474)	(50,000)	(54,435)	(55,154)	(55,590)
Other species	—	7,000	9,900	14,000	15,500
	—	(7,000)	(9,900)	(20,000)	(21,700)

Source: Ref. 3.
Note: Production is in 1,000 lb, and value, in parentheses, is in $1,000.

Table 5. U.S. Fishery Products Imports by Principal Items, 1987 and 1988

Item	1987		1988	
	Thousand pounds	Thousand dollars	Thousand pounds	Thousand dollars
	Edible fishery products			
Fresh and frozen				
Whole or eviscerated				
Cod, cusk, haddock, and flounder	105,158	65,198	85,139	52,717
Halibut	9,295	23,138	11,952	25,432
Salmon	41,902	113,008	50,144	155,173
Tuna				
Albacore	201,988	171,988	195,991	202,967
Other[a]	370,517	158,771	354,156	169,331
Other	214,628	183,022	222,696	203,399
Fillets and steaks				
Flounder	73,003	148,734	58,534	119,996
Groundfish	315,418	570,065	253,187	431,126
Other	232,564	396,324	205,988	358,766
Blocks and slabs	403,577	539,358	303,237	382,482
Shrimp	461,173	1,676,844	489,740	1,725,971
Crabmeat	12,571	67,427	10,821	59,639
Lobster				
American (includes fresh-cooked meat)	38,974	178,069	39,732	183,482
Spiny	41,949	397,854	37,806	363,195
Scallops (meats)	39,934	162,273	32,039	115,706
Analogue products with shellfish	30,539	51,197	24,516	41,569
Other fish and shellfish	98,996	169,855	92,128	163,444
Canned				
Herring, not in oil	5,617	8,726	6,541	10,264
Sardines				
In oil	27,352	35,106	22,813	30,824
Not in oil	37,670	25,470	30,546	23,154
Tuna				
In oil	329	869	318	744
Not in oil	211,356	206,051	244,186	297,922
Balls, cakes, and puddings				
Analogue products without shellfish	4,737	6,786	2,574	4,361
Other	8,797	13,130	5,650	8,376
Abalone	2,790	19,867	2,434	22,487
Clams	13,974	15,288	11,268	13,993
Crabmeat	7,967	20,626	7,720	19,622
Lobsters				
American	637	4,184	594	5,342
Spiny	136	748	52	280
Oysters	32,668	36,144	27,524	39,817
Shrimp	17,132	33,380	14,138	28,730
Other fish and shellfish	57,579	89,499	52,350	80,952
Cured				
Pickled or salted				
Cod, haddock, hake, etc	31,893	60,542	31,361	52,665
Herring	22,213	9,594	20,333	8,618
Other	9,991	22,792	11,784	27,718
Other fish and shellfish	16,108	29,306	10,966	29,119
Total edible fishery products	*3,201,132*	*5,711,233*	*2,970,958*	*5,459,383*
	Inedible fishery products			
Meal and scrap	393,730	52,508	265,310	49,567
Fish oils	30,509	18,930	27,688	9,666
Other	—	3,035,026	—	3,353,379
Total inedible fishery products	—	*3,106,464*	—	*3,412,612*
Grand total	—	*8,817,697*	—	*8,871,995*

Source: Ref. 2.

Note: Data include imports into the United States and Puerto Rico and include landings of tuna by foreign vessels at American Samoa. Statistics on imports are the weight of individual products as exported, ie, fillets, steaks, whole, headed, etc.

[a]Includes loins and disks.

Table 6. U.S. Domestic Fishery Products Exports, by Principal, 1987 and 1988

Item	1987		1988	
	Thousand pounds	Thousand dollars	Thousand pounds	Thousand dollars
	Edible fishery products			
Fresh and frozen				
Whole or eviscerated				
Cod, cusk, haddock, and flounder	105,158	65,198	85,139	52,717
Halibut	9,295	23,138	11,952	25,432
Salmon	41,902	113,008	50,144	155,173
Tuna				
Albacore	201,988	171,988	195,991	202,967
Other[c]	370,517	158,771	354,156	169,331
Other	214,628	183,022	222,696	203,399
Fillets and steaks				
Flounder	73,003	148,734	58,534	119,996
Groundfish	315,418	570,065	253,187	431,126
Other	232,564	396,324	205,988	358,766
Blocks and slabs	403,577	539,358	303,237	382,482
Shrimp	461,173	1,676,844	489,740	1,725,971
Crabmeat	12,571	67,427	10,821	59,639
Lobster				
American (includes fresh-cooked meat)	38,974	178,069	39,732	183,482
Spiny	41,949	397,854	37,806	363,195
Scallops (meats)	39,934	162,273	32,039	115,706
Analogue products with shellfish	30,539	51,197	24,516	41,569
Other fish and shellfish	98,996	169,855	92,128	163,444
Canned				
Herring, not in oil	5,617	8,726	6,541	10,264
Sardines				
In oil	27,352	35,106	22,813	30,824
Not in oil	37,670	25,470	30,546	23,154
Tuna				
In oil	329	869	318	744
Not in oil	211,356	206,051	244,186	297,922
Balls, cakes, and puddings				
Analogue products without shellfish	4,737	6,786	2,574	4,361
Other	8,797	13,130	5,650	8,376
Abalone	2,790	19,867	2,434	22,487
Clams	13,974	15,288	11,268	13,993
Crabmeat	7,967	20,626	7,720	19,622
Lobsters				
American	637	4,184	594	5,342
Spiny	136	748	52	280
Oysters	32,668	36,144	27,524	39,817
Shrimp	17,132	33,380	14,138	28,730
Other fish and shellfish	57,579	89,499	52,350	80,952
Cured				
Pickled or salted				
Cod, haddock, hake, etc	31,893	60,542	31,361	52,665
Herring	22,213	9,594	20,333	8,618
Other	9,991	22,792	11,784	27,718
Other fish and shellfish	16,108	29,306	10,966	29,119
Total edible fishery products	*3,201,132*	*5,711,233*	*2,970,958*	*5,459,383*
	Inedible fishery products			
Meal and scrap	393,730	52,508	265,310	49,567
Fish oils	30,509	18,930	27,688	9,666
Other	—	3,035,026	—	3,353,379
Total inedible fishery products	—	*3,106,464*	—	*3,412,612*
Grand total	—	*8,817,697*	—	*8,871,995*

Source: Ref. 2.

Note: Data include imports into the United States and Puerto Rico and include landings of tuna by foreign vessels at American Samoa. Statistics on imports are the weight of individual products as exported, ie, fillets, steaks, whole, headed, etc.

[a]Includes loins and disks.

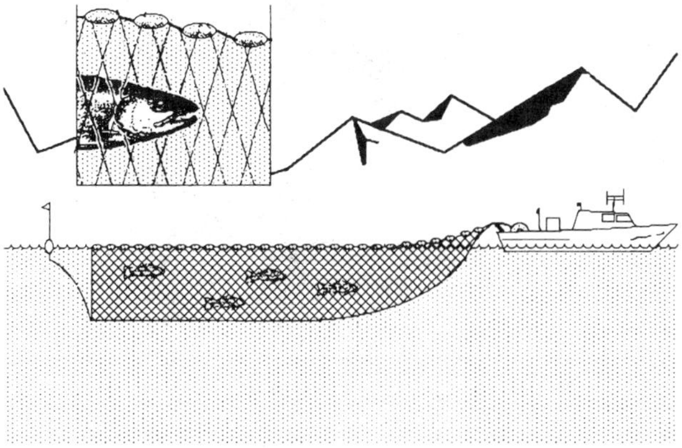

Figure 2. Static fishing gear. Gillnetting—a method of fishing in which fish swim into a suspended net and become entangled by their gills in the webbing. The net can be placed at various depths, depending on the fishery and locale. *Source:* Courtesy of *Seafood Leader*.

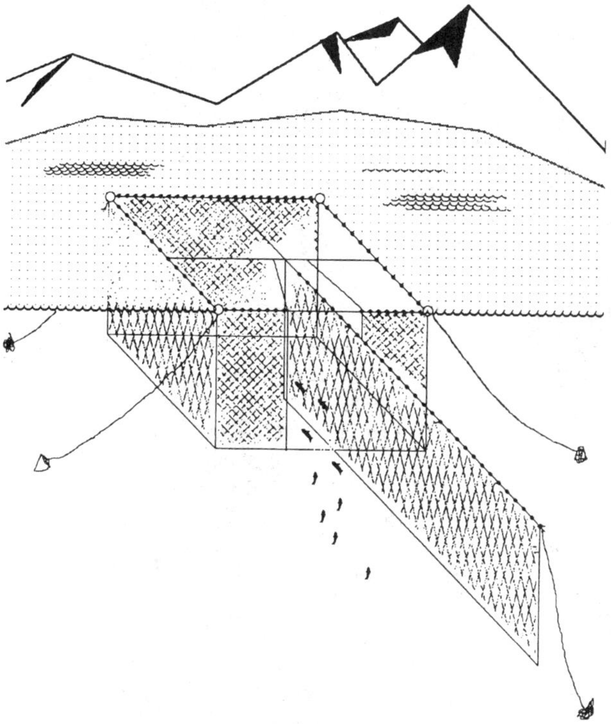

Figure 3. Static fishing gear. Cod traps—set on the ocean floor close to shore, these open-trapped box nets contain a door facing the shore, where seashore feeding cod are deterred by a net fence that directs them to the trap; once inside, they tend to swim in circles. *Source:* Courtesy of *Seafood Leader*.

cling the school of fish and bringing the end back to the vessel. Most seines (Fig. 11) have a line on the bottom that can be pursed to close the net and prevent fish from escaping. Typical fish that can be exploited by this method of harvest include anchovy, pilchard, sardine, salmon, herring, and tuna.

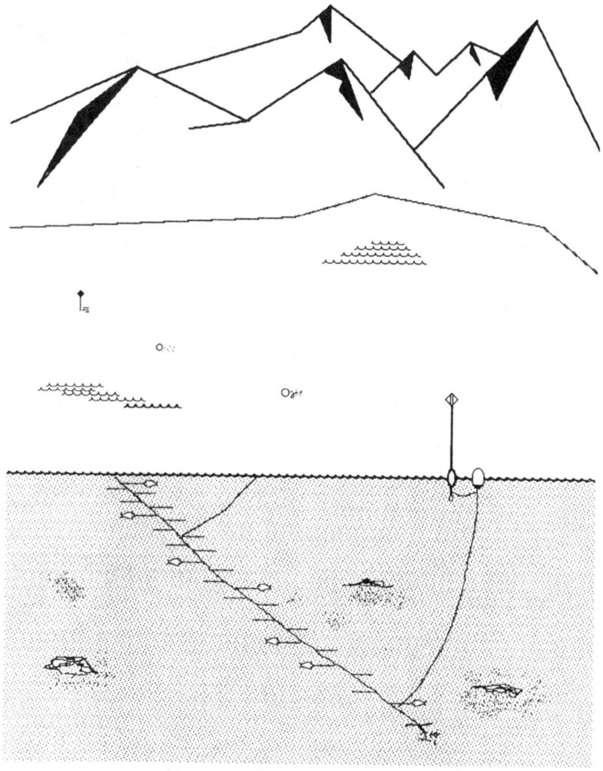

Figure 4. Static fishing gear. Longlining—a method of fishing involving one main line to which a series of shorter lines (gangions) with baited hooks are attached. Used at various depths; ie, surface longlining for pelagic species such as tuna and swordfish, bottom longlining for demersal species as halibut and cod. *Source:* Courtesy of *Seafood Leader*.

Fishing Vessels

The subject of fishing vessel design, construction, and operation is a most complex subject and will only be mentioned here. Fishing vessels range from small outboard motor boats operated by artisan fishermen to large ocean-going vessels, which require a sophisticated crew for efficient operation. The smaller vessels operate close to shore and return with the catch each day.

The length and sophistication of modern near-shore fishing vessels depends on the fishery, the fishing location and distance from shore, the distance that must be covered going to and from the fishing grounds, and the value of the catch. It is common to find vessels 20–50+ ft in length operating in the near-shore fishery and catcher-processor vessels of several hundred feet in length operating offshore on the high seas. The crews number from one to five on near-shore vessels and up to several hundred (including process crew) on the large catcher-processors. Likewise, the cost of fishing vessels varies tremendously, from about \$200,000 for a small near-shore vessel to approximately \$30,000,000 for the large offshore catcher-processors.

MAINTAINING QUALITY OF THE CATCH

Quality is a term that has many definitions, depending on the background and interests of those queried. To some,

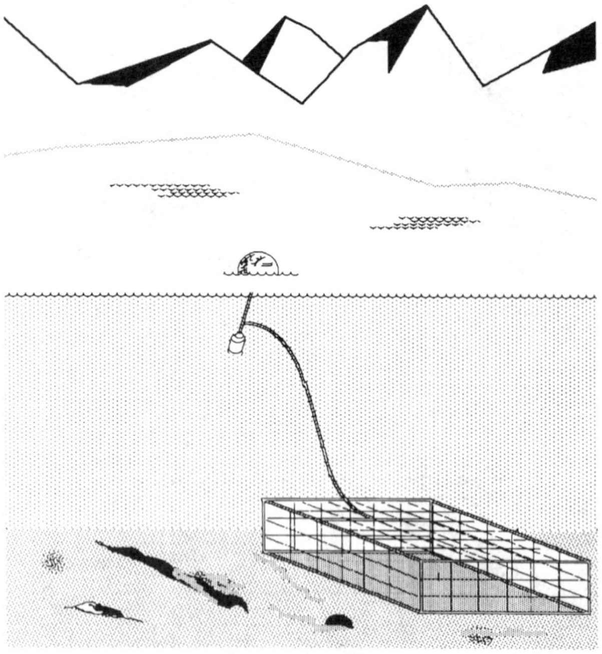

Figure 5. Static fishing gear. Crab traps—framed with iron rods and covered with polyethylene rope webbing, crab trays are usually fished on single lines. Size varies according to fishery; ie, small traps are used for blue crab, large traps are used for king crab. *Source:* Courtesy of *Seafood Leader*.

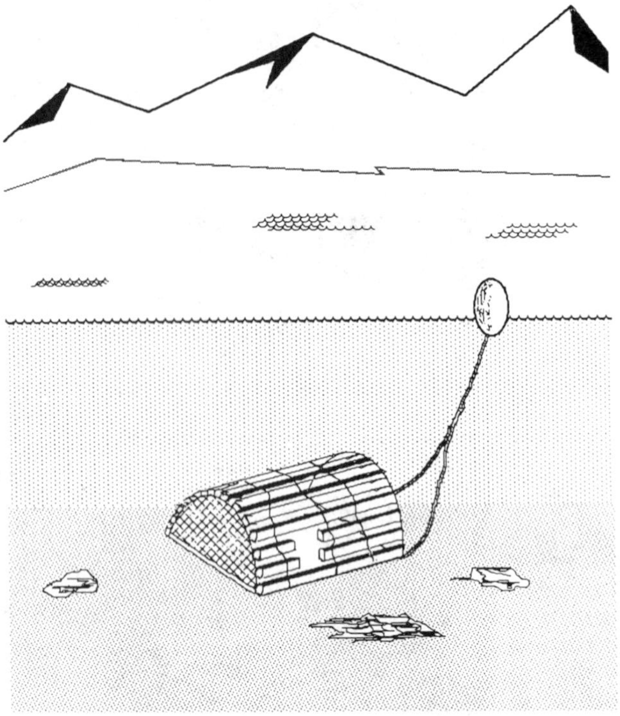

Figure 7. Static fishing gear. Inshore lobster fishing—wooden or wire traps with cotton or nylon twine are set on the ocean floor at various depths, either individually or in strings on a line, baited with either fresh or salted fish. *Source:* Courtesy of *Seafood Leader*.

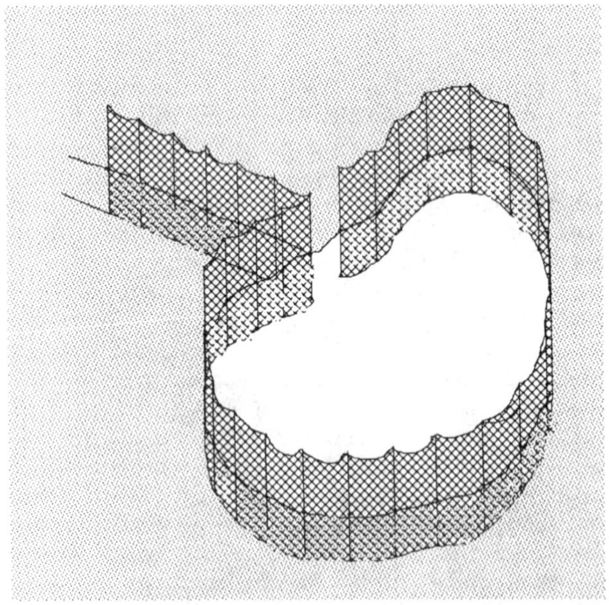

Figure 6. Static fishing gear. Weir fishing—rigid poles are driven into the mud bottom to form a heart-shaped configuration; a straight line of poles, leading from the shoreline to the weir, acts as a barrier directing the fish into the weir. Using skiffs, fishermen first seine the catch, then use a brailer or dip-net to collect them. *Source:* Courtesy of *Seafood Leader*.

the measuring of the product degradation by biological factors such as microorganisms or enzymes is the only means of determining quality. To others the aesthetic values that make a product look good are just as important in defining quality. In reality the biological factors determining safety and nutritional values and the physical factors used in a grading system (ie, size and weight uniformity, color, and blemished surfaces) are integrated to mean quality to most people.

The maintenance of quality or the fresh nature of landed seafood depends on many operations from the catching, landing, and shipboard handling to the transporting, storing, processing, and distributing. A biological specimen can only decrease in quality as it travels through the various steps of a commercial venture. Enzymes and microorganisms cause spoilage and degradation that are irreversible. Physical damage not only affects the appearance of a product, but such damage as skin ruptures allow microorganisms to invade the tissue and cause earlier deterioration. Therefore, all participants in the commercial seafood chain are important for maintaining high-quality products.

The Impact of Fishing Methods

Fishing gear is usually designed to give maximum efficiency in catching fish. The total cost of catching a given amount of fish includes vessel cost and operation, fishing gear cost and maintenance, manpower required, and other

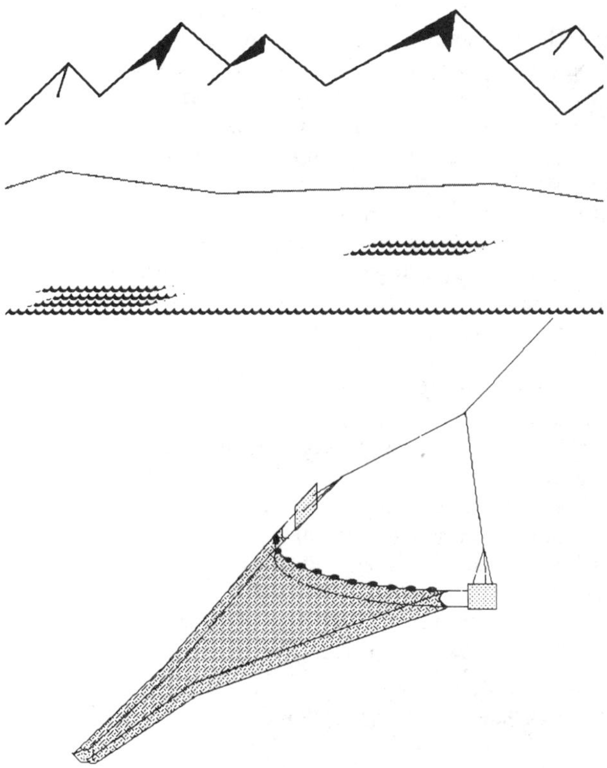

Figure 8. Towed or dragged fishing gear. Otter trawling—a method of fishing in which a large wedge-shaped net is dragged along the ocean bottom; an otter door is attached to each side of the net to hold the net open and keep it horizontal. Fish collect in the cod end (the back) of the net. *Source:* Courtesy of *Seafood Leader*.

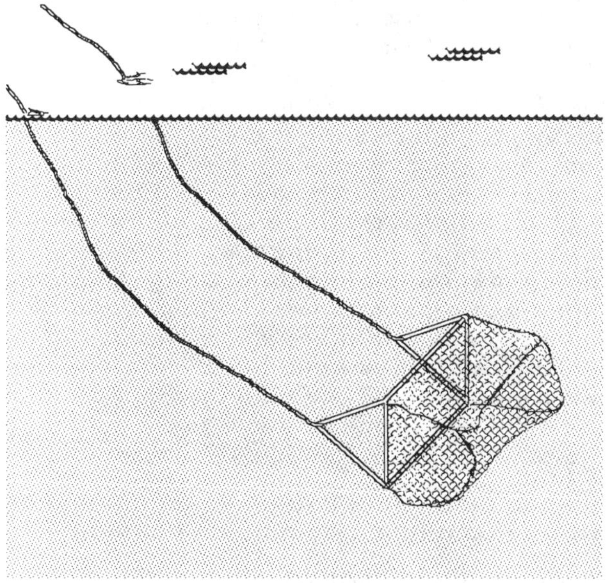

Figure 9. Towed or dragged fishing gear. Scallop dredging—a method of fishing which involves raking a metal frame with teeth and a chainmesh bag across the ocean bottom. *Source:* Courtesy of *Seafood Leader*.

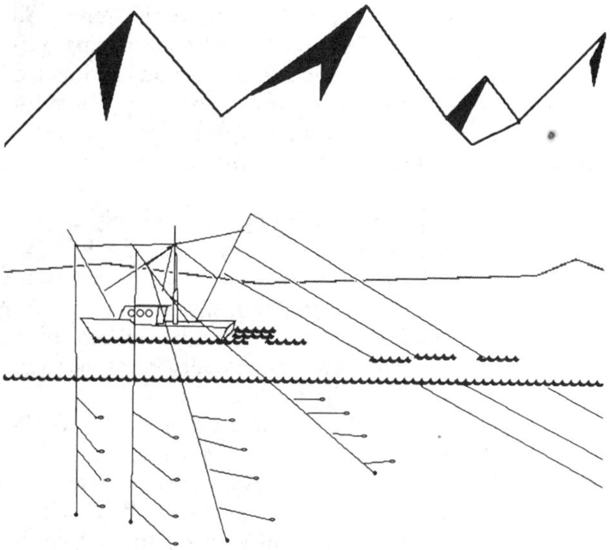

Figure 10. Towed or dragged fishing gear. Trolling—a method of fishing in which several fishing lines with numerous lures are dragged slowing through the water.

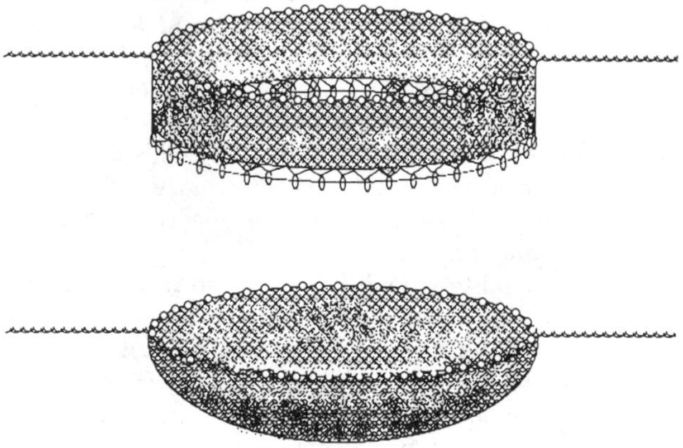

Figure 11. Encircling gear. Purse seining—a method of fishing involving a long, deep net that stands like a fence in the water, supported at the surface by floats and held down by lead lines at the bottom. A person in a skiff takes one end of the net around the school of fish and joins it at the other end, and the vessel hauls in the wire purse line strung through the bottom of the net, forming a purse under the fish. *Source:* Courtesy of *Seafood Leader*.

costs involving machinery and equipment. All factors are combined to give the cost of a given harvest known as the catch per unit effort (CUE). However, the best CUE does not insure good quality in the fish landed, and a processor is somewhat at the mercy of the fishing operation for initial quality of raw materials from the sea.

The type of gear used in fishing has a definite bearing on the quality. When a fish dies after or during vigorous exercise, such as struggling on a hook and line, metabolic activity including that of protein and lipid degrading enzymes, adversely affects the subsequent spoilage rate of the dead animal. If a fish has been feeding and has food in its stomach, the increased metabolic activity greatly accelerates the loss of quality in the slaughtered animal. The so-called soft belly of a fish is caused by this enzyme activity after the fish is dead. Because the amount of time that a hooked fish struggles is directly related to the subsequent spoilage rate, troll-caught fish are often of better quality than longline-caught fish that are allowed to die thrashing in the water.

The quality of gill net-caught fish is extremely variable depending on the length of time that the net is in the water. Hence, there can be a tremendous difference in the quality of fish taken from a given set, the last caught often being of higher quality. This factor is certainly realized by the wholesale buyers in that troll-caught fish consistently command a higher price than gill net-caught fish.

Certain visual or aesthetic factors also affect the market price of fish even though there may not be any real difference in the quality of the flesh. Marks from the web of a gill net are often caused when the fish struggles back and forth to release its entrapped gills. Unless the web causes cuts that allow bacteria to enter the flesh, these marks normally do not adversely affect the biological quality of the flesh. Hence, the word *quality* has different definitions, depending on whether it refers to the biological state of the edible portion or the visual appearance of the fish.

There are several types of fishing gear that cause abrasions and punctures during the catching and subsequent handling. Trawling, dredging, spearing or harpooning, and gillnetting all cause different degrees of damage. Trawling, which accounts for ca 40% of the world's fish catch, exerts extreme pressure on the fish as the cod end becomes full. In addition to scale abrasion, ruptures in the skin and internal portions release gut bacteria and decrease shelf life of the subsequent product.

The best quality is maintained in fish that are caught by trapping. As long as the trapping device is emptied on a reasonable cycle, the fish remains alive and is quickly killed prior to sale or processing. Some species, such as crabs, must be kept alive prior to butchering or there will be a blue color in the meat. This is due to the blood chemistry of crabs; they have a copper complex instead of the heme, or iron, complex found in most animals. If the crab is not butchered live and the blood removed prior to processing, the copper will oxidize, giving a blue color to the meat. Although the aesthetic value of the crab is impaired, the eating quality is not affected. However, as in the case of abraided fish skin, the consumer is not willing to consider blue crabmeat as anything but a low-grade, poor-quality product.

Aquaculture is somewhat akin to catching wild fish in trapping devices. The fish or shellfish are raised in an enclosed area and then removed when ready for market. As in the case of trap-caught fish, farmed fish are live when harvested.

Shipboard Handling of the Catch

The proper handling of fish during harvesting and on shipboard can minimize the adverse effects of gear. Of utmost importance when fish are first landed is that they are segregated and placed in a sanitary chilled environment. Minimizing the bacterial and enzymatic activity by fast reduction of temperature in freshly landed finfish and shellfish is probably the most important step in the entire chain of events that takes a fish from the water to the table.

Figure 12 shows the extreme variation in storage life of fresh and frozen commercial fish prepared for the market. This curve has been compiled from many published sources that give the shelf life of fish as related to the handling methods (6). It has been shown that landed high-quality fish that is chilled rapidly and carefully handled, processed, and packaged can be acceptable for up to three weeks after being caught. On the other end of the scale, a fish that has undergone poor handling on shipboard (eg, 70% of as caught quality) and subsequent marginal handling during the processing and marketing stages is inedible after four days.

Fish that has been properly handled, processed, packaged, and frozen can be held up to one year without significant deterioration. However, there are many factors that must be considered in discussing shelf life of a fresh or frozen product. The species, the oil content, the catching technique, and the state of the fish when harvested are all uncontrollable factors that have a major bearing on the shelf life of a seafood product.

PROCESSING SEAFOOD

Inspecting As-Received Seafoods

Seafoods received in the processing area of a vessel or in a shore-based plant vary tremendously in the state and form. This is the situation when batches of product from different sources or catching vessels are mixed in the received lot. A sensory inspection must be made to insure that the raw material passes the criteria specified by the buyer or processor.

A sensory evaluation utilizes touch, odor, and sight to determine the acceptability of a given lot of seafood (7). An on-site inspection should concentrate on microbial contamination, enzymatic degradation, and other chemical and physical factors that reduce the marketability of the seafood. A faint fresh, nonfishy odor; firm and elastic flesh; bright and full translucent eyes; bright pink gills; and bright and moist skin surface with no heavy deposits of mucus or slime are all properties of a good fresh fish.

In addition to microbial and enzymatic degradation that can take place in improperly handled seafood, the oil in fatty fish that have not been chilled rapidly or adequately protected from the environment are subject to oxidative rancidity. This is both an aesthetic and nutritional problem. Oxidation from the air (and sometimes autooxidation within the seafood) not only affects the odor and taste acceptability of the product but also destroys the omega-3 ($n - 3$) fatty acids in the oil that are so important for human nutrition.

Preprocessing As-Received Seafoods

Preprocessing begins on shipboard where the seafood is, at the minimum, segregated and chilled. Many fish destined for the fresh market are butchered and washed. The minimal butchering operation consists of removing the visceral portion and often the gills. The ultimate in shipboard

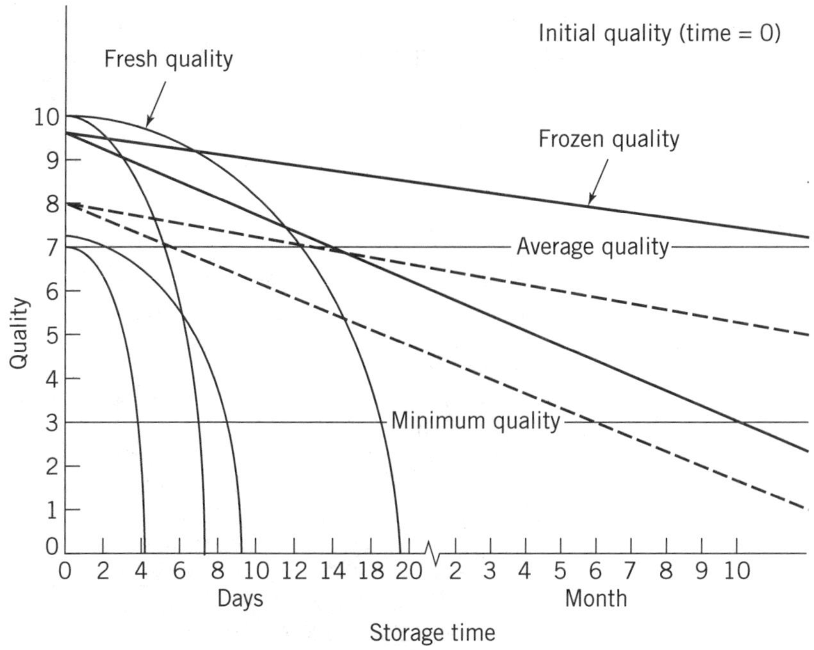

Figure 12. Shelf life of fishery products. *Source:* Ref. 12.

butchering is heading and gutting, which consists of removing the viscera, head, and often the tail and fins. Shrimp are normally iced but sometimes the head is removed on shipboard. Crab are either delivered live directly to the shore plant or are kept alive in seawater tanks until delivery. In all shipboard preprocessing operations the most important procedure is to lower the temperature of the catch as soon as possible after it is taken from the water.

Preprocessing operations not carried out on shipboard are completed in the processing plant. After a second visual inspection, fish are butchered. This primarily consists of eviscerating but also can include scaling, trimming, and further cleaning when necessary for a specific processing operation.

Finfish are portioned for processing or direct marketing by filleting, steaking, or dressing a whole fish or section for roasting or broiling. Depending on the size and sophistication of the processing plant, these operations are carried out either by hand or machine. Crustacea (eg, crayfish, lobster, and crab) and mollusks (eg, clams, oysters, and mussels) are handled and processed quite differently from finfish.

Shrimp are iced on shipboard and unloaded by basket and conveyor into the plant receiving area. In the plant they are segregated and graded as to size by machine, passed over a visual inspection table for removing substandard specimens, cooked, and then headed and peeled (normally by machine). Large prawns are sometimes headed and handled individually on shipboard or in the plant. Those destined for market in an unpealed condition are then frozen. Prawns with head off and not cooked are called headed green prawns in the trade. Shrimp, regardless of whether they are marketed cooked and peeled, green fresh or frozen, or in other forms are sold by the count to designate size. For example 21–25 count shrimp means that there are 21–25 shrimp per lb.

Crab is cooked prior to processing or marketing. It is important that the crab are alive when butchered just before being cooked whole. This is due to the high copper content in the crab blood, which oxidizes to a blue color if allowed to remain in the meat of a dead crab. Crab, depending on the species, are sold in the shell whole (eg, blue and dungeness), segregated with shell on into legs and body portions (eg, king and snow), or as leg and body meat. Meat is removed or shaken from the cracked shell portions by hand.

Bivalves must be alive when purchased and subsequently cooked in a plant, restaurant, or home. A healthy live oyster, clam, or mussel will have a tightly closed shell. Any gapers, eg, those with an open shell, must be discarded. This requirement for handling only live molluscs is important because they normally are not iced after being harvested by digging, picking, dredging, or tonging, but are delivered directly to the receiving station or plant.

Problems involving toxins (eg, paralytic shellfish poisoning) or communicable diseases being transmitted to the consumer by mollusks have become increasingly more prevalent as the coastal waters become more polluted (8). These animals are static so they are particularly vulnerable to any fluctuating environmental pollution problem that may exist. Furthermore, the conditions causing the meat to be inedible cannot be detected by simple in-plant inspection. This has resulted in an intricate system involving surveillance of mollusk-growing areas by federal and local government agencies who are responsible for closing harvest areas when there is a potential problem.

Total Utilization

There is a growing emphasis on improving the total utilization of seafood raw material (8,9). For many years, only the most desirable portion of the fish, often accounting for 20 or 30% (eg, fillets) of the fish, was used. The remainder was considered waste or a raw material for preparing cheap animal feeds. Environmental and economic considerations dictate that this gross misuse of base raw material must be stopped. Hence, the modern attitude is that there is no such thing as fish waste. The portions remaining after the initial edible portion is removed should be considered secondary raw materials that can often equal or exceed the amount of the primary edible portion. The initial preprocessing operation must consider the ultimate total utilization destination of all portions of the raw material. Some of the products that can be prepared or manufactured from secondary raw materials are outlined in Table 7.

The developing operations and markets that use minced flesh for human consumption are beginning to have an impact on the economics of operating and the market for seafood products. Minced fish is used in engineered and formulated foods, the fastest growing segment of the food industry. Sources of minced flesh include

1. Frames (remaining skeleton) from a fillet operation.
2. Industrial fish presently being used for meal and oil.
3. Small fish currently being discarded or converted into fish meal that cannot be economically filleted or otherwise processed.
4. Freshwater fish presently being underutilized.
5. Low-fat fish that do not have good keeping properties due to rapid enzyme action (eg, pollock).

Table 7. Total Utilization of Secondary Raw Materials from the Sea

	Product	
Raw material	Edible	Industrial
Finfish[a]	Steaks Roasts Fillets Minced flesh, surimi, formed foods, and extracted protein Specialty foods Refined oil Roe and milt Food additives	Meal, oil, pet foods, pharmaceutical raw materials, and leather (skins)
Industrial fish[a]	Minced flesh, surimi, formed foods, and extracted protein Specialty foods Refined oil Roe and milt Food additives	Meal, oil, pet foods, pharmaceutical raw materials, and leather (skins)
Shellfish	Whole (as caught) Portions Whole meat Minced meat Specialty products Food additives	Crustacea shell, extracted protein, chitin products, and calcium salts Bivalve shell land fill Meal and oil
Seaweed	Food Food additives	Pharmaceutical raw materials Extracted products for industry

Source: Ref. 8.

[a]Process residues are an important source of recyled high-protein ingredients for on-site preparation of aquaculture-raised fish and shellfish feed.

Minced flesh can be used in many products that require a protein base or binder. Such items include sausage (much lower saturated fat and calories than those made from pork), wieners, other cased meats, extruded meat products, nutrition-controlled foods (eg, low fat, high protein, and low sugar or carbohydrate), and other foods requiring ingredients with highly functional properties.

In the past, fish oils have been by-products from the production of fish meal. This low-grade oil has been used for industrial purposes or for making margarine. The large amount made into margarine has been manufactured in foreign countries because fish oil is not allowed for this purpose in the United States. The recent interest in fish oils for health has created a challenge for researchers to develop satisfactory refining techniques and subsequent edible products from the refined oils.

Some 30 yr ago it was recognized that fish oil has beneficial fatty acids that are active in preventing or minimizing the effects of certain cardiac problems and other diseases. However, it was not until the early 1980s that highly publicized work demonstrated that diets of Greenland Eskimos, high in marine fish and mammals, were associated with greatly reduced numbers of deaths due to ischemic heart disease when compared to populations in more developed countries consuming low fish and high animal fat diets (10). Fish oil, through the consumption of more fish and formulated foods containing fish oil are now considered a valuable contribution to a more healthful diet (11–13).

Heat Processing

Methods of Heating. Seafood, like any other food is cooked or heated to make it taste good by changing the texture and bringing out flavors and odors. If sufficient heat is added to pasteurize the food, microorganisms that can cause public health diseases and illnesses are destroyed or inactivated. Enzymes are also inactivated by heat so that protected cooked food stored under refrigeration is subjected to a minimum of degradation by hydrolysis.

The ultimate in heat processing is sterilization whereby the product is heated for a sufficient time at a given temperature. If the product is hermetically sealed so that there is no postprocessing contamination, it will have an indefinite shelf life free from degradation by microorganisms and enzymes.

Seafood is cooked and pasteurized by standard radiant energy baking ovens, infrared heating ovens, and microwave ovens. The heating in any oven is due to a combination of conduction, convection, and radiation. Food in a standard oven receives conduction heating from the pan or oven rack, convection heating from the air that is heated by the walls of the oven, and radiant heating from the exposed heating units. Infrared ovens do not have open heating elements because there is little convection heating from the air being heated by conduction.

Microwave heating is a specialized form of dielectric radiant heating that has certain advantages over other dielectric methods. Because the molecular polarizations are reversed many millions of times per second, the friction of molecules contacting each other cause heat and subsequent rapid, uniform heating of a food. Microwave heating has risen rapidly over the past decade. There are now microwave ovens in about 80% of U.S. households. Restaurants and institutional kitchens are rapidly increasing the use of this fast way of heating and cooking foods. The microwave oven with additional convection heating is becoming popular in that it combines the advantages of fast, uniform heating with the advantages of a conventional convection oven.

The growing popularity of microwave ovens has greatly increased the demand for microwavable foods. The major challenge to the food scientist developing microwavable foods is lack of radiant heat that causes the surface of a food to brown. Hence, cooked nondeep-fried potatoes, chicken, white fish, etc remain white on the surface and do not take on the normally expected desirable browning on the surface. Microwavable batter and breadings have been developed that are the color of deep-fried products. Hence, the future of microwaved foods will indeed be insured if low-fat, consumer-acceptable microwavable batter and breaded foods can be developed that take the place of deep-fried products. Fish fillets and formed patties from minced fish base stand to benefit as much from this development as any other food product. Fish sandwiches and batter and breaded fish have been growing in popularity, especially in fast-food restaurants. However, as shown in Table 8, the desirable omega-3 fatty acids suffer and the fat calories are tremendously increased when a fish is deep fried (9).

Commercial sterilization is carried out in steam retorts, which can process at temperatures above that of boiling water, normally at or above 117°C (242°F). The time and temperature required for sterilization must be sufficient to kill *Clostridium botulinum* spores that, when viable, can grow in an anaerobic (nonoxygen) atmosphere and produce lethal toxins. These spores must be held for 32 min at 110°C (230°F) to insure total destruction. Many low pH (high acidity) products such as certain fruits, vinegar-packed foods, and highly acid formulated foods prevent *C. botulinum* spores from growing. Many of these foods do not have to be sterilized at the temperatures required for high pH (low acidity). However, near neutral or high pH vacuum-packed products, such as seafood, are particularly vulnerable to anaerobic spore growth and sterilization must be insured.

To ensure a margin of safety, the sterilization requirement for canned fish is that the geometric center of a can or pouch must be held for 32 min at 116°C (240°F) (14). Each food and each different geometric form of container requires a different total processing time to accomplish sterilization. These processing times, determined by thermal death time laboratory studies, are mandatory for each processing company that is canning hermetically sealed food. Each batch of canned food must be coded and retort processing records kept to prove that the product was sterilized.

It should be emphasized that anaerobic conditions often prevail in a canned food even though an incomplete or no vacuum is drawn on the can prior to sealing. This is a result of the subsequent oxidation of components in a food by the remaining oxygen in the air.

Another precaution that must be practiced by processors is the venting of a steam retort prior to beginning the official retorting time. This is to prevent air pockets from insulating some of the cans so that there is nonuniform temperature in the retort. Of course, hydrostatic retorts that heat containers in a column of water kept above normal atmospheric boiling temperature by hydrostatic water pressure do not have a problem with entrapped air.

Effects of Heating. Fresh or frozen seafood being cooked for a meal should be heated for the minimum time required to improve the texture and the taste for the consumer. The normally dangerous microorganisms, those known as public health disease organisms are destroyed at relatively low temperatures. As shown in Table 9, the most heat resistant of the group, thermophiles, have an optimum growth at 50–60°C (122–150°F). Hence, a seafood is normally safe to eat if the geometric center has been raised above 150°F when it is actually pasteurized.

Table 8. Effect of Processing on Omega-3 ($n-3$) Fatty Acids[a]

Species	Product	Lipid, %	Fatty acid in oil EPA, %	DHA, %
Cod	Fresh	0.29	18.67	27.98
	Batter and breaded	0.47	12.53	12.25
	Batter and breaded, deep-fried in liquid vegetable oil	5.53	0.03	0.09
	Batter and breaded, deep-fried in solid vegetable fat	9.14	0.21	0.65
	Batter and breaded, deep-fried in beef shortening	7.07	0.66	1.51
Salmon, Sockeye	Fresh	—	6.25	9.73
	Canned	—	5.0	5.35
	Canned (added salmon oil)	—	5.76	8.61

Source: Ref. 9.

Table 9. Optimum Growth Temperature Range for Bacterial Group

Bacterial group	Optimum temperature range
Psychotrophs	14–20°C (58–68°F)
Mesophiles	30–37°C (86–98°F)
Faculative thermophiles	38–46°C (100–115°F)
Thermophiles	50–66°C (122–150°F)

Source: Ref. 14.

Overcooking causes heat degradation of nutrients, oxidation of vitamins and oils, and leaching of water-soluble minerals and proteins. The retention of B vitamins, zinc, and iron is particularly important for populations that consume fish as the major source of meat in their diet. In addition, overcooking causes too much water to be released and the drying effect causes flesh to become tough, thus nullifying the desired effect of texture improvement.

Refrigeration and Freezing Technology

As has already been stressed, the most important factor in handling fresh fish is to lower the temperature to just above freezing as soon as it is removed from the water. Because the condition of the harvested fish and the subsequent handling determines the shelf life of a fresh fish, it is not possible to state exact times that a seafood can be held in ice or under refrigeration and be considered a high-quality food. This is shown in the wide range of shelf life that has been published in the literature (Fig. 12). In general, fish with a high oil content or enzyme activity have greatly reduced shelf life and are often of marginal quality after a few days.

High-quality seafood, when frozen properly soon after being removed from the water, is often far superior to fresh fish available on the market. This is due to the fact that all microbial action is stopped and enzyme action is significantly reduced in frozen fish. However, the initial quality of the fish being frozen, the rate of freezing, the temperature at which the frozen seafood is held, and the uniformity of the freezing temperature are all important to maintaining a high-quality product. It is surprising how many people involved in the seafood chain do not understand some of these basic factors in insuring the high quality of a seafood. Thus the constant challenge of those involved in seafood technology is the continual education and reeducation of everyone involved in the commercial seafood chain.

Freezing Seafood. During the freezing of seafood, structural changes take place in the cells and cell walls as well as in components that are between the cells. Many of these adverse changes are caused by water crystals that expand and rupture the cell walls. This allows liquid within the cells to leak out when the flesh is thawed. Hence, free liquid, called drip, exudes from seafood when it is thawed. This loss of free water reduces the water content of the seafood causing economic loss to the seller and greatly reduces the fresh qualities of the thawed product.

As seafood is cooled above and below approximately 28°F in a constant-temperature environment there is a near linear relationship between the temperature decrease and the time. However, as the water in the flesh begins to freeze, there is a long period of time during which the temperature remains almost constant. This period is a critical range for freezing and is caused by heat (heat of fusion) being removed from the fish to freeze water rather than to lower the temperature of the flesh. This relationship is shown in Figure 13.

The longer a product remains in the critical zone, the larger the ice crystals formed in the cells will be. When water is frozen rapidly, small crystals form and do not have time to increase in size due to the nucleation characteristics of a water molecule. Experience has shown that if seafood passes through the critical zone in ca 1.5 h, there will be little damage to cell structure as a result of large crystals formed during freezing. This is demonstrated by microphotographs of cells frozen under different conditions. Figure 14 showns cells (magnified 640 ×) of a rainbow trout that was rapidly frozen and passed through the critical zone in less than 1 h (15). Note the well-defined intact cell structure is similar to that of high-quality fresh flesh. Figure 15 is a similar microphotograph of cells from the other half of the same rainbow trout, which took approximately 5 h to pass through the critical zone. In this case extreme damage to the cell structure caused by large ice crystals can be noted. This is a dramatic demonstration of why fast freezing of seafood is essential to maintaining the quality of frozen products.

There is wide variation in commercial freezing facilities available to the seafood industry, and it is most important that commercial operators carefully study their options when purchasing and installing new equipment. The most efficient methods of freezing involve both conduction and convection freezing in an immersion system. Liquid refrigerants such as the freons have been used for this purpose

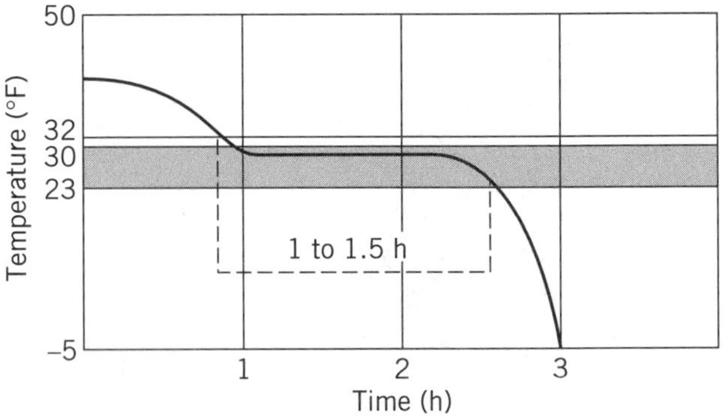

Figure 13. Idealized freezing curve for fish muscle.

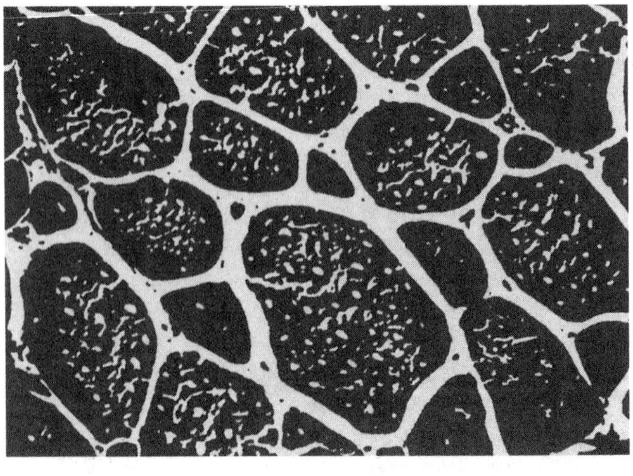

Figure 14. Cell structure of rainbow trout frozen rapidly (×640). *Source:* Ref. 15.

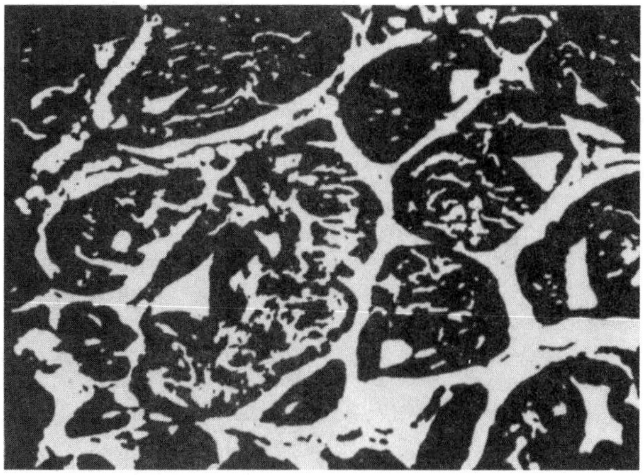

Figure 15. Cell structure of rainbow trout frozen slowly (×640). *Source:* Ref. 15.

but have proven too costly. Immersion freezing using refrigerated brine is fast and efficient but a considerable amount of salt is absorbed into the flesh during freezing and can be a problem to product marketing. Refrigerated seawater is used extensively, especially in Alaska for rapidly chilling and holding the fresh catch on shipboard.

Plate freezers rapidly freeze, by conduction, products that are in contact with the plates. Plate freezers are used to their best advantage when the plates can be brought together to contact both sides of a package. This is the case for rectangular packaged seafood portions or prepared formulations. A disadvantage is that irregular items, particularly large fish, can only contact the plate on one side and the exposed side must be frozen by convection. Another disadvantage is using plate freezers for whole fish, because the plate only contacts one side of the large irregular shaped item and causes one side of the finished product to be flat while the other side is the nice rounded shape of the live fish. However, unless the fish are suspended by hooks, this same condition prevails in any freezer where the fish are placed on shelves during freezing.

Convection blast freezing, where cold air is circulated over the products, is less efficient than conduction freezing but can be used for larger volumes. A disadvantage to blast freezing is that the rapidly moving dry air can dehydrate a product that is exposed for any length of time. This can be eliminated by proper packaging protection, but it also greatly decreases the heat transfer rate and lengthens the time of freezing and the subsequent time at which the product is in the critical range.

The problems with blast freezing can often be minimized or eliminated when a combined blast-plate-freezing system is used. This can be demonstrated by comparing Figures 16 and 17 in which chum salmon weighing 4 lb were frozen in two commercial units (16). Figure 16 is the freezing curve for a conventional blast freezer in which whole (headed and gutted) fish were placed on racks and frozen. Note that the time in the critical range was about 4 h and the total time to reduce the temperature to 0°F was about 8 h. The freezing curve shown in Figure 17 was that of a 4-lb chum salmon in a combined blast and plate freezer. The time in the critical range for this system was about 40 min, and the total freezing time to reach 0°F was slightly over 1.5 h. The fish frozen in the combined blast-plate freezer were far superior to those frozen in the blast freezer alone. Furthermore, the reduction in weight due to drip loss gave a significant economic advantage to the more rapidly frozen fish.

Holding Seafood in Cold Storage. Good freezing practices involving fast freezing and minimal dehydration of fresh seafood insures that high-quality products are delivered to the cold storage for holding. Equally important to the continuing maintenance of high quality is the environment under which the products are kept during the cold storage period. Unless frozen seafood is kept at extremely low temperatures, there is a certain amount of free water (not frozen) remaining in the product. This is a result of the antifreeze effect of soluble salts in the cell, which become more concentrated as water freezes. Depending on the specific food, the physical and chemical conditions of the food, and the composition (including water content), the point at which all of the water is frozen is in the range of −45°F. Because few, if any commercial cold storages are held at that low temperature, there is a certain amount of free water remaining in all frozen food (9). This is depicted in Figure 18, indicating that the normal commercial cold storage temperatures are well above −45°F. In fact, different commercial cold storage warehouses in the United States range in temperature from slightly above 0°F to a low that seldom is below −20°F.

Not only is the average commercial cold storage facility maintained at a temperature at which several percent of the water remain unfrozen but there is normally a significant fluctuation in the base cold storage temperature. Thus when the temperature fluctuates above and below the average, some of the water in the product is continually frozen, thawed, and refrozen. The effect of this fluctuation is to greatly increase the effect of enzyme action that essentially digests the protein in the same manner that it is digested in the gut of an animal. This is emphasized by the fact that continual thawing and refreezing is not limited to the original free water and causes the enzyme action to spread throughout the flesh. This is why the meat in seafood held for long periods can be extremely soft when thawed. This effect is even more noticeable in many home freezer units where temperature fluctuations are greater than those found in commercial facilities. Hence, holding a product at a higher but constant cold storage temperature can result in a better product than when it is held at a fluctuating lower temperature. This concept is depicted in Figure 19.

It can be important to know how much fish a given size cold storage unit will hold. The true density of a seafood ranges from 70 to 80 lb/ft^2. Because frozen fish blocks are essentially composed of solid fish flesh, they have about the same density as the natural flesh. At the other end of the spectrum, individually frozen and loosely packed fish range from 30 to 50 lb/ft^2. A well-run cold storage, allowing for the average distribution of product forms and allowing for air spaces and movement within the room, usually holds about 20–30 lb/ft^2.

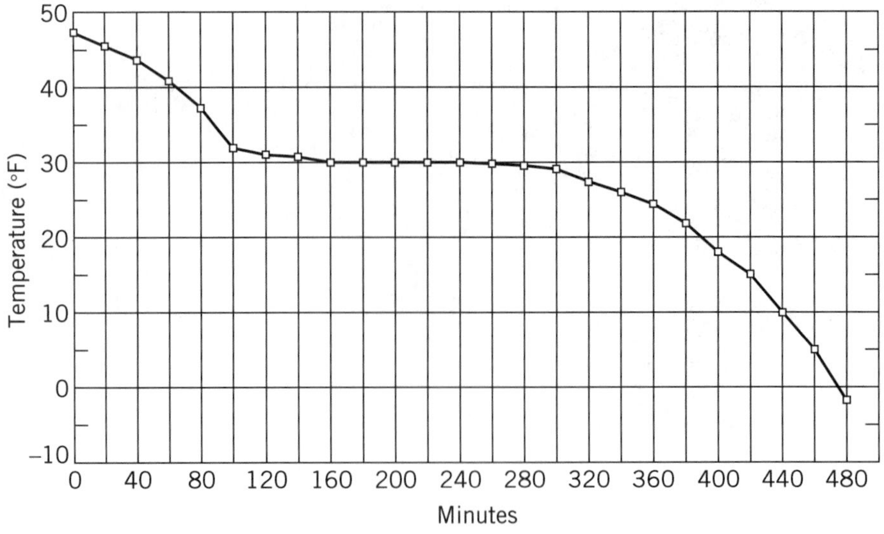

Figure 16. Freezing curve for blast freezing of chum salmon (4 lb).

Refreezing Seafood. Often it is necessary for a processing plant to receive frozen fish (either whole, butchered, or partially processed portions) for subsequent final processing or reprocessing. This occurs when fish are frozen at sea or other remote areas where total processing is not practical due to limited facilities or economic considerations. A typical example is when fish blocks are shipped to a plant for thawing, trimming, battering and breading, and refreezing. A similar situation occurs when frozen fish or preprocessed raw materials are held in cold storage until they are processed in response to a specific market demand. For example, if fresh salmon are cut into steaks and frozen, there is no opportunity to sell the fish in the fillet form if the demand or price of fillets takes a sudden increase.

It is often said that seafood cannot be thawed and refrozen. This is based on the too-often encountered situation whereby the fresh fish has been abused, the products have not been frozen rapidly, or have been held at high or fluctuating cold storage conditions. However, if the fish have been properly frozen and held in cold storage as described above, there will be little cell damage and the thawed product will have minimum water loss and reduction in quality from that of the original fresh raw material. Thus it can be thawed, reprocessed, and refrozen without significantly altering the quality of the finished product from that of the fresh fish.

The increase of high-seas processing vessels encouraged by the 200-mi limit (Fishery Conservation Zone) has greatly accelerated the final processing of frozen products prepared at sea. The future will probably bring an even greater acceleration of this trend as such products as batter and breaded (nondeep-fried) portions, whole fish, and products engineered and formulated from minced flesh or chunk portions are being developed for microwave cooking. These are inexpensive low-fat products with good sensory properties especially acceptable to the increasingly large number of nutrition-conscious consumers.

Commercial Refrigeration Systems

There is a distinct difference between selecting facilities for freezing or for cold storage for a commercial seafood operation.

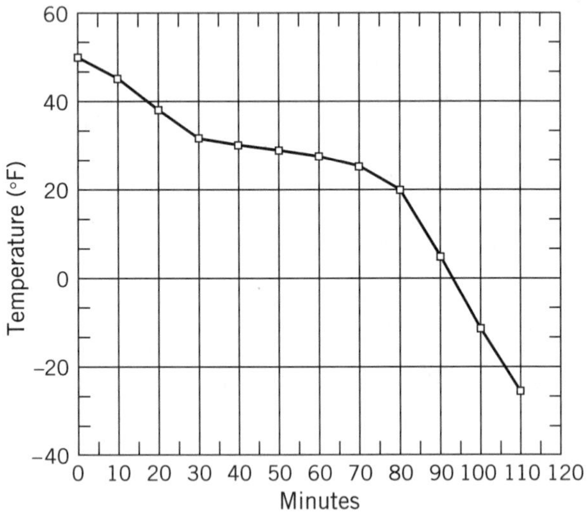

Figure 17. Freezing curve for blast-plate freezing of chum salmon (4 lb).

Freezing Facilities. The freezing facility is related to the products that will be processed, the throughput of product, and the space availability in the processing area. The differences between various freezing techniques lie in the control of the type of heat transfer between the product and the refrigerant. The types of freezer as related to the products being frozen are as follows.

Natural Convention Freezing. This facility is a room or chamber in which the product is frozen by natural convection. There are minimum problems with dehydration but the freezing rate is slow. Only products not affected by slow freezing should be frozen in this type of freezer. These would include formulated products or very small items where the water loss or cell damage through slow freezing is not a problem.

Combined Conduction and Natural Convection. The addition of freezer plates on which the product is placed

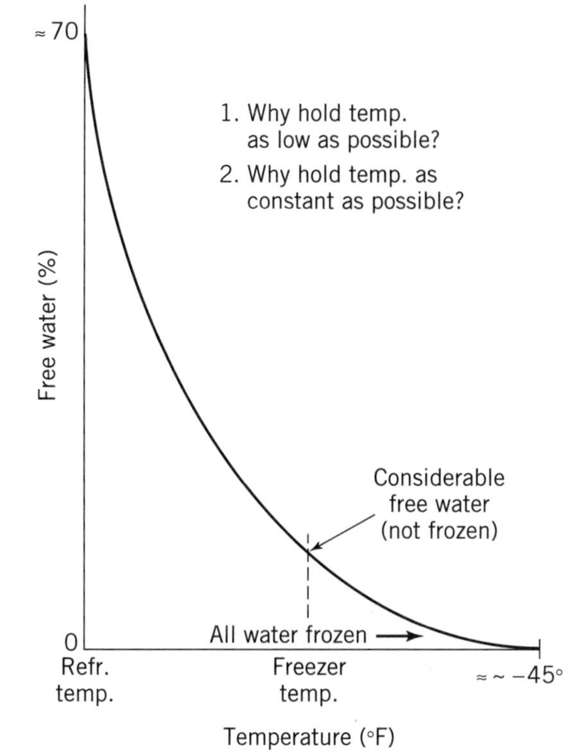

Figure 18. Remaining free water in frozen food held at different temperatures.

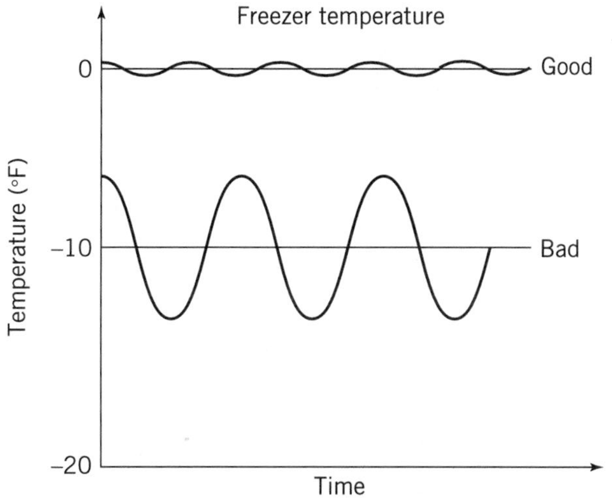

Figure 19. Time-temperature relationship of food during cold storage holding.

greatly accelerates the freezing rate and maintains the advantage of minimum hydration during freezing.

Blast Convection Freezing. Blast freezing is a popular method of freezing irregularly shaped seafood items such as whole or partially dressed fish. A major disadvantage of this facility is that considerable loss of water through hydration can cause an unsightly surface condition known as freezer burn. This hydration, if allowed to continue, can remove a considerable amount of water from the flesh, making the fish inedible. Such is the case when a product is frozen and then allowed to remain (actually to be stored) in the freezer for some length of time before being removed.

When unpackaged products are destined for storage in a cold room with fast moving air, they are glazed to protect against dehydration. This consists of immersing the frozen product in cold water and then allowing the water film to freeze on the surface. The layer of ice protects the fish, because the air removes water from the glaze rather than from the product.

Combined Conduction and Blast Convection Freezing. This facility normally consists of refrigerated plates in a forced-air cold room. This freezer has the advantage of fast conduction freezing assisted by convection air on portions of the seafood that are not in contact with the plate surface. The tremendous increase in freezing rate by this combination of freezing techniques as compared to that of a blast freezer were previously discussed and compared in Figures 16 and 17. Normally the freezing rate is high enough to eliminate the problem of surface dehydration.

Conduction Freezing. When small rectangular products or packages with two parallel surfaces are placed between two refrigerated plates in a cold room or chamber, fast freezing takes place with minimum harm to the products. These types of facility have mechanisms that allow the plates to be vertically adjusted. This allows the plates to open, or separate, during loading and then they come in contact with the product for freezing.

Immersion and Cryogenic Freezing. Immersion in a cold liquid that will not affect the safety of the product (eg, brine, liquid nitrogen, or freon) is the fastest method of freezing. Freezing takes place by both convection from the circulating fluid and conduction from being in direct contact with the liquid. Immersion freezing in extremely cold refrigerants such as nitrogen or freon is called cryogenic freezing. Brine freezing utilizes saturated salt solutions that freeze rapidly due to the combined convection and conduction but takes place at about 0°F.

Cryogenic freezing has two limiting factors for extensive use in freezing seafood products. The first is that the freezing is so rapid that it causes extreme internal tension due to the freezing of the fiberous materials at a differential rate. This causes the flesh to rupture. It can be minimized by allowing the product to temper at room temperature before further handling, packaging, and placing in cold storage. The splitting problem can be fairly well overcome for small items such as fillets and steaks but large fish cannot be satisfactorily frozen by this means. Brine, on the other hand does not have the splitting problem and is a major means of freezing whole tuna fish on high-seas catcher vessels.

The second problem with cryogenic freezing is that it is uneconomical to operate unless the freezer is used continuously, because a considerable amount of heat from the cryogen is lost each time a processing unit is shut down and restarted. Most seafood-processing plants do not operate on an extended basis so that the cost per pound for intermittent freezing limits cryogenic freezing to a few fish processing plants.

Modern cryogenic freezing no longer uses liquid immersion true freezing. It is more economical to spray the liquid refrigerant on the food as it moves along on a conveyor belt. This uses less refrigerant and still gives the same advantages as immersion. Furthermore, when extremely cold liquids are used for freezing, especially in immersion freezing, vapor bubbles are formed that insulate the liquid from the product and prevent the rapid heat transfer expected from such a large temperature difference between the product and the liquid.

Cold Storage Facilities. Cold storage facilities must meet the requirements for long-term storage of commercial seafood products ranging from large whole glazed fish to cases of packaged retail products. Of particular importance in purchasing or contracting for such facilities in a plant or choosing a public cold storage for use include the following:

1. A design with proper insulation and construction materials that will insure a minimum of heat loss through the walls, ceiling, and floor.
2. Properly designed protection to minimize heat loss through doors and other openings while the product is being taken into or out of the cold storage room.
3. Refrigeration machinery that will have sufficient capacity to hold the cold storage at the desired temperature.
4. Sufficient refrigeration capacity to insure that the temperature in the cold storage rooms does not fluctuate to the extent of adversely affecting the products being stored. This is probably the major problem encountered with commercial cold storage facilities. So often a low bid is awarded to a contractor who is the lowest due to cutting back on the amount of refrigeration machinery and thus the ability to maintain constant storage temperature during variable and seasonal outside weather conditions.

Cured and Dried Seafood Products

Control of water activity, defined as the ratio of the vapor pressure of water in a product at any given temperature divided by the vapor pressure of pure water at the same temperature, not only applies to smoking and drying but also to salting, pickling, and product formulation. Product stability, the growth of microorganisms, and chemical reactions occurring during processing and storage are all dependent on the water activity, that is, the ability of water to move and interact with other ingredients in the food. The importance of water activity (a_w) in foods is shown in Table 10. The many facets of water activity and its relation to the preservation, safety, and shelf life of foods has been summarized by a group of internationally recognized experts (17).

Dehydration. Drying or dehydration is a means of controlling water activity by reducing the water content of a product. Many dried products, such as cereal grains, legumes, and many nuts and fruits, are dried by nature in the field prior to harvesting. In the past humans dried seafood products in the sun, long before they were aware that they were controlling water activity. In fact, today many developing countries located in tropical parts of the world use the sun as a major means of drying seafood and other food products.

The principal cured fishery products produced in the United States are shown in Table 11 (2). It should be noted that these statistics do not differentiate between smoked, salted, dried, and pickled products due to the fact that all of the processes are based on controlling or reducing water content. It is often difficult to distinguish between process classifications. For example, salted and salted-dried fish have a different final moisture content but the mechanism of removing the water and stabilizing the product are the same, control of water activity.

The most efficient means of drying a seafood product is through dehydration by forced-air drying, vacuum drying, or vacuum freeze-drying. In each case the drying mechanism is a combination of adding heat to increase the temperature and vapor-driving forces between the product and the environment. The drying time is divided into two distinct periods: constant drying rate and falling drying rate. During the constant rate period, all of the heat added to the product is used to evaporate water from the surface and near surface of the product. In this case, there is free water in contact with the environment and drying occurs similarly to that of an open container of water. During the falling rate period, part of the heat energy is imparted to the product to cause water to migrate to the surface. Therefore, the product is heated during this period of the drying cycle.

Excellent highly nutritious dried formulated fish-base products can be prepared from the minced flesh of seafood. Shaped into forms such as patties and air dried, these items have a long and stable shelf life (18).

Curing. Whereas dehydration removes sufficient water to inhibit growth of microorganisms, curing consists of adding sufficient chemicals (eg, sodium chloride, sugars, and acetic acid) to prevent degradation of a product by microorganisms. Although sufficient water is not removed to accomplish this objective, the water activity is reduced to the point where growth is prevented.

Curing methods currently practiced include dry salting, where split fish is covered with salt and the brine liquor is allowed to escape, and pickling where products are immersed in a strong brine, or pickle, allowing salt to penetrate the product and water to be exuded into the brine solution. Low-fat white fish such as cod are dry salted by the heavy (hard) cure and fatty fish are cured in airtight barrels by the Gaspe (light) cure. Figure 20 shows the process for the hard cure; the last step is air drying, which results in a salted-dried product with long-term, room temperature shelf life and a water activity of between 0.75 and 0.85. The Gaspe or light-cured product remains edible only a few days in the wet-stack stage (a_w = 0.85–0.90) at room temperatures and must be pressed and mechanically dried for longer-term storage.

Smoking. The age-old practice of smoking has changed drastically over the last few decades. The process as originally practiced by Eskimos and American Indians to pre-

Table 10. Importance of Water Activity in Foods

a_w	Phenomena	Food examples
1.0		Water-rich foods (a_w = 0.90–1.0): foods with 40% sucrose or 7% NaCl, cooked sausages, bread crumbs, and kippered fish
0.95		
0.90	General lower limit for bacterial growth	Foods with 55% sucrose or 12% NaCl, dry ham, medium-age cheese, and hard-smoked fish
0.85	Lower limit for growth of most yeast	Intermediate-moisture foods (a_w = 0.55–0.90): foods with 65% sucrose or 15% NaCl, salami, old cheese, and salt fish
0.80	Lower limit for activity of most enzymes	Flour, rice (15–17% water), fruitcake, and sweetened condensed milk
0.75	Lower limit for halophilic bacteria	Foods with 26% NaCl (satd), marzipan (15–17% water), and jams
0.70	Lower limit for growth of most xerophilic (dry loving) molds	
0.65	Maximum velocity of Maillard reactions	Rolled oats (10% water)
0.60	Lower limit for growth osmphilic or xerophilic yeasts and molds	Dried fruits (15–20% water), toffees, and caramels (8% water)
0.55	DNA becomes disordered (lower limit for life to continue)	Dried foods (a_w 0–0.55)
0.50		Noodles (12% water), spices (10% water), and fish protein concentrate (10% water)
0.40	Minimum oxidation velocity	Whole egg powder (5% water)
0.30		Crackers and crusts (3–5% water)
0.25	Maximum heat resistance of bacterial spores	
0.20		Whole mild powder (2–3% water); dried vegetables (5% water), and cornflakes (5% water)
0.00	Maximum oxidation velocity	

Source: Ref. 17.

serve fish for the winter months was essentially a drying process whereby heat from a fire was used to reduce the moisture content sufficiently for extended storage. The smoke flavor was somewhat incidental to the process. In fact in some areas the fish were dried by the sun and the natural air currents and the smoke was used to prevent flies and insects from consuming and contaminating the product. This dried smoked fish, which takes days to cure, is known as hard smoked.

Today, most smoked fish is smoked for the flavor and there is relatively little loss of water in the process. The change in processing is a reaction to the consumer, who prefers a soft, moist texture rather than the tough texture of a dried product, and to the processor, who cannot afford to tie up large processing areas for longer-term smoking. Furthermore, the minimizing of moisture loss greatly improves the economics of processing and marketing smoked products. Hence, although smoked fish are considered to be processed in the same manner as fish in which water activity is altered, in reality the modern product is a partially or wholly cooked fish that has smoke added as a condiment (19).

Most of the smoked fish prepared in the United States has little shelf life stability beyond that of a fresh fish. The commercial process of smoking involves splitting and cleaning the fish, salting or brining (soaking in a brine solution) to firm the texture of the meat, draining to remove excess moisture, and smoking. Smoking is carried out as a cold smoke, the temperature of the smoke does not rise above 85°F, or as a hot smoke, the smoke is hot enough (eg, 250°F) to raise the center temperature of the fish to above 140°F. It is also common to smoke the fish with colder smoke and then to raise the smoke and air temperature during the terminal part of the smoking to pasteurize the fish.

Today, much of the smoking is carried out in commercially constructed smoking facilities, or kilns, that have smoke generators (using sawdust), controlled temperature forced air, and humidity control of the air. There are as many specific smoking procedures as there are processors in the business. Each processor has a favorite method, including the use of certain additives in the brine, controlled-temperature drying, smoking, cooking, and cooling. Modern kiln smoking has allowed precise control of the variables in smoking and has removed much of the artisan approach that was so prevalent when all smoking was carried out using open wood fires.

Specialty hard-smoked fish, known as jerky, is prepared by smoking and drying fish to a hard, chewy consistency. This product is popular in the bar and tavern trade and for hikers who wish to carry a meat product that will not spoil during a several-day outing.

Irradiation

The United States is out of step with the rest of the world when considering the use of irradiation for preserving food. During the early 1980s, the World Health Organization gave ionizing radiation its blessing after an extensive review of many years of scientific work and investigation (20). Since that time, many countries in the world have been using irradiation on a wide variety of foods (9). Al-

Table 11. U.S. Production of Principal Cured Fishery Products

Item	Thousand pounds	Thousand dollars
Salted and pickled		
Cod	2,633	2,509
Halibut	45	73
Herring		
Lake	460	185
Sea	12,190	18,102
Mullet	133	79
Sablefish	1,032	1,404
Salmon	11,386	32,024
Total	*27,819*	*54,376*
Smoked and kippered		
Carp	153	134
Chubs	2,151	3,971
Cod	210	588
Eels	184	669
Halibut	281	630
Herring		
Lake	38	52
Sea	963	1,876
King mackerel	13	52
Lake trout	84	167
Marlin	9	43
Mullet	137	212
Paddlefish	257	469
Pollock, Pacific	900	875
Sablefish	1,880	4,157
Salmon	13,071	63,532
Sturgeon	377	1,338
Trout, unclassified	242	406
Tuna	48	108
Tunalike fish, bonita, yellowtail	4	3
Whitefish	2,910	5,526
Whiting	1,028	485
Unclassified fish and crustaceans	293	746
Total	*25,233*	*86,039*
Dried		
Cod	579	971
Shrimp	368	2,064
Total	947	3,035

Source: Ref. 2.

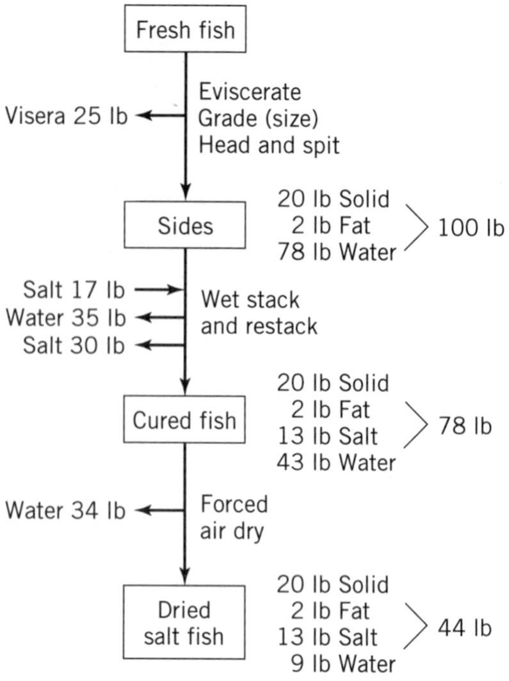

Figure 20. Process for hard curing and drying of fish.

though there has been steady progress in the United States toward approval of the process, there is still a widespread disagreement between scientific, government, and consumer groups regarding the pros and cons of using this form of energy for processing foods (21).

Two factors probably have equally contributed to this battle, namely the association in the minds of people between ionizing radiation and nuclear warfare, and the DeLaney Clause of the 1958 Food Additive Amendment to the Food, Drug and Cosmetic Law. The amendment, best known for the prohibition of a food additive that is shown to be carcinogenic in a test animal at any dose level, states that irradiation is a food additive and not a food-processing technique. It is interesting to note that the U.S. Department of Agriculture has issued an extensive list of publications that show the safety and wholesomeness of irradiated foods (22).

Some confusion exists about the units used for measuring radiation doses that a product receives. This is because the original unit of measurement, the rad, was changed to kilograys (kGy); 100,000 rad (100 krad or 1 Mrad) is equivalent to 1 kGy (100,000 Gy). As shown in Table 12, the doses for processing range from 0.75–2.5 kGy for pasteurization to 30–40 kGy for sterilization. Note that specific process terminology has been suggested for the various operations. Pasteurization is radurization, sanitization is radicidation, deinfestation is the process for destroying eggs and larvae, and sterilization is radapertization. Two radioisotopes, Co-60 and Ces-137, and electron-accelerator-generated electron beams meet the requirements for producing sufficient energy and intensity to penetrate food and accomplish the four basic processes.

The radurization of fresh seafood has an excellent potential for extending shelf life. Being in the low-dose range, this process should not meet the resistance that is found for the publicized high doses that accomplish sterilization, but result in some off-flavor (23). The extra week of shelf

Table 12. Dose Ranges for Food Irradiation Processes

Process	Dose range	
Radurization (pasteurization)	75–250 krad	0.75–2.5 kGy
Radicidation (sanitization)	250–1,000 krad	2.5–10 kGy
Destroy eggs/larvae (deinfestation)	below 100 krad	below 1 kGy
Radappertization (sterilization)	3–4 Mrad	30–40 kGy

Source: Ref. 23.

life that can be given to a fresh fish by radurization would be of economic benefit to the entire industry. The extra shelf life would eliminate product loss at the retail level and would extend the market range for delivering fresh fish.

There is every reason to believe that irradiation will eventually become an accepted and extensively used processing technique for seafood, especially seafood for the fresh market, where all parties will benefit. Perhaps during the next decade the radurization of seafoods will be commonplace. After all, 20 yr ago there was the same consternation over the use of microwave ovens, and today they are in 80% of U.S. homes.

Packaging

Proper packaging of products is necessary to protect food from adverse losses and changes in weight, texture, flavor, nutritional components, and protection against contamination and physical damage. With the growing demand for high-quality seafood, the major emphasis is on packaging for fresh and frozen products. Basic types of packaging, in order of increasing ability to protect fresh or frozen food, include paper, coated paper, fiber, foil, films, laminates, and combinations.

One of the primary considerations in selecting packaging for fresh and frozen foods is the gas and water permeability of the material, usually plastic films, plastic containers, and treated paper products. There is always a balance between the cost of the package and the degree of protection given to the seafood. A very low cost uncoated cellophane has a moisture permeability that is so high that frozen and even cold fresh products are rapidly dessicated during holding. A film such as poly(vinyl chloride) (PVC) costs more but gives moisture-barrier protection that improves the overall economics of the processing and marketing chain.

The same economic analysis must be made for gas permeability, because oxygen entering the package tends to oxidize lipids and give off-flavors while some of the desirable volatile taste and odor components may be lost from the product. The fact that some films give good vapor-barrier protection but allow moisture to pass while others act in a reverse manner has encouraged the development of laminated films for food packaging. Laminates for seafood packaging often consist of polyethylene (which is cheap, is heat sealable, is usable over a wide range of temperatures, and is a good moisture barrier) and another film that has a low permeability for gas (eg, CO_2, O_2, and N_2). Examples of laminates used in frozen seafood products include polyethylene laminated with low-gas-permeability poly(vinylidene chloride) (PVDC) or saran), polyester, cellulose, aluminum, or nylon.

Packaging not only protects the food but is also an important factor in the sales appeal of the product. Packaging that shrinks film to give better evacuation of air from the package, overwraps for multiple packages and trays, overwraps to give rigidity to a package, allows see-through to the product, uses vacuum to minimize oxidation of the product, and uses gas flushing to expel further air are all popular methods of protecting processed seafoods. Packaging materials have become quite sophisticated and the accompanying use technology has been developed to meet the requirements for economic applications, long-term storage of products (particularly frozen), and enhanced consumer appeal.

Over the last decade, the market for fresh food products has become increasingly popular. A system combining closely controlled combinations of low temperature, high humidity, and proper ventilation has allowed extensive increase of the shelf life of perishable fresh foods. This system, hypobarics, was awarded the Institute for Food Technology's Technology Industrial Achievement Award in 1979 (24). Although hypobarics is a container system rather than a package, it can be considered a package in that it is used to contain a fresh product during the plant storage and transport time required to move the product to market. Originally developed for fruit, the system was extended to all fresh food items including seafood. Although the system has not been used extensively on seafood, it has been very successful with numerous products and should certainly be mentioned as a potential use for extending the shelf life of fresh seafood.

Another important area involving packaging technology is modified-atmosphere packaging (MAP), which has been shown to increase the shelf life of many perishable foods and formulated products by reported values of from 50 to 400% (25). However, there has been considerable concern by both regulatory authorities and researchers over maintaining the safety of MAP foods as they travel through the channels of commerce. The system, consisting of modifying the gaseous environment with a package, results in various combinations of carbon dioxide, nitrogen, and oxygen. Carbon dioxide is the important factor replacing much of the oxygen and nitrogen in the air (eg, 75% CO_2, 15% N_2, and 10% O_2). Of concern is maintenance of the proper refrigeration temperatures and the gas ratios for a given product so that microbial growth does not present an undue safety hazard.

A considerable amount of research has been carried out in the MAP of seafood products (26). A significant amount of effort is being expended on developing the processor-distributor combination to improve quality and value of fresh seafoods.

SEAFOOD SAFETY

The discussion of MAP is a good place to end the packaging discussion because it leads to the subject of general safety of foods as presented to the consumer. Often packaging is a major factor involved in these concerns, because the environment created by the combination of processing and packaging often creates potential for adverse microbial growth. There are two programs, one specifically for shellfish and one for refrigerated foods, that are important in reducing the microbiological risks associated with harvesting, handling, processing, and distributing to the consumer.

The National Shellfish Sanitation Program

The National Shellfish Sanitation Program (NSSP) is designed to prevent harvesting of shellfish in polluted waters containing pathogenic organisms or other contaminants (27). The program, begun in 1925 following illnesses culminating in typhoid fever outbreaks caused by contaminated sewage, is administered cooperatively among the federal government, states, and industry through the interstate Shellfish Sanitation Conference. The Food and Drug Administration is responsible for appraising each state's shellfish program to insure that they are complying with the specified requirements. NSSP gives each state the responsibility of defining or classifying its waters from which shellfish are harvested. The waters are tested and analyzed on a continual basis and classified as follows:

1. *Approved.* These waters may be harvested for the direct marketing of shellfish at all times.
2. *Conditionally Approved.* These waters do not meet the criteria for the approved waters at all times but may be harvested when criteria are met.
3. *Restricted.* Shellfish may be harvested from restricted waters if subjected to a suitable purification process.
4. *Prohibited.* Harvest for human consumption cannot occur at any time.

The term harvest limited is used to refer to conditionally approved, restricted, or prohibited waters. A closure area is an area in which some restriction on harvest has been placed (eg, a harvest-limited area).

Hazard Analysis and Critical Control Point Principles

As the popularity and volume of refrigerated foods have been significantly increasing over the last decade, there has been an accompanying increase in the risks of foodborne disease. The overall safety record of a wide variety traditional refrigerated fresh fruits and vegetables; raw meats, poultry, and seafood; pasteurized milk; cured, ready-to-eat meats and seafood; high-moisture cheeses, yogurt, and pickles; and perishable delicatessan products has been good. However, the complexity of food-distribution systems presents risk factors that are increasing. This is especially the case of a new generation of engineered and formulated foods including frozen or restaurant dinners; frozen or delicatessan entrées; dry, frozen, or canned pasta; fresh, refrigerated salads; canned, frozen, or dry gravies; canned or dry soups; and frozen, canned, or refrigerated cooked meats and seafood (28).

In response to this increasing hazard, the National Academy of Science has studied the problem and recommended a system, the hazard analysis and critical control point (HACCP) system, which "provides a more specific and critical approach to the control of microbiological hazard in foods than that provided by traditional inspection and quality control approaches" (29). The system, similar to the critical control point system used by engineers in planning design, construction, and startup of a building or operation, include the following principles:

1. Describing the product and how the consumer will use it.
2. Preparing a flow diagram for intended manufacturing and distribution of the product.
3. Conducting risk analysis for ingredients, product, and packaging; reducing the risks by making changes to the design; and incorporating these changes into the processing and packaging schemes.
4. Selecting critical control points (CCP) and designating their location on the flow diagram and describing CCP, establishing monitoring.
5. Implementing HACCP in routine activities.

The assurance of food safety is a complex subject that impacts all phases of the commercial food chain, ending with the consumer. HACCP undoubtedly will have a strong influence on the future of many food operations. This impact will especially affect the seafood industry with its myriad safety and sanitation problems, many of which are specific to products from the marine and fresh-water environments.

FUTURE DEVELOPMENT

Fish is health food is not a marketing gimmick but truly states the merits of seafood. The demand for this highly nutritious protein food, much avowed for its healthful omega-3 fatty acids, will continue to grow in demand. This pressure on a limited resource will encourage better biological management of the natural wild stocks (including international cooperation), faster expansion of aquaculture operations, and the total utilization of the raw material.

Much of the so-called industrial fish and the waste from present processing operations, currently being reduced to cheap animal foods, can and will be upgraded to human foods. Headed and gutted industrial fish can be deboned to give a highly acceptable minced flesh for engineered and formulated foods. A fish frame from which fillets have been removed contains as much meat as the fillets. This meat can be removed by deboning, and the minced flesh, nutritionally equal to the fillet or to any other minced flesh, is a tremendous source of base protein materials for formulated foods. Modern processing machinery, growing knowledge about the technology of formulated foods, and the demand for high-quality prepared foods support the trend to use previously underutilized seafood and seafood portions for this market.

As has been discussed, the technology of handling, packaging, and transporting fresh seafoods is rapidly developing and the future will see more fresh seafood on the market. In conclusion there is an exciting future for seafood in the United States. The new generation of refrigerated and frozen seafood and formulated products, as well as better quality traditional items, will continue to encourage increased consumption of both wild and farmed fish and shellfish.

BIBLIOGRAPHY

1. S. N. Jhavari, P. S. Karakoltsidis, J. Montecalvo, Jr., and S. M. Constantinides, *Journal of Food Science* **49**, 110 (1984).

2. National Marine Fisheries Service, *Current Fishery Statistics No. 8800*, United States Department of Commerce, Washington, D.C., 1988.
3. U.S. Department of Agriculture, "Aquaculture Situation and Outlook Report," *AQUA 1* **31**, (Oct. 1988).
4. "Farm-Raised Catfish," *Aquaculture Magazine* **18**, 81 (1989).
5. L. A. Nielsen and D. L. Johnson, *Fisheries*, American Fisheries Society, Bethesda, Md., 1983.
6. K. S. Hildebrand, unpublished data, 1984.
7. G. M. Pigott, "Total Utilization of Raw Materials from the Sea," *Proceedings of the Conference on Formed Foods*, Brigham Young University, Brighton, Utah, Apr. 1–2, 1985.
8. J. J. Sullivan, M. G. Simon, and W. T. Iwaoka, "Comparison of HPLC and Mouse Bioassay Methods for Determining PSP Toxins in Shellfish," *Journal of Food Science* **48**, 1321 (1983).
9. G. M. Pigott and B. W. Tucker, *Seafood: The Effect of Technology on Nutrition*, Marcel Dekker, Inc., New York, 1990.
10. H. O. Bang and J. Dyerberg, "Plasma Lipid and Lipoprotein in Greenlandic West Coast Eskimos," *Acta Medica Scandinavica* **192**, 85–94 (1972).
11. G. M. Pigott and B. W. Tucker, "Science Opens New Horizons for Marine Lipids in Human Nutrition," *Food Reviews International* **3**, 105–138 (1987).
12. W. E. M. Lands, "Fish and Human Health: A Story Unfolding," *World Aquaculture* **20**, 59–62 (1989).
13. B. W. Tucker, "Sterols in Seafood: A Review," *World Aquaculture* **20**, 69–72 (1989).
14. *Canned Foods*, The Food Processors Institute, Washington, D.C., 1980.
15. R. A. Bello and G. M. Pigott, "Ultrastructural Study of Skeletal Fish Muscle after Freezing at Different Rates, *Journal of Food Science* **47**, 1389–1394 (1982).
16. G. M. Pigott, insert in Ref. 9.
17. L. B. Rockland and L. R. Beuchet, eds., *Water Activity: Theory and Applications to Food*, Marcel Dekker, Inc., New York, 1987.
18. R. A. Bello and G. M. Pigott, "Dried Fish Patties: Storage Stability and Economic Considerations," *Journal of Food Science* **4**, 247–260 (1980).
19. G. M. Pigott, "Smoking Fish—Special Considerations," *Proceedings of the Smoked Fish Conference Symposium*, University of Alaska and University of Washington Sea Grant, Seattle, Wash., Apr. 27–29, 1981.
20. Codex Alimentarius Commission, *Microbiological Safety of Irradiated Foods*, FAO/World Health Organization Joint Office, Rome, Italy, 1983.
21. G. Giddings, "Irradiation: Progress or Peril?" *Prepared Foods* **158**, 62–67 (1989).
22. U.S. Department of Agriculture, *Safety and Wholesomeness of Irradiated Foods*, The National Agricultural Library, Beltsville, Md., 1986.
23. G. M. Pigott, "Radurization of Aquaculture Fish: A Value-Added Processing Technology of the Future," *The Proceedings of Aquaculture International Congress and Exposition*, Vancouver, B.C., Sept. 6–9, 1988.
24. N. H. Mermelstein, "Hypobaric Transport and Storage of Fresh Meats and Produce Earns 1979 IFT Food Technology Industrial Achievement Award," *Food Technology* **33**, 32–40 (1979).
25. J. H. Hotchkiss, "Experimental Approaches to Determining the Safety of Food Packaged in Modified Atmospheres," *Food Technology* **42**, 55–64 (1988).
26. M. F. Layrisse and J. P. Matches, "Microbiological and Clinical Changes in Spotted Shrimp Stored under Modified Atmosphere," *Journal of Food Protection* **47**, 453–457 (1984).
27. D. L. Leonard, M. A. Broutman, and K. E. Harkness, *The Quality of Shellfish Growing Waters on the East Coast of the United States*, Ocean Assessments Division, NOAA, Rockville, Md., 1989.
28. D. A. Corlett, Jr., "Refrigerated Foods and Use of Hazard Analysis and Critical Control Point Principles," *Food Technology* **43**, 91–94 (1989).
29. Food Protection Committee, Subcommittee on Microbiological Criteria, *An Evaluation of the Role of Microbiological Criteria for Foods and Food Ingredients*, National Academy of Sciences, National Research Council, National Academy Press, Washington, D.C., 1985.

GEORGE M. PIGOTT
University of Washington
Seattle, Washington

FISH CAKES. See FISH AND SHELLFISH PRODUCTS.

FISHES: ANATOMY AND PHYSIOLOGY

The diversity of forms found among the species of fishes is matched only by the diversity of aquatic environments on the earth. Aside from the shared characteristics of possessing a backbone, being cold blooded, and having gills and fins, the variation in anatomy, physiology, and ecology of all the fish species is extensive indeed. In size, they range from a 1-cm goby to the 15-m whale shark. They are found in waters from nearly 4,000 m above sea level to 7,000 m below sea level as well as in water temperatures from below freezing to 44°C. They are also found in environments that are nearly distilled to those that are very salty. Fish inhabit surface waters in full sunlight as well as very deep waters in complete darkness where pressures are extremely high. While most are strict water breathers, some have adapted to breath air where oxygen in water might be lacking. It may not be surprising, therefore, that there may be more than 25,000 species of fishes, accounting for more than half of all vertebrate species in the animal kingdom. Fishes also play an important role in human life by providing food and recreation. The traditional commercial fishery and now, to an increasing degree, aquaculture provide a high-quality protein source for human consumption. The sport fishery and hobby aquarium industry are both very important sectors of the economy. Feed for agricultural animals and fish oil supplements to human diets are examples of secondary uses of fish products for human benefit.

The purpose of this article is to outline some of the fundamental aspects of the anatomy and physiology of the bony fishes. The reference list at the end of the article serves as a guide to more detailed literature on various subjects covered here. The anatomy and physiology of eight major systems in fish are briefly described. These are the digestive, renal, cardiovascular, respiratory, nervous, endocrine, reproductive, and immune systems. The information is basic and is presented here with the assumption that the reader has some knowledge about vertebrate anatomy and physiology but very little or no knowledge about fish. The intention is to present some basic information and stimulate interest in this important food animal.

GROSS ANATOMY AND ORIENTATION

The anatomy of the salmonid is emphasized here because they are one of the most widely studied genera in terms of physiology and aquaculture science. Figure 1 presents the general orientation and terminology that will be used throughout this article. It also points out the major external features of the animal.

Figure 2 illustrates the general features of the internal organs of the salmonid. The muscle mass is, by far, the largest component of the body. The muscle mass, relative to the body, is larger in fish than in other terrestrial or aerial vertebrates. The dense medium of water requires a large force for locomotion compared to air. That same medium, however, provides substantial structural support such that the fish can maintain neutral bouyancy with a swim bladder or body lipids. It is also this great mass of muscle that makes the fish uniquely desirable as a food animal. There are two types of muscle in the fish: red and white. In most species of fish less than 10% of the total muscle mass is made up of red muscle. The color-oriented terminology is based on the visual appearance of two types of muscle. The red muscle appears darker because there is up to a 10-fold greater blood capillary to muscle fiber ratio in the red muscle compared to the white muscle. Red muscle is used for sustained swimming, fueled by aerobic glycolysis and lipolysis, while the white muscle is used for burst swimming, fueled by anaerobic glycolysis. The approximate weight distribution, in percent, of the major body parts of a trout are listed below.

Liver	1.2
Spleen	0.1
Intestine	4.7
Heart	1.2
Swim bladder	0.2
Kidney	0.9
Muscle	55.9
Skin	8.7
Axial skeleton	13.5
Gills	2.8
Head	11.8

DIGESTIVE SYSTEM

The anatomy and function of the digestive system in fish is basically the same as in other vertebrates. It is made up of a combination of mechanical and chemical processes. Food is ingested, broken down, and either absorbed into the blood or remains in the gut and is eventually expelled as feces. Differences arise between fish species in feeding habits, and the environment may influence the presence, position, shape, and size of a particular organ. Fish may be divided into three categories according to the food they eat. Herbivores are specialized and eat detritus and plants. Omnivores eat both plants and animal material, and carnivores feed on larger invertebrates and other fish.

Digestive Organs

The oral cavity is made up of lips, mouth, teeth, and tongue. Fleshy lips are common to herbivores, designed for grazing and straining. Carnivore lips are generally unmodified and thin. Bottom feeders often have barbels, sensory appendages that help guide them when feeding and grazing.

Mouths are large or tubular. Large mouths are usually associated with predatory fish, such as a barracuda, whereas tubular mouths such as that of a seahorse, are better adapted for plankton eaters due to the enhanced ability to suck water and food. Mouth position can be variable depending on the types of feed normally consumed. Bottom feeders have mouths on the ventral surface, surface feeders have a dorsal mouth and bony fish (teleosts) have a terminal mouth.

Teeth are thought to have originally arisen from scales covering the lips. Fish generally have small teeth for gripping and capturing prey. There are four possible types of teeth present in fish, also depending on feeding habits: (*1*) jaw teeth, (*2*) palate teeth, (*3*) tongue teeth, and (*4*) throat teeth. There is a strong correlation between dentition, feeding habits, and food eaten. Jaw teeth are more developed in carnivores, but they are poorly developed or absent in herbivores, which usually have well-developed throat teeth attached to the gill arches.

The tongue of a fish is very rudimentary. It has no salivary glands but does bear sensory taste buds. Any lubricant fluid in the mouth originates from mucous cells scattered throughout the inner lining, not from the tongue. Plankton eaters also have well-developed gill rakers. This is a sieve-type apparatus attached to the inner edges of the gill arches and serves to filter out phytoplankton, crustaceans, and, in some cases, even diatoms.

A short, broad, and muscular esophagus connects the mouth and elastic stomach. It functions in the transport and taste of food. The musculature of the esophagus tends to be more developed in freshwater fish than in saltwater fish, as it plays a role in minimizing water intake while ingesting food.

The stomach is muscular and motile. The inner surface is lined with gastric glands that secrete digestive enzymes, necessary for the breakdown of food. The size of the stomach is related to the size of food particles and the interval between meals, ie, large food size and infrequent feeding are usually associated with large stomachs, and smaller food particles ingested more constantly are associated with fish with small or no stomachs. The pyloric valve, a ring of muscle, controls the passage of food between the stomach and intestine.

The intestine is a relatively simple tube that has digestive glands and an abundant supply of blood vessels. There are two valves, one at each end. The pyloric valve is at the anterior end and the ilioceacal valve is at the posterior end. Here is where much of the absorption of nutrients takes place. Digestive products are picked up in solution. Intes-

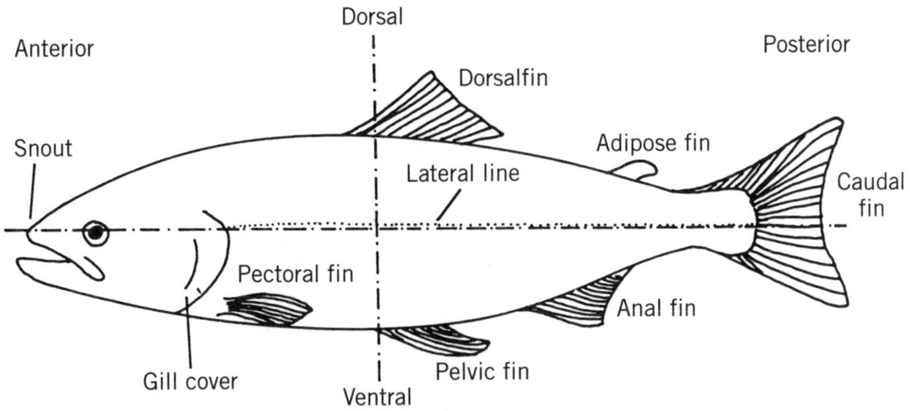

Figure 1. External anatomy.

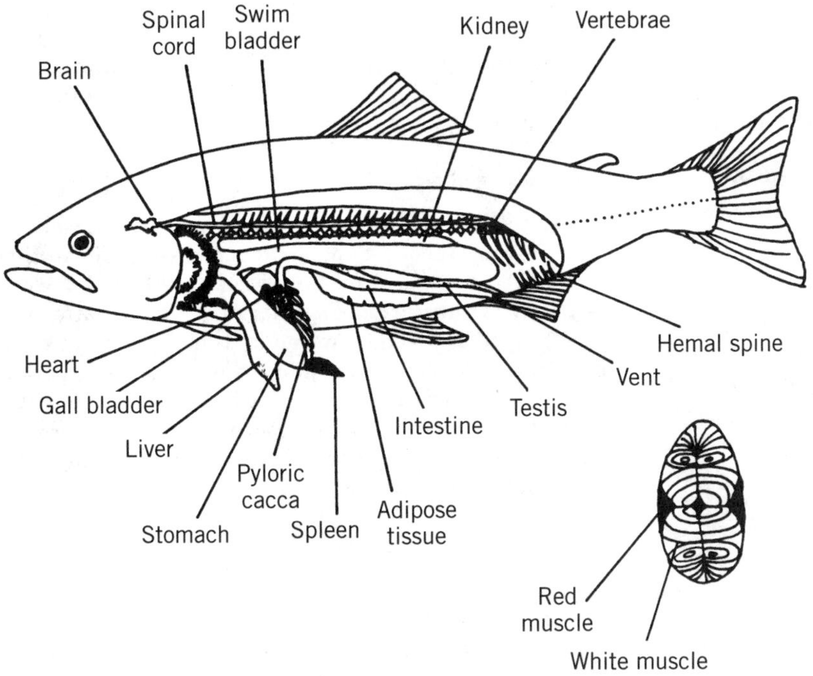

Figure 2. Internal auditing and cross section of trunk.

tine length varies with feeding habits: carnivores have short straight intestines, herbivores have long and coiled intestines. Gut length seems more correlated to the amount of indigestible material in the food rather than the nature of the food. A large surface area is realized by extensive infoldings and ridges. The intestine leads to the rectum where vascularization and secretory cells are sparse, but where more mucous cells are present. The anal opening is called a vent and is also the terminal location for the urinary and reproductive ducts. While the common collection of these ducts suggest a corollary with the avian cloaca, it is not the case. The three ducts in fish all empty to the external environment and not to a common internal chamber. Absorption of plant material is probably poor; probably about 20% or less. Hatchery diets for salmonids may be absorbed up to 80%.

Food Movement

Movement of food through the digestive system is aided by peristaltic waves of muscle contraction. Cilia also line the tract and are especially useful in stressful situations when peristalsis may become reduced or even stop. This suggests that peristalsis is under some nervous control. Mucous membranes in the mouth, esophagus, and stomach provide lubrication to aid the passage of the food along the digestive tract.

Associated Organs

The pyloric caeca is an important appendage that opens into the intestine. It has the same epithelium as the intestine and its purpose is to increase the effective surface area. It may have both an absorptive and digestive function. A two or more lobed liver, which serves several metabolic and energetic functions, also produces bile, which aids in digestion. The gallbladder, which opens into the intestine, stores excess bile, which is secreted when required. The pancreas, which is large and distinct in the elasmobranch but diffuse in teleosts, secretes insulin and glucagon in response to nutritional intake, but also secretes digestive enzymes into the intestine. In teleosts, small beads of pancreatic tissue are scattered in the mesentaries near the digestive tract and are supplied with artery, vein, nerve, and pancreatic ducts. Little is known about the pancreatic secretions and most knowledge is based on histological evidence.

Digestive Enzymes

While goblet cells secrete mucus, gland cells throughout the digestive system contribute digestive enzymes necessary for the breakdown of food. The stomach is lined with secretory cells (gastric glands) that secrete hydrochloric acid (HCl) and pepsinogen. These chemicals break down protein molecules into amino acids and are, therefore, especially important to carnivorous fish. The stomach is very acidic (pH 2–4). Pepsin activity is dependent on pH and temperature, whereas HCl secretion is dependent on temperature and meal size. Optimum pepsin activity occurs at pH values of 2 and 4. Stomach distension stimulates gastric secretions.

Both the pyloric caeca and intestinal mucosa are sources of lipase, which breaks down fats into fatty acids and glycerine. Absorption of fat may occur in the anterior of the intestine. Intestinal secretions include proteases such as trypsin, lipase, and carbohydrase. These enzymes work best at neutral to alkaline pH. Bicarbonate is secreted (possibly from the pancreas) to raise the pH of the food coming from the stomach. Bile salts, secreted from the gallbladder may also play a role in adjusting the digestive juices to the proper pH to facilitate the digestive process. Bile salts are detergentlike substances formed from the decomposition of cholesterol and other steroids that function in fat emulsification aiding in the digestion of lipids. Fat soluble vitamins (A, D, E, K) are digested by bile. Digestive proteases are also secreted from the pancreatic tissue.

Most of the food absorption takes place in the intestine and pyloric caeca with the undigested material passing on to the rectum. Factors such as temperature, fat content, and presence of indigestible material affect the passage and digestion and absorption of food.

Innervation of Digestive Organs

Innervation of the digestive organs is not well understood, although it is thought that there is sympathetic innervation from paired ganglia, lateral to the spinal cord to the stomach, intestine, and rectum. This adrenergic system has an inhibitory effect. Parasympathetic innervation is through three cranial nerves to various parts of the digestive tract. Glossopharyngeal (IX) and facial (III) nerves innervate the mouth and esophagus areas. The vagus (X) innervates all visceral portions. These cranial nerves are cholinergic and have a stimulatory effect. Intrinsic nerves form a network inside the tissue of the digestive system. These nerves do not originate in the brain or the spine. They may assist in peristaltic movement. Intestinal and pancreatic secretions appear to be also under both nervous and hormonal control.

Feeding Types

Predators have well-developed grasping and holding teeth, ie, sharks, pike, and gar, and well-defined stomachs with strong acid secretions and short intestines (relative to herbivores). They depend on their senses of vision or the lateral line sensory organ to detect their prey. Grazers browse on plants and organisms and sometimes on each other. The trout is an example of this, and in crowded ponds they often nip at the fins of the other fish. Strainers usually have well-developed gill rakers that strain out the plankton and crustaceans in the water that flows through the mouth and gills. Suckers suck in their food. They have mouths on the ventral surface of their heads but adaptive lips. An example of this type of feeder is the sturgeon. Parasites live off the body fluids of other fish. An examples of this type of feeder is the lamprey.

Stimuli for Feeding

Internal motivation is driven by factors such as season and temperature, diurnal light, light intensity, and the time and nature of the last feeding. Fish have been known to feed better at dusk and early morning. Due to patterns of growth, fish also feed in spring and through the summer much more voraciously than in the winter. The hypothalamus is also thought to be involved especially where hunger and satiation triggers are concerned. Stimuli for feeding also comes through the senses, such as smell, taste, sight, and the lateral line system.

RENAL SYSTEM

The kidneys are important in filtering undesirable materials out of the blood as well as serving a role in water and ion balance. Active transport, together with epithelial permeability are adjusted in order that ions and other materials vital to the body can be conserved and control over body water content can be exercized.

Basic Anatomy and Physiology

The kidney in salmonidae is made up of two parts running along the anterior-posterior axis between the body cavity and vertebral column. Anterior and posterior portions, also known as head and trunk, respectively, have slightly different functions but there is no visual distinction between them. The anterior kidney is associated with inter-renal tissue and chromaffin cells, which are involved in blood cell and hormone production; the posterior kidney is associated with filtration and urine production.

Kidney tissue is made up of individual units called nephrons (Fig. 3). Each nephron is made up of a renal corpuscle, consisting of the glomerulus and Bowman's capsule and a kidney tubule. The glomerulus is a network of afferent and efferent arterioles whose blood supply comes from

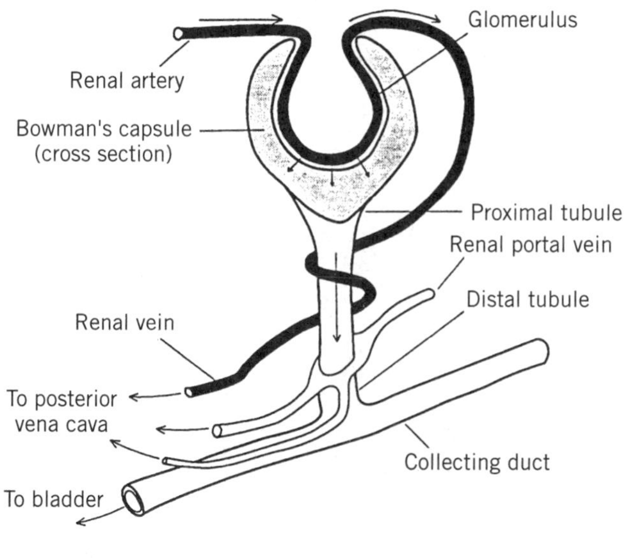

Figure 3. Kidney nephron.

the dorsal aorta or a venous supply called the renal portal system. The Bowman's capsule encapsulates the glomerulus and nonselectively filters solutes and molecules less than 70,000 mol. wt. from the blood. The filtrate is carried through the renal tubule, which is made up of the proximal, intermediate, and distal segments. It is in the tubule that reabsorption of selected electrolytes, minerals, amino acids, glucose, and other plasma organics takes place via a capillary network back into the blood. The remaining fluid containing unreabsorbed constituents flows to the collecting tubule and gathers in the urinary bladder, if present, where reabsorption of more ions may take place before being excreted out the urinary duct.

The process of nonselective filtration and subsequent selective reabsorption is an efficient process, because the body reabsorbs only what it needs. It is an adaptable system, because needs may change over time and through changing circumstances. Although efficient, this system requires energy as many ions are reabsorbed by active transport. Control of kidney function is accomplished by a variety of hormonal mechanisms and includes tissues of the thyroid, kidney, gonads, hypothalamus, and possibly the pituitary. In mammals, it is well established that a renin-angiotensin system and antidiuretic hormone (ADH) regulate renal function for ionic and water regulation. While the exact role of such a system in fishes is not well established, there is evidence that similar mechanisms are found in fishes. In this model, renin is secreted under nervous and hormonal control from a juxtaglomelular apparatus, located near the glomerulus, in response to decreased Na^+ concentration or blood pressure as well as under stressful conditions. Renin is then converted to angiotensin, which increases blood pressure through the constriction of blood vessels and causes aldosterone to be release from the adrenal cortex (of mammals). Aldosterone stimulates ionic uptake from the distal tubule of the nephron. ADH, from the pituitary, controls the amount of water leaving the nephron by controlling the permeability of the epithelium in the collecting duct. An increase in blood osmolarity causes an increase in ADH, which increases the permeability of the epithelium, resulting in the increased absorption of water out of the filtrate, back into the blood.

Freshwater fishes are hyperosmotic regulators. That means that the concentration of ions and other solutes is greater in the blood than in their surrounding water, which in most cases is very dilute. The fish will thus absorb water osmotically from their environment through all permeable epithelia such as the gills, skin, and gut. The regulatory problem is one of getting rid of the excess water and the kidney plays that important role. A large amount of urine is produced, which is dilute and contains creatine, uric acid, and some ions. The volume of urine produced must balance the quantity of water entering the body. Sodium (Na^+) and chloride (Cl^-) ions passively diffuse out of the body across permeable epithelia and are actively taken up, to a large extent, across the gill epithelium.

Electrolyte reabsorption out of the urine takes place across the renal tubule. Na^+ is actively extracted and it appears that Cl^- passively follows. Calcium (Ca^{2+}), magnesium (Mg^{2+}), and other divalent ions must also be reabsorbed, because they are normally absent in the urine of freshwater fish. The reabsorption of these ions is usually accomplished without the osmotic absorption of water. The distal segment, collecting duct and urinary bladder appears to be relatively impermeable to water. Macromolecule reabsorption, including glucose, amino acids, and other plasma organic constituents takes place in the first segment of the proximal tubule. Salts are reabsorbed in the distal segment, and any remaining in the filtrate may be reabsorbed from the urinary bladder. Only a small proportion of total organic nitrogen is excreted via the kidneys, although this appears to be an important excretory pathway for minor nitrogenous products, such as creatine and uric acid. Major nitrogenous products such as ammonia and urea are excreted via the gills.

The kidney in marine fishes plays a crucial role in hypoosmotic osmoregulation. Converse to the freshwater fish, the blood concentrations of ions and solutes in fish blood relative to the concentrated ionic environment of the marine environment is rather dilute. The osmotic problem created by such a gradient is one of dessication, which the fish counteracts by actively taking in water from the environment by drinking and reducing water loss at the kidneys. Between 60 and 80% of ingested water is absorbed through the gut, along with the monovalent ions Na^+, Cl^-, and potassium (K^+). Those excess salts are actively excreted. The gills play a major role in that function, although the kidney excretes Mg^{2+}, sulfate, and other divalent ions. Although less than 20% of the ingested divalent ions are actually absorbed most are passively eliminated via the intestines, that which is absorbed is handled by the kidney tubules.

The kidney nephron of marine fishes often lack the distal segment of the tubule and has less glomeruli. In some cases, such as the goosefish, glomeruli are absent. Because glomeruli are suited to the excretion of large volumes of water the importance of the glomerulus in marine teleosts is greatly reduced. Urine from a marine kidney thus has high osmolality and is low in volume. Much of the water entering the glomerulus reabsorbed to help prevent dehydration, a case opposite to the freshwater kidney. Nitrogenous excretion is similar to the pathway in freshwater fish, with ammonia and urea excreted through the gills and minor nitrogenous products such as creatine and uric acid excreted by the kidney.

Anadromous and catadromous fish must be capable of adjusting their osmotic balance to survive the changes in salinities that they experience in their life cycle. They have glomerular kidneys that can adjust to changing urine volumes, and possess gills and oral membranes capable of both uptake and secretion of certain ions against the prevailing diffusion gradients. The kidneys in those animals are capable of adjusting urine volume and composition on demand. When those fish, for example, are transferred or voluntarily move from fresh water to salt water, the urine composition gradually changes after a few days. Urine flow decreases and osmolality increases as divalent ions are excreted. Sodium and chloride content in the urine decreases as the chloride cells (specialized salt excreting cells on the gills) take over this function. Glomelular filtration rate may temporarily slow down and tubular reabsorption increases.

CARDIOVASCULAR SYSTEM

The cardiovascular system moves the blood around the body. It is basically composed of the heart (cardio-) and network of blood vessels (-vascular) throughout the body. The purpose of blood circulation is the transport of materials between certain locations within the body. Gases such as oxygen are taken up from the water and move from gills to the tissues and others such as carbon dioxide move from tissues back to the gills for excretion. End products of metabolism such as lactate, originate in the muscles and are transported to the liver for breakdown. Glucose made in the liver must be moved to the tissues where it will be used. Materials that are foreign to the body are excreted by gills, released in the urine, or engulfed and destroyed by specialized cells in the blood. Amino acids and other nutrients from digested materials must move from gut to the tissues. Blood cells that are produced in the anterior kidney and spleen must be distributed throughout the body. Because of the vast range of environments in which fish are found, the diversity in the form and function of the cardiovascular systems are as wide ranging. The basic conformity among salmonid fishes are emphasized here.

Heart

The teleost heart has four chambers: the sinus venosus, the atrium, the ventricle, and the bulbus (Fig. 4). The venous blood from the body is received in the sinus venosus. This is a rather flat bag anterior to the transverse septum separating the heart and head compartments from the visceral cavity. The atrium is a thin-walled bag that feeds the blood to the muscular ventricle. The ventricle is the powerhouse that contracts and propels the blood throughout the body. The bulbus is rather muscular as well and helps to dampen pressure waves as the blood pulses out of the ventricle. There are one-way valves that prevent the blood from flowing backward with each contraction. Cardiac output in teleosts ranges from 5 to 100 mL/kg per minute with a general mean about 15–30 mL/kg per minute.

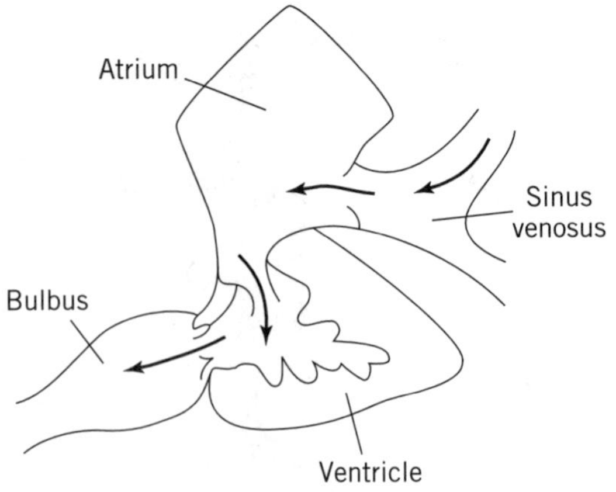

Figure 4. Cross section of a trout heart. *Source:* Ref. 1.

Blood flow is accomplished primarily by the heart, which is a pump for the system. The basic route is from the heart to the gills, where gas and ionic exchange with the environment occurs, and then to the rest of the body (Fig. 5). The blood is then pumped back to the heart through a venous network where the cycle begins over again. About 4–6% of body weight is a reasonable approximation of the blood volume. Blood pressures in teleosts are in the range of 30–70 mm Hg. Because the resistance to blood flow increases as the diameter of the blood vessels decreases, blood flow declines from the point where it leaves the heart. For example, the blood vessel network in the gills reduces the pressure found in the ventral aorta by about 40–50%. That resistance may increase with stresses such as low oxygen levels in the water, whereas other conditions such as exercise or elevated levels of adrenaline dilate the blood vessels and decrease the resistance to blood flow.

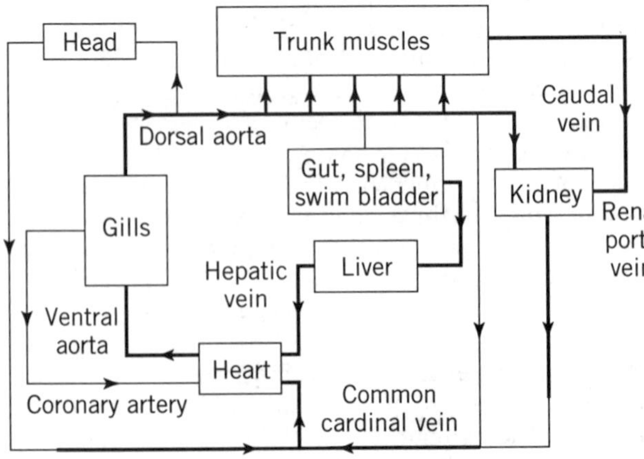

Figure 5. General flow of blood in fish.

The Secondary Circulation

The primary circulation in fish comprises the blood vessels that contain red and white blood cells. There is a secondary circulation, sometimes known as the lymphatic or venolymphatic system, that is characterized by low blood cell content and by low pressure. There are direct connections between the primary and secondary circulation through small vessels, often at right angles to the blood vessel of the primary circulation. This physical configuration causes plasma to be skimmed out of the primary circulation and into the secondary circulation. The secondary circulation feeds this fluid, low in cells as a result of the skimming, and fluid from tissue beds back to the primary circulation via the heart. Salmonids have major ducts just under each lateral line and along the dorsal midline. These may serve to collect fluid from the secondary circulation and return it to the heart. Fin and trunk movements may play an important role in pumping venous fluids back to the heart. There is a tail heart in some fishes that aids the movement of this secondary fluid back to the heart.

RESPIRATORY SYSTEM

The respiratory system is extremely varied among the 25,000 species of fishes. This, once again, reflects the wide range of habitats to which fish have managed to adapt. The primary function of the respiratory system is to transport gases between the environment and the tissues of the body that consume oxygen as well as excrete ammonia and carbon dioxide. The vital components in this system are the water flow over the gills outside the fish and the circulatory system inside the body. The consumption of oxygen (O_2) as well as the excretion of carbon dioxide (CO_2) and ammonia (NH_3 and $NH_4{}^+$) take place across the gill.

Anatomy

The fish gill has four gill arches on each side of the midline and two rows of primary or gill filaments per arch (Fig. 6). Arranged perpendicularly to the filament are rows of secondary lamellae on both side of each filament. The plates of secondary lamellae form narrow channels through which the water flows. This space is approximately 0.02–0.05 mm wide, 0.2–1.6 mm long, and 0.1–0.5 mm high. This width is particularly important in that one-half of that width is the minimum distance for gases and dissolved materials such as ions to diffuse between water and blood. The surface of the secondary lamellae is the primary surface across which gases, ions, and other dissolved materials pass between water and blood.

Water Flow over the Gills

The water flow over the gills is directed in an anterior to posterior direction. More importantly, it flows in the opposite direction to the flow of blood through the secondary lamellae in the gills. This countercurrent flow maintains the maximum gradients between blood and water for the gases throughout the transit of water through the gills. The mouth or buccal pump is driven by skeletal muscles that control the mouth of the floor and opercular covers.

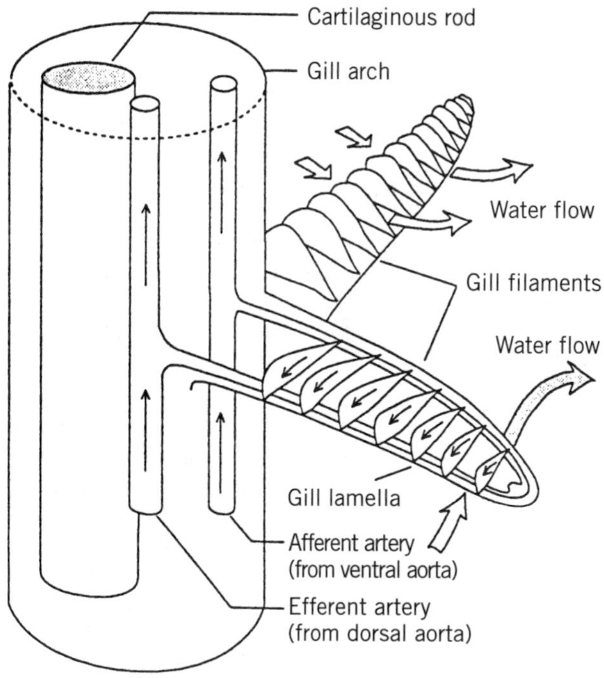

Figure 6. Fish gill with patterns of blood and water flow. *Source:* Ref. 2.

Water, therefore, flows from mouth, over gills, and out the operculum. When the floor of the mouth is lowered, negative pressure is created in the buccal cavity and water is sucked into the mouth. As the mouth is closed and the floor of the buccal cavity raised, the water is forced posteriorly, across the gills, through the opercular cavity. This cycle is repeated continuously and the result is a unidirectional flow of water through the mouth and gills.

While this rhythmic ventilation is the general rule, some fish are ram ventilators, that is, they ventilate the gills by keeping their mouths open and swimming forward through the water. Salmonids also do this at moderate to high swimming velocities. There are also other variations on how fish move water across the respiratory surface of the gill for respiration. Some of these variations such as ram ventilation may reduce the energetic costs of ventilation.

Blood

Blood is a suspension of various cells in a solution of proteins and electrolytes, which make up the plasma. It is the vehicle for transporting materials from different locations in the body, as discussed above. The volume of blood in teleosts ranges between 3 and 6% of wet body weight. Fish do not have bone marrow and the origin of the blood cells is the anterior kidney (anterior 30%) and the spleen. These are termed hematopoietic tissues. The production of blood cells is stimulated by conditions such as bleeding and inhibited by starvation. There is an erythropoesis stimulating factor, found in the plasma of fish that stimulates blood cell production. Very little is found in plasma of resting fish.

White blood cells or leukocytes, are involved in the clotting of blood and play an important part in the immune system. The red blood cell, or erythrocyte is important in the transport of oxygen in the body. The packed red cell volume, or hematocrit, is expressed in percent and varies between 20 and 35%. The erythrocyte contains the respiratory pigment, hemoglobin (Hb), which binds oxygen and enables the oxygen content of the blood to be higher than that of simple dissolved oxygen. The concentration of Hb in blood is approximately 7–12 g % (equal to grams/100 mL). The total oxygen content is about 9–14 vol %.

Blood Proteins

The proteins in the plasma serve to maintain the blood osmolarity. It is the source of colloid osmotic pressure. Some other functions are to buffer pH changes in the blood and to transport vitamins, hormones, and inorganic ions, that attach to these molecules in blood, to tissues. Another important function of special plasma proteins are the antibody proteins that play an important role in the specific immune system of the body. In carp, for example 4.15% of the blood weight are protein. Of that weight, 2.82 g is albumin, 0.79 g is globulin, and 0.23 g is fibrinogen.

Response to Environmental Change

Fish respond to environmental changes in different ways. Some fish are known as oxygen regulators because they maintain O_2 consumption at lower O_2 levels by increasing ventilation. Oxygen conformers, on the other hand, adjust O_2 consumption according to ambient O_2 levels. Salmonids are regulators. The following are some effects of environmental changes on the respiratory system and the response of fish.

Temperature increase presents a three-pronged problem for respiration. It increases the O_2 demand with increased metabolism, lowers the solubility of O_2 in water, and lowers the affinity of O_2 to the haemoglobin molecule. Both heart rate and ventilation increase in response to an increase in temperature. The resistance of the peripheral circulation also decreases.

Hypoxia, or lowered environmental O_2, causes an increase in the frequency and amplitude of ventilation in most fish. While there is a decrease in heart rate, there is an increase in the volume of each stroke of the heart (stroke volume), thus maintaining the output of blood from the heart. This results in a large increase in the ratio of water pumped over the gills to the blood perfused through the gills (ventilation-to-perfusion ratio). Oxygen uptake is, therefore, maintained in the face of lowered water O_2 concentration.

Exercise increases cardiac output due to an increase in stroke volume. Ventilation increases and there is a fourfold to eightfold increase in O_2 uptake. The ventilation-to-perfusion ratio remains relatively constant. Acidic conditions generally increase ventilation.

NERVOUS SYSTEM

The nervous system, acting with the endocrine system, is the way a body coordinates its bodily functions. There are sensory input from external as well as internal sources, and nervous output that sends coordinated commands to various parts of the body as nervous impulses and hormones, which can be stimulatory or inhibitory. Finally, there is integration between those two components at various levels including simple automatic reflexes; integration of vital processes, such as breathing; and higher levels, which include complex learning such as in the homing mechanisms of the salmon returning to its freshwater spawning site. The fish nervous system is not completely understood and it is complicated by the many diverse adaptations present. Generalization is difficult but the basics of the fish nervous system are presented here with an emphasis on the bony fish.

Central Nervous System

The central nervous system constitutes the brain and the spinal cord (Fig. 7). The configuration of the fish brain resembles that of other aquatic vertebrates. All the typical nerves and paired lobes are present. While this basic form is consistent among fish species, the relative sizes of the lobes reflect the adaptive response of the particular species to its physical and biological environment. The optic lobes, therefore, in fish with large eyes that live in dimly lit environments are larger than in fish with small eyes. The basic anatomy and function of the brain of a bony fish (including teleosts) is described here.

The Prosencephalon is made up of the telencephalon and diencephalon. The telencephalon is known as the forebrain and contains the paired olfactory lobes. The forebrain is predominantly involved with reception and conduction of smell impulses. The size of the lobes indicates the importance the role of smell has in the fish species. Elasmobranchs (sharks) and bony fishes have quite pronounced olfactory lobes. The forebrain also appears to play a role in fish behavior. The diencephalon is sometimes known as the between brain. It lies in between the olfactory lobes of the telencephalon and contains the pineal organ. There are two possible functions of the pineal organ. It has been suggested that the pineal body is either a photosensory structure or plays a secretory role. The pineal gland may react to the chemical composition of the cerebrospinal fluid or brain tissue by external or internal (endocrine) secretion. Another important component of the diencephalon is the hypothalamus, which is an important area of interaction between the nervous and endocrine systems. Nervous inputs into this area from the brain result in the production and release of many regulatory hormones of the body (more details in the endocrine section). The diencephalon appears to be an important center for incoming and outgoing messages relating to internal homeostasis.

The optic lobes lie in the Mesencephalon and are the central control for vision. The optic tectum, which is the nervous center for the optic lobes, is also found in the mesencephalon. The tectum is made up of neurons and the fibers from the optic lobe end here. The tectum correlates visual impressions with muscle reaction such as darting for prey or avoiding objects in the water. The choroid

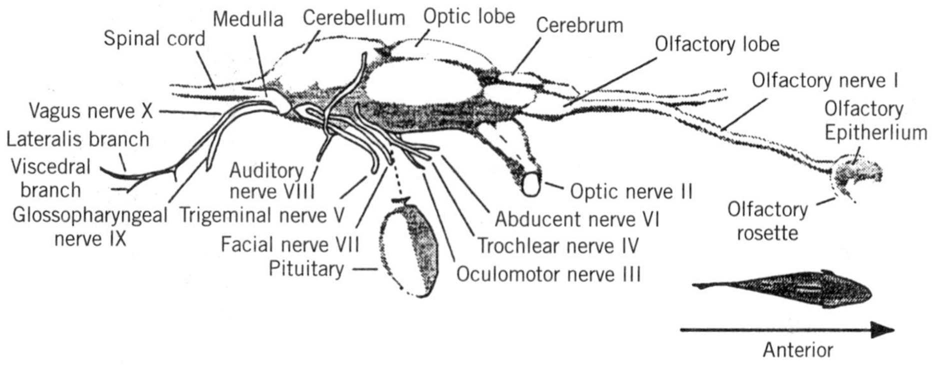

Figure 7. Brain. *Source:* Ref. 3.

plexus, which serves to nourish the central nervous system, is found in the midbrain. It is filled with cerebrospinal fluid and is richly supplied with blood vessels.

The cerebellum has a role in swimming equilibrium, muscle tone maintenance, and fish orientation. The lateral line and sense organs are associated with the cerebellum. The division between the medulla oblongata and the spinal cord is not distinct. All sensory nerves except smell (I) and sight (II) lead to this center. This acts as a relay center between the spinal cord and the higher brain areas. It also controls certain somatic and visceral functions including respiration, osmoregulation, and swimming equilibrium.

As in other vertebrates the gray matter makes up the central region of the spinal cord and consists of nerve cells. White matter, which surrounds this core, contains the myelinated nerve fibers. The spinal cord receives sensory, afferent (to the brain) fibers by way of the dorsal roots of the spinal nerves and gives off the efferent (from the brain) motor fibers to the ventral roots of the spinal nerves.

Peripheral Nervous System

The peripheral nervous system provides a means of communication from the environment where stimuli are received by sensory organs, to the central nervous system and from the central nervous system to the proper effector organs in the body, muscles, or glands. The spinal nerves are paired and carry sensory and motor fibers from and to the fish body. Ganglia act as relay centers along the spinal nerves. Fish have 10 cranial nerves (higher vertebrates have 12). They are similar to spinal nerves except they do not have a ventral or dorsal root and they emerge from the skull. Some cranial nerves are strictly sensory (afferent) and some are strictly motor (efferent).

The autonomic nervous system is part of the peripheral nervous system that innervates smooth muscle, heart muscle, and glands. Within this system is the sympathetic nervous system, made up mostly of spinal nerves, and the parasympathetic nervous system, made up of cranial nerves. The autonomic nervous system controls many of the vegetative (slow, routine, and automatic) systems of the body. As in mammals, the innumerable voluntary functions of the body are controlled in various centers of the brain.

Coordination within the nervous system involves matching the input stimuli to appropriate actions. Those actions may involve the regulation of heart rate, for example, with various feedback of important information from the cardiovascular system. It involves the coordination of gross behavior such as schooling, where inputs from visual and acoustical sensors are coordinated with gross muscular action involved in swimming, which in itself involves a balancing act involving the labyrinths, fin positions, and trunk movement.

Sensory Organs

Fish have an acute sense of smell and rely on the olfactory organ to find food, avoid predators, and to help guide migratory fish to their spawning grounds. Some fish utilize pheromones, chemical substances that can be smelled, to signal alarm to the rest of the school. Catfish, which have an excellent sense of smell, use the specific odor of the slime of individual fish to differentiate fish in a school hierarchy. As the water passes through the nostril over a transverse septum, olfactory epithelium connected to the olfactory nerve signals the smell response to the olfactory lobes in the brain.

Taste buds are the gustatory organs and can be found in the epithelium of the mouth, lips, esophagus and snout. Some fish, mainly those that are not sight feeders, may have taste buds at other places on the body. Free nerve endings from the cranial nerves appear in sensitive regions, suggesting that touch works with taste to stimulate the feeding response. External taste receptors respond to monovalent ions and, to a lesser degree, divalent ions. Internal receptors respond to amino acids, sugars, strong salts, and acids. Both taste and smell are based on organic chemoreception. It also appears that tactile, chemical, and temperature sensations are closely integrated in fish skin.

Sight is extremely important to salmonids and other fish that depend on this sense to detect movements and color changes that signal prey, predator, and mate. The eyes of most bony fishes have a spherical lens, the most powerful shape for a single lens, that in some cases can be changed in convexity and in distance from the retina, the visual layer of the eye (Fig. 8). This is how the fish focus, because very little focusing is done by the cornea. There is little difference in the refractive index, the source of light diffraction, between the cornea and water and, therefore, little focusing is necessary. The cornea does play a role in protecting the rest of the eye from physical pollutants and harm and may help prevent dehydration. The iris controls the amount of light getting through to the retina. Most fish

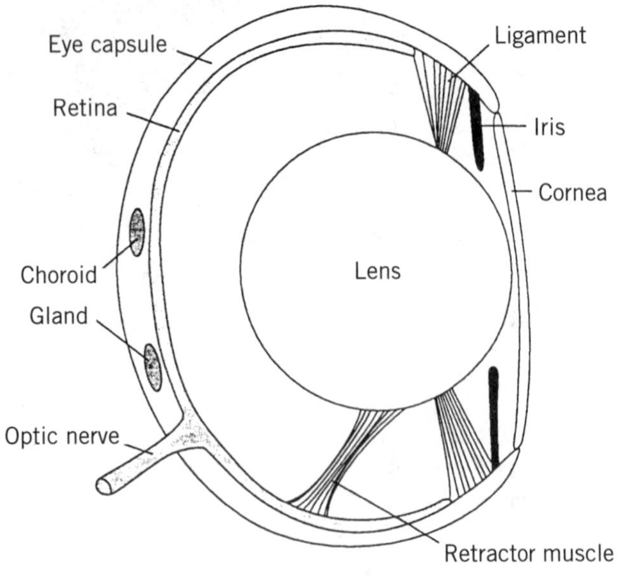

Figure 8. Cross section of a fish eye. *Source:* Ref. 2.

(eg, elasmobranchs) have a fixed iris that contains guanine and melanin. Sight feeders, such as the salmonids, have a well-developed lens muscle and focus by actually moving the lens. The eyeball is ellipsoid and lends a great depth of field, that is the sharp simultaneous focus of objects that are distant and near. This is due to a wide-angle configuration of the components: the flattened eyeball, the round lens, and the protrusion of the lens out of the iris. Light travels through the lens and is focused on the retina. Retinal layers such as the optic nerve fibers, ganglion cells, bipolar cells, and photoreceptor cells, (the cones and rods) all have a role in converting the light to sight images in the brain. The rod cells are sensitive to low light levels but not to color. The cone cells react to strong light and are color sensitive. Photochemical reactions in these cells are transmitted as electrical impulses to the optic nerve and eventually to the optic lobe in the brain. The sensitivity of the retina to particular colors of light depends on the pigments found in the rods and cones in the retina.

The small pineal organ serves a visual as well as a secretory role in fish. It is a small fingerlike organ is located just under the cranium on the dorsal midline of the brain. It serves possible functions in the detection of day length. While it is not vital to life, it may serve an important role in the adjustment of body color to the environment as well as in the synchronization of physiological changes associated with smoltification, in anadromous salmonids, with day length.

The inner ear functions in both equilibrium and hearing. It is found on either side of the medulla oblongata of the brain. The inner ear is made up of two sections, the pars superior and the pars inferior. The pars superior has three semicircular canals filled with fluid and has an ampullae and utriculus. The ampullae contains the receptor tissue, which has sensory cells with long sensory hairs similar to taste buds, and the lateral line. Displacement of the body causes the sensory cells to react and the proper righting mechanism to take place. The utriculus of the pars superior contains an otolith; a calcarious stone. This stone lies on the sensory hairs and responds to the force of gravity. This system works with the retina for maintenance of balance. The pars inferior is the structure of actual sound reception and it is composed of the two vesicles, sacculus and lagena. These vesicles, containing otoliths, are innervated by the auditory nerve. The otoliths will vibrate at a different resonance then the rest of the body when presented with sound waves. This triggers the sensory hairs to send a nervous impulse up the cranial nerve.

The lateral line is a set of sense organs found only in fishes and the aquatic stages of amphibians. The lateral line is innervated by cranial nerves and is closely associated with the auditory centers. The brain relays secondary impulses from the lateral line to the hypothalamus, thalamus, the optic tectum, and the cerebellum. The lateral line helps in the detection of prey, predators or mates by sensing hydrodynamic displacement. The lateral line receptors respond to pressure waves and low frequency vibrations and other localized disturbances.

ENDOCRINE SYSTEM

The endocrine system coordinates physiological processes with each other as well as with the external environment by way of chemical messengers called hormones. The hormones are released in response to changes such as light, temperature and stress. They ellicit and coordinate the appropriate responses from parts of the body, whether it be by a generalized stress response, a fright reaction, maturation of the gonads, smoltification, migration or spawning. Hormones thus play important roles in many physiological processes such as reproduction, osmoregulation, mineral metabolism, and growth. Endocrine glands are ductless tissues that release hormones into the blood.

The effects of hormones may be slow but are persistent in contrast to the rapid and transient action of nerves. It is important to realize the close relationship between the endocrine system and the nervous system. These two systems work together to create a communication network with each system dependent on the other for complete integration. Neurohormones are released by nerve cells, not endocrine glands, but have similar actions. They are carried by the blood to a target organ or to an endocrine gland, which can, in turn, release its own hormone. Neurohormones act as a link between the endocrine and nervous system.

Pituitary Gland

The pituitary gland lies close to the hypothalamus in the brain. It is made up of the infundibulum, the down growth of tissue from the brain, and the hypophysis, the upgrowth of tissue from the roof of the mouth. The hypophysis is differentiated into the adenohypophysis and the infundibular tissue. The adenohypophysis is the site of synthesis, storage, and release into the blood of several different peptide hormones. The infundibular, or neural, tissue forms the neurohypophysis, which is connected to the brain by the infundibular stalk. The neurohypophysis is not a discrete tissue in teleosts and appears to be a storage-release center for the materials that are synthesized in the hypothalamus and then transported to the neurohypophysis along neurosecretory axons.

Growth hormone increases appetite, improves food conversion, increases protein synthesis, decreases nitrogen loss, stimulates fat mobilization and oxidation, and stimulates insulin synthesis and release. Prolactin is necessary for electrolyte and water balance in fresh water. It reduces water absorption and reduces sodium movement through the gill, kidney and urinary bladder. It may also cause the dispersal of xanthophore pigment in the skin of some fish. It is involved in lipid metabolism fat storage and reduces thyroxine levels of the serum. Thyroid stimulating hormone (TSH) regulates thyroid function and is controlled by negative feedback from the levels of circulating thyroid hormones. Adrenocorticotropic hormone regulates the release of adrenocorticoids (steroids) from the interrenal tissue of the kidney. Melanophore stimulating hormone has a small effect on external body pigmentation in teleost. Fish also have sympathetic innervation of melanophores. Gonadotropic hormone, luteinizing hormone, and, to a minor extent, follicle stimulating hormone, have crucial roles to play in gonad development and spawning. Oxytocin is involved in reproduction: egg formation, mating, and spawning. Vasopressin affects diuresis and, therefore, assists in osmoregulation.

It is clear that the pituitary gland is the major gland in the endocrine system, exerting a large influence on processes of the body such as growth, osmoregulation, and reproduction. Target organs of the pituitary hormones are the thyroid gland, interrenal tissue, testis, and ovary. The synthesis of stimuli from the environment and subsequent hormonal release demonstrates the link between the central nervous system and the endocrine system of the body. That link is the hypothalamus–pituitary axis. The hypothalamus receives external stimuli and communicates changes to the adenohypophysis of the pituitary, in close proximity, where the release of the appropriate hormone is elicited.

Thyroid Gland

Thyroid tissue is generally scattered throughout the pharyngeal area, in the head kidney, and in some cases around the eye. The hormones secreted by the thyroid are thyroxine (T4) and triiodothyronine (T3). Thyroid action is linked closely to the pituitary, which secretes TSH. Secretion of TSH is under negative feedback control depending on the levels of circulating thyroid hormones. There also appears to be a strong relationship between thyroid activity and the lunar cycle. Peak thyroxine levels have been shown at the period of the new moon, which is also when peaks in the outward migration of coho salmon have been observed. The surge of thyroxine influences growth, migratory restlessness, and seawater preference.

Ultimo Branchial Gland

This gland is equivalent to the parathyroid gland in terrestrial mammals. It is generally located between the abdominal cavity and sinus venosus. Calcitonin is produced by this gland, which regulates calcium by controlling its conservation and mobilization.

Pancreas

The pancreas of teleost is usually diffuse and scattered around the intestine and spleen, and often extends into the liver. Insulin and glucagon are the two hormones associated with this gland. The action of the pancreas appears to be modified by different nutritional states, particularly in response to changes in protein metabolism. Insulin affects amino acid metabolism and incorporates amino acids into skeletal muscle and may not be as important to glucose homeostasis in teleosts as it is in mammals. It decreases the rate of gluconeogenesis. It may have some role in saltwater survival. Glucagon simulates gluconeogenesis from muscle amino acids.

Renal Tissue and Interrenal Tissue

A number of hormones and tissues are involved in kidney function. Those include renin, angiotensin, and antidiuretic hormone. Their functions are covered in the section on the renal system. Interrenal tissue is located in the anterior kidney tissue, or the head kidney. The primary hormones released are corticosteroids. Under control of the pituitary through the action of adrenocorticotropic hormone (ACTH), the corticosteroids are released in response to appropriate stimuli. Chromaffin cells, the functional equivalents of medullary tissue of the adrenals in mammals, secrete the catecholamines, adrenaline, and noradrenaline.

Corticosteroids such as cortisol are important hormones produced in the interrenal glands. Interrenal tissue response increases with stressful conditions, such as temperature change, exposure to toxic elements, and spawning. This elicits a subsequent increase in cortisol levels. Cortisol is important in regulation of water and electrolyte balance by affecting ATPase activity and sodium flux across the gills. It is known as the saltwater adapting hormone, because it increases renal sodium retention, reduces glomerular filtration rate, and increases permeability of urinary bladder. It is important in smoltification and subsequent saltwater acclimation. Cortisol is also involved in metabolism, by promoting gluconeogenesis and related carbohydrate and protein metabolism. It also plays a role in preventing exhaustion by mobilizing energy reserves. Corticosteroids influence oocyte maturation and seem to be influenced by pituitary gonadotropins.

Catecholamines, such as adrenaline and noradrenaline, respond immediately to stress. Catecholamines cause hyperglycemia (high blood glucose), with a reduction in liver and muscle glycogen in some cases. They also cause systemic vasoconstriction, vasodilation in the gills, and increase the force and rate of the heartbeat.

REPRODUCTIVE SYSTEM

The reproductive systems and strategies in fishes are extremely varied. Examples of the salmonid family are emphasized here. Salmonids reproduce sexually. Pacific salmon spawn only once in their life cycle, whereas the trouts and Atlantic salmon spawn repeatedly during their life span. Unlike other fish, lipid levels have little to do with initiating spawning activity or gamete development in the salmon. Lipid levels are, however, correlated to the distance that animals have to migrate up freshwater streams to reach their spawning grounds. Lipids as well as body proteins are consumed in this migration. The primary environmental factor involved in sexual maturation is day length. The changes in day length are probably more important than absolute day length in the processes that affect reproduction. Temperature also plays an important role in gonadal development in some fishes.

Both the testes and ovaries are located along the midline of the fish, ventral to the swim bladder, extending from the anterior to the posterior end, terminating at the vent. Eggs are commonly released into the peritoneal cavity and reach the outside through a funnel and ovarian duct. Sperm almost always stay inside ducts until they are released.

Egg Size and Fecundity

In general, fish that lay smaller eggs lay many of them. Cod, for instance, lay millions of eggs that incur a high mortality rate. Smaller eggs also tend to hatch in a relatively short time and the hatched young require microscopic food almost immediately because the stored nutrients in the way of yolk is limited or nonexistent. Conversely, fish that lay larger eggs lay fewer eggs that take longer to hatch and have higher survival rates. The young also can survive on the yolk for days or weeks and can ingest large food particles when feeding begins.

There is a correlation between the length of parental care and egg size. That is, the larger and fewer the eggs, the greater the parental care of the young. Herring spawn millions of eggs in kelp beds and abandon them. Salmon lay fewer, larger eggs and bury them in the gravel but still die and do not tend to their young. In the other extreme, some fish produce only a few eggs that are incubated in the body and live young are born that are more or less ready to fend for themselves. The yolk material is solid in salmonids. Their eggs sink in water because they are very dense. The yolk material influences the density and the development time. There is only about 30% water in the yolk material of salmon, whereas the herring egg has oil droplets in the yolk, thus allowing the egg to float. Oviparous fish are egg layers such as salmon, viviparous fish are placental live-bearers as mammals, and ovoviviparous fish are live-bearers but are passive parents.

Hormones

Gametogenesis, the formation of gametes, is dependent on pituitary hormones called gonadotropins. The gondadotropins are leutenizing hormone (LH) and follicle stimulating hormone (FSH). Blocking gonadotropins reduces gametogenesis. These names are taken from mammalian sources. Although both, or analogues of both, may be present in fish, it is LH that predominates. This hormone appears to regulate the maturation and release of the egg. The follicles probably produce the estrogens, estrone, and estradiol under hormonal control from the pituitary gonadotropin. These hormones control the development of secondary sex characteristics in the female fish. They also serve an important function in the production of vitellogenin in the liver. Vitellogenin is then taken up from the blood and incorporated into the yolk proteins. Progesterone, which maintains the uterine development during pregnancy in mammals, may be absent in fish. The corticosteroids testosterone and androstenodione may control the secondary sex characteristics in the male fish.

Gonads and Gamete Formation

The early development of gametes is similar to other vertebrates. The processes of spermatogenesis, (sperm development) and oogenesis (egg development) proceed in the following manner. In spermatogenesis a diploid spermatogonium, through meiosis, gives rise to haploid spermatocytes, which then become spermatids. A process called spermiogenesis is the final step of differentiation of sperm from spermatids. As Figure 9 shows, the sertoli cells serve as nutritive cells. During this process, the cysts swell and finally rupture, releasing the sperm to the lumen, which is continuous with the sperm duct. The interstitium between the cysts contain interstitial cells, fibroblasts, and blood vessels. In oogenesis, cells destined to become eggs proliferate, enlarge, and each becomes surrounded with follicular, or nutritive cells. The granulosa layer of the follicle cells is thought to give rise to the yolk. The proteins in the yolk are derived from vitellogenin as discussed above. The development of all eggs in the salmonid occurs synchronously, that is all at once.

Ovulation

Ovulation, or the release of all the eggs from the follicles, is triggered by the completion of yolk deposition. Once the ova leaves the follicle, they lose the supply of nutrients provided by the follicular cells. They depend on the ovarian

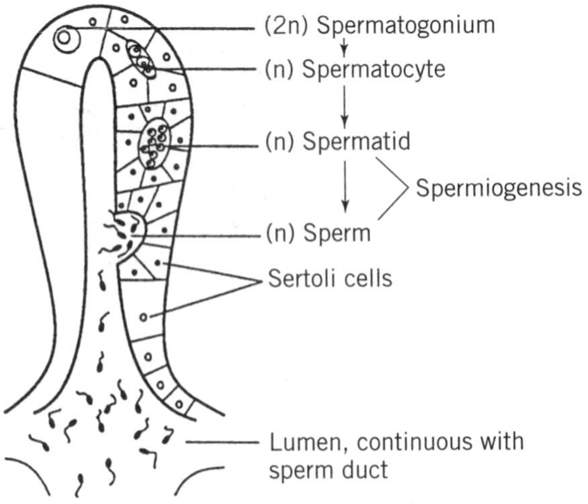

Figure 9. Spermatogenic cyst with several spermatogonial cells.

fluid for nutrients and oxygen. The origin, quantity, and turnover rate of the ovarian fluid are unknown, although it could be a dilute ultrafiltrate of plasma.

Spawning

Spawning follows soon after ovulation. The range of spawning behavior as well as the spawning itself in fish is probably wider than that in all land vertebrates put together. This event is probably controlled by a combination of hormones and environmental factors. There may be a special spawning hormone acting directly on the central nervous system. Mammalian reproduction involves both the pituitary gonadotropins, LH and FSH, affecting the gonads and the ovary, which respond by producing estrogen and progesterone, whereas the LH hormone activity in fish predominates and FSH activity is absent or weak in some species.

Fertilization

Fertilization is also varied among species. Generally, one sperm enters the egg via a micropyle, which closes after entry (Fig. 10). This prevents multiple fertilization. Thereafter, the two nuclei unite. The chorion, the tough outer shell, or protective coating, of the egg swells by the egg imbibing water. This water hardening provides the egg with a protective shell permeable to dissolved substances, gases, and water. Water hardening produces a perivitelline space between the cell membrane and the external environment. Eggs and sperm survive only a short time if fertilization does not take place.

Development

Developmental time depends on environmental conditions. The important factors are temperature, salinity, oxygen, and light. All conditions have varying effects, depending on species. The following outline shows the major steps in the development of the fish, from embryonic to adult stages (Fig. 11). It is similar in pattern to the general pattern in vertebrate development.

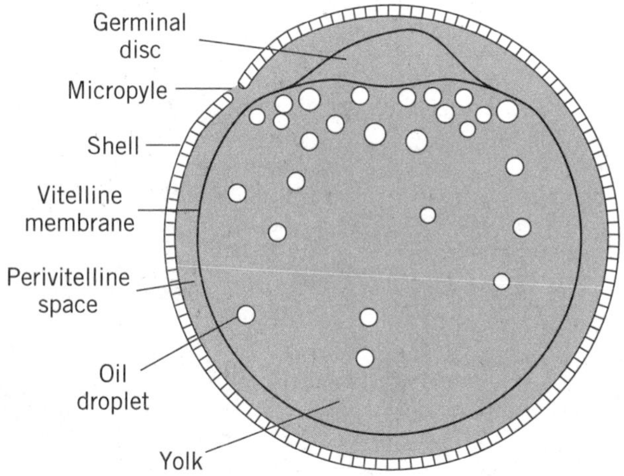

Figure 10. Cross section of a salmonid egg.

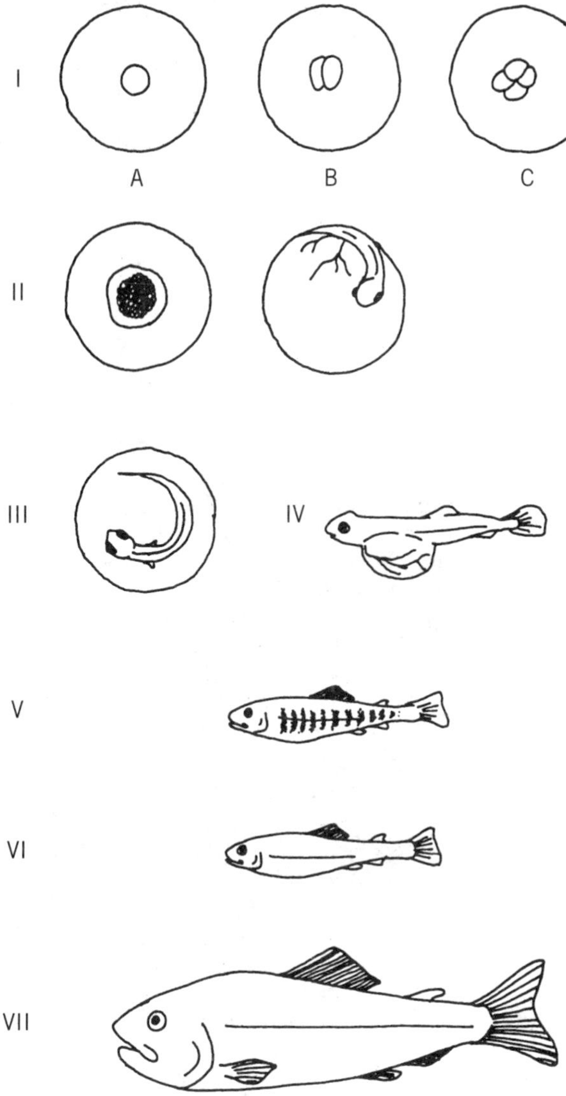

Figure 11. Development stages of a salmonid. I = blastic cleavage; cell division is shown in A, B and C. Cell division continues. II = gastrulation, III = organogenesis, IV = hatching, V = parr, VI = smolt, VII = juvenile to adult.

I. Blastodisc cleavage (2 cells)
II. Gastrulation
  Specialization of cells
  Body tissues develop and grow over the yolk
  Eyed stage
  Basic body form is evident
III. Organogenesis
  Internal organs form
  Fins appear
  Circulatory system develops connection between yolk and internal vasculature
IV. Hatching
  Hatching glands develop around head of embryo and secrete chorionase, an enzyme that dissolves

the tough chorion, or egg shell, for the embryo to escape.

V. Parr
General body form resembles that of the adult
Dark vertical bars along the lateral body wall

VI. Smolt
Parr marks disappear
Body coloration is silver
Body shape is slightly more elongate than smolt

VII. Juvenile to adult
General body shape and color is like the smolt
Size increases as it grows and matures at sea

IMMUNE SYSTEM

The immune system comprises the defense mechanisms of the body that prevent or control disease-producing organisms from entering the body and causing disease. The primary problem of the immune system is to recognize self from nonself. Self refers to all normal physical and chemical components of the body and nonself is everything else. The latter may include bacteria, parasites, viruses, and tumors.

Before discussing immune system in detail, a few key terms should be defined. Immunocompetence is the ability of a fish to mount a specific immune response against a foreign body or agent. The first stages of immunocompetence in salmonid fishes start at 1–2.5 g body weight. It is fully developed at 4 g. Immunological tolerance refers to a condition where the host does not recognize an antigen as being foreign and, therefore, tolerates its presence. Vaccination should not be conducted before immunocompetence is established, or immunotolerance may result. An antibody is a protein manufactured by white blood cells (WBC) that are designed to bind to an antigen for which it was made. An antigen is any substance, usually foreign to the body that is capable of causing the activation of an immune response.

The immune system is made up of two kinds, or lines, of defenses. The first is the non-specific defense and the second is the specific defense. The nonspecific defense mechanisms are also divided into two categories of those involved in preventing nonself substances from entering the body and those that deal with substances that manage to enter the body. Nonspecific defense mechanisms are mounted against any foreign substance, whereas the specific defense mechanisms are designed to attack very specific antigens. Usually the term immune system refers to this last category.

Nonspecific Defense Mechanisms

Physical and chemical barriers to pathogen entry include the mucus on the skin and the skin itself. Mucus is continuously produced from goblet cells and it provides both a physical and chemical barrier to pathogen entry. It sloughs away pathogens and contains antibacterial proteins such as lysozymes, agglutinins, and lysins. Both the epithelium and the scales that make up the skin provide an effective physical barrier to pathogens. Some pathogens, however, do enter the body. Once the pathogens are inside, the following components of the nonspecific immune system work to limit growth and multiplication of the pathogen. Serum transferrin is a protein that keeps free iron concentrations low. Growing and replicating bacteria need iron. C-reactive protein is a nonspecific agglutinating protein that coats the surface of the antigen and allows phagocytosis to occur more readily. Interferon is a soluble substance that acts against viruses. It prevents viral DNA replication in an infested host and cell. Cellular nonspecific defenses are performed by certain WBC such as neutrophils, macrophages and natural killer cells.

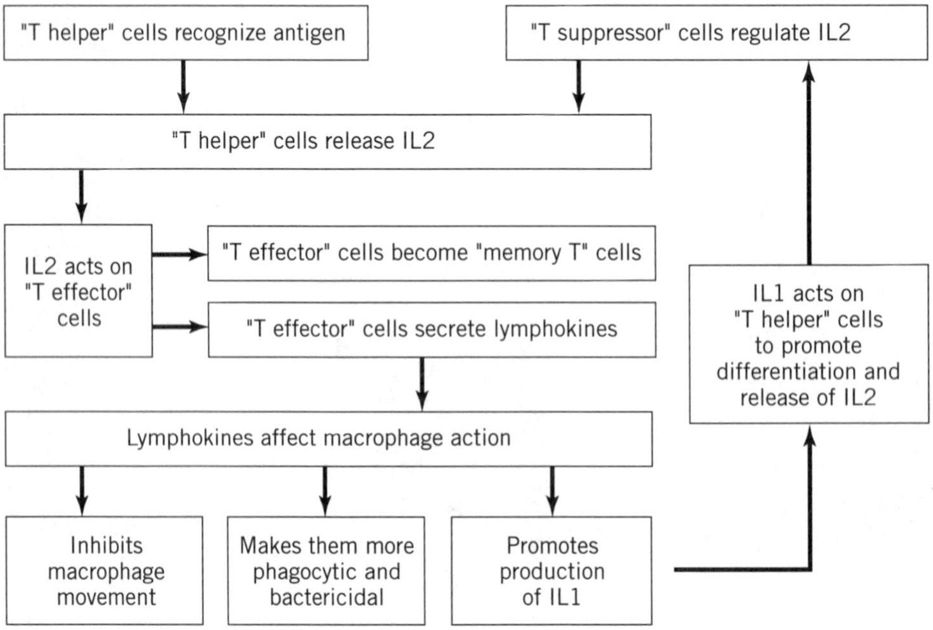

Figure 12. 1. Recognition of antigen by T helper (Th) cell. 2. Th cell releases a soluble messenger cell called interleukin (IL2). 3. Interleukin 2 acts on T effector cells. 4(a). T effector cells secrete lymphokines. 4(b). T effector cells begin differentiating to memory T cells. 5. Lymphokines mediated macrophage action, which results in antigen or antigen-host cell destruction. 6(a). Lymphokine promotes macrophage production of another interleukin (IL1). 6(b). Migration inhibition factor, a lymphokine, inhibits macrophage movement away from the infection locus where antigen was recognized. 6(c). Another lymphokine activates macrophages and makes them more rapidly phagocytic and more strongly bactericidal. 7. IL1 acts on Th cells to promote their differentiation and the release of more IL2. This action is regulated by T suppressor cells.

Specific Defense Mechanisms

There are two types of specific defense mechanism. Humoral immunity defense involves the production of specific immunoglobulins (Ig), which are globulin proteins that have antibody activity. The cellular immunity involves the stimulation of certain white blood cells to destroy specific antigens. The organs involved include the anterior kidney, thymus and spleen. The head kidney and the spleen have antibody-producing cells as well as phagocytes and macrophages. The spleen produces lymphocytes. Fishes lack bone marrow, and the above tissues act, in place, to supply the necessary elements for a functional immune system. Fish immunoglobins are tetramer molecules as opposed to the pentamers in humans and there is only one class of immunoglobins in fish compared to the two to five classes in mammals. These proteins are found in lymph, blood, mucus, and interstitial fluids.

Humoral Immunity. In this type of specific immune response, antigens are selectively removed from the circulation by the use of antibody-producing white blood cells, which are specific to particular antigen. There are two types of cell involved in humoral immunity. Some of the cells produce the antibody and some act as memory cells. Once exposed to a particular antigen, the memory cells can rapidly grow and multiply when they encounter that antigen again.

Cellular Immunity. In this type of specific immunity, the antigen causes certain groups of primed white blood cells to act specifically against the invading antigen. These WBCs can ingest or destroy the antigen as well as proliferate chemical messengers signaling the presence of nonself material. It seems that immersion-vaccination results in eliciting a particularly strong cellular immunity. This type of immunity is also referred to as cell mediated immunity. The following are various cell types involved in cell mediated immunity.

Neutrophils are the first line of defense after antigen invasion. These cells are phagocytic in humans and some fish such as catfish, but they are not in salmonids. They may release a toxic substance in the vicinity of the antigen. They are abundant in the WBC picture. Macrophages have a phagocytic function and also play a significant role as a secretory immune cell. They are more like the human neutrophil in their phagocytic and bactericidal action. Special processing macrophages process and present the antigen to the lymphocytes. The antigen is taken up by these macrophages and processed internally. The antigen is then brought to the surface for lymphocytes to destroy. T-cell lymphocytes are antigen-specific lymphocytes that send chemical messengers that enhance phagocytic and bactericidal activity of macrophages. T-cells do not engulf pathogens, although they do have receptors for antigens. They also enhance macrophage activity. The activities of T cells are restricted by the fact that the antigen must be first processed by a macrophage and presented to this lymphocyte. They are unable to respond directly to free antigen. Their response to an antigen presented to them is to differentiate into effector T-cells and memory T-cells. The coordinated action of the various components of the specific immune system is demonstrated in Figure 12.

BIBLIOGRAPHY

1. D. J. Randall, *American Zoologist* **8**, 179–189 (1968).
2. G. A. Wedemeyer, F. P. Meyer, and L. Smith, *Diseases of Fishes, Book 5: Environmental Stress and Fish Diseases*, TFH Publications, Neptune, N.J., 1976.
3. L. Smith and G. Bell, *A Practical Guide to the Anatomy and Physiology of Pacific Salmon*, Misc. special publication of Environment Canada, Fisheries and Marine Service, Ottawa, 1975.

GENERAL REFERENCES

Reference 3 is a good general reference.

P. J. Bentley, *Comparative Vertebrate Endocrinology*, Cambridge University Press, London, 1976.

P. J. Bentley, *Endocrines and Osmoregulation*, Springer-Verlag, New York, 1971.

C. E. Bond, *Biology of Fishes*, W. B. Saunders Co., Toronto, 1979.

C. P. Hickman, Jr., and B. F. Trump, *Excretion, Ionic Regulation, and Metabolism*, in W. S. Hoar and D. J. Randall, eds., *Fish Physiology*, Vol. 1, Academic Press, Inc., Orlando, Fla., 1969.

W. S. Hoar and D. J. Randall, eds., *Fish Physiology, Vol. IV: The Nervous System, Circulation and Respiration*, Academic Press, Orlando, Fla., 1970.

W. S. Hoar and D. J. Randall, eds., *Fish Physiology, Vol. II: The Endocrine System*, Academic Press, Orlando, Fla., 1969.

K. F. Lagler, J. E. Bardach, R. R. Miller, and D. R. May Passino, *Ichthyology*, John Wiley & Sons, Inc., New York, 1977.

A. J. Matty, *Fish Endocrinology*, Croom Helm, London, 1985.

P. Moyle and J. J. Cech, Jr., *Fishes: An Introduction to Ichthyology*, Prentice-Hall Press, Englewood Cliffs, N.J., 1982.

C. Pincher, *A Study of Fish*, Duell, Sloan and Pearce, New York, 1948.

G. H. Satchell, *Circulation in Fishes*, Cambridge Press, 1971.

L. Smith, *Introduction to Fish Physiology*, F. F. H. Publications, 1982.

George Katsushi Iwama
University of British Columbia
Vancouver, British Columbia
Canada

FISHES: SPECIES OF ECONOMIC IMPORTANCE

This article consists of brief descriptions of fish species of importance to either the commercial or sport fisheries in North America. The description of the each species consists of information about their physical appearance, notes about how they are caught and what food value they hold, their distribution, and their reproductive biology. The common and scientific names used in this article comply with published information (1).

In some cases, limiting the selection of species to those found in North America and of commercial or sport importance restricted the representatives of a particular genus or family. In most cases, however, the omitted species of any genus share most of the major characteristics of those species represented in this review. The amount of information given here about a particular species is largely a function of the body of published knowledge about that fish. This is the reason for any uneven treatment among species. The salmonlike fishes, for example, are probably the most intensively studied family of fishes in the world. The body of knowledge is consequently much larger than a group of fishes such as the ocean perches. Measured units such as body size, weight, fecundity, and age at sexual maturity are estimates in many cases, and are based on measurements on a few individuals. They should, therefore, be considered as rough estimations of the average value. The reference list at the end of the article contains reviews articles and books relevant to the species covered in this article.

ANCHOVY

The anchovies belong to the family Engraulidae. They belong to the same order as the herrings, order Clupeiformes. The anchovies are a schooling fish that are almost exclusively marine, except for some populations that migrate into or become landlocked in freshwater systems. The three species found in North American waters are the northern anchovy, the striped anchovy, and the silver anchovy. The northern anchovy is found off the Pacific Coast and the latter two are Atlantic species that are relatively rare in North America. The anchovies are typically found in temperate to tropical waters of Pacific, Atlantic, and Indian Oceans. They are a small and silver fish that usually measure less than 15 cm, although they have been reported as large as 50 cm. They have characteristically silvery sides with greenish blue to gray backs and a bright strip along the midlateral flank. The anchovies have no lateral line.

The northern anchovy (*Engraulis mordax mordax*) is presently of commercial importance in California. Their distribution extends from Cape San Lucas in Baja California to the Queen Charlotte Islands off the British Columbia coast. The preferred temperature range seems to be between 14.5 and 18.5°C. Although there was a sufficient abundance of northern anchovy off the British Columbia coast in the 1940s, stocks have declined to a point that a viable fishery cannot be sustained there. Populations off the California coast have been increasing in the last 20 yr, where they are captured in schools with nets for the fresh, cured, or canned markets. They have also been caught for the bait and fish meal markets.

Evidence from California, shows that the mature anchovy may spawn several times each year. Spawning takes place at temperatures between about 13 and 17.5°C near the surface and at night. Several thousand eggs are broadcast and fertilized in the open waters. The eggs are ellipsoid and measure about 1.4 mm along the long axis and slightly less than 1 mm in diameter. As the embryo develops, it hangs upside down during development. At those temperatures, the eggs hatch in about two to four days and the alevins are 2.5–3 mm long. The warm waters also enhance the utilization of yolk, taking only about three days for the yolk sac to be absorbed. The young resemble the adults by the time they are about 25 mm long. They become sexually mature when they are about 13 cm. While about half of the two- to three-year-old fish will be mature, all at four years of age are sexually mature. They may live to be seven years old. Like other coastal fishes, the anchovies move to deeper waters during the winter and migrate toward the inshore during the summer. They also spend the daylight hours near the darker bottom and rise toward the surface at night.

The striped anchovy (*Anchoa hepsetus*) is particularly abundant from Chesapeake Bay to the West Indies and Uruguay. Only strays from those areas have ended up in the northern United States and Nova Scotia, Canada. The silver anchovy (*Engraulis eurystole*) is normally found in the Atlantic Ocean from Woods Hole, Mass. to Beaufort, N.C.

STRIPED BASS

The name bass is usually used in reference to the largemouth bass (*Micropterus salmonides*), which has the reputation for being the most popular game fish in North America. The striped bass (*Morone saxatilis*), while also being an extremely popular game fish, is classified under a different family (Fig. 1). Like groups of other fishes, the basses include a number of species that are quite different from each other. The striped bass is not hermaproditic as the sea bass is and it has only two spurs on the opercular bone instead of three. It, therefore, belongs to the Percichthyidae family instead of the Serranidae to which the sea bass belong.

There is no predictable way in which this animal can be caught because of its sporadic feeding habits. The fisherman may be lucky with a particular lure or bait one day but be frustrated without a bite on the identical tackle and bait the next day. The animal is caught with a wide variety of sport fishing tackle. The most common tackle is the spinning rod with a lure that resembles the shad. The striped bass can also be caught on the fly rod with streamers or casting with bait. While it may be caught in open lakes and reservoirs, the most popular locations are at the outfalls of hydroelectric power plants or at the foot of dams. The elusive fish attracts the avid fisherman of inland waters. The present record is a 26.9 kg striper taken from the Colorado River in 1977. In addition to its large size, the popularity of the striped bass as a sport fish is due to the fact that it feeds actively in warm as well as cold temperatures. It also actively feeds in daylight as well as in the dark. This popularity, however, has added to the plight of dwindling populations of this species in Atlantic coastal waters. The effects of pollution and mechanical destruction of their free-floating eggs in dams in combination with the increasing catch by sport fishing has imposed great pressures on populations of this fish.

Although the striped bass is found on both coasts of North America, its native habitat is the Atlantic Coast. The populations in California and Oregon are decedents of a handful of animals that were transplanted from the East Coast into the San Francisco Bay area between 1879 and 1881. The construction of dams, which trapped migrating

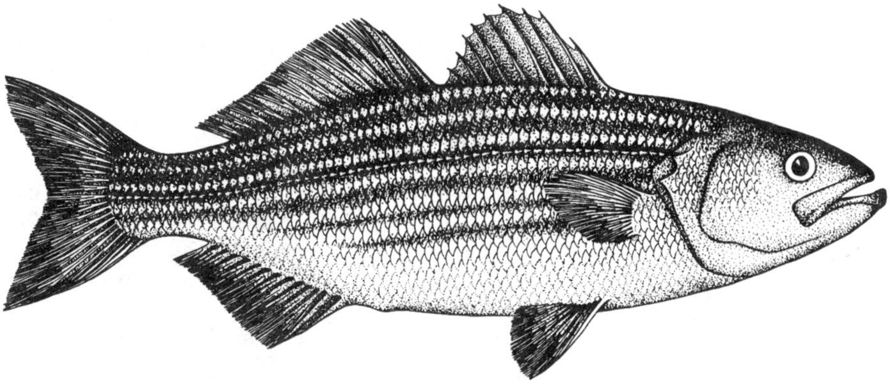

Figure 1. Striped Bass (*Morone saxatilis*). *Source:* Copyright 1990 by B. Guild-Gillespie.

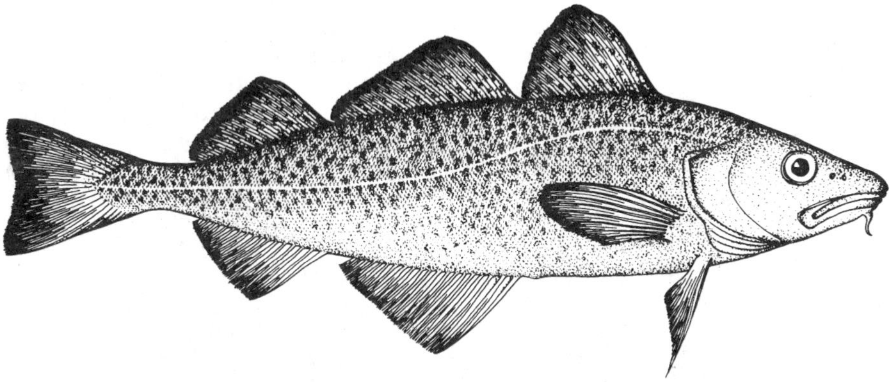

Figure 2. Atlantic cod (*Gadus morhua*). *Source:* Copyright 1990 by B. Guild-Gillespie.

populations of striped bass, demonstrated that these animals can adapt to a wholly freshwater existence. They can thrive without the saltwater phase of their life cycle. This, along with the development of successful rearing technology in hatcheries, led to the transplantation of populations to inland states throughout North America and to the establishment of successful fisheries in 17 inland states.

The striped bass is omnivorous. The juvenile fish feeds on planktonic organisms, such as copepods, and on insect larvae. After the animals are between 7.6 and 12.7 cm in length, they begin to capture and feed on smaller fishes. The adult will feed on a wider range of organisms including fish, worms, shrimps, and shellfish. The striped bass is an anadromamous fish. That means that, like the costal salmon species, it spends its adult life in salt water and migrates into freshwater streams to spawn. Unlike the salmon that migrate great distances in salt water, however, the bass's saltwater residence is confined to coastal waters.

The female produces about a million eggs, 1 mm in diameter, for every 4.5 kg of body weight. The females are always larger than the males and also live longer. The females can live up to 40 yr, reaching sizes of over 4.5 kg. All animals that are greater than 13.6 kg are almost exclusively females. With the increasing day length and warming temperatures of spring, the migration into freshwater streams occurs. Spawning generally takes place between 4-yr-old females and 2-yr-old males. The spawning takes place in open waters of the stream. The female broadcasts her eggs, which are fertilized by several males. Due to the oil content of the eggs, they are bouyant and float freely in the current of the river and later in the estuary. After a rather short incubation period of two or three days, the alevins hatch with yolk sacs. The alevins obtain nourishment from the yolk for the first four to six days, after which they resemble a small adult. Very much like the salmon young, the 5.1- to 7.6-cm young show vertical parr marks, which are eventually replaced by the characteristic horizontal stripes when they are about 6 in. in length. The juvenile then spends its growing and maturing period in the coastal marine environment until sexual maturity.

COD FISHES

The cods belong to the family Gadidae, which has three subfamilies, the Lotinae, the Phycinae, and the Gadinae. As Figure 2 shows, the cods have elongate bodies with large heads and mouths. They have well-formed caudal fins and may have two or three dorsal fins as well as one or two anal fins. They are normally dark and colored brown to gray. The cods produce free-floating eggs that do not contain oil globules. There are approximately 55 species of cods belonging to 21 genera. They are primarily marine and inhabit the cooler waters, near the bottom of northern seas. Most migrate inshore during the summer and move to deeper waters in the fall and winter. There are exceptions such as the burbot (*Lota lota*), which is a freshwater fish, and landlocked populations of the Atlantic tomcod (*Microgadus tomcod*). Furthermore, there are species that reside in southern seas. The 7 species discussed here were selected on the basis of their commercial importance as food fishes.

Atlantic Cod

The Atlantic cod (*Gadus morhua*) is one of the most important food fishes of the world. It is Canada's most important commercial species in terms of net value. Atlantic cod is caught commercially in a many ways, including trawl gear, seine, and gill nets, hand lines and jiggers. In 1984, the 463,100 tons of Atlantic cod landed in Canada was valued at $168.6 million. The Atlantic cod is also caught for sport with hook and line. The Atlantic cod cheek is a delicacy, served in the finest restaurants. By-products of the cod include fish meal, cod liver oil, and glue. Between these two extremes, the Atlantic cod is present in almost every food niche. It is sold fresh, frozen, smoked, salted, and canned. Fish and chips and fish sticks are prime market products for the cod.

The Atlantic cod is distributed throughout the cooler northern waters of the Atlantic Ocean. On the eastern shore, the Atlantic cod is found from Iceland, south to the Baltic Sea and to the Bay of Biscay. On the western side of the Atlantic, it is found in the North Atlantic from Greenland and southern Baffin Island, to Cape Hatteras, N.C. The majority of the populations are found on the continental shelf.

The Atlantic cod can be found spawning during most months of the year, from February to December, throughout their extensive range. The northern populations spawn earlier. As an example, spawning in the Grand Bank, which is one of the major fishing grounds for the Atlantic cod, begins in April, peaks in May, and ends in June. Spawning takes place in the open ocean at varying depths. While some may spawn at depths less than 110 m, other Atlantic cod populations will spawn deeper than 182 m. Both eggs and milt are broadcast into the water. Fecundity is high and increases with age. Although a fish 51 cm in length may produce 200,000 eggs, a female 140 cm may produce 12 million eggs. Like the haddock, the eggs are about 1.5 mm in diameter, spherical, transparent, and bouyant. They rise and float freely under the surface during incubation. As for all fish eggs, incubation time depends on temperature. Due to the colder waters that characterize their environment, the eggs take from 40 to 60 days to hatch (−1.5 to 1°C). The newly hatched young are about 3–6 mm long and remain pelagic until a length of about 25–30 mm. They then descend to the bottom to feed and mature to adult sizes. The average adult caught by the commercial fishery is about 5–6 yr old, 50–60 cm long, and about 1–2 kg. They can live to be over 20 yr old and the largest recorded animal was 95.9 kg.

Pacific Cod

The Pacific cod (*Gadus macrocephalus*) is known as the common or gray cod. Like its counterpart on the Atlantic Coast, it is the most important trawl-caught bottom fish in western Canada. This trawl catch increased to a peak of about 9.1 million kg in 1966. It decreased after that to about 2.3 million kg in 1970. The Pacific cod is sold fresh and frozen. The flesh is filleted, and products such as fish sticks or fillet blocks are made for the domestic and export markets.

The Pacific cod is similar in appearance to the Atlantic cod. In general, it may be more slender and may have a longer barbel under the lower jaw that equals or exceeds the diameter of the eye. The Pacific cod is found on both sides of the Pacific Ocean. Its distribution extends from Santa Monica, Calif. north to Alaska and around into the Okhotsk and Japan seas. It is also found off Korea and in the Yellow Sea to Port Arthur.

The Pacific cod may be found in shallow waters in the spring but migrates into deep water in the fall. They spawn in the winter months. Spawning females range from 40 to 60 cm in length and are two to three years old when they spawn the first time. A 60-cm female may produce about 1.2 million eggs, which measure about 1 mm in diameter. The eggs and milt are broadcast into open waters and the fertilized eggs may float near the surface, depending on the salinity. In the cold northern waters, the eggs may take four weeks to hatch at 2°C. The incubation time is reduced to about 8 days at about 10°C. At 5°C, the yolk sac is absorbed in about 10 days. The young will grow to about 20–50 cm in one year and to about 45–75 cm in about two years.

Haddock

The haddock (*Melanogrammus aeglefinus*) is a commercially valuable food fish of the cod family. Haddock are caught commercially with the otter trawl. They are sold as a fresh, frozen, smoked, and canned product. The major stocks are presently recovering from the overexploitation of the 1960s.

Like the other cods, the haddock is rather dark in color. The color fades from the purplish gray on the dorsal surface, laterally to a light pink below the lateral line, and eventually to a white on the ventral surface. There is a characteristic black blotch on the lateral side posterior to the head and just dorsal to the pectoral fins.

The haddock is found on both sides of the Atlantic. On the eastern side of the North Atlantic, it is distributed from Iceland south to the English Channel. It is found in the White, Kara, Norwegian, and North seas. On the western side of the Atlantic, the haddock is found from southwestern Greenland, south along the east coast of Canada and the United States to about Cape Cod. They migrate between the shallower (30–37 m) warm waters of the banks during the summer and the deeper (55–125 m) waters in the winter.

The haddock spawns in waters up to 90 m in depth, primarily on the Grand, Emerald, Browns, and Georges banks. They spawn between January and May, depending on location. The spawning adults are between 3 and 5 yr old. While many features of the reproductive process in the haddock are characteristic of the cods, observations of aquarium specimens suggest that the reproductive behavior of the male fish is complex, involving color changes, sound production, and courtship displays. The spawning female measures between 40 and 60 cm long and may produce between 230,000 and 1.77 million eggs, respectively.

The eggs resemble that of the Atlantic cod. Like the other cods, the eggs and milt are broadcast into the open waters. After fertilization, the eggs float to the surface and float freely until they hatch. In the cooler waters of the North Atlantic, incubation may take between one (10°C) and four (2°C) weeks. The newly hatched alevins are about 4 mm long and grow to young larvae of about 25 mm before they resemble the adult in body form. They start to descend to deeper waters between 40 and 50 mm. Although the average 5-yr-old haddock may measure about 50 cm in length, the fish can live to be over 15 yr in age and an individual 112 cm long and 16.8 kg has been recorded.

Pollock

The pollock (*Pollachius virens*) is also known by other names such as Boston bluefish, coalfish, and green cod. The flesh, which is of a darker color than the cods and haddock, has a richer flavor than its relatives. The adult weighs about 2–5 kg and is fished commercially as an important food fish. The pollock is caught commercially in a number of ways, including trawling, longline, hand line, weirs, and traps. It is sold in fresh, frozen, and smoked forms. It is also fished for sport in the shallower waters with artificial lures.

Like the haddock, the pollock are distributed throughout the North Atlantic. On the east side, they are found from Iceland, south to the Bay of Biscay. On the west side of the Atlantic, the pollock are found from southwestern Greenland to Cape Hatteras, N.C.

At maturity, the male and female are about 3 yr old and are about 50 cm long; although records show that they can live up to 14 yr, weight 70 kg, and measure over 100 cm. The pollock shares much of the reproductive features of other cod fishes. They produce about 225,000 eggs each; the eggs are about 1 mm in diameter, spherical, bouyant, and remain pelagic until hatching. Eggs and milt are released into the open waters at a depth of about 100–200 m. Unlike other cod fishes, the pollock spends more time swimming throughout the water column as opposed to being a primarily bottom fish. Its preferred depth is about 110–180 m. Like the cod fishes, however, the pollock migrates toward shore and shallow waters in the summer and offshore to deeper waters in the winter.

Walleye Pollock

The walleye pollock (*Theragra chalcogramma*) has come to be one of the most important commercial fisheries in the world. It has been described as the most productive single-species fishery in the world. While it once had a limited market for human consumption due to its soft flesh, it is fished intensively today for the roe, frozen fish block and fillets, and processed fish flesh markets. Surimi, a processed minced fish flesh product, is also one of the prime products from the flesh of walleye pollock. These fishery products that derive from the walleye pollock are taken by separate fisheries, although trawling and seining are the common methods of catch. The surimi fishery consists of large factory trawlers or large factory ships with fleets of smaller catcher vessels; walleye pollock of all sizes are taken from the bottom and midwater depths. The freezer fishery consists of large factory trawlers and smaller independent fisherman that harvest walleye pollock off the bottom, along with other species such as turbot and the Pacific cod. The spawning adults are caught for the roe fishery from the midwaters.

The walleye pollock has a typical codlike appearance with three, well-separated dorsal fins and very small or no barbels projecting from the lower jaw. It usually has an olive green to brown color on its back with irregular blotches or mottled appearance. It is lighter on its ventral surface and has silvery sides.

The walleye pollock is distributed from central California, north through the Bering Sea, around to Asian waters to the southern reaches of the Sea of Japan. It has a wide distribution vertically in the water column. It has been found from the surface to depths below 380 m.

Walleye pollock begin to mature sexually at about 2–3 yr of age, when they measure about 25 cm. By the time they are 5–6 yr old and 45 cm, more than 90% are mature. The fecundity of a 50-cm female is between about 200,000 and 220,000 eggs. Estimates of fecundity range from about 100,000 to 1 million eggs for females between 30 and 70 cm in length, and 225 to 2,000 g, respectively. While spawning occurs during one season of the year, the female will spawn a number of times within that season. The eggs develop and mature in batches over time to allow multiple spawnings. Spawnings takes place mostly between March and April. The eggs measure about 1.4 mm in diameter and float freely in the water, once they are spawned. They are found mostly in the upper 20 m of the water column. Hatching may take place between 14 (5°C) and 24 (2°C) days. The hatched young measure about 4 mm in length and resemble tadpoles, with characteristic markings. They are born with yolk sacs, which are absorbed by the time they attain about 7 mm in length. At 22 mm and about 50 days after hatching, they have all the adult fin rays and are considered juveniles. Adults can live to 15 yr, although most are between 1 and 7 yr of age.

Hake

The red hake (*Urophycis chuss*) and the white hake (*U. tenuis*) are often considered as good food fishes. The white hake grows to a larger size and is distributed over a broader range than the red hake. Characteristics of the white hake are presented here. Similar in size to other cod fishes, the hake is distinguished by the smaller head, large eyes, long dorsal and anal fins and the numerous barbels on the ventral side of the chin. They have a reddish brown color on their back, which fades to lighter shades on the side and white on the belly. The lateral line is pale. While some hake fisheries exist, this species is mostly taken incidental to other fisheries.

The hake live on mud bottoms, 200–1,000 m in depth. They are distributed on the continental slopes in the North Atlantic from Iceland south to North Carolina. Some have strayed as far south as Florida in deep water. Like the other cod fishes, they move to deeper waters in the fall and winter.

The average size of an adult at maturity is between 60 and 70 cm, although specimens over 100 cm have been recorded. Spawning takes place at different times depending on the location, although most takes place in the winter and early spring. Fecundity is high relative to other cods. A 70-cm-long female may produce about 4 million eggs, and one 90 cm long may produce 15 million eggs. The eggs, therefore, are smaller than those of the other cod fishes and measure about 0.75 mm in diameter. The eggs, which are transparent and bouyant, float about the sea until the young alevins, about 2 mm long, hatch. The young grow up to about 80 cm as pelagic fish, after which they migrate to deeper waters to feed and grow to maturity.

Whiting

The name whiting can refer to any one of three species of fish. It was a name originally given to the walleye pollock (*Theragra chalcogramma*) (1). The Pacific tomcod (*Microgadus proximus*) is also commonly called the whiting. While the Pacific tomcod is highly regarded as a tasty food fish, it is not abundant enough for a commercial fishery. The blue whiting (*Micromesistius poutassou*) on the other hand is fished in Europe for food. The blue whiting is rather rare in North America.

The blue whiting is normally found in the northeast Atlantic, off the coast of Greenland and off southern Iceland. It may also be found off Europe between Norway, south to western Mediterranean. Records of the blue whiting in North American range from Sable Island, south to Woods Hole, Mass. The blue whiting is a deep-water fish, caught by otter trawls at depths beyond 183 m off the continental slope during the spring or early summer. The scarce data on the reproductive features of this fish suggest that they resemble, verly closely, reproduction in other species of the Gadidae. It has been suggested that spawning occurs between mid March to mid May (2). Spawning probably occurs in deep waters, deeper than 1,000 m, and the eggs are about 1–1.3 mm in diameter. The few specimens inspected from the northwest Atlantic suggest that the blue whiting is smaller than other Gadids, measuring less than 40 cm.

FLATFISHES

As the name implies, the flatfishes include the fishes that are flat in the dorsoventral axis (Fig. 3). They have both eyes on one side of the body and are often found lying on the bottom of the ocean covered in layer of sand or mud. All the flatfishes are classified in the order Heterosomata and include the soles of the family Soleidae and the flounders in the family Pleuronectidae. This latter family includes the commercially important species of the true halibuts in the genus *Hippoglossus*, the turbots in the genus *Reinhardtius*, and the right-eyed flounders in the genus *Platichthys*. The commercially important species tend to be right-sided; that is the side that is visible to the observer is the right side of the fish. The left eye migrates, along with appropriate skeletal, muscular and neural rearrangements, so that both eyes end up on the right, uppermost side of the animal. Although these fishes are well known as benthic animals, fish in the juvenile stages are also pelagic. The reading of rings in the otoliths, or ear stones is sometimes used to age flatfishes. Nine commercially important species are discussed here.

Halibut

Although it seems reasonable at this time to regard both the Atlantic and Pacific halibuts as belonging to the same species *Hippoglossus hippoglossus*, the Pacific halibut is considered in current literature to be a separate species *Hippoglossus stenolepis* (Schmidt) (3,4). There were declining levels of global commercial catches during the 1960s and 1970s. The large increases in bottom trawling in the 1960s and 1970s were significant contributions to the decline. These global trends were even more exaggerated in some local fisheries such as the Atlantic landing in New England. The average annual landing was 4.83 million kg (eviscerated and heads off) in 1879. In 1930–1939 it was 802,858 kg and by the period 1970–1975, it had fallen to 75,296 kg. Regulation and stock management have curtailed this trend and Pacific stocks, for instance, seem to be recovering.

The chief way in which commercially caught halibut is preserved for storage and transport to market is freezing. The animal is eviscerated and has the head off before the intact trunk is frozen. While earlier methods included salting and chilling with ice, current methods of quick freezing and glazing or vacuum packaging in plastic provides a more reliable and consistent method.

Atlantic Halibut. The Atlantic halibut (*Hippoglossus hippoglossus*) is a highly prized food fish. The Atlantic hal-

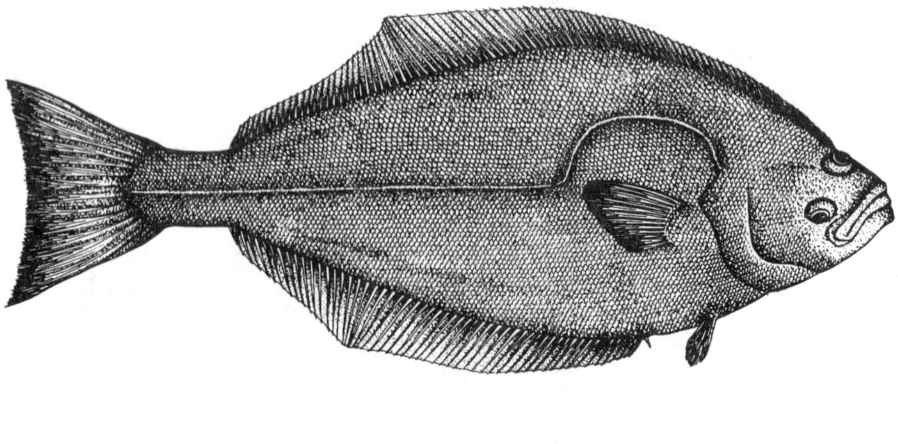

Figure 3. Halibut (*Hippoglossus hippoglossus*). *Source:* Copyright 1990 by B. Guild-Gillespie.

ibut is commercially taken by otter trawls and longlines, with smaller fish being taken by trawls and the larger fish being caught on longlines. Both sizes are taken by fisheries specifically directed at their catch or as incidental catches when trawling or long-lining for cod, haddock, or other ground fishes. The flesh is marketed fresh or frozen. The high price it commands enables fishermen to take relatively small quantities. In spite of this fact, there is evidence suggesting that certain populations are being overfished. The average length of the mature male and female halibut was 84 cm and 98 cm, respectively, in the period 1959 to 1969. Those lengths dropped to 66 cm and 70 cm, respectively, in the years 1970 to 1979. This reduction in average size at maturity is characteristic of populations subjected to heavy harvesting of the larger individuals.

The Atlantic halibut has a typical appearance of a flatfish. It has a flattened body, lying on its left side with its right side showing. The lower side is characteristically white and the right uppermost side is colored various shades of brown. The lateral line is prominent and curved over the pectoral fins. It is a very large fish, with record weights of over 100 kg. The larger fishes are between 30 and 35 yr old. It is the largest of the flatfishes. The females, in particular, have the highest growth rates as well as growth potential among the flatfishes. Females, therefore, become larger than males.

The Atlantic halibut is found on the east and west shores of the north Atlantic Ocean. The distribution on the east shore spans the Bay of Biscay to the Barents Sea. On the west side, the Atlantic halibut is found off the west coast of Greenland, south along the Canadian coast to the coast of Virginia. It prefers cooler waters between 3 and 9°C, at depths of 200–300 m.

Knowledge about reproduction in the halibut is relatively new. Atlantic halibut spawn between February and March around the distribution in Canada. They spawn in deep waters ranging from about 180 m on the western side and about 1,000 m or deeper in the northeastern Atlantic. The mature female, which weighs about 90 kg, may produce over 2 million eggs. The eggs are about 3 mm in diameter, spherical, and neutrally buoyant. They float during development at depths greater than 54 m but the hatched young sink as they develop. These free-floating eggs are most abundant at temperatures between 4.5 and 7°C and salinities between 33.8 and 35.0 ppt. Incubation takes about 16 days at 6°C. Little else is known about the details of the early life stages of the Atlantic halibut.

Pacific Halibut. The Pacific halibut (*Hippoglossus stenolepis*) has had a long commercial history on the Pacific coast of North America. The transacontinental railways across North America facilitated the Pacific halibut catches to be shipped across to eastern destinations and contributed to the near demise of the Pacific populations of this species. At the turn of the century, the total Canadian and U.S. catches was more than 4.54 million kg. Additional technological advancements in fishing vessels and gear increased harvests to a point that by 1930, increases in fishing efficiency no longer yielded greater catches. Through regulation under the International Fisheries Commission, the Pacific halibut stocks have rebuilt over time. The Pacific halibut is commercially caught mainly on hook and line or on other gear that does not involve nets of any kind. Numerous baited hooks are set on lines and rest on the bottom. Those sets can be made by dories from larger mother ships. Longlines of baited hooks can be set and retrieved by the mother ship.

The Pacific halibut is distributed throughout both western and eastern coasts of the Pacific. On the western side, it is found from southern California, north along the western coast to the Bering Sea. It is found off Anadyr, Kuril Islands, and Kamchatka, down to the northern areas of Japan. Like other flatfishes, the Pacific halibut is benthic (living on the bottom) at depths of about 1,100 m.

Reproduction and details of the early life stages in the Pacific halibut are better understood than the Atlantic halibut. Spawning takes place in the winter between November and January from 275 to 412 m. It is not entirely clear whether females spawn every year or whether they spawn every other year. Both may be possible. Pacific halibut are known to migrate as far as 1,600 km to spawn. The mature female is about between 8 and 16 yr old and about 100–150 cm long. Males are considerably smaller. The female may produce between 600,000 and 1.5 million spherical eggs of about 3–3.5 cm in diameter, which are neutrally bouyant between 100 and 200 m. The egg surface has a honeycombed appearance due to the presence of many small holes. The eggs as well as the newly hatched young are pelagic for four to five months after spawning. Like the Atlantic halibut, the newly hatched alevins start to sink to deeper waters and are found below about 200 m. The alevins are transparent, except for the eyes, and about 8–15 mm long with very large yolk sacs. The young are still symmetrical at this point. When they reach a size of about 18 mm, the left eye starts to migrate to the right side and the yolk sac is barely evident. The young appear as small adults by the time they are about 30 mm. By the time the young are about three to five months old, they rise in the water column to about 100 m. Currents carry the young inshore and they become benthic fish at about six to seven months in age. As they mature, they gradually move to deeper waters. By the time they are about 5–7 yr old, they are established at depths of about 100 m and available to the commercial fishery.

Greenland Halibut. The Greenland halibut (*Reinhardtius hippoglossoides*) is a moderately flat fish that has a rich-tasting flesh. It is also known commonly as the Greenland turbot in the marketplace. Greenland halibut are caught commercially by longlines, gill nets, and otter trawls. It is only moderately flat because both sides of the fish are equally muscled. While it is primarily sold as frozen fillets, it is also sold as a smoked product. It is a moderate-size fish that grows to about 25 kg and 120 cm in Atlantic waters; the Pacific specimens are somewhat smaller. The underside is usually white in a young fish but gray in older fish. The visible side is dark brown to black. It is distinguished by its prominent teeth along a protruding lower jaw.

The Greenland halibut is distributed in cold waters in both the Atlantic and Pacific Oceans. In the northeast Atlantic, it is found from around Iceland and the Greenland

Sea to the Arctic Ocean, and both Barents and Norwegian Seas. From there, it is found southward to the Faroe-Shetland Ridge. In the northwest Atlantic, it is distributed from western Greenland, south to Georges Bank. In the Pacific, it is found from northern California to Japan through the Bering and Okhotsk Seas, Sakhalin, and Kamchatka.

The limited knowledge about the details of spawning in this species is, no doubt, due in part to the fact that it spawns in the winter in the far north and at depths of about 650–1,000 m, at temperatures of about 0–3°C, making observation difficult. Spawning takes place in the winter or early spring, depending on location. The Greenland halibut can live to an age between 15 and 20 yr old. Until they are about 5–7 yr old, the males and the females grow at about the same rate. After that, the females grow larger and at a faster rate. All fish over 90 cm are females. The mature female may produce between 30,000 and 300,000 eggs, depending on size. The eggs are larger than the halibut eggs, measuring about 4–4.5 mm in diameter. The eggs are transparent and the newly hatched young rise to depths of about 30 m below the surface, where they live as pelagic fish until they are about 70 mm long. At that stage, they move to deeper waters but do not become benthic like the halibuts; although they are associated close to the bottom.

Yellowtail Flounder. The yellowtail flounder (*Limanda ferrunginea*) is valued as a tasty product that is often sold fresh but that may also be sold frozen. It is either fished directly with otter trawls or as an incidental catch on longlines in the American plaice fishery. While it is fished commercially, it is usually just the largest of the population that gets caught because of its small mouth. While it has the lateral line that curves over the pectoral fins, like the halibuts, the yellowtail flounder has a small mouth and the body size is much smaller. An average adult caught in the commercial fishery is about 30 cm long. This fish gets its name from the yellow markings anterior to the tail and at the bases of dorsal and anal fins. It has a white underside and the color of the visible right side may range from a reddish brown to an olive green.

The yellowtail flounder has a relatively limited distribution, compared to other flatfishes. It is found only in the western North Atlantic Ocean, from southern Labrador down to the Chesapeake Bay. The time of spawning depends on location, but mainly occurs in the spring to summer months. The female produces a large number of small eggs. Females that measure 40–45 cm may produce between 1 and 2 million eggs that are slightly less than 1 mm in diameter. The eggs and milt are deposited on, or near the bottom. Fertilization occurs there and the bouyant eggs then float to near the surface and drift during incubation. The incubation time is about five days at about 10°C. The newly hatched alevins are about 2–3.5 mm long and grow to about 11.5–16 mm before they metamorphose from the anatomically symmetrical young to a smaller version of the adult form. The yellowtail flounder may live to 12 yr of age, and individuals have been recorded up to about 60 cm. The typical fish caught commercially is between 4 and 10 yr old, depending on the location of the catch. The females grow at significantly faster rates than the males.

American Plaice. The American plaice (*Hippoglossoides platessoides*) is a white-fleshed fish that is highly enjoyed as a food fish. It is a very important commercial fish that constitutes about one-half of the commercially caught flatfish species in Canada. Most of the northwest Atlantic fishery is Canadian. American plaice is commercially caught with otter trawls, gill nets, and with the Danish seine. Most of the commercial fishing takes place along the Labrador Shelf, Grand Bank, along the southern Gulf of St. Lawrence, and on the banks of the Scotian Shelf. The flesh is normally sold as frozen fillets. It is colored white on the underside and reddish to grayish brown on the visible right side. The maximum body size rarely reaches over 60 cm.

The American plaice is found on both sides of the North Atlantic Ocean. It is distributed from Iceland, south to the English Channel, on the eastern side and from Baffin Island, south to Rhode Island on the western side. It is normally found between about 75 and 275 m of water, preferring temperatures around freezing. It is benthic and prefers the fine sand or mud bottom. It has a seasonal migratory pattern, which brings it to deeper waters in the winter and to shallower waters in the spring.

Spawning take place between early April and June, depending on location. Spawning takes place at depths of about 90–180 m, at temperatures between 0 and 2.5°C. The mature female produces many small eggs. A 40-cm-long female, 8 yr old, may produce between 250,000 and 300,000 eggs, while a 70-cm-long female has been recorded to produce 1.5 million eggs. Males mature earlier at 4–5 yr old. The eggs measure about 1.5–2.8 mm in diameter. They are bouyant, without oil droplets and float near the surface during incubation. They, therefore, drift widely. At surface temperatures of about 5°C, the young alevins of about 4–6 mm, hatch in about 11–14 days. They transform from the symmetrical young to a small fish resembling the adult form at about 18–34 mm in length. They can live to be 25 yr old, and the record size is 81.2 cm long and 6.3 kg dressed.

Petrale Sole. The petrale sole (*Eopsetta jordani*) is a large flatfish that is prized as an excellent food fish, especially along the west coast of North America. The livers of this fish are a rich source of vitamin A. It is distributed from Bering Sea, south to Baja California. Like other flatfishes, it migrates seasonally and is found at depths of around 70–130 m throughout most of the year, except during the winter, when it moves to deeper waters, around 300–460 m. The petrale sole is white on the underside and olive brown on the visible right side. It has a large mouth and lacks free spines at the origin of the dorsal fin.

The oldest female petrale sole recorded was 25 yr old, whereas the eldest male was 19 yr old. The average spawning female is about 44 cm and the male is about 38 cm. The female produces a large number of relatively large eggs, the number depending on size. Records show a 42-cm female produced 400,000 eggs and a 57-cm female produced 1.2 million eggs. The eggs measure about 1.3 mm in diameter. Spawning takes place between late winter and early spring. Spawning takes place at depths of about 350 m and the bouyant eggs float to shallower depths where they are carried by prevailing currents during incubation. The eggs hatch into 3 mm alevins in about 8–9 days at 7°C and the yolk sac is absorbed within a further 10 days. Samples that measured 22 mm in length showed metamorphosis complete. Other evidence suggests that they settle to their adult benthic existence by the time they are 1–2 yr old.

Rock Sole. The rock sole (*Lepidopsetta bilineata*) is the most commonly used food fish of the smaller flatfishes. There have been several subspecies identified in the Pacific Ocean. At 15 yr of age, the females may reach a length of 60 cm, and the males, 50 cm. It is recognized by a canal formed along the lateral line that arches, typically, over the pectoral fin. It is a right-sided flatfish. Unlike other flatfishes, it tends to inhabit shallower waters. It is commercially taken at depths up to about 200 m, although it is scarce from about 100 m and deeper. Like other flatfishes, however, it does migrate to even shallower waters during the summer. Its distribution extends to both sides of the Pacific Ocean from southern California, up along the North American west coast to the Bering and Okhotsk seas and around to Korea and the Sea of Japan.

Observation off the North American west coast show that spawnings takes place between February and April. The fecundity of females measuring 35 cm and 46 cm in length were estimated at 400,000 and 1.3 million eggs, respectively. The eggs are pigmented yellowish orange and are adhesive. They measure about 1 mm in diameter. Incubation is a function of temperature and may take between 6 and 25 days, corresponding to a temperature range of about 8 to 3°C, respectively. Hatched alevins are about 5 mm long and yolk absorption takes about 10–14 days, influenced again on ambient temperature. Fully metamorphosed young have been observed at about 20 mm in length. The oldest recorded female was 25 yr old, and the eldest male on record was 15 yr old.

Dover Sole. The Dover sole (*Microstomus pacificus*) is highly prized for its quality flesh and excellent keeping qualities in the frozen state. It was originally dismissed as a viable commercial species because of its softness and sliminess. It may be uniformly brown on the visible right side and the underside may range in color from a light to a dark gray. It is characterized by having a lateral line canal that is almost straight as well as the excessive production of slime. While the body is also extraordinarily flaccid, it is known as a very hardy fish. The Dover sole is distributed from northern Baja California, up to the Bering Sea, found mainly on soft substrates. The average body size is about 70 cm.

Spawning in California takes place from November to February. Mature females are approximately 45 cm and males mature at about 40 cm. The female produces a wide range of eggs depending on size. Samples from Oregon showed a 42.5-cm female produced 52,000 eggs, while another 57.5-cm female had 266,000 eggs. The eggs are large, measure 2–2.6 mm in diameter, and have a wrinkled surface. After hatching, the young remain bilaterally symmetrical for several months. Flatfishes usually metamorphose and settle to a near benthic existence at about 20 mm. Dover sole specimens up to 100 mm in length have been observed in a pelagic life stage.

English Sole. The English sole (*Parophrys vetulus*) is characterized by its pointed head and, like the Dover sole, a lateral line canal that is almost straight. It is a moderate-size flatfish that grows to about 60 cm as a female and up to 50 cm as a male. It is colored a uniform brown on the uppermost right side and pale yellow to white on the underside. It has had a long history of being a commercially important species in North America. Its particular iodine flavor, found in some inshore populations, has identified its place in the marketplace. It is typically fished commercially at depths shallower than about 130 m, although its distribution extends to about 300 m.

Spawning off British Columbia occurs between January and March. The range in fecundity is extreme. Records show that a 30-cm female produced 150,000 eggs, and one measuring 44 cm produced 1.9 million eggs. The size of the maturing male and female are rather similar. Mature females measure about 30 cm, whereas the males measure about 26 cm. The eggs are small and measure slightly less than 1 mm in diameter. The eggs float due to the presence of oil droplets of various sizes in the yolk of the egg, start to sink just before hatching. The surface of the egg is covered with small wrinkles and pores. Observations in California show that the incubation time is about 90 days and the new alevins are about 2.8 mm long. Due to the oil droplets, the alevins hang upside down until the yolk is absorbed, in about 10 days, at which time they can swim. They remain pelagic for 6–10 weeks, after which they metamorphose to their adult form and seek the bottom. Young English sole are found in shallow waters but as they mature, the larger fish move to deeper waters. Like the other flatfishes, it tends to inhabit deeper waters in the winter and seek the shallower waters closer to shore in the spring. Another rather peculiar characteristic of this flatfish is that it can migrate long distances. There are records of English sole traveling over 1,000 km, between Vancouver Island and California.

HERRINGS AND SARDINES

The herrings are commercially important in that they support a variety of food fish markets (Fig. 4). Until the late 1960s, the herring was reduced and utilized heavily as a source of oil and fish meal. The herring has and continues to be used for pet food. Herring as food for humans has had a long history. Today, they are sold fresh, frozen, smoked (kippers), and pickled. The smaller Atlantic herring, which are canned, is well known around the world as a sardine. The larger Atlantic herring are canned as kipper snacks and fillets. The roe from herring has recently found a market in Japan as a delicacy item known as kazunoko, as their local herring populations have declined. The herrings belong to the family Clupeidae. Both Atlantic and Pacific herrings are subspecies of the species *Clupea harengus*. The two fish look similar, characterized by a silvery body

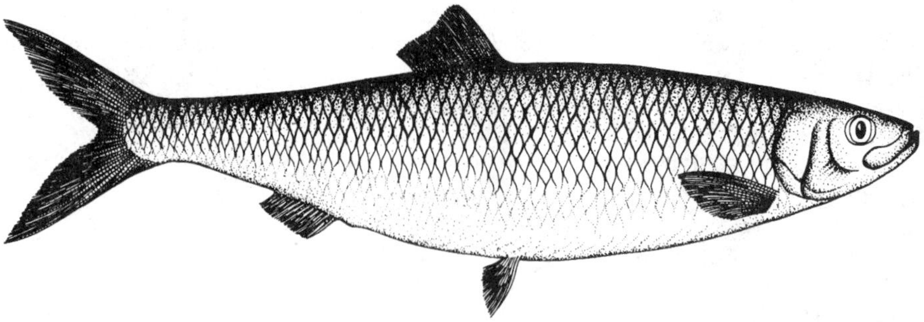

Figure 4. Herring (*Clupea harengus harengus*). *Source:* Copyright 1990 by B. Guild-Gillespie.

that is highly compressed laterally, large cycloid scales and large eyes. The pearl essence from the scales was in high demand during the 1940s for use in high-quality paints for aircrafts.

The Pacific sardine (*Sardinops sagax*) is commonly referred to as the pilchard. It is not the fish that North Americans commonly refer to as the sardine. While the pilchard has some commercial value for its oil and for fish meal, very little is canned.

Pacific Herring

The Pacific herring is fished today for its roe and for reduction purposes. Both herring and herring spawn have been fished by the North American natives since 800 BC. The commercial herring fishery that began in the late 1800s was based on the salted herring market. This was replaced by a fishery based on the reduction industry in which the herring carcasses were reduced for oil and meal for commercial feeds for poultry and fish culture. The relatively new market of herring roe for the Japanese market started in the early 1970s. Today, the herring roe goes to Japan, and the carcasses are reduced for oils and meal.

The Pacific herring (*Clupea harengus pallasi*) is distributed throughout the coastal regions of both eastern and western shores of the Pacific Ocean. On the eastern shore, it is found from northern Baja California up to the Beaufort Sea. On the western side, it is found from Korea, north to the Arctic Ocean. The adults are found at depths of 100–150 m. Juveniles are found between 150 and 200 m of water.

The Pacific herring spawn in the late winter through to April, with peak spawning occurring in March. This exclusive spring spawning distinguishes the Pacific herring from their Atlantic counterpart. Most are able to spawn by the time they are about three years of age. The spawning process is very dramatic as the fish broadcast eggs and sperm near shore. The water in which they spawn turns white with the milt from the males, covering as much as 257 km of shore line and at depths from the surface to about 10 m. The spawning area is usually in the intertidal zone, in sheltered bays or on open sand beaches, but not on exposed coastal areas. The texture of the substrate seems to be an important factor in determining the exact location of spawning.

The female produces between 9,000 and 38,000 eggs depending on her size, which can range from about 20 to about 30 cm. The relative fecundity is about 200 eggs per gram body weight per year. In extreme cases, fish can grow to 50 cm and produce over 100,000 eggs. The eggs are about 0.9 mm in diameter before fertilization but expands to about 1.2–1.5 mm in diameter after fertilization and the absorption of water. They are also very sticky once they are exposed to water. The eggs commonly adhere to aquatic plants such as eelgrasses or rockweed. The incubation time is about 10 days and the newly hatched alevins measure about 7.5 mm in length. The yolk is absorbed within the following two weeks and the fish then begin to feed on planktonic organisms. At that stage, the young do not resemble the adults. They are white, thin, and have large eyes. In about two months following yolk absorption, the young start to resemble the adult form and begin to form schools. By the late summer, the young are about 2.5–4.0 cm long and move to deeper waters in the fall. The Pacific herring can live to be 10 yrs old. During growth and maturation, they may be about 15 mm by the end of their second year, about 20 mm at the end of their fourth year, and about 23 cm at the end of the eighth year of life. They return to shallow waters close to shore as they mature sexually and prepare to spawn. They also tend to have a diurnal migration as rise to the surface in the evening and swim to deeper waters at dawn.

Atlantic Herring

The Atlantic herring (*Cuplea harengus harengus*) looks like its Pacific counterpart with iridescent blue or bluish green back and sides and a silver belly. Like the Pacific herring, its abundance has been declining steadily over time. The Canadian landings, for example, were 528,000 tons in 1968; 250,000 tons in 1975; 177,000 tons in 1980; and 147,000 tons in 1982. The advent of the highly efficient purse seine net has contributed to the harvest pressures on populations of herring. Gill nets, trap nets, and weirs have traditionally been used to catch herring. It is distributed on both sides of the North Atlantic Ocean. On the west side, it is found from Greenland south along the east coast of North America to Cape Hatteras. On the east side of the Atlantic, its distribution extends from Iceland, south to Europe between the White Sea to the Strait of Gibraltar.

Atlantic herring mature to spawn at three to five years of age. There seems to be several discrete stocks that spawn at different times of the year; there are spring, summer, and fall spawning populations. There may be stocks spawning every month between April and November throughout its distribution. Spring spawning occurs in shallower inshore areas, whereas summer and fall spawning occurs in deeper offshore waters. Like the Pacific herring, eggs and milt are broadcast into the open water where fertilization occurs. The eggs then sinks to adhere to bottom plants such as Irish moss and several algal species at depths of 1–4 m. The fecundity can range an order of magnitude from 23,000 to 261,000 eggs, depending on body size and age. The fecundity increases with body size and age up to a certain age, after which egg numbers decline with further aging. Fecundity is also a function of when spawning occurs. Spring spawners produce up to about half the egg numbers of summer and fall spawner. The eggs, however, have a larger yolk mass. This may reflect a strategy to survive the colder months of spring when food supply might be less abundant than in the summer and fall months. The small eggs, which measure from 1 to 1.5 mm in diameter, take from 10 to 30 days to hatch, depending on the ambient water temperature. The hatched alevins are about 4 mm long. They are light sensitive and avoid bright light, seeking deeper waters during the day. Unlike the Pacific herring that stay relatively close to their spawning grounds, Atlantic herring migrate offshore extensively.

MENHADEN

The Atlantic menhaden (*Brevoortia tyrannus*) is a bony, oily fish that is of commercial importance for oil, fish meal, and fertilizer products. It belongs to the family Cludeidae. It is a major protein component in commercial feed fed to cultured fishes. Livestock and poultry feed may also contain menhaden meal. The production of paints, soaps, and certain lubricants may use the menhaden oil. It is not consumed by humans because it has a lot of bones and its oily nature gives off an unpleasant odor when cooked. It is a particularly important fishery from Massachusetts to the Carolinas. It resembles the herring in appearance except for several features. The menhaden is deeper in the body, being more elliptical from a lateral view. It has a large head and lacks teeth. It also has a distinctive black spot posterior to the gill covers with more spots of irregular shapes and sizes along the ventral halves of the flanks. It has silvery sides with a back that can have a blue, brown, or green hue to it. Menhaden are harvested almost exclusively by the purse seine net. Purse seine fishing depends almost completely on the sighting of schools of menhaden at the surface, from aircraft or large vessels. The schooling behavior is an outstanding characteristic of this species, from larval to adult stages.

The menhaden is distinct in its feeding habits in that it is one of few fish that can feed on planktonic organisms. While it is rather common for juvenile fishes to feed on plankton, most fishes use feed higher up the food chain. The menhaden has very fine gill rakers that filter out phytoplankton such as diatoms as well as planktonic crustaceans.

The menhaden is euryhaline. The Atlantic menhaden is mainly an ocean fish that schools off coastal waters of the Atlantic, although it has been reported in fresh water as well. It is distributed from the Gulf of St. Lawrence to southern Florida. Freshwater populations have been reported in the St. John River, New Brunswick, Canada.

The Atlantic menhaden can live up to 12 yr, although specimens over seven yr old are rare. While a few mature at 1 yr of age, 80% are sexually mature at 2 yr of age, and all are sexually mature by the time they are 3 yr old. The body size at 3 yr is approximately 25 cm. Reproduction of the menhaden is similar to the herring. They spawn at sea or in large bays, throughout their distribution. They may also be able to spawn in the St. John River, where they are found year-round. The spawning period for Maine to Massachusetts, its northern distribution, occurs between May and October. Populations south of that area spawn during the other half of the year. The male and female broadcast eggs and sperm into the water where fertilization occurs. The fertilized eggs measure about 1.3–1.9 cm in diameter. Fecundity of females measuring 20–35 cm were 38,000–631,000. The eggs are spherical and bouyant, the latter being due to the presence of an oil globule in the yolk. The young are approximately 2.4–4.5 mm in length at hatching.

MULLET

Distributed mainly in the Atlantic Ocean, Black Seas, and the Mediterranean, the order Mullus includes several species of mullet that frequently move up to the Norwegian coast (Fig. 5). Two species, the red mullet (*Mullus barbatus*) and the striped mullet (*Mullus surmuletus*) generally spawn off the coast in the summer and reach sexual maturity in two to three years. They prefer sandy or muddy bottoms. The oil bubbles in the eggs permit them to float in the water. With the arrival of fall season, the juveniles seek greater depths. Preferring sandy or muddy ground, the red mullet has a steep forehead whereas the striped mullet, usually found above sandy ground, has a less-steep forehead. Although teeth are not present in the upper jaw, they are located on the vomer and the gums. Mullets had been an edible delight since Roman times. Although they are small, they commanded a high price and were commonly brought in the dining hall alive.

Nowadays, the commercially important mullet species is another striped mullet (*Mugil cephalus*). It is large and measures 90 cm and weighs 7 kg. This ash gray fish has a dark blue shimmer with 9–10 light longitudinal stripes on the sides of the body. The striped mullet resides in warm seas including the Mediterranean and frequents river mouths or lagoons. Young striped mullets are nourished in salt water or brackish water ponds until they have grown to acceptable size for the consumer market.

The suborder Mugiloidei, family Mugilidae includes several species of mullet that inhabit coastal waters and can acclimatize to brackish water, fresh water, or salt water. They prefer soft ground with a rich plant source located in tidal zones. The generic name *Mugil* (sucker) derives from their feeding habit. They prey on detritus and tiny organisms on the floor. Otherwise, these mullet feed on mussels, snails, planktons, and little organisms that frequent algal populations.

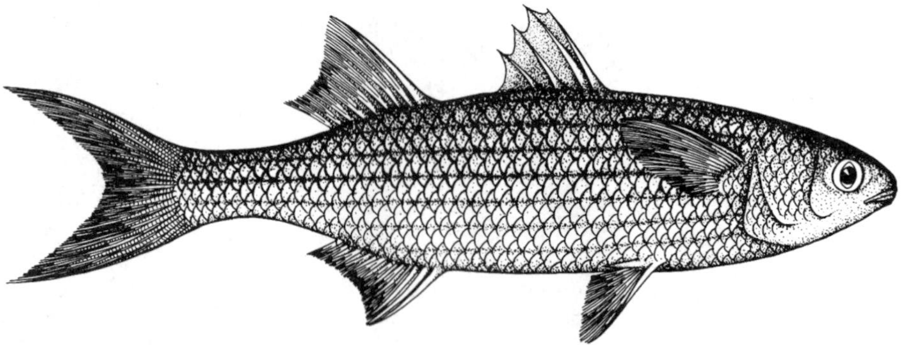

Figure 5. Striped mullet (*Mugil cephalus*). *Source:* Copyright 1990 by B. Guild-Gillespie.

Ocean Perch, Rockfish, and Redfish

The fishes, commonly known as ocean perches, rockfishes, and redfishes belong to the family Scorpaenidae, the scorpion fishes (Fig. 6). In general, these are highly valued food fishes that resemble the freshwater bass in appearance. They are found in all tropical and temperate marine environments around the world. There are 60 genera with over 300 species in this family. The representatives of the popular food fishes in North America belong to the genus *Sebastes*. Fishes in this genus are ovoviviparous, that is they bear live young with some passive nuturing from the mother. The commercially important representatives comprise two species in the Pacific and three species in Atlantic waters.

Pacific Ocean Perch

The Pacific ocean perch (*Sebastes alutus*) constitutes the major species of the Pacific *Sabastes* that is caught commercially for food. The fish may be caught by trawl or hand lines offshore and is mainly sold as fresh and frozen fillets. It can be recognized by a prominent knob, or overhanging tissue, off of the lower lip. It is colored a bright red over the entire body, including fins. The ventral side is usually lighter red. There are olive brown patches below the dorsal fin and on the dorsal side of the caudal penduncle. The Pacific ocean perch is found from southern California north to the Bering Sea. A close relative, the *Sebastes paucispinosus*, is found in the western shores of the Pacific, mainly around northern Honshu in Japan. It is usually found at depths below 130 m and fished most often between about 160 and 300 m.

The mature Pacific ocean perch can be between 30 and 50 cm in length. As for most species, the male is generally smaller, lives longer, and grows more slowly than the female. Spawning occurs in the winter months, and the young are born between January and February. Fecundity can vary an order of magnitude. Fish about 32 cm long and 7 yr old produce about 30,000 eggs, and a female 44 cm long and 20 yr old produces about 300,000 eggs. Unlike the bottom-dwelling adults, the younger are pelagic. Ocean perch have low growth rates. Because of this, the young stay pelagic until the second or third year of life before heading to the bottom to grow to maturity. Adults may live to be 30 yr old.

Yelloweye Rockfish

The yelloweye rockfish (*Sebastes ruberrimus*) are named for their brilliant yellow eyes. They have a characteristic orange-yellow color to their body with a pink tint along the back and sides. The fins are also pink with black margins. The bellies are white with black dots. They are found around reefs from Baja California up the west coast of North America to the Gulf of Alaska. These fish are mostly caught commercially by setlines with live or dead herring bait. They contribute significantly to the whitefleshed fillet market in fresh or frozen states. The yelloweye rockfish can grow to be about twice the size of the ocean perch. They can grow to be about 90 cm. They are normally found at depths between about 50–550 m. The fecundity of a 9 kg

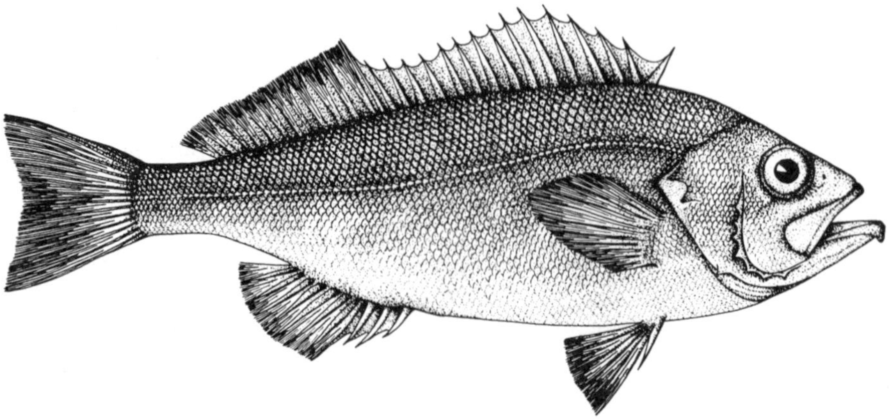

Figure 6. Golden redfish (*Sebastes marinus*). *Source:* Copyright 1990 by B. Guild-Gillespie.

female can exceed 2.5 million young. The young off the Washington coast are born in June.

Atlantic Redfishes

There are several species that belong to *Sebastes* in the Atlantic. Four occur in the North Atlantic Ocean, three in the western North Atlantic and one (*Sebastes viviparous*) in the eastern North Atlantic. The three species of Atlantic redfishes found off the North American coast are the Acadian redfish (*Sebastes fasciatus*), the golden redfish (*Sebastes marinus*), and the deepwater redfish (*Sebastes mentella*). They are all commercially valuable as food fishes, although specific fisheries for each species do not exist. They are, rather, fished as a group and marketed as fresh and frozen fillets under the name of ocean perch. The deepwater redfish probably makes up the majority of the catch off the Newfoundland-Labrador coast, while the Acadian redfish is the main species in the commercial catch of redfish off the Georges Bank-Gulf of Maine region. These fishes are caught with otter trawls.

The Acadian redfish has an orange-red body with red fins, with the pelvic and anal fins having a particularly deep red color. There are green-black blotches below the dorsal fin and on the posterior part of the gill covers. There may be green iridescent flecks on the body above the lateral line. The deepwater redfish is bright red all over. The golden redfish can be colored either orange-yellow or gold-yellow.

The distribution of the Atlantic redfishes extends from Iceland, south to about Virginia on the U.S. coast. The golden redfish is more common in the northern range of this distribution but less common in North American waters compared to the Acadian redfish or the deepwater redfish. The golden redfish is common off Greenland, Iceland, Norway, and in the southern Barents Sea. The deepwater redfish has the broadest distribution of the three redfishes and occurs throughout the North Atlantic range of this genus except the North Sea and the Gulf of Maine. It is also found farther offshore than the other two species. The Acadian redfish may be considered the North American redfish because it is most common along the Canadian continental shelf and southward, particularly along the Georges Bank and in the Bay of Fundy and Gulf of Maine waters. The Acadian redfish occupies the shallowest waters of the three species, found most often at depths between about 130 and 370 m. The golden redfish is found from about 300 to about 370 m. The deepwater redfish is found in waters about 350–700 m.

About half of the populations of the Acadian and deepwater redfishes mature at about 19 cm and 30 cm for males and females, respectively. The female golden redfish matures at about 41 cm. There is evidence suggesting the young hatch inside the female golden redfish in April and are born between April and May. The Acadian and deepwater redfishes hatch their young between March and June and release their young between March and June. A study in the Gulf of Maine showed that the Acadian redfish about 30 cm retains about 50,000 fertilized eggs and releases about 15,000 to 20,000 living young. The young of the Acadian redfish grow to be about 8 cm in the first year and grow an average of about 2.5 cm each year up to about 10 yr of age, declining in rate after that. The growth rates in males and females seem to be about the first 20 yr of life.

The Atlantic redfishes, like the Pacific *Sebastes*, grow very slowly. Sections of the otoliths have been used to age these fishes. That data show that the golden and deepwater redfishes can live to be 48 yr old. This depends on the population. The populations of Acadian redfish in the Gulf of Maine may live up to 20 yr and grow to a maximum size of about 46 cm and 1.4 kg.

SALMON

The salmon family Salmonidae comprises numerous landlocked and anadromous species (Fig. 7). They are found in most waters of the northern hemisphere and dominate the northern waters of North America, Europe, and Asia. This family is made of of three subfamilies, the Salmoninae (salmon, trout, char), the Coregoninae (whitefish), and the Thymallinae (grayling). A common feature among all species is the adipose fin located on the dorsal surface between the dorsal and caudal fins. While representatives of the Coregoninae and Thymallinae belong to this family, the term salmon is usually applied to the salmon and trout of the Salmoninae. This article deals with only those species that are commonly referred to as salmon, which includes species in the genera *Oncorhynchus* and *Salmo*. Even this restriction does not eliminate the confusion between common names and the scientific classification. The Atlantic salmon, for example, is strictly classified as a trout in the genus *Salmo*. The rainbow trout, on the other hand, has just recently been transferred from the genus *Salmo* to *Oncorhynchus*, but it continues to be called a trout. The species covered here include the Pacific salmon species, the rainbow trout, and the Atlantic salmon.

The anadromous species spawn in fresh water where the eggs incubate and where the young spend varying lengths of time. It is common for adults to return to the streambed where they were hatched. During the sexual maturation process, the typically silver fish begins to darken and takes on various colors including black, brown, orange, and red. There is usually a dramatic transformation of the shape of the head and in some instances, the whole trunk. The head of the male usually elongates and there is a pronounced development of teeth. The chum salmon, for instance, develops a large hump back during this process. Many species do not feed once they enter fresh water and begin their migration to their spawning grounds. The spawning process culminates a long journey, sometimes covering thousands of miles, and normally ends in death for the Pacific salmon. The trouts, on the other hand, usually recuperate and may return to spawn a number of times. The Pacific salmon, Atlantic salmon, and steelhead trout are anadromous. The landlocked salmonids discussed in this article include the rainbow trout and the kokanee salmon. While these are the natural life cycles, some species can be grown entirely in fresh water. For example, rainbow trout can be grown quite successfully in seawater netpens and the coho and pink salmon can spend their entire life cycle in fresh water.

Figure 7. Rainbow trout (*Oncorhynchus mykiss*). *Source:* Copyright 1990 by B. Guild-Gillespie.

Spawning behavior typically involves a high degree of territoriality and aggression between individuals as mates selection takes place. The female will usually create a depression, called a redd, in the gravel bottom by beating the rocks with the side of her tail. After eggs and milt are deposited simultaneously for fertilization to take place, the female will cover the eggs with gravel by moving upstream from the redd and making another redd, causing the displaced gravel to cover the first redd that contains the eggs. The fertilized eggs then incubate in fresh water, which percolates through the gravel, supplying oxygen and carrying away metabolic products such as ammonia and carbon dioxide. The developing embryos as well as the hatched alevins live off yolk carried in a yolk sac while they are still under the gravel. After the yolk is consumed, their mouths open and they begin to emerge from the gravel and feed on organisms in the open waters of the stream. The young of anadromous fishes spend varying lengths of time in fresh water. This fresh water residence ends when the animal undergoes the process of smoltification. This is the total of behavioral, morphological, and physiological changes that occur in the juvenile fish to prepare it for life in salt water. After the process of smoltification, the young migrate to sea to spend the rest of their life cycle growing to sexual maturity.

Pacific Salmon

Chinook Salmon. The chinook salmon (*Oncorhynchus tshawytscha*) is one of the most prized fishes on the west coast of North America. It is also one of the most important commercial species. The chinook salmon is known as the tyee or king salmon when the body size reaches greater than 13.6 kg. This fish has the largest body size of the Pacific salmons. While 13.6 kg is more common as the maximum size, the world's record is 57.3 kg. It is this characteristic combined with its high market value that has made it a popular choice for aquaculture in the northwestern states and on the west coast of Canada. There are two varieties of chinook salmon, the red chinook and the white chinook, which are named according to their flesh color. The red flesh commands the higher price. This segregation is rather unique among the salmon. Chinook salmon are caught commercially with trolling gear, purse seine or gill nets, long lines, and fish wheels. The majority are caught by trollers. Chinook salmon are sold as fresh, frozen, and canned products, with the canned products coming from the net fishery. They are caught for sport with a variety of lures including artificial lures, bait, and fiys. Dead herring serve as a desirable bait, trolled deep in the water column.

Chinook salmon adults migrate extensively and are primarily found in the Pacific Ocean from the Ventura River in southern California, north to Point Hope, Alaska. Limited evidence suggests, for instance, that the Canadian shinook migrate north to the Gulf of Alaska but remain within 160 km of that area until they return south to spawn. This species is less common in the Arctic Ocean and in the Bering and Okhotsk seas and the Sea of Japan. The largest numbers come from the largest rivers such as the Fraser River in British Columbia. Substantial numbers of chinook also come from the Yukon River. While there have been numerous attempts to transplant this species to many locations throughout North America and around the world, it seems that only success was on South Island, New Zealand, where a self-supporting population was established.

The average age of an adult returning to spawn varies from four to seven years, depending on the location. Three- to five-year-old returning adults are more common in southern streams, whereas five- and six-year-old adults are more common in northern streams. It is also quite common for river systems to have more than one stock of chinook salmon returning to it, where there may be spring, fall, and winter runs of returning adults.

The time of spawning depends on the location and the distance the adults must swim upriver to reach the spawning grounds. In the Fraser River, for example, the chinooks spawn between July and November and the adults may travel up to 965 km from the mouth of the river to their spawning grounds. The chinooks that return to the Yukon River spawn between July and August and may travel up to 1930 km to reach their spawning grounds. Other stocks that travel only a few miles, if that, in British Columbia spawn in September and in October. Chinook salmon tend to spawn in deeper waters and in larger gravel than other Pacific salmon species. Each female may carry from 4,000 to 13,000 eggs. The eggs are large for fish eggs and measure about 7 mm in diameter. The eggs hatch the following spring and the alevins spend two to three weeks in the gravel with their yolk sacs. The emergent young remain in the freshwater rivers for varying lengths of time, depending primarily on the water temperature. The young will smolt and migrate to sea after about three months in warmer southern areas such as in southern British Columbia, while chinook spend at least one year in fresh water in northern areas. Chinook juveniles may spend two years in fresh water in the Yukon River. The maturing adults may spend two to three years at sea before returning to fresh water to spawn. Spawning females are, therefore, about four to five years old.

Coho Salmon. The coho salmon (*Oncorhynchus kistuch*) is also an important commercial and sport fish in North America. It is the mainstay of the saltwater sport fishery for salmon. Most of the coho are commercially caught in August by trolling with plugs, spoons, or feathered jigs, although purse seines and gill nets are also used. The trolled catch is sold as the fresh or freshly frozen product, while the netted fish are either canned or smoked. Sport fish are caught in the late summer to early fall months of July to October. The gear used by sport anglers is similar to the commercial trolling gear except it is smaller in size. Coho are also caught on the fly with bucktail flies and with bait such as frozen or pickled herring. Because coho remain silver and continue to feed while traveling upstream to spawn, they are pursued by the sport angler in many freshwater rivers and streams.

The adult distribution extends from southern California to the Gulf of Alaska. The juveniles are found in the fresh waters from Monterey Bay, Calif., to Point Hope, Alaska. Coho salmon are also found in the Anadyr River in the former USSR and south from that point to Hokkaido, Japan. This species has been transported to other parts of the world such as Argentina and Chile, with Chile reporting some success in the establishment of naturally reproducing stocks. Coho have also been introduced into the Great Lakes by both Canadians and U.S. citizens. While there are reports of natural reproduction, the maintenance of those stocks are heavily dependent on the annual planting of cultured stocks.

The full-grown adult can weigh from 1.8 to 5.5 kg. It enters fresh waters to spawn from late September to October and may travel up to about 240 km upriver to spawn. Asian populations exhibit more of the separation of summer and fall, or fall to winter runs than those populations returning to North American river systems. Spawning takes place in October or November. The eggs are rather large, measuring about 6–7 mm in diameter. The range in fecundity is about 2,000–3,000 eggs. After the eggs have hatched, the young spends between one and two years growing in fresh water. This again, depends on the location. Coho juveniles in British Columbia spend about one year in fresh water while those in the Yukon spend two years growing in fresh water. After their freshwater residence, the young smolts migrate to sea. The saltwater residence usually lasts about 18 months, although the phenomenon of precocious sexual maturation, or jacking, is predominant in this species. While normal adults will return between three and four years of age, jacks return to spawn at only two years of age. Coho salmon normally stay within about 40 km of the coast.

Chum Salmon. While the chum salmon (*Oncorhynchus keta*), or dog salmon, is a less desirable species for the commercial or sport fisheries, it has maintained an important position in the diets of the native people of North America who capture the fish with nets and fish wheels. It is a preferred species for smoking because of its low fat content. Its pale to white flesh has the lowest fat content of the Pacific salmons. This species has been used the least for transplantations to other geographic areas, perhaps due to its relatively low commercial and sport value. Most of the chum salmon are harvested while they are migrating toward their home streams to spawn. Commercial catches are harvested mainly between August and October with purse seines and gill nets. Some are also caught by trolling gear. The trolled catch is sold fresh. Some are smoked or dry salted for preservation. The bulk of the commercial catch, however, is canned.

The marine adults are found in Pacific and Arctic oceans, in the Sea of Japan, and in the Okhotsk and Bering seas. They probably have the widest distribution among the Pacific salmons. The major spawning areas in North America lie between Puget Sound in Washington State to Kotzebue, Alaska. The geographic range for their spawning covers the west coast of North America from Oregon to the Mackenzie River. The time frame for adults arriving at the spawning grounds ranges from July, in northern British Columbia to September or even as late as January for streams further south. The location of the spawning grounds, reflecting the degree to which the chum salmon swim into fresh water to spawn, varies considerably. Most arrive at the mouths of rivers in an advanced state of sexual development and spawn relatively close to the ocean. In general, few chum salmon migrate more than 160 km from the mouth of a river to spawn. The Yukon River, however, is an exception; the population ascends over 1,930 km to spawn.

The mature adult is usually between two and four years old. While the adult chum salmon can weigh from 3.6 to 5.5 kg, fish up to 13.7 kg have been reported. The fecundity ranges from about 2,000 to 3,000 eggs and each egg measures about 5–6 mm in diameter. The hatched young spend a period in the gravel feeding off the yolk until the spring when they emerge and migrate directly to sea. Depending on the distance to the ocean, the young may take from a single night up to several months until they reach sea water. The migratory behavior of the young chum are similar to the pink salmon and the young of both species may be found making the seaward journey together. They usually travel by night and hide in the gravel during the day. They are also not dependent on schooling to make the trip, although they do school once they have reached the estuary. Once in the ocean, the young chum will usually spend a few months, until mid to late summer, in the coastal areas before migrating further out to sea. Chum salmon from North America have been captured as far as 4,180 km off the west coast in the North Pacific. Most chum salmon spend two or three years in the ocean before returning to their home streams to spawn. They begin to appear at the mouths of rivers around May or June of their final year at sea.

Pink Salmon. The pink salmon (*Oncorhynchus gorbuscha*) is the most abundant of the Pacific salmon species. It gets its name from the pink color of its flesh. Like the chum salmon, this species is not regarded as being highly desirable compared to the deep red flesh of the sockeye salmon. The size range of the adult is about 1.4–2.3 kg, although animals as large as 6.4 kg have been recorded. A unique characteristic of this species is its fixed, two-year life cycle.

There is no overlap between the reproducing stock of one year to the next. This allows two separate stocks to utilize any stream for reproduction. Where there are odd-year stocks as well as even-year stocks using the same river, one often dominates. The Fraser River pink salmon run in British Columbia, for example, is primarily an odd-year stock, while the run in the Queen Charlotte Islands is an even-year stock.

Pink salmon is caught commercially with purse seine and gill nets as well as with trolling gear, the latter contributing only a minor part of the total catch. Fish caught by trolling is sold fresh, but 95% of the commercial catch is canned. Pink salmon is also fished for sport by trolling artificial lures.

Pink salmon is found in the Pacific and Arctic oceans, the Bering and Okhotsk seas, and the Sea of Japan. Its distribution in North America extends from the Sacramento River in California, north to the delta of the Mackenzie River, being most abundant around the central area of this range. It can be found from the surface to depths of about 36.6 m. Transplantations have established self-sustaining populations of pink salmon in northern Europe and in fresh water in Lake Superior. There are transplanted populations along the Atlantic Coast of North America that seem to be self-sustaining.

Pink salmon may spawn in a wide range of locations from tidal areas of certain rivers up to 483 km upstream from the mouths of large rivers. The pink salmon are also known as humpbacks because the sexually mature males develop a large hump behind the head. The snout elongates dramatically and large teeth emerge from the jaws. The adults will appear at the mouths of rivers from June to September to begin their migration to their spawning beds. Pink salmon spawn from mid-July to late October and the eggs hatch from late December to late February. Like the chum salmon, the young emerge from the gravel after the yolk supply is consumed and move directly to sea. The emergent young, which measure about 3.8 cm in fork length, may travel 16 km in a single night to reach the ocean. Their appearance is distinctive for salmon in that they lack the vertical parr marks on their sides. They are colored blue-green on their back and have silver sides. If the journey takes more than a night, the young will hide during the day in the gravel and emerge again at night to be carried by the current to the sea.

Sockeye Salmon. The anadromous sockeye was the first Pacific salmon species to be fished commercially. The sockeye has long been the mainstay of the Pacific salmon fishery. Its deep red flesh and high oil and protein contents have always brought it the highest price among the Pacific salmon species on the market. The sockeye is mainly caught with purse seine and gill nets, although trolling is also used. The product is mainly canned, although it may be sold fresh when the fishery is open. Native people use traditional methods of nets, weirs, and gaffs to harvest the fish. Although the sockeye salmon are not an important sport fish, effective artificial lures have been developed that have served both sport and commercial trollers at the mouth of major sockeye rivers such as the Fraser in British Columbia.

The sockeye salmon (*Oncorhynchus nerka*) is represented by an anadromous form and a landlocked form known as the kokanee. Specific recognition as subspecies has been given to each. The anadromous form i *O. nerka nerka* Walbaum and the kokanee is designated *O. nerka kennerlyi* Suckley. The anatomy of the two subspecies is very similar except for the total body size. The average weight of an adult anadromous sockeye is about 2.7–3.2 kg, whereas the adult kokanee weighs less than 455 g. Adult sockeye have been recorded to be up to 6.8 kg.

The distribution of this species extends from the Kalamath River in California to Point Hope in Alaska. They are also found in Asia from northern Hokkaido, Japan, to the Anadyr River in the former USSR. The distribution of the kokanee certainly follows that of the anadromous sockeye. It occurs in lakes where the anadromous salmon must have had access at one time. Both kokanee and most anadromous sockeye return to the spawning grounds as four- or five-year-old fish. Both species spawn in the fall. The actual time depends on location. Kokanee, for example, spawn in September and October in Kootenay Lake in British Columbia but in November and December in Boulter Lake, Ontario. The major spawning grounds for this species are the watersheds draining into the Fraser, Skeena, and Nass rivers in British Columbia. Anadromous sockeye spawn between July and December, again depending on location.

Sockeye become sexually mature from three to eight years in age. While a one-year freshwater residence and a two- to three-year saltwater residence is normal, precocious males may return to spawn after only one year at sea. Fecundity is highly variable in this species. It ranges from 370 to 1760. While the kokanee egg is naturally small, the egg of the anadromous sockeye is large, measuring about 4.5–5 mm in diameter. After hatching and consuming all the yolk, the young sockeye will spend at least one year in fresh water, although some will spend two or even three years in fresh water before migrating to sea. This variability is, to some extent, responsible for the range in size of adult sockeye. These years in fresh water are spent in nursery lakes. The fry feed on planktonic organisms. Anadromous sockeye smolts will migrate to sea in the spring of their second and fifth year of life. At sea, the maturing sockeye migrate north and northwest, spreading out to the distribution outlined above.

Rainbow Trout. The rainbow trout, steelhead trout, and kamloops trout all belong to the same species, *Oncorhynchus mykiss* formerly *Salmo gairdneri*. This species represents one of the most highly prized game fishes in North America. The names rainbow and kamloops trout refer to the nonanadromous populations of this species, while the steelhead is normally used to refer to the large, anadromous variety of this species. The kamloops trout usually refers to the larger variety of the nonanadromous trout. All varieties of this species are highly prized sport fishes. The flesh is red to pink in fish that have been feeding on planktonic organisms and it tends to be more pale in fish from larger lakes where they feed on fish more than plankton. They take the bait or lure aggressively and fight the line, jumping out of the water many times in the process.

Most of the steelhead are caught in fresh water in coastal areas. The commercial catch is usually incidental to the salmon fishery and are almost exclusively canned. The commercial gear involved in this fishery, therefore, consists of gill nets.

Although the native range of this species is the eastern Pacific Ocean and for the freshwater forms, west of the Rocky Mountains from Mexico to Alaska, all forms of the rainbow trout can be found throughout North America in suitable habitats. It has also been introduced successfully in New Zealand, Australia, Tasmania, South America, Africa, Japan, southern Asia, Europe, and Hawaii. It is a very plastic species that adapts to different environments readily and may show variations in its physical and behavioral characteristics throughout its range. It is probably the most studied fish in terms of fish physiology and anatomy.

The nonanadromous rainbow and kamloops trout are naturally spring spawners. They move from the lakes to inlet or outlet streams from about mid-April to June and spawn in streams with beds of fine gravel. This is the timing for North America. Because their distribution spans the globe, the actual month of spawning will depend on when spring occurs at a particular location. The fecundity is about 1,400–2,700 eggs, but this is extremely variable; the range extends from 200 to 12,750. The eggs are 3–5 mm in diameter. The eggs usually hatch in about four to seven weeks and the yolk is consumed in about three to seven days. The young emerge from the gravel between mid June and mid August and may reside in the stream until fall of that year or they may spend as long as three years in the stream. The young of the rainbow and kamloops trout will then migrate upstream or downstream to the lake to feed and mature until they are ready to spawn.

The anadromous steelhead trout populations may spawn in the spring, fall or winter. An established population for a given stream, however, will maintain a consistent pattern from year to year. The fecundity and egg size are similar to the rainbow and kamloops trouts. Most of the young will spend two to three years in fresh water before migrating to sea as smolts in the spring, where they may spend an equal amount of time maturing in the ocean. Many will survive the spawning and return for a second or third spawning.

Atlantic Salmon

The Atlantic salmon (*Salmo salar*) has attracted both commercial and sport fisherman like no other species throughout history; so much so that populations of Atlantic salmon disappeared from the Thames in 1833 and from Lake Ontario in 1890 due to overexploitation. The Atlantic salmon are caught commercially with net and troll gear. They are caught in the sport fishery with live bait as well as on the fly and artificial lures. The intensive interest on this species has, unfortunately, eliminated entire populations of this species from certain watersheds over time. The commercial fishery for the Atlantic salmon off Greenland places a particularly heavy pressure on the Atlantic salmon population at large, because those waters represent the major feeding grounds for all the stocks in the Atlantic Ocean.

The native distribution of the Atlantic salmon is the basin of the North Atlantic Ocean, from the Arctic Circle to Portugal and from northern Quebec south to the Connecticut River. While fertilized eggs and live fish transplantations to the Pacific waters have occurred since 1905, there is no record of the establishment of self-sustaining populations. Like several of the Pacific salmons, there are established landlocked populations of this species in North America. The landlocked Atlantic salmon are sometimes called the ouananiche. There are self-sustaining populations in Sebago Lake, Maine; Lake Ontario; and Lake St. John, Quebec.

The Atlantic salmon is a typical anadromous fish. They spend 1–3 yr in fresh water as juveniles and migrate to sea as smolts where they may spend 1–2 yr growing and maturing. While mortalities in spawning are high, significant numbers of spawners return to spawn two or three times. The normal spawning time is from October to late November. The fish in the southern regions tend to spawn later. The female will produce about 1,540 eggs per kg body weight. The size of returning adults ranges from 2.7 to 6.8 kg. The returning adult spawner will normally have spent 2 yr at sea. The Atlantic salmon generally have longer life spans than the Pacific salmon. Ages up to 11 yr have been reported. While the eggs may hatch in April of the following year, the young alevins will stay in the gravel and consume their yolk until May or June, at which time they emerge into the water column. The period of freshwater residence is variable, depending on the location. The younger parr in North America will spend 2–3 yr in fresh water until they smolt, at about 15.2 cm in length. Populations in Greenland, however, have been recorded to spend as long as 4–8 yr in fresh water, attaining about the same body size.

Unlike the slow growth in fresh water, the Atlantic salmon undergo rapid growth at sea. An exception to the average size of returning adults, stated above, are the precociously matured fish that return after one year at sea. These are often males and are in the 1.4–2.7 kg range. There are many reports of individuals exceeding the normal size range. While the average Atlantic salmon caught commercially is about 4.6 kg, record sizes of 35.9 and 37.7 kg have been reported. While the average landlocked Atlantic salmon weighs about 0.9–1.8 kg, individual fish weighing 16.1 and 20.3 kg have been caught in North American lakes.

SMELT

The smelts are small, slender fishes that belong to the family Osmeridae. This group of fish probably got its name from one characteristic, their smell. The smell, which may be described as that of freshly cut cucumbers, is particularly noticeable when large numbers are caught during spawning runs. The six genera and 10 species are found in the northern hemisphere in the Atlantic, Arctic, and Pacific oceans. They are circumpolar in distribution. Some species in this family spend their entire life cycle in fresh water and others in sea water. There are species, furthermore, that are anadromous. The four species found in

North America are the pond smelt (*Hypomesus olidus*), the longfin smelt (*Spirinchus thaleichthys*), the Eulachon (*Thaleichthys pacificus*), and the rainbow smelt (*Osmerus mordax*). The latter two species are discussed here because the existence of both sport and commercial fisheries for them. The former two species play a minor part in any commercial fishery and in normal consumption.

Rainbow Smelt

As Figure 8 shows, the rainbow smelt is a slender fish that may reach lengths of 17.8–20.3 cm, although 35.6-cm fish have been recorded. As the name implies, it is a colorful silver fish with a pale green back and iridescent sides. The rainbow smelt has supported a commercial fishery for over 100 yr. It has always been abundant throughout its distribution. Most fish are caught in the spawning migration by trawl nets. The Atlantic rainbow smelt is also a popular sport fish. It is caught on hook and line through the ice as well as by dip netting or seine netting during the spawning run. It is an anadromous fish but like other anadromous fishes, it can successfully spend its entire life cycle in fresh water. The landlocked forms tend to have a darker color than the marine counterparts. As with other anadromous fishes, they lose the silver appearance and darken when they begin to spawn.

There are two subspecies of the *Osmerus mordax*: the Arctic rainbow smelt (*O. mordax mordax*) and the Atlantic rainbow smelt (*O. mordax dentex*). The Arctic rainbow smelt includes those populations that are in the Pacific Ocean and are distributed from Vancouver Island, north around the state of Alaska, and into the Arctic Ocean to Cape Bathurst. The distribution of the Atlantic smelt was originally restricted to the marine waters from New Jersey to Labrador. The Atlantic smelt today is represented by both anadromous forms and landlocked forms in the Great Lakes.

The anadromous and landlocked forms behave in a similar way in spawning. The spawning adults ascend streams to lay and fertilize their eggs. They begin their ascent in the spring when the ice cover disappears, usually in the months between March and May. Spawners can be up to six years old, but are commonly between two and three years old. Spawning, which usually occurs at night, involves two or more males to each female. While spawning normally takes place in streams, fish will spawn on the shoreline if the currents are too strong to permit migration. Eggs and milt are released into the moving current and the eggs quickly stick within seconds to anything they come in contact with. The fertilized eggs, which measure about 1 mm in diameter, eventually end up resembling balloons on stems as the sticky outer coat washes off except at the point of attachment. Most of the spawners, like many other anadromous fishes, die after spawning. Although reports of fecundity vary, about 494–530 eggs per gram of body weight are produced. The eggs incubate for about two to four weeks in water temperatures between 6 and 10°C. The hatched young, which are about 5 mm long are quickly swept down to the lake or estuary where they continue to feed and grow to a length of about 5 cm. They spend the remaining period up to spawning in the middle of the water column of lakes or in coastal waters close to shore.

Eulachon

Thaleichthys pacificus, the eulachon, means oily fish of the Pacific. Aside from the characteristic odor the oiliness of this fish is its outstanding feature. The oil content of the body is so high that the dried carcass can be lit like a candle. Another common name for this fish is candlefish, because in the days before the luxuries of candles and inexpensive oils, small strips of rags were inserted into the mouths of dried eulachons and lit for light. The rich oils lend a flavorful taste to this fish, which probably accounts for its high value as a food fish. The eulachon resembles the rainbow smelt in appearance, although it is darker in coloration and lacks the iridescence in the body walls. It is typically brown to blue-black on the dorsal surface. The coloration on the ventral surface is white and the sides are a light color, intermediate between the two extremes.

The eulachon is caught by the native Indian community as well as by the commercial fishery. The volume caught by the native fishery has exceeded, and probably still exceeds, the commercial catch. The commercial fishery utilizes drift gill nets and occurs principally in the Fraser River as the eulachon migrates to spawn. The commercial catch is used for human consumption but mostly for feed at fur farms. The native catch has always been based on the oil from the eulachon. It is used for cooking and for curing. Traditionally, the grease trails got their name because they were used to carry the extracted eulachon oil from the fishing grounds to the points of trade.

The eulachon is found only on the west coast of North America, from the Kalamath River in California up to the panhandle of Alaska and to the Pribilof Islands. Like the

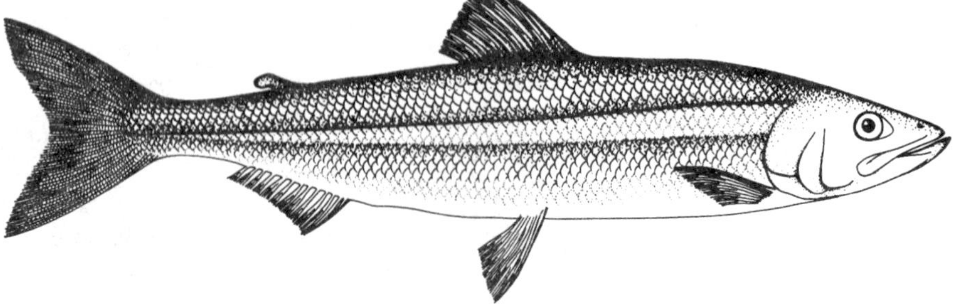

Figure 8. Rainbow smelt (*Osmerus mordax*). *Source*: Copyright 1990 by B. Guild-Gillespie.

rainbow smelt, the eulachon is an anadromous fish that migrates from the sea up freshwater streams, soon after the ice cover breaks, to spawn. This occurs from mid March to mid May. The spawning females, which measure about 15.2–17.8 cm in length, produce an average of 25,000 eggs and shed them in the free current along with the milt from the male partners. They are at least three years old when they spawn. Unlike the small rainbow smelt egg, the eulachon produces a relatively large egg of 8–10 mm in diameter, which is irregular in shape. The eulachon eggs have two outer layers. The outer, adhesive, layer breaks on spawning and everts to anchor the egg to coarse particles such as gravel. The outer and inner layers remain attached at one point. While most of the adults die after spawning, some survive to spawn a second time. Like the rainbow smelt, the eggs take about two to three weeks to hatch and the just-hatched young resemble young herring. They measure about 5 mm in length and are slender and nearly transparent. The small young are swept to sea with the current. They move to deeper waters farther from the shore as they mature. Although they normally measure 10.2–17.8 cm in length, individuals up to 30.5 cm have been recorded.

TUNA AND SWORDFISH

Tuna

The true tuna belong to the tribe Thunnini, which belongs to the family Scombridae of the suborder Scombroidei, order Perciformes (Fig. 9). The bullet, frigate, bluefin, albacore, yellowfin, skipjack, blackfin longtail, and the slender tuna are all true tunas. The other tribes in this family include the Sardini (bonitos), the Scomberomorini (seerfish), and the Scombrini (mackerel). The smallest tunas (bullet and frigate) usually weigh less than 3 kg while the largest (bluefin) tunas can exceed 700 kg. The large bluefin tunas can live to 38 yr. This family of fishes possesses remarkable limits for growth. The billfishes, such as the black marlin, grow even larger than the tuna. Commercial fisherman have reported taking black marlin in excess of 3,000 kg, no doubt among the largest fishes in all the seas.

The tunas have been fished for food for thousands of years. The skipjack, yellowfin, and bigeye tunas dominate the commercial catches. Most of the tuna harvests occur in the Pacific Ocean. About 50% of the world catch of tuna is caught by Japan and the United States. The balance is caught by almost every nation in the areas where the tuna are found. They are caught with almost every kind of gear including the pole and line method as well as giant purse seines and long-lining. About 40% of the tuna are caught by the pole and line method where large schools of fish are first baited close to the boat with live anchovies or sardines that are thrown overboard and then caught with barbless lures on single poles. This process is called chumming. The balance of the commercial catch are harvested by purse seines (30%) and by long lines (30%). The purse seiner encircles the school of fish up to about a kilometer in diameter while letting out a special net that can be gathered at the bottom, like a purse or bag. As the boat completes the circle, the purse is gathered up to the boat along with the fish. This technique had caused high mortalities in dolphins (also called porpoises by fisherman) that often congregate above schools of yellowfin tuna. Many of the dolphins die because they become entangled in the net and are prevented from surfacing to breath. After this came to public attention in the mid-1960s, purse seine nets and seining methods have been modified to prevent this entanglement and to allow the dolphins to escape from the net. The long-line vessel will let out a line up to 130 km in length with as many as 2,000 baited hooks dangling from this main line, supported by floats. In addition to the commercial catch, the tuna represent one of the most desirable sport fish in the world. It is well known for attracting avid sport fishermen who spend thousands of hours and dollars in pursuit of this challenging and exciting catch.

Tuna is a very expensive food item. It is at least as expensive as beef, relative to protein content. Over 90% of the world catch is consumed by the United States, Japan, France, Spain, Italy, and West Germany. A typical tuna fish sandwich is probably made of albacore, yellowfin, skipjack, or bigeye tuna. Certain species, such as the skipjack, are sold as a dried product in Japan. The flesh of the bluefin and bigeye tunas are valued as raw sashimi in Japan. Record prices for tuna come from Japan where the flesh of such large tuna, with a high oil content, has sold for as much as $26,000 U.S. a ton.

The tuna and the billfishes represent some of the most interesting fishes in the world in that they deviate from the norm. The tuna are found in every temperate and tropical ocean around the world. They are most commonly found in the Atlantic, Pacific, and the Indian oceans. Tuna and billfishes are not only among the largest fishes in the oceans, but are also the fastest swimmers in the world. At high speeds, their body forms represent an extremely high degree of streamlining an hydrodynamic efficiency. At

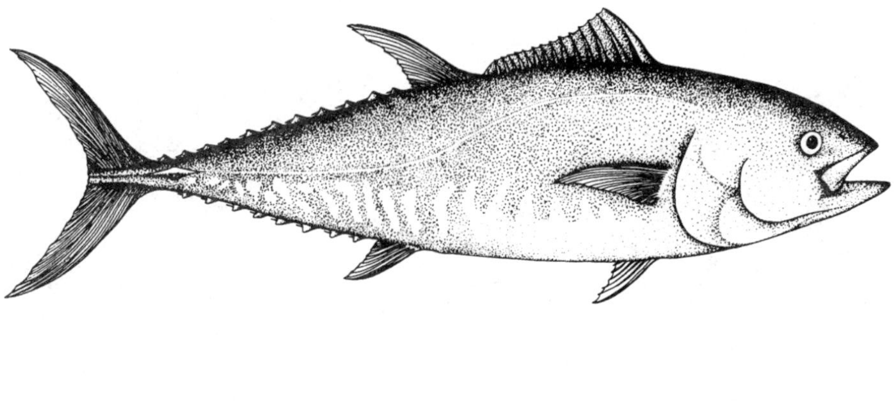

Figure 9. Tuna (*Thunnus thunnus*). *Source*: Copyright 1990 by B. Guild-Gillespie.

those speeds, their fins retract into grooves and their eyes become flush with the surface of the body. Even the slowest pace for the tuna represents the top speed any human could hope to achieve. Relatives of the tuna, the wahoo and sailfish have been recorded at swimming speeds over 75 and 100 km/h. Not only are these animals most remarkable for their speed but they are also known for the incredible distances they travel. Tagging studies, which tend to yield conservative estimates, have shown individual bluefin tuna and black marlin to travel over 10,000 km, at a rate of over 30 km/day. While most of the distance records are held by the larger bluefin tuna, the skipjack tuna has been recorded to travel over 9,500 km and the yellowfin tuna, over 5,000 km. There are numerous records of individually tagged members of this group of remarkable fish crossing every major ocean in the world.

As stated above, the general body shape of the tuna is one designed for hydrodynamic efficiency. The body is rather round in cross section and laterally oval. The tail is lunate (shaped like a crescent moon), which provides a high degree of forward thrust, with minimum drag. Like many other fishes, the upper body is dark and the ventral side is light in color to aid in camouflage. Body markings change with age and may include vertical bars or markings of different shapes. These markings may serve to identify sex, age, species, and particular stages in the reproductive cycle.

One need for the efficiency in swimming is their mode of respiration. Tuna are ram ventilators. That is, they must swim with their mouths open to pass sufficient water over their gills. One result of this swimming activity is the high energy requirements of this fish. A tuna can consume up to one-quarter of its body weight in food each day. Such a high metabolic rate naturally produces a great amount of heat. The tuna are rather unique among fishes in that four genera of the family regulate body temperature above that of the ambient water. The bluefin tuna, for example, maintains a visceral temperature 10–15°C above ambient water temperature. Maintaining a higher body temperature presumably allows quicker digestion as well as a more rapid mobilization of stored energy reserves. It may be more accurate to state that those tunas that regulate body temperature have specialized mechanisms for dissipating the heat from metabolism. Special circulatory structures enable this heat to be transferred to the external surface.

As stated above, the tuna is highly active fish. The need for high rates of oxygen uptake by the fish to feed this high metabolic rate is met by a larger gill surface area and higher hemoglobin content than most other fishes. The constant swimming motion in combination with pectoral fins that act like hydrofoils maintains the fish at a certain position in the water column. This greatly reduces the need for the swim bladder, which functions in other fishes as a buoyancy organ. The swim bladder is greatly reduced or absent in species of tuna. Unlike most other fishes, there is a greater proportion of red muscle to white muscle mass in the body. White muscle is used for short bursts of anaerobic activity while the red muscle is used for routine swimming in fishes. This makeup provides the machinery with which the tuna can accomplish the incredible journeys over long distances.

Spawning takes place in various regions, depending on the species. For example, the Atlantic mackerel spawns in the northwest Atlantic and the large bluefins spawn in the Straits of Florida and the Gulf of Mexico. Both eggs and sperm are broadcast into the water, near the surface. The eggs of the Atlantic mackerel are found mainly in the top 10 m of water. The female tuna lays about 100,000 eggs per kg body weight; there might be about 10 million eggs from a 100-kg animal. This is not unreasonable because the bluefin tuna can reach a maximum size of about 700 kg. These tiny eggs of about 1 mm float about in the upper layers of the ocean, held bouyant by an oil droplet inside the egg. While the various tuna species are found in waters of a wide range of temperatures, most eggs take only a few days to hatch. As in many other fishes, only about two of those individuals survive to maturity.

Swordfish

The swordfishes belong to the family Xiphidae (Fig. 10). The Xiphidae are related to other giant mackerellike fishes such as the tunas (Scombridae) and the spearfishes (Istiophoridae), in that they all belong to the suborder Scombroidei. The single representative of this species (*Xiphias gladius*) is known as the common swordfish or broadbill. It is highly prized as a game fish and as a food item. Like the tunas, there is evidence that the swordfish has been fished for thousands of years. Its dominant physical feature is the large sword protruding from the upper jaw. While this feature was thought to serve a predatory function, it may function more in the enhancement of swimming speed by reducing the resistance of water to the body, much like the pointed head of a rocket. There are reports, however, of many whales with broken bills of swordfish impaled in their bodies, some through their vertebrae. This fish may be referred to by sport anglers as the broadbill swordfish. The teeth and the spines of the first dorsal fin are lost with age. Its large body is scaleless.

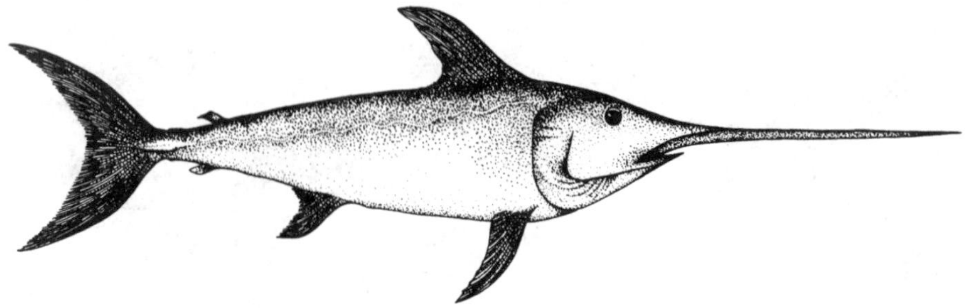

Figure 10. Swordfish (*Xiphias gladius*). *Source*: Copyright 1990 by B. Guild-Gillespie.

Although an individual weighing 537 kg was caught in 1953 off the Chilean coast, most other record sizes are closer to half that weight. This is relatively small compared to members of the other families in the Scombroidei. The billfishes also represent a minor part of the commercial catch, relative to the tunas. The average annual catch for the swordfish between 1982 and 1985 was estimated at 49,000 t, about 1.6% of the total catch for the Scombroidei.

Unlike the tunas and other billfishes that avoid deep waters, the swordfish moves widely in the water column. When it dives to great depths and returns to the surface, it is often found in semiconsciousness at the surface until it regains equilibrium. It is in this stupor that fishermen have used the harpoon to catch this prize. While this was the popular method of catch until the early 1960s, the majority of today's commercial catch of swordfish is by long line. This billfish also contributes to the specialized market of big game fishing around the world where spending $1,000/day in the pursuit of a sailfish or billfish is not uncommon. The local communities where big-game fishing ports are located may depend heavily on the economy of that sport.

The distribution of the swordfish extends throughout the temperate, subtropical, and tropical waters in all oceans of the world. They are found in coastal as well as oceanic areas. Like the related tunas and spearfishes, the swordfish are highly migratory and have body forms that are very streamlined. The basic physiological and reproductive features of the swordfish are similar to those of the tuna. Spawning is thought to occur throughout the year in the Caribbean Sea, Gulf of Mexico, and off the Florida coast. Although the swordfish egg is somewhat larger than the tuna egg, and the fecundity is slightly lower, the major features of the reproductive system and process are the same as those for the tuna. The eggs are approximately 1.7 mm in diameter and contain a large oil globule. Young hatch out at about 4 mm in length and may grow rapidly, up to 2 mm per day.

BIBLIOGRAPHY

1. *List of Common and Scientific Names from North America and Canada*, 4th ed., American Fisheries Society, 1980.
2. W. A. Clemens and G. V. Wilby, *Fishes of the Pacific Coast of Canada*, 2nd ed., Fish. Res. Board Can. Bulletin **68**, 1961.
3. D. Miller, "The Blue Whiting, *Micromesistius poutassou*, in the Western Atlantic, with Notes on Its Biology," *Copeia* **2**, 301–305 (1966).
4. F. H. Bell, *The Pacific Halibut: The Resource and the Fishery*, Alaska Northwest Publishing Co., Anchorage, 1981.
5. W. B. Scott and M. G. Scott, "Atlantic Fishes of Canada," *Can. Bull. Fish. Aquat. Sci.* **219**, 1–731 (1988).

GENERAL REFERENCES

References 1–4 are good general references.

D. L. Alverson, A. T. Pruter, and L. L. Ronholt, *A Study of Demersal Fishes and Fisheries of the Northeastern Pacific Ocean, H. R. MacMillan Lecture Series in Fisheries*, Inst. Fish. University British Columbia, 1964.

G. Godson, *Fishes of the Pacific Coast*, Stanford University Press, Stanford, Calif., 1988.

J. L. Hart, "Pacific Fishes of Canada," *Bull. Fish. Res. Board Can.* **180**, 1973.

J. G. Hunter, S. T. Leach, D. E. McAllister, and M. B. Steigerwald, "A Distributional Atlas of Records of the Marine Fishes of Arctic Canada in the National Museums of Canada and Arctic Biological Station," *Syllogeus* **52**, 1–35 (1984).

H. Kasahara, *Fisheries Resource of the North Pacific Ocean. Part 1. H. R. MacMillan Lectures in Fisheries*, Institute of Fisheries, the University of British Columbia, Vancouver, 1960.

P. A. Larkin and W. E. Ricker, "Canada's Pacific Marine Fisheries. Past Performance and Future Prospects," in *Inventory of the Natural Resources of British Columbia*, 1964, 194–268.

Y. Okada, *Fishes of Japan*, Maruzen Co. Ltd., Tokyo, 1955.

F. S. Russell, *The Eggs and Planktonic Stages of British Marine Fishes*, Academic Press, Inc., Orlando, Fla., 1976.

D. J. Scarratt, ed., "Canadian Atlantic Offshore Fishery Atlas," *Can. Spec. Publ. Fish. Aquat. Sci.* **47**, 1982.

W. B. Scott and E. J. Crossman, "Freshwater Fishes of Canada," *Bull. Fish. Res. Board Can.* **184**, 1973.

A. Wheeler, *The Fishes of the British Isles and North-west Europe*, Michigan State University Press, East Lansing, 1969.

N. J. Wilimovsky, "List of the Fishes of Alaska," *Stanford Ichthyology Bull.* **4**, 279–294 (1954).

Specific to SpeciesS. F. Hildebrand, "Family Engraulidae," in *Fishes of Western North Atlantic, Mem. Sears Found. Mar. Res.*, Vol. 3, Yale University, New Haven, Conn., 1963, pp. 151–249.

J. L. McHugh and J. E. Fitch, "An Annotated List of the Cluepoid Fishes of the Pacific Coast from Alaska to Cape San Lucas, Baja California, *Calif Fish Game* **37**, 491–495 (1951).

R. H. Boyle, *Bass*, W. W. Norton & Co., New York, 1980.

R. H. Boyle, *The Hudson River, A Natural and Unnatural History*, New York, 1979.

J. N. Cole, *Striper*, Boston, 1978.

N. Karas, *The Complete Book of the Striped Bass*, New York, 1974.

E. C. Raney, E. F. Tresselt, E. H. Hollis, V. D. Vladykov, and D. H. Wallace, *The Striped Bass.* Bulletin of the Bingham Oceanographic Collection 14, Peabody Museum of Natural History, Yale University Press, New Haven, Conn., 1952.

M. P. Fahay, "Guide to the Early Stages of Marine Fishes Occurring in the Western North Atlantic Ocean, Cape Hatteras to the Southern Scotian Shelf," *J. Northw. Atl. Fish. Sci.* **4**, 1–423 (1983).

L. S. Incze, C. M. Lynde, S. Kim, and R. Strickland, "Walleye Pollock, *Theragra chalcogramma*, in the Eastern Bering Sea," in N. J. Wilimovsky, L. S. Incze, and S. J. Westerheim, eds., *Species Synopsis: Life Histories of Selected Fish and Shellfish of the Northeast Pacific and Bering Sea*, Washington Sea Grant Program and Fisheries Research Institute, University of Washington, Seattle, 1988, pp. 55–69.

H. A. Innis, *The Cod Fisheries, The History of an International Economy*, Yale University Press, New Haven, Conn., 1940.

H. A. Innis, *The Cod Fisheries*, University of Toronto Press, Toronto, 1954, pp. 1–10.

A. C. Jensen *The Cod*, Fitzhenry and Whiteside, Toronto, 1972.

S. Kim and D. R. Gunderson, "Walleye Pollock, *Theragra chalcogramma*, in the Gulf of Alaska," in H. J. Wilimovsky, L. S. Incze, and S. J. Westrheim, eds., *Species Synopsis: Life Histories of Selected Fish and Shellfish of the Northeast Pacific and Bearing Sea*, Washington Sea Grant Program and Fisheries Research Institute, University of Washington, Seattle, 1988, pp. 70–82.

D. H. Steele, "Pollock (*Pollachius virens* (L.)) in the Bay of Fundy," *J. Fish. Res. Board Can.* **20**, 1267–1314 (1963).

C. R. Forrester, "Life History Information Some Groundfish Species," *Fish. Res. Board Can. Tech. Rep.* **105**, 1969.

F. B. Hagerman, *The Biology of the Dover Sole, Microstomus pacificus (Lockinton)*, Calif. Div. Fish. Gam Fish. Bull. **78**, 1949.

International Pacific Halibut Commission, *The Pacific Halibut: Biology, Fishery, and Management*, IPHC Technical Report No. **22**, 1987.

K. S. Parker, "Pacific Halibut, *Hippoglossus stenolepsis*, in the Gulf of Alaska," in N. J. Wilmovsky, L. S. Incze, and S. J. Westrheim, eds., *Species Synopsis: Life Histories of Selected Fish and Shellfish of the Northeast Pacific and Bering Sea*, Report of the Washington Sea Grant Program and Fisheries Research Institute University of Washington, Seattle, 1988, pp. 94–111.

J. H. S. Blaxter, "The Herring: A Successful Species," *Can. J. Fish. Aquat. Sci.* **42**(Suppl. 1), 21–30 (1985).

J. H. S. Blaxter and F. G. T. Holliday, "The Behavior and Physiology of Herring and Other Clupeoids," in F. S. Russell, ed. *Adv. Mar. Biol.*, Vol. 1, Academic Press, Orlando, Fla., 1963, pp. 261–393.

J. H. S. Blaxter and J. R. Hunter, "The Biology of the Clupeoid Fishes," in J. H. S. Blaxter, F. S. Russell, and M. Yonge, eds., *Advanced Marine Biology*, Vol. 20, Academic Press, Orlando, Fla., 1982, pp. 1–223.

D. J. Grosse and D. E. Hay, "Pacific Herring *Clupea harengus pallasi*, in the Northeast Pacific and Bering Sea," in N. J. Wilimovsky, L. S. Incze, and S. J. Westrheim, eds., *Species Synopsis: Life Histories of Selected Fish and Shellfish of the northeast Pacific and Bering Sea*. Washington Sea Grant Program and Fisheries Research Institute, University of Washington Press, Seattle, 1988, pp. 34–54.

S. Morita, "History of the Herring Fishery and Review of Artificial Propagation Techniques for Herring in Japan," *Can. J. Fish. Aquat. Sci.* **42**(Suppl 1), 22–229 (1985).

J. W. Reintjes, *Synopsis of Biological Data on the Atlantic Menhaden, Brevortia Tyrannus*, FAO Species Synop. **42**, Cir. 320, Washington, D.C. 1969.

W. Templeman, *Redfish Distribution in the North Atlantic*, Bull. Fish. Res. Board Can. **120**, 1959.

B. A. Branson, "Sockeye Salmon," *Oceans* **9**, 25–29 (1976).

R. E. Foerster, "The Sockeye Salmon, *Oncorhynchus nerka*," *Bull. Fish. Res. Board Can.* **162**, 1968.

W. S. Hoar, "The Chum and Pink Salmon Fisheries of British Columbia, 1917–1947," *Bull. Fish. Res. Board Can.* **90**, 1951.

F. Neave, "The Origin and Speciation of *Oncorhynchus*," *Trans. Roy. Soc. Can.* **52**, 25–39 (1958).

S. D. Sedgwick, *The Salmon Handbook*, Andre Deutsch Ltd, 1982.

L. Shapovalov and A. C. Taft, "The Life Histories of the Steelhead Rainbow Trout (*Salmo gairdneri gairdneri*) and Silver Salmon (*Oncorhynchus kistuch*)," *Calif. Fish Bull.* **98**, 1985.

N. J. Wilimovsky, ed., *Symposium on Pink Salmon*, Institute of Fisheries, University of British Columbia, Vancouver, 1962.

M. Fish, "A Review of the Fishes of the Genus *Osmerus* of the California Coast," *Proc. U.S. Nat. Mus.* **46**(2027), 29–297 (1913).

J. L. Hart, "Pacific Fishes of Canada," *Bull. Fish. Res. Board Can.* **180**, 1973.

J. L. Hart and J. L. McHugh, "The Smelts (*Osmeridae*) of British Columbia," *Bull. Fish. Res. Board Can.* **64**, 1944.

M. J. A. Butler, "Plight of the Bluefin Tuna," *National Geographic* **169**, 220–239 (1982).

J. Joseph, W. Klawe and P. Murphy, *Tuna and Billfish-Fish Without a Country*, Inter-American Tropical Tuna Commission, Scripps Institution of Oceanography, LaJolla, Calif. 1988.

G. D. Sharp, and A. E. Dizon, eds., *The Physiology Ecology of Tunas*, Academic Press, Inc., Orlando, Fla., 1978.

George Katsushi Iwama
University of British Columbia
Vancouver, British Columbia
Canada

FISH, MINCED. See Surimi: Science and Technology.

FLAVOR CHEMISTRY

OVERVIEW OF FLAVOR COMPOUNDS

Flavors

Flavors possess a variety of chemical groups and structures. They can be heterocyclic, carbocyclic, terpenoid, aromatic, and so on. The overall flavor of foods is due to carbohydrates, lipids, and proteins; however, specific flavors can be elicited by numerous other classes of compounds, such as alcohols, aldehydes, ketones, and various heterocyclic compounds (pyrazines, pyrroles, pyridines etc). Flavor components in food range in number from 50 to 250 compounds in fresh products such as fruits and vegetables and to more than double this number in foods subjected to heat or enzyme treatment; for example, more than 700 compounds have been reported in roast coffee aroma (1). The investigation of around 200 different food products has led to the identification of nearly 5000 compounds (2,3), the vast majority of them identified by gas chromatography/mass spectrometry (GC/MS) analysis.

Flavor sensation may be due to a single compound (4) or to a group of compounds. Single compounds are called flavor notes; examples are 4-hydroxy-3-methoxybenzaldehyde (vanillin) (Fig. 1), the character note for vanilla flavor, and 3-phenyl-2-propenal (cinnamaldehyde) (Fig. 2) for cinnamon flavor. On the other hand, a group of compounds representing a particular flavor is called the flavor profile

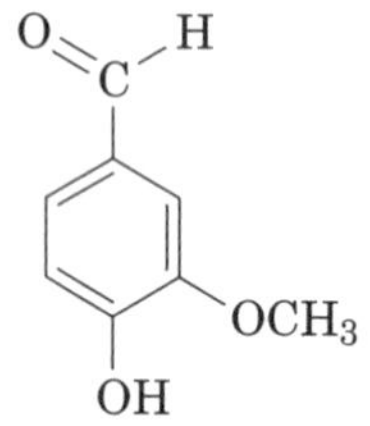

Figure 1. 4-Hydroxy-3-methoxybenzaldehyde (vanillin).

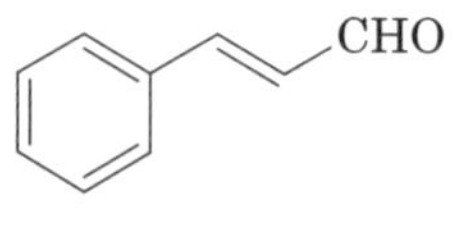

Figure 2. 3-Phenyl-2-propenal (cinnamaldehyde).

of that food product (5). Although a large number of flavor compounds are usually identifiable in a flavor profile, in general only a small number have a significant effect on the overall flavor, these compounds are often called character impact compounds. An example is 2-isobutylthiazole (Fig. 3), the character impact compound found in tomato. The relative concentration of each component in a flavor profile is crucial for its imitation. Flavors can be biosynthesized naturally in foods, such as apples, bananas, and other fruits and vegetables, or they can be produced from precursors during processing or thermal treatment such as baking, roasting, and frying. Furthermore, flavors can be generated by enzymatic modifications such as in cheese or by microbial fermentation as in butter.

The origin of natural flavors and their precursor compounds is found in animal and plant tissue, so flavor research interfaces with other disciplines in its effort to extract and identify these compounds (6,7).

Flavor Enhancers

Monosodium glutamate (MSG), 5′-ribonucleotides such as 5′-inosine monophosphate (5′-IMP) and 2-methyl-3-hydroxy-4-(4*H*)-pyrone (maltol) (Fig. 4) are called flavor enhancers. The actual mechanism of flavor enhancement is not known. These substances contribute a delicious or umami taste to foods when used at levels in excess of their detection limits and enhance flavors of food at levels below their detection thresholds. Their effects are prominent and desirable in the flavors of vegetables, dairy products, meats, poultry, and fish.

Essential Oils

Essential oils are also known as volatile oils, or essences. When exposed to air they evaporate at room temperature. They are usually complex mixtures of a wide variety of organic compounds (such as hydrocarbons, terpenes, alcohols, ketones, phenols, aldehydes, esters, etc). Essential oils are obtained by steam distillation or solvent extraction from many odorous plant sources such as clove, cinnamon, orange, lemon, jasmine, rose, and so on. The organic constituents of essential oils are synthesized by the plant during its normal growth.

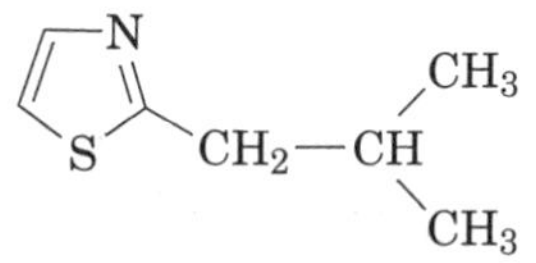

Figure 3. 2-*iso*-Butylthiazole.

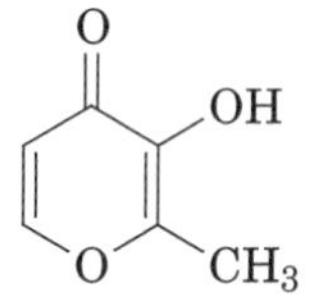

Figure 4. 2-Methyl-3-hydroxy-4(4*H*)-pyrone (maltol).

Resins

Resins are natural products that can be obtained either directly from the plant as exudates or are prepared by alcohol extraction of plants that contain resinous materials. Naturally occurring resins are solids or semisolids at room temperature. They are soluble in alcohol or basic solutions but insoluble in water. They are usually noncrystalline and soften or melt on heating. Chemically, they are complex oxidation products of terpenes. They rarely occur in nature without being mixed with gums and/or volatile oils forming oleoresins.

SENSORY BASIS OF FLAVOR

Flavor can be defined as the combined perception of taste and smell. It involves receptors in both the oral and nasal cavities. In the oral cavity the taste buds are mainly distributed on the surface of the tongue. Each taste bud consists of a barrel-like structure in which the taste cells are packed. Taste receptors, which are found on the surface of the taste cells, are linked to the brain by way of cranial nerves, which carry the nerve impulses to the brain after the neurotransmitters are released from the taste cells. This process is initiated by the formation of the taste compound–receptor complex. Olfactory cells on the other hand are situated in the upper part of the nasal cavity. Their receptors perform a similar function to that of the taste cells through their own nerve fibers, which transmit neural impulses from receptors directly to the olfactory bulb in the brain. There is a common view that there are four primary taste qualities—sweet, sour, bitter and salty—and seven primary odors—camphoraceous, musky, floral, peppermint, ethereal, pungent, and putrid (8).

Sweet Taste

Sweet taste is produced by several different classes of compounds (Fig. 5), such as sugars, polyhydric alcohols, α amino acids, proteins, and synthetic sweeteners.

Sour Taste

Sour taste results from the presence of hydrogen ions on the tongue; however, sourness and acidity (pH) are not directly related, but there is some correlation. Two acids having the same pH do not produce the same degree of response.

Salty Taste

Salt taste is stimulated by most soluble salts having low molecular weights.

Bitter Taste

Three major classes of organic compounds encountered in food materials are associated with bitterness: alkaloids, glycosides (Fig. 6), and peptides.

Astringency

A taste-related phenomena perceived as a dry feeling in the mouth along with a coarse puckering of the oral tissue,

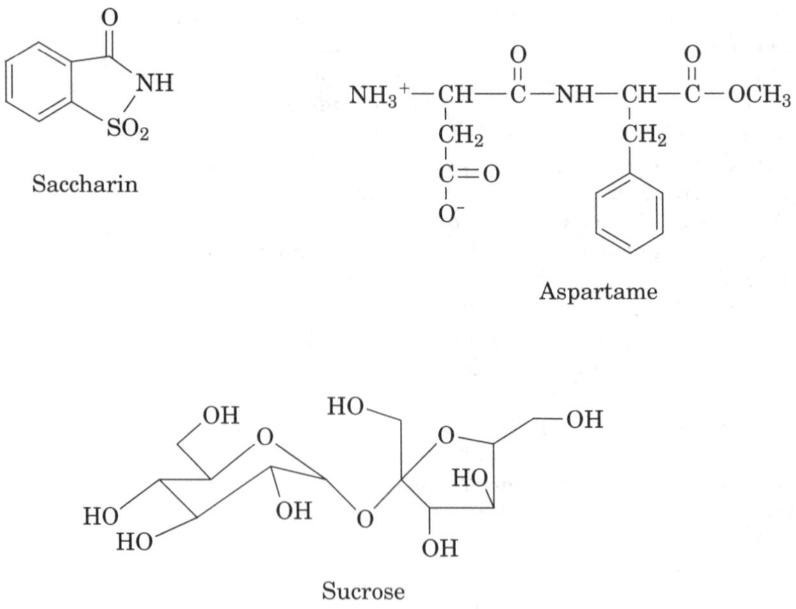

Figure 5. Examples of sweet-tasting compounds.

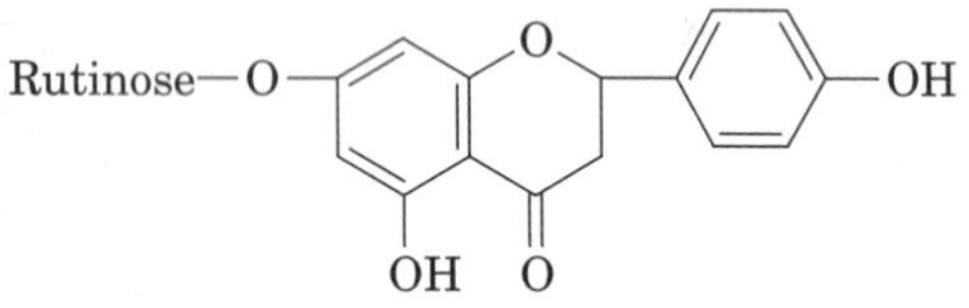

Figure 6. Naringin, a bitter-tasting glycoside.

astringency is due to tannins or polyphenols interacting with the proteins in the saliva to form precipitates or aggregates. Astringency may be a desirable flavor property, such as in tea and red wine.

Pungency

Certain compounds (Fig. 7) found in several spices and vegetables cause characteristic hot, sharp, burning, and stinging sensations that are known collectively as pungency.

Cooling Effect

Cooling sensations occur when certain chemicals (Fig. 8) contact the nasal or oral tissues and stimulate a specific receptor. This effect is most commonly associated with mintlike flavors, including peppermint, spearmint, and wintergreen.

CLASSIFICATION AND ORIGIN OF FLAVOR COMPOUNDS

Classification of Flavor Compounds Based on Their Mode of Formation: Biogenetic and Thermogenetic

One way of classifying flavor compounds is based on their mode of formation; flavors can be generated either naturally (9,10) or by heat treatment during food processing (11,12). Natural flavors are mainly the secondary metabolites of the living tissue, formed during the natural growth cycle of the organism by the action of enzymes, whereas flavors produced by heat treatment are the result of thermal degradation and oxidation of various food ingredients and their complex interactions. The initial or primary precursors of flavors are the polymers found in food such as proteins, lipids, polysaccharides, and DNA, which can undergo either enzymatic or thermal hydrolysis to produce the intermediate precursors, mainly dimers and monomers. These, in turn, can undergo different biotransformation reactions during normal metabolic growth and produce metabolites that have specific flavor qualities that remain *in situ* when the plant is harvested and produce the perceived flavor effect when consumed. On the other hand, during thermal processing of food products, the intermediate precursors formed by thermal hydrolysis can undergo complex chemical transformations, and further

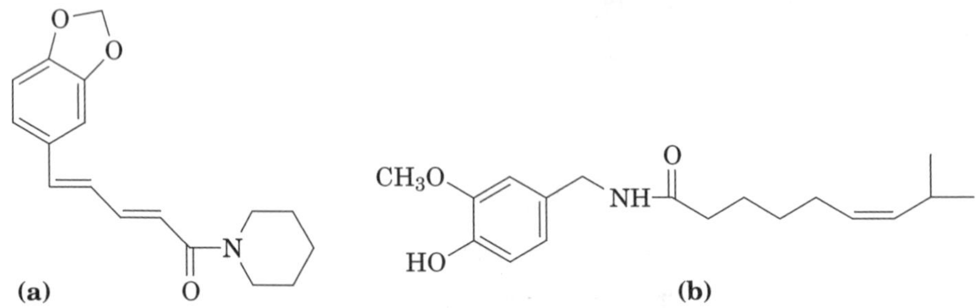

Figure 7. Examples of compounds causing pungency. (a) Piperine, component responsible for the sensation of black pepper; (b) capsaicin, the pungent principal of "hot" pepper.

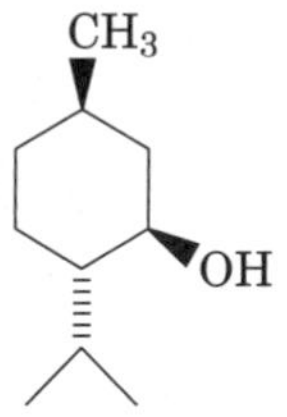

Figure 8. (−)-Menthol causes a cooling sensation.

degradations, to generate cooked flavor (Fig. 9). The type of flavor effect depends on the conditions of processing and the type of the initial precursors found in the particular food product. In terms of composition, thermally generated flavors are richer in heterocyclic compounds compared with enzymatically produced flavors.

Precursors of Flavor Compounds in Food

Meat and Related Products. Raw meat has no particular appealing flavor. However, during various processes of cooking, the initial precursors in meat produce water-soluble intermediate precursors, such as glycopeptides, free nucleotides, peptides, amino acids, amino-sugars, free sugars, fatty acids, and so on. The meat flavor, therefore, is the combined effect of the chemicals produced from the thermal degradation of these intermediates and the products formed from the amino–carbonyl interactions, such as between the reducing sugars and amino acids; this reaction is known as the Maillard reaction or nonenzymatic browning as it is responsible also for the brown color produced when foods are heated (11–13). During thermal degradation amino acids and sugars produce complex mixture of compounds. Sulfur-containing amino acids are specially important for the generation of meat aroma; cysteine, cyctine, and methionine produce sulfur-containing small molecules, such as hydrogen sulfide, 3-(methylthio)-propionaldehyde, and 2-mercaptoethylamine, that play an important role in the aroma of meat products (Fig. 10). Thermal degradation of thiamine produces furans, thiophenes, and thiazoles, important heterocyclic compounds with meatlike flavor. Thermal oxidation of unsaturated fatty acids leads to the formation of different aldehydes.

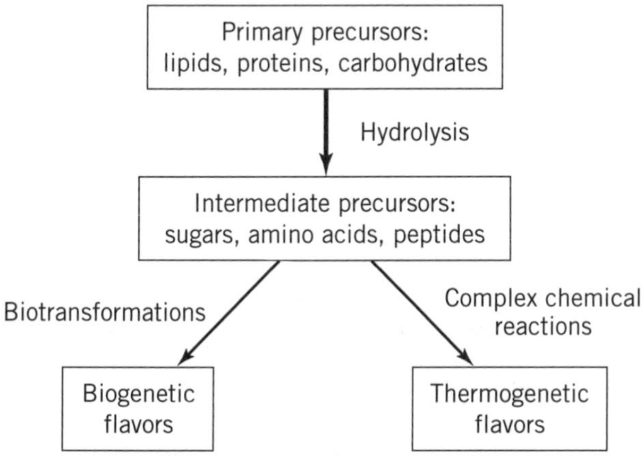

Figure 9. Origin of flavor compounds in food.

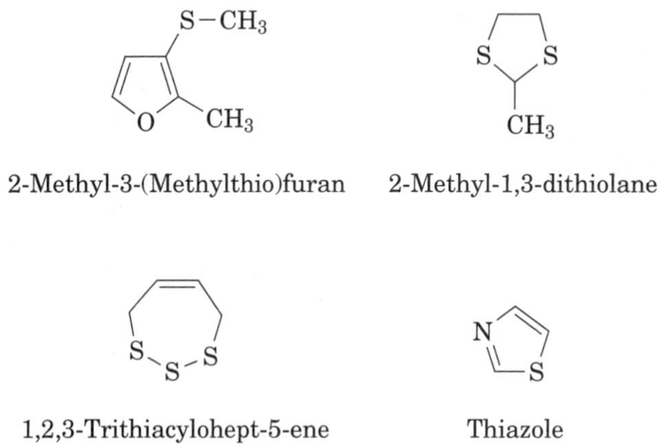

Figure 10. Aroma compounds identified in cooked beef.

Nonalcoholic Beverages: Tea and Coffee. Tea and coffee are examples of food products in which both modes of flavor formation, thermal and enzymatic, play a role in the generation of flavor. During the initial fermentation of green tea leaves, for example, important intermediate precursors are produced by the action of endogenous enzymes to produce polyphenols, carotenoids, and unsaturated fatty acids, such as lenolenic acid, which under the action of lipoxygenase enzymes can produce important flavor aldehydes such as (*Z*)-3-hexenal and (*E*)-2-hexenal. In the heat-treatment stage tea leaves are dried at 85°C and the coffee beans are roasted at >180°C to produce numerous volatile aroma compounds (Fig. 11) by thermal degradation and Maillard reaction processes. The main intermediate precursors involved during roasting of coffee are sugars, amino acids, fatty acids, peptides, amines, phenolic acids such as 3-(3′,4′-dihydroxyphenyl)propenoic acid (caffeic acid), 1-methylpyridinium-3-carboxylate (trigonelline), and so on. In addition, tea and coffee contain alkaloids such as 1,3,7-trimethylxanthine (caffeine) and 3,7-dimethylxanthine (theobromine) that impart a bitter taste to the beverage.

Alcoholic Beverages: Wine. There are three main sources of flavor compounds in wine; some are already present in the grape, some are formed during fermentation process, and some may form during aging by Maillard-type reactions. The type of wood in which the wine is stored can

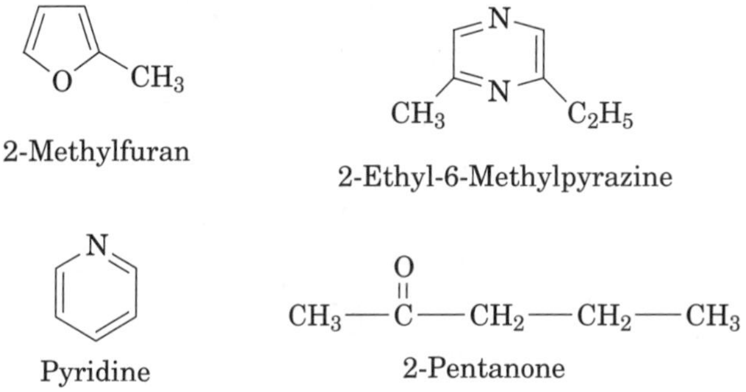

Figure 11. Some volatiles identified in coffee aroma.

also contribute to the overall flavor of the wine. Most of the 300 or so flavor compounds identified in different wines such as 2-methyl-1-propanol, 1-hexanol, 2-phenyl-ethanol, ethyl acetate, and ethyl lactate, are formed during fermentation. The natural aroma of some grape varieties results from a mixture of different terpenes such as (*E*)-3,7-dimethyl-2,6-octadien-1-ol (geraniol), 3,7-dimethyl-1,6-octadien-3-ol (linalool), (*Z*)-3,7-dimethyl-2,6-octadien-1-ol (nerol) (Fig. 12). Sugars such as arabinose, xylose, mannose, and rhamnose produce furans during the aging process by reacting with the amino acids such as proline, alanine, and asparagine by the Maillard reaction (14,15).

Fruits and Vegetables: Tomato. Generally, the aroma of all fresh fruits and vegetables is produced by the action of enzymes on the hydrolysis products of proteins, lipids, and carbohydrates. Free fatty acids in tomato and other vegetables are generated by the action of phospholipases. Then, lipoxygenase enzymes convert the unsaturated fatty acids into important flavor aldehydes such as (*Z*)-3-hexenal, hexanal and alcohols such as 3-hexenol. Amino acids can also be converted into aldehydes and alcohols by the action of enzymes for example leucine is converted into 3-methyl-butanal and 3-methyl-butanol; carotenoids in tomato can be converted into flavor ketones such as 2,6-dimethyl-2,6-undecadien-10-one.

Milk and Dairy Products. More than 200 aroma compounds have been identified in differently processed milk. Some of these compounds have their origin in the raw milk, and some of them are formed during processing due to the degradation of the main precursors found in the raw milk such as fats, carbohydrates (mainly lactose), and proteins. These precursors undergo two types of transformations: (*1*) chemical, such as oxidations, and (*2*) biochemical with the participation of the enzymes and the microbial flora contained in the milk or by exogenous bacteria and microbial flora, such as during preparation of cheese and yogurt. The most important precursor responsible for the production of milk aroma is the lipid. Milk triglycerides are hydrolyzed by bacterial lipases to produce free fatty acids, which in turn are oxidized into different carbonyl compounds, among them many methyl ketones responsible for the milk aroma. Other important carbonyl compounds include oct-1-en-3-one (responsible for the metallic aroma), (*Z*)-4-heptenal (characteristic odor of cream), and γ and δ lactones formed by the thermal oxidation of fatty acids (8).

OCCURRENCE AND ORGANOLEPTIC PROPERTIES OF HETEROCYCLIC COMPOUNDS IN FOOD

Although heterocyclic compounds are present only in minute amounts in foods, they constitute the most important character impact compounds because of their low threshold values. They are extremely important as compounding ingredients and in the development of new flavors. Almost half of the 5000 flavor compounds identified till now are heterocyclic in nature (16).

Furans

Furans are mainly associated with caramel-like, sweet, fruity, nutty, meaty, and burnt odor impressions. Because of their olfactory properties, many furans are commercially important flavoring chemicals. Furans are formed in food by thermal degradation of carbohydrates and by the Maillard reaction. They are almost present in all food aromas and essential oils. The most abundant furans are 2- and 2,5-disubstituted, such as 2-methylfuran, which is found in coffee aroma. The 2-substituted furans with aldehyde, ketone, or alcohol functional groups generally have fruity aromas with the mild flavor of caramel when added in small amounts to nonalcoholic beverages and ice creams. 3-Acetyl-2,5-dimethylfuran is used in imitation nut flavors (Fig. 13).

Thiophenes

Thiophenes can significantly contribute to the sensory properties of foods, but they are not as numerous as furans; they have been detected in boiled and canned beef, cooked chicken, asparagus, leeks roasted peanuts, popcorn, rice, bread, and coffee. Of the 46 sulfur-containing compounds identified in pressure-cooked lean ground beef, 20 were thiophenes. The most frequently found thiophenes are the parent compound itself, 2-alkylated, 2- and 3-acylated derivatives; they are described as being pungent and green-sweet. 2-Acetyl-3-methylthiophene (Fig. 14) is described in a patent as imparting a honeylike flavor to syrup bases, and 3-acetyl-2,5-dimethylthiophene is recommended in another patent for improving the aroma of tobacco and perfumes.

Pyrroles

Pyrroles are among the most widespread heterocyclic compounds found in food flavors such as coffee, roasted peanuts, popcorn, and tobacco smoke. Few pyrroles have also

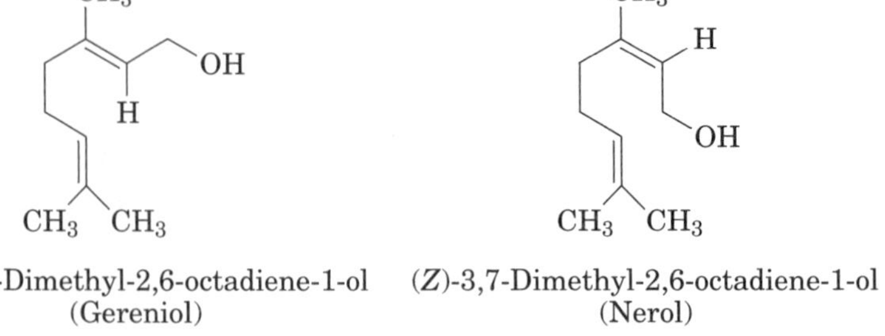

Figure 12. Terpenoids in wine.

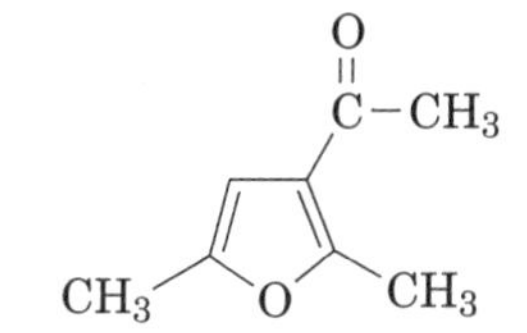

Figure 13. 3-Acetyl-2,5-dimethylfuran.

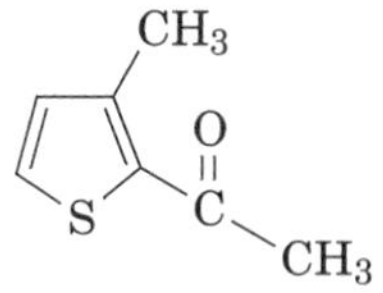

Figure 14. 2-Acetyl-3-methylthiophene.

been identified in cooked beef, cooked asparagus, leeks, and bread. They are mainly formed via the Maillard reaction during cooking. N- and 2-substituted pyrroles are more common, and they impart a burnt character note to the food product. 1-Pyrroline significantly enhances butter flavor of manufactured margarine (Fig. 15).

Thiazoles

Thiazoles possess extraordinary potent sensory properties that can be described as green, roasted, or nutty. They play an important role in the flavors of meat products, vegetables, passion fruit, roasted products, milk, coffee, rum, and whisky. Thiazoles can be formed in food products either naturally or through the Maillard reaction. 2-Isobutylthiazole (Fig. 3), the characteristic aroma of fresh tomato, when added to canned tomato can produce a more intense flavor. 2,4,5-Trisubstituted thiazoles usually have roasted and meaty flavor characteristics. 2-Hydroxymethyl-4-methylthiazole is a patented flavor compound to produce woody and burnt flavor notes.

Pyrazines

Pyrazines represent 4% of all aroma compounds used as flavoring agents. They are the most widely distributed heterocyclic compound in food flavors, found in more than 50 food products of vegetable and animal origin whether processed or not. Pyrazines are described as having nutty, roasted, green, and fruity flavors. They principally occur in cooked beef, potato, mushroom, roasted nuts, bread, cheese, coffee, and some alcoholic beverages. 2-Alkyl-3-methoxypyrazines are widespread in vegetables; the characteristic aroma of green pepper is due to 2-*i*-butyl-3-methoxypyrazine (Fig. 16). 2-Methoxy-3-ethylpyrazine imparts a potato flavor to food products; 2,5-dimethylpyrazine has the flavor of fried chicken, whereas 2-ethoxy-6-methylpyrazine gives the strong aroma and flavor of fresh pineapple.

Figure 15. 1-Pyrroline.

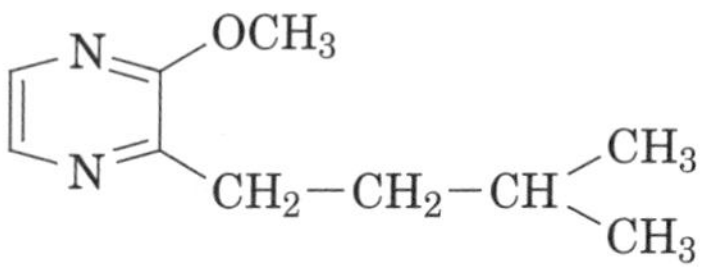

Figure 16. 2-*iso*-butyl-3-methoxypyrazine.

MAILLARD REACTION AND FORMATION OF HETEROCYCLIC FLAVOR COMPOUNDS IN FOOD

Maillard Reaction

Flavors can be produced either naturally by the action of enzymes or by thermal processing due to the interaction of different food components or their pyrolytic degradation. The interaction of reducing sugars with amino-containing components and their subsequent reactions is termed nonenzymatic browning or Maillard reaction (17–21). This reaction is considered to be the most important reaction in food chemistry because in addition to flavors, it is also responsible for the formation of color, antioxidants, carcinogens, and so on. It can also reduce the nutritional value of foods by effectively decreasing the concentration of essential amino acids. The Maillard reaction results in the formation of distinctive brown color and aroma of broiled, baked, and roasted food products. Consequently, many unpleasant-tasting raw foods can be transformed by Maillard reactions into desirable products via processes such as bread baking, coffee roasting, and chocolate manufacture. The Maillard reaction is initiated by the interaction of the open chain form of the reducing sugars with amino acids resulting in the formation of a Schiff base that exists in equilibrium with glycosylamino acid. The equilibrium constant for this reaction is unfavorable, but the glycosylamino acid slowly undergoes a rearrangement reaction to yield a relatively stable derivative. The type of derivative formed depends on the reducing sugar; aldoses undergo Amadori rearrangement (22) to produce 1-(amino acid)-1-deoxy-2-ketoses, whereas ketoses undergo Heyns rearrangement to produce 2-(amino acid)-2-deoxy-1-aldose. Both rearrangements are acid catalyzed; the carboxyl group of the amino acids provides the internal acid catalyst. The net result of this reaction is the transformation of an aldose into a ketose, and vice versa, via the formation of N-glycosides (Fig. 17). This initial stage of the Maillard reaction is well documented and understood; however, the subsequent reactions of these rearrangement products that produce flavors and colors are not very well defined. The rearrangement products themselves are colorless, nonvolatile compounds that do not impart to the food products any specific flavor qualities. However, during the thermal processing of foods they decompose to produce different reactive intermediates that interact further with other food components to produce, among others, a wide range of heterocyclic compounds. These heterocyclic compounds possess potent flavor qualities at very low concentrations

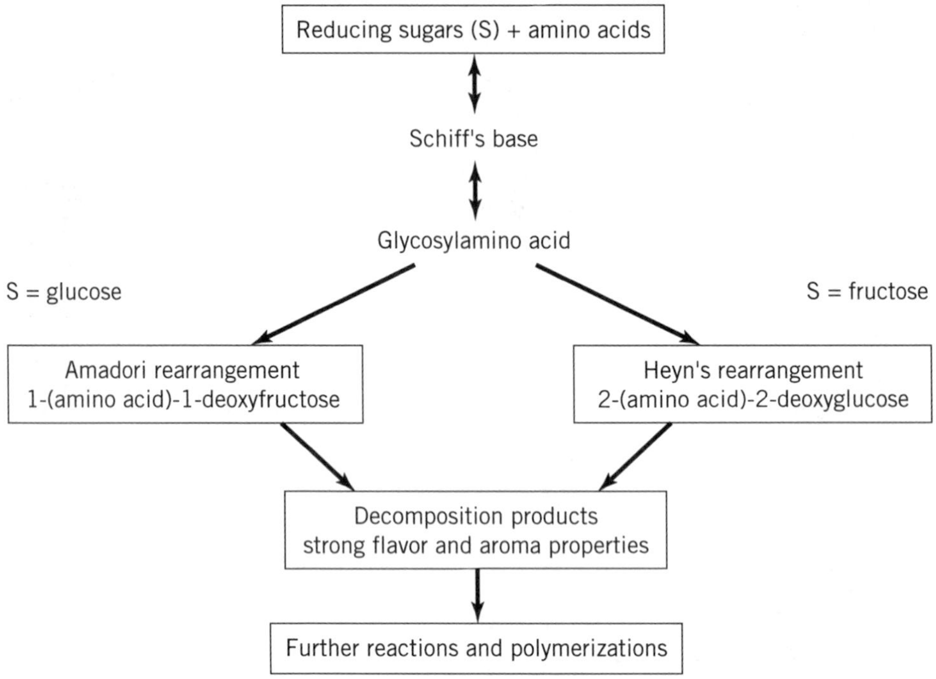

Figure 17. Maillard reaction.

that makes them important contributors to the flavor of baked, roasted, and cooked food. Eventually some of the intermediates formed will polymerize to form brown colored melanoidins that are characteristic of cooked food products.

Further Reactions of Amadori and Heyns Rearrangement Products

The mechanism of decomposition of Amadori and Heyns rearrangement products (AHRP) to produce heterocyclic flavor compounds is not well understood due to the complexity of the food matrix. However, based on the chemistry of carbohydrates, some decomposition pathways have been proposed that involve 1,2- and 2,3-enolizations of the open chain form of the AHRPs followed by β-eliminations to produce 1,2- and 2,3-dicarbonyl compounds. These reactive intermediates can undergo many reactions, including dehydrations, cyclizations to produce furans, retroaldolizations, and further reactions with nitrogen and sulfur nucleophiles to produce N- and S-heterocyclic compounds, such as pyrazines by the Strecker degradation (17).

FLAVOR FORMATION BY ENZYMES AND MICROORGANISMS

In contrast to thermally produced flavors, which comprise a large number of heterocyclic compounds, flavors generated by enzymes in vegetables and fruits consist mainly of aldehydes, ketones, esters, alcohols, terpens, terpenoids, and S-containing aliphatic and aromatic volatile compounds. They can be produced from nonvolatile precursors by the action of different enzymes in the intact tissue before harvest or due to the disruption of the cell tissue so that the compartmentalized endogenous enzymes and the substrates interact. Alternatively, flavors can be produced by the action of exogenous enzymes or microorganisms during fermentation (8).

Origins of Enzymatically Produced Flavors in Vegetables

Aldehydes. Aldehydes containing less than five carbon atoms usually are associated with the development of off-flavors. However, unsaturated alkenals containing 5 to 10 carbon atoms impart desirable flavor attributes to fresh vegetables; *cis*- and *trans*-2-hexenals are both potent aroma compounds of "green leaf" character. The melonlike odor of cucumbers are due to *cis*-2-nonenal (Fig. 18). Aliphatic aldehydes may arise and accumulate in vegetables from precursors such as esters, amino acids, and free fatty acids. Esters can be hydrolyzed by esterases and followed by enzymatic oxidation (NADP-dependent dehydrogenases) to produce aldehydes. Amino acids can be converted into α-keto acids by the action of transaminases, followed by enzymatic decarboxylation to produce flavor aldehydes. Triglycerides in plant tissue can be hydrolyzed by lipases to produce free fatty acids, which in turn can produce long-chain aldehydes by the action of lipoxygenases and lyases.

Sulfur Volatiles. Sulfur-containing volatiles are produced mainly by alliaceous (onion, garlic, leek, etc) and cruciferous (cabbage, mustard, broccoli, horseradish, etc) plants. Alliaceous vegetables produce disulfides and re-

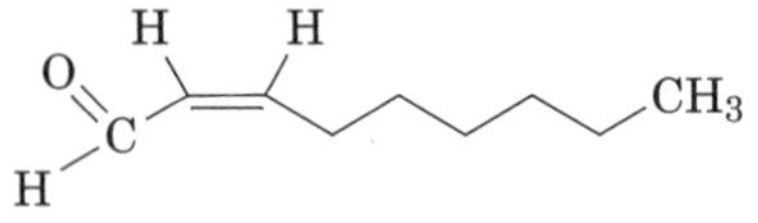

Figure 18. *cis*-2-Nonenal.

lated volatiles, whereas cruciferous vegetables generate isothiocyanates. The main precursor of disulfides in alliaceous plants are S-substituted L-cysteine sulfoxides, which on the action of allinases produce thiosulfinates (Fig. 19), the principal component responsible for the odor of fresh alliaceous plants. On heating, the thiosulfinates are converted into corresponding disulfides, which are responsible for the cooked flavor. The pungent taste of cruciferous vegetables, on the other hand, are caused by isothiocyanates (Fig. 20). The enzymes responsible for the release of these compounds are thioglucosidases, which act on different thioglucosides found in these vegetables to produce the corresponding isothiocyanates. The enzymes and the substrates in these vegetables are compartmentalized; the tissue of the plant should be disrupted to bring the substrate close to the enzyme for the generation of the flavor compounds.

Origins of Enzymatically Produced Flavors in Fruits

About 2000 distinct fruit volatiles have been isolated till now. They consist mainly of esters, aromatic aldehydes, lactones, alcohols, terpenoids, and some thioesters. Unlike fresh vegetables, the aroma is more frequently preformed and arises directly from the intact fruit.

Esters and Aromatic Aldehydes. Esters make the most important contribution to what we usually perceive as fruit flavors. They are present in higher levels than any other class of compound, and also they are present in a greater variety compared with any other class. Most of the esters contribute to the character impact of the resultant aroma. Certain esters have been associated with the aroma of specific fruits, such as methyl butyrate with apple, isopentyl acetate with banana, and ethyl butyrate with orange (Fig. 21). Esters are biosynthesized by enzymatic esterification of carboxylic acids with alcohols through the action of such enzymes as esterases and acyl CoA-alcohol transacetylase.

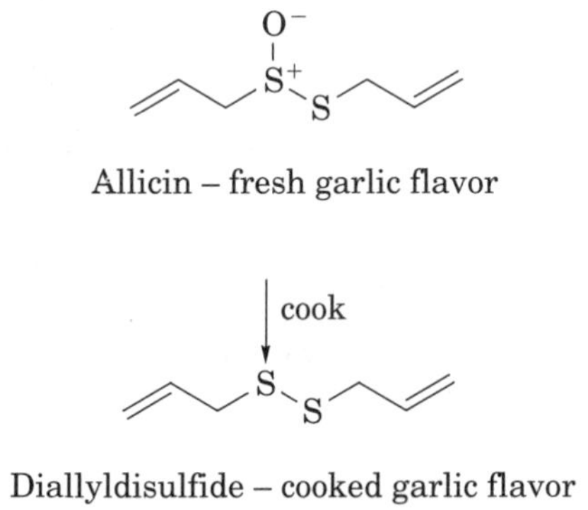

Figure 19. Components responsible for the garlic flavor.

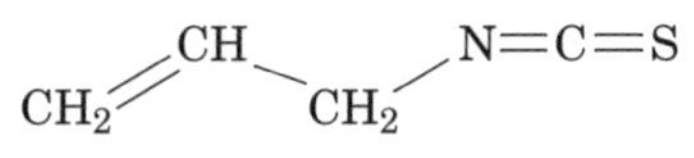

Figure 20. Allylisothiocyanate, principal "hot" component in mustard.

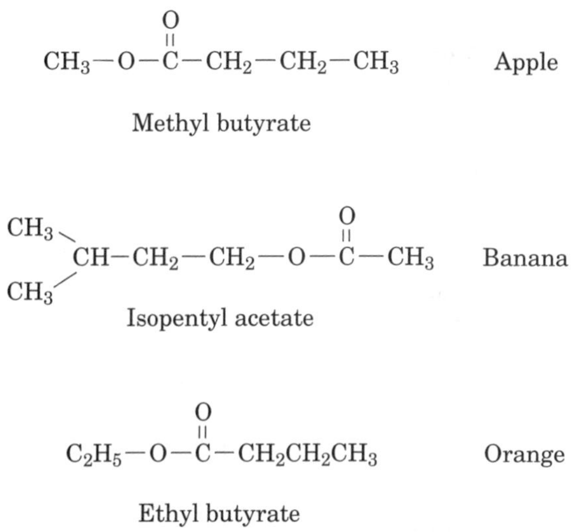

Figure 21. Esters with specific fruit flavors.

In general, the most abundant alcohol in the volatile fraction of a fruit is usually present in the most abundant ester. Aromatic aldehydes sometimes constitute the main flavor component of certain fruits. Benzaldehyde (Fig. 22), for example, is the principal flavor compound in bitter almond. It is also present in the volatiles of peaches, apricots, cherries, and plums. They are generally produced from cyanogenic glycosides by the action of β-glucosidases, which hydrolyzes the glycoside into the free sugar and (R)-mandelonitrile. The latter then undergoes an elimination reaction in the presence of hydroxynitrile lyase to produce the benzaldehyde. 4-Hydroxy-3-methoxybenzaldehyde (vanillin), another aromatic aldehyde, is also produced by the same series of enzymes.

Origins of Enzymatically Produced Flavors in Dairy Products: Milk and Cheese

In contrast to vegetables and fruits, in which the endogenous enzymes are responsible for the release of aroma compounds, dairy flavors are generated by in addition to endogenous enzymes, added microbial enzymes that play a predominant role in the formation of flavors. In general, the flavor of dairy products originates from microbial, enzymatic, and chemical transformations (mostly oxidations)—the relative importance of which is not always understood. These transformations give rise to a series of volatile and nonvolatile compounds, some of which have been shown to correlate well with some typical dairy flavor notes and some have little indication of their real contri-

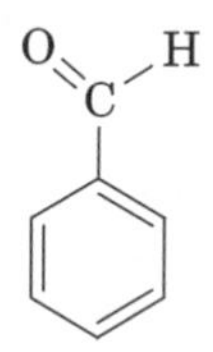

Figure 22. Benzaldehyde.

bution. For any dairy aroma, the total number of compounds identified is far less than in those foods subjected to thermal treatment. Thousands of heterocyclic compounds present in such foods are missing in fermented ones. The characteristic dairy food flavors are usually related to the occurrence of relatively few key components.

Milk furnishes the flavor precursors for all the fermented products (cheese, yogurt etc) made from it. The milk lactose, in addition to being a nutrient for the growth of the starter microorganisms, is also the principal precursor of 2,3-butadione (diacetyl), an important constituent of cultured products. It imparts characteristic creamy note to most cheeses and other dairy products. It is found in butter, unripened soft cheeses such as cottage cheese, and also in Swiss cheese. Milk proteins (caseins) are the source of bitter peptides that contribute to the aroma of different cheeses. The sweet flavor of Swiss cheese is due to a complex formed between calcium (or magnesium) and peptides. The burned note of Gruyère cheese is attributed to peptides complexed with 2,5-dimethyl-4-hydroxy-3-[2*H*]furanone; these peptides and amino acids are produced during the ripening of cheese by the action of proteolytic enzymes on the milk caseines. Some amino acids can be oxidized into corresponding carboxylic acids, such as alanine, and into propionic acid, which is responsible in part for the aroma and taste of Swiss cheese. Although carbohydrates and proteins play an important role as precursors of cheese aroma, fatty acids are the key precursors for both desirable and undesirable aroma. C4–C10 fatty acids impart well-known pungent notes to the cheese, whereas 2-methylketones, lactones, and dairy-associated aldehydes such as *cis*-4-heptenal are important precursors to other flavors. In addition, they act as a medium for the action of enzymes responsible for the formation of aroma, and they serve as a repository for all the other aroma compounds. As such, they determine the relative composition of the aroma vapor above the cheese.

ANALYSIS OF FLAVOR COMPOUNDS

The analysis of food flavors is complicated because of several factors. The flavor-active compounds are found in very low concentrations, ranging from a few 100 ppm for strongly flavored food products to less than 10 ppm for weakly flavored foods. In addition, they represent a wide range of functional groups with widely differing physical and chemical properties. Moreover, some of the components are thermally labile whereas others are highly reactive and may be lost if great care is not exercised during their isolation and concentration as well as in the subsequent analysis. A further complication that puts great demands on the analytical methodology is that the trace components, even when present at levels of ppm or less, may sometimes make a greater contribution to the flavor than components present in vastly greater amounts. The economic importance of such a low threshold value flavor compound lies in the reduced cost of production of highly flavored food products. Despite impressive data available on the chemical composition of food flavors, there is still lack of complete understanding of what determines the flavor of many foods. This may be due, for some products at least, to the inability of the analytical techniques to identify important trace components. However, major advances both in GC and MS technology—such as the introduction of inert thermostable capillary columns; gentle nonvaporizing on-column injection methods; improved interfaces between GC and MS; and sophisticated facilities for spectral acquisition, storage, data manipulation, and library search—make the modern GC/MS an analytical tool of outstanding powers, capable of acquisition of useful mass spectral data from the narrowest of GC peaks.

Establishing the chemical composition and the structure of flavor compounds found in food is important for correlating structures to sensory properties and for understanding the mechanisms by which flavors are formed from their precursors. Although analysis of food flavors is similar in many respects to analysis of any other mixtures, there are special problems associated with it, such as the low concentration of flavor components and their extreme complexity, frequently containing several hundred components of different functional groups. In some cases, trace components in an aroma have far greater sensory importance than other components present in larger amounts; hence, sample preparation and concentration is specially important for analysis of flavor compounds.

Sample Preparation

There are two different approaches to prepare samples for flavor analysis. The first approach is total volatile analysis, which attempts to isolate and concentrate from target food products all the chemicals that could possibly contribute to the flavor. This can be achieved by different distillation and extraction methods. The most appropriate method is largely dictated by the nature of the product under study and by the extent to which its flavors are affected adversely by heat. Fresh fruits and other heat-sensitive products, for example, can be distilled at low temperatures under vacuum, whereas heat-stable products can be steam distilled and simultaneously extracted into an organic solvent by using Likens- and Nickerson-type apparatuses. The second approach is headspace analysis, which aims at analysis of volatile compounds found at equilibrium in the vapor phase above the food. The second approach is faster and simpler and requires small quantities of the food product for analysis. In addition, the headspace contains volatiles in the same relative concentrations as one actually inhales. The headspace volatiles can be analyzed by injecting a small volume of the volatiles directly into a gas chromatograph (GC) or, after its concentration, by cryogenic trapping or by the use of porous polymers such as Porapak and Chromosorb or by activated charcoal (23).

Chemical Methods of Analysis

The identification of the flavor components present in a sample can be established rapidly by GC/MS analysis. However, in complex mixtures, important trace components are often either poorly separated from or totally masked by other components. Consequently, interpretation of their mass spectra may be difficult to perform. To overcome this problem the sample can be modified prior to, during, or after GC/MS analysis to simplify its complexity. Modification of the sample prior to GC/MS analysis can be achieved by preliminary fractionation of its components into polar, nonpolar, and weakly polar fractions or into acidic, basic, and neutral fractions by column chromatography using silica gel. Carbonyl compounds as a group are of special significance since they occur in a wide range of food products of plant and animal origin. As a group they can be isolated by treating the sample with an acidified solution of 2,4-dinitrophenylhydrazine and analyzed as their 2,4-dinitrophenylhydrazone derivatives by GC, HPLC, TLC, and so on. Modification of the sample injected onto a GC/MS can be performed on-line by means of chemical abstractors deposited on deactivated solid supports contained in a coiled tubing and placed before the analytical column of the GC. The function of the abstractors is to remove components bearing specific functional groups by reacting with them. The absence of a particular peak is an indication of the presence of the functional group when the resulting chromatograms are compared with those obtained in the absence of the abstractor. Alternatively, to achieve greater resolution for incompletely separated peaks, components of each chromatographic peak can be recovered by different trapping techniques and analyzed again using different columns. Trapping techniques may be very simple, such as collection of peaks in cooled glass capillary tubes inserted into the GC column or by use of porous-layer open-tubular (PLOT) glass capillaries containing a layer of diatomacous earth support permanently fused to the wall of the capillary tube (23).

Gas Chromatography/Mass Spectrometry

The combined GC/MS remains the most powerful technique available to the chemist for the separation and identification of complex mixtures, because it generates the maximum amount of structural information for the smallest amount of sample in the shortest time. For more than 20 years it has been the mainstay technique in the analysis of flavor components in food and still is, despite recent advances in complementary techniques such as GC/IR (24).

The separation of the complex mixture into its components is achieved on the gas chromatographic column, which can be divided into two types: packed and open tubular (or capillary). The separated components are then introduced through the GC/MS interface into the ionization chamber of the mass spectrometer, where the molecules are bombarded with a beam of energetic electrons (electron impact mode), causing them to ionize and fragment in a way that is characteristic of the molecule. The resulting mixture of ions are then separated on the basis of their mass/charge ratio (m/z) and their relative abundances are recorded. The results are then displayed as a plot of ion abundance versus m/z, which is called the mass spectrum. This represents the characteristic fingerprint of the molecule. The mass spectrum of an unknown compound then can be used to identify its structure by comparison to other known mass spectra; this can be done conveniently by computers.

Gas Chromatography–Olfactometry (GCO)

GC is used mainly in the quantitative analysis of flavor mixtures. However, when the instrumental detector is replaced with a human nose—through the use of a sniffing port—GC can also be used to perform sensory analysis. The technique is known as GCO (25). The main application of GCO is to establish and identify the odor-active components and odor quality of individual components present in a mixture and separated on a GC column. The most common techniques that utilize GCO include Charm-Analysis™ and Aroma Extract Dilution Analysis (AEDA). Both techniques are based on dilution analysis and produce quantitative estimates of relative potency of compounds eluting from a GC column. In dilution analysis, the aroma extracts are serially diluted, and each dilution is analyzed until no significant odor can be detected. In AEDA, the number of dilutions required to eliminate the odor is used to estimate the potency of that particular compound. In CharmAnalysis™, retention times are included in the calculation of the potency factors that are known as "charm values."

Purge and Trap–Thermal Desorption

Low concentration of aroma volatiles can be isolated and concentrated on porous polymer traps such as poly[2,6-diphenyl-*p*-phenylene oxide] (Tenax-GC) using purge and trap (P&T) apparatus (26). Subsequently, the trapped volatiles can be *thermally* desorbed into a GC column using thermal desorption (TD) devices mounted on a GC. Both solid and liquid samples can be analyzed, if proper sampling apparatus is used. In both cases, samples are purged with purified inert carrier gas for a specific period of time (20–90 min), at a flow rate in the range of 10 to 80 mL/min. The temperature, which can range from subambient to few hundred degrees Centigrade, depends on the nature of the sample. The adsorbant trap is then removed and fitted with a syringe needle and attached to the thermal desorber. The traps are thermally desorbed for a period of 3 to 10 min at temperatures depending on the nature of the sample (50–350°C). During the desorption period, the temperature of the GC column is maintained at subambient values to cryofocus the aroma volatiles. After the desorption phase, the volatiles can be separated and eventually identified by initiating the temperature programming of the column.

Solid-Phase Microextraction

Solid-phase microextraction (SPME) is a simple adsorption/desorption technique (26) that eliminates some of the problems associated with solvent extraction of aroma volatiles. The device used to perform SPME consists of a 1-cm fused silica fiber, coated with a specific stationary phase (such as polydimethylsiloxane, carboxene, etc) and bonded to a stainless steel plunger and encased in a holder. The fused silica fiber can be drawn into a hollow needle attached to the holder by using the plunger. The device can be used to adsorb aroma compounds either from a solution or from the headspace above the solution, by inserting the needle through the septum that seals the sample vial. After sample adsorption, the needle can be introduced into the GC injector where the adsorbed volatiles are thermally desorbed into the GC column and analyzed. The adsorption equilibrium is usually attained in 2 to 30 min, depending on the concentration of the sample and the thickness and the type of the coating of the fiber.

BIBLIOGRAPHY

1. S. Van Straten, F. De Vrijer, and J. C. de Beauveser, eds., *List of Volatile Compounds in Food*, 3rd ed., supplements 1–8, Central Institute for Nutrition and Food Research, TNO, Zeist, The Netherlands, 1977–1980.
2. S. Van Straten et al., eds., *Volatiles in Food, Qualitative Data*, 5th ed. supplement I, Division for Nutrition and Food Research, TNO, Zeist, The Netherlands, 1983.
3. H. Maarse, and C. A. Visscher, eds., *Volatile Compounds in Food. Qualitative Data*, 5th ed., supplement I, Division for Nutrition and Food Research, TNO, Zeist, The Netherlands, 1984.
4. P. Z. Bedoukian, *Perfumary and Flavoring Synthetics*, Allured, Wheaton, Ill., 1986.
5. L. B. Sjöström, *The Flavor Profile*, A.D. Little, Cambridge, Mass., 1972.
6. H. B. Heath, *Source Book of Flavors*, AVI, Westport, Conn., 1981.
7. H. B. Heath and G. Reineccius, *Flavor Chemistry and Technology*, AVI, Westport, Conn., 1986.
8. G. G. Birch and M. G. Lindley, eds., *Developments in Food Flavors*, Elsevier Applied Science, London, United Kingdom, 1986.
9. T. H. Parliment and R. Croteau, eds., *Bioggeneration of Aroma*, ACS Symposium Series 317, American Chemical Society, Washington, D.C., 1986.
10. P. R. Ashurst, ed., *Food Flavourings*, Blackie, United Kingdom, 1991.
11. T. H. Parliment, R. J. McGorrin, and C. T. Ho, eds., *Thermal Generation of Aromas*, ACS Symposium Series 409, American Chemical Society, Washington, D.C., 1989.
12. T. H. Parliment, M. J. Morello, and R. J. McGorrin, eds., *Thermally Generated Flavors*, (ACS Symposium Series 543), American Chemical Society, Washington, D.C., 1994.
13. W. G. Moody, "Beef Flavor—A Review," *Food Technol.* **37**, 227–232 (1983).
14. M. A. Amerine and C. S. Ough, *Wine and Must Analysis*, John Wiley and Sons, New York, 1980.
15. J. R. Piggott and A Paterson, eds., *Distilled Beverage Flavour*, VCH, Ellis Horwood Ltd., Chichester, United Kingdom, 1989.
16. G. Vernin, ed., *Chemistry of Heterocyclic Compounds in Flavours and Aromas*, Ellis Horwood Ltd., Chichester, United Kingdom, 1982.
17. C. Eriksson, ed., *Maillard Reactions in Food: Chemical, Physiological, and Technological Aspects*, Prog. Fd. Nutr. Sci. Vol. 5, Pergamon Press, Oxford, United Kingdom, 1981.
18. G. R. Waller and M. S. Feather, eds., *The Maillard Reaction in Foods and Nutrition*, ACS Symposium Series 215, American Chemical Society, Washington, D.C., 1983.
19. M. Fujimaki, M. Namiki, and H. Koto, eds., *Amino Carbonyl Reactions in Food and Biological Systems*, Dev. Food Sci. No 13, Elsevier, Amsterdam, The Netherlands, 1986.
20. P. A. Finot et al., eds., *The Maillard Reaction in Food Processing, Human Nutrition and Physiology*, Birkhauser Verlag, Basel, Switzerland 1990.
21. T. P. Labuza et al., eds., *Maillard Reaction in Chemistry, Food, and Health*, The Royal Society of Chemistry, Cambridge, U.K., 1994.
22. V. A. Yaylayan and A. Huygues-Despointes, "Chemistry of Amadori Rearrangement Products," *Crit. Rev. Food Sci. Nutr.* **34**, 321–369 (1994).
23. I. D. Morton and A. J. Macleod, eds., *Food Flavours—Part A*, Elsevier Scientific, New York, 1982.
24. J. Gilbert, ed., *Applications of Mass Spectrometry in Food Science*, Elsevier Applied Science, London, United Kingdom, 1987.
25. C-H. Ho and C. H. Manley, eds., *Flavor Measurement*, Marcel Dekker, New York, 1993.
26. Z. Zhang and J. Pawliszyn, "Headspace Solid-phase Microextraction," *Anal. Chem.* 1843–1852 (1993).

V. A. YAYLAYAN
McGill University
St. Anne de Bellevue, Quebec
Canada

FOAMS AND SILICONES IN FOOD PROCESSING

Potato chips, jams and jellies, beers and wines, fruits and vegetables. What do all these foods have in common? The manufacturing processes for all these products commonly experience unwanted foaming.

FOAMS

Foam can be defined as a gas dispersed in a liquid at a ratio such that the mixture's bulk density approaches that of a gas rather than a liquid (1). Foaming can occur for several reasons, stemming from both mechanical and chemical origins. Examples of mechanical contributors of foam formation are agitation near a gas–liquid interface, excessive vacuums or pressure gradients, or the free fall of liquids (2). The addition of certain chemicals that act as surface-active materials, or surfactants, may also produce foaming within a process. The existence of starch in a food process is an example of this. Foam formation can cause many problems within a process, including reducing the capacity of open vats, increasing the need for head space in closed systems, slowing the time needed to drain liquids or dried foods, interfering with process instruments, or it can increase housekeeping costs by introducing the need for maintenance personnel to deal with safety hazards such as slippery floors and the overflowing of process tanks (2).

Two basic actions can be taken to stop the formation of foam in a food process. The first involves mechanical means of eliminating the foam, such as heating, centrifuging, spraying, or ultrasonic vibration (3). The other option is to introduce chemical antifoams or defoamers to the process. This article will focus on the chemical method of foam control. Defoamers are materials that specialize in destroying existing foam immediately. Antifoams are materials that are designed as long-lasting means of preventing the formation of foam. Examples of foam control products range from liquid products such as hydrocarbons (kerosene, vegetable oils, organic phosphates, and acetylenic glycols), polyethers, fluorocarbons, and silicones, to solid products such as fatty amides, hydrocarbon waxes, solid fatty acids, and esters (3). The remainder of this article will focus on silicone-based antifoam products.

SILICONE ANTIFOAMS

Two theories exist to describe how antifoams work. Both are based on the fact that pure liquids do not foam. The presence of some stabilizing material is necessary in the surface layer of foam films to allow them to remain stable. These theories are also based on the premise that antifoaming agents either replace or modify these stabilizing materials. The first theory states that the antifoam disperses in the form of fine drops into the liquid film between the bubbles and spreads as a thick duplex film, ie, a film that is thick enough to have two definite surfaces. The tension created by the spreading duplex film leads to the rupture of the original liquid film (bubble wall), destroying the foam.

The second theory is similar to the first and states that the antifoam produces a less cohesive, mixed monolayer on the surface of the liquid film. Because this monolayer is of less coherence than the original film stabilizing monolayer, it will cause the destabilization of the bubble wall (3). In either case, it is important to note that antifoams must be surface active, that they must be able to spread to the entire foaming surface of the process, and that they must remain insoluble to continue to be active.

Since the discovery of polydimethylsiloxane in 1943 by scientists of what is now Dow Corning Corp., there have been many applications found for this unique substance. Applied as antifoams since shortly after their discovery, these silicone-based products are able to provide benefits that most organic defoamers cannot (4). Due to the inert nature of the polydimethylsiloxane molecule, silicone-based antifoams will not react with most process media. This inert nature also allows the antifoam to remain effective longer than reactive organic material. Silicones exhibit a low surface tension (<20 dynes/cm) (5), allowing them to spread quickly and evenly over the foaming surface, an important requirement for an antifoam already mentioned.

Both of these attributes contribute to the most important benefit of using silicone antifoams: their efficiency. Silicone-based antifoams are effective usually in amounts less than 100 parts active silicone per million parts foaming media in industrial applications and in as little as 5–10 ppm in food-grade systems.

Silicone antifoams are available as three basic types of product. First, the polydimethylsiloxane, or silicone oil, can be used as an effective antifoam for many nonaqueous-type processes. Second, the silicone oil can be compounded with hydrophobic silica to form a product that can be useful in both aqueous and nonaqueous systems. It has been theorized that the addition of this silica gives an added benefit of physically disrupting the foam interlayer. The third and most effective product for aqueous systems is an emulsion of the silicone–silica compound. An emulsion is the most effective carrier of a compound in a water environment, allowing for ease in dilution and dispersion.

Silicones can be used in many processes. There are silicone-based antifoams that are permissible for use in direct contact with food, as per the Food and Drug Administration's regulation #173.340, which allows for up to 10 parts of active silicone per million parts of food in the final product. Certain silicones also meet the U.S. Department of Agriculture's specifications for federally inspected foods. Many are also kosher approved. Silicone antifoams usually will not impart any taste or smell to a product. Because these antifoams are essentially inert and are used in such low amounts, they rarely affect process reactions.

BASIC GUIDELINES

There are several basic guidelines to follow when using silicone antifoams to be the most effective as possible. Silicone antifoams should be prediluted and mixed in with a small amount of the foaming medium prior to introduction into the system. This allows the antifoam to disperse evenly and can prevent shocking of an emulsion, which could cause the emulsion particles to coalesce, or stick together. The antifoam should be dispensed in such a manner as to allow it to disperse evenly and reach the surface of the foaming site. This could involve spraying the antifoam onto the surface, or providing slight agitation when adding the antifoam to allow it to reach the surface. The antifoam should be added prior to the point where foaming occurs. This allows the product to act as it was designed; that is, as an antifoam rather than a defoamer.

When using silicone emulsified antifoams, there are additional precautions that should be taken that involve the amount of shear that the antifoam is subjected to. Because this oil-based antifoam is emulsified using surfactants and thickeners, the product can be destroyed by disturbing the emulsion particles, causing coagulation or separation. High levels of mechanical shear can do this through turbulent agitation, violent shaking, or pumping through gear type, centrifugal pumps. Care should be taken to provide gentle agitation when prediluting, or when adding the antifoam to the foaming process.

FUTURE DEVELOPMENT

Significant advancements continue to be made in the development of silicone-based antifoams. Recent developments have focused on areas involving organic–silicone blends. Many newly commercialized products use this new technology and may impart even more efficiency to current antifoam product lines. Work is also being conducted on varying the types of delivery systems for silicone antifoams. Instead of water-based emulsions, solid media are being investigated as delivery systems for the antifoam compound to the foaming interface. This could provide benefits to separation processes using ultrafiltration. Typically, emulsifiers used in antifoam emulsions have shown a tendency to clog and even damage many types of cross-flow ultrafiltration membranes.

As the food industry continues to expand with quicker, more convenient processed foods, so too will the opportunities grow for foam control. Currently, an area experiencing growth is the microwavable foods segment. This is also an area where foaming has been a problem. Many manufacturers desire microwaved food to resemble food that has been prepared by conventional means. Although the formation of foam on food or drinks that have been microwaved does not affect the taste of these products, it is not aesthetically pleasing. The use of silicone antifoams can alleviate these types of problem.

BIBLIOGRAPHY

1. A. P. Kouloheris, "Foam: Friend and Foe," *Chemical Engineering* (1987).
2. J. B. McGee, "Selecting Chemical Defoamers and Antifoams," *Chemical Engineering* (1989).
3. M. J. Owen, "Antifoaming Agents," in *Encyclopedia of Polymer Science and Engineering*, Vol. 2, 2nd ed., John Wiley & Sons, Inc., New York.
4. C. C. Currie, "What Silicones Can Do for the Food Processor," *The Canner* (1952).
5. Internal Dow Corning Surface Tension Studies, TIS #1952-I0030-119.

GENERAL REFERENCES

F. LaBelle, "Taming Foam Problems in Formulations and Products," *Prepared Foods* **163**, 75 (1994).
W. H. B. Denner, "Food Additives: Recommendations for Harmonization and Control," *Food Control* **1**, 150–162 (1990).
P. R. Garrett, "Defoaming: Theory and Industrial Application," Marcel Dekker, New York, 1993.
E. G. Gooch, "Determination of Traces of Silicone Defoamer in Fruit Juices by Solvent Extraction/Atomic Absorption Spectroscopy," *J. AOAC Int.* **76**, 581–583 (1993).
S. Hegenbart, "Processing Aids: The Hidden Helpers," *Prepared Foods* **159**, 83–84 (1990).
M. Holland, "Effective Foam Control in Beet Processing," *International Sugar Journal* **98**, 347–348 (1996).
A. Monferrer and J. Villalta, "Practical Aspects of Frying II," *Alimentacion Equipos y Tecnologia* **12**, 85–90 (1993).
K. Yasunaga et al., "Stabilization Mechanism for Bubbles Formed in Frying Foods," *Journal of the Japanese Society for Food Science and Technology* **44**, 400–406 (1997).
A. A. Zotto, "Antifoams and Release Agents," in J. Smith, ed., *Food Additive's User's Handbook*, Blackie and Sons Ltd., Glasgow, Scotland, 1991, pp. 236–241.

CHRIS COMBS
Dow Corning Corporation
Midland, Michigan

FOOD ADDITIVES

The broadest practical definition of a food additive is any substance that becomes part of a food product either directly or indirectly during some phase of processing, storage, or packaging.

Direct food additives are those that have been intentionally added to food for functional purpose, in controlled amounts, usually at low levels (from parts per million to 1–2% by weight). Basic foodstuffs are excluded from the definition, although ingredients that are added to foods (eg, high-fructose corn syrup [HFCS], starches, and protein concentrates) are often included among food additives.

Indirect or nonintentional additives, on the other hand, are those entering into food products in small quantities as a result of growing, processing, or packaging. Examples of these are lubricating oils from processing equipment or components of a package that migrate into the food before consumption.

The legal definition of a food additive is much more complex and often the subject of controversial interpretation; it is addressed in the sections of this article dealing with government regulations.

The practice of adding chemicals (eg, salt, spices, herbs, vinegar, and smoke) to food dates back many centuries. In recent years, however, the ubiquitous presence of chemical additives in processed foods has attracted much attention and public concern over the long-term safety of additives to humans.

Although the safety issue is far from subsiding, the scientific consensus is that food additives are indispensable in the production, processing, and marketing of many food products, and that the judicious use of chemical additives—typically in the range from a few parts per million (ppm) to less than 1% by weight of the finished food—contributes to the abundance, variety, stability, microbiological safety, flavor, and appearance of the food supply. While food additives offer a major contribution to the palatability and appeal of a wide variety of foods, their level of use is relatively insignificant in the total human diet.

There is much discussion about whether a food additive or food product is natural or synthetic. This classification, in many instances, has become somewhat arbitrary. Many food additives synthesized in chemical laboratories are also found naturally occurring in normal food. Monosodium glutamate, a flavor-enhancing food additive, for instance, is the sodium salt of glutamic acid, an amino acid found in many foods such as mushrooms and tomatoes and metabolized by the human body using the same biochemical pathways of digestion. Vitamin C (ascorbic acid) and its isomer, erythorbic acid, are the same chemicals that are found in oranges. Similarly, citric acid, which is today produced commercially by enzymatic fermentation of sugars, is the same chemically as the naturally occurring chemical that has been found to make lemons and limes tart.

Direct food additives serve several major functions in foods. Many additives, in fact, are multifunctional. The basic functions are as follows:

- *Preservation.* Food preservation techniques have advanced in the past hundred years and now include thermal processing, concentration and drying, refrigeration and freezing, modified atmosphere, and irradiation. However, the use of chemical preservatives frequently augment these basic preservation techniques and represent the most economical way for food manufacturers to ensure a reasonable shelf life for the product. Antioxidants and antimicrobial agents perform some of these functions as well.
- *Processing.* Food processors are increasingly using food additives to ensure the integrity and appeal of finished products. Emulsifiers maintain mixtures and improve texture in breads, dressings, and other foods. They are used in ice cream when smoothness is desired, in breads to increase shelf life and volume and to distribute the shortening, and in cake mixes to achieve batter consistency. Sorbitol, a humectant and sweetener, is used to retain moisture and enhance flavor. With the removal of sugar from many

foods for dietetic reasons, a substitute-bulking agent is growing in importance.

- *Appeal and convenience*. The changing eating habits of the consumers, partly brought about by the large increase in the percentage of women who work outside the home, is creating a growing need for convenience foods. In many of these types of foods, it is essential that a variety of additives be used to provide the taste, color, texture, body, and general acceptability that are required. This need for convenience, while maintaining aesthetic appeal and taste, is becoming extremely important. Most food additives such as gums, flavoring agents, colorants, and sweeteners are included by food processors because consumers in the developed countries demand that food look and taste good as well as being easy and safe to serve.
- *Nutrition*. There have been tremendous advances in knowledge of human nutrition, and consumers are increasingly aware of the value of good nutrition. Vitamins, antioxidants, proteins, and minerals are added to foods and beverages as supplements in an attempt to ensure proper nutrition for those who do not eat a well-balanced diet. In addition, additives such as antioxidants are often used to prevent deterioration of natural nutrients during processing. Recently there is more importance attributed to disease prevention through proper nutrition, as well as to increasing performance through sport nutrition products. On the other hand, the medically based desire for good nutrition through a balanced diet may adversely affect consumer demand for some food additives, such as fat substitutes.

Substances that come under the general definition of direct food additives number in the thousands and include both inorganic and organic chemicals, natural products, and modified natural and synthetic or artificial materials. Most food additives have a long history of use; others are the result of recent research and development to fill particular requirements of modern food processing. Some are common chemicals of industry that are upgraded in terms of purity to allow their use in food.

Food additives comprise more than 30 segments. With well over 2300 food additives currently approved for use in the United States (and more petitioning for approval), it would be impossible in a brief article such as this to discuss each and every substance.

The U.S. Code of Federal Regulations (CFR) provides classification for food additives. In the CFR, direct food additives are divided into the following eight categories:

1. Food preservatives (eg, sodium nitrate, sorbates)
2. Coatings, films, and related substances (eg, polyacrylamide)
3. Special dietary and nutritional additives (eg, vitamins)
4. Anticaking agents (eg, sodium stearate, silicon dioxide)
5. Flavoring agents and related substances (eg, vanillin)
6. Gums, chewing gum bases, and related substances (eg, xanthan gum)
7. Other specific usage additives (eg, calcium lignosulfonate)
8. Multipurpose additives (eg, glycine)

Secondary direct food additives permitted in food for human consumption are divided into four different types in the CFR, as follows:

1. Polymer substances for food treatment (eg, acrylate acrylamide resins)
2. Enzyme preparations and microorganisms (eg, rennet, amylase)
3. Solvents, lubricants, release agents, and related substances (eg, hexane, trichloroethene)
4. Specific usage additives (eg, boiler water additives, defoaming agents)

Indirect food additives included in the CFR are divided into eight categories, as follows:

1. Components of adhesives (eg, calcium ethyl acetoacetate 1,4-butanediol modified with adipic acid)
2. Components of coatings (eg, acrylate ester copolymer coatings and polyvinyl fluoride resins)
3. As components of paper and paperboard (eg, slimicides, sodium nitrate/urea complex, and alkyl ketene dimers)
4. As basic components of single- and repeated-use food contact surfaces (eg, cellophane, ethylene-acrylic acid copolymers, isobutylene copolymers and nylon resins)
5. As components of articles intended for repeated use (eg, ultrafiltration membranes and textiles and textile fibers)
6. Controlling growth of microorganisms (eg, sanitizing solutions)
7. Antioxidants and stabilizers (eg, octyltin stabilizers in vinyl chloride plastics)
8. Certain adjuvants and production aids (eg, animal glue, hydrogenated castor oil, synthetic fatty alcohols, and petrolatum)

FUNCTIONALITY OF FOOD ADDITIVES

Adopted from the National Academy of Sciences national survey of food industries the Food and Drug Administration (FDA) uses the following terms to describe the physical or technical effects for which direct human food ingredients may be added to foods:

- *Anticaking agent or free flow agent*. Substance added to finely powdered or crystalline food product to prevent caking, lumping, or agglomeration

- *Antimicrobial agent.* Substance used to preserve food by preventing growth of microorganisms and subsequent spoilage, including fungistats, mold, and rope inhibitors; also includes antimicrobial agents, antimyotic agents, preservatives, and mold-preventing agents (indirect additives)
- *Antioxidant.* Substance used to preserve food by retarding deterioration, rancidity, or discoloration due to oxidation
- *Boiler water additive.* Substance used in a steam or boiler water system as an anticorrosion agent, to prevent scale or to effect steam purity
- *Color or coloring adjunct.* Substance used to impart, preserve, or enhance the color or shading of a food; includes color fixatives, color-retention agents, and so on
- *Curing or pickling agent.* Substance imparting a unique flavor and/or color to food, usually producing an increase in shelf-life stability
- *Dough strengthener.* Substance used to modify starch and gluten, thereby producing more stable dough
- *Drying agent.* Substance with moisture-absorbing ability; used to maintain an environment of low moisture
- *Emulsifier or emulsifier salt.* Substance that modifies surface tension in the component phase of an emulsion to establish a uniform dispersion or emulsion
- *Enzyme.* Substance used to improve food processing and the quality of finished food
- *Firming agent.* Substance added to precipitate residual pectin, thus strengthening the supporting tissue and preventing its collapse during processing
- *Flavor enhancer.* Substance added to supplement, enhance, or modify the original taste and/or aroma of a food, without imparting a characteristic taste or aroma of its own
- *Flavoring agent or adjuvant.* Substance added to impart or help impart a taste or aroma in food
- *Flour treating agent.* Substance added to milled flour at the mill to improve its color and/or baking qualities, inducing bleaching and maturing agents
- *Formulation aid.* Substance used to promote or promote or to produce a desired physical state or texture in food. Including carriers, binders, fillers, plasticizers, film formers, and tableting aids, and so on
- *Freezing or cooling agent.* Substance that reduces the temperature of food materials through direct contact
- *Fumigant.* Volatile substance used for controlling insects or pests
- *Humectant.* Hygroscopic substance incorporated in food to promote retention of moisture; includes moisture retention agents and antidusting agents
- *Leavening agent.* Substance used to produce or stimulate production of carbon oxide in baked goods to impart a light texture; includes yeast, yeast foods, and calcium salts
- *Lubricant or release agent.* Substance added to food contact surfaces to prevent ingredients and finished products from sticking to them (direct additives); includes release agents, lubricants, surface lubricants, waxes, and antiblocking agents (indirect additives)
- *Malting or fermenting aid.* Substance used to control the rate or nature of malting or fermenting process; includes microbial nutrients and suppressants and excluding acids and alkalis
- *Masticatory substance.* Substance that is responsible for the long-lasting and pliable property of chewing gum
- *Nonnutritive sweetener.* Substance having less than 2% of the caloric value of sucrose per equivalent unit of sweetener capacity
- *Nutrient supplement.* Substance necessary for the body's nutritional and metabolic process
- *Nutritive sweetener.* Substance having greater than 2% of sucrose per equivalent unit of sweetening capacity.
- *Oxidizing or reducing agent.* Substance that chemically oxidizes or reduces another food ingredient, thereby producing a more stable product.
- *pH control agent.* Substance added to change or to maintain active acidity or basicity; includes buffers, acids, alkalis, and neutralizing agents
- *Processing aid.* Substance used as a manufacturing aid to enhance the appeal or utility of a food or component; Includes clarifying and clouding agents, catalysts, flocculents, filter aids, crystallization inhibitors, and so on
- *Propellant.* Gas used to supply force to expel a product or to reduce the amount of oxygen in contact with the food in packaging
- *Sequesterant.* Substance that combines with polyvalent metal ions to form a soluble metal complex to improve the quality and stability of products.
- *Solvent or vehicle.* Substance used to extract or dissolve another substance
- *Stabilizer or thickener.* Substance used to produce viscous solutions or dispersions; to impart body, to improve consistency, or to stabilize emulsions; includes suspending and bodying agents, setting agents, and bulking agents
- *Surface-active agent.* Substance used, other than emulsifiers, to modify surface properties of liquid food components for a variety of effects; includes solubilizing agents, dispersants, detergents, wetting agents, rehydrating enhancers, foaming agents, defoaming agents, and so on
- *Surface finishing agent.* Substance used to increase palatability, preserve gloss, and inhibit discoloration of foods; includes glazes, polishes, waxes, and protective coatings
- *Synergist.* Substance used to act or react with another food ingredient to produce a total effect different from or greater than the sum of the effects produced by the individual ingredients
- *Texturizer.* Substance that affects the appearance or feel of the food.

- *Tracer*. Substance added as a food constituent (as required by regulation) so that levels of this constituent can be detected after subsequent processing and/or combination with other food materials
- *Washing or surface removal agent*. Substance used to wash or assist in the removal of unwanted surface layers from plant or animal tissues

GOVERNMENT REGULATIONS

The application of food additives is highly regulated worldwide, although regulatory philosophy, the approval of specific product, and the level of enforcement differ from country to country. Basic regulations in the United States, Western Europe, and Japan are described in this section. These three major industrial regions are the largest consumers of food additives. With only 13% of the world's population, these countries account for more than two-thirds of the food additive market.

United States

The U.S. Food and Drug Administration (FDA) is the principal U.S. regulatory body controlling the use of food additives. It does so through the 1958 Food Additives Amendment to the Food, Drug & Cosmetic (FD&C) Act of 1938. The amendment was enacted with the threefold purpose of

1. protecting public health by requiring proof of safety before a substance could be added to food,
2. advancing food technology, and
3. improving the food supply by permitting the use of substances in food that is safe at the levels of intended use.

According to the legal definition, food additives that are subject to the amendment include "any substance the intended use of which results or may reasonably be expected to result directly or indirectly in its becoming a component or otherwise affecting the characteristics of any food." This definition includes any substance used in the production, processing, treatment, packaging, transportation, or storage of food.

If a substance is added to a food for a specific purpose, it is referred to as a direct additive. For example, the low-calorie sweetener aspartame, which is used in beverages, puddings, yogurt, chewing gum, and other foods, is considered a direct additive.

Indirect food additives are those that become part of the food in trace amounts due to its packaging, storage, or other handling. For example, minute amounts of packaging substances may find their way into foods during storage. Food packaging manufacturers, therefore, must prove to the FDA that all materials coming in contact with food are safe, before they are permitted for use in such a manner.

For regulatory purposes, all food additives fall into one of three categories:

1. generally recognized as safe (GRAS) substances,
2. prior sanctioned substances, or
3. regulated direct/indirect additives.

GRAS substances (of which there are approximately 700 in all) are a group of additives regarded by qualified experts as "generally recognized as safe." These substances are considered safe because their past extensive use has not shown any harmful effects.

Prior sanctioned substances (approximately 1400) are products that already were in use in foods prior to the 1958 Food Additives Amendment to the federal FD&C Act and are therefore considered exempt from the approval process. Some prior sanctioned substances also appear on the GRAS list. This is the grandfather clause of the amendment. The FDA is involved in an ongoing review of the GRAS and prior sanctioned lists to ensure that these substances are tested by means of the latest scientific methods. Likewise, the FDA also reviews substances that are not currently included on the GRAS list to determine whether they should be added.

All other additives are regulated; that is, a specific food additive petition must be filed with the FDA, requesting approval for use of the additive in any application not previously approved. A food or color additive petition must provide convincing evidence that the proposed additive performs as it is intended. Moreover, animal studies using large doses of the additive for long periods are often necessary to show that the substance would not cause harmful effects at expected levels of human consumption.

In deciding whether an additive should be approved, the agency considers the composition and properties of the substance, the amount likely to be consumed, its probable long-term effects, and various other safety factors. Absolute safety of any substance can never be proven. Therefore, the FDA must determine if the additive is safe under the proposed conditions of use, based on the best scientific knowledge available.

In addition, FDA operates an Adverse Reaction Monitoring System (ARMS) to help serve as an ongoing safety check of all additives. The system monitors and investigates all complains by individuals or their physicians that are believed to be related to specific foods, food additives, or nutrient supplements.

All color additives are subject to the Color Additive Amendment of 1960. Colors permitted for use in foods are classified either as certified or exempt from certification. Certified colors are man-made, with each batch being tested by the manufacturer and the FDA (certified) to ensure that they meet strict specifications for purity.

Color additives that are exempt from certification include pigments derived from natural sources. However, color exempt from certification also must meet certain legal criteria for specifications and purity.

The FD&C Act also authorizes the establishment of reasonable standards of identity and quality for food products. The standards of identity specify in detail what can and cannot be packaged under a given product name. Standards of identity exist for milk, cream, cheese, frozen dessert, bologna products, cereal products, cereal flours, pasta, canned and frozen fruits and vegetables, juices, eggs, fish, nuts, nonalcoholic beverages, margarine, sweeteners, dressings, and flavorings. An approved food additive in the United States may be precluded from use in certain foods characterized by the standards of identity unless the additive is specifically required by or is listed as an optional ingredient in the standards. The standards of identity establish the ingredient composition of a given food, which can then be labeled by its common name. If the manufacturer does not adhere to the standard composition, the food must be labeled "imitation."

Flavor substances are regulated somewhat differently, and the rules are less restrictive. However, the use of aroma chemicals as flavor ingredients is regulated also under laws that may differ from country to country. Following the lead of the United States, inclusion in a positive list that spells out which chemicals are permitted for food use has become the prevalent legislation for regulating flavor chemicals worldwide. The United States has a list of flavor substances that are deemed GRAS based on history of use, review of available toxicology, and the opinion of experts. These GRAS lists (through GRAS 18 issued in 1998) have been compiled since 1977 by the expert panel of Flavor Extracts Manufacturers Association (FEMA) of the United States. Over the years, more than 1800 materials have appeared on FEMA lists. Formed in 1909, FEMA is an industry association that originally started pursuing voluntary self-regulation and later was granted quasi-official status on regulatory matters regarding flavor chemicals by the FDA. The FEMA expert panel was formed in 1960. This independent panel, composed of eminently qualified experts recruited from outside the flavor industry, has expertise in human nutrition, physiology, metabolism, and toxicology and chemical structure-activity relationships. Most industrial countries more or less follow the U.S. system.

In the United States, the FDA has primary jurisdiction over food additives, although clearance for use of additives in certain products must be obtained from other government agencies as well. For example, the U.S. Department of Agriculture (USDA) through the Meat Inspection Division (MID) exercises jurisdiction over additives and ingredients for meat and poultry, and the Bureau of Alcohol, Tobacco, and Firearms (BATF) of the U.S. Department of the Treasury controls the ingredients used in alcoholic beverages.

The Federal Insecticide, Fungicide and Rodenticide Act (FIFRA), which was issued in 1972 and amended in 1988, covers pesticides used on raw agricultural products. The FDA, however, is responsible for enforcing tolerances for pesticide residues that end up in food products, such as ethylene dibromide used in grain products.

Under FDA, USDA, and BATF regulations, the ingredients of a food or beverage must be stated on the product label in decreasing order of predominance. For many direct additive categories, chemical constituents must be identified by their common names and the purpose for which they were added.

One of the recent regulations involving the food industry as well as food additive manufacturers came with the passing of the Nutrition Labeling and Education Act of 1990 (NLEA), which amends the federal FD&C Act, to make nutrition labeling mandatory for most FDA-regulated foods. The nutrition labeling regulations issued by the FDA and the USDA's Food Safety and Inspection Service (FSIS) required compliance by August 8, 1994. The FDA found that by the end of May 1995, more than 80% of the domestic and imported food products checked were in compliance with the regulations.

The FDA's Food Additives Amendment also contains what is known as the Delaney Clause, which mandates the FDA to ban any food additive found to cause cancer in humans or animals, regardless of dose level or intended use. The clause applies not only to new food additives but also to those in use prior to 1958. The Delaney Clause is totally inflexible in that it does not recognize any threshold level below which the additive might not present a health hazard. Thus, it has caused a number of problems for the food industry and for food additives.

Certain additives, such as the sweetener cyclamate, FD&C Red #2 food color, and so on, have been banned after they were found to be potential carcinogens, although many experts believe that feeding tests in animals at massive dose levels may not bear any correlation to the potential risk to humans of chronic ingestion at very low levels. Were it not for a moratorium mandated by Congress, saccharin would have also been banned in the United States several years ago by the FDA in compliance with current food laws.

Approval Process

A new substance gains approval for food use through the successful submission of a Food Additive petition that must document the following:

- Safety, including chronic feeding studies in two species of animals
- Intended use
- Efficiency data at specific levels in specified food systems
- Manufacturing details and product specifications
- Methods for analysis of the substance in food
- Environmental impact statement

Quite frequently, this process can be lengthy—up to 10 years in the case of aspartame and Olestra™—and costly in terms of production hours and dollars. There is little doubt that every phase of the U.S. food additives business is affected by regulations and operates with a constant awareness of the importance of FDA decisions. Not only is the introduction of a new food additive impossible without FDA approval, but also the additives in use are under constant scrutiny by the regulatory agency and remain always vulnerable to new unfavorable toxicological findings. Although the barring of an additive may create opportunities for suppliers to develop new or substitute materials, the potential market is often too small to create sufficient incentive, and the potential loss of the ingredient may raise excessive havoc within affected sectors of the food industry. For example, the ban on cyclamates, followed by the close call on saccharin, almost caused the demise of a strong diet soft drink industry. The well-recognized need for alternative safe sweeteners that would command a high price undoubtedly acted as a stimulus for G. D. Searle (now Monsanto NutraSweet/Kelco division) to engage in a 10 year effort to have aspartame cleared for food use.

European Union

Food additives are regulated by Directive 89/107/EEC of December 21, 1988, on the approximation of the laws of member states concerning food additives authorized for use in foodstuffs intended for human consumption. The European Union (EU) food additives law recognizes 106 food additives. Later, several amendment and adaptations of Directive 89/107/EEC were introduced or proposed, including the following:

- A list of additives, the use of which is authorized to the exclusion of all others.
- The list of foodstuffs to which these additives may be added, the conditions under which they may be added and, where appropriate, a limit on the technological purpose of their use.
- The rules on additives used as carrier substances and solvents, including where necessary their purity criteria.
- The proposal in 1990 by the European Commission of a first specific directive relating to sweeteners and on food additives other than colors and sweeteners. The Sweetener Directive took effect in 30 July 1994. Member countries were asked to adopt the directive by December 31, 1995, which did not occur. The sweetener guidelines are expected to open up new markets for low-calorie food products and will simplify the logistical matters.

Efforts have been toward a uniform registration process so that a registration obtained in one country would be valid in all EU member countries. The new EU food additive law, however, will not prevent individual countries from asking for additional or country-specific requirements for new product registrations. At the EU level, several institutions and groups are involved in the development of food additives law, including the Scientific Committee for Food (SCF), one of the institutions of the European Commission that deals with safety issues; it includes representatives from different national professional organizations, representatives from the food industry, retailers, and so on.

The EU rules for the evaluation, marketing, and labeling of novel foods such as genetically modified foods are also being developed. The new marketing rules would also oblige manufacturers to obtain permission before placing new foods or ingredients on the market, with the exceptions for products that are substantially equivalent to existing foods. The new rules have still to be cleared by the European Parliament, which has the power to veto under the new codecision procedure introduced in 1995.

In many countries, additives must be declared in the labeling. Within the European Community, some additive groups have been uniformly codified with "E" numbers for the orientation of consumers. Some countries, such as Germany, have gone further, adopting regulations on an acceptable daily intake (ADI) basis that build on the newest toxicological knowledge. Some examples of "E" numbers are shown in Table 1.

Japan

In Japan, the Food Chemistry Division of the Ministry of Health and Welfare (MHW) has jurisdiction over food additives through the Food Sanitation Law, which was enacted in January 1948 with several amendments adopted since then. Amendments to the regulations as well as additions or deletions to Kohetisho (the Japanese Codex of Food Additives) were mostly influenced by two major objectives: (1) protection of food sanitation and customer safety, and (2) harmonization with international regulatory requirements.

Most discussions on regulating food additives in Japan have been related to defining what food additives should be under legal restriction and on labeling requirements. Very often in these discussions, differentiating "synthetic" and "natural" food additives had been at issue. In Japan,

Table 1. Selected Food Additives and their European Codes

	Antioxidants
E300	L-Ascorbic acid (vitamin C)
E307	α-Tocopherol, synthetic (vitamin E)
E311	Propyl gallate (PG)
E320	Butylhydroxyanisole (BHA)
	Colorants
E100	Curcumin
E101	Riboflavin
E102	Tartrazin (Yellow No. 5)
E150	Caramel
E160	Annatto
E160a	Beta-carotene
E162	Beetroot red (Betanin; Betanidin)
	Emulsifiers
E322	Lecithin
E471	Mono- and diglycerides of fatty acids
E475	Polyglycerol esters of fatty acids
	Preservatives
E200	Sorbic acid
E202	Potasium sorbate
E210	Benzoic acid
E211	Sodium benzoate
E212	Potassium benzoate
E282	Potassium propionate
	Thickeners and stabilizers
E401	Sodium alginate
E415	Xanthan
E420	Sorbitol
E440a	Pectin

those two generally used terms have often misled customers into a blind belief in natural food additives. However, regulatory bodies as well as the food additive industry no longer distinguish additives with these terms. The latest amendment of the Food Sanitation Law (May 24, 1995) includes deletion of the term "chemically synthesized substances" from the law. Thus, "natural" food additives are regulated under the law amended (being enacted from May 24, 1996), unless they are listed as "existing food additives."

The following short summary of the history of food additive regulations and their effect shows why a distinction had been made between synthetic and natural and why those terms are now obsolete.

It was in 1948 that the term food additive appeared in the Food Sanitation Law and a positive list of food additives was created in Japan. It was the first positive list created in the world, and it did not distinguish between synthetic or natural additives. In the Food Sanitation Law, the term additive means anything added to, mixed into, permeating, or otherwise put in or upon food for the purpose of processing or preserving it.

Since 1989 regulations such as labeling requirements cover natural additives as well, and other requirements for natural additives are planned for the future. Legal coverage of natural additives was indicated by General Agreement on Tariffs and Trade (GATT) officials as a part of the move toward international harmonization of regulations. It was in 1987, when the Japanese government gave notification to GATT officials about a new labeling regulation, that officials pointed out that the Japanese regulation only covered synthetic additives, creating a large inconsistency with regulations of other nations. Today, when new natural food additives are used in Japan, suppliers also need to report them to the MHW. However, in general, data requirements for natural additives are still not as strict as those for chemical substances. There are about 1200 items in the list of natural additives while the conventional list contained 349 compounds as of 1992.

Another important regulation that strongly affects this industry is labeling. The Enforcement Regulation of the Food Sanitation Law, Chapter II (amended on June 10, 1981 and enacted in 1983), defines the term label, and it requires that both synthetic and natural food additives be declared on the food label.

DESCRIPTION OF MAJOR FOOD ADDITIVES

Practically every food manufacturing operation depends in some degree on the use of food additives, but the range of additives necessary for the formulation varies (Table 2). Of the variety of ways food additives have been classified, most involve functional grouping. Although chemical classifications may be useful, because they place molecules of similar structure and physical-chemical properties in comparable categories, compounds of a single chemical family may perform different functions in the food industry. Therefore, the following brief description of the major food additives is based on functional classification.

Anticaking Agents

Food products that contain hygroscopic substances require the addition of an anticaking agent to inhibit formation of aggregates and lumps and thus retain the free-flowing characteristic of the products. Such chemicals as calcium aluminum silicate, calcium phosphate tribasic, magnesium carbonate, magnesium silicate, calcium stearate magnesium stearate, microcrystalline cellulose, and sodium aluminum silicate at a level of 1.0% or less have been suggested as effective anticaking agents in free-running salt and in dry powdered mixes. Colloidal silicon dioxide is also used at a level of 1.0% as an anticaking agent in salt (NaCl). The decahydrate of sodium ferrocyanide or Yellow Prussiate of soda at the extremely low level of 5 ppm causes the formation of star-shaped crystals of sodium chloride rather than the usual cubical ones. The star-shaped crystals known in the trade as dendritic salt possess nonsegregating properties; hence, they are less likely to cake.

Antifoaming (Defoaming) Agents

Antifoaming agents are substances used to reduce foaming caused by proteins or gases that may interfere with processing. Foaming may be largely suppressed or completely eliminated by the use of small quantities, generally about 10 ppm, of dimethylpolysiloxane, mono- and diglycerides, oleic acid, silicon dioxide, white mineral oil, and a number of other fatty acids. These compounds have been suggested for use in the preparation of certain comminuted meat products, bakery products, confections, dairy products, vegetable oils, alcoholic and nonalcoholic beverages, jams and jellies, molasses, soups, starches, syrups, and pickles.

Antioxidants

Antioxidants are described in the "Preservatives" section of this article.

Bleaching, Maturing, and Dough-Conditioning Agents

Because some chemicals serve as both bleaching and maturing agents, and others are referred to as dough-conditioning agents or bread improvers, it is perhaps desirable to consider all of them under one heading.

Bleaching agents are used in the production of certain cheeses, processed fruits, crude fats and oils, and meat products to neutralize color that may be present naturally. As an example, after crude oils are refined to remove impurities, they must be further treated by bleaching to remove coloring materials that are usually present in crude fats and oils. Sulfuric and metaphosphoric acids and hydrogen peroxide have been used for this purpose. Calcium hypochlorite is used in bleaching sugar syrup prior to crystallization.

Bleaching plays special importance in the flour milling and baking industries. Freshly milled wheat flour has a yellowish color, caused by the presence of small quantities of carotenoid and other pigments. Such flour also lacks the quality necessary to make elastic, stable dough. When such flour is stored and allowed to age for several months, it gradually becomes whiter because of oxidation and matures, making it acceptable for baking. For many years,

Table 2. Selected Food Additives and their Major Uses

	Processed/cured meats	Dairy products	Jams/ Jellies	Pickles/ sauces	Breakfast cereals	Cookies/ crackers	Baked goods	Mayonnaise/ dressings
Anticaking agents					X			
Antifoaming agents								X
Antioxidants	X				X			X
Bleaching/maturing agents						X	X	
Colors	X	X	X	X	X	X	X	X
Emulsifiers		X						
Enzymes		X					X	
Fat substitutes		X				X	X	X
Firming agents				X				
Flavors		X		X	X	X	X	X
Glazing compounds							X	
Humectants							X	
pH-adjusting compounds	X	X	X	X				X
Preservatives	X	X	X	X		X	X	
Release agents						X	X	
Sequesterants								X
Sweeteners		X	X	X	X	X	X	
Thickeners/stabilizers	X	X	X				X	X
Vitamins		X			X			

	Carbonated/still beverages	Snacks/chips	Dessert mixes	Frozen entrees	Fats/oils	Margarine	Candies/chewing gums
Anticaking agents			X				X
Antifoaming agents	X				X		X
Antioxidants	X	X			X		X
Bleaching/maturing agents					X		
Colors	X	X	X			X	X
Emulsifiers	X		X	X		X	X
Enzymes							
Fat substitutes		X	X	X		X	
Firming agents							
Flavors	X	X	X	X			X
Glazing compounds							
Humectants							
pH-adjusting compounds	X						
Preservatives	X	X		X			
Release agents		X					X
Sequesterants					X		
Sweeteners	X		X	X			X
Thickeners/stabilizers	X	X	X	X			
Vitamins	X		X			X	

the natural bleaching and aging was the only means millers and bakers had of producing the desired material. These natural processes were slow and costly and did not always yield a consistently satisfactory product. Then it was discovered that certain oxidizing agents incorporated into the flour in small amounts brought about rapid improvements in its color and bread-making properties. Oxidizing agents such as ammonium persulfate, potassium bromate, nitrogen peroxide, and chlorine dioxide is recommended for flour treatment.

Colorants

Colorants, both synthetic and natural derivatives, are used in many foods to enhance their appearance and thereby influence the perception of texture and taste.

Colors can be divided into two types: FD&C certified and noncertified colors. The term FD&C color is a carry-over from the federal FD&C Act of 1938. It refers to synthetic colors that were regarded as “harmless and suitable for use” in foods, drugs, and cosmetics. To market their products, U.S. producers must submit product samples from each batch of material for certification.

Certified food colorants can be divided into dyes and lakes. Dyes are water-soluble compounds that impart color to a substance through dissolution. Lakes are both water and oil insoluble and impart color through dispersion in food. Because of their high tinctorial strength, synthetic colorants of the azo and triphenyl-methane class are used at levels of few ppm to about 300 to 500 ppm, whereas naturally derived colorants require much higher levels, often a few percent as in the case of caramel color.

Noncertified colors can be either natural origins, such as vegetables (eg, carrot oil, red beet juice, paprika, etc), fruits (eg, grape skin extract, cranberry juice concentrate, etc), and insects (cochineal extract), or produced by chemical synthesis (eg, ferrous gluconate, riboflavin, synthetic carotenoid, etc). Although a great number of noncertified food colorants are used by the industry, demand is concentrated in a few products; namely, caramel, annatto, paprika, and synthetic carotenoids. Food color additives and their sources are listed in Table 3.

Emulsifiers

Emulsifiers are additives that allow normally immiscible liquids, such as oil and water, to form a stable mixture. They are widely used in foods to achieve the texture, taste, appearance, fat reduction, and shelf life desired in foods.

The most common and commercially important emulsifiers are monoglycerides and diglycerides of fatty acids and their esters (eg, glyceryl monostearate), lactylated esters (eg, sodium stearyl lactylate), propylene glycol mono- and diesters (eg, propylene glycol monostearate), lecithin, sorbitan esters (eg, sorbitan monostearate), and polysorbates (eg, polyoxyethylene sorbitan monolaurate). These products are usually formulated for specific applications so that the combination provides both enhanced performance and ease of use. A great number of other emulsifiers are also used by the food industry, but in smaller volumes (Table 4).

Enzymes

Enzymes are catalysts used during food processing to make chemical changes to the food. They are biological catalysts that make possible or greatly speed up chemical reaction by combining with the reacting chemicals and bringing them into the proper configuration for the reaction to take place. They are not affected by the reaction. All enzymes are proteins and become inactive at temperatures above 40°C or in unfavorable conditions of acidity or alkalinity. Food enzymes perform some of following functions:

- Speed up reactions
- Reduce viscosity
- Improve extractions
- Carry out bioconversions
- Enhance separations
- Develop functionality
- Create/intensify flavor

Table 3. Selected Food Color Additives

	Color	Source
Allura Red No. 40	Red	Synthetic
Amaranth, Red No. 2[a]	Red	Synthetic
Annatto extract	Yellow	Vegetable
Beet juice	Red	Vegetable
Brilliant blue FCF, Blue No. 1	Blue	Synthetic
Beets, dehydrated (beet powder)	Purple	Vegetable
Canthaxanthin	Red	Synthetic
Caramel	Brown	Semisynthetic
Apo-carotenal	Orange	Synthetic
Beta-carotene	Yellow	Synthetic
Carrot oil	Yellow	Vegetable
Chlorophyll	Green	Vegetable
Cochineal Extract, (carmine)	Red	Insect
Erythrosine, Red No. 3	Red	Synthetic
Fast Green, Green No. 3	Green	Synthetic
Fruit juice (grape and cranberry)	Red	Fruit
Grape skin extract (enocyanin)	Red	Fruit
Paprika	Red	Vegetable
Paprika oleoresin	Red	Vegetable
Ponceau 4R, (Red No. 4)[a]	Red	Synthetic
Riboflavin	Yellow	Synthetic
Saffron	Yellow	Vegetable
Sodium Indigo Disulfonate, Blue No. 2	Blue	Synthetic
Sunset Yellow, Yellow No. 6	Yellow	Synthetic
Tartrazine, Yellow No. 5	Yellow	Synthetic
Titanium dioxide	White	Synthetic
Turmeric (curcumin)	Yellow	Vegetable
Turmeric oleoresin	Yellow	Vegetable
Vegetable juice	Red	Beet and red cabbage juice

[a]Approved in Europe and Japan; banned in the United States.

Table 4. Emulsifiers and their Regulatory Status in the United States and Europe

Emulsifier	USA[a]	European no.
Mono- and diglycerides (GRAS)	182.4505	E 471
Succinyl monoglyceride	172.830	
Lactylated monoglyceride	172.852	E 472
Acetylated monoglyceride	172.828	E 472
Monoglyceride citrate	172.832	E 472
Monoglyceride phosphate (GRAS)	182.4521	
Stearyl-monoglyceride citrate	172.755	E 472
Diacetyl-tartrate ester of monoglyceride (GRAS)	182.4101	E 472
Polyoxyethylene monoglyceride	172.834	
Polyoxyethylene (8) stearate	172.838	
Propylene glycol monoester	172.854	E 477
Lactylated propylene glycol monoester	172.850	
Sorbitan monostearate	172.842	E 491
Sorbitan tristearate	172.842	
Polysorbate 60	172.836	E 435
Polysorbate 65	172.836	E 436
Polysorbate 80	172.840	E 433
Calcium stearoyl lactylate	172.844	E 482
Sodium stearoyl lactylate	172.846	E 481
Stearoyl lactylic acid	172.848	
Stearyl tartarate		E 483
Stearyl monoglyceridyl citrate	172.755	
Sodium stearyl fumarate	172.826	
Sodium lauryl sulfate	172.822	
Dioctyl sodium sulfosuccinate	172.810	
Polyglycerol esters	172.854	E 475
Sucrose esters	172.859	E 473
Sucrose glycerides		E 474
Lecithin (GRAS)	184.1400	E 322
Hydroxylated lecithin	172.814	E 322
Triethyl citrate (GRAS)	182.1911	

[a]Code of Federal Regulation (CFR) Title 21.

- Synthesize chemicals

Food enzymes are usually classified into the following categories:

- Carbohydrases (amylases are commercially the most important subgroup hydrolyzing 1,4 glycosidic bonds in carbohydrates)
- Proteases, hydrolyze peptide bonds in proteins
- Lipases, split hydrocarbons from lipid
- Pectic enzymes and cellulases, hydrolyzing the plant cell wall material
- Specialty enzymes

These enzyme categories can be divided further into around 15 to 20 subgroups. Traditional roles of enzymes in the food industry have been in the processing of bakery goods, alcoholic beverages, and starch conversion. But interest is now focused on newer and more varied applications, such as hydrolysis of lactose, the preparation of modified fats and oils, the processing of fruit juices, and other processes where newer enzymes are being identified. Important food applications of enzymes are listed in Table 5.

The commercially important food enzymes are α amylase, glucoamylase, and glucose isomerase, which are especially significant in the United States because of their widespread use in the conversion of cornstarch into high-fructose corn syrup (HFCS). Rennin, which is used in cheese making, is also of significant volume, followed by a host of other enzymes, including pectinases, invertase, lactase, and maltase (used for the modification of starches and sugars), catalase, pepsin, glucoseoxidase (an antioxidant for canned foods), and bromelin, ficin, and papain (plant proteases used for tenderizing meat and producing easily digestible foods). Enzymes are extremely specific and can act only on a single class of chemicals such as proteins, carbohydrates, or fats.

Enzymes are produced from animal tissues (eg, pancreatin, tripsin, lipase), plant tissues (eg, ficin, bromelin), and most frequently by microorganisms (eg, pectic, or starch enzymes). Microbial production from a variety of species of molds, yeast, and bacteria is increasingly becoming the predominant source of enzymes.

Application of genetic engineering to the development of enzymes has already made a significant impact. The first food ingredient produced by genetic engineering was chymosin, "Chy-Max™," a microbial rennet developed by Pfizer (now Cultor Food Science), which has been approved by the regulatory agencies in the United States, Canada, the United Kingdom, Australia, Italy, and several other countries.

Table 5. Enzyme Applications

Dairy products	Milk coagulation (rennet), milk protein modification, cheese flavor development, enzyme modified cheeses, and removal of hydrogen peroxide
Baking and cereals	Antistaling, dough improvement, improved crust color, and gluten hydrolysis
Sugar processing	Removal of starches and processing from cane sugar
Starch conversion	Starch modification, liquefaction, isomerization, saccharification, modification, and increasing yield
Oils and fats	Improving yields, interesterification, oil extraction, and lecithin production
Flavors	Synthesis of flavors, production of natural esters
Alcohol fermentation	Starch liquefaction, improving yeast growth
Brewing	Adjunct liquefaction, enhanced fermentability, filtration improvement, production of light beer, and removal of protein haze
Fruit juices, and wines	Increasing press yields, juice clarification, shelf-life extension
Coffee processing	Separation of bean, viscosity control of concentrate
Chemical processes	Biotransformation, synthesis (eg, emulsifiers)
Analytical	Tests for dietary fiber, sugars
Waste treatment	Breakdown of cellulose, lignin, oil residues, and other solid waste material

Firming Agents

Fruits and vegetables contain pectin components that are relatively insoluble and form a firm gel around the fibrous tissues of the fruit and prevent its collapse. Addition of calcium salts causes the formation of calcium pectate gel, which supports the tissues and affords protection against softening during processing. The calcium salt is sometimes added to the canned vegetable in the form of a tablet containing both sodium chloride and calcium chloride.

Canned vegetables, canned apples, frozen apples, and tomatoes are sometimes treated during processing with calcium chloride, calcium citrate, monocalcium dihydrogen phosphate, or calcium sulfate to prevent them from becoming soft and disintegrating. Suggested level of use of these calcium salts is 0.02%, calculated as calcium in the final food product. In canned potatoes, calcium chloride and calcium citrate at a level of 0.5% (calculated as calcium) are used.

Aluminum sulfate, ammonium aluminum sulfate, potassium aluminum sulfate, and sodium aluminum sulfate are used as firming agents in pickles and relishes. A more recently introduced firming agent is aluminum sulfate for canned crabmeat, lobster, salmon, shrimp, and tuna. Calcium chloride acts as a firming agent in cheddar and cottage cheese.

Flavors

The word *flavor* describes a complex sensation provided by compositions of many defined aromatic ingredients. Flavorings are concentrated preparations used to impart a specific aroma to food or beverage. Flavoring ingredients are the most numerous single group of intentional additives utilized by the food industries. Flavors should not be viewed as a single homogeneous class of food additives but as a composite of closely interrelated and somewhat overlapping sectors with differentiated characteristics, as discussed next.

Essential oils and natural extracts are usually defined as the volatile material obtained from a particular plant species by the process of distillation, expression (cold pressing), and maceration. Essential oils represent complex aroma mixtures containing as many as hundreds of chemical constituents. Included are vanilla, cocoa, cola, spice oleoresins, and so on. Essential oils may be used for imparting scent or aroma to consumer products, as raw materials for compounding flavor compositions, or as the source of isolated aroma chemicals, also used in compounding.

Aroma chemicals comprise organic compounds with a defined chemical structure that are isolated from microbial fermentation, plants, or animal sources or are produced by organic synthesis. Included are anethole, vanillin, citronellol, geraniol, and so on. Aroma chemicals may be added directly to foods and beverages or used as raw materials in flavor compositions.

Flavor compositions consist of complex mixtures of various aromatic materials from a few to 100 or more constituents. Compounded flavors may contain aroma chemicals, natural extracts, essential oils, solvents, and in some cases other functional additives (eg, antioxidants, acidulants, emulsifiers, etc).

Flavor compositions are added to foods and beverages to

- Create a totally new taste;
- Enhance, extend, round out, or increase the potency of flavors already present;
- Supplement or replace flavors to compensate for losses during processing;
- Simulate other more expensive flavors or replace unavailable flavors; and
- Mask less desirable flavors—to cover harsh or undesirable tastes naturally present in some food, other than hide spoilage.

Flavoring substances may be classified as follows:

- *Natural flavoring substance.* Obtained by physical separation, enzymatic process, or microbial process from vegetable or animal sources, either in the raw state or after processing (including drying, torrefaction, and fermentation)
- *Nature-identical flavoring substance.* Obtained by synthesis or isolated by chemical processes, chemically identical to substances naturally present in vegetable or animal sources (this classification is used in Europe, but not allowed in the United States)
- *Artificial flavoring substance.* Obtained by chemical synthesis and not found to occur in nature
- *Flavoring preparation.* Products other than natural substances, whether concentrated or not, with flavoring properties, obtained by physical separation, or enzymatic or microbial processes, from material of vegetable or animal origin, either in the raw state or after processing (including drying, torrefaction, and fermentation)
- *Process flavorings.* Products obtained by heating to a temperature not exceeding 180°C for a period not exceeding 15 min using a mixture of ingredients, not necessarily themselves having flavoring properties, of which at least one contains nitrogen (amino) and another is a reducing sugar.
- *Smoke flavorings.* Smoke extracts used in traditional foodstuff smoking processes
- *Flavor enhancers.* Some amino acids and nucleotides, as well as sodium salts (such as monosodium glutamate, sodium inositate and sodium guanylate), have only a weak taste by themselves but have the power to considerably enhance the taste sensation caused by other ingredients in savory flavors.

The types of flavor compositions, their manufacturing processes, the starting materials for manufacturing them, and the common product forms are summarized in Table 6.

Glazing and Polishing Agents

Glazes and polishes are used on coated confections to give luster to the otherwise dull coating. Chemicals that are used for this purpose include acetylated monoglycerides, beeswax, carnauba wax, gum arabic, magnesium silicate, mineral oil, petrolatum, shellac, and zein. Generally these compounds are used at levels of about 0.4%, with the exception of mineral oil and petrolatum, which are used at 0.15%, and zein at 1.0%.

Humectants

Substances that have affinity to water, with stabilizing action on the water content of a material, are termed humectant or moisturizing agents. These are hygroscopic materials that prevent loss of moisture when incorporated into foodstuffs. Ideally, a humectant maintains within a rather narrow range of moisture content caused by humidity fluctuations. Among commonly used humectants are the following substances: glycerin, mono- and diglycerides of fatty acids, propylene glycol, pectin, molasses, potassium polymethaphosphate, and sorbitol.

pH-Adjusting Agents

A large group of chemical additives that are widely used in foods might be considered under the broad heading of pH-adjusting agents. Other terms that describe these chemicals include acidulants, acids, alkalis, buffers, and neutralizers. These chemicals are used in most segments of the food processing industries, including:

- the baking industry as chemical leavening agents,
- in soft drinks to provide tartness,
- in certain dairy products to adjust the acidity,
- in cheese spreads for emulsification,

Table 6. Commercial Flavor Compositions

Type of flavor	Classification	Manufacturing process	Raw materials	Product form
Compounded flavors	Natural or synthetic	Blending, mixing	Essential oils, natural extracts, fruit juice concentrates, aroma chemicals	Liquid Spray-dried, encapsulated
Natural extracts	Natural	Extraction, enzymatic treatment	Food substrates (eg, plants, fish, meat, etc)	Liquid Paste
Reaction flavors (thermally processed)	Natural	Heating/pressure cooking	Amino acids and sugars, hydrolyzed proteins	Paste powder
Enzymatically modified flavors	Natural	Enzymatic/microbial reaction	Food substrates (eg, cheese)	Paste powder

- in confectionery products as flavoring, to control the degree of inversion of sugars, and to control the texture in the processing of chocolates,
- in jams and jellies to provide proper gel formation.

Citric acid is the most versatile and widely used of the food acidulants, and very large volumes of phosphoric acid are required for cola beverages. Other acidulants frequently used in processed foods include acetic, adipic, fumaric, hydrochloric, lactic, malic and tartaric acids, and glucono-δ-lactone.

Important alkalies used in the food field include ammonium bicarbonate, ammonium hydroxide, calcium carbonate, calcium oxide, potassium bicarbonate, potassium hydroxide, sodium bicarbonate, sodium carbonate, sodium hydroxide, and sodium sesquicarbonate.

Quite often, the pH may be difficult to adjust or to maintain after adjustment. Stability of pH can be accomplished by the addition of buffering agents that, within limits, effectively maintain the desired pH even when additional acid or alkali may be added. Examples of representative buffer solutions are as follows:

- Phosphoric acid : dibasic potassium phosphate
- Formic acid : sodium formate
- Acetic acid : sodium acetate
- Sodium bicarbonate : sodium carbonate
- Dibasic sodium phosphate : sodium hydroxide

Preservatives

Chemical preservatives play a very important role in the food industry, from manufacture through distribution to the ultimate consumer. The choice of a preservative takes into consideration the product to be preserved, the type of spoilage organism endemic to it, the pH of the product, period of shelf life, and ease of application. No one preservative can be used in every product to control all organisms, and therefore combinations are often used. In certain foods, specific preservatives have very little competition. In the concentrations used in practice, none of the preservatives discussed here is lethal to microorganisms in foods. Rather, their action is inhibitory.

Preservatives may be divided into two main groups: (1) antioxidants and (2) antimicrobials. Antioxidants are food additives that retard atmospheric oxidation and its degrading effects, thus extending the shelf life of foods. Examples of food oxidative degradation include products that contain fats and oils in which the oxidation would produce objectionable rancid odors and flavors, some of which might even be harmful. Antioxidants are also used to scavenge oxygen and prevent color, flavor, and nutrient deterioration of cut or bruised fruits and vegetables.

Food antioxidants may be divided into oil- and water-soluble compounds (Table 7) and also classified as natural or synthetic.

The most frequently used natural antioxidants are ascorbic acid (vitamin C), its stereo isomer erythorbic acid, and their sodium salts (sodium ascorbate, sodium erythorbate), plus the mixed δ- and γ-tocopherols. Vitamin C finds more major use as a nutritive supplement or in pharmaceutical preparations. Smaller amounts, however, are intentionally used for antioxidant purposes. Erythorbic acid (*iso*-ascorbic acid) is virtually devoid of vitamin C activity (only 5% that of ascorbic acid). Citric acid and tartaric acid are also natural antioxidants (and antioxidant synergists) but are predominantly added to foods as acidulants.

Synthetic antioxidants used as direct food antioxidants include butylated hydroxyanisole (BHA), butylated hy-

Table 7. List of Major Food Antioxidants

Oil-soluble antioxidants
Butylated hydroxyanisole (BHA)
Butylated hydroxytoluene (BHT)
tert-Butyl hydroquinone (TBHQ)
Propyl gallate (PG)
Tocopherols
Thiodipropionic acid
Dilauryl thiodipropionate
Ascorbyl palmitate
Ethoxyquin
Water-soluble antioxidants
Ascorbic acid
Sodium ascorbate
Erythorbic acid
Sodium erythorbate
Glucose Oxidase/catalase enzymes
Gum Guaiac
Sulfites
Rosemary extract

droxytoluene (BHT), *tert*-butyl hydroquinone (TBHQ), and propyl gallate (PG). These antioxidants are oil-soluble compounds. Their primary application is in fats and oils to retard the development of rancid taste.

To improve the performance of antioxidants, two other types of food additives, sequestrants (eg, EDTA, citric acid) and synergists (eg, mixtures of antioxidants and lecithin), are frequently used with them. Antioxidants may also be present in food packaging as indirect food additives.

Recently, definitive studies have shown and widely publicized in the news media that antioxidant nutrients such as ascorbic acid (vitamin C) and tocopherols (vitamin E) can protect against harmful cell damage and thus prevent certain human diseases. Foods formulated with antioxidants and other vitamins are now recommended to prevent and cure cancer, cardiovascular diseases, and cataracts. The same antioxidants that are used to prevent oxidative deterioration of food may be used in functional foods (also called nutraceuticals, designer foods, etc) to create products that prevent or cure certain chronic diseases.

Antimicrobials are capable of retarding or preventing growth of microorganisms such as yeast, bacteria, molds, or fungi and subsequent spoilage of foods. The principal mechanisms are reduced water availability and increased acidity. Sometimes these additives also preserve other important food characteristics such as flavor, color, texture, and nutritional value. The primary food additives used for this function are sorbic acid, potassium and sodium sorbates, calcium and sodium propionates, benzoic acid, sodium and potassium benzoates, and parabens.

In addition to these compounds, several organic acids such as citric, malic, lactic, ascorbic, phosphoric, and tartaric also act as antimicrobial agents, but because of their greater use as acidulants, they are covered elsewhere.

Sulfur dioxide and sulfites are also applied extensively for controlling undesirable microorganisms in soft drinks, juices, wine, beer, and other products. Salt, sugar, alcohol, spices, essential oils, and herbs also inhibit growth of microorganisms, but usually their primary function is different when added to food.

Major uses of antimicrobial preservatives by food industry include the following:

- Sorbates are used as mold and yeast inhibition in processed cheese and spreads, salad dressings, and dried fruits, and are effective in the acidic pH range up to pH 6.5.
- Benzoates are used in beverages, fruit juice, and pickles, and are effective in the pH range between 2.5 and 4.0.
- Propionates are used as mold and rope inhibitors in bread and baked goods.
- Parabens are effective in low-acid foods (pH greater than 5.0), such as meat and poultry products.
- Sulfur dioxide and sulfite salts (eg, potassium bisulfite, potassium metabisulfite, sodium bisulfite, sodium metabisulfite, and sodium sulfite) are the most effective inhibitors of deterioration of dried fruits and fruit juices. Sulfur compounds also used in the fermentation industry to prevent spoilage by microorganisms and as a selective inhibitor of undesirable organisms.

Release Agents

Release agents are used to prevent confectionery, and to a lesser extent baked goods, from sticking to equipment or to the container in which they are heated. They are also used to prevent pieces of confection from adhering to each other. Included are acetylated monoglycerides, calcium stearate, magnesium carbonate, magnesium silicate, magnesium stearate, mannitol, mineral oil, mono- and diglycerides, sorbitol, and stearic acid.

Sequestering Agents

Also called chelates, sequestering agents combine with polyvalent metal ions to form a soluble metal complex to improve the quality and stability of products as free metallic ions promote oxidation of food. They are used in various aspects of food production and processing chiefly to obviate undesirable properties of metal ions without the necessity of precipitating or removing these ions from solutions.

Ethylenediaminetetraacetic acid (EDTA) is the most commonly used sequestering, or metal-complexing, agent used in the food industry. This compound, as well as the disodium or calcium disodium salts of tetraacetic acid retards discoloration of dried bananas, beans, chickpeas, canned clams, pecan pie filling, frozen potatoes, and canned shrimp. Also, these compounds improve flavor retention in canned carbonated beverages, salad dressings, mayonnaise, margarine, and sauces; retard struvite formation in canned crabmeat and shrimp; and protect against rancidity in dressings, mayonnaise, sauces, and sandwich spreads. EDTA is used at concentrations of 33 to 800 ppm.

Other chemicals that may be included in this category are calcium acetate, calcium gluconate, calcium sulfate, citric acid, stearyl citrate, tartaric acid, sodium tartarate, calcium monoisopropyl citrate, sodium hexametaphosphate, phosphoric acid, potassium citrate, and various calcium, potassium, and sodium phosphates.

Sweeteners

Sweeteners are used in formulated foods for many functional reasons as well as to impart sweetness. They render certain foods palatable and mask bitterness; add flavor, body, bulk, and texture; change the freezing point and control crystallization; control viscosity, which contributes to body and texture; and prevent spoilage. Certain sweeteners bind the moisture in food that is required by detrimental microorganisms. Alternatively, some sweeteners can serve as food for fermenting organisms that produce acids that preserve the food, thus extending shelf life by retaining moisture. These auxiliary functions must be kept in mind when considering applications for artificial sweeteners.

Sweeteners may be classified in a variety of ways:

Table 8. Polyols: Their Calorie Value and Relative Sweetness to Sugar

Polyol sweetener	Relative sweetness, sucrose = 100	Calorie value (U.S. FDA allowance), Kcal/g
Erythritol	60–70	0.2
Isomalt	45–65	2.0
Lactitol	40	2.0
Maltitol	90	3.0
Mannitol	70	1.6
Sorbitol	50–70	2.6
Xylitol	100	2.4

- *Nutritive or nonnutritive*. Materials either are metabolized and provide calories or are not metabolized and thus are noncaloric.
- *Natural or synthetic*. Commercial products that are modifications of a natural product (eg, honey or crystalline fructose is considered natural; saccharin is a synthetic compound).
- *Regular or low-calorie / dietetic / high-intensity*. Although two sweeteners may have the same number of calories per gram, one may be considered low-calorie or high-intensity if less material is used for equivalent sweetness.
- *As foods*. For example, fruit juice concentrates can impart substantial sweetness.

Sweetness is a subjective perception influenced by a multitude of variables, including temperature of the food being tasted, pH, other flavors and ingredients in the food, physical characteristics of the food sweetener, concentration, rate of sweetness development, and permanence of sweetness and flavor. Sweetness is measured via sensory methods by taste panels. Results can vary depending on foods consumed prior to testing (even several hours before testing), the flavors to which the taster is accustomed, tasting experience of the panelist, time of day, and physical surroundings in the test room.

Sucrose, commonly known as table sugar (or refined sugar), is the standard against which all sweeteners are measured in terms of quality of taste and taste profile. It is consumed in the greatest volume of all sweeteners. However, sucrose, high-fructose corn syrup (HFCS), and other natural sweeteners, such as molasses, honey, maple syrup, and lactose sweeteners, are not considered additives and are not covered in this report.

Polyols (sugar alcohols) are a group of sweeteners that provide the bulk of sugars, without as many calories as sugar (Table 8). Polyols are important sugar substitutes and utilized where their different sensory, special dietary, and functional properties make them feasible. Polyols are obtained from their parent sugars by catalytic hydrogenation. In most European countries and in the United States polyols are utilized in low-calorie food formulations. Polyols are absorbed more slowly in the digestive tract than sucrose; therefore, they are useful in certain special diets. However, when consumed in large quantities some of them have a laxative effect. Polyols offer the same preservative benefit and a similar bodying effect to food than sucrose. Polyols are more resistant to either thermal breakdown or hydrolysis than sugar. Moreover most polyols are resistant to fermentation by oral bacteria, and therefore prime ingredients for tooth-friendly confectioneries (eg, "sugarless chewing gums).

High-intensity sweeteners, once used mainly for dietetic purposes, are now used as food additives in a wide variety of products. They are termed high-intensity because they are many times sweeter than sucrose and closely mimic its sweetness profile. Because of the very low use levels, however, high-intensity sweeteners cannot perform other key auxiliary functions in food and often must be used in conjunction with other additives such as low-calorie bulking agents. The relative sweetness of high-intensity sweeteners to sucrose sugar and their regulatory status are summarized in Table 9.

In 1988, the U.S. FDA approved the use of Hoechst AG's *acesulfame K* (Sunette™) for use in chewing gum, dry beverage mixes, instant coffee and tea, gelatins, puddings, and nondairy creamers. In 1998 its approval was extended to

Table 9. High Intensity Sweeteners: Their Regulatory Status and Sweetness Relative to Sugar (1998)

Sweetener	Sweetness, sucrose = 1	United States	Canada	Europe	Japan
Cyclamate, Na Salt	30	P	A	A	N
Aspartame	200	A	A	A	A
Acesulfame K	200	A	A	A	N
Saccharin	300	A	N[d]	A	A
Sucralose[a]	600	A	A	P	N
Thaumatin (Talin)	3,000	N	N	N	A
Alitame[b]	2,000	P	P	P	P
Neohesperidin DC	2,000	N	N	A	N
Stevioside	300	N	N	N	A
Glycirrhizin	50	N[c]	N	N	A

Note: A = approved; P = petition filed; N = not approved.
[a]Sucralose is approved in Australia, Russia, Brazil, New Zealand, Quasar, Romania, and Mexico.
[b]Alitame is approved in Australia, New Zealand, People's Republic of China, Indonesia, Colombia, and Mexico.
[c]Glycirrhizin is approved as a flavoring, but not as a sweetener in the United States.
[d]Saccharin in Canada is limited for use in personal care products and pharmaceutical, but it is banned in foods and beverages.

Table 10. Functions of Major Food Thickeners and Stabilizers

	Thickening	Emulsion stabilization	Suspending properties	Gelation	Crystallization control	Water ending	Mouth-feel	Foam stabilization	Flavor fixation	Protective film forming	Synergistic effect	Fat substitution
Agar				X								
Unmodified starches	X	X	X	X								
Modified starches	X	X	X	X		X	X					X
Casein						X		X		X		
Gelatin	X	X		X					X	X	X	
Carboxymethylcellulose (CMC)	X	X			X	X	X		X	X		X
Methylcellulose (MC)	X	X		X				X		X	X	
Guar gum	X	X				X					X	
Alginates	X	X	X	X		X	X	X		X	X	
Xanthan gum	X	X	X		X		X					X
Pectin	X			X			X			X		X
Locust bean gum	X	X										
Gum arabic		X	X		X			X		X		
Carageenan	X	X	X	X	X	X	X				X	X

liquid beverages. Acesulfame K has a rapidly perceptible sweet taste 200 times as potent as sucrose.

Alitame is a dipeptide made of two amino acids, L-aspartic and D-alanine. The sweetener is 2000 times as sweet as sugar with the same taste as sugar. Its potential market applications include bakery products, snack foods, candies and confectionery, ice cream, and frozen dairy products.

Aspartame was approved in the United States in 1981 for use in prepared foods and dry beverage mixes, as a tabletop sweetener, and in 1983 in liquid soft drinks. Aspartame is about 200 times as sweet as sucrose. Unlike most low-calorie sweeteners, aspartame is digested by the body to amino acids, which are metabolized normally. However, because of its intense sweetness, the amounts ingested are small enough that aspartame is generally considered noncaloric.

Cyclamate is 30 times sweeter than sugar. It has a sugarlike taste, a good shelf life, and a synergistic effect when combined with other high-intensity sweeteners. Currently it is approved in Canada and Europe, but it has been banned in the United States since 1970.

Neohesperidin dihydrochalcone (DHC) is derived from bioflavonoids of citrus fruits. It is 1500 to 2000 times sweeter than sucrose, leaves a licorice aftertaste, and gives a delayed perception of sweetness. DHC is produced from bitter Seville oranges by hydrogenation of natural neohesperidin, the main flavonoid constituent of some oranges.

Saccharin, used primarily as its sodium salt, is approximately 300 times as sweet as sucrose. Saccharin is the most widely used nonnutritive sweetener worldwide and the least expensive on a sweetness basis. It combines well with other sweeteners and has excellent shelf life. However, it has several disadvantages: it is bitter, has metallic aftertaste, and there is concern over its safety. Saccharin has been used primarily in soft drinks, but also as a tabletop sweetener and in a wide range of other beverages and foods.

Sucralose is a selectively chlorinated derivative of sucrose, the only high-intensity sweetener based on sucrose. It is approximately 600 times sweeter than sucrose. Sucralose is approved for food use in the United States, Canada, and several other countries.

Stevia (Stevia rebaudiana), a plant native to South America, is the source of the stevia extract, which is the natural sweetener. It can be used in food products that require baking or cooking because of its stability in high temperatures. Stevia is approved for use as a sweetener in Japan.

Thaumatin, a mixture of sweet-tasting proteins from the seeds of *Thaumato coccusceus daniellii*, a West African fruit, is about 2000 to 2500 times sweeter than sucrose. Its taste develops slowly and leaves a licorice aftertaste. Thaumatin acts synergistically with saccharin, acesulfame K, and stevioside; potential applications include beverages and desserts. It has been permitted in Japan as a natural food additive since 1979.

Thickeners and Stabilizers

Thickeners and stabilizers (also called as hydrocolloids, gums, or water-soluble polymers) provide a number of useful functions to food products (Table 10). The technical base for these effects results from the ability of these materials to modify the physical properties of water. Most food and beverage products largely consist of water. Thickener and stabilizer materials

- function as rheology modifiers, affecting the flow and feel (mouth) of food and beverage products,
- act as suspension agents for food products containing particulate matter,
- stabilize oil/water mixtures,
- act as binders in dry and semidry food products, and
- create both hard and soft gels in food products that require this physical form.

During the 1990s fat replacement became a major application for modified starches and gums as these additives provide unique texturing, bulking, and emulsifying properties of the displaced fat. Also, natural gums were offered as good sources of dietary fiber. Thickeners and stabilizers are generally used in very small amounts in most food products, in concentrations of 0.15% in jam, 0.35% in ice cream, and 1 to 2% in salad dressings.

Two principal classes of these materials are recognized (a third class known as "synthetic polymers," obtained from petroleum or natural gas precursors, are not used in food):

1. Natural materials obtained from plants or animals, including gum arabic, locust bean gum, guar gum, alginates, carrageenan, pectin, starches, casein, gelatin, and so on.
2. Semisynthetic materials that are manufactured by chemical derivatization of natural organic materials (generally based on a polysaccharide), or by microbial fermentation. This group includes carboxymethylcellulose (CMC) and other modified cellulose compounds, dextran, and gellan and xanthan gums.

Vitamins

Vitamins are nutritive substances required for normal growth and maintenance of life. They play an essential role in regulating metabolism, converting fat and carbohydrates into energy, and forming tissues and bones. Vitamins are typically divided into two groups: fat-soluble and water-soluble vitamins (Table 11). The fat-soluble group is

Table 11. Vitamins Consumed as Food Additives and U.S. RDI Values

Vitamin	Principal synonyms	Major market form	U.S. RDI value
		Vitamin A	
A_1	Retinol	Vitamin A acetate Vitamin A palmitate	5000 IU
A_2	Dehydroretinol		
		Vitamin B	
Niacin	Vitamin B_3	Nicotinic acid	20 mg
Thiamin	Vitamin B_1	Thiamin hydrochloride	1.5 mg
Riboflavin	Vitamin B_2	Riboflavin	1.7 mg
Pantothenic acid	Vitamin B_5	Calcium pantothenate	10 mg
Pyridoxine	Vitamin B_6	Pyridoxine hydrochloride	2 mg
Cyanocobalamin	Vitamin B_{12}		6 mcg
Folic acid, biotin	Vitamin B_c	Folate	0.4 mg, 0.3 mg
		Vitamin C	
	Ascorbic acid	Ascorbic acid Sodium ascorbate Calcium ascorbate	60 mg
		Vitamin D	
D_2	Ergocalciferol		400 IU
D_3	Cholecalciferol		
		Vitamin E	
	Tocopherols	DL α-tocopherol acetate D-α-tocopherol D-α-tocopheryl acid succinate	30 IU
		Vitamin K	
K_1	Phytonadione	Phylloquinone	65 mcg
K_3	Menadione		

Table 12. Fortified Food Groups

Food	Vitamin	Use level	Remarks
Milk	Vitamin D	420 IU/L	Optional but generally added
Beverages (noncarbonated)	Vitamin C	15–100% of US RD1 per serving	Optional; also added as an antioxidant
Cereals	Most essential vitamins	25–100% of US RDI per serving	Optional; added to 90% of cold cereals
Flour	Thiamin, riboflavin, niacin	8–15% of US RDI per 2-oz serving	Mandatory
Margarine	Vitamin A	33,100 IU per kilogram	Optional, but generally added
Miscellaneous foods (eg, instant breakfast, energy bars, etc)	Most essential vitamins		Added to position food as complete meal replacement

usually measured in International Units (IU). The water-soluble group is usually measured in units of weight.

Thirteen vitamins are recognized as essential for human health, and deficiency diseases occur if any one is lacking. Because the human body cannot synthesize most vitamins, they must be added to the diet. Most vitamins are currently consumed as pharmaceutical preparations, or over-the-counter vitamin supplements. Some, like vitamins B, C, D, and E, however, are added directly to food products (Table 12). Ready-to-eat breakfast cereals are a successful example of fortification. Because the primary use of these cereals is as a complete breakfast entree, they are commonly formulated to provide 25% or more of the Daily Value (% DV) per serving of 10 to 12 important vitamins and minerals common to cereals.

Vitamins are added to processed foods for several related reasons:

- To restore vital nutrients lost during processing; this is important with dried milk, dehydrated vegetables, canned foods, and refined and processed foods
- To standardize nutrient levels in foods when they fluctuate because of seasonal variations, soil differences, and methods of preparation
- To fortify fabricated foods that are low in nutrients and promoted as substitutes for traditional products; this includes complete breakfasts, breakfast drinks, meat extenders, and imitation products such as eggs, milk, cheese, and ice cream
- To fortify a major staple, such as bread, with a nutrient known to be in short supply
- For the preparation of designer food (nutraceuticals) containing vitamins that are shown to be useful in preventing chronic diseases.

In addition, vitamins may be used as functional ingredients in foods. Vitamin E (tocopherol) and vitamin C (ascorbic acid) protect foods by serving as antioxidants to inhibit the destructive effects of oxygen. This helps protect the nutritive value, flavor, and color of food products. In addition, ascorbic acid enhances the baking quality of breads, increases the clarity of wine and beer, and aids color development and inhibition of nitrosamine formation in cured meat products. Beta-carotene and β-apo-8′-carotenal are vitamin A precursors, which are brightly pigmented and may be added to foods such as margarine and cheese to enhance their appearance. The roles of these substances outside their nutritional functions are discussed elsewhere in this article (see the sections "Antioxidants," "Preservatives," and "Colorants").

Water-Correcting Agents

Water used in the brewing industry is often corrected to a uniform mineral salt content that corresponds to water known to give the most satisfactory final product. A wide variety of salts are used for this purpose, including mono- and diammonium phosphates, calcium chloride, calcium hydroxide, calcium oxide, calcium dihydrogen phosphate, calcium sulfate, magnesium sulfate, potassium aluminum sulfate, potassium chloride, potassium sulfate, sodium bisulfate, and mono-, di-, and trisodium phosphates. Some of the chemicals, in addition to standardizing the salt content, also control the acidity, thus providing uniform conditions for yeast fermentation.

GENERAL REFERENCES

A. T. Brannen, *Food Additives*, Marcel Dekker, New York, 1980.

Code of Federal Regulations, *Title 21—Food and Drugs. Subchapter B—Food For Human Consumption*, Parts 100–199, U.S. Government Printing Office, Washington, D.C., 1998.

J. M. Concon, *Food Toxicology, Contaminants, and Additives*, Vols. 1 and 2, Marcel Dekker, New York, 1987.

F. J. Francis, *Colorants*, American Association of Cereal Chemists, St. Paul, Minn., 1998.

T. E. Furia, *Handbook of Food Additives*, 2nd ed., CRC Press, Boca Raton, Fla., 1980.

M. Glicksman, *Food Hydrocolloids*, CRC Press, Boca Raton, Fla., 1982.

R. J. Lewis, *Food Additives Handbook*, Van Nostrand Reinhold, New York, 1989.

T. Nagodawithana and G. Reed, *Enzymes in Food Processing*, Academic Press, San Diego, Calif., 1993.

P. Newberne, "GRAS Flavoring Substances, 18th List," *Food Technol.* **52**, 65–92 (1998).

L. O'Brien Nabors and R. C. Geraldi, *Alternative Sweeteners*, 2nd ed., Marcel Dekker, New York, 1991.

G. Reineccious, *Source Book of Flavors*, 2nd ed., Chapman & Hall, New York, 1994.

L. P. Somogyi, "Direct Food Additives in Fruit Processing," in L. P. Somogyi, H. S. Ramaswamy, and Y. H. Hui, eds., *Processing Fruits, Biology, Principles, and Applications*, Technomic Publishing, Lancaster, Penn., 1996.

LASZLO P. SOMOGYI
Consulting Food Scientist
Kensington, California

FOOD ALLERGY

The ancient Roman poet and philosopher Lucretius was quoted as saying: "One man's food may be another man's poison." Although Lucretius was not referring to food allergies, his quote applies nicely to these illnesses. Most foodborne illnesses have the potential to affect everyone in the population. However, food allergies affect only a few individuals in the population. Food allergies involve an abnormal immunologic response to a particular food or food component, usually a naturally occurring protein component of the food. The same food or food component would be safe and nutritious for the vast majority of consumers.

True food allergies can occur through two different immunologic mechanisms, the antibody-mediated mechanism and the cellular mechanism. The most common type of food allergies involve mediation by immunoglobulin E, or IgE.

IGE-MEDIATED FOOD ALLERGIES

In IgE-mediated food allergies, allergen-specific IgE antibodies are produced in the body in response to exposure to a food allergen, usually a protein. These IgE antibodies are highly specific and will recognize only a specific portion of the protein that they are directed against. Occasionally, IgE antibodies produced against one particular protein in a specific food will confer sensitivity to another food either because the food is closely related or because it shares a common segment with the allergenic protein. Some food proteins are more likely to elicit IgE antibody formation than others. Although exposure to the food is critical to the development of allergen-specific IgE, exposure will not invariably result in the development of IgE antibodies even among susceptible people. Many factors, including the susceptibility of the individual; the immunogenic nature of the food and its constituent proteins; the age of exposure; and the dose, duration, and frequency of exposure, are likely to influence the formation of allergen-specific IgE antibodies.

Allergen-specific IgE antibodies are produced by plasma cells and attach themselves to the outer membrane surfaces of two types of specialized cells: mast cells, which are found in many different tissues, and basophils, which are found in the blood. In this so-called sensitization process, the mast cells and basophils become sensitized and ready to respond to subsequent exposure to that specific food allergen. However, the sensitization process itself does not result in any symptoms. No adverse reactions will occur without subsequent exposure to the specific allergenic protein or some closely related protein. Sensitization to a particular allergen distinguishes allergic individuals from nonallergic individuals.

Once the mast cells and basophils are sensitized, subsequent exposure to the allergen results in the allergen cross-linking two IgE molecules on the surface of the mast cell or basophil membrane. This interaction between the allergen and the allergen-specific IgE triggers the release of a host of mediators of allergic disease, which are either stored or formed by the mast cells and basophils. Several dozen different mediators have been identified, including histamine, prostaglandins, and leukotrienes. Histamine is one of the primary mediators of IgE-mediated allergies. Histamine is responsible for many of the early symptoms associated with allergies. Many of the other mediators are involved in the development of inflammation. The interaction of a small amount of allergen with the allergen-specific IgE antibodies results in the immediate release of comparatively large quantities of the various mediators into the bloodstream and tissues. Thus, exposure to extremely small amounts of allergens can elicit symptoms. This mechanism of IgE-mediated reactions is involved in many different types of allergies to foods, pollens, mold spores, animal danders, bee venom, and pharmaceuticals. Only the source of the allergen is different.

IgE-mediated food allergies are sometimes called immediate hypersensitivity reactions because of the short onset time (a few minutes to a few hours) between the ingestion of the offending food and the onset of symptoms. Since the mediators released from the mast cells and basophils can interact with receptors in a number of different tissues in the body, a rather wide variety of symptoms can be associated with IgE-mediated food allergies (Table 1). The most common symptoms associated with food allergies are those involving the skin and the gastrointestinal tract. Respiratory symptoms are less frequently involved with food allergies than with various inhalant allergies such as pollen and animal dander allergies. However, asthma is a very serious, though uncommon, respiratory manifestation of food allergies. Fortunately, most food-allergic individuals suffer from only a few of the many possible symptoms.

Most of the symptoms of IgE-mediated food allergies are not particularly definitive, which can make clinical diagnosis rather difficult. For example, the gastrointestinal manifestations of food allergies can also be associated with many other foodborne illnesses and a variety of other diseases as well. Additionally, there are millions of asthmatics, but only a few are allergic to foods.

Anaphylactic shock is, by far, the most serious manifestation of food allergies. Anaphylactic shock involves gastrointestinal, cutaneous, and respiratory symptoms in combination with a dramatic fall in blood pressure and cardiovascular complications. Death can ensue within minutes of the onset of anaphylactic shock. Fortunately, very few individuals with food allergies are susceptible to such severe reactions after the ingestion of the offending food.

The severity of an allergic reaction will depend to some extent on the amount of the offending food that is ingested. Severe reactions are more likely to occur when an allergic individual inadvertently ingests a large amount of the offending food, especially if that individual happens to be

Table 1. Symptoms of IgE-Mediated Food Allergies

Gastrointestinal symptoms	Cutaneous symptoms
Vomiting	Urticaria (hives)
Diarrhea	Dermatitis
Nausea	Angioedema
Respiratory symptoms	Other symptoms
Rhinitis	Anaphylactic shock
Asthma	Laryngeal edema

exquisitely sensitive. However, exposure to even trace quantities can elicit noticeable reactions due to the large release of mediators.

The most common allergenic foods are peanuts, tree nuts (almonds, walnuts, pecans, cashews, etc), soybeans, cows' milk, eggs, fish, crustacea (shrimp, crab, lobster, etc), and wheat. Peanut allergy is the most common food allergy, especially in the United States where peanuts are a popular dietary item and peanut butter is introduced at an early age. Throughout the world, cows' milk allergy is the most common food allergy among infants due to the widespread ingestion of milk during the first months of life. Any food that contains protein has the potential to elicit an allergic reaction in someone. The most common allergenic foods tend to be foods with high protein content that are frequently consumed. The exceptions are beef, pork, chicken, and turkey, which are uncommonly allergenic despite their frequent consumption and high protein content.

The prevalence of IgE-mediated food allergies is not precisely known. The overall prevalence of food allergies in the developed countries of the world ranges from 4 to 8% in infants to perhaps 1% in adults. Thus, many infants and young children outgrow their IgE-mediated food allergies. The reasons for the development of tolerance to previously allergenic foods are not understood but may involve the development of blocking antibodies of other types, especially IgG and IgA. Allergies to some foods, such as cows' milk and eggs, are more frequently outgrown than allergies to other foods, such as peanuts.

The diagnosis of food allergies is typically approached in a stepwise fashion. The diagnosis of food allergies by an allergist is often critical because parental diagnosis and self-diagnosis are often incorrect, leading to identification of the wrong incorrect foods or the identification of too many foods as allergens. Most individuals with IgE-mediated food allergies are allergic to one or two foods; only on rare occasions do coexistent allergies occur to more than three foods. Thus, the goal of medical diagnosis is to establish a cause-and-effect relationship between one or a few foods and the onset of allergic symptoms. Most physicians begin the diagnosis by taking a careful history of the patient's adverse reactions taking note of the foods eaten immediately before the onset of symptoms, the amount of various foods consumed, the type, severity, and consistency of symptoms, and the time intervals between eating and the onset of symptoms. Sometimes histories are needed from several episodes to reach a probable diagnosis. Challenge tests with the suspected food(s) can be used to establish with certainty the role of a specific food in the reaction. The double-blind, placebo-controlled food challenge (DBPCFC) is considered the most reliable procedure. In the DBPCFC, neither the patient nor the medical personnel know when the food (in capsules or disguised in another food or beverage) is going to be administered and when the placebo is to be administered. Thus, the DBPCFC is free of bias. Single-blind and open challenge tests also have value in some situations. History alone can be sufficient to make the diagnosis in some situations if the cause-and-effect relationship is particularly compelling. Challenge tests are seldom used on individuals who experience life-threatening allergic reactions for rather obvious reasons.

Once the adverse reaction has been clearly linked to the ingestion of a specific food, an assessment of the possible role of IgE must be conducted to determine if the adverse reaction has an allergic mechanism. The diagnosis of an IgE-mediated mechanism can be made with either the skin prick test (SPT) or radioallergosorbent test (RAST). The SPT is the simpler of the two procedures. In the SPT, a small amount of a food extract is applied to the patient's skin, the site is pricked with a needle to allow entry of the allergen, and the site is observed for the development of a wheal-and-flare (basically a hive) response. A wheal-and-flare response at the skin prick site demonstrates that IgE affixed to skin mast cells has reacted with some protein in the food extract releasing histamine into the surrounding tissue and resulting in the formation of the hive. The RAST is an alternative procedure that uses a small sample of the patient's blood serum. In the RAST, the binding of serum IgE to food protein bound to some solid matrix is assessed using radiolabeled or enzyme-linked antihuman IgE. While the RAST is more expensive than the SPT, it is equally reliable and can be conducted in a specialized laboratory in the absence of the patient. The RAST is preferred for patients with extreme sensitivities because of the risk associated with severe reactions to the SPT. It should be emphasized that a positive SPT or RAST in the absence of a history of allergic reactions to that particular food is probably meaningless. The SPT and RAST are the most frequently used and reliable tests to assess the role of IgE in an adverse food reaction.

The specific avoidance diet is the primary means of treatment for IgE-mediated food allergies. For example, if allergic to peanuts, don't eat peanuts. With IgE-mediated food allergies, very low amounts of the offending food can be tolerated by most allergic individuals. Thus, the construction of a safe and effective avoidance diet can be quite difficult. Food-allergic patients must have considerable knowledge of food composition. For example, casein, whey, and lactose are common food ingredients that are derived from cows' milk. These milk ingredients would likely be hazardous for milk-allergic individuals. The ingredient must contain the specific allergenic protein to be hazardous to the allergic consumer. For example, peanut oil and soybean oil, despite being derived from allergenic sources, do not contain protein and would not be hazardous for peanut-allergic or soy-allergic individuals unless the oils had become contaminated during use. The careful reading and complete understanding of food labels is critical to the implementation of a safe and effective avoidance diets. Of course, the manufacturers of packaged foods have the responsibility to ensure that the label statements on packages are accurate. Occasionally, errors are made by food processors that result in the presence of undeclared residues of allergenic foods in a packaged food. The contamination of one food with another from the use of shared food processing equipment is one of the most common errors occurring in food manufacturing. However, restaurant and other food service meals can present an even bigger challenge for food-allergic individuals. Residues of allergenic foods can arise from the use of shared food preparation equipment (utensils, cooking surfaces, pots and pans, etc). Additionally, the accurate identification of all of the ingredients in food service and restaurant meals can sometimes be quite difficult, and such foods are not labeled. As a result, many inadvertent exposures occur among allergic consumers who are attempting to avoid their offending food(s).

Cross-reactions are another perplexing issue for food-allergic consumers as they attempt to develop effective avoidance diets. Cross-reactions can occur but do not inevitably occur between closely related foods. For example, many individuals are allergic to peanuts, but most of these individuals are not allergic to other legumes such as soybeans, peas, green beans, and so on. A few of these individuals are cross-reactive with one or more other legumes. Alternatively, cross-reactions frequently occur among the various crustacea (shrimp, crab, lobster, and crayfish). Cows' milk and goats' milk invariably cross-react as do the eggs of various avian species. Cross-reactions can also occur between foods and other environmental allergens. The most common examples are the cross-reactions that occur between some fresh fruits and vegetables and certain pollen allergies in some individuals and the cross-reaction that occurs in a few individuals with allergies to natural rubber latex with several foods, including bananas and kiwis. The basis for such cross-reactions is frequently not known.

Infants born to parents with histories of allergic disease are much more likely than other infants to develop food allergies. Prevention of the development of food allergies in such infants is quite difficult. The avoidance of commonly allergenic foods such as cows' milk, eggs, and peanuts primarily through breast feeding appears to delay but not prevent the development of food allergies.

A few specialized hypoallergenic foods are available in the marketplace. These foods are intended for infants who have developed allergies to infant formula made with cows' milk. The most effective hypoallergenic infant formulae are based on extensively hydrolyzed casein. Although casein is a common milk protein and a major milk allergen, the hydrolysis of its peptide bonds renders it safe for cows' milk–allergic infants.

Other approaches to the treatment of food allergies are considered controversial. Immunotherapy (eg, allergy shots), sublingual food drops, and the use of rotation diets would be examples of such controversial approaches.

Pharmacological approaches can be used to treat the symptoms of allergic reactions. In particular, epinephrine (also known as adrenalin) is prescribed for individuals who experience life-threatening food allergies. The early administration of epinephrine after inadvertent exposure to the offending food can be life-saving for such patients. Antihistamines can also be used to treat the less-serious symptoms of food allergies.

CELL-MEDIATED FOOD ALLERGIES

Cell-mediated food allergies are sometimes called delayed hypersensitivity reactions, because the symptoms of these reactions typically appear 6 to 24 hours after consumption of the offending food. Cell-mediated allergies involve the interaction of food allergens with sensitized lymphocytes, usually in the intestinal tract. These reactions occur without the involvement of IgE or other antibodies. The interaction between the allergen and the sensitized lymphocyte results in lymphokine production and release, lymphocyte proliferation, and the generation of cytotoxic T lymphocytes. Lymphokines are soluble proteins that exert profound effects on tissues and cells resulting in localized inflammation. Lymphocyte proliferation increases the number of reactive cells thus magnifying the inflammatory process. The generation of cytotoxic or killer T cells results in the destruction of other intestinal cells including the critical absorptive epithelial cells.

The T lymphocytes responsible for cell-mediated allergies abound in the gut-associated lymphoid tissue. Intestinal T lymphocytes are likely to be very critical in food-related, delayed hypersensitivity. However, the inaccessibility of these lymphocytes has hampered experimental studies of the role of cell-mediated reactions in food allergies. As a result, the prevalence and importance of cell-mediated food allergies remains unknown.

Celiac disease, also known as gluten-sensitive enteropathy, is the one illness that seems likely to involve a cell-mediated mechanism. Celiac disease occurs in certain individuals following the ingestion of wheat, rye, barley, triticale, and perhaps oats. Although the mechanism of celiac disease is not completely understood, an immunocytotoxic reaction mediated by intestinal lymphocytes is probable. On ingestion of proteins from the offending grains, the absorptive cells of the small intestinal epithelium are damaged and the absorptive function of the small intestine is severely compromised, resulting in a malabsorption syndrome. The symptoms of celiac disease include diarrhea, bloating, weight loss, anemia from inadequate iron absorption, bone pain from impaired calcium absorption, chronic fatigue, weakness, muscle cramps, and, in children, failure to gain weight and growth retardation. The prevalence of celiac disease in the U.S. is not precisely known but is thought to be about 1 in every 2000 individuals. Celiac disease appears to occur more frequently in Europe and Australia. Celiac disease rarely occurs in individuals of Chinese or African heritage. Celiac disease may manifest at any age. Environmental factors, such as viral illness, may possibly contribute to the onset of celiac disease in some cases.

Celiac disease is triggered by the ingestion of the protein fractions, specifically the gliadin proteins, of wheat and related grains. Other grains, such as corn and rice, do not contain similar gliadin proteins. As with IgE-mediated food allergies, most consumers do not react adversely to the ingestion of these particular grain proteins. In susceptible individuals, a particular segment of the gliadin protein interacts with the sensitized T lymphocytes to elicit a cell-mediated immune response.

The most definitive diagnosis of celiac disease requires taking a small bowel biopsy. The biopsy material is examined for evidence of flattened intestinal villi, a feature that is characteristic of the disease. A normal appearance of the biopsy material is restored after avoidance of wheat and related grains. Alternatively, a blood sample from the patient can be examined for the presence of antiendomysial antibodies, which are elevated in celiac patients. Because of the invasiveness of the biopsy procedure, the diagnosis is often made tentatively on the basis of symptomatic improvement after avoidance of wheat and related grains.

Celiac disease is also treated with thorough implementation of an avoidance diet. In the case of celiac disease, the total avoidance of wheat, rye, barley, triticale, and perhaps oats is usually advocated. Some, but not all, patients can tolerate oats. Ingredients derived from wheat, rye, barley, and oats that contain protein must also be avoided. The tolerance of celiac sufferers for these grains is not precisely known, but it is thought that small amounts of these foods can provoke reactions.

CONCLUSION

Food allergies affect a small percentage of the population. The reactions can range in severity from mild to life-threatening. Food allergies are caused primarily by naturally occurring substances in foods. Food allergies can be triggered by ingestion of very small amounts of the offending food.

GENERAL REFERENCES

J. Brostoff and S. J. Challacombe, eds., *Food Allergy and Intolerance*, Bailliere Tindall, London, United Kingdom, 1987.

P. J. Lemke and S. L. Taylor, "Allergic Reactions and Food Intolerance" in F. N. Kotsonis, M. Mackey, and J. Hjelle, eds., *Nutritional Toxicology*, Raven Press, New York, pp. 117–137.

D. D. Metcalfe, H. A. Sampson, and R. A. Simon, eds., *Food Allergy-Adverse Reactions to Foods and Food Additives*, Blackwell Scientific Publications, Boston, Mass., 1991.

J. E. Perkin, ed. *Food Allergies and Adverse Reactions*, Aspen Publishers, Gaithersburg, Md., 1990.

W. Strober, "Gluten-Sensitive Enteropathy: A Nonallergic Immune Hypersensitivity of the Gastrointestinal Tract," *J. Allergy Clin. Immunol.* **78**, 202–211 (1986).

R. S. Zeiger and S. Heller, "The Development and Prediction of Atopy in High-Risk Children: Follow-up at Age Seven Years in a Prospective Randomized Study of Combined Maternal and Infant Food Allergen Avoidance," *J. Allergy Clin. Immunol.* **95**, 1179–1190 (1995).

STEVE L. TAYLOR
SUSAN L. HEFLE
University of Nebraska
Lincoln, Nebraska

FOOD ANALYSIS

Food scientists analyze foods to obtain information about composition, appearance, texture, flavor, shelf life, safety, processibility, and microstructure (1–4). Nevertheless, the term food analysis is often reserved for determination of food composition, and so this will be the emphasis of this article. Knowledge of the composition of foods is important for several reasons, including compliance with legal standards, quality assurance, determination of nutritional value, detection of adulteration, safety, processing, and research and development (2,3). An understanding of the methods of food analysis is important to food scientists working in academia, industry, and government.

The diversity and complexity of food materials mean that a large number of different analytical techniques have been developed to analyze their properties. These techniques can be conveniently divided into either classical methods (such as gravimetry, titration, distillation, and solvent extraction) or instrumental methods (such as spectroscopy, chromatography, and electrophoresis).

INFORMATION SOURCES

Food analysts often have to identify the various types of analytical methods available for determining the concentration of a specific component and then select the one that is most suitable for their particular application. A knowledge of the various sources where information about analytical methods is available is therefore extremely important to the food analyst.

Books

A number of textbooks and reference books have been written on the subject of food analysis (1–8). These books cover topics such as sampling, data analysis, principles of analytical techniques, and procedures for analyzing specific food components (eg, water, fat, protein, carbohydrates, and minerals). This type of book usually provides a general overview of analytical methods, rather than the specific details needed to actually carry out an analysis.

A variety of other types of book contain information that is also useful to food analysts (eg, monographs, conference proceedings, encyclopedias, and handbooks). Monographs are books that deal with a particular specialized subject area, such as spectroscopy or chromatography. Conference proceedings contain papers that were presented at a scientific or technical meeting on a particular topic and are usually published by the sponsoring organization, for example, the American Chemical Society. Encyclopedias provide information about a wide range of different analytical techniques, for example, *Kirk-Othmer Encyclopedia of Chemical Technology* (9). Some handbooks contain information about the physicochemical properties of materials (molecular weight, structure, density, refractive index, solubility, specific heat capacity), which are useful for the identification and determination of specific food components (8,10–12).

Official Methods

In many circumstances it is important for a food analyst to use a standard analytical method that has been approved by a professional association, such as the Association of Official Analytical Chemists, the American Association of Cereal Chemists, or the American Oil Chemists Society (13–15). These methods are compiled in volumes that are periodically updated as the methods are improved or as new methods are accepted. Standard methods have been developed after years of collaborative testing and are considered reliable and official.

Periodicals

Analytical methods developed by other scientists are often reported in scientific journals. Useful information about analytical methods may be obtained from food science journals (eg, *Food Chemistry, Food Microstructure, Food Testing and Analysis, Journal of Food Science*) or from journals published in other areas (eg, *Analytical Chemistry, Journal of the Chemical Society, Applied Spectroscopy*). For a detailed overview of many of the useful periodicals for food analysts see chapter 1 of Pomeranz and Meloan (1).

Reviews

Critical reviews are usually written by one or more experts in the field and are intended to offer a comprehensive survey of the current state of the knowledge on a specific topic. A critical review may be written about the application of a specific analytical instrument, recent developments in instrumentation, or the determination of a specific component. Critical reviews are usually published in scientific journals or book series (eg, *Advances in Analytical Chemistry and Instrumentation, Advances in Food Research, Critical Reviews in Food and Nutrition*).

Theses

Details about analytical methods can often be located in master's and doctoral theses. Since 1938 theses from U.S. universities have been processed for microfilming by University Microfilms of Ann Arbor, Michigan. Abstracts of up to 600 words of these theses are published in *Dissertation Abstracts*. An annual list of U.S. master's theses in the pure and applied sciences has been published since 1955–1956 by the Thermphysical Properties Research Center, Purdue University, Lafayette, Indiana.

Modern Information Retrieval Systems

During the past few decades, computerized systems for storing and retrieving scientific information have developed rapidly. Many of these can be accessed from a computer terminal in a library or via the Internet. An awareness of the systems available and their proper use can significantly reduce the time and effort involved in a literature search on a particular subject.

Several databases of particular interest to food analysts are available: AGRICOLA (USDA, National Agricultural Library, Beltsville, Md.), BIOSIS PREVIEWS (Biosciences Information Services, Philadelphia, Pa.), CA Search (Chemical Abstracts Service, Columbus, Ohio), CAB Abstracts (Commonwealth Agricultural Bureau, Slough, U.K.), CRIS (USDA, Washington, D.C.), Dissertation Abstracts Online (University Microfilms International, Ann Arbor, Mich.), Science Citation Index (Institute for Scientific Information, Philadelphia, Penn.), and Food Science and Technology Abstracts (International Food Information Service, Reading, U.K.).

Trade Publications

Many companies that manufacture chemicals or analytical instruments produce publications that are valuable sources of information for food analysts. These include bibliographies and abstracts of technical/scientific articles in a specific area, or detailed handbooks giving specifications, properties, and details of analytical procedures. The instruction manuals normally provided by companies with the instrumentation they sell are indispensable for installing, using, and servicing equipment. Several scientific and trade journals periodically prepare lists of major commercial providers of chemicals and instrumentation (eg, *Science, Prepared Foods, and Laboratory Equipment Directory*).

SAMPLING

Sample selection, preparation, and labeling are extremely important aspects of any food analysis procedure, which can lead to large errors if not carried out correctly. Ideally, the sample analyzed in a laboratory should have exactly the same properties as the total population it is supposed to represent. To achieve this a food analyst would have to analyze every sample, which is rarely possible because many analytical techniques are either destructive, time-consuming, expensive, or labor intensive. The analyst must therefore select a limited number of samples from the total population using a sampling plan that will ensure that their properties are a good representation of the true value. A sampling plan is a set of rules that an analyst uses to decide the number of samples of a given population to test, the location from which the samples should be selected, and the method used to collect them. The choice of a particular sampling plan depends on the purpose of the analysis, the physical property to be measured, the nature of the total population and of the individual samples, and the type of analytical technique used to characterize the samples. For certain products and types of populations, sampling plans have already been developed and documented by various organizations that authorize official methods.

Once a representative sample has been selected, it must be prepared for analysis in the laboratory. This step must be done very carefully in order to make accurate and precise measurements. Typical preparation steps involve: making the sample homogeneous, reducing the sample to a manageable size, and preventing sample deterioration (eg, due to evaporation, enzyme activity, lipid oxidation, or microbial growth). It is important to carefully label samples with information such as the nature of the sample, the place it was selected from, the date it was taken, and the person who collected and prepared it. It is also important to report the results of any analysis in a clear fashion, stating the procedures used, the number of replications performed, and the estimated reliability of the measured value.

SELECTION OF AN ANALYTICAL METHOD

A food analyst often has a large number of different analytical procedures that can be used to analyze a particular food product. Selection of the most suitable procedure is often the key to the success of an analysis. Some of the most important factors that should be considered are as follows (4):

1. *Precision*. A measure of the ability to reproduce an answer between determinations performed by the same scientist (or group of scientists) using the same equipment and experimental approach.
2. *Reproducibility*. A measure of the ability to reproduce an answer by scientists using the same experimental approach but in different laboratories using different equipment.
3. *Accuracy*. A measure of how close one can actually measure the true value of the parameter being measured.
4. *Simplicity of operation*. A measure of the ease with which the analysis may be carried out by relatively unskilled workers.
5. *Economy*. The total cost of the analysis, including the reagents, instrumentation, and time.
6. *Speed*. The time needed to complete the analysis.
7. *Sensitivity*. A measure of the lowest concentration of material that can be detected or quantified by a given technique.
8. *Specificity*. A measure of the ability to detect and quantify specific components within a food material, even in the presence of other similar components.
9. *Safety*. A measure of the potential hazards associated with reagents and procedures used in the analysis.
10. *Destructive / nondestructive*. Whether the sample is destroyed during the analysis or remains intact.
11. *Official approval*. Various international bodies have given official approval to methods that have been comprehensively studied by independent analysts and shown to be acceptable to the various organizations involved, for example, AOAC or AOCS.

ANALYTICAL METHODS

It is convenient to classify analytical methods as being either classical or instrumental (15). Classical methods include gravimetry, titration, distillation, evaporation, and solvent extraction. Instrumental methods include various spectroscopic (infrared, visible, ultraviolet, nuclear magnetic resonance, mass spectroscopy), chromatographic (thin-layer chromatography, gas chromatography, high-performance liquid chromatography), electrical (potentiometry, conductiometry, electrophoresis), and optical (polarimetry and refractive index) methods.

Classical Methods

The classical methods of food analysis were developed more than a century ago to give reliable and reproducible results using chemicals and equipment that were readily available in most laboratories (eg, balances, glassware, ovens, heaters). Typically, classical methods involve procedures such as weighing, volume determination, titration, filtering, evaporation, distillation, and solvent extraction. Although these classical methods are still widely used today, they are often being replaced by more convenient instrumental methods.

Instrumental Methods

A variety of instrumental methods have been developed to analyze the properties of food materials (1–5). These can be categorized according to the physical principles on which they are based.

Spectroscopy. Many instrumental methods utilize interactions between some form of radiation (electromagnetic, acoustic, electron beam, or neutron beam) and matter to obtain information about the properties of foods (1,2,3,16). Radiation beams may interact with materials in a variety of ways, including absorption, emission, reflection, refraction, scattering, and polarization. Instrumental methods based on each of these interactions are utilized in food analysis, although those based on absorption and emission tend to be the most common.

Atomic spectroscopy utilizes electromagnetic radiation in the ultraviolet and visible region and relies on the transition of outer-shell electrons between different energy levels. In atomic emission spectroscopy (AES), a sample is heated to atomize it and to excite the electrons to a higher energy state. The emission of radiation by the excited atoms is then measured using suitable detectors. In atomic absorption spectroscopy (AAS), the absorption of radiation when a beam of electromagnetic radiation is passed through an atomized sample is recorded. Each element has its own unique set of energy levels and so the wavelengths of emitted or absorbed radiation can be used to identify it, whereas the extent of the absorption/emission can be used to quantify it. Atomic spectroscopy is principally used to determine the concentration of mineral elements in foods.

UV-visible absorption spectrophotometry is one of the most common types of spectroscopic technique used by food analysts. Like atomic spectroscopy, it relies on transitions of outer-shell electrons. A beam of radiation is transmitted through the sample, and the reduction in its amplitude due to absorption is measured. By carrying out measurements over a range of wavelengths a spectra is obtained that contains peaks corresponding to the absorbing groups. The concentration of a substance can be determined by measuring the height of one of these absorption peaks (provided there is no interference from other molecules). A number of more sensitive and specific methods based on luminescence are available for materials that fluoresce or phosphoresce.

Infrared (IR) techniques rely on electromagnetic radiation absorbed or emitted by a sample due to the vibration or rotation of molecules. The interaction between infrared radiation and a sample leads to a spectra that contains peaks that correspond to different types of chemical groups. The location and magnitude of a peak provides information about the type and concentration of components present. IR techniques are available to simultaneously determine moisture, protein, lipids, and carbohydrates in foods.

Nuclear magnetic resonance (NMR) techniques rely on transitions of nuclei between different energy levels when they are placed in a static magnetic field. Only nuclei that have a property referred to as spin can be detected by NMR, the most common being ^{1}H. NMR is an extremely powerful analytical technique that can be used to determine concentrations, molecular structures, phase transitions, and images.

A number of analytical instruments have been developed to provide information about the composition or structure of food materials that utilize the absorption, diffraction, emission or scattering of X rays.

Optical Methods. Analytical techniques that utilize the rotation of monochromatic light (polarimeters) or the refractive index (refractometers) of a material are frequently used to determine the concentration of components in liquids. These techniques are simple to use and provide rapid measurements, so they are often used for quality control purposes.

Chromatography. Chromatography is a term used to designate a variety of analytical techniques that can separate mixtures of molecules into different fractions. It is mainly used in the food industry for two purposes: (*1*) to purify materials, and (*2*) to determine the types and amounts of specific substances present. Chromatography involves passing a solution of molecules (usually in a gas or liquid carrier) through a matrix. The carrier is called the mobile phase because it moves through the matrix, whereas the matrix is called the stationary phase because it remains stationary. Molecules are separated according to their interaction with the matrix: the stronger the interaction, the slower they move. The strength of the interaction depends on the physicochemical properties of the substance and the matrix. Molecules can be separated according to their size, polarity, electrical charge or molecular interactions. It is convenient to divide chromatographic methods into different categories according to the physicochemical basis of the separation (Table 1).

Table 1. Physicochemical Basis of Separation in Chromatographic Techniques

Basis	Physical principle	Examples
Molecular size	Differences in the size and shape of molecules	Gel filtration, size exclusion
Polarity	Differences in partition coefficient between polar and nonpolar solvent	Partition
Electrical charge	Differences in electrostatic interaction between charged groups on molecules in the mobile phase and the matrix	Ion-exchange
Molecular affinity	Specific interactions between molecules in the mobile phase and the matrix	Affinity

Chromatographic separations can be carried out using a variety of different experimental arrangements. In paper chromatography the sample is spotted on to the bottom of a sheet of paper. The paper is usually impregnated with a liquid, that acts as the stationary phase. The bottom of the paper is placed in a vessel that contains a solvent. The solvent acts as the mobile phase because it carries the sample with it as it moves up the paper due to capillary forces. The molecules are separated according to the strength of their interaction with the stationary phase: the stronger the interaction with the stationary phase, the slower they move. More-sophisticated separations can be carried out using two-dimensional paper chromatography.

Thin-layer chromatography (TLC) has largely replaced paper chromatography because it is quicker, more sensitive, has greater resolution, and has better reproducibility. The general principles of TLC are similar to paper chromatography, except that the paper is replaced by a glass plate coated with a thin layer of porous material that acts as the stationary phase.

Column Chromatography. Originally, column chromatographic techniques used glass columns with either gravity or a slight vacuum to move the mobile phase through them. Steel columns and high pressures are now widely used to speed up this process. In high-pressure (or performance) liquid chromatography (HPLC) the sample to be analyzed is dissolved in a liquid mobile phase and passed through a column containing the stationary phase (which may be either solid or liquid). In gas chromatography the sample is usually volatilized and then carried by a gaseous mobile phase through a thin tube that is coated with the stationary phase. The efficiency of the separation can be manipulated by carefully selecting the most appropriate combination of stationary and mobile phases.

Electrical Techniques. A number of analytical instruments utilize measurements of the electrical properties of foods to obtain information about their composition. Potentiometric methods rely on measurements of the potential difference between an indicator electrode and a reference electrode. The magnitude of the potential difference is related to the concentration of a specific component within a solution. Coulometric methods are based on the measurement of the electrical charge required to completely electrolyze the substance being analyzed: the greater the electrical charge, the higher the concentration of the substance. The composition of some foods can be determined by measuring their electrical conductivity. The current passing between a pair of electrodes is measured and then related to the composition of a specific component within the food.

Electrophoresis relies on differences in the migration of charged macromolecules or colloidal particles in a solution when an electrical field is applied across it. It is used to separate molecules and particles on the basis of their size, shape, or charge. A buffered solution containing the substance to be analyzed is poured onto a porous matrix (usually a strip of paper or a gel) and a voltage is applied across it. The charged substance moves through the gel in a direction that depends on the sign of their electrical charge, and at a rate that depends on the magnitude of the charge and the friction to their movement. The friction of a substance is a measure of its resistance to movement through the matrix and is largely determined by the relationship between the effective size of the substance and the size of the pores in the matrix. The smaller the size of the substance, or the larger the size of the pores in the matrix, the lower the resistance and therefore the faster the substance moves through the matrix. Matrices with different porosities can be purchased from chemical suppliers, or made up in the laboratory, so that substances with different sizes can be analyzed. Electrophoresis techniques can be carried out in two dimensions to improve resolution. Substances are separated in one direction on the basis of their size, and then in a perpendicular direction on the basis of their charge.

Biochemical Analysis. A number of biochemical techniques are available that can be used to determine the concentration or identity of specific food components. Enzymic techniques utilize specific enzyme reactions, whereas immunological techniques utilize specific interactions between an antigen and an antibody. Biochemical assays can be purchased commercially for many important food components. These kits contain the chemicals and instructions required to carry out the analysis and are usually easy to use, rapid, extremely sensitive, and specific.

Because some of the nutritional requirements of microorganisms and experimental animals are similar, it is possible to employ analytical microbiology to determine some substances that are essential constituents of living cells. Microorganisms, as reagents, have been used to determine amino acids, vitamins, nucleic acids, heavy metals, growth factors, nutritional value of proteins, and antibodies. The basic principle is that in the presence of limiting amounts of certain compounds, the amount of microbial growth is a function of the amounts of the compounds. The microorganisms used for assay are primarily bacteria, but yeasts, fungi, and protozoans also have been used. The assay methods include diffusion in a gel, turbidimetric and dilution methods, gravimetric methods, and metabolic response methods.

MAJOR FOOD COMPONENTS

A knowledge of the concentration of the major components present in a food material are important to food scientists for a variety of reasons, including legal requirements, labeling, quality assurance, processing, food safety, nutrition, and so on. In this section, methods for analyzing the major components in foods are reviewed: moisture, carbohydrates, protein, lipids, and minerals. The analytical techniques described in the previous section often form part of these methods.

Moisture

Moisture determination is one of the most important and widely used measurements in the processing and testing of foods. The moisture content ($M\% = 100m_{water}/m_{sample}$), where m_{water} is the mass of the water and m_{sample} is the mass of the whole food) is simply related to the total solids ($T\% = 100 - M\%$).

The thermal drying method relies on the evaporation of water from a sample and is one of the most commonly used procedures for determining moisture content (eg, conventional oven, forced air oven, vacuum oven, microwave oven, and infrared oven). The mass of a sample is weighed before and after it is dried, and the moisture content is then calculated. Care must be taken to ensure that the water is removed in a consistent fashion and that there are no interferences from chemical degradation or evaporation of other substances. In distillation methods, the food sample is mixed with an organic solvent and then placed in a specially designed glass vessel with a collection arm. The glass vessel is heated to evaporate the water, which is collected in the graduated collection arm, where its volume is determined. The most commonly used chemical methods for moisture determination are the Karl Fischer titration and the gas production method. These methods rely on specific chemical reactions between water and another substance that lead to some quantifiable parameter. Several instrumental methods based on differences in the physical properties of water compared with the other components are widely used to determine moisture content, including infrared, gas chromatography, NMR, and several electrical methods. These instruments are simple to use and provide rapid and reliable measurements and are therefore particularly suitable for routine quality control applications.

Carbohydrates

Carbohydrates are the most abundant and widely distributed food components. They include mono-, di-, oligo- and polysaccharides (17). The latter can be divided into structural polysaccharides (cellulose, hemicellulose, lignin) and nutrient polysaccharides (starch, glycogen). Assay methods of mono-, di- and oligosaccharides include chemical, colorimetric, chromatographic, electrophoretic, optical, and biochemical techniques. Many procedures involve preliminary separation by chromatographic or electrophoretic techniques prior to quantification by classical chemical or optical assays. The use of enzymatic assays is gaining popularity because of their high specificity, sensitivity, and ease of use.

Starch can be extracted from foods using perchloric acid or calcium chloride and precipitated with iodine, and the starch liberated from the complex can be determined colorimetrically.

The determination of structural polysaccharides is fraught with many difficulties because the materials contain many undefined polymers, varying in size and composition. In food composition tables, carbohydrates are often reported as the difference after subtracting from 100% the moisture, protein, oil, and ash content. In some cases, fiber (crude, total, and dietary) is determined separately.

Protein

Methods of determining the total protein content of foods (crude protein) are empirical in nature. Proteins are polymers of varying length and molecular structure that may contain up to 24 different kinds of amino acid (17). Amino acids can be determined, after hydrolysis of proteins, by colorimetric, enzymatic, microbiological, and chromatographic methods. The most powerful methods of amino acid analysis in protein hydrolysates are based on ion-exchange and gas chromatography.

The most common procedures for determination of crude protein rely on a measurement of a specific element or chemical group in the proteins. The protein content is then calculated using an experimentally established calibration factor. For example, the approximate protein content of a food can be estimated by measuring the nitrogen content and then multiplying by 6.25 (because on average 16% of a protein's mass is nitrogen). The elements or groups most commonly used are nitrogen, aromatic amino acids, or the peptide linkage. It is assumed that the constituent determined is present entirely in the protein fraction and that the empirical conversion factor is a constant. Determination of the protein content based on the nitrogen content can be made by the Kjeldahl method (based on digestion, neutralization, and titration) or the Dumas method (based on pyrolysis). A variety of spectrophotometric methods that rely on the presence of specific groups are also available for determining the protein content. These include the biuret method (formation of a purple complex between copper salts in alkali solutions and compounds containing at least two peptide bonds), the Lowry method (color formed after interaction of proteins with a phenol reagent and copper under alkaline conditions), direct method (measurement of absorbance at 280 nm, 210 nm, or two wavelengths), turbidimetric methods (based on scattering of light from protein aggregates formed when a certain reagent is added), and dye-binding methods (add excess dye to protein solution and measure unbound dye). A variety of other instrumental methods can also be used to determine protein content, the most popular being NMR and IR.

The concentrations of specific types of proteins in foods can be determined using liquid chromatography or electrophoretic techniques. The proteins are first separated on the basis of their molecular weight or electrical charge and are then identified by comparison with standards of known properties.

Lipids

The most common methods of lipid analysis are based on their sparing solubility in water and their considerable solubility in organic solvents. Successful lipid analysis often requires that covalent bonds between lipids and other constituents (such as proteins or carbohydrates) are broken before the solvent extraction is carried out, so that the lipids are free to be solubilized. The efficiency of extraction is usually greatest when the polarities of the lipid and solvent are similar. Ethyl ether and petroleum ether are two of the most common extraction solvents. Combinations of alcohol and ether, ternary mixtures of chloroform–methanol–water, and water-saturated butanol are other examples of effective solvents. Solvent extraction techniques may be batch (eg, an extraction funnel), semicontinuous (eg, Soxhlet), or continuous (eg, Goldfish). The fat content of milk and other dairy products is often determined by nonsolvent extraction techniques, such as the Babcock, Gerber, or detergent methods. Numerous instrumental methods are available for determining the lipid content of certain foods, including IR, NMR, densitometry, X-ray absorption, and dielectric measurements. The identity of the molecules in the lipid phase can be determined by chromatography, mass spectrometry, or NMR.

The processing and quality of food oils are often determined by physical characteristics such as their melting point, smoke point, flash point, fire point, solid fat content, and degree of oxidation. Consequently, analytical methods have been developed to characterize these quantities.

Mineral Components

Ash is the inorganic residue remaining after the incineration of organic matter. It is a measure of the total mineral components in a food. Depending on the nature of the mineral components, various ashing procedures can be employed, such as dry ashing (in a muffle furnace) or wet ashing (by acid digestion). The individual elements within an ash can be determined by titrimetric methods, colorimetric methods, or atomic spectroscopy techniques. Atomic spectroscopy techniques can provide a complete profile of the different types of element in a food material, whereas the titrimetric and colorimetric methods are usually designed to determine a particular element. In addition, atomic spectroscopy techniques have a much higher sensitivity and specificity.

BIBLIOGRAPHY

1. Y. Pomeranz and C. E. Meloan, *Food Analysis, Theory and Practice*, 3rd ed., Chapman and Hall, New York, 1994.
2. S. S. Nielsen, ed., *Introduction to the Chemical Analysis of Foods*, Jones and Bartlett, Boston, Mass., 1994.
3. D. W. Gruenwedel and J. R. Whitaker, eds., *Food Analyses, Principles and Techniques*, Vols. I–V, Marcel Dekker, New York, 1984.
4. C. S. James, *Analytical Chemistry of Foods*, Blackie Academic and Professional, London, United Kingdom, 1994.
5. D. King, ed., *Developments in Food Analysis*, Elsevier Applied Science, London, United Kingdom, 1984.
6. G. C. Birch and K. J. Parker, eds., *Control of Food Quality and Analysis*, Elsevier Applied Science, London, United Kingdom, 1984.
7. W. Diemair, ed., *Handbook of Food Chemistry, Food Analysis*, Springer-Verlag, Berlin, Germany, 1967.
8. *CRC Handbook of Chemistry and Physics*, 58th ed., CRC Press, Cleveland, Ohio, 1998.
9. J. I. Kroschwitz and M. Howe-Grant, eds., *Kirk-Othmer Encyclopedia of Chemical Technology*, 4th ed., Wiley-Interscience, New York, 1991–1998.
10. R. M. C. Dawson et al., *Data for Biochemical Research*, 3rd ed., Clarendon Press, Oxford, United Kingdom, 1986.
11. S. Budavari, *The Merck Index*, 12th ed., Merck, Whitehouse Station, N.J., 1996.
12. *Approved Methods of Analysis*, 8th ed., American Association of Cereal Chemists, St. Paul, Minn., 1983.
13. *Official Methods of Analysis*, 15th ed., Association of Official Analytical Chemists, Washington, D.C., 1990.

14. *Official Methods of Recommended Practices*, 4th ed., American Oil Chemists Society, Chicago, Ill., 1990.
15. D. A. Skoog, D. M. West, and F. J. Holler, *Analytical Chemistry: An Introduction*, 6th ed., Harcourt Brace, Philadelphia, Pa., 1994.
16. R. H. Wilson, ed., *Spectroscopic Techniques for Food Analysis*, Wiley-VCH, New York, 1994.
17. O. R. Fennema, ed., *Food Chemistry*, Marcel Dekker, New York, 1996.

JULIAN MCCLEMENTS
University of Massachusetts
Amherst, Massachusetts

FOOD AND NUTRITION SCIENCE ALLIANCE (FANSA)

The Food and Nutrition Science Alliance, FANSA, is a partnership of four professional scientific societies whose members have joined forces to speak with one voice on food and nutrition science issues. FANSA's combined membership includes more than 100,000 food, nutrition, and medical practitioners and scientists.

FANSA's statements help consumers sort through the confusing array of information about food and nutrition and offer tips for evaluating the validity of such information. For instance, FANSA alerts consumers to be cautious about claims made about dietary supplements:

1. When buying dietary supplements, caveat emptor applies.
2. If it sounds too good to be true, it probably is not true.
3. Dietary supplements encompass vitamins and minerals as well as herbals and botanicals.
4. Multivitamins may help some people.
5. Less is known about herbals and botanicals than vitamins and minerals.
6. High doses of some dietary supplements may be harmful.
7. "Natural" is not synonymous with "safe."
8. For good health, eat a variety of foods.

Other FANSA statements address the following topics:

- Interpreting nutrition research
- Making sense of scientific research about diet and health
- Making sense of risks associated with diet
- Folic acid: a reminder for women before and during pregnancy

FANSA's "10 Red Flags of Junk Science" help consumers think twice before reacting to the latest news bulletin about nutrition. Any combination of the following signs should send up a red flag of suspicion about the accuracy of the information:

1. Recommendations that promise a quick fix
2. Dire warnings of danger from a single product or regimen
3. Claims that sound too good to be true
4. Simplistic conclusions drawn from a complex study
5. Recommendations based on a single study
6. Dramatic statements that are refuted by reputable scientific organizations
7. Lists of "good" and "bad" foods
8. Recommendations made to help sell a product
9. Recommendations based on studies published without peer review
10. Recommendations from studies that ignore differences among individuals or groups

FANSA member organizations include the American Dietetic Association (ADA), American Society for Clinical Nutrition (ASCN), American Society for Nutritional Sciences (ASNS), and the Institute of Food Technologists (IFT).

As the nation's largest group of food and nutrition professionals with 68,500 members, ADA serves the public by promoting optimal nutrition, health, and well-being. With 1300 members, ASCN is the scientific society for clinical nutritionists in medicine and the health sciences. With 3400 members, ASNS (formerly American Institute of Nutrition) is the principal professional organization of nutrition research scientists in the United States. And as the society for food science and technology, IFT represents 28,000 food scientists and others in related professions, working in all facets of the food system, in academia, industry, and government (IFT's website can be accessed at *http://www.ift.org*).

ELLEN J. SULLIVAN
Institute of Food Technologists
Chicago, Illinois

FOOD AND NUTRITION SCIENCE ALLIANCE ORGANIZATIONS

The Food and Nutrition Science Alliance (FANSA) is a partnership of four professional scientific societies whose members have joined forces to speak with one voice on food and nutrition science issues. FANSA's combined membership includes more than 100,000 food, nutrition, and medical practitioners and scientists.

FANSA member organizations are as follows:

The American Dietetic Association (ADA)
American Society for Clinical Nutrition (ASCN)
American Society for Nutritional Sciences (ASNS)
Institute of Food Technologists (see the article INSTITUTE OF FOOD TECHNOLOGISTS).

THE AMERICAN DIETETIC ASSOCIATION

As the nation's largest group of food and nutrition professionals with nearly 70,000 members, ADA (founded in

1917) serves the public by promoting optimal nutrition, health, and well-being.

ADA members in the United States and abroad help shape the food choices and impact the nutritional status of the public. The membership includes dietitians, dietetic technicians, students, and others holding baccalaureate and advanced degrees in nutrition and dietetics. ADA members play a key role in shaping the public's food choices, thereby improving its nutritional status, and in treating persons with illnesses. Members offer preventive and therapeutic nutrition services in a variety of settings.

The ADA is a collective partnership of members, staff, and allied organizations. A number of networks provide the framework for member involvement, including state and district associations and 26 dietetic practice groups. The practice groups include members who specialize in public health, education, research, long-term care, food-service management, private practice, sports nutrition diet counseling for diabetes, weight management, and heart disease, among other areas.

ADA members provide expert testimony at hearings, lobby Congress and other government bodies, comments on proposed federal and state regulations, and develop position statements on critical food and nutrition issues. The Commission on Accreditation and Approval for Dietetics Education accredits and approves more than 600 entry-level dietetics education programs for entry into the profession. Through its National Center for Nutrition and Dietetics, ADA offers programs to educate consumers about the links between food and nutrition and health.

The *Journal of the American Dietetic Association* and the *ADA Courier* supply members with current food and nutrition research and practice information, as well as Association news. As an important player in the International Committee of Dietetic Associations, ADA helps promote information sharing and professional cooperation among dietetics practitioners around the world.

Further information is available at *http:www .eatright.org*.

AMERICAN SOCIETY FOR CLINICAL NUTRITION

With 1300 members, ASCN is the premier scientific society for clinical nutritionists in medicine and the health sciences.

The goals and objectives of ASCN are as follows:

- To encourage and implement undergraduate and graduate education in basic and clinical nutrition, particularly in medical schools
- To expand research and clinical training opportunities in nutrition for health professionals
- To provide opportunities for investigators to present and discuss current research in human nutrition
- To offer continuing education programs in clinical nutrition for physicians, basic scientists, and allied health professionals
- To provide a journal for publication of meritorious work in clinical and experimental nutrition
- To promote and publicize the professional application of nutrition science in health promotion, disease prevention, and patient care
- To facilitate and promote the issuing of credentials for practitioners of clinical nutrition
- To participate in developing and implementing public policies and procedures concerned with clinical nutrition
- To expand federal and private funding of clinical nutrition research and training
- To maintain integrity and ethical behavior among nutrition societies through adherence to the society's formalized code of professional responsibility

ASCN sponsors two major forums for continuing education of physicians, basic scientists, and allied health professionals: the annual meeting and the postgraduate course. The postgraduate course, held annually since 1985, addresses current nutrition research findings and their clinical applications in health and disease. ASCN also co-sponsors meetings with other nutrition societies, such as ASNS, the American Society for Parenteral and Enteral Nutrition, the Canadian Society for Nutritional Sciences, and the International Union of Nutritional Sciences, which organizes the International Nutrition Congress every four years.

The *American Journal of Clinical Nutrition* (AJCN), the society's official journal, publishes peer-reviewed original research contributed by scientists throughout the world and provides perspectives on nutrition, editorials, special articles, meeting reports, book reviews, and letters to the editor. Special supplements are devoted to symposia, workshops, and other long reports.

Government and the news media turn to ASCN members for authoritative information on issues related to clinical nutrition and its role in maintaining health and preventing and treating disease. ASCN also works with other professional societies to disseminate accurate information on current nutrition and dietary issues.

General management of the society is vested in an elected council. The executive officer, as the society's chief administrative officer, implements financial, scientific, publication, and education programs under policies established by the council. In addition, the executive officer interacts with legislators, the National Academy of Sciences, the National Institutes of Health, and other government bodies to enhance the role of ASCN as a national source of clinical nutrition expertise; it also maintains liaison with Federation of American Societies for Experimental Biology (FASEB) constituent societies and with national and international nutrition societies.

Further information is available at *http://www.faseb.org/ascn*.

AMERICAN SOCIETY FOR NUTRITIONAL SCIENCES

With 3400 members, ASNS is the principal professional organization of nutrition research scientists in the United States. It is dedicated to improving the quality of life through the science of nutrition.

The Society fulfills its mission by

- fostering and enhancing research in animal and human nutrition;
- providing opportunities for sharing, disseminating, and archiving peer-reviewed nutrition research results (at its annual meeting and in its official publication, the *Journal of Nutrition*);
- fostering quality education and training in nutrition;
- upholding standards of ethical behavior in research, the protection of human research participants, and the care and treatment of research animals;
- providing opportunities for fellowship and support among nutritionists; and
- bringing scientific knowledge to bear on nutrition issues through communication and influence in the public domain.

The ASNS Secretariat and its staff, housed in Bethesda, Maryland, administer the affairs of the Society. The executive officer interacts with the elected officers of the Society to manage the business affairs of the Society and the *Journal of Nutrition*, to maintain a relationship with the Federation of American Societies for Experimental Biology (FASEB), to coordinate public affairs activities, and to provide support for the selection and scheduling of the scientific program at the annual meeting (Experimental Biology).

Further information is available at *http://www.faseb.org/asns*.

KRISTEN MCNUTT
Consumer Choices, Inc.
Winfield, Illinois

FOOD CHEMISTRY AND BIOCHEMISTRY

The Institute of Food Technologists (IFT; *http://www.ift.org*) is one of the largest scientific societies representing professionals in food science and technology. A food scientist is defined by the IFT as a person who studies the physical, microbiological, and chemical makeup of food. In an accreditation program, the IFT has defined minimum standards for undergraduate food science degrees. These standards highlight the importance of chemistry in a food scientist's education, with a greater minimum requirement for chemistry than any other single discipline. Four introductory chemistry courses are required: two in general chemistry, one in organic chemistry, and one in biochemistry. Building on this base are a lecture and laboratory course in food chemistry and a food analysis lecture and laboratory course that utilizes chemical and microbiological techniques. Over the last few decades, however, food chemistry has evolved into a field of such scope that these courses are barely adequate to serve as an introduction to the field.

EDUCATION

One indication of the rapid evolution of food chemistry is the increase in specialized food chemistry and food chemistry analysis books. In the early 1970s there were virtually no general food chemistry text books. From the mid 1970s on, however, there was an explosive growth in such books (Table 1).

Most of the books have chapters on water, carbohydrates, lipids, amino acids and proteins, enzymes, vitamins and minerals, color, flavor, texture, toxic compounds and additives. Besides the trend to more general textbooks in the food chemistry area, there is a trend to increased complexity and proficiency in these books. Also, new food

Table 1. Some General Food Chemistry Textbooks

Author or editor	Title	Publishing company	Year
F. A. Lee	*Basic Food Chemistry*	AVI, Westport, Conn.	1975
O. R. Fennema (ed.)	*Food Chemistry* (3rd ed.)	Marcel Dekker, New York	1996 (original version, 1976)
C. Zapsalis and R. A. Beck	*Food Chemistry and Nutritional Biochemistry*	Originally John Wiley and Sons, New York, now Macmillan, London	1985
H. D. Belitz and W. Grosch	*Food Chemistry* (translation from the German version by D. Hadziyev)	Springer-Verlag, Berlin	1986
J. M. deMan	*Principles of Food Chemistry* (3rd ed.)	Aspen Publishers, Gaithersburg, Md.	1999 (original version 1980)
D. W. S. Wong	*Mechanism and Theory in Food Chemistry*	Van Nostrand Reinhold, New York	1989
D. Pearson	*The Chemical Analysis of Foods* (7th ed.)	Churchill Livingstone, New York	1976 (original version 1926)
Y. Pomeranz and C. E. Meloan	*Food Analysis: Theory and Practice*	Chapman and Hall	1978
S. S. Nielsen	*Food Analysis* (2nd ed.)	Aspen Publishers, Gaithersburg, Md.	1998 (original version 1994)
C. M. Weaver	*Food and Chemistry Laboratory*	CRC Press, New York	1996
L. M. L. Nollet	*Handbook of Food Analysis* (2 volumes)	Marcel Dekker, New York	1996

Table 2. Food Chemistry Topics

Water

Physical properties, pH
Hydrogen bonding, solubility of food components
Water activity

Carbohydrates

Basic structures and naming conventions
Chemistry of reducing and nonreducing sugars
Nonenzymatic browning
Natural homo and heteropolysaccharides, sources, biosynthesis and uses
Chemical modification of natural polysaccharides and their uses
New carbohydrate polymers from genetic engineering

Lipids

Basic structures and naming
Lipid reactions, hydrolytic and oxidative rancidity
Crystal structure and physical properties
Refining, hydrogenation, and interesterification reactions
Chemistry of artificial fats
Modified fat composition from genetic engineering

Amino acids and proteins

Basic structure and naming
Protein structure
Denaturation conditions
Protein chemical changes in processing

Enzymes

Catalytic action, stereochemical specificity
Food enzymes, action, kinetics, stability, uses, and sources
Enzymatic browning and rancidity
Enzyme immobilization
Postharvest physiology of plants
Postmortem changes in animals
Modified enzyme activity through genetic engineering

Vitamins and minerals

Chemistry and interaction of vitamins and minerals with food components under processing conditions
Biosynthesis and biodegradation, metabolism

Color

Natural food pigments and their chemistry
Artificial colors and their metabolism
Color measurement

Flavor

Biosynthesis of flavor components
Sweetening and the flavor theories
Physiology of taste and smell
Chemistry of artificial sweeteners

Texture

Food colloids
Gel formation
Emulsifiers and foaming agents
Viscosity

Table 2. Food Chemistry Topics *(continued)*

Toxic compounds

Natural microbial, plant, and animal toxins
Synthetic food toxins
Radionucleotides and toxic metals
Residues
Metabolism and excretion of toxic compounds

Food additives

Sequestrants and their chemistry
Artificial and natural sweeteners
Antioxidants
Acidulants
Bases, leavening agents
Antibiotic and probiotic agents
Gases and propellants
Other food additives

Food analysis

Analysis of any of the above food components using:
- Wet chemical and physical methods
- Chromatography techniques including high performance liquid chromatography (HPLC), gas chromatography (GC), capillary electrophoresis
- Spectroscopy techniques, visible (absorption and reflectance), ultraviolet, fluorescent, infrared, atomic absorption and emission, mass spectrometry (MS), nuclear magnetic resonance (NMR), electron spin resonance (ESR).
- Electroanalytical techniques, selective electrodes, coulometry and voltammetry, electrophoresis
- Enzymatic methods, enzymatic electrodes, enzyme immunoassays
- Thermal methods, differential scanning calorimetry, cryoscopy
- Rheological methods

chemistry analysis books have recently been emphasized. Food chemistry analysis borrows techniques from many other disciplines and has therefore become very complex. An indication of this increased complexity is the two-volume, 3000-plus-page series entitled, *Handbook of Food Analysis*. Included in the two volumes are 48 chapters with each chapter written by a group of experts in their field.

With the increase in intricacy of the discipline, increased specialization is necessary. As early as 1944 the need for specialization in food science and technology was recognized by educators in the field (1). Nowadays food chemistry itself covers such a variety of topics (see Table 2) that further specialization is an inevitable trend.

PUBLICATIONS

There is no doubt that parallel with the increase in educational material, publications in the food chemistry discipline have increased. As in the educational area there is indication that the discipline is becoming increasingly specialized and complex.

It is instructive to look at trend-setting journals in the food chemistry area. Although it is difficult to define a journal as solely dealing with food chemistry, leading journals in food science and technology have been rated by the In-

Table 3. 1996 Food Science and Technology Journals Rated By Impact Factor

Rank	Journal title	Impact factor
1	*CRC Critical Reviews in Food Science*	2.173
2	*Journal of Agricultural and Food Chemistry*	1.732
3	*Biotechnology Progress*	1.588
4	*Trends in Food Science and Technology*	1.430
5	*Food and Chemical Toxicology*	1.398
6	*International Journal of Food Microbiology*	1.387
7	*Journal of Dairy Research*	1.374
8	*Journal of Food Protection*	1.372
9	*Journal of Cereal Science*	1.354
10	*Journal of American Oil Chemists Society*	1.241
11	*Journal of Food Science*	1.225
12	*Cereal Chemistry*	1.160
13	*Journal of Dairy Science*	1.139
14	*Food Microbiology*	1.127
15	*Food Technology*	1.100
16	*Netherlands Milk and Dairy Journal*	1.056
17	*Meat Science*	1.048
18	*Food Hydrocolloids*	1.034
19	*Journal of Texture Studies*	0.985
20	*Journal of the Institute of Brewing*	0.955

Source: Ref. 2.

stitute for Scientific Information (2) on the basis of impact factor (the top 20 journals out of a list of 73 are given in Table 3). Impact factor is defined as the number of yearly citations of journal articles divided by the number of yearly articles. Therefore, an impact factor of 1.00 means that if a journal contained 50 papers in a given year, there would also be 50 citations of former journal papers for the same year.

Review journals such as *CRC Critical Reviews in Food Science, or Trends in Food Science and Technology* will always have somewhat higher impact factors because reviews are generally larger, well-referenced papers and preferentially cited over many smaller papers in a field. Also, some journals are not be considered primarily food technology journals and may not appear on the list for this reason. An example of this latter case is the *Journal of the AOAC International* (1.256 impact factor). However, impact factors do give a measure of the influence of various journals independent of the number of papers that appear in a journal (ie, large journals that have more total citations are rated on an equivalent scale to small journals because of the per-article nature of impact factors).

From Table 3 it can be seen that two leading food chemistry journals are the *Journal of Agricultural and Food Chemistry* (*JAFC*), and *Food and Chemical Toxicology*. The latter journal covers an important food chemistry specialization. The first of these is published by the American Chemical Society, which is the world's largest scientific society. *JAFC* is also rated by impact factor as the second most influential agriculture journal and the fifth most influential applied chemistry journal. Mirroring the growth of food chemistry and related fields, this journal has grown from 1548 pages in 1989 to almost 5000 pages in 1996. With this expansion the journal has developed subheadings (Table 4) to organize the research areas.

Table 4. Subheadings Used by the *Journal of Agricultural and Food Chemistry*

Analytical Methods
Biotechnology
Chemical Changes During Processing/Storage
Chemistry of Crop and Animal Protection
Composition of Foods/Feeds
Environmental Chemistry
Flavors and Aromas
Food Chemistry/Biochemistry
Nutrition Toxicology

Other leading food science and technology journals are often commodity publications such as the *Journal of Dairy Research* and the *Journal of Cereal Science*. All of these commodity publications contain a large proportion of food chemistry articles.

Food chemistry has now grown and matured to a point that it has influenced other disciplines such as nutrition and medicine. Two examples are nonenzymatic browning reactions and antioxidant protection of lipids to prevent oxidative rancidity. The former is now known to be a very general reaction that impacts conditions such as diabetes, aging, and even Alzheimer's disease (3). The nonenzymatic browning products of concern in these conditions have been called advanced glycated end products. Food antioxidants, long known to play a role in preventing off-flavors or rancidity in food, now also find an important role in human health. The search for new and better natural antioxidants is an area of intense research interest, with scientists still unsure of the best protective agents found in mostly plant foods (4). In fact, the interest in foods and ingredients that may have health benefits has developed into a large new interdisciplinary research area dealing with functional foods or nutraceuticals.

Even though food chemistry has developed a great deal over the last few decades, clearly there is still room for tremendous growth in the discipline because there is still a great deal that we do not understand about the chemistry of an increasingly diverse food supply.

BIBLIOGRAPHY

1. O. Fennema, "Educational Programs in Food Science: A Continuing Struggle for Legitimacy, Respect, and Recognition," *Food Technol.* **43**, 170–182 (1989).
2. E. Garfield, *SCI Journal Citation Reports*, Institute for Scientific Information, Philadelphia, Penn., 1996, p. 83.
3. M. P. Vitek, et al., "Advanced Glycation End Products Contribute to Amyloidosis in Alzheimer's Disease," *Proc. Natl. Acad. Sci. U.S.A.* **91**, 4766–4770 (1994).
4. B. Halliwell, "Antioxidants in Human Health and Disease," *Annu. Rev. Nutr.* **16**, 33–50 (1996).

PETER SPORNS
University of Alberta
Edmonton, Alberta
Canada

FOOD CHEMISTRY: MECHANISM AND THEORY

Food chemistry deals with the chemical identity of food components and chemical reactions governing the changes and performance of individual or interacting food components during handling, processing, and storage. Although the individual constituents may often be readily identified, the interactions of food components are extremely complex.

Food chemistry is a branch of chemistry with its foundation built on chemical principles and reaction mechanisms, and a comprehension of the subject often requires thorough understanding and application of knowledge from various chemistry and chemistry-related disciplines. In this regard, food chemistry is quite similar to biochemistry, except that the former relates chemistry to food systems and the latter to biological systems. For example, a biochemist may be interested in elucidating the biosynthetic pathways of wheat storage proteins, but for a food chemist, it is more relevant to relate the chemical structures of these proteins to functional properties, such as changes and effects in dough quality. Another example is found in the Maillard reaction. An organic chemist may investigate the chemical reactions and their mechanism in the formation of melanoidin compounds. A biochemist is likely more interested in the reactions because they are related to the aging process of certain vital proteins, such as lens crystalline, collagen, and elastin. A food chemist is also interested in the Maillard reaction, but more in linking the reaction mechanism to physical and chemical changes in food systems such as flavor development, browning, and nutritional loss.

In this article reaction mechanisms of sufficient importance in food systems and their current developments are presented. A number of good background references have been published (1–5).

CARBOHYDRATES

The Maillard Reaction

In 1912, the French chemist Louis-Camille Maillard first observed that yellow-brown pigments formed in the reaction among sugars and amino acids, peptides, and proteins in a heated solution. Food chemists have recognized the practical relevance of this reaction to many chemical and physical changes during processing and storage of food. The first review (in English) on the Maillard chemistry in food systems was published in 1951 (6). Since then, numerous reviews on this subject have appeared (eg, 7–9,15). The biological importance of this reaction has been recognized only in the last 20 years. It is now well established that the reaction is linked to glycosylated hemoglobin (HbA_{1c}) in diabetes, hardened lens crystallins in cataract disease, and a number of other aging proteins (10,11).

The Maillard reaction comprises a series of reactions: (*1*) formation of glycosylamine via a Schiff-base formation between a reducing sugar and the amino group of an amino acid, (*2*) Amadori rearrangement in which glycosylamine is converted to ketosamine, (*3*) enolization (C1–C2 or C2–C3), followed by cyclodehydration. In general, under acidic conditions, the nitrogen is protonated, and 1,2-enolization is assisted by the positively charged nitrogen acting as an electron sink. Alkali and strong basic amine favor 2,3-enolization (Fig. 1).

The actual reactions are far more complicated than those outlined here, and there are many variations in the pathway. The initial step in the Amadori rearrangement is suggested as N-protonation. However, addition of the proton to the ring oxygen has also been proposed. Most discussions on the Maillard reaction concern the monosubstituted amines, but the ketosamine formed in the reaction can also react with another molecule of an aldose resulting in disubstitution. Another variation is the Heynes rearrangement in the conversion of D-fructosylamine (a ketosamine) to 2-amino-2-deoxy-D-glucose (an aldosylamine).

A pathway that involves sugar fragmentation and free-radical formation prior to the Amadori rearrangement has been suggested (12,13) (Fig. 2). The radical has been structurally identified to be N,N′-disubstituted pyrazine cation radical and it is formed by the dimerization of a two-carbon enaminol product from the cleavage of glycosylamine.

Direct dehydration of the Amadori compound has been proposed as an alternative to enolization (14,15) (Fig. 3). In this mechanism, the Amadori compound undergoes a trans-elimination at C2–C3, followed by a second dehydration at C3–C4, to form a hydroxypyran and finally a pyrylium ion. The highly electrophilic pyrylium ion can undergo various nucleophilic additions, ring opening, and recyclization.

Maillard reaction products have very diverse structures and are involved in various secondary reactions. A compilation of 450 volatile Maillard reaction products and related compounds (16) and reviews on this subject are available (7–9,15). Dicarbonyl compounds, such as 3-deoxyglycosulose generated by the 1,2-enolization pathway and the glycosulos-3-ene formed via the 2,3-enolization are the key intermediates for subsequent degradative reactions relating to color and flavor production. One of the well-known pathways is the Strecker degradation in which the carbonyl forms a Schiff base with the α-amino group of an amino acid. Enolization, decarboxylation, and hydrolysis yield an aldehyde corresponding to the original amino acid with one fewer carbon. The aldehydes derived from this degradative pathway constitute many important flavor compounds in food systems. Compounds generated by degradation of the dicarbonyl compounds include pyrroles, pyrazines, oxazoles and derivatives, pyrrolines, pyrrolidines, pyrones, thiazole, and thiazoline.

Sulfite inhibition of nonenzymatic browning also involves the dicarbonyl intermediates reacting with sulfur oxoanion to form stable sulfonate product (17). The sulfur oxoanion may replace the C4 hydroxy group of the deoxyglycosulose or undergoes 1,4-addition to the double bond of the α,β-unsaturated glycosulos-3-ene.

Formation of melanoidins is caused by polymerization of unsaturated carbonyl compounds. Condensation between 3-deoxyglycosulose and its enamine, which is a reductone, has been suggested to be the structural unit of the polymer (18,19) (Fig. 4), although polymers of repeating units of furan or pyrrole have also been proposed.

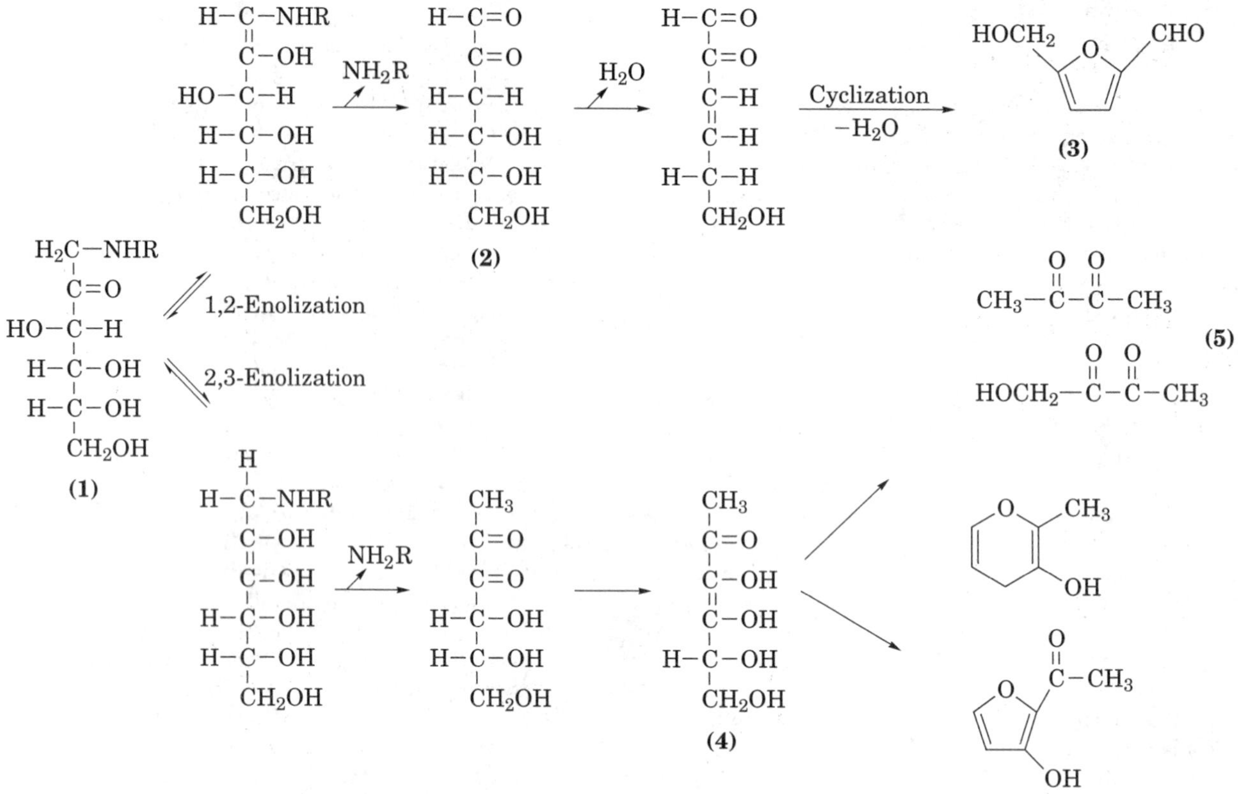

Figure 1. Reaction pathways of Amadori product: (**1**) Amadori product, (**2**) deoxyglycosulose, (**3**) furaldehyde, (**4**) glycosulos-3-ene, (**5**) reductones.

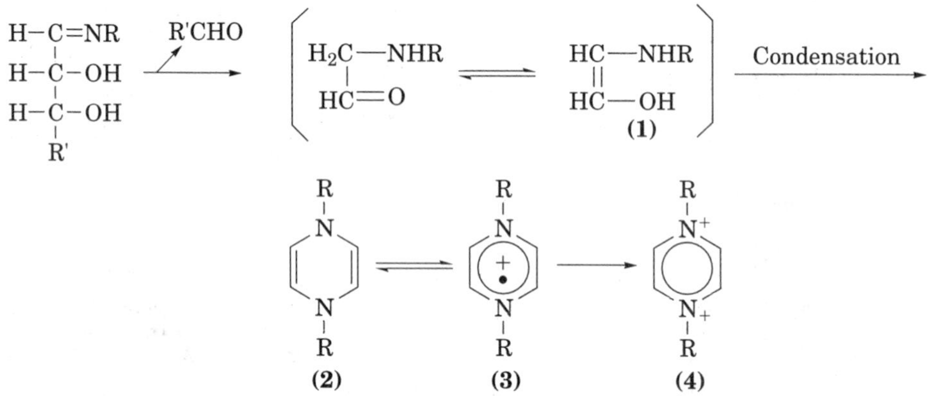

Figure 2. Fragmentation and free radical formation in the Maillard reaction: (**1**) enaminol, (**2**) dialkyldihydropyrazine, (**3**) dialkylpyrazine cation radical, (**4**) dialkylpyrazinium compound. *Source:* Ref. 12.

The Maillard reaction offers a perfect example of how basic organic chemistry—studies of the reaction mechanism between sugar and amino acids and the structural identification of possible intermediate compounds—has made it possible to explain some of the most interesting and important interactions in food systems.

Polysaccharide Polymers

The past two decades have witnessed an increasing interest in the relationship between molecular structure and functional mechanism of food polysaccharides. Growth in the knowledge of polymer chemistry has made a dramatic change in the way food scientists interpret structure and function of food polysaccharides.

One milestone in this area is the elucidation of the gelling mechanism of alginate in the early 1970s (20). The mechanism of gelation has been shown to be the interaction between calcium ions and the polyguluronate blocks of alginate polymer. Such interactions provide junctions that cross-link alginate polymers into a three-dimensional network, known as the "egg box" model (Fig. 5).

The concept of junction-zone linking long-chain polymers has since been included in explaining various kinds of gelation. For example, carrageenan polymers are known to associate by intermolecular double helices. Gelation oc-

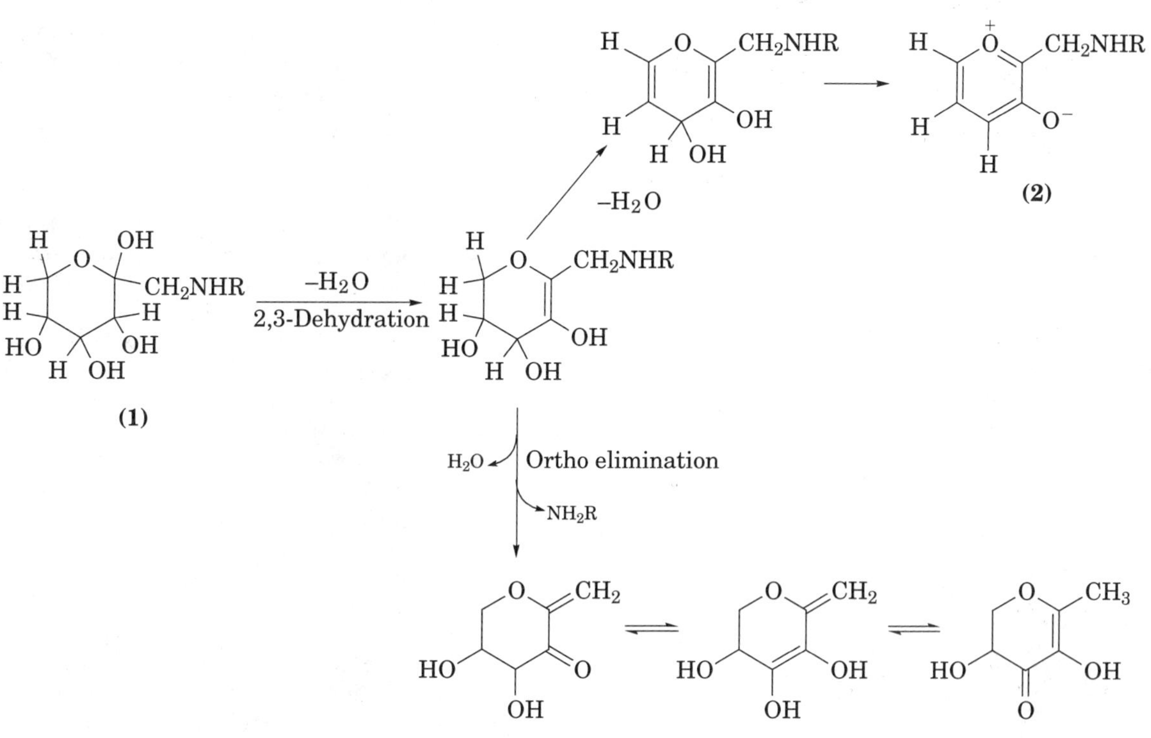

Figure 3. Direct sequential dehydration of Amadori product: (**1**) Amadori product, (**2**) pyrylium ion, (**3**) 2,3-dihydro-3,5-dihydroxy-6-methyl-4H-pyran-4-one. *Source:* Refs. 14 and 15.

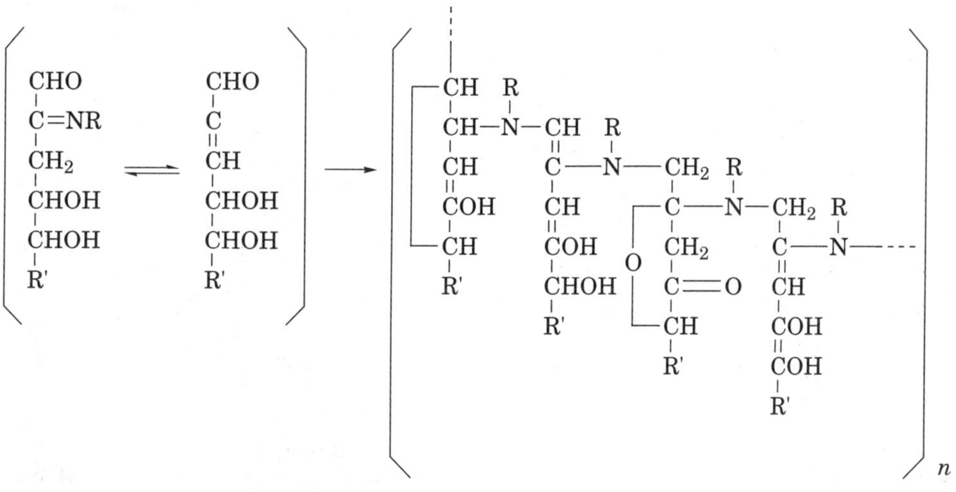

Figure 4. Repeating unit of melanoidin. R-NH_2 = amine, R′ = H or CH_2OH. *Source:* Ref. 18.

curs with the subsequent aggregation of these helices providing cross-linking junctions to build a continuous network (21,22). The number of charged sulfate groups along the polymer contributes to the degree of aggregation and the characteristics of the gel.

Carrageenan forms stable complexes with κ-casein via the interaction between the sulfate anions and the highly positively charged region of the protein. The synergistic interaction between κ-carrageenan and locust bean gum has also been interpreted on the basis of junction cross-linking.

High methyoxy pectin gels at low pH and in the presence of a cosolute such as sucrose. At low pH, the carboxyl groups are protonated, causing a decrease in electrostatic repulsion. Addition of a cosolute lowers the water solvation of the polymer. Both factors increase hydrophobic interaction and association of the polymers into cross-linking junction zones (23).

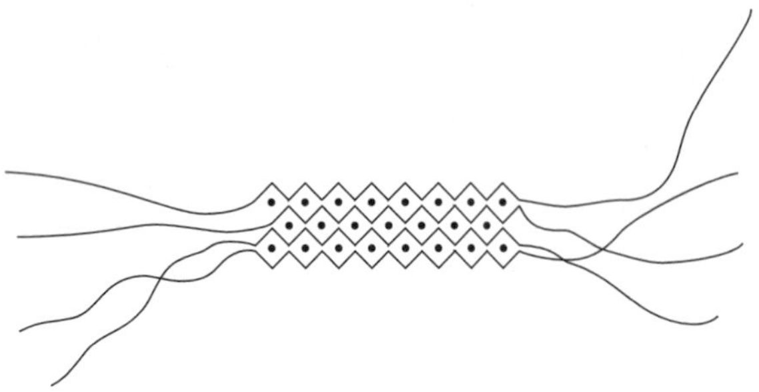

Figure 5. The "egg box" model. Line = alginate polymer; dark circle = calcium ion. *Source:* Ref. 20.

Starch gelation represents a more complex system, which has received continuous attention from various disciplines. The structure of crystalline amylose (of B-form starch) as originally elucidated by X-ray diffraction consists of two parallel strands of right-handed sixfold helices packed in an antiparallel double helix (24). More recent work has suggested a parallel packing of left-handed, parallel-stranded double helices for crystalline amylose from both A- and B-form starch (25,26). In solution, the conformation of amylose assumes a random coil containing short segments of loose and irregular helical structure (27,28).

Various models have been proposed for the structure of amylopectin. Most investigations seem to support a cluster-type model in that amylopectin is composed of clusters of oriented chains with the branching points collected together toward the reducing end (29).

Gelation occurs when starch granules in suspension are heated above the gelatinization temperature and cooled. Heating causes the granules to swell irreversibly, accompanied by the solubilization of amylose while most of the amylopectin is retained. On cooling, the starch gel formed has a composite structure of swollen amylopectin granules distributed in a matrix of amylose gel (30). Retrogradation involves crystallization of both the amylose in the gel phase and the amylopectin in the granules. The amylose molecules associate through hydrogen bonding into an insoluble precipitate. Amylopectin also exhibits interchain association. However, amylopectin molecules have an average chain length of 20 to 25 and a degree of polymerization of approximately 10^3. Interchain association between the polymers can only extend for 15 to 20 glucose units before being interrupted by the branching points. Crystallization of the amylopectin fraction is slow and results in a gradual increase in the rigidity of the granules and hence the amylose gel matrix. Reversion of a starch gel to the granular crystalline state is termed retrogradation, a process chiefly responsible for the staleness of bread.

It is important to note that in a gel the intermolecular association of polymers usually involves extensive segments of the polymeric chains held together by hydrogen bonding, electrostatic forces, hydrophobic or ionic interactions to form cross-linking junctions. Depending on the gel, a variety of rheological properties can exist. The entire field of polysaccharide gels has gone through a rapid expansion in the scope of basic research and intensified investigation in the relationship between the molecular arrangement and gelling characteristics.

LIPIDS

Lipid Oxidation

Lipid autoxidation is a free-radical chain reaction involving hydroperoxides. The mechanism has been extensively investigated since the first systematic study by Bateman and Bolland in the 1940s. In 1970, Rawls and van Santen proposed a possible role for singlet oxygen in the initiation of fatty acid autoxidation (31). In the presence of a suitable sensitizer, lipid oxidation proceeds via the "ene" reaction in which the dioxygen molecule is added to the olefinic carbon with a subsequent shift in the position of the double bond, and both conjugated and nonconjugated hydroperoxides are formed. In contrast, the classical free-radical mechanism produces only the conjugated hydroperoxides. The distribution of hydroperoxides are different from that found in the free-radical reaction.

In linoleate autoxidation, the product is composed predominantly of the 9- and 13-hydroperoxides, because the stability of the conjugated diene system favors the oxygen attack at the end position. Experimentally, however, four hydroperoxides are obtained: two with *trans, cis* and the other two with *trans, trans* stereochemistry (13-hydroperoxy-9-*cis*, 11-*trans*-octadecadienoic; 13-hydroxyperoxy-9-*trans*, 11-*trans*-octadecadienoic; 9-hydroxyperoxy-10-*trans*, 12-*cis*-octadecadienoic, and 9-hydroperoxy-10-*trans*, 12-*trans*-octadecadienoic). A unified mechanism has been proposed to account for the formation of these products (32) (Fig. 6). In this scheme the peroxy radicals initially formed have the *trans, cis* configuration and exist in two conformational isomers. The normal pathway involves the abstraction of hydrogen from another linoleate by either isomer to yield the *trans, cis* hydroperoxides. The alternative pathway involves the loss of oxygen from the conformers to give either the original pentadienyl radical or a new carbon radical in which the stereochemistry of the partial double bond is inverted. Oxygen addition to this new pentadienyl radical yields a *trans, trans*-diene peroxy radical which ultimately gives the *trans, trans*-conjugated hydroperoxide.

Because autoxidation involves free radicals, radical scavengers should effectively terminate the chain reaction. Most antioxidants are substituted phenolic compounds that act by transferring hydrogen to lipid peroxy radicals. The resulting aryloxy radical of the antioxidant is stable and unreactive in oxidative reactions. The resonance stability of the aryloxy radical depends on the substitution groups. Electron-releasing groups decrease the transition energy for the formation of the aryloxy radical. Bulky substituents stabilize the aryloxy radical but also create steric hindrance making the antioxidant less accessible to the lipid peroxy radical (33).

Physical Chemistry of Lipids

Another area of great interest to food chemists concerns the physical chemistry of lipids—polymorphism, crystal

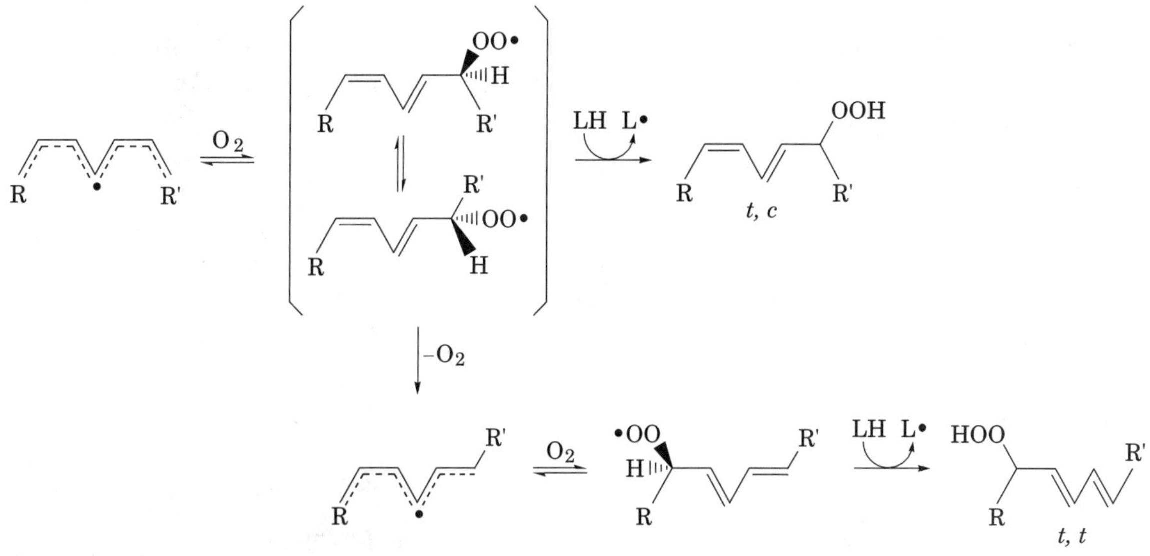

Figure 6. A unified mechanism for lipid autoxidation. *Source:* Ref. 32.

habit, emulsion, etc. Triglycerides exhibit multiple crystalline structures. For example, tristearin melts at 54.7, 64.0, and 73.3°C, representing the transition from the less stable α form to the more stable forms, β' and β. The α form has a hexagonal crystal subcell. The intermediate melting β' form is orthorhombic and the highest melting polymorph; the β form has a triclinic subcell. This multiplicity of molecular conformation and packing influences the fluidity, texture, and appearance of the product. Alpha crystals are fragile platelets of 5 μm in size. Beta crystals tend to be large and coarse with 25–50 μm in diameter. Fats in the β' form appear as tiny needles of ~1 μm in length. The β' crystals can incorporate large numbers of air bubbles providing a smooth texture to an oil product such as margarine and shortening. The fatty acid composition, as well as the position of a particular fatty acid in the glyceride, affects the crystal habit of a particular fat. Interesterification usually causes a β' to β conversion.

The current concept of emulsion stability came with the understanding of the factors controlling attractive/repulsive forces and their interaction with distance between disperse particles. Flocculation and coalescence of oil droplets in an emulsion is dependent on the balance between van der Waals attraction and electrostatic repulsion. The combination of these two gives a net potential energy vs distance curve with a potential minimum at certain intermediate ranges, where repulsion may be greater than attraction and a measure of stability exists (34). At long and short ranges, the net potential is always attraction.

An emulsifier consists of both hydrophilic and hydrophobic segments in the same molecule. Food grade emulsifiers are usually partial esters of fatty acids, polyols, and water-soluble organic acids. When an emulsifier is dispersed in water and heated, a liquid crystalline mesophase is formed (35). The mesophase assumes a lamellar, hexagonal, or cubic structure depending on the type of emulsifier and temperature. In a ternary system such as an emulsifier in an oil/water system, similar types of mesophases are formed (36). The ordered layer of this liquid-crystalline phase stabilizes the oil-in-water emulsion by forming a film at the interface. It decreases the attractive forces between the oil droplets and provides a steric barrier against coalescence between the droplets.

PROTEINS

Chemical Reactions in Processing

Because proteins contain many amino acids with reactive side-chain groups, it is expected that a variety of reactions may occur during food processing. One of the most extensively studied reactions is alkali degradation. Alkaline treatment is used in food industry for peeling, solubilization, and texturization of food proteins and manufacture of gelatin, sausage casings, and tortillas.

The most well-studied reaction is β-elimination in which the α-hydrogen of an amino acid residue is abstracted by the hydroxide ion (37) (Fig. 7). In protein-bound cysteine, the resulting product is a dehydroalanine which is an α,β-unsaturated compound. Nucleophilic side groups, such as the lysyl ϵ-amino and cysteinyl sulfur, react with dehydroalanine via Michael addition, leading to new cross-linkings in the proteins and loss of certain essential amino acids.

Alkaline treatment also causes hydrolysis of the amide groups in asparagine and glutamine, and the guanidino group in arginine. Racemization of amino acids is also detected. Protein-bound amino acids are more susceptible to α-hydrogen abstraction and hence recemization to the D-form (38). Prolonged heating results in isopeptide formation between the ϵ-amino group of lysine and the carbonyl group of aspartic or glutamic acid, or the amide groups of glutamine or asparagine.

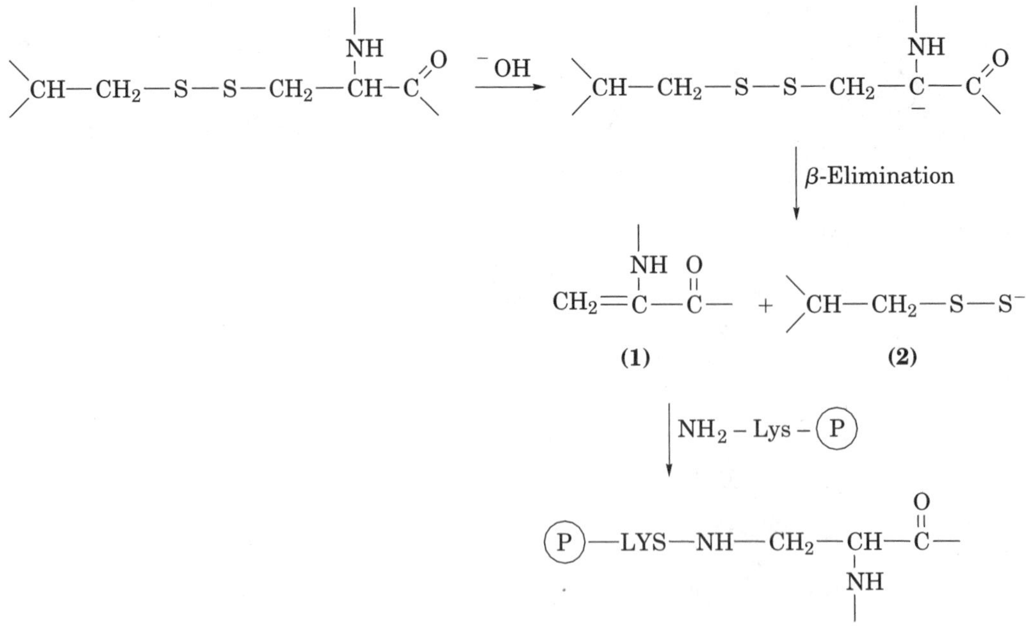

Figure 7. Reaction of β-elimination in alkali degradation of proteins: (**1**) dehydroalanine, (**2**) persulfide product, (**3**) lysinoalanine.

Protein Structure and Functionality

For a food chemist the structure of proteins is quite often viewed in the context of functionality in a food system. For example, the chemistry of muscle fibers and the mechanism of muscle contraction are related to rigor mortis and postmortem tenderness of meat (39). The onset of rigor mortis follows rapid depletion of ATP and breakdown in the regulatory system that controls the calcium level in muscle. Increasing concentration of calcium in the sarcoplasm induces contraction, whereas lack of ATP in the system stops the dissociation of the actin-myosin complex formed. The muscle loses its natural extensibility and this postmortem change is known as rigor mortis. Postmortem tenderness, however, is related to proteolysis of the muscle proteins. The acid proteases, such as the cathepsins, have received much attention in this respect. The calcium-dependent proteinase, calcium activated factor, has been linked to the causes of postmortem tenderization (40). In the last decade, there have been significant advances in our understanding of the molecular structure as well as the morphology and mechanism of muscle cytoskeletal proteins (41–43). These new developments in the basic knowledge of muscle proteins will inevitably affect the way their functionality in food systems is interpreted.

Despite the importance of cereal seed proteins, until recently, their molecular structure has been little understood. In the last several years, the complete amino acid sequence (or sequence deduced from cDNA) of gliadin (α-type) has been determined (44). The low molecular weight (LMW) glutenin subunits have been mapped by two-dimensional gel electrophoresis and sequenced (45). The high molecular weight (HMW) glutenin genes have been sequenced and expressed in *Escherichia coli* (46). Most recently, wheat transformation with genes for HMW subunits has been achieved, and the expression of one or two transgenic proteins in wheat results in stepwise increase in dough elasticity (47). Structural analysis of the amino acid sequence of the HMW glutenins is especially revealing. The protein molecule contains a large central repetitive region rich in glutamine, forming a loose spiral structure (48). Several cysteine residues are located at the α-helical region near the *N*- and *C*-terminal ends. Intermolecular disulfide bonds between the terminal cysteine residues cross-link the glutenin subunits into gluten polymers with the spiral regions in between. Hydrogen bonding could form among the side chains of glutamine residues, as well as the peptide backbone. However, attempt to relate the elastic mechanism of HMW subunits to a model consisting of spiral motifs that can extend and reform has not been entirely convincing (49). The continuing efforts on the investigation of protein sequences and structures in combination with genetic engineering studies are of immense value in providing the molecular basis for the specific role of HMW glutenins.

Bovine casein micelles exist in large spherical colloidal particles of 500 to 3000 Å in diameter and 10^7 to 3×10^{10} in particle weight. It has long been postulated that a micelle is assembled from submicelles containing a mixture of various casein molecules. However, the supramolecular structure of casein submicelles and micelles is unclear. One model suggests that submicelles are bound by electrostatic interaction via colloidal calcium phosphate through their ester phosphate groups (50). Because κ-casein is almost phosphate free, binding occurs only among the other caseins in the submicelle. Submicelles with a low level of κ-casein are oriented in the interior, and the surface of the micelle is covered entirely with submicelles having a high content of κ-casein. A more recent model depicts the structure of micelles as a protein gel without the formation of discret submicelles (51). The core of a micelle is a gel matrix with casein proteins held by microgranules of calcium phosphate. On the surface of the micelle is a "hairy" layer composed of a uniform density of macropeptide segments of κ-casein. The highly flexible and hydrated polar polypeptide chains provide steric stabilization to the micelle. Cleavage of the macropeptides by chymosin changes the surface characteristics of the micelle resulting in aggregation.

The cDNA sequences for the four major caseins (α_{s1}, α_{s2}, β, and κ) are known. Suggestions have been made to improve the functionality of casein using genetic engineering techniques (52). These include (*1*) alteration of the proportion of κ-casein to enhance the stability of casein micelles, (*2*) construction of an additional cleaving site in casein for chymosin, resulting in a change of rheological effects of proteolysis, (*3*) dephosphorylation of casein, creating additional phosphate groups for stabilization, (*4*) deletion of a polar segment from the otherwise nonpolar *N*-terminus of κ-casein, thereby enhancing its amphiphilicity.

Similar strategy has been applied to the whey protein β-lactoglobulin. It has been postulated that the thermal instability of this protein is due to the unfolding of the polypeptide segment (residues 115–125) containing the free cysteine-121, and the subsequent sulfhydryl-disulfide exchange with the disulfides (65–160 and 106–143) or with κ-casein in a milk system. Hence, deletion or substitution of the cys-121 by site-directed mutagenesis is expected to enhance the thermal stability and the functional properties of β-lactoglobulin.

COLORANTS

There are two basic categories of color compounds: conjugated polyenes and metalloporphyrins. The former includes carotenoids, annatto, anthocyanins, betanain, dyes, and lakes. The effect of conjugation is lowering the π-π^* transition energy from the highest occupied molecular orbital to the lowest unoccupied molecular orbital. Increased conjugation in the molecule shifts the absorption maximum to a higher wavelength. Substituent groups with lone pairs of electrons tend to increase π conjugation by resonance.

Carotenoids consist of a basic structure of eight repeating isoprene units that are highly conjugated. In the conjugated system, the terminal double bond has the highest electron density and is most susceptible to oxidative attack. Degradation proceeds from the end to the center of the molecule, resulting in a progressive shortening of the polyene chain (53).

Anthocyanins are flavonoid compounds characterized by the flavylium nucleus. The flavylium form is stabilized by resonance with the positive charge delocalized throughout the entire structure, making the anthocyanin molecule intensely colored. The structural transformation of anthocyanins in aqueous medium was thoroughly investigated in the early 1980s. Flavylium salts exist in equilibrium in different forms: flavylium cation (AH^+), quinoidal base (A), carbinol pseudobase (B) and chalcone (C). The equilibrium between AH^+ and A involves the transfer of proton from the C5, C7, or C4′ hydroxyl groups to a water molecule. In the hydration reaction, the water molecule is preferentially added to the C2 of the pyrylium ring of AH^+, resulting in colorless B. The conversion of B to C is a base-catalyzed tautomerization (Fig. 8). Because most isolated natural anthocyanins when placed in slightly acidic medium (pH 3–6) exist largely in the colorless forms (B and C), these fundamental studies should have practical implications. Evidence indicates that substitution pattern at various positions of the anthocyanin molecule influences the equilibrium constants of these reactions and hence the distribution of the colored (AH^+, A) and colorless (B, C) species (54).

The two best known examples of metalloporphyrins found in food are the myoglobin and chlorophylls. A porphyrin metal complex possesses 19 π-electrons in an 18-atom ring. The main effect of the metal on the transitions is the conjugation of the metal $p\pi$ orbital with the porphyrin π orbital (55). The splitting of the δ orbitals of the metal ion (due to the porphyrin) exhibits additional loss of degeneracy from the theoretically predicted octahedral symmetry. In oxymyoglobin heme complex the iron coordination positions are directed to the four porphyrin nitrogens, and in the fifth and the sixth positions, to histidine F8 and O_2 (or H_2O), respectively. The role of the protein globin is to stabilize the steric and electronic configuration of the iron heme, and to facilitate the back-bonding of electrons from the iron to the π^* orbital of the oxygen (Fig. 9). The hydrophobicity of the heme pocket excludes the binding of ionic ligands such as CN^-, OH^-, and the closely packed amino acid side chains restrict the size and orientation of the ligand (56).

Chlorophylls, unlike myoglobin which contains a transition metal ion, are porphyrins complexed with the alkali earth metal Mg^{++}. The compounds are hydrophobic because of the a long chain C_{20} phytol esterified to the propionic acid side chain at C7. The magnesium atom of chlorophyll can be readily replaced by weak acids or other metals such as copper, zinc, and iron. The free base obtained after removal of the metal is pheophytin, with the transition intensity shifted to a lower wavelength. Hydrolysis of the phytyl chain by alkali or enzyme yields the chlorophyllides, which are water-soluble and green-colored, with spectral characteristics similar to those of the chlorophylls. Removal of magnesium from the chlorophyllides yields pheophorbides, which have the same spectral properties as those of the pheophytins. These types of conversion have been frequently implicated as causes of color loss in processed green vegetables (57,58). Copper complexes of pheophytin and clorophyllide are stable to acid and are used as food colors in some European countries (59).

There is a considerable interest in colorants from natural resources. Cape jasmine (*Gardenia jasminoides*) have been investigated for the production of carotenoid and crocin. The flower is also rich in flavonoids. Another example is the mold *Monascus purpureus*, which has been used in Asian countries for centuries. The pigments produced are a mixture of red, yellow, and purple polyketides (60). Pigments extracted from algae, yeasts, and insects have also

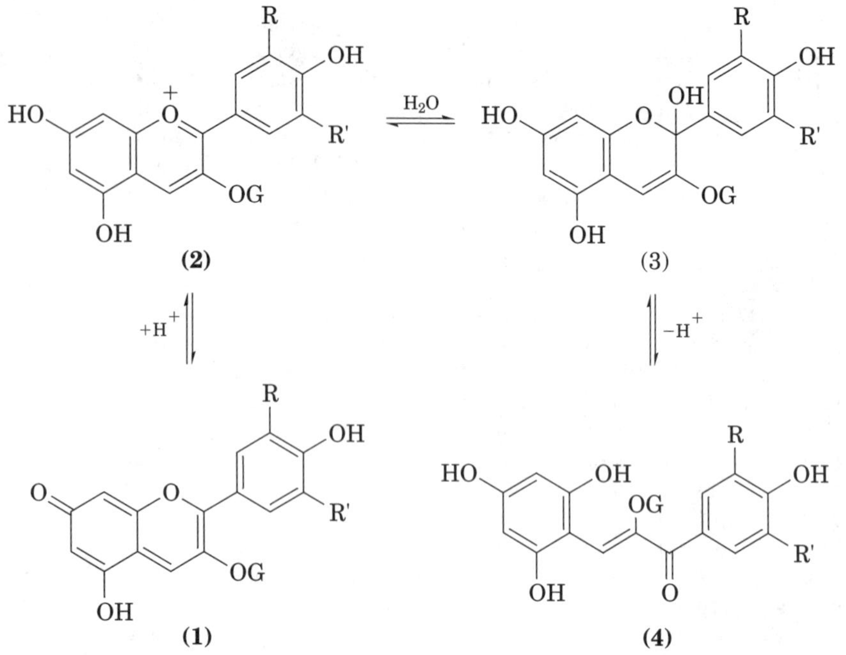

Figure 8. Resonance stabilization of flavylium cation: (**1**) flavylium cation (red), (**2**) carbinol pseudobase (colorless), (**3**) chalcone (colorless), (**4**) quinoidal base (blue). *Source:* Ref. 54.

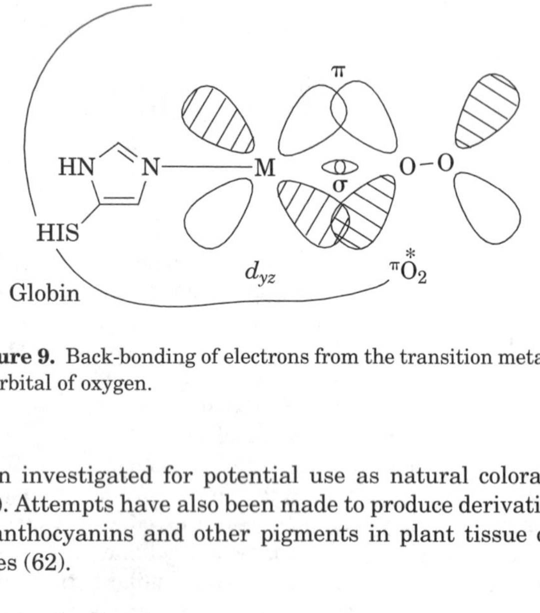

Figure 9. Back-bonding of electrons from the transition metal to π^* orbital of oxygen.

been investigated for potential use as natural colorants (61). Attempts have also been made to produce derivatives of anthocyanins and other pigments in plant tissue cultures (62).

FLAVORS

Flavor is the sensation produced by a material and perceived principally by the senses of taste and smell. Over the years, scientists have attempted to relate chemical senses to the molecular structures of various compounds in an effort to develop a coherent theory. In 1967 it was proposed the glucophore unit, which is responsible for sweetness in compounds, should consist of an AH, B hydrogen bonding system where H is an acidic proton and B is an electronegative atom or center (63). In 1972 the two-component system was extended to include a dispersion bonding component designated as γ component. The distance parameters of the resulting tripartite structure are A, B = 2.6, B, γ = 5.5, and A, γ = 3.5 Å. The orbital distance between the AH proton and B is 3.0 Å. In a three-dimensional picture, the glucophore binds to the receptor site and the sweet taste is initiated by intermolecular hydrogen bonding between the glucophore and a similar AH, B unit on the receptor. The γ component acts to align the molecule to the receptor. The locations of AH, B, γ components in many sweet compounds are known. For example, in aspartame, the protonated α-amino group and the ionized β-carboxyl group of the aspartyl residue represent the AH and B unit, respectively (Fig. 10). The phenyalanine end of the molecule represents the γ component. Only the L-L isomer of aspartame is sweet, and this type of enantiomeric effect in sweetness is also shown in several simple amino acids. To account for this fact, the receptor AH, B site is believed to consist of a spatial barrier. Hence, the L-D isomer of aspartame has the methyl ester group so positioned that the molecule cannot fit into the pocket for interactions (64). A similar concept is also applied to the aminosulfonates, such as saccharin and acesulfame K. Ring substitution experiments on acesulfame K indicate a loss of sweetness when the length of the hydrophobic group on the nitrogen exceeds 0.7 Å, suggesting that a bulky substituent would create a steric hindrance (65).

Another group of sweeteners worthy of attention are the sweet proteins, which are approximately 100,000 times sweeter than sucrose on a molar basis and several times on a weight basis. Thaumatin I, the most studied sweet protein, consists of a single polypeptide of 207 amino acids (M_W 22,000), with 8 disulfides and no histidine. The largest domain is a flattened β barrel formed by 11 antiparallel β strands, with 6 residues per strand on the average. Attached to this are domains II and III, with loops linked and stabilized by disulfide bonds (66). Monellin, the other known sweet protein, consists of subunits I and II of 50 and 44 amino acid residues, respectively. The protein consists of a 5-stranded, antiparallel β sheet (2 β strands from

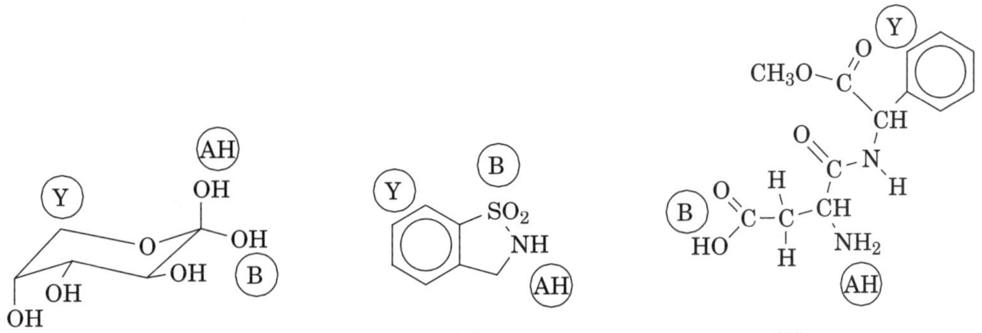

(1) (2) (3)

Figure 10. Assignment of glucophore unit in sweet compounds: (**1**) fructose, (**2**) saccharin, (**3**) aspartame. *Source:* Ref. 64.

subunit I and 3 from subunit II) where each strand contains an average of 10 residues (67). There is only one sulfhydryl group (in subunit II), and in contrast to thaumatin I, no disulfide bond. Despite the intense sweetness of both proteins, there is little similarity in their three-dimensional structures. However, antibodies raised against thaumatin also cross-react with monellin (68). Five homologous amino acid sequences are found in thaumatin I and monellin; two are topologically similar and one of these is located in a β bend protruding from the surface of the molecule. This tripeptide sequence (-Glu-Tyr-Gly-) has been postulated to be the active site responsible for the sweet taste (69,70). The structure–function relationship of these proteins is not clear.

A similar AH, B concept has been employed to explain bitterness with little success. The structural requirements for bitter taste have been proposed, and it has been shown that, in contrast to the AH/B system for sweeteners, bitter compounds need only one polar group (electrophilic or nucleophilic with a negative charge) and a hydrophobic group, with the intensity of bitter taste depending on the size and shape of the hydrophobic part of the molecule (71,72). Although the monopolar–hydrophobic concept for bitter taste requires more research, the importance of hydrophobicity to bitterness has been well established. Introduction of an additional hydrophobic group onto a sweet compound often results in modifying the taste from sweet to bitter.

Several theories have been presented over the years to correlate the molecular structure with the perceived odor quality of a chemical compound. The infrared theory has had some success in correlating odor quality with low-energy molecular vibrations (73). The site-fitting theory attempts to link the size, shape, and electronic status of a molecule to a complementary receptor site (74). It has been suggested that although the overall shape and size of an odorous compound is responsible for the sensory perception, the functional group determines the orientation and the affinity of the molecule to the receptor (75). Molecular connectivity terms have been derived to relate the quantitative description of certain aspects of molecular topology to several classes of odor compounds and found significant correlations between the two (76). Another study on the qualitative structure activity relationship has been conducted using computerized pattern recognition techniques (77). Both chemical composition and the shape of the molecule seem to be important contributing factors. In addition to the properties mentioned, it is also evident that the chirality and hydrophobicity of the compound are important for odor intensity.

Current knowledge about olfactory reception suggests the initial interaction occurs in the olfactory epithelium between the odorant with receptor proteins that extend from the dendrites of the olfactory sensory neurons (78,79). In humans, there are around 1,000 genes encoding 1,000 different odor receptors, each expressed in thousands of olfactory sensory neurons. A particular odor would bind to a characteristic set of multiple receptors rather than individual receptor cells, depending on the chemical groups of the odor molecule. Binding of an odorant to receptors activates a G protein–adenylate cyclase cascade, which results in the generation of cAMP. The cAMP causes the opening of an ion channel either directly or by phosphorylation via a cAMP-dependent protein kinase (80,81). The signals are transmitted to the localized glomeruli in the olfactory bulb, but exactly how the cortex decodes and reconstructs the signals into sensory responses is not known. There is still lack of a clear relationship between chemical structure and odor quality, and it is impossible to predict odor quality of a compound with a known structure.

NATURAL TOXICANTS

There are a large number of natural toxic chemicals in our daily diet. Toxicants may be chemical constituents of the food itself, contaminants from microbial infestation, or degradation products formed during food storage and processing. Toxic chemicals are commonly present in all plants, in which they act as protective agents against microbial attack, insects, and predators (82). Although the generally low level of these compounds together with a varied diet usually eliminate the risk of intoxification, the presence of many of these natural pesticides in our food seems to be receiving very little attention from the general public.

One well-known example of natural pesticides is the glycoalkaloids, solanine and chaconine, found in potatoes

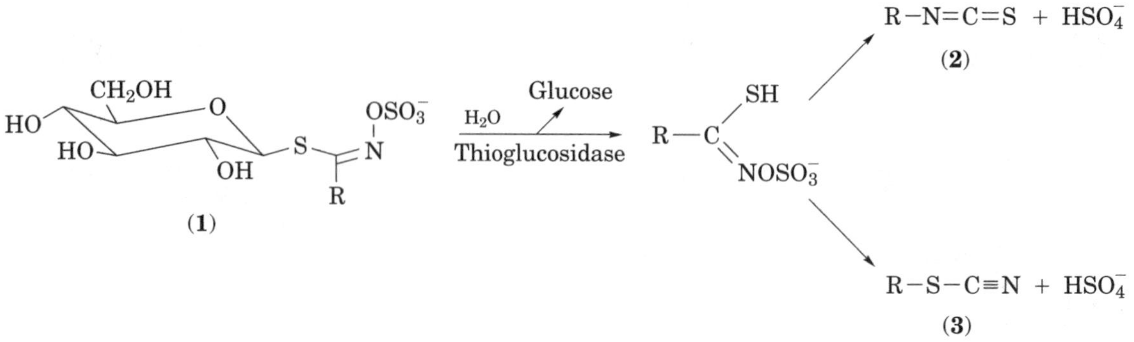

Figure 11. Degradation of glucosinolate: (**1**) glucosinolate, (**2**) isothiocyanate, (**3**) thiocyanate. *Source:* Ref. 83.

at the level of approximately 10 mg/100 g fresh weight (83). These glycoalkaloids are strong inhibitors of cholinesterase causing neurological disorder symptoms. Lethal doses for humans range from 3 to 6 mg/kg body weight, and doses of greater than 2 mg/kg are normally considered toxic.

Another class of toxicants, glucosinolates, occurs predominantly in vegetables such as cabbage, cauliflower, radish, mustard greens, brussels sprouts, and broccoli. Glucosinolates are hydrolyzed by the enzyme thioglucoside glycohydrolase when the plant tissues are crushed. The unstable aglycone produced is converted to isothiocyanate in a process similar to the Lossen rearrangement or thiocyanate (84) (Fig. 11). The latter compound is a goitrogen.

A number of amino acids, peptides, and proteins constitute an interesting class of toxicants. There are more than 250 nonprotein amino acids found in plants, many of which are structurally similar to the protein amino acids. For example, β-pyrazol-1-ylalane, pipecolic acid, homoarginie, α,β-diaminobutyric acid are analogs of histidine, proline, arginine, and lysine, respectively. Toxicity of these nonprotein amino acids is usually caused by the competition with structurally similar protein amino acids for proteins, resulting in functionally inactive products, or in interference with protein synthesis.

The notorious toxins produced by *Clostridium botulinum* are neurotoxins that block the release of acetylcholine

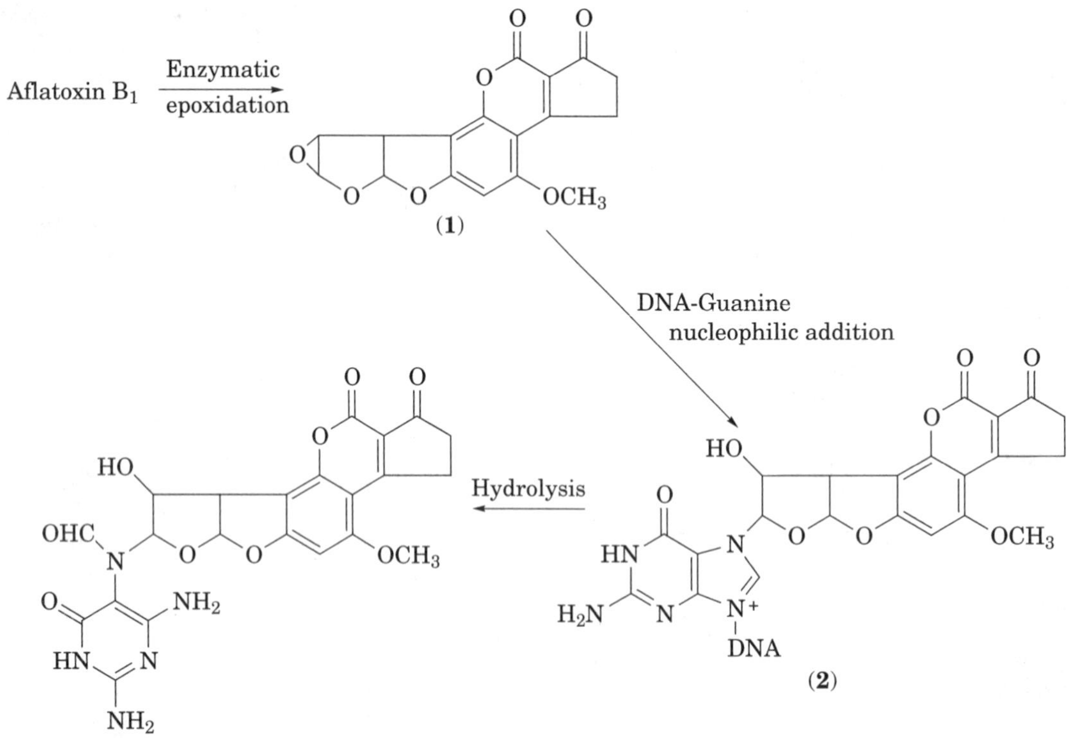

Figure 12. Formation of conjugate between aflatoxin B_1 and DNA: (**1**) AFB$_1$-8,9-oxide, (**2**) AFB$_1$-Guanine-DNA. *Source:* Ref. 85.

in the peripheral nervous system, resulting in paralysis. The lethal dose is in the range of 1 to 5 ng/kg body weight. *Botulinum* toxins are synthesized as an inactive precursor, which is a single polypeptide with a molecular weight of ~150,000 Da. The protein is cleaved by an endogenous protease into an active molecule consisting of a heavy chain and a light chain linked by a disulfide bond. The toxin initially binds to gangliosides (sialic acid–containing glycolipids) on the plasma membrane of cholinergic nerve. The toxin–ganglioside complex undergoes endocytosis, and the low pH in the vesicle activates the insertion of the toxin into the membrane (85).

Mycotoxins are toxic fungal metabolites that often contaminate peanuts, cereals, and dairy products. The most extensively studied mycotoxins are the aflatoxins produced by *Aspergillus* and *Penicillium*. The toxicity of aflatoxin B_1 is caused by their metabolism to the reactive AFB_1-8,9-oxide, which forms covalently linked conjugates with DNA and proteins (86) (Fig. 12), or undergoes hydrolysis to AFB_1-8,9-dihydrodiol. The latter product is a dialdehyde phenolate ion stabilized by resonance and forms Schiff base with proteins.

Polycyclic aromatic hydrocarbons are known carcinogens found in grilled meat products. These compounds include benzo[a]pyrene, benzo[a]fluroanthene, benzo[a]-anthracene, and chrysene. They are metabolized by enzymatic activation to form a dihydrodiolepoxide that binds DNA. Heterocyclic amines found in high temperature cooking are enzymatically converted to *O*-acyl derivative, which covalently binds DNA, resulting in mispairing in replication and transcription (87).

Many *N*-nitrosamines are found in cured meat and fish, including *N*-nitrosodimethylamine, *N*-nitrosodiethylamine, *N*-nitrosopyrrolidine, and *N*-nitrosopiperidine. Metabolically, nitrosamines undergo enzymatic α-hydroxylation, followed by C-N bond cleavage to yield alkydiazohydroxide, which further breaks down to alkyldiazonium ion. Both alkydiazohydroxide and alkyldiazonium ion react with DNA to form covalent conjugates (88).

CONCLUSIONS

This brief overview clearly indicates the dynamic interactions among various interrelated disciplines. The very fact that interactions among food components are complex reflects the scope and diversity of the underlying chemistry. It is often inadequate to treat food chemistry in a descriptive manner. Approaches must be taken to analyze the kind of chemical mechanisms involved and to follow the sequence of steps in terms of chemical equations and principles. Food chemistry encompasses reaction mechanisms involving free radicals, transition metals, coordination chemistry, stereochemistry, molecular structures, biopolymers, and colloids, in addition to common organic chemical reactions. The process of searching for a fundamental understanding of how food components react and interact is indeed an exciting challenge for food scientists.

DISCLOSURE STATEMENT

Reference to a company and/or products is only for the purposes of information and does not imply approval or recommendation of the product to the exclusion of others which may also be suitable. All programs and services of the U.S. Department of Agriculture are offered on a nondiscriminatory basis without regard to race, color, national origin, religion, sex, age, marital status, or handicap.

BIBLIOGRAPHY

1. D. W. S. Wong, *Mechanism and Theory in Food Chemistry*, Chapman & Hall, New York, 1989.
2. H. Belitz and W. Grosch, *Food Chemistry*, 2nd ed., Springer-Verlag, Berlin, Heidelberg, 1982, Translated in English by D. Hadziyev, 1987.
3. O. R. Fennema, *Food Chemistry*, 3rd ed., Marcel Dekker, New York, 1996.
4. J. M. deMan, *Principles of Food Chemistry*, 2nd ed., Van Nostrand Reinhold, New York, 1990.
5. D. W. S. Wong, *Food Enzymes: Structure and Mechanism*, Chapman & Hall, New York, 1995.
6. J. P. Danehy and W. W. Pigman, "Reactions Between Sugars and Nitrogenous Compounds and Their Relationship to Certain Food Problems," *Adv. Food Res.* **3**, 241–290 (1951).
7. E. Dworschak, "Nonenzymatic Browning and Its Effect on Protein Nutrition," *Crit. Rev. Food Sci. Nutr.* **13**, 1–40 (1980).
8. J. P. Danehy, "Maillard Reactions: Nonenzymatic Browning in Food Systems With Special Reference to the Development of Flavors," *Adv. Food Res.* **30**, 77–138 (1986).
9. M. Namiki, "Chemistry of Maillard Reactions: Recent Studies on the Browning Reaction Mechanism and the Development of Antioxidants and Mutagens," *Adv. Food Res.* **32**, 115–184 (1988).
10. A. Cerami, H. Vlassara, and M. Brownlee, "Glucose and Aging," *Sci. Am.* **256**, 90–96 (1987).
11. A. Cerami, "Aging of Proteins and Nucleic Acids: What Is the Role of Glucose?" *Trends Biochem. Sci.* **11**, 311–314 (1986).
12. M. Namiki and T. Hayashi, "A New Mechanism of the Maillard Reaction Involving Sugar Fragmentation and Free Radical Formation," in G. R. Waller and M. S. Feather, eds., *The Maillard Reaction in Foods and Nutrition*, ACS Symposium Series 215, American Chemical Society, Washington, D.C., 1983, pp. 21–46.
13. T. Hayashi, and M. Namiki, "Role of Sugar Fragmentation in an Early Stage Browning of Amino-Carbonyl Reaction of Sugar With Amino Acid." *Agric. Biol. Chem.* **50**, 1965–1970 (1986).
14. V. Yaylayan and P. Sporns, "Novel Mechanisms for the Decomposition of 1-(Amino acid)-1-deoxy-D-fructoses (Amadori Compounds): A Mass Spectrometer Approach," *Food Chem.* **26**, 283–305 (1987).
15. V. Yaylayan and A. Huyghues-Despointes, "Chemistry of Amadori Rearrangement Products: Analysis, Synthesis, Kinetics, Reactions and Spectroscopic Properties," *Crit. Rev. Food Sci. Nutr.* **34**, 321–369 (1994).
16. S. Fors, "Sensory Properties of Volatile Maillard Reaction Products and Related Compounds," in G. R. Waller and M. S. Feather, eds., *The Maillard Reaction in Foods and Nutrition*, ACS Symposium Series 215, American Chemical Society, Washington, D.C., 1983, pp. xx–xx.
17. B. L. Wedzicha and D. J. McWeeney, "Non-enzymatic Browning Reactions of Ascorbic Acid and Their Inhibition. The Production of 3-Deoxy-4-sulphopentosulose in Mixtures of Ascor-

bic Acid, Glycine and Bisulphite ion," *J. Sci. Food Agric.* **35**, 577–587 (1974).

18. H. Kato and H. Tsuchida, "Estimation of Melanoidin Structure by Pyrolysis and Oxidation," *Prog. Food Nutr. Sci.* **5**, 147–156 (1981).
19. H. Kato, S. B. Kim, and F. Hayase, "Estimation of the Partial Chemical Structures of Melanoidins by Oxidative Degradation and ^{13}C CP-MAS NMR," *Dev. Food Sci.* **13**, 215–223 (1986).
20. G. T. Grant et al., "Biological Interactions Between Polysaccharides and Divalent Cations: The Egg Box Model," *FEBS Lett.* **32**, 195–198 (1973).
21. E. R. Morris, D. A. Rees, and G. R. Robinson, "Cation-specific Aggregation of Carrageenan Helices; Domain Model of Polymer Gel Structure," *J. Mol. Biol.* **138**, 349–362 (1980).
22. G. Robinson, E. R. Morris, and D. A. Rees, "Role of Double Helices in Carrageenan Gelation: The Domain Model," *J. Chem. Soc. Chem. Comm.* **1980**, 152–153 (1980).
23. D. G. Okenfull, "The Chemistry of High Methoxyl Pectins," in R. H. Walter, ed., *The Chemistry and Technology of Pectin*, Academic Press, New York, 1991, pp. 87–108.
24. H.-C. H. Wu and A. Sarko, "The Double-helical Molecular Structure of Crystalline β-Amylase," *Carbohydr. Res.* **61**, 7–25 (1978).
25. A. Imberty and S. Perez, "A Revisit to the Three Dimensional Structure of β-Type Starch," *Biopolymers* **27**, 1205–1221 (1988).
26. A. Imberty, H. Chanzy, and S. Perez, "The Double-helical Nature of the Crystalline Part of A-starch," *J. Mol. Biol* **201**, 365–378 (1988).
27. R. C. Jordan, D. A. Brant, and A. Cesaro, "A Monte Carlo Study of the Amylose Chain Conformation," *Biopolymers* **17**, 2617–2632 (1978).
28. A. Neszmelyi and J. Hollo, "Some Aspects of the Structure of Starch—A 3-D Molecular Modelling Approach," *Starch / Starke* **41**, 1–3 (1988).
29. D. J. Manners, "Recent Development in Our Understanding of Amylopectin Structure," *Carbohydr. Polym.* **11**, 87–112 (1989).
30. M. Miles, et al., "The Role of Amylose and Amylopectin in the Gelation and Retrogradation of Starch," *Carbohydr. Res.* **135**, 271–281 (1985).
31. H. R. Rawls and P. J. Van Santen, "A Possible Role for Singlet Oxygen in the Initiation of Fatty Acid Autoxidation," *J. Am. Oil Chem. Soc.* **47**, 121–125 (1970).
32. N. A. Porter, "Mechanisms for the Autoxidation of Polyunsaturated Lipids," *Acc. Chem. Res.* **19**, 262–268 (1986).
33. G. Scott, "Antioxidants *in vitro* and *vivo*," *Chem. Br.* **21**, 648–653 (1985).
34. S. Friberg, "Emulsion Stability," in S. Friberg, ed., *Food Emulsions*, Marcel Dekker, New York, 1976, pp. 1–37.
35. E. Boyle, and B. German, "Monoglycerides in Membrane Systems," *Crit. Rev. Food Sci. Nutr.* **36**, 785–805 (1996).
36. L. Hernquist, "Polar Lipids in Emulsions and Microemulsions," in E. Dickinson, ed., *Food Emulsions and Foams*, Royal Society of Chemistry, London, 1987, pp. 110–157.
37. J. R. Whitaker and R. E. Feeney, "Chemical and Physical Modification of Proteins by the Hydroxide Ion," *Crit. Rev. Food Sci. Nutr.* **19**, 173–212 (1983).
38. R. Liardon and S. Ledermann, "Racemization Kinetics of Free and Protein-bound Amino Acids Under Moderate Alkaline Treatment," *J. Agric. Food Chem.* **34**, 557–565 (1986).
39. Y. L. Xiong, "Structure–Function Relationships of Muscle Proteins," in S. Damodoran and A. Paraf, eds., *Food Proteins and Their Applications*, Marcel Dekker, New York, 1997, pp. 341–392.
40. D. E. Goll et al., "Role of Muscle Proteinases in Maintenance of Muscle Integrity and Mass," *J. Food Biochem.* **7**, 137–177 (1983).
41. I. Rayment et al., "Structure of the Actin–Myosin Complex and Its Implications for Muscle Contraction," *Science* **261**, 58–65 (1993).
42. M. Irving et al., "Tilting of the Light-Chain Region of Myosin During Step Length Changes and Active Force Generation in Skeletal Muscle," *Nature* **375**, 688–632 (1995).
43. R. D. Vale, "Getting a Grip on Myosin," *Cell* **78**, 733–737 (1994).
44. D. D. Kasarda, et al., "Nucleic Acid (cDNA) and Amino Acid Sequences of α-Type Gliadins From Wheat (*Triticum aestivum*)," *Proc. Natl. Acad. Sci. U.S.A.* **81**, 4712–4716 (1984).
45. H. P. Tao and D. D. Kasarda, "Two-dimensional Gel Mapping and *N*-Terminal Sequencing of LMW-Glutenin Subunits," *J. Exp. Bot.* **40**, 1015–1020 (1989).
46. G. Galili, "Heterologous Expression of a Wheat High Molecular Weight Glutenin Gene in *Escherichia coli*," *Proc. Natl. Acad. Sci. U.S.A.* **86**, 7756–7760 (1989).
47. F. Barro et al., "Transformation of Wheat With High Molecular Weight Subunit Genes Results in Improved Functional Properties," *Nature Biotechnology* **15**, 1295–1299 (1997).
48. P. R. Shewry et al., "Biotechnology of Breadmaking: Unraveling and Manipulating the Multiprotein Gluten Complex." *Bio / Technology* **13**, 1185–1190 (1995).
49. D. D. Kasarda, G. King, and F. Kumosinski, "Comparison of Spiral Structures in Wheat High Molecular Weight Glutenin Subunits and Elastin by Molecular Modeling," in T. F. Kumosinski and M. N. Liebman, eds., *Molecular Modeling: From Virtual Tools to Real Problems*, American Chemical Society, Washington, D.C., 1994, pp. 209–220.
50. D. G. Schmidt, "Association of Caseins and Casein Micelle Structure," in P. F. Fox, ed., *Developments in Dairy Chemistry*, Applied Science Publishers, London and New York, 1982, pp. 61–86.
51. C. Holt, "Structure and Stability of Bovine Casein Micelles," *Adv. Protein Chem.* **43**, 63–151 (1992).
52. Y. Kang and T. Richardson, "Genetic Engineering of Caseins," *Food Technol.* **39**, 89–94 (1985).
53. C. Marty and C. Berset, "Factors Affecting the Thermal Degradation of All-*trans*-β-carotene," *J. Agric. Food Chem.* **38**, 1063–1067 (1990).
54. R. Brouillard, G. A. Iacobucci, and J. G. Sweeny, "Chemistry of Anthocyanin Pigments. 9. UV-Visible Spectrophotometric Determination of the Acidity Constants of Apigeninidin and Three Related 3-Deoxyflavylium Salts," *J. Am. Chem. Soc.* **104**, 7585–7590 (1982).
55. D. J. Livingston and W. D. Brown, "The Chemistry of Myoglobin and Its Reactions," *Food Technol.* **35**, 244–252 (1981).
56. D. Ladikos and B. L. Wedzicha, "The Chemistry and Stability of the Haem–Protein Complex in Relation to Meat," *Food Chem.* **29**, 143–155 (1988).
57. H. K. Lichtenthalev, "Chlorophylls and Carotenoids: Pigments of Photosynthetic Biomembranes," *Methods Enzymol.* **148**, 350–382 (1987).
58. S. J. Schwatz and T. V. Lorenzo, "Chlorophylls in Foods," *Crit. Rev. Food Sci. Nutr.* **29**, 1–27 (1990).
59. A. H. Humphrey, "Chlorophyll," *Food Chem.* **5**, 57–67 (1980).

60. A. J. Taylor, "Natural Colors in Food," in J. Walford, ed., *Developments in Food Colors—II*, Applied Science Publishers, London, 1984, pp. 31–49.
61. K. Spears, "Developments in Food Colourings: The Natural Alternatives," *Trends in Biotechnol.* **6**, 283–288 (1988).
62. D. K. Dougall et al., "Biosynthesis and Stability of Monoacylated Anthocyanins," *Food Technol.* **51**, 69–71 (1997).
63. R. S. Shallenberg and T. E. Acree, "Molecular Theory of Sweet Taste," *Nature* **216**, 480–482 (1967).
64. F. Lelj et al., "Interaction of α-L-Aspartyl-L-phenylalanine Methyl Ester With the Receptor Site of the Sweet Taste Bud," *J. Am. Chem. Soc.* **98**, 6669–6675 (1976).
65. F. Pautet and C. Nofre, "Correlation of Chemical Structure and Taste in the Cyclamate Series and the Steric Nature of the Chemoreceptor site," *Z. Lebensm.-Unters.-Forsch.* **166**, 167–170 (1978).
66. C. M. Ogata et al., "Crystal Structure of a Sweet Tasting Protein Thaumatin I, at 1.65 Å Resolution." *J. Mol. Biol.* **228**, 893–908 (1992).
67. J. R. Somoza et al., "Two Crystal Structures of a Potently Sweet Protein: Natural Monellin at 2.75 Å Resolution and Single-Chain Monellin at 1.7 Å Resolution," *J. Mol. Biol.* **234**, 390–404 (1993).
68. C. A. M. Hough and J. A. Edwardson, "Antibodies to Thaumatin as a Model of the Sweet Taste Receptor," *Nature* **271**, 381–383 (1978).
69. H. van der Wel, "Physiological Action and Structure Characteristics of the Sweet-tasting Proteins Thaumatin and Monellin," *Trends Biochem. Sci.* **5**, 122–123 (1980).
70. S. H. Kim, A. deVos, and C. Ogata, "Crystal Structures of Two Intensely Sweet Proteins," *Trends Biochem. Sci.* **13**, 13–15 (1988).
71. A. van der Heijden. "Sweet and Bitter tastes," in T. E. Acree and R. Teranishi, eds., *Flavor Science: Sensible Principles and Techniques*, American Chemical Society, Washington D.C., 1993, pp. 67–115.
72. H. D. Belitz et al., "QSAR of Bitter Tasting Compounds," *Chem. Ind. (London)* **1**, 23–26 (1983).
73. R. H. Wright, *The Science of Smell*, Basic Books, New York, 1964.
74. J. E. Amoore, *Molecular Basis of Odor*, Charles C. Thomas, Springfield, Ill., 1970.
75. M. G. J. Beets, "Relationship of Chemical Structure to Odor and Taste," in *Proceedings, Third International Congress on Food Science and Technology*, Institute of Food Technologists, Chicago, Ill., 1971, pp. 379–395.
76. L. M. Kier, T. D. Paolo, and L. H. Hall, "Structure–Activity Studies on Odor Molecules Using Molecular Connectivity," *J. Theor. Biol.* **67**, 585–595 (1977).
77. G. Ohloff, "Chemistry of Odor Stimuli," *Experienta* **42**, 271–279 (1986).
78. L. Buck and R. Axel, "A Novel Multigene Family May Encode Odorant Receptors: A Molecular Basis for Odor Reception," *Cell* **65**, 175–187 (1991).
79. K. Mori and Y. Yoshihara, "Molecular Recognition and Olfactory Processing in the Mammalian Olfactory System," *Progress in Neurobiology* **45**, 585–619 (1995).
80. S. H. Snyder, P. B. Sklar, and J. Pevsner, "Molecular Mechanisms of Olfaction," *J. Biol. Chem.* **263**, 13971–13974 (1988).
81. W. Pickenhagen, "Enantioselectivity in Odor Perception," in R. Teranishi, R. G. Buttery, and F. Shahidi, eds., *Flavor Chemistry, Trends and Developments*, ACS Symposium Series 388, American Chemical Society, Washington, D.C., 1989, pp. 151–157.
82. B. N. Ames, "Chemicals, Cancers, Causalities and Cautions," *CHEMTECH* **19**, 590–598 (1989).
83. S. F. Osman, "Glycoalkaloids in Potatoes," *Food Chem.* **11**, 235–247 (1983).
84. G. R. Fenwick, R. K. Heaney, and W. J. Mullin, "Glucosinolates and Their Breakdown Products in Food and Food Plants," *Crit. Rev. Food Technol.* **18**, 123–200 (1983).
85. E. Schantz and E. A. Johnson, "Properties and Use of Botulinum Toxin and Other Microbial Neurotoxins in Medicine," *Microbiol. Rev.* **56**, 80–89 (1992).
86. R. G. Croy and G. N. Wogan, "Temporal Patterns of Covalent DNA Adducts in Rat Liver After Single and Multiple Doses of Aflatoxin B_1," *Cancer Res.* **41**, 197–203 (1981).
87. Y. Hashimoto, K. Shudo, and T. Okamoto, "Activation of a Mutagen, 3-Amino-methyl-5H-pyrido[4,3-b]indole. Identification of 3-Hydroxyamino-1-methyl-5H-pyrido[4,3-b]indole and Its Reaction With DNA," *Biochem. Biophys. Res. Comm.* **96**, 355–362 (1980).
88. M. C. Archer, "Reactive Intermediates From Nitrosamines," in R. Snyder, ed., *Biological Reactive Intermediates—II, Advances in Experimental Medicine and Biology*, Vol. 136B, Plenum Press, New York, 1982, pp. 1027–1035.

DOMINIC W. S. WONG
USDA
Albany, California

FOOD CONSUMPTION SURVEYS IN THE U.S. DEPARTMENT OF AGRICULTURE

One mission of the U.S. Department of Agriculture (USDA) is to encourage the production and availability of a sufficient, safe, and nutritionally adequate supply of food for Americans. In support of this mission, USDA conducts surveys to monitor food use and food consumption patterns in the U.S. population. Early studies on food and nutrition were begun at the end of the last century. These small-scale studies aimed to help people in the working class achieve good diets at low cost. As time went on, recognition of the need for nationally representative food and nutrient intake data resulted in the development of larger surveys. The USDA's most recent national food consumption survey was the 1994–1996 Continuing Survey of Food Intakes by Individuals (CSFII), which measured the kinds and amounts of foods eaten by Americans. Conducted as a telephone follow-up to the CSFII, the 1994–1996 Diet and Health Knowledge Survey (DHKS) measured attitudes and knowledge about diet and health among Americans. The CSFII and DHKS were designed so that individuals' attitudes and knowledge about healthy eating could be linked with their food choices and nutrient intakes. Both surveys address the requirements of the National Nutrition Monitoring and Related Research Act of 1990 (P.L. 101-445) for continuous monitoring of the dietary status of the American population, including the low-income population. See the section "The National Nutrition Monitoring and Related Research Program."

USES OF THE DATA

As shown in Table 1, evaluations of diet quality and tracking changes in the diet over time have many useful federal, state, and local applications, including policy formation, program planning, and nutrition education. Users of the survey data include federal government agencies, such as the Environmental Protection Agency (EPA), the Food and Drug Administration (FDA), the Federal Trade Commission, and numerous USDA agencies; state agencies and larger county health departments; food and agricultural industries; and universities. The data are used to determine the food choices Americans make and to evaluate the content and adequacy of their diets in relationship to the Dietary Guidelines for Americans (1); other federal government statements of dietary policy, such as the year 2000 nutrition objectives (2); and the National Academy of Sciences' Recommended Dietary Allowances (3). Health interventions such as the National Cancer Institute's "5-A-Day for Better Health Program" (4) and nutrition education materials such as the Food Guide Pyramid (5) are developed and targeted based on survey results. The data are also used to assess the nutritional impact of the USDA's food assistance programs; to estimate exposure to pesticide residues, food additives, and contaminants; to develop food fortification, enrichment, and food labeling policies; and to assess the demand for agricultural products and marketing facilities.

Table 1. Uses of Data from Nationwide Food Surveys

Current nationwide food surveys conducted by USDA's Agricultural Research Service encompass two types of surveys: surveys of foods eaten by individuals both at home and away from home and surveys of attitudes and knowledge about healthy eating, about diet and health relationships, and about dietary guidance. A third type of survey that measured food used by households and the costs of those foods has been discontinued although results from those surveys are still used. Results from these three types of surveys are used for a variety of purposes.

Assessment of dietary intakes

Provide detailed benchmark data on food and nutrient intakes and eating patterns of the population
Monitor the nutritional quality of diets and determine the size and nature of populations at risk of having diets low in certain nutrients
Identify socioeconomic and attitudinal factors associated with diets
Identify changes in food and nutrient consumption that would reduce health risks

Economics of food consumption

Predict demand for agricultural products and marketing facilities
Determine the effects of socioeconomic factors on the demand for food
Determine the demand for food away from home and its effects on the nutritional quality of diets

Food policy, programs, and guidance

Determine appropriate levels of enrichment or fortification based in part on food use
Track use of food labels and their effect on dietary intakes
Monitor food security, hunger, and diet quality
Identify populations that might benefit from intervention programs
Determine the amounts of foods that are suitable to offer in food distribution programs
Identify factors affecting participation in some food programs and estimate the effect of participation on diet quality
Estimate the effect of food programs on demand for food
Develop food plans that reflect food consumption practices and meet nutritional and cost criteria
Develop food guides and dietary guidance materials that target nutritional problems in the U.S. population
Identify educational strategies to increase the knowledge of nutrition and to improve the eating habits of Americans

Food safety

Estimate exposure to pesticide residues, food additives, contaminants, and naturally occurring toxic substances
Predict food items in which a food additive can safely be permitted in specified amounts
Determine the need to modify regulations in response to changes in consumption

Historical trends

Correlate food consumption and dietary status with incidence of disease over time
Follow food consumption through the life cycle
Predict changes in food consumption and dietary status as they may be influenced by economic, technological, and other developments

RECENT USDA FOOD CONSUMPTION SURVEYS

The 1994–1996 CSFII and the 1994–1996 DHKS, popularly known as the *What We Eat in America* survey, are the most recent surveys conducted by the USDA (6). The 1994–1996 survey collected dietary data on individuals of all ages in the United States. A survey of food and nutrient intakes by children under 10 years of age was conducted in 1998 as a supplement to the CSFII 1994–1996. The Supplemental Children's Survey (SCS) was conducted in response to the 1996 Food Quality Protection Act of 1996 (P.L. 104–170), which required the Secretary of Agriculture to provide the EPA with information on food consumption patterns of a statistically valid sample of infants and children. This requirement follows a 1993 report of the National Academy of Sciences entitled *Pesticides in the Diets of Infants and Children* in which concern was raised that current food consumption data did not provide sufficient sample sizes for adequate estimation of exposure to pesticide residues in the diets of children (7).

In the CSFII/DHKS 1994–1996, a nationally representative sample of individuals of all ages was asked to provide, through in-person 24-h recall interviews, information about their food intakes on two nonconsecutive days as well as socioeconomic and health-related information. About two weeks after the food intake interview, a subsample of individuals age 20 years and over was asked to answer a series of questions about knowledge and attitudes toward dietary guidance and health. The sample design included oversampling of low-income individuals to yield a national sample of the low-income population. The final survey data set includes information on two days of food and nutrient intake data for more than 15,000 individuals and information on diet, health, and food safety issues from nearly 6000 adults who also responded to the DHKS (6). The method of data collection for the SCS is identical to that used in the 1994–1996 CSFII; the SCS includes two days of dietary intake on approximately 5000 children birth through 9 years of age.

Methodology research has always been important in planning the increasingly complex food consumption surveys at the USDA. Most surveys have contained a component to improve methods in future surveys as well as to evaluate food consumption of the sample under study. Such studies build on earlier experience and findings (8). To meet the objectives of improving the accuracy and comprehensiveness of dietary intake data and developing a more cost-effective method of data collection for national surveys of food consumption to be conducted in the year 2000 and later, the USDA's Food Surveys Research Group (part of the Beltsville Human Nutrition Research Center of the Agricultural Research Service) has a dietary survey methods research program. The objective of this research program is to develop and test new dietary intake methodology, utilizing a telephone mode of data collection. This research includes the development and testing of a computer-assisted telephone interview (CATI) for collecting dietary data and the selection and testing of food measurement aids to improve portion size estimation. A critical component of the research program is a large-scale validation study of 400 individuals to compare estimates of energy intake as determined by the 24-hour dietary recall method against estimates of actual energy expenditure by individuals as determined by the doubly labeled water technique (9). The doubly labeled water technique provides precise measures of energy expenditure in free-living individuals and may be used to validate the assessment of energy intake by other methods (10).

HISTORY OF USDA FOOD CONSUMPTION SURVEY METHODOLOGY

Household Food Consumption Surveys

The USDA has conducted household food consumption surveys for about 100 years. In 1894 Congress mandated that human nutrition investigations be conducted by the USDA Office of Experiment Stations. W. O. Atwater, the first director of the Experiment Stations, is credited with the first food consumption studies in the United States in the late 1800s. He recognized the essential links between such studies and research on food composition, nutritional requirements, and dietary guidance, and he pioneered studies in all of these areas. Atwater (11,12) sought food consumption information that would help him develop recommendations on what a working man should eat and how families could spend their food money wisely. By 1898 USDA investigators had made studies of food consumption by more than 300 families (13). In early studies, participants were "willing families." Researchers used a food inventory-record to collect data by determining the weight and cost of food used by the family from inventories of food on hand at the start and end of the survey period and records of foods brought into the home during the period (14).

Because this complex procedure was found to be too intrusive, too time-consuming, and too costly, it was replaced in the 1930s by the food list-recall (or food list). The new technique required only an interview with the household respondent (usually the homemaker) who recalled, using the food list, the quantities of listed foods used by the household during the preceding week and the amounts paid for purchased items. Although the list-recall procedure was introduced with little preliminary study, response rates for the food list-recall were later shown to be much higher than for the food inventory-record method (8).

Nationwide Household Food Consumption Surveys

The USDA began periodic nationwide surveys of households in the 1930s using the list-recall method along with statistical sampling techniques that permitted the collection of data from large numbers of households in relatively short periods of time. In the 1930s and 1940s, nationwide surveys included the 1935–1936 Consumer Purchases Study, the 1942 Spending and Saving in Wartime Survey, and a survey of food consumption of urban families in 1948. The surveys in 1935–1936, 1942, and 1948 preceded the advent of probability sampling in surveys (15) and were less than fully representative of the U.S. population (8).

Later household surveys attempted to represent the total U.S. population. Their results indicated the way U.S. food supplies were distributed among households of different sizes, income, and location. Because each food was reported as brought into the household, its cost could be obtained and an estimate made of the money value of food used in households. The 1955 Household Food Consumption Survey (HFCS) was the first of the USDA's food consumption surveys in which a commercial firm selected the sample and collected and edited the data. The USDA retained responsibility for overall planning and monitoring of all survey operations, provision of support data for analysis, and the interpretation and publication of results. The 1965–1966 HFCS was the first to cover all four seasons of the year and to collect information on food intake by individuals (see later), as well as the food used by the household as a whole.

In addition to the periodic nationwide surveys of households, the USDA has conducted smaller methodological or special-purpose surveys of food consumption. More precise methods evolved as various problems were investigated (8). For example, several studies addressed survey methodology issues such as food discard measurement, questionnaire design and wording, and interviewer training and compared the food inventory-record and food list-recall methods of data collection, confirming the decision to adopt the list-recall technique for use in future surveys (16–21). Other studies (22,23) explored techniques for collecting dietary data from individuals. The 1950s and early 1960s saw widespread concern for disadvantaged and low-income families, and the Pilot Food Stamp Program was initiated in 1961 in eight economically depressed areas. A before-and-after study of food consumption and dietary levels in an urban and a rural area showed that the Food Stamp Program increased the purchase of more nutritious foods by needy families and also expanded the market for agricultural products—a major government objective being to utilize farm surpluses (24). The Food Stamp Program became permanent with the Food Stamp Act of 1964.

Two additional nationwide household food consumption surveys were conducted—in 1977–1978 and in 1987–1988.

The 1977–1978 Nationwide Food Consumption Survey (NFCS) included five supplemental surveys: low-income households, households with elderly persons, and households in Puerto Rico, Alaska, and Hawaii. Some of these supplemental surveys facilitated studies of factors affecting food consumption and diet quality among groups eligible for some of the USDA's food assistance programs; others provided information on food consumption in locations outside the conterminous states that had not previously been studied (13). In 1979–1980, a follow-up survey to the 1977–1978 low-income survey was conducted to assess whether shifts had occurred in food consumption and dietary quality that might be associated with rising food costs and changes in the Food Stamp Program since the survey of similar households in 1977–1978.

As food supplies increased and became more varied, the number of foods on the food list-recall form increased rapidly: about 200 items in 1948, about 2300 items in 1977, and nearly 3000 items by 1987 (8). In the NFCS of 1987–1988, one innovation was the use of a laptop computer programmed to handle the burden of a growing food list. The computer was intended to improve processing time also.

The collection of two types of information—household use of food and individual intake—in the nationwide surveys of 1965–1966, 1977–1978, and 1987–1988 created large respondent burden and, in the 1987–1988 survey, low response rates. Because of these concerns and increasing national awareness of the relationship between diet and health, the collection of information on household use of food was discontinued after the 1987–1988 survey, completing the shift in emphasis from collecting household food use data to individual intake data.

Dietary Intake by Individuals

Information on food intakes by individuals is more precise than household food use data for the assessment of diet quality (13). For several reasons, household data were less than ideal for analyses of diet quality relative to the Recommended Dietary Allowances (RDA), which were the only standards available (25). Household food consumption data included food discard, resulting in overestimates of nutritional quality. To compare household intake levels with the RDAs, it was necessary to adjust for the consumption of food away from home, which was not surveyed, as well as to make various assumptions related to the apportionment of food among household members and their differing nutritional needs. Individual intake data represent foods as eaten, excluding food discard and including both food eaten at home and food away from home. Individual intakes can appropriately be compared with sex- and age-specific RDAs. The first USDA nationwide survey of food intakes by individual members of households was conducted in spring 1965 as a supplement to the 1965–1966 HFCS. One day of intake for about 14,500 individuals in the United States was collected using a one-day (24-hour) recall.

The 1965 individual intake data were found to be very useful as baseline data, resulting in many requests for enlarging their scope—more intake days per individual, all seasons, and more questions on dietary practices. Consequently, the scope of the second nationwide survey of dietary intakes by individuals was greatly expanded in 1977–1978 and the name changed from the *Household* Food Consumption Survey to the *Nationwide* Food Consumption Survey. Three consecutive days of food intake were collected in four seasons for more than 30,000 individuals in the 48 conterminous states. In the NFCS 1977–1978, individuals provided information on their own intakes, unlike in the HFCS 1965 when the household respondent had provided information for all household members. In 1987–1988, another NFCS was conducted, which included the collection of three consecutive days of food intake data from about 15,000 individuals. Most of the procedures used to obtain food intake reports were similar in the two NFCS surveys. Intake data were collected using a 24-hour recall and a two-day record. Interviewers collected the first day of data in the home and instructed the individuals in the households on filling out the two-day records. The interviewer then returned after the third intake day to answer any questions and collect the records.

In 1985 the first national USDA survey of dietary intake by individuals independent of a household food use component began. Its purpose was to collect data more frequently than every 10 years, thus providing up-to-date information on the adequacy of diets of selected population groups and early indications of dietary changes—important considerations for data that are used in planning food assistance and educational programs and in administering a variety of public programs affecting the supply, safety, and distribution of the nation's food. The CSFII was repeated in 1986. In both years the survey included two nationally representative samples—an all-income sample and a low-income sample. In both years, the all-income sample included about 1500 women 19 to 50 years of age and about 500 of their children 1 to 5 years of age. The low-income sample included about 2100 women and about 1200 children 1 to 5 years of age in 1985 and about 1300 women and 800 children in 1986. A sample of men age 19 to 50 years of age was included in 1985 only. Food intake data were collected using a panel approach: Collection from each individual took place on up to six nonconsecutive days at intervals of approximately two months over a one-year period. The first day of intake was collected using an in-person interview; subsequent days of data were collected by telephone when possible. A 24-hour dietary recall was used for all intakes.

In 1989 the panel aspect of the 1985 and 1986 CSFII was dropped and the CSFII 1989–1991 was conducted using the same methodology as the individual intake portion of the NFCS 1987–1988. Individuals who took part in the CSFII 1989–1991 were asked to provide three consecutive days of dietary data. The first day's data were collected in an in-home interview using a one-day (24-hour) dietary recall. The second and third days' data were collected using a self-administered two-day dietary record.

Diet and Health Knowledge Survey

In 1989 the DHKS was initiated to improve understanding of factors that affect food choices and provide a link between an individual's knowledge and attitudes and his or her dietary behavior. Individuals who were identified as the main meal-planners/preparers in the CSFII were contacted by telephone, if possible, about six weeks after collection of the dietary data and were asked to answer a series of questions about knowledge and attitudes toward diet, health, and food safety. In the 1994–1996 DHKS, adults 20 years and over replaced main meal-planners/preparers as the unit of observation.

An overview of USDA nationwide food surveys is provided in Table 2.

TRENDS IN FOOD CONSUMPTION

In the depression years of the 1930s, concern about the quality of American diets was high. To address this and other issues related to family economics, the USDA and the Bureau of Labor Statistics conducted several national studies (13). The comprehensive picture of household food consumption and dietary levels obtained in the Consumer Purchases Study of 1935–1936 indicated that one-third of the nation's families had diets that were poor by nutritional standards (26). These findings gave impetus to the enrichment of white flour and bread with iron and three B vitamins, establishment of the National School Lunch Program, and expansion of nutrition education and research. Also, USDA economists used results to project food consumption in the United States and to develop food budgets to help families select good diets. A later version of the least costly of these food budgets, the Thrifty Food Plan, is still used in the federal formulas for counting the nation's poor (27) and for setting benefit levels in the Food Stamp Program (28).

The 1942 Spending and Saving in Wartime Survey measured the early effects of World War II on food consumption in urban, rural, and farm families at different income levels (29). The survey found marked improvement from the 1930s in diets overall, but many families' intakes of several nutrients were low in comparison with the new standards, the RDAs, first issued in 1941. Between the 1935–1936 survey and the 1948 survey of urban areas, great strides were made in the distribution and storage of food products, most notably in home refrigeration. These changes affected the way people purchased and used food.

Between the household food consumption surveys of 1955 and 1965–1966, the availability and consumer acceptance of many new, more convenient food products changed the cooking practices in many American households. For example, the use of mixes for baked products such as cakes and muffins and the availability of ready-made baked products led to a decrease in baking "from scratch," and household consumption of flour, sugar, and other basic baking ingredients decreased.

Between 1965–1966 and 1977–1978, the proliferation of new products was especially marked. Technological changes, such as freeze-dried coffee, and the increasing variety of commercially frozen foods reflected breakthroughs in food processing and packaging. Lifestyle changes, such as increases in the proportion of women employed outside the home, may have decreased the time spent in meal preparation and increased the demand for convenience foods and fast-food restaurants.

Between 1977 and 1985, when the CSFII was initiated, substantial changes occurred in food intakes—shifts to lower fat milk, less meat eaten separately (not as part of a mixture), and more grain products. These shifts, most prominent among higher income, more educated respondents, may have reflected concerns about diet and health issues. Nutrient intakes were at least as good, if not better in some respects, in 1985 than in 1977. However, the intakes of some nutrients were still below the 1980 RDAs (30); these observations were apparent at all levels of income and in all geographic regions.

The NFCS 1987–1988 showed a continuation of the dietary trends observed between 1977 and 1985 toward lower-fat milk, less meat eaten separately, and more grain products. Total fat intakes as a percentage of calories fell from 40% in 1977–1978 to 36% in 1987–1988. More of the household food dollar was spent away from home, and fewer meals were consumed from household food supplies in 1987–1988 than in 1977–1978. These changes may have resulted from a desire for increased convenience and variety. The food industry responded in a number of ways: more and varied restaurants, more microwavable packaging, and more bakeries, delicatessens, and salad bars in supermarkets.

In 1989–1991 survey data indicated that eating habits more closely followed national dietary guidelines than in the past, but the amount of fat in the average diet was still higher than the 30% recommended by the Dietary Guidelines for Americans (1), and Americans were eating lower amounts of fruits, vegetables, grains, and low-fat dairy products than recommended. In 1989–1991, results from the DHKS showed that only about one-fourth of main meal-planners/preparers met dietary recommendations for fat and saturated fat, and that their perceptions about their diets did not always match reality.

Data from the 1994–1996 CSFII indicated a continued decline in the percentage of calories from fat—from 36% in 1987–1988 to 33% in 1994–1996. However, only about one-third of adults had fat intakes that provided 30% or less of calories. Consumption of grain-based products, especially grain mixtures such as pizza and lasagna, continued to rise between the late 1980s and the mid-1990s. Among young children, consumption of fluid milk decreased by 16% since the late 1970s, while consumption of carbonated soft drinks increased by 16%.

PYRAMID SERVINGS DATA

Along with improvements in data collection methods, another methodological advance during the 1990s involves the way information on food intakes by individuals is reported to the public. Since the 1965–1966 HFCS, when information on the intakes of food by individuals was collected for the first time, average quantities of foods consumed have been reported in grams or as the percentages of individuals consuming food from selected food groups or subgroups.

Information expressed as percentages of individuals using food or as gram amounts of food has numerous uses, including comparing food consumption over time. Also, nutrient intakes could be calculated from food intakes and then compared with standards such as the RDAs (3). These comparisons gave important information about the ade-

Table 2. Overview of USDA Nationwide Food Surveys, 1936 to 1998

Survey	Population	Sample	Type of data collected	Dietary method
1935–36 Consumer Purchases Study	Farm, village, and city households in five geographic regions	Husband and wife families, white and native born	Household food use	Seven-day list-recall; seven-day food inventory record
1942 Family Spending and Saving in Wartime	Cities, rural nonfarm areas, and farms	Housekeeping families and single persons	Household food use	Seven-day list-recall
1948 Food Consumption of Urban Families	Urban families nationwide in spring plus surveys in four cities	Housekeeping families of two or more persons	Household food use	Seven-day list-recall
1955 Food Consumption of Households	48 states plus a supplement of farm households	National, self-weighting probability sample of housekeeping households	Household food use	Seven-day list-recall
1965–66 Household Food Consumption Survey (HFCS)	48 states	Two separate samples (basic and low income); selected household members were asked to provide intake information	Household food use Individual intake	Seven-day list-recall 24-h dietary recall in spring only
1977–78 Nationwide Food Consumption Survey (NFCS)	48 states	Two separate samples (basic and low income); all household members were asked to provide intake information	Household food use Individual intake	Seven-day list-recall Three consecutive days: 24-h dietary recall and two-day diet record
1985–86 Continuing Survey of Food Intakes by Individuals (CSFII)	48 states	Two separate samples (basic and low income); women 19–50 years and their children 1–5 years in both years and men 19–50 years in 1985 only	Individual intake	Women and children: six nonconsecutive 24-h dietary recalls; day one in person and remaining days by telephone Men: day one only
1987–88 NFCS	48 states	Two separate samples (basic and low income); all household members were asked to provide intake information	Household food use Individual intake	Seven-day list-recall Three consecutive days: 24-h dietary recall and two-day diet record
1989–91 CSFII	48 states	Two separate samples (basic and low income); all household members were asked to provide intake information	Individual intake	Three consecutive days: 24-h dietary recall and two-day diet record
1989–91 Diet and Health Knowledge Survey (DHKS)	48 states	Main meal-planners/ preparers with a completed day one intake in CSFII	Dietary knowledge, behavior, and attitudes	Telephone followup to CSFII
1994–96 CSFII	50 states	Oversampling of the low-income population; only selected household members were asked to provide intake information	Individual intake	Two nonconsecutive 24-h dietary recalls
1994–96 DHKS	50 states	Adults 20 years and over with a completed day one intake in CSFII	Dietary knowledge, behavior, and attitudes	Telephone followup to CSFII
1998 Supplemental Children's Survey to CSFII 1994–96	50 states	Children 0–9 years	Individual intake	Two nonconsecutive 24-h dietary recalls

quacy of diets but were difficult to interpret to the public, especially in light of more recent dietary recommendations that specified recommendations as the number of servings from specified food groups to eat each day. These recommendations are from the Dietary Guidelines for Americans and the USDA Food Guide Pyramid (1,5). To meet the need for a more easily understood way to report on the adequacy of food intakes, USDA developed a method for converting data from the CSFII into servings (31). This method facilitates comparisons of food intakes by Americans to dietary recommendations in the Food Guide Pyramid. The Pyramid shows a recommended range of servings to eat each day from five major food groups—grain, vegetable, fruit, dairy, and meat. In general, the number of servings that is right for a person depends on calorie needs; people who need more calories should eat more servings. The Pyramid also provides guidance to help people choose diets low in fat and moderate in sugars.

The method developed by the USDA adheres to Pyramid principles, uses the serving sizes specified by the Pyramid, and strictly categorizes foods according to Pyramid criteria (5). Since many foods people eat—foods like pizza, soups, and pies—count toward more than one food group, the method separates foods into their ingredients before servings are counted. Table 3 compares the average American diet against Food Guide Pyramid recommendations. The average number of servings from the fruit, dairy, and meat groups are below minimums recommended. Those from the grain and vegetable groups are near the bottom of recommended ranges.

The Pyramid recommends that Americans limit the fat in their diets to 30% of calories. People will consume about half of this amount as nondiscretionary fat if they eat the recommended number of servings from each food group, select the lowest fat choices within each group, and add no fat to their foods in preparation or at the table. Additional fat, up to 30% of calories, is considered discretionary in that people can decide whether to get it from higher-fat food choices or from additions to their foods during preparation or at the table. Examples of discretionary fat are the fat absorbed by french-fried potatoes during preparation, the fat from margarine spread on bread at the table, and the fat in whole milk. In 1994–1996, the average fat intake of Americans, at 33% of calories, exceeded the recommendation. Only 8% of calories were from nondiscretionary fat sources. Discretionary fat accounted for 25% of calories.

The Pyramid suggests that Americans try to limit their added sugars (defined as any sugar that is not naturally occurring in the food) to 6 tsp a day if they eat about 1600 calories, 12 tsp at 2200 calories, or 18 tsp at 2800 calories. In 1994–1996, Americans consumed an average of 20 tsp of added sugars a day in a diet that provided about 2000 calories. Added sugars accounted for 16% of calories.

THE NATIONAL NUTRITION MONITORING AND RELATED RESEARCH PROGRAM

"Growing recognition of the relation of food to health and its consequent social implications has created widespread interest in estimates of the adequacy of diets of different population groups." That statement, written in 1939 by Hazel K. Stiebeling and Esther F. Phipard in the introduction to an early survey report from the USDA, still rings true (32). Although tremendous progress in evaluating dietary adequacy occurred during the first hundred years of nutrition research, many research needs remain. In contrast with W. O. Atwater's landmark work in the late 1880s and early 1890s (11), which yielded information only on dietary energy, protein, fat, and carbohydrate, national food consumption surveys at the end of the twentieth century provide information on about 40 additional nutrients and dietary components. Yet, as we face the new millennium, there are still nutrients for which the analytical methodology continues to evolve, and there are still blanks in our understanding of the relationships between diet and health.

Nutrition research encompasses many different fields, ranging from agriculture and economics to biochemistry and medicine. The missions of many different agencies of the U.S. government include responsibility for monitoring or promoting the health of different population subgroups. In the 1990s, with passage of the National Nutrition Monitoring and Related Research Act of 1990 (P.L. 101-445), the federal government set in place a comprehensive plan to link all the agencies engaged in food and nutrition monitoring activities.

The 1990 Act required the federal government to develop the National Nutrition Monitoring and Related Research Program (NNMRRP) along with a Ten-Year Comprehensive Plan for Nutrition Monitoring and Related Research. As required by law, the plan was transmitted by the President to Congress and was published in the *Federal Register* on June 11, 1993 (33). The NNMRRP in the United States is a complex system of coordinated activities that provides information about the dietary, nutritional, and related health status of Americans; the relationships between diet and health; and the factors affecting dietary and nutritional status. The Ten-Year Plan serves as the basis for planning and coordinating the activities of the more than 20 federal agencies responsible for nutrition monitoring and related research activities. The primary goals of the plan are to ensure that the agencies participating in the NNMRRP collect data that are continuous, timely, and reliable; coordinate data collection with other member agencies; use comparable methods for data collection and reporting of results; and conduct research on the issues and topics relevant to monitoring the nutritional and health status of the population and subgroups at nutritional risk.

Prior to the NNMRR Act of 1990, nutrition monitoring research was conducted as a part of the National Nutrition Monitoring System (NNMS). The history of the goals and milestones of federal nutrition monitoring in the United States can be found in Refs. 34 to 37.

Two comprehensive reports on nutrition monitoring activities conducted as part of the NNMS were published in the late 1980s. The first report, *Nutrition Monitoring in the United States: A Progress Report from the Joint Nutrition Monitoring Evaluation Committee*, was prepared by a federal advisory committee jointly sponsored by the U.S. Department of Health and Human Services (HHS) and the USDA (36). The report provided an overview of the dietary and nutritional status of the U.S. population and was intended to serve as a reference, or baseline, for subsequent reports. The two major data sources for that report were the National Health and Nutrition Examination Survey of 1976–1980 conducted by HHS and the NFCS of 1977–1978 conducted by the USDA. The second report, *Nutrition Monitoring in the United States: An Update Report on Nutrition Monitoring* (37), used data produced or released since publication of the 1986 report to provide an update on the dietary and nutritional status of the U.S. population and on selected health conditions and behaviors.

Table 3. Comparison of the Average American Diet Against Food Guide Pyramid Recommendations

Food group	Recommended range of servings	Average number of servings consumed per day, two-day average	Percentages of individual consuming servings recommended based on caloric intake, two-day average[a]
Grain group	6 to 11	6.7	38
Vegetable group	3 to 5	3.3	41
Fruit group	2 to 4	1.5	23
Dairy group	2 to 3	1.5	23
Meat group (ounces)[b]	5 to 7	4.7	32

Source: The Food Guide Pyramid (5). USDA's Continuing Survey of Food Intakes by Individuals, 1994–1996; individuals 2 years and older (6).

[a]Recommended amounts were derived from sample patterns in the Food Guide Pyramid (5). For the grain, vegetable, fruit, and meat groups, individuals consuming less than 2200 calories met the recommendation if they consumed at the bottom of the range of servings, individuals consuming 2200 up to 2,800 calories meet the recommendation if they consumed in the middle of the range, and individuals consuming 2800 calories or more met the recommendation if they consumed at the top of the range. For the dairy group, women who were pregnant or lactating and individuals 11 through 24 years of age were counted as meeting the recommendation if they consumed at least three dairy servings a day; all other individuals were counted as meeting the recommendation if they consumed at least two dairy servings a day.

[b]The Pyramid recommends that servings of the meat group should be the equivalent of 5 to 7 oz of cooked lean meat, poultry, or fish. According to the Pyramid, 1 egg, 1/2 c of tofu, 2 tb of peanut butter, 1/3 c of nuts, and 1/4 c of seeds are each equivalent to 1 oz of cooked lean meat.

A third report was published following passage of the 1990 Act. The *Third Report on Nutrition Monitoring in the United States* was based on data gathered by the NNMRRP since the 1989 report, giving special emphasis to low-income and high-risk population subgroups (38). This report concluded with a chapter on 33 recommendations spanning all the component areas. Future years will no doubt see the achievement of many of those recommendations as well as the emergence of new opportunities for the expansion of nutrition knowledge.

BIBLIOGRAPHY

1. U.S. Department of Agriculture and U.S. Department of Health and Human Services, *Nutrition and Your Health: Dietary Guidelines for Americans*, 4th ed., Home and Garden Bull. No. 232, U.S. Dept. of Agriculture, Washington, D.C., 1995.
2. U.S. Department of Health and Human Services, Public Health Service, *Healthy People 2000: National Health Promotion and Disease Prevention Objectives*, DHHS Publication No. (PHS) 91-50212, U.S. Government Printing Office, Washington, D.C., 1991.
3. Food and Nutrition Board, National Research Council, *Recommended Dietary Allowances*, 10th ed., National Academy Press, Washington, D.C., 1989.
4. A. S. Subar et al., *5-a-Day for Better Health: A Baseline Study of Americans' Fruit and Vegetable Consumption*, National Institutes of Health, National Cancer Institute, Washington, D.C., 1992.
5. *The Food Guide Pyramid*, Home and Garden Bul. No. 252, U.S. Dept. of Agriculture, Washington, D.C., 1992.
6. *The 1994–96 Continuing Survey of Food Intakes by Individuals and the 1994–96 Diet and Health Knowledge Survey*, CD-ROM, NTIS Accession No. PB98-500457, U.S. Dept. of Agriculture, Washington, D.C., 1998.
7. National Research Council, *Pesticides in the Diets of Infants and Children*, National Academy Press, Washington, D.C., 1993.
8. E. M. Pao, K. E. Sykes, and Y. S. Cypel, *USDA Methodological Research for Large-Scale Dietary Intake Surveys*, 1975–88, Home Econ. Res. Rep. No. 49, U.S. Department of Agriculture, Washington, D.C., 1989.
9. D. A. Schoeller, "Measurement of Energy Expenditure in Free-Living Humans by Using Doubly Labeled Water," *J. Nutr.* **118**, 1278–1289 (1988).
10. A. E. Black et al., "Measurements of Total Energy Expenditure Provide Insights Into the Validity of Dietary Measurements of Energy Intake," *J. Am. Dietetic Assoc.* **93**, 571–579 (1993).
11. W. O. Atwater, *Foods: Nutritive Value and Cost*, Farmers' Bul. No. 23, U.S. Dept. of Agriculture, Washington, D.C., 1894.
12. W. O. Atwater, *Methods and Results of Investigations of the Chemistry and Economy of Food*, Expt. Sta. Office Bull. No. 21, U.S. Dept. of Agriculture, Washington, D.C., 1895.
13. B. B. Peterkin, "USDA Food Consumption Research: Parade of Survey Greats," *J. Nutr.* **124**, 1836S–1841S (1994).
14. A. P. Bryant, "Some Results of Dietary Studies in the United States," in *USDA Yearbook of Agriculture 1898*, U.S. Dept. of Agriculture, Washington, D.C., 1899, pp. 439–452.
15. W. G. Cochran, *Sampling Techniques*, John Wiley and Sons, New York, 1953.
16. S. Adelson, E. Asp, and I. Noble, "Household records of food used and discarded," *J. Am. Dietetic Assoc.* **39**, 578–584 (1961).
17. S. F. Adelson and E. C. Blake, *Diets of Families in the Open Country, a Georgia and an Ohio County, Summer 1945*, Misc. Pub. No. 704, U.S. Dept. of Agriculture, Washington, D.C., 1950.
18. F. Clark and L. Fincher, *Nutritive Content of Homemakers' Meals, Four Cities, Winter 1948*, Info. Bul. No. 112, U.S. Dept. of Agriculture, Washington, D.C., 1954.
19. F. Clark et al., *Food Consumption of Urban Families in the United States . . . With an Appraisal of Methods of Analysis*, Ag. Info. Bull. No. 132, U.S. Government Printing Office, Washington, D.C., 1954.
20. B. B. Reagan and E. Grossman, *Rural Levels of Living in Lee and Jones Counties, Mississippi, 1945, and a Comparison of Two Methods of Data Collection*, Ag. Inf. Bull. No. 41, U.S. Department of Agriculture, Washington, D.C., 1951.

21. J. Murray et al., *Collection Methods in Dietary Surveys. A Comparison of the Food List and Record in Two Farming Areas of the South*, S. Coop. Series Bull. 23, S.C. Ag. Expt. Sta, Clemson, S. Car., 1952.
22. S. F. Adelson, "Some Problems in Collecting Dietary Data From Individuals," *J. Am. Dietetic Assoc.* **36**, 453–461 (1960).
23. S. F. Adelson, and A. Keys, *Diet and Some Health Characteristics of 123 Business and Professional Men and Methods Used to Obtain Dietary Information*, Ag. Res. Serv. Pub. No 62-11, U.S. Dept. of Agriculture, Washington, D.C., 1962.
24. R. B. Reese and S. F. Adelson, *Food Consumption and Dietary Levels Under the Pilot Food Stamp Program, Detroit, Michigan and Fayette County, Pennsylvania*, Ag. Econ. Rep. No. 9, U.S. Dept. of Agriculture, Washington, D.C., 1962.
25. *Dietary Levels: Households in the United States, Spring 1977. Nationwide Food Consumption Survey 1977–78*, Rep. No. H-11, U.S. Dept. of Agriculture, Washington, D.C., 1985.
26. H. Stiebeling, *Are We Well Fed?*, USDA Pub. No 430, U.S. Dept. of Agriculture, Washington, D.C., 1941.
27. M. Orshansky, "How poverty is measured," *Monthly Labor Review* **92**, 37–41 (1969).
28. B. B. Peterkin et al., *The Thrifty Food Plan, 1983*, Human Nutrition Info. Serv. Admin. Rep. No 365, U.S. Dept. of Agriculture, Washington, D.C., 1983.
29. *Family Food Consumption in the United States*, USDA Misc. Pub. No. 550, U.S. Dept. of Agriculture, Washington, D.C., 1944.
30. Food and Nutrition Board, National Research Council, *Recommended Dietary Allowances*, 9th ed., National Academy Press, Washington, D.C., 1980.
31. L. E. Cleveland et al., "A Method for Assessing Food Intakes in Terms of Servings Based on Food Guidance," *Am. J. Clin. Nutr.* **65** (Suppl), 1254S–1263S, (1997).
32. H. K. Stiebeling and E. F. Phipard, *Diets of Families of Employed Wage Earners and Clerical Workers in Cities*, USDA Circular No. 507, U.S. Dept. of Agriculture, Washington, D.C., 1939.
33. U.S. Department of Health and Human Services and U.S. Department of Agriculture, "Ten-year Comprehensive Plan for the National Nutrition Monitoring and Related Research Program," *Federal Register* **58**, 32751–32806 (1993).
34. M. F. Kuczmarski and R. J. Kuczmarski, "Nutrition Monitoring in the United States," in M. E. Shils, J. S. Olson, and M. Shike, eds, *Modern Nutrition in Health and Disease*, Vol. 2, 8th ed., Lee & Febiger, Philadelphia, Penn., 1994, pp. 1506–1516.
35. M. F. Kuczmarski, A. Moshfegh, and R. Briefel, "Update on Nutrition Monitoring Activities in the United States," *J. Am. Dietetic Assoc.* **94**, 753–760 (1994).
36. U.S. Department of Health and Human Services and U.S. Department of Agriculture, *Nutrition Monitoring in the United States—A Report From the Joint Nutrition Monitoring Evaluation Committee*, DHHS Publication No. (PHS) 86-1255, U.S. Government Printing Office, Washington, D.C., 1986.
37. Federation of American Societies for Experimental Biology, Life Sciences Research Office, *Nutrition Monitoring in the United States—An Update Report on Nutrition Monitoring*, DHHS Pub. No. (PHS) 89-1255, Public Health Service, U.S. Government Printing Office, Washington, D.C., 1989.
38. Federation of American Societies for Experimental Biology, Life Sciences Research Office, *Third Report on Nutrition Monitoring in the United States*, U.S. Government Printing Office, Washington, D.C., 1995.

Katherine S. Tippett
Cecilia Wilkinson Enns
Alanna J. Moshfegh
USDA/ARS
Beltsville, Maryland

FOOD CROPS: NONDESTRUCTIVE QUALITY EVALUATION

Quality of food crops is not a single characteristic but comprises many properties or attributes. Quality encompasses sensory properties (appearance, texture, and flavor), nutritive values, chemical constituents, mechanical properties, functional properties, and defects. It is convenient to think of *quality* as the composite of intrinsic characteristics of the commodity and *acceptability* as people's perceptions of and reactions to those characteristics. Instrumental methods for measuring quality of food crops were introduced in the 1920s, with improvements and new technologies being introduced continuously. Instrumental methods generally measure chemical constituents or physical properties of the commodity that can be related directly or indirectly to functional behavior or sensory acceptability. Measurement by some instrumental methods involves preparation that destroys the sample, which restricts use to a limited number of random samples for predicting quality of the population. Nondestructive methods leave the product intact and undamaged so it can be stored or marketed after analysis. Nondestructive methods are preferred over destructive methods for commercial purposes and for monitoring quality over time in research programs. Various methods are available for measuring specific quality attributes nondestructively; a few are used commercially, some are used in postharvest research, and others are in the developmental stages.

PRINCIPLES USED FOR NONDESTRUCTIVE INSTRUMENTAL QUALITY EVALUATION

Nondestructive instrumental methods for measuring quality involve electromagnetic (visible and near infrared light, X-ray, and magnetic resonance), electrochemical, or mechanical (force, deformation, or vibration) responses. These principles are applied in the following methods of measurement.

DESCRIPTION AND APPLICATION OF METHODS

Electromagnetic Radiation

Electromagnetic wavelengths encompass, from longest to shortest, radio waves, microwaves, light, X-ray, and gamma waves. Light here refers to ultraviolet, visible, near infrared, and infrared wavelengths. Visible light is that portion of the electromagnet spectrum that humans can see. The visible and near infrared ranges have been most used for measuring quality, but X-ray and magnetic resonance permit deeper penetration of the tissues.

Light

Visible and Near Infrared Wavelengths. Each chemical bond in the chemical constituents of the tissue absorbs energy in a characteristic manner that causes the absorption of only certain wavelengths of light. Analysis of the relative amount of each wavelength of light that is reflected or transmitted through the tissues can inform us of the concentration of pigments, sugars, oils, and certain other chemical constituents within the tissue. Simple measurements may involve the difference or ratio between two specific wavelengths of light. Some defects are presently detected by using reflectance images taken at two wavelengths where sound and damaged tissue have different relative absorbances. Recently whole-spectrum data reduction techniques using partial least squares (PLS) or similar mathematical treatments are being implemented to reduce spectral data to a small number of basic *factors* that are then used in multiple regression equations to predict components. PLS has been particularly useful for near infrared data.

Reflectance, Transmittance, and Interactance. Light can be measured in reflectance, transmittance, or interactance (body reflectance) modes. Food crops are cellular tissues that are highly light scattering. Most of the light impinging on a food crop penetrates its surface and interacts with the chemical constituents and physical structure of the tissue. The small portion of light that is reflected directly at the surface is called specular reflectance or gloss and is not generally useful for measuring quality. The portion that is reflected after very shallow penetration is usually also called reflectance but is more properly termed diffuse reflectance. Reflectance from deeper within the tissues may be called interactance or body reflectance. Energy that passes entirely through the product is called body transmittance or direct transmittance, depending on whether the material does or does not scatter the light internally ("clear" juices or oils pressed from some fruits and vegetables cause little scattering).

The principle of light absorption by specific chemical bonds applies whether the measurement is reflectance or transmittance. The measurement of transmittance requires that the illuminating light must not reach the detector except by passing through the product. Such a light seal is difficult to achieve with irregularly shaped products like fruits and vegetables, especially at high speeds. Therefore body reflectance is preferred over transmittance measurement for on-line sorting operations that must handle each sample in a fraction of a second (about 5 to 12 fruit per second is currently desired for various products). However, the distribution within the product of pigments or other constituents of interest determines whether reflectance can be used or if transmittance is required.

The reactions of visible light with pigments result in the color we perceive when we view the product: chlorophyll looks green; anthocyanins give us shades of reds, purples, and blues; carotenoids are yellow, orange, or red. The pigment contents of many fruits and vegetables change as the product ripens, so ripeness often can be estimated by measuring pigment content or color. Additionally, color is often associated with consumer acceptability of the product. Reflectance from the outer few cells of the product is used in grading lines of packing houses to sort apples, carrots, citrus, nectarines, peaches, pears, plums, potatoes, and tomatoes according to defined color categories (1). Diffuse reflectance in the 640- to 750-nm wavelength range is used to predict maturity of lettuce, oranges, and peaches (2,3) based on the chlorophyll content, which has peak absorption at 680 nm. The difference in absorbance (optical density) at 690 nm and 740 nm (noted as ΔOD [690–740 nm]) was associated with ripeness of apple (4). Maturity of tomatoes is based on chlorophyll and lycopene contents, which can be predicted by ΔOD (710–780 nm) and ΔOD (570–780 nm) values, respectively (5). Maturity of blueberries, papayas, and peaches can also be predicted by transmittance based on pigment content. The effectiveness of ripeness sorting by these methods depends on the composition of the product, interfering absorbance by other components, and the relationship of pigment content to ripeness. In peaches the density of the stone as well as the amount of anthocyanin around the stone must be considered when using transmittance measurements of chlorophyll content. With apples, watercore may affect chlorophyll readings.

Near-infrared diffuse reflectance spectra are used to predict starch, protein, oil, and moisture contents of grains and oil seeds (6) and sugars in fruit (7,8). Dry matter content of onions and potatoes, mainly carbohydrates, is determined by near infrared wavelengths (9).

The composition and therefore the absorbance characteristics of healthy tissue differ from those of damaged or decayed tissues. Internal defects that can be detected by transmittance include hollow heart in potato tubers, core breakdown in pears, watercore in apples, freeze damage in citrus, and smut content on wheat. Hollow heart detection is based on ΔOD (800–710 nm), core breakdown on ΔOD (690–740 nm), watercore on ΔOD (810–760 nm), and smut on wheat on ΔOD (890–930 nm). Surface defects that can be quantified by reflectance (R) include bruise ($R_{400-450\ nm}$, $R_{770-890\ nm}$, and others) (10,11), scald, russet, and other defects of apples; rind defects on lemons ($R_{580-650\ nm}$); mold, scab, crack, and exposed pit on dried prunes (near infrared wavelengths) (12), and blight, soft rot, and scab in potato tubers (13).

Fluorescence and Delayed Light Emission. Fluorescence is light that is instantaneously emitted by an illuminated sample at a wavelength longer (lower energy) than the illuminating wavelength. Delayed light emission (DLE) is similar to fluorescence but is emitted after a time delay that allows chemical reaction to occur. In fruits and vegetables, chlorophyll exhibits fluorescence and DLE, both with characteristic kinetics (time courses). Fluorescence emission peaks characteristic of the photosystem II antennae (686 nm), reaction center (696 nm), and photosystem I antenna (730–740 nm) are used to study chlorophyll function during the growth and ripening periods of fruit (14). Because chlorophyll content is often related to maturity in fruits and vegetables, chlorophyll fluorescence and DLE have been proposed for measuring maturity of apricots, citrus, melons, papaya, persimmons, tea leaves, and tomatoes (15–18). Total, maximum, or changes in fluorescence or DLE can be used to detect exposure to physiological stress injury such as caused by chilling temperatures, high temperatures, air pollutants, and mechanical stress (15,16,19,20). Equilibration in the dark for up to 10 min may be required to obtain stable values of chlorophyll fluorescence or DLE; however, a relatively recent chlorophyll fluorescence technique, pulse amplitude modulated (PAM) fluorometry, does not require the dark period. PAM fluorescence has been used to follow development of chilling injury (21) and responses to heat treatments (22).

A fluorescence application *not* based on chlorophyll detected mechanical injury of oranges by measuring fluorescence from oils that leaked from damaged oil cells (23).

X ray. X-ray technology has been modified to recognize rapidly internal characteristics of commodities on grading lines or on mechanical harvesters. Absorption of X rays varies directly with density and water concentration. Commercial uses include detection of hollow heart of potatoes (24) and freeze damage in citrus (25). X ray is being studied for automated detection of bruises and watercore in apples (26), split pits in peaches, presence of pits in processed stone fruits, and insect infestation in fruits and nuts (27). In research, X ray computed tomography shows density changes in tomato locules during maturation (28) and internal sprouting and ring separations due to microbial rot in onion (29).

Magnetic Resonance. Nuclear magnetic resonance (NMR) detects concentration and mobility of hydrogen nuclei (and certain other nuclei) and indicates amounts of water, oil, and sugar. Of particular interest in horticultural applications, areas of greater free water content produce a stronger NMR signal than surrounding tissues, so that disorders involving water distribution can be detected (30). Magnetic resonance imaging (MRI) has been used to show morphology, ripening, seeds or pits, voids, pathogen invasion, worm damage, bruises, and other defects, as well as changes due to ripening, heat, chilling, and freezing (30–34). The major disadvantages of MRI are cost and speed. In a less expensive adaptation, a limited volume of tissue is averaged to detect concentrations of sugars or oils within a product (35,36). Currently, MR and MRI are not practical for routine quality testing; but, like all technologies, they are becoming cheaper, faster, and more feasible for research and specialized applications. Magnetic resonance detectors have great potential for evaluating internal quality of products.

Imaging. The electromagnetic technologies—light, X ray, and MR—can be applied using sensors that integrate a limited area or volume of the object being measured or using imaging technology (machine vision). Imaging provides spatial information. Most imaging is two-dimensional, much like the image on a computer monitor or television screen; X-ray CT and MRI can be three-dimensional. The more dimensions, the more expensive is the equipment and the more computationally intensive is the measurement. At present, light imaging is being used on-line for sorting apples, peaches, citrus, and some other commodities according to color, for some surface defects, and for sizing some commodities (11,37–39). Research is directed at improving speed, accuracy, and cost.

Electrochemical

The relative concentrations of volatile compounds (nonaromatic and aromatic) within a fruit or vegetable increase during maturation and ripening. The electrical conductivity of semiconductor gas detectors, based on different polymers and metal oxides, decreases on exposure to volatiles. A battery of several detectors can produce a profile that may indicate maturity or presence of some disorders. The electronic sniffer concept (40,41) has been tested on apples, blueberries, melons, and strawberries. Further research is needed to explore the selection of semiconductors and to relate the profiles to quality categories (42).

Mechanical

Quasi-Static Force and Deformation. Most force/deformation measurements are destructive, for example, the familiar Magness-Taylor fruit firmness test (a penetrometer) or the Kramer shear test (multiblade shear widely used in the processed foods industry), or too slow for on-line use, such as the Cornell firmness tester (a creep tester) (42). A nondestructive, noncontact firmness detector was recently patented (43) that uses a laser to measure the deflection caused by a short puff of pressurized air, similar to some devices used by ophthalmologists to detect glaucoma. Under fixed air pressure, firmer products deflect less than softer ones. This appears to be a quite localized measurement; that is, a very small portion of the total fruit or vegetable is actually tested. In early tests, laser-puff readings correlated fairly well with destructive Magness-Taylor firmness values for apple, cantaloupe, kiwifruit, nectarine, orange, pear, peach, plum, and strawberry.

Dynamic Force and Deformation

Impact. The impact response of a product is directly related to its mechanical properties, mass, and shape. Impact is the rapid collision of two objects, whether one or both objects are in motion. Impact testing techniques include drop, falling mass (or impact probe), and impact ram. A number of impact parameters have been proposed to measure horticultural product firmness, including peak force and time in contact with the impacting object. Impact responses have been studied for apples, blueberries, cherries, kiwifruit, peaches, pears, potatoes, and tomatoes (44–46). One potential problem with impact tests on fruits and vegetables is that some bruising may occur.

Vibration. Sonic (or acoustic) vibrations encompass the audible frequencies from 20 Hz to about 15 kHz and ultrasonic vibrations are above that range. When an object is excited by periodic forces, such as at sonic frequencies, or by certain types of impact it vibrates. Resonance (maximum vibration) occurs at certain frequencies that depend on the mechanical properties, size, shape, and density of the product. The firmer the flesh, the higher the resonant frequency for products of the same size and shape. The traditional watermelon ripeness test is based on the acoustic principle, where one thumps the melon and listens to the pitch (frequency) of the resonance. Resonant frequency, wave propagation velocity, attenuation, and reflection are the important parameters for evaluating texture of horticultural commodities. The sonic vibration method is nondestructive and is suitable for rapid firmness measurement. Sonic measurement generally represents the mechanical properties of the entire product, unlike puncture or compression tests that sample localized tissues. Sonic measurements are excellent for following changes in individuals over time in research applications and are suitable for determining average firmness of grower lots of fruit. Sonics may not be suitable for sorting operations as they have not always proved capable of predicting firmness of individual fruit as determined with a penetrometer (47–53). Some of the fruit tested have been apples, avocados, bananas, grapes, kiwifruit, head lettuce, mangoes, melons, peaches, pears, pineapples, and tomatoes.

Ultrasonic measurement has generally not been successful for quality measurement on fruits and vegetables despite its success in medicine and animal studies. The structure and air spaces in fruits and vegetables make it difficult to transmit sufficient ultrasonic energy through them to obtain useful measurements (54). Wave propagation velocity, attenuation, and reflection are the important ultrasonic parameters for quality evaluation of horticultural commodities. Skin texture of oranges and cracks in tomatoes could be evaluated by reflectance and backscatter, respectively (54). Hollow heart in potatoes was detectable (55), but bruises in apples could not be readily detected (56). Ultrasonic measurements correlated well with firmness or ripeness of melon and avocado (57). However, a more powerful ultrasonic source is required to penetrate most fruits and vegetables.

Sensor Fusion

There is growing recognition that quality is a multifaceted attribute, leading to a need for "sensor fusion" or the combining of several measurements into a quality classification (38,58). The measurements may be from multiple sensors examining different parts of the same fruit but more likely will be from different kinds of sensors that detect different characteristics. Sorting requires high-speed, nondestructive sensing of each piece of fruit or vegetable and a statistical procedure to combine the measurements into a classification decision. Currently, automated sorting is mainly by size, color, and, for some commodities, surface defects. Fruit packinghouses in the United States require sorter speeds of 5 to 12 fruit per second. Statistical combination of measurements will increase the likelihood of predicting overall quality. However, it is important to understand what each sensor measures and to understand the relationship of that characteristic to quality, so the limitations of the sorting method are appreciated.

FUTURE DEVELOPMENTS

Engineers and horticulturists continuously develop new, nondestructive methods for evaluating quality of fruits and vegetables. Some of the methods are incorporated into grading lines to sort the individual fruit or vegetables by one or several quality attributes. Some methods, such as MRI, are too costly or too slow for commercial, on-line implementation at present, but these methods are useful for horticultural research. And new sensing and computing technologies will likely render these sensors commercially economical in the future.

BIBLIOGRAPHY

1. M. J. Delwiche et al., "Surface Color Measurement of Fruits and Vegetables," in G. Brown and Y. Sarig, eds., *Nondestructive Technologies for Quality Evaluation of Fruits and Vegetables*, American Society of Agricultural Engineers, St. Joseph, Mich., 1994, pp. 63–71.
2. E. J. Brach et al., "Lettuce Maturity Detection in the Visible (380–720 nm), Far Red (680–750) and Near Infrared (800–1950) Wavelength Band (*Lactuca sativa*)," *Agron. Sci. Prod. Veg. et de l'Environ* **2**, 685–694 (1982).
3. Y. Chuma, T. Shiga, and K. Morita, "Evaluation of Surface Color of Japanese Persimmon Fruits by Light Reflectance Mechanized Grading Systems," *J. Soc. Agric. Machinery, Japan* **42**, 15–120 (1980).
4. K. L. Olsen, H. A. Schomer, and R. D. Bartram, "Segregation of Golden Delicious Apples for Quality by Light Transmission," *Proceedings of the American Society for Horticultural Science* **91**, 821–828 (1967).
5. A. E. Watada et al., "Estimation of Chlorophyll and Carotenoid Contents of Whole Tomato by Light Absorbance Technique," *J. Food Sci.* **41**, 329–332 (1976).
6. K. H. Norris, "Instrumental Techniques for Measuring Quality of Agricultural Crops," in M. Lieberman, ed., *Postharvest Physiology and Crop Preservation*, Plenum Press, New York, 1983, pp. 471–484.
7. S. Kawano, H. Watanabe, and M. Iwamoto, "Determination of Sugar Content in Intact Peaches by Near Infrared Spectroscopy With Fiber Optics in Interactance Mode," *Journal of the Japanese Society for Horticultural Science* **61**, 445–451 (1992).
8. D. C. Slaughter, D. Barrett, and M. Boersig, "Nondestructive Determination of Soluble Solids in Tomatoes Using Near Infrared Spectroscopy," *J. Food Sci.* **61**, 695–697 (1996).
9. G. S. Birth et al., "Nondestructive Spectrophotometric Determination of Dry Matter in Onions," *J. Am. Soc. Hortic. Sci.* **110**, 297–303 (1985).
10. B. L. Upchurch et al., "Spectrophotometric Study of Bruises on Whole Red Delicious Apples," *Transactions of the American Society of Agricultural Engineers* **33**, 585–589 (1990).
11. D. J. Aneshansley, J. A. Throop, and B. L. Upchurch, "Reflectance Spectra of Surface Defects on Apples," in *Sensors for Nondestructive Testing*, Proceedings of Sensors for Nondestructive Testing International Conference, Orlando, Fla., Feb. 18–21, 1997. NRAES (Northeast Reg. Agr. Eng. Svc.), Coop. Extn., Ithaca, N.Y., 1997, pp. 143–160.
12. M. J. Delwiche, S. Tang, and J. F. Thompson, "A High-speed Sorting System for Dried Prunes," *Transactions of the American Society of Agricultural Engineers* **36**, 195–200 (1993).
13. R. L. Porteous, A. Y. Muir, and R. L. Wastie, "The Identification of Diseases and Defects in Potato Tubers From Measurements of Optical Spectral Reflectance," *Journal of Agricultural Engineering Research* **26**, 151–160 (1981).
14. J. Gross and I. Ohad, "In vivo Fluorescence Spectroscopy of Chlorophyll in Various Unripe and Ripe Fruit," *Photochemistry and Photobiology* **37**, 195–200 (1983).
15. J. A. Abbott et al., "Technologies for Nondestructive Quality Evaluation of Fruits and Vegetables," *Horticultural Reviews* **20**, 1–120 (1997).
16. J. A. Abbott, W. R. Forbus, Jr., and D. R. Massie, "Temperature Damage Measurement by Fluorescence or Delayed Light

Emission from Chlorophyll," in G. Brown and Y. Sarig, eds., *Nondestructive Technologies for Quality Evaluation of Fruits and Vegetables*, American Society of Agricultural Engineers, St. Joseph, Mich., 1994, pp. 44–49.

17. Y. Chuma, M. Ohura, and A. Tagawa, "Delayed Light Emission as a Means of Sorting Tomatoes, DLE Characteristics of Tomatoes Excited by Flash Light," *Journal of the Faculty of Agriculture, Kyushu University* **26**, 159–167 (1982).
18. W. R. Forbus, Jr., G. G. Dull, and D. A. Smittle, "Nondestructive Measurement of Canary Melon Maturity by Delayed Light Emission," *Journal of Food Quality* **15**, 119–127 (1992).
19. R. M. Smillie, and R. Nott, "Assay of Chilling Injury in Wild and Domestic Tomatoes Based on Photosystem Activity of the Chilled Leaves," *Plant Physiol.* **63**, 796–801 (1979).
20. J. A. Abbott, A. R. Miller, and T. A. Campbell, "Detection of Mechanical Injury and Physiological Breakdown of Cucumbers Using Delayed Light Emission," *J. Am. Soc. Hortic. Sci.* **116**, 52–57 (1991).
21. S. Lurie, R. Ronen, and S. Meier, "Determining Chilling Injury Induction in Green Peppers Using Nondestructive Pulse Amplitude Modulated (PAM) Fluorometry," *J. Am. Soc. Hortic. Sci.* **119**, 59–62 (1994).
22. A. B. Woolf and W. A. Laing, "Avocado Fruit Skin Fluorescence Following Hot Water Treatments and Pretreatments," *J. Am. Soc. Hortic. Sci.* **121**, 47–151 (1996).
23. J. L. Uozumi et al., "Spectrophotometric System for Quality Evaluation of Unevenly Colored Food" [in Japanese], *Nippon Shokuhin Kogyo Gakkaishi* [Journal of the Japanese Society of Food Science and Technology] **34**, 163–170 (1987).
24. E. E. Finney and K. H. Norris, "X-ray Scans for Detecting Hollow Heart in Potatoes," *American Potato Journal* **55**, 95–105 (1978).
25. B. L. Upchurch et al., "Detection of Internal Disorders," in G. Brown and Y. Sarig, eds., *Nondestructive Technologies for Quality Evaluation of Fruits and Vegetables*, American Society of Agricultural Engineers, St. Joseph, Mich., 1994, pp. 80–85.
26. E. W. Tollner, J. K. Brecht, and B. L. Upchurch, "Nondestructive Evaluation: Detection of External and Internal Attributes Frequently Associated with Quality or Damage," in S. E. Prussia and R. L. Shewfelt, eds., *Postharvest Handling: A Systems Approach*, Academic Press, New York, 1993, pp. 225–255.
27. T. F. Schatzki et al., "Defect Detection in Apples by Means of X-ray Imaging," in *Sensors for Nondestructive Testing*, Proceedings of Sensors for Nondestructive Testing International Conference, Orlando, Fla., Feb. 18–21, 1997, NRAES (Northeast Reg. Agr. Eng. Svc.), Coop. Extn., Ithaca, N.Y., 1997, pp. 161–171.
28. J. K. Brecht et al., "Using X-ray Computed Tomography (X-ray Ct) to Nondestructively Determine Maturity of Green Tomatoes," *HortScience* **26**, 45–47 (1991).
29. E. W. Tollner et al., "Nondestructive Testing for Identifying Poor Quality Onions," in J. A. DeShazer and G. E. Meyer, eds., *Optics in Agriculture, Forestry, and Biological Processing*, Proceedings of the Society of Photo-Optical Instrumentation Engineers 2345, International Society for Optical Engineering, Bellingham, Wa., 1995, pp. 392–397.
30. C. J. Clark et al., "Application of Magnetic Resonance Imaging to Pre- and Post-Harvest Studies of Fruit and Vegetables," *Postharvest Biology and Technology* **11**, 1–21 (1997).
31. M. Saltveit, Jr., "Determining Tomato Fruit Maturity with Nondestructive in Vivo Nuclear Magnetic Resonance Imaging," *Postharvest Biology and Technology* **1**, 153–159 (1991).
32. J. S. MacFall and G. A. Johnson, "The Architecture of Plant Vasculature and Transport as Seen with Magnetic Resonance Microscopy," *Canadian Journal of Botany* **72**, 1561–1573 (1994).
33. K. Akimoto, W. F. McClure, and K. Shimizu, "Non-destructive Evaluation of Vegetable and Fruit Quality by Visible Light and MRI," *Proceedings of Automation and Robotics in Bioproduction and Processing* **1**, 117–124 (1995).
34. B. Zion, M. J. McCarthy, and P. Chen, "Real-time Detection of Pits in Processed Cherries by Magnetic Resonance Projections," *Lebensm.-Wiss. Technol.* **27**, 457–462 (1994).
35. P. Chen et al., *Development of a High-speed NMR Technique for Sensing Fruit Quality*, Paper 95-3613, American Society of Agricultural Engineers, St. Joseph, Mich., 1995.
36. S. I. Cho et al., "Nondestructive Sugar Content Measurements of Intact Fruit Using Spin-spin Relaxation Times (T_2) Measurements by Pulsed ^{1}H Magnetic Resonance," *Transactions of the American Society of Agricultural Engineers* **36**, 1217–1221 (1993).
37. P. H. Heinemann et al., "Machine Vision Inspection of Golden Delicious Apples," *Applied Engineering in Agriculture* **11**, 901–906 (1995).
38. M. J. Delwiche, S. Tang, and J. F. Thompson, *Prune Defect Detection by Line-scan Imaging*, Paper 88-3024, American Society of Agricultural Engineering, St. Joseph, Mich., 1988.
39. B. K. Miller and M. J. Delwiche, *A Color Vision System for Peach Grading*, Paper 88-6025, American Society of Agricultural Engineering, place of publication 1988.
40. M. Benady et al., "Fruit Ripeness Determination by Electronic Sensing of Aromatic Volatiles," *Transactions of the American Society of Agricultural Engineers* **38**, 251–257 (1995).
41. J. E. Simon "Electronic Sensing of Aromatic Volatiles for Quality Sorting of Blueberry Fruit," *J. Food Sci.* **61**, 967–969, 972 (1996).
42. E. Kress-Rogers, ed., *Handbook of Biosensors and Electronic Noses: Medicine, Food, and the Environment*, CRC Press, Boca Raton, Fla., 1997.
43. A. R. Hamson, "Measuring Firmness of Tomatoes in a Breeding Program," *Proceedings of the American Society for Horticultural Science* **60**, 425–433 (1952).
44. U.S. Pat. 5,372,030 (Dec. 13, 1994), S. E. Prussia, J. J. Astleford, B. Hewlett, and Y. C. Hung (to University of Georgia Research Foundation).
45. M. J. Delwiche, S. Tang, and J. J. Mehlschau, "An Impact Force Response Fruit Firmness Sorter," *Transactions of the American Society of Agricultural Engineers* **32**, 321–326 (1989).
46. V. A. McGlone and P. N. Schaare, *The Application of Impact Response Analysis in the New Zealand Fruit Industry*, Paper 93-6537 American Society of Agricultural Engineers, St. Joseph, Mich., 1993.
47. J. Sugiyama, K. Otobe, and Y. Kikuchi, *A Novel Firmness Meter for Fruits and Vegetables*, Paper 94-6030, American Society of Agricultural Engineers, St. Joseph, Mich., 1994.
48. J. A. Abbott et al., "Nondestructive Sonic Firmness Measurement of Apples," *Transactions of the American Society of Agricultural Engineers* **38**, 1461–1466 (1995).
49. J. A. Abbott and D. R. Massie, "Nondestructive Sonic Measurement of Kiwifruit Firmness," *J. Am. Soc. Hortic. Sci.* **123**, 317–322 (1998).
50. P. R. Armstrong, H. R. Zapp, and G. K. Brown, "Impulsive Excitation of Acoustic Vibrations in Apples for Firmness Determination," *Transactions of the American Society of Agricultural Engineers* **33**, 1353–1359 (1990).
51. K. Peleg, U. Ben-Hanan, and S. Hinga, "Classification of Avocado by Firmness and Maturity," *J. Texture Stud.* **21**, 123–129 (1990).
52. M. L. Stone et al., *Watermelon Maturity Determination in the Field With Acoustic Impedance Techniques*, Paper 94-6024, American Society of Agricultural Engineers, St. Joseph, Mich., 1994.

53. I. Shmulevich, N. Galili, and N. Benichou, "Development of a Nondestructive Method for Measuring the Shelf-life of Mango Fruit," *Proceedings Food Processing Automation IV Conference*, Chicago, Ill., Nov. 3–5, 1995.
54. H. Yamamoto, M. Iwamoto, and S. Haginuma, "Nondestructive Acoustic Impulse Response Method for Measuring Internal Quality of Apples and Watermelons," *Journal of the Japanese Society for Horticultural Science* **50**, 247–261 (1981).
55. N. Sarkar and R. R. Wolfe, "Potential of Ultrasonic Measurements in Food Quality Evaluation," *Transactions of the American Society of Agricultural Engineers* **26**, 624–629 (1983).
56. Y. Cheng and C. G. Haugh, "Detecting Hollow Heart in Potatoes Using Ultrasound," *Transactions of the American Society of Agricultural Engineers* **37**, 217–222 (1994).
57. B. L. Upchurch et al., "Ultrasonic Measurement for Detecting Apple Bruises," *Transactions of the American Society of Agricultural Engineers* **30**, 803–809 (1987).
58. N. Galili, A. Mizrach, and G. Rosenhouse, *Ultrasonic Testing of Whole Fruit for Nondestructive Quality Evaluation*, Paper 93-6026, American Society of Agricultural Engineers, St. Joseph, Mich., 1993.
59. N. Ozer, B. Engel, and J. Simon, "Fusion Classification Techniques for Fruit Quality," *Transactions of the American Society of Agricultural Engineers* **38**, 1927–1934 (1995).

JUDITH A. ABBOTT
USDA/ARS
Beltsville, Maryland

FOOD CROPS: POSTHARVEST DETERIORATION

Harvest is a major event for any food that is derived from plants. As a plant part is severed from the plant it loses its source of supply of nutrients and its repository for metabolic waste products. Until the detached plant part undergoes conventional food processing, it continues to live, respire, transpire, and senesce, ultimately leading to death. Postharvest deterioration continues until the item is either processed or consumed. Handling techniques have been developed to slow the physiological processes to provide a product that is satisfactory to the consumer.

Food crops are categorized as agronomic or horticultural. Agronomic (field) crops are primarily the grains and oilseeds that are harvested in a dry to semidry state and tend to be relatively stable to handling and storage as long as they are protected from moisture and insects. Horticultural (garden) crops, which comprise fruits, vegetables, and nuts, tend to be much more perishable, requiring sophisticated handling systems to transport them from field to consumer.

Estimates vary widely on how much of a crop is actually consumed (1,2). Processing techniques such as canning, freezing, and drying are designed to minimize these losses and extend the length of the season they are available for consumption. Postharvest handling techniques that manipulate the storage environment extend the life of the product while keeping it in a fresh state. Minimal processes such as cutting, slicing, and dicing increase the appeal and convenience of the item frequently at the expense of greater perishability (3). Losses of edible product begin in the field during harvesting and loading; continue during transport to the processing plant, packinghouse, or market; during storage at any point in the distribution scheme; and during food preparation or even by the consumer at the point of consumption. Losses may be complete resulting from the discarding of part (removal of outer leaves) or all (discarding a rotten fruit) of a given item. Frequently losses are more subtle and less tangible such as loss of acceptability, nutritional quality, or economic value. An understanding of the scope of losses incurred for a particular fruit or vegetable requires an understanding of the complexity of the handling and distribution system for that item (4,5).

Key concepts provide insight into the perishability of most fresh horticultural crops. Physiological deterioration of a fresh item begins at harvest and continues until processing or consumption. Respiration is the metabolic breakdown of food constituents to release the necessary energy to sustain the healthy tissue. Transpiration is the release of moisture from the surface of the fruit or vegetable. Senescence is genetically programmed deterioration that leads to cell and tissue death (4). Quality refers to the properties of a particular fruit or vegetable that make it unique and influence its purchase and consumption by the consumer. Shelf life is the time period a product can be maintained at an acceptable level of quality (5).

Although most harvested products are at their peak of quality at harvest, climacteric fruits continue to ripen after detachment from the plant. These fruits (eg, apples, bananas, pears, and tomatoes) will develop color, flavor, and textural attributes during postharvest storage. In many cases climacteric fruits do not develop full flavor off the plant, but the perishability of the fully ripe fruits precludes distribution.

CAUSES OF LOSSES

Many factors contribute to losses of fresh products during postharvest handling and storage. Mechanical injury results in cuts or bruises that decrease purchase acceptability. Such damage may not be immediately evident as softening and discoloration associated with bruising takes time to develop. Thus it is frequently difficult to determine the cause of the injury and how to take corrective action.

Microbes can invade plant tissue, particularly when it has been mechanically damaged. Until recently, postharvest pathologists were primarily interested in plant pathogens present in the field or introduced during handling that lead to decay and spread from item to item within bulk containers. More recently, introduction of human pathogens by contamination from irrigation water, untreated animal manure, or association with raw meats or their exudates has led to safety concerns in fruits and vegetables that are not thoroughly washed or cooked prior to consumption (5).

As mentioned, physiological processes occur in the fruit or vegetable after detachment from the plant. Increased respiration accelerates tissue degradation and can lead to flavor development, which may be desirable in the form of ripe-fruit flavor or may be undesirable off-flavors. Excess transpiration results in shriveling, wilting, or loss of turgor causing rejection of squash, lettuce, or broccoli, and so on.

Physiological disorders can develop during storage from preharvest nutritional deficiencies, postharvest stress conditions, or an interaction of preharvest and postharvest factors leading to unacceptable quality or reduced shelf life.

Some fruits and vegetables are susceptible to chilling injury, which develops at low temperatures above the freezing point. Chilling injury produces different symptoms in different crops including abnormal ripening, surface lesions or pitting, increased susceptibility to decay, browning discoloration, and off-flavor development. Chilling-susceptible products include bananas, beans, cucumbers, grapefruit, melons, and tomatoes (6). Other physiological disorders can result from improper mineral nutrition during growth and development; during exposure to low levels of oxygen (O_2), high levels of carbondioxide (CO_2), or high levels of ethylene (C_2H_4) during storage; or an interaction of these and other factors.

As consumer demand increases for more fresh fruits and vegetables to meet nutritional concerns and provide more convenience to fit them into a busy lifestyle, more fresh items are sold in a cut or otherwise minimally processed form. Cutting tends to break apart cells, leading to loss of cell material; accelerate respiration, leading to more rapid degradation of the tissue; and increase surface area, resulting in surface evaporation. Composition of the gaseous atmosphere that is most likely to preserve quality of these products can result in the production of off-odors and flavors. Cutting appears to decrease the losses due to chilling injury in susceptible fruits and vegetables.

PREVENTION OF LOSSES

The goal of postharvest handling is to minimize the loss of product quality between harvest and either processing or consumption. The first line of defense is prevention of physical injury. Environmental conditions during storage and handling represent the next line of defense. Addition of chemical compounds can prevent losses. Finally, sophisticated packaging techniques are being developed and used that employ one or more of the preceding lines of defense.

Physical protection of products from injury is used to minimize mechanical damage. Reduction of mechanical injury can be achieved by decreasing the number and height of falls of individual items or containers during handling. When falls are unavoidable, damage due to impact can be minimized by cushioning, which can be achieved by foam padding, liquid foam, water, or even product that will be discarded. Impact damage of one fruit on another can be reduced by decreasing dumping operations and using spacer bars in conveyor lines to minimize fruit-to-fruit contact. Vibration damage tends to increase with an increase in the size of bulk handling and the distance from field (or orchard) to packing facility. Packing the product in wholesale or retail packages close to the field, reducing or eliminating tractor-drawn vehicles from field to packinghouse, and use of paved roads for transport will reduce vibration damage.

Likewise, microbial decay and insect damage can be reduced by physical protection. Since mechanical damage frequently provides a route for invasion, any means of mechanical protection will help decrease microbial and insect damage. Once endogenous protective barriers such as rind or peel are penetrated, chances for future losses are increased. A greater incidence of pests and disease in the field will be reflected in greater problems during postharvest handling. By leaving these problems in the field, similar problems during handling and storage will be reduced. Physical removal of diseased or infected items during sorting or grading to prevent the spread of pathogens reduces subsequent infections. To prevent cross-contamination from human pathogens, all areas in which fresh fruits and vegetables are handled should be kept free from insects, birds, their droppings, and any raw animal products. The relatively short handling periods of fruits and vegetables are such that insect infestation is not usually a major problem. With grain products, however, storage times are long, and insects, and the microbes they deposit during their visits, pose a more serious threat. Physical barriers are an important part of an insect-protection plan.

In addition, screening products from light affects quality. Light enhances chlorophyll breakdown and thus speeds yellowing of green vegetables. Light also enhances chlorophyll synthesis in nongreen vegetables such as potatoes. This greening is a quality defect as it is an indicator of light-catalyzed synthesis of toxic alkaloids such as solanine. Light enhances β-carotene synthesis in tomatoes, but the effect on color quality is much more significant preharvest than postharvest.

Manipulation of environmental conditions is also an important tool available to postharvest handlers. In general, lowering the temperature while maintaining high relative humidity increases the shelf life of a product by reducing the rates of respiration and transpiration. Composition of gaseous atmosphere can be either modified or controlled in the storage room, container, or consumer package to slow ripening and senescence.

Proper temperature control is the most important tool in preventing postharvest losses. As the temperature is lowered, rates of respiration and transpiration decrease. The growth of microorganisms is also slowed by lower temperatures. For most fresh products, storage at temperatures as close to freezing as possible will extend shelf life. Freezing should be avoided as inadvertent freezing and thawing of fresh items leads to breakage of cell membranes and loss of desirable texture. Quick cooling after harvest to remove field heat is imperative in items like strawberries and green vegetables that respire rapidly and perish quickly. Hydrocooling and icing are used for products that can withstand water, but water is an excellent vehicle for spreading microorganisms. Forced-air cooling is another effective method, while vacuum-cooling is used for high-value items with a large surface area like lettuce. Slower cooling such as room cooling is permissible for products being stored for a longer time such as apples, but the final temperature should be as close to optimal as possible. When calculating refrigeration requirements, it must be remembered that respiring plant material evolves heat, known as the heat of respiration.

Prevention of chilling injury can be achieved by storing susceptible commodities at temperatures above the critical storage temperature, which ranges from 4°C for snap beans to 15°C for bananas. A complete list of optimal storage temperatures is available (7). In commercial practice a compromise temperature between 5 and 10°C is frequently used to store most fresh items. At this temperature it is assumed that damage to chilling-susceptible product will be minimal while the decrease in shelf life to nonsusceptible items will not be economically significant. Ice is usually added to green vegetables to lower the temperature and increase relative humidity (RH) without changing room temperature. The success of these strategies depends on a rapid turnaround of fresh product to minimize losses.

Maintenance of a high RH lowers transpiration of heavily transpiring products. Just as each commodity has an optimal storage temperature, it also has an optimal RH. If the RH is too high, microbial growth is enhanced. If it is too low, shriveling or wilting can result. Rapid changes in temperature of a product can lead to condensation on the surface and increased susceptibility to microbial decay.

Food additives are effective agents for the protection of plant products, but they are coming under greater scrutiny as consumers become more wary of chemicals. Many currently used compounds are under regulatory review and the approval of new compounds is unlikely. Microbial inhibitors help prevent the growth of spoilage microorganisms. Fumigants have been used to disinfest products from insects. External waxes are applied to porous fruit such as citrus fruits and cucumbers to slow water loss. These waxes also enhance appearance by providing gloss.

Although not a chemical as such, food irradiation induces chemical changes similar to conventional processing and is considered a food additive by the FDA. Low-dose irradiation has been approved in many countries for the inhibition of sprouting in potatoes and onions, insect disinfestation, and shelf-life extension. In some crops like strawberries, irradiation is effective in extending shelf life, but in others damage is induced at doses lower than effective for extension. Irradiation appears to be a safer technique for disinfestation than chemical fumigation, but questions of consumer acceptance of irradiated product have limited willingness of the food industry to adopt it as a widespread technique.

Shelf life of fresh products may also be extended by modification of the composition of atmospheric gases. Respiration and other metabolic processes are slowed with a decrease in oxygen and an increase in carbon dioxide. In some crops such as apples, pears, and onions, long-term storage is enhanced by controlling the atmosphere. Other crops such as lettuce and most root crops are susceptible to CO_2. Optimal storage atmospheres for crops have been published (7). Controlled-atmosphere storage usually occurs in large storage rooms where the gaseous atmosphere is monitored and changed to maintain the desired composition. In modified-atmosphere storage the initial gaseous composition is established but changes as respiration leads to decreased O_2 and increased CO_2 in the container. Atmosphere modification is very effective at maintaining texture and appearance but can lead to the development of off-flavors. Modified atmosphere packaging (MAP) is the major technique used to preserve fresh-cut products. MAP is most effective when used in conjunction with temperature reduction. In fresh-cut vegetable products like lettuce, the consumer has shown a willingness to sacrifice some losses in fresh flavor for added convenience. It is not clear that the consumer is willing to make the same sacrifice for fresh-cut fruits.

Advances in film technology have introduced greater sophistication in packaging of fresh products. Packaging protects the product by confining it, preventing contamination, shielding from mechanical damage and pests, permitting atmosphere modification, and providing instructions for optimal handling. The type of container used and its function can change as the item moves through handling and distribution. In general, the fewer handling steps and product transfers, the less opportunity for damage.

Plastics are being widely used in the packaging of fresh fruits and vegetables. They are employed at the pallet level for containerization and prevention of moisture loss around as well as used for MAP within shipping cartons, retail packages, or even individual items. Barrier films have different transmission properties to permit or exclude specific gases such as water vapor, CO_2, O_2 and C_2H_4, depending on specific requirements of the individual items. These films permit in-package atmosphere modification, extending the advantages of the technology to the supermarket shelves. Determination of the best initial composition of gases has been limited by the variation of the individual items in response to differing atmospheres. While accumulation of CO_2 and C_2H_4 could be detrimental to product quality, absorbers of these compounds can be included in sachets enclosed in the package or imbedded in the packaging material itself (8).

A logical extension of MAP is shrinking of the film tightly around the individual produce item. Although generally considered a type of modified atmosphere, individual plastic films are really more analogous to externally applied waxes, which change the diffusion properties of the item with the external atmosphere. These films slow transpiration while modifying the internal gaseous composition. Shelf life of some products like lemons is dramatically extended while other products develop off-odors and off-flavors due to altered metabolism. Edible films such as sucrose polyesters and proteins permit moisture control and changes in respiration of whole and cut products either as a retail item or within a retail package (9).

BIBLIOGRAPHY

1. M. C. Bourne, *Post Harvest Food Losses—The Neglected Dimension in Increasing the World Food Supply*, Cornell International Agriculture Mimeograph 53, Ithaca, N.Y., 1977.
2. E. B. Pantastico and O. K. Bautista, "Postharvest Handling of Tropical Vegetable Crops," *HortScience* **11**, 122–124 (1976).
3. R. C. Wiley, ed., *Minimally Processed Refrigerated Fruits and Vegetables*, Chapman & Hall, New York, 1994.
4. A. A. Kader et al., *Postharvest Technology of Horticultural Crops*, University of California, Davis, Calif., 1992.
5. R. L. Shewfelt and S. E. Prussia, eds., *Postharvest Handling: A Systems Approach*, Academic Press, San Diego, Calif., 1993.
6. C. Y. Wang, "Chilling Injury of Tropical Horticultural Commodities," *HortScience* **29**, 986–988 (1994).
7. R. E. Hardenburg, A. E. Watada and C. Y. Wang, *The Commercial Storage of Fruits, Vegetables, and Florist and Nursery Stocks*, USDA Agriculture Handbook 66, Washington, D.C., 1986.

8. T. P. Labuza, "An Introduction to Active Packaging for Foods," *Food Technol.* **50**, 68–71 (1996).
9. J. Krotcha, E. A. Baldwin, and M. O. Nisperos-Carriedo, *Edible Coatings and Films to Improve Food Quality*, Technomic Publishing Co., Lancaster, Penn., 1994.

GENERAL REFERENCES

J. K. Brecht, "Physiology of Lightly Processed Fruits and Vegetables," *HortScience* **30**, 18–22 (1995).

A. C. Cameron, P. C. Talasila, and D. W. Joles, "Predicting Film Permeability Needs for Modified-Atmosphere Packaging of Lightly Processed Fruits and Vegetables," *HortScience* **30**, 25–34 (1995).

I. J. Church and A. L. Parsons, "Modified Atmosphere Packaging Technology—A Review," *J. Sci. Food Agric.* **67**, 143–152 (1995).

R. B. How, *Marketing Fresh Fruits and Vegetables*, Chapman and Hall, London, United Kingdom, 1991.

S. J. Kays, *Postharvest Physiology of Perishable Plant Products*, Chapman and Hall, London, United Kingdom, 1991.

B. Ooraidul and M. E. Stiles, *Modified Atmosphere Packaging of Food*, John Wiley & Sons, New York, 1991.

D. S. Smith et al., eds., *Processing Vegetables: Science and Technology*, Technomic Publishing Co., Lancaster, Penn., 1997.

C. Y. Wang, *Alleviation of Chilling Injury of Horticultural Crops*, CRC Press, Boca Raton, Fla., 1990.

ROBERT L. SHEWFELT
University of Georgia
Athens, Georgia

FOOD CROPS: STORAGE

This article describes various factors to be considered during the storage of fresh fruits and vegetables and some supplemental methods used with cold storage. Space limitations preclude detailed prescriptions of storage for specific crops. This information is available in other publications (1–4).

Tissues of fruits and vegetables after harvest are still alive and continue to respire, metabolize, and change. Successful storage of fruits and vegetables extends the marketing period of these commodities and minimizes economic losses because most of the crop is utilized. Good storage conditions will maintain crop quality, reduce deterioration, render excellent salable commodities, and avoid causing injury to tissues by chilling, freezing or ammonia fumes.

FACTORS AFFECTING STORAGE LIFE

The most important factor affecting storage life is the quality of the product at harvest. Other factors to be considered during storage are temperature, humidity, and sanitation. Optimal storage conditions for fresh food crops will change from time to time, as changes in technology and popularity of different cultivars occur.

Product Quality

The quality of produce at harvest greatly affects the length of storage. The commodity should be harvested at the recommended stage of maturity (which differs with commodity) for maximal storage and should be free of physical damage, as skin breaks and bruises can lead to increased susceptibility to water loss and to bacterial and fungal attack. The extent of incipient infections by pathogens should be determined prior to storage, because such infections can cause rot and decay during storage.

Other factors to be considered in the determination of storage life and conditions are cultivar, environmental conditions during growth, cultural practices, handling practices before storage, and transit conditions. Allowances for improper handling, unfavorable growing conditions, and poor or long-distance transit conditions should be made in determining optimal storage conditions and predicting storage life. Furthermore, different parts of plants are used as food, including roots, stems, fruits, flowers, and leaves, and these parts have different storage requirements.

Respiration

Plant tissues respire, using oxygen in the conversion of carbohydrates to carbon dioxide, water, and energy. The energy is primarily given off as heat. Respiration rates and the amount of heat released increase as storage temperature increases, about two to three times for every 10°C (18°F) rise in temperature. The storage life of a commodity is inversely proportional to the rate of respiration. Therefore, as temperature increases, respiration rate increases, and storage life decreases. Respiration rates differ with plant species, cultivar, morphological tissue, and growing conditions. Rates gradually decrease after harvest for many fruits and vegetables, but in certain crops respiration rates gradually increase as they ripen, peak to a climacteric, and then decrease (5).

Temperature

Temperature affects the rates of produce metabolism and respiration, spore germination and pathogen growth, softening, moisture loss, and undesirable growth, such as sprouting. For every 10°C increase in temperature, the rate of deterioration or undesirable growth increases two to three times. Refrigerated storage is therefore recommended for many perishable commodities to slow the changes listed above (6). For optimal storage, storage room temperatures should be held constant, with variations of less than 1–2°C. The recommended storage temperature for many crops is 0°C, while temperatures below 0°C will cause freezing damage. Crops sensitive to chilling should be stored at 12.5°C; temperatures below 12.5°C but above 0°C will injure these commodities. Storage temperatures higher than optimal will shorten the expected shelf life of the produce. Temperature fluctuations can cause condensation on the product, leading to an increase in pathogen growth and subsequent decay.

Precooling

The first step in good temperature management of perishable commodities is precooling after harvest and before shipment, storage, or processing (7–10). Prompt precooling can reduce commodity respiration, water loss, metabolism, and pathogen growth. This is done using special equipment and rooms, when rapid cooling is essential and economically beneficial. Precooling can be accomplished in refrigerated storage rooms equipped with proper control of airflow.

Precooling methods quickly transfer heat from the produce to a cooling medium, such as water, ice, or air. The rate of transfer depends on the availability of the cooling medium to the commodity, the temperature difference between the medium and the commodity, the type of medium, and the flow rate of the medium. The commercial methods used include hydrocooling, vacuum cooling, air cooling, and icing. The methods used depend on the type of produce, the availability of equipment, facilities, containers, cost, and the proximity to market. See Table 1 for a list of methods of cooling for different commodities.

Humidity

Maintenance of proper humidity around the commodity is necessary during precooling and storage. Fruits and vegetables transpire and lose water during precooling, storage, and transit. They will wilt or shrivel if humidity is too low. Humidity is generally expressed as percent relative humidity (% RH), the ratio of water vapor pressure in the air to saturation vapor pressure at a certain temperature (11). As temperature increases, the capacity of air to hold water increases. For most vegetables, a relative humidity of 98–100% are recommended to prevent wilting. Water may be sprinkled on the floor of a storage room or misted in the air if it is necessary to increase the relative humidity around a commodity. Other practices to ensure adequate humidity include using good room insulation, preventing air leaks, and providing sufficient cooling surface to keep the difference between refrigerant surface temperature and commodity temperature as small as possible.

Sanitation

Sanitary conditions must be maintained in storage facilities to prevent the growth of contaminating organisms. Under high humidity conditions, molds grow on the surfaces of packages, walls, and ceilings, and although they may not directly decay produce, they may produce unwanted substances that accelerate decay, give off-flavors to the commodity, or create favorable environments for decay organisms. Periodic and thorough cleaning of storage surfaces and good air circulation help prevent the growth of surface molds. Whitewash can be sprayed on ceilings and walls when storage rooms are empty and before a new load comes in. Surfaces may also be scrubbed with sodium hypochlorite or trisodium phosphate, with subsequent rinsing and airing before new commodities are loaded. Fungicidal paint or fumigation with a mixture of 85% carbon dioxide and 15% ethylene can also be used. If odors have developed in a room, air purification can be accomplished by absorbing the odors with 6–14 mesh-activated coconut-shell carbon or washing the air with water (12).

Table 1. Methods of Cooling for Different Commodities

Cooling method	Commodities cooled
Hydrocooling	Asparagus, cantaloupes, celery, cucumbers, green peas, peaches, peppers, radishes
Vacuum cooling	Asparagus, artichokes, broccoli, Brussels sprouts, cabbage, cauliflower, celery, endive, escarole, head lettuce, parsley, spinach, sweet corn
Air cooling	Cauliflower, cucumbers, grapes, melons, peppers, strawberries, tomatoes
Icing	Broccoli, Brussels sprouts, cantaloupes, carrots, green onions, kale, radishes

INJURIES

Certain injuries to produce can occur during refrigerated storage under suboptimal conditions and include those due to chilling, freezing, or ammonia. Chilling injury usually occurs to commodities of tropical origin at temperatures below 10–13°C (50–55.4°F), although certain temperate commodities are also susceptible (13–15). Table 2 lists some commodities susceptible to chilling injury and their symptoms. Often symptoms cannot be detected at low temperatures and only become visible several days after the commodity has been removed to warmer temperatures. The extent of damage depends on length of exposure and chilling temperature. It may occur in a commodity exposed a short amount of time to a temperature considerably below the danger zone, but not if it is exposed longer to a higher temperature in the danger zone. Also, chilling injury may be cumulative, and chilling during transit or in the field before harvest may add to the effects of storage chilling. Chilling injury has been reduced by treatments such as intermittent warming, temperature preconditioning, controlled atmosphere or hypobaric storage, waxing, and film packaging.

Freezing injury occurs when ice crystals form in produce tissues, and is manifested by a mushy, limp, and water-soaked appearance. The extent of freezing injury depends on exposure time and temperature and varies with different commodities.

Ammonia injury occurs when ammonia escaping from direct-expansion refrigeration units come in contact with the commodity. Injury is evident as brown to greenish black discoloration of the outer tissues and, in severe cases, as discoloration and softening of inner tissues (16). Daily

Table 2. Commodities Susceptible to Chilling Injury and Their Symptoms

Commodity	Lowest safe temperature (°C)	Symptoms
Avocados	4.5–13.0	Grayish brown flesh discoloration
Bananas	11.5–13.0	Dull color when ripe
Cantaloupes	2–5	Pitting, surface decay
Cucumbers	7	Pitting, decay
Eggplants	7	Surface scald, *Alternaria* rot, black seeds
Lemons	11–13	Pitting, red blotch
Peppers	7	Sheet pitting, *Alternaria* rot on pods and calyxes, seed blackening
Potatoes	10	Browning, sweetening
Tomatoes, ripe	7–10	Water soaking, softening, decay

Source: Ref. 1.

odor checks or installation of ammonia detection systems help prevent ammonia injury to produce. Fumes can be removed by aeration and washing the air of the storage room with water.

SUPPLEMENTAL METHODS TO COLD STORAGE

Although refrigeration is the most effective method for retarding decay of food crops, supplements can be used to control the growth of pathogens, slow respiration rates, control physiological disorders and sprouting, reduce moisture loss, or retard ripening and senescence. Supplemental methods in use include chemical treatments, controlled- and modified-atmosphere storage, waxes, irradiation, and protective packaging.

Chemical treatments are used if they are nontoxic to humans and the commodity, cost-effective, effective for their intended purpose, and approved by federal regulatory agencies, for example, the use of growth regulators to control sprouting of potatoes and onions during storage. Calcium chloride is used to control bitter pit and maintain firmness in apples. Fungicides and bactericides, such as chlorine and sodium *o*-phenylphenate, are often added to wash water, soak tanks, and hydrocoolers to reduce growth of pathogens prior to and during refrigerated storage (17).

In controlled-atmosphere (CA) storage, commodities are held in atmospheres of specified proportions of oxygen, carbon dioxide, or nitrogen, different from ambient air. This is accomplished by adding or scrubbing oxygen or carbon dioxide in airtight containers or rooms (18–21). In modified-atmosphere storage, storage atmospheres are different from that of ambient air, but are not controlled precisely.

Waxing is used to reduce moisture loss and improve the appearance of the commodity. Fungicides can also be mixed with the wax to retard decay. The effectiveness of the wax depends on the thickness of the coat and the uniformity of application.

Gamma irradiation has been used to disinfect papayas, mangoes, and grapefruit. Commercial use is limited because of the cost and size of necessary equipment and consumer acceptance. Dosages of 1.5–2 kilogray can control decay and sprouting in certain commodities (22), but can also cause discoloration, pitting, softening, abnormal ripening, and flavor loss (23).

Protective packaging safeguards produce against physical damage during packing, such as that acquired from dropping of containers, overfilling of containers, or movement of the commodity within containers (24). Protective packaging must be sturdy and moisture tolerant. Packaging materials include plastic foam, waxed fiberboard, film box liners, wooden pallet bins, nailed wooden boxes, wirebound veneer crates, and perforated polyethylene or polypropylene bags.

FUTURE DEVELOPMENTS

Technologies are continually being researched and developed to improve the storage life of food crops. Future developments should result in more economical and energy-effective cooling methods, more efficient transportation and distribution systems, improved shipping containers, and increased usage of controlled- and modified-atmosphere storage.

BIBLIOGRAPHY

1. R. E. Hardenburg, A. E. Watada, and C. Y. Wang, *The Commercial Storage of Fruits, Vegetables, and Florist and Nursery Crops*, USDA Agricultural Handbook, 66, rev., U.S. Government Printing Office, Washington, D.C., 1986.
2. A. A. Kader and co-workers, *Postharvest Technology of Horticultural Crops*, Cooperative Extension Publication 3311, University of California, Berkeley, 1985.
3. A. L. Ryall and W. J. Lipton, *Handling, Transportation and Storage of Fruits and Vegetables*, Vol. 1, *Vegetables and Melons*, 2nd ed., AVI Publishing Co., Inc., Westport, Conn., 1979.
4. A. L. Ryall and W. T. Pentzer, *Handling, Transportation and Storage of Fruits and Vegetables*, Vol. 2., *Fruits and Tree Nuts*, 2nd ed., AVI Publishing Co., Inc., Westport, Conn., 1982.
5. J. B. Biale and R. E. Young, "Respiration and Ripening in Fruits—Retrospect and Prospect," in J. Friend and M. J. C. Rhodes, eds., *Recent Advances in the Biochemistry of Fruit and Vegetables*, Academic Press, Inc., Orlando, Fla., 1981.
6. J. A. Bartsch and G. D. Blanpied, "Refrigerated Storage for Horticultural Crops," *Agric. Eng. Ext. Bull.* **448**, Cornell University, Ithaca, N.Y., 1984.
7. F. E. Henry, A. H. Bennett, and R. H. Segall, "Hydroaircooling: A New Concept for Precooling Pallet Loads of Vegetables," *ASHRAE Trans.* **82**(part 2), 541–547 (1976).
8. F. M. R. Isenberg, R. F. Kasmire, and J. Parson, *Vacuum Cooling Vegetables*, Cornell University Cooperative Extension Information Bulletin **186**, 1982.
9. F. G. Mitchell, R. Guillou, and R. A. Parson, *Commercial Cooling of Fruits and Vegetables*, California Agricultural Experimental Station Extension Service Manual **43**, 1972.
10. I. J. Pflug and co-workers, "Precooling of Fruits and Vegetables," *ASHRAE Symp.* SF-4-70, 1970.
11. J. J. Gaffney, "Humidity: Basic Principles and Measurement Techniques," *HortScience* **13**, 551–555 (1978).
12. H. Hansen and J. Kuprianoff, "Some Experiences with Air-Washing in Cold and Gas Stores of Pears and Apples," *Internat. Inst. Refrig. Bul. Sup.* **1**, 337–340 (1961).
13. H. M. Couey, "Chilling Injury of Crops of Tropical and Subtropical Origin," *HortScience* **17**, 162–165 (1982).
14. J. M. Lyons, "Chilling Injury in Plants," *Annual Review of Plant Physiology* **24**, 445–466 (1973).
15. L. L. Morris, "Chilling Injury of Horticultural Crops: An Overview," *HortScience* **17**, 161–162 (1982).
16. E. Brennan, I. Leone, and R. H. Daines, "Ammonia Injury to Apples and Peaches in Storage," *Plant Dis. Rptr.* **46**, 792–795 (1962).
17. J. W. Eckert, "Postharvest Diseases of Fresh Fruits and Vegetables—Etiology and Control," in N. F. Haard and D. K. Salunkhe, eds., *Symposium: Postharvest Biology and Handling of Fruits and Vegetables*, AVI Publishing Co., Inc., Westport, Conn., 1975.
18. P. E. Brecht, "Use of Controlled Atmospheres to Retard Deterioration of Produce," *Food Technology* **34**, 45–50 (1980).
19. F. M. R. Isenberg, "Controlled-Atmosphere Storage of Vegetables," *Hortic. Rev.* 1, 337–394 (1979).
20. D. G. Richardson and M. Meheriuk, eds., *Controlled Atmospheres for Storage and Transport of Perishable Agricultural Commodities, Proceedings of the Third National CA Research Conference, Beaverton, Oreg., July 1981.*

21. R. M. Smock, "Controlled Atmosphere Storage of Fruits," *Hortic. Rev.* **1**, 301–336 (1979).
22. J. H. Moy, "Potential of Gamma Irradiation of Fruits: A Review," *Journal of Food Technology* **12**, 449–457 (1977).
23. R. A. Dennison and E. M. Ahmed, "Irradiation Treatment of Fruits and Vegetables," in Ref. 17.
24. F. G. Mitchell, *Packaging Horticultural Crops*, Univ. Calif. Coop. Ext. Perishables Handling 52, 1983, pp. 2–4.

GENERAL REFERENCES

F. Artes and A. J. Eseriche, "Intermittent Warming Reduces Chilling Injury and Decay of Tomato Fruit," *J. Food Sci.* **59**, 1053–1056 (1994).

J. R. Birewar, "Development of Improved On-farm Grain Drying Facility in Nigeria," *Agricultural Mechanization in Asia and Latin America* **27**, 51–53 (1996).

J. Dincer, "Precooling of Cylindrically Shaped Grapes: Experimental and Theoretical Heat Transfer Rates," *Journal of Food Processing and Preservation* **17**, 57–71 (1994).

I. Edeogu, J. Feddes, and J. Leonard, "Comparison Between Vertical and Horizontal Air Flow for Fruit and Vegetable Precooling," *Canadian Agricultural Engineering* **39**, 102–112 (1997).

E. A. Estes, "Feasibility and Affordability Considerations in Precooling Fruits and Vegetables," *American Society for Agricultural Engineers Publication* **1–95**, 390–396 (1995).

S. L. Gillies and P. M. A. Toivonen, "Cooling Method Influences the Postharvest Quality of Broccoli," *HortScience* **30**, 313–315 (1995).

A. L. Kyung and Y. J. Yong, "Effects of Low Temperature and CA on Quality Changes and Physiological Characteristics of Chilling Injury During Storage of Squash (*Circubitai imoschatai*) *Journal of the Korean Society for Horticultural Sciences* **39**, 402–407 (1998).

S. Lurie and A. Sabehat, "Prestorage Temperature Manipulations to Reduce Chilling Injury in Tomatoes," *Postharvest Biology and Technology* **11**, 57–62 (1997).

A. Nagao, T. Indou, and H. Dohi, "Effects of Curing Condition and Storage Temperature on Postharvest Quality of Squash Fruit," *Journal of the Japanese Society for Horticultural Science* **60**, 175–181 (1991).

A. Picon et al., "Effects of Precooling, Packaging Film, Modified Atmosphere and Ethylene Absorber on the Quality of Refrigerated Chandler and Douglas Strawberries," *Food Chem.* **48**, 189–193 (1993).

A. C. Purvis, "Interaction of Waxes and Temperature in Retarding Moisture Loss From and Chilling Injury of Cucumber Fruit During Storage," *Proceedings of the Florida State Horticultural Society* **107**, 257–260 (1994).

T. B. Puttaraju and T. V. Reddy, "Effects of Precooling on the Quality of Mango (cv. Mallika)," *Journal of Food Science and Technology India* **34**, 24–27 (1997).

J. M. Sala, "Involvement of Oxidative Stress in Chilling Injury in Cold-Stored Mandarin Fruits," *Postharvest Biology and Technology* **13**, 255–261 (1998).

B. D. Shukla and R. T. Patil, "Dryers and Drying Technology for Food Crops," *Indian Journal of Agricultural Science* **62**, 579–589 (1992).

R. Turk and E. Celik, "The Effect of Vacuum Precooling on the Half Cooling Period and Quality Characteristics of Iceberg Lettuce," *Acta Horticulture* **343**, 321–324 (1993).

C. Y. Wang, "Chilling Injury of Fruits and Vegetables," *Food Rev. Int.* **5**, 209–236 (1989).

C. Y. Wang, "Heat Treatment Affects Postharvest Quality of Kale and Kollard, but not Brussel Sprouts," *HortScience* **33**, 881–883 (1998).

A. B. Woolf, "Reduction of Chilling Injury in Stored Hass Avocado Fruit by 38 Degree Water Treatment," *HortScience* **32**, 1247–1251 (1997).

A. B. Woolf et al., "Reducing External Chilling Injury in Stored Hass Avocados With Dry Heat Treatments," *J. Am. Soc. Hortic. Sci.* **120**, 1050–1056 (1995).

A. B. Woolf et al., "Reduction of Chilling Injury in Sweet Persimmon Fuyu During Storage by Dry Air Heat Treatments," *Postharvest Biology and Technology* **11**, 155–164 (1997).

CINDY B. S. TONG
ALLEY E. WATADA
USDA/ARS
Beltsville, Maryland

FOOD CROPS: VARIETAL DIFFERENCES, MATURATION, RIPENING, AND SENESCENCE

Nutritional labeling of packaged food products conveys a misleading impression that similar foods are uniform in nutrient composition. Raw agricultural crops are particularly susceptible to wide variations in nutrient content, sensory quality, and suitability as ingredients in prepared or processed food products. Differences in composition result from many factors, including cultivar, growing condition, maturation, ripening, and senescence (1). Terms critical to an understanding of this subject are defined in Table 1.

VARIETAL DIFFERENCES

Humans eat a wide variety of plant products, thanks to the large number of plant species cultivated for food. Most individual fruits, vegetables, nuts, oilseeds, and grains represent separate species. A notable exception is *Brassica oleracea*, a species comprising the crops of broccoli, brussels sprouts, cabbage, cauliflower, collards, kale, and kohlrabi. Although not as diverse as *B. oleracea*, most species have variations in cultivars that are apparent in differences in growth response, pest and disease resistance, stress response, nutritional quality, and shipping stability.

The concept of varietal differences is based on Mendel's classic study of genetics in peas. Varietal differences are evident in the growth patterns of the crop in the field, such as the height of the plant, its branching characteristics, color and morphology of flower and fruit, and yield potential for edible product at harvest. Some cultivars offer greater resistance to pests and disease at harvest. Resistance can be achieved as a result of removal of chemical compounds that attract insects or synthesis of compounds that are toxic to the disease microorganisms or pests. Unfortunately, the same compounds that are toxic to the pest may adversely affect quality or may even be allergenic or toxic to humans (4).

Cultivars adapt in different ways to changes in growing locations and environmental conditions. Some cultivars will thrive in one location and suffer in others. Cultivars of fruits and vegetables vary in their response to temperature, cultural practices, and soil conditions, such as fertility, soil compaction, and moisture content. Certain cultivars within a species may show greater resistance to

Table 1. Definition of Terms

Cultivar (cultivated variety)	Group of plants within a particular cultivated species that is distinguished by a character or group of characters and that maintains its identity when propagated either asexually or sexually (2).
Horticultural maturity	The stage of development when a plant or plant part possesses two prerequisites for use by consumers for a particular purpose (3).
Maturation	The stage of development leading to the attainment of physiological and horticultural maturity (3).
Physiological maturity	The stage of development when a plant or plant part will continue to ontogeny even if detached (3).
Ripening	The composite of the processes that occur from the latter stages of growth and development through the early stages of senescence and result in characteristic aesthetic quality and food quality, as evidenced by changes in composition, color, texture, and other sensory attributes (3).
Senescence	Those processes that follow physiological maturity or horticultural maturity and lead to death of tissue (3).
Species	A group of plants (within a genus) that often exhibit many more morphological similarities than to members of the genus (2).

water, salt, and low-temperature or high-temperature stresses. The response to stress by plants is often reflected in changes in the edible portion of the tissue (eg, tomato plants resistant to low-temperature stress tend to produce fruit resistant to low temperatures during storage). Varietal differences have also been noted in the nutritional composition of crops at harvest and in the ability of a crop to maintain marketable quality during handling, shipping, and storage.

Crops may be improved by selecting for desired traits from existing cultivars or wild strains. The most celebrated modern example of crop improvement was the development of high-yield rice cultivars by Borlaug that resulted in the Green Revolution. In addition to conventional breeding techniques, biotechnology offers powerful tools and recombinant DNA in the development of better crops. Cell culture techniques permit rapid screening of potential strains if a reliable index for desirable trait(s) is available. Genetic engineering permits the direct manipulation of genetic material within the plant, promising the development of designer plants with the characteristics desired. These techniques are effective when the trait is clearly defined and a single gene is involved, such as disease resistance (4) or high-linoleic oilseeds (5). Much less effective has been the genetic modification of plants to improve quality that involves modification of texture or flavor where the physiological determinants of the characteristic are not clearly defined.

Many examples demonstrate the importance of varietal differences in the marketplace. Differences in color and sweetness in fresh apple cultivars help satisfy regional and individual tastes. Other cultivars are preferred for specific processing or home applications. The length of the peach season in a particular growing area is the result of a succession of early-, middle-, and late-maturing cultivars; different cultivars adapt to different climatic regions. Clingstone cultivars provide a firmer peach for canning and a better appearance, but many consumers prefer the flavor of freestone peaches in the fresh market. The cultivar of choice for a grape grower depends on the intended application: fresh, juice, jelly, raisins, or wine. Supersweet cultivars of corn were developed to increase sugar content by slow hydrolysis of sucrose during postharvest handling and storage. Processing cultivars of tomatoes must have a high-solids content, commercial cultivars that are shipped across the country must be able to withstand the rigors of handling and storage, and cultivars grown in backyard gardens must have high flavor impact. Hard cultivars of wheat are higher in protein and are used in breads, whereas soft cultivars are used in other applications, such as cookies and cakes.

Mechanization of harvesting operations has been achieved for several commodities. Certain species are more amenable to mechanical harvesting than others, and cultivars for processing are more likely to have been adapted for mechanization than fresh-market cultivars. Certain cultivars are more amenable to mechanical harvest than others. Characteristics important in ease of mechanical harvest include the uniformity of size and maturity at harvest, decreased susceptibility to bruising and other damage, and adaptability to fully mechanized production and handling systems. Successful mechanization requires an integration of breeding, quality, and engineering considerations.

MATURATION

Harvest is a critical point for the quality of any product derived from plants. The optimum harvest time varies with the period of growth and development of a plant from crop to crop. For example, celery (stems), cabbage (leaves), and broccoli (inflorescences) are harvested during vegetative growth, and cucumber and beans are harvested as partially developed fruits. Maturity or stage of development at harvest can be viewed from either a biological or market perspective. Full development biologically is called physiological maturity, and satisfactory sensory quality is related to horticultural maturity (3).

Many crops present a narrow window for optimal harvest. Maturity at harvest affects yield, stability during storage and handling, susceptibility to bruising or other damage, nutrient composition, and the quality of an item at purchase and consumption. All these factors should be considered in determining the best time to harvest. With most vegetables, this decision requires a compromise between harvesting early enough to obtain tender, succulent quality and waiting to obtain better yield.

RIPENING

The desirable attributes associated with high-quality fruit develop during ripening. Nonclimacteric fruits develop these attributes while still attached to the plant and are characterized by low levels of carbon dioxide (CO_2) and ethylene (C_2H_4) evolution. Little further ripening occurs in nonclimacteric fruit after harvest. Nonclimacteric fruits should be harvested as close to the flavor peak as possible because flavor will not develop after detachment.

Climacteric fruits are characterized by a rapid respiratory burst of greatly increased CO_2 and C_2H_4 at the initiation of ripening. The climacteric rise, triggered by C_2H_4, can occur on or off the plant. The fruit will generate its own C_2H_4 from aminocyclopropane carboxylic acid (ACC) by ACC oxidase, and ACC is generated from *S*-adenosyl methionine (SAM) by ACC synthase (6). If the fruit is harvested in a physiologically mature state before the increased generation of ACC and is kept in an environment free of C_2H_4, it can be maintained in a mature, unripe state during the transportation and storage. If climacteric fruits are harvested too early, they will not develop full flavor potential during storage. If harvested too late, they will become unacceptably soft before purchase or consumption. The marketplace offers greater incentives to the grower who harvests too early than to the one who harvests too late. In crops such as peaches and tomatoes where this window between too early and too late is narrow, premium quality product is difficult to obtain. Ripening can be artificially triggered by using an external source of C_2H_4. This process permits the shipment of green bananas from the tropics to geographically distant markets where ripening is triggered close to the consumer, long-distance truck shipment of green tomatoes for ripening close to the distribution point, and controlled-atmosphere storage of apples before ripening to permit year-round consumption.

Desirable changes associated with ripening include changes in appearance, flavor, texture, and nutritive value. Color changes are the result of the degradation of chlorophyll (green), which unmasks xanthophylls and carotenes (yellow and orange), and the biosynthesis of anthocyanins (red, blue, or purple) and the carotenoid lycopene (red). The development of sweetness in fruits is the result of a disappearance of sour (acids) and bitter (including tannins) compounds and the accumulation of sweet (sugars) compounds. Flavor is the combination of taste (sweet, sour, bitter) and aroma that results from the accumulation of volatile compounds during ripening, producing characteristic flavors associated with specific fruits. Softening during ripening is the result of degradation of cell walls, primarily pectins, by a series of hydrolase enzymes (5). Vitamins (particularly A and C) accumulate during ripening, thereby enhancing the nutritional quality of the product.

SENESCENCE

Plants senesce by predictable, controlled patterns that lead to death. Senescence is distinct, although not always distinguishable, from aging (longevity). Plant organs also senesce, which is frequently enhanced by detachment. Ripening, as described earlier and defined in Table 1, is a special case of the early phase of senescence. In somewhat oversimplified terms, ripening tends to lead to improvement of quality attributes of fruits, whereas other senescence processes tend to lead to loss of quality. The primary objective of postharvest handling and storage is to control ripening and slow senescence, thus providing the consumer with a product of optimum quality at a reasonable price.

The most obvious signs of senescence in vegetables are loss of color and tissue softening. Unlike similar changes noted in ripening fruits that are considered quality improvements, yellowing and softening of green vegetables is considered detrimental. Yellowing, as in ripening, involves the unmasking of yellow pigments by chlorophyll depletion as chloroplasts are converted to chromoplasts. Likewise, softening is associated with breakdown of cell wall components, such as pectin and cellulose.

The cellular physiology of senescence is complex. Although it was previously believed that senescence resulted from the release of organic acids and hydrolases within the cell, it is now generally recognized that senescence is controlled by genetics. Senescence affects gene expression, protein and nucleic acid degradation, chemical composition and physical properties of membranes, and the structure and function of plant organelles (7). Although predominantly a degradative process, it is misleading to view senescence solely in the context of degradation. Metabolic processes function as a series of delicate balances between biosynthetic and degradative reactions. During growth and development as well as ripening, accumulation of a compound is the net result of biosynthesis exceeding degradation. Likewise, during senescence, net loss of a compound is the result of increased degradation, decreased synthesis, or a change in the balance between the two.

Early changes observed in the ultrastructure of senescing plant tissue include swelling of thylakoids in chloroplasts. Mitochondria and vacuoles are affected later in the process. Changes in the nucleus are generally observed late in senescence before disruption of the plasma membrane, leading to cell and tissue death. Of particular interest in the study of senescence are the increase in protease enzymes and loss of membrane integrity. Membranes serve as internal barriers within the cell and regulate the flow of ions and metabolites. During senescence and aging, these membranes lose their integrity, which leads to increased permeability. A current theory suggests that membrane lipids are degraded by hydrolytic and peroxidative mechanisms, resulting in increased permeability and decreased function of membrane-bound proteins.

Not all degradation of plant tissue is attributable to senescence. Mechanical damage can cause physical rupture of membranes and lead to mixing of enzymes and substrates, which results in the browning and discoloration called a bruise. Inadequate nutrition of the plant during growth and development can lead to deficiencies within a detached plant organ and result in a physiological disorder. Low temperatures and controlled atmospheres slow the respiration and senescence of many plant organs, but certain species are susceptible to physiological disorders that result from exposure to low temperatures, low O_2, high CO_2, or elevated C_2H_4. In addition, excess water loss can lead to loss of turgor pressure, resulting in undesirable textural properties.

BIBLIOGRAPHY

1. R. L. Shewfelt, "Sources of Variation in the Nutrient Content of Agricultural Commodities from the Farm to the Consumer," *Journal of Food Quality* **13**, 37–54 (1990).
2. J. Janik, *Horticultural Science*, 4th ed., W. H. Freeman, New York, 1986.
3. A. E. Watada et al., "Terminology for the Description of the Developmental Stages of Horticultural Crops," *HortScience* **19**, 20–21 (1984).
4. S. L. Franck-Oberaspach and B. Keller, "Consequences of Classical and Biotechnological Resistance Breeding for Food Toxicology and Allergenicity," *Plant Breeding* **116**, 1–17 (1997).
5. D. S. Brar, T. Ohtani and H. Uchimiya, "Genetically-Engineered Plants for Quality Improvement," *Biotechnology and Genetic Engineering Reviews* **13**, 167–179 (1996).
6. J-M. Lelievre et al., "Ethylene and Fruit Ripening," *Physiol. Plant* **101**, 727–739 (1997).
7. L. D. Nooden, "The Phenomena of Senescence and Aging," in L. D. Nooden and A. C. Leopold, eds., *Senescence and Aging in Plants*, Academic Press, Orlando Fla., 1988, pp. 1–50.

GENERAL REFERENCES

M. J. Chrispeels and D. E. Sadava, *Plants, Genes, and Agriculture*, Jones and Bartlett, Boston, Mass., 1994.

J. Gross, *Pigments in Vegetables: Chlorophylls and Carotenoids*, Chapman and Hall, London, United Kingdom, 1991.

R. C. Hoseney, *Principles of Cereal Science and Technology*, American Association of Cereal Chemists, St. Paul, Minn., 1994.

A. A. Kader, ed., *Postharvest Technology of Horticultural Crops*, University of California Press, Oakland, Calif., 1992.

S. J. Kays, *Postharvest Physiology of Perishable Plant Products*, Chapman and Hall, London, United Kingdom, 1991.

R. Lal and S. Lal, *Genetic Engineering of Plants for Crop Improvement*, CRC Press, Boca Raton, Fla., 1993

G. Mazza and E. Maniati, *Anthocyanins in Fruits, Vegetables and Grains*, CRC Press, Boca Raton, Fla., 1993.

G. B. Seymour, J. E. Taylor, and G. A. Tucker, eds., *Biochemistry of Fruit Ripening*, Chapman and Hall, London, United Kingdom, 1993.

L. P. Somogyi, H. S. Ramaswamy, and Y. H. Hui, eds., *Processing Fruits: Science and Technology*, Vol. 1, Technomic, Lancaster, Penn., 1996.

ROBERT L. SHEWFELT
University of Georgia
Athens, Georgia

FOOD ENGINEERING

The largest business activity in the world is the supplying of food to an ever-increasing population. This business includes growing and harvesting, transportation and handling, storing, processing and preservation, packing, distribution, and marketing. During the past 50 to 75 years, the various phases of the food business have grown from small family-type enterprises to gigantic, increasingly sophisticated and integrated food supply systems. The need for this change has been dictated by increasing concentrations of people in large urban areas, where the livelihood of large segments of the population depend on huge quantities of foodstuffs being readily available. Because large-volume production of foods and food materials is often remote from dense concentrations of consumers, efficient mass production and transportation of food supplies are absolutely necessary.

The vertically integrated food industry that has grown from these demands, probably more than any other human activity, requires the support of diversified, well-rounded teams of scientists, engineers, economists, and marketing specialists. Food engineering is a relatively new professional and scientific field defined in the 1950s when several engineering-educated food scientists and technologists employed by educational institutions and various segments of the food industry recognized that few engineers were educated and trained for the increasingly complex world food industry.

Foods are composed of a large variety of physically and chemically complex materials. A powerful analysis and design concept, called unit operations, originated and was extensively developed by those in the field of chemical engineering. This concept has been immensely useful in food engineering operations ranging from raw material assessment to finished product evaluation. Unit operations permits a myriad of processing steps to be seen as relatively few basic physical and chemical transformations. For example, crushing, mixing, and filtering are physical unit operations; reduction and polymerization are chemical unit operations.

Engineers in other industries are almost exclusively physical-science oriented. Functional engineers in the food industry must be knowledgeable in the biological sciences as applied to the food industry, including sanitation, spoilage, public health, environmental control, and biological process engineering, in which microorganisms are used to drive or mediate processes to produce food materials or food products.

The food requirements of the modern world can no longer be met by small, isolated, and nonintegrated food production systems, each involved in a single phase of the food industry. The logistic requirements and the complexity of feeding a world in which many countries are unable to produce sufficient food for their own populations have created a demand for a more science- and engineering-based approach. Today, and increasingly in the future, the food engineer must have the ability to play an important role in the integration of all phases of food production, preservation, and distribution into a smoothly functioning industry.

PROFESSIONAL RESPONSIBILITIES OF A FOOD ENGINEER

Some principal responsibilities of food engineering professionals, whether in industry or academia, are to develop and improve the basic steps involved in the production of food materials, process these materials into finished products, and apply this knowledge to commercial operations. Considerations during all phases must include consumer acceptability (aesthetic and sensory) and safety of the product for human or animal consumption through good manufacturing practices, as prescribed by the U.S. Food and Drug Administration.

Food engineers are involved in a wide spectrum of activities, including operations, engineering feasibility, and pre-engineering studies for updating existing facilities and planning new operations. Other responsibilities include designing process machinery and equipment layouts and integration, developing and improving unit operations and processes, ensuring food sanitation and safety, and maintaining and upgrading facilities. Planning operations also incorporate logistics of supplying, storing, and transporting raw materials and ingredients as well as finished products.

Food engineers are increasingly required to have knowledge of computers and computer applications, especially methods and techniques relating to process control in manufacturing operations. The feedback of information in closed-loop systems is particularly important because the biological state is the primary factor in controlling the quality of the final product.

EDUCATIONAL REQUIREMENTS OF A FOOD ENGINEER

The first collegiate year is about the same for all engineering curricula. The second year is likely to be similar, but with some departures for beginning specialization by field. Even in the third and fourth years, commonality for related specialities is not unusual.

The first year for all engineering students is primarily devoted to core curriculum requirements and beginning foundation courses in the natural sciences and mathematics. Most foundation science courses have decidedly engineering slants; the subject matter is presented and used quantitatively, rather than descriptively. Courses directed toward engineering specialities usually begin in the second year and increase markedly in number and complexity in the third and fourth years.

Engineering curricula tend to require more prescribed courses as well as a greater number of courses than other curricula. There are two important results from this: (*1*) it is not so unusual for engineering students to take an extra term to complete a degree, and (*2*) the options for taking nonrequired courses are limited, meaning that elective courses may have to be taken as overloads.

There is no opportunity at present to pursue a 4-year accredited collegiate program leading to a baccalaureate in food engineering in any American college or university. Furthermore, no such program will exist until a food engineering program meeting the requirements of the Engineers Council for Professional Development (ECPD) has enrolled and graduated students and then has been examined and approved by an accreditation committee of the ECPD. The procedure briefly outlined here is the one followed by more than a dozen separately identified engineering degree programs that have received accreditation.

At present, the best opportunity to prepare for a career in food engineering after a baccalaureate is to select a college or university that provides goal-compatible, ECPD-accredited programs in engineering; formal instruction for majors and nonmajors in related scientific and technical areas; and existing procedures for permitting significant student and advisor input into the planning of programs tailored to meet student aims. Goal-compatible engineering curricula might include agricultural, chemical, mechanical, civil, bio- and biochemical engineering. Related scientific and technical areas might include food science, nutrition, microbiology, sanitary engineering, pollution control, and public health engineering.

An alternative might be to pursue a program in food science, with technical electives selected from appropriate engineering courses (perhaps with advisors in food science and engineering). The minimum grounding in engineering should include fluid mechanics, mechanics of materials, heat and mass transfer, thermodynamics, unit operations and processes, process control, computer science for engineers, and applied mathematics for engineers.

The opportunities mentioned in the previous paragraphs for study in or related to food engineering might require an extra term or two of academic work. Because of this, a third alternative might be worth considering. With careful planning from both the engineering and food science viewpoints, and with institutional cooperation, it is conceivable that a B.S. degree in engineering and an M.S. degree in food science could be earned in 5 academic years.

RESEARCH AND DEVELOPMENT AREAS IN FOOD ENGINEERING

In the United States, there has been relatively little financial support given to food engineering programs for fundamental investigation of the complex composition and physical properties that must be considered in the design, construction, and operation of food processing operations. European countries have taken a world leadership position in the design, manufacturing, and marketing of food processing equipment. The support given to process engineers parallels the development of innovative new technologies, such as aseptic processing and controlled atmosphere storage and packaging.

The challenge of engineering practice is to find usefully precise and economically feasible solutions to real problems. The basic laws of physics, chemistry, and biology, expressed in the language of mathematics, show the relationships among the elements composing a system, whether it be a piece of machinery or the planet on which we live. However, because of many shortcomings of the precise knowledge of these relationships, balanced mathematical expressions are not determinable without the use of experimentally designed constants and secondary relationships that balance the equations and make them useful for application to practical situations. Hence, the application to sound engineering principles implies the use of experimentally determined information combined with scientific knowledge. These relationships can be practically applied to design a successful plant, to design and build machinery and equipment for successful and economical operations, and to complete records and information relating to production performance so that present operations can be improved and new ones developed. The problems associated with using an engineer's tools in the food industry are all too common. Food processing and food handling machinery and equipment are often based on previous practice rather than efficient functioning, easy maintenance, and meeting sanitary requirements. Modern-day food manufacturing requires reliance on sound engineering data, which often are not available or are in need of updating or improvement to evaluate and commercialize manufacturing methods and techniques. Some of the important food engineering areas in need of further study include

1. The changes in mass and composition that occur in foods during processing.
2. The pack densities, specific gravities, and compressibilities that are brought about in loading bins, storage tanks, and vats.
3. The safe loading depths for fresh or raw products during transportation and storage.
4. The flow properties for various foods in different forms and process states.
5. The properties of viscous material such as purées, pulps, slurries, etc.
6. The specific heat, heat capacity, and heat conductivity of foods in all states of preservation, especially when simultaneous mass and heat transfer occur.
7. The effect of particle size on various processes, which has a bearing on many processes such as spray drying, grinding, filtering slurries and miscella, and disintegration operations.

Both mechanical and chemical unit operations and processes are necessary in food production. Mechanical operations include washing, sizing, sorting, grinding, mixing, polishing, grading, packaging, and materials handling. Processes for preserving (and often to improve texture and flavor) include a combination of chemical and physical unit operations and processes. These include (*1*) adding heat for cooking, pasteurizing, or sterilizing; (*2*) removing heat for cooling (refrigerating), freezing, or storing in a frozen state; (*3*) adding chemicals to control water activity, aid in fermentation or other processes (eg, enzyme hydrolysis), and improve sensory characteristics and acceptance; (*4*) removing water to control water activity; and (*5*) irradiating to pasteurize or sterilize. Most food processing uses a combination of these steps. For example, drying is normally carried out by adding heat to vaporize the water, whether it be under vacuum or atmospheric pressure. Of course, the primary purpose of these operations is to control or destruct microorganisms and their metabolic products and to reduce certain detrimental chemical and physical reactions. However, many undesirable chemical and physical relationships within foods, such as oxidative rancidity, vitamin deterioration, and certain textural changes, are often caused or accelerated by processing methods.

FOOD ENGINEERING AS RELATED TO OTHER DISCIPLINES AND PROFESSIONAL SOCIETIES

Food science and food technology are related and overlapping disciplines that have similar basic science requirements, such as general and organic chemistry, physics, mathematics through calculus, and microbiology. Food science courses emphasize the relationships between basic science courses and the various aspects of food production, processing, and preservation. Separations—differentiation among career aims and goals—usually occur after college graduation and the entrance into graduate school or the industrial and commercial job market. Food scientists ordinarily concentrate on the basic chemistry and microbiology of foods; food technologists usually are more involved in the applied aspects of handling and processing foods. Many basic researchers specialize in the biological relationships of foods and must have particularly strong backgrounds in biology, biochemistry, and microbiology.

The third member of the team is the food engineer. Engineers professionally involved in the food industry must have the food science and technology background to actively and productively participate in maintaining an adequate food supply for the world's burgeoning population.

The close work and professional relationships among food scientists, food technologists, and food engineers are demonstrated by the extent to which they maintain memberships in cross-disciplinary professional societies. The Institute of Food Science and Technology (IFT) was founded in 1939. Several decades later, a food engineering division was formed in the IFT. Many of the founding members of the division were food technologists who, through experience or further training, had learned applied engineering related to food processing. Today, the food engineer has found more direct relationships with fellow engineers in professional engineering societies, including the American Society of Agricultural Engineers, the American Society of Chemical Engineers, the American Society of Mechanical Engineers, and other scientific and professional organizations. The food engineer is now established as an indispensible contributor to the further scientific and technological development and growth of our most important scientific and industrial enterprise.

J. T. CLAYTON
University of Massachusetts
Amherst, Massachusetts

FOOD, DRUG, AND COSMETIC COLORANTS.

See COLORANTS: FOOD, DRUG, AND COSMETIC COLORANTS.

FOOD FERMENTATION

DEFINITION

Food fermentation is the study of microbial activity, usually anaerobic, on suitable substrates under controlled or uncontrolled conditions resulting in the production of desirable foods or beverages that are characteristically more stable, palatable, and nutritious than the raw substrate.

HISTORY

The history of food fermentation paralleled the development of microbiology and food microbiology. Traditionally many foods were prepared by fermentation, but the reasons behind the success or failure of the processes were not known. After the work of Pasteur in the 1850s and others, who demonstrated that a specific microorganism (eg, yeast) acting on a suitable substrate (grape juice) will produce a desirable product (wine), the science of food fermentation began. Now many food fermentation principles and practices are well established, and food companies can predictably produce consistently good-quality fermented products. With the advances in genetic engineering, old processes are being improved and new ones are being discovered. Also, many indigenous fermented foods (such as some Oriental foods and African tribal foods) and their processes are not well known and are areas for future investigation.

Although the principles of fermentation in many foods are understood in a laboratory setting, the scaling-up of these processes to commercially successful operations is complicated. A detailed account of all aspects of anaerobic fermentation, including methodology of anaerobic cultivation; mutation and genetic engineering of anaerobic bacteria; industrially important strains and pathways, biochemistry, kinetics, and transport in anaerobic fermentation; bioenergetics of anaerobic processes; data collection and analysis; mixed culture interactions; and design and application of anaerobic systems has been published by Erickson and Fung (1).

MICROORGANISMS USED IN FERMENTED FOODS

The number of microbial species used in fermented foods is surprisingly small. Of the thousands of species of microorganisms in nature, only the following genera are well utilized by the food fermentation industry (2):

Bacteria: *Acetobacter, Streptococcus, Lactococcus, Leuconostoc, Pediococcus, Lactobacillus, Propionibacterium, Brevibacterium, Bacillus, Micrococcus,* and *Staphylococcus*. There has been a major renaming of some key bacteria used in fermentation as follows:

Streptococcus cremois to *Lactococcus lactis* subsp. *cremoris*

Streptococcus lactis to *Lactococcus lactis* subsp. *lactis*

Streptococcus diacetylactis to *Lactococcus lactic* subsp. *lactis biovar diacetylactis*

Leuconostoc citrovorum to *Leuconostoc mesenteroides* subsp. *cremoris*

Lactobacillus bulgaricus to *Lactobacillus delbrueckii* subsp. *bulgaricus*

Streptococcus thermophilus unchanged

Lactobacillus helvticus unchanged

Yeast: *Saccharomyces* (especially *S. cerevisiae* and *S. carlsbergensis*), *Candida, Torulopsis*, and *Hansenula*

Mold: *Aspergillus, Penicillium, Rhizopus, Mucor, Monascus*, and *Actinomucor*

In many tradition or indigenous fermented products, the cultures are not known or identified.

TYPES OF FERMENTATION PROCESSES

Single-Culture Fermentation

The key to the success of single-culture fermentation is to provide the culture with a sterile substrate and environment with no contamination during the fermentation process. Examples include wine making, beer making, bread making, production of single-culture fermented dairy products, and vinegar production. In this type of fermentation the viable cells increase in a typical growth curve sequence of lag phase, log phase, stationary phase, and death phase. Primary metabolites (alcohol, acid, etc) are made during the log phase, and secondary products (antibiotics, toxins, etc) are made after the culture reaches the stationary phase.

An ideal culture for fermentation should possess the following attributes:

1. Organisms must be pure.
2. Organisms must grow and reproduce quickly.
3. Organisms must be genetically stable yet amiable to manipulation for better performance.
4. Organisms must produce uniform product in a short time.
5. Organisms must not produce undesirable by-products.
6. Organisms must have a protective mechanism (eg, acid or bacteriocin production) against other undesirable contaminants.

Mixed Pure Culture Fermentation

Some products need a mixture of known cultures. The mixed pure cultures can be a controlled mixture of bacteria or bacteria with a combination of yeast and mold. Yogurt making and cheese making are good examples of this type of mixed fermentation. It is necessary to provide a balance of the two cultures for maximum performance. For example, in yogurt making the desired ratio of cocci (*S. thermophilus*) to rod (*L. bulgaricus*) is 5:1 in cell numbers, but by weight the ratio should be 1:1.

Mixed Natural Culture Fermentation

In many parts of the world, especially in developing countries, the flora of indigenous fermented foods are mixed natural cultures. The interactions of those microbes are exceedingly complex, and the success of such fermentations depends on following traditional processes and not on scientific principles, because many of the responsible cultures have not been isolated and studied. The principal fermentation reactions in foods are listed in Table 1. This list does not include reactions in industrial fermentation of solvents, acids, and so on.

Microbial Interactions

The interactions of mixed cultures are of great interest in studying the kinetics of population development in liquid

Table 1. The Principal Fermentation Reactions in Foods

Lactic acid fermentation

Homofermentative: $C_6H_{12}O_6$ (glucose) to $2CH_3CHOHCOOH$ (lactic acid)

Heterofermentative: $C_6H_{12}O_6$ (glucose) to $CH_3CHOHCOOH$ (lactic acid) + CO_2 + C_2H_5OH (ethyl alcohol)

Propionic acid fermentation

$3C_6H_{12}O_6$ (glucose) to $6CH_3CHOHCOOH$ (lactic acid)

$3CH_3CHOHCOOH$ (lactic acid) to $2CH_3CH_2CO_2H$ (propionic acid) + CH_3COOH (acetic acid) + CO_2 + H_2O

Citric acid fermentation

$CH_2COOHHOCCOOHCH_2COOH$ (citric acid) to $2CH_2COCOOH$ (pyruvic acid) to $CH_2COCHOHCH_3$ (acetylmethylcarbinol) + $2CO_2$

Acetylmethylcabinol can be oxidized to $CH_3COCOCH_3$ (diacetyl) or reduced to $CH_3CHOHCHOHCH_3$ (2,3, butylene glycol)

Alcoholic fermentation

$C_6H_{12}O_6$ (glucose) to $2C_2H_5OH$ (ethyl alcohol) + $2CO_2$

Butyric acid fermentation

$C_6H_{12}O_6$ (glucose) to CH_3COOH (acetic acid) + CH3CH_3CH_2COOH (butyric acid) + CH_3CH_2OH (ethyl alcohol) + $CH_3(CH_2)_2CHOH$ (butyl alcohol) + CH_3COCH_3 (acetone) + CO_2 + H_2

Gassy fermentation

$2C_6H_{12}O_6$ (glucose) + H_2O to $2CH_2CHOHCOOH$ (lactic acid) + CH_3COOH (acetic acid) + C_2H_5OH (ethyl alcohol) + $2CO_2$ + $2H_2$

Acetic acid formation (oxidative)

C_2H_5OH (ethyl alcohol) + H_2O O_2 to CH_3COOH (acetic acid)

and solid fermented foods. The following is a discussion of the study on the interactions between an important fermentative yeast (*S. cerevisiae*) and an important industrial bacterium (*Acetobacter suboxydans*) and an important environmental contaminant (*Escherichia coli*).

Figure 1 shows the interaction between *S. cerevisiae* and *A. suboxydans*. For the yeast, both direct count (under the microscope) and electronic count (Coulter Electronic Counter, which can differentiate particles sizes) increased with time. The direct count registered about 10 times more cells than the electronic count. For the bacterium, the direct count increased with time, but the electronic count increased to about 7.8 log cell/mL and then started to decrease. This decrease in number is due to the adhesion of many of the bacterial cells to yeast cells. The electronic counter cannot differentiate these two populations when they are in clumps. However, microscopic observations can ascertain the different counts. The concentration of glucose completely disappeared in the first 10 h of yeast and bacterial interactions. The pH of the medium first decreased and then stabilized at pH 3. This is due to oxidation of alcohol (produced by *S. cerevisiae*) by *A. suboxydans*.

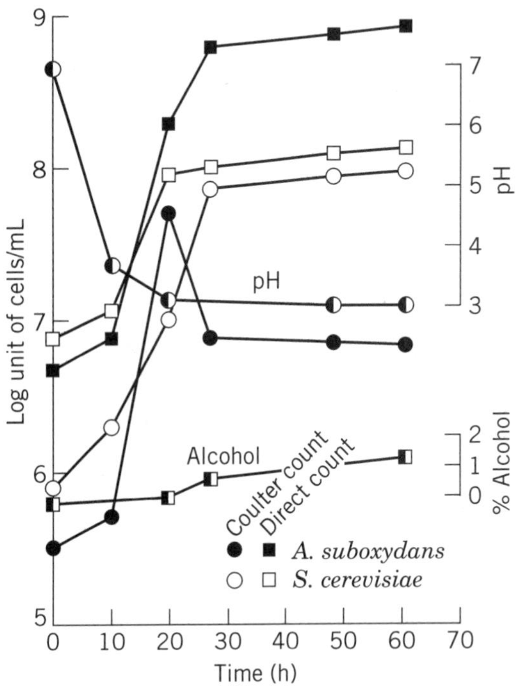

Figure 1. Interaction of mixed cultures of *S. cerevisiae* and *A. suboxydans* in terms of cell numbers monitored by direct count and electronic count as well as product formation (alcohol and acid production). *Source:* Ref. 3, used with permission.

The interactions of yeast and *E. coli* (Fig. 2) showed interesting contrasts compared with the yeast–*Acetobacter* interactions. Growth of yeast as monitored by viable cell count, direct cell count, and electronic count all increased with time. After reaching a stationary phase, viable yeast count decreased, but direct count and electronic counts registered no reduction in numbers. A likely explanation is that the latter two biomass measurements counted both live and dead cells, but the viable cell count method only registered living cells. After the stationary phase, some of the yeast cell went through the death cycle. The growth of *E. coli* showed an increase in the viable cell count and direct cell count. The electronic count reached 8 log cells/mL and then declined, exhibiting a trend similar to that observed in the yeast and *Acetobacter* interactions. Alcohol contents increased to about 2%, and the pH first dropped to around 4 and then rose to 6. This pattern differed from that obtained from the interaction of yeast and *Acetobacter*. *E. coli* cannot oxidize alcohol, and this did not create large amounts of acid to counteract the basic metabolites in the reaction vessel. This resulted in a medium that reverted to a more alkaline state.

This study indicated that different methods of estimating biomass may provide different results in mixed culture fermentation. The electronic counting method is useful when a single culture is monitored. In mixed culture interaction that involves clumping, this method is not suitable for differentiating mixed populations.

The following sections are synopses of major fermented food. Detailed treatment of these subjects can be found in books listed in the reference section.

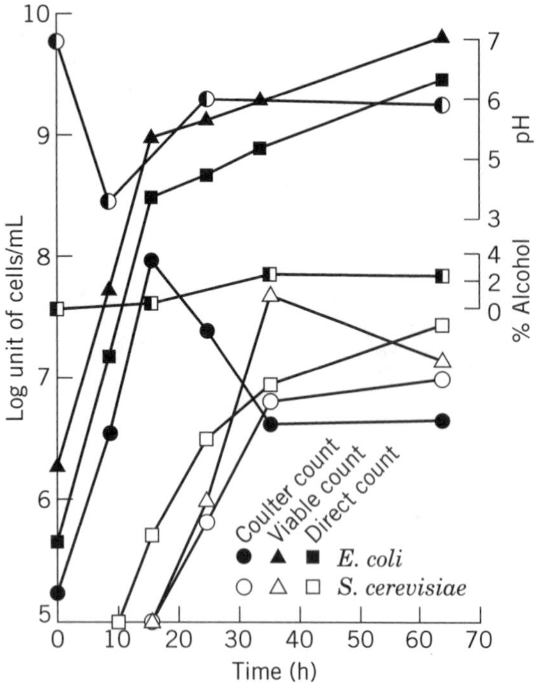

Figure 2. Interaction of mixed culture of *S. cerevisiae* and *E. coli* in terms of cell numbers monitored by direct count, electronic count, and viable cell count as well as product formation (alcohol and acid production). *Source:* Ref. 3, used with permission.

Fermented Liquid and Semisolid Dairy Products

Cultured Buttermilk. Pasteurized skim milk inoculated with 0.5% mixture of *Streptococcus (Lactococcus) lactis, S. (L.) cremoris, Leuconostoc cremoris*, or *L. dextranicum* incubated at 22°C for 14 to 18 h, results in a product with pH 4.5 acid curds with pleasant aroma and flavor. It has acid tastes.

Sour Cream. Milk with 19% milk fat and 0.2% citric acid inoculated with 1% *S. (L.) lactis* and *S. (L.) cremoris*, incubated at 22°C for 14 to 16 h results in a product with pH 4.5 that is creamy and sour and good for baked potatoes and other foods.

Acidophilus Milk. Skim milk inoculated with 5% *Lactobacillus acidophilus* incubated at 38°C for 18 to 24 h results in an acidic (1% acid) sour milk product. It is claimed to have health benefits due to large numbers of live cultures in the product.

Sweet Acidophilus Milk. Skim milk inoculated with a large percentage of *L. acidophilus* results in 5×10^6; *L. acidophilus* per mL of *live* cultures in milk. The product *is not fermented* and must be kept cold (4°C) until sold. Contrary to regular acidophilus milk, objected to by some consumers due to the high acidity, this product has no acid taste but is claimed to relieve gastrointestinal problems.

Bulgaricus Milk. Whole milk inoculated with 2% *L. bulgaricus* incubated at 38°C for 10 to 12 h results in a sour (2–4% acid) product, claimed to relieve gastrointestinal problems. It is consumed in large quantities by people in the Georgia region of Europe.

Yogurt. Milk with added 2 to 4% nonfat dry milk powder inoculated with 5% combination of *S. thermophilus* and *L. bulgaricus* (1:1 ratio, by weight, or 5:1 ratio by cocci to rod) incubated at 45°C for 3 to 6 h or 32°C for 12 to 14 h results in a smooth, viscus gel with a delicate walnutlike flavor. The product must be chilled immediately. A variety of fruits, berries, and flavors can be added to the product. This is the fastest-growing product in the dairy industry.

Kefir. Goat, sheep, or cow's milk inoculated with kefir grain (a mixture of *Saccharomyces kefir, Torula, Lactobacillus caucasicus, Leuconostoc*, etc) incubated at 22°C for 12 to 18 h results in a product with 1% alcohol and 0.8% acid. The product is strained through cheesecloth. The milk foams and fizzles like beer.

Koumiss. Mare's milk (which has five times more vitamin C than cow's milk) inoculated with *L. bulgaricus* plus *Torula* yeast incubated at 22°C for 12 to 18 h results in a drink with 1 to 2.5% alcohol and 0.7 to 1.8% acid.

THERAPEUTIC VALUE OF CULTURED MILK

In 1907 Metchnikoff wrote the book *Prolongation of Life* and claimed that drinking sour milk regularly gave people in southeast Europe long life and good health, because premature old age was a result of absorption of toxins produced by bacteria in the large intestines. Large volumes of sour milk with live cultures changed the balance of microflora in the intestines so that toxic compounds were not produced and people lived longer. At a minimum, drinking live cultures of *L. acidophilus* or *L. bulgaricus* is not harmful. For people taking antibiotics after major surgery, which tends to sterilize the intestines, these live organisms might very well be beneficial. People with gastrointestinal problems may find sour milk helpful in suppressing diarrhea. This is still a controversial area.

FERMENTED SOLID DAIRY PRODUCTS—CHEESE

Cheese is made by the coagulation or precipitation of milk protein by acid produced by starter cultures (*S. [L.] lactis* and *S. [L.] cremoris*) with the aid of rennin (from the stomach of calves) added to the milk. Milk is usually pasteurized at 62.8°C for 30 min or 72.0°C for 15 s. Raw milk has been used to make cheese but is not advisable due to the potential of survival of foodborne pathogens such as *Listeria monocytogenes*. Regulations indicate the cheese made with raw milk cannot be sold until a 60-day aging period. The assumption is that in 60 days pathogens, if present in the cheese, will be killed by the starter cultures and the acidity involved in the product. A food-grade dye, annatto, is often added to give the cheese the familiar yellow color. Some cheese makers are now not using the dye so that the cheese will not have additives in the product. When milk proteins coagulate, the resultant liquid portion is called whey. From 10 lb of milk, only 1 lb of cheese is made and 9 lb are discarded as whey. Whey utilization is an important area of research because whey has very high biological oxygen demand (BOD) (30,000–60,000 ppm) and can be an important environmental contaminant if the BOD is not reduced. Also, whey contains about 50 to 70% of the nutrients of milk, and this rich food source should be effectively utilized. After the curd is formed, it can be poured into perforated molds lined with cheesecloth, and the extra whey is drained. For harder cheese, the curd is cooked, stirred, cut, and pressed. The more whey expressed from the protein mass, the harder the resultant cheese will be. Salt is often added to the curd to give flavor as well as to prevent undesirable microorganisms from growing. Cheese is then ripened for several months or even years before consumption.

Classification of Cheese

There are literally hundreds of cheese varieties being made in the world. Most of the names are from the town or city in which the cheese originated. From a texture point of view, cheese can be classified into the following.

Soft, Unripened or Ripened Cheeses. Cottage cheese is an example of soft, unripened cheese. It can be made by acid produced by the starter cultures (*Lactococcus lactics* and *L. cremoris*) along with rennin or by direct set using food-grade acid such as mesolactide and D-glucono-δ-lactone. For the direct set method, the cottage cheese will coagulate very fast—within seconds at 4°C. For starter cultures, one can use a "short set" method by inoculating the milk with 5% starter cultures, and the process will be completed in 5 h. For the "long set" method, a 3% percent inoculation can be used, and it will take 10 to 14 h to complete the process. After setting the curd, it be cut using 1/4″ wire for small-curd cottage cheese and 3/4″ wire for large-curd cottage cheese.

Soft, Ripened Cheeses. Limburger cheese is an example of soft, ripened cheese. The soft cheese is placed on wooden shelves, and a surface bacterium (*Brevibacterium linens*) grows on the surface of the cheese and produces a brownish red surface growth. Protein is broken down into ammonia and gives the strong flavor to this cheese. Camembert cheese is another soft cheese, but it is ripened by a surface mold, *Penicillium camemberti*, with a mixture of bacteria. This famous French cheese is called the queen of cheese.

Semisoft, Ripened Cheeses. Several mold-ripened cheeses are classified under this category. They are all ripened by *Penicillium roqueforti*, a mold that grows throughout the cheese mass. They are also called blue-veined cheese. Roquefort cheese is called the king of cheese. Other cheeses in this group are Stilton, Gorgonzola, and bleu.

Hard Cheeses. These cheeses are well ripened and hard. Among the most important cheeses are cheddar, a bacterial-ripened cheese, and Swiss cheeses, which are cheeses with eyes. The characteristics eyes of Swiss cheeses are caused by CO_2 produced by *Propionibacterium shermanii* during anaerobic fermentation in the ripening stage.

Cheese Varieties and Descriptions

Principal cheese varieties and descriptions are as follows (4).

American. This term is used to identify the group that includes cheddar, Colby, and so on, popularized in the United States.

Bleu. The French name, bleu, is used for cheese similar to Roquefort but either not made in Roquefort, France, or not made from ewe's milk.

Blue. Roquefort-type cheese made in the United States and Canada is referred to as blue. Brick cheese, of American origin, is made from whole milk, with a mild but pungent and sweet flavor.

Cheddar. This is the most important cheese and accounts for 75% of cheese made in the United States. The name came from the town Cheddar in the UK where it was first made. In addition, cheddaring is the name of a step in the manufacturing process of piling and repiling of curd. This will provide better adhesiveness of the protein and also allow starter cultures to produce more acid to control coliforms. Most cheddar cheeses are ripened for 60 days and some for a year or more (sharp and very sharp cheddar).

Colby. Similar to cheddar, Colby has a softer body and more open texture.

Cottage. A soft unripened cheese, cottage cheese is made from skim milk. Flavoring materials such as peppers, olives, pimientos, or garlic may be added. When more than 4% of fat is added, it is called creamed cottage cheese.

Cream. Cream cheese is a soft, rich, unripened cheese made of cream or a mixture of cream and milk.

Edam. Made from whole milk, Edam has a mild flavor and firm body. It is usually shaped like a flattened ball and covered with red coloring or red paraffin.

Gouda. Gouda is similar to Edam, except that it contains more fat and is usually packaged like Edam.

Limburger. A soft, surface-ripened cheese, Limburger has a characteristic strong flavor and aroma.

Parmesan. A very hard cheese that will keep almost indefinitely, Parmesan is used as grated cheese on salads, soups, and with pasta.

Process Cheese. Process cheese is made by grinding and mixing together by heating and stirring one or more naturally fermented cheese of the same or different varieties, together with an emulsifying agent into a homogenous mass at very high temperature (130°C) for a few seconds. After the mixture has been properly prepared, the cheese can be made into various thickness, shapes, and forms. Also, a variety of ingredients such as bacon, spices, port wine, and so on can be incorporated into the cheese. At least one-third of all cheese marketed in the United States, except soft, unripened cheese, is process cheese.

Process Cheese Food. Process cheese food is made in the same way as process cheese, except that certain dairy products (cream, milk, skim milk, cheese whey, or whey albumin) may be added. At least 51% of the weight of the finished product must be cheese.

Process Cheese Spread. This is made in the same way as process cheese food, except that it contains more moisture and less fat and must be spreadable at a temperature of 21.1°C (70°F). Fruits, vegetables, or meats may be added.

Roquefort. A cheese made only in the Roquefort area of France from ewe's milk, Roquefort is characterized by sharp, peppery flavor and blue-green veins throughout the cheese. It is ripened principally by blue mold in the interior.

Swiss. A hard cheese with an elastic body, nutlike flavor, Swiss cheese is best known by the holes or eyes that develop as the cheese ripens.

An excellence reference set—*Cheese and Fermented Milk Foods Vol 1: Origins and Principles* and *Vol 2: Procedures and Analysis*—that covers all aspects of fermented diary foods was published by Kosikowski and Mistry (5).

ALCOHOLIC BEVERAGES

The per capita consumption of alcohol beverages in the United States is about 2.4 gal of wine, 24 gal of beer, and 1.6 gal of spirits. In 1995 alcoholic beverages accounted for $103.9 billion in sales, with beer being $62.6 billion, spirit being $29.5 billion, and wine being $11.8 billion. Alcoholic beverages can be classified as fermented, not distilled, and distilled.

Fermented, Not Distilled Beverages

Wine. By definition, wine is a fermented product from fruits. By far the most important wine is produced from grapes, although it can be produced from apples, pears, berries, and other fruits. *Vitis vinifera* and *V. labrusca* are the most important species. Grape growing is a science itself (viticulture), and wine making is called enology.

To obtain a good wine, the enologist must consider the species and varieties used, the climate and soil conditions that dictate the vintage of a good wine, and the time to pick the grapes to make the wine. Europe produces about 80% of the world's wine, with North and South America producing about 14%; Africa, 4%; Asia, 1%; and Oceania, 1%. Whereas in Europe people consume about 30 gal of wine a year per person, the U.S. population consumes 2 to 3 gal per year per person. Grapes are first picked in autumn, when the amount of sugar in the grape stabilizes, then the grapes are stemmed and crushed to separate and remove the leaves and stems. The resultant materials, juice, skin, pulp, and so on, are called must (*L. mustum*, "new wine"), which is the source of the wine.

Next SO_2 is added at 50 to 100 ppm to condition the must for its ability to inhibit undesirable organisms and its antifungal and antioxidative properties. The correct amount of sugar is added to define the final concentration of alcohol in the wine. The rule is when one wants to make a wine of X % alcohol one would add 2 X % of sugar. This is because 1 g of sugar will be converted to half a gram of alcohol and another half a gram of carbon dioxide, which is released into the environment. The conditioned must will then be inoculated with a wine yeast (usually *S. cerevisiae* var. *ellipsoideus* Montrachet strain) at 1% inoculum level. For red wine the must can be fermented on the skin, which allows extraction of the red anthocyanin by alcohol to give the red color to the wine. Alternatively, the must can be hot-pressed (the must is heated to 62.7°C, and the juice is pressed out while hot), which also extracts the red color. To make white wine, white grapes can be used or, if red grapes are used, a cold press will yield a white juice. Pressing of the wine is an important step. In a traditional wine press, when the must is poured into the press, some juice flows through the system without pressing. This "free-run" juice yields the best wine. The harder the grapes are pressed, the poorer the quality of the wine produced. At this point it is important to check the acidity of the must. The liquid should be about pH 3.6 and 0.7 to 0.9 g acid/100 mL of juice. If adjustment is needed, tartaric acid is used to acidify the must, and water is used to dilute the juice. This is the amelioration step. The juice is now ready for secondary fermentation.

Secondary fermentation involves putting the fermenting must into a closed container (cooperage) and applying an airlock to prevent oxygen from entering the system and to allow carbon dioxide to escape. Because heat is produced (exothermic reaction), the fermenting must is kept at 15.5°C to 21°C for red wine and 12.8°C to 18.3°C for white wine.

Fermentation will continue for two to six weeks depending on the amount of dryness (reduction of sugar) desired in the finished wine. Before the completion of fermentation, several racking steps may be done. Racking is a process in which the wine is separated from lees (dead yeast cells and insoluble materials at the bottom of the cooperage). The first racking is done after fermentation to desired dryness (an appropriate amount of sugar is fermented to alcohol), followed by racking after dropping (crystallizing) the insoluble tartaric in cold temperature (ca 0°C), and the final racking is done after clarification and before bottling. Clarification can be achieved by the addition of fining agents such as gelatin, casein, bentonite, and polyclar AT. Enzymes may also be added to degrade pectin. The new wine is put into a wooden cooperage for aging. The biochemistry involved in aging is complex, but at least one year of aging is necessary for a good wine to be produced. The exact time for bottling of wine is determined by the wine master of the particular winery. Wine can be pasteurized (55.5–65.5°C for a few seconds), but more often it is bottled without heat treatment, because heat tends to destroy flavor compounds. To remove any residues after aging, a filtration step can be made before bottling. Wine continues to improve in the bottle. Ten years of aging is optimum.

Classification of Wines

Hundreds of varieties of wines are produced around the world, and no classification can cover all of them. The following is a functional classification of wines.

Dry or Sweet Wines. In dry wine all fermentable sugar is converted to alcohol. Typically the wine has about 8 to 12% alcohol. Examples are Chablis, Riesling, Burgundy, and Chianti. Sweet wine has some unfermented sugar. The alcohol is about 13 to 15%. Examples are Tokay and sauterne.

Fortified or Unfortified. In fortified wine more alcohol is added to the product. Some examples are sherry, port, muscatel, and champagne. Unfortified wine is a wine in which all the alcohol resulted from fermentation.

Sparkling or Still. Sparkling wine is bottled before fermentation has ceased and liberates effervescence after the bottle is opened. Champagne made in France is the true example of fermentation in the bottle. To be able to show that the gas is generated in the bottle, a label can be printed as "fermented in this bottle." Injection of CO_2 into a wine to achieve effervescence is called crackling. In this case the manufacturer cannot claim that the wine is fermented in this bottle. Still wines are those in which fermentation ceased before bottling and no CO_2 exists.

Red or White. Fermentation of the skin of red grape will result in red wine such as claret, Burgundy, port, and chianti. Reisling is a good example of a white wine.

Generic Wine. Wines blended from several varieties of grapes are called generic wines. Examples are Burgundy, chianti, Chablis, and claret.

Variety Wine. Wines from predominately (at least 75%) one type of grape variety are variety wines. Examples are cabernet, pinot noir, zinfandel, and chardonnay.

Vintage Wine. Vintage wine must have at least 95% of the juice coming from the year claimed.

Prevention of Wine Spoilage

Oxygen is the worst enemy of wine because it promotes the growth of *Acetobacter*, which oxidizes alcohol to acetic acid. Oxygen also oxidizes wine color from white to heavy amber or from rich purple to tawny brown and affects the flavor of the wine. Dirty equipment and environment contributes to spoilage of wine directly by introducing microorganisms or indirectly due to residues of unwanted chemicals. Undesirable microorganisms during the fermentation process or in aging and storage can also contaminate wine.

Champagne Processing

Champagne processing is a fascinating subject and deserves some discussions. About 250 years ago a monk named Don Perignon in the Champagne region of France discovered the process of making champagne. He exclaimed, "Come and see I am drinking stars." Champagne making is now a highly regulated manufacturing process under the control of the French government. At the perfect time of maturity of the grapes (chardonnay, pinot meunier, and/or pinot noir), they are picked by hand. Great care is taken not to bruise the delicate skin of the grapes during the harvesting stage. The grapes are then transported to the winery to be pressed to obtain the grape juice. Four thousand kg or 8800 lb of grapes are placed in a tub called marc. Each marc is pressed several times, yielding a total of 2550 L as laid down by the rules of *appellation*, which calls for 1.6 kg (3.52 lb) of grapes per liter of juice. The first three pressings yield 2050 L of first-quality juice known as the cuvee. A final pressing then yields 500 L of taille or lesser-quality juice.

After pressing, the vin de cuvee and vin de taille are allowed to stand for the solids to settle. After that time the must will be allowed to ferment and is made into wine. Only wine made from vin de cuvee is used for champagne making. After fermentation, the cuvee will have about 10% alcohol. Into this cuvee more sugar and yeast culture will be added, and a strong cap is applied to the bottle to start a secondary fermentation in the bottle. The amount of added sugar will dictate the amount of pressure generated in the bottle. This is why champagne bottles are made with very strong and thick glass. During this fermentation, the yeast will convert the added sugar to alcohol and carbon dioxide. It is this carbon dioxide that causes champagne to bubble. An important step in champagne making is to remove the yeast cells from the enclosed bottle so that the champagne will not be cloudy. To achieve this, a very tedious procedure is used in which a worker will turn the champagne bottle one quarter time per day for two weeks (riddling). Each day after riddling, the bottle will be tilted a little more toward the neck such that yeast will slowly migrate to the neck of the bottle by gravity. After completion of the riddling process, all the yeast will be at the neck of bottle. The neck is then frozen. This will cause a column of ice to be formed, which traps the yeast. The bottle is then opened quickly, and the pressure of the gas in the bottle will expel the column of ice with the yeast. Quickly the worker will add some sterile wine into the bottle to compensate for the lost liquid, and the bottle is tightly closed. A champagne is then made. The champagne can be stored in caves for years before shipment for sale. In the Meot & Chandon company in Epernay, France, something like 5 million bottles of champagne are stored in 7.1 mile of caves in this lime country. Current research work on champagne includes: (*1*) development of disease-resistant strains of grape vine; (*2*) use of alginate pellets to immobilize yeast cell in the champagne for secondary fermentation; after fermentation, it would take only 40 s to settle all the alginate pellets with yeast to the neck and remove the yeast a frozen column—this can take the place of the tedious "riddling" process by hand; and (*3*) measurement of amount of bubble and foam by use of robots and camera to examine champagne quality.

Beer

Although beer is more complicated to produce, it was developed earlier in history than wine at around 6000 B.C. By definition beer is an alcoholic drink derived from grains. The most common source of beer is barley, with rice and corn used as adjuncts. The per capita consumption in Europe is about 32 gal of beer a year, whereas in the United States the per capita consumption is about 25 gal. Beer is made by malting barley to obtain α- and β-amylases, then by mashing, which is a process in which more grain (adjuncts) and water are added to the malt. The mixture is heated slowly to first achieve proteolysis (35°C for 1 h) and then starch hydrolysis (67–68°C for 20–30 min); finally, the enzymes are inactivated (mashing off at 75–80°C). The main purpose is to allow α- and β-amylase to degrade amylose and amylopectin to yield glucose and maltose for fermentation by yeast, because yeast cannot degrade starch directly. The liquid (which can be considered as nutrient broth for yeast) is separated from the husks of the grains by lautering and sparging. Basically these are filtration and rinsing steps to obtain clear liquid for fermentation. The resultant clear fluid is called wort. Before addition of yeast for fermentation, the wort is boiled and hops (*Humulus lupulus*) is added to give the bitter flavor of beer. After cooling to 8.8°C, yeast is added (pitching). *Saccharomyces carlsbergensis* (bottom yeast) is the yeast of choice for lager beer making. Fermentation continues for three to four days, at which time a lot of foam is developed in the fermentation vessel (Krausen formation). After five days, the Krausen collapses and after 10 to 12 days fermentation ceases. At this stage the beer is called green beer. Yeast cells can be recovered for animal feed, and CO_2 is collected and later injected into finished beer for foam formation. The green beer is aged for about three months before being bottled for sale. Beer can be sold in unpasteurized form (draft beer), which has a short shelf life. Most of the beer on the market is pasteurized at 57.2 to 60°C for 15 to 20 min in bottles and cans. Some beers are filtered and not pasteurized (draft canned beer). Beer made in the preceding manner is called lager beer, and the alcohol content is about 3.5 to 4%. Using *S. cerevisiae* (top yeast), ale can be made that has about 6% alcohol. European beers are "heavier" because they use only barley, which has more protein than barley used in the procedure in the United States, where barley is mixed with rice and corn adjuncts that decrease the total protein contents.

Light Beer. Light beer is made by using enzymes such as glucoamylase and pullulanase along with the α- and β-amylases to convert all the starch to fermentable sugar; thus, no sugar is left in the beer after fermentation. The beer has fewer (lighter) calories than the regular beer, which has residual starch. Light beer still has calories from alcohol since 1 g of alcohol provides 7 cal of energy.

Dry Beer. Dry beer was started in Japan. Japanese beer has a sweeter taste than Western beer and is not blended. By a process called *superattenuated fermentation*, special yeast strains are used for prolonged fermentation, resulting in the utilization of all the sugars. In dry beer making, no enzyme is added to reduce the carbohydrate.

Ice Beer. This is the newest entry into the beer market. Beer is first made and then quickly frozen to obtain small ice crystals. The beer is then filtered through a membrane that retains the ice crystals. The reason behind this process is that the ice crystal will capture undesirable compounds of the beer with a resultant smoother beer. Some people think that this is just another commercial effort to keep consumers entertained by having another style of beer. Because the water is removed from the beer during the freezing and filtration processes, ice beer has a higher alcohol content of about 6% compared with regular beer.

Sake Making

Sake, a pale white drink, is closely related to Sinto religion and is often called rice wine. Technically it is a beer because it is made of rice. The rice is first inoculated with spores of *Aspergillus oryzae* (Koji), which is the source of amylase to degrade starch. The Koji is then added to more rice along with yeast to achieve fermentation. The resultant liquid has a 14 to 20% alcohol content with no gas.

Beverage from Distilled Fermented Liquids

Because natural fermentation can achieve only about 12 to 15% alcohol and in extreme cases like sake, 20% drinks of higher alcoholic content are achieved by concentrating alcohol through the process of distillation.

Whiskey includes rye, bourbon, Scotch, Irish, Canadian Pisco, and vodka, which are distilled from fermented grains. Rums (Cuba, Puerto Rico, Philippines, Jamaica, Barbados, Martinique, and New England) are distilled from sugarcane and molasses. Tequila is distilled from agave. Brandies are distilled from a variety of fermented fruit juices such as grape (cognac, Armagnac, and brandy), apple (calvados and apple jack), cherry (kirsch and cherry brandy), and plum (slivovitz and Micabelle). In addition, distillation with flavored compounds provide such drinks as gins, liqueurs, absinthe, aquavit, and bitters.

Fermented Products from Vegetables

Vegetables have low buffering capacity. After harvest, vegetables have a heterogeneous population, including *Pseudomonas, Bacillus, Chromobacterium*, and a variety of enteric organisms; lactic acid bacteria (the organisms responsible for fermentation) exists in relatively small numbers. However, once the material is placed under anaerobic conditions with the addition of 2.5% salt, lactic acid bacteria quickly predominates and initiates a favorable fermentation process. Such is the case in sauerkraut making. Shredded cabbage has 3 to 6% sugar (glucose, fructose, and sucrose). The first group of organisms growing are *Enterobacter cloacae, Erwinia herbicola*, and *Leuconostoc mesenteroides* (a heterofermentative lactic). As the lactic acid increases, the enterics die off and *Leuconostoc mementeroides* predominates, producing favorable compounds for the development of *Lactobacillus plantarum* (a homofermentative lactic). This organism continues to produce lactic acid to a level of 1.5 to 2.0% and completes the fermentation process. Vegetable fermentation represents a form of natural fermentation because, regardless of the type of starter culture used, *L. plantarum* always ends up as the principal organism in a properly treated fermentation vet. A popular fermented vegetable in Korea is Kimchi. Hundreds of varieties of Kimchi are made by anaerobic fermentation of mixed vegetable along with meat, spices, and other ingredients. It is interesting to note that in the West, vegetable fermentation is usually by use of one type of vegetable, but in the Orient the rule is to use several kinds of vegetable in the same fermentation vessel.

The fermentation of cucumbers in the making of dill pickles and salt stock pickles is another example of vegetable fermentation. The microbial succession of cucumber fermentation closely follows the pattern achieved in sauerkraut fermentation. Almost any kind of vegetable can be fermented.

Fermented Products from Meats

Since ancient times many types of fermented meat product have been developed. However, scientific understanding of fermented sausage resulted only after the advancement of meat microbiology in the last 100 years. Currently most meat is consumed unfermented. However, many varieties of meat products are preserved in the form of sausages (such as frankfurters, bologna, Vienna sausage, loaves and luncheon meats, bratwurst, brockwurst, and braunschweiger). Only 6% of U.S. sausages are fermented. They are either fermented sausages (pepperoni and salami) or semidry fermented sausages (cervelat and Thuringer). The basic ingredients for fermented sausage are ground meat (one type or a mixture of a few types), salt (2.5–3.5%), sugar (1%, for fermentation), spices, nitrites, ascorbate and erythorbate, and lactic acid bacteria. The traditional process of applying cultures is to let chance contaminants start the fermentation. This process is highly unreliable because ground meat contains many types of microorganism other than lactic acid bacteria. The second process is by back-slopping, when the batter from a previous successful batch is added to a fresh sausage batch to initiate the fermentation. This method is more reliable than the traditional method but is still subject to unwanted variations. The third and most reliable process is by the addition of starter cultures. Lactic acid bacteria such as *Pediococcus* and *Lactobacillus* have been successfully used to make fermented sausages. These bacteria in the presence of sugar and the anaerobic environment provided by the casing of the sausage quickly produce lactic acid and also develop the tangy flavor typical of the taste of fermented sausage. Fermented sausages can also be smoked to add distinctive flavors.

Fermented Cereal Foods—Bread

According to many people of the world, bread is the staff of life. Bread is only one type of fermented cereal product. Other products include rolls, Danish pastry, crackers, doughnuts, pretzels, and so on. Bread and bakery products are a multibillion dollar business in the United States, the majority of which is bread. The definition of a bread is a product of moistened, kneaded, fermented, and baked meal of flour (mainly from wheat) with appropriate added ingredients. Yeast (*S. cerevisiae*) is the organism to facilitate the fermentation process to make bread. Yeast can be supplied to the baking industry in a compressed, bulk, or active dried yeast form. Bread is made by many different processes, but the most popular method is the sponge-dough method. After all the ingredients are mixed for the sponge part (65% of total flour, water, yeast, and yeast nutrients), the sponge is kept at 25°C for 4.5 h for the first fermentation. The volume increases to four to five times, and the temperature increases to 30°C due to the exothermic reaction of fermentation. The rest of the ingredients (flour, water, sweetener, fat, dairy product, crumb softener, rope and mold inhibitor, dough improver, and enrichment) are then added to make the final dough. The dough is mixed at 72 rpm to produce a smooth cohesive dough that has a glossy sheen. The mixed dough is allowed to rest for 20 to 30 min, then the dough is divided and rounded and put into molds. In the proof box (35–43 °C at 80–90% humidity for 60 min) the dough expands to the desired volume due to the fermentation of sugar with the development of CO_2, which is trapped by the elastic gluten of the dough. Baking is the last step in bread making. During baking the heat of the process expands the trapped gas in the dough matrix and causes the dough fabric to oven-spring. Enzymes are active until the dough reaches 75°C, when the gluten matrix coagulates and dough structure is set into the form of the bread. When the bread surface reaches 130 to 140°C, sugar and soluble protein react chemically to give the crust color and textures. The center of the bread does not exceed 100°C. After cooling, slicing, wrapping, and distributing, the bread reaches consumers. In many places in Europe, bread is baked a couple of times a day and consumers eat bread fresh from the oven. In the United States, breads in supermarkets are usually half a day to a day old before consumers purchase the products. One of the problems of the science of bread making is staling. After years of research, the exact cause of staling is still unresolved.

Production of Vinegar

Vinegar is literally a result of souring (aigre) of wine (vin). The origin of vinegar no doubt followed the production of wine, because a bad batch of wine will result in some form of vinegar. Vinegar has been used as a flavoring agent, food preservation agent, and even as medicine. Both Eastern and Western cultures have records of vinegar in ancient history. Although vinegar production is always treated as a part of food fermentation, it is, in reality, an oxidative process. In the presence of molecular oxygen and *Acetobacter aceti*, or related species, alcohol is oxidized to acetic acid. Historically vinegar was made by the let-alone process or the field process; poor quality wine was allowed to sit in the open air for oxidation to occur. Currently the methods to make vinegar are the trickling process and the submerge culture generator. In the trickling process, wine is trickled into a Frings-type generator where the liquid comes in contact with wood shavings, thereby exposing it to large amounts of oxygen. The alcohol is then oxidized by the *Acetobacter* in the system. At the bottom of the system, the collected liquid contains about 12% vinegar. This process takes about two weeks. The submerged culture system takes about 35 h to achieve the same concentration of vinegar. In this closed system, air is pumped into the generator containing wine, and the organisms in the mixture actively oxidize alcohol to vinegar. Since *Acetobacter* can overoxidize acetic acid to hydrogen and carbon dioxide, *Gluconobacter* is used in submerged systems because this organism does not have enzymes to further oxidize acetic acid to other compounds. Vinegar can be pasteurized to render it more stable in storage.

Mold-Modified Foods

In many parts of the world the indigenous fermented products are modified by molds along with yeast and bacteria. These products are usually the result of uncontrolled, naturally mixed culture fermentation. Much research needs to be done on these foods to ascertain the microbes involved in the process, the safety of the products, and the economical feasibility of large-scale production in modern industrial plants. Some of the more important foods in this category are listed next.

Soy Sauce. A product of the Orient, soy sauce is made from fermentation of whole or defatted soybean and soybean products along with roasted wheat, using a combination of *Aspergillus oryzae, Pediococcus halophilus, Lactobacillus delbrueckii, Torulopsis versatilus*, and *Saccharomyce rouxii*, resulting in a dark reddish liquid, salty taste, and an important flavoring agent.

Miso. This is a product of the Orient from fermentation of whole soybeans mixed with rice or barley using *Aspergillus oryzae, S. rouxii, Torulopsis etchellsii*, and *P. halophilus* to produce a dark reddish, smooth paste with a strong flavor and salty taste.

Hamanatto. A product of the Orient, hamanatto is made from fermentation of whole black soybeans and wheat flour by *Aspergillus, Streptococcus*, and *Pediococcus*. These black soft beans have a salty flavor and are used as a condiment.

Sufu. Sufu is a product of China obtained by fermentation of soybean curd (tofu) using *Actinomucor elegans* and *Mucor dispersus*, resulting in salty and strongly flavored cream cheese–type cubes.

Tempeh. Tempeh is a product of Indonesia and vicinity using whole soybeans inoculated with *Rhizopus oligosporus*, resulting in a soft bean cake bound by mycelia of the mold. It has a clean taste and can be fried, cooked, or eaten as is.

Natto. A product of Japan, natto is obtained by fermentation of whole soybeans with *Bacillus subtilis* var. *natto*, which results in beans covered with viscous, sticky glutamic acid polymers produced by the bacteria. It has a strong ammonia odor and is eaten with or without further cooking.

Other mold-modified foods include *bongkrek* (coconut cake, Indonesia), *ontjom* (peanut cake, Indonesia), *lao-chao* (glutinous rice, China and Indonesia), *ang-kak* (red rice, China and the Philippines), *idli* (rice, India), *doza* (rice, India), *trahana* (parboiled wheat, meat, and yogurt, Turkey), *injera*, (teff, Ethiopia), *kishk* (wheat and milk, Egypt and Middle East), *gari* (cassava roots, West African and Nigeria), *ogi* (maize, Benin and Nigeria), *mahewu* (maize, South Africa), *pozol* (maize, Mexico), and many others.

Other food products are also fermented, such as coffee, cacao, vanilla, tea, citron, ginger, mead, fish sauces (*nu'oc mari*), poi, olives, *pidan* (hundred-year-old egg), and so on. The field of food fermentation can even include such processes as production of single-cell protein, lactic acid, citric acid, glutamic acid, lysine, antibiotics, vitamins, lipids, ascorbic acid, and so on. Many areas of food fermentation overlap with industrial fermentation and biochemical engineering.

THE ROLE OF GENETIC ENGINEERING IN FOOD FERMENTATION

Most of the cultures used in food fermentation are naturally occurring cultures. Selection of special strains allow the manufacturers to produce special aromas, flavors, textures, and odors of foods. By selection, mutation, and crossing of genes, much improved cultures have been developed to produce better fermented food. Now with the advent of genetic engineering scientists can insert genes of almost any living things into the chromosomes or plasmids of another organism to produce desirable products. For example, the rennin gene of calf has been cloned into *E. coli* to produce rennin that is chemically and biochemically exactly like the rennin extracted from the calf, but the enzyme is much cheaper. A lot of work has been done in developing phage-resistant dairy starter cultures. These types of development will no doubt help advance the science and productivity of starter culture technology and, it is hoped, provide more delicious, nutritious, plentiful, less expensive, and healthy fermented food products.

In conclusion, food fermentation is an important field of study. Although much has been learned about some well-known fermentation processes such as beer, wine, cheese, and bread, much needs to be studied concerning mold-modified foods, indigenous foods, and improvement of existing processes. With the rapid advancement of genetic engineering, some dramatic changes in the field of food fermentation should occur in the near future.

BIBLIOGRAPHY

1. L. E. Erickson and D. Y. C. Fung, eds., *Handbook on Anaerobic Fermentations*, Marcel Dekker, New York, 1988.
2. S. E. Gilliland, *Bacterial Starter Cultures for Food*, CRC Press, Boca Raton, Fla., 1985.

3. D. Y. C. Fung et al., "Mixed Culture Interactions in Anaerobic Fermentations," in L. E. Erickson and D. Y. C. Fung, eds. *Handbook on Anaerobic Fermentations*, Marcel Dekker, New York, 1998, pp. 501–536.
4. G. Sauder, *Cheese Varieties and Descriptions*, USDA Agricultural Handbook No. 54, Washington, D.C., 1978.
5. F. V. Kosikowski, and V. V. Mistry, *Cheese and Fermented Milk Foods*, 3rd ed. Vols. 1 and 2, Edwards Brothers, Ann Arbor, Mich., 1997.

GENERAL REFERENCES

M. A. Amerine et al., *Technology of Wine Making*, 4th ed., AVI Westport, Conn., 1980.

G. J. Banwart, *Basic Food Microbiology*, 2nd ed., Chapman and Hall, New York, 1989.

L. R. Beuchat, *Food and Beverage Mycology*, AVI, Westport, Conn., 1978.

C. M. Bourgeois, J. Y. Leveau, and D. Y. C. Fung *Microbiological Control of Food and Agricultural Products* (English ed.), VCH, New York, 1995.

D. R. Buege and R. C. Cassens, *Manufacturing Summer Sausage*, Extension. No. A 3058, University of Wisconsin, Madison, Wis., 1980.

R. A. Copeland, *Enzyme*, Wiley-VCH, New York, 1996.

W. Crueger and A. Crueger, *Biotechnology: A Textbook on Industrial Microbiology*, Sinauer, Sunderland, Mass., 1984.

M. P. Doyle, L. R. Beuchat, and T. J. Montville, *Food Microbiology: Fundamentals and Frontiers*, ASM, Washington, D.C., 1997.

D. Y. C. Fung and K. Vicheinroj, *Introduction to Food Fermentation*, Kansas State University, Manhattan, Kans. 1998.

Cultures for the Manufacture of Diary Products, Chr. Hansen, Inc., Milwaukee, Wis., 1997.

ICMSF, *Microorganisms in Foods 5: Microbiological Specifications of Food Pathogens*, Blackie Academics and Professionals, New York, 1996.

J. M. Jay, *Modern Food Microbiology*, 5th ed., Van Nostrand Reinhold, New York, 1996.

P. Z. Margalith, *Flavor Microbiology*, Charles C Thomas, Springfield, Ill., 1981.

R. T. Marshall, *Standard Methods for the Examination of Dairy Products*, 16th ed., American Public Health Association, Washington D.C., 1992.

D. A. A. Mossel et al., *Essentials of the Microbiology of Foods*, John Wiley, New York, 1995.

C. S. Pederson, *Microbiology of Food Fermentation*, 2nd ed., AVI, Westport, Conn., 1979.

H. J. Peppler and D. Perlman, *Microbial Technology*, Vol. 11, Academic Press, Orlando, Fla., 1979.

R. K. Robinson, *Dairy Microbiology*, Vols. 1 and 2, Elsevier Applied Science, Barking, United Kingdom, 1976.

J. H. Silliker et al., *Microbial Ecology of Foods*, Vols. 1 and 2, Academic Press, Orlando, Fla., 1980.

P. F. Stanbury and A. Whitaker, *Principles of Fermentation Technology*, Pergamon Press, Elsmford, N.Y., 1984.

K. H. Steinkraus, *Handbook of Indigenous Fermented Foods*, Marcel Dekker, New York, 1983.

R. P. Vine, *Commercial Winemaking: Processing and Controls*, AVI, Westport, Conn., 1981.

D. I. C. Wang et al., *Fermentation and Enzyme Technology*, John Wiley, New York, 1979.

DANIEL Y. C. FUNG
Kansas State University
Manhattan, Kansas

FOOD FREEZING

Freezing has long been established as an effective method for the long-term preservation of foods. It has the potential to maintain quality and nutritional characteristics close to those of fresh foods. Although refrigerated storage has been developed to a stage where high quality can be maintained for a few weeks, when extended shelf life is required of a preservation technology, the most commonly used methods are freezing, thermal processing, and canning. It is generally accepted that of these three methods, freezing is superior (1). Properly conducted freezing retains much of the flavor, color, texture, and nutritive value of food. The activity of microorganisms is much reduced, and, while frozen, many enzymes also have reduced activity. However, the freezing process can disrupt the control systems that influence enzyme action, leading to undesirable changes during thawing. Where this is a problem, special techniques may be used to destroy or reduce enzymic activity before freezing. The consequences of thawing are mentioned because, with the exception of frozen commodities that are consumed directly (such as ice cream), normal freezing preservation requires that products be thawed before they are used.

Long before commercial application, refrigeration was used as a preservation technique. An early report is found in the *She King*, Chinese poetry written in 1100 B.C. (2). The earliest refrigeration agent was natural ice from frozen rivers and lakes and mountain snows. Natural refrigeration was used to preserve meat, game, fish, and, more recently, eggs, vegetables, and fruits. It has long been accepted that low temperatures reduce the rate of spoilage of food products compared to higher temperatures, and those who reside in colder climates have tended to expect food to store for longer times than those from hot climates. Modern progress in the refrigerated storage of foods, however, required the development of systems capable of producing low temperatures on demand.

REFRIGERATION HISTORY

The early history of the development of refrigeration has been reviewed and summarized (1–4). Table 1 presents some of the important milestones in the development of refrigeration.

The use of cool conditions for food storage dates back before the dawn of recorded history. Closer to modern times, in 1683, Robert Boyle reported the effects of refrigeration and freezing on foods and insects. Ice was reportedly used by a Massachusetts fisherman to preserve his catch until it could be brought to port and marketed (5). Natural ice, harvested in winter and stored in special facilities, was the primary source of refrigeration until the development of mechanical systems for ice making in the mid 1800s, based on the developing science of thermodynamics.

The early development of mechanical refrigeration was targeted to medical needs rather than to food preservation. John Gorrie developed a mechanical system, using air compression and expansion and the Joule–Thomson effect, to

Table 1. Important Events in the Development of the U.S. Frozen Food Industry

Year	Event
500 B.C.–1800 A.D.	Cooling achieved by use of snow, natural ice, air in cold climates, evaporative cooling of water, and radiative cooling
1820	Natural ice had come into general use as an article of commerce and was used on a large scale for food preservation.
1851	Jacob Fussell of Baltimore first sold ice cream on a significant commercial scale in the United States.
1861	Thomas S. Mort established what is believed to be the world's first cold-storage plant in Darling Harbor, Australia.
1864	Ferdinand Carre patented an ammonia compression machine in France.
1870	Ammonia compression machines were brought to a level of practicality almost simultaneously by Dr. Carl Linde in Germany and David Boyle in the United States.
ca 1880	Ammonia compression machines and insulated rooms began to be used in the United States.
ca 1916	Scientific work on the methodology of freezing foods began in earnest.
1923	The quick freezing industry began with the founding of a freezing company by Clarence Birdseye.
1929	MA Joslyn and W.V. Cruess reported on the need to blanch vegetables before freezing.
	The Birdseye organization was bought by Postum Co. (now General Foods Corp.), marking the real beginning of marketing of frozen foods through retail stores in the United States.
1930	Mechanical refrigeration had assumed an almost indispensable part of the food distribution system in the United States.
1949	Mechanically refrigerated railroad cars came into use about this time.
Early 1960s	Fluidized-bed freezers and individually quick frozen foods began to assume a position of importance in the United States
1962	Liquid nitrogen food freezers were first used commercially.
1968	Freon®-12 freezers were first used commercially.

Source: Ref. 1.

cool air in a cylinder and produce ice for application in the treatment of malaria (2). Commercial ice making began about 1876. The number of mechanical refrigerated warehouses began increase about 1890, but it was not until about 1915 that large-scale construction began. These early refrigeration systems were much too large to be installed in individual homes.

In the mid to late 1930s and the early 1940s, small-scale compressors became readily available. This allowed for the design and manufacture of relatively cheap and compact refrigeration units, which meant that home refrigeration was now practical. The dangers of ammonia refrigerant led to the development of an alternative, safer group of refrigerants, the fluorocarbons (Freon), by the DuPont company. In recent years, the undesirable environmental consequences of these refrigerants has led to continued development of more environmentally acceptable refrigerants.

In 1923, Clarence Birdseye created the first modern frozen food business, which set in motion a revolution in food preservation that has had significant consequences to the American diet. By 1930, the list of frozen foods consisted mainly of fish, small fruits, poultry, meat, eggs, and a few vegetables. Quality was variable. Joslyn and Cruess recognized the role of enzymes in quality degradation and introduced the blanching process for the destruction of enzymes. This led to significant quality improvement. During the past 60 years, the research efforts of frozen food processors, packaging manufacturers, and university and government scientists have led to further improvements in the quality and stability of frozen foods.

When describing the technologies of food freezing, it is best to consider the food freezing process from two perspectives: (*1*) the methodologies for removal of heat, that is, the cooling process, and (*2*) the effects of removal of heat on the properties of the system undergoing cooling.

THE REMOVAL OF HEAT (REFRIGERATION)

Principles of Refrigeration

Heat flows from a high temperature to a lower temperature. To remove heat from an object by convection, conduction, or radiation, there must be a heat sink at a lower temperature. If the object is at ambient temperature, then the heat sink must be at a temperature below ambient to achieve cooling. There are several ways to reduce temperature below ambient. Early cooling methods involved evaporative cooling, which is what one experiences when a cooling breeze causes the evaporation of perspiration. Evaporative cooling has long been used for cooling water with porous pots and for air conditioning with swamp coolers. The principles behind evaporative cooling will be described later. Another simple method is to use ice, but if natural ice is not used, ice must how be produced in refrigerated cooling systems capable of producing temperatures below the freezing point of ice.

Two observations are key to the understanding of refrigeration methods. First, a gas increases in temperature after being compressed with no heat removed and, similarly, reduces in temperature after expansion. Second, heat has to be supplied to a liquid in order for it to vaporize at constant temperature, and the condensation of a vapor yields heat. The two most commonly used refrigeration methods employ the *vapor compression refrigeration cycle* and the *absorption refrigeration cycle* (6,7). Mechanical energy drives the vapor compression system, which is commonly found in warehouses. Thermal energy drives the absorption cycle, which benefits from greater mechanical simplicity and the ability to use a variety of sources for thermal energy. In both cycles, the primary heat sink is a liquid boiling at low temperature and pressure, and the primary heat rejection is a vapor at higher temperature and pressure condensing to a liquid. In the vapor compression system, work is used to compress the vapor; in the absorption system, heat is used to raise the temperature of the absorbent and boil off the vapor at a higher temperature.

To better understand the operation of these refrigeration systems, it is necessary to learn the principles of thermodynamics. Thermodynamics is the name given to the study of energy, its transformations, and its relation to the states of matter. The units of measurement representing amount of heat or work are the calorie (cal), the joule (J), and the British thermal unit (Btu). These are defined as follows (8):

> One calorie is the heat required to raise the temperature of 1g of pure water from 14.5°C to 15.5°C.
>
> One joule is the work done when a force of 1 N acts through a distance of 1m.
>
> One Btu is the quantity of heat required to raise the temperature of 1lb of water from 63°F to 64°F.

Refrigeration is the process of removal of heat from an object or area. As has already been stated, heat moves from a warm to a cold body. In addition, two thermodynamic laws must always be obeyed. The first law of thermodynamics, the law of conservation of energy, states that energy cannot be created or destroyed. This law may be restated in a variety of forms, all of which imply that although energy may assume many forms, the total amount of energy must remain constant. It follows that if energy should disappear in one form it must simultaneously appear in other forms (8,9). Consequentially, different forms of energy are interconvertible and equivalent. For each possible transformation, there is a definite conversion ratio.

In one form, the second law of thermodynamics states that no system can receive heat at a given temperature and reject it to a higher temperature without the input of work from the surroundings. Because a refrigeration system performs such a transfer of heat, from a lower temperature region to a higher temperature region, this transfer necessarily requires the input of work in some form.

To properly describe a thermodynamic process, it is necessary to maintain a careful bookkeeping system that accounts for energy in all the forms that are interconverting. This permits keeping track of the transformations quantitatively. There are some standard thermodynamic functions that are used for these bookkeeping purposes. In addition, the conditions of the process must be defined to make the bookkeeping feasible. Considering equilibrium states, such characteristic thermodynamic properties as internal energy (U), entropy (S), and enthalpy (H) are defined functions of easily measurable properties, such as pressure (P), temperature (T), and volume (V). Also, for equilibrium states, there are relationships among P, PV, and T. For an ideal gas, for example, the relationship $PV = RT$ holds, where R is a constant. Real gases have pressure–volume–temperature relationships that are extensions of this simple equation. Several standard process routes can be to describe the thermodynamic relationships of a system, including:

1. *Constant temperature (isothermal) process.* In this process, temperature is held constant and volume varies with pressure.
2. *Constant pressure (isobaric) process.* In this process, pressure is held constant and volume varies with temperature.
3. *Constant entropy (adiabatic) process.* In this process, no heat transfer occurs between the system and its surroundings.

The bookkeeping can then be carried out by using some standard thermodynamic diagrams, which are discussed in the next section.

Thermodynamic Diagrams

In a thermodynamic diagram, the temperature, pressure, volume, enthalpy, and entropy relationships of substances are represented. The most frequently used diagrams are temperature–entropy and pressure–enthalpy diagrams. In these diagrams, it is important to know whether the system being described is one phase or multiphase and which phases are present. This is most readily achieved by plotting the saturation lines, which represent, in a system capable of existing as liquid or as vapor (ie, a fluid), the point at which the first bubble of vapor appears in the liquid and the point at which the last drop of liquid evaporates to vapor. Thus, the saturation lines delineate the boundary between the one-phase and the two-phase regions. The one phase may be liquid or vapor, or the two phases may be liquid and vapor in coexistence. At some temperature and pressure, known as the critical point, liquid and vapor can no longer be distinguished.

To calculate heat and work in a mechanical refrigeration system, it is necessary to have knowledge of the thermodynamic properties of the working fluid, or refrigerant. This can readily be represented by means of a pressure–enthalpy diagram. A schematic pressure–enthalpy diagram is shown in Figure 1. The line that describes the property of saturated liquid (g), the line that describes the property of saturated vapor (h), and the critical point (*) are marked. To the left of g is a single-phase region of liquid. The g describes the liquid at the point when vapor just begins to boil off. The h describes the vapor as the last drop of liquid vaporizes. Between g and h is a two-phase region of mixed liquid and vapor. Beyond h, the single phase is vapor. Vertical lines describe conditions of constant enthalpy, where the heat content of the refrigerant is unchanging. Horizontal lines describe conditions of constant pressure. The line c describes a constant temperature path. Note that in the two-phase region, this line is horizontal, that is, at constant pressure. At constant pressure, a pure liquid boils at constant temperature. As pressure increases, the boiling point increases. The line e describes a path of constant entropy. Additional lines, such as constant quality (ie, constant liquid/vapor ratio) and constant density lines can also be plotted. Diagrams such as these are readily available for all common refrigerants. Similar

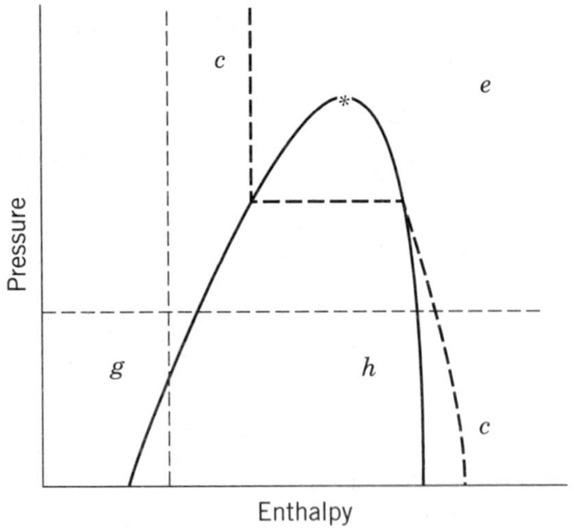

Figure 1. A schematic pressure–enthalpy diagram.

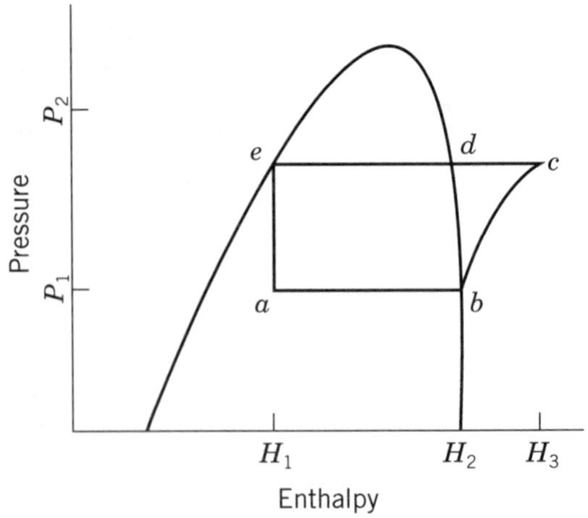

Figure 3. Pressure–enthalpy diagram for the mechanical refrigeration system shown in Figure 2.

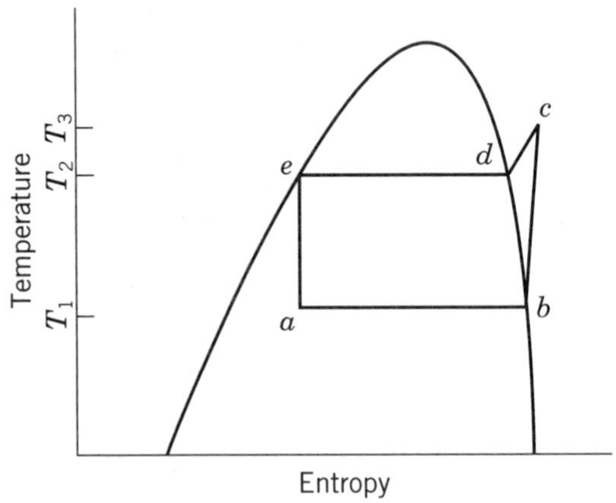

Figure 4. Temperature–entropy diagram for the mechanical refrigeration system shown in Figure 2.

diagrams for temperature–entropy relations can be drawn. The key lines for understanding refrigeration systems are those that describe the saturated liquid and the saturated vapor.

Compression Refrigeration Cycles

Figure 2 schematically shows a mechanical refrigeration system. In the evaporator, a liquid under low pressure boils as it absorbs heat entering from the surroundings. The vapor travels to a mechanical compressor, which raises its pressure. The temperature increases. The high-pressure vapor passes through a condenser, where it rejects the heat to the surroundings. As heat is lost, the vapor condenses. The resulting liquid is held in a reservoir. As required, liquid from the reservoir passes through an expansion valve into the low-pressure side and enters the evaporator, where the cycle begins again. An operating principle of such a device, which uses work to remove heat in a cyclical process, may be described by a reversed Carnot cycle. Figures 3 and 4 are the basic thermodynamic diagrams that describe the process. In Figure 3, the changes in unit mass of refrigerant fluid are as follows. Point *e* represents the saturated liquid in the reservoir. It has a heat content of H_1, pressure of P_2, and temperature of T_2. As it passes through the expansion valve, the pressure drops, and some vaporization occurs, lowering the temperature and bringing the fluid to *a*, at temperature T_1, pressure P_1 and heat content still H_1. As the refrigerant takes in heat from the

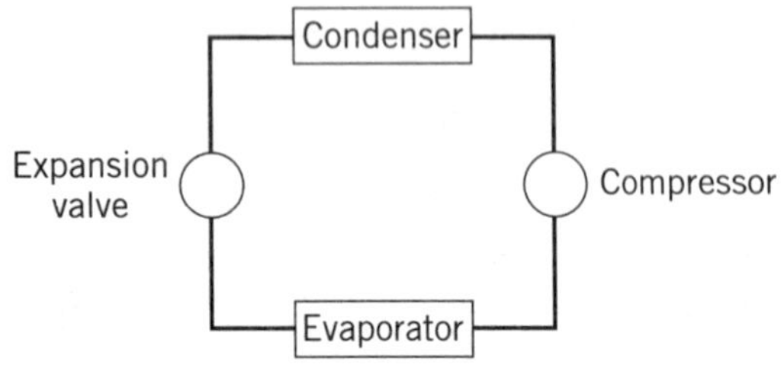

Figure 2. A mechanical refrigeration system, schematic.

surroundings (at a temperature greater than T_1), the heat content increases. At *b*, the heat content is now H_2, with the pressure still P_1 and the temperature T_1. This point is on the saturation line and describes the boiling of the last drop of the liquid. From *b* to *c* describes the compression of the initially saturated vapor, an adiabatic (or isentropic) step. The increase in heat content is the work supplied by the compressor to increase the pressure. At *c*, the temperature is T_3, the heat content is H_3, and the pressure is P_2. The vapor enters the condenser and loses heat to the surroundings. At *d*, on the saturation line, condensation just starts to occur. The pressure is still P_2, the temperature has dropped to T_2 (still above the surroundings), and the heat content is now H_2. As heat continues to leave the refrigerant, it enters the two-phase region, until the saturated liquid is reached at *e*. At *e*, the condition of the liquid reservoir, the heat content is H_1, the pressure is P_2, and the temperature is T_2. Thus, in one pass through the cycle, the unit mass of refrigerant has picked up an amount of heat, $H_2 - H_1$, in the evaporator at temperature T_1 and has rejected this heat in addition to the heat representing the work of compression, $H_3 - H_2$ in the condenser. Work is being used to enable heat to be transferred from a lower temperature to a higher temperature. A refrigerant is chosen so that the pressure required for a boiling point at or below the refrigeration temperature is not too low and so that the pressure required to have a boiling point high enough to reject heat into the surroundings of the condenser is not too high. Table 2 lists appropriate temperatures and pressures for some common refrigerants. The major components of a vapor compression system are the compressor, the condenser, the evaporator, and the controls. Any standard type of compressor is satisfactory; reciprocating compressors are commonly used, but rotary and centrifugal compressors can also be used. Condensers and evaporators are simply heat exchangers. The condenser, which serves to reject heat from the system, may use air cooling, water cooling, or evaporative cooling. Water-cooled condensers flow the water through one of three configurations: shell and tube, shell and coil, and tube in tube. An air-cooled condenser circulates the high-temperature, high-pressure refrigerant through the condenser coil and uses air moved by free convection, wind, or fans to remove the heat. An evaporative condenser transfers heat from the coil to a film of water that evaporates into an air stream passing over it. In the evaporator, the liquid refrigerant boils at low temperature and pressure within the coil, and the medium to be cooled is in contact with the exterior of the coil.

Absorption Refrigeration Cycle

Figure 5 illustrates a typical absorption refrigeration system. Once again, heat is picked up from the surroundings by the boiling of the low-pressure refrigerant liquid. As the vapor exits the evaporator, it contacts the absorbent and dissolves. The solution is pumped around to the boiler, where it is heated. The refrigerant boils off from the still-liquid absorbent and is transferred to the condenser, where it loses heat and is liquefied. The liquid refrigerant is stored in a reservoir until required, when it passes through an expansion valve into the evaporator. The absorption at low temperature and pressure, heating, and boiling off of the vapor at high temperature and pressure are the equivalent of the mechanical compression step. In each case, the refrigerant is taken from the T_1, P_1 condition to the T_2, P_2 condition. The ammonia absorption refrigeration system requires some additional components because the absorbent fluid—water—is itself volatile; therefore, there is a need to separate a significant amount of water vapor from the ammonia vapor exiting from the boiler. This is achieved by using a distillation column.

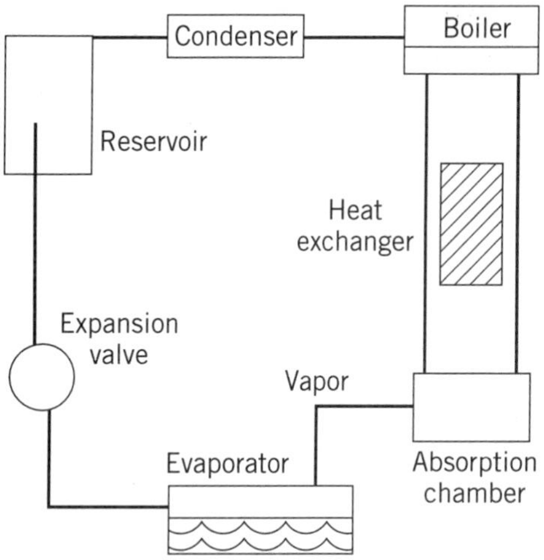

Figure 5. An absorption refrigeration system, schematic.

For the absorption system, the special components are the absorption chamber, the boiler, and the heat exchangers that transfer the heat in the returning "stripped" absorbent to the refrigerant-loaded absorbent being pumped to the boiler. As mentioned earlier, for an ammonia-water system, a distillation coil is required to separate water from ammonia in the exiting vapor stream from the boiler. The evaporator and condenser can be similar to those of the vapor compression systems, although special modifications may be required to deal with residual water in the ammonia-water system.

To compare the capabilities of different refrigeration systems, some definitions are appropriate.

Ton of refrigeration. The standard unit of refrigeration capacity is known as the ton of refrigeration. It is the

Table 2. Properties of Some Important Refrigerants

	Vapor pressure (MPa)		
Temperature (°C)	Ammonia	R22	R134
−50	0.0408	0.0299	0.0645
−30	0.119	0.164	0.085
−10	0.290	0.355	0.201
+10	0.613	0.681	0.415
+30	0.855	1.192	0.770
+50	2.028	1.943	1.318

			Properties at −40°C evaporator, +20°C condenser		
Refrigerant	Evaporator P (MPa)	Condenser P (MPa)	Compression ratio	Net refrigeration (kJ/kg)	Refrigerant circulation (kg/s)
Ammonia	0.071	0.853	11.99	1131.45	0.0031
R22	0.105	0.910	8.65	164.21	0.00609
R134	0.051	0.569	11.23	146.95	0.02392

rate at which heat would have to be removed to convert 1,000 kg of water at 0°C into ice at the same temperature in a 24-hour period. The latent heat of fusion of water is 79.68 kcal/kg. Hence, a ton of refrigeration is equal to a rate of heat removal of 3,320 kcal/h.

Enthalpy. Enthalpy, a measure of the heat quantity in a product, is a relative quantity. For most refrigerants, the enthalpy of the saturated liquid at 0°C is by convention set at 200 kcal/kg. For frozen foods, the enthalpy at −40°C is by convention set at 0 kcal/kg.

Refrigeration effect. The net refrigeration effect for a system under given operating conditions can be calculated from the pressure–enthalpy diagram as

$$RE = H_g - H_f$$

where RE is the refrigerating effect (kcal/kg), H_g is the enthalpy of vapor refrigerant leaving the evaporator, and H_f is the enthalpy of the refrigerant entering the evaporator. In the example of the mechanical refrigeration cycle in Figure 4, RE is given by $H_2 - H_1$.

Quantity of refrigerant circulated per ton of capacity. This quantity is self-defining.

Heat of compression. This is the enthalpy difference in the refrigerant between compressor discharge and inlet. In Figure 4, it is represented by $H_3 - H_2$. It can be expressed as kcal/kg.

Work of compression. This is the product of refrigerant mass flow and heat of compression. It can be expressed as kilocalories per minute. Standard conversions allow it to be expressed in other units where necessary.

Condenser heat load. This is determined by subtracting the enthalpy of the saturated liquid leaving the condenser from the enthalpy of the superheated vapor entering the condenser: $H_3 - H_1$.

Coefficient of performance. This is the ratio of output to input. The theoretical value is given by the refrigeration effect divided by the heat of compression. This, however, ignores friction losses in the compressor, which cause the heat input from the compressor (ie, the work of compression) to be less than the energy required to operate the compressor. The practical coefficient of performance is often only 60% of the theoretical value.

The choice of refrigerant is made on the basis of several criteria. These include (10):

1. The latent heat of evaporation. A high latent heat is preferred.
2. The pressure that would be experienced in the condenser at the condensing temperature. Excessively high pressures are to be avoided.
3. The freezing temperature of the refrigerant must be below the temperature of the evaporator.
4. The critical temperature of the refrigerant must be sufficiently high to give an effective working diagram for the temperatures experienced in the high-pressure side.
5. The refrigerant should ideally be nontoxic, noncorrosive, and chemically stable.
6. It should be easy to detect leaks in the system.
7. For industrial applications, where large quantities of refrigerant are used, low-cost refrigerants are preferred.

Clearly, not all these criteria can be satisfied by any one refrigerant. Ammonia has a high latent heat of evaporation, and leaks are easily detected. However, it is toxic and corrosive. On balance, the ease of leak detection renders it safe for use despite its toxicity. The fluorocarbons are also commonly used in refrigeration units. They have a wider range of applicable temperatures than ammonia, but environmental concerns are associated with their use.

Where lower temperatures are required, it is possible to use multistage refrigeration, where essentially two or more refrigeration cycles are coupled—the low stage of one is the high stage of the next. An alternative approach is to have two-stage compression (Figure 6).

To freeze a product, it must in some way be coupled to the refrigeration effect. Freezing equipment uses a variety of solutions. Where mechanical refrigeration is used, the cooling source is the evaporator. It is possible to put the product in direct contact with the evaporator, as is the case in some designs of plate freezers, where the plates themselves are machined internally as the evaporator. It is more common, however, to use some form of secondary refrigerant, which is cooled by passing through the heat exchanger of the evaporator, and then to contact this fluid, directly or indirectly, with the object. The secondary cooling medium may be a variety of fluids or air. The ability of this fluid to transport heat has an influence on the freezer design. Fluids based on aqueous solutions have a high volumetric heat capacity, so the volume of fluid required to transport a given quantity of heat is not too high. Air has a much smaller volumetric heat capacity (given the much lower density), so the volume of air required to transport the same quantity of heat is much greater. Freezers that use air as the transfer fluid need powerful fans to blow large volumes of air past the product to be frozen if heat

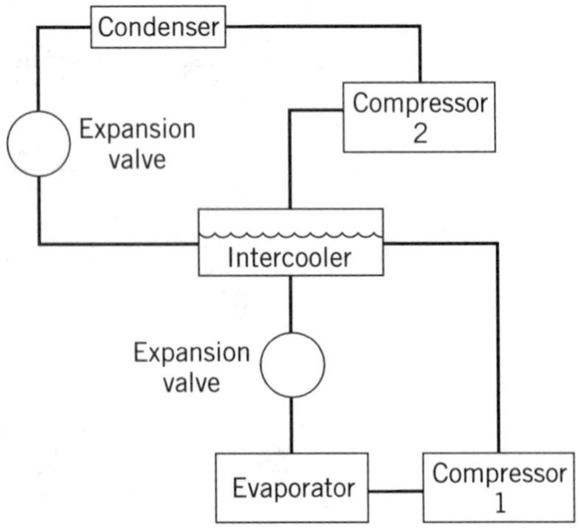

Figure 6. Two-stage refrigeration, schematic.

removal is to be rapid. A variety of freezer designs exist (11). In plate freezers, the product is in contact with metal plates, cooled either directly (ie, the plates are themselves the evaporator coil) or indirectly, by circulating a secondary refrigerant (usually an aqueous solution such as glycol solution or brine of sufficiently low freezing point) through the coolant channels. This gives effective cooling of the product if the geometry of the product results in good thermal contact with the plates, but poorer cooling occurs if air spaces interrupt the thermal contact. An alternative cooling method is direct immersion of the product in the secondary cooling fluid. Using aqueous secondary refrigerants with high volumetric heat capacity, immersion results in excellent cooling rates, but the potential for contamination of the product exists. Using air as refrigerant, the potential for contamination is much less, but the volumetric heat capacity is much lower, so the heat transfer capabilities depend on the volume of cold air that can be brought into close contact with the product in unit time. A wide variety of freezer designs exist that use air as the heat transfer fluid. Static freezers use natural air circulation, and forced air (blast) freezers use a high-velocity cold air stream that passes by the product. Many mechanisms have been evolved for ensuring product throughput, and batch and continuous processes exist. In either case, air temperature and air velocity are critical to the rate of heat removal.

The common characteristic of the mechanical refrigeration systems so far discussed is their ability to produce cooling at the throw of a switch. This characteristic is also found in thermoelectric cooling systems (12). Thermocouples are devices that produce an electrical voltage as a result of temperature difference. Thermoelectric coolers are the reverse concept—devices that produce a temperature difference on passage of an electric current. If the higher temperature node is at ambient temperature, the lower temperature node is a heat sink. Using modern semiconductor materials, effective thermoelectric cooling devices have been manufactured that can be used to provide refrigeration. These are used primarily in special situations. They do not have the high capacities of mechanical refrigeration systems.

Mechanical or thermoelectric refrigeration is not the sole means of providing a heat sink. Just as ice has traditionally been used as a cooling source, cryogenic media may be used as a cooling source. As with ice, these are not turnkey systems to produce cooling at the throw of a switch. Rather, these are systems where cooling capacity is stored in a convenient form. Special cooling processes are used to produce the cryogenic medium. In food freezing, common cryogenic coolants are liquid nitrogen or solid carbon dioxide. These are produced centrally and distributed and stored in special insulated containers before use. The properties of the cryogens are summarized in Table 3. These cryogens can be used directly, bringing them into contact with the product to be frozen. The vaporization of the cryogen is an important component of the heat sink capacity, but the heat required to raise the temperature of the cryogen is also an important source of cooling capacity. Because the three primary drivers in rates of heat transfer are the volumetric contact rate between coolant and product, the volumetric heat capacity of the coolant, and the temperature difference between the coolant and the object to be cooled, the rates of heat transfer tend to be higher than for mechanical refrigeration systems, in which the refrigeration temperature is higher, and the volumetric heat capacity of the secondary refrigerant is lower. Refrigeration systems are capable of providing a large range of cooling rate capabilities for products to be frozen. The design of a refrigeration system should take into account the length of the operating season, the refrigeration load, the amplitude of fluctuations in this load, the buildup of frost on the evaporator surfaces and the need to remove it to maintain cooling efficiency, the target temperatures, and the cooling medium of choice. Other factors to be considered are the type of refrigeration, capital cost, and ease of maintenance (13,14). The characteristics of the product often determine the most appropriate freezing system for the product. The next section describes the freezing process within a product and its dependence on the external cooling conditions and the characteristics of the product.

Table 3. Thermal Properties of Cryogenic Refrigerants

Property	Nitrogen	Carbon Dioxide
Boiling point (°C)	−195.4	−78.6 (sublimes)
Specific heat of vapor (kJ/kg K)	1.03	0.837
Heat of vaporization (kJ/kg)	199	573
Heat removal (bp to −18°C, kJ/kg)	384	623

THE FOOD FREEZING PROCESS

What Is Freezing?

At its simplest, freezing can be taken to refer to a process that reduces the temperature of a product to 0°C or below. However, it also implies that ice should form. In addition, when considering food freezing, it is normally required that the temperature be reduced to −18°C if extended storage life is to be achieved (15). The primary component of food is water. The use of the term freezing implies that some, or most, of this component enters the solid state, that is, becomes ice. In order for ice to form, heat must be removed. As heat is removed, the temperature of a food will begin to reduce. When it reaches the freezing temperature, ice does not necessarily form at first. Often, the temperature continues to reduce below the freezing point, resulting in a supercooled system. At some point, the creation of in-ice nucleus initiates ice formation. Some ice can now form very rapidly from supercooled water, as very small crystals. The heat released as this supercooled water converts to ice raises the temperature back toward the freezing point. Once the ice formation has begun, further ice crystals form in concert with heat removal. The extent of initial supercooling determines the depth to which the initial burst of crystallization penetrates. Beyond that, the rate of heat removal has the primary influence on ice crystal size. If heat removal is rapid, the propagation of a few crystals is insufficient to match the heat flux. Hence, incipient supercooling occurs, which results in an increased probability of nucleation. As a result, a larger number of ice crystals are formed (15). If the cooling rate is slower, propagation can keep pace with heat removal, and fewer larger crystals are the result. Figure 7 illustrates the relationships between concentration and ice content that govern the freezing process. As ice forms, removing water from the unfrozen phase, the concentration of solutes in the unfrozen phase increases. Figure 7 shows the concentration of the unfrozen phase as a function of temperature or, alternatively, shows the concentration dependence of freezing point. These two statements are equivalent. The two curves illustrate schematically the differences that exist for different materials. The important point to grasp is that for two systems with the same initial water content, freezing can result in different amounts of ice at the same temperature. Also, addition of more solute can result in a lower freezing point. Both of these observations are germane to the understanding of the behavior of frozen foods.

When cellular systems are frozen, ice may or may not form within cells (16,17). In freezing, there is a driving force to keep the concentration of the unfrozen phase matched to the requirements of the phase diagram (Figure 7). Within a cell, ice must form or water will translocate to the external medium for this to happen. If cooling is rapid, there may not be time for translocation, and ice will form within cells. Slow cooling may leave enough time for translocation, so that ice forms only in external locations. Product quality is better if translocation of the water has not occurred; therefore, fast freezing tends to result in higher initial quality.

Frozen food quality is influenced not only by the initial freezing conditions, but also by the conditions of storage. Chemical and physical change is possible under frozen storage conditions, at rates that depend on the temperature (17,18). In general terms, the lower the temperature, the slower the change, although it should be recognized that, depending on the storage life required, there will be a threshold temperature that is low enough. However, even brief exposure to higher temperatures, especially above −10°C, significantly reduces quality and shelf life. Rate of freezing, which is related to rate of cooling, is an important determinant of initial quality. Figure 8 shows a typical plot of the enthalpy of a product as a function of temperature. Freezing involves the removal of heat. The amount of heat to be removed to go from ambient temperature to the storage temperature can be determined from Figure 8. However, it is not only the changed heat content that will influence the rate of the freezing process. The size of the object will also influence the rate of the process. The amount of heat to be removed depends on the mass and volume of the object. Because heat removal is through the surface, the rate of heat removal depends on the surface area and the effectiveness of heat transfer between the surface and the refrigerating medium. It also depends on the internal rates of heat transport within the object. For small products, the surface heat transfer is most important, and the cooling rate is influenced primarily by the heat transfer characteristics of the freezer. For larger products, the internal characteristics of the product assume greater importance, and the heat content and internal heat transfer characteristics are the factors that have the greatest influence on the freezing rate and the rate of change of temperature. More simply, small products can be frozen slowly or rapidly, depending on the choice of freezer conditions. Large objects take much longer to freeze because the internal properties control the freezing rate, and the freezer characteristics are no longer the controlling factor. Only in the surface zone do the freezer characteristics markedly influence freezing rates. Beyond some critical surface heat transfer condition, the maximum possible surface heat transfer rate will exceed the rate of heat transfer from the interior of the product. Because of the mechanism of heat transfer, there exists a gradient in temperature between the surface of the product and the center. This temperature gradient is one determinant of the rate of internal heat transfer. As Figure 8 shows, the change in

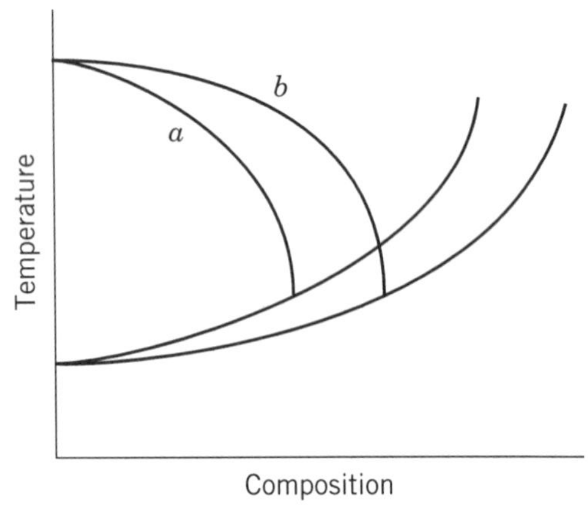

Figure 7. A typical phase diagram for freezing.

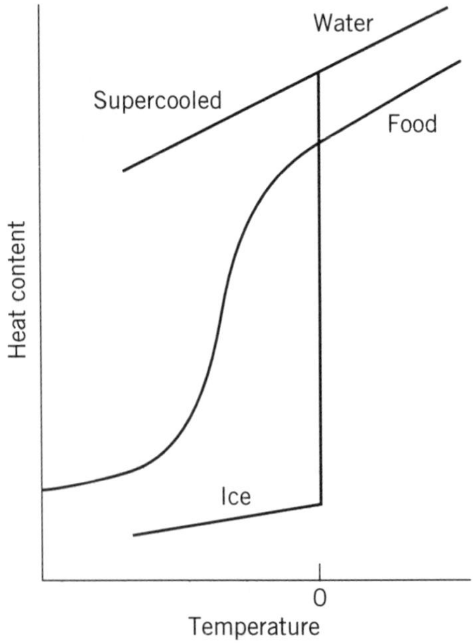

Figure 8. The enthalpy–temperature diagram for aqueous systems in the region of the freezing temperature.

heat content for a 1°C drop in temperature is dependent on temperature. This is a consequence of the two different processes that result in a change in heat content: temperature change and phase change. The heat associated with temperature change is called sensible heat (ie, heat that can be sensed). When phase change occurs in a pure system at constant pressure, there is no temperature change. The heat associated with phase change is called latent heat. In Figure 8, the slope of the plot for pure water (sensible heat) is approximately 1 kcal/kg/K. For ice, it is approximately 0.5 kcal/kg/K. The conversion of water to ice is accompanied by a change in heat content of approximately 80 kcal/kg. Thus, cooling of water to below freezing temperatures will result in a steady drop in temperature, followed by a region of steady temperature during the phase change, and then a further steady drop in temperature. Because nucleation is necessary, there is a region of supercooling where the temperature drops below the freezing point before freezing initiates, and the temperature returns to that of the freezing plateau. This is illustrated in Figure 9, which shows the cooling profile at the surface and the center of a pure water sample. In a real product, with heat content as illustrated in Figure 8, the freezing curves appear as in Figure 10. Supercooling is still apparent, but now the temperature declines slowly as ice forms. Once most of the ice has formed, the rate of temperature change increases again. The trajectory of Figure 10, except for the supercooling dip, can be estimated from the phase diagram of Figure 7, because the reason for the plateau region is the need to remove latent heat, and the amount of latent heat to be removed depends on the change in ice content as a function of temperature. It is also possible to determine Figure 8 from Figure 7, and vice versa. The enthalpy–temperature profiles of Figure 8, traversing from ambient temperature, show first the sensible heat zone. The slope

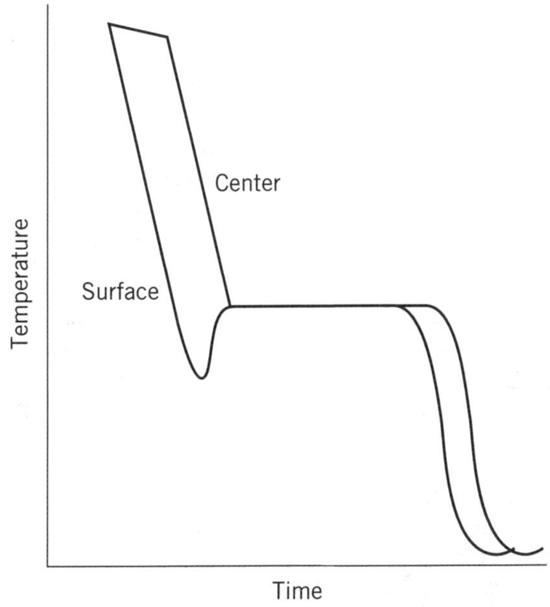

Figure 9. A temperature–time cooling diagram for a pure water sample (surface and center).

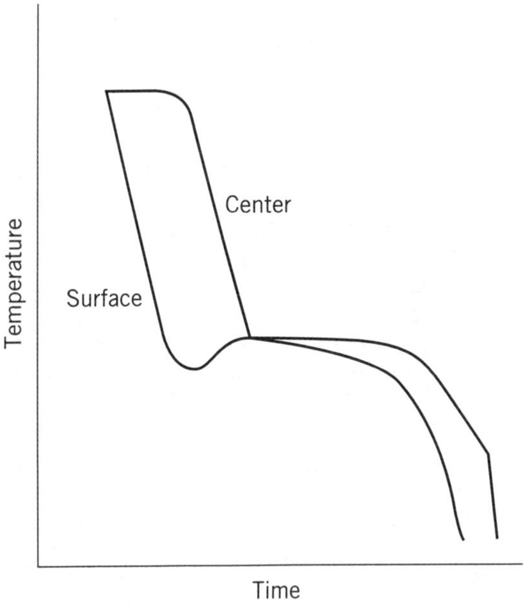

Figure 10. A temperature–time cooling diagram for a food sample (surface and center).

is the specific heat of the product and depends on water content and solids content. Water typically has a specific heat of 1 kcal/kg/K, and food solids typically have a specific heat of about 0.3 kcal/kg/K. In the freezing zone, the enthalpy change is a combination of sensible heat for the unfrozen material and the latent heat released by the formation of ice. Once most of the freezing is over, the slope is characteristic of the sensible heat of the frozen system, in part influenced by the amount of unfrozen water that remains. Figures 9 and 10 illustrate the difference in freezing profile for different positions in the sample, which are in part a consequence of the internal temperature gradients that accompany heat conduction.

Freezing is normally considered complete when the center of the product reaches a predefined temperature. To estimate the freezing time, it is necessary to factor in the external conditions and the internal product properties. Successful design of a refrigeration system requires an estimate of the freezing time, because this helps define the refrigeration requirement. The alternative International Institute of Refrigeration (IR) definitions of freezing time (19) are (*1*) nominal freezing time, the time between the surface reaching 0°C and the thermal center reaching a temperature 10K colder than the temperature of initial ice formation at the thermal center; and (*2*) effective freezing time, the time to lower the temperature of the product from its initial temperature to a given temperature at its thermal center. Because of the complex dependence of enthalpy on temperature, prediction is very complex. The anisotropic structure of food also adds complexity. Heat transfer coefficients are not well known. A variety of analytical and numerical predictive methods, are available and are discussed in the literature (20–24).

A simple method to reach a rough estimate of freezing time uses Plank's equation:

$$t_f = \frac{\rho L}{T_f - T_a}\left(P\frac{a}{h} + R\frac{a^2}{k}\right)$$

where t_f is the freezing time, L is the latent heat of the water fraction, ρ is the density, T_f is the initial freezing temperature, T_a is the temperature of the refrigerating medium, a is the characteristic dimension, h is the surface heat transfer coefficient, k is the thermal conductivity, and P and R are constants that depend on the geometry of the sample. This equation clearly illustrates the effect of size, which was discussed previously. For a small object, the relative importance of the a and a^2 terms highlights the term involving h. As a increases, a^2 increases in significance, and the term in k increases in importance. Thus, the freezing time of small objects is determined primarily by the surface heat transfer, and that for large objects is determined primarily by the thermal conductivity of the object. One purpose of estimating freezing time is to determine the heat load for the refrigeration system. If the heat to be removed and the freezing time are known, then the rate of heat removal required can be estimated. This then gives an indication of the required capacity of the refrigeration system.

Many different types of food products are preserved by the freezing process, but they do not exhibit the same characteristics in the freezing process or require the same process conditions. For example, plant tissues, on freezing, tend to undergo disruption of the cell structure (16). In fresh plant tissue, the cell wall complex can act as an effective barrier to ice propagation during freezing, with the result that ice may form only externally under slow cooling conditions. Water will tend to transfer from the cell. The resulting textural quality is often inferior. If rapid cooling is used, greater supercooling is produced in the cells and internal freezing becomes possible. Water transfer is minimized, and the textural properties are improved. Freezing in plant tissues often produces structural disruptions of the cells, which allows for unwanted enzymic activity during storage or thawing. The results of this enzyme activity significantly change flavor and other attributes and lead to lowered quality or complete unacceptability of the product. To prevent this, it is often advised to thermally treat (blanch) the material before freezing to inactivate the enzymes responsible for the unwanted changes. This procedure was first introduced by Joslyn and Cruess. Because blanching is a thermal treatment, it causes textural changes that are acceptable in products normally consumed cooked, but they are often unacceptable in products normally consumed raw. Blanching is most often used for vegetables and less often for fruits, which are frozen as is. For high-water-content plant tissues, freezing damage can cause significant textural alteration. Where texture is important, this might mean that freezing is an unacceptable process. A good example is lettuce, where the high turgor pressure, essential to the eating quality, is lost by freezing. The individual material has to be examined to determine whether freezing and thawing cause unacceptable changes and whether blanching can be used to prevent these changes. Animal tissues exhibit different characteristics from plant tissues (16). First, the resistance to internal cell freezing is much less, and so cell dehydration during freezing, with consequential drip loss on thawing, is less of a problem. Also, the effect of freezing on the enzymatic processes is less marked than for plant systems. The biochemical system most at risk from freezing is the muscle contraction system. The structural disruption of freezing causes uncoupling of the control of the contractile mechanism. If ATP is present, uncontrolled contraction can occur, resulting in extremely tough material. This is obviated by ensuring that rigor has been passed. Because rigor is a time-consuming process, methods to accelerate the depletion of the ATP reserve have been developed. These allow for the freezing process to be speeded up without causing unacceptable textural changes in the meat.

Storage Temperatures

Because the purpose of freezing is to extend the shelf life of food products, the influence of storage temperature on storage life and thawed quality is a concern. Traditionally, it has been assumed that storage at −18°C (0°F) is adequate for all products. Although it is true that storage at lower temperatures is preferred to storage at higher temperatures, −18°C does not always result in an acceptable shelf life. Some products require lower storage temperatures for extended storage. Ice cream and fish, for example, are much better stored at temperatures below −30°C if unacceptable changes in quality are to be avoided. The influence of temperature on storage life is now much better understood. At temperatures close to −5°C, change tends to be accelerated, compared to temperatures above or below. This is because of the reaction-enhancing effect of increased solute concentration in the unfrozen phase outweighing the general slowing effect of lowered temperature (18). This is of special significance in the careful handling of frozen foods. Beyond the zone of maximum ice formation, the influence of temperature is the primary determinant of reaction rates. However, it is now understood that it is the relative temperature, rather than some absolute temperature, which is significant. As a system freezes and ice forms, the properties of the unfrozen aqueous phase change. Increasing concentration and decreasing temperature lower the mobility of the solute molecules. At some characteristic temperature, solute mobility becomes severely constrained, and reaction rates decrease markedly. Although it may not be economically feasible to use temperatures below this characteristic temperature for storage, it also serves as a reference temperature against which the actual temperature of storage can be compared. For products where this temperature is lower, storage temperatures also should be lower to achieve extended shelf life.

As previously indicated, freezing is necessarily followed by thawing. The thermal properties of water and ice are such that the freezing process and the thawing process are not mirror images. Thawing follows a different profile (18). Because ice is a better conductor of heat, heat transfer into a frozen product is more rapid than heat removal from a similar unfrozen product. As ice begins to melt at the surface, a thermal barrier is formed, because liquid water is a less effective heat transfer medium. As a result, on thawing the temperature rapidly rises into the zone around

−5°C, and then heat transfer becomes less effective. The latent heat plateau takes longer to traverse than in freezing, where the heat transfer is through an external layer of ice. As a consequence of this property, exposure of frozen products to higher temperatures even for a short time can lead to significant quality loss. The thawing process, too, if not well controlled, can result in loss in quality, as a consequence of the potential for significant reaction at −5°C.

Freezing is an effective method for food preservation. The deceptively simple technology requires care to be exercised in temperature control for optimum quality of product to be maintained to consumption.

BIBLIOGRAPHY

1. O. Fennema, "The U.S. Frozen Food Industry: 1776–1976," *Food Technol.* **30**, 56–68 (1976).
2. R. V. Enochian and W. R. Woolrich, "The Rise of Frozen Foods," in N. W. Desrosier and D. K. Tressler, eds., *Fundamentals of Food Freezing*. AVI, Westport, Conn., 1977, pp. 1–24.
3. T. P. Labuza and A. E. Sloan, "Forces of Change: From Osiris to Open Dating," *Food Technol.* **35**, 34–43 (1981).
4. C. P. Mallett, "Editorial Introduction," in C. P. Mallett, ed., *Frozen Food Technology*. Blackie, London, United Kingdom, 1993, pp. xix–xxv.
5. H. T. Cook, "Refrigerated Storage," in C. W. Hall, A. W. Farrall, and A. L. Rippen, eds., *Encyclopedia of Food Engineering*. AVI, Westport, Conn., 1986.
6. W. R. Woolrich and A. F. Novak, "Refrigeration Technology," in N. W. Desrosier and D. K. Tressler, eds., *Fundamentals of Food Freezing*. AVI, Westport, Conn., 1977, pp. 25–53.
7. F. W. Sears, M. W. Zemansky, and H. D. Young. *University Physics*, 5th Ed., Addison-Wesley, Reading, Mass., 1976.
8. J. M. Smith and H. C. Van Ness, *Introduction to Chemical Engineering Thermodynamics*, 3rd ed., McGraw-Hill, New York, 1975.
9. R. P. Singh and D. R. Heldman, *Introduction to Food Engineering*, Academic Press, Orlando, Fla., 1986.
10. P. O. Persson and G. Londahl, "Freezing Technology," in C. P. Mallett, ed., *Frozen Food Technology*, Blackie, London, United Kingdom, 1993, pp. 20–57.
11. E. R. Hollowell, *Cold and Freezer Storage Manual*, 2nd ed., AVI, Westport, Conn., 1980.
12. H. J. Goldsmith, "Thermoelectric Refrigerators," in D. M. Rowe, ed., *CRC Handbook of Thermoelectrics*, CRC Press, Boca Raton, Fla., 1995.
13. *Engineered Refrigeration Systems (Industrial and Commercial)*, American Society of Heating, Refrigeration and Air Conditioning Engineers, Atlanta, Ga., 1994.
14. D. S. Reid, "Fundamental Physicochemical Aspects of Freezing," *Food Technol.* **37**, 110–115 (1983).
15. D. S. Reid, "The Freezing of Food Tissues," in B. W. W. Grout and G. J. Morris, eds., *The Effects of Low Temperatures on Biological Systems*, Arnold, London, United Kingdom, 1987, pp. 478–479.
16. D. S. Reid, "Physical Phenomena in the Freezing of Tissues," in C. P. Mallett, ed., *Frozen Food Technology*, Blackie, London, United Kingdom, 1993, pp. 1–19.
17. O. R. Fennema, W. D. Powrie, and E. H. Marth, *The Low Temperature Preservation of Food and Living Matter*, Dekker, New York, 1973.
18. International Institute of Refrigeration, *Recommendation for the Processing and Handling of Frozen Foods*, 3rd ed., International Institute of Refrigeration, Paris, France, 1986.
19. Y.-C. Hung, "Prediction of Cooling and Freezing Times," *Food Technol.* **44**, 137 (1990).
20. "Cooling and Freezing Times of Foods," in *ASHRAE Handbook Fundamentals*, American Society of Heating, Refrigeration and Air Conditioning Engineers, Atlanta, Ga., 1997.
21. A. C. Cleland, *Food Refrigeration Process-Analysis, Design and Simulation*, Elsevier Applied Science, United Kingdom, London, 1990.
22. R. Z. Plank, 1913, cited by A. J. Ede, "The Calculation of the Rate of Freezing and Thawing of Foodstuffs," *Modern Refrigeration* **52**, 52–55 (1949).
23. D. J. Cleland, A. C. Cleland, and R. L. Earle, "Prediction of Freezing and Thawing Times for Multi-Dimensional Shapes by Simple Formulae. Part 2: Irregular Shapes," *International Journal of Refrigeration* **10**, 234 (1987).
24. Q. T. Pham, "Simplified Equation for Predicting the Freezing Time of Foodstuffs," *J. Food Technol.* **21**, 209 (1986).

GENERAL REFERENCES

A. D. Althouse, S. H. Turnquist, and A. F. Bracciano, *Modern Refrigeration and Air Conditioning*, The Goodheart-Wilcox Co., South Holland, Ill., 1988.

American Society of Heating, Refrigeration and Air Conditioning Engineers (ASHRAE), *Refrigeration Systems and Applications*, ASHRAE, Atlanta, Ga., 1994.

American Society of Heating, Refrigeration and Air Conditioning Engineers (ASHRAE), *Equipment*, ASHRAE, Atlanta, Ga., 1996.

American Society of Heating, Refrigeration and Air Conditioning Engineers (ASHRAE), *Fundamentals*, ASHRAE, Atlanta, Ga., 1997.

O. R. Fennema, W. D. Powrie, and E. H. Marth, *The Low Temperature Preservation of Food and Living Matter*, Dekker, New York, 1973.

W. B. Gosney, *Principles of Refrigeration*, Cambridge University Press, Cambridge, United Kingdom, 1982.

L. E. Jeremiah, ed., *Freezing Effects on Food Quality* Marcel Dekker, New York, 1996

M. K. Karel, O. R. Fennema, and D. B. Lund, *Principles of Food Science, Part II. Physical Principles of Food Preservation*, Marcel Dekker, New York, 1975.

C. P. Mallett, ed., *Frozen Food Technology*, Blackie, London, United Kingdom, 1993.

M. R. Okas, *Physical and Chemical Properties of Food*, ASAE Publication No. Q0986, St. Joseph, Mich., 1987.

S. Rahman, *Food Properties Handbook* CRC Press, Boca Raton, Fla., 1995.

R. P. Singh and D. R. Heldman, *Introduction to Food Engineering*, Academic Press, Orlando, Fla., 1986.

R. Thevenot, *History of Refrigeration*, International Institute of Refrigeration, Paris, France, 1979.

D. S. Reid
University of California
Davis, California

FOOD MARKETING

Most of the efforts and resources in food science and technology are devoted to making food production faster, more efficient, of higher quality, and, of course, less expensive. Food marketing is concerned with a number of other important activities such as the identification of consumer needs; the design of need-satisfying food products and services; the pricing, distribution, and communications (including advertising and promotion) of food and allied products. Most people think of food marketing as synonymous with advertising, and while this is the most obvious activity, it is by no means the most important. Marketing is more than just a set of techniques to entice people to buy a product, it is a philosophic approach to doing business.

MARKETING PHILOSOPHY

The marketing philosophy has been described by numerous authors, in literally hundreds of textbooks. Some of the more classic definitions are "marketing is the performance of activities that seek to accomplish an organization's objectives by anticipating customer or client needs and directing a flow of need-satisfying goods and services from producer to customer or client" (1) and "marketing is a social and managerial process by which individuals and groups obtain what they need and want through creating and exchanging products and value with others" (2). In 1985, the American Marketing Association approved this definition: "Marketing is the process of planning and executing the conception, pricing, promotion, and distributing of ideas, goods, and services to create exchanges that satisfy individual and organizational objectives." (3). A simpler definition that captures the essence of marketing is as follows: "Marketing is the anticipation, understanding, and satisfaction of customer needs while realizing the organization's objectives." The key to all of these definitions is that marketing involves activities that begin with assessing what the consumer needs and wants and translates that into food products that can be sold for a profit. Food marketers believe that the most profitable way to sell products is to offer the consumer the products that they value the most. Food marketers understand that what they make in the factories may not be the same thing consumers are buying at the checkout line. Although not involved in the food industry, Charles Revson, of Revlon Cosmetics, summed up this concept as follows: "In the factories we make cosmetics, at the counters we sell hope." McDonald's makes hamburgers but its customers are buying quick and convenient appetite satisfaction. Food marketers must understand and focus on what consumers want to buy and not simply what food marketers make. But producing the right product is not enough. Food marketers are also concerned with moving that product to the point of consumption (distribution) and making sure that intermediate (wholesalers and retailers) as well as final (consumers) buyers know all the differential benefits derived from product usage. This differential advantage represents the primary basis for persuading intermediate and final buyers to purchase and use a particular food product. The tools available for persuasion include advertising, personal selling, consumer, and trade promotions.

Production vs Marketing vs Sales

After World War II, consumers had considerable spending power, pent-up demand, and too few goods to purchase. Firms found instant success by producing for the mass market. During the production era, food marketers focused on low-cost products to satisfy excess demand. An example is TV dinners. The earliest versions were frequently bland and cheap. Companies that made these products focused on cost reduction and just getting the product to market. However, as demand decreased and competition increased, firms shifted to a sales orientation. While the company made only minor changes to the TV dinners, they increased sales promotions and gave price deals to increase sales.

One point of confusion is the difference between sales and marketing. In the 1960s when marketing came into vogue, many companies promoted their sales managers to marketing managers. Marketing to many was just modern-day sales. Nothing could be further from the truth marketing is a way of doing business. Marketing focuses on the customer and profits, and it sets the stage so that when the sales force is asked to sell, the odds are in their favor. Marketing ensures that the product the sales force is selling is one that the customer wants, in the type of package, in the proper stores, and in the right sizes, etc. The sales force is responsible for getting the volume. Marketing makes sure that the volume is profitable.

As the TV dinner people found that increased selling activity increased sales, it also meant lower margins and profits. Food marketers began to ask the question; "What do consumers really want from frozen prepared foods?" It was not the traditional TV dinner that they had been pushing on the public in the previous years. Food marketers, not the production department, defined what should be done to make the product more attractive (and, therefore, more profitable). Today's frozen entrees and dinners are much better, and many consumers say a delicious, product choice. By taking a marketing approach, the company wins with higher profits and the consumer wins with better products.

THE MARKETING MANAGEMENT PROCESS

The marketing management process consists of analyzing market opportunities, researching and selecting target markets, developing marketing strategies, planning marketing tactics, and implementing and controlling the marketing effort.

The steps in the marketing management process are as follows:

1. Identify market opportunities.
2. Select target market(s) and position the product in the target market.
3. Develop a strategy for each target market.
4. Plan the tactics to accomplish the strategy.
5. Provide a procedure to implement the strategy.

6. Put in place a system for monitoring performance.
7. Evaluate and modify as needed.

MARKETING STRATEGY

Marketing strategy is how food marketers get from where they are to where they want to be. Many people confuse marketing strategy with simply beating the competition. However, marketing strategy is serving customers' real needs. Competitive realities are what possible strategies are tested against. Marketing strategy involves manipulating the marketing mix in the context of environmental variables to satisfy the needs of the target market. The target market is a fairly homogeneous (similar) group of customers to whom a company wishes to appeal. The marketing mix represents the controllable variables the company manipulates to satisfy the target market. The marketing mix is made up of four basic variables generally known as the four *P*'s product, place, promotion, and price.

MARKETING ENVIRONMENT

The marketing environment consists of forces that are not controllable but that influence both the consumer and the food marketer. Although the list of such environmental, uncontrollable or external forces is endless, the most common factors confronting the marketer are as follows.

1. *Competition*. Competitive analysis has been in vogue for the last five years. Understanding competitive strengths and weaknesses is a major input to marketing strategy today. A number of books have focused on just this topic (4,5).

2. *Demographics (Population Descriptors)*. Changes in demographics has also had a major impact on food marketing. Baby boomers and the aging of America are just two examples of demographic trends that will lead to new food marketing techniques. *American Demographics* is one journal that provides an excellent overview of changes in American demographics.

3. *Politics*. No area of business is unaffected by politics. The implication to the food industry is through the USDA, FDA, FTC, FCC, and more recently the DEP.

4. *The Economy*. Recession, depression, inflation, a booming economy, unemployment, etc all affect the ability of food marketers to carry out their mission. Even more important to the food business is Wall Street, which frequently determines corporate policies based on current stock prices. Companies attempting to raise stock prices by cutting costs and raising profits often find competitors taking market share from them.

5. *The Physical Environment*. This usually refers to available resources. Probably the most important variables are oil prices and availability of farmland.

6. *Technology*. Little needs to be said about technology. It has influenced every aspect of food marketing. At retail, computerized shelf space management, talking shopping carts, and in-store videos will change the nature of the shopping experience. Distribution is already taking advantage of satellite dishes on trucks to monitor the exact location and contents of the trucks. Warehouses are using scanning technology to position and control product movement. Food manufacturers have similarly benefited from technology. The *Journal of the International Food Technologists (IFT)* addresses technology in food processing.

7. *Social—Cultural Factors*. Of all factors changing food marketing, the changing American family is number one. Women leaving home to join the work force has made convenience a priority. More meals are being purchased from alternative outlets than ever before. No segment of the food industry has been unaffected by these changes (6).

Food marketers employ the controllable variables, the four *P*'s, to address the uncontrollable forces that influence the success of a firm's marketing strategy.

MARKETING MIX (THE FOUR *P*'s)

Product

In the area of food marketing, a product can be described as the need-satisfying offering of a firm. Because consumers are buying satisfaction, companies must be concerned with product quality. From a marketing perspective, quality means the ability of a product to satisfy a customer's needs or requirements. The product area of the four *P*'s is concerned with developing the right product for a specific target market. A product can either be a physical good, a service, or a blend of both. Products in the food industry are usually limited to consumer or industrial products. Consumer products are those products meant for the final consumer. Industrial products are products meant for use in producing other products.

Product Development

Product development involves offering new or improved products for a present market. In the food industry, competition is strong and dynamic, making it essential for companies to keep developing new product. New product planning is not optional, it must be done just to survive in today's changing markets. A new product is one that is new in any way for the company concerned. According to the Federal Trade Commission (FTC), a product can only be called new for six months and only if it is entirely new or changed in a functionally significant or substantial respect. The cost of new product introductions can be high. Experts estimate that consumer-product companies spend at least $20 million to introduce a new brand and 70–80% of these new brands are flops (7).

Place (Distribution)

The place component of the four *P*'s can be thought of as making products available in the right quantities and locations when the customers want them. A product is not much good to a consumer if it is not available when and where it is wanted. Place requires the selection and use of marketing specialists to provide target consumers with the product. Most consumer goods in our economy are distributed through multiple institutions, or middlemen, which are commonly referred to as marketing channels, or channels of distribution. A channel of distribution is any series of firms or individuals who participate in the flow of goods and services from producer to final user or consumer.

The issue of distribution in food marketing has historically fallen into either food retail such as supermarkets or food service such as restaurants or school lunch programs. In the old model, food was either sold as components, taken home, and made into meals for supermarkets or sold as complete meals for the food service. Customers either bought meat, potatoes, and vegetables and made dinner or went to a restaurant for dinner.

Recently this model of distribution has become less clear. Time-starved American consumers want and are willing to pay for more prepared meals. Each year the percentage of the food dollar spent on traditional supermarket food has dropped and food service has increased. In 1997 the percentage of the food dollar spent on traditional supermarket products has fallen to less than 50%, while it was as high as 70% in the 1950s and 1960s.

This consumer shift in demand for prepared foods has led to two major shifts in the distribution channel. First, a new form of food service supplier has appeared. These stores specialize in providing inexpensive complete prepared meals to be consumed either at home or at the store. Unlike the fast-food restaurants, they offer more traditional meals such as meat loaf, roasted chicken, baked ham, and macaroni and cheese. These dishes are referred to as comfort food and are sold in outlets called Boston Market, Kenny Rogers Roasters, or Chili's. These outlets look more like restaurants than supermarkets and are frequently called family restaurants.

To compete against the family restaurants, many food retailers have actually changed into a new format called a "grocerant." Grocerants look more like supermarkets than restaurants but have significantly more space dedicated to prepared meals. In many cases the grocerant will have a number of chefs in the store preparing the food as it is ordered. These meals are sold in outlets called Eatzi's of Dallas, or Zagarra's of Philadelphia. The grocerant still sells various food components and groceries, but this is just a small portion of total sales. Eatzi's reported selling 38% ready-to-heat, 30% ready-to-eat, 12% raw prepared, and 20% conventional groceries.

Both the traditional supermarket and restaurant have changed to stop the loss of sales to the two new formats. Restaurants are providing more attention to takeout, and take home. Once just a nuisance, restaurants now see that they can make larger margins and increase total sales by not being limited to the number of tables available. Many restaurants are allocating space to a take-out section and have comfortable waiting areas for the take-out patrons to wait. The second defense against the new formats is to add entertainment to the traditional dining experience. Rather than compete just for the food, the new restaurants are creating themes that draw customers for fun as well as food. These theme restaurants called "eatertainment" include Rain Forest Café, Planet Hollywood, King Henry's Feast, and many others.

Supermarkets have also tried to appeal to the time-starved consumer by providing prepared food in the stores. They are often called Meal Solutions or Home Meal Replacement and offer complete meals in the deli sections. In some cases these meals are prepared in-house, sometimes in a central commissary for the chain, and in some other cases an outside supplier is used. At the time of this writing no single format has been demonstrated to be superior.

Regardless of the end point, traditional supermarket, restaurant, grocerant, eatertainment, or new supermarket, the people who supply those companies fit into one of the following categories. Marketing intermediaries fit into one of the following categories.

1. *Merchant wholesalers* take title to (own) the goods they sell and sell primarily to other resellers (retailers), industrial, and commercial customers rather than to individual consumers.

2. *Agent middlemen*, such as manufacturer's representatives and brokers, also sell to other resellers and industrial or commercial customers, but they do not take title to the goods they sell. They usually specialize in the selling function and represent client manufacturers on a commission basis.

3. *Retailers* sell goods and services directly to final consumers for their personal, nonbusiness use.

4. *Facilitating agencies*, such as advertising agencies, marketing research firms, collection agencies, and railroads, specialize in one or more marketing functions on a fee-for-service basis to help their clients perform those functions more effectively and efficiently (8).

Choosing the correct channel of distribution is critical in getting products to the target market's place. Because in the food industry physical goods are almost always involved, place requires physical distribution (PD) decisions. Physical distribution is the transporting and storing of goods to match target customers' needs with a firm's marketing mix (9). From the customer point of view, the concern is not how the product was stored or moved, but rather what is the customer service level, how rapidly and dependably a firm can deliver what the customer wants. It is important for food marketers to understand the customer's point of view. Physical distribution is usually the invisible part of marketing and only gets the customer's attention if something goes wrong.

Promotion

Promotion is concerned with telling the target market about the right product. A promotion is a direct inducement that offers an extra value or incentive for the product to the sales force, distributors, or the consumer with the primary objective of creating an immediate sale (10). Promotions are aimed at both consumers and middlemen for the purpose of keeping goods moving smoothly through the pipeline. This is accomplished by using a combination of push and pull strategies. A push strategy provides incentives to the middlemen to buy and resell the product, therefore pushing the goods on to the next stage of the pipeline to the ultimate consumer. A pull strategy provides incentives to the consumer to pull the product out of the end of the pipeline at the retailer level. Food marketers should use a combination of both strategies so that the product will flow easily from the manufacturer to the consumer. The ultimate goal of promotions is to induce behavior. Sales promotions refer to promotional activities other than advertising, publicity, and personal selling that stimulate interest, trial or purchase by final customers or others in the channel.

Consumer Promotion. Examples of consumer promotions are coupons, sweepstakes, contests, product samples, refunds, rebates, tie-ins, premiums, bonus packs, trade-ins, and exhibitions. These promotions are directed at consumers who purchase products at the retail level and are designed to provide them with an inducement to purchase the marketer's brand. Consumer promotions are part of a promotional pull strategy and work along with advertising to encourage consumers to create a demand for a particular brand.

Trade Promotion. Trade promotions in the food industry are sometimes known as promotions to the retailer. This term may be misleading as these promotions are targeted to distributors, wholesalers, and retailers. Trade promotions are critical because product cannot be sold to consumers if it is not first sold to the middlemen. Because the average supermarket stocks more than 10,000 items, promotions need creativity to break through the clutter. Trade promotions include activities such as promotional allowances, dealer incentives, point-of-purchase displays, sales contests and sweepstakes, and trade shows. Trade promotions are designed to motivate distributors and retailers to carry a product and make an extra effort to promote it to their customers (push strategy).

One of the most commonly used trade promotions in the food industry is the promotional allowance. These are payments made by a manufacturer to resellers (often off-invoice) for merchandising its products or running in-store promotional programs such as reduced shelf prices, special displays, or in-store advertising. Another common but controversial promotion is the slotting allowance (sometimes known as a stocking allowance or street money). Slotting allowances are the fees that are often demanded by the retailers to gain admission into their stores. The position taken by the retailer is that with increased competition, a proliferation of new products, and small profit margins, they are required to ask for these fees, claiming that they will be used to promote the products, redesign shelves, and reprogram computers. Most manufacturers feel differently about how slotting fees are used. One food industry source estimated that 70% of all slotting fees go directly to the retailers' bottom line (11). Many food manufacturers view slotting allowances as little more than corporate bribery.

Traditionally, food companies have spent more of their promotional dollars on trade promotions than on consumer promotions, in an effort to win shelf space in the midst of a rising tide of new products. Recently, however, expenditures on promotions to consumers have been growing while expenditures on promotions to the trade have been declining slowly. Of the three major areas of a company's promotional spending, consumer promotions, trade promotions, and media spending, the three-year trend shown in Table 1 has appeared (12).

As an example, Kellogg Co., the world's leading cereal manufacturer, is one of the companies that intends to tilt its marketing strategy away from the trade and back toward the consumer. This comes on the heels of sluggish domestic sales. In addition to Kellogg Co., Kraft General Foods has decided to step up advertising for its Maxwell House coffee, a brand that stopped advertising in 1987 and lost its number one position in the coffee business (13) (Table 2).

Advertising

Simply stated, advertising tells the target market that the right product is available, at the right price, and at the right place and time. Advertising is defined as any paid form of nonpersonal presentation of ideas, goods, or services by an identified sponsor. It includes the use of such media as magazines, newspapers, radio and television, signs, and direct mail (14). The most common vehicles of advertising used by food companies are network and spot television, radio, and magazines. According to one source

> the food industry is expected to be the biggest ad spender in 1990 with a budget of $8.4 billion, according to a study by Schonfeld & Associates of Evanston, Ill. The 107 publicly owned restaurant and fast-food chains are forecast to spend 8% more this year, to more than $1.5 billion. Supermarket chains will raise spending by 5.6% to $1.8 billion, with Kroger Co. the biggest at $225 million. Ad spending for health and beauty aids is expected to rise by 11% to more than $11.2 billion with Bristol-Myers Co. leading the way at $1.1 billion (15).

Personal Selling

Personal selling is defined as direct communication between sellers and potential customers. Personal selling is usually face-to-face, but communication can take place over the telephone. The strength of personal selling in food

Table 1. Annual Survey of Promotional Practices

	Consumer Promotions	Trade Promotions
1987	26.1%	39.9%
1988	26.3%	39.6%
1989	27.1%	39.4%

Table 2. Biggest Ad Spenders in the Prepared Food Category

Rank	Company
1	Kellogg Company
2	General Mills Inc.
3	Philip Morris Companies, Inc.
4	Quaker Oats Company
5	Campbell Soup Company

Source: Ref. 16.
Note: For a one-year time period ending in June 1989. Total advertising expenditures for prepared foods for this time period was $1.2 billion.

marketing is its flexibility, as it provides immediate feedback that helps salespeople adapt to the current situation.

In total there are more than 11 million consumer salespersons in the United States, with another 9 million salespersons in the industrial marketing area. How important are salespersons? One study, for example, found that executives of industrial firms rated the sales function as 5 times more important than advertising in their marketing mixes; for consumer durables marketers, sales was rated 1.8 times as important as advertising, while for consumer nondurables, advertising and person selling were rated as about equally important (17).

The basic steps of personal selling are as follows; prospecting, planning sales presentations, making sales presentations, and following up after the sale. Prospecting involves following all the leads in the target market to identify potential customers. Once the prospect has been located, it is necessary to make a sales presentation that is the salesperson's effort to make a sale.

Price

The fourth *P* a food marketer must be concerned with is price. When a company sets a price, they must take into consideration the kind of competition in the target market and the cost of the whole marketing mix. Other considerations to be made in pricing decisions are customer reaction to possible prices, current practices as to markups, discounts, and other terms of sale, and finally, legal restrictions on pricing. Price is an important consideration in food marketing because if a customer will not accept the price, all of the planning decisions and efforts could be wasted.

MARKETING RESEARCH

Finally, food marketing usually encompasses marketing research. Marketing researchers represent the information-gathering arm of the firm and helps to determine exactly what products consumers want as well as how to best distribute, advertise, and promote those products. Marketing research is also used extensively to measure performance. Do consumers really understand the benefits of the product, how much is being sold and to whom, etc? Simply stated, marketing research is intended to provide timely and relevant information to improve marketing decision making.

In attempting to meet the needs of consumers, food marketers need to keep up with all of the changes taking place in their markets. Most food marketers rely on marketing research to help them make decisions about the marketing plan. The American Marketing Association adopted this definition of marketing research in 1987:

> Marketing Research is the function which links the consumer, customer, and public to the marketer through information—information used to identify and define marketing opportunities and problems; generate, refine, and evaluate marketing actions; monitor marketing performance; and improve understanding of marketing as a process. Marketing Research specifies the information required to address these issues; designs the method for collection information; manages and implements the data collection process; analyzes the results; and communicates the findings and their implications (19).

Food marketers must use marketing research to help them make good decisions. One way is to use the scientific method—a decision-making approach that focuses on being objective in testing ideas before accepting them. This way a marketer does not assume his or her intuition is correct without evidence to support it. The marketing research process is a five-step application of the scientific method. It includes

1. Defining the problem.
2. Analyzing the situation.
3. Getting problem-specific data.
4. Interpreting the data.
5. Solving the problem.

CONCLUSION

In order for the food industry to maintain its preeminence, all the parts must not only work individually, but work together. While the engineers and food technologists are developing new products, food marketers must ensure that the research and development is market directed. Making food less expensive to produce is only valuable if people want and will buy the less-expensive product. Food marketing is the discipline that brings the consumer into the corporate planning process. It ensures that the products desired will not only be available, but that consumers will know about the benefits of the product and where to get it. Food marketing is no more important than any other food business function, but it is essential in today's food business environment.

BIBLIOGRAPHY

1. E. J. McCarthy, *Basic Marketing*, Richard D. Irwin, Inc., Homewood, Ill., 1990, p. 8.
2. P. Kotler, *Principles of Marketing*, Prentice Hall, Inc., Englewood, N.J., 1989, p. 5.
3. *Marketing News*, 1 (Mar. 1, 1985).
4. A. Ries and J. Trout, *Marketing Warfare*, McGraw-Hill Inc., New York, 1983.
5. M. Porter, *Competition in Global Industries*, Harvard Business School Press, Boston, 1986.
6. J. Stanton, "2001: A Food Odyssey," *HEIB*, 1990
7. Ref. 1, p. 261.
8. H. Boyd, *Marketing Management*, Richard D. Irwin, Inc., Homewood, Ill., 1990, p. 22.
9. Ref. 1, p. 344.
10. L. Haugh, "Defining and Redefining," *Advertising Age*, M-44 (Feb. 14, 1983).
11. J. Dagnoli and L. Freeman, "Marketers Seek Slotting-Fee Truce," *Advertising Age*, 12 (Feb. 22, 1988).

12. Donnelly Marketing, *12th Annual Survey of Promotional Practices*, Stamford, Conn., 1990, p. 3.
13. R. Gibson, "Kellogg Shifts Strategy to Pull Consumers In," *The Wall Street Journal*, Jan. 22, 1990, p. 131.
14. Ref. 1, p. 366.
15. *Ad Week's Marketing Week*, July 3, 1989, p. 6.
16. B. Bagot, "1990 Industry Outlooks," *Marketing and Media Decisions*, 33 (Jan. 1990).
17. W. Wilkie, *Consumer Behavior*, John Wiley & Sons, Inc., New York, 1990, p. 492.
18. D. Lehmann, *Market Research and Analysis*, Richard D. Irwin, Inc., Homewood, Ill., 1989, p. 3.

John L. Stanton
Richard J. George
Carol A. Gallagher
Saint Joseph's University
Philadelphia, Pennsylvania

See also United States food marketing system in the Supplement section.

FOOD MICROSTRUCTURE

The microstructure (location, distribution, size, morphology, porosity, etc) of a food material is closely correlated to its texture (1), component functionality (2), and nutritional quality (3). The challenge lies in finding the appropriate methodology (or combination of such) for microstructural characterization. This review will focus mainly on the various techniques available for such characterization with special emphasis on their applications and limitations.

A basic approach for visualizing the microstructure of a sample is to obtain a magnified view of its components through microscopic techniques. Excellent examples of the relationship of microscopic observations to changes observed during processing can be found in the literature (4,5). Our attempt will be to summarize the differences of some basic microscopy techniques and their limitations and advantages.

LIGHT MICROSCOPY

The most routinely applied microscopy techniques are those collectively categorized under light microscopy. At the heart of these techniques is a microscope equipped with a light source and a series of lenses: the condenser, objective, and ocular. Figure 1a shows the basic principles of image magnification in a light microscope. A light source provides illumination, that passes through first the condenser, then the specimen, followed by the objective lens producing a reversed upside-down image, which is then further magnified by the ocular lens (6,7). The resolution of light microscopy is between 200 and 500 nm (5). The proper use and limitation of each of these components has been described elsewhere (7).

A specimen's microstructure can only be seen in the microscope when there is some form of contrast. Contrast is usually established through the staining (chemically or physically attaching specific dyes) of particular components. Some commonly used stains in food research are presented in Table 1. Vaughan (4) described in detail the use of various stains to locate specific components in a variety of commodities. An example of the relationship of microstructural changes occurring during processing looked at the effect of mixing on the structure of dough made from different quality flours by observing uranyl-acetate stained gluten fractions (8). Proper dough development required a matrix network of protein strands. For living cells, which may be harmed by stains, a phase ring and plate may be added to achieve phase contrast (7) as shown in Figure 1b.

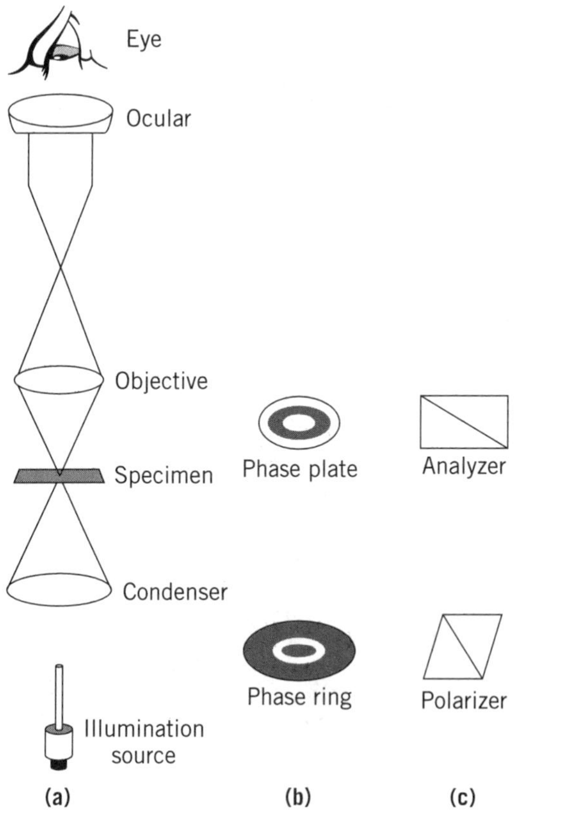

Figure 1. (**a**) Schematic of image magnification in light microscopy; (**b**) phase plates required for phase contrast microscopy (see text for details); (**c**) polarizers required for polarized light microscopy (see text for details).

Table 1. Stains of Food Components Used in Light Microscopy Work

Compound	Stain
Starch	Iodine
Protein	Fast green, acid fuchsin
Plant cell walls, muscle tissue	Toluidine Blue O
Fats	Oil Red O, Sudan III

Source: Ref. 6.

Polarizers are common modifications added to the compound microscope to view highly ordered (crystalline) material. The polarizers are placed between the illumination source and the condenser (polarizer) and between the specimen and objective (analyzer) as shown in Figure 1c. Many organic crystals are birefringent: entering light is split into two component and polarized in two perpendicular directions (7). Birefringence is the measure of the differences in refractive indices of the two components (9). Polarized microscopy has been used in food to measure the sacromere length in muscle fiber and subsequently related to meat quality (5), to identify the botanical origin of starch grains and in turn follow the specific patterns of gelatinization and recrystallization (9), and to differentiate the three polymorphic forms of triglycerides in fats, oils, and emulsions and relate each to lipid stability (5).

A powerful light microscopy technique relies on the ability of objects to fluoresce. Fluorescence is described as a process in which light energy is absorbed at a particular wavelength that excites the molecule to a higher electron energy state and higher vibrational state. Subsequently, this energy is re-emitted in the form of light in a range of wavelengths that can be quantified (10). Various food components are autofluorescent, whereas others require the attachment of specific fluoropores for them to be observed under fluorescence (Table 2). Selecting the appropriate fluorescence probes and/or filters to minimize background signals is critical for improved sensitivity of the detected fluorescence. The two types of filters used to obtain adequate fluorescence are the exciter and barrier filters. The exciter filter transmits only those wavelengths that are near the excitation maximum of the fluorescent compound while the barrier filter transmits only the wavelengths emitted by the sample (11). Vodovotz et al. (12) discussed the environmental factors that affect the efficacy of a specific probe including solvent polarity, proximity, and concentrations of quenching species and pH of the aqueous medium. Various examples of the use of fluorescence in food research include the effect of processing on starch as well as cell wall structure, enzyme activity, and nutritional quality of cereal grains (3,11).

The main drawbacks of light microscopy are limited resolution and the need to obtain thin sections for analysis. A specimen with considerable thickness yields a blurred or diffused image because the depth of field in conventional microscopy at high power is 2 to 3 μ while the resolving power is 0.2 μ (13). The confocal microscope circumvents this problem by focusing a point light source on a small volume within the specimen, rendering an image of in-focus plane only, with the out-of-focus parts appearing as black background (14). The confocal microscope functions by focusing a light source on an x-y plane at a specific depth (z). The illumination is often coherent (laser) light since its greater intensity allows for additional applications such as observing fluorescence in a specimen. The specific fluorescent probes can aid in the localization of different components depending on the number of laser lines available in the instrument. Figure 2 shows the principal parts and light path of a confocal microscope. The laser light passes through an objective lens (which acts as a condenser as well as a collector; hence, the term confocal), bounces off a diachroic mirror, and is then projected onto a pinhole (confocal aperture) in front of a detector (photomultiplier tube). To obtain more than just a pinhole image, an x-y plane is scanned (rastered), and the image is then collected and reconstructed in a computer (12,14,15).

Table 2. Fluorescent Probes and Stains Used with Food Components in Fluorescence Microscopy

Food Component	Fluorescence probe
Collagen	Autofluorescence
Elastin	Autofluorescence
Pigments (chlorophyll, carotenoids)	Autofluorescence
Phenolic compounds (lignin and ferulic acid)	Autofluorescence
Cartlidge, bone	Autofluorescence
β-D-Glucan	Calcofluor white, Congo Red
Periodate/Schiff's	Starch
Proteins	Acid fuchin, Thiazine Red R
Lipids	Nile Blue A, Sudan III

Source: Refs. 5, 6, and 12.

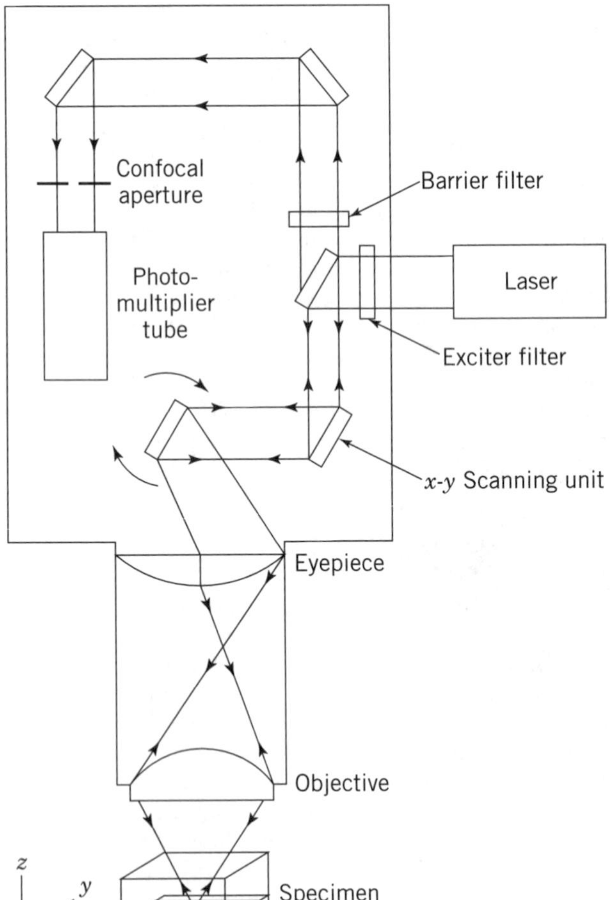

Figure 2. Schematic of the light path in a confocal microscope. *Source:* Ref. 12, used with permission.

An important application of the confocal microscope is the ability to observe a three-dimensional image without the need to physically section the sample, allowing a disturbance-free observation of the specimen (12). Although transmission electron microscopy (TEM) can provide a three-dimensional view of the specimen, the sample must be dried or frozen (or replicated with platinum and carbon) since TEM is carried out in vacuum. Additionally, confocal microscopy requires less sample preparation such as dehydration and fixation, since thin sections are obtained optically. For example, beef actomyacin solutions mixed with glutaminase were viewed directly at different time intervals as a gel network was formed (16). Dynamic processes such as dough formation (17) are also possible.

The combination of two powerful techniques such as laser scanning confocal microscopy (LSCM) and fluorescence microscopy can be used to localize and identify different components. Heertje and coworkers (18) followed the rising of dough containing fluorescein isothiocyanate solution at room temperature. They were able to distinguish various locations of gluten, starch, and air pockets. A similar study by Blonk and van Aalst (15) looked at the influence of various additives and the gluten network in dough on bread loaf volume. In their study, gluten was stained with rhodamine and was shown to have a finite expandability, whereas inclusion of an additive with surfactant capabilities aided in the formation of a continuous gluten film. No difference was detected in the location and distribution of starch (stained with periodic Acid-Schiff's reaction) and gluten (stained with FuoroLink Cy3) when comparing fresh and aged bread (19).

LSCM has been most commonly used in the fields of medicine and microbiology/biochemistry. The main benefit of LSCM is the elimination of diffuse light by the use of a pinhole. However, this requires greater illumination, which may result in photobleaching and photodamage to live specimen. To circumvent this problem, a microscope with an efficient light path from the specimen to the detector is required as well as its operation at maximal sensitivity (highest gain) (20). Some relevant applications include the study of pH imaging in cells (21,22), detection of single molecules (such as DNA) in solution (23), the determination of the intracellular 3-D (or 4-D) structure of individual cell components by microinjecting dyes into living cells (24), and observing the location and distribution of *Salmonella enteritidis* in egg shells held at various conditions (25).

ELECTRON MICROSCOPY

Electron microscopy (EM) provides better resolution and higher magnifications compared with conventional light microscopy (26) because the illumination source in EM is electrons focused with magnetic lenses rather than photons focused with glass lenses. Electrons are absorbed by air; therefore, operation is carried out under vacuum. The sample must not release any volatile substances when placed in the microscope, and thus it is dried or frozen or replicated with platinum and carbon. In the latter case the replica is examined (6).

EM can be carried out in two modes: scanning electron microscopy (SEM) and TEM. SEM is used to examine surfaces. Nonconductive materials such as food specimens need to be made electrically conductive. This can be achieved by depositing a metal layer such as gold palladium onto the sample using a spatter coater. The sample is scanned by a focused electron beam. In conventional SEM the sample needs to be completely dry because the vacuum within the microscope would cause the water to evaporate. In addition, lipids in the sample would melt under the electron beam (27). Careful specimen preparation is critical to avoid artifacts and essentially determines the value of information obtained (28). The preparation methods are dictated mainly by the type of sample. Moisture and fat content as well as structural differences should be considered. For example, fixation, dehydration, and coating may require longer time for compact structures such as low-fat cheeses compared with full-fat cheeses (6). Critical point drying after chemical fixation with osmium tetroxide and a low concentration of gluteraldehyde is an effective sample preparation method for protein systems such as gluten (29). Dry cereal grains and starch have been examined with SEM with relatively simple sample preparation, to assess relations among components and extent of damage due to milling (30,31), action of amylolytic enzymes on starch granules (32) and to relate granule size and surface morphology to the genetic background of the source plant (33). Observation of bread by SEM can provide information on gas cell size. However, the use of fixatives and removal of moisture can cause the protein matrix to shrink away from the starch granules, altering the appearance of the bread components (34,35).

High-moisture samples, as most foods are, and high-fat systems can be examined with cryo-SEM. This technique involves freezing the specimen and examining it in the SEM at a temperature at which neither water is lost nor fat is melted (36). Allan-Woajtas (37) has discussed methods to freeze samples in preparation for SEM. Cryo-SEM is very useful in the study of food materials that are easily damaged by conventional SEM, such as baked products and emulsions. A significant advantage of the technique is the ability to study dynamic phenomena by freezing samples at time intervals, during emulsion destabilization, for instance, or dough formation and baking (38). Sargent (36) has reviewed some of the numerous applications of cryo-SEM in food systems. Dairy products such as cheese, ice cream, and full- and low-fat spreads have been examined in great detail. Low-temperature SEM has helped reveal the mechanism of emulsion and foam destabilization in products such as salad dressings, meringues, and mayonnaise. Cocoa powders, butter, and chocolate have been extensively characterized as well. Structure organization in cereals and changes during heating or baking (35,39), dough formation and effect of mixing on the gluten network (28,40), and examination of cake batters (41) are some of the many applications of cryo-SEM in the study of cereal products (42).

Despite the extent of applications, cryo-SEM is not free of artifacts. The most common are due to formation of ice crystals, which can displace structural elements and destroy the initial structure (6). Some of the limitations of the conventional and cryo-SEM can be resolved by an emerging technique, the environmental SEM (ESEM) (39,43), that can produce images at ambient temperature without drying or metal coating. This technique could have great impact in the direct observation of food systems.

TEM visualizes the internal structure of food samples. Thin sections of samples embedded in epoxy resin or platinum-carbon replicas of the sample are placed in the path of the electron beam, and the electrons are transmitted through the sample (6). Cryo-fixation followed by replication produces superior TEM images of high resolution. The image quality depends on the quality of the replica, which requires high skill and experience. TEM has been used to identify protein and lipid structures in grains (30), to determine structure and stability of cake batters (41), to study the properties of microencapsulated droplets in starch/water gels (44), to assess enzyme activity on flours (45), to monitor changes in plant cell walls as a result of processing (46), and to obtain images of food-related proteins (47).

EM techniques in general provide qualitative information on the structure and microscopic organization of food products. Quantitative information can be obtained by image analysis techniques that employ sophisticated statistical analysis. Computer-aided digitization of the micrographs can be used to obtain, for instance, to study air cell size distributions (48) or even two- and three-dimensional reconstructions of protein structures (47). Image analysis will be described in more detail later.

ATOMIC FORCE MICROSCOPY

The light and electron microscopes helped to cross many boundaries in the world of molecules through elegant methods of sample preparation that have enabled microscopists to approach the theoretical resolution of their instruments. Still, to pass from the subcellular level to the molecular domain, the ability to resolve objects smaller than 1 nm is needed. This task can be accomplished by the scanning probe microscopes (49), a family of microscopes that make it possible to examine molecules or groups of molecules without the heavy metal coatings that are often necessary to produce contrast in optical and high-vacuum systems and without the many of the other constraints inherent to optical and electron microscopes.

The first probe microscope, the scanning tunneling microscope, was invented in 1986 and led to the development of the atomic force microscopy (AFM) (50). AFM is more suitable for nonconductive samples such as biological specimen (51).

The major difference between AFM and light or electron microscopy is in the way the image is created. Conventional microscopy employs lenses and relies on optics to create an image. In AFM the image is created by scanning a sharp tip, which is attached to a flexible cantilever, across a sample surface (Figure 3), much like a record stylus plays a record. When the tip is in close proximity to the sample, repulsive forces cause the cantilever to bend away from the surface. By measuring the extent of cantilever bending, a detailed topography can be recorded. The most common

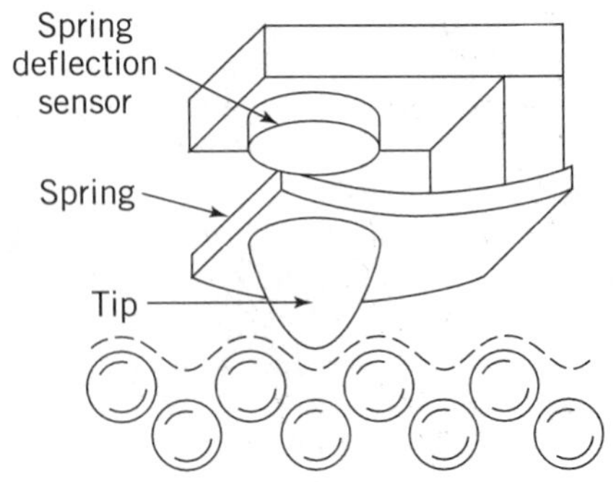

Figure 3. Schematic of the AFM cantilever-tip mechanism. *Source:* Ref. 51, used with permission.

detection system is the deflection of a laser beam (52) reflected off a mirror mounted on the back of the cantilever mechanism. This system is very sensitive and can detect cantilever deflections produced by the tip scanning over individual atoms and molecules. The force exerted by the tip on the sample is very important for precision imaging. Too low a force causes the tip to mistrack and lose detail, whereas too high a force can displace or damage the sample. To control the force exerted by the tip, the cantilever assembly is mounted on a piezoelectric device that also controls the precision with which the tip is positioned relative to the surface of the sample. The precision of this mechanism is the key to the high resolution of the AFM (53).

One of the major advantages of the AFM is the minimal sample preparation required to obtain high-resolution images. Specimens are applied to a substrate as a dilute solution or thin film. No coating, extensive drying, or other preparation is needed. The substrate can be any flat surface such as a glass slide or freshly cleaved mica.

AFM can be carried out in three modes: contact, noncontact, and tapping mode (6). In the contact mode, the tip touches the sample. As the cantilever bends in response to surface features, a variable force is exerted on the sample. This force is compared with a preset value, and the piezoelectric device moves up or down to meet that value. The whole scan is then performed at constant force. The distance that the piezo needs to travel to achieve that constant force is recorded and amplified to create the image (53). Higher scan speeds can be attained with the contact mode, which is the only mode that can obtain "atomic resolution" images. Contact mode is the technique of choice for very rough samples with extreme changes in vertical topography; for example, deep grooves or valleys. The main disadvantage is that contact mode can be damaging to very soft samples due to the high forces, a result of capillary interactions from the absorbed fluid layer on the sample surface (54).

Most of the AFM imaging in food systems has been done with the contact mode. Aggregates of β-lactoglobulin, a whey protein, were imaged and characterized after deposition on mica and graphite (55). This level of detail cannot be accomplished with electron microscopy because of the restrictions imposed by the size of the metal coatings necessary to obtain the EM image. Kirby et al. (56) obtained high-resolution images of xanthan gum that showed extreme stiffness of individual molecules entangled in a network. Images of pectin (57), another thickening agent, appeared to be a more flexible structure that can be easily distorted. This can explain the relatively lower viscosity attained by pectin as compared with xanthan gum. Round et al. (58) studied an unusual branched structure for tomato pectins that could not be predicted from enzymatic hydrolysis, showing that AFM can be used to differentiate among polysaccharides from different plant sources. Interactions between α-gliadin layers were also investigated (59). It was shown that the molecules in the protein film have a very compact conformation that was affected by the presence of additives, demonstrating the potential use of AFM in characterization of edible protein-based films. Kirby et al. (53) obtained images of cellulosic gelled structures exuded by *Acetobacter xylinum* that show a network of flat ribbons enclosing the bacteria. Baker's yeast strains along with polysaccharide molecules from the cell wall were imaged (60) with minimal sample preparation. The action of α-amylase on starch granules has been captured in real time (61), demonstrating the ability of AFM to directly observe enzymatic reactions and structure changes on the substrate (Figure 4).

In the noncontact mode the cantilever is vibrated close to its resonance frequency, and the tip does not touch the sample. The image is created by detecting the damping effect on the cantilever vibrations caused by long-range attractive forces between the atoms of the tip and the sample. The main advantage of using the noncontact mode is that the shearing forces imposed on the sample by the tip are reduced or even eliminated, making imaging of very soft specimens possible. The main disadvantage is that the noncontact mode usually only works on extremely hydrophobic samples where the absorbed fluid layer at the surface of the sample is at minimum.

The tapping mode overcomes this limitation (54). It is similar to the noncontact mode, but, in this case, the cantilever is strongly vibrated and the tip intermittently touches the sample. The major advantage of this technique is that the adhesive forces arising from absorbed water on the surface are reduced. Shearing is also minimized since the tip touches the surface only momentarily. As a result, this mode provides high-resolution images that could otherwise be obtained only with both the sample and the cantilever immersed in a liquid (53) and is very suitable for soft hydrophilic specimens. Tapping mode is gaining momentum in studying food systems. Tsoubeli and Taub (62) obtained high-resolution images of bovine myofibrils in air and followed aggregation and gel formation as a result of thermal processing.

DYNAMIC LIGHT SCATTERING

Dynamic light scattering (DLS) is an alternative method for determining food microstructure that is based on scattering of light by moving particles. Details of this noninvasive and relatively rapid method are given elsewhere (63). In the food industry, DLS is used for determining particle sizes in the range of 1 to 300 μm in such diverse products as chocolate, wheat and soy flour, confectionery sugar, spices, and mayonnaise. More specific instruments have been developed to detect particles outside this range (64). The main application of this technique has involved very dilute systems (milk and oil in water emulsions), because the concentration of scatters had to be sufficiently low that an incident photon of light is scattered only once by the sample (63). Therefore, DLS can rarely be used directly on most foods, since particulate concentration is too high and native structure is difficult to maintain upon dilution. Nevertheless, DLS has been used to look at the changes in the average size of casein micelles with addition of calcium

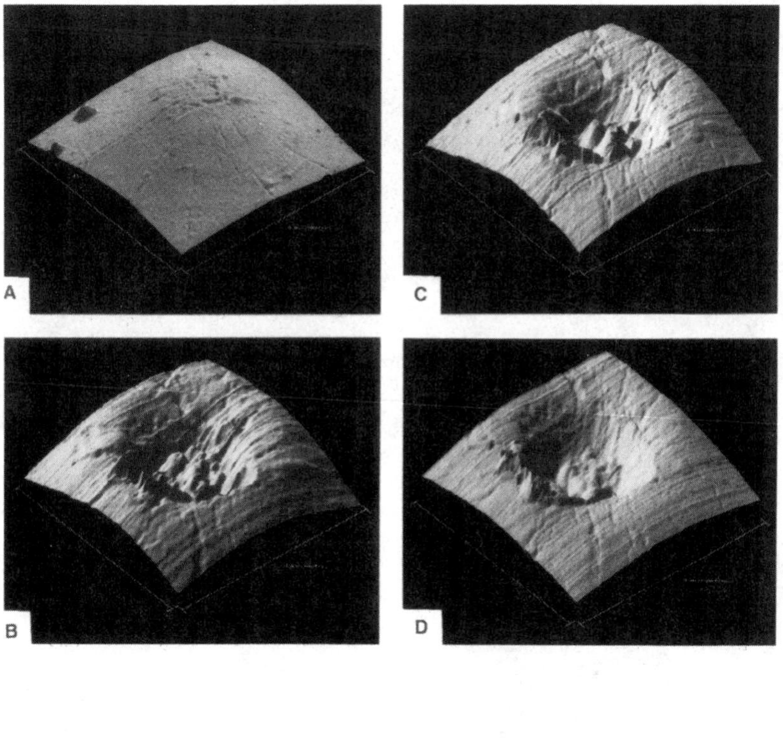

Figure 4. α-Amylase action on a single starch granule obtained by contact mode AFM. A–D: 0, 105, 420, and 735 s, respectively. *Source:* Ref. 61, used with permission.

phosphate, increased temperature, and different pHs (63). These changes (reflecting processing protocols) could be related to the functional properties of milk. Additionally, DLS has been used to detect size distribution (65) and adsorbed monolayers (66) in emulsions.

A modification of DLS, diffusion wave spectroscopy (DWS), in which incident and back-scattered light are conducted through fiber optics, can be used to provide estimates of mean particle size in concentrated suspensions. For example, the particle-size changes in rennet (67) and acid-induced (68) gelation of milk were studied by DWS. However, to obtain accurate particle sizes with DWS, the instrument needs to be calibrated with separate samples with known size scatterers, which may not have the same optical characteristics as the food samples used, making the interpretation of results difficult (63).

IMAGE ANALYSIS

To obtain microstructural information from microscopic or other imaging techniques, data need to be collected and analyzed. The first step would require a computer rendering of the image, which may be accomplished by photographing and digitizing the data. In the case of LSCM, this would be done automatically. Once microstructural data are collected, subjective analysis may not be sufficient. Quantification of objects observed in an image may be as simple as measuring its length with an eyepiece micrometer or counting individual components. These measurements, if performed in a large number of samples, could be extremely time-consuming; therefore, automated image analysis techniques may be preferable. However, the automated system used is required to recognize the desired object(s), and differentiate it (them) from artifacts such as overlapping or touching cells, debris, and dirt, so as to exclude them from the measurements (7). Regardless of the method used, analysts need to be sure that their observations are not due to artifacts arising from sample preparation, such as fixing and embedding, or instrumental limitations such as the optics used.

Various objects display irregular shapes, which are difficult to characterize quantitatively. Fractal analysis can be employed to estimate the degree of irregularity in a sample. Unlike Euclidean geometry in which dimensions are expressed in whole numbers (eg, 2 for area and 3 for volume), fractal dimensions consist of fractional numbers (such as 1.3, etc). However, for true fractal objects they must exhibit self-similarity, at least over the range being described (69). This technique can be used to follow changes in the irregularity of a sample due to processing. For example, the extent of agglomeration of coffee particles can be quantified and related to their dispersibility in a liquid (69). The difficulty with fractal analysis lies with the ability to obtain accurate data from imaging technique, such as the edge pixels in a digitized image.

These techniques represent a few, but not all, of the methods available to obtain microstructural information of food products. It is imperative to understand the limitation of each technique, especially when interpreting the results. Proper microstructural analysis can be extremely beneficial in characterizing changes occurring in food products, which manifest themselves in a macroscopic level.

BIBLIOGRAPHY

1. D. W. Stanley, "Food Texture and Microstructure," in R. Moskowitz, ed., *Food Texture: Instrumental and Sensory Measurement*, Marcel Dekker, New York, 1987, pp. 35–64.
2. A. Hermansson, "Protein Functionality and Its Relation to Food Microstructure," *Qualitas Plantarum, Plant Foods for Human Nutrition* **32**, 369–388 (1983).
3. S. H. Yiu, "Food Microscopy and the Nutritional Quality of Cereal Foods," *Food Structure* **12**, 123–133 (1993).
4. J. G. Vaughan, *Food Microscopy*, Academic Press, New York, 1979.
5. J. M. Aguillera and D. W. Stanley, *Microstructural Principles of Food Processing and Engineering*, Elsevier Science, New York, 1990.
6. M. Kalab, P. Allan-Wojtas, and S. S. Miller, "Microscopy and Other Imaging Techniques in Food Structure Analysis," *Trends Food Sci. Technol.* **6**, 177–186 (1995).
7. J. James and H. J. Tanke, *Biomedical Light Microscopy*, Kluwer Academic Publishers, Dordecht, The Netherlands, 1991.
8. D. B. Bechtel, Y. Poranz, and A. de Francisco, "Breadmaking Studied by Light and Transmission Electron Microscopy," *Cereal Chem.* **55**, 392–401 (1978).
9. E. Varriano-Marston, "Polarization Microscopy: Applications in Cereal Science," in D. B. Bechtel, ed., *New Frontiers in Food Microstructure*, American Association of Cereal Chemists, St. Paul, Minn., 1983, pp. 71–104.
10. R. G. Fulcher and P. J. Wood, "Identification of Cereal Carbohydrates by Fluorescence Microscopy," in D. B. Bechtel, ed., *New Frontiers in Food Microstructure*, American Association of Cereal Chemists, St. Paul, Minn., 1983, pp. 111–147.
11. R. G. Fulcher, P. J. Wood, and S. H. Yiu, "Insight Into Food Carbohydrates Through Fluorescence Microscopy," *Food Technol.* **1**, 101–106 (1984).
12. Y. Vodovotz et al., "Bridging the Gap: Use of Confocal Microscopy in Food Research," *Food Technol.* **6**, 74–82 (1996).
13. J. J. Lemasters et al., "Laser Scanning Confocal Microscopy of Living Cells," in B. Herman and J. J. Lemasters, eds., *Optical Microscopy: Emerging Methods and Applications*, Academic Press, San Diego, Calif., 1993.
14. H. D. Cavanagh, W. M. Petroll, and J. V. Jester, "Confocal Microscopy: Uses in Measurement of Cellular Structure and Function," *Progress in Retinal and Eye Research* **14**, 527–565 (1995).
15. J. C. G. Blonk and H. van Aalst, "Confocal Scanning Light Microscopy in Food Research," *Food Res. Int.* **26**, 297–311 (1993).
16. S. H. Kim et al., "Polymerization of Beef Actomyosin Induced by Transglutaminase," *J. Food Sci.* **58**, 473–474, 491 (1993).
17. G. J. Brakenhoff et al., "3-Dimensional Imaging of Biological Structures by High Resolution Confocal Scanning Laser Microscopy," *Scanning Microscopy* **2**, 33–40 (1988).
18. I. Heertje et al., "Confocal Scanning Laser Microscopy in Food Research: Some Observations," *Food Microstructure* **6**, 115–120 (1987).
19. Y. Vodovotz and C. Chinachoti, "Confocal Microscopy of Bread," in M. H. Tunick, S. A. Palumbo, and P. M. Fratamico, eds., *New Techniques in the Analysis of Foods*, Plenum Press, New York, 1998, pp. 9–17.

20. J. J. Lemasters et al., "Laser Scanning Confocal Microscopy of Living Cells," in B. Herman and J. J. Lemasters, eds., *Optical Microscopy, Emerging Methods and Applications*, Academic Press, San Diego, Calif., 1993, pp. xx–xx.
21. R. Sanders et al., "Quantitative pH Imaging in Cells Using Confocal Fluorescence Lifetime Imaging Microscopy," *Anal. Biochem.* **227**, 302–308 (1995).
22. A. N. Hassan et al., "Formation of Yogurt Microstructure and Three-Dimensional Visualization as Determined by Confocal Scanning Laser Microscopy," *Journal of Dairy Science* **78**, 2629–2636 (1995).
23. S. Nie, D. T. Chiu, and R. N. Zare, "Real-Time Detection of Single Molecules in Solution by Confocal Fluorescence Microscopy," *Anal. Chem.* **67**, 2849–2857 (1995).
24. S. J. Wright et al., "Introduction to Confocal Microscopy and Three-Dimensional Reconstruction," *Methods in Cell Biology* **38**, 1–45 (1993).
25. J. W. Wong Liong, J. F. Frank, and S. Bailey, "Visualization of Eggshell Membranes and Their Interaction With *Salmonella enteritidis* Using Confocal Scanning Laser Microscopy," *J. Food Prot.* **60**, 1022–1028 (1997).
26. M. T. Postek et al., *Scanning Electron Microscopy—A Student's Handbook*, Laboratory Research Industries, Burlington, Vt., 1980.
27. M. Kalab, "Electron Microscopy of Foods," in E. B. Bagley and M. Peleg, eds., *Physical Properties of Foods*, AVI, Westport, Conn., 1983, pp. 43–104.
28. K. Autio and T. Laurikainen, "Relationships Between Flour/Dough Microstructure and Dough Handling and Baking Properties," *Trends Food Sci. Technol.* 8(6), 181–185 (1997).
29. A. M. Hermansson and W. Buchheim, "Characterization of Protein Gels by Scanning and Transmission Electron Microscopy," *Colloid Interface Sci.* **81**, 519–530 (1981).
30. E. Varriano-Marston, "Intergrating Light and Electron Microscopy in Cereal Science," *Cereal Foods World* **26**, 558–561 (1981).
31. T. Otto, B. K. Baik, and Z. Czuchajowska, "Microstructure of Seeds, Flours and Starches of Legumes," *Cereal Chem.* **74**, 445–451 (1997).
32. P. M. Mathias et al., "Studies of Amylolytic Breakdown of Damaged Starch in Cereal and Non-cereal Flours Using High Performance Liquid Chromatography and Scanning Electron Microscopy," in *Starch: Structure and Functionality*, Royal Society of Chemistry, Cambridge, United Kingdom, 1997, pp. 129–140.
33. M. Obanni and J. N. BeMiller, "Ghost Microstructures of Starch From Different Botanical Sources," *Cereal Chem.* **73**, 333–337 (1996).
34. J. F. Chabot, L. F. Hood, and M. Liboff, "Effect of Scanning Electron Microscopy Preparation Methods on the Ultrastructure of White Bread," *Cereal Chem.* **56**, 462 (1979).
35. T. P. Freeman and D. R. Shelton, "Microstructure of Wheat Starch: From Kernel to Bread," *Food Technol.* **45**, 162–168 (1991).
36. J. A. Sargent, "The Application of Cold Stage Scanning Electron Microscopy to Food Research," *Food Microstructure* **7**, 123–135 (1988).
37. P. Allan-Wojtas, "A Simple Carrier for Freezing Difficult Food Samples in Preparation for Scanning Electron Microscopy," *Food Structure* **9**, 75–76 (1990).
38. B. E. Brooker, "Low-Temperature Microscopy and X-Ray Analysis of Food Systems," *Trends Food Sci. Technol.* **1**, 100–103 (1990).
39. C. M. McDonough et al., "Microstructure Changes in Wheat Flour Tortillas During Baking," *J. Food Sci.* **61**, 995–999 (1996).
40. L. G. Evans, A. M. Pearson, and G. R. Hooper, "Scanning Electron Microscopy of Flour-Water Doughs Treated With Oxidizing and Reducing Agents," *Scanning Electron Microscopy* **III**, 583–592 (1981).
41. S. I. Hsieh, E. A. Davis, and J. Gordon, "Cryofixation Freeze-Etch of Cake Batters, Cereals and Cereal Products, Electron Microscopy," *Cereal Foods World* **26**, 562–564 (1981).
42. S. H. Yiu, "Food Microscopy and the Nutritional Quality of Cereal Foods," *Food Structure* **12**, 123–133 (1993).
43. E. Doehne and D. C. Stulik, "Applications of the Environmental Scanning Microscope to Conservation Science," *Scanning Microscopy* **4**, 275–286 (1990).
44. K. Eskins et al., "Ultrastructural Studies on Microencapsulated Oil Droplets in Aqueous Gels and Dried Films of a New Starch-Oil Composite," *Carbohydr. Polym.* **29**, 233–239 (1996).
45. M. P. Oria, B. R. Hamaker, and J. M. Shull, "Resistance of Sorghum Alpha-, Beta-, and Gamma-Kafirins to Pepsin Digestion," *J. Agric. Food Chem.* **43**, 2148–2153 (1995).
46. J. T. Marle et al., "Chemical and Microscopic Characterization of Potato Cell Walls During Cooking," *J. Agric. Food Chem.* **45**, 50–58 (1997).
47. R. Y. Yada et al., "Visions in the Mist: The *Zeitgeist* of Food Protein Imaging by Electron Microscopy," *Trends Food Sci. Technol.* **6**, 265–270 (1995).
48. J. Tan, H. Zhang, and X. Cao, "SEM Image Processing for Food Structure Analysis," *J. Texture Stud.* **28**, 657–672 (1997).
49. V. J. Morris and T. J. McMaster, "Molecular Microscopy—Probing Biological Structure," *Trends Food Sci. Technol.* **2**, 80–84 (1991).
50. G. Binnig, C. F. Quate, and C. Gerber, "Atomic Force Microscope," *Phys. Rev. Lett.* **59**, 930–933 (1986).
51. P. K. Hansma et al., "Scanning Tunneling Microscopy and Atomic Force Microscopy: Application to Biology and Technology," *Science* **242**, 209–216 (1988).
52. G. Meyer and N. M. Amer, "Novel Optical Approach to Atomic Force Microscopy," *Appl. Phys. Lett.* **53**, 1045–1047 (1988).
53. A. R. Kirby, A. P. Gunning, and V. J. Morris, "Atomic Force Microscopy in Food Research: A New Technique Comes of Age," *Trends Food Sci. Technol.* **6**, 359–365 (1995).
54. *Scanning Probe Microscopy Training Notebook*, Digital Instruments, Santa Barbara, Calif., 1998.
55. M. N. Tsoubeli, E. A. Davis, and J. Gordon, "Exploring Whey Proteins at the Molecular Level With Atomic Force and Scanning Tunneling Microscopy," *Scanning 96 Abstracts* **18**, 254 (1996).
56. R. Kirby, A. P. Gunning, and V. J. Morris, "Imaging Xanthan Gum by Atomic Force Microscopy," *Carbohydr. Res.* **267**, 161–166 (1995).
57. A. R. Kirby, A. P. Gunning, and V. J. Morris, "Imaging Polysaccharides by Atomic Force Microscopy," *Biopolymers* **38**, 355–366 (1996).
58. A. N. Round et al., "Unexpected Branching in Pectin Observed by Atomic Force Microscopy," *Carbohydr. Res.* **303**, 251–253 (1997).
59. L. Wannerberger et al., "Interaction Between Alpha-Gliadin Layers," *Journal of Cereal Science* **26**, 1–13 (1997).
60. R. de S. Pereira, N. A. Parizotto, and V. Baranauskas, "Observation of Baker's Yeast Strains Used in Biotransformation by Atomic Force Microscopy," *Appl. Biochem. Biotechnol.* **59**, 135–143 (1996).

61. P. R. Shewry et al., "Scanning Probe Microscopes—Application in Cereal Science," *Cereal Chem.* **74**, 193–199 (1997).
62. M. N. Tsoubeli and I. A. Taub, "ESR and AFM Studies of Bovine Myofibrils and Restructured Meat Systems," *IFT Book of Abstracts*, 58 (1998).
63. D. G. Dalgleish and F. R. Hallett, "Dynamic Light Scattering: Applications to Food Systems," *Food Res. Int.* **28**, 181–193 (1995).
64. P. E. Plantz, "Laser Light Scattering," *Cereal Foods World* **28**, 252–253 (1983).
65. F. R. Hallett et al., "Particle Size Analysis: Number Distributions by Dynamic Light Scattering," *Canadian Journal of Spectroscopy* **34**, 63–70 (1989).
66. D. G. Dalgleish, "Conformations and Structures of Milk Proteins Adsorbed to Oil-Water Interfaces," *Food Res. Int.* **29**, 541–547 (1996).
67. D. S. Horne and C. M. Davidson, "The Use of Dynamic Light-Scattering in Monitoring Rennet Curd Formation," *Milchwissenschaft* **45**, 712–725 (1990).
68. D. G. Dalgleish and D. S. Horne, "Studies of Gelation of Acidified and Renneted Milks Using Diffusion Wave Spectroscopy," *Milchwissenschaft* **46**, 417–420 (1991).
69. M. Peleg, "Fractals and Foods," *Crit. Rev. Food Sci. Nutr.* **33**, 149–165 (1993).

Y. VODOVOTZ
NASA Johnson Space Center
Houston, Texas
M. TSOUBELI
Campbell Soup Co.
Camden, New Jersey

FOOD PLANT DESIGN AND CONSTRUCTION

Design and construction of a new food plant or a major expansion is a relatively rare event in the normal career of a food professional. When it does occur, it is important to have an understanding of critical issues and to have a well-planned approach. This article provides the nonexpert food professional with enough information to interact effectively with the engineers, architects, and constructors who are usually involved in any large project.

CRITICAL ISSUES

Nearly every food plant design must face the following issues:

- *Sanitary design.* Hazard analysis and critical control point (HACCP) plans must be accommodated to ensure human safety and product quality. Certain aspects of sanitary design are dictated by government regulation, whereas others are voluntary but highly recommended.
- *Efficient material and people flow.* Flow of material and of people dictates plant layout and thus overall design. It also strongly influences expansion opportunities, sanitation, and safety.
- *Environmental impact.* A food plant discharges wastes that can require treatment. A plant can be affected by its neighbors' discharges, and it must be a good neighbor aesthetically.
- *Minimum capital and operating cost.* Food plants must be efficient competitors. Resources must be invested wisely in light of life-cycle costs and designs that minimize such costs as maintenance and sanitation.

Other more specific issues and further details on these issues are addressed in the description of the overall approach in the balance of this article.

PROJECT APPROACH

The approach outlined in this section has proved successful in the design and construction of new or renovated production facilities. It is adapted from one used by a large equipment and engineering firm focused on the food industry. It includes three steps or phases: a preliminary study of feasibility; an engineering and design phase (which itself may have some divisions); and a construction phase. Table 1 describes the phases in outline form.

PHASE I—FEASIBILITY STUDY

The primary objective of the feasibility study is to establish a financial analysis, specifically the capital and operating costs that can be compared with the estimated benefits to test whether the proposed project meets investment criteria. The feasibility study also imposes a discipline on the various interested parties, such as production, engineering, management, and marketing. They must agree on specific assumptions and design criteria. Once established and proved satisfactory, these criteria are the basis for all subsequent designs.

Develop Process Parameters

Present and Historical Sales. The most important decisions about a new plant are what it should make and how much. These are dictated by corporate strategy, but sizing has such an influence on design and cost that it requires special analysis. The typical trade-off is line capacity against inventory levels, especially for seasonal products. Minimum inventory level consistent with efficient customer service must be established. Modern practice is to minimize inventory as much as possible, even to the extent of importing raw materials when necessary.

Projected Sales Data. The plant should satisfy the projected sales requirement for some future identified date. This forecast is based on growth rate projections for each product category. The design basis year is often designated as the fifth year after completion of construction. Expansions beyond the year used as the design basis are provided by physically expanding the building or adding production capacity.

Raw Ingredient Utilization and Procurement. Usage levels and purchase options of raw ingredients are evaluated to determine liquid and dry ingredient storage and receiving requirements. Issues include whether ingredients are

Table 1. Three Phases of the Design and Construction of a Food Plant

Phase I—Feasibility study	
Develop process parameters	
	Present and historical sales
	Projected sales data
	Raw ingredient utilization and procurement
	Operations and production
	Packaging and filling
	Automation
	Packaging material utilization
	Materials handling methods
	Finished product storage
Develop utility requirements	
	Water utilization
	Fuel utilization
	Electrical utilization
	Sewage and wastewater
	Solid wastes
Develop site parameters	
	Area needed
	Projected plant configuration
	Owner's building style
Cost estimate and schedule	
Phase II—Engineering and architectural design	
Process specifications and engineering	
	Piping and instrumentation diagram
	Process equipment arrangement
	Process piping
	Process equipment specifications
	Process installation specifications
Utility specifications and engineering	
	Mechanical flow diagrams
	Mechanical equipment arrangements
	Mechanical equipment specifications
	Mechanical piping
	Mechanical installation
	Electrical single-line diagram
	Electrical equipment arrangement
	Electrical equipment details
	Electrical equipment specifications
	Electrical equipment installation
Site and facility specifications and engineering	
	Site layout
	Architectural design
	Architectural details
	Building and zoning permits
Cost estimate and financial analysis	
Phase III—Project and construction management	
Bidding	
	Contract type
	Site
	Building
	Process equipment
	Mechanical
	Electrical
	Process installation
Site management	
	Scheduling of site activities
	Receiving storing, and releasing material and equipment
	Verification of work
	Accounting and payments
	Project documents
	Start-up and commissioning
	Close out

received in bulk, semibulk containers, or discrete packages (bags, drums, and boxes). Minimum inventory levels depend on ordering quantities, time for delivery, safety stock, and stability of the materials. These factors influence equipment and space requirements.

Operations and Production. Decisions must be made about production rates, operating days, and shift arrangements. Traditionally, food plants have operated five days per week, with two shifts of production and one for cleanup per day, but there is a growing trend toward operating plants for 24 hours per day and seven days per week as the ideal to minimize capital costs. There are significant challenges in achieving such a schedule in many food plants, including finding qualified workers willing to keep the required schedule. Allowance must also be made for swings in required production rates, and equipment must be chosen that can accommodate such swings. Many types of food processing equipment have only a 2:1 turndown ratio, meaning they can only run over a range of 50% to 100% of nominal capacity. Some types are more flexible but many are less.

An important document resulting from this step is the process flow diagram, which identifies schematically each unit operation with associated flow rates and important process conditions.

Packaging and Filling. The type and rate of commercially available filling equipment must be evaluated for present and projected capacities. There may need to be several packaging lines for some process lines. Packaging lines are typically highly automated because packaging often is labor intensive. Flexibility in packaging in the sense of package size changes, material changes, and case configurations must be considered.

Automation. The level of plant automation must be established during this phase. Plant floor control systems are typically integrated with management information systems even up to the corporate level so that production requirements can be transmitted to the factory and production costs and performance can be transmitted to accounting. This is a fast-changing area, but some assumption must be made for the purpose of cost estimating. A typical automation structure includes PLCs (programmable logic controllers) on major equipment, PCs (personal computers) as operator workstations controlling segments of a process, and somewhat larger computers at the plant level, all connected by local area networks (LANs). The plant and the corporate computers are typically connected by dedicated high-speed communication channels.

Packaging Material Utilization. The usage levels and purchasing methods for packaging must be evaluated to determine storage and receiving requirements. Unique storage requirements such as temperature or environment control must be defined. For example, some paper and polymer packaging material must be equilibrated with the relative humidity of the plant to perform well. Glass containers received during cold weather must be warmed to prevent condensation of moisture. The packaging materials storage capacity will reflect such factors as proximity to supplier, long-term purchasing agreements, and volume discounts.

Materials Handling Methods. Materials handling methods for raw materials, work-in-process, and finished goods must be established. Choices include pneumatics for bulk powders, pipe for liquids, conveyors, totes, automated vehicles, person-driven vehicles, overhead rail, and manually powdered carts. Common issues include delivery of packaging material to point of use, where to put unitizing (palletizing), and how to handle trash such as empty ingredient bags and drums. It is desirable to minimize trash generation in food processing areas. Glass and wood should not be used (except for glass jars and bottles used as packages).

Finished Product Storage. An important issue is whether the plant site is also a distribution center for a wider line of products or simply ships its output to another site. This has a significant impact on storage requirements, traffic patterns, information systems, and overall design. If the plant is not to also be a distribution center, then the requirement for on-site finished goods storage may be minimal, and costs can be saved. If the plant is also to be a distribution center, then issues of receiving from other sites, order receiving and picking, and order shipping must be faced.

Develop Utility Requirements

Critical utilities include water, fuel, electricity, wastewater treatment, and solid waste disposal. If the process involves refrigeration, uses refrigerated ingredients, or produces refrigerated products, then it will probably require a central ammonia system. Utility requirements are estimated from the process flow diagram and material and energy balances. Local providers of water, fuel, waste disposal, and power are sources of information on rates, characteristics, and limitations. It is important that capacity be sufficient for current and future requirements and that costs be stable and predictable. Disposal of liquid and solid wastes is normally best handled by municipal facilities, but in some locations, a food plant must become self-sufficient in these matters. When permitted, land disposal of liquid wastes can be satisfactory, but it requires large amounts of land. Often, local farmers will accept food solid waste as animal feed and remove it at their own cost. Disposing of packaging waste can be a challenge in some areas.

Develop Site Parameters

To develop an accurate cost estimate, it is important to have a specific site in mind, but before one can be identified, it is necessary to understand the requirements. It is not unusual for land area to be 4 to 5 times the building area and sometimes much more, such as when land disposal of liquid wastes is included. Ideal sites are generally level and well-drained, have utilities supplied or within relatively short distances, are close to desired transportation, have no undesirable neighbors, are removed from residential areas, have safe and efficient traffic patterns, and permit designs with room for expansion in the future.

Site location at a broader scale is dictated largely by marketing considerations, but it is also influenced by other factors such as labor availability and cost, utility availability and cost, taxes, regulatory environment, raw materials, transportation access (including air, for visits from headquarters), and quality of life (to help in recruiting management).

Sadly, the ideal piece of land probably does not exist, at least not in the ideal location. This means various compromises must be made, which usually have cost implications. Generally, in the feasibility phase, a time zone or region of the country is known, perhaps even a specific state or community. Rarely is a specific plot of land identified at this stage, because the feasibility is still under study. Although it is tempting to assume an ideal plot of land (large enough, solid-bearing soil, easy access to utilities, great location), it is prudent to test the sensitivity of costs to various adverse assumptions, such as the need for pilings because of weak soil, the need to pay for utility access, or the need to pave an access road. These alternatives can be identified with assumed costs as subcases of the ideal. They rarely are the sole cause of unsatisfactory cost projections, but they can contribute.

In economic evaluation, if an ideal case is not feasible, then no worse case will be. However, if an ideal case is marginal, then one or two adverse circumstances can make a difference, and the project may be shaky by corporate standards. It is important to discover this early before large sums are spent on design and construction.

Projected Plant Configuration. A preliminary plan illustrating the production facility on the assumed lot is developed. This helps identify site-related costs and assists in presenting the concept to management, municipalities, and designers.

Owner's Building Style. The final design of a process facility must suit the owner's needs, which may include using the facility in marketing. Visits by the public are relatively rare, but it is common to welcome customers to food plants, and some provision for safe touring might be appropriate. Good design can make a utilitarian facility an attractive place to work and a good neighbor in its community at very little cost.

Cost Estimate

An estimate of the capital cost of process equipment and services can be determined from the preliminary design. Operating costs depend on the quantities and costs of utilities. The facility cost is estimated from the preliminary layout and the site plan, often by applying an experience-based unit cost to the estimated area. For many food plants in the late 1990s, facility costs were about $80 to $100 per square foot. Facility costs for food plants are typically 35 to 50% of the total cost, depending on the complexity of the process involved.

The feasibility study summarizes all the assumptions, projected costs, and operating procedures as well as the parameters on which the final design will be based. The study report is a reference document for the balance of the project and also can support such requirements as environmental impact statements.

The financial analysis includes a projection of capital requirements during design, construction, and the first five years of operation to meet projected production requirements. Pro forma operating statements, staffing requirements, inventory costs, and other scheduled operating costs are the basis of operating statements as well as profit and loss calculations. Modern financial analysis computes the net present value (NPV) of future cash flows using a predetermined hurdle or discount rate, often close to 20%. If the estimated NPV is positive over the expected life of the plant, the project is feasible and should be funded. Cash flow is the sum of profits after taxes plus depreciation. Capital investments are treated as negative cash flows when they occur. Clark discusses cost estimating and economic analysis in greater detail (1).

A schedule for the project is prepared that shows the major activities and their expected duration. It is important to recognize the dependence of certain activities on others—construction cannot begin until the relevant design is complete; steel cannot be erected until foundations are laid; it takes time for structural steel to be fabricated; earthwork is difficult when the ground is frozen; obtaining permits, as mentioned later, may require a long time; and equipment cannot be installed until the building is largely complete.

PHASE II—ENGINEERING AND ARCHITECTURAL DESIGN

The new facility should be designed around the process. It should accommodate the projected growth in production and future expansions, keeping in mind that new products will be developed and existing products discontinued. Layout should consider how changes could be made with minimal disruption to existing operations.

Structural steel, foundations, and other elements should be designed to support equipment and piping and to facilitate expansion. Floor drains are important in wet areas, and floors should be sloped to drain. Layout of equipment to permit access for maintenance and easy changes is critical.

Room, floor, and ceiling finishes are dictated by process conditions and should be chosen with an understanding of life-cycle costs. Ceilings should be avoided in process areas if possible because they collect dust and moisture and are hard to clean. Heating, ventilating,and air conditioning (HVAC) requirements are also set by the process and can be stringent in food plants, which may be dusty, humid, and require extremes of heating and cooling. Meat and poultry plants, for instance, are routinely cooled to 50°F to reduce microbial growth, and some cereal processing plants are heated to 140°F on occasion to eliminate insect infestations. Employee comfort is achieved by adequate ventilation and appropriate cooling or heating.

Process Specifications and Engineering

Piping and Instrumentation Diagram. The piping and instrumentation diagram is based on the process flow diagram but adds much more detail. All equipment, piping, and instruments are shown with indications of size, operating rates, and expected conditions of temperature, flow rate, and pressure. Each device is identified with a number that is used to track it through procurement, receiving, installation, and operation. It is good practice to determine if the corporation has an existing asset accounting system and to use numbers from the beginning of the project that are compatible with the system. Thus, the new assets are easily incorporated into the corporation's records with little effort. The same identification system can be extended to a preventive maintenance system. If cleaning-in-place is involved in the process, and it often is, then there should be a separate cleaning-in-place piping and instrumentation diagram to avoid confusion. The cleaning-in-place system in some aspects nearly duplicates the process and in others simply connects with the process equipment. Trying to show both on one drawing is difficult.

Process Equipment Arrangement. The major process equipment is located on scale drawings. These often rely on vendor shop drawings that typically are prepared late in a project, yet the arrangement drawings are needed early to establish building size and layout. One solution to this dilemma is to make generous space allowances, because equipment always seems to grow; it rarely shrinks. Also, supports, platforms, chutes, conveyors, motors, and control cabinets are often overlooked early in design and added later, so additional space is always needed.

Process Piping. Typical piping details, isometrics, pipe support methods, and piping routes are identified. As with the equipment, piping seems to grow during a project as needs are refined and detail is developed. Piping can encounter conflicts with structural steel, lighting, fire protection, electrical conduit, and HVAC ducting because typically each of these engineering disciplines develops independently within a project. It falls to project management and sometimes to the owner to identify and resolve early such interferences and conflicts. One solution is to identify chases or spaces reserved for certain purposes, such as electrical, process piping, or utility piping. Even with this precaution, there will be conflicts that must be resolved. It is much less expensive to do so in the design stage than in the field.

Process Equipment Specifications. Each piece of equipment is defined by a specification that describes its performance, special features, regulatory issues if any, and desired spare parts and support, such as assistance during start-up. The specifications are used primarily for purchasing but also become an important part of the documentation of the project.

Process Installation Specifications. The primary purpose of this document is to define responsibility for procurement, receipt, and installation of the equipment. A typical issue is defining responsibility for supplying electrical power to a unit. Often, the electrical power is installed up to a distribution panel by one contractor but then must be delivered to each piece of equipment, which can only be done after the equipment is set in place. Various parties may become involved, but it is best to establish uniform standards, routes, details, and practices.

Utility Specifications and Engineering

Mechanical Flow Diagrams. The mechanical flow diagrams will indicate all mechanical equipment type, size, and operating rates. Instrumentation for mechanical control and alarm are identified. Each device is assigned a unique item number. The relationship of the mechanical equipment with the process must be shown and defined. Typical mechanical systems included in the mechanical flow diagram include steam, hot water, cold water, compressed air, fuel gas, inert gas, vacuum, chilled water, ammonia, glycol coolant, and conveyor chain lubricant. There can be several pressures of steam and several levels of ammonia refrigerant.

Mechanical Equipment Arrangements. The mechanical equipment arrangement drawings locate all mechanical equipment.

Mechanical Equipment Specifications. Each piece of mechanical equipment is defined by an equipment specification, which includes its performance, rates, features, and regulatory requirements.

Mechanical Piping. Typical mechanical piping details, pipe support methods, insulation, and mechanical piping routes are identified. The distinction between process piping and mechanical piping is that process piping often is stainless steel, sanitary tubing, whereas the mechanical piping often is carbon steel, fiberglass, copper, or other materials. Routing, hangers, and other details should be compatible for both; that is, the same sanitary design practices should be followed in food processing areas. Some utility piping may exist outside the food processing areas, and there less expensive practices may be followed. However, an area that initially is not a processing area, such as a storage area, may later be converted into a processing area, so it is wise to anticipate such possibilities when specifying utility hangers, insulation, and routing.

Mechanical Installation. The installation responsibility for each piece of mechanical equipment and piping system is defined, including responsibility for procurement, receipt, and installation.

Electrical Single-Line Diagram. Electrical distribution for all process, mechanical and facility equipment drives, and lighting is indicated on the single-line diagram.

Electrical Equipment Arrangement. The electrical equipment, switchgear, distribution, motor control centers, low-voltage panels, and lighting panels are located. Normally, these do not occupy prime space, but they should be accessible, and space for future expansion should be provided. The relationship of the switchgear to incoming power and transformers is defined. Reliability of local power must be assessed to determine if two independent sources are needed, and provisions for future requirements should be made.

Electrical Equipment Details. Typical electrical details and conduit arrangements are identified. In food plants, electrical conduit should be mounted away from walls so that the space between can be cleaned. Perforated channel (Unistrut) and all-thread rods should not be used as hangers in food processing areas and probably should not be used at all because they are hard to clean.

Electrical Equipment Specifications. Each major electrical device—transformer, switchgear, motor control center, mechanical control panel, process control panels and lighting panels—are defined for purchase and installation. It is common for corporations to establish preferred suppliers for such equipment in an attempt to standardize and simplify maintenance.

Electrical Equipment Installation. The installation responsibility for each piece of electrical equipment is defined, including responsibility for procurement, receipt, and installation.

Site and Facility Specifications and Engineering

Site Layout. A survey, topographical study, and appropriate soil borings are obtained for the specific site. These help guide placement of the building on the site and influence foundation design. The site plan or layout shows building location, access routes for trucks, visitors and employees, and provisions for expansion (typically dotted in). Location of underground utilities is also shown. Identification of bordering neighbors is common.

Architectural Design. Each floor or level of the building is shown in plan view, and elevations (side views) are prepared. Because it is common to put support equipment on the roof, a roof plan is also needed. Overall people, material, and traffic flow patterns are shown. A perspective rendering, which is a drawing showing what the building will look like when finished, is sometimes prepared for presentations, especially to management and local governments.

Architectural Details. Architectural details include materials of construction, wall finishes, floor finishes, doors, security systems, lighting (exterior and interior), and various amenities such as cafeteria, if any, lounges, offices, rest rooms, and health centers. Modern plants commonly contract for food service, but may only provide vending machines and a place to eat. Rest rooms and lockers must be convenient to work areas but separated, usually by corridors. Exterior lighting should be directed at the plant, not mounted on the walls, and should not use mercury vapor lamps because they attract insects. Mercury vapor lamps can be used on the perimeter of the property to attract insects away from the plant.

Floors, walls, and ceilings (if any) in food processing areas should be impervious, easy to clean, and resistant to moisture and strong cleaning solutions often used in food plants. There are many possibilities, but epoxy-based coatings properly selected for the service are economic choices. Epoxy-based floor coatings can be harmed by thermal shock, such as exposure to live steam, unless they are filled with the correct amount of inert material. Such fillers also provide some texture to the coating, which may improve traction on the floor. Fiberglass-reinforced epoxy panels are often used for walls and ceilings. Insulated sandwich panels are often used for refrigerated spaces and can be a sanitary and economic choice for walls and ceilings. Each architectural detail in a food plant must be selected with an understanding of the environment and hazards to which it is exposed. The environment is often wet or dusty; there are strong chemicals (acids and caustics) used for cleaning; there can be oils and abrasive materials; and the area must be protected against insect, rodent, and bird infestation. Common commercial and industrial materials and details are not adequate for a food plant.

Building and Zoning Permits. The site layout, architectural details, and layouts are used to obtain construction and building permits. These are issued by local government agencies after review of the plans for compliance with site zoning and building requirements. Normally, the procedure for obtaining permits begins early in the design phase because it can influence design. Setback requirements, for instance, affect building location on a site. Final permits require final design plans, usually signed and sealed by a registered architect and engineers licensed to practice in the state. Exceptions to some requirements can be obtained, but it is usually prudent to comply with most requirements. Obtaining permits can be time-consuming, especially if public hearings are required for zoning changes, but areas seeking economic development and investment will usually expedite the process if asked.

Cost Estimate and Financial Analysis

The cost estimate and financial analysis prepared in the feasibility phase is revised in light of the completed design and any subsequent information, such as revised sales forecasts, costs, or other developments. The cost estimate prepared at this point becomes the control budget for construction and should always be preserved as the basis for any future changes. A detailed schedule is also prepared, adding detail to those prepared earlier.

PHASE III—PROJECT AND CONSTRUCTION MANAGEMENT

The critical issues during this phase are control of the budget, schedule, and quality of construction. The budget is based on the estimate made at the end of design. As bids are received, the budget is adjusted to reflect reality. If good estimating practice is followed by experienced engineers and architects working from a reasonably complete design, the total cost should not change during bidding and construction *if the design does not change*. The biggest risk to the budgeted cost is design changes during construction. These arise for a number of reasons: new information, new people with new opinions, unforeseen circumstances and events, and changes in strategy. Good project management strives to control design changes, for whatever reason. The contingency allowance always included in any cost estimate is not intended to cover design changes; it exists to account for errors in the estimate corresponding to the original scope. Knowing that there is almost always some uncertainty about scope, some owners provide a separate allowance, called the owner's contingency, to cover such incidents. The schedule is a living document that is revised as information is received, as estimated duration of activities are converted into actual duration, and as activities are added. It is common for management to insist on a fixed completion date because of market obligations and other constraints, even in the face of delays in approval, design, and construction. Reconciling the inevitable conflicts is a significant challenge. Techniques include overlapping construction activities, starting construction before design is complete (fast-track approach), paying premium wages to work extra shifts and weekends, and, as a last resort, compromising design intent, by using more easily available materials, for example.

Maintaining quality during construction requires constant vigilance by the owner and the construction manager. There is always a temptation to cut corners on materials or procedures to save money, but once details have been specified, they must be followed unless prior approval is received for a substitute. Tests are common for concrete, coatings, electrical wire, pipe, and other materials. Experience and care are required in sanitary welding and application of coatings. Before allowing a contractor to proceed, the manager should make sure that the tradesmen are skilled, and the manager should routinely inspect the work as it proceeds. Construction can rarely improve on a design, but it can easily compromise the intent if permitted to do so.

Bidding

There are several approaches to organizing and managing construction. The entire project can be managed by a general contractor, who may also have performed the design (design–build approach), or the design may have been performed by a different firm, typically an architecture and engineering company. The owner can engage a construction manager, who may or may not have performed the design, but who advises the owner on the selection of subcontractors but does not perform any of the work himself. In the construction management approach, the owner directly engages each subcontractor, and the Construction Manager maintains the budget and schedule and monitors construction quality. Finally, the owner can serve as his or her own construction manager. Because of the relatively unique skills needed for construction, as compared with those of the typical food professional, it is a rare corporation that can perform this task on a large project successfully.

Depending on the project approach taken, there may be just one bid (or negotiated price) or many. Typically, it is easier to get competitive bids on smaller and more specialized packages than on a smaller number of larger packages. On the other hand, the more parties involved the more complex the management task. A common solution is to have bid packages for the following major components: site, building shell, mechanical, electrical, process equipment installation (millwright), and process and packaging equipment (multiple packages). The mechanical and electrical contractors often install the utility and service equipment and connect the process equipment. Some mechanical contractors install process equipment, but because much of it is highly specialized, often the vendors assume this responsibility.

Some types of process and packaging equipment have long lead times and so must be ordered early in the project to ensure delivery at the proper time. Ideally, equipment is delivered when its correct space is available so that handling is minimized.

Lead times for structural steel, process and packaging equipment, and large mechanical and electrical equipment have significant impact on the schedule. Once ordered, their progress through manufacturing must be monitored to ensure adherence to the plan, which may require a full-time person or more.

Site Management

Typically, the owner, the contractor or construction manager, and the major subcontractors have staff members on site, usually in trailers. A good practice followed by many owners is to hire or assign the future plant manager and key staff members early in a project and make them responsible during all phases of the construction project. Alternatively, project management specialists may be assigned. In any event, the following activities occur on site.

Scheduling of Site Activities. Coordinating the various trades and subcontractors is a critical, ongoing effort. It is common to have a daily meeting of all parties to report attendance, revise schedules, resolve interferences and conflicts, and predict needed material and people.

Receiving, Storing, and Releasing Material and Equipment. As equipment and material are received, they are inspected, verified against specifications and orders, and stored securely, protected against weather and unauthorized access. As needed, equipment and material are released for use. Contractors are generally responsible for their own materials, but the owner or the construction manager may handle process equipment and piping.

Verification of Work. Contractors routinely file liens against owners to ensure payment for their work. As work is completed, the owner or construction manager secures waivers of liens and approves payment, upon inspection and satisfaction. It is normal to retain a portion of payment against correction of faults, usually documented in a punch list. The punch list is maintained as a living document, and responsibility for and disposition of each item are noted.

Accounting and Payments. Accurate, fair, and timely payment for completed and accepted equipment, material, and services is critical. This requires documentation in the form of purchase orders, contracts, receipts, invoices, inspection reports, and the punch list. Typically, contractors expect to be paid as they incur expenses subject to a retained portion, often 10%, which is paid at completion and satisfaction of any defects. To document progress, a contractor may submit time sheets supporting the labor costs and receipts for materials. If the contract is a lump sum, these are compared to an estimate of the total; if the contractor is being paid for time and materials, these are the costs plus an agreed profit and overhead, if any. As costs are actually incurred for the project, the control budget is adjusted, and projections are made for amounts to complete. The owner must be kept informed of the budget status almost continuously.

Project Documents. In addition to the cost and disbursement documents, schedules, and inspection reports, there are important permanent records that accumulate during construction. These include operations and maintenance manuals for all equipment, equipment and facility drawings as designed and as built, and all the permits and certificates issued. These must be collected, cataloged, and preserved.

Start-up and Commissioning. Once process, packaging, and mechanical equipment is installed and utilities are provided, each unit can usually be tested in isolation to be sure motors rotate correctly, instruments read correctly, and controls work as expected. This testing is usually the responsibility of the party installing the device. After all components of a system, such as one process line, are in place, then it can be tested as a whole. This may require coordination of speeds, alignment of conveyors, and integration of controls. Vendors of process and packaging equipment routinely provide assistance in this phase, during which operators are trained in safe and effective operation and maintenance of the system. Commissioning occurs with the initial operation of a system using realistic raw materials and attempting to produce acceptable final product. It is wise not to expect design efficiency at this point, and in fact, any salable product should be seen as a dividend. Each process unit operation must be assessed over its expected range of operating rates and conditions and compared with process design and equipment specifications. Deficiencies, if any, must be corrected or otherwise accommodated, perhaps by lowering expectations, if necessary. The limiting components of the system must be identified. These may well be different than was intended or expected in design. Ideally, at the end of commissioning, which may take a month of effort, the new process line is operating at expected design rates and efficiencies, the operators are trained and self-sufficient, the vendor technicians have departed, and the plant or line is ready to accept orders routinely.

Close Out. It is wise to have a formal project completion conference among the owner, the construction manager, the designer, and major vendors to transfer any remaining documents, resolve any outstanding invoices, and transfer accumulated knowledge and experience. Sometimes this can correspond with dedication of a new facility and involve some modest ceremony and souvenirs; on other occasions, it can be a simple business meeting, but it is important to somehow mark the official end of a project.

BIBLIOGRAPHY

This article was adapted from A. Cukurs, *Food Plant Design and Construction*, APV Crepaco, Lake Mills, Wisc., 1989.

1. J. P. Clark, "Cost and Profitability Estimating," in K. J. Valentas, E. Rotstein, and P. Singh, ed., *Handbook of Food Engineering Practice*, CRC Press, Boca Ranton, Fla., 1997, pp. 537–557.

GENERAL REFERENCES

T. J. Imholte, *Engineering for Food Safety and Sanitation*, 2nd ed., Technical Institute of Food Safety, Crystal, Minn., 1998.

J. PETER CLARK
Oak Park, Illinois

FOOD PRESERVATION

Matching the supply of food as produced by the agricultural sector with the demand for food by consumers in both time and space necessitates the use of a variety of preservation techniques. Food is being produced in larger quantities by fewer people in rural areas often very distant from the urban consumer. The production of agricultural commodities follows cyclical patterns of supply, dictated by such things as cropping times and yearly fluctuations. The requirements on the food supply by the consumer include safety, quality, adequate shelf life, variety, and convenience. Thus, the demand for food tends to be more constant than its production, food needs to be transported from the production sector to the consuming markets, food needs to be transformed from its raw state as produced by the agricultural sector into a vast array of consumer goods, and throughout this process quality must be maintained, safety must be guaranteed, and the economics must be favorable, minimizing losses and waste. This matching of supply and demand in both space and time both defines and introduces the field of food preservation.

Food in its many forms is subject to very rapid deterioration beginning soon after harvest. The factors contributing to this process include biological deterioration and postharvest loss from bacteria, yeasts, molds, insects, and rodents, and chemical breakdown of food components catalyzed by enzymes, light, or oxygen. Preservation techniques attempt to control these deterioration processes through the destruction of microorganisms present in the food, through the manipulation of factors essential for the continued growth of microorganisms, or through the control of factors responsible for chemical deteriorations. It is difficult to define the time frame necessary to consider a food preserved. Milk can be preserved through pasteurization and refrigeration to attain a shelf life of 10 to 14 days. Milk can also be preserved through spray drying to attain a shelf life of a year or longer. Thus a food might be considered preserved when it is effectively moved from production to consumption, maintaining safety and quality, minimizing losses, and being delivered in a form which is convenient and acceptable to the consumer.

Although new preservation processes have been developed and traditional processes have been modified, food preservation has been practiced for millennia. Since the beginning of man the need has existed to gather food. Processes such as sun-drying, salting, smoking, or food fermentations are considered to be early forms of food preservation. The technology of food preservation increased dramatically with the discovery of microorganisms late in the eighteenth century and the appreciation of their role in food deterioration that followed. The reader is referred to Ref. 1 for a complete history of food preservation.

CAUSES AND MANIFESTATIONS OF FOOD DETERIORATION

Microorganisms

Microorganisms are ubiquitous living organisms that need nutrients, moisture, appropriate oxygen conditions, and favorable pH ranges to grow. Microorganisms can contaminate our food supply from the point of production or harvest until the time of consumption. Many microbial food contaminants are native to the soil or animal environment from which the food is derived, and many are added unintentionally through handling practices. The components of most foodstuffs (carbohydrates, proteins, etc) serve as ideal nutrients for the growth of microorganisms and this growth leads to potentially harmful populations of microorganisms, potentially harmful buildup of microbial metabolites, for example, toxins, enzymes, polysaccharides, or pigments, and a deterioration in food quality resulting from the metabolism of food constituents, for example, amino acid or fatty acid release from proteins or lipids. Microorganisms important to food deterioration include bacteria, yeasts, and molds. Some bacteria can produce spores, a dormant form that is capable of withstanding long periods under conditions in which the vegetative form could not grow. These spores can then germinate when conditions once again become favorable. Spores are much more heat resistant than their vegetative form, and hence during processing, if elimination of all bacteria is a goal, then the heat resistance of the spore must be targeted.

It is important to distinguish between pathogenic and nonpathogenic species. Pathogens are organisms that cause disease, and a number of human pathogens can be transmitted through food. They include *Salmonella* spp., *Staphylococcus aureus, Clostridium botulinum, Listeria monocytogenes*, and many others. Mycotoxins are fungal metabolites, some of which are potentially toxic or carcinogenic to humans, for example, aflatoxin from *Aspergillus flavus*. It is essential that all pathogenic species be controlled in preserved foods to eliminate foodborne disease. This is done by providing measures to eliminate what may be present in raw materials (eg, those of animal origin) through processing, and by providing measures to eliminate the entry of contaminating pathogens to the food supply through hygienic manufacturing practices. Most microorganisms, however, are not pathogenic but can cause food spoilage if allowed to grow, primarily as a result of off-flavors and odors from metabolic by-products. These, too, need to be controlled as they account for diminished shelf life and economic loss. In this latter case, however, it is important to recognize that spoilage organisms are not eliminated during processing, but rather kept below a critical number to provide adequate shelf life to the product. An example would be pasteurized milk, in which pasteurization eliminates all pathogens, but a population of spoilage organisms remains and limits the shelf life to 14 to 16 days in most cases. For further reading on food microbiology, the reader is referred to Refs. 2 to 5.

Enzymes

Enzymes are naturally present in most raw food commodities and a number of enzymes catalyze reactions that are detrimental to product quality. These include off-flavor production by lipoxygenases, lipases, and proteases; textural changes due to pectic enzymes and cellulases; color changes due to polyphenol oxidase, chlorophyllase, and peroxidase; and nutritional changes due to ascorbic acid oxidase or thiaminase. A mild heating process such as blanching or pasteurization is usually sufficient to destroy any harmful enzymes that may be present in a food.

Pests

Insect and rodent infestations and losses can occur readily in open fields during growth and production and in bulk stores of food exposed or partially exposed to the environment, and this is especially prevalent in developing countries where storage conditions are less than adequate. Insects can attack grains, reducing their nutritive value and imparting a sour taste to flour, peas, beans, meat, fish, cheese, and so on. Insects damage far more food than they consume due to the deposition of larvae in webbings, rendering much of the food store as waste. Rodents are also a serious concern in food storage, also damaging far more than they consume through adulteration with droppings, filth, and hair and potential transmission of disease.

Temperature

Uncontrolled heat and cold can cause detrimental reactions to occur in foods. Increased but moderate temperatures, 30 to 40°C, accelerate the rates of reactions such as oxidation, which lead to the production of off-flavors and odors, color changes, and nutrient loss and significantly decrease the generation times of microorganisms, leading to enhanced populations. Decreased temperatures, in the 0 to 15°C range, can lead to detrimental physiological reactions in certain susceptible products. These include chill injury in tomatoes or bananas, which results in color and texture changes and increase in microbial susceptibility; cold shortening in meats resulting in tough carcasses after slaughter; chill sweetening in potatoes resulting in increased glucose contents and enhanced browning during frying; and staling of bread, which occurs much more readily at these temperatures.

Moisture

The water present in a food plays a critical role in supporting microbial growth and chemical reactions. The moisture is an integral part of the structure of a food, both at the molecular level (associated with macromolecules as bound or plasticizing water) and at the structural level (eg, intracellular or extracellular). Water activity is defined as the equilibrium relative humidity of the food, or a measure of free water in the food, that water which is able to participate in reactions. It has been shown that detrimental reactions in foods are often related more to the water activity of the food than its moisture content, and water activity has been used widely over the last several decades as a measure of food stability. It is increasingly being recognized, however, that the physical state of the solids in the food and their interaction with water is also a determining factor in reaction rates. Both rates of microbial growth and chemical reactions are also influenced by water and solute diffusion. In recent years, the glass transition of both small molecules (eg, sugars) and macromolecules (eg, starches) has been recognized as a key event in the occurrence of solute diffusion-controlled deteriorative reactions, and hence another measure of food stability related specifically to water relations in the food. A good review of water activity and glass transitions related to food preservation can be found in Ref. 6. Thus the gain or loss of moisture from a food due to its conditions of storage, particularly relative humidity, may influence the water activity and solute diffusion rates and thus the microbial and chemical stability of the product. Additionally, changes in the relative humidity of storage can also affect structure and texture of the food. High relative humidities of storage can favor splitting or cracking of the skins of some fruits, while low relative humidities can favor wilting and shriveling of fruits and vegetables, leading to a loss of weight and thus economic value.

Oxygen

The oxygen requirements for both the storage of foods and for microorganisms vary considerably. Bacteria have very strict oxygen requirements and are classified accordingly as aerobic, anaerobic, or facultative organisms with wider tolerances. Oxygen is also an initiator of chemical reactions that are detrimental to various food constituents, including lipids, vitamins, pigments, and some amino acids. However, reduced oxygen contents can impair the physiological function of food tissues, such as fruits, vegetables, or meat. Hence control of oxygen contents through modified atmospheres and protective packaging can help to reduce food deterioration.

Light

Exposure to light can be detrimental to some food products. Light can promote oxidative rancidity of lipids; oxidation of milk leading to breakdown of proteins and formation of unpleasant volatiles; changes in various pigments (eg, myoglobin in red meats); breakdown of vitamins such as vitamin A, riboflavin, and ascorbic acid leading to nutrient loss; or the development of potentially toxic light-induced glycoalkyloids such as solanine in potatoes that have turned green due to increased chlorophyll content (from light exposure).

Time

Time itself plays a very great role in food deterioration. While some preservation methods such as canning are intended to produce food for years of storage, other preservation methods such as pasteurization and refrigeration produce shelf lives of days to weeks. Such processes aim to slow but not eliminate deteriorations, and recontamination becomes possible unless protective packaging is employed or after package integrity is broken. Food deteriorations described earlier will substantially reduce the shelf life of a product, but virtually all foods will become unconsumable over time.

PRESERVATION PRINCIPLES AND APPLICATIONS

Preservation techniques put into practice control mechanisms for reducing food deterioration. Often, various methods are combined to ensure safety and preservative action while maintaining maximal quality and stability. For example, fluid milk is preserved through a combination of pasteurization and subsequent refrigeration, aseptic thermal processing is usually combined with aseptic packaging, and freezing is usually accompanied by a blanching pretreatment to reduce enzymatic activity. For more detailed general information on preservation principles and techniques, the reader is referred to Refs. 7 to 15.

Physical Preservation

Thermal Methods. A number of preservation processes use heat to extend the shelf life of foods. High-temperature preservation methods performed commercially are controlled processes that include canning, aseptic processing, pasteurization, and blanching. Microorganisms differ in their heat resistance and are classified as psychrotrophs, mesophiles, or thermophiles according to their tolerance to heat. However, all microorganisms can be destroyed by the application of heat, and each organism has a certain time/temperature relationship associated with it to ensure a given reduction in its population. The most intense heating preservation process would render the food sterile. Sterilization refers to the complete destruction of all microorganisms. Complete sterility is difficult to achieve and often leads to a reduction in the quality attributes of the food, since most food components such as proteins or vitamins are also heat sensitive. Commercial sterility has been defined as the destruction of all pathogens and spoilage organisms in a food. Canning, thermal retorting in aluminum cans or flexible pouches, and aseptic processing, thermal processing prior to packaging followed by aseptic packaging techniques, target commercial sterility as their goal. Provided the food is maintained in the commercially sterile state after processing, a shelf life of two years or more can be achieved.

Canned food is typically packaged and maintained under a vacuum to eliminate oxygen and to act as an indicator for the consumer of loss of can seal integrity and hence sterility. As a result, the growth of anaerobic microorganisms, those that cannot grow in the presence of oxygen, would be favored if present. *C. botulinum* is an anaerobic, pathogenic, spore-forming microorganism that can cause severe illness and death; hence, it is essential that all spores of *C. botulinum* have been destroyed during the canning operation. *C. botulinum* is also pH sensitive, with a minimum growth requirement of pH 4.6. Thus, foods to be sterilized are characterized as either low-acid or high-acid foods, with the latter requiring a less severe (eg, boiling water) process than the former (must be processed under pressure to achieve temperatures in excess of 121°C). As the food is thermally processed after being packaged, the heat transfer characteristics of the food itself also become of importance in determining the processing time required at a specified temperature. Thus the calculations required to ensure lethality of all *C. botulinum* spores become very complex. The reader is referred to Refs. 16 to 19 for further reading on the subject of canning. Aseptic processing also targets commercial sterility but relies on continuous methods of heat treatment of the food prior to being packaged. Such heat treatment methods either use heat indirect heat exchangers, in which the food is separated from the heating medium by a barrier, or direct methods such as steam injection or steam infusion, where the heating medium and food are in direct contact. Such processes need to account for any added water from steam condensation, and hence are usually followed by a vacuum treatment for water removal, which also provides rapid cooling to the processed product. Once processed, there must be no contamination before packaging, implying sterile processing environments, and aseptic packaging must follow, to provide a package that will maintain the product in its commercially sterile condition until consumption. Details of aseptic processing techniques can be found in Refs. 20 to 23.

Pasteurization is a low-order, time- and temperature-dependent heating process that is designed to destroy all pathogens present in the food; to reduce the bacterial load in the case of milk and eggs; to reduce the yeast and mold count in the case of beer, wine, and fruit juices; and to extend the shelf life. Pasteurization can be performed prior to packaging in either batch or continuous heat exchangers, as in milk, or subsequent to packaging in either continuous or batch processes, as in bottles or cans of beer. Blanching is another low-order heat process used primarily as a pretreatment step prior to either freezing or canning. In a freezing operation, blanching is performed primarily to destroy enzymes that may continue to be active during freezing, or after thawing, when tissue rupture due to the freezing process may have allowed for enzymes and their substrates to come together, promoting fast enzymic deterioration. Prior to canning, blanching is performed primarily to cleanse and wilt tissue for ease of packaging, and to expel tissue gas that may interfere with an appropriate can vacuum.

Low-temperature methods include both refrigeration and freezing. Storage below 15°C but above freezing retards growth of microorganisms, retards metabolic activities of animal tissues postslaughter and plant tissues postharvest, retards deteriorative chemical reactions such as oxidation and enzyme-catalyzed reactions, and retards moisture loss. Unlike heating preservation, cold preservation does not destroy microorganisms, only retards their growth. The foods are still perishable and organisms will grow more rapidly once conditions become favorable. The shelf life can be extended from less than one week for highly perishable products such as raw milk or ripe tomatoes to more than six months for more durable products such as onions or smoked meats. Freezing preservation is achieved through both low temperatures that inhibit microbial growth and rates of reactions and reduction in water content as a result of ice crystallization. A reduction to subzero temperatures will cause water in the food to crystallize to ice, but ideally there should be little or no other changes to food components as a result. Shelf life of frozen foods can range from three months to a year or longer, depending on storage temperature, but is still limited by enzymatic activity, oxidation, and freeze dehydration (sublimation, "freezer burn"), especially if packaging has not been adequate to provide an oxygen or moisture barrier. Freezing will reduce the population of microorganisms that may be present, but will not eliminate them, so thawed foods again become very perishable, often more so than their fresh form due to damage of cellular and tissue integrity caused by ice formation, which may also bring enzymes in closer contact with their substrates, a process referred to as decompartmentalization. Commercial food freezing facilities utilize several methods, including air freezing through sharp (natural convection), blast, or fluidized-bed (forced convection) techniques; indirect contact systems such as plate freezers or scraped surface freezers; or direct contact freezing systems utilizing low-temperature liquids or cryogens. Packaging may be done either prior to or after a freezing process, depending on the commodity and equipment utilized. Detailed information on food freezing can be found in Refs. 24 to 26.

Reduction of Water Content. Microorganisms need favorable moisture conditions within the food, water activity greater than 0.6 to 0.8, for their growth. It is therefore possible to manipulate the water activity of the food to inhibit microbial action. Concentration of liquid food products through thermal evaporation, freeze concentration, osmotic dehydration, and membrane processes achieves lowered water contents, and provided that water activity or solute diffusion rates have been lowered sufficiently, offers preservation action. Thermal concentration removes water in the form of vapor from the liquid. It is usually performed in multiple effect falling-film vacuum evaporators designed for thermal efficiency and product quality. Freeze concentration removes water from liquid foods in the form of ice and is particularly suited to foods such as fruit juices with depressed freezing points due to high sugar concentrations within the food. Osmotic dehydration has been used for fruit slices in sugar solutions whereby water will migrate from the fruit slice into the sugar solution due to the high osmotic pressure of the solution. Membrane processes such as reverse osmosis, ultrafiltration, or microfiltration remove water from foods in the form of liquid water, due to the presence of a semipermeable membrane and the imposition of a pressure gradient. Solvent and low molecular weight solute, depending on the membrane pore size, pass through the membrane in the permeate stream while the higher molecular weight solutes are concentrated in the retentate stream.

The nearly complete removal of water through dehydration by solar, cabinet, tunnel, drum, or spray-drying methods also offers a form of preservation by reduction of available water for microbial growth or chemical or enzymatic reactions. Dehydration occurs under controlled conditions that cause minimal changes in the food properties. The food can then be consumed dried, as in some dried fruits or meats, but is more likely to be rehydrated prior to consumption, as in dried milk and eggs, instant potato flakes, or instant coffee. The reconstituted product should resemble as closely as possible the quality of the original food. In addition to preservation, the drying of foods decreases the weight and bulk of the original food and adds a measure of convenience to the product. Sun drying has been practiced for centuries and is still used for the dehydration of grains, seaweeds, raisins, and other foods, particularly in developing countries. Most commercial fruit and vegetable operations employ continuous tunnel or belt dehydration systems that utilize heated air as the drying medium. The majority of liquid foods, such as skim milk, cake and soup mixes, flavors, purees, juices or instant coffee, are dried in spray dryers that atomize the usually preconcentrated liquid product into tiny droplets that dry rapidly in the surrounding heated environment, being recovered from the air stream as fine powder particles. This spray-dried powder is capable of further conditioning to render it more soluble for easy dispersion/dissolution, a process called instantizing. Such a process is frequently conducted on powders destined for the consumer market, but rarely on powders destined for commercial processing. Freeze drying removes water from a frozen food through sublimation under vacuum and is particularly suited to thermally sensitive products, such as instant coffee or convenience-type prepared entrees. See Refs. 27 to 29 for more information on food dehydration and Ref. 30 for detailed information on freeze drying.

Oxygen Control. Because of the strict oxygen requirements for bacterial growth and the participation of oxygen in a number of chemical reactions, oxygen control can act as a means of food preservation. Controlled and modified atmospheric storage of foods are techniques to maintain gaseous atmospheres with strictly controlled oxygen contents. The controlled or modified atmosphere can be maintained in warehouses for bulk foods (eg, apples), often prior to further processing, or can be maintained at the microatmospheric level within a food package. Food packaging also offers protective barriers to food against the action of contaminating microorganisms, pests, moisture, oxygen, and light. The packaging necessary to maintain preservation is usually chosen to accompany the particular process. Examples include multilaminate flexible packaging for aseptically processed foods, or rigid aluminum cans for retorting. Information on modified atmospheres can be found in Refs. 31 to 36 and on food packaging in Refs. 37 to 41.

Radiation. The use of nuclear energy in the form of gamma radiations, short wavelengths emitted by unstable isotopes of cobalt 60, or cesium 137, to inactivate microorganisms has been a developing technology since 1945. The main goal of irradiation is to extend the shelf life of foods where heat or chemical means are unfeasible due to the nature or geographic location of the food. Major potential applications of this process include spices due to the heat-sensitive volatile flavor components; insect disinfestation of grains and fruit; extended shelf life of fruits, vegetables, fish, shellfish, and meat products; sterilized diets for military, space, and medical uses; and animal feeds and moist pet foods. The irradiation occurs in an enclosed chamber in which the product can be exposed to an even distribution of the penetrating gamma rays for the necessary time to accomplish microorganism inactivation. The safety of irradiated foods has been extensively studied and proved. Further details on food irradiation can be found in Refs. 42 to 45.

Chemical Preservation

Intermediate Moisture Foods. An intermediate moisture food (IMF) is one that can be eaten as is, without rehydration, and yet is shelf stable without refrigeration or thermal processing. Whereas most foods have water activities in the range of 0.9 to 1.0, IMF rely on water activities in the range of 0.65 to 0.85, below that required for the growth of the most tolerant organisms, for their preservation effects. The aqueous portion of such foods is also very high in viscosity, due to the high concentration of dissolved solutes present in a low amount of water, and hence water and solute diffusion is also a limiting factor in moisture availability to sustain growth and promote chemical reactions. Included in this category of foods are jams and jellies, fruit cakes, pepperoni, sweetened condensed milk, marshmallows, soft cookies, and many others. Sugar and other humectants, water-absorbing compounds such as sorbitol, glycerol, starches, or gelatin, can be used to formulate these foods. Although the technology can produce a range of products with acceptable texture, many of which have been in existence since historical times, the flavor profile created by the various humectants has been a limiting factor in new IMF product development. Another factor to consider in the preservation of IMF is the heterogeneity of the product. There are many multicomponent foods consisting of two or more distinct parts (eg, bakery product and fruit or icing, pasta and sauce). Not only does the stability of each component need to be considered separately, but the stability of the components after moisture equilibration (based on water activity or solute diffusion properties) also needs to be considered. See Refs. 6 and 46 for further information on IMF.

pH Control. Acids can be used to lower the pH of foods to below the tolerable range for microorganisms. Acid can also enhance the lethality of heating processes. *C. botulinum*, the organism of concern in commercial canning processes, will not grow at less than pH 4.5 and thus it is not necessary to thermally process high-acid foods (pH < 4.5) under the same rigid time/temperature standards as is the case with low-acid foods (pH > 4.5). The addition of acid to such foods as soft drinks and the production of acid in some food fermentations are effective controls of microbial growth. However, pH control is normally associated with some other means of preservation as well since the palatability of many foods and chemical stability of their constituents, for example, proteins, also decrease at low pH levels.

Chemical Addition. Salt can be added to foods for its contribution to the preservation of the food, for example, butter, fish, or cured meat products such as bacon. The action of salt results from the osmotic pressure created in the aqueous environment surrounding the microbial cell, in an analogous manner to the addition of sugar in intermediate moisture foods or the use of sugar syrups for osmotic dehydration of fruits. Plasmolysis, the partial dehydration of the cell, results, and the viability of the microorganism is thus destroyed. Smoke is also a type of chemical preservative and has been used since historical times to preserve foods, especially meat and fish products. The action of smoke results from the formation of small amounts of preservative chemicals, and the internal temperatures and dehydration of tissues associated with the hot-smoking process. Cold smoking at temperatures less than 30°C relies solely on the formation of bactericidal chemicals and is usually associated with other means of preservation such as salting, refrigeration, or packaging. Chemical preservatives, such as benzoic acid or sodium benzoate, sorbic acid or potassium sorbate, sodium nitrite or nitrate, and sulfur dioxide are permitted at low levels in some foods as preservative agents against microbial growth. The use of chemical preservatives and other food additives are closely regulated by governmental agencies. Chemical preservatives have been reviewed in Refs. 47 to 49.

Biological Preservation

Fermentations. Unlike the processes described previously, food fermentations have as their goal an increase in the numbers of microorganisms present in a food. The traditional foods of many countries rely on fermentation processes, and fermentation is a historical but important means of food preservation throughout the world. Fermented foods are preserved through the action of a particular organism, unique to each given commodity, on a particular substrate within the food product, primarily carbohydrates but also proteins and lipids. The conditions of fermentation favor the growth of the desirable organism, which is often added in the form of a pure culture, and cause the competitive disappearance of undesirable spoilage or pathogenic organisms. The metabolic by-products of the fermentation change conditions, such as pH or oxygen content within the food, that also act to inhibit the undesirable organisms and include lactic and other acids, ethanol and other alcohols, gases such as CO_2, and a variety of other compounds at low levels that are responsible for the unique flavor characteristics of the particular product. Examples of food fermentations include the production of alcohols by yeasts in wine, cider, and beer and the production of lactic acid by bacteria in fermented milks, sour cream, yogurt, fermented meats, pickles, sauerkraut, and vinegar. Further information of food fermentations can be found in Refs. 50 to 52.

Others

There is much interest from food technologists in developing new and novel food preservation methods, to improve all aspects of food preservation from energy efficiency to food quality and convenience. Examples of these methods include the so-called hurdle technology, which is a sequence of processing steps, each designed to address a particular deterioration source, but none used to the extent that alone it would provide sufficient preservation. There is also recent interest in the use of natural antimicrobial agents, such as nisin, produced by one group of microorganisms (lactic acid bacteria) that controls growth of another group. The use of nonthermal processes such as microfiltration, which would filter out bacteria from a liquid food without relying on heat to destroy the vegetative cell, are being utilized; such a process is being used for "cold pasteurization" of beer or milk. Nontraditional thermal methods such as microwave or ohmic heating processes are also being investigated and exploited. See Refs. 53 and 54 for further details of some of these novel food preservation methods.

BIBLIOGRAPHY

1. S. Thorne, *The History of Food Preservation*, Barnes and Noble Books, Totowa, N.J., 1986.
2. W. C. Frazier and D. C. Westhoff, *Food Microbiology*, 4th ed., McGraw-Hill, New York, 1988.
3. J. M. Jay, *Modern Food Microbiology*, 5th ed., Chapman and Hall, New York, 1996.
4. G. J. Mountney and W. A. Gould, *Practical Food Microbiology and Technology*, 3rd ed., Van Nostrand Reinhold, New York, 1988.
5. B. Ray, *Fundamental Food Microbiology*, CRC Press, Boca Raton, Fla., 1996.
6. H. Levine and L. Slade, "Beyond Water Activity: Recent Advances Based on an Alternative Approach to the Assessment of Food Quality and Safety," *Crit. Rev. Food Sci. Nutr.* **30**, 115–360 (1991).
7. P. Fellows, *Food Processing Technology: Principles and Practice*, VCH Publishers, New York, 1988.
8. G. W. Gould, ed., *Mechanisms of Action of Food Preservation Procedures*, Elsevier Applied Science, New York, 1989.
9. P. Jelen, *Introduction to Food Processing*, Reston Publishing, Reston, Va., 1985.
10. M. Karel, O. R. Fennema, and D. B. Lund, *Physical Principles of Food Preservation*, Marcel Dekker, New York, 1975.
11. E. Karmas and R. S. Harris, eds., *Nutritional Evaluation of Food Processing*, 3rd ed., Van Nostrand Reinhold, New York, 1988.
12. V. Kyzlink, *Principles of Food Preservation*, Elsevier Science, New York, 1990.
13. N. N. Potter and J. H. Hotchkiss, *Food Science*, 5th ed., Chapman and Hall, New York, 1995.
14. I. A. Taub and R. P. Singh, *Food Storage Stability*, CRC Press, Boca Raton, Fla., 1997.
15. S. Thorne, *Developments in Food Preservation*, Elsevier Applied Science, New York, 1982.
16. A. C. Hersom and E. D. Hulland, *Canned Foods: Thermal Processing and Microbiology*, 7th ed., Chemical Publishing, New York, 1980.
17. S. D. Holdsworth, *Thermal Processing of Packaged Foods*, Blackie Academic and Professional, New York, 1997.
18. J. Larousse and B. E. Brown, eds., *Food Canning Technology*, Wiley-VCH Publishers, New York, 1997.
19. A. Lopez, *A Complete Course in Canning*, 11th ed., Canning Trade, Baltimore, Md., 1981.
20. J. R. David, R. H. Graves and V. R. Carlson, *Aseptic Processing and Packaging of Food: A Food Industry Perspective*, CRC Press, Boca Raton, Fla., 1996.
21. A. C. Hersom, "Aseptic Processing and Packaging of Food," *Food Rev. Int.* **1**, 215–270 (1985).
22. S. D. Holdsdworth, *Aseptic Processing and Packaging of Food Products*, Elsevier Applied Science, New York, 1992.
23. E. M. A. Willhoft, ed., *Aseptic Processing and Packaging of Particulate Foods*, Blackie Academic and Professional, New York, 1993.
24. M. C. Erickson and Y.-C. Hung, eds., *Quality in Frozen Food*, Chapman and Hall, New York, 1997.
25. L. E. Jeremiah, ed., *Freezing Effects on Food Quality*, Marcel Dekker, New York, 1996.
26. C. P. Mallett, ed., *Frozen Food Technology*, Blackie Academic and Professional, New York, 1993.
27. G. V. Barbosa-Canovas and H. Vega-Mercado, *Dehydration of Foods*, Chapman and Hall, New York, 1996.
28. D. MacCarthy, ed., *Concentration and Drying of Foods*, Elsevier Applied Science, New York, 1986.
29. A. S. Mujumdar, *Handbook of Industrial Drying*, Marcel Dekker, New York, 1987.
30. S. H. Goldblith, L. R. Rey, and W. W. Rothmayr, eds., *Freeze Drying and Advanced Food Technology*, Academic Press, New York, 1975.
31. B. A. Blaikstone, ed., *Principles and Applications of Modified Atmosphere Packaging of Foods*, 2nd ed., Blackie Academic and Professional, New York, 1998.
32. A. L. Brody, ed., *Controlled / Modified Atmosphere / Vacuum Packaging of Foods*, Food and Nutrition Press, Trumbull, Conn., 1989.
33. M. Calderon and R. Barkai-Golan, eds., *Food Preservation by Modified Atmospheres*, CRC Press, Boca Raton, Fla., 1990.
34. J. M. Farber and K. L. Dodds, eds., *Principles of Modified-Atmosphere and Sous Vide Product Packaging*, Technomic, Lancaster, Pa., 1995.
35. B. Ooraikul and M. E. Stiles, eds., *Modified Atmosphere Packaging of Food*, E. Horwood, New York, 1991.
36. R. T. Parry, ed., *Principles and Applications of Modified Atmosphere Packaging of Foods*, Blackie Academic and Professional, New York, 1993.
37. J. F. Hanlon, *Handbook of Package Engineering*, 2nd ed., McGraw-Hill, New York, 1984.
38. T. Kadoya, ed., *Food Packaging*, Academic Press, San Diego, Calif., 1990.
39. M. Mathlouthi, ed., *Food Packaging and Preservation: Theory and Practice*, Elsevier Applied Science, New York, 1986.
40. F. A. Paine and H. Y. Paine, *Handbook of Food Packaging*, L. Hill Co., Glasgow, Scotland, 1983.
41. G. L. Robertson, *Food Packaging: Principles and Practice*, Marcel Dekker, New York, 1993.
42. Council for Agricultural Science and Technology, *Ionizing Energy in Food Processing and Pest Control. I. Wholesomeness of Food Treated with Ionizing Energy*, Task Force Report No. 109, 1986, *II. Applications*, Task Force Report No. 115, Council for Agricultural Science and Technology (CAST), Ames, Iowa, 1989.

43. P. S. Elias and A. J. Cohen, eds., *Recent Advances in Food Irradiation*, Amsterdam, Elsevier Biomedical Press, 1983.
44. J. Farkas, *Irradiation of Dry Food Ingredients*, CRC Press, Boca Raton, Fla., 1988.
45. E. S. Josephson and M. S. Peterson, eds., *Preservation of Food by Ionizing Radiation*, CRC Press, Boca Raton, Fla., 1982.
46. T. M. Hardman, ed., *Water and Food Quality*, Elsevier Applied Science, New York, 1989.
47. E. Lueck, *Antimicrobial Food Additives: Characteristics, Uses, Effects*, 2nd ed., Springer-Verlag, New York, 1997.
48. B. Ray and M. Daeschel, eds., *Food Biopreservatives of Microbial Origin*, CRC Press, Boca Raton, Fla., 1992.
49. R. H. Tilbury, ed., *Developments in Food Preservatives*, Applied Science, London, United Kingdom, 1980.
50. A. H. Rose, ed., *Fermented Foods*, Academic Press, Toronto, Ontario, 1982.
51. K. H. Steinkraus, ed., *Handbook of Indigenous Fermented Foods*, Marcel Dekker, New York, 1983.
52. B. J. B. Wood, ed., *Microbiology of Fermented Foods*, Elsevier Applied Science, New York, 1985.
53. G. V. Barbosa-Canovas et al., *Nonthermal Preservation of Foods*, Marcel Dekker, New York, 1998.
54. G. W. Gould, ed., *New Methods of Food Preservation*, Blackie Academic and Professional, London, United Kingdom, 1995.

H. Douglas Goff
University of Guelph
Guelph, Ontario
Canada

FOOD PROCESSING

Individuals, companies, and corporations in the business of food processing have three basic responsibilities: (*1*) to ensure food safety, (*2*) to enhance food stability, and (*3*) to alter the form of feedstocks and ingredients to fit consumer demands. Food safety is paramount. The hazards associated with processed foods can be microbial, physical, or chemical. Microbial hazards are from pathogenic microorganisms or viruses that either invade the consumer or produce toxins in food during growth and dormancy phases. Examples of physical hazards are broken glass, rocks, or metal pieces, which can inflict serious harm if consumed inadvertently. Chemical hazards can come from cleaning compounds or other equipment maintenance fluids or can be associated with overuse of regulated food ingredients (such as sulfites in wine that control oxidative discoloration and growth of undesirable microorganisms but that can lead to bronchospasms in some individuals). Many of the processes used in food production are specifically used to ensure food safety.

Simply ensuring food safety does not, however, guarantee stability for the product shelf life desired by consumers. For example, retorting canned low-acid food through 12 decimal reductions for *Clostridium botulinum* to prevent botulism will only result in approximately a two-decimal reduction in *Bacillus stearothermophilus*, a thermophilic organism that causes flat sour spoilage if canned foods are stored above 38°C. Therefore, canned low-acid foods generally receive a more severe heat treatment than is required for ensuring food safety. In another example, refrigeration can prevent the growth of most pathogenic microorganisms but will not prevent enzyme activity or the growth of spoilage psychrophiles (some pathogens, such as *Listeria monocytogenes* or *Yersinia entercolitica*, are now being found to be psychrophilic). Food spoilage can result from nonpathogenic microorganisms, enzyme activity, chemical deterioration (eg, oxidation of lipids), contamination from pests, and mechanical damage.

In addition to safety and stability, much of modern food processing is designed to provide food to the consumer in a form convenient for use and desirable for taste. For example, while potatoes can be stored safely for long periods, precooking, mashing, and drum drying renders them in a form that can be consumed simply by adding hot water. Addition of dried spices or cheese flavoring may further enhance the appeal of the product. Ice cream can safely be packaged in one-gallon tubs and stored for months, but an individual serving on a stick with a chocolate coating will be much more convenient and appealing.

Conventional food processing, regardless of type of food, can be divided into three classes: separation, assembly, and preservation. These can occur at harvest, at the food-processing plant, or even at the point of retail sales. Increasing emphasis is being placed on field processing as exemplified by the widespread use of mechanical harvesters fitted with cleaning, sorting, color measurement, and size-grading systems. The known food-processing operations are grouped in Tables 1, 2, and 3.

Foods can be categorized as living tissue, or raw foods, and nonliving tissue. Living tissue foods include fresh fruits, vegetables, meats, and grains. Processes for preserving living tissue are considered separately from those used for the preservation of nonliving tissue, for example, canned, frozen, and dried foods.

FOOD COMPOSITION

Chemical Composition

Virtually all foods are derived from living tissues, although individual nutrients and additives, including lipids, carbohydrates, amino acids, and vitamins, can be synthesized. A nutritionally adequate dietary regimen of pure nutrients would be neither economically nor aesthetically useful except for medical applications such as intravenous feeding. Foods consist of hundreds of compounds because they are derived from the life processes of living tissues. The components of food composition are the nutrients that are essential to sustain life processes. Information on specific compounds in foods can be obtained from the literature, as can tables listing the nutrient composition of commodities, refined food components (eg, sucrose and gelatin), and processed as well as standardized formulated foods (eg, bread and margarine). Tables of nutrient composition are available for most foods found worldwide and contain quantitative data on moisture and caloric value (kJ or kcal × 4.184) and on lipid, protein, carbohydrate, fiber, ash, mineral, and vitamin contents. Tables of food composition are useful only as a first approximation to the actual nutritive value and gross chemical composition of

Table 1. Separation Unit Operations of the Food Industry Arranged by Mode of Separation and Phases Being Separated

Phases to be separated	Mode of separation: Physical	Chemical	Mechanical
Gas–gas			
Gas–liquid	Condensation and blanching		Deaeration
Gas–solid	Blanching		Controlled atmosphere, deaeration, compression, and densifying
Liquid–liquid	Distillation and membrane filtration	Solvent extraction	Centrifugation
Liquid–solid	Foaming, coagulation, and drying	Ion exchange, solvent extraction, and flocculation	11 Unit operations[a]
Solid–solid	Dialysis, freeze concentration, and sublimation		34 Unit operations[b]

[a]Mechanical separations include churning, centrifugation, clarification, draining, expelling, filtration, flotation, pressing racking, rendering, and skimming.
[b]Mechanical separations include abrading, boning, cutting, coring, crushing, dividing, defeathering, eviscerating, flaking, finishing, filleting, grinding, harvesting, husking, hulling, inspection, milling, peeling, picking, pitting, pulverizing, slicing, sieving, sorting, shredding, sizing, sifting, scarifying, sampling, shucking, stemming, vining, and winnowing.

Table 2. Assembly Unit Operations of the Food Industry Arranged by Mode of Assembly and Phases Being Assembled

Phases to be assembled	Mode of assembly: Physical	Chemical	Mechanical
Gas–gas			
Liquid–gas			Aeration, foaming, carbonation, and whipping
Solid–gas	Baking, puffing, and agglomeration	Humidification, oxidizing, and proofing	Extrusion and aeration
Liquid–liquid		Neutralization and acidification	Emulsification, homogenization, dispensing, mixing, and pumping
Liquid–solid	Crystallization	Soaking, malting, rehydration, and acidification	Dispersing, dissolving, immersion, mixing, and pumping
Solid–solid	Braising and roasting	Aging	7 Unit operations[a]
Gas–liquid–solid		Sprouting	Weighing and blending

[a]Mechanical assembly operations for solid–solid mixtures, include coating, enrobing, filling, forming, molding, pelleting, and stuffing.

Table 3. Preservation Unit Operations of the Food Industry Arranged by Mode of Preservation and Spoilage Vectors

	Mode of preservation					
	Physical					
Spoilage vector	Heat	Cold	Dehydration	Nonthermal	Chemical	Mechanical
Microbes	Frying, boiling, pasteurizing retorting	Cooling and freezing	Dehydration, desiccation, evaporation, and lyophilization	Irradiation, ultrahigh hydrostatic pressure, and high-intensity electromagnetic and ultrasonic fields	Acidification, brining, fermentation, isomerization, pickling, and smoking	Cleaning, centrifugation, and washing
Enzymes	Blanching, boiling, and scalding	Hydrocooling		Irradiation and ultrahigh hydrostatic pressure	Acidification	
Chemical	Exhausting and deodorizing	Chilling			Hydrogenation esterification, and smoking	Cleaning, degassing, and washing
Economic pests	Pasteurization	Freezing	Dessication	Irradiation	Fumigating	Cleaning, compressing, and washing
Mechanical damage		Freezing				Aspiration, cleaning, centrifugation, sorting, and washing

specific foods. Foods derived from living tissue are of extremely variable composition. The concentration of any nutrient is affected by horticultural, genetic, harvesting, handling, storage, and distribution factors. Furthermore, chemical analytical procedures may not correlate with the biological activity of available nutrients in the food.

Functional and Conformational Data

Foods and food ingredients are selected not only on the basis of nutritional content but for their functional and conformational attributes, which include toughness and tenderness; fiber content; style of cut; color and surface appearance; odor and flavor; microbial content; defects and extraneous matter; adaptability to freezing or heat treatment; genetic and varietal factors; geographical or regional production area; date of production (vintage); portion of plant or animal used; method of preservation; key ingredients; emulsifying capacity; water-binding capacity; and foaming capacity.

FOOD LAWS AND REGULATIONS

In the United States, a number of food laws have been passed with the primary objective to prevent food adulteration, either economic or hazardous. Some of the more prominent laws are the 1938 Food, Drug, and Cosmetic Act (FDCA), the 1958 Food Additives Amendment to the FDCA (including the Delaney clause, which prohibited the use of any known human or animal carcinogen in foods), the 1966 Fair Packaging and Labeling Act, and the 1990 Nutrition Labeling and Education Act. Food regulations promulgated by the Food and Drug Administration (FDA) for the purpose of enforcing these laws are found in Title 21 of the Code of Federal Regulations, while regulations promulgated by the U.S. Department of Agriculture (USDA) (for meat, poultry, eggs, and their products) are in Title 9. In addition to these national regulations, food processors dealing in international commerce may have to follow international regulations, such as those defined by the Codex Alimentarius.

Currently, regulations exist pertinent to current good manufacturing practices (CGMPs), nutritive requirements, food standards of identity, labeling, use of Hazard Analysis and Critical Control Point (HACCP) plans, and others. The most recent and important regulations are the mandating of HACCP by the FDA in 1995 for domestic and imported fish and fishery products and in 1996 by the Food Safety and Inspection Service of the USDA (FSIS-USDA) covering domestic and imported meat and poultry products. HACCP is a systematic approach to identification, evaluation, and control of food safety hazards. The systematic approach is based on the following seven points:

1. Conduct a hazard analysis.
2. Determine the critical control points.
3. Establish critical limits.
4. Establish monitoring procedures.
5. Establish corrective actions.
6. Establish verification procedures.
7. Establish record keeping and documentation procedures.

The regulations require as a prerequisite to the HACCP systems of an individual food processor the establishment of sanitation standard operating procedures (SSOPs), which are written procedures determining how a food processor will meet sanitary conditions and practices in a food plant. SSOPs are based on the CGMP regulations.

Grades and Grading

The USDA also provides grading services for fruits, vegetables, meats, poultry, dairy products, and grains. Services may be obtained on a contractual basis for inclusive grading or for grading of selected lots of product. Grades are not mandatory as are identity standards. Grades provide a rational basis for the evaluation of the worth of processed foods for trading, financial, or contractual purposes where the buyer or seller cannot sample or inspect the products themselves. All grades and grading procedures, fees, labeling and identification procedures, and sampling plans (including those for canned, frozen, and dried fruits and vegetables) are specified. Because changes in regulations take place frequently, the regional or the Washington, D.C., office of the FDA or the USDA should be consulted for the latest regulations.

PRESERVATION OF LIVING TISSUE

The genetically controlled taste, color, nutritive value, cell structure and conformation of fresh foods are usually best at harvest. Protection of these qualities is achieved by retarding or inhibiting the detrimental action of microbes, enzymes, and other chemical degradation while maintaining the integrity of the membranes, enzyme systems, and gross structure of the food. Specifically, the following must be accomplished:

1. The growth rate of microbes normally associated with the product at harvest or during handling and transport must be inhibited or reduced.
2. Production of deteriorative enzymes or enzyme systems responsible for softening, color loss, or flavor changes must be inhibited or their rate of action must be reduced. Enzymes may be associated with the product or with incidental microbial contamination.
3. Loss of moisture and deterioration by chemical contamination must be prevented. Volatile organic compounds generated through respiratory action or present in storage often must be removed.
4. Higher life forms, insects, mites, and so on, must be inactivated or eliminated.
5. Bruising and various forms of mechanical damage must be prevented.

Processing operations required for the distribution of living tissue depend on the following factors: inherent storage potential of the tissue (this can be improved through genetic engineering, eg, by improving skin strength, resistance to microbes, or reducing respiration rate), desired storage life, degree of handling and shipping required, and intended use of the tissue (eg, further processing, food service, or retail consumption).

The intensity of processing operations is limited by the tissue itself in terms of temperature range, water activity, respiratory gas composition, mechanical stresses, and concentrations of chemicals. Temperature has the greatest effect on shelf life, because activation energies (and hence reaction rates) for microbial growth and enzyme activity are two to five times greater than those of most deteriorative chemical reactions. Except for certain fruits and vegetables, storage should be to 0°C without ice formation (Table 4). When shipping distances are great, expensive and highly perishable commodities are routinely shipped by air freight. Many fruits and vegetables are stored in atmospheres with controlled or modified O_2, CO_2, or C_2H_4 compositions.

A clear distinction is now made between processing whole, intact living tissue and minimally processed products that have arisen largely due to consumer demand in developed countries for freshlike, high-quality convenience foods. Minimally processed products include living tissues that have been cut, peeled, or shredded and generally require refrigeration in combination with other treatments, such as pH control, antioxidant addition, or chlorinated water dips, to extend the shelf life of the products. The application of "hurdle" technology to ensure safety and stability refers to synergistically applying several preserving treatments in series to minimize or eliminate microbial and enzymatic activity while preserving the freshlike nature of the products.

PRESERVATION OF NONLIVING TISSUE

Most processed foods, for example, canned, frozen, and dried products, are marketed as nonliving tissue or man-

Table 4. Storage Conditions and Approximate Useful Storage Life for Selected Living-Tissue Foods

Food	Temperature (°C)	Relative humidity (%)	Atmosphere composition (vol. %)	Useful storage (days)
Apples	0	90–95	air	100–250
			1–5% CO_2, 2–3% O_2	>180
Bananas	14	90–95	air	7–10
Beans, snap	6	95	air	7–10
Beets (topped)	0	>95	air	90–150
Broccoli	0	>95	air	10–14
	0	>95	1–5% CO_2, 3.5% O_2	40
Cabbage	0	>95	air	30–180
Carrots (topped)	0	>95	air	30–270
Celery	0	95	air	30–60
Cherries	0	>90	air	15
Corn, sweet	0	95	air	4–8
Cranberries	3	90–95	air	60–120
Cucumbers	10	90–95	air	10–14
Eggs	0	85–92	air	150–180
Grapefruit	10–15	85–90	air	40–60
Grapes	0	80–95	air	20–180
Lettuce	0	>95	air	15–20
Meat	0	95	air	5–15
Melons	7–10	90–95	air	3–15
Onions	0	60–75	air	200
Oranges	0	85–90	air	50–80
Oysters (shucked)	0	100 (wet ice)	—	7–10
Papaya	10	85–90	air	5–20
Peaches	0	90	air	15–30
Pears	0	90–95	air	60–180
Peas	0	90–95	air	5–15
Peppers (green)	10	92–95	air	15
Potatoes	4	90	air	150–250
Poultry	0	>95	air	15
Pumpkin	10–12	70–75	air	90
Radishes	0	>90	air	60–120
Raspberries	0	>90	air	2
Shrimp	0	ice	air	10–12
Spinach	0	>90	air	10
Strawberries	0	>90	air	5
Sweet potato	12–16	85–90	air	90–200
Tomatoes (green)	12–16	90	air	15–20
Turnips	0	>90	air	100–150

ufactured foods. Separation, assembly, and preservation operations determine the final quality of the product. Typical processing temperatures, pressures, and pH values are shown in Table 5. Because most preservation operations allow little or no residual microbial or enzyme activity, deterioration during storage is chemical. The relatively low activation energies of deteriorative chemical reactions compared with enzyme reactions present little advantage to low-temperature storage and make practical the storage of sterile foods at 25°C for months without appreciable loss of original nutrients and quality factors. Often foods are manufactured with a specified economic shelf life. Expiration dates are used to ensure adequate quality while minimizing inventory costs. The effect of various processing operations on the nutritive content of foods has been reported. Measurable changes in the vitamin content of foods usually indicate significant changes in flavor, color, and structure.

HEAT TREATMENT

Populations of all life forms show characteristic death rates when exposed to elevated temperatures, pressures, and concentrations of certain chemicals. Enzymes show similar inactivation kinetics. The benefits of heat preservation, for example, the high inactivation rates of microbes and deteriorative enzymes, considerably outweigh the

Table 5. Typical Temperatures, Pressures, and pH Values in Food-Processing Operations

Temperature	Example of use	Limits of applications
	Heat-transfer media	
Cryogenic freezing—liquid nitrogen	Rapid freezing to minimize moisture loss	Cost, stress cracking during freezing
Cryogenic freezing—solid carbon dioxide (−78.5°C)	Rapid freezing	Cost
Cryogenic freezing—air, plate and aqueous base freezants (−40 to −5°C)	Commercial freezing; parasite and insect destruction; storage; freeze drying	Foods not suitable for freezing
Refrigerated storage (0–10°C)	Commercial refrigerated storage	Microbial growth possible
Room temperature storage, air (20–40°C)	Canned and dry food storage	Suitable only for preserved and packaged foods
Water, air, atmospheric steam (50–100°C)	Pasteurization of milk and eggs; blanching; sterilization of acid foods (pH 4.5); air drying; cooking;	Excessive times will cause poor color, flavor, and structure
Steam (110–130°C)	Thermal sterilization of nonacid foods; destruction of antinutritive factors	Rapid thermal degradation of nutrients, pigments, structure, and flavors
Oil, steam, infrared radiation (180°C)	Frying, roasting, baking, generation of browning reaction products	Short duration surface treatments
Various infrared radiation sources (180°C)	Surface heating for flash drying, peeling	Charring, pyrolysis without precision control
	Pressure[a]	
<0.5 kPa	Freeze drying	Cost
0.5–4 kPa	Hypobaric storage, deaeration, vacuum concentration, vacuum cooling	Cost
101 kPa	Most processing operations	
0.1–1 MPa	Steam sterilization, over pressure for glass and pouch packs; carbonated and aerosol packaged food	Cost
>1 MPa	Extruders, hydraulic pressing, homogenization, microbial and enzyme inactivation at 1 GPa (10,000 atm) and higher	Cost, specialized products
	Hydrogen ion concentration	
pH 1	Acid hydrolysis	Must be neutralized
pH 2.5	Lemons, limes, vinegar, organic acids	Taste
pH 3–4.5	Fruits, acidified foods	Noncompatability of foods; protein denaturation
pH 3.5–7	Normal pH of most foods	
pH 8–9	Solubilization of certain proteins for extraction, alkali process cocoa	Must neutralize to normal pH for consumption
pH 12	Limed corn	Color, flavor, nutrient loss
33% sodium hydroxide	Peeling of fruits and vegetables	Surface treatment only, must neutralize to normal pH for consumation

[a]To convert kPa to mmHg, multiply by 7.5; to convert MPa to atm, divide by 0.101.

drawbacks, for example, heat-induced losses of desirable food nutrient, structures, colors, and flavors.

The heat treatment needed to inactivate microbes or enzymes can be calculated from the energy for activation and the known rate of inactivation at a given temperature, usually 121°C. These values depend on the food system and its pH, water activity, and chemical profile. Typical activation energies and inactivation rates at 121°C for heat-resistant microbes are 209–335 kJ/mol (50–80 kcal/mol) and 0.1 to 10 decimal reductions/min. Commercial heat preservation operations assume a starting concentration of *C. botulinum* spores (mixed varieties) of 10 to 12/g and an inactivation rate of 5 to 10 decimal reductions/min at 121°C.

Preservation heating times at any temperature can be determined by integrating the lethal temperature–time effects at the slowest heating point in the package. Thus if the geometry of the container and the thermophysical properties of the food are known, it is possible to calculate the heating time necessary to ensure a safe level of microbes or residual enzyme activity in the food and the residual concentration of desired nutrients. In general, higher life forms (eg, insect eggs, mites, etc) are killed by even mild heat treatments. Computer programs are available for determining safe heat-preservation operations when given information about the product, including its initial temperature, container characteristics, fill, and heat-exchange system (pure steam, steam–air, water, etc).

The hydrogen ion is extremely toxic to most microbes. Foods having a pH below 4.5 can be commercially sterilized by heating to 100°C with a limited holding period.

Aseptic preservation involves performing separate heat-preservation operations on the food and the containers prior to assembly. Liquid products can be heated and cooled rapidly under optimum conditions of heat transfer in specialized heat exchangers. The sterile product is packaged under sterile filling conditions into sterile unit retail packages, drums, or bulk-storage tanks.

NONTHERMAL PROCESSES YIELDING MICROBIAL INACTIVATION

Thermal preservation processes such as retorting canned foods and pasteurizing milk and eggs have the disadvantage of degrading product quality as perceived by the consumer as well as nutritional quality. In recent years, much research has been conducted with success to develop nonthermal food processes that destroy microorganisms. These processes include irradiation, ultrahigh hydrostatic pressurization, pulsed electric fields, oscillating magnetic fields, high-intensity visible and UV light, and ultrasonics (manothermosonication). In all of these processes, microbial inactivation increases with the duration and intensity of the treatment, analogous to thermal processes.

Food irradiation has gained prominence worldwide as a nonthermal method. Irradiation refers to ionizing electromagnetic radiation that inactivates microorganisms by producing high-energy electrons within the food product. Irradiation has been approved for use in the United States on a number of food products (Table 6). Specifically, the FDA has approved irradiation from sealed units of radioactive nuclei (^{131}Cs or ^{60}Co), electron beams generated from machine sources at energies not to exceed 10 MeV, or X rays generated from machine sources not to exceed 5 MeV.

The legal basis for regulation of food irradiation by the FDA is that irradiation is included in the definition of "food additive" in the 1958 Amendment to the FDCA. The primary concern is that the radiation, in the process of destroying microorganisms and enzymes, will produce toxic substances. However, the Joint FAO/IAEA/WHO Expert Committee on the Wholesomeness of Irradiated Food (JECFI) concluded in 1980 after review of extensive research that irradiation of any commodity up to an overall average dose of 10 kGy presents no toxicological hazard. Despite its safety and potential benefits for a safe food supply, food irradiation still remains a controversial food process with many food processors unwilling to install irradiation systems.

The other nonthermal preservation methods have not found widespread usage and are still under development. Ultrahigh hydrostatic pressure (UHP) has been shown to be effective in inactivating microorganisms (vegetative bacteria, parasites, yeasts, and molds) at pressures between 300 and 600 MPa (43,000 to 87,000 psi). Processes can be made continuous for liquid foods, and processing times range from one minute to one hour at room temperature. The UHP process can also be enhanced by increasing temperatures up to 90°C, to the point of inactivating enzymes and spores. The other nonthermal processes involve exposing the food to high-intensity field strengths for short durations, sometimes including up to 100 rapid pulses. The degree of microbial inactivation depends on the strength of the field, its duration, and the number of pulses. Increasing temperature during these processes also enhances inactivation.

FREEZING

The rate of loss of color, flavor, structure, and nutrients in foods is a function of temperature; thus, lower storage temperatures prolong the useful life of foods. However, below 0°C, the free water in food forms ice crystals as a function of moisture content, solute composition, and storage temperature. Ice formation is both beneficial and detrimental. Benefits include strengthening of structures and removal of free moisture, which reduces water activity. Benefits, however, are often far outweighed by the deleterious effects of ice crystal formation, the partial dehydration of the tissue surrounding the ice crystal, and the freeze concentration of potential reactants. Ice crystals disrupt cell structures mechanically, and the increased concentration of cell electrolytes can result in the chemical denaturation of proteins.

The technology of food freezing emphasizes as short a passage of time through the temperature zone of maximum ice crystal formation as possible. The formation of as small an ice crystal as possible minimizes the mechanical disruption of cells and possibly reduces the effects of solute concentration damage. Rapid freezing can only be accom-

Table 6. Approved Uses of Food Irradiation in the United States

Food product	Purpose of irradiation	Dose (kGy)	Date approved[a]
Fresh fruits and vegetables	Inhibition of growth and maturation	1 (max)	April 1986 (FDA)
Dehydrated aromatic vegetable substances (herbs, seeds, spices, teas, vegetable seasonings)	Control of foodborne pathogens	30 (max)	April 1986 (FDA)
Food	Anthropod pest control	1 (max)	April 1986 (FDA)
Enzyme preparations, dehydrated	Control of foodborne pathogens	10 (max)	April 1986 (FDA)
Pork	Control of *Trichinella*	0.3 to 1.0	January 1986 (USDA) July 1985 (FDA)
Hawaiian papaya	Fruit flies	0.15 (min)	February 1989 (USDA)
Packaged poultry products, fresh or frozen, including ground, hand-boned and skinless products	Control of foodborne pathogens	1.5 to 3.0	September 1992 (USDA) May 1990 (FDA)
Mechanically deboned poultry meat	Control of foodborne pathogens	1.5 to 3.0	September 1992 (USDA) May 1990 (FDA)
Meat, meat by-products, and certain meat food products	Control of foodborne pathogens	4.5 (max) for refrigerated products 7.0 (max) for frozen products	Proposed by the USDA in February 1999 Approved by the FDA in December 1997

[a]Final rules published in the Federal Register by the Food and Drug Administration (FDA) or the United States Department of Agriculture (USDA) unless otherwise indicated.

plished by large temperature differences and high heat-transfer coefficients.

Many processed foods and certain animal products tolerate freezing and thawing because their structures can accommodate ice crystallization, movement of water, and the related changes in solute concentrations. Starches can be modified to form gels that accept several freezing and thawing cycles without breakdown. By contrast, most fruits and vegetables lose significant structural quality on freezing because their rigid cell structures fail to accommodate ice crystal formation. However, it is not possible to store foods at temperatures low enough to ensure complete conversion of all water to ice; as a result, commercial frozen food storage temperatures represent an economic balance between storage costs (time, energy, and capital investment) and projected shelf life.

The freezing process disrupts tissue structures and allows cell contents to become mixed so that undesirable enzyme reactions can take place at significant rates even at storage of −18°C. These reactions can generate off-flavors, reduce nutrient concentrations, and cause major changes in the structure and appearance of foods. The amount of free liquid or drip found after a freeze–thaw cycle is a good indication of the structural damage.

Heat treatment (blanching) prior to freezing eliminates enzyme-mediated changes in color, flavor, and structure. Most deteriorative enzymes are inactivated by exposure to a temperature of 100°C for 1 to 5 min. The enzymatic oxidative deterioration of frozen fruits can be inhibited with sulfur dioxide, sucrose, and combinations of citric acid, sodium chloride, and ascorbic acid (preceded by vacuum removal of oxygen if heat is not used).

Most frozen foods have a useful storage life of one year at −18°C; however, foods high in fat (eg, sausage products) may become rancid in two weeks. Frozen storage can result in moisture loss from the food through a freeze-drying process, because the heat-transfer surfaces used to maintain storage temperatures are at a lower temperature than the storage area. For this reason, frozen foods must be protected against drying by a moisture barrier. In addition, foods subject to oxidative deterioration must be protected from air.

Freezing-preservation equipment can be classified by method of heat transfer. Usually air is the heat-exchange medium for freezing foods. Foods are loaded on a belt or vibrating conveyor and passed through air flowing upward at up to 5 m/s at temperatures as low as −40°C. The air is recycled through coils and fans located next to the conveyor and returned through the conveyor. Because the air has a partial water vapor pressure lower than the food, freeze drying can occur, and some of the water in the product is removed and deposited as ice on the heat-exchange coils. As a freezing medium, air has other drawbacks. The low gas–solid heat-transfer coefficient and the heat capacity of air require either low temperatures or high velocities to obtain needed high heat fluxes. Low air temperatures increase refrigeration costs and high air velocities generate additional fan heat loads. For these and other reasons, freezing by conduction or by liquid heat-transfer methods are less costly in terms of capital investment and energy.

Liquid heat-transfer media that are used for direct-immersion freezing include food-grade dichlorodifluoromethane, nitrous oxide, and water solutions of various edible salts, sugars, alcohols, acid, and esters. Liquid heat-transfer agents offer a high-heat-transfer coefficient and reduced pumping costs, eliminate product desiccation, and allow operation at high low-side equipment temperatures. Drawbacks include possible changes in food flavor and costs of processing. Although dichlorodifluoromethane offers major operating advantages as compared with other heat-transfer media, cost and environmental concerns have reduced its potential usefulness as an ideal direct-immersion freezant. There is a need for a direct-immersion liquid freezant that is safe, low cost, thermodynamically efficient, and compatible with foods with respect to flavor, color, and odor.

Conduction freezing between chilled plates is a cost-effective method of heat removal provided the product can be assembled in a geometry compatible with the plate surfaces. Packages having semiinfinite slab geometry are loaded between stacks of platens through which refrigerant is circulated. Good heat transfer is maintained by maintaining a pressure on the stack of platens.

Other freezing methods use direct immersion in liquid nitrogen, exposure to solidified carbon dioxide, and immersions of packaged products in liquid freezants, for example, sodium or calcium chloride brines, methanol, or propylene glycol solutions.

The quality of frozen food is related to storage temperature; however, because constant storage temperatures are not always feasible, the shelf life is often determined by the highest temperature and total length of time that food is exposed to that temperature before use. Maintaining a −18°C storage environment from time of freezing until use continues to be a major technical problem facing the frozen-food industry.

REDUCTION OF WATER ACTIVITY (DEHYDRATION)

Microbes require a specific minimum level of water activity (a_w), defined by the relative humidity (measured in equilibrium with the food) for growth and reproduction at a given temperature and substrate composition. Foods possess characteristic equilibrium relationships between water activity and moisture content at given temperatures, which are known as sorption isotherms. Preservation against microbial spoilage by dehydration requires a moisture content equal to a water activity below 0.65 (Fig. 1). Dehydration contrasts with food concentration where water is removed for reasons other than for effective reduction in water activity.

Living tissue, when dried to a water activity below 0.97, suffers irreversible disruption of metabolic processes. Deteriorative chemical reactions, enzyme-catalyzed or not, are generally a function of water activity and reactant concentration. Thus, nonenzymatic browning (Maillard reaction), oxidation, and internal rearrangements (eg, staling and protein cross-linking) can increase in rate as water activity is reduced because of the increase in concentration of reactants. Many reactions show a minimum rate in the range of $a_w = 0.4$.

Prior to dehydration, foods are usually heat treated to inactivate enzyme systems. Those foods susceptible to rapid nonenzymatic browning resulting from high concentrations of reducing sugars are treated with sulfur dioxide, and products subject to oxidative rancidity can be treated with antioxidants and packaged to prevent exposure to oxygen. Low-temperature storage (5°C) reduces chemical deterioration. Blends of dehydrated products must be assembled from ingredients having the same water activity.

High-quality dried foods can be obtained only if the drying system is designed to match heat penetration with the rate of release of moisture from the food. Typically, continuous-belt dryers are staged to provide three or more zones into which the product is reapplied into progressively deeper beds. Each zone operates at a dry- and wet-bulb temperature, a through-flow (upflow and downflow through a bed of materials) air velocity, and a bed depth that optimize product quality, energy use, and production rate.

Liquids and pastes are commonly dried in spray, drum, or freeze dryers. In spray drying, product is atomized through a nozzle (rotary, pneumatic, or high pressure) into a drying chamber in a continuous operation with hot air entering at typically 200 to 300°C. Evaporative cooling provides rapid cooling of the inlet air and prevents heating of the product until in a dry, stable state. Drum drying is generally a much more severe heat treatment unless done under vacuum. Freeze drying preserves original product qualities (flavor, structure, nutrients, etc) to the greatest extent, but it is a slow and costly process and therefore finds application mostly with high-value products such as spices and coffee. Particulate foods can be dried in continuous conveyor systems, fluidized beds, freeze dryers, or batch tunnel systems. Fluidized beds use pneumatic flow to fluidize particulate product and have the advantage of optimizing drying surface area for heat and mass transfer per unit volume of dryer. Some fruits, salted animal products, nuts, berries, and many field crops are sun-dried. Other food processes that produce dehydration include osmotic dehydration (in which intact fruits and vegetables are exposed to hypertonic solutions), vacuum dehydration, microwave dehydration, extrusion cooking, pneumatic dehydration, foam-mat dehydration, and gun puffing.

MECHANICAL FOOD PROCESSES ENHANCING PRESERVATION

Mechanical food-preservation methods are characterized by separation operations, which tend to affect food quality factors less than other methods, as exemplified by the production of sterile draft beer by filtration.

Packaging is the primary mechanical preservation method; barrier materials protect preserved foods from external spoilage factors during storage and distribution. Other mechanical food-processing methods include sieving, filtration, air classification, washing, sorting, ultrafiltration, microfiltration, centrifugation, extraction, stripping, flotation, and so on. Size grading, defect sorting, and washing are routinely used to reduce or eliminate microbes, enzymes, chemicals, insects, and other matter. Virtually all food materials require repetition of these operations between harvest and preservation.

Washing equipment is designed to accommodate the type of food being washed. Leafy vegetable washers, for example, use a series of rotating paddles that transfer the material along the water surface in a tank so that sand can settle to the bottom of the tank. Fruit washers have reels or belts fitted with high-pressure water sprays that mechanically loosen dirt and debris. For tomatoes, tree fruit,

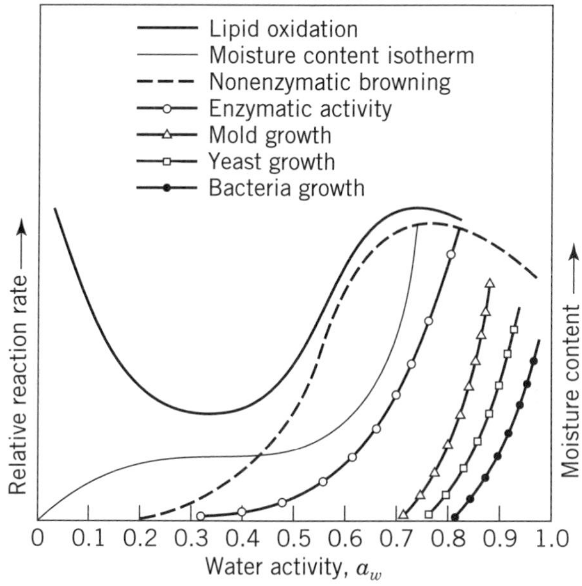

Figure 1. Relative rate of food deterioration factors as a function of water activity a_w.

and root vegetables washing units have spinning soft-rubber disks that are mounted in rows to give an intensive wiping action at the surface of the food as it rolls over the disk array. Disk washers reduce water consumption in tomato washing from 1,500 L/t (360 gal/short ton) to as little as 20 L/t (5 gal/short ton). Other food washing equipment uses combinations of sprays, brushes, and mechanical agitators. For example, froth flotation removes foreign particulate organic matter from cut corn and peas. Approved wetting agents (eg, sodium lauryl sulfate) are often incorporated in wash water to improve the efficiency of the washing operation. The proper chlorination of wash and transport water also is extremely important in maintaining low levels of microbes. Break-point chlorination ensures adequate but not excessive residual free chlorine (because >5 ppm of free chlorine can spoil flavors of some foods). This method of chlorination is implemented by adding sufficient chlorine to oxidize organic materials in the water and leave a residual concentration to ensure antimicrobial activity. Water used in the final washing of food materials processed under FDA jurisdiction must be suitable for drinking.

Liquid foods free of suspended solids may be sterilized by filtering spoilage microbes. The filtered product, for example, wine, beer, and certain fruit juices, is packaged under aseptic conditions in a sterilized container. Centrifugation can be used to reduce microbial contamination and to remove extraneous organic matter in liquid foods such as milk.

CHEMICAL PRESERVATION METHODS

Addition of chemicals usually occurs in conjunction with other preservation methods because they enhance the effectiveness of heat, refrigeration, drying, or packaging. Examples include the addition of antioxidants to fried foods to reduce the need for more expensive oxygen-impermeable flexible packaging and the use of sulfur dioxide to retard the development of brown color in frozen or dried apples.

The extreme toxicity of the hydrogen ion toward food-spoilage microbes and its tolerance by humans has made it a preferred food preservative. Studies have shown that various organic acids exert a strong inhibitory effect on the growth of microbes beyond the activity of the hydrogen ion itself. Lactic, propionic, acetic, sorbic, and benzoic acids; their sodium or potassium salts; and certain derivatives of benzoic acid find extensive use as yeast, mold, and bacterial inhibitors in bread, beverages, and sauces. Combinations of hydrogen ion, organic acids, mild heat treatment, and exclusion of oxygen have allowed the simple and safe preservation of foods not tolerant of conventional heat sterilization above 100°C, freezing, or dehydration.

Traditional food-preservation methods were based largely on drying in conjunction with preservation with locally available chemical preservatives. These included salt, organic acids generated by a lactic or acetic acid fermentation (as in pickling), wood smoke, and ethanol derived from the fermentation of substances containing sugar. Many traditionally preserved products are available in the marketplace, although modern preservation methods predominate because they are convenient.

Chemical preservatives and chemical preservation methods used in conjunction with other preservation methods for meat and poultry are controlled by USDA regulations. The FDA has jurisdiction over the use of chemical preservatives in other foods. Additives are reviewed continually for heat safety.

TOXICOLOGICAL IMPLICATION OF PROCESSED FOOD

A primary objective of food processing is to ensure a safe food supply. Certain harvested foods can contain naturally occurring poisonous substances, compounds with pharmacological effects, or compounds that can interfere with the utilization of nutrients present in other foods when they are consumed together.

Foods just harvested may contain or become contaminated with spore or vegetative forms of microbes, parasites, or the wastes of higher life-forms. Careless postharvest handling, storage, processing, and distribution operations provide additional opportunities for microbial growth or contamination. In addition to the hazard associated with pathogenic microorganisms, microbial growth can result in the release of toxins that the consumer has no way of identifying, consequently, food poisoning outbreaks occur. Fortunately, some toxins are heat sensitive and are inactivated during the cooking process. However, only good manufacturing, handling, and distribution practices ensure the prevention of microbial growth. These practices must be carried out from the time of harvest to the time of consumption.

A third source of potential health hazards in food supplies are toxic compounds formed during the chemical degradation of processed and stored foods. Oxidative breakdown products of unsaturated fatty acids have been implicated as potential carcinogens, for example, malonaldehyde.

In addition to the preceding, foods must be protected from contamination by pesticides and other chemicals used for crop protection; from exposure to hazardous chemicals and extraneous materials, for example, asbestos in air and water supplies; and radioactive isotopes. Many toxic compounds can be found as soil contaminants or as contaminants in lakes, oceans, or groundwater supplies and can be concentrated in the edible tissues of plants and animals.

FOOD-PROCESSING FACILITIES

Factors important in the selection of sites for food-processing operations include availability of raw materials, abundant potable water supply, low-cost waste disposal facilities, adequate low-cost energy supply, adequate seasonal and nonseasonal labor supply, ease of access to rail and truck transportation, proximity to consumers, and adequate storage areas. The relative importance of each of these factors depends on the length of the processing season, the ease of storage and handling of raw materials, the perishability of the finished products, and the complexity of the processing operation.

Highly perishable salad vegetables can be trimmed, inspected, bagged, and vacuum cooled using mobile equipment in or at the edge of the growing area. Processing of frozen french fries requires raw potato storage capacities of up to 10 months; thus plants tend to be located near principal potato-growing areas, for example, the northwestern United States. Continued developments in mechanical harvesting and mobile harvester separation operations for fruits and vegetables will cause further processing plant shifts toward concentrated growing areas where irrigation, climate, and level terrain make mobile processing feasible.

As processing technology improves, an increasing number of plants are relocating away from eastern U.S. population centers toward sources of raw materials. This shift is particularly noticeable in the meat industry and has occurred concurrently with the development of vacuum-packed cut beef, which provides more than a 50% reduction in storage and shipment space over conventional quarter or sides.

Certain food products are highly perishable, have marginal value, or have only regional acceptance. Dairy products, bread, soft drinks, beer, glass-packed items, and certain cured-meat products are usually manufactured from more stable raw materials or are locally produced starting materials and are distributed daily. With such distribution, production schedules are determined by local weather forecasts, by day of the week and holidays, and by seasonal factors so that returns of stale or overage products can be minimized. Considerable progress has been made in the analysis of perishable food distribution practices. A universal product code applicable to all processed foods for retail sale has further improved distribution efficiencies by allowing automatic retail checkout and continuous inventory review.

FOOD PACKAGING

Glass, woven, natural fiber, and ceramic containers have been used to store food since antiquity. However, the invention of tin-plated steel containers (ca 1800) provided the food industry with its first functional and disposable packaging system. Advances in metal container manufacturing continue to maintain the status of the cylindrical can as the primary package for heat-processed, shelf-stable foods and beverages. Notable developments include substitution of electrolytic tin plating for hot dipping, new organic linings to greatly reduce the need for tin, reduced steel thickness, and improved steel composition for better strength and corrosion resistance. Developments in coating technology have resulted in the availability of two-piece drawn aluminum and steel containers and of numerous easy-open systems. New glass containers have increased strength and have continued to reduce package weight and cost.

Rigid glass or metal containers have a major processing advantage: they can be filled and sealed at speeds in excess of 1000/min. Thus the relatively higher container cost can be recovered to some degree by higher production rates. Their rigidity (ie, sturdiness) facilitates stacking and handling during distribution and storage.

Paper, plastics, and aluminum have provided the food industry with container material that can be tailored to specific barrier needs. Ease of forming, sealing, opening, and decoration; strength; and low weight have made them preferred materials for refrigerated, dry, and frozen products.

Resistance to thermal treatment and desired barrier and mechanical qualities are obtained by laminating or coextruding appropriate combinations of paper, aluminum foil, and plastic. Lamination and coextrusion seal transparent packages against moisture and gas permeation and strengthen the packaging.

Pouches formed by heat-sealing paper–foil–plastic laminates on three sides, filling, and then heat-sealing the fourth side have found extensive use with moisture- and oxygen-sensitive dry foods as well as with high-moisture thermally processed foods. Pouches capable of withstanding saturated steam or steam–air mixtures at 121°C for several hours are finding greater use as substitutes for metal and glass containers. The advantage of pouches for heat sterilization is their slab rather than cylindrical shape. Their thinner cross section allows a shorter heat-processing time than cylindrical containers of the same capacity. Mechanical advances in forming, filling, sealing, and handling pouches will result in continued substitution of laminates for rigid metal and glass containers.

Plastics and plastic-coated papers are widely used for beverages, including carbonated drinks. Several aseptic systems are marketed for bulk packing of liquids in multiliter plastic bags.

Semirigid containers are also gaining increased acceptance, particularly as replacements for the #10 can (2.72 kg or 96 oz). Initial developments have produced a lightweight drawn-metal container measuring 24 × 30 × 6 cm with a double-seamed top. The thickness of this container (6 cm) is such that during heat sterilization, heat travels only 3 cm by conduction versus about 8 cm in the conventional #10 can. Shorter heat treatments allow the preservation of a greater variety of specialty products. The half-steam-table size eliminates cleaning of pans, because the opened unit serves as a serving tray. Other semirigid systems have been developed to replace the steel can for single-portion service. These containers generally use a heat-sealed or glued paper–foil–plastic laminated closure.

Aerosol containers have played a specialized role in the food industry as dispensers for such foods as whipped toppings (using nitrous oxide as a propellant), cake frosting, and barbecue sauces. Special filling techniques and product formulations are required because the aerosol package cannot be heat sterilized once it has been sealed.

Packaging forms an integral part of most food processing. For this reason, studies of storage conditions over a range of temperatures, relative humidities, and handling conditions must be made to determine the suitability of a package. Packaging in contact with foods must be tested for migration of packaging components under actual conditions of use, and pickup of off-flavor must be evaluated on a product-by-product basis. Because of potential migration of packaging materials into foods, current food and drug regulations should be consulted prior to marketing foods packaged in nonstandard materials.

BIBLIOGRAPHY

This article was adapted from M. Grayson, ed., *Kirk-Othmer Encyclopedia of Chemical Technology*, Vol. 11, 3rd ed., John Wiley & Sons, New York, 1979. Note: All specific citations have been removed. A user must refer to the original text to obtain specific references.

GENERAL REFERENCES

G. V. Barbosa-Cánovas and H. Vega-Mercado, *Dehydration of Foods*, Chapman and Hall, New York, 1996.

D. A. Corlett, *HACCP User's Manual*, Aspen Publishers, Gaithersburg, Md., 1998.

J. F. Diehl, *Safety of Irradiated Foods*, 2nd ed., Marcel Dekker, New York, 1995.

W. A. Gould, *CGMP's / Food Plant Sanitation*, CTI Publications, Baltimore, Md., 1994.

W. A. Gould, *Unit Operations for the Food Industries*, CTI Publications, Baltimore, Md., 1996.

D. R. Heldman and D. B. Lund, *Handbook of Food Engineering*, Marcel Dekker, New York, 1992.

S. D. Holdsworth, *Aseptic Processing and Packaging of Food Products*, Elsevier, New York, 1992.

P. Jelen, *Introduction to Food Processing*, Prentice-Hall, Englewood Cliffs, N.J., 1989.

N. G. Marriott, *Essentials of Food Sanitation*, Chapman and Hall, New York, 1997.

N. N. Potter and J. H. Hotchkiss, *Food Science*, 5th ed., Chapman and Hall, New York, 1995.

G. L. Robertson, *Food Packaging: Principles and Practice*, Marcel Dekker, New York, 1993.

R. P. Singh and D. R. Heldman, *Introduction to Food Engineering*, 2nd ed., Academic Press, New York, 1993.

K. J. Valentas, E. Rotstein, and R. P. Singh, *Handbook of Food Engineering Practice*, CRC Press, New York, 1997.

R. C. Wiley, *Minimally Processed Refrigerated Fruits and Vegetables*, Chapman and Hall, New York, 1994.

JAMES FALLER
University of Illinois
Urbana, Illinois

FOOD PROCESSING: EFFECT ON NUTRITIONAL QUALITY

The definition of *processing* is "a natural phenomenon marked by gradual changes that lead toward a particular result"; "a series of actions or operations conducing to an end"; or "a continuous operation or treatment especially in manufacture" (1). *Food processing*, therefore, refers to the series of actions involved in order to prepare and preserve a food supply by some continuous operation or treatment to achieve as a goal a safe, high-quality product with extended shelf life.

Changes that occur to food quality may be thought of as two types: deterioration and spoilage. Deterioration involves changes in quality induced by physicochemical and/or biochemical reactions taking place with or without the intervention of a physical environment (such as oxygen, carbon dioxide, water, light, heat, etc). Spoilage, on the other hand, generally refers to changes in quality due to action of biological agents such as bacteria, molds, or insects. Effects of deteriorative reactions and spoilage agents on food quality result in changes in both sensory properties (ie, appearance, flavor, texture) and nutritive value (vitamin content, protein value, etc). The extent to which these reactions can occur depends on the sensitivities and types of food products considered. For example, certain fruits such as strawberries or raspberries may be readily spoiled by the presence of molds. Green peas or beans, on the other hand, lose flavor and result in undesirable textures due to enzyme-induced reactions.

Through various types of food processing, quality factors can be maintained or extended, the extent of which depends on the food product to be considered. By canning fruits and vegetables (eg, pears, peaches, peas, beets) with proper heat sterilization and packaging, these products will not suffer spoilage, although they may deteriorate to some extent in terms of color, flavor, and texture. The extent of these changes depends on a wide array of factors, including storage conditions after packaging. Factors affecting quality of food products include: (*1*) initial quality of the raw materials and handling from harvesting to the manufacturing plant; (*2*) pretreatment (including cleaning, sanitizing, washing, aspiration, screening, filtration, chlorination, fumigation, etc); (*3*) sorting (removal of extraneous materials); (*4*) peeling, coring, dehairing, defeathering, husking, stemming, and so on; (*5*) disintegration/physical separation in cases where grinding, pulping, pressing, or expelling are needed, generally followed by screening, filtering, or centrifugation; and finally, of course (*6*) the final finished food manufacturing step. These final manufacturing steps comprise a myriad of different types of food processes including clarification/filtration; crystallization; curing/smoking; dehydration; evaporation and distillation; fermentation; foaming; mixing/shearing operations (whipping, kneading, blending); forming/shaping; heating/cooking; maturation/aging; or cooling/freezing.

To discuss the effects of food processing on nutritional quality, it is important to have an understanding of some of the basic types of food manufacturing available, their historical development and why they are used. Table 1 provides a very brief chronological sequence of events that have taken place in the development of food processing since the beginning of time. Although in a very subjective way there has been general interest in the effect of processing treatments on product integrity, most traditional processes have developed by accident over time. Any true acknowledgement of nutritional or overall quality of foods has only become more of a science and concern during the last half of the twentieth century. As knowledge expands on the effects of microorganisms and chemicals on man, the standards for purity and safety increase. The standards in the United States, for example, are largely determined by state and federal food and drug agencies as well as individual food industry standards.

In addition to introducing various types of manufacturing techniques and some of the processing parameters that influence quality of foods, it is also important to include a brief discussion on chemical kinetics, which encompasses the study of the rates at which chemical reactions proceed. It is through a basic understanding of kinetics of degradation of nutrients and general physical characteristics in combination with a thorough knowledge of how their reactions rates are affected by different processing conditions that a food technologist can optimize a food process to achieve the highest quality possible and predict a maximum shelf life for that product. The area of kinetics in food systems has received a great deal of attention in past years, primarily due to efforts to optimize or at least maximize the quality of food products during processing and storage. A good understanding of reaction kinetics provides a better idea of how to formulate or fortify food products to preserve and/or extend the existing nutrients or components in a food system or minimize the appearance of undesirable breakdown products. Although limited kinetic information is presently available for food systems or ingredients, several researchers have compiled kinetic data on various quality attributes, including enzyme and protein changes (2), flavor changes (3), physical/textural changes (4), and vitamins and pigments (5,6), in order to facilitate the development of mathematical formulas for the optimization of a given food process. The interaction of the study of kinetics and its relationship to food processing and maintaining optimum food quality will be discussed in a later section. Since a complete discussion of all types of food manufacturing and their effect on quality of foods is beyond the scope of this article, only a few selected processes that are representative of various influencing processing variables that may affect a food product will be presented.

THERMAL PROCESSING

One of the earliest commercially available and most prevalent methods of food preservation during the twentieth century has been thermal processing of foods. The term thermal processing generally refers to a process during which a food product is subjected to high temperatures with the objective of inactivating undesirable microorganisms and/or enzymes in order to comply with public health standards. It includes operations such as conventional and aseptic canning as well as pasteurization. Foods are by nature contaminated by microorganisms and contain indigenous enzymes that may cause undesirable changes in products during storage. Proper thermal processing can extend shelf life by weeks (eg, pasteurization) or months to years (eg, canning). Unfortunately, in addition to destruction of microorganisms, there is an associated undesirable degradation of heat-sensitive vitamins and other quality factors. It is, therefore, of critical importance to determine the appropriate heat treatment for individual food products when calculating process times to ensure high-quality and safe products.

Pasteurization

In the case with pasteurization, not all vegetative microorganisms are killed and thus must be stored under conditions that minimize growth. For example, with milk, pasteurization is used to kill pathogenic microorganisms; however, some vegetative spoilage microorganisms can still survive this heat treatment and, thus, milk requires refrigeration. In the case with beer, pasteurization serves mainly to kill spoilage microorganisms. Techniques used to extend shelf life include refrigeration; chemical additives that alter the microenvironment of food, reducing growth of microorganisms (eg, sweetened condensed milk, food acids in pickles and fruit juices); packaging (eg, maintain anaerobic conditions in bottled beer); and fermentations with desirable organisms (eg, cheese, yogurt, etc).

The individual heat treatment applied to a food product depends on the specific heat resistance of a particular microorganism that is used as a basis for the process. For instance, high-temperature, short-time (HTST) processing for milk is 161°F (71.7°C) for 15 s as compared with conventional pasteurization of 145°F (62.8°C) for 30 min. HTST is better due to less nutrient destruction and sensory quality changes. This is because destruction rates for nutrients and quality factors are less dependent on temperature than destruction rates for microorganisms. The process criteria for pasteurization of milk are the thermal destruction of *Coxiella burnetti*, the rickettsia organism responsible for Q fever. These process criteria vary according to food type and pH. For instance, in the case with high-acid fruits (cherries), the criteria are set for the destruction of yeasts and molds, and for fermented beverages (beer, wine), the criteria are set for the destruction of wild yeast. In the case with pasteurization of foods, since there is not established sterilization, these products need to have lowered storage temperatures.

Canning

Canning is a general term applied to processes in which a product is sealed in a hermetic container and contains no active microorganism or enzymes that could grow under conditions imposed during storage. A food product is not

Table 1. Historical Development of Processing and Preservation Techniques

Date	Event
30,000 B.C.	In-ground ovens /w/ hot stones, Europe
27,000 B.C.	Clay ovens, Czech Republic
before 15,000 B.C. (Old Stone Age)	Roasting, pounding, drying
after 15,000 B.C. (Middle Stone Age)	Dried fish, boiling, food storage, smoking, steaming
11,000 B.C.	First grain mills, North America
before 9000 B.C. (New Stone Age)	Alcoholic fermentation, acetification, salting, baking, bread making, sieving, primitive pressing, seasoning
6000–5000 B.C.	Olive oil press, Near East
4500 B.C.	Dairy produce, butter, cheese, Near East
3500 B.C. (Bronze Age)	Filtration, lactic acid fermentation, flotation, leavened bread, flavoring
1500 B.C. (Iron Age)	Refinement of flavoring, and cooking
1200 B.C.	Deep frying, Egypt
after 1000 B.C.	Ice used for refrigeration
600–400 B.C. (Roman Period)	Food adulteration common
A.D. 100s	Invention of mill /w/ vanes, Britain
A.D. 300	Water mills
A.D. 700s	Distillation of spirits, Iran
A.D. 1100	Distillation of alcoholic beverages, Italy
A.D. 1276	First official whiskey distillery, Ireland
1300s	Rice mill, Italy
end of Middle Ages	Preservation of fruits with sugar instead of honey
1500	Mechanical sieving
1588	Ramelli's grain mill
1678	Champaign, France
1795	Introduction to thermally processed foods (N. Appert)
1800–1870	Canning industry develops
1801	Beet sugar refinery, France
1835	Evaporated milk process patented
1851	High-pressure steam sterilizer patented by R. C. Appert
1850–1860	Pasteurization of wine
1853	Vacuum evaporated milk process developed (G. Borden)
1800–1850	Ice-making machines patented
1840	Roller mills for flour, Hungary
Civil War period	Expansion of canned food industry
late 1800's	Development of pure cultures for beer
1877	Laval's centrifugal cream separator
1899	First report on high-pressure preservation (B. H. Hite)
1902	Fat hydrogenation patented
1912	Discovery of vitamins, Poland
1915–1920	Large-scale refrigerated warehouses
1920	Electropure process for milk (electrical discharge for inactivation of microorganisms)
WW I	Introduction of dehydrated foods
1930s–1950s	Most work accomplished on isolation of individual vitamins
1940–1946 (WW II period)	Fortification of foods, expansion of dehydrated foods (eg, milk, eggs, vegetables)
1945	Introduction of radiation for preservation of foods
by end of 1950s	Irradiation ready for commercial processing of foods
1958	Irradiation of foods halted by FDA for three decades
1985	FDA approves irradiation for *Trichinella spiralis* control in pork
1986	FDA approves irradiation for microorganism control in dehydrated enzymes, herbs, spices, vegetable seasoning; disinfestation of vegetables and fruits; and ripening delay of fruits
1990	FDA approves irradiation for microorganism control in fresh/frozen poultry
1991	First commercially UHP-treated fruit products, Japan
1995	FDA approves irradiation for *Salmonella* control in animal feed and pet food
1997	FDA approves irradiation for microorganism control in fresh chilled/frozen red meats

necessarily sterile, however, since excessive storage temperatures (120–140°F [48.9–60°C]) could result in growth of certain microorganisms. To achieve commercial sterility, the following factors need to be considered: nature of the food (eg, pH); storage conditions following thermal processing; heat resistance of the microorganism or spore; heat-transfer characteristics of the food container and heating medium; and the initial load of microorganisms.

The nature of the food and its microenvironment during storage are extremely important for establishing guidelines for a thermal process. Since, generally, canned foods are void of oxygen, obligate aerobes do not pose health hazards or spoilage problems. Spores of anaerobic microorganisms are also less heat resistant than those of anaerobic organisms (facultative or obligative anaerobes). An exception would be the case of canned cured meat products where oxygen may not entirely be removed. Aerobes such as *Bacillus subtilis* and *Bacillus mycoides* have been identified in these cases as spoilage agents (7).

A major factor for consideration in anaerobic environments is the pH of the food system. For instance, there are extremely heat-resistant spores that may survive a typical commercial process but do not constitute a health or spoilage hazard because of the low pH of the food. For this reason, the protocol for thermal process design for foods has been set according to the pH of the food products as follows:

1. High-acid foods (pH < 3.7): spore-forming bacteria do not grow, and thermal inactivation is based on yeasts and molds
2. Acid foods with $3.7 < \text{pH} < 4.5$: processed to inactivate acid-producing spore formers (eg, *Bacillus coagulans*)
3. Low-acid foods with pH > 4.5: processed to inactivate *Clostridium botulinum*, which produces an extremely poisonous toxin and is assumed to be present in all products intended for canning; it cannot grow below pH 4.5.

Other spores that are more heat resistant than *C. botulinum* such as *Bacillus stearothermophilus*, a nonpathogenic thermophilic spore former 20 times more heat resistant but requires elevated storage temperatures (120–130°F [48.9–54.4°C]) to grow, are referred to as *flat sour* organisms. These could be potential problems if elevated storage temperatures were present. Often in qualifying a process, the organism used is another putrefactive anaerobe referred to as PA 3679 since it is less toxic, easy to assay, and exhibits a heat resistance slightly greater than *C. botulinum*.

Two main criteria must be considered in designing a thermal process to accomplish inactivation of spores or vegetative cells. The first is the understanding of thermal inactivation kinetics or rate of destruction of food-spoilage-causing organisms and the dependence of rate on temperature. The second is taking into consideration the temperature history of the product or the heat-transfer characteristics (heat penetration) that determine the temperature profiles achieved within the container during processing. Destruction of microorganisms and their spores generally obey first-order kinetics, although exceptions have been described by Stumbo (8). The rate of destruction is characterized by a first-order reaction rate constant. The general rate equation, $-dC/dt = kC$, as presented later in a short discussion on kinetics, applies to any first-order reaction process and is used here to describe the thermal inactivation of bacterial spores, where C = concentration of viable spores, k = the rate constant (t^{-1}) and t = time. After rearranging terms and integrating over time, this equation becomes $\ln [C_0/C] = kt$, where C_0 = initial concentration of viable spores. From a semilog plot of ln C versus time (t) of exposure to a constant lethal temperature, a straight line will result that intercepts the ordinate axis at $\ln C_0$ with a slope of $-k$. The temperature dependency of the rate constant is described by the Arrhenius equation. It is also an exponential function that is described by a straight line on a semilog plot ln k versus $1/T$. The equation is $\ln (k/k_0) = (E_a/R) \cdot [(T_0 - T)/T_0T]$, where T = temperature (K), k_0 = reference rate constant at a reference temperature (T_0), E_a = energy of activation (cal/mole), and R = universal gas constant (1.987 cal/mol · K).

In the canning industry, another term is used that is related to the first-order reaction rate constant, referred to as the D-value, defined as the time in minutes at a given temperature to reduce the population of microorganisms or spores by 90%, or the logarithmic order of death, using the common (base 10) logarithm. It has a similar time dependency as k, and the mathematical relationship between the two terms is $D = 2.303/k$. The temperature dependence of the rate of destruction (D-value) also follows a well-defined relationship: log (D) is linearly related to temperature (Fig. 1). It can be seen that a plot of log (D) versus temperature (°F) provides a z-value that is the reciprocal of the slope (Fig. 2). The z-value represents the temperature change (°F) necessary for the D-value to change by 90% (one log cycle). This plot is referred to as the thermal death time (TDT) curve. This z-value reflects the sensitivity of change of temperature on the destruction of the microorganism, similarly to the energy of activation (E_a). This provides the information required to equate the various time/temperature treatments in terms of thermal destruction of microorganisms and spores. In fact, it may be observed from Figure 2 that many different process designs could be selected to achieve the same thermal lethality.

The fact that thermal inactivation of microorganisms or spores follows a first-order reaction rate indicates that an

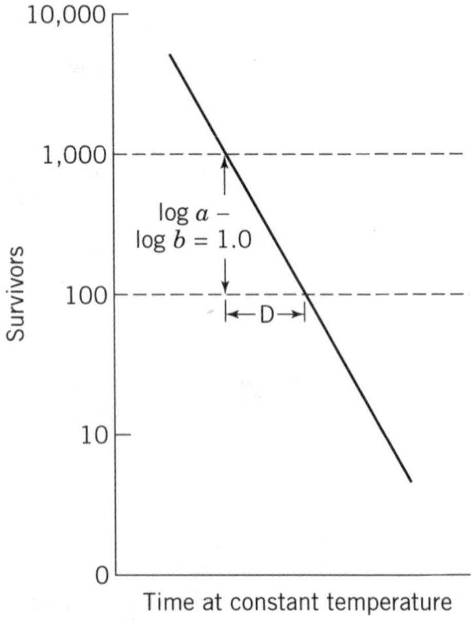

Figure 1. Graphical representation of a first-order kinetics for the destruction of bacterial spores.

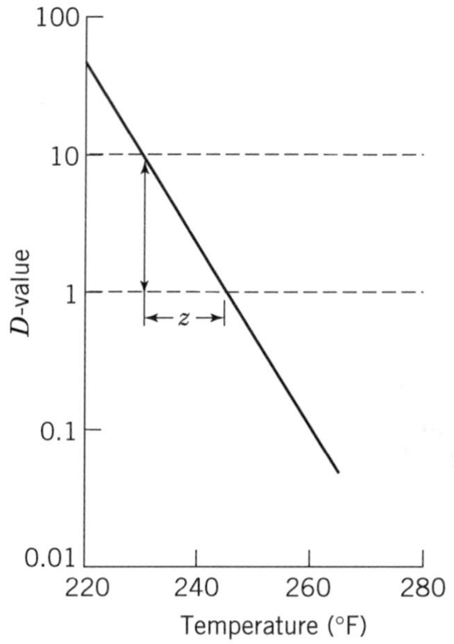

Figure 2. Graphical representation of thermal death time (TDT) for microorganisms.

absolute zero count theoretically cannot be reached. Each order of D minutes results in a reduction of 90%, which implies that 10% are left. Thus, this becomes the probability of survival and a number of D-values is used to define the order of the process. To specify a TDT curve, all that is needed is a slope (z-value) and a reference point providing a D-value at a reference temperature. For the sterilization of low-acid foods with highly resistant microorganisms, a reference temperature of 250°F (121°C) is generally specified and the order = 12D, or 12 decimal reductions. In the case of *C. botulinum*, in low-acid foods, the highest D_{250} value known for this organism is 0.21 min; therefore, the minimum process sterilizing value is F = 0.21 × 12 = 2.52 min. Lethal rate may be calculated as $1/t = 10^{(T-250)/z}$. Actually, all low-acid foods are processed far beyond the minimum botulinum cook in order to deal with spoilage bacteria of much greater heat resistance. In the case of high-acid foods or pasteurization processes in which microorganisms have lower heat resistance, a reference temperature of 212° (100°C) or 150°F (65.6°C) is used. Factors affecting thermal resistance of microorganisms or spores are pH and buffer components (eg, salts), ionic environment, water activity (particularly concentrated-type products, such as condensed soups or evaporated milk), and overall medium composition.

This information in combination with the temperature history of the sample will provide the information required to design an appropriate thermal process. Foods processed in containers require that their temperature history be determined at the slowest heating point (8,9). Heat penetration depends on: (*1*) surface heat-transfer coefficient, (*2*) physical and thermal properties of the product, (*3*) ΔT between the steam and initial product temperature, (*4*) container size (overall dimensions), and (*5*) type of heat transfer within the container (ie, convection or conduction). A good review on determining heat penetration in canned foods, which is beyond the scope of this presentation, and the use of Ball's formula for calculating thermal process times is presented by Teixeira (10).

In optimizing a thermal process, it is not only important to inactivate microorganisms or spores but also to retain sensory and nutrient quality factors. Since reaction rate constants and dependency on temperature are very different for nutrient and sensory quality factors as compared with thermal destruction of microbes, an opportunity exists to optimize quality. Aseptic canning systems have become more and more prevalent in recent years to minimize quality losses that occur in slow heated foods processed in conventional retort systems and to allow more economical packaging systems. The basic concept allows sterilizing the product using either HTST or ultrahigh temperature (UHT) processing outside the container, followed by filling aseptically into separately sterilized containers. A good example of quality improvement with such aseptic systems has been presented by Josylyn and Heid (11) showing thermal inactivation of a bacteria (F_0 = 6 min) compared with thiamin destruction curves for various levels of loss (Fig. 3). The only criteria is that a time/temperature relationship be chosen along the TDT curve. It is clear that higher processing temperatures allow greater retention of thiamin than at lower processing temperatures at longer processing times. Thiamin is one of the most thermally labile of all the vitamins and provides a good illustration of how quality can be maximized without sacrificing microbiological safety. It should be kept in mind, however, that this particular example also assumes that a first-order rate re-

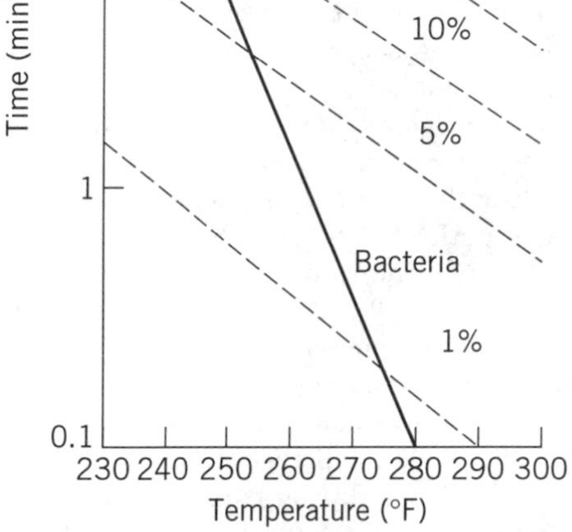

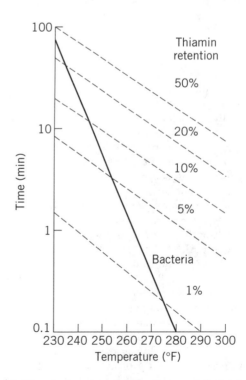

Figure 3. Time-temperature sterilization curve for bacteria (F_0 = 6 min) compared with time–temperature destruction curves for 1, 5, 10, 20, and 50% loss of thiamin. *Source:* Ref. 11.

action applies to the destruction of thiamin, which may or may not be the case with other nutrients.

Currently, most aseptically processed products are of fluid nature, although studies are constantly under way to apply aseptic processing to fluid products containing solid particulates, such as soups or stews with chunks of meat and vegetables. These particulates would be the cold spots in the process. Until recently, there has been no way to actually monitor the temperature at these points, and, thus, food engineers have relied on mathematical models to predict temperature changes. Mathematical models to account for heat-transfer coefficients have been demonstrated by Sastry (12). A joint task between the National Center for Food Safety and Technology (NCFST) and the Center for Aseptic Processing and Packaging Studies (CAPPS) resulted in establishing a protocol for use by any company that wants to produce a low-acid aseptic product containing particulates; their aseptic process filing was accepted by the FDA in 1997 (13). This milestone opened a new market for aseptically processed particulate foods, resulting in higher quality and safety, and better color, texture, and overall quality products. To further ensure safety and quality of thermally processed foods, specific guidelines have been set forth by the U.S. Code of Federal Regulations. All plants producing low-acid canned foods must be registered with the Food and Drug Administration (FDA), indicating name and place of business, location, processing method, and list of food products processed. There are four operational levels for low-acid foods: (*1*) adequacy of equipment and procedures to perform safe processing operations; (*2*) adequacy of record keeping to prove safe operations; (*3*) justification of the adequacy of time/temperature processes; and (*4*) qualifications of supervisory staff responsible for thermal processing and container closure operations.

Irradiation

Similarly to conventional thermal processing, the concept of irradiation is to temporarily raise the energy level of a system sufficient enough to result in the death of microorganisms, thus extending the shelf life of a product. In the case of irradiation, however, instead of raising the temperature of the system for a given period of time, the system is exposed to a source of radiation, resulting in ionization of individual atoms or molecules to produce an electron and a positively charged atom. The energy of radiation is often measured in electron volts (eV), where one electron volt is the energy acquired by an electron falling through a potential of 1 V (or 1/eV $= 1.602 \times 10^{-12}$ erg). The total effect of the radiation on a given material depends on the energy of each photon, its source, as well as the total number of photons impinging on the material. The relative ionization of electrons varies with the depth of absorption in a given material (absorber). Using irradiation as a technique for preservation allows the foods to be kept cold or frozen during treatment, thus potentially allowing greater stability of quality factors of the food product.

In looking at units of measurement for irradiation of foods from an historical perspective, a variety of different units have been proposed. Since the amount of energy absorbed by a material and the amount of ionization produced, as a result of the interaction of radiations with a given material, depend on a number of variables—including the number of particles passing through the material, energy of the particles, and the nature of the absorber—measurement of the radiation dose becomes a complex issue. Early references mention the roentgen (R) as a unit of measure or the amount of energy delivered by X-rays or γ-rays to a gram of dry air or 83.3 erg. Since measurements in biological systems, however, were found to range from 93 erg/g in water or biological tissue up to 150 erg/g in bone, it was decided by the International Commission on Radiological Units that the term *rad* would be used. That is the quantity of radiation that results in the absorption of 100 erg/g at the point of interest (14). Now the radiation absorbed is conventionally reported in units of *grays* (Gy) or *kilograys* (kGy), where 1 Gy = 1 J/kg (1 J = 10^7 erg).

Although the concept of using ionizing radiation for the preservation of foods has existed since about 1945 and near-commercialization since the end of the 1950s, its actual practice as a common food processing technique still has not taken a strong hold in the United States even at the close of the twentieth century. One of the reasons contributing to this delay was the passage of the Food Additives Amendment to the Food, Drug and Cosmetics Act in 1958, which classified radiation as a food additive rather than a preservation processing technique. This legislation resulted in extensive and time-consuming testing before acceptance could be considered (15). Although the first actual approval of irradiation of a food product in the United States was for insect disinfestation of wheat flour in 1963, approval for the first meat product for the control of *Trichinella spiralis* in pork was not until 1985. This was followed by approval in the United States of irradiation of chicken in 1990 and beef in 1997 (Table 1). A review by Pauli and Tarantino (16) presents the FDA's standards by which they determine safety of proposed applications of radiation. The major areas that they pursue to ensure safety include radiological, toxicological, and microbiological safety and nutritional adequacy. From a radiological point of view, there is no concern since the currently approved radiation sources are of too low an energy level to induce radioactivity. Assessment of the toxicological safety of irradiated foods was accomplished through the Bureau of Foods Irradiated Food Committee, set up by the FDA. After going through a series of assessments based on animal feeding and mutagenicity studies, the group concluded that radiation doses below 1 kGy present no evidence of possible toxicological risk. For larger radiation doses it was determined to assess the process on a case-by-case basis, particularly foods consumed in significant quantities.

There has been a great deal of controversy over irradiation of foods and its safety to both workers in the plants conducting the treatment as well as the handling and consumption of irradiated food products. To a large extent these concerns are still present, but with the education of consumers regarding safety and benefits of irradiation, opinions are gradually changing. Hashim and coworkers (17) made recommendations for increasing consumer acceptance of irradiated foods, including: (*1*) development of educational programs to increase consumer understanding of the role of irradiation of foods; (*2*) preparation of informative irradiation labels and/or posters to assure consumers of the safety of irradiated products; (*3*) conduct television shows, promoting children interaction along with pamphlets and informational brochures; and (*4*) conduct in-store sampling surveys of cooked irradiated foods. Since the FDA acceptance of irradiation to control growth of microorganisms in red meat in December of 1997, there has been a great deal of changing of public opinion over irradiation as a means to preserve foods. Resurreccion and Galvez (18) and Lusk et al. (19) have also discussed the need for consumer acceptance of irradiation as a safe processing method of preservation. At present the consumer's perception of this technique is still questionable but has been shown to greatly improve after conducting educational programs addressing the overall safety of the process.

From a microbiological safety viewpoint, some issues that are raised on the effects of irradiation of foods are whether irradiation may result in the mutation of microorganisms that may lead to more virulent pathogens and, if there is reduction of the spoilage microorganisms, whether, as a consequence, pathogens will be able to grow undetected without competition. The FDA, however, has indicated that radiation-induced mutation is not a concern with respect to increased virulence or increased heat resistance. In fact, Farkas (20) has indicated that radiation is more likely to reduce virulence of surviving pathogens. Other issues of concern involve losses of nutrients. There has been general evidence, however, that indicates conventional cooking alters the nutrient quality much more than irradiation. Macronutrients, including proteins, lipids, and carbohydrates, are not significantly affected up to doses of 10 kGy with only minor changes at sterilization doses of 50 kGy (21). There is evidence that vitamins may degrade with irradiation, which would be expected due to any process that elevates the energy level of the individual food constituents. The degree of degradation will depend on a number of factors such as the dose of radiation, the food type, the temperature at which irradiation occurs and the presence of oxygen. Generally, low-temperature radiation in the absence of oxygen reduces significantly the losses of vitamins as well as maintaining storage at low temperature and in sealed containers (22). Similarly to thermal processing, it has been found that thiamin is the most labile of the water-soluble vitamins in the presence of irradiation processing, whereas vitamin E has been shown to be the most susceptible of the oil-soluble vitamins to degradation from irradiation. The FDA requires that those vitamins most affected by irradiation are not a significant source from that particular processed food product in the overall diet.

Overall quality of irradiated foods may be affected by (*1*) radiation dose, (*2*) dose rate, (*3*) temperature and atmospheric conditions during irradiation (eg, presence of oxygen), (*4*) temperature and environmental conditions following irradiation during storage, and (*5*) development of radiolytic products (23). The presence of radiolytic products can result in oxidation of myoglobin and fat, which may result in discoloration and rancidity or other off-odor and/or off-flavor development. For instance, ozone, which is a strong oxidizer produced during irradiation in the presence of oxygen, may oxidize myoglobin, resulting in a bleached appearance. Color changes may be influenced by the packaging environment. It has been reported that irradiated vacuum-packaged meats develop a fairly stable bright pink or red color in turkey breasts, pork, or beef (24–26). This stresses the importance of elimination of oxygen before irradiation. Irradiation in the frozen state minimizes movement of free radicals to react throughout the food system and, thus, can minimize sensory quality issues. According to Kropf et al. (27) and Luchsinger et al. (28), any irradiation-induced off-odors may be removed during conventional cooking; however, studies are ongoing to investigate. Another issue that should be considered is the effect of the type of packaging that is used to prevent the evolution of hydrogen, low-molecular-weight hydrocarbons, and halogenated polymers (29). Packaging materials must be approved by the FDA according to the 21 CFR 179.45.

Flavor-transfer problems to the food could occur as a result of using conventional fresh meat overwrap (eg, polyvinylchloride). The development of detection methods for irradiation of foods is currently an active area of investigation. This is important from a regulatory compliance view point (15). With the development of these methods, there should be an acceleration of approval of irradiated foods as well as new applications and enhancement of international trade. Since irradiation involves no major chemical, physical, or sensory changes in foods, minute changes must be focused upon. Glidewell et al. (30) prepared a comprehensive review on detection methods for irradiated foods. For example, detection of hydrocarbon formation from irradiated lipid-containing foods offers a potential method of detection. Crone et al. (31) detected the formation of 2-alkyl-cyclobutanone formed from fatty acids in irradiated but not cooked foods. Other methods of detection include measurement of cell membrane damage through measurement of electrical impedance, electron spin resonance, and thermal and near-infrared analysis.

From a microbiological point of view, the predominant spoilage organisms are Gram-negative psychrotrophic microorganisms, which are very susceptible to irradiation (32). It has been shown that doses of about 1 kGy virtually eliminate Gram-negative microorganisms. However, it is not as effective on Gram-positive lactic acid–producing microorganisms. Nevertheless, refrigerated storage of meats has increased dramatically as a result of irradiation. Lambert et al. (33) found pork loin slices packaged under nitrogen and irradiated to 1 kGy had an extension of 21 days beyond the control at 5°C. As with TDT in conventional thermal processing, the death of a microorganism resulting from exposure to radiation can also be evaluated by plotting the logarithm of the surviving fraction, in this case, against dose. With thermal sterilization, the effect of kill is not solely dependent on the quantity of heat absorbed by the cell but also on the intensity factor (temperature) and on time. Radiation sterilization, on the other hand, is actually less complicated since the intensity factor is called *dose rate*, or the amount of radiation absorbed by the cell per unit time. Although dose rate has some lethal effects, it is possible to relate radiation effects to dose alone according to the following equation: $n = n_0 e^{-D/D_0}$, where n = the number of live organisms after irradiation, n_0 = the initial number of microorganisms, D = dose of radiation received, and D_0 = constant dependent on organism type and environmental factors (Fig. 4).

Depending on the dose level of radiation energy applied, foods may be pasteurized to reduce or eliminate pathogens or they may be sterilized to eliminate all microorganisms. The approved sources of radiation for food use include γ-rays (produced by radioisotopes cobalt-60 or cesium-137) and machine-generated X rays (with a maximum energy of 5×10^6 electron volts [eV]) or electrons (with a maximum energy of 10 MeV).

FOOD PRESERVATION THROUGH CONTROL OF WATER

As mentioned in the general discussion of food preservation, the most serious limitation of shelf life and nutritional quality is from microbiological activity. This section covers briefly the concept of reducing the amount of water in food systems as a means of food preservation. By lowering the amount of water available for microorganisms to remain active and reducing mobility of potentially reactive chemical species in a food system, it is possible to extend quality of a product from both a microbiological and nutritional point of view. Food preservation processes falling in this category include concentration processes such as evaporation or freeze-concentration; dehydration including air drying, spray drying, freeze drying, and so on, and methods of adding salt or sugar and thereby lowering the available water by increasing osmotic pressure. Since all of these methods depend on lowering available water, instead of describing in complete detail each type of process, which is beyond the scope of this article, the overall general relationship between availability of water and chemical reactions or microbiological activity will be discussed in this section.

One of the most important concepts in understanding the relationship between moisture removal and shelf-life extension is that of *free* versus *bound* water in foods, which are terms that grew in importance particularly during the latter half of the twentieth century. It was realized that active water is more important to the stability of a food system than the total amount of water present, or water content (34). The term water activity (a_w) is a thermodynamic property defined as the ratio of the vapor pressure of water in a food system (p) to the vapor pressure of pure water (p_0) at a given temperature or the equilibrium relative humidity (ERH) of the air surrounding the food system at a given temperature, such that $a_w = p/p_0 = \%\ \text{ERH}/100$. Water activity is currently the most common term used by researchers and industry professionals in the food processing business. A sorption isotherm is a graphical illustration of the relationship of water activity, or relative humidity of the vapor space surrounding a food system, and the equilibrium moisture content of that material. An isotherm is generally divided into three major regions (Fig. 5). Region A represents strongly bound water with an enthalpy of vaporization considerably higher than that of pure water. This region represents the first layer of water molecules sorbed at hydrophilic, charged and polar groups of food constituents (proteins, polysaccharides) and includes structural water (hydrogen-bonded water), monolayer water, and hydrophobic hydration water (35). This bound water is generally considered unfreezable and unavailable for chemical reactions or acting as a plasticizer.

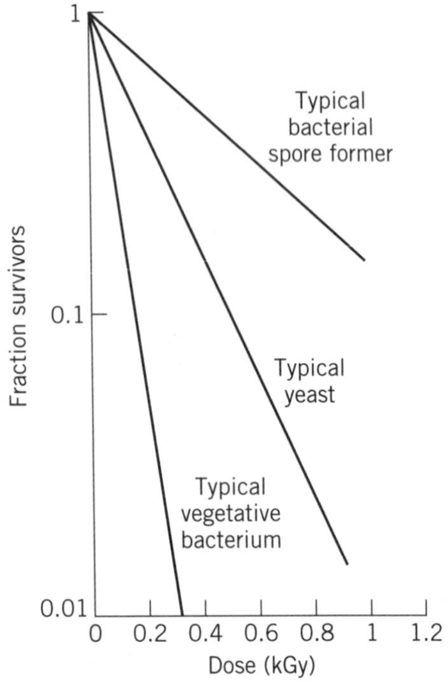

Figure 4. Representation of dose-response curves for different types of microorganisms. *Source:* Ref. 14.

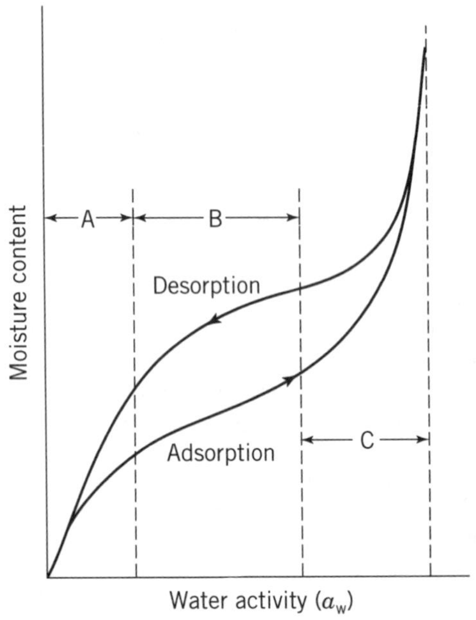

Figure 5. General type of sorption isotherm for food products. *Source:* Ref. 50.

Region B includes water molecules that are less firmly bound than those in region A with an enthalpy of vaporization only slightly more than that of pure water. This water may or may not be unfreezable and is available for low-molecular-weight solutes and some biochemical reactions. It may be considered a sort of transition zone between bound and free water. Region C consists of water molecules that have properties resembling more closely those of free water, often referred to as capillary water or that which is loosely bound with no excess heat of binding detected in this region.

Concerning the effect of temperature on water activity, as temperature increases, the amount of water adsorbed usually decreases. In fact, the effect of temperature on isotherms follows a Claussius–Clayperon relationship as depicted in an integrated form, holding moisture content constant: $\ln (a_2/a_1) = \Delta H_s/R(1/T_1 - 1/T_2)$, where ΔH_s = heat of sorption (cal/mole); $R = 1.987$ cal/mol · K; a_1 and a_2 are different water activities at their respective temperatures T_1 and T_2. ΔH_s refers to that amount of energy above and beyond that associated with the latent heat of vaporization. The average heat of sorption or the energy of binding of water (also referred to as E_b) can be determined by plotting the natural logarithm of a_w versus $1/T$ at different moisture contents. The resulting slope of the straight line is $-\Delta H_s/R$ (Fig. 6). Iglesias and Chirife (36) tabulated values for E_b for different food systems at different temperature ranges. Several techniques of measurement of water activity have been sited (37–41). Probably the most common approach to estimating the *amount* of bound water is through the Brunauer–Emmet–Teller (BET) isotherm or the "BET monolayer value" as determined from the following equation: $a/m(1 - \text{a}) = [1/m_1c] + [(c - 1)/(m_1c)]\text{a}$, where a = water activity, m = water content (g water/g solid), m_1 = monolayer value, and c = constant.

Information on sorption behavior is important in concentration and dehydration processes, particularly toward the end of a dehydration process where the total heat of vaporization (ΔH) plus heat of sorption are of concern. Because the heat of sorption depends on the partial pressure of water over the food and on the energy of water binding in the food, knowledge of these values is helpful in the design of these processes in calculating the amount of energy required for proper drying. Furthermore, since water activity affects stability, it must be brought to an appropriate level at the end of drying that is suitable for long-term storage. Different quality factors and microbiological growth all have different sensitivities to different water activities. Knowledge of the relative reaction rates in combination with heats of sorption and rates of moisture diffusivity in different types of food materials will provide a food technologist with the tools to design the most efficient process by which to remove water from a food in order for it to maximize drying and optimize quality with extended shelf life. It can be seen from Figure 7 that, depending on the type of food system, different levels of water activity will need to be achieved to best reduce degradation reactions of certain types. In general, for most foods, microbiological activity can be curtailed providing a water activity below 0.91 for bacteria and below 0.80 for most molds, although some xerophilic fungi have been found to grow as low as 0.65. On the other hand, oxidation-type reactions are minimal around 0.3 to 0.4 a_w, but nonenzymatic browning reactions are maximum around 0.4 to 0.60 a_w. It is clear that processing condition designs are highly dependent on the individual system under consideration. The importance of hysteresis should also be mentioned. By this it is meant that the upward (adsorption side) of the curve has a different path than the desorption side of the isotherm (Fig. 5). This in combination with the effect of temperature on hysteresis is also important in design and operating and controlling drying processes and reverse osmosis.

Several types of deleterious changes may take place during dehydration, including nonenzymatic browning, lipid oxidation, vitamin degradation, pigment loss, reduced solubility and textural changes as well as flavor and aroma volatilization. Through an understanding of the influences of water activity on the rates of degradation of various food components, food technologists have the ability to design and regulate the processes during the different stages of drying to best maintain the integrity of the natural components of the system. For instance, it has been shown by several researchers that browning reactions follow zero-order kinetics after an induction period (42–44), and it is most severe toward the end of the drying period at low moisture where less evaporative cooling takes place (42). Franzen et al. (45) developed a model for nonenzymatic browning in skim milk within a temperature range of 35 to 130°C as a function of time, temperature, and moisture content. Ascorbic acid, on the other hand has been shown to be particularly sensitive at high moisture contents. To optimize vitamin C retention, a product should be dried at low temperatures during the constant rate period of drying, followed by increasing the temperature as the moisture content decreases (46). Rates of ascorbic acid degradation during drying have been reported in the literature (47–49). Different methods of de-

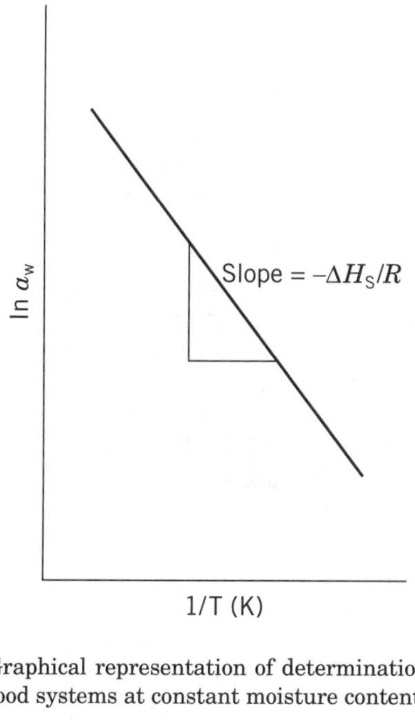

Figure 6. Graphical representation of determination of heat of sorption in food systems at constant moisture content.

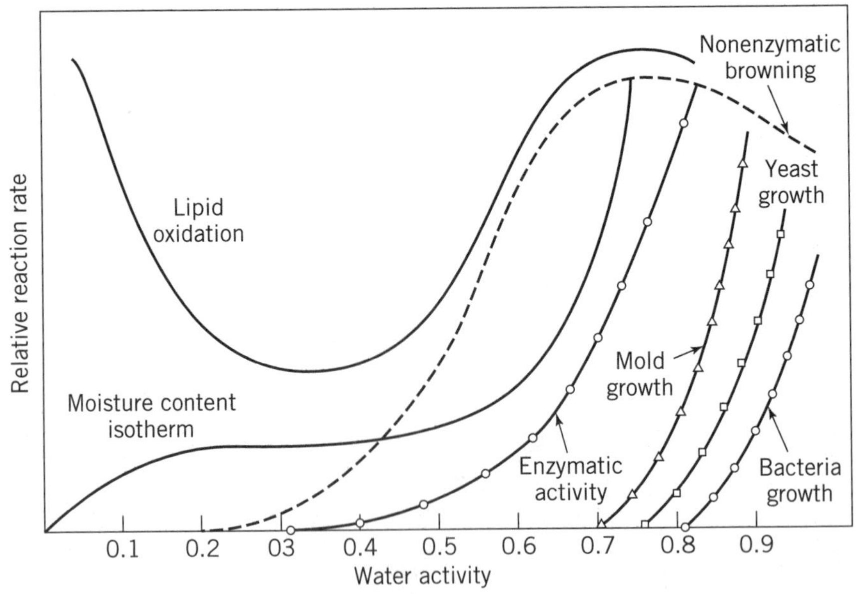

Figure 7. Relative rates of typical deteriorative reactions in foods as a function of water activity. *Source:* Ref. 99.

hydration and their major characteristics have been reviewed by Okos et al. (50).

EMERGING FOOD PROCESSING TECHNOLOGIES AND THEIR IMPLICATIONS ON QUALITY

High Pressure

As can be ascertained from the history of time in the developments made within the area of food processing and preservation and the gradual rise in concern over the nutritional benefits resulting from these processes, most advancement has occurred within the past half century. With the ever-increasing issues of a faster pace of life and the desire for more economical and better quality products for consumption, there has been greater interest in new processes to improve efficiencies. A few new technologies are beginning to emerge in the food industry that are worth mentioning. One area of recent interest is in using ultrahigh hydrostatic pressure (UHP) processing of foods. According to the Le Chatelier principle, any reaction, conformational change, or phase change that is accompanied by a decrease in volume will be favored at high pressure, while reactions involving an increase in volume will be inhibited (51). High-pressure treatments are independent of product size and geometry, and their effect is uniform and instantaneous, unlike conventional heat-transfer systems (52–54). It is somewhat difficult to predict, however, exactly the effects of increasing pressure on food systems due to their complexity. Pressure is applied via a pressure-transferring medium such as water or other fluid and can be effective at ambient temperature, thereby decreasing the amount of thermal energy needed during conventional processing (55,56). The following are considered the major effects of high pressure on food systems: (*1*) most important, the inactivation of microorganisms (57); (*2*) modification of biopolymers such as protein denaturation (58), enzyme inactivation or activation (59), or gel formation (60); (*3*) quality retention such as flavor and color (60,61); and (*4*) product functionality (eg, density changes, freezing and melting temperatures, or texture attributes (62,63). There have been reports that lipid oxidation may increase with application of high pressure in codfish and pork (64,65) but browning may decrease at elevated pressures (66). Due to the lower temperatures required for UHP treatment of foods, nutritional quality is subject to less degradation. UHP keeps covalent bond intact and affects only noncovalent ones, resulting in improvement in overall quality characteristics (61). For instance, Horie et al. (67) reported 95% retention of vitamin C in strawberry jam after UHP processing.

In high-pressure processing, foods are generally subjected to pressures in a range of 100 to 600 MPa at around room temperature. In April 1990, a Japanese company, Meidi-ya Food, introduced the first UHP-processed food, a high-acid fruit jam, and by May 1991 a variety of products including yogurts, fruit jellies, and sauces were UHP processed. The Japanese companies Pokka and Wakayama set up the first semicontinuous UHP processes for bulk treatment of citrus juices at a rate of 4000 to 6000 L/h (68). At present, however, there are no companies in the United States using this UHP process in food manufacture. The key to the success of this particular technology will be the inactivation of microorganisms with the simultaneous retention of nutritive and organoleptic characteristics of a given product. The attainment of reliable and consistent data on the destruction of microorganisms has yet to be accomplished. Great strides have been made by investigators; some have reported first-order kinetics on destruction of bacteria and yeast as a function of pressure (55,69–71). Some researchers, however, have reported a possible two-phase inactivation with the first population inactivated rapidly and the second found to be more resistant (72). It is important to determine inactivation kinetics as a function of pressure, temperature, and medium composition (73). There are many reports on the effect of UHP on reduction of microorganisms but few authors report kinetics, which as previously mentioned is essential to the prediction of destruction of pathogenic bacteria and spores. Those authors who have determined first-order kinetics have attempted to apply the traditional concepts of *D*- and *z*-values; however, other authors have shown evidence that death is not first order, indicating the invalidity of *D*- and *z*-values. As pointed out by authors Villota and Hawkes (5), a major drawback in thermal processing is the incomplete reporting of necessary conditions when collecting kinetic data. In the case of UHP processing, this is even more crucial with added variables, such as rate of pressure increase and decrease, the come-up time for the pressure, the actual isostatic pressure, and the initial population concentration. Their pressure resistance depends on the type of microorganism and suspension media composition, temperature of applied pressure, gas solubility, ionic strength, and pH. Recovery after release of pressure treatment is also an important consideration for proper assessment of the treatment.

Ohmic Heating/Pulsed Electric Field

Other areas of food processing that have received a great deal of attention include ohmic heating and pulsed electric field (PEF). Ohmic heating was one of the earliest forms of electricity applied to food pasteurization (74). Its mode of operation is by direct passage of electric current through the food product, which generates heat as a result of electrical resistance. Ohmic heating reduces the time of heating as compared with conventional heat transfer by convection or conduction. Its advantages are (*1*) rapid and uniform heating, (*2*) less thermal damage to product, (*3*) decreased operational costs, and (*4*) absence of hot surfaces and therefore reduced fouling, as occurs in UHT processing (75). Ohmic heating is of particular interest to aseptic processing, keeping in mind factors such as particle size, particle density, carrier viscosity and composition, and electrical conductivity (76). Although this technology was introduced in the early twentieth century, with the pasteurization of milk pumped between parallel plates with different voltages, the technology disappeared due to the lack of suitable electrode materials (ie, breakdown of carbon electrodes resulting in contamination of product) and adequate control systems. With the development of new electrode materials such as stainless steel, there has been renewed interest (77).

One of the earliest systems using electrical discharge for inactivation of microorganisms was the Electropure process for milk in the 1920s. This process involved passing an electric current through carbon electrodes and heating the milk to 70°C to inactivate *Mycobacterium tuberculosis* and *Escherichia coli*. An electrohydraulic treatment was introduced in the 1950s to inactivate microorganisms suspended in liquid foods. By increasing the electric field intensity, the frequency of the number of pulses and their duration, greater inactivation of microorganisms can be achieved (78–81). This, again, depends on other factors such as treatment temperature, pH, ionic strength, and conductivity of the fluid medium (ie, high-lipid-containing fluids are less conductive than aqueous media). Vega-Mercado et al. (82) have reviewed some of the most recent advances in the area of high-intensity PEFs. It is clear that this is still a developing technology but warrants mentioning as a new technology of the future in food processing. Nevertheless, a great deal of work needs to be accomplished in the area of determining the kinetics of destruction of microorganisms as a function of PEF; determination of uniform delivery of treatment; the impact of temperature, pH, moisture, and lipid content on the process; and the influence of food additives. Also, little is known about the influence of PEF on vitamin retention. It has been suggested that this methodology will have most of its potential in combination with other processing techniques.

REACTION KINETICS IN FOOD SYSTEMS

Advances in more efficient and versatile methods of food processing and preservation have been occurring exponentially over the past few decades in order to meet the continually increasing population and consumer's demands for quality foods, with particular focus on their nutritional aspects. Not only does quality of processed foods depend on the initial integrity of the raw materials, but also on the changes occurring during processing and subsequent storage. Emphasis is growing in the area of nutraceuticals and fortified high-energy foods (83–85). Not only nutritional quality is important to the food processor, however, but also the general appearance of the food, its flavor, color, and texture, many factors of which are highly dependent on the target consumer groups of interest, which are based on different cultural, geographical, and sociological backgrounds. It is, therefore, of critical importance to the food industry to minimize losses of quality in food products during processing, as well as subsequent storage. It is through the development of mathematical models to predict behavior of food components under a set of conditions and optimization of processes for maximum product quality that continued advancement can be achieved in processing techniques. To obtain these goals, extensive information is needed on the rates of destruction of quality parameters and their dependence on variables such as temperature, pH, light, oxygen, and moisture content. A food engineer can then develop new processing techniques to achieve optimum product quality based on an understanding of reaction rates and mechanisms of destruction of individual quality factors combined with the knowledge of the application of kinetics. There are three main areas of concern when dealing with reaction kinetics: (*1*) the stoichiometry, (*2*) the order and rate of reaction, and (*3*) the mechanism. For simple reactions, the stoichiometry is probably the first consideration. Once this is clarified or elucidated, the mechanisms involved in the reaction can be determined. It should be mentioned that based on actual kinetic data, our idea of the stoichiometry may change. In highly complex reactions, as is the case of many reactions occurring in food systems, a great deal of overlap exists among the three aforementioned areas. Thus, it is of critical importance to take a close and analytical look at the overall system to be able to characterize reaction pathways.

BASIC PRINCIPLES OF KINETICS

Order of Reaction

The first step in developing the basis for the kinetics of degradation of a particular nutrient is the determination of the order of the reaction. Understanding of the mechanisms involved in a reaction is important to properly obtain and report meaningful kinetic information, select reaction conditions leading to a desired end product, and/or minimize the appearance of undesirable compounds. Unfortunately, very seldom has effort been given to clearly understand the mechanisms involved in the reaction in complex systems, as in the cases with food and biological materials. Most information available has been oversimplified. In fact, most investigators have often tried to adapt fairly simple zero- or first-order reaction kinetics to complex situations without trying to understand the actual pathways involved. Although, from a practical point of view it is clear that simplifications may be taken, applicability of the information may be restricted only to the conditions encompassed by the experimental design, and thus, one may incorrectly predict trends by directly using reported information.

The reaction pathway, also called reaction mechanism, may be determined through proper experimentation. A chemical reaction may take place in a single step, as in the case of elementary reactions, or in a sequence of steps, as would be the case of most reactions occurring in food systems. Conditions such as temperature, oxygen availability, pressure, initial concentration, and the overall composition of the system may affect the mechanism of the reaction. For instance, the degradation of folic acid and ascorbic acid can be affected by the presence of oxygen, resulting in modification of the reaction pathway and thus, the type of breakdown products. Moreover, the rate at which these parent compounds disappear may be highly influenced by the presence and concentration of the breakdown products generated. It is true that the level of complexity involved in these reactions may be of such magnitude that a complete understanding of the mechanism of deterioration cannot always be easily determined or identified or, even more, may hinder the development of simple techniques to rapidly evaluate the stability of a given system. Nevertheless, it should be stressed that more reliable information is obtained when understanding of the reaction pathways is achieved. Reaction pathways for vitamin and pigment degradation kinetics have been reviewed (5).

The most common approach for the determination of the reaction order for a simple reaction, taking into consideration its initial rate, is as follows: $-dC/dt = k \cdot C^n$, where C = concentration, k = reaction rate constant, n = order of the reaction, and t = time. After taking the natural logarithm on both sides of the equation, a plot of the $\ln(-dC/dt)$ versus $\ln C$ will give a straight line, whose slope corresponds to the reaction order (n), as shown in Figure 8. Although the intercept should correspond to the reaction rate constant (k), it is normally considered that, for the sake of accuracy, this would not be the preferred approach for its estimation. Rather, once the reaction order has been determined, the rate constant can be calculated by applying the corresponding equation for that reaction order. Other methods can also be used to determine the order of the reaction with respect to the reactants and products involved in the reaction, depending on the stoichiometry (5).

Reaction Rate

A common approach in reporting reaction rates in food systems is as the change in concentration of a reactant as a function of time. The reaction rate thus provides a measurement of the reactivity and stability of a given component in a particular system. A number of factors that have been observed to influence the reaction rate include (*1*) concentration of reactants, products, and catalysts; (*2*) environmental factors such as temperature, pressure, and oxygen availability; (*3*) wavelength and intensity of light; and (*4*) physicochemical properties such as viscosity, pH, ionic strength, and conductivity. Depending on the type of reaction and the components, other factors will also be influential in controlling reaction kinetics.

Although traditionally one can apply reaction kinetics to monitor chemical changes occurring in a system, other physicochemical changes may also be described using a kinetic approach. For instance, textural and color changes occurring in food systems can be described using reaction rates. It is obvious that the numbers obtained represent the final effect caused by other complex reaction mechanisms leading to an overall result. For instance, color changes in a product containing carotenoids, may be an indication of the stability of the system, and in particular, stability of the carotenoids as related to environmental conditions. Another example is the textural changes in starch-based systems as a function of time, which may be the result of starch retrogradation mechanisms as well as lipid–amylose interactions as influenced by environmental conditions. Since reactions in food systems are normally complex and a combination of several elementary processes, additional basic information may be necessary to postulate reaction rate expressions. Identification of intermediates and previous knowledge of rate equations to fit data for other systems may provide assistance in properly characterizing a given reaction. The most commonly found reactions in food and biological systems are the zero-, first-, and second-order reactions.

Zero-Order Reactions. In zero-order reactions the rate is independent of the concentration. This may occur in two different situations: (*1*) when intrinsically the reaction rate is independent of the concentration of reactants and (*2*) when the concentration of the reacting compound is so large that the overall reaction rate appears to be independent of its concentration. Many catalyzed reactions fall in the category of zero-order reactions with respect to the reactants. On the other hand, the reaction rate may depend on the catalyst concentration or other factors unrelated to the concentration of the compound under investigation. Thus for a zero-order reaction at constant density, the overall expression would be as follows: $-dC/dt = k_0$, where C = concentration, k_0 = the zero-order reaction rate constant, and t = time. After integrating, this equation be-

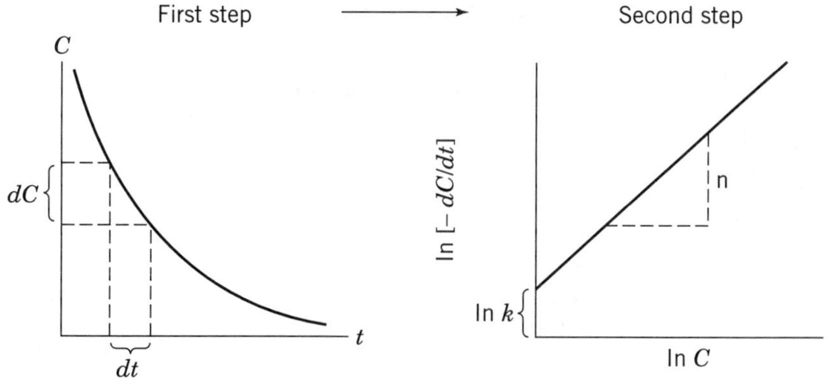

Figure 8. Graphical representation for determining reaction order (n) of a reaction.

comes $C_0 - C = k_0 t$, where C = concentration at time t, and C_0 = initial concentration. According to this mathematical expression, a distinguishing feature for this type of reaction is a linear decrease in concentration as a function of time as illustrated in Figure 9. Typical reactions that have been represented by zero-order reactions include some of the autooxidation reactions. It is clear that zero-order reactions do not appear to occur as frequently in food systems as other reaction orders. In most cases, it is evident that the most common situation for this type of reaction is when the concentration of the reactants is so large that the system appears to be independent of concentration.

First-Order Reactions. A large number of reactions occurring in food systems appears to follow a first-order reaction. A mathematical expression for this behavior would be as follows: $-dC/dt = k_1 C$, where k_1 = first-order reaction rate constant. By integration, this equation becomes $\ln[C/C_0] = k_1 t$. Thus, according to this mathematical expression ln (C) versus time will be a linear function where the slope corresponds to $-k$ as shown in Figure 10. The term half-life ($t_{1/2}$) of a reactant is commonly used and may be described as $t_{1/2} = \ln 2/k_1$. These mathematical expressions clearly indicate that the half-life and the reaction rate for a true first-order reaction are independent of the initial concentration. Although in a number of systems this may be the case, often formulated products will not follow true first-order reaction kinetics, but rather a *pseudo-first-order* reaction. In fact, in formulated systems the presence of breakdown products may strongly influence the order of the reaction. However, for only a given value of initial concentration, the reaction may follow apparent first-order kinetics. To determine if a given reaction does indeed follow a pseudo-first-order kinetics, conditions for the kinetic study can be chosen to follow the technique of *flooding*. Through this approach, all but one of the concentrations are set sufficiently high that, compared with the one reagent present at lower concentration, the others are effectively constant during the time of the experiment. Since only one of the concentrations changes appreciably during the run, the effective kinetic order is reduced to the reac-

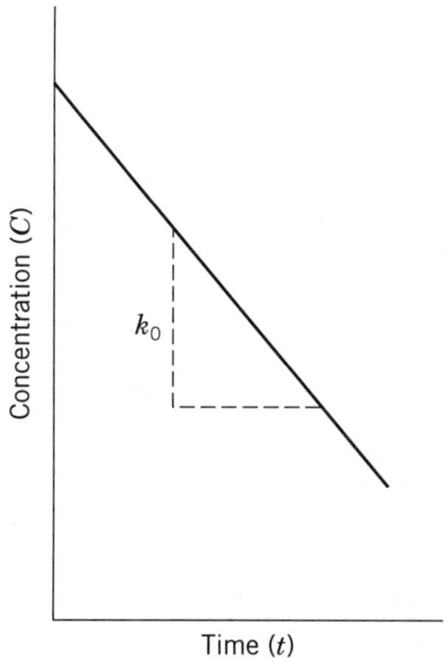

Figure 9. Graphical representation for the determination of a zero-order rate constant (k_0).

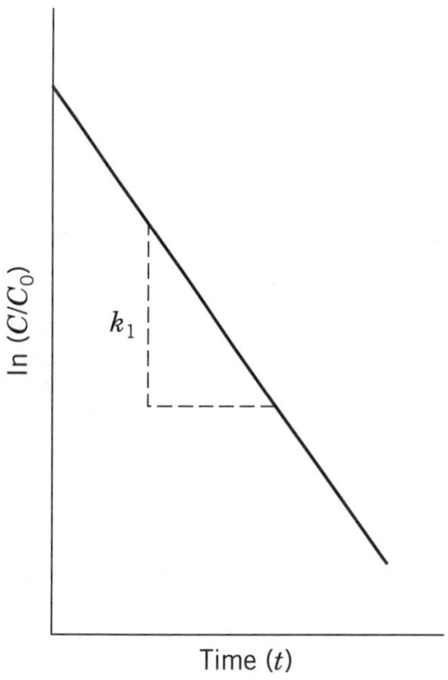

Figure 10. Graphical representation for the determination of a first-order rate constant (k_1).

tion order with respect to that one substance. If the order of the reaction is determined to be one, the reaction is said to follow a pseudo-first-order reaction. The degradation of ascorbic acid, for instance, has been primarily found to follow first-order kinetics in food systems. On the contrary, degradation of ascorbic acid in model systems has frequently been found to follow pseudo-first-order kinetics. It appears that the presence of breakdown products modifies the kinetics of deterioration of ascorbic acid, and, thus, its initial concentration will influence its rate of degradation. These factors, of course, serve to further complicate the prediction of nutrient retention.

Second-Order Reactions. Two types of second-order reaction kinetics are of importance. Type I, A + A → P, where A is a reactant and P is a product, may be mathematically described as $-dC_A/dt = k_2 \cdot C_A^2$. Type II reaction, A + B → P, where A and B are the reactants and P the product, may be mathematically described as $-dC_A/dt = k_2C_AC_B$, where C_A = concentration of reactant species (A) at time (t), C_B = concentration of reactant species (B) at time (t) and k_2 = second-order reaction rate constant. For Type I, the integrated kinetic expression yields: $[1/C_A] - [1/C_{A_0}] = k_2t$, which in terms of the half-life becomes $t_{1/2} = 1/k_2 \cdot C_{A_0}$. For Type II, the integrated form yields: $k_2t = [1/(C_{A_0} - C_{B_0})] \cdot \ln [C_{B_0} \cdot C_A/C_{A_0} \cdot C_B]$, where C_{A_0} and C_{B_0} are the respective initial concentrations and C_A and C_B are the respective concentrations at time (t). It should be stressed, however, that Type II reactions do not have to necessarily follow a second-order reaction. For instance, for the particular case where component A is present in large amounts as compared with component B, the reaction may follow first-order kinetics with respect to B. A typical plot of second-order kinetics is presented in Figure 11.

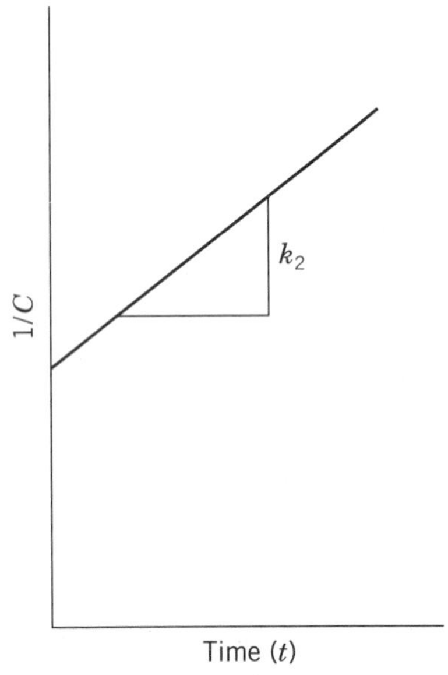

Figure 11. Graphical representation for the determination of a second-order rate constant (k_2) for a Type I reaction.

Nonelementary Reactions

Many different types of reactions fall under the category of nonelementary reactions, which will only be mentioned by name. To characterize the kinetics of nonelementary reactions, one can assume a series of individual elementary reactions taking place. In these reactions, intermediates may not be observed or quantitated, either because they are present in very small amounts or because they are unstable. Such reactions would fall under three main categories: (*1*) consecutive or series reactions, (*2*) reversible or opposing reactions that attain a finite equilibrium, and (*3*) parallel or competitive reactions. The types of intermediates postulated may fall in any one of the following categories, namely, free radicals, ions and polar substances, molecules, and transition complexes (chain reactions and nonchain reactions). It should be mentioned that a classical example of a transition complex-chain reaction is the degradation of β-carotene. It involves an autoxidation reaction involving three main periods: (*1*) induction (formation of free radicals), (*2*) propagation (free-radical chain reactions), and (*3*) termination (formation of nonradical products). A more in-depth analysis and mathematical approach to characterizing this reaction has been reported by several authors (86–89).

Another example of a transition complex reaction is that of a nonchain catalyzed type that may involve the interaction of a substrate with a catalyst to form a complex, followed by its decomposition to form a product. Upon decomposition, the catalyst is then regenerated and is capable of taking part in the reaction once again. An example of this type of transition complex reaction is that of enzyme-catalyzed reactions. It should be mentioned that most of the reactions occurring in biological systems are catalytic in nature. The basic principles for enzyme-catalyzed reactions have been presented by Michaelis–Menten, who proposed the theory of complex formation. Although the general principle of chemical kinetics may apply to enzymatic reactions, the phenomenon of saturation with substrate is unique to enzymatic reactions. In fact, at low substrate concentrations the reaction velocity is proportional to the substrate concentration, and thus the reaction is first order with respect to the substrate. As the substrate concentration increases, the reaction progressively decreases, being no longer proportional to the concentration of the substrate and deviating from any first-order kinetics. The reaction follows zero-order reaction kinetics, due to saturation with the substrate. For the particular case of enzyme kinetics, however, the cases of competitive and noncompetitive inhibition also add complications to the reaction and are thus not as easily defined by simple kinetics. A general description of enzyme-substrate kinetics can be found in most any classical biochemistry textbook (eg, 90, 91).

Temperature Effects

When considering reaction rates, it is clear that these values may be influenced by a large number of parameters, including temperature and pressure. In fact, equilibrium yields, chemical reaction rates, and product distribution may be drastically influenced by temperature. Since chemical reactions are accompanied by heat effects, if these are large enough to cause a significant change in temperature of the reaction mixture, these effects also need to be considered. This would be particularly important in reactor design. The effect of temperature for an elementary process may follow, in most cases, the Arrhenius equation: $k = k_0\exp(-E_aRT)$ where, k_0 = frequency or collision factor, E_a = activation energy (cal/mol), R = gas constant (1.987 cal/mol · K), and T = absolute temperature (K). It is obvious that if the frequency factor and the activation energy could be evaluated from molecular properties of the reactants, it would be possible to estimate the values corresponding to the reaction rate. Unfortunately, our knowledge of kinetics is limited, particularly for complex systems, as would be the case of food systems or products.

It is, however, important to mention the collision theory as an approach to deal with kinetics. In Figure 12, the energy levels involved in a reaction are illustrated. According to the collision theory, upon the collision of reactive molecules, enough energy is generated to provide the necessary activation energy. Such a theory was used as the foundation for the determination of rate expressions based on the frequency of molecular collision required to generate a minimum energy.

Another theory, the activated-complex or transition-state theory, has also been suggested. According to this approach, which still relies on reactions occurring due to collision between reactive molecules, an activated complex is formed from the reactants that eventually decomposes to generate products. The activated complex is in thermodynamic equilibrium with the reactants. Complex decomposition is, then, the limiting step. Regardless of the theories considered, they do not provide the means to rapidly and easily calculate activation energies from simple thermodynamic information. Thus in practical terms, one has to obtain basic kinetic information to be able to determine the effect of temperature as affecting reaction kinetics. Based on the Arrhenius equation, it is clear that if one plots the ln k versus $1/T$, the slope would correspond to the activation energy divided by the gas constant. Moreover, this value by itself will not provide any idea on the reactivity of a given system, only information on temperature dependence of the reaction.

Although the Arrhenius equation is commonly used to describe temperature dependence of the reaction rate in most food systems, deviations may occur as reported by several authors, including Labuza and Riboh (92). In fact, a large number of factors may contribute to deviations.

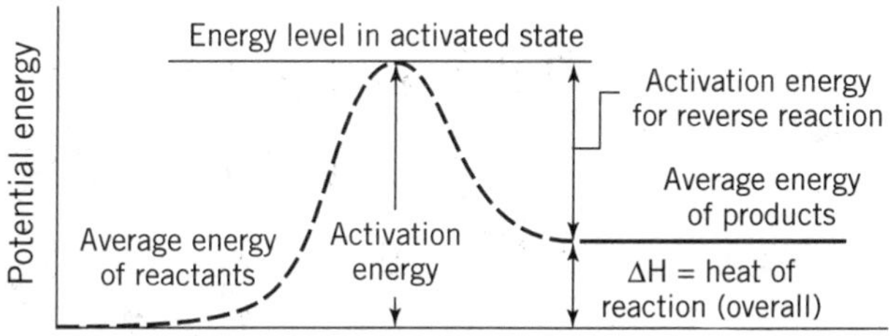

Figure 12. Representation of potential energy levels during the process of a given endothermic reaction.

Changes in reaction mechanisms may occur for a large temperature range. For instance, it is highly possible that mechanisms of deterioration may change at conditions below the freezing point due to a concentration effect. On the other hand, at high temperatures, changes in the physical state of some compounds, including fats and sugars, may occur. Lipids may change from a solid to a liquid state, while sugars may change from an amorphous to a crystalline or to a liquid state. Because of the high complexity of food systems, it is also possible that when various mechanisms of deterioration operate simultaneously, the effect of temperature may alter the rate of one, thus causing inhibition or catalysis in the other mechanisms. Finally, irreversible changes such as starch hydrolysis or protein denaturation may occur due to temperature, thus modifying the reactivity of the system. In fact, although enzyme-catalyzed reactions will have an increasing reaction rate upon an increase in temperature, a decrease will be observed beyond a certain temperature due to enzyme inactivation.

Of particular importance to food processors is the determination of any nutritional and overall quality changes that may occur as a result of processing conditions encountered in operations such as dehydration, sterilization, extrusion, and so on. It is not only important to establish the extent of these undesirable changes, but it is also crucial to know the rate at which these undesirable changes take place. It is the ultimate goal of the food manufacturer to be able to optimize the quality of the final product while still maintaining an accurate perspective of the economics of this approach. Although a variety of methodologies have been proposed for the optimization of nutrient or quality retention during processing, only a few authors have actually tested the feasibility of these methods. Due to the complex nature of food materials, complete kinetics models are not easily attainable, resulting in one of the major obstacles in optimization in the food industry. Several researchers, however, have successfully applied optimization techniques to different types of food processes such as thiamin retention during sterilization (93) and ascorbic acid retention during air drying (94) using Pontryagin's maximum principle. Mishkin et al. (95) applied the complex method for optimization of ascorbic acid in air drying and minimizing browning in the dehydration of white potatoes. Saguy (96) published a practical text on computer-aided techniques for food technologists covering model building with applications and implementation of kinetics, simulation, heat transfer, and so on. With the fast-paced advancement in computer software, computer prototyping based on mathematical models rather than actual physical models has greatly aided product and process design. Datta (97) has reviewed some different commercially available computer-aided engineering (CAE) software programs and their sources with examples of computational fluid dynamics (CFD) and heat transfer for processes such as canning of liquid and solid foods, extrusion, and continuous sterilization of liquids.

FINAL REMARKS

It is important for the food technologist to attain a complete understanding of the principal processing variables involved in a general wide array of types of food processing techniques combined with a general knowledge of the kinetics of degradation of food quality characteristics. Although not discussed due to time constraints, the effect of storage conditions, whether ambient, refrigerated, frozen, or under controlled atmosphere, depending on its mode of preprocessing, is also an area of importance in maintaining quality of processed foods. Although generally many of the deleterious reactions occurring in foods are significantly slowed down at reduced temperatures, mechanisms of reaction can be altered such that prediction and optimization of these conditions becomes even more of a challenge. It should be kept in mind that some of the new emerging technologies in food processing will require more basic studies in order to predict retention of vitamins and organoleptic properties. Understanding of the mechanisms involved in reactions leading to the destruction of a compound or quality character should facilitate development of kinetics information with wider applicability. This will facilitate the optimization process of process and storage conditions and provide guidelines to determine formulation or fortification protocols of products leading to a higher nutritional value. It will also facilitate and encourage the development of new technologies for future food processing systems needed not only to feed the ever-expanding world population but also to ensure a high-quality nutritious food supply (98,99).

BIBLIOGRAPHY

1. *Merriam-Websters Collegiate Dictionary*, 10th ed., Merriam-Webster, Inc., Springfield, Mass., 1999.
2. D. R. Thompson and J. Norwig, "Microbial Population, Enzyme and Protein Changes During Processing," in M. R. Okos, ed., *Physical and Chemical Properties of Food*, ASAE, St. Joseph, Mich., 1986, pp. 202–265.
3. D. Lund, "Kinetics of Physical Changes in Foods," in M. R. Okos, ed., *Physical and Chemical Properties of Food*, ASAE, St. Joseph, Mich., 1986, pp. 367–381.
4. L. A. Wilson, "Kinetics of Flavor Changes in Foods," in M. R. Okos, ed., *Physical and Chemical Properties of Food*, ASAE, St. Joseph, Mich., 1986, pp. 382–407.
5. R. Villota and J. G. Hawkes, "Reaction Kinetics in Food Systems," in D. R. Heldman and D. B. Lund, eds., *Handbook of Food Engineering*, Marcel Dekker, New York, 1992, pp. 39–144.
6. R. Villota and J. G. Hawkes, "Kinetics of Nutrients and Organoleptic Changes in Foods During Processing," in M. R. Okos, ed., *Physical and Chemical Properties of Food*, ASAE, St. Joseph, Mich., 1986, pp. 266–366.
7. L. B. Jensen, *Microbiology of Meats*, 2nd ed., Garrard Press, Champaign, Ill., 1945.
8. C. R. Stumbo, *Thermobacteriology in Food Processing*, 2nd ed., Academic Press, New York, 1973.
9. A. López, ed., *A Complete Course in Canning, Book 1—Basic Information on Canning*, 11th ed., The Canning Trade, Inc., Baltimore, Md., 1981.
10. A. Teixeira, "Thermal Process Calculations," in D. R. Heldman and D. B. Lund, eds., *Handbook of Food Engineering*, Marcel Dekker, New York, 1992, pp. 563–619.
11. M. A. Joslyn and J. L. Heid, *Food Processing Operation: Their Management, Machines, Materials and Methods*, Vol. II, AVI, Westport, Conn., 1963.
12. S. K. Sastry, "Mathematical Evaluation of Process Schedules for Aseptic Processing of Low-Acid Foods Containing Discrete Particles," *J. Food Sci.* **51**, 1323–1328 (1986).
13. S. Palaniappan and C. E. Sizer, "Aseptic Process Validated for Foods Containing Particulates," *Food Technol.* **51**(8), 60–68 (1997).
14. M. Karel, "Radiation Preservation of Foods," in M. Karel, O. R. Fennema, and D. B. Lund, eds., *Principles of Food Science. Part II. Physical Principles of Food Preservation*, Marcel Dekker, New York, 1975, pp. 93–130.
15. D. G. Olson, "Irradiation of Food," *Food Technol.* **52**(1), 56–62 (1998).
16. G. H. Pauli and L. M. Tarantino, "FDA Regulatory Aspects of Food Irradiation," *J. Food Protect.* **58**, 209–212 (1995).
17. I. B. Hashim, A. V. A. Ressureccion, and K. H. MacWatters, "Consumer Attitudes Toward Irradiated Poultry," *Food Technol.* **50**(3), 77–80 (1996).
18. A. V. A. Resurreccion and F. C. F. Galvez, "Will Consumers Buy Irradiated Beef?," *Food Technol.* **53**(3), 52–55 (1999).
19. J. L. Lusk, J. A. Fox, and C. L. McIlvain, "Consumer Acceptance of Irradiated Meat," *Food Technol.* **53**(3), 56–59 (1999).
20. J. Farkas, "Microbiological Safety of Irradiated Foods," *Rev. Int. J. Food Microbiol.* **9**, 1–15 (1989).
21. J. F. Diehl, ed., "Nutritional Adequacy of Irradiated Foods," in *Safety of Irradiated Foods*, 2nd ed., Marcel Dekker, New York, 1995.
22. *Safety and Nutritional Adequacy of Irradiated Food*, World Health Organization, Geneva, Switzerland, 1994.
23. D. W. Thayer, "Food Irradiation: Benefits and Concerns," *J. Food Quality* **13**, 147–169 (1990).
24. J. A. Lynch, H. J. H. MacFie, and G. C. Mead, "Effect of Irradiation and Packaging Type on Sensory Quality of Chilled-stored Turkey Breast Fillets," *J. Food Sci. Technol.* **26**, 653–668 (1991).
25. N. Lebepe et al., "Changes in Microflora and Other Characteristics of Vacuum-packaged Pork Loins Irradiated at 3.0 kGy," *J. Food Sci.* **55**, 918–924 (1990).
26. J. G. Niemand, H. J. Van der Linde, and W. H. Holzapfel, "Shelf-life Extension of Minced Beef Through Combined Treatments Involving Radurization," *J. Food Protect.* **46**, 791–796 (1983).
27. D. H. Kropf et al., "Palatability, Color, and Shelf-life of Low-dose Irradiated Beef," in *Proc. of 1995 Int. Congr. of Meat Science and Technology*, San Antonio, Tex., 1995.
28. S. E. Luchsinger et al., "Sensory Analysis and Consumer Acceptance of Irradiated Boneless Pork Chop," *J. Food Sci.* **61**, 1261–1266 (1996).
29. M. Lee et al., "Irradiation and Packaging of Fresh Meat and Poultry," *J. Food Protect.* **59**, 62–72 (1996).
30. S. M. Glidewell et al., "Detection of Irradiated Food: A Review," *J. Sci. Food Agric.* **61**, 281–300 (1993).
31. A. V. J. Crone, J. T. G. Hamilton, and M. H. Stevenson, "Effects of Storage and Cooking on the Dose Response of 2-Dodecylcyclobutanone, a Potential Marker for Irradiated Chicken," *J. Sci. Food Agric.* **58**, 249–252 (1992).
32. J. D. Monk, L. R. Beuchat, and M. P. Doyle, "Irradiation Inactivation of Food-borne Microorganisms," *J. Food Protect.* **58**, 197–208 (1995).

33. A. D. Lambert, J. P. Smith, and K. L. Dodds, "Physical, Chemical and Sensory Changes in Irradiated Fresh Pork Packaged in Modified Atmosphere," *J. Food Sci.* **57**, 1294–1299 (1992).
34. W. J. Scott, "Water Relations of Food Spoilage Microorganisms," *Adv. Food Res.* **7**, 83–127 (1957).
35. J. E. Kinsella and P. F. Fox, "Water Sorption by Proteins: Milk and Whey Proteins," *Crit. Rev. Food Sci. Nutr.* **24**, 91– (1986).
36. H. A. Iglesias and J. Chirife, *Handbook of Food Isotherms*, Academic Press, New York, 1982.
37. J. A. Troller and J. H. B. Christian, *Microbial Survival, Water Activity and Food*, Academic Press, New York, 1978.
38. M. Karel, "Water Activity and Food Preservation," in M. Karel, O. R. Fennema, and D. B. Lund, eds., *Principles of Food Science. Part II. Physical Principles of Food Preservation*, Marcel Dekker, New York, 1975, pp. 237–263.
39. T. P. Labuza, *Moisture Sorptions: Practical Aspects of Isotherm Measurement and Use*, American Association of Cereal Chemists, St. Paul, Minn., 1984.
40. S. S. H. Rizvi, "Thermodynamic Properties of Foods in Dehydration," in M. A. Rao and S. Rizvi, eds., *Engineering Properties of Foods*, 2nd ed., Marcel Dekker, New York, 1995, p. 123.
41. M. S. Rahman, *Food Properties Handbook*, CRC Press, Boca Raton, Fla., 1995.
42. C. E. Hendel, V. G. Silveira, and W. O. Harrington, "Rates of Nonenzymatic Browning of White Potato During Dehydration," *Food Technol.* **9**(9), 433–438 (1955).
43. C. Petriella et al., "Kinetics of Deteriorative Reactions in Model Food Systems of High Water Activity: Color Changes Due to Nonenzymatic Browning," *J. Food Sci.* **50**, 622–626 (1985).
44. R. K. Singh, *Kinetics and Computer Simulation of Storage Stability in Intermediate Moisture Foods*, Ph.D. Thesis, University of Wisconsin, Madison, Wis., 1983.
45. K. Franzen, R. K. Singh, and M. R. Okos, "Kinetics of Nonenzymatic Browning in Dried Skim Milk", *J. Food Eng.* **11**, 225–239 (1990).
46. M. Mishkin, I. Saguy, and M. Karel, "Optimization of Nutrient Retention During Processing: Ascorbic Acid in Potato Dehydration," *J. Food Sci.* **49**, 1262–1266 (1984).
47. R. Villota and M. Karel, "Prediction of Ascorbic Acid Retention During Drying. I. Moisture and Temperature Distribution in a Model System," *J. Food Process. Preserv.* **4**, 111–134 (1980).
48. R. Villota and M. Karel, "Prediction of Ascorbic Acid Retention During Drying. II. Simulation of Retention in a Model System," *J. Food Process. Preserv.* **4**, 141–159 (1980).
49. S. G. Haralampu and M. Karel, "Kinetic Models for Moisture Dependence of Ascorbic Acid and β-Carotene Degradation in Dehydrated Sweet Potato," *J. Food Sci.* **48**, 1872– (1983).
50. M. R. Okos et al., "Food Dehydration," in D. R. Heldman and D. B. Lund, eds., *Handbook of Food Engineering*, Marcel Dekker, New York, 1992, pp. 437–462.
51. A. Williams, "New Technologies in Food Preservation and Processing: Part II," *Nutr. Food Sci.* **1**, 16 (1994).
52. G. D. Alemán et al., "Pulsed Ultra High Pressure Treatments for Pasteurization of Pineapple Juice," *J. Food Sci.* **61**, 388 (1996).
53. D. Knorr, "Effects of High-Hydrostatic-Pressure Process on Food Safety and Quality," *Food Technol.* **47**(6), 156–161 (1993).
54. F. Zimmerman and C. Bergman, "Isostatic Pressure Equipment for Food Preservation," *Food Technol.* **47**(6), 162–163 (1993).
55. C. Hashizume, K. Kimura, and R. Hayashi, "Kinetic Analysis of Yeast Inactivation by High Pressure Treatment at Low Temperatures," *Biosci. Biotechnol. Biochem.* **59**, 1455 (1995).
56. E. Palou et al., "Combined Effect of High Hydrostatic Pressure and Water Activity on *Zygosaccharomyces bailii* Inhibition," *Lett. Appl. Microbiol.* **24**, 417 (1997).
57. D. G. Hoover et al., "Biological Effects of High Hydrostatic Pressure on Food Microorganisms," *Food Technol.* **43**(3), 99–107 (1989).
58. K. Heremans, "High Pressure Effects on Proteins and Other Biomolecules," *Ann. Rev. Bioeng.* **11**(1), 1–21 (1982).
59. E. Morild, "The Theory of Pressure Effects on Enzymes," in C. B. Anfinsen, J. T. Edsall, and F. M. Richards, eds., *Advances in Protein Chemistry*, Vol. 34, Academic Press, London, United Kingdom, 1981, pp. 93–166.
60. J.-C. Cheftel, "Applications des Hautes Pressions en Technologie Alimentaire," *Actualité des Industries Alimentaires et Agro-Alimentaires* **108**(3), 141–153 (1991).
61. R. Hayashi, "Application of High Pressure to Food Processing and Preservation: Philosophy and Development," in W. E. L. Spiess and H. Schubert, eds., *Engineering and Food*, Vol. 2, Elsevier Applied Science, London, United Kingdom, 1989, pp. 815–826.
62. D. Farr, "High Pressure Technology in the Food Industry," *Trends Food Sci. Technol.* **1**, 14–16 (1990).
63. T. Deuchi and R. Hayashi, "Pressure-Application to Thawing of Frozen Foods and to Food Preservation Under Sub-Zero Temperature," in R. Hayashi, ed., *High Pressure Science for Food*, San-ei Pub. Co., Kyoto, Japan, 1991, pp. 101–110.
64. P. B. Cheah and D. A. Ledward, "High Pressure Effects on Lipid Oxidation," *J. Am. Oil Chem. Soc.* **72**, 1059 (1995).
65. P. B. Cheah and D. A. Ledward, "High Pressure Effects on Lipid Oxidation in Minced Pork," *Meat Sci.* **43**, 123 (1996).
66. T. Tamaoka, N. Itoh, and R. Hayashi, "High Pressure Effect on Maillard Reaction," *Agric. Biol. Chem.* **55**, 2071–2074 (1991).
67. Y. Horie et al., "Jam Preparation by Pressurization," *Nippon Nogeikagaku Kaishi* **65**, 975–980 (1991).
68. B. Mertens and G. Deplace, "Engineering Aspects of High Pressure Technology in the Food Industry," *Food Technol.* **47**(6), 164–169 (1993).
69. P. Butz and H. Ludwig, "Pressure Inactivation of Microorganisms at Moderate Temperatures," *Physica* **139** & **140B**, 875 (1986).
70. A. Carlez et al., "High Pressure Inactivation of *Citrobacter freundii, Pseudomonas fluorescens* and *Listeria innocua* in Inoculated Minced Beef Muscle," *Lebensm.-Wiss. Technol.* **26**, 357 (1993).
71. J. Smelt and G. Rijke, "High Pressure Treatment as a Tool for Pasteurization of Foods," in C. Balny et al., eds., *High Pressure and Biotechnology*, Vol. 224, Colloque INSERM, John Libbey Eurotext, Montrouge, France, 1992.
72. J.-C. Cheftel, "High Pressure, Microbial Inactivation and Food Preservation," *Food Sci. Technol. Int.* **1**, 75 (1995).
73. H. Ludwig et al., "Inactivation of Microorganisms by Hydrostatic Pressure," in C. Balny et al., eds., *High Pressure and Biotechnology*, Vol. 224, Colloque INSERM, John Libbey Eurotext, Montrouge, France, 1992.
74. K. Allen, V. Eidman, and J. Kinsey, "An Economic Engineering Study of Ohmic Food Processing," *Food Technol.* **50**(5), 269–273 (1996).
75. E. Scott, "Ohmic Heating Hits Commercial Scale," *Food Technol.* **NZ30**(7), 8 (1995).

76. P. Zoltai and P. Swearingen, "Product Development Considerations for Ohmic Processing," *Food Technol.* **50**(5), 263–266 (1996).
77. S. Palaniappan, S. K. Sastry, and E. R. Richter, "Effects of Electricity on Microorganisms: A Review," *J. Food Proc. Preserv.* **14**, 393 (1990).
78. R. Benz and U. Zimmermann, "Pulse-length Dependence of the Electrical Breakdown in Lipid Bilayer Membranes," *Biochim. Biophys. Acta* **597**, 637 (1980).
79. D. Knorr et al., "Food Application of High Electric Field Pulses," *Trends Food Sci. Technol.* **5**, 71 (1994).
80. T. Y. Tsong, "Review on Electroporation of Cell Membranes and Some Related Phenomena," *Biochim. Bioeng.* **24**, 271 (1990).
81. J. W. Larkin and S. H. Spinak, "Regulatory Aspects of New/Novel Technologies," in D. I. Chandarana, ed., *New Processing Technologies Yearbook*, National Food Processors Association, Washington, D.C., 1996.
82. H. Vega-Mercado et al., "Nonthermal Preservation of Liquid Foods Using Pulsed Electric Fields," in M. S. Rahman, ed., *Handbook of Preservation*, Marcel Dekker, New York, 1999, pp. 487–520.
83. A. E. Sloan, "The New Market: Foods for the Not-So-Healthy," *Food Technol.* **53**(2), 54–60 (1999).
84. M. Molyneau and C. M. Lee, "The U.S. Market for Marine Nutraceutical Products," *Food Technol.* **52**(6), 56–57 (1998).
85. J. Giese, "Vitamin and Mineral Fortification of Foods," *Food Technol.* **49**(5), 109–122 (1995).
86. E. V. Alekseev et al., "Kinetic Principles of the Oxidation of Polyenic Hydrocarbons. Communication 1. Decomposition of β-Carotene in the Presence of Free Radical Initiators," *Bulletin of the Academy of Sciences of the USSR. Division of Chemical Sciences* (translated from *Izvestiya Akademii Nauk SSSR, Seriya Khimicheskaya*), **11**, 2342–2347 (1968).
87. A. B. Gagarina, O. T. Kasaikina, and N. M. Émanuél, "Kinetics of Autooxidation of Polyene Hydrocarbons in Aromatic Solvents," *Doklady Akademii Nauk SSSR* **195**, 387–390 (1970).
88. E. I. Finkel'shtein, É. V. Alekseev, and É. I. Kozlov, "The Kinetics of β-Carotene Solid Film Autooxidation," *Doklady Akademii Nauk SSSR* **208**, 1408–1411 (1973).
89. E. I. Finkel'shtein, É. V. Alekseev, and É. I. Kozlov, "Kinetic Relationships of the Solid State Autooxidation of β-Carotene," *Zhurnal Organicheskoi Khimii* **10**, 1027–1034 (1974).
90. A. L. Lehninger, *Biochemistry*, 2nd ed., Worth, New York, 1975.
91. M. F. Mallette et al., *Introductory Biochemistry*, Williams & Wilkins, Baltimore, Md., 1971.
92. T. P. Labuza and D. Riboh, "Theory and Application of Arrhenius Kinetics to the Prediction of Nutrient Losses in Foods," *Food Technol.* **36**(10), 66–74 (1982).
93. I. Saguy and M. Karel, "Optimal Retort Temperature Profile in Optimizing Thiamine Retention in Conduction-type Heating of Canned Foods," *J. Food Sci.* **44**, 1485 (1979).
94. M. Mishkin, M. Karel, and I. Saguy, "Applications of Optimization of Food Dehydration," *Food Technol.* **36**(7), 101–109 (1982).
95. M. Mishkin, I. Saguy, and M. Karel, "Dynamic Optimization of Dehydration Processes: Minimizing Browning in Dehydrated Potatoes," *J. Food Sci.* **48**, 1617–1621 (1983).
96. I. Saguy, ed., *Computer-Aided Techniques in Food Technology*, Marcel Dekker, New York, 1983.
97. A. K. Datta, "Computer-Aided Engineering in Food Process and Product Design," *Food Technol.* **52**(10), 44–52 (1998).
98. M. Toussaint-Samat, *History of Food* (Anthea Bell, Trans.), Barnes & Noble, New York, 1998.
99. J. M. Flink, "Intermediate Moisture Food Products in the American Marketplace," *J. Food Proc. Preserv.* **1**, 324–339 (1977).

JAMES G. HAWKES
Continental Colloids, Inc.
Chicago, Illinois

FOOD PROCESSING: STANDARD INDUSTRIAL CLASSIFICATION

When one is in a grocery store, one is faced with an incredible selection of processed foods. For food manufacturers, this illustrates only one point: the amount of research and development that precedes their marketing. For regulatory agencies such as the Food and Drug Administration and the Department of Agriculture, this confirms the amount of work needed to assure the safety and wholesomeness of these products. However, for some other federal agencies, this poses an entirely different headache. For example, the Department of Labor (DOL) and the Environmental Protection Agency (EPA) must classify and give a name to each establishment that manufactures each product in order to enforce their legal mandates. To them it is a huge logistic nightmare. For example, the DOL must have access to a record of the number of employees suffering injuries in each type or category of food processing plants. To fill this need, the executive branch of the government has developed the Standard Industrial Classification (SIC) for food processing plants. The *SIC Manual* is one of the most useful federal publications available to both government agencies and the industries. One example of its usefulness is in assisting the allocation of resources. For example, a government agency can plan its budget accordingly by using the SIC to determine the number of food processing plants it has to inspect annually. Another application is the use of SIC to determine the employment status of the food processing profession. Such classification applies only to food processing establishments in the United States and this article discusses the SIC in details.

STANDARD INDUSTRIAL CLASSIFICATION MANUAL

The SIC Manual is issued by the Office of Management and Budget (OMB) of the Executive Office of the President of the United States. The manual divides industries in the United States into major groups. One major group 20 covers food and kindred products. This group includes establishments manufacturing or processing foods and beverages for human consumption, and certain related products, such as manufactured ice, chewing gum, vegetable and animal fats and oils, and prepared feeds for animals and fowls. Products described as dietetic are classified in the same manner as nondietetic products, eg, as candy, canned fruits, and cookies. Chemical sweeteners are classified in Major Group 28.

Major Group 20 is subdivided according to the following criteria (Table 1).

The following discusses each industry under food and kindred products. The abbreviations used are: * = The

Table 1. Standard Industrial Classification (SIC) and Subdivisions for Major Group 20: Food and Kindred Products

Industrial group no.: Industrial group name Industrial no.: Industrial name
201: Meat products
2011: Meat-packing plants
2013: Sausages and other prepared meat products
2015: Poultry slaughtering and processing
202: Dairy products
2021: Creamery butter
2022: Natural, processed, and imitation cheese
2023: Dry, condensed, and evaporated dairy products
2024: Ice cream and frozen desserts
2026: Fluid milk
203: Canned, frozen, and preserved fruits, vegetables, and food specialties
2032: Canned specialties
2033: Canned fruits, vegetables, preserves, jams, and jellies
2034: Dried and dehydrated fruits, vegetables, and soup mixes
2035: Pickled fruits and vegetables, vegetable sauces and seasonings, and salad dressings
2037: Frozen fruits, fruit juices, and vegetables
2038: Frozen specialties, not elsewhere classified
204: Grain mill products
2041: Flour and other grain mill products
2043: Cereal breakfast foods
2044: Rice milling
2045: Prepared flour mixes and doughs
2046: Wet corn milling
2047: Dog and cat food
2048: Prepared feeds and feed ingredients for animals and fowls, except dogs and cats
205: Bakery products
2051: Bread and other bakery products, except cookies and crackers
2052: Cookies and crackers
2053: Frozen bakery products, except bread
206: Sugars and confectionary products
2061: Cane sugar, except refining
2062: Cane sugar refining
2063: Beet sugar
2064: Candy and other confectionery products
2066: Chocolate and cocoa products
2067: Chewing gum
2068: Salted and roasted nuts and seeds
207: Fats and oils
2074: Cottonseed oil mills
2075: Soybean oil mills
2076: Vegetable oil mills, except corn, cottonseed, and soybean
2077: Animal and marine fats and oils
2079: Shortening, table oils, margarine, and other edible fats and oils, not elsewhere classified
208: Beverages
2082: Malt beverages
2083: Malt
2084: Wines, brandy, and brandy spirits
2085: Distilled and blended liquors
2086: Bottled and canned soft drinks and carbonated waters
2087: Flavoring extracts and flavoring syrups, not elsewhere classified
209: Miscellaneous food preparations and kindred products
2091: Canned and cured fish and seafoods
2092: Prepared fresh or frozen fish and seafoods
2095: Roasted coffee
2096: Potato chips, corn chips, and similar snacks
2097: Manufactured ice
2098: Macaroni, spaghetti, vermicelli, and noodles
2099: Food preparations, not elsewhere classified

item has been made in the same establishment as the basic material; ** = The item has been made from purchased materials or materials transferred from another establishment; IGN = Industry Group No; IN = Industry No.

MEAT PRODUCTS (IGN 201)

Meat Packing Plants (IG 2011)

Establishments primarily engaged in the slaughtering, for their own account or on a contract basis for the trade, of cattle, hogs, sheep, lambs, and calves for meat to be sold or to be used on the same premises in canning, cooking, curing, and freezing, and in making sausage, lard, and other products. Also included in this industry are establishments primarily engaged in slaughtering horses for human consumption. Establishments primarily engaged in slaughtering, dressing, and packing poultry, rabbits, and other small game are classified in Industry 2015; and those primarily engaged in slaughtering and processing animals not for human consumption are classified in Industry 2048. Establishments primarily engaged in manufacturing sausages and meat specialties from purchased meats are classified in Industry 2013; and establishments primarily engaged in canning meat for baby food are classified in Industry 2032.

1. Bacon, slab and sliced*
2. Beef*
3. Blood meal
4. Boxed beef*
5. Canned meats, except baby foods and animals feeds*
6. Corned beef*
7. Cured meats*
8. Dried meats*
9. Frankfurters, except poultry*
10. Hams, except poultry*
11. Hides and skins, cured or uncured
12. Horse meat for human consumption*
13. Lamb*
14. Lard*
15. Luncheon meat, except poultry*
16. Meat extracts*
17. Meat packing plants
18. Meat*
19. Mutton*
20. Pork*
21. Sausages*
22. Slaughtering plants: except animals not for human consumption
23. Variety meats edible organs*
24. Veal*

Sausages and Other Prepared Meat Products (IN = 2013)

Establishments primarily engaged in manufacturing sausages, cured meats, smoked meats, canned meats, frozen meats, and other prepared meats and meat specialties, from purchased carcasses and other materials. Prepared meat plants operated by packing houses as separate establishments are also included in this industry. Establishments primarily engaged in canning or otherwise processing poultry, rabbits, and other small game are classified in Industry 2015. Establishments primarily engaged in canning meat for baby food are classified in Industry 2032. Establishments primarily engaged in the cutting up and resale of purchased fresh carcasses, for the trade, (including boxed beef) are classified in Wholesale Trade, Industry 5147.

1. Bacon, slab and sliced**
2. Beef**
3. Bologna**
4. Calf's-foot jelly
5. Canned meats, except baby foods and animal feeds**
6. Corned beef**
7. Corned meats**
8. Cured meats: brined, dried, and salted**
9. Dried meats**
10. Frankfurters, except poultry**
11. Hams, except poultry**
12. Headcheese**
13. Lard**
14. Luncheon meat, except poultry**
15. Meat extracts**
16. Meat products: cooked, cured, frozen, smoked, and spiced**
17. Pastrami**
18. Pigs' feet, cooked and pickled**
19. Pork: pickled, cured, salted, or smoked**
20. Potted meats**
21. Puddings, meat**
22. Sandwich spreads, meat**
23. Sausage casings, collage
24. Sausages**
25. Scrapple**
26. Smoked meats**
27. Spreads, sandwich: meat**
28. Stew, beef and lamb**
29. Tripe**
30. Vienna sausage**

Poultry Slaughtering and Processing (IN 2015)

Establishments primarily engaged in slaughtering, dressing, packing, freezing, and canning poultry, rabbits, and other small game, or in manufacturing products from such meats, for their own account or on a contract basis for the trade. This industry also includes the drying, freezing, and breaking of eggs. Establishments primarily engaged in cleaning, oil treating, packing, and grading of eggs are classified in Wholesale Trade, Industry 5144; and those engaged in the cutting up and resale of purchased fresh carcasses are classified in Wholesale and Retail Trade.

1. Chickens, processed: fresh, frozen, canned, or cooked
2. Chickens: slaughtering and dressing
3. Ducks, processed: fresh, frozen, canned, or cooked
4. Ducks: slaughtering and dressing
5. Egg albumen
6. Egg substitutes made from eggs
7. Eggs: canned, dehydrated, desiccated, frozen, and processed
8. Eggs: drying, freezing, and breaking
9. Frankfurters, poultry
10. Game, small: fresh, frozen, canned, or cooked
11. Game, small: slaughtering and dressing
12. Geese, processed: fresh, frozen, canned, or cooked
13. Geese: slaughtering and dressing
14. Ham, poultry
15. Luncheon meat, poultry
16. Poultry, processed: fresh, frozen, canned, or cooked
17. Poultry: slaughtering and dressing
18. Rabbits, processed: fresh, frozen, canned, or cooked
19. Rabbits, slaughtering and dressing
20. Turkeys, processed: fresh, frozen, canned, or cooked
21. Turkeys: slaughtering and dressing

DAIRY PRODUCTS (IGN 202)

This industry group includes establishments primarily engaged in:

1. Manufacturing creamery butter; natural, processed, and imitation cheese; dry, condensed, and evaporated milk; ice cream and frozen dairy desserts; and special dairy products, such as yogurt and malted milk; and
2. Processing (pasteurizing, homogenizing, vitaminizing, bottling) fluid milk and cream for wholesale or retail distribution.

Independently operated milk-receiving stations primarily engaged in the assembly and reshipment of bulk milk for use in manufacturing or processing plants are classified in Industry 5143.

Creamery Butter (IN 2021)

Establishments primarily engaged in manufacturing creamery butter.

1. Anhydrous butterfat
2. Butter oil
3. Butter powder
4. Butter, creamery and whey
5. Butterfat, anhydrous

Natural, Processed, and Imitation Cheese (IN 2022)

Establishments primarily engaged in manufacturing natural cheese (except cottage cheese), processed cheese, cheese foods, cheese spreads, and cheese analogs (imitations and substitutes). These establishments also produce byproducts, such as raw liquid whey. Establishments primarily engaged in manufacturing cottage cheese are classified in Industry 2026, and those manufacturing cheese-based salad dressings are classified in Industry 2035.

1. Cheese analogs
2. Cheese products, imitation or substitutes
3. Cheese spreads, pastes, and cheeselike preparations
4. Cheese, except cottage cheese
5. Cheese, imitation or substitutes
6. Cheese, processed
7. Dips, cheese based
8. Processed cheese
9. Sandwich spreads, cheese
10. Whey, raw: liquid

Dry, Condensed, and Evaporated Dairy Products (IN 2023)

Establishments primarily engaged in manufacturing dry, condensed, and evaporated dairy products. Included in this industry are establishments primarily engaged in manufacturing mixes for the preparation of frozen ice cream and ice milk and dairy- and nondairy-based cream substitutes and dietary supplements.

1. Baby formula: fresh, processed, and bottled
2. Buttermilk: concentrated, condensed, dried, evaporated, and powdered
3. Casein, dry and wet
4. Cream substitutes
5. Cream: dried, powdered, and canned
6. Dietary supplements, dairy and nondairy based
7. Dry milk products: whole milk, nonfat milk, buttermilk, whey, and cream
8. Eggnog, canned: nonalcoholic
9. Ice cream mix, unfrozen: liquid or dry
10. Ice milk mix, unfrozen: liquid or dry
11. Lactose, edible
12. Malted milk
13. Milk, whole: canned
14. Milk: concentrated, condensed, dried, evaporated, and powdered
15. Milkshake mix
16. Skim milk: concentrated, dried, and powdered
17. Sugar of mix
18. Whey: concentrated, condensed, dried, evaporated, and powdered
19. Whipped topping, dry mix
20. Yogurt mix

Ice Cream and Frozen Desserts (IN 2024)

Establishments primarily engaged in manufacturing ice cream and other frozen desserts. Establishments primarily engaged in manufacturing frozen bakery products, such as cakes and pies, are classified in Industry 2053.

1. Custard, frozen
2. Desserts, frozen: except bakery
3. Fruit pops, frozen
4. Ice cream: eg, bulk, packaged, molded, on sticks
5. Ice milk: eg, bulk, packaged, molded, on sticks
6. Ices and sherbets
7. Juice pops, frozen
8. Millorine
9. Parfait
10. Pops, dessert: frozen-flavored ice, fruit, pudding, and gelatin
11. Pudding pops, frozen
12. Sherbets and ices
13. Spumoni
14. Tofu frozen desserts
15. Yogurt, frozen

Fluid Milk

Establishments primarily engaged in processing (eg, pasteurizing, homogenizing, vitaminizing, bottling) fluid milk and cream, and related products, including cottage cheese, yogurt (except frozen), and other fermented milk. Establishments primarily engaged in manufacturing dry mix whipped toppings are classified in Industry 2023; those producing frozen whipped toppings are classified in Industry 2038; and those producing frozen yogurt are classified in Industry 2024.

1. Buttermilk, cultured
2. Chocolate milk
3. Cottage cheese, including pot, bakers', and farmers' cheese
4. Cream, aerated
5. Cream, bottled
6. Cream, sour
7. Dips, sour cream based
8. Eggnog, fresh: nonalcoholic
9. Flavored milk drinks
10. Half and half
11. Milk processing (pasteurizing, homogenizing, vitaminizing, bottling)
12. Milk production, except farm
13. Milk, acidophilus
14. Milk, bottled
15. Milk, flavored
16. Milk, reconstituted
17. Milk, ultrahigh temperature
18. Sour cream

19. Whipped cream
20. Whipped topping, except frozen or dry mix
21. Yogurt, except frozen

CANNED, FROZEN, AND PRESERVED FRUITS, VEGETABLES, AND FOOD SPECIALTIES (IGN 203)

The canned products of this industry group are distinguished by their processing rather than by the container. The products may be shipped in bulk or in individual cans, bottles, retort pouch packages, or other containers.

Canned Specialties (IG 2032)

Establishments primarily engaged in canning specialty products, such as baby foods, nationality speciality foods, and soups, except seafood. Establishments primarily engaged in canning seafoods are classified in Industry 2091.

1. Baby foods (including meats), canned
2. Bean sprouts, canned
3. Beans, baked: with or without meat—canned
4. Broth, except seafood: canned
5. Chicken broth and soup, canned
6. Chili con carne, canned
7. Chinese foods, canned
8. Chop suey, canned
9. Chow mein, canned
10. Enchiladas, canned
11. Food specialties, canned
12. Italian foods, canned
13. Macaroni, canned
14. Mexican foods, canned
15. Mincemeat, canned
16. Nationality specialty foods, canned
17. Native foods, canned
18. Pasta, canned
19. Puddings, except meat: canned
20. Ravioli, canned
21. Soups, except seafood: canned
22. Ravioli, canned
23. Soups, except seafood: canned
24. Spaghetti, canned
25. Spanish foods, canned
26. Tamales, canned
27. Tortillas, canned

Canned Fruits, Vegetables, Preserves, Jams, and Jellies (IN 2033)

Establishments primarily engaged in canning fruits, vegetables, and fruit and vegetable juices; and in manufacturing catsup and similar tomato sauces, or natural and imitation preserves, jams, and jellies. Establishments primarily engaged in canning seafoods are classified in Industry 2091; and those manufacturing canned specialties, such as baby foods and soups, except seafood, are classified in Industry 2032.

1. Artichokes in olive oil, canned
2. Barbecue sauce
3. Catsup
4. Cherries, maraschino
5. Chili sauce, tomato
6. Fruit butters
7. Fruit pie mixes
8. Fruits, canned
9. Hominy, canned
10. Jams, including imitation
11. Jellies, edible: including imitation
12. Juice, fruit: concentrated-hot pack
13. Juices, fresh: fruit or vegetable
14. Juices, fruit and vegetable: canned or fresh
15. Ketchup
16. Marmalade
17. Mushrooms, canned
18. Nectars, fruit
19. Olives, including stuffed: canned
20. Pastes, fruit and vegetable
21. Preserves, including imitation
22. Purees, fruit and vegetable
23. Sauces, tomato based
24. Sauerkraut, canned
25. Seasonings (prepared sauces), tomato
26. Spaghetti sauce
27. Tomato juice and cocktails, canned
28. Tomato paste
29. Tomato sauce
30. Vegetable pie mixes
31. Vegetables, canned

Dried and Dehydrated Fruits, Vegetables, and Soup Mixes (IN 2034)

Establishments primarily engaged in sun drying or artificially dehydrating fruits and vegetables, or in manufacturing packaged soup mixes from dehydrated ingredients. Establishments primarily engaged in the grading and marketing of farm dried fruits, such as prunes and raisins, are classified in Wholesale Trade, Industry 5149.

1. Dates, dried
2. Dehydrated fruits, vegetables, and soups
3. Fruit flour, meal, and powders
4. Fruits, sulphured
5. Olives, dried
6. Potato flakes, granules, and other dehydrated potato products
7. Prunes, dried
8. Raisins

9. Soup mixes
10. Soup powders
11. Vegetable flour, meal, and powders
12. Vegetables, sulphured

Pickled Fruits and Vegetables, Vegetable Sauces and Seasonings, and Salad Dressings (IN 2035)

Establishments primarily engaged in pickling and bringing fruits and vegetables and in manufacturing salad dressings, vegetable relishes, sauces, and seasonings. Establishments primarily engaged in manufacturing catsup and similar tomato sauces are classified in Industry 2033, and those packing purchased pickles and olives are classified in Wholesale or Retail Trade. Establishments primarily engaged in manufacturing dry salad dressing and dry sauce mixes are classified in Industry 2099.

1. Blue cheese dressing
2. Brining of fruits and vegetables
3. Cherries, brined
4. French dressing
5. Fruits, pickled and brined
6. Horseradish, prepared
7. Mayonnaise
8. Mustard, prepared (wet)
9. Olives, brined: bulk
10. Onions, pickled
11. Pickles and pickle salting
12. Relishes, fruit and vegetable
13. Russian dressing
14. Salad dressings, except dry mixes
15. Sandwich spreads, salad dressing base
16. Sauces, meat (seasoning): except tomato and dry
17. Sauces, seafood: except tomato and dry
18. Sauerkraut, bulk
19. Seasonings (prepared sauces), vegetable: except tomato and dry
20. Soy sauce
21. Thousand Island dressing
22. Vegetable sauces, except tomato
23. Vegetables, pickled and brined
24. Vinegar pickles and relishes
25. Worcestershire sauce

Frozen Fruits, Fruit Juices, and Vegetables (IN 2037)

Establishments primarily engaged in freezing fruits, fruit juices, and vegetables. These establishments also produce important byproducts such as fresh or dried citrus pulp.

1. Concentrates, frozen fruit juice
2. Dried citrus pulp
3. Frozen fruits, fruit juices, and vegetables
4. Fruit juices, frozen
5. Fruits, quick frozen and cold pack (frozen)
6. Vegetables, quick frozen and cold pack (frozen)

Frozen Specialties Not Elsewhere Classified (IN 2038)

Establishments primarily engaged in manufacturing frozen food specialties, not elsewhere classified, such as frozen dinners and frozen pizza. The manufacture of some important frozen foods and specialties is classified elsewhere. For example, establishments primarily engaged in manufacturing frozen dairy specialties are classified in Industry Group 202, those manufacturing frozen bakery products are classified in Industry Group 205, those manufacturing frozen fruits and vegetables are classified in Industry group 205, those manufacturing frozen fruits and vegetables are classified in Industry 2037, and those manufacturing frozen fish and seafood specialties are classified in Industry 2092.

1. Dinners, frozen: packaged
2. French toast, frozen
3. Frozen dinners, packaged
4. Meals, frozen
5. Native foods, frozen
6. Pizza, frozen
7. Soups, frozen: except seafood
8. Spaghetti and meatballs, frozen
9. Waffles, frozen
10. Whipped topping, frozen

GRAIN MILL PRODUCTS (IGN 204)

Flour and Other Grain Mill Products (IN 2041)

Establishments primarily engaged in milling flour or meal from grain, except rice. The products of flour mills may be sold plain or in the form of prepared mixes or doughs for specific purposes. Establishments primarily engaged in manufacturing prepared flour mixes or doughs from purchased ingredients are classified in Industry 2045, and those milling rice are classified in Industry 2044.

1. Bran and middlings, except rice
2. Bread and bread-type roll mixes*
3. Buckwheat flour
4. Cake flour*
5. Cereals, cracked grain*
6. Corn grits and lakes for brewers' use
7. Dough, biscuit*
8. Doughs, refrigerated or frozen*
9. Durum flour
10. Farina, except breakfast food*
11. Flour mills, cereals: except rice
12. Flour mixes*
13. Flour: buckwheat, corn, graham, rye, and wheat
14. Frozen doughs*
15. Graham flour
16. Granular wheat flour

17. Grits and flakes, corn: for brewers' use
18. Hominy grits, except breakfast food
19. Meal, corn
20. Milling of grains, dry, except rice
21. Mixes, flour: eg, pancake, cake, biscuit, doughnut*
22. Pancake batter, refrigerated or frozen*
23. Pizza mixes and prepared dough*
24. Semolina (flour)
25. Wheat germ
26. Wheat mill feed

Cereal Breakfast Foods (IN 2043)

Establishments primarily engaged in manufacturing cereal breakfast foods and related preparations, except breakfast bars. Establishments primarily engaged in manufacturing granola bars and other types of breakfast bars are classified in Industry 2064.

1. Breakfast foods, cereal
2. Coffee substitutes made from grain
3. Corn flakes
4. Corn hulled (cereal breakfast food)
5. Farina, cereal breakfast food
6. Granola, except bars and clusters
7. Hominy grits prepared as cereal breakfast food
8. Infants' foods, cereal type
9. Oatmeal (cereal breakfast food)
10. Oats, rolled (cereal breakfast food)
11. Rice breakfast foods
12. Wheat flakes

Rice Milling (IN 2044)

Establishments primarily engaged in cleaning and polishing rice, and in manufacturing rice flour or meal. Other important products of this industry include brown rice, milled rice (including polished rice), rice polish, and rice bran.

1. Flour, rice
2. Milling of rice
3. Polishing of rice
4. Rice bran, flour, and meal
5. Rice cleaning and polishing
6. Rice polish
7. Rice, brewers'
8. Rice, brown
9. Rice vitamin and mineral enriched

Prepared Flour Mixes and Doughs (IN 2045)

Establishments primarily engaged in preparing flour mixes or doughs from purchased flour. Establishments primarily engaged in milling flour from grain and producing mixes or doughs are classified in Industry 2091.

1. Biscuit mixes and doughs**
2. Bread and bread-type roll mixes**
3. Cake flour**
4. Cake mixes**
5. Dough, biscuit**
6. Doughnut mixes**
7. Doughs, refrigerated or frozen**
8. Flour: blended or self-rising**
9. Frozen doughs**
10. Gingerbread mixes**
11. Mixes, flour: eg, pancake, cake, biscuit, doughnut**
12. Pancake batter, refrigerated or frozen**
13. Pancake mixes**
14. Pizza mixes and doughs**

Wet Corn Milling (IN 2046)

Establishments primarily engaged in milling corn or sorghum grain (milo) by the wet process, and producing starch, syrup, oil, sugar, and byproducts, such as gluten feed and meal. Also included in this industry are establishments primarily engaged in manufacturing starch from other vegetable sources (eg, potatoes, wheat). Establishments primarily engaged in manufacturing table syrups from corn syrup and other ingredients, and those manufacturing starch-based dessert powders, are classified in Industry 2099.

1. Corn oil cake and meal
2. Corn starch
3. Corn syrup (including dried), unmixed
4. Dextrine
5. Dextrose
6. Feed, gluten
7. Fructose
8. Glucose
9. High fructose syrup
10. Hydrol
11. Meal, gluten
12. Oil, corn: crude and refined
13. Potato starch
14. Rice starch
15. Starch, instant
16. Starch, liquid
17. Starches, edible and industrial
18. Steep water concentrate
19. Sugar, corn
20. Tapioca
21. Wheat gluten
22. Wheat starch

Dog and Cat Food (IN 2047)

Establishments primarily engaged in manufacturing dog and cat food from cereal, meat, and other ingredients.

These preparations may be canned, frozen, or dry. Establishments primarily engaged in manufacturing feed for animals other than dogs and cats are classified in Industry 2048.

Prepared Feeds and Feed Ingredients for Animals and Fowls, Except Dogs and Cats (IN 2048)

Establishments primarily engaged in manufacturing prepared feeds and feed ingredients and adjuncts for animals and fowls, except dogs and cats. Included in this industry are poultry and livestock feed and feed ingredients, such as alfalfa meal, feed supplements, and feed concentrates and feed premixes. Also included are establishments primarily engaged in slaughtering animals for animal feed. Establishments primarily engaged in slaughtering animals for human consumption are classified in Industry Group 201. Establishments primarily engaged in manufacturing dog and cat foods are classified in Industry 2047.

1. Alfalfa, cubed
2. Alfalfa, prepared as feed for animals
3. Animal feeds, prepared: except dogs and cats
4. Bird food, prepared
5. Buttermilk emulsion for animal food
6. Chicken feeds, prepared
7. Citrus seed meal
8. Earthworm food and bedding
9. Feed concentrates
10. Feed premixes
11. Feed supplements
12. Feeds, prepared (including mineral): for animals and fowls—except dogs and cats
13. Feeds, specialty: mice, guinea pigs, minks, etc
14. Fish food
15. Hay, cubed
16. Horsemeat, except for human consumption
17. Kelp mean and pellets
18. Livestock feeds, supplements, and concentrates
19. Meal, bone: prepared as feed for animals and fowls
20. Mineral feed supplements
21. Oats: crimped, pulverized, and rolled: except breakfast food
22. Oyster shells, ground: used as feed for animals and fowls
23. Pet food, except dog and cat: canned, frozen, and dry
24. Poultry feeds, supplements, and concentrates
25. Shell crushing for feed
26. Slaughtering of animals, except for human consumption
27. Stock feeds, dry

BAKERY PRODUCTS (IGN 205)

Bread and Other Bakery Products, Except Cookies and Crackers (IN 2051)

Establishments primarily engaged in manufacturing fresh or frozen bread and bread-type rolls and fresh cakes, pies, pastries and other similar perishable bakery products. Establishments primarily engaged in producing dry bakery products, such as biscuits, crackers, and cookies, are classified in Industry 2052. Establishments primarily engaged in manufacturing frozen bakery products, except bread and bread-type rolls, are classified in Industry 2053. Establishments producing bakery products primarily for direct sale on the premises to household consumers are classified in Retail Trade, Industry 5461.

1. Bagels
2. Bakery products, fresh: bread, cakes, doughnuts, and pastries
3. Bakery products, partially cooked: except frozen
4. Biscuits, baked: baking powder and raised
5. Bread, brown: Boston and other—canned
6. Bread, including frozen
7. Buns, bread-type (eg, hamburger, hot dog), including frozen
8. Buns, sweet, except frozen
9. Cakes, bakery, except frozen
10. Charlotte Russe (bakery product), except frozen
11. Croissants, except frozen
12. Crullers, except frozen
13. Doughnuts, except frozen
14. Frozen bread and bread-type rolls
15. Knishes, except frozen
16. Pastries, except frozen: eg, Danish, French
17. Pies, bakery, except frozen
18. Rolls, bread-type, including frozen
19. Rolls, sweet, except frozen
20. Sponge goods, bakery, except frozen
21. Sweet yeast goods, except frozen

Cookies and Crackers (IN 2052)

Establishments primarily engaged in manufacturing fresh cookies, crackers, pretzels, and similar dry bakery products. Establishments primarily engaged in producing other fresh bakery products are classified in Industry 2051.

1. Bakery products, dry: eg, biscuits, crackers, pretzels
2. Biscuits, baked: dry, except baking powder and raised
3. Communion wafers
4. Cones, ice cream
5. Cookies
6. Cracker meal and crumbs
7. Crackers: eg, graham, soda

8. Matzoths
9. Pretzels
10. Rusk
11. Saltines
12. Zwieback

Frozen Bakery Products, Except Bread (IN 2053)

Establishments primarily engaged in manufacturing frozen bakery products, except bread and bread-type rolls. Establishments primarily engaged in manufacturing frozen bread and bread-type rolls are classified in Industry 2051.

1. Bakery products, frozen: except bread and bread-type rolls
2. Cakes, frozen, pound, layer, and cheese
3. Croissants, frozen
4. Doughnuts, frozen
5. Pies, bakery, frozen
6. Sweet yeast goods, frozen

SUGAR AND CONFECTIONERY PRODUCTS (IGN 206)

Cane Sugar, except Refining (IN 2061)

Establishments primarily engaged in manufacturing raw sugar, syrup, or finished (granulated or clarified) cane sugar from sugar cane. Establishments primarily engaged in refining sugar from purchased raw cane sugar or sugar syrup are classified in Industry 2062.

1. Cane sugar, made from sugarcane
2. Molasses, blackstrap: made from sugarcane
3. Molasses, made from sugarcane
4. Sugar, granulated: made from sugarcane
5. Sugar, invert: made from sugarcane
6. Sugar, powdered: made from sugarcane
7. Sugar, raw: made from sugarcane
8. Syrup, cane: made from sugarcane

Cane Sugar Refining (IN 2062)

Establishments primarily engaged in refining purchased raw cane sugar and sugar syrup.

1. Molasses, blackstrap: made from purchased raw cane sugar or sugar syrup
2. Refineries, cane sugar
3. Sugar, granulated: made from purchased raw cane sugar or sugar syrup
4. Sugar, invert: made from purchased raw cane sugar or sugar syrup
5. Sugar, powdered: made from purchased raw cane sugar or sugar syrup
6. Sugar, refined: made from purchased raw can sugar or sugar syrup
7. Syrup, made from purchased raw can sugar or sugar syrup

Beet Sugar (IN 2063)

Establishments primarily engaged in manufacturing sugar from sugar beets.

1. Beet pulp, dried
2. Beet sugar, made from sugar beets
3. Molasses beet pulp
4. Molasses, made from sugar beets
5. Sugar, granulated: made from sugar beets
6. Sugar, invert: made from sugar beets
7. Sugar, liquid: made from sugar beets
8. Sugar, powdered: made from sugar beets
9. Syrup, made from sugar beets

Candy and Other Confectionery Products (IN 2064)

Establishments primarily engaged in manufacturing candy, including chocolate candy, other confections, and related products. Establishments primarily engaged in manufacturing solid chocolate bars from cacao beans are classified in Industry 2066, those manufacturing chewing gum are classified in Industry 2067, and those primarily engaged in roasting and salting nuts are classified in Industry 2068. Establishments primarily engaged in manufacturing confectionery for direct sale on the premises to household consumers are classified in Retail Trade, Industry 5441.

1. Bars, candy: including chocolate-covered bars
2. Breakfast bars
3. Cake ornaments, confectionery
4. Candy, except solid chocolate
5. Chewing candy, except chewing gum
6. Chocolate bars, from purchased cocoa or chocolate
7. Chocolate candy, except solid chocolate
8. Confectionery
9. Cough drops, except pharmaceutical preparations
10. Dates: chocolate covered, sugared, and stuffed
11. Fruit peel products: candied, glazed glace, and crystallized
12. Fruits: candied, glazed, and crystallized
13. Fudge (candy)
14. Granola bars and clusters
15. Halvah (candy)
16. Licorice candy
17. Lozenges, candy: nonmedicated
18. Marshmallows
19. Marzipan (candy)
20. Nuts, candy covered
21. Nuts, glace
22. Popcorn balls and candy-covered popcorn products

Chocolate and Cocoa Products (IN 2066)

Establishments primarily engaged in shelling, roasting, and grinding cacao beans for the purpose of making chocolate liquor, from which cocoa powder and cocoa butter are derived, and in the further manufacturing of solid chocolate bars, chocolate coatings, and other chocolate and cocoa products. Also included is the manufacture of similar products, except candy, from purchased chocolate or cocoa. Establishments primarily engaged in manufacturing candy from purchased cocoa products are classified in Industry 2064.

1. Baking chocolate
2. Bars, candy: solid chocolate
3. Cacao bean products: chocolate, cocoa butter, and cocoa
4. Cacao beans: shelling, roasting, and grinding for making chocolate liquor
5. Candy, solid chocolate
6. Chocolate bars, solid: from cacao beans
7. Chocolate coatings and syrups
8. Chocolate liquor
9. Chocolate syrup
10. Chocolate, instant
11. Chocolate, sweetened or unsweetened
12. Chocolate butter
13. Chocolate mix, instant
14. Chocolate powdered: mixed with other substances

Chewing Gum (IN 2067)

Establishments primarily engaged in manufacturing salted, roasted, dried, cooked, or canned nuts or in processing grains or seeds in a similar manner for snack purposes. Establishments primarily engaged in manufacturing confectionery-coated nuts are classified in Industry 2064, and those manufacturing peanut butter are classified in Industry 2099.

1. Nuts, dehydrated or dried
2. Nuts: salted, roasted, cooked, or canned
3. Seeds: salted, roasted, cooked, or canned

FATS AND OILS (IGN 207)

Cottonseed Oil Mills (IN 2074)

Establishments primarily engaged in manufacturing cottonseed oil, cake, meal, and linters, or in processing purchased cottonseed oil other than into edible cooking oils. Establishments primarily engaged in refining cottonseed oil into edible cooking oils are classified in Industry 2079.

1. Cottonseed oil, cake, and meal: made in cottonseed oil mills
2. Cottonseed oil, deodorized
3. Lecithin, cottonseed

Soybean Oil Mills (IN 2075)

Establishments primarily engaged in manufacturing soybean oil, cake, and meal, and soybean protein isolates and concentrates, or in processing purchased soybean oil other than into edible cooking oils. Establishments primarily engaged in refining soybean oil into edible cooking oils are classified in Industry 2079.

1. Lecithin, soybean
2. Soybean flour and grits
3. Soybean oil, cake, and meal
4. Soybean oil, deodorized
5. Soybean protein concentrates
6. Soybean protein isolates

Vegetable Oil Mills, Except Corn, Cottonseed, and Soybean (IN 2076)

Establishments primarily engaged in manufacturing vegetable oils, cake and meal, except corn, cottonseed, and soybean, or in processing similar purchased oils other than into edible cooking oils. Establishments primarily engaged in manufacturing corn oil and its byproducts are classified in Industry 2046, those that are refining vegetable oils into edible cooking oils are classified in Industry 2079, and those refining these oils for medicinal purposes are classified in Industry 2833.

1. Castor oil and pomace
2. Coconut oil
3. Linseed oil, cake, and meal
4. Oils, vegetable: except corn, cottonseed, and soybean
5. Oiticica oil
6. Palm kernel oil
7. Peanut oil, cake, and meal
8. Safflower oil
9. Sunflower seed oil
10. Tallow, vegetable
11. Tung oil
12. Walnut oil

Animal and Marine Fats and Oils (IN 2077)

Establishments primarily engaged in manufacturing animal oils, including fish oil and other marine animal oils, and fish and animal meal; and those rendering inedible stearin, grease, and tallow from animal fat, bones, and meat scraps. Establishments primarily engaged in manufacturing lard and edible tallow and stearin are classified in Industry Group 201; those refining marine animal oils for medicinal purposes are classified in Industry 2833; and those manufacturing fatty acids are classified in Industry 2899.

1. Feather meal
2. Fish liver oils, crude
3. Fish meal

4. Fish oil and fish oil meal
5. Grease rendering, inedible
6. Meal, meat and bone: not prepared as feed
7. Meat and bone meal and tankage
8. Oils, animal
9. Oils, fish and marine animal: eg, herring, menhaden, whale (refined), sardine
10. Rendering plants, inedible grease and tallow
11. Stearin, animal: inedible
12. Tallow rendering, inedible

Shortening, Table Oils, Margarine, and Other Edible Fats and Oils, Not Elsewhere Classified (IN 2079)

Establishments primarily engaged in manufacturing shortening, table oils, margarine, and other edible fats and oils, not elsewhere classified. Establishments primarily engaged in producing corn oil are classified in Industry 2046.

1. Baking and frying fats (shortening)
2. Cottonseed cooking and salad oil
3. Margarine oil, except corn
4. Margarine, including imitation
5. Margarine-butter blend
6. Nut margarine
7. Oil, hydrogenated: edible
8. Oil, partially hydrogenated: edible
9. Oil, vegetable winter stearin
10. Olive oil
11. Peanut cooking and salad oil
12. Shortenings, compound and vegetable
13. Soybean cooking and salad oil
14. Vegetable cooking and salad oils, except corn oil: refined

BEVERAGES (IGN 208)

Malt Beverages (IN 2082)

Establishments primarily engaged in manufacturing malt beverages. Establishments primarily engaged in bottling purchased malt beverages are classified in Industry 5181.

1. Ale
2. Beer (alcoholic beverage)
3. Breweries
4. Brewers' grain
5. Liquors, malt
6. Malt extract, liquors, and syrups
7. Near beer
8. Porter (alcoholic beverage)
9. Stout (alcoholic beverage)

Malt (IN 2083)

Establishments primarily engaged in manufacturing malt or malt byproducts from barley or other grains.

1. Malt byproducts
2. Malt: barley, rye, wheat, and corn
3. Malthouses
4. Sprouts, made in malthouses

Wines, Brandy, and Brandy Spirits (IN 2084)

Establishments primarily engaged in manufacturing wines, brandy, and brandy spirits. This industry also includes bonded wine cellars that are engaged in blending wines. Establishments primarily bottling purchased wines, brandy, and brandy spirits but that do not manufacture wines and brandy, are classified in Wholesale Trade, Industry 5182.

1. Brandy
2. Brandy spirits
3. Wine cellars, bonded: engaged in blending wines
4. Wine coolers (beverages)
5. Wines

Distilled and Blended Liquors (IN 2085)

Establishments primarily engaged in manufacturing alcoholic liquors by distillation, and in manufacturing cordials and alcoholic cocktails by blending processes or by mixing liquors and other ingredients. Establishments primarily engaged in manufacturing industrial alcohol are classified in Industry 2869, and those only bottling purchased liquors are classified in Wholesale Trade, Industry 5182.

1. Applejack
2. Cocktails, alcoholic
3. Cordials, alcoholic
4. Distillers' dried grains and solubles
5. Eggnog, alcoholic
6. Ethyl alcohol for medicinal and beverage purposes
7. Gin (alcoholic beverage)
8. Grain alcohol for medicinal and beverage purposes
9. Liquors: distilled and blended—except brandy
10. Rum
11. Spirits, neutral, except fruit—for beverage purposes
12. Vodka
13. Whiskey: bourbon, rye, scotch type, and corn

Bottles and Canned Soft Drinks and Carbonated Waters (IN 2086)

Establishments primarily engaged in manufacturing soft drinks and carbonated waters. Establishments primarily engaged in manufacturing fruit and vegetable juices are classified in Industry Group 203; those manufacturing fruit syrups for flavoring are classified in Industry 2087; and those manufacturing nonalcoholic cider are classified in Industry 2099. Establishments primarily engaged in bottling natural spring waters are classified in Wholesale Trade, Industry 5149.

1. Beer, birch and root: bottled or canned
2. Carbonated beverages, nonalcoholic: bottled or canned
3. Drinks, fruit: bottled, canned, or fresh
4. Ginger ale, bottled or canned
5. Iced tea, bottled or canned
6. Lemonade: bottled, canned, or fresh
7. Mineral water, carbonated: bottled or canned
8. Soft drinks, bottled or canned
9. Tea, iced: bottled or canned
10. Water, pasteurized: bottled or canned

Flavoring Extracts and Flavoring Syrups, Not Elsewhere Classified (IN 2087)

Establishments primarily engaged in manufacturing flavoring extracts, syrups, powders, and related products, not elsewhere classified, for soda fountain use or for the manufacture of soft drinks, and colors for bakers' and confectioners' use. Establishments primarily engaged in manufacturing chocolate syrup are classified in Industry 2066.

1. Beverage bases
2. Bitters (flavoring concentrates)
3. Burnt sugar (food color)
4. Cocktail mixes, nonalcoholic
5. Coffee flavorings and syrups
6. Colors for bakers' and confectioners' use, except synthetic
7. Cordials, nonalcoholic
8. Drink powders and concentrates
9. Flavoring concentrates
10. Flavoring extracts, pastes, powders, and syrups
11. Food colorings, except synthetic
12. Food glace, for glazing foods
13. Fruit juices, concentrated: for fountain use
14. Fruits, crushed: for soda fountain use

MISCELLANEOUS FOOD PREPARATIONS AND KINDRED PRODUCTS (IGN 209)

Canned and Cured Fish and Seafoods (IN 2091)

Establishments primarily engaged in cooking and canning fish, shrimp, oysters, clams, crabs, and other seafoods, including soups: and those engaged in smoking, salting, drying, or otherwise curing fish and other seafoods for the trade. Establishments primarily engaged in shucking and packing fresh oysters in nonsealed containers, or in freezing or preparing fresh fish, are classified in Industry 2092.

1. Canned fish, crustacea, and mollusks
2. Caviar, canned
3. Chowders, fish and seafood: canned
4. Clam bouillon, broth, chowder, juice: bottled or canned
5. Codfish: smoked, salted, dried, and pickled
6. Crab meat, canned and cured
7. Finnan haddie (smoked haddock)
8. Fish and seafood cakes: canned
9. Fish egg bait, canned
10. Fish, canned and cured
11. Fish: cured, dried, pickled, salted, and smoked
12. Herring: smoked, salted, dried, and pickled
13. Oysters, canned and cured
14. Salmon: smoked, salted, dried, canned, and pickled
15. Sardines, canned
16. Seafood products, canned and cured
17. Shellfish, canned and cured
18. Shrimp, canned and cured
19. Soups, fish and seafood: canned
20. Stews, fish and seafood: canned
21. Tuna fish, canned

Prepared Fresh or Frozen Fish and Seafoods (IN 2092)

Establishments primarily engaged in preparing fresh and raw or cooked frozen fish and other seafoods and seafood preparations, such as soups, stews, chowders, fishcakes, crabcakes, and shrimpcakes. Prepared fresh fish are eviscerated or processed by removal of heads, fins, or scales. This industry also includes establishments primarily engaged in the shucking and packing of fresh oysters in nonsealed containers.

1. Chowders, fish and seafood: frozen
2. Crabcakes, frozen
3. Crabmeat picking
4. Crabmeat, fresh: packed in nonsealed containers
5. Fish and seafood cakes, frozen
6. Fish fillets
7. Fish sticks
8. Fish: fresh and frozen, prepared
9. Oysters, fresh: shucking and packing nonsealed containers
10. Seafoods, fresh and frozen
11. Shellfish, fresh and frozen
12. Shellfish, fresh: shucked, picked, or packed
13. Shrimp, fresh and frozen
14. Soups, fish and seafood: frozen
15. Stews, fish and seafood: frozen

Roasted Coffee (IN 2095)

Establishments primarily engaged in roasting coffee, and in manufacturing coffee concentrates and extracts in powdered, liquid, or frozen form, including freeze dried. Coffee roasting by wholesale grocers is classified in Wholesale Trade, Industry 5149.

1. Coffee extracts
2. Coffee roasting, except by wholesale grocers

3. Coffee, ground: mixed with grain or chicory
4. Coffee, instant and freeze dried

Potato Chips, Corn Chips, and Similar Snacks (IN 2096)

Establishments primarily engaged in manufacturing potato chips, corn chips, and similar snacks. Establishments primarily engaged in manufacturing pretzels and crackers are classified in Industry 2052; those manufacturing candy covered popcorn are classified in Industry 2064; those manufacturing salted, roasted, cooked or canned nuts and seeds are classified in Industry 2068; and those manufacturing packaged unpopped popcorn are classified in Industry 2099.

1. Cheese curls and puffs
2. Corn chips and related corn snacks
3. Popcorn, popped: except candy covered
4. Pork rinds
5. Potato chips and related corn snacks
6. Potato sticks

Manufactured Ice (IN 2097)

Establishments primarily engaged in manufacturing ice for sale. Establishments primarily engaged in manufacturing dry ice are classified in Industry 2813.

1. Block ice
2. Ice cubes
3. Ice plants, operated by public utilities
4. Ice, manufactured or artificial: except dry ice

Macaroni, Spaghetti, Vermicelli, and Noodles (IN 2098)

Establishments primarily engaged in manufacturing dry macaroni, spaghetti, vermicelli, and noodles. Establishments primarily engaged in manufacturing canned macaroni and spaghetti are classified in Industry 2032, and those manufacturing fried noodles, such as Chinese noodles, are classified in Industry 2099.

Macaroni and products, dry: eg, alphabets, rings, seashells

1. Noodles: egg, plain, and water
2. Spaghetti, dry
3. Vermicelli

Food Preparations, not Elsewhere Classified (IN 2099)

Establishments primarily engaged in manufacturing prepared foods and miscellaneous food specialties, not elsewhere classified, such as baking powder, yeast, and other leavening compounds; peanut butter; packaged tea, including instant; ground spices; and vinegar and cider. Also included in this industry are establishments primarily engaged in manufacturing dry preparations, except flour mixes, consisting of pasta, rice, potatoes, textured vegetable protein, and similar products that are packaged with other ingredients to be prepared and cooked by the consumer. Establishments primarily engaged in manufacturing flour mixes are classified in Industry Group 204.

1. Almond pastes
2. Baking powder
3. Bouillon cubes
4. Box lunches for sale off premises
5. Bread crumbs, not made in bakeries
6. Butter, renovated and processed
7. Cake frosting mixes, dry
8. Chicory root, dried
9. Chili pepper or powder
10. Chinese noodles
11. Cider, nonalcoholic
12. Coconut, desiccated and shredded
13. Cole slaw in bulk
14. Cracker sandwiches made from purchased crackers
15. Desserts, ready to mix
16. Dips, except cheese and sour cream based
17. Emulsifiers, food
18. Fillings, cake or pie: except fruits, vegetables, and meat
19. Frosting, prepared
20. Gelatin dessert preparations
21. Gravy mixes, dry
22. Honey, strained and bottled
23. Jelly, corncob (gelatin)
24. Leavening compounds, prepared
25. Marshmallow creme
26. Meat seasonings, except sauces
27. Molasses, mixed or blended**
28. Noodles, fried (eg, Chinese)
29. Noodles, uncooked: packaged with other ingredients
30. Pancake syrup, blended and mixed
31. Pasta, uncooked: packaged with other ingredients
32. Peanut butter
33. Pectin
34. Pepper
35. Pizza, refrigerated: not frozen
36. Popcorn, packaged: except popped
37. Potatoes, dried: packaged with other ingredients
38. Potatoes, peeled for the trade
39. Rice, uncooked: packaged with other ingredients
40. Salad dressing mixes, dry
41. Salads, fresh or refrigerated
42. Sandwiches, assembled and packaged: for wholesale market
43. Sauce mixes, dry
44. Sorghum, including custom refining
45. Spices, including grinding
46. Sugar grinding

47. Sugar, industrial maple: made in plants producing maple syrup
48. Sugar, powdered**
49. Syrups, sweetening: honey, maple syrup, sorghum
50. Tea blending
51. Tofu, except frozen desserts
52. Tortillas, fresh or refrigerated
53. Vegetables peeled for the trade
54. Vinegar
55. Yeast

BIBLIOGRAPHY

"Standard Industrial Classification Manual," Office of Management and Budget, Executive Office of the President. U.S. Government Printing Office, Washington, D.C., 1987.

Y. H. Hui
American Food and Nutrition Center
Cutten, California

FOOD PROCESSING: TECHNOLOGY, ENGINEERING, AND MANAGEMENT

The technology of food is intertwined with the business of selling food. Although this is not unique to food technology, having insight into the food business will provide some insight into the motivation it provides for the development of new technologies and the maintenance of old ones. Decision making in the food industry requires knowledge of both the business of selling food and the technology of producing, holding, and delivering food. The right information about the technology and the business, coupled with organized approaches to decision making, can yield important benefits to those responsible for making decisions in the industry.

Some generalizations about the food business are that it is high volume, low margin, multiple product, transportation intensive, and end-user marketing intensive. Because of the need for food to be ubiquitous, the business requires multiple distribution points and complicated distribution networks. The nature of food as material and as a perception of the consumer both defines and constrains the food business. The intent is to deliver safe, palatable, and profitable product to the purchaser. Technology constrains the business by specifying what can and cannot be done with food materials while maintaining a viable product. The reality is that the technology and the business of selling food are inseparable.

Because profit per sale is usually low, food businesses must make many sales to gain a reasonable profit. This emphasis on volume requires food technology to provide speed in production. This makes the production facility challenge one of turning out salable foodstuffs by the ton. In the food industry, great value is placed on technological innovation that enables automated handling of materials and scale-up of processes and procedures from small to large volumes. In such a high-volume business, small improvements in efficiency of the technology or distribution or the decision making can have important consequences for the profit (or even survival) of an enterprise.

Most food companies see opportunities for growth in the introduction of new products. This has led to varieties of products from single companies and much product differentiation within categories of products. The technology to develop, manufacture, and distribute these new product entries must be invented and applied in order for such growth to happen. This perception of how growth is attained, and the perception that growth is important, provide important spurs for technological development.

That food must be ubiquitous leads to situations in which the business and the technology cannot be separated. Although lengthening shelf life is a regular goal and concern of the food technologist, the distribution system must be designed around what the technology and the consumer will bear. Some fresh foodstuffs are flown to market, whereas other products are shipped by train or barge. The network of warehouses, transportation corridors, vehicles, and storage facilities all must accommodate to the realities of the food as material and to the requirement that consumer's expectations are matched in food products once delivered. The business accommodates to the realities of the product, whereas the technologist tries to change the product to achieve new advantages for the business. Because food must be everywhere there are people, the marketing of food must be ubiquitous—ultimately, food products are consumed teaspoon by teaspoon, and each individual consumer must be sold on the product.

The other technological reality that affects the food business is the need for uniform products when the inputs are variable. The biological systems that generate the inputs to food products require modification to yield uniform outputs. The modification is usually called processing. Besides turning out food by the ton, food technology must also provide a means to control the performance of the inputs so that the consumer's perception of the product is the same container to container and bite by bite.

The nature of the business of food has important consequences for the technologist, and the technology of food has important consequences for the businessperson. Decisions, whether they are viewed as business oriented or technology oriented, should not be seen in isolation from either arena.

CONTROLLING FOOD QUALITIES

Much of the application of food technology is for the purpose of producing uniform output from variable input. This is a significant difference from fabricators who assemble parts into a product. Such assembly presumes interchangeable inputs (parts). By contrast food processors assume that the inputs are variable. Milk from farm A is different from milk from farm B. Peas harvested on Monday are different from peas harvested on Tuesday. The challenge and opportunity is to control the quality of the output product by controlling the processing steps applied to the inputs.

Consistently reproducing a food quality implies that this quality is somehow measurable. From a consumer perspective this is usually not the case. Even articulating the nature of a food quality may be difficult for a consumer. Although there are words like *sweetness* and *saltiness*, more often the expressions describing quality are much more general—"it tastes good" or "it doesn't taste right." This problem for quality measurement is further complicated by the variability among consumers of what is considered a good quality for food to have. Preferences vary. For control to be possible the consumer's perception of quality must be separated from the measurable qualities of food materials. The food technologist seeks desirable effects through variables that are controllable. The underlying assumption is that controlling ingredient amounts, processing times and temperatures, holding times, and the like will produce a food product that meets the consumer's quality expectations. The connection between processing variables that can be measured and replicated and the consumer's perception of the qualities of a food product is real, but it is indirect. Many intervening factors influence this connection and most of them are not measurable.

For a given product the best hope of consistent qualities is to control what can be measured as carefully as possible. Keeping in mind that the inputs are intrinsically variable and that processes also vary, "control" must be defined statistically. If a quality measure of an ice cream product is its gross weight, then the ideal is that the gross weight of every container comes from the same distribution of such weights. Because we cannot eliminate variability entirely, the best we can do is to make sure that the underlying distribution does not change. In these situations a distribution is said to have "changed" if its center (mean) or its dispersion (standard deviation) changes. A manufacturing step is said to be "in control" if the measurements taken are scattered according to the same distribution.

Deciding whether an operation is in control or not can be done by routinely sampling the output from the operation and measuring the target quality for each item in the sample. Sample statistics can be computed, including typically the mean of the sample, the range of the sample, and the standard deviation of the sample. These numbers have their own distributions, but these distributions are related to the target quality measurement in well-known ways. The easiest connection to describe is the one between the mean of the measurements and the mean of the sample. The mean of the sample means has the same value as the mean of the target measurements, and the standard deviation of the sample means (also called the standard error of the mean) is the standard deviation of the measurements divided by the square root of the sample size. Furthermore, the distribution of the sample means is distributed approximately normally, provided the sample size is big enough (in practice big enough typically means four or more items in the sample). This fact is true even if the distribution of the measurements is not normal. This enables a method for making the in control decision.

W. A. Shewhart (1) developed a strategy and a graphic that made this decision making feasible for a wide variety of situations. He proposed that if the sample mean fell more than three standard deviations from the empirically established mean of the process (established when the process was in control), this was evidence that the operation producing the measurement was not in control. If the sample mean was less than three standard deviations from the established process mean, the operation was deemed in control. Because the distribution of the sample means is approximately normal, the probability of the sample mean falling beyond three standard deviations by chance (ie, without the distribution of the individual measurements changing) can be computed and is quite small (less than 0.003).

The graphic that Shewhart developed is called a control chart (Fig. 1). This consists of three equally spaced parallel lines (or sometimes two lines for the range chart). The centerline corresponds to the process mean that is also the mean of the sample distribution; the upper and lower lines correspond to the mean plus three standard deviations and the mean minus three standard deviations of the sample distribution, respectively. Sample means and sample

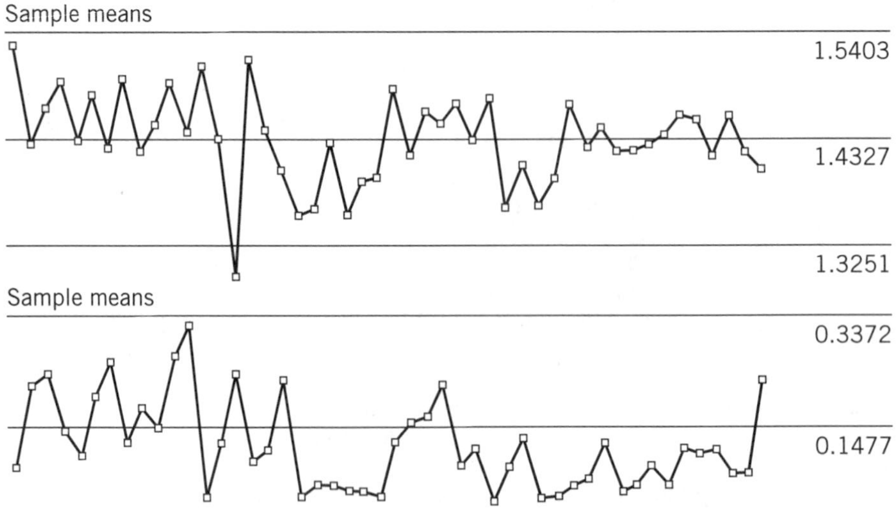

Figure 1. An example control chart for means and ranges of sample size four for gross weights of quarts of ice cream. There is one out-of-control point in the sample means chart.

ranges can be plotted successively from left to right to indicate a time sequence. Any time a plotted point falls outside the three standard deviation limits, the operation is assumed to be out of control.

Many of Shewhart's followers have made refinements of this device for decision making, but the essence of this method has remained the same since originally suggested. The main idea is that unusual values for the sample statistics indicate that the underlying distribution of measurements has changed. It is a signal that the cause of this change should be sought and found. Making decisions based on this method can produce two types of errors. On the average, three times in a thousand samples the sample mean will be beyond the "three sigma limits" even though the underlying distribution has not changed. Or the plotted point will fall between the control limits even though the distribution has changed. Such errors can be anticipated, and the use of this device can be adjusted to take into account the importance of these errors.

Control charts have been used to help propel the quality revolution. Emphasis on good quality products and on continuous quality improvement has had large impact on some industries. In automotive and electronic industries the commitment to constant quality improvement has indeed made products more reliable and convenient. The food industry has benefited less from the quality emphasis for two reasons. First, it is difficult to reduce the intrinsic variability in viable biological systems. The amount of this variability is unlikely to be reducible because genetic diversity is a strength of biological systems. This often leads to a level of variability in food materials that is hard to reduce. Second, distinguishing between a successful product and a failure is not clear-cut. This makes it difficult to define and detect quality improvement. For example, if a stereo or one if its components fail, the symptom is usually obvious to most users. In a food product, the wrong amount of an ingredient may be a failure, but the result may not be detectable by the consumer. Similarly, a food manufacturer may choose to change a product. While some customers may judge the altered product a failure, others may like it better.

This brief article cannot deal with the details of developing and using control charts, but there are many texts that can assist (2). The basic underlying idea is that patterns in variation of measurements can be used to determine the state of control (or lack of it) of the qualities of food products.

THE OPTIMIZING DECISION MAKER

Those responsible for the policy and assets of a food business are (usually without knowing it) optimizers. They try to achieve the best results in the business situation they face while recognizing the technological situation. This means conscientious managers will attempt to use resources efficiently. The usual procedure when optimizing is to array the alternatives, eliminate the infeasible ones from consideration, and then measure the remaining alternatives until a best one is determined (3). Sometimes this is an informal procedure, with little information gathering and minor mathematics, and sometimes it is very formal with structured information requirements and complex mathematics. The science associated with food materials often structures information requirements, the technology provides numbers and determines what is feasible, and the business objectives provide the measuring stick to choose among alternatives. This kind of decision making links the business to the technology in important and useful ways. Not only is the approach structured and systematic, but also for many kinds of decisions there are formal (mathematical) methods that can assist the decision maker.

Decisions that affect the efficiency of a food industry enterprise are made at many levels in the business. Sometimes they are onetime occurrences (should we invest in a new production process?) and sometimes they are recurring (how long should we process that product and at what temperature?). Because some decisions become routine, they are sometimes dismissed as trivial or unimportant. This is not always the case, as routine decisions about food formulations, delivery, processing parameters, production scheduling, and the like have important impacts on costs, profits, and the safety of the food product. If a recurring decision is not optimal, the result is regular losses to the business. In many cases, formal models can be built for these decision situations, and formal methods can be applied to give the decision maker good information to act on. Although the optimization opportunity itself is constrained by the quality of the information available to the user, it still will be the best the food manager can do in the situation he or she faces.

An important piece of optimizing is the objective measure (4). This device allows the user to choose between alternatives—providing a measure of which one is better. There is sometimes a perceived conflict between the objectives of the food technologist and the food industry manager. If the technologist insists on some kind of quality objective—best product, highest-quality product—this may come in conflict with profit objectives of the business. The inputs required to achieve such extreme quality objectives are often so expensive that the necessary product price becomes more than the market will bear. The contribution of food technology in this business context must be to specify what is feasible. Otherwise, the decision will be driven by two (or more) conflicting objectives and will never be optimal.

EXAMPLES

Allocation of Milk Resources in Cheese Making

In a step in the manufacture of cheddar cheese, a variety of milk resources are blended in a cheese vat with other ingredients (5). Making the blend is called standardizing, and the resulting blend is called cheese milk. The blend is allowed to coagulate before other processing steps are taken. The milk resources used are valuable; the cheesemaker pays to acquire them. Furthermore, the cheese yield and cheese quality are affected by which combination and proportions of resources are used in the blend. In the United States, the composition of the resulting product (which is determined by the composition of the cheese milk) is also constrained by government regulations and definitions. A recurring decision that the cheesemaker faces is determining the constituent amounts for each of the inputs to the cheese milk. The decision is constrained in many ways, but it is made regularly with the ordinary business objective of making a profit.

Technological investigation has determined a cheese yield formula (6) that allows the cheesemaker to predict the cheese yield if enough detail is known about the casein, fat, salt retention, and water retention of this step in the process. This technological model requires that the casein-to-fat ratio of the cheese milk be restricted to a small range. If this is done, the formula can be used to predict pounds of output for a given set of inputs. The potential inputs to the cheese milk include milk, condensed skim milk, nonfat dry milk, whey protein concentrate, cream, frozen cream, water, and other ingredients. Almost all the inputs affect the casein-to-fat ratio and the cheese yield. Because each of these resources will have different prices at different times, the cheesemaker can optimize the cheese milk by making the optimal choice of resources for the current set of price and availability circumstances.

This optimization opportunity may be modeled as a linear program (7). The variables are the inputs to the cheese milk; the constraints may be constructed out of these variables to restrict the casein-to-fat ratio and to reflect government regulations, company policies, resource availability, and other restrictions on the cheese product. An objective measure may be constructed from the variables and corresponding costs coupled with the sales value of the resulting product. The advantage of casting this as a linear program is that the resulting model can then be solved by computer to quickly determine the optimal constituents of the cheese milk. The result will simultaneously reflect the technological, legal, and economic realities facing the cheesemaker. Solutions generated this way have been shown to have a profound impact on the profitability of the cheesemaker. Without such a model and the computer support, cheesemakers can make good guesses, but they cannot be assured of regularly choosing optimal inputs to the cheese vat.

In this example we have a high-volume product (the vats may hold 20,000 lb of cheese milk) and variable inputs (the fat and solids content of incoming milk varies from farm to farm and animal to animal) and the recurring decision of how to standardize the cheese milk to achieve uniform outputs. Because of its recurring nature, this decision has important economic consequences for the cheesemaker. Investigation of cheese technology provided the model (the cheese yield formula) that enables the application of optimization methods to assist with the decision of how to standardize the cheese milk.

Formulating a Cheese Topping

When a new product is fielded, there is the opportunity to establish its formulation. This decision affects the economics associated with the product and the consumer's perception of the product. The challenge is to devise a formulation that is technologically feasible and commercially viable. Because there are many alternatives to any formulation, some strategy has to be adopted to determine the desired formulation. An optimization strategy provides a good way to sift through alternatives and come up with an attractive formulation.

Cheese toppings may be manufactured by making processed cheese in such a way that the substance has many potential consistencies and textures. Such material may be mixed with other ingredients such as onions, chives, bacon bits, red pepper, and many more alternatives. The resulting product may be used by consumers as a topping for vegetables, potatoes, and salads. Properly formulated, it may be used hot or cold in a variety of ways. Details about the formulation of the plain cheese topping may be found in Hanrez-LaGrange (8). For this illustration we will consider the problem of determining what proportions of the noncheese ingredients added to the plain cheese toppings will make the best product. Preliminary studies have guided the decision makers to the point of determining what combination of red pepper pieces and nacho flavor will make the best cheese topping.

With a new product in the laboratory or pilot stage, determination of an acceptable objective measure for optimization is difficult. This measure will be used to compare alternative formulations, but it should indicate the performance of the product when it is ultimately marketed. Because this amounts to a forecast, the measure is beset with intrinsic variability problems of all forecasts. A procedure is to ask a sample of consumers to taste and then rate the product. Presuming this sample represents the intended market, the formulator may use the rating as a guide for adjusting the formulation. The connection between the formulator and the consumer panelist is a ballot requiring the panelist's responses. The questions used to solicit these responses can be used to fashion an objective measure. Although there are many ways to phrase such questions, care should be taken that the responses somehow relate to the potential market for the product. One possibility is ask the consumer to rate the product as acceptable or not acceptable. If the sample of tasters represents the target market for the product, the set of purchasers will be a subset of the set of acceptors. The bigger the set of acceptors, the larger the potential market. Actual sales will depend on many factors not controllable by manipulating the cheese topping formulation, such as advertising, distribution, price, and the competition. This acceptor set size measure can be taken for any potential formulation and becomes a way to compare formulations in order to discard the less desirable ones. Other measures are possible, but for this example, the acceptor set size will be used as the objective measure.

The strategy, then, is to ask a sample of consumers whether various formulations of this product with varying amounts of red peppers and nacho flavoring are acceptable. The objective is to find a formulation that will maximize the acceptor set size when the product is marketed. The acceptor set size is presumed to be a function of the formulation. Because the intention is that the product is to be sold in large volume, this function may be presumed to be continuous and differentiable over a reasonable range of values for input amounts of red pepper and nacho flavor. For such a function the mathematical construct called the gradient exists and may be used as a guide in seeking a better (bigger acceptor set size) formulation. The gradient of a function is a vector of partial derivatives of the function with respect to the variables. It has the property that it always points in the direction in which the function is increasing most rapidly. The advantage here is that we may estimate the partial derivatives of this function without knowing an explicit formula for the acceptor set size function. Thus, the gradient may be estimated and improvements to the acceptor set size function may be made by determining the gradient, taking a step in that direction, and retesting to determine how to adjust the formulation further. In symbols:

$$A = f(x, y) \text{ is the acceptor set size function grad}(A) = (f_x, f_y)$$

where x is the amount of nacho flavor, and y is the amount of red pepper

If (x_0, y_0) is the initial formulation, then

$$(x_1, y_1) = (x_0, y_0) + k \times \text{grad}(A)(x_0, y_0)$$

where k is the step size in the gradient direction. This process of estimating the gradient and determining the next test formulation ends when there is no more improvement possible (the gradient is at or near (0, 0)) or when the noise in the data overpowers the information in the estimate of the gradient.

Partial derivatives are measures of rates of change in the direction parallel to the axes of the chosen variable. To estimate the value of the partial derivative at a given formulation requires that functional estimates be generated at higher and lower values of the chosen variable while the other variables are held constant. In this example, the initial formulation was $x_0 = 3\%$ and $y_0 = 13.75\%$. To obtain estimates of the partial derivative required functional estimates for 2%, 3%, and 4% nacho flavor while holding the amount of red pepper in the mix at a constant 13.75%. To get the other partial derivative required holding nacho flavor at a constant 3% and getting functional estimates of the acceptor set size with red pepper at 8.75%, 13.75%, and 18.75%. The results of the initial test are shown in Figure 2.

The gradient vector for this test was estimated as:

$$\text{grad}(A)I(x_0, y_0) = (0.50, -0.38)$$

which implies that a better formulation is possible with more nacho flavor and less red pepper. Choosing the step size depends on the product and ingredient circumstances and the judgment of the investigator. The next step in the optimization process was to test with the formulation set at

$$(x_1, y_1) = (4.35\%, 12.67\%)$$

The functional estimates for the acceptor set size are shown in Figure 3.

Further adjustments in this formulation are possible providing that big enough samples of consumers can be

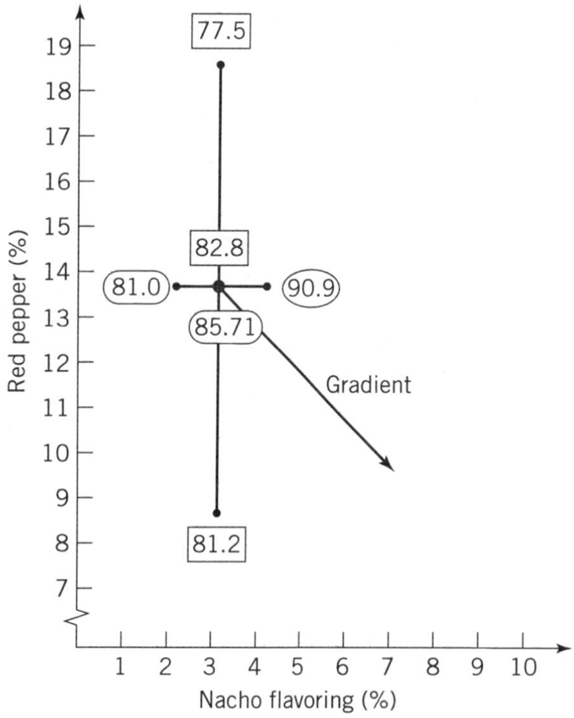

Figure 2. Determining the initial gradient in the ingredient space. Note that the gradient in the figure suggests the objective value will increase if the amount of red pepper decreases and the amount of nacho flavor increases.

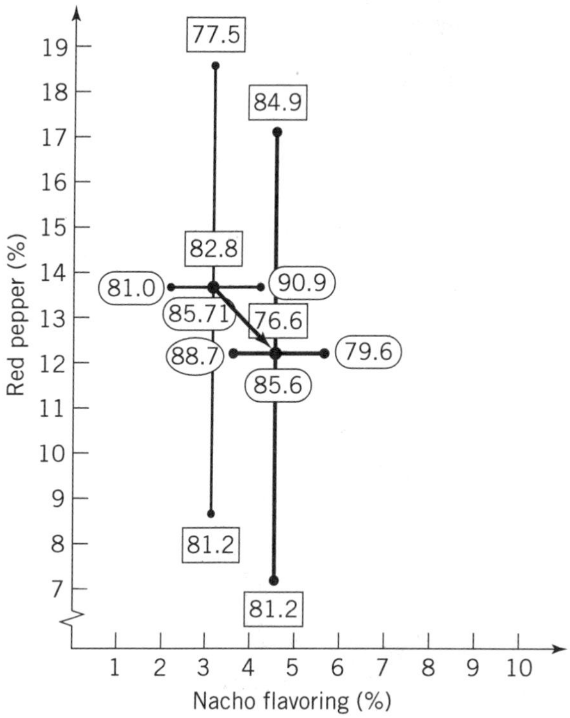

Figure 3. The values after one step in the gradient search process.

used to generate the accuracy necessary to make the results useful. This approach can be used for more than two variables, but the number of formulations for which data must be gathered also grows.

This sketch of a gradient search method leaves out many details of how the estimates are generated and used to determine the next formulation. More information may be found in Hanrez-LaGrange (8) and Hanrez-LaGrange and Norback (9). This approach has some similarity to response surface methods (RSM). The main difference is that in RSM, a formula for the objective measure must be generated, and this estimation of the objective measure then may be searched for a maximizing value. Because this function is maximized over the entire range of feasible formulations, it usually requires more expensive data collection than direct estimates of the gradient for a specific formulation and is of little value beyond its use in finding a maximum. Besides finding the maximum, the gradient search approach shows the impact of different components on the objective measure.

The weaknesses in this method are in the determination of the step size at each iteration, in the noise intrinsic to sampling from a population of consumers, and the fact that it may be suboptimal if the objective measure is multimodal. The latter difficulty can be avoided by careful prescreening of the product and its potential formulations followed by restriction of the feasible formulations so that only one maximum may occur over the range of ingredient values being considered. In every test case, the procedure has yielded good solutions to formulation problems in two or three iterations. The gradient search procedure may be used for any functional objective measure, not just for the acceptor set size measure.

Making and adjusting formulations will have important economic consequences in high-volume situations. In this case, the food technology constrained the formulation, whereas the business objective of making the acceptor set size largest adjusted the formulation among the feasible alternatives. The situation could also be constrained by the cost of the inputs to the product, by company policy, or by government regulation. Decision support methods (gradient search) were applied to assist in determining the right formulation for the product. Such methods could be applied again if the product were to be reformulated for a substitute ingredient. Small variations in a formulation can have important consequences to the consumer's response to the product and to its profitability. With infinitely many formulations to choose from, these methods provide an organized means of choosing a good one.

Extending the Shelf Life of Fresh Seafood

An ongoing effort by the purveyors of fresh seafood is the opening of inland markets. This can be done by speedier transportation, but this turns out to be costly and sometimes hard to arrange if these products are to enter ordinary marketing channels. This business objective can also be advanced by improving the holding technologies for fresh seafood products (4). With such technologies the product can come through shipping and handling and still be desirable to the consumer. Perhaps the most common of these technologies is refrigeration, but this too has limits, which still leaves large markets untapped.

Two emerging technologies have application here: modified atmosphere packaging and the use of a sorbate dip on the product. With modified atmosphere packaging, the product is enclosed by a barrier material and the atmosphere over the product is replaced by one without oxygen. The permeability of the barrier material to oxygen is a controllable variable—the packaging material must be chosen (and paid for) by the seafood seller. The sorbate dip technology requires that the product be coated with a sorbate solution of variable concentration. Because these two technologies can be used simultaneously, a reasonable question is, what combination of both would be most beneficial to the seafood seller?

This was investigated by gradient search methods. Consumer response data were taken by asking consumers to taste and then to rate the sample tasted as acceptable or unacceptable. (Other data were taken as well.) The purpose was to determine the consumer response to applications of combinations of these technologies by estimating the acceptor set size over time. The system was constrained by the limits on the sorbate concentration that seemed acceptable and by available barrier materials. The acceptor set size function was assumed to be differentiable and the data taken were used to estimate the gradient of this function. The gradient of any differentiable function has the property that it points in the direction of the steepest ascent of the function. Thus, after each completed test, new information was available that indicated what adjustment in the technologies could be made to make the combined application better. This approach is an example of a gradient search method and is a decision-making tool that is broadly applied in optimization, often as part of a computer algorithm. The novelty here is that data collection must be done before each iteration of the algorithm, making the whole procedure considerably slower than most algorithms.

A starting permeability of the barrier film and concentration of sorbate is chosen. Tests are run for higher and lower permeabilities at this sorbate concentration and for higher and lower sorbate concentrations at this permeability. The resulting data allows the estimation of the gradient vector at the starting point. If the gradient vector is at or near zero, little improvement is possible, otherwise a step in the direction of the gradient will improve the performance of the combined technologies. New values for permeability and concentration are chosen and the test is repeated.

The application of this procedure portrayed consumer response through time to different combinations of technologies. An important question not addressed is, is the extra expense of implementing and using the technology worth the resulting extension in shelf life? Addressing this question would mean a more complicated model would have to be developed and perhaps different computational methods would be required.

Managing Material Flow and Batching

Production scheduling, production quality control and inventory management in the food industry lead to a focus on the flow of materials through the production facility. Many important decisions depend on accurate and timely information regarding the amount and quality of materials that are undergoing some step in the manufacturing sequence. This information is used by the manager to determine product costs, procurement, or harvest requirements, and to project output product amounts and to assist in setting prices. Product safety, good product qualities, and product profitability require that the manufacturing sequence conform to the requirements of the food as material and that the processing facilities be managed efficiently. Because of the typical low margin per sale, the throughput for such facilities must be large. This means that small improvements in efficiency will have important consequences for profitability. In many food production facilities, the manufacturing sequence is further complicated because either the technology or the economics require batch processing. Although the ideal may be "continuous processing," the technology may not be available as in the manufacture of cheddar cheese, the blending of processed cheese, or the processing of low-acid foods. In other cases, the continuous technology is too expensive or otherwise infeasible to be adopted as in the retort processing of low-acid foods. This leads to complications in achieving target productions, especially in cases where batches of intermediate materials are used in multiple products. The decision making surrounding procurement, production scheduling, and quality control is strongly connected to the technology of the food being processed. This technology once again determines the manufacturing sequence and constrains the amount of throughput.

The flow of materials in canning requires that incoming vegetables be washed, sorted by grades, placed in a container that is partially filled with brine, then retort processed before being cooled, labeled, and placed in cases. Although not all of these steps are used in every manufacturing sequence, brining and retort processing are both batch-processing steps. The flow of materials provides a structure to organize information regarding this manufacturing sequence. Such information can be fed forward to control subsequent steps in the processing. For example, the size of peas (as well as the variety of peas and the formulation of the brining solution) will affect the time and temperature settings for the subsequent retort processing. Amounts of materials and corresponding costs may be kept in this same flow of materials structure as well as information required by government regulation or company policy.

Using a flow of materials structure provides great organizational advantage. Although it is possible to do the arithmetic and track the resources by hand for a few products, with many products that use many intermediate products the mass of detail that accumulates is soon overwhelming. Using the flow of materials as a structure provides more than organizational convenience, however. The development of the so-called Gozinto matrices can be done from the flow of materials structure (10,11), and their connection to materials requirements planning (MRP) means that accurate projections of materials use, costs, and inventory amounts can be conveniently made. Although the time phasing of inventory and production is not as important a problem in the food industry as it is in fabrication industries, tracking the use of intermediate products, product costing, and batching are.

A Gozinto matrix is created through a procedure that organizes production information into a lower triangular, invertible matrix. The rows and columns of this matrix correspond to products and inputs to products, organized so that no row corresponding to an input to a product occurs above the row corresponding to the product in the matrix. This means that the bottom rows in the matrix correspond to ingredients purchased or harvested from outside the production facility, whereas the top rows correspond to the output products of the facility. The rows in between correspond to intermediate products that may find there way into many output products. More details may be found in Mize and coworkers (11).

Inverting this matrix and doing some elementary matrix algebra provides a means of anticipating and tracking inventories and production. For specified target production, the precise input needs of each ingredient or intermediate product may be determined. Inventories would be kept to a minimum if we could procure or manufacture exactly these quantities of input materials. But getting exactly the quantity of intermediate products required is complicated by batch processing and by the use of batch output in multiple products. The amount of brine needed for a target number of cases of canned vegetables is rarely a whole number of batches of brine. More likely, the target will require 12.37 batches of brine, or some other in between number. Is it better in these cases to make 12 batches or 13 batches or to make partial batches (ie, 0.37 batches)?

If we make more batches than the target, we will have excess quantity of batch output, which must be stored or discarded, or used in excess production of final product, which then must be stored. If we make less than the target production, we may not meet demand (causing stock outs) and opportunity costs associated with lost sales. If we make partial batches, we face the same labor and fixed costs associated with full batches but get less output. Furthermore, partial batches often imply important technological changes–especially in the food industry—since batch inputs may not scale linearly. Salt and spices are good examples of inputs for which it is not good enough to simply cut the amounts in half when the batch is half the size. If this technological information is not available, product quality will vary when the batch is rescaled.

In these circumstances, what is the best the production manager can do? If the costs just described are understood and can be estimated, an optimization model can be built and used to determine how many batches of input are best for target production amounts. The difficult part is the construction of an objective measure, which simultaneously takes into account the costs of the different alternatives as well as potential revenues from the products. Measuring profits for a production run will require determining the revenue from each of the products made and subtracting ingredient costs, batching costs, and the costs from overproduction and underproduction. The resulting detailed model can be found in Chung (12). For purposes of this review suffice it to say that it is possible and profitable to apply optimization in batching situations. The goal of the model is to find not only a product mix but also a batch mix that maximize the penaltied profits.

Although other models are possible, this example shows how such a model might be constructed. Such models may require that certain values be integers but still may be solved by computer methods. The result will be optimal choices regarding the number of batches produced that reflect the costs and revenues that the production facility faces, rather than the intuition of the manager.

CONCLUSION

The examples given here give some ideas how food technology may be incorporated in the structure of decision-making models. When there are many alternatives to choose from, the effort required to search for a best one often leads to more profit. The business of food technology and the food technologist in this decision-making realm is to constrain the set of alternatives. The objectives of the organization can then drive the decision-making process to the choice of an appropriate alternative. This optimization philosophy helps keep all participants in food industry decision making pulling in the same overall direction, even though interests and expertise of the individuals may be vastly different.

What is seldom investigated is this connection between the technology of food and the decision making done by food industry managers. If this connection is managed properly, important consequences for the efficiency and profitability of the business can be realized. The procurement of materials and the production, marketing, and distribution of food products all depend on the nature of food as material and the consumer's perception of that material as each spoonful is consumed. Approaching this from an optimization perspective—arraying the alternatives, measuring the alternatives, and choosing a desirable one—is a useful and profitable way to manage.

BIBLIOGRAPHY

1. W. A. Shewhart, *Economic Control of Quality of Manufactured Product*, D. Van Nostrand, Princeton, N.J., 1931. Republished by American Society for Quality Control, Milwaukee, Wis., 1980.
2. E. L. Grant and R. S. Leavenworth, *Statistical Quality Control*, 7th ed., McGraw-Hill, New York, 1996.
3. N. M. Gordon and J. P. Norback, "Choosing Objective Measures When Using Sensory Methods for Optimization and Product Positioning," *Food Technol.* **39**, 96, 98–101 (1985).
4. W. F. Sharp, J. P. Norback, and D. A. Stuiber, "Using a New Measure to Define Shelf Life of Fresh Whitefish," *J. Food Sci.* **51**, 936–939, 959 (1986).
5. G. L. Kerrigan and J. P. Norback, "Linear Programming in the Allocation of Milk Resources for Cheese Making," *Journal of Dairy Science* **69**, 1432–1440 (1986).
6. L. L. Van Slyke and W. V. Price, *Cheese*, Ridgeview, Independence, Ohio, 1979.
7. J. P. Norback and S. R. Evans, "Optimization and Food Formulation," *Food Technol.* **37**, 73–76, 78, 80 (1983).
8. V. Hanrez-LaGrange, *An Optimization Procedure with Sensory Analysis Data in the Development of Process Cheese Toppings*, Master's Thesis, University of Wisconsin, Madison, Wis., 1987.
9. V. Hanrez-LaGrange and J. P. Norback, "Product Optimization and Acceptor Set Size," *Journal of Sensory Studies* **2**, 119–136 (1987).
10. A. Vazsonyi, *Scientific Programming in Business and Industry*, John Wiley, New York, 1958.
11. J. H. Mize, C. R. White, and G. H. Brooks, *Operations Planning and Control*, Prentice-Hall, Englewood Cliffs, N.J. 1971.
12. H. K. Chung, *Planning and Optimization for Logistics Management in the Food Industry*, Ph.D. Thesis, University of Wisconsin, Madison, Wis., 1999.

John Norback
University of Wisconsin
Madison, Wisconsin

FOOD REGULATIONS: INTERNATIONAL, CODEX ALIMENTARIUS

International food regulations govern the import and export of both raw agricultural commodities and processed food products to ensure food safety, control unwanted release of exotic species of plants and animals, and limit the spread of both plant and animal diseases. International food regulations also include standards for trade, including units of measure and minimum quality standards.

Until recently, there were essentially no unified international food regulations. Each nation negotiated with specific trading partners in cooperative agreements. This was a process driven by the standards of the importing country, since the exporter was responsible for ensuring that the product in question complied with the regulations of the importing country. Because national food regulations varied widely from nation to nation, this proved to be an inefficient process. This process also allowed countries to use food regulations as de facto barriers to foreign competition. Although there are still no legally binding international food regulations, increasing international trade and broadening participation under the World Trade Organization (WTO) is now forcing a rapid harmonization of international standards. A set of unified international food standards has emerged, with the force of the world's largest international trade organization behind it. More importantly, the WTO has established a process for the adoption of future standards, as well as for settling trade disputes involving food standards.

HISTORY

The General Agreement of Tariffs and Trade (GATT) was created in 1947 in an effort to foster international trade. From 1947 to 1994, the GATT served as the umbrella organization for international trade, establishing standards and rules to allow countries to trade with each other in an open and fair manner. Despite major success at fostering trade in nonfarm products, the GATT contained few effective rules for agricultural trade.

Since 1948 national food safety and animal and plant health measures that affect trade were subject to GATT rules. These rules required nondiscriminatory treatment of imported agricultural products from different foreign suppliers, and, in theory, imported products were supposed to be treated no less favorably than domestically produced agricultural goods. However, the agreement allowed that for purposes of protecting human, animal, or plant health, governments could impose more stringent requirements on imported products than they required of domestic goods. This clause was often intentionally misused to keep imported farm products out of a country to protect domestic producers. In 1979 the Tokyo Round of trade negotiations attempted to limit abuse of this clause and produced the Agreement on Technical Barriers to Trade. In this agreement, governments agreed to use relevant international food standards—except when they considered that these standards would not adequately protect health. Although the agreement established the precedent of using international standards, there remained a large loophole for using sanitary and phytosanitary measures as barriers to trade.

In April 1994 the GATT completed the final act of a series of international trade negotiations, commonly referred to as the Uruguay Round, including reforms of international food regulations. The negotiations culminated in a treaty that created the WTO (which superseded the GATT as the umbrella organization for international trade) and contained the WTO Agreement on Sanitary and Phytosanitary (SPS) Measures concerning the application of food safety and animal and plant health regulations.

The SPS agreement, now the centerpiece for international food regulations, stated that the WTO and its member countries would adopt the existing international standards, guidelines, and recommendations developed by specific international organizations as the basis for its food standards. The SPS agreement recognizes the standards of the Codex Alimentarius (Codex), the International Office of Epizootics (OIE), and the International Plant Protection Convention (IPPC). Importantly, the SPS agreement contains specific guidelines for the continual creation and adaptation of food standards to respond to new developments and knowledge in food safety and plant and animal health. The SPS also outlines a specific dispute settlement procedure, with effective timetables and requirements. In effect, the SPS agreement, administered through the WTO and based on the recommendations of independent organizations, is the first effective set of unified international food regulations.

The SPS agreement is an interesting and important success story. It demonstrates that voluntary, market-oriented collaboration can achieve difficult goals, and it endorses the existence and work of international technical advisory bodies. The three organizations to which the WTO defers for its food standards were created independently and prior to the creation of the WTO. None of these organizations had legal authority to impose their recommendations on any nation. But the recommendations and guidelines set by these organizations were well founded, as they represented the consensus opinions of scientists and experts from around the globe.

Although the adoption of the standards of these organizations created a much needed baseline for international trade, it is important to note that these standards do not supersede the ability of nations to maintain their own food safety standards. First, these rules apply only to member nations of the WTO. But even WTO member nations maintain control of their food safety standards. No member nation is required to adopt any international food standard as its own. However, if a member nation maintains a food standard stricter than an established international standard that results in restrictions on trade, WTO rules require that the nation demonstrate the need for the stricter standard using sound scientific principles. The SPS agreement also stipulates: "Members shall accept the sanitary or phytosanitary measures of other Members as equivalent, even if these measures differ from their own . . . if the exporting Member objectively demonstrates to the importing Member that its measures achieve the importing Member's appropriate level of sanitary or phytosanitary protection" (1). In other words, each country is free to devise its own sanitary and phytosanitary measures as long as those measures achieve the proper level of sanitary/phytosanitary protection for its exported product. The SPS agreement also states that sanitary and phytosanitary measures "shall not be applied in a manner which would constitute a disguised restriction on international trade" and standards must be applied "only to the extent necessary to protect human, animal or plant life or health, is based on scientific principles and is not maintained without sufficient scientific evidence" (1).

To ensure that the standards-setting process remains relevant to WTO member countries, the SPS agreement stipulates that "members shall play a full part, within the limits of their resources, in the relevant international organizations and their subsidiary bodies," with particular reference to the Codex, OIE, and IPPC "to promote within these organizations the development and periodic review of standards, guidelines and recommendations with respect to all aspects of sanitary and phytosanitary measures" (1). The SPS also created a Committee on Sanitary and Phytosanitary Measures to monitor the process of international harmonization and coordinate efforts in this regard with the relevant international organizations.

The Codex

The Codex Alimentarius (Codex), to which the WTO defers for food safety standards, was created in 1962 by the United Nation's Food and Agriculture Organization (UNFAO) and the World Health Organization (WHO) under the direction of the FAO/WHO Food Standards Program. The Codex Alimentarius, which means "food code" in Latin, was created "to guide and promote the elaboration and establishment of definitions and requirements for foods, to assist in their harmonization and, in doing so, to facilitate international trade" (2). The Codex is composed of more than 160 member countries, each represented by delegates who participate through the various committees that comprise the Codex (see Fig. 1).

There are several committee categories within the Codex: the executive committee, general subject committees

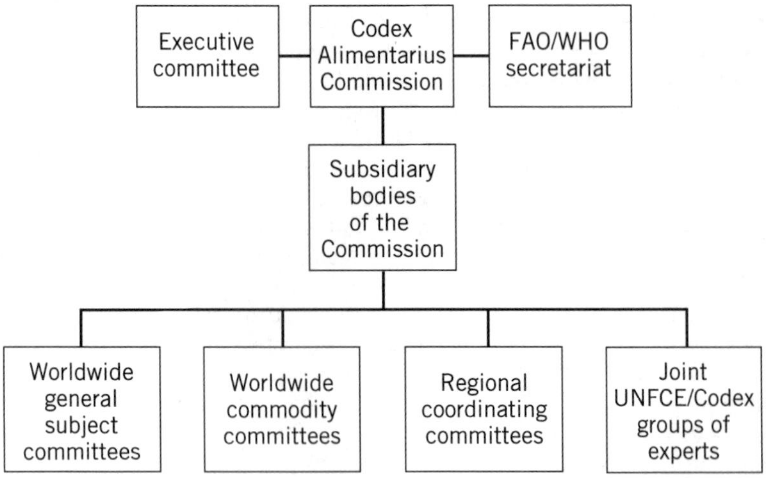

Figure 1. Subsidiary bodies of the Codex Alimentarius Commission.

and commodity committees. The executive committee is the executive organ of the Commission; that is, it is responsible for its functioning. It is comprised of a chair and three vice-chairpersons, as well as one member representing each of six geographic regions of the globe: Africa, Asia, Europe, Latin America and the Caribbean, North America, and the southwest Pacific. Actual standards and food safety recommendations are established either through a general subject committee, such as food labeling, or commodity committees, such as milk and milk products (see Table 1). There are 8 general subject committees and 14 commodity committees. As the Codex has operated for more than 30 years and has been establishing standards and guidelines throughout this time, most committees meet only on an irregular basis when developments require new or revised standards.

Table 1. Codex Commodity and General Subject Committees

Worldwide Codex Commodity Committees (and their host countries)
Cocoa products and chocolate (Switzerland)
Sugars (United Kingdom)
Processed fruits and vegetables (United States)
Fats and oils (United Kingdom)
Soups and broths (Switzerland)
Fresh fruits and vegetables (Mexico)
Milk and milk product (New Zealand)
Cereals, pulses, and legumes (United States)
Processed meat and poultry products (Denmark)
Meat hygiene (New Zealand)
Vegetable proteins (Canada)
Fish and fishery products (Norway)
Nutrition and foods for special dietary uses (Germany)
Natural mineral waters (Switzerland)
Worldwide Codex General Subject Committees
Food labeling (Canada)
Food additives and contaminants (Netherlands)
Food hygiene (United States)
Pesticide residues (Netherlands)
Veterinary drugs in foods (United States)
Methods of analysis and sampling (Hungary)
Food import and export inspection and certification systems (Australia)
General principles (France)

Examples of standards established by the Codex general committees include limits for allowable levels of pesticide residues in food products, guidelines for hygienic handling of food products, limits on food additives and contaminants (ie, preservatives, coloring agents, etc), veterinary drug residues, and even analysis and sampling methods.

Examples of commodity committees are cereals, pulses, and legumes; milk and milk products; cocoa products and chocolate; sugars; and natural mineral waters.

The Codex works closely with both its parent organizations, the UNFAO and the WHO, as well as the WTO to ensure that its work addresses the relevant concerns of these organizations.

The OIE

The OIE, the organization responsible for animal health, was created in 1924. The OIE is an intergovernmental organization headquartered in Paris, France. It consists of more than 150 member countries who, as with the Codex, are represented by delegates participating through various commissions and committees. The OIE is organized into three types of commissions: an administrative commission, regional commissions, and specialist commissions, as well as working groups.

The International Committee, the highest authority of the OIE, meets once a year. Its main function is to coordinate the work of the OIE commissions and adopt international standards for animal health, especially for international trade, as well as to adopt resolutions on the control of major animal diseases. Due to the geographic nature of animal health problems, regional commissions were created to address the specific animal health concerns for the five major geographic areas of the world: Africa, the Americas, Asia, Europe, and the Middle East.

To facilitate its role in harmonizing regulations for trade in animals and animal products, the OIE maintains the *International Animal Health Code*, prepared by the

International Animal Health Code Commission, as well as a *Manual of Standards for Diagnostic Tests and Vaccines* prepared by the OIE's Standards Commission. These publications establish minimum standards to protect animal health (both domesticated and farm animal as well as wild animals) and to prevent the spread of animal diseases that might result from the trade in animals or animal products.

Examples of standards established by the OIE are guidelines for assessing the health of animals before, during, and after transport of animals in international trade, guidelines for destroying pathogens in animal products, and recommended diagnostic procedures.

The International Plant Protection Convention

The IPPC is the phytosanitary (plant health) standard-setting organization of the SPS agreement and is the basis on which countries collaborate to prevent the spread and introduction of plant pests and plant products. The purpose of the Convention according to the IPPC preamble is "international cooperation in controlling pests of plants and plant products and in preventing their international spread, and especially their introduction into endangered areas." The scope of the Convention extends to the protection of both cultivated and natural flora and includes both direct and indirect damage by pests.

From its creation in 1951 until 1992, the IPPC existed as an international agreement administered through the FAO and implemented primarily through the cooperation of regional and national plant protection organizations. However, the FAO recognized the role that the IPPC would play in the then-developing WTO/SPS agreement and created an IPPC Secretariat in 1992. The FAO also established a standard-setting process and formed the Committee of Experts on Phytosanitary Measures (CEPM) in 1993. The IPPC currently has more than 100 member countries and is administered through the UNFAO's Plant Protection Service.

In addition to the Secretariat and CEPM, amendments to the Convention resulted in the creation of the Commission on Phytosanitary Measures to review the state of plant protection in the world, provide direction to the work of the IPPC Secretariat, and approve international standards.

The Secretariat is responsible for coordinating the work of the IPPC and, most importantly, the elaboration of International Standards for Phytosanitary Measures (ISPMs). These standards are developed by a process that begins with submission of proposed standards from the Secretariat, by expert groups organized by the Secretariat, or by regional or national plant protection organizations. Draft standards are reviewed by the CEPM and are sent to member governments for consultation before being submitted to the Commission for adoption.

IPPC standards fall into three categories: reference standards, concept standards, and specific standards. To date, the IPPC has produced primarily reference and concept standards to support the subsequent specific standards.

The IPPC also has a dispute settlement clause to help resolve international trade disputes, although this process is separate from the WTO dispute settlement process and is not legally binding.

CONCLUSION

The demand for agricultural products is expected to more than double worldwide over the next 40 years, through both population growth and dietary improvement in developing country populations. Many of these countries lack adequate land and agricultural resources to meet their expected food needs domestically without significant destruction of wildlife habitat, elevating the importance of international agricultural trade. Consumers throughout the world are also raising their food safety expectations, thus the need for viable international food safety standards.

BIBLIOGRAPHY

1. *Agreement on the Application of Sanitary and Phytosanitary Measures*, World Trade Organization, Geneva, Switzerland, 1995.
2. *Codex Alimentarius: FAO/WHO Food Standards*, United Nations FAO, Rome, Italy, 1996.

GENERAL REFERENCES

"Agreement on the Application of Sanitary and Phytosanitary Measures," available from the World Trade Organization at *www.wto.org/wto/goods/spsagr.htm.*

"Understanding the WTO Agreement on Sanitary and Phytosanitary (SPS) Measures," available from the World Trade Organization at *www.wto.org/wto/goods/spsund.htm.*

All background material used to prepare this article came from and is available at the following World Wide Web addresses:

World Trade Organization, www.wto.org
Codex Alimentarius, www.fao.org/waicent/faoinfo/economic/esn/codex/ and www.fsis.usda.gov/OA/codex/
OIE, www.oie.int
IPPC, www.fao.org/ag/agp/agpp/pq/

ALEX A. AVERY
DENNIS T. AVERY
Hudson Institute
Churchville, Virginia

FOOD SAFETY AND RISK COMMUNICATION

Foods commonly contain a number of microorganisms (bacteria, fungi, viruses, etc) as well as potential chemical contaminants (food additives, pesticide residues, veterinary drug residues, metals, etc) that may pose some level of risk to consumers. Although the magnitudes of the risks may vary dramatically among the various risk agents, it is clear that significant public attention focuses on virtually all types of food safety risks, both large and small. Effective communication of such food safety risks to the general population as well as to lawmakers is an important goal of health scientists, government agencies, and food industry representatives.

The process of risk communication was defined by the National Research Council (NRC) to include an interactive process of exchange of information including information on the nature of the risk and other messages regarding concerns, opinions, or reactions to the risk messages (1). With respect to communication of food safety risks, common risk communication methods have been mostly one-way and technocratic and frequently involve regulatory agencies, industry, or government officials transmitting the results of risk assessments with the aim that the public accept the risk messages and act accordingly (2). In the last decade this view has been challenged, and alternative risk communication methods have been developed to improve dialogue with the public and affected or interested parties. It has been suggested that such a dialogue should provide effective public involvement in the decision-making process and not merely reflect transference of food safety risk information from experts to nonexperts.

RISK MESSAGES

Risk messages are only a part of the interactive risk communication process and are often designed to inform the public about food safety risks. Due to the highly technical nature of the subject matter, the task of formulating food safety risk messages that are clear, accurate, and not misleading is a challenging one. Complex language and technical jargon are commonly used in risk messages, leading to information that is often incomprehensible to the public. Effective risk messages should be developed using clear and plain language and should transmit relevant information in addition to recommending practical actions that the public can take (1).

Effective risk communication requires communicators to recognize and overcome several obstacles that are rooted in the limitations of scientific risk assessment and in public understanding. Technical barriers to effective risk communication include the need to make assumptions and subjective judgments in the risk assessment process. In the risk assessment of a food chemical there is a need to consider any possible gaps in knowledge that will lead to uncertainties in the risk estimates (3). Uncertainties frequently arise from the use of animal models in the risk assessment process. Typical human health risk assessments rely on results from laboratory animal studies that must be extrapolated to humans. Evaluating results from toxicology studies where chemicals are administered to laboratory animals at a very high dose requires extrapolation to humans exposed to a low dose, sometimes over a lifetime, generating different levels of uncertainty. According to Young (4), the greater the uncertainty about a given effect, the more likely it is to be overestimated. Experts often adopt measures based on conservative assumptions, which provides a tendency to dramatically overestimate the predicted levels of risk (5).

Risk messages should present a clear explanation of the scientific evidence, including data gaps, significance of uncertainties, and possible differences in opinion among experts. In formulating risk messages it is important to be truthful and to discuss practical factors to be considered for risk management/control. Risk communicators should explain steps to be followed for risk reduction or, when applicable, address possible consequences (health, economic) resulting from risk control measures. The common views of experts should be expressed, and consumer concerns need to be considered (6). Nonetheless, consumer reaction to food safety risk messages may often involve misunderstanding of the scientific facts and may lead to general distrust of those providing the messages.

RISK COMPARISONS

The sheer complexity and uncertainty inherent in risk assessment provides a significant barrier to public understanding and appreciation of the magnitude of risks. One method of explaining risk information is to make comparisons to other risks. Some feel that comparing and contrasting risks with similar assessments of commonplace situations help the consumer better understand risks (7), because comparisons are more meaningful to the public than unfamiliar magnitudes (probabilities) associated with risks. Examples of such an approach include comparing the risks of cigarette smoking with the risks of motor vehicle accidents, home accidents, background radiation levels, or hang gliding.

Although risk comparisons can be helpful in communicating the magnitude of risks, they are not by themselves adequate determinants of decisions. Despite the appeal of using risk comparisons to put results of risk assessments in perspective, such risk comparison practices have been subject to criticism, because they ignore critical elements concerning public values and acceptability of different types of risks. Moreover, some comparisons involving different types of risk often ignore different levels of uncertainty inherent in the risk estimates.

With regard to chemical hazards, Ames et al. (8) and Gold et al. (9) ranked the potential human carcinogenic risks of exposures to a variety of environmental pollutants, synthetic pesticide residues, naturally occurring toxins, and pharmaceutical products using an index that relates predicted human exposure levels for carcinogens to their carcinogenic potency in rodents. Their results indicated that the risks posed by residues of synthetic pesticides or environmental pollutants ranked low in comparison to risks of naturally occurring carcinogens.

RISK AND CONSUMER PERCEPTIONS

Perception of risks among members of society is very complex and is determined by both biological and nonbiological factors. Due to the often highly technological content of risk messages, such messages can generate frustration on the part of the public who ultimately complain of the inaccessibility of the scientific knowledge expressed in those messages. On the other hand, if risk messages are oversimplified, they can also result in negative reactions from the public, such as distrust or unnecessary fear. While many experts define a food safety risk as a probability of an adverse health effect caused by exposure to some potentially hazardous agent in food, perceived risk may be defined as

Table 1. Qualitative Factors Affecting Risk Perception and Evaluation

Factor	Conditions associated with increased public concern	Conditions associated with decreased public concern
Voluntariness of exposure	Involuntary	Voluntary
Controllability	Uncontrollable	Controllable
Fairness	Inequitable distribution of risks and benefits	Equitable distribution of risks and benefits
Familiarity	Unfamiliar	Familiar
Origin	Caused by human actions or failures	Caused by acts of nature
Memorability	Memorable	Not memorable
Dread	Dreaded effects	Effects not dreaded
Catastrophic potential	Fatalities and injuries grouped in time and space	Fatalities and injuries scattered and random
Benefits	Effects irreversible	Effects reversible
Effects manifestation	Delayed effects	Immediate effects
Effects on children	Children specifically at risk	Children not specifically at risk
Victim identity	Identifiable victims	Statistical victims
Trust in institutions	Lack of trust in responsible institutions	Trust in responsible institutions

Source: Ref. 3.

the sum of hazard, derived from the risk assessment process, and outrage, which involves nonbiological, qualitative factors (10). Table 1 lists a variety of factors related to public concern and conditions that provoke high or low levels of outrage among consumers. It has been pointed out that the public pays too little attention to hazard while the scientific experts pay little or no attention to outrage, which explains the common differences between public and expert opinions concerning risks. Some risks possessing high hazard but low outrage may be of less public concern than those with low hazard but high outrage (10). The acceptability of a risk depends on those outrage factors. Many major food safety–related disputes relate more to risk acceptability than to the actual risks calculated. It is common that a voluntary risk is more acceptable than one perceived as being involuntarily inflicted on consumers. Slovic (11) provides many examples of how hazard and outrage influence public perception of risks.

Pesticide residues in foods provide a good example of how outrage factors may significantly influence public perception of risks. Consumer attitude surveys revealed that between 72 and 82% of the public deem pesticide residues as a major concern in food safety (12), although this contrasts greatly with the opinion of most health scientists and regulatory agencies who contend that the dietary risks from exposure to pesticides in foods are minimal (13). Pesticide residues are associated with several conditions of increased public concern, such as being involuntary, (largely) uncontrollable, unfamiliar, and inequitable (it is commonly assumed that consumers may be the ones taking the risks while food producers may be reaping the benefits from pesticide use). Significant concern is also expressed by the public with regard to new technologies such as food irradiation, biotechnology, engineered foods, and the use of growth hormones given to cows with the aim of boosting milk production. The scientific processes of risk assessment and risk management provide information concerning the risks, costs, and benefits of policy choices, but the ultimate management of the risks is an issue of social policy that requires decisions to be made on the basis of value choices (6).

BIBLIOGRAPHY

1. National Research Council, *Improving Risk Communication*, National Academy Press, Washington, D.C., 1989.
2. C. W. Scherer, "Strategies for Communicating Risks to the Public," *Food Technol.* **45**, 110–116 (1991).
3. C. K. Winter and F. J. Francis, "Assessing, Managing and Communicating Chemical Food Risks," *Food Technol.* **51**, 85–92 (1997).
4. F. E. Young, "Weighing Food Safety Risks," *FDA Consumer* **23**, 8–13 (1989).
5. P. H. Abelson, "Health Risk Assessment," *Reg. Toxicol. Pharmacol.* **17**, 219–223 (1993).
6. E. Groth, "Communicating With Consumers About Food Safety and Risk Issues," *Food Technol.* **45**, 248–253 (1991).
7. R. Wilson and E. A. C. Crouch, "Risk Assessment and Comparisons: An Introduction," *Science* **23**, 267–270 (1987).
8. B. N. Ames and L. S. Gold, "Too Many Rodent Carcinogens: Mitogenesis Increases Mutagenesis," *Science* **249**, 970–971 (1990).
9. L. S. Gold et al., "Rodent Carcinogens: Setting Priorities," *Science* **258**, 261–265 (1992).
10. P. M. Sandman, "Risk Communication: Facing Public Outrage," *EPA Journal* **13**, 21–22 (1987).
11. P. Slovic, "Informing and Educating the Public About Risk," *Risk Analysis* **6**, 403–415 (1986).
12. C. M. Bruhn et al., "Consumer Response to Pesticide/Food Safety Risk Statements: Implications For Consumer Education," *Dairy, Food and Environmental Sanitation* **18**, 278–287 (1998).
13. S. O. Archibald and C. K. Winter, "Pesticides in Our Food: Assessing the Risks," in C. K. Winter, J. N. Seiber, and C. F. Nuckton, eds., *Chemicals in the Human Food Chain*, Van Nostrand Reinhold, New York, 1990, pp. 1–50.

Elisabeth L. Garcia
Carl K. Winter
University of California
Davis, California

See also Food safety and risk management; Toxicology and risk assessment.

FOOD SAFETY AND RISK MANAGEMENT

The emerging field of food safety risk analysis is composed of three dependent yet separate fields: risk assessment, risk management and risk communication. This trinity of risk-related factors forms the basis for the decisions about chemicals requiring regulation and the type of regulation needed. The ultimate goal of the process is food safety, the prevention of adverse health effects due to exposure to hazardous substances in food. Many different chemicals enter the food supply, including pesticides, natural toxins (such as aflatoxins), metals, and intentionally used food additives.

Risk management is a process to determine what level of risk is significant and to identify, select, and implement options to reduce or minimize levels of risk (1). Probabilistic risk information obtained through the risk assessment process is provided to risk managers who determine what action, if any, should be taken to manage (control) the risk in question. Representatives of both the government and industry sectors are frequently involved in risk management activities.

Risk management provides a regulatory decision-making process that involves consideration of political, social, economic, and technological information in addition to simply considering the results from the risk assessment process. As such, risk management requires the use of value judgments on issues such as acceptability of risk and the reasonableness of the costs of control (2).

Although risk assessment and risk management activities must be coordinated, these entities must also maintain some independence to guarantee the scientific integrity of the risk assessment process and to minimize potential conflicts of interest between risk assessors and risk managers (3,4). Nevertheless, it has been suggested that the separation between risk assessment and risk management may not be advantageous or possible, and some authors propose an integrated model, which combines elements of risk assessment and management (5).

Risk management considers risk characterization results, searches for viable options or strategies to minimize or reduce the potential risk, and identifies risk control strategies, if needed, to reduce the risks to acceptable levels. Ideally, a monitoring step should follow to allow an evaluation of the impact of regulatory decision on public health.

Since risk management considers political, economic, and social issues in addition to risk characterization results, optimal risk management decisions should include input from a variety of stakeholders that may be impacted by such decisions. Stakeholders may represent either the public or private sectors and include consumer and environmental organizations, the food industry (producers, processors, distributors, retailers, and restaurants), regulatory bodies, the legislative sector, and trade associations.

CONSIDERATION OF RISK CHARACTERIZATION

Once the risk assessment process has been completed to provide a mathematical characterization of the risk and a description of the various uncertainties inherent in the calculation of the risk, risk managers will consider this risk estimate as the starting point from which to develop an appropriate risk mitigation strategy, if it is deemed that the risk should be reduced. It should be emphasized that optimal risk management efforts require scientifically appropriate risk assessments.

An important aspect in the consideration of the risk characterization is the need for experienced judgment. Inherent in the risk assessment process are large gaps in knowledge that require many choices to be made among competing models and assumptions; this introduces considerable uncertainty into the risk estimates that must be appreciated in the risk management process. Optimally, risk managers should be allowed the flexibility to make risk-related decisions using a weight-of-evidence approach that allows for the consideration of all available valid scientific data. It has been suggested, however, that risk assessments are often conducted using a strength-of-evidence approach in which experiments demonstrating positive toxicological effects are given more weight than any number of negative experiments of equal quality (6). This may be particularly true in the case of carcinogen risk assessment, where conservative assumptions may exaggerate risks greatly and, therefore, may distort regulatory practices (7).

RISK, COSTS, AND BENEFITS

In food safety decision-making processes, benefit–cost analysis is frequently adopted to reduce risk to a single dimension (such as the risk of developing cancer, nervous system disorder, or death), to avoid subjectivity in the decision making, and as an aid in setting priorities. An important role of the risk manager is to identify various risk-control options and to assess the risks and benefits associated with each option. Costs and benefits may be calculated on both economic and health scales; such a process is highly controversial as it often assigns monetary values to human life, deaths, illnesses, and health issues. As an alternative to conventional cost–benefit analysis, other techniques have been developed to provide assistance in evaluating various options where costs and benefits may be expressed on nonmonetary scales. It should be emphasized that although benefit–cost analysis can be of some help in the decision-making process, it is fundamental to ensure that important qualitative factors are not underplayed by quantitative factors (2).

PREVENTION, INTERVENTION, AND CONTROL ACTIONS

Chemical hazards can be introduced into the food supply at all stages along the food chain. Standards for food safety are set, and the food industries are entrusted to comply with them. In addition to regulatory standards, consumer demands and purchasing habits may also require food industry representatives to consider implementing various risk control measures.

In the process of preparing a strategy to reduce potential risks, it is clear that considerable improvements can be achieved by appropriate education, training, and prevention. Implementation of Good Manufacturing Practices and Hazard Analysis Critical Control Point (HACCP) programs have led to the widespread adoption of several risk prevention measures in the production and processing of food products.

Legislative mandate largely determines the flexibility afforded risk managers in interpreting the results of risk assessments and in considering other factors before making regulatory decisions. A variety of models are often considered in the food safety risk management process and their uses are often prescribed by law. Some common models include:

Zero risk, which applies to food additives shown to induce cancer in humans or animals. For chemicals regulated in such a manner there is a zero tolerance.

Reasonable certainty of no harm or negligible risk, which typically deems acceptable a cancer risk of no more than one excess cancer above background per million people exposed, as calculated using conservative (risk-magnifying) risk assessment methods.

Risk balancing, which specifically requires that risk and benefits should be considered together before making a decision.

Technical feasibility, which considers both the level of risk and the availability of current technological methods to control the risk.

UNITED STATES REGULATIONS

Food additives are regulated by the U.S. Food and Drug Administration (FDA). Before an additive is approved for food use it must be demonstrated that it is safe under its intended conditions of use. Some food ingredients have been reviewed by qualified personnel and are granted the status of Generally Recognized as Safe (GRAS). Food additives that have been demonstrated to be noncarcinogenic are allowed for use if the exposure estimates are below the Acceptable Daily Intake (ADI), which is derived by applying conservative uncertainty factors to exposure levels that typically do not cause observed effects in laboratory animals.

Within the Federal Food, and Drug and Cosmetic Act, the Delaney Clause, adopted in 1958, states that no substance which has been shown to induce cancer in humans or animals can be used as a food additive (zero risk basis). As such, potentially carcinogenic chemicals are not allowed as food additives regardless of the expected level of exposure.

The FDA has applied the concept of a negligible risk to veterinary drugs while taking negligible risk and risk balancing approaches to regulate specific carcinogenic food contaminants, such as polychorinated biphenyls (PCBs) in fish and aflatoxins in peanuts and other products (8).

The U.S. Environmental Protection Agency (EPA) regulates drinking water contaminants under provisions of the Safe Drinking Water Act. For noncarcinogenic drinking water contaminants, allowable levels are set to ensure that a fraction of the ADI is not exceeded. For carcinogenic drinking water contaminants, it has been recognized that zero risk is not technologically attainable. As an alternative, maximum containment levels are established at the lowest technologically feasible levels; these typically result in lifetime cancer risks in the order of 1 in 100,000 or lower, but risks for some chemicals at the maximum containment level exceed 1 in 100,000 (8).

The major law regulating pesticides, the Federal Insecticide, Fungicide, and Rodenticide Act, provides the EPA with the authority to permit specific pesticide uses when it has been determined that the potential benefits of the uses of the pesticides outweigh their potential risks. Some benefits of pesticides may be directly related to health; an example concerns the use of a fungicide that may result in food residues yet may prevent the formation of naturally occurring fungal toxins of potentially greater health risk. Substitution effects are also important, since elimination of the use of a specific pesticide that may leave food residues could lead to an increase in the use of less-effective pesticides, resulting in greater potential for environmental disruption and worker-safety concerns. Another benefit considered by EPA is the pesticide's ability to produce an abundant, available, and affordable food supply by increasing crop yields and reducing production costs and consumer prices.

In August 1996 new legislation repealed the Delaney Clause with respect to pesticide residues. The legislation limited risk-balancing provisions concerning pesticide residues in foods and instituted a "reasonable certainty of no harm" standard. In addition, regulatory practices were prescribed to consider exposures and sensitivities of specific population subgroups such as infants and children, to consider the aggregate exposure of pesticides from dietary, water, and residential exposure, to consider the cumulative risks posed from exposure of families of pesticides that possess common toxicological mechanisms of action, and to consider other types of toxicological effects such as endocrine disruption.

As noted in the previous discussion, a variety of different risk management practices are prescribed and the choice of which practices to use depends on the classification of the chemical (eg, pesticide, food additive, veterinary drug) and the medium from which humans are exposed (eg, food, water). As such, an acceptable level of risk for one type of food chemical may differ greatly from what is considered legally acceptable for another type of food chemical, and the use of practices such as risk balancing (comparing risks with benefits and/or economic impact) and technical feasibility may be allowed under some laws and not allowed under others. A summary of the various regulatory models used in the United States is provided in Table 1. In essence, it is possible for the same chemical to be subjected to different allowable levels of risk depending on whether consumers eat it, breathe it, or drink it. It is common that higher risks are tolerated for workers who are exposed to chemicals during their job (such as pesticide manufacturers and applicators) than for the general population. In general, such risks are justified because the general population includes children, pregnant women, elderly, and ill people, whereas the workforce is on average healthier and frequently subjected to medical surveillance, and their work environment, in many cases, is monitored (8).

Table 1. U.S. Federal Laws Regulating Chemicals in Food and Water

Law	Regulatory agency	Regulated products	Regulatory model
Food, Drug and Cosmetic Act (1906, 1938, amended 1958, 1960, 1962, 1968)	FDA	Food and feed additives Veterinary drugs Natural toxins	Zero risk Negligible risk Negligible risk/risk balancing
Federal Insecticide, Fungicide, and Rodenticide Act (1948, amended 1972, 1975, 1978)	EPA	Pesticides	Reasonable certainty of no harm/limited risk balancing
Safe Drinking Water Act (1974, amended 1977, 1996)	EPA	Drinking water contaminants	Technical feasibility

Source: Ref. 9.

INTERNATIONAL REGULATIONS

International food standards are set by the Codex Alimentarius Commission, which was established by the Food and Agriculture Organization (FAO) and World Health Organization (WHO) Assembly in 1961–1962. The Codex Alimentarius Commission membership is open to all member nations and associate members of FAO and/or WHO, and currently comprises more than 160 members. The overall objective of the Codex is to ensure consumer protection and to facilitate international trade.

Scientific risk assessment provides the basis for Codex risk management decisions. Two FAO/WHO scientific advisory bodies, the Joint Expert Committee on Food Additives (JECFA) and the Joint Meeting on Pesticide Residues (JMPR) focus their activities primarily on risk assessment. These committees are independent of the Codex system and advise FAO and WHO and their members. Within Codex, risk management is defined as the process of considering risk assessment results, weighing policy alternatives, and, if required, selecting and implementing appropriate control options including regulatory measures. The outcome of the risk management process is the development of standards, guidelines, and other recommendations, which are developed by the Codex subsidiary bodies, such as the Committee on Food Additives and Contaminants (CCFAC), the Committee on Pesticide Residues (CCPR), and the Committee on Residues of Veterinary Drugs in Food (CCRVDF). In the case of food contaminants, the normal approach is for CCFAC to set levels of contaminants that are as low as reasonably achievable. The CCRVDF recommends maximum residue limits (MRLs) for residues of veterinary drugs in food, and the CCPR sets ADIs and MRLs for pesticide residues. Monitoring of both the effectiveness of the control measure (through periodic evaluation of the decision) and its impact on risk to the exposed population is expected following implementation of the control measure (3).

BIBLIOGRAPHY

1. V. T. Covello and M. W. Merkhofer, *Risk Assessment Methods: Approaches for Assessing Health and Environmental Risks*, Plenum, New York, 1993.
2. National Research Council, *Improving Risk Communication*, National Academy Press, Washington, D.C., 1989.
3. United Nations Food and Agricultural Organization, *Risk Management and Food Safety*, FAO Food and Nutrition Paper 65, FAO/WHO, Rome, Italy, 1997.
4. U.S. Food and Drug Administration, "Food Safety From Farm to Table . . . A National Food Safety Initiative Report to the President," *Dairy, Food and Environmental Sanitation* **17**, 555–574 (1997).
5. M. J. DiBartolomeis, "Risk Assessment and Risk Management," in A. M. Fan and L. W. Chang, eds., *Toxicology and Risk Assessment*, Marcel Dekker, New York, 1996, pp. 775–776.
6. G. M. Gray, *Key Issues in Environmental Risk Comparisons: Removing Distortions and Insuring Fairness*, Reason Foundation, Los Angeles, Calif., 1996.
7. A. L. Nichols and R. J. Zeckhauser, "The Perils of Prudence: How Conservative Risk Assessments Distort Regulation," *Reg. Toxicol. Pharmacol.* **8**, 61–75 (1988).
8. J. V. Rodricks, *Calculated Risks: The Toxicity and Human Health Risks of Chemicals in Our Environment*, Cambridge University Press, New York, 1992.
9. C. K. Winter and F. J. Francis, "Assessing, Managing, and Communicating Chemical Food Risks," *Food Technol.* **51**, 85–92 (1997).

ELISABETH L. GARCIA
CARL K. WINTER
University of California
Davis, California

See also FOOD SAFETY AND RISK COMMUNICATION; TOXICOLOGY AND RISK ASSESSMENT.

FOOD SAFETY EDUCATION: CONSUMER

SOURCES OF CONSUMER INFORMATION

Consumers indicate that their primary source of food safety information is the media (1–4). Television newspaper stories alert the public to current food safety issues with magazines frequently providing more in-depth information. Labels and cookbooks also serve as sources of information.

Food industry and consumer groups, government agencies, educational institutions, and professional organizations provide information to the consumer through the media or directly via educational flyers, articles in magazines or newsletters, and meetings.

The Institute of Food Technologists (IFT), a professional society of food and nutrition scientists, has identified about 80 academic members who serve as food science communicators. An IFT media guide with food science communicator contact information organized by topic and geographic areas is distributed to major news services across the country. Communicators initiate or respond to almost 1000 media contacts a year.

The American Dietetic Association (ADA) has organized a similar media outreach activity. Although ADA spokespersons primarily address nutritional issues, food safety is also an area of inquiry. The association has published position papers on a variety of issues with safety components, such as food irradiation, biotechnology, and fat substitutes.

Other professional societies have developed food safety material for the public. For example, *Before Disaster Strikes, A Guide to Food Safety in the Home* can be ordered through the International Association of Milk, Food and Environmental Sanitarians Web site, *www.iamfes.org*.

The International Food Information Council (IFIC) Foundation communicates science-based information on nutrition and food safety issues to health professionals, media, educators, and government officials. The Foundation is supported primarily by the broad-based food, beverages, and agricultural industries. IFIC Foundation activities include publishing a bimonthly newsletter, "Food Insight"; placing science-based information on the Internet (www.ificinfo.health.org); sponsoring educational events and roundtables for opinion leaders; writing scientific white papers, consumer brochures, and school curricula; and preparing educational exhibits at scientific and professional meetings. IFIC Foundation has also prepared a comprehensive media guide with background and consumer publications on current issues and lists of scientific experts.

The American Council on Science and Health, Inc. (ACSH) is a consumer educational consortium concerned with food safety and other issues. The ACSH board, composed of 250 physicians, scientists, and policy advisers in a variety of fields, reviews the Council's reports and participates in ACSH educational activities. ACSH produces a wide range of publications (*www.acsh.org*), including peer-reviewed reports on important health topics. ACSH representatives appear regularly on television and radio, in public debates and in other forums, and hosts media seminars and press conferences. Publications related to food safety include: *Eating Safely: Avoiding Foodborne Illness; Irradiated Foods; Feeding Baby Safely: Facts, Fads and Fallacies; An Unhappy Anniversary: The Alar "Scare" Ten Years Later;* and *ACSH Holiday Dinner Menu*.

The U.S. Food Safety and Inspection Service established the U.S. Department of Agriculture (USDA) hot line to reach consumers with information on safe handling of meat and poultry. Since its inception the hot line has responded to more than one million calls for information on meat and poultry safety with more than 100,000 calls received on the toll-free line from January through December 1996 (5). Information is also available through a fast fax service and at the Web page *http: / / www.usda.gov.fsis*. In 1996 over 80% of hot line calls were from consumers, with the remainder coming from businesses and professional clients including educators, communicators, health and human service professionals, students, teachers, and the media. Through their professional capacity, many of these callers also extend information to the public.

The *FDA Consumer*, published 10 times per year and available electronically at *www.fda.gov / fdac /*, contains consumer-oriented articles on various health topics including food safety.

The USDA provides information to food and health educators and the land grant university cooperative extension system through material such as "Food Safety Focus" and "Food Safety Educator." The cooperative extension system in each state uses these materials or others developed within the state to provide food safety information to the public. Many of these are listed in the National Food Safety Database.

The National Food Safety Database serves as a management system of U.S. food safety materials used by the Cooperative Extension Service (CES), consumers, industry, and other public health organizations. The Web site *www.foodsafety.org* lists many of the food safety resources developed for various audiences, ranging from producer to consumer. In addition to printed resources, a growing amount of information now is accessible in electronic format, and on the Internet.

Food industry and consumer groups prepare consumer material on proper food handling or address areas of food safety concern. The Center for Science and the Public Interest reaches the public through its newsletter and interaction with the media. In recent years *Consumer Reports*, published by Consumer's Union, has addressed food safety and safe food handling.

CONSUMER CONCERNS

Better documentation of the consequences of microbiological-based foodborne illness and increased media coverage have probably led to increased consumer concern about microbiological hazards. Nationwide surveys indicate that more people volunteer concerns about microbiological hazards than any other potential food safety issue. Survey responses regarding volunteered concern about microbiological safety increased from 36% in 1992 to 69% in 1997 (6). When the question of concern about food contamination by bacteria or germs was specifically asked, 82% of respondents classified it as a serious hazard. This response indicates more concern about this type of food hazard than any other food safety risk.

Although pesticide residues continue to generate concern among a large segment of the population, only 65% ranked it a serious hazard in 1997, a significant decreased from 82% in 1989 (6). Concern about other food safety areas has also decreased. Those expressing serious concern with antibiotics and hormones used in poultry or livestock decreased from 61% in 1989 to 43% in 1997. Those rating the use of nitrites in food as a serious hazard decreased from 44% in 1989 to 28% in 1997. Those rating use of additives and preservatives as a serious hazard decreased from 30% in 1989 to 21% in 1997.

The USDA hot line calls indicate the food safety areas related to animal products where people sought information. In 1996 most calls, 74%, were inquiries about safe storage, handling, and preparation of meat and poultry products (5). Twelve percent of inquiries dealt with marketing and inspection issues, 6% addressed questionable products and practices, and 4% related to labeling, product dating, and basic nutrition. More than 6000 inquiries related to handling of turkey, followed by beef and veal with almost 5000 calls.

CONSUMER SAFE-HANDLING KNOWLEDGE

Although consumers recognized the potential seriousness of foodborne bacteria and believed they were well informed (1), many were unaware of safe-handling and storage practices, with those age 35 or younger demonstrating the lowest level of knowledge (7–8). A 1998 nationwide poll indicated more than 90% of consumers had heard of *Salmonella*, yet critical gaps existed in food safety knowledge (9).

Many factors have contributed to consumers' lack of familiarity with safe food handling. Increased participation in the paid labor force has resulted in less in-home training of young people in food handling; few schools offer or require food preparation classes; partially prepared foods may have different, less familiar handling requirements (7). Additionally, some historic food safety recommendations related to temperature and acidity do not eliminate risks from some pathogens.

Food safety experts have identified the most common food-handling problem by consumers at home. The top factors attributed to mishandling include contaminated raw food, inadequate cooking or heat processing, obtaining food from unsafe sources, improper cooling, intervals of 12 hours or more between preparation and eating, and colonized person handling implicated food or poor hygiene (10). Mishandling associated with specific pathogens included the same contributing factors (11).

In a nationwide mail survey, Williamson et al. (8) found specific safe-handling practices were not practiced by 15 to 30% of respondents. People failed to rapidly cool cooked food, with 29% indicating they would let roasted chicken sit on the counter until it cooled completely before refrigerating. Only 32% indicated they would use small shallow containers to refrigerate leftovers. People were not aware that failure to refrigerate may jeopardize safety, with 18% not concerned or uncertain about the safety of cooked meat and 14% about poultry left unrefrigerated for more than four hours. The need for sanitation was not recognized, with only 54% indicating they would wash a cutting board with soap and water between using it to cut raw meat and chop vegetables.

Similarly, in a California survey, only 63% indicated they clean the food preparation area with soap and water. The importance of temperature control was not fully understood, with 50% indicating they refrigerate leftovers in large containers. Of particular concern, more than half of consumers always or sometimes tasted leftovers to check whether they are still safe (1).

Even consumers who knew they were being watched and evaluated made mistakes. An audit of consumer food-handling practices among 106 U.S. and Canadian households found 96% had at least one critical violation. Households average 2.8 critical and 5.8 major violations per household (12). A critical violation was defined as one that, by itself, can potentially lead to a foodborne illness, whereas major violations are unlikely to cause foodborne illness but are frequently cited as contributing factors.

CONSUMER PERCEPTION OF THE SOURCE OF FOODBORNE ILLNESS

Many consumers misperceive the nature of foodborne illness and the most likely source of the pathogen (13). Consumer belief about the type of food responsible for foodborne illness—meat, poultry, seafood, eggs—is consistent with expert opinions. However, consumers believe foodborne disease is a minor illness without fever that occurs within a day of eating a contaminated food. Infections caused by *Salmonella* and *Campylobacter*, the most common foodborne illnesses in the United States (14), are not consistent with the symptoms consumers described, because of the longer latency and fever-causing properties.

Most consumers believe their illness is caused by food prepared somewhere other than the home. Williamson et al. (8) found about one-third of consumers thought food safety problems most likely occurred at food manufacturing facilities, and one-third blamed unsafe practices at restaurants. Only 16% thought the home was the most likely place for mishandling. Fein et al. (13) found 65% attributed foodborne illness to food prepared at a restaurant, with 17% believing the mishandling occurred at the supermarket and 17% at home. Similarly, a nationwide 1998 survey found 39% of consumers thought safety problems occurred at the food processor or manufacturer, 20% believed they occurred at the restaurant, and only 15% thought mishandling occurred in the home (15). In contrast, food safety experts believe sporadic cases and small outbreaks in home are far more common than those cases constituting recognized outbreaks (7).

If consumers misperceive the nature and origin of foodborne illness, they underestimate the frequency of serious consequences and thus are less motivated to change. Schafer et al. (16) found that motivation to practice safe food-handling behavior requires the belief that someone could be harmed by not following guidelines, and that new behavior could prevent foodborne disease. The failure to associate at-home food-handling practices with foodborne illness is a serious impediment to convincing people to discontinue potentially hazardous food-handling behavior (13).

U.S. NATIONAL FOOD SAFETY PROGRAM

A coalition of national industry, government, and consumer groups formed the Partnership for Food Safety Education to provide safe food-handling information to the public. In 1998 the Partnership launched the *Fight BAC* (bacteria) education campaign. The initial goal of this campaign was to covey four key principles of food safety: Clean: wash hands and cooking utensils; Separate: prevent cross-contamination; Cook: heat foods to proper temperature; and Chill: store foods correctly. The program encourages grassroots education in local communities through dissemination of a *Fight BAC* kit with implementation ideas and reproducible materials. Although it is too early to judge the success of the program, materials have been widely disseminated and *Fight BAC* symbols and flyers are in major supermarkets.

BIBLIOGRAPHY

1. C. M. Bruhn and H. G. Schutz, "Consumer Food Safety Knowledge and Practices," *J. Food Safety* **19**, 73–87 (1999).
2. T. Hoban, *Consumer Awareness and Acceptance of Bovine Somatotropin*, Grocery Manufacturers of America, Washington, D.C., 1994.
3. T. Hoban and P. A. Kendall, *Consumer Attitudes About the Use of Biotechnology in Agriculture and Food Production*, North Carolina State University, Raleigh, N.C., 1993.
4. J. C. Busby and C. Ready, "Do Consumers Trust Food Safety Information," *Food Rev.* **19**, 46–49 (1996).
5. *Making the Connection, USDA's Meat and Poultry Hotline, 1996*. U.S. Dept. of Agriculture, Washington, D.C., 1998.
6. Abt Associates, *Trends in the United States, Consumer Attitude and the Supermarket 1997*, Food Marketing Institute, Washington, D.C., 1997.
7. Institute of Food Technologists' Expert Panel on Food Safety and Nutrition, "Scientific Status Summary, Food Borne Illness: Role of Home Food Handling Practices," *Food Technol.* **49**, 119–131 (1995).
8. D. M. Williamson, R. B. Gravani, and H. T. Lawless, "Correlating Food Safety Knowledge With Home Food-Preparation Practices," *Food Technol.* **46**, 94–100 (1992).
9. Food Safety and Inspection Service, "Consumers Are Changing," *Food Safety Educator* **3**, 2 (1998).
10. F. Bryan, "Risks Associated With Vehicles of Food Borne Pathogens and Processes That Lead to Outbreaks of Food Borne Diseases," *J. Food Protect.* **51**, 663–673 (1988).
11. N. H. Bean and P. M. Griffin, "Food Borne Disease Outbreaks in the United States, 1973–1987; Pathogens, Vehicles, and Trends," *J. Food Protect.* **53**, 804–817 (1990).
12. R. Daniels, "Home Food Safety," *Food Technol.* **52**, 54–56 (1998).
13. S. B. Fein, C. T. Jordan, and A. S. Levy, "Food Borne Illness: Perceptions, Experience, and Preventive Behaviors in the United States," *J. Food Protect.* **58**, 1405–1411 (1995).
14. R. V. Tauxe, "Epidemiology of *Campylobacter jejuni* Infections in the United States and Other Industrialized Nations, in J. Nachamkin, M. J. Blaser, and L. S. Tompkins, eds., *Campylobacter jejuni: Current Status and Future Trends*, American Society for Microbiology, Washington, D.C., 1992, pp. 9–19.
15. Abt Associates, *Trends in the United States, Consumer Attitude and the Supermarket 1998*, Food Marketing Institute, Washington, D.C., 1998.
16. R. B. Schafer et al., "Food Safety: An Application of the Health Belief Model," *J. Nutr. Educ.* **25**, 17–24 (1993).

CHRISTINE M. BRUHN
University of California
Davis, California

FOOD SCIENCE AND TECHNOLOGY: DEFINITION AND DEVELOPMENT

Food science and technology have been defined in various ways. All definitions basically specify that food technology is the application of the principles and facts of science, engineering, and mathematics to the processing, preservation, storage, and utilization of food. Food science, on the other hand, deals chiefly with the acquiring of new knowledge to elucidate the course of reactions or changes occurring in foods whether natural or induced by handling procedures. With respect to "natural," nearly all our foods are of biological origin. Foods consist of chemicals, contain enzymes of intrinsic origin, and generally are subject to the action of enzymes of extrinsic origin and to microorganisms. Furthermore, various physical phenomena affect their state. Foods are subject to a myriad of forces, many of which are known; a far greater number of influences are still unknown. There is thus a great need for food science, which, among its functions, is the acquiring of new knowledge. Although in theory food science deals with the acquiring of new knowledge without there being any intended benefit from the new knowledge, in practice there is usually some hoped-for benefit such as greater stability of the food, a gain in acceptability or nutritional value, or a reduction in the likelihood of spoilage.

Food science and technology arose as separate disciplines through the coalescing of those portions of other sciences that dealt with foods. Traditionally, chemists, microbiologists, and other kinds of scientists studied or taught about foods from their particular points of view, but food processing transcends any one of the traditional disciplines. The distinction between chemists and microbiologists who study foods and food scientists and technologists is that the former devote their efforts chiefly to determining chemical composition, or the microbial state of foods. They are concerned with only one aspect of food. There is a need for those whose vision and competency cuts across discipline lines. When food processing is involved, knowledge of only one branch of science is not enough. There is a need for scientists and technologists whose knowledge of the different fields germane to various aspects of food preservation enables them to couple principles from the traditional fields with principles of food preservation to effect the most suitable solution to a food processing problem. The same applies to food engineering. There are principles that apply to the solution of engineering problem. Sometimes the most efficient engineering solution is not the most suitable or, more serious, not even an acceptable solution from the point of view of handling a food.

There have been instances of having an individual trained in one of the traditional branches of engineering assigned to the task of devising a new procedure for the handling of food. The process designed often was efficient from an engineering standpoint but a disaster from a microbiological point of view because allowance was not made for proper cleaning of the equipment. Food retained in the system could act as an inoculum of spoilage organisms for portions of food subsequently passing through the equipment. An engineer trained as a food technologist would recognize that possibility and select the next best option, one less efficient from an engineering point of view but one least likely to permit spoilage organisms to flourish.

In effect, a food scientist or technologist is a generalist in science rather than being an adherent of only one branch of science. The same applies to a food engineer. Food engineers augment their knowledge of engineering principles with the same kind of multidisciplinary knowledge that characterizes the food scientist.

Because foods themselves are intrinsically complex and subject to various forms of change (biological, chemical, nutritional, physical, and sensory) it was quite logical that a science arose that was designed to pull together the wide range of knowledge that impinges in some way on food and food processing.

FOOD RESEARCH BEFORE THE TERM "FOOD SCIENCE" AROSE

Food science as a distinct discipline is not quite a century old, but aspects of it have existed for centuries. Appert's development in 1810 of the process of canning (1) was an epochal event. The process wasn't called canning then, and Appert himself did not really know the principle upon which his process depended, but canning has had a major impact on food preservation ever since its development. It was the first of the purposely invented processes. Other methods developed earlier, such as drying and fermentation, go back to antiquity and were a result of the evolution of procedures over centuries rather than the purposeful application of the scientific method. Pasteur's study on the spoilage of wine and his description in 1864 (2,3) of how to avoid such spoilage persists not only because of the scientific importance of his findings but also because the term, pasteurization, is so much a part of our vocabulary. There were other early studies on food spoilage. In 1897 Prescott and Underwood published a seminal study of the spoilage of canned clams and lobsters (4). There was also a 1898 paper of theirs that dealt with the spoilage of canned corn. Russell (5) carried out what is now believed to be the first scientific work on spoilage of canned foods. He studied the amount of heat needed to preserve canned peas. His study, published in 1895, was little noticed for some time because it was buried in a report of the Wisconsin Agricultural Experiment Station (5).

A study initiated in 1920 that led to a truly scientific breakthrough was that of the National Canners Association (NCA; now called the National Food Processors Association). The NCA study (6) was the first to result in sound mathematical expressions for the amount of thermal energy needed to destroy microbial spores, the amount of heat put into a can and removed from it during heating and cooling, and a means of calculating the least amount of heat needed to destroy some specified number of spores in the can. Actually, there is no such thing as attaining a complete kill. One has to content oneself with the probability of there being one surviving spore. The usual level specified is 12D, or the probability of one container in 100 billion containing a viable spore.

The NCA study is a good illustration of fundamental scientific knowledge being discovered as an outgrowth of an investigation carried on for a very practical reason: to avert food spoilage and the risk of a *Clostridium botulinum* spore being able to grow out later and cause a very serious form of food intoxication (poisoning). Throughout the food field, the pharmaceutical industry, and others where sterility must be attained, the NCA study is the source of the knowledge needed.

THE FIRST OF THE TEACHING DEPARTMENTS

Just a little bit earlier than investigations such as that of the NCA, universities began to teach phases of food preservation that would be called food science today. Dairy processing and meats processing were taught during the nineteenth century, but the subject matter of each tended to be taught from the point of view of that commodity alone. In the 1910s things changed. At four different universities, predecessor departments to food science and technology arose. Like dairy manufacturers, their first departmental names were descriptive of the commodity area that they represented, but the pioneer teachers of food preservation recognized that the principles and facts that applied to their products had application to other commodities; thus, they tended to broaden their approach by teaching food preservation across commodity and discipline lines. At the University of Massachusetts, the departmental designation was Horticultural Manufacturers; at the University of California, it was Fruit Products; at Oregon State, the departmental title, like that at Massachusetts, related to Horticulture; at the Massachusetts Institute of Technology (MIT), the first courses were taught in Biology, since that was the school in which Prescott was located.

Among those who instituted formal teaching of that which is called food science today were Walter Chenoweth (G. E. Livingston, unpublished) of the University of Massachusetts; William V. Cruess, at California; Ernest Weigand, at Oregon State University; and Samuel C. Prescott (7), at MIT. The early teachers found that they could not truly teach food preservation unless they looked at the food and processes from the points of view of more than one discipline or commodity, for example, bacteriology in Prescott's case (4). That kind of thinking led to the departments becoming the first of the present food science and technology departments.

In defense of the dairy departments and other commodity-oriented departments of that time, it must be pointed out that in the last century there was less of what would be called further processing than today, when one type of commodity is incorporated into another commodity to produce still a third form of food; consequently, there was less reason to look across commodity lines.

Before discussing the evolution of the terms food science and technology, two other pioneers merit special mention. One of those is Dr. Carl R. Fellers (G. E. Livingston, unpublished). He was one of the most influential of the early teachers and researchers (8). He spent most of his career at the University of Massachusetts. From the early 1920s on, he evinced both in this teaching and research the concept that science and technology as they relate to food are one, that there are particular facts associated with particular foods, but the same principles and methods apply whether the food be apples, confections, fish, tortillas, veal, or zucchini. Another individual of the same sort was Dr. Bernard E. Proctor (G. E. Livingston, unpublished), a professor at MIT. The book *Food Technology*, published in 1937 by Prescott and Proctor (9), was among the first to use that term. Incidentally, Prescott, Cruess, Fellers, and Proctor all were early presidents of the Institute of Food Technologists (IFT).

FOOD SCIENCE AND TECHNOLOGY AS A TITLE

The founding of the IFT in 1939 (10) led to general use of the term food technologist. Reference was made in the last section to Prescott and Proctor's book for its use of the term food technology; another who did so in the same year, 1937, was Bitting (11). He called himself a food technologist on the title page of his book. By 1945, the original four departments that had taught the subject under different titles had changed their departmental names to food science, food science and technology, food engineering, or some variant such as food science and nutrition. Several other food science departments were organized in the 15 years after World War II. A few of the departments started ex nihilo; most had their origins in some commodity department or represented amalgamations where phases of food science taught in various departments were transferred and combined to form a new department. A factor that also solidified the use of food technology was development of the IFT's curricular minimum standard for food science and technology (12). That subject may be found in the article INSTITUTE OF FOOD TECHNOLOGISTS: HISTORY AND PERSPECTIVES.

The coalescing of the interests of those who conduct research on foods and food processing came somewhat late in the history of science. Not until after World War II did the term food science/technology come into widespread use. The two disciplines are so interlinked they function as one and are generally taught that way. Food engineering has generally been taught under the rubric food technology. Sometimes it is taught in an agricultural engineering department. Today one thus finds departments designated as food science, food science and technology, and sometimes food science and nutrition; if nutrition is a part of the title, however, the designation is generally food science and human nutrition. The reason for specifying "human" is that nutrition is taught in animal science courses and in foods and nutrition departments, and a distinction had to be made between the types of nutrition taught in these older kinds of departments. Nutrition as taught in foods and nutrition departments often deals with the clinical side of nutrition, whereas that taught in food science departments is more likely to delve into the relations between the retention of nutrients as influenced by various processing techniques. Both types of department naturally have common ground in elucidating the functions the various nutrients perform, interrelations among nutrients, and the well-being of humans when foods are used wisely.

POSITIONS FILLED BY FOOD SCIENTISTS/ TECHNOLOGISTS

The types of industrial jobs available to food scientists and technologists cover the gamut of responsibilities from examination of raw products to final sales. Positions of responsibility are also available in government (the Food and Drug Administration, the U.S. Department of Agriculture, the Department of Defense, and among others, the Public Health Service). The same applies to state and local governments such as state departments of agriculture and state and local health departments. In the academic world, teachers and researchers find many opportunities. Entrepreneurs such as food consultants often establish a remunerative niche for themselves. The roles food scientists and technologists play will be more fully set forth in the article FOOD SCIENCE AND TECHNOLOGY: THE PROFESSION. So too will developments as far as food processes are concerned. The next topic here will deal with changes taking place in the makeup of members of the profession.

Mermelstein (13) prepared in 1998 an article demographic as to changes taking place in the profession and as to employment, salaries, and the locations of food scientists/technologists both geographically and by employment field. He pointed out that at present males constitute 61% of IFT members, whereas females constitute 39%. Predictive of the future is the fact that female members tend to be younger than males. That 39% of IFT members are females is significant in itself, for 30 years ago they constituted a very small part of IFT membership. Since they tend to be younger today than their male counterparts, this suggests that in not too many years they will be the dominant portion of the membership.

The ascendancy of women in the profession can be seen in the colleges, too. Three decades ago there were few women among food science/technology graduates. Today that situation has changed dramatically. Female students outnumber male majors in many food science departments. The same applies to the Student Division of the IFT. Female student members are as numerous as male students, and often they assume the dominant roles in running the various student chapters by being elected chair of their university's IFT student chapter. There is little question that women are occupying roles, some of which were closed to them in years past, and at far higher levels than heretofore. Throughout industry, there are several female vice presidents. In the academic world, professors are as likely to be female as male, and several of the teaching departments have female heads or chairs.

An interesting facet of the survey Mermelstein reported is that a far higher proportion of those who work in the research and development (R&D)/scientific/technical category are women. Overall the salaries for women were lower than for men, the median salary being $68,500 for men versus $50,000 for women, but part of that discrepancy is a reflection of the fact the women are younger and thus have worked for fewer years. For those with a bachelor's or master's degree and less than one year's experience, the discrepancy was substantial, that is, $4300 and $5500, respectively. At the Ph.D. level, the discrepancy was only $400.

For those in industry with all years of experience combined, those with the B.S., M.S., Ph.D. and MBA degrees earned median salaries of $54,000; $60,000; $72,000; and $75,000, respectively. For educational institutions, the salaries were $33,682; $34,150; and $64,400; the MBA degree was not listed since it is outside of food science. For the first three categories, government employees averaged $49,825; 50,000; and $70,500. For scientific/trade organizations, the median salaries were $55,000; $57,000; $76,351; and $75,500. At the Ph.D. level and that of the MBA, median salaries by form of employment were comparable.

In summary, food science became defined only in the fore part of the twentieth century, and the term did not come into common use until nearly midway in this century. It differs from many other sciences in the scope of its outlook. As most sciences have developed, they have looked inward toward greater and greater specialization. Notwithstanding that, most practitioners within a field must today keep abreast of developments elsewhere in the great area in which their science belongs. A microbiologist, for example, generally attempts to stay knowledgeable of development elsewhere in the biological and biochemical fields. The scientist thus looks outward to other areas of his/her science. Food scientists and technologists must look outward with exceptional vigor. They cannot confine themselves to one major area such as the biological, biochemical, or physical sciences or engineering. Because food and food processing encompass all these areas, the food scientist or technologist must stay abreast of developments in several other diverse areas to practice the profession most effectively. The profession is thus one of unusual challenge. Because so many sciences impinge on food processes and since food is processed in an increasingly number of different ways today, there is an even greater need for food science and technology than there was when the two were first constituted as a new and separate field. Food scientists and technologists are the individuals most fit by training to be able to recognize, study, use, and interpret the myriad of interrelated processes that occur whenever food is preserved, stored, or utilized.

BIBLIOGRAPHY

1. S. A. Goldblith, "The Science and Technology of Thermal Processing. Part I," *Food Technol.* **25**, 1256–1262 (1971).
2. M. Frobisher, Jr., *Fundamentals of Bacteriology*, W. B. Saunders Co., Philadelphia, Penn., 1937, pp. 443–448.
3. J. M. Jay, *Modern Food Microbiology*, 5th ed., Chapman and Hall, New York, 1996, pp. 5, 7.
4. S. A. Goldblith, "The Science and Technology of Thermal Processing, Part 2," *Food Technol.* **26**, 64–69 (1972).
5. H. L. Russell, *Gaseous Fermentation in the Canning Industry, 12th Annual Report*, Wisconsin Agricultural Experiment Station, Madison, Wis., 1895, p. 227.
6. W. D. Bigelow et al., *Heat Penetration in Processing Canned Food*, Bulletin 16-L, National Canners Association, Berkeley, Calif., 1920.
7. S. A. Goldblith, *Pioneers in Food Science*, Vol. 1, Food and Nutrition Press, Trumbull, Conn., 1993.
8. R. L. Hall, "Pioneers in Food Science and Technology: 'Giants in the Earth'," *Food Technol.* **43**, 192 (1989).
9. S. C. Prescott and B. E. Proctor, *Food Technology*, McGraw-Hill, New York, 1937.
10. N. H. Mermelstein, "History of the Institute of Food Technologists: The First 50 Years," *Food Technol.* **12**, 14–18, 35–52 (1989).
11. A. W. Bitting, *Appertizing or the Art of Canning: Its History and Development*, The Trade Pressroom, San Francisco, Calif., 1937.
12. R. M. Schaffner, "What Training Should a Four-Year Food Technology Student Recieve," *Food Technol.* **12**, 7–14 (1958).
13. N. H. Mermelstein, "1997 IFT Membership Employment & Salary Survey," *Food Technol.* **52**, 63–73 (1998).

JOHN POWERS
University of Georgia
Athens, Georgia

See also FOOD SCIENCE AND TECHNOLOGY: THE PROFESSION.

FOOD SCIENCE AND TECHNOLOGY: THE PROFESSION

Food scientists and technologists are employed in academia, by government and industry, by consulting firms, in entrepreneurial endeavors of various sorts and as private consultants. Crucial to almost any profession is the education its members receive prior to entering into that profession. Food science and technology is no exception. For that reason, the development of the Institute of Food Technologists' (IFT) minimum undergraduate curriculum will be described first.

In 1958 the IFT sponsored a conference between industrial leaders and university personnel to devise an undergraduate curriculum that would answer industry's needs for graduates educated in food science (1). A model curriculum was devised. In essence it was a consensus between the course work industry thought graduates coming into their fields should have versus the very practical matter that there is a limit to the number of courses a curriculum can have and still permit students to graduate in four years.

In 1962 and again in 1965 additional conferences were held (2). They resulted in the IFT adopting its minimum undergraduate curriculum (2). It was revised again in 1977 (3). The eventual outcome was that the gradual refining of thought as to courses that are essential for a would-be food scientist to take became the IFT's minimum undergraduate standard curriculum.

The reason the specified curriculum became a standard arises from the fact the IFT put teeth into a related program. The IFT awards approximately 145 scholarships (4). IFT scholarships can go only to universities that have a curriculum conforming to IFT's minimum requirements, and, furthermore, not only must the scholarship holder be at a university conforming to the IFT's minimum undergraduate curriculum, s/he must be following that curriculum. Many food science and technology departments have more than one curriculum. At the University of Georgia, for example, there is an environmental public health program administered by the Food Science and Technology

Department. A student in that program is not eligible for an IFT scholarship because the public health curriculum does not specify some of the courses the IFT considers to be critical in the training of a food scientist.

In the United States there are approximately 60 universities claiming they teach food science and technology. Of these, 43 meet the minimum standard (4). The standard recommended is just that. It is not accreditation. Accreditation does exist for some sciences and certain other fields, but the IFT has not sought to foster that kind of program conformity. In general, universities are not in favor of all types of programs being accredited because that tends to tie the hands of the university. Sometimes a university must supply funds to an accredited department to enable it to keep its accreditation. Loss of accreditation reflects unfavorably upon a university. To avoid that happening, a university may supply the funds to the accredited department even though the funds are needed more critically elsewhere.

For the purpose of awarding scholarships, the IFT's minimum standard applies throughout North America. There are seven Canadian universities and one Mexican university, the curricula of which meet the IFT's minimum standard. Students in those countries are eligible to apply for the IFT scholarships. Not every year, but often, a student in Canada or Mexico receives one of the IFT's scholarships. The Canadian Institute of Food Science and Technology has a scholarship program of its own. Students at some Canadian universities are fortunate in that they can seek to obtain a scholarship under either of these two North American programs.

The purpose of the minimum standard curriculum is to ensure that individuals graduating from a food science program bring to the job the kind of qualifications employers have a right to expect and the kind of education most conducive to the long-term interests of graduates in food science.

The most recent revision of the undergraduate minimum standard curriculum occurred in 1992 (5). The IFT reexamines the curriculum at intervals not greater than 10 years to take into account changes in the industry, use of new tools such as students developing computer literacy, and the institution of new kinds of courses. For example, a "capstone" course was not originally required. Today the IFT considers it important that students at the senior level take a course that incorporates and unifies principles the students should have learned in food chemistry, microbiology, and engineering; food processing; nutrition; sensory analysis; and statistics. How that unification is fostered is up to the faculty. The course can be one of food processing, product development, some combination thereof, or a wholly different form of instruction that requires the students to plan and think for themselves in carrying out a project.

Not only does the Education Committee revise the standard curriculum at intervals, it monitors conformance to the curriculum by evaluating each university's curriculum at intervals not greater than five years to be sure the university is still in compliance with the minimum standard.

Aside from the capstone course, the curriculum currently in place (5,6) requires one or more courses in food chemistry, food analysis, food microbiology, food processing, food engineering, written and oral communication skills, computer literacy, and courses that call for "critical thinking skills." These are courses in food science. Additional courses in food science depend on the student's interest, but allowance has to be made in the student's program so that required courses in basic science and mathematics may be taken. In fact, they *must* be taken. Among these required courses are two courses in general chemistry plus one course each in organic chemistry and biochemistry; two courses in calculas, one in general physics, one in statistics, two courses in communication skills (normally prior to the food science courses calling for development of such skills), and not less than four courses in history, ethics, economics, literature, art, sociology, philosophy, psychology, and foreign languages. The intent is that students, no matter their specializations, should have enough "general" education to be a well-rounded citizen. The necessity for the science and mathematics courses is obvious. If students are to secure the maximum benefit from instruction in food science, they must have a sufficient and sound background in the basic sciences underlying each of the different types of course. Food science courses can then be taught at the appropriate level of comprehensiveness and depth.

Additional requirements such as the food science and technology unit should preferably be administratively independent and have an adequate budget. The faculty members must be trained in the subjects they teach; they should be encouraged to participate in professional associations and to engage in research. Teaching loads should not be so high as to interfere with adequate preparation for teaching.

Not only has the Education Committee of the IFT been involved, but other individuals and groups have examined various aspects of the educational process (7–9). The International Union of Food Science and Technology (IUFoST) has considered the types of educational programs available, including a listing of food science and technology programs in various countries (10). The listing is quite extensive. It is not, however, complete. The University of Pretoria and the University of Stellenbosch in South Africa were not listed. Both universities have well-developed food science programs. In 1991 a symposium, covering food science teaching worldwide, was held in Europe (11). Educational systems—as is well known—are different there. The same applies to the developing countries. Netto (12) commented that in most developing countries, education does not come first, that in both industrialized and developing countries the expertise of food scientists and technologists is needed to win the fight against food-borne diseases and contaminants and to search for safer and more nutritious food products.

THE PROFESSION

Once trained, the question facing any professional is: Where can I use my education and what will be the opportunities? Those in food science and technology are fortunate. The food industry in the United States is large, sophisticated, and highly efficient. Since the end of World War II disposable income required for food purchases has declined precipitously. It was once above 22%. In 1985 food expenditures as a proportion of household disposable income had declined to 18% (13). In 1995 the percentage was approximately 11% (14). Much of the change came about because of the efficiency of American agriculture and that of the food industry. Roberts et al. (15) claim that the food industry, allowing for stimulation of other industries (transportation, packaging, printing, etc) is a $1.8 trillion industry.

Since the end of World War II the value-added part of the food industry has increased substantially. In 1980 it surpassed agriculture's contribution. The overtaking of agriculture is significant, for U.S. agriculture is known worldwide for its efficiency. Only about 2.5% of the U.S. population is required to produce the food we consume plus that exported. Developments in efficiency and in the value-added side of the food industry are largely a result of the efforts of food scientists and technologists. They are not always the originators of new ideas. Marketeers regularly initiate requests for new, value-added products. Whether or not food scientists or technologists originate the idea, they are the ones who bring hoped-for results to reality.

STRENGTHENING VALUE-ADDED COMPONENTS

This next immediate description involves a few illustrations of ways food scientists and technologists have strengthened the value-added components of the American food industry and point out in particular that all fundamental knowledge does not come from scientific projects and that applied projects generally have to have a scientific component before they come to full fruition.

In the companion entry, "Food Science and Technology: Definition and Development," reference was made to the National Canners Association (NCA) research on thermal processing of food. The reason the project was undertaken was because canned foods had acquired a bad reputation in the 1910s. They were the cause of botulism outbreaks. The outbreaks were the result of failure to heat sterilize the canned product adequately, but the underlying cause was the lack of fundamental knowledge to determine just how much heat needed to be applied. The food scientists, technologists, engineers, and a mathematician on the research team brought into being the knowledge required to prevent such outbreaks.

The NCA study illustrates the gamut of technological, scientific, and mathematical knowledge often brought to bear on a technological problem and the fact that basic scientific knowledge is frequently one of the outcomes of a very practical project. The mathematician on the research team, Ball, devised the mathematical procedure allowing safe thermal processing to be attained (16,17). The point to going into detail is to highlight the fact that the food science profession is many faceted. Science, engineering, and mathematics all may need to be brought into play to solve the problems at hand. In the case here, a trade organization was the agency that developed scientific knowledge so fundamental that its reach has been well beyond the food industry.

A second benefit attained from the NCA study is that knowledge of proper thermal processing methods enabled canned foods to shake off a reputation as an unsafe food product and to acquire one of almost complete safety. That in turn has been a major factor in canned foods, a value-added product, assuming a prominent role in the U.S. economy. Aside from convenience and being amenable to long-term storage, canned foods are "safe." Food technologists are the ones who have made that so.

Two other illustrations will be given to demonstrate that the profession works in various ways. Clarence Birdseye, an engineer and an inventor working on his own, brought about change because of innovative thought. He reasoned that food frozen quickly enough would retain its original qualities. He was close to being correct but not entirely so. The process worked for seafoods, but later on when freezing was extended to vegetable products, trouble arose. The flavor, color, and texture of the vegetables usually deteriorated within a month or so. Blanching had to be inserted as a processing step to inactivate the enzymes naturally present in the vegetables (18). Even at freezing temperatures, enzymes can cause changes deleterious to food quality. Researchers at the University of California demonstrated that peroxidase had to be destroyed to be sure that it and other enzymes would be inactivated (17). Here science had to step in to bring an application to fulfillment. The frozen foods industry has most certainly affected the eating habits of Americans and made a major contribution to the value-added portion of the food industry.

The third illustration will be that of research at a state government laboratory. Rare today is the homemaker who squeezes oranges for juice at breakfast. At the Florida Citrus Experiment Station scientists using fundamental knowledge reasoned that if orange juice were concentrated under a vacuum, the temperature would be low enough to avoid the heat-induced taste to which canned orange juice is so susceptible. Furthermore, they reasoned that if a small amount of fresh juice were added back to the concentrate, the top flavor notes could be restored to the concentrate. The process was successful until the volume produced began to cause the orange concentrate to be on the market for an appreciable time; then trouble arose as it had with frozen foods. As the juice was concentrated, calcium compounds in the juice were likewise concentrated. Upon storage even at freezing temperatures, the calcium compounds reacted with low methoxyl pectin produced by enzymes in the orange concentrate to cause curdling and gelation. Mild heating—not enough to affect the sensory quality—had to be added as a pasteurization step to inactivate the pectic enzymes (19,20). As with frozen foods, fruit and vegetable concentrates increased greatly in volume and went on to add materially to the value-added part of the food industry. In this instance, governmental scientists brought about a new industry, first by using scientific reasoning to devise a new application, then by using scientific research to overcome a limitation not anticipated.

NEWER PROCESSES

Enough illustrations have been given in detail to show that the profession functions in various settings: industry, trade organizations, universities, the government, and entrepreneurial enterprises. Within recent years, newer forms of processing have arisen. Among them is extrusion processing (21), controlled atmosphere and modified atmosphere storage of fruits and vegetables; controlled atmosphere packaging of food products (22,23); minimal processing coupled with refrigeration and some forms of packaging to permit the marketing of fruits, vegetables, and other kinds of foods (24–25); and edible packaging of various sorts (26–28). Not identical to the fresh produce but generally close to it, minimal processed foods have made a decided niche for themselves. Aseptic processing has unquestionably developed for itself a major role in food preservation (29), and ohmic heating is coming to the fore (30).

Irradiation of foods is being considered anew. Outbreaks of food infections or intoxication from organisms on fresh produce has revived the idea of irradiating fruits to destroy pathogenic microorganisms (31).

An area where a major change is taking place in the food field is that of biotechnology or "genetic engineering" as it is sometime called (32–35). Katz (33) points out that the movement toward improved vegetable oils will be helped by the New Plant Genome Initiatives signed into law by President Clinton in 1997. Biotechnology research and other innovations are not without objection (36) or the risk of development of allergenicity (37).

NUTRACEUTICALS

A phase of food science coming to the fore is that of so-called nutraceuticals (38–41). They are often also referred to a phytochemicals. In many respects, this phase of science is at the same stage as was research into vitamins 70 to 80 years ago. Nutraceuticals are chemicals in the food that either upon theoretical grounds or by observation appear to possess physiological properties. The questions are (*1*) do nonnutritive compounds affect in any way the action or use of demonstrated nutritional substances; if so, how and to what extent, and (*2*) are such actions beneficial or detrimental to health? The task is to relate the presence of one or more of the nutraceuticals to specific effects. Today, of course, isolation and identification of unknown compounds is far more feasible than it was when the first of the vitamins began to be unequivocally identified. Testing for effects however is no less arduous.

GOVERNMENTAL ACTIVITIES

Very brief mention was made of a state governmental laboratory when the development of citrus concentrate was referred to, but that research was a very small part of the effect governmental agencies have on food science and technology. The research laboratories of the U.S. Department of Agriculture (USDA) employ many food scientists and technologists. Aside from the five major regional laboratories and the USDA laboratories in Beltsville, Maryland, there are numerous area laboratories such as the Southern Area Laboratory in Lane, Oklahoma, and the Tree Fruit Research Laboratory in Wenachee, Washington, which carry on research. The research is often oriented toward raw products, and livestock or poultry production, but at many of these laboratories, major effort today goes into food processing or utilization. There is also the joint state–federal program of the Agricultural Research Service and the Land-Grant Experiment Stations.

The Department of Defense (DOD) likewise employs a good number of food scientists and technologists. The RD&E Center of the U.S. Army Natick Facility conducts research on the stability of military rations, demographic acceptability of different kinds of food, and nutritional studies. The National Marine Fishery Service, Department of Commerce, also employs food scientists and technologists, though its operations are much smaller than those of the USDA or DOD.

One of the things that has added stature to the profession is that the *U.S. Federal Government's Directory of Occupational Titles* began to list both food science and food technologists as professional fields rather than lumping food scientists under some older category such as microbiologist or chemists (42).

GOVERNMENTAL REGULATION

A major employer of food scientists and technologists is the Food and Drug Administration (FDA). In spite of the matter of occupational titles just referred to, they may not be called food scientists or technologists because their duties are generally that of an analyst or an inspector. The same applies to the USDA and its Food Inspection Service. Of increasing importance today has been the institution of Hazard Analysis at Critical Control Points (HACCP) and Standard Operating Procedures (SOP) for sanitation, which are programs mandated by the federal government but implemented by the food firm according to specifications set down by the government (43) where the company's own quality control personnel are responsible for inspecting of the food at points along the processing line. HACCPs arise where there is the possibility of microbial organisms contaminating the food or increasing appreciably in number or where other forms of contamination might be the cause of a regulatory complaint.

Major effort is being directed toward solutions to the problem of pathogenic organisms infecting the food supply. Among the pathogens are protozoa (44), viruses (45); nor is the home exempt from such problems (46). In assessing dangers, recognition has to be taken of the fact that not all fecal coliforms are necessarily of public health significance (47).

Miller (48) pointed out that an integration of food science, toxicology, microbiology, nutrition, and genetics is needed to solve problems of food safety. Hutt (49) described several of the modifications made in the FDA Modernization Act of 1997. In *Food Technology* (50) the proceedings of a symposium to honor Peter Baron Hutt are described. Hutt himself explained and criticized the approval system for food additives. Articles on food additives have not been cited here. They really need to be considered in the light of risk assessment (51).

SUMMARY

The food industry is, as O'Conner said, an industrial powerhouse (13). It adds greatly to the gross domestic production of the United States. For many of the other industrial countries of the world, the same is true. Food production is a key part of the economy. Developments originated by food scientists, technologists, and engineers, while not the sole source of advances made by the food industry, have been among the foremost in bringing the United States to the enviable position it occupies. The roles food scientists and technologists assume have been illustrated here mostly by example. The very nature of the profession requires that its practitioners be individuals with a strong, broad background in basic science and mathematics and with a sufficiently broad knowledge of food science and engineering to be able to fit within anyone of the several areas the food industry encompasses. That is the reason the IFT promulgated an undergraduate minimum standard curriculum only. There is a general consensus that specialization should not commence until graduate studies are undertaken. The survey reported by Mermelstein (52) teems with information as to the demographics of job opportunities, degree level helpful, gender, types of employment, and salaries received by those in various fields of employment.

The food industry and food science are on the move in other countries, too (53). In a synopsis of a publication by the National Science Board (54), academic universities are stated to be the "manufacturing plants for new knowledge." They are the largest producers of basic research, but it is pointed out that Europe now leads North America in science and engineering doctoral production. Western and Eastern European universities now graduate about 60% more doctoral students than does North America. The report states further, contrary to popular opinion, that most doctoral graduates do not work in academia, only 28.4% do. There has been a rapid rise in R&D spending outside the United States, but there has also been a rapid rise in spending within the United States by foreign firms. The United States possesses 46% of the world's industrial R&D capacity.

BIBLIOGRAPHY

1. R. M. Schaffner, "What Training Should a Four-Year Food Technology Student Receive?" *Food Technol.* **12**, 7–14 (1958).
2. "Report of IFT Council," *Food Technol.* **20**, 1567–1599 (1966).
3. "Discussion Papers and IFT Undergraduate Curriculum Minimum Standard—1977 Revision," *Food Technol.* **44**, 32–61 (1990).
4. *IFT Administered Scholarship / Fellowship Program for the 1998–99 School Year*, No. 25, Institute of Food Technologists, Chicago, Ill. 1997.
5. L. D. Satterlee, "Introduction to the 1992 IFT Undergraduate Minimum Standards," *Food Technol.* **46**, 155 (1992).
6. "IFT Undergraduate Curriculum. Minimum Standards for Degrees in Food Science (1992 Revision)," *Food Technol.* **46**, 156–157 (1992).
7. O. Fennema, "Education Programs in Food Science: A Continuing Struggle for Legitimacy, Respect and Recognition," *Food Technol.* **43**, 170–182 (1989).
8. T. P. Labuza, "Mission 2000: IFT's Future," *Food Technol.* **43**, 68–84 (1989).
9. T. P. Labuza and D. R. Lineback, "The University-Industrial Relationship in Food Science and Technology," *Food Technol.* **41**, 74–91 (1987).
10. International Union of Food Science & Technology, *Directory of Courses and Professional Organizations in Food Science and Technology*, 2nd ed. rev., CSIRO Division of Food Research, North Ryde, Australia, 1980
11. I. D. Morton and J. Lengee, eds., *Education and Training in Food Science, a Changing Scene*, Ellis Horwood-Simon and Schuster International Group, Hemel, Hempstead, Hertfordshire, England, 1992.
12. A. G. Netto, "Present and Future of Food Science and Technology in Developing Countries," *Food Technol.* **43**, 148–168 (1989).
13. J. M. O'Connor, *Food Processing: An Industrial Powerhouse in Transition*, D.C. Heath, Lexington, Mass., 1988.
14. A. Manchester and A. Clauson, "Spending for Food up Slightly in 1995," *Food Review* **19**, 2–5 (1996).
15. T. A. Roberts, C. R. Dillon, and T. J. Siebenmorgen, "The Impact of Food Processing on the U.S. Economy," *Food Technol.* **50**, 65–68 (1996).
16. C. O. Ball, "Mathematical Solution of Problems on Thermal Processing of Canned Foods," *University of California Publications in Public Health* **1**, 1–244 (1928).
17. C. O. Ball, "Thermal Process Time for Canned Food," *Bulletin of the National Research Council* **7**, Part 1, 5–76 (1923).
18. A. L. Arighi, M. A. Joslyn, and G. L. Marsh, "Enzyme Activity in Frozen Pack Vegetables," *Ind. Eng. Chem.* **26**, 595–598 (1936).
19. E. L. Moore, R. L. Huggart, and E. C. Hill, "Storage Changes in Frozen Concentrated Citrus Juices—Preliminary Report," *Proc. Florida State Hort. Soc.*, 63rd Annual Meeting, p. 165 (1950).
20. F. W. Wenzel et al., "Gelation and Clarification in Concentrated Citrus Juice. 1. Introduction and Present Status," *Food Technol.* **5**, 454–457 (1951).
21. S. Ilo and E. Berghofer, "Kinetics of Thermomechanical Destruction of Thiamin During Extrusion Cooking," *J. Food Sci.* **63**, 312–316 (1998).
22. M. A. Cliff, O. L. Lau, and M. C. King, "Sensory Characteristics of Controlled Atmosphere- and Air-Stored 'Gala' Apples," *Journal of Food Quality* **21**, 239–249 (1998).
23. M. M. Wall and R. D. Berghage, "Prolonging the Shelf-Life of Fresh Green Chile Peppers Through Modified Atmosphere Packaging and Low Temperature Storage," *Journal of Food Quality* **19**, 467–477 (1996).
24. D. G. Hoover, "Minimally Processed Fruits and Vegetables: Reducing Microbial Load by Nonthermal Physical Treatments," *Food Technol.* **51**, 66–71 (1997).
25. G. Gunes and C. Y. Lee, "Color of Minimally Processed Potatoes as Affected by Modified Atmosphere Packaging and Antibrowning Agents," *J. Food Sci.* **62**, 572–575 (1997).
26. E. A. Baldwin et al., "Use of Lipids in Coatings for Food Products," *Food Technol.* **51**, 56–62, 64 (1997).
27. J. M. Krochta and C. De Mulder-Johnson, "Scientific Status Summary. Edible and Biodegradable Polymer Films: Challenges and Opportunities," *Food Technol.* **51**, 61–74 (1997).

28. T. D. Cai and K. C. Chang, "The Effect of Coatings and Processing Methods on Dehydrated Precooked Whole Pinto Beans Quality," *Journal of Food Quality* **20**, 315–328 (1997).

29. D. M. Dignan et al., "Safety Considerations in Establishing Ascptic Processes for Low Acid Foods Containing Particulates," *Food Technol.* **43**, 118–121, 131 (1989).

30. "Overview of Articles on Ohmic Heating," *Food Technol.* **50**, 241–273 (1996).

31. D. G. Olson, "Status Summary, Irradiation of Food," *Food Technol.* **52**, 56–62 (1998).

32. K. Liu and E. A. Brown, "Enhancing Vegetable Oil Quality Though Plant Breeding and Genetic Engineering," *Food Technol.* **50**, 67–71 (1996).

33. F. Katz, "The Move Towards Genetically Improved Oils," *Food Technol.* **51**, 66(1997).

34. J. Q. Wilkinson, "Biotech Plants: From Lab Bench to Supermarket Shelf," *Food Technol.* **51**, 37–42 (1997).

35. F. Katz, "Biotechnology—Newer Tools in Food Technology's Toolbox," *Food Technol.* **50**, 63–65 (1996).

36. D. E. Pszczola, "Activists Target Transgenic Soybeans," *Food Technol.* **50**, 29 (1996).

37. R. L. Fuchs and J. D. Astwood, "Allergenicity Assessment of Foods Derived from Genetically Modified Plants," *Food Technol.* **50**, 83–88 (1996).

38. "Articles on Nutraceuticals," *Food Technol.* **52**, 44–57 (1998).

39. M-T Huang et al., *Food Phytochemicals for Cancer Prevention. I Fruits and Vegetables*, ACS Symposium Series 546, American Chemical Society, Washington, D.C. 1994.

40. C-T Ho et al., *Food Phytochemicals for Cancer Prevention. II Teas, Spices and Herbs*, ACS Symposium Series 547, American Chemical Society, Washington, D.C. 1994.

41. P. A. LaChance, *Nutraceuticals: Designer Foods. III Garlic, Soy and Licorice*, Food and Nutrition Press, Trumbull, Conn. 1997.

42. *Dictionary of Occupational Titles*, 4th ed., U.S. Government Printing Office, Washington, D.C. 1991.

43. H. R. Cross, "HACCP: Pivotal Changes for the Meat Industry," *Food Technol.* **50**, 236 (1996).

44. G. J. Jackson et al., "Cyclospora—Still Another New Foodborne Pathogen," *Food Technol.* **51**, 120 (1997).

45. D. O. Cliver, "Scientific Status Summary: Virus Transmission Via Food," *Food Technol.* **51**, 71–78 (1997).

46. R. W. Daniels, "Home Food Safety," *Food Technol.* **52**, 54–56 (1998).

47. M. P. Doyle, "Fecal Coliforms in Tea: What's the Problem?" *Food Technol.* **50**, 104 (1996).

48. S. A. Miller, "Developing a New Food Wholesomeness Science to Ensure Food Safety," *Food Technol.* **51**, 62–65 (1997).

49. P B. Hutt, "A Guide to the FDA Modernization Act of 1997," *Food Technol.* **52**, 54–60 (1998).

50. "Food Regulatory Policy: A Symposium to Honor Peter Barton Hutt," *Food Technol.* **50**, 101–128 (1996).

51. C. R. Winter and F. J. Francis, "Assessing, Managing, and Communicating Chemical Food Risks," *Food Technol.* **51**, 85–92 (1997).

52. N. H. Mermelstein, "1997 IFT Membership Employment and Salary Survey," *Food Technol.* **52**, 63–73 (1998).

53. "Setting the Course for Global Food Markets," *Food Review* **19** (1996).

54. "NSF Data Show Shifts in R&D," *Chem. Eng. News* **76**, 44–52 (1998).

JOHN POWERS
University of Georgia
Athens, Georgia

See also FOOD SCIENCE AND TECHNOLOGY: DEFINITION AND DEVELOPMENT.

FOOD SPOILAGE

Food is considered spoiled when an undesirable change in the color, flavor, odor, or texture of the product has occurred. Microbial activity, endogenous chemical activity, insect/rodent damage, or physical injury such as bruising and freezing are responsible for spoilage. Spoilage classifications will vary with the consumer; that is, one individual may consider a food spoiled where another does not. A perfect example is fermented food. To some individuals, the food may be considered delicious but to other individuals who have had limited exposure to that type of food, the unfamiliar odors associated with the product would be considered spoiled. Individuals also differ in their thresholds to various odors and flavors, which would affect their judgment of the food's quality. Less obvious changes in food composition may also classify the food as spoiled. For example, nutrient losses, presence of harmful bacteria, or the formation of toxic substances within the food would all render the product spoiled. In these and other cases where food has been labeled as spoiled, the food is removed from distribution and labeled unfit for human consumption. The significance of this action is that there are economic losses associated with removal of the products from distribution and consumption, and the producer, processor, or consumer must assume these losses.

ORGANOLEPTIC DESCRIPTION OF SPOILAGE

Changes in the visual, tactile, olfactory, and flavor characteristics of food constitute spoilage, symptoms varying with the commodity. As an aid to summarize the types of spoilage occurring in many of the major food commodities, Table 1 lists some of these spoilage descriptors and the associated factor/mechanism for their development. Examples of visual changes would include the surface darkening of red meats, greening of cooked and cured meats, surface darkening of cut fruits, and fuzzy areas being generated on the surface of fruit, bread, or cheese. Other visual changes that also overlap tactile perceptions include sliminess on meats and vegetables and ropiness and curdling in fluid milk products. Off-odors and off-flavors associated with spoiled foods may be described by the terms putrid, rancid, bitter, soapy, sour, stale, and earthy. These off-odors and off-flavors vary not only between food categories but may also vary within a food category. For instance, the spoilage of marine temperate water fish is characterized by development of fishy, rotten, and hydrogen sulfide notes, whereas tropical fish and freshwater fish

Table 1. Selected Spoilage Indicators of Various Food Commodities.

Spoilage descriptor	Spoilage factor
Meat	
Off-odors: fruity, sweet smelling esters, followed by the formation of putrid sulfur compounds (indole, methanethiol, dimethyl disulfide and ammonia)	*Pseudomonas* spp.
Slimy or tackiness: masses of bacterial growth and the softening or loosening of meat structural proteins	*Pseudomonas* spp.
Greening: discoloration of meat pigments in frankfurters	Heterofermentative species of lactobacilli and *Leuconostoc*
Discoloration: surface browning	Pigment oxidation
Fish	
Off-odors: hydrogen sulfide; methyl mercaptan; dimethyl sulfide; trimethylamine	*Pseudomonas* spp.
Softening/mushiness: liquefaction of muscle tissue in Pacific whiting	Proteolytic enzyme activity
Discoloration: blackening of tissue in shrimp	Enzymic oxidation of polyphenols
Eggs	
Runny yolk and off-odors: H_2S and musty odors	*P. graveolens*
Discolored yolks	*Pseudomonas, Aeromonas, Acinetobacter, Proteus*
Pinspots: mycelial growth on inside of eggs	Molds—*Penicillium* and *Cladosporium*
Milk	
Off-flavors and odors: Bitter and occasionally putrid	*Pseudomonas* spp.
Clotting or curdling	Lactic acid bacteria
Butter	
Off-odors: putrid odors due to certain organic acids (ie, isovaleric acid)	*P. putrefaciens*
Off-odors: rancid odors from short-chain fatty acids hydrolyzed from fats	*P. fragi*
Cheese	
Holes or gas pockets	Gas-producing bacteria
Fuzzy areas	Molds
Vegetables	
Soft roll: initially sunken areas, followed by sunken area becoming soft, and wet and spreading to encompass the entire vegetable	*Erwinia* spp. and *Pseudomonas* spp.
Gray mold rot: Gray mycelium	*Botrytis cinerea*
Black semisoft blemishes	*Alternaria* spp.
Black rot: in potatoes	*Ceratocystis fimbriata*
Chilling injury (pitting and russeting; surface browning; improper ripening; decay)	Membrane lipid oxidation
Shriveling	Moisture loss
Citrus fruit	
Green rot: dense olive-colored layer of mycelium and powdery conidia	*Penicillium digitatum*
Noncitrus fruits	
Brown rot: brown discolorations in peaches, plums, and apples	*Monilinia fructicola* and *Penicillium expansum*
Downy mildew: in grapes	*Plasmapara viticola*
Gray mold rot: gray mycelium (fuzz) in strawberries	*Botrytis*
Legumes	
Reddish-brown spots/streaks	*Uromyces*
Off-flavors: beany, grassy, and rancid	Chemical lipid oxidation
Grains	
Toxic compound generation: mycotoxins	Molds

Table 1. Selected Spoilage Indicators of Various Food Commodities.

Spoilage descriptor	Spoilage factor
Bread	
Fuzzy areas: mycelium	*Rhizopus stolonifer*
Nuts	
Fuzzy areas: mycelium	Molds
Canned foods	
Swelling of can	Gas-producing organisms
Wine	
Off-flavors: grassy (*cis*-hexen-3-al; *trans*-hexen-2-al); cork taint (2,4,6-trichloroanisole); vinegary (acetic acid)	*Acetobacter* spp.
Beer	
Off-odor: buttery or honeylike due to presence of diacetyl	*Pediococcus cerevisiae*
Ropiness: liquid becomes viscous and pours as an "oily" stream	*Acetobacter, Lactobacillus, Pediococcus cerevisiae,* and *Gluconobacter oxydans*
Off-flavors: sourness due to oxidation of ethanol to acetic acid	*Acetobacter* spp.
Turbidity	*Zymomonas anaerobia* and *Saccharomyces* spp.
Browning and haze formation	Chemical oxidation of polyphenols
Chocolate	
Surface whitening	*Chrysosporium* spp. or fat recrystallization
Oils	
Off-flavors: rancid	Chemical lipid oxidation
Potatoes	
Sprouting: toxic formation of solanine	Physiological

spoilage is characterized by fruity and sulfhydryl notes. Development of spoilage flavors and odors are seldom abrupt but occur in successive steps. In the case of fish, the initial accumulation of inosine monophosphate (IMP) contributes to the pleasant flavor of the fish while subsequent breakdown of IMP results in loss of such flavor. Further storage leads to the microbial or chemical generation of putrid or rancid compounds, respectively, that are responsible for making the product unpalatable. It is important to note that oftentimes no specific volatile compound may be held responsible for the spoilage aromas. Instead the concentration of a class of volatile compounds, such as the sulfur class of volatiles, and the relative concentration of that class to other classes of volatiles may better describe differences in spoilage aromas.

FOOD SPOILAGE MECHANISMS

Microbiological Degradation

Organisms and Factors Affecting Growth. The types of microorganisms that are responsible for spoiling foods include bacteria, molds, and yeasts. Like other living things, microorganisms require nutrients, water, and minerals to survive and grow, and to accomplish these activities, the microorganisms degrade the foods. Many factors affect the growth and biochemical activities of microorganisms in foods and these are the chemical composition of the food, the conditions of processing and storage, and the inherent properties of the microbial species present. With regard to the latter point, different types of microorganisms have growth habitats for which they are best suited. For example, bacteria are adapted to fast growth in nonacidic (pH 5.0–8.0) and moist conditions, whereas molds grow more slowly but are better adapted to growth in relatively acidic (pH 2.5) or dry conditions. Yeasts are similar to molds, but many also tolerate conditions containing high concentrations of sugar.

The microbial flora in a food will be initially dictated by the product's environment or by contaminants picked up during processing, handling, packaging, and storage of that product. For example, freshly caught fish from warm or tropical waters carry a microbial population primarily of mesophilic bacteria (bacteria that grow best from 15 to 40°C) such as *Micrococcus coryneforms* and *Bacillus*. In contrast, cold and temperate water fish species harbor predominantly psychrotrophic bacteria (bacteria that grow at refrigeration temperatures) that includes the genera *Moraxella/Acinetobacter, Pseudomonas, Flavobacterium,* and *Vibrio*. Upon refrigerated storage, however, the psychrotrophic bacteria in both the warm and cold water fish become the predominant species as a result of natural competition. Since a greater concentration of psychrotrophic organisms are initially found in seafood harvested from temperate waters, their spoilage could be expected to be quicker than seafood harvested from tropical waters. Such conjecture has at least been substantiated with organoleptic evaluations of tropical and temperate shrimp stored in ice (1). Another important intrinsic factor in fish that has played an important role in determining the microbial flora during storage is the pH. Most fish contain only very little carbohydrate ($<0.5\%$) in the muscle tissue, and thus only small amounts of lactic acid are produced postmortem. As a result, fish muscle contains a high postmortem pH (>6.0), and under these conditions the pH-sensitive spoilage bacteria *Shewanella putrefaciens* may grow. In the case of bivalves such as oysters and clams that store energy in their tissues as glycogen, lactic acid may be produced in much greater quantities. As a result, the reduction in pH favors multiplication by fermentative-type microorganisms such as Lactobacilli, Streptococci, and yeasts.

In other meat items stored at refrigeration temperatures, only approximately 10% of the initial bacteria present are able to grow, and of those microorganisms able to grow, the fraction causing spoilage is even lower. Hence, it is important to identify these "specific spoilage organisms" in each food so treatments can be optimized to minimize their growth. Influencing the composition of those specific spoilage organisms on meat, however, will be the available moisture. In a high humidity environment, bacterial spoilage in fresh meats is dominant, with coalescence of colonies producing a slime layer on the surface. When the surface of the meat is too dry, as it is with many dried and fermented meat products, mold tends to predominate as the spoilage organisms. On the other hand, the application of vacuum or modified atmosphere packaging to cooked meat products generates conditions conducive to dominance by lactic acid bacteria. Producing acids such as lactic acid, acetic acid, and formic acid, the bacteria generate spoiled products described as sour and acid. When these products are stored aerobically following their anaerobic storage, additional obnoxious odors may be produced that are described as slightly sweet and cheesy.

Molds are able to grow on many other kinds of food in addition to meat. These include cereals/grains, milk, fruit, vegetables, and nuts. As a result of the mold growth, several kinds of food spoilage may arise: off-flavors, toxins, discolorations, and rotting. Of these, the most important spoilage problem associated with molds is the formation of mycotoxins. Mycotoxins are secondary metabolites of molds that are toxic to vertebrate animals in small amounts when introduced via a natural route. The most important toxic effects are different kinds of cancer and immune suppression. While some mycotoxins are only present in the mold, most of them are excreted in the foods. The high resistance of mycotoxins to physical and chemical treatments makes them difficult to remove from foods once they are present. Hence, it is important to limit mold growth on food items.

The well-accepted conclusion that molds are mostly responsible for the spoilage of fruit is based on the visible presence of fungal mycelium and spores on the surface of these products at the time of rotting. Yeasts, however, commonly occur on the surface of freshly harvested fruits at populations of 10^3 to 10^5 cells per square centimeter (2). These yeasts in most circumstances are considered to remain relatively inactive as they do not produce the appropriate enzymes to degrade the skin of the fruit and establish infection. Physical damage of the skin, however, by overripening, mechanical injury, or fungal attack exposes the fruit tissue upon which yeast can rapidly grow to produce secondary spoilage.

In products, such as beer and wine, that are dependent on microbial action, spoilage may also ensue when growth of undesirable microorganisms occurs. Even in chocolate-covered confectionery products, spoilage may result from growth of foodborne *Chrysosporium* species. Growth of these xerophilic fungi has produced a white surface discoloration that for all purposes was similar in appearance to the effects observed from recrystallization of cocoa butter (3).

Enzymatic/Biochemical Activity of Microorganisms. During their growth, microorganisms utilize food constituents and generate metabolic end products. As a consequence, the physical, chemical, and sensory properties of the food are substantially changed. Generally, the nature of these changes is not well described, making it difficult to explain spoilage on a biochemical basis. These biochemical activities, however, often include enzyme activity of a pectinolytic, proteolytic, and/or lipolytic nature.

In fruits and vegetables, pectolytic fluorescent pseudomonads, mainly *P. fluorescens* and *P. viridiflava*, account for substantial proportions of postharvest rot in cold storage and at wholesale and retail markets. The ability of these pseudomonads to cause maceration of plant tissues is primarily due to their ability to produce an extracellular pectate lyase capable of degrading pectin components of plant cell walls.

Grain is also subject to breakdown by carbohydrases from spoilage fungi. These fungi have been shown through *in vitro* studies to produce cellulase, polygalacturonase, pectin methyl esterase, 1-4-β-glucanase, β-glucosidase, and β-xylosidase (4). Environmental conditions such as water activity (A_w) and temperature, however, influence the production of these enzymes.

A key metabolic reaction of most yeasts when cultured under favorable anaerobic conditions is the fermentation of sugars such as glucose, fructose, sucrose, and maltose to produce ethanol and carbon dioxide. In addition to the loss of sweetness, this activity produces gassy products with a distinctive alcoholic aroma and flavor. Many hundreds of other secondary end products are formed during fermentation, and these also have a significant impact on the sensory properties. Such products include higher alcohols, organic acids, esters, aldehydes, and ketogenic substances that, although produced in small concentrations, have very low flavor and aroma thresholds. Unfortunately, with the exception of *Saccharomyces cerevisiae*, the production and sensory relevance of these secondary metabolites by yeasts are not well known (2).

Fragmentary knowledge exists on yeast metabolism of nitrogen compounds. As one example where this type of biochemical activity is of significance, metabolism of lysine during the growth of the yeast *Brettanomyces intermedius* in wine produced mousy taints due to the production of substituted tetrahydropyridines (2). It is also well documented in the brewing and wine literature that some, but not all, strains of *S. cerevisiae* can produce objectionable concentrations of hydrogen sulfide and sulfur dioxide and that these properties are linked to their metabolism of amino acids and inorganic sulfur compounds in the growth medium.

Yeast can both synthesize and metabolize organic acids. These activities change the acidities and flavor profiles of the product. In addition, when oxidative utilization of organic acids by yeast occurs via the tricarboxylic acid cycle, the pH of the product may rise to values that allow the growth of spoilage bacteria.

In foods, microbial proteolysis is another key spoilage reaction. Initially, it leads to the development of bitter flavors, followed by the production of strong ammonia odors and putrefaction. These proteases, secreted into the food to break down the proteins, include both endoproteases and exoproteases (aminopeptidases and carboxypeptidases). Thus, volatile fatty acids originate from proteins by deamination of amino acids while volatile bases arise from decarboxylation of amino acids and deamination of adenylates in the food. The degradation products, in turn, serve as a nutrient source for the microorganisms. When ample concentrations of low molecular weight nitrogen constituents (free amino acids, creatine, taurine, etc) are already present in the food product, however, as they are in fresh fish, protease synthesis and excretion is inhibited. In later stages of spoilage when most of the free amino acids have been depleted, then proteolysis becomes important. Even so, it appears that some level of low molecular weight peptides must be present to activate the synthesis of the extracellular proteases in *P. fragi*, an active spoiler of muscle food (5). Other factors in the food medium have also been shown to affect synthesis and excretion. For example, in milk supplemented with iron, the size of the *Pseudomonas* population when extracellular enzymes were produced was 10 times larger in the supplemented milk than in the nonsupplemented milk (6). The types of proteases produced by a microorganism, however, will vary. Thus, *P. aeruginosa* secretes two proteases, *V. alginolyticus* secretes a collagenase and five alkaline serine proteases, and *Aeromonas hydrophila* secretes two proteinases and one aminopeptidase. In general, these proteases have molecular weights ranging from 20 to 50 kDa, consist of only a single polypeptide, and lack the cysteine amino acid. The latter two characteristics are especially important because they contribute to the high heat stability of the extracellular proteases. Hence, even though the vegetative cells may be killed by a heat treatment, extracellular microbial enzymes continue to exert their activity. To reduce the possibility of this type of activity in milk, thresholds have been set in raw milk at approximately 2×10^6 colony forming units/mL (7). Levels above this threshold would thus generate pasteurized milk that would spoil quickly despite bacterial counts being reduced greatly. Product deterioration by proteolytic enzymes generated from psychrotrophic sporeformers, such as *Bacillus* spp., however, is also receiving attention in the dairy industry. When in the spore state, these microorganisms easily survive the typical range of pasteurization conditions with subsequent germination and outgrowth of vegetative cells. Consequently, during their growth, activity by degradative enzymes, including proteases, leads to gelation of ultra-high temperature (UHT) milk, sweet curdling of milk, bitter and unclean off-flavors in cheese, decreases in cheese yield, and textural and body defects such as "wheying off" in cultured dairy products. Interestingly, it has been noted that the optimum temperature for production of degradative enzymes by these microorganisms is usually lower than the optimum temperature for cell division (8). Thus, it is possible for milk held at refrigeration temperatures for extended periods of time to develop off-flavors through microbially produced enzymes even though the observed microbial population might remain below that normally associated with formation of microbial defects.

Lipases are another group of hydrolytic enzymes released by microorganisms to break down the food. In this case, the enzymes cleave fatty acids from triglycerides (a molecule containing three fatty acids esterified to a glycerol backbone). For those lipases found in milk, a preference occurs for cleavage at the 3-position in the triglyceride where butyric and caproic fatty acids reside. Upon hydrolysis, these volatile fatty acids at very small concentrations (1.5 mEq/100 g of fat) have distinctly unpleasant smells and tastes, which is termed hydrolytic rancidity. Also, in milk, some thermoduric psychrotrophs have been shown to produce phospholipases, particularly phospholipase C, that hydrolyse the ester linkage between the glycerol backbone and the phosphoryl group of a phospholipid. In milk, this attack takes place at the fat globule membrane where phospholipids are found. The lipolysis at the fat globule membrane results in an increased susceptibility of the exposed milkfat to the subsequent action of lipases. It has also been suggested that the degradation of the fat globule membrane by phospholipase C results in the aggregation of fat globules that leads to the bitty cream defect frequently observed in cream products.

Not all lipolytic action results directly in flavor defects. Only when the free fatty acids (FFAs) are short to medium range (C4 to C12) do they contribute directly. In many cases, FFAs react with protein constituents in the food producing textural defects in the product, but rarely are these defects of a severity to render the product spoiled. In cases where the FFAs are of an unsaturated nature (contain one or more double bonds in their structure), they may increase or decrease the susceptibility of the food to undergo lipid oxidation. When the FFA is cleaved from a membrane phospholipid early in storage, it will increase the oxidative stability of the product, but release of FFA from membranes later in storage will decrease oxidative stability. FFAs released from storage triglycerides almost always are more susceptible to lipid oxidation than they were when esterified. This process of lipid oxidation, in turn, leads to the presence of off-flavors and off-odors in the product. Lipid oxidation will be described in more detail in the next section.

The last group of microbial enzymes to be discussed in this section is one that leads to trimethylamine oxide (TMAO) degradation. In marine fish, TMAO is believed to be involved in the fish's osmotic regulation, and levels in the fish will be dictated by the levels required for counteracting the osmotic pressure of seawater as well as amounts present in the food. Degradation of this compound by psychrotrophic bacteria, such as *Achromobacter*, apparently involves both a triamineoxidase for activation of the TMAO substrate and a dehydrogenase for reduction of the compound to trimethylamine (TMA). In turn, when TMA reacts with the fat in the muscle, it creates a characteristic fishy odor. Odors appear at TMA levels as low as 2.9 mM of muscle extract, while definite odors have been observed at levels of 7.2 mM in the muscle extract (9). Significant amounts of TMA, however, will not be observed until after the bacterial lag phase that extends from the onset of rigor to its resolution, due to a postulated bacteriostatic effect of rigor mortis.

Lipid Oxidation

Oxidative rancidity has long been recognized as a major cause of spoilage in food. In many products (potato chips, nuts, oil, cereals, etc) stored for varying lengths of time, the off-flavors and off-odors generated as a result of lipid oxidation are described as rancid, cardboardy, or painty. In meat that has been cooked, refrigerated, and then rewarmed, the off-flavors are described as warmed-over. In stored fish, the off-odors may be described as either rancid or fishy.

Lipid oxidation is a radical-driven process that begins with an activated oxygen species abstracting a hydrogen atom from an unsaturated fatty acid. The active oxygen species can be produced by light; or by interaction of the oxygen with enzymes such as lipoxygenase, peroxidase, or microsomal enzymes; or with transition metal ions, particularly iron. The oxygenated fatty acid, in turn, forms a hydroperoxide (primary reaction product) by abstracting a hydrogen atom from an adjacent fatty acid, which, in turn, can continue the process of abstracting hydrogen from a neighboring fatty acid. Hydroperoxides do not directly cause an adverse effect on the flavor and aroma of the food; rather, it is their breakdown to aldehydes and ketones (secondary reaction products). During this breakdown, additional free radical compounds may be generated that, in turn, could start another sequence of lipid oxidation reactions.

The susceptibility of a food to oxidize is dependent on both its chemical composition and its physical environment. The major chemical component in the food that will affect oxidation is the degree of unsaturation that exists in the fatty acids. The more double bonds in the fatty acid, the faster it will oxidize. For example, fish contain unsaturated fatty acids with one to six double bonds, and they will oxidize faster than chicken, which contains unsaturated fatty acids with one to four double bonds. Concentrations of catalysts in the food system are also important in determining the susceptibility of the food to oxidize. Of these catalysts, the nonenzymic catalysts (transition metal ions) play a key role, as they not only participate in the initiation process but also participate in the breakdown of hydroperoxides to low molecular weight off-flavors and off-odors. The fact that most foods, even immediately after harvest and slaughter, are believed to have hydroperoxides present, would lend credence to a dominant initiation role being played by transition metal-catalyzed hydroperoxide breakdown. Other compositional factors affecting the food's susceptibility to oxidize include pro-oxidants and antioxidants (tocopherol, flavonoids), as well as the microbial contamination. Illustrating this latter point, for instance, lipid oxidation was reduced in pork muscle stored at 2°C when lipolytic enzyme-producing bacteria were present (10). As for the physical environment, oxygen availability, temperature, and water activity play a role in the susceptibility of a food's lipid to oxidize.

Biochemical/Physiological Degradation

Chilling Injury. Chilling injury is a term used to describe the physiological damage that occurs in many plants and plant products, particularly those of tropical and subtropical origin, as a result of their exposure to low but nonfreezing temperatures. Visual symptoms vary with the product but may include surface pitting, water rot, poor color development or inability to ripen, and general loss of structural integrity. Appearance of these symptoms has been found to depend on the time and temperature of exposure to the chilling temperature as well as the maturity of the plant product. It primarily occurs only after transfer to nonchilling temperatures.

One of the most important visual symptoms of chilling injury is pitting, a general collapse of the fruit tissue induced by low-temperature dehydration. It is more evident in fruits such as limes, grapefruit, mango, or avocado in which the outermost covering is harder and thicker than the adjacent layer. In fruits with thin and/or soft peels such as peppers, eggplant, and tomato, water soaking, surface discoloration, and sheet pitting are more common. More specifically in eggplant, pitting begins with the deformation and browning of the cells located several layers beneath the epidermis followed by deformation and browning of the pericarp cells. In broccoli, chilling injury deterioration is manifested by chlorophyll loss and yellowing of flower buds.

Physiological and biochemical investigations into the nature of chilling injury have focused on membrane-induced damage in chilling-susceptible fruits and vegetables as a result of their inability to increase protective enzyme systems that would reduce and scavenge free radicals attacking the membranes. As evidence for this line of thought, it was found that while superoxide dismutase activity increased during cold storage in both chilling-sensitive and -tolerant cultivars, catalase, ascorbate oxidase, and glutathione reductase activities were higher in tolerant cultivars than nontolerant cultivars at low temperature (11). Further support for the free-radical membrane-damage hypothesis has been in the observation that free-radical-mediated membrane damage precedes the visual symptoms (12). Tolerance to low temperatures may be increased, however, by exposing the sensitive fruits to nonchilling low temperatures prior to refrigerated storage. The acclimation period apparently enables the material to activate and produce the necessary enzymes required for protection during low-temperature storage. Inactivation of these enzymes, on the other hand, would most likely occur during minimal processing (cutting, slicing, coring, peeling, trimming, or sectioning of agricultural produce), and this inactivation would account for the observation that minimally processed vegetables are more sensitive to chilling injury than the intact produce. In fact, spoilage as a result of chilling injury occurred sooner in minimally processed vegetables than microbial degradation (13), which demonstrates that changes in the processing and handling of the food will often shift what type of spoilage is observed.

Senescence. Senescence is an ill-defined collective term used by plant physiologists to describe deteriorative reactions occurring during the final stage of the life of a plant, its constituent tissues, or its cells. Although the senescence of different organs can vary in rate and with organ function, several consistent features are found in a range of senescence situations. For physical changes, there may be loss of green color or synthesis of new pigments such as carotenoids or flavonoids. In terms of texture, there may be softening, wilting, or drying. Increased development of infections and lesions are also signs of senescence that indicate the material has lost its resistance to pathogens. Bear in mind that fruits and vegetables continue to live, respire, and change after harvest, and although nonclimacteric fruits and vegetables, such as citrus fruits and beans, do not ripen or improve after harvest, climateric fruits and vegetables, such as tomatoes and bananas, continue to ripen after harvest. Thus, nonclimacteric fruits and vegetables would be considered to immediately start the process of senescence following harvest, whereas climacteric fruits and vegetables only undergo senescence during storage after development of desirable sensory characteristics. Studies designed to elucidate the mechanism of senescence point again to the role of membrane lipid oxidation. It appears that during storage, protective systems that exist in the tissue become disabled and thus provide increased opportunities for free-radical reactions to inactivate key metabolic pathways. Physiological changes accompanying senescence include: loss of chlorophyll; disassembly of chloroplast structure; degradation of cell walls; altered membrane composition; loss of membrane fluidity; loss of cellular compartmentation; release of vacuolar contents; altered sugar content; a switch to alternative substrates for respiration; net loss of RNA; increased protease activity; net loss of protein; and altered amino acid content. In simplified terms, however, these changes could be described as arising from a change in the balance between anabolic and catabolic processes.

Endogenous Enzymatic Degradation. Activity of enzymes endogenously present in the food product may also lead to quality deterioration and spoilage during storage. Examples of these types of enzymes include phenolases, lipoxygenases, proteases, dimethylases, and peroxidases.

Postmortem, polyphenoloxidases present in shrimp and lobsters catalyze oxidation of tyrosine to *o*-quinones, which then condense with other quinones or with amino acids or proteins to form melanin. Although tyrosine is not colored, melanins are dark brown to black pigments, and their formation in shrimp and lobster is described as blackspot development.

Tenderization or flesh softening is considered to be associated with the disappearance of Z-disks, dissociation of actomyosin complex, destruction of connectin, and general denaturation of collagenous tissue. Of the proteases indigenous to skeletal muscle, Ca^{2+}-dependent proteases and the lysosomal cathepsins have been reported to be the best candidates for bringing about the textural changes during postmortem storage. Whereas these activities are desirable, activities associated with proteases invading from an animal's abdominal cavity are not. For example, herring, mackerel, and capelin caught during periods of heavy feeding are very susceptible to autolytic tissue degradation, and it is highly recommended to gut these fish as soon after capture as possible (9).

Breakdown of trimethylamine N-oxide (TMAO) by the enzyme TMAO demethylase in gadoid fish species generates dimethylamine and formaldehyde (FA). The FA, in turn, is extremely reactive and cross-links with myofibrillar proteins, causing a toughening of the texture and a loss of water-holding capacity that may be described as a "cottony" and "spongy" texture. Generally this reaction is evident during frozen storage of the gadoid species; however, if fresh fish are stored for long periods of time, such as when the fish are treated with chemicals and/or modified atmospheres, the consequences of this enzyme activity may also be evident.

For those fish species where red pigmentation is desirable (eg, salmon), postharvest bleaching would be considered undesirable. Lipoxygenases associated with fish skin, however, are capable of bleaching carotenoid pigments in fish skin and/or flesh.

A common occurrence that foods undergo during harvesting and processing is tissue disruption. With fruits and vegetables, this action could be either intentional (cutting broccoli, harvesting asparagus) or unintentional (fruit bruising, nicking of potatoes). For grains and oilseeds, grinding is often applied to abstract the desirable fractions, whereas with muscle foods, it could be anything from cutting into portions to mincing or deboning the tissue. These actions have important consequences, since during tissue disruption, a loss of an essential control system, enzyme decompartmentation, occurs. Enzyme and substrate not normally in contact with each other are thrown together, and the resulting activity leads to deterioration in quality. Harvesting asparagus, for example, initiates a lignification process that can render the stalk tough and inedible in a short period. This reaction is thought to proceed by peroxidase-catalyzed polymerization of phenolic compounds that then couple to cell wall constituents in vascular tissue, creating a rigid network (14). In minced fish tissue, a more extreme example of tissue disruption, lipolytic enzymes are released from lysosomes and proceed to attack the muscle lipids, rendering changes in texture and water-holding capacity of the product.

Physical Degradation

Insects. During growth, harvesting, and storage, insects may invade and feed on plant products. In some cases, this may directly lead to extensive losses in the product. In other cases, the appearance of the product would be considered unacceptable to the consumer and in this sense could be considered spoiled. Cured meats like country-style hams may also be damaged when insects lay their eggs in the meat. The larvae arising from the eggs burrow and feed on the meat causing a corky appearance. Insects may also serve as the vector for subsequent spoilage of the product. For example, the common fruit fly will transfer the mold that causes soft rot in vegetables (15).

Dehydration. The control of water in foods during storage is essential for fresh and dry foods. In the case of many types of fresh produce, they must be wrapped to prevent transpiration that would cause wilting and shriveling. On the other hand, it is important that the water vapor transfer rate of the wrapping not be too low so as to cause condensation. At 100% relative humidity, an acceptable water vapor transfer rate for packaging of most produce would be between 50 and 100 $g/m^2/day$ (16). When the relative humidity in the packaging is less than 80 to 95% of saturation, fruit and vegetables lose moisture, and reduction in quality occurs when as little as 3 to 6% of the produce moisture is lost (17).

Dehydration or loss of moisture is also a problem for food stored in the frozen state. In meats, such moisture loss is often evidenced as freezer burn, a glassy appearance produced by the presence of tiny cavities left behind by sublimated ice. Apart from appearance of the product, loss of moisture will also affect the food's juiciness and texture. Oftentimes this moisture is recrystallized and leads to an unattractive formation of ice within the package.

Absorption of Odors. Many foods high in fat and improperly packaged have the potential to absorb odors from other products. For instance, milk left opened in the refrigerator will pick up odors from other highly volatile opened products in the refrigerator. The presence of odors not normally associated with the product may thus be justification for it being labeled as spoiled.

TOOLS FOR DETECTION OF SPOILAGE

Sensory Evaluations

Since quality and acceptability in most cases is ultimately determined by the consumer, sensory evaluation employing humans is one of the standard criteria used for judgment of spoilage. Depending on the product and the projected method of spoilage in that product, participants may be asked to judge visual and/or organoleptic characteristics. Many different approaches (tests) may be taken during sensory evaluation, but, in general, there may be difference tests (panelist asked to judge which of several samples is different) or hedonic scaling tests (panelist asked to rate a characteristic on an arbitrary scale with defined end points). The difficulty with using these type of analyses on a routine basis is the expense and time involved to conduct such analyses.

Microbial Evaluations

When microbial degradation is the major source of spoilage in the food, it is common practice to analyze for total numbers of bacteria in the food. The food may be rinsed to determine surface count or it may be ground/mixed with a specially formulated liquid medium of which diluted portions are taken and spread onto an agar plate. One of the drawbacks with regard to this type of analysis is that it takes a minimum of one to two days for the bacteria to grow on the agar and generate visible colonies for counting. The second even more important drawback is the realization that total count of bacteria does not always correlate to degree of spoilage. When spoilage is evident as visible growth (mold, pigmented or nonpigmented, slimy bacterial colonies), there is a direct relationship between total numbers of microorganisms and degree of spoilage. More often, though, spoilage is a result of the production of off-odors and off-flavors caused by bacterial metabolism. In these cases, there is no correlation between total numbers of bacteria and spoilage since only a fraction of the total flora participates in the spoilage. Thus, it becomes advantageous to identify which organisms are responsible for the spoilage and concentrate on enumerating their populations by employing selective media. No specific spoilage bacteria, however, have been identified in the case of vacuum-packaged beef or with vacuum-packaged cold-smoked salami (18). Even when the specific spoilage organisms are well established, there are instances where their enumeration is of limited value. For example, spoilage occurs at lower cell densities on high pH (>6.0) meat than on normal pH (<5.8) meat (19). Therefore, in those cases, it has been more appropriate to concentrate on identifying and quantifying the metabolic products responsible for the spoilage.

Other more advanced methods exist to identify spoilage organisms, and these rely on the immunochemical and genetic characteristics of the microorganisms. For detecting molds in grains, for example, antibodies to specific fungi may be linked directly or indirectly to a fluorescent dye. This technique is suitable for work on individual grains, but difficulties arise in assessment of molds in large samples of grain and in quantifying levels of molds present. Genetic detection techniques, on the other hand, are generally based on the hybridization of genomic DNA with a specific DNA probe. Hybridization may be accomplished by a variety of methods, but the most popular is the polymerase chain reaction (PCR). Such tools have proven useful in the identification of wild yeast belonging to the genus *Saccharomyces* that are biochemically and physiologically very similar to the culture yeast used in beer. By and far, the largest number of PCR applications have been directed at specific food pathogens, where it is essential to confirm the causative agent from a public health standpoint. Spoilage organisms, however, are of concern primarily from an economic significance, hence it is not necessary to detect them individually in foods. Consequently, primers may be designed for a conserved region of the DNA that allows detection of the majority of spoilage organisms as a group for a given commodity. Using such an approach for the PCR-ELISA assay, a detection threshold of 10^2 cfu/cm_2 was found (20). The advantage to using PCR rather than traditional plating is that the detection of the microorganisms is not dependent on the organisms' state of growth or on environmental influences. Limitations of PCR, however, are the inability to differentiate between viable and nonviable microorganisms, and substances in the food variably affect the activity of the DNA polymerase.

Chemical Evaluations

The chemical methods used for evaluating spoilage basically embody the assumption that as foods undergo spoilage, some new product or products are created by spoilage. Some of the specific components for which analyses have been undertaken and related to spoilage include: chitin, ergosterol, D-amino acids, D-lactate, tyramine, volatile basic nitrogen, trimethylamine, hydroperoxides, thiobarbituric acid-reactive substances, and headspace volatiles.

Chitin (N-acetyl-glucosamine) is an important component of fungal cell walls and is assessed by hydrolysis of chitin to glucosamine, followed by deamination to an aldehyde that is measured colorimetrically. Chitin, unfortunately, is also a major cuticular component of grain storage insects. Because insect debris is common in stored grain, the levels of chitin may not always be related to the level of spoilage.

Ergosterol (ergosta-5-7, 22-trienol) is another predominant sterol found in most fungi belonging to the classes Ascomycetes and Deuteromycetes. The amount of ergosterol produced by the fungi, however, is influenced by substrate composition, extent of aeration, and growth phase of mycelium; thus, it may not always be a true indicator of the extent of spoilage.

D-amino acids are important components of bacterial peptidoglycans. The presence of >1 ppm D-alanine in a variety of fruit juices indicated bacterial as opposed to yeast contamination (21). Thus, the potential exists for this compound to be used as a spoilage indicator. Similarly the D-isomer of lactate is not found in fresh foods and can be a diagnostic feature for bacterial spoilage.

Biogenic amines (putrescine, spermine, spermidine, histamine, tyramine) are produced by a number of bacteria and represent a hazard due to their psychoactive or vasoactive effects. In terms of their usefulness as spoilage indicators, only tyramine has been detected prior to the appearance of a faint putrid smell (initial stage of putrefaction). Using a sensor, composed of a tyramine oxidase-immobilized column and an oxygen electrode, tyramine levels have been found useful for estimating bacterial spoilage in aging beef. (22)

Both trimethylamine (TMA) and total volatile basic nitrogen (TVBN) levels have been routinely measured during iced and refrigerated storage of fish products. A recent comprehensive study, however, using 115 specimens of marine teleosteans and examination after three, five, and eight days of chilled storage revealed a strong relationship of organoleptically detected spoilage with TMA content, but not with TVBN content (23).

Food spoilage due to lipid oxidation has commonly been monitored by analyzing foods for the presence of hydroperoxides (primary reaction products) and for thiobarbituric acid-reactive substances (TBARS, secondary reaction products). Which compound is targeted depends on the food being analyzed. For instance, in frying oils, hydroperoxides are the compound of choice, whereas in muscle foods, TBARS has been the most common substance measured. The latter compound is preferred in muscle because hydroperoxides tend to break down quickly from interaction with the iron in the muscle.

At the root of many of the spoilage problems in food is the presence of off-flavors and off-odors. It therefore stands to reason that these compounds would be the target for identification and quantification. One of the most common means to measure the food for off-flavors and off-odors has been to sample the headspace of the food for its volatile content and subject it to gas chromatography (GC), oftentimes in combination with mass spectrometry (MS). In this manner, the dominant volatiles, 3-methyl-1-butanol, 1-octen-3-ol and 3-octanone, have been used as indicators of spoilage in stored grain (24). Another food item where GC has been used to correctly identify spoilage samples has been with shrimp. In this case, the data from 9 to 11 volatiles were required for identification of rancidity in shrimp meat (25). Similarly, GC/MS in combination with principal component analysis of the data was proven useful in rapid differentiation of control reduced-fat milk samples from reduced-fat milk samples abused by light, heat, copper, and microbial (*P. fluorescens, P. aureofaciens,* or *P. putrefaciens*) contamination (26).

Another instrumental method to analyze volatiles that just came on the scene in the early to mid 1990s is the electronic nose. This instrument consists of an array of nonspecific electronic chemical sensors that react differentially with the headspace volatiles of the food. Similar to the human nose, the electronic nose recognizes odor patterns rather than specific patterns. While the human brain processes the signals from our nose, a pattern recognition routine analysis program analyzes the sensor responses in the electronic nose. Thus, in conjunction with sensory panels, the instrument can be trained to recognize products that have changed significantly in aroma and flavor. The potential of electronic noses to serve as spoilage indicators has been demonstrated with products such as grain, ground meat, vacuum-packaged beef, and freeze-stored chicken (27). To further the use of this instrument for spoilage measurements, however, improvements will need to be made in the sensors' stability, susceptibility to humidity, and lifetime.

TOOLS FOR PREDICTION OF SPOILAGE

Although it is all well and good to be able to detect a spoiled food, it is even better if we have the tools to predict when spoilage will occur. With that information, preventative measures may in some cases be applied. More often, the information is used to determine the distribution chain that should be taken by the product in order that it be bought and consumed before spoilage takes place.

Two approaches can be taken for prediction of spoilage. In one approach, the food is subjected to conditions that would accelerate its degradation. The time it takes to reach a defined level of deterioration is then related to the time it would take for the food to spoil under common storage practices. The inherent assumption to this approach, however, is that the conditions used for acceleration (temperature, oxidative catalysts, etc) would lead mechanistically to the same type of reactions as would have occurred without the activation. Such is not always the case.

In the second approach for prediction of spoilage, the state of the food at some earlier time period is related to the time at which spoilage would occur. In this approach, either microbial or chemical attributes may be used to describe the state of the food. For example, it has been shown that a chicken carcass with a significantly higher number of *P. fluorescens* would not continue to be hygienically acceptable as long as one that had a fewer number of spoilage organisms at day of processing (28). On the other hand, headspace GC data for milk that underwent a preincubation (24°C, 18 h) was related to the shelf life of milk stored at refrigeration temperatures (29). In both these examples, however, the predictions would only be valid for the product and storage conditions that existed for the study developing the model. In contrast, predictive microbiology aims to take into account the variability that occurs in product and storage conditions. It is based on the premise that the responses of populations of microorganisms to environmental factors are reproducible and that by considering environments in terms of identifiable dominating constraints it is possible, from past observations, to predict the responses of those microorganisms. The necessary components to predictive microbiology therefore include determining the specific spoilage microorganisms, their spoilage domains, and their growth kinetics for which a mathematical model is derived and validated. Those factors that should be considered during the development of the models are the pH, water activity, temperature, atmosphere, preservatives, and interactions among microorganisms. Many of the models currently developed have been incorporated into a commercial software program (Food Micromodel Ltd., Leatherhead, UK) where they may be applied to broad categories of food.

MEASURES TO PREVENT SPOILAGE

To ensure that food items reach the consumer before spoilage takes place, many cost-effective treatments may be applied to delay spoilage. These treatments concentrate on modifying those factors known to affect the rate at which the changes contributing to spoilage would take place and may involve processing or storage operations. They may be grouped into treatments where temperature is modified, pressure is applied, additives are added, or atmospheres are modified.

High temperatures have traditionally been applied to a number of foods to inactivate microorganisms and enzymes. In applying such treatments, however, the temperatures used should not decrease the desirable properties of the food. Also, many high-temperature treatments do not absolutely ensure a stable product. As alluded to previously, some microorganisms (thermoduric and thermophilic) may be able to survive the heat treatment, and extracellular microbial hydrolytic enzymes are also extremely heat resistant.

Refrigeration and frozen storage is another common technique used to extend shelf life. In addition to inhibiting microorganisms, low temperatures work by slowing down metabolic and chemical reactions in the product. Unfortunately, some fruits and vegetables are sensitive to refrigeration temperatures in that they exhibit chilling-injury symptoms. For these items, the most successful means to avoid chilling injury is to simply refrain from exposing susceptible commodities to temperatures below their critical threshold. Alternatively, the severity of chilling injury in a few commodities may be reduced when temperature conditioning (lowering temperatures to slightly above the chilling range for various periods) is applied.

High-pressure processing has been a recent treatment for delaying spoilage in food in that vegetative cells, including yeast and molds, can be inactivated by pressures of ~300 to 600 MPa (30). An advantage of the new technology is that flavor and vitamins are unaffected or only minimally altered by high-pressure processing at room temperature. On the other hand, bacterial spores are highly pressure resistant, since pressures exceeding 1200 MPa may be needed for their inactivation.

The use of chemicals may also be used to prevent or delay spoilage. Examples include antimicrobials, antioxidants, and calcium, which inhibit, respectively, microbial, oxidative, and chilling-injury spoilage. For any chemical additive, however, it must be legally approved for use in the product of concern.

Modified atmospheres containing carbon dioxide, oxygen, and nitrogen in varying proportions are used commercially to delay or prevent the growth of spoilage organisms in foods. As the concentration of carbon dioxide increases in the atmosphere, the growth rate of most bacteria is more or less inhibited. Excessive concentrations of carbon dioxide, however, may cause physiological damage in fresh produce such as uneven ripening in tomatoes, surface pitting in asparagus, and discoloration in celery. Therefore, the usual concentrations of carbon dioxide used for extending shelf life are in the range of 5 to 20%.

Another variation of modifying the atmosphere of a food is to apply edible coatings. Such coatings are made of edible materials that are used to enrobe the food and provide a semipermeable barrier to gases and water vapor. These coatings have the potential to reduce moisture loss, restrict oxygen entrance, lower respiration, retard ethylene production, seal in flavor volatiles, and carry additives that retard discoloration and microbial growth.

The measures just listed to reduce or prevent spoilage represent only a handful of those available; however, these and other techniques are increasingly used together in combination with preservative or hurdle technologies. These are applied on the basis that organisms develop homeostatic mechanisms to survive environments and that by attacking different elements of the survival response, increased microbial inactivation will occur with application of milder treatments. Keep in mind that removal of one spoilage mechanism does not imply that the product would be spoilage-free. More often than not, other spoilage mechanisms would surface with the extension of storage that has been incurred. Whether the mechanisms are of concern, however, would depend on the typical length of time that the product takes to get to the consumer. Thus, while spoilage may be viewed as a very complex set of activities, it must ultimately be viewed in the context of the distribution chain and the point at which consumption would take place.

BIBLIOGRAPHY

1. I. N. A. Ashie, J. P. Smith, and B. K. Simpson, "Spoilage and Shelf-Life Extension of Fresh Fish and Shellfish," *Crit. Rev. Food Sci. Nutr.* **36**, 87–121 (1996).
2. G. Fleet, "Spoilage Yeasts," *Crit. Rev. Biotechnol.* **12**, 1–44 (1992).
3. J. L. Kinderlerer, "*Chrysosporium* Species, Potential Spoilage Organisms of Chocolate," *J. Appl. Microbiol.* **83**, 771–778 (1997).
4. N. Magan, "Early Detection of Fungi in Stored Grain," *International Biodeterioration & Biodegradation* **32**, 145–160 (1993).
5. V. Venugopal, "Extracellular Proteases of Contaminant Bacteria in Fish Spoilage: A Review," *J. Food Prot.* **53**, 341–350 (1990).
6. L. Fernandez et al., "Proteolytic and Lipolytic Activities of *Pseudomonas fluorescens* Grown in Raw Milk With Variable Iron Content," *Milchwissenschaft* **47**, 160–163 (1992).
7. D. D. Muir, "The Shelf-life of Dairy Products: 4. Intermediate and Long Life Dairy Products," *Journal of the Society of Dairy Technology* **39**, 119–124 (1996).
8. R. R. Meer et al., "Psychrotrophic *Bacillus* spp. in Fluid Milk Products: A Review," *J. Food Prot.* **54**, 969–979 (1991).
9. A. Pedrosa-Menabrito and J. M. Regenstein, "Shelf-Life Extension of Fresh Fish—A Review. Part I—Spoilage of Fish," *Journal of Food Quality* **11**, 117–127 (1988).
10. Y. J. Chung-Wang, M. E. Bailey, and R. T. Marshall, "Reduced Oxidation of Fresh Pork in the Presence of Exogenous Hydrolases and Bacteria at 2°C," *J. Appl. Microbiol.* **82**, 317–324 (1997).
11. J. M. Sala, "Involvement of Oxidative Stress in Chilling Injury In Cold-Stored Mandarin Fruits," *Postharvest Biology and Technology* **13**, 255–261 (1998).
12. H. Zhuang, D. F. Hildebrand, and M. M. Barth, "Senescence of Broccoli Buds is Related to Changes in Lipid Peroxidation," *J. Agric. Food Chem.* **43**, 2585–2591 (1995).
13. J. S. Kang and D. S. Lee, "Susceptibility of Minimally Processed Green Pepper and Cucumber to Chilling Injury as Observed by Apparent Respiration Rate," *International Journal of Food Science and Technology* **32**, 421–426 (1997).
14. D. W. Stanley, "Biological Membrane Deterioration and Associated Quality Losses in Food Tissues," *Crit. Rev. Food Sci. Nutr.* **30**, 487–553 (1991).
15. R. E. Brackett, "Vegetables and Related Products," in L. R. Beuchat, ed., *Food and Beverage Mycology*, 2nd ed., Van Nostrand, New York, 1982, pp. 129–154.
16. C. H. Mannheim, "Control of Water in Foods During Storage," *Journal of Food Engineering* **22**, 509–532 (1994).
17. C. A. Phillips, "Review: Modified Atmosphere Packaging and Its Effects on the Microbiological Quality and Safety of Produce," *International Journal of Food Science and Technology* **31**, 463–479 (1996).
18. Y. Blixt and E. Borch, "Using an Electronic Nose for Determining the Spoilage of Vacuum-Packaged Beef," *International Journal of Food Microbiology* **46**, 123–134 (1999).
19. R. H. Dainty, "Chemical/Biochemical Detection of Spoilage," *International Journal of Food Microbiology* **33**, 19–33 (1996).
20. R. Gutiérrez et al., "Quantitative Detection of Meat Spoilage Bacteria by Using the Polymerase Chain Reaction (PCR) and an Enzyme Linked Immunosorbent Assay (ELISA)," *Lett. Appl. Microbiol.* **26**, 372–376 (1998).
21. I. Gandolfi et al., "D-Alanine in Fruit Juices: A Molecular Marker of Bacterial Activity, Heat Treatments and Shelf-Life," *J. Food Sci.* **59**, 152–154 (1994).
22. Y. Yano et al., "Changes in the Concentration of Biogenic Amines and Application of Tyramine Sensor During Storage of Beef," *Food Chem.* **54**, 155–159 (1995).
23. T. Civera et al., "Sensory and Chemical Assessment of Marine Teleosteans—Relationship Between Total Volatile Basic Nitrogen, Trimethylamine and Sensory Characteristics," *Sciences Des Aliments* **13**, 109–117 (1993).
24. T. Borjesson, U. Stollman, and J. Schnurer, "Volatile Metabolites Produced by Six Fungal Species Compared With Other Indicators of Fungal Growth on Cereal Grains," *Applied and Environmental Biology* **58**, 2599–2605 (1992).
25. L. S. Bak, L. Jacobsen, and S. S. Jorgensen, "Characterisation of Qualitative Changes in Frozen, Unpeeled Cold-Water Shrimp (*Pandalus borealis*) by Static Headspace Gas Chromatography and Multivariate Data Analysis," *Z. Lebensm. Unters.-Forsch.* **208**, 10–16 (1999).
26. R. T. Marsili, "SPME-MS-MVA as an Electronic Nose for the Study of Off-Flavors in Milk," *J. Agric. Food Chem.* **47**, 648–654 (1999).
27. J-E. Haugen and K. Kvaal, "Electronic Nose and Artificial Neural Network," *Meat Science* **49**, S273–S286 (1998).
28. S. M. Russell, "A Rapid Method for Predicting the Potential Shelf Life of Fresh Broiler Chicken Carcasses," *J. Food Prot.* **60**, 148–152 (1997).
29. B. Vallejo-Cordoba and S. Nakai, "Keeping-Quality Assessment of Pasteurized Milk by Multivariate Analysis of Dynamic Headspace Gas Chromatographic Data. 1. Shelf-Life Prediction by Principal Component Regression," *J. Agric. Food Chem.* **42**, 989–993 (1994).
30. M. Hendrickx et al., "Effects of High Pressure on Enzymes Related to Food Quality," *Trends Food Sci. Technol.* **9**, 197–203 (1998).

MARILYN ERICKSON
University of Georgia
Griffin, Georgia

FOOD SURFACE SANITATION

Of paramount importance in food manufacture is the freedom of microbial (spoilage and pathogenic microorganisms) and foreign body contamination in the final product. Such contamination may arise from the constituent raw materials or the processing environment, which includes food contact surfaces, the air, people, and pests. Failure to control these factors may lead to product recalls, loss of sales or profits, adverse publicity, and, if regulatory requirements have been infringed, fines, sanctions, or ultimately site closure or loss of production/export license. For example, incomplete sanitation of a meat grinder was responsible for a large outbreak of *Salmonella typhirium* infected from ingestion of ground beef (1). Many food service operations are integrating hazard analysis and critical control point (HACCP) along with sanitation to ensure food safety (2).

The sanitation of surfaces, when undertaken correctly, is cost effective, easy to manage, and, if diligently applied, can reduce the risk of microbial or foreign body contamination of product. When incidents of such contamination occur, it is often not easy to trace and rectify the exact source and as surface sanitation is relatively cheap, it can provide management with a good tool to reduce this risk. In this context, surface sanitation:

1. Removes microorganisms, or material conducive to microbial growth. This reduces the chance of contamination by pathogens and extends the shelf life of some products.
2. Removes materials that could lead to foreign body contamination or could provide food or shelter for pests.

Surface sanitation is also implemented in food processing to provide a wide range of additional benefits including:

1. Reduces waste and improves the appearance and quality of product by removing product left on lines that may deteriorate and reenter subsequent production runs.
2. Increases process performance in some areas (eg, plate and scrape surface heat exchangers).
3. Extends the life of, and prevents damage to, equipment and services.
4. Provides a safe and clean working environment for employees and thus increases morale and productivity.
5. Presents a favorable image to customers and the public.

In this article, food surfaces are defined as both food contact and environmental surfaces. Environmental surfaces are included as, although they are not in direct contact with the product, contamination may be transferred from them to the product by people, pests, the air, or cleaning procedures. Only hard surfaces are considered eg, equipment, floors, walls, and utensils as other surfaces, eg, protective clothing or skin, would be traditionally dealt with under personal hygiene.

It is further assumed that the surfaces addressed have been designed hygienically. Poor hygienic design will restrict the efficiency of even the most effective cleaning procedure and may vitiate any subsequent disinfection programs. The principles of hygienic design are comprehensively described elsewhere (3–5). The basic hygiene design of equipment for open food processing has been reviewed recently (6).

FOOD SOILS

Debris will build up on surfaces throughout the production period; this will require subsequent removal by cleaning. This debris may be the result of normal production, spillages, line jams, maintenance, packaging, or general dust and dirt. Such undesirable material, which may include food residues, microorganisms, and foreign matter, is referred to as soil. In practical terms, a soil is anything in the wrong place at the wrong time; peas on a conveyor during production are product but after production or on the floor are soil.

Product soils are usually easy to visualize and are characterized primarily by the product type eg, protein, fat or carbohydrate. The process however is also important as a given product may present a variety of cleaning problems, depending on whether it is dry, wet, heat treated, frozen or the length of time it is left prior to cleaning.

Microbial soils cannot generally be observed by the eye but require microscopic examination. Bacterial attachment to surfaces is well documented (7–9), and the influence of bacterial growth on surfaces, termed biofilms, in the food industry has been discussed (10,11). Examination of stainless steel coupons attached to production lines, by epifluorescent microscopy, has been used to assess microbial levels in a range of food soils (12). An example of a heavy biofilm buildup over 16 h is shown in Figure 1 for a baked bean soil. After 12 h or so, this biofilm appeared to be of a brown vegetable appearance, but the photographs show clearly that it is bacterial in nature. A thorough understanding of a soil's characteristics is therefore required to ensure a successful and economic sanitation program.

FUNDAMENTALS

The process of sanitation is intended to remove all undesirable material (food residues, microorganisms, foreign bodies, and cleaning chemicals) from surfaces in an economical manner, to a level at which any residues remaining are of minimal risk to the quality or safety of the product. Sanitation is divided into two broad areas, the cleaning of open and closed surfaces, though the principles are essentially the same. Open surface sanitation refers to all equipment and environmental surfaces that are readily cleaned manually. Closed surface sanitation, generally referred to as clean-in-place or CIP, is undertaken where manual cleaning is difficult, impractical or impossible (e.g., tanks, homogenizers and pipelines).

Sanitation programs are a combination of four major factors:

- Mechanical or kinetic energy
- Chemical energy
- Temperature or thermal energy
- Time

Mechanical or kinetic energy is employed to physically remove soils and may include manual brushing (physical abrasion), pressure jet washing (fluid abrasion), or the circulation of fluid in CIP systems (turbulent flow). Of these

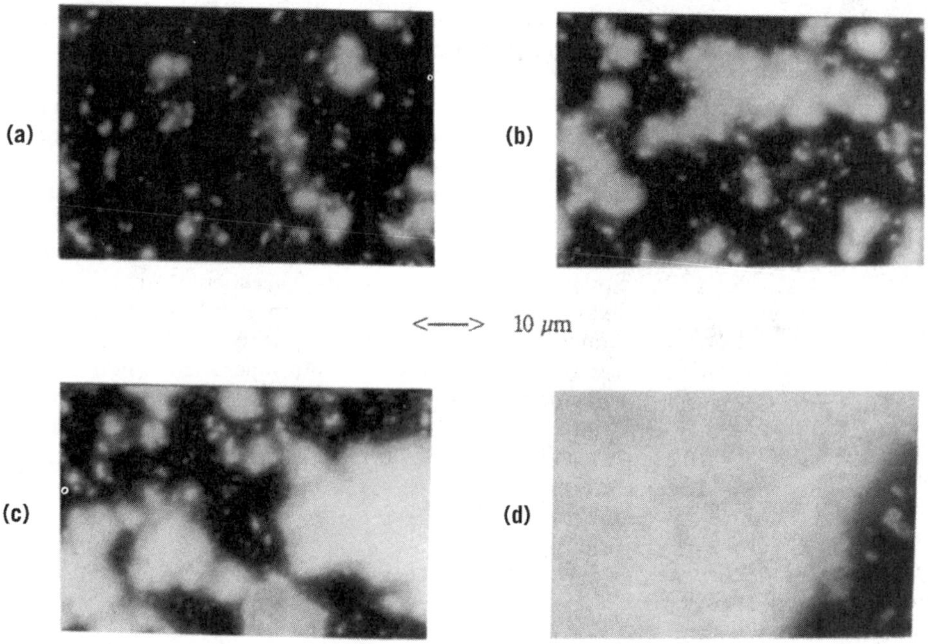

Figure 1. Build up of a bacterial biofilm on a baked bean production line. Photographs were taken using epiflourescent microscopy such that bacteria fluoresced orange while the background remained dark. (**a**) 4 h, (**b**) 8 h, (**c**) 12 h, (**d**) 16 h.

methods, physical abrasion is the most efficient in terms of energy transfer (13); for turbulent flow, a mean velocity of 1.5 m/s should be achieved (14).

Chemical energy is fundamental to both the cleaning and disinfection elements of sanitation. In cleaning, chemicals are used to break down soils so that the soils are less tenacious and to suspend them in solution to allow them to be rinsed away. In disinfection, chemicals are used to reduce the viability of microorganisms remaining on surfaces after cleaning.

Temperature or thermal energy is important for several reasons. Cleaning and disinfection chemical effects increase with temperature linearly and approximately double for every 10°C rise. Temperatures above the melting point of fatty or oily soils are used to break down and emulsify these deposits and high temperatures, particularly in CIP systems, have a disinfection effect in their own right.

Time is a factor that is often overlooked in sanitation systems. It is an essential prerequisite for the previously discussed energy forms and generally the longer the time period employed, the more efficient the process. Time can also be used to reduce the degree of energy input required from other sources when precleaning soaking is undertaken.

The combinations of these four factors varies for different cleaning systems such that if one energy source is restricted, this shortfall may be compensated for by utilizing greater inputs from the others. For example, in CIP cleaning, the energy that can be derived from mechanical energy is low but much higher temperatures and chemical concentrations are possible than can be safely used in open surface cleaning. The influence of chemical (detergents), temperature, and mechanical energy (pressure washing) has been described for open surfaces (15–18) and for CIP systems (19).

Soil removal from surfaces has been shown to basically follow first-order reaction kinetics (20,21) such that the decrease in the log of the mass of soil per unit area remaining is linear with respect to cleaning time (Fig. 2). This approximation is only valid in the central portion of the plot; it has been reported (22) that in practice, soil removal is initially faster and ultimately slower than that which a first-order reaction predicts (dotted line in Fig. 2). The reasons for this are unclear although initially, unadhered, gross soil is usually easily removed whereas ultimately, soils held within surface imperfections or otherwise shadowed from cleaning effects would be more difficult to remove.

As routine cleaning operations are therefore not 100% efficient over multiple soiling/cleaning cycles, soil deposits will accumulate on surfaces. During this phase, cleaning will become less efficient and attached microbial numbers will increase. This situation is usually controlled by the

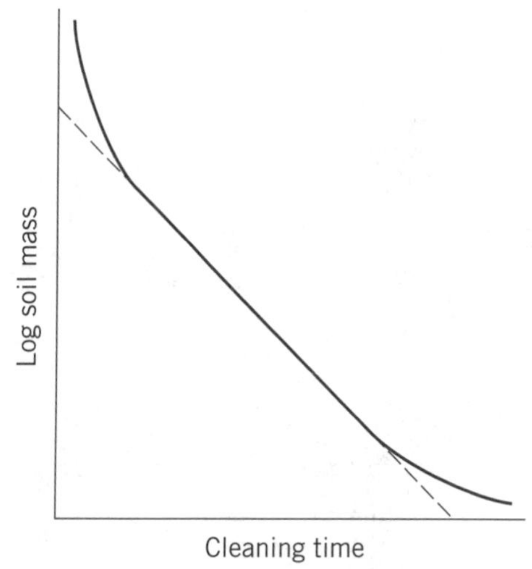

Figure 2. Removal of soil from surfaces with cleaning time.

application of a periodic clean (23), the object of which is to periodically return the surface bound soil accumulation to an acceptable base level (Fig. 3). This is achieved by increasing cleaning time and/or energy input (eg, higher temperatures, alternative chemicals, or manual scrubbing) and is the basis of many food processors weekend clean down.

SANITATION PROCEDURE

The principal stages involved in a typical sanitation procedure are as follows:

1. Preparation. Dismantle equipment as far as is practicable or necessary and/or remove unwanted utensils/equipment. Protect electric or other sensitive systems and/or screen off other lines or areas to prevent transfer of debris by the sanitation process.
2. Gross soil removal. Where appropriate, remove all loose or gross soil by eg, brushing, shoveling, scraping, or vacuuming. Wherever possible, soil on floors and walls should be picked up rather than washed to drains.
3. Prerinse. Rinse with low pressure cold water to remove loose small debris. Hot water can be used for fatty soils, but too high a temperature may coagulate proteins.
4. Cleaning. Apply cleaning chemicals, temperature, and mechanical energy to remove adhered soils.
5. Interrinse. Rinse with low pressure cold water to remove soil detached by cleaning operations and cleaning chemical residues.
6. Disinfection. Apply chemical disinfectants to remove or reduce the viability of remaining microorganisms to as low a level as possible.
7. Postrinse. Rinse with low pressure cold water to remove disinfectant residues if required.
8. Intercycle conditions. Remove excess water and/or do everything necessary to prevent the growth of microorganisms in the period up until the next production process or the next use of the equipment/area.

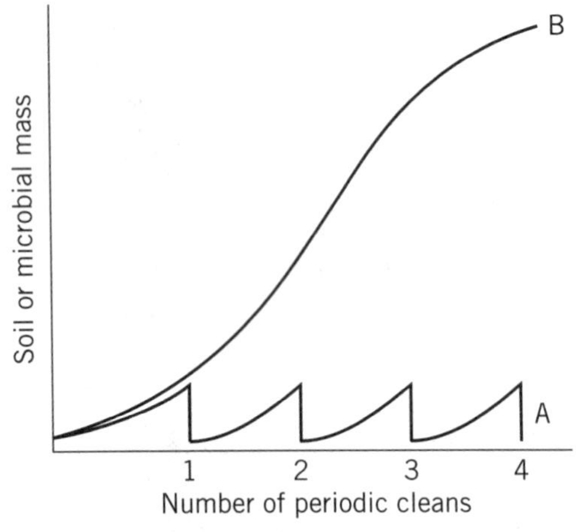

Figure 3. Build up of soil and/or microorganisms A, with periodic cleans and B, without periodic cleans (19).

Although broadly similar for all general sanitation procedures, alternatives may be used. In CIP systems a second cleaning cycle may be added if both an alkaline and acidic phase are incorporated. With only light soiling to be removed, it may be appropriate to combine stages 4–6 by using a detergent-sanitizer, a chemical with both cleaning and antimicrobial properties. This should only be used for light soils, however, as with normal or heavy soils the antimicrobial properties will be quickly lost. Stages 3, 5, and 7 may be omitted in dry cleaning operations where organic solvents replace water in the cleaning and, where appropriate, the disinfection stages. Because of the general nature of this article, specialist dry cleaning is not discussed further.

CHEMICALS

This section gives a brief and general introduction to the chemicals used in sanitation. For further information, readers are directed to the specific articles on detergents and disinfectants in this encyclopedia. Sanitation chemicals, because of the procedure outlined in the previous section, are usually employed as cleaning or disinfection agents. This section is therefore divided into cleaning and disinfection subsections. Some chemicals routinely used, such as quaternary ammonium compounds (QUATS) or iodophores, have both cleaning and biocidal properties, although in this section they are described for their primary function only.

Cleaning

Unfortunately no single cleaning agent is able to perform all the functions necessary to facilitate an efficient cleaning program, so a cleaning solution, or detergent, is blended from some of the following characteristic components: water, surface active agents inorganic alkalies inorganic and organic acid, and sequestering agents.

Water. Water is the basic ingredient of most cleaning systems as it provides the cheapest readily available transport medium for removing soils. It also has dissolving powers to remove ionic and water soluble compounds such as salts and sugars, will help emulsify fats at temperatures above their melting point, and can be used as an abrasive agent when high pressure washing. It is, however, a poor wetting agent and cannot dissolve nonionic compounds.

Surface Active Agents. Surface active agents, wetting agents or surfactants are composed of a long, nonpolar (hydrophobic or lyophilic) chain or tail and a polar (hydrophilic or lyophobic) head as illustrated in Figure 4. Surfactants may be anionic, cationic, or nonionic, depending on their ionic charge in solution. The polar end is able to penetrate into water as the ionic charges are greater in magnitude than the weaker hydrogen bonding between the water molecules. The nonpolar end is unable to easily break apart the water molecules' hydrogen bonding but can enter

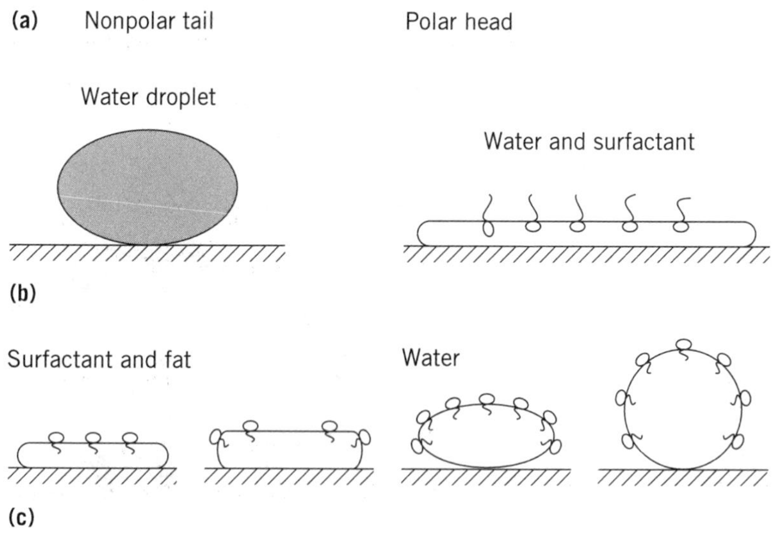

Figure 4. Schematic diagrams showing: (**a**) structure of surfactant molecule, (**b**) surfactant improving the wettability of water, and (**c**) surfactant removing fat from a surface and dispersing it in suspension.

other nonpolar compounds such as fats and oils. This aids cleaning in two ways: it reduces water surface tension and removes and suspends fats.

Surface tension is reduced as the polar head enters the water and breaks down hydrogen bonding. If a surfactant is added to a drop of water on a surface, the surface tension of the water is reduced and the drop collapses to wet the surface (Fig. 4). Increased wettability allows greater water penetration into soils and hence better cleaning action. Fats and oils are removed from surfaces as the polar head of the surfactant molecule dissolves in the water, whereas the hydrophobic end dissolves in the fat. Due to the forces acting on the fat-water interface, the fat particle tends to form a sphere, as this has the lowest surface area for a given volume. In so doing, the fat particle will roll up, become detached from the surface, and remain in suspension (Fig. 4).

Inorganic Alkalies. Alkalies are useful cleaning agents because they are cheap, break down proteins, saponify fats, and may be bactericidal. A typical saponification reaction is shown below in which a nonpolar R group is changed to a soluble form that has good surfactant properties.

$$\begin{array}{llll} CH_2COO\text{—}R & & CH_2OH & \\ | & & | & \\ CHCOO\text{—}R & + NaOH \rightarrow & CHOH & + 3R\text{—}COO^{-}Na^{+} \\ | & & | & \\ CH_2COO\text{—}R & & CH_2OH & \\ \text{Triglyceride fat} & \text{Sodium hydroxide} & \text{Glycerol} & \text{Soap} \end{array}$$

Generally, the stronger the alkali the greater the degree of saponification, although this is a compromise, as corrosiveness also increases with alkali strength. Alkaline compounds can also precipitate scum, a reaction of the soap with water hardness ions (Ca^{2+} and Mg^{2+}), and may have poor rinsability. For some applications, alkaline detergents are chlorinated, as at high pH chlorine increases peptizing of proteins and may reduce mineral deposition. Chlorine at high pH, however, is also very corrosive.

Inorganic and Organic Acids. Acids are not generally used in the food industry as they have little dissolving power for fats, oils, and proteins. They are useful, however, in making soluble mineral scales such as hard-water deposits, beer stone, and milk stone. A typical reaction would be as follows:

$$\underset{\text{Insoluble}}{CaCO_3} + 2HCl \rightarrow \underset{\text{Soluble}}{CaCl_2} + H_2O + CO_2$$

As with alkalies, the stronger the acid the more corrosive it is. Strong acids (mineral acids) include hydrochloric, nitric, and phosphoric acid, whereas weaker organic acids typically used include citric, lactic, and acetic acid.

Sequestering Agents. Sequestering agents are employed to sequester or chelate mineral salts by forming soluble complexes with them. Their primary use is in the control of water hardness ions although they are also useful in maintenance of alkaline conditions by buffering, emulsification of oils, and fats and increasing rinsability. Sequestrants can be both organic and inorganic, organic sequestrants being usually based on polyphosphates and inorganic chelating agents commonly being the potassium or sodium salts of ethylene diamine tetraacetic acid (EDTA).

A general-purpose food detergent may contain a strong alkali to saponify fats, weaker alkali builders to aid corrosion resistance, surfactants to improve wetting, dispersion, and rinsability and sequestrants to control hard water ions. Ideally, the detergent should also be safe, nontainting, stable, noncorrosive, biodegradable, and cheap. The detergent chosen for a particular application will depend on the soil to be removed; the solubility characteristics and cleaning procedure recommended for a range of food soils is shown in Table 1.

Table 1. Solubility Characteristics and Cleaning Procedures Recommended for a Range of Solid Types

Food (or soil)	Solubility characteristics	Cleaning procedure recommended
Sugars, organic acids, salt	Water soluble	Mildly alkaline detergent
High-protein foods (meat, poultry, fish)	Water soluble Alkali soluble Slight acid soluble	Chlorinated alkaline detergent
Fatty foods (fat meat butter, margarine, oils)	Water insoluble Alkali soluble	Mildly alkaline detergent; if ineffective, use strong alkali
Stone-forming foods, mineral scale (milk products, beer, spinach)	Water insoluble Acid soluble Alkali insoluble	Chlorinated cleaner or mildly alkaline cleaner, alternate with acid cleaner on each 5th day
Heat-precipitated water hardness	Water insoluble Alkali insoluble Acid soluble	Acid cleaner
Starch foods, tomatoes, fruits, vegetables	Partly water soluble Alkali soluble	Mildly alkaline detergent

Source: Ref. 24.

Disinfection

The cleaning portion of the sanitation cycle has been shown to reduce bacterial numbers on surfaces by up to 99.9% or three log orders (25). Work in our own laboratories has shown that with detergent soaks and pressure washing, up to 4–5 log orders may be removed. However, given that bacterial numbers on surfaces could be between 10^7 and 10^{10} organisms/cm (2,25), viable bacteria are likely to be present on surfaces after cleaning. The aim of disinfection procedures is to remove or reduce the viability of these remaining microorganisms.

If possible, temperature is used as a disinfectant as it penetrates into surfaces, is noncorrosive, is nonselective to microbial types, and is easily measured and leaves no residue (20). Whereas high temperatures are often used as disinfectants in CIP systems, their use on open surfaces is usually uneconomic, hazardous, or impossible. In such cases, chemical biocides are employed and, as with cleaning chemicals, no single disinfectant satisfies all the performance requirements. Biocides are rarely mixed, however, so a choice has to be made from a limited number of disinfectant types. Universally used biocides include chlorine releasing components, quaternary ammonium compounds, iodine compounds, amphoterics, and peracetic acid.

Chlorine is the most widespread and cheapest disinfectant used in the food industry and is available in fast-acting (chlorine gas, hypochlorites) or slow-releasing forms (e.g., chloramines, dichlorodimethylhydantoin). Quaternary ammonium compounds (QUATS or QAC's) are based on ammonium salts with substituted hydrogen atoms and a chlorine or bromine anion, whereas iodophores are soluble complexes between elemental iodine (active ingredient) and nonionic surface active agents. Amphoterics are based on the amino acid glycine, often incorporating an imidazole group. Peracetic acid may be used by itself or formulated with hydrogen peroxide. A range of characteristics for examples of these disinfectant types is shown in Table 2. Other disinfectants used to a limited extent include biguanides, formaldehyde, ozone, chlorine dioxide, and bromine; however, biocides successfully used in other industries, eg, phenolics or metal ion-based products, are not used for food applications due to safety or taint problems. Frank and Chmielewski (26) confirmed the effectiveness of quaternary ammonium compounds and chlorine in reducing *S. aureus* 1,000-fold on stainless steel and domestic food preparation surfaces.

Disinfectant concentration and contact time are important considerations in the reduction of microbial viability. The relationship between death and concentration is not linear but follows a sigmoidal curve dependent on the resistance of organisms within the population. Disinfectants do not, therefore, necessarily kill all microorganisms in a population, and increasing concentration may not enhance this effect. For disinfectants to be effective, they must find, bind to, and transverse microbial cell envelopes before they reach their target site (27). Chemical disinfectants are also effective in preventing cross-contamination from food contact surfaces (28). Sufficient contact time is therefore critical to give good results. Amphoterics and QUATS may be left on surfaces for extended periods between production runs without rinsing, as they are FDA approved as indirect food additives at low concentration.

The performance of the cleaning procedure may influence disinfection efficiency. Any soil or cleaning chemical residues remaining may protect microorganisms from disinfectant penetration or may react with the disinfectant and destroy its antimicrobial abilities. Biocides are best used within their specified pH range, although performance can generally be increased by increasing the temperature. A study by Best et al. (29) examined the efficacy of 14 disinfectants against *Listeria* spp. Their effectiveness varied depending on whether the test organism was dried on a steel surface, in suspension, or in the presence of organic material. Consequently, it is important to select the appropriate disinfectant for the particular contaminated surface, especially if organic material is present.

Table 2. Characteristics of Some Universal Disinfectants

Property		Chlorine	QUAT	Iodophore	Amphoteric	Peracetic acid
Microorganism control	Gram positive	+ +	+ +	+ +	+ +	− +
	Gram negative	+ +	−	+ −	+ +	− +
	Spores	+	−	+ / −	−	− +
	Yeast	+ +	− −	+ −	+ +	+ +
Developed microbial resistance		−	−	−	+	−
Inactivation by	Organic matter	+ +	+	+	+	+
	Water hardness	−	+	−	−	−
Detergency properties		−	+ +	+	+	−
Residual film formation		−	− +	+ / −	+ +	−
Foaming potential		−	+ +	+	+ +	−
Problems with	Taints	+ / −	−	+ / −	−	+ / −
	Stability	+ / −	−	−	−	+ / −
	Corrosion	+	−	+	−	−
	Safety	+	−	+	−	+ +
	Other chemicals	−	+	+	−	−
Cost		−	+ +	+	+ +	+

Note: − No effect (or problem).
+ Effect.
+ + Large effect.

METHODS

Open Surfaces

After gross soil removal, open surfaces can be cleaned and disinfected in their normal position or dismantled and/or transferred to a separate area (cleaned-out-of-place, COP). Sanitation procedures can be undertaken by hand using simple tools eg, brushes or cloths (manual cleaning) or by using specialized equipment designed to cover larger areas more rapidly.

Manual cleaning is often undertaken because no specialist equipment is required and high levels of mechanical energy input can be used at exactly the right point. Sanitation procedures can also be undertaken quickly (over small surface areas) and in places that alternative techniques could not reach or may damage. Due to operator safety, however, only low levels of temperature and chemical energy can be applied and it is costly in time and labour to cover large surface areas. Only tools specifically designed for food sanitation should be used, eg, plastic-handled brushes with stiff, colored nylon bristles, and they should be regularly cleaned and stored in a disinfectant solution to reduce cross-contamination in use.

Manual cleaning of smaller, dismantled items can be improved using COP techniques. When soak tanks are used, items can be subjected to higher temperatures and chemical concentrations for an extended time period prior to manual cleaning and disinfection. For an automated approach, specialized equipment can be used that also applies the mechanical energy input, eg, tray washers and tunnel dish washers.

For large areas of open surface, sanitation equipment is designed to disperse chemicals and/or provide mechanical energy. Chemicals may be applied as mists, foams, or gels, whereas mechanical energy is provided by water jets or scrubbing actions. The use of high temperature is rarely used for cleaning or disinfections, because of the excessive energy requirements needed to heat soil and/or surfaces to suitable temperatures.

Mist spraying of chemicals onto surfaces is undertaken using small, hand-pumped containers, knapsack sprayers (Fig. 5) or pressure washing systems at low pressure. As for the other chemical application methods, pressure washing systems can be mobile units (Fig. 5), wall mounted serving one or more outlets, or centralized where one unit may supply many outlets via a ring main. Chem-

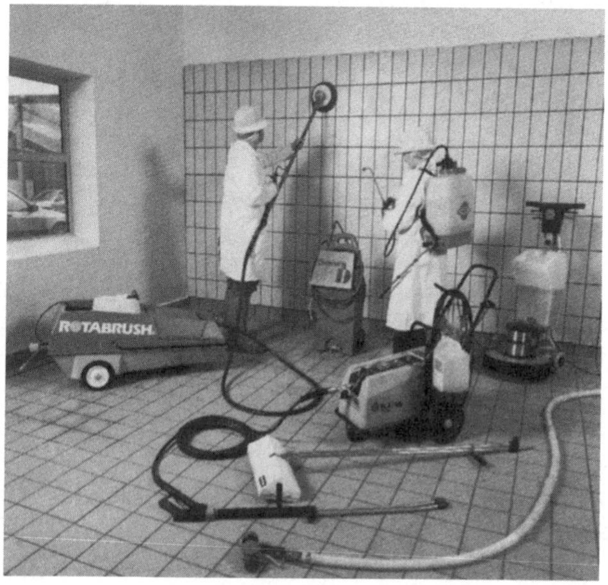

Figure 5. A range of cleaning equipment suitable for use in food hygiene. Clockwise from left: electrically driven scrubber brush, gel applicator, detergent or disinfectant mist sprayer, floor scrubber, mobile high-pressure/low-volume washer with standard lance connected, shrouded lance to reduce aerosol spread, low-pressure/high-volume gun.

icals are added, usually by venturi injectors, either at the pumps of mobile or wall-mounted units, at the outlets of high pressure water ring main systems or via separate low pressure chemical ring mains. Misting is only able to wet vertical smooth surfaces and therefore only small quantities can be applied. Only weak chemicals can be applied, as the technique tends to form aerosols, so for cleaning purposes, misting is only useful for light soiling. The detergent and loosened soil are removed via high or low pressure water rinsing. Once the surfaces have been cleaned, however, misting is a very useful and most widely applied technique for applying disinfectants.

If more tenacious soiling is to be removed, detergents may be applied as a foam. Foams are more viscous than mists, are not as prone to aerosol formation, and remain on vertical surfaces for around 5–20 min. This allows the use of more concentrated detergents, longer contact times, and because of the nature of the foam, a more consistent application of chemicals as it is easier to spot areas that have been missed. Foams can be generated and applied by the entrapment of air in high pressure systems or by the addition of compressed air in low pressure equipment, and are rinsed away by high or low pressure water. Although the visible nature of the foam should aid complete removal from surfaces, its foaming capabilities can make rinsing difficult.

The use of thixotropic gels has recently been introduced to further extend contact time over foams and improve rinsability. Gels are typically fluid at high and low concentrations but become thick and gelatinous at a concentration of around 5–10%. They are easily mixed and applied through high and low pressure systems, foaming equipment, or portable electrically pumped units (Fig. 5), remain on surfaces for extended periods (hours) and are easily diluted and rinsed away with high or low pressure water. Gels are usually more expensive than foams however.

The proportions of input from each energy source for a range of open surface cleaning techniques are shown in Figure 6. The figure illustrates the various methods' abilities to satisfactorily remove both light and heavy soiling from surfaces.

High pressure-low volume water systems, in which water is typically pumped at pressures up to 100 bar through a 15° nozzle, are widely used in the food industry. Water jets confer high mechanical energy, can be used on a wide range of equipment and environmental surfaces, are not limited to flat surfaces, will penetrate into surface irregularities, and can mix and apply sanitation chemicals. Soil removal does not necessarily increase linearly with pressure, as illustrated for a bacterial soil in Table 3, and may decrease. This is related to the water droplet size emerging from the nozzle, for which there will be an optimum size and hence impingement, for each soil type (24). Droplet size reduces with increasing pressure and pressures of around 50 bar are satisfactory for many operations. Care must be taken when using pressure washers as they are able to transfer soil and bacteria from one surface to another over large distances and may damage electrical installations or other delicate equipment.

For the cleaning of floors and other predominantly flat surfaces eg, walls, equipment that produces a mechanical scrubbing action may be used. Such equipment may include water-driven attachments to high pressure systems, electrically operated small diameter (23 cm) brushes, and floor scrubbers (Fig. 5). With these techniques, high mechanical input is combined with chemical energy after which surfaces may be rinsed with low pressure water. Larger and more sophisticated scrubber dryers, or floor automats, have a squeegee and vacuum unit mounted behind the scrubbing brush(es) to suck up the resulting soil and detergent emulsion. These devices provide a rapid and effective clean, and produce a dry surface that is safer and less conducive to microbial growth but can only be used efficiently in food processing areas designed for their use. All mechanical scrubbing units should be cleaned and disinfected after use to avoid cross-contamination.

Enclosed Surfaces, CIP

This section gives a brief and general introduction to CIP and readers are directed to the specific article on CIP in

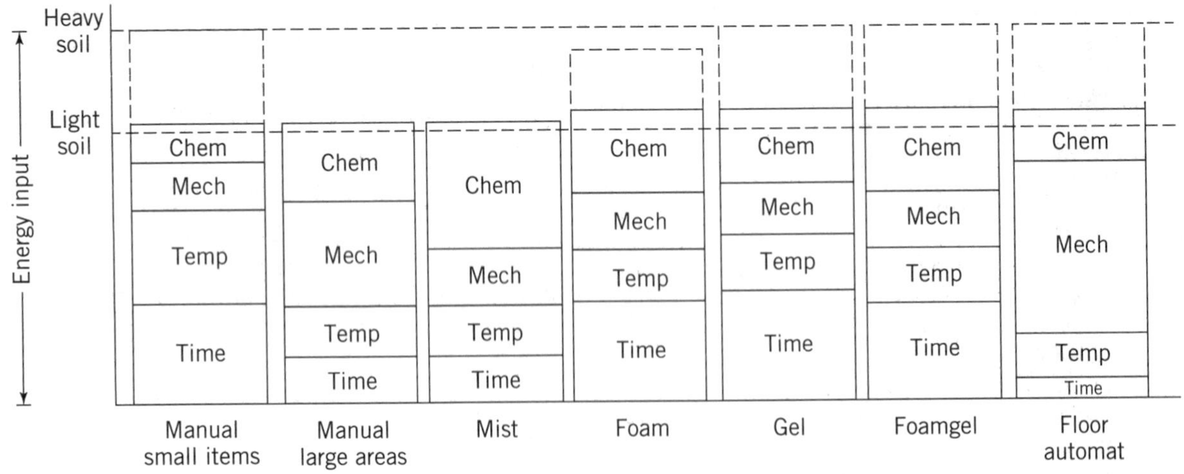

Figure 6. Relative energy inputs for a range of cleaning techniques. The dotted lines above each technique represent their abilities to cope with heavy soiling (13).

Table 3. The Effect of Water Pressure on Spray Cleaning Efficiency

Pressure (bar)	2	25	50	75	98
Microorganisms remaining cm^2 (as measured by DEM)	2.9×10^3	3.1×10^4	1.9×10^4	7.7×10^3	3.3×10^4

this encyclopedia, for further information. As a technique, CIP is well established, especially in liquid food processing, and is achieved by the sequential circulation of water, detergents, and disinfectants through processing equipment that remains assembled. Mechanical or kinetic energy is provided by turbulent flow in pipes and jet impingement in tanks and vessels. CIP systems are best employed to clean process equipment that has been designed specifically for CIP cleaning and should supply detergents at the correct temperature, concentration, and velocity in the right place for the required contact time.

It is fundamental with enclosed systems that hygienic design and product safety from sanitation chemicals is addressed. Attention must be paid to hygienic design so that areas where product and microorganisms can lodge, out of the flow of cleaning chemicals and disinfectants, are eliminated (3). Special devices should be built into the system, eg, key pieces, flow plates, swing bends, and electrical interlocks to ensure that product contamination by CIP solutions cannot occur, and a single valve must never be relied upon to safeguard product (30).

There are three main types of CIP systems: single use, reuse, and multiuse. Single-use systems are the most basic and are so called because the detergent is only used once and then drained. A simple system is shown in Figure 7. The system can be located close to the equipment to be cleaned, to eliminate long pipe runs, and is low on capital cost. It may be operated automatically or manually and the final rinse may be collected to use as the initial rinse in

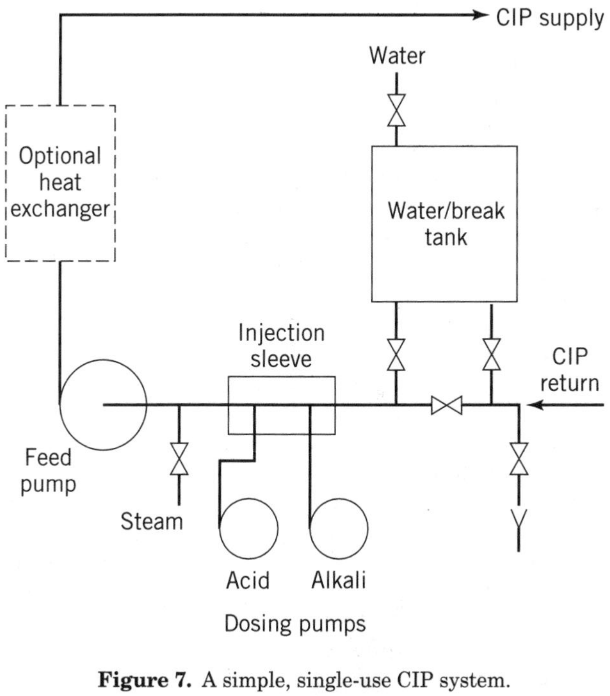

Figure 7. A simple, single-use CIP system.

the following cleaning cycle. Although it is expensive on detergent, a single-use system may be the only system applicable if the product soil to be removed is such that detergent contamination occurs to a degree that detergent recovery is impractical.

Reuse systems are generally more complex and may be used to clean more than one process. They are designed to recover the cleaning fluids for reuse and are best employed where product soil is light or where much of the soil can be removed with the prerinse. Reuse systems are normally run automatically with programmed cleaning cycles for each circuit and there is, therefore, the need for timing, temperature and dosing control equipment, filters, storage vessels, and recording instrumentation. With large installations, the CIP unit can be placed out of the process area, or centralized, which may be advantageous for some applications. This can, however, lead to extensive pipe runs, which require large solution volumes and are prone to heat loss and expansion problems. This can be partially overcome in decentralized systems in which the detergent solution is mixed, heated, and distributed centrally, but rinse waters and other chemicals are heated and/or mixed and supplied close to the process equipment to be cleaned.

Multiuse systems combine the best design features of both single-use and reuse systems and are used where versatility in temperature, cleaning fluid type, concentration, and the ability to recover or drain cleaning fluids are required.

As a cleaning technique, CIP has many advantages. Cleaning schedules are adhered to automatically and use optimum cleaning fluids, concentrations, and temperatures to help achieve consistently satisfactory results. Processes can be cleaned as soon as production finishes to reduce down time and the life of equipment is extended due to reduced dismantling. Manual input can be reduced and operator safety increased through less contact with hazardous materials or work places. The major disadvantages are initial capital cost and the inflexibility of some systems if manufacturing processes are altered. Russel (31) reviewed some of the difficulties in cleaning processing equipment and emphasized the importance of designing safe, easy-to-use cleaning equipment as well as safer sanitation chemicals. More automated CIP systems will probably be introduced to reduce human error.

EVALUATION OF EFFECTIVENESS OF SANITATION SYSTEMS

As with other important aspects of safe and wholesome food production, sanitation programs require regular evaluation of their effectiveness as part of a specified quality assurance system. Traditionally, this has been undertaken on two levels—an immediate assessment of the performance of a sanitation program by sensory evaluation and a historical measurement of microbial surface populations if surfaces are visibly clean. Sensory evaluations are used as a process control to rectify immediately obviously poor sanitation. Bacterial evaluations may be used to optimize sanitation procedures and ensure compliance with microbial standards and in hygiene inspection and troubleshooting exercises. Recently, rapid methods have been developed to assess microbial surface populations in a time relevant to process control and in this context, only techniques that provide estimates in less than 15–20 min are considered.

As with the inherent faults of end product analysis in describing production quality, reliance should not be placed on assessment techniques that can only sample a very small proportion of plant surfaces. Assessment techniques should, rather, be designed to monitor the effectiveness of an integrated HACCP-type (Hazard Analysis and Critical Control Point) approach to sanitation in which such critical control points may include detergent and disinfection concentration, solution temperature, application procedures, and chemical stock rotation.

Sensory evaluation involves visual inspection of surfaces under good lighting, feeling for greasy or encrusted surfaces, and smelling for product and/or offensive odors. If these assessments indicate the presence of product residues, no further analysis is required and microbiological examination will be misleading. If no product residues are detected, microbiological techniques may be applied. All of the microbiological techniques appropriate for food factory use involve the sampling of microorganisms from contaminated surfaces and their culture using standard agar plating methods. Microorganisms may be collected via cotton or alginate swabs (after which they are resuspended by vortex mixing or dissolution), water rinses for larger enclosed areas, or directly using agar contact plates. Traditional techniques have been reviewed (32,34) since which time few improvements have occurred other than with contact methods which are now available commercially as premade sampling units.

Traditional microbiological techniques all require a minimum of 24–48 h to provide results and these only provide a historical view of the efficiency of sanitation programs. Rapid methodology was devised to ascertain sanitation efficiency and for some products, when to clean, in a time relevant to process control. The two most common techniques are epifluorescent microscopy and the ATP technique. With epifluorescent microscopy, microorganisms may be sampled by swabbing and filtering the resuspension media or filtering rinses (direct epifluorescent filter technique, DEFT) or by enumerating the surface of coupons attached to product contact surfaces (direct epifluorescent microscopy, DEM). After swabbing, estimates may be made of soil and/or microorganisms by analysis of total or microbial adenosine triphosphate (ATP) levels. ATP bioluminescence provides a reliable and rapid alternative to traditional microbiological methods (35). The efficiency and flexibility of this method makes it a unique hygiene-monitoring system for food plants (36). A recent study by Green et al. (37) showed that cleaning agents and sanitizers can affect ATP bioluminescence measurements differently. Consequently, it is important to select carefully the type and concentration of chemical cleaner or sanitizer used to clean the processing equipment when using this method to monitor hygiene. In the majority of practical applications, where an assessment of cleaning is required, analysis for total ATP is preferred on the assumption that any residues, soil or microorganisms, should have been removed. The use of DEM and DEFT has been described for assessing open surface hygiene (12,38), rinses (39), and the use of ATP (40,41).

The accuracy of a range of surface hygiene assessment techniques as compared to the most accurate method, DEM, has been described (42), and is shown in Table 4. The table shows the ability of a number of commercial contact, self-prepared (DIY) contact, traditional swabbing and rapid techniques (ATP and DEFT) to predict a DEM count of 10^4, 10^6 and 10^8 bacteria/cm^2. The results indicate that only above 10^6 bacteria/cm^2 are these methods accurate and they are therefore useful for indicating gross contamination only. For cotton swabs, the most widely used method in practice, a result of zero is more likely than a result of 10^4 when assessing surface populations of 10^4 bacteria/cm^2. The rapid methods available are at least as accurate as traditional methods and as shown in Table 5, which compiles a range of attributes for typical assessment methods, the choice between rapid and traditional techniques is a balance of cost and speed of result.

EVALUATION OF THE EFFECTIVENESS OF DISINFECTION

The effectiveness of a disinfectant can only be undertaken by assessing viable microbial levels before and after exposure. Ideally, this should be carried out in the field, but due to problems with reproducibility and scale (assessing only a small area of the total plant), this may take many weeks before satisfactory results are achieved. Disinfectant testing is, therefore, usually undertaken in the laboratory under strictly defined conditions to enable reproducible and rapid results.

At their simplest, disinfectant test methods consist of adding a known concentration of disinfectant to a known inoculum of bacteria for a defined time and temperature period and assessing surviving bacterial numbers. Microorganism survival may be assessed by inoculating a sample of the disinfectant/microorganism mixture into traditional culture medium and enumerating any microcolonies produced or by the use of a disinfectant concentration/microorganism mixture dilution range and recording the disinfectant dilution that just prevents microbial growth. These disinfectant tests, referred to as suspension tests, are simple to undertake but can only be used to screen potential products for disinfectant action or to confirm the biocidal abilities of known disinfectants as a quality control procedure.

More extensive suspension test methodologies are available that simulate some of the conditions under which disinfectants may have to work. Such conditions may include the presence of soil, different contact times, alternative temperatures, a variety of bacterial types and the degree of water hardness (43). The results of suspension tests have been used to ascertain recommended in-use dis-

Table 4. Accuracy in the Estimation of Surface Microbial Populations for Various Hygiene Assessment Methods as Compared to DEM

		Prediction of a DEM count (cm^2)			
Method used	Repeats	10^4	10^6	10^8	$2SD^2$
Commercial contact A	58	3.42	6.18	8.94	±1.11
DEFT swab	108	2.06	4.46	6.86	±1.22
Commercial contact B	30	3.57	6.01	8.45	±1.77
ATP swab	31	2.36	4.66	6.96	±2.22
DIY contact	30	4.76	6.15	7.55	±2.33
Commercial contact/swab	17	1.50	3.04	4.59	±2.81
Cotton swab	70	1.47	4.41	7.35	±2.88
Alginate swab	70	0.61	3.55	6.49	±4.06
Commercial contact C	46	2.81	5.35	7.89	±5.54

Table 5. Attributes of a Range of Surface Hygiene Assessment Techniques

Attribute commercial	DEM	DEFT Swab	ATP Swab	TVC Swab	CONTACT DIY	CONTACT
Analysis for	Total count	Viable/total	Viable/soil	Viable count	Viable count	Viable count
Preparation time	5 min	15 min	10 min	1 h	1h	1 min
Test time	5–10 min	15 min	2 min	25 min	1 min	1 min
Surface requirement	Coupon	Any	Any	Any	Flat only	Flat only
Field application	Research Yes	Yes	Yes	Yes	Yes	Yes
Degree of competence	Micro training	Micro training	Basic training	Basic micro	Basic micro	Little/none
Facilities required	UV microscope	UV microscope	Luminometer (portable)	Basic lab	Basic lab	Incubator
Speed of result	<5 min	<5 min	<1 min	>1–2 days	>1–2 days	>1–2 days
Test cost	50p/$1	50p/$1	£1/$2	50p/$1	50/$1	£1–4/$2–8
Initial cost	£2–5,000 $4–10,000	£3–15,000 $6–30,000	£2.5–10,000 $5–20,000	Low	Low	Very low

infectant concentrations, but in reality they only highlight conditions under which disinfectants that have failed should not be used.

Through the auspices of national standard organizations, many countries have standard disinfectant testing procedures, including Australia, New Zealand, South Africa, France, Germany, The Netherlands, the UK, and the United States (44). There is little standardization between countries, however, and to address this issue, with the planned harmonization of Europe in 1992, the European Suspension Test (EST) (45) was published with some agreement between European countries. To pass the EST, a disinfectant must reduce the viability of 5 test organisms by 5 log orders in 5 min at both low (0.03% bovine albumin) and high (1.0% bovine albumin) soil levels. Results for two examples each of a range of disinfectant types are shown in Table 6. These results illustrate a number of points including: the relative resistance of microbial species, the variation of disinfectant action within a single microbial species, the influence of organic load and hence the necessity to thoroughly clean prior to disinfection in practice, and the range of biocidal action between and within product types. In its early stages of development, a collaborative trial in 10 European countries found the EST to be sufficiently repeatable and reproducible to be adopted internationally (46).

Microorganisms are not primarily found in the food production environment in suspension. In an attempt to more closely simulate an in-use conditions surface, tests have been developed in which microorganisms are dried onto surfaces prior to disinfection (47–49). Although these tests provide a stronger challenge to the disinfectant (50), there are problems with the tests in that the drying process may alter microbial resistance, it is difficult to remove a consistent proportion of microorganisms for enumeration after disinfection and again, microorganisms are rarely dried onto surfaces in practice. A surface test has recently been described (51) in which disinfectants are tested against bacterial biofilms developed on stainless steel samples and in which the viability of the biofilm was assessed, while surface bound using a Malthus microbiological growth analyser (Malthus Instruments Ltd. (Radiometer), UK). The relative resistance of bacteria when attached to surfaces as compared to in suspension is shown in Table 7. A range of disinfectants was tested under the conditions of the EST at their manufacturers' recommended concentration (MRC) for bacteria in suspension and at the MRC and 10 and 100 times this concentration, when attached to a surface. The results showed that bacteria were 10 to 100 times more resistant to biocides when attached to a surface and that biocides that performed well in suspension did not necessarily perform well on surfaces. Care should therefore be exercised in transferring the results from suspension tests to in-use applications.

FUTURE DEVELOPMENT

Sanitation is likely to feature much more prominantly in the eyes of management in the future due to their desire

Table 6. European Suspension Test Results for a Range of Commonly Used Disinfectants Tested at their Manufacturers Recommended Concentrations (MRC)

			Log reduction					
Product	MRC(%)	Organic Load (%)	*Pseudomonas aeruginosa*	*Proteus mirabilis*	*Staphylococcus aureus*	*Enterococcus faecium*	*Saccharomyces cerevisiae*	Result
Peracid	0.1	0.03	7, 6	7, 7	7, 7	6, 5	5, 1, 1	Fail
		1.0	7, 1, 5	7, 6, 7	7, 1, 6	6, 0, 5	5, 0, 0	Fail
Peracid	0.1	0.03	5, 5	7, 7	7, 7	7, 7	>4, 5	Pass
		1.0	5, 5	7, 7	6, 7	7, 7	>4, 5	Pass
HOCl	0.25	0.03	7, 7	7, 7	7, 6, 6	5, 6	4, 5, 2	Fail
		1.0	4, 3	2, 2	2, 2	0, 0	3, 5, 3	Fail
HOCl	0.3	0.03	6, 6	5, 4, 4	6, 5	6, 6	>4, 3	Fail
		1.0	3, 3	0, 3, 2	1, 4	2, 2	0, 0	Fail
QUAT	0.2	0.03	2, 2	3, 5, 4	7, 7	6, 5	5, 3, 4	Fail
		1.0	2, 2	3, 1	7, 6	5, 5	4, 1	Fail
QUAT	1.0	0.03	5, 6	6, 5	5, 7	5, 5	5, 5	Pass
		1.0	5, 7	5, 6	7, 7	4, 6, 6	4, 5, 5	Pass
Biguanide	0.5	0.03	2, 5, 5	2, 5, 5	2, 6, 7	3, 6, 7	5, 4	Pass
		1.0	2, 1	2, 1	2, 2	1, 1	2, 1	Fail
Biguanide	0.5	0.03	6, 6	5, 6	7, 5	7, 7	1, 2	Fail
		1.0	6, 6	4, 3	4, 4	7, 7	1, 1	Fail
Iodophore	0.2	0.03	7, 7	7, 7	7, 6	7, 6	4, 4	Fail
		1.0	1, 1	3, 2	4, 2	1, 1	0, 0	Fail
Iodophore	1.0	0.03	4, 4	7, 7	5, 5	3, 4	2, 3	Fail
		1.0	2, 3	3, 5, 5	5, 5	3, 3	1, 1	Fail
Amphoteric	1.0	0.03	5, 5	4, 7, 7	7, 7	6, 7	4, 5, 5	Pass
		1.0	4, 5, 7	6, 4, 7	7, 7	4, 7, 6	2, 5, 3	Fail
Amphoteric	1.0	0.03	7, 6	7, 7	7, 7	6, 6	5, 5	Pass
		1.0	7, 6	6, 6	7, 7	6, 6	5, 5	Pass

Table 7. The Relative Resistance of Bacteria Attached to a Surface (Malthus) and in Suspension (EST)

		Result			
		EST		Malthus	
Product type	Conc. (%)	0.03	1.0	0.03	1.0
Peracid	0.1	Pass	Pass	Pass	Fail
	1.0			Pass	Pass
	10.0			Pass	Pass
Iodophore	0.2	Pass	Fail	Fail	Fail
	2.0			Pass	Pass
	20.0			Pass	Pass
Biguanide	0.5	Pass	Fail	Fail	Fail
	5.0			Pass	Fail
	50.0			Pass	Pass
HOCl	0.25	Pass	Pass	Fail	Fail
	2.25			Fail	Fail
	25.0			Pass	Pass
Amphoteric	1.0	Pass	Pass	Fail	Fail
	10.0			Pass	Fail
QUAT	1.0	Pass	Pass	Fail	Fail
	10.0			Fail	Fail

Note: For both the EST and Malthus tests, three organisms were used—*Pseudomonas aeruginosa, Proteus mirabilis,* and *Staphylococcus aureus;* a pass criteria of a 5 log reduction in 5 min for all three organisms was used.

to seek increased standards of hygiene. These have arisen from:

1. The realization of the importance of environmental routes of infection to product for both microbial and foreign body contaminants.
2. Pressure from customers and consumers for higher standards of hygiene.
3. Trends in food production towards short shelf-life products that intrinsically demand higher standards of hygiene.
4. The ability to sell product of a low microbiological load at a premium.
5. Increased plant size, where single mistakes will incur large financial losses.
6. Legislation.

Demand for increased standards of hygiene could lead to further advances in rapid methodology to determine surface hygiene in a time relevant to process control, chemical testing schemes that more closely simulate in-use conditions, improved application techniques that are more efficient and reduce cross-contamination via cleaning methods, the design of equipment and processing areas with more thought to ease of cleanability, and the increased use of automation and/or robotics for both production and sanitation operations.

Other trends may be influenced by environmental and energy restrictions. This will lead to the use of chemicals that leave no undesirable environmental residues. For example, the formation of chemical by-products when chlorine is used as a disinfectant remains a great concern. Chlorination of drinking water has been shown to produce a number of cancer-causing chemicals (52). Consequently, there is a definite trend to increase the use of alternative disinfectants such as ozone, chlorine chloride, and chloramine for treating drinking water as they have fewer chlorinated disinfectant by-products (53). Further trends will include a reduction in water usage, and an associated move towards dry cleaning methods such as ring-main vacuum lines, and as for hygiene reasons, more efficient and less manual sanitation application techniques.

BIBLIOGRAPHY

1. T. H. Roels et al., "Incomplete Sanitation of a Meat Grinder and Ingestion of Raw Ground Beef: Contributing Factors to a Large Outbreak of *Salmonella typhyrium* Infection," *Epidemiology and Infection* **119**, 127–134 (1997).
2. M. Setiabuhdi, M. Theis, and J. Norback, "Integrating Hazard Analysis and Critical Control Point (HACCP) and Sanitation for Verifiable Food Safety," *Journal of the American Dietetic Association* **97**, 889–891 (1997).
3. "Hygienic Design of Equipment for Open Food Processing," *Trends Food Sci. Technol.* **6**, 305–310 (1995).
4. Technical Manual No. 7, *Hygienic Design of Food Processing Equipment*, Campden Food and Drink Research Association, Chipping Campden, UK, 1983.
5. Technical Manual No. 8, *Hygienic Design of Post Process Can Handling Equipment*, Campden Food and Drink Research Association, Chipping Campden, UK, 1985.
6. Technical Manual No. 17, *Hygienic Design of Liquid Handling Equipment for the Food Industry*, Campden Food and Drink Research Association, Chipping Campden, UK, 1987.
7. R. C. W. Berkeley, J. M. Lynch, P. R. Rutter, and B. Vincent, *Microbial Adhesion to Surface*, Ellis Horwood Ltd., Chichester, UK, 1980.
8. K. C. M. Marshall, *Microbial Adhesion and Aggregation*, UK, Springer-Verlag, Berlin, 1984.
9. C. K. Bower, J. McGuire, and M. A. Daeschel, "The Adhesion and Detachment of Bacteria and Spores on Food-contact Surfaces," *Trends Food Sci. Technol.* **7**, 152–157 (1996).
10. S. Bouman, D. B. Lund, F. M. Driessen, and D. G. Schmidt, "Growth of Thermoresistant *Streptococci* and Deposition of Milk Constituents on Plates of Heat Exchangers during Long Operating Times," *Journal of Food Protection* **45**, 806–812 (1982).
11. S. J. Lewis and A. Gilmour, "Microflora Associated with the Internal Surfaces of Rubber and Stainless Steel Milk Transfer Pipeline," *Journal of Applied Bacteriology* **62**, 327–333 (1987).
12. J. T. Holah, R. P. Betts and R. H. Thorpe, "The Use of Epifluorescence Microscopy to Determine Surface Hygiene," *International Biodeterioration* **25**, 147–153 (1989).
13. M. T. Offiler, "Open Plant Cleaning: Equipment and Methods," in *Proceedings of Hygiene for the 90s* Campden Food and Drink Research Association, Chipping Campden, UK, November 7–8, 1990.
14. D. A. Timperley, "The Effect of Reynolds Number and Mean Velocity of Flow on the Cleaning-in-Place of Pipelines," in *Proceedings of Fundamentals and Applications of Surface Phenomena Associated with Fouling and Cleaning in Food Processing*, Lund University, Lund, Sweden, April 6–9, 1981.
15. D. G. Dunsmore, "Bacteriological Control of Food Equipment Surfaces by Cleaning Systems. 1. Detergent Effects," *Journal of Food Protection* **44**, 15–20 (1981).
16. W. L. Shupe, J. S. Bailey, W. K. Whitehead, and J. E. Thompson, "Cleaning Poultry Fat from Stainless Steel Flat Plates," *Transactions of the American Society of Agricultural Engineers* **25**, 1446–1449 (1982).
17. M. E. Anderson, H. E. Huff, and R. T. Marshall, "Removal of Animal Fat from Food Grade Belting as Affected by Pressure and Temperature of Sprayed Water," *Journal of Food Protection* **44**, 246–248 (1985).
18. N. E. Middlemiss, C. A. Nunes, J. E. Sorensen, and G. Paquette, "Effect of a Water Rinse and a Detergent Wash on Milkfat and Milk Protein Soils," *Journal of Food Protection* **48**, 257–260 (1985).
19. D. A. Timperley and C. N. M. Smeulders, "Cleaning of Dairy HTST Plate Heat Exchangers: Optimisation of the Single-Stage Procedure," *Journal of the Society of Dairy Technology* **41**, 4–7 (1965).
20. W. G. Jennings, "Theory and Practice of Hard-Surface Cleaning," *Advances in Food Research* **14**, 325–459 (1965).
21. H. J. Schlussler, "Zur Kinetik von Reinigungsvorgangen an festen Oberflachen," Symposium uber *Reinigen und Desinfizieren lebensmittel verarbeitender Anlagen*, Karlsruhe, West Germany, 1975.
22. M. Loncin, "Modelling in Cleaning, Disinfection and Rinsing," in *Proceedings of Mathematical Modelling in Food Processing*, Lund Institute of Technology, Lund, UK, 7–9 September, 1977.
23. D. G. Dunsmore, A. Twomey, W. G. Whittlestone, and H. W. Morgan, "Design and Performance of Systems for Cleaning Product Contact Surfaces of Food Equipment: A Review," *Journal of Food Protection* **44**, 220–240 (1981).
24. R. P. Elliott, "Cleaning and Sanitation," in *Principals of Food Processing Sanitation*, A. M. Katsuyama, ed., The Food Processors Institute, USA, 1980.
25. H. Mrozek, "Development Trends with Disinfection in the Food Industry," *Deutsche Molkerei-Zeitung* **12**, 348–352 (1982).
26. J. E. Frank and R. A. N. Chmielewski, "Effectiveness of Sanitation with Quaternary Ammonium Compound or Chlorine on Stainless Steel and Other Domestic Food-Preparation Surfaces," *J. Food Prot.* **60**, 43–47 (1997).
27. R. Klemperer, "Tests for Disinfectants: Principles and Problems," in *Disinfectants: Their Assessment and Industrial Use*, Scientific Symposia Ltd., London, 1982.
28. S. F. Bloomfield and E. Scott, "Cross-Contamination and Infection in the Domestic Environment and the Role of Disinfectants," *J. Appl. Microbiol.* **83**, 1–9 (1997).
29. M. Best, M. E. Kennedy, and F. Coates, "Efficacy of a Variety of Disinfectants Against *Listeria* spp.," *Appl. Environ. Microbiol.* **56**, 377–380 (1990).
30. D. A. Timperley, "Cleaning in Place (CIP)," *Journal of the Society of Dairy Technology* **42**, 32–33 (1989).
31. M. Russell, "Coming Clean on Sanitation," *Food Engineering* **66**, 104–106 (1994).
32. J. T. Patterson, "Microbiological Assessment of Surfaces," *Journal of Food Technology* **6**, 63–72 (1971).
33. J. D. Baldock, "Microbiological Monitoring of the Food Plant: Methods to Assess Bacterial Contamination on Surfaces," *Journal of Milk Food Technology* **37**, 361–368 (1974).
34. S. M. Kulkarni, R. B. Maxy, and R. G. Arnold, "Evaluation of Soil Deposition and Removal Processes: An Interpretive View," *Journal of Dairy Science* **58**, 1922–1936 (1975).

35. J. M. Hawronskyj and J. Holah, "ATP: A Universal Hygiene Monitor," *Trends Food Sci. Technol.* **8**, 79–84 (1997).
36. L. S. Madl, "Sanitation in a Flash, ATP Bioluminescence Technology Makes its Way into Food Plants," *Food Processing* **58**, 100 (1997).
37. T. A. Green, S. M. Russell, and D. L. Fletcher, "Effect of Chemical Cleaning Agents and Commercial Sanitizers on ATP Bioluminescence Measurements," *J. Food Prot.* **62**, 86–90 (1999).
38. J. T. Holah, R. P. Betts, and R. H. Thorpe, "The Use of Direct Epifluorescent Microscopy (DEM) and the Direct Epifluorescent Filter Technique (DEFT) to Assess Microbial Populations on Food Contact Surfaces," *Journal of Applied Bacteriology* **65**, 215–221 (1988).
39. C. H. McKinon and R. Mansell, "Rapid Counting of Bacteria in Rinses of Milking Equipment by a Membrane Filtration Epifluorescent Microscopy Technique," *Journal of Applied Bacteriology* **51**, 363–367 (1981).
40. P. Thompson, "Rapid Hygiene Analysis Using ATP Bioluminescence," *European Food and Drink Review*, 42–48 (Spring 1989).
41. W. J. Simpson, "Instant Assessment of Brewery Hygiene Using ATP Bioluminescence," *Brewers Guardian* **118**, 20–22 (1989).
42. J. T. Holah, "Monitoring the Hygienic Status of Surfaces," in *Proceedings of Hygiene for the 90s*, Campden Food and Drink Research Association, Chipping Campden, UK, 1990.
43. A. D. Russel, "Factors Influencing the Efficiency of Antimicrobial Agents," in A. D. Russell, W. B. Hugo, and G. A. J. Ayliffe, eds., *Principles and Practice of Disinfection*, Blackwell Scientific Publications, London, 1982.
44. *Review of Worldwide Disinfectant Test Methods*. British Association for Chemical Specialties, Sutton, UK, 1989.
45. Test Methods for the Antimicrobial Activity of Disinfectants in Food Hygiene. Council of Europe, Strasbourg, 1987.
46. B. Van Klingeren, A. B. Leussink, and W. Pullen, "A European Collaborative Study on the Repeatability and the Reproducibility of the Standard Suspension Test for the Evaluation of Disinfectants in Food Hygiene," *Report of the National Institute of Public Health and Environmental Hygiene*, No. 35901001, 1981.
47. *Tube Test for the Evaluation of Detergent-Disinfectants for Dairy Equipment*. International Standard FIL-1DF44. Brussels. International Dairy Federation, 1967.
48. "AOAC Use-Dilution Method," in S. Williams, ed., *Official Methods of Analysis of the Association of Official Analytical Chemists*, 14th Edition, Association of Official Analytical Chemists, Inc., Arlington, Va., 1984.
49. *AFNOR T72-190 Germ-Carrier Method*. Paris: l'Association Française de Normalisation, 1986.
50. S. F. Bloomfield, "Disinfectant Testing in Relation to the Food Industry," in *Proceedings of Hygiene for the 90s*, Campden Food and Drink Research Association, Chipping, Campden, UK, 1990.
51. J. T. Holah, C. Higgs, S. Robinson, D. Worthington, and H. Spenceley, "A Conductance-based Surface Disinfectant Test for Food Hygiene," *Letters in Applied Bacteriology* **11**, 255–259 (1990).
52. S. D. Richardson, "Drinking Water Disinfection By-products," in R. A. Meyers, ed., *The Encyclopedia of Environmental Analysis and Remediation*, Vol. 3, John Wiley & Sons, New York, 1998, pp. 1398–1421.
53. S. D. Richardson et al., "Chemical By-products of Chlorine and Alternative Disinfectants," *Food Technol.* **52**, 58–61 (1998).

GENERAL REFERENCES

P. R. Hayes, *Food Microbiology and Hygiene*, Elsevier Applied Science Publishers, New York, 1985.
A. M. Katsuyama, *Principles of Food Processing Sanitation*, The Food Processors Institute, Washington, D.C., 1980.
N. G. Marion, *Essentials of Food Sanitation*, Chapman and Hall, London, 1997.
N. G. Marriot, *Principles of Food Sanitation*, AVI Publishing Co., Westport, Conn., 1985.
A. J. D. Romney, *Cleaning in Place*, The Society of Dairy Technology, Huntingdon, Cambridgeshire, U.K., 1990.

FOOD TOXICOLOGY

Food toxicology is a specialized area of the discipline of toxicology that deals with toxic substances in foods. It is the study of the nature, sources, and formation of toxic substances in foods, their deleterious effects on consumers, the manifestations and mechanisms of these effects, and the identification of the limits of the safety and thus the regulation of these substances. Because the public is concerned about the quality and safety of foods, it is important to understand the principles of food toxicology and food safety. In this article, the general principles of toxicology are presented before the discussion of food toxicology.

TOXICOLOGY: PRINCIPLES, DEFINITIONS, AND SCOPE

Toxicology is the study of the adverse effects of substances on living organisms. It is a multidisciplinary field of study dealing with the detection, occurrence, properties, effects, and regulations of toxic compounds. It therefore involves an understanding of chemical reactions as well as biological mechanisms of toxic actions.

A toxicological study usually consists of four elements: (*1*) a chemical agent capable of causing a deleterious response; (*2*) a biological system with which the chemical agent may interact to produce the deleterious response; (*3*) a means by which the chemical agent and the biological system are permitted to interact; and (*4*) a response that can be used to quantitate the deleterious effect on the biological system.

Two aspects of interaction between substances and living organisms are of importance: the influence of the substances on the living organism and the influences of the organism on the substances. The chemical agent capable of causing a deleterious effect in the organism is defined as a poison or toxicant. A toxicant will exert toxicity, which is defined as the capacity to produce toxic injury to cells or tissues, only at appropriate conditions when the biological system is exposed to a certain dose of the toxicant. In effect, Paraceisus (1493–1541) noted, "All substances are poisons; there is none which is not a poison. The right dose differentiates a poison and a remedy." Thus, as a rule, a substance is a toxicant only in toxic doses; virtually any substance, even pure water or sugar, is poisonous when taken in great excess. The capability to detect subtoxic levels of toxicants in biological system, such as in the plasma or urine, is of particular importance because, once known, further exposure can be avoided. In toxicology, exposure is the total amount of toxicants received by the biological system of interest. It can be expressed as the product of concentration and duration. A chemical agent does not produce toxic effects in a biological system unless that agent or its biotransformation products reach appropriate sites in the body at a concentration and for a length of time sufficient to produce the toxic manifestation.

The route or site of exposure affects the toxicity of a chemical agent to the biological system. In food toxicology, the route of exposure is through the gastrointestinal tract (ingestion). The duration of animal exposure to toxicants is usually divided into four categories: acute, subacute, subchronic, and chronic. Acute exposure is defined as exposure to a toxic chemical for less than 24 hours. Although acute exposure usually refers to a single administration, repeated exposures may be given within a 24-hour period for some slightly toxic or practically nontoxic chemicals. Subacute exposure refers to repeated exposure to a toxic chemical for 1 month or less; subchronic exposure, to repeated exposure for 1 to 3 months; and chronic exposure, to repeated exposure for more than 3 months.

The LD_{50} of a compound is commonly reported as a measure of the toxicity of that compound. The LD_{50} can be defined as the dose of a compound that causes 50% mortality in a population. However, limitations of the LD_{50} as a measure of toxicity are often not recognized. The LD_{50} is an indicator of acute toxicity, using death as the end point, and therefore is not indicative of the effect of the compound under low dose or long-term exposure, which is usually the case for food toxicants. Chronic toxicity studies are needed to establish the effects of long-term or repeated administration of a compound. Chronic toxicity studies are used to determine the no observed effect level (NOEL) or no observed adverse effect level (NOAEL). From these studies, the reference dose (RfD) can be established. The RfD is defined as the maximum dose (mg/kg body weight) of an agent that is assumed to be without an adverse noncancer health impact on the human population (1).

The biological systems used for toxicity testing can be whole animals, including humans, dogs, and rodents, or they can be tissues or organs in culture, cell cultures, cell-free systems, eukaryotes (such as yeast and *Aspergillus*) and prokaryotes (such as *Salmonella typhimurium* and *Escherichia coli*), and plants. The age, sex, strain, and nutritional and disease status of the animal species all affect the outcome of the toxic effect. The use of enzyme inducers or inhibitors to modulate drug metabolism enzyme systems also affects animal susceptibility and toxicity to environmental chemicals. Whole animals are used to determine LD_{50} and organ target toxicity. They are also used to assess whether a compound is carcinogenic (ie, capable of inducing cancers) or teratogenic (ie, capable of inducing defects in the developing embryo). Insects are used in conjunction with pesticide development and for the detection of environmental mutagens (ie, causing a change in the DNA). Eukaryotes and prokaryotes are now widely used for the determination of mutagenic and carcinogenic potencies of environmental toxicants, drug impurities, and compounds present in foods. Currently, there is a trend in toxicology to use cell cultures as an alternative model system to animals. In this case, the TC_{50} (the toxic concentration that will induce poisonous effects to 50% of the cell population) is used to indicate the potential toxicity of the test compound. Cell-free systems are used to study the biochemical mechanism of toxicity.

Thus, the occurrence of a toxic response is dependent on the chemical and physical properties of the agent, the duration of exposure, and the susceptibility of the biological system or subject.

The single-most important factor that determines the potential harmfulness of a toxicant is the relationship between the concentration of the toxic agent and the effect produced in the biological system. This is referred to as the dose–response relationship. Toxic responses will not occur unless the chemical interacts with the target site(s). The degree of the response is related to the concentration of the agent at the reactive site, which in turn is related to the dose administered. In addition, the toxic response should be quantifiable.

A graphic expression of the typical dose–response relationship is shown in Figure 1. A sigmoidal response curve is obtained when the dosage is plotted on a logarithmic scale. The response may be applied to an individual, a system, or a fraction of a population; it ranges from 0 to 100%. The lowest dose of any toxicant that evokes a stated all-or-none response is called the threshold dose. Below this dose, there is no response. It is through the use of this dose–response relationship that the toxicologist is able to obtain the LD_{50} of a toxicant if mortality is used as an end point. See Reference 2 for a discussion of dose–response and LD_{50}.

After administration to a test animal, toxicants usually undergo a series of complex processes, including absorption, distribution, metabolism, and excretion, before they exert their toxic effects (Fig. 2). An overview of these processes can be found in References 3 and 4. During the ex-

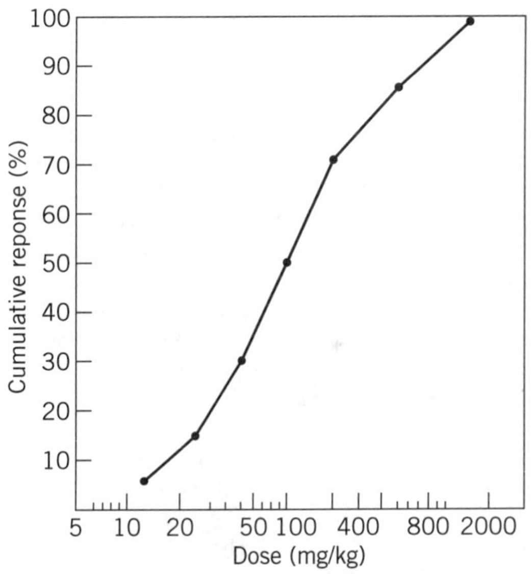

Figure 1. A typical sigmoid form of the dose-response relationship. Dosage is most often expressed as mg/kg and plotted on a log scale.

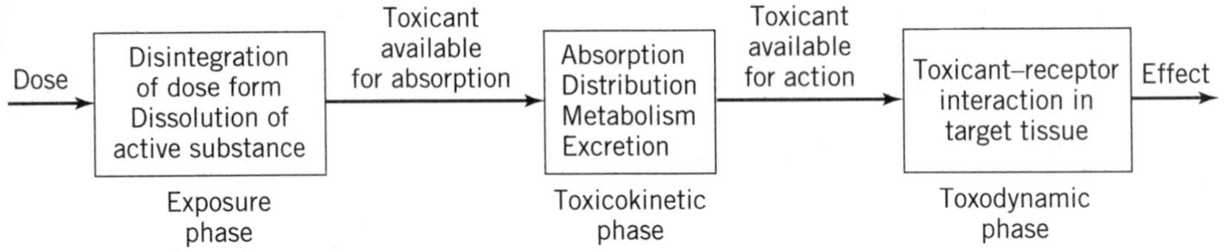

Figure 2. The three phases of toxicant action: the exposure phase, the toxico-kinetic phase, and the toxodynamic phase.

posure phase, the toxicant may undergo chemical alteration to compounds that may be more or less toxic than the parent compound. For example, hydrolysis of esters can take place in the gastrointestinal tract through the action of intestinal microflora. In this connection, azo compounds can be reduced to the more toxic aromatic amines.

The toxicokinetic phase includes all the processes involved in the relationship between the effective dose of a toxicant and the concentration present at the various body fluid compartments and the target tissue. During the toxicokinetic phase, two types of processes play an important role:

1. *Distribution processes that involve absorption, distribution to the organs, and excretion.* Toxicants are transported and then may bind to protein carriers or tissue components. The principles of pharmacokinetics apply to this distribution process of toxicants.
2. *Biotransformation of toxicants.* This usually involves the bioactivation of the toxic agent. This metabolic biotransformation is accompanied by changes in chemical properties such as hydrophilicity and lipophilicity, which in turn affect their distribution in the organisms, the binding to macromolecules (such as proteins and DNA), and excretion.

Metabolism of toxicants mainly occurs in the liver but may also occur in other tissues, such as the lung, kidney, skin, and gonads. Through enzymatic biotransformation processes, the lipophilic compounds are converted to more water-soluble metabolites. Two types of enzymatic reactions are involved in toxicant metabolism: phase I reactions, which involve oxidation, reduction, and hydrolysis; and phase II reactions, which consist of conjugation reactions. Phase I reactions generally convert compounds to derivatives that are more water soluble than the parent molecule. The reactions occur mainly via two oxidative enzyme systems, the cytochrome P-450 system (the mixed-function oxygenase) and the mixed-function amine oxidase. More important than these particular conversions is that these two systems also add or expose functional groups such as —OH, —SH, —NH_2, and —COOH, which promote the compound's covalent conjugation with endogenous moieties such as glucuronic acid, sulfate, and amino acids through the actions of phase II reaction enzymes. These conjugated secondary metabolites possess increased water solubility and significant ionization properties at physiologic pH that in turn facilitate their secretion or transfer across hepatic, renal, and intestinal membranes.

The toxodynamic phase comprises the action of the toxicant molecules on the specific sites of action and the expression of the observed toxic effect. The target organ on which the toxicant acts and the effector organ in which the effect is induced, or on which the effect is observed, need not be identical. The concentration of the active toxicant metabolite reached in the target determines to what degree a biological action will be elicited. The toxic effect observed in the biological system can be the result of interference with the normal function of the enzyme systems; blockade of the oxygen transport by hemoglobin; interference with the general functions of the cell; interference with DNA, RNA, and protein synthesis; hypersensitivity reactions; and direct chemical irritation of tissues.

Many carcinogens undergo enzymatic activation to reactive ultimate carcinogens that are electrophilic and are capable of covalent interaction with cellular macromolecules, including DNA. In addition to these secondary carcinogens, there are also primary carcinogens that are reactive and do not require metabolic activation. If the damaged DNA is not repaired, the genome lesions are expressed in replicated cells that later will transform into abnormal cells. For a review of carcinogen metabolism, DNA adduct formation, and DNA repair, see reference 5. These abnormally altered or initiated cells may be removed through the process of programmed cell death (apoptosis) or may undergo proliferation to form preneoplastic lesions. The growth and progression of preneoplastic lesions into a neoplasm or cancer depends on the presence of promoting or inhibiting compounds or conditions in the animal's environment. For many types of cancers, dietary factors have been shown to play a major role in the promotion or inhibition of tumor development. For detailed discussions on carcinogenesis mechanisms, refer to References 5 and 6. The role of dietary factors in cancer development was summarized in Reference 7.

FOOD TOXICOLOGY

Food contains hundreds of thousands of substances, and most of these have not been characterized or tested. The recognition of the safety of these substances, at the level consumed in the diet, is based on a history of consumption of those foods. Substances added to foods that do not have this history, such as new food additives, must undergo specific toxicological testing to determine that the federal Food Drug and Cosmetic Act (FD&C Act) standard of "reasonable certainty of no harm" is met.

The toxicants present in food can be of biological origin, a direct or indirect food additive, a contaminant, or produced during cooking or processing. Exogenous factors, such as the nature of the compounds, the dose, the frequency of exposure, the route of exposure, the presence of other nutrients or drugs, and various environmental factors, affect the toxicity of food toxicants. Endogenous factors, such as the physiology and morphology of the gastrointestinal tract, the nature of the gastrointestinal microflora, and the metabolic activity of the body, can also affect toxicity. The intestinal microflora play important roles in inducing the formation of some toxicants, such as nitrosamines and cyanogenic glycosides. The toxicology of food compounds has been the subject of several books (8,9).

Foodborne Hazards of Microbial Origin

Foodborne disease agents are characterized by their diversity. Some produce their effects through toxic metabolites resulting from the growth of microorganisms in the food before ingestion and are classified as foodborne intoxications (eg, staphylococcal food poisoning and botulism). Others produce adverse effects through ingestion of living microorganisms and thus are called foodborne infections (eg, *Salmonella, Vibrio parahemolyticus, Listeria monocytogenes*, and *Clostridium perfringens* poisoning). A detailed description of these bacteria, their toxins, prevalence, and toxicity can be found in a review by Concon (10). Other pathogens are emerging as important public health threats, including *E. coli* O157:H7, *Campylobacter jejuni*, and *Yersinla* (11). The severity of the toxic effects ranges from temporary discomfort to the acute lethality of botulism. Infants, the elderly, and persons on immunosuppressive or chemotherapeutic drugs are thought to be more susceptible to the toxic effects of these microbial agents. The main source of these hazards may be on the farm, during food processing, or more likely, during food service preparation or preparation at home. The changing consumer lifestyle, including increased number of women in the workforce, limited time for food preparation, and an increase in the number of single heads of households, may also impact the emergence and reemergence of foodborne pathogens (12).

Environmental Contaminants

Toxicants included in this category are the trace elements and organometallic compounds (eg, mercury, lead, and cadmium) as well as a variety of organic compounds (eg, polychlorinated biphenyls [PCBs]). These toxicants tend to have common behavioral characteristics, although they are quite different in chemical structure. Environmental toxicants tend to be stable and thus persistent in the environment. They tend to bioaccumulate in the food chain and can be biotransformed with increasing toxicity in humans. Lead and mercury have been shown to cause major toxic effects in infants and young children because of the greater absorption of the compounds, increased sensitivity of developing systems, and their greater frequency of exposure to the toxicants.

The major source of lead exposure through food is in fruits, vegetables, and grains, which is primarily the result of deposition of environmental lead into the plants. The elimination of leaded gasoline, lead in soldered food cans, and lead in pottery glazes have all contributed to a decrease in the levels of lead in food. Fish, plants, and animals take up cadmium from the environment, and shellfish are the major source of cadmium in the diet. Mercury exposure via food most often occurs when fish and seafood are contaminated with mercury. The Food and Drug Administration (FDA) limits the amount of heavy metals in our foods. For example, cadmium in food colors is limited to 15 parts of cadmium per million parts of food color (15 ppm), and the maximum acceptable level of mercury in fish and seafood is set at 0.5 ppm.

Because PCBs tend to be stored in the fat tissues of animals, the FDA has established maximum allowed levels of between 0.2 and 3 ppm PCBs in milk, eggs, animal fats, and shellfish. The used of PCBs was put under regulatory control in 1972, and production was ceased in the United States in 1977. As a result, the amount of PCBs in the U.S. diet has dramatically dropped since 1971. A possible role of PCBs in the development of breast cancer has been hypothesized. However, findings from epidemiological studies have been inconclusive. For a review of this issue as well as other health issues related to exposure to chlorinated contaminates, see the work by Holland (13). The toxicity of metals (14) and other food contaminants (8) has been reviewed.

Naturally Occurring Toxicants

Naturally occurring toxicants are products of the metabolic processes of animals, plants, and microorganisms from which the food products and nutrients are derived. Humans may be exposed to these naturally occurring toxicants through direct consumption of the foods or through secondary exposure from the edible by-products of food-producing animals. Most toxicants of animal origin come from fish and shellfish, including saxitoxin (paralytic shellfish poisoning), tetrodotoxin (puffer fish poisoning), ciguatoxin (ciguatera poisoning), and histamine (scombroid poisoning). The toxins produced by fungus on infected grains and other foods are also considered naturally occurring toxicants. They include aflatoxins, ergot toxins, ochratoxin, zearalenone, and the trichothecene toxins. Many of these mycotoxins (toxins produced by fungal cultures) are carcinogenic, mutagenic, and teratogenic. Many mushrooms also produce toxins with varying types of effects. See references 9 and 10 for further details on the toxicity of mycotoxins and toxic mushrooms.

Plants contain hundreds of biologically active compounds, which recently have been termed phytochemicals. Interest in phytochemicals has grown dramatically in recent years as a result of reports that they may offer potential health benefits (including cancer prevention) and agricultural benefits (increased resistance to plant pathogens). However, in many cases, these compounds have also been demonstrated to have toxic effects. Examples of toxic compounds found in plants consumed by humans are listed in Table 1 (9,10).

The concept of compounds having both beneficial and harmful effects, depending on the dose, has been well es-

Table 1. Examples of Toxic Compounds Found in Plants Consumed by Humans

Type of toxicity	Source of compounds involved
Cyanide poisoning	Cyanogenic glycosides in almonds, lima beans, cassava root, and sorghum
Goitrogens	Glucosinolates, isothiocyanates in the *Brassica* family (cabbages, brussels sprouts)
Favism	Fava beans
Lathyrism	Chick peas or garbanzos
Hemaglutinins	A wide variety of legumes
Enzyme inhibitors	Legumes, potatoes, eggplant, tomatoes
Carcinogenic	Pyrolizidine alkaloids in some herbs
Vasoactive amines	Bananas, avocado, cheeses
Interferes with hormone action	Phytoestrogens in soybeans
Antinutrient compounds	Phytic acid in cereals and legumes, oxalates in spinach and tea
Stimulants and other psychoactive effects	Caffeine in teas and coffee

tablished for various nutrients and pharmaceuticals. The phenomenon is called hormesis and is defined as the production of beneficial effects in a population at low exposures and adverse effects at high exposures to a given chemical. Data demonstrating that hormesis may occur with compounds that have previously been considered only as toxins, such as pesticides and heavy metals, has recently been reviewed (13). It appears likely that this holds true for many plant compounds as well. For example, tannins represent a group of natural plant compounds that have been shown to have both beneficial and toxic effects (15).

Further investigations into the mechanisms of action of these biologically active plant compounds are needed. In many cases, there is little information on the toxicity of these compounds, especially when compared to the amount of toxicology information needed for food additives and pesticides (16). Increased consumption of these compounds is likely because of the current development of "functional foods" that often contain various phytochemicals with the goal to improve health. However, the optimum levels of consumption for healthful effects and the levels that would result in toxicity are currently unknown for most phytochemicals.

Concerns regarding the safety of genetically engineered foods include the potential development of increased levels of plant toxins or of new allergens in these foods. In the United States, it is the responsibility of FDA to ensure that genetically engineered foods are safe. See reference 17 for a discussion of FDA's policy for foods developed by biotechnology.

Reaction Products

Toxic reaction products may be produced after cooking, processing, or charcoal broiling of foods. Carcinogenic nitrosamines are produced in cured meats such as bacon during cooking from the reaction of nitrite with secondary amines. Potent mutagenic and carcinogenic heterocyclic amines such as PhIP (2-amino-1-methyl-6-phenylimidazo[4,5-*b*]pyridine), IQ (2-amino-3-methylimidazo[4,5-*f*]quinoline), MeIQ(2-amino-3,4-dimethylimidazo[4,5-*f*]-quinoline), MeIQx(2-amino-3,8-dimethylimidazo[4,5-*f*]-quinoxaline) are found in cooked fish and meat owing to pyrolysis of amino acids, peptides, and proteins (18). Polycyclic aromatic hydrocarbons such as the carcinogen benzo(a)pyrene accumulates in foods during charcoal broiling and smoking. The production of reaction products from food irradiation, a method of food preservation, has been investigated in numerous studies. The safety of irradiated food produced in accordance with good manufacturing practices has been acknowledged by the World Health Organization (19). An evaluation of the safety of food irradiation was required for recent approval of food irradiation as a method of preservation of meat (20). See reference 21 for further information on the safety of irradiated foods.

Residues

Residues of chemicals occur in foods as a result of the use of pesticides, drugs in food-producing animals, and food packaging materials. To ensure a safe food supply, the Environmental Protection Agency (EPA) regulates the safety of food by setting safety standards to limit the amount of pesticide residues that legally may remain in or on food or animal feed that is sold in the United States. The FDA and the U.S. Department of Agriculture (USDA) ensure compliance with these safety standards by monitoring domestic and imported foods. Annual reports summarizing the findings of pesticide monitoring programs are available from the USDA. The regulation of pesticides in food production is currently undergoing major reform because of the passage of the Food Quality Protection Act (FQPA) in August 1996. Major changes in pesticide residue safety standards resulting from the FQPA include a new safety standard based on a "reasonable certainty of no harm"; consideration of aggregate exposure from all sources of pesticides (drinking water, residential, and dietary exposure); and consideration of all pesticides with a common mechanism of toxicity as a group to determine cumulative exposure. The EPA must consider children's special sensitivity and exposure to pesticides. Under the new law, the EPA is required to develop and use a screening and testing program for chemicals with the potential to disrupt the endocrine process (endocrine disruptors). The impact of this new law on the use of pesticides has yet to be determined because the regulations, which will enforce this law, are still under development.

Various drugs and antibiotics are used in the production of animal food products. The use of these drugs and the amount of residues allowed in the animal tissues are regulated by FDA. In most cases, tolerances range from 0 to

1 ppm. The unintentional transfer of compounds from packaging materials into foods is another source of food residues. Food residues from pesticides, drugs, and packaging materials are called unintentional food additives. Animal drug residues and food contact substances, which "may reasonably be expected to become a component of food," must be shown to be safe for humans under similar standards as food additives. Although public concern regarding food residues has always been the subject of the lay press, there is little evidence of health hazards from the small amount of residues present in foods produced in the United States (22). The use of pesticides in other countries may not be as well regulated, and therefore the safety of pesticide residues in imported foods continues to be a concern.

Food Additives

A food additive is defined as a substance or mixture of substances, other than a basic foodstuff, that is present in a food as a result of any aspect of production, processing, storage, or packaging. Today, more than 2,500 different additives are intentionally added to produce desired effects. They include preservatives, antioxidants, sweeteners, nutritional additives, flavoring agents, coloring agents, bleaching agents, texturizing agents, and miscellaneous additives. A comprehensive summary of food additive toxicology has recently been published (23). The majority of direct food additives, predominantly spices and flavors, are generally recognized as safe (GRAS) substances that have been used for many years and have been found to present no significant hazard with normal human food uses. However, the use of some food additives, such as cyclamates, red no. 2 and violet no. 1, is banned owing to their potential carcinogenic activity.

When a manufacturer submits a petition for a food additive approval, the FDA requires information on the chemical identity of the additive, the purpose and amount proposed for use, functionality testing to demonstrate effectiveness of the additive, methods of analysis in food, and safety data for the additive. Safety evaluation is discussed in the next section. The FDA operates an adverse reaction monitoring system (ARMS) to help serve as an ongoing safety check of all food additives. The system monitors and investigates all complaints by individuals or physicians that are believed to be related to specific foods, food and color additives, or vitamin and mineral supplements. The ARMS computerized database helps officials decide whether reported adverse reactions represent a real public health hazard associated with food, so that appropriate action can be taken. However, it is possible that postmarket surveillance may not be sensitive enough to identify a small subsection of the population that has severe adverse effects to a particular additive. This concern has been raised with the recent approval of the fat replacement Olestra. The purported benefit of fat replacers is possible prevention of obesity through calorie reduction. However, adverse gastrointestinal effects in some individuals have raised concern and resulted in a label requirement on foods containing this additive.

Dietary Supplements

A dietary supplement is any product that is taken by mouth and intended to supplement the diet and is labeled as a dietary supplement. Dietary supplements may contain vitamins, minerals, herbs, and amino acids as well as substances such as enzymes, organ tissues, metabolites, extracts, or concentrates. The Dietary Supplement Health and Education Act (DSHEA) of 1994 established new guidelines for the safety requirements of dietary supplements and dietary ingredients of supplements. One requirement is that "the dietary supplement contains only dietary ingredients which have been present in the food supply as an article used for food in a form in which the food has not been chemically altered." The alternative requirement is that there is a history of use or other evidence of safety establishing that the dietary ingredient when used under the conditions recommended or suggested in the labeling of the dietary supplement "will reasonably be expected to be safe." As a result, dietary ingredients used in dietary supplements are no longer subject to the premarket safety evaluations required of other new food ingredients.

The ARMS program for food additive postmarket surveillance also serves to monitor safety of dietary supplements. Recently, ARMS reports of adverse health effects associated with dietary supplements containing ephedrine alkaloids resulted in proposed safety measures by the FDA, including limits on the amount of ephedrine per serving and label warning statements for ephedrine-containing supplements.

SAFETY EVALUATION AND RISK ASSESSMENT

A number of different laws—the Food, Drug, and Cosmetic Act (FD&C Act); the Federal Insecticide, Fungicide, Rodenticide Act (FIFRA); the recent FQPA and DSHEA; the Meat Inspection Act; and the Poultry Products Inspection Act—are routinely employed by the FDA, USDA, and EPA to govern the safety of food. For chemical food safety issues, a risk assessment process is used to determine the levels of a compound that will be allowed in foods. There are four steps in risk assessment: (*1*) hazard identification, (*2*) dose–response assessment, (*3*) exposure assessment, and (*4*) risk characterization. See reference 1 for an overview of risk assessment.

Hazard Identification

Hazard identification is the determination of the known or potential health effects of consumption of the compound. In some cases, the need for toxicological testing can be predicted from the chemical structure based on known structure–activity relationships. Several chemical structures, such as *n*-nitroso or aromatic amine groups, are potentially carcinogenic. In many cases, manufacturers use this information to decide whether to go forward with expensive animal testing of their products. *In vitro* and short-term animal tests are also used before investing in long-term animal testing. These can include tests for mutagenicity, carcinogenecity, developmental toxicity, reproductive toxicity, neurotoxicity, and immunotoxicity (3). However, for final approval of pesticides and food additives, the long-term animal bioassay is a key component in the hazard identification process. Table 2 lists the types of toxicological tests that are conducted using the animal model system. For food additives, the safety evaluation involves assigning a concern level to the additive, based on structural information and an estimate of exposure. The minimum toxicological testing required for each concern level is listed in Table 3.

Epidemiological studies, in which a positive association between exposure and disease has been observed, and case reports of accidental poisonings have provided convincing evidence of the human risk of various compounds. This has been the basis for the identification of many naturally occurring toxins, including ergot alkaloids, potato alkaloids, and toxic mushrooms.

Dose–Response Assessment

The dose–response assessment examines the relationship between the magnitude of the exposure and the probability of adverse effects. There are two approaches to dose–response assessments: a threshold approach for noncancer end points and a nonthreshold approach for cancer end points. The threshold is the level of exposure below which no adverse effects result. This approach is based on establishing a NOAEL, which is the highest dose in milligrams/kilogram body weight that results in no detectable damage to the animal during chronic or subchronic testing. Safety factors are applied to the NOAEL to determine the RfD and the acceptable daily intake (ADI). The ADI values are used by the World Health Organization for pesticides and food additives to define "the daily intake of a chemical, which during an entire lifetime appears to be without appreciable risk on the basis of all known facts at that time." The safety factors, also called uncertainty factors, allow for differences in species and individual susceptibility to toxicity. A factor of 10 is used for each extrapolation (ie, from animals to humans, from a small number of animals to a population, from a short-term study to a chronic study) to result in a total safety factor of 100 to 1,000. The FQPA requires an additional 10-fold safety factor for pesticides if adverse effects on infants and children are unknown or increased compared to adults. Therefore, the RfD or ADI = NOAEL ÷ safety factors.

Dose–response models for cancer, which is considered to be a nonthreshold end point, can be categorized into two types. One type is the statistical or probability model, such as the log–probit and logit models. The other type is a mechanistic-based model, such as the one-hit, multistage, and linearized multihit models. A discussion of these models and the use of pharmacokinetic models to improve cancer risk assessment can be found in reference 25.

Table 2. Types of Toxicological Tests

Acute tests (single exposure or dose)
Determination of median lethal dose (LD_{50})
Acute physiologic changes (blood pressure, pupil dilation, etc)
Subacute tests (continuous exposure or daily doses)
Three-month duration
Two or more species (one nonrodent)
Three dose levels (minimum)
Administration by intended or likely route
Health evaluation, including body weight, complete physical examination, blood chemistry, hematology, urinalysis, and performance tests
Complete autopsy and histopathology on all animals to determine the presence of any tissue damage or cancerous tissues
Chronic tests (continuous exposure or daily doses)
Two-year duration (minimum)
Two species selected for sensitivity from previous tests
Two dose levels (minimum)
Administered by likely route of exposure
Health evaluation including body weight, complete physical examination, blood chemistry, hematology, urinalysis, and performance tests
Complete autopsy and histopathology on all animals to determine the presence of any tissue damage or cancerous tissues
Special tests
Teratogenicity
Multigenerational reproduction feeding study (all aspects other than teratogenicity)
Skin and eye effects
Behavioral effects

Exposure Assessment

For exposure assessment, estimates of the likely levels of human consumption of the particular food substance or the estimate of daily intake (EDI) must be made. It is necessary to know the following: (*1*) the amount of substance in the foods; (*2*) daily intake of each food containing the substance by consumers; and (*3*) the portion of the population that are consumers. The amount of a direct food additive in the foods will be based on the amount needed to obtain the desired effects of the additive and this will have been determined by the manufacturer petitioning for approval of the additive. For pesticides, tests are conducted to determine the level of pesticide that remains on the agricultural commodities after the pesticide is used in the production of that commodity. Food consumption surveys, food disappearance figures, and market basket surveys can be used to determine the EDI. However, food consumption surveys are most often used. The anticipated use and the toxicity of the compound will affect whether values for average consumers or heavy consumers (at or above 95th percentile for food intake) and values for the whole population or consumers only are used in the EDI. The daily intake of food consumed (kilograms per day) multiplied by the amount of the compound in the food (milligrams/kilogram) provides an estimate of the amount of the compound (milligrams per day) that will be consumed. Guidelines on the estimations of exposure to food additives and chemical contaminants have been published by the U.S. FDA (26).

Table 3. Minimum Toxicological Testing of Food Additives Required by the U.S. Food and Drug Administration

	Indirect food additives
Virtually nil exposure (<0.05 ppm)	Acute oral study—rodent
Insignificant exposure (>0.05 ppm)	Acute oral study—rodent[c]
	Subchronic feeding study (90-day)—rodent with *in utero* exposure
	Subchronic feeding study (90-day)—nonrodent
	Multigenerational reproduction feeding study (minimum of 2 generations) with teratology phase—rodent[a]
	Teratology study
	Short-term tests for carcinogenic potential[b]
Significant exposure (>~1 ppm)	Acute oral study—rodent[c]
	Subchronic feeding study (90-day)—rodent[c]
	Lifetime feeding study (about 2 y)—rodent with *in utero* exposure for carcinogenesis and chronic toxicity
	Lifetime feeding study (about 2 y)—rodent for carcinogenesis
	Short-term feeding study (at least 1 year)—nonrodent
	Multigenerational reproduction feeding study (minimum of 2 generations) with teratology phase—rodent
	Teratology study[a]
	Short-term tests for carcinogenic potential[b]
	Metabolism studies[b]
	Direct food additives
Concern level I (lowest concern)	Short-term feeding study (at least 28 days)—rodent
	Short-term tests for carcinogenic potential
Concern level II	Acute oral study—rodent[c]
	Subchronic feeding study (90-day)—rodent
	Subchronic feeding study (90-day)—nonrodent
	Multigenerational reproduction feeding study (minimum of 2 generations) with teratology phase—rodent
	Teratology study[a]
	Short-term tests for carcinogenic potential
Concern level III (highest concern)	Acute oral study—rodent[c]
	Subchronic feeding study (90-day)—rodent[c]
	Lifetime feeding study (about 2 y)—rodent with *in utero* exposure for carcinogenesis and chronic toxicity
	Lifetime feeding study (about 2 y)—rodent for carcinogenesis
	Short-term feeding study (at least 1 year)—nonrodent
	Multigenerational reproduction feeding study (minimum of 2 generations) with teratology phase—rodent
	Teratology study[a]
	Short-term tests for carcinogenic potential
	Metabolism studies[b]

Notes: If carcinogenicity is suspected as a problem, carcinogenicity studies will be done no matter what the circumstances. References to current guides on these toxicology studies are contained in the "Redbook" (24).

[a]If indicated by available data or information.

[b]Suggested.

[c]If needed as preliminary to further study.

Risk Characterization

Risk characterization is the estimate of the incidence of health effects that are likely to occur under specific conditions of exposure. This is dependent on the exposure levels and the toxicity of the compound. Simply, risk = toxicity × exposure. Risk characterization requires a judgment of the applicability of the science to the potential human exposure conditions.

For noncarcinogens, the approach is to ensure that the estimated intake or exposure does not exceed the RfD. Intakes at levels below this level are considered to be without appreciable risk. The amount of additive or pesticide residue that is allowed in foods (ie, the tolerance) is based on the lowest amount needed for efficacy, not safety. As a result, the tolerance may be many folds lower than the safe level based on the RfD. See Table 4 for an example of the determination of tolerance levels.

For carcinogens, the risk assessment is based on the assumption that there is no threshold for carcinogenesis and that the dose–response relationship observed at high doses is similar to that at low doses. An acceptable level of risk has been set at a consumption level estimated to produce no more than one cancer in a million individuals. The insensitivity of the animal bioassay for detecting carcinogens has led to the use of high doses in order to generate a sufficiently high frequency of response to be statistically significant. Results from these studies then need to be ex-

Table 4. Example of Determination of Maximum Permissible Levels

1. The synthetic food color additive Rosy Red was tested in rats and dogs in a chronic feeding study. A NOAEL of 1,000 mg/kg body weight per day (mg/kg/day) was determined.
2. Safety factor = 10 (extrapolation to humans) × 10 (individual variation) = 100
3. RfD = 1,000 mg/kg/day ÷ 100 = 10 mg/kg/day
4. Rosy Red is being developed for use in licorice candy. Using food consumption survey data, it was found that the person with the highest intake would consume 100 g of candy per day.
5. The maximal permissible intake per day (MPI) = RfD × 60 kg (adult body weight) = 10 mg/kg × 60 kg = 600 mg per day.
6. Assuming that Rosy Red will not be added to any other foods, the maximum permissible level (MPL) (also called tolerance level) that would be approved would be MPL = MPI/food factor. Food factor represents an estimation of the amount of that food in the diet. MPL = 600 mg ÷ 100 g licorice = 6 mg Rosy Red/g licorice. However, if only 1 mg Rosy Red/g candy was needed for the desired color, the MPL would be 1 mg/g.
7. The RfD is usually well above the EDI of a food additive.

trapolated to the low-dose range that represents the human exposure levels. A very large number of animals would be required to obtain a statistically valid response level if low doses were used, resulting in increased testing costs. Therefore, the dose used may be up to the maximum tolerated dose (MTD), which is the highest dose that does not produce overt toxic responses in subchronic studies. Concerns with the use of the MTD are that this dose may result in depressed food intake and may saturate other physiological processes, such as activation and inactivation by enzyme systems, active transport, and DNA repair. Cytotoxicity caused by high doses may result in mitogenesis or increased cell proliferation, enhancing the carcinogenesis process. Thus, use of the MTD may result in the ability of a compound to induce cancer through mechanisms that do not occur at low doses.

In part to address this issue, the EPA proposed new guidelines for cancer risk assessment in 1996 (27). The 1986 cancer guidelines classified compounds as A, B1, B2, C, D, or E carcinogens. Classification as an A carcinogen was based on "sufficient evidence of carcinogenecity from human studies," and an E classification was based on "evidence of noncarcinogenecity." Tumor findings in animals or humans were the dominant components of the classification. The 1996 proposed guidelines will use a narrative description of the likelihood and conditions of human hazard, such as "not likely to be a human carcinogen, because animal data shown to be not relevant to humans." Therefore, consideration of the route of exposure, the dose–response relationship, and mode of action information are changes in the proposed guidelines. Further discussion on the proposed guidelines can be found in the Science Advisory Board review (28).

A major difference in the cancer risk assessment process for food additives and for pesticides is that the Delaney clause still applies to food additives, but has been removed for pesticides by the FQPA. The Delaney clause of the FD&C Act states that "no additive shall be deemed to be safe if it is found to induce cancer when ingested by man or animal, or if it is found, to induce cancer in man or animal" (Food Additive Amendment of 1958, section 409). The Delaney clause therefore does not allow for consideration of dose or mechanism of action. Therefore, even if it is clearly demonstrated that the mechanism of cancer induction in animals is not relevant to humans, the Delaney clause would prohibit the use of the product in foods. This is one reason why the new FQPA eliminated the Delaney clause from pesticide regulations.

CONCLUSIONS

The safety of the food supply affects the entire population of a country. Current laws and regulations in the United States have resulted in the provision of one of the safest food supplies in the world. However, as production and processing methods of food continue to change, new challenges and issues in food toxicology continue to arise.

The level of exposure to naturally occurring toxins may increase in the near future as a result of the increased development of functional foods and pest-resistant crops. Functional foods or nutraceuticals represent a rapidly growing segment of the food industry. Functional foods are defined as foods that contain components with anticipated health benefits. Examples include beverages with added herbal extracts, cereals with added fiber, and vegetables and fruits bred to contain higher levels of antioxidants or other phytochemicals. Concerns regarding environmental and health impacts of pesticide use have resulted in the drive to lower pesticide use through the development of biological pesticides and pest-resistant crops. In many cases, pest resistance is achieved through increasing the plant's naturally occurring defense mechanisms, which may involve naturally occurring toxicants. In both cases, the goal is to improve overall health. However, as mentioned earlier, our current level of understanding of the dose–response curves for naturally occurring compounds is very limited, making it extremely difficult to determine when levels of exposure cease being beneficial and start to become harmful.

Understanding of the interactions between the many compounds in the diet and the interaction between diet and other environmental compounds is another area of needed research. The requirement of the FQPA to consider all pesticides with a common mechanism of action is one step toward recognizing the need to consider the impact of exposure to a variety of compounds. This is a very complex and difficult question to address, but research on this topic is beginning to increase.

Consumer understanding of the dose–response concept is critical to their understanding of how low levels of food additives and pesticides in foods can pose little to no risk, even though these compounds may have toxic effects at high doses. Consumer acceptance of pesticide residues and processes such as food irradiation has been shown to increase after educational programs to develop this type of understanding. The commonly held beliefs "natural is safe" and that "synthetic chemicals pose a greater risk than natural compounds" also need to be addressed through public educational programs. Greater understanding of these issues will prevent undue concern and overreaction to the presence of small amounts of synthetic chemicals and may prevent poisonings and adverse health effects caused by overzealous consumption of "natural" compounds with promised health benefits.

BIBLIOGRAPHY

1. E. M. Faustman and G. S. Omenn, "Risk Assessment," in C. D. Klaassen, ed. *Casarett and Doull's Toxicology: The Basic Science of Poisons*, 5th ed., McGraw-Hill, New York, 1996, pp. 75–88.
2. D. L. Eaton and C. D. Klaassen, "Principles of Toxicology," in C. D. Klaassen, ed. *Casarett and Doull's Toxicology: The Basic Science of Poisons*, 5th ed., McGraw-Hill, New York, 1996, pp. 13–33.
3. C. D. Klaassen, *Casarett and Doull's Toxicology: The Basic Science of Poisons*, 5th ed., McGraw-Hill, New York, 1996.
4. A. W. Hayes, *Principles and Methods of Toxicology*, 3rd ed., Raven Press, New York, 1994.
5. H. C. Pitot and Y. P. Dragan, "Chemical Carcinogenesis," in C. D. Klaassen, ed. *Casarett and Doull's Toxicology: The Basic Science of Poisons*, 5th ed., McGraw-Hill, New York, 1996, pp. 201–267.
6. R. W. Rudon, *Cancer Biology*, 3rd ed., Oxford University Press, New York, 1995.
7. *Food, Nutrition and the Prevention of Cancer: A Global Perspective*, World Health Cancer Fund and American Institute for Cancer Research, Washington, D.C., 1997.
8. T. Shibamoto and L. F. Bjeidanes, *Introduction to Food Toxicology*, Academic Press, San Diego, Calif., 1993.
9. J. M. Concon, *Food Toxicology: Principles and Concepts*, Vol. 1, Marcel Dekker, New York, 1988.
10. J. M. Concon, *Food Toxicology: Contaminants and Additives*, Vol. 2, Marcel Dekker, New York, 1988.
11. L. Slutsker, S. F. Altekruse, and D. L. Swerdlow, "Foodborne Diseases. Emerging Pathogens and Trends," *Infect. Dis. Clin. North Am.* **12**, 199–216, 1998.
12. J. E. Collins, "Impact of Changing Consumer Lifestyles on the Emergence/Reemergence of Foodborne Pathogens," *Emerg. Infect. Dis* **3**, 471–479 (1997).
13. C. D. Holland, *Chemical Hormesis: Beneficial Effects at Low Exposures Adverse Effects at High Exposures*, Texas Institute for Advancement of Chemical Technology, Bryan, Tex., 1998.
14. F. K. Ennever, "Metals," in A. W. Hayes, ed. *Principles and Methods of Toxicology*, Raven Press, New York, 1994, pp. 417–446.
15. K. T. Chung, C. I. Wei, and M. G. Johnson, "Are Tannins a Double-Edged Sword in Biology and Health," *Trends Food Sci. Technol.* **9**, 168–175 (1998).
16. G. J. A. Speijers, "Toxicological Data Needed for Safety Evaluation and Regulation on Inherent Plant Toxins," *Natural Toxins* **3**, 222–226 (1995).
17. J. H. Maryanski, "FDA's Policy for Foods Developed by Biotechnology," in K. Engel, G. R. Takeoka, and R. Teranishi, eds., *Genetically Modified Foods: Safety Issues*, American Chemical Society, Washington, D.C., 1995, pp. 12–22.
18. H. A. J. Schut and E. G. Snyderwine, "DNA Adducts of Heterocyclic Amine Food Mutagens: Implications for Mutagenesis and Carcinogenesis," *Carcinogenesis* **20**, 353–368 (1999).
19. *Safety and Nutritional Adequacy of Irradiated Foods*, World Health Organization, Geneva, Switzerland, 1994.
20. U.S. Food and Drug Administration, "Irradiation in the Production, Processing and Handling of Food," *Federal Register* **62**, 64107–64121 (1997).
21. J. F. Diehl, *Safety of Irradiated Foods*, Marcel Dekker, New York, 1995.
22. National Research Council, *Carcinogens and Anticarcinogens in the Human Diet*, National Academy Press, Washington, D.C., 1996.
23. J. A. Maga and A. T. Tu, eds., *Food Additive Toxicology*, Marcel Decker, New York, 1995.
24. U.S. Food and Drug Administration, *Toxicological Principles for the Safety Assessment of Direct Food Additives and Color Additives Used in Food*, U.S. Food and Drug Administration, Bureau of Foods (now Center for Food Safety and Applied Nutrition), Washington, D.C., 1982.
25. D. C. Rees and D. Hattis, "Developing Quantitative Strategies for Animal to Human Extrapolation," in A. W. Hayes, ed., *Principles and Methods of Toxicology*, Raven Press, New York, 1994, pp. 275–315.
26. M. J. DiNovi and P. M. Kuznesof, *Estimating Exposure to Direct Food Additives and Chemical Contaminants in the Diet*, U.S. Food and Drug Administration Center for Food Safety and Applied Nutrition, Office of Premarket Approval, Washington, D.C., 1995.
27. U.S. Environmental Protection Agency, *Proposed Guidelines for Carcinogen Risk Assessment*, Washington, D.C., 1996.
28. Science Advisory Board, *An SAB Report: Guidelines for Cancer Risk Assessment*, U.S. Environmental Protection Agency, Washington, D.C., 1997.

BERNADENE MAGNUSON
University of Idaho
Moscow, Idaho

See also TOXICOLOGY AND RISK ASSESSMENT.

FOOD UTILIZATION

Most people take the process of eating for granted, knowing that the body will take care of itself. But between eating and the cellular utilization of dietary nutrients, hundreds of thousands of metabolic processes take place. Food must first be digested, or broken down into particles of a size and chemical composition that the body can readily absorb. Absorption takes place mostly in the small intestine, where specialized cells transfer digested nutrients to the blood and lymph vessels. In some cases, special changes are needed so that the nutrients can be transported to the cells where they are to be used or further processed. Within the cells, the nutrients are either stored or metabolized, that is, broken down into simpler components for energy or excretion (catabolism), or used to synthesize new materials for cellular growth, maintenance, or repair (anabolism).

The complicated processes involved are different for each nutrient, although the paths that certain nutrients take intersect at various points. The chemical details of metabolism belong to the disciplines of biochemistry and physiology; standard texts in these subjects can be consulted for detailed information. This article provides a brief overview of what happens in the body to the foods that are eaten, with particular attention to the three major nutrients: carbohydrates, proteins, and fats.

THE ALIMENTARY SYSTEM

The alimentary or digestive system is a long tube that consists of the mouth, esophagus, stomach, small intestine, colon, rectum, and anus. Some important accessory organs connected to the digestive tract are the salivary glands, gallbladder, pancreas, and liver. Along this tract, foods are broken down into smaller units, both physically and chemically, and then absorbed for use by the body. Figure 1 shows the general outline of the entire human digestive system.

The Food Path

The food placed in the mouth is chewed, softened, and swallowed. In the stomach, it is churned and then propelled into the small intestine where it is mixed with the bile from the gallbladder and digestive enzymes from the intestinal walls. The products of this digestion are partly or completely absorbed into either the portal vein or the lacteal system.

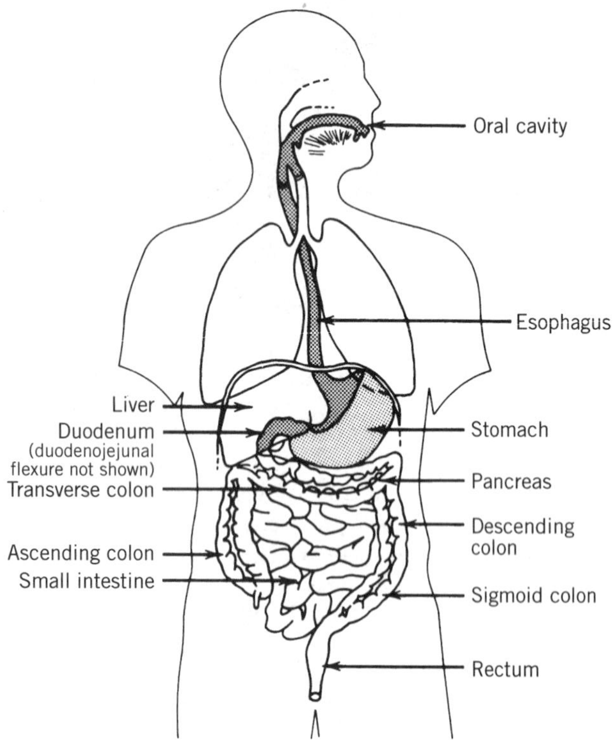

Figure 1. The digestive system of the human body.

In the mouth, chewing (mastication) reduces large food lumps into smaller pieces and mixes them with saliva. This wetting and homogenizing facilitates later digestion. In clinical dietetics, edentulous (toothless) patients or those with reduced saliva secretion have trouble eating dry foods and require a soft, moist diet. Saliva facilitates swallowing and movements of the tongue and lips, keeps the mouth moist and clean, serves as a solvent for taste bud stimulants, acts as an oral bugger, provides some antibiotic activity, and inhibits loss of calcium from the teeth by maintaining a neutral pH. Saliva contains ptyalin (salivary amylase, a digestive enzyme) and mucin (a glycoprotein). Mucin lubricates food and ptyalin digests carbohydrates to a small extent. Each day the salivary gland makes about 1,500 mL of saliva (ca 5.5–6 cups).

The bolus of food is propelled forward by rhythmic contractions of the entire intestinal system. These peristaltic waves move the food from the mouth, through the esophagus, and into the stomach. Certain individuals, especially nervous people, tend to swallow air when eating. When part of the air is expelled through the mouth, belching results; the remaining air is expelled as flatus. If too much air is swallowed, there will be abdominal discomfort.

From the mouth, food travels through the esophagus, the stomach cardia, the stomach body, the greater curvature, the pylorus, and the duodenum. These are all parts of the stomach, where food is well mixed. Figure 2 shows the general structure of this organ and the site of specific secretions. The acid, mucus, and pepsin cause partial digestion, and peristalsis mixes up the food. The food is then released gradually through the pylorus into the duodenum.

The gastrointestinal (GI) system, the stomach and intestines, breaks down complex carbohydrates, proteins, and fats into absorbable units, mainly in the small intestine. Vitamins, minerals, fluids, and most nonessential nutrients are also digested and absorbed to varying degrees. Foods are digested by enzymes secreted by different parts of the GI system. Table 1 summarizes the major digestive enzymes and their actions. After the digestive process is complete, nutrients are ready for absorption, which occurs mainly at the small intestine. The absorption of each nutrient is discussed later.

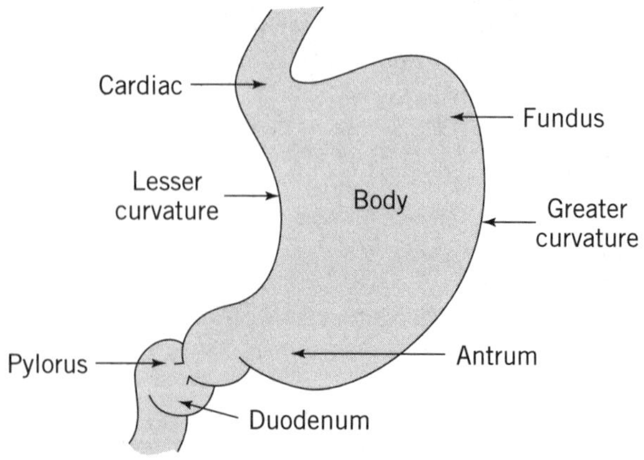

Figure 2. General structure of the stomach.

After the nutrients have been absorbed, they enter the circulation in two ways. Most fat-soluble nutrients enter the lacteal or lymphatic system, which eventually joins the systemic blood circulation at the thoracic duct. Other nutrients enter the hepatic portal vein and are received by the liver, which eventually releases them to the bloodstream.

Enzymes and Coenzymes

After digestion and absorption, the nutrients exist as hexoses (mainly glucose and fructose), fatty acids, glycerols, and amino acids and are then metabolized in various fashions. Many of the metabolic processes require the presence of a catalyst, a substance that can facilitate a chemical reaction. Although participating in the process, it may or may not undergo physical, chemical, or other modification itself. Nonetheless, the catalyst usually returns to its original form after the reaction.

In the body, most biological reactions require a special class of catalysts: the protein catalysts or enzymes. Each enzyme catalyzes only one or a small number of reactions. There are many enzymes, each with a specific responsibility. Without enzymes, most biological reactions would proceed at a very slow speed. Coenzymes are accessory substances that facilitate the working of an enzyme, mainly by acting as carriers for products of the reaction. In this case, the enzyme is composed of two parts: a protein (apoenzyme) and a nonprotein (cofactor or coenzyme). Many coenzymes contain vitamins or slightly modified vitamins as the major ingredient. A coenzyme can catalyze many types of reaction. Some coenzymes transfer hydrogens; others transfer groups other than hydrogens. Table 2 describes the characteristics of the former; Table 3, those of the latter. Because most of the metabolic reactions discussed in this article involve coenzymes, a knowledge of the biochemical role of vitamins is important.

CARBOHYDRATES

Starch, cellulose, and their derivatives are the only polysaccharides consumed to any extent by humans. The major simple sugars ingested include the monosaccharides (such as glucose and fructose in honey and fruit juices) and disaccharides (such as maltose in beer, lactose in milk, and sucrose in table sugar). Also ingested are dextrins, sugar alcohol, and trisaccharides and tetrasaccharides, although in very small quantities. The sections that follow discuss briefly the digestion and absorption pathways of these carbohydrates, the involvement of carbohydrates in energy formation and storage, the specific processes by which carbohydrates are broken down or synthesized by the body, and the regulation of glucose levels in the blood.

Digestion and Absorption

Starch is partially hydrolyzed by ptyalin in the mouth (Table 1 and Fig. 3). The short stay in the oral cavity permits only dextrines and small polysaccharide fragments to break off from the starch molecules, and the action of ptyalin is terminated by the acid in the stomach. In the small intestine, all digestible carbohydrates are reduced to monosaccharides, namely, glucose, fructose, galactose, mannose, and pentoses. Currently, it is believed that the final hydrolysis, or digestion, of disaccharides to monosaccharides occurs in the intestinal mucosal walls. In humans, all nondigestible carbohydrates such as lignin, hemicellulose, and cellulose are passed into the colon, where they are mainly fermented to release gas. By contrast, ruminants (animals such as cattle, sheep, and goats) have the ability to digest fiber.

Most of the monosaccharides are absorbed before the food residue reaches the end of the ileum. Absorption is principally carried out by active transport (requiring energy), although diffusion (passive movement) also occurs. Glucose and galactose enter the blood five times faster than mannose and pentoses, whereas fructose is absorbed about two to three times faster. Glucose entry into the blood may be as high as 129 g/h.

The monosaccharides traverse the intestinal and portal veins to reach the liver, from which they may eventually be released into blood circulation. In the liver, most of the fructose and galactose are converted to glucose, which is the main simple sugar in the blood, although fructose and galactose may be present throughout the bloodstream if a person consumes a large amount of them. An actively nursing mother also has some lactose in the blood, because this disaccharide is manufactured by the active mammary tissues. In the liver, part of the glucose is released to the circulation, part converted to glycogen for storage, part changed to other essential substances required by the body, and part oxidized to energy. Normally, a peak plasma glucose level of 120 to 140 mg/100 mL is reached within 60 min after a mixed meal.

After glucose has reached the bloodstream, some enters cells to give energy, and some is converted to glycogen in tissues such as muscle. Glycogen is found in many organs in the body, but the liver and muscle are the major storage sites. Most of the glucose is used to provide energy through a three-stage process: glycolysis, the citric acid cycle, and the respiratory chain.

Energy Formation and Storage

Everything a body does requires energy. Nature has provided the animal body with a wide spectrum of methods that permit energy either to be released to be stored and released to satisfy its energy need. There are five different energy systems known to operate in animal cells.

Direct Release of Energy (Mainly as Heat)

$$\text{glucose (or fat or protein)} + \text{oxygen} = \text{carbon dioxide} + \text{water} + \text{energy to be stored} + \text{heat.}$$

Energy Stored as Adenosine Triphosphate (ATP). One technique of storing part of the energy released from the oxidation of foodstuffs is incorporating it into ester bonds between certain organic compounds and phosphoric acid groups. The resulting substances are called high-energy phosphate compounds, the most important of which is probably adenosine triphosphate. This ubiquitous mole-

Table 1. Characteristics of the Enzymatic System of Digestion

Location	Food or substrate	Products of digestion	Enzyme(s) involved in digestion		
			Name(s)	Source(s)	Active in acid-base (pH)
Mouth	Starch	Maltose, dextrins, disaccharides, monosaccharides, branched oligosaccharides	Ptyalin, or salivary amylase	Salivary glands	Slightly acidic (6.7)
Stomach	Protein	Proteoses, peptones, polypeptides, dipeptides, amino acids	Pepsin	Peptic or chief cells of stomach	Acidic (1.6–2.4)
	Milk casein	Milk coagulation	Rennin	Stomach mucosa	Acidic (4.0); requires calcium for activity
	Fat	Triglycerides; some monoglycerides and diglycerides, glycerol, fatty acids	Gastric lipase	Stomach mucosa	Acidic
Small intestine (mainly duodenum and jejunum)[a]	Protein proteoses, peptones, etc	Polypeptides, dipeptides, etc	Trypsin (activated trypsinogen)	Exocrine gland of pancreas	Alkaline (7.9)
	Proteoses, peptones, etc	Polypeptides, dipeptides, etc	Chymotrypsin (activated chymotrypsinogen)	Exocrine gland of pancreas	Alkaline (8.0)
	Polypeptides with free carboxyl groups	Lower peptides, free amino acids	Carboxypeptidase	Exocrine gland of pancreas	
	Fibrous protein	Peptides, amino acids	Elastase	Exocrine gland of pancreas	
	Carbohydrate Starch, dextrins	Maltose, isomaltose, monosaccharides, dextrins	α-Amylase (amylopsin)	Exocrine gland of pancrease	Slightly alkaline (7.1)
	Fat Triglycerides	Monoglycerides and diglycerides, glycerol, fatty acids	Lipase (steapsin)	Exocrine gland of pancreas	Alkaline (8.0)

	Cholesterol	Cholesterol esters	Cholesterol esterase	Exocrine gland of pancreas	
	Nucleic acids	Nucleotides			
	Ribonucleic acid	Ribonucleotides	Ribonuclease	Exocrine gland of pancreas	
	Deoxyribonucleic acid	Deoxyribonucleotides	Deoxyribonuclease	Exocrine gland of pancreas	
	Protein				
Small intestine (mainly jejunum and ileum)	Polypeptides	Amino acids	Carboxypeptidase, aminopeptidase, dipeptidase	Brush border of the small intestine	
	Carbohydrate				
	Sucrose	Glucose, fructose	Sucrase	Brush border of the small intestine	Acidic-alkaline (5.0–7.0)
	Dextrin (isomaltose)	Glucose	α-Dextrinase (isomaltase)	Brush border of the small intestine	
	Maltose	Glucose	Maltase	Brush border of the small intestine	Acidic (5.8–6.2)
	Lactose	Glucose, galactose	Lactase	Brush border of the small intestine	Acidic (5.4–6.0)
	Fat				
	Monoglycerides	Glycerol, fatty acids	Lipase (enteric)	Brush border of the small intestine	
	Lecithin	Glycerol, fatty acids	Lecithinase	Brush border of the small intestine	
	Nucleotides	Nucleosides, phosphate	Nucleotidase	Brush border of the small intestine	
	Nucleosides	Purines, pyrimides, pentose	Nucleosidase	Brush border of the small intestine	
	Organic phosphates	Free phosphates	Phosphatase	Brush border of the small intestine	Alkaline (8.6)

[a]The food is not grouped together (eg, all fat, all proteins, etc), instead the food is placed in an order that follows the sequence of digestion along the duodenum to jejunum. This attempts to present the digestive enzymes in their expected sequence of action.

Table 2. Characteristics of Coenzymes That Transfer Hydrogens

Enzyme system	Coenzyme	Vitamin component	Nonvitamin component
Dehydrogenase	Flavin adenine dinucleotide (FAD)	Riboflavin (vitamin B_2)	Adenine, ribose, phosphate
Dehydrogenase	Nicotinamide adenine dinucleotide (NAD)	Niacin	Adenine, ribose, phosphate
Part of dehydrogenase	Lipoic acid (thiotic acid)	None	Lipoic acid
Respiratory chain	Coenzyme Q	None	Quinone (vitamin E-related substance)

Table 3. Characteristics of Coenzymes That Transfer Nonhydrogen Groups

Enzyme system	Coenzyme	Vitamin component	Nonvitamin component
Transaminases, decarboxylase	Pyridoxal phosphate	Vitamin B_6	None
Part of dehydrogenase	Lipoic acid (lipoamide)	None	Lipolic acid
Dehydrogenase	Coenzyme A	Pantothenic acid	β-Mercaptoethylamine, adenine, ribose, phosphate
Cocarboxylase (decarboxylase, transketolase)	Thiamin pyrophosphate	Thiamin	Phosphate
Methyl transferase	5-Methyltetrahydrofolate	Folic acid	None
Transmethylase	Coenzyme B_{12}	Vitamin B_{12}	Adenine, ribose
Carboxylase	Carboxyl-biotin complex	Biotin	None

cule is considered the energy powerhouse of the body. It releases its energy in the following reactions:

$$\text{ATP} + H_2O = \text{ADP} + \text{P} + 7.5 \text{ kcal}$$
$$\text{ADP} + H_2O = \text{AMP} + \text{P} + 7.5 \text{ kcal}$$

where ATP = adenosine diphosphate, P = inorganic phosphate, and AMP = adenosine monophosphate. Theoretically, the conversion of 1 mol of ATP to AMP can produce 15 kcal. However, within the body ATP is normally changed only to ADP when energy is needed. The energy released from this process can be used for such work as organ building, heartbeat, transportation across cell membranes, and muscle contraction. Sometimes these compounds are called active phosphate carriers and dischargers. Figure 4 summarizes the role of ATP in body energy and metabolism.

Energy Stored in Creatine Phosphate, or Phosphocreatine. Creatine phosphate is another energy-rich phosphate compound found in muscle. It can contribute to muscle energy metabolism in two ways:

$$\text{creatine phosphate} + H_2O = \text{creatine} + \text{P} + 7.5 \text{ kcal}$$

$$\text{creatine phosphate} + H_2O \xrightarrow{\text{ADP} \;\to\; \text{ATP}} \text{creatine}$$

In the first reaction the energy is released directly. In the second, the energy is transferred to ADP and later released for muscular or other work. Creatine phosphate is sometimes called an active phosphate carrier. Similar substances are 1,3-diphosphoglyceric acid and phosphoenolpyruvic acid.

Energy Stored in Active Acetate. The active acetate is the substance acetyl-CoA, which participates in intermediate metabolism. In terms of energy, formation of 1 mol of acetyl-CoA is equivalent to that of 1 mol of ATP.

Low-Energy Phosphate Compounds. Not all organic phosphates are of the high-energy type. Substances such as glucose-6-phosphate are phosphate compounds carrying a small amount of stored energy, such as 2–3 kcal/mol. If the above energy systems are summarized the oxidation or complete metabolism of glucose will yield the following:

$$C_6H_{12}O_6 \text{ (glucose)} + 6O_2 = 6CO_2 + 6H_2O + \text{heat} + 38 \text{ ATP}$$

Cellular Metabolism Processes

Within the cells, a series of complex biochemical processes is needed to degrade glucose (carbohydrate) to release the energy needed by the body. Three biological processes are involved: glycolysis, the citric acid cycle, and the respiratory chain. The glucose involved derives from food and internal production, which occurs mainly in the liver. In the cells of the liver, stored glycogen may be converted to glucose by the process of glycogenolysis; in the process of gluconeogenesis, glucose is synthesized from noncarbohydrate sources. However, as will be seen later, the muscles can also indirectly contribute the energy from cellular metabolism by participating in these processes. These processes are briefly defined in Table 4 and discussed in the following sections.

Glycolysis. The first step of carbohydrate metabolism is glycolysis, in which the six-carbon glucose is converted to a three-carbon substance (pyruvic or lactic acid), as indicated in Figure 5. Figure 6 illustrates the intermediate metabolic steps during the transformation of glucose to pyruvic acid. The interconversion between pyruvic and lactic acid occurs mainly in the muscle and will be discussed later. During the process of glycolysis, the conversion of 1

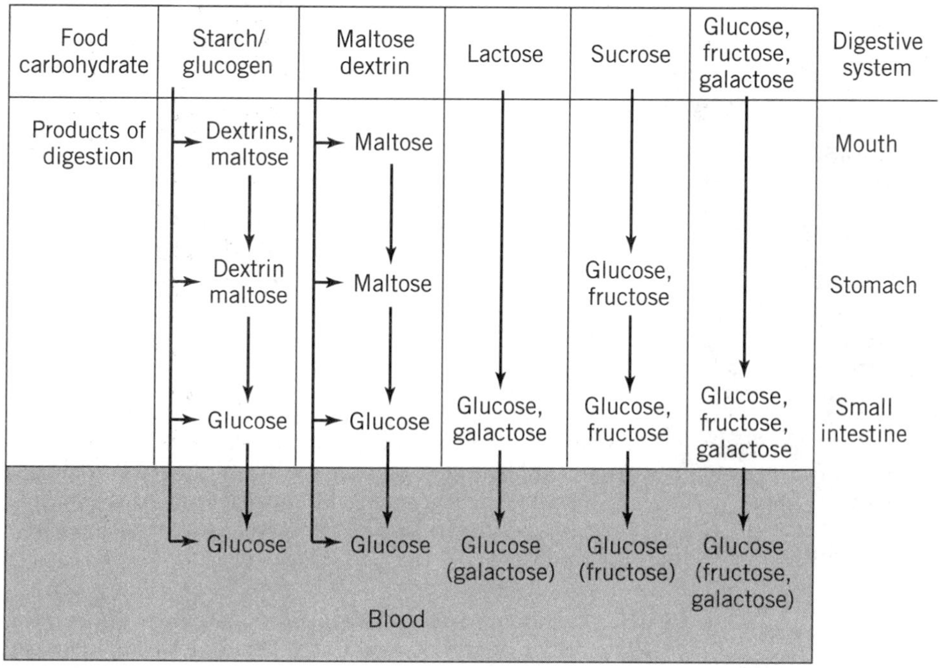

Figure 3. Carbohydrate digestion.

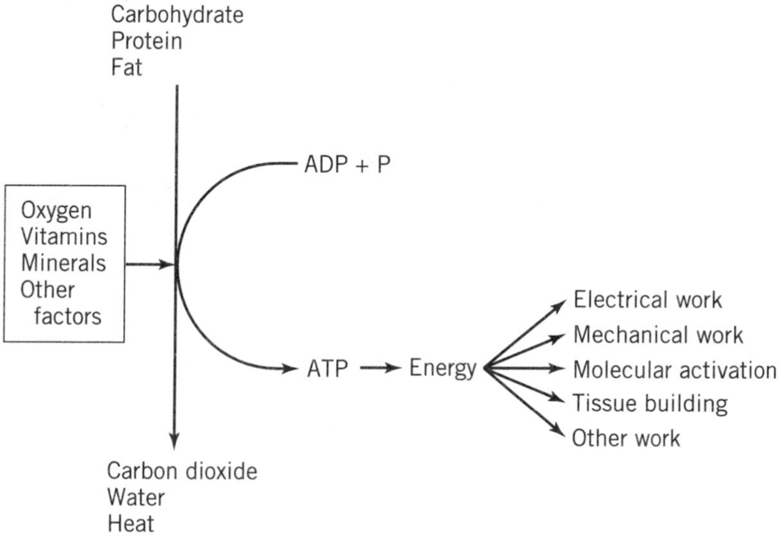

Figure 4. The role of ATP in body energy metabolism.

mol of glucose to 2 mol of pyruvic acid generates four hydrogen atoms and 8 mol of ATP. The hydrogen atoms released are eventually converted to water; the details are discussed below. Before the pyruvic acid can be converted to carbon dioxide, water, and energy, it must be transformed into a highly versatile metabolite, the two-carbon substance acetyl-CoA (Fig. 6). This transformation is irreversible.

Citric Acid Cycle. The citric acid cycle is a series of chemical reactions that metabolizes acetyl-CoA to carbon dioxide and hydrogen atoms, as indicated in Figures 7 and 8. In each cycle, two carbon dioxide molecules and four pairs of hydrogen atoms are put through the respiratory chain, together with the electrons, to generate 12 mol of ATP and 4 mol of water from oxygen. This cycle is the major link, or common path, in the transformation of carbohydrate, fat, and protein to carbon dioxide and water. Metabolites of the three nutrients enter the cycle at different strategic points. Because the cycle requires the respiratory chain to complete its work, it will not function in the absence of oxygen (anaerobically).

As indicated above, pyruvic acid may be removed from the glycolysis process by being converted to lactic acid. If so, there must be a source of hydrogen atoms, which are normally obtained from the production of phosphoglyceraldhyde (Fig. 6). In this case, glucose metabolism and energy (ATP) production can continue for a while without oxygen (that is, without going through the citric acid cycle). This anaerobic respiration occurs in muscle where an oc-

Table 4. Definitions of Some Metabolic Terms

Term	Definition
Glycolysis	The breaking down of hexoses (six carbon sugars), mainly glucose, into three-carbon substances (pyruvic or lactic acid); the process is sometimes termed the Embden-Meyerhof pathway
Glycogenesis	The formation of glycogen from glucose
Glycogenolysis	The breaking down of glycogen into glucose and its metabolites
Gluconeogenesis	The synthesis of glucose (and thus glycogen) from noncarbohydrate sources, such as lactate, glycerol, and amino acids
Citric acid cycle	Also termed the Krebs Cycle or tricarboxylic acid cycle; the process whereby carbohydrate, fat, and protein is completely oxidized to carbon dioxide, water, and energy; accomplished with the assistance of the respiratory chain
Respiratory chain	The transport of hydrogen atoms from biological oxidation for acceptance by oxygen atoms to form water molecules

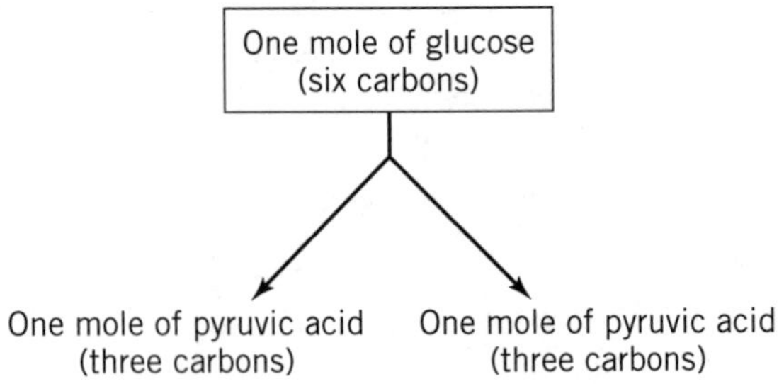

Figure 5. The overall result of glycolysis.

casional burst of energy is needed. The lactic acid that accumulates is converted back to pyruvic acid when the oxygen supply is restored, in which case the citric acid cycle is reactivated. The soreness of muscle from heavy work or exercise results from the presence of a large amount of lactic acid.

Respiratory Chain. Biological oxidation, or the respiratory chain, is a very complicated process whereby hydrogen atoms released by substances through oxidation are transported by a number of intermediates until the hydrogen atoms are accepted by oxygen to produce water. The respiratory chain involves both oxidation and reduction. Ox-

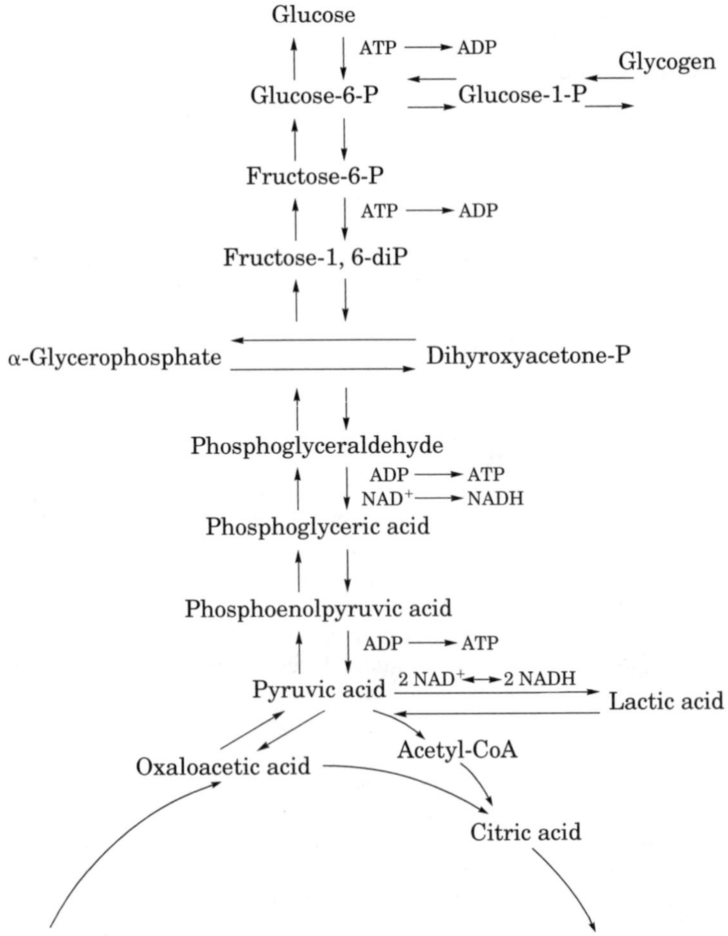

Figure 6. Reactions involved in glycolysis.

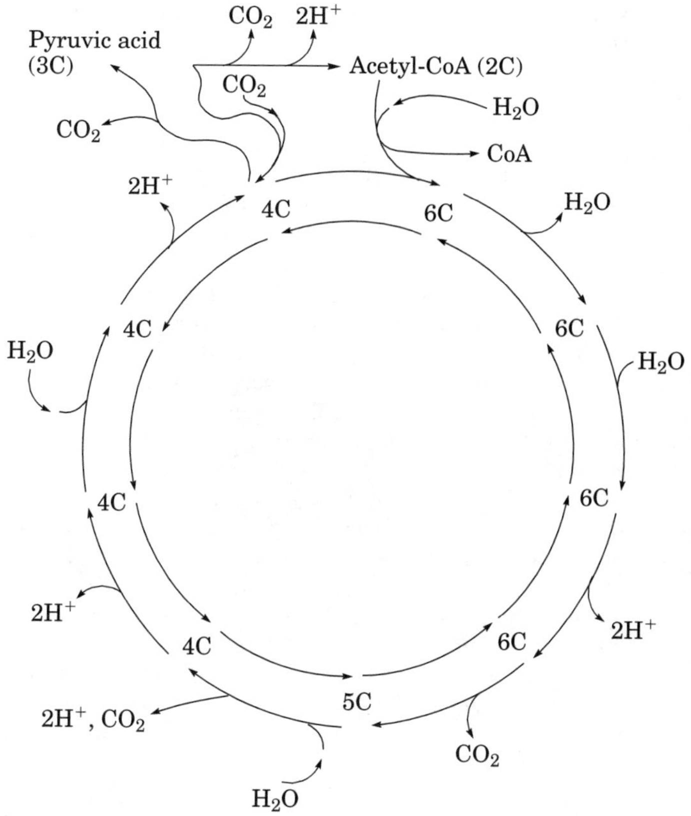

Figure 7. Outline of the citric acid cycle (C refers to a carbon atom).

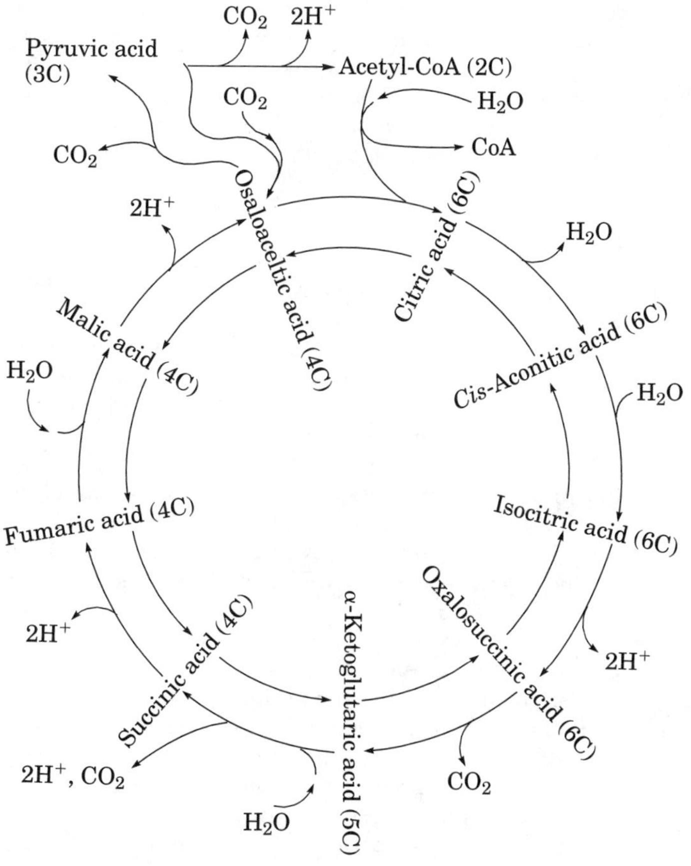

Figure 8. Components and products of the citric acid cycle.

idation is the process whereby a substrate either takes up oxygen or loses hydrogen. The substrate is oxidized, while the source substance that provides the oxygen or accepts hydrogen is the oxidizing agent. Reduction is the reverse process, whereby a substrate loses oxygen or accepts hydrogen. This substrate is reduced and the source substance that gains oxygen or loses hydrogen is the reducing agent.

The respiratory chain involves coenzymes (Tables 2 and 3) and consists of the following series of events:

1. Hydrogen atoms are released from a substrate, a process requiring an energy source.
2. The coenzyme NAD accepts hydrogen (NAD to $NADH_2$).
3. The hydrogen in NAD is accepted by FAD (FAD to $FADH_2$).
4. The hydrogen in $FADH_2$ is accepted by coenzyme Q.
5. The hydrogen in $CoQH_2$ is released as hydrogen by losing an electron, and a chain of cytochromes becomes reduced by accepting the electron.
6. The electron is transferred to molecular oxygen (O_2).
7. The negatively charged oxygen (O_2) reacts with two protons (H^-) to form water.

The entire process of the respiratory chain, including the formation of ATPs, is illustrated in Figure 9. The complete conversion of glucose to carbon dioxide, water, and energy is shown in Figure 10. The amount of energy used or stored is shown in Tables 5 and 6.

Glycogenesis and Glycogenolysis. In plants, carbohydrate is stored as starch, a polysaccharide of glucose. In animals, carbohydrate is stored in the form of glycogen, also a polysaccharide of glucose. The amount of glycogen stored in the body depends on the diet and the physiological status of the animal. In man, glycogen is produced and stored mainly in the liver and muscle. The process, known as glycogenesis, occurs readily when adequate glucose is present.

However, when the glucose concentration in the liver and muscle decreases, their glycogen content must be broken down to provide glucose for energy. This reverse process of glycogenesis is glycogenolysis. Its occurrence in the muscle is slightly different from that in the liver. In the liver, the glycogen can be directly degraded to glucose. However, in the muscle, the enzyme for the final step is missing so that glycogen is degraded only to glucose-6-phosphate, which has to be converted to pyruvic and lactic acids instead (Fig. 6). These can be converted to glucose via the citric cycle.

Gluconeogenesis. Gluconeogenesis is the synthesis of glucose (and thus glycogen) from noncarbohydrate sources such as amino acids, fatty acids, glycerol, and lactic and pyruvic acids. Figure 11 shows how muscle glycogen can be converted to glucose in the liver in spite of the muscle's lack of the appropriate enzyme. Figures 11 and 12 show how protein and fat can be converted to glucose. The interrelationship among the three nutrients, carbohydrate, protein, and fat, will be discussed at length in later sections.

Regulation of Blood Glucose

Another important aspect of what happens to carbohydrates in the body is the balancing of blood glucose (or blood sugar) levels. In a normal person, blood glucose fluctuates within narrow limits: between 70 and 100 mg/100 mL of blood. This is achieved by a balance between the supply and removal of blood glucose. If blood glucose drops below the norm, hypoglycemia occurs. In a healthy individual, the blood sugar is restored to normal by the provision of glucose from three sources. Simply eating additional carbohydrates increases the absorption of monosaccharides, and the liver can then release more glucose. Second, the glycogen in liver and muscle may be degraded (glycogenolysis) to form more glucose. Third, protein and fat may be degraded to provide glucose (gluconeogenesis).

If a person's blood glucose rises above the norm, hyperglycemia occurs. If the person is in normal health, the body spontaneously lowers the blood glucose levels in one or more of the following ways: (*1*) more insulin is released to drive glucose into cells for oxidation, (*2*) more glycogen is formed (glycogenesis) in the liver and muscle, (*3*) more glucose is changed to fat (lipogenesis) in fat cells, and (*4*) more glucose is excreted in the urine (glucosuria).

PROTEINS

Dietary protein exists in three forms. The major portion is conjugated with other substances, a small fraction is associated with fats and carbohydrates, and only a very small part exists as free protein, such as that in egg white. To be used by the cells, all proteins must be broken down into their constituent amino acids. In the sections that follow, the paths of protein digestion, absorption, and metabolism will be traced.

Digestion

Digestion of protein begins in the stomach (Table 1 and Fig. 13) where acid activates the pepsinogen (an enzyme) to release pepsin (another enzyme). The pepsin cleaves peptide linkages in the protein to produce polypeptides, each with two or more amino acids. When the stomach content reaches the duodenum, the pH is raised to about 6.5 by the presence of alkaline pancreatic juice. In the small intestine, chymotrypsin and trypsin hydrolyze most of the protein molecules to form small polypeptides and dipeptides. These are further digested to form free amino acids by the pancreatic enzyme carboxypeptidase and the intestinal enzymes aminopeptidase and dipeptidase. Those small peptides not split into individual amino acids may gain enter to mucosal cells to be digested later, for body tissues can utilize only amino acids.

Because protein molecules also contain nonpeptide linkages, they must be denatured first (by heat or stomach acid) before the appropriate enzymes can digest them. The denaturation process exposes the protein molecules, providing more surface area for enzymatic action. On the

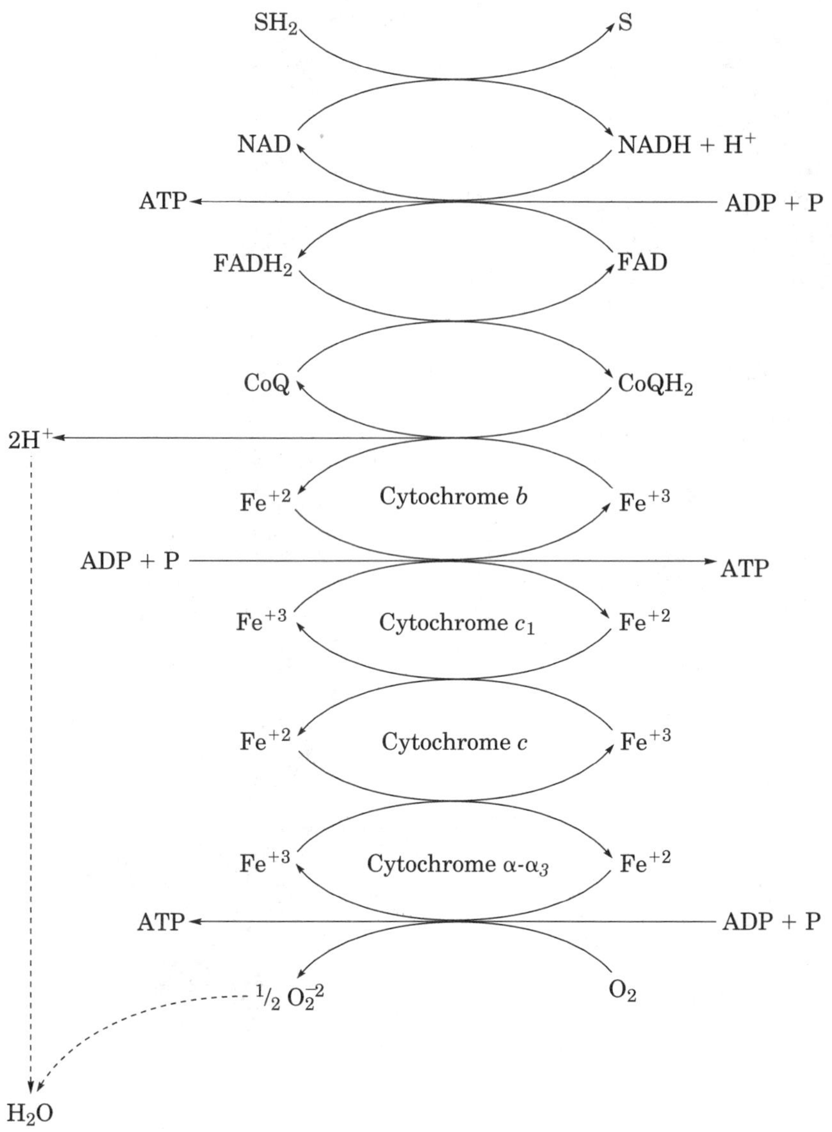

Figure 9. The respiratory chain or electron transport system (S = substrate).

other hand, excess heating or cooking can also reform some other linkages, making digestion more difficult.

Absorption

Normally only amino acids (and certain small peptides) are absorbed. In the small intestine all free amino acids, whether ingested or digested products, are absorbed via the mucosal cells into the hepatic portal vein. The D-amino acids are not absorbed as well as the L forms. The L-amino acids, the biologically active ones, are absorbed by active transport (a process requiring energy). Absorption occurs along the entire small intestine, with the slowest rate along the ileum. The stomach and colon may also absorb some amino acids. About 20–30% of ingested proteins are unabsorbed and excreted in the stools.

Although normally only amino acids are absorbed, it is well known that in small infants some undigested proteins are also absorbed. The subsequent antigen–antibody interaction causes the child to develop an allergic reaction when ingesting the same protein foods later. This explains the high allergic incidence among infants to foods such as eggs and cereals. If adults show allergy to ingested protein foods, they are still probably capable of absorbing whole protein molecules. For the majority of the population, this ability disappears with age.

Amino acids seem to be utilized best when they are absorbed in accordance with the body need for growth and function. Any excess absorbed will not be stored but excreted in the urine or metabolized to ammonia. Studies have shown that absorption and utilization of proteins are optimized when adults evenly distribute their protein intake throughout the day.

In some cases, erratic patterns of protein absorption may be due to stomach irregularities rather than irregular protein ingestion, for the stomach regulates the emptying of the nutrient into the small intestine. In patients with partially or completely removed stomachs, the rate of protein emptying is so disturbed that much of this valuable nutrient is lost in the fecal waste or degraded by intestinal

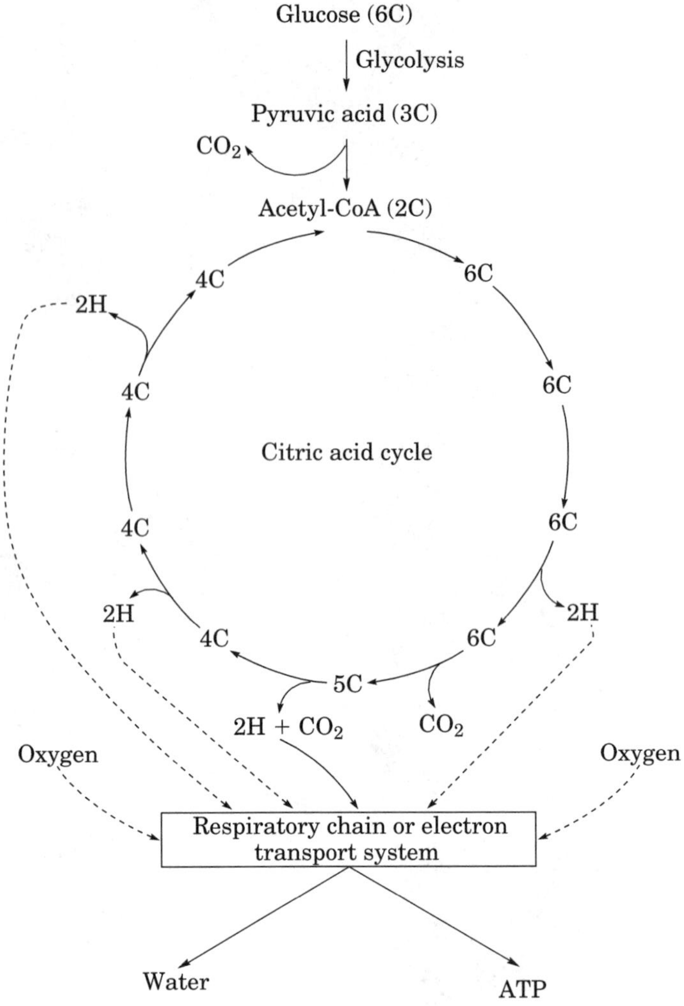

Figure 10. The three stages of converting glucose to carbon dioxide, water, and ATP.

Table 5. Energy Production during the Conversion of 1 Mol of Glucose to Carbon Dioxide and Water in the Presence of Oxygen

		Phosphorylation[a]			ATP	
Sequence	Product of specific reaction	SD	SP	OP	Loss	Gain
Glycolysis	Glucose-6-P	x			1	
	Fructose-1,6-diP	x			1	
	Phosphoglyceric acid		x			2
	Phosphoglyceraldehyde			x		6
	Pyruvic acid		x			2
Citric acid cycle	Acetyl-CoA			x		6
	Oxalosuccinic acid			x		6
	Succinyl-CoA			x		6
	Succinic acid		x			2
	Fumarate			x		4
	Oxaloacetate			x		6
Total number of ATP gained during the oxidation of 1 Mol of glucose under aerobic conditions						38

[a]SD = substrate dephosphorylation; SP = substrate phosphorylation; OP = oxidative phosphorylation.

Table 6. Energy Production during the Conversion of 1 Mol of Glucose to Carbon Dioxide and Water in the Presence of Oxygen

		Phosphorylation[a]			ATP		
Sequence	Product of specific reaction	SD	SP	OP	Loss	Gain	Hydrogen atoms
Glycolysis	Glucose-6-P	x			1		
	Fructose-1,6-diP	x			1		
	Phosphoglyceraldehyde			x			Released
	Phosphoglyceric acid		x			2	
	Pyruvic acid		x			2	
	Lactic acid						Accepted

[a]SD = substrate dephosphorylation; SP = substrate phosphorylation; OP = oxidative phosphorylation.

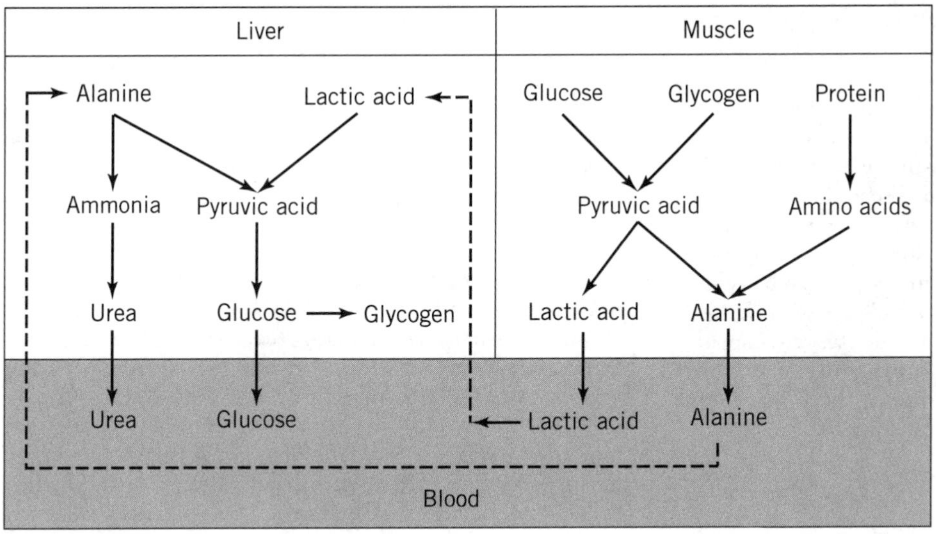

Figure 11. The role of liver and muscle in gluconeogenesis.

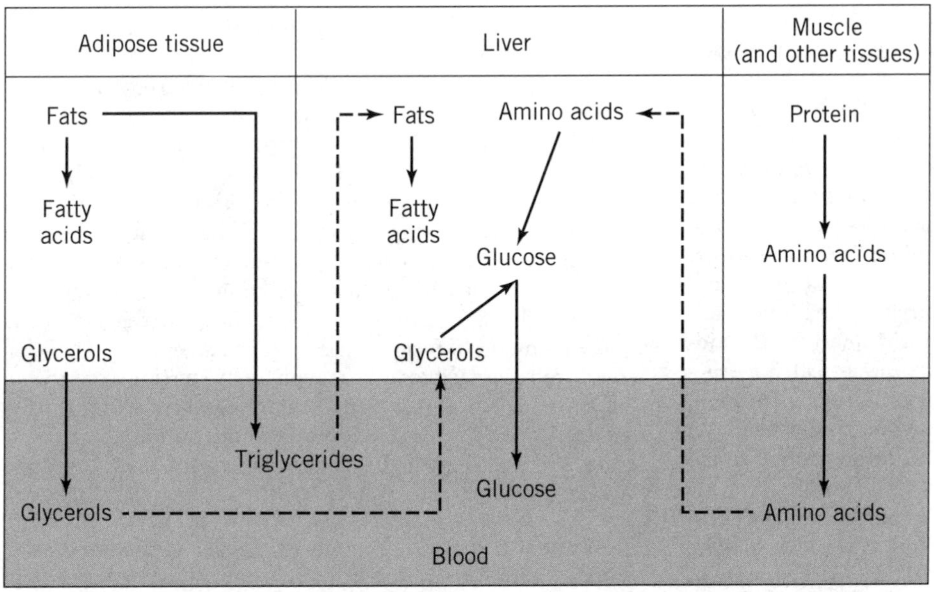

Figure 12. An overview of glucose formation from fat and protein.

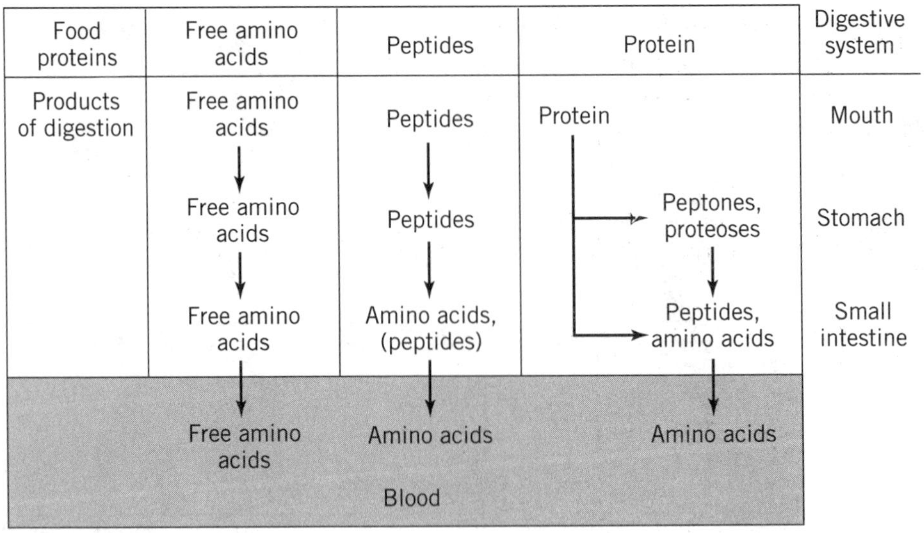

Figure 13. Protein digestion.

flora. When undigested amino acids are decarboxylated by intestinal bacteria, the important chemicals such as histamine, tyramine, ammonia, and similar substances are formed, many of which are absorbed. These substances are undesirable in excessive quantities. The ammonia formed by deamination plays a critical role in certain pathological conditions, for the absorbed ammonia can cause body deterioration. The absorption of amino acids is also impaired if the intestinal mucosa is damaged, as occurs in sprue, ulcerative colitis, and resection of a moderate amount of the small intestine. With time, however, there is a functional adaptation of the intestinal mucosa.

Metabolism

The liver may release some absorbed amino acids for body metabolism. Body amino acids may be catabolized or used in body repair, building, and maintenance. The importance of protein to the body is illustrated by its numerous functions and the complexity of its metabolism. If a normal person has a regular diet, three major factors determine the direction of protein metabolism: the quality and quantity of protein consumed, the amount of calories ingested, and the physiological and nutritional status of the body.

Protein metabolism revolves around a body pool of amino acids that are continuously released by protein hydrolysis and resynthesized. The amino acids ingested are the same as those released in the body. Body tissues can utilize only amino acids, not small peptides. About 50–100 g of body protein turns over daily, ranging from the slowest rate in the collagen to the fastest rate in the intestinal mucosa. Any amino acids filtered through the kidneys are reabsorbed, although certain congenital defects in the kidney tubules may interfere with this process. During pregnancy, infancy, childhood, and other conditions of growth, protein synthesis exceeds degradation. Individuals in these categories, therefore, require a large pool of amino acids.

Figure 14 illustrates the general catabolism, or degradation, of protein to amino acids and other metabolites. Figure 15 gives the general outline of protein formation in the body, and Figure 16 provides an overview of protein metabolism. The sections that follow discuss certain aspects of protein metabolism: degradation, protein synthesis, nitrogen balance, and metabolism of creatine and creatinine.

Protein Degradation. Protein is degraded to its individual amino acids in the muscle and other tissues, but the major site of actual destruction (catabolism) of each amino acid is the liver. The amino acids released from all other organs, especially the muscles, reach the blood and are diverted to the liver for degradation (Figure 14). Under normal circumstances, the catabolism of protein is balanced by its formation, although during stresses such as starvation and disease destruction can outstrip synthesis.

The procedure for the oxidation of an amino acid begins with deamination, a process in which the ammonia or amino group of the acid is removed so that only the carbon skeleton of the amino acid is left. Two metabolites are formed: the keto acids and ammonia (Fig. 14). The ammonia is changed to urea, which is released to the blood and eventually excreted in the kidney. Normally, we excrete about 20–30 g of urea in the urine each day. Because urea forms exclusively in the liver, advanced liver disorders raise the blood urea nitrogen level.

The keto acids formed from deamination enter the citric acid cycle to be oxidized. The exact process whereby keto acids and ammonia are formed varies with each individual amino acid, although the goal is the same. Figure 17 shows the points at which each of the acids and their keto acids enter the citric acid cycle or are transformed to pyruvate to form glucose. Within the citric acid cycle, the keto acids may be made to form glucose (gluconeogenesis). The keto acids are also tied to fat metabolism by the interconversion of keto acids and fatty acids. In sum, when protein is degraded amino acids are formed. Some amino acids circulate, some are oxidized, some are converted to glucose, and some are directed to other paths.

Protein Synthesis. Protein synthesis is the process of linking different amino acids by their amino groups (pep-

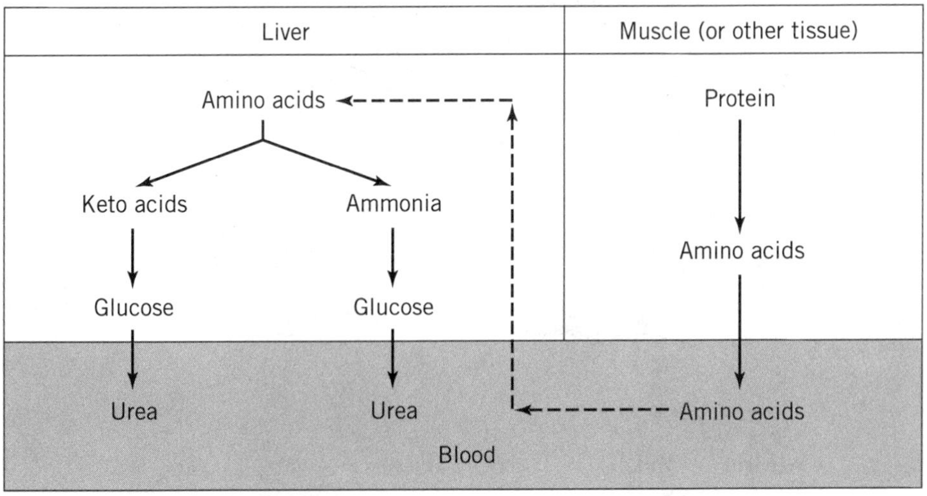

Figure 14. An overview of protein catabolism (degradation).

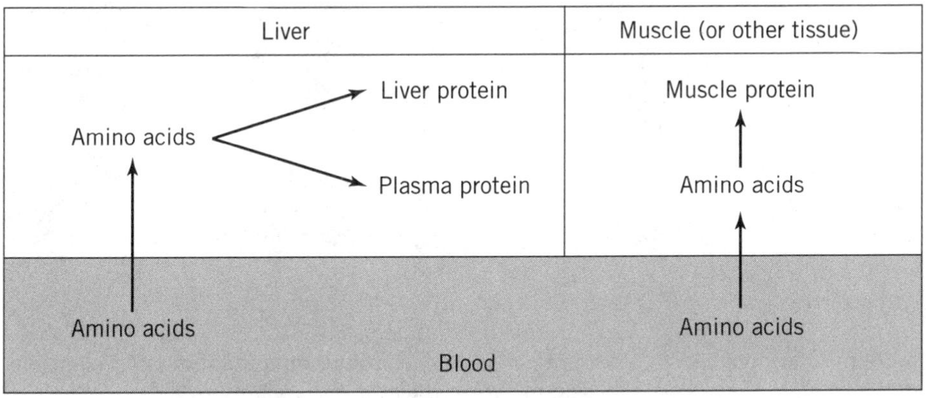

Figure 15. An overview of protein anabolism (synthesis).

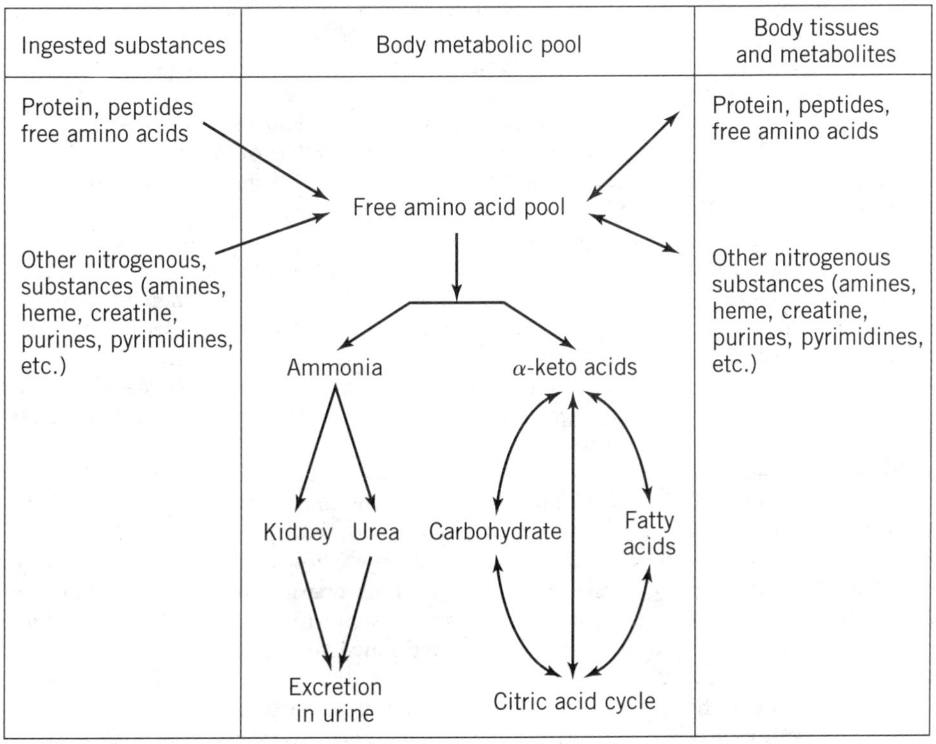

Figure 16. An overview of protein metabolism including nonprotein nitrogenous substances.

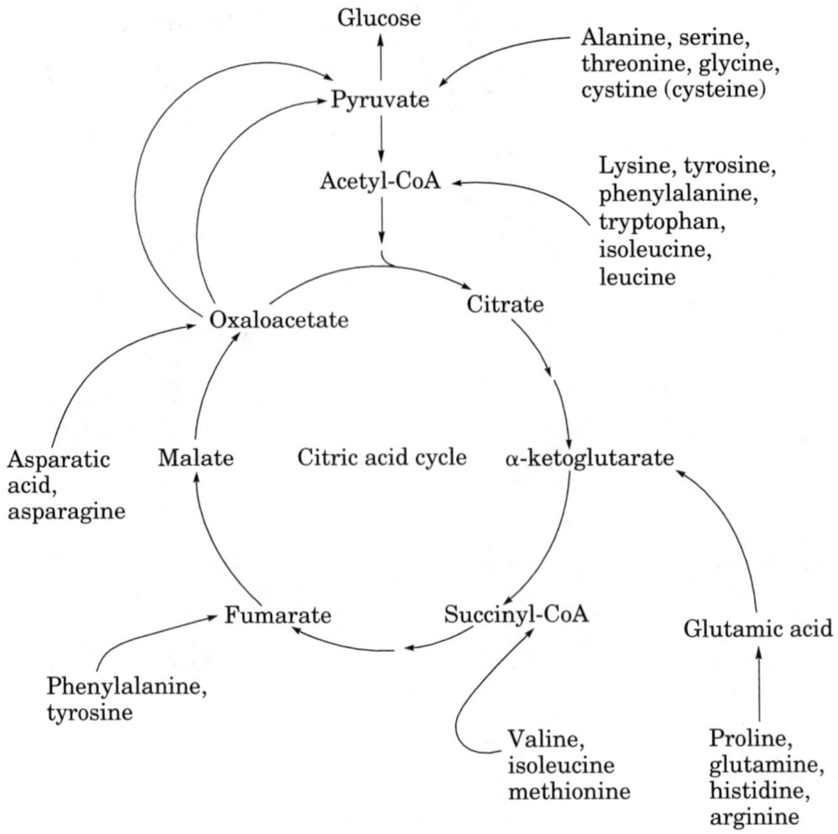

Figure 17. Introduction of the carbon structures of individual amino acids into the citric acid cycle.

tide linkages) to form a molecule of protein. This molecule must contain the right number of amino acids in the appropriate pattern and sequence needed for a specific body structure, such as hair, skin, muscle, tears, saliva, enzymes, or hormones. The nucleic acids DNA and RNA control this process, which takes place in cell nucleus cytoplasm. The formation of each specific protein requires a highly specific genetic code, the details of which may be obtained from a standard biochemistry or molecular biology text.

Although it occurs in nearly all cells, protein synthesis in some is more frequent and intense than in others. The organs that seem to synthesize the most protein are the liver, muscle, and those cells, tissues, and organs that manufacture or secrete enzymes, hormones, and other protein substances. One exception is the synthesis of plasma proteins by the liver. This occurs because red blood cells have no nuclei and are thus unable to synthesize protein themselves.

For a protein to be synthesized, the appropriate type and number of amino acids must be available. There are 8–10 amino acids whose carbon skeletons cannot be synthesized or manufactured by the body. They are called the essential amino acids and must be supplied in the diet. In addition, there are 10–15 amino acids that the body can manufacture (including their carbon skeletons), which are the nonessential amino acids. If the appropriate carbon skeleton of an amino acid is present, the body can add, subtract, and transfer the amino group until the right amino acid is formed. Protein synthesis is normally preceded by massive deamination, amination, and transamination to obtain the appropriate amino acids.

Nitrogen Balance. Of the three major nutrients protein is the most important in the sense that it makes up lean body mass. Another factor that makes protein important is that body protein turnover is tremendous. For example, the formation of hair, skin, saliva, and sweat all involve large losses of protein that must be constantly replaced. One acceptable technique to determine if there is enough protein in the diet is to measure the ingestion and excretion of nitrogen (nitrogen balance), because all protein has a relatively constant content of this element. Theoretically, if a young adult is in normal health and a proper stage of development, the amount of nitrogen consumed should be equal to the amount excreted. This is nitrogen equilibrium. However, depending on age, physiological condition, dietary intake, and other factors, some people are in positive nitrogen balance and others are in negative nitrogen balance.

Metabolism of Creatine and Creatinine. Body muscles perform work and may occasionally be required to provide a sudden burst of energy. They achieve this by the hydrolysis of two unique high-energy bonds. One is ATP and the other is phosphocreatine (or creatine phosphate), both of which were mentioned above. The body, mainly the muscles, contains about 100–150 g of phosphocreatine and creatine. Although not an amino acid, creatine is a unique nitrogenous chemical derived from three amino acids:

arginine, glycine, and methionine. Creatine, which is water soluble, is found in meat and meat products (such as extracts, soups, and gravies). Food contains little or no phosphocreatine, because of its easy degradation to creatine (or creatinine, another metabolite) and phosphoric acid (Fig. 18).

Phosphocreatine is a high-energy compound that provides instant energy when the muscles need it. If it is depleted, ATP is then used. Phosphocreatine seems to help maintain the ATP levels. After exertion and after all stored energy is used, creatine is rephosphorylated to form phosphocreatine.

Although an important body constituent, creatine is usually not excreted as such. Instead, the spontaneous, nonenzymatic dephosphorylation of phosphocreatine produces creatinine, in an irreversible process. Every healthy individual excretes a constant amount of creatinine, which is thus considered to be a normal waste product. The amount in the urine also reflects the amount of active muscle mass in the host. A woman excretes about 15–22 mg/kg, and a man about 20–26 mg/kg. Because creatinine is normally excreted rapidly, any increase in its level in the blood is a sign of kidney malfunction. Physicians attempting to make sure of the proper collection of urine frequently exploit the constancy of creatinine excretion. The volume of urine collected should reflect a 24-h excretion if the level of creatinine is within the normal range.

Although creatine is normally not excreted in the urine, children and women occasionally do dispose of the chemical in this manner. The excretion rate for women is especially high during and after pregnancy. The urinary level of creatine is also high in patients with diabetes, hyperthyroidism, and fever and those experiencing malnutrition or simple starvation. This reflects the degradation of musculature.

FATS

The digestion, absorption, transportation, and cellular metabolism of fats follow yet a third set of chemical and biological pathways in the body. However, they intersect the pathways of carbohydrates and proteins at several points. The following sections briefly describe the general fate of fats in the body.

Digestion, Absorption, and Transportation

Ingested fat meets its first significant digestive enzyme in the duodenum, where the exocrine gland of the pancreas provides the most important lipase (Table 1 and Fig. 19). The lipases in the saliva, stomach, and small intestine have only a small effect on fat digestion, as shown by the tremendous reduction in fat digestion when the pancreas is disabled. When the exocrine gland of the pancreas is not working properly, undigested and unabsorbed fat causes steatorrhea (bulky, clay-colored, fatty stools). The combined detergent actions of bile salts (from the gallbladder),

(a) Creatine synthesis

Glycine + amidine group from arginine $\xrightarrow{\text{Kidney}}$ Glycocyamine (guanidoacetic acid)

Liver | Methyl group from methionine or choline

↓

Creatine

(b) Phosphocreatine metabolism

ATP + Creatine $\xleftrightarrow{\text{Skeletal muscle}}$ ADP + Phosphocreatine + H_2O

Phosphocreatine + H_2O \longrightarrow Creatine + P + 7.5 kcal

(c) Creatinine formation

Physiological medium

Phosphocreatine $\xrightarrow[\text{Spontaneous, nonenzymatic}]{P}$ Creatinine

Laboratory conditions

Creatine $\xrightleftharpoons[\text{Alkali}]{\text{Acid}}$ Creatine

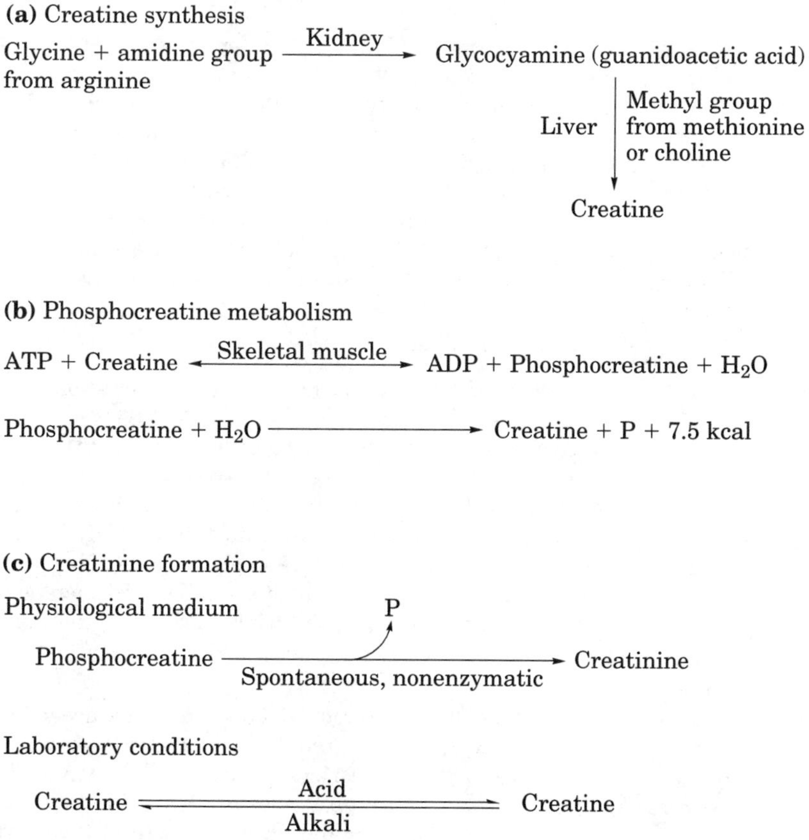

Figure 18. Metabolism of creatine and creatinine.

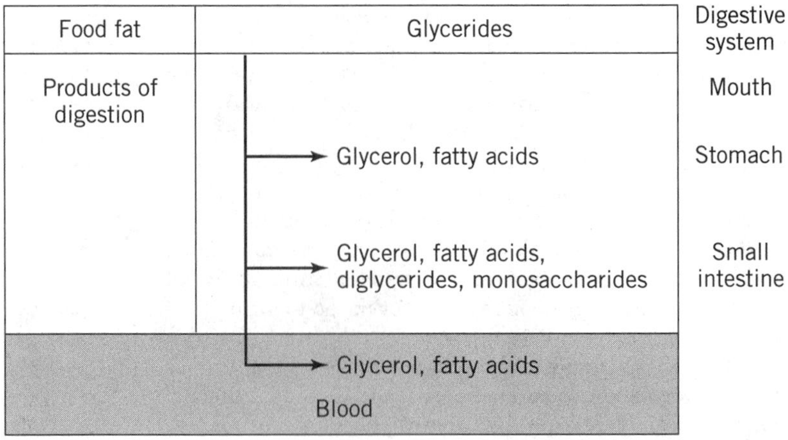

Figure 19. Triglyceride digestion.

fatty acids, and glycerides emulsify fat, thus facilitating its digestion by lipase.

In the small intestine, half of the ingested triglycerides are hydrolyzed by lipase to form free fatty acids and glycerols. The rest are changed to monoglycerides and a small amount of diglycerides. The monoglycerides are absorbed into the intestinal mucosa, where they are further hydrolyzed to glycerols and free fatty acids.

Free fatty acids in the intestine are absorbed in two ways. Fatty acids with less than 10–12 carbon atoms pass directly from the intestinal lumen, through mucosal cells, into the portal vein, and to the liver. Here some are released into circulation as free fatty acids, some are converted to triglycerides for deposition, and some are circulated in the blood as glycerides or as fatty acids that reside within the complex of lipoproteins. Fatty acids with more than 10–21 carbon atoms are absorbed into the mucosal cells, where they are regrouped with glycerols to form triglycerides. The triglycerides attach themselves to very low density lipoproteins to form chylomicrons, which enter the systemic circulation via the lymph and thoracic duct. Chylomicrons are fat globules 1 μm in diameter and visible under the microscope.

Other fatty substances are absorbed to varying degrees. For example, animal sterols are absorbed easier than plant sterols. Pancreatic secretion, fatty acids, and bile salts, which together emulsify and esterify cholesterol, are necessary for cholesterol absorption of cholesterol. It is currently believed that cholesterol is absorbed mainly in the ileum. Like the triglycerides, absorbed cholesterol is incorporated into the chylomicrons, which reach the systemic circulation.

Within two to three hours after the ingestion of food containing short-chain fatty acids, the blood level of chylomicrons remains unchanged, although it may rise sharply if the meal contains long-chain fatty acids. Normally, after a mixed meal, the plasma develops a milky appearance because of the presence of chylomicrons in the blood. This is sometimes known as lipemia. In the presence of the enzyme lipoprotein lipas, these plasma chylomicrons are cleared and their contents diverted to the liver and adipose tissue.

Blood plasma, therefore, contains fat in the following forms: fatty acids, glycerol, glycerides, cholesterol, cholesterol esters, and phospholipids. These forms are bound to the albumin, α-globulin, and β-globulin fractions of the plasma proteins. The resulting lipid-protein complexes have varying densities. The highest densities occur in those with the most protein and least lipid; the lowest densities occur in those with the least protein and most lipid. Consequently, the complexes are classified into high-density, low-density, and very low density lipoproteins. In general, very low density lipoproteins carry mainly triglycerides; low-density lipoproteins carry mainly cholesterol; and α-lipoproteins carry phospholipids, albumin, and free fatty acids. For a normal person, about 95% of ingested fat is absorbed, mainly in the duodenum and jejunum, with some absorption by the ileum. About 5% of fecal waste is fat, which comes from the diet, cell debris, and bacterial synthesis.

Although most of the fats are emptied into the lymphatic system after absorption and eventually reach the systemic circulation, the bile salts separate from the fats and travel through the portal vein into the liver. There they are reincorporated into the bile. Bile salts are thus cycled through the enterohepatic circulation (the liver, gallbladder, intestinal lumen, portal vein, and back to the liver). About 80–90% of bile salts in the intestinal lumen are reabsorbed in this way; the rest are lost in the stool.

Cellular Metabolism

The adult body distributes fats to two main locations: the membranes and other structural parts of cells (commonly called structural fats) and the fat cells (neutral fats), which are mainly white. Infants have some brown fat cells, which can regulate body temperature by producing heat to support the baby's higher metabolic rate. Neutral body fat contains mainly triglycerides, plus small amounts of diglycerides and monoglycerides, which are important metabolic intermediates. Consequently, triglycerides are the main form of stored energy.

Fat Degradation. Stored fat is degraded as needed to provide energy. Fat degradation occurs in two major stages: hydrolysis of glycerides and oxidation of fatty acids. In the adipose tissues, glycerides are hydrolyzed by a lipase to form fatty acids and glycerols. Both of these are released into the circulation for transport to the liver, where further hydrolysis may occur. When triglycerides are hydrolyzed, the released glycerols can be converted to phosphoglyceraldehyde in the liver (Fig. 6 and 20). This compound can in turn be converted to either carbon dioxide and water or glucose.

The process of oxidizing the fatty acids to carbon dioxide, water, and energy is called β-oxidation, or alternate oxidation. It occurs mainly in the mitochondria of liver cells. The carbon chain is broken down by the successive removal of two-carbon fragments from the carboxyl end to form acetic acids. These can combine with CoA to form acetyl-CoA, which can enter the citric acid cycle to be oxidized (Figs. 8, 17, and 20). When the fatty acids are reduced to acetyl-CoA, hydrogen atoms are also released, which can be passed on to the respiratory chain. When fatty acids are completely oxidized, they generate more ATP than the molecular equivalent of carbohydrate because less oxygen is present. This explains why fat has a higher caloric value. However, unsaturated fatty acids generate less energy than the molecular equivalent of saturated fatty acids because less hydrogen is present in the former.

Most of the naturally occurring fatty acids are even numbered, and thus their oxidation always produces acetyl-CoA at the end. However, if the fatty acids happen to be odd chained, propionyl-CoA is formed instead. Propionyl-CoA can also enter the citric acid cycle if the coenzyme with vitamin B_{12} is available.

Fat Synthesis. Fat synthesis takes place in two major stages: the formation of fatty acids and the formation of triglycerides. Fatty acids synthesis is achieved in two places: the mitochondria and the cytoplasm. Within the mitochondria, β-oxidation is reversed and two-carbon units are added until the appropriate fatty acids are formed. Outside the micochondria, in the cytoplasm, another form of fatty acid synthesis occurs. Here the starting compound is acetyl-CoA which serves as the end of the fatty acid molecule. The remaining carbons are incorporated as two-carbon units derived from the malonyl group. The incorporation is accompanied by simultaneous recarboxylation. The fatty acids formed are mainly 12–14 carbons long and rarely more than 16. The body can synthesize unsaturated fatty acids from the saturated ones by removing hydrogen, although it is unable to synthesize the essential ones.

In the adipose tissues, fatty acids combine with glycerol to form triglycerides, or neutral fats. This reaction occurs in the mitochondria. Figure 20 summarizes the information on fat synthesis.

As indicated earlier, glycerol can be converted to glucose (gluconeogenesis). However, acetyl-CoA cannot be converted to pyruvic acid (Fig. 10). Although keto acids can enter the citric acid cycle, there is very little net conversion

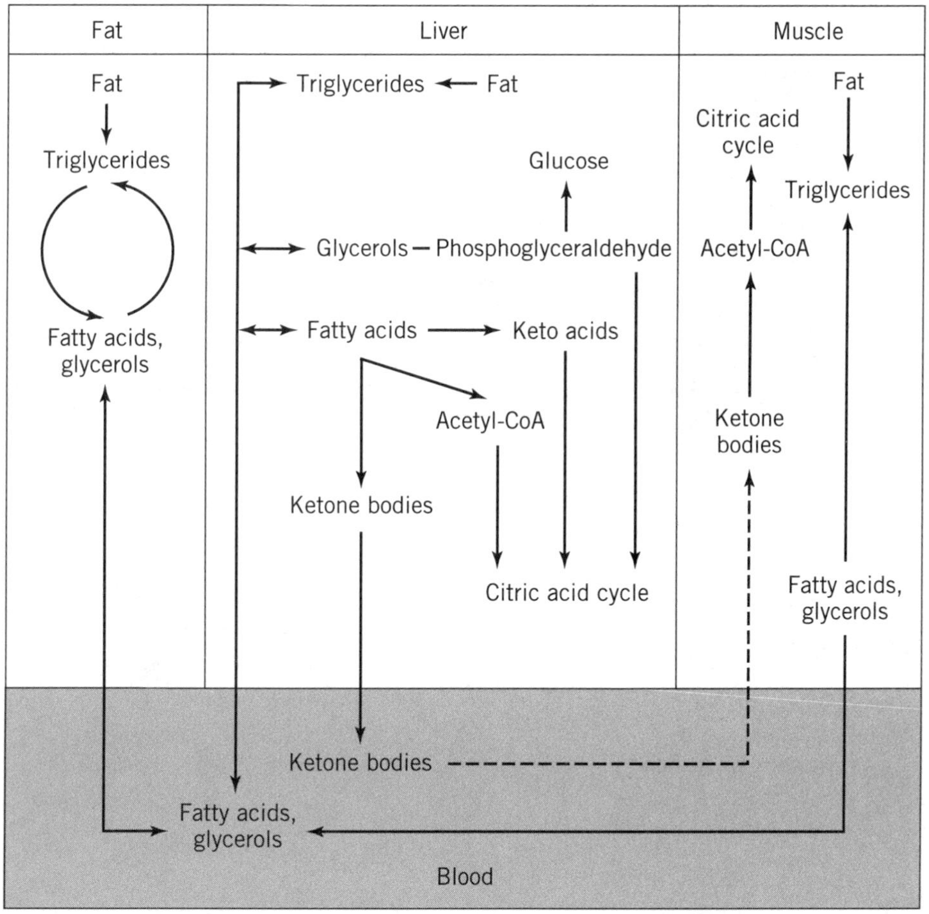

Figure 20. An overview of fat metabolism.

of fat to carbohydrate in the body, with the exception of the small amount of phosphogyceraldehyde formed from glycerol.

Ketone Bodies. During the normal process of β-oxidation of fatty acids, the liver has the appropriate enzyme to remove the CoA from acetoacetyl-CoA to form acetoacetic acid. Acetones and β-hydroxybutyric acids can be formed from acetoacetic acids. The last three compounds are collectively called ketone bodies. The small amount of ketone bodies normally made by the liver is transported by the circulation to the muscle for conversion to acetyl-CoA, which is put through the citric acid cycle (Fig. 20). Acetone is eliminated via urination and respiration. Because under normal circumstances the ketone bodies are metabolized as soon as they are formed, a person rarely excretes more than 1 mg of ketone each day, and blood levels are usually less than 1 mg/100 mL.

However, the ketones can accumulate under certain conditions, and the resulting clinical condition is known as ketosis. The main cause of ketosis is the accumulation of acetyl-CoA because the citric acid cycle in the liver is not operating at its normal or optimal efficiency. The most common cause is a sequence of events called intracellular carbohydrate starvation. First, decreased supply of glucose leads to a reduction in pyruvic acid, acetyl-CoA, and cellular energy supply. Second, for compensation, fatty acid oxidation is increased to provide energy with an accumulation of acetyl-CoA. Third, the oversupply of acetyl-CoA leads to the formation of ketone bodies.

Glucose supply to cells is reduced in people with diabetes mellitus and people who undergo dietary alterations

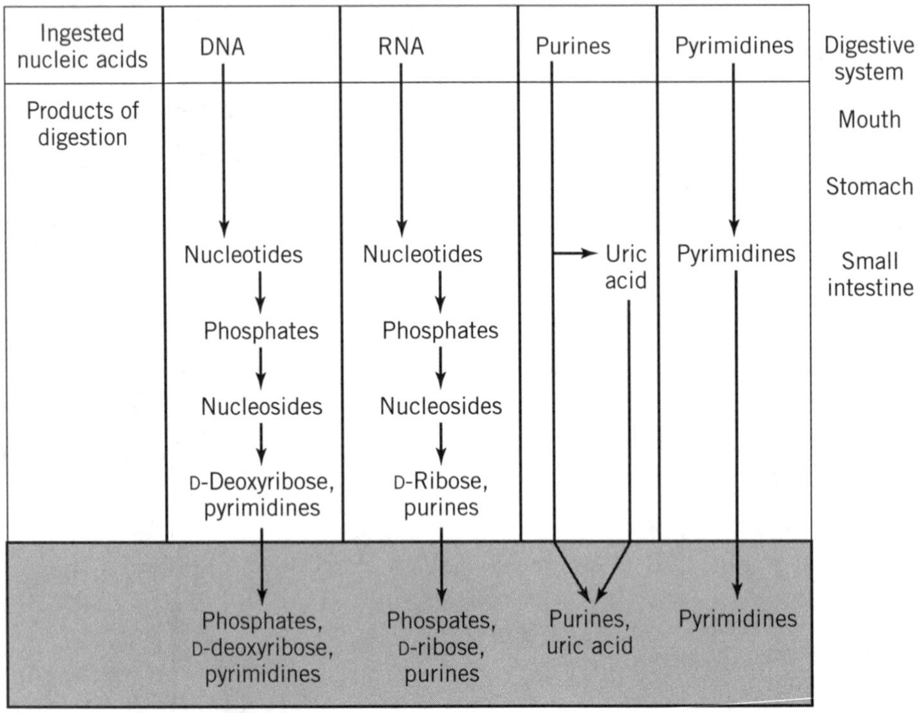

Figure 21. Digestion of nucleic acids.

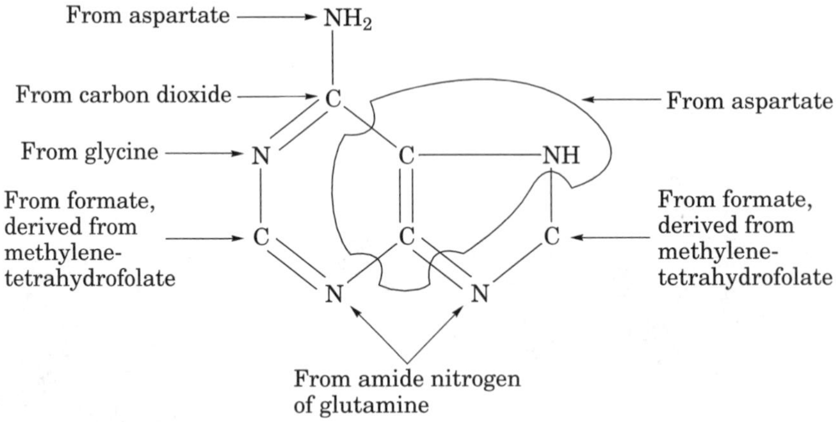

Figure 22. Origins of the atoms in purine.

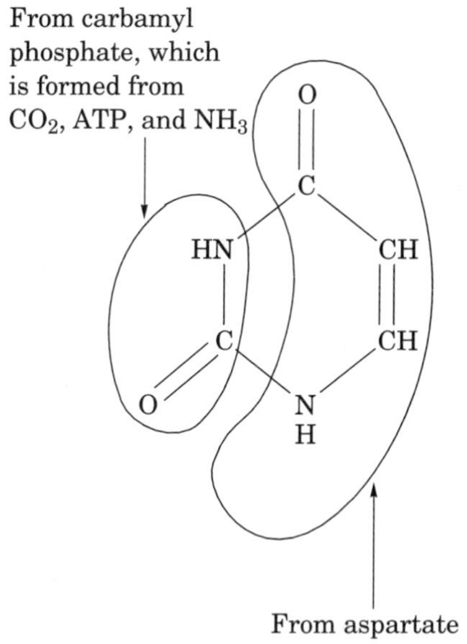

Figure 23. Origins of the atoms in pyrimidine.

such as high-fat—low-carbohydrate intake or simple starvation. In a diabetic patient, the lack of insulin prevents glucose from entering cells. When a persons's diet is low in calories, high in fat, or low in carbohydrate, a similar metabolic pattern takes place. The inadequate intake of carbohydrate means a low supply of glucose to cells, and ketosis may develop. However, an intravenous introduction of glucose counteracts ketosis, which is why carbohydrate is an antiketogenic agent.

Cholesterol Metabolism. Body metabolism of cholesterol is of special concern to nutritional scientists. Dietary cholesterol comes mainly from animal products such as fats, eggs, and organ meats. Sterols are obtained from plant foods, although they are not absorbed by the human digestive system. Ingested cholesterol is readily absorbed via the lymphatic system after esterification in the intestinal mucosa (with fatty acids). The body can also synthesize cholesterol, mainly in the intestinal mucosa and liver. It is currently believed that the amount of cholesterol synthesized by the body is inversely related to the quantity consumed. However, the problem of regulating serum cholesterol level by reducing dietary intake is a much debated issue.

Synthesis and degradation of cholesterol occur simultaneously and continuously. The body removes cholesterol by conjugating it with taurine or glycine in the liver and excreting it in the bile, although the enterohepatic circulation makes sure that some cholesterol is reabsorbed.

NUCLEIC ACIDS

Practically everything humans eat contains nucleic acids, which occur in cell chromosomes; nucleic acids are responsible for heredity. During digestion, ingested nucleic acids are initially cleaved into nucleotides by pancreatic nucleases (Fig. 21 and Table 1). Next, the small intestine secretes nucleotidase, which hydrolyzes nucleotides to form phosphoric acid and nucleosides. The latter are split by intestinal nucleosidases to form sugars, purine, and pyrimidine, all of which are absorbed by active transport. Undigested large molecules of nucleic acids are excreted in the stool.

Nucleic acids can also be synthesized. In the body, purines, pyrimidines, and sugars are put together to form ribonucleic acid (RNA), deoxyribonucleic acid (DNA), nicotinamide adenine dinucleotide (NAD), and other related substances. However, the liver can also make pyrimidines and purines. Figure 22 shows the origins of atoms in synthesized purine, and Figure 23 those of pyrimidine.

Within a cell, although DNA is stable throughout life, RNA is in constant equilibrium with a metabolic pool. Purines and pyrimidines may be excreted as such in the urine or be metabolized to uric acid (purines) or carbon dioxide and ammonia (pyrimidines).

Uric acid in the body comes from two sources: synthesis from glycine and degradation of purines. In humans, uric acid is excreted in the urine, although in most other mammals it is converted to allantoin before excretion. The normal blood level of uric acid is 4 mg/100 mL. The kidney reabsorbs much of the filtered uric acid, but the body excretes about 1 g of uric acid in 24 h. A standard reference text should be consulted for additional information on DNA, RNA, and molecular genetics.

WATER, VITAMINS, AND MINERALS

From the stomach to the colon, water passes freely and reversibly between the intestinal lumen and body compartments, although less so in the stomach than elsewhere in the gastrointestinal tract. Water moves in or out of the intestinal lumen to insure osmotic equilibrium on the two sides. In general, the osmolality of the contents of the small intestine resembles that of the plasma. After nutrients in the intestine have been absorbed, the excess water in the lumen is passed out with fecal waste to maintain osmotic equilibrium. Sodium moves freely according to the concentration gradient on the two sides of the mucosal cells. In the colon, sodium moves from body to luman in accordance with the osmotic gradient.

All water-soluble vitamins are absorbed along the small intestine. Except for vitamin B_{12}, a healthy person can absorb these vitamins rapidly. All fat-soluble vitamins require the presence of pancreatic enzyme, bile salts, glycerides, and fatty acids for absorption, as does fat itself. In a healthy person, all essential minerals are absorbed easily by the body, although the extent varies with individual minerals.

BIBLIOGRAPHY

This article has been adapted from Y. H. Hui, *Human Nutrition and Diet Therapy*. Jones and Bartlett, Boston, 1983. Used with permission.

GENERAL REFERENCES

D. A. Bender, *Amino Acid Metabolism*, 2nd ed., John Wiley & Sons, Inc., New York, 1985.

E. Braunwall and co-workers, *Harrison's Principles of Internal Medicine*, 11th ed., McGraw-Hill, Inc., New York, 1987.
M. D. Cashman, "Principles of Digestive Physiology for Clinical Nutrition," *Nutr. Clin. Pract.* **1**, 241 (1986).
J. J. Cunningham, *Introduction to Nutritional Physiology*, G. F. Stickley, Philadelphia, 1983.
G. L. Eastwood. *Core Textbook of Gastroenterology*, Lippincott, Philadelphia, 1984.
S. I. Fox, *Human Physiology*, 2nd ed., W. C. Brown, Dubuque, Iowa, 1987.
E. Goldberger, *A Primer of Water, Electrolyte and Acid-Base Syndromes*, 7th ed., Lea & Febiger, Philadelphia, 1986.
A. Guyton, *Textbook of Medical Physiology*, 3rd ed., Saunders, Philadelphia, 1986.
R. V. Heatley and co-workers, eds., *Clinical Nutrition in Gastroenterology*, Churchill, London, 1986.
Y. H. Hui, *Human Nutrition and Diet Therapy*, Jones and Bartlett Publishers, Boston, 1983.
J. O. Hunter and V. A. Jones, eds., *Food and the Gut*, Bailliere Tindall, London, 1985.
M. C. Linder, *Nutritional Biochemistry in Metabolism*. Elsevier Science Publishing Co., Inc., New York, 1985.
R. K. Murray and co-workers, *Harper's Biochemistry*. Appleton & Lange, Norwalk, Conn., 1988.
R. Pike and M. Brown, *Nutrition: An Integrated Approach*, 3rd ed., John Wiley & Sons, Inc., New York, 1985.
W. K. Stephenson, *Concepts in Biochemistry*, 3rd ed., John Wiley & Sons, Inc., New York, 1988.
L. Stryer, *Biochemistry*, 3rd ed., W. H. Freeman, New York, 1988.
G. V. Vahouny and co-workers, eds., *Dietary Fiber: Basic and Clinical Aspects*, Plenum Press, New York, 1986.
J. West, *Best and Taylor's Physiological Basis of Medical Practice*, 11th ed., Williams & Wilkins, Baltimore, Md., 1985.
P. C. Wilson and H. L. Greene, "The Gastrointestinal Tract: Portal to Nutrient Utilization," in M. E. Shils and V. R. Young, eds., *Modern Nutrition in Health and Disease*, 7th ed., Lea & Febiger, Philadelphia, 1988.

Y. H. Hui
American Food and Nutrition Center
Cutten, California

FOODBORNE DISEASES

All through history, human beings no doubt were affected by a great variety of foodborne diseases through consumption of water and food. No one knows for certain the number of cases of foodborne intoxications and infections occurring annually in the world. In the United States the Centers for Disease Control and Prevention reported that between 1988 and 1992 a total of 2423 outbreaks of foodborne diseases with 77,373 cases of illness. Among outbreaks for which the etiology is known, bacterial pathogens caused 79% of the outbreaks and 90% of the cases. Since many outbreaks and cases are not reported, the estimation of foodborne cases is between 6 and 33 million cases per year and from 525 to 9000 deaths. The total cost of productivity losses from a few major pathogens ranges from $5 to $6 billion per year. Class 1 recalls of food products for life-threatening bacteria increased from 79 recalls in 1988 to 378 recalls in 1995. In countries with poor sanitation, one can only surmise that the number of foodborne disease cases is much higher. There is a heightened awareness of the role of foodborne diseases by consumers in the United States due to some sensational outbreaks of foodborne diseases affecting a large number of people and the deaths of children after the consumption of undercooked hamburger or contaminated cheeses. As a result, the government has implemented greater monitoring programs and Hazard Analysis Critical Control Point (HACCP) procedures in the food industry to attempt to curtail the outbreaks and protect the safety of consumers. Consumers are much more aware of the great potential for large-scale foodborne outbreaks and demand a safer food supply. At the same time consumers are also demanding more fresh foods, minimally processed food, and organic foods where control of foodborne pathogens are more difficult. There is also a drastic demographic change in the society where more and more people live longer and thus have lower resistance to diseases in general, and more people are at risk due to immunocompromised diseases and conditions that make them more susceptible to foodborne diseases. Food distribution systems also have been greatly improved, and thus production of food in one location can be transported to hundreds and thousands of miles in a short time. When a problem occurs, the amount of food involved can be astronomical, such as a case of recalling 25 million pounds of ground beef due to one contamination source. The company involved is no longer in existence as a major player in food supply. Another important development is international trade. Vast amounts of food are regularly shipped from one country to another with minimal monitoring of the microbial safety of the food involved. To complicate matters further, there are actually now microorganisms that are emerging or reemerging in the food supply and make tracking and controlling of these organisms more difficult. Fortunately, with new developments in microbial detection methods and systems these microorganisms are being detected more frequently with better accuracy and rate (see the article Rapid methods of microbial analysis). Also there are better and more efficient intervention strategies and methods in food processing to control unwanted microorganisms (see the articles Food fermentation and Microbiology of foods). Thus food microbiologists, food scientists, epidemiologists, medical personnel, public health workers, and consumer educators are charged with the responsibility of studying the occurrence, enumeration, isolation, detection, characterization, prevention, reporting, education about, and control of foodborne microorganisms in and from food, water, and the environment nationally and internationally.

DEFINITIONS

Food intoxication is the ingestion of toxic compounds in foods, from chemical contamination or preformed by toxigenic microorganisms, by susceptible persons who later became ill.

Food infection is the ingestion of large numbers of viable, pathogenic microorganisms in food by susceptible persons who later became ill. Usually the number is around one million live organisms per gram of food, but there are cases that at low as 100 cells can become infectious.

Food poisoning is the ingestion of contaminated food containing either chemical preformed toxins or live microbes by susceptible persons who later became ill.

Foodborne outbreak is the consumption of contaminated food from one source by two or many people who later became ill. In the case of botulism one affected person will constitute one outbreak.

Foodborne disease case is the consumption of contaminated food by one susceptible person who later became ill. A foodborne outbreak can have 2 cases or 100,000 cases, with the exception of botulism, when one case is considered one outbreak.

Endemic is the usual cases of a particular illness in a community.

Epidemic is an unusual, large number of cases of a particular illness from a single source in a community.

Pandemic is a disease affecting the entire world.

Epidemiology is the study of diseases in a population using statistical methods. An epidemiologist studies patterns of diseases and their causative agents in terms of a population, whereas a physician treats individual patients.

Etiologic agent is the agent that caused a specific disease.

FOOD INTOXICATIONS

Chemical intoxications are usually the result of accidents. People have been poisoned by inorganic compounds such as antimony, arsenic, cyanide, cadmium, lead, selenium, and mercury. The symptoms usually occur rapidly (a few minutes or hours), and reactions are usually violent in cases of ingestion of large doses of the toxic compounds. Immediate medical assistance is essential for the victims in such cases.

Long-term chemical intoxication is also possible by the ingestion of a small amount of toxins in food or water over many years and later the person became ill. There are also many naturally occurring chemical toxins in foods as well as unintentionally added chemical toxins such as pesticides and chemical residuals from packaging materials and the environment.

BACTERIAL AND MICROBIAL INTOXICATION

This form of intoxication is the result of consumption of preformed toxic compounds in the food by susceptible persons. The effects can be rapid (within hours) but usually are longer than chemical intoxication.

Clostridium botulinum

The first recorded outbreak of botulism was in 1793 involving sausages (*botulus*) in Germany. Since that time many outbreaks all over the world have been reported. From 1899 to 1977 there were 766 outbreaks involving 1961 cases and 999 deaths. In the United States between 1971 and 1985, three outbreaks were recorded with 485 cases and 55 deaths. In 1990 there were 12 outbreaks, 22 cases, and 5 deaths and in 1994 there were 42 cases of foodborne botulism, 86 cases of infant botulism, and 11 cases of wound botulism with no fatality. The general public usually have panic in the report of botulism due to its high fatality rate. The organism is a Gram-positive, anaerobic, spore-forming rod that can grow at temperatures from 3.3°C to as high as 50°C. Most strains will grow well at 30°C, with optimum temperature at 37°C. To control the growth of the organism the pH must be below 4.6, salt content of 10% and above, and a water activity of less than 0.94. The vegetative cells of this organism are easily killed by heat, but the spores formed by this cell are far more resistant to heat, cold, acid and basic chemicals, radiation, and other forms of preservation methods. Thus control of botulism is geared to the destruction of the spores. Time and temperature combinations of canning of all foods are designed to kill the most heat-resistant spores of *Clostridium botulinum*. The spores formed by this organism can reside in soil, water, and the environment and can be transmitted to foods. Foods involved in botulism cases usually are improperly home-canned medium- or low-acid foods. Information since 1899 indicates that about 70% of the outbreaks can be traced to improperly processed home-canned foods and 9% to commercially processed food, with the other outbreaks from unknown sources. Symptoms develop from 18 to 96 h after ingestion of toxic foods. They include vomiting, nausea, fatigue, dizziness, vertigo, headache, dryness of mouth, muscle paralysis, and death by asphyxiation. Since the toxin affects peripheral nerves the patient is alert until the moment of death. There are several types of botulin toxins (types A, B, C_1, C_2, D, E, F, and G). These are large molecular weight proteins (about 1 million dalton). The important toxins affecting human beings are toxins A, B, and E, and rarely F. These toxins are among the most toxic materials produced by a biologic system. It was estimated that one pure ounce of toxin can kill 200 million people. Treatment is by administration of monovalent E, bivalent AB, trivalent ABE, or polyvalent ABCDEF antisera. Fortunately, the toxins are heat sensitive. Boiling of the toxin for 10 min will destroy it. The toxins can be detected by animal tests using mice as well as immunologic tests using specific antibodies (gel diffusion tests, ELISA, RIA tests, etc). Recently a rapid polymerase chain reaction (PCR) method has been employed to detect the botulin gene harbored by cultures isolated from foods. The information, however, does not directly imply that the food is toxigenic and harbors the toxin. The key to preventing botulism is to know the composition (pH, A_w oxidation-reduction potential, presence of inhibitory compounds, etc) of the food and to utilize proper time and temperature for processing as well as correct packaging and storage of the processed food. All high-moisture, low-acid foods processed and then stored under anaerobic conditions either in cans, glass bottles, or pouches should be subject to close scrutiny to avoid the possibility of *C. botulinum* surviving and later germinating and producing the toxins. Since there are proteolytic and nonproteolytic strains of *C. botulinum* the absence of off-odor of a suspected canned food cannot guarantee the safety of the food.

Never taste a suspected food or use food from swollen, dented, or deformed cans. When in doubt, always boil the suspected food for 10 min before discarding it.

Staphylococcus aureus

Staphylococcus aureus is a Gram-positive facultative anaerobic coccus occurring in clusters. The organisms is ubiquitous and can be found in human skin, nose, hair, and many food items. The organisms, when allowed to grow in food, may produce a class of low-molecular-weight (ca 30,000 daltons) protein toxins called staphylococcal enterotoxins (A, B, C_1, C_2, C_3, D, and E). These toxins, when ingested by a susceptible person, will cause severe nausea, vomiting, abdominal cramps, diarrhea, and prostration about 4 to 6 h after consumption. Recovery is about 24 to 72 h: Victims will not die, but may wish they had, as the reactions are very violent. Also there is no immunity against the toxin, thus a person can have staphylococcal intoxication repeatedly. Along with *Salmonella* and *Clostridium perfringens*, staphylococcal intoxication ranks among the top three agents of foodborne disease in the past 25 years. In 1992 there were six outbreaks involving 206 cases with no death. Since this is a nonreportable disease, many more outbreaks and cases occur regularly without being reported to public health officials. These toxins are heat stable. Once formed in food the toxin is very hard to destroy. Heating the toxins at 80°C for 5 h will not destroy the toxin. Boiling for 3 h will destroy the toxin and cooking under pressure (121°C) will inactive the toxin in 30 min. For practical purposes these toxins are not inactivated by normal cooking procedures.

Due to the fact that the enterotoxins are heat stable, detection of live *S. aureus* in foods has only limited value in terms of assessment of the potential of the food to cause staphylococcal food intoxication. For example, assume a food is contaminated with a toxigenic strain of *S. aureus* and the organisms grew to 1 million cells and released large amount of enterotoxins (in micrograms) in the food and the food is subsequently cooked. Even though live *S. aureus* cannot be found in this food item, the food is still capable of causing a case of staphylococcal intoxication due to the heat-stable toxins in the food. Several years ago there was a big concern regarding canned mushrooms that were shipped from a foreign country to the United States. Certainly no live *S. aureus* were found in canned mushrooms, yet the preformed enterotoxins in the canned mushrooms caused many cases of food intoxication.

The value of monitoring live *S. aureus* is to ascertain the hygienic quality of the food and the potential of the live organisms to grow and produce the enterotoxins in foods. Detection of staphylococcal enterotoxins has been a subject of much research in the past 25 years. Monkey, cats, and other animals have been used to detect toxins but are not practical for routine testing. Immunological methods such as the ELISA test, Latex agglutination tests, and gel diffusion tests are used to detect the toxin in foods. The commercial kits can detect enterotoxin A, B, C, D, and/or E either singularly or in combination.

Fortunately *S. aureus* is not a good competitor compared with other spoilage organisms (eg, *Pseudomonas*) in raw foods such as ground beef and fish, and so on. However, in the absence of competitors, such as in salty food (eg, ham) or processed foods (eg, processed cheese), the organism can grow and produce heat-stable toxins. Enough toxins can be produced in 4 h at room temperature to cause a problem. That is why this intoxication is called picnic food poisoning; during a picnic, food may be left nonrefrigerated for hours before consumption by partygoers after various activities. It is therefore essential to prevent *S. aureus* from growing in the food by proper refrigeration (4°C) of food or keeping hot food hot (60°C). This advice is applicable to all subsequent discussions on food intoxication and infections.

Aspergillus

Aspergillus flavus and *A. parasiticus* are molds that can produce a group of carcinogenic toxins called aflatoxins (*A-fla*-toxin). In 1960 in England, 100,000 turkeys died of unknown causes, and the disease was called Turkey X disease. After much work, the contaminant was found to have originated from peanut meals from Brazil. The organisms responsible for producing the toxic compounds were identified as *A. flavus*. Later, *A. parasiticus* was also found to be able to produce the toxin. Recently *A. nomius* has also been added to the list of cultures producing aflatoxin. The mold can grow between 7.5°C and 40°C, with optimal temperature at 24 to 28°C. The minimal water activity for growth is 0.82 and the optimal is 0.99. pH range of growth is from 2 to near 11. Production of the toxins generally parallel the growth of the organisms. Sporulation of the cultures seems to be a prerequisite to toxin production. In one to three days of growth, the organism can produce the toxins. The primary toxins are B_1, B_2, G_1, and G_2. B and G indicate that the toxins fluoresce blue or green under ultraviolet light, respectively. When cows consume B_1 and B_2 toxins, they can modify the toxins and excrete the toxins as M_1 and M_2 in milk.

Spores of these molds are ubiquitous; the organisms have been found to grow in rice, sorghum, peanut, corn, wheat, and soybean crops, as well as in animal feed. Human food shown to support growth and toxin production of this mold include peanut, peanut butter, pecans, beans, dried fruits, fish, and even cheese. Because the toxins are carcinogenic, they are under strict government scrutiny since the Delaney Clause of 1958 prohibits the presence of carcinogenic compounds in foods. Currently the allowed limit is 20 ppb for animal feed and all foods except milk, which has an action level of 0.5 aflatoxin M_1. Although no direct food-related aflatoxin cases have been reported in the United States, there are concerns that aflatoxin can affect the immune systems of patients. There were aflatoxin cases reported in Southeast Asia when people consumed food heavily contaminated with molds. The toxins can be detected by animal tests using ducklings or chick embryos. Thin-layer chromatography and high-performance liquid chromatography can also be used to detect these toxins. Recently, monoclonal antibodies have been employed to detect these toxins with great rapidity (10 to 30 min) and sensitivity (1 ppb and lower). Attempts to detoxify aflatoxin by ozone, peroxides, and ammonia have been met with limited success. Thus, the best preventive measure is not to allow the mold to contaminate the food and feed and to keep these commodities in a dry environment unfavorable for mold growth.

Exotoxins Versus Endotoxins

It is necessary to differentiate these toxins before a discussion on foodborne infections. *Exotoxins* are toxins produced by an organism and later released into the environment. The cell remains alive and intact. Ingestion of these preformed toxins causes foodborne intoxication. These toxins are protein toxins mainly produced by Gram-positive organisms. Because they are proteins, they can be neutralized by corresponding antibodies and detected by a variety of immunologic methods. These toxins are relatively heat sensitive (except the staphylococcal enterotoxins as described earlier). These toxins also have distinct pharmacology. Examples of exotoxins are staphylococcal enterotoxins (affecting the intestinal tracts) and botulinum neurotoxins (affecting the nervous system).

Endotoxins are part of the cell wall material of Gram-negative cells. Every Gram-negative bacterium examined has endotoxins. These are complex molecules containing protein, carbohydrate, and lipid. The protein moiety determines antigenicity, the carbohydrate moiety determines immunologic specificity, and the lipid moiety causes toxicity. Unlike exotoxins, antibodies will not neutralize toxicity because the toxic part is the lipid. All endotoxins have the same action and are released when the Gram-negative bacterium lysogenizes. These endotoxins cause fever by acting as *exogenous pyrogens*. The exogenous pyrogen, when absorbed into the bloodstream, causes injury to the leukocytes, which in turn releases an *endogenous pyrogen*. This endogenous pyrogen stimulates the thermoregulatory center of the brain at the hypothalamus and causes fever. Therefore, fever production by a patient is indicative of a foodborne infection case. Endotoxins can be detected by the limulus amebocyte lysate (LAL) test. In the presence of endotoxins, the LAL will form a gel. The reaction takes about 1 h. Currently, hospital materials should be pyrogen-free, and LAL is the standard test for pyrogens in the hospital supplies and environment. Because endotoxins are released upon lysis of the cell, it is advisable in certain cases that antibiotics not be administered in mild food infection cases. Lysis of cells by antibiotics such as *Escherichia coli* O157:H7 may allow the release of other harmful toxins in the intestinal tract and may case a more severe infection case.

BACTERIAL INFECTION

Clostridium perfringens

C. perfringens occupies an interesting position as being both a foodborne infection agent as well as a foodborne intoxication agent. On the one hand, the susceptible person has to ingest large numbers of viable *C. perfringens* before coming down with food poisoning, and on the other hand, the organism produces an enterotoxin to cause the illness. In 1990 there were 11 outbreaks and 1240 cases with no deaths in the United States. The estimate, however, is 652,000 cases with an average of 7.6 deaths per year at an annual cost of $123 million to the U.S. economy. *C. perfringens* is a Gram-positive anaerobic spore-forming rod and produces at least 13 different toxins that can cause diseases such as gas gangrene. One of the toxins is named *C. perfringens* enterotoxin (CPE), which is released in the intestinal tract and causes the infection/intoxication by this organism. The generation time (time for doubling of a population of cells) of *C. perfringens* in ideal conditions is as short as 7 min, making it the fastest-growing organism known. Spores of the organism distribute widely in nature and can easily contaminate foods. Most of the incidences of *C. perfringens* food poisoning involve meats prepared in large quantities one day and consumed the next day while the food is held at lukewarm temperatures. In such conditions, most vegetative cells of competitors die off while the spores of *C. perfringens* have a chance to survive, germinate, and grow into large numbers. The organisms in large numbers (about 10,000 to 1 million per gram), when ingested by a susceptible person, will start to sporulate in the small intestine due to the favorable anaerobic environment. The gene coded for sporulation also controls the release of an enterotoxin that is responsible for the diarrhea characteristics of *C. perfringens* food poisoning. It is noteworthy that *C. perfringens* does not sporulate in foods and therefore the CPE is not preformed in the food to cause the food poisoning cases. Symptoms occur between 8 and 20 h after ingestion of a large number of viable *C. perfringens* and include acute abdominal pain, diarrhea, and nausea, with rare vomiting. The symptoms are milder than those caused by *Salmonella*. Detection of this organism is by anaerobic cultivation of food using differential anaerobic agar such as tryptose sulfite cycloserine agar. *C. perfringens* forms black colonies in this agar medium. Immunologic methods such as reverse-passive agglutination assay and ELISA test have been developed to detect the CPE in food, culture fluid, and feces.

Salmonella

Salmonella is the classic example of foodborne infection. *Salmonella enteritidis* was isolated in 1884 and still is an important foodborne organism. In 1992 there were 80 outbreaks with 2834 cases and four deaths due to *Salmonella* in the United States. The organism is a Gram-negative, facultative anaerobic, non-spore-forming rod, motile by peritrichous flagella. It does not ferment lactose and sucrose but ferments dulcitol, mannitol, and glucose. There are exceptions to the general characteristics. For example, lactose-positive cultures have been found, and nonmotile species such as *S. pullorum* and *S. gallinarum* exist. The organism is heat sensitive but can tolerate a variety of chemicals, such as brilliant green, sodium lauryl sulfite, selenite, and tetrathionate. These compounds have been used for the selective isolation of this organism from food and water. To confirm the isolate as a *Salmonella*, one must perform serologic tests by use of polyvalent anti-O antiserum (against cell surface antigens) or polyvalent anti-H antiserum (against flagella antigens). This genus went through several revisions in classification and taxonomy of species and subspecies in the past 20 years due to advancement of genetic typing systems.

Currently there are two species, namely, *S. enterica* with six subspecies and 2356 serovars and *S. bongori* with 19 serovars. Each serovar is potentially pathogenic. In the literature, many scientists still use the traditional genus and species nomenclature such as *S. typhimurium, S. typhosa*, and so on. *Salmonella* has been found in water, ice, milk, dairy products, shellfish, poultry, poultry meat products, eggs and egg products, animal feed, pets, and so on. Human beings can be healthy carriers of this organism. It has been estimated that 4% of the general public carries this organism. There are actually three types of diseases caused by *Salmonella*: (*1*) enteric fever caused by *S. typhosa* (typhoid fever), in which the organism, ingested along with food, finds its way into the bloodstream and disseminates to the kidney and is excreted in the stools; (*2*) septicemia caused by *S. cholerasuis*, in which the organism causes blood poisoning; and (*3*) gastroenteritis caused by *S. typhimurium* and *S. enteritidis*, a true foodborne infection. In the last case, large numbers of live *Salmonella* are ingested with food and, in one to three days, liberate the endotoxins that cause localized violent irritation of the mucous membrane with no invasion of the bloodstream and no distribution to other organs. Symptoms of salmonellosis occur 12 to 24 h after ingestion of food containing 1 to 10 million *Salmonella* per gram and include nausea, vomiting, headache, chills, diarrhea, and fever. The illness lasts for two to three days. Most patients recover; however, death can occur in the very old, the very young, and those with compromised immune systems.

Since no *Salmonella* is allowed in cooked food for interstate commerce and international trade, the detection of *Salmonella* has been a subject of much research and development. Detection of *Salmonella* by the classical method includes preenrichment of culture from food sample, enrichment or selective enrichment of the liquid culture, plating of liquid on selective agar to isolate cultures, biochemical tests of suspect colonies, and confirmation of isolates with typical biochemical profiles by serological tests. These procedures may take up to five days for completion. Recently a variety of methods and procedures have been developed and implemented for the effective isolation, enumeration, detection, identification, and characterization of *Salmonella*. Improvement of preenrichment and enrichment procedures have been made by the manipulation of incubation temperature (use 42°C instead of 37°C), addition of various stimulation compounds such as Oxyrase enzyme, and concentration of cells by immunomagnetic separation (Dynal system). A large number of biochemical diagnostic kits, such as API, MicroID, Enterotube, Biolog, Vitek, and so on, have been developed and marketed to conveniently and automatically identify isolates. Manual and automated sandwich ELISA tests, by EIA Assurance test, VIDAS, Tecra, and so on, and lateral immunomigration tests kits by BioControl VIP system and Neogen have been used widely. For the genetic tests, DNA/RNA probes system by Genetrak, PCR test by Perkin-Elmer, Probelia, and BAX system and ribotyping by Qualicon system and others are finding their ways into food microbiology laboratories (see the article RAPID METHODS OF MICROBIOLOGICAL ANALYSIS).

Because *Salmonella* is heat sensitive, proper cooking will destroy the organism. Also, proper chilling, refrigeration, and good sanitation will minimize the problem. *Salmonella* remains one of the most important food pathogens in our food supply.

Shigella

Shigella is a Gram-negative, facultative anaerobic non-spore-forming rod quite often confused with *Salmonella* in the bacteriologic diagnostic process. It is nonmotile and hydrogen sulfide negative. The colonies are smaller than *Salmonella*. In terms of foodborne infection, *Shigella* is not as prevalent as *Salmonella*, but this organism is very important in waterborne diseases, especially in tropical and subtropical countries where sanitation conditions are poor. The organism is transmitted by water, food, humans, and animals. The "4 Fs" involved in the transmission of *Shigella* are food, finger, feces, and flies. One to four days after ingestion of the organisms, there will be an inflammation of walls of the large intestines and ileum. Invasion of the blood is rare. Bloody stool will occur, owing to superficial ulceration. The cell wall of *Shigella*, when lysed, will release endotoxins. In addition, *S. dysenteriae* produces an exotoxin that is a highly toxic neurotoxin. This toxin can be neutralized by specific antibody. Mortality rate of shigellosis is higher than that of salmonellosis. Prevention of shigellosis can be achieved by sanitation, good hygiene, treatment of water, prevention of contamination, detection of carriers, and isolation of patients from the general public.

Vibrio cholerae

Vibrio cholerae was a very important disease-causing organism in the late nineteenth century and early twentieth century worldwide. The organism is under control in many industrialized countries; however, it is still a very important waterborne disease in places with poor sanitary conditions. The classical work of John Snow in 1854 showed the transmission of *V. cholerae* through poorly designed water delivery systems in London. His work led to the development of much improved water delivery systems and water treatment systems by public health official and environmental engineers in developed countries around the world. In the United States in 1990 there was one outbreak involving 26 cases and one death. The appearance of *V. cholerae* in industrialized countries often causes panic because this organism has the potential to start a pandemic infection. It is a Gram-negative, curved rod that looks like a comma under the microscope, thus the original name of *V. comma*. No spore is formed. *V. cholerae* grows well in alkaline medium and is actively motile by a single polar flagellum. The organism is endemic in India and Southeast Asia and is spread by person-to-person contact, water, milk, food, and insects. The organism produces enterotoxins and endotoxins in the intestines and causes severe irritation to the mucous membranes with resultant outflow of fluid and salts and impairs the sodium pump of mammalian cells, thus causing severe diarrhea, dehydration, acidosis, shock, and even death. The mortality rate may be as high as 25 to 50%. The most effective therapy is replacement of water and electrolytes to correct severe dehydration and salt depletion. *V. cholerae* remains a dreaded communicable disease in many parts of the world, and much education and public health work need to be done to reduce human suffering from this organism. Besides conventional biochemical tests, currently there are immunologic and DNA probes and PCR methods for rapid detection of this organism.

Vibrio parahemolyticus

Vibrio parahemolyticus is an organism that has caused many cases of foodborne disease in Japan for many years. This is because citizens in Japan like to consume raw or undercooked seafood, which may be contaminated with the organisms, especially in the summer months when the water is warm in Northern Hemisphere. Most of the original reports and research works were in Japanese and not readily understandable or available to microbiologists in the West. U.S. scientists started working on the organism in earnest around 1969. In 1971 three outbreaks of this organism occurred in the United States. Since U.S. citizens do not regularly consume raw seafood, the sources of the illness were probably recontamination of cooked foods. In 1990 there were four outbreaks and 21 cases reported. The fact that no outbreaks were reported in 1988, 1989, 1991, and 1992 indicates that this organism is not a source of common foodborne infections in the United States.

The organism is a Gram-negative curved rod and is halophilic (salt loving); it grows best in 3 to 4% salt medium and can grow in 8% salt also. The growth temperature range is 15 to 40°C, and the pH range is 5 to 9.6. The organism is sensitive to streptomycin, tetracycline, chloramphenicol, and novobiocin but resistant to polymyxin and colistin. The Kanagawa positive strains hemolyze human blood. Environmental strains are negative for this test. The organism is distributed in fish and shellfish from seawater as well as from fresh water. Most of the outbreaks are recorded in the summer months when the water is warm in the Northern Hemisphere. Symptoms of the disease occur about 12 h after ingestion of a large number of viable cells (10^5/g) and include abdominal pain, diarrhea, vomiting, mild chills, and headache. The symptoms are similar to those of salmonellosis but more severe. It has been described that salmonellosis affects the abdomen of the patient, whereas *V. parahemolyticus* infection affects the stomach of the patient. Detection of the organism is best achieved by good selective medium such as BTB-salt-Teepol agar. Prevention of the occurrence of infection by this organism is by adequate cooking of seafood.

Vibrio vulnificus

This organism can be considered as an emerging pathogen. It causes more than 90% of all seafood-related deaths in the United States. The organism is widespread in estuarine environments and has been isolated from waters around the world. Consumption of raw oysters contaminated with this organism may lead to septicemia and death. Also the organism may invade wounds of victims as the persons wade or work in the contaminate water. The organism is a typical vibrio-shaped organism and is classified as biotype 1 and biotype 2. The organism grows well in common bacteriological agars such as MacConkey agar and blood agar. The incubation time of the illness is from one to seven days. The disease involves fever; chills; nausea and, to a lesser extent, vomiting; abdominal pain; and diarrhea. Development of secondary lesions can be serious and may result in vasculitis and necrotizing fasciitis necessitating surgical removal of tissues or even limb amputation. Since the organism is killed by common cooking practices, the problem is the consumption of raw seafood, especially raw oysters. People with liver damage and with immunocompromised conditions should definitely avoid eating raw seafood. There was one outbreak with two cases and one death reported in the United States in 1990.

Bacillus cereus

Bacillus cereus and other *Bacillus* species have been implicated in foodborne diseases only in recent years, although these organisms have been suspected as agents of foodborne illness for a long time. In 1991 there were five outbreaks, 253 cases, and no deaths reported in the United States. These are Gram-positive, aerobic, spore-forming rods occurring widely in nature and contaminating foods easily. Because of the general resistance of spores of this organism and the prolific biochemical activity of the vegetative cells, it can be considered one of the most important environmental bacterial contaminants of foods. Two distinct clinical symptoms are caused by this organism. The diarrheal syndrome occurs after 12 to 24 h of ingestion of large numbers (about 1 million) of viable *B. cereus* and includes abdominal pain, watery diarrhea, rectal tenesmus, and nausea without vomiting. The diarrheal enterotoxin is formed in the intestine of the host and causes the disease. The diarrheal syndrome is the result of consuming proteinaceous foods, pudding, milk and milk products, sauces, and vegetables. The emetic syndrome causes illness almost exclusively associated with cooked rice and noddles and is characterized by a rapid onset (1–5 h) with nausea, vomiting, and malaise. The toxin is preformed in the food by large numbers of *B. cereus* (1 to 10 million cells). A large number of viable *B. cereus* found in food indicates poor food handling and storage practices. To truly assess the foodborne illness potential, one has to detect the toxins involved. Currently no diagnostic kits are available for the detection of emetic toxin, but two kits available for diarrheal enterotoxin, one by OXOID that utilizes reverse passive latex agglutination tests and another by Tecra that uses the ELISA format.

Other *Bacillus* suspected of causing foodborne diseases include *B. licheniformis* and *B. subtilis*, in which large numbers (10^5–10^6 organisms/gram of food) of these organisms are ingested by susceptible persons. It should be noted that *B. subtilis var natto* is used to ferment a popular food item in Japan called Natto in which the organisms produce a polymer of glutamic acids as well as other flavor compounds. Control of *Bacillus* food poisoning is complicated by the ubiquitous nature of this organism. The best measures are to prevent the spore from germinating and to prevent multiplication of vegetative cells in cooked and ready-to-eat foods. Freshly cooked food eaten hot immediately after cooking should not be a problem. However, slow reheating of previously cooked rice products should be treated with caution. Refrigeration of leftover cooked rice products is highly recommended as a preventive measure.

Campylobacter jejuni

Campylobacter jejuni, recognized as an emerging pathogen in the past 10 years, has been reported as the most common bacterial cause of gastrointestinal infection in humans, even surpassing rates of illness caused by *Salmonella* and *Shigella*. In 1992 there were six outbreaks, 138 cases, and two deaths reported in the United States. *Campylobacter* was originally called *Vibrio fetus*, because it was first recognized as an agent of infertility and abortion in sheep and cattle. The organism is a member of the family *Spirillaceae* because of the physiologic and morphologic similarities to *Spirillum*. The organism is a Gram-negative, slender, curved bacteria that is motile by a single polar flagellum. It neither ferments nor oxidizes carbohydrates, is oxidase positive, reduces nitrates, but will not hydrolyze gelatin or urea and is methyl red and Voges-Proskauer reaction negative. It will grow between 25 and 43°C. The organism is an obligate microaerophile that grows optimally in 5% oxygen. This attribute has been used for isolation of the organism by applying appropriate gas mixtures into the headspace of cultivation media. Recently in D. Fung's laboratory the enzyme Oxyrase was found to greatly stimulate the growth of this organism, even in the absence of special gas mixtures, thus facilitating rapid and convenient detection and isolation of this organism. The incubation time of *C. jejuni* food poisoning ranges from two to five days; the duration of the sickness may be up to 10 days. The patient will exhibit enteritis, fever, malaise, abdominal pain, and headache. The stools become liquid and foul smelling. Blood, bile, and mucus discharge may occur in serious cases. The organism has a worldwide distribution, with outbreaks related to milk, poultry, eggs, red meat, pork, and water reported. It has been isolated from 50 to almost 100% of poultry carcasses in several studies. Competitive exclusion protocols have been devised to prevent the attachment and growth of *C. jejuni* by inoculating large numbers of natural intestine microorganisms in newly hatched chicks. Detection of this organism is by suitable liquid and solid growth media designed for the organism and rapid tests involving ELISA, PCR, ribotyping, and so on. One complication in studying this organism is the presence of a viable but nonculturable population of *C. jejuni* in the environment. Proper food processing techniques (heating, cooling, chemical treatment of foods, etc) will control this fragile organism. Its prevalence as a foodborne pathogen can be attributed to post-processing contaminations of food. Again, good sanitation and hygiene should reduce the incidence of this organism in our food supplies. Because of increased outbreaks and cases related to this organism much research worldwide is being conducted to monitor the organism. *C. jejuni* may be the next major foodborne disease-causing organism to be faced by food microbiologists nationally and internationally.

Escherichia coli

Escherichia coli is one of the most common bacteria in our environment. Most people do not think of *E. coli* as a foodborne pathogen; however, recent research and information indicates that some strains of *E. coli* can indeed cause severe foodborne diseases. The sensational outbreak of *E. coli* O157:H7 involving hundreds of people and four deaths of children by the consumption of undercooked hamburger served by a fast-food chain in the United States in 1993 awakened the general public to the importance of food safety. Food industries, academic communities, regulatory agencies, and consumer groups have been actively working on solving the problem of *E. coli* O157:H7 ever since that outbreak. Although much has been learned about this organism, far more needs to be done to determine the habitat, the detection, and the control of this organism. *E. coli* is a Gram-negative, facultative anaerobic, non-spore-forming rod that occurs widely in nature as well as in intestines of humans and animals. It is glucose and lactose positive, indole and methyl red positive, but Voges-Proskauer and citrate negative. The most useful way to classify the species is by serotyping, using antibodies against O, H, and K antigens of various strains of *E. coli*. Most *E. coli* isolated from the environment are not pathogenic. However, there are six classes of pathogenic and diarrheagenic *E. coli*. They are enterohemorrhagic (EHEC), enterotoxigenic (ETEC), enteroinvasive (EIEC), enteroaggregative (EaggEC), enteropathogenic (EPEC), and diffusely adherent (DAEC) *E. coli*.

EHEC, or enterohemorrhagic *E. coli*, was first identified as a human pathogen in 1982. The most important serotype is O157:H7. Other serotypes in this group are O26:H11, O103, O104, O111, and others. *E. coli* O157:H7 causes the most concern worldwide because of its unusual cultural characteristics and pathogenicity. Unlike most *E. coli*, this serotype does not ferment sorbitol within 24 h, does not possess β-glucuronidase activity, and does not have the ability to hydrolyze 4-methylumbelliferyl-β-D-glucuronide (MUG), which is an important diagnostic characteristic of most other *E. coli* strains. Because of these differences in routine microbiological manipulations, *E. coli* O157:H7 has been excluded in the protocol for common *E. coli*. The organism produces one or more Shiga-like toxins (SLT; also known as verotoxin, VT) and it possesses an attaching and effacing gene (*eae* gene) and a large plasmid (60 MDA). The organism causes severe illness, especially in children and immunocompromised patients. There are three manifestations of the disease: hemorrhagic colitis (HC), hemolytic uremic syndrome (HUS), or thrombotic thrombocytogenic purpura (TTP). Symptoms of HC occur, within one to two days after consuming contaminated foods. The initial symptoms are mild, nonbloody diarrhea followed by severe abdominal pain, and a short fever or no fever. The watery diarrhea will last for 24 to 48 h followed by 4 to 10 days of bloody diarrhea, severe abdominal pain, and dehydration. Patients with HC may develop more severe life-threatening complications such as HUS or TTP.

HUS symptoms are characterized by microangiopathic hemolytic anemia (pallor, intravascular destruction of red blood cells), thrombocytopenia (depressed platelet counts), and acute renal failure and may lead to death. TTP affects mostly adults and is a rare syndrome of *E. coli* O157:H7 infection. It causes neurological abnormalities such as nervous system deterioration, seizures, and strokes. Patients will often develop blood clots in the brain and may die. The infectious dose of *E. coli* O157:H7 is between 2 to 200 cells. Adhesion of the organisms to the intestinal wells is important, but it does not enter the circulatory system. The organism colonizes the intestinal tract where toxins are produced and then become active in the colon. For this reason much research is being conducted to have competitive exclusion by nonpathogenic organisms to attach to the epitheleal cells before *E. coli* O157:H7 can have a chance to interact with the epitheleal cells. Research in this area is in its infant stage.

Much information is now known about the characteristics of this organism. It grows well at 37°C but poorly at 44 to 45°C, a temperature usually used to isolate *E. coli*. It can grow between 8 and 10°C and can survive in ground beef at −20°C for nine months. It grows in neutral pH ranges of 5.5 to 7.5 but can grow in the pH 4.0 to 4.5 range, and more recent data indicated that it can survive in apple cider in the range of 3.6 to 4.0. The organism is heat sensitive. Proper cooking temperature of 160°F or 71°C will destroy the organism in foods. The organism is quite salt tolerant with the ability to grow at 8% NaCl at 37°C; however, at the lower incubation temperature of 10°C, growth was inhibited at 4 to 6%. This organism grows well in the water activity range of 0.99 with a minimum at 0.95.

Outbreaks of *E. coli* O157:H7 have been reported from water, meat, poultry, dairy products, salad, apple cider, and even fermented meats and mayonnaise. Detection methods include conventional culture procedures designed specifically for *E. coli* O157:H7, a variety of diagnostic kits, serologic tests, ELISA, PCR, ribotyping, and so on. The aim is to accurately and rapidly screen the presence or absence of the organisms in 25 g of food. A 24-h negative screening protocol is now available. One commercial company even suggested an 8-h protocol. This organism will continue to be very important in food microbiology for the foreseeable future.

ETEC, or enterotoxigenic *E. coli*, is a the major cause of infantile diarrhea in developing countries and is most frequently responsible for traveler's diarrhea. The serotypes involved include O8, O15, O20, O25, and others.

EIEC, or enteroinvasive *E. coli*, strains cause nonbloody diarrhea and dysentery by invading and multiplying within colonic epithelial cells. Serotypes include O28ac, O112, O124, and others.

EAggEC, enteroaggregative *E. coli*, affects infants and children with persistent diarrhea. These cells have the characteristic pattern of aggregative adherence on Hep-2 cells.

EPEC, or enterotoxigenic *E. coli*, has been defined as "diarrheagenic *E. coli* belonging to serogroups epidemiologically incriminated as pathogens, but whose pathogenic mechanisms have not been proven to be related either to heat-liable enterotoxins (LT) or heat-stable enterotoxins (ST) or to *Shigella*-like invasiveness." The serotypes included in EPEC are O55, O86, O111ab, O119, O125ac, O126, and others.

DAEC, or diffusely adherent *E. coli*, have been associated with diarrhea in children in Mexico and can produce mild diarrhea without blood or fecal leukocytes.

A comprehensive treatment of *E. coli* O157:H7 and other *E. coli* strains as well as many foodborne pathogens can be found in the book by Doyle, Beuchat, and Montville (1997). Prevention and control of pathogenic *E. coli* is best done by education of food handlers, who should adhere to strict hygienic practices. Fecal and other waste materials from humans and animals should be decontaminated and not allowed in contact with water and food supplies. Kalamaki, Price, and Fung (1997) summarized screening and identification test kits for *E. coli* (see Table 1). Similar tables for the detection of *Enterobacteriaceae, Campylobacter, Salmonella, Listeria*, Rotavirus, *Staphylococcus aureus, Vibrio cholerae*, and *V. vulnificus* are presented in the same publication. Due to these developments it is possible now to have a negative screening of *E. coli* O157:H7 in about one day. However, when a food sample showed a positive screening result, the conventional methods must be used to confirm the presence or absence of *E. coli* O157:H7.

Yersinia enterocolitica

Yersinia enterocolitica is a Gram-negative, facultative anaerobic, non-spore-forming bacterium, sucrose positive, rhamnose negative, indole positive, motile at 20°C but not at 37°C, and highly virulent to mice. Serotyping is very important in separating this organism from other closely related Gram-negative bacteria. Although *Y. enterocolitica* has an optimal growth temperature at around 32 to 34°C, it is often isolated on enteric agars at 22 to 25°C. It grows slowly in simple glucose-salts medium but grows much better with supplements such as methionine or cysteine and thiamine. One important aspect of this organism is that it can grow in vacuum-packaged meat under refrigeration because it is a facultative anaerobe and is a psychrotroph. After ingestion of large numbers of this organism, the susceptible person can develop fever, abdominal pain, and diarrhea, with nausea and vomiting occurring less frequently. More serious intestinal disorders include enteritis, terminal ileitis, and mesenteric lymphadenitis. Extraintestinal infections of *Y. enterocolitica* have been reported, including septicemia, arthritis, erythema nodosum, sarcoidosis, skin infection, and eye infection.

Foods suspected of being a source of yersiniosis in the United States include chocolate milk, milk powder, chow mein, tofu, and pasteurized milk. Pork products have also been suspected.

Isolation of this organism typically goes through an enrichment step using nutrient broth or Rappaport broth and then through a plating medium using an enteric agar (SS, XLD, DCL, etc). The CIN agar (cefsulodin–irgasan–novobiocin agar) is commercially available for the isolation of this pathogen; however, many other organisms, such as *Salmonella* and *Serratia*, also grow on this agar. In D. Fung's laboratory a new agar named KV202 has been de-

Table 1. *E. coli* Screening and Identification Test Kits

Test kit	Supplier	Assay principle	Limit of detection	Sensitivity	Specificity	% Agreement	False negative rate	False positive rate	Total time	Cost per assay
ANTI *E. coli* O157	Dynal, Inc., Lake Success, NY	Immunomagnetic separation	2 cells/g	NR[a]	NR	NR	NR	NR	30 h	$2.60
Assurance EIA EHEC[b]	BioControl	ELISA	<0.003–0.093 cfu/g	100%	98.4% (a) 99.5% (b) 100% (c)	100 (b) 99.4 (c)	0% (b) 1% (c)	1.1% (b) 0% (c)	<20 h	$5.15
Bactident *E. coli*	Merk	β-D-glucoronidase activity assay		96.30%	100%	NR	85%[a]	100%[d]	30 min for test only	NR
COLI ST EIA:ETEC STA	Denkaseiken Co Ltd, Tokyo, Japan	ELISA	NR	100%	100%	NR	0%	0%	24 h	NR
E. coli O157:H7 Kit	AMPCOR (Neogen Corp.)	Immunoprecipitation	10^4–10^3 cfu/mL	97.90% (c)	95% (a) 90.4% (c)	94.1 efficiency (c)	<1% (a) 2.1% (c)	<1% (a) 9.6% (c)	6–28 h	$9.50
E. coli O157:H7 VIA	TECRA	ELISA	1–5 cfu/25g	93%	>96%	NR	7%	<4%	20 h	$6.00
E. coli O157 Latex Test	Unipath Ltd.	Latex agglutination	Pure colony	100%	79%	NR	0%	21%	2 min for test only	NR
EHEC-Tek	Organon Tecknika	ELISA	<1 cell/g	76.5% (c)[e]	NR	NR	NR	NR	24–26 h	$9.60
EZ COLI	DIPCO	ELISA	10^4 cfu/mL	100%	95.7%	97.8 accuracy	0%	<2%	24 h	$9.48
GENE-TRAK	GENE-TRAK	Nucleic acid hybridization	10^4 cfu/mL	97.40% (a)	100% (a) 100% (b)	NR	1.2% (a)	0% (a)	24–48 h	$7.00
HEC O157 ELISA[f]	3M Company	Blot ELISA	<1 cell/g	84.5% (e)	95.2% (e)	89.9% efficiency (e)	0% (c) 15.5% (e)	2% (c) 4.8% (e)	26–28 h	$5.00
Premier EHEC	Meridian Diagnostics	ELISA	Verotoxin detection	100%	99.6%	99.7	0%	0.4%	NR	NR
Rapidec coli[a]	bioMerieux	β-D-glucoronidase activity assay	Pure colony	91.30%	100%	NR	70.80%[c]	100%[d]	2–6 h for test only	NR
Rhabdebact ETEC-LT	Karo BioDiagnostics AB Hudding, Sweden	Immunofluorescence	Pure colony	96%	100%	59.20	4%	0%	25.5 h	NR
VIDAS	bioMerieux Vitek	ELFA	10^3 cfu/mL	NR	NR	NR	11.10%	0% (b) 1.7% (a)	24–48 h	$8.00
VIP EHEC[b]	BioControl	Immunoprecipitation	<0.003–2.4 cells/g	100% (b) 99.6% (c)	100% (b) 94.3% (c)	99.4 (a)	0% (b) 0.3% (c)	0% (b) 5.6% (c)	18 h	$8.50
VET-RPLA[h]	Unipath Ltd.	Latex Agglutination	titre 20–128	100%	100%	88.50	0%	0%	37 h	$11.04

[a]No reported.
[b]AOAC first action.
[c]Negative predictive value.
[d]Positive predictive value.
[e]Increases with IMS step.
[f]FSIS method.
[g]Not available in North America.
[h]Heat labile enterotoxin detection.
Source: Ref. 1, used with permission.

veloped that separates *Salmonella* and *Serratia* colonies from *Yersinia* by the development of black colonies. Control of yersiniosis is through proper handling of raw and cooked food of all types, especially pork products, and of water for food processing. There has been no reported outbreaks of *Y. enterocolitica* between 1988 and 1992.

Listeria monocytogenes

Listeria monocytogenes has developed into a very important food pathogen in the past 10 years from the standpoint of economic and public health impact. The organism is a small, short, Gram-positive non-spore-forming rod. It is motile by a characteristic tumbling motion or slightly rotating fashion. The organism grows on simple laboratory media in the pH range between 5 and 9. On solid agar, the colonies are translucent, dewdrop-like, and bluish when viewed by 45° incident transmitted light (Henry's illumination step). Biochemically, this organism can be confused with such organisms as *Lactobacillus, Brochothrix, Erysipelohrix*, and *Kurthia*. A variety of biochemical tests have been devised to separate *L. monocytogenes* from other *Listeria* species, such as *L. innocua, L. welshimeri*, and *L. murrayi*. Serotyping is also important in the identification of this organism, the most important ones being 1/2a, 1/2b, 1/2c, 3a, 3b, 3c, and 4b. *Listeria* is a psychrotroph capable of growing at temperatures as low as 2.5°C and as high as 44°C. Because dairy products have been implicated in outbreaks of listeriosis, much research has been directed toward cheese and milk products. The organism has been found to survive the processing of cottage cheese, cheddar cheese, and Colby cheese. A question of great concern is whether *L. monocytogenes* can survive the current pasteurization temperature of milk (ie, 63°C for 30 min or 72°C for 15 s). Data on this issue are still inconclusive, and research on this topic is still ongoing. It is important to note that, at present, the time and temperature regulation for pasteurization of milk has not been affected by the possible heat resistance of *L. monocytogenes*. The disease starts with infection of the intestine—the infective dose is not known at this point. Patients may develop transitory flu-like symptoms such as malaise, diarrhea, and mild fever. In severe cases, virulent strains are capable of multiplying in macrophages and later producing septicemia. When this occurs, the bacteria can affect the central nervous system, the heart, the eyes, and may invade the fetus of pregnant women and result in abortion, stillbirth, or neonatal sepsis.

Several well-documented cases of listeriosis have been reported in Nova Scotia (1981), Massachusetts (1983), and the most well known one involving Mexican-style soft cheese in Southern California (1985). Due to concerted effort by the food industry and government agencies, the outbreaks of *L. monocytogenes* seemed to have subsided for about 10 years. Between 1998 and 1992 only one outbreak, involving two cases and one death, was recorded for *L. monocytogenes*, and it appeared that the problem of *L. monocytogenes* was under control. However, recently in 1998 and 1999 the organism was found in surfaces and equipment of production lines of frankfurters, air condition lines, hot dogs, lunch meat, and turkey breasts and caused many outbreaks and recalls. One company recalled 30 million lb of ready-to-eat products due to *L. monocytogenes*. Another company recalled 15 million lb of hot dog and deli meat products due to an outbreak of *L. monocytogenes* that included 20 deaths—14 adults and 6 miscarriages/stillbirths—and at least 97 illnesses in 22 states. There is a resurgence of concern about this organism due to the pathogenic nature of the illness involving miscarriages and stillbirth of fetuses.

L. monocytogenes has been isolated in a variety of commodities, including poultry carcasses, meat and chopped beef, dry sausages, milk and milk products, cheese, vegetables, and surface water. Control measures include eliminating occurrence of the organism in the raw food materials, transporting vehicles, and food-processing plants (especially in controlling cross-contamination of raw and finished products); in practicing good general sanitation of the entire food-processing environment; in regular monitoring of the occurrence of this organism in the food-processing facilities; and in preventing pregnant females from working in and around an environment that might have the possibility of exposing them to *L. monocytogenes*. Because the organism is killed by heat and is susceptible to sanitizing agents, proper cooking of food and decontamination of the food preparation environment will also help reduce risks. Much research has been devoted to the rapid isolation, enumeration, and identification of this organism. Many diagnostic kits, immunological systems, and genetic systems have been developed to rapidly screen this organism in the food supply.

Aeromonas hydrophia

Aeromonas hydrophia has been associated with foodborne infection, although the evidence is not conclusive. The organism is a facultative anaerobic, Gram-negative, motile rod. Biochemically, it is similar to *E. coli* and *Klebsiella*. The optimal temperature for growth is 28°C and the maximum is 42°C. Many strains can grow at 5°C, which is a temperature usually considered adequate to prevent growth of foodborne pathogens.

Diseases caused by *A. hydrophila* include gastroenteritis (cholera-like illness and dysentery-like illness) and extraintestinal infections such as septicemia and meningitis. This organism has been isolated from fish, shrimp, crabs, scallops, oysters, red meats, poultry, raw milk, vacuum-packaged pork and beef, and even bottled mineral water.

Because the organism is a psychrotroph, cold storage is not an adequate preventive measure. Proper heating of food offers sufficient protection against this organism. Consumption of undercooked food or raw food such as raw shellfish is discouraged.

Plesiomonas shigelloides

Plesiomonas shigelloides has been a suspect in foodborne disease cases. The organism is Gram-negative, facultative anaerobic, catalase negative, and fermentative. It is oxidase positive and can be differentiated from bacteria in the family *Enterobacteriaceae* by this test, since the latter is oxidase negative. The organism also resembles *Shigella* but can be differentiated from *Shigella* by being motile. It is capable of producing many diseases, ranging from enteritis to meningitis.

Gastroenteritis by *P. shigellosides* is characterized by diarrhea, abdominal pain, nausea, chills, fever, headache, and vomiting after an incubation time of one to two days. Symptoms last for a week or longer. All reported food involved with cases of gastroenteritis were from aquatic origin (salted fish, crabs, and oysters). The organism can be isolated from a variety of sources, including humans, birds, fish, reptiles, and crustaceans. The true nature of this organism as a foodborne agent is not fully known because the organism has not been well studied to date.

Miscellaneous Bacterial Foodborne Pathogens

Many other microbes are suspected of being foodborne pathogens. However, they are not currently being labeled as true foodborne pathogens owing to a lack of reports of these organisms, as well as a lack of isolation methods and research on these organisms. Many of these organisms may be identified as foodborne pathogens in the future. Among these organisms are the Gram-negative bacteria *Citrobacter, Edwardsiella, Enterobacter, Klebsiella, Hafnia, Kluyvera, Proteus, Providencia, Morganella, Serratia, Vibrios*, and *Pseudomonas* and the Gram-positive bacteria *Corynebacterium, Streptococcus*, and other species of *Bacillus* and *Clostridium*. Miscellaneous organisms include *Brucella, Mycobacterium* (T, B), *Coxiella burnetii* (Q-fever), and *Leptospirosis, Erysipelas*, and *Tularemia*.

Foodborne Viruses

Foodborne viruses are much less studied by food microbiologists than are bacteria and fungi owing to the difficulty of cultivating these entities, as conventional bacteriologic media will not allow these particles to grow. There are, no doubt, many foodborne outbreaks and cases caused by a variety of viruses, but scientists in many cases are not able to identify the sources of the infection. Viruses that have been incriminated in foodborne diseases include hepatitis A virus (oysters, clams, doughnuts, sandwiches, and salad), Norwalk virus (oysters), polio virus (milk and oysters), ECHO virus (oysters), enteroviruses (oysters), and coxsackievirus (oysters). Much more research needs to be done in the field of food virology to help reduce the incidences of foodborne diseases caused by viruses.

Protozoa and Related Organisms

Protozoans such as *Cryptosporidium, Cyclospora, Toxoplasma, Giardiasis, Entamoeba, Balantidium*, and others can also cause human foodborne diseases. The most sensational outbreak was the one involving *Cryptosporidium parvum*, which affected 400,000 people and caused several deaths in Milwaukee in 1993. *Cyclospora cayetanensis* also was in the news due to imported fruits and caused a foodborne outbreak. These organisms have complex life cycles and are studied by specialists in this area. Recently a organism named *Pfiesteria piscicida* was responsible for killing million of fish in the Eastern shores of the United States. The organism has 24 life stages ranging from a cyst stage to a toxic zoospore phase to an amoeba stage. People in contact with water infected with this organism complained of vomiting and liver problems, but no conclusive data are available on the pathogenicity of this organism on human.

Nonmicrobial Foodborne Disease Agents

Consumption of food containing other living organisms can directly and indirectly cause foodborne diseases as well. Among nonmicrobial foodborne disease agents are scombroid fish (associated with high level of histamine), cestodes (flatworms such as *Taenia saginata, T. solium*, and *Diphyllobothrium latum*), nematodes (hookworm such as *Trichinella spiralis*), trematodes (fluke such as *Clonorchis sinensis*), shellfish (indirectly toxin from the dinoflagellate *Gonyaulax catenella*), ciguatera (from eating fish such as barracudas, groupers, and sea basses that feed on toxic algae), and other poisonous fish (such as puffer fish and moray eel).

SUMMARY

Food safety is everybody's responsibility. Scientists are charged with identifying the agents causing foodborne infections and intoxications and studying the mechanisms of the intoxication and infection as well as working on the isolation, enumeration, characterization, and identification of the causative agents and their control by developing intervention strategies and preservation methods. The food industry uses this basic knowledge and applies it to good manufacturing practices to produce wholesome, nutritious, and safe foods by utilizing modern equipments, systems, processing techniques, and distribution systems. Government agents are charged with the responsibility of monitoring the safety of food supplies and enforcing regulation to ensure the production, distribution, and sales of wholesome foods. The consumer must also be educated in the handling of raw and cooked food at the point of purchase, as well as preparation of the food and final consumption. All parties are responsible for the food safety of all involved. The delightful book *Safe Eating* by Acheson and Levinson (1998) details the problems involving food safety and solutions to protect consumers in laypeople's terms and yet provides much scientific information about the entire issue of food safety and consumer protection. It is a book worth reading and studying by consumers concerned about food safety.

BIBLIOGRAPHY

1. M. Kalamaki, R. Price, and D. Y. C. Fung, "Rapid Methods for Identifying Seafood Microbial Pathogens and Toxins," *J. of Rapid Methods and Automation in Microbiol.* **5**, 87–138 (1997).

GENERAL REFERENCES

M. R. Adams and M. O. Moss, *Food Microbiology*, The Royal Society of Chemistry, London, United Kingdom, 1995.

G. J. Banwart, *Basic Food Microbiology*, AVI, Westport, Conn., 1981.

C. M. Bourgeois, J. Y. Levean, and D. Y. C. Fung, *Microbiological Control of Foods and Agricultural Products*, English ed., VCH, New York, 1995.

J. E. L. Corry, D. Roberts, and F. A. Skinner, *Isolation and Identification Methods for Food Poisoning Organisms*, Academic Press, New York, 1982.

M. P. Doyle, *Food Borne Bacterial Pathogens*, Marcel Dekker, New York, 1989.

M. P. Doyle, L. R. Beuchat, and T. J. Montville, *Food Microbiology: Fundamentals and Frontiers*, ASM Press, Washington, D.C., 1997.

U.S. Food and Drug Administration, *Bacteriology Analytical Manual*, 8th ed., AOAC INTERNATIONAL, Arlington, Va., 1995.

D. Y. C. Fung, "Types of Microorganisms," in F. E. Cunningham and N. A. Cox, eds., *Microbiology of Poultry Meat Products*, Academic Press, New York, 1987, pp. 5–27.

D. Y. C. Fung, "What's Needed in Rapid Detection of Foodborne Pathogens," *Food Technol.* **49**, 64–67 (1995).

D. Y. C. Fung, "Overview of Rapid Methods of Microbiological Analysis," in M. C. Tortorello and S. M. Gendel, eds., *Food Microbiology Analysis: New Technologies*, Marcel Dekker, New York, 1997, pp. 1–25.

D. Y. C. Fung, *Handbook of Rapid Methods and Automation in Microbiology Workshop*, Kansas State University, Manhattan, Kans., 1998.

D. Y. C. Fung et al., "Novel Methods to Stimulate Growth of Food Pathogens by Oxyrase and Related Membrane Fractions," in R. C. Spencer, E. P. Wright, and S. W. B. Newsome, eds., *Rapid Methods and Automation in Microbiology*, Intercept Limited, Andover, United Kingdom, 1994, pp. 313–326.

J. M. Jay, *Modern Food Microbiology*, 5th ed., Van Nostrand Reinhold, New York, 1996.

R. T. Marshall, ed., *Standard Methods for the Examination of Dairy Products*, 16th ed., American Public Health Association, Washington, D.C., 1992.

M. D. Pierson and N. J. Stern, *Foodborne Microorganisms and Their Toxins: Developing Methodology*, Marcel Dekker, New York, 1985.

B. Ray, *Fundamental Food Microbiology*, CRC Press, Boca Raton, Fla., 1996.

A. N. Sharpe, *Food Microbiology: A Framework for the Future*, Charles C. Thomas, Springfield, Ill., 1980.

J. H. Silliker et al., *Microbial Ecology of Foods*, Vols. 1 and 2, Academic Press, New York, 1980.

C. Vanderzant and D. Splittstoesser, *Compendium of Methods for the Microbiological Examination of Foods*, American Public Health Association, Washington, D.C., 1991.

DANIEL Y. C. FUNG
Kansas State University
Manhattan, Kansas

FOODBORNE MICROORGANISMS: DETECTION AND IDENTIFICATION

The many methods that are used to enumerate microorganisms or to detect their toxic products in foods can be placed in two groups. One group includes those methods that require the organisms to be viable, while the other includes methods that detect cells that may be nonliving, or parts and products of cells, and the two groups are listed in Table 1 along with the reported minimum detectable numbers of cells or products for each. In both groups are methods that are used primarily for the detection and enumeration or microorganisms, whereas some are used primarily to identify microorganisms.

Table 1. Examples of the Two Broad Categories of Enumeration/Detection Methods and their Minimum Response Cell Numbers

	Minimum numbers
A. Viable / respiring cells required	
1. Standard plate count	1
2. Most probable number (MPN)	<1
3. Dye reductions	$\sim 10^{5a}$
4. Hydrophobic grid membrane filter (HGMF)	<10
5. Microcolony-DEFT	10^3
6. Electrical impedance	10^6–10^7
7. Radiometry	10^4–10^5
8. Catalasemeter	$\sim 10^4$
9. Microcalorimetry	$\sim 10^4$
B. Viable cells are not needed	
1. Direct microscopic count (DMC)	10^4
2. Direct epifluorescent filter technique (DEFT)	10^3–10^4
3. Fluorescent antibody	10^6–10^7
4. ATP assay	10^5–10^6
5. *Limulus* amoebocyte lysate test (LAL)	$\sim 10^2$
6. Radioimmunoassay	$<10^3$
7. ELISA, EIA	10^5–10^6
8. Thermostable nuclease (TNase)	10^7
9. DNA probes	10^6–10^7

[a]Four hours for resazurin reduction.

DETECTION METHODS THAT REQUIRE VIABLE CELLS

This group of methods includes some of the oldest in use along with some that were developed only in the last two decades. Because they all depend on dividing cells, their speed is dependant directly upon the growth rate of the organisms of interest. Their chief value is that they may be used to determine the number of viable or colony-forming units (cfu) in a food product, and the number of organisms found depends on the factors of culture media employed, temperature and time of incubation, incubation environment (aerobic or anaerobic), pH, and a_w of culture medium. Each of those listed in Table 1 is further discussed as follows.

Standard/Aerobic Plate Count (SPC, APC)

Along with direct microscopic count, this is one of the oldest methods for enumerating bacteria in foods and food specimens. Official methods for determining cfus are described in standard references (1,2) and no further details are presented here.

There are two basic ways to conduct an SPC: pour and surface plating of samples or diluents. By the former, diluent is placed in empty petri dishes followed by pouring of cooled molten agar and mixing with diluted sample. With surface plating, diluents are spread over the surface of dried agar media, typically with bent glass rods (hockey sticks). The following factors may be considered in deciding whether to use the pour or surface plate method, assuming the same time and conditions of incubation.

1. Size of colonies: smaller on pour, larger on surface.
2. Spreading of colonies: less pronounced on pour.
3. Enumeration of psychrotrophs: better with surface.
4. Crowding of colonies on plate: more with surface.
5. Use of selective/differential media: better with surface.
6. Colonial features of colonies: better with surface.
7. Microaerophilic organisms: better with pour.
8. Strict aerobes: better with surface.
9. Colony pigmentation: better observed with surface.
10. Strict anaerobes: better with pour.
11. Subculturing of colonies: better with surface.

Reproducibility: Generally Better With Pour

Although surface plating has some advantages over pour plating, the latter method is much more widely used. The way in which food specimens are homogenized for cfu determinations is an important decision that affects the numbers of organisms found. Brisk shaking by hand or by use of mechanical shakers provides suitable results when comminuted meats or powdered food specimens are used, but reproducibility is not always good. Homogenizing with a Waring blender is a widely used method and this device allows one to homogenize chunks of food as well as comminuted products. One drawback to using this method is the heat build up that occurs in the container when blending is carried out for more than about 2 min. Other drawbacks to the use of the Waring blender have been discussed elsewhere (3). The Stomacher has emerged as the method of choice for food homogenizations. Developed by Sharpe and Jackson (4), this device has been compared to the Waring blender and other methods and found to give better results overall (5–7).

Spiral Plater. The use of the Spiral plating device is another way to achieve surface plating, and a large number of investigators have shown that the proper use of this device achieves results comparable to those by the more traditional methods (8–11). Among the advantages it offers is the use of only one petri dish to effect the enumeration of cfus over a wide range, and dilution blanks are not needed. The plating of solid foods can present problems unless care is taken to prevent particles from clogging the Spiral plater stylus.

Dry Film. A more recent surface-plating methodology is the use of dry film such as Petrifilm where an inoculum is spread over the surface of a prescribed area impregnated with culture medium. The colonies that develop are smaller than for the classical surface plate method, but their enumeration is aided by color development from tetrazolium in the film. This method has been shown to give results comparable to those by pour or surface plating in petri dishes (12,13).

Other Methods. Other variations of methods that allow one to enumerate cfus include roll tube methods that are especially useful for enumerating strict anaerobes, and Rodac or contact plates that may be used to determine cfus by their direct application to food plant surfaces. A common problem with contact plates employing nonselective media is the spreading of colony growth and also overgrowth by molds, but this can be minimized by using selective culture media for specific groups of organisms (14), or using properly dried contact plates when nonselective media are used.

Most Probable Numbers (MPN)

This is a statistical method that can be used to enumerate all viable cells in a food product or specific indicators, pathogens, or fungi, depending on the media employed. MPN is run either as a 3- or 5-tube method and results are generally higher than by either pour or surface Plating methods. Details of MPN methodology are presented in *Standard Methods* (2). Some of the problems encountered with plate count methods are avoided by the use of MPN. For example, problems presented by spreading colonies as well as those of inaccurate counts are avoided. On the down side, no information is provided on colony features or the types of organisms when nonselective media are employed. As noted above, MPN counts are generally higher than plate counts; and large quantities of glassware are needed (3). MPN is the method of choice for enumerating coliforms or *Escherichia coli* in foods and waters. This method has recently been reported for *Listeria* spp. (15) and *Salmonella* spp. (16) in raw meat and poultry.

Dye Reductions

Although these methods cannot be used to make precise determinations of microbial numbers, they can be used to make estimates of certain organisms within number ranges. The dyes most commonly used are methylene blue, resazurin, and tetrazolium, and detailed procedures for their use are described in *Standard Methods* (2).

Dye reduction methods are based on the fact that respiring microorganisms effect their reduction resulting in color changes. Methylene blue changes from blue to white, resazurin from slate blue to pink or white, and tetrazolium from colorless to its red formazan. In general, the color changes are essentially linear in the $\log_{10}$ 5 to 8 cfu/g range, although not all organisms are equal in their reductive abilities.

Dye reduction methods are normally employed as screens. Resazurin has been shown to be an excellent screen for $\log_{10}$ 5.0 cfu/g for raw milk (17,18), meats (19–21) poultry (22), and frozen shrimps (23). Methylene blue has a long history of use as a screen for raw milk (17) where the time (in hours) it takes for the blue color to be reduced to its colorless form is used as an approximation of overall microbial load. When compared to nitroblue tetrazolium and indophenyl nitrophenyl tetrazolium, resazurin produced faster results (20). Using surface samples from sheep carcasses, resazurin was reduced in 300 min by 18,000 cfu/cm^2, nitroblue in 600 min by 21,000 cfu/cm^2 whereas with indophenyl nitrophenyl tetrazolium 18,000 cfu/cm^2 effected reduction in 660 min (20).

These methods are both simple to use and inexpensive, but as noted above, some groups of organisms are more efficient in reducing the oxidized forms than others. Their use has been described (1) and discussed elsewhere (3).

Hydrophobic Grid Membrane Filter (HGMF)

A hydrophobic grid membrane filter consists of 1600 wax grids on a single membrane filter that restricts growth and colony size to individual grids (24–26). On one HGMF, from 10 to 9×10^4 cells can be enumerated by an MPN procedure, and enumeration can be automated. The method can detect as few as 10 cells/g and results can be achieved in 24 h or so, and it can be used to enumerate all cfu's or specific groups such as indicator organisms (27–29), yeasts and molds (30), or pathogens such as salmonellae (31). This method has been automated (30) and given AOAC approval (27). Hydrophobic grid membrane filter systems have been reported for enumerating *Escherichia coli* (32) and *Listeria monocytogenes* (33,34).

A typical application of HGMF consists of filtering 1 mL of a 1:10 homogenized sample through a HGMF membrane and placing the membrane on a suitable medium for incubation overnight to allow colonies to develop. The grids that contain colonies are enumerated and MPN is calculated.

Microcolony Direct Epifluorescent Filter Technique (Microcolony-DEFT)

The DEFT, described shortly is a method for determining viable and nonviable cells, whereas microcolony-DEFT is a variation used to determine viable cells only. Typically, food homogenates are filtered through DEFT membranes and the membranes are then placed on the surface of appropriate media and incubated for 3 to 6 h (3 h for gram-negative and 6 h for gram-positive bacteria). The microcolonies on the filters are stained and enumerated (35). The latter authors showed this method to be satisfactory for enumerating coliforms, pseudomonads, and staphylococci, and that as few as 10^3/g could be detected with results within 8 h.

In another variation of DEFT, Rodrigues and Kroll (36) devised a microcolony epifluorescence microscopy method that combines DEFT with the hydrophobic grid membrane filter (HGMF). By this method, nonenzyme detergent-treated samples are filtered through Nucleopore polycarbonate membranes. The membranes are transferred to the surface of selective agar and incubated for 3 h for gram-negative and 6 h for gram-positive bacteria. The membranes are then stained with acridine orange and the microcolonies are enumerated by epifluorescence microscopy. Results were achieved in less than 6 h without resuscitation for injured organisms, and in about 12 h when a resuscitation step was employed.

Electrical Impedance

Impedance does not lend itself to the enumeration of microbial numbers per se, but it can be used to indicate the presence of respiring cells in the range of about 10^4/g or above. The method is based on the measurement of impedance decrease caused by respiring microorganisms in suitable culture substrates, and it lends itself to the screening of food products for the presence or absence of a certain minimum number of cells (37,38). The lowest number of viable bacterial cells that can elicit an impedance response is in the 10^6–10^7/mL range. If a product contains a lower number of cells/milliliter, say, 10^3/g, the time it takes this number to reach the threshold level is the impedance detection time (IDT).

As a screen, impedance measurement may be used to determine how long it takes (in hours) for IDT where the longer the time the lower the number of cells in the food sample. By use of known numbers of organisms in appropriate substrates, one can determine cutoff times to reflect, eg, 10^5 cells/g. All aerobic bacteria (total numbers) may be detected as well as specific groups of organisms by using the appropriate growth media (39). Results can be obtained within hours, depending on initial numbers, and the method lends itself to automated data collection. This method has been used to detect coliforms in foods (40–42), to estimate the flora of frozen orange juice concentrate (43), raw meats (44–46), and milk (47). The potential food and nonfood applications for the technology were recently reviewed (48).

Radiometry

This method employs the use of ^{14}C-labeled fermentable substrates with the radiolabel on a carbon atom that respiring organisms can release as $^{14}CO_2$. In general, the larger the number of cells the more $^{14}CO_2$ released, and approximate numbers of cells are determined by the amount of radioactive carbon dioxide released. It normally requires 10^4–10^5 coliforms to release enough $^{14}CO_2$ from lactose to be detected.

Although radiometry is not widely used, it has been shown to be applicable to the detection of the microbial flora of foods (49,50) and to the specific detection of salmonellae (51).

Radiometry is used as a screen in much the same way as impedance and dye reduction methods, where the time required to produce a threshold quantity of $^{14}CO_2$ is a function of the number of cells that can release CO_2 from the labeled substrate. The radiolabel and the necessity for radioactivity counters make this method unpopular in the food industry.

Catalasemeter

This disk flotation device was designed by Gagnon et al. (52) and further developed at the University of Quebec (53) to detect the volume of oxygen released when microbially produced catalase acts on hydrogen peroxide. The catalasemeter consists of the preparation of food extracts that contain catalase (formed by catalase-positive organisms) followed by soaking a filter paper disk with the enzyme preparation. The filter paper disk is immediately placed in a tube containing 3% hydrogen peroxide. The disk-containing tube is placed in a photometer and the time in seconds that elapses between the initial fall of the disk to the bottom of the tube and its subsequent rise, due to the buoyancy caused by bubbles of oxygen at its surface, is recorded as flotation time. When large numbers of catalase-positive organisms are in the food preparation, more catalase is present to produce more oxygen from the hydrogen peroxide and subsequently the shorter the flotation time.

The Catalasemeter as described above was employed by Dodds et al. (54) to make rapid assessments of the microbial quality of vacuum-packaged cooked turkey products, and cfus of 10^4/g or higher could be detected accurately in 300 sec, with longer and less accurate flotation times required with lower numbers. Overall, the test correlated well with cfu's of catalase-positive Enterobacteriaceae with a correlation coefficient of 0.804.

Microcalorimetry

This method has received only limited study, but it offers potential as a screen for ca 10^4 microorganisms/g in canned foods as well as in ground beef. It is based on the fact that respiring cells emit heat and that accurate measurements of the heat correspond to cell numbers. Either a batch or flow type microcalorimeter is used, and some of these instruments are sensitive to a heat flow of 0.01 cal/h (55). The organisms that grow in canned foods can be detected by observing and measuring the thermogram produced when sensors are placed on the outside of cans (56). About 10^4 cells/g are necessary to give a detectable response (3), and specific groups of organisms, such as lactic acid bacteria (57) and yeasts (58,59) can be detected or characterized under appropriate conditions.

DETECTION OF ORGANISMS AND/OR THEIR PRODUCTS BY METHODS THAT DO NOT REQUIRE VIABLE CELLS

Unlike some of the enumeration methods that can be used to make rather precise determinations of microorganisms in foods, most of those that do not require cell viability can be used only to determine numbers at or above their respective threshold ranges. In addition to the latter, some of these methods lend themselves to the identification of certain groups of organisms, and they are discussed further in this regard in a later section. Their application for enumeration of cells or the detection of cell products is discussed as follows.

Direct Microscopic Count (DMC) Methods

In addition to being the oldest ways to determine the numbers of microorganisms in foods and other specimens, DMCs are the fastest. For food use, the two oldest and most widely used DMC methods are the Breed method for bacteria, and the Howard mold count slide for molds.

The Breed method consists of adding 0.01 ml of a sample to a 1-cm^2 area on a special microscope slide followed by drying, fixing, staining, and viewing with the oil immersion objective of a calibrated compound microscope (1). The method is widely used for raw milk, and it can be used for other food products such as powdered eggs. Results can be obtained in about 5 min. but a minimum of around 10^4 cells/mL are needed. Results obtained by this method are generally always higher than for viable cell methods such as APC since no distinctions can be made between viable and nonviable in cells (3). In spite of its drawbacks, this method is valuable when one wishes to know in the shortest possible time if a given level of microorganisms exists in a food product.

A slide method that may be used to detect and enumerate only viable cells was developed by Betts et al (60). Key to the method is the use of the tetrazolium salt 2-(*p*-iodophenyl-3-(*p*-nitrophenyl)5-phenyl tetrazolium chloride (INT). Cells are exposed to filter-sterilized INT for 10 min at 37°C in a water bath followed by filtration on 0.45 μm membranes. Following drying of membranes for 10 min at 50°C, the special membranes are mounted in cotton seed oil and viewed with coverslip in place. The method was found to be workable for pure cultures of bacteria and yeasts. Betts et al. found that the INT method underestimated APC by 1–1.5 log cycles when compared to APC using milk. By use of fluorescence microscopy and Viablue (modified aniline blue fluochrome), viable yeast cells could be differentiated from nonviable cells (61,62).

The Howard mold count is a slide method specifically developed for monitoring tomato products for molds (63). A similar method has been developed for *Geotrichum candidum* in fruits and vegetable products. These methods can be used to assess the prevalence of fungal contamination by observing their mycelia or mycelial fragments, but the values obtained rarely correspond to viable mold counts when plating methods are employed (3).

Direct Epifluorescent Filter Technique (DEFT)

DEFT is a more recent modification of the classical DMC method. A good review of this technique was published by Pettipher (64). By the basic method, a diluted food homogenate is filtered through a 5-μm nylon filter. The filtrate is collected and treated with 2 mL Triton X-100 and 0.5 mL trypsin. After incubation, the treated filtrate is passed through a 0.6-μm Nucleopore polycarbonate membrane and the filter is stained with acridine orange. Following drying, the stained cells are enumerated by epifluorescence microscopy and the number of cells/g calculated by multiplying the average number/field by the microscope factor (65–67). Results can be obtained in 25–30 min, and numbers as low as around 6,000 cfu/g can be obtained from meats and milk products.

DEFT has been used successively not only with milk but to estimate numbers of microorganisms on meat and poultry (67) and on food contact surfaces (68). It has been adapted to enumerate viable gram-negative and all gram-positive bacteria in milk in about 10 min (69). The further adaptation of DEFT to the enumeration of viable cells was previously discussed. A shortened procedure for DEFT developed by Champagne et al. (70) was used to assay the quality of raw milk in tanker deliveries.

Fluorescent Antibody (FA)

This is a microscope slide method that finds its widest use in food microbiology in the examination of foods for salmonellae (71,72), although the method can be used for any organisms to which an antibody can be made. By this method, an antibody is made to the organism of interest and made fluorescent by coupling it to a fluorescent compound such as fluorescein isocyanate. Following the preparation of slide smears, drying, and fixing, the coupled antibody–fluorescent dye is added. If the organism to which the antibody is made is present, it will react with the cells and evidence for the reaction is ascertained by observing for fluorescence under a microscope with oil immersion objective. The FA technique can be run as a direct or an indirect method, with the latter being preferred by many. By the latter, the homologous antibody is not coupled with fluorescent label but instead, an antibody to this antibody is prepared and coupled. This method obviates the need to prepare antibody for each organism of interest. More recent modifications of FA include the use of monoclonal antibodies, which appear to reduce false positive reactions (3). A modified antibody–DEFT technique has recently been reported by a number of researchers to provide a rapid and direct method for enumerating *E. coli* 0157:H7 in beef (73,74) and *Listeria* in vegetables (75).

This method is rapid but requires the presence of about 10^6–10^7 cells/ml for success in finding positive cells on the slide. It is official by AOAC (76), and it is more satisfactory in the hands of some than others. False positive results may run as high as 8–10% and false negatives may be as high as 1–3%, depending on the experience of the microscopist. The FA technique is more of a screen for the presence of targeted organisms in the 10^6–10^7/mL range than an enumeration method per se. It can be automated (77) and viable salmonellae cells can be enumerated when combined with a microcolony technique (78). Its utility as a method to identify given organisms is obvious.

ATP Assay

This method is based on the fact that all cells contain ATP and that the quantity detected in a specimen is referable to a given number of cells. The presence and amount of ATP are detected by use of the firefly luciferin-luciferase assay with the use of a photometer to measure emitted light (79,80). The problems that must be dealt with are that (1) the quantity of ATP varies for different bacterial cells, with the amount ranging from 0.1 to 4.0 fg/bacterial cell and for yeast cells from 13 to 100 fg/cell and (2) background ATP from food substances must be excluded from the analysis. By one commercially available system, background ATP can be eliminated and ATP assays can be run in an automated manner.

It generally requires 10^5–10^6 cells/g in order to have enough ATP to emit consistent detectable light, but results can be obtained within 1 h depending on the initial number of cells. The method has been shown to be comparable to APC for the aerobic flora of ground beef (81) over the APC range of 10^5–10^6 cells/g and to be applicable to seafoods (82), and yeasts in beverages (83). Sensitivity can be increased for beverages by filtration and concentration of cells. Recent studies by Ellerbroek et al. (84) found that the ATP assay was suitable for monitoring the hygiene of poultry carcasses in slaughterhouses by enumerating psychrotrophic and mesotrophic bacteria present. This method was also reported for assessing the microbial quality of milk samples (85).

Limulus Amoebocyte Lysate (LAL) Test

The LAL method is applicable to the determination of lipopolysaccharides (LPS) or endotoxins in foods, and as such the detection of gram-negative bacteria, both viable and nonviable. It employs the lysate from the amoebocytes of the horseshoe crab (*Limulus polyphemous*), which is the most specific substance known for detecting LPS (86). Because all gram-negative bacteria produce LPS while gram-positives do not, LAL is a test for gram-negative bacteria.

The LAL test can be run in several ways, and these have been presented and discussed elsewhere (87). The most commonly used is a tube gelation method where the food specimen is diluted in pyrogen-free water and 0.1 or 0.2 mL amounts are added to similar quantities of the LAL reagent in pyrogen-free tubes (all glassware and reagents must be pyrogen-free for use by this method). Following incubation at 37°C for 1 h, positive tests are indicated by the presence of firm gels. The chromogenic substrate method consists of using synthetic substrates that produce a color change in the presence of LPS and the LAL reagent, and the color change is read spectrophotometrically. The latter method can be run in 30 min, and it lends itself to automation (87,88).

Commercially available LAL reagents can detect 1.0 pg of LPS, and since a typical *Escherichia coli* cell contains about 3.0 fg of LPS, the LAL test should be able to respond to about 300 cells. Studies with pseudomonads from meats have shown the test to be capable of detecting 10^2 cfu/mL (89), and it has been shown to be effective for testing meats (90), raw milk (91–93), seafoods (94), sugars and some processed meats (54).

Since the total gram-negative flora of a food product can be determined within 1 h by use of LAL, this test has been adapted to the approximation of the total bacterial count using fresh ground beef (87,95). By this procedure, the gram-negative bacteria are represented by LAL values, whereas gram-positives are estimated by multiplying respective gram-negative numbers by predetermined ratio values of gram negatives to gram-positives. Estimated total bacteria by this procedure have been shown to compare favorably to the total count by use of APC methods (89,95).

Since the LAL test can be conducted in 1 h or less, it can be used to make rapid assessments of the microbial load or overall sanitation of a food product. In this application, low LAL values are more meaningful than high values since the latter could represent nonviable cells.

Radioimmunoassay (RIA)

This method does not lend itself to the enumeration of microbial cells per se, but it can be used to detect/quantitate toxic products of microbial cells. Briefly, RIA consists of adding a radioactive label (eg, ^{125}I) to a soluble antigen followed by reaction of the antigen with its homologous antibody bound to a solid such as polystyrene. After washing, the adherence of the radiolabel to the polystyrene is assessed by use of a radioactivity counter.

RIA is one of the most sensitive methods known for detecting staphylococcal enterotoxins (96,97), enterotoxins of *E coli* (98), *Vibrio cholerae* (99), mycotoxins (100,101) and other similar products (3). It is sensitive to 0.1 ng of staphylococcal enterotoxin A, 0.5 ng of aflatoxin B_1, and 20 ppb ochratoxin A. The most serious drawbacks to this method is the need for radioactive compounds and the consequent need for a radioactivity counter.

Enzyme Immunoassay, Enzyme-linked Immunoabsorbent Assay (EIA, ELISA)

In its simplest form, ELISA can be viewed as RIA without a radioactive element. Instead of the latter, an enzyme substrate reaction is used to indicate an homologous antigen—antibody reaction. Because of the absence of radioactivity, ELISA is much more widely used to detect microbial toxins in foods, and it has been used to detect salmonellae cells in food products (102–104). An automated ELISA test was recently reported for detecting *Salmonella* in chocolate (105). The ELISA test gave 15% fewer false negatives compared to conventional culture methods. For the latter, 10^4–10^5 cells/mL may be detected but the general minimum is around 10^6/mL. It can detect 0.1 ng/ml of *Clostridium perfringens* enterotoxin, and 10 pg/mL or less of aflatoxin M_1 (106) and other mycotoxins (107–109). ELISA tests to detect botulinal toxins Type A (110), Type E (111), and Type G (112) have been developed as well as tests for staphylococcal enterotoxins (113–115).

Although ELISA methods can be run with either polyclonal or monoclonal antibodies, the latter tend to give better results (116). A commercial method for detecting *Listeria monocytogenes* employing monoclonal antibodies is available (117). A collaborative study on the ELISA test proved it effective for detecting *L. monocytogenes* and other *Listeria* spp. in foods (118). This method was subsequently adopted by AOAC International for detecting *L. monocytogenes* in dairy, seafood, and meat products. A monoclonal method for salmonellae has been show capable of detecting as few as 10^6 cells/mL (119).

Thermostable Nuclease (TNase)

This test is specifically designed for *Staphylococcus aureus* and it can be run in 1–2 h (120). All strains of *S. aureus* that produce enterotoxins also produce this heat-stable nuclease, although many TNase-positive strains do not produce enterotoxins (121).

The microslide method consists of a layer of agar containing toluidine blue O and DNA into which holes are made for test samples. After heating to 100°C, food or food homogenates are placed in the holes and the slide is incubated at 37°C for 1–2 h. If TNase is present, the DNA is degraded and this is manifested by a change in color of the dye (122). It generally takes about 5×10^7 *S. aureus* cells to produce detectable amounts of TNase and this enzyme is generally produced before enterotoxins. Thus, this test can be used to screen foods for the presence of enterotoxins even in the absence of viable *S. aureus* cells (123,124). Occasional false positive tests may occur from *Streptococcus faecalis* strains that also produce TNase.

DNA Probes

Although DNA probes are most often used to detect specific organisms, their use may be viewed as screens for the organism in question since 10^6–10^7 cells/mL are needed for positive tests. In a typical application, one selects a unique DNA sequence of the organism of interest and tags a single strand with a radioisotope. DNA fragments of unknown organisms are prepared by use of restriction endonucleases. After separating the fragment strands, they are transferred to cellulose nitrate filters and hybridized to the radioactive probe. After gentle washing to remove unreacted probe DNA, the presence of the radiolabel is detected by autoradiography or by use of a radioactivity counter.

A number of investigators have developed and tested DNA probes for salmonella (125–127), listeriae (128–131), and other food-borne pathogens and found the method to be satisfactory for detecting the presence of target organisms in mixed culture. Commercial methods exist for salmonellae and listeriae, but they generally require 44+ h when used on products that contain low numbers of target organisms.

DNA probes are used in colony hybridization methods where micro- or macrocolonies of the target organism are allowed to develop directly on membranes that are incubated on suitable agar plates. Following treatment of the colonies to separate and fix DNA strands, the DNA probe is applied to the membrane and the radiolabel is tested for. Colony hybridizations have been employed with success to detect *Listeria monocytogenes*, enterotoxigenic strains of *E. coli* (132,133), and *Yersinia enterocolitica*. These developments have resulted in the polymerase chain reaction (PCR) technique which has found considerable application for detecting foodborne pathogens (134).

One of the drawbacks to the use of DNA probes in the food industry is the presence of radioactive elements. One of the most promising nonisotope probes employs biotin and its detection using a streptavidine-alkaline phosphatase conjugate to produce an insoluble color precipitate in the presence of a dye. One such method was developed (135) and shown to be workable for *E. coli* and to detect 2×10^7 cells/g in about 30 h. The nonisotopic detection system required about 3 h for results.

BIBLIOGRAPHY

1. E. H. Marth, ed., *Standard Methods for the Examination of Dairy Products*, 14th ed., American Public Health Association, Washington, D.C., 1978.
2. M. L. Speck, ed., *Compendium of Methods for the Microbiological Examination of Foods*, 2nd ed., American Public Health Association, Washington, D.C., 1984.
3. J. M. Jay, *Modern Food Microbiology*, 3rd ed., Van Nostrand Reinhold, New York, 1986.
4. A. N. Sharpe and A. K. Jackson, "Stomaching: A New Concept in Bacteriological Sample Preparation," *Applied Microbiology* **24**, 175–178 (1978).
5. B. S. Emswiler, C. J. Pierson, and A. W. Kotula, "Stomaching vs Blending. A Comparison of Two Techniques for the Homogenization of Meat Samples for Microbiological Analysis," *Food Technology* **31**, 40–42 (1977).
6. W. H. Andrews, C. R. Wilson, P. L. Poelma, A. Romero, R. A. Ruce, A. P. Duran, P. D. McClure, and D. E. Gentile, "Usefulness of the Stomacher in a Microbiological Regulatory Laboratory," *Applied Environmental Microbiology* **35**, 89–93 (1978).
7. J. M. Jay and S. Margitic, "Comparison of Homogenizing, Shaking, and Blending on the Recovery of Microorganisms and Endotoxins from Fresh and Frozen Ground Beef as As-

sessed by Plate Counts and the *Limulus* Amoebocyte Lysate Test," *Applied Environmental Microbiology* **38**, 879–884 (1979).

8. J. E. Gilchrist, J. E. Campbell, C. B. Donnelly, J. T. Peeler, and J. M. Delaney, "Spiral Plate Method for Bacterial Determination," *Applied Microbiology* **25**, 244–252 (1973).
9. C. B. Donnelly, J. E. Gilchrist, J. T. Peeler, and J. E. Campbell, "Spiral Plate Count Method for the Examination of Raw and Pasteurized Milk," *Applied Environmental Microbiology* **32**, 21–27 (1976).
10. B. Jarvis, V. H. Lach, and J. M. Wood, "Evaluation of the Spiral Plate Maker for the Enumeration of Micro-organisms in Foods," *Journal of Applied Bacteriology* **43**, 149–157 (1977).
11. Association of Official Analytical Chemists "Spiral Plate Method for Bacterial Count: Official First Action," *Journal of the Association of Official Analytical Chemists* **60**, 493–494 (1977).
12. C. L. Nelson, T. L. Fox, and F. F. Busta, "Evaluation of Dry Medium Film (Petrifilm VRB) for Coliform Enumeration," *Journal of Food Protection* **47**, 520–525 (1984).
13. L. B. Smith, T. L. Fox, and F. F. Busta, "Comparison of a Dry Medium Culture Plate (Petrifilm SM plates) Method to the Aerobic Plate Count Method for Enumeration of Mesophilic Aerobic Colony-forming Units in Fresh Ground Beef," *Journal of Food Protection* **48**, 1044–1045 (1985).
14. M. P. deFigueiredo and J. M. Jay, "Coliforms, Enterococci, and Other Microbial Indicators," in M. P. deFigueiredo and D. F. Splittstoesser, eds., *Food Microbiology: Public Health and Spoilage Aspects*, Avi Publishing, Westport, Conn., 1976.
15. L. S. L. Yu, R. K. Prasai, and D. Y. C. Fung, "Most Probable Numbers of *Listeria* Species in Raw Meats Detected by Selective Motility Enrichment," *Journal of Food Protection* **58**, 943–945 (1995).
16. E. S. Humbert et al., "Miniaturized Most Probable Number and Enrichment Serology Technique for the Enumeration of *Salmonella* spp. on Poultry Carcasses," *Journal of Food Protection* **60**, 1306–1311 (1997).
17. C. K. Johns, "Place of the Methylene Blue and Resazurin Reduction Tests in a Milk Control Program," *American Journal of Public Health* **29**, 239–247 (1939).
18. R. Dabbah, W. A. Moats, S. R. Tatini, and J. C. Olson, Jr., "Evaluation of the Resazurin Reduction One-Hour Test for Grading Milk Intended for Manufacturing Purposes," *Journal of Milk Food Technology* **32**, 44–48 (1969).
19. P. J. Dodsworth and A. G. Kempton, "Rapid Measurement of Meat Quality by Resazurin Reduction. II. Industrial Application," *Canadian Institute of Food Science Technology Journal* **10**, 158–160 (1977).
20. D. N. Rao and V. S. Murthy, "Rapid Dye Reduction Tests for the Determination of Microbiological Quality of Meat," *Journal of Food Technology* **21**, 151–157 (1986).
21. R. A. Holley, S. M. Smith, and A. G. Kempton, "Rapid Measurement of Meat Quality by Resazurin Reduction. I. Factors Affecting Test Validity," *Canadian Institute of Food Science Technology Journal* **10**, 153–157 (1977).
22. H. W. Walker, W. J. Coffin, and J. C. Ayres, "A Resazurin Reduction Test for Determination of Microbiological Quality of Processed Poultry," *Food Technology* **13**, 578–581 (1959).
23. R. Kummerlin, "Technical Note: Resazurin Test for Microbiological Control of Deep-Frozen Shrimps," *Journal of Food Technology* **17**, 513–515 (1982).
24. A. N. Sharpe and G. L. Michaud, "Hydrophobic Grid-Membrane Filters: New Approach to Microbiological Enumeration," *Applied Microbiology* **28**, 223–225 (1974).
25. A. N. Sharpe and G. L. Michaud, "Enumeration of High Numbers of Bacteria Using Hydrophobic Grid-Membrane Filters," *Applied Microbiology* **30**, 519–524 (1975).
26. A. N. Sharpe, M. P. Diotte, I. Dudas, S. Malcolm, and P. I. Peterkin, "Colony Counting on Hydrophobic Grid-Membrane Filters" *Canadian Journal of Microbiology* **29**, 797–802 (1983).
27. A. O. A. C., "Enumeration of Coliforms in Selected Foods, Hydrophobic Grid Membrane Filter Method, Official First Action," *Journal of the Association of Official Analytical Chemists* **66**, 547–548 (1983).
28. P. Entis, "Enumeration of Coliforms in Non-Fat Dry Milk and Canned Custard by Hydrophobic Grid Membrane Filter Method: Collaborative Study," *Journal of the Association of Official Analytical Chemists* **66**, 897–904 (1983).
29. M. H. Brodsky, P. Entis, A. N. Sharpe, and G. A. Jarvis "Enumeration of Indicator Organisms in Foods Using the Automated Hydrophobic Grid Membrane Filter Technique," *Journal of Food Protection* **45**, 292–296 (1982).
30. M. H. Brodsky, P. Entis, M. P. Entis, A. N. Sharpe, and G. A. Jarvis, "Determination of Aerobic Plate and Yeast and Mold Counts in Foods Using an Automated Hydrophobic Grid Membrane Filter Technique," *Journal of Food Protection* **45**, 301–304 (1982).
31. P. Entis, "Rapid Hydrophobic Grid Membrane Filter Method for *Salmonella* Detection in Selected Foods," *Journal of the Association of Official Analytical Chemists* **68**, 555 (1985).
32. J. R. Sage and S. C. Ingham, "Survival of *Escherichia coli* 0157:H7 After Freezing and Thawing Ground Beef Patties," *Journal of Food Protection* **61**, 1181–1183 (1998).
33. P. I. Peterkin, E. S. Idziak, and A. V. Sharpe, "Use of Hydrophobic Grid-Membrane Filter DNA Probe to Detect *Listeria monocytogenes* in Artificially-Contaminated Food," *Food Microbiology* **9**, 155–160 (1992).
34. W. Yan et al., "Comparison of the Hydrophobic-Grid Membrane Filter DNA Probe Method and the Health Protection Standard Method for the Detection of *Listeria monocytogenes* in Food," *International Journal of Food Microbiology* **30**, 379–384 (1996).
35. U. M. Rodrigues and R. G. Kroll, "Rapid Selective Enumeration of Bacteria in Foods Using a Micro-Colony Epifluorescence Microscopy Technique," *Journal of Applied Bacteriology* **64**, 65–78 (1988).
36. U. M. Rodrigues and R. G. Kroll, "Microcolony Epifluorescence Microscopy for Selective Enumeration of Injured Bacteria in Frozen and Heat-Treated Foods," *Applied Environmental Microbiology* **55**, 778–787 (1989).
37. D. Hardy, S. J. Kraeger, S. W. Dufour, and P. Cady, "Rapid Detection of Microbial Contamination in Frozen Vegetables by Automated Impedance Measurements," *Applied Environmental Microbiology* **34**, 14–17 (1977).
38. P. Cady, D. Hardy, S. Martins, S. W. Dufour, and S. J. Kraeger, "Automated Impedance Measurements for Rapid Screening of Milk Microbial Content," *Journal of Food Protection* **41**, 277–283 (1978).
39. J. M. Wood, V. Lach, and B. Jarvis, "Detection of Food-Associated Microbes Using Electrical Impedance Measurements," *Journal of Applied Bacteriology* **43**, xiv–xv (1977).
40. M. P. Silverman and E. F. Munoz, "Automated Electrical Impedance Technique for Rapid Enumeration of Fecal Coli-

forms in Effluents From Sewage Treatment Plants," *Applied Environmental Microbiology* **37**, 521–526 (1979).

41. S. B. Martins and M. J. Selby, "Evaluation of a Rapid Method for the Quantitative Estimation of Coliforms in Meat by Impedimetric Procedures," *Applied Environmental Microbiology* **39**, 518–524 (1980).
42. R. Firstenberg-Eden and C. S. Klein, "Evaluation of a Rapid Impedimetric Procedure for the Quantitative Estimation of Coli-Forms" *Journal of Food Science* **48**, 1307–1311 (1983).
43. J. L. Weihe, S. L. Seist, and W. S. Hatcher, Jr., "Estimation of Microbial Populations in Frozen Concentrated Orange Juice Using Automated Impedance Measurements," *Journal of Food Science* **49**, 243–245 (1984).
44. R. Firstenberg-Eden, "Rapid Estimation of the Number of Microorganisms in Raw Meat by Impedance Measurement," *Food Technology* **37**, 64–70 (1983).
45. R. Andreine, "Automated Microbiological Techniques," *Latte* **19**, 456–463 (1994).
46. M. Campanini, S. Barbuti, and G. Camorali, "Quick Evaluation of Microbial Contamination of Fresh Meat by an Impediometric Procedure," *Industria-Conserve* **64**, 121–125 (1989).
47. R. Firstenberg-Eden and M. K. Tricarico, "Impedimetric Determination of Total, Mesophilic and Psychotrophic Counts in Raw Milk," *Journal of Food Science* **48**, 1750–1754 (1983).
48. P. Silley and S. Forsythe, "Impedance Microbiology—A Rapid Change for Microbiologists," *Journal of Applied Bacteriology* **80**, 233–243 (1996).
49. R. A. Lampi, D. A. Mikelson, D. B. Rowley, J. J. Previte, and R. E. Wells, "Radiometry and Microcalorimetry—Techniques for the Rapid Detection of Foodborne Microorganisms," *Food Technology* **28**, 52–55 (1974).
50. D. B. Rowley, J. J. Previte, and H. P. Srinivasa, "A Radiometric Method for Rapid Screening of Cooked Foods for Microbial Acceptability," *Journal of Food Science* **43**, 1720–1722 (1978).
51. B. J. Stewart, M. J. Eyles, and W. G. Murrell, "Rapid Radiometric Method for Detection of *Salmonella* in Foods," *Applied Environmental Microbiology* **40**, 223–230 (1980).
52. M. Gagnon, W. M. Hunting, and B. Esselen, "New Method for Catalase Determination," *Analytical Chemistry* **31**, 144–146 (1959).
53. R. Charbonneau, J. Therrien, and M. Gagnon, "Detection and Measurement of Bacterial Catalase by the Disk-Flotation Method Using the Catalasemeter," *Canadian Journal of Microbiology* **21**, 580–582 (1975).
54. K. L. Dodds, R. A. Holley, and A. G. Kempton, "Evaluation of the Catalase and *Limulus* Amoebocyte Lysate Tests for Rapid Determination of the Microbial Quality of Vacuum-Packed Cooked Turkey," *Canadian Institute of Food Science Technology Journal* **16**, 167–172 (1983).
55. W. W. Forrest, "Microcalorimetry," *Methods in Microbiology* **6B**, 285–318 (1972).
56. L. E. Sacks and E. Menefee, "Thermal Detection of Spoilage in Canned Foods," *Journal of Food Science* **37**, 928–931 (1972).
57. T. Fugita, P. R. Mond, and I. Wadso, "Calorimetric Identification of Several Strains of Lactic Acid Bacteria," *Journal of Dairy Research* **45**, 457–463 (1978).
58. B. F. Perry, A. E. Beezer, and R. J. Miles, "Flow Microcalorimetric Studies of Yeast growth: Fundamental Aspects," *Journal of Applied Bacteriology* **47**, 527–537 (1979).
59. B. F. Perry, A. E. Beezer, and R. J. Miles, "Characterization of Commercial Yeast Strains By Flow Microcalorimetry," *Journal of Applied Bacteriology* **54**, 183–189 (1983).
60. R. P. Betts, P. Bankes, and J. G. Board, "Rapid Enumeration of Viable Micro-Organisms by Staining and Direct Microscopy," *Letters in Applied Microbiology* **9**, 199–202 (1989).
61. T. C. Hutcheson, T. McKay, L. Farr, and B. Seddon, "Evaluation of the Strain Viablue for the Rapid Estimation of Viable Yeast Cells," *Letters in Applied Microbiology* **6**, 85–88 (1988).
62. H. A. Koch, R. Bandler, and R. R. Gibson, "Fluorescence Microscopy Procedure for Quantification of Yeasts in Beverages," *Applied Environmental Microbiology* **52**, 599–601 (1986).
63. Association of Official Analytical Chemists, *Official Methods of Analysis*, 15th ed., Vol. 1, AOAC, Arlington, Va., 1990.
64. G. L. Pettipher, "The Direct Epifluourescent Filter Technique," in M. R. Adams and C. F. A. Hope, eds., *Progress in Industrial Microbiology*, Vol. 26, Elsevier, Amsterdam, 1989, pp. 19–26.
65. J. E. Hobbie, R. J. Daley, and S. Jasper, "Use of Nucleopore Filters for Counting Bacteria by Fluorescence Microscopy," *Applied Environmental Microbiology* **33**, 1225–1228 (1977).
66. G. L. Pettipher, R. Mansell, C. H. McKinnon, and C. M. Cousins, "Rapid Membrane Filtration-Epifluorescent Microscopy Technique for Direct Enumeration of Bacteria in Raw Milk," *Applied Environmental Microbiology* **39**, 423–429 (1980).
67. B. G. Shaw, C. D. Harding, W. H. Hudson, and L. Farr, "Rapid Estimation of Microbial Numbers on Meat and Poultry by Direct Epifluorescent Filter Technique," *Journal of Food Protection* **50**, 652–657 (1987).
68. J. T. Holah, R. P. Betts, and R. H. Thorpe, "The Use of Direct Epifluorescent Microscopy (DEM) and the Direct Epifluorescent Filter Technique (DEFT) to Assess Microbial Populations on Food Contact Surfaces," *Journal of Applied Bacteriology* **65**, 215–221 (1988).
69. U. M. Rodrigues and R. G. Kroll, "The Direct Epifluorescent Filter Technique (DEFT): Increased Selectivity, Sensitivity and Rapidity," *Journal of Applied Bacteriology* **59**, 493–499 (1985).
70. C. P. Champagne et al., "Determination of Viable Bacterial Populations in Raw Milk Within 20 Minutes by Using a Direct Epifluorescent Filter Technique," *Journal of Food Protection* **60**, 874–876 (1997).
71. J. M. Goepfert, M. E. Mann, and R. Hicks, "One-day Fluorescent-Antibody Procedure for Detecting Salmonellae in Frozen and Dried Foods," *Applied Microbiology* **20**, 977–983 (1970).
72. B. M. Thomason, "Rapid Detection of *Salmonella* Microcolonies by Fluorescent Antibody," *Applied Microbiology* **22**, 1064–1069 (1971).
73. L. Restaino et al., "A 5-h Screening and 24-h Confirmation Procedure for Detecting *Escherichia coli* 0157:H7 in Beef Using Direct Epifluourescent Microscopy and Immunomagnetic Separation," *Letters in Applied Microbiology* **24**, 401–404 (1997).
74. M. L. Tortorello and D. S. Stewart, "Antibody-Direct Epifluourescent Filter Technique for Rapid Direct Enumeration of *Escherichia coli* 0157:H7 in Beef," *Applied and Environmental Microbiology* **60**, 3553–3559 (1994).
75. M. L. Tortorello, K. F. Reinecke, and D. S. Stewart, "Comparison of Antibody-Direct Epifluourescent Filter Technique With Most Probable Number Procedure for Rapid Enumeration of *Listeria* in Fresh Vegetables," *Journal of AOAC International* **80**, 1208–1214 (1997).

76. L. D. Fantasia, J. P. Schrade, J. F. Yager, and D. Debler, "Fluorescent Antibody Method for the Detection of *Salmonella*: Development, Evaluation, and Collaborative Study," *Journal of the Association of Official Analytical Chemists* **58**, 828–844 (1975).
77. T. E. Munson, J. P. Schrade, N. B. Bisciello, Jr., L. D. Fantasia, W. R. Hartung, and J. J. O'Connor, "Evaluation of an Automated Fluorescence Antibody Procedure for Detection of *Salmonella* in Foods and Feeds," *Applied Environmental Microbiology* **31**, 514–521 (1976).
78. U. M. Rodrigues and R. G. Kroll, "Rapid Detection of Salmonellas in Raw Meats using a Fluorescent Antibody-Microcolony technique," *Journal of Applied Bacteriology* **68**, 213–223 (1990).
79. A. N. Sharpe, M. N. Woodrow, and A. K. Jackson, "Adenosinetriphosphate (ATP) Levels in Foods Contaminated by Bacteria," *Journal of Applied Bacteriology* **33**, 758–767 (1970).
80. G. A. Kimmich, J. Randles, and J. S. Brand, "Assay of Picomole Amounts of ATP, ADP, and AMP Using the Luciferase Enzyme System," *Analytical Biochemistry* **69**, 187–206 (1975).
81. C. J. Stannard and J. M. Wood, "The Rapid Estimation of Microbial Contamination of Raw Meat by Measurement of Adenosinetriphosphate (ATP)," *Journal of Applied Bacteriology* **55**, 429–438 (1983).
82. D. R. Ward, K. A. LaRocco, and D. J. Hopson, "Adenosine Triphosphate Bioluminescent Assay to Enumerate Bacterial Numbers on Fresh Fish," *Journal of Food Protection* **49**, 647–650 (1986).
83. K. A. LaRocco, P. Galligan, K. J. Littel, and A. Spurgash, "A Rapid Bioluminescent ATP Method for Determining Yeast Contamination in a Carbonated Beverage," *Food Technology* **39**, 49–52 (1985).
84. L. Ellerbroek et al., "Hygiene Monitoring in Poultry Meat Production, Use of Bioluminescence ATP Assay," *Fleischwirtschaft* **7**, 486–487, 489 (1998).
85. S. Girotti et al., "Determination of Microbial Contamination in Milk by ATP Assay," *Czech Journal of Food Science* **15**, 241–248 (1997).
86. J. D. Sullivan, Jr., F. W. Valois, and S. W. Watson, "Endotoxins: The Limulus Amoebocyte Lystae System, in A. W. Bernheimer, ed., *Mechanisms in Bacterial Toxinology*, Wiley, New York, 1976.
87. J. M. Jay, "The *Limulus* Amoebocyte Lysate (LAL) Test, in M. R. Adams and C. F. A. Hope, eds., *Progress in Industrial Microbiology: Rapid Methods in Food Microbiology*, Elsevier, Amsterdam, 1989.
88. K. Tsuji, P. A. Martin, and D. M. Bussey, "Automation of Chromogenic Substrate *Limulus*Amebocyte Lysate Assay Method for Endotoxin by Robotic System." *Applied Environmental Microbiology* **48**, 550–555 (1984).
89. H. J. Fallowfield and J. T. Patterson, "Potential Value of the *Limulus* Lysate Assay for the Measurement of Meat Spoilage," *Journal of Food Technology* **20**, 467–479 (1985).
90. J. M. Jay, S. Margitic, A. L. Shereda, and H. V. Covington, "Determining Endotoxin Content of Ground Beef by the *Limulus* Amoebocyte Lysate Test as a Rapid Indicator of Microbial Quality," *Applied Environmental Microbiology* **38**, 885–890 (1979).
91. K.-J. Zaadhof and G. Terplan, "Der *Limulus*-test—ein Verfahren zur Buerteilung der Mikrobiologischen Qualität von Milch und Milchprodukten," *Deutsch Molkerzeit* **34**, 1094–1098 (1981).
92. J. Mottar et al., "Routine Limulus Amoebocyte Lysate (LAL) Test for Endotoxin Determination in Milk Using a Toxinometer ET-201," *Journal of Dairy Science* **60**, 223–228 (1993).
93. P. C. Vasavada, "Rapid Methods and Automation in Dairy Microbiology," *Journal of Dairy Science* **76**, 3101–3113 (1993).
94. J. D. Sullivan, Jr., P. E. Ellis, R. G. Lee, W. S. Combs, Jr., and S. W. Watson, "Comparison of the *Limulus* Amoebocyte Lysate Test with Plate Counts and Chemical Analyses for Assessment of the Quality of Lean Fish," *Applied Environmental Microbiology* **45**, 720–722 (1983).
95. J. M. Jay, "Rapid Estimation of Microbial Numbers in Fresh Ground Beef By Use of the *Limulus* Test," *Journal of Food Protection* **44**, 275–278 (1981).
96. B. A. Miller, R. F. Reiser, and M. S. Bergdoll, "Detection of Staphylococcal Enterotoxins A, B, C, D, and E in Foods By Radioimmunoassay, Using Staphylococcal Cells Containing Protein A as Immunoadsorbent," *Applied Environmental Microbiology* **36**, 421–426 (1978).
97. H. Robern, M. Dighton, Y. Yano, and N. Dickie, "Double-Antibody Radioimmunoassay for Staphylococcal Enterotoxin C_2," *Applied Microbiology* **30**, 525–529 (1975).
98. R. A. Giannella, K. W. Drake, and M. Luttrell, "Development of a Radioimmunoassay for *Escherichia coli* Heat-Stable Enterotoxin: Comparison With the Suckling Mouse Bioassay," *Infectious Immunology* **33**, 186–192 (1981).
99. D. B. Shah, P. E. Kauffman, B. K. Boutin, and C. H. Johnson, "Detection of Heat-Labile-Enterotoxin-Producing Colonies of *Escherichia coli* and *Vibrio cholerae* by Solid-Phase Sandwich Radioimmunoassay," *Journal of Clinical Microbiology* **16**, 504–508 (1982).
100. O. Aalund, K. Brunfeldt, B. Hald, P. Krogh, and K. Poulsen, "A Radioimmunoassay for Ochratoxin A: A Preliminary Investigation," *Acta Pathologic Microbiologica Scandinavica Section* **83**, 390–392 (1975).
101. J. J. Pestka, V. Li, W. O. Harder, and F. S. Chu, "Comparison of Radioimmunoassay and Enzyme-Linked Immunosorbent Assay for Determining Aflatoxin M_1 in Milk," *Journal of the Association of Official Analytical Chemists* **65**, 294–301 (1981).
102. S. A. Minnich, P. A. Hartman, and R. C. Heimsch, "Enzyme Immunoassay for Detection of Salmonellae in Foods," *Applied Environmental Microbiology* **43**, 877–883 (1982).
103. B. J. Robison, C. I. Pretzman, and J. A. Mattingly, "Enzyme Immunoassay in Which a Myeloma Protein is Used for Detection of Salmonellae," *Applied Environmental Microbiology* **45**, 1816–1821 (1983).
104. R. S. Flowers, K.-H. Chen, B. J. Robison, J. A. Mattingly, D. A. Gabis, and J. H. Silliker, "Comparison of *Salmonella* Bio-EnzaBead Immunoassay Method and Conventional Culture Procedure for Detection of *Salmonella* in Crustaceans," *Journal of Food Protection* **50**, 386–389 (1987).
105. B. Pacher-Beck, "Automated ELISA Test for Detection of *Salmonella* in Chocolate—A Note," *Lebensmittel & Biotechnologie* **13**, 59 (1996).
106. W. J. Hu, N. Woychik, and F. S. Chu. "ELISA of Picogram Quantities of Aflatoxin M_1 in Urine and Milk," *Journal of Food Protection* **47**, 126–127 (1984).
107. J. J. Pestka, P. K. Gaur, and F. S. Chu, "Quantitation of Aflatoxin B_1 and Aflatoxin B_1 Antibody by an Enzyme-Linked Immunosorbent Microassay," *Applied Environmental Microbiology* **40**, 1027–1031 (1980).

108. J. J. Pestka, B. W. Steinert, and F. S. Chu, "Enzyme-Linked Immunosorbent Assay for Detection of Ochratoxin A," *Applied Environmental Microbiology* **41**, 1472–1474 (1981).
109. J. J. Pestka and F. S. Chu, "Enzyme-Linked Immunosorbent Assay of Mycotoxins Using Nylon Bead and Terasaki Plate Solid Phases," *Journal of Food Protection* **47**, 305–308 (1984).
110. S. Notermans, J. Dufrenne, and M. van Schothorst, "Enzyme-Linked Immunosorbent Assay for Detection of *Clostridium botulinum* Type A," *Japanese Journal of Medical Science Biology* **31**, 81–85 (1978).
111. S. Notermans, J. Dufrenne, and S. Kozaki, "Enzyme-Linked Immunosorbent Assay for Detection of *Clostridium botulinum* Type E Toxin," *Applied Environmental Microbiology* **37**, 1173–1175 (1979).
112. G. E. Lewis, Jr., S. S. Kulinski, D. W. Reichard, and J. F. Metzger, "Detection of *Clostridium botulinum* Type G Toxin by Enzyme-Linked Immunosorbent Assay," *Applied Environmental Microbiology* **42**, 1018–1022 (1981).
113. G. C. Saunders and M. L. Bartlett, "Double-Antibody Solid-Phase Enzyme Immunoassay for the Detection of Staphylococcal Enterotoxin A," *Applied Environmental Microbiology* **34**, 518–522 (1977).
114. G. Stiffler-Rosenberg and H. Fey, "Simple Assay for Staphylococcal Enterotoxins A, B, and C: Modification of Enzyme-Linked Immunosorbent Assay," *Journal of Clinical Microbiology* **8**, 473–479 (1978).
115. S. Notermans, R. Boot, P. D. Tips, and M. P. DeNooij, "Extraction of Staphylococcal Enterotoxins (SE) from minced meat and subsequent detection of SE with Enzyme-Linked Immunosorbent Assay (ELISA)," *Journal of Food Protection* **46**, 238–241 (1983).
116. J. M. Farber and J. I. Speirs, "Monoclonal Antibodies Directed Against the Flagella Antigens of *Listeria* Species and Their Potential in EIA-Based Methods," *Journal of Food Protection* **50**, 479–484 (1987).
117. B. T. Butnam, M. C. Plank, R. J. Durham, and J. M. Mattingly, "Monoclonal Antibodies Which Identify a Genus-Specific *Listeria* Antigen," *Applied Environmental Microbiology* **54**, 1564–1569 (1988).
118. M. S. Curiale, W. Lepper, and B. Robinson, "Enzyme-Linked Immunoassay for Detection of *Listeria monocytogenes* in Dairy Products, Seafoods and Meats: A Collaborative Study," *Journal of the AOAC International* **77**, 1472–1489 (1994).
119. J. A. Mattingly, "An Enzyme Immunoassay for the Detection of All *Salmonella* Using a Combination of Myeloma Protein as a Hybridoma Antibody," *Journal of Immunological Methods* **73**, 147–156 (1984).
120. B. M. Hill, "The Thermostable Nuclease Test as a Method for Identifying *Staphylococcus aureus*," *Australian Journal of Dairy Technology* **38**, 95–96 (1983).
121. C. E. Park, H. B. El Derea, and M. K. Rayman, "Evaluation of Staphylococcal Thermonuclease (TNase) Assay as a Means of Screening Foods for Growth of Staphylococci and Possible Enterotoxin Production," *Canadian Journal of Microbiology* **24**, 1135–1139 (1978).
122. A. Koupal and R. H. Deibel, "Rapid Qualitative Method for Detecting Staphylococcal Nuclease in Foods," *Applied Environmental Microbiology* **35**, 1193–1197 (1978).
123. J. F. Kamman and S. R. Tatini, "Optimal Conditions for Assay of Staphylococcal Nuclease," *Journal of Food Science* **42**, 421–424 (1977).
124. B. S. Emswiler-Rose, R. W. Johnston, M. E. Harris, and W. H. Lee, "Rapid Detection of Staphylococcal Thermonuclease on Casings of Naturally Contaminated Fermented Sausages," *Applied Environmental Microbiology* **40**, 13–18 (1980).
125. R. Fitts, M. L. Diamond, C. Hamilton, and M. Neri, "DNA-DNA Hybridization Assay for Detection of *Salmonella* Spp. in Foods," *Applied Environmental Microbiology* **46**, 1146–1151 (1983).
126. R. S. Flowers, M. A. Mozzola, M. S. Curiale, D. A. Gabis, and J. H. Silliker, "Comparative Study of a DNA Hybridization Method and the Conventional Culture Procedure for Detection of *Salmonella* in Foods," *Journal of Food Science* **52**, 781–785 (1987).
127. J. R. Bandekar, K. K. Ussuf, and P. M. Nair, "Detection of *Salmonella* in Chicken and Fish Samples Using DNA Probes," *Journal of Food Safety* **15**, 11–19 (1995).
128. S. Notermans, T. Chakraborty, M. Leimeister-Wächter, J. Dufrenne, K. J. Heuvelman, H. Maas, W. Jansen, K. Wernars, and P. Guinee, "Specific Gene Probe for Detection of Biotyped and Serotyped *Listeria* Strains," *Applied Environmental Microbiology* **55**, 902–906 (1989).
129. A. R. Datta, B. A. Wentz, D. Shook, and M. W. Trucksess, "Synthetic Oligodeoxyribonucleotide Probes for Detection of *Listeria Monocytogenes*," *Applied Environmental Microbiology* **54**, 2933–2937 (1988).
130. R. F. Duvall and A. D. Hitchins, "Pooling Noncollaborative Multilaboratory Data for Evaluation of the Use of DNA Probe Test Kits in Identifying *Listeria monocytogenes* Strains," *Journal of Food Protection* **60**, 995–997 (1997).
131. L. Herman and H. de Ridder, "Evaluation of a DNA Probe Assay for the Identification of *Listeria monocytogenes*," *Michwissenschaft* **48**, 126–128 (1993).
132. S. L. Moseley, P. Echeverria, J. Seriwatana, C. Tirapat, W. Chaicumpa, T. Sakuldaipeara, and S. Falkow, "Identification of Enterotoxigenic *Escherichia coli* by Colony Hybridization Using Three Enterotoxin Gene Probes," *Journal of Infectious Disease* **145**, 863–869 (1982).
133. W. E. Hill, J. M. Madden, B. A. McCardell, D. B. Shah, J. A. Jagow, W. L. Payne, and B. K. Boutin, "Foodborne Enterotoxigenic *Escherichia coli*: Detection and Enumeration by DNA Colony Hybridization," *Applied Environmental Microbiology* **45**, 1324–1330 (1983).
134. W. E. Hill, "The Polymerase Chain Reaction: Application for the Detection of Foodborne Pathogens," *CRC Reviews in Food Science and Nutrition* **36**, 123–173 (1996).
135. S. Dovey and K. J. Towner, "A Biotinylated DNA Probe to Detect Bacterial Cells in Artificially Contaminated Foodstuffs," *Journal of Applied Bacteriology* **66**, 43–47 (1989).

FOODSERVICE SYSTEMS

OVERVIEW

The foodservice industry is defined as the industry responsible for preparing food or meals outside the home. In the last part of the 20th century, the food dollar has moved away from traditional food shopping. In the late 1990s, approximately 50% of food dollars were spent for food prepared away from the home. It is predicted that by 2005, the foodservice industry will have more points of distribution, many new selections, and fresher foods.

In the 1960s, the development of new food technologies in food production, packaging, transportation, changes in lifestyles, increased mobility of the population, and development of computers made possible the growth of modern foodservice. Higher expenditure on food away from home is the result of higher incomes, several paychecks in a family, and busier lifestyles. Any foodservice operation that supplies prepared food to a customer, such as institutions, fast-food operations, restaurants, and espresso stands, can be classified as a system.

The external environment energizes this dynamic foodservice system. Without this external energy, the system fails. We can approach large or small foodservice systems as interrelated activities that make up the total system. As in any system, the basic operating activities influence the total system.

Inputs in the foodservice system are the operational, physical, and people required to produce the operation objectives, a food(s) product. The activity, procurement, and production of food are the finished product; the quality of food and service are the output of the system. An internal control is the menu. The external controls include the federal and state laws and regulations governing the operation. The responses from internal and external environments include the customer comments, plate waste, cost, and frequency of selection of a food item. All external feedback provides the energy that keeps the system going.

We can divide the foodservice industry into three groups: commercial, institutional, and military feeding. The commercial feeding is the largest segment of the foodservice industry. This industry includes those operations that are open to the public and supply meal service daily for profit, such as, on-site feeding from table service, amusement and ball park concessions, airlines, cruise ships, espresso stands, food courts, and casual theme eateries. The institutional noncommercial feeding groups include business, education, correctional institutions, and health care facilities that operate their own foodservice. Many of these establishments do not operate for profit but must stay within a break-even budget. Some institutional foodservice operations do make a profit, but profit is not the primary goal. For many of these organizations, nutritious food, nutrition education, and customer satisfaction are the primary goals. Military feeding includes the foodservice to the troops, noncommissioned officers and officers clubs, and military exchange foodservice. The military is now emphasizing healthy entrees and nutrition education. Menus are modified to reduce fat, salt, and cholesterol and to increase the use of fresh fruits and vegetables.

Foodservice systems are designed for efficient purchasing, storage, preparation, delivery of product, service, and transportation. Recipes are standardized for each type of operation. In some facilities, methods of preparation are controlled using assembly-line techniques in back-of-the-house and front-of-the-house operations to speed customer service. Customers in the 1990s do not have the patience or time to stand in line. Sophisticated, computerized equipment has added to the efficiency of many foodservice operations. Microwave ovens are used to reheat or reconstitute many food items in institutional and commercial operations.

Restaurants and large supermarkets prepare cooked-to-order food or use prepared food already portioned and shaped (eg, ready-cut potatoes for frying). They may also choose to use meat and desserts proportioned and ready to serve. Computer technology has increased efficiency of the institutional and commercial foodservice operations when ordering food, either from a central commissary or direct from a manufacturer. The larger restaurants and fast-food chains order food from a central location where whole meals are prepared and packaged so that they remain fresh. The preprepared food or meals are distributed to other establishments, sometimes at a great distance from the preparation facility. Preservation methods include fast freezing, precooking, and dehydrating. New food technologies and food irradiation add shelf life to food products, which increases efficiency and decreases cost.

Many successful foodservice operations have identified a particular segment of the population and focus on specific consumer desires. The take-out phenomenon of foodservice that grew from recognizing the needs of a particular consumer population is an example. In the past decade, more gourmet take-out foods from large supermarkets, specialty take-out, and fast food eateries have developed. The larger franchise chain operations use marketing research and advertising strategies to identify and increase customer sales.

Commercial foodservice includes eating and drinking operations, food contractors, recreation and sports centers, stores and gas station foodservice, and lodging establishments. The franchise foodservice provides more theme or specialty foodservice. Fast-food foodservice includes specialties in sandwiches or hamburgers, seafood, and ethnic foods. Local operators or fully franchised operations may own some fast-food operations. Hotels may have an in-house managed foodservice operation or contract with a franchised operation. A trend in the hotel foodservice operations is to have independent coffee or espresso kiosks at customer traffic areas. Some large department stores combine a gourmet menu and take-out service with their restaurant foodservice.

In-house management or a food contractor may operate the foodservice operations in health care institutions. Increasing costs, competition, and change of population have provided challenges for the health care industry. The number of older patients who require complicated dietary regiments and modified diets, operating expenses of hospitals, convalescent homes, and long-term care facilities continues to increase. The philosophical change in feeding the older population in convalescent homes and long-term care facilities to provide a more liberalized diet has helped to lower the variety of foods prepared. This trend, along with new technologies, has helped the health care foodservice industry to become more efficient. Feeding facilities for employees, the public, and catering are potential revenue sources. Health care foodservice directors are encouraged to meet their customer's needs and have become very entrepreneurial.

Changes in the public's eating behaviors have influenced changes in college and university feeding. College costs, fewer residential colleges and universities, and more diverse college populations have had an impact on the foodservice operation. Nontraditional board plans, à la carte programs, cash operations, 24-hour service, take-out service, espresso kiosks, and a variety of menus are innovations that have developed. Foodservice is no longer obtained in just one location for those living in residence halls. Specialty dining halls for the faculty and guests and catering activities are other types of foodservice that may be provided.

Centralization of various functions has become a stable operational function in college and university foodservice operations. They may centralize food purchasing, storage, preparation, and some production. Foodservice employees may then transport the prepared item or raw material to a satellite unit. Recipe and menu standardization, preprepared purchased food items, and computerized operations have made college and university foodservice operations more efficient and cost effective.

School lunch in the United States has a long history. The rejection of young men for military service in World War II because of poor nutritional status led the impetus to the development of the National School Lunch Program. The National School Lunch Act in 1946 provided more than 23.5 million children with a school lunch, with approximately 50% receiving reduced price or free lunches. The National School Lunch Act had a dual purpose: "to safeguard the health and well-being of the nation's children and to encourage the domestic consumption of nutritious agricultural products." Federal regulations control the meal pattern and provide for joint administration of the program through a federal-state-local school district relationship.

The Federal Government has changed the Child Nutrition Act often since 1946. Legislation and amendments have changed the funding policy and circumstances for offering free or reduced-price meals to children. Legislation has reshaped the commodity distribution plan, provided for supplemental feeding programs, and authorized the School Breakfast Program. One of the biggest impacts on the federal school lunch program was the Omnibus Reconciliation Act of 1980. This legislation reduced the reimbursement rate to schools and changed the income eligibility standard for students who could receive free or reduced-price meals. Other legislation modifications were made in the 1990s to help achieve reductions in federal spending. School foodservice personnel work creatively to meet these legislation changes while maintaining a nutritious and satisfying meal program that will appeal to the student population.

School foodservice programs vary in type of organization and management from the one unit, one manager operation to a centrally managed unit in a large city school system. Many large school systems have centralized food preparation centers that distribute food in bulk or preplated, hot or refrigerated service to students in schools with satellite service centers. Since early 1970, program regulations have changed, allowing schools to negotiate with food contractors to provide meals.

Commercial and industrial organizations also offer employee feeding programs. The increase in white collar workers and suburban living increased foodservice sales in office buildings. The use of food and beverage vending machines with microwave ovens has simplified and reduced the cost of these foodservice operations.

Food contractors use marketing strategies to encourage sales, such as on-site bakeries, espresso coffee centers, and take-home food items. A variety of settings are used, from elegant executive dining rooms to the vending-machine cafeteria.

In the transportation segment of foodservice, food contractors provide most of the in-transit airline feeding. They adapt menus and food items to specifications of the various airline carriers. Sales in the transportation feeding for passenger and cargo liners and railroads continue to decline. The major thrust in transportation foodservice has been efficient productivity.

MENU SYSTEM

The foodservice operation begins and ends with the menu, a detailed list of foods served. Creative menu planning uses originality, imagination, and knowledge. The menu is a guide for the kitchen and wait person and a printed list for customers. The menu determines the foods purchased, equipment needed for production, personnel required for production and service, and work schedules of the employees. It is a marketing tool and an important factor in controlling costs.

The menu must be consistent with the goals of the organization and meet the food preferences of the identified customer. Considerations to use in menu planning for commercial foodservice include the goals of the organization, equipment and physical facilities, skill of employees, money available, and type of service. Most institutional foodservice operations include the consideration of serving nutritious foods. Commercial foodservice has moved to include healthy food and the caloric value of foods served on their menus. In the 1990s, many restaurants have included caloric values or notations identifying "healthy heart" food items. The public's avoidance of many of these healthy foods has led some foodservice establishments to discontinue the nutrition information menu. Managers planning menus for any foodservice operation must be aware of the food preferences, food habits, and cultural make-up of their customers. Monitoring current trends in eating behaviors is necessary. Balancing these food preferences of the clientele with the food budget, the constraints of the facility, and employees skills are important tasks for all foodservice managers.

To determine effectiveness of the menu, many foodservice operations prepare a method to evaluate menus. This may be a record of food purchased, customer preference surveys, plate waste checks, or verbal opinion from customers. Production employees can contribute to the evaluation of menus. Providing a routine employee menu evaluation allows the employees some ownership in the production of the menus.

PURCHASING, RECEIVING, AND STORAGE SYSTEMS

To maintain an effective and cost-efficient foodservice operation, the purchasing and inventory system must be carefully planned. Selecting the establishment's suppliers on an objective basis and developing specifications and formal buying procedures contribute to efficient management. Computers have made this part of the foodservice system more efficient and effective.

Effective purchasing procedures include the knowledge of availability of food, number of suppliers in the area, skill of production employees, organization's food and nonfood items budget, and the quality and value of food items. The purchasing manager must use appropriate buying methods, establish good ordering schedules, and develop an effective system of communicating needs from the production to the buyer. The purchasing manager must have a thorough knowledge of the menu, organization mission, food and nonfood specifications, and personnel resources.

Specifications for food and nonfood items should be detailed to ensure the continuous quality of products. Several federal agencies have established quality grades for the safety of many food products. Determination of the quality of a food or nonfood item to buy and whether to prepare or buy prepared foods will be governed by the organization's mission, menu, and identified customer.

Computer technology is used to develop an effective system for purchasing and recording the delivery of food and nonfood items. Employees must be trained to receive foods and store them at appropriate temperatures. All preplanning and savings can be lost by careless employees in the storage and receiving area.

Designing effective food-receiving procedures provides control and predictability. Receiving procedures that involve daily reports, a tagging system, and invoice stamps are appropriate management tools. Providing a checking system to evaluate the accuracy of the foodservice receiving system is important.

Adequate procedures for storage can reduce spoilage and product theft. Storage areas, including dry, refrigerated, and frozen, should be checked regularly. Procedures for checking dry, refrigerated, and frozen food items are essential for inventory and product control. Physical inventory systems are essential. Computer software has been developed for day-to-day inventories interfaced with ordering and purchasing. These systems give the foodservice organization a painless, perpetual inventory system.

PRODUCTION SYSTEM

Basic preparation standards and procedures are necessary in any foodservice system. A standard product can be developed consistently, and waste can be avoided. Using preprepared foods will reduce the need for skilled labor and reduce energy requirements. Because the cost of operating a foodservice establishment is still climbing, the production system has the potential to be the most expensive of all the systems. Skilled labor is expensive, but so is untrained, unskilled labor. Therefore, the use of standardized recipes, good quality ingredients, training, and proper supervision of production is essential.

The menu is the input and control, where production planning begins. The production system transforms the raw product into the salable output of the foodservice system. Accurate forecasting is the prediction of the food needs for the meal, day, or defined period based on established trends. Forecasting is a procedure that can limit food waste and maintain cost controls. Past sales trends can be used along with present and future developments. Food sales performance determines the quantities of food and portion size to be prepared. This information is included on the production schedule, along with special instructions for preparation and work assignments. New computer technologies and modern equipment can help in making the production equipment more efficient. To maintain quality and consistent foodservice, employees must follow the standardized recipes, procedures, and portion control. Quality preprepared and preplated food items have provided automation and decreased personnel expense in this area.

To maintain food quality, the purchase of quality food is the first step. Using standardized recipes, supervision of food production, and critical tasting techniques are other factor in producing quality items. The food needs to be safe, wholesome, and pleasing to the targeted customer for the foodservice operation to be successful.

THE SERVICE SYSTEM

The wait person or server communicates the establishment's attitude toward the customer. Some foodservice operations use the term guest to emphasize the value placed on their clientele. The server spends more time with the customer and may be the first and last person with whom the customer has contact. A good culture of service occurs when servers are empowered to respond positively to all customer requests. Management must promote this "culture of service" to all employees that have contact with the customer. Good servers are aware of the customer's discomfort and dissatisfaction. Having a positive regard for the customer is training by example. Continuing education for managers and employees is necessary.

A full-service restaurant practices one of the four basic service styles or a combinations of styles. Family-style service uses serving dishes placed on the table, and the customers serve themselves. This is also known as English service. Smorgasbords and self-serve salad bars are forms of the family-style service. Plate service, or American service, is generally used by most restaurants. All the food is put on plates in the kitchen and served in the dining room. Bread and butter may be served on a bread and butter plate. French service, or table-side service, is another style. This service is characterized by the service of food from a heated cart by a waiter and assistant. The waiter and assistant prepare finishing touches to the food on the cart at the table and serve the food to the customer. During platter service, or Russian service, the food is fully prepared and cut into portions in the kitchen. The waiter serves the food on platters directly to the customer. Both French and Russian types of service allow for showmanship and drama at the customer's table. More equipment, utensils, and skilled waiters are needed for this type of service. Portion control may also be a problem.

In the 1990, most full-service restaurants added another type of service: the take-out and delivery service. Take-out foods have become a vital part of the American's way of life. Restaurants and food retailers have become kitchens-on-the-go. A consumer survey revealed that approximately three-quarters (78%) of U.S. households make one carry-out or delivery purchase in a typical month. The take-out customer is "in a hurry," "tired," or has "no other place to go." Therefore, the customer is looking for speedy service, accurately filled orders, easy access, reasonable prices, and sometimes portion sizes that provide enough for leftovers. The restaurant's take-out contact person may be the only employee the customer sees. A good service culture is important.

HUMAN RESOURCE SYSTEM

Foodservice operations have had a negative production record in the past, but the trend is to reverse this image. The foodservice industry is learning to maximize its best resource—the people they employ. Increasing productivity and decreasing employee turnover should be objectives of a good foodservice manager. Effective recruitment, selection, orientation, and training will enhance these objectives.

A plan for recruiting and selecting new employees communicates the knowledge and skill needed for employment. Employee orientation and training programs developed by the fast-food chains have emphasized the need for training programs in all foodservice operations. Managers need to be trained for their positions and to understand the principles of learning to maximize training efforts. Managers who are not trained to use effective investigative methods, active listening, and conflict-resolution techniques when small employee problems occur may discover a larger employee problem. Failure to resolve small problems may contribute to an unhappy working environment and frequent employee changes. Unresolved employee problems open the door to labor unions to represent the employees better.

Government legislation affects human resources planning. Since the 1960s, several pieces of legislation have had significant impact on the hiring, disciplining, promotion, demotion, internal organization environment, and separation of employees.

Developing standards of performance to evaluate employee and staff can increase the employee's involvement and motivation. Management by objective can be used for staff personnel and is an effective personal motivator. "Management by walking around" can be an effective way to "catch" an employee doing something right. Management by walking around adds to more positive employee performance evaluations. Using this technique will allow the manager to correct inappropriate behavior or work procedures on the spot. An employee manual with the foodservice operation's policies and procedures is necessary for good employee–employer relationships. A positive approach, humor, and a good self-image can be valuable assets for a foodservice manager. Providing a positive work environment and paying attention to the employee and the employee's contributions to the organization's goals can be the most effective motivator in the industry.

EQUIPMENT AND ENERGY SYSTEM

Knowledge of equipment and energy is necessary to provide efficient transformation of the raw food to quality food products. Training of personnel in the care and use of the equipment is important for preserving the equipment. Today's current use of computerized equipment requires better equipment care to reduce the cost of repair. On the other hand, many foods are purchased preprepared or precut and require less equipment and skill to prepare and cook. When purchasing a piece of equipment, consideration should be given to the initial cost, installation, maintenance, personnel using the equipment, and energy cost. Calculations can be used to determine these answers and the equipment cost over time.

In energy management, a planned and organized program is necessary for an efficient organization. An equipment maintenance checklist can provide a starting point. Many equipment manufacturers and utility companies have tips for energy conservation.

QUALITY ASSURANCE CONTROL SYSTEM

Quality assurance programs allow foodservice operations to be proactive rather than reactive. Satisfaction of the customer is the goal of the operation. Quality assurance programs ensure the attainment of that goal. The use of the hazard analysis critical control point concept has been recommended as a preventive approach to quality control in foodservice systems. Quality audits are an important component of a quality assurance program and provide feedback. These audits allow management to control the output of the system.

GENERAL REFERENCES

I. Obenaeur, "Foodservice Trends," *Restaurants USA* **18**, 39–46, 1998.

J. Payne-Palacio and M. Theis, *West and Wood's Introduction to Foodservice*, 8th ed., Prentice-Hall, Englewood Cliffs, N.J., 1997.

A. E. Sloan, "Food Industry Forecast: Consumer Trends to 2020 and Beyond," *Food Technol.* **52**, 37–44 (1998).

DOROTHY POND-SMITH
Washington State University
Pullman, Washington

FREEZE CONCENTRATION

Freeze concentration is a process in which aqueous solutions are concentrated by partially freezing them and separating the ice produced from residual unfrozen solution. Because nonaqueous volatile components (eg, volatile flavors and aromas in foods) are not removed, freeze concentration provides more selective water removal than evaporation. It is also gentler because no heat exposure is involved. Freeze concentrated liquid foods often taste measurably better than evaporative concentrates. Unfortunately, the taste improvement does not always translate into improved sales. In spite of process improvements and cost reductions achieved in recent years, freeze concentration remains much more costly than evaporative concentration.

PRODUCTS

Coffee extract, fruit juices, milk, beer, wine, and vinegar have been concentrated by freeze concentration. Freeze concentration is also used to recover potable water in the form of meltable ice from sea water and brackish water. A less-selective analogous process, partial solidification, is used to separate high-melting fats from low-melting fats.

Freeze concentration works best for clear solutions; any insoluble solids present are likely to reduce cleanness of separation and be trapped and lost when the concentrate is separated from the ice. Therefore, insoluble solids should be removed before freeze concentration begins. They can, if desired, be added back to the concentrate after the process is complete. Freeze concentration also does not work well if solutes in the solution precipitate as concentration occurs.

FREEZING POINT DEPRESSION

Solutes depress the freezing point of water. Therefore, as ice forms and solute concentrations increase, the freezing point decreases. Figure 1 shows how the freezing points of wine and some juices depend on concentration (1). The freezing point depression is roughly inversely proportional to the solute's molecular weight. Thus, at equal concentrations, freezing point depressions for juices (eg, apple juice) in which the main solutes are fructose and glucose are greater than those for juices (eg, orange juice) containing greater amounts of sucrose. Depending on the fruit juice involved, solute concentrations in the 50% range can be obtained by cooling juice to −8 to −14°C. Cooling to −13 to −18°C is required to obtain 60% concentration.

When simple binary solutions (eg, NaCl in water) undergo freezing, the solute involved will precipitate at a certain temperature and corresponding concentration, the eutectic temperature and concentration. As the solute precipitates, the remaining water simultaneously freezes. It is extremely difficult to separate the components of the resulting solid mixture. For such solutions, freeze concentration beyond the eutectic concentration is impossible. Liquid foods contain complex mixtures of solutes that inhibit each others' crystallization. Therefore, eutectics rarely

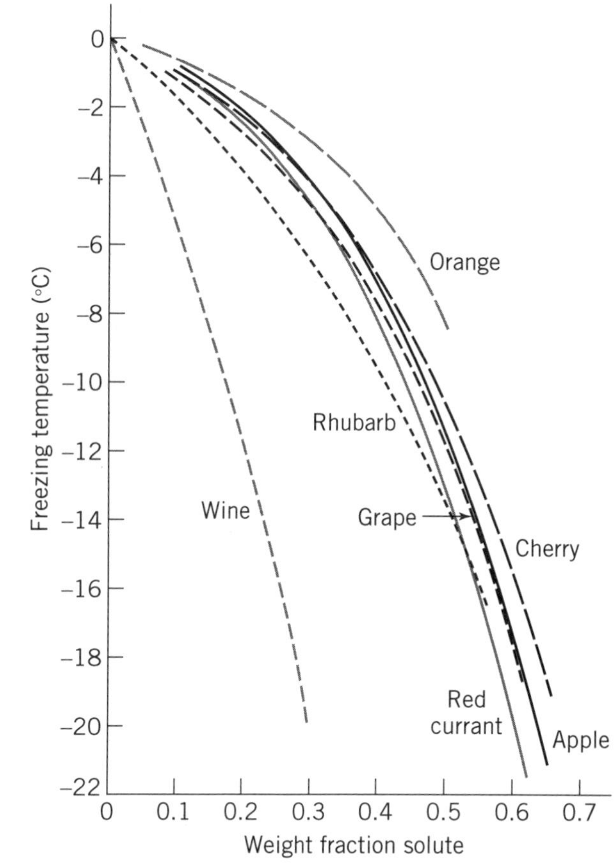

Figure 1. Freezing point versus concentration for wine and various fruit juices. *Source:* Ref. 1. Courtesy of Marcel Dekker.

form when liquid foods freeze. Around 65% solute concentration, the remaining unfrozen water in foods so strongly associates with the solutes, or is otherwise prevented from freezing, that freezing stops or radically slows down. Further, at such concentrations, concentrates become so viscous that they cannot be readily separated from ice. Therefore, 65% represents the upper concentration limit for freeze concentration of liquid foods. Final concentrations obtained in most food freeze concentration systems are lower (45 or 50%). In some cases, only a modest increase in concentration is provided, such as when low-sugar-content wine grape juices are freeze concentrated to permit production of wine that contains adequate amounts of alcohol. Large increases in concentration are used when wine grape juice is freeze concentrated so that it can be stored for subsequent use in wine production.

CLEANNESS OF SEPARATION

Apple jack has been freeze concentrated on farms by slowly freezing it in a milk can and augering out the concentrate that accumulates along the can's axis. A great deal of solute is left behind, trapped in ice. Similar loss-prone procedures can be used to freeze concentrate small amounts of aqueous solution in laboratories. In commercial freeze concentration, solute losses must be minimized; nearly all

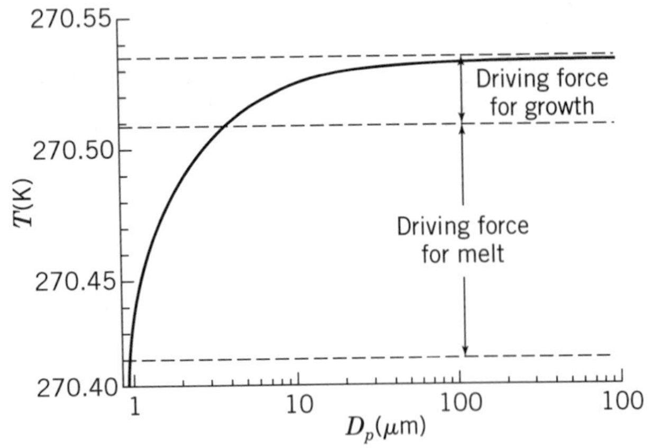

Figure 2. Freezing point versus crystal diameter for ice in a 30% sucrose solution. *Source:* Reprinted courtesy of Niro Process Technology B.V.

solute-containing liquid has to be cleanly separated from ice. This requires production of well-rounded, suitably large crystals of ice and efficient means of separation. Solute losses can be kept below 0.01% when food liquids are separated from suitably large ice crystals (200 to 250 μm diameter) in wash columns (the most efficient separation devices currently available). When centrifugation is used for ice-concentrate separation, losses are in the 1 to 3% range. For pressing, losses are around 5%. Losses are also bad for filtration. Losses for these less-efficient ice-concentrate separation systems can be reduced by using reverse osmosis to recover solutes from melted ice.

ICE CRYSTAL FORMATION AND GROWTH

Ice-free aqueous solutions have to be cooled below their equilibrium freezing point before ice crystal nuclei form and freezing starts. Ice can nucleate in several ways. Heterogeneous nucleation usually predominates in ice-free solutions and rapidly occurs roughly 6°C below the initial

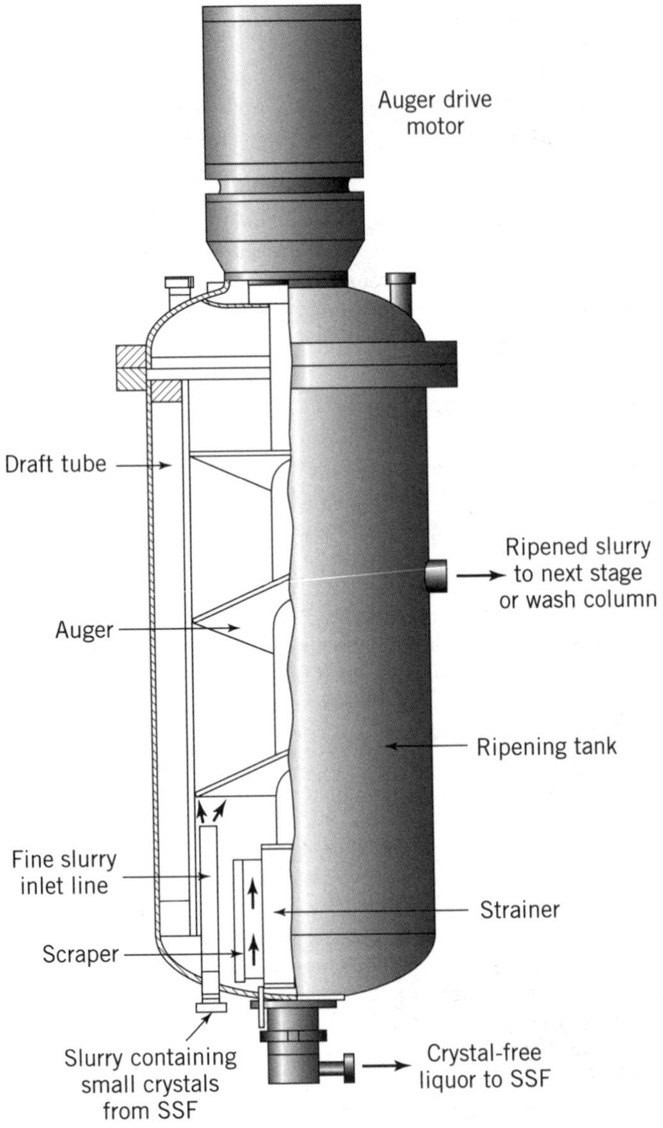

Figure 3. A ripening tank. SSF = scraped-surface freezer. *Source:* Reprinted courtesy of Niro Process Technology B.V.

equilibrium freezing point of the solution. Once nucleation occurs, the solution's temperature rises and closely approaches its equilibrium freezing point at its current concentration. Ice nuclei are very small and have to be converted into much larger ice crystals for efficient separation. This can be done by bulk supercooling that causes existing crystals to grow by maintaining the solution below its equilibrium freezing point. The greater the supercooling, the greater the rate of growth. However, if the supercooling is too large, highly branched crystals from which solution does not separate cleanly will form. Further, new nuclei will continually form on cooling surfaces if temperature differences larger than 6°C are used to provide heat removal.

Therefore, it is preferable to form small ice crystals on sharply chilled, rapidly scraped surfaces and mix those crystals with larger crystals in tanks, in which the small crystals sacrificially melt and cause large-crystal growth. The process occurs because ice crystals with small radii have slightly lower equilibrium freezing points than crystals with larger radii (Fig. 2). Slurries containing mixtures of smaller and larger crystals assume a temperature between the respective equilibrium freezing temperatures of the small and large crystals. The small crystals tend to melt, and the large crystals tend to grow, a process known as Ostwald ripening (2). The temperature differences involved are very small, but the process is fairly rapid in stirred systems because the crystals provide an enormous amount of heat-transfer and mass-transfer surface.

Ripening in freeze concentration systems is frequently carried out in a large, enclosed, vertical, cylindrical tank fitted with a helical agitator and draft tube (Fig. 3). Liquid withdrawn from the mixing tank through a scraped screen at its bottom is fed to freezers and returns to the tank as an ice-containing slurry. The process runs best when the returning slurry contains very small crystals (10 μm diameter), and the bulk of the crystals in the tank are much larger (120 to 250 μm diameter).

Table 1. Local Conditions in a Four-Stage Freeze-Concentration System Converting 5,000 kg/h of a 10% Solute Feed into a 50% Solute Concentrate

Stage	1	2	3	4
Concentration (% s.s.)	16.0	26.1	36.4	50.0
Temperature (°C)	−2.0	−3.7	−6.5	−9.6
Ice-formation rate (kg/hr)	2,100	1,200	900	600
Mean ice crystal diameter (μm)	240	198	163	120

ICE PRODUCTION

Roughly 75% of a feed by volume may be converted into ice in a typical food freeze concentration system. Stirring is needed to provide rapid ice crystal ripening and growth, but ice-concentrate mixtures containing so much ice cannot be readily stirred or pumped. Therefore, enough concentrate is retained in the system to keep ice volume percentages in the system within the stirrable range (below 28% by volume).

FREEZE-CONCENTRATION SYSTEMS

The basic components in most freeze concentration systems are freezers, ripening tanks, and separators (2). Originally, single-stage systems,—systems containing one ripening tank or, in some cases, a bulk-growth tank—were

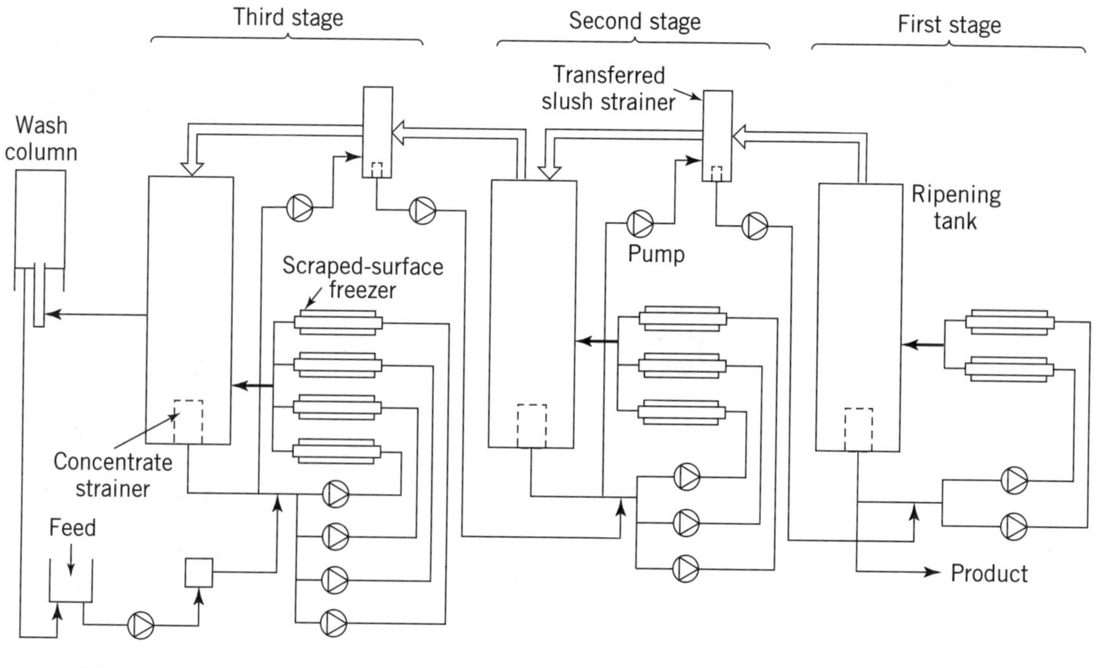

Figure 4. A three-stage freeze-concentration system. *Source:* Reprinted courtesy of Niro Process Technology B.V.

used and are still used when little concentration is required. Countercurrent multistage systems that use wash columns as separators are used to provide high degree of concentration. Compared to single-stage operation, countercurrent multistage operation provides faster, cleaner separation; greater thermodynamic efficiency (reduced energy consumption); faster ripening and crystal growth; and higher concentrations. Figure 4 shows a modern three-stage countercurrent freeze-concentration system. Systems containing up to five stages are used. Each stage consists of a ripening tank served by several scraped-surface freezers operating in parallel (3). Clear liquid, drawn out of the tank through a scraped strainer, is pumped through the associated freezers and returns to the tank as an ice-containing slurry. Although high scraping rates and strong chilling are used, the liquid passes through the freezers so quickly that only 7.5 to 8% of the stream is converted into ice, and 5- to 10-μm diameter crystals are produced.

Flow within the system is arranged so that the liquid being concentrated and the ice produced move countercurrent with respect to each other in an overall sense. Part of the liquid drawn through the ice strainer at the bottom of a ripening tank is sent to the next tank in terms of concentrate flow, that is, to the tank with the next higher concentration. Ice-concentrate mix continuously transfers out of each stage. Most of the concentrate in the mix is mechanically expelled; much of what remains is displaced by more dilute concentrate from the stage the ice is about to enter. The ice enters the next stage in the direction of ice flow, which is the next-more-dilute stage. Ice-crystal size progressively increases as ice moves from the most concentrated stage to the most dilute stage.

Strained, ice-free product concentrate discharges from the most concentrated stage. Ice is removed in slurry form at the opposite end, the dilute end of the system, where separation is easiest. The discharged mixture of ice and dilute stage liquor is sent to a wash column, where it is propelled upward and slightly mechanically compressed. The compression expels most of the accompanying liquor. The ice is melted near the top of the column, forming melted water that displaces remaining solute-containing liquor from the pores in the advancing bed of ice. If the rate of ice advance is suitably slow, very clean displacement is obtained, and, as previously mentioned, less than 0.01% solute is lost. If it is too fast, very large amounts of solute are lost. The maximum rate of ice advance that can be safely used is roughly proportional to the ice crystal diameter squared and inversely proportional to the difference between viscosity of the displaced liquor and the viscosity of the water carrying out the displacement. Concentrations, temperatures, ice-formation rates, and mean ice crystal diameters in the stages of a four-stage freeze concentration system processing 5,000 kg/h of liquid feed are listed in Table 1.

The total rate of ice production, 4,800 kg/h, exceeds the net rate of ice removal, 4,000 kg/h, because some ice melts because of stirring-induced friction in the system. Roughly two hours of ice-holdup time in a multistage, freeze-concentration system may be required to achieve adequate ice crystal size growth. The average solute holdup is usually roughly twice as long. Less than a few minutes of holdup time are used in juice-concentration evaporators.

BIBLIOGRAPHY

1. H. G. Schwartzberg, "Food Freeze Concentration," in H. G. Schwartzberg and M. A. Rao, eds., *Biotechnology and Food Process Engineering*, Marcel Dekker, New York, 1990, pp. 127–202.
2. H. A. C. Thijssen, "Current Developments in the Freeze Concentration of Liquid Foods," in S. A. Goldblith, L. Rey, and W. W. Rothmayr, eds., *Freeze Drying and Advanced Food Technology*, Academic Press, London, United Kingdom, 1975, pp. 481–501.
3. U.S. Patent 4,430,104 (Feb. 7, 1984). W. Van Pelt and J. Roodenrijs (to Grasso's Koniklijke Machine Fabrieken N.V.).

HENRY SCHWARTZBERG
University of Massachusetts
Amherst, Massachusetts

FREEZE DRYING

Freeze drying (lyophilization) is the drying of material in the frozen state. It is usually carried out under vacuum, at absolute pressures that readily permit ice to sublime (change directly from solid to vapor). Absolute pressures used in food freeze drying range between 50 μm Hg and 1,500 μm Hg.

Moist foods contain water that freezes (ie, forms ice) when suitably cooled, and smaller amounts of bound water that do not, even at -60°C, but instead form part of a solute-rich matrix that becomes extremely viscous and stiffens as temperature drops. In frozen foods, ice is embedded in that matrix. Except for clear juices and extracts, frozen foods also contain other solid components, such as cell walls, membranes, and subcellular particles. As noted in the article FREEZE CONCENTRATION, food solutes depress the freezing point of water. The initial equilibrium–freezing points of foods usually lie between -0.5°C and -2°C. As foods freeze, the solutes they contain become more concentrated, and the freezing point depression increases. Therefore, successive portions of a food's water content freeze at lower and lower temperatures. Freezing is complete when the last freezable water changes into ice. In many foods, this effectively occurs between -20°C and -30°C.

During food freeze drying, ice sublimes and bound water desorbs as vapor. Most bound water remains when the last ice has sublimed; almost all has to be removed to provide stable, freeze-dried food. To do this, the product temperatures are raised, and added vacuum drying time is provided after the last ice sublimes. Ice-containing regions in foods remain fully frozen during freeze drying; temperatures well above freezing exist in parts of regions whose ice content has already sublimed.

PRODUCTS

Soluble coffee is the material freeze-dried in greatest volume. Vegetables for dried soup mixes, mushrooms, herbs, spices, cheese starter cultures, shrimp, fruits for ready-to-eat breakfast cereals and vegetables, meats, fish, and fruits for military, camping, and space-travel rations are or have been freeze-dried commercially. Freeze drying is used in making highly compressed, dry, military rations that expand to near-normal size and shape when rehydrated. Patents (1) or other technical literature describe the freeze drying of eggs, dairy products, powdered fish and shellfish, gelatin, bacon, cooked rice, emulsified peanut butter, soluble tea, puddings, jellies, pie fillings, salads coated with dressings, berries, avocado powder, potato-based products, fruit juices, and yeasts. Laboratory-scale freeze drying is used to preserve food samples.

ADVANTAGES AND DISADVANTAGES

Advantages of freeze drying include (*1*) low thermal damage; (*2*) good retention of volatile flavors (2); (*3*) good vitamin retention with minor exceptions (eg, loss of pantothenic acid in pork) (3); (*4*) rapid product rehydration (sometimes this requires slow freezing before drying); (*5*) low product shrinkage; (*6*) long product storage life if products are suitably packed; and (*7*) retention of biological activity (particularly if cryoprotectants are used during freezing). Disadvantages of freeze drying include (*1*) freeze drying costs much more than conventional drying (4); (*2*) certain products are damaged by freezing before drying; (*3*) freeze-dried products rapidly pick up moisture unless packed and maintained at low humidity; (*4*) freeze-dried products are very friable; and (*5*) some freeze-dried products (eg, carrots) bleach when stored (this can be prevented by suitable pretreatment).

SYSTEM COMPONENTS

Figure 1 shows basic components of a simple vacuum freeze dryer: a vacuum chamber containing a set of parallel heating plates, between which is a tray holding the frozen food to be dried; a vacuum pump for evacuating the chamber and removing inleaking air; a condenser where water vapor generated during drying condenses as ice; refrigeration for cooling the condenser; and a system that supplies hot heat-transfer fluid to the plates. Other elements not shown include a pressure-tight door that permits insertion and removal of loads; instruments for sensing plate, condenser and drying food temperatures, and for sensing pressure in the chamber; and the controller that regulates plate temperatures.

Figure 1 shows a single pair of heating plates and one tray of frozen food. Plant-scale freeze dryers contain many plates and many rows of trays holding many tray-loads of product. Large-scale freeze dryers are described later in this article. Reference 5 provides design details about such dryers.

Small-scale freeze dryers may contain up to five heating plates. They can process up to four rows of product, have a transparent plastic door that permits products to be observed during drying, and may be fitted with a balance or load cells that weigh food loads as they dry. Drying rates and temperatures measured in suitably instrumented small freeze dryers are used to help select operating conditions used in production-scale drying.

PREPARATION

Foods have to be suitably prepared (cleaned, peeled, trimmed, sliced, diced, and blanched or sulfited where necessary) and properly frozen before drying. To improve permeability, holes are punched in the skins of some foods, such as blueberries and peas. Meat is trimmed to remove fat (which interferes with drying) and, when sliced, is cut across the grain to maximize heat transfer and ease of vapor escape during drying. Liquid feeds may be evaporatively concentrated or freeze concentrated to reduce water removal loads in drying.

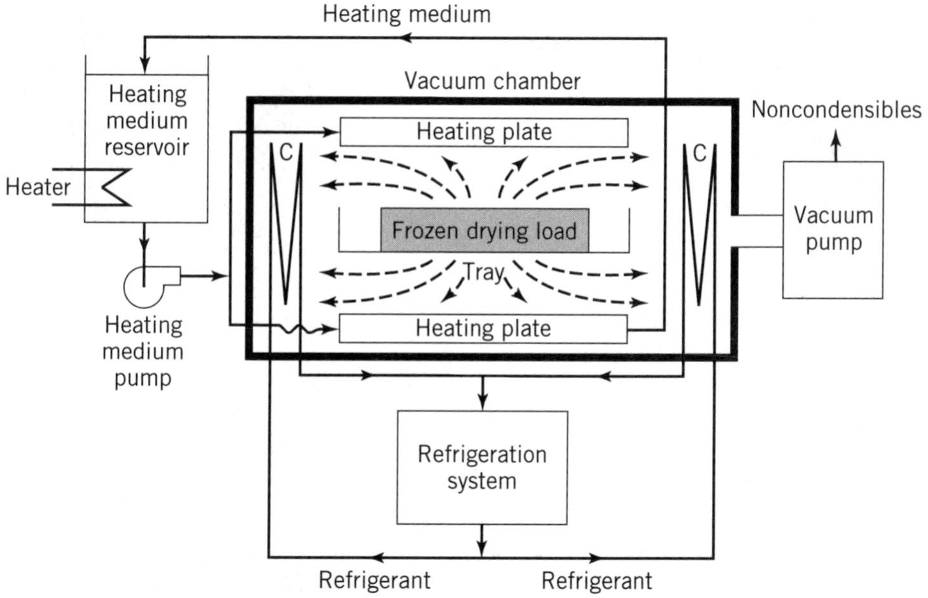

Figure 1. Basic components of a simple one-shelf vacuum freeze drier. C = condenser.

Freeze drying and the properties of freeze-dried foods are greatly affected by how foods are frozen. Slow freezing is used to produce freeze-dried coffee dark enough to be accepted by consumers and to improve the rehydratability of freeze-dried fruits. Rapid freezing permits more rapid desorption of bound water, but also causes greater resistance to vapor flow during sublimation.

BASIC PROCESS

Contact-based conductive heating has been used in some commercial freeze dryers. Microwave heating has been tested experimentally (6). In modern freeze dryers, radiant heating is by far the most commonly used method of providing heat input. Therefore, only drying based on its use will be discussed. Let us examine what happens when a slab of frozen food placed on a perforated tray is freeze dried in a simple dryer, like that shown in Figure 1. After the dryer door is closed and sealed, the refrigeration is turned on, cooling the condenser. The vacuum pump is turned on, and, as air leaves the dryer, pressure drops. When the pressure closely approaches the vapor pressure of ice at the set condenser temperature (eg, 97 μm Hg absolute for a -40°C condenser), hot fluid is pumped through the heating plates.

The plate temperatures rise until they reach a desired set value (eg, 120°C). Heat transfers radiantly from the plates to the slab, causing sublimation on both sides of the slab. Vapor produced at the bottom of the slab escapes through the perforations in the tray. Sublimation interfaces form and recede into the slab, leaving behind ice-free pores in the matrix containing the bound water. Thereafter, heat, radiantly transferred to outer surfaces of the slab, transfers conductively across the ice-free layers, causing further sublimation. Small amounts of bound water also evaporate. Vapor produced by both processes flows through the pores in the ice-free layers, causing a pressure drop that may be a large fraction of the absolute pressure in the dryer. The vapor leaves the slab and flows to the condenser, where it freezes.

The ice-free layers deepen as sublimation proceeds. Consequently, heat- and vapor-flow resistances increase, and the sublimation rate progressively decreases. After the last ice sublimes, that is, the sublimation interfaces meet near the center of the slab, the center temperature rises and bound water desorbs more rapidly. To prevent overdrying and overheating of outer parts of the slab, the plate temperature is reduced as desorption proceeds. At the end of drying, the plate temperature may be 50°C, and the center product temperature may be 30°C.

After drying is complete, the vacuum is broken by admitting air or, preferably, dry nitrogen. The product is quickly removed, rapidly transferred to a low humidity room, and quickly packed and sealed in an impermeable container. After all product is discharged from the dryer, ice is removed from the condenser by passing hot refrigerant gas through it or by spraying it with hot water. The melt produced drains from the dryer through a closeable outlet at its bottom.

HEAT TRANSFER

Heat that transfers across ice-free layers in a drying food causes a proportional amount of sublimation. Therefore, dZ/dt, the time rate of increase in the depths of ice-free layers in a slab, can be expressed in terms of the rate of heat transfer (7):

$$\frac{dZ}{dt} = \frac{k_t(T_o - T_s)}{\Lambda \rho X Z} \tag{1}$$

where T_o is the outer-surface temperature, T_s is the temperature at the sublimation interface, Z is the thickness of the ice-free layer, k_t is its thermal conductivity, Λ is the latent heat of sublimation (roughly 2,836 kJ/kg ice), and ρ and X are the initial density and weight fraction of ice in the slab, respectively.

The k_t is roughly proportional to the food's initial solids content and also depends on pressure. At absolute pressures greater than roughly 100 mm Hg, k_t is the same as at atmospheric pressure, that is, roughly 0.032 to 0.17 W $\cdot$ m^{-1} $\cdot$ °C^{-1}. At lower pressures, k_t decreases sigmoidally as pressure decreases, and roughly around 100 μm Hg levels off at roughly one half to one quarter of its atmospheric pressure value. The respective pressures at which k_t shifts start and finish depend on pore diameter. In most cases, freeze drying is carried out at or close to the low-pressure k_t limit. Thermal conductivities encountered in freeze-drying foods are substantially lower than those of many thermal insulators.

PRESSURE DROP

The pressure drop caused by vapor flow across the ice-free layers in a slab is proportional to dZ/dt. Consequently, it can be shown that

$$(P_s - P_o) = \frac{k_t(T_o - T_s)}{\Lambda K} \tag{2}$$

where P_s is the pressure at the sublimation interface, P_o the pressure at the outer surface of the slab, and K is the permeability of the porous ice-free layers. K is proportional to the volume fraction of ice initially present, roughly inversely proportional to ice-crystal diameter, increases somewhat as the mean partial pressure of water in the pores increases, and decreases if excessive amounts of noncondensibles are present. Typical K range between 0.7×10^{-9} and 10×10^{-9} kg $\cdot$ m^{-1} $\cdot$ s^{-1} $\cdot$ (μm Hg)$^{-1}$ (5×10^{-9} to 70×10^{-9} s^{-1} in SI units).

SUBLIMATION TEMPERATURE

The vapor pressure versus temperature relationship for ice, shown in Figure 2, fixes P_s in terms of T_s. In lightly loaded freeze dryers with wide vapor-transfer spaces, P_o equals the vapor pressure of the ice on the condenser. Figure 2 also depicts equation 2 solved to determine T_s and P_s for the freeze drying of a slab of coffee extract for which

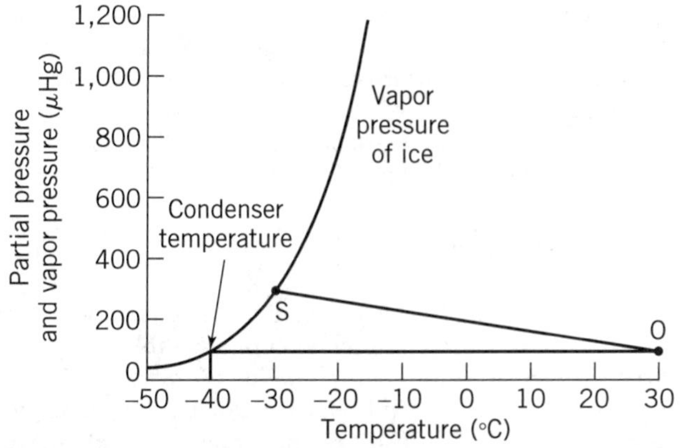

Figure 2. Vapor pressure of ice versus temperature and pressure and temperature conditions at the outer surface (O) and sublimation interface (S) of a slab of food undergoing freeze drying in a chamber where the condenser temperature is −40°C.

$k_t = 0.035\ \text{W} \cdot \text{m}^{-1} \cdot {}^\circ\text{C}^{-1}, K = 3.5 \times 10^{-9}\ \text{kg} \cdot \text{m}^{-1} \cdot \text{s}^{-1} \cdot (\mu\text{m Hg})^{-1}$, and $T_o = 30°\text{C}$. The condensed ice temperature is −40°C. Therefore, as shown by the dotted lines, $P_o = 97\ \mu\text{m Hg}$. T_o and P_o define point O. A line, drawn through O with slope $-k_t/(K \cdot \Lambda)$, intersects the ice vapor pressure curve at S, providing P_s and T_s, which respectively are 301 μm Hg and −29.4°C.

PRODUCT SURFACE TEMPERATURE

T_o is greatly affected by T_p, the heating plate temperature; higher T_p cause higher T_o. Usually, T_p is rapidly ramped up to a set value and then held there until sublimation is complete. If so, T_o initially will be only slightly higher then the condenser temperature. Then, as drying proceeds and the sublimation interfaces recede, that is, Z increases, T_o progressively rises. T_o can be calculated as a function of T_p during sublimation by relating the rate of radiant heat absorption by the food to the rate at which heat is conductively transferred through the ice free layers. Equation 2 and the vapor pressure versus temperature relationship for ice also have to be used in carrying out this calculation.

To prevent thermal damage, T_o must be kept below a specified upper limit or limits that depend on the product being dried. A fixed upper limit, such as 40°C for coffee extract, is usually used. Upper limits that depend on exposure time are used occasionally; higher T_o is allowed if the exposure time is short. Plate temperatures and drying rates must be reduced if excessively high T_o develop as Z increases. During the desorption phase of drying, plate temperatures are usually progressively reduced to prevent T_o from rising excessively.

DRYING TIME

Equation 1 can be rearranged and integrated, as shown below, to provide t_s, the time required to complete sublimation:

$$t_s = \int_{Z=0}^{L} \frac{(\Lambda \rho X)Z dZ}{k_t(T_o - T_s)} \tag{3}$$

where L is the final depth of the ice-free layer, that is, half the slab thickness for slabs where sublimation occurs at equal rates on both sides. T_o and, to a lesser extent, T_s vary as Z changes during drying. Therefore, equation 3 usually must be numerically integrated. The integrand in equation 3 is proportional to ZdZ. Therefore, t_s will be proportional to L^2 regardless of the particular way T_s and T_o change during drying. It appears that required desorption times and required total drying times are also proportional to L^2.

It usually takes 14 to 18 hours to freeze dry a 12.5-mm (0.5-in.) thick slab of food that initially contains 75 to 85% water. In freeze-drying a 12.5-mm thick slab of coffee extract containing 74% water, 60% of the drying time was used for sublimation, 40% for desorption. Concentrated extracts with lower water contents usually dry more rapidly, but as concentration increases, a larger proportion of the drying time is used for desorption. Less information is available about the portion of drying time used for desorption for other products.

COLLAPSE

During sublimation, T_s must be kept low enough to prevent pore structure collapse caused by partial melting (8). When collapse occurs, vapor escape is cut off, the product foams, and drying rapidly fails. Collapse will occur at T_s greater than −23°C for coffee extract. Collapse can occur at T_s as low as −46°C (for Concord grape juice). At the other extreme, T_s as high as −1.5°C can be safely used for potatoes. Collapse can occur if flow-pressure drop is excessive or if P_o is too large, but it can be prevented, if need be, by reducing the condenser operating temperature or by reducing the sublimation rate by decreasing T_p.

PARTICULATE PRODUCTS

Most foods are freeze-dried in particulate form. Heat and mass transfer during the freeze drying of particles is more complex than for the freeze drying of slabs. Sublimation occurs beneath all exposed surfaces, but is most rapid where temperature is greatest—at ice–vapor interfaces closest to the heating plate. Although sublimation takes place over a greater depth of product, the required drying time remains proportional to L^2 (the product bed depth squared). Because vapor flows readily through spaces between particles, trays with solid bottoms can be used without interfering with drying in lower parts of the load. Because t_s depends on L^2, thicker parts of particulate loads can take markedly longer to dry than parts of average thickness. Therefore, care must taken to load trays uniformly.

Excessive amounts of fines were produced when slabs of freeze-dried extract were ground to provide granules for consumer use. To prevent this, frozen coffee extract is ground to produce granules of desired size before drying and freeze-dried in granule form in shallow beds in trays. Fines produced during grinding are removed by screening and recycled by mixing them with extract about to be frozen. Vapor-flow resistance during freeze drying of extract granules is much smaller than for slabs, but heat-transfer resistance is greater. In drying extract granules, heating plates are often maintained at 130°C while sublimation occurs. Yet, because of low pressure drop, T_s is only slightly higher than the condenser temperature during this period. After sublimation is complete and desorption alone occurs, plate temperatures are gradually reduced to 50°C.

Excessive product loss caused by granule entrainment in high-velocity vapor used to occur when coffee extract granules were freeze-dried, but now, in modern, properly operated, well-designed dryers, losses are now less than 0.1%.

FOAMED PRODUCTS

Gas is injected or beaten into some liquid feeds to foam them before they are frozen and dried. This reduces bulk density, improves vapor permeability during drying, or permits creation of novel foamed products. Freeze-dried soluble coffee made from concentrated extract, in foamed, frozen, granular form has the same bulk density as freeze-dried granular soluble coffee made from unfoamed, normal concentration extract. Without foaming, it would be denser and less acceptable. At regularly used continuous production conditions, foamed granules dry in roughly 7 hours.

CONDENSATION

The largest energy cost for freeze drying is that for the refrigeration used to condense vapor as ice. This cost can be reduced by removing water before drying and by operating the condenser at the highest temperature that still reliably prevents collapse. Higher condenser temperatures can be safely used when the pressure drop caused by vapor flow is low, as occurs when extract granules are dried.

In addition to removing heat of condensation (roughly the same as the heat of sublimation), the refrigerator also removes heat the vapor picks up from the product and the heating plate and heat transferred to the condenser by radiation. Heat shields or condenser enclosures that do not significantly interfere with vapor flow are used to reduce transfer of radiant heat to the condenser.

The temperature at the outer surface of deposited condensed ice is higher than at the condenser wall itself. When air inleakage is small, transparent ice forms on the condenser and the temperature difference across the ice is small. If inleakage is moderate, milky ice forms, and the temperature difference is greater. When inleakage is excessive, porous frost that conducts heat poorly forms and the temperature on the outside of the frost is markedly higher than in the condenser. When this occurs, the condenser operating temperature may have to be reduced to prevent P_o from rising enough to cause collapse. This raises refrigeration costs. Accumulation of air around condensers is minimized by having the vapor contacting the condenser efficiently sweep air toward the vacuum pump.

The drier shown in Figure 3 has two condensers, an online one that condenses vapor as ice, and an off-line one undergoing de-icing. The two condensers cyclically change roles with the aid of a pressure-tight slide gate. Thus, de-icing takes place almost continually while drying occurs. Ice buildup on the active condenser is 2 to 3 mm, at most. Up to 17 mm of ice may deposit on condensers subjected to de-icing only at the end of drying.

BATCH FREEZE DRIERS

Plant-scale, batch, freeze drying of foods is usually carried out in cylindrical vacuum chambers available in various sizes. A very large unit may provide 90 m^2 of shelf area and up to 227 kg/h of sublimation capacity. In some systems, trays of frozen food are loaded on a rack of shelves that is pushed into a drier on an overhead rail. There, the tray-laden shelves intermesh with built-in heating plates, as shown in Figure 3. Seven pairs of shelves are shown; 15 pairs are usually used. In other driers, loaded trays sit on built-in guide rails that lie between built-in heating plates. In still another system, both shelves and heating plates form part of a cart that is pushed into the drier. Heating fluid flows through the plates through connections made after the cart enters the drier. In each system, the sequence of operations used and basic processes involved are, in most respects, essentially the same as those described for a one-shelf drier.

Groups of batch freeze driers operating on a staggered schedule are used to efficiently freeze-dry food on a sustained basis. Staggered operation evens out refrigeration, heating, and vacuum pumping loads; reduces manpower needs; and permits sharing of refrigeration, heating, and vacuum pumping systems.

CONTINUOUS FREEZE DRIERS

Freeze-dried foods produced year-round in large volume are often freeze-dried continuously in cylindrical chambers, up to 3.7-m wide and up to 25-m long, that process 15 rows of trays at a time (9). Generated vapor is removed by using multiple pairs of switchable condensers, similar to those shown in Figure 3. Trays of frozen food enter such driers one-at-a-time through a double-valve lock, without loss of vacuum. The injected trays are automatically elevated, arrayed in a vertical set, and then pushed forward as a set. The pushed trays move through the drier in separate rows between heated plates, whose local temperatures are programmed to provide heat input that varies properly with axial tray position. At the discharge end of the drier, trays and their now-dry contents are individually automatically lowered and exit, one-at-a-time, through a double-valve lock. Pairs of locks are used at each end of large driers. In smaller driers, individual locks are used at each end. Depending on drier size and feed moisture content, available continuous freeze driers can process 5,000 to 27,500 kg of feed per day.

ALTERNATIVE FREEZE DRYING METHODS

U.S. food freeze drying patents issued between 1961 and 1976, an important period of development, have been re-

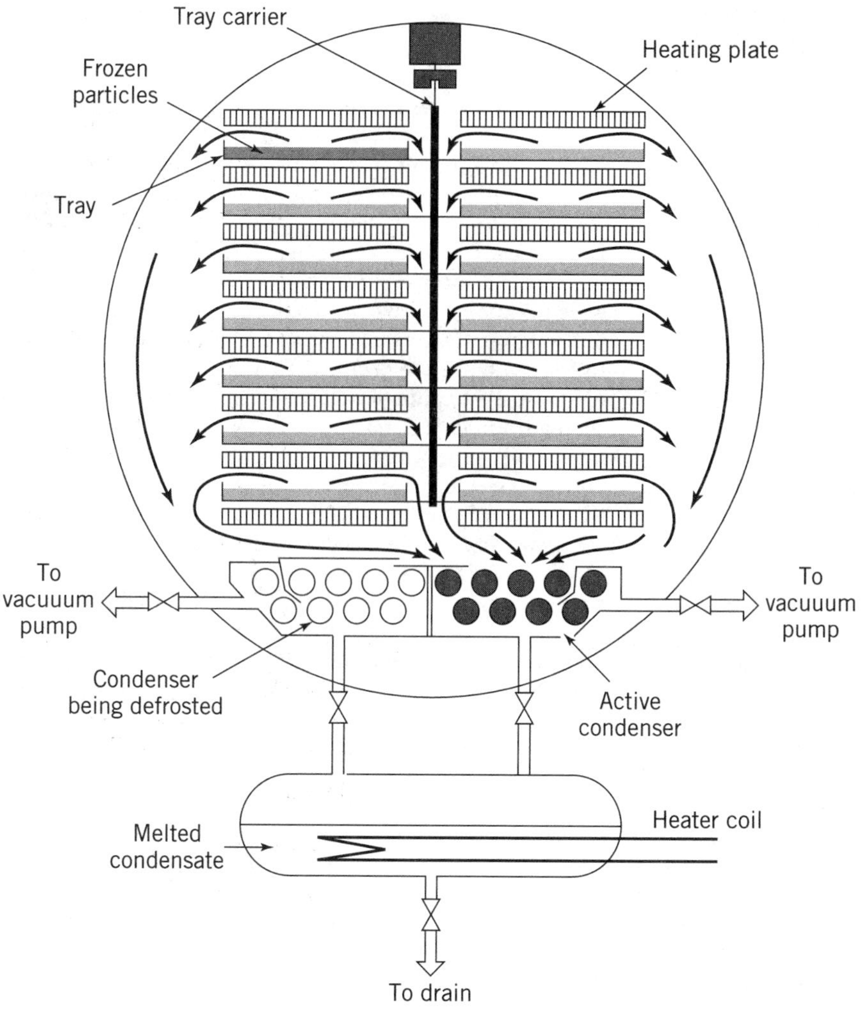

Figure 3. A batch freeze drier in which dual, switchable condensers are used and in which trays enter on a rack of shelves supported on an overhead track. *Source:* Reproduced with permission of Atlas-Stord Inc.

viewed by Gutcho (1). Some of these patents describe now-standard freeze drying methods and equipment, methods for freeze drying particular products, or methods for preparing such products for drying. Others describe freeze drying methods that for one reason or another have been abandoned or are rarely or infrequently used today. Some describe infrequently used heating methods: direct-contact heating; microwave heating; infrared heating; induction heating; use of layers of expanded metal mesh or of ribbed trays to facilitate heat transfer; or the use of heat pumping to couple heating and condensation. Others describe spray freeze drying, various types of atmospheric freeze drying, circulation of low-pressure gas through the drying material, or vapor removal by adsorption instead of condensation. Still other patents describe use of moving belts, vibrating conveyers, helical conveyors or vertical stacks of scraped trays as particle conveying devices in continuous freeze driers. Food freeze drying patents and literature for the years 1977 to 1998 were reviewed in preparing this article. Aside from development of the previously described continuous freeze drier, innovation in the field has slowed down. Atmospheric freeze drying in sorbent-containing fluidized beds and use of microwave heating in freeze drying are still of interest, and articles dealing with these methods continue to appear from time to time.

BIBLIOGRAPHY

1. M. H. Gutcho, *Freeze Drying Processes for the Food Industry*, Noyes Data Corp., Park Ridge, N.J., 1977.
2. M. Karel, "Fundamentals of Dehydration Processes," in A. Spicer, ed., *Advances in Preconcentration and Dehydration*, Elsevier Applied Science, London, United Kingdom, 1974, pp. 45–94.
3. J. M. Flink, "Effect of Processing on Nutritive Value of Food: Freeze-Drying," in M. Recheigl, ed., *Handbook of the Nutritive Value of Processed Food*, Vol. 1, CRC Press, Boca Raton, Fla., 1982, pp. 45–62.
4. J. Rolfgaard, "Freeze Drying: Processing, Costs and Applications," in A. Turner, ed., *Food Technology International Europe*, Sterling, London, United Kingdom, 1984, pp. 47–49.
5. J. Lorentzen, "Freeze Drying: The Process, Equipment and Products," in S. Thorne, ed., *Developments in Food Preservation*, Vol. 1, Elsevier Applied Science, London, United Kingdom, 1981, pp. 153–175.
6. U. Rosenberg and W. Bogl, "Microwave Thawing, Drying and Baking in the Food Industry," *Food Technol.* **41**, 85–91 (1987).
7. M. Karel, "Dehydration of Foods," in O. R. Fennema, ed., *Principles of Food Science, Part 2, Physical Principles of Food Preservation*, Marcel Dekker, New York, 1975, pp. 359–395.

8. R. J. Bellows and C. J. King, "Freeze Drying of Aqueous Solutions: Maximum Allowable Operating Temperatures," *Cryobiology* **9**, 559 (1972).
9. H. G. A. Unger, "Revolution in Freeze Drying," *Food Processing Industry* **51**, 20 (April 1982).

HENRY SCHWARTZBERG
University of Massachusetts
Amherst, Massachusetts

FREEZING SYSTEMS FOR THE FOOD INDUSTRY

FUNDAMENTALS

During storage foods are subjected to changes that affect food quality and that sooner or later will lead to severe deterioration and eventually spoilage of the foods. These changes are caused by microorganisms and chemical and physical reactions. Often a combination or interaction of different reactions will cause changes, lowering the quality primarily by changing the sensory properties of the product.

The wide variety of circumstances, including type of food and ingredients, will determine the type of changes that will dominate. In meat and fish products or other foods rich in protein, changes caused by microorganisms will dominate while food rich in fat is more susceptible to chemical and biochemical change. Physical changes occur in all types of food.

The purpose of all food preservation methods is to inhibit or decrease the speed of reaction responsible for the deterioration. All of these reactions are among other factors influenced by temperature. The speed of reaction is decreased at lower temperatures. Cooling and chill storage therefore are perhaps the most important methods of enhancing the storage life of most food products. But even at temperatures near the freezing point some reactions, including growth of many microorganisms, continue at a rate that will limit the preservability to a relatively short period of time.

When the food temperature is $<-10°C$ microbiologic growth will cease. Chemical, biochemical, and physical reactions will still continue at very low temperatures but at a slow pace. Storage life is substantially enhanced as compared to storage at chilled or ambient temperature.

HISTORY OF REFRIGERATION

Refrigeration applied both above and below the freezing point of foods has been used for thousands of years to preserve foods and to increase comfort. Historians estimate that caves were used for food storage about 100,000 yr ago (1).

Temperature inside the caves is naturally low as a result of the evaporization of water, which is often present.

An Egyptian frieze from 3000 B.C. shows a slave waving a fan in front of a clay pot. The cooling effect from vaporization of water was also utilized in this case. Egyptians also utilized terrestrial radiation toward space at night under a clear sky to produce ice.

Ice was used locally when available. Later it was harvested during wintertime and stored for the summer season. In 1100 B.C. a Chinese poem mentioned ice houses. Ice and snow were transported over great distances, eg, from the Apennines to Rome, and caravans were transporting ice and snow from Lebanon to the palaces of the Califs of Damascus, Baghdad, and the sultans in Cairo.

In many countries ice was believed to be a gift of the gods to humans. This was confirmed as late as during the mid-nineteenth century, when the American John Goorie in 1844 managed to produce ice with an air compressor but did not dare to publish the invention under his own name. Instead, under a pseudonym he wrote a scientific article describing an ice machine as a future possibility. The New York newspaper, *The Globe*, shortly after published an article titled "Some lunatic in Florida believes his machine can make ice equally good as the All Mighty."

Obviously hunters and gatherers living in a cold climate did use the natural freezing of their food products in order to preserve them over long periods of time.

Exactly when the temperature-decreasing effect—with the addition of certain salts to water—was discovered is not known. There is, however, reason to believe that the method was used in India in the fourth century A.D. During the fourteenth and fifteenth century a number of European scientists were working with salt solutions and managed to achieve temperatures as low as -15 to $-20°C$. With those salt mixtures the stage between natural and artificial cooling was passed.

It was not until 1755, however, that the first apparatus for making ice was constructed by William Cullen. Vaporization of water at reduced pressure was utilized.

During the first half of the nineteenth century four events of fundamental importance for the refrigeration industry took place: (1) the systematic work on the liquefaction of gases, (2) the genesis of thermodynamics originated by Nicolas Carnot in 1824 on the invention of the refrigeration machine using compression of a liquefiable gas (often referred to as the Carnot engine), (3) Jacob Perkins's work in mechanical refrigeration in 1834, and (4) the air cycle machine (Goorie, 1844); the latter two inventions remained undeveloped for two decades.

The start of industrial freezing of food is often set at around 1880, even if the first industrial installations were made some 20 years earlier. During 1870–1880 frozen meat was transported from the southern hemisphere to Europe. Initially those endeavors were unsuccessful. In beef cargo from Buenos Aires to Rouen (port on Seine River) in France as much as 25% in weight was lost by sublimation and the quality was unacceptable. The breakthrough came in 1877 when a shipment of frozen meat was brought in from Buenos Aires to Marseilles on the steamer *Paraguay* (2). This shipment was followed by another one from South America to New York in 1879, and in 1880 the steamer *Strathleven* made the journey from Sydney and Melbourne to London.

As compared to the quality achieved today, much was to be wished for. The freezing was carried out very slowly and the storage temperatures were high compared to the storage temperatures used today. In 1915 the German scientist Rudolf Plank showed the importance of rapid freezing in experiments on fish. This knowledge was soon applied for other products.

Although fish, meat, poultry, and berries have long been subjected to preservation by freezing, frozen-state deliveries to consumers did not take place until the development of "quick freezing" in the mid-1920s. The date is often set at the October 14, 1924, the day when Clarence Birdseye received a patent for a plate freezer. This apparatus was revolutionary, as it was now possible to quick freeze packaged foods for the retail outlets. In 1929 the American company Postum/General Foods bought Birdseye's Company, and in the following year the first consumer package of frozen foods were marketed in Springfield, Massachusetts (3).

In 1938 frozen foods were introduced to Europe and the year after, British companies were licensed to produce quick-frozen foods. After World War II the market expanded rapidly throughout all industrialized countries.

FUNDAMENTAL CHANGES DURING FREEZING

The importance of early freezing is related to the need to decrease the rate of the deterioration processes caused by chemical, biochemical, and physical reactions as well as microbiological activity. Chemical and biochemical reactions influence the product quality not during the freezing process, but during subsequent storage. The importance of quick freezing is some times argued from a sensory point of view (4). It must not, however, be forgotten that quick freezing is most important from a technical–economical–operational point of view. With reference to quality, the rate of freezing determines the size of weight loss and in some cases also the microbiologic quality of the product. The drip loss or loss of product juice on thawing is determined by the rate of freezing as well.

The freezing process may be seen as a lowering of the product temperature from its original value to the storage temperature in question. However, from a technical–economical–operational point of view a more strict definition is needed.

Ice Crystallization

The major component of most foods is water, which in the freezing process is transferred from a liquid to a solid state. This transfer obviously result in numerous changes.

Most food products consist of or contain animal and/or vegetable cells forming biological tissues. The water solution of the tissue is contained between the cells—intercellular fluid—and within the cells—intracellular fluid.

When the food product is cooled below 0°C ice begins to form at the initial freezing point. The temperature at which freezing starts depends on the concentration of dissolved substances—salts and other solubles—present. The concentration is higher within than outside the cells. The cell membrane acts as an osmotic barrier and maintains the difference in concentration.

When the product is frozen the first ice crystals are formed outside the cell since the freezing point is higher because of the more diluted fluid here than inside the cell. Once started, the rate of ice crystallization is a function of the speed of heat removal as well as the diffusion of water from within the cell to the intercellular space. If the freezing rate is low, few crystallization centers—nucelli—are formed in the intercellular space. During the freezing process the cell looses water by diffusion through the membrane, and this water crystallizes to ice on the surface on the crystals already formed outside the cell. As in slow freezing, there are few nucleins formed; those existing as crystals grow to a relatively large size.

As the cells loose water, the remaining solution within the cells becomes increasingly concentrated and the cell volume shrinks causing the cell wall to partly or entirely collapse. The large ice crystals formed outside the cell wall occupy a larger volume than does the corresponding amount of water, and therefore they will exert a physical pressure on the cell wall. In some cases this pressure can be sufficient to damage the cell wall and contribute to an increased drip loss on thawing.

By increasing freezing rates a larger number of ice crystallization nuclei are formed, which results in a much smaller size of the final crystals as compared to slow freezing. However, even in the case of high freezing rates most of the crystals are formed outside the cells. Only at extremely high freezing rates not obtainable in commercial freezing of food products, small crystals are formed uniformly throughout the tissue both externally and internally with respect to the cell.

As the food products are cooled down below the initial freezing point, an increasing amount of water is turned into ice and the residual solutions become more concentrated. The relation of water frozen out as ice and the concentration of the remaining solution have an impact on the preservability of a number of food products (5).

The size of the ice crystals has long been regarded as crucial for the quality of the frozen product. It appears, however, from experience as well as a number of investigations that the differences in ice crystal size and distribution have little effect on the sensory properties of the food product when presented to the consumer, provided up-to-date equipment and good commercial practice have been used (6).

The freezing rate is not negligible, however. On the contrary, in good commercial practice the freezing time must be determined for each product in order to safeguard against microbiologic growth, which is most important from a safety point of view and low controlled weight losses, which is important from an economic point of view.

The practical result of different freezing rates—size and locations of ice crystals—can be seen as a difference in drip loss of water or "product juice" when a product is thawed. Loss of juice results in a more or less pronounced loss of texture, flavors, and—in most cases—nutrients. For this reason the drip loss is often used as an indication of the quality loss during freezing and subsequent storage. The relation between speed of freezing and drip loss for two different foods is illustrated in Figure 1.

For strawberries it is rather obvious that the consumer will not readily accept the product with a 20% drip loss if a product with only 8–10% drip loss is available. The latter

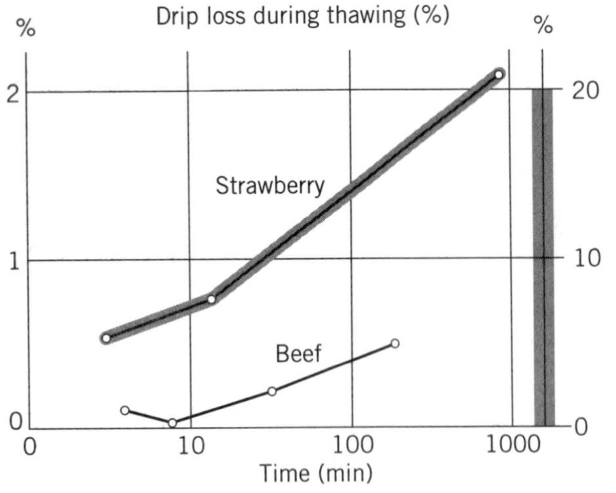

Figure 1. Drip loss during thawing of sliced beef and strawberries, frozen at various rates, to an equalization temperature of −18°C. *Source:* Ref. 6.

level of drip loss can be achieved in modern freezers usually employing the fluidization technique.

When comparing slices of beef even very slow freezing rates gave a small drip loss hardly noticeable to the consumer. The small improvement achieved by quick freezing is normally not observed.

Microbiology

With reference to temperature requirements, bacterias are divided into four basic groups according to type of growth: thermophilic, mesophilic, psychrophilic, and psychrotrophic. Of those the two latter groups are of special interest in food spoilage at low temperatures. Psychrophilics are often an important cause of spoilage of protein-rich foods such as meat, fish, poultry; psychrophilic bacteria grow well at temperatures above 0°C.

The optimum temperature for growth, however, is much higher. Psychrotrophics are also able to grow close to the freezing point but the optimum temperature is higher than that for psychrophilics. Lowering the temperature will slow down the growth rate of all bacteria, and at temperatures used in commercial storage of frozen foods all microbiologic growth ceases completely. The time to decrease the temperature to below the freezing point is critical. A product temperature of −10°C is normally considered safe with regard to microbiologic growth.

During freezing and frozen storage some bacteria are impaired and even destroyed. Under certain circumstances the death of bacteria will cause a considerable decrease in the total number of viable cells in the frozen products. Since some species are more susceptible to freezing injuries than others there may also be a change in the relation between various species (7).

From a microbiologic point of view the total flow from the production to the consumption must be regarded. During this flow, food products are subjected to various temperatures and other growth-affecting factors. Large variations occur from product to product; meat and meat products have been chosen as examples in the following discussion. The general concept is obviously valid for most food products.

During slaughter and subsequent handling the surface of the meat is infected by microorganisms originating from the animal itself and from the environment.

The number of organisms present is dependent on the hygienic conditions, but the flora consists to a great extent of spoilage organisms that thrive on the meat surface and multiply.

Rapid cooling reduces the rate of growth substantially, but many of the psychrophilic and psychrotropic organisms will grow even at chill temperatures. These organisms depend on free oxygen for their metabolism, and since oxygen is available in the surface layer only, no growth will occur in the interior of the meat.

Freezing of carcass meat therefore normally will not cause any serious problems from a microbiologic point of view. In commercial freezing the freezing rate is fast enough to stop the growth at the surface. As the microorganisms cannot develop under the surface layer, the freezing rate is of less importance microbiologically.

Processing, like cutting and mincing, increases the microbiologic contamination as the surface/volume ratio increases. The freezing rate becomes more critical. A common pack in the industry today is the 30-kg carton normally frozen in a traditional air-blast tunnel. It is then essential that a good air circulation be provided in order not to prolong the freezing time.

Freezing times for cartons subjected to both adequate and inadequate airflow are compared in Figure 2.

Inadequate airflow has been achieved by placing the meat cartons directly on pallets with only a small wooden spacer (30 mm) between each layer instead of being placed on freezing racks or with rigid layer separators with a minimum height of 50 nm.

If spacers used as in the improper airflow system do not cover the total carton area, some cartons may collapse, which will prevent airflow through the different layers.

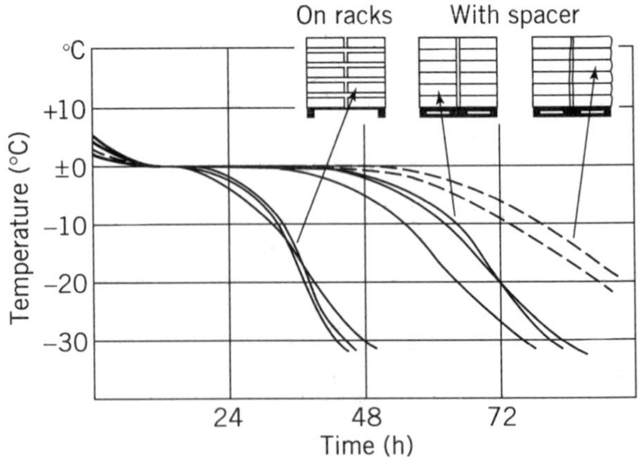

Figure 2. Recorded temperature decrease during the freezing of cartons in an air-blast tunnel (air temperature −38°C, front air velocity 1.5 m/s, size of cartons 160 × 400 × 600 mm). *Source:* Ref. 7.

The freezing time is obviously prolonged. There is also a risk that cartons are removed from the air-blast tunnel before the freezing is completed if blocking of the air channels is not visible from the outside. In the latter case there may be substantial growth of microorganisms during subsequent storage, ie, until a sufficiently low temperature has been reached.

In most cases a freezing time of 24–36 h down to − 10°C in the center of wholesale cut meat will not cause any microbiologic problems. If the degree of cutting is increased to smaller cuts, such a long freezing time may become a major problem. Those products should preferably be frozen integrated in the processing line before packaging or in very small packages that allow for a much faster freezing.

As most prepared foods involve a high degree of processing as well as mixing of different ingredients the freezing becomes very important. The general pattern of the growth of microorganisms in the production of prepared foods is illustrated in Figure 3.

During storage of raw material as well as during handling and preparation, microorganism growth will take place. If the preparation is followed by heat treatment, the total number of microorganisms will be reduced.

At this point the product could be handled in two different ways: either packaged and frozen in batch-operated equipment or frozen in-line and then packed. If the products are placed on racks and then transported to a freezing tunnel for freezing, there may be a time lapse, resulting in a marked growth of microorganisms. If the product has been heat treated, it will pass through the temperature zone of optimum microbiologic activity.

A chilling operation immediately after the heat treatment—which means that the products are cooled down to below − 10°C, then packaged and frozen—will reduce the growth of bacteria. However also in this case microbiologic growth can be recorded. If the products are instead frozen in-line immediately after the heat treatment and then packaged, there will be almost no increase in the number of bacteria present.

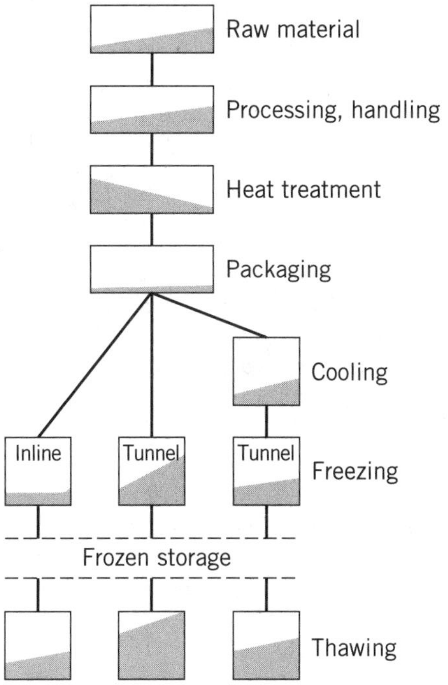

Figure 3. Growth pattern of bacteria during different steps in the processing of precooked frozen foods. *Source:* Ref. 7.

In-line freezing is, of course, even more essential when prepared foods are processed without heat treatment prior to the freezing process.

A low microbiologic load in the frozen state will directly influence the preservability of the food product after thawing. All measures taken to arrest growth of microorganisms prior to freezing are therefore beneficial.

Compared to batch freezing in a tunnel, the modern in-line freezer provides much faster freezing, and even more important, the in-line process itself minimizes the delays in product flow from preparation to freezing and through the temperature zone, which is critical from a microbiologic point of view.

Desiccation

During the freezing process it is unavoidable that a certain evaporation of water from the surface takes place, resulting in both a quality loss and a weight loss. Only if the product is tightly enclosed in a water vapor inpermeable packaging material, evaporation can be completely avoided. If there are small spaces between the product and the packaging material ice is deposited in these.

Freezers that are poorly designed and improperly used may cause a weight loss in the order of 5–7%, while properly designed and used equipment will cause no more than 0.5–1.5% weight loss. As the total freezing cost often is no greater than 3–5% of the product value, it is obvious that dehydration losses are of great importance in any comparison of different freezing methods.

There is a definite correlation between the degree of dehydration loss and the rate of freezing. In Figure 4, the temperature gradient in a product with a thickness of 2 b and in the air surrounding the product is plotted against the partial pressure of water vapor in moist air.

The size of the dehydration is influenced by a number of factors related to the biological materials as well as to physical handling, temperature of the heat-transfer medium, dimensions of the product, etc. The rate of evaporation from the product is determined by the conditions at the surface.

In a simplified consideration the evaporation rate may be regarded as proportional to the vapor pressure Δp^2. The curve corresponds to the "relative" humidity of the product surface. This concept is introduced in order to represent the diffusion resistance that may exists in cell walls, etc.

During infinitely slow freezing the surface temperature approaches the average temperature t_3. Then the evaporation rate increases to the value Δp_3. Hence it is clear that $t_3 - t_2$ should be as great as possible in order to obtain low dehydration losses. Three principal factors contribute to this:

- b large. In thick products the surface temperature will be low during most of the freezing process.

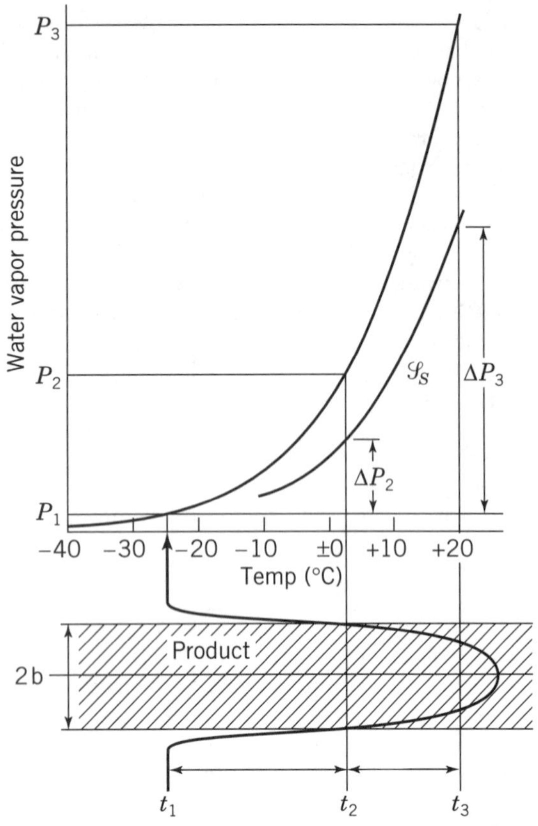

Figure 4. Dehydration mechanism during freezing. *Source:* Ref. 8.

- t_1 low. The lower the temperature of the ambient air, the more curved the temperature gradient in the product will be.
- High heat-transfer rate. A low air temperature has no appreciable effect if the heat-transfer rate is too low. Therefore, it is extremely important that the heat-transfer conditions be favorable. This is more important for thinner products.

Wet products generate water vapor at a rate proportional to the difference between the vapor pressure at the surface of the product and that of the surrounding air. In a product that has a more or less dry surface there is a resistance in the cell walls against diffusion of vapor from the interior of the product to the surface and the air. This results in a reduction of vapor pressure at the surface of the product.

The water vapor pressure decreases rapidly when the temperature is reduced, which means that the dehydration will be less the colder the heat-transfer medium is. The importance of a low temperature during freezing is well illustrated in Figure 5, where the accumulated weight loss during different freezing tests with a temperature varying from −13 to −35°C is plotted against the core temperature of the product. The diagram illustrates both the importance of the latent heat zone and the drastic influence of the air temperature.

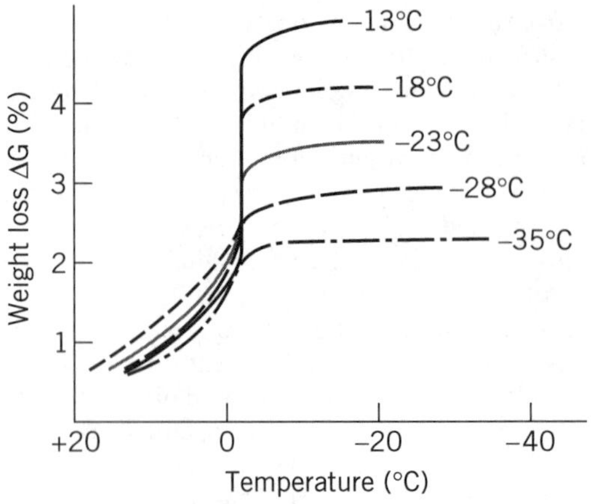

Figure 5. Accumulated dehydration during freezing at different air temperatures. *Source:* Ref. 9.

Glazing for Protection of Product Quality. In order to improve shelf life by preventing desiccation and oxidative changes, it has become standard practice to glaze certain individually frozen products, eg, shrimp, after freezing. The product quality is greatly improved as the thin ice layer prevents the product from the changes mentioned above. The glazing is carried out by spraying the frozen product with cold water, which immediately freezes on the surface. Even if the product leaves the freezer at a low outfeed temperature, the product temperature may increase considerably after the glazing operation. This high-end temperature may cause clumping of the product when packed, and the slow decrease in temperature during subsequent storage and distribution may lead to noticeable texture changes on thawing.

To avoid these problems special equipment has been developed—GLAZoFREEZE—that is designed to lower the product temperature of the glazed product immediately after the glazing operation. The equipment utilizes the fluidization technique. In GLAZoFREEZE the fluidization must be specially gentle to the product in order to avoid damage of the glazed surface.

DEFINITIONS

Definitions of the freezing process, freezing time, freezing rate, and speed of freezing are most useful in comparisons of different systems and equipment. The International Institute of Refrigeration gives the following definitions (10).

Freezing Process

During the freezing process different parts of a product will pass through the various stages at different times. If a particular location in the product is considered, three stages of temperature changes can be defined.

Prefreezing Stage

This is the interval between the time at which a high-temperature product is subjected to a freezing process and the time at which the water starts to crystallize.

Freezing Stage

The period during which the temperature at the considered location is almost constant because the heat that being extracted is causing the majority of water to change phase into ice.

Reduction to Storage Temperature

This is the period during which the temperature is reduced from the temperature at which most of the freezable water has been converted to ice to the intended final temperature. The final temperature can result when the storage temperature is reached in any part of the product, including the thermal center or the equalization temperature. The equalization temperature is the temperature achieved under adiabatic conditions—without heat exchange with the environment.

Freezing Time

The freezing time is defined as the time elapsed from the start of the prefreezing stage until the final temperature is reached. The freezing time depends not only on the initial and final temperature of the product and the quantity of heat removed but also on the dimensions—especially the thickness—and shape of the product as well as on the heat-transfer process and its temperature.

Freezing Rate

For a product or a package the freezing rate (degrees Celsius per hour) is the difference between the initial and the final temperatures divided by the freezing time. In a given point of a product the local freezing rate is equal to the difference between the initial temperature and the desired temperature divided by the time elapsed until the moment at which the latter temperature is achieved in this particular point.

Speed of Ice Front Movement

The freezing rate may be evaluated by the speed of movement of the ice (in centimeters per hour) through a product. This speed is faster near the surface and slower toward the center. As a result, reported freezing rates from different sources are not necessarily comparable.

Practical Freezing Time

Besides the above-mentioned definitions the industry has developed a common definition of freezing time for practical commercial purposes.

From a practical point of view the freezing time is defined as the time required to lower the temperature of a product to an equalization temperature of −18°C under adiabatic conditions. This definition determines the commercial capacity of the freezing equipment.

The time the product is held in the freezer is known as standard freezing time (SFT) or holding time. In Figure 6 ET stands for equalization time. In determining capacity of equipment there will be an equalization to −18°C. However, in practical life this temperature equalization takes place during subsequent handling and storage and the equalization temperature will be that of the surrounding environment. From a theoretical point of view the total freezing time (TFT) for a product is SFT + ET.

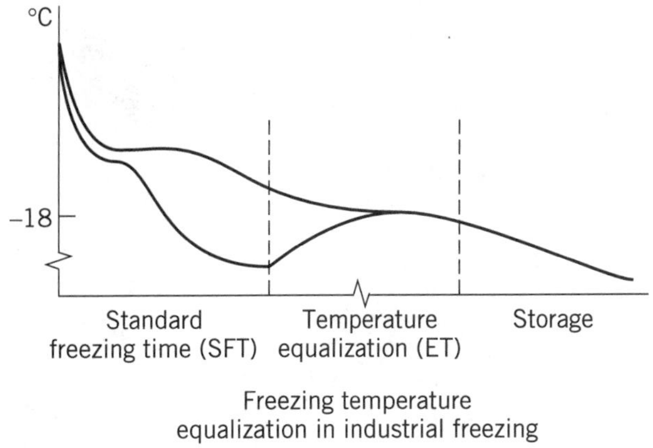

Figure 6. Practical commercial definition of the freezing process.

Important is that contrary to what is commonly believed the product temperature is not brought down to −18°C in the center of the product in the freezer. This temperature or preferably a lower temperature is achieved during subsequent handling, packaging, and storage.

Freezer Capacity

The investment in any freezing equipment is based on requirements to freeze a certain quantity of food per hour. As a principle the following relation is valid for any type of freezer:

$$C = Q/F = Vq/F$$

where C is the capacity of the freezer expressed as tons per hour, Q is the quantity, in tons, of the product that can be accommodated in the freezer, F is the holding time, in hours, of the product *in the particular freezer*, V is the volume, in cubic meters, of the product that can be accommodated in the freezer, and q is the volume weight, in tons per cubic meter, of the product.

The holding time refers to the temperature change that is required, ie, usually from incoming temperature to −18°C equalization temperature.

It is important to understand the fundamental importance of the holding time for the capacity of a freezer. Every product has different holding times. The amount of heat to be extracted per kilogram of product usually varies only little within a group of products, eg, vegetables, but the dimensions of each product particle have a drastic influence (see Fig. 7).

As all heat must be extracted from the product through its surface, the relation between surface and weight, the specific surface in square meters per kilogram, is of great interest. Volume and weight are entirely proportional to each other and therefore can substitute each other in this discussion. In Figure 7 can be seen that the freezing time is inversely proportional to the specific surface for particles.

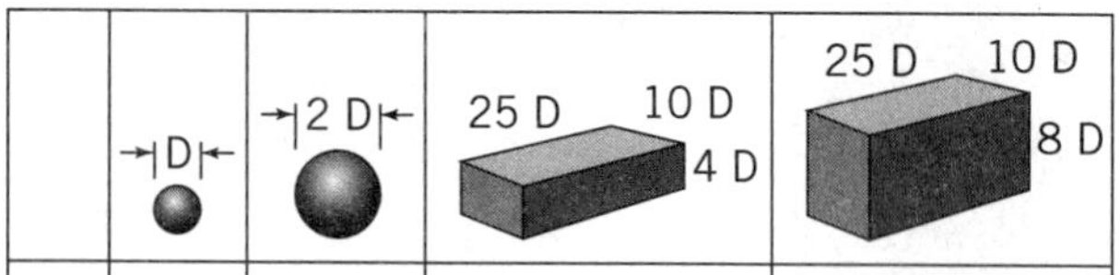

V =	$\frac{\pi \cdot D^3}{6}$	$8 \cdot \frac{\pi \cdot D^3}{6}$	$1000 \cdot 0{,}56\ D^3$	$2000 \cdot 0{,}56\ D^3$
F =	πD^2	$4\pi D^2$	$780\ D^2$	$1060\ D^2$
$\frac{F}{V}$ =	$6 \cdot \frac{1}{D}$	$3 \cdot \frac{1}{D}$	$1{,}40 \cdot \frac{1}{D}$	$0{,}94 \cdot \frac{1}{D}$
I =	1	2	4,3	6,4
T =	6 min	12 min	2,5h	5,5h

Figure 7. Heat to be extracted during freezing of different package sizes (where V = volume, F = surface, I = inverted specific surface, T = practical holding time, F/V = specific surface).

As a practical example a simple rack tunnel with a capacity of one ton per hour of spinach purée in 45-mm-thick packages can be taken. It may have the following capacities:

Spinach, packaged, 45-mm layer	1 ton/h
Peas, unpackaged, 30 mm layer	1.5 ton/h
Peas, packaged, 45 mm layer	0.6 ton/s
Parsley, packaged, 45 mm layer	0.4 ton/h
TV-dinner, packaged, 35 mm layer	0.2 ton/h

In the case of parsley and TV dinners the capacity is reduced primarily because of the low volume weight and secondarily because of the poor heat transfer inside the packages.

Design Capacity—Working Capacity. For an in-line operation with short freezing time, (eg, 6–10 min), it is important to distinguish between design capacity and working capacity.

Design capacity refers to temperature reduction in volume per hour, according to specifications, if the product feed is entirely steady and continuous. In reality the load will be fluctuating. Because the temperature is influenced by fluctuations as short as 3–5 min, the product feed will have to be cut back until the peaks do not exceed 100% of the design capacity. This means that the average product flow will be only 80–90% of the design capacity.

Other reductions are required since every minute of every working hour cannot be utilized. When the operation is started in the morning it takes time to build up the product flow to 100%. If there are any production breaks for lunch or shift changes, the product feed will have to be stopped some time in advance.

Finally there may be stoppages in the product flow because of breakdowns anywhere along the processing line. In total, the working capacity may be as low as 70% of design capacity with respect to one and the same product. To obtain maximum production out of a line all links in the chain must meet required capacity.

Frequently a freezer may be used for several different products. For such products the above working capacity will have to be multiplied with the capacity factor, which relates the capacity of that particular product to the capacity of the product stated in the design specification. For products with very low capacity factor, the operating conditions usually differ so much that a separate working capacity must be established.

Precooling. Precooling is defined as the cooling of the product before it enters the freezer. The process of precooling has a positive effect on the energy required to freeze a product and generally reduces the frost load on the coils as well. Precooling can be done by blowing ambient air over the product, blowing ambient air that has been evaporatively cooled by water sprays, immersion in cold water (refrigerated or nonrefrigerated), or blowing refrigerated air over the product.

Sanitation during precooling processes is an important consideration because the conveyor system could be at a temperature for all or part of the cycle that can allow bacterial buildup.

Precooling reduces the load on the low-temperature refrigeration system, which reduces the power required by the low-temperature system. If a refrigerated precooler is used, the precooling heat is removed by a refrigeration system that consumes less power per unit of heat removed from the product, and a net power savings is achieved.

FREEZING EQUIPMENT

Today's freezing equipment can be divided into two main groups: integrated in the processing line and operating in batches.

According to the heat-transfer method, there are basically three main types of equipment:

- Air-blast freezers, which use air for heat transfer. Because air is the most common freezing media, this method of heat transfer has probably the largest range of designs.
- Contact freezers. Heat transfers occurs through conduction. A refrigerated surface is placed in direct contact with the product or package to carry away the heat. Alternatively, the product is immersed in a cold liquid—brine.
- Cryogenic freezers. These freezers use liquefiable gases, nitrogen or carbon dioxide to produce vapors that precool and freeze the products.

Combinations of these heat-transfer methods can be seen in special designs.

The freezing equipment can also be divided into two main groups with reference to the product. Individually quick-frozen (IQF) and packaged products.

All of these methods are used in the food processing industry; however, the preferred systems are those that can be operated in-line, integrated with the processing and packaging operations.

The freezing equipment must be designed to accommodate the three stages of the freezing process and should ideally optimize the total process.

The following design criteria are of special interest (see below under "Major Considerations in Freezer Design"):

Product quality
Hygiene
Minimum product losses
Reliable and simple operation
Simple maintenance
Economy, freezing cost

In the following sections the freezing equipment will be discussed with the basis on the method of heat transfer.

Air-Blast Freezers

Air is the most common freezing medium, and for that reason a number of designs can be found. Even if a storage room should never be considered as a piece of freezing equipment, it is sometimes used for this purpose. However, freezing in a storage room involves so many disadvantages that it should be used only in exceptional cases. The freezing is so slow that the quality of almost all products will be affected adversely.

Sharp Freezer, Blast Room

Basically, a sharp freezer or blast room is a cold storage room that has been especially constructed and equipped to operate at low temperatures for freezing. Even if this room is equipped with extra refrigeration capacity as well as fans for air circulation, there is normally no controlled airflow over the products, and for that reason freezing is normally slow. This type of equipment is, however, still used sometimes for bulk products such as beef quarters, but not for processed food products.

Tunnel Freezers

In tunnel freezers refrigerated air is circulated over the product, which is placed on trays or special spacers that stand in or pass through the tunnel in racks or trolleys.

The racks or spacers are arranged to provide an air space between each layer of trays. The racks or trolleys can be moved in and out of the freezer manually or by a fork-lift truck pushed through the tunnel by a pushing mechanism or slide-through. Tunnels are also used for freezing hanging meat carcasses carried on a suspension conveyor or in especially designed racks.

Practically all products can be frozen in a tunnel freezer. Whole, diced, and sliced vegetables may be frozen in cartons or unpacked in a 30–40-mm-deep layer on trays. Spinach, broccoli, meat patties, fish fillets, and prepared foods are frozen in packages in this type of equipment. It is, however, important to recognize that both refrigeration capacity and arrangements for air circulation have been designed for a specific product range. This means that if the tunnel freezer is designed for freezing meat carcasses or bulk meat cartons, the tunnel design is not appropriate for handling unpacked products such as vegetables. The result is often a increase in weight loss that influences both the sensory properties of the product and the economics of the freezing operation.

The flexibility of this type of equipment is balanced by high manpower requirements and a considerable weight loss if improperly used. The high manpower is caused by the handling, releasing, cleaning, and transportation of the trays and of the racks.

The tunnel should always be filled with product in such way that uniform airflow is obtained over all products to be frozen (see Fig. 8).

Mechanized Freezing Tunnels

A certain degree of mechanization is achieved when the racks are fitted with casters or wheels. The racks or trolleys are usually moved on rails by a pushing mechanism often powered hydraulically. Mechanized tunnel freezers are known as push-through tunnels, carrier freezers, and sliding-tray freezers. A typical design is shown in Figure 9.

As in any other tunnel freezer, the products are placed on trays that, in turn, are stacked on trolleys. The tunnel contains one or two rows of trolleys that are pushed forward stepwise by each other on rails in line with the product line. When a trolley leaves the freezer, the frozen prod-

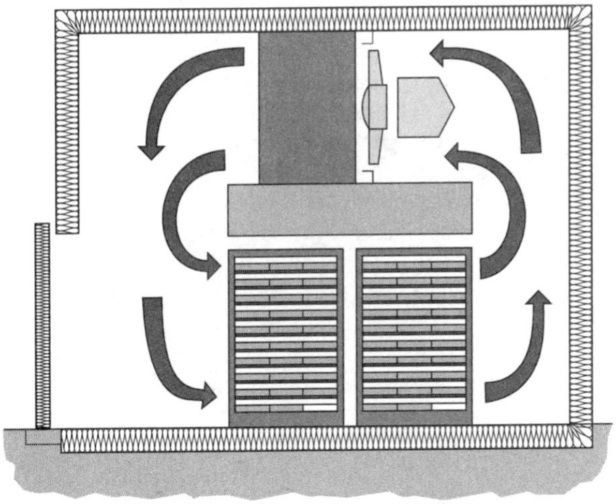

Figure 8. Stationary freezing tunnel.

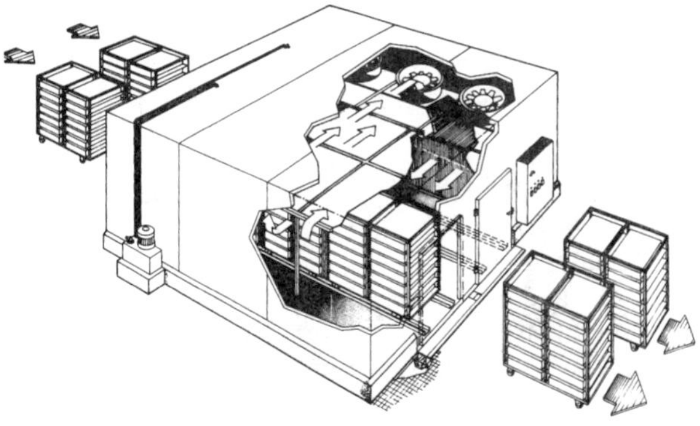

Figure 9. Mechanized freezing tunnel.

ucts are removed from the trays and the trolleys are returned to the loading station.

The design and construction of the freezer is common to most tunnel freezers. The evaporation coils are placed on a steel frame standing on the insulated floor. They have a thin spacing, which varies with the depth of the coils. The spacing is wide at the air inlet and narrow at the outlet. This ensures an even frost buildup on the coils without detriment to the airflow. The coils are furnished with liquid and suction headers and arranged for pump circulation of the refrigerant. In some installations arrangements are also made for gravity feed. Standard defrosting is effected by hot gas or by leaving the doors open and running the fans during the night.

The fans circulate the air down between the coil and the wall through the coil, through the trolleys and the products, and then the air is deflected up along the wall and back to the fans above the sheeting.

Tunnel freezers are built for capacities varying from a few hundred kilograms to several tons per hour. The freezing time in this type of equipment varies considerably from a few hours when freezing unpackaged vegetables in a thin layer to 48 h freezing meat carcasses.

The sliding-tray freezer is basically a tunnel consisting of one huge rack accommodating many large trays on each tier. At one end of the construction there is an elevating mechanism that lifts the arriving tray to the top tier, where it is pushed in, forcing all the other trays on this tier to advance one step. The tray at the far end is pushed onto an elevator that brings it down one tier where the tray is entered. Consequently, on every odd tier the trays will be advancing, and on every even tier they will be returning. For each tray that enters, all trays will advance one step.

All mechanisms are usually hydraulically powered. Outside the freezer enclosure automatic loading and unloading of the trays may be arranged.

The freezing trays are in the traveling-tray freezer connected to one or two sturdy roller chains at each end. These are arranged to move the trays forward to a set of sprockets elevating the trays one tier while maintaining their horizontal position. This type is constructed as large as those mentioned previously, but requires more space because the minimum pitch between the tiers is decided by the sprockets, which are 200–300 mm in diameter.

The carrier freezer may be regarded as two push-through tunnels, one on top of the other. At the stop section a row of carriers is pushed forward, while it is returned in the lower section. At both ends there are elevating mechanisms. A carrier is similar to a bookcase with shelves. When it is indexed up at the loading end of the freezer, the products on one tier at a time are pushed off the shelf onto a discharge conveyor.

When the carrier is indexed up the next time, this shelf is level with the loading belt from which new products are transferred to the carrier (see Fig. 10).

The carriers may be designed for almost any pitch between the tiers and for any length and width, giving maximum compactness. The loading and unloading may be manual or fully automatic.

Another automatic freezing tunnel is the reciprocating spiral freezer, which consists of two parallel sets of rails, one of which is fixed. In between there is a set of movable rails. Initially the products rest on the fixed set. The movable set lifts the products clear of the fixed set, advances one stroke, descends to leave the products resting on the fixed set again, and then returns to the initial position.

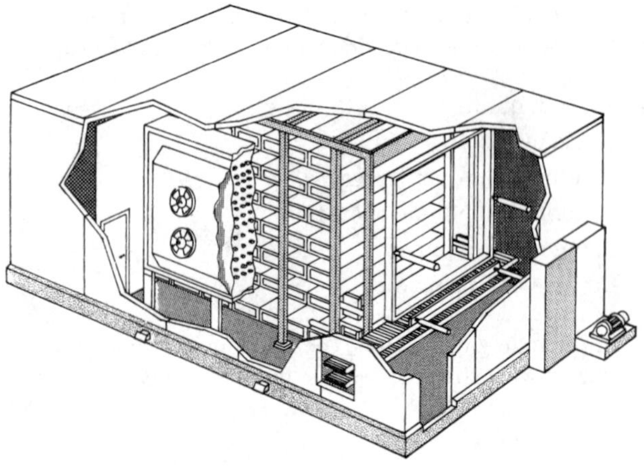

Figure 10. Carrier freezer.

Both sets are arranged to form a large spiral, the fixed set supported from an external steel structure and the movable set fitted to a central cylinder. This provides a reciprocating motion around the vertical axis as well as up and down.

The infeed can be arranged very easily for a range of carton sizes provided the utilization of the conveyor area is limited. Small items must be placed on trays that are loaded, unloaded, and transferred from outfeed to infeed separately.

As the total product load is accelerated and decelerated for each stroke, a relatively slow motion is necessitated, making this type of freezer suitable mainly for intermediate and large package sizes.

All freezer designs described above are intended primarily for packaged products. Attempts to freeze fish fillets, meat patties, etc individually on trays invariably have been only moderate successful because of a number of problems, primarily the following:

1. Products stick to the trays. This causes damage and weight losses if products are removed mechanically. An alternative is to heat the trays to release the products. This requires complicated equipment and causes reduced capacity.
2. Hygiene. The trays must be washed after removal of the frozen products if acceptable hygiene is to be obtained.
3. The handling of trays from outfeed to infeed is costly whether it is manual or automatic.

The principal advantages of automatic air-blast freezers in comparison with automatic plate freezers, which are usually the alternative, are the following:

4. Products of much varied thickness may be frozen simultaneously or immediately after another.
5. Products do not need to be square in shape.
6. Higher capacities per unit are possible.

The automatic or mechanized tunnel freezer generally has the same advantages and disadvantages as the classical tunnel except that it is slightly better suited as an inline freezer.

Labor costs can be reduced and the flexibility is somewhat better as different products can be handled at the same time by different tracks having different dwell times.

Belt Freezers

The first belt freezers basically consisted of a wire mesh belt conveyor in a blast room, which satisfied the need for a continuous product flow. In addition to the disadvantage of poor heat transfer in a blast room, many mechanical problems arose.

Modern belt freezers normally utilize vertical airflow, forcing the air through the product layer, which creates good contact with all product particles. A condition is, however, that the product be evenly distributed over the entire belt area. Where the product layer is thin or nonexisting, there is less resistance to the air, which will concentrate to these areas and bypass the thicker product layer. This phenomenon, called channeling, may result in poorly frozen products and thus must be avoided by careful and even spreading of the product across total belt width under all operating conditions. Single-belt freezers designed for freezing unpacked products can be designed to achieve a fluidized freezing.

In order to decrease the necessary floor space of a single-belt freezer the belts can be stacked above each other as in a multitier belt freezer or a spiral belt freezer. The latter being the most important modern belt freezer equipment. A typical multitier freezer is shown in Figure 11.

A multitier freezer can be used where the factory layout requires a straight-through product flow and where available space is narrow. A freezer consists of three conveyor systems positioned one above the other with fans and coils positioned above the top belt supported by the same steel structure that carries the conveyor system.

Products are fed onto the top belt and transported through the cooling zone into the freezing zone and to the opposite end of the freezer. Here the products are transferred from the top belt to the second via a stainless-steel transfer chute. On a second belt they are conveyed back through the freezing zone to the infeed, where they are transferred to the third belt. The third belt takes the products through the freezing zone and delivers the frozen products to the outfeed, from which the products leave the freezer. This arrangement has the advantage that the product, after being surface frozen on the first belt, may be stacked in a rather deep bed on the lower belts. Thus the total belt area required can be reduced.

By vertically stacking the belt in tiers a minimum of floor space is occupied by the freezer. This is the case in the modern spiral belt freezer (Fig. 12), which maximizes the belt surface area in a given floor space. This is achieved by using a belt that can be bend laterally around a rotating drum. The belt in supported by rails and driven by the friction against the drum. The most advanced and refined spiral freezers operate with a low tension drive system.

The continuous-belt design eliminates product transfer points where product damage can occur within the freezing system. The products are placed on the belt outside the freezer where they can be monitored and will remain in the same spot until leaving the freezer. The flexibility of the belt used allows for more than one infeed and outfeed with one and the same belt, and infeed and outfeed may be arranged in any direction desired to suit the layout of the processing line.

In the most modern version of spiral belt freezers a self-stacking belt—FRIGoBELT—where each tier rests directly on the vertical side links of the tier beneath is used. This construction eliminates the need for rails and runners and allows more tiers of belt to be installed in a given space. The whole stack turns as a unit. Products cannot roll and blow out of the closed freezing zone and cannot become stuck within freezer because there are no stationary structural parts to snag them. Equally important is the improvements of the hygienic conditions of the enclosed product zone. The belt in contact with the product is regularly cleaned and dried in an external washing unit outside the freezer.

The side links of the belt will also serve to channel the vertical airflow in the freezing zone. The air is blown down over the upward-moving products in a countercurrent flow, which is a very efficient form of heat transfer.

The spiral belt freezer provides great flexibility with regard to the product range to be handled. Both unpackaged and packaged products are frozen and typical products are meat patties, fish cakes, fish fillets, and bakery products, which all may be frozen raw or prepared.

Fluidized-Bed Freezers

Previously freezing of vegetables took place in a plate freezer or tunnel freezer, and the result was more or less a block frozen product that was hard to thaw and rather inconvenient in handling. The use of "cluster busters" in order to obtain a more free-flowing product caused considerable mechanical damage. Belt freezers were introduced soon after World War II, but in order to meet the high freezing demands those freezers became rather huge. In the early 1960s fluidized freezing was introduced after years of experiments and tests, and it was possible for the first time to quick-freeze vegetables individually very fast in a commercial application.

Fluidization occurs when particles of fairly uniform shape and size are subjected to an upward airstream. At a certain air velocity the particles will float in the airstream, each one separated from the other but surrounded by air and free to move. In this state the mass of particles can be compared to a fluid.

If a mass is held in a container that is fed in one end and the other end is lower, the mass will move toward the lower end as long as more products are added. By utilizing low-temperature air to achieve the fluidization the products are frozen and simultaneously conveyed by the same air without the aid of a conveyor (see Fig. 13).

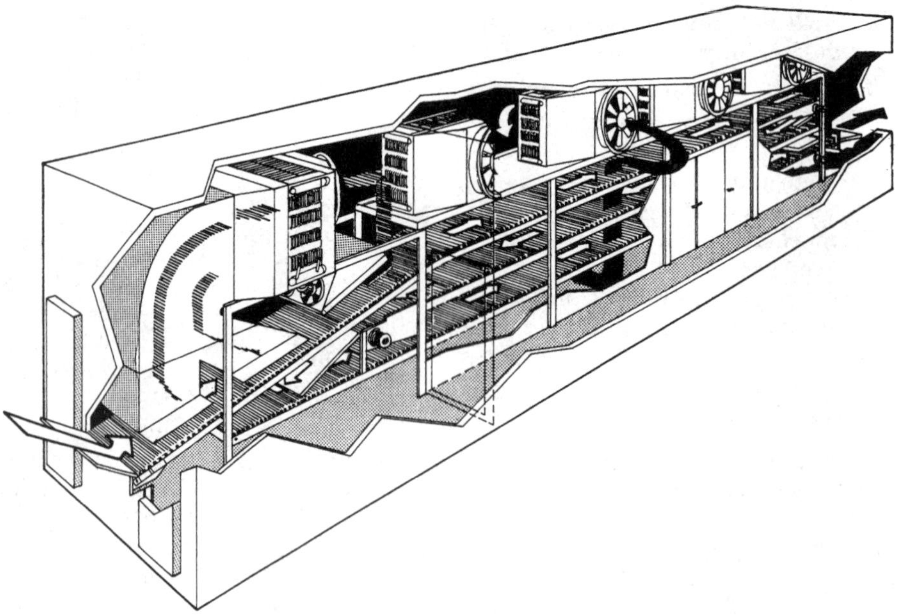

Figure 11. Multitier freezer (TRIoFREEZE).

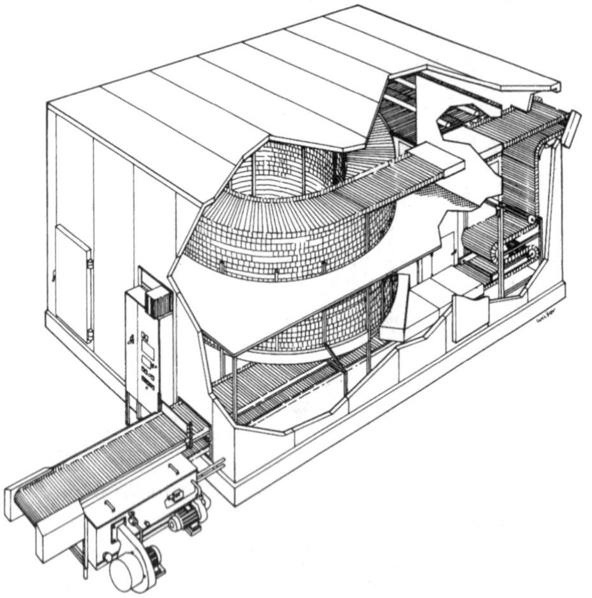

Figure 12. Spiral freezer (GYROCOMPACT).

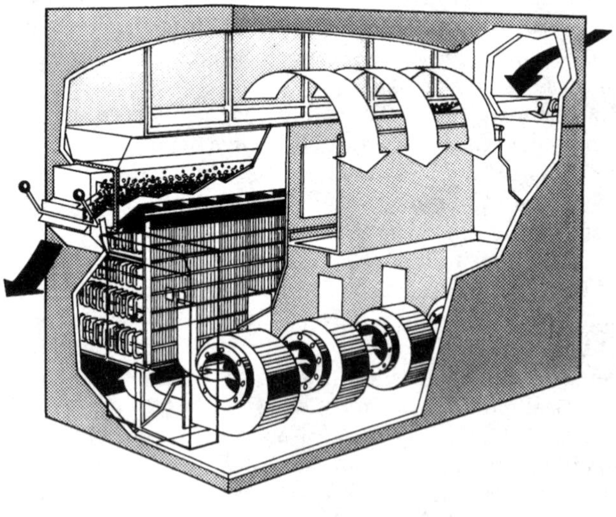

Figure 13. Fluidized-bed freezer (FLoFREEZE).

The use of the fluidization principle gives a number of advantages in comparison with the use of a belt freezer. The product is always individually quick-frozen (IQF); this applies also to products with a tendency to stick together, eg, French-style green beans, sliced carrots, and sliced cucumber.

The freezer is totally independent of fluctuations in load. If the freezer is partly loaded, the air distribution can be the same as for the full load, ie, with no hazard of channeling. The variability of freezing with products is greatly improved because a deep fluidized bed can accept products with more surface water. Consequently, there is no hazard of belt damage if a dewatering screen is broken down temporarily.

One type of fluidized-bed freezer combines the fluidized bed with a conveyor belt. The freezer operates in two stages: a crust freezing zone and a finishing freezing zone. In the former the product is carried on a fluidized bed that guarantees an efficient heat transfer, a quick crust freezing, and particle separation. The crust frozen product is then conveyed on a belt through the second freezing zone.

The fluidization technique achieves a very efficient air-product contact, which gives heat-transfer rates which are much higher than those for conventional air-blast freezing tunnels or belt freezers. The efficiency of the heat removal can also be seen in the physical dimensions of the equipment, which are generally one-third of the comparable belt freezer.

The fluidized-bed freezer is in-line equipment suitable not only for vegetables, berries, and other fruits but also for processed products such as French fried potatoes, peeled cooked shrimps, diced meat, and meatballs.

Contact Freezers

In a contact freezer the product is either in direct contact with the freezing media—immersed—or indirectly by being in contact with a belt or plate containing the freezing media.

Immersion Freezers

The immersion freezers consists of a tank with a cooled freezing medium, such as salt, sugar, or alcohol solution in water or other nontoxic mixtures of water and solutes. The product is immersed in this brine or sprayed while being conveyed through the tank.

This type of equipment has been quite commonly used for surface freezing of turkeys and other poultry on markets where a light color is demanded. Final freezing is accomplished in a separate blast tunnel or during cold storage. The latter, however, may jeopardize quality because of slow core freezing. It is necessary to protect the product from contact with the brine by using high-quality packaging materials with absolutely tight seal. Brine residues on the packages are washed off with water at the freezer exit.

A sodium chloride brine was earlier sometimes used in direct contact with the product in the fishing industry. For freezing tuna fish, for example, it is still used in some places.

Indirect-Contact Freezers

The most commonly contact freezer is the plate freezer, where the product is pressed between hollow metal plates that are positioned horizontally or vertically, with a refrigerant circulating through them.

Another type of freezer utilizes two belts with the refrigerant circulating outside the belts or alternatively placing the product on a single belt. All these arrangements provide a very good heat transfer, which is reflected in short freezing times, provided the product itself is a good heat conductor.

The advantage of good heat transfer at the surface is gradually reduced with increasing product thickness. For this reason it is often limited to a maximum of 50–60 mm. It is further important that the packages are well filled and if metal trays are used to carry the packs that these are not distorted.

The pressure from the plates or belts maintain throughout the freezing process practically eliminates what is known as "bulging" and the frozen packs will maintain their regular shape within very close tolerances.

Plate Freezers

There are two main types of plate freezer: the horizontal and the vertical plate freezer. Either type can be manual or automatic. The typical manual horizontal plate freezer contains 15–20 plates. The product is placed on metal trays or in other systems metal frames and transported to the freezer where they are manually loaded between the plates.

In order to obtain automatic operation of the horizontal plate freezer the whole battery of plates is moved up and down in an elevator system. At the loading conveyor level the plates are separated, and a row of packages that have accumulated on a transport conveyor from the processing and packaging are pushed in between the open plates, simultaneously discharging a row of frozen products at the opposite side. This cycle is repeated until all frozen packages at this level has been replaced. Then the space between the two plates are closed, all plates are indexed up, and the next set of plates are opened for loading and unloading. A typical automatic plate freezer is shown in Figure 14.

The vertical plate freezer is used mainly for freezing products in blocks weighing 10–15 kg and has been specially developed for freezing fish at sea. The freezer consists of a number of vertical freezing plates forming partitions in a container with an open top. The product is simply fed in from the top and the blocks after freezing is discharged either to the side, upwards, or down through the bottom. Usually this mechanism is automized. The discharge of the products is enhanced by a short period of gas defrost at the end of the freezing cycle and the use of compressed air or a hydraulic system to force out the product.

Band Freezers

Single-band and double-band freezers are designed to freeze thin product layers. The freezers can either be straight forward bands as shown in Figure 15 or as a drum (Fig. 16).

The band freezer illustrated in Figure 16 is designed to freeze and form liquids and semiliquids into individual pellets as an in-line operation.

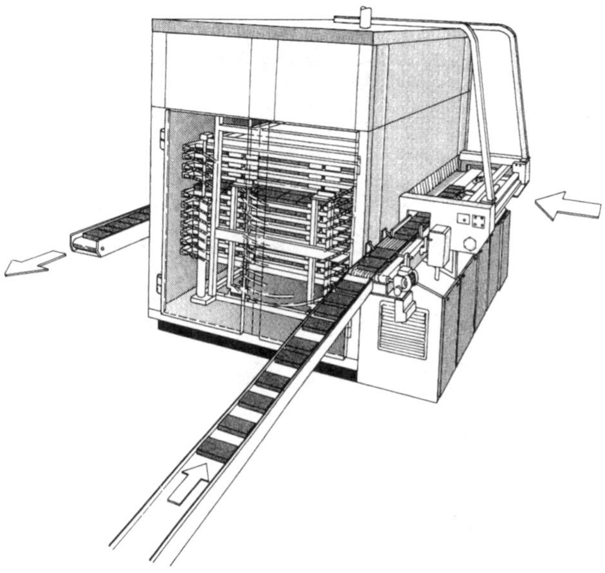

Figure 14. Automatic plate freezer (AUToPLATE).

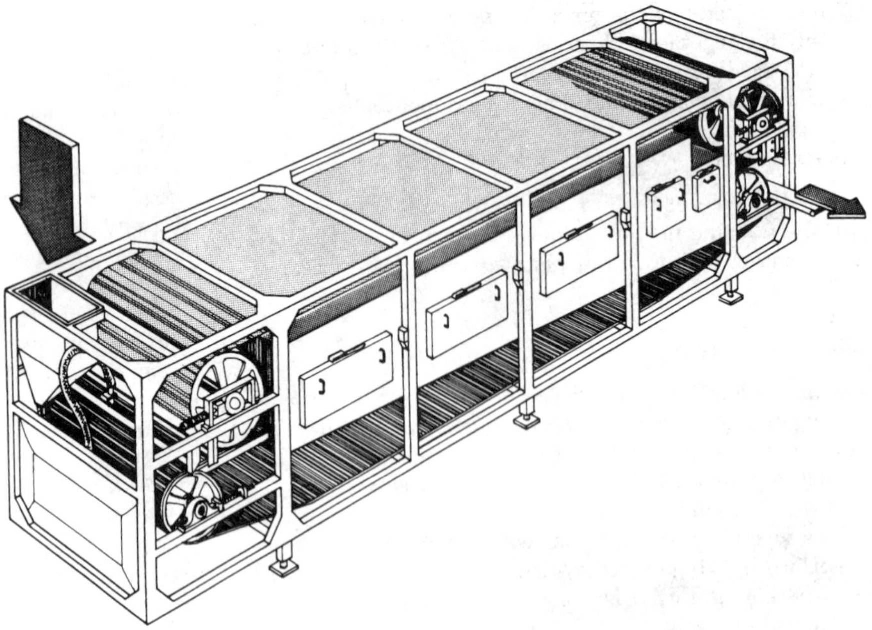

Figure 15. Contact band freezer (PELLoFREEZE).

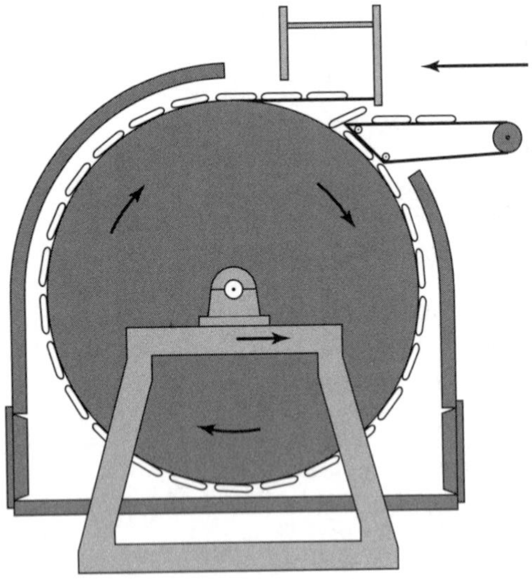

Figure 16. Direct-contact band freezeR—drum freezer.

The product is formed and frozen between two endless stainless-steel bands, of which the top band is flat and the lower band is corrugated with flexible seals on each side. The product is supplied on the corrugated band by a spreading device, after which the flat band is brought in contact with the product, thus totally enclosing it. After the freezing-forming zone the two bands are separated. The liquid is now frozen to a mat and passes through the final forming operation. The product finally enters the outfeed conveyor in an IQF form.

A monopropylene glycol–water brine is used as an intermediate freezing medium. The brine is circulated by pumps from a sump below the freezing pump via a brine cooler to the freezing zone, where it passes over the outer surfaces of the band with high velocity.

Typical products frozen in band freezers are chopped spinach purée, fruits pulps, egg yoke, sauces, and soups.

The drum freezer can be viewed as a more compact band freezer. Distinction is made between both vertical and horizontal drums. This type of freezer is also known as a rotary freezer.

Cryogenic Freezers

Cryogenic freezers differ from all other types of freezers in one fundamental respect: they are not connected to a refrigeration plant. The heat-transfer medium is nitrogen or carbon dioxide liquefied in large industrial installations and shipped to the food freezing factory in low-temperature well-insulated pressure vessels.

The design of cryogenic freezers has improved significantly in recent years. As for all types of freezers also this type can be found as straight-belt, multitier, and spiral belt as well as immersion designs. In principle the same basic equipment can be used for both gases, but with slight modifications. The size and mobility of cryogenic freezers allow for flexibility in design and redesigning a processing plant.

Key attributes of the equipment are high heat-transfer rates, low investment costs, and rapid installation and startup. Especially interesting applications are those for chilling, firming, or crusting products.

A typical belt liquid nitrogen (LIN) freezer is shown in Figure 17. Liquid nitrogen at -196°C is sprayed into the freezer in which the atmosphere is circulated by small fans. The freezant or liquid nitrogen partially evaporates immediately leaving the spray nozzles and on contact with the products. The cold gas is circulated by fans toward the infeed end precooling the products entering the freezer and thereafter extracted by an exhaust fan.

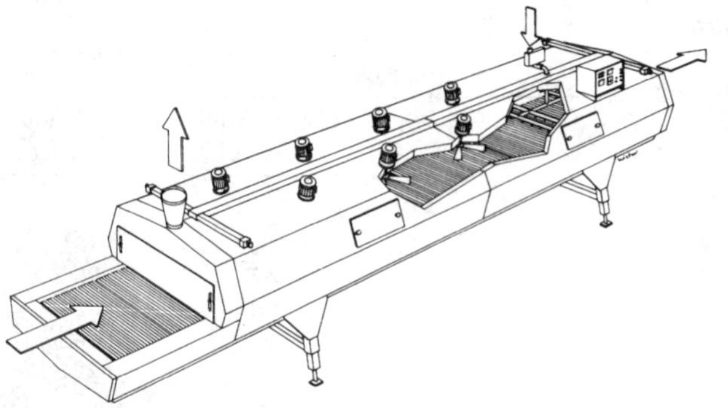

Figure 17. Cryogenic freezer (AGA FREEZE).

The freezant thus passes in countercurrent to the movement of products on the belt and giving a high heat-transfer efficiency, which is an advantage in terms of quality for some special products.

However, the quick freezing may also result in cracking of the product surface if sufficient precautions are not taken. The freezant consumption is in the range of 1.2–2.0 kg per kilogram of product. The capacity can vary from 150 to 1000 kg/h, and typical products are meat cuts, fish fillets, seafood, fruit, berries, pies, and pastries.

Prefreezer

A recent development utilizing liquid nitrogen as freezing medium is an apparatus designed for quick crust freezing of extremely wet, sticky, or sensitive products that can then be easily handled in a spiral belt freezer or a fluidized-bed freezer for completion of the freezing process without deformation or breakage. The freezers also offer a possibility of freezing products that are difficult to freeze in conventional systems (see Fig. 18).

The products are frozen by means of direct immersion in liquid nitrogen, which gives an almost instantaneous freezing of the surface. The product is fed vertically into an IQF tank with a continuous flow of liquid nitrogen.

In this stream the products are gently received and separated at the same time as a very thin layer of the product surface is frozen. From this first step the products drop down on a belt and are fed into a bath of liquid nitrogen that together with a spray completes the crust freezing. The liquid nitrogen is then separated from the product and collected in a sump from where it is recirculated by means of a specially designed pump.

Carbon Dioxide Freezer

Liquid carbon dioxide is normally stored under high pressure. At atmospheric pressure it exists only as a solid or a gas. When the liquid is released to the atmosphere, 50% of the liquid becomes dry-ice snow and 50% vapor both at −79°C. Because of these unusual properties, carbon dioxide freezer designs vary widely.

In a LIN freezer the cold gas phase is used to precool the product before it is exposed to the nitrogen spray. As liquid carbon dioxide forms snow that needs time for sub-

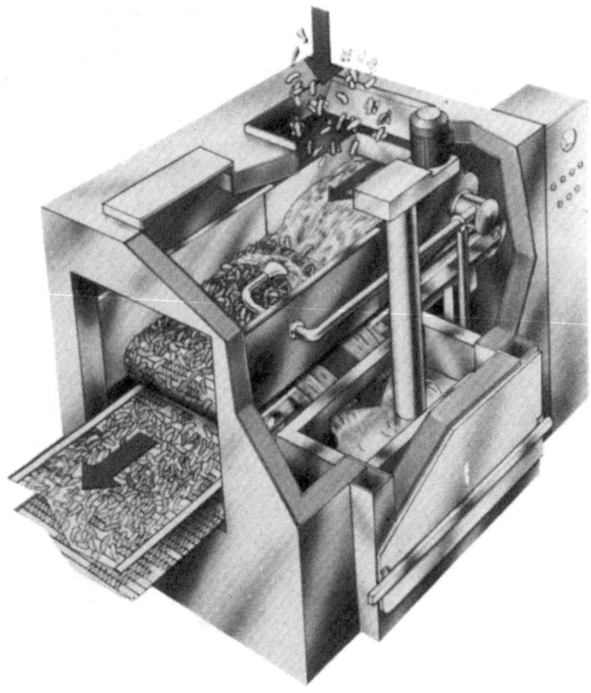

Figure 18. Prefreezer (CRUSToFREEZE).

limation, the injection is moved closer to the product infeed as compared to the LIN freezer.

THERMODYNAMIC PROPERTIES

An important consideration in all designs of food processing involving heat exchange is the thermal properties of the food products to be handled in the system. The variable composition and structure of foods influence those properties. During storage the chemical and physical properties, the time, and the temperatures are the most important factors. The composition of the food product varies with species, growth condition, age, feed, harvest, slaughter, catch, handling, and processing, as well as storage conditions. All of these factors influence the thermal properties.

For these reasons it is easily understood that the values of thermal properties are not exact values but most often estimates. The more detailed a description of the food product measured the more accurate the values given. Another important consideration when using experimentally found thermal properties is the difference in the methods of measurement, which may place limitations on the value of the data (4). Computer programs have been developed to estimate the thermal physical properties from the knowledge of the specifications of the products, such as chemical composition, temperature, and density.

Values of specific heat and latent heat of fusion are often calculated directly from the water content of the product. This is the case for the values given in Table 1.

The values of water content in Table 1 are averaged for the product. The water content for fruit and vegetables

Table 1. Thermal and Related Properties of Food and Food Materials

Food or food material	Water content % (mass)[a]	Highest freezing point, °C[c]	Specific heat[b] Above freezing, kJ/kg · °C	Specific heat[b] Below freezing, kJ/kg °C	Latent heat of fusion[d] kJ/kg
		Vegetables			
Artichokes, Globe	84	−1.2	3.78	1.90	281
Artichokes, Jerusalem	80	−2.5	3.68	1.85	268
Asparagus	93	−0.6	4.00	2.01	312
Beans, snap	89	−0.7	3.90	1.96	298
Beans, lima	67	−0.6	3.35	1.68	224
Beans, dried	11	—	1.95	0.98	37
Beets, roots	88	−1.1	3.88	1.95	295
Broccoli	90	−0.6	3.93	1.97	302
Brussels sprouts	85	−0.8	3.80	1.91	285
Cabbage, late	92	−0.9	3.98	2.00	308
Carrots, roots	88	−1.4	3.88	1.95	295
Cauliflower	92	−0.8	3.98	2.00	308
Celeriac	88	−0.9	3.88	1.95	295
Celery	94	−0.5	4.03	2.02	315
Collards	87	−0.8	3.85	1.94	291
Corn, sweet	74	−0.6	3.53	1.77	248
Cucumbers	96	−0.5	4.08	2.05	322
Eggplant	93	−0.8	4.00	2.01	312
Endive (escarole)	93	−0.1	4.00	2.01	312
Garlic	61	−0.8	3.20	1.61	204
Ginger, rhizomes	87	—	3.85	1.94	291
Horseradish	75	−1.8	3.55	1.79	251
Kale	87	−0.5	3.85	1.94	291
Kohlrabi	90	−1.0	3.93	1.97	302
Leeks	85	−0.7	3.80	1.91	285
Lettuce	95	−0.2	4.06	2.04	318
Mushrooms	91	−0.9	3.95	1.99	305
Okra	90	−1.8	3.93	1.97	302
Onions, green	89	−0.9	3.90	1.96	298
Onions, dry	88	−0.8	3.88	1.95	295
Parsely	85	−1.1	3.80	1.91	285
Parsnips	79	−0.9	3.65	1.84	265
Peas, green	74	−0.6	3.53	1.77	248
Peas, dried	12	—	1.97	0.99	40
Peppers, dried	12	—	1.97	0.99	40
Peppers, sweet	92	−0.7	3.98	2.00	308
Potatoes, early	81	−0.6	3.70	1.86	271
Potatoes, main crop	78	−0.6	3.63	1.82	261
Potatoes, sweet	69	−1.3	3.40	1.71	231
Pumpkins	91	−0.8	3.95	1.99	305
Radishes	95	−0.7	4.06	2.04	318
Rhubarb	95	−0.9	4.06	2.04	318
Rutabagas	89	−1.1	3.90	1.96	298
Salsify	79	−1.1	3.65	1.84	265
Spinach	93	−0.3	4.00	2.01	312
Squash, summer	94	−0.5	4.03	2.02	315
Squash, winter	85	−0.8	3.80	1.91	285
Tomatoes, mature green	93	−0.6	4.00	2.01	312
Tomatoes, ripe	94	−0.5	4.03	2.02	315
Turnip greens	90	−0.2	3.93	1.97	302
Turnip	92	−1.1	3.98	2.00	308
Watercress	93	−0.3	4.00	2.01	312
Yams	74	—	3.53	1.77	248

Table 1. Thermal and Related Properties of Food and Food Materials *(continued)*

Food or food material	Water content % (mass)[a]	Highest freezing point, °C[c]	Specific heat[b] Above freezing, kJ/kg · °C	Specific heat[b] Below freezing, kJ/kg °C	Latent heat of fusion[d] kJ/kg
		Fruits			
Apples, fresh	84	−1.1	3.78	1.90	281
Apples, dried	24	—	2.27	1.14	80
Apricots	85	−1.1	3.80	1.91	285
Avocados	65	−0.3	3.30	1.66	218
Bananas	75	−0.8	3.55	1.79	251
Blackberries	85	−0.8	3.80	1.91	285
Blueberries	82	−1.6	3.73	1.87	275
Cantaloupes	92	−1.2	3.98	2.00	308
Cherries, sour	84	−1.7	3.78	1.90	281
Cherries, sweet	80	−1.8	3.68	1.85	268
Cranberries	87	−0.9	3.85	1.94	291
Currants	85	−1.0	3.80	1.91	285
Dates, cured	20	−15.7	2.17	1.09	67
Figs, fresh	78	−2.4	3.63	1.82	261
Figs, dried	23	—	2.25	1.13	77
Gooseberries	89	−1.1	3.90	1.96	298
Grapefruit	89	−1.1	3.90	1.96	298
Grapes, American	82	−1.6	3.73	1.87	275
Grapes, Vinifera	82	−2.1	3.73	1.87	275
Lemons	89	−1.4	3.90	1.96	298
Limes	86	−1.6	3.83	1.92	288
Mangoes	81	−0.9	3.70	1.86	271
Melons, Casaba	93	−1.1	4.00	2.01	312
Melons, Crenshaw	93	−1.1	4.00	2.01	312
Melons, honeydew	93	−0.9	4.00	2.01	312
Melons, Persian	93	−0.8	4.00	2.01	312
Melons, watermelon	93	−0.4	4.00	2.01	312
Nectarines	82	−0.9	3.73	1.87	275
Olives	75	−1.4	3.55	1.79	251
Oranges	87	−0.8	3.85	1.94	292
Peaches, fresh	89	−0.9	3.90	1.96	298
Peaches, dried	25	—	2.30	1.16	84
Pears	83	−1.6	3.75	1.89	278
Persimmons	78	−2.2	3.63	1.82	261
Pineapples	85	−1.0	3.80	1.91	285
Plums	86	−0.8	3.83	1.92	288
Pomegranates	82	−3.0	3.73	1.87	275
Prunes	28	—	2.37	1.19	94
Quinces	85	−2.0	3.80	1.91	285
Raisins	18	—	2.12	1.07	60
Raspberries	81	−0.6	3.70	1.86	271
Strawberries	90	−0.8	3.93	1.97	302
Tangerines	87	−1.1	3.85	1.94	291
		Whole fish			
Haddock, cod	78	−2.2	3.63	1.82	261
Halibut	75	−2.2	3.55	1.79	251
Herring, kippered	70	−2.2	3.43	1.72	235
Herring, smoked	64	−2.2	3.28	1.65	214
Menhaden	62	−2.2	3.23	1.62	208
Salmon	64	−2.2	3.28	1.65	214
Tuna	70	−2.2	3.43	1.72	235
		Fish fillets or steaks			
Haddock, cod, perch	80	−2.2	3.68	1.85	268
Hake, whiting	82	−2.2	3.73	1.87	275
Pollock	79	−2.2	3.65	1.84	265
Mackerel	57	−2.2	3.10	1.56	191

Table 1. Thermal and Related Properties of Food and Food Materials *(continued)*

Food or food material	Water content % (mass)[a]	Highest freezing point, °C[c]	Specific heat[b]		Latent heat of fusion[d] kJ/kg
			Above freezing, kJ/kg · °C	Below freezing, kJ/kg °C	
		Shellfish			
Scallop, meat	80	−2.2	3.68	1.85	268
Shrimp	83	−2.2	3.75	1.89	278
Lobster, American	79	−2.2	3.65	1.84	265
Oysters, clams, meat, and liquor	87	−2.2	3.85	1.94	291
Oyster in shell	80	−2.8	3.68	1.85	268
		Beef			
Carcass (60% lean)	49	−1.7	2.90	1.46	164
Carcass (54% lean)	45	−2.2	2.80	1.41	151
Sirloin, retail cut	56	—	3.08	1.55	188
Round, retail cut	67	—	3.35	1.68	224
Dried, chipped	48	—	2.88	1.44	161
Liver	70	−1.7	3.43	1.72	235
Veal, carcass (81% lean)	66	—	3.33	1.67	221
		Pork			
Bacon	19	—	2.15	1.08	64
Ham, light cure	57	—	3.10	1.56	191
Ham, country cure	42	—	2.72	1.37	141
Carcass (47% lean)	37	—	2.60	1.31	124
Bellies (33% lean)	30	—	2.42	1.22	101
Backfat (100% fat)	8	—	1.87	0.94	27
Shoulder (67% lean)	49	−2.2	2.90	1.46	164
Ham (74% lean)	56	−1.7	3.08	1.55	188
Sausage, links or bulk	38	—	2.62	1.32	1.27
Sausage, country style, smoked	50	−3.9	2.93	1.47	168
Sausage, frankfurters	56	−1.7	3.08	1.55	188
Sausage, Polish style	54	—	3.03	1.52	181
		Lamb			
Composite of cuts (67% lean)	61	−1.9	3.20	1.61	204
Leg (83% lean)	65	—	3.30	1.66	218
		Dairy products			
Butter	16	—	2.07	1.04	54
Cheese, Camembert	52	—	2.98	1.50	174
Cheese, Cheddar	37	−12.9	2.60	1.31	124
Cheese, cottage (uncreamed)	79	−1.2	3.65	1.84	265
Cheese, cream	51	—	2.95	1.48	171
Cheese, Limburger	45	−7.4	2.80	1.41	151
Cheese, Roquefort	40	−16.3	2.67	1.34	134
Cheese, Swiss	39	−10.0	2.65	1.33	131
Cheese, processed American	40	−6.9	2.68	1.34	134
Cream, half-and-half	80	—	3.68	1.85	268
Cream, table	72	−2.2	3.48	1.75	241
Cream, whipping, heavy	57	—	3.10	1.56	191
Ice cream, (10% fat)	63	−5.6	3.25	1.63	211
Milk, canned, condensed, sweetened	27	−15.0	2.35	1.18	90
Milk, evaporated, unsweetened	74	−1.4	3.53	1.77	248
Milk, dried (whole)	2	—	1.72	0.87	7
Milk, dried (nonfat)	3	—	1.75	0.88	10
Milk, fluid (3.7% fat)	87	−0.6	3.85	1.94	291
Milk, fluid (skim)	91	—	3.95	1.99	305
Whey, dried	5	—	1.80	0.90	17

Table 1. Thermal and Related Properties of Food and Food Materials *(continued)*

Food or food material	Water content % (mass)[a]	Highest freezing point, °C[c]	Specific heat[b] Above freezing, kJ/kg · °C	Below freezing, kJ/kg °C	Latent heat of fusion[d] kJ/kg
		Poultry products			
Egg, whole (fresh)	74	−0.6	3.53	1.77	247
Eggs, white	88	−0.6	3.88	1.95	295
Eggs, yolks	51	−0.6	2.95	1.48	171
Eggs, yolks (sugared)	51	−3.9	2.95	1.48	171
Eggs, yolks (salted)	50	−17.2	2.93	1.47	168
Eggs, dried (whole)	4	—	1.77	0.89	13
Eggs, dried (white)	9	—	1.90	0.95	30
Chicken	74	−2.8	3.53	1.77	248
Turkey	64	—	3.28	1.65	214
Duck	69	—	3.40	1.71	231
		Miscellaneous			
Honey	17	—	2.10	1.68	57
Maple syrup	33	—	2.50	1.26	111
Popcorn, unpopped	10	—	1.92	0.97	34
Yeast, baker's compressed	71	—	3.45	1.73	238
		Candy			
Milk chocolate	1	—	1.70	0.35	3
Peanut brittle	2	—	1.72	0.87	7
Fudge, vanilla	10	—	1.92	0.97	34
Marshmallows	17	—	2.10	1.05	57
		Nuts, shelled			
Peanuts (with skins)	6	—	1.82	0.92	20
Peanuts (with skins, roasted)	2	—	1.72	0.87	7
Pecans	3	—	1.75	0.88	10
Almonds	5	—	1.80	0.90	17
Walnuts, english	4	—	1.78	0.89	13
Filberts	6	—	1.82	0.92	20

[a]Water contents of fruits and vegetables are from Lutz and Hardenburg (1968) except for Jerusalem artichokes; dried beans; and peas, yams, dried apples, figs, peaches, prunes, and raisins; the latter are from Watt and Merrill (1963). Water contents of meats, dairy, and poultry products, miscellaneous candy, and nuts are also from Watt and Merrill; water contents of eggs (yolks, salted) and fish are from ASHRAE (1972, 1974, and 1978).
[b]Freezing points of fruits and vegetables are from Whiteman (1957), and average freezing points of other foods are from ASHRAE (1972, 1974, and 1978).
[c]Specific heat was calculated from Siebel's formulas (1892).
[d]Latent heat of fusion was obtained by multiplying water content expressed in decimal form by 144, the heat of fusion of water in Btu/lb.
Source: Reprinted with permission of the American Society of Heating; Refrigerating and Air-Conditioning Engineers from the 1989 *ASHRAE Handbook—Fundamentals.* Ref. 11.

varies with the stage of development or maturity when harvested and also with type of species, growing conditions, and the storage conditions after harvest. The values given in the table apply to mature product shortly after harvest. For meat the water content values are for the time of slaughter or after the aging period. In reality the water content varies considerably, not only between different animals but also between different muscles from the same animal. For processed products the water content depends on the specific process used.

The freezing points given in Table 1 are based on experiments in which the product has been cooled slowly until freezing occurred. For fruits and vegetables the highest temperature at which the product freeze are given and for other foods average freezing temperature is shown.

With reference to specific heat it should be observed that this is a function of temperature. The value given in Table 1 are from 0°C. In a unfrozen product the specific heat will be slightly lower as temperature rise, and in frozen foods there is a large change in specific heat as temperature decreases. The latter is, of course, related to the changes in composition, in particular the water content. When calculating specific heat of a frozen product it is assumed that the water is frozen to ice and that the specific heat involved is that of ice. This assumption is not totally correct. The freezing of most foods that is transferring from liquid water to ice is a gradual process that occurs over a wide temperature range.

With reference to the latent heat of fusion given in Table 1, these values are also subject to error because they do not consider the chemical composition other than water content. They are, in other words, the product of the heat of fusional water and the water content.

The variations that occur in values given on thermal properties should be taken into account in practical applications. Today a comprehensive work is carried out at several research institutes and universities, eg, at Campden Food and Drink Research Association, UK (12) to compare different data. Besides Table 1, a compilation of thermal properties presented by Polley and co-workers is of interest (13).

The definitions of the thermal properties normally listed are

Water content—the mass of water in the product divided by the total mass expressed in percentage.

Average freezing point—the temperature at which the liquid and solid state of a product are in equilibrium expressed in Celsius.

Latent heat—the quantity of heat necessary to change 1 kilogram of liquid to solid without change of temperature measured as kJ/kg.

Specific heat—the amount of heat needed to rise the temperature of 1 kg of a food product 1° measured as kJ/kg, °C.

Heat of respiration—the quality of heat generated per 24 h measured as kJ/24 h, kg.

The large variations that occur for thermal properties of food products explain why theoretical calculations on freezing time are difficult and should not be used to determine the requirements for a specific freezer. Those requirements should be determined by practical tests.

DETERMINATION OF HOLDING TIME

For calculation of the capacity of any freezer, the necessary holding time is essential. For bulk products such as peas, green beans, French fried potatoes, and fish sticks the capacity of standard equipment is usually specified separately. This is also the case with packaged products that are homogeneous such as spinach purée and fish fillet blocks. For other products holding time must be determined before capacity can be stated.

In the literature many different formulas can be found for accurate determination of holding times. In reality, however, they are of little use, because products differ so much in composition and form that the work to transform the characteristics of a product into mathematical terms is a lot more time-consuming than test freezing of the very product itself. This can usually be done in less time than it takes to analyze the composition of the product.

Of course, such test freezing should be carried out under controlled conditions that do correspond to those that can actually be achieved in production. A suitable pilot freezer for different airflow directions is shown in Figure 19.

MAJOR CONSIDERATIONS IN FREEZER DESIGN

Safety of Personnel

The safety of personnel that operate, clean, and service a freezer should be a main consideration in the freezer design. In too many instances, serious injuries have been the result of unsafe design and operation of freezing machinery.

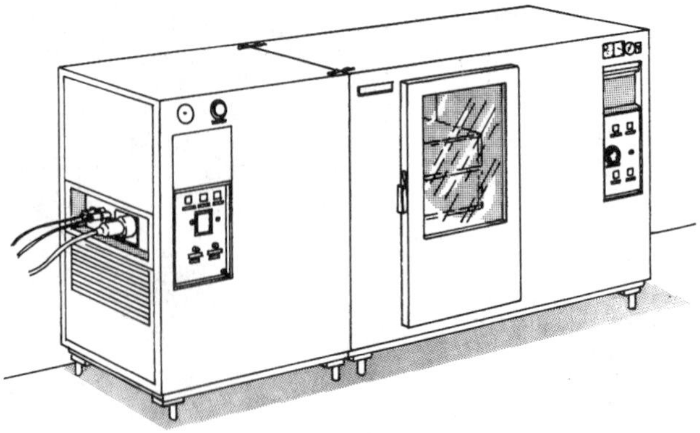

Figure 19. Test freezer.

Mechanical hazards exist in the conveyor drive systems, fans, and other areas. The machinery must be designed so that all areas of a freezer can be easily cleaned and inspected while ensuring that personnel are suitably protected. All drives and fans must be fully guarded so that no worker individual or worker's clothing can reach any part of the machine and get caught or crushed. Emergency stop switches need to be located throughout the machine so that if someone does get caught in the machinery the machine can be stopped quickly to minimize the injury. Fan guards must be designed sufficiently open to remain unrestricted by frost buildup while still preventing personnel from getting their hands or clothing into the fan.

The extreme cold found in most modern freezers represents a hazard to personnel in the form of hypothermia and frostbite. The high air velocities found in efficient freezers that use air as the product heat-transfer medium greatly increases heat transfer and the wind-chill factor, which will quickly freeze exposed skin and draw heat from the body at a very high rate. Exposure to these low temperatures must be limited to a tolerable period, and the fans must be shut off whenever a person needs to enter such a freezer.

Cryogenic freezers must not be entered during operation; even a very brief exposure to the cryogen can quickly cause frostbite as a result of the extremely high heat transfer between the boiling liquid or sublimating solid and warm flesh.

Cryogenic freezers, which generally employ either liquid nitrogen or carbon dioxide in snow form, must be evacuated and refilled with air prior to entry by personnel. The freezers purge all the air out during operation, which results in an oxygen content insufficient to sustain consciousness or life. Such machines must be locked out during operation to prevent access and risk of asphyxiation.

Noise can reach levels that damage hearing in some freezers. Care must be taken to avoid exposure to noise levels above 95 dB by wearing a suitable form of hearing protection such as earplugs or earmuffs.

Noise can be reduced to safe levels by proper selection of fans. Excessive noise is generally a result of using a fan at a pressure above which it is designed to operate, causing cavitation.

Machinery Protection

The hostile environment in which freezers operate render human surveillance of their operation almost impossible. The machinery therefore must operate in an extremely dependable manner and have a number of detection devices mounted and operational so that if something does go wrong, the damage will be minimized.

The most common problems encountered in freezers are caused by ice accumulation, product jams, and operational errors. Ice accumulation can occur as a result of poor defrost procedure, excessive moisture on the conveyor carrying the product into the freezer, or poor startup and shutdown procedures. Ice can accumulate in locations in the machine where it impedes the safe operation of the freezer by jamming a part of by forcing the conveyor or other parts out of their normal operating position. Product jams can occur because the products to be frozen is improperly loaded onto the conveyor or the conveyor can be blocked by an obstruction in the product path. In either case good operation of the freezer and sound design can alleviate the problem. Operational errors are symptomized by a wide variety of problems. Typical symptoms are frozen conveyors due to improper startup, unbalanced fans, or an ice-plugged coil due to improper defrost.

Excessive pressure in the refrigeration piping and the coils can lead to a failure and subsequent loss of refrigerant and potential safety hazard. During normal operation the refrigeration piping in the freezer is under low pressure, so the risk of a ruptured pipe is small at that time. When a freezer is defrosted, the temperature of the coil is raised in order to melt the frost of the coil. If the refrigeration piping is not properly designed and installed, potentially dangerous pressures can develop in the coil. It is imperative that suitable relief devices be installed in the coil so that if the operator makes an error the pressure of the refrigerant in the coil cannot go above a safe level.

Product Safety

In a typical freezer application the value of the product that passes through the freezer in a period of just a few weeks can exceed the cost of the freezer. It is therefore sound practice to ensure that the product is not damaged or contaminated by the freezer.

Contamination can occur as a result of improper cleaning, debris such as surface coatings, or particles created by wear. Even if the contaminant is not harmful or toxic, its presence can render the product unsalable.

Product contamination can be minimized by sound freezer design in which wear debris cannot be generated in a location where it can get into the product as it is being conveyed through the freezer. Typical sources of contamination are wear debris created between the conveyor belt and the wear strip supporting the belt, dripping of condensed water at entries and discharges from freezers, flaking coatings, and leaking fluids such as hydraulic oil.

Product handling within the freezer can physically damage the product. Symptoms of products that have been damaged within the freezer include clumping together of pieces intended to be separate and free-flowing (IQF), pieces of the product torn from the conveying device as a result of the product being frozen to the conveyor, collision damage as a result of the product itself, and buildup of ice on the product.

Clumping together of product is generally a result of improper handling by the freezer as a result of poor design or improper operation. It is important that relative motion between the pieces to be frozen is maintained while the surface of the product freezes. The relative motion between the particles can be achieved by fluidization, mechanical agitation, or immersion in a boiling liquid that boils at a temperature well below the product freezing point.

In order to avoid damage to a product as it is removed from the conveyor that is transporting it through the freezer, it is essential that the product be solidly frozen and that the product does not adhere to the conveyor excessively.

Excessive adhesion can be avoided by careful selection of the conveyor material or by transporting the product in a manner such that it does not rest on a solid conveyor during the period that the surface of the product is being frozen.

Collision damage is generally a result of excessive loading of product onto the conveyor, which can result in the product colliding with other products inside the freezer or colliding with stationary structures within the freezer. Such collisions frequently result in product jams, which can destroy a considerable amount of product at each occurrence. To avoid such problems even feed equipment should be used.

Ice buildup on products can result in an insightly product. This condition can occur when excessive free moisture is transported into the freezer and is transferred from unfrozen to frozen product in the early stages of freezing. This can be controlled by minimizing the free moisture going into the freezer with the product and by controlling the motion of the product once it is inside the freezer.

Hygiene in the Freezer Environment

Proper hygiene in the freezer environment must be defined by the user. Factors such as the sensitivity of the product to various degrees of contamination must be considered in setting sanitary standards. For example, if a cooked product is frozen that requires no additional cooking prior to consumption, extreme hygienic procedures and very sanitary equipment must be employed because of the high risks to the consumer. Contamination of the product can assume many forms. Typical forms of contamination that must be considered are bacterial, wear debris, pieces of a different product, and foreign debris.

Bacterial contamination is generally the result of bacterially contaminated food coming in contact with the product and attaching itself. This can occur if the conveying system has not been adequately cleaned and sanitized.

Wear debris will be created in all freezing and processing machinery. Whether or not the product is considered contaminated is a function of how obvious the debris is and what the composition is. Stainless steel rubbing on either a plastic such as polyethelyne or directly on another piece of stainless steel will generate a significant amount of dark-gray powder. If this gray powder is deposited on the product such that it is visible macroscopically, the product is considered contaminated.

Contamination can also occur if previously frozen food or any food particle is deposited onto a product that is being frozen. This can be the result of inadequate cleaning between changes in product changes on the freezer or production line. A clean design and thorough cleaning can alleviate this problem.

Automatic cleaning of freezing machinery permits cleaning with less labor and reduces the need to access difficult areas and dangerous locations while still providing adequate cleaning. Additionally, automatic cleaning reduces labor and the necessary cleaning time (see Fig. 20).

Cleaning solutions must be selected carefully to adequately clean the machinery without damaging the materials or surface coatings used on the freezer. After cleaning, a sanitizing agent is sometimes employed to sterilize the machine.

Materials for Freezing Machinery

Hygienic standards demand that all food machinery including freezers be constructed of nontoxic materials that permit use of aggressive cleaning agents such as mild caustics without corroding. Typical materials used presently are 300 series stainless steel, galvanized steel, aluminum, and various food-grade plastics.

Special considerations must be taken into account in selection of materials for specific functions in the freezer.

Surfaces that come in contact with the product must be smooth and totally noncorrosive and not adhere to the product, either frozen or thawed. Stainless steel and plastics are generally used for product contact.

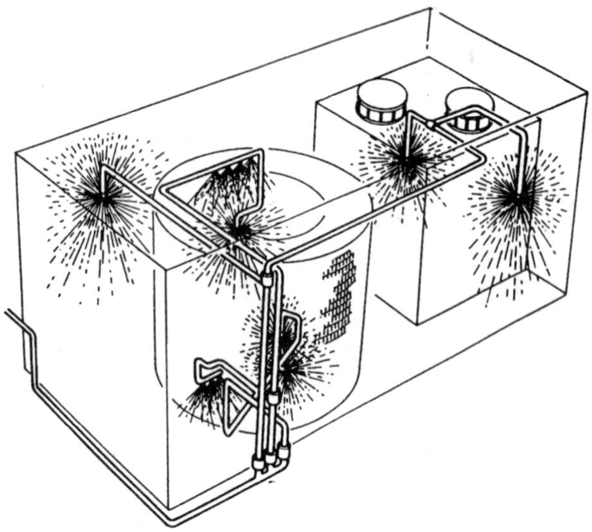

Figure 20. Cleaning in place (CIP).

Materials through which heat must pass must have a high thermal conductivity. Applications that fall under this category are heat-transfer coils, flat metal belts used on contact freezers, and plates used on horizontal and vertical plate freezers.

High-insulation properties are required in practically all freezers to separate the cold environment from the ambient air as well as to prevent the formation of condensation on the warm external walls. Generally the insulating walls are of a panel construction with either metal or fiberglass skins and a plastic low thermal conductivity core. The panel is generally bonded together to give suitable mechanical properties and also to prevent the ingress of moisture. If moisture were to enter the panel, deterioration of the panels thermal and mechanical properties could result because of the freeze-thaw cycles encountered in a typical freezer.

Lubricants selected for food freezer applications must be selected on the basis of both their low-temperature properties and their toxicity. Gearboxes in the freezer must be located so that any leakage of lubricant cannot contaminate the product. Greases and oils used in close proximity to the product or product-carrying surface must be edible as insurance against accidental contamination over the life of the freezer is impossible. Lubricants must remain viscous enough over the entire operating range they are to be exposed (even when the freezer is held at extremely low temperatures without product) to permit dependable operation of the machinery. This generally requires the use of synthetic lubricants at temperatures below 7°C. Lubricants should also have the ability to maintain their properties with considerable moisture content as the cleanup and thawing will result in significant water accumulation in the lubricant. Lubricants must generally be changed at short intervals as a result of water contamination.

Coatings used in freezers must be permanent, or contamination of the product can result. Painted coatings requiring meticulous preparation in application are frequently fragile and require routine maintenance. Metallic coatings such as galvanizing, flame spraying, and plating are generally durable but are not as durable as a component made entirely of noncorrosive material. When coating a material with zinc, consideration should be given the thickness of the coating desired as the zinc will be consumed over time depending on its thickness and the environment. Zinc may not be in contact with the food product.

Materials at freezer temperatures generally undergo large changes in their physical properties, such as brittleness and size. The property changes can result in large stress-induced distortions and breakage unless adequately accounted for.

Mechanical Efficiency

The mechanical efficiency of the freezer is a measure of how much work goes into the mechanical devices within the freezer. The mechanical devices typically consists of conveyor drives, fans, and other powered devices. It is important to realize that any energy put into the freezer either by the work performed in the freezer or by removing heat from the product must be removed by either the refrigeration system or the cryogen. Therefore, a significant multiplier must be applied to the cost of adding extra energy to the freezer environment.

The quantity, types, and power consumption of energy-consuming devices in a given freezer vary greatly between freezer types. Fans are generally the largest power-consuming component in fan-equipped freezers. Fan efficiencies are defined as the ratio of useful fan work to the actual power consumed by the fan motor and can vary from 40 to 70% depending on the selection of the fan. Conveyor drives are generally relatively small power consumers.

Coils

Coils are employed in almost all types of freezers that transfer heat by circulating cold air over the product and serve the function of transferring the heat from the air to the refrigerant. The efficiency of such a freezer is greatly influenced by the design of the coil as the design determines the difference between the air and refrigerant temperatures, which has a large bearing on the initial and operating cost of the refrigeration system. Coil efficiency is a function of materials of construction, configuration of the surface and refrigerant pipes, air velocity, and refrigerant circulation.

In addition to heat-transfer considerations, freezer coils must be designed with sanitation, corrosion, frost buildup, and speed of defrosting in mind.

In order to remain operational as long as possible, coils must be designed to accept frost without becoming plugged up prematurely. This is generally done by careful selection of the spaces between the heat-transfer surfaces or by continuous defrosting.

Continuous defrosting can be accomplished by installing multiple banks of coils and shutting off one bank and defrosting it while the others remain operational, blowing the frost from the coils with compressed air during operation, or rinsing the frost from an operating coil with a mixture of glycol and water. In such cases a glycol concentration is used to remove the water. A continuous (automatic) defrosting system is illustrated in Figure 21.

Fans

Fans for freezers should be selected to provide the highest possible economy over the life of the freezer taking into account dependability, efficiency, and initial costs.

Fans positioned such that the air flows over them before the coil are subject to large frost buildup. Frost buildup on a fan will result in imbalance, which will impose large stresses on the fan wheel and the motor. The fan must be designed to accept the resultant imbalance without failing.

Fan efficiency is determined by how the fan is selected and applied. Inlet and discharge effects can greatly alter a fan's performance. The air pressure and volume of air that pass through the fan are normally presented graphically by the fan producer. A fan must be selected to give the correct quantity of air over a fully normal operating range, which can vary significantly from the time when a freezer has been recently defrosted and lightly loaded with product to when the freezer is frosted up and heavily loaded with product.

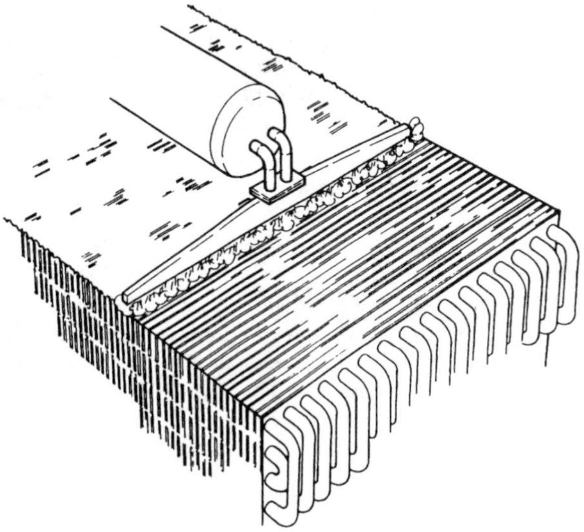

Figure 21. Automatic defrosting (ADF).

Fan motors must be selected and designed to take the mechanical stresses imposed on them as well as to power the fan wheel over a wide operating range. Fan motor bearings should be specially lubricated to run freely over a full range of operating temperatures. The bearings need to be adequate to tolerate a significant amount unbalance as discussed before.

Electrical Considerations

Dependable and safe electrical installations require special attention in a freezer environment because of the high moisture and large temperature changes that cause water to migrate and condensate on electrical wiring and components. Electrical devices and wiring need to be protected from moisture by ensuring that they always remain above the dewpoint under all operating conditions and by protecting them from water entry during defrosting and cleaning. Electrical panels external to the freezer need to be protected from the cleanup conditions found in most plants by taking special precautions in sealing and ventilation for cooling.

Airflow

In freezers employing air as the heat-transfer medium the design of the airflow has a major influence on the performance of the freezer and on the food product quality. The significant factors to consider in regard to airflow are quantity, evenness of distribution over the product, and power consumed by the fans.

The quantity of the air circulated influences the performance of the coils, the change in air temperatures through the freezer, and heat-transfer rates between the air and the product, all assuming that the areas through which the air flows remain constant. An efficient airflow is illustrated in Figure 22.

The distribution of the air over the product should be carefully controlled to give the designed freezing rate uniformly over all the product passing through the freezer.

The deterioration of the airflow through the freezer with increased coil frosting and heavy product loading needs to be taken into account when designing the air system so that the freezer maintains full performance under the full range of operating conditions.

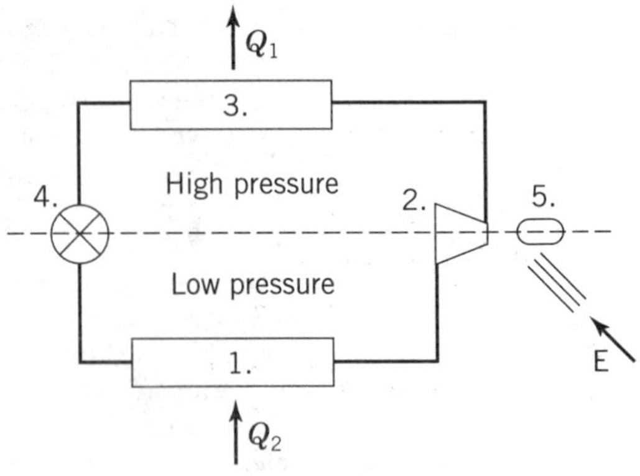

Figure 23. Mechanical refrigeration system: (1) evaporator; (2) compressor; (3) condenser; (4) expansion valve; (5) motor.

REFRIGERATION SYSTEMS

The Second Law of Thermodynamics states that heat can be raised from a low to a high temperature level only by expenditure of work. This means that energy inherently flows only from high to low temperatures. From the long list of refrigeration processes, only two are normally of importance for the food industry: (1) a closed mechanical refrigeration system containing a compressor, a condenser, an expansion valve, and an evaporator and (2) an open cryogenic refrigeration system using either liquid nitrogen (N_2), or carbon dioxide (CO_2).

Mechanical Refrigeration

The main elements in a mechanical refrigeration system are shown in Figure 23. The closed system is filled with a refrigerant like Freon or ammonia. Refrigerant as a gas is pulled from the evaporator (1) to the compressor (2) driven by a motor (5). The compressor discharge the gas to the condensing pressure and the gas condensates to liquid in the condenser as heat is removed from the condensor to the surrounding air or to water circulating through the condenser.

From the condenser the liquid goes through the expansion valve. The expansion valve is regulating the flow of refrigerant being evaporated in the evaporator and also keeping the pressure between high and low levels.

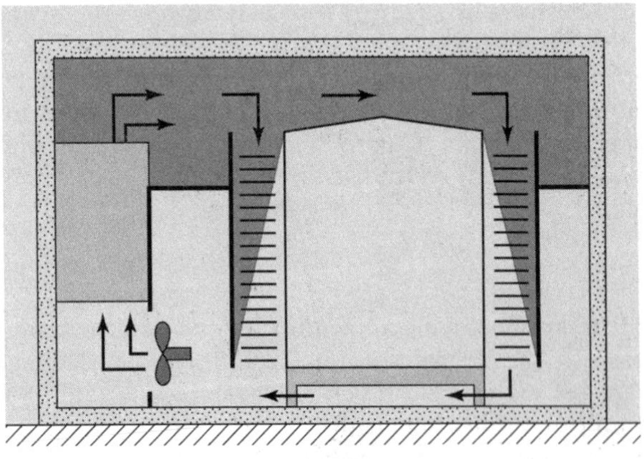

Figure 22. Example of efficient airflow in a freezer.

Energy picked up in the evaporator (Q_2) plus energy introduced to the compressor shaft (E) equals the energy delivered from the condensor (Q_1):

$$Q_2 + E = Q_1$$

The system described is the same as used in heat pumps. The main difference between the two systems is that in a refrigeration application the heat pickup in the evaporator is the objective, while in a heat pump the main goal is to produce heat in the condensor.

Heat recovery from refrigeration systems will be of greater importance as energy costs rice and environmental questions require an increasing amount of attention. Whenever a demand for heat exists near a refrigeration plant, one form or another of heat recovery from the refrigeration plant should be considered.

One- or Two-Stage Systems

The simple arrangement illustrated in Figure 23 is generally used for all installations where the purpose is to keep the temperature around the freezing point or higher.

For storage temperatures around $-30°C$ and for freezing equipment it is normally economical to use a two-stage system. Such a system contains, in principle, two subsystems, as shown in Figure 23, working together. The condensing pressure for the low-stage compressor is about the same as the evaporation pressure for the high-stage compressor.

The vapor pressure (in megapascals) for refrigerants R12, R22, R134A, and R717 (ammonia, NH_3) as a function of the temperature in Celsius is shown in Table 2. Assuming that ammonia R717 is used, an evaporation temperature of $-30°C$ corresponds to a pressure of 0.119 MPa and a condensing temperature of $+35°C$ corresponds to a pressure of 1.36 MPa. This gives a compression ratio of 11.4 which is uneconomically high. For economical operation the ratio should not be more than 8–10; therefore, in this case a two-stage system should be used.

Table 2. Saturated Vapor Pressure of Some Refrigerants, MPa

Temperature, °C	Refrigerant[a]			
	R12	R22	R134A	R717
−50	0.0391	0.0645	0.0299	0.0408
−40	0.0642	0.105	0.0516	0.0715
−30	0.100	0.164	0.0847	0.119
−20	0.152	0.245	0.133	0.190
−10	0.219	0.355	0.201	0.290
0	0.309	0.498	0.293	0.428
+10	0.424	0.681	0.415	0.613
+20	0.567	0.910	0.572	0.855
+30	0.745	1.192	0.770	1.164
+40	0.959	1.534	1.016	1.551
+50	1.217	1.943	1.318	2.028

[a]R12 = dichlorodifluoromethane, CCl_2F_2, R22 = dichlorodifluoromethane, $CHClF_2$, R134A = tetrafluoroethane, CF_3CH_2F, R717 = ammonia, NH_3.
Source: Reprinted with permission of ASHRAE from the 1990 *ASHRAE Handbook-Refrigeration*. Ref. 14.

The intermediate pressure is selected so that the two compressors will work with the same compression ratio. This means in this case that the intermediate pressure should be about 0.44 MPa, corresponding to a temperature of 0°C.

The Refrigerant

Refrigerants are the vital working fluid in a refrigeration system. They absorb heat from one area and dissipate heat into another. The design of the refrigeration equipment is influenced by the properties of the refrigerant selected.

A refrigerant must satisfy many requirements, some of which do not directly relate to its ability to transfer heat. The environmental consequences of a refrigerant that leaks from a system must also be considered. Because of their stability, fully halogenated compounds, called CFCs, persist in the atmosphere for many years and eventually diffuse into the statosphere. According to several scientists the existence of CFCs in the stratosphere has a detrimental effect on the ozone layer. The ozone layer protects the earth from too much ultraviolet radiation. A depletion of ozone might mean a dangerous increase of uv radiation on earth. This fact, coupled with the CFC contribution to the greenhouse effect, has led most nations to urge the industry to minimize the leakage of CFCs into the atmosphere and gradually to phase out their use.

Several of the large chemical companies are at present heavily engaged in finding good substitutes for the CFCs. One such substitute is refrigerant 134a, which is listed in Table 2.

The present available substitutes do not have the same properties as, for example, R12. The users therefore have to accept the fact that a plant designed for the use of R12 will have some limitations when operated with, eg, R134a, and in most cases modifications are necessary.

Most likely the use of R717 (ammonia) will be more and more frequent, particularly for large industrial applications.

For commercial installations such as those in department stores most likely we will see a central refrigeration plant using ammonia chilling brine being pump-circulated to the various locations where refrigeration is needed. By using a brine the risk of ammonia gas leaking and coming in contact with the public will be minimized.

Compressors

Piston compressors of various sizes dominated in the refrigeration industry until about 25 years ago, when screw compressors were first used also for refrigeration applications.

Today screw compressors are commonly used for industrial applications, while for small systems such as refrigerators and home freezers the use of hermetic piston compressors is still widespread.

Large compressors are often equipped with a capacity-control mechanism, which makes it possible to maintain constant evaporation temperatures even when the demand varies. It should, however, be noted that the efficiency generally decreases when the compressor works with reduced capacity.

The efficiency of refrigeration systems is influenced by many factors other than the compressor. Only when all details are well designed and tuned together can an optimal result be achieved.

Engine Room as Package Unit or Site-Built

In the early days of the refrigeration industry all engine rooms were site-built. Today, however, it is more common for the refrigeration plant to be delivered as a package unit (see Fig. 24). This offers the benefit of quick deliveries with very little work on-site. If the package unit is designed by an experienced supplier, it also means less risk for functional problems during startup. Finally it offers a greater flexibility than a site-built installation, which tends to be more important as the industry must accommodate changing demands.

Economical Operation Conditions

Two basic facts are important for anyone responsible for the operation of an industrial refrigeration plant.

Out of an existing refrigeration installation more refrigeration capacity is achieved if the evaporation temperature is elevated. This is clearly illustrated in Figure 25. Therefore, it is important to make sure that the evaporator surfaces are kept as clean as possible from ice and that good air circulation is obtained over the whole surface. It is also important that the refrigerant be in good contact with every part of the evaporator.

The condition of the condensor is also of importance for a good economy. As can be seen from Figure 26 the electrical power consumption goes down with lowering the condensing temperature. For efficient operation the evaporation temperature should be kept high and the condensing temperature low.

Figure 24. Package refrigeration unit. Refrigeration capacity = 400 kW at −40°C evaporation temperature and 35°C condensation temperature. Length = 12 m; width = 2.1 m; height = 3 m.

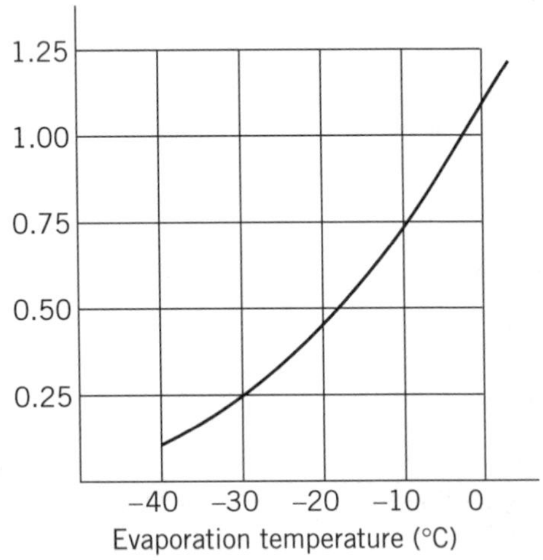

Figure 25. Refrigeration capacity as a function of evaporation temperature.

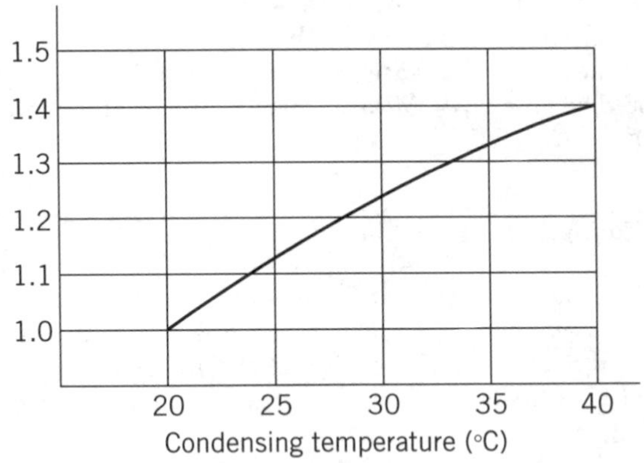

Figure 26. Electric power consumption as function of condensing temperature. Constant condensing temperature.

With the help of an automatic process control system it is possible to improve the efficiency of a refrigeration installation, including the operation of the freezing equipment.

Figure 27 illustrates a typical example of how a process line is shown on a data screen being part of an automatic process control system.

Cryogenic Refrigeration Systems

The second refrigeration process of importance for the food industry is the cryogenic system. In contrast to the mechanical refrigeration system described above, the cryogenic system is open, which means that the refrigerant is consumed and not recirculated. Liquid nitrogen or liquid or solid carbon dioxide (CO_2) are normally used as refrigerants.

In a cryogenic system the refrigeration machinery for production of LIN or liquid CO_2 are large industrial units normally located at a distance remote from the food processing factory. With help of special transport equipment the cryogenics are transported from the place of manufacturing to storage tanks at the food factories.

From an economic viewpoint the principal difference between a mechanical refrigeration system and a cryogenic system is that for the former the investment cost for the food producer is fairly high while the running cost is low. The opposite goes for cryogenic systems. This fact is illustrated in Figure 28.

THE ECONOMICS OF FREEZING

The freezer is usually the largest investment in a processing line. The freezer's impact on operating costs as well

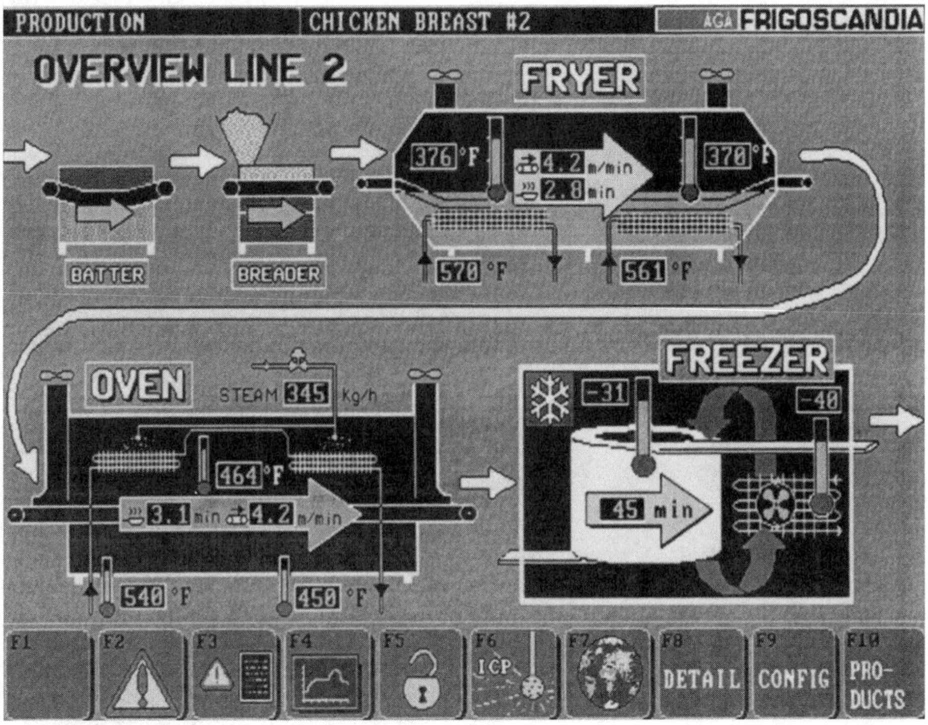

Figure 27. Screen illustration of part of a processing line being controlled automatically.

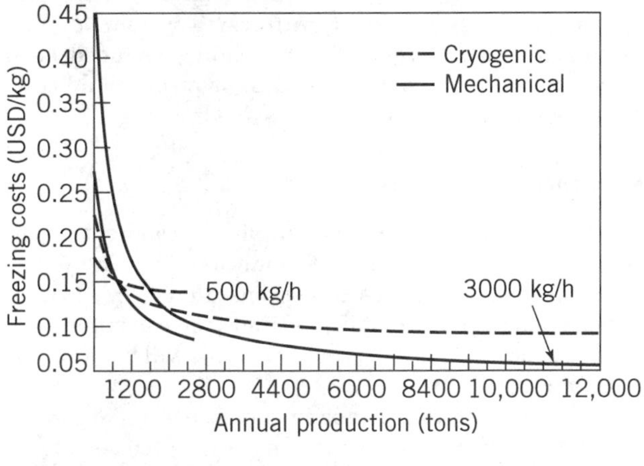

Figure 28. Freezing cost as function of utilization per year for mechanical freezing and cryogenic freezing. Constant evaporating temperature.

as on the final product quality and cost is similarly high. For the investor, it is essential to ensure that the design of the freezing system allows it to be properly integrated into the process and consistently operated at an optimal level.

Reliability

The value of the food products that pass through a freezer in a few weeks' time is often much higher than the investment cost of the freezer itself. This makes reliability a crucial consideration for the food processor.

Not only is the time the freezer is in operation important, but also the amount of time it is working to full capacity. A breakdown of any part of the line stops the whole line, and product that moves through a line without freezing properly may loose much or most of its market value.

With some systems, product blows out of the freezing zone, lodging in and freezing to other parts of the equipment. This can jam up the line and eventually cause damage.

Dehydration

Dehydration losses will always be present in any freezing system. The evaporation of water vapor from unpacked products during freezing becomes evident as frost builds up on evaporation surfaces. This frost is also caused by excessive infiltration of warm, moist air into the freezer. Still air inside the diffusion-tight carton often creates larger dehydration losses than does the unpackaged products frozen in an IQF freezer. Heat transfer is poor because no air circulation occurs within the package. The result is an evaporation of moisture that can be significant; however, the frost may stay inside the carton.

A poorly designed freezing tunnel may operate with dehydration losses of 3–4%, while a well-designed tunnel can be built to operate with losses of 0.5–1.5%. Liquid nitrogen tunnels normally operate with a dehydration loss of about 0.2–1.25%. This loss occurs when the nitrogen gas is circulated over the product at the infeed end of the freezer.

Freezing Cost Comparison

The cost of a freezing system comes both from the purchase price and the true cost of operation over the lifetime of the system. High operating costs can offset the advantages of a low initial purchase price. Consequently, it is essential to consider all the factors before making an investment, ie, to consider a freezer's "all-in economy."

In making preinvestment analyses, a comparison should be made of all pertinent factors involved, eg, capital, power, operation, cleaning, and maintenance, dehydration, and downtime costs. Figure 29 shows the relative costs when the real-life data from a GYRoCOMPACT were compared with the data from a conventional spiral freezer.

The graphs in this cost study (based on freezing of 1500 kg/h of hamburgers) clearly illustrates that the real cost of a freezer is not merely a question of the initial capital outlay. All the other factors involved in the economics of freezing contribute to the all-in cost picture. Figure 30 shows the effect of an efficient operation over a 5-yr period.

With all the crucial factors taken into account, the somewhat higher purchase price of one freezer was more than offset by the unique module design concept of the same freezer, including a particular belt design indicating major overall cost savings. The extra investment has a payback time of about 1 yr. Thereafter, the continued cost savings means profit for the food processor.

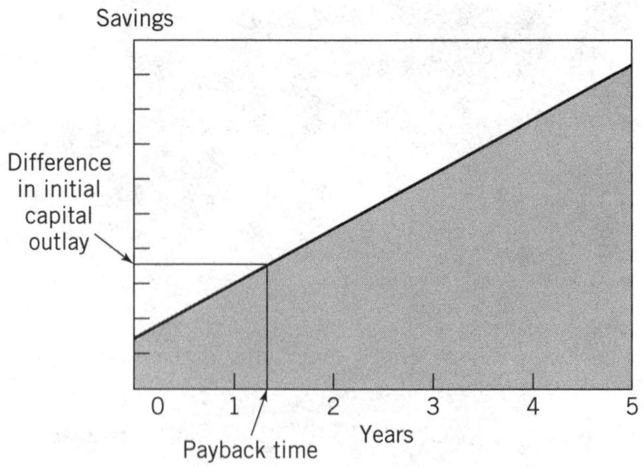

Figure 30. Accumulated annual savings.

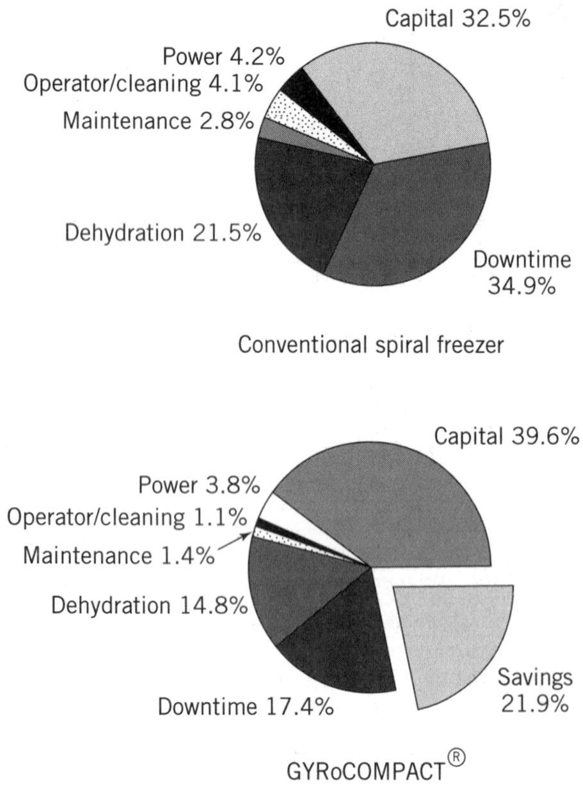

Figure 29. Relative freezing cost.

Flexibility and Upgradability

With the competitiveness of today's consumer market, timing is of decisive importance. A system that is easy to modify will make it possible for the processor to respond quickly to the requirements of a new market and to expand or relocate the system easily. A freezing system built as a permanent fixture that cannot be readily expanded, easily removed, or conveniently traded in or resold could cost a great deal more at a later date as needs change.

Maintenance and Service Support

Freezers create an extremely harsh environment for mechanical parts. Optimum performance from even the most efficient freezer requires proper maintenance. Attention must be paid to how complicated and time-consuming such maintenance will be, and what effect this will have on valuable processing time.

Another important factor in the economics of freezing is backup support and a wide range of related services. In judging a freezer supplier's support capability, consideration should be given to the level of service offered in the planning and installation stages as well as after the deal is complete.

The processor will also need to consider where the supplier's service centers are located, and what commitment of service resources they are prepared to make. This will determine how quickly a service technician can respond in an emergency.

BIBLIOGRAPHY

1. R. Thévenot, *A History of Refrigeration throughout the World*, International Institute of Refrigeration, Paris, 1979.
2. W. R. Woolrich, *The Men Who Created Cold*, Exposition Press, New York, 1967.

3. O. E. Anderson, Jr., *Refrigeration in America*, Princeton University Press, Princeton, N.J., 1953.
4. M. Jul, *The Quality of Frozen Foods*, Academic Press, Orlando, Fla., 1984.
5. F. Lindelöv, "Chemical Reactions in Frozen Cured Meat Products. TTT—Examination of Cured and Uncured Pork Bellies," in *Proceedings of Commissions C1 and C2 Meeting* (Karlsruhe, W. Germany, International Institute of Refrigeration, Paris, 1977.
6. S. Åström and G. Löndahl, "Air Blast In-line Freezing versus Ultra Rapid Freezing. A Comparison of Freezing Results with Some Various Vegetables and Prepared Foods," in *Proceedings of Commissions 4 and 5 Meeting* (Budapest), International Institute of Refrigeration, Paris 1969.
7. G. Löndahl and T. Nilsson, "Microbiological Aspects of Freezing of Foods," *International Journal of Refrigeration* **1**, 1978.
8. S. Åström, "Freezing Equipment Influence on Weight Losses," in *SOS / 70 Proceedings, 3rd International Congress of Food Science and Technology* (Washington, D.C., 1970), Institute of Food Technologists, Chicago, 1971.
9. G. Lundgren and K. Nilsson, "Vattenförluster vid infrysning av livsmedel" ("Water Loss during Freezing of Foods"), *Livsmedelsteknik*, No. 9, 1969 (Sweden).
10. *Recommendations for the Processing and Handling of Frozen Foods*, International Institute of Refrigeration, Paris, 1986.
11. American Society of Heating, Refrigeration and Air-Conditioning Engineers, *ASHRAE Handbook—Fundamentals*, 1989.
12. S. D. Holdsworth, CFDRA Technical Memorandum No. 452, Campden Food & Drink Research Association, Chipping Camden, UK, 1979.
13. S. L. Polley, O. P. Snyder, and P. Kotnour, "A Compilatision of Thermal Properties of Foods," *Food Technology* (Nov. 1980).
14. American Society of Heating, Refrigeration and Air-Conditioning Engineers, *ASHRAE Handbook—Refrigeration*, 1990.

JIM HEBER
GORAN LÖNDAHL
PER OSKAR PERSSON
LEIF RYNNEL
Helsingborg, Sweden

FRESH-CUT FRUITS AND VEGETABLES: MODIFIED ATMOSPHERE PACKAGING

Fresh and fresh-cut fruits and vegetables are living tissue that undergoes catabolic metabolism including respiration (1). Unlike their fresh, intact counterparts, fresh-cut or "minimally processed" fruits and vegetables are subjected to specific processing and preservation steps. Fresh-cut fruits and vegetables are prepared by one or more unit operations such as peeling, slicing, shredding, juicing, and so on. They also are subjected to a partial but not end-point preservation treatment that may include use of minimal heat, antioxidants and other preservatives, radiation, pH control, chlorinated water dips, or a combination of these treatments (2). The initial preparation and preservation treatments are usually followed by a controlled/modified atmosphere, vacuum-packaging step. The packaged product should be maintained at a reduced temperature above the freezing point (~1°C) throughout storage, distribution, and marketing up until the time of preparation for consumption.

Fresh-cut produce is a perishable food that is included in Brody's (3) minimally processed/reduced-oxygen-packaged/chilled-distribution category of commercial foods. A major advantage is that it offers freshlike, ready-to-eat convenience and quality for both institutional and retail markets.

The composition of the earth's atmosphere (commonly called air), exclusive of water vapor and trace contaminants, is 78.08% nitrogen, 20.94% oxygen, 0.93% argon, 0.34% carbon dioxide, and trace amounts of several other noble gases including neon and helium (4). In nature all fresh, unpackaged food not immersed in water is surrounded, and to some extent permeated, by air. Chemical interactions routinely occur between fresh, unpackaged food and the air's oxygen and carbon dioxide, which are highly reactive gases. These reactions result in significant short- and long-term changes in the chemical and physical characteristics of food that can have positive or negative impacts on a food's utility, quality, storability, and nutrient status.

Fresh-cut produce items in the marketplace include peeled, sliced carrot sticks; sliced/diced onions; celery sticks; peeled, sliced/diced Irish potatoes; melon balls; peeled, cored/sliced/diced pineapple; apple slices; cole slaw; sliced tomatoes; destemmed grapes; decapped, sliced strawberries; shredded lettuce; mixed salad greens; and even prepared Waldorf salad with sliced apples, raisins, diced celery, and walnut pieces.

PARTIAL PROCESSING OF FRESH PRODUCE

Postharvest processing operations on fresh produce are essentially limited to those that take place prior to blanching. Thus, enzyme systems remain functional, and abundant microflora are present on minimally processed produce at the time it is packaged. Food processing operations and treatments applied to fresh-cut produce include chilling, washing, peeling, coring, slicing/cutting, acidified chlorine solution immersion, applications of antioxidants/firming agents/antimicrobials, and dewatering via centrifugation. Obviously many of these operations cause stress or damage to the plant tissue, shortening its shelf life. Furthermore, despite the sanitizing treatments carried out during fresh-cut processing, surviving psychotrophic microorganisms can thrive at damaged tissue sites even under refrigeration.

THE RATIONALE FOR MODIFIED ATMOSPHERE PACKAGING OF FRESH-CUT PRODUCE

Damaged plant tissue usually demonstrates marked increases in aerobic respiration and catabolic metabolism and is accompanied by increased moisture loss and a decline in sensory and nutritional quality. Two primary means that are used to delay postprocessing quality degradation are refrigerated storage/distribution and alteration of the atmosphere surrounding the fresh-cut produce. Modified atmosphere packaging (MAP) of fresh-cut produce results in a reduced oxygen concentration and an increased carbon dioxide concentration surrounding the product.

Living plant tissue takes up oxygen and liberates carbon dioxide during aerobic respiration. When this tissue is hermetically sealed inside a package that is selectively permeable to oxygen and carbon dioxide, the initial rate of oxygen uptake by the tissue exceeds the rate at which oxygen permeates by diffusion into the package from the surrounding air. When the oxygen concentration inside the package falls to around 12%, the respiration rate of the tissue begins to slow (5). At an 8% oxygen concentration, metabolic activity such as ripening of fruit tissue declines significantly. Aerobic respiration, and the rate of oxygen uptake by the tissue, reach a minimum when the in-package oxygen concentration attains a 1 to 5% level. If the package material is appropriately permeable, the minimum oxygen concentration is maintained at a steady state; that is, the rate of oxygen uptake by the tissue equals the rate of oxygen permeation into the package from surrounding air. At this minimum respiration rate, tissue quality loss due to catabolic metabolism is minimized and the onset of senescence is delayed.

Since the plant tissue also liberates carbon dioxide during respiration, the carbon dioxide concentration inside the package rises simultaneously as the oxygen level declines. Therefore, the MAP material must be sufficiently permeable to carbon dioxide to prevent excessively high levels from accumulating inside the package, which would injure the plant tissue. After the package is sealed, the interior relative humidity rapidly increases to nearly 100%. The package material is a good barrier to water vapor, so the rate of moisture vapor transmission through the package is much less than the rate at which moisture vapor accumulates inside the package via evapotranspiration from the plant tissue. Thus, shortly after the produce is sealed inside the package the moisture vapor pressure deficit between the tissue and the in-package atmosphere attains zero and further tissue moisture loss is arrested.

PROBLEMS AND CONCERNS ASSOCIATED WITH MAP OF FRESH-CUT PRODUCE

Considerable care must be taken in designing MAP for fresh-cut produce. Polymeric packaging materials having appropriate permeabilities for oxygen and carbon dioxide must be carefully selected to prevent the development of an anaerobic atmosphere within the package after filling and sealing. When the in-package atmospheric oxygen concentration declines to a low level (usually somewhere between <1 and 3%), the tissue aerobic respiration rate reaches a minimum termed the extinction point. If the oxygen level falls below the extinction point, anoxia ensues and the tissue begins anaerobic respiration, which is fermentative in nature. This event triggers the rapid destruction of the tissue's like-fresh sensory quality, generating off-flavors and negatively impacting texture and appearance; in short, the product becomes unmarketable.

The optimum atmosphere within the modified atmosphere (MA) package, with respect to oxygen and carbon dioxide concentration, depends on the variety of produce, the extent of tissue damage sustained during processing, and the product temperature. Maximum quality retention is achieved when product temperature is maintained either just above its freezing point or, for some items, just above the chill-injury temperature. Lowering produce temperature reduces respiration by a factor of two- to threefold for each 10°C decline ($Q10 = 2$–3), and use of an appropriate MA package can result in an additional fourfold reduction in respiration rate (6).

When the temperature is sufficiently cool and an appropriate MA package results in an in-package oxygen concentration slightly above the extinction point, aerobic microbial growth also is slowed, further retarding tissue decay. However, if the package is temperature abused or improperly designed or filled, causing the oxygen concentration to fall below the extinction point, the anoxic atmosphere can encourage the growth of anaerobic and/or facultative anaerobic, psychotrophic microorganisms (2), some of which are pathogenic. This potential for pathogenic microorganisms to develop within MA-packaged fresh-cut products is of considerable concern to food science professionals and regulatory authorities.

In many cases the steady-state (equilibrium) oxygen content in fresh-cut MA packages is 2 to 5%, and there is a potential that carbon dioxide will increase to as high as 16 to 19% (7). Maximum tolerable carbon dioxide concentrations for many fresh-cut or intact items, including cauliflower, lettuce, celery, kiwifruit, cabbage, radish, sweet pepper, banana, and carrot, are in the 2 to 5% range, and physiological damage can occur at levels in excess of 2 to 6% (5). Thus, the polymeric materials selected for use in MAP need to be four to six times more permeable to carbon dioxide than to oxygen. Sometimes it is necessary to use package parameters for fresh-cut produce that provide a compromise, with the equilibrium atmosphere somewhat higher in oxygen concentration than ideal, in order to avoid excessive accumulation of carbon dioxide. The package material selected also should have a moderate to high permeability to moisture vapor in order to minimize the chance of moisture condensation on the inside surface of the package due to minor temperature fluctuations during storage and distribution.

MODES OF MAP

There are two modes of MAP—passive and active (7). Passive MAP involves placing the plant tissue in a selectively gas-permeable polymeric package, hermetically sealing the package, and then allowing tissue aerobic respiration to reduce oxygen and increase carbon dioxide concentrations inside the package to a desired steady-state equilibrium. Active MAP involves placing the plant tissue in a selectively gas-permeable package, evacuating the package atmosphere, and replacing it by flushing the unsealed package with a specific mixture of oxygen, carbon dioxide, and nitrogen gases (8) followed by rapid sealing of the package. The flushing gas composition is usually selected to provide optimum levels of oxygen and carbon dioxide (with a balance of nitrogen) to immediately diminish aerobic respiration rate of the particular tissue system being packaged. Active MAP may also include the utilization of absorbers or adsorbers placed inside the package to scavenge oxygen, carbon dioxide, ethylene (5), and water vapor as well as the use of antimicrobial agents such as carbon monoxide. The goal of passive and active MAP design is to achieve a balance between the enclosed plant tissue respiration rate and package gas permeability to attain or maintain an acceptable equilibrium atmosphere within the package; that is, a MA that will delay ripening/senescence and thereby extend product shelf life.

SHELF-LIFE REQUIREMENTS OF MAP FRESH-CUT PRODUCTS

In the United States fresh-cut products are processed and packaged either at the source of production or at regional or local processing facilities. The postprocessing shelf life needed at each of these processing sites is 14 to 21 days, 7 days, and a short time (12–24 hours), respectively (9). The use of vacuum and gas flushing techniques in conjunction with differentially permeable packages (primarily flexible polymeric film bags) is essential for attaining a 21-day shelf life and desirable for achieving 7 days of postprocessing shelf life. Simpler packaging materials, combined with vacuum or gas flushing, are usually sufficient for attaining a short postpackaging shelf life.

SELECTION OF MAP MATERIALS

Determining the optimal atmosphere/packaging material requirements for MAP of fresh-cut produce is a complex task. Various researchers have worked to develop mathematical models to aid package material selection. Information about the various mathematical models can be obtained by referring to references provided by Zagory and Kader (7). However, use of these models to select MA package design criteria cannot be relied upon exclusively. Empirical testing of selected packaging is absolutely essential for ultimate successful use of MAP (10).

MAP of fresh-cut produce requires careful selection of polymeric packaging materials especially with regard to their permeability to oxygen, carbon dioxide, and water vapor. It is necessary that the polymeric material selected be three to six or more times more permeable to carbon dioxide than to oxygen. Polymeric materials potentially suitable for MAP fresh-cut produce include low-density polyethylene (LDPE), linear low-density polyethylene (LLDPE), medium-density polyethylene (MDPE), high-density polyethylene (HDPE), polypropylene (PP), polyvinyl chloride (PVC), polystyrene (PS), ethylene vinylacetate (EVA) copolymers, ionomers, and polybutylene (10).

SAFETY CONSIDERATIONS

The probability of a microbial foodborne disease resulting from consumption of MA-packaged fresh-cut produce is relatively low. Only a few pathogens are capable of growing at refrigerated temperatures, and normal spoilage organisms which are usually psychotrophic have a competitive advantage over most pathogens that may be present (11). Moreover, many fruits are sufficiently acidic to prevent pathogen growth. Nevertheless, foodborne disease does occur with fruits and vegetables. Some pathogens that have potential to develop in MA-packaged fresh-cut products are *Escherichia coli, Aeromonas hydrophila, Clostridium botulinum, Listeria monocytogenes, Salmonella* sp., and parasites including *Entamoeba histolytica* and *Giardia lamblia* can be present (11). Therefore, an intense commitment to sanitation and adherence to U.S. Food and Drug Administration (FDA) Good Manufacturing Practices (GMPs) as well as implementation of a Hazard Analysis Critical Control Point (HACCP) program should be a part of fresh-cut food processors quality assurance efforts (12).

Another important food safety consideration is the potential for migration of harmful/toxic chemicals from packaging materials in immediate contact with the fresh-cut produce into it. Commercial plastic packaging materials usually include nonplastic components that might migrate to contained foods. Thus, polymeric MAP materials must comply with all governmental regulations related to indirect food additives, and packaging suppliers must be able to certify such compliance (10).

BIBLIOGRAPHY

1. Institute of Food Technologists, "Quality of Fruits and Vegetables—A Scientific Summary by IFT's Expert Panel on Food Safety and Nutrition," *Food Technol.* **44**(6), 99–106 (1990).
2. R. C. Wiley, "Introduction to Minimally Processed Refrigerated Fruits and Vegetables," in R. C. Wiley, ed., *Minimally Processed Refrigerated Fruits and Vegetables*, Chapman and Hall, New York, 1994, pp. 1–14.
3. A. L. Brody, "Minimally Processed Foods Demand Maximum Research and Education," *Food Technol.* **52**(5), 62–66, 204–206 (1998).
4. "Components of Atmospheric Air," in R. C. Weast and S. M. Selby, eds., *Handbook of Chemistry and Physics*, 48th ed., The Chemical Rubber Co., Cleveland, OH, 1967–1968, p. F-142.
5. A. A. Kader, D. Zagory, and E. L. Kerbel, "Modified Atmosphere Packaging of Fruits and Vegetables," *Crit. Rev. Food Sci. Nutr.* **28**, 1–30 (1989).
6. D. O'Beirne, "Modified Atmosphere Packaging of Fruits and Vegetables," in T. R. Gormely, ed., *Chilled Foods the State of the Art*, Elsevier Applied Science, New York, 1990, pp. 183–199.
7. D. Zagory and A. A. Kader, "Modified Atmosphere Packaging of Fresh Produce," *Food Technol.* **42**(9), 70–77 (1988).
8. J. P. Smith, H. S. Ramaswamy, and B. K. Simpson, "Developments in Food Packaging Technology, Part II: Storage Aspects," *Trends Food Sci. Technol.* **1**(5), 111–118 (1990).
9. M. Cantwell, "Postharvest Handling Systems: Minimally Processed Fruits and Vegetables," in A. A. Kader, ed., *Postharvest Technology of Horticultural Crops*, 2nd ed., University of California Division of Agriculture and Natural Resources, Oakland, Calif. 1992, pp. 277–281.
10. D. V. Schlimme and M. L. Rooney, "Packaging of Minimally Processed Fruits and Vegetables," in R. C. Wiley, ed., *Minimally Processed Refrigerated Fruits and Vegetables*, Chapman and Hall, New York, 1994, pp. 135–182.

11. R. E. Brackett, "Microbiological Spoilage and Pathogens in Minimally Processed Refrigerated Fruits and Vegetables," in R. C. Wiley, ed., *Minimally Processed Refrigerated Fruits and Vegetables*, Chapman and Hall, New York, 1994, pp. 269–312.
12. D. M. Dignan, "Regulatory Issues Associated with Minimally Processed Refrigerated Foods," in R. C. Wiley, ed., *Minimally Processed Refrigerated Fruits and Vegetables*, Chapman and Hall, New York, 1994, pp. 327–353.

DONALD SCHLIMME
MARK KANTOR
University of Maryland
College Park, Maryland

See also CONTROLLED ATMOSPHERES FOR FRESH FRUITS AND VEGETABLES.

FRUIT DEHYDRATION

Dehydration, or drying, is a preservation process that involves the reversible removal of water from the fruit tissues, thereby extending the storage life because microorganisms and the native enzymes of fruits are deprived of the moisture necessary for their activity. However, it is not a sterilizing process, and means must be provided to preserve the low-moisture equilibrium and prevent the fruit tissue from regaining moisture until such time as deliberate reconstitution is desired. Although preservation is usually the principal reason for dehydration, other considerations are often important; for example, significant reductions in the weight and bulk of fruits for economical transport and storage.

Drying fruits is one of the oldest techniques of food preservation; Persians, Greeks, and Egyptians have used it since ancient times. Sun drying was the method applied in early times, and it still accounts for a significant part of the dried fruit consumed in the world today. However, since the latter part of the nineteenth century mechanically dehydrated fruits are produced in increasingly larger quantities.

Today the food dehydration industry is large and extends to all countries throughout the world. The following countries are highly important dried fruit producers: Australia, Argentina, Chile, France, Greece, Portugal, Spain, South Africa, Turkey, and the United States (Table 1). In the United States more than 90% of the dried fruit output is produced in California, where apples, apricots, dates, figs, peaches, pears, prunes, and raisins represent the significant volume of dried fruits; raisins account for over 61% of the total volume (Table 2). Dried fruits, particularly apples, are also produced in Washington and Oregon. Consumption of dried fruits in the United States has remained almost constant during the past 20 years. Annual per capita consumption was 2.47 lb (1121 g) in 1977 and 2.94 lb (1335 g) in 1997 (Fig. 1). The small increase in consumption is largely attributed to the higher use of raisins. Recently the development of an increasing variety of convenience foods, such as instant beverages, breakfast cereal, healthy fruit snacks, and so on, often depend on the availability of high-quality dehydrated fruits. Development of new forms of dried fruit and successful preparation of intermediate-moisture products have led to products that are stable to store and pleasant to eat directly.

Table 1. Major Dried Fruit Producing Countries

Dried Fruit	Principal Countries of Origin
Apples	Chile, Italy, Spain, U.S.
Apricots	South Africa, Turkey, U.S.
Cranberries	U.S.
Currants	Australia, Greece, U.S.
Dates	Egypt, Iran, Iraq, Saudi Arabia, U.S.
Figs	Greece, Turkey, U.S.
Peaches	Australia, Chile, South Africa, U.S.
Pears	Australia, Chile, South Africa, U.S.
Prunes	Argentina, Chile, Croatia, South Africa, U.S.
Raisins	Afganistan, South Africa, U.S.

PRINCIPLES

Dehydration involves simultaneous heat transfer and moisture diffusion (mass transfer). The conversion of liquid (or solid in the case of freeze drying) to vapor demands the supply of latent heat to the product; this can be achieved by a variety of methods: conduction by contact with a heated metal plate, convection from heated air, radiation from an infrared source, or microwave energy. The process may be accelerated by the application of vacuum.

Dehydration Terms

Dried is the term applied to all dried products, regardless of the method of drying. Dehydration refers to the use of mechanical equipment and artificial methods under carefully controlled conditions of temperature, humidity, and airflow. Although the term dehydrated does not refer to any specific moisture content in the finished product, it is usually considered implying virtually complete water removal to a range of 1 to 5% moisture. Products with such low water content can be stored in a moisture-proof package at room temperature for periods well in excess of two years with no detectable change in quality.

Evaporation refers to the use of the sun or forced-air dryers to evaporate moisture from fruit to a fairly stable product. The moisture level of evaporated fruit is approximately 15 to 25%. In general, sun drying will not lower the moisture content of fruit below 15%. The shelf life of such fruit products does not exceed one year, unless they are held in cold storage. For the extended shelf life of most evaporated fruits, sulfur treatment (except for prunes, unbleached raisins, and dates) or the use of a chemical mold inhibitor is necessary.

Vacuum drying is a method of drying in a vacuum chamber under reduced atmospheric pressure to remove water from the fruit at less than the boiling point of water under ambient conditions.

Freeze drying is a method of drying in which the fruit is frozen and then dried by sublimation in a vacuum chamber under high vacuum (an absolute pressure of less than

Table 2. Production Volume and Value of Dried Fruits in California Five Year Averages, 1992–1996

Commodity	Drying ratio: fresh to dry fruit[a]	Average annual production (tons)	Average farm value (dollars)
Apples	7–10 to 1	2,720	3,168,800
Apricots	5.5–8.5 to 1	2,200	4,840,00
Dates		19,900	19,900,000
Figs	3 to 1	14,350	13,919,00
Peaches, Freestone	6–8.75 to 1	1,420	1,377,400
Pears, Bartlett	6–7 to 1	970	834,000
Prunes	2.75–3.25 to 1	174,730	167,740,800
Raisins	4–5 to 1	342,750	332,467,500
Total		*559,040*	*544,247,500*

[a]Pounds of fresh fruit required to yield one pound of dehydrated product.
Source: Dried Fruit Association of California, Sacramento, California.

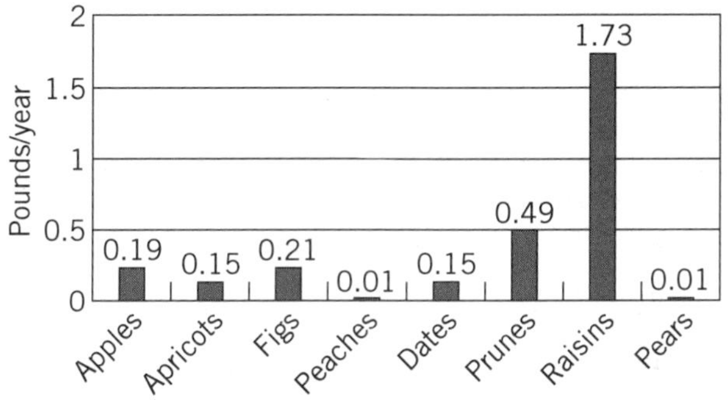

Figure 1. Per capita consumption of dried fruits in the United States, 1997.

4.6 mmHg, the vapor pressure of the ice and water at 0°C, is necessary) to around 2% moisture.

Intermediate-moisture fruits (IMF) are dried on the basis of water activity level rather than on the percentage of moisture. Restraint of the water molecules to a degree that prohibits spoilage by microorganisms occurs at different moisture contents depending on the amount and nature of the dissolved material present and to some degree on the insoluble components. Food may be classified as an IMF if it has water activity level greater than a common low-moisture fruit (0.2) and less than that of most fresh fruit (0.85). In practice, most IMF has a water activity in the range of 0.65 to 0.85 and contains 15 to 30% moisture.

Water activity (a_w) is defined as the ratio of the water vapor pressure in equilibrium with a food to the vapor pressure of water at the same temperature. Bacteria will not grow if a_w is below 0.9, and yeast and molds, are inhibited below a_w of 0.7.

Dehydrofreezing is a hybrid process that combines the best features of both drying and freezing. The process consists of drying fruit to about 50% of its original weight, then this intermediate material is quick-frozen by standard freezing techniques. The quality of dehydrofrozen fruit is equal to that of frozen products, as the drying process is discontinued at a stage where quality impairment usually does not occur. The advantages of the process are a 50% reduction in storage and freight charges and even greater savings in packaging costs in comparison with traditional frozen products.

DRYING METHODS

Quality requirements, raw material characteristics, and economic factors determine selection of a drying method as well as the operating conditions for a given fruit. Several drying methods are commercially used, each better suited for a particular situation.

The selection of drying methods depends on the following factors:

- Physical form of raw material: whole fruit (eg, cranberry, grape, plum, etc), sliced form (eg, banana, mango, kiwi, etc), liquid, paste, slurry, pulp, thick liquid, and large versus small aggregates
- Properties of raw material: sensitive to oxidation, sensitive to temperature damage, thermoplastic residues, and so on
- Susceptibility to microbial attack
- Sugar content
- Presence of a "skin" that acts as a barrier for water removal (eg, grapes, blueberries)
- Desired product characteristics: powder, instant solubility, retention of shape, and so on
- Value of the finished product: low, medium, or high price

The following three basic types of drying process are used for fruits:

1. Sun drying and solar drying are practiced for certain fruits, such as prunes, grapes, and dates.
2. Atmospheric dehydration processes, including:
   a. stationary, or batch, processes (using kiln, tower, and cabinet dryers)
   b. continuous processes (eg, tunnel, belt trough, conveyor, and fluidized bed)
   c. spray dryers, suitable for making powders from fruit juice concentrates, and drum, or roller, dryers, useful for drying fruit juice concentrates, slurries, or pastes
3. Subatmospheric dehydration techniques, including:
   a. vacuum dehydration processes, which are useful for the processing of low-moisture fruits with high

sugar content, such as peaches, pears, and apricots

b. freeze drying, which ensures high flavor retention and minimal damage to product structure and nutritional value; the finished product with open structure permits fast and nearly complete rehydration

More than 200 types of dryers are available commercially to process the diverse physical forms of foods and required product characteristics. For drying fruits, conventional types of dryers are commonly used because of their simplicity of construction and operation, as well as low cost. Most products are seasonal in nature, which means that dryers are operated only over a fraction of the year. Thus, the units must have low capital costs for economic reasons.

Sun Drying

Sun drying of fruit crops is still practiced unchanged from ancient times. This method is limited to climates with hot sun and dry atmosphere, and to certain fruits such as prunes, grapes, dates, figs, apricots, and pears. These crops are processed without much technical aid by simply spreading fruits on the ground, on racks, or roofs and exposing them to the sun until dry. Because sun-dried products generally have moisture levels no lower than 15 to 20%, they have a limited shelf life. It is a slow process, unsuitable for producing high-quality products, and dust, dirt, and insects often contaminate the finished products.

Solar Drying

In recent years considerable interest has been focused on the use of solar energy because of the rapid increase of fuel costs. In this method, solar energy is used alone or may be supplemented by an auxiliary energy source. A simple method of accelerating the sun-drying rate of fruit is to paint the trays black. A solar trough application of mirrors to increase solar energy and indirect solar dryers applying solar collectors are also recommended to improve the efficiency of this process.

Kiln Drying

The natural draft from rising heated air used for drying fruit is the oldest type of dehydration equipment still in commercial use. Kiln dryers generally have two levels: gas burners on the lower floor provide heat, and the warm air rises through a slotted floor to the upper level. Food materials such as apple slices are spread out on the slotted floor in a layer about 25 cm deep and turned over periodically by fork until they reach the desired moisture, usually between 16 and 24%.

Cabinet Drying

Cabinet dryers are arranged for batch operation and are usually held at constant temperature, though humidity may decrease during the drying process. They consist of an enclosed chamber fitted with a variable heater and a fan for air stirring, together with deflectors for airflow adjustment, outlet air louvers, and adjustable inlet air louvers. Cabinet drying is a particularly useful research tool for establishing the drying characteristics of a new product, prior to a large-scale commercial run.

Tunnel Drying

The equipment is similar to a cabinet dryer except that it allows a continuous operation along a rectangular tunnel through which move tray-loaded trucks. The tunnel dryer provides rapid drying without injury to fruits and permits a uniform drying process; therefore, it is widely used in drying fruit.

Tunnel dryers are classified by the direction in which the air traverses the product. In a parallel flow unit, the fresh material encounters the driest and warmest air initially, and leaves the drier at the coldest end; in a countercurrent flow unit, the air direction is opposite to the movement of the product, so the dry product leaving the drier encounters hot dry air. Multistage dryers consisting of three, four, or five dryer stages are also used. Such systems are flexible and can achieve close to optimum drying conditions for a wide variety of products.

Conveyor Drying

Conveyor dryers are continuous processing equipment and consist of an endless belt that carries the material to be dried through a tunnel of warm circulating air. The speed of the conveyor is variable to suit both the product and the heat conditions. Furthermore, process conditions are usually controlled by designing the system in sections, thus allowing different flow rates, humidities, and temperatures to be set in each section, and by rotating the product when it moves from one section of the belt to the next.

This drying method has the advantage of essentially automatic operation, which minimizes labor requirements. The conveyor dryer is best adapted to the large-scale drying of a single commodity for the whole operating season. It is not well suited to operations in which the raw material or the drying conditions are changed frequently, because extensive hours of start-up and shutdowns make it difficult to produce satisfactory products.

Spray Drying

Spray drying involves the dispersion of liquid or slurry in a stream of heated air, followed by the collection of the dried particles after their separation from the air. The process is widely used to dehydrate fruit juices.

The general construction of a spray dryer incorporates four main features:

1. a heater and at least one fan to produce air at the required temperature and velocity;
2. an atomizer or jet to produce liquid particles of the required size;
3. a chamber in which the liquid droplets are brought into intimate contact with the hot air; and
4. a means of removing the product from the air stream.

The final product is delivered as a free-flowing powder.

Drum Drying

In drum drying, which is suitable for a wide range of liquid, slurried, and pureed products, a thin layer of product is applied to the surface of a slowly revolving heated drum. In the course of about 300 degrees of one revolution, the moisture in the product is flashed off, and the dried material is scraped off the drum by a stationary or reciprocating blade at some point on the periphery. The residence time of the product in the drier is on the order of two to few minutes.

Drum drying is an inexpensive method; however, its commercial application is limited to less heat sensitive products. Its usefulness for fruit dehydration is quite limited because the high temperature required, usually above 120°C, imparts a cooked flavor and off-color to most fruit products. Also, the high sugar content of most fruit juices makes them difficult to remove from the drum dryers because of their high thermoplasticity.

Vacuum Drying

Dehydration under vacuum has special merits for certain fruits in terms of final quality. Drying can be carried out at lower temperatures than with air drying, and heat damage is minimized. Furthermore, as drying is carried out with virtual absence of air, oxidation of the fruit is virtually eliminated.

Vacuum-drying systems have the following components: (*1*) a vacuum chamber, (*2*) a heat supply, (*3*) a vacuum-producing unit, and (*4*) a device to collect water vapor as it evaporates from the food. The vacuum shelf dryer, the simplest type of vacuum drier, is used to process a wide range of fruit products, including liquid, pastes, discrete particles, chunks, slices, and wedges. The equipment consists of a vacuum chamber containing a number of shelves arranged to supply heat to the product and to support the trays on which the product is loaded into the chamber. The shelves may be heated electrically or, more often, by circulating heated fluid through them. The vacuum chambers connect to suitable vacuum-producing equipment, located outside the vacuum chamber, which may be a vacuum pump or a steam injector.

Another essential part of a vacuum dryer that has a vacuum pump is a condenser, which collects water vapor to prevent it from entering the pump. A steam injector, which is often used to create vacuum, is an aspirator in which high-velocity steam jetting past an opening draws air and water vapor from the vacuum chamber.

Because of the high installation cost and operating cost of vacuum driers, they are used only for high-value raw materials or products requiring reduction to extremely low levels of moisture without damage.

Freeze Drying

In freeze drying (also called lyophilization) the moisture is removed from the fruit by sublimation, that is, converting ice directly into water vapor. Therefore, no transfer of liquid occurs from the center of the mass to the surface. As drying proceeds, the ice layer gradually recedes toward the center, leaving vacant spaces formerly occupied by ice crystals.

The processing involves two basic steps: (*1*) the raw fruit is first frozen in the conventional manner followed by (*2*) drying to around 2% moisture in a vacuum chamber while still frozen. The most common type of freeze-drying equipment is a batch chamber system similar to a vacuum shelf dryer but with special features to meet the needs of the freeze-drying process.

Freeze dehydration produces the highest-quality fruit products obtainable by any drying method. The porous nonshrunken structure of the freeze-dried product facilitates rapid and nearly complete rehydration when water is added. The low processing temperatures and the rapid transition minimize the extent of various degradative reactions such as nonenzymatic browning and also help to reduce the loss of flavor substances.

The industrial application of freeze drying to a wide range of fruits has been limited because of the requirements of high capital investment, high processing costs, and the need for special packaging. This process is suitable only for high-value products.

PREDRYING TREATMENTS

Preparation of raw fruit for drying is similar to that of canning and freezing. It includes washing, sorting for size and maturity, peeling, and cutting into halves, wedges, slices, cubes, nuggets, and so on. Alkali dipping is used for raisins and prunes. Blanching is used for some fruits.

Sulfur is applied to help preserve the color of dried fruits such as apples, pears, peaches, and apricots. The FDA generally recognizes several sulfite salts and SO_2 gas as safe (GRAS) for use in foods. Both enzymatic and nonenzymatic browning of fruits could be inhibited by application of sulfites. However, sulfites have been associated with severe allergy-like reactions in some asthmatics, prompting the FDA to limit sulfites to certain categories of food products. The food regulation in the United States requires that the presence of sulfites must be declared on the label when the sulfiting agents have a functional effect or are present at a detectable level, defined as 10 ppm or more.

Although sulfites are allowed in dry fruits, their use is unpopular to many consumers. Therefore, alternative treatments to retard enzymatic browning and other oxidative reactions during drying have been investigated. Probably the best-known alternative to sulfite is ascorbic acid and its isomer erythorbic acid. These compounds are added to syrups or applied by dipping the fruit in solutions containing the browning inhibitor, sometimes in combination with an organic acid such as citric acid and a calcium salt. Browning inhibitor penetration may be enhanced by vacuum infiltration treatment, which also removes air from the fruit tissues.

Current alternatives to sulfites are not equivalent to sulfites in effectiveness, cost, or functionality. Because sulfites serve multiple functions, for example, inhibiting enzymatic and nonenzymatic browning and controlling growth of microorganisms, a sulfite replacement is likely to contain several components having different and complementary functions.

POSTDRYING TREATMENTS

Treatments of finished fruits vary with the kind of fruit and intended use of the products. The addition of an anticaking agent, usually calcium stearate at less than 0.1% concentration, is necessary to prevent caking of most dehydrated fruits.

Sweating of dehydrated fruits is a treatment to equalize moisture of the batch. It is usually accomplished by keeping the dried fruit in bins or boxes. Bins are also used for secondary drying to reduce moisture levels of particulate fruits from 10–15% to 3–5%, a range at which drying rates are limited by slow diffusion of water. Temperatures of about 40°C and airflow provided by a blower fan, of about 33 m/min, suit the nearly dry product. These conditions minimize the risk of heat damage at a stage when fruit products are most susceptible to degradation.

Screening is often required to remove the unwanted size portion of the dried product. Removal of unwanted size pieces, or fines, is usually accomplished by passing the dry product over a vibrating wire cloth or perforated metal screens and collecting the fractions separately. Fines may be used in other products, or sometimes represent a loss.

The acceptable fraction passes onto the final inspection operation to remove foreign materials, discolored pieces, or other imperfections such as skin, carpel, or stem particles. Manual and visual selection of most dehydrated products is necessary and is carried out by inspectors while the product is moving along on a continuous belt. In addition to inspectors, magnetic devices are usually installed over the belt to remove ferrous metal contaminants.

PACKAGING

The shelf life of a dehydrated fruit product is influenced to a large extent by its packaging, which must protect the dehydrated product against moisture, light, air, dust, microflora, foreign odor, insects, and rodents. To avoid caking, air-conditioned and dehumidified rooms (below 30% RH) are required to package low-moisture fruits with high sugar content.

Freeze-dried fruits must be packed in inert gas to ensure storage stability. Nitrogen gas is commonly used to extend the storage stability of oxygen-sensitive products. In inert gas packaging, a headspace oxygen level of 1 to 2% is targeted.

PROCEDURES FOR SELECTED COMMODITIES

Apples

Apples are either dried immediately after harvesting or after being stored in cold or controlled-atmosphere (CA) storage until a convenient processing time occurs. The best varieties for drying are Gravenstein, Pippin, and Golden or Red Delicious apples. Two types of dried apple products are recognized under U.S. standards: (*1*) evaporated apples and (*2*) dehydrated (low-moisture) apples.

Evaporated apples, also called regular moisture or dried, are cut to desired size and dried to average, not more than 24% moisture by weight. Unsulfured apples however, must average not more than 20% moisture. Only artificial dryers such as kiln, tunnel, or continuous belt dryers are commonly used to produce dried apples. The process involves sizing, coring, and slicing the fruit pieces to 0.95 to 1.3 cm (3/8–1.5 in.) in thickness. The fruit is usually dipped into a sodium sulfite solution, then dried in a kiln, tunnel, or continuous belt dryer to approximately 16 to 25% moisture. The fruit slices during drying are exposed to the fumes of burning sulfur. Drying in the kiln requires 14 to 18 h at 65 to 74°C (150–165°F).

If a tunnel dryer is used, the air entering the tunnel is at 74°C (165°F) with a relative humidity of 35%. At the outlet, the air is at about 54°C (130°F) with a relative humidity of 35%.

Evaporated apples can be stored for short periods of time, less than three months, at ambient room temperatures in a dry atmosphere. For prolonged storage, 7°C (45°F) or less is required. Unsulfured evaporated apples require 4 to 5°C (39–40°F) cold storage. Evaporated apples will generally reconstitute with one part apples in five parts water by weight.

If the product is processed to a low-moisture state (less than 5% moisture), the dried apples are cut to the desired size, usually 0.64 and 0.85 cm (1/4 and 3/8 in.), dice. Frequently the fruit pieces are instantized by compression or perforation prior to dehydration. Dehydrated (low-moisture) apples are processed in forced-air dryers, such as the continuous belt dryer, using evaporated fruit as raw material. Some apples are vacuum dried for snack or other specialty product applications.

Only 300 ppm SO_2 is necessary to prevent browning. To prevent caking, 0.5% maximum calcium stearate may be added. Packaging is generally in fiberboard boxes with a net weight of 15 to 40 lb (6.8–18 kg), depending on the product destiny.

Approximately 100 kg of fresh unpeeled fruit will yield about 10 to 11 kg of dehydrated apples. Dehydrated apples will generally fully reconstitute with one part of apple in six parts water by weight. The maximum allowable SO_2 level in dried apple products in the United States is 1000 ppm and a maximum of 500 ppm in the European Community.

Apricots

Apricots are picked for drying from mid-June to early July when they are fully tree-ripened. The best varieties for drying are Royal, Blenheim, and Moorparks. California produces more than 90% of the U.S. crop and 10% of the world crop.

Apricots are either sun-dried or dried in tunnel driers. The fruit is halved, the pit is removed, and the fruit is placed cup up on a flat wooden tray. The filled trays are exposed to sulfur dioxide fumes for about 12 h. After sulfuring, the trays are transferred to a field, exposing the fruit to full sun. Apricots are allowed to dry in this manner for one day, then the individual trays are transferred to a shady area and stacked 3 to 4 ft high. They are allowed to dry in the stack for approximately one week, then they are removed from the trays, placed into boxes or bins, and ultimately delivered to a packing plant.

Berries

Berry flavor and color degrades with exposure to prolonged elevated temperatures. Also, the juices of both strawberries and raspberries tend to drip during heating. Therefore, conventional air drying techniques, such as continuous belt driers are not used for drying berries.

Freeze drying has been the typical method for drying strawberry and raspberry pieces. The berries are individually quick-frozen and then spread onto drying trays. The freeze-dry chamber is filled with the trays, and the shelf temperatures during the cycle may be varied to maintain an optimal product temperature.

The packaging must be a high-moisture barrier film, such as foil-paper laminate or foil polyethylene laminate, because with exposure to air freeze-dried berries absorb moisture very quickly. Cereal and confection products often contain freeze-dried berries.

Drum dryers are also used to dry strawberry and raspberry purees. Because of high temperatures (>220°F, >105°C) for a time period of 1 to 5 min, the color is often browner and flavor is more caramelized than those of freeze-dried products.

Cranberries are being dehydrated and sold as a snack food and are being used in mixtures of other dehydrated fruit as a snack. The Ocean Spray Company has developed *Craisin*, which is a sugar-induced cranberry that is dried and used in baked and cereal products or sold as a snack in its own package.

Blueberries are dried by tunnel dryers and freeze-dryers. For both processes individually frozen berries are used as the raw material. The use of dried blueberries is gaining acceptance in cereals and baked products. The drying of currants and gooseberries is mainly a home industry due to lack of volume production of these fruits.

Dates

Dates are almost exclusively sun-dried, as a favorable dry climate exists in the areas where date palms are grown. Such areas as Southern Iran, Iraq, and Egypt have a very high daily temperature during the growing and harvesting seasons and thus are ideal for natural sun drying of dates. In Arizona and Southern California, where dates are also grown, they are seldom dehydrated except in the preparation of low-moisture date products. In such cases, the dates are cut to small pieces or extruded to paste and dried in vacuum dryers.

Figs

The principal varieties grown for drying are Calimyrna, Mission, Adriatic, and Kadota. Figs are usually allowed to dry partially on the trees. In some cases, the trees are lightly shaken at intervals. Figs are usually mechanically gathered from the ground and are typically dry enough to be loosely packed in boxes or bins. Sometimes they are further dried on trays in the sun to a moisture content of approximately 17 to 18%.

After being transported to the plant in sweat boxes or bins, the figs are normally screened, graded for size, and then inspected to remove insect-damaged fruits. The screening is necessary to divide the figs into the required finished product styles. After the first sorting and grading operation, the figs to be stored for later processing are packed in boxes and placed in an airtight chamber and fumigated. This operation is repeated several times over a two-week holding period.

The figs to be processed are conveyed through a cold-water reel washer to remove dust and foreign material. They are then directed to a processing unit where they are immersed in hot water for 5 to 10 min. Soak time depends on the size and variety of fruit being processed. The figs at this point have adsorbed some water and, because of increased susceptibility to mold, are sprayed with potassium sorbate. They are conveyed, typically, over a dewatering belt where they may either be put into small plastic tubs to equilibrate or placed into retorts directly. The retorting process includes exposure to live steam for 2 or 3 min, which further softens the figs. The fruit is air-cooled and packaged.

Peaches and Nectarines

The primary peach for drying is the Lovell, a highly colored freestone variety. Faye Elberta is also dried. Sungrade and Le Grande nectarines are preferred for drying; both are freestone fruits.

Freestone peaches and nectarines are harvested and processed similarly to apricots. After sulfuring, the fruits are placed in full sun for two to three days or longer, depending on weather conditions, at which time they are transferred to shady stack storage and dried for several more weeks. Finally, they are removed from the trays, transferred to boxes or bins, and delivered to the packing plant.

Dehydrated (low-moisture) apricots and peaches are processed from the sun-dried (evaporated) fruit to a limited extent. Because of the high sugar content and the sensitivity of the yellow pigment to heat, a vacuum process is necessary to dehydrate these fruits to less than 5% moisture. Vacuum shelf dryers are used for the process. The evaporated fruit halves are sliced or diced before loading on the drying trays.

Pears

Pears that are to be dried are allowed to ripen on the tree. The summer Bartlett variety is used for drying in California. The fruits are handpicked and transported to cutting sheds where they are cored and halved. Placed cup up on wooden racks, they are stored overnight in sulfur houses where they are exposed to burning sulfur to prevent browning. The pear halves are removed from the sulfur house, dried in the sun for four to eight days, and then transferred to stacked storage for an additional two to three weeks.

Once dried, the fruit is delivered to the packing plant, where it is usually processed to fill orders. The dried fruit from the field may sometimes be stored as long as several years before being repacked. Pears are little known as a dried fruit and most commonly appear in packages of mixed dry fruits.

Prunes

Although plums are one of the most widely distributed fruits in the United States, only one type of plum, the French type, is designated a prune. The La Petite d'Agen variety brought to the Santa Clara Valley of California by a French nurseryman in 1856, known as the California French prune, is used for drying and is grown almost entirely in California. Prunes are the second most important dried fruit crops. The California production accounts for 100% of the U.S. production and 70% of the world's supply.

Most of California's plums for prune production are harvested mostly in late August by machine. The soluble solids content of juice must reach 22% prior to harvest. Immediately after harvesting, the orchard-ripened fruit is taken to the dehydrator yard where it is washed, placed on large wooden trays, and dehydrated to about 18% moisture in forced-draft tunnel dehydrators. The drying process requires 24 to 36 h, depending on the size and solids content of the fruit. Three pounds of plum will yield 1 lb of prune.

Dried prunes are processed through a series of screening, grading, and washing steps. Grading involves separation according to size, ranging from 23 to 150 prunes per pound. Hand sorting for cull removal follows, after which the prunes are conveyed to a blancher where they are held from 8 to 20 min to inactivate enzymes and preserve color and flavor. Potassium sorbate and fresh water are then sprayed onto the prunes to maintain proper moisture content and to add a preservative. Fruit to be pitted is sent through automatic pitting machines that either squeeze the pit out with mechanical fingers or punch it out. The pitted or unpitted prunes are again handsorted for rejects, automatically weighed into boxes or sacks, sprayed with potassium sorbate preservative, and sealed. Other popular prune products include prune juice (a water leachate of the prune) and prune paste, used in baking and confections.

Raisins

The United States and Turkey produce over two-thirds of the world's supply of raisins. Approximately 25% of the U.S. grape production is made into raisins, processed almost entirely in the San Joaquin Valley of California. Raisins are the most important dried fruit crops. Cultivation for raisin grapes is little different than for fresh market grapes.

Thompson Seedless, Muscat, Black Corinth, and Sultan grapes are the principal varieties processed for raisins. The fruit is handpicked in August from the vine in bunches and set on paper to be dried in the sun for 7 to 10 days. At this point, fruit dehydration has not been uniform. Thus, turning of the berries is done to speed the drying process and allow raisins to dry more uniformly. Without turning, some raisins will be too dry, while others will be too moist, requiring a long curing period later on. Within approximately five to seven days after turning, the raisin trays are ready to be retrieved and loaded into sweat boxes to equalize and then to be shipped to the processing plant. The maximum moisture allowed at the packinghouse for incoming raisins is 16%, but usually they are in the 9 to 12% moisture range.

Golden Bleach raisins are dipped in 0.25% hot NaOH (lye) solution after harvest for 3 to 6 s, sulfured for 4 to 6 h, and dried in the sun or in forced-air dryers. Soda-dipped raisins are dipped for 30 to 60 s in a 4% solution of soda ash and Na_2CO_3 at 35 to 38°C and then processed in the same manner as Golden Bleach, but they are not sulfured. From this point, the processing is the same as for natural-dried raisins.

Raisins are processed by a series of screening, destemming, and air separations. These processes are repeated until lightweight particles are removed. Small raisins may then be utilized in distilled alcohol products or in cereal or bakery products.

The raisins are next washed and sent through a dewatering operation, which removes excess surface water. They go through another destemming operation, and their moisture content is adjusted to 18% by water sprays. The fruits are sorted by a mechanical sorter and packaged for storage or shipment. The packages used for raisins are 14-g (0.5-oz) miniboxes, 500-g (1.1-lb) canisters, and 1-kg (2.2-lb) bags, and bulk size packages that contain from 13 to 500 kg (28.6 to 1100 lb) each. Raisins are also sold to industrial users as raisin paste and raisin juice concentrate (70° Brix).

BIBLIOGRAPHY

1. J. G. Brennan, *Food Dehydration*, Technomic Publishing Co., Lancaster, Pa., 1994.
2. L. Calvin, *Fruit and Tree Nuts. Situation and Outlook Report*, USDA Economic Research Service, Washington, D.C., 1996.
3. A. I. Liapis, "Freeze Drying," in A. S. Mujundar, ed., *Handbook of Industrial Drying*, 2nd ed., Marcel Dekker, New York, 1995, pp. 295–326.
4. M. R. Okos et al., "Food Dehydration," in D. R. Heldman and D. B. Lund, eds., *Handbook of Food Engineering*, Marcel Dekker, New York, 1992, pp. 437–562.
5. G. M. Sapers, "Browning of Foods: Control by Sulfites, Antioxidants, and Other Means," *Food Technol.* **47**, 75–84 1993.
6. L. P. Somogyi, D. M. Barrett, and Y. H. Hui, *Processing Fruits, Major Processed Products*, Technomic Publishing Co., Lancaster, Pa., 1996.
7. L. P. Somogyi and B. S. Luh, "Dehydration of Fruits," in J. G. Woodroof and B. S. Luh, eds., *Commercial Fruit Processing*, 2nd ed., AVI Publishing Co., Westport, Conn., 1986, pp. 353–405.
8. L. P. Somogyi, H. S. Ramaswamy, and Y. H. Hui, *Processing Fruits, Biology, Principles, and Applications*, Technomic Publishing Co., Lancaster, Pa., 1996.

LASZLO P. SOMOGYI
Consulting Food Scientist
Kensington, California

FRUIT PRESERVES AND JELLIES

The fruit preserve and jelly categories are certainly a major milestone in food evolution. Their history dates back to ancient times, when a confection or a dessert has been documented as a part of the meal. The use of sugar widened the possibilities for preserving fruits. In fact, during colonial times, a jam sometimes formed while fruit was being boiled. This jam or gel formed when the correct proportions of pectin, sugars, and acids occurred. Jam and jelly production was once considered an art; now it is a science. More is known about the components necessary to produce this kind of gel. This knowledge has led to other applications such as stabilized fruit fillings and sauces, processed fruit juices, canned fruits, and frozen desserts and confections.

DEFINITIONS AND STANDARDS

The name preserves covers a broad range of products, including jams, butters, marmalades, and conserves, as well as ordinary preserves. Preserves contain the largest fruit pieces, whereas jams contain smaller pieces that are crushed or chopped with added acid. Fruit butters are made of fruit pulp cooked to a smooth consistency. They are pressed through a coarse strainer and are more concentrated than jams. Scorching can be a problem because of their high viscosity. Marmalades have the characteristics of both jellies and preserves. They contain thin citrus peel or fruit pieces and are chiefly made from citrus fruits, alone or in combination with other fruits. Conserves are similar to jams, except that two or more fruits are cooked together and raisins and nuts can be added. Jellies are in a class by themselves. They are clear sparkling spreads in which fruit juices as the source of flavor, and, in some cases, the thickening agent.

The Federal Standards and Definitions do not differentiate between preserves and jams (1,2). A preserve is minimally 45 parts prepared fruit with 55 parts of sugar and is concentrated to 65% or higher solids, resulting in a semisolid product. Jellies are similar to preserves, with 45 parts of clarified fruit juice and 55 parts of sugar, resulting in a minimum of 65% solids. Both categories can utilize a maximum of 25% corn syrup for sweetness as well as pectin and acid to achieve the gelling texture required. Fruit butters are prepared from mixtures containing not less than five parts by weight of fruit to two parts of sugar.

GELATION-PECTIN MECHANISM

Typical Gel Formations

Gelation, the formation of the polymer network that gives commercial fruit preserves and jellies their texture, depends on four essential ingredients—pectin, sugar, acid, and water—added in the correct proportions. A pectin gel is a system resembling a sponge filled with water. This polymer is in a partially dissolved, partially precipitated state. The chain molecules are locally joined by limited crystallization, forming a three-dimensional network in which water, sugar, and other solutes are held. Some fruits such as tart apples, red and black raspberries, oranges, and cranberries have enough pectin and acid present. Still others, such as ripe apples and plums, contain sufficient pectin but lack enough acid. Pectin or acid must be added when using most fruits. Sugar is always needed when high-methoxyl pectin is used.

Since fruits vary widely with regard to maturity, climatic conditions, and storage, it can be difficult to ensure the proper composition. Fruit should be picked just before processing to ensure taste and texture. It should be picked as ripe fruit in the early morning to ensure quality. Overripe fruit will have reduced sugar quality, and the pectin will suffer molecular breakdown from enzyme activity. If fresh fruit is not available, frozen, cold-pack, or canned fruit can be used for jams and preserves.

The juice of grapes, currants, lemons, sour oranges, and grapefruits contains sufficient pectin and acid for jelly manufacture. Strawberries, rhubarb, and apricots usually contain sufficient acid but may lack pectin. On the other hand, sweet cherries and quinces may lack acid yet have enough pectin. Commercial pectin, either liquid or powder, can be added as a supplement. The viscosity of a fruit juice is an index of its gelling power.

Pectin is found in the flesh, skins, and seeds of most fruits. It can be extracted when fruit is boiled. Pectin is a complex carbohydrate consisting of polygalacturonic acid chains having a wide variety of molecular weights. The chains contain some carboxyl groups that are partially methylated, forming the ester known as pectin.

Generally, a degree of methylation (DM) of 50% divides commercial pectin into two main groups: high-methoxyl (HM) pectins and low-methoxyl (LM) pectins. The LM-pectin group includes both conventional and amidated versions. The HM pectins are the predominate choice for the standard jellies and preserves. The LM pectins can be utilized in low-sugar fruit spreads.

The degree of methylation of pectin has a critical influence on the solution and gelation characteristics of preserves and jellies. The highest DM that can be achieved by extraction of the natural raw material is about 75%. Pectins of DM ranging from 0 to 70% are produced by demethoxylation in the manufacturing process.

The DM of HM pectins controls their relative speed of gelation; hence, the terms slow-set and rapid-set HM pectin. If a higher degree of methylation of pectin is used, the higher will be the pH required for a fast set. A fast set is necessary to suspend fruit pieces and prevent fruit flotation or sinking. A slow set is necessary for a clear jelly, so that air bubbles are removed. The pectin's quality is standardized on the basis of its 150° standard, which means that, under controlled conditions, 1 lb of 150-grade pectin will gel 150 lb of sugar. This method is known as SAG.

Another method for testing pectin gels is the Voland-Stevens LFRA texture, analyzer in which the gel's elasticity is exceeded. The Tarr-Baker gelometer was the original jelly strength tester. However, after it was discontinued in 1965, the Voland-Stevens analyzer was declared an acceptable replacement because of its operational similarity, portability, compactness, reproducibility, and ease of use and calibration (3). This flexible instrument can be attached to a printer to measure elasticity and other attributes of a gel. This method is comparable to the SAG method.

Jellies are usually produced at a pH of 3.1 and jams at 3.3. HM pectin can gel sugar solutions with a minimum of about 55% soluble solids within a pH range of approximately 2.0 to 3.4. For any soluble solids with a value above 55%, there is a pH value at which gelation is optimal for a particular HM pectin and a pH range within which gelation can be controlled.

Sugars have a general dehydration effect on HM-pectin solubility. At higher solids values, there is less water available to act as a solvent for the pectin. Hence, there is an increased tendency to gel. Because gelation relies on the proper balance of soluble solids and pH in the medium, it is possible to compensate for a reduction in soluble solids by reducing the pH. Any HM pectin can gel rapidly or slowly and the rate can be controlled by the soluble solids and pH.

An attempt was made to quantify fruit content in jams by combining chemical composition data, particularly the inorganic elements that are stable to processing such as ash, magnesium, and potassium, with the rheological forces such as yield stress and flow index in a regression analysis that could explain 90% of the variability in fruit content (4).

Novel Gel Systems

The typical manufacturing methods for jams and jellies use the four necessary components: fruit, pectin, sugar, and acid. Some combination pectin sources have been devised. In one (5), an emulsifier is added to the sugar surface and blended with the bulk of the sugar. A very fine pectin is then mixed with acid and combined with the emulsifier–sugar complex. The emulsifier acts not only as a glue, causing the fine pectin particles to adhere to the sugar surface, but it also acts as an antifoaming agent and a dispersing agent in the final gel production.

Another convenient product, one-step pectin gelling composition, has been developed (6). Pectin particles are mixed with moistened coarse sugar particles. Acid may be added in a dry form and will adhere to the sugar, or it can be predissolved in water and sprayed onto the mix. Much less pectin can be used in this method as compared with dry blending fine pectin, acid, and sugar. This is due to the use of larger sugar particles and smaller pectin particles, so that the pectin dissolves faster while the concentration of dissolved sugar solids is retarded.

Instant pectins are technologically optimized pectins with better dispersion and solubility properties. They dissolve rapidly even at low temperatures and work into jam mixes without prior mixing with sucrose or corn syrup (7).

The LM pectin offers another gel formation method for preserves and jellies. However, the gel formed does not conform to the federal standard of identity. The use of this pectin does result in reduced-sugar jams, jellies, and preserves, the LM-pectin group includes both conventional (acid demethuylated) and amidated types. These pectins require calcium ions for gelation, not sugar or acid.

The gelation of LM pectin is controlled primarily by the reaction of a divalent cation with the acid groups of the pectin chains. LM pectin can be used at solids levels as low as 10%. The pH range for LM pectin is 3.0 to 6.0 because the role of acid is minimized. For successful gelation to occur, 50 to 100% of the acid groups must be complexed with calcium. The amidated pectin, which has fewer free acid groups, requires less calcium for gelation and relies on hydrogen bonding between the amide and free acid groups. Amidated pectins form gels that are more rigid than those formed with conventional pectins. Conventional pectins produce a thickening effect and are aptly used for jams and preserves; amidated pectins are used for jellies.

CARBOHYDRATE SWEETENERS

In HM-pectin systems, sugar accounts for more than 50% of the total weight and 80% of the total solids in a jam. It contributes solids; maintains microbiological shelf life; provides sweetness, body, and mouth-feel; contributes to gelation; and adds color and shine to the jam.

Other sugars that can be used are glucose syrup, dextrose, invert sugar syrup, and honey. When other sugars are substituted for sucrose in jam, the effects on the HM-pectin gelation are as follows:

- Inversion of sucrose reduces gel strength and lowers the gelling temperature.
- Glucose syrup usually reduces the gel strength. High-dextrose-equivalent (DE) glucose syrups decrease gelling temperature; regular-DE syrups increase it.
- Sugar alcohols such as sorbitol and xylitol are used in dietary products. Sorbitol jams can be made with HM pectin, soluble solids of 65%, and a pH of 3.0 Xylitol has limited solubility. At the 39% limit of solubility, gelatin with HM pectin can be obtained when the pH is lowered to 2.7.
- Maltitol syrups are used in the manufacture of sugar-free jellies. Suitable selection of the maltitol content needs to complement the acidified and nonacidified gelatin to deliver the appropriate applications (8).

A study was developed to compare some effects of gelling agents and sweeteners in high- and low-sugar-content carbohydrate gels (9). HM pectin, LM pectin, carrageenan, and alginate gels were the gelling agents, while sucrose and high-fructose corn syrup (HFCS) were the sweeteners. Soluble solids ranged from 35 to 65% with polydextrose as the bulking agent. The properties compared were bound water, water activity, syneresis (35% soluble solids), closely simulated HM-pectin gels because of the comparable spreadability properties. The water-binding property of sucrose exceeded HFCS with most gel systems, except where a combination of LM pectin and carrageenan was used. The water binding served as an index for predicting syneresis, or weeping; spreadability; and shear. However, a synergy of carrageenan with pectin or alginate with pectin resulted in increased bound water compared with individual gums.

Another study to improve nonsugar jam systems used the addition of xylitol or saccharin to veltol to improve the color and taste of apricot jam with minimal changes during storage (10). The use of aspartame in fruit spreads was minimally documented until its stability and effectiveness were measured in 1986 (11). A high-performance liquid chromatographic method was the quality control tool for monitoring levels of aspartame. This study predicted an average half-life of 168 days for aspartame in a fruit spread kept at 25°C. It occurs only at higher temperatures.

Gel strength and gelling temperatures of both amide and conventional LM pectins are influenced by the type of sugar used in the gel. Gels prepared with HFCSs have significantly lower gel strengths at all calcium levels than gels prepared with sucrose. However the use of 42- and 62-DE corn syrups give higher gel strengths than sucrose in LM-pectin formulations.

PROCESSING TECHNIQUES

Traditional Process

The traditional process used for preserve and jelly manufacture is the open kettle, batch boiling technique. The boiling process, in addition to removing excess water, also partially inverts the sugar, develops the flavor and texture, and destroys yeast and mold. In jelly manufacture, the fruit is boiled to extract the pectin and destroy the pectin-hydrolyzing enzymes. The juice is then separated by straining or pressing, and the press cake is boiled with more water to obtain more pectin. Pectin deficiency is remedied by the addition of commercial pectin. Added pectin needs to be dispersed with sugar to ensure uniform distribution. Either liquid or dry sugar is added.

A second boiling step is necessary to concentrate the juice to the critical point for gel formation of the particular pectin–sugar–acid system being used (12). Extended boiling causes acid volatilization, pectin breakdown, and losses in flavor and color. Vacuum concentration (50–60°C) produces a higher-quality jelly than atmospheric pressure boiling (105°C). A refractive index reading indicating soluble-solids content is the point at which the concentration stops. A flow diagram for these processes can be found in Reference 13.

The pH of the jelly will determine the set temperature of the pectin. The setting temperature of a jelly at pH 3.0 can be lowered approximately 10°C with rapid-set pectin or 20°C with slow-set pectin by decreasing the acidity to pH 3.25 (14).

For preserves and jams, the same procedure is used except that the fruit pulp is not strained. Rapid-set pectin is preferred to suspend the fruit more evenly and to minimize settling out.

Both products are packed hot, at about 85°C, into containers that are then sealed. Hot sterilized jars with hot sterilized lids and caps can also be used. Once filled, the jars are turned over to heat the lids and then returned to the upright position. The hot water bath technique can be a water bath and boiled from 5 to 15 min, depending on the fruit. This is a better method for deterring mold growth for fruit preserves and jellies. The U.S. Department of Agriculture (USDA) sanctions the use of paraffin or a two-piece metal lid and screwband for sealing jellies, but it highly recommends processing them for extra safety precaution.

The continuous process utilizes a premix for its efficiency. The APV system uses a plate evaporator for jellies. The Alfa-Laval system uses a scraped-surface heat exchanger for preserves because of the fruit pieces involved. The soluble-solids content or Brix degree is monitored by either an in-line unit or an automatic unit with electrical feedback to control the evaporator (15,16).

The filling temperature for these processes should be 85 to 95°C. This will ensure proper setting, fruit distribution, and a sterile product. A rotary multiple-piston displacement machine is used for filling. Speeds range from 100 to 600 jars per minute. Jars are washed and preheated before filling. Capping occurs immediately afterward, ensuring a vacuum seal.

The pack will be sterile in most cases if it is filled at not less than 85°C and capped using a steam flow closure. If a steam flow closure is not used, the sterile pack will require the use of a steam-sterilizing unit to cool it. Jars can be cooled continuously using water sprays of about 60°C to avoid thermal shock. Subsequently, 20°C water is used to finish this process. The jar temperature should be above ambient. The vacuum seal is check by a nonvacuum detector to ensure a hermetic seal. Jars are passed through a visual inspection point to locate and remove jars with unfavorable attributes such as foreign material, floating fruit pieces, and bubble formation. The jars are then passed to labeling machines, packed in cases or trays, and shrink-wrapped.

The order of addition of ingredients is very important. As in vacuum cooking, a slow-set pectin is preferred to limit the change of preset with the pectin. Most manufacturers use pectin solutions that are easily prepared and dissolve much more effectively than powder. The pectin solution can be added before or after concentration of the batch. Addition of the pectin after concentration results in a faster cooking rate due to the lower viscosity of the batch during concentration. For jelly manufacture, it is best to add the pectin solution before cooking is completed. The addition of 15 to 25% corn syrup deters crystallization from occurring due to sugar inversion resulting from low-temperature vacuum cooking (17). Low-sugar jams require less cooking than jellies and can use larger quantities of HM pectin to improve gel quality.

Modern Processing Techniques

A process to replace canning has been developed in Sweden by Alfastar (16). This multitherm process is said to preserve food for several months without chemicals as well as to achieve a fresher-tasting product. The process is rapid, with even heating through the product. The processing temperature is 150°C and can be reached in less than a minute.

The use of enzymes has made the process more efficient. Serum separation can be reduced by using an effective amount of pectin esterase to the aqueous phase and incubating it in the presence of divalent cations (18).

Another modern technique being used today includes extraction of pectin by pretreating the fruit in an electromagnetic field of superhigh frequency (19).

The microwave oven is the latest method for processing jams and jellies for the homemaker (13). An oversized container must be used for this process to avoid boiling over. Fruit, sugar, and some butter are mixed and allowed to stand for 30 min. The butter will help to deter the frothiness that may develop. The mixture is microwaved until it boils, with frequent stirring. It is then cooked for 10 to 13 min more in the microwave. Jams produced from this mixture keep well in the refrigerator for several months. They can also be canned for greater safety.

The no-cooked freezer jams are by far the easiest of these processing techniques. The fruit is mixed with an appropriate amount of sugar. This technique does not rely on pectin as much, because pectin is not heated to form the gel bonding that cooking at high temperatures creates. Lemon juice, if any, is added to the pectin, and then stirred into the sugar–fruit mix. The mixture is placed in sterilized containers and covered with two-piece metal lids and kept at room temperature for 24 h before placing in the freezer. Once opened, it can be stored up to three weeks in the refrigerator.

QUALITY PARAMETERS

The overall quality of fruit preserves and jellies has increased because of improvements in processing techniques, increased knowledge of fruit characteristics, and competitive situations. Fruit quality control is most important because it affects flavor, odor and color of the preserves and jellies.

The following criteria are important for manufacturing a quality product:

- Fruit appearance, ripeness, and solids must be optimized.
- Fruit juice must be clarified properly to ensure a clear jelly.
- Appropriate pectin grade, 120 to 200, must be used.
- Sugar assay and appearance must be appropriate.
- Corn syrup buffering capacity, solids, and appearance must meet minimum specifications.

Processing must be monitored in the areas of appearance, flavor, color, viscosity, pH, and solids. Powder pectin use results in a dark, rich color and stiff gel. In contrast, liquid pectin, which is less concentrated, makes a less stiff product. Typically, high-DM pectin must be conditioned to increase its set time and to optimize the DM. This pectin follows first-order kinetics in its stability.

Jelly quality attributes are similar to those of jam except that jelly is clear and bright, does not contain fruit pieces, holds its shape when unmolded, and cuts easily with a spoon. A stiff jelly is so firm that the mold will retain the mold shape.

Some problems that can occur are the following:

Cloudy jelly. Unclarified juice, underripe fruit, or pouring so slowly into containers that gelling occurs can result in a cloudy jelly.

Color changes. Darkening at the top of the jars can be caused by storing them in too warm a place or by an imperfect jar seal.

Color fading. Fading can occur with red fruits if they are stored in too warm and too bright areas or stored too long. The natural colorants in the fruit are highly susceptible to high temperatures and light. Another possible cause of color fading could be that the processing was not sufficient to either destroy the enzymes that can affect color, or that the processing time elevated the temperature, causing color destruction. Trapped air bubbles can also contribute to the chemical changes caused by oxidation.

Crystal formation. An excess of sugar can "seed" the jelly when HM pectin is used. This excess sugar comes from overcooking, too little acid, or from undercooking the recipe. Tartrate crystals can form in grape jelly, if juice is left to stand in the cold for several hours before being used. Moreover, if the glass interior is scratched, seeding can occur.

Floating fruit. This can result either from undercooking or from not driving off enough water to create the viscous gel necessary to maintain even fruit distribution. Fruit pieces not properly cut or not ripe enough can also lead to floating fruit.

Gummy jelly. Gummy jelly can result from overcooking and creating invert sugar.

Mold. The appearance of mold can be the result of imperfectly sealed jars and airborne contamination, if the full sugar complement was not used. The water activity created makes a favorable environment for contamination from the jars if they were not properly sterilized or simply underprocessed. A change in the appearance and off-odor or fermented smell will not necessarily occur. However, mold is often seen before taste is affected.

Weeping jelly. Syneresis in jelly can be overcome by not overcooking, not storing in a warm place, and using the appropriate amount of pectin or acid.

Stiff or tough jelly. Overcooking or using too much added pectin delivers a tough jelly.

Jelly failures. An improper balance has occurred when a gel is not formed. Inaccurate measurement, insufficient cooking, overcooking, or increasing the recipe prevents the pectin from building its network.

In a study (20), reduced boiling time improved both the aroma and flavor of fruit preserves. However, it was noted that the retention of flavor and color can be protected during the shelf life by means of modified packaging and appropriate storage practices, thus eliminating light and oxygen and storing at 15°C.

CURRENT TRENDS WORLDWIDE

The jam, jelly, and preserve market is expected to reach $1.5 billion by 2000, with the upscale market showing the greatest potential for growth. Gourmet fruit spreads, preserves, jams, and jellies, including more imports and exotic flavors, are the new products. According to *Food & Beverage Marketing*, the estimated growth rate for these products is 3% per year. The nutritional and health benefits take the forms of less sugar and more fruit; no sugar is added because high-sugar fruit juice is used instead of 100% fruit (18,21). A more convenient packaged powdered pectin has sugar added to it to be used as a sugar and pectin mix for preparing jams and jellies at home. Argentina has a line of dietetic jams, the Netherlands markets lower-calorie jams and preserves, and Japan produces a jelly drink.

The packaging revolution has also affected this market in the form of squeezable plastic containers for convenience, trays and containers of layered foil materials that add longer shelf life to products, and contemporary plastic jelly jars and lids featuring dinosaurs.

SUMMARY

Fruit preserves and jellies have become segmented over the last few years and will continue to grow via the gourmet marketplace. They have grown right along with the new processes, continuous operation, aseptic processing, and microwaving. There is still opportunity for growth in refining current gel systems and developing more convenient ones for either commercial or home use.

BIBLIOGRAPHY

1. *United States Standards for Grades of Fruit Preserves (or Jams)*, 4th ed., Dept. of Agriculture Food Safety and Quality Service, Washington, D.C., Jan. 4, 1980.
2. *United States Standards for Grades of Fruit Jelly*, 2nd ed., Dept. of Agriculture, Food Safety and Quality Service, Washington, D.C., Sept. 3, 1979.
3. S. A. Angalet, "Evaluation of Voland Stevens LFRA Texture Analyzer for Measuring the Strength of Pectin Sugar Jellies," *J. Texture Stud.* **17**, 87–96 (1986).
4. E. Costell, E. Carbonell, and L. Duran, "Chemical Composition and Rheological Behavior of Strawberry Jams," *Acta Alimentaria* **16**, 319–330 (1987).
5. U.S. Pat. 4,686,106 (Aug. 11, 1987), R. Ehrlich and R. Cox (to General Foods Corporation).
6. U.S. Pat. 4,800,096 (Jan. 24, 1989), D. DiGiovacchino, R. Carlson, R. Jonas, and S. Marion (to General Foods Corporation).
7. D. Hesse, "Instant Pectins for Use in Jams and Marmalades," *Food Marketing and Technology* **8**, 12–14 (1994).
8. "Sugar Free Jellies and Gums," *Kennedy's Confection* **2**, 9 (1995).
9. D. L. Gerdes, E. E. Burns, and L. S. Harrow, "Some Effects of Gelling Agents and Sweetners on High and Low Sugar Content Carbohydrate Gels," *Lebensm.-Wiss. Technol.* **20**, 282–285 (1987).
10. M. Ragab, "Characteristics of Apricot Jam Sweetened With Saccharin and Xylitol," *Food Chem.* **23**, 55–64 (1987).
11. M. C. Dever et al., "Measurement and Stability of Aspartame in Fruit Spread," *Canadian Institute of Food Science Technology Journal* **19**, 86–88 (1986).
12. M. Glocksman, "Pectins," in *Gum Technology in the Food Industry*, Academic Press, New York, 1969, pp. 159–190.
13. R. W. Broomfield, "Preserves," in *Food Industries Manual*, 22nd ed., AVI, Westport, Conn., 1988, pp. 335–355.
14. R. M. Ehrlich, "Controlling Gel Quality by Choice and Proper Use of Pectin," *Food Product Development* **2**, 36–42 (1968).
15. *Continuous Processing Systems*, Chilton Food Engineering, Chilton Co., Chicago, Ill., 1984, pp. 90–92.
16. U.S. Pat. 4,562,085 (Dec. 31, 1985), F. Ruggiero (to Alfa Laval, Inc.).
17. D. Tressier, "Jams, Jellies, Marmalades, and Preserves, Candied and Glaced Fruits, Fruit Syrups and Sauces," in *Fruit and Nut Products*, Vol. 3, AVI, Westport, Conn., 1976, pp. 76–98.
18. E. C. Nwanekezi and M. Kpolulu, "Characterization of Pectin Substances From Selected Tropical Fruits," *Journal of Food Science and Technology India* **31**, 159–161 (1994).
19. D. O'Belme and S. Egan, "Some Effects of Reduced Boiling Time on the Quality of Fruit Preserves," *Lebensm.-Wiss. Technol.* **20**, 241–244 (1987).
20. M. Kratchanova et al., "Extraction of Pectin From Fruit Materials Pretreated in an Electromagnetic Field of Super High Frequency," *Carbohydr. Polym.* **25**, 141–144 (1994).
21. E. Kratz, "Jams, Jellies, Marmalades" *Food Marketing and Technology* **7**, 5–6, 8, 10, 13 (1993).

GENERAL REFERENCES

R. Baker, N. Berry, and Y. Hui, "Fruit Preserves and Jams," in L. P. Somogyi, D. M. Barrett, and Y. H. Hui, eds., *Processing Fruits: Science and Technology*, Technomic, Lancaster, Pa., 1996, pp. 117–133.

S. A. El-Nawawi and Y. A. Helkel, "Factors Affecting Gelations of High Ester Citrus Pectin," *Process-Biochemistry* **32**, 381–385 (1997).

B. R. Thakur and A. K. Handa, "Chemistry and Uses of Pectin—A Review", *Crit. Rev. Food Sci. Nutr.* **37**, 47–73, 229 (1997).

R. Whistler and J. Be Miller, *Carbohydrate Chemistry for Food Scientists*, Eagen Press, St. Paul, Minn., 1997, pp. 171–210, 217–231.

Marnie L. DeGregorio
Kraft Foods U.S.A.
Tarrytown, New York

FRUITS, SEMI-TROPICAL

ACEROLA

Acerola (*Malpighia glabra*, Barbadas cherry, or West Indian cherry, of the family Malpighiaceae) is native to the West Indies and South America. There are also large plantations in Puerto Rico. The fruit is grown on a large shrub, 10 to 15 ft in height, and resembles cherries. Each bright-red fruit contains three seeds and several vertical furrows. The fruit is the richest known natural source of vitamin C, containing 1000 to 4000 mg/100 g, over 20 times as much as oranges. The juice is often used to enrich the vitamin C content of other fruit products. The tart, fruity fruit and juice are used to make jams, jellies, preserves, and the like.

BAEL

Bael (*Aegle marmelos*, bel, or Bengal quince, of the family Rutaceae) is native to India and is also grown in Southeast Asia. The fruit is grown on a tree up to 40 ft tall and resembles a citrus fruit with a thick rind. The pulp is yellow and mucilaginous when fresh and becomes reddish when dried. The aromatic, pleasant-tasting fruit is eaten fresh or dried and in juice preparations. It is also used in the treatment of diarrhea and dysentary (1).

CALAMONDIN

Calamondin (*Citrus madurensis*, of the family Rutaceae) looks like a small orange. It is grown on a small evergreen shrub often planted as an ornamental. *Hortus III* (2) lists calamondin (*Citrofortunella mitis*) as a hybrid of *Citrus* and *Fortunella*. The fruit is very acidic and rarely consumed fresh, but it makes a desirable, flavorful ingredient for marmalades, jams, jellies, and the like.

CAROB

Carob (*Ceritonia siliqua*, John's bread, locust, or algaroba bean, of the family Leguminosae) is native to the eastern part of the Mediterranean area. The tree bears pods up to 1 ft in length that have a thick, juicy pulp containing up to 50% dry weight of sugar. The large carob seeds are contained in the pulp. The fruit is eaten raw, discarding the seeds, or the seeds can be dried and ground to make a flour for baking. Carob seeds also serve as a source of gum for the food industry.

CHIRONJA

Chironjas are a citrus fruit of the family Rutaceae. Chironja is believed to be a natural hybrid of the orange, *C. sinensis*, and the grapefruit, *C. paradisi*. It was found growing wild in the mountains of Puerto Rico in 1969. It is an excellent fresh fruit and can also be preserved in a manner similar to oranges and grapefruit (3).

FIG

Figs (*Ficus carica*, of the very large family Moraceae) are native to Asia Minor and spread to the Mediterranean area more than 4000 years ago. The fruit is a multiple fruit developing from a whole inflorescence inside a protective coating. The best-known cultivated varieties are the common, or Adriatic, fig and the Smyrna fig. The former produces fruit by parthenocarpy and does not need to be fertilized. The Smyrna fig is fertilized by a small wasp that enters the fruit through a small hole in the end. In the United States, the Smyrna fig is known as the Kalimyrna, and the Italian Dottato type is known as the Kadota. Figs may be consumed fresh or dried, canned, or frozen. They are used in conserves, jams, confections, and baked goods.

GRAPEFRUIT

The origin of the grapefruit (*C. paradisi*, of the family Rutaceae) is uncertain, but it is believed to have developed in the West Indies from a cross between the sweet orange, *C. sinensis*, and the pummelo or shaddock, *C. grandis*. Shaddock is native to Thailand and Malaysia and was brought to the Barbados by a ship captain named Shaddock. Grapefruits became an important import to Europe between the two world wars.

Grapefruit is an important crop, with world production estimated at about 3.5 million metric tons in 1992; the United States produced about half (4). Florida grows more than 80% of the U.S. crop and more than 90% of the export volume. About half of the crop is processed into a variety of products. Grapefruits are divided into two major groups, the white grapefruit, with a pale yellow flesh, and the pink, or pigmented, group. The red color is due to the presence of the carotenoid lycopene, which has a very appealing pink color in fresh fruit or juice but tends to turn brownish on heat processing. Juice is the major processed product from grapefruit. Grapefruit juice sometimes contains high concentrations of limonin, which produces a bitter taste. The content of limonin is often reduced to a desirable level by a solvent-extraction debittering step, but some limonin is retained because at low levels it imparts a desirable flavor. Oil recovered from the outer layer of the skin is an important product. Grapefruit is classified along with lemons and limes as yellow citrus, as opposed to the orange citrus group. Oil from all three is used as a flavoring additive in juice products. Canned grapefruit sections are a minor product compared with juice, but they are still very popular on the consumer market. Pectin may be produced from the peels. The peels are also dried and sold as cattle feed. The peels are usually treated with lime to break down the pectin before pressing and drying. The press juice may be concentrated into citrus molasses and sold as cattle feed.

KUMQUAT

Kumquats (of the family Rutinaceae) are considered by some to be a citrus fruit even though they belong to the genus *Fortunella* not *Citrus*. The two best-known species,

F. japonica (the marumi kumquat) and *F. margarita* (the nagami kumquat) are native to China and cultivated in Japan and China. They are grown in the United States mainly as an ornamental plant. The small fruits (about 1 in. in diameter) have a fruity, acidic flavor and are often used in marmalades, jams, jellies, and preserves.

LEMON

The lemon (*C. limon*, of the family Rutaceae) is the most diversified and most widely grown of the citrus species. Worldwide production in 1992 was estimated at about 3.4 million metric tons (4), and the United States produced about 680,000 tons. The major product from lemons is juice, which is produced in a manner similar to orange and grapefruit juice. Lemon oil from the peels is considered to be superior as a flavoring agent to the other citrus oils because of its high content of aldehydes. The prominent aldehyde is citral, which is a mixture of nerol and geraniol, as compared with the decanol of oranges and the nootkatone of grapefruit. Lemon oil is highly prized as a flavorant for many juice drinks and other food products as well a fragrance for cosmetics.

LITCHI

Litchis (leechee, lichi, or lychee) belong to several species of the family Sapindaceae; they are native to China and also grown in the Philippines and India. The litchi, *Nephelium litchi*, is a true nut, with the inner nut (about 1 in. in diameter) surrounded by an edible white, fleshy aril layer and a red spiny outer coat. The term *litchi nut* usually refers to the whole nut with the dried aril inside; it tolerates shipment very well and is often found in European markets. The edible aril layer is usually eaten fresh, dried, or canned. The inner nut is not eaten (1).

LONGAN

Longans (*Nephelian longan*, of the family Sapindaceae), native to China, are also grown in India and the Malay Archipelago. The fruit resembles the litchi in size, structure, and appearance, except that the outer coat is brown. The edible white, fleshy aril layer is usually eaten fresh, dried, or canned. The inner nut has no value as a food.

LOQUAT

The loquat (*Eriobotrya japonica*, Japanese plum, or Japanese medlar, of the family Rosaceae) is native to China. It is widely grown in Japan and many other parts of the world. The fruit is a pear-shaped yellow pome, up to 3 in. long, with a five-lobed calyx and two to four black stony seeds. The fruit is eaten fresh or in the form of jams, jellies, sauces, or alcoholic beverages. The seeds are also used in cooking.

MANDARIN

Mandarins (*C. reticulata* or *C. nobilis*, mandarin orange, or satsuma orange or tangerine, of the family Rutaceae), native to southern China, is now grown in many parts of the world. The many varieties may be divided into five groups: *C. reticulata* Blanco (common tangerines), *C. unshiu* Marc (satsuma tangerines), *C. deliciosa* Tenore (Mediterranean tangerines), *C. nobilis* Lourerio (King tangerines), and *C. madurensis* Lourerio (calamondin tangerines). The satsumas, with yellow-orange flesh, are the largest and hardiest and are common in Japan and Europe. The mandarins are similar but have orange flesh. The tangerines are smaller, with red-orange flesh. One cultivar that originated in Algeria is being sold under the name *Clementine. Mandarin orange* is the preferred name for all groups, and the names *mandarin* and *tangerine* are often used interchangeably. Some important hybrids associated with tangerines are the tangelo (tangerine × pummelo) and the tangor (tangerine × orange). Tangerines are a major crop, with worldwide production in 1992 estimated at 6 million metric tons and U.S. production at 300,000 tons (5). The fruit is mainly eaten fresh or canned or processed into juice.

MEDLAR

Medlar (*Mespilus germanica*, of the family Rosaceae), native to Persia, is a very old fruit and was well-known by the Romans. The fruit is less appealing than other pomes and is usually used in jams (1).

ORANGE

Citrus fruits, of the family Rutaceae, date back over 4000 years and are believed to have developed on the slopes of the Himalayas in northeastern India. They soon spread throughout Europe and were introduced to the Americas by Columbus. Citrus juices constitute a major portion of the worldwide fruit products and are second only to grapes in volume. Worldwide production of oranges was estimated in 1992 at 36 million metric tons. Japan was the leading producer, at 1.9 million tons, with U.S. production at 310,000 tons (5).

Commercial oranges are divided into two groups: *C. sinensis*, the sweet orange, and *C. aurantium*, the sour, or Seville, orange. The sweet oranges can be subdivided into four groups: common, navel, blood, and acidless, or sugar, oranges. The dominant variety in the common group is the Valencia cultivar, and it is the main citrus variety grown in all the world. The navel orange dominates the second group, and it is usually grown for the fresh trade. Blood oranges are grown primarily in the Mediterranean area and are claimed to be the most delicious of all the citrus fruits. The red color is due to the presence of anthocyanin pigments in addition to the yellow-orange carotenoid pigments found in all the other citrus fruits. The acidless oranges produce fruit that is very low in acid and less desirable for the juice trade. The sour orange group is dominated by the Seville cultivar, which is grown primarily for its peel and used in the manufacture of marmalade. The juice by-product from the manufacture of marmalade can be debittered and added to 100% juice products at the 5% level in the United States.

Oranges for the fresh market are a major outlet, but most growers combine a fresh fruit operation with a processing capability to utilize the fruit that does not have outer skin of the desirable appearance for marketing as fresh fruit. For juice production, the fruit is washed, sorted, and conveyed to a juice extractor. The fruit is conveyed to upper and lower cups that have sharp edges that cut a 1-in. hole in the bottom of the fruit; the inner contents of the fruit are then pushed through. In one pass the fruit is separated into an oil emulsion, juice, peel, and core material. The oil emulsion is washed and centrifuged to collect the oil. The remaining liquid is then recycled through the process. The juice is filtered to remove some of the pulp and sent to the packaging units. The pulp is washed and filtered, and the pulp wash is added back to the juice. In many areas that grow citrus, varieties more suitable for the fresh trade are grown, and they sometimes produce a bitter juice due to the presence of limonin. The limonin content can be lowered by passing the juice through an ion exchange column. In 1990 the U.S. Food and Drug Administration (FDA) approved the debittering process for frozen concentrated orange juice and concentrated juice for manufacturing, thus permitting more efficient use of navel oranges. Citrus oil is an important by-product of citrus processing. The major component (90%) of orange oil is *d*-limonene, which contributes little to the flavor but is a carrier of other flavor components. The most prominent flavor components are the aldehydes, primarily decanol and ethyl butyrate. The excess pulp, core, peels, seeds, and so forth can be dried and sold as cattle feed. Frozen concentrated orange juice is the major processed product, but specialty products such as molasses, alcoholic beverages, and candied peel are also produced.

POMEGRANATE

Pomegranates (*Punica granatum*, of the family Punicaceae) are native to Iran and are grown in large quantities in India and the Far East. The pomegranate was held in great esteem in ancient times but is less popular today. The fruit is round, about 2 to 3 in. in diameter, and has a leathery, red outer skin. The seeds are surrounded individually by a reddish-purple, highly flavorful pulp. The fruit is eaten fresh or, more usually, as a flavorant in ice cream, sherbet, juice drinks, and the like.

PUMMELO

Pummelos (*C. grandis*, pomelo, or shaddock, of the family Rutaceae) are large citrus fruits up to 10 in. in diameter native to Thailand and Malaysia. Grapefruit is believed to be a cross between the pummelo and the sweet orange.

SOUR ORANGE

See *Orange*.

SUGAR APPLE

Sugar apples (*Annona squamosa*, custard apple, or sweetsop, of the family Annonaceae) are grown in the lowland areas of South America. The yellowish-green fruits, up to 3 in. in diameter, are consumed fresh or in juice drinks.

TANGERINE

See *Mandarin*.

UGLI

Ugli fruit, of the family Rutaceae, is a large citrus hybrid with a thick rind produced by a cross between *C. reticulata* and *C. paradisi*. It is cultivated in Jamaica.

BIBLIOGRAPHY

1. B. Brouk, *Plants Consumed by Man*, Academic Press, New York, 1975.
2. The L. H. Bailey Hortorium, *Hortus III*, Cornell University, Ithaca, N.Y., 1976.
3. C. E. Bueso, Soursop, Tamarind, and Chironja, in S. Nagy and P. E. Shaw, eds., *Tropical and Subtropical Fruits*, AVI, Westport, Conn., 1980, pp. 375–406.
4. D. Kimball, "Grapefruits, Lemons, and Limes," in L. P. Somogyi, D. M. Barrett, and Y. H. Hui, eds., *Processing Fruits: Science and Technology. Major Processed Products*, Technomic, Lancaster, Pa., 1996, pp. 305–336.
5. D. Kimball, "Oranges and Tangerines," in L. P. Somogyi, D. M. Barrett, and Y. H. Hui, eds., *Processing Fruits: Science and Technology. Major Processed Products*, Technomic, Lancaster, Pa., 1996, pp. 265–304.

F. J. FRANCIS
Editor-in-Chief
University of Massachusetts
Amherst, Massachusetts

FRUITS, TEMPERATE

APPLE

Apples have been known since the beginning of recorded history. The fruit referred to in the Bible by Adam and Eve is thought to be an apple. Apples were very popular in ancient Rome and Greece. Modern apples developed in southwestern Asia in the area from the Caspian to the Black Sea. The Stone Age lake dwellers of central Europe learned to preserve apples by drying them in the sun. The apple was brought to America by the colonists and soon spread across the new continent. The life of John Chapman, better known as Johnny Appleseed, born in Leominster, Massachusetts, in June 1776, is an American legend. He carried apple seeds with him on his travels west and planted them wherever he went. Apples are the most widely planted of all fruits and are found in nearly all temperate zones around the world (1).

The genus *Malus*, of the family Rosaceae, contains a number of species and literally hundreds of cultivars of both the edible and ornamental varieties. Large apples are decendants of the species *M. pumila*, which originated in southwestern Asia, and hybrids with *M. sylvestris* of Europe. Edible crab apples are derived from *M. baccata*. Ornamental apples are hybrids of the edible species. The taxonomy of modern cultivated apples is so obscured by centuries of breeding and selection by humans that it is very difficult to assign modern apples to any one species of *Malus* (2). Botanically, the apple is a pome fruit developed from an inferior ovary and is derived from the ovary wall and the floral tube. The fleshy mesocarp constitutes the main edible portion. The five cavities each contain two seeds.

World production of apples in 1993 was about 40 million metric tons, with the United States contributing about 5 million tons (3). Production is increasing because production in China, Russia, Korea, Poland, and Romania is believed to be underreported. In the United States, about 55% of the apple crop was marketed fresh and 45% processed. Of the processed portion, about 44% was utilized in juice, 26% was canned, 6% was dried, 4% was frozen, and 3% was used in miscellaneous products such as jelly, wine, and vinegar. All of the current cultivars are used to some extent for processing, and some cultivars are grown exclusively for processing; however, most apples used for processing are salvaged from the fresh market.

Apple juice is sold in many forms. Fresh apple juice, or sweet cider, is produced from sound, ripe fruit that has been pressed and bottled. No form of preservation is used other than refrigeration. In the United States, apple cider refers to the fresh juice, but worldwide, it usually means apple juice that has been fermented. Shelf-stable apple juice has been treated with some form of preservation, usually heat treatment. The processed juice can be in several forms: crushed, with a high pulp content; unfiltered, with a lower pulp content; or clarified. The most popular product in the United States has been treated with ascorbic or erythorbic acid to produce a lighter color, depectinized with a pectinase enzyme, and filtered before being pasteurized to produce a clear juice. The manufacture of applesauce is a relatively simple procedure. Apples are washed, sorted, and chopped. Sugar is added, the mixture is cooked, and the skins and seeds are removed with a screen extractor. The puree is then canned or bottled. For sliced apples the fruit is washed, graded, peeled, cored, and sliced. The slices are placed in a container, and a vacuum is applied to remove the air from the slices. The vacuum is broken by injecting a solution of water, salt, ascorbic acid, and/or sugar. The slices are then put into containers, sugar syrup is added, and the containers are steam vacuumed, closed, and thermally processed. For frozen slices, the apple slices are vacuum treated, blanched, put into 35-lb containers, and frozen. For dried apple slices, the slices are prepared as noted previously and treated with sulfur dioxide, or one of its salts, to maintain a light color and to minimize enzymatic activity. Two types are recognized in the United States: Evaporated apple slices have not more than 24% moisture, and dehydrated slices contain not more than 3.5% moisture. A number of specialty products are produced from apples, such as glazed apples, spiced crab apples, apple butter, apple jelly, apple vinegar, and baked apples. Apple butter is similar to applesauce except that it is produced with slower heating and the final product is darker in color, more caramelized, and thicker. Apple jelly is made from concentrated apple juice.

APRICOT

Apricots are native to China and have been cultivated for more than 4000 years. They were well-known in ancient Greece and Rome, having followed the trade routes westward. From Europe, the apricot was introduced to America on the ships of the early explorers and settlers. Worldwide production in 1990 was estimated at nearly 700,000 metric tons (4). U.S. production was estimated at about 123,000 metric tons, with 97% produced in California.

The apricot belongs to the genus *Prunus* of the family Rosaceae. It is a stone fruit, in which the seed is encased within a hard, lignified endocarp referred to as the stone. Most commercial apricots belong to *P. armeniaca*, but several other varieties (*P. siberica* and *P. mandshurica*) are known. *P. mume*, known as the winter-flowering plum in Asia, has several cultivars. There is considerable genetic diversity in the apricot, but nearly all commercially important cultivars in the United States are derived from *P. Armeniaca*.

Before World War II, most apricots in the United States were sun dried, but production of canned apricots currently exceeds that of dried products. About 16% of apricots are marketed fresh and about 10% frozen. For canned apricots, Patterson and Tilton are the varieties of choice. The apricots are either picked by hand or shaken off the trees mechanically and transported to the cannery. The fruits are washed in chlorinated water; extraneous matter such as leaves, sticks, immature fruit, and the like are removed; and the fruit is conveyed to the cutters. The fruit is cut along the sutures, the pits are removed, and the fruit is placed in containers. Syrup consisting of light or heavy sugar syrup, apricot juice, or pear juice is added, and the containers are thermally processed. For sun-dried apricots, Blenheim is the cultivar of choice because of its superior flavor, but Patterson and Tilton are also used. Sun drying is a method of preservation limited to climates with hot sun and dry atmospheres. There are six basic steps: (*1*) select and sort fresh fruits, (*2*) wash, (*3*) cut into halves and remove pits, (*4*) place fruit cut side up on drying trays, (*5*) treat with burning sulfur or gaseous sulfur dioxide, and (*6*) place trays in the drying yard in full sun. The sulfur dioxide preserves the color of the dried fruit by minimizing enzymatic browning and reduces degradation of carotene and ascorbic acid. The fruits usually take 5 to 10 days to dry to a moisture content of 15 to 20%. The trays are then removed from direct sunlight and stacked to allow the moisture content to rise to about 27%. The fruits are then placed in boxes for 2 to 3 weeks and allowed to equilibrate to the desired moisture content. Once cured, the apricots are graded and packaged for sale. Frozen apricot slices are produced by washing, grading, cutting, slicing, and freezing the fresh fruit.

BILBERRY

Bilberries (*Vaccinium myrtillus*, of the family Ericaceae) are native to northern parts of Eurasia and are also called whortleberries or blaeberries. The small (8 mm in diameter), dark blue fruits are closely related to blueberries and are eaten raw or as jams, jellies, and preserves. They are also used in bakery goods.

BLACKBERRY

Blackberries, of the family Rosaceae, belong to the genus *Rubus*, which, particularly its subgenus *Eubatis*, is highly genetically heterogeneous. A series of species and subspecies makes it difficult to identify the heritage of many of the current cultivars. The cultivated blackberries of North America are divided into five groups: (*1*) the erect or nearly erect types of the eastern United States, (*2*) the eastern trailing types, (*3*) the southeastern trailing types found along the Atlantic and Gulf Coasts, (*4*) the trailing types of the Pacific Coast, and (*5*) the semitrailing types of the Pacific Coast. Trailing blackberries are also called dewberries, running blackberries, or ground blackberries. A number of thornless blackberry cultivars have been introduced and are sometimes called boysenberries, loganberries, or youngberries. All blackberries consist of a collection of drupelets that are partly fleshy and partly hard (the seed). Blackberries can be distinguished from raspberries because when a raspberry is picked the core stays on the bush. When a blackberry is picked, the core remains with the berry and becomes part of the edible fruit. A tayberry is a cross between a blackberry and a red raspberry

Approximately 90% of the blackberry crop is processed, perhaps because the fresh fruit has a storage life of only 4 to 5 days. In earlier years, blackberries were more popular as fresh fruit rather than being processed because of the difficulty in handpicking canes with abundant thorns. The introduction of mechanical picking machines changed the economics of harvesting, and blackberry products became much more popular. Canning and freezing are the main methods of preservation. Juice production is increasing along with the national increase in consumption of fruit juices.

BLUEBERRY

Blueberries belong to the family Ericaceae, subfamily Vacciniaceae. True blueberries belong to the ancient genus *Vaccinium*, subgenus *Cyanococcus*. Blueberries are divided into two major groups, lowbush and highbush. The lowbush type generally grows wild over large areas and is the source of a large quantity of fruit on the commercial market. Northern lowbush blueberries consist of several species: *V. myrtilloides, V. angustifolium, V. lamarkii*, and *V. vacillans. V. angustifolium* is gradually replacing the others in areas where rotational burning is used for weed control and pruning. The northern highbush blueberry is *V. corymbosum*, and the southern type is *V. australe*. The rabbit-eye blueberry is *V. ashei*, so named because when the fruit begins to ripen the pink color resembles a rabbit's eye. It is grown mainly in the southern states.

Blueberry production in the United States was about 85,000 tons in 1993 and is increasing rapidly. Production increased 50% between 1992 and 1993 alone. About 60% of the blueberries were processed and 40% sold as fresh berries. The rapid increase was due to the addition of blueberry products to a wide variety of foods, including breakfast cereals, dessert toppings and fillings, pies, cakes, cheesecakes, salads, muffins, fruit cocktails, yogurts, ice creams, breads, bagels, pastries, frozen muffin and cookie doughs, and the like. The forms of blueberries used in these products are fresh, frozen, puree, juice, dehydrated, and concentrate.

CHERRY

Cherries, of the family Rosaceae, are believed to have originated in the Caspian and Black Sea area, and wild trees inhabit all of Europe (5). They were domesticated in Greece as early as 300 B.C. The colonists brought them to America. Cherries are drupe fruits with a stone center. The two main types of cherries are sweet (*P. avium*) and pie, tart, or sour (*P. cerasus*). U.S. production in 1993 was estimated at 170,000 tons for sweet cherries and 160,000 tons for tart cherries. Several preservation methods are available. Traditionally cherries have been canned. Canning involves soaking the cherries in cold water to firm them for the pitting operation. After pitting, the cherries are put into cans or jars, syrup is added, and the containers are thermally processed. Tart cherries are usually pitted, but most sweet cherries are not. Frozen cherries are usually pitted and blanched before filling and freezing. Brining involves soaking the cherries in a 1% solution of sulfur dioxide, or one of its salts, which inhibits microbial growth and enzyme activity. It also bleaches the cherries to a pale yellow color. Brined cherries can be stored for a long time and used for jam and jelly production. The red color returns after removal of the sulfur dioxide. In the past, this was a very popular method of preservation, but it has been replaced by freezing because the frozen product offers a superior flavor in the final product.

Maraschino cherries are produced from brined cherries by soaking the cherries in a 0.5% solution of calcium chloride to firm the fruit, soaking in a sodium sulfite solution for further bleaching, and neutralizing with a sodium bicarbonate solution. After addition of a red colorant, usually erythrosine, the cherries are soaked in a solution of citric acid to fix the color. The cherries are then put into containers, a syrup flavored with benzaldehyde is added, and the containers are thermally processed. Cherries are also pressed to produce single-strength or concentrated juice, which has been well received as a colorful and flavorful food ingredient. Cherries are also dehydrated to make a desirable component of bakery products. Cherries lend themselves to many specialty products such as pie fillings, glazed products, wines, sauces, juice powders, spreads, candies, and flavors (5).

CHOKEBERRY

The black chokeberry, also known as chokecherry (*Aronia melanocarpa*), is a member of the family Rosaceae. It is a native North American plant introduced into Europe in the late eighteenth century. It produces clusters of small stone berries that, particularly in Poland, have been used to make jams, jellies, juices, and wines. Chokeberry juice is very astringent and has been used to improve the taste of apple products. Its astringency is not unlike that of cranberries, and it may find a place as an adjunct for flavoring products that now use cranberries. Chokeberries have also been suggested as a source of anthocyanin pigments for applications as natural colorants (6). Chokeberries are a very minor crop in the United States.

CRANBERRY

The cultivated cranberry (*V. macrocarpon*, of the family Ericaceae) is native to peat bogs in many northern states from Massachusetts to Minnesota and some Canadian provinces. The wild cranberry, *V. oxycoccus*, is native to the Pacific states but not the eastern states. It is not grown commercially because of its small berry size. Another species, *V. vitis-idea*, lingonberry, is grown on the upland areas of Scandinavia and Alaska. It is sometimes called mountain cranberry, European cranberry, partridgeberry, or foxberry and is used locally for jams and jellies. These are not to be confused with the highbush cranberry, *Viburnum trilobum* or *V. opulus*; it is primarily grown as an ornamental shrub, but the berries are sometimes used for jellies. North America produces nearly all of the world crop of cranberries, and production in 1998 was estimated at 300,000 tons, of which 95% was processed (M.S. Starr, personal communication, 1999).

Cranberries are one of the three fruits native to North America, the others being blueberries and the Concord grape. Cranberries were part of the American diet long before recorded history in the United States. Native Americans made pemmican from dried meat, fat, and cranberries. The cranberries probably contributed a pleasant taste and appearance to a food that must not have been very palatable. They also probably acted as a preservative by raising the acidity of the product, and the benzoic acid content may have had some antimicrobial action. Pemmican may have been the original trail food.

The cranberry is a low-growing, trailing, woody, broadleaf, evergreen plant that is grown in bogs and swamp areas generally unsuited for other types of agriculture. A plentiful supply of water is required for irrigation and flooding to prevent frost damage before harvest. In earlier years, harvesting was accomplished by hand rakes to collect the berries. Today, mechanical harvesters are used on both dry and flooded bogs. Flooding the bogs before shaking the berries off the vines produces a bigger yield because the berries float and can be skimmed by harvesters. Dry harvesting is usually reserved for the fresh trade, and wet berries are usually sent to freezers for later use. North America currently supplies nearly all of the world's cranberries, but test plantings are being developed in Chile, the Ukraine, Poland, and central Europe.

Cranberries are used to produce three major products. Cranberry sauce is a mixture of cranberries, sugar, and water. The mixture is cooked, passed through a strainer to remove the skins and seeds, poured into cans, and thermally processed. Cranberry whole sauce is a similar product, but it contains the skins and seeds. Both products form a firm gel because of the high pectin content. For cranberry juice, frozen cranberries are thawed and pressed to collect the juice. Commercial juice, as marketed for consumption, contains about 20 to 30% juice, sugar, and water. The press juice has to be diluted because full-strength juice is too astringent and does not make a palatable product. Cranberry juice and a number of combinations with other juices such as raspberry, orange, apple, blueberry, and prune have proved to be very popular in the recent rapid increase in fruit juice consumption by the American public. The popularity of the fruit juices has led to increased demand for cranberry juice concentrate, which is made by filtering the press run juice and passing it through an evaporator. Cranberry juice concentrate also provides a very concentrated source of red anthocyanin pigments, which make a very attractive red color (7). Fresh and dried cranberries have been added to muffins, cakes, cookies, breads, fillings, toppings, and a variety of specialty products. One cranberry product, Craisins, is similar to raisins.

CURRANT

Cultivated currants, of the family Saxifragaceae, are currant-bearing shrubs grown in the temperate and cold regions of both North and South America. Currants are extremely cold hardy, and their culture extends nearly to the Arctic Circle. Red and white currants are *Ribes sativum*, and black currants are *R. nigrum*. The flowering currant is *R. odoratum*. The black currant produces a more vigorous plant with a higher yield and stronger flavor than the other types and is probably more important commercially; however, it is a minor crop. A hybrid of black currant and gooseberry named Jostaberry is even more vigorous and produces fruit that looks like a black currant except larger Jostaberries are attracting the interest of home gardeners. Currants are used to make jams and jellies and as a very flavorful juice concentrate for addition to other juices. Dried currants are not from the *Ribes* genus but are dried grapes from the Black Corinth cultivar of *V. vinifera*. They are sometimes called Zante currants.

GOOSEBERRY

Gooseberries, of the family Saxifragaceae, are closely related to currants and require essentially the same cultural conditions. Gooseberries of American origin are *R. hirtellum*, and those of European origin are *R. uva-crispa*. The fruit may be white, yellow, green, or red and may have a prickly, hairy, or smooth surface. Gooseberries have an acidic and astringent flavor and are used to make jams and jellies. They are a minor crop in the United States.

GRAPE

Grapes, of the family Vitaceae, are the largest fruit crop in the world. The Food and Agriculture Organization of the United Nations estimated the grape crop in 1989 to be about 60 million metric tons, of which about 80% was used in wine production. Grape growing in the United States is divided into four areas: (*1*) the European type (*Vitis vinifera*), grown in California and Arizona; (*2*) the American bunch type (*V. labrusca*), grown in the Great Lakes area (New York, Pennsylvania, Michigan, Ohio, and Ontario, Canada), the Pacific Northwest (Washington, Oregon, and British Columbia), and the Midwest (Arkansas, Missouri, Iowa, Illinois, Indiana, Kansas, and Nebraska); (*3*) French hybrids (*V. vinifera*), grown in Ontario, New York, and British Columbia, and (*4*) muscadine (*V. rotundifolia*), grown in the South Atlantic and Gulf states. Actually, with the rise in popularity of wines, nearly every U.S. state has a grape and wine production area. The French hybrids have an interesting background, particularly with regard to wine. They originated in France as a group of cultivars produced by crossing *V. vinifera* with certain wild American grapes, mainly *V. lincecumii* (post-oak or turkey grape) and *V. rupestris* (sand, sugar, or rock grape). The French breeders hoped to obtain cultivars that combined the desirable qualities of the European cultivars with the hardiness and disease resistance of the American species. The breeding program has been going on for more than 100 years, and a number of cultivars are available, mainly for the wine trade (8). There are literally thousands of grape cultivars available for the grape-growing regions around the world.

Grapes for fresh consumption are grown in many geographical areas, but most are in California. Cultivars of *V. vinifera* dominate the production of table grapes. The most prominent cultivar for juice production is the Concord (*V. labrusca*), New York and Washington led the United States in production of Concord grapes in 1993 with 300,000 metric tons (9). Grapes for table consumption are usually picked by hand to preserve the integrity of the grape bunches. Grapes for juice production are usually mechanically harvested; a number of different machine designs are available, but all of them use beaters or rods to shake the grapes from the vines. The grapes are collected, transferred to bulk containers, and transported immediately to the factory. The grapes are dumped into a holding tank and then transferred by auger to a stemmer-crusher that separates the fruit from the stems. The mixture is transferred to a holding tank, heated to 140°F, and a pectolytic enzyme is added to break down the pectin. After 30 min, a filter aid, such as rice hulls or paper pulp, is added, and the mixture is pressed to remove the juice. The filtered juice is allowed to stand to precipitate the argols (crude tartar). The juice is then racked off, filtered, poured into containers, and thermally processed. Grape juice concentrate can be made from the juice. Cold pressing is similar to the method just mentioned except that the process is conducted at ambient temperature. Grape juice, prior to the tartar removal step, can be pasteurized and stored in large tanks of up to 100,000 gal for future manufacture into jams jellies, toppings, fillings, and the like.

Production of grape juice and grape juice concentrate has recently increased markedly as a replacement for sugar in many drinks on the market. Some of the concentrates are bland in flavor and mix very well with some of the more prominent flavors inherent in other fruits. A large industry has developed for grape concentrate with up to eight times the normal red pigment content of grape juice; this product provides a colorant as well as a desirable mild, fruity flavor. The colorant concentrate is made from selected grape cultivars with a very high pigment content. Both red and yellow concentrates are available (10). Grape colorants (under the generic name enocyanin), prepared from the press cake left over from the wine-making process, have been marketed for more than 120 years (11). The preparation and use of colorants from grapes were described by Francis (12).

HUCKLEBERRY

Huckleberries (*Gaylussacia baccata*, of the family Ericaceae) grow wild on shrubs in North America. The small black fruits grow in clusters and resemble blueberries; they are not true berries but drupes that contain 10 stones. The fruit is used raw or frozen.

KIWI

Kiwi, also called kiwifruit (*Actinidia chinensis*, of the family Actinidiaceae), was developed in New Zealand, where it was known as the Chinese gooseberry. This name did not lend itself to marketing, so the name was changed to Kiwi; it began to appear in U.S. stores in the 1980s. The fruit is oval, about 2 in. in length, with brown fuzzy skin, green flesh, and a refreshing, tart flavor. It is consumed fresh or made into jam, jelly, or wine. It is sometimes dried and used as a meat tenderizer. Kiwi grows on a vigorous, deciduous vine, and hardier types are finding favor in home gardens in the United States. The hardy kiwi, *A. arguta*, produces fruit about 1 in. in length, with a pleasant, tart, slightly phenolic flavor. The hardy kiwi requires both a male and a female plant for fruit production, but another cultivar, Issae, is self-fertile.

LOGANBERRY

Loganberries (*R. loganobaccus*, of the family Rosaceae) are believed to be a cross between an American variety of blackberry and a raspberry. The fruit is eaten raw or in jams and preserves.

MULBERRY

Mulberry trees are well-known in Asia because silkworms feed on their leaves. The mulberry trees (*Morus alba*, of the family Moraceae) grown in the northern United States are large and grown for ornamental purposes, as a windbreak, and also for their fruit, which is very attractive to birds. The trees are prolific in fruit production. The more tender black mulberry (*M. nigra* and *M. rubra*) is grown in the south. The fruit resembles a blackberry and has a fruity but rather insipid flavor. The fruit is used for juice, jam, jelly, toppings, and the like.

PEACH

The peach is considered to be the queen of temperate-zone fruits and is second only to the apple as the world's most widely grown tree fruit. Peaches are native to China and have been grown for over 4000 years. They were well-known by the ancient Greeks and Romans and came to America on the ships of the colonists. Peaches are grown on all continents, but most of the peaches in world commerce are grown in the United States. World production in 1990 was estimated at about 7 million metric tons (4).

All commercial peaches are *P. persica*, of the family Rosaceae. Breeding and selection down through the ages have resulted in numerous cultivars designed for fruit quality, size, yield, and horticulturally desirable characteristics. Canning is the most popular method of preservation for both freestone and clingstone types of peaches. Sound fruit is washed, peeled, pitted, cut, and put into cans or jars. Syrup is added, and the containers are thermally processed. Frozen peaches are prepared the same way except that the containers are frozen. Single-strength juice and concentrate can be prepared by pressing the fruit. Peach nectar is juice and sweetener with enough fruit pulp to produce a more viscous liquid. Dried peaches are prepared by sun-drying the peach halves or by dehydration, in the same manner as apricots.

PEAR

Pears, of the family Rosaceae, are native to Europe and Asia and were well-known by the ancient Greeks and Romans. Breeders in France 400 years ago developed a number of cultivars, and France is still the largest grower of pears in the world. The pear was introduced to England and then to America and soon became an important crop in many countries (13). The largest producers are France, Germany, the United States, Australia, New Zealand, South Africa, Argentina, and Japan. In North America, pears are of three botanical groups: (*1*) the European pear, *Pyrus communis*, which includes all the old standard cultivars; (*2*) the Asian, Oriental, or sand pear, *P. pyrifolia* or *P. serotind*; and (*3*) the Eurasian pear, *P. lecontei*, a hybrid between the previous two. The pear is a pome fruit closely related to the apple, but it never quite attained the popularity of apples, possibly because it does not store as well.

The major processed product is canned pears prepared in a manner similar to canned peaches and apples. Dried pears are prepared as described for peaches and apples. Pear juice, pear juice concentrate, and nectar are prepared by crushing and pressing the ripe pears.

PERSIMMON

Persimmons (*Diospyros kaki*, of the family Ebenaceae) are native to the Far East and were originally cultivated in China and Japan. Today persimmons are cultivated in Europe, the Mediterranean, the United States, and many other warm parts of the world. The fruit resembles a tomato in size and shape and has a fruity, tart, astringent flavor. Another variety, *D. lotus*, cultivated in Asia, is also known as the Chinese date plum. In the United States, another native species, *D. virginiana*, occurs, but its fruit is much smaller and inferior to *D. kaki*. All commercial plantings in the United States are based on *D. kaki* and are known as the Oriental or Japanese persimmon. They are usually eaten raw or processed into juice or nectar.

PLUM

Plums are fleshy stone fruits of the family Rosaceae. The large-fruited European type, *P. domestica*, is the most important type. It originated in the Caucacus Mountains, probably from a doubling of the chromosomes of a cross between *P. cerasifera* and *P. spinosa*. The small-fruited European type, *P. insititia*, includes the purple damsons and the yellow mirabelles. The Japanese plum, *P. salicina*, is next in importance. The Simon plum, *P. simonii*, was introduced from China. A number of plums are native to America, including *P. americana, P. americana nigra* (from Canada), *P. munsoniana* (the wild goose plum), and *P. hortulana. P. angustifolia*, the Chicasaw or sand plum, is native to the southern United States. *P. maritima*, the beach plum, is abundant in the Cape Cod area. *P. subcordata* grows wild in the western United States. Plums have been extensively hybridized; more than 2000 varieties exist, but only a few are commercially important. The large European plums are largely consumed fresh, but many of the other types are canned, frozen, dried, or processed into jams, jellies, preserves, nectars, and so forth. Yugoslavia is the largest producer, with 90% of its crop processed into slivovitz brandy. Germany is second and the United States third, with about 300 tons. California produces 90% of the U.S. crop.

PRUNE

Prunes are dried plums. All prunes are plums, but not all plums are prunes. One group of plums, *P. domestica*, is known as the prune type; it is a major source of prunes, but prune plums are also eaten fresh. Plum designates a variety primarily for uses other than drying such as for fresh consumption, canning, freezing, jams, and jellies. Most plum varieties ferment if dried with the pit; however, if they are dried after removal of the pit, they are called dried plums rather than prunes (14). World production of prunes in 1994 was about 800,000 tons; California produced 70% of the world crop and 98% of the U.S. crop in 1994.

In previous years, prunes were produced by sun drying, but today nearly all prunes are produced by dehydration. Prune juice can be produced by simply leaching prunes with hot water, but recently the disintegration process has become the method of choice. Prunes are cooked in a pressure cooker with agitation to break up the fruit. The juice is then separated by centrifugation or pressing and filtered. The clear juice is then concentrated under vacuum to 19 to 20°Brix and poured into cans or bottles. Prune juice concentrate is made by treating the prune juice with a pectolytic enzyme to reduce the viscosity and concentrating to the desired solids level. At 70°Brix for domestic use and 72°Brix for export, the product is shelf stable and does not require freezing or a chemical preservative. The bulk concentrate in 5-gal pails, 55-gal drums, or tank cars is often used to reconstitute single-strength juice.

QUINCE

Quinces (*Cydonia oblongata*, of the family Rosaceae), native to western Asia, were well-known by the ancient Romans. The fruit is closely related to the genus *Pyrus* and resembles an apple. The fruit is very firm, orange, and up to 5 in. in diameter. Six varieties are known: *C. sinensis, C. lusitanica, C. malformis, C. marmorata, C. pyriformis*, and *C. pyramidalis*. The pomological quinces also include four species of the Asiatic genus *Chaenomeles* such as *C. speciosa*. Most of these are grown as ornamentals because the fruit is unpalatable. The fruit of *C. oblongata* is the most popular and is often added to other fruit because of its high pectin content or used to make quince jam and jelly.

RAISIN

Raisins are a very old product and have been known since biblical times. Raisin production spread from the Middle East to many other parts of the world. The United States and Turkey produce two-thirds of the world production of about 1 million metric tons. Raisins are produced from varieties of grapes (*Vitis vinifera*) that have a high sugar content. The sultanas produced from the Thompson seedless variety of grapes are the most well-known. Grapes to be dried into raisins are picked by hand and spread on paper trays between the rows of grapevines. The grapes on the trays have to be turned to promote even drying. They are then transported to the packing shed for cleaning, grading, adjustment of moisture content, and packaging. Raisins are usually added to bakery products, breakfast cereals, confections, chocolate bars, and the like. Raisins can also be used to make brandy, such as the aniseed-flavored ouzo in Greece or raki in Turkey.

ROSE HIP

Rose hip is the name given to the fruit of a number of *Rosa* spp. in the family Rosaceae, a very large family with more than 100 genera and 2000 species. Nearly all rose hips are very high in vitamin C, with some species (eg, *R. laxa*) containing up to 10% dry weight of vitamin C. This is 200 times that found in orange juice, so rose hips have been promoted as a source of vitamin C. The hips are seldom eaten raw and usually made into juice extracts, jams, jellies, and preserves.

STRAWBERRIES

Strawberries are a major crop with worldwide production estimated at 1,260,000 tons (15). The U.S. produces about 700,000 tons with about 80% of the production in California. Strawberries are not true berries in the botanical sense but are aggregate or multiple fruits. Strawberries belong to the family Rosaceae and the genus *Fragaria*. Many domesticated cultivars are crosses between *F. chiloensis* and *F. virginiana*, also known as *F. ananassa*. The plants are planted one year, bear fruit the second year, and sometimes are allowed to produce for a third year before replanting.

Strawberries are picked by hand and transported in flats for the fresh trade or for processing. The major processed products are frozen purees and puree concentrates for use in jams, preserves, and nectars. Strawberries are also frozen as whole or sliced fruit individually or in blocks, with or without sugar. Strawberries are also canned or made into juice concentrate. Dehydrated berries, fruit preparations, fillings, syrups, toppings, beverages, and wines are other products. Combinations of strawberry juice with many other fruit juices are increasing in popularity due to the trend for increased consumption of fruit juices.

BIBLIOGRAPHY

1. B. J. E. Teskey, and J. S. Shoemaker, *Tree Fruit Production*, AVI, Westport, Conn., 1972.
2. The L. H. Bailey Hortorium, *Hortus III*, Cornell University, Ithaca, N.Y., 1976.
3. W. H. Root, "Apples and Apple Processing," in L. P. Somogyi, D. M. Barrett, and Y. H. Hui, eds., *Processing Fruits: Science and Technology. Major Processed Products*, Technomic, Lancaster, Pa, 1996, pp. 1–36.
4. R. Scorza and Y. H. Hui, "Apricots and Peaches," in L. P. Somogyi, D. M. Barrett and Y. H. Hui, eds., *Processing Fruits: Science and Technology. Major Processed Products*, Technomic, Lancaster, Pa, 1996, pp. 37–76.
5. M. R. McLellan, "Cherry and Sour Cherry Processing," in L. P. Somogyi, D. M. Barrett, and Y. H. Hui, eds., *Processing Fruits: Science and Technology. Major Processed Products*. Technomic, Lancaster, Pa., 1996, pp. 77–94.
6. G. Mazza and E. Miniati, *Anthocyanins in Fruits, Vegetables and Grains*, CRC Press, Boca Raton, Fla., 1993.
7. F. J. Francis, "Cranberries. Effect of Production and Processing on Sensory Properties," in H. E. Pattee, ed., *Evaluation of Quality of Fruits and Vegetables*, AVI, Westport, Conn., 1986, pp. 199–216.
8. J. S. Shoemaker, *Small Fruit Culture*, AVI, Westport, Conn., 1975.
9. J. R. Morris and K. Striegler, "Grape Juice: Factors that Influence Quality, Processing Technology and Economics," in L. P. Somogyi, D. M. Barrett, and Y. H. Hui, eds. *Processing Fruits: Science and Technology. Major Processed Products*, Technomic, Lancaster, Pa., 1996, pp. 197–234.
10. F. J. Francis, "Concentrates as Colorants," *World of Ingredients*, 26–27 (1995).
11. P. Markakis, *Anthocyanins as Food Colors*, Academic Press, New York, 1992.
12. F. J. Francis, *Handbook of Food Colorants*, Eagen Press, St. Paul, Minn., 1999.
13. B. Brouk, *Plants Consumed by Man*, Academic Press, New York, 1975.
14. L. P. Somogyi, "Plums and Prunes," in L. P. Somogyi, D. M. Barrett and Y. H. Hui, eds., *Processing Fruits: Science and Technology, Major Processed Products*, Lancaster, Pa., 1996, pp. 95–116.
15. C. L. Duell, "Strawberries and Raspberries," in L. P. Somogyi, D. M. Barrett, and Y. H. Hui, eds., *Processing Fruits: Science and Technology, Major Processed Products*, Technomic, Lancaster, Pa., 1996, pp. 117–158.

F. J. FRANCIS
Editor-in-Chief
University of Massachusetts
Amherst, Massachusetts

FRUITS, TROPICAL

AVOCADO

Avocados (*Persea americana*, avocado pears, alligator pears, of the family Lauraceae) are native to Central America and are grown worldwide, but they became an important commercial crop in the United States and Mexico in the twentieth century. Worldwide production in 1992 was estimated to be about 1.4 million metric tons, of which Mexico produced about 785,000 tons and the United States 155,000 tons (1).

Avocados are classified into three types:

1. The West Indies type (*P. americana* var. *americana*) is grown in tropical regions. The fruit is large and round, with a thick skin and a yellowish-green pulp color. The oil content is less than 8%.
2. The Mexican type (*P. americana*, var. *drymifolia*) is grown in both tropical and subtropical regions. The small fruit has a thin green skin, and the green pulp has an oil content up to 30% by weight.
3. The Guatemalan type (*P. americana* var. *guatemalensis*) is grown in both tropical and subtropical regions. The green fruit has a thick skin, a high oil content, and a nutty flavor.

Avocados have traditionally been marketed fresh, but a large processing industry developed in the 1960s primarily to produce a spiced puree known as guacamole. The avocados are washed, sorted, sanitized in a hypochlorite solution, and cut to remove the large seed; the flesh is then passed through an extractor to remove the skin. The resultant puree is packed, frozen, and sold to both the consumer and institutional trade for the manufacture of guacamole. Oil can be extracted from the fruit by solvent extraction of the dried fruit, by hydraulic pressing of the dried slices, or by centrifugation of the fruit puree.

Avocados as purchased are usually very firm and have to be allowed to ripen before consumption. They do not ripen on the tree. After harvesting the rate of ripening depends on the maturity of the fruit, the temperature, the levels of oxygen and carbon dioxide, and the humidity. The processors require a uniform supply of ripe fruit, so the fruit is usually treated with ethylene gas in ripening rooms to ensure uniform ripening. For the fresh fruit trade, the fruit may be preconditioned with a short ethylene treatment to ensure more rapid ripening in supermarkets and restaurants. Discoloration after cutting is a major problem for the processors, but it has been minimized by rapid handling and proper selection of cultivars. Avocados are unique in having the highest protein and oil content of any fruit. The U.S. processed market has been estimated at 23,000 tons and will probably increase as Mexican foods become more popular (1).

BANANA

Bananas (*Musa paradisiaca* var. *sapientum*, of the family Musaceae) are native to India and Malaysia and are grown in every wet tropical area in the world. The cultivated varieties are derived from the wild species *M. acuminata* and *M. balbisiana*, and a number of other *Musa* spp. are grown as ornamental plants. The only edible species that thrives outside the tropics, where the temperature never falls below 18°C, is *M. cavendishii*, which despite its delicious flavor is not popular because its small brown fruit is less desirable in appearance. Bananas are the most widely consumed and most highly consumed (in terms of volume) of any fresh fruit in the world, partially due to their ease of shipping and year-round availability.

Musa spp. are large herbaceous plants that grow from an underground rhizome. The shoots, which may be up to 30 ft tall, produce an inflorescence at the top that develops into a stem of bananas. The weight of the bananas, up to 120 lb, causes the stem to bend over. After harvesting, the stem is cut off and left in the field. A new planting will produce a stem in about 18 months, and a daughter stem from the same rhizome will produce a stem in about 12 months. A third stem from the same rhizome may produce a stem in about 8 months, after which the field is usually replanted. The stem is cut while the fruit is immature and green in color, and the fruit ripens during and after shipping. This provides a 2- to 3-week period for shipping, which is very convenient for worldwide delivery. Most of the banana production is centered in Central America, South America, and the Caribbean.

The large fresh fruit industry provides the raw material for the processed industry because fruit that is unsuitable for the fresh trade or is in oversupply is usually processed. The green fruit is transported to the holding sheds, treated with ethylene gas to ensure uniform ripening, and processed. The major product is banana puree, which is made by mechanically peeling the fruit, pureeing the flesh, and deseeding the puree if desired. Actually, bananas have no seeds; the black specks are atrophied seed bags, but they are commonly called seeds. The seeds can be removed with a screen. The puree is then deaerated, which also removes some of the flavor essence. The flavor components are collected and added back to the puree. The puree can be packed as single strength or (more often) concentrated and thermally treated or frozen before packaging. Both single-strength and concentrated puree are used extensively in the baby food industry, as juice components, and in many baking and confectionery products. Banana flakes are prepared by drum-drying the puree. Banana powder is made by grinding banana flakes. Banana "figs" are made by drying slices of ripe bananas. Banana chips are made by slicing green cooking bananas and deep-frying them in oil. A number of other minor products are also made.

BREADFRUIT

Breadfruit (*Artocarpus altilis*, of the family Moraceae) is native to Polynesia. The breadfruit tree was discovered by Captain Cook in Tahiti, and it was thought that the fruit would be a good source of food for slaves in the West Indies. Accordingly, Captain Bligh, on the Bounty, was commissioned to bring seedlings to the West Indies. A mutiny thwarted his first try, but he succeeded on his second trip. He even planted one himself in the botanic garden of St. Vincent in 1793, and apparently it is still there today (2).

The fruits are large, up to 10 in. in diameter, and borne on a large tree up to 60 ft tall. The edible portion is a thick, fleshy layer between the rind and the core. It is usually eaten immature when the flesh is still white and is boiled, baked, roasted, or fried but never eaten raw. Because of its high starch content, the fruit can be dried, ground into flour, and baked into bread, cakes, and a fermented dough product known as mahe. The usual breadfruit has no seeds and is propagated by cuttings, but some cultivars produce seeds, which are called bread nuts. The seeds are eaten boiled or roasted. Breadfruit has been introduced into all of the wet tropical countries, but it never attained the importance envisioned by Captain Cook.

CARAMBOLA

Carambolas (*Averrhoa carambola*, caramba, bliming, or country gooseberry, of the family Oxalidaceae) are native to Indonesia but are grown in all tropical countries. The yellow fruit is up to 5 in. long and 3 in. wide, with five pronounced ribs and a star-shaped cross section. The flesh is translucent, yellow, tart, and juicy and contains large brown seeds. The fruit is eaten raw or made into juices, jams, jellies, and tarts. A related species, *A. belimbi*, known as Belimbi, produces green acidic fruit that is used to make pickles, curries, and preserves.

CASHEW APPLE

The cashew tree (*Anacardium occidentale*, of the family Anacardiaceae) is native to tropical America, but today it is grown chiefly in India, Tanzania, and Mozambique. The tree produces three products of commercial interest. The cashew apple is a pear-shaped fruit about 4 in. long with a very acidic taste. The fruit is eaten raw in the local markets or made into fresh and fermented juices. A protuberance on the end of the apple contains the cashew nut, which is highly prized on the worldwide market. The outer coating of the cashew nut, which is removed before shipping, contains a very corrosive, black oil that is recovered and sold on the industrial market. Cashew nuts can be pressed to produce cashew nut oil, which is a prized flavorant and should not be confused with the oil from the seed coat.

CERIMAN

Ceriman (*Monstera deliciosa*, or monster plant) is native to Mexico and Guatemala and belongs to the family Aracea. This is a very large family with about 15 genera and 2000 species. Several aroids are grown in the tropics for their edible tubers (taro, dasheen, tannia, malonga, and ocumo) and are an important starch staple of the diet. *M. deliciosa* is a robust climbing vine that produces tiny fruit on a spadix up to 10 in. long. The spadix is eaten raw and is reputed to have a delicious flavor resembling that of pineapple.

CHERIMOYA

Cherimoyas (*Annona cherimola*, of the family Annonaceae) are native to South America. The fruit is heart shaped or conical and can weigh up to 1 lb. The soft white flesh contains many beanlike seeds, and the flavor somewhat resembles that of a pineapple. The fruit is eaten raw or processed into juice.

CUSTARD APPLE

Custard apples (*A. squamosa*, of the family Annonaceae), also known as sweetsops or sugar apples, are native to South America. The yellowish-green fruits, up to 4 in. in diameter, have a scalelike covering that resembles a coniferous cone. The edible custardlike, granular pulp is white, and the seeds are embedded in it. The pulp is very sweet, with up to 18% sugar. The fruit is very perishable and hence little known outside the producing regions. The fruit is eaten raw.

DATE

Dates (*Phoenix dactilifera*, of the family Palmae) probably are native to India despite the fact that they have been grown in the Arab countries for many thousands of years. The date palm is grown in all hot, dry areas of the world but primarily in the desert areas of the Middle East and the Sahara. Dates are also grown in California, Arizona, and Mexico. The date palm is a dioecious plant, meaning that both male and female plants must be grown to ensure pollination. The fruit is borne at the top of a very tall (up to 90 ft) tree and is pollinated by suspending a small branch of the male flowers over the female inflorescence. Each individual fruit contains a hard seed called a stone. The seed is surrounded by an edible fleshy endocarp that when dried contains up to 70% sugar. The high sugar content ensures that the fruit will not spoil during storage or shipping. Dry dates are consumed fresh or incorporated into a wide variety of bakery products, confections, dairy products, preserves, spreads, beverages, and other products. The Arabs are believed to have over 800 different uses for the date, perhaps because it is almost the only plant available in the desert areas (2).

DURIAN

Durians (*Durio zibethicus*, of the family Bombacaceae) are native to Malaysia. The fruits, weighing up to 10 lb, are borne on a very large tree up to 90 ft tall. The fruit has a spiny outer covering surrounding cream-colored flesh with a large seed. When ripe, the fruit drops to the ground and soon turns rancid. The durian is noted for the foul odor of the ripe fruit, and only people who can get past the odor can enjoy the delicious taste of the fruit.

FEIJOA

Feijoa (*Feijoa sellowiana*, or pineapple guava, of the family Myrtaceae) grows on a small tree native to South America. The green fruit tinged with red is 2 to 3 in. long. The white flesh is sweet and consumed fresh or made into juices, jams, jellies, and preserves.

GUAVA

Guavas (*Psidium guajava*, of the family Myrtaceae) are native to tropical America and are grown in nearly all tropical areas in the world. The fruit of the best-known guava, *P. guajava* var. *pyriferum*, varies greatly in size, up to 0.5 lb, and color, from white to yellow or red. The fruit from *P. guajava* var. *pomiferum* is smaller and redder. *P. cattleianum* or *P. littorale*, native to Brazil, has small, round, purple fruit. The small purple strawberry guava, *P. littorale* var. *longipes*, is also widely cultivated. The fruit may be consumed raw, canned, frozen, or in the form of juice. It is widely used for jams, jellies, preserves, and confections. The fruit contains about 240 mg/100 g of vitamin C and has been promoted as a good source of vitamin C.

ILAMA

Ilamas (*A. diversifolia*, of the family Annonaceae) are native to South America. The fruit is a medium-sized berry formed by the union of the pistils and the receptacle. The fruit is consumed fresh or in juices, jellies, sherbets, or fruit desserts.

JABOTICA

Jaboticas (*Myrciaria cauliflora*, of the family Myrtaceae) are native to South America and grow on small evergreen shrubs. The grapelike berries may be consumed fresh or as juice, jam, jelly, or preserves.

JACKFRUIT

Jackfruit (*Artocarpus heterophyllus* [Syn. *A. integrifolia* or *A. integra*]) belongs to the same family (Moraceae) as breadfruit. It is a native of India and is grown in almost every tropical area. The trees are very large, up to 70 ft tall, and bear the largest known fruit. The fruit is up to 4 ft long, 1 ft in diameter, and weighs up to 100 lb. It is a compound fruit developed from the entire female inflorescence, with a green coating covered with hexagonal spines. The pulp is composed of many fruits, each surrounding a large seed. The edible part is the red pulp surrounding the seeds, and it has a very sweet but rather insipid flavor. The pulp is consumed raw or as juice, jam, or jelly. The seeds are also edible after boiling or roasting. The immature fruit is consumed as a vegetable, either boiled or roasted.

JAVA PLUM

Java plums (*Syzygium cumini*, of the family Myrtaceae) are native to Asia. The small purplish-red fruit is eaten fresh.

JUJUBE

Jujubes (*Ziziphus jujuba* or *Z. mauritana*, Chinese jujube, or Chinese date, of the family Rhamnaceae) are native to the Mediterranean area. The fruit is small, mealy, deep yellow, and borne in clusters. It is used in sweet pickles, stews, preserves, and confections. It is sometimes called lotus in Greek mythology, not to be confused with the sacred lotus, an aquatic plant, *Nelumbo nucifera* or *Nelumbian nuciferum*, grown in India and China. In Homer's *Odyssey*, lotus probably refers to *Z. jujuba*, but it should be called a pseudocereal or pseudobeverage rather than a fruit (2).

LIMES

Limes (of the family Rutaceae), native to the East Indian Archipelago, are the least cold hardy of the citrus fruits and are grown in tropical areas throughout the world. Three lime species are recognized: the small-fruited *Citrus aurantifolia*, the large-fruited *C. latifolia*, and the sweet lime, *C. limettoides*. Limes are often described by their area of production. The most popular small-fruited type is called the West Indian, Mexican, or Key lime. The large-fruited types are called Tahitian or Bearss and are grown in Australia, Brazil, California, and Florida. The Indian type is the favorite sweet lime and is grown in India and Pakistan. U.S. production of limes in 1992 was estimated at 1.6 million boxes (3).

Limes are very acidic and are seldom eaten fresh. They are much in demand as a flavorant for other products such as juices, jams, jellies, baked goods, and alcoholic beverages. British sailors became known as limeys because they were fed limes for the prevention of scurvy. Despite this folklore, limes, at 37 mg/100 g, have less vitamin C than lemons (2).

MAMEY

Mameys (*Mammea americana*, or Mammey apple) belong to the family Guttiferae and are native to Central America and the West Indies. The fruit is a drupe the size and shape of an orange and with a brown russeted skin. The yellow, edible flesh surrounds a large seed. The fruit is consumed fresh or as jams and preserves.

MANGO

Mangos (*Mangifera indica*, of the family Anacardiaceae) are native to southern and Southeast Asia. It is one of the most important tropical and subtropical crops in terms of production and popularity. The country with the largest production is India, where it has been cultivated for over 4000 years, but it is now grown in almost all tropical areas. India produces 60% of the worldwide crop of about 17 million tons (4).

Mangos grow very large, handsome trees up to 90 ft tall with a spread of 125 ft. The fruit is ovoid or heart shaped, about 3 to 5 in. long, with a skin color varying from green to yellow to red. When ripe, the flesh is yellow-orange and fibrous, with a large central seed. The flavor is fruity with a terpene flavor. One irreverent British visitor described the wild mango as having a flavor resembling that of a piece of cotton batting dipped in turpentine. But centuries of breeding have resulted in hundreds of commercial cultivars with truly delicious flavors. The major processed product is mango puree, which is used as an ingredient for many other products such as juices, nectars, jams, preserves, confections, and baked goods. Mango slices, canned or frozen, are available. Immature, green frozen slices are also popular. Mango leather is an ethnic food in India. It is prepared from mango puree by removing some of the fiber and drying in the sun. The unripe fruit is used in chutneys and pickles.

MANGOSTEEN

Mangosteens (*Garcinia mangostana*, of the family Guttiferae) are native to Malaysia and are found in Java, Sumatra, the Philippines and Sri Lanka. The fruit is a reddish-purple berry about the size of a small apple with segments similar to citrus fruits. The edible portion is the white flesh in the arils surrounding the seeds.

MIRACLE FRUIT

Miracle fruit (*Synsepalum dulcificum*, of the family Sapotaceae) is native to tropical West Africa. The fruit is a small red berry with white flesh. It is consumed fresh with other foods. Its major quality is the ability to make sour foods taste sweet. This effect on the taste buds received much research attention, but the fruit was not commercialized because of lack of U.S. Food and Drug Administration (FDA) approval.

PAPAYA

Papayas (*Carica papaya*, or pawpaw, of the family Caricaceae) are native to tropical America. Brazil is the major producer, with worldwide production in 1992 estimated at 4 million metric tons (4). The papaya plant is a giant herb, rather than a tree, because it lacks woody tissue. The stem grows 20 to 30 ft tall and bears fruit near the stem. The fruit may be oblong or elliptical, up to 20 lb, and orange. The flesh may be orange to red and surrounds a large seed cavity containing many seeds. The fruit is consumed fresh or as puree, which is used to produce juices, nectars, jams, jellies, sherbets, dairy products, and confections Papaya leather is made by drying the puree. Papaya slices or chunks are canned or frozen and used in fruit salads, ice creams, sherbets, and confections.

Another plant is also called PaPaw. This is *Asimina triloba* (Michigan banana, of the family Annonaceae), a small ornamental tree that grows in temperate zones. The fleshy, oblong fruit is about 3 to 5 in. long and has a banana-like flavor. It is eaten fresh or as an ingredient in fruit salads, baked goods, and confections.

PASSION FRUIT

Passion fruit (*Passiflora edulis*, of the family Passifloraceae) is native to Brazil and is grown in all tropical areas around the world. A number of species are grown for their edible fruit: purple granadilla (*P. edulis*), yellow granadilla (*P. laurifolia*), sweet granadilla (*P. ligularis*), sweet calabash (*P. maliformis*), curaba (*P. mollissima*), and giant granadilla (*P. quadrangularis*). The round berry, grown on a woody vine, is about 2 in. in diameter and contains a juicy pulp with many seeds. Single-strength or concentrated juice, frozen or canned, is the major processed product. The flavor is very susceptible to damage by heat; therefore, the frozen product is preferred over the canned product, but recent advances in aseptic packaging have produced superior products. Passion fruit juice is very popular for blending with other juices. The fruit is also used in fruit salads, jams, jellies, and confections.

PINEAPPLE

Pineapples (*Ananas comosus*, of the family Bromeliaceae) are native to tropical South America and are grown in nearly all tropical areas. The plant is a rosette of long, stiff, fleshy leaves that produces a short flower stem. The individual fruitlets fuse to produce a compound fruit. Despite its economic importance, only one cultivar is of commercial importance. The cultivar Smooth Cayenne comprises 95% of the worldwide crop. The major processed products are solid pack (slices, chunks, and tidbits), crushed (dices), and juice. The fruits are handled by a machine called a ginaca, which processes about 100 pineapples a minute into eight fractions: the cylinder, core, skin, eradicator meat from the skin, crown end, eradicator meat from the crown end, butt end, and eradicator meat from the butt end. The cylinder is used to make the high-value products: slices, chunks, tidbits, and crushed pineapple. The eradicator fractions and the core are used to make juice. The skin, crown end, and butt end are used to make by-products. The mixture may be pressed to make mill juice, which can be added to the solid pack or used to dissolve sugar to make a syrup for addition to the solid pack. The residue from the pressing operation is sold as cattle feed. The mill juice can also be used as a sugar source for the production of alcohol or vinegar. Recent research indicates that mill juice can be purified sufficiently to allow the product to compete with grape juice as a sugar base for fruit beverages. Solid pack pineapple, crushed pineapple, and pineapple juice are usually preserved by thermal treatment (5). Pineapples are often consumed raw; many people consider the pineapple to be the most palatable fruit of all (2). They are often used as flavorful ingredients for juice products, jams, jellies, preserves, dairy products, and confections. The pineapple plant is also the source of the proteolytic enzyme bromelin.

PLANTAIN

Plantains (*M. paradisiaca*, or cooking bananas, of the family Musaceae) are closely related to bananas. They are grown in most tropical countries. They have a higher starch content than bananas and are eaten raw or as a cooked vegetable. A flour is often prepared from plantains and added, as a thickener, to a number of foods. The flour itself can be used to prepare baked goods. The fruit can be french fried to make chips or fermented to make alcoholic beverages.

PRICKLY PEAR

Prickly pears (*Opuntia ficus-indica*, tuna, or Indian fig), native to the dry areas of North and South America, belong to the Cactaceae or cactus family. Six other species of *Opuntia* also produce edible fruit. The red to purple fruit, 3 to 4 in. long, is covered with small prickles and is consumed fresh or dried. The fruit is also used in jams, jellies, preserves, and candy.

PULASAN

Pulasans (*Nephelium mutabile*, of the family Sapindaceae), native to Southeast Asia, are another arillate fruit similar to litchis and rambutans.

RAMBUTAN

Rambutans (*N. lappaceum*, of the family Sapindaceae) are native to Malaysia. The rambutan is similar to the litchi in appearance and structure except that rambutans are covered with soft spines. The fruit is eaten raw or canned, primarily for export from Thailand.

ROSE APPLE

Rose apples (*S. jambos*, of the family Myrtaceae) are native to Asia. The small yellow fruits are eaten fresh, dried, or in confections. The species *S. malaccense* is closely related, with larger fruit up to 2 in. in diameter. The fruit is also called Malay apple and is eaten raw, cooked, or in jams, preserves, and wine.

SANTOL

Santols (*Sandoricum koetjape*) are native to Asia. The medium-sized, spherical, yellow fruit has a fruity, sour taste and is consumed fresh or as jams, jellies, and preserves.

SAPODILLA

Sapodillas (*Manikara achras* [Syn. *M. zapotilla* or *Achras zapota*], sapotilla, sapota, zapotle, chiku, or naseberry, of the family Sapotaceae) are native to Mexico and South America and are grown in nearly all tropical areas. The fruit is a fleshy berry, up to 3 in. in diameter, with a rusty brown skin. The pulp is yellowish-brown to red in color, with a mild aroma and a slightly astringent, pleasant taste. The fruit is usually eaten fresh because processing degrades the delicate flavor (6).

SOURSOP

Soursops (*A. muricata*, guanabana, or sapote agrio, of the family Annonaceae) are native to North and South America. The oval fruit weighs up to 3 lb and has a dark green, leathery skin covered with many short, curved, fleshy spines. The white cottony pulp contains many shiny dark brown seeds and has a distinctive flavor and aroma. The fruit can be eaten fresh, but more often the pulp is extracted and used in beverages, sherbets, ice creams, and syrups. The pulp and the puree can be canned or frozen (6).

STAR APPLE

Star apples (*Chrysophyllum cainito*, or caimito, of the family Sapotaceae) are native to North and South America. The fruits are globular, up to 4 in. in diameter, and star shaped in cross section, with a light green to purple skin. The white, translucent pulp contains three to eight seeds, and the fruit is usually eaten fresh.

TAMARIND

Tamarinds (*Tamarindus indica*, tamarin, tamarindo, or imli, of the family Leguminosae) are probably native to East Africa but were introduced into India long ago. The tamarind was well-known by the Egyptians and Greeks of the fourth century. The tamarind grows on a large evergreen tree of the family Leguminosae. The fruit is an irregularly shaped pod about 8 in. long. As the pod matures it turns into a brown, brittle shell containing a dry sticky pulp that surrounds red-brown seeds. Tender shoots and flowers are consumed as vegetables in salads, soups, and curries. Immature pods are used as seasoning for cooked rice, fish, and meats. Pulp from the mature pods is pressed into cakes and is valued as highly as dates and figs. Tamarind pulp is an important ingredient in barbecue sauces and as a fruit base in juices, jams, jellies, preserves, sherbets, dairy products, and alcoholic beverages (6).

TREE TOMATO

Tree tomatoes (*Cyphomandra betaceae*, or tamarillo, of the family Solanaceae) are native to tropical America. The round, juicy, orange-red fruit, about 2 in. in diameter, grows on a 10-ft shrub and has a mild sour taste. The fruit can be eaten fresh but is usually made into jams, jellies, and preserves.

WAX APPLE

Wax apples (*S. samarangense*, java apple, or jambosa, of the family Myrtaceae) are native to Asia. The fruit is a whitish-red berry about 1 in. long with a mild insipid flavor. It is usually consumed fresh.

BIBLIOGRAPHY

1. A. Kurlaender, "Avocadoes," in L. P. Somogyi, D. M. Barrett, and Y. H. Hui, eds., *Processing Fruits: Science and Technology. Major Processed Products*, Technomic, Lancaster, Pa., 1996, pp. 445–458.
2. B. Brouk, *Plants Consumed by Man*, Academic Press, New York, 1975.

3. D. Kimball, "Grapefruits, Lemons and Limes," in L. P. Somogyi, D. M. Barrett, and Y. H. Hui, eds., *Processing Fruits: Science and Technology. Major Processed Products*, Technomic, Lancaster, Pa., 1996, pp. 305–336.
4. J. S. Wu and M. J. Scheu, "Tropical Fruits," in L. P. Somogyi, D. M. Barrett, and Y. H. Hui, eds., *Processing Fruits: Science and Technology. Major Processed Products*, Technomic, Lancaster, Pa., 1996, pp. 387–418.
5. C. E. Mumaw, "Pineapples," in L. P. Somogyi, D. M. Barrett, and Y. H. Hui, eds., *Processing Fruits: Science and Technology, Major Processed Products*, Technomic, Lancaster, Pa., 1996, pp. 337–360.
6. S. Lakshminarayana, "Sapodilla and Prickly Pear," in S. Nagy and P. E. Shaw, *Tropical and Subtropical Fruits*, AVI, Westport, Conn., 1980, pp. 415–441.

F. J. FRANCIS
Editor-in-Chief
University of Massachusetts
Amherst, Massachusetts

FRYING TECHNOLOGY

Frying is a process of dehydrating food from the surface inward. The process typically uses triglyceride-based oil (lipid) from an animal or vegetable origin to transfer thermal energy from a heat source to food immersed in the oil. The efficiency of heat transfer is mediated by surfactant chemical species (wetting agents) included or formed in the oil that control the contact time between hydrophobic oils and aqueous foods. Heat capacity of the oil is primarily determined by the ratio of triglyceride to polymer formed in the oil.

A dynamic balance occurs between water movement from and oil movement into the frying food at a given temperature (typically 150–190°C) and food immersion time. The dynamic balance and the kinetics of the process are further mediated by the state of degradation of the oil as influenced by its exposure to primarily heat, oxygen, water, and chemicals and particles from the frying food. Fatty acid and smaller organic molecules formed in the oil generally increase thermal conductivity of the oil at the food surface.

Three surveys of frying topics covering chemical, physical, engineering, sensory, and nutritional aspects of frying oils and fried food are currently available (1–3).

PROCESS ANALYSIS

Frying is a unit operation composed of multiple subsystems. A fryer is a process reactor wherein chemical and physical interactions are induced to occur between oil and food. Thermal energy is supplied to the reactor by heating the oil in the fryer vat through direct or indirect means, and waste energy escapes to the environment by conduction through the reactor walls and by water evaporation. Raw food containing an excess of water (and sometimes solids) enters the reactor in immersed batches (basketloads) or immersed or submerged continuous (conveyor) streams, and the food emerges in a cooked state separated from waste materials: water (liquid and vapor), organic volatiles, and particulate solids.

Oil in the reactor vat is replenished to make up for losses incurred by adsorption and absorption by the food. Oil is also filtered (cleaned) to remove particulates, some oil degradation chemicals, and interaction by-products formed between oil and food degradation products such as emulsified water. Oil filtration and/or treatment materials are discarded from the oil cleaning and/or filtering apparatus, and the oil is recycled into the fryer until the oil is out of specification for food manufacture; it is then discarded from the process. The process reactor when drained of oil also requires cleaning and sanitation to prevent fouling and to provide a contaminant-free process environment; this step also generates a waste stream of chemicals, particulates, and water.

Process monitoring and control involve thermal and other measurements of the reactor's input and output streams, sampling oil and food to judge product quality and potential shelf life, and sampling waste streams to suggest the extent of needed treatment.

ENGINEERING PRINCIPLES

It is possible to calculate the many aspects of heat and mass transfer most important to the frying process based on the expected loss of water and the mass of food to be dehydrated. Unfortunately, these calculations do not take into account the constant changes occurring in a degrading heat transfer medium (the oil) and the accumulation of surfactant species in the oil due to both food and process influences. Numerical modeling has not yet advanced beyond examining fresh oil model systems.

There are five stages in the life of frying oil that produce, in sequence, raw, cooked, and overcooked food. The following is a description of how frying proceeds and why surfactant chemicals control the kinetics and dynamics of the frying process. From the perspective of physical chemists and process specialists, the cooking of food in an oil can be reduced to simple engineering principles with parallel simple measurement and control procedures. This is a new paradigm of frying and is different from that of the paradigm of organic chemists and food scientists, who initially studied the complexities of frying oil and food chemistry to develop databases of results and observations.

The model of understanding frying in terms of physical chemistry and engineering leads to the belief that the foremost way to judge frying and frying oils is by evaluating the physical properties of fried foods. Only the process variables affecting the physical properties of fried foods can be controlled in the engineering sense. Temperature profiles, water loss from food, and oil absorption into food are amenable to process control. On rare occasions, the pressure over the frying oil is also controlled.

Factors such as flavor development and typical finished food color are not primarily controlled by the process. Rather, they are dependent on the source of the oil, the content and type of surfactants, the type and composition of food fried, and a range of organic reactions, only some of which depend directly on process variables.

Two traditional means of controlling the transfer of thermal energy to frying food are heater temperature control (at designed energy flux) and residence time of the food in the heated oil bath. Overheated oil at the surface of the heaters is reduced to carbon deposits (coke) on the heater surfaces. As the carbon layer builds and becomes an insulating jacket, the heater cycle on times and temperatures are increased to keep the oil at frying temperature. A new frying oil somewhat resists this process, and heat transfer rates and ratios are essentially a static system. As the oil degrades, however, and products such as thermally formed and food-formed surfactants increase, the dynamics of the heat transfer system accelerates. Oil makes successively better contact with the food exterior and thus excessively dehydrates that layer. Thermal energy is expended to convert more and more surface water to steam. This ever-deepening dehydration of the crust phenomenon robs energy that otherwise would go to heat and cook the interior of the food.

The time food spends in heated oil can be varied to achieve a particular degree of cooking. Times vary from blanching for 30 s to cooking a chicken for 20 min. Manual placement and removal of basketloads of food control residence time in smaller fryers. Larger, or continuous, fryers often use conveyor systems to control frying food residence time. Unfortunately, heater designs and residence time variations leave out considerations of the chemical changes that take place in the oil with time. Such changes have been described in depth by analytical and organic chemists, but physicochemical changes that affect heat and mass transfer of the oil with use have been largely, although not completely, ignored. Although electromechanical devices for controlling fryer heaters and residence time are well developed (based on the incomplete engineering concept that the thermal properties of oils are constant), further analysis of frying systems has ignored available information related to chemical changes taking place in the frying oil.

FRYERS

Fryers are basically oil vats with heaters. The heaters transfer thermal energy into frying oil. The frying oil then transfers heat energy to the food's surface. Heat not used up at the surface is conducted into the interior of frying food. Heaters are turned on and then off as predetermined low and high set points of temperature (energy) in the oil are reached.

The heaters operate at temperatures considerably higher than the maximum set point temperature of the oil. The energy density (flux) of the heater's surface is high to heat the oil quickly to operating temperature regardless of energy losses to the environment. Cycle rates are adjusted by a set point thermocouple to counter overall heat loss at the heater surface, which includes energy dissipated to oil, fryer machinery, and also to further heat loss in the oil due to food loading.

Engineering of heater configuration is usually based on thermal calculations supposing that frying is really a food dehydration process. Heater design calculations entail determining the specific heat of the dry mass of the food, the percentage of water, the change of state of water from ice to water and water to steam, and the specific heat of the frying oil (figured as a constant representing new oil). For production fryers, heater placement and energy density in the oil vat are designed to compensate for food-loading points (high density) and take-away points (low density) and to compensate for heat sinks such as conveyers, crumb collectors, and filtering systems.

FOOD AND OIL QUALITY INTERACTIONS

A broad model describing all the qualities of frying oils and foods prepared in these oils is a continuum of frying oil degradation changes that could be associated with fried food qualities. Five stages of frying oil degradation are identified.

Figure 1 suggests that all frying oils behave similarly in the frying environment and that their bulk effects change slowly, even though the minor products of degradation may have a strong relationship to consumer acceptance of foods. Furthermore, each oil degradation stage can have different associated analytical values depending on the food being cooked. For instance, the degraded oil stage end points for potato chips, breaded vegetables, and fast food are found at 13%, 17%, and 25% total polar materials (TPM) content, respectively.

Within narrow windows, these TPM chemical index values seem relatively invariant across the food industry and are closely linked with the heating history of the oils. However, specific chemical markers used as process measurement end point specification values such as percentage of free fatty acids (FFAs) and other relatively minor degradation products vary with the mix of foods being fried. The percentage of FFAs end point values for the degraded stage are 0.5%, 2.0%, and 5.0%, respectively, for snack food, processed meat food, and fast-food industry segments. A generalized table for fast foods is shown in Table 1.

FRYING OIL DEGRADATION

The simple hydrolysis of oils (triglycerides) due to heat and moisture at ordinary frying temperatures forms quantities of FFAs. The FFAs then react with oxygen, each other, food juices, and a variety of intermediate degradation products

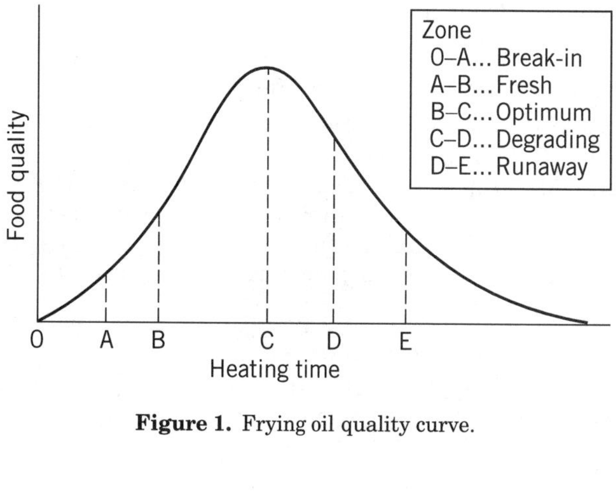

Figure 1. Frying oil quality curve.

Table 1. Values for Fast-Food Degradation by Stages

Parameter	0	Break-in	Fresh	Optimum	Degraded	Runaway
TPM (%)	<4	10	15	20	25	35
FFAs (%)	0.02	0.5	1.0	3.0	5.0	8.0
Mono- and diglycerides (%)	2	6	7	3	1	0.5
Surfactants	0–7	10	35	65	>150	>200
Dimer and polymer (%)	0.5	2	5	12	17	25
Oxidized fatty acids (OFA) (%)	0.01	0.08	0.2	0.7	1.0	2.0
Metals (ppm)	1	5	10	15	25	50

in the oil. These new chemicals in frying oil are either volatile (and leave) or are soluble (and some are even suspended colloids) in the triglycerides that comprise the oil. Monoglycerides and diglycerides formed after hydrolytic splitting off of fatty acids are generally more reactive than the original triglycerides. Acrolein, formed by the dehydration reaction of glycerol, may be an important toxicant or irritant in the oil and frying vapor, but it is rarely investigated.

The degradation of frying oil is also accelerated by the soluble chemicals that are either left in the oil as residues, additives, and artifacts of refining or that are introduced into the oil by frying food. As plant and animal cells (whole food) explode during cooking, the watery (plasma) juices of the cells introduce foreign chemicals into the oil. A similar process occurs when watery systems in formed and fabricated foods are introduced into hot frying oil.

Exposure to heat, light, crumbs, reactive metals, and food juices causes frying oil to react more and more rapidly with its own degradation products, with oxygen from the air, and with food-introduced contaminants. The ongoing combination of chemical changes finally causes a fry chef or line operator to dispose of a batch of oil because poor oil makes poor-quality fried foods.

Frying oils cook poorly when degraded by physical (thermal) stress and chemical changes. With use, oil changes its ability to hold heat. Heat capacity of oil increases with an increasing load of polymer (it takes more energy to attain a given temperature), and oil increases its rate of transferring heat (contact time increases with increasing levels of surfactants). Fresh oil, by comparison, holds heat well and acts like a thermal reservoir. Fresh oil also slowly releases heat to the surface of frying food.

Following are the most common chemical tests performed by inspectors in food production facilities, restaurants, and laboratories to indicate the approach of the specified discard point of a used frying oil.

Total Polar Materials (Chemical Index)

As oil breaks down, a group of polar materials includes most of the nontriglycerides. In the laboratory, a column chromatography method allows the determination of the percentage of polar materials in frying oils. The determination may also be made with a colorimetric quick test onsite; the quick test is then correlated to the standard method.

Polymer in Oil (Chemical Marker)

The formation of polymer in frying oils is directly related to the time oil is exposed to frying temperatures. At elevated temperatures, unsaturated fatty acids in triglycerides and also fatty acids free in solution combine with oxygen from the air. Some of the oxidized fatty acids cross-link to form high molecular weight polar compounds known as dimers. These dimeric triglycerides and fatty acids again cross-link during further exposure to frying conditions and form polar trimers, tetramers, and oligomers. Observation supports the hypothesis that no later than at about 50 h of heating and frying the oil becomes saturated with polymer, and any excess polymer plates out onto the walls of the fryer and onto the fry basket.

Free Fatty Acids

The measurement of FFAs as an indicator of oil quality is a widespread practice because it is easy to do and the percentage of FFAs rises with increasing use of a single batch of frying oil. However, the measurement is nearly meaningless because the rate of formation is not uniform, and many of the FFAs are continually being changed to volatile and nonacidic, nonvolatile decomposition products. The measured percentage of FFAs (titratable acidity) just documents what is left over from other degradation processes. In addition, when fresh oil is used to top up a fryer, the FFAs in the original batch are diluted to further confuse the actual relationship between oil quality and oil use.

An accepted reason to measure percentage of FFAs in the laboratory is to be able to predict the approximate resistance to oxidation of the oil on a potato chip distributed in a plastic bag. The FFAs are readily oxidized to yield a rancid odor. Colorimetric quick tests for percentage of FFAs are available for laboratory and for at-process control of frying oils.

Surfactants

Surfactants (wetting agents) in frying oil control the rate and amount of water removed from the surface of cooking food, overall oil pickup, and, by indirect (conductive and perhaps convective) forces, the degree of cooking of the interior. Although fatty acids are not surfactants, degradation products of fatty acids can be surfactants. For example, the soaps formed by combination of fatty acids with metals leached from food cells and food ingredients are surfactants. These parts per million of soaps along with other surfactants, such as monoglycerides, diglycerides, phospholipids, oxidized fatty acids, and oxypolymer fractions, control how (and even if) a food will fry in oils. The efficiency of heat transfer from heater to oil to food, the oil soakage into food, and the foaming tendency of the oil (oxygen incorporation route) are mediated by traces of surfactants. A colorimetric quick test for surfactants in frying oil is available.

Color in Oil

The often intense color of an oil is unrelated to the production of lightly colored, delicious food. The color bodies in oil do not seem to enter into the chemistry and physics of frying. They are inert with respect to the cooking process. The color of oil is not related to the color of food fried because the layer of oil on a food's surface is so thin it does not contribute significant color to the food. Removing color from oil is not equivalent to restoring cooking quality to the oil. Regardless of applicability, color matching and spectrophotometric techniques are available.

FOOD QUALITY

Very fresh oil does not cook well (food remains relatively raw) until it is broken in by frying a few portions of food or by adding a small amount of old oil to the new. Food fried in moderately used oil develops a well-cooked interior, a crisp exterior with a minimum of oil soakage, and typical golden colors and tempting flavors. Food fried in degraded oil does not cook completely in the interior, develops an excessively hard exterior, and suffers from oil soakage, dark coloration, and off-flavor.

Food fried in fresher oil is crisp, tasty, and well received. Food fried in degraded oil is limp and rejected. The quality of fried food from differently aged oil is often adjusted for by varying heater temperature and food residence time. These latter adjustments, however, never fully compensate for changes in the cooking properties of frying oil. Attainment of a crisp crust differentiated from a well-cooked interior is the goal of many groups that use frying to process food. The structure and chemical composition of fried crust is not yet understood. Dehydration produces hard structures, and oil incorporation (and perhaps interaction) produces an elastic component. This lack of knowledge keeps product development focused more on ingredient technology than on scientific principles.

Oil in Prefried Foods

The coatings of battered and breaded foods introduce soluble contaminants into the oil. When prefried foods are refried, the aged oil in the food comingles with the new frying oil and again introduces foreign chemicals. This latter case is especially important to food service operators and is difficult to guard against. Some prefried food vendors even tout the fact that their products are heavy with oil and will provide the food service fryer with extra free oil. This free oil is often loaded with soluble contaminant chemicals such as surfactants. Often a food service operation complains to its frying oil vendor that a new batch of oil was no good because it failed early in use. Actually, the prefried food was the culprit and the wrong vendor was blamed. An opportunity always exists to better control the production and increase the quality of fried foods by identifying the causative agent of variation in oil and its relationship to the process of frying.

NUTRITION AND TOXICOLOGY

The volatile and nonvolatile (possibly toxic) organic decomposition products of frying oils that have been extensively studied and commented on amount to, at most, a small percentage of the total mass of heated frying oil. Other, less studied components can be present in used oils at large percentages (20% or more for polymer) and have severe operating and economic consequences for fryer operators and consumers. It seems appropriate at this time to study the larger chemical variables and their consequences in frying technology. Fried food may be sterile on the surface as it emerges from the hot frying oil, but the interior bulk may still contain viable bacteria due to energy balance deficiencies in the process. This is especially worrisome in products such as coated meats, poultry, fish, and cheese and vegetable roll-ups.

REGULATORY

Recently, regulatory agencies and fast-food chains have shown an interest in having a simple means to determine when oil should be discarded. Many regulatory officials and nutritionists believe that it would be a wise practice to discard frying oils from food production before their TPM content has risen to 30% by weight of the oil. This attitude is nearly universal even though abused cooking oil consumed with fried food has not been shown to cause tumorigenesis or chronic toxicity in humans.

When a frying oil contains about 25% polar materials, numerous inspectors and restaurant operators believe the oil is abused and that such oil incorporated into food causes the food to be out of quality specification. The TPM content of oil included in food can then be regarded as a chemical index of fried-food quality. A food can be considered adulterated with respect to label declaration when the listed oil is only about 70 to 75% pure.

FRYING RESEARCH

Frying research has traditionally focused on the volatile and nonvolatile degradation chemicals formed in oils as a consequence of heating the oils in air. Summaries of important areas of study at an earlier time have been published (4,5). The degradation pathways and products of heated frying oils have been described (6). Important areas of frying oil analysis have been suggested (7). Diverse tools have been used to attempt to find relationships between different breakdown products in frying oils (8). Facets of oil use influencing fried-food quality have been studied (9,10). A simple surfactant formed in frying oils is related to oil soakage into food (11). Oil flavor chemistry has been described (12). Early evidence of a toxic material formed in abused frying oils has been presented (13). A summary of the nutrition and toxicology of frying oils has been published (14).

No researcher, however, has been able to specifically assign some element or elements in the complex mix of degradation chemicals in oils to the overall physical and sensory quality of food produced in aging frying oils. There is no textbook statement that if this element is examined in oil and controlled, the physical and sensory quality of fried food can be controlled. Some control of texture may be attainable if the chemistry causing crispness can be elucidated. The fractal nature of crust may be controlled by understanding the relationship between oil and starch interactions in potatoes, for example. Only general texts are available in this discipline (15). Flavor and taste components and rates and origins of organic reactions have been intensely studied, but the public as often as not still eats poor-quality fried food.

PROCESS OPTIMIZATION

The dynamics of heat and mass transfer while oil is undergoing chemical changes have not been well studied. Selective absorption of deleterious materials from frying oil solution has been proposed by the industry as a way to lengthen the service life of oils. Both filtration and infiltration of oil into frying food can be improved by studying aspects of fluid mechanics. Again, only general texts are available (16). Optimization of the frying process needs to be reduced to principles and rules for computerization in expert systems.

FRYING OIL SUBSTITUTES

To obtain either formulation compatibilities or reduced calories per serving, the food industry has frequently funded research on synthetic oils to be used either as replacers, extenders, or sensory enhancers. The group of discovered oil mimetics copy some sensory properties but cannot replace the functionality of frying oils in resistance to heat and oxidation. Oil replacers are created to substitute for natural triglyceride oils functionally but are not bioavailable and so do not contribute to nutrition. The lack of bioavailability is often touted as yielding low-calorie oil, but a consequence of replacers to date has been to remove fat-soluble materials, such as some vitamins, from the gut as the replacer passes through unchanged. Fortification of the replacers with fat-soluble vitamins is now suggested. The similarity between synthetics and the nonabsorbable polymer formed in abused oil is striking.

FRYING OIL CONTROVERSIES

Frying research has generally been conducted in ways that illustrate the dynamics and kinetics of oil degradation in nearly fresh oils or nearly abused oils. The middle ground, which is of most concern to industry and commerce, has been scarcely touched. Thus the emphasis has been on the onset of degradation or on thoroughly degraded systems that are not typical of food service or food processing by frying. Problems with processed oils such as flavor and color reversion, hydrogenation flavor, color associated with bitterness, high solids content, *trans*-fatty acids, unusual oxidation products, and processing residues and additives are rarely discussed. The importance of saturates and cholesterol and the potentially negative aspects of highly polyunsaturated oils, animal-source oils, saturated tropical oils, and heavily hydrogenated oils are not yet resolved.

BIBLIOGRAPHY

1. G. Varela, A. E. Bender, and I. D. Morton, eds., *Frying of Food*, VCH, New York, 1988.
2. E. G. Perkins and M. D. Erikson, eds., *Deep Frying: Chemistry, Nutrition and Practical Applications*, AOCS Press, Champaign, Ill., 1996.
3. M. M. Blumenthal, "Frying Technology," in Y. H. Hui, ed., *Bailey's Industrial Oil and Fat Products*, 5th ed., Vol. 3, John Wiley & Sons, New York, 1996, pp. 429–481.
4. H. Roth and S. P. Rock, *Bakers Digest* **46**(5), 38 (1972).
5. H. Roth and S. P. Rock, *Bakers Digest* **46**(6), 38 (1972).
6. C. W. Fritsch, "Measurements of Frying Fat Deterioration: A Brief Review," *J. Am. Oil Chem. Soc.* **58**, 272 (1981).
7. S. G. Stevenson, M. Vaisey-Genser, and N. A. M. Eskin, "Quality Control in the Use of Deep Frying Oils," *J. Am. Oil Chem. Soc.* **61**, 1102 (1984).
8. P. Wu and W. W. Nawar, "A Technique for Monitoring the Quality of Used Frying Oils," *J. Am. Oil Chem. Soc.* **63**, 1363 (1986).
9. L. M. Smith et al., "Changes in Physical and Chemical Properties of Shortenings Used for Commercial Deep-fat Frying," *J. Am. Oil Chem. Soc.* **63**, 1017 (1986).
10. J. Mancini-Filbo et al., "Effects of Selected Chemical Treatments on Quality of Fats Used for Deep Frying" *J. Am. Oil Chem. Soc.* **63**, 1452 (1986).
11. M. M. Blumenthal and J. R. Stockler, "Isolation and Detection of Alkaline Contaminant Materials (ACM) in Used Frying Oils," *J. Am. Oil Chem. Soc.* **63**, 687 (1986).
12. S. S. Chang, R. J. Peterson, and C. T. Ho, "Chemical Reactions Involved in the Deep-Fat Frying of Foods," *J. Am. Oil Chem. Soc.* **55**, 718 (1978).
13. D. Firestone et al., "Heated Fats. I. Studies of the Effects of Heating on the Chemical Nature of Cottonseed Oil," *J. Am. Oil Chem. Soc.* **38**, 253 (1961).
14. E. G. Perkins and W. J. Visek, eds., *Dietary Fats and Health*, American Oil Chemists' Society, Champaign, Ill., 1983.
15. A. Harrison, *Fractals in Chemistry*, Oxford University Press, Oxford, United Kingdom, 1995.
16. E. L. Cussler *Diffusion, Mass Transfer in Fluid Systems*, 2nd ed., Cambridge University Press, Cambridge, United Kingdom, 1997.

MICHAEL M. BLUMENTHAL
Libra Technologies, Inc.
Metuchen, New Jersey

FUMIGANTS

Fumigants have been used for centuries as a method of controlling stored insect pests. Because some insects deposit their eggs inside the grain kernel, fumigants must possess certain properties so that they exist in a gaseous state to penetrate the kernel to effectively control the infestation. In recent years the number of fumigants used worldwide has decreased to only two, phosphide and methyl bromide (1). The industry must take greater care in the use of the available products and exercise greater caution so that misuse cannot occur. Improper sealing of an area to be fumigated, poor application techniques coupled with wrong dosages, and inadequate fumigation time always result in failure. One of the principal concerns with poor fumigations is the development of resistant insects.

Newer government regulations require that applicators become more knowledgeable in regard to fumigant chemicals and their application. Safety of the applicator is a prime consideration for any fumigation, both during application of the chemical and the subsequent aeration of the fumigated commodity. The total procedure will only be successful if the space to be fumigated has been properly sealed so that the gas concentration can be maintained for the required time period to ensure total control. Improper sealing will result in loss of gas during the fumigation, so it is recommended that the applicator take gas readings during the fumigation to determine gas release and concentration. It is always surprising to hear how much time was taken for a fumigation that failed only to hear that some area was not sealed and that gas readings were not taken. Applications of this nature can only result in the development of resistance because generations of insects become exposed to nonlethal concentrations. The industry still has sufficient means to control stored product pests, but these methods must be used effectively and with proper techniques.

Each fumigation is different and, therefore, the selection of the appropriate fumigant to be used is crucial. It is especially significant that the properties of each chemical is considered for both success and to determine how best to apply the product so that all areas of the space to be fumigated have sufficient gas concentration during the treatment period. Fumigant products vary in mode of action depending on many factors including temperature, relative humidity, sorption, diffusion, and penetration.

A good fumigation will require prefumigation planning. All aspects of the fumigation must be decided so that success will be achieved. Among the factors to be considered are safety of the applicator, proper sealing of the structure, properties of the fumigant to be used, application equipment or techniques, gas readings, placement of warning signs, posting of guards, and notification of emergency personnel in the area in the event of an episode. The proper method of aeration of the treated commodity must also be analyzed and planned so that the release of the fumigant will not result in exposure or risk to anyone. Once all aspects of the fumigation have been established, it then is necessary to choose which fumigant will be used from among those available.

AVAILABLE FUMIGANTS

Metal Phosphides

There are two metal phosphide products that are presently used as a source of hydrogen phosphide: aluminum phosphide and magnesium phosphide.

Aluminum Phosphide. This product is manufactured as a tablet or pellet and is packaged in a formulation in a bag or sachet or as a tablet or pellet prepac. The paraffin-coated, 3-g tablets will produce 1 g of hydrogen phosphide gas when exposed to certain heat and humidity conditions. The pellets are one-fifth the size of the tablets and produce 200 mg of hydrogen phosphide gas. This gas has a density of 1.18 as compared to air and, as a result, has excellent diffusing and penetrating properties. The tableted product usually contains ammonium carbamate, which will break down into ammonia and carbon dioxide when exposed to heat. The gases escape from the tablet or pellet by causing ruptures in the paraffin coating. The ammonia also acts as a warning gas to the applicator. Atmospheric moisture then penetrates the ruptured coating to react with the aluminum phosphide resulting in the release of hydrogen phosphide gas, which is the resultant fumigant. Tablet and pellet prepacs release hydrogen phosphide in an identical manner. These formulations cause a delayed release until appreciable gas concentrations are reached and act as a safeguard to the applicator. Bag or sachet formulations also have a delayed release; atmospheric moisture must penetrate the paper or cloth packaging material before any toxic gas is released. It is important to follow the manufacturer's applicator's manual for use of aluminum phosphide because low temperature and low relative humidity can result in only partial reaction and a minimum of gas release. The result will be low concentration of gas, a good chance at fumigation failure, and possible problems on deactivation of the unreacted metal phosphide. When temperatures are lower or relative humidity is lower, then it is wise to consider the use of a magnesium phosphide product.

Magnesium Phosphide. These products are produced and marketed as either tablets or pellets or as magnesium phosphide impregnated in a plastic matrix. No bag or sachet formulation of this chemical is available. There is a magnesium phosphide pellet prepac product that is used only as a spot fumigant for plant processing equipment. The reactivity of magnesium phosphide with atmospheric moisture is so much faster than aluminum phosphide, it can be used in conditions of high temperatures, low relative humidity or low temperatures and low to medium relative humidity.

PROPERTIES OF METAL PHOSPHIDES

There are several properties of metal phosphide fumigant products that must be considered in any application.

1. Its density is such that any leakages will result in loss of gas.
2. Hydrogen phosphide gas can self-ignite at a concentration of 17,900 ppm if there is an ignition source available. The reaction between a metal phosphide and atmospheric moisture, which yields hydrogen phosphide, is exothermic and, therefore, the piling of tablets during fumigation is to be avoided as the heat generated from this exothermic reaction can be the ignition source.
3. Hydrogen phosphide reacts with certain metals, especially with copper, so care must be taken during fumigation to prevent corrosion of electrical and electronic equipment and instruments.
4. Note that 1 g of hydrogen phosphide gas per 1,000 ft^3 will result in a concentration of 25 ppm. There-

fore, if a dose of 20 tablets/1,000 ft^3 is applied, the theoretical maximum concentration would be 500 ppm (20 g $\times$ 25 = 500 ppm), which is equivalent to a dose of 0.7 g/m^3.

5. Aluminum or magnesium phosphide tablets and pellets that contain ammonium carbamate also release ammonia and, therefore, can be phytotoxic to living plants if sufficient concentration is present as well as darkening some nut meats when direct contact is made with the product. The darkening of nut meat is usually reversible. It would be better to use the magnesium phosphide impregnated in a plastic matrix (Fumi-Cel/Fumi-Strip) for some of these fumigations because it contains no ammonium salts.
6. Hydrogen phosphide has no effect on germination and, therefore, can safely be used to fumigate various seeds.
7. Hydrogen phosphide has no real CT product and fumigations must be no less than three days for complete control.
8. Do not fumigate if commodity temperature is below 5°C (40°F).
9. A fumigated commodity or space must be aerated until there is 0.3 ppm or less of hydrogen phosphide before reentry is allowed without respiratory protection.
10. Repeated fumigation with hydrogen phosphide will not result in a buildup of residues if the product is properly aerated following fumigation.

Finally, it must also be noted that hydrogen phosphide produced from metal phosphide product has a garlic or carbidelike odor. One word of caution in this regard is to emphasize that pure hydrogen phosphide has no odor, and the odor noted during a fumigation is an impurity and can be sorbed on a commodity, therefore, the lack of odor cannot be considered as a key for the lack of hydrogen phosphide being present. A measuring device must be used to determine concentration.

Metal phosphide products can be applied directly to raw agricultural commodity and animal feeds and have a tolerance of 0.1 ppm for this purpose. No direct contact can be made with these products on processed food so packaged metal phosphide products such as prepacs, bags, sachets, or Fumi-Cel/Fumi-Strip are used for these applications, and the approved tolerance is 0.01 ppm. Care should always be taken when deactivating residual materials as fumigation conditions could have resulted in having some unreacted phosphides in the residuals. The manufacturer's applicator's manual should be consulted prior to initiating this activity.

The half-life of hydrogen phosphide in the atmosphere is short and depends on climatic condition; it is in the range of 5–28 h. Hydrogen phosphide is known to be oxidized to phosphate both in the atmosphere and during fumigation.

Toxicology

Hydrogen phosphide has been shown to be an acute toxic material and has no chronic effects. Effects shown by experimental animals as a result of subchronic exposure were shown to be reversible. This substance also had no teratological or mutagenic action when these studies were completed.

Aluminum phosphide, a fumigant used extensively in India, is extremely toxic to humans with a fatal dose reported to be 1.5 g (2). Gupta and Ahlawat (3) attributed aluminum phosphide poisoning to the release of a cytotoxic phosphine gas to the major organs of the body including heart, lungs, and kidneys.

Respiratory Protection

If the application is made inside a structure, then gas readings must be taken and a gas mask with approved canister must be worn if the levels exceed 0.3 ppm. An air pack must be worn when the levels exceed 15 ppm. These same requirements also apply to aeration.

METHYL BROMIDE

Methyl bromide is packaged as a liquid under pressure in either small cans or in various sizes of steel cylinders. It has a density of approximately 3.3 times that of air and, as a result, should be used with fans to prevent stratification, resulting in fumigation failures in the higher portion of the area. To prevent some of this phenomenon from occurring, it is usually recommended that the product be released near top of the site to be fumigated whether it is a silo or mill. If fans are not employed, there is a possibility that the upper portion of the site will not be successfully fumigated. If bagged commodities are to be fumigated under a tarpaulin or in a truck or container, the entry tube should be leakproof and a pan should be placed under the end of the tube so that no liquid methyl bromide is allowed to come into contact with the bagged commodity, which can result in staining.

The properties of methyl bromide that must be considered prior to fumigation are as follows.

1. Its density is such that fans should be used to recirculate the gas for even distribution.
2. A heat exchanger may be necessary to properly volatize the methyl bromide.
3. Methyl bromide is nonflammable but can result in the formulation of hydrobromic acid when open flames are present. Hydrobromic acid can be corrosive.
4. Methyl bromide can be sorbed under certain condition and can react chemically with certain compounds resulting in damage or bad odors. These materials include certain foodstuffs, rubber goods, furs, leather goods, woolens, rayons, various paper products, photographic chemicals, cinder blocks, charcoal or any material that contains active sulfur compounds.
5. Methyl bromide can effect germination of seeds and should be used for this purpose with caution.
6. Methyl bromide does have a true CT product, and fumigation usually can be completed in 12–24 h depending on dosage.

7. Normally fumigation with methyl bromide is not recommended when the commodity temperature is lower than 15.5°C (60°F).
8. Following fumigation, aeration must continue until the level of methyl bromide is below 5 ppm. If the level exceeds this amount, then proper respiratory equipment must be worn.
9. Repeated fumigations with methyl bromide can result in residues exceeding tolerances established by federal agencies for various commodities.

Finally, it should be noted that methyl bromide is odorless and proper measuring instrumentation is required to measure concentration. The halide detector is not sensitive enough to be used to determine levels suitable for reentry into an aerated facility. Aeration may have to be prolonged due to sorption of methyl bromide to commodities or materials in the structure. Fans will help in the aeration process. The half-life of methyl bromide can be 15–18 months depending on climatic conditions.

Toxicology

Methyl bromide is known to be a chronic toxic chemical, that is, the effects due to exposure are cumulative and are not reversible. Major studies such as chronic toxicity, teratology, pharmacokinetics, and mutagenicity are being conducted to fill data gaps for regulatory purposes.

Methyl bromide is known to be a chronic toxic chemical, as the effects due to exposure are cumulative and not reversible. Consumption of methyl bromide from fumigated food has been estimated at 0.00125 mg/kg/day in the average diet. Wilson et al. (4) fed dogs a diet containing 0.27–0.28 mg/kg/day methyl bromide over a period of a year. No adverse effects were observed, indicating it did not present a significant health hazard to humans. Another recent feeding trial on rats and rabbits conducted by Kaneda et al. (5) concluded that methyl bromide was not fetotoxic or teratogenic to rat and rabbit fetuses up to dose levels of 30 and 10 mg/kg/day, respectively. Nevertheless, the use of methyl bromide as a fumigant was banned by the Montreal Protocol (an international treaty) as of 2001 because it is an ozone depleter. This could have a significant effect on Florida's horticultural industry unless alternative methods are found (6,7).

Respiratory Protection

If application is made inside a structure, then an air pack must be worn for any level above 5 ppm. This same restriction applies to aeration.

Alternatives to Methyl Bromide

A number of alternative methods are being explored for possible implementation prior to the ban on methyl bromide in 2001. Warner (8) suggested a combination of moist heat and controlled atmosphere as an alternative quarantine treatment for codling moth in fruit. Subjecting apples, pears, and cherries to 113–117°F (45–47°C) would be adequate to destroy the codling moth larvae. Storing the fruit in a controlled atmosphere of 0.5 % oxygen and 15% carbon dioxide shortened the heating time at 117°F needed to kill the larvae by as much as one half or two-thirds compared to just heating alone. Warner (9) also recommended irradiation as an alternative quarantine treatment to fumigation with methyl bromide for cherries. Irradiation up to 50 krad did not have a detrimental effect on the quality of the fruit. A recent study by Sholberg (10) used short chain organic acids (acetic, formic, and proprionic) as fumigants against postharvest decay by fungal pathogens. All three acids reduced decay in citrus by *Penicillium digitatum* from 86% to 11% and decay in pome fruit by *P. expansum* from 98% to 14%, 4%, and 8%, respectively. It is clear that alternative methods to fumigants will soon be implemented before the ban on methyl bromide comes into force.

CONTROLLED ATMOSPHERES

The gases primarily used for controlled-atmosphere fumigations are carbon dioxide, nitrogen, or a mixture of these gases. These gases, in effect, are applied to reduce the oxygen content in a storage.

Carbon Dioxide

The addition of this gas not only reduces the oxygen content but also acts directly on stored pests by acting as a dessicant on the insect body fluids. Carbon dioxide is usually applied from trucks or tanks that contain liquid carbon dioxide. Appropriate vaporizers and regulators are then used to supply gaseous carbon dioxide to the structure to be fumigated. The gas is usually applied at the top of a silo and because it is denser than air the gas will move down through the grain mass. The silo or warehouse to be fumigated must be tightly sealed, and carbon dioxide must be added daily to maintain a 60–70% concentration to be efficient. A system must also be used in the structure that will not allow the pressure inside to build up during application. Usually a vent is placed in the roof or a hatch or portion of polyethylene used to seal roof vents is opened during application.

Nitrogen

Nitrogen can also be used as an inert gas to control oxygen concentrations. A nitrogen generator is required, and normally the oxygen concentration must be reduced to approximately 1%. Fumigation time can vary but usually requires approximately 10 days for good control.

Inert Atmosphere Generators

There are several inert atmosphere generators that result in a mixture of nitrogen and carbon dioxide being applied to the structure to be fumigated.

To use a controlled atmosphere for fumigation, the following factors must be considered.

1. Temperature of the commodity and within the storage area should be 21°C (70°F) or higher to have an effective fumigation. Lower temperature will effect the mortality of the insect present in the structure or commodity.

2. A vent or pressure relief system must be present in the warehouse or silo to prevent structural damage due to pressure build up.
3. Application time can take up to 12 h with daily recharging, which will normally take 3–4 h.
4. Total fumigation time can be from 4 to 10 days depending on temperature and gas tightness of the structure.
5. Even though these gases are considered to be nontoxic, an air pack must be used when entering a facility that has low oxygen content during fumigation.

MIXTURES OF GASES

Research is being conducted by several groups to try to determine if mixtures of gases can result in more favorable fumigation conditions with a shorter duration. Mixtures such as hydrogen phosphide and carbon dioxide, methyl bromide and carbon dioxide, or increased temperature and carbon dioxide are being tested. Early studies show that some mixtures may be promising, but additional studies will have to be performed. Elevated temperature and carbon dioxide combinations have not been successful to date.

BIBLIOGRAPHY

1. R. W. D. Taylor, "The Future of Postharvest Fumigants," *Postharvest News and Information* **8**, 26N–31N (1997).
2. S. Singh et al., "Aluminum Phosphide Ingestion—A Clinico-Pathologic Study," *Journal of Toxicology and Clinical Toxicology* **54**, 703–706 (1996).
3. A. Gupta and S. K. Ahlawat, "Aluminum Phosphide Poisoning—A Review," *Journal of Toxicology and Clinical Toxicology* **33**, 19–24 (1995).
4. N. H. Wilson et al., "Methyl Bromide: 1-Year Dietary Study in Dogs," *Food and Chemical Toxicology* **36**, 575–584 (1998).
5. M. Kaneda et al., "Oral Teratogenicity Studies of Methyl Bromide in Rats and Rabbits," *Food and Chemical Toxicology* **36**, 421–427 (1998).
6. J. Noling and J. Gilreath, "USDA Evaluates Alternatives to Methyl Bromide in Florida," *Citrus and Vegetable Magazine* **62**, 21, 24 (1998).
7. J. J. VanSickle and T. Spreen, "Florida Without Methyl Bromide: The Potential Result," *Citrus and Vegetable Magazine* **58**, 26–28 (1994).
8. G. Warner, "Heat and CA Treatment May Replace Fumigation," *Good Fruit Grower* **46**, 8–9 (1995).
9. G. Warner, "Irradiation is Kinder to Fruit than Fumigation," *Good Fruit Grower* **47**, 9 (1996).
10. P. L. Sholberg, "Fumigation of Fruit with Short Chain Organic Acids to Reduce the Potential of Postharvest Decay," *Plant Disease* **82**, 689–693 (1998).

GENERAL REFERENCES

E. J. Bond, *Manual of Fumigation for Insect Control*, Food and Agriculture Organization of the United Nations, Rome, 1984.

Truman's Scientific Guide to Pest Control Operations, Purdue University/Edgell Communications.

FUNCTIONAL FOODS

Functional foods usually refer to foods containing significant levels of biologically active components that impart health benefits beyond basic nutrition. These components are often known as phytochemicals—meaning plant chemicals. Several terms are used to describe the many natural products currently being developed for health benefit. These include nutraceutical, functional food, pharmafood, designer food, vitafood, phytochemical, and foodaceutical. Functional food may be the most internationally recognized term for the category of enhanced foods with potentially strong preventive or therapeutic properties and is used widely in Japan and Europe. In North America, the terms functional food and nutraceutical have been used interchangeably, and a number of definitions of what are included under these all-encompassing headings have been advanced. According to Health Canada, a functional food is similar in appearance to conventional food, is consumed as part of the usual diet, and has demonstrated physiological benefits and/or reduces the risk of chronic disease beyond basic nutritional functions. A nutraceutical is a product produced from foods but sold in pills, powders (potions), and other medicinal forms not generally associated with food and demonstrated to have a physiological benefit or provides protection against chronic disease (1).

The following statement provides the definition for functional food endorsed by the European expert committee organized by the International Life Science Institute (LSI): "A food can be regarded as functional if it is adequately demonstrated to beneficially affect one or more target functions in the body, beyond normal nutritional effects, in a way which is relevant to either the state of well being and health and/or the reduction of the risk of a disease." Definitions are also based on, and relevant for, target groups. In Germany, for example, functional foods are targeted for the general population, while the term *nutraceuticals* is used for products aimed at specific age categories, and *medical foods* is the term for products designed for patients (2). Currently in Japan, the functional foods approved as Foods for Specified Health Use (FOSHU) regulation established in 1991 by the Japanese Ministry of Health and Welfare are defined as everyday foodstuffs expected to have a specific effect on health due to relevant constituent(s). This regulation allows claims regarding the foods' beneficial effect on health.

In the United States, the movement toward functional foods began during the late 1970s when Lee Wattenberg of the University of Minnesota extolled the health benefits of components of crucifers and cabbage. As the next step, the National Academy of Sciences' report on "Diet and Cancer" in 1982 not only made a connection between fat and fiber in the diet but also nonnutritive components that might be important in the diet. The term designer foods coined by Herbert F. Pierson, Jr., in 1989, was publicized in the popular press in the United States several years ago when the National Cancer Institute (NCI) announced a $20 million five-year research program to study the anticarcinogenic properties of components of citrus, flax, aged garlic extract, licorice extract, soybean meal, and umbelliferous vegetable juice beverage. In 1989, because of the regulatory confusion concerning foods versus drugs, Stephen L. DeFelice, chair of the Foundation for Innovation in Medicine, introduced the term nutraceuticals to include isolated nutrients, dietary supplements, genetically engineered foods, and herbal products. The term nutraceuticals has received considerable press and carries a high-tech image that fits comfortably with pharmaceutical companies.

The concept of functional foods is based on evidence that certain vegetables, herbs, fruit, and grains naturally rich in various phytochemicals have a protective effect against diseases such as cancer. There are at least 14 classes of phytochemicals known or believed to possess cancer-preventive properties (3). These include sulfides, phytates, flavonoids, glucarates, carotenoids, coumarins, monoterpenes, triterpenes, lignans, phenolic acids, indoles, isothiocyanates, phthalides, and polyacetylenes. A variety of these phytochemicals are abundantly present in garlic, green tea, soybeans, cereal grains, licorice roots, flaxseed, and plants from the cruciferous, umbelliferous, citrus, solanaceous, and cucurbitaceous family (Table 1).

Increased intakes of nonnutrient phytochemicals with potential health benefits can be improved by increased intake of a food rich in products such as garlic extract; by conventional plant breeding to increase the concentration in a crop; through biotechnology; by selective processing, such as milling and extraction, to enhance the concentration of one or more chemicals in a food; or by adding a phytochemical to a food. Food processing and preparation procedures also affect the physiological consequences of food. Fermented products, including dairy products, have long been recognized to alter gastrointestinal flora and even reduce circulating cholesterol. While heating tomatoes may improve lycopene availability and thus improve its antioxidant potential, heating of unpeeled garlic reduces its anticancer potential. Ingestion of functional foods represents an effective strategy to maximize health and reduce the risk of diseases. Interest in the health benefits of foods is propelled by rising health care costs, legislative changes (ie, the Nutritional Labelling and Education Act [NLEA] and Dietary Supplement Health and Education Act [DSHEA]) that permit claims for foods and associated components, and by new and exciting scientific discoveries. Functional food is one of today's most prevalent trends for food product development. According to Datamonitor, Inc., a global market research firm, the U.S. market for nutraceuticals has grown from $11 billion to $16.7 billion, representing a growth of 10.9% over the past four years. The firm's survey, "U.S. Nutraceuticals 1997," found that a general trend toward healthy eating (represented by low-fat, low-calorie foods) has shifted toward more active prevention of disease and maintenance of health.

FUNCTIONAL FOODS FROM PLANT SOURCES

Epidemiological studies have shown that plant-based diet can reduce the risk of chronic disease, particularly cancer and cardiovascular disease. This is supported by the low incidence of cardiovascular disease in vegetarians and reduced cancer risk in people consuming diets high in fruits and vegetables. Biologically active components of plants

Table 1. Phytochemicals from Major Plant Food

	Phenolic acids	Flavonoids	Coumarins	Carotenoids	Triterpenes	Monoterpenes	Glucarates	Lignans	Sulfides	Phytates	Isothiocyanates
Cereals	✓	✓	✓	✓	✓		✓			✓	
Soybeans	✓	✓	✓	✓	✓			✓		✓	
Flaxseed	✓	✓	✓					✓			
Solanaceous	✓	✓	✓	✓	✓	✓	✓				
Garlic	✓				✓	✓			✓		
Cruciferous	✓	✓	✓	✓	✓	✓	✓		✓		✓
Citrus	✓	✓	✓	✓	✓	✓	✓				
Umbelliferous	✓	✓	✓	✓	✓	✓					
Green tea	✓	✓	✓				✓				

Source: Ref. 3.

with demonstrated physiological benefits that may enhance health by reducing or alleviating the risk and/or incidence of diseases have been identified in plants commonly grown in commercial agriculture.

Cereals

Cereals such as wheat and rice are major constituents of the diet of a vast number of the world's population and therefore can play an important role in the nutritional quality of the diet and human health. The role of cereal grains in gastrointestinal health has long been recognized. Dietary fiber, particularly from cereals, has been postulated to be the major preventive factor against a number of cancers including colorectal carcinogenesis. However, a recent report (4) has challenged the connection between fiber consumption and colorectal cancer. Increased dietary fiber intake, especially soluble fiber, has also been linked to reduced risk of coronary heart disease. Oat is one of the few foods that have been rigorously tested as a dietary source of the cholesterol-lowering soluble fiber β-glucan in clinical settings among free-living individuals. Consumption of diets high in soluble fiber β-glucan has been associated with reduction in total and low-density lipoprotein (LDL) cholesterol, thereby reducing the risk of coronary heart disease (CHD). For this, the Food and Drug Administration (FDA) awarded the first food-specific health claim under the NLEA in January 1997 (5), in response to a petition submitted by the Quaker Oats Company. A food bearing the health claim must contain 13 g of oat bran or 20 g oatmeal and provide without fortification at least 1.0 g of β-glucan per serving. In February of 1998, the soluble fiber health claim was extended to include psyllium fiber. There is great expectation that the health claim for the soluble fiber β-glucan of oats will expand to other grains, including barley. In this regard, the prowash barley variety from Montana State University is said to contain twice the level of soluble cholesterol-lowering β-glucan fiber found in ordinary barley. Recently, stabilized full-fat rice bran performed similarly to oat bran in reducing hypercholesterolemia in humans (6). Wheat foods, the major source of dietary fiber in the North American diet, generally accepted in promoting regularity, is also being touted as having protective effects against colon and breast cancers. Wheat bran, even in processed forms, can offer protection against the development of colon adenocarcinoma even when the carcinogenic process is promoted by high-fat low-calcium high-risk Western-style diets (7). In this regard, the FDA has under consideration a petition from the Kellogg Company seeking approval for the health claim that eating products containing insoluble fiber from wheat bran can help reduce the risk of colon cancer.

Legumes

The Leguminosae plant family, of which soy is of the utmost economic importance, is most abundant in phytoestrogens. The low mortality due to breast, colon, and prostate cancer of Asians, especially Japanese and Chinese, compared with Westerners has been attributed mainly to extensive use of soy foods. Several classes of anticarcinogens have been identified in soybeans, including protease inhibitors; phytosterols; saponins; phenolic acids; phytic acid; the precursors of mammalian lignans, especially seicoisolariciresinol; and isoflavones, a class of phytoestrogens. The two primary isoflavones, genistein, reported to be the most biologically active phytoestrogen, and diadzein, present in high concentration in soybeans, are known to modulate hormone-related conditions, such as heart disease, osteoporosis, breast and prostate cancers, and menopausal symptoms. Since isoflavones are structurally similar to the estrogenic steroids, they may act as antiestrogens by competing with the more potent, naturally occurring endogenous estrogens for binding to the estrogen receptor. In this regard, an isoflavone extracted from soybeans, 17β-estradiol (Estrace®, Roberts Pharmaceuticals Canada Inc.), is being marketed as a prescriptive oral estrogen for effective management of menopause. The cholesterol-lowering effect of soy is the most well documented physiological effect and is attributed mainly to soy isoflavones. However, very recent studies suggest that the combination of both soy protein and isoflavones are necessary to produce the greatest cholesterol-lowering effect. In addition to being a high-quality protein, soybean protein is now thought to play a preventive and therapeutic role in cardiovascular disease. In this regard, the FDA has proposed to authorize a health claim regarding the association between soy protein and reduced risk of coronary heart disease as petitioned by Protein Technologies International, Inc. Soy protein in the amount of 6.25 g with a minimum of 12.5 mg of total isoflavones (equivalent to 25 g of soy) per day is required to qualify an individual food to bear the health claim (8). Recently, the hypocholesterolemic effects of protein concentrates from lesser known legume seeds have been demonstrated (9).

Oilseeds

Flaxseed is the most prominent oilseed studied to date as functional food since it is a leading source of the omega-3 fatty acid, α-linolenic acid (ALA) (52% of total fatty acids), and phenolic compounds known as lignans (>500 μ/g, as is basis). Dietary omega-3-rich oil moderates the recognized current American excess of omega-6 intake that leads to eicosanoid-mediated disorders (eg, thrombotic heart attack and stroke, cardiac arrhythmia, atherogenesis, arthritis, asthma, osteoporosis, tumor metastases, etc). In this context, flaxseed has been fed to poultry, cattle, and pigs to increase omega-3 fatty acids in eggs, milk, and meat. Omega-3 fatty acids, precursors of long-chain polyunsaturated fatty acids, have been shown to be important for visual and brain development, especially for termed infants. The lignan secoisolariciresinol diglycoside (SDG), extracted from flaxseed, is the precursor of the mammalian lignan enterodiol and its oxidation product, enterolactone. Because enterodiol and enterolactone are structurally similar to estrogens, they may play a role in the prevention of estrogen-dependent cancers. Flaxseed has a protective effect against both mammary and colon cancer. Long-term studies confirm the protective effect of SDG on tumor growth due to the biological properties of mammalian lignans derived from SDG. Dietary flaxseed also provides a significant benefit in preventing the decline in renal function and has been found to be a suitable adjunct to therapy in patients with lupus nephritis and other forms of progressive renal disease. Flaxseed is also a good source of dietary fiber known to improve glucose metabolism as well as other phytochemicals associated with the decreased risk of developing colorectal cancer (10). Consumption of flaxseed has been shown to reduce total and LDL cholesterol and platelet aggregation. Other oilseed crops are predominantly linoleic-acid-rich oils and have the advantage of being the richest natural dietary sources of vitamin E, in addition to providing the essential fatty acids. Recent trends for commercial vegetable oil have been the reduction of polyunsaturates, as in the case of sunflower (Nusun), to overcome the presence of undesirable *trans* fatty acids due to hydrogenation (11).

Horticultural Crops

Tomatoes. Lycopene, the primary carotenoid found in tomatoes, has been implicated in risk reduction of prostate cancer as well as other cancers, including breast, digestive tract, cervix, bladder, skin, and lung, whose risks have been inversely associated with serum or tissue levels of lycopene. Recently, intake of lycopene has also been inversely associated with risk of myocardial infarction, probably due to modulation of cholesterol metabolism (12).

Garlic. The purported health benefits of garlic include cancer chemopreventive, antibiotic, antihypertensive, and cholesterol-lowering effects. Allicin, produced when garlic cloves are crushed, and responsible for the characteristic odor of fresh garlic, spontaneously decomposes to form sulfur-containing compounds with chemopreventive activity.

Garlic has been historically used as a spice and remedy for common ailments. The reported medicinal properties of garlic such as antibacterial activity, antifungal activity, inhibition of platelet aggregation, and reduction of serum cholesterol levels are due to biologically active sulfur compounds. Attention has been directed to garlic and garlic extracts as potential cardiovascular and anticancer agents. Several epidemiological studies demonstrate the effectiveness of garlic in reducing human cancer risk, especially cancer of the stomach, colon, and gastrointestinal tract. Garlic has also been advocated for the prevention of cardiovascular diseases, possibly through antihypersensitive and cholesterol-lowering effects. Plants of the genus *Allium*—including garlic, onion, and leek—are also good sources of sapogenins, bioactive compounds that are relatively stable within the food during and after processing.

Cruciferous vegetables. Epidemiological studies indicate reduced risk of breast, colon, gastric, and prostate cancers on frequent consumption of cruciferous vegetables. The anticarcinogenic properties of cruciferous vegetables are attributed to their relatively high content of glucosinolates and their derivatives, of which the allyl isothiocynate and indoles are most potent. Indole-3 carbinol (I3C) is currently under investigation for its cancer chemopreventive properties, particularly of the mammary gland, and may be a novel approach for reducing the risk of breast cancer. Besides glucosinolates and its derivatives, several groups of compounds including dithiolthiones, sulfonates, and vitamins from cruciferous vegetables have been reported to alleviate diseases and promote health.

Fruits. Flavonoids are the main constituents in most fruits known to protect against a variety of human degenerative diseases. These and other fruit components are major sources of dietary antioxidants that increase the plasma antioxidant capacity resulting in inhibition of atherosclerosis-related diseases in humans with increased consumption of fruits (13). Flavonoid intake from various sources including fruits and vegetables has been inversely related to coronary heart mortality (14). The limonoids, a group of chemically related triterpene derivatives, present in large quantities in citrus fruits has been suggested as the preventive agent against a variety of human cancers. Cranberry juice has been recognized for a long time as efficacious in the treatment of urinary tract infections. The inhibition of the adherence of *Escherichia coli* to uroepithelial cells has also been attributed to fructose and a nondialyzable polymeric compound from cranberry juice (15). Flavonoids and other phenolics present in grape and grape products have been shown to possess anticarcinogenic, anti-inflammatory, antihepatotoxic, antibacterial, antiviral, antiallergic, antithrombic, and antioxidative effects. The ability of phenolic substances in prevention of atherosclerosis and risk reduction of cardiovascular diseases has been aptly demonstrated by the benefits of red wine consumption, often referred to as the French Paradox. The health benefits of grape phenolics in inhibiting the oxidation of LDL can also be obtained from commercial grape juice (16). Red wine is also a significant source of *trans*-resveratrol, a phytoalexin with estrogenic and anticancer properties (17).

Polyphenols comprising up to 30% of the total dry weight of fresh tea leaves have been associated with the cancer chemopreventive effects particularly from green tea. Decrease recurrence of breast cancer in Japanese women and substantial reduction in risk of cardiovascular diseases have been associated with tea consumption (18). Phenol antioxidants from tea with lipoprotein-bound antioxidant activity constitute a major source of antioxidants in the North American diet.

Blueberry fruits, particularly *Vaccinium myrtillus* L., also known as bilberries are a rich source of anthocyanins and other flavonoids, which have demonstrated a wide range of biochemical and pharmacological effects, including anticarcinogenic, antiatherogenic, anti-inflammatory, antimicrobial, and antioxidant activities. Studies carried out in small animals or on a small number of humans, indicate that concentrated extracts of anthocyanins obtained from bilberries may benefit visual acuity, as well as provide protection against macular degeneration, glaucoma, and cataracts.

FUNCTIONAL FOODS FROM ANIMAL SOURCES

Dairy Products

Prebiotic- and probiotic-containing dairy products lead the functional food market in most of Europe and Japan. The term prebiotics has been applied to products such as oligosaccharides, used as bifidogenic factors in fermented dairy products to promote the growth of beneficial organisms. Natural sources of these bifidogenic oligosaccharides include chicory, Jerusalem artichokes, and other plants. Besides prebiotic effect, the major fructo-oligosaccharide inulin has physiological benefits in absorption of minerals, carbohydrate and lipid metabolism, immune functions, modulation of serum lipids, and reduction of cancer risks (19). The generally accepted definition of probiotics is that they are live microbial food or feed supplements that benefit humans and animals by improving the microbial balance in the intestine. Several health benefits have been attributed to probiotics, including anticarcinogenic, hypocholesterolemic, and antagonistic effects against enteric pathogens and other intestinal organisms. The role of probiotics in cancer risk reduction has been evident, particularly in colon cancer (20). Products that contain both prebiotics and probiotics, referred to as synbiotics, are being successful marketed in Europe. Traditional dairy products are also good sources of calcium and other milk-based bioactive compounds, such as casein-based bioactive peptides, whey proteins, and conjugated linoleic acid (CLA), with anticarcinogenic and immunogenic properties (21).

Beef

The mixture of positional and geometric isomers of linoleic acid with conjugated double bonds referred to as conjugated linoleic acid is most abundant in fat from ruminant animals. CLA, first isolated from grilled beef (22), has been shown to be effective in suppressing forestomach tumors and mammary carcinogenesis. Beef fat contains 3 to 8.5 mg CLA/g fat, and diets containing only 0.1 to 1% CLA have been shown to have anticarcinogenic effect in a mammary tumor model (23). CLA has also been suggested as a weight reduction agent since it can change body composition. Recent studies indicate that CLA has protective effects in cases of human breast cancer, chronic renal failures, and reduction in the severity of preexisting aortic atherosclerosis (23). It has been suggested that supplementation of CLA in diets alters the incorporation of n-6 fatty acids and CLA isomers into adipose tissue and enhances certain anticancer immune defenses.

Fish

Seafood lipids are rich in long-chain omega-3 polyunsaturated fatty acids (PUFA). The beneficial health effects of

Table 2. Selected Functional Foods and Bioactive Food Ingredients and Their Health Benefits

Functional food	Benefit
β-glucan derived from oats	Reduces risk of coronary heart disease
Canola oil with reduced saturated fats	Reduces risk of coronary heart disease
Dairy products with probiotics and prebiotics	Maintains a healthy digestive tract
Flaxseed lignans	Reduces risk of breast cancer and improves glucose metabolism
Flaxseed oil—high in omega-3 fatty acids	Decreases triglycerides and blood cholesterol
Fructo-oligosaccharides (especially for people with irritable bowel syndrome)	Restores colonic and intestinal health and reduces colonic precancerous lesions
Green tea extracts and beverages	Aids relaxation and reduces risk of cardiovascular disease and certain cancers
Herbal extracts	
Ginseng	Antistress, immunostimulant
Gingko biloba	Improves cognitive performance
St. John's wort	Controls mild to moderate depression
Echinaceae	Boosts immune system
Milk thistle	Supports healthy liver function
Valerian	Promotes restful sleep
High laurate canola oil	Lowers cholesterol
Insoluble fiber from cereals	Reduces risk of colon cancer
Oil rich in decosahexanoic acid	Enhances neurological and visual developments in infants
Oils engineered without *trans* fatty acids	Lowers heart disease risk
Omega-3 enriched eggs	Reduces serum triglycerides
Products made from blueberries, cranberries, bilberries and lingonberries (*Vaccinium genus*)	Reduces risk of inflammation and peripheral vascular disease, controls diabetes and improves vision
Resistant starch from grains	Promotes gastrointestinal health and lowers risk of colon disease
Shiitake mushrooms	Enhances immune activity, stimulates immune function and inhibits tumor growth
Soluble fiber from cereals	Aids in control of blood glucose
Soluble fiber from konjac glucomannan	Maintains intestinal health and suppresses appetite
Soy isoflavones	Reduces the negative effects of estrogen and lowers risk of several cancers
Soy protein	Reduces risk of coronary heart disease
Sunflower oil enriched with conjugated linoleic acid	Enhances anticancer immune defenses
Tomato with enhanced lycopene content	Reduces risk of specific types of cancer (eg, prostate)

omega-3 PUFA have been attributed to their ability to lower serum triglycerol and cholesterol. Consumption of 35 g or more of fish daily has been shown to reduce the risk of death from nonsudden myocardial infarction (24), and as little as one serving of fish per week was associated with a significantly reduced risk of total cardiovascular mortality (25). In addition, omega-3 PUFA are essential for normal growth and development and may also play a role in prevention and treatment of hypertension, arthritis, other inflammatory and autoimmune disorders, and cancer (26). In this context, several countries, including Canada, the United Kingdom, and Denmark, have dietary guidelines for omega-3 PUFA and requirements for its addition in infant formula.

CURRENT TRENDS

Consumer demand for foods capable of promoting good health and preventing or alleviating diseases is growing. This demand is likely to increase as baby boomers strive to manage the chronic health problems associated with aging and seek cost-effective alternatives to drugs. Also, with the ongoing legislative changes that permit claims for foods and associated components, and with the new and exciting scientific discoveries, many more companies, including multinationals, are and will continue to put more resources into the development and marketing of functional foods (27). Table 2 lists selected functional foods and bioactive food ingredients currently under active development in North America and around the world. In the process of developing new and improved functional products, a better understanding of food, nutrition, and health will likely emerge. Nonetheless, to gain wide acceptance, functional foods must offer health benefits that consumers can understand. Although terminology used in the functional food area can be confusing, it is generally agreed that the primary goal is to develop foods with added health benefits, rather than to rely on supplements. Any claim about the benefits of foods, however, must be based on sound and accurate scientific information. The development of functional foods also requires careful attention to safety, labeling, and claims as well as to nutritional and physiological rationale, cost, and sensory quality and should depend heavily on scientific substantiation.

BIBLIOGRAPHY

1. F. W. Scott et al., *Recommendations for Defining and Dealing With Functional Foods. A Discussion Paper*, Bureau of Nutritional Sciences, Food Directorate, Health Canada, Ottawa, 1996.
2. B. F. Haumann, "Health Benefits May Be Key for Foods of the Future," *Int. News Fats, Oils Rel. Mat.* (INFORM) **10**, 18–22 (1999).
3. A. B. Caragay, "Cancer-Preventive Foods and Ingredients," *Food Technol.* **46**, 65–68 (1992).
4. C. S. Fuchs et al., "Dietary Fiber and the Risk of Colorectal Cancer and Adenoma in Women," *N. Engl. J. Med.* **340**, 169–176 (1999).
5. U.S. Dept. Health and Human Services/Food and Drug Administration, "Food Labelling: Health Claims; Oats and Coronary Heart Disease," *Federal Register* **62**, 3584–3601 (1997).
6. A. L. Gerhardt and N. B. Gallo, "Full-Fat Rice Bran and Oat Bran Similarly Reduce Hypercholesterolemia in Humans," *J. Nutr.* **128**, 865–869 (1998).
7. O. Alabaster, Z. Tang, and N. Shivapurkar, "Inhibition by Wheat Bran Cereals of the Development of Aberrant Crypt Foci and Colon Tumors," *Food Chem. Toxicol.* **35**, 517–522 (1997).
8. C. M. Hasler, "Functional Foods: Their Role in Disease Prevention and Health Promotion," *Food Technol.* **52**, 63–70 (1998).
9. C-F. Chan, P. C. K. Cheung, and Y-S. Wong, "Hypocholesterolemic Effects of Protein Concentrates From Three Chinese Indigenous Legume Seeds," *J. Agric. Food Chem.* **46**, 3698–3701 (1998).
10. B. D. Oomah and G. Mazza, "Flaxseed Products for Disease Prevention," in G. Mazza, ed., *Functional Foods Biochemical and Processing Aspects*, Technomic, Lancaster, Pa., 1998, pp. 91–138.
11. M. K. Gupta, "NuSun—The Future Generation of Oils," *Int. News Fats, Oils Rel. Mat.* (INFORM) **9**, 1150–1154 (1998).
12. S. K. Clinton, "Lycopene: Chemistry, Biology, and Implications for Human Health and Disease," *Nutr. Rev.* **56**, 35–51 (1998).
13. G. Cao et al., "Increases in Human Plasma Antioxidant Capacity After Consumption of Controlled Diets High in Fruit and Vegetables," *Am. J. Clin. Nutr.* **68**, 1081–1087 (1998).
14. H. M. Meltzer and K. E. Malterud, "Can Dietary Flavonoids Influence the Development of Coronary Heart Disease?" *Scand. J. Nutr.* **41**, 50–57 (1997).
15. D. R. Schmidt and A. E. Sobota, "An Examination of the Anti-adherence Activity of Cranberry Juice on Urinary and Nonurinary Bacterial Isolates," *Microbios.* **55**, 171–181 (1998).
16. A. P. Day et al., "Effects of Concentrated Red Grape Juice Consumption on Serum Antioxidant Capacity and Low-Density Lipoprotein Oxidation," *Ann. Nutr. Metab.* **41**, 353–357 (1998).
17. M. Jang et al., "Cancer Chemopreventive Activity of Resveratrol, A Natural Product Derived From Grapes," *Science* **275**, 218–220 (1997).
18. K. Nakachi et al., "Influence of Drinking Green Tea on Breast Cancer Malignancy Among Japanese Patients," *J. Jap. Cancer Res.* **89**, 254–261 (1998).
19. R. Klont and P. Mannion, "Inulin and oligofructose," *World of Ingredients*, 34–37 (Sept. 1998).
20. B. K. Mital and S. K. Garg, "Anticarcinogenic, Hypocholesterolemic, and Antagonistic Activities of *Lactobacilus acidophilus*," *Crit. Rev. Microbiol.* **21**, 175–214 (1995).
21. P. Jelen and S. Lutz, "Functional Milk and Dairy Products," in G. Mazza, ed., *Functional Foods Biochemical and Processing Aspects*, Technomic, Lancaster, Pa., 1998, pp. 357–380.
22. Y. L. Ha, N. K. Grimm, and M. W. Pariza, "Anticarcinogens From Fried Ground Beef: Health-Altered Derivatives of Linoleic Acid," *Carcinogenesis* **8**, 1881–1887 (1987).
23. E. Doyle, "Scientific Forum Explores CLA Knowledge," *Int. News Fats, Oils Rel. Mat.* (INFORM) **9**, 69–72 (1998).

24. M. L. Daviglus et al., "Fish Consumption and 30-Year Risk of Fatal Myocardial Infarction," *N. Eng. J. Med.* **336**, 1046–1053 (1997).
25. C. M. Albert et al., "Fish Consumption and Risk of Sudden Cardiac Death," *JAMA, J. Am. Med. Assoc.* **279**, 23–28 (1998).
26. F. Shahidi, "Functional Seafood Lipids and Proteins," in G. Mazza, ed., *Functional Foods Biochemical and Processing Aspects*, Technomic, Lancaster, Pa., 1998, pp. 381–401.
27. V. Brower, "Nutraceuticals: Poised for a Healthy Slice of the Healthcare Market?" *Nature Biotech.* **16**, 728–731 (1998).

B. Dave Oomah
G. Mazza
Pacific Agri-Food Research Centre
Summerland, British Columbia
Canada

G

GELATIN

Gelatin is a substantially pure protein food ingredient, obtained by the thermal denaturation of collagen (1), which is the structural mainstay and most common protein in the animal kingdom. Today, gelatin is usually available in granular powder form, although in Europe sheet gelatin is still available.

There are two main types of gelatin. Type A, with isoionic point of 7 to 9, is derived from collagen with exclusively acid pretreatment. Type B, with isoionic point of 4.8 to 5.2, is the result of an alkaline pretreatment of the collagen. However, gelatin is sold with a wide range of special properties, such as gel strength, to suit particular applications.

Gelatin (2) forms thermally reversible gels with water, and the gel melting temperature (<35°C) is below body temperature, which gives gelatin products unique organoleptic properties and flavor release. The disadvantage of gelatin is that it is derived from animal hide or bone (not from horses as is a common perception), so there are problems with regard to kosher and Halal status. Vegetarians also have objections to its use. Competitive gelling agents, such as starch, alginate, pectin, agar, carrageenan, are all carbohydrates from vegetable sources, but their gels lack the melt-in-the-mouth and elastic properties of gelatin gels.

CHEMISTRY AND BIOCHEMISTRY

Gelatin is an amphoteric protein with isoionic point between 5 and 9 depending on raw material and method of manufacture. Like its parent protein, collagen (3), it is unique in that it contains 14% hydroxyproline, 16% proline, and 26% glycine. The only other animal product containing hydroxyproline is elastin, at a very much lower concentration, so hydroxyproline is used to determine the collagen or gelatin content of foods. The protein is made up of peptide triplets, glycine-X-Y, where X and Y can be any one of the amino acids but proline has a preference for the X position and hydroxyproline the Y position (3). Approximately 1,050 amino acids produce an α-chain with the left-handed proline helix conformation. Collagen exists in many different forms, but gelatin is only derived from sources rich in Type I collagen, which contains no cystine. However, hide or skin contains some Type III collagen, which can be the source of traces of cystine found in some gelatins. Although Type I collagen contains no cystine, the α-procollagen chains excreted by the cell do contain cystine at the C-terminal end of the protein that is thought to be the site of assembly of three α-chains. The three chains then spontaneously (4) coil together, zipper fashion, to form a right-handed helix. After spontaneous helix formation, cross-links between chains are formed in the region of the N-terminal telopeptides (globular tail portion of the chains), and then the telopeptides (containing the cystine and tyrosine of procollagen) are shed, leaving the rodlike, about 3,150-amino-acid-containing triple helix. These collagen rods assemble together with a quarter-stagger to form the collagen fiber, and the fibers are stabilized by further cross-links.

Gelatin is the product of denaturation or disintegration of collagen. Initially, the α-chains of collagen are held together with several different but easily reducible cross-links. As the collagen matures, the cross-links become stabilized (3). As time progresses, the eta-amino groups of lysine become linked to arginine by glucose molecules (Maillard reaction) to form the pentosidine type cross-links that are extremely stable (5). Hence, when the alkaline processing is used on young animal skin, the alkali breaks one of the initial (pyridinoline) cross-links and as a result, on heating, the collagen releases mainly denatured α-chains into solution (5). Once the pentosidine cross-links of the mature animal have formed in the collagen, the main process of denaturation has to be thermal hydrolysis of peptide bonds, resulting in protein fragments of various molecular weights (polydisperse protein fragments). With the acid process, the collagen denaturation is limited to the thermal hydrolysis of peptide bonds with a small amount of α-chain material from acid-soluble collagen in evidence (6).

Nutritionally, gelatin is not a complete protein food because the essential amino acid tryptophan is missing, and methionine is present at a low level.

Type A gelatin (dry and ash-free) contains 18.5% nitrogen, but because of the loss of amide groups, Type B gelatin contains only about 18% nitrogen (7). Gelatin is abnormally stable, and a special catalyst has to be used to obtain the correct Kjeldahl nitrogen content.

The amino acid analysis of gelatin (8) is variable, particularly for the minor constituents, depending on raw material and process used, but proximate values by weight are glycine, 21%; proline, 12%; hydroxyproline, 12%; glutamic acid, 10%; alanine 9%; arginine, 8%; aspartic acid, 6%; lysine, 4%; serine, 4%; leucine, 3%; valine, 2%; phenylalanine, 2%; threonine, 2%; isoleucine, 1%; hydroxylysine, 1%; methionine and histidine, <1%; and tyrosine, <0.5%. The peptide bond has considerable aromatic character, so gelatin shows an absorption maximum at about 230 nm.

Collagen is resistant to most proteases and requires special collagenases for its enzymic hydrolysis. Gelatin, however, is susceptible to most proteases, but they do not break gelatin down into peptides containing much less than 20 amino acids.

The cross-linking of gelatin with aldehydes is being used to extend the uses of gelatin. In particular, treatment of gelatin films with glutaraldehyde is receiving considerable study to improve thermal resistance, decrease solubility in water, and improve mechanical properties. In Japan and Brazil, the cross-linking of gelatin using the enzyme *trans*-glutaminase and its use in joining gelatin to other proteins is approved for food use. An occasional phenomenon is the loss of gelatin solubility after storage in a new kitchen cupboard where the residual formaldehyde vapor from the adhesives used causes cross-linking of the gelatin. This reaction has been used to make gelatin adhesives water resistant. Furthermore, the smokes used in food preservation are rich in aldehydes and thus can have unwanted reactions with gelatin.

GELATIN MANUFACTURE

There are a large number of unit processes used in the manufacture of gelatin. The raw materials from which it is derived are demineralized bone (called ossein), pigskin, cow hide, and fish skin. In China, donkey hide is used quite extensively. In theory, there is no reason for excluding any collagen source from the manufacture of gelatin, but the raw materials listed are currently commercially available. Interestingly, in countries where pork is sold with its skin intact, there is no pigskin available for gelatin manufacture.

As mentioned earlier, there are two processes by which collagen is processed to gelatin: the acid process and the alkali process. The acid process (studied in detail by Reich et al. [9]) is mainly used with pigskin and fish skin and sometimes bone raw materials. It is basically one in which the collagen is acidified to about pH 4 and then heated stepwise from 50°C to boiling to denature and solubilize the collagen. Thereafter, the denatured collagen or gelatin solution has to be defatted, filtered to high clarity, concentrated by vacuum evaporation or membrane ultrafiltration treatment to a reasonably high concentration, and then dried by passing dry air over the gel. The final process is one of grinding and blending to customer requirements and packaging. The resulting gelatin has an isoionic point of 7 to 9 based on the severity and duration of the acid processing of the collagen, which causes limited hydrolysis of the asparagine and glutamine amino acid side chains.

The alkali process (studied in detail by Cole [10]) is used on bovine hide and collagen sources where the animals are relatively old at slaughter. The process is one in which collagen is submitted to a caustic soda or lengthy liming process before extraction. The alkali hydrolyzes the asparagine and glutamine side chains to glutamic and aspartic acid relatively quickly (11), with the result that the gelatin has a traditional isoionic point of 4.8 to 5.2. However, with shortened (7 days or less) alkali treatment, isoionic points as high as 6 are produced. After the alkali processing, the collagen is washed free of alkali and treated with acid to the desired extraction pH (which has a marked effect on the gel strength to viscosity ratio of the final product). The collagen is then denatured and converted to gelatin by heating, as with the acid process. Because of the alkali treatment, it is often necessary to demineralize the gelatin solution to remove excessive amounts of salts using ion–exchange or ultrafiltration. Thereafter, the process is the same as for the acid process—vacuum evaporation, filtration, gelation, drying, grinding, and blending.

Although gelatin is often considered a commodity like sugar, the descriptions of the processes and raw materials above should indicate that gelatin has the potential for being a variable product. Users must ensure that they are using the best product for each particular application. In the past, little emphasis has been placed on the animal age of the raw material, particularly in the case of gelatins from bovines; however, it is now known that this factor plays a significant role in the molecular structure of the derived gelatin. The role of liming in the alkali process used to be considered one of progressive alkali hydrolysis of the collagen, which made it possible to denature the collagen at lower temperatures and thus maximize the yield of top-quality gelatin. Recently, however, it has been shown that the role of liming is limited to the hydrolysis of one collagen cross-link that fluoresces at 290 to 380 nm and that liming has increasing less effect on extractability in older animals. The result is that alkali treatment times have been greatly reduced. One of the less well-recognized effects of alkali treatment is the "opening up" of the hide collagen, as it is called in leather manufacture, or the destruction of the proteoglycans associated with the collagen fibrils. This probably results in a more pure gelatin via the alkali process, as is indicated by electrophoresis of the gelatin proteins (6).

At present, enormous developments are being made in the understanding of the structure of collagen and the changes occurring with senescence, and these developments are bound to have an impact on the appreciation of the variables in gelatin, particularly at the molecular level.

GELATIN SAFETY

Gelatin is regarded as a food ingredient rather than an additive, and it is generally regarded as safe (GRAS). In 1993, the Food and Drug Administration (FDA) reiterated the GRAS status of gelatin and stated that there was no objection to the use of gelatin from any source and any country, provided that the hide from animals showing signs of neurological disease were excluded and that specified raw materials were excluded from the manufacturing process. At the beginning of the bovine spongiform encephalopathy (BSE) scare in Europe, the popular media brought suspicion on all products of bovine origin as being possible transmitters of the disease to humans as Creutzfeldt-Jakob disease. This was a thoroughly unscientific assessment of the dangers of spreading infection. It is now recognized that BSE is a neurological and brain problem that is not associated with the hide of the animal. It is also recognized that the processes of manufacturing gelatin make it virtually impossible for the survival of a defective prion, if it were present in the first place.

Detailed and unbiased information on BSE is available from the Institute of Food Science and Technology website. Gelatin retains its GRAS status, and the Joint Expert Commission on Food Additives (JECFA) placed no limit on the use of gelatin in 1970.

Gelatin is an excellent growth medium for most bacteria, so considerable care needs to be taken during manufacture to avoid contamination. This care is evidenced by the use of documented hazard analysis of critical control points (HACCP) programs by manufacturers. In the same way, to ensure product reproducibility, most companies are implementing ISO 9000 quality management systems.

GELATIN PROPERTIES AND USES

Solubility in Water

Gelatin is only partially soluble in cold water, but dry gelatin swells or hydrates when stirred into water. Such mixtures should generally not exceed 34% gelatin. On warming to about 40°C, gelatin that has been allowed to hydrate for about 30 minutes melts to give a uniform solution. Alternatively, dry gelatin can be dissolved by stirring into hot water, but stirring must be continued until solution is complete. This method is normally only used for dilute solutions of gelatin.

If gelatin solutions are spray-dried or drum-dried from the sol state, the resulting gelatin is cold-water soluble, and such gelatins gel quickly when stirred into cold water. These gels are generally not clear, so the use of this form of gelatin is limited to milk puddings and other products where solution clarity is not required.

The compatibility of gelatin in aqueous solution with polyhydric alcohols, such as glycerol, propylene glycol, and sorbitol, is virtually unlimited, and they are used to modify the hardness of gelatin films.

Adhesive Properties

Possibly the oldest use of gelatin was as animal glue. For adhesion to take place, a warm gelatin solution must be used, and the gelatin must not have gelled before the surfaces to be glued are brought together. An example of this use of gelatin is in pharmaceutical or confectionery tableting and in licorice allsorts where it can be used to join the layers.

Gelling Properties

The most common use of gelatin is for its thermally reversible gelling properties with water, such as in the production of table jellies. Gelatin is also used in aspic to add flavor to meat products, and after gelling it also provides a pleasing shiny appearance to the product. In some cases, gelling is known as its water-absorbing property. For example, in canned hams, gelatin can be added to the can before cooking. After cooking, the exudate from the meat is absorbed by the gelatin and appears as a gel when the can is opened.

In confectionery, gelatin is used as the gelling binder in gummy products and wine gums. In the manufacture of these products, gelatin is combined with sugar and glucose syrups. Incompatibility between gelatin and glucose syrup can occur (12) and is a function of the concentration of glucose polymers containing more than 2 glucose units in the syrup. Competition between gelatin and glucose polymers for water in low-water content products can result in, at worst, precipitation of the gelatin and, at best, a marked loss in gelling properties or hardness of the product. It is also known that different gelatins with similar properties in water can have very different properties in confectionery formulations.

Some raw fruits, such as pineapple and papaya, contain proteolytic enzymes (bromelain) that hydrolyse gelatin and destroy its gelling ability. In such cases, it is essential that the fruit is cooked to destroy the protease before the fruit is added to gelatin solutions.

In general, one can say that the lower the mean molecular weight (MW) of a gelatin, the lower the gel strength and viscosity of its solution. However, the collagen chain (MW 100 kD; gel strength, 364 g Bloom) is the main contributor of gel strength (13), and higher MW components (β-chain, MW 200 kD; γ-chain, MW 300 kD; and microgel, MW > 300 kD) make a relatively low contribution to gel strength but a high contribution to viscosity.

Foaming Properties

Gelatin is a very efficient foam stabilizer, and this property is exploited in the manufacture of marshmallows. Different gelatins have different foam stabilizing properties, and gelatin for this use needs to be carefully selected. However, the foaming properties can be standardized by the use of sodium lauryl sulfate (14), if this is permitted by local food additive regulations. In marshmallows, the gelatin's film-forming properties are also used to stabilize the foam on cooling, and because the product is normally not acidified, it has to have a much lower moisture content (>85% solids) than gummy products (76% solids) to avoid mold growth in storage.

Protective Colloid and Crystal Habit-Modifying Properties

If a gelled jelly is frozen, the product will suffer from syneresis, and after thawing, the clear jelly will disintegrate with much exuded water. However, if water containing 0.5% gelatin is frozen, the water will freeze as millions of small discrete crystals, instead of forming a single solid block of ice. This effect is most desirable in frozen ice treats (popsicles or ice lollies). Gelatin is also used in ice cream manufacturing to obtain a smooth product with small ice crystals and to ensure that any lactose precipitates as fine crystals, which avoids the development of graininess with time.

Film-Forming Properties

Gelatin's film-forming properties are used in the manufacture of hard and soft (pharmaceutical) capsules. Gelatin films shrink with great force on drying; hence, such uses usually involve the addition of polyhydric alcohols to modify the adhesion and flexibility of the dry film. Also, for film forming, a gelatin with a high viscosity is preferred to one with a low viscosity. For hard capsules and in photography, ossein gelatin is preferred and commands a premium price.

Emulsifying Properties

The amphoteric character and the hydrophobic areas on the peptide chain give gelatin limited emulsifying and emulsion-stabilizing properties used in the manufacture of toffees and water-in-oil emulsions, such as lowfat margarine.

Stability

Dry gelatin has an almost infinite shelf life as long as the moisture content is such as to ensure that the product is stored below the glass-transition temperature.

The stability of gelatin in solution depends on temperature and pH. Generally, to minimize loss of gel strength and viscosity with time, the pH of the solution should be in the range 5 to 7, and the temperature should be kept as low as possible, consistent with the avoidance of gelation and the suitability of the solution viscosity to the particular application. Often, the cause of degradation or hydrolysis of gelatin in solution is microbial proliferation, so gelatin solutions should not be stored for longer than is absolutely necessary. After addition of the acid to confectionery formulations, the solution should be used and cooled or gelled with minimal delays.

Microencapsulation and Mixed Film-Forming Properties

Besides being precipitated by polymers competing for water, gelatin is amphoteric—it has positive and negative charges on the molecule (and no net charge at the isoionic point). Hence, at a pH where the basic side chains do not carry a charge, acid groups, for example, from gum arabic, can react with the basic groups of gelatin to form an insoluble gelatin–arabate complex that can be precipitated around emulsified oil droplets, forming microencapsulated oil. The microcapsules are hardened with formaldehyde or glutaraldehyde before harvesting and drying. In this application, the pI of the gelatin is critical. This process has been used in the food industry for encapsulating flavors.

Milk and Food Stabilizing Properties

Gelatin is used as a stabilizer, particularly in yogurt, where the addition of 0.3 to 0.5% acts to prevent syneresis, thus allowing the production of stirred and fruit-containing products. In this application, the gelatin reacts with the casein in the milk to reduce its tendency to separate water from the curd. Gelatin can also be used in cheese manufacturing to improve yield and in the stabilization of thickened cream.

Fruit Juice Clarifying Properties

In fining applications, gelatin reacts with polyphenols (tannins) and proteins in fruit juices, forming a precipitate that settles and leaving a supernatant that is stable to further cloud formation with storage time. In wine, usage levels are about 1 to 3 g/hL, and excess usage, which would lead to protein instability, needs to be avoided. Traditionally, low-Bloom-strength gelatins have been used, but high-Bloom-strength gelatins are equally effective (15). However, from the practical point of view, the use of low-Bloom-strength gelatin is cheaper and makes it easier to mix the gelatin into the bulk of the cold juice before gelation can occur. In this regard, it has become common practice to treat cold grapes, during the initial crushing process, with gelatin that has been hydrolyzed to the extent that it can no longer gel.

Texturizing Properties

Gelatin is used in dried soups to provide the appropriate mouth feel (viscosity) to the final product.

Nutritional Properties

As stated earlier, gelatin is not a complete protein source because it is deficient in tryptophan and low in methionine. However, the digestibility is excellent, and it is often used in feeding invalids. The high level of lysine (4%) is noteworthy. Although controversial, studies have shown that the consumption of 7 to 10 g/day can significantly improve nail growth rate and strength (16), and it also promotes hair growth (17). Gelatin has also been shown to benefit arthritis patients in a large proportion of cases (18).

Corrosive Properties

Although 304 stainless steel (s/s) can be used with milk, gelatin attacks 304 s/s, and tubing can be perforated after a few months of continuous usage. With gelatin, it is essential to use 316 s/s. If heat exchanger plates are involved, the use of 316 s/s with the minimum specified molybdenum content of only 2% can be unacceptable.

Fish Skin Gelatin

Fish skin gelatin is available commercially and can be produced for kosher use provided that the appropriate conditions are met (such as the use of fish with scales). Fish gelatin with normal gel strength has a normal hydroxyproline content (19) and is made from fish from warmer waters and not necessarily from fresh water, although this is normally the case. Fish gelatin with low or no gel strength (20) has a low hydroxyproline content (7) and is produced from cold water species that are obtained typically from the sea.

The low-gel-strength gelatin has been used to emulsify vitamin A before spray drying to give another type of microencapsulated product using gelatin.

GELATIN TESTING METHODS

The best published sources of gelatin testing methods are *British Standard 757* (21) or *Standard Methods for Sampling and Testing Gelatin*, published by the Gelatin Manufacturers of America (22) or the *Pharmacopoeias*. Many of the methods used in laboratories need to be modified to suit the peculiarities of gelatin.

Identification

Gelatin gives the normal positive trichloroacetic acid, biuret, and ninhydrin tests for protein. The precipitate with 5% tannic acid is a particularly sensitive test for very dilute solutions of gelatin. In addition, the thermally reversible gelation of a 6% solution in water between 10° and 60°C is unique for this protein.

Gel Strength

The most important attribute of gelatin is its gel strength, and when determined by the standard method (21), this is called the Bloom strength. Bloom strength is the force in grams required to press a 12.5-mm diameter plunger 4 mm into 112 g of a standard 6 2/3% w/v gelatin gel at 10°C.

Several penetrometer-type instruments have been adapted to determine Bloom strength.

A frequent question is how to substitute gelatin of one Bloom strength for a gelatin of another. As a guide one can say

$$C \times \sqrt{B} = k$$

or

$$C_1(\sqrt{B_1})/(\sqrt{B_2}) = C_2$$

where C = concentration, B = Bloom strength, and k = constant (11). However, there are other considerations besides gel strength that can invalidate such a substitution calculation. For example, in a gummy formulation, the texture using 250-Bloom gelatin is far less elastic and sticky than when 180-Bloom gelatin is used.

Viscosity

From the point of view of functionality, the solution viscosity of gelatin is probably the second-most important parameter. The standard method calls for the viscosity of a 6-2/3% solution at 60°C. Low viscosity (and a high gel strength) is required for poured confectionery, and high viscosity is required for film-forming applications.

In viscosity calculations, usually $C \log V = k$, but the model is not as good as the mathematical model for Bloom calculations.

Color and Clarity

Solution color and turbidity or clarity are attributes that may or may not be important depending on the application. Poor clarity markedly affects the ability to measure color (23), and at this stage, there are no internationally accepted methods for determining these attributes. However, if clarity is good, then gelatin color obeys Beer's law.

pH

Solution pH (1%) is usually about pH 5 but can vary considerably. At this pH, the viscosity of type B gelatin is minimal and the gel strength is maximal; hence, from the manufacturer's point of view, it is advantageous to manufacture gelatin at this pH. However, because of the strong buffering capacity of gelatin, this pH may not be the most advantageous for the customer.

Moisture

The moisture content of gelatin may be as high as 16%, but normally it is about 10 to 13%. At 13.0% moisture content, the glass-transition temperature (24) of gelatin is about 64°C, which allows particle size reduction to be a simple operation. In addition, at 13% moisture content and 25°C gelatin is close to equilibrium with ambient air moisture contents of about 46% relative humidity. Gelatin with a moisture content of 6 to 8% is very hygroscopic, and it becomes difficult to determine the physical attributes with accuracy.

Because of the variable granule size of gelatin, the rate of moisture loss at 105°C can be variable. Hence, it is normal to add water to the gelatin powder before placing the sample in the drying oven. This means that the gelatin melts and water is lost from a uniform thin film of protein. Metal dishes must be used because after drying, the film of gelatin shrinks and breaks glass or ceramic containers.

Finally, the drying of gelatin to very low moisture contents results in cross-linking and loss of solubility. It is thus difficult to distinguish between free and bound water in gelatin.

Ash

The gelatin ash content is determined by pyrolysis at 550°C. Ash contents up to 2.5% can usually be accepted in food applications. However, the nature of the ash can be important. For example, 2% $CaSO_4$ in gelatin can have excellent clarity in spite of the solubility product of the ash being exceeded (because of the crystal-habit modifying effect of gelatin). However, after dilution of the gelatin in a confectionery formulation, the ash can precipitate. Furthermore, ammonia is often used as a pH modifier in gelatin preparation, and salts such as NH_4Cl are not determinable by pyrolysis.

Sulfur Dioxide Content

Sulfur dioxide is used as a biocide and bleach in gelatin manufacture. The nationally permitted level of residual SO_2 in gelatin is variable, and the methods for its determination can give a great variation in results. It is known that gelatin promotes oscillating redox reactions (25,26), and the control of this contaminant is not easy. Hydrogen peroxide is often used to control the SO_2 content of gelatin, and sometimes the permitted level of this contaminant is also specified. It is interesting to note that both H_2O_2 and SO_2 can be shown to coexist in gelatin.

Heavy Metal Content

The determination of heavy metals in gelatin can be a problem. It is difficult to completely degrade gelatin, and the main component of the ash in gelatin can be of low solubility, such as calcium sulfate, and have a variable ability to absorb traces of heavy metals. Internal standards must be used wherever possible.

Isoionic Point

The isoionic point of gelatin (11) is best determined by passing a 1% solution of the gelatin at 40°C through a mixed-bed column of ion–exchange resin (Rohm & Haas MB3) at a flow rate of not more than 10 bed volumes per hour and measuring the pH of the eluate. After cooling, isoionic gelatin has poor clarity, and the conductivity should be between 1 and 5 μs/cm for type B gelatin.

Microbiological Properties

Gelatin is an excellent nutrient for most bacteria, so the manufacturing processes have to carefully avoid contamination. Most countries have microbiological specifications for gelatin, but generally they are not very onerous. Total mesophyllic plate counts of 1,000 are generally accepted with various countries limiting the presence of coliforms, *Escherichia coli*, *Salmonella*, clostridial spores, staphylococci, and sometimes even pseudomonads.

BIBLIOGRAPHY

1. A. J. Bailey and R. G. Paul, *Journal of the Society of Leather Technologists and Chemists* **82**, 104–110 (1998).
2. M. Glicksman, *Gum Technology in the Food Industry*, Academic Press, New York, 1969, pp. 359–397.
3. A. J. Bailey and N. D. Light, "Genes, Biosynthesis and Degradation of Collagen," in A. J. Bailey and N. D. Light, eds., *Connective Tissue in Meat and Meat Products*, Elsevier Applied Science, London, United Kingdom, 1989, pp. 225–247.
4. D. J. Prockop, *Matrix Biol.* **16**, 519–528 (1998).
5. C. G. B. Cole and J. J. Roberts, *Proceedings of the International Union of Leather Technologists and Chemists Societies Congress*, 57–64 (1997).
6. C. G. B. Cole and J. J. Roberts, *Journal of the Society of Leather Technologists and Chemists* **80**, 136–141 (1996).
7. J. E. Eastoe and A. A. Leach, A Survey of Recent Work on the Amino Acid Composition of Vertebrate Collagen and Gelatin, in G. Stainsby, ed., *Recent Advances in Gelatin and Glue Research*, Pergamon Press, New York, 1958, pp. 173–178.
8. P. V. Stevens, *Food Australia* **44**, 320–324 (1992).
9. G. Reich, S. Walther, and F. Stather, *Deutsche Lederinstitut* **18**, 15–23 (1962).
10. C. G. B. Cole, "The Occurrence of Dark Coloured Gelatin," in *Occurrence, Measurement and Origins of Gelatin Colour as Determined by Fluorescence and Electrophoresis*, University of Pretoria, Pretoria, South Africa, 1995, pp. 19–155.
11. A. Veis, *The Macromolecular Chemistry of Gelatin*, Academic Press, New York, 1964, pp. 107–113, 196, 392–396.
12. W. M. Marrs, "Gelatin/Carbohydrate Interactions and Their Effect on the Structure and Texture of Confectionery Gels," in G. O. Phillips, P. A. Williams, and D. J. Wedlok, eds., *Progress in Food Science and Nutrition 6*, Pergamon Press, New York, 1982, pp. 259–268.
13. E. Heidemann et al., *Proceedings of the 5th IAG Conference: Photographic Gelatin*, International Arbeitsgem. Photogelatine, Fribourg, Switzerland, 1989.
14. *Federal Register*, May 15, 1964, p. 6383.
15. W. Bestbier, *Wynboer* **621**, 6–62 (1983).
16. M. Schwimmer and M. G. Mulinos, *Antibiotic Medicine and Clinical Therapy* **IV**, 403–407 (1957).
17. United States Patent 4,749,684 (June 7, 1988), B. Silvestrini (to Bruno Silvestrini).
18. M. Adam, *Therapiewoche* **38**, 2456–2461 (1991).
19. European Patent 0 436 266 A1 (Published 10.07.91), S. Grossman (to Bar Ilan University).
20. A. G. Ward, "Conversion of Collagen to Gelatin, and Chemical Composition," in G. Stainsby, ed., *Recent Advances in Gelatin and Glue Research*, Pergamon Press, New York, 1958, pp. 137–139.
21. *Methods for sampling and testing gelatine. BS 757 : 1975. Gr8*, British Standards Institution, London, United Kingdom, 1975.
22. *Standard Methods for Sampling and Testing of Gelatin*, Gelatin Manufacturers of America, New York, 1986.
23. C. G. B. Cole and J. J. Roberts, *Meat Science* **45**, 23–31 (1997).
24. M. H. McCormick-Goodhart, "Research Techniques in Photographic Conservation," *Proceedings of the Copenhagen Conference* 65–70 (May 1995).
25. C. R. Chinake and R. H. Simoyi, *S. Afr. J. Chem.* **48**, 1–7 (1995).
26. Z. Melichova, A. Olexova, and L. Treindel, *Z. Phys. Chem.* (Munich) **191**, 259–264 (1995).

GENERAL REFERENCES

A. J. Bailey and N. D. Light, *Connective Tissue in Meat and Meat Products*, Elsevier Applied Science, London, United Kingdom, 1989.

M. Glicksman, *Gum Technology in the Food Industry*, Academic Press, New York 1969.

G. Stainsby, *Recent Advances in Gelatin and Glue Research*, Pergamon Press, New York, 1958.

A. Veis, *The Macromolecular Chemistry of Gelatin*, Academic Press, New York, 1964.

A. G. Ward and A. Courts, *The Science and Technology of Gelatin*, Academic Press, New York, 1977.

BERNARD COLE
Leiner David Gelatin
Krugersdorp, South Africa

GENETIC ENGINEERING: ANIMALS. See GENETIC ENGINEERING: ANIMALS in the Supplement section.

GENETIC ENGINEERING: FOOD FLAVORS

Traditional food biotechnologies are frequently characterized by the formation of ethanol, acetic acid, propanoic acid, and lactic acid. The resulting products, though different from the original fruit must, flour paste, milk, or meat in many ways, were experienced not only as having no adverse effects on well-being (if consumed in moderation), but they even possessed new, attractive sensory properties. Based on these very roots of biotechnology (1), a great number of nonvolatile flavor compounds, such as acidulants, amino acids, 5′-nucleotides, and certain carbohydrates are now manufactured on an industrial scale (2). Although the enzymatic origins of the volatile flavor fraction of many foods are clearly recognized (3), aroma-producing microorganisms and enzymes thereof have, for a long time, been rather regarded as laboratory oddities (4). Today, the increasing number of papers and patents on the biotechnology of aromas indicates a strong trend in both academic and industrial flavor research (5–26). Increasing confidence in the future of biotechnology (27) and in the so-called soft chemistry approach has merged together with "all-natural" marketing claims into the recent development of pilot and production scale bioprocesses for aroma chemicals (Table 1).

Table 1. Industry Driving Forces

Technical Push
High selectivities or specificity (chemo-, regio-, stereo-)
High reaction rate at low molar fractions
Environmentally compatible, renewable substrates and mild reaction conditions
Multistep syntheses including cofactor regeneration (whole cell)
Business Pull
Health- and nutrition-conscious lifestyles demand "natural" products
Classical plant sources suffer from microbial or insect infestation, sociopolitical instability and depend on seasonal variation, fertilizers, etc
Character impact components may possess additional bioactivities and, therefore, often occur in traces only in their plant sources

Source: Ref. 28.

Volatile flavors and fragrances command a growing (currently ca $10 billion) market (29). The market of food flavors in the United States alone is estimated at over $1 billion and is continuously increasing (21). The resulting demand of the industry for natural ingredients suggests that the agricultural sources with their inherent instabilities will no longer satisfy the need. The physiologically most active constituents of natural flavors, the so-called character impact components, are often present in plants in traces and in bound forms, an additional obstacle for an economic isolation. Recent advances in cell biology and in bioengineering—and particularly the ability to alter the biocatalyst's properties genetically—possess great potential for the flavor industry, as the legal definitions of the Food and Drug Administration and of the Council of the European Communities classify fermentation flavors (with few restrictions) as "natural" compounds. From a chemical viewpoint, the use of a chiral (bio)catalyst would offer the most adequate synthetic approach anyway: The sensory properties of many flavors strongly depend on the chiral structure of the molecules, and the enantiomeric forms of many terpenes, lactones, and so on may exhibit different, sometimes even opposite, sensory characteristics (30).

An estimated number of less than 100 flavor compounds and building blocks are now available on the market, among them carboxylic fatty acids, esters and lactones derived thereof, cheese and yeast flavors, methyl ketones, benzaldehyde, vanillin, cinnamates, and some other character impact components. Particular progress has been made using lipases and certain other hydrolases, whereas the improvement of cell based processes appears more difficult. According to the increasing structural complexity of the biocatalyst, this article highlights some recent achievements, but also current problems of the biotechnology of volatile flavors using enzymes, microbial and fungal cells, and plant cells.

ENZYME MIXTURES AND PURIFIED ENZYMES

Microbial enzymes, products of bioprocesses themselves, are an integrated part of modern food technology. Their long-established use by the food industry leaves no doubt that biocatalysts are particularly advantageous for the safe production of food and pure food ingredients. In view of regiospecifity and enantioselectivity, substrate specifity, and ability to function at statistically cold temperatures, it would seem obvious to use enzymes for the generation of chemically often sensitive, volatile molecules.

Indirect Action on Food Flavor

Enzymes of main group three (hydrolases) possess a share of about 80% of the industrial enzyme market. The application of pure or combined lipases, amylases, glucoamylases, cellulases, invertases, proteases, or ribonucleases may affect the overall flavor of food in various ways: (*1*) by direct accumulation of monomer flavor compounds from polymer or conjugated precursors; (*2*) by increasing the pool of precursors for a subsequent thermal flavor generation (reaction flavors); and, (*3*) by liberating previously adsorbed compounds from the hydrolyzed polymer carrier molecules.

Phosphatases and glycosidases were applied to liberate volatile flavors directly from their odorless transport or storage forms (Table 2). These cofactor independent enzymes are, like all hydrolases, conveniently applied to practical processes, as the cosubstrate water is abundant in most food systems. Acyl sugar esters represent a novel class of non-volatile precursors of volatiles. 1-*O*-*trans*-cinnamoyl-β-D-glucopyranose and related ester compounds may be the natural progenitors of methyl and ethyl cinnamate, widely occurring volatiles in various fruits (32). Comparable to CoA-dependent activation of acyl moieties, 1-*O*-sugar esters are thought to be activated conjugates. In the intact plant cell specific transferases may convert them into new esters, while a spontaneous chemical alkylation was observed under slightly alkaline conditions in the presence of methanol or ethanol. As these hydrolases and transferases may not be easily separated from their producer cells, the plant gene may be isolated, linked to a suitable construct, and transferred into a microbial host, to finally overexpress the gene coding for the target enzyme. The rDNA approach has been suggested for endo-β-

Table 2. Gycoconjugated Flavors Liberated by Hydrolases

Source	Organ	Flavor products
Camellia sinensis (tea)	Leaf and aerial part	(3*Z*)-Hexenol, 2-phenylethanol
Carica papaya	Fruit	Benzaldehyde, benzyl isothiocyanate
Cydonia oblonga (quince)	Fruit	Norisoprenoids
Fragaria ananassa (strawberry)	Fruit	Furaneol
Malus domestica (apple)	Fruit and leaf	Alkanols, eugenol, octadiols, norisoprenoids
Vanilla planifolia	Fruit	Vanillin
Vitis vinifera (grape)	Fruit	Geraniol, nerol, linalool, norisoprenoids

Source: Ref. 31.

glucosidases from *Aspergillus niger* or *A. oryzae*. Some of these fungal hydrolases are distinguished by a superior activity at elevated processing temperature and at high concentrations of ethanol; their application to wine should enhance the flavor.

Looking at the second field of application, rhamnosidase could be the target of a genetic approach. This enzyme is present as an impurity in commercial pectinases. Its action on *Citrus* wastes containing bitter glycosides results in the liberation of a glucose/rhamnose mixture. Glucose oxidase from *Gluconobacter suboxydans* is then used subsequently to remove any remaining glucose. The residual pure rhamnose is an important precursor of the widely used flavor, 2,5-dimethyl-4-hydroxy-2,3-dihydrofuran-3-one (Furaneol) by reacting the methylpentose with an amino acid. The product resulting from the oxidation of glucose, 5-keto-gluconic acid, may serve as a precursor of the savory flavor 4-hydroxy-5-methyl-3(2*H*)-furanone (26).

The third approach is exemplified by the treatment of extracted vanilla pods with an exo-enzyme mixture of *A. niger* that led to the additional formation of flavor (33). This idea has been transferred to other flavor-bearing plant sources with a rigid morphology, such as *Iris* rhizom. The addition of enzymes for the removal or prevention of off-flavors, for example, the enzymatic debittering of *Citrus* products, may also be interpreted as an indirect action of enzymes on the final sensory quality of the food product.

Classical Fermented Food

Numerous enzymes of various microorganisms must act together in a well-balanced mode to bring about the typical sensory characteristics of the complex substrates of fermented food. In an approach to rationalize the traditional technology, enzyme and substrate mixtures, or purified lipases and proteases together with defined substrates, such as butter oil, are used. Various microbial lipases preferably cleave the ester bond between glycerol and short- to medium-chain fatty acids. The liberated free fatty acids are the sensory backbone of many dairy flavors. Butter and cheese flavor are now generated by the action of lipases or combinations of enzymes on curd or milk fractions, and some of the flavor-enriched products are traded as "Enzyme Modified Cheese."

Some meat products, such as salami-type sausages, were also described to benefit from added enzymes: lipase, protease, and collagenase enzymes improved taste and accelerated ripening (34–36). Proteolytic, lipolytic, and further biopolymer degrading activities can be isolated from yeast cultures and combined with the inactivated producer cells to obtain food flavors with savory notes. The claimed gain in flavor intensity is seldomly quantified, which makes the success of certain bioprocesses difficult to assess. If, in a food-type substrate, the addition of a single enzyme caused a fortified flavor, the traditionally used starter culture strain should benefit from a transmission of the corresponding gene or from the use of gene dosage effects. Several recent reviews deal with rDNA techniques for the elucidation and improvement of relevant properties of starter cultures (eg, 37,38), but their application to improve the flavor formation properties is still somewhat neglected (39).

Concerted Enzymatic Production of Flavor

Lipase and esterase preparations catalyze reverse hydrolysis (esterification, lactonization) and transesterification reactions in one-phase, two-phase, microemulsion, and even in supercritical fluid systems (40). Elegant methods perform the reaction in one of the substrates as the solvent and apply chemical or physical sinks to trap the reaction water. Although essential for the hydration of the enzyme, water is detrimental in excess and will lead to incomplete conversion. Various techniques for water limitation such as the use of molecular sieve, distillation, condensation on a cold surface, and reduced pressure were described. Flavor applications included bacterial, fungal, and mammalian sources of enzyme for the production of chiral aliphatic esters, terpenol esters, and lactones with fruity, flowery, nutty, and creamy odors (Table 3). Similarly, proteases were used for the production of aspartam, a dipeptide sweetener, and for peptides possessing *umami* or sodium glutamate-like flavors (44). Carbohydrate esters may find an industrial use as surfactants to stabilize emulsion-type composed *Citrus* flavors, for example, in nonalcoholic beverages (45).

Vice versa, the selective hydrolysis of racemic amides or esters yields chiral amines, alcohols, or acids. Since the early seventies important industrial applications of hydrolases are the enzymatic resolution of racemic amino acids, and the asymmetric hydrolysis of menthyl esters to produce optically pure L-menthol (candies, sweets, toothpaste, tobacco, ointments, etc), one of the bulk compounds of the flavor industry. More than 50 lipases of different origin have been characterized with respect to their efficiency and selectivity. Only tertiary alcohols and branched-chain and benzoic acids were usually excluded from the wide range of substrates. Subjects of recent work were the effects of solvent, pH, temperature, surfactants, water content, immobilization of the enzyme, pretreatment of the solid support, mass transfer-reaction interactions, isozymes, and chemical modification of the enzyme. Few researchers have been interested in the unambiguous identification of the catalytically active constituent of the often crude industrial enzyme mixtures. A closer look at commercial porcine pancreas extract showed that the N-terminal end of the active protein was homologous to a known cholesterol esterase sequence (47). This enzyme

Table 3. Compounds from Enzymatic Hydrolysis or Reverse Hydrolysis

Substrates	Products	Ref.
Racemic carboxylic acids	Chiral alcohols	41
Acetocarboxylic acid esters	Chiral hydroxy esters	42
Acid plus alcohol/ esters	Geranyl butanoate etc.	43
Amino acids	Oligopeptides	44
Carbohydrate and fatty acid	Fructose stearate, linoleate, etc	45
Racemic lactones	Jasmolactone, massoia lactone, tuberolactone	46

catalyzed the selective esterification of 2-pentanol with butanoic acid in *n*-heptane. Proofed protocols for flavor ester synthesis can be obtained from some enzyme suppliers.

The step from hydrolytic/reverse hydrolytic reactions toward serial enzymatic reactions is bound to require cofactor-dependent enzymes (Table 4). With an ADH/NAD^+/CH_3CHO-system for the production of geranial (lemon flavor constituent), 1500 cycles of regeneration of the cofactor were achieved by coimmobilization of NAD. The bienzymatic method for cofactor regeneration instead of the coupled substrate method was suggested to convert ethanol to acetaldehyde, cinnamic alcohol to cinnamaldehyde, and leucine to 3-methylbutanal (malty odor) and 3-methylbutanol (fusel oil odor). In the latter case (50), *Streptococcus* enzymes were conencapsulated with substrate, NAD, and enzymes of *Gluconobacter oxidans*, which oxidize ethanol to acetic acid in a milk fat coat. Cheese that contained the enzyme capsules exhibited a stronger taste than controls. Such coimmobilizates of biocatalysts of different origin or of biocatalyst and substrate belong to a second generation of immobilized biocatalysts; they seem to be especially promising in the area of flavor production with its often poorly water soluble substrates and products.

An oxidoreductase from *Candida parapsilopsis* reduced the carbonyl function of keto esters; aliphatic, aromatic, and alicyclic ketones; aldehydes; and ketoacetals with high conversion rates (51). The products may be flavors or useful for flavor ester syntheses. To demonstrate its preparative value, methyl (*S*)-(+)-3-hydroxybutanoate, a versatile chiral building block, was synthesized with coupled coenzyme regeneration. Prenylation of small, polar compounds helps to increase their volatility and, thus, their odor activity. Prenylated odorants were found in roasted coffee and believed to contribute to the overall flavor (52). A transfer of the prenyl moiety onto benzoic acid derivatives was achieved using a membrane bound transferase from overproducing strains of *Escherichia coli* (53). This example documented particularly well the inherent advantages of enzyme catalysis: The substitution proceeded regioselectively, and no isomerization of the substituent's double bonds occurred.

MICROBIAL CELLS

Microbial cells may be regarded as a very natural form of immobilized enzymes that provide self-regeneration of catalytic activities and of all cofactors, active transport systems for substrates, and an optimal spatial arrangement of enzyme chains. Empirical biotechnologies rely on *Schizophyta* and *Protoascomycotina* to improve keeping quality, digestibility, and, as a welcome side effect, flavor of foods and beverages.

Food Strains

Fatty acids and simple carbonyl compounds are the predominant volatile compounds produced by *Streptococci* and *Lactobacilli* (54). *Penicillium* strains used in the manufacture of surface-molded cheese and raw sausages add methylketones and, if the cells were injured by mechanical stress, C_8-compounds to the spectrum. Typical yeast volatiles are esters, thio-compounds, and reduced carbonyls. The fermentation of food strains on classical or on chemically defined substrates usually resulted in the production of these common flavor molecules. More interesting compounds were rarely formed or formed in low amounts (<mg L^{-1}) only.

However, some recent applications demonstrate that industry is willing to exploit the capabilities and the operational advantages of microbial cells, despite their obvious genetic limitations. The examples selected (Table 5) reflect some of the actual trends: new substrates and old strains, for example, for the malolactic fermentation of fruit and vegetable juices; better defined, often continuously operated systems for better understanding of biochemical regulation phenomena; cell recycle and immobilized systems for reuse of the biocatalyst (55,56). It has been observed that both quality and quantity of volatiles produced by mutants or mixed cultures with enhanced or complementary metabolic abilities were superior to products fermented with a pure standard strain. The cell counts in mixed cultures often run through transient maxima for the single strains resulting in fluctuations of formation of volatiles and their precursors. Sometimes the success of the entire process is jeopardized. Genetic engineers are now in the position to step forward from combining strains to recombinant strains with a more stable and complex physiology.

The reported generation of Furaneol (2,5-dimethyl-4-hydroxy-3(2*H*)-furanone; "Enhansol" in the United States) in an aerated culture of *Lactobacillus helveticus* deserves special attention (57). This heterocyclic character impact compound with an intense fruity/pineapple to caramel/burnt note (depending on its actual concentration) belongs to the most potent flavor compounds, but it is rather unstable in aqueous fermentation media. Special precautions

Table 4. Cofactor Coupled Reactions Generating Volatile Flavor

System components	Product	Ref.
Geraniol/Acetaldehyde/NAD^+/ADH	Geranial	48
CinnamylOH/Octanal/NAD^+/ADH/Lip-DH/DCIP	Cinnamaldehyde	49
L-Leucine/EtOH/NAD^+/Enzyme mixture	3-Methylbutanal/ol	50
Oxo compound/NADH	Chiral alcohol	51

Table 5. Food Strains and Flavor Formation

Substrate	Catalyst	Flavor	Ref.
Citrate	*Streptococcus lactis*	Diacetyl, etc	55
Whey plus amino acids	*Lactobacillus helveticus*	Furaneol, etc	57
Milk fat plus methionine	*Brevibacterium linens*	Methanethiol	56
Koji	*Pediococcus plus Yeast*	"Soy sauce"	58
Octanoic acid	*Penicillium roquefortii*	2-Heptanone	59
Citronellol	*Botrytis cinerea*	Terpenols, etc	60

were taken for its quantitative recovery from a submerged culture by ion-exchange resins. This should sharpen the view for a possibly concurrent or even predominant chemical formation of volatiles in the often complex fermentation broths. Furaneol was also detected in a casein peptone medium supplemented with L-rhamnose and grown with *Pichia capsulata* (61). However, Furaneol was formed from a hypothetical intermediate of rhamnose generated during heat sterilization of the medium, as Furaneol was neither detected in sterile filtered medium nor after separate heat sterilization of rhamnose and the peptone. Capillary gas chromatography coupled to combustion/isotope-ratio mass spectrometry confirmed that the carbons of Furaneol were actually derived from the rhamnose skeleton.

The closely related sotolone (4,5-dimethyl-3-hydroxy-2(5*H*)-furanone), another powerful odorant (lovage odor) was prepared using an amino acid oxidase of resting cells of the bacterium *Morganella morganii* and subsequent lactonization (62). The same compound is known to be developed during the aging of wine (63). Under experimental conditions, however, sotolone was formed by a purely chemical mechanism from α-ketobutanoic acid and acetaldehyde by an aldol addition followed by cyclization. Another strong odorant among this group of chemicals is 5-ethyl-4-hydroxy-2-methyl-3(2*H*)-furanone. The efficiency of numerous sugars and sugar phosphates to act as a precursor in the formation of this heterocycle by the shoyu yeast *Zygosaccharomyces rouxii* was investigated (64). Biosynthesis proceeded under conditions of an operating pentose-phosphate cycle in the yeast. The same potent flavor enhancer has also been found in Swiss cheese, an indication that another yeast-independent way of formation may exist (65).

A fermentation of soy sauce from koji with immobilized whole cells of lactic acid bacteria of the genus *Pediococcus* and yeast cells to produce a soy sauce flavor was reported by a Japanese group. The calcium alginate entrapped cells were kept in three column reactors with a working volume of 280 L. Thereby, the ecological situation of the fermenting microorganisms was simulated (58). Substrates of *Penicillia* in blue cheese flavor formation were pure fatty acids, ripened curd, or lipolyzed milk fat. Experiments to stabilize the active fungal spores in continuous culture by entrapment into alginate gel were successful (59). The reproducible manufacture of a high-quality fermented sausage is still a challenge to the producer, because choice of ingredients and processing conditions may severely alter product characteristics (66). An earlier drawback, the involvement of an incidental indigenous or contaminating microflora, has been overcome by the use of defined starter cultures; comprehensive investigations have also been devoted to the surface molds of salami (67,68). The key role of microorganisms in flavor formation is now widely recognized, but problems of isolating the lipophilic flavor from the fatty matrix have impeded a sound correlation of volatile compounds and the underlying microbial pathways. Because mixed populations are no longer regarded as the antithesis of good experimental work, new perspectives have opened up. The biodegradation of complex polyester, polyglycosidic, polyamidic, or polyaromatic structures usually leads to the formation of flavor intermediates. In all of these cases a stimulating effect of the various forms of mutualism or commensalism within microbial communities may be expected.

Nonfood Organisms

The range of accessible volatile flavors is considerably expanded by using nonfood species. The need to provide all those pathways with energy and substrates that are not directly required for an intended conversion reaction may be compensated by the cell's self-regeneration of active biocatalyst. Just when cofactor-dependent steps are aimed at, intact cells still appear to be the catalysts of choice.

A good example is the oxidation of fusel oil alkanols to carboxylic acids such as 2-methylbutanoic acid by *Enterobacteria*. A following conversion of this acid to its ethyl ester by a reverse hydrolysis step will yield one of the most potent fruit flavor compounds (13). Proliferating nonfood yeasts substituted purified, immobilized lipase or alcohol dehydrogenase in the production of carboxylic acid esters or aliphatic aldehydes including acetaldehyde, an impact component of citrus and yogurt flavors (69); and strains of *Penicillium, Pseudomonas*, and *A. niger* substituted the action of plant enzymes in degrading terpenoid structures such as carotenoids, resulting in, for example, tea and tobacco flavor constituents. Genetically altered *Pseudomonas* strains with special plasmids were described for effecting this catabolism (20). The formation of geosmin (4,8-dimethyl-decahydro-4-naphthalinol) with its strong, earthy smell needs to be controlled in food and drinking water. Bacteria, fungi, and some plants are able to synthesize the compound. Parameters that affects its accumulation in *Streptomyces tendae* and in *Penicillium expansum* were studied (70), and a hypothetical pathway of formation was suggested based on structural properties of some byproducts (71). Soil bacteria with their sometimes extensive secondary metabolism have also proven useful for the rapid degradation of ferulic acid to yield 11.5 g of vanillin per liter (72). The patent describes the use of *Amycolatopsis* species of the genus *Pseudonocardia* and adsorbent resins such as Amberlite, XAD-types, or Lewatites for product accumulation. Another character impact component, 2-acetyl-1-pyrroline (bread crust, popcorn), was isolated from cocoa fermentation boxes naturally contaminated with *Bacillus cereus*. A pathway of formation was made likely using ^{13}C- and ^{15}N-labeled precursors fed to various *Bacillus* species and an atomic emission detector coupled with the chromatographic separation of volatiles (73).

Protoplast fusion, another prominent novel technique to recombine genetic information, was used to overcome unstable cocultivations and to produce single compounds such as 4-ethylguaiacol, a seasoning agent, and to improve wine flavor (74). Problems encountered with hybrid cells are the sometimes difficult selection and insufficient genome stability of fusant strains. In a strict sense, the first transfer of flavor-related genes from one procaryotic species into another one was reported in 1995 only (75): An expression of genes coding for the transformation of limonene to α-terpineol (lilac odor) of *B. stearothermophilus* in *E. coli* was achieved. Compared with strains of *Ps. gladioli* and *P. digitatum* that perform the same reaction, the yields were still too low for a biotechnological application.

Higher Fungi

Eumycetes, taxonomically located between less-organized microorganisms with limited potential for flavor formation and the complex seed plants, appear to be a kind of golden mean for the biogeneration of flavors. They grow well in certain synthetic liquid media and offer a tremendous metabolic diversity. Some producer strains are related to edible species that should facilitate approval of the products by the authorities.

Not only mushroom-like compounds such as 1-octene-3-ol, but also fruity, floral, spicy, and even chocolate-like (due to pyrazines) odors emanated from fungal cell cultures (Table 6). Volatiles of almost every chemical class were synthesized de novo or via bioconversions by submerged cultured mycelia. Depending on strain and composition of the nutrient medium, frequently found volatiles were phenolics, terpenes, and lactones. With respect to the vast number of strains, their metabolic flexibility under changing conditions, and the broad bioengineering experience, the potential of fungal cells to produce commercial food flavors can hardly be overestimated. Previously reported low yields are not an inevitable feature of fungal cells but were rather due to suboptimal culture conditions.

The example of *Polyporus durus* demonstrated that these saprophytic organisms are very adaptable to changing culture conditions; thus, they are amenable to a stepwise optimization of product yields (Table 7; 76). The presence of a reesterified coconut oil fraction strongly stimulated the formation of volatiles. More than 60 constituents, among them almost 20 lactones, were accumulated in the culture broth. A total concentration of volatiles of about 800 mg L^{-1} was reached. Capillary gas chromatography on chiral and achiral phases showed a high enantiomeric purity of some of the chiral lactones (>98% ee). After the pioneering work using castor oil as a substrate for the fungal production of 4-decanolide (fruity-fatty odor) had ended up with yields of less than 5 g L^{-1} (77), the race for the best process was opened. During the past 10 years, every major flavor house started to develop its own lactone process. When, for example, species of the *Mucor* fungi were fed under aerated conditions with fatty acids or fatty acid esters, 4- and 5-alkanolides with the respective carbon numbers were produced in good yields (78). Similar effects of triacylglycerol substrates were reported for many other microorganisms.

Table 7. Stepwise Improvement of 4-Octanolide Production by *Polyporus durus*

Parameter	Productivity [mg L^{-1} d^{-1}]
Standard medium	Traces
N-source	0.2
Transfer to submerged cultivation	0.6
C-Source, pO_2	3
Addition of precursor and surfactant	16
Preparation and size of inoculum	31
Semicontinuous cultivation	46

Source: Ref. 74.

Table 6. Volatiles from Cultured Fungi

Species	Product(s)
Agaricus bisporus, etc	Benzaldehyde, phenyl acetaldehyde
Ascoidea hylecoeti	2-Phenyl ethanol
Bjerkandera adusta	Anisaldehyde, veratraldehyde
Hebeloma sacchariolens	2-Amino benzaldehyde
Hyanellum suaveolens	Coumarins
Inocybe sp.	Methyl cinnamate
Ischnoderma benzoinum	Benzaldehyde, anisaldehyde
Lentinus sp.	Benzyl acetate, methyl anisate, methyl cinnamate
Hycoacia uda	Methyl acetophenone, methyl benzylalcohol, *p*-tolualdehyde
Nidula sp.	Raspberry ketone, cinnamic acid derivatives
Phanerochaete chrysosporium	Veratraldehyde
Phellinus sp.	Methyl benzoate, salicylate
Pycnoporus cinnabarius	Methyl anthranilate, vanillin
Sirodesmium diversum	*p*-Hydroxybenzaldehyde
Sparassis ramosa	Oak moss impacts (orcine derivatives)
Trametes sp.	Anisaldehyde, methyl phenylacetate
Tyromyces sambuceus	Alkanolides, benzaldehyde, ethyl benzoate

Source: Ref. 24.

Based on the universality of biochemistry, microbial models for studying flavor synthesis have widely replaced the inconvenient plant cell systems. One should, however, be aware that nature frequently provides more than just one pathway for a certain flavor molecule. This obviously restricts an unseen extrapolation of results. Genetic engineering should therefore be considered as a second step, after a biogenetic route has been fully elucidated and selected for gene transfer. Microbial versus plant formation of 4-dodecanolide (peachy-fatty odor; odor threshold in the $\mu g\ L^{-1}$ range) may illustrate the situation: A chiral 10-hydroxystearic acid was obtained by bacterial conversion of oleic acid, and a baker's yeast concluded the sequence by β-oxidation and lactonization (79). Data obtained with fruits of strawberry and peach, however, support a sequence starting with an epoxy fatty acid and proceeding through a β-oxidative chain shortening of the ω-3,4-dihydroxy fatty acid. Elimination of water from the hydroxylactone followed by reduction of the enlactone concluded the biosynthesis (80). Both routes led to the same optically pure (*R*)-form of the lactone.

Only about 0.2% (equivalent to 20 to 30 tons) of the vanillin consumed worldwide is isolated from the botanical source, whereas the remainder is synthetic. This disproportion has stimulated research for a biotechnological substitution, despite some arguing about the details of its biosynthesis (20). Until now, neither plant cell cultures of *Vanilla* nor bacteria or fungi yielded more than trace amounts of the target compound. Again, as mentioned for the preceding bacterium (72), feeding of precursors (here lignin, eugenol, ferulic acid, curcumin, or benzoe siam resin) presented a solution. Different ^{13}C-isotope ratios of the vanillins allowed to assign the flavor compound to a specific geographical or biochemical source (81).

Benzaldehyde, the second most important flavor molecule after vanillin, is obtained from amygdalin in peach or apricot kernels and is a character impact ingredient in stone fruit flavors, such as cherry coke. Equimolar amounts of hydrocyanic acid are formed concurrently and cause major safety problems. Among the biotechnological alternatives is the microbial degradation of natural L-phenylalanine. This amino acid has become easily available (from biotechnology), as it is an intermediate of the synthesis of the high intensity sweetener, aspartame (26). Strains of the basidiomycete *Ischnoderma benzoinum* afforded significant yields of benzaldehyde upon phenylalanine feeding (82). The metabolic pathways were elucidated using submerged cultures of this fungus and ring-labeled deutero-L-phenylalanine as a precursor (83). Depending on the phase of the cell cycle operating, phenylalanine was almost completely converted to benzaldehyde or to 3-phenylpropanol (flowery, roselike odor).

Generally, the supplementation of cell cultures with suitable (natural) precursor substrates, customary in numerous bioprocesses for pharmaceuticals, is an efficient tool to further increase the yields of volatiles. While high yields will continue to be key to the economy of a bioprocess, a more critical evaluation of terms such as "optimization" or "best performance" has started: Optimization should not be regarded equivalent to the maximization of any one single parameter, but depends on the definition of a frame of reference to which the developmental activities are directed. A first step into this direction is the replacement of the term "yield" by time- and biomass-related productivity data.

PLANT CATALYSTS

Most of the volatile flavors processed by the food industry are directly or indirectly derived from plant metabolism. Pressing, distillation, extraction, and chromatography are the common techniques for obtaining raw materials of natural flavors from intact plants.

Crude Enzymes and Homogenates

Crude enzyme preparations from fruits and vegetables were able to partly restore the fresh odor impression in processed food by converting nonvolatile odorless precursors to volatile flavors. Developed in the late fifties, this so-called flavorese-enzyme concept failed from economic and biogenetic reasons. Successors used by-products of food processing as a less expensive source of enzymes. Following up earlier work with apple peel, a Hungarian group developed a preparative scale generation of apple flavor (84). A vigorous formation of the volatile C_6-compounds typical of plant lipoxygenase/hydroperoxide lyase systems occurred in homogenates of some apple cultivars and grass species. The yield of the main volatile compound, (2*E*)-hexenal (pleasant fresh-green note), reached some 100 mg $kg^{-1}\,h^{-1}$ (85,86).

Progress toward understanding how fatty acid peroxides are obtained from lipoxygenase action (87), how the products can be analyzed (88), and how these precursors are further transformed by a cytochrome-type lyase (89) has culminated in the development of immobilized biocatalysts (90,91). Patents describe, for example, the generation of (3*Z*)-hexenol by subjecting linolenic acid and watermelon foliage to mechanical shearing in the presence of a yeast (92), or the generation of several green note compounds using a lyase from guava (93). This biosynthetic principle has been extended to the cooxidation of carrot oleoresin (rich in carotenoids) by fatty acid hydroperoxides to obtain oxidative breakdown products, such as ionones (fruity-violet odors) or dihydroactinidiolide (fragrant-woody notes) (94). The genetic engineering approach holds a lot of promise for these bioprocesses, because the single enzymes involved are well known. The operational instability of plant lyases may be improved by enzyme engineering in the future.

Callus and Suspension Culture

Derived from a wounding-induced tissue (primary callus), a continuously growing sterile culture of plant cells can be established by mechanical separation and further subculturing on a synthetic medium (callus culture). After transfer to a liquid medium, cells of a macroorganism are finally reduced to the single-cell level (suspension culture). According to its origin, the somatic cell is provided with a complete genom and able to develop into a whole plant again. This totipotency was translated into the idea that all genetic functions, including, for example, essential oil and aroma formation, would be expressed in such cells (95). However, numerous disappointing experiments have caused the industry to almost completely refrain from further investments into this kind of plant biotechnology. The absence of flavor compounds in cell cultures of flavor-yielding plants has been attributed to a lack of morphologically differentiated structures (eg, oil glands, resin ducts, etc), or to a rapid catabolism of the presumed intermediates. Detailed studies on the synthesis of lower terpenes in callus cultures of essential oil bearing plants showed all species to possess high prenyltransferase and pyrophosphate isomerase activities, and most of the cultures to possess further activities of the mevalonate pathway, although a significant accumulation of products did not occur.

A physical stimulus to secondary metabolism in plant cells has been neglected for a long time: light. Although not always sufficient, the cytodifferentiation from heterotrophic to phototrophic cells induced the formation of leaf-typical constituents, such as pigments, quinones, and essential oils. Chlorophyll formation in the presence of light is tied to a preceding formation of phytol; this diterpenoid alcohol in turn can only be formed if the initial steps of the mevalonate pathway are operating. The accumulation of the intermediate lower terpenoids along this route is then to be expected. In addition to light a fine-tuning of the phytoeffector concentrations in the growth medium is mandatory to support both growth and aroma formation, as was shown with illuminated callus cultures of *Citrus* (96). These cells accumulated a full spectrum of volatile terpenes, and the best yields were about 5% of the volatiles isolated from a mature tissue of the mother explant (grapefruit peel; Table 8). Addition of precursors resulted in con-

Table 8. Some Volatile Constituents of *in vitro* cells of *Citrus paradisi*

Compound, mg kg^{-1} fresh wt	Low phytoeffector conc	High phytoeffector conc
Chlorophyll	193	310
(+)-Limonene	48	177
α-Pinene	0.09	0.58
β-Pinene	0.22	0.67
n-Octanal	1.17	0.41
Citronellal	0.95	0.59

Source: Ref. 96.
24-h photoperiod and 3,000 lux

version to and stable accumulation of volatiles, especially in oxyfunctionalization reactions. One example of economical relevance could be the regioselective biotransformation of valencene into the sesquiterpene ketone nootkatone. After six hours of incubation, almost 70% of the substrate was transformed with no other detectable volatile by-products. Natural nootkatone with its grapefruit-like odor and bitter taste is a sought-after flavor compound with limited availability. As was found repeatedly in earlier work with plant cell cultures, flavor formation was favored by slow growth, an indication that flavor syntheses may be a general characteristic of a heterostatic physiology of the cell (26). The preceding findings disagree with early statements that either the cytotoxicity of the once formed volatiles or their instability in the surrounding medium compartment would prevent flavors from being accumulated in elevated concentration.

The addition of heat-inactivated microbial homogenates to growing plant cell cultures was discussed as a means of inducing metabolic cascades involved in chemical defense. As some flavor compounds interfere with bacterial or fungal growth, their synthesis was thought to be triggered by simulating a respective microbial attack. This so-called elicitation was demonstrated by the addition of autoclaved fungal mycelium to *Petroselinum* cells that responded with the formation of volatile phthalides, coumarins, and phenolics (97). A phototrophic state of the cells was required, because only the plastid differentiated cells tolerated the higher cytokinin concentrations and the fungal homogenate. This kind of ecological stress typically leads to the secretion of the newly formed compounds into the surrounding nutrient medium, where they unfold their bioactivity, and from where they can be easily recovered. The use of lipophilic traps (solvent or adsorbent) simulating a natural morpholicical accumulation site can be combined with the elicitor approach.

Cell immobilization is an expanding area of biotechnology. The characteristics of continuous operation, reuse of the biocatalyst, ease of process control, and improved biocatalyst stability should prove especially beneficial for cultured plant cells, as immobilized cells are similar in many respects to the tissues of intact plants. Aggregated cells face heterogeneous microenvironments, release intracellular products, and divide slower. These and other factors may redirect plant secondary metabolism. The use of immobilized plant cells in column or membrane reactors may offer better perspectives for a future application on an industrial scale. If single enzymes were rate limiting, as particularly in biotransformations, it should be possible to select for higher yields by making use of the somaclonal variation that builds up in plant cell cultures, or by enhanced mutagenesis. Increased levels of essential oils were reported for mutant plants of *Pelargonium* and other species regenerated from UV or ethyl methane sulfonate treated callus cultures; this type of approach is somewhat restricted by the laborious procedures and assays required (98).

In spite of the achievements it should not be overlooked that de novo syntheses as well as biotransformations using plant cells will always have to compete not only with field plants, but also with microbial and enzyme technologies. Plant cells are not very adaptable to extreme conditions of cultivation, and the logical consequence would rather be to use *in vitro* plant cells of suitable cultivars as a convenient source of genes and to link these plant genes to microbial operators in future rDNA experiments.

Specialized Plant Cells

The biotechnological ideal of an immobilized plant cell with active flavor metabolism, available in any amount, was created by nature: mature fruits. It has been known for a long time that aged fruit tissues are able to take up exogenous substrates, metabolizing them to flavor compounds. The same holds true for intact fruits and volatile substrates. A procedure was developed to expose fruits during storage to vapors of precursors of volatile flavors. By analogy with controlled atmosphere (CA) storage, this biotechnological concept was termed PA (for precursor atmosphere) storage. One example refers to the drying of banana tissue (99). Although many compounds of banana flavor showed a good retention in the starchy matrix of the tissue, the concentrations of some highly volatile carboxylic esters decreased by 50 to 80% during freeze drying. When bananas were submitted to PA-storage in 3-methylbutanol vapors before dehydration, the fruits rapidly accumulated volatile impact components. By managing endogenous organized enzymes and their substrates, a dried fruit with an amount of 3-methylbutyl esters comparable to that of the genuine fresh fruit was finally obtained (Table 9). As a result, a preceding PA storage can compensate for flavor deficiencies caused by one-sided breeding, by improper transport and storage, or by physical losses during thermal processing operations.

Table 9. Effect of a Preceding Precursor Atmosphere (PA) Storage on Impact Volatiles of Freeze-Dried Banana Slices

	Fruit		
		Freeze-dried	
3-Methylbutyl ester, μg 100 g^{-1} fresh wt	Fresh	Without	With PA-storage
Acetate	7460	2860	6990
2-Methylpropanoate	70	27	60
Butanoate	600	250	960

Source: Ref. 99.

COMMERCIAL BIOFLAVORS

Novel bioprocesses are now an accepted alternative source of a wide range of high-prized volatile flavors. A couple of companies advertise 100% natural compounds. Whether these were obtained by conventional physical treatments of plant (or other natural) sources, or whether they in fact originated from a bioreactor, is sometimes difficult to assess. True products of concerted bioprocesses may be diacetyl (butter note, from starter cultures), cheese flavors, yeast products (meaty and savory notes), fatty acids and related alcohols and esters (from lipase technology or intact microorganisms), C_6- and C_8-alcohols and related carbonyls (green and mushroom notes, from plant and fungal homogenates), and some specialties such as lactones, vanillin, and nootkatone from complex biosystems.

The legal definitions (and the average consumer) clearly discriminate artificial flavors against natural ones; this fact has undoubtedly contributed to the recent surge of activity in bioflavor research. The legal discrimination itself, however, must be regarded artificial, because "no compounds are made on earth other than those permitted by the laws of nature" (100). The existing legal restrictions might be challenged less by the public with its unfounded reservations than by problems to analytically distinguish natural from synthetic. Sophisticated analytical tools comprise multidimensional chiral gas chromatography, high resolution NMR-spectroscopy, and isotope ratio mass spectroscopy (30,81). Despite this selection of analytical options, some volatile compounds are still difficult to differentiate.

Regardless of the actual legal situation in a country, the superiority of biocatalysts to generate complex, sensitive, and chiral molecules is more and more recognized in organic synthesis. The classical field of food biotechnology will continue to profit from modern developments in gene recombination and delivery, bioreactor and sensor design, on-line control, and related techniques. The improvement of fungal strains and plant cultivars is still restricted by several fundamental scientific and applied hurdles. The use of eucaryotic and mitochondrial genome fragments and the chromosomal integration of rDNA characterize lines of current research. Special emphasis should be put on safety testing of food generated by genetically manipulated strains, particularly if these are to remain in the consumed products (101).

BIBLIOGRAPHY

1. R. H. Michel, P. E. McGovern, and V. R. Badler, "Chemical Evidence for Ancient Beer," *Nature* **360**, 24 (1992).
2. B. H. Lee, *Fundamentals of Food Biotechnology*, VCH, New York, 1996.
3. P. Schreier, *Chromatographic Studies of Biogenesis of Plant Volatiles*, Hüthig, Heidelberg, 1984.
4. E. Sprecher and H.-P. Hansen, "Recent Trends in the Research on Flavours Produced by Fungi," in R. G. Berger, S. Nitz, and P. Schreier, eds., *Topics in Flavour Research*, Eichhorn, Hangenham, 1985, pp. 387–403.
5. F. H. Sharpell Jr., "Microbial Flavors and Fragrances", in M. M. Young, ed., *Comprehensive Biotechnology*, vol. 3, Pergamon Press, Oxford, United Kingdom, 1985, pp. 965–981.
6. T. H. Parliment and R. Croteau, eds., *Biogeneration of Aromas*, ACS Symposium Series 317, American Chemical Society, Washington, D.C., 1986.
7. P. Schreier, ed., *Bioflavour '87*, deGruyter, Berlin, 1988.
8. P. Schreier, "Aspects of the Biotechnological Production of Food Flavours," *Food Rev. Int.* **5**, 289–315 (1989).
9. F. W. Welsh, W. D. Murray, and R. E. Williams, "Microbiological and Enzymic Production of Flavor and Fragrance Chemicals," *Crit. Rev. Biotechnol.* **9**, 105–169 (1989).
10. E. W. Seitz, "Flavor Building Blocks," in I. Goldberg and R. Williams, eds., *Biotechnology and Food Ingredients*, Van Nostrand Reinhold, New York, 1991, pp. 375–391.
11. D. A. Romero, "Bacteria as Potential Sources of Flavor Metabolites," *Food Technol.* **46**, 122–126 (1992).
12. M. I. Farbood, "Micro-organisms as a Novel Source of Flavour Compounds," *Biochem. Soc. Trans.* **19**, 690–694 (1992).
13. L. Janssens et al., "Production of Flavours by Microorganisms," *Proc. Biochem.* **27**, 195–215 (1992).
14. R. L. S. Patterson et al., eds., *Bioformation of Flavours*, Royal Society of Chemistry, Cambridge, United Kingdom, 1992.
15. J. M. Belin, M. Bensoussan, and L. Serrano-Carreon, "Microbial Biosynthesis for the Production of Food Flavours," *Trends Food Sci. Technol.* **3**, 11–14 (1992).
16. P. Winterhalter and P. Schreier, "Biotechnology: Challenge for the Flavor Industry," in T. E. Acree and R. Teranishi, eds., *Flavor Science*, American Chemical Society, Washington, D.C., 1993, pp. 225–258.
17. P. Schreier and P. Winterhalter, eds., *Progress in Flavour Precursor Studies*, Allured, Carol Stream, Ill., 1993.
18. S. Hagedorn and B. Kaphammer, "Biocatalysis in the Generation of Flavors and Fragrances," *Annu. Rev. Microbiol.* **48**, 773–800 (1994).
19. A. Gabelman, ed., *Bioprocess Production of Flavor, Fragrance, and Color Ingredients*, John Wiley & Sons, New York, 1994.
20. R. G. Berger, *Aroma Biotechnology*, Springer, Berlin, 1995.
21. K. Kevin, "A Brave New World: Capturing the Flavor Bug," *Food Proc.* 66–70 (1995).
22. M. Tyrell, "Advances in Natural Flavors and Materials," *Perf. Flavorist* **20**, 13–21 (1995).
23. P. Étiévant and P. Schreier, eds., *Bioflavour 95*, INRA, Paris, 1995.
24. U. Krings, B. Abraham, and R. G. Berger, "Plant Impact Volatiles—A Biotechnological Perspective," *Perf. Flavorist* **20**, 79–86 (1995).
25. G. Feron, P. Bonnarme, and A. Durand, "Prospects for the Microbial Production of Food Flavours," *Trends Food Sci. Technol.* **7**, 285–293 (1996).
26. R. G. Berger, ed., *Biotechnology of Aroma Compounds*, Adv. Biochem. Engin. Biotechnol. 55, Springer, Berlin, 1996.
27. G. R. Takeoka et al., eds., *Biotechnology for Improved Foods and Flavors*, ACS Symposium Series 637, American Chemical Society, Washington, D.C., 1996.
28. P. J. S. Cheetham, "The Use of Biotransformations for the Production of Flavours and Fragrances," *Trends Biotechnol.* **35**, 478–488 (1993).
29. L. Somogyi, "The Flavour and Fragrance Industry: Serving a Global Market," *Chem. Ind.* (London) **4**, 170–173 (1996).

30. A. Mosandl, "Enantioselective Capillary Gas Chromatography and Stable Isotope Ratio Mass Spectrometry in the Authenticity Control of Flavors and Essential Oils," *Food Rev. Int.* **11**, 597–664 (1995).
31. P. Winterhalter and G. K. Skouroumounis, "Glycoconjugated Aroma Compounds: Occurrence, Role and Biotechnological Transformation," in R. G. Berger, ed., *Biotechnology of Aroma Compounds*, Adv. Biochem. Engin. Biotechnol. 55, Springer, Berlin, 1996, pp. 73–105.
32. S. Latza, D. Ganßer, and R. G. Berger, "Carbohydrate esters of cinnamic acid from fruits of *Physalis peruviana, Psidium guajava and Vaccinium vitis-idaea*," *Phytochemistry* **43**, 481–485 (1996).
33. M. P. Pouget, A. Pourrat, and H. Pourrat, "Recovery of Flavor Enhancers From Vanilla Pod Residues by Fermentation," *Food Sci. Technol.* **23**, 1–3 (1990).
34. I. Zalacain et al., "Dry Fermented Sausages Elaborated With Lipase From *Candida cylindracea*," *Meat Science* **40**, 55–61 (1995)
35. B. F. Hagen et al., "Bacterial Proteinase Reduces Maturation Time of Dry Fermented Sausages," *J. Food Sci.* **61**, 1024–1029 (1996).
36. K. Marggrander, "Collageneous Proteins as Excipients for the Improvement of the Technological and Sensory Properties of Meat Products and Ready Meals," *Fleischwirtschaft* **76**, 725–726 (1996).
37. H. Teuber, "The Use of Genetically Manipulated Microorganisms in Food: Opportunities and Limitations," *Eur. Congr. Biotechnol.* **4**, 383–391 (1987).
38. D. R. Berry, "Manipulation of Flavour Production by Yeast: Physiological and Genetic Approaches," in J. R. Piggott and A. Paterson, eds., *Distilled Beverage Flavour*, Ellis Horwood, Chichester, United Kingdom, 1989, pp. 299–307.
39. P. Christen and A. López-Munguía, "Enzymes and Food Flavor—A Review," *Food Biotechnol.* **8**, 167–190 (1995).
40. A. Ballesteros et al., "Enzymes in Non-conventional Phases," *Biocatalysis and Biotransformation* **13**, 1–42 (1995).
41. K. Laumen, D. Breitgoff, and M. P. Schneider, "Enzymic Preparation of Enantiomerically Pure Secondary Alcohols," *J. Chem. Soc. Chem. Commun.* **22**, 1459–1461 (1988).
42. I. Glänzer et al., "Enantioselective Hydrolysis of Esters of Secondary Alcohols Using Lyophylized Baker's Yeast," *Enzyme Microb. Technol.* **10**, 744–749 (1988).
43. G. Langrand, C. Triantaphylides, and J. Baratti, "Lipase Catalyzed Formation of Flavor Esters," *Biotechnol. Lett.* **10**, 549–554 (1988).
44. K. Aso, T. Uemura, and Y. Shiokawa, "Protease Catalyzed Synthesis of Oligo-L-Glutamic Acid from L-Glutamic Acid Diethyl Ester," *Agric. Biol. Chem.* **52**, 2443–2449 (1988).
45. M. A. Ku and Y. D. Hang, "Enzymatic Synthesis of Esters in Organic Medium With Lipase From *Byssochlamys fulva*," *Biotechnol. Lett.* **17**, 1081–1084 (1995).
46. WO 94/07887 (Sept. 21, 1993), J.-P. Bourdineaud, C. Ehret, and M. Petrzilka (to Givaudan-Roure Intl.).
47. K. Sostmann and P. Schreier, "Esterification in Non-aqueous Solvents: Cholesterol Esterase as a Selective Biocatalysator From Porcine Pancreas," *Z. Lebensm. Unters. Forsch.* **200**, 428–431 (1995).
48. M. D. Legoy, H. S. Kim, and D. Thomas, "Use of ADH for Flavor Aldehyde Production," *Process Biochem.* **20**, 145–148 (1985).
49. J. S. Deetz and J. D. Rozzell, "Enzyme-Catalyzed Reactions in Non-aqueous Media," *Trends Biotechnol.* **6**, 15–19 (1988).
50. S. D. Braun and N. F. Olsen, "Microencapsulation of Cell-free Extracts to Demonstrate the Feasibility of Heterogeneous Enzyme Systems and Cofactor Recycling for Development of Flavor in Cheese," *J. Dairy Sci.* **69**, 1202–1208 (1986).
51. J. Peters, T. Minuth, and M.-R. Kula, "A Novel NADH-dependent Carbonyl Reductase With an Extremely Broad Substrate Range From *Candida parapsilopsis*: Purification and Characterization," *Enzyme Microb. Technol.* **15**, 950–958 (1993).
52. W. Holscher, O. G. Vitzthum, and H. Steinhart, "Prenyl Alcohol, Source for Odorants in Roasted Coffee," *J. Agric. Food Chem.* **40**, 655–658 (1992).
53. L. A. Wessjohann and B. Sontag, "Prenylierung von Benzoesäurederivaten katalysiert durch eine Transferase aus *E. coli* Überproduzenten," *Angew. Chemie Intl. Ed. Engl.* **35**, 1697–1699 (1996).
54. R. Imhof, H. Glättli and J. O. Bosset, "Volatile Organic Aroma Compounds Produced by Thermophilic and Mesophilic Mixed Strain Dairy Cultures," *Food Sci. Technol.* **27**, 442–449 (1994).
55. P. Schmitt et al., "Citrate Utilization by Free and Immobilized *Streptococcus lactis* ssp. *diacetylactis* in Continuous Culture," *Appl. Microbiol. Biotechnol.* **29**, 430–436 (1988).
56. S. C. Kim and N. F. Olsen, "Production of Methanethiol in Milk Fat-Coated Microcapsules Containing *Brevibacterium linens* and Methionine," *J. Dairy Sci.* **56**, 799–811 (1989).
57. J. Kowalewska et al., "Isolation of Aroma Bearing Material From *Lactobacillus helveticus* Culture and Cheese," *J. Dairy Sci.* **68**, 2165–2171 (1985).
58. K. Osaki et al., "Fermentation of Soy Sauce With Immobilized Whole Cells," *J. Food Sci.* **50**, 1289–1292 (1985).
59. C. Larroche and J.-B. Gros, "Special Transformation Processes Using Fungal Spores and Immobilized Cells," in R. G. Berger, ed., *Biotechnology of Aroma Compounds*, Adv. Biochem. Engin. Biotechnol. 55, Springer, Berlin, 1996, pp. 179–220.
60. P. Brunerie et al., "Bioconversion of Citronellol by *Botrytis cinerea*," *Appl. Microbiol. Biotechnol.* **27**, 6–10 (1988).
61. R. Roscher et al., "I-Rhamnose: Progenitor of 2,5-dimethyl-4-hydroxy-3(2H)-furanone Formation by *Pichia capsulata*?" *Z. Lebensm. Unters. Forsch.* **204**, 198–201 (1997).
62. K. Lerch and M. Ambühl, "Biotechnological Production of 4,5-dimethyl-3-hydroxy-2(5H)-furanone," in P. Étiévant and P. Schreier, eds., *Bioflavour 95*, INRA, Paris, 1995, pp. 381–384.
63. T. T. Pham et al., "Optimal Conditions for the Formation of Sotolone from α-Ketobutyric Acid in the French *'Vin Jaune'* ", *J. Agric. Food Chem.* **43**, 2616–2619 (1995).
64. M. Sasaki, N. Nunomura, and T. Matsudo, "Biosynthesis of 4-hydroxy-2(or 5)-ethyl-5(or 2)-methyl-3(2H)-furanone by Yeasts," *J. Agric. Food Chem.* **39**, 934–938 (1991).
65. M. Preininger, M. Rychlik, and W. Grosch, "Potent Odorants of the Neutral Volatile Fraction of Swiss Cheese (Emmentaler)," in H. Maarse and D. G. van der Heij, eds., *Trends in Flavour Research*, Elsevier, Amsterdam, The Netherlands, 1994, pp. 267–270.
66. W. M. Park et al., "Effects of Lactic Acid Bacteria Isolated From Fermented Foods on the Microbiological Properties of Fermented Sausages," *Food Biotechnol.* **6**, 145–148 (1997).
67. W. P. Hammes, "Fermentation of Non-dairy Food," *Food Biotechnol.* **5**, 293–303 (1991).

68. H. J. Hwang, R. F. Vogel, and W. P. Hammes, "Entwicklung von Schimmelpilzkulturen für die Rohwurstherstellung," *Fleischwirtschaft* **73**, 327–332 (1993).
69. L. Janssens et al., "Fusel Oil as a Precursor for the Microbiological Production of Fruity Flavours," *Med. Fac. Landbouww. Rijksuniv. Gent* **54**, 1387–1392 (1989).
70. C. P. Dionigi and D. A. Ingram, "Effects of Temperature and Oxygen Concentration on Geosmin Production by *Streptomyces tendae* and *Penicillium expansum*," *J. Agric. Food Chem.* **42**, 143–145 (1994).
71. F. Pollak and R. G. Berger, "Geosmin and Related Volatiles in Bioreactor Cultured *Streptomyces citreus* CBS 109.60," *Appl. Environ. Microbiol.* **62**, 1295–1299 (1996).
72. EP 761817 (March 12, 1997), J. Rabenhorst and R. Hopp (to Haarmann & Reimer GmbH).
73. L. J. Romanczyk, Jr. et al., "Formation of 2-Acetyl-1-pyrroline by Several *Bacillus cereus* Strains Isolated From Cocoa Fermentation Boxes," *J. Agric. Food Chem.* **43**, 469–545 (1995).
74. S. Lino and M. Watanabe, "Wine-making by Killer Wine Yeast" [in Japanese], *After CA* **113**, 13754 (1989).
75. H. C. Chang, D. A. Gage, and P. J. Oriel, "Cloning and Expression of a Limonene Degradation Pathway From *Bacillus stearothermophilus* in *Escherichia coli*," *J. Food Sci.* **60**, 551–553 (1995).
76. R. G. Berger, K. Neuhäuser, and F. Drawert, "Odorous Constituents of *Polyporus durus* (Basidiomycetes)," *Z. Naturforsch.* **41c**, 963–970 (1986).
77. U.S. Patent 4,560,656 (Dec. 24, 1985), M. I. Farbood and B. J. Willis (to Fritzsche Dodge & Olcott, Inc.).
78. WO 89/12104 (Dec. 14, 1989), G. V. Page and G. R. Eilerman (to BASF Corp.).
79. S. Gocho et al., "Biotransformation of Oleic Acid to Optically Active γ-Dodecalactone," *Biosci. Biotech. Biochem.* **59**, 1571–1572 (1995).
80. M. Schöttler and W. Boland, "Über die Biosynthese von γ-Dodecanolacton in reifenden Früchten: Aroma-Komponenten der Erdbeere (*Fragaria ananassa*) und des Pfirsichs (*Prunus persica*)," *Helv. Chim. Acta* **78**, 847–856 (1995).
81. I. Benz and A. Muheim, "Biotechnological Production of Vanillin," in A. J. Taylor and D. S. Mottram, eds., *Flavour Science—Recent Developments*, Royal Society of Chemistry, Cambridge, United Kingdom, 1996, pp. 111–117.
82. C. E. Fabre, P. J. Blanc, and G. Goma, "Production of Benzaldehyde by Several Strains of *Ischnoderma benzoinum*," *Sci. Aliments* **16**, 61–68 (1996).
83. U. Krings, M. Hinz, and R. G. Berger, "Degradation of [^{2}H] Phenylalanine by the Basidiomycete *Ischnoderma benzoinum*," *J. Biotechnol.* **51**, 123–129 (1996).
84. L. Vámos-Vigyázó, N. Kiss-Kutz, and A. Hersiczky, "Preparative Scale Generation of Apple Flavour from By-Products of Processing," in *3rd International Flavour Symposium*, Kekssemet, 1984, pp. 34–35.
85. R. G. Berger, A. Kler, and F. Drawert, "The C_6-Aldehyde Forming System in *Agropyron repens*," *Biochim. Biophys. Acta* **883**, 523–530 (1986).
86. F. Drawert, A. Kler, and R. G. Berger, "Optimierung der Ausbeuten von (2*E*)-Hexenal bei pflanzlichen Gewebehomogenater," *Lebensm.-Wiss. Technol.* **19**, 426–431 (1986).
87. H. W. Gardner, "Lipoxygenase as a Versatile Biocatalyst," *J. Am. Oil Chem. Soc.* **73**, 1347–1357 (1996).
88. C. Schneider, P. Schreier, and M. Herderich, "Analysis of Lipoxygenase-derived Fatty Acid Hydroperoxides by Electrospray Ionization Tandem Mass Spectrometry," *Lipids* **32**, 331–336 (1997).
89. K. Matsui et al., "Bell Pepper Fruit Fatty Acid Hydroperoxide Lyase Is a Cytochrome P450 (CYP74B)," *FEBS Lett.* **394**, 21–24 (1996).
90. A.-F. Hsu, T. A. Foglia, and G. J. Piazza, "Immobilization of Lipoxygenase in an Alginate-Silicate Solgel Matrix: Formation of Fatty Acid Hydroperoxides," *Biotechnol. Lett.* **19**, 71–74 (1997).
91. A. Nunez et al., "Immobilization of Hydroperoxide Lyase From *Chlorella*," *Biotechnol. Appl. Biochem.* **25**, 75–80 (1997).
92. WO 95/26413 (March 8, 1994), R. B. Holtz, M. J. McCulloch, and S. J. Gargar (to Reynolds Technologies, Inc.).
93. WO 93/24644 (May 3, 1993), B. Muller, A. Gautier, C. Dean, and J.-C. Kuhn (to Firmenich, S.A.).
94. WO 94/08028 (Sept. 27, 1993), J.-M. Belin, B. Dumont, and F. Ropert (to BFA Laboratories).
95. T. Mulder-Krieger et al., "Production of Essential Oils and Flavors in Plant Cell and Tissue Culture," *Plant Cell Tissue Organ Cult.* **13**, 85–154 (1988).
96. G. Reil and R. G. Berger, "Accumulation of Chlorophyll and Essential Oil in Photomixotrophic Cell Cultures of *Citrus* sp.," *Z. Naturforsch.* **51c**, 657–666 (1996).
97. G. Reil and R. G. Berger, "Elicitation of Volatile Compounds in Photomixotrophic Cell Cultures of *Petroselinum crispum*," *Plant Cell Tissue Organ Cult.* **46**, 131–136 (1996).
98. A. H. Scragg, "The Production of Aromas by Plant Cell Cultures," in R. G. Berger, ed., *Biotechnology of Aroma Compounds*, Adv. Biochem. Engin. Biotechnol. 55, Springer, Berlin, 1996, pp. 239–263.
99. R. G. Berger, F. Drawert, and H. Kollmannsberger, "PA-Lagerung zur Kompensation von Aromaverlusten bei der Gefriertrocknung von Bananenscheiben," *Z. Lebensm. Unters.-Forsch.* **183**, 169–171 (1986).
100. R. Teranishi, "New Trends and Developments in Flavor Chemistry," in R. Teranishi, R. G. Buttery, and F. Shahidi, eds., *Flavor Chemistry*, ACS Symposium Series 388, American Chemical Society, Washington, D.C., 1989, p. 2.
101. D. Anderson and W. F. J. Cuthbertson, "Safety Testing of Novel Food Products Generated by Biotechnology and Genetic Manipulation," *Biotechnol. Genet. Eng. Rev.* **5**, 369–395 (1987).

R. G. BERGER
Institut für Lebensmittelchemie der Universität Hannover
Hannover, Germany

GENETIC ENGINEERING: PRINCIPLES AND APPLICATIONS

Genetic engineering is an important and widespread enabling technology in the development of the discipline of food science and technology and our notions of food, from production to processing. A decade ago, Douglas McCormick, the editor of *Bio / Technology*, wrote: "By most standards, all of this [biotechnology] is still brand new. Yet biotechnology is the product of intellectual earthquakes that change things so completely that it is difficult, after the event, to remember that once the landscape was different. It is easy to forget that the change is recent, and more change is just around the corner" (1). McCormick's statement has been true so far and will hold true for some time. In this article we describe the vocabulary, concepts, tools, techniques, and applications of genetic engineering to food science and technology.

HISTORY

The discovery of DNA ligase in 1967 followed by the DNA restriction endonucleases and RNA reverse transcriptase in 1970 was the beginning of era of genetic engineering. Also, in 1970 the first record for total synthesis of a gene *in vitro* was produced. In 1972 the first *in vitro* recombinant DNA molecule was generated, and a year later the first plasmid vector-mediated transformation was demonstrated. In 1975 the methodology for Southern blot was published. The blotting system (Fig. 1) allows for isolation and identification of DNA fragments by gel electrophoresis and their transfer (or blotting) onto a membrane filter for direct hybridization to a single-stranded radioactively labeled or biotin-linked probe. Four years later, the discovery of introns and exons in eukaryotic genes changed our concepts of gene transcripts, and the same year rapid DNA sequencing methods became available. In the decades following these outstanding achievements we have seen a rise in the commercial application of genetic engineering and production of human, plant, and microbial processes or products (2). It is amazing how rapidly genetic engineering has grown and what in the future it holds.

PRINCIPLES OF GENETICS

The hereditary material of most cellular organisms is DNA, although in many bacteriophages and plant or animal viruses it could be either DNA or RNA. DNA is largely the basis for the preservation of the instructions for the organization, structure, and functioning of living cells. Historically, it was realized that the passage of hereditary traits to one's offspring depended on those that were found

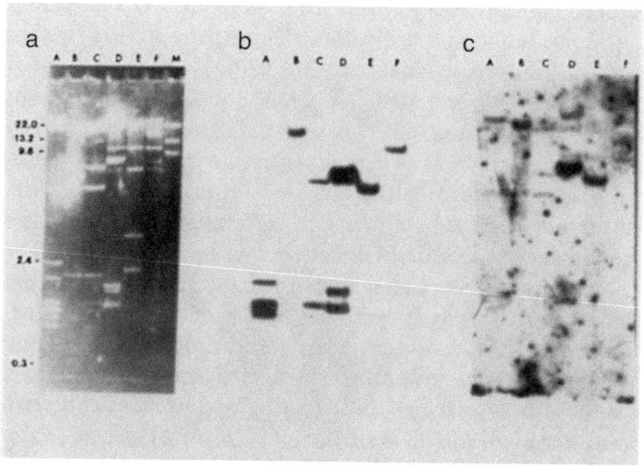

Figure 1. DNA obtained from human genome (**a**) has been hybridized by either (**b**) DNA or (**c**) RNA by Southern blotting method. *Source:* Photograph courtesy of J. M. Mac Pherson.

in the parents. In the early 1860s the Austrian monk Gregor Mendel not only experimentally confirmed this in pea plants, but also established the principal rules governing the appearance of specific traits, that is, the parental factors (subsequently these became known as genes) and their assortment into individual progeny. He also made the distinction between the external appearance of such traits (later named phenotypes) and those that are reflections of the composition (later named genotypes). Mendel also advanced the idea of the expression of traits when they were in single copies from those that required two copies (later named dominant and recessive genes). The advent of microscopy and the discovery of dyes that could differentially stain cellular parts permitted the examination of the anatomical features of cells including the nucleus. The hereditary material, housed in distinct and organized structures, were named chromosomes. The number of chromosomes per cell are constant in any species. In most cells, the chromosomes are found in pairs of homologous structures, varying in number from a few two to three dozen. The concept of ploidy was developed to reflect the presence of haploid (N) and diploid (2N) cells, such as those found in bacteria and gametes or sex cells and in nonsomatic or fertilized cells, respectively. The N was defined to be the number of nonhomologous chromosomes found in a cell.

Analysis of Hereditary Materials and Exchanges

In the 1940 to 1950s the nature of DNA was elucidated by the use of quantitative genetic experiments, application of radioisotopes to analytical biochemistry, and the discovery of the electron microscope. These tools permitted the visualization of DNA and demonstrated its many activities: replication, repair, exchange, and recombination.

The structures of prokaryotic (eg, bacterial) and eukaryotic (eg, yeasts, fungi, plant, and animal) cells, which are both of significance in food science, and their genetic apparati have been investigated in great detail (3,4). The best-studied prokaryotic cell, the workhorse of many genetic engineering technologies and a causative of certain foodborne illnesses, is that of the bacterium *Escherichia coli*. This bacterial cell is cylindrical in shape, 1×0.5 μm in size (Fig. 2), and can grow and double in number every 20 min. The *E. coli* chromosome is in a covalently closed circular (ccc) form about 1 mm long and is made of 4,600,000 base pairs (4600 kilobase pairs, kb). The chromosome of this bacterium is mapped and completely sequenced.

We are beginning to realize that the structure of the genome in bacteria is more dynamic and diverse than once thought and accepted. Analysis of more than 150 bacterial species and isolates for genome size indicates a range, 580 kbp for *Mycoplasma genitalium* and 9200 kbp for *Myxococcus xanthus*, the smallest and the largest known genomes, respectively (3). These two organisms carry a gene number of 470 and more than 10,000, respectively, representing the lifestyles of an obligate parasite within living hosts and a metabolic generalist, undergoing sporulation, mycelial growth, and differentiation. The geometry of chromosomal DNA is found in a double-stranded circular, lin-

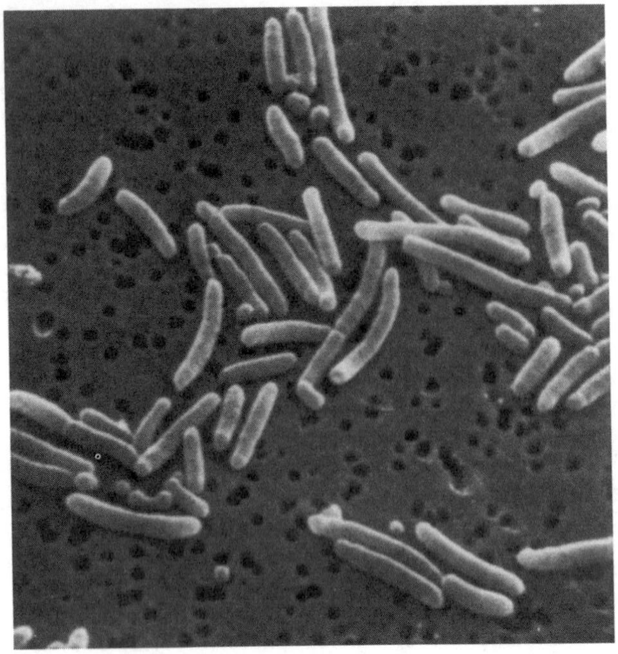

Figure 2. The simple binary fission of *Escherichia coli* cells along with the production of DNA-less minicells. *Source:* Ref. 30.

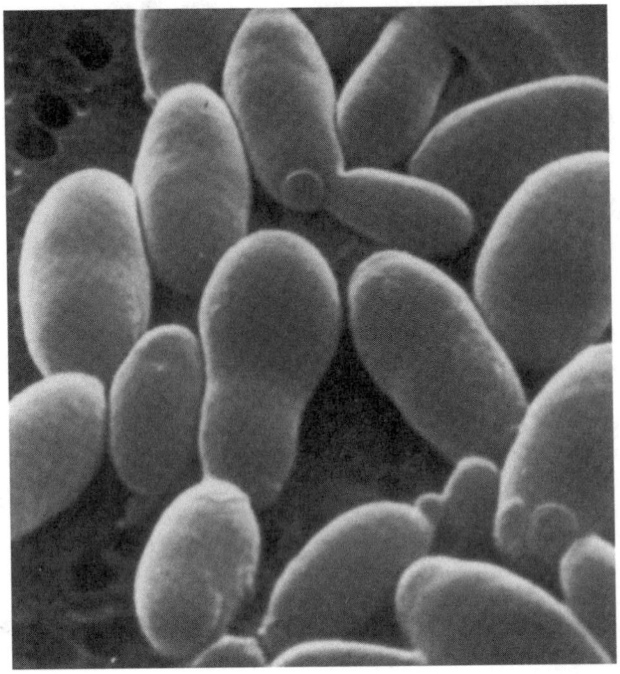

Figure 3. The yeast *Kluyveromyces marxianus* showing budding of a daughter cell and the location of previous bud scars.

ear, and folded configuration called the nucleoid or folded chromosome. Some bacteria have more than one set of chromosomes; for example, *Deinococcus radiodurans* has four chromosomes per cell in its stationary phase, suggesting that the classification of bacteria as haploid is an oversimplification (3). The current view of bacterial chromosome replication is that the DNA replication machinery is held at a relatively fixed position (at inner cell membrane) through which chromosomal DNA is threaded to produce two daughter chromosomes (5).

Often DNA of extra chromosomal origin, called plasmids, which are autonomously replicating pieces of DNA about 1/100 of the chromosome size, are found in the cytoplasm. These plasmids frequently contain genes for antibiotic resistance, conjugation, and production of proteins generally deemed nonessential to normal cell functioning. Some bacterial plasmids are nonconjugative, yet others are conjugative and permit their and/or other plasmids' transfer within the species or a broad range of hosts from different genera. Certain conjugative plasmids can transfer DNA between kingdoms; for example, the Ti plasmids of the bacterium *Agrobacterium tumefaciens* can be transferred to dicotyledonous and monocotyledenous plants or to the yeast *Saccharomyces cerevisiae*. In all cases, transfer of plasmids requires cell-to-cell contact and presence of transfer and mobilizing genes.

Cells of the yeasts *S. cerevisiae* and *Kluyveromyces marxianus* (Fig. 3), which are used in fermentation, bread making, and food-grade enzyme production, are quasi-spherical, 3 to 5 μm in diameter, and can double hourly. Yeasts may be haploid or diploid, depending on their life-cycle stage. All eukaryotic cell chromosomes contain some basic proteins or histones, which wrap around the DNA to form nucleosomes. *S. cerevisiae* contains 17 linear chromosomes ranging from 150 to 2500 kb. The collaboration of more than 600 laboratories in the United States, Canada, Europe, and Japan to sequence 12 billion bases and arrange 6000 genes of this yeast has been one of the largest decentralized experiments in molecular biology to date. A whole issue of the magazine *Nature* was devoted in 1997 to the yeast genome. DNA duplication in eukaryotic cells occurs during the S-phase of the cell cycle. Here the nuclear DNA will contain more than 100 putative replication factories, each with 300 or so replication points (forks) (6). Individual chromosomes can be separated by pulsed-field gel electrophoresis technique. Eukaryotic cells may also contain membrane-bound intracellular organelles such as mitochondria (Mtc), endoplasmic reticulum (Er), or chloroplasts (Chl). In Chl and Mtc (Fig. 4) a ccc DNA is compartmentalized and expressed. mtDNA are believed to have been derived from some prokaryotic cell genome. During cell division, all DNA, whether organellic, plasmid borne, or chromosomal, are divided equally and partitioned between the two daughter cells.

Bacterial cells are able to accept DNA from another parent or donor and undergo *in vivo* recombination. This was studied initially in the 1950s through the 1970s. These studies used whole cells and relied on natural exchange such as mating, transformation, and recombination occurring *in vivo* between donor DNA and recipient cells. During the 1960 to 1980s, after many of the requirements of DNA metabolism were discovered, it became possible to perform *in vitro* the synthesis, breakage, and joining the DNA from homologous or heterologous origins. These discoveries gave birth to the science and production of *in vitro* recombinant DNA (rDNA) molecules (7), which when transferred into living cells could express new traits.

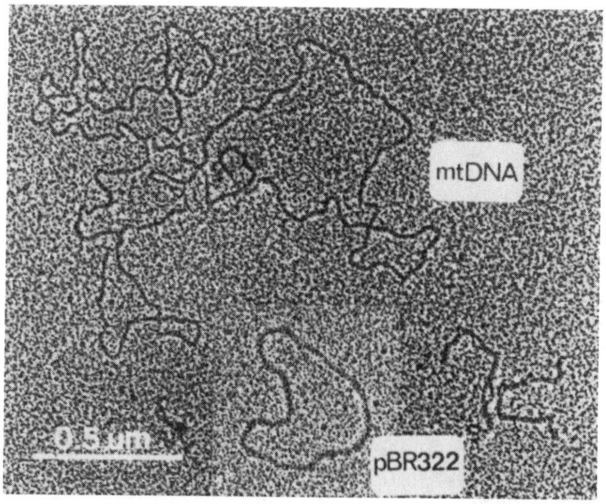

Figure 4. Covalently closed circular DNA molecules from the mitochondrion of a fungus and *E. coli* plasmid pBR322.

The terms transformation, transduction, transfection, and conjugation are used to indicate "natural" gene transfer into a cell (4). Transformation and transfection involve transfer of donor DNA into a recipient cell, whereas conjugation and transduction require the presence of a donor cell and a bacteriophage, respectively. Both natural and $CaCl_{2+}$ heat-induced transformation of many microorganisms by rDNA are now possible. Natural transformation of certain bacteria (eg, *Bacillus subtilis*) by linear duplex chromosomal DNA, and induced transformation of others (eg, *E. coli*) using ccc-plasmid DNA, has become common practice in genetic engineering (4). DNA can also be introduced into a cell through one of many laboratory-devised techniques of gene transfer, such as electroporation, particle gun or biolistics, microinjection, microlaser technique, and liposome fusion (see later). Transformation of many animal and plant cells is also possible, although the underlying mechanisms are quite different from those in bacteria. Transformation of cells can occur by intergeneric fusion of two individual cells from, in the order of their discovery, plant, microbial, and animal origins (8). Microbial and plant-cell fusion requires the production of protoplasts or cells without external surface layer(s). Protoplasts are osmotically unstable; in hypotonic environments they lyse, but in the presence of stabilizers (sugars, salts) they remain intact. When two protoplasts are brought into contact in the presence of a fusogenic substance (eg, Polyethylene glycol 6000) their membranes fuse, causing cytoplasmic and nuclear mixing events to occur. A transient fusant contains chromosomes from both parents; subsequently, karyogamy and recombination of nuclear materials and chromosomes can take place. With animal cells the fusogen could be an animal virus (8).

Analysis of the DNA as Genetic Code and Its Functions

The flow of genetic information is in general from DNA through transcription to RNA (messenger, transfer, and ribosomal RNA) and through translation of mRNA to proteins (Fig. 5). This was known as the central dogma of molecular biology until the discovery of the enzyme reverse transcriptase, which could synthesize a complementary DNA (cDNA) from an mRNA molecule. DNA and RNA molecules contain four bases: two purines, adenine (A) and guanine (G); and two pyrimidines, cytosine (C) and thymine (T) in DNA and uracil (U) in RNA. These bases are connected to the sugar deoxyribose (DNA) or ribose (RNA) to form a deoxynucleoside or nucleoside and are phosphorylated to form deoxyribonucleotides (DNA) or ribonucleotides (RNA), respectively (Fig. 6). The double helical or Watson-Crick form of base pairing for A:T and G:C in DNA was the first to be discovered (Fig. 7). Other types of Watson-Crick base pairing called k and π with a hydrogen bonding pattern (9) have increased the genetic alphabet from four to six letters. Many of the natural bases can be modified by addition of organic groups, for example, methylcytosine. Nucleotides are linked through phosphodiester linkage, and each polymer of DNA has a 5′ to 3′ polarity. The ratio of (G + C)/(A + T) is known by the designation (G + C) content or % (G + C) and reflects taxonomic relatedness and molecular characteristics. The (G + C) content of two different DNA molecules determines their separation during buoyant density-gradient centrifugation. Additionally, double-stranded (ds) DNA of higher %(G + C) will have a higher melting temperature (Tm) than that with a higher %(A + T). Through melting, dsDNA can be dissociated into two single-stranded (ss)DNAs, and through cooling down, they will regain the complementary ds structure. This process forms the basis for several DNA:DNA or DNA:RNA hybridization techniques. *In vitro*, every DNA molecule has a topological feature. The open-ended DNA molecule is either in ss or ds form and can be in rod shape. DNA molecules can bend; for example, a small polymer of 242 base pairs (bp) or larger can go from a linear (open-ended) to a circular (ccc) form by the joining of its open ends with ligase. More complex forms of dsDNA are also found. DNA molecules can assume many forms, including winding or unwinding, depending on their physical and enzymatic environments, for example, presence of enzymes called topoisomerases (for twisting) and endonucleases (for nicking and relaxing the twisted DNA). In addition, the sequence of bases within a ssDNA could create structures of their own. When a sequence of -AAAAAAAGCTTTTTTT- from a DNA duplex is allowed to separate into two ss polymers, each can generate a hairpinlike structure through base pairing of all A and T residues. Such sequences are thought to be recognition segments within DNA for its interaction with regulatory proteins. RNA structures also have unique organizations such as folding on itself and forming of a hairpinlike structure (eg, tRNA).

All DNA synthesis is enzymatic and utilizes the four deoxyribonucleotides. The pathways for the biosynthesis of deoxyribonucleotide triphosphates and degradation of DNA are well known. The enzyme responsible for DNA biosynthesis, DNA polymerase, appears in the forms I, II, and III in prokaryotic and α, β, and γ in eukaryotic cells and, along with several ancillary proteins, comprise the DNA synthesis machinery (6,10–12). DNA polymerase is used either for repair, in a 3′ to 5′ direction (which could also be due to exonuclease activity) or replication in a 5′ to

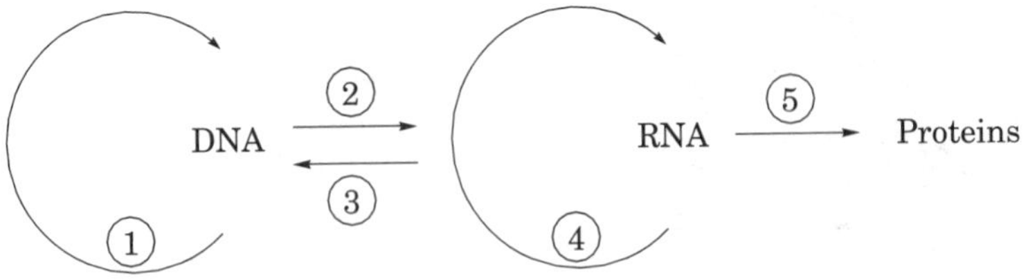

Figure 5. Flow of genetic information as perceived in molecular biology: (1) DNA replicates through DNA polymerase system; (2) transcription of DNA into RNA occurs via RNA polymerase system; (3) RNA is copied into DNA via reverse transcriptase action; (4) RNA as an enzyme can act on itself; (5) RNA being translated into proteins by the translational system.

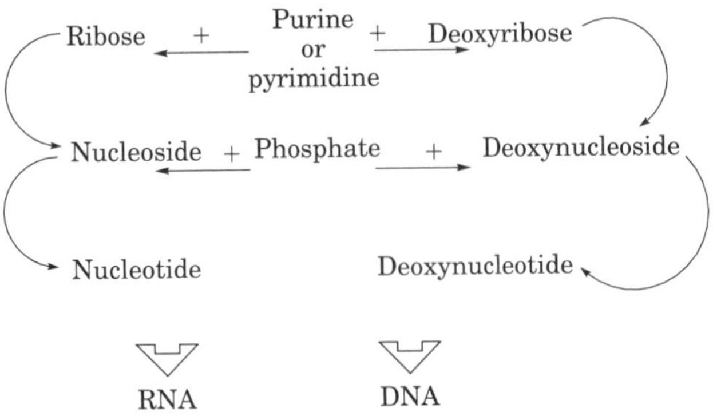

Figure 6. The biosynthesis of nucleosides and nucleotides for their use in DNA or RNA.

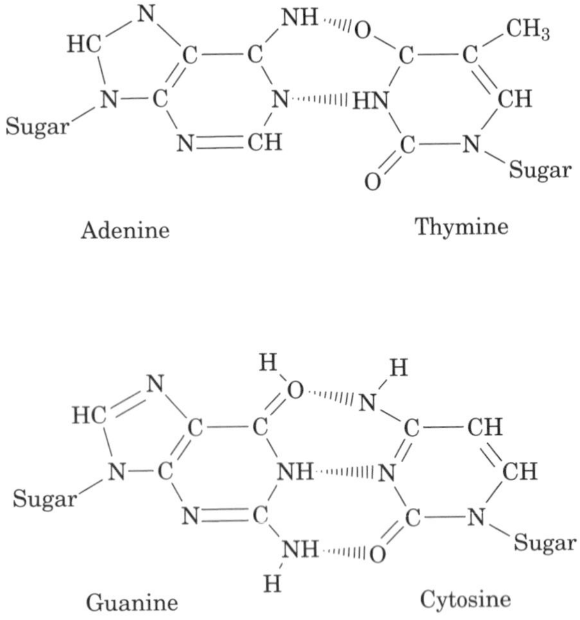

Figure 7. The base pairing between A:T and G:C in DNA.

3′ direction. DNA helicases are essential motor proteins that function to unwind double-stranded (duplex) DNA by an active, rolling mechanism to yield the transient ssDNA intermediates for replication, recombination and repair (11). During the synthesis of DNA each deoxynucleotide is joined to the previous one through a phosphodiester bond. DNA polymerization occurs at a maximum speed of 100 and 1000 nucleotides per second for bacterial and animal cells, respectively. The open ends of the DNA at 3′-hydroxyl and 5′-phosphoryl position could be joined by DNA ligase (12). In certain mammalian cells, there are polynucleotide kinases that possess both 5-phosphotransferase and a 3-phosphatase activity that can restore DNA strand breaks, with 5-hydroxyl termini or 3-phosphate termini, or both, to a form that supports the subsequent action of DNA repair polymerases and DNA ligases, that is, 5′-phosphate and 3-hydroxyl termini (13).

A gene is a segment of DNA or, as in certain viruses, RNA made of a stretch of bases that, respectively, code for an RNA and a polypeptide molecule. Synthesis of a polypeptide from the transcript of a gene has been visualized (Fig. 8). It requires translational machinery to synthesize proteins from N-terminal to C-terminal by joining amino acids through their carboxy-terminals. In some viruses and eukaryotic organisms some genes and hence their RNAs are made of translatable and intervening sequences (known as exons and introns, respectively). The genetic code is read as a codon of three bases at a time. There are 64 codons, of which 61 code for amino acids and 3 for termination. There is some degeneracy in certain codons; for example, the amino acid glycine is coded for by the triplets GGA, GGG, GGU, and GGC. Different organisms have a bias in codon usage; for example, they may prefer to use one set of these over the other for the synthesis of polypeptides. Polypeptides in their nascent form are self-assembled into functional structures through the physicochemical properties of the constituent amino acids. Other polypeptides require a special class of proteins called nucleoplasmins, heat-shock proteins, and chaperon-

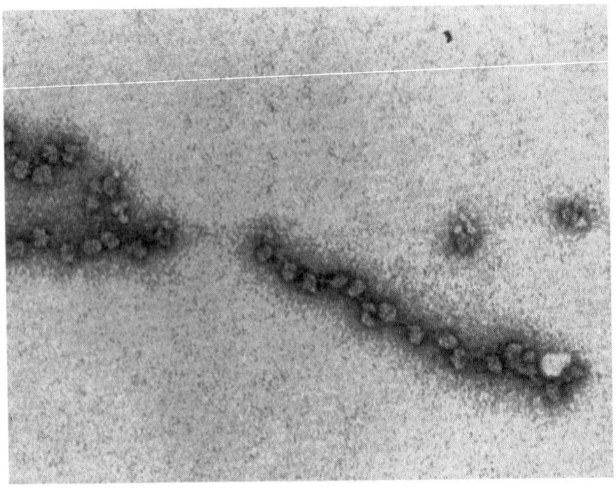

Figure 8. Transcription of bacterial DNA into an mRNA and its translation by polysomes into polypeptide. *Source:* Photograph courtesy of B. A. Hamakalo.

ins for assembly. Chaperonins are multifunctional proteins that catalyze the correct folding of other proteins by preventing side reactions such as aggregation (14). They are not themselves components of the final functional proteins. Different members of the chaperonin family assist in folding in a concerted manner. Some, but not all, molecular chaperonins are heat-shock or -stress proteins. Chaperonins act in the assembly of other proteins where they are not components of the final structure (14). Protein structure, folding, and enzymatic stability are dictated by covalent, disulfide linkages, and by noncovalent forces and interactions (15). Mutational replacement of amino acids within certain regions of a protein could affect its conformation and, hence, function, for example, thermostability and reaction rates (15).

Mutations and Selections

Most heritable variation of phenotypes is explained by mutations in genotypes. Improved mutant organisms have been used in industrial microbiology and food and fermentation technology. The initial attempts at strain improvements for food technology were through selection for spontaneously occurring variants from the original strain (16). Subsequently, fundamental studies on the occurrence and induction of mutations, enrichment and isolation of mutants, site-specific mutagenesis, and use of genetic recombination have made this process more manageable (15,16). *In vivo* mutations in microorganisms can occur spontaneously or with the mediation of mutagens at frequencies in the $10^{-(7-9)}$ or $10^{-(4-6)}$ range, respectively. Mutations alter DNA base sequences, by adding, deleting, or substituting base(s) and often affect the structure, and hence the function, of proteins. Mutagenic agents fall into three classes: physical (UV and X-ray irradiation), chemical (nitrous acid or nitrosoguanidine), and biological (transposons and mutagenic bacteriophages) (16).

A special class of genetic elements that are capable of insertion into any DNA are known as insertion sequences (ISs) and transposons (TNs). Insertion sequences constitute an important component of most genomes. More than 500 individual ISs have been described (17). The organization of a typical IS includes discrete DNA sequences of 0.8 to 1.4 kbp with a short DNA sequence at each terminus or 30 to 40 bp inverted repeats (IRs). Transposons are a class of genetic elements that contain at least one gene such as gene(s) for metabolism, antibiotic resistance, or synthesis of toxins (class I TNs), and in addition the transposase (class II TNs) (18). Insertion of a TN into the genome of a host requires the presence in the recipient DNA of ISs. Two functions are associated with a TN DNA once inserted into DNA molecules in a host cell. First, TN through its own insertion causes disruption of the sequence of a gene leading to a deletion-type mutation (18,19). Second, TN can affect the regulation of the expression of genes in the neighborhood of their integration site; for example, by inactivation of one gene and increased transcription of another adjacent gene (20). Another example is the deactivation of a regulatory phenotype by TN insertions in the *lac*A gene, encoding an endogenous β-galactosidase of *B. subtilis*, leading to inactivation of its negative regulator, *lac*R (21). The mechanism of such phenotypic change is due to effects involving one or several elements internal to the TN. On the other hand, the loss and restoration of mutator TN activity in maize has shown evidence against dominant-negative regulator associated with loss of activity (22).

As expected the exact removal of a TN DNA should restore the gene's function. This form of mutagenesis is called transposon mutagenesis and has wide application in genetic engineering (23). The end result of *in vivo* mutagenesis is alterations in DNA sequence(s), which must be fixed, replicated, and segregated. The rare mutant clone is then screened, enriched, purified, and characterized. One problem with *in vivo* mutagenesis is the difficulty in targeting specific genes. Certain industrially useful mutants are hard to grow or maintain because of the multiple or deleterious mutations they acquire during induced, but random mutagenesis. These problems can be eliminated by using *in vitro* mutagenesis. Here, a gene sequence is isolated, cloned into a suitable vector, and usually treated with a chemical mutagen, to specifically alter a base. More commonly, an oligonucleotide with a modified sequence is synthesized and then inserted into a gene (15,23). Through these approaches it is now possible to generate gene- or site-specific mutations. Recombination between two mutants with different characteristics can generate new organisms for industrial exploitation (6).

Elements and Regulation of Gene Expression

Gene expression requires template DNA, which is divided into regulatory and structural regions, and RNA transcription to produce the intermediate RNA (mRNA, rRNA, and tRNA). Although the primary function of mRNA is to serve as a blueprint for translation of a genetic message into a polypeptide (23), certain RNA molecules, called ribozymes, can function as self-splicing catalysts (24). The discovery of ribozymes has changed our dogmatic views of catalysis being solely the domain of protein enzymes.

To regulate their growth and metabolism, organisms must determine when and how much cellular constituents are needed. This task is achieved by controlling gene expression (3,23). The following elements regulate and control prokaryotic gene expression (Fig. 9). The operator is a DNA sequence where a repressor protein binds and pre-

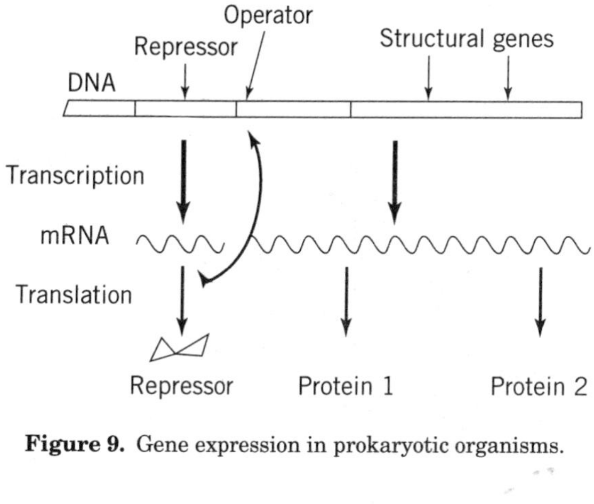

Figure 9. Gene expression in prokaryotic organisms.

vents transcriptional occurrence. Repressor-operator recognition involves a complex code of DNA base-pair sequences. Induction refers to expression of a gene subsequent to the removal of the repressor, usually by the addition of specific compound(s) called inducers. The promoter is located immediately in front of the gene and is the RNA polymerase recognition and transcriptional initiation site. Natural promoter efficiency varies, some are stronger than others. During the transcription of prokaryotic genes, RNA polymerase identifies in the 5′ to 3′ direction of the sense strand of the promoter before the initiation codon, a recognition sequence of bases, TTGACA or -35bps sequence, and a TATAAT (the Pribnow box) sequence at -10bps for its subsequent binding. Mutationally induced base changes in these two sequences can have mild or severe effects on transcription. Through its specificity for DNA transcriptional initiation sites, binding of RNA polymerase generates a localized melting and separation of the two strands of a small segment of the DNA leading to the initiation of RNA synthesis usually starting with GTP or ATP. DNA topoisomerases contribute to this process by relaxing DNA tension (25). There is continuous polymerization of ribonucleotides into an RNA transcript until the terminus of the gene is reached. At the terminus, either a sequence of six uridine residues and a hairpinlike structure or a sequence lacking such bases but requiring a termination factor, rho, to signal the stoppage of transcription. Some structural genes contain a leader and a trailer sequence, respectively, immediately after the promoter and before the terminator regions. At the end of transcription, RNA polymerase is freed to repeat the cycle.

The control of gene expression occur through transcriptional and translational control. One mechanism of control of transcription of genes is through induction of specific mRNA synthesis when an inducing substance is present (to turn on gene expression) and its repression (or turning off) once such inducers are removed. For example, when lactose is present in the growth medium of *E. coli*, the enzyme β-galactosidase is synthesized and, conversely, when it is absent, the repression of the enzyme synthesis occurs through a repressor protein. This modality of regulation is called negative control. In addition, several positive regulators of RNA polymerase are involved in control. For example, intracellular regulator cyclic AMP (cAMP) through its binding to cAMP binding protein (CAP) creates a complex, cAMP-CAP, which binds to specific promoters to activate certain operons. Finally, through a process called attenuation, the biosynthesis of the amino acid tryptophan is controlled. In this case, the transcription of the biosynthetic genes begins in the usual manner. However, instead of a complete transcript of the genes, an incomplete transcript of relatively short mRNA (162 bases) responsible for the tryptophan leader protein is made. The leader protein is tryptophan rich. When intracellular tryptophan supply is high, the mRNA makes a structure to block further continuation of transcription, by stalling (attenuating) the translation. With low tryptophan levels, a different structure in the stem-loop region of mRNA develops, which allows the continuation with transcription to the end of the operon. Therefore, in the case of tryptophan operon, the removal of repressor and attenuation generates a 70-fold (derepression) and another 8- to 10-fold (attenuation) or an overall of 600- to 700-fold increase in the expression (23).

In certain eukaryotic genes and hence their mRNAs (Fig. 10) there are some sequences of about 1000 bp inserted (introns) between the coding sequences (exons). Transcription begins at the promoter (5′) of a gene and proceeds through the exons and introns (if present) to the 3′ terminus, producing a primary mRNA transcript or heterogeneous nuclear RNA. In eukaryotic cells, several processing steps must occur to change the primary mRNA transcript into functional mRNA and allow its translation into proteins. Any introns present are subsequently removed or spliced out. The actual ending of the transcription in some cases could be several hundred base pairs beyond the polyadenylation site of the 3′ end of a mRNA. The third step required to change a primary mRNA transcript to a functional mRNA is the addition of a 5′ cap.

Eukaryotic promoters are recognized by specific DNA-binding proteins of less than 100 amino acid residues in size that bind to DNA and activate or repress gene transcription. Most bacterial RNA polymerases are large multimeric proteins; for example, *E. coli* enzyme contains two α, one β, and one β′ subunits. This holoenzyme has additional subunits, the ∂ 70, ω, and σ subunits, which are integral to its functioning. In sporulating bacteria (eg, *B. subtilis*), RNA polymerase subunits are responsible for the selection of sporulation promoters. The RNA polymerases of eukaryotic organisms differ from the preceding both in molecular size and number of enzyme species. As many as two to four different types of RNA polymerases can be found in eukaryotic cells, and in several cases their β and β′ subunits have conserved sequences. In higher eukaryotes, RNA polymerase I initiates transcription for rRNA genes from nucleolus, and RNA polymerases II and III transcribe mRNA, tRNA, and 5S rRNA from nuclear matrix (26). The transcriptional machinery of eukaryotic—as compared with the prokaryotic—cells is much more com-

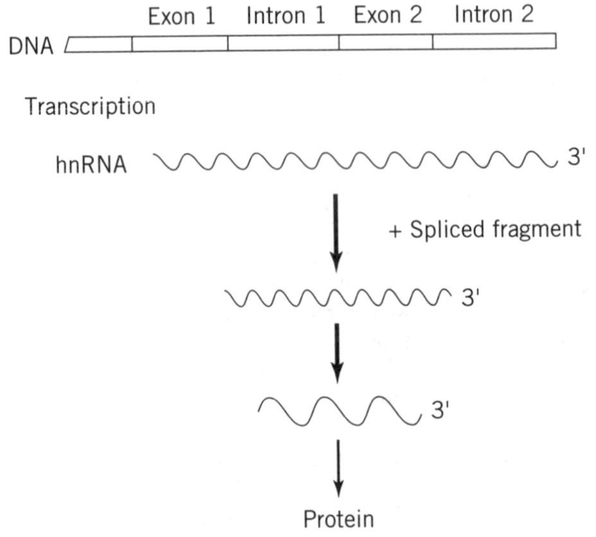

Figure 10. Transcription of an eukaryotic gene containing introns and exons.

plex (27) and can involve allosteric control by DNA in the case of selective gene transcription (28). Functions of RNA can be regulated by antisense RNA, which is an RNA complementary to the sequence in mRNA that through complementary, stable base pairing, prevents its translation (29). Antisense RNA also regulates or even prevents gene expression in prokaryotic, plant, and animal cells.

Whether prokaryotic or eukaryotic, the mRNAs are translated into proteins using ribosomes, charged tRNA molecules, and auxiliary proteins, beginning with codons ATG or GTG (23). The ribosome is a large multifunctional complex composed of both RNA and proteins. Accumulating evidence suggests that rRNAs play a central role in the critical ribosomal functions of tRNA selection, and binding, translocation, and peptidyl transferase (30). Usually several ribosomes (or a polysome) are present on a mRNA. In prokaryotic cells transcription and translation occur concurrently within the cytoplasm; but in eukaryotic cells, translation occurs outside the nucleus. Translational control includes the ribosome-binding sequence (Shine-Delgarno sequences) and attenuation, which is prevention of translation of mRNA because of its folding back on itself. After its transcription, the mRNA is degraded. Messenger stability, among other things, depends on the presence of AU-rich sequences at 3′-end, resulting in shorter half-lives than those lacking such sequences. If a nonsense codon is reached either at the end or through mutations within a gene, the translation is halted and a partial polypeptide results. Finally, protein stability and half-life depends on proteases that degrade not only normal but also for certain heterologous and abnormal (mutant) proteins. In eukaryotic cells protein stability depends on the small protein ubiquitin, which when conjugated to proteins leads to their rapid degradation.

Genetic Exchange and Recombination in Vivo

Prior to the discovery of genetic engineering *in vitro*, it was known that genetic information can be exchanged *in vivo* among various organisms (3,4). The simplest forms of genetic exchange—transformation, transduction, and conjugation—are those found among bacteria. Only conjugation requires cell-to-cell contact. Conjugation was initially discovered in the common enteric bacterium *E. coli*, but it occurs in many genera of bacteria and between members of two kingdoms, for example, *A. tumefaciens* and many plants species (31), as well as *E. coli* and *S. cerevisiae* (32). In certain bacteria such as streptococci, mating depends on the presence of special mating pheromones, which are small polypeptides promoting mating aggregation. In fungi, mating occurs when two haploid cells of different mating types fuse to allow genetic exchange to occur. In yeasts this requires special mating control locus (MAT) and switching loci (HML or HMR), which are silent cassettes that introduce the a and α mating-cell types through direct transposition of required genetic cassettes. In the absence of sexual cycle, certain fungi use parasexual mating events. In transformation, whether natural or induced, naked DNA or DNA contained in membranous vesicles called transformosomes are transferred into a recipient cell (4). In generalized and specialized transduction, DNA packaged into a bacteriophage (Fig. 11) is injected into a bacterial cell. The latter two modes of genetic exchange are used in recombinant DNA technology for packaging of rDNA (eg, bacteriophage λ) or transformation of commercially available competent *E. coli* cells. Once inside a cell, homologous DNA can undergo recombination using recombinational enzymes to generate hybrid DNA molecules. All mechanisms of genetic exchange are currently used in food research.

GENETIC ENGINEERING TECHNOLOGIES

Instrumentation and Tools

Central to and fundamental in the process of genetic engineering is the isolation, manipulation, insertion into, and expression of DNA in a host cell (33). The instrumentation and tools needed for the technical tasks are shown in Table 1.

Organisms

Microorganisms, plants, and animals are used organisms in food production and processing technologies. While the production of microbial cells through fermentation technology has been a long-established art, that of plant cells in cultures is rather new. In the 1970s several groups studied the ability of plant cells and tissues to grow in liquid nutritive media. Cultures from diverse origins such as (*1*) organs (roots, flowers, anthers), (*2*) meristems (shoot, leaf), (*3*) callus (undifferentiated cell mass), (*4*) cells (homogenized tissues), and (*5*) protoplasts were found to grow on defined salts media containing a carbon source, vitamins, plant hormones, and various other nitrogenous substances. Plant cell biotechnology and economic value-added or value-derived products have matured more rapidly with advances in plant cell culture systems, construction of transgenic plants, and other applications of biotechnology (see later).

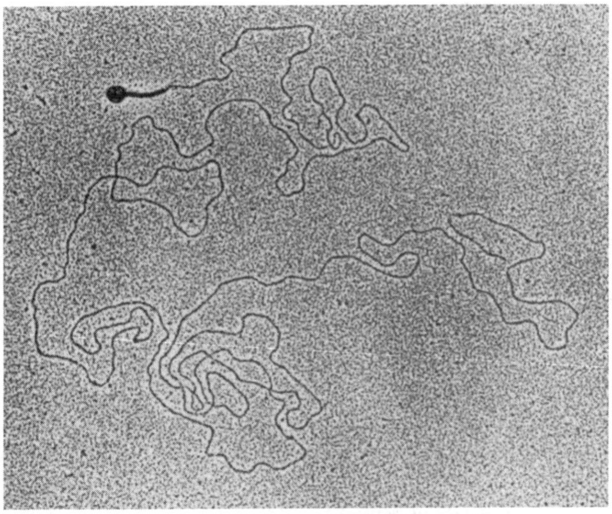

Figure 11. The bacteriophage lambda of *E. coli* ejecting its DNA.

Table 1. Basic Steps in a Simple Genetic Cloning

Step	Tools or systems needed
1. Propagation of organisms	Growth chambers, incubators, fermenters, greenhouses
2. DNA extraction and purification	French pressure cell, cell lytic enzymes, spectrophotometer, preparative centrifugation, cesium chloride density ultra centrifugation
3. DNA library preparation	Restriction enzymes
4. DNA size separation	Ultracentrifugation or electrophoresis
5. Vector and probe DNA	Polymerase chain reaction or cDNA, preparation technology, DNA manipulative enzymes, electrophoresis
6. DNA sequencing	Maxam-Gilbert, Sanger's and/automated sequencing methods
7. Introduction of rDNA into a host cell	Electroporation, bio-bolistics host transformation
8. Identification of transformed cells	Selective media, reporter genes, Southern blots, Northern blots, Western blots
9. Gene location	OFAGE, CHEF, various blots

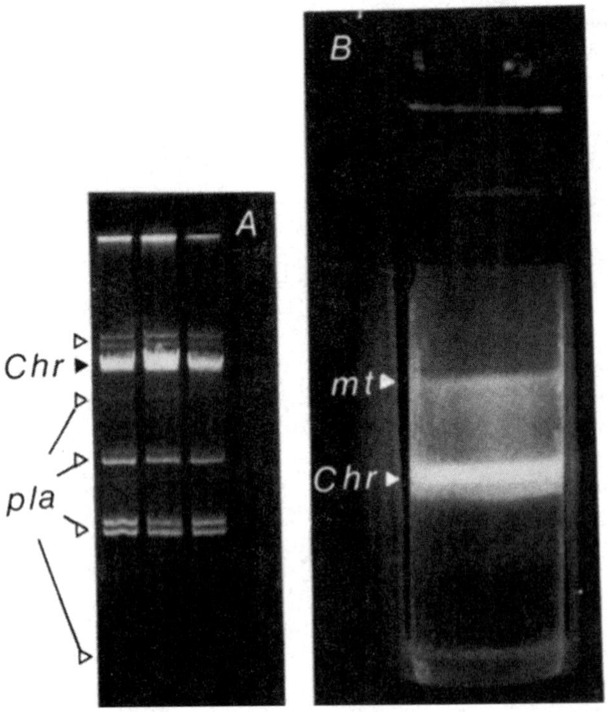

Figure 12. Separation of chromosomal and plasmid DNA from the bacterium *Bacillus thuringiensis* HD-1 by agarose gel electrophoresis (**a**) and chromosomal and mitochondrial DNA from the fungus *Beauveria bassiana* by cesium chloride bisbenzimide gradient centrifugation (**b**).

Isolation and Preparation of DNA

DNA from most cellular organisms can be isolated through the disruption of cell membranes and/or walls by using lytic enzymes or other physicochemical forces (osmotic pressure, shear forces, and ultrasound). DNA of high (chromosomal) and low molecular weights (from plasmids, mitochondria, and chloroplasts) can be extracted. Because cellular lysates are rich in DNA-degrading enzymes, such reactions are performed in the presence of chelating agents or cooler temperature. DNA is then purified from proteins and other cellular constituents by extraction with a mixture of phenol-chloroform and is recovered by ethyl alcohol or isopropanol precipitation. To purify further, high-speed dye (ethydium bromide or bisbenzimide) buoyant density ultracentrifugation is performed and to characterize the plasmid or other types of DNA molecules, agarose gel electrophoresis is employed (Fig. 12). Although DNA isolated by this method is of high quality, the procedure is a lengthy one requiring expensive equipment. Nowadays, commercial DNA purification kits that utilize silica-based resins and anion exchange have become available. These kits allow for the isolation of DNA in purity equivalent or superior to that obtained by two successive rounds of cesium chloride gradient centrifugation (34). Such plasmid DNA is now routinely used in such applications as transfection, microinjection, automated and manual sequencing, restriction analysis and *in vitro* transcription.

In electrophoresis, DNA molecules migrate according to their molecular weight from the negative to the positive electrode; that is, the smaller molecules move further from the origin. If such DNAs are cut by any of some 600 restriction endonucleases, enzymes that cleave phosphodiester linkages at specific sequences within DNA, the resultant fragments can be separated in an electrophoretic separation gel, according to their sizes (Fig. 13). DNA fragments after electrophoresis can be transferred permanently by blotting onto specific membranes. Often the electrophoretic separation of DNA followed by the ethidium bromide staining is used for visualization of DNA molecules, although the same can be achieved by electron microscopic analysis for size (Figs. 4 and 11) or heteroduplex analysis. Table 2 lists most commonly used enzymes for cell lysis, DNA extraction, and subsequent rDNA construction. While all DNA-sequencing protocols are based on the Sanger's chemical method (Fig. 14), other subsets of methodology by enzymatic (Klenow polymerase, Taq polymerase, and polymerase chain reaction, or PCR) systems are also used. Sequencing tasks are now automated and commercially available both for synthesis and sequencing (33).

Polymerase Chain Reaction

The PCR developed by Cetus (Emeryville, California) scientists in 1985 (35) is a powerful *in vitro* method for amplifying a segment of DNA that lies between two regions of known sequence, defined by a set of primers. The basic steps of PCR are denaturation of the DNA, annealing the primers to complementary sequences, and extension of the annealed primer with a thermostable DNA (eg, *taq*) polymerase. Together these steps represent one cycle of amplification.

As illustrated in Figure 15 the double-stranded DNA template is first denatured by heating in the presence of a

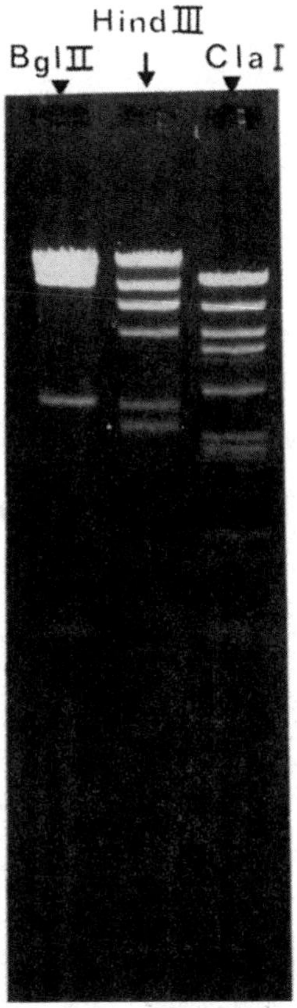

Figure 13. Restriction endonuclease digestion of bacteriophage lambda DNA into fragments. The names of enzymes are indicated on the top of each track.

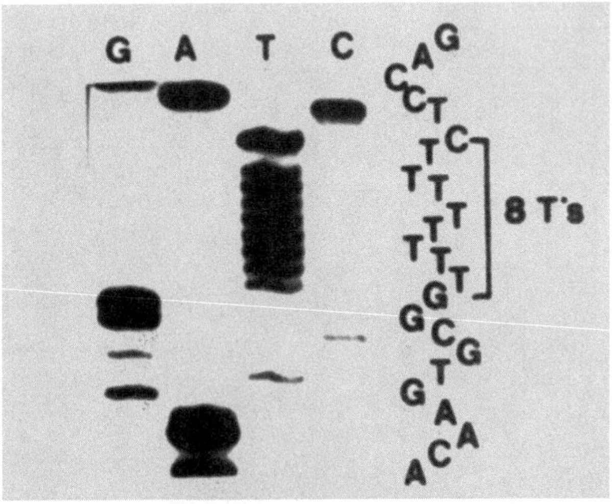

Figure 14. Sample autoradiogram of DNA sequencing reaction pattern by the Sanger method. *Source:* Photography courtesy of J. M. MacPherson.

large molar excess of two specific oligonucleotides and four dNTPs. The reaction is then allowed to cool to a temperature that allows for the annealing of pairs of oligonucleotides to their target sequences. Once annealing has occurred DNA polymerase mediates the 5′ to 3′ extension of the primer-template complex. These three steps are repeated several times and the major product of this reaction is a significant amplification of the desired segment of double-stranded DNA whose termini are defined by the 5′ termini of the oligonucleotide primers.

The first experiments in PCR used the Klenow fragment of *E. coli* DNA polymerase I, which has a temperature optimum of 37°C. However, since the Klenow fragment is inactivated at temperatures required to denature DNA, it was necessary to add a fresh aliquot of the enzyme to the reaction repeatedly. Unfortunately, these reactions tended to only work well for the amplification of DNA fragments less than 200 bp. For larger fragments it was found that the yields were poorer and the products were often heterogeneous in size (36). These problems were solved with the discovery of the thermostable *taq* DNA polymerase, isolated from the cells of thermophilic bacterium *Thermus aquaticus*. *Taq* polymerase can survive extended incubation at 95°C; therefore, all the components can be added at the start of the reaction without any further replenishment. Also, since annealing and extension can be carried

Table 2. Enzymological Aspects of rDNA Work

Purpose	Enzyme	Function
Cell lysis	Lysozyme, cellulase, mutanolysin	Removes cell walls
Proteolysis	Pronase, protease K	Removes proteins
RNA degradation	RNase H	Elimination of RNA
Nick translation	*E. coli* DNA Polymerase I	Synthesize DNA
	Klenow fragment	
	T4 DNA polymerase	
cDNA synthesis	Reverse transcriptase	(cDNA) from mRNA
Process DNA	Nuclease Bal31	
	Exonuclease digestion	
	Mung-bean nuclease,	
	Restriction endonucleases	Endonuclease cuts
	DNA methylases	Methylated bases
	Phosphatases	Remove 5′ phosphate
Joining of polynucleotides	Polynucleotide ligase	Join ends of DNA or RNA
Amplifying DNA	Taq polymerase	Produce oligonucleotides by polymerase chain reaction

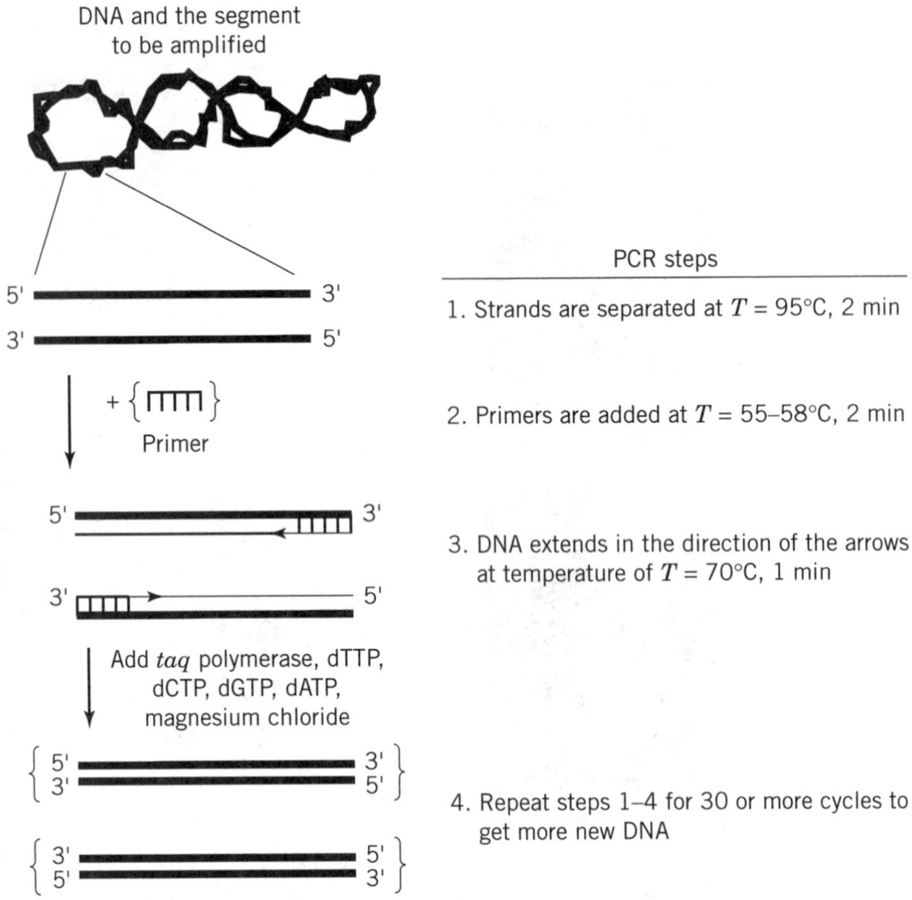

Figure 15. PCR the double-stranded DNA template is first denatured by heating in the presence of a large molar excess of two specific oligonucleotides and four dNTPs.

out at elevated temperatures, mispriming is reduced, thus resulting in improvements in the specificity and yield of the amplification reaction.

PCR amplification is currently used in a variety of needs in molecular cloning and analysis of DNA, for example, probes, generation of large amounts of DNA for sequencing, chromosome crawling, creation of mutant sequences, and the generation of cDNA from small amounts of mRNA. PCR is also extensively being used for (*1*) the specific amplification of cellular protein coding genes by differential or global expression, (*2*) detection of nucleic acid sequences of genetic modified plants or plant products, (*3*) pathogenic organisms in food and clinical samples, (*4*) the diagnosis of genetic disorders, and (*5*) in forensic cases.

Several modified versions of the PCR method have been developed. Compared with the original use of PCR, that is, to amplify segments of DNA located between two specific primer hybridization sites, a single-sided PCR method has been developed that initially requires specification of only one primer hybridization side. The second site is then defined by the ligation-based addition of a unique DNA linker. This method, referred to as ligase chain reaction (LCR), allows for exponential amplification of any fragment of DNA. Another modification of the basic PCR has led to the development of anchored PCR. By using anchored PCR it is possible to amplify full-length mRNA when only a small amount of the sequence information is available. PCR-based techniques such as Random Amplified Polymorphic DNA (RAPD), Amplified Fragment Length Polymorphism (AFLP), DNA Amplification Fingerprinting (DAF), and microsatellite/PCR have been developed and used for identifying DNA markers.

In many cases it becomes essential to have a PCR detection system that can identify desired gene sequences quickly, with high specificity and in large volumes. The best analytical tool available to do this is mass spectrometry (37). A mass spectrometer works by vaporizing DNA and then accelerating the molecules through a vacuum chamber with the help of an electric field. Tiny differences in the time that it takes the fragments to reach the detector reveal small differences in their mass and hence their sequence. This technique is known as matrix-assisted laser desorption ionization time-of-flight mass spectrometry, or MALDI-TOF MS, and has the capability to analyze hundreds of DNA samples for a variety of assays for point mutations and polymorphism analysis (38). Although originally used over a decade ago for protein analysis, it was not available for DNA analysis until 1993, when various matrices were developed that would work with DNA fragments as long as 100 base pairs. However, for practical sequencing MALDI-TOF would have to work with DNA fragments much longer than the current 100 base pair capacity. Currently, new matrices are being studied that could extend MALDI-TOF reach to 1000 bases, and if this works, then this technique would be a major breakthrough for high-throughput sequencing.

Cloned Genes in Vitro

Gene cloning is the use of experimental techniques that generate rDNA molecules *in vitro* with the desire for its incorporation and expression in a cell. With the use of rDNA technology one can produce a genomic DNA or cDNA library and hence a fragment containing any gene(s) from any source (39). Cloned genes are essential foundations of biotechnology. To clone a gene (Fig. 16), the genetic information, whether DNA or RNA, is processed by restriction endonucleases or reverse transcriptase or amplified by PCR, as the case may be. The steps involved in cloning foreign DNA into host cells include: (*1*) isolating RNA or DNA to be cloned; (*2*) choosing a suitable vector; (*3*) processing the DNA or RNA by restriction endonucleases or reverse transcriptase, respectively; (*4*) inserting the DNA fragments into the vector; (*5*) transferring of DNA into the desired host cell; (*6*) identifying those cells which have taken up the DNA, and (*7*) confirming that the clones are carrying the desired DNA fragment. These are the general steps within each of which several options are available. Figure 16 depicts the strategies involved in a gene cloning.

Several criteria are needed in deciding on a choice for cloning vector. Depending on fragment size to be cloned, specific vectors must be employed. Thus, for fragments of DNA with sizes of <4 kb, <10 kb, <23 kb, and <46 kb, the recommended vectors are phage M13, bacterial plasmids (pEMBL, pBR, etc), λ phage (Fig. 11), and cosmids (33). Cosmids, which are a combination of a plasmid and a bacteriophage (lambda) based vector, can be made to contain large DNA fragments from the genomic library of an organism. The number of clones needed for fragments of 35 kb from *E. coli* genome would be 340; from *S. cerevisiae*, 6000; and from tomato plant cell, 60,000. Often the cloning strategies, for example, choice of available restriction sites, reporter genes (to indicate the expression of cloned gene), presence of one or more selectable phenotypes, autonomous replication in two or more hosts (shuttle vectors), nature of the experiment (eg, sequencing, preparation of probes, study of gene regulation), and exploitation and safety consideration (biohazard consideration, placement of nonconjugability functions, suicide vector system) decide the particular details of the system. Finally, in deciding about the choice for the vector, the question of cloned gene copy number (low or high) and ssDNA (used for heteroduplexing and sequence analysis) versus dsDNA phages should be made in advance. In recent years more advanced vectors have been produced, for example, expression vectors and cassette vectors. They contain a promoter, a terminator, and ribosomal-binding sites, with the additional feature of a restriction site where a desired structural gene will be inserted. Presence of strong regulatory sequences aid significantly in the expression of the inserted gene in such a tailor-made expression vector system. This system offers great advantages for the production of rDNA-derived foreign products from cells. Additional uses of cassette vectors occur in those cases where partial removal of the reading frame from a structural gene has been made. After the start, the chimeric DNA, containing fragments of two different organisms, is inserted. The foreign gene product is recovered as a single (stand-alone) or fused (chimeric) polypeptides.

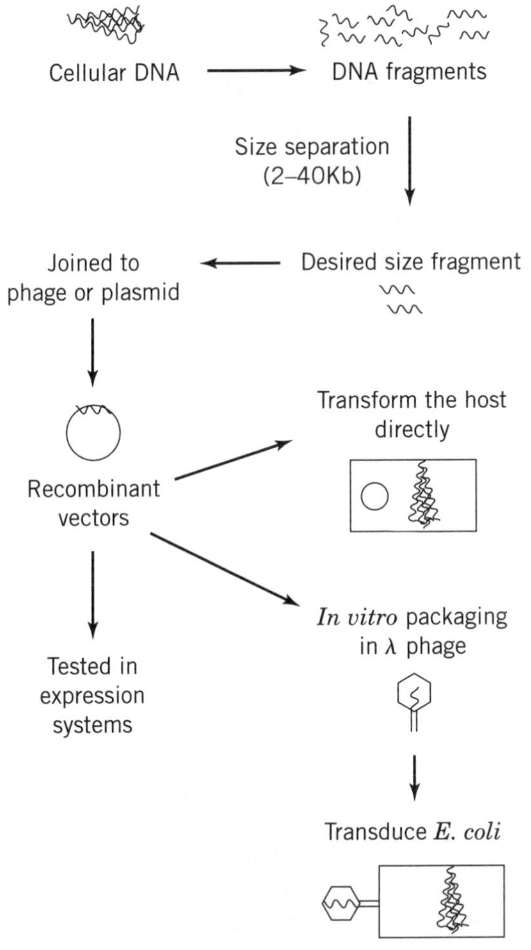

Figure 16. A summary of the basic gene cloning experiment using *E. coli* as the ultimate host.

To examine the polypeptide products from rDNA, several *in vitro* (various transcription and translation systems) and *in vivo* (whole bacterial cell, minicell, and maxicell systems) can be used (40,41). Minicells are unique in that they are polar buds from rod-shaped bacteria (Fig. 2) lacking chromosomal DNA and therefore permit the screening of only those proteins that have been derived from the rDNA. After a recombinant DNA is generated, it is introduced into the new host through one of many techniques of gene transfer, such as electroporation, particle gun or biolistics, microinjection, microlaser technique, and liposome fusion (33,41,42). The initial choice of a bacterial host for rDNA work reflects the specific usage, for example, laboratory versus industrial; however, for food usage the GRAS (generally regarded as safe) status is important and strains of microorganisms with particular genetic markings are available. The transformed cell is then identified by the expression of the recombinant gene product under some selection or growth condition or by using variety of DNA, RNA, enzyme function related, fluorescent (eg, the green fluorescent protein), and immunological probes for colony or plaque hybridization (Fig. 17) and detection (33).

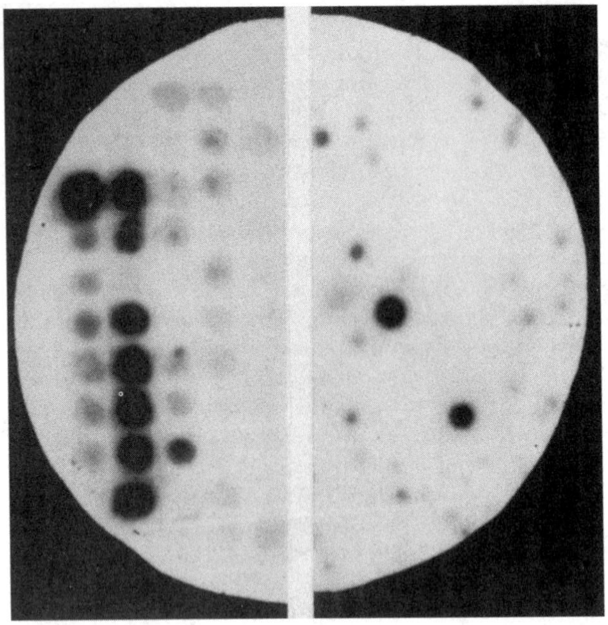

Figure 17. Autoradiogram of two halves of plates showing colony lift (left half) and plaque lift (right half). *Source:* Photograph courtesy of J. M. MacPherson.

APPLICATIONS OF GENETIC ENGINEERING IN FOOD SCIENCE

There are numerous applications of genetic engineering in food science and technology. The obvious use of this technology is in aiding the detection food contaminants or pathogens and safe foods (43). Additionally, however, we can change the starting materials in a food and add valuable attributes to the ingredients that are or get into the food system. In the following paragraphs, selected examples are given, more from a generic rather than commodity-specific points of view.

Genetic engineering tools have deciphered the total genomic sequences of more than a dozen microorganisms and tens of thousands of genes. Many molecular biology database systems have been established to deposit and acquire sequences, for example, EMBL (European Molecular Biology Laboratory), GeneBank (Genetic Sequence Data Bank), and NBRF (National Biomedical Research Foundation). These libraries can be accessed and searched through software tools and programs such as BLAST and FASTA (Lipman-Pearson search algorithms). The availability of many gene sequences has permitted inspection and identification of gene regulatory element-binding sites. One key element is the open reading frame (ORF) and the other is unidentified reading frame (URF). ORFs can locate ribosome attachment sites and, hence, through *in vitro* mutagenesis, base sequence changes within ORFs can be made without much need for formal genetic analysis (ie, mutants). Naturally occurring and engineered genes could be utilized from such gene banks.

The main source for the enzymes used in general, and the food industry in particular, are those enzymes that in terms of quantity of enzymes used come from microbial, plant, and animal sources, respectively. Enzymes not only transform food ingredients but also play a crucial role in restoring the lost flavor and aroma that can occur in certain instances of food processing. Various lipolytic, proteolytic, and multicomponent hydrolytic enzymes are being used for the addition of flavors to food. Thus, single-flavor compounds such as fatty acids, ketones, diacetyl, acetaldehyde, lactones, esters, pyrazines, and mustard oils and flavor enhancers such as nucleotides, amino acids, and terpens can all be produced by microbial enzymes for aroma and flavoring needs. In addition to the desirable flavors, enzymes such as lipoxygenases contribute to the off-flavoring of corn oil, while exerting a positive effect on the flavor of tomatoes (44,45). Genetic engineering, without creating a dilemma for the internal functions of a cell, could control the undesirable effects of many enzymatic reactions that affect postharvest or postprocessed foods. Through the knowledge of the DNA and protein sequence for many enzymes, the genetic analysis for the physical requirements of an enzyme (eg, active, catalytic, and ligand binding sites), and use of computer modeling the production of ideal enzymes should be possible (15). Production of a functional enzyme of minimal size or with improved thermotolerance, whether through solid-state peptide synthesis or synthetic oligonucleotides and rDNA technology, should be possible (15).

Recent production of transgenic animals and plants by such diverse methods as microinjection (plants and animal cells), conjugal transfer (microbial and plant cells), electroporation (microbial and plant cells), biolistics or microprojectile, and other less commonly known techniques has confirmed genetic engineering to be a strong and viable technology. Currently there are many fruits, vegetables, and seeds constructed through the tools of genetic engineering, which have produced transgenic plants containing viral, bacterial, and other eukaryotic genes presenting desirable traits such as resistance to disease, insects, herbicides, and tolerance to certain environmental factors (Table 3). Through an extension of this technology it should be possible for a majority of edible crops and other commercial plants to contain various nutritional and health-related qualities (45). Through combined knowledge of flowering-plant pigment and hormone biosynthesis, it is now possible to isolate, manipulate, and transfer genes for floral shape, and color. The extension of this knowledge toward derivation of food coloring and flavoring substances should be an exciting possibility. Use of antisense RNA in the area of genetic engineering of flowering plants and ripening of climacteric fruit and vegetables has been achieved.

Animal embryo transfer technology and production of transgenic animals by microinjection of foreign genes into eggs (transgenesis) to increase animal productivity and product value has become possible. Biotechnologists have produced transgenic sheep ("Dolly") and injected about 1 million copies of rat growth hormone gene into fish eggs to show about 5% of fish hatched were transgenic (with rat growth hormone). In sheep, goats, pigs, and cattle, presence of growth hormone produces leaner meat. In Edinburgh, Scotland, researchers have had success in creating transgenic sheep that produced milk with human anti-

Table 3. Examples of Transgenic Food Plants

Genetic modification	Plant
Amino acids	Canola/rapeseed
Antioxidant enzyme	Tomato
Carbohydrate metabolism	Soybean
Development	Soybean
Disease resistance	Apple, grapes, pear, potato, rice, wheat
Fertility	Corn
Fiber strength	Cotton
Flowering time	Apple
Herbicide tolerance	Beet, canola/rapeseed, corn, rice, soybean
Insect resistance	Alfalfa, corn, potato, rice, soybean, sunflower
Lysine content	Soybean
Male sterility	Corn
Maturity	Corn
Oil profile/quality	Canola/rapeseed, corn, soybean
Pectin esterase	Tomato
Protein quality	Barley, corn, soybean,
Reduced disulfides	Barley
Ripening	Melon, cantaloupe
Seed composition	Soybean
Shelf life	Mango, papaya
Size	Pear
Stalk strength	Corn
Vaccines	Bananas
Virus resistance	Grapes, papaya, plums, potato, soybean, tomato
Yield	Corn

hemophilic substance, factor IX, and have obtained sheep β-lactoglobulin from transgenic mice. In Australia through an ingenious injection of epidermal growth factor (Bioclip) in sheep and fitting of the animal with a net, the fleece falls within a week of injection, which is caught in a net. The commercial impact of the latter example is in saving spring shearing of some 150 million animals every year in Australia alone.

Genetic engineering, as described here, is a subject about which courses are given and textbooks are written. What is clear from this article is that which was said at the beginning: these changes are recent and indeed more change is just around the corner. Needless to say, the consumer education and buy-in position can ultimately determine how far this technology can go or stop: the bench top, the field, the factor, and the food menu.

BIBLIOGRAPHY

1. D. McCormick, "Genome Know-How," *Bio / Technology* **8**, 5 (1990).
2. Congress of the United States: Office of Technology Assessment, *Impact of Applied Genetics: Microorganisms, Plants and Animals*, U.S. Government Printing Office, Washington, D.C., 1981.
3. S. Casjens, "The Diverse and Dynamic Structure of Bacterial Genomes," *Annual Reviews of Genetics* **32**, 339–377 (1998).
4. E. A. Birge, *Bacterial and Bacteriophage Genetics*, Springer-Verlag, New York, 1994.
5. K. R. Lemon and A. D. Grossman, "Localization of Bacterial DNA Polymerase: Evidence for a Factory Model of Replication," *Science* **282**, 1516–1519 (1998).
6. S. Waga and B. Stillman, "The DNA Replication Fork in Eukaryotic Cells," *Annu. Rev. Biochem.* **67**, 721–751 (1998).
7. S. Cohen et al., "Construction of Biologically Functional Bacterial Plasmids *in vitro*," *Proc. Natl. Acad. Sci. U.S.A.* **70**, 3240–3244 (1973).
8. G. Poste and G. L. Nicholson, *Membrane Fusion*, Elsevier/North Holland, Amsterdam, The Netherlands, 1978.
9. J. A. Piccirilli et al., "Enzymatic Incorporation of a New Base Pair into DNA and RNA Extends the Genetic Alphabet," *Nature* **343**, 33–37 (1990).
10. Z. Kelman and M. O'Donnell, "DNA Polymerase III Holoenzyme. Structure and Function of a Chromosomal Replicating Machinery," *Annu. Rev. Biochem.* **64**, 171–200 (1995).
11. T. M. Lohman and K. P. Bjornson, "Mechanism of Helicase Catalyzed DNA Unwinding," *Annu. Rev. Biochem.* **65**, 169–214 (1996).
12. A. Kornberg, *DNA Replication*, W. H. Freeman, San Francisco, Calif., 1995.
13. U. K. Busheri-Feridoun et al., "Repair of DNA Strand Gaps and Nicks Containing 3′-Phosphate and 5′hydroxyl Termini by Purified Mammalian Enzymes," *Nucleic Acids Res.* **6**, 4395–4400 (1998).
14. P. B. Sigler et al., "Structure Function in Gro EL-Mediated Protein Folding," *Annu. Rev. Biochem.* **67**, 581–608 (1998).
15. R. L. Jackman, T. J. Cottrell, and L. J. Harris, "Protein Engineering," in Y. H. Hui and G. G. Khachatourians, eds., *Food Biotechnology: Microorganisms*, VCH, New York, 1995, pp. 181–236.
16. A. L. Demain and N. A. Solomon, eds., *Manual of Industrial Microbiology and Biotechnology*, American Society for Microbiology, Washington, D.C., 1986.
17. J. Mahillon and M. Chandler, "Insertion Sequences," *Microbiology and Molecular Biology Reviews* **62**, 725–774 (1998).
18. D. Berg and M. M. Howe, eds., *Mobile DNA*, American Society for Microbiology, Washington D.C., 1989.
19. N. L. Craig, "Target Site Selection in Transposition," *Annu. Rev. Biochem.* **65**, 427–474 (1996).
20. D. leCoq, S. Aymerich, and M. Steinmetz, "Dual effect of a transposon Tn917 insertion into the *Bacillus subtilis sacX gene*," *J. Gen. Microbiol.* **137**, 101–106 (1991).
21. R. A. Daniel et al., "Isolation and Characterization of the *lac*A Gene Encoding Beta-Galactosidase in *Bacillus subtilis* and a Regulator Gene, *lac*R," *J. Bacteriol.* **179**, 5636–5638 (1997).
22. J. Brown and V. Sundaresan, "Genetic Study of the Loss and Restoration of Mutator Transposon Activity in Maize: Evidence Against Dominant-Negative Regulator Associated With Loss of Activity," *Genetics* **130**, 889–898 (1992).
23. J. D. Watson, J. Tooze, and D. T. Kurtz, *Recombinant DNA: A Short Course*, W. H. Freeman, New York, 1992.
24. T. R. Cech, "The Chemistry of Self-Splicing RNA and RNA Enzymes," *Science* **236**, 1532–1539 (1987).
25. J. C. Wang, "DNA Topoisomerase," *Annu. Rev. Biochem.* **65**, 635–692 (1996).
26. N. J. Proudfoot, "How RNA Polymerase II Terminates Transcription in Higher Eukaryotes," *Trends in Biochemistry* **14**, 105–109 (1989).
27. J. A. Lefstin and K. R. Yamamoto, "Allosteric Effects of DNA on Transcriptional Regulators," *Nature* **392**, 885–888 (1998).

28. J. M. MacPherson and G. G. Khachatourians, "Biotechnology and Genetics: Concepts and Applications," in P. N. Cheremisinoff and L. M. Ferrante, eds., *Biotechnology Current Progress*, Technomic, Lancaster, Pa., 1991, pp. 21–37.
29. P. J. Green, O. Pines, and M. Inouye, "The Role of Antisense RNA in Gene Regulation," *Annu. Rev. Biochem.* **55**, 569–597 (1986).
30. R. Green, and H. F. Noller, "Ribosomes and Translation," *Annu. Rev. Biochem.* **66**, 5679–5716 (1997).
31. D. M. Raineri et al., "*Agrobacterium*-mediated Transformation of Rice (*Oryza sativa* L.)," *Bio/Technology* **8**, 33–37 (1990).
32. J. A. Heinemann and G. F. Sprague, Jr., "Bacterial Conjugative Plasmid Mobilize DNA Transfer Between Bacteria and Yeast," *Nature* **340**, 205–209 (1989).
33. T. Manniatis, E. F. Fritsch, and J. Sambrook, *Molecular Cloning: Laboratory Manual*, Cold Spring Harbor Laboratory, Cold Spring Harbor, N.Y., 1989.
34. M. Schleef and P. Heimann, "Cesium Chloride or Column Preparation? An Electron Microscopical View of Plasmid Preparations," *BioTechniques* **14**, 544 (1993).
35. K. B. Mullis et al., "Specific Enzymatic Amplification of DNA *in vitro*: The Polymerase Chain Reaction," *Cold Spring Harbor Symposium Quantitative Biology* **51**, 263–273 (1986).
36. S. J. Scharf, G. T. Horn, and H. A. Erlich, "Direct Cloning and Sequence Analysis of Enzymatically Amplified Genomic Sequences," *Science* **233**, 1076–1080 (1986).
37. J. Alper, "Weighing DNA for Fast Genetic Diagnosis," *Science* **279**, 2044–2045 (1998).
38. P. Ross et al., "High Level Multiplex Genotyping by MALDI-TOF Mass Spectrometry," *Nature Biotechnology* **16**, 1347–1351 (1998).
39. E. Jay, J. Rommens, and G. Jay, "Synthesis of Mammalian Proteins in Bacteria," in P. N. Cheremisinoff and R. P. Ouellete, eds., *Biotechnology Handbook*, Technomic, Lancaster, Pa., 1985, pp. 388–400.
40. G. G. Khachatourians and C. M. S. Berezowsky, "Expression of Recombinant DNA Functional Products in *Escherichia coli* Anucleate Minicells," *Biotechnology Advances* **4**, 75–93 (1986).
41. G. G. Khachatourians, "The Use of Anucleated Minicells in Biotechnology: An Overview," in P. N. Cheremisinoff and R. P. Ouellette, eds., *Biotechnology Handbook*, Technomic Publishing, Lancaster, Pa., 1985, pp. 308–318.
42. G. G. Khachatourians and A. R. McCurdy, "Biotechnology: Applications of Genetics to Food Production," in D. Knorr, ed., *Impact of Biotechnology on Food Production and Processing*," Marcel Dekker, New York, 1987, pp. 1–19.
43. G. G. Khachatourians and D. K. Arora, "Biochemical Identification Techniques—Modern Techniques," in R. K. Robinson, C. A. Batt, and P. Patel, eds., *Encyclopedia of Food Microbiology*, Academic Press, San Diego, Calif., 1999.
44. Y. H. Hui and G. G. Khachatourians, eds., *Food Biotechnology: Microorganisms*, VCH, New York, 1995.
45. Y. H. Hui et al., eds., *The Handbook of Transgenic Food Plants*, Marcel Dekker, New York, 2000.

GEORGE G. KHACHATOURIANS
ADRIENNE E. WOYTOWICH
University of Saskatchewan
Saskatoon, Saskatchewan
Canada

GRAINS AND PROTECTANTS

Several insecticides are available for postharvest application to grains in the United States: malathion, chlorpyrifos-methyl (Reldan), pirimiphos-methyl (Actellic), methoprene (Diacon), *Bacillus thuringiensis* (Dipel), synergized pyrethrins, and diatomaceous earth (Table 1). Malathion, Reldan, and Actellic are most the commonly used insecticides; however, Reldan and Actellic can be applied only to specific grains.

Malathion, an organophosphate (OP) insecticide, has been labeled for application to grain since 1957. It can be applied to all major grain commodities in addition to some minor crops (1–4). The labeled rate for application to stored grains is 10.4 ppm and the established Environmental Protection Agency (EPA) tolerance is 8.0 ppm. In addition to its use as a grain protectant, malathion has been widely used in control programs for production agriculture and disease vector control. However, many companies have voluntarily withdrawn their formulations of malathion that were registered on stored grains, and there are doubts as to whether this product will be reregistered. Many insect pests of stored grain have developed resistance to malathion (5–13), which is also affecting decisions regarding registration.

In 1985, almost 30 years after the introduction of OPs as protectants, Reldan (chlorpyrifos-methyl) received a registration for direct application to grain (14). Reldan can be applied to barley, oats, sorghum, rice, and wheat, but not to corn. In addition, Reldan can be used as a prebin spray only if wheat, sorghum, rice, barley, and oats will be stored in the bin. Pirimiphos-methyl (Actellic) was registered in 1986 (1–4). It is labeled for corn and sorghum only and does not have a label for use as a prebin spray.

Acceptable daily intake (ADI) calculations are used by the EPA to regulate nononcogenic compounds. The ADI is an estimate of the daily exposure dose that can be consumed without potential for any harmful effects, and the intake of commodities treated with a particular pesticide is represented as a percentage of the ADI. Although in future actions by the EPA, ADI will be referred to as reference dose (rfd), it was not used at this writing because no EPA documents have used this terminology for pesticide registrations. The total ADI for all registered uses of an active ingredient should not exceed 100%. The calculated ADI for malathion is about 507% for the general population (1–4). In addition, in the most susceptible age group (nonnursing infants aged one to six years) the ADI exceeds 1100%. Malathion has broad use patterns and was registered prior to the EPA's use of ADI as a regulatory tool. Neither Reldan nor Actellic exceed the 100% calculated ADI value.

Synergized pyrethrins have been registered as a protectant for many years but have not been used extensively due to the limited availability and high cost. Synergized pyrethrins have a high acute toxicity to insects but low toxicity to mammals and are generally used as preventative surface applications instead of treatments to the entire grain mass. The base pyrethrum compounds are extracted from chrysanthemum flowers, and while the natural products give quick knockdown, insects often re-

Table 1. Protectants and Grain Surface Treatments, United States, Rar Agricultural Commodities

Product	Crops	Label rate (ppm)	Tolerance (ppm)	% ADI[a]
Malathion	Barley, corn, oats, wheat, sorghum, rice, rye, sunflowers, and almonds	10.4	8.0	507
Chlorpyriros-methyl (Reldan)	Wheat, milo (sorghum), rice, barley, and oats	6.0	6.0	<100
Pirimiphos-methyl (Actellic)	Corn and sorghum (milo)	6.0–8.0	8.0	<100
Methoprene (Diacon)	Corn, wheat, sorghum (milo), barley, rice, oats, and peanuts	5.0	5.0	<100
Bacillus thuringiensis (Dipel)	Grain, soybean, seed, popcorn, birdseed, herbs and spices	3.17	Exempt	N/A
Synergized pyrethrins (pyrethrins + piperonyl butoxide)	Wheat, corn, rye, and sorghum (milo)	3–20	3–20	<100
Diatomaceous earth	Grains	N/A	N/A	N/A

[a]ADI, the average daily intake, is an estimate of the daily exposure dose to a pesticide, and the intake of commodities treated with a pesticide is represented as a percentage of the ADI.

cover from this initial knockdown (15). Addition of the synergist piperonyl butoxide increases the toxicity of the compounds.

Diatomaceous earth is a silicaceous earth composed of the cell walls of diatoms. This insecticide kills insects by removing the waterproof layer, or exoskeleton, of the insect, causing a continuous loss of lipids from the exoskeleton (16). It is available in dust formulations, and many older products were physically irritating to exposed workers and were suspected of causing pulmonary fibrosis. Many new formulations have been registered in recent years that are not as irritating to humans as the older compounds.

A biological compound, Dipel (*B. thuringiensis*), was registered for use on grain and other stored commodities in 1977. *B. thuringiensis* is a naturally occurring pathogen isolated from insects and is exempt from an EPA tolerance and ADI consideration (1–4). The current application rate calculates to 3.17 ppm with respect to traditional pesticide tolerances, which equals 46.02 billion International Units of *B. theringiensis* per million pounds of grain. The grain surface application is for Lepidopteran pests (eg, Indianmeal moth, *Plodia interpunctella*, Angoumois grain moth, *Sitotroga cerealla*, Mediterranean flour moth, *Ephestia kuehiniella*), and almond moth (Cuda cautella) because this is the only group of grain pests affected by *B. thuringinensis*. Indianmeal moths can develop resistance to *B. thuringiensis* (17).

Diacon (methoprene) is an insect growth regulator (IGR) that was registered in 1988 (1–4). It affects insects by interrupting the molting process. Diacon has a low mammalian toxicity and a tolerance of 5 ppm on all of the major grain crops as well as uses in vector and household insect control. As indicated, Diacon affects insect molting and egg hatch in some species (Fig. 1). This product does not eliminate existing adults, but has a secondary effect on the F_1 generation by limiting population explosions of most stored grain insects. The rice weevil, *Sitophilus oryzae*, granary weevil, *Sitophilus granarius*, and lesser grain borer, *Rhyzopertha dominica*, are difficult to control with methoprene because the larvae develop inside the grain kernel. In countries other than the United States, additional compounds labeled as grain protectants include resmethrin, bromophos, fenitrothion, bioresmethrin, deltamethrin, and permethrin (18).

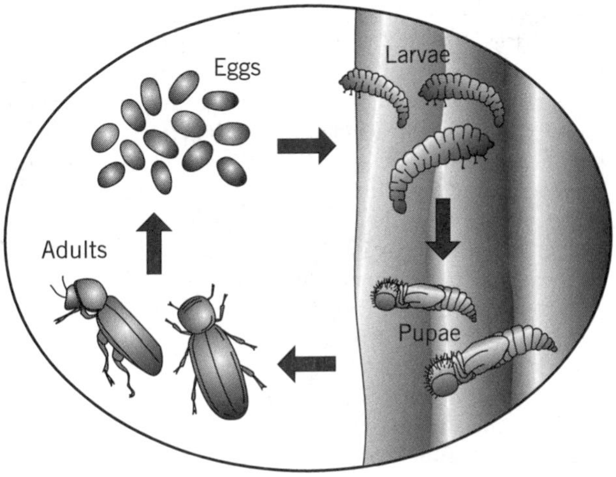

Figure 1. Effects of insect growth regulators (IGRs) on stored product insects (effected stages shown in shaded areas).

In addition to those protectants and grain surface treatments labeled in the United States, several chemicals are labeled for application to grain storage structures (19). Methoxychlor is a chlorinated hydrocarbon that has been registered for more than 30 years. As mentioned earlier, Reldan can be used as a prebinning treatment if wheat, oats, sorghum, rice and barley will be stored in the bin. Actellic is not labeled as a prebinning spray. Cyfluthrin (Tempo) is a pyrethroid insecticide that can be used as a residual spray treatment for all storage structures, including grain bins.

Currently, protectants formulated for direct application to grain are emulsifiable concentrates (EC) or ready to apply as dry material. The ECs are diluted in water or FDA-approved mineral or soybean oil. Liquid solutions are applied through gravity-flow or pressurized systems. In commercial facilities pressurized pump systems are used to treat from 30 t/h to 60,000 t/h (1,000 to 200,000 bu/h) (Fig. 2). The dry formulations are metered into the auger with mechanical applicators or dispersed on the grain surface, or cut into the grain with a scoop before unloading from a truck. As the grain is augured to the bin, distribution throughout the grain is adequate for intended protection. Only one application of a protectant is recommended

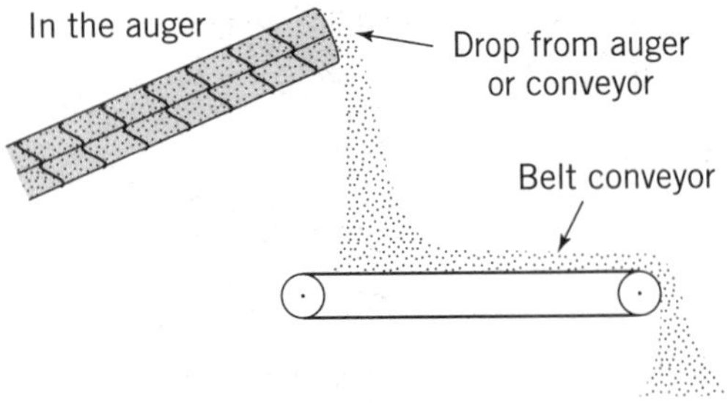

Figure 2. Suggested points at which chemical should be applied.

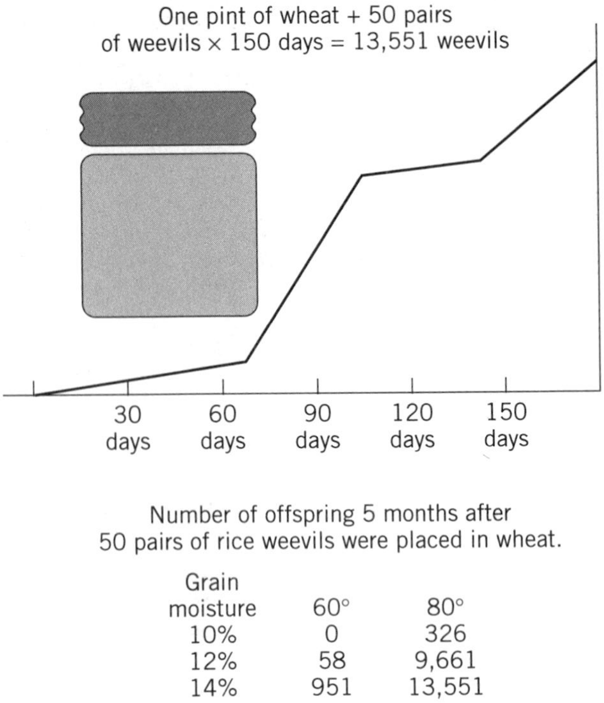

Figure 3. Number of offspring five months after 50 pairs of rice weevils were placed in wheat. *Source:* Ref. 21.

to ensure that residues remain below allowable tolerances. Protectants should be applied as the grain is loaded into the bin (Fig. 2). Organophosphate insecticides degrade more rapidly as temperature and moisture content increase (20), but evaluations for protectant registrations are made at the full application rate, which helps minimize the risks of exposure associated with protectant insecticides.

To illustrate the challenge of maintaining grain quality, a study on rice weevils (Fig. 3) shows how 50 adults can multiply to more than 13,551 adults in five months (21). Low moisture has a negative effect on weevil populations. The reproductive potential of several important species is shown in Table 2 (22). Studies conducted at the Gustafson Seed and Grain Technology Laboratory (Fig. 4) show that the percentage of insect-damaged kernels (IDK) is directly related to the insect population-to-grain ratio. The percentage of IDK was 18 and 97% with 2 and 50 insects per 200 g of wheat, respectively.

In May 1988, revised Federal Grain Inspection Service (FGIS) grading standards went into effect (23). These standards were more stringent; in wheat, rye, and triticale, one live weevil or lesser grain borer and one or more other live insects (OLI) injurious to stored grain in a 1-kg sample can cause that load of grain to be graded infested (Table 3). In corn, barley, oats, sorghum, soybeans, sunflower seed, or mixed grain, the infested grade is assigned when two or more live weevils or lesser grain borers, or one live weevil or lesser grain borer and five or more OLI injurious to stored grain, or ten or more OLI injurious to stored grain (Table 3).

An additional FGIS revision established an IDK limit of 32 damaged kernels per 100 g of wheat. Prior to the FGIS revisions when IDK was not included as part of the grade, a load of wheat could have significant damage and still have graded U.S. #1. Under the new standards, if a sample (100 g) reveals 32 or more IDK, the load is graded U.S. Sample Grade wheat. The grading agency then notifies the FDA and the load or lot is declared unfit for human consumption, forcing it to be marketed as feed grain or for alcohol production. This market can be from $0.20 to $1.50 per bushel less than food-grade wheat.

Given the reproductive potential of insects, the complex life cycles, and the fact that several different species of insects are capable of damaging stored grain, management plans that utilize multiple control strategies are used for stored grains. Many programs include but are not limited to the following:

1. Emphasis on maintaining high grain quality standards (recent FGIS changes have increased the awareness of the need for better management)
2. Sanitation and cleaning around storage facilities to prevent insects from establishing resident populations

Table 2. Stored-Grain Insects—Life Cycles and Characteristics

Insect	Life cycle (days), egg to egg	Adult life (days)	Number of eggs	Penetrate packages	Fly
Confused flour beetle	40	365	450	Yes (weak)	No
Flat grain beetle	45	365	240	Yes (weak)	Yes
Lesser grain borer	30	180	300–500	Yes	Yes
Rice weevil	30	90–185	300–400	Yes (weak)	Yes
Granary weevil	30–50	210–240	30–50	Yes (weak)	No
Sawtoothed grain beetle	20–80	135–300	280	Yes (weak)	No

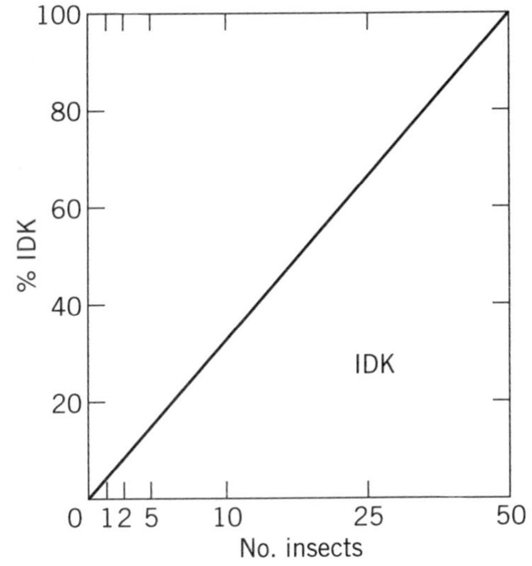

Figure 4. Relationship between numbers of rice weevils and percentage of insect-damaged kernels (IDK).

Table 3. Insect Infestation Tolerance Comparison

Grain	Insect standard
Wheat, rye, triticale	One live weevil[a] and one or more other live insects[b]
Corn, sorghum, soybeans, barley, oats, sunflowers, mixed grain	Two or more live weevils,[a] or one weevil and five or more other live insects,[b] or ten or more other live insects

Note: This table illustrates the number of live insects that, if detected in a 1-kg sample, will result in an infested grade under the FGIS insect-infestation standards.
[a]Weevils are rice, granary, maize, and cowpea weevils, as well as the lesser grain borer.
[b]Other live insects include the grain and flour beetles, moths, and vetch bruchids.

3. Application of grain protectants
4. The use of aeration to cool the grain mass at selected times during storage
5. Regular inspection and monitoring of the grain
6. Control of insects with fumigants if populations become excessive

Insect attractants (pheromones) and new methods of trapping insects have been developed to detect insects in stored grain (24). Traps are more efficient than traditional grain trier and sieve monitoring methods because they often can detect insects that would otherwise go undetected. There has been increased interest in recent years in new programs to sample insects with traps and to use this information in control programs. The grain storage industry is becoming increasingly more involved in Integrated Pest Management (IPM) programs by taking steps to manage insect populations, instead of reacting after large populations are detected and IDK are present.

BIBLIOGRAPHY

1. Environmental Protection Agency, "*Bacillus thuringiensis Berliner*, Viable Spores for Residues," *U.S. Code of Federal Regulations*, section 40, Part 180.1011, 1988, p. 456.
2. Environmental Protection Agency, "Established Tolerances for Residue of the Insecticide Methoprene," *Federal Register* **53**, 27391–27392 (1988).
3. Environmental Protection Agency, "Guidance for the Reregistration of Pesticide Products Containing Malathion as the Active Ingredient," Case-248, p. 315; "Malathion Tolerances for Residues," *U.S. Code of Federal Regulations*, Section 40, Part 180.111, 1988 pp. 33–335.
4. Environmental Protection Agency, "Pirimiphos-methyl, Final Rule," *Federal Register* **51**, 28223–28229 (1987).
5. P. C. Bansode and W. V. Campbell, "Evaluation of North Carolina Field Strains of the Red Flour Beetle for Resistance to Malathion and Other Organophosphorus Compounds," *Journal of Economic Entomology* **72**, 331–333 (1979).
6. R. W. Beeman, W. E. Speirs, and B. A. Schmidt, "Malathion Resistance in Indianmeal Moths Infesting Stored Corn and Wheat in the Northcentral United States," *Journal of Economic Entomology* **75**, 950–954 (1982).
7. J. L. Zettler, "Pesticide Resistance in *Tribolium castaneum* and *T. confusum* (Coleoptera: Tenebrionidae) From Flour Mills in the United States," *Journal of Economic Entomology* **84**, 763–767 (1991).
8. J. L. Zettler and G. W. Cuperus, "Pesticide Resistance in *Tribolium castaneum* (Coleoptera: Tenebrionidae) and *Rhyzopertha dominica* (Coleoptera: Tenebrionidae) in Wheat," *Journal of Economic Entomology* **83**, 1677–1681 (1990).
9. F. H. Arthur, J. L. Zettler, and W. R. Halliday, "Insecticide Resistance Among Populations of Almond Moth and Indianmeal Moth in Stored Peanuts," *Journal of Economic Entomology* **81**, 1283–1287 (1988).
10. C. L. Storey, D. B. Sauer, and D. Walker, "Insects and Fungi in Wheat, Corn and Oats Stored on the Farm," *Journal of Economic Entomology* **77**, 784–788 (1984).
11. B. H. Subramanyam and P. K. Harein, "Status of Malathion and Pirimiphosmethyl Resistance in Adults of Red Flour Beetle and Saw-toothed Grain Beetle Infesting Farm-Stored Corn in Minnesota," *Journal of Agricultural Entomology* **7**, 127–136 (1990).
12. B. H. Subranamyam and D. W. Hagstrum, "Resistance Measurement and Management," in B. H. Subramanyam and D. W. Hagstrum, eds., *Integrated Management of Insects in Stored Products*, Marcel Decker, New York, 1996, pp. 331–398.
13. J. L. Zettler, "Insecticide Resistance in Selected Stored-Product Insects Infesting Peanuts in the Southeastern United States," *Journal of Economic Entomology* **75**, 359–362 (1982).
14. Environmental Protection Agency, "Chlorpyrifos-Methyl Tolerances for Residues," *U.S. Code of Federal Regulations*, Section 40, Part 180.419, pp. 426–427.
15. R. D. Obrien, *Insecticide Actions and Metabolism; Pyrethroids*, Academic Press, Orlando, Fla., 1967, pp. 168–171.
16. P. Golub, "Current Status and Future Perspectives for Inert Dusts for Control of Stored Product Insects," *Journal of Stored Product Research* **33**, 69–79 (1997).
17. W. H. McGaughey and D. E. Johnson, "Indianmeal moth (*Lepidoptera: Pyralidae*) Resistance to Different Strains and Mixtures of *Bacillus thringiensis*," *Journal of Economic Entomology* **85**, 1594–1600 (1992).

18. N. D. G. White and J. G. Leesch, "Chemical Control," "Resistance Measurement and Management," in B. H. Subramanyam and D. W. Hagstrum, eds., *Integrated Management of Insects in Stored Products*, Marcel Decker, New York, 1996, pp. 287–330.
19. J. G. Touhey, "Future Development and Availability of Pesticides for Grain: How Pesticides are Registered by EPA," *UDAGIS Grain Insect Interagency Task Force Meeting*, Washington, D.C., July 21, 1988.
20. F. H. Arthur, J. E. Throne, and R. A. Simonaitis, "Degradation and Biological Efficacy of Chlorpyrifos-Methyl on Wheat Stored at Five Temperatures and Three Moisture Contents," *Journal of Economic Entomology* **85**, 1994–2002 (1982).
21. R. A. Higgins, *Management of Stored Grain Insects, Part 1*, Kansas State University MF-726, Manhattan, Kans., 1984.
22. F. J. Bauer, *Insect Management for Food Storage and Processing*, American Association of Cereal Chemists, Minneapolis, Minn., 1984, pp. 54–67.
23. U.S. Department of Agriculture, Federal Grain Inspection Service, "Grain Standards: Official U.S. Standards, Handling Practices and Insect Infestation, Final Rules," *Federal Register* **52**, 2424–2442 (1988).
24. T. W. Phillips, "Semiochemicals of Stored-Product Insects: Research and Applications," *Journal of Stored Product Research* **33**, 17–30 (1997).

J. TERRY PITTS
Gustafson, Inc.
Plano, Texas

FRANK H. ARTHUR
Agricultural Research Service, United States Department of Agriculture
Manhattan, Kansas

GRAPEFRUIT. See FRUITS: SEMI-TROPICAL.

GUMS

Food gums are water-soluble or water-dispersable polysaccharides (glycans) and their derivatives (1–6) and gelatin (3) (see the article CARBOHYDRATES: CLASSIFICATION, CHEMISTRY, LABELING). In general, they thicken or gel aqueous systems at low concentration. Although starches, flours, and modified food starches are polysaccharide materials and have similar properties, they are usually considered separately as they are in this encyclopedia (see the article STARCH). Polysaccharide food gums can be classified by source (Table 1) or by structure (Table 2).

The usefulness of gums is based on their physical properties, in particular their capacity to thicken and/or gel aqueous systems and otherwise to control water. Because all gums modify the flow of aqueous solutions, dispersions, and suspensions, the choice of which gum to use for a particular application often depends on its other characteristics, characteristics that are responsible for their utilization as binders, bodying agents, bulking agents, crystallization inhibitors, clarifying agents, cloud agents, emulsifying agents, emulsion stabilizers, encapsulating agents, film formers, flocculating agents, foam stabilizers, gelling materials, mold release agents, protective colloids, suspending agents, suspension stabilizers, swelling agents, syneresis inhibitors, texturing agents, and whipping agents, in coatings, and for water absorption and binding. Most gums are not true emulsification agents but stabilize emulsions and suspensions by increasing the viscosity and the yield value of the aqueous phase. Some also exhibit protective colloid action.

Table 1. Classification of Polysaccharide Food Gums by Source

Class	Examples
Seed gums	Guar gum, locust bean gum, tara gum
Tuber and root gums	Konjac mannan
Seawood extracts	Algins, carrageenans, agar, furcellaran
Plant extracts	Pectins
Exudate gums	Gum arabics, gum tragacanth
Fermentation/microbial gums	Xanthan gums, gellan gums, curdlan
Derived gums	CMCs, hydroxypropylcelluloses, HPMCs, MCs

Food gums are tasteless, odorless, colorless, and nontoxic. All (except the starches, starch derivatives, and gelatin) are essentially noncaloric and are classified as soluble dietary fiber (see the article FIBER, DIETARY).

Gums, in general, do not form true solutions because of their large molecular size and intermolecular interactions. Hence, the term hydrocolloid is often used interchangeably with gum. The rheology (flow characteristics and gel properties) of gum solutions is a function of the size, shape, flexibility, solvation, and ease of deformation of particles, and the presence and magnitude of charges on them. Particles may be hydrated molecules and/or molecular aggregates. Many gum solutions exhibit shear thinning; that is, they are pseudoplastic, occasionally thixotropic. The variables that affect their rheology are polymer structure, molecular weight, concentration, shear rate, temperature, pH, and the concentration of salts, other solutes, and sequestrants. Factors that affect dispersion and dissolution are pH, presence of salts, presence of other solutes, gum type, particle size, physical form of particles, shear rate, and method of dispersion (mixing efficiency).

Most gums are available in a range of viscosity grades, which are produced by depolymerization (reducing the molecular weight) of the native polysaccharide. If viscosity is the goal, a high-viscosity grade at low solids concentration is used; if binding or protective colloid action, for example, is the goal, a low-viscosity grade at high solids concentration is used.

In general, gels made with food gums are composed of 95.0 to 99.5% water and 0.5 to 5.0% gum. Important characteristics of gels are means of gelation (chemical gelation, thermal gelation), reversibility, texture (brittle, elastic, plastic), rigidity (rigid or firm, soft or mushy), tendency for syneresis, and whether they are cutable or spreadable.

Gums, excluding native and modified starches, that are used in foods in the United States are described later. Not included is curdlan, which was only recently introduced in the United States, inulin, or the seldom used gum ghatti and gum karaya. Uses of curdlan and inulin are being ex-

Table 2. Classification of Polysaccharidic Food Gums by Structure

Classification schemes	Examples
By shape	
Linear	Algins, carrageenans, cellulosics, furcellaran, konjac mannan, pectins
Branched	
Short branches on a linear backbone	Galactomannans (guar gum, locust bean gum, tara gum, enzymically modified guar gum), xanthan gums
Branch-on-branch structures	Gum arabics, gum tragacanth (tragacathin)
By monomeric units[a]	
Homoglycans	Cellulosics, curdlan
Diheteroglycans	Agarose, algins, carrageenans, furcellaran, galactomannans, konjac mannan, pectins
Triheteroglucans	Gellan gums, xanthan gums
Tetraheteroglycans	Gum arabics
Pentaheteroglycans	Gum tragacanth (tragacanthin)
By charge	
Neutral	Agarose, HPMCs, MCs, galactomannans, konjac mannan
Anionic (acidic)[b]	Algins, CMCs, carrageenans, furcellaran, gellan gums, gum arabics, gum tragacanth (tragacanthin), pectins, xanthan gum

[a]Considers only the basic monosaccharide units. A derivatized monosaccharide unit, such as D-galactopyranosyl 6-sulfate unit, is not considered as a unit separate from a D-galactopyranosyl unit, for example.
[b]From the presence of uronic acid, sulfate half-ester, or pyruvyl cyclic acetal groups.

plored. Also not included are the polysaccharides that occur naturally in many foodstuffs, that may be extracted during processing, and that impart the same functionalities as gums added as ingredients; examples of such polysaccharides are the arabinoxylans (pentosans) of cereal flours, xyloglucans of many cell walls, arabinogalactans and arabinogalactan-proteins of legumes, and psyllium seed gum. For β-glucan, which has the properties of a gum, but which is not isolated for commercial use, see the article FIBER, DIETARY.

ALGINS

Algins (alginates) (2,4–6) are extracted from brown algae (*Phaeophyceae*). Alginates are salts (generally sodium) and the propylene glycol ester of alginic acid. Alginic acid is a generic term for polymers of D-mannuronic acid and L-guluronic acid (see the article CARBOHYDRATES: CLASSIFICATION, CHEMISTRY, LABELING). In alginic acid molecules, at least three different types of polymer segments exist: poly(β-D-mannopyranosyluronic acid) segments, poly(α-L-gulopyranosyluronic acid) segments, and segments with alternating sugar units. Ratios of the constituent monomers and ratios of chain segments vary with the source. Alginates are linear polymers. The degree of polymerization (molecular weight) is controlled and varied in commercial products.

Specific properties of algins depend on the percentage of each type of building block. An important and useful property of sodium, potassium, and ammonium alginates is their ability to form gels on reaction with calcium ions. Different types of gels are formed with alginates from different sources. Alginates with a higher percentage of poly(guluronic acid) segments form gels more readily and form the more rigid, more brittle gels that tend to undergo syneresis. Alginates with a higher percentage of poly(mannuronic acid) segments form the more elastic, more deformable gels that have a lesser tendency to undergo syneresis.

Algins are most often used as thickeners and are available in a range of viscosity grades. Algin solutions are pseudoplastic, that is, exhibit shear thinning. Solutions of propylene glycol alginate (PGA, a partial propylene glycol ester) are somewhat thixotropic and much less sensitive to pH and polyvalent cations. The specific properties exhibited by an algin solution depend on the ratio of monomeric units, the concentration and type of cations in solution, the temperature, and the degree of polymerization.

Primary products using sodium alginate are breakfast and cereal bars, fruit fillings, dry mixes for reconstitution with water or milk, and frozen products. Calcium alginate gels are found in structured foods, such as fruit pieces, onion rings, and pimiento strips for Spanish olives. Primary products in which PGA (which can be labeled algin derivative) is used are pourable salad dressings, buttered pancake syrups, sauces, and beverages. It is also used as a beer foam stabilizer. Alginic acid, which swells in water, is used as a tablet disintegrant.

CARBOXYMETHYLCELLULOSES

Carboxymethylcellulose (CMC) (2,4–6) is the sodium salt of the carboxymethyl ether of cellulose (see the article CARBOHYDRATES: CLASSIFICATION, CHEMISTRY, LABELING). In an ingredient label, it can be designated cellulose gum, CMC, sodium CMC, sodium carboxymethyl cellulose, or carboxymethyl cellulose.

Modified polysaccharides, such as CMC, have a degree of substitution (DS). There are, on average, three hydroxyl groups per polysaccharide hexopyranosyl (six-carbon-atom sugar) unit. In the case of cellulose, each β-D-glucopyranosyl unit has three hydroxyl groups (see the article CARBOHYDRATES: CLASSIFICATION, CHEMISTRY, LABELING). The average number of hydroxyl groups derivatized (with a carboxymethyl ether group in the case of CMCs) per monomeric unit is the DS; the maximum DS of a cellulose derivative is 3.0.

Table 3. Some Properties of Food Gums

Gum	Solubility: RT[a] water or milk	Solubility: Hot water or milk	Thickener[b]	Gel former[c]	Surface tension reduction	Emulsion/ suspension stabilizer	Interactions and synergisms
Agar		+		+			LBG, konjac mannan
Algins							
Na^+ alginate	+	+	+	Ca^{2+}			Proteins
PGA	+	+	+		+	+	
Cellulosics							
CMCs	+	+	+			+ (certain types)	Proteins, guar gum (weak)
MCs	+		+	Heat	+	+	Xanthan gums
HPMCs	+		+	Heat	+	+	Xanthan gums
Carrageenans							
κ-type	Na^+ only	+		K^+, Ca^{2+}, LBG			Milk proteins, LBG, konjac mannan, xanthan gums, guar gum (weak)
ι-type	Na^+ only	+		Ca^{2+}			Milk proteins, LBG, starches
λ-type	+	+					Milk proteins, starches (weak)
Furcellaran	+		+				LBG, guar gum (slight)
Galactomannans							
Guar gum	+	+	+				Xanthan, κ-carrageenan, furcellaran, CMC (all weak)
LBG	Partial	+	+	Xanthan gums, κ-carrageenan			Xanthan gums, κ-carrageenan, ι-carrageenan, agar, furcellaran
Gelatin		+		+	+	+	Agar, algins, carrageenans, gum arabics, LM pectins
Gellan gums		+		Cations			
Gum arabics	+	+		+	+		
Gum tragacanth	+	+	+		+	+	
Konjac mannan	Partial	+	+	OH^-, κ-carrageenan			Agar, κ-carrageenan, xanthan gums, starches
Pectins							
HM		+		H^+, sugar			
LM		+		Ca^{2+}			
Amidated LM		+		Ca^{2+}			
Xanthan gum	+	+	+	LBG		+	LBG, konjac mannan, κ-carrageenan, MCs, HPMCs, guar gum (weak)

Note: Blanks are negative indicators.
[a]RT = room temperature
[b]Commonly used to thicken.
[c]Used to form gels. Listing of another ingredient or ion signifies a requirement for gelation.

Solution characteristics of CMCs are primarily determined by the DS, the average chain length/degree of polymerization (DP), and the uniformity of substitution. The most widely used types have a DS of 0.7 or an average of 7 carboxymethyl ether groups per 10 β-D-glucopyranosyl units (Fig. 1). Many viscosity grades of CMC are manufactured.

CMCs hydrate rapidly and form clear, stable solutions. Viscosity building is their most important property. Solutions can be either pseudoplastic or thixotropic, depending on the type of product used, but are most often pseudoplastic. Like solutions of hydroxypropylcelluloses, meth-

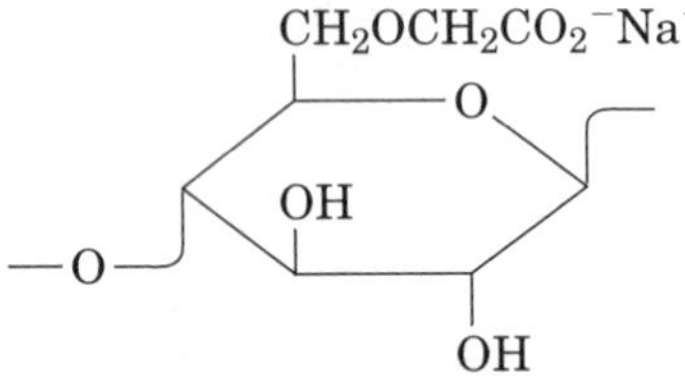

Figure 1. A representative unit of sodium carboxymethylcellulose, that is, a representative β-D-glucopyranosyl unit containing a carboxymethyl ether group.

ylcelluloses, and hydroxypropylmethylcelluloses, CMC solutions are Newtonian at low shear rates and become pseudoplastic as the shear rate is increased. The shear rate at which solutions change from being Newtonian to being pseudoplastic increases with increasing molecular weight and decreases with increasing concentration. Also, the higher-molecular-weight products are more affected by shear. Solutions are stable over a wide pH range (pH 4–10).

CMCs help to solubilize various proteins and to stabilize their solutions. They have a synergistic viscosity building effect when used with casein and soy protein.

Examples of CMC use are as a retarder of ice crystal growth in ice creams and related products; as a physiologically inert and noncaloric thickening and bodying agent in dietetic foods; in cake and related mixes to hold moisture, enhance organoleptic properties, and extend shelf life; as an extrusion aid; in batters for viscosity control, adhesion, and suspension stabilization; in icings, frostings, and glazes to hold moisture and retard sugar crystallization; as a thickener in pancake syrup; as a suspending agent and suspension stabilizer in hot cocoa and fruit drink mixes; as a water binder in slightly moist pet foods; as a casein stabilizer in dips and sour cream; and in frozen poultry sticks and nuggets.

CARRAGEENANS, AGAR, AND FURCELLARAN

Carrageenan is a generic term applied to polysaccharides extracted from a number of closely related species of red seaweeds (2,4–6). Agar and furcellaran are also red seaweed extracts and members of the same larger family. All polysaccharides in this family are derivatives of linear galactans, and all have alternating monosaccharide units and linkages. In agar, the monosaccharide units are D-galactopyranosyl and 3,6-anhydro-L-galactopyranosyl units. In κ-carrageenan and ι-carrageenan, they are D-galactopyranosyl and 3,6-anhydro-D-galactopyranosyl units. λ-Carrageenan contains only D-galactopyranosyl units (Fig. 2). Agar contains little or no sulfate half-ester groups. Other members of the family are sulfated: in increasing order, κ-carrageenan ($\sim$25%), ι-carrageenan ($\sim$30%), and λ-carrageenan ($\sim$35%). Each polymer type is heterogeneous; that is, none contains an exact repeating unit structure. Agar is the least soluble of this family of polysaccharides and λ-carrageenan the most soluble. Agar forms the strongest gels. λ-Carrageenan does not gel.

Commercial carrageenans are composed primarily of three types of polymer: κ-, ι-, and λ-carrageenans. The composition of an extract of a carrageenophyte with respect to each of the three types of polymers and its properties depends on the species collected, growth conditions, and treatment during production. Products are blended and standardized with respect to any of several properties.

As pH values decrease below 6, carrageenan solutions become increasingly unstable when heated. The loss of viscosity is due to polymer chain cleavage and, hence, is irreversible.

Carrageenans form complexes with proteins. Carrageenans are used extensively as gelling, thickening, and suspending agents in milk-based products. Preparations are blended to provide products that will form a variety of gels: clear gels and turbid gels, rigid gels and elastic gels, tough gels and tender gels, heat-stable gels and thermally reversible gels, gels that undergo syneresis and gels that do not. Carrageenan gels do not require refrigeration because they do not melt at room temperature.

Carrageenans react with milk proteins. The thickening effect of a κ-type carrageenan in milk is 5 to 10 times greater than it is in water; at a concentration of 0.025% in milk, a weak thixotropic gel is formed via interaction of κ-carrageenan with κ-casein micelles. This property finds use in the preparation of chocolate milk (to keep the cocoa suspended) and ice cream (to prevent whey-off). Other (stabilizing) interactions are important in producing evaporated milk, infant formulas, and whipped cream required to be freeze–thaw stable.

There is a synergistic effect between κ-carrageenan and locust bean gum; the two gums together produce a much more elastic gel with markedly greater gel strength and less syneresis. Such gels find use in canned pet foods, fruit gels, and processed meats and seafoods. The κ-carrageenan–locust bean gum mixture is also used to provide body and fruit suspension in yogurt, to stabilize low-fat yogurt, and in other dairy products such as cheese spreads, cottage and cream cheese, dips, ice creams, other frozen desserts and novelties, and whipped toppings.

ι-Type carrageenans form elastic, syneresis-free, thermally reversible gels that are stable to repeated freeze–thaw cycling. They are used in ready-to-eat milk gels. Blending is a common practice, and blends of κ- and ι-type carrageenans are used to prepare water dessert gels (where refrigeration is unavailable), whipped toppings, instant whipped desserts, and eggless custards and flans. ι-Type carrageenans exhibit a synergistic interaction with starches.

λ-Type carrageenans are nongelling and are used as emulsion stabilizers in such products as whipped cream, instant breakfast drinks, milkshakes, nondairy coffee creamers, and dry mix sauce systems.

Agar (agar–agar) is the least soluble of this class of polysaccharides. Agar is composed of a gelling polysaccharide, agarose, and a nongelling polysaccharide, agaropectin. Agar can be dispersed only at temperatures above 100°C (212°F). When its dispersions are cooled, strong, brittle, turbid gels form. Agar gels remelt when heated, undergo syneresis, and are unstable to freeze–thaw cycles. Agar is used primarily in bakery icings, glazes, and (in Asia) dessert gels. A form of agar that dissolves at lower temperature (about 65°C/150°F) is made by drying agar with sugars/or maltodextrins, producing a so-called agglomerated form.

Furcellaran (Danish agar) is a less-sulfated κ-type carrageenan, placing it between agar and κ-carrageenan in properties. It is not always considered as a separate entity, often being included in the carrageenan family. Furcellaran is used to prepare milk-based puddings, flan and related jellies, and low-sugar jams and other fruit preserves and elsewhere where κ-type carrageenans might also be used.

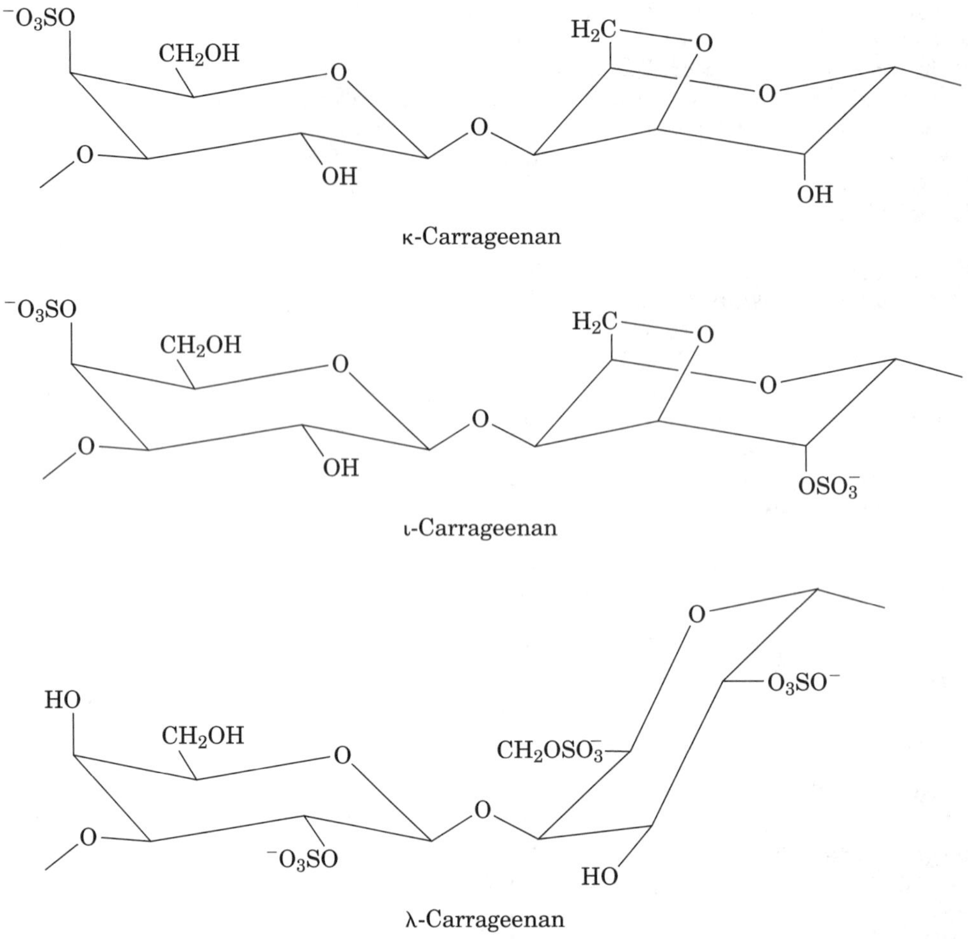

Figure 2. Idealized structures of the three types of carrageenan.

GELATIN

Gelatin is a protein rather than a polysaccharide. It is produced by acid (type A) or alkali (type B) treatment of collagen obtained from pig skin, cattle hides, and bones. While gelatin contains 18 different amino acids, about 95% of the structure is composed of 7. The amino acid sequence of most of the chain is a tripeptide repeating unit of (glycine)–(proline or hydroxyproline)–(alanine, glutamic acid, aspartic acid, or arginine). Gelatin contains no tryptophan.

The important functional properties in food applications of gelatin (3) are gel strength, gelling temperature, setting rate, pH, and isoelectric pH. Typical values for some of these properties are as follows:

Type	Gel strength, g Bloom	pH	Isoelectric pH
A	75–300	4.5–6.0	8.0–9.5
B	75–275	5.0–7.5	4.7–5.4

The presence of ash, sulfur dioxide, and peroxides can affect product appearance, that is, can cause gel cloudiness and/or discoloration, but do not affect functional properties. The gelling temperature of gelatin solutions is related to concentration. Below pH 5, types A and B gelatins have a net positive charge and will interact with negatively charged polymers such as agar, carrageenans, algins, low-methoxyl (LM) pectins, and gum arabic.

Although gelatin could be used in a variety of food products, its use is essentially limited to preparation of meltable-in-the-mouth water-dessert gels (gelatin desserts), meat products, and marshmallows and other confectioneries. It is also used to stabilize emulsions; as a stabilizer for ice cream and other dairy products; as a clarifying agent for wines, vinegar, and juices; and in bakery fillings and icings. It is not normally used as a thickener but is used to thicken low-fat spreads where the gel structure is broken down by processing and tempered by hydrophobic ingredients such as oils and emulsifiers.

GELLAN GUM

Gellan, known commercially as gellan gum, is an extracellular microbial polysaccharide produced by the organism *Sphingomonas elodea* (5). It is composed of a linear tetrasaccharide repeating unit of D-glucose–D-glucuronic acid–D-glucose–L-rhamnose. Two forms of gellan gum are available—the native acylated form and a deacylated form. Cooling of hot solutions of the low-acyl gellan (gum) forms firm and brittle gels of texture similar to that of agar and κ-carrageenan gels.

Gellan gum requires cations for gelation; but unlike other ion-sensitive gelling polysaccharides, it gels with a wide variety of cations, including hydrogen ions. Divalent ions such as Ca^{2+} are, however, much more efficient at forming gels than are monovalent ions such as Na^+ and K^+. Since divalent cations decrease solubility, sequestering agents are required if such cations are present initially. Like agar gels, gellan gum gels exhibit marked setting and melting temperature hysteresis; and by selection of the ion and its concentration, both temperatures can be controlled.

Gellan gum is functional in milk, although a calcium ion-sequestering agent is normally required. It can be used in confectioneries; structured foods; pie and bakery fillings; bakery icings, frostings, and glazes; dairy products; water-based gels; and films and coatings. Generally, gellan gum is blended with other gums or ingredients to modify its performance.

GUAR AND LOCUST BEAN (CAROB) GUMS

Commercial guar gum (2,4–6) is the ground endosperm of seeds of the guar plant. The guar plant resembles the soy plant, and guar seeds are produced much as are soy beans. Commercial guar gum usually contains 80 to 85% polysaccharide, 10 to 14% moisture, 3 to 5% protein, 1 to 2% fiber, 0.5 to 1.0% ash, and 0.4 to 1.0% lipid.

Guaran is the purified polysaccharide from guar gum. It is composed of D-galactose and D-mannose and is, therefore, a galactomannan. It has a linear backbone chain of β-D-mannopyranosyl units, approximately 56% of which (on average) are substituted with an α-D-galactopyranosyl unit, giving a comblike structure (Fig. 3). The mannan chain is rather evenly substituted with D-galactopyranosyl units, but in a somewhat random manner.

Like guaran and the endosperm polysaccharides of other legumes, locust bean (carob) gum (LBG) is also a galactomannan. Like guaran, it has a linear mannan backbone. However, in locust bean gum, on average only approximately 20% of the β-D-mannopyranosyl units are substituted with an α-D-galactopyranosyl unit and the branched units are clustered. The locust bean gum molecule contains "smooth" regions that contain no α-D-galactopyranosyl side-chain units and "hairy" regions in which most main-chain units contain single-unit α-D-galactopyranosyl branches.

Commercial locust bean gum is the ground endosperm of the seeds of the locust (carob) tree. The tree, which grows primarily in the Mediterranean region, is slow to mature and does not begin to bear until it is about 15 years old. Hence, the supply of locust bean seeds is limited and is not expected to increase. Locust bean gum usually has about the same composition (polysaccharide, moisture, protein, fiber, ash, lipid content) as does guar gum.

Guar gum forms very high viscosity, slightly pseudoplastic solutions at low concentrations. Because commercial guar gum contains protein, fiber, and lipids, its solutions are cloudy. Particle size is important in the use of guar gums. Coarse granulations are used for rapid, easy dispersion; fine granulations give rapid hydration. All forms will develop some additional viscosity after their dispersions have been heated. Guar gum is nonionic, so the viscosity of its solutions is not greatly affected by pH. It hydrates most rapidly at pH 6 to 9.

Locust bean gum has low cold-water solubility and is generally used when delayed viscosity development is desired. Only when dispersions of locust bean gum are heated (eg, to 85°C [185°F]) and cooled is high viscosity obtained. The general properties of locust bean gum are similar to those of guar gum. A difference is its synergism with κ-carrageenan, agar, and xanthan gum, with which it forms gels. Guar gum has only a weak synergistic effect on the viscosity of xanthan gum solutions and an even weaker effect on carrageenan and CMC solutions.

Guar gum is used in foods as a thickener and a binder of water. It is often used in combination with other gums, for example, with carrageenan and/or xanthan plus locust bean gum, particularly in dairy products such as ice creams, frozen novelties, whipped toppings, sour cream, cottage cheese, and low-fat yogurts. It is used with xanthan gum in pickle relish and sauces such as pizza and Sloppy Joe sauce to bind and control water. Both guar gum and locust bean (carob) gum are used (often in combination with carrageenan or xanthan gum) in stabilizers for ice cream. Other applications of these blends are in the manufacture of canned frostings, dips, cream cheese, cottage cheese dressings, and other processed cheese products. Guar gum is used in instant hot cereals and dry soup and other mixes, egg substitutes, dipping batters, pimiento strips for Spanish olives, sauces, condiments, and pet foods.

Two forms of enzymically modified guar gum are produced. One is a depolymerized guar gum designed to be a noncaloric bulking agent and a less viscous source of dietary fiber. The other is a guar gum that has had some of the D-galactosyl units removed to give it properties more like those of the more expensive locust bean gum.

Locust bean gum is almost always used in combination with one or more other gums in a variety of dairy products. In combination with xanthan gum and starch, it is used to make chewy fruit confections. Locust bean gum–carrageenan blends are used in pet food gels.

Tara gum is a galactomannan obtained from seeds of the tara shrub, which grows in northern regions of South America and Africa. It has a D-galactosyl unit:D-mannosyl unit ratio between those of guar gum and locust bean gum,

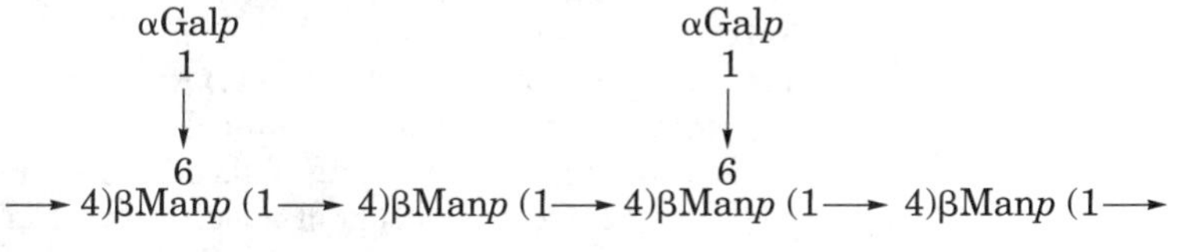

Figure 3. Idealized structure (shorthand) of guaran.

giving it properties most like those of a locust bean gum with a high D-galactose content. Only relatively small amounts of tara gum are currently used.

GUM ARABICS

Of the several gums that are dried, gummy exudations collected by hand from various trees and shrubs, only gum arabic, also called gum acacia and acacia gum, is still in significant use (2,4–6).

Gum arabics are collected by hand from various species of *Acacia*, a small tree. Tears of gum exude from wounds in trunks and branches. Wounds can be natural, but trees often are cut deliberately. Dried tears are sorted into lots based on clarity, color, and gross impurities. Representative samples of each lot are analyzed for ash content, water-insoluble impurities, and viscosity. The lots are then cleaned, converted into grains or powders of varying mesh sizes, and blended. The best gum comes from the region just south of the Sahara Desert. The most highly purified grade is produced by spray drying a clarified solution.

Gum arabic preparations are mixtures of highly branched, branch-on-branch, acidic polysaccharides, and protein-polysaccharides that occur as mixed salts. Their specific compositions and structures can vary with species, season, and climate.

Gum arabic is unique among gums because of its high solubility and the low viscosity and Newtonian flow of its solutions. While most other gums form highly viscous solutions at 1 to 2% concentration, 20% solutions of gum arabic resemble a thin sugar syrup in body and flow properties.

Gum arabic also has the unique property of being an effective emulsifier of flavor oils plus an effective stabilizer of the resulting emulsions. It is often used in the preparation of baker's emulsions of citrus and other essential oils. Another major application is in the preparation of dry fixed-flavor powders, which are prepared by adding citrus oils and other fruit or imitation flavors to gum arabic solutions and spray drying the resulting emulsions. Some gum arabic is also used in the preparation of confectioneries and lozenges.

GUM TRAGACANTH

Gum tragacanth (2,4–6), is the dried exudate of *Astragalus* species. It is produced from natural or deliberately made wounds in the roots, trunks, and/or branches of the small bushes, comes for the most part from Iran and Turkey, and is relatively expensive. Its use, therefore, is quite limited.

Gum tragacanth contains two polysaccharides. One (60–70%), termed both tragacanthic acid and bassorin, only swells in water, forming a gel. Tragacanthic acid contains a highly branched, acidic galactan covalently bound to protein. The minor polysaccharide is a neutral arabinogalactan in which L-arabinose is the predominant monosaccharide; it most probably consists of a core galactan chain to which highly branched arabinan chains are attached.

The most important physical properties of gum tragacanth are its relative acid stability, ability to lower surface and interfacial tensions, and hydration to a gel. The primary use of gum tragacanth is as an emulsifying agent and water controller in low pH foods. Gum tragacanth is found in some pourable salad dressings and pickle relishes and in certain bakery products.

HYDROXYPROPYLCELLULOSES

Hydroxypropylcelluloses (HP) (2,4–6) are cellulose ethers prepared by reacting cellulose with propylene oxide. The products are characterized in terms of moles of substitution (MS) rather than DS (see "Carboxymethylcelluloses"). MS is used because the reaction of a propylene oxide molecule with cellulose leads to the formation of a new hydroxyl group with which another alkylene oxide molecule can react to form an oligomeric side chain. Therefore, there is no limit to the moles of substituent that can be added to each D-glucopyranosyl unit. MS denotes the average number of moles of alkylene oxide that has reacted per D-glucopyranosyl unit (Fig. 4.)

In general, the MS controls the solubility of a hydroxypropylcellulose. Commercially available hydroxypropylcelluloses are insoluble in hot water, soluble in cold water, and compatible with several oils. HP is produced in a wide range of viscosity grades and forms clear, smooth, uniform solutions with pseudoplastic rheology similar to that of other cellulosics (see "Carboxymethylcelluloses").

Because they are nonionic gums, hydroxypropylcelluloses are unaffected by pH. Because of their ability to reduce surface and interfacial tension, they are used as emulsifying agents, emulsion stabilizers, and whipping aids. Flexible, nontacky, heat-sealable packaging films and sheets can be produced from hydroxypropylcelluloses by conventional extrusion techniques. Because hydroxypropylcelluloses form strong, edible films that provide a barrier to oxygen and water vapor, they are used to coat nuts and confections.

KONJAC MANNAN

Konjac mannan is the principal component of konjac flour, the product of commerce. Konjac flour, the powdered tuber of *Amorphophallus konjac*, is used primarily in Japan to make a gelled food product, in noodles, and as a fat replacer.

Konjac mannan is a glucomannan. It is a linear polymer of β-D-mannopyranosyl and β-D-glucopyranosyl (ratio 1.6:1.0) units all linked (1→4). Removal of some of the naturally occurring acetate ester groups with alkali is required for gelation.

Konjac flour is cold-water swelling but hydrates only slowly in room-temperature water, so unlike many other gum products, it disperses easily in water. As the particle size is reduced, the time required for hydration and dissolution is reduced. When fully hydrated, it forms solutions of high viscosity and pseudoplasticity.

Konjac mannan, by itself, when deacetylated, forms elastic gels that are stable in boiling water. In fact, gel strength increases reversibly on heating. This accounts for its traditional use in the stabilization of noodles that can

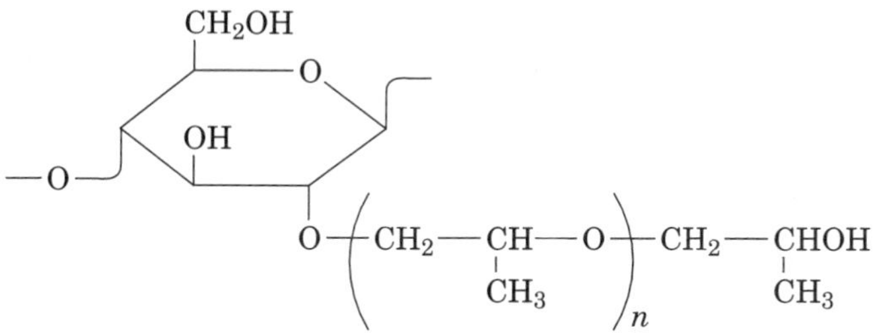

Figure 4. A representative β-D-glucopyranosyl unit of cellulose containing a hydroxypropyl group or a poly(propylene oxide) chain ($n = 0, 1, 2, 3$, etc).

be cooked in an autoclave. In addition, konjac mannan exhibits a strong synergism with agar and κ-carrageenan, forming elastic, thermally reversible gels after heating and cooling. It also will form a gel via synergistic interaction with xanthan and interacts synergistically with starches to increase viscosity and form heat-stable gels. Strong edible films can be made with konjac mannan as the base.

METHYLCELLULOSES AND HYDROXYPROPYLMETHYLCELLULOSES

Methylcelluloses (MCs) contain methoxyl groups in place of some of the hydroxyl groups along the cellulose molecule; hydroxypropylmethylcelluloses (HPMCs) contain, in addition to methyl ether groups, hydroxypropyl ether groups along the cellulose chain (2,4–6). The properties of MCs and HPMCs are primarily a function of the amount of each type of substituent group and the molecular-weight distribution.

MCs are made by reacting cellulose with methyl chloride until the DS (see "Carboxymethylcelluloses") reaches 1.1 to 2.2. HPMCs are made by using propylene oxide in addition to methyl chloride in the reaction; hydroxypropyl group MS (see "Hydroxypropylcelluloses") levels in commercial products are 0.02 to 0.3.

Members of this family of gums are cold-water soluble. Conversion of some of the hydroxyl groups of cellulose molecules into methyl ether groups increases the water solubility of the polymer and reduces its ability to aggregate (ie, reduces intermolecular interactions). Solubility and solution stability is increased even more when hydroxypropyl groups are added to MC.

The most interesting property of these nonionic products is thermal gelation. Solutions of members of this family of gums decrease in viscosity when heated, as do solutions of most other polysaccharides. However, unlike solutions of other gums, when a certain temperature is reached (depending on the specific product), the viscosity will increase rapidly and the solution will gel. Gelation can occur at various temperatures from 45 to 90°C (115 to 195°F), depending on the viscosity type, DS/MS, and proportions of methyl and hydroxypropyl substituent groups. The thermal gelation is reversible; that is, the gel will revert to a fluid upon cooling.

These gums reduce surface and interfacial tensions and can, therefore, be used to stabilize emulsions and make foams. They will form high-strength films that are clear, water-soluble, and oil- and grease-resistant and have low oxygen and moisture vapor transmission rates. They are used in the preparation of dietetic foods, in baked goods for their glutenlike properties, in dipping batters (where film formation, low oil migration, moisture retention, thermal gelation, and tackiness are important properties), in whipped toppings, in frozen desserts and novelties, in canned fruit juice and fruit drink mixes (as a bodying agent, drying aid, emulsion stabilizer, and/or cloud agent), and as a binder and lubricator in extrusion processes. Their solutions exhibit pseudoplastic rheology similar to that of other cellulosics (see "Carboxymethylcellulose").

PECTINS

Pectins are mixtures of polysaccharides that originate from plants, contain pectinic acids as major components, are water soluble, and whose solutions gel under suitable conditions (2,4–7). Pectinic acids are galacturonoglycans [poly(α-D-galactopyranosyluronic acids)] with various, but greater than negligible, contents of methyl ester groups. Pectinic acids may have varying degrees of neutralization. Salts of pectinic acids are pectinates. Pectic acids are galacturonoglycans without, or with only a negligible content of, methyl ester groups. Pectic acids may have varying degrees of neutralization. Salts of pectic acids are pectates. The principal and key feature of all these molecules is a linear chain of α-D-galactopyranosyluronic acid units. In all commercial pectins, some of the carboxyl groups are in the methyl ester form; some or all of the remaining carboxylic acid groups may be in a carboxylate salt form (Fig. 5).

Pectins are subdivided according to their degree of esterification (DE), a designation of the percentage of carboxyl groups esterified with methanol. Pectins with DE values >50% are high-methoxyl (HM) pectins; those with DE values <50% are low-methoxyl (LM) pectins. The DE strongly influences the solubility, gel-forming ability, con-

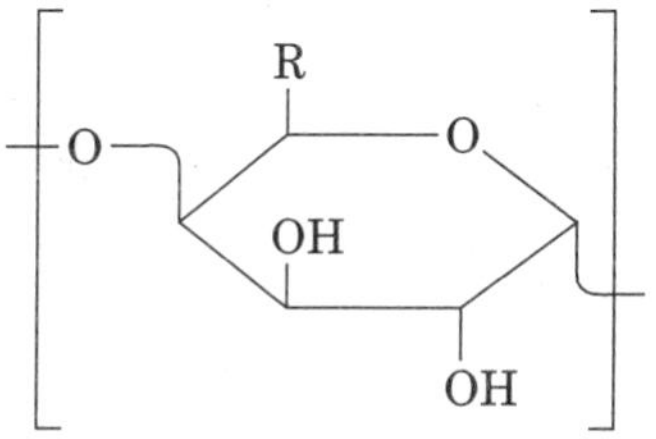

$R = -CO_2CH_3$, $-CO_2^-$, or $-CO_2H$

Figure 5. Monomer units of pectins.

ditions required for gelation, gelling temperature, and gel properties of the preparation. In some LM pectins, termed amidated pectins, some carboxyl groups have been converted into carboxamide groups. The degree of amidation (DA) indicates the percentage of carboxyl groups in the amide form.

Pectins are soluble in hot water. The importance of pectin is predominately the result of its unique ability to form spreadable gels when a hot solution is cooled. HM-pectin gels are formed by the addition of sugar (at least 55%, but normally ~65%, soluble solids) to a hot, acidic (pH ~3) solution of pectin in a fruit juice. LM pectins will gel only in the presence of calcium ions and do not require soluble solids to gel. Increasing the concentration of calcium ions increases the gelling temperature and gel strength.

Any system containing pectin at potential gelling conditions (ie, necessary concentration of an appropriate pectin, pH, concentration of cosolutes, and concentration of divalent cations) must be prepared at a temperature above the gelling temperature. The temperature at which structure is formed upon cooling is the gelling temperature. LM pectins, particularly amidated pectins, are sensitive to divalent cations, such as calcium ions, and require use of a sequestering agent for dissolution if they are present.

The primary use of pectins is in the preparation of spreadable gels (jams, jellies, marmalades, and preserves, including the preserves in fruit yogurt and fillings for chocolates) from fruit juices or whole fruits, with or without added sugar. Pectins are also used in preparation of chewable fruit candies, fruit roll ups, canned fruit juices, cheese spreads, and icings and frostings.

XANTHAN GUMS

Xanthan (2,4–6) (also known as xanthan gum) is widely and extensively used as a food gum. It is an extracellular microbial polysaccharide produced by fermentation. Its characteristics vary somewhat with variations in the strain of the organism and the fermentation conditions used.

Xanthan gum has a linear main chain that has the same structure as cellulose (see the article CARBOHYDRATES: CLASSIFICATION, CHEMISTRY, LABELING). In xanthan gum, every second β-D-glucopyranosyl unit of the cellulosic backbone is substituted with a trisaccharide unit. About half of the trisaccharide side-chain units carry a terminal pyruvic acid cyclic acetal group.

Xanthan gum solutions are extremely pseudoplastic and have a high at-rest viscosity. These properties make xanthan gum almost ideal for the stabilization of aqueous dispersions, suspensions, and emulsions. Whereas other polysaccharide solutions decrease in viscosity when they are heated, xanthan solutions change little in viscosity over the temperature range 0 to 95°C (32 to 205°F). Although xanthan gum is anionic, pH has almost no effect on the viscosity of its solutions. A synergistic viscosity increase results from the interaction of xanthan gum with κ-type carrageenans, MCs, locust bean gum, and konjac mannan. The latter two combinations form thermally reversible gels when hot solutions of these polysaccharides are cooled.

The wide range of properties that make xanthan gum so widely and extensively used in foods includes its temperature stability, pH stability, ability to improve gloss (sheen), ability to hold moisture, and bland flavor; the high at-rest viscosity, stability to salts, high pseudoplasticity, and stable viscosity of its solutions from the freezing temperature to the boiling temperature of water; the extraordinary increase in solution viscosity it imparts with increases in concentration; and its synergistic interactions. Xanthan gum is sometimes used in combination with modified food starch. It is used in almost all pourable salad dressings, very frequently in combination with propylene glycol alginate. Among other products in which it is found are spoonable dressings, including reduced-calorie mayonnaise; bakery products; cereal bars; condiments; dairy products; egg substitutes; frozen foods; meat products; drink mixes; sauces; spreads; syrups; and toppings.

Blends of xanthan and locust bean gum and/or guar gum are excellent ice cream stabilizers and are often used in combination with a CMC and a carrageenan. Other applications of these blends are in the preparation of canned frostings, dips, cream cheese, cottage cheese dressings, other processed cheese products, low or nonfat dairy-based systems, sauces such as pizza and Sloppy Joe sauce, and starch-based gum candies.

BIBLIOGRAPHY

1. A.-C. Eliasson, ed., *Carbohydrates in Food*, Marcel Dekker, New York, 1996.
2. M. Glicksman, ed., *Food Hydrocolloids*, CRC Press, Boca Raton, Fla., 1982 (Vol. 1), 1983 (Vol. 2), 1986 (Vol. 3).
3. J. R. Mitchell and D. A. Ledward, eds., *Functional Properties of Food Macromolecules*, Elsevier Applied Science Publishers, London, 1986.
4. A. M. Stephen, ed., *Food Polysaccharides and Their Applications*, Marcel Dekker, New York, 1995.
5. R. L. Whistler and J. N. BeMiller, eds., *Industrial Gums*, 3rd ed., Academic Press, New York, 1993.
6. R. L. Whistler and J. N. BeMiller, *Carbohydrate Chemistry for Food Scientists*, Eagen Press, St. Paul, Minn., 1997.
7. M. L. Fishman and J. Jen, eds., *Chemistry and Function of Pectins*, American Chemical Society Symposium Series, Vol. 310, 1986.

J. N. BEMILLER
Purdue University
West Lafayette, Indiana

H

HAZARD ANALYSIS AND CRITICAL CONTROL POINTS (HACCP).

See POSTHARVEST INTEGRATED PEST MANAGEMENT in the Supplement section.

HEAT

When two bodies are at different temperatures, there is a transfer of heat energy from the body having the higher temperature t_1 to the body having the lower temperature t_2. Hence, the state of energy of the colder body is increased and that of the warmer body decreased. This situation of unsteady-state heat transfer continues with the driving force $t_1 - t_2$, or δt, decreasing until both bodies are at the same temperature and there is no unbalanced state to give a driving force. Such is the case when a hot or warm food is placed in a container of cold water to decrease the temperature of the food.

Steady state heat transfer occurs when the temperature driving force Δt remains constant and the rate of heat transfer between two bodies is constant. An example of this type of steady-state transfer is boiling of water on a hot element of a stove. The element temperature is kept constant by a continuous flow of electrical energy being converted to heat and the boiling water stays constant at the boiling point of water (100°C at standard atmospheric pressure). On the other hand, if the heating element is kept constant and a food is being heated below the boiling point, there is an unsteady-state heat transfer, which results in a temperature rise of the food.

The last two examples also illustrate that there are different types of heat within a substance. Sensible heat is the amount of heat that can be added or removed from a given mass of product between two temperatures of the product (t_1 to t_2) without changing the state of the body (eg, heating or cooling a food by cooking or refrigerating). Latent heat refers to the amount of heat necessary to change a given mass from one state to another (eg, boiling water, freezing food products). Sensible heat results in a temperature rise within a body, whereas latent heat is the heat (at constant temperature) necessary to change the state.

There are numerous systems used in the world to quantify mass and energy, including the English (Imperial) system and several systems using metric units. This can be well demonstrated by the measurement of sensible heat. Each food or material requires a different amount of heat, called the specific heat (heat capacity), to raise or lower a given mass to a given temperature. Through laboratory research, specific heats have been determined and given in different unit systems. A British thermal unit (English system) is defined as the amount of heat required to raise the temperature of 1 lb of water 1°F. A kilocalorie in the metric system is the amount of heat required to raise 1 kg of water 1°C.

Several international organizations have attempted to standardize the unit systems, symbols, and quantities to prevent the confusion that often exists (1,2). The result is the Systeme International d'Unites, or the SI system, in which base metric units have been defined. Table 1 gives the SI system unit definitions along with factors for converting from the Imperial system to SI units.

In the SI system, the basic unit of force is the newton (N), the force that gives a mass of 1 kg an acceleration of 1 m/s. The basic unit of energy is the joule (J), the work done when the point of application of 1 N is displaced by a distance of 1 m. Hence specific heat c is expressed in the three principal systems as:

$$c = 1\ \text{Btu/lb°F} = 1\ \text{cal/g°C} = 4.1865\ \text{J/g K}$$

The mathematical relationship between heat, temperature, and specific heat in a food can be expressed in SI units as:

$$Q = M \int_{t_1}^{t_2} c dt$$

where Q = heat (heat gained or lost in KJ), M = mass (kg), c = specific heat (kJ/kg°C), and dt = temperature change (°C).

If the specific heat is at constant pressure it is designated as c_p. If the process is carried out at constant volume (eg, a container of compressed gas), it is designated as c_v. For liquid and solid food, the difference between c_p and c_v is negligible. Because over the range of temperatures for food processes the c_p is essentially constant, heat in a system is normally calculated as:

$$Q = (M)(c_p)(t_1 - t_2) = Mc_p\Delta t$$

Thus if 100 kg of apples with a specific heat of 3.6 kJ/kg/K) are cooled from tree temperature of 18°C to 5°C, the amount of heat removed would be

$$Q = (100\ \text{kg})(3.6\ \text{kJ/kg} \cdot \text{K})(18\text{–}5\ \text{K}) = 4{,}680\ \text{kJ}$$

As a point of reference, 1 Btu equals 1.055 kJ; therefore, this would be equal to 4,436 Btu.

When learning the SI system, which will eventually be the world standard, it is useful to remember some basic approximate conversion factors to visualize the relationships between values. This is especially true for the United States where the English system has become so well entrenched. It is helpful to remember that 1 ft is approximately 0.3 m, 1 Btu is approximately 1 kJ, 1 lbf is approximately 4.5 N, 1 Btu/h is approximately equal to 0.3 W, and 1 Btu/lb is approximately 2.33 kJ/kg. If the equivalent English and SI values are visualized when working a problem in SI units, it will become easier to think in both systems.

Table 1. Base Units of SI System (Metric) and Conversion from Imperial to SI units

Measurable quantity	SI base unit	SI symbol	Imperial base unit	Imperial base symbol	Conversion factor (imperial × factor = SI)
Length	Meter	m	Foot	ft	0.30480
Mass	Kilogram	kg	Pound mass	lb	0.453592
Time	Second	s	Second	s	1.0
Temperature	Degree Kelvin	K	Degree Rankine	R	0.55556
Electric current	Ampere	A	Ampere	A	1.0
Amount of substance	Mole	mol	Mole	mol	1.0

PROCESSING FOODS BY ADDING OR REMOVING HEAT

The addition or removal of heat from a food increases shelf life and ensures the safety of the product. Heat is added to make a food more acceptable (improving sensory characteristics), to reduce microorganism populations, to lessen enzyme activity (pasteurization or cooking), or to kill microorganism and completely inactivate the enzymes (sterilizing). Heat is also added during dehydration processes to remove water through vaporization and thus lower the water activity a. Heat is removed to slow the growth of microorganisms and the action of enzymes (cooling or refrigerating) or to change the state (freezing), which prevents growth of microorganisms and further reduces enzyme action.

CHANGING THE STATE OF A FOOD PRODUCT

There are three states in which a food can exist, namely solid, liquid, or vapor. Because most natural foods are in some state of equilibrium with water, most of the significant changes of state in food products involve water. Food processes involving a change in state include freezing, heating to the point of evaporating a component (eg, vaporization or dehydration), condensing vapors to liquids (eg, solvent extraction), and subliming water directly from the frozen state to vapor.

Freezing Food

Freezing foods involves removing heat, resulting in the changing of liquid water in the food to ice. This has been demonstrated in the frozen fish industry where sensible heat is removed until the temperature reaches approximately −1°C (30°F). It then takes about 45 min during the so-called critical period to remove the latent heat of fusion from the water. After this period, the temperature drops rapidly as sensible heat is again removed (see FISH AND SHELLFISH PRODUCTS). It should be noted that foods are not pure substances so do not have precise temperatures at which there is a change in phase. This is because water has dissolved materials that increase or decrease in amount as a food is being processed and some of the water is bound to components of the food rather than being free to flow and act like pure water.

Vaporization Processes

The three vaporization processes involve the change in basic characteristics of a food or material by applying heat or other forms of energy that is converted to heat within the product.

Evaporation. Evaporization is the concentration of a liquid by supplying sufficient energy to vaporize the more volatile component or components. The process is used to concentrate solutions for stability or prior to further processing and to recover solvent from an extraction system.

Drying or Dehydrating. Drying or dehydration involves adding energy, usually heat, to vaporize a liquid, usually water, from a solid food. Foods are dried by supplying sufficient heat energy to vaporize water. This (1) preserves the product by reducing water activity; (2) reduces the cost or difficulty of packaging, handling, storing and shipment; and (3) produces convenience items (eg, instant coffee). In certain cases energy is supplied through other energy forms and then converted to heat in the product. For example, in processing by microwave heating, the energy is supplied to the system in the form of hertzian waves that are absorbed by the food, the resulting friction between vibrating molecules converts the wave energy to heat energy.

There are two distinct periods of drying in which different mechanisms control the rate of drying. During the first period, the constant rate period, the heat transfer predominates; all of the heat added is directly used in evaporating water and the rate of drying is independent of the nature of the food. In this case the moisture movement near the surface is rapid enough to maintain a saturated condition at the surface and the temperature of the food remains constant.

Each food has a critical moisture content at which the moisture can not migrate to the surface by diffusion rapidly enough to utilize all of the heat for evaporation. At this point the second phase, or falling rate period, begins and the food begins to absorb heat and rise in temperature.

A special form of drying involves sublimation of water directly from a frozen product to vapor, that is without passing through the liquid state. This is accomplished by placing a frozen food in a chamber under high vacuum, if the partial pressure of water vapor in the chamber is maintained below that of the ice at 0°C or 4.58 mm Hg, the water does not thaw prior to becoming a vapor.

Distillation. Distillation is the separation of two or more liquids through vaporizing the more volatile component or components. A principal use of this process in the food industry is in the recovery of a product after solvent extraction. In the case of organic solvent extraction of oilseeds, the solvent is recovered for reuse and the oil is retained as a pure food product.

Heat energy is certainly the most important factor involved in the processing of food products. In dealing with the science and engineering aspects of the food industry it is necessary to be constantly aware of the units involved in measuring the amount of energy being added to or removed from a food product. Unit equations (those showing the units as well as the numerical values) ensure the consistency of units and dimensions.

BIBLIOGRAPHY

1. R. P. Singh and D. R. Heldman, *Introduction to Food Engineering*. Academic Press, Inc., Orlando, Fla., 1984.
2. H. Wolf, *Heat Transfer*, Harper & Row, New York, 1983.

GEORGE M. PIGOTT
University of Washington
Seattle, Washington

HEAT EXCHANGERS

HEAT EXCHANGERS

Previous discussions on heat and the mechanisms for heating or cooling foods have emphasized that virtually every food-processing operation depends on transferring heat. The operations involving heat transfer use a wide variety of heat exchangers for heating or cooling products and for operational aspects of auxiliary equipment. Although there are many types of heat-exchanger equipment used in the food industry, there are relatively few principles that govern the heat transfer and operation of the equipment.

There are two basic classifications of heat exchangers. One is the contact type in which there is direct physical contact between the food product and the heating or cooling medium. The other is the noncontact type in which the heat is transferred through a body that separates the product from the heating or cooling source. Within these two categories there are many proprietary designs and models of heat exchangers depending on the specific requirement for transferring heat to or from a given type of food product.

Capital investment, safety, and economics of processing are important operational factors to consider when purchasing or installing exchanger equipment. Of equal value in determining the type or model of equipment to install is the effect on the physical and chemical properties of the end products. Maintaining or improving nutritional value, aesthetics, safety of the products, and sensory attributes affect the marketability.

Paramount to the successful long-term operation of heat exchangers in meeting the above goals is the efficiency and ease of sanitizing the entire equipment. This is especially important for closed systems that can not be dismantled while cleaning. The operation of cleaning in place (CEP) requires sanitary design that ensures complete sanitation when cleaning liquids are circulated through the exchangers. Hence, engineering design is the controlling factor in insuring successful heat-transfer operations involving food processing.

FOOD-PROCESSING OPERATIONS REQUIRING HEAT EXCHANGERS

It is common to think of all heat exchangers as being composed of two adjoining compartments through which fluids are flowing, each cooling or heating the other. This concept is natural because the basic engineering approach to studying heat transfer is to discuss each type of heat exchange operation as a specific process (eg, product freezing, refrigeration systems, batch heating, nonsteady-state operations, etc) with steady-state, liquid-liquid heat exchangers being studied under the subject of heat exchangers. However, when processing food, it is important to think of heat exchangers as being the total range of equipment that is used for cooking, blanching, pasteurizing, sterilizing, cooling, freezing, and cold-storage holding. In addition, there are many processes, such as drying and extraction, in which the transfer of heat is important to accomplish the basic goal of the process. Food products that are heated or cooled during processing, including the heat transfer medium, cover a wide range of vapors, liquids, and solids. Table 1 indicates how machinery and equipment must be used for the transfer of heat in the many different food-processing operations.

In addition to the type of heat transfer taking place during processing, the physical and chemical condition of the food before, during, and after heating or cooling must be considered. Products requiring different approaches to heat transfer during the food-processing operation include

1. Dense, hard, solid foods (eg, potatoes) that can stand considerable mechanical abuse during processing but become somewhat fragile during the final phase of heating.
2. Soft, solid foods (eg, tomatoes) that cannot stand any mechanical abuse during or after cooking.
3. Purées or soft items that flow like a liquid but congeal or gel to a solid when heated (eg, surimi being processed into seafood analogues and extruded cereals).
4. Liquids that range from heat stable to extremely heat labile.
5. Foods, particularly those high in protein, that congeal with heat and cause baked-on deposits on the heat-transfer surfaces.

Table 1. The State of a Food Product as Related to Classification of Heat Exchanger

Food form	Heat transfer media	Classification	Example
Vapor	Liquid	Noncontact	Condensing vapors
Liquid	Vapor	Contact	Steam infusion, steam injection
		Noncontact	Heating by steam (condensing), refrigerant cooling
	Liquid	Noncontact	Transfering heat between liquids
	Solid	Contact	Melting ice
		Noncontact	Heating on metal hot plate in a container
Solid	Vapor	Noncontact	Extrusion (heating by steam jacket)
		Contact	Drying, cooking
	Liquid	Contact	Blanching, immersion freezing, deep-fat frying, poaching, steeping
		Noncontact	Extrusion (heating by liquid jacket)
	Solid	Contact	Dry ice cooling, freezing on plates
		Noncontact	Cooking on stove
Any food	None	Noncontact	Radiant heating, irradiation

Processing requirements for these types of food range from individual batch cooking to large-scale continuous sterilizing. Many products must also be heated and then cooled to ensure maximum retention of nutrients and desired product form.

HEAT-EXCHANGER SYSTEMS AND PRINCIPLES OF OPERATION

Heat is transferred to or from a food in batch and continuous systems. Batch systems involve unsteady-state heat transfer whereby the food being heated or cooled begins at a given temperature and increases or decreases until the desired temperature is reached. The heat-transfer medium can vary in temperature (eg, a hot surface or liquid that changes temperature as it gives or receives heat) or can be at steady state (eg, condensing steam).

Batch steady-state systems involve a series of batch systems that give the overall effect of steady state processing. For example, liquid-filled tanks in series can be stepwise heated by batch heating but can be connected so that the end result of the overall heating is a steady-state emission of constant-temperature liquid continuously flowing at the same rate as the cold liquid entering the first tank.

A true steady-state system is found in flowing liquids or viscous solids whereby mass flow rate, temperature, pressure, and physical properties of the food and the heat transfer medium are constant at any given cross section.

Heating and Cooling Liquids

Batch Heating. Typical batch heating of a liquid food takes place in a steam-jacketed kettle. While the food is being heated, the system is under unsteady-state conditions. After the food reaches the desired processing temperature and is heated at a constant temperature, steady-state conditions are reached. Assuming that the final temperature of the food is to be maintained somewhat below that of the condensing steam, steady-state conditions prevail as the steam flow rate is adjusted to maintain the processing conditions. As is the practice for all commercial heat-exchanger equipment, the outside of the kettle is lagged to minimize heat loss to the surroundings and isolate the environment of the unit operation.

Large kettles (tanks) being used for heating a liquid often have coils in the tank to transfer heat. This greatly improves the heat-transfer efficiency by increasing the heat-transfer surface above that of the outside wall receiving heat from the jacketed heating (or cooling) source. An additional improvement of the heat transfer in tanks can be realized by installing mechanical stirring equipment, which increases heat transfer above that of natural convection.

Tubular Heat Exchangers. The simplest continuous heat exchange occurs when two fluids of different temperature are flowing through concentric pipes or tubes. The flow in steady-state heat exchangers can be either cocurrent or cocurrent (parallel) flow. During countercurrent flow, one stream (liquid or vapor) is introduced at the opposite end of the unit. By controlling the flow rates, it is possible to heat the cold liquid above the outlet temperature of the entering hot stream. Conversely, when the two liquids are introduced at the same point, the stream being heated can never leave at a temperature above that of the stream being cooled. This cocurrent system is normally less efficient than a countercurrent system because the temperature difference driving force can become quite small as the temperatures of the two streams meet. However, there are circumstances whereby cocurrent flow can be used to ensure that a heat-sensitive material does not rise above a certain temperature during processing. In the case of using steam to heat a flowing liquid, the food is heated while the condensing steam is maintained at the saturation temperature of the steam. As in the case of a steam kettle cooker, the most common and efficient heating medium is condensing steam. Shell-and-tube heat exchangers are essentially improved tubular heat exchangers where a few to many tubes replace the single concentric inner tube.

Whereas the heat transfer coefficient is constant under these steady-state conditions, the temperature driving force (ΔT) is calculated as the log mean temperature difference

$$\Delta T_M = \frac{(T_S - T_{f_1}) - (T_S - T_{f_2})}{\ln \dfrac{(T_S - T_{f_1})}{(T_S - T_{f_2})}} \qquad (1)$$

and the steady-state heat transfer is calculated as

$$Q = UA\Delta T_M \quad (2)$$

where T_S = temperature of condensing steam, T_f = temperature of liquid food being heated, U = overall heat transfer coefficient, and ΔT_M = log mean temperature driving force.

Plate Heat Exchangers. Plate heat exchangers solve one of the principal processing problems encountered with tubular and shell-and-tube heat exchangers, that of sanitation. It is virtually impossible to thoroughly clean and sanitize closed system exchangers when a food liquid or slurry is passed through the larger diameter or the shell side where velocity is low. The basic units of plate exchangers are stainless steel (for sanitation and corrosion resistance) plates that are pressed, machined, or formed to accomplish several special design features. The contour of a stack of the plates is such that, when a formed gasket is placed between each plate, a heat exchanger allowing two liquid streams to flow between plates is formed. In practical operation, the plates are suspended from horizontal rails or pipes that allow them to be brought together and tightly compressed during operation or separated for cleaning and maintenance.

These types of heat exchanger are used extensively in the dairy industry. The exchanger is highly efficient due to the high turbulence and minimum volume flowing between the channels. Capacity can be increased to any flow rate desired (eg, 10,000 kg/h) by increasing the number of plates in the frame. When operated at the proper flow rate, the velocity of liquid through the small channels decreases the tenacious baked-on deposits that are caused when colloidal suspended components (eg, proteins in milk) contact hot surfaces. Also, the plates can be separated for thorough cleaning and sanitizing during the maintenance periods.

The disadvantages of plate heat exchangers is the high initial cost compared to tubular type exchangers and the high cleaning and maintenance cost of taking the exchanger apart for each cleaning period. Also, the maximum velocity is limited by the small cross-sectional area and the pressure limitations of the gasketed plates. The minimum velocity is determined by the varying cross section, which allows dead spots of low velocity and subsequent bake-on of the suspended or dissolved solids.

Direct Steam Heating. Direct contact between steam and a food is the most efficient means of heating by directly transferring the latent heat of vaporization to the food. However, the product must be able to sustain the dilution effect caused by the added water resulting from the condensed steam that remains in the heated product. Furthermore, special consideration must be given to producing steam that is safe for human consumption.

When steam is added to a product, the process is known as steam injection. When the product is sprayed into a chamber of steam, thus adding the product to the steam, the process is called steam infusion.

Falling Film Evaporators. When a film of liquid is allowed to flow down a heated wall, the transfer of heat is extremely rapid. This type of unit is used for rapid heating of a fluid that is being evaporated. However, due to the limitation of flow rate as compared to other types of heat exchanger, this method is not often used solely for heating a product.

Heating and Cooling Solids

Batch Heating. 'Batch heating of solids is carried out in the type of heat-exchange facilities normally associated with cooking foods. Broiling, roasting, or baking operations are accomplished by a combination of radiant heat, convection, and conduction, the predominant mechanism of heat transfer depending on the specific commercial equipment. Ovens are heated by elements emitting radiation at a wavelength of 5×10^{-6} m at a temperature of 250–400°C. A more effective radiant heating occurs in chambers heated by an electric bulb infrared source. In this case the air is not heated by the source so that little heating of the food is due to convection or conduction heating.

Microwave heating is a specialized form of dielectric radiant heating that has many advantages over other dielectric methods because there is no requirement for critical spacing between the food and the capacitor plates. Heating is accomplished by the friction of excited molecules rubbing against each other as the strong alternating energy reverses the polarization of the molecules in the food many millions of times per second. The advantages of microwave heating include extremely rapid heating of the food, uniform distribution of the heat throughout the entire food mass, high efficiency, and good control of the energy being added to the food.

Many products are cooked and heated by convection heating when immersed in hot vegetable oil (deep frying) or poached in hot or boiling water. Batches of products are often heated by placing a container of the product on a hot surface (eg, hot plate or element of a stove) or in a steam environment. Steam retort canning of foods is a good example of batch heating by steam. Hermetically sealed cans are placed in baskets and placed in a steam chamber that can be closed and pressurized, normally at about 10 psi (117°C or 242°F). Condensing steam transfers energy to the outside of the can by conduction and convection, whereby the heat is conducted through the can by conduction and then to the food. Solid packs with no free liquid are heated in the can by conduction while a pack with free water transfers heat by both conduction and convection. After being held at the required temperature and time to accomplish sterilization, the cans are removed and air cooled. Sensitive products that tend to scorch or decrease in nutritional value during long air cooling periods are often pressure cooled with water prior to being removed from the retort.

Continuous Heating. More efficient production control, increased product throughput, and improved processed food quality can be accomplished when processes are upgraded from batch to continuous. Modern processing of solid foods involves large continuous production lines utilizing continuous baking ovens (eg, bakery products); deep-frying tanks (eg, french-fried potatoes), microwave ovens, radiant heat chambers, and steam chambers.

Batch Cooling or Freezing. Removing heat from foods, with the exception of cooling in tubular type exchangers, requires considerably different types of facilities than heating. This is due to the nature of the recycling refrigerant or the cryogenic liquids used to remove heat from a food. An additional factor is the psychrometric properties of air that are often recycled near the humidity saturation point in refrigeration facilities in which natural or forced convection is involved in the process.

The basic unit involved with refrigeration cooling of foods is a heat exchanger in which a refrigerant is introduced through an expansion valve into the coils that are located in the cooling or freezing chamber. Thus the cooling or freezing heat exchanger has a cold gaseous refrigerant on one side of the coil wall and the food or heat-transfer medium on the other. Air or liquid brines are the normal mediums used for transferring heat by convection from the food to the refrigeration coils. When the food product is in direct contact with the refrigeration coils, heat is transferred directly by conduction.

Blast Cooling and Freezing Facilities. Forced-air convection is used in blast refrigeration to transfer heat from refrigerated coils to the product. A blast freezer or cooler is actually a double heat exchanger. One exchange takes place between the product and direct contact with air flowing past, and the second is the cooling of the recirculating air as it passes over the freezing coils. During the first phase of this cycle, cold air is circulated over the product and through the freezing chamber, which results in an increase in humidity due to the humidity driving force H between the product environment and the colder refrigerant coil. Thus the air removes water from the freezing chamber and deposits it as ice on the refrigeration coils during the second phase when the air is cooled below the saturation point. If the product being frozen is not completely protected from the air, desiccation will occur in the product. Also, if the doors to the chamber are not completely sealed when closed, moist warm air will enter and further complicate the moisture transfer problem. Of course, as the moisture, in the form of ice, builds up on the coil it acts as a heat-transfer barrier and greatly reduces the efficiency of the system. Hence, the advantage of blast refrigeration is the relatively simple facility required while the main disadvantage is desiccation of the food and buildup of ice deposits that must be removed from the refrigeration coils. The ice greatly reduces the efficiency of heat transfer and increases the cost of operation.

Contact Plate Freezers. Many solid food products are frozen by conduction on freezer shelves, called plates, that contain circulating refrigerant. Thus plate freezing involves heat exchange between a solid food and a vapor refrigerant. Efficient plate freezing is limited to foods and food packages that have flat surfaces (eg, rectangular packages of vegetable) because irregular geometries (eg, turkeys) cannot contact the flat freezer plate. The efficiency of freezing suitable packages is further increased by plate freezer systems in which the plates can be adjusted after loading to contact both the top and the bottom of the package.

Immersion Freezing in Brine. Saturated brine solutions have freezing points well below the freezing point of water and can be used efficiently to freeze products, particularly irregularly shaped items such as turkeys and fish. The product is immersed in a cold brine solution that is maintained at the low temperature by freezing coils. As in the case of blast freezing, there is a two-phase heat transfer. The refrigerant takes heat from the brine and the brine removes heat from the food by conduction and convection. This heat exchange is between a solid food and a liquid.

Cryogenic Freezing. Immersion of a solid food in a liquid refrigerant is similar to freezing in a brine except that there is a much higher temperature driving force between the liquid (eg, liquid ammonia or freon) and the product. Thus the freezing is rapid. Batch freezing by this method is not ordinarily carried out commercially because the cost is prohibitive. Due to the extremely fast freezing, a cryogen immersion frozen product must be carefully tempered before further handling and processing or the internal stresses produced will cause cracking of the frozen item.

Continuous Cooling or Freezing. As in the case of heating a food, continuous freezing is much more efficient than batch freezing. Many modern freezing operations involve continuous lines whereby conveyors carry a food through a freezing apparatus. This includes blast freezing and cryogenic freezing. One improvement over immersion cryogenic freezing is the continuous freezing in tunnels in which a liquid cryogenic is flowed over the product. In this operation the liquid expands to a vapor through nozzles directed toward the moving product line. This freezing of a solid by direct contact with a vapor greatly reduces the amount of refrigerant used during processing. However, the cost of the liquid cryogens used (ammonia, freon, carbon dioxide, and nitrogen) are such that only continuous processing lines operating long hours can be justified economically.

There are many different types of heat exchanger and auxiliary equipment available to the food processor. The length of the processing season, the type of food or food product, the value of the raw materials, the cost of utilities at the processing location, environmental factors, and common sense are all factors that must enter into the plans for food-processing operations. Judicial selection of the equipment and facilities for a given food and a given process are necessary to insure that the highest quality product is produced efficiently and economically.

GEORGE M. PIGOTT
University of Washington
Seattle, Washington

HEAT EXCHANGERS: FOULING

THE FOULING FACTOR

In view of its complexity and variability and the need to carry out experimental work on a long-term basis under actual operating conditions, fouling remains a somewhat neglected issue among the technical aspects of heat transfer. Still, the importance of carefully predicting fouling resistance in both tubular and plate heat-exchanger calculations cannot be overstressed. This is well illustrated in Tables 1 and 2.

Note that for a typical water–water duty in a plate heat exchanger, it would be necessary to double the size of the unit if a fouling factor of 0.0005 were used on each side of the plate (ie, a total fouling of 0.001).

Although fouling is of great importance, there are relatively little accurate data available and the rather conservative figures quoted in Kern (*Process Heat Transfer*) are used all too frequently. It also may be said that many of the high fouling resistances quoted have been obtained from poorly operated plants. If a clean exchanger, for example, is started and run at the designed inlet water temperature, it will exceed its duty. To overcome this, plant personnel tends to turn down the cooling-water flow rate and thereby reduce turbulence in the exchanger. This encourages fouling and even though the water flow rate eventually is turned up to design, the damage will have been done. It is probable that if the design flow rate had been maintained from the onset, the ultimate fouling resistance would have been lower. A similar effect can happen if the cooling-water inlet temperature falls below the design figure and the flow rate is again turned down.

SIX TYPES OF FOULING

Generally speaking, the types of fouling experienced in most CPI (cleaning in place) operations can be divided into six fairly distinct categories. First is crystallization—the most common type of fouling that occurs in many process streams, particularly cooling-tower water. Frequently superimposed with crystallization is sedimentation, which usually is caused by deposits of particulate matter such as clay, sand, or rust. From chemical reaction and polymerization often comes a buildup of organic products and polymers. The surface temperature and presence of reactants, particularly oxygen, can have a very significant effect. Coking occurs on high-temperature surfaces and is the result of hydrocarbon deposits. Organic material growth usually is superimposed with crystallization and sedimentation and is common to sea water systems. And corrosion of the heat-transfer surface itself produces an added thermal resistance as well as a surface roughness.

In the design of the plate heat exchanger, fouling due to coking is of no significance since the unit cannot be used at such high temperatures. Corrosion also is irrelevant since the metals used in these units are noncorrosive. The other four types of fouling, however, are most important. With certain fluids such as cooling-tower water, fouling can result from a combination of crystallization, sedimentation, and organic material growth.

Table 1. Typical Water/Water Tubular Design. Clean Overall Coefficient 500 Btu/h · ft² · °F

Fouling resistance (h · ft² · °F/Btu)	Dirty coefficient (Btu/h · ft² · °F)	Extra surface, required, %
0.0002	455	10
0.0005	400	25
0.001	333	50
0.002	250	100

Table 2. R405 Paraflow, Water/Water Duty, Overall Coefficient 1000 Btu/h · ft² · °F; Single Pass-Pressure Loss 9 psig

Fouling Resistance (h · ft² · °F/Btu)	Dirty coefficient (Btu/h · ft² · °F)	Extra surface, required, %
0.0002	833	20
0.0005	666	50
0.001	500	100
0.002	333	200

A FUNCTION OF TIME

From Figure 1, it is apparent that the fouling process is time-dependant with zero fouling initially. The fouling then builds up quite rapidly and in most cases and levels off at a certain time to an asymptomatic value as represented by curve A. At this point, the rate of deposition is equal to that of removal. Not all fouling levels off, however, and curve B shows that at a certain time the exchanger would have to be taken off line for cleaning. It should be noted that a Paraflow is a particularly useful exchanger for this type of duty because of the ease of access to the plates and the simplicity of cleaning.

In the case of crystallization and suspended solid fouling, the process usually is of the type A. However, when the fouling is of the crystallization type with a pure compound crystallizing out, the fouling approaches type B and

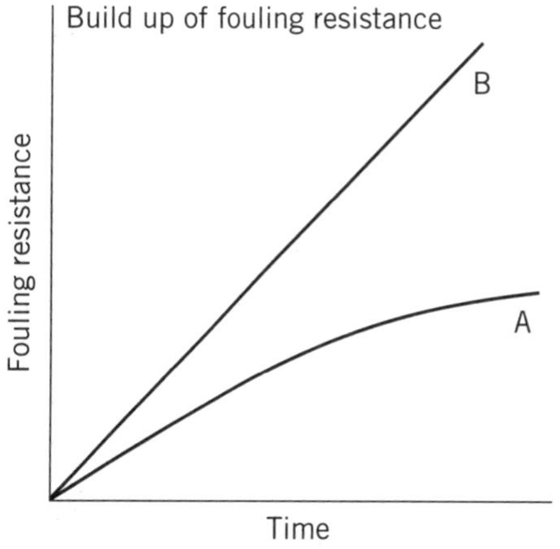

Figure 1. Buildup of fouling resistance.

the equipment must be cleaned at frequent intervals. In one particularly severe fouling application, three Series HMB Paraflows are on a 4-1/2 hour cycle and the units are cleaned in place for 1-1/2 h in each cycle.

Biological growth can present a potentially hazardous fouling since it can provide a more sticky surface with which to bond other foulants. In many cases, however, treatment of the fluid can reduce the amount of biological growth. The use of germicides or poisons to kill bacteria can help.

LOWER RESISTANCE

It generally is considered that resistance due to fouling is lower with Paraflow plate heat exchangers than with tubular units. This is the result of five Paraflow advantages:

1. There is a high degree of turbulence, which increases the rate of foulant removal and results in a lower asymptotic value of fouling resistance.
2. The velocity profile across a plate is good. There are no zones of low velocity compared with certain areas on the shell side of tubular exchangers.
3. Corrosion is maintained at an absolute minimum.
4. A smooth heat-transfer surface can be obtained.
5. In certain cooling duties using water to cool organics, the very high water film coefficient maintains a moderately low metal surface temperature, which helps prevent crystallization growth of the inversely soluble compounds in the cooling water.

The most important of these is turbulence. HTRI (Heat Transfer Research Incorporated) has shown that for tubular heat exchangers, fouling is a function of flow velocity and friction factor. Although flow velocities are low with the plate heat exchanger, friction factors are very high and this results in lower fouling resistance. The effect of velocity and turbulence is plotted in Figure 2.

Marriot of Alfa Laval has produced a table showing values of fouling for a number of plate heat-exchanger duties.

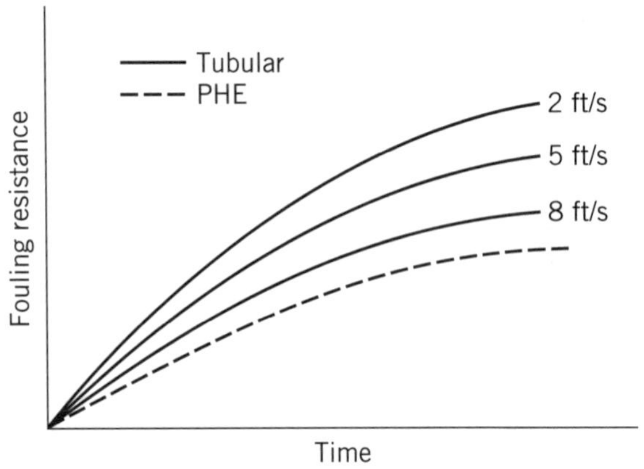

Figure 2. Effect of velocity and turbulence.

Fluid	Fouling resistance (hr · ft^2 · °F/Btu)
Water	
Demineralized or distilled	0.00005
Soft	0.00010
Hard	0.00025
Cooling tower (treated)	0.00020
Sea (coastal) or estuary	0.00025
Sea (ocean)	0.00015
River, canal	0.00025
Engine jacket	0.00030
Oils, lubricating	0.00010–0.00025
Oils, vegetable	0.00010–0.00030
Solvents, organic	0.00005–0.00015
Steam	0.00005
Process fluid, general	0.00005–0.00030

These probably represent about one half to one-fifth of the figures used for tubulars as quoted in Kern, but it must be noted that the Kern figures probably are conservative, even for tubular exchangers.

APV, meanwhile, has carried out test work that tends to confirm that fouling varies for different plates with the more turbulent type of plate providing the lower fouling resistances. In testing an R405 heating a multicomponent aqueous solution containing inverse solubility salts (ie, salts whose water-solubility decreases with increasing temperature), it was learned that the rate of fouling in the Paraflow was substantially less than that inside the tubes of a tubular exchanger. The tubular unit had to be cleaned every 3 or 4 days while the Paraflow required cleaning about once a month.

Additional tests on cooling water fouling were sponsored by APV at Heat Transfer Research, Inc. (HTRI) with the test fluid being a typical treated cooling tower water. Results of these experiments follow.

OBJECTIVES

In many streams, fouling is an unavoidable by-product of the heat-transfer process. Fouling deposits can assume numerous types such as crystallization, sedimentation, corrosion, and polymerization. Systematic research on fouling is relatively recent and extremely limited (1–3) and almost exclusively concentrated on water—the most common fluid with fouling tendencies. The few research data that exist usually are proprietary or obtained from qualitative observations in plants. However, it is generally recognized that the prime variables affecting fouling buildup are flow velocity, surface temperature, and surface material. In the case of water, the water quality and treatment must also be considered.

The importance of fouling on the design of heat exchangers can be seen from the rate equation

$$\frac{1}{U} = \frac{1}{h_1} + \frac{1}{h_2} + R_f$$

where U is the overall heat transfer coefficient, Btu/h · ft^2 · °F, h_1, h_2 is the film coefficients of the two heat-transferring fluids, Btu/h · ft^2 · °F, and R_f is the fouling resistance, h · ft^2 · °F/Btu. It is obvious from equation 1 that the higher the film coefficients, the greater effect the fouling resistance will have on the overall coefficient and therefore on the size of the exchanger.

In tubular heat exchangers, water-side heat-transfer coefficients in the order of magnitude of 1000 Btu/h · ft^2 · °F are quite common. In plate exchangers, the coefficients are substantially higher, typically around 2000 Btu/h · ft^2 · °F. Assuming that both types of equipment operate with water—water systems, overall clean coefficients of 500 and 1000 Btu/h · ft^2 · °F respectively are obtained. Using a typical fouling resistance of 0.001 h · ft^2 · °F/Btu (equal to a coefficient of 1000 Btu/h · ft^2 · °F), the inclusion of the fouling will cause the tubular exchanger size to increase by a factor of 4, while for the plate exchanger, the corresponding factor is 7.

This example clearly demonstrates the crucial importance of fouling, especially in plate exchangers. Yet, fouling resistances that are unrealistically high often are specified and invariably have been based on experiences derived from tubular equipment. The common source of water fouling resistances in TEMA which recommends R_f values spanning a tremendous range between 0.0015 and 0.005 h · ft^2 · °F/Btu.

Flow velocity, as mentioned earlier, is a crucial operating parameter that influences the fouling behavior. For flow inside the tubes, the definition of flow velocity and the velocity profile is straightforward. But in plate exchangers, flow velocity is characterized by constant fluctuations as the fluid passes over the corrugations. It is postulated that this induces turbulence that is superimposed on the flow velocity as a factor that diminishes fouling tendencies. This has been observed qualitatively in practical applications and confirmed by unpublished APV research.

TEST APPARATUS AND CONDITIONS

The plate heat exchanger (PHE) tested was an APV Model 405 using APV Type R40 stainless-steel plates 45 in. high and 18 in. wide. The plate heat-transfer area was 4 ft^2 with a nominal gap between plates of 0.12 in. with the plate corrugations, the maximum gap was 0.24 in. Seven plates were used, creating three countercurrent passages each of the cooling water and the heating medium. A schematic diagram of the installation is shown in Figure 3.

The PHE was mounted on the HTRI Shellside Fouling Research Unit (SFRU) together with two small stainless-steel shell-and-tube exchangers. The PHE was heated using hot steam condensate, and the fouling was determined from the degradation of the overall heat-transfer coefficient. Simultaneously, tests also were run on an HTRI Portable Fouling Research Unit (PFRU), which uses electrically heated rods with the cooling water flowing in an annulus. The SFRU and PFRU are described (5) elsewhere.

The fouling tests were conducted at a major petrochemical plant in the Houston, Texas area. The test units were installed near the cooling-tower basin on the plant. The cooling-tower operating characteristics are summarized in Table 3. The 140,000-gpm system had two 1600-gpm sidestream filters. Filter backwash accounted for most of the blowdown. The makeup water for the cooling-water system came from several sources and was clarified with alum outside the plant. The water treatment used is as follows:

- Chromate-zinc-based inhibitor for corrosion control (20–25 ppm chromate)
- Organic phosphonate and polymer combination as dispersant (2 ppm organic phosphonate)
- Polyphosphate as anodic passivator (5–6 ppm total inorganic phosphate)
- Chlorine for biological control (continuous feeding of 386 lb/day)
- Biocide for biological control (15 ppm biocide once per week)
- Sulfuric acid for pH control (pH 6–6.5)

TEST PROCEDURE

The velocity for the PHE is defined by

$$V = \frac{w}{\rho A_c}$$

where V is the velocity, w is the mass flow rate, ρ is the fluid density, and A_c is the cross-sectional flow area. The cross-sectional flow area is based on the compressed gasket spacing. In other words, the flow are is computed as if there were no corrugations but smooth plates instead. The surface temperature is defined as the fluid–deposit interface temperature. Since the PHE was heated by constant temperature steam condensate, the metal temperature remained constant and the surface temperature decreased as fouling built up. In addition, because of the counterflow nature of the PHE, the surface temperature was not constant from inlet to outlet. The surface temperature reported here is the initial surface temperature at the midpoint of the plate.

Conditions of operation were selected so that all the SFRU test exchangers operated roughly at the same nominal flow velocity and surface temperature. During the operation, weekly water samples were taken for chemical analysis. At the end of each test, deposits were photographed and sampled for chemical analysis. The exchanger plates then were cleaned and the unit reassembled and new test conditions established.

TEST AND RESULT DESCRIPTION

Five test series were performed testing several velocities and surface temperatures (eg, see test 2 in Fig. 4). Typical fouling time histories are shown in Figure 5 along with the operating conditions. Notice that the PHE fouling resistances establish a stable asymptotic value after about 600–900 h of operation. The results of the other tests are shown in Figure 6 only as the values of the asymptotic

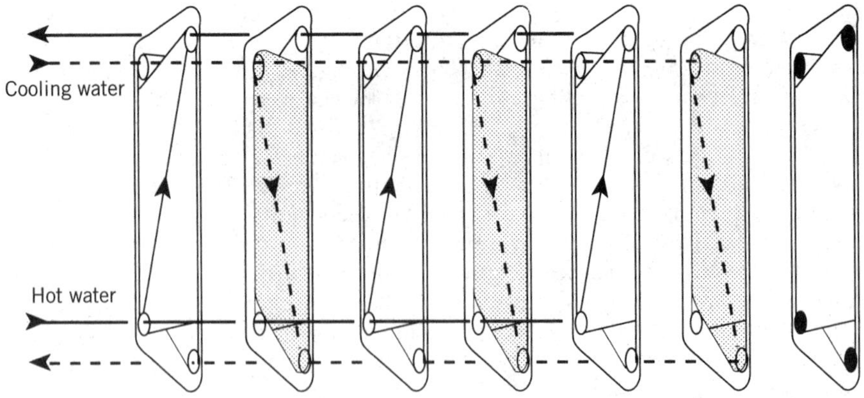

Figure 3. APV plate heat exchanger with three parallel countercurrent parallel passages for both the cooling water and the hot water.

Table 3. Cooling-Water System Characteristics

	Tower and circulation system	
	Circulation rate	140,000 gpm
	Temperature difference across tower	23.4°F
	Number of sidestream filters	2
	Corrosion rate from carbon steel coupon tests	
	<1 mil/yr	

Water description	Cooling-tower water	Makeup water
Treatment	10 cycle concentration Blowdown from coolers and sidestream filters	Unclarified San Jacinto River 1000 gpm Clarified San Jacinto River 2400 gpm Lissie Sand Well 475 gpm
	Typical composition	
Total hardness as $CaCO_3$, ppm	520	48
Calcium as $CaCO_3$, ppm	420	40
Magnesium as $CaCO_3$, ppm	120	8
Methyl Orange alkalinity as $CaCO_3$, ppm	20	34
Sulfate as SO_4, ppm	1600	34
Chloride as Cl, ppm	800	48
Silica as SiO_2, ppm	150	18
Total inorganic phosphate as PO_4, ppm	10	
Orthophosphate as PO_4, ppm	7	
pH	6	7.5
Specific conductance, micromhos, 18C	5000	300
Chromate as CrO_4, ppm	25	
Chromium as Cr, ppm	0.5	
Soluble zinc as Zn, ppm	2.5	
Total iron as Fe, ppm		0.8
Suspended solids, ppm	100	10

fouling resistances, as these are the data required for design.

The chemical analysis of the fouling deposit was made after each test and the results are summarized in Table 4. The primary elements found were phosphorus, zinc, and chrome from the water treatment and silicon from suspended solids in the water.

The photograph of the plates of the PHE shown in Figure 4 at the termination of test 2 indicates that fouling occurred only in the upper third of the plates. This is the region near the hot-water inlet and cold-water outlet, region of high surface temperature. Figure 5 shows a surface temperature profile for test 2 conditions of 2.8 ft/s and an inlet bulk temperature of 90°F. From the water chemistry parameters, the saturation temperature for calcium phosphate above which precipitation is expected was calculated to be 163°F (6). The shaded region on Figure 7 indicates the expected precipitation region and corresponds closely to the fouled region seen in Figure 4.

The surface temperatures during some of the tests were higher than those experienced in normal PHE operation. This is the result of the setup condition criteria that the

Figure 4. Fouled plates, test 2.

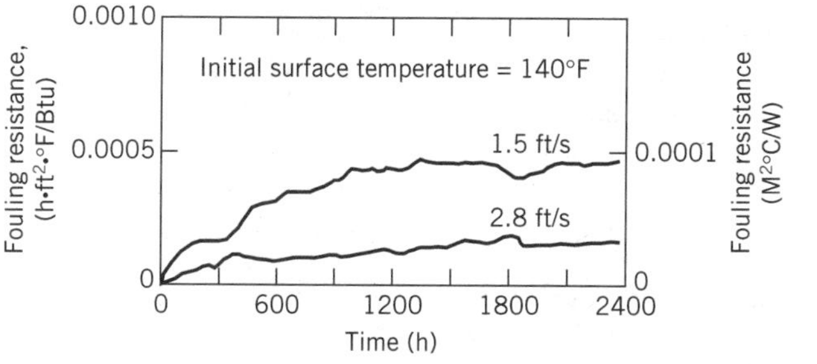

Figure 5. Typical PHE fouling curves.

Table 4. Fouling Deposit Analysis, %

	Series 2	Series 3	Series 4	Series 5	Series 6
Loss of ignition	—	25	25	17	21
Phosphorus	19	24	13	12	22
Aluminum	2	2	8	6	4
Silicon	1	4	18	16	13
Calcium	4	8	3	15	9
Chrome	5	13	15	15	9
Iron	2	7	4	5	13
Zinc	2	20	12	13	7
Sodium	—	1	1	—	1
Magnesium	—	—	1	—	1

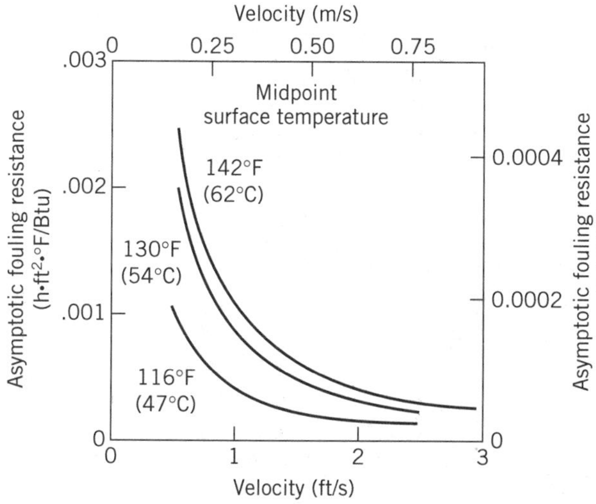

Figure 6. Asymptotic fouling resistance versus velocity with surface temperature as a parameter.

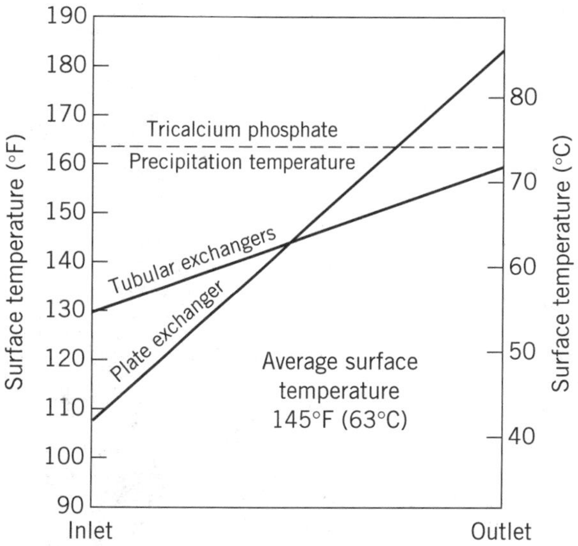

Figure 7. Surface temperature profile with indicated calcium phosphate precipitation.

three exchangers on the SFRU operate at the same midpoint surface temperature. Since the ratio of flow rate to surface area for the shell side of the shell and tube exchangers is twice that for the PHE, the cooling water had a longer residence time in the PHE than in the tubular exchanger. As a result, the lower flow rate to surface area ratio yields a larger temperature rise of the cooling water. The result is the steeper surface temperature profile shown in Figure 7. Although cocurrent operation was investigated as a correction for this problem, the required higher hot-water flow rate was not available. Consequently, the PHE data in some of these tests was slightly penalized because part of the surface is above the critical temperature for tricalcium phosphate precipitation, a condition that normally would not be encountered in industrial operations.

1. The loss of ignition (LOI) is performed by drying the sample to constant weight at 220°F (105°C) and then combusting at 1470°F (800°C). LOI represents bound water and organics present in the deposit. An organic test on samples of test series 5 showed that about 1/3 of the LOI was organic.
2. Aluminum, silicon, and iron come from the suspended solids. Aluminum comes from operation of the clarifier. The silicon is not from magnesium silicate since little magnesium is in the analyses.
3. Chrome, zinc, and phosphorus are derived from the water treatment.
4. Low values of calcium indicate little or no tricalcium phosphate. Most of the calcium comes from association with zinc and polyphosphates in the treatment.

It is difficult to make meaningful comparisons between the fouling tendencies of different types of heat exchangers. Unless carefully done, the results are misleading. This is particularly true when attempting to compare the fouling experience in the PHE to that of tubeside operation as in the PFRU. The geometries, surface area, and operational characteristics are very different even though the cooling water may be the same. Such comparisons often are attempted using TEMA recommended fouling resistances.

TEMA recommends a fouling resistance of 0.002 h · ft^2 · °F/Btu for the water system used. However, the maximum fouling resistance measured in the PHE was less than 0.0005 h · ft^2 · °F/Btu-only 25% of the TEMA recommendation. This confirms the earlier assumption that applying TEMA-recommended fouling resistances for a shell and tube exchangers to the PHE seriously handicaps the performance ratings. Because of the inherently high heat-transfer coefficients, the effects of fouling resistances are more pronounced. Consider a PHE operating at 1.5 ft/s. From Table 5, the fouled overall coefficient using a TEMA fouling resistance of 0.002 h · ft^2 · °F/Btu is about one half of the overall coefficient using the measured fouling resistance of this investigation. This example illustrates the need for caution in using TEMA fouling recommendations for equipment other than shell-and-tube exchangers.

Figure 8 compares the performance of the PHE with typical tube-side data and with the TEMA-recommended fouling resistance.

A direct comparison of the PHE and the shell-side exchangers can be made through the deposit analyses. A sample analysis is given in Table 6. The amounts of phosphates, calcium, chrome, and zinc are somewhat higher for

Table 5. Effects of Fouling Resistance on PHE Performance

	Overall heat-transfer coefficient		
		Fouling coefficient	
Operating velocity (both sides)	Clean	TEMA 0.002 h · ft^2 · °F/Btu (0.00035 m^2°C/W)	Present study 0.0005 h · ft^2 · °F/Btu (0.00009 m^2°C/W)
1.5 ft/s	1078 Btu/h · ft^2 · °F	341 Btu/h · ft^2 · °F	700 Btu/h · ft^2 · °F
0.45 m/s	6121 W/m^2 · °C	1936 W/m^2 · °C	3975 W/m^2 · °C

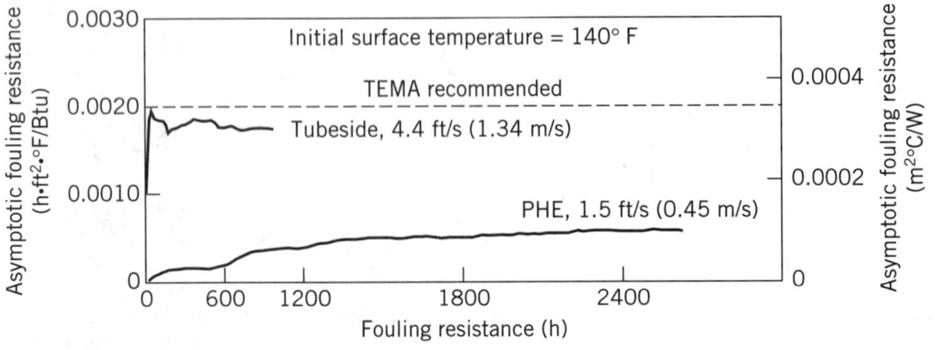

Figure 8. Comparison of tube-side and PHE fouling.

Table 6. Comparison of Deposit Analyses from Plate Exchanger and Shell-Side Test Exchanger for Series 5

	Fouling deposit analysis, %	
	Plate	Shell-side
Loss on ignition	17	24
Phosphorus	12	9
Aluminum	6	7
Silicon	16	33
Calcium	15	7
Chrome	15	9
Iron	5	4
Zinc	13	8

the PHE. The operation of the PHE at a higher surface temperature than the shell-side units resulted in more crystallization fouling. However, the amount of silicon for the shell-side unit is much higher than for the PHE. The high silicon concentration is due to sedimentation in the baffle-shell corners of the bundle, while the high turbulence promoted by the corrugated plates in the PHE minimizes sedimentation fouling. In heavily sediment-laden waters, the PHE would be especially superior.

CONCLUSIONS

Although the tests were performed on one water system only experience with tubular data indicates that the overall trends generally are valid; ie, plate exchangers should be designed to substantially lower values of fouling than would be used on tubular equipment for the same stream. However, the comparison is not always that straightforward, as flow velocity itself is not a valid criterion. The wall shear stress in a PHE operating at 2.8 ft/s is equivalent to that in a tube-side exchanger operating at 8.2 ft/s. On the basis of typical industrial velocities, it was found that the PHE at 1.5 ft/s fouls about one-half as much as a tube-side exchanger operating at 5.9 ft/s, Furthermore, because of the high-heat-transfer coefficients typical to the PHE, surface temperature may differ from those in tubular equipment and should be carefully watched.

The turbulence inducing corrugation pattern prevalent in heat-exchanger plates produces very high local velocities. This results in high friction factors and therefore high shearing. This high shear, in turn, results in less fouling in plate exchangers than in tubular units. For cooling-water duties, Heat Transfer Research Incorporated (HTRI) has shown that the plate exchanger fouls at a much lower rate than either the tube side or the shell side of tubulars. TEMA recommendations for fouling therefore should never be used for plate units since they probably are five times the value found in practice. In particular, the plate exchanger is far less susceptible to fouling with heavy sedimented waters. This is contrary to popular belief. It must be noted, however, that all particular matter must be significantly smaller than the plate gap and generally particles over 0.1 in. in diameter cannot be handled in any plate heat exchanger.

It has been shown that quoting a high fouling resistance can negate a plate heat exchanger design by adding large amounts of surface and thereby overriding the benefits of the high coefficients. Fouling design resistance, therefore, should be chosen with care, keeping in mind that with a Paraflow unit, it always is possible to add or subtract surface to meet exact fouling conditions.

DISCLOSURE STATEMENT

As a general policy, HTRI as the project operator disclaims responsibility for any calculations or designs resulting from the use of the date presented.

BIBLIOGRAPHY

Adapted from:

J. D. Usher and A. Cooper, "Paraflow Seminar, Part Four," *CPI Digest* **3**, 6–9 (1974).

J. D. Usher and A. Cooper, "Cooling Water Fouling in Plate Heat Exchangers," *CPI Digest* **5**, 2–9 (1979).

1. J. Taborek, T. Aoki, R. B. Ritter, J. W. Palen, and J. G. Knudsen, "Fouling—The Major Unresolved Problem in Heat Transfer, Parts I and II," *CEP* **68**, (1972).
2. J. W. Suitor, W. J. Marner, and R. B. Ritter, "The History and Status of Research in Fouling of Heat Exchangers in Cooling Water Service," Paper No. 76-CSME/CSChE—19 presented at the 16th National Heat Transfer Conference, St. Louis, Missouri, August 8–11, 1976.
3. R. W. Morse and J. G. Knudsen, "Effect of Alkalinity of the Scaling of Simulated Cooling Tower Water," Paper No. 76-CSME/CSChE—24 presented at the 16th National Heat Transfer Conference, St. Louis, Missouri, August 8–11, 1976.
4. *Standards of Tubular Exchanger Manufacturers Association*, 5th Ed., New York, 1968.
5. P. Fisher, J. W. Suitor, and R. B. Ritter, "Fouling Measurement Techniques and Apparatus," *CEP* **71**, 66 (1975).
6. J. Green and J. A. Holmes, "Calculation of the pH of Saturation of Tricalcium Phosphate," *Journal of the American Water Works Association* **39**, (1947).

APV CREPACO, INC.
Lake Mills, Wisconsin

HEAT EXCHANGERS: PARAFLOW

PRINCIPLES AND APPLICATIONS

The Paraflow is the original plate-type heat exchanger designed by APV to provide maximum efficiency and cost-effectiveness in handling thermal duties while minimizing maintenance downtime and floor space requirements.

Frame, Plates, and Gaskets

The Paraflow plate heat exchanger as shown in Figure 1 consists of a stationary head and end support connected

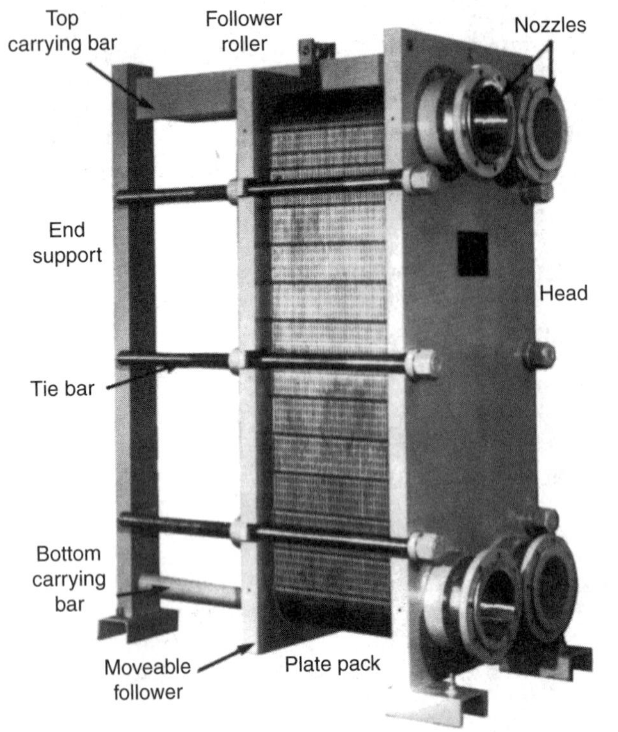

Figure 1. The APV CREPACO Paraflow plate heat exchanger.

by a top carrying bar and bottom guide rail. These form a rigid frame that supports the plates and moveable follower. In most units, plates are securely compressed between the head the follower by means of tie bars on either side of the exchanger. In a few models, central tightening spindles working against a reinforced end support are used for compression. When Paraflows are opened, the follower moves easily along the top bar with the aid of a bearing supported roller to allow full access to each individual plate.

With the exception of some sanitary models that are clad with stainless steel, Paraflow frames are fabricated of carbon steel and are finished in chemical-resistant epoxy paint. Frame ports accept bushings of stainless steel or alternative metals which, with various types of flanged or sanitary connections, form the inlet and outlet nozzles. By using intermediate connector plates as shown in Figure 2, units can be divided into separate sections to accommodate multiple duties within a single frame.

The closely spaced metal heat-transfer plates have troughs or corrugations that induce turbulence to the liquids flowing as a thin stream between the plates (Fig. 3).

The plates have corner ports which in the complete plate pack form a manifold for even fluid distribution to the individual plate passages (Fig. 4).

The seal between the plates is established by a peripheral gasket that also separates the thruport and flow areas with a double barrier. The interspace is vented to atmosphere to prevent cross-contamination in the rare event of leakage (Fig. 5).

As an exclusive feature, Paraflow heat-exchanger plates have interlocking gaskets in which upstanding lugs and scallops are sited intermittently around the outside edges. These scallops ensure that there are no unsupported portions of the gaskets and, in combination with the patented form of pressed groove, provide mechanical plate-to-plate support for the sealing system. The upstanding lugs (Fig. 6) maintain plate alignment in the Paraflow during pack closure and operation. The groove form provides 100% peripheral support of the gasket, leaving none of the material exposed to the outside. In addition, the gasket-groove design minimizes gasket exposure to the process liquid.

Plate Arrangement

Comparison of Paraflow plate arrangement to the tube and shell-side arrangement in a shell and tube exchanger is charted in Fig. 7. Essentially, the number of passes on the tube side of a tubular unit can be compared with the number of passes on a plate heat exchanger. The number of tubes per pass also can be equated with the number of passages per pass for the Paraflow. However, the comparison with the shell side usually is more difficult since with a Paraflow, the total number of passages available for the flow of one fluid must equal those available for the other fluid to within ± 1. The number of cross passes

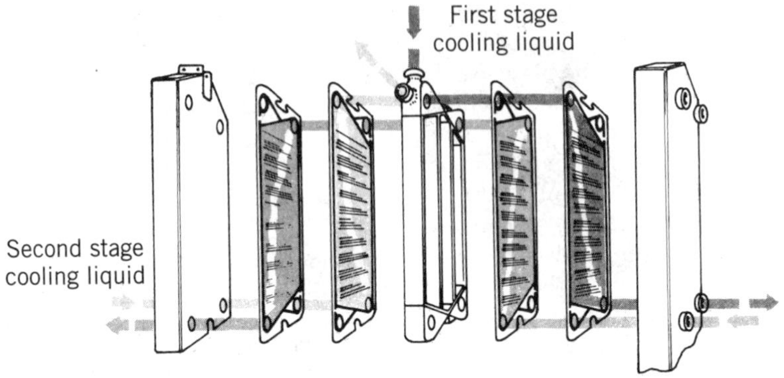

Figure 2. Two-section Paraflow with connector plate.

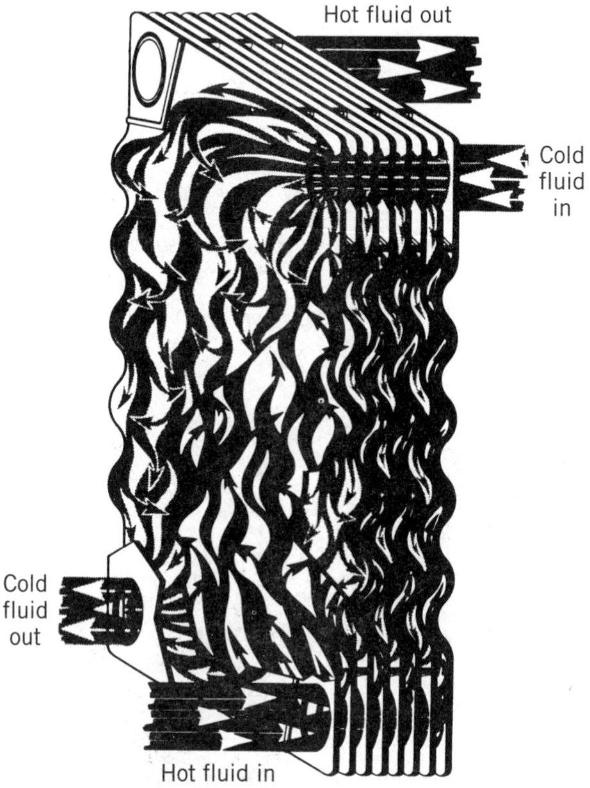

Figure 3. Cutaway of Paraflow plate shows turbulence during passage of product and service liquids.

on a shell, however, can be related to the number of plate passes and since the number of passages/pass for a plate is an indication of the flow area, this can be equated to the shell diameter. This is not a perfect comparison but it does show the relative parameters for each exchanger.

With regard to flow patterns, the Paraflow advantage over shell and tube designs is the ability to have equal passes on each side in full countercurrent flow, thus obtaining maximum utilization of the temperature difference between the two fluids. This feature is particularly important in heat-recovery processes with close temperature approaches and even in cases with temperature crossovers.

Whenever the thermal duty permits, it is desirable to use single-pass, countercurrent flow for an extremely efficient performance. Since the flow is pure counterflow, correction factors required on the LMTD approach unity. Furthermore, with all connections located at the head, the follower is easily moved and plates are more readily accessible.

Plate Construction

Depending on type, some plates employ diagonal flow while others are designed for vertical flow (Fig. 8). Plates are pressed in thicknesses between 0.020 and 0.036 in. (0.5–0.9 mm) and the degree of mechanical loading is important. The most severe case occurs when one process liquid is operating at the highest working pressure and the other at zero pressure. The maximum pressure differential is applied across the plate and results in a considerable unbalanced load that tends to close the typical 0.1–0.2-in. gap. It is essential, therefore, that some form of interplate support be provided to maintain the gap and this is done by two different plate forms.

One method is to press pips into a plate with deep washboard corrugations to provide contact points for about every 1–3 $in.^2$ of heat-transfer surface (Fig. 9). Another is the chevron plate of relatively shallow corrugations with support maintained by the peak-peak contact (Fig. 10). Alternate plates are arranged so that corrugations cross to provide a contact point for every 0.2–1 $in.^2$ of area. The plate then can handle a large differential pressure, and the cross pattern forms a tortuous path that promotes substantial liquid turbulence and thus a very high heat transfer coefficient.

Mixing and Variable-Length Plates

To obtain optimum thermal and pressure drop performance while using a minimum number of heat-exchanger plates, mixing and variable-length plates are available for several APV Paraflow plate heat-exchanger models. These

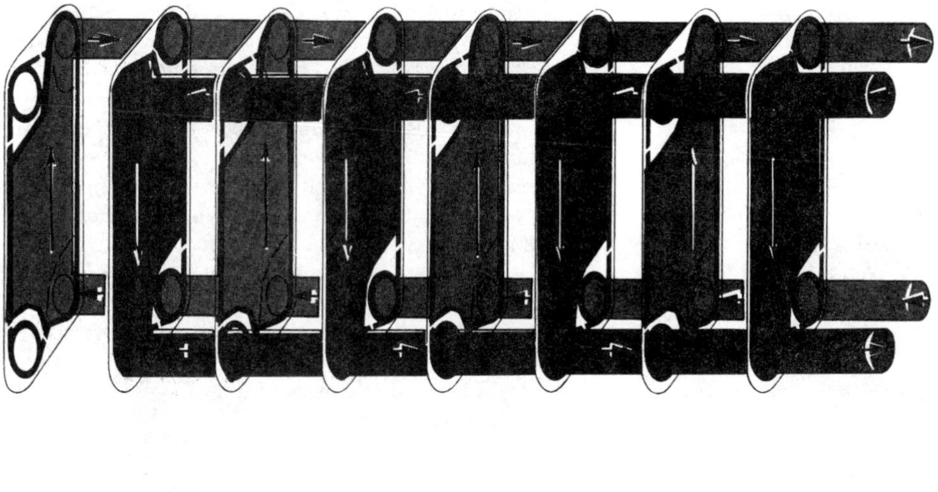

Figure 4. Single-pass countercurrent flow.

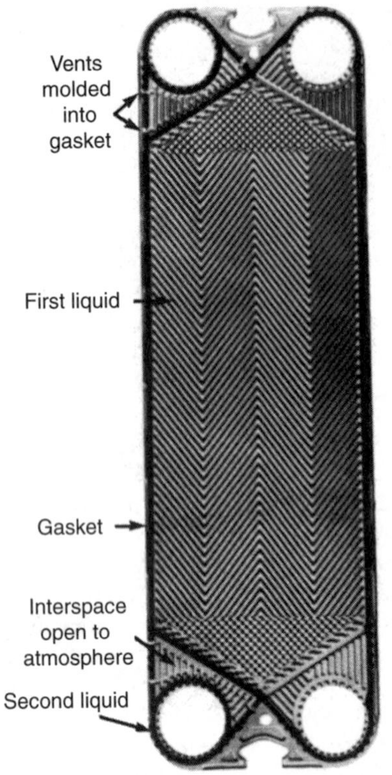

Figure 5. Gasket showing separation of throughport and flow areas.

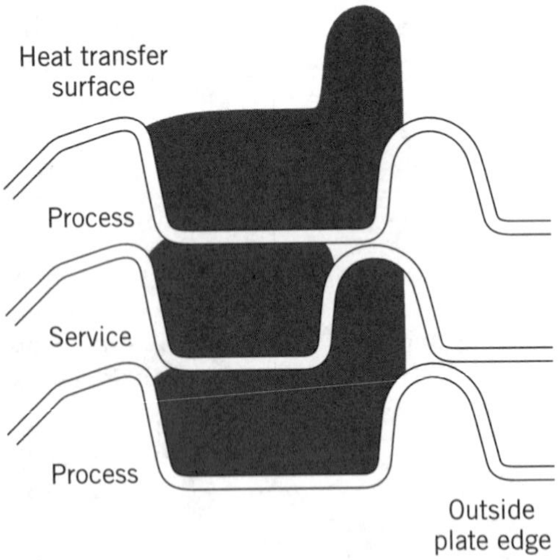

Figure 6. Exclusive interlocking gasket.

	Shell and Tube	Plate Equivalent	
Tube side	No. of passes	No. of passes	Side 1
Shell side	No. of tubes/pass	No. of passages/pass	Side 2
	No. of cross passes (No. of baffles + 1)	No. of passes	
	Shell diameter	No. of passages/pass	

Figure 7. Pass arrangement comparison: plate versus tubular.

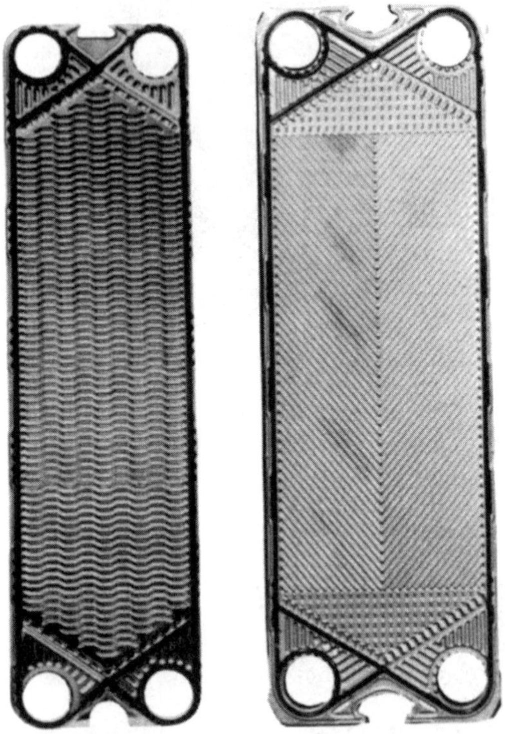

Figure 8. Diagonal and vertical flow patterns.

plates are manufactured to the standard widths specified for the particular heat exchanger involved but are offered in different corrugation patterns and plate lengths.

Since each type of plate has its own predictable performance characteristics, it is possible to calculate heat-transfer surface that more precisely matches the required thermal duty without oversizing the exchanger. This results in the use of fewer plates and a smaller, less expensive exchanger frame.

To achieve mixing, plates that have been pressed with different corrugation angles are combined within a single heat-exchanger frame. This results in flow passages that differ significantly in their flow characteristics and thus heat-transfer capability from passages created by using plates that have the same corrugation pattern.

For example, a plate pack (Fig. 11) of standard plates that have a typical 50° corrugation angle (to horizontal) develops a fixed level of thermal performance (HTU) per unit length. As plates of 0° angle (Fig. 12) are substituted into the plate pack up to a maximum of 50% of the total number of plates, the thermal performance progressively increases to a level that typically is twice that of a pack containing only 50° angle plates.

Figure 9. Corrugations pressed into plates are perpendicular to the liquid flow.

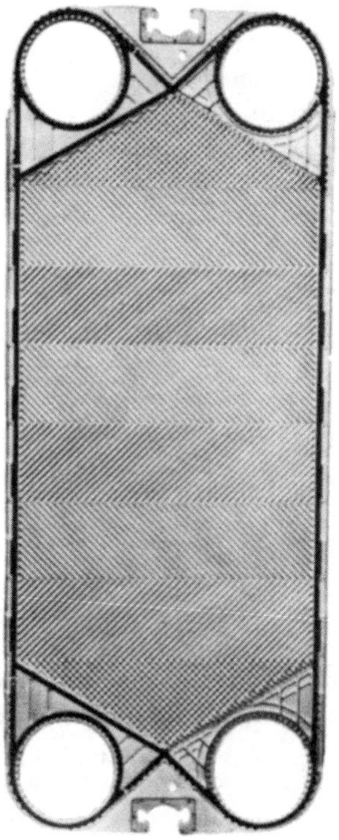

Figure 10. Troughs are formed at opposite angles to the centerline in adjacent plates.

Figure 11. Low HTU passage.

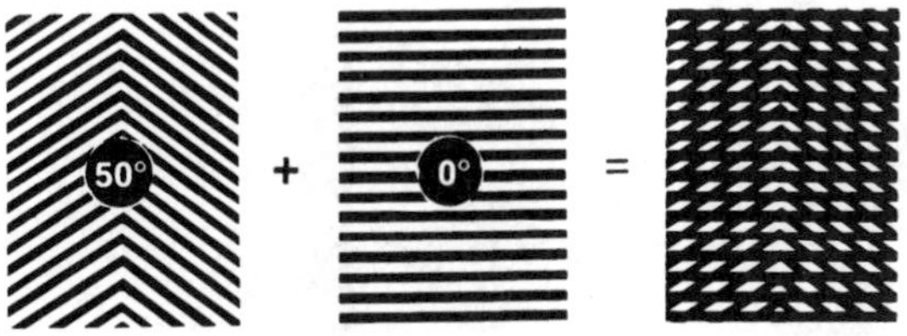

Figure 12. High HTU passage.

Thus, it is possible for a given plate length to fine-tune the Paraflow design in a single or even multiple-pass arrangement exactly to the thermal and pressure drop requirements of the application.

Of more recent development are plates of fixed width with variable lengths that extend the range of heat-transfer performance in terms of HTU. This is proportional to the effective length of the plate and typically, provides a range of 3–1 from the longest to the shortest plate in the series. As shown in Figure 13, mixing also is available in plates of varied lengths and further increases the performance range of the variable-length plate by a factor of approximately 2.

This extreme flexibility of combining mixing and variable-length plates allows more duties to be handled by a single-pass design, maintaining all connections on the stationary head of the exchanger to simplify piping and unit maintenance.

Plate Size and Frame Capacity

Paraflow plates are available with effective heat-transfer area from 0.28 to 50 ft^2, and up to 600 of any one size can be contained in a single standard frame. The largest Paraflow can provide in excess of 30,000 ft^2 of surface area. Flow ports are sized in proportion to the plate area and control the maximum permissible liquid throughput Figure 14. Flow capacity of the individual Paraflow, based on a maximum port velocity of 20-ft/s ranges from 15 gpm in the "junior" to 11,000 gpm in the Model SR235. This velocity is at first sight somewhat high compared to conventional pipework practice. However, the high fluid velocity is very localized in the exchanger and progressively is reduced as distribution into the flow passages occurs from the port manifold. If pipe runs are long, it is not uncommon to see reducers fitted in the piping at the inlet and exit connections of high-throughput machines.

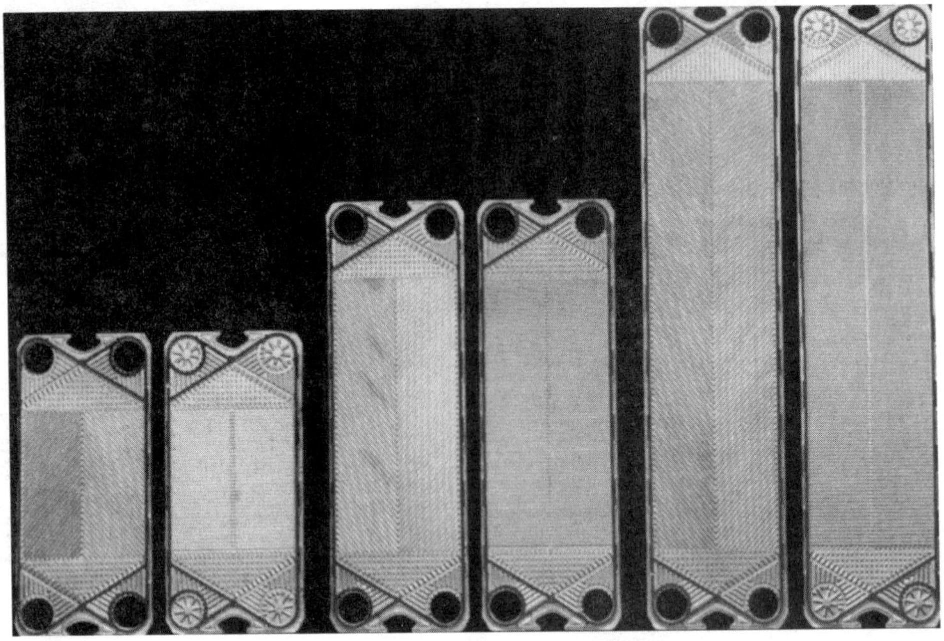

Figure 13. Variable-length plates with mixing options.

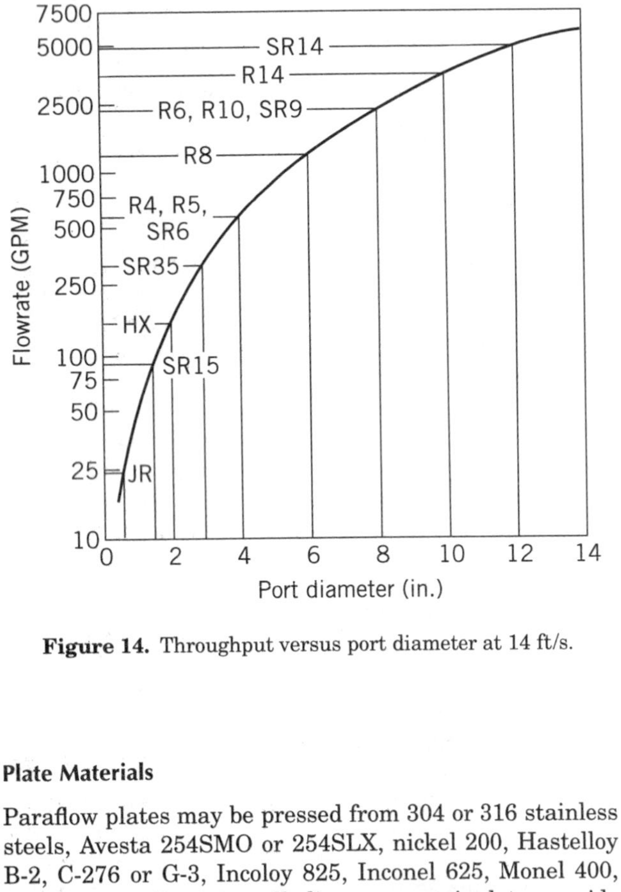

Figure 14. Throughput versus port diameter at 14 ft/s.

Plate Materials

Paraflow plates may be pressed from 304 or 316 stainless steels, Avesta 254SMO or 254SLX, nickel 200, Hastelloy B-2, C-276 or G-3, Incoloy 825, Inconel 625, Monel 400, titanium or titanium-palladium as required to provide suitable corrosion resistance to the streams being handled.

Gasket Materials

As detailed in Figure 15, various gasket materials are available as standard that have chemical and temperature resistance coupled with excellent sealing properties. These qualities are achieved by specifically compounding and molding the elastomers for long-term performance in the APV Paraflow.

Since the temperatures shown are not absolute, gasket material selection must take into consideration the chemical composition of the streams involved as well as the operating cycles.

THERMAL PERFORMANCE

The Paraflow plate heat exchanger is used most extensively in liquid—liquid duties under turbulent flow conditions. In addition, it is particularly effective for laminar flow heat transfer and is used in condensing, gas cooling, and evaporating applications.

Turbulent Flow

For plate heat transfer in turbulent flow, thermal performance can best be exemplified by a Dittus—Boelter-type equation:

$$\mathrm{Nu} = (C)(\mathrm{Re})^n(\mathrm{Pr})^m\left(\frac{\mu}{\mu_w}\right)^m$$

where Nu is the Nusselt number hD_e/k, Re is the Reynolds number vD_e/μ, Pr is the Prandtl number $Cp\mu/k$, D_e is the equivalent diameter (2× average plate gap), (μ/μ_w) is the Sieder—Tate correction factor, and reported values of the constant and exponents are

$$C = 0.15\text{–}0.40, \quad m = 0.30\text{–}0.45$$
$$n = 0.65\text{–}0.85, \quad x = 0.05\text{–}0.20$$

Typical velocities in plate heat exchangers for waterlike fluids in turbulent flow are 1–3 ft/s (0.3–0.9 m/s), but true

Gasket material	Approximate maximum operating temperature	Application
Paracril (medium nitrile)	275°F (135°C)	General aqueous service, aliphatic hydrocarbons
Paratemp (EPDM)	300°F (150°F)	High temperature resistance for a wide range of chemicals and steam
Paradur (fluoroelastomer)	400°F (205°C)	Mineral oils, fuels, vegetable and animal oils
Paraflor (fluoroelastomer)	400°F (205°C)	Steam, sulfuric acids

Figure 15. Paraflow gasket guide.

velocities in certain regions will be higher by a factor of ≧4 due to the effect of the corrugations. All heat-transfer and pressure drop relationships are, however, based on either a velocity calculated from the average plate gap or on the flow rate per passage.

Figure 16 illustrates the effect of velocity for water at 60°F on heat-transfer coefficients. This graph also plots pressure drop against velocity under the same conditions. The film coefficients are very high and can be obtained for a moderate pressure drop.

One particularly important feature of the Paraflow is that the turbulence induced by the troughs reduces the Reynolds number at which the flow becomes laminar. If the characteristic length dimension in the Reynolds number is taken at twice the average gap between plates, the Re number at which the flow becomes laminar varies from about 100 to 400 according to the type of plate.

To achieve these high coefficients, it is necessary to expend energy. With the plate unit, the friction factors normally encountered are in the range of 10–400 times those inside a tube for the same Reynolds number. However, nominal velocities are low and plate lengths do not exceed 7.5 ft so that the term $(V^2)L/(2g)$ in the pressure drop equation is much smaller than one normally would encounter in tubulars. In addition, single-pass operation will achieve

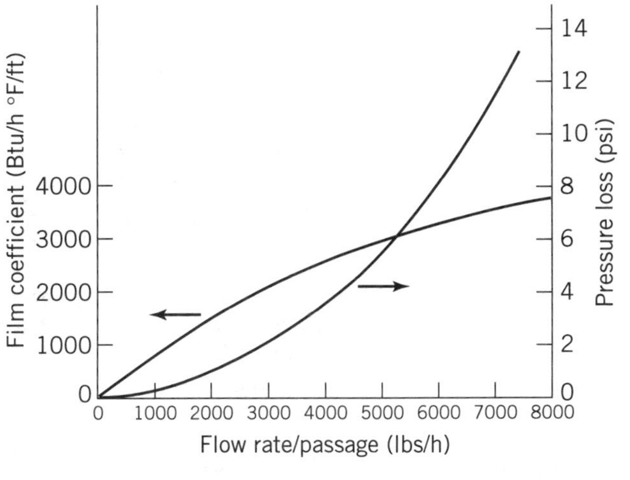

Figure 16. Performance details: Series HX Paraflow.

many duties so that the pressure drop is efficiently used and not wasted on losses due to flow direction changes.

The friction factor is correlated with the equations:

$$f = \frac{B}{(\mathrm{Re})^y}, \quad \Delta p = \frac{fL_pV^2}{2gd}$$

where y varies from 0.1 to 0.4 according to the plate and B is a constant characteristic of the plate.

If the overall heat-transfer equation $Q = UA\Delta T$ is used to calculate the heat duty, it is necessary to know the overall coefficient U, the surface area A, and the mean temperature difference ΔT.

The overall coefficient U can be calculated from

$$\frac{1}{U} = r_{fh} + r_{fc} + r_w + r_{dh} + r_{dc}$$

The values of r_{fh} and r_{fc} (the film resistances for the hot and cold fluids, respectively) can be calculated from the Dittus—Boelter equations described previously and the wall metal resistance r_w can be calculated from the average metal thickness and thermal conductivity. The fouling resistances of the hot and cold fluids r_{dh} and r_{dc} often are based on experience.

The value taken for A is the developed area after pressing. That is the total area available for heat transfer and, because of the corrugations, will be greater than the projected area of the plate, ie, 1.81 ft^2 versus 1.45 ft^2 for an HX plate.

The value of ΔT is calculated from the logarithmic mean temperature difference multiplied by a correction factor. With single pass operation, this factor is about 1 except for plate packs of less than 20 when the end effect has a significant bearing on the calculation. This is due to the fact that the passage at either end of the plate pack only transfers heat from one side and therefore the heat load is reduced.

When the plate unit is arranged for multiple-pass use, a further correction factor must be applied. Even when two passes are countercurrent to two other passes, at least one of them must experience cocurrent flow. This correction factor is shown in Figure 17 against a number of heat

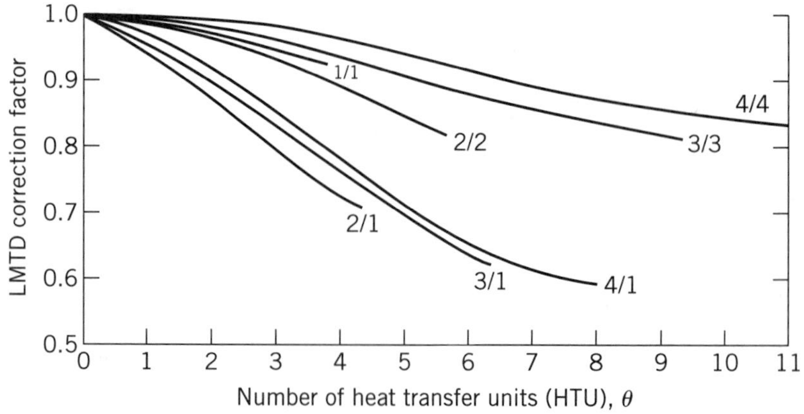

Figure 17. LMTD correction factor.

transfer units (HTU = temperature rise of the process fluid divided by the mean temperature difference). As indicated, whenever unequal passes are used, the correction factor calls for a considerable increase in area. This is particularly important when unequal flow conditions are handled. If high and low flow rates are to be handled, the necessary velocities must be maintained with the low fluid flow rate by using an increased number of passes. Although the plate unit is most efficient when the flow ratio between two fluids is in the range of 0.7–1.4, other ratios can be handled with unequal passes. This is done, however, at the expense of the LMTD factor.

The issue of how to specify a fouling resistance for a plate heat exchanger is difficult to resolve. Manufacturers generally specify 5% excess HTU for low fouling duties, 10% for moderate fouling, and 15–20% excess for high fouling. The allowed excess surface almost always is sufficient even though in many cases, it represents a low absolute value of fouling in terms of $(\text{Btu/h} \cdot \text{ft}^2 \cdot {}^\circ\text{F})^{-1}$. If a high fouling resistance is specified, extra plates have to be added, usually in parallel. This results in lower velocities, more extreme temperatures during startup, and the probability of higher fouling rates. Because of the narrow plate gaps and, in particular, because of the small entrance and exit flow areas, fouling in a plate usually causes more problems by the increase in pressure drop and/or the lowering of flow rates than by causing large reductions in heat-transfer performance. Increasing the number of plates does not increase the gap or throat area, and a plate unit sometimes can foul more quickly when oversurfaced.

The customer who does not have considerable experience with both the process and the plate heat exchanger should allow the equipment manufacturer to advise on fouling and the minimum velocity at which the exchanger should operate.

Laminar Flow

The other area suitable for the plate heat exchanger is that of laminar flow heat transfer. It has been pointed out already that the Paraflow can save surface by handling fairly viscous fluids in turbulent flow because the critical Reynolds number is low. Once the viscosity exceeds 20–50 cP, however, most plate heat exchanger designs fall into the viscous flow range. Considering only Newtonian fluids since most chemical duties fall into this category, in laminar ducted flow the flow can be said to be one of three types: (1) fully developed velocity and temperature profiles (ie, the limiting Nusselt case), (2) fully developed velocity profile with developing temperature profile (ie, the thermal entrance region), or (3) the simultaneous development of the velocity and temperature profiles.

The first type is of interest only when considering fluids of low Prandtl number, and this seldom exists with normal plate heat-exchanger applications. The third is relevant only for fluids such as gases that have a Prandtl number of about 1. Therefore, consider type 2.

As a rough guide for plate heat exchangers, the ratio of the hydrodynamic entrance length to the corresponding thermal entrance length is given by

$$\frac{L_{\text{th}}}{L_{\text{hyd}}} = 1.7\ \text{Pr}$$

Plate the heat transfer for laminar flow follows the Dittus—Boelter equation in this form:

$$\text{Nu} = c\left(\frac{\text{Re} \cdot \text{Pr} \cdot \text{De}}{L}\right)^{1/3}\left(\frac{\mu}{\mu_w}\right)^x$$

where L is the nominal plate length, c is the constant for each plate (usually in the range 1.86–4.50), and x is the exponent varying from 0.1 to 0.2 depending on plate type. For pressure loss, the friction factor can be taken as $f = a/\text{Re}$ where a is a constant characteristic of the plate.

It can be seen that for heat transfer, the plate heat exchanger is ideal because the value of d is small and the film coefficients are proportional to $d^{-2/3}$. Unfortunately, however, the pressure loss is proportional to d^{-4} and the pressure drop is sacrificed to achieve the heat transfer.

From these correlations, it is possible to calculate the film heat-transfer coefficient and the pressure loss for laminar flow. This coefficient combined with the metal coefficient and the calculated coefficient for the service fluid together with the fouling resistance then are used to produce the overall coefficient. As with turbulent flow, an allowance has to be made to use the LMTD to allow for either end effect correction for small plate packs and/or concurrency caused by having concurrent flow in some passages. This is particularly important for laminar flow since these exchangers usually have more than one pass.

Beyond Liquid–Liquid

Over many years, APV has built up considerable experience in the design and use of Paraflow plate heat exchangers for process applications that fall outside the normal turbulent flow that is common in chemical operations. The Paraflow, for example, can be used in laminar flow duties, for the evaporation of fluids with relatively high viscosities, for cooling various gases, and for condensing applications where pressure drop parameters are not overly restrictive.

Condensing

One of the most important heat-transfer processes if possible, is the condensation of vapors—a duty that often is carried out on the shell side of a tubular exchanger but is entirely feasible in the plate-type unit. Generally speaking, the determining factor is pressure drop.

For those condensing duties where permissible pressure loss is less than one pound per square inch, there is no doubt but that the tubular unit is most efficient. Under such pressure drop conditions, only a portion of the length of a Paraflow plate would be used and substantial surface area would be wasted. However, when less restrictive pressure drops are available the plate heat exchanger becomes an excellent condenser since very high heat-transfer coefficients are obtained and the condensation can be carried out in a single pass across the plate.

The pressure drop of condensing steam in the passages of plate heat exchangers has been investigated experimentally for a series of different Paraflow plates. As indicated in Figure 18, which provides data for a typical unit, the drop obtained is plotted against steam flow rate per passage for a number of inlet steam pressures.

It is interesting to note that for a set steam flow rate and a given duty, the steam pressure drop is higher when the liquid and steam are in countercurrent rather than cocurrent flow. This is due to differences in temperature profile.

Figure 19 shows that for equal duties and flows, the temperature difference for countercurrent flow is lower at the steam inlet than at the outlet with most of the steam condensation taking place in the lower half of the plate. The reverse holds true for cocurrent flow. In this case, most of the steam condenses in the top half of the plate, the mean vapor velocity is lower, and a reduction in pressure drop of 10–40% occurs. This difference in pressure drop becomes lower for duties where the final approach temperature between the steam and process fluid becomes larger.

The pressure drop of condensing steam therefore is a function of steam flow rate, pressure, and temperature difference. Since the steam pressure drop affects the saturation temperature of the steam, the mean temperature difference in turn becomes a function of steam pressure drop. This is particularly important when vacuum steam is being used since small changes in steam pressure can give significant changes in the temperature at which the steam condenses.

By using an APV computer program and a Martinelli—Lockhart-type approach to the problem, it has been possible to correlate the pressure loss to a high degree of accuracy. Figure 20 cites a typical performance of a steam heated Series R4 Paraflow. From this experimental run during which the exchanger was equipped with only a small number of plates, it can be seen that for a 4–5-psi pressure drop and above, the plate is completely used. Below that figure, however, there is insufficient pressure drop available to fully use the entire plate and part of the surface therefore is flooded to reduce the pressure loss. At only one psi allowable pressure drop, only 60% of the plate is used for heat transfer, which is not particularly economic.

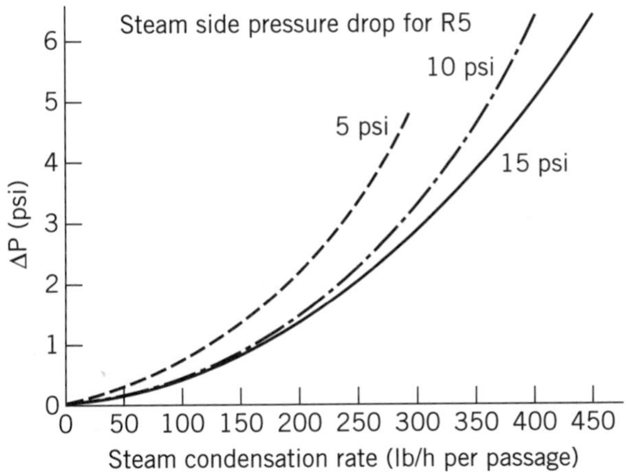

Figure 18. Steam-side pressure drop for R5.

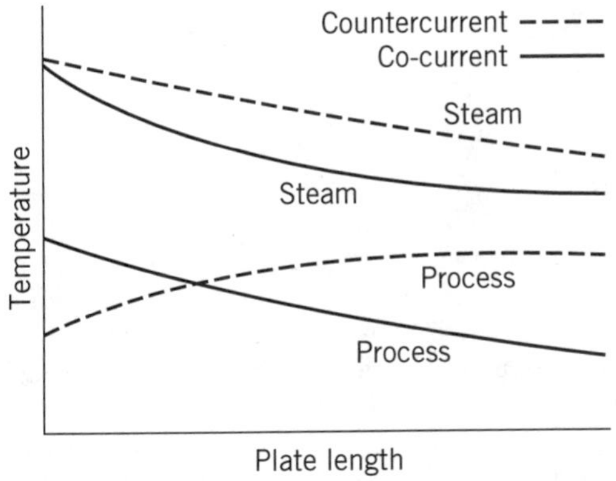

Figure 19. Temperature profile during condensation of steam.

Available pressure loss, psi	1	2	3	4
Total duty, Btu/h	207,000	256,000	320,000	333,000
Fraction of plate flooded	40	30	4	0
Effective overall heat-transfer coefficient, clean	445	520	725	770
Pressure loss, psi	1	2	4	4.5

Figure 20. Steam heating in an R4 Paraflow (water flow rate 16,000 lb/h, inlet water temperature 216°F, inlet steam temperature 250°F, total number of plates 7).

This example, however, well illustrates the application of a plate heat exchanger to condensing duties. If sufficient pressure loss is available, then the plate type unit is a good condenser. The overall coefficient of 770 Btu/h · ft^2 · °F for 4–5-psi pressure loss is much higher than a coefficient of 450–500 Btu/h · ft^2 · °F, which could be expected in a tubular exchanger for this type of duty. However, the tubular design would, for shell-side condensation, be less dependent on available pressure loss and for a 1-psi drop, a 450–500 Btu overall coefficient still could be obtained. With the plate, the calculated coefficient at this pressure is 746 Btu but the effective coefficient based on total area is only 60% of that figure or 445 Btu/h · ft^2 · °F.

Gas Cooling

Plate heat exchangers also are used for gas cooling with units in service for cooling moist air, hydrogen, and chlorine. The problems are similar to those of steam heating since the gas velocity changes along the length of the plate due either to condensation or pressure fluctuations. Designs usually are restricted by pressure drop, so machines with low pressure drop plates are recommended. A typical allowable pressure loss would be 0.5 psi with rather low gas velocities giving overall heat transfer coefficients in the region of 50 Btu/h · ft^2 · °F.

Evaporation

The plate heat exchanger also can be used for evaporation of highly viscous fluids when as a Paravap the evaporation occurs in the plate or as a Paraflash the liquid flashes after leaving the plate. Applications generally have been restricted to the soap and food industries. The advantage of these units is their ability to concentrate viscous fluids of up to 5000 cP.

CONCLUSION

It has been shown that the plate heat exchanger is a relatively simple machine on which to carry out a thermal design. Unlike the shell side of a tubular exchanger where predicting performance depends on baffle—shell leakage, baffle—tube leakage, and leakage around the bundle, it is not possible to have bypass streams on a Paraflow. The only major problem is that the pressure loss through the ports can cause unequal distribution in the plate pack. This is overcome by limiting the port velocity and by using a port pressure loss correlation in the design to allow for the effect of unequal distribution.

The flow in a plate also is far more uniform than on the shell side. Furthermore, there are no problems in calculation of heat transfer in the window, across the bundle, or allowing for dead spots as is the case with tubular exchangers. As a result, the prediction of performance is simple and very reliable once the initial correlations have been established.

BIBLIOGRAPHY

Adapted from:
J. D. Usher and A. Cooper, "Paraflow Seminar, Part One," *CPI Digest* **2**, 2–7 (1973). Copyrighted APV Crepaco, Inc. Used with permission.

APV CREPACO, INC.
Lake Mills, Wisconsin

HEAT EXCHANGERS: PLATE VERSUS TUBULAR

SELECTION

In forming a comparison between plate and tubular heat exchangers, there are a number of guidelines that will generally assist in the selection of the optimum exchanger for any application. In summary, these are

1. For liquid—liquid duties, the plate heat exchanger usually has a higher overall heat-transfer coefficient and often the required pressure loss will be no higher.
2. The effective mean temperature difference will usually be higher with the plate heat exchanger.
3. Although the tube is the best shape of flow conduit for withstanding pressure, it is entirely the wrong shape for optimum heat-transfer performance since it has the smallest surface area per unit of cross sectional flow area.
4. Because of the restrictions in the flow area of the ports on plate units, it is usually difficult, unless a moderate pressure loss is available, to produce economic designs when it is necessary to handle large quantities of low density fluids such as vapors and gases.
5. A plate heat exchanger will usually occupy considerably less floor space than a tubular for the same duty.

6. From a mechanical viewpoint, the plate passage is not the optimum and gasketed plate units are not made for operating pressures much in excess of 300 psig.
7. For most materials of construction, sheet metal for plates is less expensive per unit area than tube of the same thickness.
8. When materials other than mild steel are required, the plate will usually be more economical than the tube for the application.
9. When mild steel construction is acceptable and when a close temperature approach is not required, the tubular heat exchanger will often be the most economic solution since the plate heat exchanger is rarely made in mild steel.
10. Plate heat exchangers are limited by the necessity that the gasket be elastomeric. Even compressed asbestos fiber gaskets contain about 6% rubber. The maximum operating temperature therefore is usually limited to 500°F.

Heat-Transfer Coefficients

Higher overall heat-transfer coefficients are obtained with the plate heat exchanger compared with a tubular for a similar loss of pressure because the shell side of a tubular exchanger is basically a poor design from a thermal point of view. Considerable pressure drop is used without much benefit in heat transfer because of the turbulence in the separated region at the rear of the tube. Additionally, large areas of tubes even in a well designed tubular unit are partially bypassed by liquid and low heat-transfer areas thus are created.

Bypassing in a plate-type exchanger is less of a problem, and more use is made of the flow separation that occurs over the plate troughs since the reattachment point on the plate gives rise to an area of very high heat transfer.

For most duties, the fluids have to make fewer passes across the plates than would be required through tubes or in passes across the shell. Since a plate unit can carry out the duty with one pass for both fluids in many cases, the reduction in the number of required passes means less pressure lost due to entrance and exit losses and consequently, more effective use of the pressure.

Mean Temperature Difference

A further advantage of the plate heat exchanger is that the effective mean temperature difference is usually higher than with the tubular unit. Since the tubular is always a mixture of cross and contraflow in multipass arrangements, substantial correction factors have to be applied to the log mean temperature difference. In the plate heat exchanger where both fluids take the same number of passes through the unit the LMTD correction factor is usually in excess of 0.95. As is illustrated in Figure 1, this factor is particularly important when a close or relatively close temperature approach is required.

In practice, it is probable that the sea water flow rate would have been increased to reduce the number of shells in series if a tubular had to be designed for this duty. While this would reduce the cost of the tubular unit, it would result in increased operating costs.

Design Case Studies

Figure 2 covering a number of case studies on plate versus tubular design demonstrates the remarkable heat-transfer performance that can be obtained from Paraflow units. Even for low to moderate available pressure loss, the plate heat exchanger usually will be smaller than a corresponding tubular.

Because of the high heat-transfer rates, the controlling resistance usually is fouling, so an allowance of 20–50% extra surface has been made based on APV experience.

One limitation of the plate heat exchanger is that it is rarely made in mild steel; the most inexpensive material of construction is stainless. Therefore, even when the plate surface requirement is much lower than the tubular on some heat-transfer duties, the tubular will be less expensive when mild steel construction is acceptable. If a close temperature approach is required, however, the plate unit always will cost less and where stainless steel or more exotic materials are required for process reasons, the plate unit usually will cost less.

Physical Size

One important advantage of the plate over a tubular unit is that for a particular duty, the plate heat exchanger will be physically smaller and require far less floor space. This

	Tubular	Plate
Size	3 shells in series: 20-ft tubes, 1 shell-side pass, 4 tube passes, 2238 tubes, 12,100-ft^2 area	R10 Paraflow with 313 plates arranged for 3 process passes and 3 service passes, 3630 ft^2 of area
Overall coefficient	270 Btu/h · ft^2 · °F	700 Btu/h · ft^2 · °F
MTD	7.72°F	9.0°F
Pressure loss	10–19 psi	10.8–8.8 psi

Figure 1. Duty: demineralized water/seawater. To cool 864,000 lb/h of water from 107°F to 82.4°F using 773,000 lb/h of 72°F sea water.

Case study	Fluid A (lb/h)	Fluid B (lb/h)	Duty (°F)	Tubular design pressure loss			Plate design pressure loss			
				A	B	ft^2	A	B	ft^2	type
A	13,380 hydrocarbon	53,383 water	320 A 120 120 B 92	0.42	3.05	301	0.3	2.2	92	HX
B	864,000 water	77,300 sea water	107 A 82.4 99.1 B 71.6	19	19	12,100	10.8	8.8	3,630	R10
C	17,500 solvent	194,000 water	140 A 104 95 B 79	2.9	4.5	1,830	3.0	3.7	445	R5
D	148,650 desalter effluent	472,500 salt water	222 A 100 106 B 68	10.8	8.0	1,500	4.5	10	380	R4

Figure 2. Case studies of plate heat exchangers and tubular designs. All tubular designs were carried out with the aid of the HTRI program ST3.

is shown most graphically in Figure 3, where a Series R405 Paraflow is being installed next to tubular units that need twice the amount of space for the identical duty. It is further illustrated in the volumetric comparisons of Figure 4.

If the tightened plate pack of a Paraflow is regarded as a rectangular box, each cubic foot contains 50–100 ft^2 of heat-transfer area according to the type of plate used. Allowing for metal thickness, the contained liquid is some 80% of this volume. Thus, the total of both heating and cooling media is about 5 gal. Expressed in another way, the liquid hold up per square foot for each stream varies according to plate type from about 0.06 gal down to 0.03 gal.

By comparison, one cubic foot of tubular exchanger of equilateral triangular pitch with a tube pitch-tube diameter ratio of 1.5 has a surface area of 10 ft^2 for 2-in.-OD tubes or 40 ft^2 for 1/2-in.-OD tubes. The average contained liquid is proportionately 0.27 gal/ft^2 down to 0.07 gal/ft^2 of heat-exchanger area with no allowance for the headers. If the heat-transfer coefficient ratio between plate and tubular is conservatively taken as 2, the plate exchanger volume to meet a given duty varies from 1/10 to 1/3 of that of the tubular. For a lower tube pitch-tube diameter ratio of 1.25, the comparison becomes 1/7 to 1/5. These facts demonstrate why the Paraflow plate heat exchanger is referred to as "compact."

Figure 3. At the Glidden Durkee Baltimore plant, a Series R405 Paraflow is being readied to cool 1000 gpm of deionized water. This unit requires only half the space of a tubular exchanger for the same duty.

Thermal Limitations

While it would appear offhand that the plate heat exchanger always provides a better performance at usually a lower price than the tubular exchanger, consideration must be given to the thermal as well as mechanical limitations of the plate-type machine. These usually are based on allowable pressure loss.

For single-phase liquid—liquid duties, the plate heat exchanger can be designed for moderately low pressure loss. However, if the pressure loss across any plate passage that has liquid flowing downward is lower than the available liquid static head, the plate will not run full and performance therefore will be reduced. This is termed low plate rate. Use of a plate below the minimum plate rate is inadvisable since it causes a wastage of surface area and results in unreliable operation. It is, however, possible to function below the minimum plate rate in a single-pass arrangement by making sure that the low plate rate is operated with a climbing liquid flow.

These problems are not quite so severe with a tubular exchanger and therefore, operation at a moderately lower available pressure loss is possible.

Conclusion

To summarize, the gasketed plate heat exchanger generally will be the most economical heat exchanger for liquid—liquid duties provided the material of construction is not mild steel and provided the operating temperatures and pressures are below 500°F and 300 psig respectively. For other types of duty such as gas cooling, condensation or boiling, the plate heat exchanger can be a very econom-

Plate			Tube			
Plate pitch (in.)	Heat-transfer area per ft^3 of exchanger (ft^2)	Liquid contained per ft^2 of heat-transfer area (gal)	Tube diameter (in.)	Ratio tube pitch tube diameter	Heat-transfer area per ft^3 of exchanger (ft^2)	Liquid containment per ft^2 of heat-transfer area (average of both sides) (gal)
1/4	50	0.06	2	1.5	10	0.32
1/8	100	0.03	1/2	1.5	40	0.085

Figure 4. Volumetric comparison, plate versus tube.

ical type of unit if the pressure loss allowed is sufficient to utilize the very high heat-transfer performance characteristics of the plate.

RECOVERY OF PROCESS HEAT

Since it is quite clear that never again will energy be as inexpensive as in the past, it therefore is necessary to conserve this natural resource—to recover more of the process heat that currently is dissipated to waterways and atmosphere. Some of this heat can be recovered with the aid of high-efficiency heat exchangers, which can operate economically with a close temperature approach at relatively low pumping powers. One type of unit that is particularly suited for this duty is the APV Paraflow plate heat exchanger. For many applications, this equipment can transfer heat with almost true countercurrent flow coupled with high coefficients to provide efficient and inexpensive heat transfer.

Unfortunately, the plate heat exchanger has been considered by many chemical engineers to be suitable only for hygienic heat-transfer duties. This, of course, is not so. Nowadays, many more plate type units are sold for chemical and industrial use than are sold for hygienic applications. A further mistake is the claim that the plate heat exchanger can be used only for duties when the volumetric flows of the two fluids are similar. Again, this is not true, although it must be stated that the plate heat exchanger is at its most efficient when flows are similar.

Basic Considerations

Since plate and tubular heat exchangers are the most widely used types of heat-transfer equipment, it is well to draw a brief comparison of their respective heat-recovery capabilities for the energy-conscious plant manager.

While the plate heat exchanger does have mechanical limitations with regard to withstanding high operating pressures above 300 psig, it is more efficient thermally than shell and tube units, especially for liquid—liquid duties. In many waste heat recovery applications, however, both pressure and temperature generally are moderate and the plate-type unit is an excellent choice since its thermal performance advantage becomes very significant for low-temperature approach duties. Higher overall-heat transfer coefficients are obtained with the plate unit for a similar loss of pressure because the shell side of a tubular basically is a poor design from a thermal point of view. Pressure drop is used without much benefit in heat transfer on the shell side due to the flow reversing direction after each cross pass. In addition, even in a well-designed tubular heat exchanger, large areas of tubes are partially bypassed by liquid and areas of low heat transfer thus are created. Conversely, bypassing of the heat transfer area is far less of a problem in a plate unit. The pressure loss is used more efficiently in producing heat transfer since the fluid flows at low velocity but with high turbulence in thin streams between the plates.

For most duties, the fluids also have to make fewer passes across the plates than would be required either through tubes or in passes across the shell. In many cases, the plate heat exchanger can carry out the duty with one pass for both fluids. Since there are fewer passes, less pressure is lost as a result of entrance and exit losses and the pressure is used more effectively.

A further advantage of the plate heat exchanger is that the effective mean temperature difference usually is higher than with the tubular. Since in multipass arrangements the tubular is always a mixture of cross and contraflow, substantial correction factors have to be applied to the log mean temperature difference. In the plate unit for applications where both fluids take the same number of passes through the exchanger, the LMTD correction factor approaches unity. This is particularly important when a close or even relatively close temperature approach is required.

Thermal Performance Data

Although the plate heat exchanger now is widely used throughout industry, precise thermal performance characteristics are proprietary and thus unavailable. It is possible, however, to size a unit approximately for turbulent flow liquid—liquid duties by use of generalized correlations that apply to a typical plate heat exchanger. The basis of this method is to calculate the heat-exchanger area required for a given duty by assuming that all the available pressure loss is consumed and that any size unit is available to provide this surface area.

For a typical plate heat exchanger, the heat transfer can be predicted in turbulent flow by the following equation

$$\frac{hD_e}{k} = 0.28\left(\frac{GD_e}{\mu}\right)^{0.65}\left(\frac{Cp\mu}{k}\right)^{0.4} \tag{1}$$

$$f = 2.5\left(\frac{GD_e}{\mu}\right)^{-0.3} \tag{2}$$

$$\Delta P = \frac{2fG^2L}{g\rho D_e} \tag{3}$$

The pressure loss can be predicted from equations 2 and 3.

Obviously, equation 1 cannot represent accurately the performance of the many different types of plate heat exchangers that are manufactured. However, plates that have higher or lower heat-transfer performance than given in equation 1 usually will give correspondingly higher or lower friction factors in equation 2. Experience indicates that the relationship for pressure loss and heat transfer is reasonably consistent for well-designed plates. In the Appendix, equations A1–A3 are further developed to show that it is possible for a given duty and allowable pressure loss to predict the required surface area. This technique has been used for a number of years to provide an approximate starting point for design purposes and has given answers to ±20%. For accurate designs, however, it is necessary to consult the manufacturer.

Heat-Recovery Duties

In any heat-recovery application, it always is necessary to consider the savings in the cost of heat against the cost of the heat exchanger and the pumping of fluids. Each case must be treated individually since costs for heat, electricity, pumps, etc will vary from location to location.

One point is obvious. Any increase in heat recovery, and thus heat load, results in a decrease in LMTD and considering a constant heat-transfer coefficient, subsequently in the cost of the exchanger. This effect is tabulated in Figure 5. Because the cost of an exchanger increases considerably for relatively small gains in recovered heat above the 90% level, such applications, even with the plate heat exchanger, must be studied closely to verify economic gain. The economic break-even point is far lower for a tubular exchanger. Situations where it is advantageous to go above 90% recovery usually involve duties where higher heat recovery reduces subsequent heating or cooling of the process stream. High steam or refrigeration costs therefore justify these higher heat recoveries.

As shown in Figure 5, the cost of increasing heat recovery from 85 to 90% at a constant pressure loss of 12 lb/in.^{2} is $2600. From a practical standpoint, going from 90 to 95% requires a significantly higher pressure loss and nearly doubles the exchanger cost. However, even with this 95% heat recovery and assuming steam costs at $6.00/1000#, payback on the plate heat exchanger would take 530 hs.

The plate heat exchanger thus provides a most eco-Xnomic solution for recovering heat (Fig. 6). This degree of heat recovery cannot be achieved economically in a tubular exchanger since the presence of cross flow and multipass on the tube side causes the LMTD correction factors to become very small or, alternately, requires more than one shell in series. This is shown in the Figure 7 comparison.

As detailed, this example illustrates that the plate heat exchanger has considerable thermal and therefore price advantage over the tubular exchanger for a heat recovery of 70%. Since the overall heat-transfer coefficient and the effective mean temperature difference both are much higher for the plate unit, reduced surface area is needed. Furthermore, because of cross-flow temperature difference problems in the tubular, three shells in series were needed to handle the duty within the surface area quoted. Using only two shells would have resulted in a further 40% increase in surface area.

The small size of the plate heat exchanger also results in a saving of space and a lower liquid holdup. For this type of heat recovery duty, a stainless-steel plate heat exchanger almost always will be less expensive than a mild steel tubular unit. Although the tubular exchanger physically will be capable of withstanding higher temperatures and pressures, there is a considerable and for the most part

			12 lb/in.^{2} Pressure loss					25 lb/in.^{2} Pressure loss				
Heat recovered (Btu/h)	%	Temperature (°F)	LMTD	Factor	Actual MTD	Area (ft^2)	Relative price ($)	LMTD	Factor	Actual MTD	Area, (ft^2)	Relative price ($)
3,600,000	60	200 → 140 160 ← 100	40	0.985	39.4	85	5,500	40	0.985	39.4	74	5,400
4,500,000	75	200 → 125 175 ← 100	25	0.975	24.4	197	6,450	25	0.980	24.5	172	6,250
5,100,000	85	200 → 115 185 ← 100	15	0.965	14.4	425	8,400	15	0.965	14.4	362	7,850
5,400,000	90	200 → 110 190 ← 100	10	0.95	9.5	739	11,000	10	0.92	9.2	629	10,100
5,700,000	95	200 → 105 195 ← 100	5					5	0.83	4.16	1580	18,100

Hot liquid 60,000 lb/h of water at 200°F; cold liquid 60,000 lb/h of water at 60°F.

Figure 5. Effect of percentage heat recovery on PHE cost.

Figure 6. Two shrouded Paraflow units provide an optimum return on investment with energy and cost saving regeneration of 88 and 81%.

Duty: To heat 1,300,000 lb/h of water from 73 to 97°F using 1,300,000 lb/h of water at 107°F.
Available pressure loss 15 lb/in.^{2} for both streams.

Heat recovery = 70.5%
Number of heat-transfer units (HTU) = 2.4

	Plate	Tubular	
Heat-transfer area	4820	13,100	ft^2
Heat-transfer coefficient (clean)	734	386	Btu/h · ft^2 · °F
Heat-transfer coefficient (dirty)	641	271	Btu/h · ft^2 · °R
Effective mean temperature difference	10.0	8.7	°F
Pressure drop: hot fluid/cold fluid	4.6/4.7	15/15	lb/in^2
Fouling resistance: hot fluid/cold fluid	0.0001/0.0001	0.0005/0.0005	(Btu/h · ft^2 · °F)$^{-1}$
Pass arrangement	1/1	4 tube side 3 shells in series	baffled
Approximate price	$95,000 Stainless steel	$135,000 Mild steel	

This example demonstrates that quite high heat-transfer coefficients can be obtained from a PHE with only a moderate pressure loss.

Figure 7. Heat recovery—comparison between plate and tubular heat exchangers.

unnecessary penalty to pay for these features both in price and size.

For heat recovery duties in excess of 70%, the plate heat exchanger will become increasingly more economical than the tubular.

Typical Applications

One of the more common uses of regeneration is found in many of the nation's breweries, where it is necessary to cool huge amounts of hot wort before it is discharged to fermentation tanks. Typical in scope is an operation where 850 barrels per hour of wort (220,000 lb/h) are cooled from 200 to 50°F by means of 242,000 lb/h (R = 1.1:1) of water entering at 35°F and being heated to 165°F. As a result; approximately 33,000,000 Btu/h are saved, there is an excellent water balance for use throughout the brewery, and no water is discharged to the sewer.

For a dairy, regeneration usually involves the transfer of heat from pasteurized milk to cooler raw milk entering the system. After initially heating 100,000 lb/hr of milk from 40 to 170°F by high temperature, short time pasteurization, 90% regeneration permits cooling of the milk back down to 40°F with a savings of 11,700,000 Btu/h. Only 10%

of the total heat or cooling must be supplied, and no cooling medium such as city water is used and discarded.

Chemically, there are many and varied regenerative applications. For desalination, APV has supplied a number of Paraflows, which are virtually perpetual motion machines. These units achieve 95% regeneration in heating 87,000 lb/h (175 gpm) from 70 to 197°F while cooling a secondary stream flowing at 78,000 lb/h from 214 to 73°F. Savings in the Btu load in this case are 11,050,000 per hour.

While flow rates in hot oil applications are quite low in comparison, temperatures are very high. It is possible to cool 400 gph of vegetable oil from 446 to 226°F while heating 400 GPH of oil from 200 to 400°F with 80% regeneration.

In the production of caustic soda, where very corrosive product streams are encountered, Paraflows with nickel plates are being used to cool 10,000 lb/h of 72% NaOH from 292 to 210°F while heating 14,000 lb/h of 50% NaOH from 120 to 169°F.

Cost and Efficiency

To examine a hypothetical case of Paraflow regeneration from the viewpoint of efficiency and dollar savings, consider the following process duty:

Duty Heat 100 GPM of fluid 1 from 50°F to 200°F while cooling 100 gpm of fluid 2 from 200 to 50°F (see Tables 1–3)

Under ordinary conditions, steam required to heat fluid 1 would be in the nature of 7500 lb/h with an equivalent cooling requirement for the second stream of 625 tons of refrigeration. Using 85% regeneration, however, the energy needs are drastically reduced.

At this point in the process, fluid 2 in a conventional system has been cooled to only 90°F by means of 85°F city water and will require supplemental refrigeration for final cooling to 50°F. At the same time, fluid 2 in an APV regenerative system has been cooled to 72.5°F by means of 85% regeneration and must be cooled further to 50°F.

Duty	Cool from 90 to 50°F	Cool from 72.5 to 50°F
Supplemental refrigeration required	$\frac{50{,}000 \times 40}{12{,}000}$ = 166.5 tons	93.5 tons
Assuming average electrical cost of 4¢/kWh	166.5 × 3.5 = 583.8 kWh	93.5 × 3.5 = 327.25 kWh
Annual supplemental refrigeration cost	$\frac{4}{100} \times 583.8 \times 24 \times 365$ = $204,571	$\frac{4}{100} \times 327.25 \times 24 \times 365$ = $114,669

Definition

The number of heat-transfer units is defined as the temperature rise of fluid one divided by the mean temperature difference. A further term for the HTU is the temperature ratio (TR), ie,

$$\text{HTU} = \frac{t_1 - t_2}{\text{MTD}}$$

where MTD = mean temperature difference.

Table 1. Energy Needs, Fluid 1

	Conventional	85% Regeneration
	Heat 100 GPM 50 → 200°F	Heat 100 GPM 50 → 177.5°F by cooling fluid 2 from 200 to 72.5°F
Steam required	$\frac{50{,}000 \times 150}{10_3}$ = 7500 lb/h to heat to 200°F	$\frac{50{,}000 \times 22.5}{10^3}$ = 1125 lb/h suppemental heat to raise temperature to 200°F
Assuming average steam cost: $6.00 per 1000#		
Steam cost/h	7.5 × $6.00 = $45.00/h	1.125 × $6.00 = $6.75/h
Annual steam cost	$45.00 × 24 × 365 = $394,200	$6.75 × 24 × 365 = $59,130

Table 2. Energy Needs, Fluid 2

	Conventional	85% Regeneration
	Cool 100 GPM 200 → 50°F	Cool 100 GPM 200 → 72.5°F by heating fluid 1 from 50 to 177.5°F
Equivalent refrigeration required	$\frac{7{,}500{,}000}{12{,}000}$ = 625 tons to cool to 50°F	.15 × 625 = 93.5 tons supplemental cooling to lower temperature to 50°F
Using available 85°F cooling-tower water	200 gpm to cool 200 → 90°F	None
	200 × 60 = 12,000 GPH	None
Assuming average water cost of $0.05 per 1000 gal		
Water cost/h	12 × .05 = $.60/h	None
Annual water cost	$.60 × 24 × 365 = $5250	None

Table 3. Cost Comparison

	Original Cost, $	Regenerative System Cost, $	Regenerative Savings, $
Steam	394,200	59,130	335,070 annually
Tower cooling water	5,250	0	5,250 annually
Refrigeration	204,571	114,669	89,902 annually
	604,021	173,799	430,222

Note: Capital cost of a Paraflow plate heat exchanger for the above duty would be approximately $17,000.

APPENDIX

For heat transfer in a typical PHE with turbulent flow one can write the heat transfer performance in terms of a dimensionless Dittus—Boelter equation

$$\frac{hD_e}{k} = 0.28\left(\frac{GD_e}{\mu}\right)^{0.65}\left(\frac{Cp\mu}{k}\right)^{0.4}\left(\frac{\mu}{\mu_w}\right)^{0.14} \qquad \text{(A1)}$$

For applications in turbulent flow it is usually sufficiently accurate to omit the Sieder—Tate viscosity ratio and therefore the equation reduces to

$$\frac{hD_e}{k} = 0.28\left(\frac{GD_e}{\mu}\right)^{0.65}\left(\frac{Cp\mu}{k}\right)^{0.4} \qquad \text{(A2)}$$

The pressure drop can be predicted in a similar exchanger by equations A3 and A4.

$$f = 2.5\left(\frac{GD_e}{\mu}\right)^{-0.3} \qquad \text{(A3)}$$

$$\Delta P = \frac{2fG^2L}{g\rho D_e} \qquad \text{(A4)}$$

To solve these equations for a particular duty it is necessary to know G, L, and D_e, and it is shown below how these can be eliminated to produce a general equation.

For any plate,

$$G = \frac{m}{A_f} \qquad \text{(A5)}$$

where m is the total mass flow rate and A_f the total flow area.

D_e is defined by APV as four times the flow area in a plate divided by the wetted perimeter. Since the plate gap is small compared with the width, then:

$$D_e = \frac{4 \times A_f}{\text{wetted perimeter}} \qquad \text{(A6)}$$

but since $A_s = L \times$ wetted perimeter where A_s is the total surface area:

$$L = \frac{A_s}{4A_f} D_e \qquad \text{(A7)}$$

and it is possible to eliminate L from equation A4.

Similarly, by substituting G using equation A5 in equations A2, A3, and A4 and then by rearranging the equations to eliminate A_f, it is possible to arrive at equation A8 for the film heat-transfer coefficient.

$$h = \frac{J_2}{J_1^{0.241}}\left(\frac{m}{A_s}\right)^{0.241} D_e^{-0.28} \qquad \text{(A8)}$$

where

$$J_1 = \frac{2.5\,\mu^{0.3}}{2g\rho\Delta P}$$

and

$$J_2 = 0.28\left(\frac{Cp\mu}{k}\right)^{0.4} k\mu^{-0.65}$$

That is, the film heat-transfer coefficient is expressed only in terms of the surface area and equivalent diameter. A computer program solves equation A8 using a constant value of D_e and using an empirical factor Z to account for port pressure loss and other deviations from equations A2–A4. The equation is solved using the HTU approach where

$$\text{HTU} = \frac{(t_1 - t_2)}{\text{MTD}} = \frac{UA_s}{m_1c_1}$$

That is, the number of heat-transfer units is the temperature rise of fluid 1 divided by the mean temperature difference.

U is defined as

$$\frac{1}{U} = \frac{1}{h_1} + \frac{1}{h_2} + \text{metal resistance}$$

where h_1 and h_2 are calculated from an empirical modification of the constants in equation A8. The powers in equation A8 are not modified.

BIBLIOGRAPHY

Adapted from:

J. D. Usher and A. Cooper, "Paraflow Seminar, Part Two," *CPI Digest* **2**, 6–10 (1974).

J. D. Usher and A. Cooper, "Thermal Performance," *CPI Digest* **2**, 10–14 (1974).

J. D. Usher and A. Cooper, "Recovery of Process Heat," *CPI Digest* **3**, 2–7 (1974).

APV Crepaco, Inc.
Lake Mills, Wisconsin

HEAT EXCHANGERS: SCRAPED SURFACE

BASIC CONSIDERATIONS

The scraped-surface heat exchanger (SSHE) is a specialized piece of heat-transfer equipment that was patented around 1926 by Clarence Vogt in an effort to develop a more efficient freezer for making ice cream. The design incorporated a scraping action to prevent buildup of frozen ice cream on heat-transfer surfaces. The concept was successful. Thermal efficiency was improved and production capacity greatly increased.

Since that time, the SSHE has become essential for numerous processes in the dairy and food industries. Many of these applications are similar to the ice cream problem in that, as the product is heated or cooled, it fouls heat-transfer surfaces and reduces efficiency. Such applications, for example, include cooling peanut butter or plasticizing shortening and margarine. In each instance, fat crystals that form are like the ice crystals found in ice cream. Without the scraping of the heat exchanger sweeping the solidified fats away from the surface, heat-exchange efficiency would fall off very rapidly. Conversely, in heating applications where a product tends to "burn on," the scraping action removes product from the surface before fouling can occur. Typical heating examples are aseptic cheese sauces and aseptic puddings where processing temperatures approach 300°F (150°C).

The modern SSHE also is capable of processing products containing particulates. Currently, standard units can handle 1/2-in. (13-mm) particles without excessive breakage while special designs are available to accommodate 1-in. (26-mm) particulates.

While more expensive than other types of heat exchanger, the SSHE is the best and only thermal approach in hundreds of applications where high viscosity, large particulates, crystallization, and burn-on are problems that must be considered.

A scraped-surface heat exchanger (Fig. 1) basically consists of a jacketed cylinder fitted with a rotating shaft on which scraper blades are mounted. Product is pumped through the cylinder while a heating or cooling medium is circulated in the annular space between the cylinder and the jacket. The blades are fixed to pins that allow them to swing freely. No springs are necessary since centrifugal force holds the blades in position against the inside of the cylinder wall as product constantly is swept away from the heat-transfer surface and new product exposed to treatment.

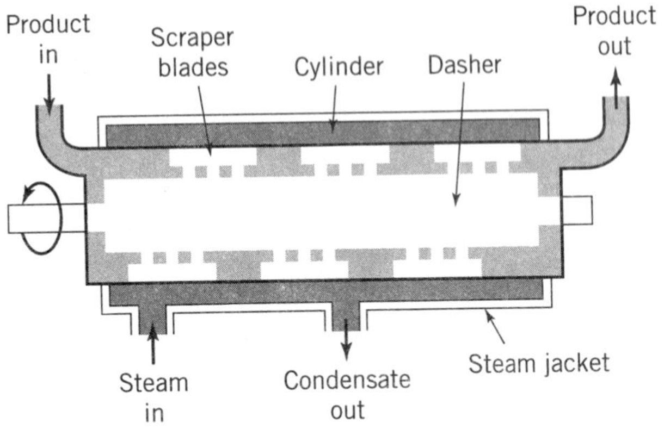

Figure 1. Cutaway of horizontal SSHE.

The standard horizontal exchanger generally has from one to three independently functioning jacketed cylinders mounted on a heavy steel base (Fig. 2, 3). A stainless-steel casing covers the cylinders, base and drives to form a completely enclosed system. Vertical units also are available for use when floor space is at a premium (Fig. 4).

Many processes require more than one unit, and in such cases where multiple cylinders are used, product always should be piped in series. Heat transfer will be higher because of the higher flow through each cylinder. Parallel

Figure 2. Model 2HD-648 SSHE with water flush seals.

Figure 3. A pair of scraped—surface heat exchangers provide an added dimension to an aseptic system by permitting the processing of fluids containing particulates.

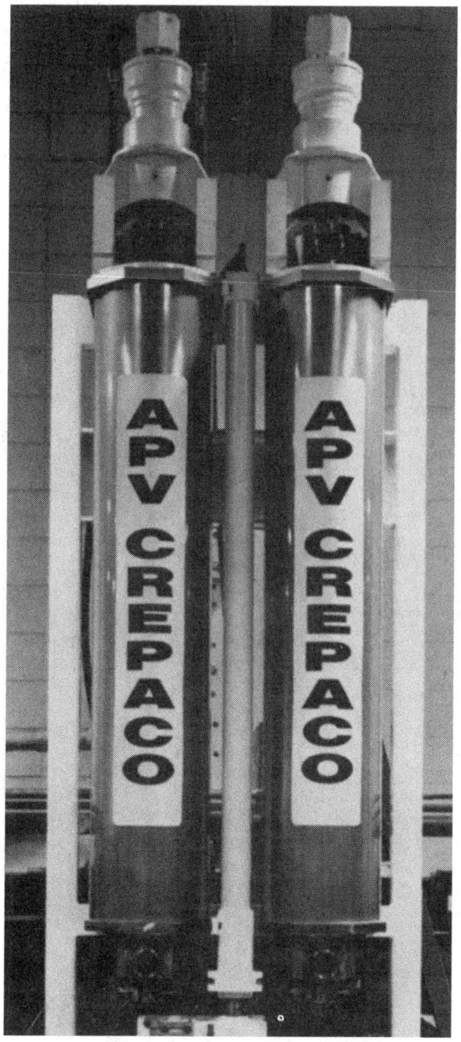

Figure 4. Vertical Model VExHD-884 SSHE.

flow arrangements fail because there is no way to ensure equal flow to each cylinder unless individual pumps are used for each circuit. It should be noted that a scraped surface heat exchanger does not do any pumping. A pump is required to move product through the unit.

Dashers

The shaft which carries the scraper blades is called a dasher by some manufacturers and a mutator by others—the "dasher" terminology being derived from early days when the exchanger initially was used as an ice cream freezer.

Dashers are engineered to achieve high heat-transfer coefficients with minimum power consumption and are supported by heavy-duty bearings located outside the product contact zone. Three standard designs (Fig. 5a, b, 6) provide product flow spaces of different sizes to accommodate different product viscosities, dwell time, level of blending, and size of particulates. For margarine or plasticizing applications, dashers with internal water circulation reduce adhesion of product to the dasher surface.

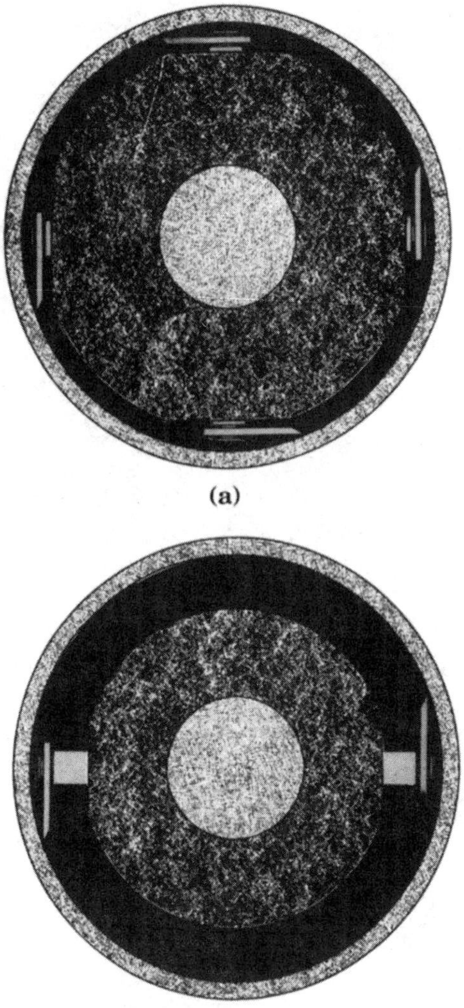

(a)

(b)

Figure 5. Dasher—blade assemblies cutaway. (**a**) Series 55; (**b**) Series 45.

Typical dasher speeds vary from 60 to 420 rpm with standard motors providing a choice of three drive methods—direct-driven hydraulic, belt-driven electric, or direct-driven gearhead.

Scraper Blades

Designed to promote the rapid removal of product from cylinder walls while enhancing product agitation and mixing, scraper blades are available in a selection of materials and configurations. Most common materials are stainless steel and plastics since the blade is the wearing part and as such, must be softer than the cylinder wall or lining.

Blade selection generally is determined by product temperature, pressure, and formulation as well as by the cylinder material and service media being used.

Heat-Exchange Cylinders

To provide optimum performance and economy of operation, heat-exchange cylinders are available in a selection of sizes and materials of construction.

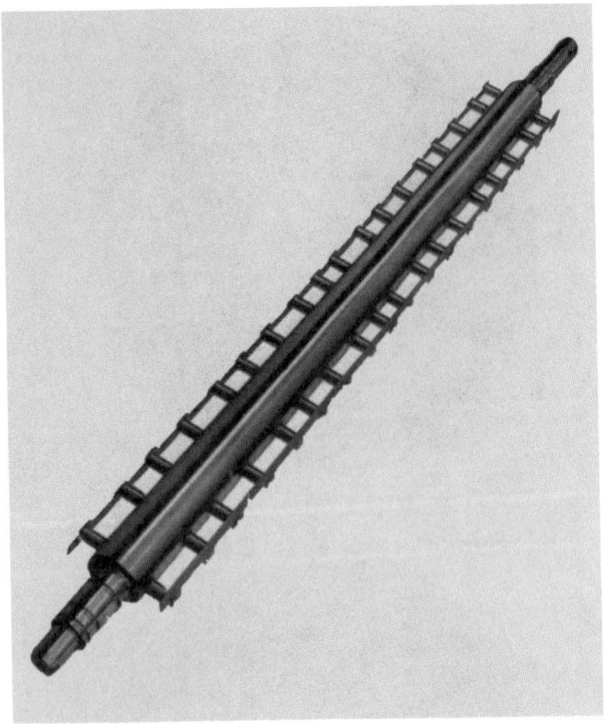

Figure 6. Series 30 dasher for viscous products or particulates.

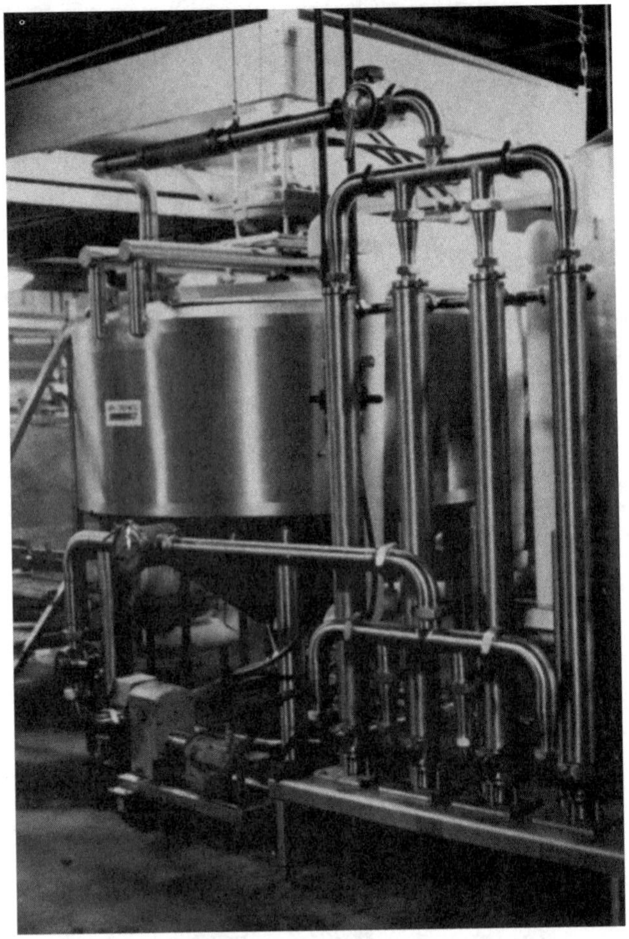

Figure 7. Four-barrel vertical scraped-surface heat exchanger is designed to cool meat gravies from 190 to 70°F. Unit capacity is 5000 lb/h.

The most common diameter for the scraped SSHE cylinder is 6 in. (152 mm) with lengths established at 48 in. (1220 mm) and 72 in. (1830 mm). There are, however, other sizes available. Some are 4 in. (102 mm) in diameter with lengths of 60 in. (1520 mm) (Fig. 7) and 120 in. (305 mm) while others are of 8 in. (203-mm) diameter and lengths to 84 in. (2130 mm).

The most common materials of construction are stainless steel, nickel with or without chrome plating, and, more recently, a bimetallic combination.

Since the cost per square foot of heat-transfer surface is higher for a SSHE than other types of heat exchangers, it is essential that cylinder material with the highest feasible heat-transfer coefficient be selected. This selection, however, must be tempered by consideration of the compatibility between cylinder and scraper blade materials and the susceptibility of these materials to acid attack and corrosion.

Table 1. Heat Transfer Coefficients of SSHE Cylinder Metals

Material	K, Btu/h · °F · in.
Nickel	52
Chrome nickel	52
Bimetallic	31
Stainless steel	9.4

As charted in Table 1, nickel exhibits the best thermal conductivity while stainless steel is the least conductive. Even so, nickel is not suitable for all applications. It is relatively soft when compared to typical metal scraper blade materials and would wear rapidly if this combination were to be used. Since SSHE always are designed to make the blades rather than the more expensive cylinders the wearing part, plastic blades must be used with nickel cylinders if only for reasons of economics and reduced downtime. For the same reasons, only plastic blades are run on stainless-steel cylinders.

To retain the superior heat-transfer characteristics of nickel while benefiting from the extended durability of steel blades, nickel cylinders may be chrome-plated. The scraping surface then becomes the hardest material in use and can accommodate any of the common blade materials. The sole drawback is that the chrome is sensitive to salt and acid attack.

Bimetallic cylinders offer some advantages over those of any other material, whether nickel, chromed nickel, or stainless. Produced from two different materials, the bimetallic cylinder usually is made by centrifugally casting a hard, corrosion-resistant alloy inside a tube that has high tensile strength and thermal conductivity approaching nickel. While the lining is not quite as hard as chrome, it is hard enough to withstand abrasion from both stainless-steel and plastic blades. One disadvantage other than a slight loss in heat transfer is that the inner alloy coating is susceptible to strong acid corrosion.

The selection of a cylinder with a compatible blade material is compounded by the products to be processed and the cleaning procedures used. For example, on slush freezing applications or where media temperature is below freezing, plastic blades can be abraded within hours as hard ice is scraped away. And as already noted, despite being the hardest cylinder material in use, chrome plated nickel is subject to attack by common acid CIP solutions and does not stand up well to salty products over extended periods of time. Table 2 shows the difference in corrosion for various materials and acids, and it is worth noting that it is the cleaning regimen rather than the product that causes problems in some cases. A change in CIP procedures may well extend the life or allow the use of nickel or chrome nickel cylinder material. In effect, much thought and experience must go into selecting the best cylinder and blade combinations—a selection best left up to the equipment manufacturer.

Cylinder Jackets—Media

The three common media used in a SSHE are water, steam, and refrigerants such as ammonia and Freon. Each requires a slightly different cylinder jacket design for optimum performance.

The water jacket is designed with a small spacing between the jacket and heat exchange cylinder. This induces high liquid media velocities that cause turbulence, which in turn, improves heat-transfer efficiency. It also minimizes fouling on the media side. Normally, countercurrent flow is recommended between product and media. In many cases, however, because of the high flow rate of media over product and the high temperature differences between media and product, the difference in performance between countercurrent and cocurrent flow is small.

On steam units, media enters the jacket via a header that distributes the steam over the entire length of the cylinder. Condensate runs to the bottom of the jacket, where it is collected and removed by a steam trap. Direction of product flow is immaterial.

When ammonia or Freon is used, the refrigerant is handled differentially by various manufacturers. Most exchanger designs are of the flooded type with media being fed at the bottom of the jacket and boiling occurring within the jacket. Since the key to efficient heat transfer is to maintain a wetted surface on the cylinder, not all of the refrigerant is allowed to evaporate. The combination of gas and liquid is carried over into a surge tank, where the phases are separated and the liquid recycled to the bottom of the jacket. Liquid level is maintained by a valve that allows more liquid to enter the surge tank to replace that which has been evaporated.

One advantage of the flooded arrangement as illustrated in Figure 8 is the incorporation of a quick shutoff valve for instant stop/instant start. Although the valve is open during normal operation, it may be closed when a freeze-up is imminent or the operator want to stop cooling. Since the vapor cannot go up and out, it expands and rapidly pushes the liquid refrigerant down and away from the cylinder. When liquid no longer touches the cylinder and the surface no longer is wetted, all freezing stops. To restart the process, the valve is opened and with liquid quickly filling the jacket area, freezing begins again.

The primary advantage of the SSHE over other types of heat-transfer equipment is its ability to accommodate product viscosities of 500,000 cP and higher and to handle particulates up to 1 inch (26 mm) in size provided that they are not shear-sensitive. Where the product undergoes a major change of state, ie, liquid to gel, or of viscosity as occurs during cooling, the SSHE often is used as a finisher for either the heating or cooling mode in conjunction with other methods of heat transfer.

Table 2. Corrosion of SSHE Cylinder Metals

Acid	Common use/source	Material corrosion at 140°F (maximum in./yr) Nickel	Chrome	316 SS
Nitric	Cleaning	0.050	nr[a]	0.002
Citric	Product	0.020	nr	0.002
Acetic	Product	0.020	nr	0.002
Phosphoric	Cleaning	0.050	nr	0.002
Malic	Product	0.020	nr	0.020
NaCl	Product	0.020	nr	0.020

[a]nr = not recommended.

THEORY AND CALCULATIONS

The operating principle of scraped-surface heat exchange is based on the constant movement of a product away from the heat-exchange surface in order to minimize the formation of films that resist heat transfer.

In considering the scraped-surface heat exchanger, four types of thermal exchange are taken into account:

- *Sensible Heat*. The heat produced by the increase or decrease in temperature of a product (without change of state).
- *Latent Heat*. Heat exchange associated with a physical change in the material being processed.
- *Heat of Reaction*. The heat given off (exothermic) or taken up (endothermic) when two or more chemicals react.
- *Mechanical Heat*. Power is consumed in turning the dasher of a SSHE. Most of this energy is absorbed as heat energy into the product within the heat exchanger.

The ideal form of heat transfer occurs when one product at an elevated temperature is brought into direct contact with another material at a lower temperature. The warmer product gives up its heat without loss of energy and at a rate equivalent to the ability to mix or disperse the two materials.

In practice, however, it is rare that two materials may be brought into direct contact. As a general rule, there is an intervening heat-transfer surface such as a tank or tube

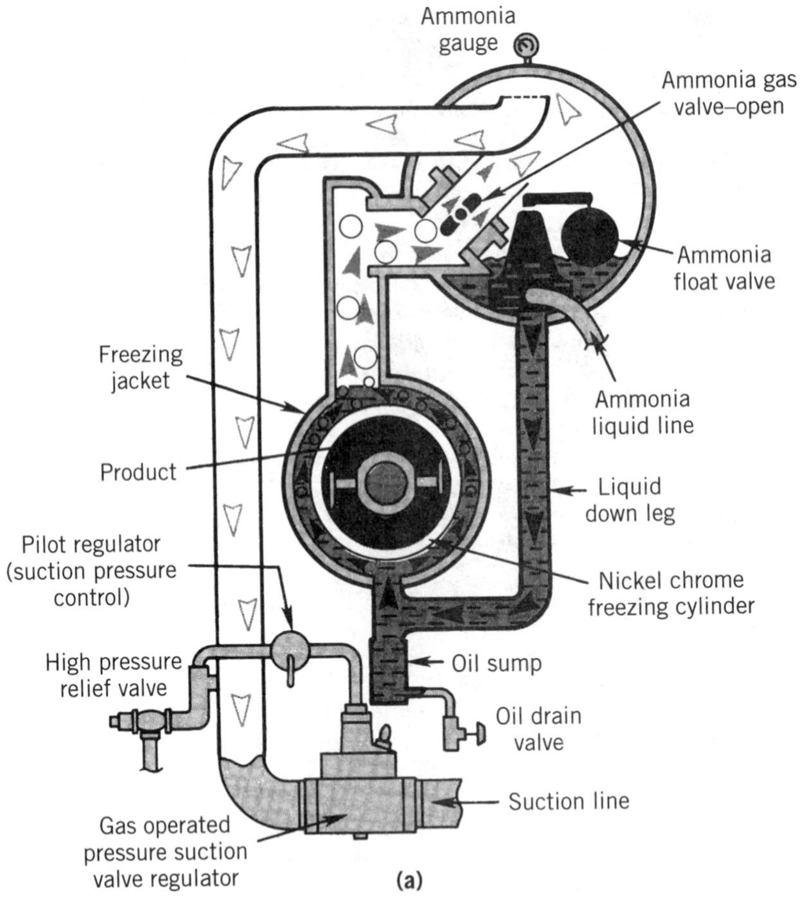

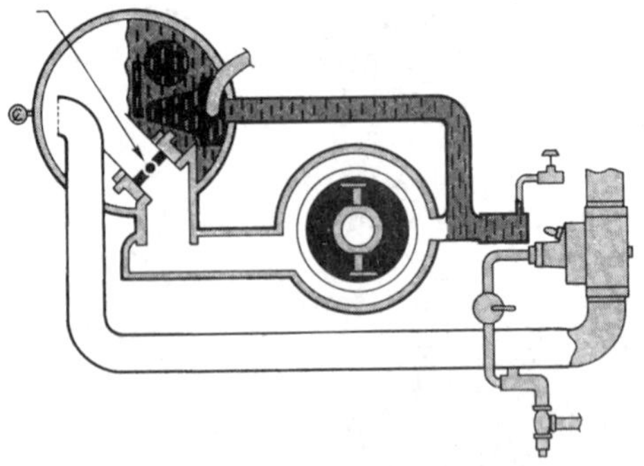

Figure 8. Full flooded SSHE refrigeration system (**a**) shown in operation, (**b**) shown shut down.

wall. This surface presents problems to heat transfer because it resists the passage of heat. It further induces resistances related to hydraulic drag and the buildup of deposits or other films that further retard the passage of heat. Figure 9 illustrates the various resistances that may be encountered.

Since these resistances are in series, they are additive. The adverse effect, however, may be reduced by agitating the fluids on both sides of the wall. Since the SSHE eliminates buildup of deposits on the product side of the tube wall, the resistance to heat transfer is minimized.

The total resistance to heat transfer then is

$$R = R_1 + R_2 + R_3$$

The difference in temperatures between the product and the media is the driving force to push across this resistance. Therefore, the equation for heat transfer is

$$\text{heat-transfer rate, Btu/h} \cdot \text{ft}^2 = \frac{\text{temperature, °F}}{\text{resistance}}$$

This equation can be developed into a useful form by letting the temperature difference be ΔT and resistance be R. Therefore

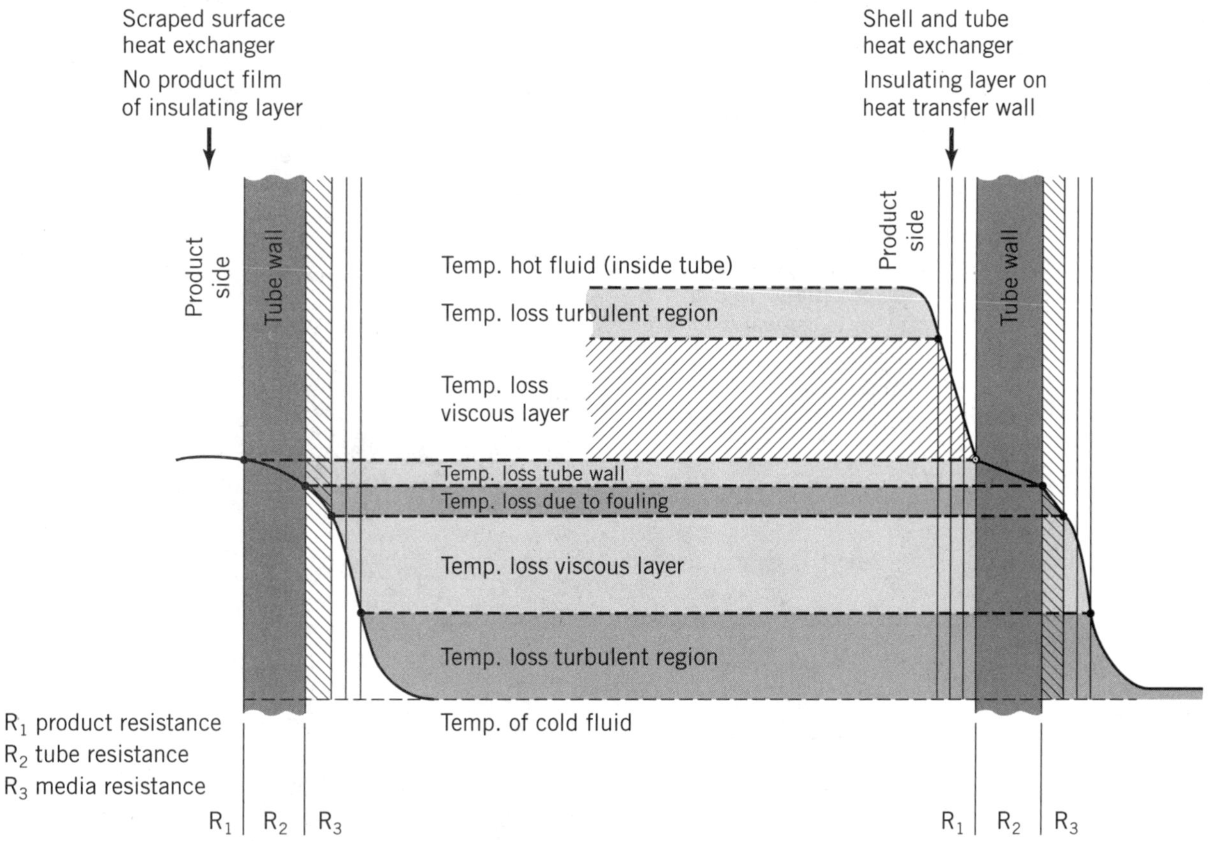

Figure 9. Comparison of scraped surface and shell and tube heat transfer profiles.

$$\text{heat-transfer rate} = \Delta T/R$$

Furthermore, by taking the inverse of resistance, which is conductance, $1/R = U$. The heat-transfer equation now is heat-transfer rate $= U\Delta T$.

Heat-exchange U values in a scraped-surface heat exchanger consist of three parts as shown in this formula:

$$U = \frac{1}{R_1 + R_2 + R_3} = \frac{1}{1/h_p + t/k + 1/h_m}$$

where h_p is the heat-transfer coefficient for the product and ranges from 200 to 800 Btu/h · ft^2 · °F, t/k is the heat-transfer coefficient through the cylinder wall (t usually is between 1/4 and 1/8 in. and k varies depending on the material, and h_m is the heat-transfer coefficient for the media and ranges from 800 (water media) to 2000 (steam or ammonia) Btu/h · ft^2 · °F. The overall U value then varies from 150 to 420 Btu/h · ft^2 · °F. Unfortunately, the h_p is difficult to determine theoretically and, because h_p tends to influence the overall U value much more than h_m and t/k, manufacturers size the heat exchangers using a U that is determined from lab test data or from production runs. Attempts have been made to predetermine the expected U value, but many factors influence the heat-transfer rate:

SSHE geometry: an SSHE can be equipped with different diameter dashers. The larger the dasher diameter, the smaller the annular space in which the product travels. A small annular spacing will improve heat transfer.

Viscosity: SSHE units normally can handle viscosities of up to 1,000,000 cP. The greater the viscosity, the lower the U value. Products with viscosities that vary greatly at different temperatures; eg, peanut butter and cooking—cooling starch for salad dressings, usually produce lower U values. This is due to mass rotation. It normally is encountered in cooling applications where the low-viscosity warm product short-circuits through the heat-exchange cylinder while the cooler, thicker product rotates with the dasher.

Dasher speed: high dasher speeds improve heat transfer but heat generated by high rpm is counter-productive in cooling applications.

Number of scraper blades: dashers typically are equipped with two or four rows of scraper blades. Units with four rows of blades provide higher heat-transfer rates at any given dasher rpm. When supplying two rows of blades, many manufacturers compensate by increasing the dasher speed.

Specific heat: the lower the specific heat, the lower the U values.

Thermal conductivity: the lower the thermal conductivity, the lower the U value.

Flow rate: the higher the product flow rate, the higher the U value. Flow rates below 5 gpm are considered low. See Figure 10.

Applications: heating applications produce higher U values than do cooling applications.

Cylinder materials: the cylinder materials, as mentioned previously, can be nickel, stainless steel or bimetallic. Figure 11 charts the differences in overall U values when cylinder thickness or materials are changed. Nickel is the most efficient and is used as the basis to compare the other materials commonly available.

Once a U value for an application has been determined, a calculation to size the heat exchanger can be made.

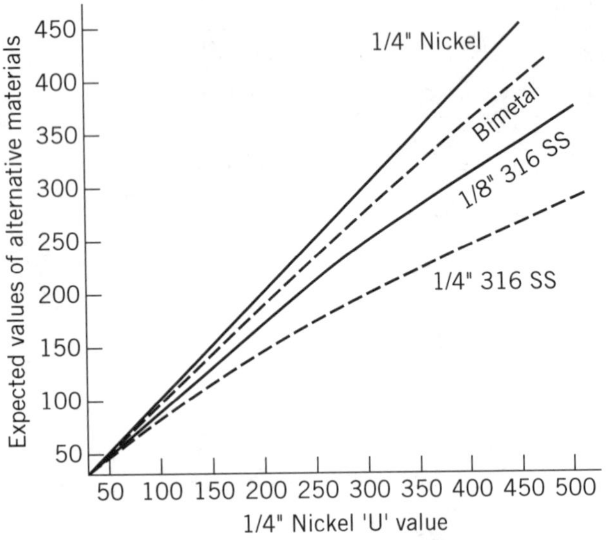

Figure 10. Comparison of nickel and other materials.

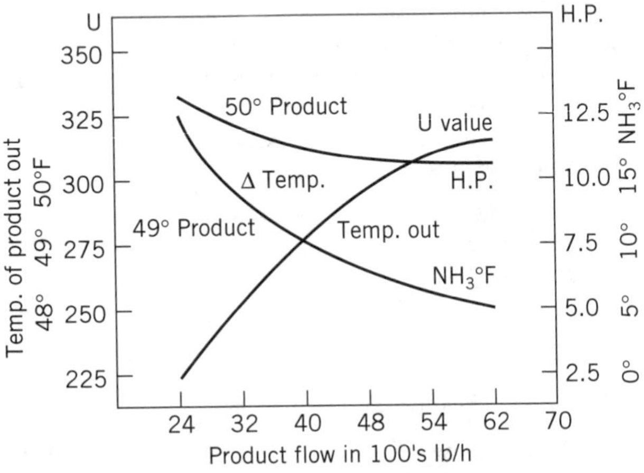

Figure 11. Curves represent a 6-ft² SSHE margarine cooler with 4-1/2-in.-diameter dasher. Discharge temperature 49–50°F flow rate increased from 2400 to 6200 lb/h; U value increases with flow rate and eventually levels off. Ammonia temperature is lowered to maintain proper exit temperature.

A Practical Formula

Deriving the practical scraped-surface heat exchange formula is achieved as follows:

$$\text{heat-transfer rate} = U\Delta T$$

Product heat load Q (heat load being transferred from the product to the media) will be equal to the heat-transfer rate times the heat transfer area A. Thus

$$Q = UA\Delta T$$

where A is the area, ft². The product heat load is equated

$$Q = \text{(product) lb/h} \times \text{temperature change} \times \text{SpHt} + \text{latent heat} + \text{heat of reaction} + \text{mechanical heat}$$

In the majority of applications, heat of reaction does not occur. Therefore, by eliminating this factor and combining equations, the generally accepted formula for calculating scraped-surface heat exchangers is

$$\text{area} = [\text{product flow} \times (T_{\text{in}} - T_{\text{out}}) \times \text{SpHt} + \text{latent heat} + \text{hp} \times 2545\ \text{Btu/h/hp}] \div U \times \text{LMTD}$$

$$\text{LMTD} = \frac{(T_{\text{in}} - T_{\text{media out}}) - (T_{\text{out}} - T_{\text{media in}})}{\ln\left[\dfrac{(T_{\text{in}} - T_{\text{media out}})}{(T_{\text{out}} - T_{\text{media in}})}\right]}$$

For steam, ammonia, and Freon units,

$$T_{\text{media in}} = T_{\text{media out}}$$

Area: square feet of surface required
Flow: pounds per hour of product
T_{in}: temperature of product in (°F)
T_{out}: temperature of product out (°F)
$T_{\text{media in}}$: temperature of media in (°F)
$T_{\text{media out}}$: temperature of media out (°F)
Latent heat: heat removed from ice formation or fat crystallization Btu/h
hp: dasher drive horsepower. Use only on cooling applications; disregard on heating. Note that horsepower is converted to Btu by the conversion factor 2545 Btu/h/HP
U: U value Btu/h · ft² × °F
LMTD: log mean temperature difference

Sample Calculation

Product: gravy
Specific heat: 0.8
Application: cool from 200 to 60°F using 5°F ammonia
U value: 280
Flow rate: 5000 lb/h

Assume 15 hp is required.

$$\text{LMTD} = \frac{(200 - 5) - (60 - 5)}{\ln[(200 - 5)/(60 - 5)]} = \frac{140}{1.266} = 110$$

$$\text{area} = \frac{5000 \text{ lb/h} \times (200 - 60°) \times 0.8 + 15 \text{ hp} \times 2545}{110 \times 280}$$

$$= \frac{598{,}175}{30{,}800} = 19.4 \text{ ft}^2 \text{ required}$$

SELECTION OF COMPONENTS

The scraped-surface heat exchanger is used on a wide range of diverse applications, many requiring different materials of construction and component design. Choices are based on a number of criteria involving both the compatibility of materials with the product to be processed and compatibility between various heat-exchanger components and materials. To better understand the reasoning behind component and material selection, it is necessary to know the equipment options available, the properties of the product, and the process temperatures. The four major areas of concern are cylinder materials, blade materials, seal selection, and dasher sizing.

Selecting Cylinder Materials

Since the per square foot cost of SSHE heat-transfer surface is relatively expensive, it is desirable to select heat-transfer cylinder material that provides the highest heat transfer coefficient. This will maximize the overall thermal performance for any given application. The commonly used materials are stainless steel, nickel with or without chrome plating, and bimetallic combinations. Nickel offers the highest HTC with a coefficient five times that of stainless steel.

To contain internal product pressures and prevent implosion by media pressure, cylinders generally are designed with a quarter inch wall thickness. For lower-pressure applications, however, an 1/8-in. wall occasionally is used to improve the performance of stainless-steel cylinders.

Since nickel has such a high heat-transfer coefficient, it might be expected that this material would have universal application. Unfortunately, however, nickel is susceptible to acid corrosion, so stainless-steel cylinders must be used with high acid type products. Furthermore, nickel is relatively soft and wears rapidly if subjected to constant abrasion by metal blades. Since simple economics dictate that the blades rather than the more expensive cylinders accept the wear, plastic blades must be used with nickel cylinders. Also, because most metal blades are of a stainless alloy and there is not enough difference in hardness between blade and cylinder, plastic blades are used with stainless steel cylinders.

To benefit from both the heat transfer characteristics of nickel and the durability of steel blades, nickel cylinders may be chrome-plated. This provides a scraping surface harder than any of the common blade materials but still cannot be used for all applications since the chrome also is subject to salt and acid attack.

An alternative to nickel, chromed nickel or stainless is the bimetallic cylinder (Fig. 12). With a hard corrosion-resistant alloy centrifugally cast within a tube having high tensile strength and thermal conductivity higher than stainless and approaching that of nickel, the bimetallic cylinder provides an acceptable compromise. While slightly softer than chrome, the inner liner is hard enough for use with any blade material. One limitation on a bimetallic cylinder is that it can be damaged by strong CIP acids, especially nitric acid.

Selecting Blade Materials

While scraper blades are available in plastic or metal (Fig. 13), the determination of which blade will provide optimum service is influenced by the cylinder material, by the media being used, and by product temperature and formulation.

Generally speaking, plastic scraper blades can be used in conjunction with all types of cylinder materials while metal blades are run only against chrome plated surfaces

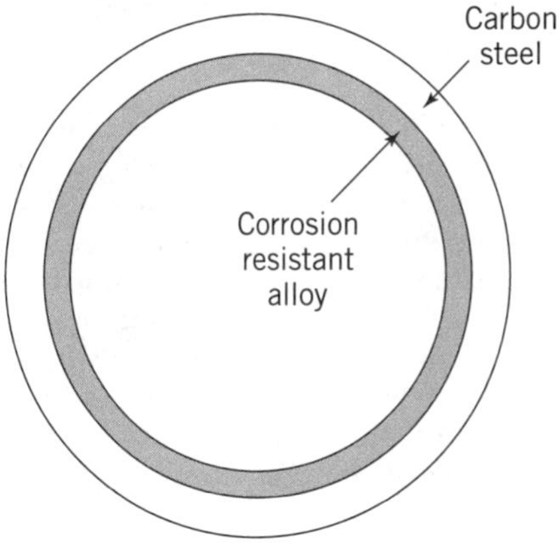

Figure 12. Bimetal cylinder cutaway.

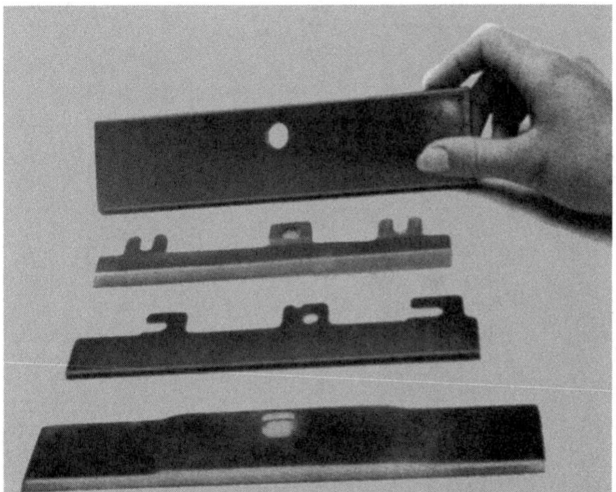

Figure 13. Available scraper blades.

or bimetallic cylinders. Limitations on plastic blades are wear life, temperature, and certain solvents. Metal blades, on the other hand, provide a longer wear life, can be sharpened to extend their use, and are not affected by temperature and solvents.

Plastic Blades

While the most common material for plastic blades is fiber reinforced phenolic resin, new proprietary injected molded plastic materials also are available.

As may be seen from the curves plotted in Fig. 14, injected molded plastic blades can withstand higher temperatures than can those of phenolic resin. Note, however, that there is a break point at about 180°F where temperature becomes the important factor to wear. Below 180°F, abrasion caused by the rubbing of plastic on metal is the prime wear related factor. Since the plastic blade is a wearing part regardless of the application, it must be replaced periodically to preclude breakage and possible product contamination by plastic fragments.

As stated previously, metal blades are not affected by processing temperatures. They are made of any number of tempered stainless alloys such as 17-4 PH or 410 stainless steel, and this hardening improves the wear life of the blades.

This is especially important for slush freezing applications where there is an added abrasive action caused by the scraping of ice from the heat-transfer surface. Since ice impacts the scraping edge as the blade sweeps the surface, it is important that the blade be maintained with a sharp edge (Fig. 15). For ice cream freezing applications, resharpening normally is done on a monthly basis to ensure optimum performance and the blades are replaced every one or two years. In other applications, metal blades may be sharpened every 3–12 months and replaced every 1–3 yr.

Selecting Seals and Materials

Shaft seals tend to be the most troublesome maintenance area on a scraped-surface heat exchanger. SSHE applications differ widely, and no one seal can handle the variety of temperatures, pressures, and product characteristics constantly encountered.

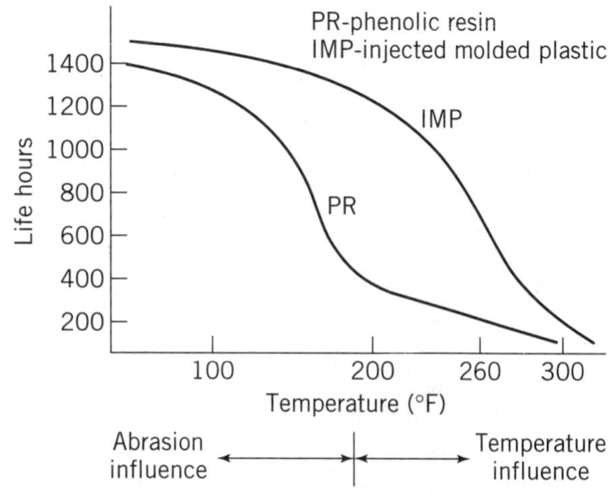

Figure 14. Temperature—wear relationship.

Most manufacturers offer a number of shaft seals, each designed to satisfy the demands of certain applications. The simplest and least expensive seal to maintain is the O-ring seal (Fig. 16a). This type of seal is recommended for use with products having high lubricity, eg, margarine, lard, shortening, ice cream mix, certain meat products, oils, and fats. Normally these products are not processed at temperatures above 140°F. High temperatures above 180°F shorten the effective life of an O-ring by causing it to lose elasticity. The sealing action of an O-ring seal depends on its ability to be sufficiently flexible to fill the gap between the rotating shafts and the static O-ring groove.

While the advantage of the O-ring is that it is inexpensive to replace when worn, the disadvantage is that it wears rapidly. Although daily replacement is not uncommon, this is not a problem when frequent equipment inspection is scheduled. The material of the O-ring depends on the fats or oils in the product to be processed. Typical materials are Buna N, Viton, and EPDM.

Rotary face seals are for all other applications (Fig. 16b). These can handle high temperatures and abrasive products, and offer longer wear life than O-rings. Rotary seals can be broken down into two groups—those with and without a seal water chamber.

The seal water chamber allows water or steam to flow around the edges of the seal faces and is recommended for use when handling products that tend to crystallize or dry out in the sealing area. Examples would be liquid sugars, tomato products, candies, fondants, and processed cheese. Water flushing increases the useful life of the seal. Steam is injected into the chamber for aseptic applications since FDA standards require a sterile barrier media on all rotating shaft seals in low acid aseptic processes. The steam, a sterial medium, prevents sterilized products from being contaminated should there be a seal leak. Many manufacturers build an integral water chamber into their equipment with the chamber to be used only as required.

Critical to the life of a seal is the materials used for the seal faces. The typical seal material is hardened stainless steel on carbon. Alternates for the carbon can be phenolic resins, plastics, silica carbide, or ceramics. Other more exotic seal materials to replace the hardened stainless include tungsten carbide and chrome oxide. Each has advantages and disadvantages, such as wear life, and cost. Normally, the manufacturer recommends the seal type and seal materials based on test work and field experience.

Selecting Dashers

Dashers or shafts that carry the scraper blades are available in diameters from 1-1/2 to 6-1/2 in. and are mounted within heat-exchanger cylinders that, in turn, are manufactured in diameters from 4 to 8 in. Since the most widely used cylinder has an ID of 6 in., the following is based on that model.

The best heat-transfer efficiency is provided when the annular space between a dasher and a cylinder is small. For example (Fig. 17), a 5-1/2-in. dasher within a 6-in. cyl-

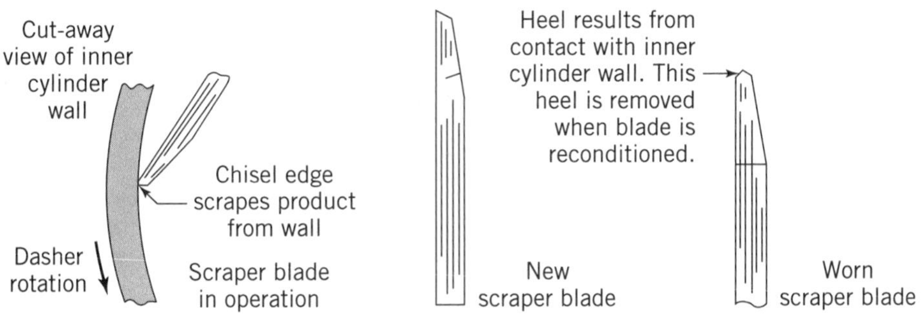

Figure 15. Wear pattern on metal scraper blades.

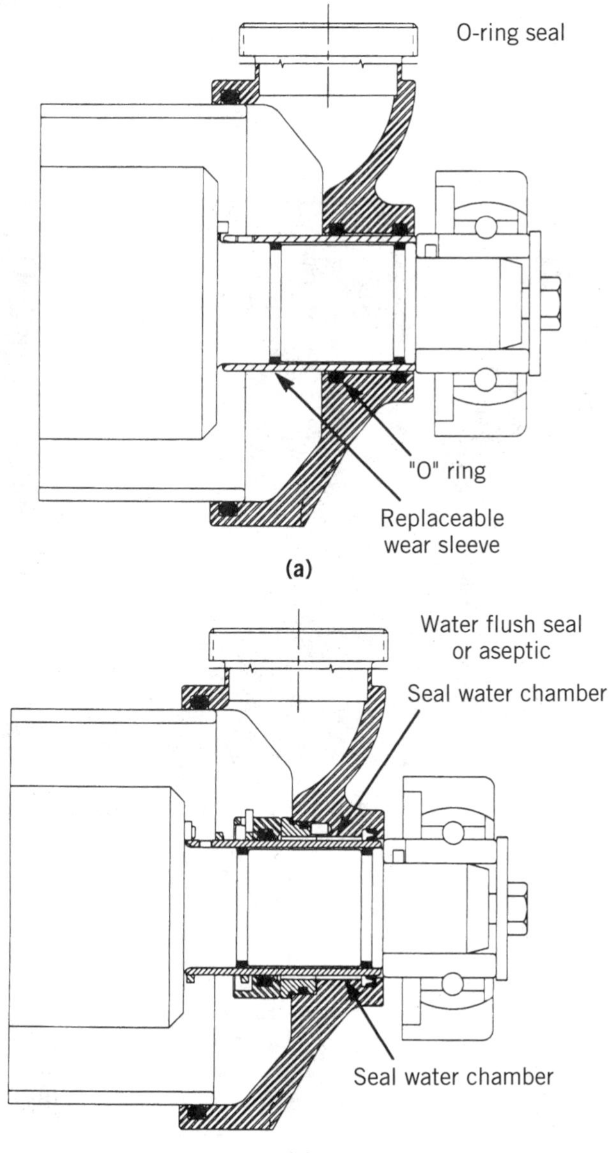

Figure 16. Shaft seals: (**a**) simple O-ring seal; (**b**) rotary mechanical seal with water flush or stream aseptic outer chamber.

inder provides a quarter inch annular product flow space, which causes a high axial flow velocity, induces turbulence, and offers a short product residence time. This is particularly useful in crystallizing applications such as for processing shortening and margarine since these products essentially are subcooled and allowed to crystallize after leaving the heat exchanger. Dashers for these products are water circulated.

The disadvantages of the 5-1/2-in. dasher are that it cannot handle viscous products because of high product pressure drop and cannot accommodate products containing particulates over a 1/4 in. in size. In such cases, the 4-1/2-in. dasher with its 3/4-in. annular product flow space is recommended. The smaller the dasher diameter, the larger the annular space, and the larger the particulates that can be processed. As the dasher diameter decreases, so does the effective U value for a given dasher speed.

Dasher speed is another design consideration. Typical dasher speeds vary from 60 to 420 rpm, and, in general, the higher the dasher speed, the higher the U value (Fig. 18). While this is desirable, increased dasher speed also increases the motor load, which can cause problems in cooling applications. Recalling the heat-transfer formula given previously, heat removed when cooling comes from both the product and the motor load required to turn the dasher. In some applications, the motor load actually can equal the product heat load. Therefore, in some cases a slower dasher speed nets a more beneficial end result.

The graph in Figure 19 compares horsepower and rpm involving various products. All are cooling applications performed on a 6-in. diameter by 72-in.-length unit. Note how small changes in dasher speed on marshmallow change the horsepower requirement significantly while the power requirements vary little while cooling cheese sauces at different speeds. It should be pointed out that horsepower requirements do not necessarily follow viscosity of the product. Rather, the term tenacity would best describe the measurement that would equate to the horsepower—rpm relationship. A good example is corn syrup versus a cooked starch slurry base for salad dressings. At the same viscosity, the corn syrup has much more tenacity in retarding dasher rotation than does the cooked starch.

To reduce motor loads while processing tenacious products, slow dasher speeds are required but the dasher cannot be run so slowly that the U value is reduced significantly. The properties of such products normally generate a low U value at any dasher speed, usually in the range of 120–180 Btu/h · ft$^{2\circ}$F. At very slow speeds, the U value is

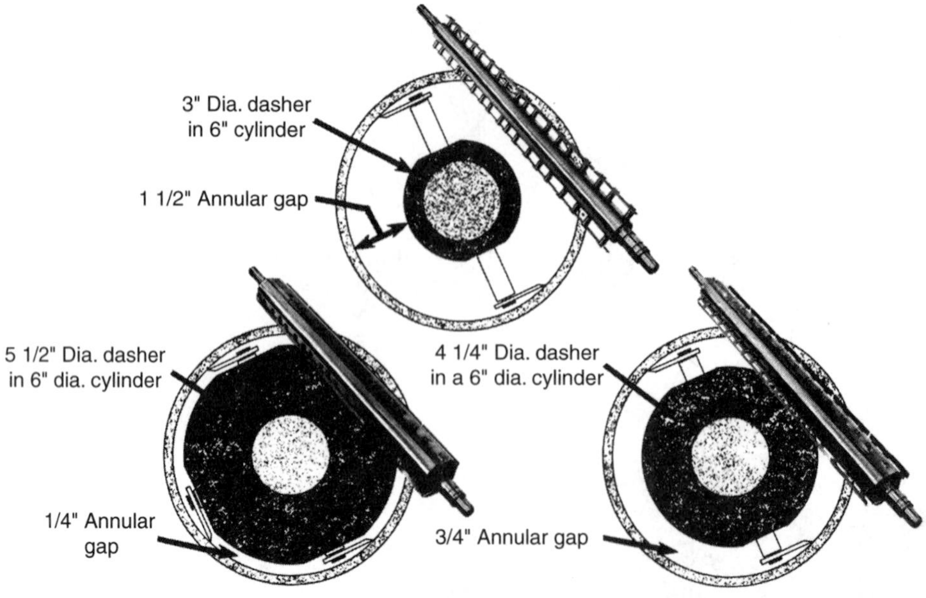

Figure 17. Dasher comparisons.

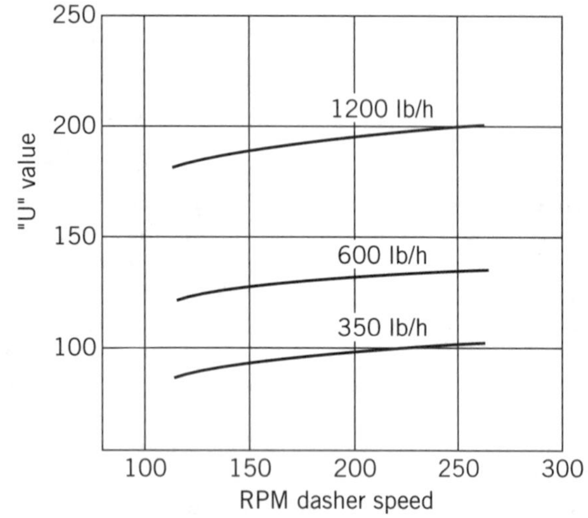

Figure 18. *U* value versus dasher speed and product flow.

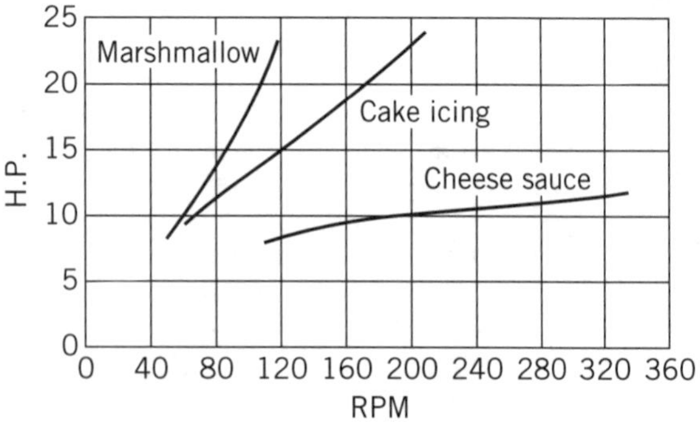

Figure 19. Dasher horsepower versus rpm.

Figure 20. Three-cylinder SSHE uses ammonia to cool 6000 lb/h of MDM turkey from 50 to 35°F.

further reduced from these already low levels and an economic sizing is not possible. It also should be noted that a certain minimum dasher speed is required to centrifugally "throw" and hold the scraper blades against the heat transfer cylinder. Therefore, there is a definite optimum dasher rpm that maximizes the net capacity of the heat exchanger for tenacious-type products in cooling applications.

Another reason for using low dasher speeds is retention of product identity. When large particulates are being processed with a sauce or gravy, fast-turning dashers tend to cause breakage of the particulates. The dasher speed

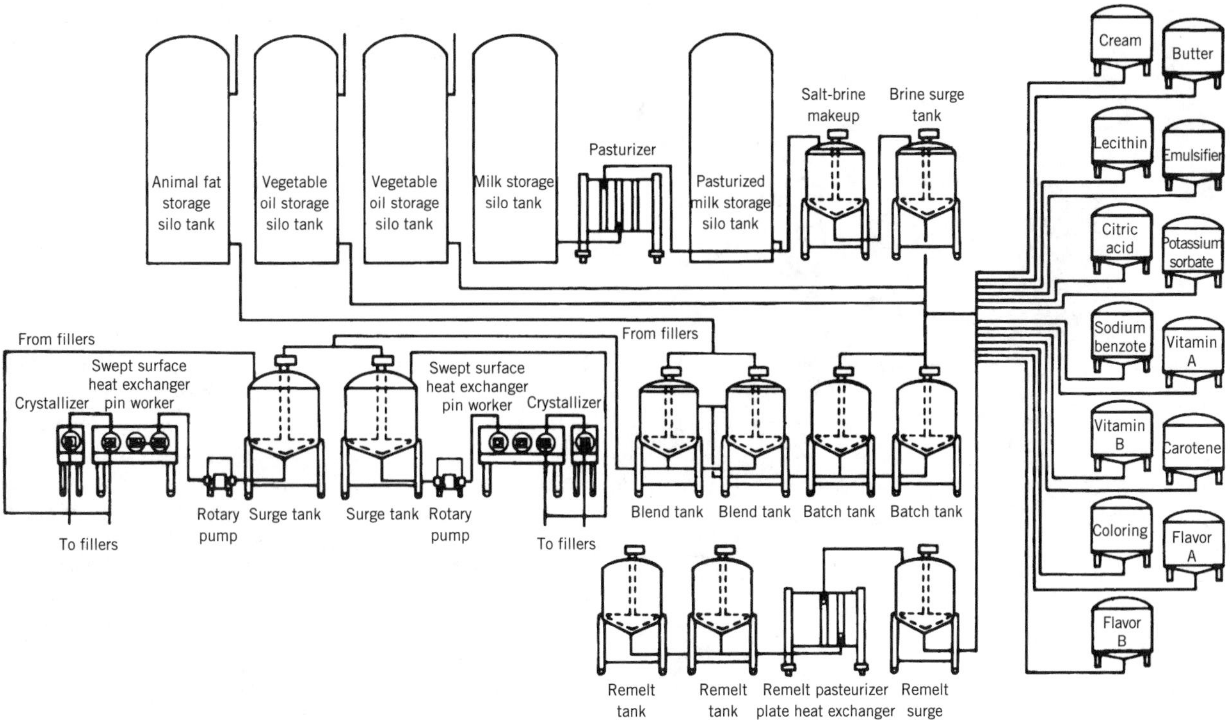

Figure 21. Margarine manufacture.

Table 3. SFI Values for Typical Shortenings

Shortening	Plastic range	Melting point °C	Melting point °F	SFI value 10°C 50°F	21.1°C 70°F	26.7°C 80°F	33.3°C 92°F	37.8°C 100°F
High stability[a]	Narrow	43	109	44	28	22	11	5
All-purpose[a]	Wide	51	124	28	23	22	18	13
14% hardfat in cottonseed oil[b]	Wide	51	124	16	14	14	12	11

Source: [a]Ref. 1 and [b]Ref. 2.

Figure 22. 10,000 lb/h system. Control panel, SSHE with water circulated dashers, pinworker, pumping unit with surge tank.

therefore is not always determined by the desired net heat-transfer efficiency but by limitations of the product.

In both the tenacious and particulate-type products, field measurements and/or lab test data are required to determine optimum dasher speeds.

SYSTEMS APPLICATIONS

As new products emerge from development labs across the country, many processors have determined that the scraped-surface heat exchanger is essential to effective, cost-efficient operations. In some cases, this type of heat exchanger is used for continuous and uniform heating and cooling of highly viscous pastes, fillings that contain up to 80% total solids, or products carrying discrete particles of up to 1 in. in size. In others, the SSHE has been selected for texturizing, gelling, whipping, plasticizing, and crys-

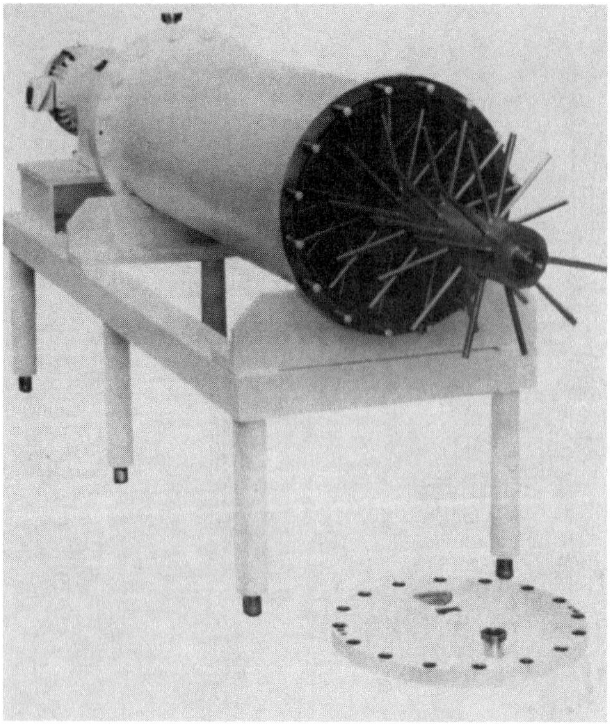

Figure 23. Pinworker with cover removed to expose shaft and pins.

Figure 24. Plasticizer with combination scraped-surface heat exchanger and pinworker.

tallizing duties. A few of the more interesting applications are detailed on the following pages.

Mechanical Deboning

In the mechanical deboning of turkey, chicken, and red meats, the heat that is generated in the deboning process must be dissipated to prevent product spoilage. This is done best by means of a scraped-surface heat exchanger.

The mechanical deboning machine (MDM) works on the principle of separating hard material such as bone, gristle, sinew, and cartilage from soft, whole meat. Following prebaking or grinding to reduce large bones to 3/8-in. size, the meat and bones are compressed by an auger inside the deboner and are forced under high pressure into a perforated tube. The soft meat passes through the many small holes, thus separating it from the bones. The meat is emulsionlike in consistency and drops into a hoppered pump. This deboning action requires considerable mechanical working of the meat and, in doing so, increases product temperature by 10–20°F. A scraped-surface heat exchanger (Fig. 20) is used to cool the meat from 55 to 35°F since a temperature of 40°F or less is required to prevent rapid bacterial growth and subsequent spoilage.

When used as a meat emulsion chiller, the SSHE operates on ammonia and is equipped with chrome nickel cylinders and a dasher and blade assembly specifically designed for viscous or particulate products. The typical flow rate of product through the chiller ranges from 2500 to 6000 lb/h. Since turkey and red meat are lower in moisture than MDM chicken, they are more viscous. A chiller for these products, in general, operates at two-thirds the capacity of a unit processing chicken and requires a larger dasher drive motor.

Margarine

While a typical system generally consists of a series of storage and blending tanks combined with a crystallizer, high-pressure pumps, and both plate and scraped-surface heat exchangers, the latter is the key to the production of high-quality margarine.

As indicated by the schematic, (Fig. 21) basic oils and fats are drawn from storage and blended with water, salts,

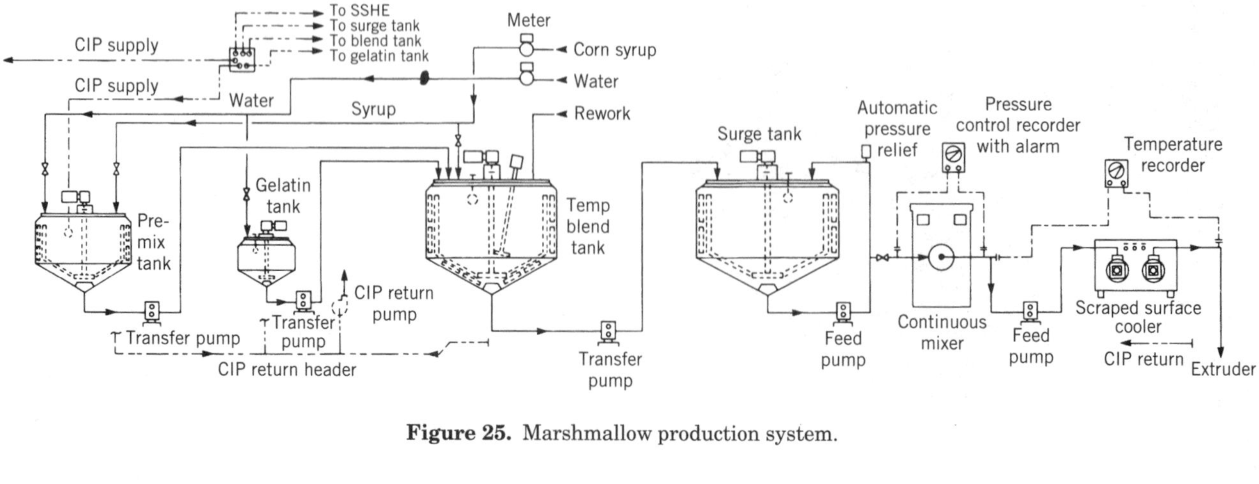

Figure 25. Marshmallow production system.

Figure 26. Scraped-surface heat exchanger cooks and cools ground beef for frozen food entrees. Ground beef is pumped into the rear of the left hand cylinder. Steam is used to head from 45 to 185°F. Cooked product flows out the front, through a holder tube, and into the rear of the right hand cylinder where ammonia at −40°F cools it to below 50°F.

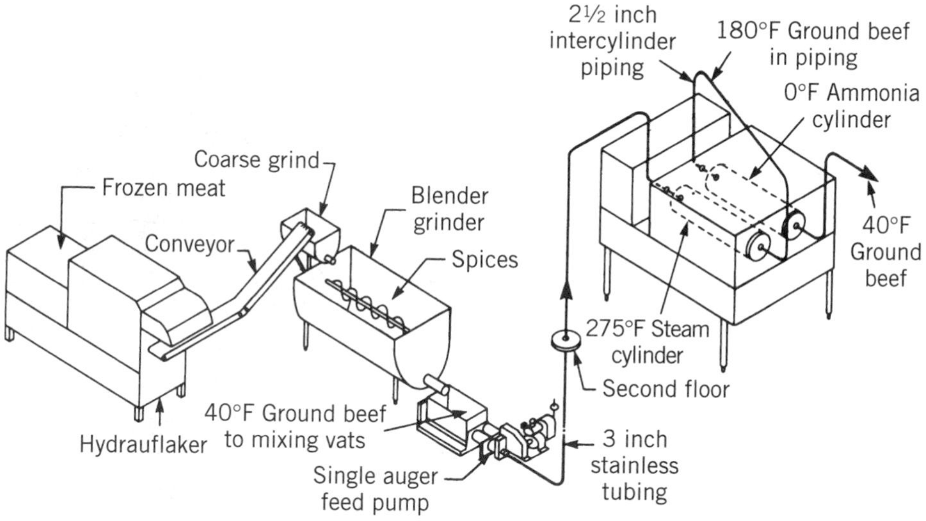

Figure 27. Ground beef cooking and cooling.

Figure 28. Three-cylinder SSHE aseptically heats and cools filling with 1/4–1/2 pieces of fruit.

emulsifiers, coloring, preservatives, and flavorings for containment in surge tanks at temperatures ranging from 110 to 130°F. If milk ingredients are involved, a Paraflow plate heat exchanger is used for legal pasteurization and cooling before the blending phase. The final mixture then is pumped through an SSHE unit for cooling to approximately 45°F.

While the scraped surface exchanger generally operates on ammonia and is equipped with chromed nickel cylinders and metal blades, the special feature for margarine production is the water-circulated dasher. Since margarine is a crystallizing application, product crystals would tend to stick and collect on the surface of a standard dasher. Such a buildup would increase not only product pressure drop through the exchanger but also the motor load and accelerate overall wear on seals and bearing components. By

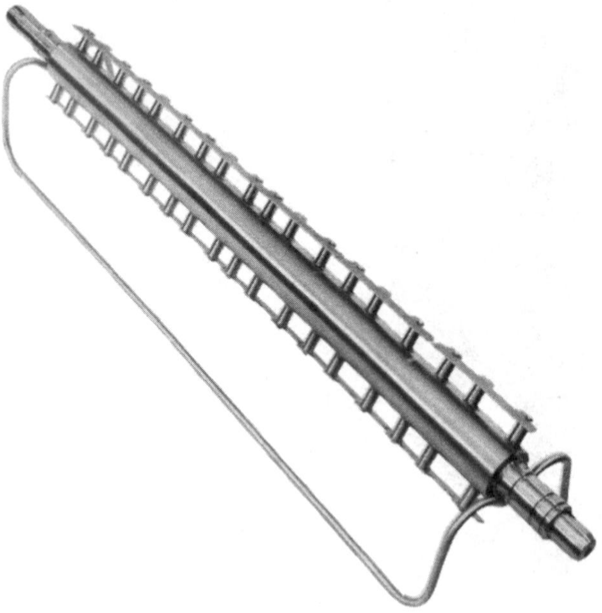

Figure 29. To minimize product damage, clearance between the blade and the dasher shaft should be equal or greater than the size of the particle that is being processed.

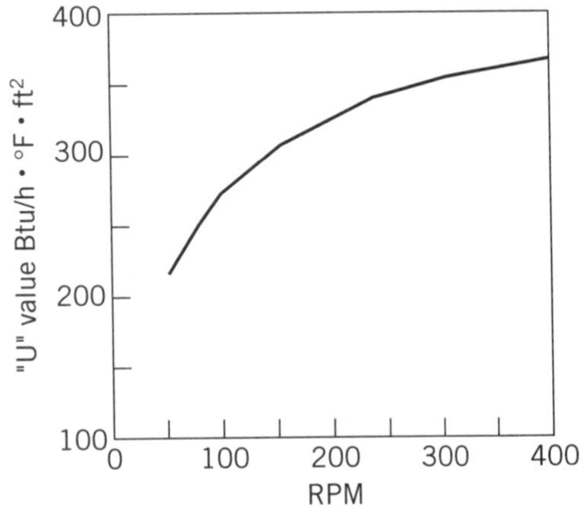

Figure 30. Dasher speed versus heat transfer (meat gravy).

passing water at about 100°F through the hollow dasher, enough heat is provided to prevent crystal formation while still not contributing significantly to the overall heat load.

Following the cooling cycle, the viscous product is pumped from the scraped-surface chiller through a static crystallizer for table-grade, stick-type margarine or to a pinworker crystallizer for tub, soft table, or bakery product. When an acceptable crystal structure has been achieved, the product is formed and packaged and allowed to gain its final viscosity during a quiescent period.

Plasticizing Shortenings

In the manufacture of solid shortening, one primary objective is to make a product that remains plastic and work-

Table 4. Partial List of Products Processed

Product	Heat	Cool
Candies, frostings, and fillings		
Cake frosting		X
Cookie filling		X
Fondant		X
Marshmallow		X
Dairy products		
Concentrated milk	X	
Cottage cheese curd		X
Cream cheese		X
Ice cream mix	X	
Melting butter	X	
Processed cheese	X	
Quark	X	X
Whey concentrate	X	X
Whipped butter		X
Yogurt	X	X
Fats and oils		
Margarine	X	X
Shortening		
Plasticizing Bulk		X
In-house		X
Vegetable oil	X	X
Fruit: Concentrates, juices, and fillings		
Aseptic applesauce	X	X
Aseptic bananas	X	X
Aseptic fruit fillings	X	X
Aseptic fruit purees	X	X
Citrus pulp		X
Jelly and jam preserves	X	X
Juice concentrate		X
Pumpkin and squash	X	X
Pumpkin pie filling		X
Meat products		
Cooking bacon	X	
Gelatin concentration		X
Lard		X
Low-temperature rendered meat		X
Meat pie fillings	X	X
Mechanically deboned		
Chicken		X
Red meat		X
Turkey		X
Sausage and hamburger	X	X
Sauces, starches, gravies and puddings		
Aseptic cheese sauce	X	X
Aseptic puddings	X	X
Chili	X	X
Enchilada filling		X
Gravies and sauces		X
Salad dressing starch	X	X

Table 4. Partial List of Products Processed *(continued)*

Product	Heat	Cool
Slush freezing for freeze concentration		
Coffee or tea extract		X
Grape juice		X
Orange juice		X
Various aroma extracts		X
Vinegar		X
Wine		X
Tomato products		
Barbecue sauce	X	X
Tomato paste	X	X
Tomato sauces	X	X
1/4-in. diced tomatoes	X	
Miscellaneous products		
Acetone—deep cooling		X
Adhesives and glues		X
Aseptic diced peppers	X	X
Autolyzed yeast	X	
Baby foods	X	X
Candle wax		X
Hand cream		X
Honey		X
Ink		X
Lipstick	X	X
Liquid antacid	X	X
Lithium grease		X
Paints	X	X
Peanut butter		X
Photographic gelatin		X
Pickle relish	X	X
Poi	X	
Resins	X	X
Shampoo		X
Shaving cream		X
Soup	X	X
Synthetic polymers	X	X
Toothpaste		X
Vanilla wafer batter		X
Wallpaper paste	X	X

able over a wide temperature range, specifically between 60 and 90°F. To do this, hydrogenated or hard fats are added to formulations to extend the plastic range. Solid shortening actually is a suspension of high-melting-point fats in liquid fat or oil. Different fats will melt or solidify as is shown in the solid fats index (SFI) charted in Table 3. At room temperatures of 68°F (20°C), the typical shortening is only about 30% solid.

Two methods commonly are used to achieve desired crystal structure, size, and dispersion when producing plasticized shortening. One is for full-scale production (Fig. 22); the other, for limited "in house" applications.

For normal production, an A or scraped-surface heat exchanger is used to cool oil from 140°F (60°C) (no solids) to 60–70°F (16–21°C). This subcooling is done rapidly, and only about 5% of the fats are crystallized. The shortening then enters a B or pinworker unit (Fig. 23), which acts as an agitated holding tube of about 3-min duration. During this time, the remaining crystals form with the latent heat of crystallization causing the shortening to heat up. Typically, if the shortening enters at 65°F (18°C), it will exit at 80°F (27°C). The now plasticized shortening is ready for filling into 50-lb cubes. Most plasticizing systems usually include a shell-and-tube exchanger to precool product from 140°F (60°C) to 120°F (49°C) and reduce the heat load on the SSHE. Nitrogen also is usually added before the SSHE to give the final product a white color.

For bakeries that plasticize shortening for in-house use, a single unit combining the features of A and B machines commonly is used. As shown in Figure 24, this involves an open type dasher shaft. The dasher is hollow and has wide slots at regular intervals. A set of beater bars acts as the pinworker, holding and working the shortening to a high degree in order to produce a soft, readily flowable texture that can be handled in pump lines and tanks downstream.

Marshmallow

Marshmallow is an aerated food product typically formulated from 45% sucrose, 27% corn syrup, 23% water, gelatin, invert sugar, and flavoring. (See Fig. 25.)

As shown in the flow schematic, a production system generally employs multiple tanks, an aerating mixer, and a scraped-surface heat exchanger. A premix tank first is used to blend and heat sugars to 245°F before the solution is transferred to another blending tank for cooling to 155°F. Gelatin and 150°F tempered water, meanwhile, are thoroughly mixed by means of a high shear agitator and blended with the sugar solution. The resulting syrup is charged to a surge tank and pumped as required to a high shear aerating mixer where it is beaten to a weight of between 35 and 48 oz/gal (specific gravity of 0.26–0.36). The 155°F marshmallow mix then flows to an SSHE for chilled water cooling to 90°F.

Marshmallow is a difficult duty for a scraped-surface heat exchanger because of the extremely high viscosity and the low heat-transfer properties of the product. Operating pressures in the mixer and heat-exchanger range from 150 to 300 psi to handle product viscosity and to force the marshmallow through an extrusion head to a belt where it is shaped and dusted with starch. The low product density results in a low U value and because of viscosity, a high torque is required to turn the dasher. Note in the Figure 19 dasher horsepower—rpm chart how dramatically the dasher hp changes relative speed. Too high an rpm results in excessive motor load that makes cooling more difficult because the motor heat has to be removed by the heat exchanger. Since even the optimum dasher speed results in about one half of the cooling load coming from the motor input, selecting the proper dasher speed for this application is very important. Care also must be given to the selection of shaft seals. Since marshmallow is mostly sugar, seals with abrasive resistant, high hardness qualities should be used and the seals should be water flushed to prevent crystals from building up on the seal faces.

Cooking–Cooling Ground Meat

While ground beef generally is cooked in steam-jacketed kettles for moderate size production runs, the tendency is to install scraped-surface heat exchangers for both cooking and cooling duties when production requirements exceed 10,000 lb/day (Fig. 26). Not only does the SSHE provide faster cooking; it also improves product yield since moisture and fat loss is far less than the 15–20% decline in product volume that occurs with the open-kettle method (See Fig. 27.)

The typical meat cooking system consists of a single auger feed hopper with a leveling ribbon followed in line by a two-cylinder SSHE, one cylinder with 275°F steam and the other with 0°F ammonia. Both the auger feeder and SSHE are hydraulically driven from a central power source. After meat is ground, spiced, and blended, it is fed to the heat exchanger by means of a single-discharge auger designed specifically to move viscous, sticky products directly into a rotary pump. The auger maintains a constant stuffing pressure at the rotary pump inlet by varying the operating speed while a leveling ribbon maintains a uniform meat level in the feed hopper. The scraped surface blade and dasher design promote rapid removal of product from the cylinder walls and enhance agitation and mixing during both cooking and cooling phases. Assembly and disassembly for blade inspection and/or replacement is quick and easy. Meat remaining in the auger feed pump, pipelines, or the scraped surface heat exchanger is easily recovered and, once clear of product, the closed system and piping are easily cleaned with a pump circulation loop.

Products with Particulates

When processing sauces, gravies, or juices containing pieces of meat, vegetables, or fruit, one of the key elements is maintaining product identity by reducing or eliminating particulate breakage. This becomes more of a concern if the particulates exceed 1/4 in. in size and the carrier is a thin liquid such as broth or soup. Thick sauces, on the other hand, seem to have a cushioning effect with less product damage occurring during passage through the system.

While the dominant type of heat exchanger used for particulate-type products is the scraped-surface unit (Fig. 28), each case still must be considered individually. Many processes require multiple SSHE cylinders that should be piped in series rather than in parallel. This eliminates the need for individual product pumps while ensuring higher heat-transfer rates and an equal flow to each cylinder.

To reduce particle breakage, three design areas must be considered—the gap between the dasher and scraper blade, the sizing of SSHE inlets and outlets, and the dasher operating speed.

The distance between the underside of the blade and the dasher shaft (Fig. 29) should be equal or greater than the largest particle to be processed. This allows passage without particle damage as the blade sweeps by. The disadvantage of a wide annular space between the dasher and the cylinder wall is a longer product retention period and a reduction in exchanger efficiency. The faster the product travels through the cylinder, the higher the U value attained. This results in improved exchanger performance with lower residence time and less product damage.

The sizing of inlets and outlets is directly related to product quality. Using the largest possible ports reduces not only pressure drop through the system but also shearing effect. Furthermore, passage of product is eased and the possibility of contact with inlet walls is minimized.

Finally, there is the matter of dasher speed. As shown by the curve in Figure 30, while higher dasher speed results in a higher U value, it also can cause increased particulate damage. Therefore, higher dasher rpm and the corresponding high heat-transfer efficiency will apply only to products with small particulates. When products contain delicate particulates of 1/2–3/4 in., a variable-speed drive is used to reduce dasher revolutions to about 120 rpm. While this protects product quality, it also lowers exchanger efficiency. Consequently, striking a balance between performance and quality should be determined by lab runs so that optimum operation at full scale becomes a matter of a proper dasher speed setting.

Peanut Butter

For the production of creamy or chunky peanut butter, a process system will consist of surge tanks, a deaerator, scraped-surface heat exchanger, ingredient feeder with in-line blender, and transfer pumps.

After mixing roasted nuts with a stabilizer, salt, and sugar to the desired formula, the product is ground and discharged at a temperature of about 150–200°F. Deaeration follows to eliminate air pockets, which initiate oil separation. During cooling from an SSHE inlet temperature of 140–190°F to a discharge temperature of 85–95° · F, the stabilizer is solidified in finely divided crystalline form and uniformly distributed throughout the mixture. At the proper temperature, the peanut butter becomes a viscous, extrudable mass. Crystal change continues and further solidification occurs after filling. For chunky style, a chunk feeder is located between the SSHE and filler or the transfer pump and deaerator.

Table 4 gives a listing of other products routinely processed using SSHEs.

BIBLIOGRAPHY

Adapted from: SSHE Handbook, SSH-1087. APV Crepaco, Inc. Copyrighted APV Crepaco, Inc. Used with permission.

APV Crepaco, Inc.
Lake Mills, Wisconsin

HEAT TRANSFER

BACKGROUND

Almost all food processes depend on or are affected by heat being added or removed at some stage in the operation. The efficient and effective utilization of heat results in economic savings, minimum adverse affects on nutrient components, higher quality consumer-ready products, and minimum effect on the many environmental factors associated with food processing. This efficient and effective use of heat depends on knowledge and subsequent application of the heat transfer mechanisms involved in heating, cooling, and changing the state of foods and food products. The rate at which heat is transferred depends on the type of product, the condition of the product, and the type of heat transfer by which heat enters or is removed from the product.

Other than maintaining a food under handling and storage conditions that ensure the highest quality product, the only option involving heat transfer is the control of processing conditions to give the desired rate of heat transfer. This means considering the three basic mechanisms that control the rate of heat transfer and selecting processing conditions and facilities that optimize the process. Hence a thorough knowledge of heat transfer mechanisms and the relationship between heat transfer rates and other physical and chemical factors is probably the most important consideration involved in the processing of foods. This is further emphasized by the fact that the control of mass and energy balances and fluid flow mechanics, the basics for designing, constructing, and operating the machinery, equipment, and facilities involved in food handling, storing, and processing, is dependent on considerations involving the mechanisms and rate of heat transfer.

Heat is transferred from one body to another by three different mechanisms, namely conduction, convection, and radiation. In fact, most processing facilities utilize two or all three of these means of transferring heat from or to a product. Before considering the mathematical relationships that describe heat transfer and allow the control of the amount and rate of transfer, it is well to have a visual understanding of how each functions.

Conduction heat transfer takes place when two bodies at different temperatures are in contact with each other. Through this direct contact, heat energy is transferred from particle to particle between the solid bodies with no bulk movement of material. When one places a hand on an object having a temperature different from body temperature there is a flow of heat. Hence, when the object is a blackboard in the classroom at room temperature, there is a sensation of the blackboard being cold. Conversely, if one touches a warm element on a stove, there is an immediate sensation of heat being transferred to the hand. The difference in temperatures between the two objects, as well as the characteristics of the conducting body, determines the rate of transfer. In the case of the blackboard, the temperature driving force was probably not more than 20°F (11°C), while the driving force between a warm stove-element and the hand is considerably higher. Hence the sensation of heat will be detected and create considerably more reaction than placing a hand on the blackboard. Other things to consider include the nature of the material to which heat is being transferred. This determines how fast the heat will penetrate as well as the temperature gradient within the material. In general, conduction is highly desired for food being frozen, especially those forms having flat surfaces that can insure maximum contact with plate freezers. In most other cases of freezing irregularly shaped items and heating, cooking, or sterilizing, a combination of conduction and other transfer mechanisms is more efficient and controllable.

Convection heat transfer depends on the bulk movement and mixing of liquids that are initially at different mass temperatures, or on the contact of a solid with a moving liquid stream of a different temperature. These two basic types of convection must be considered together in most heat transfer involved with processing food products. Regardless of whether hot liquids are being mixed in a batch or continuous basis, the temperature of the final combined liquid mixture reaches an equilibrium temperature somewhere between the original temperatures of the two liquids. If a solid is involved, the temperature equilibrium occurs between the bulk of the liquid and the surface of the solid. The velocity of a liquid flowing past a solid affects the rate of transfer between the two materials. Natural or free convection is caused by density gradients (thermal expansion) formed when a liquid is changing in temperature. These can not be controlled and are often a detriment to processing since the rate of transfer is at the mercy of the naturally rising or mixing streams. Forced convection, the pumping or blowing of a liquid or gas over a surface, can be controlled and the heat transfer rates can be predicted. Hence most food processes utilizing convection heat transfer depend on facilities and equipment that use forced convection. This can be demonstrated by considering the wind chill factor, by which a wind causes a person to feel colder than expected at a given outside temperature. For example, in the winter one might bear a weather report in which the outside temperature is given as 0°C and the wind chill factor makes it feel as if the temperature were −15°C. This is because the wind is a forced convection whereas the 0°C temperature is measured under shielded or still conditions.

Many food processing operations include a combination of conduction and convection heat transfer. This is demonstrated in cooling and freezing curves in which blast freezing or convection is the method of heat transfer (see Figure 9 in FISH AND SHELLFISH PRODUCTS) and a plate or conduction heat transfer is combined with blast freezing or convection (see Figure 9 in FISH AND SHELLFISH PRODUCTS). The critical period during which heat of fusion is being removed between about 30 to 22°F (−2 to −6°C) is 220 min in Figure 10 of FISH AND SHELLFISH PRODUCTS. The total time for cooling, freezing, and dropping from 50 to 0°F (10 to −18°C) is 472 min in Figure 10 and 93 min in Figure 9. As has been discussed in FISH AND SHELLFISH PRODUCTS, the more rapid freezing as shown in Figure 9 resulted in high quality frozen fish, similar to the fresh product. Conversely, the slower freezing in Figure 10 resulted in considerable cell degradation that greatly reduced the quality of the end product.

The third type of heat transfer used for heat processing of foods is radiation, the transmission of electromagnetic energy through space. Whereas conduction and convection are dependent on a physical medium through which to transfer heat energy, radiation requires no carrier to transfer wave energy from a surface of one body to another. The amount of energy transmitted is dependent on area and the nature of the exposed surface and the temperature of the body. The amount of radiation absorbed or deflected also depends on the nature of the absorbing body and the body temperature. Hence each body receives radiation from every other body, the amount depending on how much the bodies can "see" of each other and the ability of the body to radiate or deflect and absorb radiation (emissivity or absorptivity).

There are several different types of radiation used in the food industry, each used for a specific reason. The radiation heating from a warm or hot body to a food is akin to that of a person standing in the shade or the sun on a hot day. The ambient air temperature may be the same in both positions; however, a person becomes much warmer in the sun where the direct radiation is being added to the body in addition to the convection heating occurring from the surrounding air. The radiating body for heating a food by radiation can be the hot element in an oven or grill (roasting, baking, broiling). In this case the surface receives energy so rapidly that it can not be conducted into the food fast enough to prevent the desired browning (actually scorching and/or Maillard browning reaction) of the surface.

Other types of radiant energy are used in processing food. Some of these are becoming as prevalent as the conventional concept of cooking or processing by exposure to a hot element or heat source. When an alternating current is passed through a conductor, energy is emitted in the form of waves having a specific wavelength and frequency. Microwave heating is a good example of this process, whereby the energy absorbed by a food is converted into heat due to the friction of moving molecules or atoms.

Ionizing radiation from x rays and gamma rays, while not heating a food during a normal exposure time, can destroy microorganisms and thus accomplish the aim of pasteurizing or sterilizing a food.

PROCESSING FOODS BY CONDUCTION HEAT TRANSFER

The relationship between the factors involving heat transfer by conduction can be intuitively realized by considering the flow of heat energy. One would reason that heat flow (q) perpendicular to the direction of flow would be proportional to the area (A) of contact or flow and the temperature difference (dt) along the path of flow. Likewise, the thickness of the food or path through which the heat is flowing (dx) would be inversely proportional to this flow. Although modern mathematics allow one to derive such relationships, the basic conductive heat transfer relationship, known as Fourier's law, was derived empirically and confirmed by experimentation.

In addition to the factors, discussed above, the physical and chemical properties of a material have a significant effect on heat flow and can be represented by an experimentally determined proportionality constant called the thermal conductivity (k). The thermal conductivity is related to the number of free electrons in a material, varying from being high in metals to low in gases. It can be considered the flow of heat per unit time (watts or Btu per hour) through a given area (square meters or square feet) per unit thickness (meters or feet). The mathematical relationships involving conduction heat transfer can be represented by Fourier's law:

$$q = -kA\,dt/dx \qquad (1)$$

The units of each item in SI and English units are shown in Table 1. It is conventional to place a negative sign in front of the equation to signify a positive heat flow from the higher to the lower temperature.

Although the complete mathematical analysis of heat transfer can become extremely complicated when it comes to considering complex three-dimensional flow and integrations over total volumes, most problems involving conduction in food processing systems can be greatly simplified. In most steady-state cases, considering the relatively short range of temperature changes, the uniformity in area over the distance of heat conduction, and the uniformity in thermal conductivity over these temperature ranges, equation 1 can be simplified to

$$q = -kA(t_1 - t_2)/\Delta x - kA\,\frac{\Delta t}{\Delta x} \qquad (2)$$

Equation 2 is typical of many problems in nature that involve a driving force analogy. In the case of heat transfer, the rate of transferring some discrete quantity of energy q is equal to the driving force DF that makes the movement happen divided by the resistance R to this movement, or

$$q = \mathrm{DF}/R \qquad (3)$$

A common use of this relationship is in the transfer of electrical energy, where the current flowing (I, amperes) is equal to the driving force (E, volts) divided by the resistance (R, ohms), or

$$I = E/R \qquad (4)$$

The resistance in the heat transfer equation (2) equals the thickness (Δx), which is directly proportional to resistance to heat transfer divided by the area and thermal conductivity, both of which decrease the resistance,

$$R = \Delta x/kA \qquad (5)$$

The reciprocal of the resistance is the conductance (C), where

$$C = 1/R = kA/\Delta x \qquad (6)$$

Since the total heat being transferred through each of a series of materials in contact is the same for each material,

Table 1. Selected Coefficients of Thermal Conductivity

Property	SI units	English units
q is the heat flow	watts (W)	Btu heat flow Btu/h
h is the thermal conductivity	(W)(m)/(K)(m^2)	Btu/h of ft^2/ft
A is the area	m^2	ft^2
dt is the temperature difference	°C	°F
dx is the thickness	m	ft

$$q = q_1 = q_2 = q_x = -\frac{k_1A_1\Delta t}{x_1} = -\frac{k_2A_2\Delta t_2}{x_2} = -\frac{k_xA_x\Delta t_x}{x_x} \quad (7)$$

and considering that

$$q = -\frac{\Delta t}{R} = -\frac{\Delta t}{\Delta x/kA} \quad (8)$$

then

$$q = q_1 = q_2 = q_x = -\frac{t_1 - t_2}{x_1/k_1A_1} = -\frac{t_2 - t_3}{x_2/k_2A_2} = -\frac{t_3 - t_x}{x_x/k_xA_x} \quad (9)$$

Combining to obtain the overall resistance

$$R = (x_1/k_1A_1) + (x_2/k_2A_2) + (x_x/k_xA_x) \quad (10)$$

and the overall driving force

$$\Delta t = t_1 - t_x$$

the overall heat transfer equation becomes

$$q = -\frac{\Delta t}{R} = -\frac{t_1 - t_x}{[(x_1/k_1A_1) + (x_2/k_2A_2) + (x_x/k_xA_x)]} \quad (11)$$

When, as is often the case, $A_1 = A_2 = A_x$, the equation reduces to

$$q = -\frac{\Delta t}{R} = -\frac{t_1 - t_x}{(1/A)[(x_1/k_1) + (x_2/k_2) + (x_x/k_x)]} \quad (12)$$

EXAMPLE 1

Using the conductance heat transfer equation, calculate the amount of heat lost through 2 m^2 (21.528 ft^2) of a cold storage wall that is composed of three layers of material. The wall consists of 4-in. (0.1016-m) thick concrete block (x_c) on the outside, 6-in. (0.1524-m) thick polystyrene insulation (x_p), and 2-in. (0.0508-m) thick fir wood (x_f) on the inside. The inside wall of the cold room is −40°F (−40°C) and the outside wall is 68°F 20°C). From Table 2, the thermal conductivities are

$$k_c = 0.76 \text{ W} \cdot \text{m/m}^2 \cdot \text{K} = 5.27 \text{ (Btu)(in.)/ft}^2\text{)(h)(°F)}$$

$$k_p = 0.038 \text{ W} \cdot \text{m/m}^2 \cdot \text{K} = 0.263 \text{ (Btu)(in.)/(ft}^2\text{)(h)(°F)}$$

$$k_f = 0.11 \text{ W} \cdot \text{m/m}^2 \cdot \text{K} = 0.76 \text{ (Btu)(in.)/(ft}^2\text{)(h)(°F)}$$

Using equation 12 where the area is constant (Fig. 1),

$$q = -\frac{t_i - t_o}{(1/A)[(x_c/k_c) + (x_p/k_p) + (x_f/k_f)]}$$

In SI units:

Table 2. Thermal Conductivity of Selected Foods and Processing Facility Materials

Product	Temperature, °C	Thermal conductivity, k^a (W)(m)/(m^2)(K)	Thermal conductivity, k^a (Btu)(in.)/(ft^2)(h)(°F)
Apple	2–36	0.393	2.72
Aluminum	0	230	1594
Beef, lean	7	0.476	3.20
perpendicular to fiber	62	0.485	3.36
parallel to	8	0.431	2.99
fiber	61	0.447	3.10
Butter	46	0.197	1.37
Concrete, cinder	23	0.78	5.27
Fish muscle	0–10	0.557	3.86
Ice	0	2.3	15.9
Milk	37	0.530	3.67
Olive oil	15	0.189	1.31
	100	0.163	1.13
Polystyrene	24	0.038	0.263
Potato, raw	1–32	0.554	3.84
White fir	23	0.11	0.76
White pine	15	0.15	1.04

Source: Refs. 1 and 2.
[a] 1 (Btu)(in.)(ft.^{2})(h)(°F) = 0.144228 (W)(m)(m^2)(K).

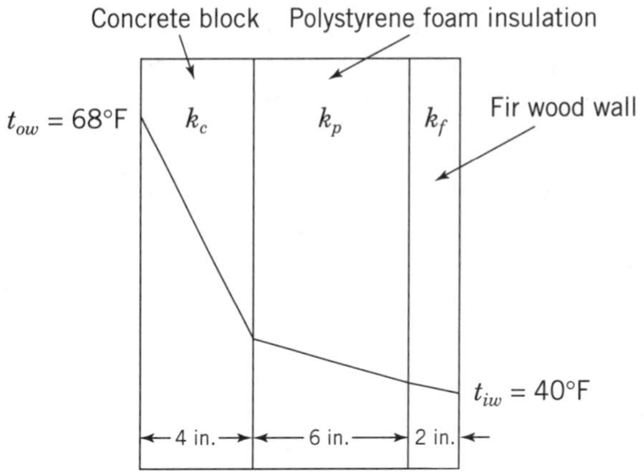

Figure 1. Temperature profile of wall in cold storage room.

$$q = -\frac{(-40) - (20)}{(1/2)[0.1016/0.76 + 0.1524/0.038 + 0.0508/0.11]}$$

carrying the unit with the numerical equation

$$q = \frac{\text{°C or K}}{(1/\text{m})[(\text{m/Wm/m}^2\text{K}) + (\text{m/Wm/m}^2\text{K}) + (\text{m/Wm/m}^2\text{K})} = \text{watts}$$

$$q = 26.05 \text{ W lost through 2 m}^2 \text{ of the wall}$$

Likewise, in English units:

$$q = \frac{(-40) - (68)}{(1/21.528)[4/5.27 + 6/0.263 + 2/0.76]}$$

carrying the unit with the numerical equation

$$q = \frac{°F}{1/ft[\Sigma(in./Btu)(in.)/ft^2(h)(°F)}= Btu/h$$

$$q = 88.74 \text{ Btu/h lost through 2 m}^2 \text{ of the wall}$$

Checking units by converting English to SI:

$$q = (88.74 \text{ Btu/h})(0.29307 \text{ W-h/Btu}) = 26 \text{ W}$$

PROCESSING FOODS BY CONVECTION HEAT TRANSFER

Convection heat transfer is concerned with the mixing of fluids or the transfer of heat from a fluid to a surface. Therefore there is no solid-body thickness through which heat must be transferred, as in conduction. In normal food processing operations the concern is usually to transfer heat between a liquid and the surface of a solid or a fluid. Newton's law of cooling, which also applies to heating, defines the driving force as the temperature difference between the surface temperature (t_s) and the bulk temperature (t_f) of the fluid medium:

$$q = -hA(t_s - t_f) \tag{13}$$

In this case the temperature of the surface is higher than the bulk temperature of the liquid and h is the convective heat transfer coefficient (commonly called the film coefficient), the experimentally determined proportionality constant that takes into consideration the flowing characteristics and liquid—solid interface effects. The units are $W/m^2 \cdot K$ or $Btu/(h)(ft^2)(°F)$. Note that there is no dx or thickness factor so that the resistance R corresponding to conduction heat transfer is

$$R = 1/hA \tag{14}$$

Figure 2 depicts the type of temperature profile that occurs when a hot fluid is transferring heat to a solid (or liquid) surface. The equilibrium is between the bulk temperature of the liquid (t_f) and the surface of the solid (t_s). Note that the temperature drops in a nonlinear manner. This is caused by the very thin layer of fluid that is almost stagnant because of the friction of the moving fluid, often referred to as the edge effect. This layer is flowing in the streamlined region and, since there is essentially no mixing with the main liquid stream, the heat transfer through this layer is actually by conduction but without a thickness (dx) to consider. The magnitude of the edge or fictitious film effect varies widely depending on the solid surface and explains why this coefficient must be experimentally determined for any given condition. Some select convection film coefficients are given in Table 3. The wide range of film coefficient values is due to the surface and geometry over which the fluid is passing and the turbulence (Reynolds number) of the flowing fluid.

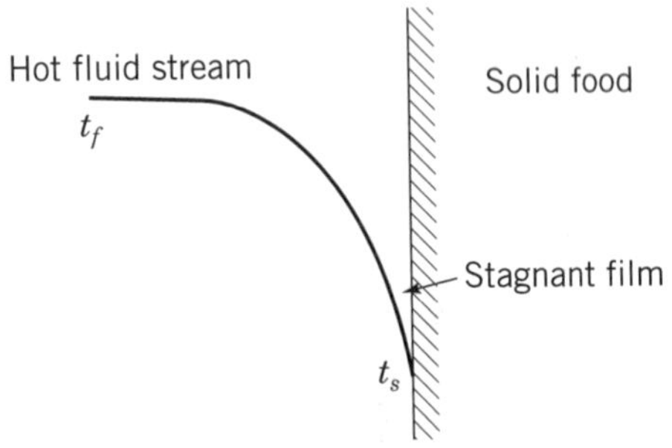

Figure 2. Temperature profile of hot flowing liquid near the surface of solid being heated.

Over a period of time, scale deposits build up on the walls of vessels and pipes. The effects of these films or deposits are determined experimentally and expressed in units of the film coefficient h. Therefore films and scale-deposit film coefficients can be handled in the same manner as the thermal conductance of a flowing liquid.

Steady-state heat transfer processes occur in the operations that involve continuous heating or cooling of foods. These include continuous freezing tunnels, deep frying, and pasteurizing and sterilizing liquids. Many processing procedures in which a solid is being heated in a continuous moving system (eg, freezing or cooking a product on a moving belt) can be treated as a total steady-state system by using specific locations on the belt as base points for heat balances. In most of these cases steam, hot water or oil, or a refrigerant is the source for adding or removing heat, so that convection heat transfer is an important factor in these processes.

The only procedures that must be handled as batch, unsteady-state systems are those in which a product is being heated or cooled in place by a steady-state or non-steady-state source (eg, cooking in an oven, retorting canned foods, and freezing a product on a freezer plate). Normally the information required to calculate heat transfer in these situations is collected experimentally and then applied. For example, the thermal death time to ensure the sterilization of canned food is experimentally determined for each size of container and product and then applied to the commercial facilities. The actual retorting time is then increased by a sufficient time to insure that there is no error in obtaining sterilization. In the case of sterilizing products, the National Food Processors Association (in coordination with the Food and Drug Administration) provides

Table 3. Approximate Range of Convection Heat Transfer Coefficients

	Conduction heat transfer film coefficient, h	
Medium	$W/m^2 \cdot K$	$Btu/(ft^2)(h)(°F)$[a]
Gases		
natural convection	2.8–28	0.9–9
forced convection	11–110	3.5–35
Viscous liquids,		
forced convection	56–560	17–178
Water, forced convection	560–5600	180–1,800
Boiling water	1,700–28,400	540–9,000
Condensing steam	5,600–1.1 × 10	1,800–3.5 × 10

Source: Refs. 1 and 2.
[a] $1 \text{ Btu/(Ft}^2)(h)(°F) = 3.15459 \text{ W/m}^2 \cdot K$.

sterilizing time requirements to the processors. The processors must then record and maintain the records of retort time and temperature for each batch processed to insure compliance with the retorting requirements. These records are available in legal cases involving undercooking.

COMBINED CONVECTION AND CONDUCTION HEAT TRANSFER

Conduction and convection have been discussed separately in order to emphasize the different mechanisms that control the heating and cooling of food products. However, seldom is a food processing operation carried out in which the heat transfer is not a combination of conduction and convection. Although radiation is also present in most of these operations, the effect is usually quite insignificant unless the process is specifically a radiation process.

Figure 3 extends the transfer of heat from a liquid to a solid surface, through the solid, and then to another liquid. This could be the situation occurring in a heat exchanger where a hot liquid is being used to warm another liquid, with the solid being the wall of the heat exchanger. The resistances to heat flow through the hot fluid (R_h), solid pipe wall (R_w), and fluid being heated (R_c) can be summed to give a total resistance

$$R = R_h + R_w + R_c \tag{15}$$

Although only three media are used in the example, there is no limit to the number of resistances that can be summed for a given problem.

An overall heat transfer coefficient U combining conduction and convection heat transfer coefficients now can be defined from equation 15, where, although only three media are used in the example, there is no limit to the summing of as many resistances as are necessary in a given situation

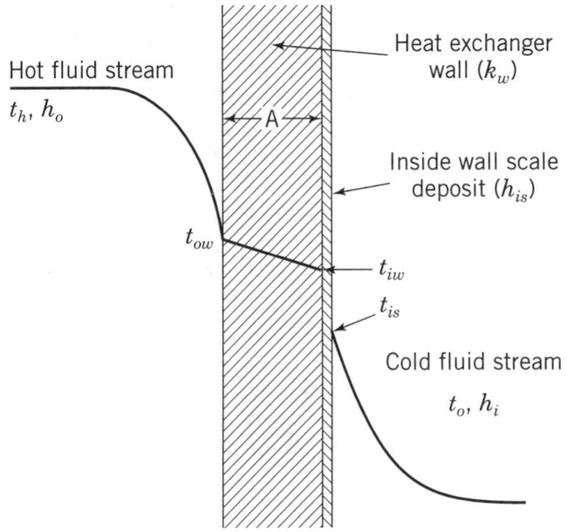

Figure 3. Temperature profile of hot liquid (t_h) transferring heat through heat exchanger wall to another colder liquid (t_e).

$$1/UA = 1/h_hA_k + x_w/k_wA_w + 1/h_cA_c \tag{16}$$

or

$$UA = \frac{1}{1/h_hA_h + x_w/k_wA_w + 1/h_cA_c} \tag{17}$$

Thus the overall heat transfer equation becomes

$$q = -UA\Delta t \tag{18}$$

As noted, U and A are not independent since, if the areas of the heat transfer surfaces vary (eg, a pipe heat exchanger), A must be based on a mean area (A_m). For example, in the process of transferring heat through a pipe that has an inside scale deposit with a determined h_{is}, the overall conductance can be calculated as

$$UA = \frac{1}{1/h_oA_{ow} + x_w/k_wA_m + 1/h_{\mathrm{is}}A_{\mathrm{is}} + 1/h_{\mathrm{i}}A_{iw}} \tag{19}$$

where

$$A_{\mathrm{iw}} = A_{\mathrm{is}} = \pi DL$$

$$A_{\mathrm{ow}} = A_{\mathrm{os}} = \pi DL$$

$$A_m = (A_{\mathrm{ow}} - A_{\mathrm{iw}})/\ln(A_{\mathrm{ow}}/A_{\mathrm{iw}})$$

multiplying the denominator by A_m/A_m gives

$$UA = \frac{1}{(1/A_m)[A_m/h_oA_{\mathrm{ow}} + x_w/k_{\mathrm{w}} + A_m/h_{\mathrm{is}}A_{\mathrm{io}} + A_m/h_{\mathrm{i}}A_{\mathrm{iw}}]} \tag{20}$$

and rearranging gives an expression for UA in a circular cross section where heat is being transferred:

$$UA = \frac{A_m}{A_m/h_oA_{\mathrm{ow}} + x_w/k_w + A_m/A_{\mathrm{iw}}h_{\mathrm{is}} + A_m/A_{\mathrm{iw}}h_{\mathrm{iw}}} \tag{21}$$

In this example, the effect of the scale deposit, h_{iw}, is on the inside wall of the pipe. There are commonly five resistances to heat transfer caused by the (1) heat transfer wall and (2) inside (i) and (3) outside (o) of the heat transfer wall, (4) 2 fluids (h_o and h_i), and (5) 2 fouling scales (h_{is} and h_{os}). Often there is a fouling scale on the surface of a transfer surface caused by oxidation or burn-on during operation. Protein foods are particularly bad in that the heat causes denaturation and subsequent adherence to the surface. When gas comes into contact with heat transfer surfaces (eg, a fire tube boiler) heavy scales can be formed. In fact, the maintenance and cleaning of heat exchanger surfaces is a continuous chore, usually accomplished by chemical cleaners. However, in extreme cases (especially in boiler tubes), a periodic rodding out is necessary to maintain the efficiency of the heat transfer.

Boilers are particularly important to the food processing industry since they are the source of steam for many heat transfer systems. The boiler is a heat exchanger that converts a liquid to a vapor. The vapor, after being condensed in a food processing heat exchanger, is normally collected and recycled. Often there are fouling scales or film deposits on both sides of a pipe or plate heat exchanger. This must be recognized in the calculation of the overall heat transfer coefficient.

EXAMPLE 2

In determining the heat transfer characteristics of a heat exchanger, it is important to have a clear picture of the physical exchanger prior to calculations. For example, in a pipe heat exchanger, it is common to determine the *UA* for a unit length of pipe. An exchanger can then be sized to meet the heat transfer requirements. Consider a heat exchanger of stainless steel tubes with the cross section shown in Figure 4. Each tube is 1-in. OD with the following dimensions:

$$D_1 = \text{inside diameter} = 0.782 \text{ in. } (0.0199 \text{ m})$$

$$D_o = \text{outside diameter} = 1.000 \text{ in. } (0.0254 \text{ m})$$

$$x_w = \text{wall thickness} = 0.109 \text{ in. } (0.00277)$$

The heat transfer coefficients are

inside: $h_i = 120$ Btu/(h)(ft^2)(°F) or 681 W/m$^2 \cdot$ K
inside fouling: $h_{is} = 660$ Btu/(h)(ft^2)(°F) or 3748 W/m$^2 \cdot$ K
outside fouling: $h_{os} = 1000$ Btu/(h)(ft^2)(°F) or 5678 W/m$^2 \cdot$ K
outside: $h_o = 1500$ Btu/(h)(ft^2)(°F) or 8517 W/m$^2 \cdot$ K

and the conductivity of the stainless steel tube is

$$k = 9.4 \text{ (Btu)(ft)/(h)(ft}^2\text{)(°F) or 16.3 W-m/m}^2\text{-K}$$

Determine *UA* for the tubes in this heat exchanger. Basis:

1. Consider a 1-ft length (L) of tube.
2. Consider a 1-m length (L) of tube.

Since the area of the tube cross section is expanding perpendicularly to the heat transfer, a mean area (A_m) value must be sued in determining *UA*, and the basic equation becomes

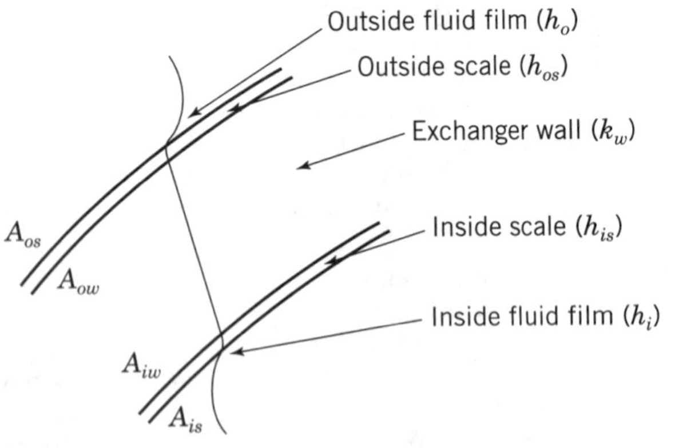

Figure 4. Temperature profile for heat exchanger in which heat transfer area is not constant.

$$UA = \frac{A_m}{(A_m/h_o A_{ow}) + (A_m/h_{os} A_{ow}) + (x_w/k_w) + (A_m/h_{is} A_{iw}) + (A_m/h_i A_{iw})}$$

Calculation of areas in English units:

$$A_{ow} = A_{os} = \pi DL = \pi(1)(1)/12 = 0.262 \text{ ft}^2$$

$$A_{iw} = A_{is} = \pi \text{DL} = \pi(0.782)(1)/12 = 0.205 \text{ ft}^2$$

$$A = \frac{A_{ow} - A_{iw}}{\ln(A_{ow}/A_{iw})} = \frac{0.262 - 0.205}{\ln(0.262/0.205)} = 0.232 \text{ ft}^2$$

Calculation of areas in SI units:

$$A_{ow} = A_{os} = \pi DL = \pi(0.0254)(1) = 0.0798 \text{ m}^2$$

$$A_{iw} = A_{is} = \pi DL = \pi(0.0199)(1) = 0.0625 \text{ m}^2$$

$$A = \frac{A_{ow} - A_{iw}}{\ln(A_{ow}/A_{iw})} = \frac{0.0798 - 0.0625}{\ln(0.0798/0.0625)} = 0.0708 \text{ m}^2$$

UA in English units:

$$\begin{aligned} UA &= \frac{A_m}{[\Sigma A/hA + x/k]} \\ &= 0.232 \Big/ \left(\frac{0.232}{(1500)(0.262)} + \frac{0.232}{(1000)(0.262)} + \frac{0.109}{(12)(9.4)} \right. \\ &\quad \left. + \frac{0.232}{(660)(0.205)} + \frac{0.232}{(120)(0.205)} \right) \\ &= 17.1 \text{ Btu/(h)(°F)(ft length)} \end{aligned}$$

UA in SI units:

$$\begin{aligned} UA &= 0.0708 \Big/ \left(\frac{0.0708}{(8517)(0.0798)} + \frac{0.0708}{(5678)(0.0798)} + \frac{0.00277}{(16.3)} \right. \\ &\quad \left. + \frac{0.0708}{(3748)(0.0625)} + \frac{0.0708}{(681)(0.0625)} \right) \\ &= 29.55 \text{ W/(K)(m length)} \end{aligned}$$

RADIATION HEAT TRANSFER

Since all bodies continuously emit thermal radiation and absorb radiation from other bodies around them, both the received and emitted energy (combining emitted, received, and reflected energies) must be considered when determining the temperature state of a body. This total energy interchange between bodies is significantly affected by the nature of the bodies and the conditions in which these bodies exist. The ability of a body to absorb and emit energy is measured by its emissivity, the fraction absorbed or emitted compared to an ideal perfect radiator or so-called blackbody that has an emissivity (ϵ) of 1.0. It has been determined that the amount of energy emission from a body is proportional to the fourth power of the absolute temperature. A heat transfer equation can be balanced by the experimentally determined proportionality constant known as the Stefan-Boltzmann constant (σ), hence, for a blackbody

$$q = \sigma T^4 \tag{22}$$

where

$$q = W/m^2 \text{ or Btu/(h)(ft}^2) = 5.6697 \times 10^{-8}\ \text{W/m}^2\text{-K}^4$$
$$= 0.173 \times 10^{-8}\ \text{Btu/(h)(ft}^2)(°\text{R})^4$$
$$T = K \text{ or } °\text{R}$$

In reality, there is no such thing as a blackbody that has an emissivity of 1.0, so the practical equation must take into account the experimentally determined emissivity (fraction of 1.0), resulting in the usable equation for radiation

$$q = \sigma \in T^4 \quad (23)$$

EXAMPLE 3

A person wearing a white shirt and pants ($\epsilon = 0.9$) at room temperature will radiate energy.

In SI units:

$$q = (5.6697)(10^{-8})(0.9)(\text{W/m}^2\text{K}^4)(293)^4(\text{K})^4$$
$$= 376\ \text{W/m}^2$$

In English units:

$$q = (0.1713)(10^{-8})(0.9)[\text{Btu/(h)(ft}^2)(°\text{R})^4](528)^4(°\text{R})^4$$
$$= 119.8\ \text{Btu/(h)(ft}^2)$$

Checking units:

$$q = (119.8)(3.1546) = 377\ \text{W/m}^2$$

where

$$1\ \text{Btu/(h)(ft}^2) = 3.1546\ \text{W/m}^2$$

Since there is an interchange between two bodies, the net amount of radiation received from a source is the radiation received minus the radiation returned or emitted. The combination of the heat being absorbed by a food and the amount being emitted gives an equation or the net heat transfer to a food from a radiation source:

$$q = \sigma\{[(\varepsilon_S\alpha_R F_{S\text{-}R})(T_S)^4] - [(\varepsilon_R\alpha_S F_{R\text{-}S})(T_R)^4]\} \quad (24)$$

where σ is the Stefan-Botzmann constant, ϵ_S is the source emissivity, α_R is the receiver absorptivity, $F_{S\text{-}R}$ is the view factor, source to receiver, T_S is the temperature of source, ϵ_R is the receiver emissivity, α_S is the source absorptivity, $F_{R\text{-}S}$ is the view factor, receiver to source, and T_R is the temperature of receiver.

The amount of energy absorbed from a body depends on how much of the radiating body can be "seen" by the receiver. In most situations involving the processing of food by radiant heat transfer, the view factor is equal to unity since the source is directly below or adjacent to the receiver. Also, the emissivity is essentially the same as the absorptivity in a food processing operation. The emissivities for many surfaces have been determined experimentally and are available in standard data tables. The emissivity of many foods is approximately 0.9, while hot metal (as in the case of oven heating units) has a wide range emissivity values, from as low as 0.1 to 0.6 or higher.

EXAMPLE 4

Bread is being baked on a stainless steel sheet. The top is cooking nicely but the bottom is not getting done. The oven (ceiling, wall, and floor) temperature is 400°F (204.4°C) and the emissivity and absorptivity are 0.90

1. The bottom of the stainless steel sheet has emissivity absorptivity of 0.25. Assuming a view factor of 1.0, what is the net rate of radiant heat transferred to the bottom of the sheet? Assume that the surface of the bread while cooking and vaporizing water is 212°F (100°C) and that the conduction heat transfer to the stainless steel sheet is to a surface at the same temperature.
2. What is the net rate of heat transfer to the top of the bread, which has emissivity—absorptivity equal to 0.90?

Solution. (1) By substituting the values in equation 22, the net heat that is being transferred to the bottom of the sheet is:

In English units:

$$q = (0.1713)(10)^{-8}[(0.9)(0.25)(1.0)(860)^4 - (0.25)(0.9)(1.0)(672)^4]$$
$$= 132\ \text{Btu/(h)(ft}^2)$$

In SI units:

$$q = (5.6697)(10)^{-8}[(0.9)(0.25)(1.0)(478)^4 - (0.25)(0.9)(1.0)(373)^4]$$
$$= 417\ \text{W/m}^2$$

Checking units:

$$(132) \times (3.1546) = 417\ \text{W/m}^2$$
$$\frac{\text{Btu}}{(\text{h})(\text{ft}^2)}\,\frac{(\text{W})(\text{hr})(\text{ft}^2)}{(\text{m}^2)\text{Btu}} = \text{W/m}^2$$

(2) By substituting the values in equation 22, the net heat that is being transferred to the top of the bread is:

In English units:

$$q = (0.1713)(10)^{-8}[(0.9)(0.9)(1.0)(860)^4 - (0.9)(0.9)(1.0)(672)^4]$$
$$= 476\ \text{Btu/(h)(ft}^2)$$

In SI units:

$$q = (5.6697)(10)^{-8}[(0.9)(0.9)(1.0)(478)^4 - (0.9)(0.9)(1.0)(373)^4]$$
$$= 1508\ \text{W/m}^2$$

It is clear that the top portion of the bread is receiving more than 3.5 times the amount of heat being received on the bottom where it is in contact with the stainless steel sheet. This explains why the top, the portion receiving the most heat, is baking at a faster rate than the bottom. It should be noted that often cookies or items being baked in an oven burn on the bottom and are obviously being heated at a faster rate than the top of the item. This is due to the fact that often the metal container is in contact with a hot shelf and is thus receiving heat by conduction.

TRANSFERRING HEAT IN FOODS

Heat transfer was introduced as being an important unit operation in almost every food process. In fact, the transfer of heat to and from a food is certainly the most important factor to consider during processing, whether it occurs in a food processing operation or through natural temperature changes during holding and storage. The type of heat transfer and the rate at which the transfer occurs affect the nutritional value, product safety, and sensory properties of every food that is consumed by humans and animals.

The processes of canning, pasteurizing, and cooking depend on transferring heat into a product to elevate the temperature to a level that will destroy microorganisms. Heat sterilization of a product (eg, pouch packaging and canning) under anaerobic conditions requires closely controlled time and temperature process conditions to ensure that spores resistant to heat (eg, *Clostridium botulinum*) do not survive. Pasteurization temperature and time are just as important to ensure that food-borne diseases are not spread by the food. Cooking a food not only improves sensory acceptance but reduces the microbial content.

Combined heating and cooling operations are commonly carried out during processing or handling of foods. The cooling normally follows the required heating time and temperature period to minimize damage to nutritional and sensory attributes after the heat has accomplished the intended purpose. For example, some fruits and vegetables are blanched prior to freezing to inactivate enzymes that cause degradation and greatly reduced shelf life. Pasteurization to kill microorganisms that might cause public health problems leaves many other microorganisms that will continue to grow. Therefore, pasteurized products (eg, milk) must be held at refrigeration temperatures to ensure a reasonable shelf life. Cooked foods that are not consumed immediately after cooking must be stored in the refrigeration for the same reason.

Many food processes require that heat be added to a food for a purpose other than reducing microorganism contamination or enzyme activity. Vaporization processes such as evaporation, dehydration, and distillation utilize heat energy to vaporize a component, usually water, to change the characteristics of the food. Hence a concentrated juice or a dried cereal product may be subjected to an elevated temperature during the processing.

Often during food processing mechanical energy is converted to heat energy by friction from the particles of food as they rub together. The resulting energy release from such operations as cutting, grinding, screening, and extruding can significantly increase the temperature of a product.

Nature is responsible for adding the heat that is necessary for drying grains in the field to the extent that they can be harvested and stored for long periods of time before being finally processed and consumed. During extremely hot harvest times many fruits and vegetables can be severely damaged during the time it takes to harvest and transport them to a processing plant.

In all cases, whether the heat be intentionally added for safety and to improve texture or unavoidably added as a result of handling or processing conditions, a certain amount of nutritional damage occurs during the heat processing of a food. Depending on the food water content, pH, and oxygen exposure, heat can severely degrade many vitamins, denature proteins, cause free radicals to form, and adversely affect quality attributes such as texture and flavor.

In summary, the transfer of heat directly or indirectly effects the acceptance, quality, and safety of a food. A major responsibility of the food scientist and engineer working with food processes is to optimize the time and temperature conditions of processing and minimize the adverse side affects of heat and other processing conditions to insure that the consumer receives the highest quality food at the minimum cost.

BIBLIOGRAPHY

1. E. L. Watson and J. C. Harper, *Elements of Food Engineering*, 2nd ed., Van Nostrand Reinhold Company, New York, 1988.
2. R. P. Singh and D. R. Heldman, *Introduction to Food Engineering*, Academic Press, New York, 1984.

GENERAL REFERENCE

Heat Transfer, H. Wolf, Harper & Row, Publishers, New York, 1983.

GEORGE M. PIGOTT
University of Washington
Seattle, Washington

HIGH-PRESSURE PROCESSING

Consumers are shifting their food purchases from heat-processed to fresh-tasting, minimally processed foods since excessive heat treatment can reduce the perceived freshness of foods. Salads, fruits, vegetables, nuts, spices, oysters, and selected cheeses are traditionally consumed without heat treatment. However, the incidence of pathogenic microbes is increasing in minimally processed and in raw foods previously considered safe. *Escherichia coli* 0157:H7 has been found in fresh apple and orange juice. This has caused the U.S. Food and Drug Administration (FDA) to recommend pasteurization. High-pressure processing (HPP) is one alternative to heat to achieve the pasteurization or commercial sterility of many fresh and freshlike processed foods. An advantage of HPP is that the treatment does not break covalent bonds. As a result, HPP does not change the flavor, color, or nutrient content of the food. Heat and ionizing radiation break covalent bonds as a function of dose and thus have the potential for changing the chemistry of a treated food. An additional advantage of HPP over other food preservation methods is that the effect of HPP is instantaneous and uniform throughout the mass of food being treated. The process is completely independent of the volume, geometry, chemical composition, and physical structure of the food. The capabilities of HPP to yield a pathogen-free, minimally processed food, with an extended shelf life due to inactivation of vegetative forms of spoilage microbes (but not spores or viruses), have attracted the attention of food processors. HPP-treated foods are appearing in the marketplace in Japan, North America, and Europe.

TECHNOLOGY OF FOOD PRESERVATION BY HIGH PRESSURE

Mechanism of Action

Research during the past 100 years has demonstrated that pressures above 100 MPa (15,000 psi) can affect the activity of enzymes and the structure of proteins. The unfolding of a protein polymer and subsequent irreversible denaturation of a protein at pressures above 300 to 400 MPa (45,000 to 60,000 psi) can affect the ability of living cells to control the flux of water and ions across their cellular membranes. Electron micrographs of pressure-treated microbes indicate a loss of cell turgor and collapse of cell structures, indicating damage to the cell membranes. It is believed that this is the mechanism that allows high pressure to inactivate pathogenic and spoilage vegetative forms of microbes in foods. Bacteria, yeast, and molds are primarily single-cell plants. Cell membrane damage can be fatal to their survival. Similarly, parasites, insects, and insect eggs can be inactivated by high pressures in the range of 200 to 680 MPa (30,000 to 100,000 psi). Food products generally consist of large integrated masses of cells. Although high pressure does denature cell membranes and can cause the gelation of starches, many foods—including cooked meats, raw and cooked vegetables, firm fruits such as peaches, and cooked pasta and rice—show little change in structure and texture after treatment at pressures in the range of 680 MPa (100,000 psi). High-pressure inactivation rates of vegetative microbes have been observed to be a function of pressure and time at any given temperature. Experimental data have shown that the rate of inactivation and the shape of the inactivation curve may be influenced by the media in which the microbes are suspended. Generally a lowering of the pH will decrease pressure resistance of a given strain of microbe. By contrast, as the water activity of the suspending media is reduced, the vegetative microbes in the media become more pressure resistant. Also, the strains of any given species of microbe can display a wide range of pressure resistance, and food materials can increase the pressure resistance of a microbe when compared with its rate of inactivation in a phosphate buffer of the same pH. Classic inactivation kinetics based on a first-order reaction may not be followed. The rate of inactivation of a microbe as a function of pressure and time may not be proportional to the remaining concentration of viable microbes. An induction period may be observed, and tailing may also be observed as low numbers of survivors are reached. These findings require that any proposed HPP food-preservation treatment must be challenged with pathogens and spoilage microbes normally associated with the product. Recommended levels of challenge are 10^6 pathogenic microbes and up to 10^8 spoilage microbes per gram of product.

Shelf-stable (commercially sterile) acidic foods may be produced by high-pressure treatment. A pH below 4.5 will block the germination of spore-forming pathogens such as proteolytic and nonproteolytic *Clostridium botulinum*. High pressure may be used to produce pressure-pasteurized, extended-shelf-life, low-acid, refrigerated foods. A water activity above about 0.98 is essential for high-pressure treatment to be effective in the inactivation of vegetative microbial cells. High-pressure treatments have limited applications in the pasteurization of dry foods such as spices, dry fruits, vegetables, and meat products. Because parasites in animal tissue can be inactivated at pressures (200 MPa, 30,000 psi) that have little effect on fresh food quality factors (such as protein denaturation), pressure may be useful in treating fresh animal foods normally infected with parasites. Insect and insect egg inactivation is another possible application of HPP.

Processing Systems for Food Preservation by High Pressure

High-pressure food treatment systems are available for semicontinuous processing of liquid foods and batch processing of packaged foods. Liquid foods processed in semicontinuous systems may be packaged in aseptic or clean-fill packaging equipment after pressure treatment. The semicontinuous treatment of liquid foods requires the following steps. A conventional pump is used to transfer liquid food from a hold tank through an inlet port into a stainless steel high-pressure vessel treatment chamber. The chamber is fitted with a free piston that is displaced from the filling port end toward the opposite end of the chamber where high-pressure water is introduced to pressurize the chamber when the filling step is complete. The filling port is closed at the end of the filling step, and high-pressure water is pumped into the chamber to drive the piston against the food. Because water is compressed to 85% of its original volume at 680 MPa (100,000 psi), the piston is displaced up to 15% of the distance down the chamber. After an appropriate hold time at the desired process pressure, usually in the range of 30 s to 5 min, the pressure is released, and a sterile discharge port is opened. The pressure treated liquid is pushed out of the pressure vessel into a sterile hold tank by low-pressure water moving the free piston to the product inlet end of the vessel. The cycle is repeated under the direction of an automated control system that can monitor and record pressures, times, piston positions, product levels in tanks, process temperatures, and the status of critical seals and valves. A single high-pressure water pump may be used to pressurize several pressure vessels under the direction of the automated control system. A semicontinuous system is manufactured by Flow International (Kent, Washington).

The high-pressure treatment of packaged food is similar to a batch steam retort process. The food is vacuum packed in packages capable of withstanding compression of up to 15% of their volume. The packaged products may be loaded into a carrier that is placed in the pressure vessel. The vessel is sealed, and water is pumped into the vessel until the desired treatment pressure is reached. After the appropriate hold time, the pressure is released, the vessel opened, and the carrier removed. The treated product is then ready for distribution. The batch process may be automated using a control system similar to that used in a semicontinuous system. A single high-pressure pump may be used to pressurize several pressure vessels.

Research on the effect of temperature on the rate of microbial inactivation has demonstrated that temperatures in the range of 50°C (122°F) can increase the rate of microbial inactivation during pressure treatment. The use of controlled temperature conditions during semicontinuous or batch pressure treatment requires that all product must be at the desired process temperature for the time and pressure of the treatment. High-pressure treatment raises the temperature of high-water-content foods about 3°C (5.4°F) per 100 MPa (15,000 psi) of imposed pressure. This temperature increase is lost upon decompression. The specification of a food process temperature in conjunction with a high-pressure treatment could subject the proposed process to FDA regulation.

Systems for the high-pressure treatment of foods require four components. These are a pressure vessel and closure system, a low-pressure pump, an intensifier to deliver high-pressure water, and a control system. Safe, reliable, pressure vessels and closures are now available commercially with capacities of 25 to 250 L (0.88 to 8.8 cubic ft). These vessels are capable of operating at pressures of 680 MPa (100,000 psi) for more than 100,000 cycles. The extreme operating pressures and rapid cycling required for food processing applications has favored wire wound pressure vessels (Flow International, Kent, Washington) and other approaches to ensure safe, "leak before break," designs. Each time a pressure vessel is cycled, the inner wall is subjected to potential crack formation. Successful pressure vessel designs require that the inner wall be under some compression even when the vessel is at full pressure so that potential cracks cannot propagate. Pressure vessels are made from high-tensile-strength steels with several concentric cylinders assembled to ensure compression of the inner wall of the vessel. Wire winding, in which miles of steel wire are wound around a cylindrical liner to ensure compression, can be used as an alternative technology to obtain cycle lives well in excess of 100,000 cycles. The reader is referred to the general references for detailed descriptions of vessel construction technologies. High pressures are achieved by pumping oil, at pressures up to 34 MPa (5000 psi), to an intensifier. The oil works against a piston in the intensifier with an area perhaps 20 times larger than the piston delivering the high-pressure water. Thus the water pressure in this case is exactly 20 times greater than the oil pressure. Pressure vessels are available with built-in intensifiers, and systems can be built with external single-stroke intensifiers. The advantage of a single-stroke intensifier is reduction in system wear and ease of decompression. The control system use in the HPP treatment of foods must be able to deliver a record of the pressure and time of each cycle. A record of the food temperature is needed if temperature is a specified component of the process.

PACKAGING FOODS FOR HIGH-PRESSURE TREATMENT

A major consideration in the packaging of foods for pressure treatment is the effect of compression on the structure and barrier properties of the package. If rigid structures are used, means must be incorporated to allow for up to 15% reduction in volume of the product during pressure treatment at 680 MPa (100,000 psi). Otherwise, flexible packages such as pouches or semiridged structures, such as trays or bowls with easy open peel-seals, can be used. Foods packaged for high-pressure treatment must be vacuum treated to remove as much dissolved and occluded gas as possible. They must be packaged without a headspace because any gas will be forced into the food during compression. Compression of gas represents extra compression work and the presence of dissolved oxygen in the food may lead to accelerated development of oxidized flavors. Ultimately the product-package system must be treated, stored, and evaluated for product quality after the desired storage time at expected storage temperatures. As with all packaging, the food must be held in the proposed package to determine if undesirable flavors transfer from the package to the food.

PRODUCT SAFETY AND REGULATORY ISSUES

Regulations covering the manufacturing and sale of pressure-treated foods are under development at this date. Pressure-treated foods distributed as products labeled "requires refrigeration" are on the market in the United States. These types of products must be free of pathogens, must be manufactured under good manufacturing practices (GMP), and must remain free of spoilage microbes until their "use by" date. Naturally acidic and directly acidified products are covered by existing regulations for these products. Their safety requirements are the same as for refrigerated foods. Currently product safety and regulatory requirements for shelf-stable, low-acid foods, preserved by high pressure, and packed in hermetically sealed containers have not been developed in the United States. Good manufacturing practices (GMP) can be expected to be similar to those for heat-processed foods. It is expected that an HPP process will be required to provide the equivalent of a 12-fold decimal reduction of a pressure-resistant strain of *C. botulinum*.

Because spores are pressure resistant, all commercially sterile high-pressure preserved foods to date have been acidic products with a water activity very close to 1.0. Examples are yogurt, fruit preserves, fruit juices, and pourable salad dressings. Viable microbial counts may be obtained when commercially sterile pressure-treated acidic foods are plated on neutral pH media. For this reason all pressure-preserved products should be stored at an appropriate storage temperature for twice their proposed shelf life to assay for possible regeneration of microbes.

ECONOMIC CONSIDERATIONS

Although high-pressure treatment systems are inherently expensive to build and maintain, construction and maintenance costs are expected to decrease as commercial use expands. For any installed high-pressure treatment system, the cost per unit of food treated is a function of the amount of product treated per cycle (batch or semicontinuous operation) and the number of cycles run per year. The shape of the package is an important factor in determining the amount of product that can be treated each cycle. Cylindrical containers may occupy 30% of a pressure vessel while product packed in bags for food service may occupy up to 90% of the available vessel volume. Cycle time is a function of the required treatment time at pressure and the time needed to move the food in and out of the pressure vessel, pressurize, and decompress the vessel. Semicontinuous systems processing liquids with a 3-min hold time at 680 MPa (100,000 psi) can cycle up to 15 times per hour and utilize 100% of the available process vessel volume. A batch system treating packaged foods may require 5 min to load, close and seal, pressurize, decompress, open and unload, in addition to the 3-min process time. This will result in seven to eight cycles per hour. Another major cost factor is system availability; that is, the percentage of time the unit is not down for maintenance and repairs and is available for production. System availability is related to the reliability of the pressure vessel closures, valves, piping, pumps, intensifier, control systems, and material handling equipment. The food industry will need a system reliability in excess of 95% exclusive of programmed maintenance.

GENERAL REFERENCES

D. A. Ledward et al., eds., *High Pressure Processing of Foods*, Nottingham University Press, Loughborough, United Kingdom, 1995.

T. Ohlsson, ed., *High Pressure Processing of Foods and Food Components—A Literature Survey and Bibliography*, Swedish Institute for Food and Biotechnology, Göteborg, Sweden, 1996.

M. W. Peck, "*Clostridium botulinum* and the Safety of Refrigerated Processed Foods of Extended Durability" *Trends Food Sci. Tech.* **8**, 186–192 (1997).

DANIEL F. FARKAS
Oregon State University
Corvallis, Oregon

HISTORY OF FOODS

Writing a history of food is a formidable task that involves identifying, describing, and discussing what people ate where and at what time, where it came from, and how it was obtained. This article attempts to sketch a broad picture of the human quest for food through time and, primarily, will be an archaeological odyssey for reasons discussed below. It treats the history of food as the general history of the human subsistence economy.

Food is an integral part of the human experience, and because the human experience is long, so must be the story of food. The greater part of the human experience occurred long before the domestication of plants and animals. The invention of writing came considerably after the development of agriculture. For some two million years humanity wrote nothing that would aid us in reconstructing its subsistence. Even after writing was invented some 5500 yrs ago, very little was written about food for a long time. Therefore the history of food is for the most part based on archaeological materials.

A history of food must have a beginning. An anthropologist would choose to begin the story in the dim, distant ages of human prehistory some two million years ago with the earliest fossils human paleontologists place in the genus *Homo*. From the two-million-year point, our evolving ancestors and their food quests and habits will be traced to the present time. Important aspects of their lifestyles that bear on the evolution of their diets and general subsistence economies will also be briefly sketched. This approach takes us first, into the realms of physical anthropology and archaeology (paleoanthropology); second, into ancient history; and third, into history. We move from the earliest foraging and scavanging subsistence economies through gathering and hunting to the invention of agriculture and beyond. Finally, some modern cuisines will be briefly addressed in order to demonstrate the varied world origins of ingredients in given dishes and meals.

The earliest known writing appeared in the Near East about 5500 yr ago. All human experience before that, there and in the rest of the world, falls under prehistory. Prehistory is the province of the prehistorian. Most prehistorians are archaeologists. All research is archaeological and based on the few imperishable remains left to us. For example, for early times we must rely on stone, bone and antler tools, bones from animals that were consumed, charcoal and ash from campfires, traces of camps, and so on. Much of what people did and ate was not recorded or preserved (see below). Because our data are limited, there are massive holes in our story. Consequently, caution is necessary when presenting it.

For roughly two million years, humans foraged, gathered, and hunted for wild food. As time passed, we grew more efficient in exploiting our planet's wild resources. This was accomplished through the gradual development of more efficient and sophisticated tool technology based on stone, bone, wood, antler, and such and through developments in social and political organization and probably ideology, which allowed greater cooperation within social groups. These developments were accompanied by increasing brain size and intelligence. This long episode in human history is called the Paleolithic or Old Stone Age. The chronological subdivisions of the Paleolithic are used below for ease of presentation of the history of food (Fig. 1).

About 10,000 or 11,000 years ago in various areas of the Old World, strides were taken that resulted in the domestication of some plants and animals. The "agricultural revolution" had begun. By 9000 yr ago, the revolution was in full swing. There was still no writing. Neither were there cities and states, only small farming villages. Some groups in many world areas still hunted and gathered for their livelihood. The so-called invention of agriculture is prehistoric. In the Near East (sometimes called the Fertile Crescent and including southern Turkey, western Iran, Iraq, Syria, Jordan, Lebanon, and Israel), this earliest agricultural period is called the Neolithic (Fig. 1), and was truly revolutionary for much of humanity. While the invention of agriculture was not simultaneous in all Old World areas, developments in India and the Far East (Southeast Asia and China) were roughly concordant. For none of these

Period		Years before present	Subsistence economics
Neolithic			Agricultural subsistence economy in the Near East
		9,000	
Mesolithic			Incipient agriculture in some world areas: beginning domestication of some plants and animals
		12,000	
Paleolithic	Upper		
		40,000	*Homo sapiens* gathers wild plants and hunts small and large game
	Middle		
		100,000	
	Lower	200,000	*Homo erectus* gathers wild plants, hunts small and large game; and scavenges carnivore kills
		1.5 mya	
		2.5 mya	*Homo habilis* gathers wild plants, forages for small animals, and scavenges carnivore kills

Figure 1. Simplified chronological chart not to scale.

areas do we have a complete picture of the development of agriculture.

In the New World, agriculture began independently a few thousand years later than it did in the Near East. There were dramatic developments in Mexico, tropical Central and South America, and in the South American Andes Mountain region. Like those of the Old World, New World domesticates were to have worldwide consequences. Also like that of the Old World, the picture is frustratingly incomplete.

The development of agriculture was revolutionary in several senses: (1) it eliminated dependence on wild resources; (2) in its areas of origin and in areas that ultimately adopted it, it permitted population growth well beyond what wild resources could support; and (3) it laid the economic foundation for the eventual development of civilization. Without agriculture, there would be no cities and nation-states with their centralized government bureaucracies, monumental public works, social class systems and inequities, exact and predictive sciences, and all other civilizational characteristics such as taxes, welfare systems, and terrorists. Humanity could not afford them.

As indicated earlier, the major part of this presentation is based on archaeological research. It is essentially an archaeological odyssey through time and around the globe; thus, and because the reader is most likely not an archaeologist, it is necessary to present a brief discussion of the problems and methods involved in reconstructing the prehistory and history of food. A glossary of technical terms may be found at the end of the article.

BACKGROUND

Reconstructing the history of food is fraught with dangers and difficulties. Much of what is written, especially that for prehistoric times, is based on a number of assumptions, some warranted, and some not. Those based on human biological necessities probably have some validity. Those based on archaeological research are subject to two major problems. First, not everything preserves. Second, archaeological interpretation of what does remain is difficult and has itself been based on assumptions. We are all familiar with the image of "man the hunter" and the portrayal of Stone Age humans as mighty, intrepid hunters of massive, dangerous, and cunning game like mammoths and cave bears. This image comes to us through the magnificent Late Stone Age paintings of western Europe, from food bones found in caves and other living sites and from kill and butchery sites found throughout the Stone Age world. Because plant foods rarely preserve and are even more rarely portrayed, incautious interpretation by some has left the impression that prehistoric humans ate mostly meat. Except for environments like the arctic and the subarctic, where edible plants are rare, this is unlikely to be the case for a variety of reasons.

Humans are primates. Most primates are herbivorous and only accidentally omnivorous, and some are omnivorous (1,2). Higher primates tend to be plant eaters, although a few, such as chimpanzees, will deliberately hunt and eat meat (2). Humans are omnivorous today, and it is likely that our hominid ancestors, as primates, were also omnivorous. Early hominid teeth, like those of modern humans, are those of omnivores (2). The greater majority of modern humans cannot extract sufficient vitamins and minerals from meat alone to survive. Eskimos are an exception, but even they eat the stomach contents of the herbivorous animals they kill and take advantage of plant foods during the short growing season (1). Nor can we live on plants alone without bringing together the right combination of plants to supply us with complete proteins. Nutritionists tell us that humans need a diet balanced so that we receive all the vitamins and minerals the body demands for efficient functioning. This means a combination of plant and animal foods with heavy emphasis on plants as opposed to animal protein.

Anthropological study of historic and modern gatherer-hunters shows that, with the exception of the Siberian and North American Eskimos of the subarctic, the ratio of wild plant food gathered is almost always greater than that of hunted or fished animal protein. The actual ratio depends on both available environmental resources and cultural factors and differs between groups. An additional factor affecting the ratio is the seasonality of resources. In some seasons plant foods are more available than in others, such as winter in the temperate northern hemisphere, when, in the absence of plant preservation techniques or even with them, more meat will be eaten—or perhaps, less food will be consumed in general.

In a given environment various small animals are foraged, such as rodents, birds, reptiles, and insects. While larger game may be preferred, it is not always present or present in sufficient quantity to appease human appetites for animal protein.

In short, gatherer-hunters adapt to the environment in which they live and the ratio of plant to animal foods will reflect that adaptation. It will also reflect cultural factors such as religious taboos and the cultural definition of what is and is not food. We also recognize that hunting is high risk and not always successful. Gathering provides an ensured food supply.

Archaeologically, little trace of the plants consumed by recent gatherer-hunters will remain. Without the ethnographic documentation of these living people, we would never have complete knowledge of their subsistence economies. This is because plant material is notoriously perishable and subject to selective preservation. In some dry desert areas, plant remains do preserve. So do human coprolites. Ancient coprolites contain seeds, pollens, and husks or shell parts that aid in reconstructing paleodiet. But human coprolites are rarely found. In wet areas, such as bogs, plants often preserve, as do human bodies. Perhaps one of the most famous and scientifically valuable examples of preservation is Tollund Man, in whose preserved stomach were found the remains of his last meal, a porridge prepared from a wide variety of seeds of domesticated and wild plants (3). A very recent technique of extracting collagen from human bone and conducting trace-element and isotopic analysis enables us to determine whether and what kinds of cereal grasses were being consumed, for example, maize. But, of course, the bone itself must be preserved and unfossilized (4). Fossilized bone is completely mineralized and contains no organic material.

Animal bone is also subject to selective preservation. Large bones survive better than do small bones, and not all bone fossilizes. Acid soil destroys bone. Insect parts rarely preserve. In sum, little of what was originally deposited at a site by human subsistence activity preserves. There is no real way of reconstructing the ratio of plant to animal food consumed by our remote ancestors or even those closest to us in time.

We are left on shaky ground. Our assumptions are: (*1*) both presapiens and sapiens hominids were omnivorous; (*2*) they adapted to the varied environments in which they lived and exploited their edible resources using the then current technology and sociopolitical organization; (*3*) gathering plants and foraging small animals provided them with a more reliable food supply than did hunting large game; and (*4*) by the Late Stone Age, fully human cultural factors surfaced that led to food selection among the edibles when people could afford it.

Figure 1 is provided as a convenient guide through the time periods discussed below. The chart is highly simplified and applies only to Europe, northern Africa, and the Near East. Although data from other world areas are discussed within the time divisions, it is fully acknowledged that the terms Paleolithic (Old Stone Age), Mesolithic (Middle Stone Age), and Neolithic (New Stone Age) do not apply well to Subsaharan Africa, the Far East and North and South America. With this in mind, the periods used and their finer subdivisions are briefly defined below.

The Paleolithic or Old Stone Age began about 2,500,000 years ago (2.5 mya) and is subdivided into three major chronological stages on the basis of advances in stone tool technology: Lower or Early Paleolithic (2.5 mya–100,000 yr ago), Middle Paleolithic (100,000–40,000 yr ago), and Upper or Late Paleolithic (40,000–12,000 yr ago) (Fig. 1). The Paleolithic ended with the end of the last ice age, called the Wurm or Weichsel in the Old World and the Wisconsin in the New World.

The Mesolithic or Middle Stone Age was a short period that began with the final retreat of the last glacier about 12,000 years ago (Fig. 1). No definitive end date that applies to the planet can be given because the proper end of the period is with the beginning of agriculture. Agriculture began at different times in different parts of the globe. In the Near East, the period fades into what is sometimes called the Proto-Neolithic. The Proto-Neolithic (about 11,000–9000 yr ago) is called such because it was the era of incipient agriculture. Various animals and plants were brought under domestication, but wild animals and plants still constituted an important part of subsistence. There appear to have been some similar developments in Thailand dating to this period.

The Neolithic began in the Near East about 9000 years ago, when human subsistence was based fully on agriculture (Fig. 1). Fully agricultural economies were later in Europe, the Far East, and the New World.

PALEOLITHIC

Lower Paleolithic (Early Stone Age): 2.5 mya–100,000 Years Ago

East Africa; Olduval Gorge, Tanzania between 1.85 and 1.5 mya; Hominid Form: *Homo habilis*. A fully bipedal form with a manipulative hand and a complex brain, the upper range of which was more than half the size of the present human brain, *H. habilis* was probably an omnivore. Although no primate is a carnivore, *H. habilis* looks like one because of the nature of what has been found on its habitation sites (living floors). Isolated and fragmentary remains of almost every conceivable available animal from mice and turtles to gazelles and saber-toothed tigers have been found. Although some paleoanthropologists argue about just how these bones got there, the consensus is that *H. habilis* was a scavanger of carnivore kills who brought pieces back to eat raw. In some cases, as at Olduvai Gorge, Tanzania, and Koobi Fora, Kenya, it camped at large animal carcasses such as elephant and hippopotamus and ate them on the spot, bringing fragments of other scavanged animals to the same place (5). Although there is no way of knowing, it is possible that the meat was not always fresh. The smaller animals could have been hunted and/or foraged, but given the hominid's size (about 4 ft) and rudimentary stone tool technology, it is doubtful the large and more dangerous animals were hunted (2,5). Like historic Australian aborigines and Amazonian Tukanoan Indians, they may also have consumed grubs and insects (6,7). One assumes they would have eaten any eggs they found.

It seems obvious that *H. habilis* sought meat. Did it also eat plants? There are no remains, but the likelihood is that it did and probably in a higher proportion than meat. Its primate physiology would have demanded this. Seeds, nuts, edible leaves and twigs, and fruits were available.

Europe, Africa, China, and the General Far East; 1.5 mya–300,000–200,000 years ago. *Homo erectus*, the presumed descendant of *Homo habilis*, took the stage around 1.5 million years ago. Although the size of the *H. erectus* brain was not yet equal to that of the *H. sapiens sapiens* brain, *H. erectus* was otherwise fully human physically. Taller, heavier, more muscular, and with a brain the upper range of which was within the lower range of ours, *H. erectus* was more intrepid than *H. habilis* and accomplished more, such as controlling fire. *H. erectus* preferred warm to temperate cold climates. The latter would be characterized by seasonality of resources and in some cases, seasonal low availability of plant foods. It is likely that in the colder climates a high degree of nomadism was required in pursuit of plant foods. Of course, in the colder environments, more meat could and would have been eaten.

True to form, *H. erectus* ate everything that was available and edible, including it seems, itself (8). *Homo erectus* more than likely continued the practice of scavanging carnivore kills and foraging for small animals. However, we know that large game was hunted. Abundant evidence from Spain (elephant and horse) (9), France (10), and the Zhoukoudien Caves in northern China indicate this (5,8). No stone spear points are yet known, but wooden spears with fire-hardened tips would have served well. Evidence from China indicates that at least some meat was cooked, including *H. erectus* (8).

Evidence from France (10), East Africa (11), and China (8), although scarce, indicates that plants were eaten.

Middle Paleolithic (Middle Stone Age): 100,000–40,000 Years Ago

When this period began depends on what is being emphasized. For present purposes the discussion here begins at 100,000 yr ago and ends 40,000 yr ago. This period is characterized by the emergence of *Homo sapiens*. The form most commonly known is Neanderthal, but there were other varieties who are given other taxonomic names.

Homo sapiens lived in a variety of environments from the Ice Age northern European subarctic tundra to tropical forests and therefore evolved a variety of environmental adaptations specific to each. The ratio of plant to animal consumption probably reflected differential environmental adaptations. In cold and temperate climates mobility was the main pattern owing to the seasonality of resources. Hunting emerged as an important pattern around the then inhabited world and stone spear points are known. Massive game such as woolly mammoth and woolly rhinoceros were taken in cold climates, but smaller game including such large cold-climate animals as reindeer and warm-climate camel and gazelle and small game such as birds and rabbits were also hunted. Harpoon heads and fish bones indicate that some fishing was done on the tundra of northern Europe (9). Opportunistic scavanging probably continued. The diet appears to have been broad-spectrum and unspecialized. Meat was probably roasted.

While no plant food remains have been found except for Kalambo Falls, Zambia (11), digging stick tips from northern Europe (9) and tooth wear patterns indicate plant food consumption. How plant foods were prepared and whether they were cooked remains a mystery, as does the existence of methods of storing them for winter. The proportion of plant to animal foods would have depended on the environment.

Upper Paleolithic (Late Stone Age): 40,000–12,000 Years Ago

The modern human form called *Homo sapiens sapiens* characterizes this period. During these 30,000 years of the latter part of the last Ice Age, humans expanded into all inhabitable environments from the tropical forests of Africa to the subarctic tundra of Siberia and into Australia and North and South America. A wide variety of macroenvironmental and microenvironmental adaptations were necessary. Humanity—through its flexible mode of adaptation anthropology refers to as culture and society—adapted. The result was the emergence of a great deal of cultural diversity. If we were to emphasize the diversity, a discussion of the history of food would become immediately unmanageable. Therefore, only the broadest picture will be drawn.

Northern peoples made cold climate adaptations that led them to specialize in hunting certain large game. That is not to say they did not take smaller game. It seems, however, that the habits of the animals on which they depended dictated their basic lifestyle. The degree of nomadism a people pursued depended on the migratory patterns of the main animal. Large game ranged from woolly mammoth and woolly rhinoceros to bison, horse, and reindeer (9). The decimation and extinction of horse and bison herds in Europe suggest that meat was stored for the winter. Fishing was carried out, especially of salmon. In southern coastal areas people not only fished for salmon but also gathered large quantities of mollusks and shellfish (12). There are no plant food remains or indisputable evidence for the storage of plants for winter (13). It is possible that more meat was consumed during the ferocious winters. The ratio of plant to animal food is impossible to determine, but at least in some areas, a wide variety of berries and other fruits, roots, tubers, and nuts would have been available in summer and autumn. Given the intelligence, knowledge, and experience of these people, one would assume they had a variety of preservation and storage techniques for at least some plant foods.

Of course, one can never say anything about the consumption of eggs and honey, or even insects, since these have no way of preserving, but their consumption is likely throughout these time periods.

During the last Ice Age, northern Africa and the Middle East enjoyed more temperate climates. The massive cold climate game did not exist. Although temperate cold, temperatures were warmer and seasons more varied than in northern Europe. As a consequence, the ratios of plant to animal food probably reflected that difference. But little can be stated with certainty. Further south, one would expect a high ratio of plant to animal food and in desert regions, the highest. Today, large game is almost absent in most desert regions and, when present, is sparse. Small game such as rodents, birds, and reptiles and a variety of insects are characteristic fauna. The ratio of plant to animal consumption is very heavily in favor of plants.

For late Ice Age North and South America, depending on latitude and altitude, the same picture may be painted. There is a dearth of information on plant food consumption and a great deal bearing on the hunting of massive, large and small game.

Mesolithic: 12,000 Yr Ago to a Regionally Variable End Date

The Mesolithic or Middle Stone Age (sometimes called the Epipaleolithic) began with the final retreat of the last glacier about 12,000 yr ago. The period's end date is variable depending on when agriculture came into being in a given area.

With the end of the last glaciation, the climatic and vegetational picture characteristic of the modern world took over. Consequently, and especially in northern latitudes, there was considerable environmental change. Humanity had to respond and did. The emerging cultural adaptations were more complex than before and an even greater amount of cultural diversity based on, among other things, microenvironmental adaptations came into being. Again, only the broadest picture will be given.

Throughout the world, the Ice Age megafauna disappeared. Mammoths, cave bears, giant sloths, gigantic deer, and their like became extinct. Smaller game, fish and other seafood became the focus of humanity's search for protein. People adapted to forest, grassland, desert, lake, riverine, and seacoast niches (9). Most people exploited several adjacent niches, such as forest and lake, and in most areas, because of the seasonality of resources, were nomadic in their lifestyles. The ratio of plant to animal food would have varied considerably depending on the adaptive strategy in a given set of niches.

One of the consequences of environmental change was the establishment of vast stands of wild wheat and barley extending in appropriate niches from east of the Caspian Sea west and south into the Near East (14). Concurrent with this was the spread of herds of wild sheep and goats in the same region (15).

During what is sometimes called the Proto-Neolithic of the Near East, gatherer-hunters took full advantage of these abundant wild resources. Some even settled down into small, permanent communities and built permanent houses. They were able to gather sufficient wild wheat to do so. Jack Harlan conducted an experiment in the same area (the ancestral wild wheat still grows there) and estimates that in 3 weeks a quantity sufficient to last a full year could be gathered (16). It is likely that it is this that permitted sedentary living. People supplemented these plant resources with hunting sheep, goats, deer, pigs, cattle, and so forth. At some point, perhaps 11000 yr ago, they began managing herds of wild sheep and/or goats and animal husbandry was born (17,18). Somewhere around 9000 yr ago, wheat and barley were domesticated. Shortly thereafter, peas and lentils, the wild varieties of which grow in the same area, were added to the list of domesticates. So were pigs. Agriculture was born in the Near East around 9000 yr ago (7000 BC) (see Tables 1 and 2).

Neolithic (New Stone Age): 7000 BC to an End-Date Variable Depending on Other Cultural Developments

In the Near East, the plant/animal food ratio appears to have been greatly in favor of plants. Sickles, grinding stones, storage pits, and ovens are highly characteristic artifacts of the period. Impressions of wheat and barley kernels appear in mud bricks, hearth materials, and pottery. Bread and porridge probably formed the main part of meals. Sheep and goat herds seem to have been husbanded mainly for wool, hides, and milk. Cheese may have been invented and added to the diet. Animal slaughter was selective by age and sex. Meat, therefore, was not consumed as a daily ration. As in ancient historic times, domesticated sheep and perhaps goats may have been religious feast foods.

True to form, humans remained omnivorous, but in the Near East, at least, dairy foods in the form of milk and perhaps, cheese took the place of meat as a daily food. For a while, some hunting was done that would have placed meat on at least some tables. People may also have traded

Table 1. Origins of Major Domesticated Food Animals

Animal	Region of Original Domestication
Cattle	Near East
Chicken	India
Goat	Near East
Pig	Near East
Sheep	Near East
Turkey	Mexico

Table 2. Origins of Selected Domesticated Food Plants

Plant	Region of Original Domestication
	Grains
Amaranth	Mexico
Barley	Near East
Maize (corn)	Mexico
Millet	Asia
Oats	Europe
Quinoa	South American Andes
Rice	China and Southeast Asia
Rye	Turkey
Sorghum	India
Wheat	Near East
	Legumes
Common beans	Mexico and South America depending on species
Lentils	Near East
Peas	Near East
Peanuts	Bolivia
Soybeans	Far East
	Sugar
Sugarcane	Indonesia or New Guinea
Sugarbeet	Europe
	Root crops
Beet	Europe
Potato	South America
Sweet Potato	South America
	Vegetables[a]
Avocado	Mexico
Cabbage (cole) family	Mediterranean
Chives	Near East
Eggplant	India
Garlic	Near East
Leeks	Near East
Olive	Mediterranean
Onion	Near East
Peppers	Mexico
Squash	Mexico
Tomato	Mexico
	Fruits[a]
Apple	Western Europe and Asia depending on variety
Banana	Southeast Asia
Grape	Near East
Kiwi	China
Lemon	Southeast Asia
Lime	Southeast Asia
Mango	India
Orange	Southeast Asia
Pear	Western Europe and Asia depending on species
Pineapple	Brazil
	Nuts and edible seeds
Almond	Mediterranean
Brazil nut	Brazil
Cashew	Brazil
Hazelnut	Europe
Pecan	United States
Pistachio	Mediterranean
Sunflower	United States
Walnut	Iran
	Spices
Black pepper	India
Chili pepper	Mexico
Mustard seed	Mediterranean
Nutmeg	Malay Peninsula
Vanilla	Tropical America
	Beverages
Chocolate	Mexico
Coffee	Ethiopia
Tea	India and China

[a]Common, not botanical classificaton.

for meat with nonagriculturalists such as hunters and pastoralists, but it is likely, as mentioned above, that it was primarily a feast-day item.

Since meat would have been scarce, salt would have become necessary. Without the salt naturally present in meat, humans cannot survive. A substitution is necessary. Salt production and trade became important.

Ultimately, certain fruits such as figs, dates, and grapes came out of the Near East (see Table 2).

Let us look at the development of agriculture elsewhere. Many Near Eastern crops and animals made their ways into Europe and together with native plants (eg, perhaps cherry and plums) ended gathering and hunting there as the major subsistence mode about 6000 BC (19). Northern China domesticated millet, which eventually was adopted by eastern Africa. Rice was domesticated in Southeast Asia and has made its way all over the world. South America gave the world the potato, peanuts, manioc for tapioca, and the sweet potato, among other crops. Regarding animals, southern Europe and/or Turkey gave us cattle, the jungles of India the chicken, and Mexico the turkey (20) (see Tables 1 and 2). Charles Heiser's book, *Seed to Civilization: The Story of Food* (20), is a valuable source for the geographic origins of domesticated plants and animals.

Mexico has given much to the world. It was one of the planet's primary centers of domestication. It took approximately four thousand years for Mexico to domesticate the wealth of foods it has given (21). Some 21 domesticated plants have their origins in Mexico (22). Maize—corn, as we call it—is the result of 4000 yr of painfully slow selection and hybridization that began around 6000 BC. Corn is a staple food in parts of Africa and South America as well as in Middle America. Mexico has also given us a large variety of beans, chili peppers, and squashes, to say nothing of a variety of tree and cactus fruits. It also gave us the tomato, avocado, and chocolate. With the limitations of space here, it is impossible to list its valuable and varied contributions to the world's diet (see Table 2).

The development of agriculture in Mexico lasted from at least 6000 to 2000 BC, and there were several centers of domestication, some humid tropical, and some semiarid desert. Throughout most of its prehistory and history, plant foods have superseded animal foods in dietary importance. In the desert areas of the southern part of the state of Tamaulipas and in the Tehuacan Valley we have the plant preservation to prove this (21). We also have the historical records from conquest times. Some people rarely, if ever, saw meat prior to the arrival of the Spaniards, and this is true even now in the most poverty-stricken rural areas. Traditionally in Mexico and over much of Middle America, meat is a fiesta food, a special-occasion food eaten only a few times a year. (This has changed among the more affluent populace.) How, then, eating plant foods almost exclusively, did these people survive?

The fortuitous congruence of maize and beans, provided they are eaten together, provides complete protein. Therefore, a meal of tortillas and beans, in terms of protein, is healthful. The lime water in which the dried maize is soaked prior to grinding into flour for tortillas provides calcium. A few chilis and other vegetables or fruits added to the meal provides additional vitamins and minerals. The use of salt provides sodium. Adults do well on such a diet. Small children, however, are often malnourished because they need more animal protein. Infants who are weaned too soon die. Infant mortality is still very high in rural areas of Mexico and Guatemala.

Of course, before the arrival of the Spaniards, the population was not completely without sources of animal protein. Turkey eggs, ant larvae, other insects, and grubs were eaten. Eggs of other birds were probably gathered. In the coastal areas marine resources, including iguana, turtle and turtle eggs, fish, crustaceans, and mollusks were consumed, although perhaps some were not allowed to everyone in these socially stratified societies. In the tropical rainforests reptiles, monkeys, deer, wild pig, and other animals existed and were hunted. But again, the consumption of some of these was probably strictly controlled. Deer, wild pig, and small game such as badger, squirrels of various kinds, rabbit, and hare abounded in the central Mexican highlands, and it is probable that the consumption of the small game was not regulated. Throughout the Middle American region, the common people probably made do with beans and corn as their main source of protein and supplemented these with small game, insects, and eggs and in the region around modern Mexico City, with fish and larvae from lakes now drained.

It is of interest to note that the beginnings of agriculture in the Near East, Far East, and Middle and South America appear to have been roughly concurrent. This is not likely to have been due to any communication of ideas. Why agriculture began is subject to much debate, and there are no clear answers. Environmental change, population pressure, religion, and the human propensity to experiment have all been invoked (1,16,17,20). It is beyond the scope of this article to discuss these debates.

Another interesting fact is that in the probably unintentional search for complete protein in the virtual absence of meat consumption, humanity in various parts of the globe domesticated plants that, when consumed together, provide complete protein. The case for Middle America has been briefly discussed. Rice and beans and other combinations also provide complete proteins (1).

A final fact that should be added is that in all world areas, the development of agriculture was a long, slow process. In a given area, not all plants were domesticated and brought under cultivation at the same time. Nor were all animals domesticated simultaneously. It took centuries and, in some cases, millennia to bring together the agricultural complexes that are prehistorically and historically known.

CONCLUSION

It would appear that for most of humanity's history, we were omnivores like many other primates and that, like other primates, we consumed more plant food than animal. Under special circumstances the ratio might swing in favor of animal over plant, but this is rare and due to special environmental circumstances. Even then, plant food is consumed, even if this means consuming the stomach contents of dead herbivores. Also rare is the society that does not seek some form of animal protein even if it means eating insects, grubs, or raw bird eggs.

The ratio of plant to animal food consumption is probably still in favor of plants around the planet, even for those of us who like our meat so much that we jokingly call ourselves carnivores. If we look at the modern Western diet, we find that even fast foods like hamburgers and pizza are heavy on the plant end. The hamburger comes on a bun made from flour and is accompanied by french fried potatoes, potato chips, or potato salad, is relished by plant products such as tomato catsup, mustard (seed, vinegar), pickles or pickle relish, onions, tomatoes, lettuce, and sometimes mushrooms. The basic pizza is, of course, made from flour and tomato sauce, regardless of the topping on the pizza. If we look at the Far Eastern diets of China, Southeast Asia, and Japan, we find that while meat or seafood may form part of the meal, the major constituent is vegetables and other plan foods like rice or noodles.

We have reached a point in the world where the history of any given dish or meal is utterly fascinating. The ancient northern Chinese somehow received wheat and the idea of flour from the Near East and ultimately invented the noodle. In the thirteenth century, Marco Polo went to China and allegedly, inter alia, encountered the noodle and found it to taste good. He took it back to Italy, where, ultimately, a variety of pastas were developed, including spaghetti. A moderately well-off Mexican sits down to a several-course meal. The first dish is *arroz à la mexicana* (Mexican-style rice). It consists of rice (Southeast Asia) toasted raw in lard (a pork product originating from the Near East) or vegetable oil (an eastern Mediterranean idea) seasoned with a sauce made from tomatoes (Mexico), onion, and garlic (both Near East) and cooked in chicken broth (ultimately India). It is served with corn tortillas (Mexico) or perhaps, flour tortillas (made from wheat flour which has its ultimate origins in the Near East). The next course is *carne de cerdo con salsa verde* (pork in green sauce). The pork is ultimately Near Eastern. The green sauce is made from *tomates* (*tomatillos* or green husk tomatoes) and chilis (Mexico), onion (Near East), and *cilantro* (Near East). The final course is *frijoles*, which, if done right, are cooked with lard or oil and also, perhaps, a little onion. Without the fats and onion, the *frijoles* are thoroughly Mexican. All may be accompanied by beer (a thoroughly European idea incorporating many Near Eastern ingredients) or carbonated soft drinks (a U.S. invention). Many of the "foreign" ingredients in the Mexican cuisine can, of course, be attributed to the sixteenth-century Spanish conquest, one of the results of which was a revolution in cooking. A glance at the history of Spain would illuminate how some of these ingredients got into its cuisine!

The cross-fertilization in our various cuisines is enormous and universal. A simple plate of bacon and hen's eggs served with orange juice, coffee, toast, and jam has its ultimate origins in the Near East, India, Southeast Asia, Ethiopia, and Europe or North America (if the jam is raspberry). If you add hashbrowns or grits, you must acknowledge South America or Mexico. A bowl of vegetable beef soup or chicken and vegetable soup contains ingredients from around the world. A vegan's vegetarian plate is a veritable travelog of time and space.

The history of food and cuisine is, among other needs and things, very much a product of trial and error experimentation and the human propensity to travel, trade, and even conquer and to bring home new and delightful products and ideas and to modify and invent new products, new combinations, and new dishes. Humans are indeed omnivores in every sense of the word.

GLOSSARY

Anthropology. The field that studies humans as biological and cultural beings. Such study includes human culture, biology, language, and history. See also *Culture*.

Archaeology. The subfield of anthropology that studies and reconstructs past cultures and societies.

Coprolites. Naturally dried fecal matter.

Culture. The patterned thought and behavior—such as social, political, economic, and religious—that individuals learn and are taught as members of social groups and that is transmitted from one generation to the next.

Ethnography. The anthropological study and description of the cultures of living groups: their social, political, economic, and ideological systems and all the behavior that relates to these.

Food bones. The bones of animals consumed by humans.

Foraging. The collection of edible wild plants and small animals, such as birds, rodents, reptiles, and insects.

Hominid. The common name for those primates referred to in the taxonomic family Hominidae: modern humans and their nearest evolutionary ancestors.

Homo. The genus to which modern humans and their nearest evolutionary ancestors belong: *Homo habilis* (handy human), *Homo erectus* (upright human), *Homo sapiens* (knowing human), and *Homo sapiens sapiens* (modern human).

Kill and butchery sites. Locations where an animal or animals were killed by humans and butchered on the same spot.

Living floors. Locations where concentrations of living debris are found. Such debris may be tools, debris from tool manufacture, food bones, and other evidence for human occupation.

Lower Paleolithic. Dating from 2.5 mya to about 100,000 yr ago. The beginning of the period is marked by the presence of the earliest known stone tools and probably, the first appearance of the *Homo* genus. The period is characterized by increasing human brain size, and capacity, intelligence, and the evolution toward greater complexity of human technology and culture.

Middle America. Mexico and Central America.

Mesolithic. A term that designates immediately preagricultural societies in the Old World. A diagnostic technological characteristic is the presence of microliths, small stone blades set into bone or wood. In the Near East, sickels used for the harvesting of wild grains were made using this technique.

Middle Paleolithic. Dating from about 100,000 to 40,000 yr ago. The period is marked by increased sophistication in stone tool technology, such as the making of tools from prepared cores, and by the presence of *Homo sapiens*. Human culture became more complex, particularly with regard to the ideological system. Religious ritual, as manifested archaeologically in the burial of the dead with grave goods, came into being.

mya. Million years ago.

Neolithic. A stage in cultural evolution generally marked by the appearance of ground stone tool technology (bowls, adzes, axes, etc) and frequently by domesticated plants and animals and permanent villages.

Paleoanthropology. The multidisciplinary approach to the study of human biological and cultural evolution. It includes physical anthropology, archaeology, geology, ecology, and many other fields.

Paleolithic. Dating from about 2.5 mya to 10,000 BC. The period during which stone tools were produced by percussion flaking (chipping). This period is characterized by the origin and evolution of modern humans and culture.

Physical anthropology. The subfield of anthropology that studies human biology and evolution.

Presapiens. Members of the genus *Homo* who lived before the appearance of the *sapiens* species: *Homo habilis* and *Homo erectus*.

Primates. The order of mammals that includes prosimians, Old and New World monkeys, apes, and humans.

Sapiens. The species to which Middle and Upper Paleolithic and modern humans belong.

Scavanging. The procurement of meat from animals killed by carnivores.

Site. A confined geographic area or location of interest to archaeologists in which the remains of earlier human activity are concentrated.

Subsistence economy. A term referring to food resources and their modes of procurement: foraging, gathering, and hunting for wild plants and animals, agriculture, marketing, etc.

Upper Paleolithic. Dating from about 40,000 to 12,000 yr ago. The final stage of the Paleolithic. It is characterized by the prevalence of modern humans with more sophisticated culture. Characteristic remains in western Europe are polychrome cave paintings, sculpture, engraving, and stone tools made from blades.

BIBLIOGRAPHY

1. P. Farb and G. Armelagos, *Consuming Passions: The Anthropology of Eating*, Houghton-Mifflin, Boston, 1980.
2. R. Jurmain, H. Nelson, and W. A. Turnbaugh, *Understanding Physical Anthropology and Archaeology*, 4th ed., West Publishing Co., St. Paul, Minn., 1990.
3. P. V. Glob, *The Bog People: Iron Age Man Preserved*, Ballentine Books, New York, 1969.
4. A. Sillen, J. C. Sealy, and N. J. Van der Merwe, "Chemistry and Paleodietary Research: No More Easy Answers," *American Antiquity* **54**, 504–512 (1989).
5. B. G. Campbell, *Humankind Emerging*, Scott, Foresman and Co., Boston, 1988.
6. N. Tindale, *Aboriginal Tribes of Australia*, University of California Press, Berkeley, 1974.
7. D. L. Dufour, "Insects as Food: A Case Study from the Northwest Amazon," *American Anthropologist* **89**, 383–387 (1987).
8. K. C. Chang, *The Archaeology of Ancient China*, 4th ed., Yale University Press, New Haven, Conn., 1986.
9. K. W. Butzer, *Environment and Archaeology: An Ecological Approach to Prehistory*, 2nd ed., Aldine-Atherton, Chicago, 1971.
10. H. de Lumley, "A Paleolithic Camp at Nice," *Scientific American* **220**, 42–50 (1969).
11. D. J. Clark, *The Prehistory of Africa*, Praeger, New York, 1970.
12. L. G. Straus, G. A. Clark, J. Altuna, and J. A. Ortega, "Ice Age Subsistence in Northern Spain," *Scientific American* **242**, 142–152 (1980).
13. R. Dennell, *European Economic Prehistory: A New Approach*, Academic Press, Orlando, Fla., 1985.
14. J. R. Harlan and D. Zohary, "Distribution of Wild Wheats and Barley," *Science* **153**, 1074–1080 (1966).
15. J. R. Harlan, "Plant and Animal Distribution in Relation to Domestication," *Philosophical Transactions of the Royal Society of London* **275**, 13–25 (1976).
16. J. R. Harlan, "A Wild Wheat Harvest in Turkey," *Archaeology* **20**, 197–201 (1967).
17. C. Redman, *The Rise of Civilization: From Early Farmers to Urban Society in the Ancient Near East*, Freeman, San Francisco, 1978.
18. H. J. Nissen, *The Early History of the Ancient Near East 9000–2000 B.C.*, University of Chicago Press, Chicago, 1988.
19. A. Whittle, *Neolithic Europe: A Survey*, Cambridge University Press, Cambridge, 1985.
20. C. B. Heiser, Jr., *Seed to Civilization: The Story of Food*, Harvard University Press, Cambridge, 1990.
21. R. S. MacNeish and D. S. Byers, *The Prehistory of the Tehuacan Valley*, Vol. 1, *Environment and Subsistence*, University of Texas Press, Austin, 1967.
22. R. C. West and J. P. Augeili, *Middle America: Its Lands and Peoples*, Prentice Hall, Englewood Cliffs, N.J., 1989.

DARLENA K. BLUCHER
Humboldt State University
Arcata, California

See also BIOLOGICALLY STABLE INTERMEDIATES.

HOMOGENIZERS

The homogenizer, which is used today in many food and dairy applications, was invented in the 1890s. Although many changes and modifications have been made to the machine over the years, the basic components of the homogenizer are the same as those early machines. Before describing the homogenizer in detail, it is important to distinguish the type of homogenizer used in food and dairy processing from the more generic term homogenizer. Today, the term homogenizer is often applied to any piece of equipment that disperses or emulsifies. This equipment may include a turbine blade mixer, an ultrasonic probe, a high shear mixer, a colloid mill, a blender, or even a mortar and pestle. The more precise definition of a homogenizer is a machine consisting of a positive displacement pump and a homogenizing valve that forms a restricted orifice through which product flows.

EARLY HISTORY

The first homogenizers were invented at the turn of the century and were used for making artificial butter (1). Gaulin invented and patented his homogenizer for the processing of milk and first showed his machine to the public at the 1900 World's Fair in Paris (2,3). The early literature attributes the term homogenizing, or homogenization, to Gaulin (1). The homogenization of milk, that is, reducing the milk fat globules in size to retard separation and the resulting cream layer, was at least 26 years ahead of its time, because the pasteurization of milk had not been perfected and public acceptance of this product was yet to be realized. Although very little homogenized milk was produced in the early years, homogenizers were sold for ice cream and evaporated milk. Few changes were made to the homogenizer design from 1900 to 1930, but after that, improvements were made to make machines more cleanable and sanitary (1). Homogenized milk became more popular in the 1940s. Some of the benefits of homogenized milk that helped to sell it were reduction of curd tension (which made milk more digestible, especially for infants), uniformity of fat throughout the product, and improvement in the appearance and palatability of the milk (4,5). Today, of course, homogenized milk is universally accepted. The dairy industry is one of the largest users of homogenizers, but homogenizers are also used for other food products.

DESCRIPTION

Figure 1 shows a modern dairy homogenizer. As previously mentioned, the homogenizer consists of a pump and homogenizing valve. The pump is usually a reciprocating positive-displacement pump, which delivers a relatively constant flow rate despite the pressure or restriction to flow made by the homogenizing valve. The positive displacement pump has a power end and a liquid end. The power, or drive, end converts rotating motion to reciprocating motion. In most machines, an electric motor is connected to a drive shaft by V-belts and sheaves. The drive shaft turns an eccentric shaft by means of gears. In other cases, the motor is connected directly to the eccentric shaft by V-belts and sheaves.

The eccentric shaft has cams that drive the plungers or pistons by means of connecting rods and crossheads. The crosshead couples the connecting rod to the plungers or pistons. A piston and a plunger are both solid cylinders that displace fluid; however, for a piston the sealing elements are moving with the piston, and for a plunger the sealing elements are stationary. The sealing elements are called plunger packing (6).

The liquid end includes the pump chamber, or block, and all its components. Figure 2 shows a typical pumping chamber. The plungers move back and forth in the chamber. The eccentric shaft cams that drive the plungers are offset to provide a steady flow. If the pump has three plungers, they would be offset by 120 degrees. This means that the flow profile is represented by the summation of three overlapping sine curves producing some peaks and valleys, but, in general, a continuous flow will be produced. Some homogenizers have five, six, or seven plungers depending on the model.

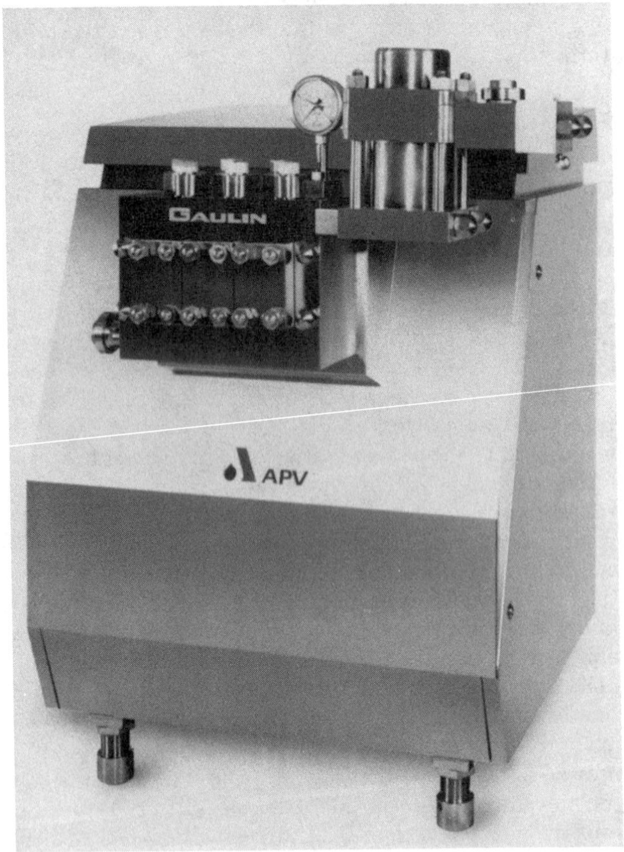

Figure 1. Modern dairy homogenizer.

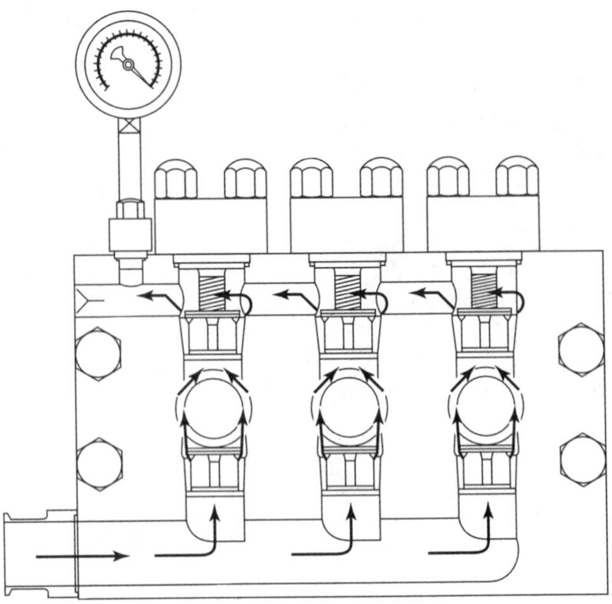

Figure 2. Pumping chamber of a dairy homogenizer.

In addition to the reciprocating plungers, the pumping chamber has suction and discharge valves. On the rearward motion of the plunger, the suction valve opens and liquid is drawn in while the discharge valve closes. On the forward discharge stroke the suction valve closes and the discharge valve opens. The plunger then displaces or pushes the liquid out of the chamber and through the discharge valve. These pump valves can be simply a ball sitting on a seat or a guided valve with a pilot aligning the valve to the seat. The piloted valve is called a poppet valve and the other a ball valve. The liquid end of the pump, especially components in contact with the liquid, is usually made of ceramics, stainless steel, or special alloys. These materials include 17-4, 15-5, 304, and 316.

Homogenizers can cover a wide range of flow rates and pressures. The range of flow rates for different machine sizes can go from 7.6 L (2 gal) per hour for a laboratory machine up to 52,996 L (14,000 gal) per hour for a large production machine. The maximum operating pressure can be 6.90 MPa (1,000 psi) up to 150 MPa (21,756 psi), but the higher the pressure rating, the lower the capacity for a given size machine. This is due to the limitation of force allowed on the drive end of the machine. For a given size machine, increasing capacity usually requires larger diameter plungers or pistons with a corresponding decrease in operating pressure due to the limiting thrust loading. Because the operating pressure is exerting force against the face of the plunger or piston, the larger diameter means that the power stroke requires greater force. Therefore, the size of the plunger or piston is limited by the maximum tolerable force on the drive end. Of course, as the pressure and capacity increase, the motor power needed also increases because the required motor power is directly proportional to the flow rate times the operating pressure.

HOMOGENIZING VALVE

The positive-displacement pump delivers fluid to the homogenizing valve that is contained in a valve body attached to the pumping chamber of the machine. Figure 3 shows a simple homogenizing valve assembly and its designated parts. The fluid to be processed flows into the homogenizing valve seat and pushes against the face of the homogenizing valve. After flowing out through the restricted orifice between the valve and seat, the liquid impinges on the impact or wear ring. Without the impact ring the fluid would cut into the body of the homogenizing valve assembly, eventually requiring replacement of this expensive part. When an impact ring is used, the worn ring can be replaced at a low cost. Figure 4 shows a two-stage homogenizing valve assembly. The actuation of the valve requires applying force on the valve to counteract the force pushing the valve open due to the pressure of the fluid. This pressure is caused by the reduction of flow area when the valve is pushed toward the seat, while the pump is delivering a constant flow rate. The force exerted by the liquid is equal to the area of the valve in contact with the liquid times the pressure generated. For example, a valve with a contact diameter of 6.35 mm (0.25 in) at 68.95 MPa (10,000 psi) needs a counteracting force of 2184 N (491 lb). A valve with a contact diameter of 25.4 mm (1 in) at 68.95 MPa requires 34,936 N (7,854 lb) of counteracting force.

If the counteracting forces are large, then a spring-loaded handwheel (manual operation) will not deliver enough mechanical advantage to achieve high pressures and some other means of actuation such as a hydraulic or pneumatic system must be used. For some products, a two-stage homogenizing valve consisting of two valves in series is used, and the valves can be manual, hydraulic, or pneumatic depending on the valve size and pressure. The homogenizing valve, seat, and impact ring are usually made of special wear-resistant materials because of the high fluid velocities in the valve assembly and because some products contain suspended solids.

As previously described, Figure 3 shows the flow through the homogenizing valve. Although this flow profile may look simple, the fluid dynamics occurring are profound. Intense energy changes occur in the homogenizing valve as the liquid goes from high pressure and low velocity to low pressure and high velocity. The best way to understand these changes is to consider an example. A homogenizer operating at 13.8 MPa (2,000 psi) and 11,356 L (3,000 gal) per hour with a conventional homogenizing valve would, typically, have a calculated gap of about 0.152 mm (0.006 in) between the valve and seat. The pressure before the homogenizing valve seat is 13.8 MPa (2,000 psi) and the velocity is about 6.1 m (20 ft) per second. As the fluid enters the gap, the pressure drops and the velocity increases to 122 m (400 ft) per second. In this case 2.1 L (0.56 gal) per second are flowing through an opening of 25.2 mm^2 (0.039 in^2) in one second. These hydrodynamic changes occur over distances less than 0.254 mm (0.1 in) in less than 0.000002 s (7–9). Homogenization is completed shortly after the liquid leaves the gap area between the valve and seat. Therefore, the potential energy of the liquid stored in the pump chamber is rapidly converted to kinetic energy producing the homogenization effect. Walstra suggests that the energy density in the fluid during homogenization is as high as 10^{12} W/m^3 (10). Only a portion of this energy is used for emulsification, but these intense flow conditions are necessary to trigger homogenization.

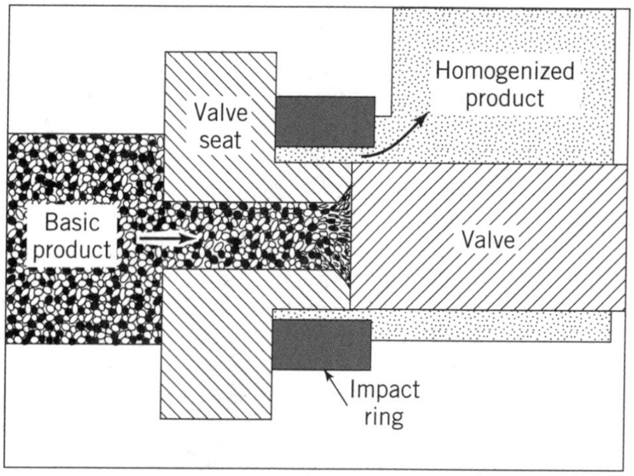

Figure 3. Simple homogenizing valve assembly.

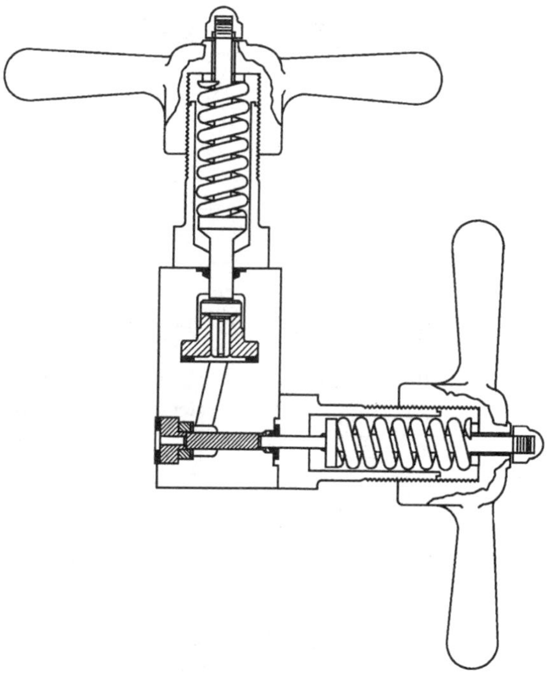

Figure 4. Two-stage manually actuated homogenizing valve assembly.

THEORIES OF HOMOGENIZATION

Because of the dynamic forces occurring in the homogenizing valve, there has been some interest in determining the mechanism of homogenization. Many researchers have suggested theories to account for homogenization by relating the presumed flow condition in the homogenizing valve to known concepts of fluid dynamics. Homogenization includes the formation of emulsions (one immiscible liquid uniformly distributed into another) or dispersions (solid particles dispersed throughout a liquid), but most theories of homogenization only consider the formation of emulsions. Therefore, these theories usually relate to the mechanism by which dispersed oil droplets are disrupted into smaller droplets and distributed throughout a continuous water phase. This approach is a consequence of the fact that most researchers were investigating the homogenization of milk when they developed their theories.

Theories of homogenization include shear, impact, wire drawing, acceleration and deceleration, homogenizing valve vibration, turbulence, and cavitation (1,11–13). Many of these theories have been discounted over the years. For example, impact of the dispersed droplets on the impact ring is not the cause of homogenization because the required velocities at impact are not large enough to disrupt the droplet, and increasing the distance from the valve seat to the impact ring does not significantly affect homogenization of an emulsion. Wire drawing is the elongation and subsequent disruption of the thinned droplet, but this is unlikely to occur due to the flow conditions in the valve. Shear is commonly mentioned as a mechanism, but the velocity gradients in the valve gap boundary layer do not appear to be adequate for emulsification. Also, the successful homogenization of oils with viscosities greater than that allowed for shearing action to occur indicate a mechanism other than shear (14).

Two other prominent mechanisms of homogenization involve turbulence and cavitation (15,16). The intense turbulent eddies generated in the liquid at the instant of energy conversion (potential to kinetic energy) produce significant local velocity gradients that disrupt the droplets. Cavitation theory suggests that the extreme pressure drop in the homogenizing valve generates cavitation bubbles, and when these bubbles collapse, the shock waves in the fluid cause the droplets to break apart. Actual measurements of cavitation noise in the homogenizing valve have suggested that the greater the intensity of cavitation, the greater is the homogenizing effect (16). However, the fact that cavitation is present does not prove that it is the actual mechanism for homogenization. Published research has shown that if cavitation is dampened or suppressed, homogenization still occurs, suggesting that turbulence may be the predominant mechanism (17). Actual visualization of the emulsification process in a homogenizing valve using a Nd:YAG laser with a pulse duration of 10 ns has shown that homogenization occurs after discharge from the gap of the valve and that emulsification is produced by the intense turbulent mixing zone in the exit region. No homogenization occurs in the region between the valve and the seat (18).

HOMOGENIZING VALVE DESIGN

One consequence of understanding the mechanism of homogenization is the ability to improve the efficiency of homogenization. The geometry of the homogenizing valve is important in regard to the quality of the finished product. Gaulin realized this, and he experimented with different valve designs at the turn of the century (1). The objective of changing valve designs is to find one that gives the best product possible at the lowest pressure (19). Some basic designs include the plug valve (Fig. 3), the piloted valve, and a grooved valve. The piloted valve has a guide on the valve that fits into the seat to align the two. The grooved valve has concentric grooves on the valve and seat. These grooves are machined so that the peaks on the valve fit into the grooves on the seat. Different configurations of these grooves have been tried over the years. Variations of the plug and piloted valve include a knife edge on the seat having a short land or travel distance, replaceable screens on the faces of the valve and seat, and a cone-shaped valve fitting into a hollowed seat. There is even a valve consisting of a tightly compressed wire bundle through which the product flows. For large flow rates, the Micro-Gap® valve is used, and this valve consists of stacked valve plates that split the flow into equal parts for optimization of homogenization (19,20). These designs seek to make efficient use of the available homogenization energy. Along with the geometry of the valves it is important to consider the size of the valves, because as the flow rate increases, valve size also increases to maintain efficient homogenization. A laboratory homogenizer operating at 114 L (30 gal) per hour may have a valve diameter of 9.53 mm (0.375 in), while a production-size homogenizer operating at 15,142 L (4,000 gal) per hour might have a valve diameter of 76.2 mm (3 in). A homogenizer at 37,854 L (10,000 gal) per hour with a Micro-Gap® valve would have the equivalent diameter (by summing the stacked valves) of a valve 787.4 mm (31 in) across. Of course, as the size of the homogenizing valve increases, the required actuating force becomes larger and this must be considered when homogenizing valves are designed.

APPLICATIONS

The homogenizer is used in the processing of many food and dairy products. Figure 5 illustrates how homogenization of milk reduces the droplet size of fat globules. The particle size distribution was measured with an AccuSizer 770 (Particle Sizing Systems, Santa Barbara, California) on samples of unhomogenized and homogenized milk (equivalent to a Micro-Gap® valve at 12 MPa) demonstrating the change in mean diameter and size distribution achieved by homogenization. The reduction in size distribution produces an increase in total surface area for the emulsion illustrating the importance of surface stabilization for fine emulsions. For example, from the size distribution curves the unhomogenized sample has a calculated surface area of 1.73 m^2 per gram of fat and the homogenized sample has a surface area of 7.30 m^2 per gram of fat. Therefore, the surface area has increased by a factor of

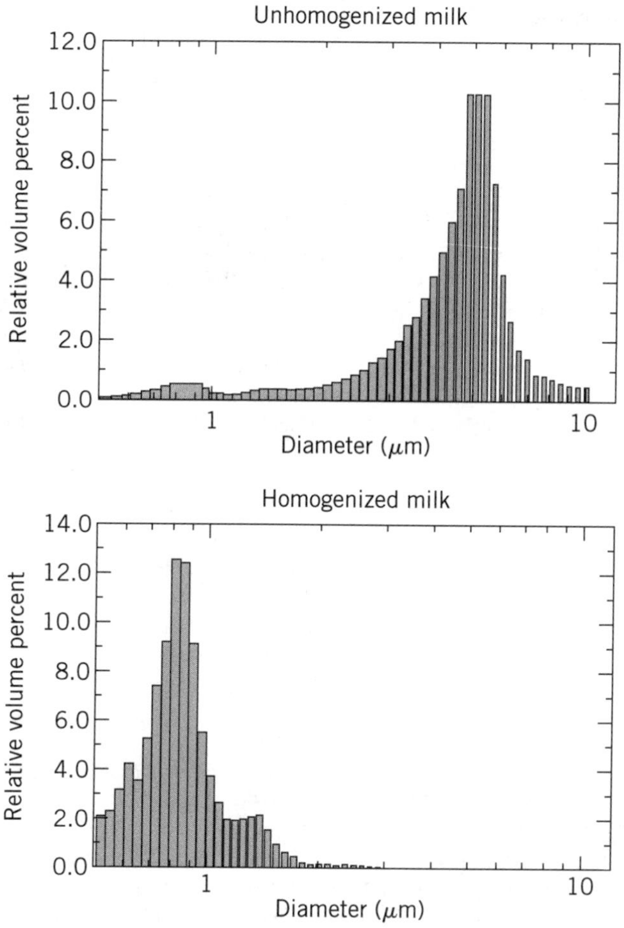

Figure 5. Particle size distribution for unhomogenized and homogenized milk.

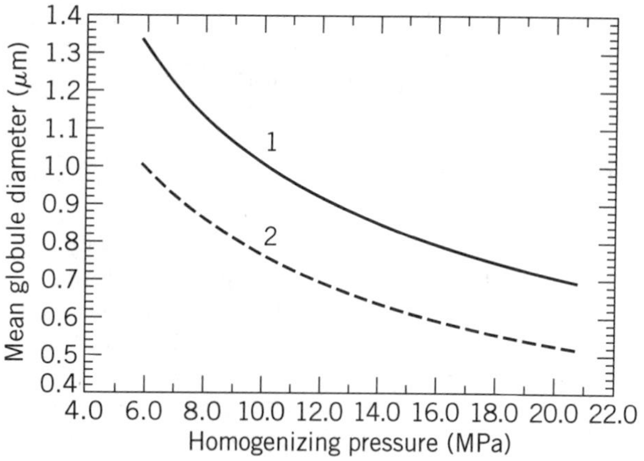

Figure 6. Mean fat globule diameter versus homogenizing pressure. Curve 1 is typical for a standard homogenizing valve at 23,000 liters per hour. Curve 2 is typical for a Micro-Gap® homogenizing valve at all flow rates. Mean diameters are measured using the spectroturbidity method. *Source:* Data from APV Americas-Homogenizers.

4.22 because of homogenization. For milk, the increased surface area means that the original milk fat globule membrane is insufficient to cover the newly formed fat surfaces, and plasma proteins are adsorbed from the milk serum to compensate for this extended surface area (21).

Figure 6 shows typical globule size reduction for milk homogenized with a standard homogenizing valve (curve 1) and a Micro-Gap® homogenizing valve (curve 2) illustrating the reduction in fat globule size with increasing homogenizing pressure.

Table 1 lists some common applications of the homogenizer. The pressure ranges given are for conventional homogenizing valves, but the actual pressures can vary depending on the product formulation, the required shelf life, and the product specifications of each processor. Also, some of these applications might require more than one pass through the homogenizer or the use of a two-stage homogenizing valve. When using the two-stage valve, it has been determined that the second stage pressure should be in the range of 10 to 15% of the total homogenizing pressure (12,19). The second stage valve eliminates cavitation in the first stage valve. This improves homogenization efficiency for an emulsion while maintaining the same total homog-

Table 1. Common Applications of the Homogenizer

Product	Pressure range (MPa)	Pressure range (psi)
	Dairy	
Evaporated milk	16.55–20.68	2400–3000
Half-and-half	12.41–13.79	1800–2000
Ice cream mix	13.79–17.24	2000–2500
Ice milk	12.41–15.17	1800–2200
Light cream	10.34–13.79	1500–2000
Pasteurized milk	12.41–15.17	1800–2200
Soft cheese	6.89–20.68	1000–3000
Sour cream	13.79–17.24	2000–2500
UHT milk	20.68–27.57	3000–4000
Yogurt base	17.24–20.68	2500–3000
	Food	
Baby foods	13.79–34.47	2000–5000
Coffee whiteners	13.79–34.47	2000–5000
Cream-base cocktails	20.68–27.58	3000–4000
Cream soup base	13.79–20.68	2000–3000
Flavor emulsions	20.68–34.47	3000–5000
Frozen whipped toppings	34.47–55.16	5000–8000
Fruit nectar	20.68–34.47	3000–5000
Infant formula	20.68–34.47	3000–5000
Liquid egg	6.89–20.68	1000–3000
Orange juice concentrate	20.68–34.47	3000–5000
Peanut butter	34.47–68.95	5000–10000
Puddings	13.79–34.47	2000–5000
Salad dressings	6.89–20.68	1000–3000
Soy beverages	20.68–34.47	3000–5000
Tomato ketchup	20.68–31.03	3000–4500
Tomato juice	3.45–6.89	500–1000
Tomato sauce	20.68–27.58	3000–4000
Toppings (hot fudge, etc)	20.68–34.47	3000–5000

enizing pressure of single stage operation. However, in some applications a single stage valve produces acceptable product, and two stages are not used.

OPERATION

Certain requirements must be met for successful operation of the homogenizer. The first of these is adequate in-feed pressure. A positive-displacement pump must have a positive in-feed pressure so the pump is not starved. If it is starved, transient high-pressure shock loading can occur, eventually resulting in severe damage to the pump (6). The required in-feed pressure will depend on the characteristics of the product and the size of the pump. Excessive amounts of entrained air in a product will cause a similar type of shock loading, and some products require deaeration to eliminate large amounts of entrained air. The correct type of pump valves should be used depending on the nature of the product. Viscous products may require ball valves. Abrasive products may require ball valves and special wear-resistant materials for the pump-valve seats and homogenizing valve. The homogenizer should be sanitary. Parts should be easily removable for cleaning or replacement, and the machine should have no flow areas where product can be trapped. Of course, the homogenizer must be rugged and reliable; for example, some dairies may require the homogenizer to operate more than 16 h a day, six days a week.

TESTING

Many methods are used to test a product for homogenization quality. These methods vary with the type of product and quality control requirements. A homogenized substance might be checked for viscosity change, either an increase or a decrease. The appearance of a product may be important; for example, texture, color, smoothness, graininess, or pulpiness. A dispersion might be checked microscopically for changes in the size of dispersed solids. An emulsion, such as milk, can be examined microscopically for the size of the milk fat globules, or more-sophisticated spectroturbidimetric methods may be used to determine the average diameter and size distribution of the fat globules (22).

SUMMARY

The homogenizer is a unique piece of equipment that was invented at the turn of the century for the processing of artificial butter and for milk but now is used for many applications in the food and dairy industries. The homogenizer consists of a pump and homogenizing valve. The pump is a positive displacement pump, and the homogenizing valve can be made in different sizes and configurations. Even after more than 90 years, the mechanism of homogenization is not completely understood, but homogenization has improved many consumer products.

BIBLIOGRAPHY

1. G. M. Trout, *Homogenized Milk*, Michigan State College Press, East Lansing, Mich., 1950.
2. W. Clayton, *The Theory of Emulsions and Their Technical Treatment*, 2nd ed., J. & A. Churchill, London, United Kingdom, 1928.
3. U.S. Pat. 756,953 (April 12, 1904), Auguste Gaulin.
4. F. J. Doan, "Changes in the Physico-Chemical Characteristics of Milk as a Result of Homogenization," *American Milk Review* 54 (June 1954).
5. F. J. Doan and C. H. Minster, "The Homogenization of Milk and Cream," *Pennsylvania State College Bull.* **287**, *School of Agriculture and Experiment Station*, 1933.
6. T. L. Henshaw, "Reciprocating Pumps," *Chemical Eng.* 104 (Sept. 1981).
7. C. C. Loo and W. M. Carleton, "Further Studies of Cavitation in the Homogenization of Milk Products," *J. Dairy Sci.* **36**, 64 (1953).
8. A. A. McKillop et al., "The Cavitation Theory of Homogenization," *J. Dairy Sci.* **38**, 273 (1955).
9. L. W. Phipps, "Action of the High Pressure Homogenizer," *Bienn. Rev. Natl. Inst. Res. Dairy*, 61 (1976).
10. P. Walstra, "Formation of Emulsions," in P. Becher, ed., *Encyclopedia of Emulsion Technology*, Vol. 1, Marcel Dekker, New York, 1983, p. 67.
11. H. Mulder and P. Walstra, *The Milk Fat Globule*, PUDOC (Centre for Agricultural Publishing and Documentation), Wageningen, The Netherlands, 1974.
12. H. G. Kessler, *Food Engineering and Dairy Technology*, Verlag A. Kessler, Freising, Federal Republic of Germany, 1981.
13. L. H. Rees, "The Theory and Practical Application of Homogenization," *J. Soc. Dairy Technol.* **21**, 172 (1968).
14. W. D. Pandolfe, "Effect of Dispersed and Continuous Phase Viscosity on Droplet Size of Emulsions Generated by Homogenization," *J. Disp. Sci. and Technol.* **2**, 459 (1981).
15. P. Walstra, "Preliminary Note on the Mechanism of Homogenization," *Neth. Milk Dairy J.* **23**, 290 (1969).
16. H. A. Kurzhals, "Untersuchungen Über die Physikalisch-Technischen Vorgänge Biem Homogenisieren von Milch in Hochdruck-Homogenisier-Maschinen," Ph.D. dissertation, Technical University Hannover, Federal Republic of Germany, 1977.
17. L. W. Phipps, "Cavitation and Separated Flow in a Single Homogenizing Valve and Their Influence on the Break-up of Fat Globules in Milk," *J. Dairy Res.* **41**, 1–8 (1974).
18. U.S. Pat. 5,749,650 (May 12, 1998), R. R. Kinney, W. D. Pandolfe and R. D. Ferguson (to APV Homogenizer Group).
19. W. D. Pandolfe, "Development of the New Gaulin Micro-Gap Homogenizing Valve," *J. Dairy Sci.* **65**, 2035 (1982).
20. U.S. Pat. 4,383,769 (May 17, 1983), W. D. Pandolfe and H. Graglia (to APV Gaulin, Inc.).
21. M. E. Cano-Ruiz and R. L. Richter, "Effect of Homogenization Pressure on the Milk Fat Globule Membrane Proteins," *J. Dairy Sci.* **80**, 2732–2739 (1997).
22. W. D. Pandolfe and S. F. Masucci, "Rapid Spectroturbidimetric Analysis," *American Laboratory* 40 (Aug. 1984).

WILLIAM D. PANDOLFE
APV Americas-Homogenizers
Wilmington, Massachusetts

HUNGER, FOOD DEPRIVATION, AND NUTRITIONAL DISORDERS

Hunger, food deprivation, and nutritional disorders remain devastating problems facing many millions, especially the poor and disadvantaged (1,2). The word *hunger* has two customary meanings. The first is the "strong desire or need for food," and the second describes the "discomfort, weakness, or pain caused by a prolonged lack of food." However, to suffer hunger is not necessarily to be hungry. Estimates by various authorities over the last three decades for the numbers of the hungry or undernourished have ranged from about 12% to more than 50% of the populations of developing countries. More recent estimates approximate to 20% (1).

ASSESSMENT OF NUTRITIONAL STATUS

Major methods (Table 1) for nutritional assessment can be divided into two categories: those that are food related and those that are people related. For monitoring world food security, food supply data are used (3) in the form of food balance sheets (FBSs). FBSs show estimates of per capita food availability derived from national production, import, and export data. These data can then be aggregated into economic or geographic groups. Such data do not show actual consumption, nor do they consider uneven distribution of the available supplies.

GLOBAL NUTRITION SITUATION

Food Energy Availability

For developing countries, in general, there was good progress in food energy availability until the mid-1980s (Table 2). However, several countries, especially many in Africa, failed to make progress. The average food energy availability for developing countries as a group is projected to increase from the 1990–1992 value of 2520 kcal to 2770 kcal per person per day by the year 2010. Many of the poorest countries, however, are likely to show only minimal gains.

In addition to the quantity of food, the quality of the diet changes as countries become richer (Table 3). With increasing wealth (gross national product [GNP]), total protein and animal protein availability increase while the dependence on cereals as a source of protein declines. Cereals are a poor source of the essential amino acid lysine, whereas animal foods are a good source of lysine. An indicator, therefore, of the overall value of the diet can be mean daily lysine availability (4). Intake of lysine ranges from 2400 mg/day in countries with a GNP less than $500 per capita per year up to 6500 mg/day for those with a GNP greater than $10,000 per capita per year. At the same time, there are also large differences in average mortality rates for children less than 5 years of age. This mortality rate is only 9 per thousand in the wealthy countries but reaches 171 per thousand live births in the poorest.

Food Production

Many scientific innovations to increase food productivity have been proposed and practiced, most especially the green revolution. More attention is now paid to environmental issues and problems such as water distribution, energy photosynthesis, land limitations and degradation, salinity and drought resistance, reduced postharvest losses, and the role of biotechnology (5,6). Nevertheless, improved nutrition for many is probably more related to uneven distribution of people's rights to land and to water; unequal access to labor, credit, and tools; and inequalities in the control over the results of food production (7). Poverty remains the major determinant of food insecurity and poor health (2). Thus agriculture can have a twofold role for improving nutrition; the first is obvious and follows from the production of food of the desired quality and quantity. The second role, highly significant in low-income countries, is in providing employment and income for the poor.

Developed and Developing Country Profiles

International agencies have grouped countries by their economies into developing and developed (industrialized) nations. The majority of the world's population lives in the developing countries. The least developed countries (LDCs) are the poorest of the developing countries. The developing countries and especially the LDCs have high child mortality rates, but the situation improved between 1960 and 1996. The 1996 rate for the LDCs (Table 4) is, however, still twice the rate for the world as a whole and is nearly five times that occurring in the industrialized countries in 1960. The maternal mortality rate is also much higher in the LDCs, some 85 times that of the industrialized countries. This enormous difference reflects the lack of basic infrastructure and medical services. Safe drinking water and sanitation are also both almost universal in industrialized nations but are far less available in poorer countries.

The Undernourished

With the exception of sub-Saharan Africa, the percentages defined as undernourished in the developing countries

Table 1. Major Methods for Assessment of Nutritional Status

	Food-related data		Health and people-related data	
	Country	Individual	Country	Individual
Method	Food balance sheets	Diet surveys	Vital and health statistics	Anthropometry (biochemical, clinical)
Outcome	Average food and nutrient availability	Approximate nutrient intakes compared with requirements	Morbidity and mortality data (risk in community)	Effect of nutrition on development function, and clinical abnormalities

Table 2. Per Capita Food Supplies for Direct Human Consumption, Historical and Projected

	kcal/capita/day				
	1961–1963	1969–1971	1979–1981	1990–1992	2010[c]
Developing countries[a]	1960	2130	2320	2520	2770
Sub-Saharan Africa	2100	2140	2080	2040	2280
Near East and North Africa	2220	2380	2840	2960	3010
Eastern Asia	1750	2050	2360	2670	3030
Southern Asia	2030	2060	2070	2290	2520
Latin America and the Caribbean	2360	2510	2720	2740	3090
Developed countries	3020	3190	3280	3350	3390
Former CPEs[b]	3130	3330	3400	3230	3380
Others	2980	3120	3220	3410	3400
World	2300	2440	2580	2720	2900

Source: Ref. 1.
[a]Ninety three developing countries accounting for 98.5% of the total population of the developing countries
[b]CPEs, centrally planned economies
[c]Projections of aggregate food availabilities for direct human consumption divided by population projections

Table 3. Total Population, Nutrient Availability, Mortality Rate, and Life Expectancy for World Economic Classes (Data from 122 Countries)

GNP class (U.S. $/capita/year)	Countries (*N*)	Total population (millions)	Food energy (kcal/day)	Total protein (g/day)	Animal protein (%)	Cereal protein (%)	Lysine (mg/day)[a]	Under-5 MR[b]	Life expectancy (years)
<500	37	2990	2070	51	20	53	2405	171	52
500–2000	41	862	2570	65	31	50	3270	76	64
2000–10,000	21	548	2913	78	45	39	4484	39	69
>10,000	23	806	3335	101	61	24	6555	9	77

[a]Lysine values calculated from protein availability data.
[b]Under-5 mortality rate: deaths at less than 5 years of age per 1000 live births.
Source: Ref. 3.

declined between 1969–1971 and 1990–1992 (1). Because of population growth, however, there were increases in the absolute numbers of the undernourished in both sub-Saharan Africa and in southern Asia. For 1990–1992 these numbered 215 (43%) and 255 million (22%), respectively. Proportions are lower in eastern Asia (16%), Latin America and the Caribbean (15%), and the Middle East and North Africa (12%). When viewed on an agroecological basis, food deprivation has tended to be highest in the arid regions of the world. Poverty is a central characteristic of the food insecure. Although the numbers of undernourished are likely to decline, only slowly and insufficiently, from the current 839 million to possibly 680 million (1) by 2010, this would represent a significant decline from 21% to 12% of the total population of the developing countries.

MALNUTRITION AND NUTRITIONAL DISORDERS

Causation of Malnutrition

A complex set of factors determines the prevalence of hunger and malnutrition. The lack of food and/or nutrients acts in combination with other environmental factors, in particular with infectious diseases, to produce a spectrum of health problems with a whole range of intensities and outcomes. Determinants of malnutrition exist at three key levels of causation (8,9): the immediate, the underlying, and the basic (Table 5). Although many food and nutrition activities continue to operate at the immediate level, basic level changes are essential to any significant global progress in eliminating malnutrition.

Diet and disease interact in a mutually reinforcing way so that it is usually impossible to distinguish either as the primary cause of malnutrition (10). Infection can result in a loss of appetite and hence initiate food deficiency; it can also result in the depletion of body stores of specific nutrients and thus cause malnutrition. Through effects on the immune system, malnutrition can make infection more likely and increase its severity. Because poverty influences both the availability of food and the quality of public health, these interactions are of major worldwide significance.

Consequences of Malnutrition

Malnutrition results in a variety of conditions such as weight loss, growth failure, anemia, learning disabilities, lower activity levels and work capacity, increased susceptibility to other diseases, blindness, and various chronic conditions. These conditions translate into social and economic costs that no country can afford (1). Malnutrition is both a symptom of broader poverty and underdevelopment problems and a cause of these same problems (2). Recently the United Nations (UN) has explicitly stated that wide-

Table 4. Profile of the Industrialized, Developing, and Least Developed Countries (Data for 1995)

Variable	Industrialized countries	Developing countries[a]	LDCs[b]	World
Number of countries	23	110	37	150
Total population (millions)	830	4526	586	5696
Population growth (%)	0.6	2.0	2.6	1.7
Total fertility rate[c]	1.7	3.4	5.6	3.1
Population <18 years (%)	23	40	50	37
Population <5 years (%)	6	12	17	11
GNP per capital (U.S.$)	24,300	1023	233	4498
Health (%)[d]	12	4	5	10
Education (%)[d]	4	11	12	6
Defenses (%)[d]	10	13	19	10
GNP growth (%)	1.9	2.9	0.1	1.9
Life expectancy (years)[e]	77	62	52	64
U5MR[f]				
1960	37	216	283	191
1980	15	137	221	122
1995	8	99	173	90
IMR[a]	7	67	109	61
LBW (%)[h]	6	19	23	18
Maternal MR[i]	13	477	1052	428
Births attended[j]	99	53	29	57
Contraceptive prevalence[k]	72	54	18	57
Urban population (%)[l]	77	37	22	45
Safe water access (%)	—	71	55	71
Sanitation (%)	—	40	35	40
Primary school enrollment, males (%)	—	86	56	88
Primary school enrollment, females (%)	—	81	45	84
Adult literacy rate, males[m]	—	79	59	81
Adult literacy rate, females[n]	—	62	38	66

[a]Developing countries include the LDCs.
[b]LDCs, Least developed countries.
[c]Total fertility rate: The number of children that would be born per woman if she were to live to the end of her childbearing years and bear the children at each age in accordance with prevailing age-specific fertility rates.
[d]Percentages of government expenditure.
[e]Life expectancy at birth: The number of years newborns would live if subject to the mortality risks prevailing for the cross section of population at the time of their birth.
[f]Under-5 mortality rate: Probability of dying between birth and exactly 5 years expressed per 1000 live births.
[g]Infant mortality rate: Probability of dying between birth and exactly 1 year of age expressed per 1000 live births.
[h]Low birth weight: Weight of less than 2500 g.
[i]Maternal mortality rate: Annual number of deaths of women from pregnancy-related causes per 100,000 live births.
[j]Births attended: Percentage of births attended by physicians, nurses, midwives, or primary health care workers trained in midwifery skills.
[k]Contraceptive prevalence: Percentage of married women aged 15–49 years currently using contraception.
[l]Urban population: Percentage of population living in urban areas as defined according to the national definition used in the most recent population census.
[m]Adult literacy rate: Percentage of persons aged 15 and over who can read or write.
Source: Ref. 2.

Table 5. Levels of Determinants of Malnutrition

Immediate causes
Food intake and health
Underlying causes
Income, land, education, water, fuel, and health service availability
Basic causes
Resources, economics, politics

Source: Ref. 9.

spread malnutrition is a clear violation of human rights in general and children's rights in particular (2,11).

People-Related Nutritional Assessment

Health- and people-related assessment tools include clinical-biochemical determinations and anthropometry (Table 1). Clinical assessment involves direct observations on criteria such as changes in the body (eg, the skin, eyes, bones, and hair). Biochemical indicators are mainly measurements in blood and urine for the levels of components such as hemoglobin, specific enzymes, and metabolites. The values are compared with standards based on well-fed, healthy individuals.

Anthropometry uses objective measurements of body dimensions as a proxy indicator of nutritional status. The most commonly used measurements assess growth and development in children and body composition of adults (12). Child and adult nutritional status is presented in terms of ratios of weight and height (eg, weight for height and, in the case of children, weight for age and height for age), and these indicators are compared with age- and gender-specific points of reference (13). In this approach, deficits in the ratio of height to age measure chronic malnutrition (stunting), whereas deficits in weight to height indicate acute malnutrition (wasting). Finally, weight for age, a composite indicator, can signify the degree of undernutrition. Stunting may be a better cumulative indicator of well-being for children than underweight (2,14). For adults, body mass index (BMI; weight/height^2) is the most suitable anthropometric indicator of under- or overnutrition. Adults with low BMI (<18.5) generally have reduced work capacity and limited social activity (15). They also have lower incomes, more sickness, and, for females, more low-birth-weight (LBW) babies. In contrast, those with high BMIs (>30.0) are considered to be at greatly increased risk for several chronic diseases including diabetes and coronary vascular disease.

Malnourished Children. The numbers of wasted, stunted, and underweight children under 5 years of age in developing countries are shown in Table 6. Southern Asia and sub-Saharan Africa have the highest percentages of undernourished children in all three categories. By far the greatest numbers of undernourished children are in southern Asia because the total population there is so large. Globally some 180 million children are underweight, 215 million children are chronically malnourished (stunted), and nearly 50 million children are acutely malnourished (wasted). For all categories, not unexpectedly, the problem is far more severe in low-income countries than in middle- to high-income developing countries.

Low Birth Weight. LBW, a major health problem in developing countries, is defined internationally as birth weight less than 2500 g and includes both premature infants and those of small size at full term. Globally, about 18% of newborn babies are of low birth weight (see Table 4). In industrialized countries only some 6% are of low birth weight, and nearly all of these infants are premature. As many as 23% may be affected in the LDCs. In contrast, these are mostly full term. If LBW infants survive, they are likely to become malnourished children and underweight adults. Poor maternal health and inadequate nutrition, both associated with poverty, are key causative factors.

Protein-Energy Malnutrition

The term *protein-energy malnutrition (PEM)* has been used to describe a range of disorders, primarily characterized by growth failure or retardation of growth in children. Mild forms are assessed by anthropometry. Extreme clinical forms of PEM associated with large weight deficits are called marasmus and kwashiorkor. Marasmus describes a wasted and emaciated child, whereas kwashiorkor is characterized by water accumulation (edema) in the body, an enlarged belly, and often severe hair and skin changes together with reduced levels of serum protein. The major cause of nutritional marasmus is inadequate food intake often induced by infection. A common hypothesis for the causation of kwashiorkor combines a deficit of protein in the diet with an adequate intake of food energy. The disease is then precipitated by an infective episode. Infants and young children are the most severely affected by PEM because of their high energy and protein needs relative to body weight and their particular vulnerability to infection.

Adaptation to Undernutrition

When food energy intake is consistently below requirements, activity and/or body size is reduced. A new balance may be established because a body that is smaller and less active needs less food. The concept of adaptation has generated passionate dispute between different schools of thought in the field of nutrition, especially concerning children who are apparently healthy but are small for their age. Some have considered such adaptations to be benign (16), but others believe that they are of major long-term health significance (17,18).

Micronutrient Deficiency Disorders

Globally, the most widespread of the micronutrient deficiencies are lack of iron, with more than 2 billion affected;

Table 6. Estimated of the Number of Wasted, Stunted, and Underweight Children under 5 Years of Age in Developing Countries, 1990

	Wasted		Stunted		Underweight	
Region or economic class	(%)	(millions)	(%)	(millions)	(%)	(millions)
Region						
Sub-Saharan Africa	7.0	6.1	38.8	33.7	30.2	26.2
Near East and North Africa	8.8	4.4	32.4	16.0	25.3	12.5
Southern Asia	17.1	26.6	59.5	92.7	58.3	90.7
Eastern and Southeast Asia	5.2	9.4	33.3	59.8	23.6	42.5
Latin America	2.6	1.5	22.7	12.7	12.0	6.7
Economic class						
Low-income countries	10.3	45.2	45.2	174.4	38.2	147.6
Middle- to high-income countries	5.6	8.0	28.7	40.8	22.0	31.2
Total	9.1	47.9	40.7	215.2	33.9	178.8

Source: Ref. 1.

iodine, with more than 1 billion at risk; and vitamin A, with 40 million affected. Many millions are also affected by the dietary lack of zinc, selenium, and other trace elements (19,20). Zinc deficiency is most common in areas where the diet consists mainly of unprocessed cereals and must now be considered a potentially major public health problem (2,20).

Vitamin A. Vitamin A is present as retinol in animal foods and as carotenoids, especially beta-carotene, in plant foods. Beta-carotene is less efficiently absorbed than retinol but is the primary source of vitamin A in the diets of most in the developing world (20). The major dietary sources and the major features of vitamin A deficiency are listed in Table 7. Deficiency of vitamin A is the most common cause of preventable childhood blindness. It occurs primarily in areas where animal foods are rarely available and fruit and vegetable consumption is low (20,21). Deficiency is caused not only by the low content in the food supply but also by poor absorption and utilization (21). Vitamin A deficient children exhibit decreased resistance to infections, decreased physical growth (22), and increased morbidity and mortality due to diarrhea, respiratory conditions, and measles (23).

At least 190 million children live in areas where vitamin A intake is low. Of these, 13 million demonstrate clinical signs of deficiency, and each year 250,000 to 500,000 children become partially or totally blind as a result (12). Two-thirds of these children die within a few months of becoming blind. Programs of vitamin A supplementation have been highly successful in reducing blindness and overall mortality rates (2,20).

Table 7. Deficiency Disorders of Vitamin A

Dietary sources

Retinol or vitamin A: foods of animal origin, liver, butter, milk, and egg yolks; beta-carotene: dark green leafy vegetables, yellow and orange fruits and vegetables

Causation

Inadequate sources of retinol or beta-carotene in diet

Most at risk

Preschool children in developing areas with insufficient intake of fruits and vegetables

Major features

Inflammation of squamous cells; softening, desiccation, and ulceration of cornea; depressed immunity; poor night vision; severe eye lesions; Bitot's spots

Consequences

Night blindness or even permanent blindness

Areas of incidence

India, Bangladesh, Indonesia, and Brazil

Measures to eliminate

Provision of diet rich in fruits, vegetables, and animal foods and fortification of dairy products or providing supplements

Iron. Iron deficiency anemia is widespread worldwide, especially in women and children. Iron in the diet is present as either heme (ie, associated with the hemoglobin of blood) or nonheme. Heme iron absorption is high and ranges between 20 and 30%, whereas that for nonheme iron can be as low as 2% (24). In the diets of individuals in developing countries heme iron constitutes only a small proportion of intake because only animal foods, especially meat products, provide heme iron. The main dietary sources of iron are shown in Table 8. Additional factors such as the amount of vitamin C in the diet further influence iron status because such reducing agents can enhance the utilization of nonheme iron. Iron nutritional status is thus a function of the amount of dietary iron and the bioavailability of that iron (20,24). In adults the most common causes of iron deficiency are defective absorption of the mineral or blood losses from various causes including parasites.

Anemia in children and infants may retard physical and cognitive development (25). In adults it can cause fatigue, reduced work capacity, and impair reproductive functions (Table 8). In pregnancy anemia is associated with retarded

Table 8. Iron Deficiency Disorder

Dietary sources

Red meats, whole and enriched grain products, and dark green leafy vegetables

Deficiency name

Iron deficiency anemia

Causation

Inadequate diet, impaired absorption, blood loss, and repeated pregnancies

Most at risk

Infants, women of childbearing age, adolescent girls, and premature infants

Major features

Weakness, fatigue, pallor, difficulty in breathing on exertion, palpitation, and vague gastrointestinal complaints

Consequences

Impaired work tolerance, impaired body temperature regulation, growth retardation, impairment in behavioral and intellectual performance, immunity and resistance to infections, and adverse pregnancy outcomes

Areas of incidence

Worldwide, especially in areas with high risk of infections, poverty, and occurrence of multiparity and premature infants

Measures to eliminate

Oral iron supplementation therapy

fetal growth, LBW, and greater perinatal mortality, 20% of maternal deaths may be associated with anemia because it predisposes mothers to hemorrhage and infections (12).

Zinc. Human deficiency of zinc was described nearly three decades ago among the poor of the Middle East (25). Major features are shown in Table 9. Deficiency symptoms include severe growth retardation, with delayed sexual maturation, and pregnancy complications. Other manifestations include suppressed immunity, poor healing, dermatitis, and impairments in neuropsychological functions (20,25,26). Zinc deficiency is found to be associated with high-phytate, high-fiber, and low-protein diets, which are common in poorer regions of the Middle East. Conditions such as hookworm and schistosomiasis, which lead to blood loss, increase the incidence of both iron and zinc deficiency (26). Animal foods are major sources of zinc. Cereals are high in fiber and phytate, which makes much of the zinc poorly available. Dietary zinc supplementation often improves growth, especially for those recovering from malnutrition and diarrhea (20,22).

Table 9. Zinc Deficiency Disorder

Dietary sources
Shellfish, red meats, and whole grain cereals (low availability)
Deficiency name
No specific name
Causation
Insufficient intake with poor digestibility and absorption caused by the effects of malnutrition on gastrointestinal function
Most at risk
Growing infants and children, pregnant and lactating women in developing countries, patients fed total parenteral nutrition solutions lacking in zinc
Major features
Reduced growth; disturbance in nucleic acid metabolism and protein synthesis; reduction in circulating proteins, growth hormones, gonadotropins, sex hormones, prolactin, and thyroid hormones, corneal lesions; skin lesions; changes in hair; delayed wound and burn healing; and diarrhea in patients fed via total parenteral nutrition
Consequences
Growth retardation, delayed sexual maturation and impotence, hypogonadism and hypospermia, immune dysfunction, patchy loss of hair and decreased pigmentation, behavioral disturbances, night blindness, impaired taste, and lack of adaptation to darkness
Areas of incidence
Areas with high consumption of cereals in the diet, such as the Middle East
Measures to eliminate
Zinc supplementation or increased consumption of animal foods

Iodine. Iodine deficiency is endemic in regions of the world where the surface soil has been leached of iodine due to glaciation, snow, and rain (Table 10). It is the most common cause of preventable mental defects (2,20). More than 200 million people have goiters, 26 million have mental defects, and 6 million have the condition termed cretinism (characterized by mental retardation and stunted growth), all of which are associated with iodine deficiency. Fetal iodine deficiency is associated with increased incidence of stillbirths, spontaneous abortions, and congenital abnormalities (25). Iodine deficiency increases the risk of LBW and early death; in older children and adults it is characterized by goiter and its complications (27,28). Linear growth retardation is a common feature in the iodine deficient child. Fetal iodine deficiency has been hypothesized as exacerbating later growth retardation (22).

Other Micronutrient Deficiencies. The vitamin deficiency disease pellagra (niacin deficiency) occurred widely in the American South in the 1800s and beriberi (thiamine deficiency) was common until recently in Asia. These deficiencies are now rare (23,25), but they may still occur together with scurvy (vitamin C deficiency) in refugee camps and other deprived populations (12). Rickets, associated with lack of vitamin D and poor calcification of the

Table 10. Iodine Deficiency Disorders

Dietary sources
Seafood and plants grown in iodine-rich soil
Deficiency name
Iodine deficiency disorders
Causation
Low-iodine soils lead to lack of iodine in diet
Most at risk
Fetuses, neonates, children, adolescents, and adults with low-iodine diets
Major features
Reduced growth and development at all ages, especially in the fetal, neonatal, and infancy stages; impaired thyroid function; impaired mental function; goiter is the primary feature
Consequences
Spontaneous abortions, stillbirths and congenital anomalies; increased perinatal and infant mortality rates; increased neurological cretinism; and higher rates of neonatal, juvenile, and adult hypothyroidism
Areas of incidence
The Himalayas, the Andes, and the European Alps; elevated regions subject to glaciation, higher rainfall, and runoff into rivers; flooded river valleys such as the Ganges
Measures to eliminate
Administration of iodized salt; iodized oil (oral or injection) is used for moderate and severe deficiency cases

bones, continues to affect many children (23,25). Scurvy has great historical significance not only because it influenced geographical exploration but also because its eventual cure and prevention followed one of the first controlled nutrition experiments (29). Deficiencies of folic acid and vitamin B_{12} are still important and affect the formation of red blood cells. Folic acid deficiency is of special significance in pregnancy (20). Pyridoxine, biotin, and pantothenic acid deficiencies have all been described but are rare.

Osteoporosis (bone mineral loss) is a major problem for postmenopausal women in developed countries. In developing countries, despite apparently lower calcium intakes, osteoporosis is much less common (20). High protein intakes and low levels of physical activity may increase risk. Copper, selenium, chromium, and fluoride deficiencies have all been described (23,25). Fluoride excess, causing fluorosis, may also be a problem in certain oasis areas. Although deficiencies of a number of other minerals such as boron, manganese, molybdenum, and vanadium can be demonstrated in specialized animal experimentation, they are either extremely rare or nonexistent in human populations.

Prevention of Micronutrient Deficiencies

Because of the severe social consequences of many micronutrient deficiencies, prevention is of major international significance (20). The Food and Agriculture Organization (FAO) of the United Nations and the World Health Organization (WHO) (1,12) and United Nations Administrative Committee on Coordination, Sub Committee on Nutrition (ACC-SCN) (20) have outlined a number of approaches toward tackling these issues. Strategies include improving dietary diversity, food fortification, and nutrient supplementation.

Vitamin A deficiency can be treated or prevented by supplementing with high-dose capsules and also by injection (2,20). Iodine deficiency disorders can be prevented or treated by the administration of iodized oil (given orally or by injection). Oral iron supplementation programs are the most common public health intervention against iron deficiency. Food fortification involves the addition of micronutrients, particularly potassium iodate, vitamin A, and iron, to common foods (30).

The growth retardation seen in many children in developing countries may be a consequence of multiple deficiencies rather than the lack of a single nutrient (22; P. L. Pellett and S. Ghosh, unpublished data). When diets high in cereal products are consumed there may also be deficiencies of the key amino acid lysine. Dietary diversity can often improve micronutrient status. Animal foods contain high levels of both lysine and bioavailable micronutrients, and even small quantities can transform poor cereal-based diets (4). Finally, public health measures to address critical environmental factors such as water quality, sanitation, and food hygiene are also essential.

Other Nutritional Disorders

The problems of undernutrition are paralleled by extensive and growing chronic disease problems with obesity and the associated conditions of diabetes, coronary vascular diseases, and certain cancers. These conditions are occurring not only in rich industrialized countries but also in low- and middle-income countries, especially in urban areas. The nutritional relationships of these diseases are discussed in the article entitled AFFLUENCE, FOOD EXCESS, AND NUTRITIONAL DISORDERS. Anorexia nervosa, with persistent intentional weight loss, is generally classified along with bulimia among the psychogenic eating disorders (25).

THE FUTURE

The green revolution, which began in the 1960s, was a global technological achievement. The introduction of improved varieties, irrigation, pesticides, and mineral fertilizers for key commodity crops, accompanied by investment in institutional infrastructure and ongoing research programs, raised food production and productivity on a wide scale. The productivity gains in rice and wheat in Asia were especially significant. With population growth and a diminishing land area to produce food, further increases in productivity are still needed in the poorer, food-insecure countries (1,5).

Science and technology have provided tools for increasing food production. More varied crops and animals, plant breeding for nutritional enhancement, improved cropping systems, and greater emphasis on integrated pest management and plant nutrition are increasingly being used. In addition, ecoregional approaches to research are being adopted to reflect prevailing biological and physical constraints such as low labor productivity, dysfunctional markets, and limited access to mechanization and energy sources. The role of biotechnology remains the subject of intense international debate concerning ethics, safety, and intellectual property rights. Despite major developments in the technologically advanced countries, it may be another 10 to 20 years before significant benefits are realized by farmers in developing countries (1).

In many countries collaboration between governments, nongovernmental organizations, and UN agencies has produced highly significant gains in the field of child survival. The well-tried paths of the GOBI strategy (growth monitoring, oral rehydration, breast feeding, and immunization programs) have been mainly responsible. More widespread recognition that malnutrition is an abuse of children's rights (2,11) and greater awareness of the importance of nutritional care are other major signs of progress. A more equitable distribution of benefits and resources to the lower-income, food-insecure groups is urgently required. Experience has shown that science and technology are essential (1,5) but cannot by themselves solve the health and food security problems of developing countries. Appropriate political, social, economic, and institutional reforms are needed not only to make advances but also to maintain what has been accomplished so far.

BIBLIOGRAPHY

1. *World Food Summit. Technical Background Documents 1–5*, Food and Agriculture Organization of the United Nations, New York, 1996.

2. UNICEF, *The State of the World's Children 1998*, Oxford University Press, New York, 1998.
3. FAOSTAT, *Food Balance Sheets 1961-199, FAOSTAT-PC: FAO Computerized Information Series Version 3*, Food and Agricultural Organization of the United Nations, Rome, Italy, 1996.
4. P. L. Pellett, *Food Nutrition Bulletin* **17**, 204–234 (1996).
5. J. C. Waterlow et al., eds., *Feeding a World Population of More Than Eight Billion People: A Challenge to Science*, Oxford University Press and Rank Prize Funds, New York, 1998.
6. United Nations Environment Programme, *Global Environmental Outlook, "For Life on Earth,"* Oxford University Press, New York, 1997.
7. B. Harriss-White, "Introduction," in B. Harriss and Sir R. Hoffenberg, eds., *Food: Multidisciplinary Perspectives*, Blackwell Publishers, Oxford, United Kingdom, 1994, pp. 1–26.
8. UNICEF, *Strategy for Improved Nutrition of Children and Women in Developing Countries*, UNICEF, New York, 1990.
9. U. Jonnson, *Food and Nutrition Bulletin* **16**, 102–111 (1995).
10. N. S. Scrimshaw, C. E. Taylor, and J. E. Gordon, *Interactions of Nutrition and Infection*, WHO Monograph Series No. 57, World Health Organization, Geneva, Switzerland, 1968.
11. United Nations Administrative Committee on Coordination, Sub-Committee on Nutrition, "The Nutrition Challenge in the Twenty-first Century," *SCN News* **14**, 3–9 (1997).
12. FAO/WHO, *International Conference on Nutrition: Nutrition and Development—A Global Assessment*, Food and Agriculture Organization, Rome, Italy, 1992.
13. *Physical Status: The Use and Interpretation of Anthropometry. Report of a WHO Expert Committee, Technical Series 854*, World Health Organization, Geneva, Switzerland, 1995.
14. United Nations Administrative Committee on Coordination, Sub Committee on Nutrition, "Stunting and Young Child Development," in *The Third Report on the World Nutrition Situation*, World Health Organization, Geneva, Switzerland, 1997, pp. 3–18.
15. FAO, Body Mass Index, A Measure of Chronic Energy Deficiency in Adults, FAO Food and Nutrition Paper No. 56, Rome, Italy, 1994.
16. D. Seckler, *Western Journal of Agricultural Economics* **5**, 219–227 (1980).
17. R. Martorell, *Human Organization* **48**, 15–20 (1989).
18. G. H. Pelto and P. J. Pelto, *Human Organization* **48**, 11–15 (1989).
19. United Nations Administrative Committee on Coordination, Sub Committee on Nutrition, "Focus on Micronutrients," *SCN News* **9**, 1993.
20. United Nations Administrative Committee on Coordination, Sub Committee on Nutrition, "Micronutrients" *The Third Report on the World Nutrition Situation*, World Health Organization, Geneva, Switzerland, 1997, pp. 19–52.
21. J. A. Olson, "Vitamin A, Retinoids & Carotenoids," in M. E. Shils, J. A. Olson, and M. Shike, eds., *Modern Nutrition in Health and Disease*, 8th ed., Vol. 1, Lea & Febiger, Philadelphia, Pa., 1994, pp. 287–307.
22. L. H. Allen, *European Journal of Clinical Nutrition* **48**, S75–S89 (1994).
23. T. K. Basu and J. W. Dickerson, *Vitamins in Human Health and Disease*, CAB International, Wallingford, United Kingdom, 1996.
24. J. L. Beard and H. D. Dawson, "Iron," in B. L. O'Dell and R. A. Sunde, eds., *Handbook of Nutritionally Essential Mineral Elements*, Marcel Dekker, New York, 1997.
25. M. E. Shils, J. A. Olson, and M. Shike, *Modern Nutrition in Health and Disease*, 8th ed., Vol. 1, Lea & Febiger, Philadelphia, Pa., 1994.
26. H. H. Sanstead, *American Journal of Diseases of Childhood* **145**, 853–859 (1991).
27. B. S. Hetzel and M. L. Wellby, "Iodine," in B. L. O'Dell and R. A. Sunde, eds., *Handbook of Nutritionally Essential Mineral Elements*, Marcel Dekker, New York, 1997, pp. 557–582.
28. B. A. Hetzel and G. A. Clugston, "Iodine," in M. E. Shils, J. A. Olson, and M. Shike, eds., *Modern Nutrition in Health and Disease*, 8th ed., Vol. 1, Lea & Febiger, Philadelphia, Pa., 1994, pp. 264–268.
29. R. A. Jacob, "Vitamin C," in M. E. Shils, J. A. Olson, and M. Shike, eds., *Modern Nutrition in Health and Disease*, 8th ed., Vol. 1, Lea & Febiger, Philadelphia, Pa., 1994, pp. 432–448.
30. A. Nilson and J. Piza, *Food and Nutrition Bulletin* **19**, 49–60 (1998).

PETER L. PELLETT
SHIBANI A. GHOSH
University of Massachusetts
Amherst, Massachusetts

HURDLE TECHNOLOGY

The microbial stability and safety as well as the sensory and nutritive quality of most preserved foods are based on a combination of several empirically applied preservative factors (called *hurdles*), and more recently on knowingly employed hurdle technology. Deliberate and intelligent application of hurdle technology allows a gentle but efficient preservation of foods and is advancing worldwide. Various expressions are used for the same concept in different languages: Hürden-Technologie in German, hurdle technology in English, *technologie des barrières* in French, *barjernaja technologija* in Russian, *technologia degli ostacoli* in Italian, *technologia de obstaculos* in Spanish, *shogai gijutsu* in Japanese, and *zanglangishu* in Chinese. The most important hurdles for the preservation of foods are temperature (high or low), water activity (a_w), acidity (pH), redox potential (Eh), preservatives (nitrite, sorbate, sulfite, etc), and competitive microorganisms (eg, lactic acid bacteria). However, more than 60 hurdles that influence the preservation and/or the quality of foods have been already described, and the list of possible hurdles is by no means complete. In this contribution the principles of hurdle technology, basic aspects, and the applications of hurdle technology for food preservation in industrialized as well as developing countries are discussed.

PRINCIPLES OF HURDLE TECHNOLOGY

The microorganisms present in a food should not be able to overcome (ie, leap over) the hurdles inherent in this food. This is illustrated by the so-called hurdle effect (1), which is of fundamental importance for the preservation of intermediate-moisture foods (2) and high-moisture foods (3), since the hurdles in a stable product control microbial spoilage and food poisoning as well as desired fermentation processes. Leistner and coworkers acknowledged that the hurdle effect illustrates only the well-known fact that complex interactions of temperature, water activity, acidity, redox potential, preservatives, and so on are significant for the microbial stability and safety of most foods (3). From an understanding of the hurdle effect, hurdle technology has been derived (4), which allows improvements of the microbial stability and safety of foods by deliberate and intentional combinations of hurdles. However, the sensory quality of a food also is determined by positive and negative hurdles. To secure the total quality of a food, the safety and the quality hurdles in a food must be kept in the optimal range (5). By an intelligent combination of hurdles, the microbial stability and safety as well as the sensory, nutritive, and economic properties of a food are secured. For the economy of a food item, it is for example, important how much water in a product is compatible with the microbial stability of this food.

Some examples will facilitate the understanding of the hurdle effect and the application of hurdle technology in food preservation. Figure 1 gives eight examples.

Example 1 represents a food that contains six hurdles: high temperature during processing (F value), low temperature during storage (t value), water activity (a_w), acidity (pH), redox potential (Eh), and preservatives (pres.). The microorganisms present cannot overcome these hurdles, and thus the food is microbiologically stable and safe. However, example 1 is only a theoretical case, because all hurdles are of the same height, that is, have the same intensity, and this rarely occurs.

A more likely situation is presented in example 2, since the microbial stability of this product is based on hurdles of different intensity. In this particular product the main hurdles are a_w and preservatives, whereas other less important hurdles are storage temperature, pH, and redox potential. These five hurdles are sufficient to inhibit the usual types and numbers of microorganisms associated with such a product.

If there are only a few microorganisms present ("at the start"), then a few or low hurdles are sufficient for the stability of the product (example 3). The superclean or aseptic packaging of perishable foods is based on this principle. On the other hand, as in example 4 if, due to bad hygienic conditions, too many undesirable microorganisms are initially present, even the usual hurdles inherent to a product may be unable to prevent spoilage or food poisoning. Example 5 is a food rich in nutrients and vitamins, which might foster the growth of microorganisms (this is called the booster effect or trampoline effect), and thus the hurdles in such a product must be enhanced, otherwise they will be overcome.

Example 6 illustrates the behavior of sublethally damaged organisms in food. If, for instance, bacterial spores in a food are damaged sublethally by heat, then vegetative cells derived from such spores lack "vitality" and therefore are inhibited by fewer or lower hurdles. In some foods the stability is achieved during processing by a sequence of hurdles, which are important in different stages of a fermentation or ripening process and lead to a stable final product. A sequence of hurdles operates in fermented sausages (example 7), and probably also in ripened cheeses or fermented vegetables, and so on. Finally, example 8 illustrates the possible synergistic effect of hurdles, which probably relates to a multitarget disturbance of the homeostasis of the microorganisms in foods, which will be discussed subsequently.

BASIC ASPECTS

Food preservation implies putting microorganisms in a hostile environment, to inhibit their growth, shorten their survival, or cause their death. The feasible response of microorganisms to such a hostile environment determines whether they may grow or die. In view of these responses more research is needed; however, recent advances have been made by considering the homeostasis, metabolic exhaustion, and stress reactions of microorganisms in relation to hurdle technology, as well as by introducing the concept of multitarget preservation for a gentle but effective preservation of foods (6,7).

Homeostasis

Homeostasis is the tendency to uniformity and stability in the internal status (internal environment) of organisms. For instance, the maintenance of a defined pH in narrow limits is a prerequisite and feature of all living cells, and this applies to higher organisms as well as microorganisms (8). In food preservation the homeostasis of microorganisms is a key phenomenon that deserves much attention, because if the homeostasis of these organisms is disturbed by preservative factors (hurdles) in foods, they will not multiply; that is, they remain in the lag phase or even die, before their homeostasis is reestablished ("repaired"). Thus, food preservation is achieved by disturbing the homeostasis of microorganisms in a food temporarily or permanently. Gentle food preservation means using an intelligent mix of hurdles that secures the safety and stability as well as the quality of foods (6). Gould drew attention to the interference by the food with the homeostasis of the microorganisms present in this food (9). During their evolution, a wide range of more or less rapidly acting mechanisms (eg, osmoregulation to counterbalance a hostile water activity of foods) has developed in microorganisms, that act to keep important physiological systems operating, in balance and unperturbed even when the environment around them is greatly perturbed (10). In most foods the microorganisms are operating homeostatically, in order to react to the environmental stresses imposed by the preservation procedures applied, and the most useful procedures employed to preserve foods are effective in overcoming the various homeostatic mechanisms that microorganisms have evolved to survive extreme environmental stresses (10). The repair of a disturbed homeostasis demands much energy of the microorganisms, and thus restriction of energy supply inhibits repair mechanisms of

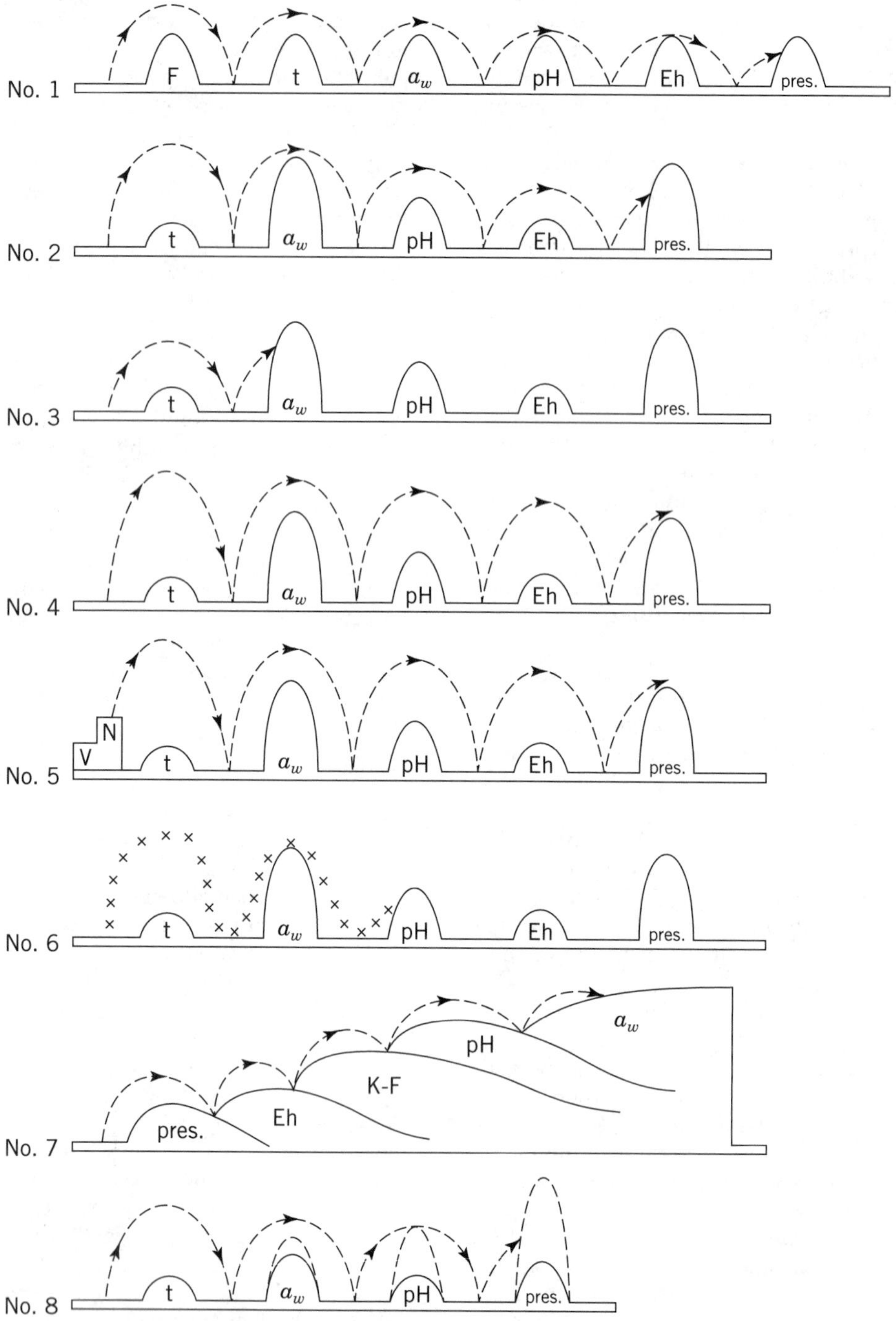

Figure 1. Eight examples of the hurdle effect that facilitate understanding of the application of hurdle technology in food preservation. See text for details. Symbols have the following meaning: F, heating; t, chilling; a_w, water activity; pH, acidification; Eh, redox potential; pres., preservatives; K-F, competitive flora; V, vitamins; N, nutrients.

the microbial cells and leads to a synergistic effect of preservative factors (hurdles). Energy restrictions for microorganisms are caused, for example, by anaerobic conditions, such as in modified atmosphere or vacuum packaging of foods (10). The interference with the homeostasis of microorganisms or entire microbial populations forms an attractive and logical focus for further improvements in food preservation techniques (10).

Metabolic Exhaustion

Another phenomenon of certainly practical importance is the metabolic exhaustion of microorganisms, which could lead to an "autosterilization" of foods. This was first observed in experiments with mildly heated (95°C core temperature) liver sausage adjusted to different water activities, inoculated with *Clostridium sporogenes*, and stored at

37°C. Clostridial spores that survived the heat treatment vanished in the product during storage (11). The most likely explanation is that bacterial spores that survive the heat treatment are able to germinate in these foods under less favorable conditions than those under which vegetative bacteria are able to multiply (12). Therefore, the spore counts in stable hurdle-technology foods actually decrease during storage, especially in unrefrigerated foods. A similar behavior has been observed with vegetative microorganisms in studies with high-moisture fruit products (HMFPs) (13,14). The counts of a variety of bacteria, yeasts, and molds that survived the mild heat treatment decreased quite fast in these products during unrefrigerated storage, since the hurdles applied (pH, a_w, sorbate, sulfite) did not allow growth. A general explanation for this behavior might be that vegetative microorganisms that cannot grow will die, and they die more quickly if the stability is close to the threshold for growth, storage temperature is elevated, antimicrobial substances are present, and the organisms are sublethally injured (eg, by heat). Apparently, microorganisms in stable hurdle-technology foods strain every possible repair mechanism to overcome the hostile environment. By doing this they completely use up their energy and die, if they become metabolically exhausted, which leads to an autosterilization of such foods. Thus, due to autosterilization the hurdle-technology foods, which are microbiologically stable, become even more safe during storage, especially at ambient temperatures (6).

Stress Reactions

A limitation to the success of hurdle technology could be stress reactions of microorganisms, because some bacteria become more resistant (eg, toward heat) or even more virulent under stress, since they generate stress shock proteins. Synthesis of these protective proteins is induced by heat, pH, a_w, ethanol, and so on, as well as by starvation. This response of microorganisms under stress might hamper food preservation and could turn out to be problematic for the application of hurdle technology. On the other hand, the switch on of genes for the synthesis of stress shock proteins, which help the organisms to cope with stress situations, should become more difficult if various stresses are received at the same time. This is because countering different stresses simultaneously will demand energy-consuming synthesis of several or at least much more protective stress shock proteins, which the microorganisms cannot deliver since they become metabolically exhausted. Therefore, multitarget preservation of foods could be the answer to avoid synthesis of stress shock proteins, which would make microorganisms in hurdle-technology foods more resistant (7).

Multitarget Preservation

The multitarget preservation of foods should be the ultimate goal for a gentle but most effective preservation of foods. For foods preserved by hurdle technology, it has been suspected for some time that different hurdles in a food could not just have an additive effect on microbial stability, but could act synergistically (1). Example 8 in Figure 1 illustrates this phenomenon. A synergistic effect could become true if the hurdles in a food hit, at the same time, different targets (eg, cell membrane, DNA, enzyme systems, pH, a_w, Eh) within the microbial cell, because then the repair of the homeostasis of the microorganisms as well as the switch on of stress shock proteins become more difficult. Therefore, employing different hurdles in the preservation of a particular food should have advantages, because optimal microbial stability could be achieved with an intelligent mix of gentle hurdles. In practical terms, this could mean that it is, for example, more effective to use different preservatives in small amounts in a food than only one preservative in larger amounts, because different preservatives might hit different targets within the microbial cell, and thus act synergistically (6). It is anticipated that the targets in microorganisms of different preservative factors (hurdles) for foods will be elucidated, and then the hurdles could be grouped in classes according to their targets within the microbial cells. A mild and effective preservation of foods, that is, a synergistic effect of hurdles, is likely if the preservation measures are based on an intelligent selection and combination of hurdles taken from different "target classes." The "multiple-drug attack" has proved successful in the medical field to fight bacterial infections (eg, tuberculosis) as well as viral infections (eg, AIDS), and thus a multitarget attack of microorganisms should be a promising approach in food microbiology, too (7).

APPLICATIONS OF HURDLE TECHNOLOGY

The intelligent application of hurdle technology for a mild but efficient food preservation is advancing worldwide in industrialized as well as in developing countries. Some examples and trends are presented next.

Industrialized Countries

Deliberate and intelligent hurdle technology for food preservation started about 20 years ago in Germany with meat products. First it was used for the gentle preservation of mildly heated, freshlike meats storable without refrigeration (4). In the meantime, several categories of these shelf-stable meat products have evolved, which are in large quantities on the German market and have caused no problems related to spoilage or food poisoning. In the manufacturing plants, processing these shelf-stable meats requires no microbiological tests to be carried out; however, other process parameters have to be strictly controlled; these are time, temperature, pH, and a_w (15). Furthermore, a better understanding of the sequence of hurdles that leads to microbial stability of fermented sausages (salami) has improved the safety and quality of these products (16). In fermented sausages, the microstructure of the products, which was studied by electron microscopy, also turned out to be an important hurdle related to the behavior of pathogens as well as starter cultures in salami (17).

More recent is the application of hurdle technology for microbial stabilization of novel healthful foods derived from meat, poultry, or fish, which contain less fat and/or salt and therefore are more prone to spoil or cause food poisoning. The reduction of salt and fat as well as the substitutes and replacers of these traditional ingredients for muscle foods diminish the microbial stability, since several hurdles (a_w, pH, preservatives, and possibly Eh and microstructure) will change. Compensation could be achieved by an intelligent application of hurdle technology (18). The advantages of hurdle technology are most obvious in high-moisture foods, which are shelf stable at ambient temperature due to an intelligent application of combined methods for preservation. However, the use of hurdle technology is appropriate for chilled foods, too, because in the case of temperature abuse, which can easily happen during food distribution, the stability and safety of chilled foods could break down, especially if low-temperature storage is the only hurdle. Therefore, it is advisable to incorporate into chilled foods (eg, sous vide dishes, salads, fresh-cut vegetables) some additional hurdles (eg, modified atmosphere packaging) that will act as a backup in case of temperature abuse. This type of safety precaution for chilled foods is called invisible technology, implying that additional hurdles act as safeguards in chilled foods, ensuring that they remain microbiologically stable and safe during storage in retail outlets as well as in the home (19).

Packaging is an important hurdle for most foods, since it supports the microbial stability and safety as well as the sensory quality of food products. Industrialized countries have the tendency to overpackage foods. This is especially true for Japan, where "active" packaging (using scavengers, absorbers, emmiters, antimicrobial or antioxidative packaging materials, etc.) has been developed to perfection. These "smart" packaging systems are very sophisticated, but wasteful, too. Therefore, Japanese experts are aiming now for less packaging of foods (20). Future packaging shall provide only necessary information and some convenience to the consumer; however, the required shelf life of the products should not come from the packaging but should be based (*1*) on superclean packaging combined with just-in-time delivery or (*2*) on the development of hurdle-technology foods that are stable and safe in spite of minimal packaging.

Finally, it also should be mentioned that novel emerging technologies for food preservation, that is, nonthermal preservation methods (high hydrostatic pressure, pulsed electric fields, oscillating magnetic fields, light pulses, etc), are often most efficient in combination with traditional food preservation methods. Thus, even for the application of futuristic food preservation methods, hurdle technology is essential (21).

Developing Countries

Most of the food in developing countries, which preferably must be storable without refrigeration since electricity is expensive and not continuously available, are based on empiric use of hurdle technology. However, recently such foods have been optimized by the intentional application of hurdles. Relevant examples are optimized meat products of China and Taiwan as well as dairy and meat products of India. Moreover, in several countries of Latin America (ie, Argentina, Mexico, Venezuela), by the application of hurdle technology, high-moisture fruit products (HMFPs) have been developed. In spite of a high water activity (a_w 0.98 to 0.93), HMFPs are storable in freshlike condition for several months at ambient temperatures and even become sterile during storage due to metabolic exhaustion of the bacteria, yeasts, and molds originally present in these products. There is a general trend in developing countries to move gradually away from the traditional intermediate-moisture foods because they are often too salty or too sweet and have a less appealing texture and appearance than high-moisture foods; this goal is achieved by the application of intentional hurdle technology. The progress made in the application of intelligent hurdle technology in developing countries of Latin America, China, India, and Africa has recently been reviewed (22).

FOOD DESIGN

Hurdle technology as a concept has proved useful in the optimization of traditional foods as well as in the development of novel products. However, it is beneficial to combine hurdle technology in food design with Hazard Analysis Critical Control Point (HACCP) and predictive microbiology. For the proper design of hurdle-technology foods a 10-step procedure has been suggested, which comprises hurdle technology, predictive microbiology, and HACCP (or GMP guidelines). This procedure has proved suitable for solving real development tasks in the food industry (15).

BIBLIOGRAPHY

1. L. Leistner, "Hurdle Effect and Energy Savings," in W. K. Downey, ed., *Food Quality and Nutrition*, Applied Science Publishers, London, United Kingdom, 1978, pp. 553–557.
2. L. Leistner and W. Rödel, "The Stability of Intermediate Moisture Foods With Respect to Micro-organisms," in R. Davies, G. G., Birch and K. J. Parker, eds., *Intermediate Moisture Foods*, Applied Science Publishers, London, United Kingdom, 1976, pp. 120–137.
3. L. Leistner, W. Rödel, and K. Krispien, "Microbiology of Meat and Meat Products in High- and Intermediate-Moisture Ranges," in L. B. Rockland and G. F. Stewart, eds., *Water Activity: Influences on Food Quality*, Academic Press, New York, 1981, pp. 855–916.
4. L. Leistner, "Hurdle Technology Applied to Meat Products of the Shelf Stable Product and Intermediate Moisture Food Types," in D. Simatos and J. L. Multon, eds., *Properties of Water in Foods in Relation to Quality and Stability*, Martinus Nijhoff Publishers, Dordrecht, The Netherlands, 1985, pp. 309–329.
5. L. Leistner, *J. Food Eng.* **22**, 421–432 (1994).
6. L. Leistner, "Principles and Applications of Hurdle Technology," in G. W. Gould, ed., *New Methods of Food Preservation*, Blackie Academic and Professional, London, United Kingdom, 1995, pp. 1–21.
7. L. Leistner, *Emerging Concepts for Food Safety*, Proceedings 41st ICoMST, San Antonio, Tex., 1995.
8. D. Häussinger, *pH Homeostasis, Mechanisms and Control*, Academic Press, London, United Kingdom, 1988.

9. G. W. Gould, "Interference With Homeostasis—Food," in R. Wittenbury et al., eds., *Homeostatic Mechanisms in Microorganisms*, Bath University Press, Bath, United Kingdom, 1988, pp. 220–228.
10. G. W. Gould, "Homeostatic Mechanisms During Food Preservation by Combined Methods," in G. V. Barbosa-Cánovas and J. Welti-Chanes, eds., *Food Preservation by Moisture Control, Fundamentals and Applications*, Technomic, Lancaster, Pa., 1995, pp. 397–410.
11. L. Leistner and S. Karan-Djurdjić, *Fleischwirtschaft* **50**, 1547–1549 (1970).
12. L. Leistner, *Food Res. Int.* **25**, 151–158 (1992).
13. S. M. Alzamora et al., *Food Res. Int.* **26**, 125–130 (1993).
14. M. S. Tapia de Daza et al., "Microbial Stability Assessment in High and Intermediate Moisture Foods: Special Emphasis on Fruit Products," in G. V. Barbosa-Cánovas and J. Welti-Chanes, eds., *Food Preservation by Moisture Control, Fundamentals and Applications*, Technomic, Lancaster, Pa., 1995, pp. 575–601.
15. L. Leistner, *Food Design by Hurdle Technology and HACCP*, Adalbert Raps Foundation, Kulmbach, Germany, 1994.
16. L. Leistner, "Stable and Safe Fermented Sausages Worldwide," in G. Campbell-Platt and P. E. Cook, eds., *Fermented Meats*, Blackie Academic and Professional, London, United Kingdom, 1995, pp. 160–175.
17. K. Katsaras and L. Leistner, *Biofouling* **5**, 115–124 (1991).
18. L. Leistner, "Microbial Stability and Safety of Healthy Meat, Poultry and Fish Products," in A. M. Pearson and T. R. Dutson, eds., *Production and Processing of Healthy Meat, Poultry and Fish Products*, Blackie Academic and Professional, London, United Kingdom, 1997, pp. 347–360.
19. L. Leistner, "Combined Methods for Food Preservation," in M. Shafiur Rahman, ed., *Handbook of Food Preservation*, Marcel Dekker, New York, 1999, pp. 457–485.
20. K. Ono, *Packaging Design and Innovation*, Snow Brand, Tokyo, Japan, 1994.
21. G. V. Barbosa-Cánovas et al., *Nonthermal Preservation of Foods*, Marcel Dekker, New York, 1998.
22. L. Leistner, "Use of Combined Preservative Factors in Foods of Developing Countries," in B. M. Lund, A. C. Baird-Parker, and G. W. Gould, eds., *The Microbiological Safety of Foods*, Aspen Publishers, Gaithersburg, Md., 1999.

LOTHAR LEISTNER
Federal Centre for Meat Research
Kulmbach, Germany

HYDROGEN-ION ACTIVITY (pH)

The effective concentration of hydrogen ion in solution is expressed in terms of pH, which is the negative logarithm of the hydrogen-ion activity:

$$\mathrm{pH} = -\log_{10} a_{\mathrm{H}^+} \tag{1}$$

The relationship between activity and concentration is

$$a = \gamma c \tag{2}$$

where the activity coefficient γ is a function of the ionic strength of the solution and approaches unity as the ionic strength decreases; that is, the difference between the activity and the concentration of hydrogen ion diminishes as the solution becomes more dilute. The pH of a solution may have little relationship to the titratable acidity of a solution that contains weak acids or buffering substances; the pH of a solution indicates only the activity of unassociated hydrogen ions. If total acid concentration is to be determined, an acid–base titration must be performed.

Thermodynamically, the activity of a single ionic species is an inexact quantity, and a conventional pH scale has been adopted that is defined by reference solutions with assigned pH values. These reference solutions, in conjunction with equation 3, define the pH.

$$\mathrm{pH}(X) = \mathrm{pH}(S) - \frac{(E_x - E_s)\mathrm{F}}{2.303\mathrm{RT}} \tag{3}$$

E_{S} is the electromotive force (emf) of the cell:

reference electrode | $\mathrm{KCl}(\geq 3.5M)$ || solution S | $\mathrm{H_2(g)}$, Pt

and E_{x} is the emf of the same cell when the reference buffer solution S is replaced by the sample solution X. The quantities R, T, and F are the gas constant, the thermodynamic temperature, and the Faraday constant, respectively. For routine pH measurements, the hydrogen gas electrode [$\mathrm{H_2(g)}$, Pt] usually is replaced by a glass membrane electrode.

The availability of multiple pH reference solutions makes possible an alternative definition of pH:

$$\mathrm{pH}(X) = \mathrm{pH}(S_1) + [\mathrm{pH}(S_2) - \mathrm{pH}(S_1)]\frac{(E_{x2} - E_{s1})}{(E_{s2} - E_{s1})} \tag{4}$$

where E_{s1} and E_{s2} are the measured cell potentials when the sample solution X is replaced in the cell by the two reference solutions S_1 and S_2 such that the values E_{s1} and E_{s2} are on either side of, and as near as possible to, E_{X}. Equation 4 assumes linearity of the pH versus E response between the two reference solutions, whereas equation 3 assumes both linearity and ideal Nernstian response of the pH electrode. The two-point calibration procedure is recommended if a pH electrode, other than the hydrogen gas electrode, is used for the measurements.

pH DETERMINATION

Two methods are used to measure pH: electrometric and chemical indicator (1–6). The most common is electrometric and uses the commercial pH meter with a glass electrode. This procedure is based on the measurement of the difference between the pH of an unknown or test solution and that of a standard solution. The instrument measures the emf developed between the glass electrode and a reference electrode of constant potential. The difference in emf when the electrodes are removed from the standard solution and placed in the test solution is converted to a difference in pH. Electrodes based on metal–metal oxides (eg, antimony–antimony oxide) have also found use as pH sensors, especially for industrial applications where their superior mechanical stability is needed. However, because of the presence of the metallic element, these electrodes suffer from interferences by reduction-oxidation (redox) systems in the test solution. Nonglass pH electrodes have also been described in which synthetic organic ionophores, selective for hydrogen ions, entrapped in plasticized polymeric membranes have shown excellent pH-response behavior. More recently, the pH ion-sensitive field-effect transistor (ISFET) has shown utility for pH measurements, especially where solid-state ruggedness is desired (7). These devices are based on replacing the metal gate of an ordinary field-effect transistor by a pH-sensitive layer such as silicon nitride. These sensors exhibit ideal Nernstian response over pH ranges comparable to conventional glass pH electrodes.

The second method, which has more limited applications, is the indicator method. The success of this procedure depends on matching the color that is produced by the addition of a suitable pH-sensitive indicator dye to a portion of the unknown solution with the color produced by adding the same quantity of the same dye to a series of standard solutions of known pH. Alternatively, the color is matched against a color comparison chart for the particular dye. Because of the limited color resolution, the results obtained by the indicator method are less accurate relative to those obtained using a pH meter and also suffer from errors when used in highly colored solutions or those containing reactive substances such as bleaches. The indicator method, however, is simple to apply and inexpensive. In addition to being used by direct addition to the test solution, indicator dyes can be immobilized onto paper strips (eg, litmus paper) or, more recently, onto the distal end of fiber-optic probes that, when combined with spectrophotometric readout, provide more quantitative indicator-dye pH determinations.

Reference Buffer Solutions

The uncertainties introduced by the reference electrode liquid junction that exist in conventional electrochemical cells can be avoided by using a cell without transference, for example,

$$\text{Ag, AgCl} \mid \text{KCl}(m), \text{ solution } S \mid \text{H}_2(\text{g}), \text{ Pt}$$

Potassium chloride, KCl, of molality m is added to each reference solution. If the standard potential, $E°$, of the cell and the molality of the chloride ion, m_{Cl} are known, emf measurements yield values of the acidity function $\text{p}(a_{\text{H}}\gamma_{\text{Cl}})$, as shown by the following equation:

$$\text{p}(a_{\text{H}}\gamma_{\text{Cl}}) = -\log(m_{\text{H}}\gamma_{\text{H}}\gamma_{\text{Cl}}) = \frac{(E° - E)\text{F}}{2.303\text{RT}} + \log m_{\text{Cl}} \quad (5)$$

To eliminate the effect of the added KCl on the acidity of the buffer solution, $\text{p}(a_{\text{H}}\gamma_{\text{Cl}})$ is determined for three or more portions of the buffer solution that contain different amounts of added chloride. The limiting value of the acidity function is obtained by extrapolation to zero molality of chloride ion. If the single-ion activity coefficient of chloride ion in the buffer solution could be obtained, the activity of hydrogen ion would be readily accessible.

$$\text{p}a_{\text{H}} = -\log(m_{\text{H}}\gamma_{\text{H}}) = \text{p}(a_{\text{H}}\gamma_{\text{Cl}}) + \log\gamma_{\text{Cl}} \quad (6)$$

To establish a conventional scale of hydrogen-ion activity, it has been suggested (8) that the activity coefficient of chloride ion in selected reference buffer solutions having ionic strengths of ≦0.1 be defined as

$$-\log\gamma_{\text{Cl}} = \frac{\text{A}\sqrt{I}}{1 + 1.5\sqrt{I}} \quad (7)$$

where A is a constant of the Debye-Hückel theory and I is the ionic strength. The $\text{p}a_{\text{H}}$ for these selected reference solutions is identified with pH(S) in the operational definition:

$$\text{p}a_{\text{H}} = \text{pH}(S) \quad (8)$$

The pH(S) values at 25°C of the primary and secondary reference buffer solutions certified by the U.S. National Institute of Standards and Technology (NIST) are listed in Table 1 (2). Of particular note, the pH 6.86 and 7.41 phosphate buffers have long been accepted as the primary reference standards for blood pH measurements, with the knowledge that the ionic strength of these buffers is significantly different from that of blood, thus biasing (however reproducibly) the pH measurement in blood due to the residual liquid-junction potential. Recently, NIST certified two concentrations each of two zwitterionic buffer systems (HEPES/HEPESate and MOPSO/MOPSOate) as pH buffers. These secondary standards have been certified at ionic strengths comparable to that of blood and conse-

Table 1. pH Standards, Molality Scale

Solution composition	pH(S) at 25°C
Primary standards	
Potassium hydrogen tartrate (saturated at 25°C)	3.557
0.05 m Potassium dihydrogen citrate	3.776
0.05 m Potassium hydrogen phthalate	4.006
0.025 m KH_2PO_4 + 0.025 m Na_2HPO_4	6.863
0.008695 m KH_2PO_4 + 0.03043 m Na_2HPO_4	7.410
0.01 m $Na_2B_4O_7 \cdot 10H_2O$	9.180
0.025 m $NaHCO_3$ + 0.025 m Na_2CO_3	10.011
Secondary standards	
0.05 m Potassium tetroxalate · $2H_2O$	1.681
0.05 m HEPES[a] + 0.05 m NaHEPESate	7.503
0.08 m HEPES + 0.08 m NaHEPESate	7.516
0.05 m MOPSO[a] + 0.05 m NaMOPSOate	6.867
0.08 m MOPSO + 0.08 m NaMOPSOate	6.865
0.01667 m TRIS[c] + 0.05 m TRIS · HCl	7.699
$Ca(OH)_2$ (saturated at 25°C)	12.454

[a]HEPES = N-2-hydroxyethylpiperazine-N'-2-ethanesulfonic acid.
[b]MOPSO = 3-(N-morpholino)-2-hydroxypropanesulfonic acid.
[c]TRIS = tris(hydroxymethyl)aminomethane.
Source: Ref. 2, and NIST, private communication.

quently should minimize the residual liquid-junction potential. The International Union of Pure and Applied Chemistry recommends the NIST primary standards plus a series of operational standards, measured versus the phthalate reference value standard in a cell with a well-defined liquid junction, for the definition of the pH scale (9). However, this recommendation is currently under review, although significant change in the present scale is not anticipated other than its extension to higher ionic strength solutions.

Accuracy and Interpretation of Measured pH Values

The acidity function $p(a_H\gamma_{Cl})$, which is the experimental basis for the assignment of pH(S), is reproducible within about 0.003 pH unit from 10 to 40°C. If the ionic strength is known and ≤ 0.1, the assignment of numerical values to the activity coefficient of chloride ion does not add to the uncertainty. However, errors in the standard potential of the cell, in the composition of the buffer materials, and in the preparation of the solutions may raise the uncertainty to 0.005 pH unit.

The reproducibility of the practical scale that has been defined using the seven primary standards includes the possible inconsistencies introduced in the standardization of the instrument with seven different standards of different composition and concentration. These inconsistencies are the result of variations in the liquid-junction potential when one solution is replaced by another and are unavoidable. The accuracy of the practical scale from 10 to 40°C therefore appears to be in the range from 0.008 to 0.01 pH unit.

Variations in the liquid-junction potential may be increased when the standard solutions are replaced by test solutions that do not closely match the standards with respect to the composition and concentrations of solutes, or to the solvent composition, for example, nonaqueous and mixed solvents. Under these circumstances, the pH remains a reproducible number, but it may have little or no meaning in terms of the conventional hydrogen-ion activity of the medium. The use of experimental pH numbers as a measure of the extent of acid–base reactions or to obtain thermodynamic equilibrium constants is justified only when the pH of the medium is between 2.5 and 11.5 and when the mixture is an aqueous solution of simple solutes in total concentration of approximately $\leq 0.2\ M$.

Sources of Error

Although subject to fewer interferences and other types of error than most potentiometric ionic-activity sensors, that is, ion-selective electrodes (qv), pH electrodes must be used with an awareness of their particular response characteristics as well as the potential sources of error that may affect the other components of the measurement system, especially the reference electrode (see also "pH Measurement System Electrodes"). Several common causes of measurement problems are electrode interferences and/or fouling of the pH sensor, sample matrix effects, reference electrode instability, and improper calibration of the measurement system (10).

In general, the potential of an electrochemical cell, E_{cell}, is the sum of three potential terms:

$$E_{cell} = E_{pH} - E_{ref} + E_{lj} \tag{9}$$

where E_{pH} and E_{ref} are the potentials of the pH and reference electrodes, respectively, and E_{lj} is the ubiquitous liquid-junction potential. After substitution of the Nernst equation for the pH electrode potential term in equation 9,

$$E_{cell} = E^{\circ}_{pH} - \frac{RT}{F}\ln a_H - E_{ref} + E_{ij} \tag{10}$$

it can be calculated that a 1 mV error in any of the potential terms corresponds to an error of about 4% in the hydrogen-ion activity. Under carefully controlled experimental conditions, the potential of a pH cell can be measured with an uncertainty as small as 0.3 mV, which corresponds to a ±0.005 pH unit uncertainty.

The measurement of pH using the operational cell assumes that no residual liquid-junction potential is present when a standard buffer is compared with a solution of unknown pH. Although this may never strictly be true, especially with complex matrices, the residual liquid-junction potential can be minimized by the appropriate choice of a salt-bridge solution and calibration buffer solutions.

pH MEASUREMENT SYSTEM ELECTRODES

Glass Electrodes

The glass electrode is the hydrogen-ion sensor in most pH-measurement systems. The pH-responsive surface of the glass electrode consists of a thin membrane formed from a special glass that, after suitable conditioning, develops a surface potential that is an accurate index of the acidity of the solution in which the electrode is immersed. To permit changes in the potential of the active surface of the glass membrane to be measured, an inner reference electrode of constant potential is placed in the internal compartment of the glass membrane. The inner reference compartment contains a solution that has a stable hydrogen-ion concentration and counter ions to which the inner electrode is reversible. The choice of the inner cell components has a bearing on the temperature coefficient of the emf of the pH assembly. The inner cell commonly consists of a silver–silver chloride electrode or calomel electrode in a buffered chloride solution.

Immersion electrodes are the most common glass electrodes. These are roughly cylindrical and consist of a barrel or stem of inert glass that is sealed at the lower end to a tip, which is often hemispherical, of special pH-responsive glass. The tip is completely immersed in the solution during measurements. Miniature and microelectrodes are also commercially available and used widely, particularly in physiological studies. Capillary electrodes permit the use of small samples and provide protection from exposure to air during the measurements, for example, for the determination of blood pH. This type of electrode may be provided with a water jacket for temperature control.

The membrane of pH-responsive glass usually is made as thin as is consistent with adequate mechanical strength; nevertheless, its electrical resistance is high, for example, 10 to 250 MΩ. Therefore, an electronic amplifier must be used to obtain adequate accuracy in the measurement of the surface potential of a glass electrode. The versatility of the glass electrode results from its mechanism of operation, which is one of proton exchange rather than electron transfer; hence, oxidizing and reducing agents in the solution do not affect the pH response.

Most modern electrode glasses contain mixtures of silicon dioxide, either sodium or lithium oxide, and either calcium, barium, cesium, or lanthanum oxide. The latter oxides are added to reduce spurious response to alkali metal ions in high pH solutions. The composition of the glass has a profound effect on the electrical resistance, the chemical durability of the pH-sensitive surface, and the accuracy of the pH response in alkaline solutions. Both the electrical and the chemical resistance of the electrode glasses decrease rapidly with a rise in temperature. Therefore, it is difficult to design an electrode that is sufficiently durable for extended use at high temperatures and yet, when used at room temperature, free from the sluggish response often characteristic of pH cells of excessively high resistance. Most manufacturers use different glass compositions for electrodes, depending on their intended use.

The mechanism of the glass electrode response is not entirely understood. It is clear, however, that when a freshly blown membrane of pH-responsive glass is first conditioned in water, the sodium or lithium ions that occupy the interstices of the silicon–oxygen network in the glass surface are exchanged for protons from the water. The protons find stable sites in the conditioned gel layer of the glass surface. Exchange of the labile protons between these sites and the solution phase appears to be the mechanism by which the surface potential reflects changes in the hydrogen ion activity of the external solution. When the glass electrode and the hydrogen gas electrode are immersed in the same solution, their potentials usually differ by a constant amount, even though the pH of the medium is raised from 1 to 10 or greater. In this range, the potential E_g of a glass electrode may be written

$$E_g = E_g^\circ + \frac{RT}{F} \ln a_{H+} \tag{11}$$

where E_g° is the "standard" (or formal) potential of that particular glass electrode on the hydrogen scale.

Departures from the ideal behavior expressed by equation 11 usually are found in alkaline solutions containing alkali metal ions in appreciable concentration, and often in solutions of strong acids. The supposition that the alkaline error is associated with the development of an imperfect response to alkali metal ions is substantiated by the successful design of cation-sensitive glass electrodes that are used to determine sodium, silver, and other monovalent cations (3).

The advantage of the lithium glasses over the sodium glasses in the reduction of alkaline error is attributed to the smaller size of the proton sites remaining after elution of the lithium ions from the glass surface. This interpretation is consistent with the relative magnitudes of the alkaline errors for various cations. These errors decrease rapidly as the diameter of the cation becomes larger. The error observed in concentrated solutions of the strong acids is characterized by a marked drift of potential with time, which is thought to result from the penetration of acid anions, as well as protons, into the glass surface (11).

The immersion of glass electrodes in strongly dehydrating media should be avoided. If the electrode is used in solvents of low water activity, frequent conditioning in water is advisable, as dehydration of the gel layer of the surface causes a progressive alteration in the electrode potential with a consequent drift of the measured pH. Slow dissolution of the pH-sensitive membrane is unavoidable, and it eventually leads to mechanical failure. Standardization of the electrode with two buffer solutions is the best means of early detection of incipient electrode failure.

Fouling of the pH sensor may occur in solutions containing surface-active constituents that coat the electrode surface and may result in sluggish response and drift of the pH reading. Prolonged measurements in blood, sludges, and various industrial process materials and wastes can cause such drift; therefore, it is necessary to clean the membrane mechanically or chemically at intervals that are consistent with the magnitude of the effect and the precision of the results required.

Reference Electrodes and Liquid Junctions

The electrical circuit of the pH cell is completed through a salt bridge that usually consists of a concentrated solution of potassium chloride. The solution makes contact at one end with the test solution and at the other with a reference electrode of constant potential. The liquid junction is formed at the area of contact between the salt bridge and test solutions. The mercury–mercurous chloride electrode—the calomel electrode—provides a highly reproducible potential in the potassium chloride bridge solution and is the most widely used reference electrode. However, mercurous chloride is converted readily into mercuric ion and mercury when in contact with concentrated potassium chloride solutions above 80°C. This disproportionation reaction causes an unstable potential with calomel electrodes. Therefore, the silver–silver chloride electrode and the thallium amalgam–thallous chloride electrode often are preferred for measurements above 80°C. However, because silver chloride is relatively soluble in concentrated solutions of potassium chloride, the solution in the electrode chamber must be saturated with silver chloride to avoid dissolution of the electrode coating.

The commercially used reference electrode–salt bridge combination usually is of the immersion type. The salt–bridge chamber usually surrounds the electrode element. Some provision is made to allow a slow leakage of the bridge solution out of the tip of the electrode to establish a stable liquid junction with the standard solution or test solution in the pH cell. An opening is usually provided through which the electrode chamber may be refilled with the salt-bridge solution. Various devices are used to constrain the outflow of bridge solution, for example, fibers, porous ceramics, capillaries, ground-glass joints, and controlled cracks. Such commercial electrodes normally give very satisfactory results, but there is some evidence that the type and structure of the junction may affect the reference potential when measurements are made at very low pH and, possibly, at high alkalinities.

Combination electrodes have increased in use and are a consolidation of the glass and reference electrodes in a single probe, usually in a concentric arrangement, with the reference electrode compartment surrounding the pH sensor. The advantages of combination electrodes include the convenience of using a single probe and the ability to measure small volumes of sample solution or in restricted-access containers, for example, test tubes and narrow-neck flasks. A disadvantage of this arrangement is that if one of the electrodes becomes defective, the entire combination assembly must be discarded.

Theoretical considerations favor liquid junctions in which cylindrical symmetry and a steady state of ionic diffusion are achieved. Special cells in which a stable junction can be achieved are not difficult to construct and are available commercially.

A solution of potassium chloride that is saturated at room temperature usually is used for the salt bridge. It has been shown that the higher the concentration of the potassium chloride solution, the more effective the bridge solution is in reducing the liquid-junction potential (12). Also, the saturated potassium chloride calomel and silver–silver chloride reference electrodes are stable, reproducible, and easy to prepare. However, after long periods and with temperature lowering, the salt-bridge chamber may become filled with large crystals of potassium chloride that block the flow of bridge solution and thereby impair the reproducibility of the junction potential and raise the resistance of the cell. A slightly undersaturated (eg, 3.5 *M*) solution of potassium chloride is preferred. The calomel electrode has the added disadvantage that it shows a marked potential hysteresis with changes of temperature.

Samples that contain suspended matter are among the most difficult types from which to obtain accurate pH readings because of the so-called suspension effect, that is, the suspended particles produce abnormal liquid-junction potentials at the reference electrode (13). This effect is especially noticeable with soil slurries, pastes, and other types of colloidal suspensions. In the case of a slurry that separates into two layers, pH differences of several units may result, depending on the placement of the electrodes in the layers. Internal consistency is achieved by pH measurement using carefully prescribed measurement protocols, as has been used in the determination of soil pH (14).

Another problem that may result in spurious pH readings is caused by streaming potentials. Presumably, these are attributable to changes in the reference electrode liquid junction that are caused by variations in the flow rate of the sample solution. Factors that affect the observed pH include the magnitude of the flow-rate changes, the geometry of the electrode system, and the concentration of the salt-bridge electrolyte; therefore, this problem may be avoided by maintaining constant flow and geometry characteristics and calibrating the system under operating conditions that are identical to those of the sample measurement.

pH INSTRUMENTATION

The pH meter is an electronic voltmeter that provides a direct conversion of voltage differences to differences of pH at the measurement temperature (15). One class of instruments is the direct-reading analogue, which has a deflection meter with a large scale calibrated in mV and pH units. Most modern direct-reading meters have digital displays of the emf or pH. The types range from very inexpensive meters that read to the nearest 0.1 pH unit to the research models capable of measuring pH with a precision of 0.001 pH unit and drifting less than 0.003 pH unit over 24 h; however, it should be noted that the fundamental meaning of these measured values is considerably less certain than the precision of the measurement.

Because of the very large resistance of the glass membrane in a conventional pH electrode, an input amplifier of high impedance (usually 10^{12}–10^{14} Ω) is required to avoid errors in the pH (or mV) readings. Most pH meters have field-effect transistor amplifiers that typically exhibit bias currents of only a picoampere (10^{-12} ampere), which, for an electrode resistance of 100 MΩ, results in an emf error of only 0.1 mV (0.002 pH unit).

In addition, most of these devices provide operator control of settings for temperature and/or response slope, isopotential point, zero or standardization, and function (pH, mV, or monovalent/divalent cation/anion). Microprocessors are incorporated in advanced-design meters to facilitate calibration, calculation of measurement parameters, and automatic temperature compensation. Furthermore, pH meters are provided with output connectors for continuous readout via a strip-chart recorder and often with binary-coded decimal output for computer interconnections or connection to a printer. Although the accuracy of the measurement is not increased by the use of a recorder, the readability of the displayed pH (on analogue models) can be expanded, and recording provides a permanent record with information on response and equilibration times during measurement (5).

TEMPERATURE EFFECTS

The emf, E, of a pH cell may be written

$$E = E_g^{\circ\prime} - k\text{pH} \tag{12}$$

where k is the Nernst factor $(2.303RT)/F$, and $E_g^{\circ\prime}$ includes the liquid-junction potential and the half-cell emf on the reference side of the glass membrane. Changes of temperature alter the scale slope because k is proportional to T. The scale position also is changed because the "standard" potential is temperature dependent: $E_g^{\circ\prime}$ is usually a quadratic function of the temperature.

The objective of temperature compensation in a pH meter is to nullify changes in emf from any source except changes in the true pH of the test solution. Nearly all pH meters provide automatic or manual adjustment for the change of k with T. If correction is not made for the change of standard potential, however, the instrument must always be standardized at the temperature at which the pH is to be determined. In industrial pH control, standardization of the assembly at the temperature of the measurements is not always possible, and compensation for shift of the scale position, though imperfect, is useful. If the value of $E_g^{\circ\prime}$ were a linear function of T, it would be easy to show that the straight lines representing the variation of E and pH at different temperatures would intersect at a point, the isopotential point or pH_i. Even though $E_g^{\circ\prime}$ does not usually vary linearly with T, these plots intersect at about pH_i when the range of temperatures is narrow. By providing a temperature-dependent bias potential of kpH_i, an approximate correction for the change of the standard potential with temperature can be applied automatically (1,5).

NONAQUEOUS SOLVENTS

The activity of the hydrogen ion is affected by the properties of the solvent in which it is measured. Scales of pH only apply to the medium (single solvent or mixed solvents, eg, water-alcohol) for which they are developed. The comparison of the pH values of a buffer in aqueous solution to one in a nonaqueous solvent has neither direct quantitative nor thermodynamic significance. Consequently, operational pH scales must be developed for the individual solvent systems. In certain cases, correlation to the aqueous pH scale can be made but, in others, pH values are used only as relative indicators of the hydrogen-ion activity.

Other difficulties of measuring pH in nonaqueous solvents are the complications that result from dehydration of the glass pH membrane, increased sample resistance and large liquid-junction potentials. These effects are complex and are highly dependent on the type of solvent or mixture used (1,5).

INDICATOR pH MEASUREMENTS

The indicator method is especially convenient when the pH of a well-buffered colorless solution must be measured at room temperature with an accuracy no greater than 0.5 pH unit. Under optimum conditions an accuracy of 0.2 pH unit may be obtainable. A list of representative acid–base indicators is given in Table 2 with their corresponding transformation ranges. A more complete listing, including the theory of the indicator color change and of the salt effect, is given in Reference 1.

Because they are weak acids or bases, the indicators may affect the pH of the sample, especially in the case of a poorly buffered solution. Variations in the ionic strength or solvent composition, or both, also can produce large uncertainties in pH measurements, presumably caused by changes in the equilibria of the indicator species. Specific chemical reactions also may occur between solutes in the sample and the indicator species to produce appreciable pH measurement errors. Examples of such interferences include binding of the indicator species by proteins and colloidal substances and direct reaction with sample components, for example, oxidizing agents and heavy metal ions.

Table 2. Acid–Base Indicators

Indicator	pH range	Acid color	Base color
Acid cresol red	0.2–1.8	red	yellow
Methyl violet	0.5–1.5	yellow	blue
Acid thymol blue	1.2–2.8	red	yellow
Bromophenol blue	3.0–4.6	yellow	blue
Methyl orange	3.2–4.4	red	yellow
Bromocresol green	3.8–5.4	yellow	blue
Methyl red	4.4–6.2	red	yellow
Bromocresol purple	5.2–6.8	yellow	purple
Bromothymol blue	6.0–7.6	yellow	blue
Phenol red	6.6–8.2	yellow	red
Cresol red	7.2–8.8	yellow	red
Thymol blue	8.0–9.6	yellow	blue
Phenolphthalein	8.2–9.8	colorless	red
Tolyl red	10.0–11.6	red	yellow
Parazo orange	11.0–12.6	yellow	orange
Acyl blue	12.0–13.6	red	blue

INDUSTRIAL PROCESS CONTROL

Specialized equipment for industrial measurements and automatic control have been developed (16) (see "pH Instrumentation"). In general, the pH of an industrial process need not be controlled with great accuracy. Consequently, frequent standardization of the cell assembly may be unnecessary. On the other hand, the ambient conditions, for example, temperature and humidity, under which the industrial control measurements are made may be such that the pH meter must be much more robust than those intended for laboratory use. To avoid costly downtime for repairs, pH instruments may be constructed of modular units, permitting rapid removal and replacement of a defective subassembly.

The pH meter usually is coupled to a data recording device and often to a pneumatic or electric controller. The controller governs the addition of reagent so that the pH of the process stream is maintained at the desired level.

Immersion-cell assemblies are designed for continuous pH measurement in tanks, troughs, or other vessels containing process solutions at different levels under various conditions of agitation and pressure. The electrodes are protected from mechanical damage and sometimes are provided with devices to remove surface deposits as they accumulate. Process flow chambers are designed to introduce the pH electrodes directly into piped sample streams or into bypass sample loops that may be pressurized. Electrode chambers of both types usually contain a temperature-sensing element that controls the temperature-compensating circuits of the measuring instrument.

Glass electrodes for process control do not differ materially from those used for pH measurements in the laboratory, but the emphasis in industrial application is on rugged construction to withstand both mechanical stresses and high pressures. Pressurized salt bridges, which ensure slow leakage of bridge solution into the process stream even under very high ambient pressures, have been developed. For less-severe process-monitoring conditions, reference electrodes are available with no-flow polymeric or gel-filled junctions that can be used without external pressurization.

BIBLIOGRAPHY

1. R. G. Bates, *Determination of pH, Theory and Practice*, 2nd ed., Wiley-Interscience, New York, 1973.
2. Y. C. Wu, W. F. Koch, and R. A. Durst, *Standardization of pH Measurements*, National Bureau of Standards Special Publication 260-53, U.S. Government Printing Office, Washington, D.C., 1988.
3. G. Eisenman, ed., *Glass Electrodes for Hydrogen and Other Cations*, Marcel Dekker, New York, 1967.
4. G. Mattock, *pH Measurement and Titration*, Macmillan, New York, 1961.
5. C. C. Westcott, *pH Measurements*, Academic Press, New York, 1978.
6. "pH of Aqueous Solutions with the Glass Electrode," *ASTM Method E 70-77*, American Society for Testing and Materials, Philadelphia, Pa., 1977.
7. J. Janata, *Principles of Chemical Sensors*, Plenum Press, New York, 1989.
8. R. G. Bates and E. A. Guggenheim, *Pure Appl. Chem.* **1**, 163 (1960).
9. A. K. Covington, R. G. Bates, and R. A. Durst, *Pure Appl. Chem.* **57**, 531 (1985).
10. R. A. Durst, "Sources of Error in Ion-Selective Electrode Potentiometry" in H. Freiser, ed., *Ion-Selective Electrodes in Analytical Chemistry*, Plenum Press, New York, 1978, pp. 311–338.
11. K. Schwabe, "pH Measurements and Their Applications" in H. W. Nürnberg, ed., *Electroanalytical Chemistry*, John Wiley & Sons, New York, 1974, pp. 495–586.
12. E. A. Guggenheim, *J. Am. Chem. Soc.* **52**, 1315 (1930).
13. H. Jenny et al., *Science* **112**, 164 (1950).
14. A. M. Pommer, "Glass Electrodes for Soil Waters and Soil Suspensions" in G. Eisenman, ed., *Glass Electrodes for Hydrogen and Other Cations*, Marcel Dekker, New York, 1967, pp. 362–411.
15. A. Wilson, *pH Meters*, Barnes and Noble, New York, 1970.
16. F. G. Shinskey, *pH and pIon: Control in Process and Waste Streams*, John Wiley & Sons, New York, 1973.

GENERAL REFERENCES

Refs. 1–6 are also general references.

RICHARD A. DURST
Cornell University
Geneva, New York

ROGER G. BATES
University of Florida
Gainesville, Florida

HYDROGENATION

Hydrogenation is unique among the unit processes used in edible oils and fats processing in that it alters the molecular structure and composition of the glycerol esters comprising the naturally occurring triglycerides of vegetable, meat, and marine origin. It is a chemical reaction in which hydrogen is added to the ethylenic linkages (double bonds). It requires the presence of a catalyst—historically almost always nickel. It normally is performed batchwise in a hydrogen-pressurized stirred reactor. Agitation is necessary to suspend the catalyst and to effect solubilization of the hydrogen, thus enabling it to react on the surface of the catalyst with a double bond in the oil.

Hydrogenation is employed worldwide on a vast scale in the edible oils and fats industry. Its principal objectives are to convert liquid oils into plastic fats usable in a wide variety of cooking and baking applications, and to improve flavor stability by retarding deterioration through oxidation. Hydrogenation has also made a major contribution to the present high degree of interchangeability among a wide variety of oils and fats.

REACTION MECHANISM

The basic chemical equation for hydrogenation of an unsaturated carbon–carbon double bond is shown in equation 1(1).

$$\text{-CH=CH-} + H_2 \xrightarrow{\text{catalyst}} \text{-CH}_2 - \text{CH}_2\text{-} \quad (1)$$

As the equation indicates, and as alluded to earlier, hydrogenation takes place when the carbon to carbon double bonds in the liquid oil, the solid catalyst, and the gaseous hydrogen have been brought together—usually in a heated stirred reactor. The hydrogen must be dissolved in the oil since only dissolved hydrogen is available for reaction.

Although the hydrogenation reaction may appear straightforward, it is in actuality very complicated. The esters may contain one, two, three, or more unsaturated bonds, and each double bond may be hydrogenated at a different rate, depending on its position in the molecule. There may also be simultaneous isomerization of the unsaturated bonds. Isomerization is principally geometrical, but some positional isomerization can also occur. In addition, the position of the ester on the glycerol also has some effect in determining the physical properties of the molecule.

The complexity of the reaction is further illustrated by noting that the partial hydrogenation of soybean oil results in the production of a minimum of 30 different linolenic, linoleic, and oleic esters whose *cis* and *trans* forms could produce more than 4000 different triglycerides. The marvel is that producers have learned to not only control the reaction but actually utilize its many-faceted complexity to produce a great variety of oils, shortenings and margarines, each designed to have physical properties (functional characteristics) making them specifically desirable for a particular application. In recent years there has been an additional strong impetus to also make the final products nutritionally desirable.

SELECTIVITY

The term *selectivity*, as commonly used in the edible oils and fats industry, relates to equation 2 (2).

$$\text{Linolenic} \xrightarrow{K_1} \text{linoleic} \xrightarrow{K_2} \text{oleic} \xrightarrow{K_3} \text{stearic} \qquad (2)$$

With some simplification, linolenic selectivity is defined as the ratio (Ln SR) of the rate of reaction of K_1 to K_2. With the same simplification, linoleic selectivity is defined as the ratio (Lo SR) of the rate of reaction of K_2 to K_3. Because of its much greater utility and its ability to be manipulated, when the term selectivity is used in the edible oil industry, the speaker or writer is referring to linoleic selectivity unless explicitly indicated otherwise. Another way of defining linoleic selectivity is as the degree of preferential conversion of dienes to monoenes compared with the conversion of monoenes to saturates.

Although few *trans* isomers occur in nature, during partial hydrogenation of oils the double bond may be either saturated or isomerized while it is being adsorbed on the catalyst surface. Some positional and many geometric isomers are formed, basically by way of the Horiute-Polanyi mechanism. Elaidic, the *trans* form of octadecanoic (oleic) acid, is the most common *trans* isomer. Its historic utility in margarine formulation is based on its having a significantly higher melting point (43.7°C) than the *cis* form (16.3°C), while still being considerably lower melting than the completely saturated stearic (69.6°C). Since the solids content of a fat at any given temperature depends on the distribution of all the esters of a triglyceride, the ratio of *trans* to *cis* is the principal determinant of the slope of the melting curve in the important refrigerator to body temperature range.

The preferential selectivity of nickel catalysts for the hydrogenation of a linolenic ester in a triglyceride compared with a linoleic (Ln SR) has been shown to be about 1.8 to 2.3. Since changes in process conditions do not significantly change this ratio, evidently the diene and the triene are hydrogenated by the same mechanism. Although it would seem the presence of the third double bond should double the chance of hydrogenation of one of the double bonds in the triene compared with the diene, this has not been shown to be the case. However, when copper catalysts are used to hydrogenate soybean oil, as will be discussed later, linolenic selectivity ratios of 8 to 12 are found. Copper catalyst obviously operates somewhat differently from nickel. It is surmised that copper catalyst causes conjugation of the linolenate triene on the catalyst surface. Since conjugated trienes have been shown to react about 200 times faster than nonconjugated ones, they do not accumulate in the product—evidently hydrogenating to a conjugated diene before desorbing from the catalyst surface. The conjugated dienes are then further reduced to monoenes. The monoenes are not reduced to saturates by the catalyst since hydrogenation with copper catalyst must involve two or more double bonds.

EFFECTS OF PROCESS CONDITIONS

When hydrogenating a specific oil with a chosen catalyst, the reaction parameters are temperature, pressure, catalyst concentration, and agitation (mixing). Because each parameter influences both reaction rate and selectivity, and because all are interrelated, it is not possible to specifically predict what their combined effect will be in a particular instance. Therefore, each of the parameters will be discussed separately, as they affect reaction rate and as they affect selectivity.

Effects of Temperature

Hydrogenation, like other chemical reactions, is accelerated by increasing temperature. Both preferential selectivity and isomerization are also greater with increasing temperature. Since there is an increasing tendency toward hydrolysis above 400°F (204°C), processors limit their upper temperature accordingly, with 450°F (232°C) being a maximum. Since some investigators claim that nutritionally undesirable positional isomers are also formed at high temperatures, a limitation as low as 400°F (204°C) is sometimes observed.

Effects of Pressure

Again, as with most chemical reactions, increasing pressure increases reaction rate. Commercial hydrogenation of edible triglyceride oils is usually performed under hydrogen pressure of 7 to 50 psig (0.5–3.5 atm)—at the lower end for partial hardening, which constitutes most of the commercial hydrogenation of edible oils such as soybean, canola, and so on. Within this lower range, even modest change in pressure has a significant effect on the inverse relationship of increasing reaction rate/decreasing selectivity—both preferential and *trans* isomer.

Effects of Catalyst Concentration

Increasing the catalyst concentration in a given system increases reaction rate up to the point where hydrogen availability becomes the limiting parameter. It has a minor effect on selectivity. The catalyst concentration employed in commercial partial hydrogenation of edible oils is chosen so that the converter reaction time fits into the other batch cycle steps of heat exchange and filtration for catalyst removal.

Effects of Agitation

Agitation (mixing) is of great importance in determining both the rate and selectivities (preferential and *trans*-isomeric) of edible oil hydrogenation. In general, better mixing increases activity and decreases selectivity. Agitation must accomplish the following: (*1*) distribute heat or cooling for temperature control; (*2*) keep the solid catalyst suspended; (*3*) solubilize and maintain the solution of hydrogen in the oil.

Whereas (*1*) and (*2*) are straightforward and quite easily achieved, (*3*) is complex. In conventional batch-operated tank-type hydrogenation reactors (Fig. 1), hydrogen is bubbled into the liquid through a spider-type gas distributor at the bottom of the converter. While the hydrogen to be solubilized comes principally from bubbles absorbed during their passage up through the oil from the spider, it partially comes from gas in the headspace that is continually being stirred back into the oil. Mixing in such a converter as shown in Figure 1 is provided by two or more turbine blades attached to a central shaft. Individual

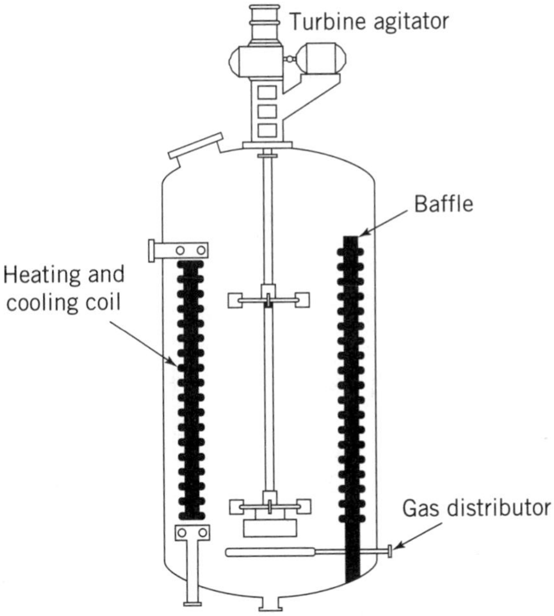

Figure 1. Conventional hydrogenation reactor. *Source:* Ref. 1.

blades may be flat/perpendicular, flat/canted, or foil shaped, depending on whether their purpose is to break the hydrogen into small bubbles, retard the flow of the bubbles upward, or create a vortex to reincorporate headspace hydrogen into the oil. Positioning of the blades on the shaft is very important. The rotation speed of the agitators, design and location of baffles on the side of the tank, and placement of heating/cooling coils also affect the flow of hydrogen bubbles passing through the oil and, consequently, the absorption of hydrogen into the oil. Mixers specially designed to better reincorporate hydrogen from the headspace into the body of the oil are gaining acceptance. They provide a more constant supply of hydrogen at the catalyst site, thus increasing the rate of reaction and also making it less variable. This is a significant aid in achieving uniform selectivity, which results in greater product uniformity. Three agitation systems designed to achieve this are depicted in Figures 2, 3, and 4.

The activity and selectivity effects of varying hydrogenation process conditions are summarized in Table 1.

CATALYSIS

According to the classic definition of Ostwald, a catalyst is a substance that alters the rate of a chemical reaction without affecting the energy factors of the reaction or being consumed in the reaction. Thus, the catalyst enters into the reaction over and over again, and a relatively small amount may be capable of transforming very large amounts of feedstock into reacted product. In the hydrogenation of vegetable oils, the concentration of nickel employed as a catalyst does not exceed a few hundredths of a percent. Furthermore, one catalyst may differ from another in its relative effect on alternate reaction rates. The hydrogenation of fats and oils furnishes examples of such specificity of catalyst action. Thus, the addition of 1 mol of hydrogen to a linoleic acid chain in a glyceride molecule may yield either a normal oleic ester or isomeric forms of it. Some nickel catalysts are more inclined than others to produce the isomeric forms.

Table 1. Activity and Selectivity Effects of Varying Hydrogenation Process Conditions

Increase in	Rate	Lo SR[a]	Isomerization
Temperature	+ + + +	+ + + +	+ + + +
Pressure	+ + +	− − −	− − −
Agitation	+ + + +	− − − −	− − − −
Catalyst concentration	+ +	−	−

[a]Lo SR, linoleic selectivity ratio.

Heterogeneous catalysts are the most important in industry generally. In edible oil hydrogenation they are used exclusively. In heterogeneous catalysis it is assumed the reaction proceeds through the formation of unstable intermediate compounds or adsorption complexes, in which the catalyst is temporarily combined with one or more of the reactants. If such compounds do exist, it is probable that in most cases they are not definite chemical combinations but consist merely of strongly bound molecules of the reactant held to the catalyst surface by secondary valence forces, or by complexing. In any event, it is essential that they be unstable. That is, capable of being either decomposed or desorbed, to permit reaction to proceed according to the scheme shown in equation 3.

$$\text{catalyst} + \text{reactants} \xrightarrow{\text{adsorption}} \text{catalyst-reactant complex} \xrightarrow{\text{desorption}} \text{reaction products} + \text{regenerated catalyst} \quad (3)$$

Although homogeneous catalysts for fats and oils hydrogenation have been studied, there is no current commercial interest in pursuing them. This is principally because of legal/environmental reasons. For homogeneous catalysts to be legally permitted in edible oil hydrogenation, they would need to achieve generally regarded as safe (GRAS) status. This would require either proof of their complete removal after use, or long-term human feeding tests demonstrating their harmlessness. In either case, the cost is too high and the potential liability too great to be considered seriously by either oil processors or catalyst manufacturers. Since enzyme catalysts might have the possibility of more easily being recognized as GRAS, they could be a future possibility.

Commercial hydrogenation catalysts are made by first combining nickel with other elements, such as in nickel oxide, nickel hydroxide, nickel carbonate, nickel formate, or nickel-aluminum alloy, and then reducing the compound to regain a portion of the nickel (now in a catalytic state) in metallic form. Considerable evidence indicates that the hydrogenation of an ethylenic compound must be preceded by two-point adsorption of the carbon atoms on either side of the double bond. This requirement imposes certain dimensional limitations on the space lattice of any catalytically active metal. Actually, the metals that are at all effective in the hydrogenation of double bonds, such as

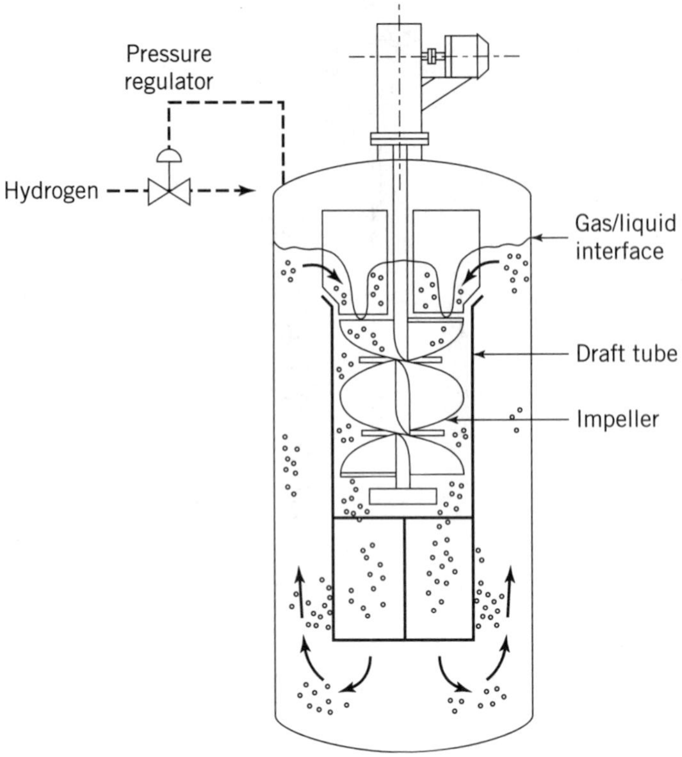

Figure 2. Helical screw mixer. *Source:* Ref. 1.

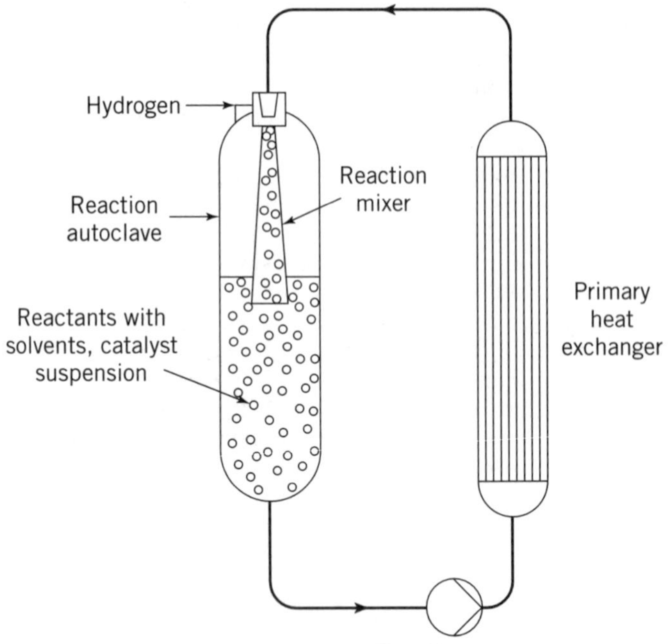

Figure 3. Loop Venturi reactor. *Source:* Ref. 1.

nickel, cobalt, iron, copper, platinum, and palladium, all have interatomic spacings close to that (2.73 Å) calculated as optimum for such two-point adsorption.

Catalysts are quite sensitive to heat and may be inactivated by temperatures considerably below the fusion point of massive nickel. This inactivation is apparently the result of a sintering process, which causes the active projecting nickel atoms to assume more stable positions on the catalyst surface. This sensitivity to heat makes their reduction to metallic form critical with respect to temperature since there is sometimes only a small interval between the temperature at which reduction becomes rapid and that at which sintering begins. Precipitated nickel catalysts are usually reduced at temperatures between about

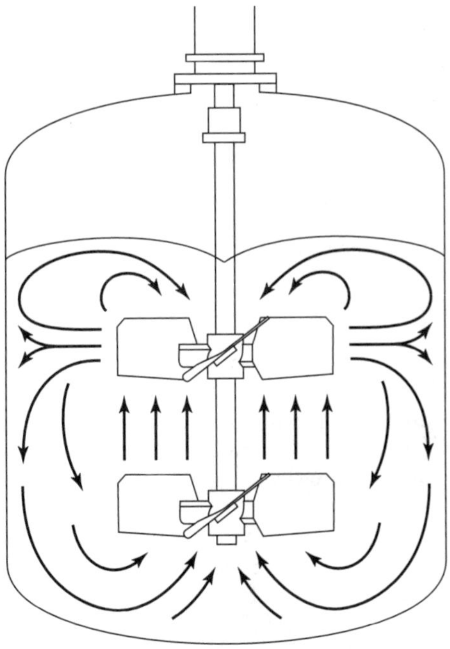

Figure 4. A340 Hydrofoil Mixer. *Source:* Courtesy of Lightnin.

800°F (425°C) and 900°F (480°C), well below nickel's sintering temperature, which begins at about 1000°F (540°C).

Freshly reduced catalysts are highly pyrophoric because of the reactive nature of their zero valence nickel atoms and their retention of hydrogen adsorbed during reduction. If exposed to air (oxygen) at this stage, they become inactivated as the atoms revert to their oxide form. Their pyrophoric properties and tendency to become inactivated through oxidation can be essentially eliminated by coating (protecting) the catalyst surface with an essentially inert material, usually a highly saturated fat. An alternate method of stabilization is by a slight back oxidation after reduction. Although this takes the "edge" off the catalyst's activity, it readily reactivates *in situ* under the reducing conditions of hydrogenation. It is a common and necessary practice to use reduced and stabilized catalysts for the hydrogenation of feedstocks to produce products in which even the small amount of protecting oil would be a serious contaminant. The conversion of fatty nitriles to amines is an example.

CATALYST POISONS

General

Since the activity of a catalyst depends on the presence of a relatively few metallic atoms of an unusually high degree of reactivity, it is to be expected that these atoms will display a marked avidity for many substances, in addition to hydrogen, if such substances are present as impurities in the reacting system. Substances that are thus able to cause catalysts to become inactive are generically termed catalyst *poisons*. They may be troublesome when present, even in traces, since the amount of catalyst is always small in relation to the amount of reactants, and the active portion of the catalyst is small in comparison with the total catalyst.

Impurities that have a negative effect on catalysts can be more specifically broken down into poisons, inhibitors, and deactivators, with each group acting according to a different mechanism. True poisons, such as sulfur and bromine, chemically react with the nickel in an irreversible manner. Inhibitors, such as phosphorous and alkali soaps, prevent hydrogenation from occurring by blocking the entrances to the catalyst pores. Deactivators, such as water and carbon dioxide, do not necessarily permanently affect the reaction since they can be removed under vacuum at elevated temperature.

Among the worst poisons for nickel catalysts are the gaseous sulfur compounds: hydrogen sulfide, carbon disulfide, sulfur dioxide, carbon oxysulfide, and so on. Historically, these compounds were of considerable concern because they may occur as impurities in hydrogen prepared by the steam–iron, water gas–catalytic, or hydrocarbon reforming processes. They are rapidly adsorbed by nickel catalysts and poison the catalyst irreversibly. The virtual disappearance of plants producing hydrogen by the steam–iron and water gas–catalytic processes, and the addition of purification steps to hydrocarbon reforming plants, have essentially eliminated gaseous sulfur compounds as being of significance in present-day edible oil hydrogenation.

Besides the sulfur compounds just mentioned, the gaseous catalyst poison that historically has been most likely to cause trouble in the hydrogenation of fats and oils is carbon monoxide. As with gaseous sulfur, the discontinuance of some methods of manufacture and the addition of purifiers, principally pressure swing adsorbers (PSAs), have also essentially eliminated carbon monoxide as a catalyst poison found in hydrogen.

Other gaseous impurities that can occur in hydrogen include carbon dioxide, nitrogen, and methane. Although these gases are not catalyst poisons per se, in a converter of the "dead-end" type they accumulate in the headspace as the reaction proceeds. By dilution, this reduces the hydrogen partial pressure and the reaction slows. Historically, this was a significant problem, particularly after the oil's iodine value had been lowered considerably. It was dealt with by venting, either occasionally or continuously. However, as with the other gaseous impurities, PSA systems have eliminated the problem. Small amounts of water vapor in the hydrogen (or the feedstock) have negligible poisoning effect on the catalyst but do cause formation of free fatty acids by hydrolysis of the triglyceride. These acids, in turn, react with the catalytic nickel to form noncatalytic nickel soaps. This phenomenon not only inactivates a portion of the catalyst but the resulting soaps slow posthydrogenation filtration.

Hydrogen produced by electrolysis contains a small amount of oxygen. Although oxygen is not a catalyst poison, its potential to cause oil oxidation makes it suspect as deleterious to product quality. It can also be readily reduced to a negligible level by a purification step.

Feedstock Poisons

All naturally occurring oils and fats contain constituents that can inhibit hydrogenation. They vary greatly, depending on the source—whether vegetable, meat, or marine. They also vary widely within those categories. For instance, in the case of vegetable oils, soybean and rapeseed contain considerable phosphorus, whereas palm has virtually none. Rapeseed oils contain sufficient sulfur to have a significant effect on hydrogenation; soybean has essentially none. Even a specific oil, such as rapeseed, can vary greatly in its content of sulfur, phosphorous, and chlorophyll. This variation may be caused by seed variety, weather, or handling. Whatever the cause or source of the hydrogenation-inhibiting substance, a principal function of the processing steps prior to hydrogenation is its reduction or elimination. This is successful to a greater or lesser degree depending on the crude material and on the effort expended.

Very significant progress has been made in oil purification. A principal impetus has been the realization by processors that the costs of purification are usually less than the downstream cost of larger catalyst usage, lower yield, and inferior product quality. The availability of improved processing controls and better analytical instruments have contributed substantially to this effort. For instance, instruments such as the inductively coupled plasma spectrophotometer have enabled processors to rapidly and accurately measure and control phosphorous and other metals.

CATALYSTS

Nickel Catalysts

Nickel alloy and wet-reduced nickel formate catalysts were historically used for edible oil hydrogenation, but they have been completely replaced by dry-reduced precipitated nickel catalysts. Catalysts of this type are prepared by precipitating a nickel salt, usually sulfate or chloride, on an inert support, such as silica or alumina or a combination thereof. The precipitate is filtered, washed, and dried. The resulting powder is reduced with hydrogen at 800 to 900°F (480–540°C). It is then coated (protected) with a triglyceride fat, chilled, and formed into a particle (pastule or flake) for packaging and shipment.

The performance characteristics (activity, selectivity, and filterability) of dry-reduced precipitated catalysts depend to a very great extent on the precipitation step. Such operating variables as the strengths of the solutions, the temperature at which precipitation takes place, the agitation of the solutions, their pH, the temperature to which the suspended precipitate is subjected, and so on, are all important and are rigidly controlled by the catalyst producer.

To a significant degree, the manufacture of dry-reduced precipitated nickel catalysts can be manipulated to produce products having various combinations of activity, selectivity, and filterability. This can be important and useful, depending on the feedstocks and the finished products to be manufactured.

Promoted Catalysts

A catalyst promoter is classically defined as a metal or other substance that enhances performance (usually activity) without actually being a catalyst for the reaction. The promotion of catalysts is much practiced in industries other than the hydrogenation of fats. In these cases, the employment of catalysts having two, three, four, or even more components is not unusual. In such complex systems the action of some of the components may more accurately be called synergistic rather than promotive, since more than one of the components may possess catalytic activity by itself.

A number of mechanisms have been proposed to explain the phenomenon of promotion. It has been supposed that the promoter acts as a secondary catalyst, accelerating the formation or decomposition of intermediate compounds, assisting in the adsorption of the reactants, or protecting the catalyst from poisons. With catalysts for the hydrogenation of fats, however, it appears more reasonable to assume that the function of a promoter would be simply structural, that is, to permit development of larger numbers of active centers on the catalyst surface.

Metals referred to in the patent literature as useful promoters for nickel catalysts used in the hydrogenation of edible oils, either as such or in the form of oxides, include chromium, cobalt, thorium, zirconium, copper, titanium, and silver. Practical use of some of them (eg, chromium) can be ruled out based on current environmental and/or nutritional laws and regulations. Most have not shown any cost-benefit performance advantage. Copper will be discussed separately, not as a promoter but as an alternate to nickel.

Nickel Sulfur Catalysts

The addition of sulfur to a nickel catalyst formulation has two effects. One is the diminishing of the nickel's hydrogenation activity and the second is an increase in *trans*-isomer formation. The increase in *trans*-isomer formation has spawned a widespread use of nickel sulfur catalysts to convert vegetable oils into products acceptable as cocoa butter substitutes useful in the manufacture of confectionery/coating fats.

Other Metal Catalysts

Although precious metals have not been used for commercial hydrogenation of fats and oils, they have been investigated quite extensively. They have several characteristics that make them attractive. These include satisfactory reactivity at about 60°C and an activity so powerful that only a few parts per million are required. However, because of the high price of the metal, to be cost-effective the metal in the spent catalyst would need to be at least 98% captured and reclaimed. Although that is feasible using current techniques, it would require considerable change in practice by both the hydrogenator and the catalyst manufacturer. Since neither has perceived any appreciable advantage in making such a change, it has not received serious attention. However, in the case of some unanticipated future development, such as the outlawing of nickel for edible oil hydrogenation, platinum or palladium would be viable substitutes.

Copper catalyst, because of its high linolenic selectivity (Ln SR), has been used to a limited extent for the hydrogenation of the linolenic ester in soybean oil. Soybean oil "brush" hydrogenated with copper catalyst, although much more linolenic preferential than nickel, still requires winterization to pass a cold test. The deleterious effect of even traces of residual copper on oxidative stability of the product is also a negative aspect. While copper is not currently used commercially, 1960s research on the possibility of doing so added considerably to understanding the selectivity of polyunsaturated ester hydrogenation.

HYDROGENATION IN PRACTICE

Batch Hydrogenation

Figure 5 is a flowsheet sketch depicting the unit operations of an industrial vegetable oil hydrogenation facility, commonly referred to as a hardening plant. The feedstock oil from upstream purification operations is brought to reaction temperature in a heat exchange with hydrogenated oil coming from the converter by way of the converter's drop tank. It is held in a hot oil measuring tank awaiting transfer to the converter. The cooled hydrogenated oil passes through a filter press where the catalyst is removed.

With state-of-the-art instrumentation and properly sized valves, pumps, and so on, the converter can easily be cycled 12 to 15 times per 24 h with even greater turnover if needed.

Hydrogen is supplied under pressure to the converter, either from on-site generation or as purchased from commercial suppliers. Catalyst is purchased from one of the three or four commercial manufacturers who supply it worldwide. It is added to the converter either as a liquid by pumping or by dry aspiration utilizing the converter's vacuum system.

A single-stage steam ejector system produces sufficient vacuum for the converter operations of evacuating air from the headspace, assisting in maintaining low-moisture feedstock, and aspirating catalyst. The converter is serviced by electricity for motors that drive the agitator and turn the centrifugal pumps that move the feedstock and hydrogenated product through the system. Although low-pressure heating steam is ordinarily available, except for startup, it is not needed in energy-efficient plants. Rather, most of the heat generated by the hydrogenation reaction exotherm is exchanged with the incoming feedstock, as already discussed, to bring it to reaction temperature. The excess calories, in the form of steam generated from the vaporization of converter cooling water, are used as a heat source for other plant operations.

Meter measurement of hydrogen usage can be a good method of determining how far the reaction has proceeded. It requires knowledge of the iodine value of a known quantity of the specific oil being hydrogenated in a reactor of known volume. Assuming no venting, the system, though indirect, is quite accurate.

An alternate method of determining the degree of hydrogenation is by measuring the refractive index of the oil and converting it to iodine value using the known relationship between them. Instruments are available to do this on-line and *in situ*. They utilize probes positioned closely together within the converter. One measures the oil's refractive index and the other measures its temperature.

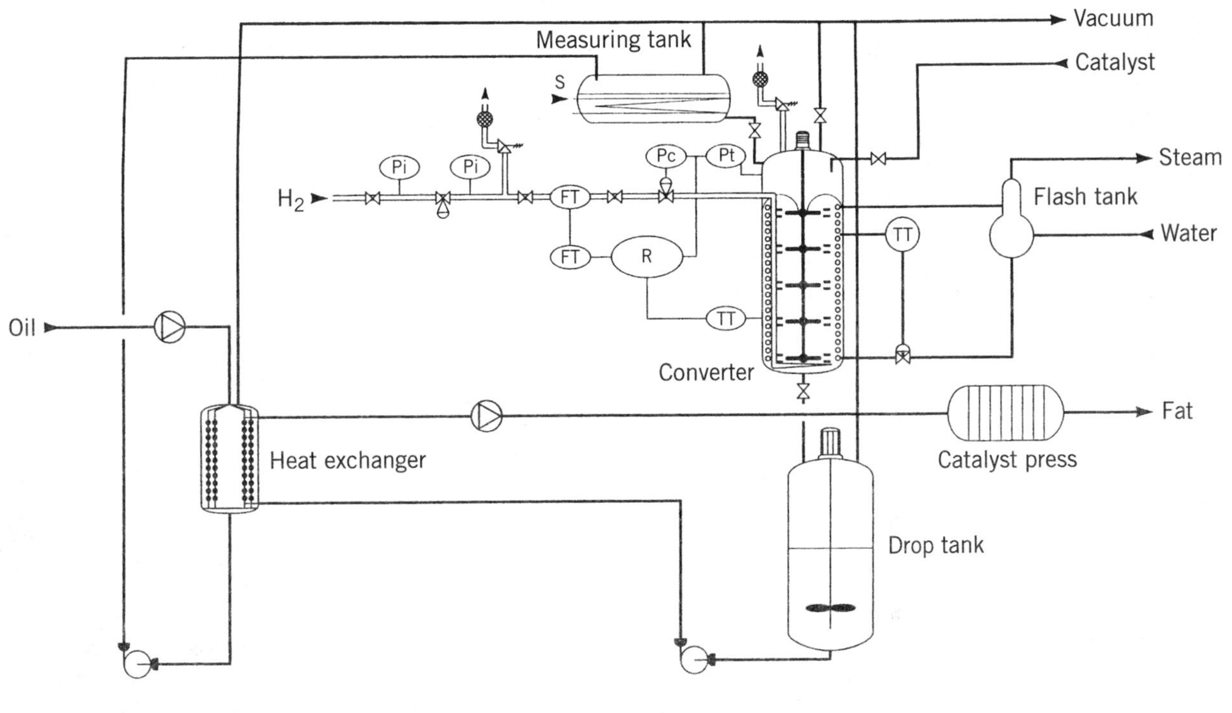

Figure 5. The hardening plant. *Source:* Ref. 1.

The probes are tied together in an external computer where software converts their measurements to iodine value at a specific reference temperature—usually 60°C.

Continuous Systems

Slurry. Employing a continuous system for hydrogenating edible oils has a number of operational advantages. A principal advantage is elimination of the numerous opportunities for equipment malfunction and human error that are inevitable in cycling operations. Since the temperature profile of the reaction is essentially constant in a continuous system, there is significantly more uniform selectivity of the hydrogenation reaction. In spite of these advantages, continuous slurry hydrogenation has achieved only very limited success in edible oil hydrogenation. The reason usually cited for its lack of acceptance is the contamination that inevitably occurs when switching from one product to another. Maintaining only a few base stocks can minimize the seriousness of this problem. Utilization of an on-line instrument to monitor refractive index (iodine value), as described previously, could also be helpful.

Fixed Bed. The fixed-bed mode may have a long-term future in continuous partial hydrogenation of triglycerides. With it, a relatively small amount of feedstock makes a single pass through a comparatively large quantity of formed catalyst. A filtration step is not needed to remove the catalyst. This simplifies the operation and enhances used catalyst disposition, since the spent will usually have a relatively high nickel content. The major disadvantage of fixed-bed systems is the gradual poisoning and/or inactivation of the catalyst caused by impurities in the feedstock accumulating on the catalyst. This affects the selectivity characteristics of the reaction and is reflected in a gradually changing melting profile of the hydrogenated product. Activity of the catalyst declines similarly. The catalyst is replaced when its activity and/or selectivity has declined to the point where either throughput or product quality becomes unacceptable. Some change in selectivity characteristics can be accommodated by manipulating the operational variables.

Impurity levels in feedstocks, and to some extent in hydrogen, have historically made the use of fixed-bed catalysts economically unattractive. More recently, perceiving the positive cost effect of lower catalyst usage, refiners have significantly reduced impurity levels in hydrogenation feedstocks. The availability of better and faster analytical techniques has been of help in this regard. Likewise, hydrogen purity has improved to the extent where it is no longer a factor in catalyst efficiency.

Based on significant lessening of the historical reasons for limited catalyst life in a fixed-bed mode, as just outlined, there may be a resurgence of interest in this approach. If nutritional advantages for the use of precious metal catalysts are verified, it would be a significant impetus for their use becoming attractive. If so, it would most likely be in a fixed-bed mode because of the much greater feasibility of precious metal reclamation from formed catalysts.

Catalyst Usage

The only correct manner for measuring catalyst usage efficiency is to relate it to quantity of catalyst consumed per quantity of feedstock hydrogenated. For accounting purposes it is translated into monetary units, for example, dollars per ton.

Although the price of catalyst has increased substantially over time, its usage efficiency cost has remained relatively constant. There have been two principal reasons for this: (*1*) a continual improvement in commercial catalyst performance, both in activity and uniformity, and (*2*) a continual improvement in feedstock purity. Purity improvements have resulted from better processing practices, from the farmers' field through oil separation, refining, and bleaching. Better and faster analytical techniques have enabled it to happen.

Some processors prefer to use catalyst once and then discard it. The principal reason they cite for this practice is catalytic performance predictability, both in activity and selectivity. Others have extensive reuse procedures. Generally, these break down into one of three alternate practices. The first is simple reuse; that is, employing the same catalyst over and over until its selectivity characteristics preclude making product specifications, or its activity extends reaction time to an unacceptable point. The second practice is to program catalyst reuse according to the type of hardening needed. This approach endeavors to take advantage of the phenomenon of preferential and *trans*-isomeric selectivities changing through use. These changes reflect the effect of feedstock poisons accumulating on the catalyst. Such a usage sequence might be a progression from brush hydrogenation for salad oil, through shortening base stock, through margarine base stock and finally, to complete hydrogenation to low iodine value stearine. A third reuse option of continually removing a portion of the catalyst and replacing it with an equal portion of fresh also has its advocates. While this option is uncomplicated, its downside is the acceptance of performance characteristics less than attainable with the other two approaches.

As referred to earlier, catalyst is often added to the converter by dissolving it in feedstock in a small side tank and then pumping into the reaction vessel. When adding new (dry) catalyst, a better way is to aspirate a weighed amount into the headspace of the converter, by utilizing the reactor's vacuum system.

Employing reused catalyst requires quite elaborate tank and measuring systems plus extensive record keeping and close monitoring. Over the years there has been a trend toward single use, particularly in the large new refineries where operations are computer controlled. A modified approach is to employ single use for all partial hardening and simple reuse for complete hydrogenation, where selectivity is not a consideration.

Filtration and Spent Catalyst Disposition

Utilizing plate and frame filter presses for removing catalyst from hydrogenated oil was standard for many years, and they are still sometimes used. They do an excellent job. Their shortcomings are high worker cost of operation and employee aversion to the hot and disagreeable job of cleaning them.

Metal screen pressure leaf filters have become standard. They are usually of the vertical plate type for ease of cleaning and used catalyst retrieval. Cleaning is by vibration. Filter aid is employed to precoat the screens to achieve rapid filtration with essentially complete removal of the catalyst. Through proper choice of filter aid and its minimal use, a level approaching 15% nickel can be achieved in the filter cake. This facilitates catalyst reuse, if desired. It also minimizes product loss from wetting of the filter aid. In addition, and most important, the high nickel content obtains the highest price for spent catalyst disposition. Because of potential liability, based on environmental concerns, less and less catalyst is being disposed of by landfill.

GENERAL REFERENCES

J. R. C. Hastert, "Hydrogenation," in Y. H. Hui, ed., *Bailey's Industrial Oil and Fat Products*, 5th ed., vol. 4, John Wiley & Sons, New York, 1996, pp. 213–300.

R. R. Allen, "Hydrogenation," in Daniel Swern, ed., *Bailey's Industrial Oil and Fat Products*, 4th ed., Vol. 2, John Wiley & Sons, New York, 1982, pp. 1–95.

D. R. Erickson and M. D. Erickson, "Hydrogenation and Base Stock Formulation Procedures," in D. R. Erickson, ed., *Practical Handbook of Soybean Processing and Utilization*, AOCS Press, Champaign, Ill. 1995, pp. 218–238.

H. B. W. Patterson, *Hydrogenation of Fats and Oils: Theory and Practice*, AOCS Press, Champaign, Ill., 1994.

ROBERT C. HASTERT
Hastech Corporation
Omaha, Nebraska

HYDROPHOBICITY IN FOOD PROTEIN SYSTEMS

BACKGROUND

Definition

The term "lyophobic" is used to describe a solute that has little or no affinity for the solvent medium in which it is placed. When the medium is composed of water or an aqueous solution, the more specific term "hydrophobic" is used as the descriptor. Similarly, the terms "lyophilic" and "hydrophilic" are used to describe affinity of a solute for solvent or water. In the case of proteins, the widely varying hydrophobicities of amino acid side chains has led to the use of a more neutral term, "hydropathy," to describe their relative preferences for aqueous and nonpolar environments (1).

The association of nonpolar solutes or moieties of a macromolecule is referred to as the hydrophobic effect, or hydrophobic interactions, rather than hydrophobic bonding. This association of nonpolar groups arises from enthalpically favorable interactions (mainly London forces between polarizable groups) in conjunction with the entropically driven tendency to minimize contact of the nonpolar groups with water. Furthermore, it is difficult to assign an absolute value of hydrophobicity to a solute or functional group. Commonly, hydrophobicity is described by an empirically measured or calculated parameter, which is related to the strength of hydrophobic interaction compared with a standard or with respect to an arbitrary scale.

Importance of Hydrophobic Interactions in Food Protein Systems

Almost all biological systems are composed of molecules that involve at least some nonpolar groups. Because water is a common constituent in these systems, it has been suggested that hydrophobic interactions must be ubiquitous in nature, playing a key role in both equilibrium and kinetic phenomena. Originally, interest in the hydrophobic effect was focused on its contribution to the stabilization of the structure of native protein molecules and on the mechanisms of protein folding. Although it was originally presumed that nonpolar or hydrophobic groups should be restricted to the interior of folded protein molecules, where they would be buried (ie, not exposed to the solvent water molecules), crystallographic analysis of the three-dimensional structures of many proteins has indicated that many hydrophobic groups are at least partly exposed on the surface of the protein molecules. The resulting hydrophobic patches play a key role in intermolecular interactions, such as the binding of small ligands or the association with other macromolecules, including protein–protein or protein–lipid interactions in biological membranes and micellar systems. A model has been proposed for elucidating quantitative structure–activity relationships by the correlation of biological properties with three structural parameters, namely, hydrophobic or lipophilic, electronic, and steric parameters (2).

In food systems, the role of the hydrophobic effect in stabilizing native protein molecules and structural networks is exemplified by the existence of stable casein micelles in milk. However, processing of foods leads to a variety of changes in both intramolecular and intermolecular interactions. The transformation of milk into cheese by enzymatic treatment and heating is one example of the dramatic effect of processing. Hydrophobic, electronic, and steric effects must all be taken into consideration to explain not only the biological properties of the native protein molecules in foods, such as enzymatic activity, but also the functional properties that give the food its characteristic texture and form. These properties include solubility, viscosity, gellability or coagulability, emulsifying, and foaming properties. For example, food products such as meringue and angel food cake rely on the unique properties of egg white proteins, including their excellent foaming properties and ability to form thermostable foams, as well as their capability to yield strong yet elastic gels on heating. The understanding of the basic molecular mechanisms responsible for these properties is essential to control the various aspects of food processing and ingredient selection.

THEORY AND MEASUREMENT OF PROTEIN HYDROPHOBICITY

Theory

The existence of hydrophobic interactions between solutes in an aqueous medium may be explained by two hypotheses. One approach is to explain the hydrophobic effect as simply arising from the attractive forces between nonpolar moieties and the phobia experienced by those groups when placed in a medium of water molecules (3). The other approach emphasizes the effect of nonpolar solutes on the structure of water molecules (4).

It is generally well recognized that water molecules exhibit strong attractive interactions with each other through hydrogen bonding to produce a continuous yet dynamic network. The entropy-driven hypothesis for hydrophobic interactions proposes that when nonpolar solutes are introduced into water, the hydrophobic effect results from disruption of the dynamic water structure, rather than an actual repulsion or phobia of the nonpolar groups by the water molecules. The ordering of water molecules around the nonpolar solutes is reflected in a large negative partial molar entropy change, that is, a decrease in entropy or randomness. To minimize this unfavorable change, nonpolar solutes or moieties associate to reduce their total surface area of contact with the water molecules. However, this hypothesis emphasizes only the disruption of water structure and does not explain the observation that the enthalpy and free energy of transfer of solutes such as phenol from an organic solvent to water differs considerably depending on the solvent, for example, octane versus toluene. Thus the contribution of favorable attractive forces between nonpolar groups to the net changes in enthalpy cannot be neglected.

In summary, hydrophobic interactions result from the tendency of nonpolar groups to interact with each other rather than with water. The hydrophobic effect is weakened by decreasing temperature, due to the loss of entropy resulting from increased ordering of water molecules into clathratelike structures around the nonpolar groups. Conversely at higher temperatures, the water-ordering effect is minimized and the hydrophobic interactions are maximized. The hydrophobic effect thus culminates from both the tendency to reduce the entropically unfavorable contact between nonpolar groups with water, as well as the need to form enthalpically favorable noncovalent associations, including those interactions broadly classified as van der Waals forces, which are the net effect of attractive London interactions, repulsive electron cloud overlap, and inducible dipole orientation and induction effects.

The attractive forces between nonpolar solutes can be calculated if the solutes are assumed to be approximately spherical bodies (5). Repulsion arising from the interaction of the diffuse double layers surrounding these molecules can also be explained by the Derjaguin–Landau–Verwey–Overbeek (DLVO) theory (6). The overall attraction or repulsion potential resulting from the net effect of these forces can thus be assessed using these theories. The underlying assumptions, principles, and mathematics for this calculation have been clearly explained in standard textbooks and will not be elaborated here. However, it must be noted that in real food protein systems, much of the theory no longer holds true because ideal dilute solutions of spherical and uniform molecules are rarely encountered. Thus while the theory enables us to understand the phenomenon of attraction or repulsion between solute and solvent molecules, it is not useful for quantitative assessment of the strength of interaction of protein molecules in food.

Empirical Assessment of Hydrophobicity

Hydrophobicity Scales. Numerous scales have been proposed to assess the relative hydrophobicities of the amino acids that constitute proteins. Broadly speaking, these scales can be classified into two categories: (*1*) those that are based on the solubility behavior in solvents of different polarity, and (*2*) those that are calculated using crystallographic or other data showing the location of amino acid residues in the molecular structure, assuming that hydrophobic residues will locate in the interior of the molecule. The former scales include those based on the free energy of transfer ($\Delta G_{\text{transfer}}$) of the amino acid residues or their derivatives from water to an organic solvent or vapor, or the partition coefficients measured as a solubility ratio between water and a nonpolar immiscible organic solvent (1,7,8), as well as those based on relative retention times of amino acids and peptides during reverse-phase liquid chromatography (9). The latter scales include those based on the accessible or buried area of the residues (10,11), or on the location of amino acid residues in proteins assessed either in terms of the distance from the protein center of mass and average orientation of the side chain (12) or in terms of the average surroundings of residue types (13). A hydropathy scale has been reported based on both the average extent of buriedness and the $\Delta G_{\text{transfer}}$ from water to vapor (14). Calculation of hydrophobicity values based on the hydrophobic fragmental constant of constituent fragments has also been proposed (15). Table 1 shows some of the values of hydrophobicity for the 20 amino acids that have been reported in the literature; the different scales have been compared (16–18).

Using the hydrophobicity scales of the constituent amino acids, various approaches have been taken to calculate values for proteins. Based on Tanford's scale for $\Delta G_{\text{transfer}}$ of amino acid side chains from an organic to an aqueous environment, Bigelow's average hydrophobicity ($H\O_{\text{avg}}$) values are calculated using only information on the amino acid composition of the protein (19). In contrast, most other values require knowledge of the amino acid sequence or primary structure of the proteins as well. Computer programs are commonly used to calculate histograms or plots showing hydrophobicity profiles or hydropathy profiles (16,20) or amphipathic helix patterns (21).

A significant drawback to the use of hydrophobicity scales of amino acid residues to calculate the corresponding hydrophobicity values or profiles for proteins is the lack of consideration of the effect of the three-dimensional structure of each protein on the extent of exposure of the residues. The inadequacy of this approach is especially true for calculations using scales developed based on the

Table 1. Scales of Hydrophobicity for Amino Acid Residues in Proteins

	$\Delta G_{transfer}$[a] Kcal/mol								
	water → cyclohexane side-chain analogues	octanol → water calculated	Σf from RP-HPLC[b]	Molar fraction buried,[c] percent	Average surrounding hydrophobicity,[d] Kcal	Hydrophobicity index,[d] Kcal	Hydropathy index[e]	Rekker's fragmental constant[f]	Average area buried,[g] Å²
Ala	−0.87	−0.39	−0.3	11.2	12.97	0.87	1.8	0.53	86.6
Arg	15.86	3.95	−1.1	0.5	11.72	0.85	−4.5	−0.82	162.2
Asn	7.58	1.91	−0.2	2.9	11.42	0.09	−3.5	−1.05	103.3
Asp	9.66	3.81	−1.4	2.9	10.85	0.66	−3.5	−0.02	97.8
Cys/2	−0.34	−0.25	—	—	—	—	2.5	0.93	—
Cys	—	—	6.3	4.1	14.63	1.52	—	1.11	132.3
Gln	6.48	1.30	−0.2	1.6	11.76	0.00	−3.5	−1.09	119.2
Glu	7.75	2.91	0.0	1.8	11.89	0.67	−3.5	−0.07	113.9
Gly	0	0	1.2	11.8	12.43	0.10	−0.4	0.00	62.9
His	5.60	0.64	−1.3	2.0	12.16	0.87	−3.2	−0.23	155.8
Ile	−3.98	−1.82	4.3	8.6	15.67	3.15	4.5	1.99	158.0
Leu	−3.98	−1.82	6.6	11.7	14.90	2.17	3.8	1.99	164.1
Lys	6.49	2.77	−3.6	0.5	11.36	1.64	−3.9	0.52	115.5
Met	−1.41	−0.96	2.5	1.9	14.39	1.67	1.9	1.08	172.9
Phe	−2.04	−2.27	7.5	5.1	14.00	2.87	2.8	2.24	194.1
Pro	—	−0.99	2.2	2.7	11.37	2.77	−1.6	1.01	92.9
Ser	4.34	1.24	−0.6	8.0	11.23	0.07	−0.8	−0.56	85.6
Thr	3.51	1.00	−2.2	4.9	11.69	0.07	−0.7	−0.26	106.5
Trp	−1.39	−2.13	7.9	2.2	13.93	3.77	−0.9	2.31	224.6
Tyr	1.08	−1.47	7.1	2.6	13.42	2.67	−1.3	1.70	177.7
Val	−3.10	−1.30	5.9	12.9	15.71	1.87	4.2	1.46	141.0

Source: [a]Ref. 1 (normalized to zero for glycine).
[b]Ref. 9.
[c]Ref. 10.
[d]Ref. 13.
[e]Ref. 14.
[f]Ref. 15.
[g]Ref. 16.

behavior of amino acids or small peptides. Scales that were formulated on the basis of location or buriedness of residues measured in different proteins attempt to address this problem but are still limited in universality of application, due to the need to extrapolate the behavior of residues to proteins other than those for which data are available. Some researchers (18) have proposed that different hydrophobicity coefficients should be computed for each of four structural classes of proteins ($\alpha\alpha$, $\beta\beta$, $\alpha + \beta$, and α/β). In the case of food systems, the problem is even more complex, due to the simultaneous occurrence of many different proteins. Determination of a net or average hydrophobicity value for complex systems requires not only the calculation of a hydrophobicity value for each protein, but also some knowledge of how these might be changed through possible interactions between the proteins. Furthermore, calculated values fail to take into account the effects of processing on buriedness or surface exposure of the residues. For these reasons, various methods of measuring parameters that may relate to the hydrophobicity of complex food protein systems are usually favored over calculation of average values or profiles based on the constituent amino acids. Examples of such methods are illustrated in the following sections.

Partition in Aqueous Two-Phase Systems. An indication of the relative hydrophobicity of proteins may be given by their partition coefficients measured as the solubility ratio between a polar and an immiscible nonpolar solvent. Due to the virtual insolubility of most proteins in organic solvents, two-phase aqueous systems containing dextran and poly(ethylene glycol), or PEG, are used (22,23). Esterification of PEG with a fatty acid such as palmitate is used to alter nonpolarity of the system. The difference in partition of proteins in phase systems with and without the fatty acyl hydrocarbon group bound to PEG is taken as a measure of hydrophobic interaction. Length of the hydrocarbon chain may be varied to investigate the effects on the partition behavior of different proteins. The method has been shown to yield useful data (22–24) but suffers from the tedious nature of the procedure and difficulties in solubilizing certain proteins.

Reverse-Phase and Hydrophobic Interaction Chromatography. The relative retention times of solutes during chromatography indicate their relative solubility in or affinity for a nonpolar stationary phase versus a mobile phase of differing polarity. According to the solvophobic effect theory (25), the retention time of peptides depends mainly on their nonpolar and polar surfaces, and thus may be a good measure of hydrophobicity. In fact, this hypothesis has proved true for small peptides (9). However, chromatographic behavior of proteins is not as straightforward. Possible denaturation of proteins under the harsh solvent conditions often required in reverse-phase chromatography has led to recommendations to use the milder conditions of hydrophobic-interaction chromatography, but the relevance of these retention data to hydrophobicity of native protein molecules has still come under question. The hydrophobicity of the stationary phase is an important parameter in maintaining native structure of the proteins upon elution (26). Nevertheless, differences in chromatographic retention between proteins have been demonstrated, and it has been suggested that the differences in binding to aliphatic versus aromatic types of adsorbent (27) may confirm the need to differentiate between these types of interactions in the hydrophobic effect (28–31).

Binding Methods. Various methods have been proposed for quantitating the binding of a nonpolar or hydrophobic ligand to proteins as a measure of the protein hydrophobicity. The ligands used have included aliphatic and aromatic hydrocarbons (32,33), sodium dodecylsulfate (34), Tween 80 (35), simple triglycerides (36), and corn oil (37). Commonly, the mixture of protein solution and ligand are incubated for a specified time to allow interaction, followed by removal of unbound or free ligand by dialysis, extraction, microfiltration, or other such techniques. The protein-bound ligand may then be quantitated by various methods, including gas chromatographic (32,33,36) or radioactive count (36) analysis, colorimetric reaction (34), or by a fluorescence probe method (37). Although the procedures are time-consuming, these binding methods can be useful in determining empirical parameters that may be relevant in assessing interaction of proteins with various nonpolar components of foods, such as flavor compounds, vitamins, pigments, fatty acids, or triglycerides.

Contact Angle Measurement. The quantitative determination of the Lifshitz–van der Waals (LW) and electron donor–acceptor or Lewis acid–base (AB) interactions that contribute to surface tension has been extended to the case of proteins. The method is based on Young's equation describing the relationship of the advancing contact angle of drops of a liquid on a flat solid surface (38). The solid surface in this case consists of protein sample prepared as flat layers. The advancing contact angles of droplets of three different well-characterized liquids (eg, water, glycerol, and α-bromonaphthalene) on the protein surface are measured. Based on Young's equation and the known surface tensions of the three liquids, the contact angles are then used to calculate the contributions of LW and AB interactions to the surface tension of the liquids on the layered protein. This approach has been used to obtain values for native hydrated proteins using only one liquid, namely drops of saline water. These values are correlated with the hydrophobicity of proteins measured as relative retention on hydrophobic chromatography (39,40). However, much higher surface tension values are obtained by contact angle measurement of a drop of protein solution on a solid polymer surface, compared with the values obtained by measuring the angle for the air-dried protein layer, or by other measurements for surface tension at an air–liquid interface such as platinum ring tensiometry, Wilhelmy plate, and pendant drop shape (39,40). This suggests that protein exposed at an interface may reorient to a much more hydrophobic configuration than found in its native hydrated state.

SPECTROSCOPIC METHODS: INTRINSIC FLUORESCENCE.

The intrinsic fluorescence spectra of proteins is primarily attributed to the aromatic amino acid residues of tryptophan, tyrosine, and phenylalanine. In practice, fluorescence from tryptophan is the most commonly studied aspect of the spectrum, because phenylalanine has a low quantum yield while tyrosine fluorescence is frequently weakened due to quenching by tryptophan residues and the protein backbone itself (41). The fluorescence of both tryptophan and tyrosine residues depends substantially on their environment, with the magnitude of fluorescence intensity as well as the wavelength of maximum fluorescence emission being sensitive to the polarity of the environment. Three spectral classes of protein tryptophan residues have been reported: the residues that are completely buried in nonpolar regions of the molecule, those that are completely exposed to the surrounding water, and those that have limited contact with water and are probably immobilized at the protein surface. The typical wavelengths of maximum emission for these three groups of tryptophan residues are 330 to 332 nm, 350 to 353 nm and 340 to 342 nm, respectively (42). More detailed information may be obtained by recording intrinsic emission spectra at different excitation wavelengths (43). Measurements of intrinsic fluorescence can give information on buriedness of aromatic residues and on the effect of interactions with other molecules (44). However, it is often difficult to directly relate this information to hydrophobic interactions as the fluorescence characteristics may also be altered by general changes in conformation.

Spectroscopic Methods: Derivative Spectroscopy. The ultraviolet absorption spectrum of proteins depends on the chromophoric properties of the constituent aromatic amino acids. Solvent perturbation and folded–unfolded difference spectra can be used to monitor the number of exposed tyrosine and tryptophan residues (41). Although the spectra of the individual chromophores are different, it is difficult to resolve quantitatively their contributions in the resulting broad absorption spectrum. Derivative spectroscopy, particularly of the second order ($d^2A/d\lambda^2$) and fourth order ($d^4A/d\lambda^4$), has the ability to resolve overlapping bands in the original spectrum (45–47). Moderately turbid samples can be analyzed by derivative spectroscopy since a horizontal baseline is obtained even with appreciable nonselective light scattering (45). In addition to quantitative determination of the aromatic amino acid contents, derivative spectroscopy has been proposed as a means to determine changes in polarity of the microenvironment around the chromophores. The maximum spectral shift observed in near ultraviolet second derivative spectra between solvent-exposed model compounds and side chains completely buried in an apolar protein core was found to be 5, 4, and 2 nm for tyrosine, tryptophan, and phenylalanine, respectively (46). However, for all three aromatic residues in proteins, there was no consistent correlation between absolute spectral band positions and average solvent accessibility, implying the influence of other local effects such as electrostatic interactions on the near ultraviolet spectra of proteins.

Spectroscopic Methods: Fluorescence Probes. Compounds whose quantum yields of fluorescence and wavelength of maximum emission depend on the polarity of their environment have been used to probe the hydrophobic or nonpolar nature of proteins. Figure 1 shows the chemical structures of several such fluorescence probes. The most popular types used include anionic probes of the aromatic sulfonic class, such as the amphiphilic 1-anilinonaphthalene-8-sulfonate (ANS), or its dimeric form, bis-ANS. These probes have high quantum yields of fluorescence in organic solvents but not in water; they thus fluoresce when bound to membranes or hydrophobic cavities in proteins (48,49). However, ANS fluorescence cannot always be directly correlated with hydrophobicity, as enhancement of fluorescence and a blue shift of the emission maximum have been observed in strong aqueous $MgCl_2$ solutions (50). It has been suggested that solvents or environments that are not necessarily nonpolar but that favor the rigid, planar configuration of the ANS molecule may influence fluorescence (51).

Another category of anionic fluorescence probes is the fatty acid analogue type, including *cis*-parinaric acid (CPA), which has been used as a probe for proteins and biological membranes (52–54). The parinaric acids are among the few nonaromatic fluorophores known. Their similarity to natural fatty acids, nonfluorescence in water, and good Stokes shift characteristics are among the advantages of their use as probes for hydrophobic regions that may be relevant to protein–lipid interactions in food systems. Good correlation was obtained between the relative hydrophobicity values of proteins determined by CPA fluorescence, and properties related to protein–lipid interactions such as interfacial tension and emulsifying activity (55).

Titration of protein solutions with increasing concentrations of the fluorescence probe can provide information on both the number and binding constants of the hydrophobic sites (56,57). Alternatively, proteins may be compared by calculation of an index of surface hydrophobicity S_0 based on the initial slope of the measured relative fluorescence intensity of excess fluorescence probe as a function of increasing concentrations of protein (55). Transfer efficiency of the excitation energy from aromatic chromophores to bound probe at an adjacent hydrophobic site has also been proposed as an index of surface hydrophobicity (55).

Limitations in using anionic probes such as ANS and CPA include the possibility that electrostatic as well as hydrophobic interactions may contribute to the binding of the probes. The use of neutral or uncharged probes such as diphenylhexatriene (37), 6-propionyl-2-(dimethylamino)naphthalene or PRODAN (58) and Nile Red (59) may circumvent this problem.

Other Spectroscopic Methods. Most of the spectroscopic methods outlined previously provide information on the contribution to hydrophobic interactions from aromatic amino acids. Only a few methods have been proposed that can assess the involvement of aliphatic amino acids in hydrophobic interactions, including the CPA probe assay or ligand binding involving aliphatic hydrocarbons. A number of other spectroscopic methods can be used to investigate protein structure and environment of constituent amino acid residues in general. Each of these methods has particular advantages and limitations. However, consid-

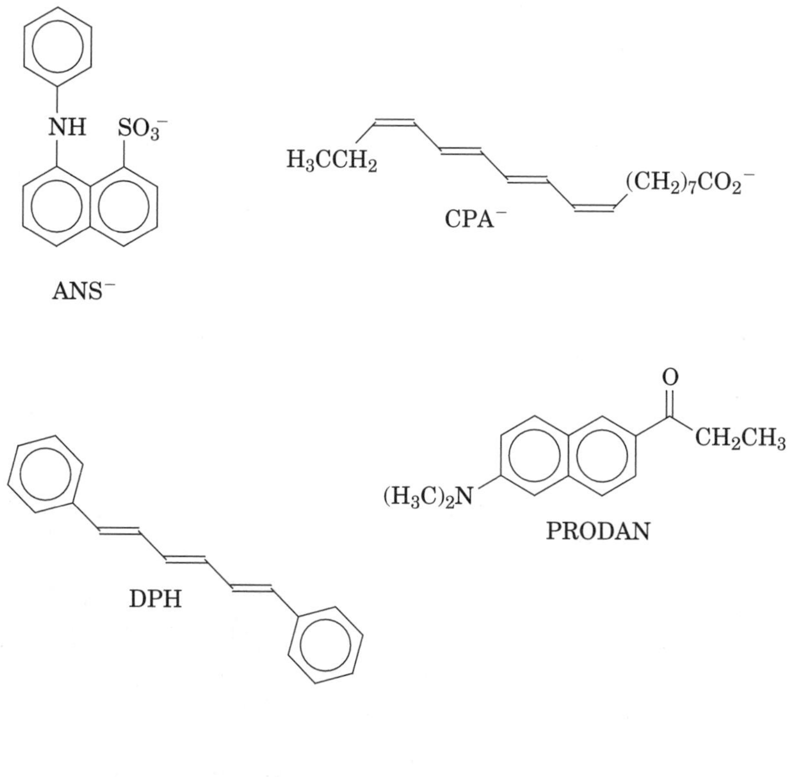

Figure 1. Chemical structures of some fluorescence probes that may be used for determination of protein hydrophobicity. ANS = 1-anilinonaphthalene-8-sulfonic acid; CPA = *cis*-parinaric acid; DPH = diphenylhexatriene; PRODAN = 6-propionyl-2-(dimethylamino)naphthalene. The two anionic probes (ANS and CPA) are shown in the dissociated acid form. *Source:* Courtesy of Dr. C. A. Haskard.

ering the need to study proteins that are typically more than 20 kilodaltons in molecular weight and in a turbid solution or solid phase, the most promising techniques to study hydrophobicity in food protein systems are nuclear magnetic resonance (60), infrared (60), and Raman spectroscopy (60,61). The reader is referred to more specialized volumes on these techniques.

HYDROPHOBICITY OF FOOD PROTEINS

Table 2 is a compilation of the relative hydrophobicity values reported for some proteins of relevance in food systems. The hydrophobicity values were either calculated from the amino acid composition of the proteins or measured by some of the empirical methods outlined earlier. Unfortunately, there is still no consensus on a standard method for quantitative measurement of hydrophobicity. In comparing the values shown in Table 2, one should bear in mind that the methodologies differ in whether they are measuring average, total, or surface-exposed hydrophobicity. Furthermore, variability in published data may arise from differences in sources or purity of protein, as well as in the conditions during measurement, including the temperature, pH, ionic strength, or buffer composition.

Simple methods such as those based on fluorescence probes are the most widely used for the study of food protein systems, since these methods are based on monitoring the probe-accessible groups in nonpolar environments, which may correlate to the hydrophobic sites that are actually available for participation in functionality. The fluorescence probe methods are relatively simple and rapid to perform. Furthermore, empirical methods can be applied to study heterogeneous protein systems and to monitor the effects of processing. Typical hydrophobicity values measured using an aromatic (ANS) and an aliphatic (CPA) probe are listed in Table 3 for some food proteins, including complex systems composed of several molecular species.

IMPORTANCE OF HYDROPHOBICITY IN FUNCTIONALITY OF FOOD PROTEINS

Functional properties of food proteins have been defined as those characteristics, other than nutritional ones, that affect their utilization. This includes a diverse range of functionality, such as solubility, viscosity, water- and fat-binding properties, emulsification, foaming, film formation, gellability or coagulability, elasticity, and flavor. Functionality depends on the interaction of the protein molecules with other components in the food system, including water; macromolecules such as other proteins, complex carbohydrates, and lipids; and small molecules such as salts, simple sugars, and flavor compounds. These interactions may take place in the bulk phase, or at a surface or interface. The interplay of electrostatic, hydrophobic, and steric parameters, important in predicting biological activity of molecules (2), also holds true for the elucidation of the relationship between structure and functionality of food proteins (17,62). Computer-assisted studies for quantitative analysis of the structure–activity relationship (QSAR) of small molecules are widely applied (63), but progress in QSAR for functionality of food protein analyses has been much slower (62). Although the quali-

Table 2. Relative Hydrophobicity Values of Some Proteins Measured by Different Methods

		Fluorescence probes			DPH probe	Chromatography		Binding			Partition
Proteins	$HØ_{one}$[a]	S_oANS	S_oCPA	S_eCPA	TG binding	Rpc	Hic	Heptane	SDS	TG	ΔlogK
Albumin, bovine	1120	1000	100	100	100	100	100	100	100	100	100
Albumin, chicken	1110	7	2	96	15	0	9–86	2	14	3	1.25
Casein, α-	1200	57[b]	6–30	—	150	400	—	—	—	160	—
Casein, β-	1320	60[b]	50	107	—	315	—	—	—	—	—
Casein, κ-	1210	83	13–93	89	—	340	—	—	30	—	—
Chymotrypsin, α-	1030	0	3	—	41	—	57–139	9	—	3	9–33
Globulin, soy 7S	1090	47	9–77	—	50	165	—	—	21	—	—
Globulin, soy 11S	950	27	2–17	—	25	150	—	—	—	—	—
Lactalbumin, α-	1150	33	9–54	—	98	225	10	—	—	—	—
Lactoglobulin, β-	1230	13	54–146	80	49	0	67–102	100	62	130	41–78
Lysozyme	970	0.7	7	—	—	—	42–113	0	19	1	0
Ovomucoid, chicken	920	—	0	22	—	—	—	24	18	108	—
Pepsin	1063	0.7	2	57	32	—	92	—	—	95	—
Ribonuclease A	870	—	1	24	—	—	4–71	17	18	2	—
Transferrin, chicken	1080	—	4	108	—	—	16–31	29	20	118	9–20
Trypsin	940	8	3	34	—	—	—	—	13	—	9–27
Trypsin inhibitor	1040	0	0	—	1	0	—	—	—	—	—

Note: Data were compiled from Refs. 17, 18, 65, references cited therein, and unpublished data. With the exception of $HØ_{one}$ and S_o ANS, the data are expressed relative to 100 for bovine albumin to facilitate comparison of values. Where varying values were reported by investigators using essentially the same method, a range of values is presented.

[a]Abbreviations: $Hø_{one}$ = average hydrophobicity calculated by Bigelow's method (19), S_o = initial slope of relative fluorescence intensity versus protein concentration plot, using native protein; S_e = S_o measured for protein solutions after treatment with 1.5% SDS at 100°C for 10 min; ANS = 1-anilino-8-naphthalene sulfonic acid; CPA = *cis*-parinaric acid; DPH = 1,6-diphenyl-1,3,5-hexatriene; TG = triglyceride; Rpc = reverse-phase chromatography; Hic = hydrophobic-interaction chromatography; SDS = sodium dodecyl sulfate.

[b]At the isoelectric pH, 530 and 800 for α- and β-casein, respectively.

Table 3. Hydrophobicity Values of Some Food Proteins Measured by Fluorescence Probe Method

Protein[a]	S_oANS	S_oCPA
Albumin, bovine serum	1600	2750
Canola isolate	110	370
Casein, bovine	150	400
Egg albumen	40	135
Gelatin	0	0
Lactalbumen	145	1700
Lactalbumin, α-	90	280
Lactoglobulin, β-	40	7000
Lysozyme, hen	1	15
Muscle, chicken breast	75	350
Myosin, chicken breast	100	—
Ovalbumin, hen	10	40
Pea protein isolate	280	825
Soy protein isolate	250	1000
Sunflower protein isolate	155	910
Whey protein concentrate	70	3600
Zein	410	390

Note: See Table 2 for abbreviations.

[a]Measurements for S_o determination were carried out for protein solutions in 0.01 M sodium phosphate buffer, pH 7.0–7.4; 0.6 M NaCl and 0.3 M NaCl were included in the buffer for salt extracts of chicken breast muscle and for isolated chicken myosin, respectively.

tative importance of the physicochemical properties in functionality is well recognized, quantitative prediction based on QSAR analysis of food proteins is still not a facile task, being limited by the need to measure relevant parameters, including hydrophobic interactions, which can take into account the effects of processing for incorporation into structure–functionality models. The following highlight the role of hydrophobic interactions in explaining some of the important functional properties of food proteins.

Solubility

Solubility has long been considered a critical property because of its effect on many other functional properties, such as emulsifying, foaming, and gelling properties. Charge frequency and hydrophobicity have been proposed to be the two major factors affecting protein solubility (17). Generally speaking, proteins are soluble in aqueous media when electrostatic repulsive forces between protein molecules are greater than the driving forces for hydrophobic interactions. Insolubility is more likely when the repulsive forces are at a minimum, near the isoelectric pH of the protein. However, proteins with relatively few exposed hydrophobic regions on the molecular surface may remain soluble even at pH near their isoelectric point. For example, the isoelectric precipitation of casein proteins during acidification of milk is attributed to their high surface hydrophobicity. On the other hand, the globular proteins in the whey fraction remain soluble. Denaturation of whey proteins, such as may occur by heat treatment, leads to exposure of previously buried hydrophobic groups and consequently to insolubilization upon acidification.

Aromatic amino acids have been suggested to play a greater role than aliphatic amino acids in contributing to the hydrophobic interactions affecting insolubilization (28). At pH values where charge effects are minimized (zero zeta potential), the insolubility of some food proteins has been correlated to hydrophobicity parameters measured using the aromatic fluorescence probe ANS or by reverse phase chromatographic behavior on phenyl Sepharose (PSC), but not to hydrophobicity measured using the aliphatic probe CPA. At other pH values, charge frequency measured as zeta potential and hydrophobicity measured by ANS or PSC are both significant parameters to explain the insolubility of proteins. Although calculations of hydrophobicity scale values based on free energy of transfer from organic solvent to water indicate that aromatic amino acids are more hydrophobic than aliphatic amino acids, statistical scales based on frequency of location of amino acid residues at the surface versus the interior of protein molecules indicate that aromatic amino acids are often considered less hydrophobic than aliphatic ones. While inherently more hydrophobic, the bulky nature of the aromatic residues may discourage their effective burial in the protein interior (29). These surface-exposed residues determine the strength of hydrophobic interactions that affect properties such as solubility.

The effects of various compounds on protein solubility can be explained on the basis of their opposite effects on electrostatic and hydrophobic interactions of the proteins. Salts may be ranked in terms of a lyotropic series describing their water structure-making or -breaking effects based on molal surface tension values (25). At high ionic strength, ammonium sulfate and sodium chloride increase the surface tension of water, whereas tetraethylammonium chloride and guanidinium chloride reduce it. The salting-in or salting-out of proteins by different salts can thus be explained theoretically by the lyotropic series. The salting-out constant of a protein is a function of its surface hydrophobicity, calculated as the hydrophobic contribution of each amino acid (7) and the fraction of exposed hydrophobic residues (10). Similarly, the stabilizing effects of sugars and polyols such as glycerol may be related to their water structure-enhancing effects, which intensifies the intramolecular hydrophobic interactions that stabilize protein structure. The effect of urea in solubilizing proteins has been explained by solvation of the hydrophobic moieties in urea, as well as by water structure-breaking effects. Precipitation by trichloroacetic or sulfosalicylic acid has been postulated to arise from an increase of accessible hydrophobicity of peptide moieties (64).

Emulsifying and Foaming Properties

Emulsion and foam-related properties depend on interactions of protein molecules at the oil–water interface and air–water surface, respectively. Although protein solubility is a key determinant in these functionalities, additional important considerations are molecular flexibility and hydrophobic interactions of the protein at the surface or interface, which result in formation and stabilization of the emulsion or foam. It has been reported that the decrease in interfacial tension and improvement in emulsifying activity are not related to hydrophobicity values calculated from the total content of hydrophobic amino acids, but rather to the effective or surface hydrophobicity of proteins measured by hydrophobic partition, hydrophobic interaction chromatography, or fluorescence probes (24,55). The extent of unfolding to expose areas for hydrophobic interactions is expected to be greater for foaming than emulsifying properties, which may be related to the higher tension at an air–water surface than an oil–water interface. The values of exposed hydrophobicity, measured by fluorescence probe assay after denaturation of protein molecules by heating in the presence of sodium dodecyl sulfate, were found to be correlated with foaming capacity (65). An example of the industrial relevance of hydrophobic interactions in surface properties is demonstrated in the positive correlation between the content of hydrophobic protein fractions separated from beer by hydrophobic interaction chromatography and beer foam stability (66).

In analogy to the hydrophile–lipophile balance (HLB) concept originally developed for nonprotein, synthetic emulsifiers, the amphiphilic nature of proteins is vital to their function as surface active agents. A balance in accessible hydrophobic and hydrophilic areas is required for optimal functionality. Solubility of the protein molecules facilitates diffusion to the surface or interface. However, once there, the ability of the protein to interact with oil or other protein molecules to form an interfacial layer depends on the flexibility and accessibility of surface groups, especially through hydrophobic interactions. The importance of hydrophobic interactions is illustrated in the observation that functionality can often be improved by mild denaturing treatments that increase surface hydrophobicity without impairing solubility (55,67,68). When solubility is a constant parameter, emulsifying properties are improved by increasing surface hydrophobicity, up to some critical point. However, as predicted by the HLB concept, excessive hydrophobicity is detrimental to functionality, and emulsifying properties are poor when proteins have excessively high values of hydrophobicity (17,62).

Thermal Functional Properties

The importance of intramolecular hydrophobic interactions in the stability of proteins to thermally induced denaturation has been assumed, due to the increasing strength of these interactions with increasing temperature up to 60 to 70°C. However, extensive comparison of hydrophobic indices in thermophilic versus mesophilic proteins has not demonstrated any definitive relationship between hydrophobicity and thermal stability. Stabilization is promoted by greater internal or intramolecular hydrophobic interactions and lower external or surface hydrophobicity. Because aliphatic residues have a greater tendency to be located in the protein interior than bulky aromatic side chains, it has been suggested that an aliphatic index may be more relevant to protein stability than a general hydrophobicity index (29).

Thermally induced gelation or coagulation are important in affecting textural characteristics of foods. Coagulation-type (concentration-dependent) and gelation-type (concentration-independent) proteins have been differentiated depending on molecular weight and relative content of hydrophobic amino acid residues. One hypothesis proposes that proteins having over 31.5 mol % of hydrophobic (Val, Pro, Leu, Ile, Phe, and Trp) residues would be of the coagulation type, whereas those with less mol % of those residues would form translucent-type gels (69). However, such generalizations assume that denaturing treatments prior to gelation result in total exposure of the hydrophobic residues. The balance of electrostatic repulsive forces and hydrophobic attractive forces on the molecular surface determine the formation of particulate, fine-stranded, or mixed structure gels, with distinctive textural characteristics and appearance (70,71). A striking example of the involvement of hydrophobic interactions is the setting phenomenon observed in particular fish species, wherein an elastic gel forms at relatively mild heating temperatures. Introduction of aromatic or aliphatic hydrophobic groups through chemical modification can induce nonsetting species to behave like easily setting fish species, exhibiting characteristics of increased viscosity and gelling on low temperature heating (72,73).

Hydrophobic interactions may be involved in the initial intramolecular stage of unfolding as well as the later stage of network or aggregate formation through intermolecular interactions. Both surface and exposed hydrophobicity may be important in the nature of the resulting product. It has been suggested that a low value of surface hydrophobicity of the native protein, coupled with a high value of exposed hydrophobicity after thermal denaturation, may lead to the formation of strong gels (74). The ability to unfold, or molecular flexibility, may be hindered by intramolecular stabilization through noncovalent interactions as well as covalent bonds, particularly of the disulfide type. The involvement of intermolecular sulfhydryl–disulfide reactions usually appears in later stages of these thermally induced phenomena and may be concentration dependent. At temperatures below 70°C, sulfhydryl reactivity is not typically involved and hydrophobic interactions may be the major driving force in network formation, with hydrogen bond formation during the cooling phase contributing to strengthening of the final gel structure (17).

Flavor

Although most proteins do not have a strong intrinsic flavor, they do influence flavor due to their ability to bind flavor compounds. This property can play an important role in the transmission of undesirable off-flavors, for example, in some foods containing soy proteins; conversely, binding of some flavor components may lead to a reduction in the perceived desirable flavor (75). The nonpolar nature of many flavor components such as aldehydes and ketones promotes hydrophobic interactions that may be critical in their binding to proteins. The binding affinity of aliphatic ketones with β-lactoglobulin increases proportionally with chain length of the ketone, suggesting that the association is primarily hydrophobic in nature (75). Influences of processing such as heat denaturation may be expected by consideration of their impact on the hydrophobicity of the protein and consequently, the flavor binding sites. Oddly, very little interest appears to have been devoted to the study of flavor retention and release, compared with the bulk of work on molecular interactions (76). Systematic study in this area should allow optimal design of flavors for new formulated foods, elimination of transmitted off-flavors and development of efficient flavor carrier systems (75).

Hydrolysis of proteins to form peptides also has a great influence on flavor. Formation of bitter peptides from casein and soy proteins, for example, is detrimental to the flavor acceptability of hydrolysates. Bitterness appears to be related to the average hydrophobicity (HQ) of the peptides, with those peptides having HQ values above 1400 cal/mol residue being bitter. However, the extent of exposure of hydrophobic residues and their incorporation into peptides in contrast to their existence as free amino acids also affects bitterness. Thus bitterness depends not only on the content of hydrophobic amino acids in the original protein, but also on the degree of hydrolysis and the location of the hydrophobic residue on the peptide sequence (77).

FUTURE TRENDS

Long-term, systematic research is required to establish quantitative, empirical rules to understand the role of hydrophobicity in functionality of food protein systems. This basic research is required to explain at the molecular basis, various well-known but not clearly understood phenomena such as the thermostability of egg white foams, elasticity of egg white gels, coagulation of casein to form cheese, binding and texture formation of muscle proteins, elasticity and extensibility of wheat gluten dough, and so on.

Although only one of the physicochemical parameters involved in elucidation of structure–function relationships, hydrophobic interactions often play a key role. Improvement of functionality including whipping and emulsifying properties by modifying the hydrophobic–hydrophilic balance of proteins is exemplified in the preparation of proteinaceous surfactants with different HLB values by attachment of hydrophobic amino acid alkyl esters to various food proteins (78). Development of novel functionality such as antifreeze or cryoprotectant properties has also been described using this approach (78). Another rapidly developing area is the enhancement of hydrophobic nature of enzymes through attachment of amphiphilic groups such as poly(ethylene glycol), coupled with selection of organic solvent media to alter the hydrophobic environment. This area has interesting applications such as the possible resolution of alcohols and acids by lipase-catalyzed esterification, protease-catalyzed synthesis of proteins, and lipase-catalyzed interesterification of fats and oils (79). The potential of "ultrahydrophobic sequences" and beta-sheet structures of corn, wheat, and other proteins in film-forming preparations and adhesives (80) may be of value in the development of edible films. The growing market for reduced- and low-fat foods, which often have high protein content, suggests that more research should be undertaken to study the role of hydrophobic interactions in the binding and release of flavor compounds by food proteins.

Site-specific modifications of proteins can now be realized through molecular biology and genetic engineering techniques. By establishing the relationship between food protein structure and function, systematic and predictable enzyme and protein engineering for tailoring of specific biological and functional properties should become a reality.

BIBLIOGRAPHY

1. T. E. Creighton, "Physical Interactions that Determine the Properties of Proteins," *Proteins. Structures and Molecular Properties*, 2nd ed., W. H. Freeman and Company, New York, 1993.
2. C. Hansch and A. J. Leo, *Substituent Constants for Correlation Analysis in Chemistry and Biology*, John Wiley & Sons, New York, 1979.
3. C. Tanford, *The Hydrophobic Effect: Formation of Micelles and Biological Membranes*, 2nd ed., John Wiley & Sons, New York, 1980.
4. C. J. van Oss, "Hydrophilic and 'Hydrophobic' Interactions," in H. Visser, ed., *Protein Interactions*, VCH, New York, 1992, pp. 25–55.
5. H. C. Hamaker, "The London-van der Waals Attraction between Spherical Particles," *Physica* **4**, 1058–1072 (1937).
6. E. Dickinson, *An Introduction to Food Colloids*, Oxford University Press, Oxford, United Kingdom, 1992, pp. 94–97, 181–184.
7. Y. Nozaki and C. Tanford, "The Solubility of Amino Acids and Two Glycine Peptides in Aqueous Ethanol and Dioxane Solutions. Establishment of a Hydrophobicity Scale," *J. Biol. Chem.* **246**, 2211–2217 (1971).
8. L. M. Yunger and R. D. Cramer III, "Measurement and Correlation of Partition Coefficients of Polar Amino Acids," *Mol. Pharmacol.* **20**, 602–608 (1981).
9. K. J. Wilson et al., "The Behavior of Peptides on Reverse-Phase Supports During High Pressure Liquid Chromatography," *Biochem. J.* **199**, 31–41 (1981).
10. C. Chothia, "Principles That Determine the Structure of Proteins," *Annu. Rev. Biochem.* **54**, 537–572 (1984).
11. G. J. Lesser and G. D. Rose, "Hydrophobicity of Amino Acid Subgroups in Proteins," *Proteins: Struct., Funct., Genet.* **8**, 6–13 (1990).
12. H. Meirovitch, S. Rackovsky, and H. A. Scheraga, "Empirical Studies of Hydrophobicity. I. Effect of Protein Size on the Hydrophobic Behavior of Amino Acids," *Macromolecules* **13**, 1398–1405 (1980).
13. P. Manavalan and P. K. Ponnuswamy, "Hydrophobic Character of Amino Acid Residues in Globular Proteins," *Nature* **275**, 673–674 (1978).
14. J. Kyte and R. F. Doolittle, "A Simple Method for Displaying the Hydropathic Character of a Protein," *J. Mol. Biol.* **157**, 105–132 (1982).
15. R. F. Rekker, *The Hydrophobic Fragmental Constant*, Elsevier Science, New York, 1977.
16. G. D. Rose, L. M. Gierasch, and J. A. Smith, "Turns in Peptides and Proteins," *Adv. Protein Chem.* **37**, 1–109 (1985).
17. S. Nakai and E. Li-Chan, *Hydrophobic Interactions in Food Systems*, CRC Press, Boca Raton, Fla., 1988.
18. H. Cid et al., "Hydrophobicity and Structural Classes in Proteins," *Protein Eng.* **5**, 373–375 (1992).
19. C. C. Bigelow, "On the Average Hydrophobicity of Proteins and the Relation Between It and Protein Structure," *J. Theor. Biol.* **16**, 187–211 (1967).
20. S. R. Krystek, Jr., and T. T. Anderson, "A Program for Hydropathy and Antigenicity Analysis of Protein Sequences," *Endocrinology* **117**, 1118–1124 (1985).

21. J. P. Segrest and R. J. Feldman, "Amphipathic Helices and Plasma Lipoproteins: A Computer Study," *Biopolymers* **16**, 2053–2065 (1977).
22. H. Walter and G. Johansson, eds., *Aqueous Two-Phase Systems*, Methods in Enzymology Vol. 228, Academic Press, New York, 1994.
23. P.-A. Albertsson, *Partition of Cell Particles and Macromolecules*, 3rd ed., Wiley Interscience, New York, 1986.
24. E. Keshavarz and S. Nakai, "The Relationship Between Hydrophobicity and Interfacial Tension of Proteins," *Biochim. Biophys. Acta* **576**, 269–279 (1979).
25. W. Melander and C. Horvath, "Salt Effects on Hydrophobic Interactions in Precipitation and Chromatography of Proteins: An Interpretation of the Lyotropic Series," *Arch. Biochem. Biophys.* **183**, 200–215 (1977).
26. L. Szepesy and G. Rippel, "Comparison and Evaluation of HIC Columns of Different Hydrophobicity," *Chromatographia* **34**, 391–397 (1992).
27. B. H. Hofstee and N. F. Otillio, "Modifying Factors in Hydrophobic Protein Binding by Substituted Agaroses," *J. Chromatogr.* **161**, 183–215 (1978).
28. S. Hayakawa and S. Nakai, "Relationships of Hydrophobicity and Net Charge to the Solubility of Milk and Soy Proteins," *J. Food Sci.* **50**, 486–491 (1985).
29. B. B. Mozhaev and K. Martinek, "Structure-Stability Relationships in Proteins: New Approaches to Stabilizing Enzymes," *Enzyme Microb. Technol.* **6**, 50–59 (1984).
30. S. K. Burley and G. A. Petsko, "Aromatic-Aromatic Interaction: A Mechanism of Protein Structure Stabilization," *Science* **229**, 23–28 (1985).
31. S. K. Burley and G. A. Putsko, "Weakly Polar Interactions in Proteins," *Adv. Protein Chem.* **39**, 125–189 (1988).
32. A. Mohammadzadeh-K., R. E. Feeney, and L. M. Smith, "Hydrophobic Binding of Hydrocarbons by Proteins. I. Relationship of Hydrocarbon Structure," *Biochim. Biophys. Acta* **194**, 246–255 (1969).
33. A. Mohammadzadeh-K., R. E. Feeney, and L. M. Smith, "Hydrophobic Binding of Hydrocarbons by Proteins. II. Relationship of Protein Structure," *Biochim. Biophys. Acta* **194**, 256–264 (1969).
34. A. Kato et al., "Determination of Protein Hydrophobicity Using a Sodium Dodecyl Sulfate Binding Method," *J. Agric. Food Chem.* **32**, 284–288 (1984).
35. B. Lieske and G. A. Konrad, "A New Approach to Estimate Surface Hydrophobicity of Proteins," *Milchwissenschaft* **49**, 663–666 (1994).
36. L. M. Smith, P. Fantozzi, and R. K. Creveling, "Study of Triglyceride-Protein Interaction Using a Microemulsion-Filtration Method," *J. Am. Oil Chem. Soc.* **60**, 960–967 (1983).
37. T. Tsutsui, E. Li-Chan, and S. Nakai, "A Simple Fluorometric Method for Fat-Binding Capacity as an Index of Hydrophobicity of Proteins," *J. Food Sci.* **51**, 1268–1272 (1986).
38. C. J. van Oss, "Energetics of Cell-Cell and Cell-Biopolymer Interactions," *Cell Biophysics* **14**, 1–16 (1989).
39. C. J. van Oss et al., "Determination of the Surface Tension of Proteins. I. Surface Tension of Native Serum Proteins in Aqueous Media," *Biochim. Biophys. Acta* **670**, 64–73 (1981).
40. C. J. van Oss, et al., "Determination of the Surface Tension of Proteins. II. Surface Tension of Serum Albumin, Altered at the Protein-Air Interface," *Biochim. Biophys. Acta* **670**, 74–78 (1981).
41. R. A. Copeland, "Spectroscopic Probes of Protein Structure," *Methods for Protein Analysis*, Chapman & Hall, New York, 1994.
42. E. A. Burstein, N. S. Vedenkina, and M. N. Ivkova, "Fluorescence and the Location of Tryptophan Residues in Protein Molecules," *Photochem. Photobiol.* **18**, 263–279 (1973).
43. C. Myers, "Functional Attributes of Protein Isolates," in F. Felix, ed., *Characterization of Proteins*, Humana Press, Clifton, N.J., 1988, pp. 491–549.
44. M. R. Eftink, "Fluorescence Techniques for Studying Protein Structure," in C. H. Suelter, ed. *Methods of Biochemical Analysis*, Vol. 35, John Wiley & Sons, New York, 1991, pp. 127–205.
45. E. Padros et al., "Fourth-Derivative Spectrophotometry of Proteins," *Trends Biochem. Sci.* **9**, 58–510 (1984).
46. H. Mach and C. R. Middaugh, "Simultaneous Monitoring of the Environment of Tryptophan, Tyrosine and Phenylalanine Residues in Proteins by Near-Ultraviolet Second-Derivative Spectroscopy," *Anal. Biochem.* **222**, 323–331 (1994).
47. H. Mach et al., "Ultraviolet Absorption Spectroscopy," *Methods Mol. Biol.* **40**, 91–115 (1995).
48. L. Stryer, "The Interaction of a Naphthalene Dye With Apomyoglobin and Apohemoglobin. A Fluorescent Probe of Nonpolar Binding Sites," *J. Mol. Biol.* **13**, 482–495 (1965).
49. G. Weber and L. B. Young, "Fragmentation of Bovine Serum Albumin by Pepsin. 1. The Origin of the Acid Expansion of the Albumin Molecules," *J. Biol. Chem.* **239**, 1415–1423 (1964).
50. G. Penzer, "1-Anilinonapthalene-8-Sulfonate. The Dependence of Emission Spectra on Molecular Formation Studied by Fluorescence and Proton Magnetic Resonance," *Eur. J. Biochem.* **25**, 218–228 (1972).
51. S. Ainsworth and M. T. Flanagan, "The Effects That the Environment Exerts on the Spectroscopic Properties of Certain Dyes That Are Bound by Bovine Serum Albumin," *Biochim. Biophys. Acta* **194**, 213–221 (1969).
52. L. A. Sklar, B. S. Hudson, and R. D. Simoni, "Conjugated Polyene Fatty Acids as Membrane Probes: Preliminary Characterization," *Proc. Nat. Acad. Sci. U.S.A.* **72**, 1649–1653 (1975).
53. L. A. Sklar, B. S. Hudson, and R. D. Simoni, "Model System Studies," *J. Supramolecular Structure* **4**, 449–465 (1976).
54. L. A. Sklar, B. S. Hudson, and R. D. Simoni, "Binding to Bovine Serum Albumin," *Biochemistry* **16**, 5100–5108 (1977).
55. A. Kato and S. Nakai, "Hydrophobicity Determined by a Fluorescence Probe Method and Its Correlation With Surface Properties of Proteins," *Biochim. Biophys. Acta* **624**, 13–20 (1980).
56. M. Cardamone and N. K. Puri, "Spectrofluorimetric Assessment of the Surface Hydrophobicity of Proteins," *Biochem. J.* **282**, 589–593 (1992).
57. J. Slavik, *Fluorescent Probes in Cellular and Molecular Biology*, CRC Press, Boca Raton, Fla., 1994, pp. 125–137.
58. G. Weber and F. J. Farris, "Synthesis and Spectral Properties of a Hydrophobic Fluorescent Probe: 6-Propionyl-2-(dimethylamino)naphthalene," *Biochemistry* **18**, 3075–3078 (1979).
59. D. L. Sackett and J. Wolff, "Nile Red as a Polarity-Sensitive Fluorescent Probe of Hydrophobic Protein Surfaces," *Anal. Biochem.* **167**, 228–234 (1987).
60. P. S. Belton, "New Methods for Monitoring Changes in Proteins," *Food Rev. Int.* **9**, 551–573 (1993).
61. E. Li-Chan, S. Nakai, and M. Hirotsuka, "Raman Spectroscopy as a Probe of Protein Structure in Food Systems," in R. Y. Yada, R. L. Jackman, and J. L. Smith, eds., *Protein Structure-Function Relationships in Foods*, Blackie Academic and Professional, London, United Kingdom, 1994, pp. 163–197.

62. S. Nakai and E. Li-Chan, "Recent Advances in Structure and Function of Food Proteins: QSAR Approach," *Crit. Rev. Food Sci. Nutr.* **33**, 477–499 (1993).
63. C. Hansch and T. E. Klein, "Quantitative Structure-Activity Relationships and Molecular Graphics in Evaluation of Enzyme-Ligand Interactions," *Methods Enzymol.* **202**, 512–543 (1991).
64. M. Yvon, C. Chabanet, and J.-P. Pelissier, "Solubility of Peptides in Trichloroacetic Acid (TCA) Solutions. Hypothesis on the Precipitation Mechanism," *Int. J. Pept. Protein Res.* **34**, 166–176 (1989).
65. A.-A. Townsend and S. Nakai, "Relationships Between Hydrophobicity and Foaming Characteristics of Food Proteins," *J. Food Sci.* **48**, 588–594 (1983).
66. S. Yokoi et al., "Hydrophobic Beer Proteins and Their Function in Beer Foam," *Journal of the American Society of Brewing Chemists* **52**, 123–126 (1994).
67. A. Kato et al., "Effect of Partial Denaturation on Surface Properties of Ovalbumin and Lysozyme," *Agric. Biol. Chem.* **45**, 2755–2760 (1981).
68. I. Nir et al., "Surface Properties and Emulsification Behavior of Denatured Soy Proteins," *J. Food Sci.* **59**, 606–610 (1994).
69. K. Shimada and S. Matsushita, "Relationship Between Thermocoagulation of Proteins and Amino Acid Compositions," *J. Agric. Food Chem.* **28**, 413–417 (1980).
70. S. Barbut, "Protein Gel Ultrastructure and Functionality," in N. S. Hettiarachchy and G. R. Ziegler, eds., *Protein Functionality in Food Systems*, Marcel Dekker, New York, 1994, pp. 383–433.
71. E. C. Y. Li-Chan, "Macromolecular Interactions of Food Proteins Studied by Raman Spectroscopy," in N. Parris et al., eds., *Macromolecular Interactions in Food Technology*, American Chemical Society, Washington, D.C., 1996, pp. 15–36.
72. E. Niwa, T. Hakayma, and I. Hamada, "Arylsulfonyl Chloride Induced Setting of Dolphinfish Flesh Sol," *Bulletin of the Japanese Society of Scientific Fisheries* **47**, 179–182 (1981).
73. E. Niwa et al., "Setting of Flesh Sol Induced by Ethylsulfonation," *Bulletin of the Japanese Society of Scientific Fisheries* **47**, 915–919 (1981).
74. E. Li-Chan, S. Nakai, and D. F. Wood, " Muscle Protein Structure-Function Relationships and Discrimination of Functionality by Multivariate Analysis," *J. Food Sci.* **52**, 31–41 (1987).
75. T. E. O'Neill, "Flavor Binding by Food Proteins: An Overview," in R. J. McGorrin and J. V. Leland, eds., *Flavor-Food Interactions*, American Chemical Society, Washington, D.C., 1996, pp. 59–74.
76. N. Boudaud and J.-P. Doumont, "Interaction Between Flavor Components and β-Lactoglobulin," in R. J. McGorrin and J. V. Leland, eds., *Flavor-Food Interactions*, American Chemical Society, Washington, D.C., 1996, pp. 90–97.
77. J. Adler-Nissen, *Enzymic Hydrolysis of Food Proteins*, Elsevier Science, New York, 1986.
78. S. Arai, M. Watanabe, and N. Hirao, "Modification to Change Physical and Functional Properties of Food Proteins," in R. E. Feeney and J. R. Whitaker, eds., *Protein Tailoring for Food and Medical Uses*, Marcel Dekker, New York, 1986, pp. 75–95.
79. H. W. Blanch and A. M. Klibanov, eds., *Enzyme Engineering 9*, Vol. 542, Annals of the New York Academy of Sciences, New York, 1988.
80. J. A. Rothfus, "Potential Beta-Sheets Surfaces of Corn and Wheat Proteins," *J. Agric. Food Chem.* **44**, 3143–3152 (1996).

GENERAL REFERENCES

References 1, 3, 15, and 17 are good sources of general references on protein hydrophobicity. The following are recommended for further reading on hydrophobic interactions in general, and on the relationship with food protein functionality.

A. Ben-Naim, *Hydrophobic Interactions*, Plenum Press, New York, 1980.

A. Ben-Naim, "Hydrophobic Interactions in Biological Systems," in A. S. V. Burgen and G. C. K. Roberts, eds., *Topics in Molecular Pharmacology*, Vol. 2, Elsevier Science, New York, 1983, pp. 1–52.

N. S. Hettiarachchy and G. R. Ziegler, eds., *Protein Functionality in Food Systems*, Marcel Dekker, New York, 1994.

S. Nakai and H. W. Modler, eds., *Food Proteins, Properties and Characterization*, VCH, New York, 1996.

EUNICE C. Y. LI-CHAN
University of British Columbia
Vancouver, British Columbia
Canada

I

ICE CREAM AND FROZEN DESSERT

BACKGROUND AND DEFINITIONS

Ice cream and frozen desserts, which include sherbets, ices (sorbets), frozen yogurt, dynamically and quiescently frozen novelties, and mellorine, play very important nutritional and social roles in our daily diet. Although debate exists as to whether ice cream originated in ancient China or Rome, there is little doubt that ice cream is America's favorite dessert. In 1996 the total retail sales of ice cream and other frozen related products reached a high of $10.8 billion (1). Ice cream eating occasions are frequently associated with fond memories of childhood, lazy summer days, and family get-togethers. In addition to its role in complementing various social occasions, ice cream provides important nutrients, particularly to infants and adolescents who consume large quantities of ice cream and frozen desserts. A 4-fl-oz (1/2 cup) serving of a typical ice cream provides 8% of U.S. RDA calcium and 5% of U.S. RDA protein as well as other key nutrients (eg, vitamin A, thiamine, riboflavin) in significant quantities.

Although overall ice cream consumption growth has decreased in recent years owing to changing demographic and dietary patterns, a substantial growth in dollar sales has been achieved through products that satisfy the consumers' craving for high-quality, indulgent, convenient, and healthy alternatives for different occasions. The premium and superpremium ice cream category, which satisfies consumers' craving for quality and indulgence, was pioneered in the early 1960s by Reuben Mattus who created Häagen-Dazs ice cream, using all top-quality natural ingredients. The wide popularity of this type of rich, clean-tasting ice cream with no artificial stabilizers or other additives is proven over the years by the success of the premium and superpremium ice creams in the United States as well as overseas. The success enjoyed by Häagen-Dazs and others in Japan and Europe demonstrates that people across different cultures have equal appreciation for ice cream.

Smooth, creamy ice cream as we know it today was developed in the United States during the early part of this century. Two significant technological breakthroughs contributed to this development (1). First, homogenization to reduce fat particle size and to create a smooth texture was developed by August Gaulin. Second, continuous freezing was made commercially feasible by Clarence Vogt to produce consistent ice crystal structure with adequate throughput. Following these initial technical breakthroughs, most of the advances in ice cream technology have consisted of refinements in formulation, stabilizer systems, and process systems.

Ice cream and frozen desserts are governed under Food and Drug Administration (FDA) and state health department regulations. Most states have adopted standards defined in the FDA *Code of Federal Regulations*, as presented in Ref. 2.

Ice cream is a food produced by dynamically freezing a pasteurized mixture of milk, cream, nonfat milk solids, sugars, and stabilizers combined with flavorings that may be added before or after pasteurization. Fruits, nuts, candies, syrups, and fudges are often added to semifrozen ice cream to create different flavors and taste impacts. Ice cream containing more than 1.4% egg yolk solids is called French ice cream or frozen custard. Ice cream contains at least 1.6 lb solids per gallon and weighs at least 4.5 lb per gallon. The minimum butterfat content for ice cream and French ice cream is 10% with a minimum of 20% total milk solids.

Sherbet is a frozen dessert containing small amounts of dairy solids (1–2% milkfat, 2–5% dairy solids) and not less than 0.35% titratable acidity (calculated as lactic acid) for fruit flavors. Sherbets may be flavored with fruit or other flavors, such as chocolate.

Water ices (sorbets) contain fruits, fruit flavors, or flavors other than fruit. Typically, these products are quiescently frozen and weigh at least 6 lb/gal. These products contain no dairy or egg ingredients, other than egg white.

Frozen yogurt is a frozen dessert with the same ingredients as ice cream that also contains yogurt bacteria of the species *Lactobacillus bulgaricus* and *Streptococcus thermophilus*. It contains at least 3.25% milkfat and 8.25% nonfat milk solids (NMS). Frozen yogurt weighs not less than 5 lb/gal. At this time, there are no federal standards for this product; however, states may regulate.

Quiescently frozen dairy confections (eg, stick novelties) are made much like water ices except that these products contain at least 13% by weight of milk solids and 33% by weight of total solids. *Quiescently frozen confections* may or may not contain milk solids.

Mellorine is made in much the same way as ice cream except that butterfat is partially or totally replaced with vegetable fat. The product should contain at least 6% vegetable fat and 2.7% protein. The addition of vitamin A must be such that 40 I.U. are available per gram of vegetable fat.

As of 1993, the National Labeling and Education Act (NLEA) created four new labels for ice cream and related products: reduced fat, light, lowfat, and nonfat ice creams (3). *Reduced fat* is made with 25% less fat than the reference product. *Light* or *Lite* products are made with 50% less fat or one-third fewer calories than the reference product (as long as <50% of the calories come from fat). *Lowfat* ice cream contains not more than 3 g milkfat/serving. *Nonfat* ice creams contain less than 0.5 g of fat per serving. With the creation of the NLEA, the FDA in 1995 revoked the standard of identity for *ice milk*.

ICE CREAM COMPOSITION AND PHYSICAL PROPERTIES

Ice cream mix contains a minimum of 10% milkfat and 20% total milk solids, except when fruit, nuts, chocolate, or other bulky flavors are added. If bulky flavors are added, the minimum milkfat is 8.8% and 16.6% milk solids. According to U.S. Federal Standards, ice cream must weigh at least 4.5 lb/gal. The minimum gallon weight permits the incorporation of air sufficient to roughly double the volume (or 100% overrun). Percent overrun is defined as:

$$\frac{\text{weight of 1 gal mix} - \text{weight of 1 gal ice cream}}{\text{weight of 1 gal ice cream}} \times 100$$

Superpremium ice creams typically have a range of 10 to 40% overrun (premium, 60–75%; economy, 75–90%).

Ice cream mix normally contains 10 to 15% sucrose, 5 to 7% corn sweetener, 0.2 to 0.3% stabilizer, 0.1% emulsifier, and natural or artificial flavors and colors. According to U.S. Federal Standards, ice cream may not contain more than 0.5% stabilizer or 0.3% emulsifier.

Physically, ice cream is an emulsion, a dispersion of ice crystals and a partly frozen foam. The ice cream mix is an emulsion in which the aqueous phases contain solutions of soluble proteins, lactose, mineral salts, and added sugars. The nonaqueous phases consist of dispersed solids such as stabilizers, proteins, and fat globules. During the freezing process and foam formation, the emulsion is partially destabilized. This emulsion destabilization allows the air cells to be stabilized by clusters of intact fat globules, which agglomerate by free oil or protein/protein interactions (4–7). The percentage of water changed into ice is dependent on the production temperature and concentrations of sugars and milk salts in solution.

Butterfat

Butterfat contributes to the rich and creamy flavor and texture of ice cream. Fat particles, which impart a smooth mouth-feel, are dispersed throughout the ice cream mix. Ice cream with high fat content tends to have smaller ice crystals and a slower rate of air incorporation. Butterfat can be obtained from a variety of sources, but fresh sweet cream is by far the most popular and desirable source. Sweet cream is about 40% butterfat. Because of its perishability, for best flavor, cream must be refrigerated and promptly used. Cream can be frozen to preserve the quality; however, extra care must be taken during handling to protect the flavors. Another source of butterfat is unsalted butter, which is about 82.5% fat and can replace 50 to 75% of the sweet cream fat. Other special fat sources sometimes used in ice cream are plastic cream (80% fat), anhydrous butter oil (99% fat), concentrated sweetened cream, and dried cream (65% fat) (3).

Milk Proteins

Milk proteins are important to the formation of ice cream. Their role in the overall structure is to help stabilize the foam or air cells and to contribute to emulsification of the fat. This is due to hydration during mixing, destabilization and complexing during heating, and attraction to the fat globules surface after homogenization. As a result, the finished ice cream is more stable and smooth.

Two main types of milk proteins are found in ice cream. These are whey proteins and caseins. Casein comprises approximately 80% of the proteins found in milk. Numerous sources of milk proteins are available for use in ice cream. Condensed skim milk at about 25 to 35% solids is widely used throughout the industry. However, it is perishable and must be used fresh. Dried skim milk, or nonfat dried milk, is also widely used and has a much longer storage life. Many types of powders (low, medium, and high heat-treated) are available, each with specific functional properties (3). Superheated condensed skim milk is sometimes used as a protein source. Heating the condensed skim milk denatures more protein, thereby imparting increased stability and whipping ability when used in ice cream. However, cooked off-flavors may result. Lactose reduced (LR) skim milk may be used to decrease the lactose concentration in ice cream to discourage the formation of lactose crystals that impart an undesirable grainy, gritty, or sandy ice cream texture.

Specialty products are also available as additional protein sources. These include sodium caseinate and dry whey, which both improve mix-whipping properties by increasing overrun and dryness. Dry whey contains high portions of both lactose and mineral salts and can replace up to 25% of the milk solids nonfat (MSNF). At higher levels, saltiness and sandiness may result. Milk protein concentrate (ultrafiltered skim milk, MPC) has potential use in the ice cream industry, having a higher protein concentration and lower lactose and soluble mineral content than NFDM. MPC is most beneficial due to its clean, milklike flavor (8).

Stabilizers and Emulsifiers

Although excellent ice cream products can be made with only the natural stabilizing and emulsifying materials present in milk (milk protein and phosphates), additional stabilizers and emulsifiers also have potential benefits.

Fluctuations in temperature during normal distribution cause ice crystals to melt and refreeze into larger ice crystals, resulting in negative textural changes in the ice cream. This phenomenon, exposure to either large temperature swings or high temperature for a period of time, is referred to as heat shock. Hydrocolloidal stabilizers function to physically bind the water formed by melting and, therefore, to help prevent the formation of large ice crystals. The amount and type of stabilizer needed is governed by the composition of the ice cream mix, processing condition, processing temperatures, storage times and conditions, and other factors (3,9–11). Most ice cream mixes are made with 0.2 to 0.5% added stabilizer. A stabilizer is capable of binding a large amount of water. A small amount is effective in producing a product with good texture and more resistance to temperature abuse (heat shock). Too much stabilizer will result in product that is gummy.

Many stabilizers used in ice cream are natural. Sodium alginates are widely used as ice cream stabilizers. Algins improve whipping qualities of the ice cream, help prevent heat shock, and give the product good eating qualities. Gelatins, which are derived from animal sources, are also used to stabilize ice cream products. Their structure and affinity for water help prevent the formation of large ice crystals and contribute to texture. Locust bean gum (carob) is normally used in conjunction with carrageenan (Irish moss) as a stabilizer. Locust bean gum has excellent water-binding properties and imparts superior heat shock resistance. Its tendency to cause wheying off, curdling of milk proteins, is prevented when it is used in combination with carrageenan. Guar gum is a complex carbohydrate that functions as a stabilizer and is usually used in conjunction with carrageenan. Guar has water-binding properties similar to that of locust bean gum and imparts similar heat-shock resistance. It solubilizes easily in cold mix, making it a good choice for mixes undergoing high-temperature short-time (HTST) pasteurization. A newer gum, similar to guar and locust bean, is tara gum, which is cold water soluble and protective to heat shock. Tara gum imparts a buttery mouthfeel and can be used at levels 20 to 25% less than locust bean gum. Sodium carboxymethyl cellulose (CMC) is also used as a stabilizer. It has excellent water-holding properties and dissolves easily in mix. It possesses some emulsifying properties. CMC provides the best results when used in conjunction with another stabilizer (4). Milk proteins also have stabilizing properties, depending on the heat treatment they have received.

The primary effect of emulsifiers in ice cream is related to their ability to de-emulsify the fat globule membrane formed during homogenization. This de-emulsification, arising from the disruption of the fat globule membranes during freezing, facilitates the agglomeration and coalescence of the fat globules, leading to partial churning out of the fat phase. The agglomerated fat globules stabilize air cells (3). Thus, emulsifiers are used to improve the whipping qualities of ice cream mix by producing smaller ice crystals and smaller air cells, resulting in a smoother ice cream texture and a drier, stiffer ice cream. Generally, a mixture of high and low hydrophile-lipophile balance (HLB) emulsifiers, such as mono- and diglycerides, and polysorbate 80 is used. Milk contains some naturally emulsifying compounds that aid in the manufacturing of ice cream. These include milk proteins, phosphates, and citrates. Egg yolks may also be used as an emulsifier as they are high in lecithin.

Fat Reduction

To achieve fat reduction in frozen desserts, one needs to replace both the functionality of the fat and the actual quantity of fat removed (formula percentage). Reducing fat in a frozen dairy product generally requires multiple approaches. Solids replacement is critical for stability (ice crystal growth) and the perception of warmth/coldness in the mouth. For example, in nonfat products there is proportionally more ice that must melt at least partially in the mouth so the product will seem "colder" than a high fat ice cream when both products are served at the same temperature. Various fat substitutes may be chosen to provide "body" or a mouth-feel that simulates the lubricity of fat. Flavor levels or flavor character may be changed to modify the flavor impact and release.

Dairy processors worldwide are continuing to adjust their fat-reduced ice cream formulas to duplicate the qualities of full-fat products. Ingredient suppliers are broadening the range and quality of fat substitutes, which significantly help ice cream manufacturers in their quest for the perfect low-fat and nonfat desserts. A recent market survey of low-fat and fat-free frozen desserts in the United States showed that 38% contained microcrystalline cellulose, 31% contained polydextrose, 31% contained maltodextrin, 25% contained modified food starch, and 50% contained added protein, for example, whey protein or egg white.

Fat substitutes can be classified as proteins, carbohydrates, or lipids. In addition to fat substitutes, a low-fat or nonfat frozen dessert may require flavor addition to compensate for the lack of fat and its effect on flavor release. When butterfat is reduced in a product with a standard of identity such as ice cream ("low-fat ice cream"), a lost nutrient, such as fat-soluble vitamin A, must also be replaced.

Protein-Based Fat Substitutes. Most of the protein-based fat mimetics function by emulating fat globules. Simplesse® is a microparticulated egg white and/or milk protein from The NutraSweet Kelco Company, which is said to mimic fat by simulating the particle size of fat globules. Dairy-Lo® or Dairy-Lite®, a heat-treated whey protein concentrate from Cultor Food Science, is another available protein-based fat substitute.

Carbohydrate-Based Fat Substitutes. Carbohydrates are generally used in ice creams to replace fat solids. Higher molecular weight carbohydrates with low sweetness are usually chosen since they will be used in addition to the carbohydrates (sweeteners) already included in the formula. One such carbohydrate is Oatrim®, a fat substitute that is produced by partially hydrolyzing the starch in oat bran or oat flour into maltodextrins. Modified starches, polydextrose, polyfructose, and pectin, such as Slendid from Hercules, are also commonly used as fat replacers. Microcrystalline cellulose is used to enhance the mouthfeel of frozen dairy products. Gelled carbohydrate solutions or suspensions provide some viscosity and lubricity but do not have the globular mouth-feel characteristics of fat. Carbohydrate usage levels may be limited by increased mix viscosity.

Lipid-Based Fat Substitutes. Lipid-based fat substitutes are the so-called designer fats that can fully replace butterfat functionality at a lower calorie level because they are only partially digested. They generally don't require other formulation changes, while protein and carbohydrates behave as fat mimetics and require formulation changes. These fat "substitutes" are chemically fats, but rendering the fat partially indigestible reduces the caloric content. Several lower-calorie fats are commercially available. Caprennin® from Procter & Gamble is an interesterified triglyceride with long-chain fatty acids, which contribute to its lower degree of digestibility. Salatrim®, a 5 calories per gram triglyceride from Cultor Food Science marketed as Benefat®, is another interesterification product between glycerol esters of acetic, propionic, and butyric acids and regular fats. FDA has approved Olestra®, a sucrose polyester with very low HLB from Procter & Gamble, for salty snacks, but not yet for frozen desserts. Emulsifiers can augment the functionality of fat.

Flavor Delivery Systems. There are two potential flavor issues associated with low-fat products: the fat replacer itself might have an off-flavor, and the loss of fat and replacement by the fat substitute might result in an unbalanced flavor release. Flavor delivery issues exist because fat replacers do not dissolve flavor compounds in the same way as fat. When fat melts, flavors are released in a characteristic sequence. In low-fat and fat-free formulations the flavor release can be faster, stronger, and less pleasurable. Some advances have been made by flavor encapsulation and controlled-release mechanisms. One example is Singer's (12) patented method for the delivery of fat-soluble flavor compounds into nonfat and low-fat food products in which fat components have been replaced by nonlipid fat substitutes. Hatchwell (13) discussed a flavor delivery system composed of fat globules into which elevated levels of fat-soluble flavor compounds have been loaded or incorporated into nonfat and low-fat food products, so that fat-soluble flavor compounds will be released in a more natural and familiar sequence.

Cholesterol Reduction. As consumer demand for low-fat cholesterol-free products has increased innovative technologies to remove cholesterol from butterfat or egg yolk have been developed to different scales of commercialization. The techniques include vacuum steam distillation, short path distillation, melt crystallization, supercritical fluid extraction, and chelation with β-cyclodextrin or quillaja saponins (14,15).

Sugars and Sweeteners

Sugars contribute to the sweet flavor of ice cream. The concentration of sugars in the mix will also determine the freezing point of ice cream. The greater the amount of sweetener contained in an ice cream mix, the lower the freezing point. Many kinds of sweeteners can be used in ice cream. They include cane and beet sugar, corn sweeteners, honey, invert sugar, lactose, fructose, and refiner's syrup.

Sucrose, the most widely used sugar, is processed from cane sugar and beet sugar. Most ice creams are made with a combination of corn syrup and sucrose. Corn syrup solids are made from a hydrolysis of cornstarch. Corn syrups are classified by dextrose equivalence (DE), which indicates its degree of hydrolysis. The higher the DE, the sweeter the corn syrup will be. The intermediate range (48–68 DE) and high conversion (58–68 DE) corn syrups are used to produce a product that will be very close in texture to that of a product made entirely with sucrose. Corn syrup imparts a denser body to the finished product and may improve its shelf life (3,9,11). Honey is used in ice cream primarily to make honey-flavored ice cream. Honey usually does not blend well with other ice cream flavors and therefore is rarely used as a sweetener.

A significant part of the MSNF is lactose, which also imparts sweetness. Lactose is a sugar found only in milk. It is less sweet and less soluble than sucrose. The concentration of lactose in mix is limited because it may separate out of solution in large crystals and produce an undesirable sandy feeling in the mouth. The maximum amount of lactose to avoid sandiness may vary depending on the freezing and storage conditions of the product.

Sucrose Reduction

For some diabetics, no-sugar-added products provide opportunities to consume frozen dairy desserts. However, it is the total amount of carbohydrate, not just the simple sugars, that affects the blood glucose response. The technologies for sucrose replacement have been reviewed by Deis (16). Traditionally, polyols such as glycerol, xylitol, sorbitol, lactitol, maltitol, palatinit, and hydrogenated starch hydrolysates are used. Polydextrose, with 1 calorie per gram, is another commonly used bulking agent with high intensity artificial sweeteners. Buzzanell et al. (17) reviewed the status and application of various high-intensity sweeteners including aspartame, saccharin, Acesulfame-K, cyclamate, sucralose, and Alitame®. Sucralose® has recently been approved in the United States in April of 1998. A current U.S. market survey of no-sugar-added products shows that the majority contain maltodextrin, polydextrose, sorbitol, and a combination of Acesulfame-K and aspartame.

A novel process for a no-sugar-added product uses lactase to hydrolyze the lactose in condensed skim to increase the mix sweetness and permit use of a higher concentration of milk without risking lactose recrystallization (sandiness) (18). This allows the complete replacement of sugar with a high potency sweetener (aspartame) without the use of bulking agents.

Flavorings

Hundreds of varieties of flavoring substances are used to flavor ice creams. There are two main characteristics to each flavoring, type and intensity. Both influence how well consumers like the finished products. Serving temperature and overrun also influence flavor.

The U.S. Federal standards of identity place flavors into the following three categories (2):

Category I. Pure extracts (no artificial flavor)

Category II. Pure extracts with synthetic or artificial components (natural flavor dominates)

Category III. Artificial flavors (or natural and artificial with artificial flavor predominate)

U.S. Federal laws also regulate how these various flavoring categories are to be labeled on the cartons. The majority of the ice creams are made with category II flavors, whereas most premium and superpremium brands use category I or pure flavors.

Vanilla is by far the most widely used flavor. Vanilla flavors are available from each of the three categories I, II, or III; pure vanilla flavor is a product of vanilla bean fermentation. Most vanilla flavors used are of the category II, pure vanilla plus some artificial components. Chocolate products are the second most popular flavorings. They come in many forms; cocoa powders, chocolate syrups, and chocolate liquor are the most predominant. Flavor intensity, color, and texture can be manipulated by using different types of cocoa products. Fruits and fruit extracts are also very popular as flavorings. They can be added fresh, dried, candied, as concentrated juices, or as fruit essences. Nuts, spices, candies, cookies, and sugars such as honey are also added to ice cream to provide a wide variety of flavors. Other substances are swirled throughout the ice cream to produce a rippled effect, or a variegated product. These include flavors such as fudge, butterscotch, marshmallow, caramel, and fruits. Liqueur flavorings such as fruit brandy distillates or fruit liqueurs are also used in some ice creams.

Other Ingredients

Any FDA-certified natural and artificial coloring may be added to ice cream. These must be declared on the label. Annatto is a very popular yellow vegetable color used to color ice cream and to add visual richness.

Egg yolk solids are an optional ingredient. They are added for their emulsifying properties. They also impart a characteristic subtle flavor. Custard-type ice creams, by definition, must include egg yolks.

ICE CREAM PROCESSING—MIX MAKING, FREEZING, AND HARDENING

After the processor has selected the formula, the ingredients are blended together to produce the ice cream mix. The basic steps inherent in almost all processing of this mix into a frozen product are blending ingredients, pasteurization, homogenization, cooling, mix storage, flavoring, freezing, packaging, hardening, and finished product storage and distribution.

Mix Blending

To begin mix processing, the fluid ingredients are pumped from their storage vessel into a blending vat. The amount of these ingredients, as determined by the formula, may be either metered or weighed into the blending vat with the use of load cells. Some blend systems contain both meters and load cells to use one to check against the other in an effort to improve the accuracy of ingredient addition.

Dry ingredients such as cocoa powders, nonfat dry milk, corn syrup solids, and stabilizers are added into a blend vat, using a high-shear mixer. A portion of the bulk fluid ingredients is circulated through by the mixer to aid incorporation into the blend vat.

Pasteurization

Pasteurization is the process of heating the ice cream mix to a required time and temperature for the primary purpose of destroying pathogenic microorganisms. Two primary methods are recognized for achieving the time and temperature pasteurization requirements, batch also known as low-temperature long-time (LTLT) or continuous. Continuous pasteurization can be accomplished at different combinations of temperature and time: high-temperature short-time (HTST), higher-heat shorter-time (HHST), and ultra high temperature (UHT) methods. The U.S. Public Health Service (USPHS) standards for the pasteurization of mix are (19):

LTLT: 155°F for 30 min
HTST: 175°F for 25 s
HHST: 191°F for 1.0 s or 212°F for 0.01 s
UHT: 280°F for a minimum of 2 s

To achieve a higher rate of microbe destruction and to impart the characteristic flavors, most processors will pasteurize at slightly higher temperatures and/or longer times.

Today, the most popular method is the continuous HTST procedure using plate heat exchangers. Using HTST, the processor benefits from (*1*) improved process control, (*2*) less stabilizer requirements, (*3*) savings of time and space, and (*4*) increased capacity. Other continuous heat exchanger designs include steam injection and infusion, tubular, and swept surface types.

Batch pasteurization may be accomplished by two methods. The classic method is when the batch pasteurizer vessel also acts as the blending vat. After the required ingredients per formulation have been added to the blending vat, hot water is circulated through the jacket of the vat. Depending on the temperature of the hot water and the flow rate through the jacket, the mix arrives at the required batch pasteurization temperature. After the legal hold period, cool water is circulated through the jacket and, when sufficiently cooled, is pumped to a cooling system similar to that for the HTST system. The second method used for batch pasteurization is a modified system employing the regeneration section and the heater section of the HTST system; however, the temperature is raised only to the required batch pasteurization temperatures. The mix is then held in holding tanks for 30 min and then consequently pumped to the regenerator and cooling system.

Homogenization

The purpose of homogenization is to reduce the size of the fat globules so that 90% of the globules are less than 2 μm. Decreasing the fat globule size to smaller globules increases the surface area. This allows for a more uniform and consistent product, smoother in texture, which resists churn out in the freezer.

Acceptable regulatory placements of homogenizers are (*1*) in front of the heater section, (*2*) after the heater section and in front of the holding tube, and (*3*) after the flow diversion device and before the regeneration section. The homogenizer is a positive displacement pump that may act as the timing pump in many HTST systems.

Cooling

After pasteurization and homogenization, the mix is passed through a cooling unit. The cooling medium may be a refrigerated glycol solution or brine solution (a concentrated food-grade salt solution). The method employed for cooling is a plate heat exchange system similar to the HTST or a surface cooler. The heat exchange system is sufficiently sized to decrease the temperature of the mix to 45°F or less. The cold mix then flows to mix holding tanks for storage.

Mix Storage

Although legal standards only require cooling mix to 45°F or less, it is advantageous for product safety and quality to cool the mix to 34 to 36°F. With respect to aging of the mix, it was believed that 24 h was necessary for the physical changes in protein structure and fat crystallization to create more consistent mix that can be processed. However, recent studies do indicate that as little as 4 h accomplish this goal.

Freezing

During the freezing process, refrigerated ice cream mix is introduced into the ice cream barrel. The barrel is a cylindrical scraped-surface heat exchanger. Ammonia or freon is used as the cooling medium. Part of the water content of the mix is frozen into small ice crystals. At the same time, air is incorporated into the emulsion with agitation. The mix is transformed into a 20 to 25°F flowable mixture that can be filled/shaped into its final form. In commercial freezers, stainless-steel blades attached to the dasher (or mutator) scrape the crystals as they are being formed on the wall surface. The dasher controls the flow dynamics of the mix. Dashers come in many designs that determine the texture of the product. Closed or partial open dashers are used for ice creams. Many freezer manufacturers have specially designed their own dashers to meet their customers needs (20–22).

There are two main types of freezers, batch freezers and continuous freezers. Originally all freezing was done with batch freezers. In the batch process, a certain volume of mix is placed into the freezing barrel. The mix is frozen as the dasher is whipping in air. Batch freezing relies heavily on operator experience.

Currently, most manufacturing plants use continuous freezers. Continuous freezers provide greater process efficiency and product consistency. In continuous freezing, ice cream mix is metered into the freezing barrel at a constant rate. As the mix passes through the cylinder, air is whipped in by the dasher. Particulates (fruits, nuts, cookies) are added through a fruit and ingredient feeder after freezing. Continuous freezers are now available fully automated. Computers check the main control points of the operation. These control points may include overrun, dasher speed, flow rates, viscosity, pressure, and temperature control. These freezers also include automatic start-up, CIP pumps, and freeze-up prevention.

Soft-serve freezers are a scaled-down version of the batch and continuous freezers (23). Mix is added to a refrigerated hopper and flows by gravity into the freezing barrel. Mix is frozen, and air is incorporated in the barrel. The finished product is drawn out at 18 to 20°F onto a cup or cone. Because the product may have to stay in the freezing barrel for hours, the compressor and dasher are cycled on/off to maintain a consistent texture (20–22).

Cleaning-in-Place (CIP)

Modern ice cream plants include as much CIP cleaning as possible. CIP refers to the cleaning and sanitizing of process equipment in its assembled condition. The detergent solutions and sanitizers are recirculated through the process pipelines (pasteurizing system, mix tanks, freezer, etc). For effective soil removal, the detergent solution must have sufficient concentration, temperature, velocity (5 ft/s), and time. Manual or clean-out-of-place (COP) cleaning is significantly more time-consuming, more expensive, and generally less effective.

The four primary steps of an effective cleaning and sanitizing program for the CIP circuit are:

1. Prerinse: Flushing with water until the water runs clear.
2. Recirculation of detergent solution at 150 to 180°F for 15 to 30 min, depending on the type of equipment to be cleaned.
3. Postrinse for 10 to 18 min to remove all traces of detergent and soil.
4. Recirculation of sanitizers using chlorine or acid sanitizing agents, or heat sanitization. Heat sanitization is the circulating of hot water, 180°F minimum, for 15 min.

Ice Cream Hardening and Storage

Most ice creams, with the exception of quiescently frozen novelties, must be passed through a final freezing process after packaging called hardening. Hardening completes the ice crystallization that began in the continuous or batch ice cream freezer, enables the ice cream to withstand storage and transportation, and aids in final textural development.

Ice cream hardening must be accomplished through the rapid extraction of heat to ensure that the small ice crystals previously formed in the ice cream freezers are kept intact, giving the ice cream its familiar cool and creamy mouth-feel. Formation of large ice crystals would lead to an undesirable defect of iciness or a coarse mouth-feel. Efficient heat extraction is achieved through the use of hardening tunnels (insulated cold boxes in which the product is continuously conveyed through an extremely cold environment).

The actual heat transfer occurs either by a cold air blast (−30°F or lower) circulated around the ice cream packages or through hollow stainless steel plates that act as both the conveyor and the heat transfer medium. Brine, ammonia, or glycol chilled to −20°F flows in the inside of the hollow plates, discharging the heat gained from the ice cream to the plant refrigeration system.

Hardening tunnels are designed to bring the center of a package down to 0°F. The tunnel conveyor system has adjustable speeds to accommodate the various packages that a normal plant will produce. Hardening times for a −40°F blast tunnel are typically 90 min for pint packages to 4 h or more for 2-1/2-ga bulk packages.

Automatic or manual unloading of the hardening tunnel into a palletizing area completes the plants responsibilities for the product. The product will typically be warehoused at −20°F. It is then distributed to the retailers where the product should be held at 0°F or colder to survive its trip to the consumer's home.

ICE CREAM DEFECTS

High-quality ingredients are a requirement in the production of a high-quality ice cream product. Adherence to formulation, industry processing standards, and proper storage requirements also is critical. Incorporation of inferior dairy ingredients, sweeteners, or bulky ingredients always will result in a sub-standard product. Table 1 summarizes the characteristics and causes of predominant flavor and texture defects found in poor-quality ice cream (3,4).

Cryostabilization

Cryostabilization is a new approach to stabilizing a frozen product over its shelf life (23–26). It is an applied technology developed from the concept of food polymer science (FPS), the application to foods of the concept of polymer science. Since the late 1980s this field has been actively researched and applied to food product and technology development. Cryostabilization is based on fundamental understanding of the critical physicochemical and thermomechanical structure-property relationships that underlie the behavior of water in all nonequilibrium food systems at subzero temperatures (27,28).

A food structure can be in different states (liquid, rubber, or glass) depending on composition and temperature. The state of a material has enormous effects on the rate of change (eg, deformation, diffusion, or chemical reaction) within the material. Since ice cream defects such as iciness and sandiness are diffusion-controlled properties, cryostabilization can be an important tool for improving product quality. An ideally stable product will not change significantly during shelf life, so factors affecting the rate of change are critical to understanding and controlling product stability.

When ice cream is frozen, ice separates out from the system and the concentration of the unfrozen matrix (unfrozen water plus solutes such as sweeteners, electrolytes, proteins, and gums) between the ice crystals increases. As the temperature decreases further, more ice crystallizes until the viscosity of the unfrozen matrix is so high that there will be no more observable ice recrystallization in the time frame of the measurement. That temperature is called Tg′, the glass transition of the maximally concentrated unfrozen matrix (23–28). The Tg′ of a typical ice cream, as measured by DSC (differential scanning calorimetry), is approximately −32°C. It was agreed during the ISOPOW 6 meeting that "there was a lower temperature glass transition and a mobility transformation at Tg′ which coincided with the onset of ice dissolution into solution phase and correlated well with product stability" (29).

The Tg′ of a product will predict its stability. In distribution and in consumers' home freezers ice cream will not be held at temperatures below −32°C. However, the smaller the differential is between the temperature at which a product is stored and Tg′, the greater the stability

Table 1. Characteristics and Causes of Ice Cream Flavor and Texture Defects

Defect	Characteristics	Cause
Acid, sour	Tingly taste sensation	Uncontrolled bacterial activity
Cooked, eggy, custard	A cooked milk, condensed milk, caramel-like flavor	Excessive heat treatment Use of "cooked" cream
Lacks freshness, stale	Lacks clean, delicate, balanced flavor	General flavor deterioration of mix during storage Marginal quality dairy ingredients Absorption of odors from equipment
Oxidized, cardboardy, metallic	Astringent, metallic, papery, painty, fishy, tallowy	Old product Poor storage conditions Marginal quality dairy ingredients
Coarse, icy, grainy	Large ice particles distributed throughout the product or localized layerlike ice crystals	Slow, inadequate freezing; insufficient stabilizer Temperature fluctuations during storage
Sandy, gritty	Slow-melting crystals that remain on tongue or stick in throat after ice cream melts	High milk solids nonfat (lactose concentration) High total solids Temperature fluctuations during storage Added particulates such as nuts
Crumbly, fluffy, brittle	Tendency of ice cream to fall apart when scooped	Excessive overrun Low total solids Inadequate stabilization
Gummy, pasty, elastic	Opposite of crumbly Ice cream at consumption temperature is like taffy	Excessive stabilization Excessive corn syrup High sugar solids
Buttery, churned, greasy	Butter particles or greasy mouth-feel after ice cream has melted	Inadequate homogenization Excessive fat destabilization High milkfat content Overemulsification of mix

of the product as it is closer to the stable glassy state. This temperature differential can be reduced by increasing the Tg′ or lowering the storage temperature. One patented example of cryostabilization uses 20 to 33% 2–6DE maltodextrin to raise the Tg′ of frozen novelties to −13°C., reducing shrinkage and deformation (30). For the greatest product stability it is important to control the product's storage temperature as well as to formulate the product to create a higher Tg′, resulting in a reduced molecular diffusion rate.

NOVELTIES

Unlike packaged ice cream that normally is scooped or soft-served, novelties are unique ice cream products that are produced in a variety of shapes and sizes. Packaged in multi-pack or popular individual servings, early novelties included ice pops, Fudgesicles®, and ice cream sandwiches. Original marketing efforts were targeted toward children. Recently, new novelties have emerged and are capturing a more mature, sophisticated consumer niche. This resurgence has opened another exciting avenue for the ice cream industry.

Novelties can be made by two different processes, dynamically frozen and extruded or statically (quiescently frozen) hardened in brine. The dynamically frozen and extruded process involves freezing ice cream to as low a temperature as possible so that it retains its shape after extrusion through the nozzle. The product is wire cut, dropped onto a plate conveyor, and transferred into a hardening tunnel. For stick novelties, sticks are inserted before the hardening tunnel. After the hardener, the product is released from the plate and may be enrobed or dipped in coating and then packaged. Shapes are easy to alter by changing the nozzle configuration.

Quiescently frozen novelties are formed by hardening mix in metal molds. The molds travel through a two-section refrigerated brine tank. In the first section, mix is filled, sticks are inserted, and product is hardened. The molds then pass through a defrost section where the product is extracted from the molds, coated, and packaged.

SHERBETS AND SORBETS (ICES)

Sherbets, as defined by the *Code of Federal Regulations* (2), contain 1 to 2% butterfat and 2 to 5% total milk-derived solids by weight of the finished product. Because of their typically low total solid contents, sherbets rely on hydrocolloidal stabilizers to physically bind water and add viscosity. This aids in distribution stability and quality. Sherbets can be flavored by both fruit and nonfruit flavors. Fruit-flavored sherbets should have a titratable acidity of at least 0.85 (calculated as lactic acid). Formulas for sherbets are comparable to ice cream and ice milk.

Sorbets (ices) contain neither milk-derived ingredients nor egg ingredients other than egg white. Unlike ice creams and sherbets, ices can be made from unpasteurized mix owing to their typically high acidity formulation. Ices may be either dynamically or statically frozen. Water ices have enjoyed a tremendous resurgence in recent years as premium sorbets in single-serve novelties.

OTHER FROZEN DESSERTS

Frozen Yogurt

Frozen yogurt is a cultured dairy product produced using cultures of bacteria: *S. thermophilus* and/or *L. bulgaricus*. Generally accepted industry standards indicate that the finished product cannot be less than 8.25% nonfat milk solids and not less than 0.5% titratable acidity (expressed as lactic acid), and finished product has to weigh at least 5 lb/gal (3). Frozen yogurt butterfat content is similar to whole milk (3.25%). Frozen yogurt can be classified as low-fat or nonfat yogurt. Low fat contains no less than 0.5% and no more than 2% butterfat, whereas nonfat contains no more than 0.5% butterfat.

An acceptable frozen yogurt should be smooth in texture and have a pleasant flavor. It will handle and freeze much the same as sherbet or reduced or low-fat ice cream. The amounts and types of milk solids and sugars present will affect the body and texture of the finished product. Not all ice cream stabilizers can be used for frozen yogurt because stabilizers used in frozen yogurts must be acid stable. Gelatin-based stabilizers perform best with cultured products. Emulsifiers may be used to improve overrun, but can also react negatively in the culturing process. They may reduce the activity of the starter cultures, resulting in longer fermentation.

Fruit flavors will work well with the acid flavor profile of yogurt. Vanilla and chocolate have also been successful. When using chocolate as a flavoring ingredient, the yogurt should not be fully fermented to ensure a better flavor balance.

Mellorine

Mellorine products are similar in composition to ice cream except that the butterfat has been replaced by a combination of vegetable and animal fat. The types of vegetable fats used may include coconut, soybean, cottonseed, and other plant fats. Hydrogenated vegetable fats will improve flavor stability and texture because they closely resemble the melting properties of milk fat. Standards of identity require that all mellorine-type products must contain not less than 6% fat, and not less than 3.5% protein (the biological value of the protein in the product must be at least equivalent to that of whole milk protein). Finished product must weigh no less than 4.5 lb/gal and contain no less than 1.6 lb of total solids/gal (2). The processing, handling, and freezing of mellorine are similar to that of ice cream. More emulsifier may need to be used in mellorine products. The maximum stabilizer allowed is 1%.

Parevine

Parevine is a nondairy frozen dessert. It is consumed mainly by those who, for religious reasons, need to avoid mixing meat and dairy products. Parevine is made from a combination of fats, water, one or more protein or carbohydrate food ingredient (from other than milk or meat sources), and nutritive sweetening ingredients other than lactose. It may also contain eggs or egg products, flavoring, coloring, and added vegetable stabilizer or vegetable emulsifier. Standards of identity require that the finished product must contain at least 10% fat and total solids not less than 1.3 lb/gal of finished product. When bulky ingredients (such as nuts, cocoa, or chocolate) are added, the fat content can be reduced to, but to no less than 8% (2).

Parevine can only be sold in properly labeled factory-filled containers that cannot be larger than 1 gal. When parevine is sold on the premises, a sign must be conspicuously displayed stating "Parevine Sold Here" in letters that can be easily read by the consumer.

Other Nondairy Frozen Desserts

Other options exist for vegetarians, those who simply prefer to avoid dairy products, or lactose intolerant persons. These options include rice- and soy-based frozen desserts and novelties. These desserts rely on added stabilizers and emulsifiers to produce a creamy, smooth product similar to dairy counterparts.

FUTURE TRENDS

The frozen dessert segment is an approximately $10 billion per year industry. Nearly 100% of the U.S. population enjoys frozen desserts as part of their diet. Current trends suggest a modest industry growth rate.

The Nutrition Label Education Act (NLEA) significantly reshaped the landscape of frozen dessert standards. One result of the NLEA is that "ice milk" is no longer a legally recognized product in the United States. Along with the NLEA, consumer health concerns related to fat and calorie intake have driven the growth of new segments for *Reduced Fat, Low-Fat, and Fat-Free* ice creams, as well as *No Sugar Added* products. After initial explosive growth when good-quality low-fat products were no longer stigmatized with the "ice milk" label, these product segments shifted to more niche segments.

Historically, fat reduction has been a key driver in this segment's development. In response to increasing consumer awareness of health and wellness issues, efforts to reduce caloric content are expected to be a new motivator in the development of healthier products. Sorbet and water ices have enjoyed a significant growth in popularity. Other segments that will emerge include organic ice cream and nutritionally-fortified frozen desserts.

Fat replacers such as certain proteins or starches as well as fat mimetics (eg, Olestra) will create new opportunities for high-quality "Better for You" products.

Starch technology advances will continue to create greater opportunities for fat replacement and specific product textures.

Milk protein concentrates, lactose-reduced skim milk, and the lactase enzyme will be keys to developing ice creams that are free of the sandy texture caused by lactose crystal formation.

Antifreeze proteins (ice crystal growth-controlling proteins found in cold climate plants and animals) have the potential to revolutionize the frozen dessert industry. They can dramatically reduce the effects of heat-shock. Their future use is dependent on the development of a protein that is both economical and label-friendly.

In the 1980s *Listeria monocytogenes* was a major food safety concern in ice cream and frozen dessert manufacture. While contamination by pathogens such as *Listeria* continues to be an industry concern, Good Manufacturing Practices (GMPs), the application of hazard analysis and critical control points (HACCP), and improved equipment design have kept this potential issue in check. Control of food allergens in frozen desserts and in other foods has emerged as a critical food safety concern. The industry will again need to rely on HACCP in both ice cream manufacturing plants and suppliers' ingredient plants.

New forms of ice cream and ice cream novelties continue to be developed to meet specific consumer needs. The trend toward targeting specific consumer segments, as well as specific product use occasions, will be a key driver of future new products.

BIBLIOGRAPHY

1. *The Latest Scoop . . . Facts and Figures on Ice Cream and Related Products*, International Ice Cream Association, Washington, D.C., 1997.
2. *U.S. Code of Federal Regulations*, Part 21, Sections 135.110.
3. R. T. Marshall and W. S. Arbuckle, *Ice Cream*, 5th ed., International Thomson, New York, 1996.
4. K. G. Berger, "Ice Cream," in S. Friberg and K. Larsson, eds., *Food Emulsions*, 3rd ed., Marcel Dekker, New York, 1997, pp. 413–490.
5. K. B. Caldwell, H. D. Goff, and D. W. Stanley, "A Low-Temperature SEM Study of Ice Cream. I. Techniques and General Microstructure," *Food Structure* **11**, 1–9 (1992).
6. J. L. Gelin et al., "Structural Changes in Oil-in-Water Emulsions During the Manufacture of Ice Cream," *Food Hydrocolloids* **8**, 299–308 (1994).
7. S. Kokubo et al., "Agglomeration of Fat Globules During the Freezing Process of Ice Cream Manufacturing," *Milchwissenschaft* **53**, 206–209 (1998).
8. A. Kilara, "Formulating Frozen Desserts," *Dairy Foods* **94**, 69–70 (1993).
9. S. A. Wittinger and D. E. Smith, "Effects of Sweeteners and Stabilizers on Selected Sensory Attributes and Shelf Life of Ice Cream," *J. Food Sci.* **51**, 1463–1466, 1470 (1986).
10. C. Sharma, "Gums and Hydrocolloids in Oil-Water Emulsions," *Food Technol.* **35**, 59 (1981).
11. T. Miller-Livney and R. Hartel, "Ice Recrystallization in Ice Cream: Interactions Between Sweeteners and Stabilizers," *J. Dairy Sci.* **80**, 447–456 (1997).
12. U.S. Patent 5,202,146 (April 13, 1993), N. S. Singer, S. Pookoter, L. C. Hatchwell, G. Anderson, A. G. Shazer, and B. J. Booth, (to The Nutrasweet Company).
13. L. C. Hatchwell, "Overcoming Flavor Challenges in Low-Fat Frozen Desserts," *Food Technol.* **2**, 98–102 (1994).
14. U.S. Patent 5,326,579 (July 5, 1994), T. Richardson and R. Jimenez-Flores, (to the Regents of the University of California).
15. E. Sunfeld, J. M. Krochta, and T. Richardon, "Separation of Cholesterol from Butteroil Using Quilaja Saponins. I. Effect of pH, Contact Time and Adsorbent," *Journal of Food Process Engineering* **16**, 191–205 (1993).
16. R. C. Dels, "Adding Bulk Without Adding Sucrose," *Cereal Foods World* **39**, 93–97 (1994).

17. P. Buzzanell and F. Gray, "Have High-Intensity Sweeteners Reached Their Peak?" *Food Review* **16**, 44–50 (1993).
18. S. F. Keller et al., "Formulation of Aspartame-Sweetened Frozen Dairy Without Bulking Agents," *Food Technol.* **45**, 102, 104, 106 (1991).
19. U.S. Public Health and Human Services, *Milk Pasteurization Controls and Tests*, 6th ed., U.S. Government Printing Office, Washington, D.C., 1997.
20. *Food Equipment and Refrigeration*, Crepaco, Inc., Chicago, Ill.
21. *Ice Cream Plant Information Manual*, O. G. Hoyer, Inc., Lake Geneva, Wisc.
22. *Taylor Operator's Manual*, Taylor, Inc., Rockton, Ill.
23. H. Levine and L. Slade, "Thermomechanical Properties of Small Carbohydrate-Water Glasses and Rubbers: Kinetically-Metastable Systems at Subzero Temperatures," *J. Chem. Soc. Faraday Trans. I* **84**, 2619–2633 (1988).
24. H. Levine and L. Slade, "A Food Polymer Science Approach to the Practice of Cryostabilization Technology," *Comments Agric. Food Chem.* **1**, 315 (1989).
25. H. Levine and L. Slade, "Principles of Cryostabilization Technology from Structure/Property Relationships of Water-Soluble Food Carbohydrates—A Review," *Cryo-Letter* **9**, 21–63 (1988).
26. V. Huang and S. Platt, "The Latest Developments in Ice Cream Technology," *Chem. Ind. (London)* **2**, 37–72 (1995).
27. F. Franks, "The Properties of Aqueous Solutions at Subzero Temperatures," in F. Franks, ed., *Water: A Comprehensive Treatise*, Vol. 7, New York, Plenum Press, 1982, pp. 215–338.
28. F. Franks, "Complex Aqueous Systems at Subzero Temperatures," in D. Simatos and J. L. Multon, eds., *Properties of Water in Foods*, Dordrecht, The Netherlands, Martinus Nilhoff, 1985, pp. 497–509.
29. D. S. Reid, *The Properties of Water in Foods: ISOPOW 6*, Blackie Academic & Professional; New York, 1998, pp. xix–xx.
30. U.S. Patent 5,486,373 (January 23, 1996), C. B. Holt and J. H. Telford, (to Good Humor Corporation).

John Clemmings
Paul Fazio
Victor Huang
Diane Rosenwald
Leigh Hadden White
Häagen-Dazs Research and Development
Edina, Minnesota

IMMOBILIZED ENZYMES

In the processing of foods, enzymes have distinct advantages over chemical catalysts of which most notable are substrate specificity and activity under mild conditions of temperature and pH. However, the cost of using soluble enzymes is a drawback so there is much interest in the use of immobilized enzymes and cells. Immobilized enzymes have been defined as "enzymes that are physically confined or localized in a certain defined region of space with retention of their catalytic activities, and that can be used repeatedly and continuously" (1). Immobilization often results in the enzyme becoming water insoluble.

Compared with processing with soluble, unconstrained enzymes, immobilization offers several advantages, including:

- Reuse or continuous use of the catalyst, thereby reducing both capital and recurrent process costs
- Absence of the enzyme from the product, thus potentially allowing for a wider range of enzymes than those normally permitted in foods
- Ease of terminating the reaction without drastic measures such as heat denaturation or extreme pH
- Greater (sometimes) thermal and pH stability and prevention of self-digestion by proteases
- Less product inhibition, and more substrate depletion with continuous processes, giving faster conversion

The main disadvantages are the cost of producing the immobilized enzyme, including the cost of the support, and altered reaction kinetics, which often result from diffusional restrictions.

METHODS FOR IMMOBILIZATION

General methods for immobilization of enzymyes and cells can be classified as follows: (*1*) adsorption onto a solid carrier, (*2*) covalent attachment to a solid carrier, (*3*) cross-linking to form insoluble aggregates (*4*) adsorption followed by cross-linking, and (*5*) entrapment in gels or encapsulation by membranes.

Adsorption onto a support surface is the simplest method of immobilization. Binding of the enzyme to the substrate is typically through electrostatic binding, hydrogen bonding, hydrophobic interactions, or van der Waals forces. Biospecific interactions between enzyme and ligand, or enzyme and antibody, are also possible. Binding forces (except for biospecific interactions) are rather weak and usually readily reversible. This makes adsorption an attractive process commercially since loading, desorption following decay of activity, and reloading are relatively simple. The main problems are leakage of enzyme from the support and changed activity characteristics caused by intimate association of the enzyme with the support matrix.

Covalent attachment provides the most secure method of immobilizing enzymes. Since enzymes are proteins, the amino acid side chains contain several functional groups that can be used for coupling to a support. The amino group of lysine is widely used, but the sulfhydryl of cysteine, hydroxyl of tyrosine or serine, and carboxyl groups of glutamic and aspartic acids are also used for coupling. However, the attachment must be via some nonessential group in the protein surface or activity will be lost. Prior to coupling, the support matrix is treated with a reagent that activates some functional groups on its surface; then the enzyme is mixed with the activated support and coupling takes place. Although many chemical reagents can be used for coupling, safety restrictions on food processes mean that only glutaraldehyde is used in commercial applications at the present time.

Cross-linking involves formation of covalent bonds between enzyme molecules, by bifunctional reagents, to produce large insoluble aggregates. Glutaraldehyde is a typical cross-linking reagent. Cross-linking may lead to large increases in stability, however, the reaction is difficult to control and the gelatinous physical nature of the products is a major impediment in many applications. Cross-linking *is* a very useful technique to minimize leakage of enzymes already immobilized by adsorption.

Entrapment involves placing the enzyme within a polymer matrix (eg, a gel) or behind a membrane to allow penetration of substrate and products but to prevent release of the enzyme protein. This restricts the application to processes that involve small substrates and products. Specific techniques include: entrapment in gels such as polyacrylamide and calcium alginate, entrapment in hollow fibers or within the microcavities of synthetic fibers, and microencapsulation of the enzyme within tiny spheres having semipermeable membranes. Entrapment has the advantage of minimal alteration of enzyme properties since there is no significant interaction with the "support." Drawbacks of the method include: enzyme leakage, mass transfer/diffusional limitations due to passage of substrates and products across a barrier, and the lack of stabilizing effects often associated with a solid support.

An extraordinary variety of supports have been used for enzyme immobilization, ranging from feathers to stainless steel spheres. For most food processes, the best supports are physically strong, chemically inert except for those groups involved in linkage, and not subject to microbial attack. A good example is glass beads. Once immobilized, the enzyme can be used in a variety of batch or continuous reactor systems, including stirred-tank, plugged-flow, and fluidized-bed systems.

Immobilization can cause substantial alterations in the activity of enzymes. There are many reasons for this, but three major influences have been identified: mass transfer/diffusion, pH shifts, and partitioning.

When an enzyme is bound to a solid surface, mass-transfer limitations often occur due to the stagnant liquid film close to the surface through which the substrates and products diffuse. This can lead to concentration differences between the bulk phase and the microenvironment of the enzyme. Diffusional restrictions are also noticeable for enzymes entrapped in gels or particles.

Shifts in pH optimum are noted when enzymes are immobilized to charged supports. A negatively charged support abstracts protons leading to a more acidic microenvironment and a shift to a higher optimum pH in bulk solution. The opposite effect is observed with positively charged supports, leading to a lower pH optimum.

Partitioning of substrates and products between the bulk liquid and the immobilized enzyme system also leads to differences in concentrations. This partitioning can be caused by charge interactions or by hydrophobic/hydrophilic interactions between the substrate/product and the enzyme support system. The kinetic consequences of these influences have been carefully explored (2,3).

Immobilized Cells

An alternative to extracting enzymes is to use immobilized cells. These can be readily entrapped in gel media such as carrageenan or calcium alginate. The major advantages of such preparations compared with immobilized enzymes include: avoidance of expensive enzyme isolation procedures, enhanced catalyst stability due to protection by the cell membrane and the local microenvironment, and the ability to catalyze multistep enzymatic reactions, particularly those using cofactors that are otherwise difficult to regenerate. Disadvantages include the following: the possibility of side reactions and hence less pure products; and both substrate and product must be small enough to pass through the gel particles. Currently, immobilized cells are used either to catalyze simple conversions such as isomerization or as growing immobilized cells to produce primary metabolites, for example, amino acids. Immobilized living cells can also be used to detoxify waste materials, for example, to remove nitrates from drinking water and to remove phenols from industrial wastewater (4).

Limitations

Despite the advantages cited, the current use of immobilized enzymes in industrial-scale food processing is quite limited. In some cases the reasons are purely economic—the soluble enzyme is so cheap that there is no cost advantage in an immobilized enzyme process. Other limitations are directly related to processing in the reactor. A major constraint is that applications are restricted to liquid systems. Even then, in complex liquid foods such as milk, enzyme particles are subject to fouling by components such as proteins. Another problem with foods is to prevent microbial buildup in the reactor. Accordingly, reactors may need to be periodically sanitized by compounds such as H_2O_2 or quaternary ammonium compounds. Chlorine-based compounds and iodophors are effective sanitizers but can lead to rapid loss of activity with some enzymes (5).

CURRENT APPLICATIONS IN FOOD PROCESSING

Glucose Isomerization

In terms of throughput, isomerization of glucose to the sweeter sugar fructose represents the largest application of immobilized enzyme technology. The glucose substrate is generated from hydrolysis of starch, and the product—high-fructose corn syrup—contains about 42% fructose, where it is isosweet with sucrose on a solids basis. The product is used extensively in the soft drinks industry. The enzyme used—glucose isomerase—is actually a xylose isomerase. It can be obtained from several microbial sources (notably *Streptomyces* spp.) and immobilized by most of the methods already described. Some processes use enzyme adsorbed onto DEAE-cellulose. However, since the enzyme is intracellular and extraction is inefficient, several preparations use whole cells. These are usually heated prior to immobilization both to prevent enzyme release after cell lysis and to inactivate other enzymes that could cause undesirable side reactions. Generally, higher activity per unit volume is obtained using enzyme adsorbed onto supports, but this is offset by the cost of the support and, if necessary, its recovery.

Lactose Hydrolysis

Hydrolyzing lactose has two major benefits: (*1*) milk becomes easier to digest for those who are lactose intolerant and (2) sweet syrups can be produced from whey/whey permeate, thereby enhancing its value and increasing utilization. Treatment of milk requires a neutral pH enzyme. Immobilization of neutral lactases has been largely unsuccessful with one notable exception: yeast lactase has been entrapped in cellulose acetate fibers and used in Italy to process sterilized skim milk in a batch reactor at 5°C.

Immobilization of acid-pH lactase for processing acid whey is easier, and two preparations have been used on a major scale. The enzyme from *Aspergillus niger* has been covalently linked to glass beads and has been employed in pilot-scale plants by Corning Glass. Demineralized, deproteinized whey was used as a substrate at 35 to 50°C, the temperature slowly increasing to compensate for activity losses. In Finland, acid lactase has been immobilized by adsorption onto a phenol formaldehyde resin and used on a commercial scale by the Valio Dairy to process acid whey.

Sugar Refining

Raffinose, an α-galactoside present in sugar beet interferes with crystallization of sucrose from the molasses and depresses the yield. α-Galactosidase hydrolyzes raffinose into sucrose and galactose and is produced by several fungi—notably *Mortierella vinaceae*. Since extraction of the enzyme is costly, mycelial pellets of *M. vinaceae* containing α-galactosidase but lacking invertase are used in a continuous process. Molasses is pumped through chambers of suspended pellets and about 65% of the raffinose is converted to sucrose. The results of processing are better sucrose yields and a reduction in waste molasses.

Aspartic Acid Production

L-aspartic acid can be enzymically synthesized by aspartase-catalyzed addition of ammonia to fumaric acid. This has been done commercially in Japan since 1973 using the aspartase in *Escherichia coli* cells immobilized in acrylamide or carrageenan. Before entrapment, the cells are heat shocked to inactivate fumarase and other unwanted enzymes. About 95% conversion is obtained using 1*M* ammonium fumarate. The immobilized cells have better operational stability than either immobilized enzyme or free cells that were previously used in batch processes. Production costs are about half of those using the earlier methods.

Resolution of L and D Amino Acids

Chemical synthesis of amino acids results in a racemic mixture of D and L forms; however, only the L form is biologically available and used for amino acid supplementation of foods. Resolution of the mixture can be achieved by acetylation followed by deacetylation with amino acylase, which acts only on the L form. The L-amino acid can then be crystallized out of the mixture. A commercial process has been developed in Japan using amino acylase immobilized on DEAE-Sephadex in a packed column reactor. This process has been in operation since 1969 and represents the first commercial-scale use of immobilized enzyme technology. Operating cost savings are about 40% compared with using the soluble enzyme.

Glucose Syrup Production

In the conversion of starch to glucose, glucoamylase is used to hydrolyze dextrins formed by α-amylase action. Traditionally the enzyme is used in the soluble form. Although it is readily immobilized, it has proven difficult to achieve satisfactory yields of glucose due to formation of side products, such as isomaltose. However, Tate and Lyle (London) has developed a satisfactory preparation in which the enzyme is immobilized in a gel surrounding bone char as an inert support. This preparation is beginning to be used in commercial operations.

Production of Aspartame

Aspartame (α-L-aspartic acid-L-phenylalanine-methyl ester) is a dipeptide sweetner widely used in diet soft drinks. It is about 180 times as sweet as sucrose. It can be synthesized chemically from its constituent amino acids or enzymatically using the protease thermolysin working in reverse to synthesize the peptide bond. The enzyme can be immobilized by adsorption to Celite or cross-linked polyacrylic ester, followed by cross-linking with glutaraldehyde. The synthesis reaction is then carried out in buffer saturated with ethyl acetate. The reaction takes place in the aqueous phase, but the product is effectively partitioned into the organic phase, thereby disturbing the equilibrium to favor synthesis.

Isomaltulose Production

The disaccharide isomaltulose is a novel sugar with noncariogenic properties that make it useful in confectionery. It is produced by the enzyme isomaltulose synthetase, which is immobilized as a constituent of whole cells of organisms such as *Erwinia rhapontici, Protaminobacter rubrum*, or *Serratia plymuthica*.

FOOD ANALYSIS

Enzymes are ideal for quantitative analysis of foods because of their specificity and sensitivity. The advantages of using immobilized enzymes for analysis include: reuse of the catalyst, stability, and predictable decay rates. There are two major approaches. The enzyme can be used in a reactor that produces a product that can be readily detected, for example, colorimetrically. Alternatively, the enzyme can be immobilized onto a sensing device capable of giving a continuous output, such as an electrode. The resulting *enzyme electrode* can then be inserted into foods to measure substrate depletion or product accumulation or some associated change such as pH. Enzyme electrodes have been developed that are capable of detecting a wide range of food components, including sugars, alcohols, amino acids, organic acids, and even cholesterol. They are particularly useful in automatic closed-system flow analysis, for example, as in-line monitors.

Where the enthalpy change of the reaction is sufficient, substrates can be assayed calorimetrically with enzyme thermistors. The heat of the reaction in a small column of immobilized enzyme is measured as a temperature change by a thermistor in the column eluent and indicates the amount of substrate converted. Enzyme thermistors can be used for continuous monitoring and control of bioconversions, for example, to sense changes in glucose concentration and control the flow of whey pumped to an immobilized lactase reactor (6).

POTENTIAL APPLICATIONS AND FUTURE DEVELOPMENTS

Application of immobilized enzymes seems promising in several areas. These include the following: use of sulfhydryl oxidase to remove the cooked flavor in UHT milk, use of trypsin to control the development of an oxidized flavor in stored milk, and the dibittering of citrus juices by enzymes that alter the bitter component limonene. Immobilized proteases could be used for chill-proofing beer, for limited hydrolysis of soy proteins to improve functionality, and for continuous production of curd for cheese making. Immobilized yeasts could be used for brewing beer and immobilized bacteria for production of vinegar.

One area that is especially promising is the use of immobilized lipases for the chemical modification of fats. Interesterification with free fatty acids or with other triglycerides using regiospecific lipases allows upgrading of cheap fats into more valuable fats such as cocoa butter substitutes. The immobilized enzyme could also be used to replace conventional chemical catalysts used in random interesterification. These processes have been well developed on the pilot scale, but the extent of commercial production is uncertain.

Primary obstacles to implementation of these applications are process economics and unfavorable enzyme characteristics such as inadequate stability. However, developments in molecular biology coupled with protein engineering promise to provide a second generation of enzymes whose properties will be designed with a particular process in mind. Greater stability toward heat, pH, and chemical denaturants is likely, as is altered specificity. The ability of enzymes to work in organic solvents will undoubtedly be exploited for further transformation of lipids and synthesis of food additives. Use of multienzyme systems and of enzymes that require cofactor regeneration should also expand the list of potential applications. At the reactor level, a major breakthrough would result from the discovery of general methods for regenerating activity of immobilized enzymes.

BIBLIOGRAPHY

1. J. F. Kennedy and J. M. S. Cabral, "Enzyme Immobilization," in H. J. Rehm and G. Reed, eds., *Biotechnology*, Vol. 72, VCH, Weinheim, Germany, 1987.
2. M. Griffin, E. J. Hammonds, and C. K. Leach, "Enzyme Immobilization," *Technological Applications of Biocatalysts*, Butterworth-Heineman, Oxford, United Kingdom, 1993, pp. 75–117.
3. L. Goldstein, "Kinetic Behaviour of Immobilized Enzyme Systems," *Methods Enzymol.* **44**, 397–443 (1976).
4. I. Chibata and T. Tosa, "Immobilized Microbial Cells and their Applications," *Trends Biochem. Sci.* **5**, 88–90 (1980).
5. J. F. Rolands, "Requirements Unique to the Food and Beverage Industry," in W. H. Pitcher, Jr., ed., *Immobilized Enzymes for Food Processing*, CRC Press, Boca Raton, Fl., 1980, pp. 55–80.
6. B. Danielson et al., "The Use of an Enzyme Thermistor in Continuous Measurements and Enzyme Reactor Control," *Biotechnol. Bioeng.* **21**, 1749–1766 (1979).

GENERAL REFERENCES

R. J. D. Barker, ed., *Technological Application of Biocatalysts*, Butterworth-Heinerman, Oxford, United Kingdom, 1993.

R. D. King and P. S. J. Cheetham, eds., *Food Biotechnology–2*, Elsevier Applied Science, London, United Kingdom, 1988.

T. Nagodawithana and G. Reed, eds., *Enzymes in Food Processing*, 3rd ed., Academic Press, San Diego, Calif., 1993.

H. E. Swaisgood, "Immobilized Enzymes: Applications to Bioprocessing of Food," in P. F. Fox, ed., *Food Enzymology*, Elsevier Applied Science, London, United Kingdom, 1991, pp. 309–341.

A. Tanaka, T. Tosa, and T. Kobayashi, eds., *Industrial Application of Immobilized Biocatalysts*, Marcel Dekker, New York, 1993.

RAYMOND R. MAHONEY
University of Massachusetts
Amherst, Massachusetts

IMMUNOLOGICAL METHODOLOGY

Since the late 1970s, application of immunoassays (IAs) has moved from the clinical field into the areas of agricultural residue and food analysis. Development of these immunoassays, and enzyme-linked immunosorbent assays (ELISAs) in particular, began when the attractive features of the assays used in the clinical diagnostic field were identified. These assays are capable of being more rapid, more sensitive, less expensive, and less laborious than many of the classical methods used for detecting compounds of importance in agriculture and the food industry. These compounds may be naturally occurring compounds such as mycotoxins, natural toxicants, or microbial contaminants. They may also be substances added to animal feeds (eg, antimicrobial agents) or applied to crops (eg, pesticides), which occur as potentially harmful residues in raw materials and finished food products. The ability to identify food constituents is also important to ensure that adulteration has not occurred.

Because of the explosion of research and development of ias in this area and the limited scope and length of this article, it is impossible to list and describe every compound of interest, whether it is a food component or a contaminant. This article will provide a fairly detailed summary of ias used in analysis of food proteins (meat and nonmeat proteins), beginning with the older classical methods (eg, agglutination and precipitin techniques), and then focus on the development of the newer, more innovative immunoassays (eg, ELISAs). The later sections, whether they describe ias designed for detecting naturally occurring substances or those designed for detecting residues resulting from application or administration of drugs and other chemicals, will focus on ELISAs.

BACKGROUND

Chromatographic methods such as thin layer chromatography (TLC), high-performance liquid chromatography (HPLC), and gas chromatography (GC) are commonly used to determine levels of many agricultural residues (1–10). They are limited because chromatography is only a method of separation, not identification. Spectral identification is necessary if structural confirmation is desired (11).

Microbiological assay methods have traditionally been used to detect a wide variety of antibiotics, regardless of the structures and sample matrix involved. The basic protocols were established in 1945 (12) and serve as the basis for the analysis involved both for inhibition and diffusion methods (13). The overall drawback to either the turbidimetric or diffusion procedures is that they are nonspecific (14). The routine procedures only achieve purification by dilution before analysis. Second, the organisms used in the assays respond to more than one antibiotic family in addition to other nonspecific materials. Therefore, sources of analytical bias and inaccurate results are inherent in the microbiological assay systems. The sensitivity of the organisms to the antibiotic being detected is usually sufficient enough to minimize some of these problems. However, the lack of specificity remains the key flaw to these assays. Fairly elaborate separations are required to separate the various antibiotic families prior to analysis. When the formulation being analyzed is known this is not a problem. However, when unknown samples or materials of uncertain origin are to be analyzed, this can pose a serious problem (14).

Detection of food-borne pathogens in foods has relied on culture procedures that allow growth, selection, isolation, and identification of isolated organisms. The major disadvantages of these procedures is the time required to obtain a negative reading. The laborious, time-consuming cultural procedures hinder prompt therapy for patients suffering from food poisoning and prompt removal of any contaminated food product from the marketplace.

Due to the drawbacks mentioned with conventional chemical and microbiological methods attention turned to employing antibody-based assays for detection of residues and organisms. Immunobased assays use antibodies that are developed to recognize a particular compound (or look-alike). Antibodies are developed using a number of domestic animals including mice, rabbits, guinea pigs, sheep, and goats. Antibody production currently takes two forms; either polyclonal or monoclonal antibodies are produced for immunoassay development.

These different types of antibody have been used to develop isotopic immunoassays such as radioimmunoassay (RIA), nonisotopic immunoassays that use a labeled antigen or antibody such as fluoroimmunoassay (FIA) and enzyme immunoassay (EIA). Agglutination, immunodiffusion, and quantitative precipitin techniques have also been used as well as affinity columns for separation and detection of haptens (such as pesticides, drugs, mycotoxins, and other potentially harmful chemicals) and food-borne pathogens and food proteins (meat and nonmeat protein additives).

Definitions

Before proceeding with a further discussion, it is necessary to provide definitions of frequently used terms and procedures.

Antigen. Antigens are any substance that when introduced into an animal's body, are recognized as foreign and elicit an immune response. The ability of an antigen to stimulate an immune response is called its immunogenicity, whereas its ability to combine specifically with an antibody is called its antigenicity.

Hapten. Hapten is a small molecule with molecular weight of less than 1,000. It is nonimmunogenic in its own right and must be chemically linked to proteins (*in vivo* or *in vitro*) to produce antibodies. It reacts with antibodies of appropriate specificity. Many of the compounds that require testing in the food industry and agriculture are small molecules. Bovine serum albumin (BSA), keyhole limpet hemocyanin (KLH), and bovine thyroglobulin (BTG) are commonly used proteins. These antigens, composed of a large protein with attached haptenic groups, are called conjugate antigens.

Antibodies. Proteins that are formed in response to an antigen and react specifically with the antigen are antibodies. All antibodies belong to a family of proteins called immunoglobulins. Although the definition states that antibodies are formed only in response to an antigen, antisera may contain immunoglobulins that recognize and bind to certain antigens regardless of known exposure. These immunoglobulins are called natural antibodies (14).

Polyclonal Antibodies. To produce antibodies, a properly selected conjugate antigen, or large protein is injected into the host animal. The host's immune system will respond to the antigen because it is recognized as being a foreign substance. The resulting antibodies are a mixture. If the antigen is a large protein or a particulate antigen, with a number of repeating antigenic determinants or with one repeating determinant, some of the antibodies will bind to parts of the protein. If the antigen is a conjugate antigen, some antibodies will bind to parts of the carrier protein, while others will bind to the small molecule. Therefore, there are a number of different antigenic determinants to which the immune cells respond, resulting in a mixture of antibodies to those determinants. If a successful preparation is made, this mixed population will contain some antibodies that are very strong, that is their affinity and avidity for the foreign protein will be great.

Monoclonal Antibodies. These antibodies comprise a homogeneous population, all with the same specificity. They are produced and secreted by hybridomas, which are cells created by fusing hyperimmune spleen cells to myeloma cells. Mice are commonly used for the spleen cell donors as well as the species of origin of the myeloma cells. The spleen cells become immortalized (do not have a finite life span in culture) by fusion with the myeloma cells and can live forever if treated properly. (15) All of the cells in a population have originated from one cell separated and identified during cloning so that they all produce a homogeneous population of antibodies.

Immunoassay. A test format that employs antibodies that are specific for a certain antigen is called an immunoassay. They are used for detection of either antigens or antibodies. Where an assay is designed for detection of an antibody in a serum sample, another antibody is produced using the antibody of interest as the immunogen. In this case, the antibody that is to be detected is the antigen.

Isotopic Immunoassays. An assay in which hapten or antigen can be measured is an isotopic immunoassay. The assay is based on competition for antibody between a radioactive indicator antigen and its unlabeled counterpart in the test sample. As the amount of unlabeled antigen in the test sample increases, less labeled antigen is bound. The concentration of antigen in the test sample can be determined from comparison with a standard calibration curve prepared with known concentrations of the purified antigen (16).

The sensitivity of RIA assay is primarily limited by the amount of radioactivity that can be introduced into the radiolabeled antigen (17). Levels as low as 1 ng can be detected when carrier-free radioactive iodine ^{125}I is used as an extrinsic label. Precipitates are usually not evident, because extremely low concentrations are used. There are several procedures that can be used to separate free and bound indicator antigen. In the general method complexes of antibody bound to radiolabeled antigen are precipitated with antiserum prepared against the antibody moiety.

There are also a number of solid-phase assays that eliminate the requirement for the second antispecies antibody. In one method antibodies to the antigen are absorbed to the walls of plastic (eg, polystyrene) tubes. Radiolabeled antigen binds specifically to the adsorbed antigens and can be counted. When competing unlabeled antigen is also present, less radiolabeled antigen is proportionally bound. The remaining unbound fraction can be decanted. The method is inexpensive, rapid, and highly sensitive. For example, it is possible to detect less than 0.01 μg of antigen when a tube is coated with 1 μg of antibody (17).

Nonisotopic Immunoassays. These assays differ from isotopic immunoassays in terms of the type of label used, the means of end-point detection, and the possibility of bypassing a separation step. Fluoroimmunoassays and enzyme immunoassays are the two types of nonisotopic immunoassay.

Fluorescein and rhodamine are commonly used for labeling molecules. There are nonseparation fluoroimmunoassays that require no separation step of bound from unbound product. There are polarization fluoroimmunoassays that depend on antibody binding to enhance the signal, and quenching fluoroimmunoassays that depend on a decrease in signal from the bound fraction, which is attributed to the antibody's ability to impair the excitation, or the emission from the labeled hapten (18). A method has been developed for several aminoglycosides that enables the use of a simple fluorimeter (19).

Eias use enzyme labels and are grouped into two categories: homogeneous and heterogeneous. The difference between these two types of eias is that homogeneous assays requires no separation of unreacted reagents because the immune reaction affects the enzyme activity. Heterogeneous assays have separation steps; ELISA, a type of heterogeneous assay, requires washing between each step to remove unbound reagents (20). Commonly used enzyme labels include alkaline phosphatase (21), glucose oxidase (22), and horseradish peroxidase (23). These enzymes catalyze reactions that cause substrates to degrade and form a colored product that can be read spectrophotometrically or visually by eye. Depending on format, either antibody or antigen is adsorbed onto a solid phase, which can be polystyrene tubes, polystyrene microtiter wells, or membranes (ie, nitrocellulose and nylon).

Competitive Direct ELISA. This is an enzyme immunoassay where free hapten competes with an enzyme-labeled hapten for a number of limited antibody sites attached onto a solid phase. Unbound reactants are washed before substrate is added. Following substrate addition the color produced is indirectly related to the amount of hapten in the test sample.

Sandwich ELISA. This is an ELISA that is suitable for detecting large antigens (bacterial, viral, and other large proteins). Two preparations of antibodies are used, one to coat the solid phase and one onto which the enzyme is attached. These antibodies can have the same specificity or can be directed against separate antigenic determinants. The color development is directly related to the amount of antigen present. The assay derives its name from the position the antigen occupies in the test. It is sandwiched between the unlabeled antibody attached to the solid phase and the enzyme-labeled antibody, which is added following addition of the antigen. A washing is done between each step to remove unbound reactants. An indirect double sandwich elisa may also be developed. In this method the solid phase is coated with antibodies specific to the large protein produced in species A (eg, rabbit). Serial dilutions of standards and sample(s) are added and incubated for a specified time. Unbound antigen is removed by washing and a fixed amount of specific antibody from species (eg, mouse) is added. Following incubation and washing an antispecies antibody labeled with an enzyme (eg, goat antimouse IgG labeled with horseradish peroxidase) is added. The labeled antibody is incubated for a specified time and then the excess is removed by washing. Substrate is added and color development is directly related to the amount of antigen present (24).

Agglutination. The principle of the passive hemagglutination test is based on the least amount of soluble antigen required to inhibit agglutination of red blood cells. This is the amount of antigen in the last tube that will give a wide ring agglutination pattern (known as a mat). This concentration is the amount in the last tube that will still cause a mat to be formed (25).

Methods based on the agglutination are semiquantitative procedures. It is typical to use a twofold dilution scheme. Therefore, the procedures can only yield results that reflect the dilution sequence. If the interval between antigen concentration were greater, then the inherence error would be greater. On the other hand, the narrower the range, the more accurate the assay (14).

Hemagglutination assays are easy to use and possess the desired sensitivity. However, they require large quantities of antiserum and some antibody preparations do not react with red blood cells. The biggest potential drawback to their use is there may be nonspecific interactions due to other proteins that are capable of causing agglutination (24). Latex particles are also used for agglutination assays.

Affinity Column. Columns are packed with gel that has antibodies attached to the beads. These antibodies are often monoclonal antibodies. An extract is poured over the column and the substance that the antibodies are prepared against binds to the antibodies. The desired substance is then eluted from the column and analyzed. If the compound is fluorescent, such as the mycotoxin aflatoxin B_1, the solution can be placed into a cuvette and the fluorescence measured in a fluorometer. The sample eluted may also be analyzed by HPLC.

Immunodiffusion

Single Radial Immunodiffusion. The polyclonal antiserum prepared against the desired antigen is uniformly dispensed in an agar gel (26). Standards of known concentration of the antigen and sample extracts are placed in wells cut into the agar. Diffusion of the standards and samples into the agar cause formation of precipitin rings. At any given time the diameter of the rings is proportional to the initial antigen concentrations in the wells. Microgram quantities of antigen may be detected but the method requires large quantities of antiserum. The assay does not lend itself for analysis of sparingly soluble antigens. However, gels that contain urea do enable analysis of such proteins (eg, gluten).

Double Diffusion. Although double diffusion (27) methods are not quantitative, they provide the ability to gain information about the immunochemical relationships of several antigens. An extract may be screened against several different antisera by placing the extract in the center well and the antisera in peripheral wells.

Immunoelectrophoresis. In this procedure the antigen is electrophoretically separated, to provide resolution of different antigenic components in the mixture, prior to immunodiffusion against the antiserum (28). The antigen is electrophoresed in a gel, then a trough is cut parallel to the direction of the separation. The antiserum is placed in the trough and the separated antigens diffuse toward one another, which results in formation of precipitates.

Rocket Immunoelectrophoresis. In this method (29) the antiserum is dispersed into the agarose gel. The antigen standards and sample extracts are added to the small wells cut into one end of the gel. The antigens are then electrophoresed onto the antibody containing gel. The assay is set up so that the antigen migrates with little or no antibody migration. The rocket-shaped precipitates that form in the gel have heights that are dependent on antigen concentration.

Quantitative Precipitin Techniques. For quantitative precipitin techniques (30) an insoluble complex forms following interaction between a soluble antigen and specific antibodies in solution. The amount of precipitate is then analyzed by protein assay. The method is fairly sensitive but the assay requires long times to obtain maximal precipitate formation. Biphasic antigen concentration versus precipitate curves also may occur.

Nonimmunoassay

Bacterial Receptor Assay. Bacteria have surface receptors that will bind various antimicrobial agents. An assay can be designed in which a hapten (eg, sulfonamides or aminoglycosides) radiolabeled with ^{14}C or ^{3}H (tritiated thymidine) competes with the unlabeled counterpart in a sample for the bacterial receptors. The assay provides the desired sensitivity for detection of low levels of the haptens. However, the test lacks the ability to detect specific drugs within a particular family. For example, presence of any one or all of the family of β-lactams will be detected (penicillin G, ampicillin, cloxacillin, and cephalopsorins). A confirmatory test would have to be performed to identify which specific drug or which combination is involved in the reaction. Furthermore, the test employs radiolabeled reagents, which may present a health hazard. This is not an immunoassay format.

FOOD CONSTITUENTS

Nonmeat Protein Additives

A variety of nonmeat proteins can be added to processed meats and other foods. The most commonly used include soy, milk, and egg proteins. The proteins are called binders when they are added up to 3.5% of the total protein weight and extenders when they are added up to 10%. Fillers, emulsifiers, and stabilizers are other terms used in the literature.

Nonmeat proteins may be added to foods for any of the following reasons: (*1*) to reduce product costs by adding extenders; (*2*) to prevent water or fat loss thereby improving cooking yield, (*3*) to act as a binder and improve meat slicing, (*4*) to enhance flavor (egg and milk proteins are commonly added for this use); (*5*) to improve water or oil binding as well as to improve the stability of water soil emulsions, and (*6*) to increase or maintain the protein content of a meat product, while adding water or salt to it (31).

Since the late 1970s use of extenders has significantly increased, but the amount of meat substitutes that would be consumed is not nearly as high as the amount predicted. This is probably due to consumers' lack of desire to purchase artificial foods as well as a lack of cost advantage of product on these meat substitutes in numerous countries.

Development of immunoassays for detection and quantification of nonmeat proteins in foods is necessary both for routine surveillance of food products for adherence to government regulations and for control of diets of individuals suffering from food allergies and intolerance. For example, in the United States, vegetable and milk proteins are not permitted in traditionally named products. However, in foods with proper labeling they are allowed. For example, milk and vegetable proteins are permitted up to 3.5%. Products containing greater than this amount must be classified as imitation. Use of nonmeat proteins has gained acceptance in the U.S. diet with up to 30% hydrated vegetable protein being added to the meats for school lunch programs and the military. Soy ground beef patties are sold in supermarkets (31,32).

Rapid detection methods for nonmeat proteins can find widespread use for testing by individuals who suffer from food allergies and food intolerance who must avoid the offending food(s). Symptoms of food allergy may be manifested in the skin, gastrointestinal tract, and cardiovascular system. IgE antibodies that are specific to food proteins are found in the blood serum and on the surface of mast cells. Following binding of the protein allergen to the mast cells, allergic mediators are released that cause the allergy symptoms (33).

Grain and legumes such as soy, milk, and eggs can cause both food allergies and intolerance (34). Gastrointestinal disturbances are common to both food intolerance and hypersensitivity. Other symptoms include anemia, behavioral changes, and protein-losing enteropathy. Gluten-sensitive enteropathy is the symptom of food intolerance suffered by celiacs.

Nonimmunological methods for detection of nonmeat protein additives include microscopic and histological methods (32) and electrophoretic and chemical methods, based on amino acid analyses of peptides or analyses of nonprotein components characteristic of particular vegetable sources (35). These methods require expensive equipment and extensive sample preparation, so that immunological methods have been developed for detection of nonmeat protein additives.

Immunological methods other than RIA and EIA have been developed for detection of food proteins (Table 1). Agglutination assays exist for detection of food antigens. Hemagglutination inhibition assays are used for analysis of meat extracts and nonmeat proteins (36–38). These tests are quick and fairly sensitive but are only semiquantitative and may have nonspecific interactions due to nonprotein components in the food extracts.

Hemagglutination methods have been developed for detection of soy in meat products (39–42). The major problem with applying this methodology is that the need exists for establishing complex calibration curves to quantitate soy proteins in meat products. For example, good results were obtained with this method when meat products were heated only to 75°C but not when products were heated to temperatures exceeding 100°C (39). One group was able to quantify soy protein in meats cooked at 15°C (41) and another used indirect hemagglutination to detect <1% soy protein present in cooked sausages (42).

Hemagglutination methods have been used for quantifying milk protein in meat products. The method was only qualitative when autoclaved meats were tested (43). A similar method was developed with antigen-coated latex particles instead of red blood cells (44).

Milk proteins are considered easier to extract than other proteins (ie, gluten). However, some harsh solutions have been used for extraction prior to analysis using an immunochemical technique. Some of these extractants are 7 *m* urea (45) urea—mercaptoethanol (46,47) formic acid-methanol (48), and performic acid (49). Urea, TRIS, tricine, and performic acid treatment have been found to reduce the artifacts attributed to liver constituents in sausages

Table 1. Agglutination Gel Immunodiffusion Assays for Detection of Food Proteins

Assay	Food protein	References
	I. Agglutination	
Hemagglutination inhibition	Soy protein	40
	Soy protein in meat products	39,41
	Soy protein in cooked sausages	42
Hemagglutination	Milk protein in autoclaved meats	43
Latex agglutination	Milk protein in autoclaved meats	44
Hemagglutination inhibition	Milk protein in liver sausages	50
Indirect hemagglutination	Egg proteins in meats	43
	II. Quantitative	
Precipitin technique	Wheat seed protein	54
	Whey proteins in nonfat dry milk and buttermilk	55
	Casein and whey in foods	53
	III. Immunodiffusion	
Single radial immunodiffusion	Soy, casein egg white	56,57
	Soy in meat products	58,59
Double diffusion	Soy in	56,60–64
	Soy and wheat protein in meat	65
	Soy in meats, grits, flours, concentrates, isolates, and extrudates	66
	Milk proteins	59,64,67
	Whole egg or egg whites in meat	41,59,68
	Gliadins	68–71
Immunoelectrophoresis	Soy in meat products	63,72,73
	Milk protein in meat samples	74,75
	Origin of milk in	76

when hemagglutination is used for determining milk protein in liver sausages (50).

Another immunological method used for food analysis is the quantitative precipitin technique. An insoluble complex forms when soluble food antigens react with specific antibodies in solution. The precipitate is further analyzed by protein assay (30). Methods based on the turbidimetric or nephelometric analysis of suspended antigen–antibody complexes have become more popular (51–55). These methods are also automated for clinical use (31).

A large number of immunodiffusion assays have been developed for analysis of nonmeat proteins. A detailed list of references is provided in Table 1. It should be noted that although these methods have found widespread use for soy analysis there are a number of disadvantages encountered (77). (1) The diffusion in gel method is slow and is only suitable for limited numbers of samples. (2) Low recoveries of protein antigens are common because mild sample extraction conditions are used to decrease loss of protein antigenicity. (3) None of the antisera developed are able to recognize all of the soy protein derivatives. In addition, due to the low aqueous solubility and low molecular charge of gluten proteins, rocket immunoelectrophoretic methods have not found wide application in gluten detection. One qualitative method has been described (78). The classical methods based on antigen diffusion in gels are designed for optimal use with water-soluble proteins. Gluten and related cereal storage proteins are insoluble in aqueous buffers. Acidic buffer conditions have been used to analyze gluten extracts by double immunodiffusion (79) and 3 *M* urea containing gels have been used for gluten analysis by immunoelectrophoresis (80). Isotopic and nonisotopic ias have been described for nonmeat protein additives (Table 2).

An immunofluorescent method was described for detecting soy protein in meats. It was labor intensive with respect to sample preparation drying and fixing (81). The first RIA and EIA for soy protein determination were elucidated in the early 1980s (82,83). Another RIA was developed that detected native and formaldehyde-treated soy protein (83). There have been a number of EIAs developed for soy detection (Table 2). For example, a two-step antigen competition immunoassay was developed (83). The same investigators then analyzed a set of model beef burgers that contained known quantities of particular soy isolate (86). The effectiveness of the method has been tested in two collaborative studies (116,117). When compared to SDS–polyacrylamide gel electrophoresis (SDS–PAGE), neither method gave any false positives and similar interlaboratory variations were observed. More accurate determinations were obtained with the EIA with a range of flours, concentrates, isolates, and textured products, while the electrophoretic method showed less intralaboratory variation.

Table 2. Isotopic and Nonisotopic Immunoassays for Detection of Nonmeat Proteins

Assay	Sample	Ref.
Fluoroimmunoassay	Soy in cooked meats	81
Radioimmunoassay	Soy, gliadins	82–84
Enzymoimmunoassay	Soy in meat products	85–93
	Casein	94
	Hydrolyzed milk proteins in sausages	95,96
	Ovalbumin	97
	Gliadins	98–115

A great deal of information about the advantages of using ELISA for soy analysis can be gathered from reading the various references listed in Table 2. These articles describe the time saved when using ELISA over other methods, the extraction methods devised to overcome poor recoveries and immunoreactivity differences observed with different forms of soy proteins, attempts to develop a simpler ELISA than the two-step antigen competition immunoassay, assays designed on nitrocellulose solid phase, and the details of collaborative studies.

EIAs have been successfully developed for whole caseins (94) as well as those for detection of hydrolyzed milk protein in cooked meat products (95,96). Detection of ovalbumin in canned mushrooms was achieved by indirect EIA and SDS–PAGE immunoblotting techniques (97). The protein is widely used in canned mushrooms because it enhances water retention and drained weight of the product.

Wheat gluten is one of the least expensive sources of edible protein. It is also able to bind meat fragments in processed meats and is used as an extender in foods. Due to the large number of individuals who cannot tolerate wheat gluten and related proteins from rye and barley a great deal of effort has been spent on developing ELISA for rapid detection of gluten and related cereal proteins. Application of elisa to gluten analysis has become extensive. Gluten is the protein-rich mass left after a wheat flour dough is washed to remove starch, water, salt, and soluble proteins. It is composed of several dozen protein components that total over 80% of the flour protein. There are two major fractions, gliadin (mol wt 30,000–75,000) and glutenin (mol wt 20,000–100,000). The rhelogical properties in wheaten dough come from gluten (118). Much of the immunological work has been carried out on gliadin due to its greater extractability and solubility than gluten (Table 2).

Because gluten exhibits low aqueous solubility and low molecular charge, electroimmunodiffusion techniques have not found wide applicability. There have been numerous descriptions of RIAs and EIAs for gliadin detection (Table 2). A major problem with the EIA and immunodiffusion tests is their unsuitability for analysis of cooked or processed foods (31). The high-mobility gliadin components, α- and β-gliadin, are probably the most toxic gliadins in gluten intolerance, and their extractability and immunoreactivity is decreased with heating (105). However, cooked foods containing gluten retain their toxicity (31). Monoclonal antibodies have been produced to Ω-gliadins (101). These antibodies enabled detection of gliadin in foods after cooking (100°C for 100 min).

The need for improved extraction methods has been addressed. Seventy percent ethanol has been used to extract gluten from uncooked foods (106,109,110). This solvent is fairly selective for gliadin and may help to reduce matrix affects from other food proteins. Extraction of protein with urea has been found to provide more sensitive gliadin detection than aqueous alcohols (101). The appropriate solvent for each test format must be determined at the time of assay development.

With the trend toward governmental regulations and industry controls on food components and quality, it is possible that the amount of growth observed in antibody-based medical diagnostics that has taken place within the last decade will be equaled in the next decade in antibody-based food analysis. Food analysis has traditionally been viewed as a discipline derived from analytical chemistry. Now with the capabilities presented by EIA, food analysis will use commercially available kits for routine testing of large samples (31).

Meat Protein Additives

There have been IA developed for detection of meat proteins. Meat is the portion of animal flesh ingested as a food source. Biochemical and biophysical changes occur in the muscles following the death of the animal. These edible tissues are composed primarily of proteins that are also the most important constituents of meat. The major meat proteins are the myofibrillar, sarcoplasmic, and connective tissue proteins (24). There are similar proteins in all the muscles of all species with the interspecies differences being extremely small. Therefore, the ability to differentiate meat from different species by visual, histological, and empirical chemical analyses is nearly impossible to do (24).

At slaughterhouses and meat-cutting plants it is standard procedure to do hot deboning on the carcasses; which are then packaged as primal joints, chilled, aged, and possibly frozen prior to shipment to retail outlets. This process has been touted as a major cost-saving measure with respect to energy costs, chiller space, and weight loss resulting from the evaporation. These savings produce greater yields of salable meat (119). Using these practices the anatomical differences in the skeletons do not permit identification of the species of origin of the meat (24).

The need exists for reliable analytical methods to monitor meat speciation to insure fair trading and for consumer protection. It is a well-known fact that eating habits differ from one society to the next largely due to cultural and religious practices. A food item considered a delicacy in one culture may be viewed as a taboo by another. Therefore, it is possible that low-cost meats may be added as substitutes to frozen meat that are not considered food by some societies. The methods must be able to identify both fresh and heat-treated meat and meat products. Immunoassays satisfy these requirements (24).

A number of immunoassays have been developed for detection of meat proteins in fresh meats (Table 3). Immunoelectrophoresis has been used to identify the presence of beef and pork in fresh meat products (120). Different species of fish have also been identified using this electrophoretic method (121). This technique provides excellent resolution of antigenic components due to their differential migration in an electric field.

Immunodiffusion (139) methods have found wide application in analysis of fresh meat (122–128). Hemagglutination inhibition tests were developed for detection of adulterants in meat (129,130). This method was also used to test insect blood meats and identify the species of animals that these hematophagous insects had fed on (131).

Table 3. Immunoassays for Detection of Free-Meat Proteins

Assay	Sample	Ref.
I. Gel Immunodiffusion		
Immunoelectrophoresis	Beef, pork, fish	120,121
Double immunodiffusion	Meat (cow, pig, sheep, horse)	122–128
II. Agglutination		
Hemagglutination inhibition	Meats, insect blood meals	129–131
III. Labeled Immunoassays		
Radioimmunoassay	Cattle meat	132
Enzymoimmunoassay		
Indirect competitive	Horse meat	133
Indirect noncompetitive	Horse meat, beef	134–136
Sandwich elisa	Meat specification	137,138

Rias have not been extensively used for meat protein detection due to the short shelf life of the reagents, expensive equipment required for analysis, potential health hazards of exposure to radioisotopes, and governmental regulations pertaining to radioisotope use. A RIA was developed that could distinguish cattle meat from sheep, donkey, pig, horse, and kangaroo meats. Contamination as low as 5% could be determined (132).

A number of different eias have been developed for fresh meat speciation. Current slaughter methods leave residual blood, as much as 1.5–2.0% in the muscle tissue (140). Some of the blood proteins, namely the albumins and immunoglobulins, have served as immunogens for producing antibody-based meat species testing kits (133,141). An indirect competitive ELISA for meat speciation that enabled detection of 0–40% horse meat in beef has been developed (133).

Competitive ELISAs have been employed for quantitative determinations, whereas qualitative species identification of fresh meats have been obtained with indirect noncompetitive ELISAs (133–136,142,143). Horse meat was differentiated from beef and mutton with an indirect ELISA in which an antibody raised against horse serum albumin was used. Adulteration of beef with goat and kangaroo was also elucidated with an indirect ELISA.

The ELISA variants are not applied for quantification because they have relied on the presence of serum antigens that can vary greatly in meat and are not indicative of the meat content (24). Stunning and bleeding methods greatly affect the content of serum proteins in the muscles (142). The stability of the antigen with time is also a factor (143). Therefore, antigens that are genuine muscle components that are indicative of the meat content must be developed (124).

Sandwich ELISAs also permit speciation of meat (137,138). Adulteration of raw meat mixtures with pig meat was identified (129), and improved speciation of raw meat was achieved using this ELISA format. Levels as low as 0.5% (137) and 17% adulteration have been detected (138).

The task of identifying the species of origin of cooked or heat-processed meats becomes more difficult than speciation of fresh meat. Heating denatures or changes the species-specific epitopes that are used for identifying the species of fresh meat (24). When meat is heated at 45°C denaturation of myofibrillar and sarcoplasmic proteins occurs (144), and at 60°C collagen shortens and changes into a more soluble form (145). As the temperature is raised from 60 to 90°C, the percentage of collagen solubilized increases gradually (146). Therefore, to develop an immunoassay that permits identification of heat-treated meat, antigens must be isolated that will retain their species specificity following the heating process. They must not be lost in cooking juices and must be antigenic and immunogenic.

The first EIA developed (147) enabled detection of origin of heat-treated meats. Kangaroo meat was detected in frankfurter-type sausages and dry sausages. Antisera produced against thermostable muscle antigens (TMAs) were used to develop an indirect ELISA that enabled the detection of origin of heat-treated meats from closely related species (148). Immunodiffusion tests have demonstrated that TMAs are species, organ, and tissue specific (149–151).

Monoclonal antibodies were used to achieve speciation of heat-treated meats because cross-reactions could be eliminated. Monoclonal antibodies were produced against heat-treated kangaroo and cattle meats, which differentiated the two kinds of meat in an elisa (152). Use of monoclonal antibodies has the advantage of being specific without requiring absorptions needed when polyclonal antisera are used; therefore, the future trend would be to make use of monoclonal antibodies for speciation of meats from closely related species.

NATURAL CONTAMINANTS

Natural Toxicants

There are natural components of food that can potentially be toxic to consumers; however, attention has been largely focused on detecting food additives and contaminants and not these natural toxicants. It is true that many of the plant and animal portions that might cause toxic effects have over time been avoided by humans or are deemed safe for consumption when they are prepared in a certain manner. When normal practices are followed the naturally occurring toxins do not present acute problems to the consumer. It is when these normal practices are changed that problems can arise (153).

Until recently a dearth of information existed concerning toxin occurrence, toxicology, and metabolism following ingestion of foodstuffs containing natural toxins. One of the biggest problems has been the dilemma presented by analysis. Chemical methods (HPLC and mass spectroscopy) have helped improve toxin analysis. Immunoassays have greatly furthered this effort. They can be highly specific, which is desirable, because compounds with very similar structures will exhibit widely different toxicity potential. The high specificity will eliminate chances of misclassification of compounds. Immunoassays can also be very sensitive, which would permit detection of toxin metabolism following ingestion. The ability to quickly test large numbers of samples is also desirable. It is common for natural toxicants to exhibit nonhomogeneous patterns of distributions so that multiple subsamples must be tested to obtain data about overall concentrations. The multiple samples can also be tested at low cost (153).

A microtiter-well elisa for quinine detection in soft drinks has been developed (154). It is, of course, added as a bittering agent. Although the risk from consumption of soft drinks containing it is low, quality-control procedures for its analysis are important to the soft drink industry. The ability to test easily for the compound permits the industry to provide greater control on a potentially dangerous substance, which will boost consumer confidence in the product containing the substance.

A number of immunoassays have been described for caffeine which is found in a variety of food products, namely those containing teas, coffee, and chocolate. A radioimmunoassay and a substrate-labeled fluoroimmunoassay for caffeine detection have been described (155,156). Assays have been developed for analysis of naringin and limonin, which are bitter substances isolated from citrus fruits. The ria for naringin (157) and the elisa for limonin (158) permit the analysis of the distribution and metabolism of the compounds in the fruit. Assays of this nature may also be of use to citrus growers for quality-control testing of both raw materials and finished products.

Many toxic compounds are found in the potato, *Solanum tuberosum*. One group is collectively known as the potato glycoalkaloids. Ninety-five percent or more of the tuber is normally composed of α-chaconine, α-solanine, and derivatives of solanidine (159).

A ria, the first immunoassay described for glycoalkaloid analysis (160), employed an antiserum that did not recognize α-chaconine and α-solanine. A more useful elisa was developed (161) that enabled detection of all three compounds. It has been shown to have good correlation with conventional chemical methods (162,163). Immunoassays have been developed for testing glycoalkaloid concentrations in body fluids. Solanidine can be detected in serum (164,165) and saliva using a RIA.

Immunoassays that permit lycoalkaloid detection can be used by plant breeders to screen tubers of all potential new varieties. Testing can be performed on the retail level as well to prevent glycoalkaloid concentrations from exceeding 20 mg/1.0 g (166).

Mycotoxins

Mycotoxins are a chemically diverse group of secondary fungal metabolites that are toxic following ingestion or environmental exposure (167). Mycotoxins are haptens with molecular weights of 300–400. Many of these molecules are nonpolar and exhibit low solubility in water. The fungi that produce the mycotoxins are ubiquitous. They can grow and elaborate the toxins in numerous agriculture commodities either in the field or following harvest during storage and processing. The factors that influence toxin elaboration are temperature, relative humidity, and moisture content of the substrate. As a result of their toxicity, mycotoxins can cause severe economic losses to farmers and livestock producers as well as pose a threat to humans consuming contaminated foods. Affected agricultural commodities include corn, wheat, peanuts, cottonseed, rice, sorghum, almonds, walnuts, and milk (168).

The target tissues that can be affected by mycotoxins are the liver, kidney, spleen, gastrointestinal tract, lymphoid tissue, reproductive organs, skin, and nervous system. In the United States the most important toxins are the aflatoxins, zearalenone, the trichothecenes and to a lesser extent ochratoxin. These toxins are produced by species of *Aspergillus* and *Fusarium* and cause a variety of health problems to livestock and humans (Table 4).

Classical methods of analysis of mycotoxins include TLC, HPLC, GC, and mass spectroscopy (MS). TLC methods have found wide use because they are easy to perform, are inexpensive, and are relatively sensitive. Many of the mycotoxins naturally, fluoresce and absorb in the UV region, so that TLC methods (169–171) while TLC determination of mycotoxins are frequently only semiquantitative and require elaborate sample cleanup. GC and HPLC methods offer sensitive quantitative determinations of mycotoxins (172–174). Sample cleanup is also extensive and labor intensive prior to analysis by these methods. Due to these drawbacks, development of immunochemical methods for mycotoxin detection have become popular.

The Aflatoxins. The majority of work on mycotoxins has focused on the aflatoxins, the family of difurancoumarins, which act as hepatotoxins and hepatocarcinogens (175). Aflatoxin B_1 (AFB_1) is reported to be the most prevalent and most potent carcinogen. Results from the modified salmonella mutagenesis test show that AFB_1 is about 30 times more mutagenic than AFG_1 or AFM_1, which are more mutagenic than AFB_2, AFG_2, or AFM_2 (Fig. 1) (176). Similar patterns exist for hepatotoxicity and carcinogenicity. The biological effects of the toxin are attributed to the difuran ring with 8:9 vinyl ether, while the coumarin ring absorbs at 360 nm and fluoresces. The Food and Drug Administration (FDA) has set an action level of 20 ppb total aflatoxins in foods or feeds.

AFB_1 can be metabolized by hepatic enzymes (*in vivo* or *in vitro*) to different metabolite (173,174). The carcinogenicity results from the reaction of the 8,9 epoxide with cellular nucleophiles (RNA, DNA, and proteins). Other detoxification pathways involve reduction or hydroxylation, which form more polar compounds that are then removed from the animal. One hydroxylated metabolite, AFM_1, is also hepatocarcinogenic and will be found in milk of lactating animals that have ingested AFB_1 contaminated feed (177). The action set by the FDA for AFM_1 in dairy products is 0.5 ppb.

The Trichothecenes. Many of the mycotoxins produced by *Fusarium* species and other fungi (*Cephalosporium*, *Verticimonosporium*, *Myrothecium*, and *Stachybotrys*) are called trichothecenes (178). These mycotoxins are tetracyclic sesquiterpenoid molecules, all possessing an epoxy group. T-2 toxin and deoxynivalenol (DOH), also known as vomitoxin (Fig. 2), are frequently identified in *Fusarium* culture filtrates. The toxins cause a strong dermatitis reaction when applied to the skin of shaved guinea pigs, characterized by severe local irritation, inflammation, desquamation, and necrosis (175).

There have been more than 148 trichothecenes identified, but only four have been unambiguously identified in naturally contaminated food samples. Trichothecenes are difficult to analyze especially because, unlike the aflatoxins, they do not possess any useful spectroscopic characteristics. Analysis of the trichothecenes have been accomplished using TLC (179), HPLC (180,181), GC (182), and MS (183,184).

The trichothecenes are believed to function as protein synthesis inhibitors. The initiation or elongation termination steps of protein synthesis are believed to be affected.

The Zearalenones. *Fusarium* species, primarily *F. Sraminearium*, produce zearalenones which are resorcylic lactones (Fig. 3). The parent compound is zearalenone, which does not exhibit acute toxicity. However, high levels of zearalenone can interrupt normal estrus in sows and cause vulvar prolapse and enlargement of the uterus. In young males, the toxin causes testicular atrophy and mammary gland hyperplasia. Physiological response in swine occur when the zearalenone level in corn used for feeds exceeds about 1 ppm (185). Rats, mice, sheep, monkeys, and humans can be affected by zearalenone (168).

The Ochratoxins. Ochratoxin contamination of grains and food products is less of a concern in the United States than contamination of products with aflatoxins, zearalenones, and trichothecenes. It is frequently detected in Scandinavian and Balkan countries and occasionally in the United States in barley, corn, wheat, oats, rye, peanuts, hay, and green coffee beans (186). Ochratoxins are a group of structurally related metabolites produced by *Aspergillus ochraceus* and related species as well as *Penicillium viri-*

Table 4. Abbreviated List of Major Mycotoxins and the Potential Health Hazards Following Exposure

Mycotoxin	Fungi	Potential Health Risk
Aflatoxins	*Aspergillus flavus, A. parasiticus*	Mutagenic, teratogenic, hepatotoxic, and carcinogenic
Trichothecenes	*Fusarium* sp.	Feed refusal, emesis, immunotoxicity, gastrointestinal tract hemorrhaging, alimentary toxic aleukia
Zearalenones	*Fusarium roseum* and other *Fusarium* sp.	Estrogenic, infertility, and reproductive problems in cattle, swine, and poultry
Ochratoxins	*Aspergillus* sp. and *Penicillium* sp.	Nephrotoxic, teratogenic, and immunotoxic

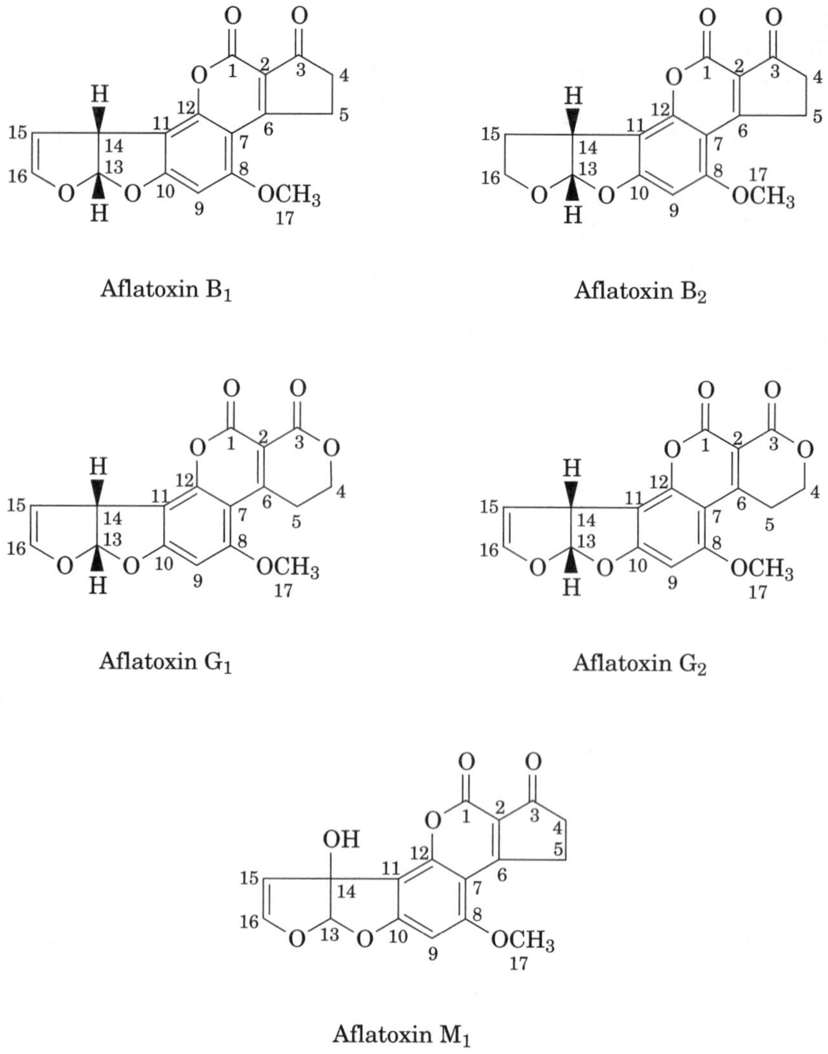

Figure 1. Structures of important aflatoxins.

dicatum and other *Penicillium* species. Ochratoxin A (OA) is the major mycotoxin in this group (Fig. 4). It has believed to cause endemic kidney disease in swine and poultry in Denmark and Sweden. When the hogs were fed diets containing 200 ppb OA, their kidneys became pale and swollen. The proximal tubules and interstitial cortical fibrils were observed to undergo atrophy (187). The toxin has been implicated in causing the human disease known as Balkan endemic nephropathy (187,188). Occurrence and toxicology of the toxin were recently reviewed (179,189).

Mycotoxin Occurrence in the Food Supply. It is possible for feeds and foods to become contaminated with mycotoxins prior to harvest, during the interval between harvest and drying and during storage. There are some mycotoxins that are only produced in the field (ie, ergot). Mycotoxins can, therefore, be considered to occur naturally in raw agricultural products, processed foods and imported products (190).

There are a number of documented occurrences of mycotoxins in raw products (190–194), processed foods (195–202) and imported products (191,192,199). Most of the information available is for the aflatoxins with less information available for ochratoxin A, trichothecenes, and zearalenone. An extensive review was published that described the effects of food processing on selected toxins (200). The stability of the mycotoxins during food processing is affected by a number of factors such as the type of food, the type of processing used, the additives, moisture content, and the amount and type of contamination (189). The mycotoxins display different degrees of variability in foods under processing conditions. It has been demonstrated that the aflatoxins are stable to moderately stable in the majority of processes. However, they are unstable when alkaline conditions and oxidizing steps are used. DON is stable during bread baking, while ergot alkaloids have been shown to be partially destroyed (190).

Although a number of mycotoxins could be identified in imported foods and feeds, the FDA only routinely tests for aflatoxins. Under the FDA compliance programs, the FDA tests about 300 samples of imported foods (eg, almonds, brazil nuts, peanuts and peanut products, sesame seeds, sunflower seeds, nutmeg, coffee beans, marzipan, cornmeal and flour, cocoa products) and about 200 feed samples (eg, corn, mixed feeds, cottonseed meal, copra pellets, and sorghum). Summaries of analyses from 1982 to 1996 are

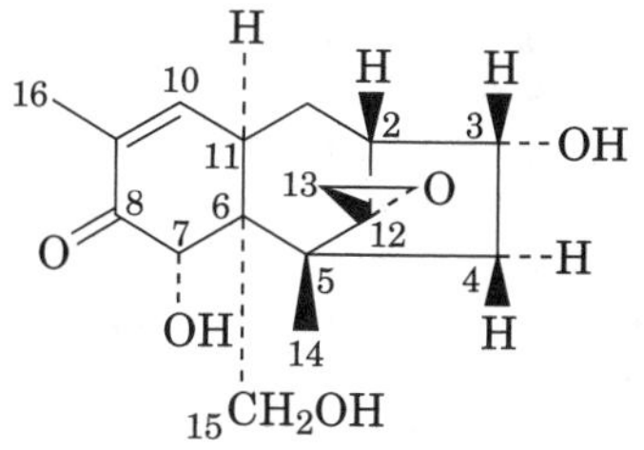

Deoxynivalenol

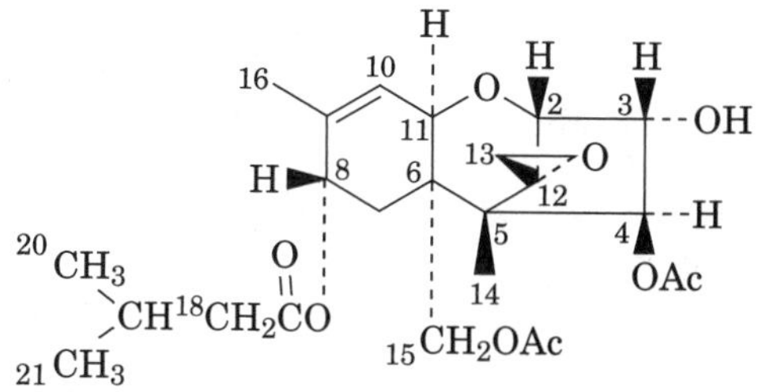

T-2

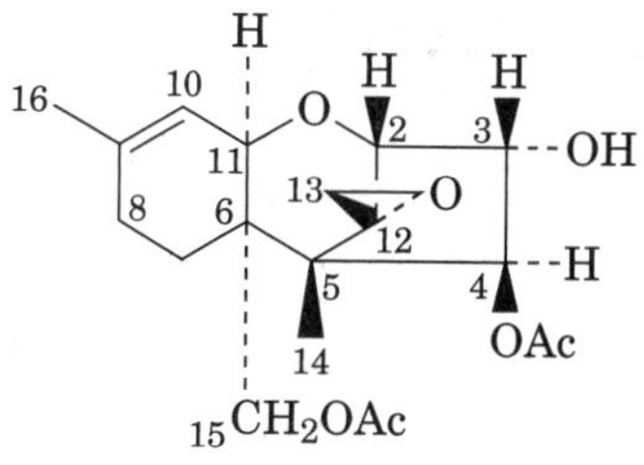

Diacetoxyscirpenol

Figure 2. Structures of three trichothecanes.

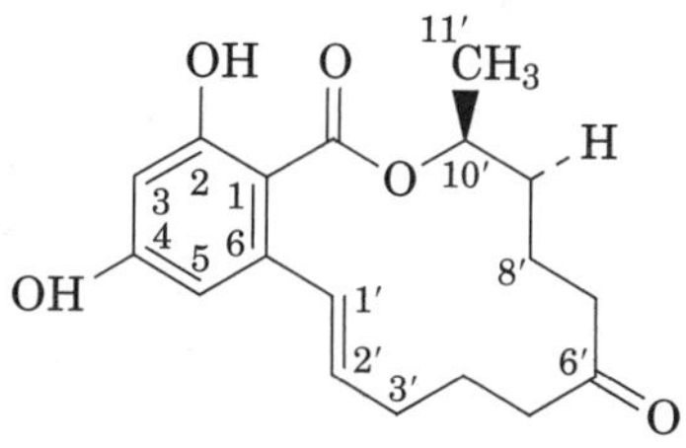

Figure 3. Zearalenone.

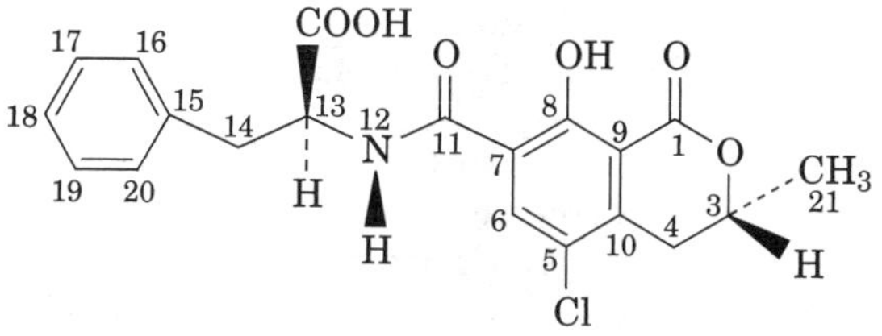

Figure 4. Ochratoxin.

provided (198), and information is also available on worldwide occurrence of mycotoxins. The data were collected from the FAO/WHO/UNMEP food contamination monitoring program in addition to other sources (198).

Regulatory Control of Mycotoxins. Although a wide range of mycotoxins have been found in the environment associated with human food and animal feeds, there is only formal regulation of aflatoxins by the FDA. The FDA does not regulate the other known mycotoxins based on the observed levels, incidence, estimated consumption, and toxicological profiles (190). The aflatoxins are regulated under the Food, Drug and Cosmetic Act Section 402(a)(1). In fact, in the United States when good manufacturing practices (GMPs) are followed, aflatoxins are regarded as unavoidable contaminants in food and feeds (191). When food is slated for human consumption, the action level of 20 ppb total aflatoxins is enforced, while for milk the level is much lower, 0.5 ppb AFM_1. The action level for aflatoxins in feeds is also 20 ppb. However, levels of 100 ppb are permitted in feeds slated for consumption by breeding cattle and swine and mature poultry. Levels of 200 ppb are allowed in food corn destined for finishing swine ($\leq$100 lb) and 300 ppb are allowed for aflatoxins in cottonseed meal incorporated in feeds and also for corn fed to feedlot beef cattle (190).

In Europe, regulatory control of mycotoxins is governed by the European Economic Communities (EEC) a body composed of 12 countries (Belgium, France, Germany, Italy, Luxembourg, the Netherlands, Denmark, the UK, Ireland, Spain, Portugal, and Greece). The formation of the EEC began with the Treaty of Rome in 1957. In 1987 the treaty was modified due to the establishment of the Single European Act. It described the stepwise establishment of the Common European Market by the end of 1992. Harmonization of food laws must be accomplished by the EEC to achieve this objective. The control of contaminants is one issue that must be addressed (203–207).

There have been increasing numbers of publications of immunochemical methods for detection of the various mycotoxins. It is now common practice to use ELISA to screen commodities for presence of mycotoxins (208). RIAs are described in some of the earlier publications. The majority of tests have been developed for aflatoxin detection (Table 5). The most attention has been focused on the aflatoxins due to their strict regulation by governmental agencies in the U.S. and other countries. There is a wealth of information provided in the references listed in Table 5. There are papers that describe assays that employ monoclonal antibodies (211,216,221) as well as those that employ polyclonal antibodies (209,210,212,215,217–220). There are descriptions of the chemistry involved in conjugate antigen production (208–224). The way in which the aflatoxin was derivatized and coupled to the large carrier molecule influences the specificity of the antibody (the antibody's ability to recognize and bind to only aflatoxin B_1 if other metabolites are bound as well). There are papers that provide comparisons of ELISAs and chemical methods when certain commodities are analyzed for aflatoxins (228) as well as comparisons of various commercial kits (229). There are also reports from different collaborative studies (225–227).

Table 5. List of Selected Immunochemical Methods for Mycotoxin Detection

Toxin	Method	References
Aflatoxin B_1	RIA	209–212
	ELISA	212–233
	Affinity column	231,234
Aflatoxin M_1	RIA	235,236
	ELISA	221,237–240
	Affinity column	241,242
Zearalenone	RIA	243
	ELISA	219,244–249
	Affinity column	249
T-2 toxin	RIA	250,251
	ELISA	230,252–258
Deoxynivalenol	RIA	259
	ELISA	260,261
Ochratoxin A	RIA	262,263
	ELISA	264–268

Both monoclonal (211,237) and polyclonal antibodies were produced against aflatoxin M-1. RIAs (235,236), ELISAs (221,237–240), and an affinity column cleanup using monoclonal antibodies (234,242) followed by direct fluorescence measurement of aflatoxin M-1 in raw milk (240) have been described. In some cases milk could be tested without any cleanup (221,235), whereas, some methods required extraction of the samples prior to analysis (235,238–242).

Zearalenone can be detected by RIA and ELISA (243,249) (Table 5). An affinity column was developed with a monoclonal antibody (247,249) and used to clean up milk samples prior to analysis by ELISA. The monoclonal and polyclonal antibodies described in these references (243,249) exhibit varying degrees of cross-reactivity to α- and β-Zearalenol. Zearalenone is metabolized to these analogues and they have been detected in trace levels in bovine milk after cows were fed feed containing high levels of zearalenone (269,270). Detection of these analogue using the antibodies is an attractive feature.

There have been immunoassays published for detection of various trichothecenes (Table 5). Both RIA and ELISA have been described for T-2, some of which employ monoclonal antibodies (230,252–255) and some employ polyclonal antibodies (256–258). A recent report was made of an immunochromatographic method (271) for isolation of group A trichothecenes of which T-2 is a member. These various antibodies exhibit differing degrees of cross-reactivity to 3′-OHT-2 and other metabolites. Detection of these metabolites in biological systems would be useful, because they are found in significant amounts when T-2 toxicosis occurs (272–274).

To date there have been few reports of antibodies produced to DON and their application to RIA or ELISA (Table 5). Two investigators have demonstrated repeated success at developing IAs for mycotoxin detection. Chu (University of Wisconsin at Madison, Food Institute) is regarded as one pioneer in mycotoxin analysis with emphasis in the area of immunoassay development. Another is Pestka (Michigan State University, Department of Food Science and Human Nutrition) a former postdoctoral associate in Chu's laboratory. The successes in developing antibodies to DON have been reported from their laboratories. Successful immunogens were prepared for immunization of mice (261) and rabbits (259,260) which enabled detection of DON by RIA and ELISA. Table 6 also provides a list of IAs developed for ochratoxin A detection. Ochratoxin A can be detected in barley (265) and wheat (266).

A number of companies sell kits that enable detection of various mycotoxins. The majority of commercial kits sold are for aflatoxin testing. The different companies claim differing cross-reactivities of their antibodies employed in their aflatoxin B_1 detection kits. Aflatoxin G_1, G_2, and B_2 are also detected in many of the kits. Aflatoxin B_1 detection ELISA kits are sold by the following companies: Neogen Corp. (Lansing, Mich.), International Diagnostics (St. Joseph, Mich.), Idexx (Portland, Maine), Environmental Diagnostics Corp. (Burlington, N.C.), and Transia (Lyon, France). Affinity columns are sold by VICAM (Somerville, Mass.) and Oxoid (Columbia, Md.). Neogen and Idexx sell AFM-1 ELISAs. Aflatoxin M_1 affinity columns are sold by VICAM and Oxoid. Environmental Diagnostics Corp. and Ube (distributor for the Japanese company is Wako Chemicals, Dallas, Tex.) market ELISA kits for ochratoxin A detection. Neogen Corp. and Environmental Diagnostics market elisa for zearalenone and T-2 toxin testing. Neogen Corp. the only company that markets an elisa kit that enables the user to test for the presence of DON in samples.

MICROBIAL CONTAMINANTS

Consumer in recent years have expressed considerable concern over the safety of the food supply. Concern has been aimed at the presence of hormones and drugs in meats, preservatives found in food, pesticides, microbial contaminants, and food additives. Reports last year of cyanide in grapes and Alar in apples have fueled the level of concern. This issue of the public's concern over the quality of food was addressed at a conference (the International Conference on Issues in Food Safety and Toxicology) held in the spring of 1990 at Michigan State University (MSU) East Lansing, Mich. The conference was sponsored by the U.S. Department of Agriculture (USDA) and MSU's Center for Environmental Toxicology. From various presentations it became apparent that the consumers' perceptions of risks from chemical agents in food differs drastically from the experts' opinions. The experts rank the risk from chemicals below microbial risk or nutritional considerations. The general public perceives the risks to be greatest from the chemical agents. It would be a gross omission to prepare a publication of this nature without providing a list of references for those individuals interested in reading more about IAs developed for detection of pathogenic microbes, particularly bacterial pathogens found in food. Detection of *Staphylococcus aureus* enterotoxins, *Clostridium perfringens* type A toxin, *Clostridium botulinum* toxins, *Bacillus cereus* enterotoxins, *Yersinia enterocolitica* toxin, *Escherichia coli* toxins, *Campylbacter jejuni, Aeromonas* toxin, and some streptococcol toxins would be desirable with IAs. *Salmonella* and *Listeria monocytogenes* can also be detected in foods using IA. Table 6 provides a list of

Table 6. Selected Immunoassays for Toxin Detection

	Toxin	References
	Immunodiffusion	
Single-gel diffusion	*Staphylococcus aureus*	275,276
	Clostridium perfringens	277
	Clostridium botulinum	278
Single-radial immunodiffusion	*S. aureus*	279
Double-gel diffusion	*S. aureus*	275,277,280–283
	Electroimmuno diffusion	
	S. aureus	284
	Agglutination	
Hemagglutination inhibition	*S. aureus*	285–287
Reverse passive hemagglutination	*C. bolutinum*	288
Latex agglutination	*S. aureus*	289
	C. perfingens	290
	Radioimmunoassay	
	S. aureus	291–305
	Escherichia coli	306,307
	C. botulinum	308,309
	C. perfingens	310
	ELISA	
	S. aureus	311–327
	C. botulinum	328–338
	C. perfringen	339–344
	E. coli	345–350

selected publications of IAs for detection of bacterial toxins. The table lists the type of assay, the species of bacterium that the assay was developed to detect, and references. It does not describe which *S. aureus* toxin or toxins are detected in the assays. Some assays only detect enterotoxin B, while others detect A and B or A, B, C, and D. To provide information on other agents that are considered to be endangering the food supply only a general list of sources has been provided. Information on risks associated with vehicles of food-borne pathogens and toxins is available (353).

To detect *Salmonella* in foods, investigators have relied on culture procedures that enable growth, selection, isolation, and identification of isolated organisms (354). The length of time needed to obtain a negative reading has always posed a problem. To overcome this disadvantage, numerous rapid testing methods have been proposed. They have not found wide acceptance due to problems with sensitivity and specificity (355). Two reports indicated the potential for use of EIA for screening for *Salmonella* (356,357) in foods. Since then there have been other reports of *Salmonella* detection using EIAs (358–363).

Listeria detection using an immunobased test requires highly specific antibodies because *Listeria* species have been shown to be antigenically related to other gram-positive organisms, namely *Staphylococcus, Streptococcus, Erisipelothrix*, and *Bacillus* species (364–368). Several groups have been successful at producing monoclonal antibodies to a genus-specific antigen (369–371).

There are several commercial kits available for toxin and pathogen detection. Oxoid Ltd. (Hampshire, UK) sells revised passive latex agglutination assays for detection of *S. aureus* toxins A, B, C, and D; *Clostridium perfringens* enterotoxin; *Vibrio chlorae* and *E. coli* heat-labile enterotoxin; and *S. aureus* toxic shock syndrome toxin. Organon Teknika Corp. (Durham, N.C.) sells Salmonella-Tek and Listeria-Tek which are ELISAs.

CHEMICAL AGENTS

Pesticides

Pesticides are used to produce high-quality, inexpensive food. With the use of pesticides, agricultural residues can be expected to occur in foods. Regulation of pesticides is shared by three federal agencies. The U.S. Environmental protection Agency (EPA) is responsible for registering or approving the use of pesticides and to setting tolerance, if residues will occur in food. The USDA is responsible for pesticides used on meat and poultry, and the FDA is responsible for enforcing tolerances for foods shipped in interstate commerce (372). Many states, including California, Florida, and Michigan, have pesticide analytical programs.

The FDA carries out a large-scale monitoring program, the objective of which is to prevent foods that contain illegal residues from entering interstate commerce. Testing of raw agricultural commodities is the focus of the monitoring. To date about 15,000 samples, 8,000 imported and 7,000 domestic (372), are collected and analyzed by the 21 field offices.

There are 300 pesticides and active ingredients with tolerances in or on foods in addition to other pesticides and related chemicals that can exist as residues (373). The EPA has supplied data showing that 53 of these pesticides have active ingredients that have been shown to be oncogenic or potentially oncogenic (351).

Gas chromatography has been used for analysis of pesticides (352,374–376). These methods were developed by the early 1960s. The use of immunoassays has spread into the area of pesticide analysis. A brief summary of chemicals that control weeds or pests in agricultural crops will be provided. There have been IAs described in the literature since the mid-1970s (Table 7). Prior to 1982 the majority of IAs described for pesticide analysis were rias and since that date there have been reports of EIAs. Much less attention has been given to development of fluoroimmunoassays for pesticide analysis.

With a few exceptions, the majority of tests developed for pesticide analysis employ polyclonal antibodies raised in rabbits. There have been some monoclonal antibodies produced against pentachlorophenol (406), surflan (407), malic hydrazide (399), paraquat (403), and paraoxon (401).

Table 7. Immunochemical Methods for Detection of Pesticides

Assay	Pesticide	Ref.
Radioimmunoassay	Aldrin	377
	Dieldrin	377
	Benomyl	378
	S-Bioallethrin	379
	2,4-D	380
	2,4,5T	381
	Diflubenzuron	382
	Parathion	383
	Paraquat	384,385
Fluoroimmunoassay	Benomyl	386
	Diclofop-methyl	386
Enzyme immunoassay	3,4-D	387
	Alachlor	388
	Aldicarb	389
	Atrazine	390
	S-Bioallethrin	391
	Blasticidin S	392
	Chlorsulfuron	393
	Cyanazine	394
	Dichlofopmethyl	395
	Diflubenzuron	396
	Fenpropimorph	397
	Irpodione	398
	Malic hydrazide	399
	Metalaxyl	400
	Paraoxon	401
	Paraquat	402–405
	Pentachlorophenol	406
	Surflan	407
	Terbutryn	408
	Triadimefon	409
	Triazines	410
Affinity chromatography	Kepone	411

For the majority of tests listed in Table 7 the limit of detection is 1 ng of pesticide, which corresponds to 1–10 ppb in a sample solution. RIA and EIA offer comparable sensitivity. The detection limit of RIA is more readily varied because it depends largely on the specific activity of the radioisotope label (17). Use of RIA is restricted to sophisticated laboratories with substantial investment in the equipment. EIA, with ELISA in particular, offers more versatility in analysis than RIA.

There are a number of pesticide detection kits sold commercially. Immuno Systems, Inc. (Biddeford, Maine) sells ELISA kits for triazine (atrazine, simazine, etc) detection as well as one for the cyclodiene insecticides (chlordane, heptachlor, etc). ELISAs for parathion and paraquat are available from Environmental Diagnostics, Inc. (Burlington, N.C.). An assay for pentachlorophenol detection is available from Westinghouse Bio-Analytical Systems (Rockville, Md.). Ohmicron Corp. (Newtown, Pa.) markets pesticide detection kits for aldicarb, atrazine, benomyl, alachlor, captan–captofol, and carbafuran that employ magnetic particle technology.

There are factors that make applications of immunoassays to pesticide analysis complex. It is often difficult to obtain representative samples of agricultural commodities. Commodities are localized temporally and geographically. Local practices can vary leading to complicated research efforts toward assay development. Agricultural commodities are often composed of mixtures of the pesticide and its metabolites at harvest. The high specificity of antibodies used in assays may be disadvantageous because metabolites may not be detected. Also there is fragmentation within the potential pesticide market, potential customers or users could be farmers, consumers, pesticide manufacturers, regulatory agency personnel, food processors, wholesalers, and retailers. The requirement for assay features can vary drastically among these users. Therefore, it is not easy for companies to identify customer needs and project the economic potential of the kits (373).

Officials at regulatory agencies have expressed concern about adopting these qualitative or quantitative immunochemical assays for routine screening of samples. Concern stems first from the need for familiarization and experience with the technology involved in the test systems. Once the testers feel confident that public health protection is not jeopardized by implementing this technology, they must address other concerns. It must be demonstrated for each method, presented in kit form, that there is no appreciable variability from lot to lot production (373).

Many of these immunochemical methods are more sensitive than the confirmatory tests and traditional methods. Therefore, qualitative screening results using IAs cannot be confirmed without effort given to increase the sensitivity of the confirmatory methods (373). As the concerns are addressed, the technology will continue to develop and practical applications of IAs for routine pesticide analysis will occur.

Drug Residues

There is an increasing number of publications that describe the development of ELISAs for drug residue analysis. One antimicrobial agent, sulfamethazine, has received a great deal of attention in the scientific literature as well as the media. There have been numerous reports of the use, or perhaps the abuse, of sulfamethazine in the livestock industry. It will be used as an example of how ELISAs have been developed for detection of the drug in place of the classical chemical methods.

Sulfonamides, namely sulfamethazine, have been incorporated into animal feeds since the 1950s. They are added as growth promotants and for control of certain diseases (412,413). The drugs are retained in the tissues of the animals eating medicated feeds. Some individuals who ingest the contaminated tissues will experience hypersensitivity allergic reactions. Furthermore, there will be preferential selection of bacterial mutants that are resistant to the drugs, which are also used in treatment of human diseases (414).

The residues are cleared from the tissues if the proper 15 day withdrawal period is followed by the producer (415). If it is not adhered to there will be a good chance for the carcasses to contain violative levels of the drug at the time of slaughter (>0.1 ppm).

Sulfonamide residues can also occur in milk due to a number of reasons, including use of the drugs in mastitis therapy, deliberate feeding, inadvertent feeding, or use of sulfamethazine-containing boluses to prevent infection in cows that have calved. One study (416) showed the 64% of milk sampled in the New York City area, central New Jersey, and eastern Pennsylvania contained one or more antibiotic–antimicrobial residues. Sulfonamide residues appeared in 42% of the samples. Other studies showed similar frequencies of the residues in milk (417,418). The FDA maintains a zero-effect tolerance. However, because approved methods cannot accurately measure extremely low levels of the sulfonamides, the tolerance level is set at 10 ppb.

The concern over the presence of sulfamethazine in meat, namely pork, and dairy products heightened when the drug was found to cause neoplasms in mice and rats (419). The need for a rapid screening method became apparent. Current methods for sulfonamide analysis include gas chromatography (1,3,420,421), gas chromatography–mass spectrophotometric (419), liquid chromatography (1,421–423), colorimetry based on the Bratton–Marshall reaction, and thin-layer chromatography (421,425–427) as well as tandem mass spectroscopy–mass spectroscopy (428,429). These methods are labor intensive. They require extensive sample cleanup or preparation and they often measure at the tolerance level. They do not lend themselves to screening samples in the field. A bacterial receptor assay (430) has been developed for detection of numerous sulfonamides. It can be used for screening samples; however, it lacks the specificity that an antibody-based test provides. The Charm Assay is marketed by Penicillin Assays, Inc. (Malden, Mass.).

The Food Safety and Inspection Service (FSIS) branch of the USDA began a residue avoidance program (RAP) designed to check animals before they are slaughtered (431). Sulfamethazine has been shown to cause the most violations in swine tissue, approximately 95% of the sulfonamide tissue violations (432). A correlation has been established for the concentrations of sulfamethazine present in the liver and edible tissues as well as how much is present in the urine and serum (432–435). Noninvasive samples can be collected and tested prior to or at the time of slaughter and tested to determine the concentration of the drug in the liver and edible tissues.

The early ELISA developed for detection of sulfamethazine in swine blood (436) required extraction of the drug from the sample and required fairly long times to perform the assay. They were not suitable for use at slaughterhouses. More recent publications describe assays with shorter times and serum or plasma samples can be used directly without extraction (437,438). Assays have also been described for detection of sulfamethazine in milk (439), feed, and tissue (440). A recent paper describes the use of high-performance immunoaffinity chromatography for drug residue analysis (441).

A number of biotechnology companies market inmunodiagnostics for sulfamethazine detection in dairy products (Neogen Corp., Lansing, Mich.; Idetek, San Bruno, Calif.; and Idexx, Portland, Maine), feed (Neogen Corp., Idetek, and Environmental Diagnostics, Inc., Burlington, N.C.), and urine and tissue (Idetek and Environmental Diagnostics).

CONCLUSION

This article has only begun to scratch the surface of all the research concerning immunoassay development for topics of interest in agriculture and the food industry. This is a day and age where consumers express concern over the quality of the food. With an ever-increasing awareness of the importance of a balanced diet and perceived dangers from additives and contaminants, as well as increased legislation, the need for the ability to test large numbers of samples routinely for the presence of potentially dangerous or offending substances becomes highly apparent. The selectivity and sensitivity of immunoassays make them excellent candidates as tools for routine analysis. ELISAs are highly appropriate assays for analyzing food and agricultural residues because they can be developed to qualitatively screen for the desired substance or they may be semiquantitative or quantitative. These assays can be cost effective, require little or no expensive equipment, permit rapid throughput of many samples, and can be performed by relatively unskilled personnel. These tests provide excellent accuracy and reproducibility, as well as sensitivity and specificity.

Over the next decade advances in immunoassays will benefit food analysts. It appears that those test formats that are the most user friendly for the end user will gain a great deal of ground. The less complex a test is to perform the more attractive it will be to nonlaboratory personnel (farmer, grain inspector, etc). Over the next decade it is likely that ELISAs will be routinely used for screening foods and feeds at governmental agencies as well as by quality-control chemists in the private sector. It is truly an exciting time to work in an area where so much evolution in analytical technology is occurring.

BIBLIOGRAPHY

1. M. C. Allred and D. L. Dunmire, *J. Chromatogr. Sci.* **16**, 553–557 (1978).
2. A. Fravolini and A. Begliomini, *J. Assoc. Off. Anal. Chem.* **52**, 767–769 (1969).
3. R. J. Daun, *J. Assoc. Off. Anal. Chem.* **54**, 1277–1288 (1971).
4. "Section 976.22," *Official Methods of Analysis*, 15th ed., Association of Official Analytical Chemists, 1990, pp. 1211–1212.
5. "Sections 986.17 and 986.18," in Ref. 4, pp. 1205–1206.
6. "Sections 980.20 and 989.06," in Ref. 4, pp. 1192–1195.
7. R. W. Beaver, *J. Assoc. Off. Anal. Chem.* **73**, 69–71 (1990).
8. M. Omura, K. Hashimoto, K. Ohta, T. Tio, S. Ueda, K. Ando, H. Hiraide, and N. Kinae, *J. Ass. Offic. Anal. Chem.* **73**, 300–305 (1990).
9. R. G. Nash, *J. Assoc. Off. Anal. Chem.* **73**, 438–442 (1990).
10. L. Q. Huang and J. J. Pignatello, *J. Assoc. Off. Anal. Chem.* **73**, 443–446 (1990).
11. M. F. Delaney, *Liquid Chromatogr.* **2**, 85–86 (1984).
12. F. Kavanaugh, *Analytical Microbiology*, Vol. 2, Academic Press, Inc., Orlando, Fla., 1972.
13. L. F. Knudsen and W. A. Randall, *J. Bacteriol.* **50**, 187–200 (1945).
14. D. E. Dixon, S. J. Steiner, and S. E. Katz, in A. Azalos, ed., *Modern Analysis of Antibiotics*, Marcel Dekker, Inc., New York, 1986, pp. 415–431.
15. J. W. Goding, *J. Immunol. Methods* **39**, 285–308 (1980).
16. R. Luft and R. S. Yalow, *Radioimmunoassay Methodology and Application in Physiology and in Clinical Studies*, George Thieme Verlag, Stuttgard, 1974.
17. H. Eisen, *Immunology: An Introduction to Molecular and Cellular Principles of the Immune Responses*, Harper & Row, New York, 1980.
18. A. J. Munro, J. Landon, and E. J. Shaw, *J. Antimicrob. Chemother.* **9**, 423–424 (1982).
19. E. J. Shaw, R. A. A. Watson, J. Landon, and D. S. Smith, *J. Clin. Pathol.* **30**, 526–531 (1977).
20. E. Engvall and P. Perlman, *Biochem. Biophys. Acta.* **251**, (1971).
21. S. Avrameas, *Immunochemistry* **6**, 43–53 (1968).
22. R. Maiolin, B. Ferrua, and R. Masseyeff, *J. Immunol. Methods* **6**, 355–362 (1975).
23. E. Engvall and P. Perlmann *Immunochemistry* **8**, 871–874 (1971).
24. K. E. Kang'ethe, "Use of Immunoassays in Monitoring Meat Protein Additives," in J. H. Rittenburg, ed., *Development and Application of Immunoassay for Food Analysis*, Elsevier Applied Science, Publishers, Ltd., Barking, UK, 1990, pp. 127–139.
25. S. J. Steiner, *The Development of a Model Immunological Assay System for the Detection of Antibiotic Residues, Ph.D. dissertation, Rutgers University New Brunswick, N.J., 1981.*
26. G. Mancini, A. O. Carbonara, and J. F. Heremans, *Immunochemistry* **2**, 235–254 (1965).
27. O. Ouchterlony, *Acta Pathol. Microbiol. Scand.* **26**, 507–515 (1949).
28. P. Grabar and C. A. Williams, *Biochim. Biophys. Acta* **10**, 193–194 (1953).
29. C. B. Laurell, *Anal. Biochem.* **15**, 45–52 (1966).
30. M. Heidelberger and F. E. Kendall, *J. Exp. Med.* **55**, 555–561 (1932).
31. J. H. Skerrit, "Immunoassays of Non-Meat Protein Additives in Food," in Ref. 24, pp. 81–125.
32. S. K. Raghavan, C. T. Ho, and H. Daun, *J. Chromatogr.* **351**, 195–202 (1986).
33. E. Bleumink, *Proc. Nutr. Soc.* **42**, 219–231 (1983).
34. S. L. Taylor, *Food Allergies Food Technol.* **39**, 98–105 (1985).
35. I. Kloczko and A. Rutkowski, *Nahrung* **28**, 9–13 (1984).
36. E. Degenkolb and M. Hingerle, *Arch. Lebensm. Hyg.* **18**, 241–247 (1967).
37. L. Kotter and C. Hermann, *Arch. Lebensm. Hyg.* **19**, 267–270 (1968).
38. C. Hermann, C. Merkle, and L. Kotter, *Fleischwirtschaft* **51**, 249–251 (1973).
39. H. Kruger and D. Grossklaus, *Fleischwirtschaft* **51**, 315–320 (1971).
40. E. J. Menzel, *Z. Ernahrungswiss.* **20**, 55–68 (1981).
41. C. Hermann and G. Wagenstaller, *Arch. Lebensm. Hyg.* **24**, 131–134 (1973).
42. C. Ring and F. Sacher, *Fleischwirtschaft* **64**, 355–357 (1984).
43. C. Hermann, C. Merkl, and L. Kotter, *Ann. Nutr. Aliment.* **31**, 153–155 (1977).
44. G. Fromm, *Fleisch. Konserven. Arch. Hyg. Bakkt.* **151**, 702–708 (1967).
45. H. J. Sinell, and I. Mentz, *Folio Vet. Lat.* **7**, 41–54 (1977).
46. J. H. Fernandez and I. S. Lopez, *Proc. Eur. Mtg. Meat Res. Workers* **28**, 434–435 (1982).
47. M. I. Santillana, E. Ruiz, E. Casado, J. Huescar, and R. Jimenez, *Bol. Cent. Nac. Aliment. Nutr.* **10**, 10–12 (1982).
48. H. Klostermeyer and S. Offt, *Z. Lebensm. Unters-Forsch.* **167**, 158–161 (1978).
49. M. Bellatti and G. Parolari, *Ind. Conserv.* **54**, 3–5 (1979).
50. H. J. Sinell, *Proc. Eur. Mtg. Meat Res. Workers* **27**, 5590–5593 (1981).
51. E. Gombocz, E. Hellwig, and F. Petuely. *Proc. Eur. Mtg. Meat Res. Workers* **27**, 594–597 (1981).
52. E. Gombocz, E. Hellwig, and F. Petuely, *Z. Lebensm. Unters. Forsch.* **172**, 355–361 (1981).
53. E. Gombocz and F. Petuely, *Z. Lebensm. Unters.-Forsch.* **177**, 2–7 (1983).
54. G. N. Festenstein, F. C. Hay, B. J. Miflin, and P. R. Shewry, *Planta* **164**, 135–141 (1985).
55. S. P. Greiner, G. J. Kellen, and D. E. Carpenter, *J. Food Sci.* **50**, 1106–1109 (1985).
56. J. Kraack, *Fleischwirtschaft* **53**, 697–698 (1973); **53**, 701–702 (1973).
57. J. Daussant and D. Bureau, "Immunochemical Methods in Food Analysis," in R. D. King, ed., *Development in Food Analysis Techniques 3*, Elsevier Applied Science Publishers, Ltd., Barking, UK, 1984, pp. 175–210.
58. M. Peter, *Arch. Lebensm. Hyg.* **21**, 220–222 (1970).
59. E. Hauser, J. Bicanova, and W. Kuenzler, *Mitt. Geb. Lebensm. Hyg.* **65**, 82–89 (1974).
60. L. A. Hanson, *Veterinaer* **16**, 201–205 (1964).
61. O. Wyler and J. J. Siegrist, *Mitt. Geb. Lebensm. Hyg.* **56**, 299–303 (1965).
62. H. Guenther, *Bestimmung Fremdeweiss Arch. Lebensm. Hyg.* **20**, 97–106 (1969).
63. H. J. Guenther, *Bestimmung Fremdeiweiss Fleuschwaren. Arch. Lebensm. Hyg.* **20**, 128–131 (1969).

64. O. D. Wyler, *Fleischwirtschaft* **51**, 1743–1745 (1971).
65. H. O. Guenther, *Ann. Nutr. Aliment.* **31**, 179–181 (1977).
66. J. C. Hammont, I. C. Cohen, and B. Flaherty, *J. Assoc. Publ. Analysis* **14**, 119–126 (1976).
67. E. Hauser, J. Bicanova, and W. Kuenzler, *Fleischwirtschaft* **54**, 240–242 (1979).
68. M. J. Escrebiano and P. Grabar, *Ann. Inst. Pasteur Paris* **110** (Suppl. 3), 84–88 (1966).
69. C. C. Nummo and M. T. O'Sullivan, *Cereal Chem.* **44**, 584–591 (1967).
70. M. R. Booth and J. A. D. Ewart, *J. Sci. Food Agr.* **21**, 187–192 (1970).
71. J. Zareba, *Pol. Tyg. Lek.* **23**, 1962–1963 (1968).
72. H. J. Sinell, *Lebensm. Arch. Lebensm. Hyg.* **19**, 121–125 (1968).
73. L. A. Appelguist, *SIK Rapport* **371**, 1–23 (1975).
74. R. Kluge-Wilm and H. J. Sinell, *Milchwissenschaft* **27**, 160–165 (1972).
75. H. O. Guenther, *Lebensm. Unters.-Forsch.* **149**, 98–99 (1972).
76. H. Elbertzhagen and E. Wenzel, *Z. Lebensm. Unters.-Forsch.* **175**, 15–16 (1982).
77. J. W. Llewellyn and R. Sawyer, *Ann. Nutr. Aliment.* **31**, 157–159 (1977).
78. S. Baudner, *Getreide Mehl Brot.* **32**, 330–337 (1978).
79. G. A. H. Elton and J. A. D. Ewart, *J. Sci. Food Agr.* **14**, 750–758 (1963).
80. M. Benhamou-Glynn, M. J. Escribano, and P. Grabar, *Bull. Soc. Chim. Biol.* **47**, 141–156 (1965).
81. J. Heitmann, *Fleischwirtschaft* **67**, 621–622 (1987).
82. E. J. Menzel and F. Glatz, *Z. Lebensm. Unters.-Forsch.* **172**, 12–19 (1981).
83. J. E. Menzel and H. Hagemeister, *Z. Lebensm. Unters.-Forsch.* **175**, 211–214 (1982).
84. P. J. Ciclitira and E. S. Lennox, *Clin. Sci.* **64**, 655–659 (1983).
85. C. H. S. Hitchcock, F. J. Bailey, A. A. Crimes, D. A. G. Dean, and P. J. Davis, *J. Sci. Food Agr.* **37**, 157–165 (1981).
86. N. M. Griffiths, M. J. Billington, A. A. Crimes, C. H. S. Hitchcock, *J. Sci. Food Agr.* **35**, 1255–1260 (1984).
87. J. H. Rittenburg, A. Adams, J. Palmer, and J. C. Allen, *J. Assoc. Off. Anal. Chem.* **70**, 582–587 (1987).
88. D. B. Berkowitz and D. W. Webert, *J. Assoc. Off. Anal. Chem.* **70**, 85–90 (1987).
89. F. W. Janssen, G. Voortman, and J. A. deBaaij, *Z. Lebensm. Unters.-Forsch.* **182**, 479–483 (1986).
90. P. Ravestein and R. A. Driedonks, *J. Food Technol.* **21**, 19–32 (1986).
91. F. W. Janssen, G. Voortman, and J. A. deBaaij, *J. Agr. Food Chem.* **35**, 563–567 (1987).
92. C. C. Hall, C. H. S. Hitchcock, and R. Wood, *J. Assoc. Publ. Analysts* **25**, 1–27 (1987).
93. Eur. Pat. 111,762 (1984), P. J. Davis, and P. Poter.
94. D. Lifier and J. C. Callin, *LeLait* **62**, 541–548 (1982).
95. P. Teufel and V. Sacher, *Fleischwertschaft* **62**, 1474–1476 (1982).
96. C. Staak and U. Kaempe, *Fleischwertschaft* **62**, 1477–1478 (1982).
97. C. Breton, L. Phan Thanh, and A. Paraf, *J. Food Sci.* **53**, 226–230 (1988).
98. J. H. Skerritt and R. A. Smith, *J. Sci. Food Agr.* **36**, 980–986 (1985).
99. J. H. Skerritt, *Food Technol. Aust.* **37**, 570–572 (1985).
100. J. H. Skerritt, *J. Sci. Food Agric.* **36**, 987–994 (1985).
101. J. H. Skerritt, J. A. Diment, and C. W. Wrigley, *J. Sci. Food Agr.* **36**, 995–1003 (1985).
102. R. Troncone, M. Vitale, A. Donatello, E. Farris, G. Rossi, and S. Aurrichio, *J. Immunol. Methods* **92**, 21–23 (1986).
103. H. Windemann, F. Fritschy, and E. Baumgartner, *Biochim. Biophys. Acta* **709**, 110–121 (1982).
104. P. J. Ciclitira, H. J. Ellis, D. J. Evans, E. S. Lennox, and R. H. Dowling, *Gut* **24**, A487 (1983).
105. J. H. Skerritt, R. A. Smith, C. W. Wrigley, and P. A. Underwood, *J. Cereal Sci.* **2**, 215–224 (1984).
106. F. Fitschy, H. Wendemann, and E. Baumgartner, *Z. Lebensm. Unters.-Forsch.* **181**, 379–385 (1985).
107. P. J. Gosling, D. F. McKellog, and P. F. Fottrel, "The Measurement of Wheat Gliadin by Enzyme Immunoassay," in *Protein Evaluation in Cereals and Legumes. Proceedings of the EEC Workshop, Thessoboniki, Greece*, Oct. 1985.
108. J. H. Skerritt and O. Martinuzzi, *J. Immunol. Methods* **88**, 217–224 (1986).
109. A. R. Freedman, G. Galfre, E. Gal, H. J. Ellis, and P. J. Ciclitera, *J. Immunol. Methods* **98**, 123–127 (1987).
110. A. R. Freedman, G. Galfre, E. Gal, H. J. Ellis and P. J. Ciclitera, *Clin. Chim. Acta* **166**, 323–328 (1987).
111. J. H. Skerritt, K. L. Jenkins, and A. S. Hill, *Food Agr. Immunol.* **1**, 161–172 (1989).
112. A. S. Hill and J. H. Skerritt, *Food Agr. Immunol.* **1**, 147–160 (1989).
113. J. H. Skerritt and A. S. Hill, *J. Cereal Sci.* **11**, 103–122 (1990).
114. J. H. Skerritt and P. Y. Lew, *J. Cereal Sci.* **11**, 103–122 (1990).
115. R. B. Johnson, J. T. Labrody, and J. H. Skerritt, *Clin. Exp. Immunol.* **79**, 135–140 (1990).
116. A. A. Crimes, C. H. S. Hitchcock, and R. Wood, *J. Assoc. Publ. Analysts* **22**, 59–78 (1984).
117. W. J. Olsman, S. Dobbelaere, and C. H. S. Hitchcock, *J. Sci. Food Agr.* **36**, 499–507 (1985).
118. B. J. Miflin, J. M. Field, and P. R. Shewry, in J. Daussant, J. Vaughan, and J. Mosse, eds., *Cereal Storage Proteins and their Effects on Technological Properties in Seed Proteins*, Academic Press, Inc., Orlando, Fla., 1985, pp. 255–319.
119. A. Cuthbertson, "Hot Processing Meat. A Review of the Rationale and Economic Implications," in *Development in Meat Science*, Vol. 1, Allied Publishers, New Delhi, 1980, pp. 66–88.
120. C. Casas, J. Tormo, P. E. Hernandez, and B. Sanz, *J. Food Technol.* **19**, 283–287 (1984); *J. Sci. Food Agr.* **35**, 793–796 (1984).
121. J. G. Sutton, J. Goodwin, G. Horscroft, R. E. Stockdale, and A. Frake, *J. Assoc. Off. Anal. Chem.* **66**, 1164–1174 (1983).
122. K. S. Swarts and C. R. Wilks, *Aust. Vet. J.* **59**, 21 (1982).
123. F. D. Shaw, E. M. Deane, and D. W. Cooper, *Aust. Vet. J.* **60**, 25 (1982).
124. L. J. Glesson, W. J. Slattery, and A. J. Sinclair, *Aust. Vet. J.* **60**, 127 (1983).
125. I. G. S. Furminger, *Nature* **202**, 1332 (1964).
126. A. R. Hayden, *J. Food Sci.* **42**, 1189–1192 (1977).
127. A. R. Hayden, *J. Food Sci.* **43**, 476–478 (1977).
128. R. P. Mageau, M. E. Cutrufelli, B. Schwab, and R. W. Johnston, *J. Assoc. Off. Anal. Chem.* **67**, 949–954 (1984).

129. T. Kamiyama, Y. Katsube, and K. Imaizumi, *Jpn. J. Vet. Sci.* **40**, 653–661 (1978).
130. T. Kamiyama, Y. Katsube, and K. Imaizumi, *Jpn. J. Vet. Sci.* **40**, 663–669 (1978).
131. K. J. Lindqvist, J. M. Gathuma, and H. F. A. Kaburia, "Analysis of Blood Meals of Haemotophagous Insects by Haemagglutination Inhibition and Enzyme Immunoassay," in P. M. Tukei and R. M. Njogu, eds., *Proceedings of 3rd Medical Scientific Conference Current Medical Research in Eastern Africa*, Africa-Science International Publishers, Ltd., Nairobi, Kenya, 1982, pp. 122–133.
132. L. A. Y. Johnston, P. Tracey-Patte, R. A. Donaldson, and B. Parkinson, *Aust. Vet. J.* **59**, 59 (1982).
133. E. K. Kang'ethe, S. J. Jones, and R. L. S. Patterson, *Meat Sci.* **7**, 229–240 (1982).
134. M. R. Patterson and T. L. Spenser, *Meat Sci.* **15**, 119–123 (1985).
135. R. G. Whittaker, T. L. Spencer, and J. W. Copland, *Aust. Vet. J.* **59**, 125 (1982).
136. R. G. Whittaker, T. L. Spencer, J. W. Copland, *J. Sci. Food Agr.* **34**, 1143–1148 (1983).
137. S. J. Jones and R. L. S. Patterson, *Meat Sci.* **15**, 1–13 (1985).
138. R. M. Patterson, R. F. Whittaker, and T. L. Spencer, *J. Sci. Food Agr.* **35**, 1018–1023 (1984).
139. O. Ouchterlony, *Acta Pathol. Microbiol. Scand.* **25**, 186–191 (1948).
140. P. D. Warriss, *J. Sci. Food. Agr.* **28**, 457–462 (1977).
141. I. R. Tizard, N. A. Fish, and F. Cadi, *J. Food Prot.* **45**, 353–355 (1982).
142. E. K. Kang'ethe, *Master's of science thesis*, University of Bristol, UK, 1981.
143. N. M. Griffiths and M. J. Billington, *J. Sci. Food Agr.* **35**, 909–914 (1984).
144. R. A. Lawrie, *Meat Science* 3rd ed., Pergamon Press, Oxford, UK, p. 77.
145. S. M. Machlik and H. N. Daudt, *J. Food Sci.* **28**, 711–718 (1963).
146. J. M. C. K. Snowden and J. F. Weidemann, *Meat Sci.* **2**, 1–18 (1978).
147. J. Manz, *Fleischwirtschaft* **63**, 1767–1769 (1983).
148. E. K. Kang'ethe and J. M. Gathuma, *Meat Sci.* **19**, 265–270 (1987).
149. E. K. Kang'ethe, K. J. Lindgvist, and J. M. Gathuma, "Immunological Reactions of Thermostable Muscle Antigens and Then Possible Use in Speciation of Cooked and Fresh Meats," in R. L. S. Patterson, ed., *Biochemical Identification of Meat Species*, Elsevier Applied Science Publishers, Ltd., Barking, UK, 1985, pp. 128–144.
150. E. K. Kang'ethe, J. M. Gathuma, and K. J. Lindqvist, *J. Sci. Food Agric.* **39**, 179–184 (1986).
151. E. K. Kang'ethe and K. J. Lindquist, "Species Identification of Internal Organs Using Antisera to Thermostable Muscle Antigens," in E. Peteaja, ed., *Proceedings of 33rd International Congress of Meat Science and Technology*, Helsinki, Finland, pp. 377–378.
152. R. Goerlich and E. Greuel, *Arch. Lebensmi.* **37**, 87–90 (1986).
153. M. R. A. Morgan, and H. A. Lee, "Mycotoxins and Natural Food Toxicants," in Ref. 24, pp. 143–170.
154. C. M. Ward and M. R. A. Morgan, *Food Addit. Contam.* **5**, 555–561 (1988).
155. C. E. Cook, C. R. Tallent, E. W. Amerson, M. W. Myers, J. A. Kepler, G. F. Taylor, and H. D. Christensen, *J. Pharm. Exp. Ther.* **199**, 679–686 (1976).
156. S. Pearson, J. M. Smith, and V. Marks, *Ann. Clin. Biochem.* **21**, 208–212 (1984).
157. P. S. Jourdan, E. W. Weiler, and R. L. Mansell, *J. Agr. Food Chem.* **31**, 1249–1255 (1983).
158. P. S. Jourdan, R. L. Mansell, D. G. Oliver, and E. W. Weiber, *Anal. Biochem.* **138**, 19–24 (1984).
159. V. Paseshnickenko and A. R. Guseva, *Biochemistry (USSR)* **21**, 606–611 (1956).
160. R. P. Vallejo and C. D. Ercegovich, *JNBS (U.S.)* **519** (Spec. Publ), 330–340 (1979).
161. M. R. A. Morgan, R. M. C. Nerney, J. A. Matthew, D. T. Coxon, and H. W.-S. Chan, *J. Sci. Food Agr.* **34**, 593–598 (1982).
162. M. R. A. Morgan, D. T. Coxon, S. Bramham, H. W.-S. Chan, W. M. H. VanGelder, and M. J. Allison, *J. Sci. Food. Agr.* **36**, 282–288 (1985).
163. K-E. Hellenas, *J. Sci. Food Agr.* **37**, 776–782 (1986).
164. J. A. Matthew, M. R. A. Morgan, R. McNerney, H. W-S. Chan, and D. T. Coxon, *Food Chem. Toxicol.* **21**, 637–640 (1983).
165. M. H. Harvey, M. McMillan, M. R. A. Morgan, and H. W-S. Chan, *Hum. Toxicol.* **4**, 187–194 (1985).
166. S. L. Sinden and R. E. Webb, *Am. Potato J.* **49**, 334–338 (1972).
167. J. J. Pestka, "Fungi and Mycotoxins in Meats," in A. M. Pearson and J. Dawson, eds., *Advances in Meat Research*, Vol. 2, AVI Publishing Co., Inc., Westport, Conn., 1986, pp. 277–302.
168. J. J. Pestka and D. E. Dixon-Holland, "Mycotoxin Detection by Immunoassay and Application of Hybridoma Technology," in B. Swaminathan and G. Prakash, eds., *Nucleic Acid and Monoclonal Antibody Probes. Applications in Diagnostic Microbiology*, Marcel Dekker, Inc., New York, 1989, pp. 657–678.
169. H. P. Van Egmonnd, W. E. Paulsch, and P. L. Schuyller, *J. Assoc. Off. Anal. Chem.* **61**, 809–812 (1978).
170. H. P. Van Egmond, W. E. Paulsch, E. Deyll, and P. L. Schuller, *J. Assoc. Off. Anal. Chem.* **63**, 110–114 (1980).
171. S. Nesheim, N. F. Hardin, O. J. Francis, and W. S. Langham, *J. Assoc. Off. Anal. Chem.* **56**, 817–821 (1973).
172. O. C. Hunt, A. T. Boudon, P. J. Wild, and N. T. Crosby, *J. Sci. Food Agr.* **29**, 234–238 (1978).
173. J. F. Gregory and D. Manley, *J. Assoc. Off. Anal. Chem.* **64**, 144–151 (1981).
174. P. M. Scott, *J. Assoc. Off. Anal. Chem.* **65**, 876–883 (1982).
175. W. F. Busby and G. F. Wogan, "Mycotoxins and Mycotoxicoses," in H. Remann and F. L. Bryan, eds., *Food-Borne Infections and Intoxications*, 2nd ed., Academic Press, Inc., Orlando, Fla., 1979, pp. 519–610.
176. J. J. Wong and D. D. Hsieh, *Proc. Natl. Acad. Sci. USA* **73**, 2211–2244 (1976).
177. L. Stoloff, *J. Prot.* **43**, 226–230 (1980).
178. J. F. Grove, *Nat. Prod. Rep.* **5**, 187–190 (1988).
179. R. R. Marquardt, A. Frohlich, and D. Abramson, *Can. J. Physiol. Pharmacol.* **68**, 991–999 (1990).
180. M. W. Trucksess, M. T. Flood, M. M. Mossoboa, and S. W. Page, *J. Agr. Food Chem.* **35**, 445–448 (1987).
181. K. C. Erhlich, L. S. Lee, and A. Ciegler, *J. Liquid Chromatogr.* **6**, 833–843 (1983).
182. U. S. Lee, H. S. Jang, T. Tanaka, Y. Joh, C. M. Cho, and Y. Ueno, *J. Agr. Food Chem.* **35**, 126–129 (1987).

183. C. J. Mirocha, S. V. Pathre, B. Schanerhamer, and C. M. Christensen, *Appl. Environ. Microbiol.* **32**, 553–556.
184. R. D. Platter and G. A. Bennett, *J. Assoc. Off. Anal. Chem.* **66**, 1470–1477 (1983).
185. H. J. Kurtz and C. J. Mirocha, "Zearalenone (F2) induced Estrogenic Syndrome in Swine," in T. D. Syelie and L. G. Morehouse, eds., *Mycotoxic Fungi, Mycotoxins and Mycotoxicoses*, Vol 2, Marcel Dekker, Inc., New York, pp. 1256–1264.
186. CAST, *Aflatoxins and Other Mycotoxins: An Agricultural Perspective*, Report No. 80, Council for Agricultural Science and Technology, Ames, Iowa 1979.
187. P. Krough, "Ochratoxins," in J. V. Rodricks, C. W. Hasseltine, and M. A. Mehlman, eds., *Mycotoxins in Human and Animal Health*, Patotox Publishers, Park Forest South, Ill., 1977, pp. 489–498.
188. J. E. Smith and M. O. Moss, *Mycotoxins: Formation, Analysis and Significance*, John Wiley & Sons, Inc., New York, 1985, p. 148.
189. S. Neshiem, L. Friedman, and A. E. Pohland, "Ochratoxin," *J. Am. Oil Chem. Soc.* **66**, 432 (1989).
190. CAST, *Mycotoxins Economic and Health Risks*, Report No. 116, Council for Agricultural Science and Technology, Ames, Iowa, 1989.
191. C. F. Jelinek, A. E. Pohland, and G. E. Wood, *J. Assoc. Off. Anal. Chem.* **72**, 223–230 (1989).
192. L. G. Morehouse, "Mycotoxins of Veterinary Importance in the United States," in J. Lacey, ed., *Trichothecenes and Other Mycotoxins. Proceedings of the International Mycotoxin Symposis, Sydney, Australia, 1984*, John Wiley & Sons, Inc., New York, 1985, pp. 383–410.
193. P. Krogh, B. Hald, and E. J. Pederson, *APMIS* **81** (Sect. B.), 681–683 (1973).
194. P. Krogh, B. Hald, L. England, L. Rutquist, and O. Swohn, *APMIS* **82** (Sect. B), 701–702 (1974).
195. P. Lafont and J. Lafont, *Experientia* **26**, 807–808 (1971).
196. U. L. Diener, "Unwanted Biological Substances in Foods: Aflatoxins," in J. C. Ayres and J. C. Kushman, eds., *Impact of Toxicology on Food Processing*, AVI Publishing Co., Inc., Westport, Conn., 1981, pp. 122–150.
197. L. B. Bullerman, *J. Dairy Sci.* **64**, 2439–2452 (1981).
198. C. F. Jelinek, *Distribution of Mycotoxin—An Analysis of World Wide Commodities Data, Including Data from FAO / WHO / UNEP Food Contamination Monitoring Programme*, Joint FAO/WHO/UNEP Second International Conference on Mycotoxins, Bangkok, Thailand, Sept. 28–Oct. 3, 1987.
199. G. E. Wood, *J. Assoc. Off. Anal. Chem.* **72**, 543–548 (1989).
200. P. M. Scott, "The Occurrence of Vomitoxin (Deoxynivalenol, DON) in Canadian Grains," in H. Kuratha and Y. Ueno, eds., *Toxic Fungi—Their Toxins and Health Hazards*, Elsevier, Science Publishing Co., Inc., New York, 1984, pp. 182–189.
201. T. Tanaka, A. Hasegawa, Y. Marsuki, and Y. Ueno, *Food Addit. Contam.* **2**, 259–265 (1985).
202. Leistner, "Toxigenic Penicillia Occurring in Feeds and Foods," in Ref. 199, pp. 162–171.
203. AIN Symposium of 10.4.1978, "Principal Hazards in Food Safety and Their Assessment," *Fed. Proceed.* **37**, 2575–2597.
204. "Completion of the Internal Market: Community Legislation Foodstuffs," EEC Commission Document COM (85)603 Final, 1985.
205. "White Paper on Completing the Internal Market," EEC Commission Document COM (85)310 Final, 1985.
206. "Communication on the Free Movement of Food Stuffs within the Community," EEC Commission Document 89/C271/03, 1989.
207. "On the Approximation of the Laws of Member States Concerning Food Additives Authorized For Use in Foodstuffs Intended for Human Consumption," Council Directive, 89/107/EEC, 1989, p. 27.
208. J. J. Pestka, *J. Assoc. Off. Anal. Chem.* **71**, 1075–1081 (1988).
209. J. J. Langone and H. VanVunakis, *J. Natl. Cancer Inst.* **56**, 591–595 (1976).
210. F. S. Chu and I. Ueno, *Appl. Environ. Microbiol.* **33**, 1125–1128 (1977).
211. J. D. Groopman, P. R. Donahue, A. Marshak-Rothstein, and G. N. Wagan, *Proc. Natl. Acad. Sci. USA* **81**, 7728–7731 (1984).
212. O. El-Nakib, J. J. Pestka, and F. S. Chu, *J. Assoc. Off. Anal. Chem.* **64**, 1077–1082 (1981).
213. D. W. Lawellin, D. W. Grant, B. K. Joyce, *Appl. Environ. Microbiol.* **34**, 94–96 (1977).
214. J. J. Pestka, P. K. Gaur, and F. S. Chu, *Appl. Environ. Microbiol.* **40**, 1027–1031 (1980).
215. T. S. L. Fan and F. S. Chu, *J. Food Prof.* **47**, 964–968 (1984).
216. A. A. G. Candlish, W. H. Stimson, and J. E. Smith, *Lett. Appl. Microbiol.* **1**, 57–59 (1985).
217. B. P. Ram, L. P. Hart, O. L. Shotwell, and J. J. Pestka, *J. Assoc. Off. Anal. Chem.* **69**, 904–907 (1986).
218. B. P. Ram, L. P. Hart, R. J. Cole, and J. J. Pestka, *J. Food Prot.* **49**, 792–795 (1986).
219. D. N. Mortimer, M. J. Shepard, J. Gilbert, and M. R. A. Morgan, *Food Addit. Contam.* **5**, 127–132 (1987).
220. R. L. Warner and J. J. Pestka, *J. Food Prot.* **50**, 502–503 (1987).
221. D. E. Dixon-Holland, J. J. Pestka, B. A. Bidigare, W. L. Casale, R. L. Warner, B. P. Ram, and L. P. Hart, *J. Food Prot.* **51**, 201–204 (1988).
222. R. B. Sashidhar and B. S. N. Rao, *J. Exp. Biol.* **26**, 984–989 (1988).
223. D. N. Mortimer, M. J. Shepard, J. Gilbert, and C. Clark, *Food Addit. Contam.* **5**, 601–608.
224. N. Reichert, S. Steinmeyer, and R. Weber, *Z. Lebensm. Unters.-Frosch.* **186**, 505–508 (1988).
225. D. L. Park, B. M. Miller, L. P. Hart, G. Yang, J. McVey, S. W. Page, J. Pestka, and L. H. Brown, *J. Assoc. Off. Anal. Chem.* **72**, 326–332 (1989).
226. D. L. Park, B. M. Miller, S. Nesheim, M. W. Trucksess, A. Vekich, B. Bidigare, J. L. McVey, and L. H. Brown, *J. Assoc. Off. Anal. Chem.* **72**, 638–643 (1989).
227. M. W. Trucksess, M. E. Stack, S. Neshiem, D. L. Park, and A. E. Pohland, *J. Assoc. Off. Anal. Chem.* **72**, 957–962 (1989).
228. J. W. Dorner and R. J. Cole, *J. Ass. Off. Anal. Chem.* **72**, 962–964 (1989).
229. A. L. Patey, M. Sharman, R. Wood, and J. Gilbert, *J. Assoc. Off. Anal. Chem.* **72**, 965–969 (1989).
230. N. Ramakrishna, J. Lacy, A. A. G. Candlish, J. E. Smith, and I. A. Goodbrand, *J. Assoc. Off. Anal. Chem.* **73**, 71–76 (1990).
231. M. W. Trucksess, K. Young, K. F. Donahue, D. K. Morris, and E. Lewis, *J. Assoc. Anal. Chem.*, 425–428 (1990).
232. L. S. Lee, J. H. Wall, P. J. Cotly, and P. Bayman, *J. Assoc. Off. Anal. Chem.* **73**, 581–583 (1990).
233. Koeltzow and S. N. Tanner, *J. Assoc. Off. Anal. Chem.* **73**, 584–589 (1990).

234. J. D. Groopman and K. Donahue, *J. Assoc. Off. Anal. Chem.* **71**, 861–867 (1988).
235. W. O. Harder and F. S. Chu, *Experientia* **35**, 1104–1105 (1976).
236. J. J. Pestka and Y. K. Li, W. O. Harder, and F. S. Chu, *J. Assoc. Off. Anal. Chem.* **64**, 294–301 (1981).
237. N. A. Woychik, R. D. Hinsdill, and F. S. Chu, *Appl. Environ. Microbiol.* **48**, 1096–1099 (1984).
238. J. M. Fremy and F. S. Chu, *J. Assoc. Off. Anal. Chem.* **67**, 1098–1101 (1984).
239. W. J. Hu, N. Woychik, and F. S. Chu, *J. Food Prot.* **47**, 126–127 (1984).
240. S. Wu, G. Yang, and T. Sun. *Chin Joncol.* **5**, 81–84 (1983).
241. T. J. Hansen, *J. Food Prot.* **53**, 75–77 (1990).
242. D. N. Mortimer, J. Gilbert, and M. J. Shepard, *Chromotogr.*, 393–398 (1987).
243. D. Thouvenot and R. F. Morfin. *Appl. Environ. Microbiol.* **45**, 16–23 (1983).
244. J. J. Pestka, M-T. Liu, B. Knudson, and M. Hogberg, *J. Food Prot.* **48**, 953–957 (1985).
245. M-T. Liu, B. P. Ram, L. P. Hart, and J. J. Pestka, *Appl. Environ. Microbiol.* **50**, 332–336 (1985).
246. R. L. Warner, B. P. Ram, L. P. Hart and J. J. Pestka, *J. Agr. Food Chem.* **34**, 714–717 (1986).
247. D. E. Dixon, R. L. Warner, B. P. Ram, L. P. Hart, and J. J. Pestka, *J. Agr. Food Chem.* **353**, 122–126 (1987).
248. O. A. MacDouggald, A. J. Thulin, and J. J. Pestka, *J. Assoc. Off. Anal. Chem.* **73**, 65–68 (1990).
249. J. I. Azcona, M. M. Abouzied, and J. J. Pestka, *J. Food Prot.* **53**, 577–580 (1990).
250. F. S. Chu, S. Grossman, R. D. Wei, and C. J. Mirocha, *Appl. Environ. Microbiol.* **37**, 104–108 (1979).
251. R. D. Wei, W. Bischoff, and F. S. Chu, *J. Food Prot.* **49**, 267–271 (1986).
252. E. H. Gendloff, J. J. Pestka, D. E. Dixon, and L. P. Hart, *Phytopathology* **77**, 57–59 (1987).
253. K. W. Hunter, A. A. Brimfield, F. D. Finkelman, and F. S. Chu, *Appl. Environ. Microbiol.* **49**, 168–172 (1985).
254. I. A. Goodbrand, W. H. Stimson, and J. E. Smith, "A Monoclonal Antibody Based ELISA for T-2 Toxin," in B. A. Morris, M. N. Clifford, and R. Jackman, eds., *Immunoassays for Veterinary and Food Analysis*, Elsevier Applied Science Publishers, Ltd., Barking, UK, 1984, pp. 355–358.
255. J. Chiba, O. Kawamura, H. Kayii, K. Ohtami, S. Nagayama, and Y. Ueno, *Food Addit. Contam.* **5**, 629–640 (1988).
256. J. J. Pestka, S. S. Lee, H. P. Lau, and F. S. Chu, *J. Am. Oil Chem. Soc.* **58**, 940a–944a (1981).
257. E. H. Gendloff, J. J. Pestka, S. P. Swanson, and L. P. Hart, *Appl. Environ. Microbiol.* **47**, 1161–1163 (1984).
258. T. S. L. Fan, G. S. Zhang, and F. S. Chu, *J. Food Prot.* **47**, 964–967 (1984).
259. Y. C. Xu, G. S. Zhang, and F. S. Chu, *J. Assoc. Off. Anal. Chem.* **69**, 967–970 (1986).
260. Y. C. Xu, G. S. Zhang, and F. S. Chu, *J. Assoc. Off. Anal. Chem.* **71**, 945–949 (1988).
261. W. L. Casale, J. J. Pestka, and L. P. Hart, *J. Agr. Food Chem.* **38**, 663–668 (1988).
262. O. Aalund, K. Brundeldt, B. Hald, P. Kroglo, and K. Paulsen, *APMIS* **83** (Sect. C), 390–392 (1975).
263. F. S. Chu, F. C. C. Chang, and R. D. Hindsdill, *Appl. Environ. Microbiol.* **31**, 831–835 (1976).
264. J. J. Pestka, B. W. Steinart, and F. S. Chu, *Appl. Environ. Microbiol.* **41**, 1472–1474 (1981).
265. M. R. A. Morgan, R. McNerney, and H. W. S. Chan, *J. Assoc. Off. Anal. Chem.* **66**, 1481–1484 (1983).
266. S. Lee and F. S. Chu, *J. Assoc. Off. Anal. Chem.* **67**, 45–49 (1984).
267. W. J. Sidwell, H. W-S. Chan, M. R. A. Morgan, *Food Agr. Immunol.* **1**, 111–118 (1989).
268. O. Kawamuna, S. Sato, H. Kajii, S. Nagayame, K. Ohtani, J. Chiba, and Y. Ueno, *Toxicon* **27**, 887–898 (1989).
269. W. M. Hagler, G. Y. Danko, L. Horvath, M. Palyusik, and C. J. Mirocha, *Acta Vet. Hung.* **28**, 209–216 (1980).
270. C. J. Mirocha, S. V. Pathre, and T. S. Robison, *Food Cosmet. Toxicol.* **19**, 25–30 (1981).
271. F. S. Chu and R. C. Lee, *Food Agr. Immunol.* **1**, 127–136 (1989).
272. T. Yoshizawa, C. J. Mirocha, J. C. Behreno, and S. P. Swanson, *Food Cosmet. Toxicol.* **19**, 31–39 (1981).
273. T. Yoshizawa, T. Sakamoto, Y. Ayano, and C. J. Mirocha, *Agr. Biol. Chem.* **46**, 2613–2615 (1982).
274. T. Yoshizawa, T. Sakamoto, and K. Okamoto, *Appl. Environ. Microbiol.* **47**, 130–134 (1984).
275. R. B. Read, W. L. Pritchard, J. Bradshaw and L. A. Black, *J. Dairy Sci.* **48**, 411–419 (1965).
276. D. Y. C. Fung and J. Wagner, *Appl. Microbiol.* **21**, 559–561 (1971).
277. C. Genigeoris, G. Sakaguchi, and H. Rieman, *Appl. Microbiol.* **26**, 111–115 (1973).
278. B. L. Vermilyea, H. M. Walker, and J. C. Ayres, *Appl. Microbiol.* **16**, 21–24 (1968).
279. R. F. Meyer and M. J. Palmieri, *Appl. Environ.* **40**, 1080–1085 (1980).
280. R. B. Read, J. Bradshaw, W. L. Pritchard, and L. A. Black, *J. Dairy Sci.* **48**, 420–424 (1965).
281. R. Robbins, S. Gould, and M. Bergdoll, *Appl. Microbiol.* **28**, 946–950 (1974).
282. M. S. Bergdoll and R. W. Bennett, "Staphylococcal Enterotoxins," in M. L. Speck, ed., *Compendium of Methods for the Microbiological Examination of Foods*, 2nd ed., American Public Health Association, Washington, D.C., 1984, pp. 428–457.
283. V. L. Zehren and V. F. Zehren, *J. Dairy Sci.* **51**, 635–644 (1968).
284. J. L. Smith and M. M. Bencivengo, *J. Food Safety* **7**, 83–100 (1985).
285. S. A. Morse and R. A. Mah, *J. Appl. Microbiol.* **15**, 58–61 (1967).
286. S. J. Silverman, A. R. Knott, and M. Howard, *Appl. Microbiol.* **16**, 1019–1023 (1968).
287. S. Yameda, H. I. Garishi, and T. Teryama, *Microbiol. Immunol.* **21**, 675–682 (1977).
288. G. M. Evancho, D. H. Ashton, E. J. Briskey, and E. J. Schantz, *J. Food Sci.* **38**, 764–767 (1973).
289. L. L. Salomon and R. W. Tew, *Proc. Soc. Exp. Biol. Med.* **129**, 539–542 (1968).
290. P. R. Berry, M. F. Stringer, and T. Uemura, *Lett. Appl. Microbiol.* **2**, 101–102 (1986).
291. H. M. Johnson, J. A. Bukovic, P. E. Kauffman, and J. T. Peller, *Appl. Microbiol.* **22**, 837–841 (1971).
292. W. S. Collins, A. D. Johnson, J. F. Metzger, and R. W. Bennett, *Appl. Microbiol.* **25**, 774–777 (1973).

293. H. M. Johnson, J. A. Bukovic, and P. E. Kauffman, *Appl. Microbiol.* **26**, 309–313 (1973).

294. J. A. Bukovic and H. M. Johnson, *Appl. Microbiol.* **30**, 700–701.

295. H. Robern, M. Dighton, Y. Yano, and N. Dickie, *Appl. Microbiol.* **30**, 525–529 (1975).

296. H. Robern, T. M. Gleeson, and R. A. Szabo, *Can. J. Microbiol.* **24**, 436–439 (1978).

297. H. Robern and T. M. Gleeson, *Can. J. Microbiol.* **4**, 137–143 (1978).

298. S. Lindroth, and A. Niskanen, *Eur. J. Appl. Microbiol.* **4**, 137–143 (1977).

299. N. Niyomvit, K. E. Stevenson, and R. F. McFeeters, *J. Food Sci.* **43**, 735–739 (1978).

300. J. F. Metzger, and A. D. Johnson, *Abst. Annu. Methods Am. Soc. Microbiol.*, 259 (1977).

301. B. A. Miller, R. F. Reiser, and M. S. Bergdoll, *Appl. Environ. Microbiol.* **36**, 421–426 (1978).

302. P. D. W. Areson, S. E. Charm, and B. L. Wong, *J. Food Sci.* **45**, 400–401 (1980).

303. G. F. Ibrahim, H. M. Radford, and J. R. Fell, *Appl. Environ. Microbiol.* **39**, 1134–1137 (1980).

304. D. S. Orth, *Appl. Environ. Microbiol.* **34**, 710–714 (1977).

305. N. Dickie and S. H. Akhtar, *J. Assoc. Off. Anal. Chem.* **65**, 180–184 (1982).

306. H. B. Greenberg, D. A. Sack, W. Rodriquez, R. B. Sack, R. G. Wyatt, A. R. Kalica, K. L. Horswood, R. M. Chanock, and A. Z. Kapikian, *Infect. Immun.* **17**, 541–545 (1977).

307. M. Ciska, "Methods," in J. J. Langone and H. van Vunakis, eds., *Enzymology*, Vol. 84, Academic Press, Inc., Orlando, Fla., 1982, pp. 238–253.

308. D. A. Boroff and G. Shu-Chen, *Fed. Proceed.* **32**, 1032 (1973).

309. A. C. Ashton, J. S. Crowther, and J. O. Dolly, *Toxicon.* **23**, 235–246 (1985).

310. G. N. Stelma, J. C. Wimsatt, P. E. Kauffman and D. Shah, *J. Food Prot.* **46**, 1069–1073 (1983).

311. G. Stiffler-Rosenberg, and H. Fey, *J. Clin. Microbiol.* **8**, 473–479 (1978).

312. P. E. Kauffman, *J. Assoc. Off. Anal. Chem.* **63**, 1138–1143 (1980).

313. J. K. S. Kuo and G. J. Silverman, *J. Food Prot.* **43**, 404–407 (1980).

314. W. Lenz, R. Thelen, P. Pickerhahn, and H. Brandis, *Zentrabl. Bakteriol. Infectionskr. Hyg. Abt. A.* **253**, 466–475 (1983).

315. H. Fey, H. Pfister, and O. Ruegg, *J. Clin. Microbiol.* **19**, 34–38 (1984).

316. P. D. Patel and P. A. Gibbs, *Biochem. Soc. Trans.* **12**, 264–265 (1984).

317. G. C. Saunders and M. L. Bartlett, *Appl. Environ. Microbiol.* **34**, 518–522 (1977).

318. J. W. Koper, A. M. Hagenaars, and S. Notermans, *J. Food Safety* **2**, 35–45 (1980).

319. H. Buning-Pfaue, P. Timmermans, and S. Notermans, *Z. Lebensm. Unters.-Forsch.* **173**, 351–355 (1981).

320. B. Berdal, P. Olsvik, and T. Omland, *APMIS Microbiol. Scand.* **89** (Sect. B), 411–415 (1981).

321. R. C. Freed, M. L. Everson, R. F. Reiser, and M. S. Bergdoll, *Appl. Environ. Microbiol.* **44**, 1349–1355.

322. S. Notermans, R. Boot, P. D. Tips, and M. P. DeNooij, *J. Food Prot.* **46**, 238–241 (1983).

323. A. A. Wieneke and R. J. Gilbert, *Int. J. Food Microbiol.* **4**, 135–143 (1987).

324. I. F. Hahn, P. Pickenhahn, W. Lenz, and H. Brandis, *J. Immunol. Methods* **92**, 25–29 (1986).

325. G. Terplan, M. Lohneis, and K. H. Jaschke, "Immunoluminometric Assay (ILMA), a New Ion-Isotopic Assay for Detection of Staphylococcal Enterotoxins," in *Proceedings of the Second World Congress on Foodborne Infections and Intoxications*, Institute of Veterinarian Medicine/Robert Von Ostertag Institute, Berlin, 1986, pp. 448–451.

326. P. Boerlin, H. Pfister, and H. Fey, "An Immuno-Dot Assay for the Detection of Staphylococcal Enterotoxins in Cultures," in Ref. 323, pp. 444–447.

327. H. Fey and H. Pfister, "A Diagnostic Kit for the Detection of Staphylococcal Enterotoxins (SET) A, B, C and D," in S. Avrameas, P. Druet, R. Masseyeff, and G. Feldman, eds., *Immunoenzymatic Techniques*, Elsevier, Amsterdam, The Netherlands, 1983, p. 345.

328. S. Notermans, J. Dufrenne, and M. Van Schotharst, *Jpn. J. Med. Sci. Biol.* **31**, 81–85 (1978).

329. S. Notermans, J. Dufrenne, and S. Kozaki, *Appl. Environ. Microbiol.* **37**, 1173–1175 (1979).

330. S. Notermans, A. M. Hagenaars, and S. Kozaki, "The Enzyme-Linked Immunosorbent Assay (ELISA) for the Detection and Determination of *Clostridium botulinum* Toxins A, B and E," in Ref. 306, 1982. pp. 223–238.

331. S. Notermans, J. Dufrenne, and S. Kozaki, *Jpn. J. Med. Sci. Biol.* **35**, 203–211 (1982).

332. S. Notermans, S. Kozaki, Y. Kamata, and G. Sakaguchi, *Jpn. J. Med. Sci. Biol.* **37**, 137–140 (1984).

333. G. E. Lewis, S. S. Kulinski, D. W. Reischard, and J. F. Metzger, *Appl. Environ. Microbiol.* **42**, 1018–1022 (1981).

334. M. Dezfulian and J. G. Bartlett, *J. Clin. Microbiol.* **19**, 645–648 (1984).

335. M. Dezfulian and J. G. Bartlett, *Diagn. Microbiol. Infect. Dis.* **3**, 105–112 (1985).

336. C. Shone, P. Wilton-Smith, M. Appleton, P. Hambleton, N. K. Modi, S. Gatley, and J. Elling, *Appl. Environ. Microbiol.* **50**, 63–67 (1985).

337. N. K. Modi, C. G. Shone, P. Hambleton, and J. Melling, "Monoclonal Antibody Based Amplified Enzyme-Linked Immunosorbent Assays for *Clostridium botulinum* Toxin Types A and B," in Ref. 323, pp. 1184–1188.

338. J. Ligieza, M. Michalik, J. Reiss, and J. Grzybowski, *Arch. Immunol. Ther. Exp.* **34**, 189–195 (1986).

339. G. N. Stelma, C. H. Johnson, and D. B. Shah, *J. Food Prot.* **48**, 227–231 (1985).

340. B. A. McClane and R. J. Strouse, *J. Clin. Microbiol.* **19**, 112–115 (1984).

341. O. Olsvik, P. E. Granum, and B. P. Berdal, *APMIS* **90**, 445–447 (1982).

342. K. G. Narayan, C. Genigeorgis, and D. Behymer, *Int. J. Zoonoses.* **10**, 105–110 (1983).

343. S. Notermans, C. Heuvelman, H. Beckers, and T. Uemura, *Zentralbl. Bakteriol. Mikrobiol. Hyg.* **179**, 225–234 (1984).

344. B. A. Bartholomew, M. F. Stringer, G. N. Watson, and R. J. Gilbert, *J. Clin. Pathol.* **38**, 222–228 (1985).

345. R. H. Yolken, H. B. Greenberg, M. H. Merson, R. B. Sack, and A. Z. Kapikain, *J. Clin. Microbiol.* **6**, 439–444 (1977).

346. I. Ketyi and A. S. Pacsa, *Acta Microbiol. Hung.* **27**, 89–97 (1980).

347. E. Back, A.-M. Svennerholm, J. Holmgren, and R. Bollby, *J. Clin. Microbiol.* **10**, 791–805 (1979).

348. A.-M. Svennerholm and G. Wirklund, *J. Clin. Microbiol.* **17**, 596–600 (1983).

349. L. Beutin, L. Bode, T. Richter, G. Pettre and R. Stephan, *J. Clin. Microbiol.* **19**, 371–375 (1984).

350. S. L. Mosely, I. Huq, M.S.O.M. Samadpour-Matalebi, and S. Falkow, *J. Infect. Dis.* **142**, 892–898 (1980).

351. National Academy of Sciences, *Regulatory Pesticides in Foods*, Committee on Scientific and Regulatory Issues Underlying Pesticide Use Patterns and Agricultural Innovation, Board of Agriculture, National Research Council, Washington, D.C., 1987.

352. J. E. Lovelock and S. R. Lipsky, *J. Am. Chem. Soc.* **82**, 431 (1960).

353. F. L. Bryan, *J. Food. Prot.* **51**, 498–508 (1988).

354. W. H. Andrews, *Food Technol.* **39**, 77–82 (1985).

355. J. H. Silliker, *Food Technol.* **36**, 65–70 (1982).

356. B. Swaminathan, J. Ag Alexio, and S. A. Minnich, *Food Technol.* **39**, 83–89 (1985).

357. J. A. Mattingly, B. J. Robinson, A. Boehm and W. D. Gehle, *Food Technol.* **39**, 90–94 (1985).

358. R. S. Flowers, K. Eckner, D. A. Gabis, B. J. Robinson, J. A. Mattingly and J. H. Silliker, *J. Assoc. of Anal. Chem.* **69**, 786–798 (1986).

359. R. S. Flowers, M. J. Klatt, B. J. Robinson, J. A. Mattingly, D. A. Gabis and J. H. Silliker, *J. Assoc. Off. Anal. Chem.* **70**, 530–535 (1987).

360. R. S. Flowers, M. J. Klatt, B. J. Robinson, J. A. Mattingly, *J. Assoc. Off. Anal. Chem.* **71**, 341–343 (1988).

361. C. Lindhardt, P. Bielefeld and M. Wilken, *J. Appl. Bacteriol.* **67**, 1989–1993 (1989).

362. G. M. Wyatt, H. A. Lee and M. R. A. Morgan, *Br. Poult. Sci.* **30**, 979–981 (1989).

363. A. L. Gorelova, G. A. Levina and S. V. Prozorovskii, *Mikrobiol. Epidemiol. Immunobiol.* **8**, 60–64 (1989).

364. R. L. Hopfer, R. Pinzon, M. Wenglar, and K. V. I. Rolston, *J. Clin. Microbiol.* **22**, 677–679 (1985).

365. M. A. Khan, A. Seaman, and M. Woodbine, *Zentralbl. Bakteriol. Parasitenk. Infektionskr. Hyg. Abt. 1* **239**, 62–69 (1977).

366. E. Neter, H. Azai, and E. A. Gorynski, *Proc. Soc. Exp. Biol. Med.* **105**, 131–134 (1960).

367. P. E. Pease, L. Nicholls, and M. R. Stuart, *J. Gen. Microbiol.* **73**, 567–569 (1972).

368. P. A. Wentworth and H. K. Ziegler, *J. Immunol.* **138**, 2671–2678 (1987).

369. B. T. Butman, M. C. Plank, R. J. Durham, and J. A. Mattingty, *Appl. Environ. Microbiol.* **54**, 1564–1569 (1988).

370. J. A. Mattingly, B. T. Butman, M. C. Plank, and R. J. Durham, *J. Assoc. Off. Anal. Chem.* **71**, 679–681 (1988).

371. R. R. Beumer and E. Brinkman, *Food Microbiol.* **6**, 171–178 (1989).

372. P. Lombardo, *J. Assoc. Off. Anal. Chem.* **72**, 518–520 (1989).

373. R. L. Ellis, *J. Assoc. Off. Anal. Chem.* **72**, 521–524 (1989).

374. W. P. McKinley and J. H. Mahon, *J. Assoc. Off. Agric. Chem.* **42**, 725–734 (1959).

375. A. Major, Jr., and H. C. Barry, *J. Assoc. Off. Agric. Chem.* **44**, 202–207 (1961).

376. J. P. Minyard and E. R. Jackson, *J. Assoc. Off. Agric. Chem.* **46**, 843–859 (1963).

377. J. J. Langone and H. VanVunakis, *Res. Commun. Chem. Pathol. Pharmacol.* **10**, 163–171 (1975).

378. W. H. Newsome and J. B. Shields, *J. Agr. Food Chem.* **29**, 220–222 (1981).

379. K. D. Wing, B. D. Hammock, and D. A. Wustner, *J. Agr. Food Chem.* **26**, 1328–1332 (1978).

380. D. Knopp, P. Nuhn, and H. J. Dobberkau, *Arch. Toxicol.* **58**, 27–32 (1985).

381. D. F. Rinder, and J. R. Fleeker, *Bull. Environ. Contam. Toxicol.* **26**, 375–380 (1981).

382. S. I. Wie, A. P. Sylwester, K. D. Wing, and B. D. Hammock, *J. Agr. Food Chem.* **30**, 943–948 (1982).

383. C. D. Ercegovich, R. P. Vallejo, R. R. Gehig, L. Woods, E. R. Bogus, *J. Agr. Food Chem.* **29**, 559–563 (1981).

384. T. Levitt, *Lancet* **2**, 358 (1977).

385. D. Fatori and W. M. Hunber, *Clin. Chim. Acta* **100**, 81–90 (1980).

386. H. R. Lukens, C. B. Williams, S. A. Levison, W. B. Dendliker, D. Murayama, and R. L. Baron, *Environ. Sci. Technol.* **11**, 292–297 (1977).

387. M. Schwalbe, E. Dorn, and K. Beyerman, *J. Agr. Food Chem.* **32**, 734–741 (1984).

388. B. S. Ferguson, in *Abstracts of Papers, The Sixth International Congress of Pesticide Chemistry, Ottawa, Canada*, IUPAC, 1986, Abst. 5C-09.

389. S. J. Wratten and P. C. Feng "Pesticide Analysis by Immunoassay," in Ref. 24, pp. 201–220.

390. J. F. Brady, J. R. Fleeker, R. A. Wilson, and R. O. Mumma, "Enzyme Immunoassay for Aidicarb," in R. G. W. Wang, C. A. Franklin, R. C. Honeyalt, and J. C. Reinert, eds., *Biological Monitoring for Pesticide Exposure*, American Chemical Society, Washington, D.C., 1989, pp. 262–284.

391. S. J. Huber, *Chemosphere* **14**, 1795–1803 (1985).

392. J. M. Van Emon, in Ref. 386, Abst. 5C-01.

393. T. Kitagawa, T. Kawasaki, and H. Munechika, *J. Biochem.* **92**, 585–590 (1982).

394. M. M. Kelley, E. W. Zahnow, W. C. Petersen, and S. T. Toy, *J. Agr. Food Chem.* **33**, 962–965 (1985).

395. K. M. Robotti, in *Abstracts of Papers, 192nd ACS National Meeting, Anaheim, Calif.*, American Chemical Society, Washington, D.C. 1986, Abst. 42.

396. S. I. Wie and B. D. Hammock, *J. Agr. Food Chem.* **30**, 949–957 (1982).

397. F. Gung, H. H. D. Meyer, and R. T. Hamm, *J. Agr. Food Chem.* **37**, 1183–1187 (1989).

398. W. H. Newsome and P. G. Collins, in Ref. 386, Abst. 5C-13.

399. R. O. Harrison, A. A. Brimfield, K. W. Hunter, and J. O. Nelson, in Ref. 386, Abst. 5C-12.

400. W. H. Newsome, *J. Agr. Food Chem.* **33**, 528–530 (1985).

401. A. A. Brimfield, D. E. Lemz, C. Graham, and K. W. Hunter, *J. Agr. Food Chem.* **33**, 1237–1242 (1985).

402. Z. Niewola, S. T. Walsh, and G. E. Davies, *Int. J. Immunopharmacol.* **5**, 211–218.

403. Z. Niewola, C. Hayward, B. A. Symington, and R. T. Robson, *Clin. Chim. Acta* **148**, 149–156 (1985).

404. J. M. Van Emon, J. N. Seiber, and B. D. Hammock, in P. A. Hedin, ed., *Bioregulations for Pesticide Control*, ACS Symp. Ser. No. **276**, American Chemical Society, Washington, D.C., 1985, pp. 307–316.

405. J. Van Emon, B. Hammock, and J. N. Seiber, *Anal. Chem.* **58**, 1866–1873 (1986).

406. A. A. Brimfield and K. W. Hunter, in *Abstracts of Papers. The 100th Annual Meeting of AOAC, Scottsdale, Ariz.*, 1986, Abst. 110.

407. K. H. Kuniyaki and S. McCarthy, in Ref. 386, Abst. 5C-11.
408. S. J. Huber and B. Hock, *J. Plant Dis. Prot.* **92**, 147–156 (1985).
409. W. H. Newsome, *Bull. Environ. Contam. Toxicol.* **36**, 9–14 (1986).
410. B. Dunbar, B. Riggle, and G. Niswender, *J. Agr. Food Chem.* **38**, 433–437 (1990).
411. R. B. Koch, T. N. Patil, B. Glick, R. S. Stinson, and E. A. Lewis, *Pest. Biochem. Physiol.* **12**, 130–140 (1979).
412. D. D. VanHouweling and F. J. Kingma, *Am. J. Vet. Res.* **155**, 2197–2202 (1969).
413. R. P. Lehman, *J. Anim. Sci.* **35**, 1340–1345 (1972).
414. D. H. Mercer, *Antimicrobial Drugs in Food Producing Animals. Veterinary Clinics of North America*, Vol. 5, W. D. Saunders, Philadelphia, 1975.
415. *Fed. Regist.* **42**, 62211 (1977).
416. M. S. Brady, and S. E. Katz, *J. Food Prot.* **51**, 8–11 (1988).
417. D. L. Thompson, D. Wood, and I. Thomson, *J. Food Prot.* **31**, 512–515 (1988).
418. H. W. Wehr, *The Incidence of Antibiotics Other Than Penicillin in Producer Raw and Finished Milk Products*, paper presented at the Annual Meeting of the IAMFES, Anaheim, Calif., 1987.
419. National Center for Toxicological Research, *Chronic Toxicology and Carcionenicity Studies of Sulfamethazine in B6C3F1 Mice*, tech. rep., NCTR, Jefferson, Ark., 1988.
420. D. Barnes, and O. H. Rigglemar, *J. Assoc. Off. Anal. Chem.* **54**, 1195–11999 (1971).
421. W. Horwitz, *J. Assoc. Off. Anal. Chem.* **64**, 104–130 (1981).
422. G. W. Peng, A. F. Gadalla, and W. L. Chori, *Res. Commun. Pathol. Pharmacol.* **18**, 233–246 (1977).
423. K. Lanbeck and B. Lindstrom, *J. Chromatogr.* **154**, 321–324 (1978).
424. A. J. Manvel and W. A. Steller, *J. Assoc. Off. Anal. Chem.* **64**, 794–799 (1988).
425. A. C. Bratton, E. K. Marshall, D. Babbitt, and A. R. Hendrickson, *J. Biol. Chem.* **28**, 537–550 (1939).
426. F. Fisler, J. L. Sutter, J. N. Bathish, and H. E. Hagma, *J. Agr. Food Chem.* **48**, 278–279 (1968).
427. J. Reider, *Chemotherapy* **17**, 1–21 (1972).
428. L. R. Alexander and E. R. Stanley, *J. Assoc. Off. Anal. Chem.* **48**, 278–279 (1965).
429. E. M. H. Finlay, D. E. Games, J. R. Startin, and J. Gilbert, *Biomed. Environ. Mass Spectrom.* **13**, 633–639 (1986).
430. S. E. Charm and R. Chi, *J. Assoc. Off. Anal. Chem.* **71**, 304–316 (1988).
431. Residues Avoidance Program, "FSIS 20," U.S. Department of Agriculture, Food Safety and Inspection Service, Washington, D.C., 1982.
432. R. B. Ashworth, R. L. Epstein, M. H. Thomas, and L. T. Frobish, *J. Vet. Res.* **47**, 2596–2603 (1985).
433. V. W. Randecker, J. A. Reagan, R. E. Engel, D. L. Soderberg, and J. E. McNeal, *J. Food Prot.* **50**, 115–122 (1987).
434. D. W. A. Bourne, R. F. Beville, R. M. Sharma, R. F. Gural, and L. W. Dittert, *J. Vet. Res.* **38**, 967–972 (1977).
435. R. F. Beville, K. M. Schemske, H. G. Luther, E. A. Ozierzak, M. Limpoka, and D. R. Felt, *J. Agr. Food Chem.* **26**, 1201–1203 (1978).
436. J. R. Fleeker, and L. J. Lovett, *J. Assoc. Off. Anal. Chem.* **68**, 172–174 (1985).
437. D. E. Dixon-Holland and S. E. Katz, *J. Assoc. Off. Anal. Chem.* **71**, 1137–1140 (1988).
438. P. Singh, B. P. Ram, and N. H. Sharkov, *J. Agr. Food Chem.* **37**, 109–114 (1989).
439. D. E. Dixon-Holland and S. E. Katz, *J. Assoc. Off. Anal. Chem.* **72**, 447–450.
440. D. E. Dixon-Holland and S. E. Katz, *J. Assoc. Off Anal. Chem.* (submitted).
441. S. E. Katz and M. S. Brady, *J. Assoc. Off. Anal. Chem.* **73**, 557–560 (1990).

GENERAL REFERENCES

G. Anguita et al., "Detection of Bovine Casein in Ovine Cheese Using Digoxigenated Monoclonal Antibodies and a Sandwich-ELISA," *Milchwissenschaft* **52**, 511–513 (1997).

D. Atkins and J. Norman, "Mycotoxins and Food Safety," *Nutrition and Food Science* No. 4/5, 260–265 (1998).

H. L. Beasley et al., "Development of a Panel of Immunoassays for Monitoring DDT, Its Metabolites and Analogues in Food and Environmental Matrixes," *J. Agric. Food Chem.* **46**, 3339–3352 (1998).

G. A. Bennet, T. C. Nelsen, and B. M. Miller, "Enzyme-Linked Immunosorbent Assay for Detection of Zearalenone in Corn, Wheat, and Pig Feed: A Collaborative Study," *J. AOAC Int.* **77**, 1500–1509 (1994).

B. W. Blais et al., "Polymacron Enzyme Immunoassay System for Detection of Naturally Contaminating *Salmonella* in Foods, Feeds and Environmental Samples," *J. Food Prot.* **61**, 1187–1190 (1998).

K. Byongki, "Quantification of Castor Bean Allergen in Castor Meal by Rocket Immunoelectrophoresis," *Foods and Biotechnology* **4**, 93–97 (1995).

N. J. Cook et al., "Radioimmunoassay for Cortisol in Pig Saliva and Serum," *J. Agric. Food Chem.* **45**, 395–399 (1997).

A. Dankwardt and B. Hock, "Enzyme Immunoassays for Analysis of Pesticides in Water and Food," *Food Technology and Biotechnology* **35**, 165–174 (1997).

H. P. van Egmond, "Analytical Methodology and Regulations for Ochratoxin A," *Food Additives and Contaminants* **13** (Suppl.), 11–13 (1996).

S. A. Eremin et al., "Polarization Fluoroimmunoassay Determination of the Pesticide 2,4,5-trichlorophenoxyacetic Acid," *Zh. Anal. Khim.* **50**, 215–218 (1995).

M. Franeck et al., "Development of a Microcolumn Radioimmunoassay for Screening of Polychlorinated Biphenyls in Milk and in Animal Fats," *J. Agric. Food Chem.* **40**, 1559–1565 (1992).

A. A. Frolich, R. R. Marequardt, and J. R. Clarke, "Enzymatic and Immunological Approaches for the Quantitation and Confirmation of Ochratoxin A in Swine Kidneys," *J. Food Prot.* **60**, 172–176 (1997).

T. Holzhauser et al., "Rocket Immunoelectrophoresis (RIE) for Determination of Potentially Allergenic Peanut Proteins in Processed Foods as a Simple Means of Quality Assurance and Food Safety," *Z. Lebensm. Unters. Forsch.* **206**, 1–8 (1998).

V. Issert, P. Grenier, and V. Maurel Bellon, "Emerging Analytical Rapid Methods for Pesticide Residues Monitoring in Foods: Immunoassays and Biosensors," *Sciences des Aliments* **17**, 131–144 (1997).

K. Jorgensen, "Survey of Pork, Poultry, Coffee, Beer and Pulses for Ochratoxin A," *Food Additives and Contaminants* **15**, 550–554 (1998).

J. P. Joung and Schu Fun, "Assessment of Immunochemical Methods for the Analysis of Tricothene Mycotoxins in Naturally Occurring Moldy Corn," *J. AOAC Int.* **79**, 465–471 (1996).

T. Kakui et al., "Development of Monoclonal Antibody Sandwich-ELISA for Determination of Beer Foam-Active Proteins," *Journal of the American Society of Brewing Chemists* **56**, 43–46 (1998).

K. E. Kerdahi and P. E. Istafanos, "Comparative Study of Colorimetric and Fully Automated Enzyme-linked Immunoassay System for Rapid Screening of *Listeria* spp. in Foods," *J. AOAC Int.* **80**, 1139–1142 (1997).

D. A. Kurtz, J. H. Skerritt, and Stanker, *New Frontiers in Agrochemical Immunoassay*, AOAC International, Arlington, Va. 1996.

R. Martin et al., "Monoclonal Antibody Sandwich-ELISA for the Potential Detection of Chicken Meat in Mixtures of Raw Beef and Pork," *Meat Science* **30**, 23–31 (1991).

E. P. Meulenberg, "Immunochemical Detection of Environmental and Food Contaminants: Development, Validation and Application," *Food Technology and Biotechnology* **35**, 153–163 (1997).

T. V. Milanez, M. B. Atui, and F. A. Lazzari, "Comparison Between Immunoassay and Thin-Layer Chromatography for Determination of Aflatoxins, Ochrotoxin A and Zearalenone in Corn and Corn Meal," *Revista do Istituto Adolfo Lutz* **57**, 65–71 (1998).

J. M. Mitchell et al., "Antimicrobial Drug Residues in Milk and Meat: Causes, Concerns, Prevalence, Regulations, Tests, and Test Performance," *J. Food Prot.* **61**, 742–756 (1998).

H. Pichler et al., "An Enzyme-Immunoassay for the Detection of the Mycotoxin Zearalenone by Use of Yolk Antibodies," *Fresenius J. Anal. Chem.* **362**, 176–177 (1998).

B. Roberts, B. A. Morris, and M. N. Clifford, "Comparison of Radioimmunoassay and Spectrophotometric Analysis for the Quantitation of Hypoxanthine in Fish Muscle," *Food Chem.* **42**, 1–17 (1991).

E. Rodrigues et al., "Sandwich-ELISA for Detection of Goat's Milk in Ewes' Milk and Cheese," *Food Agric. Immunol.* **6**, 195–111 (1994).

F. Sanchez Garcia et al., "Polarization Fluoroimmunoassay of the Herbicide Dichloroprop," *J. Agric. Food Chem.* **41**, 2215–2219 (1993).

R. Schuhmacher et al., "Interlaboratory Comparison Study for the Determination of the Fusarium Mycotoxins Deoxynivlenol in Wheat and Zearalenone in Maize Using Different Methods," *Fresenius J. Anal. Chem.* **359**, 510–515 (1997).

A. Szekacs, "Enzyme-Linked Immunosorbent Assay for Monitoring the *Fusarium* Toxin Zearalenone," *Food Technology and Biotechnology* **36**, 105–110 (1998).

L. S. Shiow and F. M. You, "Determination of Aflatoxins in Liquor Products by Immuno-Affinity Column With Fluorometry and HPLC," *Journal of Food and Drug Analysis* **5**, 161–170 (1997).

E. Simon et al., "Development of an Enzyme Immunoassay for Metsulfuron-Methyl," *Food Agric. Immunol.* **10**, 105–120 (1998).

T. G. Sokari and S. O. Anozie, "Modified Single Radial Immunodiffusion Method for Screening Staphylococcal Isolates for Enterotoxin," *Food Microbiology* **6**, 45–48 (1989).

H. Tian et al., "Rapid Detection of *Salmonella* sp. in Foods by Combination of a New Selective Enrichment and a Sandwich-ELISA Using Two Monoclonal Antibodies Against Dulcitol 1-Phosphate Dehydrogenase," *J. Food Prot.* **59**, 1158–1163 (1996).

I. A. Toscano, G. S. Nunes, and D. Barcelo, "Immunoassays for Pesticide Analysis in Environmental and Food Matrixes," *Food Technology and Biotechnology* **36**, 245–255 (1998).

M. W. Trucksess et al., "Survey of Deoxynivalenol in U.S. in 1993 Wheat and Barley Crops by Enzyme-Linked Immunosorbent Assay," *J. AOAC Int.* **78**, 631–636 (1995).

M. Tuomola et al., "Time-Resolved Fluoroimmunoassay for Measurement of Androsterone in Porcine Serum and Fat Samples," *J. Agric. Food Chem.* **45**, 3529–3534 (1997).

Y. Ueno et al., "A 4-year Study of Plasma Ochratoxin A in a Selected Population in Tokyo by Immunoassay and Immunoaffinity Column-Linked HPLC," *Food and Chemical Toxicology* **36**, 445–449 (1998).

Y. Wanjun and S. Fun, "Improved Direct Competitive Enzyme-Linked Immunosorbent Assay for Cyclopiazonic Acid in Corn, Peanuts and Mixed Feed," *J. Agric. Food Chem.* **46**, 1012–1017 (1998).

A. K. Wintero, P. D. Thomsen, and W. Davies, "A Comparison of DNA-Hybridization, Immunodiffusion, Countercurrent Immunoelectrophoresis and Isoelectric Focusing for Detecting the Admixture of Pork to Beef," *Meat Science* **27**, 75–85 (1990).

G. M. Wood et al., "Ochratoxin A in Wheat: A Second Intercomparison of Procedures," *Food Additives and Contaminants* **13**, 519–539 (1996).

J. M. Yeung and P. G. Collins, "Determination of Soy Proteins in Food Products by Enzyme Immunoassay," *Food Technology and Biotechnology* **35**, 209–214 (1997).

X. Zhao et al., "Radioimmunoassay for Insulin-like Growth Factor-I in Bovine Milk," *Canadian Journal of Animal Science* **71**, 669–674 (1991).

D. E. DIXON-HOLLAND
Neogen Corporation
Lansing, Michigan

INFANT FOODS

INFANT NUTRITION

Infants are defined, somewhat arbitrarily, as less than 1 year of age. Infant foods and feeding practices have been adapted to encompass infants and young children to allow for feeding during the first 2 to 3 years.

The infant has unique nutritional needs, substantially different from those of an adult. Energy, protein, iron, and other needs are greater than at any other period in life. Food must be relatively concentrated, and with a controlled consistency, because the volume capability, chewing, and manipulation of food are limited. At no other point in life is the growth rate of infancy duplicated. Infants typically double their birth weight in 3 to 4 months and triple it by 1 year, and the rapid growth rate places nutritional demands on the food supply. To accompany the rapid growth rate, infants have a very high basal metabolic rate. Complicating this is a limited stomach capacity, requiring an appropriate concentration of nutrients. The limited renal capability must be considered to ensure that electrolytes and protein are not present in deleterious quantities. In addition, the digestive system is immature, which requires that foods be selected that are relatively easy to digest. The most obvious of the characteristics is the limited ability to chew foods; therefore the foods must be appropriate in consistency and texture.

Breast feeding is recognized as the preferred method for feeding very young infants, and breast milk should be the exclusive food for the first 4 to 6 months unless circumstances (mother unable to breast-feed, breast milk inadequate, etc) do not allow exclusive breast feeding. Any food other than breast milk that is introduced should supplement rather than replace breast feeding.

For these reasons, a specific and well-defined set of foods have been developed for infants and young children. Infant feeding practices vary considerably in different parts of the world depending on the availability of food and custom. Definitions and standards have been developed by Codex Alimentarius but are, for the most part, utilized as guidelines.

HISTORY OF FOODS AND FEEDING PRACTICE

Infant foods have evolved largely parallel to the development of food science and technology, and the implementation of food safety and food processing practices has dramatically influenced the availability of safe foods. Before 1800 about two-thirds of children died before 5 years of age, and that statistic has dramatically changed. Wet nurses (women hired to suckle others' children) were widely used when breast feeding was impossible. Otherwise, supplementary foods were of poor quality and unclean; as a result, food was the most frequent cause of serious illness. A first major step was heat treatment of milk; pasturization was first, followed by the development of sweetened condensed milk, which provided a much safer source of milk for infants than the raw or certified milk available previously. Early efforts to prepare foods specifically for infants and young children included meat juices and cereal gruels (Mellon's Food), malted milk (Horlick's Malted Milk), and sweetened condensed milk (Borden). The milk-based foods combined with hydrolyzed complex carbohydrates (dextrins and maltodextrins) evolved to what is now considered infant formula, and foods derived from other sources have been termed baby foods and infant cereal.

INFANT FORMULA

Human milk serves as the nutritional model for infant formula. The original infant formula was derived from cow milk; however, the concentration and nutrient distribution of cow milk are not appropriate for infants. The protein content is too high, the casein is not well digested, the electrolyte concentration is too great, the fat is largely saturated and poorly absorbed, and the carbohydrate concentration is too low. Adjustments have been made using cow milk as a base, and other components have been formulated to simulate or "humanize" milk. Later in the evolution of infant formula, alternate protein sources were used; initially meat and later, during the 1960s, soy-based formulas were refined. Currently, approximately one-fourth of infant formula utilized in the United States is soy protein based.

Regulatory Aspects

Infant formula is unique in food regulation because it is regulated by specific legislative mandate in the United States. The Infant Formula Act of 1980 and subsequent amendments in 1986 specify nutrient requirements, manufacturing practices, and quality control. This specificity is considered essential because infant formula may provide the sole source of nutrition for an infant for an extended period during critical rapid growth and development.

Functionality

Infant formula has evolved as a nutritional entity to provide all the requirements for an infant during the first 6 months. After 6 months infant formula, like human milk, continues to provide a significant contribution to the nutrition of infants. The adequacy of infant formula to support growth and development is established by clinically controlled feeding trials.

Technologies

Several technologies have been utilized to ensure the safety and nutritional quality of infant formula. Concentration and controlled processing techniques ensure the sterility, keeping quality, and nutrient retention for undiluted infant formula. Drying techniques have been adapted to provide powdered infant formula with quality equivalent to liquid forms. Alternate protein sources predominately soy protein, have been refined to provide adequate protein quality for the normal growth and development of infants. Cow milk protein has been modified to imitate the casein to whey protein distribution of human milk. The carbohydrate and fat components have been adjusted to effectively imitate those of human milk. Clinical trials are used to confirm the nutritional adequacy of infant formula. Formulas based on extensively hydrolyzed protein are effectively used for infants with food protein allergies.

BABY FOODS

Over 300 varieties of baby foods are available in North America to fulfill specific needs during the development of the infant. Recommendations by the American Academy of Pediatrics and other involved organizations are that baby foods be introduced at 4 to 6 months to supplement human milk or infant formula.

Several terms are used, somewhat interchangeably, to define baby foods: supplementary foods, biekost, and solid food. Other accurate classifications include strained fruit juices, infant cereal, fruits, vegetables, and numerous combination foods with fanciful names.

The first foods introduced are generally infant cereals. Infant cereal is a specific food: precooked, partially hydrolyzed, dried cereal-based food. In North America, infant cereal is fortified with iron, vitamins, and minerals. Infant cereal is fed in a diluted form with water, human milk, or infant formula used as a diluent. Ready-to-use cereal is also available, diluted and mixed with fruit for convenient and safe feeding.

Fruits and vegetables are essentially the pureed version of the food sterilized in a container to provide an appropriate serving size. Their container (serving) sizes range from approximately 70 to 135 g (2.5–4.5 oz) for subsequent forms. Fruit juices are strained, single-ingredient and combination juices provided in appropriate-sized containers, fortified with vitamin C at a uniform level.

Meat and poultry are provided in pureed form and sterilized in small containers. Numerous combination foods are prepared, largely for consumer convenience. The terminology used for combination foods is manufacturer specific and intended to be descriptive for the consumer.

Functionality

The functionality of infant foods generally refers to the consistency (as either a puree, liquid, or slurry). Consistency relates to the physical limitations of the infant and young child, but other characteristics are equally important to establish the appropriateness of baby foods for consumption. Among these are nutritional adequacy, particularly calories and protein, and concentration. If the concentration is too high, the food may stress the infant's developing renal system. If too dilute, the food may satisfy the volume requirements for the infant digestive system without provision of adequate nutrients.

Technologies

The technologies for baby foods have been adapted from food processing techniques, with specific modifications that allow the foods to fulfill the specific requirements for the intended purpose. The processes have evolved, and specific improvements in food science and food processing have been applied. Baby foods, as presently recognized, were developed between 1920 and 1930 by numerous manufacturers. Originally, the distribution for baby foods was entirely through pharmacies; after 1933 convenience was enhanced by distribution through food outlets, first grocery stores and then supermarkets. The technology originated in North America and did not expand to other parts of the world until after World War II.

Infant cereal was developed in 1933 as a partially hydrolyzed, sterilized food prepared from flour and dried in a flake form to facilitate rapid dispersion. Pablum was the first widespread product and was the generic reference for infant cereal for many years. Supplemented with iron and water-soluble vitamins, it is a well-tolerated food used to supplement milk-based foods.

Canned baby foods are based on fruits, vegetables, and meats finely divided, or pureed. Baby foods were originally sterilized in 4- to 4.5-oz metal cans; conversion to glass containers was made in the period between 1955 and 1960. Combinations of fruit, vegetables, cereal, and meat have been made to provide convenience to the consumer.

CONTROVERSIES

Numerous controversies surrounded preparation and use of baby foods, and many were paralleled by similar discussions concerning the safety and suitability of the entire food supply. Discussions were initiated during a White House conference on nutrition in 1969, with allegations and questions that had not been comprehensively treated in earlier scientific and safety testing. Since 1969 most of the controversies have been resolved. Thorough evaluation, more extensive knowledge, and effective communications have provided consumers and scientists with assurance of the safety and suitability of the foods and ingredients used.

Salt

The intake of salt by older infants has been of interest and concern to nutrition scientists for the past 25 years. The concern was initiated based on the possibility that consumption of salt in infancy could influence the salt intake in later life, possibly contributing to the development of hypertension. The subject has been reviewed by the National Academy of Sciences and the American Academy of Pediatrics, which concluded that no advantage could be served by addition of salt to infant foods. Salt has not been an ingredient in infant foods since 1977.

Sugar

Sugar has been criticized as an ingredient in baby foods for two basic reasons: it adds calories without other nutrients and it may, if improperly used, cause tooth decay (infant dental caries). Both of the allegations or hypotheses were only partially substantiated, but most baby foods are prepared without added sugar.

Monosodium Glutamate

Monosodium glutamate (MSG) was used as a flavor enhancer in some meat-containing foods before 1969. As a result of safety questions and lack of clinical safety information, MSG is not used as an ingredient in baby foods.

Modified Food Starch

Modified food starch (MFS) has been used extensively in many foods as a digestible stabilizing system. Baby foods are an ideal use, because small amounts of MFS (1–2%) effectively suspend the particles and prevent age-associated separation. The safety, at levels of use, has been established by the National Academy of Sciences and the American Academy of Pediatrics. The use remains a source of controversy because critics of MFS consider it an extender rather than a stabilizer.

Artificial Colors and Flavors

Artificial colors, flavors, and preservatives are not used in baby foods primarily because they are not required and many consumers have an aversion to their use.

CONSUMPTION PATTERNS

Changes in Usage

From the time of their development in the 1950s and 1960s until 1972, the use of infant formulas increased substantially to a point that approximately 70% of infants were fed formula. As a result of concerted efforts by feeding advisers, primarily pediatricians, the practice of breast feeding significantly increased in both incidence and duration.

By the late 1980s the ratio had changed to a point that more than 70% of infants were breast-fed, and that ratio prevails to the present.

The consumption of baby foods increased considerably from the time of initiation in the 1920s until a maximum level of usage was reached, almost 800 jars per baby, in 1972. The advantages and the role were recognized: to provide a supplementary quantity of energy, a variety of foods, and valuable nutrients, particularly protein. Several recommendations were made during this period by the American Academy of Pediatrics in 1958, 1976, 1980, and 1986. During the decade after 1972, the usage decreased by approximately 20% as a result of a number of factors. Consumption of supplementary foods was perceived to interfere with breast feeding, questions arose concerning overfeeding, which resulted in infant obesity, and alternatives became available. Consumption has been stable during the past decade.

ALTERNATIVES

Numerous alternative foods and feeding practices were involved in the reduced consumption of baby foods. Foods were often prepared from family foods—home preparation—or were specifically prepared from selected ingredients. The use of milk-based foods, sometimes referred to as follow-on foods, was more extensive. In some instances breast feeding or infant formula was continued for a more extensive period. Consumer concerns, economics, and availability of foods were all contributing factors to the changes and use of alternatives.

GENERAL REFERENCES

American Academy of Pediatrics, *Pediatric Nutrition Handbook*, 3rd ed., American Academy of Pediatrics, Elk Grove Village, Ill., 1993.

American Academy of Pediatrics Committee on Nutrition, "Sodium Intake by Infants in the United States," *Pediatrics*, **68**, 444–445 (1981).

J. T. Bond et al., eds., *Infant and Child Feeding*, Academic Press, New York, 1981.

National Research Council Food Protection Committee, Food and Nutrition Board, *Report of the Subcommittee on Safety and Suitability of MSG and Other Substances in Baby Foods*, National Academy of Sciences, Washington, D.C., 1970.

G. A. Purvis and S. J. Bartholmey, "Infant Feeding Practices: Commercially Prepared Baby Foods," in R. C. Tsang and B. L. Nichols, eds., *Nutrition During Infancy*, Henly & Belfus, Inc., Philadelphia, Pa., 1988.

George Purvis
Purvis Consulting, Inc.
Fremont, Michigan

INSTITUTE OF FOOD TECHNOLOGISTS (IFT)

The Institute of Food Technologists (IFT) was founded July 1, 1939, by unanimous vote of attendees at the closing session at the Second Food Technology Conference held at the Massachusetts Institute of Technology (MIT), Cambridge, Massachusetts. The first conference had been organized by MIT and was chaired by Samuel Cate Prescott, Dean of Science and head of the biology department at MIT. The concept of the Institute of Food Technologists had been described at meetings at the New York State Agricultural Experiment Station in Geneva, New York, and G. J. Hucker was asked to seek opinion of scientists working in the food field, regarding the need for an organization dedicated to the use of sound science in the business of food.

The purpose of that early form of IFT was "to facilitate interchange of ideas among its members; to stimulate scientific investigations into technical problems dealing with the manufacture and distribution of foods; to promulgate the results of research in food technology; to offer a medium for the discussion of these results; and to plan, organize, and administer such projects for the advancement and application of science insofar as it is fundamental to wider knowledge of foods."

An early suggestion regarding a name for the Institute was "the Society of Food Engineers." The name "Institute of Food Technologists" was favored because it permitted a broader base of professionals. Various other names have been suggested from time to time, but the name has not been changed since the inception of the organization.

Samuel Cate Prescott was IFT's first president. He predicted, in 1941, that IFT would eventually have 1000 members. He would have been surprised at the present membership—approaching 29,000. Currently, the average member is male, about 42, and working for a food or beverage manufacturer. About 20% of members are employed by academic institutions, and a significant number of governmental agency employees are members. About 8% of total members are from countries outside the United States.

IFT operates with a paid staff of about 50 professionals, from its headquarters at 221 N. LaSalle Street, Chicago, IL 60601. The Institute uses a comprehensive committee system to coordinate its 24 divisions (which serve as centers of excellence), 55-plus regional sections, affiliate organizations, and student associations. The Executive Committee, made up of the current president, past president, president elect, treasurer, assistant treasurer, and executive director plus membership representatives, councilor representatives, and a student representative, receives recommendations for action from various standing committees and acts on many of them. Certain recommendations are made to the Council, elected by members on a yearly ballot, for action. The "three presidents" model has been developed to provide for easy transition from year to year.

The Institute has an annual meeting, during which the business of the Institute is conducted at the council meeting, and subsequent executive committee meeting and some 3000 technical presentations are given. At the same time as the annual meeting, Food Expo, the largest exposition of new ingredients, instruments, equipment, and services in the world is held, with about 1800 exhibitors covering several thousand square feet of floor space. Recently, IFT has added a virtual trade show, called

WorldFoodNet, which includes all exhibitors in the exposition, available for those who didn't attend and those who forgot what they might have seen.

The Institute is a professional society, and as such, does not lobby. The emphasis on all interaction with government is the dissemination of science-based information. The Science Communication department of IFT is devoted to the development of solid science for use by the industry, to consumers, and to the media. A project that is relatively recent is the support of a congressional fellow, who acts as a science adviser to members of Congress. This activity is supported by the IFT Foundation, which raises money to be used in important new directions for the Institute, as well as to support a number of scholarships for young people at the university level.

There is a strong educational component to IFT, both with food science departments that meet standards that have been developed by IFT, and in continuing education. Each year, 20-plus programs are given under the guidance of the Continuing Education department, to help food technologists keep up to date with new science and changing regulatory and nutritional emphasis.

Publications from IFT include *Food Technology*, a monthly magazine that includes news about Institute members and the associated academic, industrial, and regulatory components. The *Journal of Food Science* publishes technical papers, six issues per year. There is a membership directory, and Expo-associated publications. The Institute hosts a popular website at www.ift.org where more information is available.

Frances Katz
Institute of Food Technologists
Chicago, Illinois

INSTITUTE OF FOOD TECHNOLOGISTS: AWARDS

The Institute of Food Technologists (IFT) Committee on Awards seeks your nominations for IFT Achievement Awards described in the following. A single nomination form is provided. Election as an IFT Fellow requires a different form, which may be requested from IFT headquarters, by using the IFT E-XPRESS fax-on-demand service by dialing 800-234-0270 in the United States and Canada (elsewhere by dialing 913-495-2551) and requesting document number 3510, or by visiting the IFT Web site at www.ift.org. The Marcel Loncin Research Prize is given in even-numbered years, and instructions for applying for it are available on the IFT Web site or by using IFT-Express (request document 3530).

Rules for All IFT Achievement Awards

1. The nominee must not have received another IFT Achievement Award or the Marcel Loncin Research Prize (but may have been elected an IFT Fellow) within the previous 5 years. A list of prior award recipients appears in the IFT Membership Directory.
2. Any nonstudent IFT member can make a nomination with four exceptions: (*1*) nominations or letters may not come from a juror serving on the same jury, (*2*) a person cannot self-nominate, (*3*) a person may nominate his or her own company for the Industrial Achievement Award, and (*4*) student members can nominate for the Cruess Award.
3. The nomination must be received by IFT by December 1.
4. The nomination is handled in confidence. The identity of the awards jury members is confidential.
5. Unsuccessful nominations will be returned and may be resubmitted in subsequent years.
6. All material in excess of specified page limits will not be given to the jury. Letters of support from sections and divisions are included in these limits and should be addressed to the Awards Committee.
7. In any year in which none of the nominees meet the criteria, an award may not be presented.
8. When two or more people nominate the same person for the same award, the nominators will be contacted and asked to combine their nominations and submit a single nomination to the jury. If there is not enough time to do this, the first nomination received will be sent to the jury for consideration.

NICHOLAS APPERT AWARD

Award. $5,000 honorarium from IFT and a bronze medal from the Chicago Section of IFT.

Purpose. To honor an IFT member or nonmember for preeminence in and contributions to the field of food technology.

Eligibility. The nominee must have made consistent—and essentially lifetime—contributions to food science and technology. Factors of personality or achievement outside the field of food technology cannot be considered. For a detailed description and history of the award and the accomplishments of past winners, see the article "The Nicholas Appert Medalists—A Reflection of the Growth of Food Science and Technology," by Alina Surmacka Szczesniak, *Food Technology*, September 1992, pp. 144–152.

Special instructions. The nomination statement must not be more than four typed pages. Do not include a curriculum vitae or lists of publications and patents.

BABCOCK-HART AWARD

Award. $3,000 honorarium from the International Life Sciences Institute North America and a plaque from IFT.

Purpose. To honor an IFT member who has attained distinction by contributions to food technology that result in improved public health through nutrition or more nutritious food.

Eligibility. The nominee must have made practical technological contributions resulting in improved public health, as distinguished from contributions to clinical nutrition or purely scientific research, unless it has led directly to technological developments. The technological contributions, or the technological development resulting from scientific research, must be in actual, large-scale production leading to better consumer nutrition. Such contribution could be made in the area of food production, food processing, or food packaging, storage, or distribution. An example of a contribution in processing might be the development of equipment that would process food in a way to conserve nutrients or prevent wastage; another might be the development of a process that would inhibit or prevent bacterial spoilage, oxidation, desiccation, or enzymatic deterioration of food, providing such contribution eventually influences the nutritive well-being of the consumer. An example of a contribution in food production might be genetic studies to develop improved strains of cereal, fruits, or vegetables that would provide greater yield or increased nutritive content; another might be studies on animal nutrition that improve efficiency of production or studies on animal diseases that result in decreased losses. Contributions could also be made in packaging, storage, or distribution conditions that would eventually affect nutritive value.

Special instructions. The nomination statement must not be more than four typed pages. Do not include a curriculum vitae or lists of publications and patents.

SAMUEL CATE PRESCOTT AWARD

Award. $3,000 honorarium and a plaque from IFT.

Purpose. To honor an IFT member who has shown outstanding ability in research in some area of food science and technology.

Eligibility. The nominee must, by July 1 of the year of the presentation, (*1*) be younger than 36 years of age or (*2*) have received his or her highest degree within the previous 10 years. Special attention is given to contributions in methodology, competence shown, and effects of the research on advances in food science.

Special instructions. The nomination statement must not be more than four typed pages. Attach a numbered list of nominee's publications and patents and, to save space, refer to them in the nomination statement using these numbers.

INTERNATIONAL AWARD

Award. $3,000 honorarium and a plaque from IFT.

Purpose. To honor an IFT member or an institution whose outstanding efforts result in one or more of the following: (*1*) international exchange of ideas in the field of food technology, (*2*) better international understanding in the field of food technology, and/or (*3*) practical successful transfer of food technology to an economically depressed area in a developing or developed nation.

Eligibility. A balance between technology transfer and international exchange of ideas is desirable. *International exchange of ideas* may be defined as one-on-one interactions, including sustained consulting or advising in several countries. This may include educating and training students, researchers, plant workers, and others who return to foreign lands. *Better international understanding* may be defined as continued interaction through international seminars, symposia, or working groups. Organizing responsibility is an important component. *Technology transfer* is a hands-on operation involving successful one-to-one transfer of food technology from those who have the knowledge to those who need and use it. The result must be a new, commercially viable process, packaging system, postharvest preservation system, or other application of food science and technology of direct benefit to the local economy. Although such work could start with joint assessment of the need and potential market and involve training in the donor's laboratories or pilot plant, emphasis is on on-site training and development of sufficient duration (probably at least 6 months) so the process continues to function and benefit consumers long after the donor has returned home.

Special instructions. The nomination statement must not be more than four typed pages. Do not include a curriculum vitae or lists of publications and patents.

WILLIAM V. CRUESS AWARD

Award. $3,000 honorarium from IFT and a bronze medal from the Northern California Section of IFT.

Purpose. To honor an IFT member who has achieved excellence in teaching food science and technology.

Eligibility. Any IFT member who has at least 5 years of experience in university teaching in a department offering a bachelor's degree in food science, food technology, or the equivalent is eligible. Courses taught by the nominee must deal directly with some aspect of food science or food technology.

Special instructions. Members of the IFT Student Association may nominate individuals for this award. A nomination statement is not required but, if provided, must not be more than one typed page. The nomination should be accompanied by up to eight letters signed by students, former students, or groups of students who have taken courses taught by the nominee. These letters should discuss one or more of the following criteria: (*1*) mastery of subject matter, (*2*) ability to communicate effectively the principles and practices of food processing and preservation, (*3*) enthusiasm for the field of food science and technology as a profession, (*4*) ability to stimulate students to select a professional career in food science and technology, and (*5*) ability to counsel effectively. Letters from up to four professional colleagues and administrators may also be offered but will not be used as primary evidence.

CARL R. FELLERS AWARD

Award. $3,000 honorarium from Phi Tau Sigma and a plaque from IFT.

Purpose. To honor a member of IFT and Phi Tau Sigma who has brought honor and recognition to the profession of food science and technology through a distinguished career in that profession displaying exemplary leadership, service, and communication skills that enhance the effectiveness of all food scientists in serving society.

Eligibility. Nominee must have been a professional member of IFT for at least 15 years before the nomination deadline of December 1. Nominees who are not members of Phi Tau Sigma shall be simultaneously elected to Phi Tau Sigma membership by that organization upon selection as recipient of the Carl R. Fellers Award. In considering the further qualifications of nominees, priority shall be given to a record of achievement in communicating to the profession, to governmental and international agencies, and/or to the public scientific information relative to the quality, wholesomeness, safety, nutritive value, and other aspects of food and the food supply; in bringing individuals into the profession; and/or in propagating knowledge of the importance of the profession to the public, governmental agencies, and scientific and academic communities.

Special instructions. The nomination statement must not be more than four typed pages. Do not include a curriculum vitae or lists of publications and patents.

FOOD TECHNOLOGY INDUSTRIAL ACHIEVEMENT AWARD

Award. A bronze plaque from IFT.

Purpose. To honor a company or organization for developing an outstanding food process and/or food product that represents a significant advance in the application of food science and technology to food production.

Eligibility. The process or product must have been successfully applied in actual commercial operation at least 6 months but no more than 7 years before December 1 of the year in which the nomination is submitted.

Special instructions. The nomination statement should describe, as much as possible, the development itself and its public health, scientific, technological, and economic importance and must not be more than four typed pages. To assist in judging the nomination, you may include 10 copies of each of the following: reprints describing the product or process, sample of the product, and videotape not to exceed 5 minutes. Reprints should be limited to a maximum of 4 different articles, or a total of 20 pages.

CALVERT L. WILLEY DISTINGUISHED SERVICE AWARD

Award. $3,000 honorarium a plaque from IFT.

Purpose. To honor an individual who has provided continuing, meritorious, and imaginative service to IFT. The award was first presented to Calvert L. Willey, then executive director of IFT, in 1987 and initiated on the fiftieth anniversary of IFT in 1989.

Eligibility. The nominee must have been a professional member of IFT for 15 years or more, a member of IFT for 25 years or more, or on the IFT staff for 25 years or more. The nominee must not have received any other IFT Achievement Award. This does not include election as an IFT Fellow or the Marcel Loncin Research Prize. The primary consideration is continuing service to the institute rather than specific contributions to research, education, or development. Qualification is based on dedication, unique service, and supreme effort on behalf of IFT.

Special instructions. The nomination statement must not be more than four typed pages. Do not include a curriculum vitae or lists of publications and patents.

STEPHEN S. CHANG AWARD FOR LIPID OR FLAVOR SCIENCE

Award. $3,000 honorarium and a Steuben crystal sculpture.

Purpose. To honor an IFT member food scientist or technologist who has made significant contributions to lipid or flavor science.

Eligibility. The contributions can be on any aspect of lipid or flavor science, ranging from basic chemistry to applied technology, but must have had some impact on commercial operations. The award applies to all disciplines involved in lipid or flavor science. For example, the nominee (*1*) studies the basic chemistry and mechanism of autoxidation of lipids and the knowledge is used by one or more companies to improve the stability of their products; (*2*) publishes papers concerning the chemistry and processes for the removal, from fats and oils, of impurities that might be harmful to health, and the knowledge is used by one or more companies to improve the wholesomeness of their products; (*3*) studies basic chemistry and the process of deep-fat frying, and this information is used by one or more companies to improve their deep-fat-fried foods; (*4*) significantly contributes to the understanding of the chemical synthetic or biosynthetic process for the optimization of production of new, or nature-alike, flavor chemicals; (*5*) identifies the chemical profiles of potential flavor sources and pinpoints the importance of key flavor chemicals by correlating them with physiological flavor response; or (*6*) makes significant advances in the development of analytical methodology in flavor chemistry.

Special instructions. The nomination statement must not be more than four typed pages. Do not include a curriculum vitae or lists of publications and patents.

INDUSTRIAL SCIENTIST AWARD

Award. $3,000 honorarium and a plaque from IFT (if a team wins, each member of the team will receive a plaque and an equal share of the $3,000).

Purpose. To honor an IFT member industrial scientist or team of industrial scientists who have made a major technical contribution to the advancement of the food industry.

Eligibility. All team members must be IFT members. Team nominations should identify those persons primarily responsible for the achievement. Team size is limited to a maximum of five persons. Team nominations should identify the role of each team member in accomplishment. Once a nomination is submitted, no additional team members may be added.

The nominees must have made one or more major technical contributions to the advancement of the food industry while a member of an industrial food organization. The nominees' contribution must have had a significant, measurable impact in areas such as production; development of new products, processes, or packaging systems; improvement of food safety and nutrition; and so forth. Food research in industry is generally a team approach, making it difficult to assign credit for an outstanding technical accomplishment to one person or a small group. Nevertheless, the nomination should identify the person(s) most responsible for making the project(s) a success. Part of this success is the direct benefit of the scientists' contribution to consumers and to the profession of food science and technology.

Special instructions. The nomination statement must not be more than four typed pages. Do not include a curriculum vitae or lists of publications and patents.

RESEARCH AND DEVELOPMENT AWARD

Award. $3,000 honorarium and a plaque from IFT (if a team wins, each member of the team will receive a plaque and an equal share of the $3,000).

Purpose. To honor an IFT member or team of members who have made a recent, significant research and development contribution to the understanding of food science, food technology, or nutrition.

Eligibility. All team members must be IFT members. Team nominations should identify those persons primarily responsible for the achievement. Team size is limited to a maximum of five persons. Team nominations should identify the role of each team member in accomplishment. Once a nomination is submitted, no additional team members may be added.

The nominees must be primarily responsible for an achievement within the past 5 years in a research and development program. The achievement must significantly advance the discipline of food science, food technology, or nutrition. The contribution may be basic or applied in nature and must advance science or improve the human condition. The nominees may have worked in any field (including academic, industrial, government, consulting, self-employment, or other endeavors).

Special instructions. The nomination statement must not be more than four typed pages. Do not include a curriculum vitae or lists of publications but cite relevant papers, patents, or articles that show the significance of the work.

ELIZABETH FLEMING STIER AWARD

Award. $3,000 honorarium from the New York Section of IFT and a plaque from IFT.

Purpose. To honor an IFT member for pursuit of humanitarian ideals and unselfish dedication that have resulted in significant contributions to the well-being of the food industry, academia, students, or the general public.

Eligibility. The nominee must have been an IFT member for the past 10 years and active in IFT at both the section and national levels. In instances where there is no IFT section level, activity in other food industry–related organizations may be considered. The nominee must not have received any other IFT Achievement Award but may have been elected an IFT Fellow.

The nominee must have made measurable contributions to the well-being of the food industry, academia, students in food science or closely related disciplines, and/or the general public. Such contributions must have demonstrated compassion and unselfish caring for the world and its people, measured not necessarily in terms of published papers or grants received but rather in terms of the number of lives enhanced by contact with the nominee or by the achievements of individuals as a result of such contact.

Examples include but are not limited to outstanding activities in (*1*) academic advising or mentoring; (*2*) service to IFT or the industry, institutions, governments, or publics it serves; or (*3*) service to groups or individuals in need (such as work in hunger programs or in the delivery of education to persons without other means of acquiring it).

Special instructions. The nomination statement must not be more than four typed pages. Do not include a curriculum vitae or lists of publications and patents.

BERNARD L. OSER FOOD INGREDIENT SAFETY AWARD

Award. $3,000 honorarium and plaque from the Bernard L. Oser Endowment Fund.

Purpose. To honor an IFT member for his or her contribution to the scientific knowledge of food ingredient safety or for leadership in establishing principles for food ingredient safety evaluation or regulation.

Eligibility. The nominee must have been a professional in the area of food additive or GRAS ingredient safety in an academic, industrial, and/or government setting for at least 10 years. The area of professional involvement for this award can include toxicological research, analytical research, safety/regulatory involvement, and associated scientific work in support of food additive and GRAS ingredient safety evaluation and policy development. Publications and examples of leadership qualities in the area of safety evaluation will be the criteria for final selection.

Special Instructions. Must not be more than four typed pages. Attach a numbered list of nominee's publications, and, to save space, refer to them in the nomination statement using these numbers.

PATTI PAGLIUCO
Institute of Food Technologists
Chicago, Illinois

INSTITUTE OF FOOD TECHNOLOGISTS: HISTORY AND PERSPECTIVES

The Institute of Food Technologists (IFT) is one of the premier societies for food science and technology in the world. Indeed it may be *the* premier society. Founded in 1939, IFT has grown to a membership of approximately 29,000 (1).

It has 56 regional sections, some of which have subsections in adjoining geographical areas where a number of IFT members live but distance is too great for them to attend regularly meetings of the regional section. There are 24 discipline or commodity divisions such as Carbohydrate, Food Chemistry, Food Engineering, Fruit and Vegetable Products, and Sensory and a very strong student association. The latter has recently petitioned to be designated as a division. That change would be more a matter of title than of increased stature, for there are already provisions for a student to be a voting member of the IFT's Executive Committee, its Finance Subcommittee, and several other committees.

FORMATION OF THE INSTITUTE

The formation of the Institute in 1939 was the result of the efforts of several food technologists and engineers who felt that there should be a discipline for a subject as important as the production, nutrition, and safety of foods. In 1937 a food technology conference was held at the Massachusetts Institute of Technology (MIT). The conference was chaired by Dr. Samuel C. Prescott, Dean of Biology at MIT. At the conclusion of the conference, Dr. Bernard E. Proctor, also of the MIT faculty, expressed the hope that similar gatherings would be held to function as clearinghouses of information for the scientific and industrial communities. Concurrently at the New York Agricultural Experiment Station in Geneva, New York, George J. Hucker was fostering the idea of an organization of food-related societies.

BEYOND BROACHING AN IDEA

In January 1939 representatives of the two groups met in New York City to discuss the details of such an organization. Lawrence V. Burton, then the editor of *Food Industries*, came prepared with a proposed draft for such an organization, suggested a definition of food technology, and proposed that the organization be called the Society of Food Engineers. The group selected instead the name Institute of Food Technologists, because several individuals realized that the organization should not restrict itself to engineers.

The purpose of the organization was specified as follows:

To facilitate interchange of ideas among its members

To stimulate scientific investigations into technical problems dealing with the manufacture and distribution of foods

To promulgate the results of research in food technology

To offer a medium for the discussion of those results

To plan, organize, and administer such projects for the advancement and application of science insofar as it is fundamental to wider knowledge of foods

The Second Food Technology Conference was held at MIT, June 29 to July 1, 1939, under the direction of Dr. Proctor. At that meeting it was formally decided to establish the IFT. Prescott was elected president, Roy C. Newton of Swift & Co. vice president, and Hucker, secretary-treasurer. (Eventually, Burton, Hucker, and Proctor all were elected presidents of the IFT.) The first team of officials is the only one that served for a second year. (However, today the president does serve one year as president-elect and one year as immediate past president, having several official duties as president-elect and some as immediate past president.) There were 600+ scientists and engineers in attendance at the meeting in 1939.

ANNUAL MEETING THEREAFTER

In Pittsburgh in October 1940, the Council (ie, the governing body of the Institute) held its first meeting, elected members, and scheduled the first IFT conference to be held in Chicago the following June. There were 839 founder and charter members. (In 1997 there were 9031 professional members, 14,855 members, 1709 emeritus members, and 2642 student members.) The Institute has had an annual meeting and food exposition every year since 1941 with the exception of 1945 when the meeting was not convened in order to aid the war effort by avoiding unnecessary travel.

EARLY–CURRENT OFFICERS

The Institute was served for five years by Hucker as its voluntary secretary-treasurer, then for five years by Carl R. Fellers. In 1949 Charles S. Lawrence, former commandant of the Quartermaster Food and Container Institute for the armed services was appointed secretary-treasurer on a one-half time basis. Lawrence set the IFT office up in Chicago where it still is though the Food and Container Institute was later moved to Natick, Massachusetts, and is now a part of the U.S. Army Natick RD&E Center. Lawrence served as secretary-treasurer until 1961. Calvert L. Willey was hired as a full-time executive secretary; the title was later changed to executive director. He retired in 1987. That year Howard W. Mattson, who had been director of public affairs, became the executive director. He resigned for health reasons in 1991. Daniel E. Weber, who had been director of meetings and expositions for several years, became executive director upon the resignation of Howard Mattson. In 1999, his title was changed to executive vice president.

IFT ORGANIZATIONAL STRUCTURE

The Council of the IFT is the policymaking body of the organization. Aside from policy, any changes in the Constitution must be acted upon by the Council. The Council is composed of one or more councillors elected by each of the regional sections and one councillor from each of the divisions. The number of councillors a regional section has is determined by the number of IFT members within its region. In essence, the Council represents the members, for each councillor is elected by the members of his or her regional section.

The executive committee is the administrative body for the members. It is composed of six members elected from among the councillors; six elected from among the members at large; the president-elect, the president, and the immediate past president; plus, as ex officio members without a vote, the treasurer of the Institute, the assistant treasurer, and the executive vice president of the Institute. The Finance subcommittee of the executive committee prepares the budget, but approval or disapproval rests in the hands of the executive committee. The executive committee is the agency that provides direction to the staff and assignment of committee responsibilities, and other matters relating to running the organization. The executive committee answers to the Council on matters relating to policy; for matters such as establishing the budget, it merely informs the Council of its decision.

By far the bulk of members' input is by way of the committee structure. There are 25 committees, plus some of the committees have subcommittees. The executive committee has four subcommittees: finance, publications, scientific editor, and information systems. The IFT has representatives on five other organizations such as International Union of Food Science and Technology (IUFoST) and Codex Alimentarius, and it has liaisons with 11 other organizations such as the American Association for the Advancement of Science, American Oil Chemists' Society, Society of Toxicology, and the American Dietetic Association. Committee members are appointed by the president-elect for a three-year term with one-third of the committee members rotating off the committee each year.

The staff of the IFT consists of the executive vice president and six other directors of such functions as finance and administrative services, science communications, information services and field services, and continuing education. In all, the staff consists of approximately 58 employees.

MEMBERSHIP CLASSES

The four membership classes have already been referred to briefly. To become a professional member, one must first have been a member for five years. Members must have a bachelor's degree or higher degree in food science or a branch of science or engineering associated with food technology. Two-thirds of the members hold degrees in food science itself or a related basic or applied science such as nutrition. Approximately 35% of the members hold a doctorate. Those not qualified to become professional members are individuals who are active in the food field, have substantial experience in it, and, generally, have a degree, but in some field other than science or engineering.

PROGRAMS OF THE INSTITUTE

The IFT has had a strong, active awards program since 1942, when the first of the Nicholas Appert Awards was given, and a scholarship/fellowship program since 1954 (1). In 1946 the Marcel Loncin Research Price carrying an award of $10,000 was initiated (2). Aside from the Nicholas Appert Award, considered the most prestigious of the awards, there are 11 other major awards for such things as excellence in teaching, significant contributions made to improved public health through nutrition, significant contributions to lipid or flavor science, international award for transfer of knowledge, and an award for technical contributions toward advancement of the food industry. Carrying no monetary award but considered a signal honor is election as a Fellow of the Institute.

In 1997–1998, the IFT awarded 17 freshman scholarships varying from $1000 to $1500, 15 sophomore scholarships in the same award range, 63 scholarships to juniors or seniors with the range of $1000 to $2000 each, and 38 fellowships to graduate students with the range of $1000 to $2500.

SCIENCE COMMUNICATION PROGRAM

In 1985 the IFT established its Office of Scientific Public Affairs (OSPA) (1). Under the direction of the director of OSPA, regional communicators were appointed in nearly all the states in the United States. The responsibility of these regional communicators was to serve as resource personnel whenever a food editor or anyone else wanted objective information on some aspect of public health, food additives, nutrition, or any other query having some relation to foods and the food industry. A few years later the name of the function was changed to science communicators. That program still functions by way of science communicators, but their roles have been expanded. By E-mail, fax, and other means the science communicators are given advance information of impending changes or thought, especially that coming from federal agencies or Congress, so that they can be well prepared not only to answer questions but also to take the lead by disseminating information without having to wait for someone seeking information to approach them. The Science Communication office is also responsible for coordinating the writing of Scientific Status Summaries. The staff does not do all the writing—that is mostly done by an expert in the particular subject-matter being written about—but either the Science Communications Office or the editor of *Food Technology* provides input to make the Status Summary as readable as complex subject-matter permits. The idea is to provide objective, factual information for all types of people—those in related science fields, media specialists of various types, and consumers themselves.

THE IFT FOUNDATION

The newest of the IFT's endeavors was the establishment of a foundation to provide financial support for IFT programs such as scholarships and the Science Communications Program, to augment funds going into the Career Guidance Program, and, most recently, to underwrite the expenses of the IFT funding a Science Congressional Fellow. So far, the Congressional Science Fellows have spent one year each as an assistant to a senator or a representative. The foundation supports jointly with the IFT 129 scholarships/fellowships and recently provided most of the funding to distribute 5312 career videos to schools. During the past year the science communicators responded to more than 1000 inquiries from the media. More than 20 continuing education courses were given, and a food science course has been established on the World Wide Web.

The foundation was first chartered in 1985. In 1993 the foundation initiated a fund drive that brought in approximately $2.7 million, which has permitted it to expand its activities greatly. The foundation was not dormant from 1985 to 1993. It just took time to work out the means to be able to fund IFT programs while ensuring sufficient financial support to be able to pay for the costs of a fund drive.

The foundation was chartered to serve as a fund-raising arm for the IFT with the objective of acquiring resources to reach the highest level of achievement in the food science profession. The foundation board elects six individuals to serve on the board. All need not be IFT members. The IFT executive committee appoints six members. Members are appointed for three-year terms. The treasurer of the IFT, the IFT's executive vice president, and the president of the IFT are ex offico members of the Board. In 1997 the board hired a director of development for the foundation.

IFT PUBLICATIONS

The IFT publishes two journals. One is the *Journal of Food Science*. Approximately 50% of the articles it publishes come from outside the United States, which attests to the journal's high repute. In 1950 the IFT purchased the journal *Food Research* from Garrard Press and changed its name to the *Journal of Food Science (JFS)*. The first issue of *Food Technology* was published in January 1947. *JFS* of course is peer-reviewed; so too are many of the articles in *Food Technology*. The latter contributes to the reputation of the IFT, for approximately 17.6% of IFT members reside abroad, and all members and professional members receive *Food Technology* as a part of their dues. *Food Technology* features some of the outstanding papers given at the annual meeting, various didactic or expository articles, and each Scientific Status Summary. Illustration of the last are "Assessing, Managing, and Communicating Chemical Food Risks" (3), "Foodborne Disease Significance of *Escherichia coli* O157:H7" (4), "Irradiation of Food" (5), and "Glass Transition in Low Moisture and Frozen Foods" (6).

PERSPECTIVES

The preceding subjects demonstrate that the IFT is active on several fronts in bringing food science to a high state of development. For a long time it has had an international presence too. Reference was made above to the number of articles published in *JFS* by others in the global community. For many years, the IFT has had four regional sections outside the country: England, British Columbia (Canada), Japan, and Mexico. Some years ago the IFT decided not to charter any more regional sections abroad. It did so to avoid any possible conflict with existing food science societies in other countries or with the formation of societies in countries presenting lacking such an organization. The IFT has, however, maintained affiliation with several societies abroad. At present there are 22 "affiliates" of the IFT. Some countries, such as Spain, have more than one affiliate. There currently is a recommendation that the IFT change the name of the relation from "affiliate" to "allied" to be sure that the relation is understood to be between equals.

IFT AND IUFOST

Within the past three years the IFT has turned toward instituting a greater international presence. The number of international companies exhibiting products, equipment, and services at the annual meeting of the IFT as well as the number of attendees and papers given by those from outside the United States attests to its strong international presence. In 1998 the Canadian Institute of Food Science and Technology, the Mexican Association for Food Science and Technology, and the IFT held their annual meetings or conferences at the same time in Atlanta. Within the past five years, the IFT has stepped up its supporting role in IUFoST.

IUFoST is an organization made up of member countries, each of which has official delegates to IUFoST, the number being different according to the dues paid to IUFoST. No country, however, can have more than six delegates. The United States, for example, has six accredited delegates. At four-year intervals IUFoST hosts a meeting. The last one was held in 1995 in Hungary, and the next one is to be held in Australia in 1999. Currently, the United States, actually the IFT, pays its yearly dues of $6000, but it also underwrites some of the administrative expenses of IUFoST by providing services to it. During 1996–1997, the IFT and IUFoST developed a plan, called the International Action Plan, to strengthen formally each other's endeavors to facilitate cooperation among the member bodies. The IFT and IUFoST envision that there are many areas where exhibitions, short courses, other types of training courses, and various forms of cooperation could be carried on more effectively and expeditiously, by joint efforts, without compromising in any way the interests of national bodies in other countries, if the IFT and IUFoST were to take the lead in funding and supporting such endeavors. At the present time, the IFT will do most of the funding, for IUFoST has always been handicapped by the limited resources it has been able to generate. Many of its member countries are among the poorest in the world.

Based on its record to date, the perspectives for the IFT are that it will become even more the knowledgeable and influential organization that it is in the United States. Because of its current commitment to bring its organizational skill and willingness to share its knowledge and resources with food science organizations elsewhere, the prospects are that the IFT will assume in the years ahead international roles even more beneficial to food science and technology and thus to the world's food supply and the well-being of humans.

BIBLIOGRAPHY

1. N. H. Mermelstein, "History of the Institute of Food Technologists *Food Technol.* **43**, 14–52 (1989).
2. F. R. Katz, *The 1997–1998 IFT Membership Directory*, Institute of Food Technologists, Chicago, Ill., 1997, p. 14.
3. C. K. Winter and F. J. Francis, "Assessing, Managing, and Communicating Chemical Food Risks," *Food Technol.* **51**, 85–92 (1997).
4. R. L. Buchanan and M. P. Doyle, "Foodborne Diseases Significance of Escherichia coli 0157:H7." *Food Technol.* **51**, 69–76 (1997).
5. D. G. Olson, "Irradiation of Food," *Food Technol.* **52**, 56–62 (1998).
6. Y. H. Roos, M. Karel, and J. J. Kokini, "Glass Transition in Low Moisture and Frozen Foods," *Food Technol.* **50**, 95–108 (1996).

JOHN POWERS
University of Georgia
Athens, Georgia